Stantec's Water Treatment

Stantec's Water Treatment

Principles and Design

Updated Third Edition

John C. Crittenden Ph.D., P.E., BCEE, NAE
Hightower Chair and Georgia Research Alliance Eminent Scholar
Director of the Brook Byers Institute for Sustainable Systems
Georgia Institute of Technology

R. Rhodes Trussell Ph.D., P.E., BCEE, NAE
Founder and Chairman
Trussell Technologies, Inc.

David W. Hand Ph.D., BCEEM
Professor Emeritus of Civil, Environmental, and Geospatial Engineering
Michigan Technological University

Kerry J. Howe Ph.D., P.E., BCEE
Distinguished Professor of Civil, Construction, and Environmental Engineering
University of New Mexico

George Tchobanoglous Ph.D., P.E., BCEE, NAE
Professor Emeritus of Civil and Environmental Engineering
University of California at Davis

With Contributions By:

Bill Ward, J.D., P.E.
Senior Advisor, Stantec Institute for Water Technology & Policy
Director of Intellectual Property, Stantec Consulting Services, Inc.

James H. Borchardt, P.E.
Vice President
Stantec Consulting Services, Inc.

WILEY

Stantec

This book is printed on acid-free paper.

Published by John Wiley & Sons, Inc., Hoboken, New Jersey

Published simultaneously in Canada

Library of Congress Cataloging-in-Publication Data:
Names: Crittenden, John C. (John Charles), 1949- author. | Trussell, R. Rhodes, author. | Hand, David W., author. | Howe, Kerry J., author. | Tchobanoglous, George, author.
Title: Stantec's water treatment : principles and design / John C. Crittenden Ph.D., P.E., BCEE, NAE, Hightower Chair and Georgia Research Alliance Eminent Scholar, Director of the Brook Byers Institute for Sustainable Systems, Georgia Institute of Technology, R. Rhodes Trussell Ph.D., P.E., BCEE, NAE, Founder and Chairman, Trussell Technologies, Inc., David W. Hand Ph.D., BCEEM, Professor Emeritus of Civil, Environmental, and Geospatial Engineering, Michigan Technological University, Kerry J. Howe Ph.D., P.E., BCEE, Distinguished Professor of Civil, Construction, and Environmental Engineering, University of New Mexico, George Tchobanoglous Ph.D., P.E., BCEE, NAE, Professor Emeritus of Civil and Environmental Engineering, University of California at Davis.
Description: Updated third edition. | Hoboken, New Jersey : John Wiley & Sons, Inc., [2023] | Includes bibliographical references and index.
Identifiers: LCCN 2022012613 (print) | LCCN 2022012614 (ebook) | ISBN 9781119819967 (hardback ; acid-free paper) | ISBN 9781119820079 (pdf) | ISBN 9781119820062 (epub) | ISBN 9781119820086 (ebook)
Subjects: LCSH: Water–Purification.
Classification: LCC TD430 .C75 2023 (print) | LCC TD430 (ebook) | DDC 628.1/62–dc23/eng/20220511
LC record available at https://lccn.loc.gov/2022012613
LC ebook record available at https://lccn.loc.gov/2022012614

Cover image: Courtesy of William A. ward, Jr.
Cover design: Wiley

Set in 10/12pt NewBaskervilleStd by Integra Software Services Pvt. Ltd, Pondicherry, India

Printed in Singapore
M004351_080922

Contents

Preface

Making water safe to drink is complicated. A myriad of constituents can be present in natural waters, with different concentrations, different physical and chemical properties, and representing different hazards. Removing them requires integrating unit processes into a process train, in which each process may remove specific constituents or help another process work more effectively. This book has a rich history of helping students and professional engineers understand the fundamental principles of water treatment and provides a pathway to successful treatment plant design.

Some early practitioners of water treatment engineering relied on trial-and-error, and then replicating designs that were found to be successful. Others recognized that separation processes obey fundamental scientific principles. One in the latter camp was James M. Montgomery, who founded James M. Montgomery, Consulting Engineers, Inc. (JMM) in 1945. With the underpinning of Mr. Montgomery's scientific and engineering talents, the firm grew to be one of the leading environmental engineering firms in the United States and was widely respected as an innovator of the art, science, design, and engineering of water treatment systems.

Recognizing the exceptional expertise of the firm and wanting to honor their founder's devotion to training other engineers, the company's leadership set about to write a textbook that could serve as a resource to engineers throughout the rapidly growing – and rapidly changing – field of water treatment engineering. Over a period of several years, more than 50 employees contributed as authors or reviewers, led by Dr. Michael C. Kavanaugh, Dr. Susumu Kawamura, Dr. Carol Tate, Dr. R. Rhodes Trussell, and William W. Aultman. The first edition of *Water Treatment Principles and Design* was published in 1985 and quickly became the most prominent and respected resource for water treatment engineering among practitioners and students alike.

The 1980s and 1990s were a period of rapid change following the passage of the Safe Drinking Water Act in 1974. Within a decade, portions of the first edition were becoming dated and processes such as ozonation, adsorption, advanced oxidation, and reverse osmosis were becoming strategies for achieving more complex water treatment goals. In the mid-1990s, recognizing that an update was necessary, the company (then named Montgomery Watson, following a 1990 merger with the firm Watson Hawksley from the United Kingdom) embarked on writing a revision. Unfortunately, the challenges of engaging many experts within the company drove up the cost and dragged out the schedule, and after several years the effort was abandoned.

With efforts to write a revision with internal experts stymied and fearing the book would become obsolete, Dr. R. Rhodes Trussell convinced Montgomery Watson's leadership to engage a select group of outside experts to assist him in updating the original text. Drs. John C. Crittenden, George Tchobanoglous, David W. Hand, and Kerry J. Howe (a former JMM employee) were retained. By then, more time had passed and more new processes needed to be incorporated into the text. In the end, the new authors completely restructured the contents, and the second edition of *Water Treatment Principles and Design* emerged as an entirely new book while retaining the original book's (and James Montgomery's) commitment to grounding water treatment plant design on fundamental principles. The second edition was published in 2005, under the corporate logo of Montgomery Watson Harza (MWH), following the 2001 merger between Montgomery Watson and Harza Engineering Company.

As the water treatment industry continued to evolve, a third edition quickly became necessary. The authors, with support from James H. Borchardt of MWH, completed *MWH's Water Treatment Principles and Design, 3rd. edition* in 2012. Simultaneous with the 3rd edition, an abridged version of the text was prepared, trimmed to include only the essential principles that could be covered in a one-semester college course in water treatment. *Principles of Water Treatment,* weighing in at only 650 pages compared to the larger book's 1900 pages, was also published in 2012.

In 2016, MWH Global (which had grown to 6,800 employees) was acquired by Stantec, a Canadian engineering firm started by environmental engineer Dr. Don Stanley in 1954. The combined firm, with 22,000 employees, offers top-tier design services to water clients around the world from offices in over 400 locations. During the time of the acquisition, important changes were taking place in the water treatment industry and the world. In the industry, water scarcity was becoming an important driver of water treatment projects, and many communities began exploring alternative water sources like seawater desalination and potable reuse. Although desalination was already covered in the 3rd edition, the complexities of advanced treatment trains for potable reuse now warrant significant coverage. Another trend was the growing prominence of online resources and content as a way of learning and sharing information. Having acquired the legacy of *MWH's Water Treatment Principles and Design* in the merger, Stantec wanted to make sure that the book remained relevant in a changing world. Bill Ward of Stantec initiated discussions with the authors of the 3rd edition to update the book. Mr. Ward suggested the addition of QR codes in the book margins that link to online content and identified materials for the book's website (hosted on Stantec.com), while the 3rd edition authors wrote a new chapter on potable reuse. This book—*Stantec's Water Treatment: Principles and Design, updated 3rd. edition*—is the product of that latest collaboration. We hope you, the reader, will use it to further your knowledge in the exciting field of water treatment.

Important Features of This Book

This book is written to serve several purposes: (1) an undergraduate textbook appropriate for elective classes in water treatment, (2) a graduate-level textbook appropriate for teaching water treatment, groundwater remediation, and physical chemical treatment, and (3) a reference book for engineers who are designing or operating water treatment plants.

Errata- For corrections to the printed text, use this link: http://www.stantec.com/wt3r/Preface-1.html

To convey ideas and concepts more clearly, the book contains the following important elements: (1) 170 example problems worked out in detail with units, (2) 409 homework problems, designed to develop students understanding of the subject matter, (3) 247 tables that contain physical properties of chemicals, design data, and thermodynamic properties of chemicals, to name a few, and (4) 458 illustrations and photographs. Metric SI and U.S. customary units are given throughout the book. Instructors will find the example problems, illustrations, and photographs useful in introducing students to fundamental concepts and practical design issues. In addition, an instructor's solutions manual is available from the publisher.

Acknowledgments

The authors are indebted to everyone who assisted with the preparation of this latest edition of this book. Most importantly, we thank Mr. Bill Ward, PE, Director at Stantec, for being the driving force to start this revision, advocating for its publication within Stantec, serving as a liaison between Stantec and the authors, and bringing creative ideas such as the QR codes linking to online content to the project. Without his initiative, this edition would not have been published. Thanks to Dr. Amir Alansari; Dr. Laith Furatian, P.Eng.; Dr. Sanaz Imen, PE; and Dr. Ayman Shawwa, PE, BCEE, PMP, all from Stantec, and Mr. Reed Warburton of Bowdoin College for reviewing the page proofs to help ensure the accuracy of this edition.

Most of the content of this edition is held over from the 3rd edition, and those who contributed to that edition are gratefully acknowledged. Dr. Harold Leverenz prepared most of the figures, and figures for several chapters were prepared by Mr. James Howe. Dr. Zhonming Lu and Dr. Daisuke Minakata contributed to writing and revising Chapters 15 and 18, respectively. Mr. Osvaldo A. Broesicke assisted with the preparation of Figure 6-20. Mr. Carson O. Lee and Mr. Daniel Birdsell reviewed and checked many of the chapters, including the figure, table, and equation numbers, the math in example problems, and the references at the end of the chapters. Mr. James Borchardt, PE, Vice President at Stantec, served as a liaison to Stantec and coordinated technical input from Stantec staff.

For the 3rd edition, several MWH employees provided technical input, prepared case studies, gathered technical information on MWH projects, prepared graphics and photos, and provided administrative support. These include: Ms. Donna M. Arcaro; Dr. Jamal Awad, PE, BCEE; Mr. Charles O. Bromley, PE, BCEE; Dr. Arturo A. Burbano, PE, BCEE; Mr. Ronald M. Cass, PE; Mr. Harry E. Dunham, PE; Mr. Frieder H. Ehrlich, C Eng, MAIChemE; Mr. Andrew S. Findlay, PE; Mr. Mark R. Graham, PE; Mr. Jude D. Grounds, PE; Ms. Stefani O. Harrison, PE; Dr. Joseph G. Jacangelo, REHS; Ms. Karla J. Kinser, PE; Mr. Peter H. Kreft, PE; Mr. Stewart E. Lehman, PE; Mr. Richard Lin, PE; Mr. William H. Moser, PE; Mr. Michael A. One-by, PE; Mr. Michael L. Price, PE; Mr. Nigel S. Read, C Eng; Mr. Matthieu F. Roussillon, PE; Ms. Stephanie J. Sansom, PE; Mr. Gerardus J. Schers, PE; Ms. Jackie M. Silber; Mr. William A. Taplin, PE; and Dr. Timothy A. Wolfe, PE, BCEE.

We gratefully acknowledge the support and help of the Wiley staff, particularly Ms. Kalli Schultea and Ms. Amy Odum.

John C. Crittenden
R. Rhodes Trussell
David W. Hand
Kerry J. Howe
George Tchobanoglous

Foreword

Looking back on this and the prior editions of "Water Treatment Principles and Design," it is apparent that the authors have been innovating, developing, and adapting continuously to a world that is always changing. The first edition of this book was published in 1985, the second edition in 2005 and the third edition in 2012. Each edition represents a significant step forward in applying good science to the pressing issues of the day. Refinement of proven technologies and introduction of new techniques have enabled practitioners to consistently deliver safe and reliable water to people, even in the face of changing climate, emerging contaminants, water scarcity, aging infrastructure, and ever tightening budgets.

As we all work to deliver drinking water to ever more people, our governments, scientists, and engineers face challenges not seen by their predecessors. Huge populations have moved to water-limited geographies; changes in weather patterns coupled with demand have caused some supplies to evaporate before our eyes; many new and emerging chemicals are already prevalent in our environment and known to be persistent, dangerous, and difficult to treat; and every year our investment in water treatment and distribution infrastructure falls farther behind. Innovation, emerging technologies, and more effective application of best engineering practices can help us bridge these gaps.

With that in mind, our goal for this revised third edition is to collaborate with the esteemed authors to update students and practitioners on the topic of potable reuse, to deliver a focused multimedia library of supplemental information via the internet, and to address some increasingly important contaminants of emerging concern.

We all owe a debt of gratitude to the principal authors: Dr. Kerry Howe of the University of New Mexico, Dr. George Tchobanoglous of the University of California at Davis, Dr. John Crittenden of the Georgia Institute of Technology, Dr. Rhodes Trussell of Trussell Technologies, Inc., and Dr. David Hand of the Michigan Technological University for collaborating to carefully develop this invaluable encyclopedic text. Thanks also to James

Borchardt, P.E. of Stantec, who contributed to the Third Edition of the text. Thank you also, Kalli Schultea of Wiley, who helped us bring this edition forward.

And we are all grateful to you, our reader, for learning more about the science of drinking water treatment. You each contribute incrementally to a noble effort that has delivered safe and reliable drinking water to large portions of humanity, contributing to a collective work that enables so many of us to enjoy a quality of life that was not available to those who came before the field developed. With your help, may we do the same for the billions of people our profession has not yet served.

Bill Ward, J.D., P.E.
Senior Advisor
Stantec Institute for Water Technology & Policy

About the Authors

Dr. John C. Crittenden is the director of the Brook Byers Institute for Sustainable Systems and a professor in Civil and Environmental Engineering at the Georgia Institute of Technology. He holds the Hightower Chair and is a Georgia Research Alliance Eminent Scholar in Environmental Technologies. Dr. Crittenden was elected to the US National Academy of Engineering in 2002, the Chinese Academy of Engineering in 2013, and the European Academy of Sciences in 2019. He has received the NWRI Clarke Prize, the ASCE Simon W. Freese Award, and the Friendship Medal from China. He is the co-holder of five patents and is the author more than 415 articles in refereed journals, more than 100 book chapters, reports, and symposia, and has more than 30,000 citations. Dr. Crittenden has a B.S. in chemical engineering and M.S. and Ph.D. degrees in civil engineering from the University of Michigan.

Dr. R. Rhodes Trussell is a registered civil and corrosion engineer in the State of California with 50 years of water treatment experience. After working for 33 years for MWH, Inc. (now part of Stantec), he founded Trussell Technologies, Inc., a consulting firm specializing in the application of science to engineering. He was elected to the National Academy of Engineering in 1995. Dr. Trussell has been awarded the Boyd Award by AMWA, the Pohland Award by AEESP, the AP Black Award by AWWA, the Global Water Award by IWA, the Clarke Prize by NWRI, and the Simon W. Freese Award by ASCE. He has authored more than 200 publications, including several chapters in all editions of Stantec's Water Treatment: Principles and Design. He has B.S., M.S., and Ph.D. degrees in environmental engineering from the University of California at Berkeley.

Dr. David W. Hand is a professor emeritus of civil, environmental, and geospatial engineering at Michigan Technological University. His teaching and research focus on water and wastewater treatment engineering with emphasis on physicochemical treatment processes. He received the ASCE Rudolf Hering Medal, the AEESP Award for Outstanding Teaching in Environmental Engineering and Science, and a publication award from AWWA. He is a Board Certified Environmental Engineering Member of AAEES. He has authored or coauthored over 130 technical publications including six textbooks, two patents, and eight copyrighted software programs. He has B.S., M.S. and Ph.D. degrees in engineering and civil engineering from the Michigan Technological University.

Dr. Kerry J. Howe is a distinguished professor in Civil, Construction, and Environmental Engineering at the University of New Mexico, the director of the Center for Water and the Environment, and has over 35 years of water treatment experience. He worked for over 10 years as a design engineer at MWH, Inc., now part of Stantec. Dr. Howe is a registered professional engineer in New Mexico and Wisconsin, and a Board Certified Environmental Engineer by the AAEES. He has authored or coauthored over 180 presentations, technical reports, and peer-reviewed journal publications. He has a B.S. degree from the University of Wisconsin-Madison, an M.S. degree from the University of Texas at Austin, and a Ph.D. degree from the University of Illinois at Urbana-Champaign.

Dr. George Tchobanoglous is professor emeritus in the Department of Civil and Environmental Engineering at the University of California, Davis. He is a member of the US National Academy of Engineering, the European Academy of Sciences and Arts, and a corresponding member of the Academy of Athens. He has received Honorary Doctor of Engineering degrees from the Colorado School of Mines, the Aristotle University of Thessaloniki, and the Technical University of Crete, Greece. Dr. Tchobanoglous has been awarded the Clarke Prize by NWRI and the Pohland Award by AEESP. He has authored or coauthored over 600 publications, including 23 textbooks and 8 reference books, and has given more than 625 presentations. Dr. Tchobanoglous received a B.S. in civil engineering from the University of the Pacific, an M.S. in sanitary engineering from the University of California, Berkeley, and a Ph.D. in environmental engineering from Stanford University.

1 Introduction

Securing and maintaining an adequate supply of water has been one of the essential factors in the development of human settlements. The earliest developments were primarily concerned with the quantity of water available. Increasing population, however, has exerted more pressure on limited high-quality surface sources, and the contamination of water with municipal, agricultural, and industrial wastes has led to a deterioration of water quality in many other sources. At the same time, water quality regulations have become more rigorous, analytical capabilities for detecting contaminants have become more sensitive, and the general public has become both more knowledgeable and more discriminating about water quality.

Stantec's Water Treatment: Principles and Design, Updated Third Edition. John C. Crittenden, R. Rhodes Trussell, David W. Hand, Kerry J. Howe, and George Tchobanoglous.

Thus, the quality of a water source cannot be overlooked in water supply development. In fact, virtually all sources of water require some form of treatment before potable use.

Water treatment can be defined as the processing of water to achieve a water quality that meets specified goals or standards set by the end user or a community through its regulatory agencies. Goals and standards can include the requirements of regulatory agencies, additional requirements set by a local community, and requirements associated with specific industrial processes. The evolution of water treatment practice has a rich history of empirical and scientific developments and challenges met and overcome.

The primary focus of this book is the application of water treatment for the production of potable, or drinking, water on a municipal level. Water treatment, however, encompasses a much wider range of problems and ultimate uses, including home treatment units, community treatment plants, and facilities for industrial water treatment with a wide variety of water quality requirements that depend on the specific industry. Water treatment processes are also applicable to remediation of contaminated groundwater and other water sources and wastewater treatment when the treated wastewater is to be recycled for new uses. The issues and processes covered in this book are relevant to all of these applications.

This book thoroughly covers a full range of topics associated with water treatment, starting in Chaps. 2 and 3 with an in-depth exploration of the physical, chemical, and microbiological aspects that affect water quality. Chapter 4 presents an overview of factors that must be considered when selecting a treatment strategy. Chapters 5 through 8 explain background concepts necessary for understanding the principles of water treatment, including fundamentals of chemical reactions, chemical reactors, mass transfer, and oxidation/reduction reactions. Chapters 9 through 18 are the heart of the book, presenting in-depth material on each of the principal unit processes used in municipal water treatment. Chapters 19 through 22 present supplementary material that is essential to an overall treatment system, including issues related to disinfection by-products, treatment strategies for specific contaminants, processing of treatment residuals, and corrosion in water distribution systems. The final chapter, Chap. 23, synthesizes all the previous material through a series of case studies.

The purpose of this introductory chapter is to provide some perspective on the (1) historical development of water treatment, (2) health concerns, (3) constituents of emerging concern, (4) evolution of water treatment technology, and (5) selection of water treatment processes. The material presented in this chapter is meant to serve as an introduction to the chapters that follow in which these and other topics are examined in greater detail.

1-1 History of the Development of Water Treatment

Some of the major events and developments that contributed to our understanding of the importance of water quality and the need to provide some means of improving the quality of natural waters are presented in Table 1-1. As reported in Table 1-1, one of the earliest water treatment techniques (boiling of water) was primarily conducted in containers in the households using the water. From the sixteenth century onward, however, it became increasingly clear that some form of treatment of large quantities of water was essential to maintaining the water supply in large human settlements.

1-2 Health and Environmental Concerns

The health concerns from drinking water have evolved over time. While references to filtration as a way to clarify water go back thousands of years, the relationship between water quality and health was not well understood or appreciated. Treatment in those days had as much to do with the aesthetic qualities of water (clarity, taste, etc.) as it did on preventing disease. The relationship between water quality and health became clear in the nineteenth century, and for the first 100 years of the profession of water treatment engineering, treatment was focused on preventing waterborne disease outbreaks. Since 1970, however, treatment objectives have become much more complex as public health concerns shifted from acute illnesses to the chronic health effects of trace quantities of anthropogenic (manmade) contaminants.

Nineteenth Century

In the middle of the nineteenth century it was a common belief that diseases such as cholera and typhoid fever were primarily transmitted by breathing miasma, vapors emanating from a decaying victim and drifting through the night. This view began to change in the last half of that century. In 1854, Dr. John Snow demonstrated that an important cholera epidemic in London was the result of water contamination (Snow, 1855). Ten years later, Dr. Louis Pasteur articulated the germ theory of disease. Over the next several decades, a number of doctors, scientists, and engineers began to make sense of the empirical observations from previous disease outbreaks. By the late 1880s, it was clear that some important epidemic diseases were often waterborne, including cholera, typhoid fever, and amoebic dysentery (Olsztynski, 1988). As the nineteenth century ended, methods such as the coliform test were being developed to assess the presence of sewage contamination in a water supply (Smith, 1893), and the conventional water treatment process (coagulation/flocculation/sedimentation/filtration) was being developed as a robust way of removing contamination from municipal water supplies (Fuller, 1898).

Table 1-1
Historical events and developments that have been precursors to development of modern water supply and treatment systems

Period	Event
4000 B.C.	Ancient Sanskrit and Greek writings recommend water treatment methods. In the Sanskrit Ousruta Sanghita it is noted that "impure water should be purified by being boiled over a fire, or being heated in the sun, or by dipping a heated iron into it, or it may be purified by filtration through sand and coarse gravel and then allowed to cool."
3000 to 1500 B.C.	Minoan civilization in Crete develops technologies so advanced they can only be compared to modern urban water systems developed in Europe and North America in the second half of the nineteenth century. Technology is exported to Mediterranean region.
1500 B.C.	Egyptians reportedly use the chemical alum to cause suspended particles to settle out of water. Pictures of clarifying devices were depicted on the wall of the tomb of Amenophis II at Thebes and later in the tomb of Ramses II.
Fifth century B.C.	Hippocrates, the father of medicine, notes that rainwater should be boiled and strained. He invents the "Hippocrates sleeve," a cloth bag to strain rainwater.
Third century B.C.	Public water supply systems are developed at the end of the third century B.C. in Rome, Greece, Carthage, and Egypt.
340 B.C. to 225 A.D.	Roman engineers create a water supply system that delivers water [490 megaliters per day (130 million gallons per day)] to Rome through aqueducts.
1676	Anton van Leeuwenhoek first observes microorganisms under the microscope.
1703	French scientist La Hire presents a plan to French Academy of Science proposing that every household have a sand filter and rainwater cistern.
1746	French scientist Joseph Amy is granted the first patent for a filter design. By 1750 filters composed of sponge, charcoal, and wool could be purchased for home use.
1804	The first municipal water treatment plant is installed in Paisley, Scotland. The filtered water is distributed by a horse and cart.
1807	Glasgow, Scotland, is one of the first cities to pipe treated water to consumers.
1829	Installation of slow sand filters in London, England.
1835	Dr. Robley Dunlingsen, in his book *Public Health*, recommends adding a small quantity of chlorine to make contaminated water potable.
1846	Ignaz Semmelweiss (in Vienna) recommends that chlorine be used to disinfect the hands of physicians between each visit to a patient. Patient mortality drops from 18 to 1 percent as a result of this action.
1854	John Snow shows that a terrible epidemic of Asiatic cholera can be traced to water at the Broad Street Well, which has been contaminated by the cesspool of a cholera victim recently returned from India. Snow, who does not know about bacteria, suspects an agent that replicates itself in the sick individuals in great numbers and exits through the gastrointestinal tract, and is transported by the water supply to new victims.
1854	Dr. Falipo Pacini, in Italy, identifies the organism that causes Asiatic cholera, but his discovery goes largely unnoticed.

Table 1-1 *(Continued)*

Period	Event
1856	Thomas Hawksley, civil engineer, advocates continuously pressurized water systems as a strategy to prevent external contamination.
1864	Louis Pasteur articulates the germ theory of disease.
1874	Slow sand filters are installed in Poughkeepsie and Hudson, New York.
1880	Karl Eberth isolates the organism (*Salmonella typhosa*) that causes typhoid fever.
1881	Robert Koch demonstrates in the laboratory that chlorine will inactivate bacteria.
1883	Carl Zeiss markets the first commercial research microscope.
1884	Professor Escherich isolates organisms from the stools of a cholera patient that he initially thought were the cause of cholera. Later it is found that similar organisms are also present in the intestinal tracts of every healthy individual as well. Organism eventually named for him (*Escherichia coli*).
1884	Robert Koch proves that Asiatic cholera is due to a bacterium, *Vibrio cholerea*, which he calls the comma bacillus because of its comma-like shape.
1892	A cholera epidemic strikes Hamburg, Germany, while its neighboring city, Altona, which treats its water using slow sand filtration, escapes the epidemic. Since that time, the value of granular media filtration has been widely recognized.
1892	The New York State Board of Health uses the fermentation tube method developed by Theobald Smith for the detection of *E. coli* to demonstrate the connection between sewage contamination of the Mohawk River and the spread of typhoid fever.
1893	First sand filter built in America for the express purpose of reducing the death rate of the population supplied is constructed at Lawrence, Massachusetts. To this end, the filter proves to be a great success.
1897	G. W. Fuller studies rapid sand filtration [5 cubic meters per square meter per day (2 gallons per square foot per day)] and finds that bacterial removals are much better when filtration is preceded by good coagulation and sedimentation.
1902	The first drinking water supply is chlorinated in Middelkerke, Belgium. Process is actually the "Ferrochlor" process where in calcium hypochlorite and ferric chloride are mixed, resulting in both coagulation and disinfection.
1903	The iron and lime process of treating water (softening) is applied to the Mississippi River water supplied to St Louis, Missouri.
1906	First use of ozone as a disinfectant in Nice, France. First use of ozone in the United States occurs some four decades later.
1908	George Johnson, a member of Fuller's consulting firm, helps install continuous chlorination in Jersey City, New Jersey.
1911	Johnson publishes "Hypochlorite Treatment of Public Water Supplies" in which he demonstrates that filtration alone is not enough for contaminated supplies. Adding chlorination to the process of water treatment greatly reduces the risk of bacterial contamination.

(Continued)

Table 1-1 *(Continued)*

Period	Event
1914	U.S. Public Health Service (U.S. PHS) uses Smith's fermentation test for coliform to set standards for the bacteriological quality of drinking water. The standards applied only to water systems that provided drinking water to interstate carriers such as ships and trains.
1941	Eighty-five percent of the water supplies in the United States are chlorinated, based on a survey conducted by U.S. PHS.
1942	U.S. PHS adopts the first comprehensive set of drinking water standards.
1974	Dutch and American studies demonstrate that chlorination of water forms trihalomethanes.
1974	Passage of the Safe Drinking Water Act (SDWA).

Source: Adapted from AWWA (1971), Baker (1948), Baker and Taras (1981), Blake (1956), Hazen (1909), Salvato (1992), and Smith (1893).

Twentieth Century

The twentieth century began with the development of continuous chlorination as a means for bacteriological control, and in the first four decades the focus was on the implementation of conventional water treatment and chlorine disinfection of surface water supplies. By 1940, the vast majority of water supplies in developed countries had "complete treatment" and were considered microbiologically safe. In fact, during the 1940s and 1950s, having a microbiologically safe water supply became one of the principal signposts of an advanced civilization. The success of filtration and disinfection practices led to the virtual elimination of the most deadly waterborne diseases in developed countries, particularly typhoid fever and cholera.

FROM BACTERIA TO VIRUSES

The indicator systems and the treatment technologies for water treatment focused on bacteria as a cause of waterborne illness. However, scientists demonstrated that there were some infectious agents much smaller than bacteria (viruses) that could also cause disease. Beginning in the early 1940s and continuing into the 1960s, it became clear that viruses were also responsible for some of the diseases of the fecal–oral route, and traditional bacterial tests could not be relied upon to establish their presence or absence.

ANTHROPOGENIC CHEMICALS AND COMPOUNDS

Concern also began to build about the potential harm that anthropogenic chemicals in water supplies might have on public health. In the 1960s, the U.S. PHS developed some relatively simple tests using carbon adsorption and extraction in an attempt to assess the total mass of anthropogenic compounds in water. Then in the mid-1970s, with the development of the gas chromatograph/mass spectrometer, it became possible to detect these compounds at much lower levels. The concern about the potential harm of man-made organic compounds in water coupled with improving analytical capabilities has led to a vast array of regulations designed to address these risks. New issues with anthropogenic chemicals will continue to emerge as new chemicals are synthesized, analytical techniques improve, and increasing population density impacts the quality of water sources.

DISINFECTION BY-PRODUCTS

A class of anthropogenic chemicals of particular interest in water treatment is chemical by-products of the disinfection process itself (disinfection by products, or DBPs). DBPs are formed when disinfectants react with species naturally present in the water, most notably natural organic matter and some inorganic species such as bromide. The formation of DBPs increases as the dose of disinfectants or contact time with the water increases. Reducing disinfectant use to minimize DBP formation, however, has direct implications for increasing the risk of illness from microbial contamination. Thus, a trade-off has emerged between using disinfection to control microbiological risks and preventing the formation of undesirable man-made chemicals caused by disinfectants. Managing this trade-off has been one of the biggest challenges of the water treatment industry over the last 30 years.

MODERN WATERBORNE DISEASE OUTBREAKS

While severe waterborne disease has been virtually eliminated in developed countries, new sources of microbiological contamination of drinking water have surfaced in recent decades. Specifically, pathogenic protozoa have been identified that are zoonotic in origin, meaning that they can pass from animal to human. These protozoan organisms are capable of forming resistant, encysted forms in the environment, which exhibit a high level of resistance to treatment. The resistance of these organisms has further complicated the interrelationship between the requirements of disinfection and the need to control DBPs. In fact, it has become clear that processes that provide better physical removal of pathogens are required in addition to more efficient processes for disinfection.

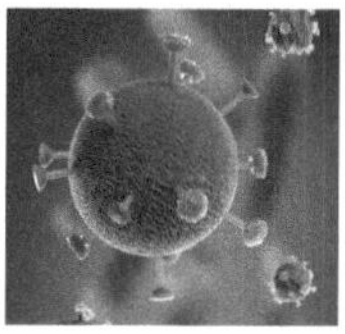

Water engineers must be knowledgeable about potential risks from new pathogens.
http://www.stantec.com/wt3r/1-1.html

The significance of these new sources of microbiological contamination has become evident in recent waterborne disease outbreaks, such as the outbreaks in Milwaukee, Wisconsin, in 1993 and Walkerton, Ontario, in 2000. In Milwaukee, severe storms caused contamination of the water supply and inadequate treatment allowed *Cryptosporidium* to enter the water distribution system, leading to over 400,000 cases of gastrointestinal illness and over 50 deaths (Fox and Lytle, 1996). The Walkerton incident was caused by contamination of a well in the local water system by a nearby farm. During the outbreak, estimates are that more than 2300 persons became ill due to *E. coli* O157:H7 and *Campylobacter* species (Clark et al., 2003). Of the 1346 cases that were reported, 1304 (97 percent) were considered to be directly due to the drinking water. Sixty-five persons were hospitalized, 27 developed hemolytic uremic syndrome, and 6 people died.

Another challenge associated with microbial contamination is that the portion of the world's population that is immunocompromised is increasing over time, due to increased life spans and improved medical care. The immunocompromised portion of the population is more susceptible to health risks, including those associated with drinking water.

Looking to the Future

As the twenty-first century begins, the challenges of water treatment have become more complex. Issues include the identification of new pathogens such as *Helicobacter pylori* and the noroviruses, new disinfection by-products such as *N*-nitrosodimethylamine (NDMA), and a myriad of chemicals, including personal care products, detergent by-products, and other consumer products. As analytical techniques improve, it is likely that these issues will grow, and the water quality engineer will face ever-increasing challenges.

1-3 Constituents of Emerging Concern

Contaminants and pathogens of emerging concern are by their very nature unregulated constituents that may pose a serious threat to human health. Consequently, they pose a serious obstacle to delivering the quality and quantity of water that the public demands. Furthermore, emerging contaminants threaten the development of more environmentally responsible water resources that do not rely on large water projects involving reservoirs and dams in more pristine environments. Creating acceptable water from water resources that are of lower quality because of contaminants of emerging concern is more expensive, and there is resistance to increased spending for public water supply projects (NRC, 1999).

Number of Possible Contaminants

The sheer number of possible contaminants is staggering. The CAS (Chemical Abstracts Service, a division of the American Chemical Society) Registry lists more than 55 million unique organic and inorganic chemicals (CAS, 2010a). In the United States, about 70,000 chemicals are used commercially and about 3300 are considered by the U.S. Environmental Protection Agency (EPA) to be high-volume production chemicals [i.e., are produced at a level greater than or equal to 454,000 kg/yr (1,000,000 lb/yr)]. The CAS also maintains CHEMLIST, a database of chemical substances that are the target of regulatory activity someplace in the world; this list currently contains more than 248,000 substances (CAS, 2010b).

Pharmaceuticals and Personal Care Products

Increasing interconnectedness between surface waters used for discharge of treated wastewater and as a source for potable water systems has created concern about whether trace contaminants can pass through the wastewater treatment system and enter the water supply. Many recent investigations have found evidence of low concentrations of pharmaceuticals and personal care products (PPCPs) and endocrine disrupting compounds (EDCs) in the source water for many communities throughout the United States and other developed nations.

The list of CECs is continually changing. Follow this link to see some recent chemicals of concern.
http://www.stantec.com/wt3r/1-2.html

Pharmaceuticals can enter the wastewater system by being excreted with human waste after medication is ingested or because of the common practice of flushing unused medication down the toilet. Pharmaceuticals include antibiotics, analgesics [painkillers such as aspirin, ibuprofen (Advil), acetaminophen (Tylenol)], lipid regulators (e.g., atorvastatin, the active ingredient in Lipitor), mood regulators (e.g., fluoxetine, the active ingredient in Prozac), antiepileptics (e.g., carbamazepine, the active ingredient in many epilepsy and bipolar disorder medications), and hundreds of other medications. Personal care products, which include cosmetics and fragrances, acne medications, insect repellants, lotions, detergents, and other products, can be washed from the skin and hair during washing or showering. Endocrine disrupting chemicals are chemicals that have the capability to interfere with the function of human hormones. EDCs include actual hormones, such as estrogens excreted by females after use of birth-control pills, or other compounds that mimic the function of hormones, such as bisphenol A. Studies have shown that some of these compounds are effectively removed by modern wastewater treatment processes, but others are not. Although the compounds are present at very low concentrations when they are detected, the public is concerned about the potential presence of these compounds in drinking water.

Nanoparticles

The manufacture of nanoparticles is a new and rapidly growing field. Nanoparticles are very small particles ranging from 1 to 100 nanometers (nm) used for applications such as the delivery of pharmaceuticals across the blood-brain barrier. Because nanomaterials are relatively new and the current market is small, a knowledge base of the potential health risks and environmental impacts of nanomaterials is lacking. As the manufacture of nanomaterials increases, along with the potential for discharge to the environment, more research to establish health risks and environmental impacts may be appropriate.

Other Constituents of Emerging Concern

In addition to the constituents listed above, other constituents of emerging concern include (1) fuel oxygenates (e.g., methyl *tert*-butyl ether, MTBE), (2) *N*-nitrosodimethylamine (NDMA), (3) perchlorate, (4) chromate, and (5) veterinary medications that originate from concentrated animal-feeding operations.

1-4 Evolution of Water Treatment Technology

To understand how the treatment methods discussed in this book developed, it is appropriate to consider their evolution. Most of the methods in use at the beginning of the twentieth century evolved out of physical observations (e.g., if turbid water is allowed to stand, a clarified liquid will develop as the particles settle) and the relatively recent (less than 120 years) recognition of the relationship between microorganisms in contaminated water and disease. A list of plausible methods for treating water at the beginning of the twentieth century was presented in a book by Hazen (1909) and is summarized in Table 1-2. It is interesting to note that all of the treatment methods reported in Table 1-2 are still in use today. The most important modern technological development in the field of water treatment not reflected in Table 1-2 is the use of membrane technology.

Table 1-2
Summary of methods used for water treatment early in the twentieth century

Treatment Method	Agent/Objectives
I. Mechanical separation	❑ By gravity—sedimentation
	❑ By screening—screens, scrubbers, filters
	❑ By adhesion—scrubbers, filters
II. Coagulation	❑ By chemical treatment resulting in drawing matters together into groups, thereby making them more susceptible to removal by mechanical separation but without any significant chemical change in the water
III. Chemical purification	❑ Softening—by use of lime
	❑ Iron removal
	❑ Neutralization of objectionable acids
IV. Poisoning processes	❑ Ozone
(now known as	❑ Sulfate of copper
disinfection processes)	❑ The object of these processes is to poison and kill objectionable organisms without at the same time adding substances objectionable or poisonous to the users of the water
V. Biological processes	❑ Oxidation of organic matter by its use as food for organisms that thereby effect its destruction
	❑ Death of objectionable organisms, resulting from the production of unfavorable conditions, such as absence of food (removed by the purification processes) and killing by antagonistic organisms
VI. Aeration	❑ Evaporation of gases held in solution that are the cause of objectionable tastes and odors
	❑ Evaporation of carbonic acid, a food supply for some kinds of growths
	❑ Supplying oxygen necessary for certain chemical purifications and especially necessary to support growth of water-purifying organisms
VII. Boiling	❑ Best household method of protection from disease-carrying waters

Source: Adapted from Hazen (1909).

Traditional Technologies

For the 100 years following the work of Fuller's team in Louisville in the late 1880s (see Table 1-1), the focus in the development of water treatment technology was on the further refinement of the technologies previously developed, namely coagulation, sedimentation, filtration, and disinfection with chlorine (see Fig. 1-1). There were numerous developments during that period, among them improvements in the coagulants available, improved understanding of the role of the flocculation process and the optimization of its design, improvements in the design of sedimentation basins, improvements in the design of filter media and in the filter rates that can be safely achieved, and improvements in the control of chlorination and chlorine residuals. These technologies have also been widely deployed, to the point where the vast majority of surface water supplies have treatment of this kind.

Introduction of Additional Treatment Technologies

A variety of new treatment technologies were introduced at various times during the twentieth century in response to more complex treatment goals. Ion exchange and reverse osmosis are processes that are able to remove a wide variety of inorganic species. A typical use for ion exchange is the removal of hardness ions (calcium and magnesium). Although ion exchange is typically expensive to implement at the municipal scale, the first large U.S. ion exchange facility was a 75.7 megaliter per day (75.7 ML/d) [20 million gallons per day (20 mgd)] softening plant constructed by the Metropolitan Water District of Southern California in 1946. The first commercial reverse osmosis plant provided potable water to Coalinga, California, in 1965 and had a capacity of 0.019 ML/d (0.005 mgd).

Aeration is accomplished by forcing intimate contact between air and water, most simply done by spraying water into the air, allowing the water to splash down a series of steps or platforms, or bubbling air into a tank of water. Early in the history of water treatment, aeration was employed to control tastes and odors associated with anaerobic conditions. The number and type of aeration systems have grown as more source waters have been contaminated with volatile organic chemicals.

Organic chemicals can be effectively removed by adsorption onto activated carbon. Adsorption using granular activated carbon was introduced in Hamm, Germany, in 1929 and Bay City, Michigan, in 1930. Powdered activated carbon was used as an adsorbent in New Milford, New Jersey, in 1930. During this time and the next few decades, the use of activated carbon as an adsorbent was primarily related to taste and odor control. In the mid-1970s, however, the increasing concern about contamination of source waters by industrial wastes, agricultural chemicals, and municipal discharges promoted the interest in adsorption for control of anthropogenic contaminants.

Satellite and aerial imagery can assist water treatment plant operators estimate TDS and TOC based on source water reflectivity

http://www.stantec.com/wt3r/1-3.html

Developments Requiring New Approaches and Technologies

During the last three decades of the twentieth century, three developments took place requiring new approaches to treatment. Two of these changes were rooted in new discoveries concerning water quality, and one was the development of a new technology that portends to cause dramatic change

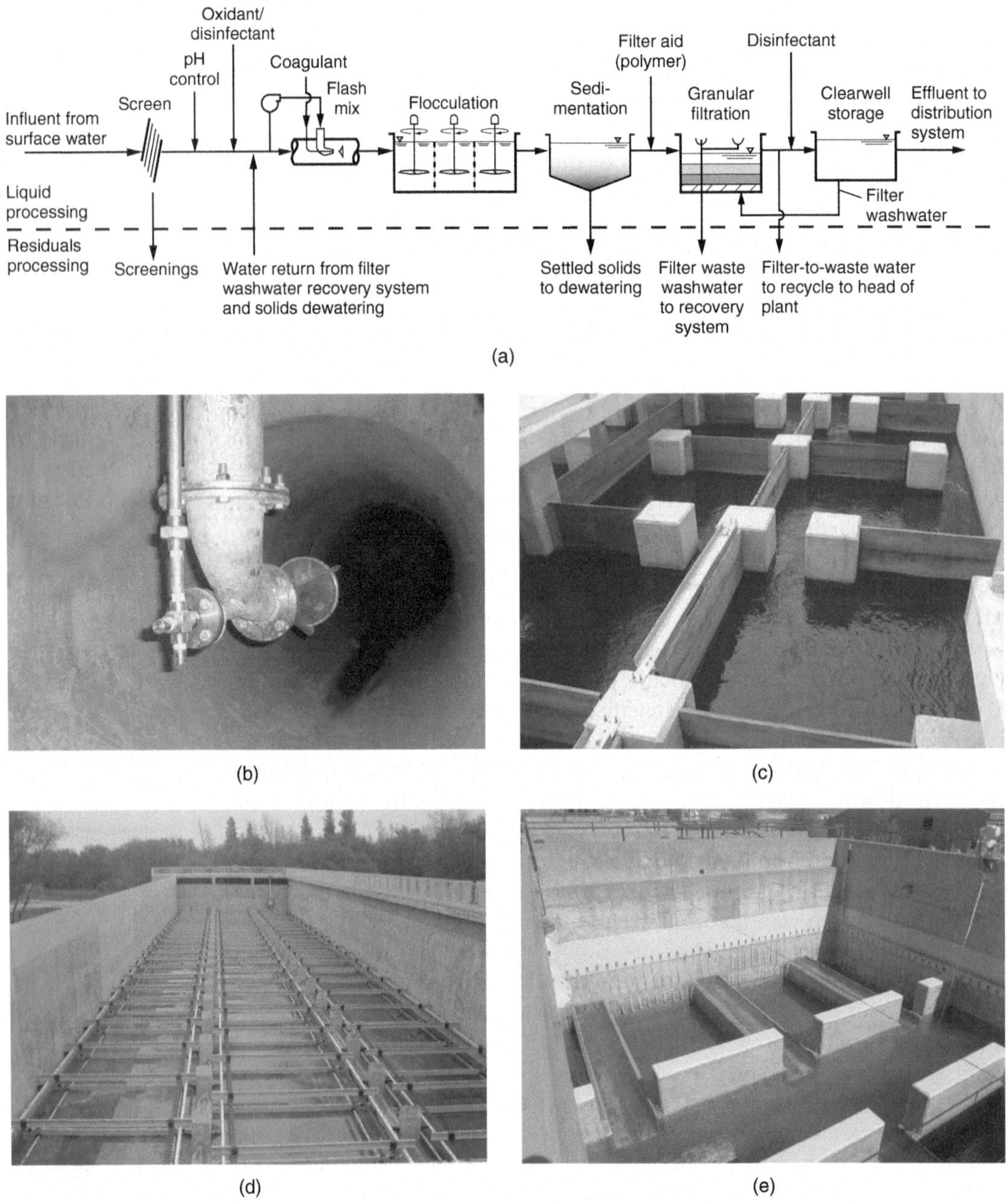

Figure 1-1
Views of conventional treatment technologies: (a) schematic flow diagram used for the treatment of surface water, (b) pumped diffusion flash mixer for chemical addition, (c) flocculation basin, (d) empty sedimentation basin, and (e) granular media filter.

in the effectiveness of water treatment. The first discovery concerning water quality was that the oxidants used for disinfecting water, particularly chlorine, react with the natural organic matter in the water supply to form chemical by-products, some of which are suspected carcinogens. The second discovery was that certain pathogenic microorganisms, namely *Giardia* and *Cryptosporidium*, can be of zoonotic origin and, therefore, can occur in a water supply that is completely free of wastewater contamination. The final and perhaps most significant change was the development of membrane filtration technologies suitable for the treatment of water on the scale required for domestic supply. Membrane technologies have the potential to completely reject pathogens by size exclusion, a possibility that could substantially improve the safety of drinking water. Further development and refinement of membrane technologies will be required before they reach their full potential.

Revolution Brought about by Use of Membrane Filtration

The first membranes were developed near the middle of the twentieth century but initially were only used in limited applications. In the late 1950s membranes began to be used in laboratory applications, most notably as an improvement in the coliform test. By the mid-1960s membrane filtration was widely used for beverages, as a replacement for heat pasteurization as a method of purification and microbiological stabilization. In virtually all of these applications the membranes were treated as disposable items. The idea of treating large volumes of drinking water in this manner seemed untenable. In the mid-1980s, researchers in both Australia and France began to pursue the idea of membrane filtration fibers that could be backwashed after each use, so that the membrane need not be disposed of but could be used on a continuous basis for a prolonged period of time. In the last decade of the twentieth century these products were commercialized, and by the turn of the twenty-first century there were numerous manufacturers of commercial membrane filtration systems and municipal water plants as large as 300 ML/d (80 mgd) were under construction (see Fig. 1-2). Membranes are arguably the most important development in the treatment of drinking water since the year 1900 because they offer the potential for complete and continuous rejection of microbiological contaminants on the basis of size exclusion.

1-5 Selection of Water Treatment Processes

To produce water that is safe to drink and aesthetically pleasing, treatment processes must be selected that, when grouped together, can be used to remove specific constituents. The most critical determinants in the selection of water treatment processes are the quality of the water source and the intended use of the treated water. The two principal water sources are groundwater and surface water. Depending on the hydrogeology of a basin, the levels of human activity in the vicinity of the source, and other

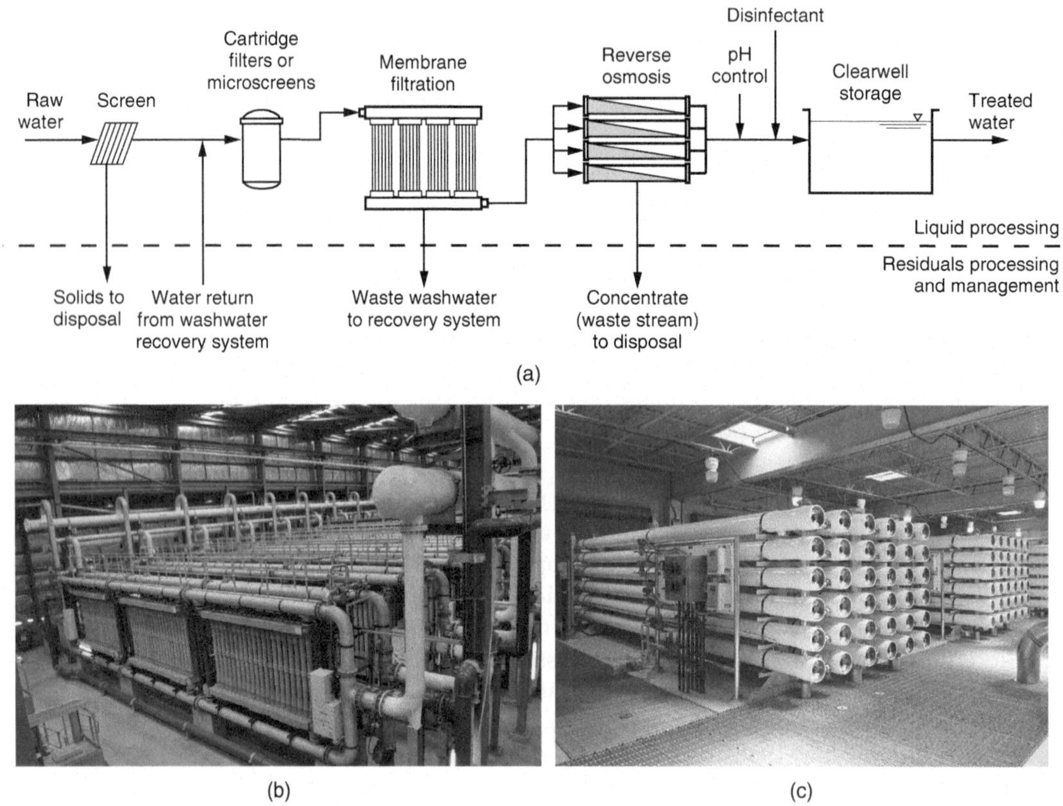

Figure 1-2
Views of membrane facilities for water treatment: (a) schematic flow diagram for a brackish water desalting plant using membrane filtration and reverse osmosis, (b) membrane filtration system, and (c) reverse osmosis system.

factors, a wide range of water qualities can be encountered. Surface waters typically have higher concentrations of particulate matter than groundwater, and groundwater often has increased concentrations of dissolved minerals due to the long contact times between subsurface water with rocks and minerals. Surface water may have more opportunity for exposure to anthropogenic chemicals.

Another major distinction is based on the level of dissolved salts or total dissolved solids (TDS) present in the water source. Water containing TDS less than 1000 mg/L is considered to be freshwater, and water with TDS between 1000 and 10,000 mg/L is considered to be brackish water. Freshwater is the most easily used for drinking water purposes, and brackish water can be used under specific circumstances with adequate treatment. Finally, the most abundant water source, the ocean, contains approximately 35,000 mg/L TDS and requires demineralization prior to use. Each of the predominant types of water sources, including natural or man-made lakes and rivers, requires a different treatment strategy.

(a) (b)

Figure 1-3
Views of pilot plant test installations: (a) test facilities for evaluation of a proprietary process (the MIEX process; see Chap.16) for the removal of natural organic matter before coagulation, flocculation, sedimentation, and filtration, and (b) reverse osmosis for the removal of dissolved constituents.

The steps that are typically involved in the selection and implementation of water treatment plants are

1. Characterization of the source water quality and definition of the treated water quality goals or standards
2. Predesign studies, including pilot plant testing (see Fig. 1-3), process selection, and development of design criteria
3. Detailed design of the selected alternative;
4. Construction
5. Operation and maintenance of the completed facility

These five steps may be performed as discrete steps or in combination and require input from a wide range of disciplines, including engineering, chemistry, microbiology, geology, architecture, and financial analysis. Each discipline plays an important role at various stages in the process. The predominant role, however, rests with professional engineers who carry the responsibility for the success of the water treatment process.

References

AWWA (1971) *Water Quality and Treatment: Handbook of Public Water Supply*, American Water Works Association, Denver, CO.

Baker, M. N. (1948) *The Quest for Pure Water*, American Water Works Association, New York.

Baker, M. N., and Taras, M. J. (1981) *The Quest for Pure Water: The History of the Twentieth Century*, Vols. 1 and 2, American Water Works Association, Denver, CO.

Blake, N. M. (1956) *Water for the Cities*, Syracuse University Press, Syracuse, NY.

CAS (2010a) http://www.cas.org/expertise/cascontent/registry/index.html.

CAS (2010b) http://www.cas.org/expertise/cascontent/regulated/index.html.

Clark, G. L., Price, L., Ahmed, R., Woodward, D. L., Melito, P. L., Rodgers, F. G., Jamieson, F., Ciebin, B., Li, A., and Ellis, A. (2003) "Characterization of Waterborne Outbreak-Associated *Campylobacter jejuni*, Walkerton, Ontario," Emerging Infect. Dis., 9, 10, 1232–1241.

Fox, K. R., Lytle, D. A. (1996) "Milwaukee's Crypto Outbreak: Investigation and Recommendations," Journal AWWA, 88, 9, 87–94.

Fuller, G. (1898) *Report on the Investigation into Purification of the Water of the Ohio River at Louisville, Kentucky*, D. Van Nostrand Co., New York.

Hazen, A. (1909) *Clean Water and How to Get It*, John Wiley & Sons, New York.

NRC (1999) *Identifying Future Drinking Water Contaminants*, Water Science and Technology Board, National Research Council, National Academy Press, Washington, DC.

Olsztynski, J. (1988) "Plagues and Epidemics," *Plumbing Mechanical Mag.*, 5, 5, 42–56.

Salvato, J. A. (1992) *Engineering and Sanitation*, 4th ed., John Wiley & Sons, New York.

Smith, T. (1893) A New Method for Determining Quantitatively the Pollution of Water by Fecal Bacteria, pp. 712–722 in *Thirteenth Annual Report for the Year 1892*, New York State Board of Health, Albany, NY.

Snow, J. (1855) *On the Mode of Communication of Cholera*, 2nd ed., J. Churchill, London.

2 Physical and Chemical Quality of Water

Stantec's Water Treatment: Principles and Design, Updated Third Edition. John C. Crittenden, R. Rhodes Trussell, David W. Hand, Kerry J. Howe, and George Tchobanoglous.

Terminology for Physical and Chemical Quality of Water

Term	Definition
Absorbance	Amount of light absorbed by the constituents in a solution.
Aggregate water quality indicators	Measured parameter values caused by a number of individual constituents.
Alkalinity	Measure of the ability of a water to resist changes in pH.
Colloids	Particles smaller than about 1 μm in size; although definitions vary, they are generally distinguished because they will not settle out of solution naturally.
Color	Reduction in clarity of water caused by the absorption of visible light by dissolved substances, including organic compounds (fulvic acid, humic acid) and inorganic compounds (iron, manganese).
Conductivity	Measure of the concentration of dissolved constituents based on their ability to conduct electrical charge.
Hydrogen bonding	Attractive interaction between a hydrogen atom of one water molecule and the unshared electrons of the oxygen atom in another water molecule.
Natural organic matter (NOM)	Complex matrix of organic chemicals present in all water bodies, originating from natural sources such as biological activity, secretions from the metabolic activity, and excretions from fish or other aquatic organisms.
Particles	Constituents in water larger than molecules that exist as a separate phase (i.e., as solids). Water with particles is a suspension, not a solution. Particles include silt, clay, algae, bacteria, and other microorganisms.
pH	Parameter describing the acid–base properties of a solution.

Term	Definition
Radionuclides	Unstable atoms that are transformed through the process of radioactive decay.
Suspended solids	See: particles
Synthetic organic compounds (SOCs)	Man-made (anthropogenic) organic synthetic chemicals. Some SOCs are volatile; others tend to stay dissolved in water instead of evaporating.
Total dissolved solids (TDS)	Total amount of ions in solution, analyzed by filtering out the suspended material, evaporating the filtrate, and weighing the remaining residue.
Total organic halogen	Total mass concentration of organically bound halogen atoms (X = Cl, Br, or I) present in water.
Trace constituents	Constituents (inorganic and organic) of natural waters found in the parts-per-billion to parts-per-trillion range.
Transmittance	Measure of the amount of light, expressed as a percentage, that passes through a solution. The percent transmittance effects the performance of ultraviolet (UV) disinfection processes.
Trihalomethane (THM)	One of a family of organic compounds named as derivative of methane. THMs are generally by-products of chlorination of drinking water that contains organic material.
Trihalomethane (THM) formation potential	Maximum tendency of the organic compounds in a given water supply to form THMs upon disinfection.
Turbidity	Reduction in clarity of water caused by the scattering of visible light by particles.

Naturally occurring water is a solution containing not only water molecules but also chemical matter such as inorganic ions, dissolved gases, and dissolved organics; solid matter such as colloids, silts, and suspended solids; and biological matter such as bacteria and viruses. The structure of water, while inherently simple, has unique physicochemical properties. These properties have practical significance for water supply, water quality, and water treatment engineers. The purpose of this chapter is to present background information on the physical and chemical properties of water, the units used to express the results of physical and chemical analyses, and the constituents found in water and the methods used to quantify them. Topics considered in this chapter include (1) the fundamental and engineering properties of water, (2) units of expression for chemical concentrations, (3) the physical aggregate characteristics of water, (4) the inorganic chemical constituents found in water, (5) the organic chemical constituents found in water, (6) taste and odor, (7) the gases found in water, and (8) the radionuclides found in water. All of the topics introduced in this chapter are expanded upon in the subsequent chapters as applied to the treatment of water.

2-1 Fundamental and Engineering Properties of Water

The fundamental and engineering properties of water are introduced in this section. The fundamental properties relate to the basic composition and structure of water in its various forms. The engineering properties of water are used in day-to-day engineering calculations.

Fundamental Properties of Water

The fundamental properties of water include its composition, dimensions, polarity, hydrogen bonding, and structural forms. Because of their importance in treatment process theory and design, polarity and hydrogen bonding are considered in the following discussion. Details on the other properties may be found in books on water chemistry and on a detailed website dedicated to water science and structure (Chapin, 2010).

POLARITY

The asymmetric water molecule contains an unequal distribution of electrons. Oxygen, which is highly electronegative, exerts a stronger pull on the shared electrons than hydrogen; also, the oxygen contains two unshared electron pairs. The net result is a slight separation of charges or dipole, with the slightly negative charge (δ^-) on the oxygen end and the slightly positive charge (δ^+) on the hydrogen end. Attractive forces exist between one polar molecule and another such that the water molecules tend to orient themselves with the hydrogen end of one directed toward the oxygen end of another.

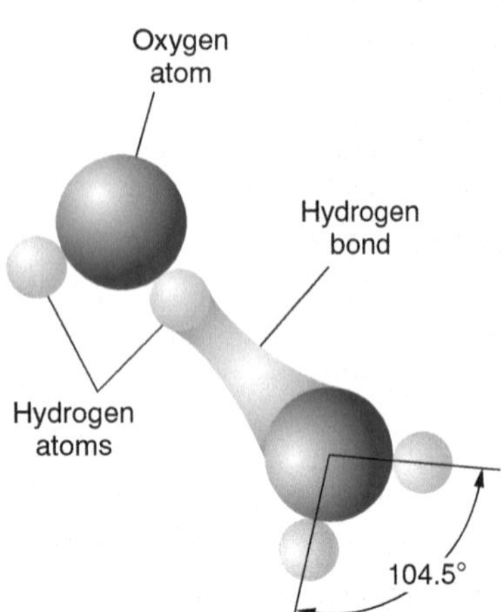

Figure 2-1
Hydrogen bonding between water molecules.

HYDROGEN BONDING

The attractive interaction between a hydrogen atom of one water molecule and the unshared electrons of the oxygen atom in another water molecule is known as a hydrogen bond, represented schematically on Fig. 2-1. Estimates of hydrogen bond energy between molecules range from 10 to 40 kJ/mol, which is approximately 1 to 4 percent of the covalent O–H bond energy within a single molecule (McMurry and Fay, 2003). Hydrogen bonding causes stronger attractive forces between water molecules than the molecules of most other liquids and is responsible for many of the unique properties of water.

Engineering Properties of Water

Compared to other species of similar molecular weight, water has higher melting and boiling points, making it a liquid rather than a gas under ambient conditions. Hydrogen bonding, as described above, can be used to explain the unique properties of water including density, high heat capacity, heat of formation, heat of fusion, surface tension, and viscosity of water. Examples of the unique properties of water include its capacity to dissolve a variety of materials, its effectiveness as a heat exchange fluid, its high density and pumping energy requirements, and its viscosity. In dissolving or suspending materials, water gains characteristics of biological, health-related, and aesthetic importance. The type, magnitude, and interactions of these materials affect the properties of water, such as its potability, corrosivity, taste, and odor. As will be demonstrated in subsequent chapters, technology now exists to remove essentially all of the dissolved and suspended components of water. The principal engineering properties encountered in environmental engineering and used throughout this book are reported in Table 2-1. The typical numerical values given in Table 2-1 are to provide a frame of reference for the values that are reported in the literature.

Table 2-1
Engineering properties of water

		Unit		Value[a]		
Property	Symbol	SI	U.S. Customary	SI	U.S. Customary	Definition/Notes
Boiling point	bp	°C	°F	100	212	Temperature at which vapor pressure equals 1 atm; high value for water keeps it in liquid state at ambient temperature.
Conductivity	κ	μS/m	μS/m	5.5	5.5	Pure water is not a good conductor of electricity; dissolved ions increase conductivity.
Density	ρ	kg/m³	slug/ft³	998.2	1.936	
Dielectric constant	ε_r	unitless	unitless	80.2	80.2	Measure of the ability of a solvent to maintain a separation of charges; high value for water indicates it is a very good solvent.
Dipole moment	p	C·m	D (debye)	6.186×10^{-30}	1.855	Measure of the separation of charge within a molecule; high value for water indicates it is very polar.

(Continued)

Table 2-1 (*Continued*)

Property	Symbol	Unit		Value[a]		Definition/Notes
		SI	U.S. Customary	SI	U.S. Customary	
Enthalpy of formation	ΔH_f	kJ/mol	btu/lb_m	−286.5	−6836	Energy associated with the formation of a substance from the elements.
Enthalpy of fusion[b]	ΔH_{fus}	kJ/mol	btu/lb_m	6.017	143.6	Energy associated with the conversion of a substance between the solid and liquid states (i.e., freezing or melting).
Enthalpy of vaporization[c]	ΔH_v	kJ/mol	btu/lb_m	40.66	970.3	Energy associated with the conversion of a substance between the liquid and gaseous states (i.e., vaporizing or condensing); high value for water makes distillation very energy intensive.
Heat capacity[d]	c_p	J/mol•°C	btu/lb_m • °F	75.34	0.999	Energy associated with raising the temperature of water by one degree; high value for water makes it impractical to heat or cool water for municipal treatment purposes.
Melting point	mp	°C	°F	0	32	
Molecular weight	MW	g/mol[e]	g/mol[e]	18.016	18.016	Also known as molar mass.
Specific weight	γ	kN/m^3	lb_f/ft^3	9.789	62.37	
Surface tension	σ	N/m	lb_f/ft	0.0728	0.00499	
Vapor pressure	p_v	kN/m^2	lb_f/in^2	2.339	0.34	
Viscosity, dynamic	μ	$N \bullet s/m^2$	$lb_f \bullet s/ft^2$	1.002×10^{-3}	2.089×10^{-5}	
Viscosity, kinematic	υ	m^2/s	ft^2/s	1.004×10^{-6}	1.081×10^{-5}	

[a]All values for pure water at 20°C (68°F) and 1 atm pressure unless noted otherwise.
[b]At the melting point (0°C).
[c]At the boiling point (100°C).
[d]Often called the molar heat capacity when expressed in units of J/mol •°C and specific heat capacity or specific heat when expressed in units of J/g •°C.
[e]Molecular weight has units of Daltons (Da) or atomic mass units (AMU) when expressed for a single molecule (i.e., one mole of carbon-12 atoms has a mass of 12 g and a single carbon-12 atom has a mass of 12 Da or 12 AMU).

2-2 Units of Expression for Chemical Concentrations

Water quality characteristics are often classified as physical, chemical (organic and inorganic), or biological and then further classified as health related or aesthetic. To characterize water effectively, appropriate sampling and analytical procedures must be established. The purpose of this section is to review briefly the units used for expressing the physical and chemical characteristics of water. The basic relationships presented in this section will be illustrated and expanded upon in subsequent chapters. Additional details on the subject of sampling, sample handling, and analyses may be found in Standard Methods (2005).

Commonly used units for the amount or concentration of constituents in water are as follows:

1. *Mole*:

$$6.02214 \times 10^{23} \text{ elementary entities (molecules, atoms, etc.) of a substance}$$

$$1.0 \text{ mole of compound} = \text{molecular weight of compound, g} \quad (2\text{-}1)$$

2. *Mole fraction*: The ratio of the amount (in moles) of a given solute to the total amount (in moles) of all components in solution (including the solvent) is expressed as

$$x_B = \frac{n_B}{n_A + n_B + n_C + \cdots + n_N} \quad (2\text{-}2)$$

where x_B = mole fraction of solute B
n_A = moles of solute A
n_B = moles of solute B
n_C = moles of solute C
⋮
n_N = moles of solvent N

The application of Eq. 2-2 is illustrated in Example 2-1.

3. *Molarity (M)*:

$$M, \text{ mol/L} = \frac{\text{mass of solute, g}}{(\text{molecular weight of solute, g/mol})(\text{volume of solution, L})} \quad (2\text{-}3)$$

4. *Molality (m)*:

$$m, \text{ mol/kg} = \frac{\text{mass of solute, g}}{(\text{molecular weight of solute, g/mol})(\text{mass of solution, kg})} \quad (2\text{-}4)$$

Example 2-1 Determination of molarity and mole fractions

Determine the molarity and the mole fraction of a 1-L solution containing 20 g sodium chloride (NaCl) at 20°C. From the periodic table and reference books, it can be found that the molar mass of NaCl is 58.45 g/mol and the density of a 20 g/L NaCl solution is 1.0125 kg/L.

Solution

1. The molarity of the NaCl solution is computed using Eq. 2-3

$$[\text{NaCl}] = \frac{20\ \text{g}}{(58.45\ \text{g/mol})(1.0\ \text{L})} = 0.342\ \text{mol/L} = 0.342\ \text{M}$$

2. The mole fraction of the NaCl solution is computed using Eq. 2-2
   a. The amount of NaCl (in moles) is

$$n_{\text{NaCl}} = \frac{20\text{g}}{58.45\text{g/mol}} = 0.342\ \text{mol}$$

   b. From the given solution density, the total mass of the solution is 1012.5 g, so the mass of the water in the solution is 1012.5g–20 g = 992.5 g and the amount of water (in moles) is

$$n_{H_2O} = \frac{992.5\text{g}}{18.02\ \text{g/mol}} = 55.08\ \text{mol}$$

   c. The mole fraction of NaCl in the solution is

$$x_{\text{NaCl}} = \frac{n_{\text{NaCl}}}{n_{\text{NaCl}} + n_{H_2O}} = \frac{0.342\text{mol}}{0.342\text{mol} + 55.07\text{mol}} = 6.17 \times 10^{-3}$$

Comment

The molar concentration of pure water is calculated by dividing the density of water by the MW of water; i.e., 1000 g/L divided by 18 g/mol equals 55.56 mol/L. Because the amount of water is so much larger than the combined values of the other constituents found in most waters, the mole fraction of constituent A is often approximated as $x_A \approx (n_A/55.56)$. If this approximation had been applied in this example, the mole fraction of NaCl in the solution would have been computed as 6.16×10^{-3}.

5. *Mass concentration:*

$$\text{Concentration, g/m}^3 = \frac{\text{mass of solute, g}}{\text{volume of solution, m}^3} \tag{2-5}$$

Note that $1.0\ \text{g/m}^3 = 1.0\ \text{mg/L}$.

6. *Normality (N)*:

$$N, \text{eq} / \text{L} = \frac{\text{mass of solute, g}}{(\text{equivalent weight of solute, g} / \text{eq})\ (\text{volume of solution, L})} \tag{2-6}$$

where

$$\text{Equivalent weight of solute, g} / \text{eq} = \frac{\text{molecular weight of solute, g} / \text{mol}}{Z, \text{eq} / \text{mol}} \tag{2-7}$$

For most compounds, Z is equal to the number of replaceable hydrogen atoms or their equivalent; for oxidation–reduction reactions, Z is equal to the change in valence. Also note that 1.0eq/m^3 = 1.0meq/L.

7. *Parts per million* (ppm):

$$\text{ppm} = \frac{\text{mass of solute, g}}{10^6 \text{ g of solution}} \tag{2-8}$$

Also,

$$\text{ppm} = \frac{\text{concentration of solute, g} / \text{m}^3}{\substack{\text{specific gravity of solution} \\ \text{(density of solution divided by density of water)}}} \tag{2-9}$$

8. *Other units*:

ppm_m = parts per million by mass (for water ppm_m = g/m^3 = mg/L)
ppm_v = parts per million by volume
ppb = parts per billion
ppt = parts per trillion

$$\text{Also, } 1\ \text{g (gram)} = 1 \times 10^3\ \text{mg (milligram)} = 1 \times 10^6\ \mu\text{g (microgram)}$$
$$= 1 \times 10^9\ \text{ng (nanogram)} = 1 \times 10^{12}\ \text{pg (picogram)}.$$

2-3 Physical Aggregate Characteristics of Water

Most first impressions of water quality are based on physical rather than chemical or biological characteristics. Water is expected to be clear, colorless, and odorless (Tchobanoglous and Schroeder, 1985). Most natural waters will contain some material in suspension typically comprised of inorganic soil components and a variety of organic materials derived from nature. Natural waters are also colored by exposure to decaying organic material. Water from slow-moving streams or eutrophic water bodies will often contain colors and odors. These physical parameters are known as *aggregate characteristics* because the measured value is caused by a number of individual constituents. Parameters commonly used to quantify the aggregate physical characteristics include (1) absorption/transmittance,

(2) turbidity, (3) number and type of particles, (4) color, and (5) temperature. Taste and odor, sometimes identified as physical characteristics, are considered in Sec. 2-6.

Absorbance and Transmittance

The absorbance of a solution is a measure of the amount of light that is absorbed by the constituents in a solution at a specified wavelength. According to the Beer–Lambert law, the amount of light absorbed by water is proportional to the concentration of light-absorbing molecules and the path length the light takes in passing through water, regardless of the intensity of the incident light. Because even pure water will absorb incident light, a sample blank (usually distilled water) is used as a reference. Absorbance is given by the relationship

$$\log\left(\frac{I}{I_0}\right) = -\varepsilon(\lambda)Cx = -k_A(\lambda)x = -A(\lambda) \qquad (2\text{-}10)$$

where
- I = intensity of light after passing through a solution of known depth containing constituents of interest at wavelength λ, mW/cm^2
- I_0 = intensity of incident light after passing through a blank solution (i.e., distilled water) of known depth (typically 1.0 cm) at wavelength λ, mW/cm^2
- λ = wavelength, nm
- $\varepsilon(\lambda)$ = molar absorptivity of light-absorbing solute at a wavelength λ, L/mol · cm
- C = concentration of light-absorbing solute, mol/L
- x = length of light path, cm
- $k_A(\lambda) = \varepsilon(\lambda)\ C$ = absorptivity at wavelength λ, cm^{-1}
- $A(\lambda) = \varepsilon(\lambda)\ Cx$ = absorbance at wavelength λ, dimensionless

If the left-hand side of Eq. 2-10 is expressed as a natural logarithm, then the right-hand side of the equation must be multiplied by 2.303 because the absorbance coefficient (also known as the extinction coefficient) is determined in base 10. Absorbance is measured using a spectrophotometer, as illustrated on Fig. 2-2. Typically, a fixed sample path length of 1.0 cm is used. The absorbance $A(\lambda)$ is unitless but is often reported in units of reciprocal centimeters, which corresponds to absorptivity $k_A(\lambda)$. If the

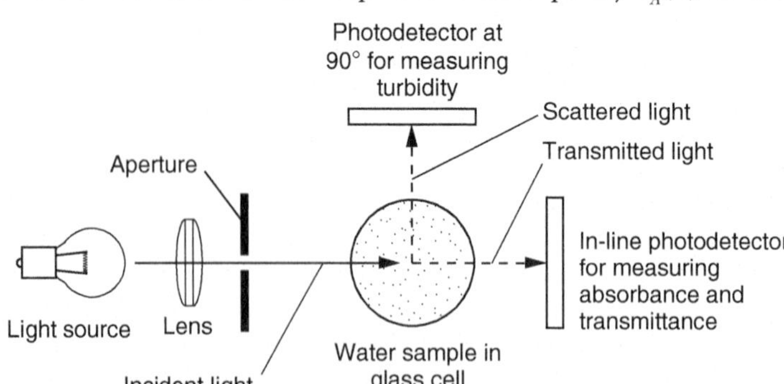

Figure 2-2
Schematic of a spectrophotometer used to measure absorbance and turbidity.

length of the light path is 1 cm, absorptivity is equal to the absorbance. The absorbance of water is typically measured at a wavelength of 254 nm. Typical absorbance values for various waters at $\lambda = 254$ are given in Table 13-10. The application of Eq. 2-10 is illustrated in the following example.

Example 2-2 Determine average UV intensity

If the intensity of the UV radiation measured at the water surface in a Petri dish is 15 mW/cm^2, determine the average UV intensity to which a sample will be exposed if the depth of water in the Petri dish is 12 mm (1.2 cm). Assume the absorptivity $k_A(\lambda) = 0.1$/cm.

Solution

1. Develop the equation to determine the average intensity.
   a. The definition sketch for this problem is given below.

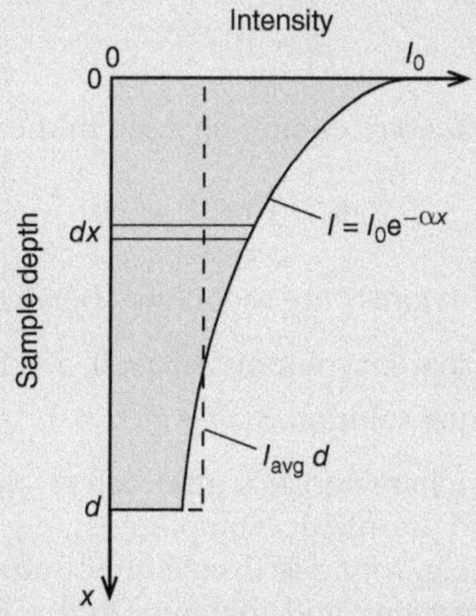

where

$$\alpha = 2.303 k_A(\lambda)$$

   b. Develop the required equation:

$$I_{avg} = \int_0^d I_0 e^{-\alpha x} dx = -\frac{I_0}{\alpha} e^{-\alpha x}\Big|_0^d$$

$$= -\frac{I_0}{d\alpha} e^{\alpha d} + \frac{I_0}{\alpha} = \frac{I_0}{\alpha}\left(1 - e^{-\alpha d}\right)$$

$$I_{avg} = \frac{I_0}{\alpha d}\left(1 - e^{-\alpha d}\right)$$

2. Compute the average intensity for a depth of 12 mm (1.2 cm):
   a. Assume $k_A(\lambda) = 0.1/\text{cm}$
   b. $\alpha = 2.303\ k_A(\lambda) = 2.303\ (0.1/\text{cm}) = 0.2303/\text{cm}$
   c. Solve for I_{avg}

$$I_{avg} = \frac{I_0}{\alpha d}\left(1 - e^{-\alpha d}\right) = \frac{15\ \text{mW}/\text{cm}^2}{(0.2303/\text{cm})(1.2\ \text{cm})}\left[1 - e^{-(0.2303)(1.2)}\right]$$
$$= 13.1\ \text{mW}/\text{cm}^2$$

The transmittance of a solution is defined as

$$\text{Transmittance, } T,\% = \left(\frac{I}{I_0}\right) \times 100 \tag{2-11}$$

Thus, the transmittance at a given wavelength can also be derived from absorbance measurements using the relationship

$$T = 10^{-A(\lambda)} \tag{2-12}$$

The term *percent transmittance*, commonly used in the literature, is given as

$$T,\% = 10^{-A(\lambda)} \times 100 \tag{2-13}$$

The extreme values of A and T are as follows (Delahay, 1957):

For a perfectly transparent solution $A(\lambda) = 0$, $T = 1$.

For a perfectly opaque solution $A(\lambda) \rightarrow \infty$, $T = 0$.

The principal water characteristics that affect the percent transmittance include selected inorganic compounds (e.g., copper and iron), organic compounds (e.g., organic dyes, humic substances, and aromatic compounds such as benzene and toluene), and small colloidal particles (≤0.45μm). If samples contain particles larger than 0.45 μm, the sample should be filtered before transmittance measurements are made. Of the inorganic compounds that affect transmittance, iron is considered to be the most important with respect to UV light absorbance because dissolved iron can absorb UV light directly. Organic compounds containing double bonds and aromatic functional groups can also absorb UV light. Absorbance values for a variety of compounds are given in the on-line resources for this text at the URL listed in App. E. The reduction in transmittance observed in surface waters during storm events is often ascribed to the presence of humic substances and particles from runoff, wave action, and stormwater flows (Tchobanoglous et al., 2003).

Turbidity

Turbidity in water is caused by the presence of suspended particles that reduce the clarity of the water. Turbidity is defined as "an expression of the optical property that causes light to be scattered and absorbed rather than transmitted with no change in direction or flux level through the sample" (Standard Methods, 2005). Turbidity measurements require a light source (incandescent or light-emitting diode) and a sensor to measure the scattered light. As shown on Fig. 2-2, the scattered light sensor is located at 90° to the light source. The measured turbidity increases as the intensity of the scattered light increases. Turbidity is expressed in nephelometric turbidity units (NTU).

It is important to note that the scattering of light caused by suspended particles will vary with the size, shape, refractive index, and composition of the particles. Also, as the number of particles increases beyond a given level, multiple scattering occurs, and the absorption of incident light is increased, causing the measured turbidity to decrease (Hach, 2008). The spatial distribution and intensity of the scattered light, as illustrated on Fig. 2-3, will depend on the size of the particle relative to the wavelength of the light source. For particles less than one-tenth of the wavelength of the incident light, the scattering of light is fairly symmetrical. As the particle size increases relative to the wavelength of the incident light, the light reflected from different parts of the particle creates interference patterns that are additive in the forward direction (Hach, 2008). Also, the intensity of the scattered light will vary with the wavelength of the incident light. For example, blue light will be scattered more than red light. Based on these considerations, turbidity measurements tend to be more sensitive to

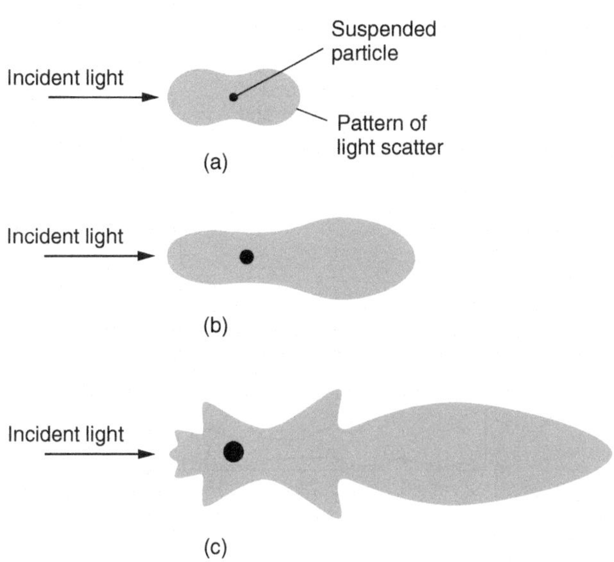

Figure 2-3
Light-scattering patterns for different particle sizes that occur when measuring turbidity. (Adapted from Hach, 2008).

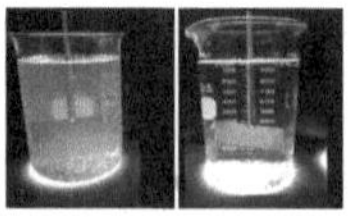

Turbidity is an essential parameter for safe water.

http://www.stantec.com/wt3r/2-1.html

particles in the size range of the incident-light wavelength (0.3 to 0.7 μm for visible light). A further complication with turbidity measurements is that some particles such as carbon black will essentially absorb most of the light and only scatter a minimal amount of the incident light.

Depending on the water source, turbidity can be the most variable of the water quality parameters of concern in drinking water supplies. Turbidity measurements are useful for comparing different water sources or treatment facilities and are used for process control and regulatory compliance. Increases in turbidity measurements are often used as an indicator for increased concentrations of water constituents, such as bacteria, *Giardia* cysts, and *Cryptosporidium* oocysts.

In lakes or reservoirs, turbidity is frequently stable over time and ranges from about 1 to 20 NTU, excluding storm events. Turbidity in rivers is more variable due to storm events, runoff, and changes in flow rate in the river. Turbidity in rivers can range from under 10 to over 4000 NTU. Streams and rivers where the turbidity can change by several hundred NTU in a matter of hours (see Fig. 2-4) are often described as "flashing" because of the rapid change in the turbidity. In such rivers, careful turbidity monitoring is critical for successful process control. The regulatory standard for turbidity in finished water is 0.3 NTU, and many water treatment facilities have a treatment goal of <0.1 NTU, which is near the detection limit for turbidity meters.

Particles

Particles are defined as finely divided solids larger than molecules but generally not distinguishable individually by the unaided eye, although clumps

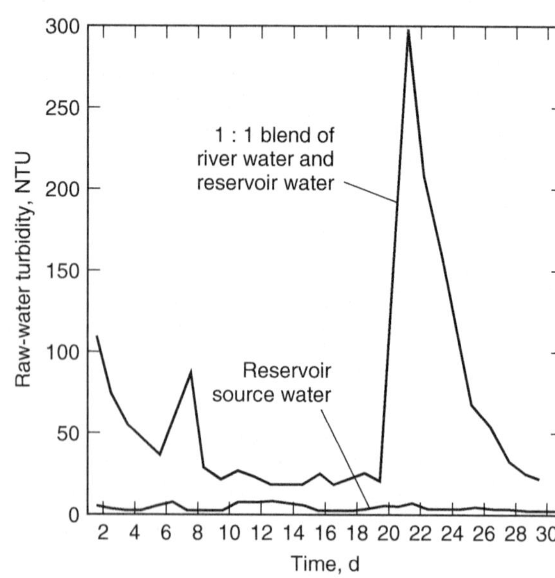

Figure 2-4
Observed variation in raw-water turbidity values. (Adapted from James M. Montgomery, 1981).

of particles are often encountered. It should be noted that with 20–20 vision it is possible to resolve a particle size of about 37 μm at a distance of 0.3 m. Particles in water are important for a variety of reasons, including their impact on treatment processes and the potential health impacts of pathogen-associated particles. Particles in water may be classified according to their source, size, chemical structure, electrical charge characteristics, and water-solid interface characteristics. The source, size, shape, number and distribution, and quantification of particles is considered in the following discussion. The electrical properties of particles and particle interactions are considered in Chap. 9. The impact of particles in water on key water treatment processes, that is, coagulation, sedimentation, granular filtration, membrane filtration, and disinfection, is considered in Chaps. 9, 10, 11, 12, and 13, respectively.

SOURCE OF PARTICLES IN WATER

The sources of particles in water are summarized in Table 2-2, along with the sources of chemical constituents and gases. As reported in Table 2-2, the principal natural sources of particles in water are soil-weathering processes and biological activity. Clays and silts are produced by weathering. Algae, bacteria, and other higher microorganisms are the predominant types of particles produced biologically. Some particles have both natural and anthropogenic sources, a notable example being asbestos fibers. Industrial and agricultural activities tend to augment these natural sources by increasing areas of runoff through cultural eutrophication, the increase in the rate of natural eutrophication as a result of human activity, or direct pollution with industrial residues. Particles may be transported into water through direct erosion from terrestrial environments, be suspended due to turbulence and mixing in water, or form in the water column during biological activity or chemical precipitation or through atmospheric deposition.

SIZE CLASSIFICATION OF PARTICLES

The size of particles in water considered in this text is typically in the range of 0.001 to 100 μm. Suspended particles are generally larger than 1.0 μm. The size of colloidal particles will vary from about 0.001 to 1 μm depending on the method of quantification. It should be noted that some researchers have classified the size range for colloidal particles as varying from 0.0001 or less to 1 μm. In practice, the distinction between colloidal and suspended particles is blurred because the suspended particles that can be removed by gravity settling will depend on the design of the sedimentation facilities. Some standard analytical procedures operationally define dissolved material as that which will pass through a 0.45 μm filter. In practice, however, colloids as small as 0.001 μm can behave as particles and affect water quality and treatment processes as particles rather than dissolved substances. A suspension comprised of particles of one size is

Table 2-2
Summary of important particulate, chemical, and biological constituents found in water according to their source

	Particulate constituents		Ionic and Dissolved Constituents		
Source	**Colloidal**	**Suspended**	**Positive ions**	**Negative ions**	**Gases and Neutral Species**
Contact of water with minerals, rocks, and soil (e.g., weathering)	Clay Silica (SiO_2) Ferric oxide (Fe_2O_3) Aluminum oxide (Al_2O_3) Manganese dioxide (MnO_2)	Clay, silt, sand, and other inorganic soils	Calcium (Ca^{2+}) Iron (Fe^{2+}) Magnesium (Mg^{2+}) Manganese (Mn^{2+}) Potassium (K^+) Sodium (Na^+) Zinc (Zn^{2+})	Bicarbonate (HCO^-) Borate ($H_2BO_3{}^-$) Carbonate ($CO_3{}^{2-}$) Chloride (Cl^-) Fluoride (F^-) Hydroxide (OH^-) Nitrate ($NO_3{}^-$) Phosphate ($PO_4{}^{3-}$) Sulfate ($SO_4{}^{2-}$)	Carbon dioxide (CO_2) Silicate (H_4SiO_4)
Rain in contact with atmosphere			Hydrogen (H^+)	Bicarbonate (HCO^-) Chloride (Cl^-) Sulfate ($SO_4{}^{2-}$)	Carbon dioxide (CO_2) Nitrogen (N_2) Oxygen (O_2) Sulfur dioxide (SO_2)
Decomposition of organic matter in environment	Various organic polymers	Cell fragments	Ammonium ($NH_4{}^+$) Hydrogen (H^+) Sodium (Na^+)	Bicarbonate (HCO–) Chloride (Cl^-) Hydroxide (OH^-) Nitrate ($NO_3{}^-$) Nitrite ($NO_2{}^-$) Sulfide (HS^-) Sulfate ($SO_4{}^{2-}$)	Ammonia (NH_3) Carbon dioxide (CO_2) Hydrogen sulfide (H_2S) Hydrogen (H_2) Methane (CH_4) Nitrogen (N_2) Oxygen (O_2) Silicate (H_4SiO_4)

Living organisms	Bacteria, algae, viruses, etc.	Algae, diatoms, minute animals, fish, etc.	—	—	Ammonia (NH_3) Carbon dioxide (CO_2) Hydrogen sulfide (H_2S) Hydrogen (H_2) Methane (CH_4) Nitrogen (N_2) Oxygen (O_2)
Municipal, industrial, and agricultural sources and other human activity	Inorganic and organic solids, constituents causing color, chlorinated organic compounds, bacteria, worms, viruses, etc.	Clay, silt, grit, and other inorganic solids; organic compounds; oil; corrosion products; etc.	Inorganic ions, including a variety of anthropogenic compounds and heavy metals	Inorganic ions, including a variety of anthropogenic compounds, organic molecules, color, etc.	Chlorine (Cl_2) Sulfur dioxide (SO_2)

Source: Adapted, in part, from Tchobanoglous and Schroeder (1985).

called monodispersed and a suspension with a variety of particle sizes is called heterodispersed (typical of natural waters).

Many water treatment processes are designed to remove particles based on sedimentation and size exclusion. The type and size of various waterborne particles and processes used for measurement and removal are presented on Fig. 2-5. As shown on Fig. 2-5, conventional treatment processes such as sedimentation and depth filtration alone are not sufficient for the removal of all water constituents; however, with the addition of coagulation and flocculation, the effective range of these treatment processes is greatly extended.

PARTICLE SHAPE

Particle shapes found in water can be described as spherical, semispherical, ellipsoids of various shapes (e.g., prolate and oblate), rods of various length and diameter, disk and disk-like, strings of various lengths, and random coils. Inorganic particles are typically defined by the dimensions of their long, intermediate, and short axes and the ratio of the intermediate-to-long and the short-to-intermediate diameters. Because of the many different particle shapes, the nominal or equivalent particle diameter is used (Dallavalle, 1948). Large organic molecules are often found in the form of coils that may be compressed, uncoiled, or almost linear. The shape of some larger particles is often described as fractal. The particle shape will vary depending on the characteristics of the source water.

PARTICLE QUANTIFICATION

Methods used for the quantification and analysis of particulate material include gravimetric techniques, electronic particle size counting, and microscopic observation. Although regulations concerning particle concentrations are typically based on turbidity measurements, monitoring particle counts throughout a treatment process can aid in understanding and controlling the process. Also, as noted above, turbidity measurements cannot be correlated to any quantifiable particle characteristics. While particle quantification may be useful for evaluating a treatment process, except for microscopic observation, these methods cannot be used reliably for determining the source or type of particle (e.g., distinguish between a viable cyst and a colloid). In addition, due to the limitations of particle analysis methods, the use of more than one method is recommended when assessing water quality data.

Gravimetric techniques

The total mass of particles may be estimated by filtering a volume of water through a membrane of known weight and pore size. Filtration of the same water sample through a series of membranes with incrementally decreasing pore sizes is known as serial filtration. Serial filtration may be used to determine an approximate particle size distribution (Levine et al., 1985).

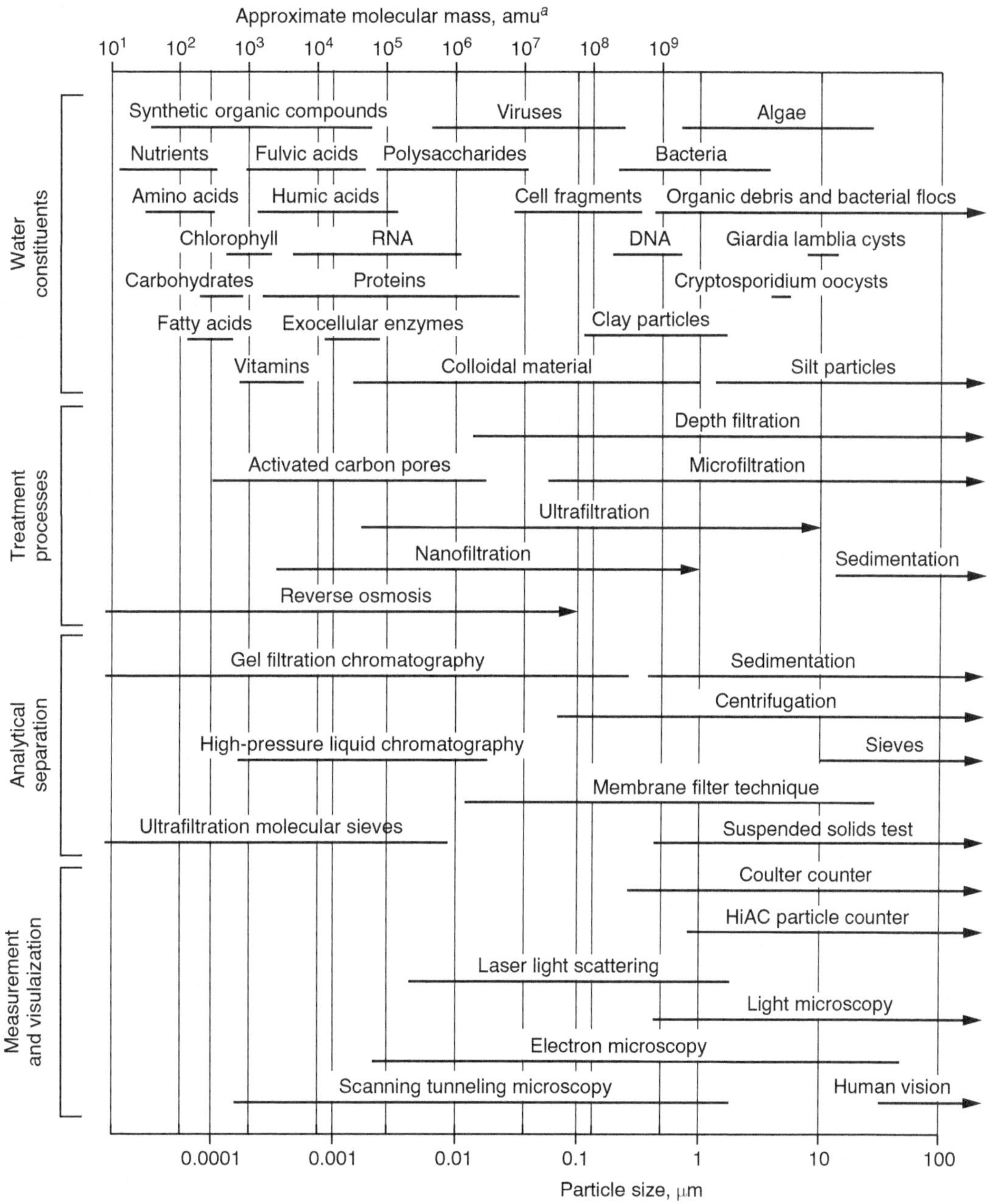

[a]An amu is an atomic mass unit (also known as a dalton, Da) and is equal to 1.66054×10^{-24} g.

Figure 2-5
Characterization of particulate matter in natural water by type and size, appropriate treatment methods, analytical separation methods, and measurement techniques. (Adapted from Tchobanoglous et al., 2003).

Particle size distribution may also be measured using electronic particle-counting devices, as discussed below.

Electronic particle size counting
Particle concentration measurements provide more specific information about the size and number of particles in a water sample. Electronic particle size counters estimate the particle size concentration by either (1) passing a water sample through a calibrated orifice and measuring the change in conductivity (see Fig. 2-6) or (2) passing the sample through a laser beam and measuring the change in intensity due to light scattering. The change in conductivity or light intensity is correlated to the diameter of an equivalent sphere. Particle counters have sensors available in different size ranges, such as 1.0 to 60 μm or 2.5 to 150 μm, depending on the manufacturer and application. Particle counts are typically measured and recorded in about 10 to 20 subranges of the sensor range. Typical particle size counters are shown on Fig. 2-7. A comparison of analytical techniques used for particle size analysis is presented in Table 2-3. Particle counts may also be used as an indicator of *Giardia* and *Cryptosporidium* cysts from water (LeChevallier and Norton, 1992, 1995).

Microscopic observation
The use of microscopic observation allows for the determination of particle size counts and, in some cases, for more rigorous identification of a particle's origin than is possible with other analysis techniques. A measured

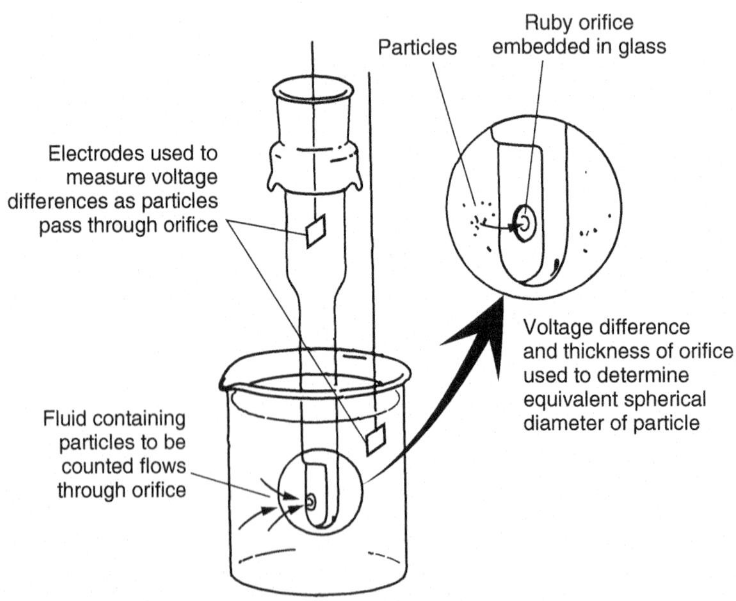

Figure 2-6
Typical particle-counting chamber used to enumerate particles in water using voltage difference to determine the size of an equivalent spherical particle. (Adapted from Tchobanoglous et al., 2003).

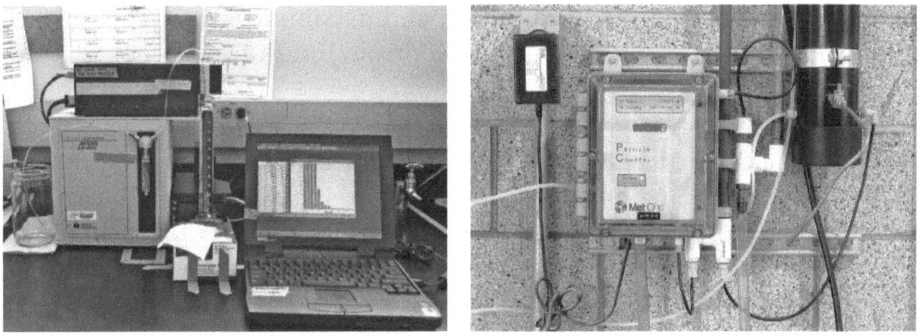

(a) (b)

Figure 2-7
Typical examples of particle size counters are (a) laboratory type connected to a computer (the sample to be analyzed is being withdrawn from the graduated cylinder) and (b) field type used to monitor the particle size distribution from a microfiltration plant.

Table 2-3
Analytical techniques used for analysis of particles in water

Technique	Typical Size Range, μm
Microscopy	
Light	0.2–>100
Transmission electron	0.0002–>0.1
Scanning electron	0.002–50
Image analysis	0.2–>100
Particle counting	
Conductivity difference	0.2–>100
Dynamic light scattering	0.0003–5
Equivalent light scattering	0.005–>100
Light obstruction (blockage)	0.2–>100
Light diffraction	0.3–>100
Separation	
Centrifugation	0.08–>100
Field flow fractionation	0.09–>100
Gel filtration chromatography	<0.0001–>100
Gravitation photosedimentation	0.1 –>100
Sedimentation	0.05–>100
Membrane filtration	0.0001–1

Source: Adapted from Levine et al. (1985).

volume of sample is placed in a particle-counting cell and the individual particles may be counted, often with the use of a stain to enhance the particle contrast. Optical imaging software may also be used to obtain a more quantitative assessment of particle characteristics. Images of water particles are obtained with a digital camera attached to a microscope and sent to a computer for imaging analysis. The imaging software typically allows for the determination of minimum, mean, and maximum size, shape, surface area, aspect ratio, circumference, and centroid location.

PARTICLE NUMBER AND DISTRIBUTION

The number of particles in raw surface water can vary from 100 to over 10,000/mL depending on the time of year and location where the sample is taken (e.g., a river or storage reservoir). The number of particles, as will be discussed later, is of importance with respect to the method to be used for their removal. The size distribution of particles in natural waters may be defined on the basis of particle number, particle mass, particle diameter, particle surface area, or particle volume. In water treatment design and operation, particle size distributions are most often determined using a particle size counter, as discussed above. In most particle size counters, the detected particles of a given size are counted and grouped with other particles within specified size ranges (e.g., 1 to 2 μm, 5 to 10 μm). When the counting is completed, the number of particles in each bin is totaled.

The particle number frequency distribution $F(d)$ can be expressed as the number concentration of particles, dN, with respect to the incremental change in particle size, $d(d_p)$, represented by the bin size:

$$F(d_p) = \frac{dN}{d(d_p)} \tag{2-14}$$

where $F(d_p)$ = function defining frequency distribution of particles d_1, d_2, d_3

dN = particle number concentration with respect to incremental change in particle diameter $d(d_p)$

$d(d_p)$ = incremental change in particle diameter (bin size)

Because of the wide particle size ranges encountered in natural waters, it is common practice to plot the frequency function $dF(d)$ against the logarithm of size, log d_p:

$$2.303(d_p)F(d) = \frac{dN}{d(\log d_p)} \tag{2-15}$$

Similar relationships can be derived based on particle surface area and volume (Dallavalle, 1948; O'Melia, 1978).

It has also been observed that in natural waters the number of particles increases with decreasing particle diameter and that the frequency distribution typically follows a power law distribution of the form

$$\frac{dN}{d(d_p)} = A(d_p)^{-\beta} \approx \frac{\Delta N}{\Delta(d_p)} \tag{2-16}$$

where A = power law density coefficient
d_p = particle diameter, μm
β = power law slope coefficient

Taking the log of both sides of Eq. 2-16 results in the following expression, which can be plotted to determine the unknown coefficients A and β:

$$\log\left[\Delta N / \Delta\left(d_p\right)\right] = \log A - \beta \log\left(d_p\right) \tag{2-17}$$

The value of A is determined when $d_p = 1$ μm. As the value of A increases, the total number of particles in each size range increases. The slope β is a measure of the relative number of particles in each size range. Thus, if $\beta < 1$, the particle size distribution is dominated by large particles; if $\beta = 1$, all particle sizes are represented equally; and if $\beta > 1$, the particle size distribution is dominated by small particles (Trussell and Tate, 1979). The value of the coefficient for most natural waters varies between 2 and 5 (O'Melia, 1978; Trussell and Tate, 1979). Typical plots of particle size data determined using a particle size counter for various waters are given on Fig. 2-8. On Fig. 2-8a, the effect of flocculation in producing large particles is evident by comparing the β values for the unflocculated versus the flocculated influent(4.1 versus 2.1). As shown on Fig. 2-8b, the removal of all particle sizes by filtration is very similar, because the slopes of the two plots are nearly identical. The analysis of data obtained from a particle size counter is shown in Example 2-3.

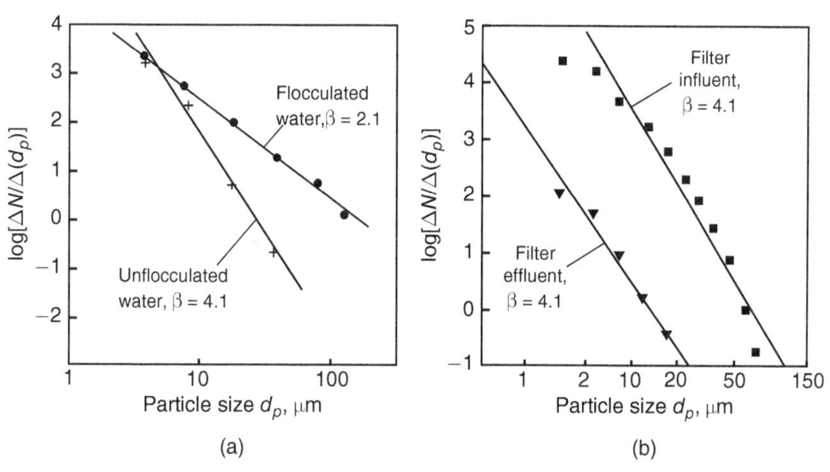

Figure 2-8
Typical examples of particle size distributions: (a) unflocculated and flocculated and (b) filter influent and effluent. (Adapted from Trussell and Tate, 1979).

Example 2-3 Analysis of particle size information

Determine the slope and density coefficients A and β in Eq. 2-17 for the following particle size data obtained from settled water during a pilot study.

Channel (Bin)	Particle size range, μm	Number of Particles, #/mL
1	1–3	1785
2	3–5	243
3	5–7	145
4	7–12	186
5	12–32	132
6	32–120	2.9
	Total	2493.9

Solution

1. Calculate the necessary values for the first data channel.
   a. Mean particle diameter:
   $$d_p = \frac{1}{2}(1\ \mu\text{m} + 3\ \mu\text{m}) = 2\ \mu\text{m}$$
   b. Log of the mean particle diameter:
   $$\log(d_p) = \log(2\ \mu\text{m}) = 0.301$$
   c. Particle diameter range:
   $$\Delta(d_p) = 3\ \mu\text{m} - 1\ \mu\text{m} = 2\ \mu\text{m}$$
   d. Number of particles:
   $$\Delta N = 1785/\text{mL}$$
   e. Log of the particle size distribution function:
   $$\log\left[\frac{\Delta N}{\Delta(d_p)}\right] = \log\left(\frac{1785/\text{mL}}{2\ \mu\text{m}}\right) = 2.95$$
2. Calculate the necessary values for the remaining data channels. The results are tabulated below.

Channel	(A) d_p	(B) $\log(d_p)$	(C) $\Delta(d_p)$	(D) ΔN	(E) $\log[\Delta N/\Delta(d_p)]$
1	2	0.301	2	1785	2.95
2	4	0.602	2	243	2.08
3	6	0.778	2	145	1.86
4	9	0.978	5	186	1.57
5	22	1.342	20	132	0.82
6	76	1.881	88	2.9	−1.48

3. Prepare a plot of $\log[\Delta N/\Delta(d_p)]$ versus $\log(d_p)$ draw a linear trendline and display the trendline equation and r^2 value on the chart. The resulting chart is shown below.

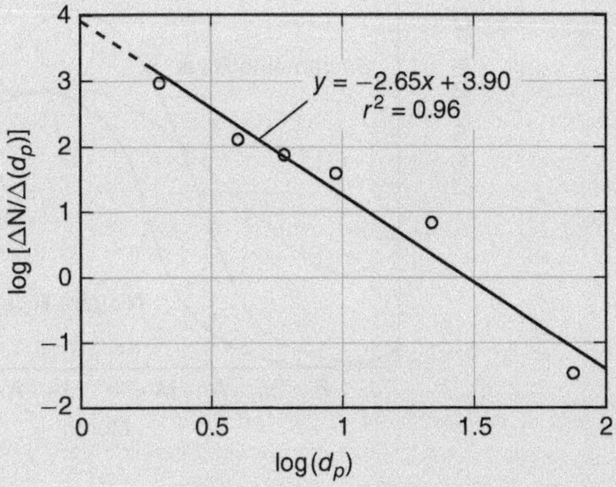

4. Determine A and β in Eq. 2-17 from the line of best fit in the above plot.
   a. When $\log(d_p) = 0$, the intercept value is equal to $\log(A)$. Thus, $A = 7{,}940$.
   b. The slope of the line of best fit is equal to $-\beta$. Thus, $\beta = 2.65$.

Color

The color of a water is an indication of the organic content, including humic and fulvic acids, the presence of natural metallic ions such as iron and manganese, and turbidity. Apparent color is measured on unfiltered samples and true color is measured in filtered samples (0.45-μm filter). Turbidity increases the apparent color of water, while the true color is caused by dissolved species and is used to define the aesthetic quality of water. The color of potable waters is typically assessed by visually comparing a water sample to known color solutions made from serial dilutions or concentrations of a standard platinum–cobalt solution. The platinum–cobalt standard is related to the color-producing substance in the water only by hue.

The presence of color is reported in color units (c.u.) at the pH of the solution. In water treatment, one of the difficulties with the comparison method is that at low levels of color it is difficult to differentiate between low values (e.g., 2 versus 5 c.u.). If the water sample contains constituents (e.g., industrial wastes) that produce unusual colors or hues that do not match the platinum–cobalt standards, then instrumental methods must be

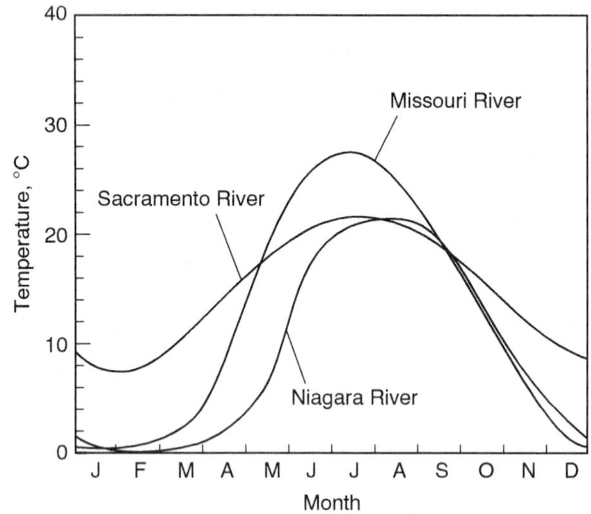

Figure 2-9
Generalized monthly variations in temperature in the Missouri River near Blair, Nebraska; in the Niagara River at Buffalo, New York; and in the Sacramento River near Sacramento, California. (Adapted from Tchobanoglous and Schroeder, 1985).

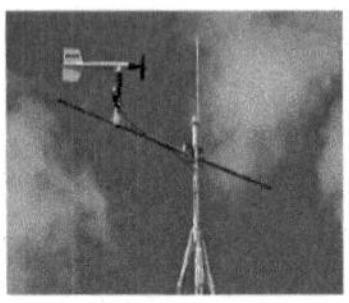

Weather Station.
http://www.stantec.com/wt3r/2-2.html

used. Instrumental methods are used to determine (1) the hue (red, green, yellow, etc.), (2) the luminance (brightness), and (3) the saturation (pale, deep, etc.) of a solution. In turn, these three parameters can be related to the *chromaticity*. It should be noted that the results obtained with the two methods are not comparable.

Temperature

Water temperature is of importance because it affects many parameters that impact engineering designs. These parameters include density, viscosity, vapor pressure, surface tension, solubility, the saturation value of gases dissolved in water, and the rates of chemical, biochemical, and biological activity. As the heat capacity of water is much greater than that of air, water temperature changes much more slowly than air temperature. Depending on the geographic location, the mean annual temperature of river water in the United States varies from about 0.5 to 3°C in the winter to 23 to 27°C in the summer (see Fig. 2-9). In small slow-moving streams, summer temperatures may exceed 30°C. Lakes, reservoirs, ponds, and other impoundments are also subject to temperature changes. Extremely wide temperature variations can occur in shallow impoundments. Typical groundwater temperatures are as shown on Fig. 2-10. In general, groundwater temperatures are not as variable as surface water temperatures.

2-4 Inorganic Chemical Constituents

Water in the environment can contain a variety of colloidal and suspended solids inorganic and organic ionic and dissolved constituents and

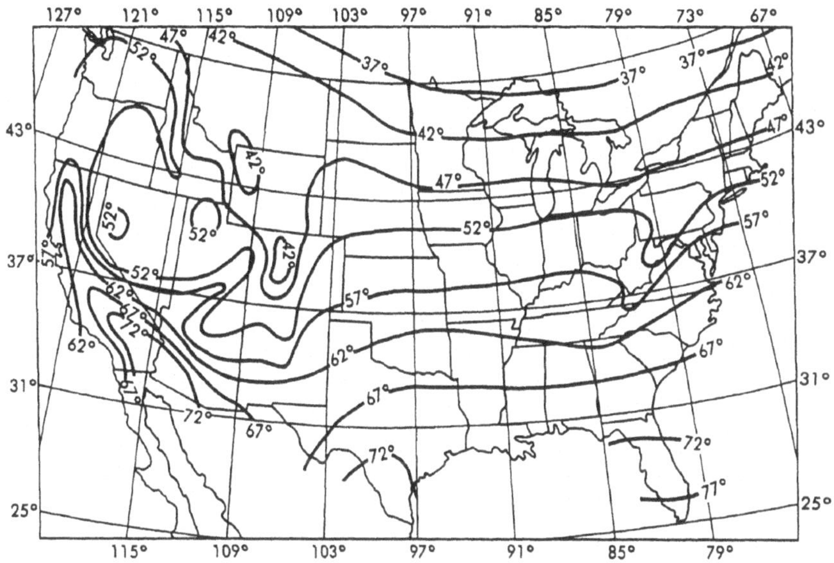

Figure 2-10 Approximate temperature of groundwater from nonthermal wells at depths varying from 10 to 20 m. Note temperatures are given in degrees Fahrenheit.

compounds, and gases (see Table 2-2). The sources of particulate (both colloidal and suspended) constituents in water were discussed previously in Sec. 2-3. The focus of this section is on the ionic and dissolved inorganic constituents found in most natural waters as identified in Table 2-2. Specific topics include (1) the major inorganic chemical constituents in natural water, (2) the minor inorganic constituents found in natural waters, and (3) the principal inorganic water quality indicators. Organic constituents are considered in Sec. 2-5.

Major Inorganic Constituents

Inorganic chemical constituents commonly found in water in significant quantities (1.0 to 1000 mg/L) include calcium, magnesium, sodium, potassium, bicarbonate, chloride, sulfate, and nitrate. Inorganic constituents that are generally present in lesser amounts (0.01 to 10 mg/L) include iron, lead, copper, arsenic, and manganese. The range of concentrations found for individual inorganic constituents in a survey of natural waters is shown on Fig. 2-11. The plotted lines for each constituent represent the percent of the samples in which each constituent was found to be equal to or less than a specified concentration. For example, potassium occurred over a range of 0.4 to 15 mg/L, and samples from 80 percent of the natural waters in this survey had potassium concentrations below 5 mg/L. Additional details on the major inorganic constituents found in natural waters are presented in Table 2-4.

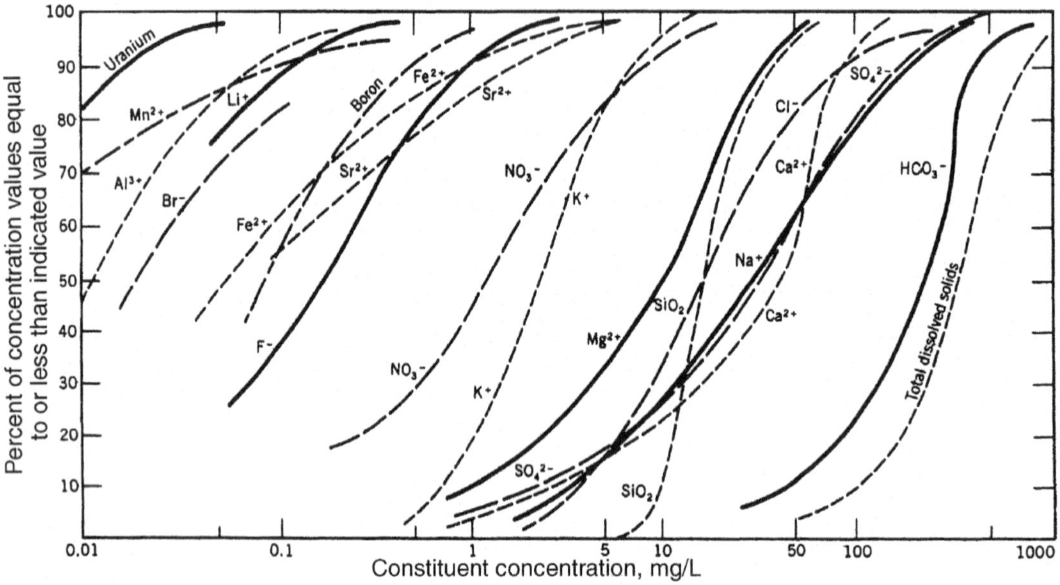

Figure 2-11
Cumulative curves showing frequency distribution of various constituents in terrestrial water. Data are mostly from the United States from various sources. (Adapted from Davies and DeWiest, 1966).

Minor and Trace Inorganic Constituents

Constituents of natural waters found in the parts-per-billion to parts-per-trillion range may still be of significant health or water quality concern. Constituents of concern include a number of inorganics and numerous trace organics, as discussed in the following section. Information on the water quality significance of several inorganic trace constituents is presented in Table 2-5. As shown, the trace constituents have been grouped under four categories: (1) alkali metals, (2) alkaline metals, (3) other metallic elements depending on their properties, and (4) nonmetals.

Inorganic Water Quality Indicators

Several chemically related quality measures are utilized to characterize the properties of a water supply including (1) the hydrogen ion concentration (pH), (2) polyvalent cation content (hardness), (3) total dissolved solids (TDS), and (4) electrical conductivity.

pH (HYDROGEN ION CONCENTRATION)

pH is a measurement of the acid-base properties of a solution. pH is an important parameter in water treatment as it directly influences the dosages of chemicals added to reduce hardness and coagulate particles. pH is measured as the negative logarithm of the concentration of hydrogen ions:

$$pH = -\log_{10}\left[H^+\right] \tag{2-18}$$

Table 2-4
Summary of the major cations and anions in natural water[a]

Ion	Description
	Cations
Calcium (Ca^{2+})	Calcium is generally among the most prevalent three or four ions in groundwaters. Common mineral forms of calcium are calcite, also known as aragonite ($CaCO_3$), gypsum ($CaSO_4{\cdot}2H_2O$), anhydrite ($CaSO_4$), and fluorite (CaF_2). Calcium is generally present as the free ion, Ca^{2+}, in natural waters and adsorbed onto soil particles. Along with magnesium and other multivalent ions Ca^{2+} is responsible for the hardness of a water as discussed later in this section.
Iron ($Fe^{2+,\ 3+}$)	Iron is found in rocks, soils, and waters in a variety of forms and oxidation states. Common mineral sources (deposits) of iron include ferric oxides and hydroxides such as hematite (Fe_2O_3) and ferric hydroxide [$Fe(OH)_3$], which gives rocks and soils their red and yellow color. In oxygenated surface waters (pH 5–8), typical concentrations of total iron are around 0.05–0.2 mg/L. In groundwater, the occurrence of iron at concentrations of 1.0–10 mg/L is common, and higher concentrations (up to 50 mg/L) are possible in low-bicarbonate and low-oxygen waters.
Magnesium (Mg^{2+})	Magnesium salts are more soluble than calcium, but they are less abundant in rocks and therefore less available for weathering reactions. Concentrations of magnesium are typically below 10–20 mg/L in surface waters and below 30–40 mg/L in groundwaters. Taken together, calcium and magnesium comprise most natural water hardness.
Manganese (Mn^{2+})	Manganese is abundant in rocks and soils, typically in the form of manganese oxides and hydroxides in association with other metallic cations. At low and neutral pH values, the predominant dissolved form of manganese is the divalent cation Mn^{2+}. Concentrations on the order of 0.1–1 mg/L are common, although in low-pH waters higher concentrations can occur. Manganese often is present with iron in groundwaters and, like iron, may cause aesthetic problems such as laundry and fixture staining.
Potassium (K^+)	Although a common element of the earth's crust, the concentration of potassium in natural waters is much lower than sodium. Potassium occurs in nature only in ionic or molecular form and has many properties that are similar to sodium, so it occasionally replaces sodium in industrial applications.
Sodium (Na^+)	Sodium compounds comprise almost 3% of the earth's crust, and a significant amount is found in rock and soil. Sodium is transported into water from rocks through weathering and soil through ion exchange reactions. In natural waters, sodium is generally present as the free ion Na^+. Several complexes and ion pairs may occur in natural waters, including sodium carbonate ($NaCO_3^-$), sodium bicarbonate ($NaHCO_3$), sodium sulfate ($NaSO_4^-$), and sodium chloride (NaCl).

(Continued)

Table 2-4 *(Continued)*

Ion	Description
	Anions and neutral species
Bicarbonate (HCO_3^-)	The carbonate–bicarbonate system in natural water performs important functions in acid–base chemistry, buffer capacity, metal complexation, solids formation, and biological metabolism. Species comprising the carbonate system include CO_2, H_2CO_3, HCO_3^-, and CO_3^{2-}. The dominant role of the carbonate system in acid–base chemistry of natural waters is well documented, although exceptions occur in waters with very high concentrations of dissolved organics or in high-sulfate groundwaters.
Chloride (Cl^-)	Chloride is present in water supplies almost exclusively as the chloride ion (Cl^-), although hydrolysis products of chlorine (HOCl and OCl^-) exist temporarily where chlorine has been added as a disinfectant. In typical surface waters the concentration of chloride ion is less than 10 mg/L; however, in areas subject to seawater intrusion or hot-spring inflows or where evaporation greatly exceeds precipitation, the chloride concentrations can approach seawater levels.
Flouride (F^-)	Although the amount of fluoride in crustal rocks is much greater than chloride, fluoride remains bound in minerals to a much greater degree. Fluorite (CaF_2) is a common fluoride mineral and fluorapatite [$Ca_5F(PO_4)_3$] also commonly contains fluoride. In natural waters, fluoride is present primarily as the F^- ion or as a complex with aluminum, beryllium, or ferric iron. In waters with TDS < 1000 mg/L, fluoride is typically <1 mg/L, although ground waters affected by volcanic activity are found with levels higher than 10 mg/L.
Nitrogen (N)	The most common and important forms of nitrogen in water and their corresponding oxidation state in the water/soil environment are ammonia gas (NH_3, –III), ammonium (NH_4^+, –III), nitrogen gas (N_2, 0), nitrite ion (NO_2^-, + III), and nitrate ion (NO_3^-, + V). The oxidation state of nitrogen in most organic compounds is –III. The oxidation states of nitrogen range from –3 to +5 and are summarized below (Sawyer et al., 2003): $\overset{-III}{NH_3} - \overset{0}{N_2} - \overset{I}{N_2O} - \overset{II}{NO} - \overset{III}{N_2O_3} - \overset{IV}{NO_2} - \overset{V}{N_2O_5}$ Other forms of nitrogen in water include organic compounds such as urea (NH_2CONH_2), amino acids and their breakdown products, ammonia (NH_3), ammonium ion (NH_4^+), hydroxylamine (NH_2OH), nitrogen gas (N_2), and nitrite (NO_2^-). Ammonia, ammonium ion, and protein by-products are all reduced species, N_2 gas is in the zero oxidation state, nitrite is at +3, and nitrate is at +5. Transformation from one state to another is closely tied to biological activity, the influx of domestic wastes, and the local use of nitrogen fertilizers.
Silica (SiO_2)	Silica is present in almost all rocks, soils, and natural waters. In water, silica is hydrated as H_4SiO_4 or $Si(OH)_4$, although water analyses commonly represent dissolved silica as SiO_2. The concentration of silica most commonly found in natural waters is between 1 and 30 mg/L. The solubility of silica is complex, but temperature is a critical factor. Sodium silicates have been used as coagulants in water treatment and as corrosion inhibitors on iron pipes.

Table 2-4 *(Continued)*

Ion	Description
	Anions and neutral species
Sulfur (S)	Sulfur occurs in natural waters as sulfate (e.g., SO_4^{2-}) and sulfides (e.g., H_2S, HS^-, $Na_2S_2O_3$). The primary sources of sulfates are evaporite rocks, which are formed by water evaporation and mineral precipitation, such as gypsum ($CaSO_4{\cdot}2H_2O$) and anhydrite ($CaSO_4$), sedimentary rock such as pyrite (FeS_2), rainfall, and bacterial metabolism. The concentration of sulfate in oxidized waters typically range from 5 to 30 mg/L.

[a]Cations and anions are arranged alphabetically

Table 2-5
Minor and trace elements found in natural waters[a]

Constituent	Concentration in Natural Waters, μg/L	Significance in Water Supplies
		Alkali Metals
Cesium	0.05–0.02[b]	
Lithium	0.001–0.3	Potentially toxic to plants, but not at concentrations likely to be encountered in irrigation waters
Rubidium	0.0015	
		Alkaline Earth Metals
Barium	0.043 (median public water)	Ingestion of soluble barium salts can be fatal. Normal water concentrations have no effect.
Beryllium	0.001–1	Highly toxic, but occurs at very low concentration.
Strontium	0.6 (median river water), 0.11 (median public water)	Concentration in natural water is less than solubility.
		Other Metallic Elements
Cadmium	ND–10	Toxic. Presence may indicate industrial contamination.
Chromium	5.8 (median river water), 0.43 (median public water)	Industrial pollutant.
Cobalt	ND–1.0	Essential in nutrition in small quantities.
Copper	10	Utilized in water treatment and metal fabrication; used to inhibit algae growth in reservoirs; essential for nutrition of flora and fauna.
Gold	ND–trace	—
Lead	1–10	Older plumbing systems contain lead, which may dissolve at low pH.

(Continued)

Table 2-5 *(Continued)*

Constituent	Concentration in Natural Waters, μg/L	Significance in Water Supplies
Mercury	ND–<10	Highly toxic. Presence indicates pollution from mining, industry, or metallurgical works.
Molybdenum	0.35 (median river water); 1.4 (median public water)	Accumulated by vegetation. Forage crops may become toxic.
Nickel	10	—
Silver	0.1–0.3	Has been used as disinfectant.
Titanium	8.6 (median river water); <1.5 (median public waters)	—
Vanadium	<70	May concentrate in vegetation.
Zinc	10	Widely found in industry wastes; found in wastes dissolved from galvanized pipes, cooling-water treatment, etc.
Arsenic	0–1000	Used in industry in some herbicides and pesticides; lethal in animals above 44 mg/kg. Long-term ingestion of 0.21 mg/L reported to be poisonous.
Bromine	20	May react with disinfectants and form brominated species, which are suspected carcinogens.
Iodine	0.2–2	Essential nutrient in higher animals; has been used to seed clouds.
Selenium	0.2	Taken up by vegetation.

[a]Values presented are approximate and represent one or more author's best estimate. ND = not detected. Public water refers to drinking water.
[b]Values observed in six analyses of rivers in Japan.
Sources bases on: NAS (1977), Livingstone (1963), Turekian (1971), and Hem (1971).

The hydrogen ion concentration in water is connected closely with the extent to which water molecules dissociate. Water will dissociate into hydrogen and hydroxide ions as follows:

$$H_2O \rightleftarrows H^+ + OH^- \tag{2-19}$$

Applying the law of mass action (see discussion in Chap. 5) to Eq. 2-19 yields

$$\frac{[H^+][OH^-]}{[H_2O]} = K \tag{2-20}$$

where the brackets indicate concentration of the constituents in moles per liter. Because the concentration of water in a dilute aqueous system is essentially constant, this concentration can be incorporated into the equilibrium constant K to give

$$[H^+][OH^-] = k_w \tag{2-21}$$

where K_w is known as the ionization constant or ion product of water and is approximately equal to 1×10^{-14} at a temperature of 25°C. Equation 2-21 can be used to calculate the hydroxide ion concentration when the hydrogen ion concentration is known, and vice versa.

With pOH, which is defined as the negative logarithm of the hydroxyl ion concentration, for water at 25°C, the following relation is used:

$$pH + pOH = 14 \tag{2-22}$$

The pH of aqueous systems typically is measured with a pH-sensing electrode. Various pH papers and indicator solutions that change color at definite pH values are also used. When using pH paper or indicator solution, pH is determined by comparing the color of the paper or solution to a series of color standards.

HARDNESS

Multivalent cations, particularly magnesium and calcium, are often present at significant concentrations in natural waters. These ions are easily precipitated and in particular react with soap to form a difficult-to-remove scum. Hardness is an important parameter to industry as an indicator of potential (interfering) precipitation, such as with carbonates in cooling towers or boilers, with soaps and dyes in cleaning and textile industries, and with emulsifiers in photographic development. For most practical purposes, hardness of water can be represented as the sum of the calcium and magnesium concentrations, given in milliequivalents per liter:

$$\text{Hardness, eq / L} = 2[Ca^{2+}] + 2[Mg^{2+}] \tag{2-23}$$

In Eq. 2-23, the concentrations of Ca and Mg are given in mol/L, and the coefficient 2 reflects the divalent nature of both ions, i.e., both have 2 equivalents per mole. Two general types of hardness are of interest: carbonate hardness, associated with HCO_3^- and CO_3^{2-}, and noncarbonate hardness, associated with other anions, particularly Cl^- and SO_4^{2-}. The balance between carbonate and noncarbonate hardness is important in water softening (hardness removal) and in scale formation. Because HCO_3^- dissociates at high temperatures, the result of heating hard water is scale formation due to $CaCO_3$ precipitation:

$$Ca^{2+} + 2HCO_3^- \rightleftarrows CaCO_3 + CO_2 + H_2O \tag{2-24}$$

Scale formation plugs pipes, decreases heat transfer coefficients, and changes the frictional resistance to flow in pipes. Hardness is also of concern to consumers due to the occurrence of scaling on fixtures and water-related appliances. With respect to hardness, waters are typically classified as follows:

Soft	0 to <50 mg/L as $CaCO_3$
Moderately hard	50 to <100 mg/L as $CaCO_3$
Hard	100 to <150 mg/L as $CaCO_3$
Very hard	>150 mg/L as $CaCO_3$

Another range of values that may be encountered in the literature for the same classifications are 0 to <60, 60 to <120, 120 to <180, and >180 as $CaCO_3$.

ALKALINITY

Alkalinity is a measure of the ability of a water to resist changes in pH. Alkalinity in water is due to the presence of weak acid systems that consume hydrogen ions produced by other reactions or produce hydrogen ions when they are needed by other reactions, allowing chemical or biological activities to take place within a water without changing the pH. The primary source of alkalinity is the carbonate system, although phosphates, silicates, borates, carboxylates, and other weak acid systems can also contribute. Alkalinity is determined by titrating with acid, and the results are expressed in concentrations of meq/L or as concentration of calcium carbonate (mg/L as $CaCO_3$). When the individual species are expressed as molar concentrations, alkalinity is calculated as

$$\text{Alkalinity, eq / L} = [HCO_3^-] + 2[CO_3^{2-}] + [OH^-] - [H^+] \quad (2\text{-}25)$$

where the coefficient on carbonate (CO_3^{2-}) is necessary because carbonate is divalent (2 eq/mol) and the other species are monovalent (1 eq/mol). When the individual species are expressed in concentrations of meq/L, alkalinity is calculated as

$$\text{Alkalinity, meq / L} = (HCO_3^-) + (CO_3^{2-}) + (OH^-) - (H^+) \quad (2\text{-}26)$$

In practice, alkalinity is expressed in terms of mass concentration as calcium carbonate. To convert from meq/L to mg/L as $CaCO_3$, it is helpful to remember that

$$\begin{aligned}\text{Millequivalent mass of } CaCO_3 &= \frac{100\,\text{mg / mmol}}{2\,\text{meq / mmol}}\\ &= 50\,\text{mg / meq}\end{aligned}$$

Thus 3 meq/L of alkalinity would be expressed as 150 mg/L as CaCO3:

$$\begin{aligned}\text{Alkalinity as } CaCO_3 &= (3.0\,\text{meq /L})\ (50\ \text{mg/meg } CaCO_3)\\ &= 50\,\text{mg/L as } CaCO_3\end{aligned}$$

TOTAL DISSOLVED SOLIDS

Total dissolved solids (TDS) is a measure of the total ions in solution, analyzed by filtering out the suspended material, evaporating the filtrate, and weighing the remaining residue. Local TDS concentrations in arid regions or in waters subjected to pollution runoff can be high. For example, Colorado River water, after reaching southern California, has a TDS content in the range of 700 to 800 mg/L. The TDS of seawater is about 35,000 mg/L.

CONDUCTIVITY

A parameter related to TDS is electrical conductivity (EC) or specific conductance. Electrical conductivity is actually a measure [in microsiemens per centimeter (μS/cm) or micromhos per centimeter (μ℧/cm)] of the ionic activity of a solution in terms of its capacity to transmit current. In dilute solutions, the two measures are reasonably comparable; that is, TDS = 0.5 × EC. However, as the solution becomes more concentrated (TDS > 1000 mg/L, EC > 2000 μS/cm), the proximity of the solution ions to each other depresses their activity and consequently their ability to transmit current, although the physical amount of dissolved solids is not affected. At high TDS values, the ratio of TDS to EC increases and the relationship tends toward TDS = 0.9 (slope of line) × EC. Thus, the slope for any one sample can fall between 0.5 and 0.9, but for several samples having the same TDS the slope will also vary; therefore, each water sample should be characterized separately.

2-5 Organic Chemical Constituents

A variety of organic compounds that can affect water quality are found in drinking water supplies. Several types of organic chemicals cause disagreeable tastes and odors in drinking water, and other types are known to be toxic. Many organic contaminants are known to be carcinogenic or are classified as cancer-suspect agents. Organic compounds in water are derived from natural and anthropogenic sources. Anthropogenic contaminants are generally present at extremely low concentrations and might not pose an immediate health hazard. However, a number of long-term research studies have been focused on the question "at what level do trace organic contaminants exert an impact on human health?" Based on the results to date it seems likely that the answers to this question will continue to be pursued.

Topics discussed in this section are (1) a brief review of organic compounds and their properties, (2) the potential sources of organic compounds and their introduction to drinking water and drinking water supplies, (3) the characteristics of the natural organic matter found in water, (4) organic compounds originating from human activity, (5) organic compounds formed during disinfection, (6) organic compounds added during treatment, (7) surrogate measures for organic water quality indicators, and (8) the analysis of trace organics.

Definition and Classification

The term *organics* refers to the general class of chemicals composed of carbon (C) and one or more of the following elements: hydrogen (H), nitrogen (N), and oxygen (O). The term organic dates to early studies of chemistry when substances were categorized as inorganic when they were obtained from mineral sources and as organic when they were derived from living organisms. Today, many organic compounds are derived from sources other than biological activity. A wide variety of materials are synthesized by the chemicals industry. The molecular structure of these synthesized compounds may also contain atoms of sulfur (S), phosphorus (P), and/or one or more of the halogens, that is, fluorine (F), chlorine (Cl), bromine (Br), and iodine (I), as well as a variety of other elements. Many naturally occurring compounds may also contain these atoms as well, but they are found to a lesser degree. There are many chemical species that are commonly considered to be inorganic in spite of having C, H, O, and N within their structure. Examples of such compounds include carbon monoxide (CO), carbon dioxide (CO_2), carbonate (CO_3^{2-}), bicarbonate (HCO_3^-), and cyanide (CN^-). The principal structural feature that distinguishes organic compounds from inorganic substances is the existence of strong carbon–carbon bonds.

CLASSIFICATION ACCORDING TO SIZE AND MOLECULAR WEIGHT

From an environmental standpoint, it is especially convenient to classify organic substances into groups according to their chemical or physical properties. Knowledge of these properties facilitates the selection of appropriate methods for the analysis and treatment of these materials in water. One important property of organic compounds is molecular weight. The molecular weight of organic compounds ranges from 16 g/mol for methane (CH_4) to values approaching one million (10^6) grams per mole for polymeric materials. The dimension of organic molecules varies from less than 1 nm for simple compounds such as chloroform ($CHCl_3$) to approximately 0.1 µm for complex organic polymers. The relative size of some organic molecules as compared to microorganisms and other material commonly found in aquatic systems was illustrated previously on Fig. 2-5.

OTHER METHODS OF CLASSIFICATION

The polarity of an organic substance can also be used to define the degree to which one segment of a molecule is either positively or negatively charged with respect to another part of the molecular structure (McMurry and Fay, 2003). A frequently used measure of the polarity of a compound is given by the dipole moment. The dipole moment of organic substances can vary from a value of 0 D (debye) for molecules such as carbon tetrachloride (CCl_4), which have a highly symmetric spatial distribution of electron density about their bonding structures, to approximately 1.87 D for chloromethane (CH_3Cl) (McMurry and Fay, 2003). The volatility of an organic substance is generally reflected by its boiling point or vapor pressure. At ambient atmospheric pressure (1 atm, 760 mm Hg), the boiling points of organic contaminants may vary from as low as –13.4°C for highly volatile compounds such as vinyl chloride to temperatures in excess of 400°C for nonvolatile polycyclic aromatic hydrocarbons.

Sources of Organic Compounds in Drinking Water

There are four major sources from which organics may be introduced to drinking water:

1. Natural organic material
2. Compounds originating from human activities
3. Compounds formed through chemical reactions that occur during disinfection
4. Compounds added or formed during the treatment and transmission of water

Each of these sources is considered in the following discussion.

Natural Organic Matter

Natural organic matter (NOM) is the term used to describe the complex matrix of organic chemicals originating from natural sources that are present in all water bodies. Natural organic matter originates from a water body due to biological activity, including secretions from the metabolic activity of algae, protozoa, microorganisms, and higher life-forms; decay of organic matter by bacteria; and excretions from fish or other aquatic organisms. The bodies and cellular material of aquatic plants and animals contribute to NOM. Natural organic matter can also be washed into a watercourse from land, originating from many of the same biological activities but undergoing different reactions due to the presence of soil and different organisms.

Historically, the significance of NOM in drinking water was related to its impact on aesthetic quality, as NOM imparts a yellowish tinge to water that many people find unpalatable. More recently, concern about NOM has focused on its ability to react with chlorine and form disinfection by-products, which are often carcinogenic. The presence of NOM affects many water quality parameters and processes. A summary of some important impacts of NOM is provided in Table 2-6.

In drinking water supplies, NOM is measured most commonly using total organic carbon (TOC) as a surrogate measure. Typical TOC concentrations for a variety of waters are reported on Fig. 2-12. The TOC concentrations of ground and surface waters often fall in the ranges of 0.1 to 2 and 1 to 20 mg/L, respectively. By contrast, the TOC levels of highly colored waters found in swamps can be in the range of 100 to 200 mg/L.

Table 2-6
Effect of NOM on water quality parameters and processes

Parameter	Effect of NOM
	Water Quality Parameters
Color	NOM can impart an unpalatable yellowish tinge to water at high concentrations.
Disinfection by-products	NOM reacts with chemical disinfectants, forming disinfection by-products. Many of these by-products have been demonstrated to be carcinogenic or have other adverse public health effects.
Metals/synthetic organics	NOM can complex with metals and hydrophobic organic chemicals (such as pesticides), making them more soluble. Once these chemicals are soluble, they can be transported more easily in the aquatic environment and are more difficult to remove during treatment.
	Water Treatment Processes
Disinfection	NOM reacts with and consumes disinfectants, so that the required dose to achieve effective disinfection is much higher than it would be in the absence of NOM.
Coagulation	NOM reacts with and consumes coagulants, so that the required dose to achieve effective turbidity removal is much higher than it would be in the absence of NOM.
Adsorption	NOM adsorbs to activated carbon, rapidly depleting the adsorption capacity of the carbon. Adsorption isotherms are much harder to predict in the presence of NOM.
Membranes	NOM adsorbs to membranes, clogging membrane pores and fouling surfaces, leading to a rapid decline in flux through the membrane.
Distribution	NOM can be biodegradable, leading to corrosion and slime growth in distribution systems (especially when oxidants are used during treatment).

CHEMISTRY OF NOM

Biological matter is composed primarily of four basic classes of organic compounds: carbohydrates, lipids, amino acids, and nucleic acids. Natural organic matter is composed of these chemicals and the products of biotic and abiotic chemical reactions between NOM molecules or between NOM and inorganic constituents of water. The wide array of biological activity in the environment leads to the production of thousands of different chemicals, so NOM is a complex mixture of different compounds with varying chemical properties, which may vary significantly from one water body to another as a result of local soil, climate, and hydrologic conditions. This complexity makes the characterization of the basic chemistry of NOM (such as functional groups or physical and chemical properties) difficult

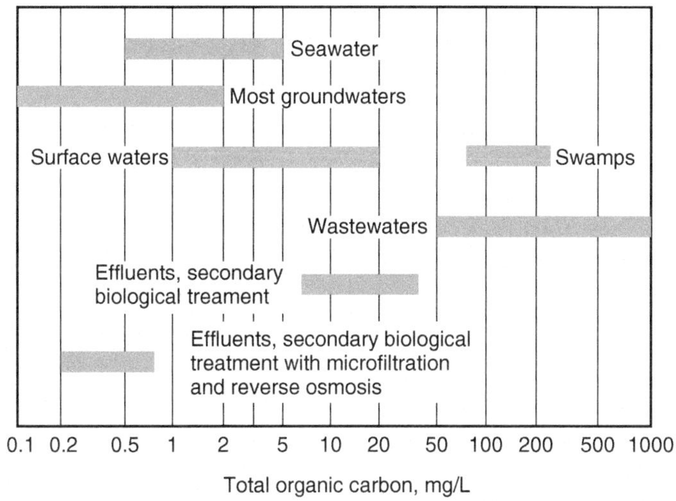

Figure 2-12
Ranges of TOC reported in a variety of waters. (Adapted from Kavanaugh, 1978).

and causes NOM from different water bodies to have different effects when subjected to water treatment processes.

Natural organic matter is not volatile. It is fairly soluble and can be concentrated to greater than 1000 mg/L without precipitating. Most NOM molecules are negatively charged and many have multiple anionic functional groups, making them polyelectrolytic. The NOM molecules have a distribution of molecular weights, with about 90 percent of NOM between about 500 and 3000 Da. The elemental composition of NOM is about 45 to 60 percent carbon, 4 to 5 percent hydrogen, 35 to 40 percent oxygen, and 1 percent nitrogen (Thurman, 1985).

Based on their solubility in acid and alkali, aquatic humates are usually divided into two principal components: humic acid (HA), which is soluble in dilute alkaline solutions but is precipitated upon acidification, and fulvic acid (FA), which remains in solution at low pH. The structural features of HA and FA are similar, but the two fractions differ considerably in molecular weight and functional group composition. The molecular weight of fulvic acid varies from 200 to 1000 g/mol, whereas the molecular weight of HA ranges up to 200,000 g/mol. The FA fraction also possesses a higher content of oxygen-containing constituents per unit weight than humic acid. However, due to the arbitrary nature of this classification scheme, the term humic material is frequently used in reference to an aggregate of FA and HA.

MEASUREMENT AND CLASSIFICATION OF NOM

The complexity of NOM makes it impractical to routinely measure individual compounds. Instead, NOM is typically quantified using bulk parameters. The most common parameters in water treatment are TOC, dissolved organic carbon (DOC), biodegradable dissolved organic carbon (BDOC), assimilable organic carbon (AOC), UV_{254} absorbance, and specific UV absorbance (SUVA). SUVA is calculated as

$$SUVA = \frac{UV_{254}}{DOC} \times 100 \tag{2-27}$$

where SUVA = specific UV absorbance, L/mg·m
UV_{254} = UV absorbance at 254 nm, cm^{-1}
DOC = dissolved organic carbon concentration, mg/L

Methods that have been used to quantify and characterize NOM are described in Table 2-7. For more detailed research, NOM can be characterized by separation into discrete fractions based on properties such as hydrophobicity, polarity, or molecular weight (Croue et al., 2000; Owen et al., 1993, 1995).

Organic Compounds from Human Activities

Organic chemicals from industry, agriculture, and municipal effluents are routinely found in water supply sources and in trace amounts in many water supplies. Surface waters are especially vulnerable to these types of contaminants, but groundwater systems can also become contaminated. Contaminants that originate from a specific site are defined as *point source pollutants*, whereas substances that enter the aquatic environment over a broad area are referred to as *non–point source pollutants*. Groundwaters are most commonly subjected to point source contamination. By contrast, large-scale surface water basins may contain organic chemicals such as trichloroethene that cannot be traced to a single site.

USE OF SYNTHETIC ORGANIC COMPOUNDS

Industries that utilize large quantities of chemicals in manufacturing processes are major sources of organic pollutants. The vast majority of organic compounds used in industry are synthesized. Synthetic organic compounds (SOCs) comprise an extremely diverse group of compounds. A general classification according to polarity and volatility is illustrated on Fig. 2-13a. Typical compounds in each of the categories identified on Fig. 2-13a are presented on Fig. 2-13b. While generally found at very low concentrations in water, many SOCs are of significant health concern. Among the sources of these compounds are the industrial and commercial facilities located in close proximity to major bodies of surface water. For example, the Mississippi and Ohio Rivers provide a plentiful supply of process and cooling water for a large fraction of the industries in the United States. Consequently, effluents from these activities can introduce a broad range of chemical contaminants to these river systems, depending on the nature of the materials being processed at each facility.

Table 2-7
Methods for quantifying and characterizing NOM

Parameter	Description
	Aggregate Properties
Total organic carbon (TOC)	NOM is oxidized completely to CO_2, which is stripped from the sample and measured in the gas phase. TOC is not equal to the NOM concentration but is a surrogate that gives an indication of the NOM concentration as long as the elemental composition does not change. The NOM concentration is typically 2 times the TOC concentration (based on the elemental composition).
Dissolved organic carbon (DOC)	The sample is analyzed identically to TOC after filtration through a 0.45-μm filter. The DOC concentration is typically 80–90% of the TOC concentration.
Biodegradable dissolved organic carbon (BDOC)	Dissolved organic carbon that can be assimilated biologically. Final value depends on the specific test procedure employed. Important in assessing the potential for regrowth of microorganisms after disinfection in the distribution system.
Assimilable organic carbon (AOC)	The fraction of the BDOC that can be readily assimilated biologically as opposed to the total, which can be biodegraded over a longer period of time. In general, the methods used to determine the BDOC and AOC will yield different results.
UV_{254} absorbance	The sample is filtered and the absorbance of UV light at a wavelength λ of 254 nm is measured with a spectrophotometer. Like TOC and DOC, UV_{254} absorbance is a surrogate for the NOM concentration. Specific molecular structures (chromophores) within NOM molecules absorb UV light, so the relationship between UV254 absorbance and NOM concentration can vary between water bodies or seasonally because of differences in NOM composition. In addition, UV_{254} absorbance may not be representative of NOM removal in a treatment process if the process removes molecules with chromophores differently than molecules without chromophores.
Specific UV absorbance (SUVA)	SUVA is calculated as the ratio of UV_{254} absorbance to the DOC (TOC has also been used). SUVA has been correlated to the hydrophobic fraction of NOM and has been used as a guide for the treatability of NOM by some processes. For instance, water with a low SUVA value may not be amenable to enhanced coagulation.
	Specific Compound Classes and Individual Constituents
Molecular weight distribution	The molecular weight distribution of NOM can be determined by serial ultrafiltration or chromatographic methods. The most advanced method is high-performance size exclusion chromatography.
Hydrophobic and ionic fractions	NOM is separated into hydrophobic, hydrophilic, cationic, neutral, and anionic fractions by retention or passage through resin columns. The effect of these specific properties is then evaluated with respect to treatment processes.
Fluorescence	Fluorescence is strongly correlated with the molecular weight of NOM.

(*Continued*)

Table 2-7 *(Continued)*

Parameter	Description
Relative polarity	Compounds can be separated based on polarity by reverse-phase high-pressure liquid chromatography (RP-HPLC) or other chromatographic methods and compared to the polarity of a standard compound. Polarity affects the reactivity and fate of NOM in many environmental processes.
Compound class identification	Assays can be performed to measure the total protein or carbohydrate concentration in samples of water containing NOM.
Spectrometry	Spectrometric methods, such as nuclear magnetic resonance (NMR), Fourier transform infrared (FT-IR) spectrometry, solid-state cross-polarization magic-angle spinning (CPMAS), ^{13}C nuclear magnetic resonance spectrometry, electrospray ionization/mass spectrometry, and pyrolysis gas chromatography mass spectrometry (pyr-GC-MS) can be used to identify the primary functional groups or compound classes present in NOM.

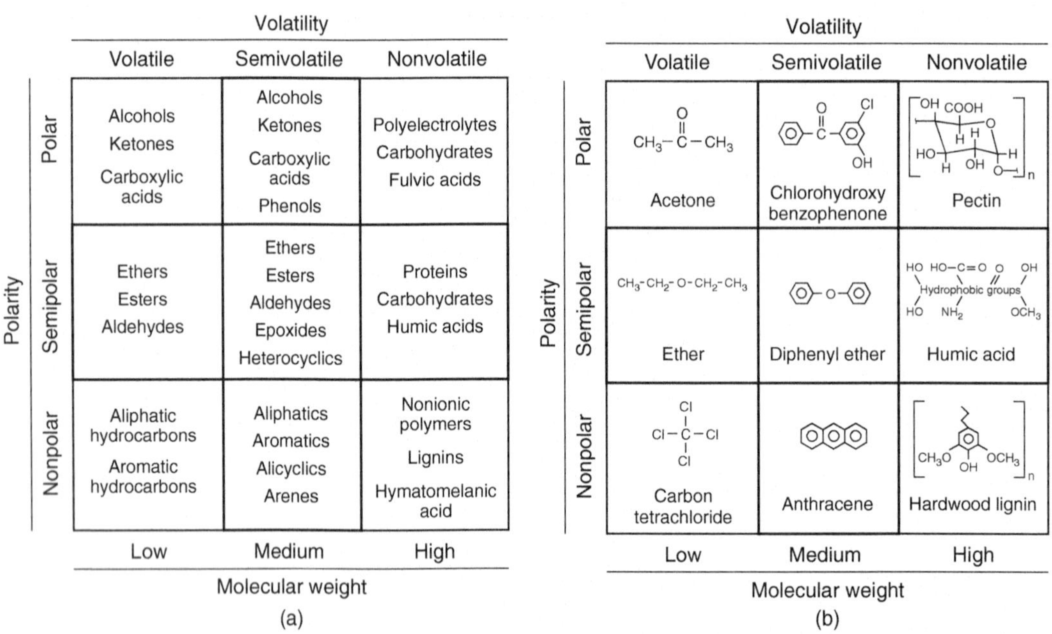

Figure 2-13
Organic compounds found in water: (a) classification based on molecular weight, polarity, and volatility and (b) representative examples of compounds in each classification. (Adapted from Trussell and Umphres, 1978).

AGRICULTURAL PESTICIDES AND HERBICIDES

The quantity of agricultural pesticides used annually in the United States is extremely large. In California alone, over 4000 tonnes of chemicals is applied each year. The vast majority of these substances are organic chemicals. In general, pesticide treatments are distributed evenly over a large acreage. Modern agricultural practice has been directed toward the use of nonrefractory pesticides, such as organophosphates, that degrade rapidly in the environment following application. Use of nonrefractory pesticides has helped to minimize the risk of water contamination. Nevertheless, the use of such large quantities of agricultural chemicals requires that programs be developed to monitor water supplies subject to agricultural runoff.

MUNICIPAL WASTEWATER DISCHARGES

Municipal wastewater treatment plants are also a major point source of organic contamination. Even with effective secondary treatment, an ever-increasing number of organic compounds is being found in the effluent from treatment plants. The U.S. Geological Survey (USGS) has identified a number of compounds termed *emerging organic compounds* that are now being found in stream waters (USGS, 2000). Many of the emerging compounds are derived from veterinary and human antibiotics, human prescription and nonprescription drugs, and industrial and household wastewater products.

Organic Compounds Formed During Water Disinfection

The processing of water for commercial applications and human consumption introduces a variety of organic compounds. More specifically, a variety of organic compounds can be formed through chemical transformations of NOM during water disinfection. For example, chlorine can efficiently convert humic substances (NOM) to trihalomethanes (THMs) and other organohalogen oxidation products under the reaction conditions encountered in water treatment systems. The formation and treatment of compounds formed during disinfection are considered in detail in Chaps. 13 and 19.

Surrogate Measures for Aggregate Organic Water Quality Indicators

A variety of measures have been developed or adapted for the quantification of the array of synthetic and naturally occurring aquatic organic material. Two types of measures are in common use: (1) those measures that are used to quantify organic matter that is composed of an aggregate (nonspecific) of constituents with similar characteristics and (2) those measures that are used to quantify individual organic constituents (specific) from within the total organic compounds present. Aggregate measures are intended to quantify part or all of the organic content of a water. They include UV absorbance, TOC, total organic halogen (TOX), and trihalomethane formation potential (THMFP). The use of some of these measures for NOM was reported previously in Table 2-7.

ULTRAVIOLET ABSORBANCE

Organic substances absorb UV light, which is light that is beyond the visible spectrum at the violet end,generally defined as having a wavelength between 100 and 400 nm. Specific organic materials show definitive UV absorbance bands reflecting their particular unsaturation pattern and/or aromatic components. Such configurations desorb the short-wavelength/ high-energy excitation of UV radiation, corresponding to excitation of electrons; increasingly shorter wavelengths are required to excite more stable molecules. Thus, simple aliphatic molecules will not tend to absorb UV light, whereas the complex multiaromatic, multiconjugated humic substances would be expected to absorb UV light very strongly. Ultraviolet absorbance at a wavelength λ of 254 nm is used as a surrogate measurement for the concentration of NOM, as described in Table 2-7. In some cases, UV absorbance at a wavelength λ of 285nm has also been measured. In reporting the absorbance of a solution, the pH must be noted. The SUVA (see Table 2-7) is another measure that has been used to quantify the NOM in water.

TOTAL ORGANIC CARBON AND DISSOLVED ORGANIC CARBON

The TOC analysis is used to quantify the total amount of organic carbon contained in a sample by converting the dissolved organic compounds to a single chemical form while excluding inorganic carbon compounds from the analysis (see Fig. 2-14). Total organic carbon is a useful measurement because it provides an assessment of organic contamination and may be correlated to the amount of disinfection by-products (DBPs) that are produced during chlorination on a case-by-case basis. Dissolved organic carbon is the fraction of the TOC that passes through a 0.45-μm filter, and

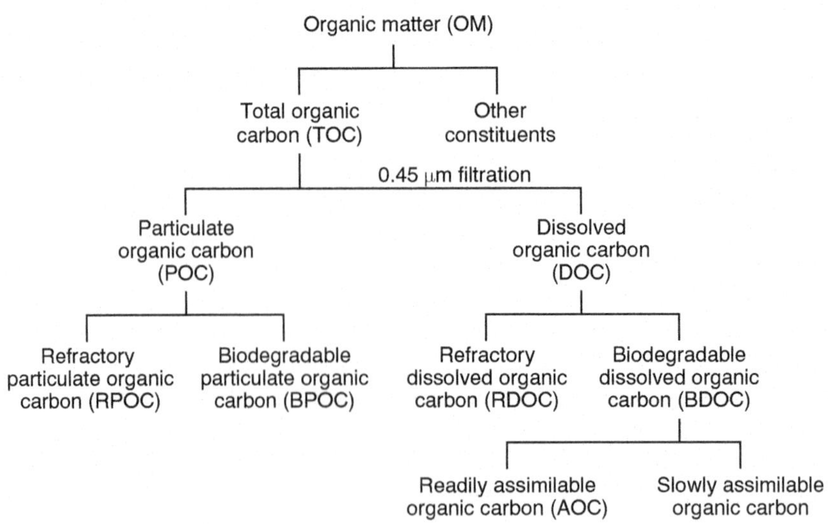

Figure 2-14 Classification of organic matter based on TOC method of analysis for total, particulate, and dissolved organic carbon. When different analytical methods of analysis are used, the term DOM is used in place of DOC. Refer to Table 2-7 for descriptions of these quantification methods.

the TOC of the material retained on the filter is defined as particulate TOC. As noted previously, the definition of DOC is operational, as a considerable amount of colloidal nondissolved material can pass through a 0.45-μm filter.

TOTAL ORGANIC HALOGEN

Total organic halogen refers to the total mass concentration of organically bound halogen atoms (X = Cl, Br, or I) present in water. From the standpoint of water quality, TOX is especially significant because it accounts not only for volatile halogen-containing compounds such as the THMs, trichloroethene, and tetrachloroethene but also includes the contribution of halogenated organic substances of high molecular weight that are also suspected health hazards. One commonly used method for TOX analysis involves the adsorption of organohalide solutes onto activated carbon (Dressman and Stevens, 1983). The particles of carbon are then washed to displace inorganic halides (predominantly Cl^-). After treatment with nitrate, the carbon adsorbent is subjected to pyrohydrolysis, which converts the organically bound halogen to hydrogen halides (HX) and hypohalous acids (HOX). The aqueous effluent from the pyrohydrolysis step (pyrohydrozylate) can be analyzed for halide ion using a specific ion probe or by direct injection of the sample into a microcoulometric titration cell.

TRIHALOMETHANE FORMATION POTENTIAL

The THMFP is employed to assess the maximum tendency of the organic compounds in a given water supply to form THMs upon disinfection. Water supply sources with low THMFP values are considered to be superior when it becomes necessary to choose between alternative sources of water. The subject of THMFP is considered further in Chap. 19.

2-6 Taste and Odor

The human senses of taste and odor (smell) are stimulated by a myriad of chemical compounds, both inorganic and organic. Certain ones of these compounds are found occasionally in domestic water supplies and, more than any other factor, influence the palatability of the product. Many water treatment plants include facilities for the feeding of taste and odor control chemicals, adsorbents, or both. Additionally, some agencies employ preventative and control measures in raw-water reservoirs, lakes, and rivers. It is impossible to estimate accurately the annual expenditure, nationally, on taste and odor control measures. Recommended limits on odors are set by the U.S. EPA in the National Secondary Drinking Water Regulations. The purpose of this section is to (1) identify the sources of tastes and odors in water supplies and (2) outline means to control their development or to remove them once they have appeared.

Sources of Tastes and Odors in Water Supplies

Tastes and odors in water supplies can generally be attributed to two different causative elements: natural forces within the environment and the actions of human beings upon the aquatic environment. These sources of tastes and odors are not unrelated. For example, odors due to biological degradation of algae and their waste products may sometimes be traced to an upstream nutrient input of human origin. Sources directly responsible for taste and odor production in groundwater and surface water supplies are considered in the following discussions. The examination of these sources is essential when attempting to identify particular tastes or odors.

TASTES AND ODORS IN SURFACE WATERS

Taste and odor problems are proportionally more common in surface waters than in groundwaters largely because of the presence of algae. In addition, direct organic inputs such as autumnal leaf fall, stormwater runoff, and agricultural drainage provide ample nutrients for microorganisms that can often generate taste- and odor-producing compounds. Decaying vegetation from leaf fall and other sources may result in brown-colored, sweet-smelling water. These effects are due to suspended and dissolved glucosides, such as tannin, that originate in vegetative matter. Other suspended particulates, such as colloidal silts and clays, may render a water unpalatable if not removed in treatment.

TASTES AND ODORS IN GROUNDWATER

Most tastes and odors in groundwater supplies are natural in origin. For example, tastes and odors are caused by bacterial actions within the groundwater aquifers or the dissolution of salts and minerals as groundwater percolates and flows through geologic deposits. Intrusion of salt or mineralbearing waters (such as seawater) may also result in taste or odor problems. Recently, tastes and odors in some groundwaters have been attributed to human sources, such as landfill leachate.

One of the most common odor problems in groundwater supplies is hydrogen sulfide (H_2S). Hydrogen sulfide is frequently characterized as a rotten-egg odor, but at low concentrations it may also impart a swampy, musty odor. The odor threshold concentration of H_2S in water is less than 100 ng/L (0.0001 mg/L), and odors from waters containing 0.1 to 0.5 mg/L or greater are offensive (Lochrane, 1979; Pomeroy and Cruze, 1969). Sulfides in groundwater result from anaerobic bacterial action on organic sulfur, elemental sulfur, sulfates, and sulfites.

Reduced iron and manganese may also pose taste problems in groundwater. Although tastes due to dissolved iron or manganese are not particularly noxious, they can render a water unpalatable and cause problems in pipelines, water services, and laundry facilities.

High salt content, as characterized by TDS or conductivity, can result in taste problems but does not usually result in objectionable odors. In general, consumers prefer waters with lower TDS content. The current widespread use of bottled mineral-bearing waters, however, may indicate that other psychophysical effects may affect taste preference.

Human-induced tastes and odors in groundwater occur as a result of chemical dumping, landfill disposal, mining and agricultural activities, or industrial waste disposal. A variety of synthetic organic chemicals have been identified in groundwater supplies. Examples include trichloroethylene (TCE), which has been found at objectionable concentrations in wells throughout the country.

Prevention and Control of Tastes and Odors at the Source

Taste and odor prevention and control may be accomplished at the source, in the treatment plant, and to a certain extent in the distribution system. Ideally, the most satisfactory site for control in surface supplies is at the source. Source control generally involves controlling the growth of algae and related organisms. For groundwater supplies, source control must be accomplished through watershed management—a difficult task. For surface reservoirs, algaecides, destratification/aeration, and watershed management are used as control methods. Purveyors using continuous draft intakes with negligible raw-water storage or detention most often address taste and odor problems in-plant rather than at the source. Taste and odor can be treated by oxidation (Chap. 8) or adsorption (Chap. 15).

2-7 Gases in Water

Gases commonly found in water, as reported in Table 2-2, include nitrogen (N_2), oxygen (O_2), carbon dioxide (CO_2), hydrogen sulfide (H_2S), ammonia (NH_3), and methane (CH_4). The first three are common gases of the atmosphere and are found in all waters exposed to the atmosphere. The latter three are derived from the bacterial decomposition of the organic matter present in water. Although not found in untreated water, other gases with which the environmental engineer must be familiar include chlorine (Cl_2) and ozone (O_3), which are used for oxidation, disinfection, and odor control.

Gases in water can form bubbles, which may interfere with sedimentation processes, as the bubbles carry particles up through the water column and filtration, as gases accumulate and disrupt flow through the filter. Gas bubbles in water can also interfere with water quality measurements such as dissolved oxygen, ions measured with electrodes, and turbidity. The quantity of a gas present in solution is governed by (1) solubility of the gas, (2) partial pressure of the gas in the atmosphere, (3) temperature, and (4) concentration of the impurities in the water (e.g., salinity, suspended solids). A discussion of the ideal gas law is presented below. The solubility of gases in water and Henry's law as applied to the gases of interest may be found in Chap. 14.

Ideal Gas Law

The ideal gas law, derived from a consideration of Boyle's law (volume of a gas is inversely proportional to pressure at constant temperature) and Charles' law (volume of a gas is directly proportional to temperature at constant pressure) is

$$PV = nRT \tag{2-28}$$

where P = absolute pressure, atm
V = volume occupied by gas, L, m^3
n = amount of gas, mol
R = universal gas law constant, 0.082056 atm/(mol/L) · K
T = temperature, K (273.15 + °C)

Using the universal gas law, it can be shown that the volume of gas occupied by 1 mole of a gas at standard temperature (0°C, 32°F) and pressure (1.0 atm) is equal to 22.414 L:

$$V = \frac{nRT}{P}$$

$$= \frac{(1\ \text{mole})[0.082056\ \text{atm}/(\text{mol}/\text{L})\cdot \text{K}][(273+0)\ \text{K}]}{1\ \text{atm}} = 22.414\ \text{L}$$

The following relationship, based on the ideal gas law, is used to convert between gas concentrations expressed in ppm_v and $\mu g/m^3$:

$$\mu\text{g}/\text{m}^3 = \frac{(\text{concentration, ppm}_v)(\text{MW, g}/\text{mol of gas})\left(10^6\ \mu\text{g}/\text{g}\right)}{22.414\times10^{-3}\ \text{m}^3/\text{mol of gas}} \tag{2-29}$$

The application of the Eq. 2-29 is illustrated in the following example.

Naturally Occurring Gases

Gases that are commonly found in untreated water include nitrogen, oxygen, carbon dioxide, ammonia, hydrogen sulfide, and methane. Ammonia, hydrogen sulfide, and methane are typically formed during the anaerobic decomposition of organic matter (see Table 2-2). Dissolved nitrogen, oxygen, and carbon dioxide are generally present in natural waters from equilibration with the atmosphere; however, these gases also have biological origins, from processes such as atmospheric nitrogen fixation, photosynthesis, and respiration, respectively.

Example 2-4 Conversion of gas concentration units

The gas released from a natural seep was found to contain 20 ppm_v (by volume) of hydrogen sulfide (H_2S). Determine the concentration in mg/m^3 and in mg/L at standard conditions (0°C, 101.325 kPa).

Solution

1. Compute the concentration in mg/L using Eq. 2-29.
   The molecular weight of H_2S = 34.08 g/mol [2(1.01)+32.06].

$$20\text{ppm}_v = \left(\frac{20\ \text{m}^3}{10^6\ \text{m}^3}\right)\left(\frac{34.08\ \text{g/mol}\ \text{H}_2\text{S}}{22.4\times10^{-3}\ \text{m}^3/\text{mol of}\ \text{H}_2\text{S}}\right)\left(\frac{10^6\ \mu\text{g}}{\text{g}}\right)$$
$$= 30{,}429\mu\text{g/m}^3$$

2. The concentration in mg/L is

$$30{,}429\mu\text{g/m}^3 = \left(\frac{30{,}429\ \mu\text{g}}{\text{m}^3}\right)\left(\frac{\text{mg}}{10^3\ \mu\text{g}}\right)\left(\frac{\text{m}^3}{10^3\ \text{L}}\right)$$
$$= 0.0304\text{mg/L}$$

Comment

If gas measurements, expressed in mg/L, are made at other than standard conditions, the concentration must be corrected to standard conditions, using the ideal gas law, before converting to ppm.

2-8 Radionuclides in Water

Radionuclides are unstable atoms that are transformed through the process of radioactive decay. Radioactive decay results in the release of radioactive particles (radiation). Radionuclides are of interest because of the health effects resulting from exposure to radioactive particles and their occurrence in natural waters. A brief review of the fundamental properties of atoms, types of radiation, and units of expression is presented in this section.

Fundamental Properties of Atoms

An atom is composed of three basic subatomic constituents: protons (positive charge, located in the nucleus), neutrons (no charge, located in the nucleus), and electrons (negative charge, located in the outer shell or orbitals surrounding the nucleus). An element is defined by its atomic number, which is equal to the number of protons in its nucleus. Elements with the same number of protons and variable number of neutrons are known as isotopes. Radium, for example, has six isotopes, ^{223}Ra, ^{224}Ra, ^{225}Ra, ^{226}Ra, ^{227}Ra, and ^{228}Ra, all of which have an atomic number of 88 (88 protons) and atomic mass of 223 to 228 (88 protons, 135 to 141 neutrons). The isotope that decays is known as the parent, and the resulting element is known as the progeny or daughter. Radioactive decay is the spontaneous disintegration of an element, resulting in greater atomic stability through change of electron orbits or release of radioactive particles or radiation.

Types of Radiation

The primary forms of radioactive decay are (1) alpha (particle) radiation, (2) beta (particle) radiation, and (3) gamma (ray) radiation. The release of alpha and beta particles transforms an isotope into a different element, while the release of gamma radiation reduces the energy of the element. Alpha, beta, and gamma radiations are known as ionizing radiation because of their ability to free electrons from their orbit in adjacent atoms.

Alpha particles are large, positively charged helium nuclei (two protons and two neutrons) released by certain isotopes during radioactive decay. Alpha particles are relatively slow and massive and are the least penetrating (may be stopped by the skin); however, when ingested, these particles can be very damaging to internal tissue and may cause cell mutation and possibly cancer. When an element emits an alpha particle, the element's atomic mass is reduced by 4 and its atomic number is reduced by 2. Beta particles are high-energy negatively charged particles released by certain elements during radioactive decay. Beta particles have smaller mass than alpha particles, which allows greater speed and penetration but creates less damage. The release of beta particles is characterized by the transformation of a neutron to a proton in the nucleus of an element and results in an increase of the atomic number. Gamma-ray emission, consisting of high- energy short-wave electromagnetic radiation (similar to x-rays) emitted from a nucleus, has tremendous penetrating power but has limited effect at low levels.

Units of Expression

The units used to quantify radionuclides in water include expressions for activity, exposure/dose, and rate of decay. Activity refers to the amount of radiation being emitted from a radioactive agent. Exposure is a function of the activity, type of radiation, and pathway of human contact, while the dose is used to express the bodily uptake of radioactivity from a given exposure scenario. The life span of a radionuclide is estimated by its rate of decay, or half-life. Activity, adsorbed dose, and dose equivalent are described below.

ACTIVITY

Radionuclides have unique properties that require units other than milligrams or moles per liter. Because the emission of radioactivity is not dependent on the mass of the element, units that quantify the activity of the element must be used. In the International System (SI) of units, the becquerel (Bq), equivalent to one disintegration or nuclear transformation (radioactive emission) per second, is the unit of radioactivity. In U.S. customary units, radiation is expressed in curies (Ci), 1 Ci is equivalent to 3.7×10^{10} disintegrations per second (37×10^9 Bq).

ADSORBED DOSE AND DOSE EQUIVALENT

Exposure to radionuclides through ingestion results in damage to internal organs as the element disintegrates. The amount of radiation that is imparted to the tissue is dependent on the number of particles emitted and is known as the absorbed dose. The SI unit for absorbed dose is the gray (Gy), where one gray equals one joule of radiation energy per kilogram of absorbing material. The corresponding U.S. customary unit is the radiation adsorbed dose (rad); 1 Gy is equal to 100 rad. Exposure to alpha, beta, and gamma radiation has different biological effects, so an exposure term known as the "dose equivalent" is used to quantify radiation that produces the same biological effect regardless of the type of radiation involved. The dose equivalent is determined by multiplying the adsorbed dose (in Gy or rad) by a quality factor. The quality factor is 1 for x-rays, gamma rays, and beta particles, and 20 for alpha particles. The units for dose equivalent is the sievert (Sv) in SI units and the Röntgen equivalent man (rem) in U.S. customary units; 1 Sv is equivalent to 100 rem.

Problems and Discussion Topics

2-1 Given the following test results, determine the mole fraction of calcium (Ca^{2+}).

Cation	Concentration, mg/L	Anion	Concentration, mg/L
Ca^{2+}	40.0	HCO_3^-	91.5
Mg^{2+}	12.2	SO_4^{2-}	72
Na^+	15.1	Cl^-	22.9
K^+	5.1	NO_3^-	5.0

2-2 Determine the mole fraction of magnesium (Mg^{2+})for the water given in Problem 2-1.

2-3 Determine the mole fraction of sulfate (SO_4^{2-}) for the water given in Problem 2-1.

2-4 Commercial-grade sulfuric acid is about 95 percent H_2SO_4 by mass. If the specific gravity is 1.85, determine the molarity, mole fraction, and normality of the sulfuric acid.

2-5 If the UV intensity measured at the surface of a water sample is 180 mW/cm^2, estimate the average intensity in a Petri dish with an average depth of 15mm (used to study the inactivation of microor-ganisms after exposure to UV light, as discussed in Chap. 13). Assume the absorptivity of the water, $k_A(\lambda)$ at $\lambda = 254$ nm, is 0.10 cm^{-1} and that the following form of the Beer–Lambert law applies:

$$\ln\left(\frac{I}{I_0}\right) = -2.303 k_A(\lambda) x$$

2-6 If the average UV intensity in a Petri dish containing water at a depth of 10 mm is 120 mW/cm^2, what is the UV intensity at the surface of the water sample? Assume the absorptivity of the water, $k_A(\lambda)$ at $\lambda = 254$nm, is 0.125 cm^{-1} and that the equation given in Problem 2-5 applies.

2-7 If the transmittance is 92 percent and a photo cell with a 12-mm path length was used, what is the absorptivity?

2-8 Given the following data obtained on two water supply sources, determine the constants in Eq. 2-16 (power law density and slope coefficients) and estimate the number of particles in the size range between 2.1 and 5. Also, comment on the nature of the particle size distributions.

	Particle Count	
Bin Size, μm	Water A	Water B
5.1–10	2500	110
10.1–15	850	80
15.1–20	500	55
20.1–30	250	36
30.1–40	80	25
40.1–50	60	20
50.1–75	28	15
75.1–100	10	10

2-9 The following particle size data were obtained for the influent and effluent from a granular medium filter. Determine the constants in Eq. 2-16 (power law density and slope coefficients) and assess the effect of the filter in removing particles.

	Particle Count	
Bin Size, μm	Influent	Effluent
2.51–5	20000	101
5.1–10	8000	32
10.1–20	2000	6
20.1–40	800	3.2
40.1–80	400	1.2
80.1–160	85	0.34
160.1–320	40	0.12

2-10 Determine the alkalinity and hardness in milligrams per liter as $CaCO_3$ for the water sample in Problem 2-1.

2-11 Given the following incomplete water analysis, determine the unknown values if the alkalinity and noncarbonate hardness are 50 and 150 mg/L as $CaCO_3$, respectively:

Ion	Concentration, mg/L
Ca^{2+}	42.0
Mg^{2+}	?
Na^{+}	?
K^{+}	29.5
HCO_3^-	?
SO_4^{2-}	144.0
Cl^-	35.5
NO_3^-	4.0

2-12 Given the following incomplete water analysis measured at 25°C, determine the unknown values if the alkalinity and noncarbonate hardness are 40 and 180 mg/L as $CaCO_3$:

Ion	Concentration, mg/L
Ca^{2+}	55.0
Mg^{2+}	?
Na^{+}	23.0
K^{+}	2.0
HCO_3^-	?
SO_4^{2-}	48.0
Cl^-	?
CO_2	4.0

2-13 Review the current literature and cite three articles in which the SUVA (specific UV absorbance) measurements were made. Prepare a summary table of the reported values. Can any conclusions be drawn from the data in the summary table you have prepared?

2-14 Review the current literature and prepare a brief synopsis of two articles in which the DOM (dissolved organic matter) was measured. What if any conclusions can be drawn from these articles about the utility of DOM measurements.

2-15 Determine the concentration in $\mu g/m^3$ of 10 ppm_v (by volume) of trichloroethylene (TCE) (C_2HCl_3) atstandard conditions (0°C and 1 atm).

2-16 If the concentration of TCE at standard conditions (0°C and 1 atm) is 15 $\mu g/m^3$, what is the corresponding concentration in ppm_v (by volume)?

References

Chapin, M. (2010) Water Science and Structure. Available at: http://www.lsbu.ac.uk/water/; accessed on Dec. 13, 2010.

Croue, J. P., Korshin, G. V., Benjamin, M. M., and AWWA Research Foundation (2000) *Characterization of Natural Organic Matter in Drinking Water*, AWWA Research Foundation and American Water Works Association, Denver, CO.

Dallavalle, J. M. (1948) *Micromeritics: The Technology of Fine Particles*, 2nd ed., Pitman Publishing, New York.

Davies, S. N., and DeWiest, R. J. M. (1966) *Hydrogeology*, John Wiley & Sons, New York.

Delahay, P. (1957) *Instrumental Analysis*, Macmillan, New York.

Dressman, R. C., and Stevens, A. (1983) "Analysis of Organohalides in Water—An Evaluation Update," J. AWWA, **75**, 8, 431-434.

Hach (2008) *Hach Water Analysis Handbook*, 5th ed., Hach Company, Loveland, CO.

Hem, J. D. (1971) *Study and Interpretation of the Chemical Characteristics of Natural Water*, Geological Survey Water Supply, Paper 1473, U.S. Government Printing Office, Washington, DC.

James M. Montgomery, Consulting Engineers, Inc. (1981) "Ute Water Conservancy District, Western Engineers Pilot Studies for Ute Water Treatment Plant Expansion."

Kavanaugh, M. C. (1978) "Modified Coagulation for Improved Removal of Trihalomethane Precursors," *J. AWWA*, **70**, 11, 613–620.

LeChevallier, M. W., and Norton, W. D. (1992) "Examining Relationships between Particle Counts and *Giardia, Cryptosporidium*, and Turbidity," J. AWWA, **84**, 12, 54–60.

LeChevallier, M. W., and Norton, W. D. (1995) "*Giardia* and *Cryptosporidium* in Raw and Finished Water," J. AWWA, **87**, 9, 54–68.

Levine, A. D., Tchobanoglous, G., and Asano, T. (1985) "Characterization of the Size Distribution of Contaminants in Wastewater: Treatment and Reuse Implications," J. WPCF, **57**, 7, 205–216.

Livingstone, D. A. (1963) *Chemical Composition of Rivers and Lakes, Data of Geochemistry*, 6th ed., Professional Paper 440-G, U.S. Geological Survey, Washington, DC.

Lochrane, T. G. (1979) "Ridding Groundwater of Hydrogen Sulfide," Water Sewage Works, Part 1, **126**, 2, 48 and Part 2, **126**, 4, 66.

McMurry J., and Fay, R. C. (2003) *Chemistry*, 4th ed., Prentice-Hall, Upper Saddle River, NJ.

NAS (1977) *Drinking Water and Health*, National Academy of Sciences Safe Drinking Water Committee, National Academy of Sciences, Washington, DC.

O'Melia, C. R. (1978) Coagulation in Wastewater Treatment, in K. J. Ives (ed.), *Scientific Basis of Flocculation*, Noordhoff International, Leyden, The Netherlands.

Owen, D. M., Amy, G. L., Chowdhury, Z. K., and AWWA Research Foundation (1993) *Characterization of Natural Organic Matter and Its Relationship to Treatability*, Foundation and American Water Works Association, Denver, CO.

Owen, D. M., Amy, G. L., Chowdhury, Z. K., Paode, R., McCoy, G., and Viscosil, K. (1995) "NOM Characterization and Treatability," J. AWWA, **87**, 1, 46–63.

Pomeroy, R., and Cruze, H. (1969) "Hydrogen Sulfide Odor Threshold," J. AWWA, **61**, 12, 677.

Sawyer, C. N., McCarty, P. L., and Parkin, G. F. (2003) *Chemistry for Environmental Engineering*, 5th ed., McGraw-Hill, Inc., New York.

Standard Methods (2005) *Standard Methods for the Examination of Water and Waste Water*, 21st ed., American Public Health Association (APHA), American Water Works Association (AWWA), and Water Environment Federation (WEF), Washington, DC.

Tchobanoglous, G., Burton, F. L., and Stensel, H. D. (2003) *Wastewater Engineering: Treatment, and Reuse*, 4th ed., McGraw-Hill, New York.

Tchobanoglous, G., and Schroeder, E. D. (1985) *Water Quality: Characteristics, Modeling, Modification*, Addison-Wesley, Reading, MA.

Thurman, E. M. (1985) *Organic Geochemistry of Natural Waters*, Martinus Nijhoff/Dr. W. Junk Publishers, Dordrecht, The Netherlands.

Trussell, A. R., and Umphres, M. D. (1978) "An Overview of the Analysis of Trace Organics in Water," J. AWWA, **70**, 11, 595–603.

Trussell, R. R., and Tate, C. H. (1979) Measurement of Particle Size Distribution in Water Treatment, in *Proceedings Advances in Laboratory Techniques for Water Quality Control*, American Water Works Association, Philadelphia, PA.

Turekian, K. K., (1971) "Rivers, Tributaries and Estuaries," in D. W. Hood (ed.), *Impingement of Man on the Ocean*, John Wiley & Sons, New York.

USGS (2000) National Reconnaissance of Emerging Contaminants in the Nations Stream Waters, U.S. Geological Survey. Available at: http://toxics.usgs.gov/regional/contaminants.html.

3 Microbiological Quality of Water

Stantec's Water Treatment: Principles and Design, Updated Third Edition. John C. Crittenden, R. Rhodes Trussell, David W. Hand, Kerry J. Howe, and George Tchobanoglous.

Terminology for Microbiological Quality of Water

Term	Definition
Aerobic	Metabolic process carried out in the presence of free oxygen, where oxygen serves as the terminal electron acceptor.
Anaerobic	Metabolic process carried out in a reduced environment in the absence of free oxygen, where compounds such as SO_4^{2-} and CO_2 serve as terminal electron acceptors.
Anoxic	Metabolic process carried out in a partially reduced environment in the absence of free oxygen, where compounds such as NO_3^-, NO_2^-, and Fe(III) serve as terminal electron acceptors.
Autotrophs	Organisms that produce complex organic compounds (such as carbohydrates, fats, and proteins) from simple inorganic molecules (e.g., CO_2) using energy from light (photosynthesis) or inorganic chemical reactions (chemosynthesis).
Cyst	Resting, nonmotile, encysted stage of amoeba or flagellate (e.g., *Entamoeba hystolytica,Giardia lamblia*).
Egg	Encysted form of helminths (e.g., *Ascaris*, Schistosomes).
Endemic	Condition where a disease is normally present in the population without external inputs.
Endopore	Dormant, highly resistant structure produced by some gram-positive bacteria (e.g., *Bacillus, Clostridium*).
Enteric	In the gastrointestinal system (mouth, throat, stomach, doudenum, small intestine, large intestine, anus).
Epidemic	Condition where a disease is rapidly spreading in the population.
Facultative	Organisms that have metabolic processes that allow them to operate under aerobic, anoxic, and anaerobic conditions.

Term	Definition
Fecal-oral	Route of disease transmission from one person to another route where a pathogen present in one person's feces is transmitted to another through the mouth (usually in food or water).
Helminths	Parasitic worms.
Heterotrophs	Organisms that use organic chemicals as carbon source; can be aerobic, facultative, or anaerobic.
Metabolism	Biochemical cellular reactions involved in maintaining cell viability. Metabolism may occur under aerobic, anaerobic, anoxic, or facultative conditions.
Morbidity ratio	Fraction of infected persons who exhibit the symptoms of the disease.
Mortality ratio	Fraction of the persons who exhibit symptoms of the disease who ultimately die from it.
Oocyst	Encysted form resulting from fertilization during life cycle of sporozoa; indicates sexual reproduction (e.g., *Cryptosporidium*).
Pandemic	Condition where an epidemic is expanding in several countries.
Parasite	Organism that draws its nutrients from another. A parasite cannot live independently.
Pathogens	Microorganisms capable of causing disease in humans, including bacteria, viruses, protozoa, helminths (worms), and algae.
Reproduction	Act of replication, either sexually or asexually. Some organisms reproduce by only one method, others can utilize either method. Microorganisms can reproduce sexually and asexually.
Virulence (pathogens)	Measure of the severity of the damage to the host.

The material on microbial quality presented in this chapter is designed as background material for subsequent chapters on treatment and disinfection. The specific objectives of this chapter are fourfold: (1) to provide an overview of the microbial world including the types of microorganisms and their characteristics, especially as they affect drinking water treatment, (2) an introduction to pathogens in drinking water and the routes of transmission of enteric disease, (3) a comprehensive discussion of bacteria, viruses, protozoa, helminths (worms), and algae of concern in drinking water, and (4) a discussion of the issues related to bacterial monitoring of water supply sources. The following discussion presumes the reader, while trained in the sciences, does not specialize in microbiology. Additional information can be obtained from literature devoted to this subject.

3-1 Overview of the Microbial World

Since the middle of the nineteenth century, biologists have assumed that an evolutionary tree of life could be constructed. It was believed that the appearance and capabilities of organisms could be traced, in a linear way, up and down that evolutionary tree as organisms evolve from one to another. Until the introduction of genetic analytical techniques late in the twentieth century, the tree was constructed mostly on the basis of the morphology and behavior of organisms relative to other organisms. At the time, there were thought to be two fundamental branches to the tree, prokaryotic organisms and eukaryotic organisms, prokaryotes having only one outer cell membrane and eukaryotes being more formally structured.

In 1977, microbiologist Carl Woese constructed an evolutionary tree using genetic information and demonstrated a third domain, the archaea (Woese and Fox, 1977). The archaea are similar to bacteria, but their genetic makeup is different. Like bacteria, archaea lack a true nucleus, usually have one deoxyribonucleic acid (DNA) molecule suspended in the cell's cytoplasm contained within a cell membrane, and most also have a rigid outer cell wall. The name *archaea* is taken from the Greek word for "ancient" because these organisms do best under extreme environmental conditions, such as those when the earth was young, for example, the conditions found in hydrothermal vents. A modern phylogenetic tree of life is shown on Fig. 3-1. As shown

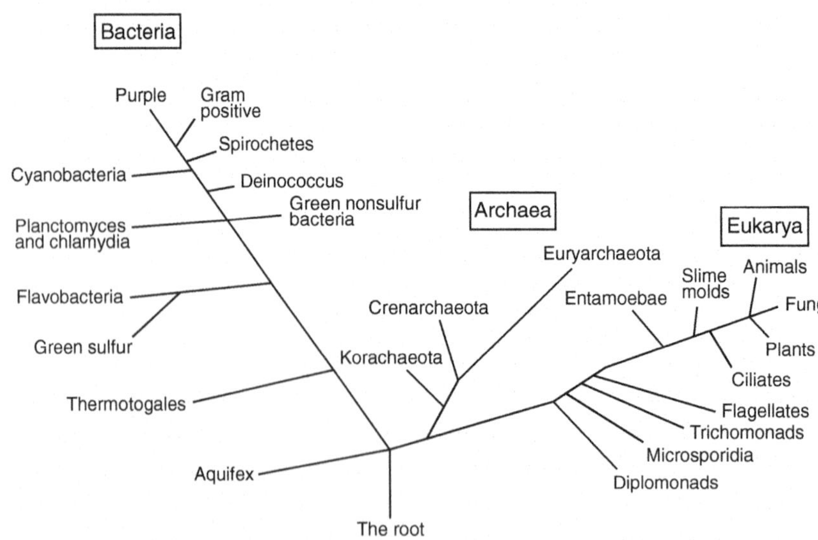

Figure 3-1
Universal phylogenetic tree of life. (Adapted from Rittman and McCarty, 2001).

on Fig. 3-1, the tree of life has three principal branches: bacteria, archaea, and eukarya. There are no known human pathogens among the archaea.

The field of aquatic microbiology encompasses diverse organisms. Where drinking water is concerned, the groups of interest include bacteria, viruses, algae, protozoa, and helminths (worms). Each of these groups is discussed in this chapter. Some aspects of each group that are important in understanding aquatic microbiology are summarized in Table 3-1. These characteristics include size, surface charge, shape, oxygen requirements, carbon and energy requirements, motility, and the environmentally resistant stage for each.

Bacteria are single-celled organisms ranging in size from about 0.1 to 10 µm. Even though their sizes range over two orders of magnitude, all bacteria have a relatively simple structure and composition. Bacteria have one key membrane structure - the membrane bounding the cell itself. The bacterial interior contains two major regions: the cytoplasm and the nuclear region, both relatively uniform in appearance. Cells of this simple type are termed prokaryotic. Cyanobacteria (also known as blue-green algae) are also prokaryotic. Other cells of algae, fungi, protozoa, plants, and animals are eukaryotic. Eukaryotic cells contain membrane-bound regions that are distinctive both morphologically and physiologically.

The simple appearance of the bacteria is deceiving; physiologically they are more diverse than any other biological group. Structural uniformity and physiological diversity make the classification of bacteria difficult. Typically, a bacterial genus is identified based on a combination of morphology and metabolic responses. For example, the genus *Escherichia* (including the species *Escherichia coli*) is described as unicellular, nonphotosynthetic, non-spore forming, straight, rod shaped, less than 2 µm wide, gram negative, aerobic, heterotrophic, acid and gas producing from glucose and lactose within 48 h, methyl red positive, Voges-Proskauer negative, and catalase positive, with four species differentiated on the basis of pigmentation, utilization of citrate, and production of H_2S. As might be suspected, the classification of bacteria is constantly changing; for example, *E. coli* was previously assigned to the genus *Bacterium* as *Bacterium coli* and to the genus *Bacillus* as *Bacillus coli*. More recent classifications have been developed that are primarily based on the organism's genome. Genetic analysis is having a revolutionary impact on our understanding of the evolution of members of the microbial community.

Physical Characteristics of Microorganisms

The size, shape, motility, and surface charge of microorganisms can have important influences on their removal by water treatment processes. As discussed in Chaps. 9, 10, and 11, particles in the range of 1 to 10 µm in diameter are more difficult to remove in conventional processes. Unfortunately, many important pathogens fall in this range, including most bacteria and the protozoan *Cryptosporidium parvum*. Size is a particularly important consideration in the treatment as well as in the detection of microorganisms. To put the microbiological world and some of its inhabitants in context, the relative sizes of a number of organisms are displayed on Fig. 3-2.

Table 3-1
Important characteristics of classes of aquatic microorganisms[a]

Organism	Size, μm	Surface Charge	Shape	Oxygen Requirement	Carbon and Energy Requirement	Motility	Environmentally Resistant Stage
Viruses	$10^{-2}-10^{-1}$	Negative	Variable, rod	NA	NA	None	Viron
Bacteria	$10^{-1}-10$	Negative	Coccoid, spiral, comma	Aerobic, anaerobic, facultative	Chemoautotroph,[b] chemoheterotroph,[c] photoautotroph, photoheterotroph	Motile, nonmotile	Spores, cystlike
Blue-green algae	1	Negative	Coccoid filamentous	Aerobic	Photoautotroph	Gliding	Cysts
Green algae	$1-10^{2}$	Negative	Colloid pennatic	Aerobic	Photoautotroph	Motile, nonmotile	Spores, cysts
Protozoa	$1-10^{2}$	Negative	Variable	Aerobic, anaerobic	Chemoheterotroph[c]	Motile, nonmotile	Cysts, oocysts
Fungi	$1-10^{2}$	Negative	Filamentous coccoid	Aerobic, anaerobic	Chemoheterotroph[c]	Nonmotile	Spores
Helminths	$1-10^{5}$	Negative	Variable	Aerobic	Chemoheterotroph[c]	Motile	Eggs

[a]NA = not applicable.
[b]Typically obtain energy from inorganic compounds (also known as Chemolithotrophic autotroph).
[c]Typically obtain energy from organic compounds (aiso known as Chemoorganotrophic heterotroph).

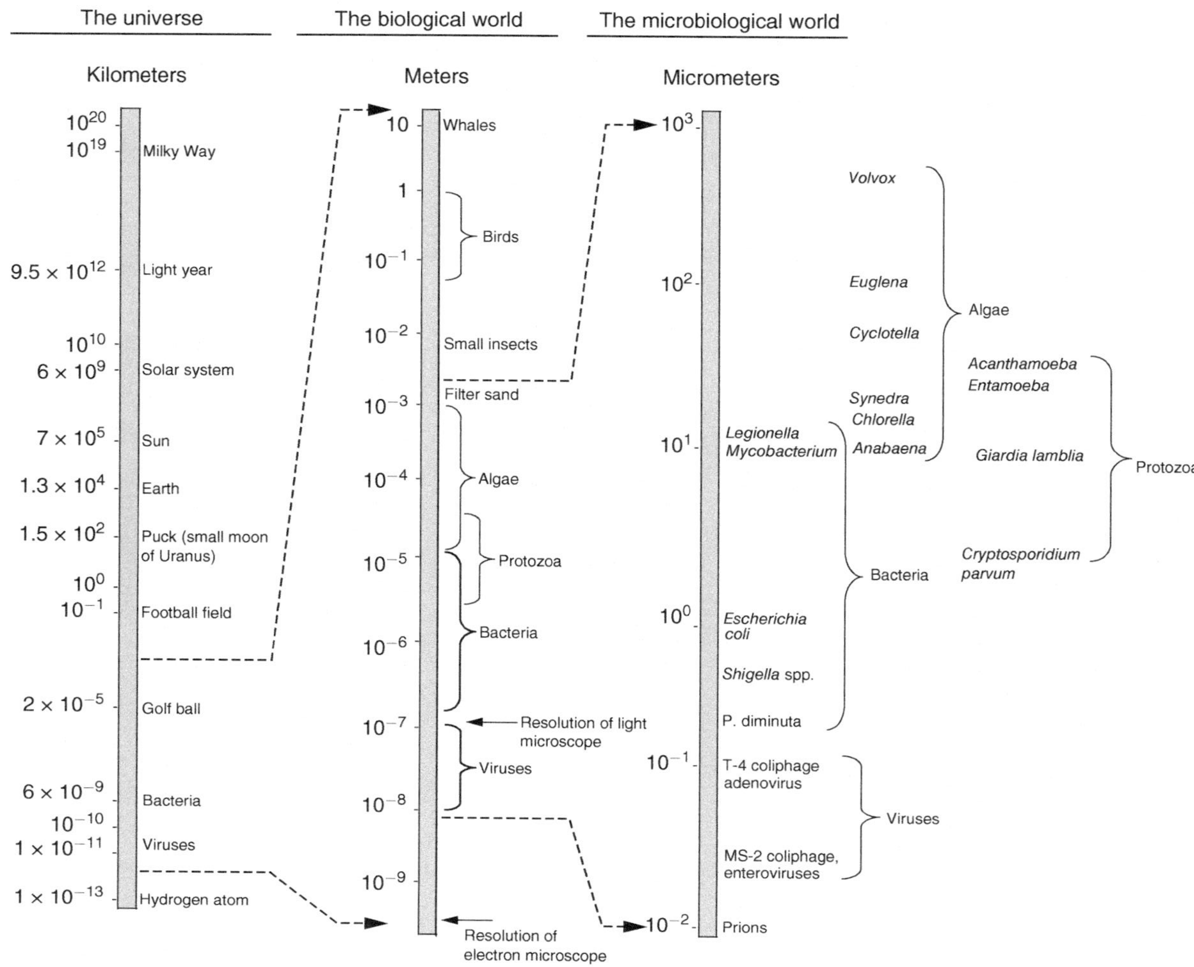

Figure 3-2
Relative sizes of universe, biological world, and microbiological world. On the left the biological world as a whole is compared with the rest of the universe, ranging from the hydrogen atom to the Milky Way. On the right the microbiological world is compared with the rest of the biological world.

The smallest inhabitants of the microbiological world are the prions, protein molecules of approximately 230 to 240 base pairs in length. *Volvox*, a species of algae formed by hundreds of cells, is among the largest members of the microbiological world.

Many algae and helminths are visible to the unaided eye, while most bacteria can only be observed through a microscope. In 1683, Anton van Leeuwenhoek reported to the Royal Society of London on the observation of bacteria found in the plaque on his teeth using the microscope he had developed. However, serious work with bacteria and disease did not develop until Robert Koch had access to one of the innovative microscopes Carl Zeiss introduced into the market in 1883. Viruses are the smallest of waterborne agents, typically ranging in size from about 0.02 to 0.2 μm. Viruses are too small to be seen with a light microscope but can be observed with an electron microscope.

Most, if not all, microbiological particles exhibit a negative surface charge over the range of pH of interest in water treatment, and so do most other aquatic particles and filter media. As a result, sedimentation and granular media filtration do not effectively remove pathogens unless a coagulant is used to address the forces of repulsion. Some organisms are motile, and experience has shown that these microorganisms are even more resistant to effective removal by granular media filtration. For these microorganisms, disinfection is first required to eliminate motility before coagulation and filtration can be effective.

Oxidation-Reduction Potential (E_H) and pH

The response of a microorganism to oxidation-reduction (redox) and pH conditions provides insight as to environments where the organism might be found. To survive under any conditions, a microorganism needs an electron acceptor and donor. When free hydrogen and oxygen are available, reducing the oxygen to water results in a transfer of electrons. As redox potential drops, other electron acceptors become important. Some of the more important conversions are the reduction of nitrate to nitrogen gas, manganese dioxide to Mn(II), ferric iron to ferrous iron, and sulfate to sulfide. Methane gas is generated under conditions of extremely low redox otential (E_H). The stoichiometry of some of these reactions is given below:

$$O_2 + 4H^+ + 4e^- \rightarrow 2H_2O \tag{3-1}$$

$$2NO_2^- + 8H^+ + 6e^- \rightarrow N_2 + 4H_2O \tag{3-2}$$

$$MnO_2 + 4H^+ + 2e^- \rightarrow Mn^{2+} + 2H_2O \tag{3-3}$$

$$Fe_2O_3 + 6H^+ + 2e^- \rightarrow 2Fe^{2+} + 3H_2O \tag{3-4}$$

$$SO_4^{2-} + 10H^+ + 8e^- \rightarrow H_2S + 4H_2O \tag{3-5}$$

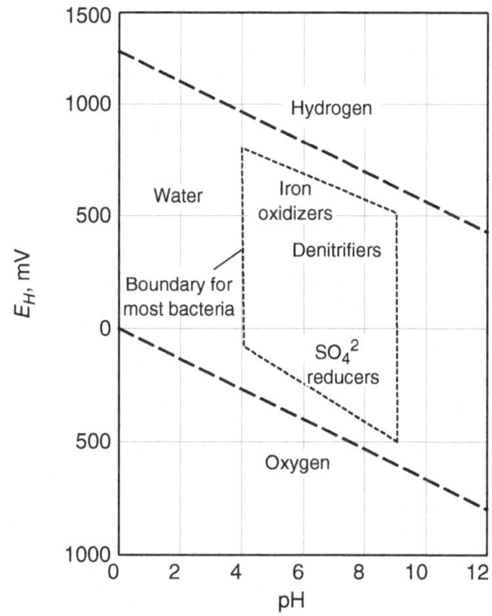

Figure 3-3
Typical range for microorganisms as a function of E_H and pH. (Adapted from Rheinheimer, 1991).

Most bacteria operate in an E_H range where the transformations they are designed for are easily accomplished. Thus, iron bacteria prefer a low pH but high E_H. Sulfate reducers (e.g., *Desulfovibrio*) prefer a higher pH and a very low E_H. Denitrifiers operate well at a neutral pH and under mildly anoxic conditions. The availability of these electron acceptors to microorganisms is influenced not only by the redox potential but also by the pH. The regions of optimal redox conditions for many microorganisms on an E_H – pH plane are shown on Fig. 3-3. Most bacteria can operate in a pH range of 4 to 9 and function best in the range of pH 6.5 to 8.5, while some can tolerate pH values outside these boundaries. Thiobacilli can tolerate pH below 1.0 and function best between pH of 2 and 3.

Biomolecular Revolution

During the past decade, significant advancements in the understanding of the chemistry of life have occurred. This revolution in understanding has its roots in the discovery of DNA, ribonucleic acid (RNA), and the role these chemicals play in the structure and growth of all living organisms.

Deoxyribonucleic acid is a double-stranded sugar-phosphate backbone with lattices made up of pairs of four nucleic acids. The sugar in the backbone is deoxyribose and the four nucleic acids are adenine, guanine, cytosine, and thymine. Ribonucleic acid is a single-stranded sugar-phosphate backbone whose lattices are also made up of pairs of four nucleic acids. The sugar in the backbone of RNA is ribose and three of the nucleic acids are the same as those in DNA, with uracil substituted for thymine. The DNA and RNA structures are shown, along with their key components

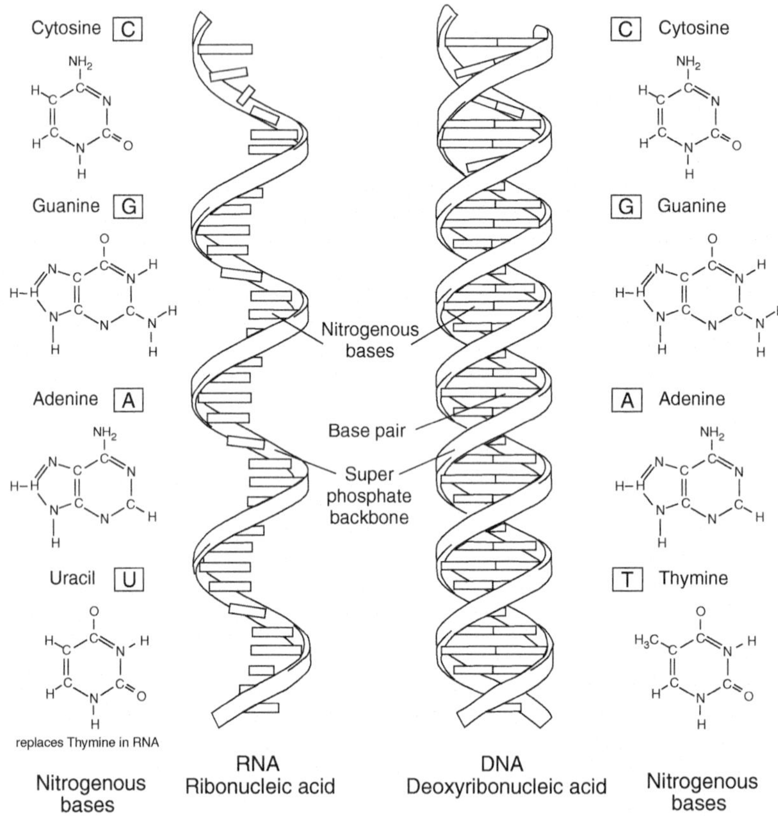

Figure 3-4
Basic structure of DNA and RNA and corresponding nitrogenous bases.

on Fig. 3-4. The DNA serves as the library of life, the database of genetic information for the cell. The role of RNA is to govern metabolic processes.

The fact that DNA played an essential role in life's genetic structure was known from the early 1940s. It was Linus Pauling who proposed that DNA had a helical structure in 1948, but it was Watson and Crick with help from Wrinkle and Franklin who developed the correct details of the DNA structure in 1953.

During recent years, the technology required to support the determination of the genetic fingerprint of a given organism has undergone dramatic improvements, and the entire genomes of numerous organisms are rapidly appearing in the public domain. The availability of this information is already beginning to have fundamental impacts on the understanding of the relationship between organisms and their behavior. For example, in addition to the slow, vertical evolutionary changes up the tree of life that were introduced earlier, microorganisms (including pathogens) also have considerable capacity for conducting more rapid horizontal exchanges in genetic material to facilitate useful evolutionary patterns.

Facilitation of Rapid Evolution among Bacterial Pathogens

Bacteria reproduce by binary fission, and in this process the chromosome in the original cell is duplicated in an identical daughter cell. For evolutionary changes to occur, changes to the chromosome in individual bacteria must take place, and these changes must then be passed on to future generations by the binary fission just described. The principal events and means that result in inheritable changes in the genome are (a) through the lysing and release of DNA from one organism, which when taken up by another organism can result in genetic transfer and recombination, (b) the direct transfer of genetic material through the exchange of plasmids from one organism to another, and (c) by mobile segments of DNA (often identified as *transposons*) that can move around to different positions in the genome of a single cell and thereby cause mutations.

Genetic transfer has been shown to occur when one cell dies and breaks up and another cell takes on a portion of its DNA. Such genetic transfer can also occur through the action of phage. Plasmids are small, circular, self-replicating units of genetic material that normally exist in the bacterial cell, but outside the chromosome itself. Some cells possess the ability to transfer their plasmids to other cells, transmitting genetic material in the process. Transposons are genetic elements that can move within the organism, from one plasmid to another and from plasmids to chromosomes.

All of these mechanisms of gene transfer have been demonstrated, but plasmid exchange is probably the best understood at the moment. Plasmids have been shown to transmit drug resistance (e.g., ampicillin, tetracycline, kanamycin, and chloramphenicol), resistance to UV light, resistance to metals (e.g., mercury, cobalt, magnesium, copper, and silver), resistance to bacteriocin (proteinaceous toxins given off by bacteria to inhibit the growth of similar bacterial strains), toxin production (both exotoxins and enterotoxins), the ability to metabolize specific sugars and hydrocarbons, and the ability to create tumors. The exchange of plasmids among bacteria is analogous to sexual processes. Some bacteria possess appendages called pilli, and through these pilli they are able to transmit plasmids to bacteria that do not have them.

3-2 Pathogens in Drinking Water

From the beginning, the practice of drinking water treatment has been rooted in the need for removal of pathogens. This goal remains among the highest priorities today. Paul Ewald, a notable evolutionary microbiologist, recently said, "No other single intervention in the history of medicine has saved as many lives and reduced as much suffering as the provisioning of uncontaminated water" (Ewald, 1994; Ewald et al., 1998).

Discovery of Waterborne Disease

It is well documented that waterborne contamination can cause disease of epidemic proportions. However, there was a time when it was a common belief that diseases such as cholera and typhoid fever were primarily transmitted by other means, such as breathing miasma, vapors emanating from the decaying victim and transported through the open air during the night.

Studying the cause of London epidemics of Asiatic cholera in 1849 and 1853, Dr. John Snow demonstrated that the second epidemic, which killed nearly 500 people in a span of 10 days (victims often died within 36 h of exposure), was associated with contamination of the water at the Broad Street Well by water from a nearby cesspool. Snow was unable to identify the specific agent in the water, but he suspected that it was microbiological in character and that it somehow replicated itself in great numbers, exiting the victim's gastrointestinal tract (Snow, 1855). In that same year, Falipo Pacini identified the organism that caused cholera (Pacini, 1854; Bentivoglio and Pacini, 1995), but his discovery went largely unknown. A short time later, William Budd (1856) demonstrated that typhoid fever could be transmitted by the same means. In 1864 a Frenchman, Dr. Louis Pasteur, articulated the germ theory of disease. In 1880, Karl Eberth (1883) isolated the organism that caused typhoid fever (*Salmonella typhi*), and in 1883, Robert Koch, one of the most important figures in the history of public health microbiology, while working in Egypt and then in India, identified the comma-shaped bacillus responsible for Asiatic cholera (*Vibrio cholerae*) (Howard-Jones, 1984). It soon became clear that a number of important epidemic diseases were often waterborne—cholera, typhoid fever, and amoebic dysentery, among them.

In the nineteenth century, the term *cholera* was used to describe many warm-weather intestinal diseases associated with pain, vomiting, diarrhea, fever, and prostration. The term *Asiatic cholera* was used to distinguish a variety of cholera, usually fatal, that had remained endemic to Asia until around 1817 when the first of several *pandemics* resulted in its spread around the world. Because the disease was thought to originate in Asia and because it was so potent, it was called Asiatic cholera.

Role of Water in Transmitting Disease

Perhaps the most unique aspect of water as a vehicle for the transmission of disease is that a contaminated water supply can rapidly expose a large number of people. When food is contaminated with a pathogen, tens to hundreds of persons are commonly infected. If a large, centralized food-packaging facility is involved, thousands might be infected. However, when drinking water is contaminated with a pathogen, hundreds of persons are typically infected, and occasionally the number of persons infected can rise to the hundreds of thousands. For example, in the 1993 Milwaukee *Cryptosporidium* incident, it is estimated that 500,000 people became ill from contaminated drinking water (MacKenzie et al., 1994).

Another important consideration is that waterborne transmission can have an impact on the virulence of the pathogens themselves. *Virulence* is a term used to describe the severity of the damage to the host (Casadevall and Pirofski, 1999). A pathogen that is more virulent causes more damage. Under normal conditions, pathogens are limited in the virulence they can exhibit because they depend on the mobility of their host for transmission.

The principal mechanisms for the transmission of enteric (intestinal) diseases are shown on Fig. 3-5. Suppose that, while it is infecting an adult, a pathogen evolves to a form that causes a severe, debilitating form of an enteric disease that immobilizes and seriously injures the infected person. The route of transmission can be analyzed using Fig. 3-5. If an adult with severe illness is too debilitated to prepare food, the organism cannot get into the food supply. However, the organism does get in the sewer, even if the sick person cannot get out of bed. Once in the sewer, the organism is then transported to the wastewater treatment plant. If the organism is not removed or inactivated at the wastewater treatment plant, it enters the receiving watercourse. If that watercourse serves as a water supply, then the organism has entered a water supply. If water treatment does not remove or inactivate the organism, both healthy toddlers and adults who drink the water are exposed and may get infected. A smaller number of individuals may also be exposed via contact with the infected persons. Thus, the entire population drinking the water supply is potentially exposed to the disease-causing agent. Under these conditions, an organism can successfully reproduce even if it causes a severe disease from which the host rarely recovers. According to some historical accounts, the classic form of Asiatic cholera that appeared in the middle of the nineteenth century behaved in this way.

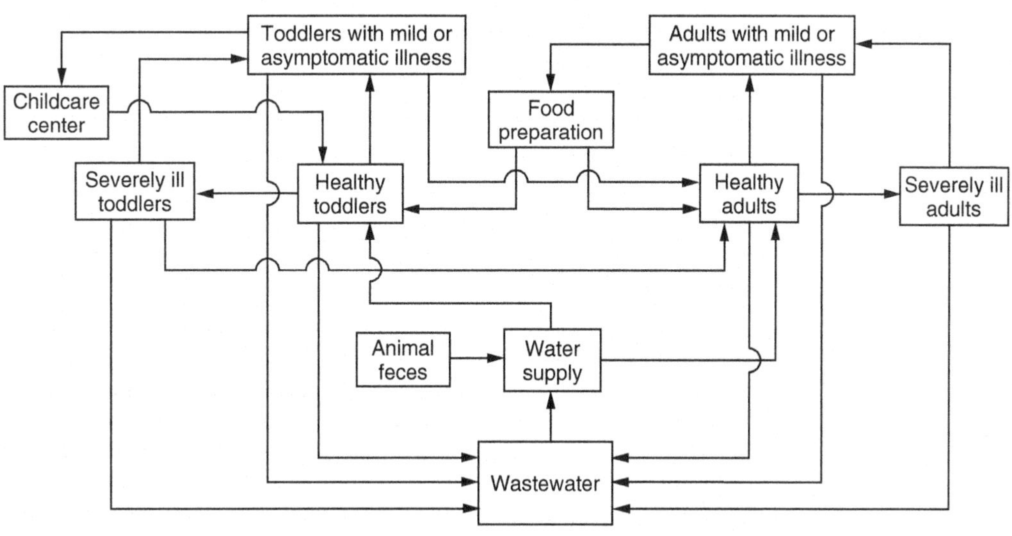

Figure 3-5
Schematic of routes of transmission for enteric disease.

The route of transmission can be interrupted by removing or inactivating the organism from the water in two places:

1. at the drinking water treatment plant
2. at the wastewater treatment plant

Organisms that evolve to cause mild disease or asymptomatic infections have different characteristics, as shown on Fig. 3-5. Examining the water route, again, as the disease is enteric, the organism will get into the wastewater. If it is not removed from the wastewater, it will also be in the water supply. And if it is not removed in water treatment either, it will also be in the drinking water—exposing both toddlers and adults. Some of these will then become mild or asymptomatic carriers. The infected toddlers will spread the disease at their childcare centers and directly to their adult caregivers if the caregivers do not exercise sufficient care. Either water treatment or wastewater treatment can intervene and stop the spread via this route.

Figure 3-5 can also be used to consider the spread of the asymptomatic disease via the food route. Adults with the disease, if they do not use adequate hygiene, may contaminate food when they prepare it. Both toddlers and adults who eat the contaminated food may then get infected. Some of those who get infected will be asymptomatic; others may exhibit mild symptoms. Infected adults may again prepare and contaminate food, and some infected toddlers will go to childcare centers. Toddlers in childcare centers will expose other toddlers who attend there. Again adult caregivers, who do not exercises sufficient hygiene, will also expose themselves while handling the sick toddlers. The drinking water has no connection to this route of communication, so treating the drinking water will not stop it. Training and regulating people working in food handling can largely eliminate transmission of the disease from an adult to others, but it seems likely that transmission among toddlers in childcare centers will be difficult to eliminate. The impact of a water treatment intervention is much greater where severe, debilitating disease is concerned.

It is not possible to eliminate diseases that are transmitted by the fecal-to-oral route. Such diseases will continue to be spread through the contamination of food, activities at daycare centers, and other endeavors of asymptomatic carriers. Eliminating the transport of pathogens through the drinking water route will result in the elimination of large-scale waterborne outbreaks. It will also make it much more difficult for organisms to survive if they produce symptoms that damage the host to the point where the host is forced to stay at home soon after contracting the disease.

An opportunistic pathogen is a microorganism that is not ordinarily able to overcome the natural defenses of a healthy human host. Under certain circumstances, however, such organisms are able to cause infection resulting in serious damage to the host. There are two circumstances when opportunistic pathogens are more successful: (a) when the immune response of the host has been compromised [e.g., persons with human immunodeficiency virus (HIV), persons on drugs that suppress the immune system, the very elderly] or (b) when the host is exposed to such high levels of the organism in question that the infection becomes overwhelming before the body can develop a suitable immune response.

Water treatment and good sanitation practices are also necessary for successful application of antibiotic and vaccine therapies for gastrointestinal (GI) diseases. In developing countries with poor sanitation and inadequate water treatment, pathogens have wide and easy access to their human hosts. Under these circumstances, pathogenic bacteria can more successfully evolve to gain antibiotic resistance, and some pathogenic viruses have even evolved the capability to evade vaccine therapies.

Pathogens of Importance to Water Treatment

It is widely known that some of humankind's worst scourges have been caused by waterborne disease. Until the last few decades of the twentieth century, water treatment technology focused on controlling diseases that were spread from one person to another by the fecal-oral route via drinking water. Diseases of this kind continue to occur throughout the world, and inadequate water treatment almost always plays a role. In parts of the world where the infrastructure for water and sewage is poorly developed, classic examples of these diseases (e.g., cholera and typhoid) still cause devastating epidemics. Even in more developed parts of the world, milder forms of gastroenteritis can become widespread when this infrastructure fails even for a short time. Clearly, preventing the spread of waterborne disease must remain the highest priority in water treatment.

New diseases have also come to the attention of the water treatment community that are not spread from one person to another via the drinking water. Most important among these are zoonotic diseases, which humans can contract from other animals. Also important are diseases caused by opportunistic pathogens, most of which are not associated with fecal contamination but instead live in various aquatic environments. While many of these diseases have come to our attention through the efforts of the medical profession to address the needs of individuals with suppressed immune systems, it has now become clear that some are important diseases for healthy individuals as well.

The most prominent zoonotic diseases of interest in water treatment at the present time are giardiasis and cryptosporidiosis. Both of these diseases are transmitted by the traditional fecal-oral route between humans but also from animals to humans. Zoonotic microorganisms of the gastrointestinal tract, such as *Salmonella* spp., infect animals as well as humans and cause a variety of human diseases and are responsible for the origins of new strains of pathogens, such as the flu (influenza viruses). It has been argued that most epidemic diseases of humans may originate from living in close community with animals (Diamond, 1999).

Pathogenicity

The microorganisms that are of most concern are the pathogens. The scientific community has not always used consistent terminology when discussing the concept of pathogenicity, partly because the behavior of microorganisms is extremely complex. One microorganism may colonize humans without causing disease or even have beneficial effects (commensalism). Another microorganism may colonize some humans without seeming to cause disease (asymptomatic infection) and yet colonize other humans and result in disease symptoms (symptomatic infection or infection resulting in disease). Casadevall and Pirofski (1999) discussed pathogenicity in depth, arguing that the unique property of pathogens is that they *damage* the host organism (i.e., cause disease).

The number of organisms that a human must be exposed to before infection varies a great deal from one pathogen to the next and also from one human to the next. The public health community uses the median infectious dose, N_{50}, as a measure of the "typical" dose required for infection in human beings. The methods for conducting such dose-response assessments are beyond the scope of this discussion but may be found in Haas et al. (1999). The wide variation in median dose from one pathogen to the next to bring about a response is illustrated on Fig. 3-6. On a mass basis, the pathogenic dose can be extremely low, as demonstrated in Example 3-1.

INFECTION OUTCOMES

Pathogenicity is a complex phenomenon. Infection by a pathogen does not automatically translate to damage to the host. Infection means that the pathogen is successfully reproducing in the host. Depending on the pathogen and the individual host involved, such an infection can result in a variety of outcomes, including (1) asymptomatic infection (no symptoms or very mild symptoms—no damage to the host), (2) mild illness (mild symptoms—no permanent damage to the host), (3) acute illness (severe symptoms—often some permanent damage to host), or (4) death. The possible infection outcomes in a conceptual framework are summarized on Fig. 3-7a. Often the fraction of infections that result in asymptomatic infection is not well known because asymptomatic individuals do not seek medical care and special studies are required to determine if an individual not showing symptoms is infected. An organism, which often causes asymptomatic infection, is likely to have many routes of transmission beyond drinking water.

Hepatitis A virus is better understood than many other pathogens. The approximate likelihood of various outcomes with this pathogen is illustrated on Fig. 3-7b. Three out of four adults infected with the hepatitis A virus are asymptomatic. Persons who contract the disease experience the sudden onset of a flulike illness. After a few days of muscle aches, headache, anorexia, abdominal discomfort, fevers, and malaise, jaundice may set in. Full recovery usually takes 2 months, but 10 to 15 percent of the

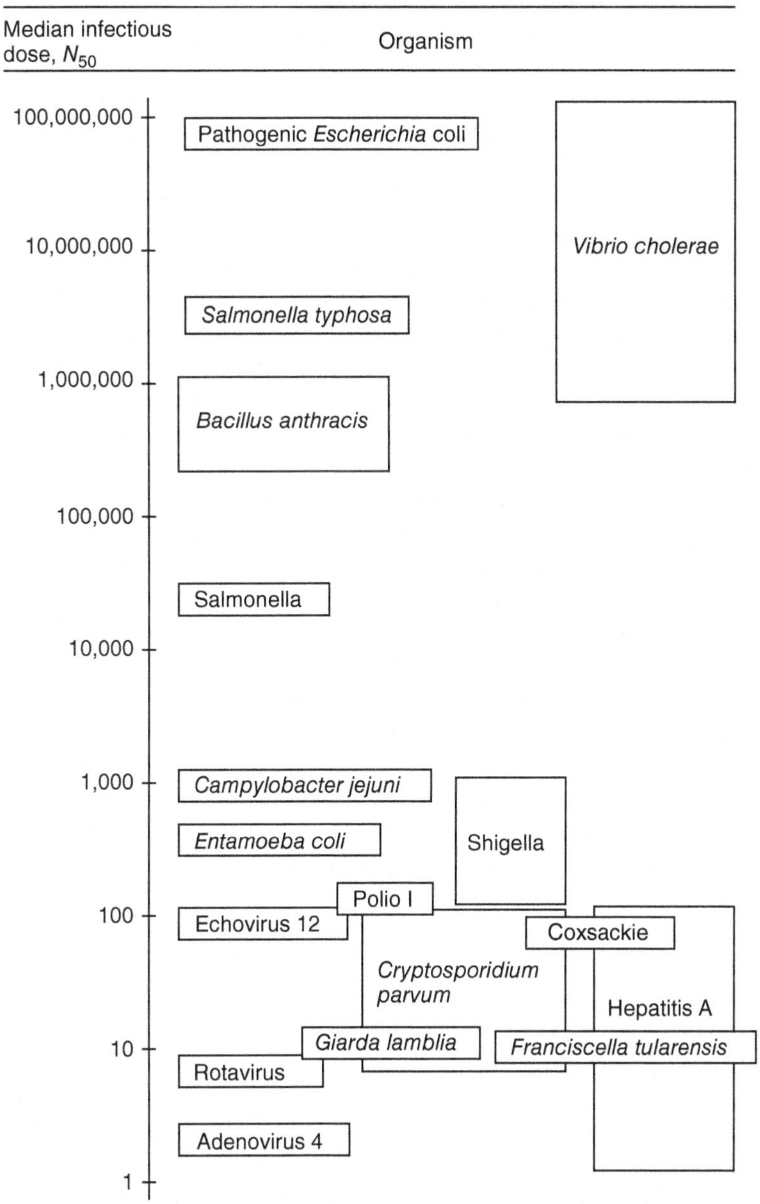

Figure 3-6
Median dose of organisms required in drinking water to cause infection.

people who contract the disease have a prolonged, relapsing course lasting up to 6 months (serious illness). Infection by the hepatitis A virus is fatal for about 0.075 to 0.15 percent of the persons who contract the disease, as demonstrated in Example 3-2.

Example 3-1 Using Infectious Dose to Assess Mass-Based Toxicity of Pathogens

Estimate the mass dose in micrograms that will result in infection in 50 percent of the persons exposed for (a) pathogenic *E. coli* and (b) the pathogenic virus hepatitis A by using the N_{50} values from Fig. 3-6. In the case of hepatitis A, use the lower range of infectious doses reported. Assume that both organisms have a specific gravity of 1.1. The diameter of *E. coli* is approximately 1 μm, and the diameter of hepatitis A is 25 nm. Assume both organisms are spherical in shape. Use 998 kg/m³ for the density of water.

Solution: Part A—Pathogenic *E. coli*

1. Estimate the number of organisms. From Fig. 3-6, N_{50} is approximately $10^{7.9}$ organisms.
2. Determine the dose on a mass basis: Assuming each organism is a sphere 1 μm in diameter with a specific gravity of 1.1, the mass corresponding to the N_{50} is

$$\text{Mass dose} = 10^{7.9}\left[\frac{4}{3}\pi\left(\frac{1.0\times10^{-6}\,\text{m}}{2}\right)^3\right](1.1)(998\ \text{kg/m}^3)(10^9\ \mu\text{g/kg})$$

$$= 46\ \mu\text{g}$$

Solution: Part B—Pathogenic Virus Hepatitis A

3. Estimate the number of organisms. From Fig. 3-6, the lower range of N_{50} is approximately two organisms.
4. Determine the dose on a mass basis. Assuming each organism is a sphere 25 nm in diameter with a specific gravity of 1.1, the mass corresponding to the N_{50} is

$$\text{Mass dose} = 2\left[\frac{4}{3}\pi\left(\frac{0.025\times10^{-6}\,\text{m}}{2}\right)^3\right](1.1)(998\ \text{kg/m}^3)(10^9\ \mu\text{g/kg})$$

$$= 1.8\times10^{-11}\ \mu\text{g}$$

Comment

These are truly low mass doses, especially when a single exposure to this level of organisms will cause infection in 50 percent of the persons exposed.

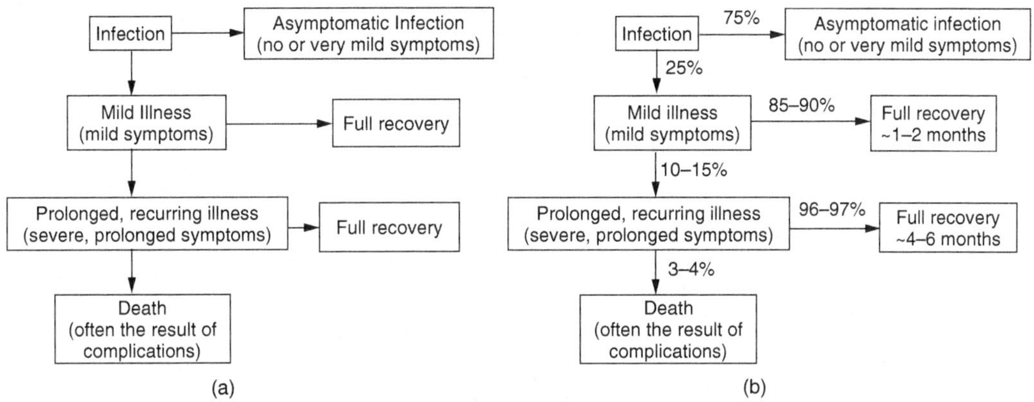

Figure 3-7
Alternative outcomes from pathogenic infection: (a) generalized representation and (b) approximate quantification of outcomes for pathogen hepatitis A.

Example 3-2 Consequences of a hepatitis A epidemic

Due to a one-time contamination incident, hepatitis A enters into a water supply serving 10,000 persons at a concentration high enough to result in infections in 50 percent of the population. Estimate the number of persons for whom the infection is asymptomatic (no symptoms), the number who become mildly ill, the number who develop a prolonged illness, the number who recover from the prolonged illness, and the number who do not survive. Use the data presented on Fig.3-7b to make the estimate. Use the upper bound for the ranges reported on Fig. 3-7b to estimate illness. Finally, what was the risk of dying from the disease to members of the community exposed?

Solution

1. Number of individuals who get infected:

$$10{,}000 \times 50\% = 5000$$

2. Number of individuals who are asymptomatic:

$$5000 \times 75\% = 3750$$

3. Number of individuals mildly ill:

$$5000 \times 25\% = 1250$$

4. Number of individuals with prolonged, recurring illness:

$$1250 \times 15\% = 188$$

5. Number of individuals who do not survive:

$$188 \times 4\% = 8$$

6. Risk of dying for members of community:

$$8/10{,}000 = 8 \times 10^{-4}$$

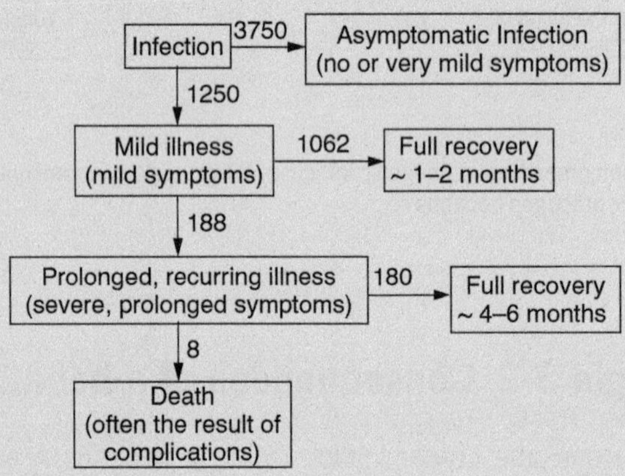

MORTALITY RATIO

Pathogens with a high probability of a fatal outcome are particularly important. A pathogen's effectiveness in causing damage to its host is referred to as virulence. The most common measure of virulence is the mortality ratio, which is the fraction of the persons who get ill to those who do not survive the disease. The mortality ratio for several pathogens is displayed on Fig. 3-8. Among the classic waterborne pathogens, the *V. cholerea* biotype 01 (Asiatic cholera) had one of the highest mortality ratios. There is evidence that contemporary strains of cholera (the El Tor and Bengal strains) are somewhat less virulent due to their dependence on asymptomatic carriers for transmission (Ewald et al., 1998). The other two organisms with the highest mortality ratio on Fig. 3-8 are *Franciscella tularensis* and *Bacillus anthracis*, organisms primarily known as agents of biological warfare.

Forms of Gastroenteritis

Gastroenteritis is the most important form of waterborne illness. Other forms are discussed in the following section. Gastroenteritis may be divided into three classes, as shown in Table 3-2: (1) noninflammatory gastroenteritis, (2) inflammatory gastroenteritis, and (3) invasive gastroenteritis. Noninflammatory gastroenteritis generally results from an organism or toxin that does not draw an immune response, such as in cases of food

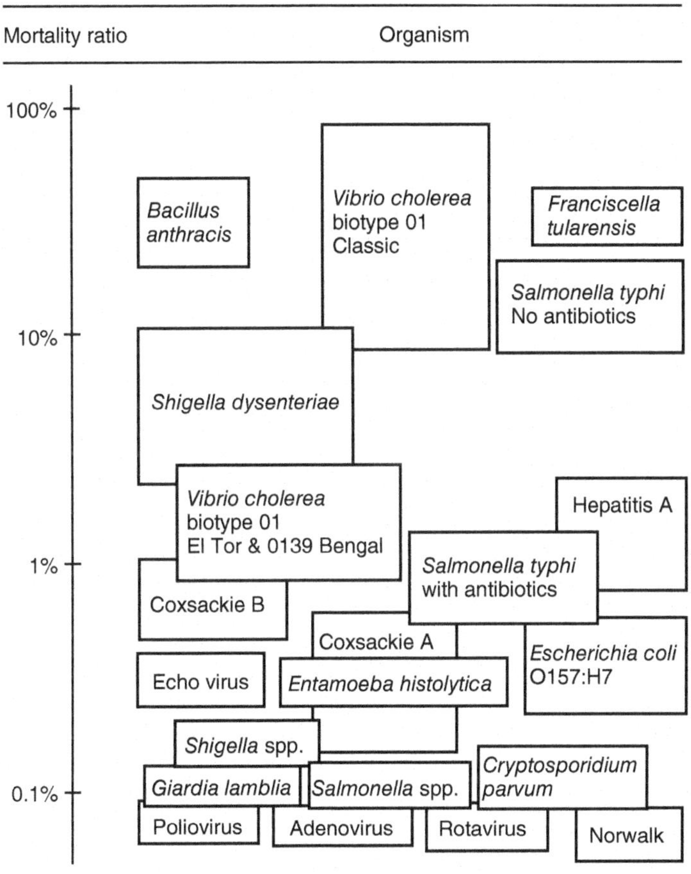

Figure 3-8
Mortality ratios for variety of pathogens.

poisoning or from protozoan infections. Food poisoning of this kind can result from the ingestion of mushrooms or food contaminated with a bacterial toxin, as is the case with *Staphylococcus aureus, Bacillus cereus, Clostridium perfringens*, or *Clostridium botulinum.* These forms of gastroenteritis are not accompanied by a fever but most result in diarrhea and/or vomiting.

Inflammatory gastroenteritis results in redness, swelling, pain, and a feeling of heat in affected areas. In these circumstances, the organism may invade the epithelial layer of the GI tract and sometimes the lymph nodes but does not venture further. The immune response is designed to address the organism and its activities but results in no permanent damage to the host. Inflammatory gastroenteritis is associated with a number of bacteria, two viruses, and one protozoan. Symptoms include fever, diarrhea, and/or vomiting and fecal leukocytes (colorless white cells with a nucleus that are associated with the lymph system and blood).

Invasive gastroenteritis occurs when the organism responsible for the infection travels beyond the epithelial layer of the GI tract, invading other

Table 3-2
Association of pathogens with different forms of gastroenteritis

Gastroenteritis	Symptoms	Responsible Organisms
Noninflammatory gastroenteritis	Diarrhea and/or vomiting, no fecal leukocytes, no blood in stool, usually no fever.	Bacteria: *S. aureus*,[a] *B. cereus*,[a] *C. perfringens*,[a] *C botulinum*[a] Viruses: noroviruses Protozoa: *Giardia lamblia (intestinalis)*, *Cryptosporidium parvum* Algae: *Pfiesteria* spp.[a]
Inflammatory gastroenteritis	Diarrhea and/or vomiting, fecal leukocytes present, usually severe fever, no blood in stool.	Bacteria: *V. cholerae*,[b] enteropathogenic *E. coli* (EPEC), *enteroaggregative E. coli* (EAggEC), *Clostridium difficile*, *Shigella* spp., enterotoxigenic *E. coli* (ETEC) Viruses: rotavirus, caliciviruses[b] Protozoa: *Entamoeba dispar*
Invasive gastroenteritis	Invasion past epithelial layer of GI tract, may not have any diarrhea or vomiting, dysentery may be present (mucus containing bloody feces), fecal leukocytes present, fever; may not have any GI tract problems but instead severe systemic problems.	Bacteria: *Salmonella* spp., *Campylobacter jejuni*, enteroinvasive *E. coli* (EIEC), enterohemorrhagic *E. coli* (EHEC), *Vibrio vulnificus*, *Yersinia* spp., *F. tularensis*, *B. anthracis*, *Helicobacter pylori* Viruses: unknown Protozoa: *E. histolytica*

[a]These microorganisms grow on food or in the environment and produce toxins that, when ingested, cause gastroenteritis a few hours later (only *Pfiesteria* spp. is of concern to drinking water).
[b]Often cited as not causing a fever.

organs and even the blood system itself. As a result, significant damage often results when these diseases are allowed to progress too far. Several bacteria and one protozoan are known to cause invasive gastroenteritis.

Waterborne pathogens include bacteria, viruses, protozoa, and helminths. Some of the more significant of each of these will be discussed in the following sections.

3-3 Bacteria of Concern in Drinking Water

Bacteria within water supplies are broadly classified into two groups: the autochthonous group (*autochthon* is from the Greek word for "one sprung from the land itself") and the allochthonous group (*allos* from the Greek for "other"). Autochthonous bacteria thrive in natural water supplies, while allochthonous bacteria do not. Allochthonous bacteria end up in the water as a result of contamination, runoff, and rainfall. Allochthonous bacteria normally have a limited life span in the natural water environment. Many bacteria of concern from the public health standpoint are classified as allochthonous. Allochthonous bacteria are more comfortable in the intestines of warm-blooded animals, and their presence in water is indicative of wastewater or fecal contamination. Outside of the warm, nutritionally rich environment of a human or animal gut, they die off rapidly. However, bacterial die-off is not an all or none response, so the rates of die-off and the initial bacterial concentrations in contaminated water are of interest.

Bacterial disease has always maintained a high profile where public health and water supply are concerned, and the situation is likely to remain that way for the foreseeable future. To provide some structure to the discussion of various bacterial pathogens, the bacteria have been divided into several broad categories. Classic waterborne pathogens, those that have plagued humans since the early understanding of waterborne illness such as cholera and typhoid fever are discussed first. Modern waterborne pathogens are those that have been the focus of health professional in more recent years. Pathogens that have been implicated in recent outbreaks but about which less is known, are considered to be pathogens of emerging concern. A final category of interest are pathogens that might be used to sabotage water systems.

Classic Waterborne Bacterial Pathogens

Classic waterborne pathogens are those that have been associated with drinking water since the early development of water treatment practices. Cholera and typhoid fever are acute epidemic diseases that can, when transmitted via the water supply, infect a significant fraction of the community overnight. Typhoid fever is caused by *S. typhi*, a bacteria of the *Salmonella* genus. These and a variety of historic bacteria of interest in drinking water are characterized in Table 3-3. Characteristics of bacteria, the symptoms of the diseases they cause, and their relevance to water treatment are discussed below.

VIBRIO CHOLERAE

Vibrio organisms are facultative anaerobic bacteria that are metabolically similar to Enterobacteriaceae and are one of the most common organisms in surface waters. *Vibrio cholerae* is the most ferocious of waterborne infectious pathogens because it is noninvasive and attacks the small intestine through secretion of an enterotoxin. The original classic *V. cholerae* (serotype 01) of the nineteenth century is thought to have been indigenous in India for many centuries. People exposed to this classic *V. cholerae* experienced explosive diarrhea and vomiting without fever. These initial symptoms usually occur 2 to 3 days or less after exposure. If left untreated, an infected individual with severe symptoms experiences dehydration, abnormally low blood pressure and temperature, muscle cramps, shock, coma, and eventually death. Often these symptoms occur in less than 18 h. Untreated cholera frequently results in death for 50 to 60 percent of those infected.

Table 3-3
Bacteria that cause classical waterborne diseases

Organism	Size, μm	Motile	Normal Habitat	Health Effects in Normal Persons	Evidence is Waterborne	Culturable
Vibrio cholerae 01 classical	0.5 × 1–2	Flagella	Human stomach and intestines	Classic cholera	Classic studies	Classic studies
V. cholerae 01 El tor	0.5 × 1–2	Flagella	Human stomach and intestines[a]	Milder but serious form of classic cholera	Several modern studies	Several modern studies
V. cholerae 0139 Bengal	0.5 × 1–2	Flagella	Human stomach and intestines[a]	Milder but serious form of classic cholera	Several modern studies	Several modern studies
Salmonella typhi	0.6	Yes	Human stomach and intestines	Typhoid fever: enteric fever	Classic studies	Classic studies
Salmonella paratyphi A	0.6	Yes	Intestines of warm-blooded animals	Enteric fever	Classic studies	Classic studies
Salmonella Schottmeulleri	0.6	Yes	Intestines of warm-blooded animals	Enteric fever	Classic studies	Classic studies
Salmonella Choleraesuis	0.6	Yes	Intestines of warm-blooded animals	Gastroenteritis, *Salmonella* septicemia	Classic studies	Classic studies

Salmonella typhimurium	0.6	Yes	Intestines of warm-blooded animals	Gastroenteritis	Classic studies	Classic studies
Salmonella enteritidis	0.6	Yes	Intestines of warm-blooded animals	Gastroenteritis	Classic studies	Classic studies
Shigella dysenteriae	0.4	No	Human stomach and intestines	Bacillary dysentery: abdominal pain, cramps, diarrhea, fever, vomiting, blood and mucus in stools	Classic studies	Classic studies
Shigella sonnei	0.4	No	Human stomach and intestines	Bacillary dysentery: abdominal pain, cramps, diarrhea, fever, vomiting, blood and mucus in stools	Classic studies	Classic studies
Shigella flexneri	0.4	No	Human stomach and intestines	Bacillary dysentery: Abdominal pain, cramps, diarrhea, fever, vomiting, blood and mucus in stools	Classic studies	Classic studies
Shigella boydii	0.4	No	Human stomach and intestines	Bacillary dysentery: Abdominal pain, cramps diarrhea, fever, vomiting, blood and mucus in stools	Classic studies	Classic studies

[a]Also shown to grow in estuarial conditions.

Vibrio cholerae's rapid impact on its victims results from the activity of the cholera enterotoxin, which activates an enzyme in the intestinal cells, causing them to extract water and electrolytes from blood and tissue and move it into the intestine. The watery diarrhea that results is speckled with flakes of mucus and epithelial cells and contains enormous numbers of *V. cholerae.* Because of its grainy appearance, it is referred to as a "rice-water stool."

In 1817, *V. cholerae* moved beyond India, causing epidemics in several other countries. Thus began the first of six pandemics, starting in the nineteenth century and continuing into the first two decades of the twentieth century. During these pandemics, the disease moved around the world killing hundreds of thousands of people in each series of events. The last of these six classic epidemics was arrested in 1923. Between 1923 and 1960, it was believed that cholera would not return in pandemic form because water supplies had been improved worldwide.

Then in 1961 a new biotype of the classic serotype of *V. cholerae* (*V. cholerae* 01 also known as El Tor) emerged, producing a major epidemic in the Philippines and initiating a seventh pandemic. El Tor is now active on six continents (Ewald et al., 1998). The early development of this new pandemic is shown on the map on Fig. 3-9.

The evolutionary biologist Paul Ewald (Ewald et al., 1998) argues that, when faced with changes in their environment (such as better water treatment), pathogens will evolve to regain their survival advantage. There are three characteristics that El Tor has developed that allow it to succeed in this new environment: (1) a higher fraction of infections result in asymptomatic carriers (99.7 to 99 percent for El Tor vs. 50 to 75 percent for classic cholera), (2) a longer period of shedding, and (3) better survival outside the

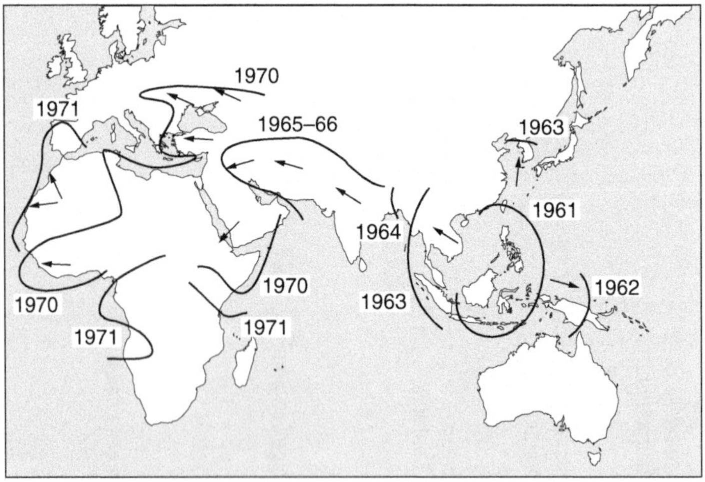

Figure 3-9
Geographical development of the first 10 years of the seventh cholera pandemic.

gut. Ewald argues that El Tor is largely successful where the classic cholera is not because it is a fundamentally milder disease with a much larger fraction of asymptomatic infections allowing for transmission of the disease by means other than water contamination (Ewald et al., 1998). Between 1969 and 1974, El Tor replaced the classic strains of endemic cholera originating from the Ganges River Delta of India. There were more than one million cases of El Tor in the Western Hemisphere in 1994.

In December 1992, a large epidemic of cholera began in Bangladesh, and large numbers of people have been infected. The organism has been characterized as *V. cholerae* serotype O139 "Bengal." It is derived genetically from the El Tor pandemic strain, but it has changed its antigenic structure such that there is no existing immunity and all ages, even in endemic areas, are susceptible. The epidemic has continued to spread and Bengal has affected at least 11 countries in southern Asia. The emergence of this new strain illustrates the ability of this organism to evolve, as it must to succeed in a changing environment.

SALMONELLA SPP

Salmonella bacteria are rod shaped, motile, non-spore-forming, and gram-negative bacteria (*S. gallinarum* and *S. pullorum* are not motile). There is a widespread occurrence of nontyphi *Salmonella* in animals, especially in poultry and swine. Several species of *Salmonella* cause gastrointestinal illness, but, of these, *S. typhi*, which causes typhoid fever, is the most notorious. Evidence suggests that typhoid fever has been with humankind from the beginning—so long that part of the population has developed resistance to the disease's most damaging effects. A Roman physician, Antonius Musa, achieved fame 2000 years ago by treating Emperor Augustus with cold baths when he fell ill with typhoid. It was Dr. William Budd who, in 1856, first demonstrated that typhoid could be waterborne and Karl Eberth who isolated the organism responsible for the disease from the organs of its victims. The incubation period for typhoid fever is usually 10 to 14 days. Early symptoms are fever [the temperature rises to 39 to 40°C (103 to 104°F)], headaches, malaise, and abdominal pain. Diarrhea is common, but many patients experience constipation. As the disease advances, *S. typhi* invades the GI cells and infects the bloodstream (many individuals of European descent have a genetic trait that interferes with the organism's effort to conduct this invasion). The bacteria then concentrate in the lymph nodes, the liver, the spleen, and the gall bladder. The bacteria often show up on the surface of the skin concentrated in "rose spots," approximately a dozen dark red spots on the abdomen and chest about 1 mm in diameter.

Asymptomatic carriers are common and have been known to shed *S. typhi* from their gall bladder for years. Typhoid Mary, the infamous food handler and restaurant worker who infected hundreds if not thousands of patrons in New York City, was a notable example of this malady. Other species of

Salmonella have similar but less severe effects. Probably the most well known *Salmonella* spp. causing an enteric fever similar to *S. typhi* is *S. paratyphi* (paratyphoid fever). Other *Salmonella* species cause acute gastroenteritis without fever, which occurs within a few days of exposure and usually lasts several days to a week. Until 1948, when the antibiotic chloramphenicol was introduced, little could be offered to typhoid victims but supportive care. Without antibiotics the mortality rate can be as high as 15 percent. Through the use of antibiotics the mortality rate can be reduced to about 1 percent, but only if recognized and treated early in the infection.

It is important to understand that, in the long term, antibiotics are no substitute for good sanitation. In countries with poor sanitation, drug resistance to *Salmonella* began to emerge in the 1970s, and by the late 1970s the rate of drug resistance among strains of *Salmonella* had reached 75 percent in Vietnam, for example. By comparison, the resistance in industrialized countries was on the order of 5 percent. Drug resistance among strains of enteric pathogens continues to be a problem in countries with poor sanitation (Isenbarger et al., 2002). Once again, these observations are consistent with the theories of evolutionary biology. In environments where pathogens have easy access to victims and constant exposure to drug therapy, they can afford the less efficient metabolic machinery required for drug resistance (Ewald et al., 1998). Good water treatment is an essential element of controlling this family of pathogens. Sanitary practices provide effective intervention, making the drug therapies more effective.

SHIGELLA SPP.

Shigella spp. organisms are gram-negative, nonmotile, non-spore-forming, rod-shaped bacteria (bacilli). Some of the more important species are *Shigella dysenteriae, S. sonnei, S. boydii,* and *S. flexneri. Shigella* is named after the Japanese scientist, Kiyoshi Shiga, who identified the organisms in 1897. The species *S. dysenteriae* is responsible for bacillary dysentery, a disease often associated with crowded, unsanitary conditions. Other species of *Shigella* usually produce milder forms of diarrheal disease. Most people who are infected with *Shigella* develop diarrhea, fever, and stomach cramps starting a day or two after exposure. The diarrhea is often bloody, the distinguishing characteristic of dysentery. *Shigella dysenteriae* is the only cause of epidemic bacterial dysentery.

Shigella, like *Salmonella* and *E. coli,* is a member of the family Enterobacteriaceae. *Shigella* differs from other members of that family by having genes in a large "virulence" plasmid that codes for epithelial cell invasion and the formation of enterotoxins. In contrast to *S. flexneri, S. sonnei,* and *S. boydii, S. dysenteriae* can also cause cell death by the production of Shiga toxin.

Shigella spp. organisms are primarily human pathogens, although they can be found in some primates such as monkeys and chimpanzees but rarely occur in other animals. *Shigella* normally affects the distal ileum or colon and is usually confined to the mucosa, but it can spread through the intestinal wall and lead to perforation. The disease is caused when virulent *Shigella* organisms attach to and penetrate epithelial cells of the intestinal mucosa. After invasion, they multiply by spreading from one epithelial cell to the next, resulting in tissue destruction. Unlike *Salmonella*, *Shigella* is acid tolerant. As a consequence, gastric acidity provides little protection against infection. Effective natural defenses include the normal flora, secretory immunoglobulin A (antibodies found in the saliva, sweat, and tears), and phagocytes. In contrast to the invasive *S. typhi*, *Shigella* rarely invades beyond the intestinal mucosa or local lymph nodes.

Contamination is usually through the fecal–oral route. Fecally contaminated water and unsanitary practices of food handlers are the most common causes of this contamination. The most commonly associated foods are salads (potato, tuna, shrimp, macaroni, and chicken), raw vegetables, milk and dairy products, and poultry. The milder forms of *Shigella* spp. infections are also spread in childcare centers in developed countries.

Shigella dysenteriae has caused epidemics of dysentery throughout the world. *Shigella dysenteriae*, *S. flexneri*, and *S. boydii* are the species most commonly found in developing countries. *Shigella sonnei* is more common in developed countries, and in these countries *S. dysenteriae* is quite rare. *Shigella dysenteriae* caused a 4-year epidemic in Central America beginning in 1968 that resulted in an estimated 500,000 cases and 20,000 deaths. No epidemics have occurred in the region since then, but *S. dysenteriea* continues to occur sporadically in the Western Hemisphere. *Shigella dysenteriae* also spread down the African continent as a major pandemic. It was first reported in eastern Zaire in 1981 and since then has been reported in Rwanda, Burundi, and Tanzania. It was first identified in the province of KwaZulu-Natal, South Africa, in March 1994 and simultaneously noted in the northern part of South Africa. Since then it has spread throughout sub-Saharan Africa. An estimated 300,000 cases of shigellosis occur annually in the United States, but these are largely infections of *S. sonnei* and some *S. flexneri*, both spread by food and oral sex. Worldwide, it is estimated that *Shigella* spp. organisms are the cause of more than 600,000 deaths annually.

As with *Salmonella*, drinking water treatment and sanitation play a major role in the ecology of *Shigella* in two ways. In countries with poor sanitation, pandemics of the most harmful species, *S. dysenteriea*, occur beyond control, whereas in more developed countries milder species such as *S. sonnei* dominate. Another related observation is the development of resistance to antibiotics among *Shigella* spp. in countries with poor sanitation. *Shigella dysenteriea* has shown that it can quickly develop antibiotic resistance. In a country with poor sanitation, antibiotics are often effective against *S. dysenteriea* for only 1 or 2 years after being introduced; resistance has even been observed to develop during the course of a single epidemic.

Modern Waterborne Bacterial Pathogens

The classical bacterial pathogens in the preceding discussion have been associated with waterborne disease for the public health community for some time. In recent years, progress in science has brought new waterborne bacterial pathogens to the forefront. Bacterial pathogens that have been shown to cause waterborne disease in recent years are summarized in Table 3-4. Most of these bacteria cause a form of diarrhea, and none presents a threat to community comparable to the cholera epidemics 100 years ago; however, they do sometimes have serious health consequences. Any serious diarrhea can be life threatening to some members of the population, especially in a poor country, and some invasive forms of these diseases have other important health consequences as well.

PATHOGENIC *E. COLI*

The largest group of bacteria discussed in Table 3-4 is the pathogenic *E. coli. Escherichia coli* belongs to the family Enterobacteriaceae and is almost invariably enteric. *Escherichia coli* is a facultative anaerobe, gram-negative rod that lives in the intestinal tracts of warm-blooded animals. A number of genera in Enterobacteriaceae are human intestinal pathogens (e.g., *Salmonella, Shigella,* and *Yersinia*). Several others are normal inhabitants of the human gastrointestinal tract (e.g., *Escherichia, Enterobacter,* and *Klebsiella*), but these bacteria may be associated with disease if they harbor virulence factors that make them capable of causing a disease.

Escherichia coli colonizes the GI tract of warm-blooded animals during the first few days after birth. The human bowel usually shows evidence of colonization in less than 4 days. *Escherichia coli* can adhere to the intestine or to the mucus overlying it. Once a strain of *E. coli* resides in an animal's colon, it remains there, undisturbed for weeks, months, or even years until disturbed by an enteric infection or antimicrobial drugs. Normally, *E. coli* serves a useful function in the body by suppressing the growth of harmful bacterial species and by synthesizing appreciable amounts of vitamins. The entire DNA sequence of the *E. coli* genome has been determined for two strains of this bacterial species and is available to the public.

While there is wide agreement that certain varieties of *E. coli* are pathogenic and that these play an important role in waterborne disease, information on these organisms continues to develop and, consequently, their classification is evolving as well. The most widely recognized groupings are the enterotoxigenic *E. coli* (ETEC), enteroinvasive *E. coli* (EIEC), and enterohemorrhagic *E. coli* (EHEC). The EHEC group consists of the well- known bacteria *E. coli* O157:H7. Newer but less widely accepted groupings are the enteropathogenic *E. coli* (EPEC) and the enteroaggregative *E. coli* (EaggEC). These organisms are implicated in waterborne disease outbreaks, and all of them have unique methods for adhering to the mucus and/or wall of the intestines of warm-blooded animals. Exactly how they adhere is related to their pathogenic nature, which is determined by which virulence genes they possess.

Table 3-4
Bacteria recently associated with waterborne disease

Organism or Bacteria	Size, µm	Motile	Health Effects in Healthy Persons	Evidence of Waterborne Pathway	Culturable
Enteropathogenic *Escherichia coli* (EPEC)	0.3 – 0.5 × 1 – 2	No	Traveler's diarrhea	Numerous waterborne outbreaks	*E. coli* test
Enteroaggregative *E. coli.* (EaggEC)	0.3 – 0.5 × 1 – 2	No	Childhood diarrhea and diarrhea among immunocompromised	Numerous waterborne outbreaks	*E. coli* test
Enteroinvasive *E. coli* (EIEC)	0.3 – 0.5 × 1 – 2	No	Childhood diarrhea	Numerous waterborne outbreaks	*E. coli* test
Enterohemorrhagic *E. coli* (EHEC, e.g., *E. coli* 0157)	0.3 – 0.5 × 1 – 2	No	Bloody diarrhea, occasionally hemolytic uremic syndrome (HUS)	Six waterborne outbreaks, notably Cabool, MI, and Walkerton, Ontario	*E. coli* test at 37.5°C
Enterotoxigenic *E. coli* (ETEC)	0.3 – 0.5 × 1 – 2	No	Traveler's diarrhea	Numerous waterborne outbreaks	*E. coli* test
Helicobacter pylori	0.5 – 1 × 2.5 – 4	Yes	Dominant cause of peptic and duodenal ulcers; associated with gastric carcinoma	Some indirect evidence	No
Yersinia enterocolitica	0.3 – 0.5 × 1 – 2	No	Fever, abdominal pain, gastroenteritis with diarrhea and vomiting	Outbreaks associated with contaminated spring water	Yes
Campylobacter jejuni	0.3 × 1 – 5	Yes	Diarrhea, abdominal pain, nausea, fever, malaise	Numerous outbreaks	Yes, but methods are difficult and insensitive

Both ETEC and EPEC are thought to be a major cause of endemic and traveler's diarrhea in underdeveloped countries. Both result in abdominal cramps and a watery diarrhea. Enterotoxigenic *E. coli* does so through the generation of toxins that lead to secretion without invading the epithelial wall and hence causes only low-grade fever. Enteropathogenic *E. coli* causes abdominal discomfort through the production of a "Shiga-like" toxin, but also by invading the epithelial wall. Its invasive character results in significant inflammation and fever.

Enteroinvasive *E. coli* does not produce toxins, but it does adhere to the colon wall and then penetrates and grows in the epithelial cells of the colon. The result is severe inflammation, fever, and bacillary dysentery, much like the symptoms of *S. dysenteriae.* The disease is also found in the same conditions of poor sanitation associated with *Shigella.* In many genetic and physiological respects, *S. dysenteriae* and EIEC are very similar bacteria.

Enteroaggregative *E. coli* is not as well studied as some of the others. The distinguishing feature of EaggEC is its ability to attach to tissue culture cells in an aggregative manner. Enteroaggregative *E. coli* is associated with persistent diarrhea in young children and immunocompromised patients. Like ETEC, EaggEC adheres to the intestinal mucosa and causes watery diarrhea without inflammation or fever. The role of EaggEC strains in human disease is less clear than the other pathogenic *E. coli* discussed in this section.

Enterohemorrhagic *E. coli,* of which *E. coli* O157:H7 (the "O" is a letter) is a main member, is a unique variety of *E. coli* that produces a Shiga-like toxin (verotoxin) that is closely related or identical to the toxin produced by *S. dysenteriae.* As a result, infection with EHEC produces hemorrhagic colitis, an illness characterized by severe cramping and diarrhea that is initially watery but becomes bloody, occasionally vomiting, and fever that is either low grade or absent. The illness is usually self-limited and lasts for an average of 8 days. Some victims develop hemolytic uremic syndrome (HUS), a rare condition affecting mostly children, characterized by destruction of red blood cells, damage to the lining of blood vessel walls, and in 10 percent of the cases kidney failure. It is uniquely but not exclusively connected to *E. coli* O157:H7. In the elderly, thrombotic thrombocytopenic purpura (TTP) can occur (HUS plus high fever and neurological symptoms), and under these conditions, TTP can have a mortality rate as high as 50 percent.

Enterohemorrhagic *E. coli* was identified as the cause of a significant waterborne outbreak in Walkerton, Ontario, in May and June 2000. The incident was caused by contamination of a well in the local water system by EHEC, and *C. jejuni* contained in manure leached into the well from a nearby farm. The EHEC was found in the manure on the farm, in the water distribution system, and in the stools of infected patients (Bruce-Grey-Owen Sound Health Unit, 2000). During the outbreak, estimates are that more than 2300 persons were ill, as shown graphically on Fig. 3-10. Of the

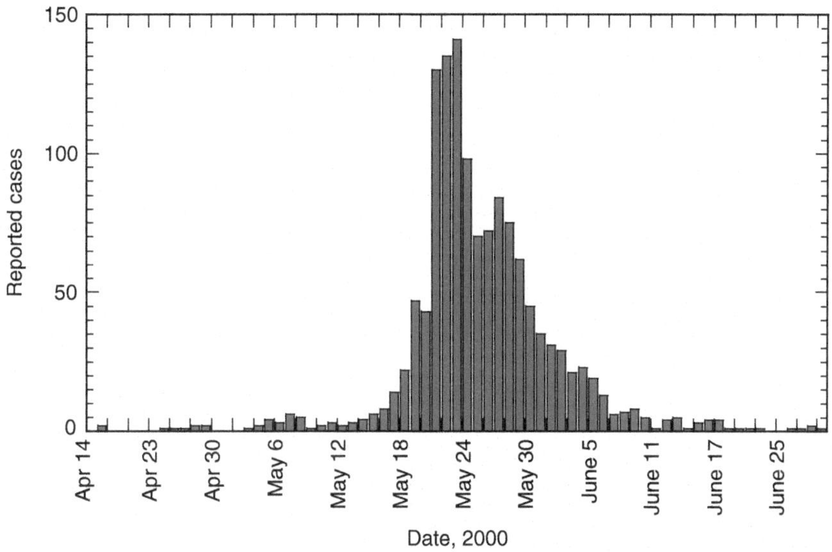

Figure 3-10
Number of illness cases reported during Walkerton, Ontario, disease outbreak of 2000. (Data from Bruce–Grey-Owen Sound Health Unit, 2000).

1346 cases that were reported, 1304 (97 percent) were considered to be directly due to the drinking water. Based on stool samples, about an equal number were infected with *E. coli* O157:H7 and *C. jejuni.* Sixty-five persons were hospitalized, 27 developed HUS, and 6 died.

CAMPYLOBACTER JEJUNI

Campylobacter jejuni is a gram-negative, slender, curved, and motile rod-shaped bacterium. Cell motility is achieved through polar flagella that emanate from a spiral-shaped bacterium. The unique shape of the cell and flagella are helpful in Gram stain identification. *Campylobacter jejuni* prefers conditions with reduced levels of oxygen (microaerophilic conditions).

Campylobacter is found in natural water sources throughout the year. The presence of *Campylobacter* is not clearly correlated with indicator organisms for fecal contamination (i.e., *E. coli*). The organism survives substantially better in a cold environment. It has been reported that, when stressed, *Campylobacter* enters a "viable but nonculturable state" but can still be transmitted to animals.

Campylobacter has been known a long time, but its significance to public health has changed over time. As early as 1886, Escherich observed organisms resembling campylobacters in stool samples of children with diarrhea. In 1913, McFaydean and Stockman identified *Campylobacter* (called related *Vibrio*) in fetal tissues of aborted sheep. In 1957 King described the isolation of related *Vibrio* from blood samples of children with diarrhea, and in 1972 clinical microbiologists in Belgium first isolated *Campylobacter* from stool samples of patients with diarrhea (Kist, 1985).

The development of selective growth media in the 1970s permitted more laboratories to test stool specimens for *Campylobacter.* Soon, *Campylobacter* spp. was established as a common human pathogen. *Campylobacter jejuni* infections are now the leading cause of bacterial gastroenteritis reported in the United States (Blaser et al., 1983). In 1996, 46 percent of laboratory-confirmed cases of bacterial gastroenteritis reported in the Centers for Disease Control and Prevention (CDC)/U.S. Department of Agriculture/Food and Drug Administration Collaborating Sites Foodborne Disease Active Surveillance Network were caused by *Campylobacter* species. Campylobacteriosis was followed in prevalence by salmonellosis (28 percent), shigellosis (17 percent), and *E. coli* O157:H7 infection (5 percent), as shown on Fig.3-11.

Campylobacter jejuni is commonly present in the GI tract of healthy cattle, pigs, chickens, turkeys, ducks, and geese and may occasionally be isolated from streams, lakes, and ponds. The closely related species *Campylobacter coli* and *Campylobacter upsaliensis* may also cause disease in humans.

Typical symptoms of *C. jejuni* illness include severe abdominal pain, diarrhea, fever, nausea, headache, and muscle pain. A majority of cases are mild and do not require hospitalization or treatment with antibiotics; however, *C. jejuni* infection can be severe. The disease is rarely fatal, less than 1 death per 1000 cases. Children under the age of 5 and young adults aged 15 to 29 are the age groups most frequently affected. The incubation period is typically 2 to 5 days, but onset times of up to 11 days have been reported. The illness usually lasts about 1 week, with severe cases persisting for up to 3 weeks. However, *Campylobacter* infection may be followed by Guillain-Barre syndrome (GBS), a rare but serious form of neuromuscular paralysis in a small proportion of cases. Many of those suffering from GBS have antibodies to *Campylobacter*, indicating a recent infection. *Campylobacter*

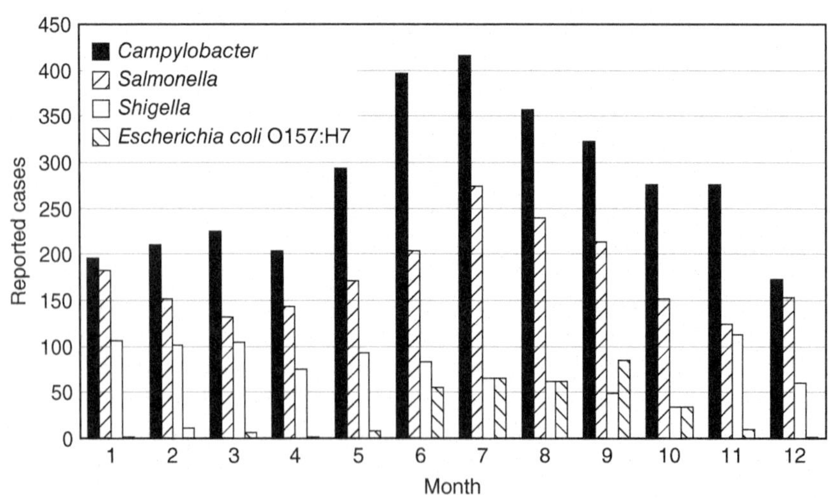

Figure 3-11 Cases of foodborne gastroenteritis in the United States in 1996. (Adapted from Altekruse et al., 1999).

has also been isolated from the stools of patients stricken with GBS. Although not everyone who is infected with *Campylobacter* will develop GBS, the association between *Campylobacter* and GBS makes this bacterium more significant.

Infections caused by strains of *C. jejuni* that are resistant to antibiotics make clinical management more difficult. Antibiotic resistance can compromise treatment of patients with bacteremia. Once again, antibiotic resistance of *C. jejuni* fits Ewald's theory of microbial evolution (Ewald et al., 1998). Resistance is highest in the developing world, where sanitation is poor, the bacteria have easy access to their hosts, and the use of antibiotics is the principal means employed to address disease. A 1994 study found that most clinical isolates of *C. jejuni* from U.S. troops in Thailand were resistant, even to ciprofloxacin, which is one of the antibiotics considered a last line of defense. Additionally, nearly one-third of isolates from U.S. troops located in Hat Yai were resistant to azithromycin (Murphy et al., 1996). Evidence also demonstrates that fluoroquinolone-susceptible *C. jejuni* becomes drug resistant in chickens receiving drug therapy.

HELICOBACTER PYLORI

Helicobacter pylori is a spiral-shaped, gram-negative rod approximately 0.5 × 3.0 μm in size, with sheathed flagella attached to one pole that allow for motility. *Helicobacter pylori* is linked to GI disease in humans. It was first identified in Australia by Dr. Barry Marshall in 1983 as a potential cause of peptic ulcers. *Helicobacter pylori* lives in the stomach and duodenum (section of intestine just below the stomach). Today it is understood to be the most common cause of gastritis in humans.

Helicobacter pylori makes its home in the mucous lining of the stomach, using its flagella and spiral shape to drill through the mucous layer in the stomach. *Helicobacter pylori* produces adhesions that bind to membrane-associated lipids and carbohydrates. The adherence of *H. pylori* to plasma membranes of surface epithelial cells has been shown using electron microscopy. *Helicobacter pylori* protects itself from the stomach's acidity by releasing the enzyme urease, which converts the urea in the stomach to bicarbonate and ammonia, creating a pH-neutral microenvironment around the cell, a benefit to the spiral bacteria but not the epithelial cells to which it is attached.

The organism's location against the stomach lining also leads to isolation from the immune system, which cannot penetrate the lining itself. Nevertheless the immune system can sense the presence of *H. pylori* and, over time, tries to respond. The immune response results in the development of a peptic ulcer. Greater than 90 percent of duodenal ulcer patients are infected with *H. pylori*. Ulcer patients without *H. pylori* infection are typically those who have taken nonsteroid anti-inflammatory drugs, such as aspirin and ibuprofen, which can commonly cause ulcers. Eradication of the bacterium from a person greatly reduces the recurrence of ulcers. There is also a strong correlation with *H. pylori*, peptic ulcers, and stomach cancer. It is estimated that *H. pylori* infection increases the risk of gastric cancer sixfold.

Helicobacter pylori is believed to be transmitted orally by means of fecal matter through the ingestion of waste-tainted food and/or water. As a result, the organism is considered a potential waterborne pathogen. Although the actual mode of transmission has not been proven, the ability to culture the bacterium from stool is supporting evidence for a fecal–oral mode of transmission, hence its consideration as a potential waterborne pathogen. Drug therapies are available to eliminate the infection.

YERSINIA ENTEROCOLITICA

Yersinia pestis was the cause of the plague, but that disease was transmitted by fleas, not by contaminated water. Its relative, *Y. enterocolitica,* also a small rod-shaped, gram-negative bacterium, is often isolated from clinical specimens such as wounds, feces, sputum, and lymph nodes near the GI tract. However, it is not part of the normal human flora. Another cousin, *Yersinia pseudotuberculosis,* has been isolated from the diseased appendix of humans. Both *Y. pseudotuberculosis* and *Y. entercolitica* are associated with diseases of the GI tract, but only *Y. entercolitica* has been connected with ingestion of contaminated water. *Yersinia enterocolitica* can be identified through its ability to grow in cold temperatures and its motility at room temperature. *Yersinia enterocolitica* is also most comfortable at neutral to alkaline pH. As a result, the use of antacids is contraindicated.

Yersinia enterocolitica is an invasive pathogen that can penetrate the intestinal lining and enter the lymph nodes, causing a systemic infection. *Yersinia entercolitica* causes intestinal inflammation called yersiniosis via the release of enterotoxins that can cause severe pain similar to that found in patients with appendicitis.

Fever and abdominal pain are the hallmark symptoms of *Y. entercolitica* infections. Four syndromes are also associated with *Y. entercolitica*: (1) enterocolitis—diarrhea, low-grade fever, abdominal cramps; (2) pseudoappendicitis—pain in the same area as the appendix along with leukocytosis; (3) extraintestinal focal infection—infections in the urinary tract, pneumonia, pharyngitis, and so on; and (4) bacteremia—entrance of organisms into the bloodstream. Enterocolitis is common in young children, pseudoappendicitis in young adults, and bacteremia in diabetics, alcoholics, and others with compromised conditions. Most infections are uncomplicated and resolve completely. Occasionally, some persons develop joint pain, most commonly in the knees, ankles, or wrists. These joint pains usually develop about 1 month after the initial episode of diarrhea and generally resolve after 1 to 6 months. A skin rash called erythema nodosum may also appear on the legs and trunk. *Yersinia enterocolitica* has been associated with reactive arthritis, which may occur even in the absence of obvious symptoms. The frequency of such postenteritis arthritic conditions is about 2 to

3 percent. Nevertheless, infections of *Y. enterocolitica* are relatively rare, and the organism is an infrequent cause of diarrhea and abdominal pain.

The *Y. enterocolitica* strains that cause human illness also reside in pigs, but other strains are found in other animals, including rodents, rabbits, sheep, cattle, horses, dogs, and cats. In pigs, the bacteria are most likely to be found on the tonsils. *Yersinia enterocolitica* has been detected in environmental and food sources, such as ponds, lakes, meats, ice cream, and milk. Isolates from environmental samples are not always pathogenic. Poor sanitation and improper techniques by food handlers, including improper storage, cannot be overlooked as contributing to food contamination.

Bacterial Pathogens of Emerging Concern

A number of pathogens have emerged in recent years that are of concern in drinking water because they are transmitted by the fecal–oral route or have been implicated in recent waterborne outbreaks. Some of these are opportunistic pathogens that grow and thrive in natural waters and in the soil environment. Four of the more significant emerging pathogens are shown in Table 3-5.

LEGIONELLA PNEUMOPHILA

Legionella pneumophila is a motile, rod-shaped, gram-negative, aerobic bacterium—a "facultative parasite." In 1976 it was discovered as the cause of a mysterious pneumonia-like disease among a group of elderly men attending an American Legion Convention in Philadelphia, Pennsylvania (hence the name). *Legionella* does not grow well in the laboratory with typical culture methods. Special media and pretreatment methods are required for its cultivation. *Legionella* has many species and serogroups. The strain responsible for Legionnaire's disease is *L. pneumophila* serogroup 1, or the *Pontiac strain.*

Legionella thrives in warm aquatic environments with rust, algae, and organic particles. It has been known to survive in tap water at room temperature for over a year. *Legionella* prefers to grow inside other organisms; for example, *L. pneumophila* has been associated with protozoa, *Hartmanella vermiformis, Tetrahymena thermophila,* and *Acanthamoeba castellani. Legionella* grows inside these amoebae and multiplies intracellularly using the ameoba as a resource for reproduction.

Legionella bacteria are transmitted to humans via aerosols, such as those generated by air-conditioning cooling towers. Successful transport is favored by a high relative humidity because it enables them to survive while airborne.

The most susceptible hosts are elderly men, smokers, and persons with suppressed immune systems (diabetics, HIV patients, organ transplant recipients, alcoholics, kidney patients, patients on steroids for allergy control, etc.).

Infection begins with the inhalation of *L. pneumophila.* In the lung, the *L. pneumophila* comes in contact with alveolar macrophage. The probability of

Table 3-5
Bacterial pathogens of emerging concern in drinking water

Organism	Size, μm	Motile	Normal Habitat	Health Effects in Healthy Persons	Modes of Transmission	Evidence of Waterborne Pathway	Culturable
Legionella pneumophila	0.3 – 0.9 × 2 – 20	Yes	Warm water	Legionaire's disease, Pontiac fever	Aerosols	Outbreaks associated with warm-water aerosols	Routine media
Aeromonas hydrophila	0.3 – 1.0 × 1.0 – 3.5	Yes	All freshwaters	Gastroenteritis (controversial)	Water exposure	Drinking contaminated water; water in open wounds	Yes on agar medium with sheep's blood
Mycobacterium avium intercellulare	0.2 – 0.6 × 10	No	Soil, dust, water, and animals	Lung infection, fatigue	Inhalation or ingestion	Increasing incidence of GI in AIDS patients	Yes
Pseudomonas aeruginosa	0.5 – 0.8 × 1.5 – 3	Polar flagellum	Inhabitant soil and water; opportunistic human pathogen	Infections of urinary tract, respiratory system, and soft tissue; dermatitis, bacteremia, and systemic infections	Contact with compromised tissue	Only for immunocompromised patients	Yes

success in this venture increases with the dose and decreases with the competence of the host immune system. At this point, the macrophage takes the bacteria into a food vacuole inside the cell where *L. pneumophila* takes over, directing processes in the macrophage to produce more of its own kind (*L. pneumophila*), eventually lysing the cell and moving on to others. *Legionella pneumophila's* ability to succeed in this mission depends on a number of special properties the organism has developed that allow it to move into a host cell, take over, reproduce itself, and then lyse the cell and seek other targets. As this process goes forward, it results in lung damage in a number of complicated ways.

Infection by *L. pneumophila* results mainly in Legionnaire's disease. However, some strains have also been known to manifest a disorder called Pontiac fever, a mild infection with influenza-like symptoms. Incidents of Legionnaire's disease have increased over the past decade, partly because of better diagnosis and partly because there are more environments that favor its occurrence. The first symptoms of Legionnaire's disease occur between 2 and 10 days, typically 5 or 6 days, postinfection. The primary indication is pneumonia; however, anorexia, malaise, sore muscles, and headaches are some early indicators. The majority of patients that develop Legionnaire's disease eventually become delirious. The major effect of Legionnaire's disease is respiratory failure. Other complications are acute renal failure, hypotension, and shock. Fifteen percent of hospitalized cases are terminal.

It might seem that keeping *L. pneumophila* out of water distribution systems would be an effective method of preventing the spread of this disease. In this regard, the current practice of maintaining a distribution system residual is helpful. Chlorine residual minimizes the spread of the organism through showerheads and similar home equipment, but keeping a residual in the distribution system is not enough. Incidences of *L. pneumophila* infection are generally associated with exposure to air-conditioning equipment, institutional hot-water supplies, and other hot-water environments that, by design, create conditions that are ideal for *L. pneumophila* development. Moreover, there are plenty of opportunities for equipment of this kind to get its *L. pneumophila* seed directly from the air supply. Taking special care in the design and operation of this kind of equipment so that the growth of *L. pneumophila* can be avoided is more likely to have a material impact. For example, regular use of oxidizing disinfectants could prevent the formation of heavy slime layers that support *L. pneumophila*, and the use of materials that do not corrode and that minimize access to nutrients can also be of help.

AEROMONAS HYDROPHILIA

Aeromonas hydrophilia and other *Aeromonas* species are straight, gram-negative, motile, non–spore-forming, facultatively anaerobic rods or coccobacilli. They appear singly and are motile with a single polar flagellum. The organisms are heat sensitive, being easily destroyed by pasteurization or equivalent heat processes.

Aeromonas causes illness through the use of a number of virulence factors, including heat-sensitive enterotoxins. The enterotoxin *A. hydrophilia* produces an effect similar to cholera toxin. Some other *Aeromonas* spp. strains also produce cytotoxins. As a result, patients infected with *A. hydrophilia* sometimes have watery diarrhea similar to that caused by *V. cholera,* although, on occasion, the diarrhea may contain mucus and blood. In addition to diarrheal illness, *Aeromonas* spp. has been associated with a number of severe and persistent infections, with symptoms varying depending on the site of infection. Severe inflammation is a common complaint.

Aeromonas hydrophilia has long been recognized as an opportunistic pathogen in immunocompromised hosts (patients on drug therapy, elderly, young). The illness can be severe, especially in the immunocompromised. Its connection with disease in normal hosts is more recent, and although *A. hydrophilia* has been associated with some chronic disease, it is not yet clear how significant *A. hydrophilia* is among the various causes of gastroenteritis.

Aeromonas spp. organisms are water loving and inhabit fresh or seawater. They are widely distributed in nature, but water is their main reservoir or source. They can be found in sinks, taps, or drainpipes. *Aeromonas* spp. is easily found in source water and also has been found in treated drinking water. There is evidence that *Aeromonas* is present in biofilms in water distribution systems. *Aeromonas hydrophilia* is found widely in fresh and brackish waters and may be present in the stools of healthy individuals. *Aeromonas hydrophilia* is usually transmitted through drinking contaminated water but may also be transmitted in foods that are in contact with contaminated water.

MYCOBACTERIUM AVIUM COMPLEX

Mycobacterium avium complex (*M. avium* and *M. intracellulare*) is an acid-fast, rod-shaped, aerobic, non–spore-forming, nonmotile family of bacilli. *Mycobacterium avium* complex (MAC) is also sometimes called MAI, which stands for *M. avium intracellulare.* These organisms are ubiquitous in soil, food, and water. *Mycobacterium avium* complex colonizes water systems as well, although it is found at much higher levels in soil (LeChevallier et al., 1991).

Mycobacterium avium complex organisms are opportunistic pathogens. In healthy individuals, when MAC causes infection, it usually causes attacks in the respiratory tract. For immunocompromised individuals, MAC is a serious bacterial infection related to tuberculosis. Symptoms can include weight loss, fevers, chills, night sweats, swollen glands, abdominal pains, diarrhea, and overall weakness. In these patients, MAC usually affects the intestines and inner organs first, causing liver test results to be high and swelling and inflammation to occur. Frequently with these patients, MAC infections also become disseminated. When this occurs, almost any organ system can be involved, especially those with many mononuclear phagocytes (e.g., the liver, spleen, and bone marrow). Signs and symptoms of disseminated MAC are generally nonspecific, such as fever, night sweats, weight loss, weakness, and anorexia. Diarrhea, malabsorption, and abdominal pain may indicate GI involvement; enlargement of the liver and spleen is common. The exact nature of the transmission of the disease is not well established, but because GI infection is often present, introduction through food or drink must be considered.

Members of the MAC are able to grow in water samples without any additional substrate and are resistant to chlorination. They can also grow over a wide range of temperatures and salinities. The cell walls of these bacteria contain high levels of lipid (waxy) material; hence they are hydrophobic and find it easy to colonize the wet surfaces in water systems (LeChevallier et al., 1991).

Some connections with water supply have also been more directly implicated. *Mycobacterium avium* strains were found in water samples (e.g., ice machines, faucets, toilets, sinks) taken from patient care sites in some Boston area hospitals. These isolates were serologically similar to clinical isolates at those same patient care sites (DuMoulin and Stottmeier, 1986). In Los Angeles, *M. avium* strains from infected patients were shown to be genetically related to isolates recovered from patients exposed to the water through drinking or bathing (Glover et al., 1994; Von Reyn et al., 1994). An epidemiological study of 290 HIV patients in San Francisco found *M. avium* complex in 4 of 528 water samples, 1 of 397 food samples, and 55 of 157 soil samples taken from potted plants (Yajko et al., 1995).

PSEUDOMONAS AERUGINOSA

Pseudomonas aeruginosa is a gram-negative, aerobic rod belonging to the bacterial family Pseudomonadaceae. *Pseudomonas aeruginosa* is also motile by means of a single polar flagellum. In fact, *Pseudomonas* is one of the most vigorous, fast-swimming bacteria seen in hay infusions and pond water samples. *Pseudomonas aeruginosa* can live in a sessile biofilm form or in a planktonic form as a free-swimming cell. *Pseudomonas* can usually be found in soil and water, showing up regularly on the surfaces of plants and occasionally on the surfaces of animals. *Pseudomonas* is better known to microbiologists as pathogens of plants rather than animals.

Pseudomonas aeruginosa is the epitome of an opportunistic human pathogen. The bacterium almost always infects tissue that has been compromised; however, there is hardly any tissue that it cannot infect once a compromise has taken place.

Pseudomonas aeruginosa typically produces three colony types; isolates from soil or water usually produce a small, rough colony. Clinical samples yield one of two smooth colony types. One type has an elevated fried- egg appearance, with large, smooth, flat edges. The other type has a mucoid appearance characteristic of urinary tract infections, and its shape is attributed to the production of alginate slime. The smooth and mucoid colonies are presumed to be due to factors that play a role in colonization and virulence.

Pseudomonas aeruginosa is primarily a pathogen acquired during hospitalization. According to the CDC, the overall incidence of *P. aeruginosa* infections in U.S. hospitals averages about 0.4 percent, and the bacterium is the fourth most commonly isolated nosocomial pathogen (a pathogen acquired during hospitalization), accounting for 10 percent of all hospital-acquired infections. For an opportunistic pathogen such as *P. aeruginosa*, the disease process begins with some alteration or circumvention of normal host defenses. The pathogenesis of *Pseudomonas* infections is multifactorial, as suggested by the number and wide array of virulence determinants possessed by the bacterium. Multiple and diverse determinants of virulence are, in part, responsible for the wide range of diseases it causes. Most *Pseudomonas* infections are both invasive and toxinogenic.

The ultimate *Pseudomonas* infection may be seen as composed of three distinct stages: (1) bacterial attachment and colonization, (2) local invasion, and (3) disseminated systemic disease. However, the disease process may stop at any stage. Particular bacterial determinants of virulence mediate each of these stages and are ultimately responsible for the characteristic syndromes that accompany the disease.

The fimbriae of *Pseudomonas* (the small hairlike protrusions surrounding the cell) adhere to the epithelial cells of the upper respiratory tract and, by inference, to other epithelial cells as well. These adhesions appear to bind to specific receptors on epithelial cells. Colonization of the respiratory tract by *Pseudomonas* requires fimbrial adherence and may be aided by production of a protease enzyme that degrades fibronectin to expose the underlying fimbrial receptors on the epithelial cell surface. Tissue injury may also play a role in colonization of the respiratory tract since *P. aeruginosa* will adhere to tracheal epithelial cells of mice infected with influenza virus but not to normal tracheal epithelium (opportunistic adherence), and it may be an important step in *Pseudomonas* inflammation of cornea and urinary tract infections as well as infections of the respiratory tract. *Pseudomonas aeruginosa* also produces a mucoid exopolysaccharide referred to as alginate. Alginate slime forms the matrix of the *Pseudomonas* biofilm that anchors the cells to their environment and, in medical situations, protects the bacteria from the host defenses, such as lymphocytes, phagocytes, the ciliary action of the respiratory tract, antibodies, and so on. Once a *P. aeruginosa* strain takes on the mucoid form, it is less susceptible to antibiotics than in its planktonic state. Mucoid strains are also associated with cystic fibrosis.

Pseudomonas aeruginosa can produce disease in any part of the GI tract from the oropharynx to the rectum. As in other forms of *Pseudomonas* disease, those involving the GI tract occur primarily in immunocompromised individuals. The organism has been implicated in perirectal infections, pediatric diarrhea, typical gastroenteritis, and necrotizing enterocolitis. The GI tract is also an important portal of entry in *Pseudomonas* septicemia.

The versatile pathogen can also be responsible for endocarditis, respiratory infections, bacteremia, central nervous system infections, ear infections, eye infections, bone and joint infections, and urinary tract infections.

Pseudomonas aeruginosa is notorious for its resistance to antibiotics, and this makes it a particularly dangerous pathogen. The bacterium is naturally resistant to many antibiotics due to the permeability barrier afforded by its outer membrane. Also, its tendency to colonize surfaces in a biofilm form makes the cells impervious to therapeutic concentrations of antibiotics. Because its natural habitat is the soil, living in association with actinomycetes and molds, it has developed resistance to a variety of their naturally occurring antibiotics. Moreover, *Pseudomonas* maintains antibiotic resistance plasmids and resistance transfer factors (Todar, 2003), and it is able to transfer these genes by means of the bacterial processes of transduction and conjugation.

It is clear that *P. aeruginosa* is an effective pathogen, and there is little doubt that it can be waterborne. What remains to be established is if drinking water is an important means of transmission for any disease associated with this organism. The organism is not only versatile but virtually omnipresent. It is not clear that removing the organism from drinking water would significantly reduce the exposure of the human population to its aggressive activity.

Bacteria and Terrorism in Water Supplies

Since the terrorist attacks on the United States on September 11, 2001, water utilities have a special interest in understanding organisms that might be considered in attempts to sabotage water systems, particularly when the objective is to create fear in the population. Although most serious bioterrorism efforts have focused on aerosols as delivery systems, there is also reason for concern that all, or a portion of, the municipal water system could be used to affect a large part of the population. The ideal pathogen for this use would be one that has an extremely low infectious dose (see Fig. 3-6), a very high mortality rate (see Fig. 3-8), and a high resistance to disinfectants. Organisms with a high mortality ratio include some of the classic waterborne pathogens, such as *V. clolerae* and *S. typli*, but also some organisms less commonly associated with water, such as *B. antlracis* (anthrax) and *F. tularensis* (rabbit fever). None of these is outstanding where infectious dose is concerned. Some of their characteristics are summarized in Table 3-6 and a brief discussion of *B. anthracis* and *F. tularensis* follows.

BACILLUS ANTHRACIS

Bacillus anthracis is a gram-positive, nonmotile bacillus typically having squared ends. *Bacillus anthracis* produces an endospore, easily seen via methylene blue or India ink stains. Capsule formation differentiates *B. anthracis* and other nonpathogenic bacilli. The endospores are ellipsoidal,

Table 3-6
Bacterial pathogens of interest in terrorism

Bacterium	Size, μm	Motile	Normal Habitat	Health Effects in Healthy Persons	Modes of Transmission	Evidence of Waterborne Pathway	Culturable
Bacillus anthracis	1 – 1.2 × 3 – 5	No	Found in soil habitats	Anthrax	Inhalation or ingestion of spores	Only of interest in bioterrorism	Ordinary nutrient medium
Francisella tularensis A	Small	No	Small mammals	Tularemia (rabbit fever)	Insect bites	Only of interest in bioterrorism	Difficult on standard media; diagnosis by serology

located centrally in the sporangium. The spores refract light and resist staining. *Bacillus anthracis* is about 1 μm wide and 3 μm long and is usually straight. Anthrax bacilli form long chains and appear similar to streptobacilli in culture. Anthrax in the spore stage can exist indefinitely in the environment. Optimal growth conditions result in a vegetative phase and bacterial multiplication.

Bacillus anthracis forms a capsule that consists of a poly-D-glutamate polypeptide. The capsule protects the organism against the bactericidal components of serum and phagocytes and also against engulfment by the phagocytes themselves. The capsule is important when infection is being established. It is less important during the terminal phase of the disease, which is dominated by the anthrax toxins. Anthrax has been around a long time and is described in the early literature of the Greeks, Romans, and Hindus. The fifth plague described in the book of Genesis may be an early description of anthrax.

The toxic nature of *B. anthracis* was not understood until the mid-1950s. Before then, it was assumed that death was due to blockage of the capillaries, popularly known as the "log-jam" theory. This theory came about because so many bacilli were found in the blood of animals dying of the disease (10^9 bacteria/mL of blood has been observed). Since then it has been demonstrated that, on the order of 10^6 cells/mL can result in death of an animal, and the blood of animals dying from anthrax was found to contain a toxin. Finally, injection of the toxin into guinea pigs caused anthrax symptoms. Thus, it was understood that anthrax toxin is important to pathogenesis. The lethal mode of action of anthrax toxin is not completely understood at this time. Anthrax victims appear to die as a result of oxygen depletion, secondary shock, increased vascular permeability, respiratory failure, and cardiac failure. Death frequently occurs suddenly and unexpectedly. The level of the lethal toxin in the blood supply increases rapidly quite late in the disease. It seems to closely parallel the concentration of organisms.

Susceptibility to anthrax varies considerably among animal species. Limited data are available on the dose required for infection. For inhalation, the infectious dose is thought to be on the order of 10^5 organisms. There is reason to believe that the infectious dose for the GI tract would have to be substantially higher, but no quantitative information is available. As shown on Fig. 3-6, the infectious dose has been set between 2×10^5 and 10^6 organisms. There is no evidence of person-to-person transmission of anthrax.

Anthrax ordinarily attacks herbivores (e.g., cattle, sheep, goats, and horses). Pigs are more resistant, as are dogs, cats, and rats. Birds usually are very resistant to anthrax. Buzzards and vultures are not vulnerable to anthrax, but they are thought to transmit the spores. Humans are relatively resistant to cutaneous invasion by *B. anthracis*, but the organisms can infest tears in the skin.

There are two manifestations of the disease that are relevant to exposure through ingestion: oropharyngeal anthrax, a disease where the mouth and throat are infested, and intestinal anthrax, where lesions are found throughout the GI tract.

Oropharyngeal anthrax typically occurs 2 to 7 days after exposure. The lesion at the site of entry into the oropharynx resembles a cutaneous ulcer. Patients with oropharyngeal anthrax may complain of a sore throat and difficulty swallowing. If the disease is allowed to progress unabated, death may result from asphyxiation due to neck edema and/or toxemia.

Intestinal anthrax occurs 2 to 5 days following ingestion. Patients with intestinal anthrax complain of nausea, vomiting, malaise, anorexia, abdominal pain, hematemesis, and bloody diarrhea. The disease is accompanied by a fever.

Primary intestinal anthrax predominantly affects the saclike first section of the large intestine, just beyond the point at which the lower part of the small intestine (ileum) joins the large intestine and produces a local lesion similar to that which results from cutaneous anthrax. In advanced cases, multiple ulcerative lesions are found throughout the GI tract secondary to hematogenous spread. Intestinal anthrax is difficult to recognize, and shock and death may occur 2 to 5 days after onset if it is not properly addressed.

With either disease a malignant pustule develops at the site of the infection. This pustule is a central area of coagulation necrosis (ulcer) surrounded by a rim of vesicles filled with bloody or clear fluid. A hard, black plaque covering an ulcer implying extensive tissue necrosis forms at the ulcer site. Extensive swelling surrounds the lesion. The organisms multiply locally and may spread to the bloodstream or other organs (e.g., the spleen). Dissemination from the liver, spleen, and kidneys back into the bloodstream can result in bacteremia. In bacteremic anthrax, hemorrhagic lesions may develop anywhere on the body. Usually, bacteremia is associated only with inhalation anthrax. Intestinal anthrax results in death in 25 to 60 percent of the cases.

FRANCISCELLA TULARENSIS

Franciscella tularensis is a small, nonmotile, gram-negative, strictly aerobic, non–spore-forming coccobacilli. *Franciscella tularensis* has been divided into two subspecies: (1) *F. tularensis* biotype tularensis (Type A) and (2) *F. tularensis* biotype palaearctica (Type B). Organisms isolated in North America are usually Type A. They are virulent in both animals and humans. In Europe and Asia, human tularemia isolates are usually Type B and much less virulent.

Tularemia is a zoonosis, also known as rabbit fever or deerfly fever. Natural reservoirs include small mammals such as rabbits, mice, ground hogs (woodchucks), ground squirrels, tree squirrels, beavers, coyotes, muskrats, opossums, sheep, and various game birds. Human-to-human transmission has not been documented. Human infections normally occur through a variety of mechanisms, such as bites of infected arthropods; handling infected animal tissues or fluids; direct contact or ingestion of contaminated water, food, or soil; and inhalation of contaminated aerosols. The infectious dose of *F. tularensis* is so low that examining an open culture plate can cause infection (see Fig. 3-6). The disease, tularemia, was first described in Japan in 1837. Its name relates to the description in 1911 of a plaguelike illness in ground squirrels in Tulare county, California (hence the name tularemia), and the extensive work subsequently done by Dr. Edward Francis, after whom it was subsequently named *(Franciscella tularensis).*

Symptoms of tularemia vary, depending on the route of introduction. In those cases where a person becomes infected from handling an animal carcass, symptoms can include a slow-growing ulcer at the site where the bacteria entered the skin (usually on the hand) and swollen lymph nodes. If the bacteria are inhaled, a pneumonia-like illness can follow. Those who ingest the bacteria may report a sore throat, abdominal pain, diarrhea, and vomiting. Advanced tularemia is accompanied with high fever, acute septicemia, and toxemia. Oral infection results in typhoidlike symptoms. In bites, a local abscess at the site of infection is followed by septicemia followed by rapid spread to the liver and spleen. Thirty percent of untreated tularemia patients die. Symptoms can appear between 1 and 14 days after exposure, typically in 3 to 5 days.

Franciscella tularensis is a particularly notorious candidate for biological warfare. During World War II, the Japanese and the United States and its allies all worked on its use for this purpose. The U.S. military stockpiled *F. tularensis* in the late 1960s and then destroyed it in 1973. The Soviet Union continued the production of biological weapons production into the early 1990s, including antibiotic-resistant strains of *F. tularensis.*

3-4 Viruses of Concern in Drinking Water

Although the existence of viruses has been known for some time and some classic diseases such as smallpox and rabies are known to be of viral origin, less is known about these pathogens when compared to bacteria. Viruses have a simpler structure than other organisms. All viruses are true parasites in that they are totally dependent on their host for the resources required for survival. Viruses can survive outside of the host, in some instances, longer than bacteria because they do not have metabolic requirements. Viruses of various kinds use animals, plants, bacteria, fungi, and algae as their hosts. The size of the viruses relative to other organisms was shown on Fig. 3-2. A basic virus consists of a core of nucleic acid (either DNA or RNA) surrounded by a protein coat. Some have protective lipid envelopes. Despite this straightforward morphology or, in some ways, because of it, viruses are more host specific. Animal viruses do not infect bacteria and vice versa. Many viruses can infect only one species of another organism.

Stantec Water Webinar-Impact of COVID-19 on Water & Wastewater Utilities

http://www.stantec.com/wt3r/3-1.html

Because of their small size, viruses could not be detected until the advent of the electron microscope (EM) in 1931. Their presence prior to that time had, however, been postulated, and they were referred to as filterable agents because they would pass through standard filters used to retain bacteria.

Even with the use of the advanced electron microscopic techniques currently available, viruses are still difficult to detect in environmental samples. Except in raw or partially treated wastewaters where their numbers may exceed 10,000 virus units per liter, large-volume sampling and subsequent concentration techniques are required. Even in untreated supplies used for drinking water, where their numbers may reach 1 to 100 units/L, samples of 100 to 1000 L must be processed to achieve significant recovery. Further sophistication is required to identify viruses collected. Given these problems, it is not surprising that definitive association of waterborne viruses with specific disease occurrences is not common. In fact, until recently infectious hepatitis was the only epidemiologically established waterborne viral disease (Grabow, 1968). However, the circumstantial evidence linking viruses to various disease outbreaks is increasingly persuasive. As discussed below, enteric viruses are regularly shed in animal and human feces; they are ubiquitous in untreated surface waters around the world, and although water treatment plants are typically effective in removing incoming viruses, inadequate treatment or temporary breaks in treatment effectiveness can allow some to survive.

Some of the more important characteristics of viruses that are associated with waterborne disease are summarized in Table 3-7. As mentioned earlier, the viruses that cause poliomyelitis and heptatitis A are probably the ones that have been documented to be associated with waterborne transmission for the longest time. But most viruses that infect the GI system are now known to be or are strong candidates for waterborne transmission. These include the viruses just mentioned as well as all the others shown in Table 3-7. For example, poliovirus, coxsackie, echo, and other enteroviruses have been known to be present in wastewater effluent for some time. More recently, rotaviruses, caliciviruses (Norwalk-like viruses or human caliciviruses), and adenoviruses have also been shown to have this

Table 3-7
Characteristics of viruses of interest in water

Species	Type or Strain	Size, nm	Shape	Nucleic Acid[a]	Genome Length, kb[a]
Coxsackie	A1–A22	28–30	Round	ssRNA	7.4
	A24	28–30	Round	ssRNA	7.4
	B1–B6	28–30	Round	ssRNA	7.4
Echovirus	1–7	28–30	Round	ssRNA	7.4
	9	28–30	Round	ssRNA	7.4
	11–27	28–30	Round	ssRNA	7.4
	29–33	28–30	Round	ssRNA	7.4
Enterovirus	68–71	28–30	Round	ssRNA	7.4
Poliovirus	1	28–30	Round	ssRNA	7.4
	2	28–30	Round	ssRNA	7.4
	3	28–30	Round	ssRNA	7.4
Hepatitis	A	27	Round	ssRNA	7.4
Calicivirus	Hawaii strain	35–39	Round	ssRNA	7.7
	Norwalk virus	35–39	Round	ssRNA	7.7
	Snow mountain strain	35–39	Round	ssRNA	7.7
	Southhampton strain	35–39	Round	ssRNA	7.7
	Taunton strain	35–39	Round	ssRNA	7.7
	Other strains	35–39	Round	ssRNA	7.7
Hepatitis	E	35–39	Round	ssRNA	7.7
Rotavirus	A	80	Round	dsRNA	16.5–21
	B	80	Round	dsRNA	16.5–21
	C	80	Round	dsRNA	16.5–21
Hepatitis	B	40–48	Round and irregular	ssDNA	1.7–3.3
	C	40–60	Polyhedral	ssRNA	9.5–12.5
	Delta	36	Spherical	ssRNA	1.7
Human adenovirus	Type 2	70–90	12 vertices	ds RNA	45.4
	Types 1 –47	70–90	12 vertices	dsRNA	45.4
	Types 1 –5	27–30	Starlike with six points	ssRNA	68–79

[a]ss = single-strand; ds = double-strand.
[b]kb = kilobase pairs.

association. Finally, the hepatitis E virus is now known to be an important agent of waterborne disease.

Nongastrointestinal Viruses

POLIOVIRUS

Fifty years ago poliomyelitis was one of the most feared diseases in the developed world. To Americans that lived through the age of polio epidemics, the iron lung is perhaps the most profound symbol.

Poliovirus is an enterovirus. The virus appears in three serotypes, of which Type I is the most common cause of the epidemic disease. Humans are the only natural hosts for poliovirus Type I. The infection occurs through oral–fecal contact and the virus is highly contagious.

Today, extensive use of vaccines has greatly reduced the incidence of the disease, even in developing countries. Because the virus has no animal host, it may be possible to eliminate polioviruses entirely through the use of vaccines, as was done for smallpox in 1980. Consequently, the World Health Organization (WHO) has a Global Polio Eradication Initiative underway to accomplish that goal.

The clinical forms of poliomyelitis vary, but the two basic patterns are the minor illness and the major illness (paralytic myelitis). The minor illness, which accounts for 80 to 90 percent of infections, mostly affects young children. The disease is mild and does not involve the central nervous system. Symptoms are a lot like the flu—a slight fever, malaise, headache, sore throat, and vomiting, which develop 3 to 5 days after exposure. Recovery typically occurs within 24 to 72 h.

Symptoms of the major illness include fever, severe headache, stiff neck and stiff back, deep muscle pain, and, occasionally, increased sensitivity of the skin or the sensation that limbs are falling asleep. Often the disease stops at this point, and complete recovery is still the probable outcome.

In other cases, the disease continues, becoming paralytic myelitis, causing loss of reflex in selected locations, weakness on one side, or paralysis of muscle groups, including those required for breathing (hence the iron lung). The major illness sometimes follows the minor illness after a few days of recovery, but, more commonly, the disease proceeds directly, particularly in older children and adults.

In paralytic poliomyelitis, slightly less than 25 percent of victims suffer severe permanent disability, about 25 percent have mild disabilities, and a little more than 50 percent eventually recover with no residual paralysis. The greatest return of muscle function occurs in the first 6 months, but improvements may continue for 2 years.

The virus enters the body through the mouth and begins to multiply in the intestinal tract. Oral transmission is the most common means of contracting the disease, but it can also be contracted by direct contact with a person who is carrying the disease. Ironically, serious poliomyelitis epidemics are a phenomenon of developed countries. Near the turn of the last century, the countries with the highest standards of hygiene suffered most (e.g., United States, Denmark, Australia, Great Britain). Today transmission is most intense where population density is high and sanitation levels are low.

Recognizing that polio is transmitted by the fecal–oral route, waterborne transmission has been suspected for some time. Nonetheless, the proposition was hard to prove at the time when polio was an important epidemic disease in the United States, and definitive results came years after the Salk vaccine (1954) and the Sabin vaccine (1958). The virus was first found in wastewater in the early 1940s (Kling et al., 1942; Paul et al., 1940; Trask and Paul, 1942). But a clear demonstration of waterborne transmission was accomplished much later (Mosley, 1967). Recently, it has been argued that better understanding of the viruses and their behavior would lead to different conclusions about earlier epidemics as well (Knolle, 1995).

Polio has been eradicated in North and South America, in most northern and southern African countries, in almost all of Eastern and Western European countries. The Western Pacific and China are nearly free of polio. Eradication of the disease in Europe is expected by the time of the publication of this book.

Nevertheless, the disease continues to strike in Africa, southeast Asia, the Indian subcontinent, and the Near East. On the order of 30 countries still report outbreaks with 15 countries reporting a total of 784 confirmed polio cases in 2003. Reservoir countries include Bangladesh, the Democratic Republic of Congo, Ethiopia, India, Nigeria, and Pakistan. A high risk of infection also exists in war-ravaged countries such as Angola, Somalia, and Sudan.

The global effort to eradicate polio stands little chance of success through the use of vaccination alone. Safe water and good sanitation and hygiene practices will be required as well. In an environment where access to new hosts is an uncommon event, the only forms of the virus that will be able to survive are those that do not cause profound symptoms in the host and are able to stay with the host for a long time. In the environments present in today's reservoir countries, opportunities for the virus to gain exposure to new hosts are myriad. In this environment, a seemingly harmless form of the virus loses no advantage in its evolutionary competition by evolving into a form that can cause serious disease. Indeed, there is some evidence that this sort of evolution is taking place in some vaccine strains being used in these countries today (Anonymous, 2001; Cherkasova et al., 2002; Kew et al., 2002; Landaverde et al., 2001).

HEPATITIS

Hepatitis is an inflammation of the liver sometimes accompanied by jaundice, a yellowish color change of skin and mucous membranes. Hippocrates first described epidemic jaundice. By the eighth century it was known to be infectious, and outbreaks were reported in military and civilian populations in the seventeenth, eighteenth, and nineteenth centuries. Early on it was proposed that the infectious jaundice was caused by a virus. Although there had been some evidence to the contrary, until World War II the disease was thought to be infectious in nature, but, during the war, a number of new outbreaks occurred as a result of vaccinations for yellow fever and measles. As a result, in 1947 MacCallum proposed that hepatitis be classified into two types: Type A—infectious hepatitis—and Type B—serum hepatitis (MacCallum, 1947). The former could be contracted from one individual to the other via the fecal–oral route, and the latter could be contracted only through direct exposure of the blood serum (e.g., blood transfusion, contaminated vaccines, contaminated drug needles, or anal sex). In the mid-1960s Blumberg found an antigen in Australian aborigines that later turned out to be the antigen for serum hepatitis (Blumberg et al., 1965). Since that time, three more viruses causing serum hepatitis have been identified. They are hepatitis virus Types C, D, and G (Choo et al., 1989; Linnen et al., 1996; Rizzetto, 1977). A second virus, Type E, causing infectious hepatitis was also identified in connection with epidemics occurring in India (Balayan et al., 1983; Tandon et al., 1985). This virus was also connected to contaminated water (Belabbers et al., 1985) and was found in wastewater (Jothikumar et al., 1993).

In summary, several different viruses cause hepatitis. They are the hepatitis virus Types A, B, C, D, E, and G (at this point a consensus has not been developed on the F candidate). All of these viruses cause acute, or short-term, viral hepatitis. The hepatitis B, C, D, and G viruses can also cause chronic hepatitis, in which the infection is prolonged, sometimes lifelong. This response is also consistent with the theory espoused by evolutionary biologists who would argue that serum hepatitis viruses must develop the ability to persist in the host for long periods because the opportunities for transfer from one host to the other are more limited. The virus causing the infectious disease, on the other hand, depends on its transmission characteristics in the environment for survival and would find long persistence in the host a more limited advantage (Ewald et al., 1998).

There have been a number of outbreaks of infectious hepatitis that have been classified serologically as non–Type A. Often these are presumed to have been outbreaks of hepatitis Type E. On the other hand, even using more up-to-date technology, some cases of viral hepatitis still cannot be attributed to the hepatitis A, B, C, D, or E viruses. This "non-A-through-E" hepatitis is sometimes called hepatitis X. The virus causing hepatitis X has not yet been identified, and it is not clear if this form of hepatitis is infectious or not.

Hepatitis A is a well-documented waterborne disease. The first study demonstrating it as a waterborne disease was conducted in 1945 (Neefe and Strokes, 1945), and newer studies using more up-to-date technologies have confirmed those conclusions in subsequent outbreaks (Sobsey et al., 1985). About 23,000 cases of hepatitis A are reported annually in the United States. These represent just under 40 percent of all hepatitis cases reported. Infectious hepatitis usually resolves on its own over a period of several weeks, but some people are asymptomatic (do not show the symptoms of the hepatitis) until the disease is advanced. Hepatitis A is often a mild disease, which explains why it is so widely spread, even in developed countries with safe water and a high standard of living. Hepatitis A has a worldwide distribution occurring in both epidemic and endemic fashion.

Due to the number of asymptomatic carriers, the mild character the disease often exhibits, and the activity of the disease in developed countries, it is clear that good water treatment will not be enough to control this disease. Nevertheless, it does appear that safe water combined with good sanitation and hygiene can keep the disease at very low endemic levels.

Hepatitis E occurs in both epidemic and sporadic-endemic forms, usually associated with contaminated drinking water. Major waterborne epidemics have occurred in Asia and North and East Africa. To date no U.S. outbreaks have been reported for this type of hepatitis. The disease is most often seen in young to middle-aged adults (15 to 40 years old). Pregnant women appear to be prone to severe disease, and high mortality (20 percent) has been reported in this group. Major waterborne epidemics have occurred in India (1955 and 1975 to 1976), USSR (1955 to 1956), Nepal (1973), Burma (1976 to 1977), Algeria (1980 to 1981), Ivory Coast (1983 to 1984), and Borneo (1987). While no outbreak has occurred in the United States, imported cases were identified in Los Angeles in 1987. Serological tests suggest that most U.S. residents have not acquired immunity to hepatitis E, but about 1 to 3 percent of the U.S. population is positive for antibodies. While hepatitis E transmission via contaminated water is well established, transmission from person to person is not common. Considering this and other data, it seems likely that the lower levels in the United States are due to the fact that this virus can be controlled by levels of sanitation infrastructure that are not successful in controlling hepatitis A (Mast and Krawczynski, 1996). Effective water treatment may be the critical component in controlling the spread of this disease.

Viral Gastroenteritis

Diarrhea remains one of the major causes of death in young children. This is especially so in Asia, Africa, and Latin America, where it causes millions of deaths in children under 4 years of age. The main long-term driving forces for high incidence and mortality are unsafe water or inadequate sanitation. The development and operation of much better physical and institutional infrastructure will be required to eliminate these driving forces. The short term causes are a variety of pathogenic microorganisms. A number of bacteria and protozoa that cause diarrhea have been identified: enterotoxigenic *E. coli*, *Salmonella*, *Shigella*, cholera, other *Vibrio* bacteria, *Giardia* and *Cryptosporidium*, for example. These pathogens account for a large part, but not all of the investigated cases and outbreaks (Lee et al., 2002).

It was not until the seventh decade of the twentieth century that technology was developed that would allow the detection and identification of viruses that cause diarrhea. As a result, an etiological agent could be found for only a limited fraction of gastroenteritis patients. Once these technologies began to develop, researchers began to recognize that a number of viruses were also etiological agents, particularly the rotaviruses and the Norwalk and Norwalk-like viruses (NLVs). Even with some of the causative agents identified, early techniques, which largely involved the use of electron microscopy, were too expensive for routine use when examining stool specimens collected from outbreaks.

In the following decade more efficient techniques were developed using immunological assays. These techniques were more effective for identifying causative agents so they brought with them new information and new understanding. On the other hand, the materials required to conduct the immunological assay are difficult to obtain. Consequently, outbreaks were still not fully investigated, and a substantial number of cases were still labeled as being of unknown etiology. One of the things the immonological assays did show is that the diversity of NLVs was substantial, and this of course only made the problem more difficult to study and characterize. To some extent, the same is true of the rotaviruses.

During the last decade of the twentieth century, breakthroughs in cloning and sequencing of diarrhea-causing viruses led to the development of sensitive molecular assays, and now information is being gathered at a rapid pace. It now appears that NLVs and rotaviruses cause most outbreaks of viral diarrhea. The rotavirus group appears to cause the majority of incidents in poor countries and the very young, and the NLVs appear to be responsible for the bulk of incidents in the United States and other developed countries (MMWR, 2001), where illness occurs in both adults and children. In an underdeveloped country either of these diseases can be life threatening, primarily as a result of lack of adequate treatment for serious cases.

ROTAVIRUS

Considering the globe as a whole and the severity of the disease that results, the most important viruses causing diarrhea are probably the rotavirus family. Rotaviruses have been estimated to cause 30 to 50 percent of all cases of severe diarrheal disease in humans. There are several groups of rotaviruses (A through G), but only three groups are of interest in human gastroenteritis: groups A, B, and C. The group A subtypes 1, 2, 3, and 4 are the leading cause of severe diarrhea among infants and children and account for about half of the cases requiring hospitalization. Group A rotaviruses are endemic worldwide. Group B normally infects pigs and rats, but group B rotavirus, also called adult diarrhea rotavirus, has caused major epidemics of severe diarrhea, affecting thousands of people in Asia, the Indian subcontinent, and North Africa. Group C rotavirus has been associated with sporadic cases of diarrhea in children in the United Kingdom and Japan.

The rotaviruses are round particles about 80 nm in diameter having the appearance of little wheels when viewed by electron microscopy (EM) with negative staining—hence the name *rota*. The virus particles enclose double-stranded RNA in two concentric protein shells (capsids). The human virus has been difficult to culture in vitro, although those belonging to group A are culturable with specialized techniques.

Human rotaviruses were once thought to be limited to the group A rotaviruses, whereas other groups (B to E) were thought to be strictly zoonotic. In the early 1980s, millions of people in China became infected in an epidemic of group B rotaviruses (Hopkins et al., 1984). Subsequent to that event, smaller group B outbreaks have occurred. Studies of immunoglobulin pools from Shanghai suggest that the Chinese population had been exposed to this pathogen in the past (Hung et al., 1987). Because group B rotavirus is a common diarrheal pathogen for swine and because all rotaviruses have a segmented genome, they may be somewhat like the influenza virus, capable of antigenic changes through reassortment of genes. Some have suggested that this human group B epidemic came about through changes that allowed the swine virus to reconfigure itself so that it could propagate in the human gut (Bridger, 1988).

The gastroenteritis caused by rotavirus exhibits the symptoms of vomiting, watery diarrhea, and low-grade fever. The disease can range from mild to serious. People are sometimes hospitalized, and in countries with limited infrastructure, it can be fatal, particularly in infants. The incidence of disease peaks among children aged less than 36 months. Children at this age are also at the greatest risk for severe disease, requiring hospitalization. The disease is normally self-limiting, meaning there is no cure beyond the human immune system. Persons with the disease often excrete large numbers of viruses (10^8 to 10^{10} infectious units/mL of feces) and the infective dose is on the order of 10 to 100 infectious units. As a result, the disease is easily transmitted via contaminated hands, food, water, or utensils.

A significant fraction of the population infected with rotavirus is asymptomatic but still excrete infective particles. As a result of this and the virus's low infectious dose, rotavirus infections are nearly ubiquitous. In both developed and undeveloped countries, 95 percent of children are infected by 5 years of age. Although water treatment can help eliminate the most lethal strains, endemic disease appears to continue even in the presence of a full complement of modern sanitation and hygiene practices. Rotavirus persistence may be related to the high level of virus shedding by infected persons (as much as a trillion viruses per gram of feces), the many other routes of fecal–oral transmission, the many different rotavirus strains that infect infants and young children, and the lack of long-lasting immunity resulting from infection. The natural immunity, which appears in the older part of the population, suggests that vaccination may also be a useful tool to achieve effective control of this disease. Vaccine development is well underway but has been set back by the occurrence of unacceptably high rates of bowel obstruction as a complication resulting from immunization (Parashar et al., 1998).

HUMAN CALICIVIRUSES

The Norwalk and Norwalk-like agents are members of the family Caliciviridae, containing several hundred viruses that have been connected with waterborne and foodborne outbreaks and share certain common characteristics. Norwalk is the pathogen that was associated with the cruise ship epidemics during the winter of 2002 to 2003. Generally they are small, round structured viruses about 35 to 39 nm in diameter, with about 7.7 kb (kilobase pairs) of single-stranded RNA and a capsid containing one major protein of about 60,000 Da. The taxonomy of the caliciviruses has been systematized recently and now uses a standard nomenclature consisting of A/B/C/D/X/Y/Z. The NLVs consist of two genogroups (GI and GII), each of which contains several subgroups (clusters or clades) of closely related but not identical viruses. The prevalence of NLVs varies in time and place among the two main genogroups and among the clades within them. Genogroup II viruses have predominated in the last several years. Another group of human caliciviruses, the Sapporo-like viruses (SLVs), is genetically distinct and has a lower prevalence than the NLVs. The serological or antigenic relationships among the NLVs remain unclear, but new antigens expressed from cloned capsid genes of these viruses are helping to establish antigenic relationships and the role of immunity in susceptibility to infection. The hepatitis E virus discussed earlier is structurally similar to the members of Caliciviridae, but it is placed in a separate virus family.

The Norwalk and NLVs cause viral gastroenteritis, acute nonbacterial gastroenteritis, food poisoning, and so-called winter vomiting disease. Symptoms are nausea, vomiting, diarrhea (not bloody), and abdominal cramps. Headache and low-grade fever may also occur, but diarrhea and vomiting are relatively prevalent among adults, whereas a higher proportion of children experience vomiting. The infectious dose is low, with perhaps 10 to 100 virus particles constituting a 50 percent infectious dose. The disease has an incubation time of 1 to 2 days and is usually self-limiting in less than 4 days. However, virus shedding can continue for 2 weeks postinfection and after symptoms of illness have disappeared. In contrast to the rotaviruses, this disease primarily affects adults and older children. Children younger than 2 years are affected, although prevalence rates are still uncertain. In the United States, illness is most commonly reported among persons of school age and older. Persons exposed to the disease develop immunity but not a long-lasting one.

The NLVs were first identified in the stool filtrates of a patient with diarrhea using EM in 1972 (Kapikian et al., 1972). Since that time they have been widely associated with both waterborne and foodborne disease. However, some instances associated with contaminated food have also been shown to be related to contaminated drinking water as well (Beller et al., 1997). Norwalk-like agents probably create a low background level of infection in a community until an infected individual contaminates a common source and an outbreak occurs. Although secondary cases can multiply the number of persons affected, outbreaks are generally limited to 1 to 2 weeks.

Like rotavirus A, the Norwalk-like agents are endemic even in the most developed countries probably due to the relatively mild character of the disease and to the large number of asymptomatic carriers, including those who just had the disease and still shed the virus particles, both of which result in more spread of the disease by personal contact.

OTHER VIRUSES IMPLICATED IN GASTROENTERITIS

Both the rotaviruses and the NLVs have been demonstrated to appear in the feces of patients with gastroenteritis; have been found in wastewater, raw water, and drinking water; and have been tied to specific waterborne outbreaks. There are also other viruses associated with gastroenteritis for which the waterborne connection, though likely, has not yet been clearly proven. Notable among these are the astroviruses and adenoviruses.

Astroviruses

The astroviruses are members of the family Astroviridae and the genus *Astrovirus.* There are several species, of which seven serotypes have been identified as human astroviruses. Astroviruses are spherical particles about 28 to 30 nm in diameter surrounded by a protein capsid but with no envelope. The name "astro" derives from the fact that the virus particles take on the appearance of a five- or six-pointed star when seen under the electron microscope. Astroviruses have a positive, single-stranded RNA genome with a length of about 7.5 kb. They are fastidious but have been successfully grown in cell culture.

Human astrovirus infections are not yet well understood. The cooccurrence of diarrhea and shedding of astrovirus in feces and the identification of astrovirus in the epithelial cells of diarrhea patients suggest that replication occurs in the human intestine (Phillips et al., 1982).

Low incidence rates of astrovirus infection among young children with diarrhea were verified with early EM studies (Kapikian et al., 1972). The development of enzyme-linked immunosorbent assay (ELISA) and reverse-transcriptase polymerase chain reaction (RT-PCR) have subsequently revealed that astroviruses are strongly associated with viral gastroenteritis in children worldwide (Cruz et al., 1992; Glass et al., 1996; Herrmann et al., 1991; Lew et al., 1991). Human astrovirus serotype 1 is the most predominant serotype worldwide, while serotypes 2, 3, and 4 also appear to be fairly common and serotypes 5, 6, and 7 appear more infrequently.

After exposure, astroviruses typically exhibit a 1- to 4-day incubation period before symptoms occur. The main symptom is typically watery diarrhea, and the disease is usually seen in young children 6 months to 2 years of age. It can also be associated with anorexia, fever, vomiting, and abdominal pain. Astrovirus infections do not normally result in serious dehydration or hospitalization, but individuals experiencing poor nutrition, immunodeficiency, severe mixed infections, or another underlying gastrointestinal disease may experience complications.

Immunity to astrovirus infection is not well understood. Young children and the institutionalized elderly are usually the populations that develop the disease, suggesting that an antibody acquired early in childhood provides protection through adult life and wanes late in life (Glass et al., 1996). Data from the limited studies that have examined the age distribution of astrovirus infection show that the majority of people acquire antibodies by the time they are 5 years of age.

At temperate latitudes, astrovirus infections are more common in the winter, while at tropical latitudes the infections tend to occur during the rainy season. This seasonal pattern of infection is similar to that of rotavirus.

Electron microscopy has long been the method of choice for identification of these viruses, but the technique is too costly and too insensitive for environmental samples. Astroviruses were first observed by EM in stool specimens from infants with gastroenteritis (Appleton and Higgins, 1975; Madeley and Cosgrove, 1975). Using cell culture and more recent molecular techniques, the viruses are frequently found in environmental samples. In data collected for the Information Collection Rule, astroviruses were found in 15 of 29 surface water samples examined using a RT-PCR-nested protocol (Chapron et al., 2000).

Astroviruses are transmitted from person to person by the fecal–oral route. Fecal–oral transmission has been verified by volunteer studies (Kurtz et al., 1979; Midthun et al., 1993). Although most transmission is probably person to person among children, contaminated water and shellfish have given rise to outbreaks. Asymptomatic shedding has also been demonstrated. To date, there has been no clear demonstration of transmission by the water route in the United States, although it is probable that such a demonstration will occur in the future.

Adenoviruses

Adenoviruses are large, nonenveloped viral particles. They are the only human enteric viruses that contain double-stranded DNA rather than single- or double-stranded RNA. Adenoviruses are widely recognized causes of infections in the lungs, eyes, urinary tract, and genitals, but adenoviruses 40 and 41 are thought to affect the intestines.

Adenovirus particles have been shown to be more common among patients with gastroenteritis than with the general population (Wadell et al., 1986). However, in numerous studies examining the victims of gastroenteritis for the presence of adenovirus types 40 and 41, the viruses have been identified in only a minority of the stools examined (Cruz et al., 1990; De Jong et al., 1993; Grimwood et al., 1995; Jarecki-Kahn et al., 1993; Wadell et al., 1986). Peak incidence of adenovirus particles in stools of diarrhea patients occurs among children less than 2 years of age, but older children and adults may be infected. Infections occur throughout the year with no clear peaks. Incubation is between 3 and 10 days, with shedding lasting longer than or equal to 1 week. Long-term immunity is thought to be acquired during childhood infection.

Person-to-person transmission is presumably the principal mechanism for the spread of infection. Neither food nor water has been demonstrated as a means of transmission, although adenoviruses have been reported as possible agents in waterborne outbreaks (Kukkula et al., 1997) and adenoviruses have been found in raw and finished drinking water (Chapron et al., 2000; Murphy et al., 1983).

Other Viruses Associated with Fecal–Oral Route

Stantec White Paper: What treatment professionals need to know about coronavirus and the water cycle.

http://www.stantec.com/wt3r/3-2.html

As the name implies, the enteroviruses are regularly found in feces and in wastewater, and all are considered capable of infecting the human gut. There is also evidence of many of these appearing in surface water and groundwater supplies. However, with the possible exception of the poliovirus, there is little conclusive evidence of waterborne transmission based on documented outbreaks. There are a number of possible reasons for this situation. For example, many of these viruses may cause asymptomatic effects or chronic rather than acute disease, where the association of the organism with disease is more difficult to make. They may also be responsible for mild diseases that do not reach the attention of the medical community. In any case, as they are residents of the gut, they are certainly candidates for waterborne transmission. Even if waterborne transmission is not their primary vehicle of transmission, their ultimate elimination will not be accomplished without effective water treatment playing a role. Other possible enteric viruses are parvoviruses, which are small, round, featureless viruses containing DNA. They have been implicated in foodborne illness in the United Kingdom, but their role as waterborne pathogens is uncertain.

3-5 Protozoa of Concern in Drinking Water

The protozoans are a group of unicellular, nonphotosynthetic organisms probably derived from various groups of unicellular algae. They are motile, moving via use of flagella (flagellates), amoeboid locomotion (amoeba), or cilia (ciliates). Two other groups of protozoa, the coccidians and the microsporidia, are nonmotile. Several protozoa are parasites transmitted by the fecal–oral route. Protozoan organisms and a description of their associated diseases are summarized in Table 3-8.

Inside the host, parasitic protozoa excystate, meaning "living stages," such as trophozoites (*Giardia*) and sporozoites (*Cryptosporidium*) are released. Only the hardy resting stages, such as cysts, oocysts, and spores (of microsporidia), can survive outside the host. For some parasites, such as *Giardia*, the life cycle is simple and involves primarily cell division (binary fission) and the formation of cysts. For *Cryptosporidium*, the life cycle of the organism is more complex and includes both asexual and sexual stages.

Table 3-8
Parasitic protozoa that infect humans and that may be of concern in drinking water

Protozoa	Organism	Reservoir	Disease	Transmission	Occurrence
Acanthamoeba Castellani	Trophozoite is amoeba 25–40 μm with thorn-like radiations, aerobic, mobile, sluggish, polydirectional. Cyst double-walled 15–28 μm polygon.	Warm stagnant water, ponds, power plant cooling ponds	Causes *Acanthamoeba* keratitis in soft contact lens wearers. Can produce a rare but serious form of encephalitis in immunosuppressed.	Gains entry through abrasions, ulcers, and as secondary invader during other infections.	Ubiquitous phagotroph in aquatic habitat. Rare endoparasite in eye or central nervous system.
Balatidium coli	Trophozoite: Ciliated amoeba (50–80 × 40–60 μm) with two nuclei, a macronucleus and a micronucleus. Cyst: round, 55 μm.	Pigs (primarily), primates, humans	Most asymptomatic. Symptoms, when present: diarrhea, occasionally dysentery, abdominal pain, weight loss. Symptoms can be severe in debilitated persons.	Waterborne transmission tied to poor sanitation or contamination with swine feces; human infection rate is low.	Worldwide. Infections occur more frequently in areas where pigs are raised.
Entamoeba histolytica	Trophozoite: 15–60 μm, avg. 25-μm amoeba with distinct nuclear envelope and small central endosome. Cyst: 10–20 μm, avg. 12 μm, four nuclei, central endosomes.	Humans	Most asymptomatic, occasionally dysentery. Usually limited to GI tract, occasional invasions with serious complications.	Fecal–oral. Fecally contaminated food or water is the primary mode of transmission. Vegetables are a problem in endemic areas.	Worldwide distribution with a higher prevalence in tropical and subtropical countries.

(*Continued*)

Table 3-8 (*Continued*)

Protozoa	Organism	Reservoir	Disease	Transmission	Occurrence
Entamoeba dispar	Trophozoite: 15–60 μm, avg. 25-μm amoeba with distinct nuclear envelope and small central endosome. Cyst: 10–20 μm, avg. 12 μm, four nuclei, central endosomes.	Humans	Most asymptomatic, mild diarrhea.	Fecal–oral. Fecally contaminated food or water is the primary mode of transmission. Vegetables are a problem in areas.	Worldwide distribution with a higher prevalence in tropical and subtropical countries.
Giardia lamblia	Trophozoite a single-celled flagellated protozoa (9–21 μm long, 5–15 μm wide, and 2–4 μm thick), "teardrop" shape, contains two nuclei at anterior and five flagella, tumbling motility; forms environmentally resistant cysts in the colon (13 μm long, oval shape, and two nuclei).	Beavers, bears, muskrats, dogs, cats, humans, maybe others	Giardiasis—asymptomatic in most; in some, sudden diarrhea with foul-smelling, greasy-looking stool with no mucus or blood; often abdominal cramps, bloating, fatigue, and weight loss; restricted to upper small intestine with no invasion; normally illness lasts 1–2 weeks; chronic infections can last months to years.	Person-to-person, fecal–oral route; infected food handlers; contaminated water or food.	Worldwide; especially areas of poor sanitation; outbreaks more common in children; waterborne outbreaks common where unfiltered waters are contaminated; childcare centers; traveler's diarrhea; most frequent cause of nonbacterial diarrhea in North America.

Naegleria fowleri	Three forms: amoeba, cyst, and flagellate. Trophozoites: 7–20-μm amoeba with single nucleus, large karyosome, and no peripheral chromatin. Cysts have a single nucleus that is almost identical to that seen in the trophozoite. Morphologicially resembles *Acanthamoeba* spp.	Soil water, decaying vegetation	Primary amoebic meningoencephalitis.	Nasal inhalation with subsequent penetration of the nasopharynx by flagellate form, then transformation to ameoba in brain; exposure from swimming in freshwater lakes.	Worldwide. On the soil, in warm, stagnant bodies of water, lakes, rivers, hot springs, unchlorinated pools, and power plant pools.
Cryptosporidium parvum type 1	Oocyst: ellipsoidal 3–5 μm, sporozoite and merozoites: wormlike 10 × 1.5 μm; shizont, microgametes (biflagellated, 5 μm), macrogametes.	Humans	Often asymptomatic; symptoms: cryptosporodiasis— diarrhea, abdominal pain, nausea or vomiting, and fever.	Person-to-person, fecal–oral route; common in childcare centers, infected food handlers; contaminated water or food, especially water and vegetables.	Worldwide.
C. parvum type 2	Oocyst: ellipsoidal 3–5 μm, sporozoite and merozoites: wormlike 10 × 1.5 μm; shizont, microgametes (biflagellated, 5 μm), macrogametes.	Cows, goats, sheep, pigs, horses, dogs, mule deer, humans	Often asymptomatic, symptoms: cryptosporodiasis— diarrhea, abdominal pain, nausea or vomiting, and fever.	Person-to-person, fecal–oral route; common in childcare centers, infected food handlers; contaminated water or food, especially water and vegetables.	Worldwide.

From the standpoint of waterborne disease the organisms of greatest importance are *Entamoeba histolytica, Entamoeba dispar, Giardia lamblia,* and *Cryptosporidium parvum.*

Entamoeba

Entamoebas are single-celled parasitic amoeboid protozoa. Trophozoites range in size from 20 to 40 μm in diameter. Sporozoites are spherical cysts measuring 10 to 16 μm in diameter with four nuclei, on rare occasions as many as eight. The morphology of the nuclei is similar in both the trophozoites and the sporozoites. Locomotion is rapid gliding, by means of a single well-defined pseudopodium, often extended explosively, without conspicuous differentiation between ecto- and endoplasm. It is fairly difficult to distinguish between the various species of *Entamoeba,* although such speciation is important because only some species are pathogens.

Humans can be host to at least six species of *Entamoeba* in addition to several amoebae belonging to other genera. However, only one species of *Entamoeba* infecting humans is known to cause a serious disease, *E. histolytica.* If untreated, infections of *Entamoeba* can last for years. Infection is often asymptomatic. The most common symptoms to occur are vague GI distress or dysentery (hence the name amoebic dysentery). Most infections occur in the digestive tract, but other tissues may be invaded. The amoebae's enzymes help it to penetrate and digest human tissues, and it can also secrete toxic substances. Complications include ulcerative and abscess pain and, rarely, intestinal blockage. Onset time is highly variable. Severe ulceration of the GI mucosal surfaces occurs in less than 16 percent of cases. In fewer cases, the parasite invades the soft tissues, most commonly the liver. Only rarely are masses formed (amoebomas) that lead to intestinal obstruction. Fatalities are infrequent.

The life cycle of *Entamoeba* spp. is illustrated on Fig. 3-12. Infection by *Entamoeba* spp. begins with the ingestion of mature cysts from fecally contaminated food, water, or hands. Once they reach the small intestine, the cysts excyst, releasing trophozoites, which migrate to the large intestine. The trophozoites multiply by binary fission and produce cysts with protective walls, which are passed in the feces. The cysts can survive a long time in

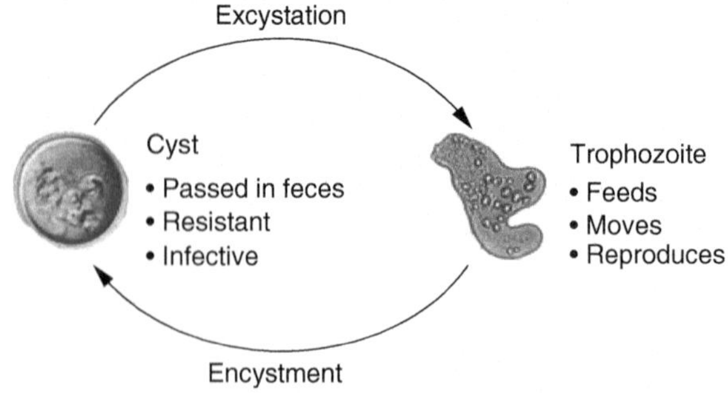

Figure 3-12
Basic model of life cycle of protozoan parasites transmitted by oral–fecal route.

the external environment and are responsible for transmission. In many cases, the trophozoites remain confined to the intestinal lumen (a noninvasive infection), which is typical of *E. dispar*. Individuals with infections of this kind have no symptoms (asymptomatic carriers); however, cysts are still passed in their stool.

In some patients the trophozoites invade the intestinal mucosa (intestinal disease), resulting in amoebic dysentery. In other patients the trophozoites also travel through the bloodstream to extraintestinal sites such as the liver, brain, and lungs (extraintestinal disease), resulting in serious complications.

In recent years invasive and noninvasive forms of the disease have been associated with two separate species, *E. histolytica* and *E. dispar*. The two species are morphologically indistinguishable.

In 1857, Löch demonstrated that amoeba can cause dysentery, naming the organism *Amoebae coli*. In 1903, Schaudinn renamed the species that caused the disease Löch studied to *E. histolytica*. It soon became evident that the vast majority of persons infected with *Entamoeba* spp., did not show the more serious conditions Löch spoke of, and in 1925 Brumpt suggested that two species be considered, one that caused pathogenic infections (*E. histolytica*) and another whose infections were not pathogenic (*E. dispar*). In the last three decades of the twentieth century, a number of developments made it clear that *E. histolytica* and *E. dispar* are very similar organisms but with different virulence factors (Bhattacharya et al., 2000; Diamond and Clark, 1993; Gilchrist and Petri, 1999; Jackson and Ravdin, 1996).

In recent years a new understanding of this organism has led to the recognition that there are in fact two species within what has previously been known as *E. histolytica*. Of these two organisms, *E. histolytica* is the cause of all invasive disease while the other, *E. dispar*, is not capable of invading tissue. These organisms were previously known as *pathogenic E. histolytica* and *nonpathogenic E. histolytica*, respectively, and these names will be encountered in the literature. The relative prevalence of these two species is not yet fully known, but it is clear that in most parts of the world *E. dispar* is easily the more common of the two. The two species are morphologically identical, and differentiating between the two is accomplished with relatively sophisticated methods: isoenzyme, antigen, and/or DNA analyses.

Entamoeba infects predominantly humans and other primates. Other mammals such as dogs and cats can become infected but usually do not shed cysts (the environmental survival form of the organism) with their feces and therefore do not constitute a significant reservoir leading to transmission.

The disease has a worldwide distribution with a higher prevalence in tropical and subtropical countries. Ten percent of the world population carries either *E. histolytica* or *E. dispar*. The infection is not unusual in the tropics and arctics or in any crowded situation with inadequate water treatment and sanitation/hygiene infrastructure in temperate zone urban environments. *Entamoeba* spp. is also frequently diagnosed among homosexual men.

Amebiasis is transmitted by fecal contamination of drinking water and foods but also by direct contact with dirty hands or objects as well as by sexual contact. Perhaps the most dramatic incident in the United States was the Chicago World's Fair outbreak in 1933 caused by contaminated drinking water. Defective plumbing allowed wastewater to contaminate the drinking water. There were 1000 cases and 58 cases were ultimately fatal. In recent times, food handlers are suspected of causing many scattered infections, but there has been no single large outbreak. In October 1983, the Los Angeles County (California) Department of Health Services was notified by a local medical laboratory of a large increase in the laboratory's diagnoses of intestinal amebiasis (*E. histolytica* infection). Thirty-eight cases were identified from August to October. The laboratory staff estimated that, before August, they had diagnosed approximately one *E. histolytica* infection per month. A preliminary investigation failed to identify a common source of the infection.

Giardia lamblia

A pathogenic flagellated protozoa, *G. lamblia*, is one of the most primitive eukaryotic, or nonbacterial, organisms. It does not have a typical endoplasmic reticulum or Golgi apparatus. The organism has no mitochondria. This lack of mitochondria has been taken as evidence that these organisms existed before the endosymbiosis event. The endosymbiosis event is an important landmark in evolution, which is thought to have occurred 100 million years ago and to have created mitochondria. Organisms created before that event do not have mitochondria. Although *Giardia* lacks these organelles, it still has ways of accomplishing many of the functions these organelles normally serve. *Giardia* has bilateral symmetry, a unique feature among organisms at its evolutionary level.

Giardia can exist either as a sporozoite (a cyst) or a trophozoite. As a trophozoite, it takes on a pear-shaped look from the front and a spoon-shaped look from the side. It is about 10 to 15 μm long and 4 to 8 μm wide. It is motile through the use of its flagellum, and its motion is similar to that of a leaf falling or rolling around. The trophozoite has two nuclei and eight flagella (four lateral, two ventral, and two caudal). The nuclei are so placed that, along with the other characteristics of the organism, they give it the appearance of a "monkey face." Another distinctive feature is the special adhesion surface the organism uses to attach to the wall of the intestine (like a suction cup). This adhesive disc is a sophisticated structure with microtubules and microribbons that allow it to firmly attach to the host organism.

As cysts, *Giardia* organisms take on an oval, ellipsoidal, or spherical shape about 11 to 14 μm long and 7 to 10 μm wide. Cysts are not motile but do include flagellar axonemes that lie diagonally across their long axis. When first formed, the cysts exhibit two nuclei, like the trophozoite. Mature cysts include four nuclei, all displaced to one pole. The life cycle of *G. lamblia* may also be described as shown on Fig. 3-12, except that it is not associated with either of the invasive stages of infection associated with *E. histolytica.*

Members of the genus *Giardia* are found in colonies attached to the intestinal linings of animals (trophozoite), in the feces of infected individuals (cyst form), or in contaminated water (cyst form). They can survive temperature extremes ranging from the internal temperature of an animal body to freshwater down to 4°C.

The two stages in the life cycle of *G. lamblia* are well adapted to survival in the environment. Exposure of cysts to gastric acid during their passage through the stomach activates excystation, although the trophozoite does not emerge from the cyst until it passes into the small intestine. The act of emergence is triggered by the milder pH in the small ileum, the small intestine just downstream of the bile ducts. The emerging parasite quickly divides into two equivalent binucleate trophozoites that use their adhesion surface or foot to attach to and colonize the small intestine. Trophozoites use their four pairs of flagella to swim in the intestinal fluid and also to assist in adhering to mucous strands. They also penetrate the mucous layer to attach to intestinal epithelial cells via their unique adhesive disc. Few other microbes normally colonize the complex and variable environment at the upper end of the small intestine. Trophozoites may continue in the small intestine for weeks to years; however, when they are carried downstream by the flow of intestinal fluid, they take the form of a cyst, as they cannot survive outside the host as a trophozoite.

Infection with *G. lamblia* causes diarrhea and abdominal pain and is implicated in a chronic fatigue syndrome that is difficult to diagnose. Giardiasis is an important contributor to the burden of human diarrheal disease. Nonetheless, the basic biology of this parasite is not well understood. *Giardia lamblia* trophozoites are not invasive and secrete no known toxin; nevertheless, they are capable of causing severe and protracted diarrhea. On the other hand, about half the persons infected with *G. lamblia* do not exhibit any symptoms (are asymptomatic). Also, in normal persons, a *G. lamblia* infection usually resolves spontaneously. Thus, both the duration and symptoms of giardiasis vary in normal people.

There are many species of *Giardia*, and they are named based on their normal host. *Giardia lamblia* (also called *Giardia duodenalis*) inhabits humans, *Giardia muris* inhabits rodents, and *Giardia ardeae* inhabits birds. Only *G. lamblia* can inhabit humans. Parasites from one animal typically exhibit a high degree of host specificity; that is, they cannot infect other animal hosts. For example, humans are not infected by *G. muris.* However, some strains appear to be able to infect more than one host. Specifically, there is evidence to suggest that *G. lamblia* can infect beavers and possibly muskrats as well as humans, making all of these a possible reservoir for the organism.

Other reservoirs of *Giardia* capable of infecting humans are calves and other agricultural animals.

Giardia lamblia causes intestinal infection throughout the world. In both industrialized and developing countries, endemic giardiasis is an important cause of illness in children and adults. *Giardia lamblia* cysts have been identified in persons with gastroenteritis numerous times, they are always present in wastewater and, despite the poor quality of analytical procedures, are regularly found in surface water supplies as well. *Giardia lamblia* has also been tied to several outbreaks of waterborne disease around the world. Clearly *G. lamblia* is a pathogen of concern in drinking water treatment. As mentioned earlier, the organism was the principal target of the U.S. EPA's original Surface Water Treatment Rule (U.S. EPA, 1989). However, *G. lamblia* is endemic through both the developed and the developing world and water treatment, while a necessary measure for controlling major outbreaks, will not eliminate the disease entirely. Based on endemic levels in both animal and human populations, this disease organism is likely to be around for a long time.

Cryptosporidium

Cryptosporidium parvum is a single-celled protozoan and an obligate intracellular parasite. It infects the epithelial cells on the intestinal wall and travels from one host to the next as an oocyst, excreted in the feces. The infectious agent, or oocyst, is about 3 to 7 μm in diameter, about half the size of a normal red blood cell.

The life cycle of *C. parvum* is illustrated on Fig. 3-13. It is much more complex than the life cycles of the protozoa discussed in the preceding sections because it involves both sexual and asexual developmental stages. Development begins with the ingestion of the sporulated oocyst, the resistant stage found in the environment. Each oocyst contains four infective sporozoites. Sporozoites exit from a suture located along one side of the oocyst. The preferred site of infection is the ileum, the small intestine just downstream of the duodenum and the bile ducts. Sporozoites penetrate individual epithelial cells in this region. The parasites were once thought to just attach to the outside surface of the epithelial layer, but it is now known that they establish themselves inside a membrane containing a thin layer of host cell cytoplasm on the surface of epithelial cells. A unique attachment/feeder organelle, plus accessory foldings of the parasite membranes, develops at the interface between the parasite proper and the host cell cytoplasm. This organanelle is called a mermont. Multiple fission occurs, resulting in the formation of eight merozoites within the mermont. Once the mermonts reach maturity, they rupture open, freeing the merozoites. The merozoites are very similar in appearance to sporozoites but represent a different stage of development. Like the sporozoites before them, the merozoites penetrate new cells where they also form additional mermonts. In fact, some merozoites develop into mermonts of exactly the same type as the sporozoites, with the same outcome. Others form a mermont that

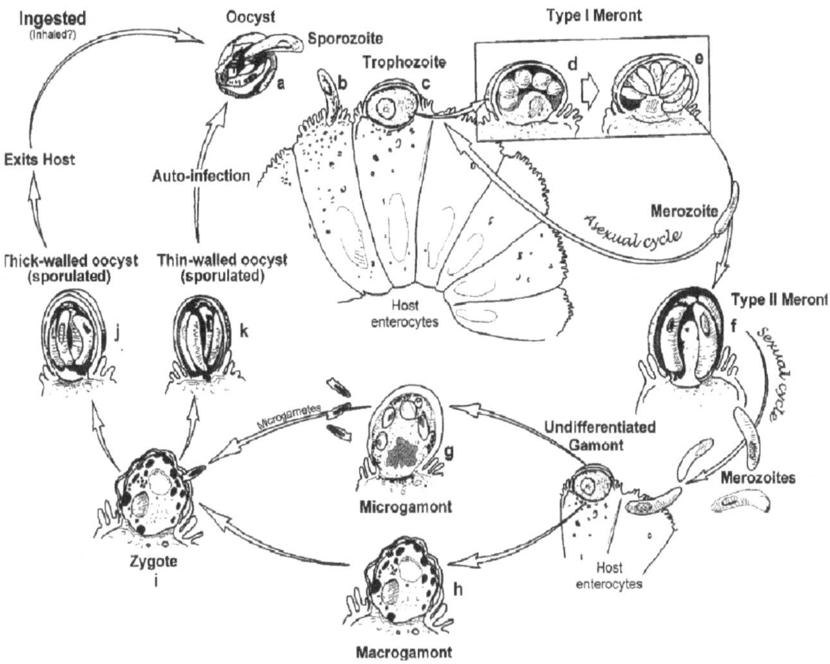

Figure 3-13
Life cycle of *C. parvum*. (Courtesy of the National Centers for Disease Control and Prevention, National Center for Infectious Diseases).

matures in a different way. This second type of mermont (Type II mermont) contains only four merozoites, unlike the Type I mermonts, which contain eight. When these four Type II merozoites are released, they form two different sexes. Some enter cells, enlarge, and form macrogametes. Others enter cells, undergo multiple fission, and form microgametocytes, which mature to contain 16 nonflagellated microgametes. When they reach maturity, the microgametocytes rupture, releasing microgametes that then seek out and penetrate the macrogametes, mating and forming a zygote. As the zygote matures, a resistant oocyst wall forms around them. Some form a thin wall and others a very thick protective wall. Inside the zygote, meiosis occurs, and four sporozoites are formed in the process. Most of the thin-walled oocysts excyst and continue the cycle in the gut. The thick-walled oocysts are passed into the environment in the feces.

The genus *Cryptosporidium* is a member of the family Cryptosporidiidae, which is assigned to the order Eucoccidiorida and to the phylum Apicomplexa. The parasite infects a wide range of vertebrates, including humans. The species *C. muris* infects the gastric glands of rodents and several other mammalian species but, under normal circumstances, not humans. The species *C. parvum*, however, infects the small intestine of an unusually wide range of mammals, including humans. It is a zoonotic species and is the agent that causes cryptosporidiosis in humans. Although numerous other species of the genus have been described over the years, the majority of those reported from mammals are synonyms of *C. parvum*. There do appear to be a number of genotypes of *C. parvum* that are more specialized.

Two genotypes of the *C. parvum* species are of special interest. So far genotype 1 (sometimes labeled genotype H for "human") has only been found to infect humans. Genotype 2 (or genotype C for "calf") occurs in a wide range of animals, including humans. Genotype 1 tends to be more aggressive in humans. Genetic markers on different chromosomes reveal there is little or no mixing between the genotypes, suggesting that these may actually be two separate species that possess the same morphology. The two isolates have a very different host range, but, at the present time, it is not known if they cause different pathology. Either genotype may cause an outbreak; however, preliminary information suggests that genotype 1 is predominant in human infections, including all seven isolates examined in the 1993 outbreak in Milwaukee.

Although there has been recent success in culturing genotype 1 in piglets and genotype 2 in vitro, virtually all the available data on infectious dose, disinfection resistance, and removal requirements are based on genotype 2 grown in calves. The inability to continuously propagate *C. parvum* in vitro is probably the most significant obstacle to developing a better understanding of the organism. Although genotype 2 isolates can be and have been maintained in calves, in vivo propagation is costly and the separation of the oocysts from the rest of the fecal matter is labor intensive. As a result, information on *C. parvum* in general and of the genotype 1 subgroup in particular is limited.

Symptoms of cryptosporodiosis include diarrhea, a loose or watery stool, stomach cramps, an upset stomach, and a slight fever. A number of persons who are infected have no symptoms (i.e., are asymptomatic). *Cryptosporidium* is a common cause of human diarrhea, although less common than diarrhea caused by human viruses. Severe infections are more common in the young, both in animals and in humans. However, people of all ages are susceptible. Childcare centers, with a large susceptible population, frequently report outbreaks.

Symptoms generally begin between 2 and 10 days after being infected. The intestinal disease is self-limiting in most healthy individuals, with watery diarrhea lasting between 2 and 4 days, the rest of the symptoms lasting no more than 2 weeks. In some outbreaks at childcare centers, diarrhea can persist for nearly a month. Occasionally the patient appears to improve for a few days, then gets worse again, before the illness finally ends.

The widespread use of highly active antiviral therapy (HAART) for patients with HIV, in particular combination therapy with protease inhibitors, has led to an overall decline in the incidence of opportunistic illnesses in persons with HIV infection. Despite this decline, cryptosporidiosis in persons with acquired immunodeficiency syndrome (AIDS) is still common. It is unlikely that HAART will be available to the vast majority of persons with HIV infection who live in countries that cannot afford these expensive therapies. In the end, a healthy immune system is the only truly effective means of eliminating the parasite.

Some data is available on infectious dose of *C. parvum.* So far these studies have been conducted on the Type 2 organism. The range of the dose to cause infection in 50 percent of the subjects (ID_{50}) reported in these studies is from 9 to 1042 oocysts (Okhuysen et al., 1999). Considering that such a high fraction of the isolates from humans have been Type 1, it seems likely that the infectious dose for this organism is on the low end of this range (i.e., *C. parvum* is an effective infectious agent). A conclusion that is consistent with the organism's weak immune defense is explained in the following paragraph.

Cryptosporidium appears to make little effort to evade the immune system of the host. Many of the surface proteins, glycoproteins, and phospholipids of *Cryptosporidium* are of such a nature that they stimulate the immune system, and many molecules on the surface of both sporozoites and merozoites will react with two or more different immune responses. To succeed, *Crytposporidium* must develop rapidly and swamp the intestine with oocysts. In fact, the infectious nature of this parasite is such that, if its stay in the body were prolonged, it would kill the host through dehydration and electrolyte imbalance. This sort of activity might eliminate the host species. Given *Cryptosporidium's* excellent ability to survive outside the host, evolution favors the strains of *Cryptosporidium* that can get in and get out again quickly, generating a lot of oocysts in the environment without seriously damaging the host.

As the first case of cryptosporidiosis in humans was reported in 1976, *Cryptosporidium* can be considered an emerging pathogen. Since then it has been recognized as a major cause of GI illness worldwide, especially in developing countries and among the immunocompromised. Outbreaks of cryptosporidiosis have been reported in several countries, the most remarkable being a waterborne outbreak in Milwaukee in 1993 that affected more than 400,000 people (Edwards, 1993). Enough surveys have been conducted to gain some idea about prevalence of the parasite among different portions of the human population. Somewhere around 0.4 percent of the U.S. population appears to be passing oocysts in their feces at any one time. Of those patients admitted to U.S. hospitals for diarrhea, 2 to 2.5 percent are reported to pass oocysts. However, the seroprevalence in the population appears to be much higher. Between 30 and 50 percent of the U.S. population have antibodies to *C. parvum.* In developing countries, this measure is higher, between 60 and 85 percent of people in these regions of the world testing positive for *C. parvum* antibodies.

The ability of *C. parvum* to infect such a broad spectrum of warm-blooded animals gives it a tremendous advantage in evolutionary competition. Consider that *C. parvum* Type 2 infects not only humans but also cows, goats, sheep, pigs, horses, dogs, and cats—even wild animals, such as the mule deer. A high environmental concentration of oocysts is expected, given the survival characteristics and broad range of hosts.

However, the numbers of *Cryptosporidium* oocysts reported by various groups from public water samples are not reliable (neither precise nor accurate). Concentration techniques for oocysts in environmental samples are poor, resulting in underestimates, and detection methods often cross-react with algae or other debris, potentially resulting in overestimates. Numerous species and strains of *Cryptosporidium* incapable of infecting humans occur in the environment and may cross-react in diagnostic tests. As a result, laboratories performing diagnostic testing on *Cryptosporidium* oocysts in water have widely varying degrees of reliability (precision and accuracy). Finally, today's methods are not capable of estimating oocyst viability. Many oocysts detected are probably not viable due to age, freezing, sunlight, or other environmental effects that have inactivated them.

Cryptosporidium is primarily transmitted from one individual to the other by the fecal–oral route. Oocysts may be found in soil, food, water, or surfaces that have been contaminated with the feces from infected humans or animals. The disease is not spread by contact with blood. As demonstrated by the experience in Milwaukee, but also in many other instances, cryptosporidiosis can be a waterborne disease. Eliminating this route of exposure is important because other measures for controlling the disease cannot succeed if the drinking water route is available. Moreover, as with other organisms, ready transport through the drinking water enables more debilitating forms of the disease to spread more effectively. For example, as better water treatment becomes available, it may be found that Type 1 *C. parvum* is isolated less frequently. However, cryptosporidiosis is the sort of disease that will never be controlled by water treatment alone for the following reasons: (1) unlike many waterborne diseases, it has a large reservoir of infections in the nonhuman mammal population; (2) *C. parvum* has a low infectious dose; it develops a number of asymptomatic carriers who can expose the rest of the population during their period of patency, the period of time when the infected person sheds oocysts in his or her feces; and (3) it is already clear that there are essentially permanent reservoirs in the human population, such as childcare centers, that water treatment will not address.

Other Protozoa of Concern in Water Supply

Entamoeba histolytica, G. lamblia, and *C. parvum* are well-established causes of waterborne disease, and all three are a potential risk throughout much of the globe. *Balantidium coli* is also a waterborne pathogen of the conventional fecal–oral route. It is normally associated with exposure to pig wastes contaminating the water supply and is thus a zoonotic disease. The disease is often asymptomatic or takes the form of a mild amoebic dysentery. In immunosuppressed persons, the disease is likely to be more serious. Waterborne outbreaks are rare, but effective water treatment can play an important role in reducing the risks from this route.

Acanthamoeba castellani and *Naegeria fowleri* are both opportunistic pathogens that are virtually ubiquitous in the aquatic environment. These are not pathogens of the fecal–oral route and the GI system is not their primary target. In both cases, serious infections can occur that involve the central nervous system. *Acanthamoeba*, which is ubiquitous in water, soil, dust, and so on, can result in serious infections for persons using contact lenses but not practicing adequate hygiene. It is not clear that drinking water treatment has much of a role to play in controlling either of these two organisms.

3-6 Helminths of Concern in Drinking Water

Parasitic worm infections are not a widespread problem in the United States. However, some types do still regularly occur, particularly in southeastern United States. Ascariasis, an infection of the small intestine, occurs worldwide and is still relatively common, especially among children in the southern United States. In more tropical countries and/or in countries with less developed wastewater treatment systems or sanitation practices, certain types of helminth infections are endemic. Schistosomiasis, a disease caused by *Schistosoma* infections of the liver or urinary system, is endemic in parts of Africa and also occurs in the Arabian peninsula, South America, the Middle East, India, and Asia. Fortunately, human schistosomiasis has been eliminated in the continental United States.

Transmission of helminths occurs in a variety of ways: via contaminated drinking water or vegetables and through body contact with contaminated irrigation water, biosolids, and/or soils. Information on some of the common helminths in the United States and worldwide is summarized in Table 3-9. The primary mode of control is through effective treatment and disposal of wastewater or human feces. If contamination of a drinking water supply does occur, conventional treatment will generally remove the helminth eggs, which are denser than water and larger than filter pores. However, some of the adult worm stages can pass through the pores of filters. For schistosomiasis, control of the intermediate snail host population can be effective.

3-7 Algae of Concern in Drinking Water

Algae play an important role in the rivers, reservoirs, and lakes that serve as surface water supplies, playing a role in the cycling of nutrients, serving as part of the food chain, and influencing the water's flavor. A few algae are pathogenic to humans, producing endotoxins that can cause gastroenteritis. Still others interfere with treatment plant operations, specifically with filter operations.

Discussion of algal roles in nutrient cycling and as the basis of food chains is outside the framework of this text. Complete descriptions of these roles can be found in limnology texts such as Horne and Goldman (1994)

Table 3-9
Characteristics of helminths that infect humans

Agent	Hosts	Disease	Transmission	Occurrence
Ascaris lumbricoides (intestinal roundworm)	Humans.	Ascariasis: moderate infections cause digestive and nutritional problems, abdominal pain, and vomiting; live worms passed in stools or vomited; serious cases involving liver can cause death.	Ingestion of infected eggs from soil, salads, and vegetable contaminated with eggs from human feces.	Worldwide: especially in moist tropical areas, where prevalence can exceed 50%; in the United States disease is most common in the south.
Schistosoma mansoni, S. haematobium, S. intercalatum, S. japonicum	Humans, domestic animals, and rats serve as primary hosts; snails act as necessary intermediate hosts.	Schistosomiasis: a debilitating infection where worms inhabit veins of host, chronic infection affects liver or urinary system.	Water infected with larvae that develop in snails; penetration through human skin; eggs excreted via urine or feces and larvae develop in water and reinfect snails.	Africa, Arabian peninsula, South America, Middle East, the Orient, parts of India; in the United States immigrants from Middle East may carry disease.
Various schistosomes	Birds and rodents; humans are nonnormal hosts.	Schistosome dermatitis (swimmer's itch): local skin dermatitis caused by penetration of larvae; larvae die in skin.	Same as previous entry.	Widely distributed but only locally endemic.
Necator americanus or *Ancylostoma duodenale* (hookworm)	Humans.	Ancylostomiasis: hookworm disease; debilitating disease associated with anemia; heavy infestations can result in retardation; slight infections produce few effects.	Eggs from deposited feces develop into larvae that penetrate the skin; *Ancyclostoma* can be acquired orally.	Widely endemic in moist tropical and subtropical areas where disposal of human feces not adequate.
Strongyloides stercoralis	Humans and possibly dogs.	Strongyloidiatis: intestinal infection causing cramps, nausea, weight loss, vomiting, and weakness; rarely results in death.	Larvae in moist soils with fecal contamination penetrate skin and reach digestive system via venous and respiratory system.	Similar distribution to hookworm.
Trichuris trichiura (whipworm)	Humans.	Trichuriasis: a nematode infection of the large intestine; often without symptoms. Heavy infestations result in abdominal pain, weight loss, and diarrhea.	Ingestion of eggs in soil and/or vegetables contaminated with fecal material.	Worldwide: especially in warm, moist environments.

and Wetzel (2001). The following sections provide an introduction to algal ecology and nomenclature and discuss algae impacts on lake trophic status, taste, and odor and summarize what is currently known about algal endotoxins and indicate some of the problems algae may cause in treatment plant filter operations.

Algae Ecology and Nomenclature

Algae have limited ability to move themselves. Some use flagella or buoyancy mechanisms to achieve limited mobility, a few filamentous forms inhabit shorelines, and some others attach to bottom substrates. However, most are free floating but tend to sink because they are slightly denser than water.

Certain groups of algae, so identified in the past, have been reclassified, based on modern methods of taxonomic analysis. For example, blue-green algae are now classified as cyanobacteria (see Fig. 3-1). Some of the flagellated algae are grouped with the protozoa. All algae, however, are photoautotrophic, using photosynthesis as their primary mode of nutrition and as the basis for synthesis of new organic matter.

The most common algal groups and various members of interest to the water quality engineer are summarized in Table 3-10.

Algal and Lake Trophic Status

The *trophic level*, or fertility, of a body of water refers to the amounts of nutrients and organic matter being cycled through it. An *oligotrophic* lake is one with a low level of nutrients and organic matter. Often the water in such lakes appears clear and free of plant life. *Mesotrophic* refers to a moderate amount of nutrient input with correspondingly moderate amounts of plant and animal life. A *eutrophic* lake is one through which large amounts of nutrients and organic matter are being cycled and that supports substantial plant life.

Progression of a lake from oligotrophic to eutrophic is a natural occurrence. In principle, every lake eventually becomes filled with organic sediments and plants and becomes dry land. With little external input, such a progression occurs over a long time frame. When nutrient inputs are higher, eutrophication will occur more rapidly.

Algae speciation at these differing trophic levels is largely dictated by each strain's nutrient uptake capabilities and requirements. Some of the major algae groups that might be expected to occur under a given trophic level and set of water characteristics are described in Table 3-11. Seasonal variation in populations, micronutrient levels, and predator-grazing intensity will create shifts in these general trends, but they are useful as a baseline.

It is important to recognize that algae are not responsible for a lake's trophic status. They are merely an indication of the degree of fertilization that has been occurring. Thus, applying copper sulfate to reservoirs to control algae blooms, while it may work in the near term, does not arrest the long-term trend. However, controlling inputs of nutrients such as nitrogen and phosphorus can act to slow the eutrophication process.

Table 3-10
Common algal groups and cyanobacteria and representative members

Group Name	Common Designation	Representative Members
Bacillariophyta	Diatoms	*Asterionella, Diatoma cyclotella, Fragilaria navicula, Melosira, Synedra, Tabellaria*
Chlorophyta	Green algae	*Chlamydomonas, Oocystis Scenedesmus, Selenastrum, Sphaerocystic, Ulothrix, Volvax*
Chrysophyta	Golden algae	*Chrysosphaerella, Dinobryon*
Cynanobacteria	Blue-green algae[a]	*Anabaena, Anacystis (Microcystis), Aphanitomenon, Oscillatoria, Spirulina*
Dinophyta	Dinoflagellates	*Ceratium, Gonyaulax, Noctiluca, Peridinium, Pfiseteria*
Euglenophyta	Euglenas	*Euglena, Trachelomonas*
Phaeophyta[b]	Brown algae	*Macrocystis pyrifera, Sargassum, Turbinaria*
Rhodophyta[b]	Red algae	*Chondrus, Corallina, Polysiphonia, Porhyra*

[a]Commonly known and identified as blue-green algae, these organisms are now classified as cyanobacteria, based on their taxonomic characteristics. They are included in the above listing because of common usage.
[b]Primarily marine algae.
Source: Adapted in part from Wetzel (2001), Hutchinson (1957), and Internet sources.

Various measures, including analyses of nitrogen and phosphorus, algae activity or productivity, organic carbon, and so on, are used to gauge lake or reservoir trophic status. The ranges of values likely to be cited in identifying such status are summarized in Table 3-12. The first four measures presented in Table 3-12 provide an overall indication of algal activity. Identifying the dominant phytoplankton groups may provide an indication of trophic status. The total phosphorous and total nitrogen measures provide a fairly direct indication of lake productivity, while the remaining three measures, light extinction, TOC, and inorganic solids, are measures that correlate with lake productivity as well as other factors.

Harmful Algal Blooms

Algal blooms are localized proliferations of algae that occur during periods of optimal growth and reduced grazing pressure. When certain microalgae (especially cyanobacteria) reach high abundance in these blooms, the bloom can take on a harmful character. These microalgal species are a small proportion of nature's repertoire, but they are capable of producing noxious or toxic substances that can cause a variety of adverse effects, including food web disruption, animal mortality, and significant human

Table 3-11
Characteristics of common major algal associations of phytoplankton in relation to increasing lake fertility

General Lake Trophy	Water Characteristics	Dominant Algae	Other Commonly Occurring Algae
Oligotrophic	Slightly acidic; very low salinity	Desmids, *Staurodesmus*, *Straurastrum*	*Sphaerocystis*, *Gloeocystis*, *Rhizosolenia*, *Tabellaria*
	Neutral to slightly alkaline; nutrient-poor lakes	Diatoms, especially *Cyclotella* and *Tabellaria*	Some *Asterionella* spp., some *Melosira* s pp., *Dinobryon*
	Neutral to slightly alkaline; nutrient-poor lakes or more productive lakes at seasons of nutrient reduction	Chrysophycean algae, especially *Dinobryon*, some *Mallomonas*	Other chrysophyceans, e.g., *Synura*, *Uroglena*; diatom *Tabellaria*
	Neutral to slightly alkaline, nutrient-poor lakes	Chlorococcal *Oocystis* or Chrysophycean-*Botryoccocus*	Oligotrophic diatoms
	Neutral to slightly alkaline, generally nutrient poor, common in shallow Arctic lakes	Dinoflagellates, some *Peridinium*, and *Ceratium* spp.	Small chrysophytes, cryptophytes, and diatoms
Mesotrophic or entrophic	Neutral to slightly alkaline, annual dominants or in eutrophic lakes at certain seasons	Dinoflagellates, some *Peridinium*, and *Ceratium* spp.	*Glenodinium* and many other algae
Eutrophic	Usually alkaline lakes with nutrient enrichment	Diatoms much of the year, especially *Asterionella* spp., *Fragilaria crotonensis*, *Synedra*, *Stephanodiscus*, and *Melosira granulata*	Many other algae, especially greens and blue-greens during warmer periods of year; desmids if dissolved organic matter is fairly high
	Usually alkaline, nutrient enriched, common in warmer periods of temperate lakes or perennially in enriched tropical lakes	Blue-green algae, especially *Anacystis (Microcystis)*, *Aphanizomenon*, *Anabaena*[a]	Other blue-green algae[a]; euglenophytes if organically enriched or polluted

[a]Blue-green algae are now classified as cyanobacteria.
Source: Hutchinson (1957).

Table 3-12
General ranges of primary productivity of phytoplankton and related characteristics of lakes of different trophic categories

Trophic Type	Mean Primary Productivity mg C/m² · d	Phytoplankton Density cm³/m³	Phytoplankton Biomass mg C/m³	Chlorophyl *a* mg/m³	Dominant Phytoplankton	Total µg/L	Total µg/L	Light Extinction Coefficients 1/µm	Total Organic; Carbon mg/L	Total Inorganic Solid mg/L
Ultraoligotrophic	<50	<1	<50	0.01–0.5	—	<1–5	<1–250	0.03–0.8	—	2–15
Oligotrophic	50–300	—	20–100	0.3–3	Chrysophyceae, Crypthyceae	—	—	0.05–1.0	<1–3	—
Oligomesotrophic	—	1–3	—	—	Dinophyceae, Bacillariophyceae	5–10	250–600	—	—	10–200
Mesotrophic	250–1000	—	100–300	2–15	—	—	—	0.1–2.0	<1–5	—
Mesoeutrophic	—	3–5	—	—	—	10–30	500–1100	—	—	100–500
Eutrophic	>1000	—	>300	10–500	Bacillariophyceae, Cyanophyceae	—	—	0.5–4.0	5–30	—
Hypereutrophic	—	>10	—	—	Chlorophyceae, Euglenophyceae	30-5000	500-15,000	—	—	400–600
Dystrophic	<50–500	—	<50–200	0.1–10	—	<1–10	<1–500	1.0–4.0	3–30	5–200

[a]Referring to approximately net primary productivity, such as measured by the 14C method.

Source: Wetzel (2001).

health risk through the consumption of contaminated food and, in at least one case, direct exposure to water or aerosols containing them (Caron et al., 2010). Additionally, in the ocean, the algal biomass itself can be a significant desalination issue, impacting the pretreatment systems and forcing treatment plants to be taken off-line. Blooms of this kind are referred to as harmful algal blooms (HABs). An example of the significance of HABs was the poisoning of cattle and wildlife or contamination of drinking water supplies by blue-green algal toxins from *Nodularia spumigena* (brackish water) and *Anabaena circinalis* and *Microcystis aeruginosa* (freshwater) (Hallegraeff, 1992).

HABs can have severe impacts on water treatment plant operations

http://www.stantec.com/wt3r/3-3.html

For marine waters, the level of concern regarding HABs has reached a point in the United States where legislation was passed to support research into the problem (HABHRCA, 1998), research that contributed to the development of policies to protect from the adverse effects of coastal HABs (HABHRCA, 2004). No similar strategy exists for freshwater HABs (Hudnell, 2010), but a special White House paper was developed examining the problems of freshwater HABs (Lopez et al., 2008). A comprehensive understanding of the role harmful algal blooms may play in drinking water supplies is not yet available. Information on their health significance can be obtained from the CDC.

Algae and Filter Clogging

Algae that pass through preliminary treatment processes and become trapped among the spaces in a filter bed can cause gradual or rapid loss of head. Although effective coagulation and sedimentation can remove up to 90 or 95 percent of the incoming algae, the remainder may be sufficient to significantly shorten filter runs, even to the extent that the amount of water required to backwash the filter is greater than the amount of filtered water produced.

Nuisance filter-clogging algae include diatoms whose rigid cell walls prevent easy passage through filter media. Common problem diatoms include *Asterionella, Fragillaria, Tabellaria,* and *Synedra.* Various members of the blue-green algae, the green algae, and the golden browns also cause problems. *Palmella,* one of the green algae, forms copious mucilaginous material around its cells and literally gums up the filter bed. Some of the more common filter-clogging algae are illustrated on Fig. 3-14.

Enumeration of Algae in Water Supplies

Enumeration of algae within a water supply can be useful in a variety of contexts. Maintaining regular records of numbers of taste and odor algae can aid a water supply manager in determining when potential problems are likely to occur. The total area, or volume, of algae, especially of diatoms, can aid in establishing a relationship between phytoplankton and length of filter runs. For recreational reservoirs, keeping track of the trophic status via measurement of indicator algae or other parameters (see Table 3-12) can help determine appropriate levels of usage for the reservoir. Finally, scanning for wastewater organisms can identify problems of contamination. Over the years, methods for measuring algal concentrations have changed. Before 1970, the standard areal unit was the dominant method for presenting algae concentrations. Since that time, total count

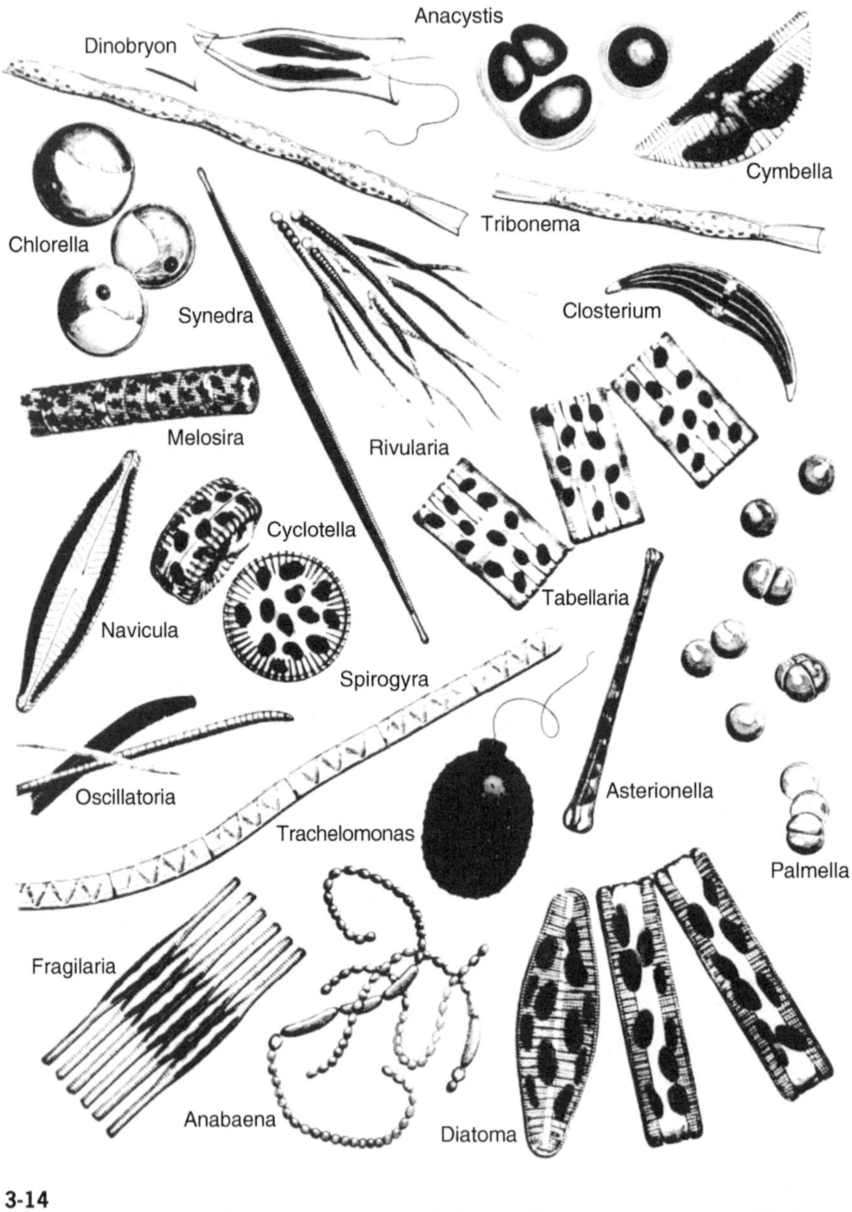

Figure 3-14
Filter-clogging algae (linear magnification in parentheses) (Palmer, 1959): *Anabaena flos-aquae* (500); *Anacystic dimidiata* (1000); *Asterionella formosa* (1500); *Chlorella pyrenoidosa* (5000); *Closterium moniliferum* (250); *Cyclotella meneghiniana* (1500); *Cymbella ventricosa* (1500); *Diatoma vulgare* (1500); *Dinobryonsertularia* (1500); *Fragilaria crotonensis* (1000); *Melosira granulata* (1000); *Navicula graciloides* (1500); *Oscillatoria princeps* (top) (250), O. *chalybea* (middle) (250), *O. splendida* (bottom) (500); *Palmella crebea* (1500); *Tribonema bombycinum* (500).

and measurements of chlorophyll a have become the dominant methods used. Of these, chlorophyll a is probably the most widely supported. Unfortunately, the methods for measuring chlorophyll a are not well standardized. While it is important to use standard methods so that results can be compared with other utilities to benchmark experience, where algae are concerned, it is perhaps even more important to use consistent methods from season to season or from year to year so that useful comparisons can be made with past experience. Utilities considering changing their methods for algae enumeration should run both the old and new methods in parallel for several seasons or until there is an adequate tie with past experience.

As summarized by Palmer (1959), an adequate routine procedure for treatment plants using surface water supplies would include regular inspection of the raw-water supply, treatment plant, and distribution system for attached growths, floating mats, and blooms. In visible growths, identification and enumeration of dominant organisms should be made. Regular plankton analyses of water samples from these areas should also be made. When supplemented with physical and chemical water quality data, a useful record of algal activity and/or problems can be established.

3-8 Assessing the Presence of Pathogens in Source Water

One of the most important elements of assessing the viability of a potential water source is collecting and analyzing water samples to gain an understanding of the potential presence or absence of pathogens. Over the past 150 years, our understanding of waterborne pathogens has grown and so has our ability to monitor for indicators of their presence. While progress has been made and new techniques continue to become available, our ability to discern the presence or absence of pathogens still fall short of what we would like to achieve, hence conservative treatment measures are often applied. Coliform bacteria are one of the most widely used indicators of microbiological contamination of source waters, but have a number of limitations. More recently, efforts have been made to include monitoring for *Cryptosporidium* into the framework of assessing the treatment requirements for a particular source water. These topics, as well as the impact of viable but not culturable bacteria, are discussed in this section.

Use of Coliform as an Indicator of the Presence of Wastewater

For the first hundred years following Snow's discovery, the public health community placed particular emphasis on controlling the transmission of waterborne bacterial enteric diseases through the drinking water supply, especially cholera and typhoid fever. As high as the infectious dose is for these organisms (see Fig. 3-6), it is still low enough that directly monitoring for their presence or absence in drinking water was unlikely to achieve a satisfactory outcome. As a result, efforts were made to develop an index of contamination. The index developed was based on *E. coli*, a small bacillus discovered by Escherich (1885) of Germany. *Escherichia coli* is plentiful in the feces of warm-blooded animals, approximately 1 billion per gram. Not long thereafter, Theobald Smith, with the Department of Health for the State of New York, used fermentation culture tubes to develop a presumptive test for the coliform group, of which *E. coli* is an important member (Smith, 1893). Shortly thereafter, the State of New York employed the technique to demonstrate that fecal contamination of the Mohawk River, a tributary of the Hudson River, had caused typhoid fever in persons drinking water from the Hudson downstream of the confluence of the two rivers (Mason, 1891).

In 1914, the U.S. PHS used Smith's fermentation tube test to set a standard requiring that drinking waters show evidence that members of the coliform group were not present. Technically speaking, the U.S. PHS standard was only applied to waters transported across state boundaries, but before long the test became a standard across the United States [American Public Health Association (APHA), 1965]. In the early 1940s, researchers at the U.S. PHS published two landmark articles demonstrating that *E. coli* densities in wastewater were correlated with those of waterborne pathogens (Kerr, 1943) and that *E. coli* is more resistant to disinfection and environmental exposure than several other pathogens (Wattie and Butterfield, 1944). This work greatly solidified the role of the coliform index as a means of confirming that a raw-water supply was not impacted by wastewater and also for determining if treatment had been successful.

As soon as the germ theory of disease became widely accepted and scientists began to use light microscopes to identify the bacteria causing disease, it became evident that certain diseases were caused by organisms that were not visible, even to the light microscope. For example, it had long been recognized that poliomyelitis was transmitted via the fecal–oral route. In the 1940s, several investigators confirmed that the virus responsible for this disease could be found in wastewater (Kling et al., 1942; Melnick, 1947; Paul et al., 1940; Trask and Paul, 1942). As early as 1945, an epidemic of infectious hepatitis was attributed to contaminated drinking water (Neefe and Strokes, 1945). Though the proposition that viruses are a cause of waterborne disease was not widely accepted at the time, it gradually became so, and by the mid-1960s a review of waterborne outbreaks of disease cited 50 outbreaks of infectious hepatitis and 8 outbreaks of polio during the period of 1946 through 1960 (Weibel et al., 1964). About that same time it also became clear that, outside of the host, viruses are not living organisms and do not have metabolic activity. This supported earlier observations that, under certain conditions, viruses can survive in the environment much longer than members of the coliform group (Neefe and Strokes, 1945). This observation raised serious questions about the suitability of coliform as an indicator. The presence of coliforms could still be taken as an indicator of contamination, but the absence of coliforms could not be taken as assurance that the water was uncontaminated.

During the last two decades of the twentieth century, the protozoa *G. lamblia* and *C. parvum* were also identified as waterborne pathogens. These organisms form cysts (*Giardia*) or oocysts (*Cryptosporidium*) that survive much longer in the environment and show much greater resistance to virtually all chemical disinfectants, particularly free and combined chlorine. Some strains of these organisms are also zoonotic; that is, they are pathogenic to humans and can proliferate in other animals as well. Because *G. lamblia, C. parvum,* and other zoonotic pathogens can be present in source waters that have no exposure to wastewater, these organisms further undermine the utility of the coliform index. Today, the measurement of coliforms is still valuable, but it must be supplemented by other measures of microbiological safety.

From the beginning it was recognized that the coliform group included organisms that are not of fecal origin. As a result, methods continuously evolved to be more specific for the original target organism. A couple of the most significant developments were the fecal coliform test (Geldreich, 1966) and the MUG (a growth medium, 4-methylumbelliferyl-β-D-glucuronide) test (Shadix and Rice, 1991). The fecal coliform test uses higher temperatures to select for thermotolerant members of the coliform group, eliminating those that cannot survive in the warm conditions of the mammalian gut. The MUG test specifically identifies *E. coli* itself, based on the action of the enzyme β-glucuronidase.

Finally, even the suitability of *E. coli* has been questioned in certain tropical climates where natural environmental conditions are occasionally adequate to support the organism's growth in the natural environment (Bermudez and Hazen, 1988; Fujioka et al., 1999; Hardina and Fujioka, 1991; Hazen et al., 1987).

The U.S. National Academy of Sciences (NAS) recently reviewed the indicator issue (NAS, 2004), and it concluded that no one indicator organism could be found that is suitable for all the purposes for which the coliform organism has been employed in the past. Rather the NAS recommended that a phased approach of flexible design be employed where a separate indicator or indicator system would be used for each circumstance. For example, *E. coli* might still be used in temperate climates as an indicator of contamination with fecal bacteria, but in tropical climates the results should be corroborated with another indicator such as *C. perfringens* (Fujioka, 2001; Fujioka et al., 1999). Moreover, the absence of *E. coli* cannot be taken as proof that source water is safe. Other indicators must be used to confirm the absence of more long-lived contaminants such as the viruses, cysts, and spores. In a similar fashion, whereas the presence of *E. coli* can be taken as an indication of fecal contamination in groundwaters, coliphage (a virus that infects *E. coli*) should also be sampled so that viruses, which are known to have longer survival in groundwater, can be detected. Whereas the absence of *E. coli* can be taken as confirmation of the performance of advanced wastewater treatment in removing bacteria, the absence of coliphage could also be a useful component of an indicator system designed to evaluate the removal of viruses by processes such as membrane filtration or reverse osmosis (Adham et al., 1999). The NAS also discussed the application of biomolecular methods and raised the possibility that, with further development, these could lead to direct monitoring of the pathogens themselves. These monitoring systems show the promise of significantly reducing the response time required for obtaining an answer, but at the present time most are only suitable for analyzing a very small sample size. A great deal of research remains to be done on the question of concentrating samples so that pathogens can be detected at low concentrations.

Viable But Not Culturable Bacteria

When van Leeuwenhoek first observed microorganisms in the seventeenth century, those that were in motion were obviously alive. For those that did not move, more information was required. Since the work of Pasteur and Koch in the late nineteenth century, growth-in-culture techniques have been the gold standard for evaluating the presence of viable organisms. In recent years, however, there has been increasing discussion of the concept of viable but nonculturable bacteria. The term, "viable but nonculturable" (VBNC) was first introduced by Oliver (Oliver, 1993; Oliver et al., 1991) to describe organisms that are alive but cannot be cultured in a laboratory using current techniques. A long-standing need in microbiology is a technique for a broad-spectrum measure of the total microbiological burden in the water (i.e., a total count of viable organisms).

When faced with the challenge of finding the total count of viable organisms, traditional culture-based methods have a serious, long-recognized shortcoming. The media used to support growth are inherently selective, that is, only a subset of the population of "viable" organisms will grow on any one media, at any one temperature. In fact, some culture-based methods are specifically designed to be highly selective, targeting one particular organism or group of organisms (e.g., the MUG method for *E. coli*) while others are designed to target a broad group of organisms (e.g. the methods for heterotrophic plate count). But no single culture method is suitable for a total viable count and some viable organisms have never been successfully grown in culture.

Others also argue that some bacteria can be present in the environment in a dormant or vegetative state and, though they may be sufficiently viable to reproduce over adequate time under the right conditions (e.g., in a host), they may not respond in culture, even when the culture techniques are those that are commonly used to detect their presence.

There is little controversy on the first point, that culture-based techniques are inherently selective and, as a result, they are not a suitable approach for determining the total count of viable bacteria in a water sample. Thus, research aimed at finding a means to obtain a "total viable count" will continue.

The second point, that organisms for which culture methods are available may be present in the environment in a vegetative state from which they can recover (in a host, but not in a culture), is more controversial. Some researchers argue that organisms that do not respond in culture can be viable, while others argue that they are dead. Nevertheless, credible work has been done to support the viable-but-not-culturable idea (Oliver, 1993).

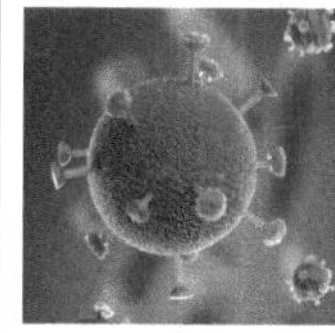

Stantec White Paper: Introduction to Coronaviruses for Water and Wastewater Utilities

http://www.stantec.com/wt3r/3-4.html

Most research directed toward developing a technique for a total viable count has explored direct methods using microscopic observations. These techniques have been labeled "direct viable count" (Rozak and Colwell, 1987; Yokomaku et al., 2000). However, there has also been criticism of these methods (NAS, 2004; Yokomaku et al., 2000). The fundamental problem is the one faced by van Leeuwenhoek so many years ago, that it is difficult to distinguish between viable and nonviable organisms through a microscope. Thus, direct count techniques often count debris (Zweifel and Hagstrom, 1995). Nevertheless, increasingly sophisticated techniques for direct viable count are becoming available, using new methods such as polymerase chain reaction (PCR), reverse transcriptase PCR (RT-PCR), and nucleic acid sequence-based amplification (NASBA) to label organisms so that the observer (often with a flow cytometer or an automated size analyzer) can better distinguish between viable and nonviable organisms. A recently developed direct bacterial count (DBC) technique using epifluorescence and a new nucleic acid stain SYBR Green I (Noble and Fuhrman, 1998) has been widely used to examine VBNC in marine samples.

Small VBNC bacteria may be particularly relevant in seawater because few nutrients are present, particularly in the Pacific (MacDonell and Hood, 1982), and some evidence suggests that many organisms, particularly marine organisms, enter into a viable but nonculturable state as a defense to starvation conditions—similar to other organisms forming spores or mammals hibernating. *Vibrio vulnificus* and *V. cholerae* are often cited for this behavior (Oliver et al., 1991). Some evidence indicates that, when they are in a starvation state, these organisms can also be extremely small, 0.2 μm and smaller (MacDonell and Hood, 1982). At this size, they may be difficult to remove by filtration, even membrane filtration (Ghayeni et al., 1999). Once exposed to a suitable environment, they may recover and form a biofilm (Winters, 2006).

Problems and Discussion Topics

3-1 Based on your reading of this chapter provide brief answers to the following questions: (a) For what contribution is Carl Woese known? (b) What are the distinguishing features of the three domains in the phylogenetic tree of life? (c) What sequence of events led to the first observation of a bacteria through the microscope?

3-2 What is the unique property of a pathogen?

3-3 Estimate the mass-based infectious dose for *B. anthracis* and adenovirus 4. Use Fig. 3-6 to estimate the median infectious dose and Tables 3-6 and 3-7 to estimate the size of the organism. Assume the specific gravity of both organisms is 1.1.

3-4 Construct a diagram similar to Fig. 3-7b for pathogenic *E. coli.* For purposes of constructing the diagram, assume that 50 percent of the persons infected are asymptomatic, 6 percent of those who get ill develop a prolonged illness, and 10 percent of those with a prolonged illness do not survive. Suppose that a town of 10,000 persons is exposed to a dose of this pathogen through the water supply and half the population is infected as a result. How many persons are likely to die as a result?

3-5 A town of 10,000 persons has an incident where *S. typhi* is present in its water supply at a level that results in infection of 70 percent of the population. Estimate the number of persons that will die from the disease. Assume that 60 percent of the infected persons are asymptomatic (i.e., 40 percent get ill). Estimate the mortality ratio using Fig. 3-8. Repeat the estimate assuming that the incident occurred before antibiotics were available.

3-6 A new pathogen evolves in an adult human being that has unusual invasive properties as well as the ability to produce high levels of a serious toxicant. After a week in bed with serious gastroenteritis the patient dies. Using Fig. 3-5, trace the transmission of the pathogen throughout the rest of the population. Consider two cases: (1) No removal occurs at either the wastewater treatment plant or the water treatment plant, and (2) the water treatment plant completely removes the pathogen.

3-7 A new pathogen that evolves in a toddler who attends a childcare center generally results in mild or asymptomatic illness—occasional diarrhea. The disease is mild enough that it escapes the notice of the child's caregivers. Using Fig. 3-5, trace the transmission of the pathogen throughout the rest of the population. Consider two cases: (1) No removal occurs at either the wastewater treatment plant or the water treatment plant, and (2) the water treatment plant completely removes the pathogen.

3-8 Explain the role of plasmids in the evolution of bacterial pathogens.

3-9 What was Dr. John Snow's contribution to our understanding of waterborne disease?

3-10 Name and describe the three different forms of gastroenteritis and at least two pathogens associated with each type.

3-11 What is the significance of zoonotic disease to water treatment?

3-12 Name the three classic waterborne pathogens and describe their effects.

3-13 Discuss the different forms of pathogenic *E. coli*. What sort of diseases do they cause?

3-14 Besides pathogenic *E. coli*, what was the principal pathogen active in the Walkerton outbreak?

3-15 Discuss the emerging pathogens listed in Table 3-5 and the role water treatment can play in preventing the diseases they cause.

3-16 Why are *F. tularensis* A and *B. anthracis* considered important candidates for waterborne terrorism?

3-17 Draw a diagram similar to Fig. 3-7b for poliomyelitis.

3-18 How many types of hepatitis have been identified. Which types are of concern in water treatment and why?

3-19 Which viruses are most likely responsible for viral gastroenteritis?

3-20 Which species of *Entamoeba* is more invasive?

3-21 Which protozoan waterborne diseases have the most serious consequences?

3-22 What is the stage in the life cycle when protozoan pathogens are in the environment?

3-23 In developed countries highly active antiviral therapy (HAART) is widely available to patients with HIV. What effect has this had on the significance of *Cryptosporidium* as a pathogen for these patients?

3-24 Which species of algae are associated with eutrophic water bodies and with oligotrophic water bodies?

3-25 What methods are normally used for algae enumeration?

3-26 Why was the coliform organism chosen as an index of fecal contamination?

3-27 What are the shortcomings of *E. coli* as an index of the presence or absence of pathogens?

References

Adham, S., Gagliardo, P., Chambers, Y., Sobsey, M., and Trussell, R. (1999) Monitoring the Reliability of an Advanced Water Treatment System for Water Repurification Using Indigenous Coliphage, in *Proceedings Specialised Conference on Rapid Microbiological Monitoring Methods*, Warrington, U.K., February 23–24, 1999, IWA, London.

Altekruse, S., Stern, N., Fields, P., and Swerdlow, D. (1999) "*Campylobacter jejuni*—An Emerging Food Borne Pathogen," *Emerging Infect. Dis.*, **5**,1, 28–35.

Anonymous (2001) "Acute Flaccid Paralysis Associated with Circulating Vaccine-Derived Poliovirus—Philippines," *MMWR Morb. Mortal. Wkly. Rep.*, **50**, 40, 874–875.

APHA (1965) *Standard Methods for the Examination of Water and Wastewater*, 12th ed., American Public Health Association, Washington, DC.

Appleton, H., and Higgins, P. G. (1975) "Viruses and Gastroenteritis in Infants," *Letter, Lancet*, 1, 1297.

Balayan, M., Andjaparidze, A., Savinskaya, S., Ketiladze, E., Braginsky, D., Savinov, A., and Poleschuk, V. (1983) "Evidence for a Virus in Non-A, Non-B Hepatitis Transmitted via the Fecal-Oral Route," *Intervirology*, **20**, 23–31.

Belabbers, E., Boughermouh, A., Benatallah, A., and Illoul, G. (1985) "Epidemic Non-A, Non-B Viral Hepatitis in Algeria. Strong Evidence for Its Spreading By Water," *J. Med. Virol.*, **16**, 3, 257–264.

Beller, M., Ellis, A., Lee, S., Drebot, M., Jenkerson, S., Funk, E., Sobsey, M., Simmons, O. 3rd, Monroe, S., Ando, T., Noel, J., Petric, M., Middaugh, J., and Spika, J. (1997) "Outbreak of Viral Gastroenteritis due to a Contaminated Well—International Consequences," *JAMA*, **7**, 7, 563–568.

Bentivoglio, M., and Pacini, P. (1995) "Filippo Pacini: A Determined Observer," *Brain Res. Bull.*, **38**, 2, 161–165.

Bermudez, M., and Hazen, T. (1988) "Phenotypic and Genotypic Comparison of *Escherichia coli* from Pristine Tropical Waters," *Appl. Environ. Microbiol.*, **54**, 3, 979–983.

Bhattacharya, A., Arya, R., Clark, C., and Ackers, J. (2000) "Absence of Lipophosphoglycan-Like Glycoconjugates in *Entamoeba dispar*," *Parasitology*, **120**, 31–35.

Blaser, M. J., Wells, J. G., Feldman, R. A., Pollard, R. A., and Allen J. R. (1983) "The Collaborative Diarrheal Disease Study Group. *Campylobacter enteritis* in the United States: A Multicenter Study," *Ann. Intern. Med.*, **98**, 360–365.

Blumberg, B., Alter, H., and Visnich, S. (1965) "A New Antigen in Leukemia Sera," *JAMA*, **191**, 7, 541–546.

Bridger, J. C. (1988) Non-Group-A Rotaviruses. In Viruses and the Gut, pp. 79–81 in *Proceedings of the Ninth BSG-SK&F International Workshop*, Smith Kline & French Laboratories, Welwyn Garden City, UK.

Bruce-Grey-Owen Sound Health Unit (2000) *The Investigative Report of the Walkerton Outbreak of Waterborne Gastroenteritis May–June, 2000*, Ontario Ministry of Health, Toronto, Canada.

Budd, W. (1856) "The Fever at Clergy Oprhan Asylum," *Lancet*, **2**, 617–619.

Caron, D., Garneau, M., Seubert, E., Howard, M., Darjany, L., Schnetzer, A., Cetinic, I., Filteau, G., Lauri, P., Jones, B., and Trussell, S., 2010, Harmful Algae and Their Potential Impacts on Desalination Operations off Southern California, *Water Res.*, **44**, 385–416.

Casadevall, A., and Pirofski, L. (1999) "Host-Pathogen Interactions: Redefining the Basic Concepts of Virulence and Pathogenicity," *Infect. Immun.*, **68**, 8, 3703–3713.

Chapron, C. D., Ballester, N. A., Fontaine, J. H., Frades C. N., and Margolin, A. B. (2000) "Detection of Astroviruses, Enteroviruses, and Adenovirus Types 40 and 41 in Surface Waters Collected and Evaluated by the Information Collection

Rule and an Integrated Cell Culture-Nested PCR Procedure," *Appl. Environ. Microbiol.*, **66**, 6, 2520–2525.

Cherkasova, E., Korotkova, E., Yakovenko, M., Ivanova, O., Eremeeva, T., Chumakov, K., and Agol, V. (2002) "Long-Term Circulation of Vaccine-Derived Poliovirus That Causes Paralytic Disease," *J. Virol.*, **76**, 13, 6791–6799.

Choo, Q., Kuo, G., Weiner, A., Overby, L., Bradley, D., and Houghton, M. (1989) "Isolation of cDNA Clone Derived from a Blood-Borne Non-A, Non-B Viral Hepatitis Genome," *Science*, **244**, 359–362.

Cruz, J., Caceres, P., Cano, F., Flores, J., Bartlett, A., and Torun, B. (1990) "Adenovirus Types 40 and 41 and Rotaviruses Associated with Diarrhea in Children from Guatemala," *J. Clin. Microbiol.*, **28**, 1780–1784.

Cruz, J.R., Bartlett, A.V., Herrmann, J.E., Caceres, P.,Blacklow, N.R., and Cano, F. (1992) "Astrovirus-Associated Diarrhea among Guatemalan Ambulatory Rural Children," *J. Clin. Microbiol.*, **30**, 1140–1144.

De Jong, J., Bijlsma, K., Wermenbol, A., Verweij-Uijterwaal, M., van der Avoort, H., Wood, D., Bailey, A., and Osterhaus, A. (1993) "Detection, Typing, and Subtyping of Enteric Adenoviruses 40 and 41 from Fecal Samples and Observation of Changing Incidences of Infections with These Types and Subtypes," *J. Clin. Microbiol.*, **31**, 6, 1562–1569.

Diamond, J. (1999) *Guns, Germs and Steel: The Fates of Human Societies*, W. Norton & Co., New York.

Diamond, L., and Clark, C. (1993) "A Redescription of *Entamoeba histolytica* Schaudinn, 1903 (Emended Walker, 1911) Separating It from *Entamoeba dispar* Brumpt, 1925," *J. Euk. Microbiol.*, **40**, 340–344.

DuMoulin, G., and Stottmeier, K. (1986) "Waterborne Mycobacteria: An Increasing Threat to Health," *Am. Soc. Microbiol. News*, **52**, 525–529.

Eberth, K. (1883) "Der Typhusbacillus und die intestinale Infection," *Sammlung KlinischerVortrage/Innere Medizin*, **66**, 18.

Edwards, D. (1993) "Troubled Waters in Milwaukee," *ASM News*, **59**, 7, 342–345.

Escherich, T. (1885) *Die Darmbakterien des Neugeborenen und Säuglings*, Fortschritte der Medizin, *München*, **3**, 515–522.

Ewald, P. (1994) *Evolution of Infectious Disease*, Oxford University Press, Oxford, England.

Ewald, P., Sussman, J., Distler, M., Libel, C., Chammas, W., Dirita, V., Salles, C., Vicente, A., Heitmann, I., and Cabello, F. (1998) "Evolutionary Control of Infectious Disease: Prospects for Vectorborne and Waterborne Pathogens," *Mem. Inst. Oswaldo Cruz, Rio deJaneiro*, **93**, 5, 567–576.

Fujioka, R. (2001) "Monitoring Coastal Marine Waters for Spore-Forming Bacteria of Fecal and Soil Origin to Determine Point from Non-Point Source Pollution," *Water Sci. Technol.*, **44**, 7, 181–188.

Fujioka, R., Stan-Denton, C., Borja, M., Castro, J., and Morphew, K. (1999) "Soil the Environmental Source of *Escherichia coli* and Enterocci in Guam's Streams," *J. Appl. Microbiol. Symp. Suppl.*, **85**, 83S–89S.

Geldreich, E. (1966) *Sanitary Significance of Fecal Coliforms in the Environment, Water Pollution Research Series. Publ. 2-3*, Federal Water Pollution Control Agency and US Department of Interior, Cincinnati, OH.

Ghayeni, S., Beatson. P., Fane, A., and Schneider, R. (1999) "Bacterial Passage through Microfiltration Membranes in Wastewater Applications, *J. Membr. Sci.*, **153**, 71 –82.

Gilchrist, C., and Petri, W. (1999) "Virulence Factors of *Entamoeba histolytica*," *Curr. Opin. Microbiol.*, **2**, 433–437.

Glass, R. I., Noel, J., Mitchell, D., Herrmann, J. E., Blacklow, N. R., Pickering, L. K., Dennehy, P., Ruiz-Palacios, G., de Guerrero, M. L., and Monroe, S. S. (1996) "The Changing Epidemiology of Astrovirus-Associated Gastroenteritis—A Review," *Arch. Virol. Suppl.*, **12**, 287–300.

Glover, N., Holtzman, A., Aronson, T., Froman, S., Berlin, O., Dominguez, P., Kunkel, K., Overturf, G., Stelma, G., Smith, C., and Yakrus, M. (1994) "The Isolation and Identification of *Mycobacterium avium* Complex (MAC) Recovered from Los Angeles Potable Water, a Possible Source of Infection in AIDS Patients," *Int. J. Environ. Health Res.*, **4**, 63–72.

Grabow, W. O. K. (1968) "The Virology of Waste Water Treatment," *Water Res.*, **2**, 675–701.

Grimwood, K., Carzino, R., Barnes, G., and Bishop, R. (1995) "Patients with Enteric Adenovirus Gastroenteritis Admitted to an Australian Pediatric Teaching Hospital from 1981 to 1992," *J. Clin. Microbiol.*, **33**, 1, 131–136.

Haas, C., Rose, J., and Gerba, C. (1999) *Quantitative Microbial Risk Assessment*, John Wiley & Sons, New York.

HABHRCA (1998) Harmful Algal Blooms and Hypoxia Research and Control Act of 1998. U.S. Congress, Washington, DC. Available at: http://www.cop.noaa.gov/stressors/extremeevents/hab/habhrca; accessed January 28, 2009.

HABHRCA. (2004) Harmful Algal Blooms and Hypoxia Amendments Act of 2004. U.S. Congress, Washington, DC. Available at: http://www.cop.noaa.gov/stressors/extremeevents/hab/habhrca; accessed January 28, 2009.

Hallegraeff, G. (1992) "Harmful Algal Blooms in the Australian Region," *Marine Poll. Bull.*, **25**, N5–8, 186–190.

Hardina, C., and Fujioka, R. (1991) "Soil, the Environmental Source of *E. coli* and Enterocci in Hawaii's Streams," *Environ. Toxicol.*, **6**, 185–195.

Hazen, T., Santiago-Mercado, J., Toranzos, G., and Bermudez, M. (1987) "What Does the Presence of Fecal Coliforms Indicate in the Waters of Puerto Rico? A Review," *Bol. Puerto Rico Med. Assoc.*, **79**, 189–193.

Herrmann, J. E., Taylor, D. N., Echeverria, P., and Blacklow, N. R. (1991) "Astroviruses as a Cause of Viral Gastroenteritis in Children," *N. Engl. J. Med.*, **324**, 25, 1757–1760.

Hopkins, R. S., Gaspard, G. B., Williams, F. P. J., Karlin, R. J., Cukor, G., and Blacklow, N. R. (1984) "A Community Waterborne Gastroenteritis Outbreak: Evidence for Rotavirus as the Agent," *Am. J. Public Health*, **74**, 3, 263–265.

Horne, A., and Goldman, C. (1994) *Limnology*, 2nd ed., McGraw-Hill, New York.

Howard-Jones, N. (1984) "Robert Koch and the *Cholera vibrio*: A Centenary," *Br. Med. J.*, **288**, 6405, 379–381.

Hudnell, H. (2010) "The state of U.S. Freshwater Harmful Algal Blooms Assessments, Policy and Legislation," *Toxicon*, **55**, 5, 1024–1034.

Hung, T., Chen, G., and Wang, C. (1987) Seroepidemiology and Molecular Epidemiology of the Chinese Rotavirus, pp. 49–62, in G. Bock and J. Whelan (eds.), *Novel Diarrhoea Viruses (Ciba Foundation Symposium; 128)*, John Wiley & Sons, Chichester.

Hutchinson, G. E. (1957) *A Treatise on Limnology*, John Wiley & Sons, New York.

Isenbarger, D., Hoge, C., Srijan, A., Pitarangsi, C., Vithayasai, N., Bodhidatta, L., Hickey, K., and Cam, P. (2002) "Comparative Antibiotic Resistance of Diarrheal Pathogens from Vietnam and Thailand, 1996–1999," *Emerging Infect. Dis.*, **8**, 2, 175–180.

Jackson, T., and Ravdin, J. (1996) "Differentiation of *Entamoeba histolytica* and *Entamoeba dispar* infections," *Parasitol. Today*, **12**, 18, 406–409.

Jarecki-Kahn, K.,Tzipori, S., and Unicomb, L. (1993) "Enteric Adenovirus Infection among Infants with Diarrhea in Rural Bangladesh," *J. Clin. Microbiol.*, **31**,3, 484–489.

Jothikumar, N., Aparna, K., Kamatchiammal, S., Paulmurugan, R., Saravanadevi, S., and Khanna, P. (1993) "Detection of Hepatitis E Virus in Raw and Treated Wastewater with the Polymerase Chain Reaction," *Appl. Environ. Microbiol.*, **59**, 8, 2558–2562.

Kapikian, A., Wyatt, R., Dolin, R., Thornhill, T., Kalica, R., and Chanock, R., (1972) "Visualization by Immune Electron Microscopy of a 27 nm Particle Associated with Acute Infectious Nonbacterial Gastroenteritis," *J. Virol.*, **46**, 2, 1075–1081.

Kerr, R. (1943) "Notes on the Relation between Coliforms and Enteric Pathogens," *Publ. Health Rep.*, **58**, 15, 589.

Kew, O., Morris-Glasgow, V., Landaverde, M., Burns, C., Shaw, J., Garib, Z., Andre, J., Blackman, E., Freeman, C. J., Jorba, J., Sutter, R., Tambini, G., Venczel, L., Pedreira, C., Laender, F., Shimizu, H., Yoneyama, T., Miyamura, T., van Der Avoort, H., Oberste, M. S., Kilpatrick, D., Cochi, S., Pallansch, M., and de Quadros, C. (2002, Apr. 12) "Outbreak of Poliomyelitis in Hispaniola Associated with Circulating Type 1 Vaccine-Derived Poliovirus," *Science*, **296**, 5566, 356–359.

Kist,M. (1985) The Historical Background of *Campylobacter* Infection: New Aspects, in A. D. Pearson (ed.), *Proceedings of the Third International Workshop on Campylobacter Infections*, Ottawa, Canada.

Kling, C., Olin, G., Fahraeus, J., and Norlin, G. (1942) "Sewage as a Carrier and Disseminator of Poliomyelitis Virus," *Acta Med. Scand.*, **112**, 217–249.

Knolle, H. (1995) "Transmission of Poliomyelitis by Drinking Water and the Problem of Prevention," *Gesundheitswesen*, **57**, 6, 349–354.

Kukkula, M., Arstila P., Klossner, M. L., Maunula, L., Bonsdorff, C. H., and Jaatinen, P. (1997) "Waterborne Outbreak of Viral Gastroenteritis," *Scand. J. Infect. Dis.*, **29**, 4, 415–418.

Kurtz, J. B., Lee, T. W., Craig, J. W., and Reed, S. E. (1979) "Astrovirus Infection in Volunteers," *J. Med. Virol.*, **3**, 3, 221–230.

Landaverde, M., Venczel, L., and de Quadros, C. (2001) "Poliomyelitis Outbreak Caused by Vaccine-Derived Virus in Haiti and the Dominican Republic," *Rev. Panam. Salud. Publ.*, **9**, 4, 272–274.

LeChevallier, M., Abbaszdegan, M., Camper, A., Izaguirre, G., Stewart, M., Naumovitz, D., Marshal, M., Sterling, C., Payment, P., Rice, E., Hurst, C., Schaub, S., Slifko, T., Rose, J., Smith, H., and Smith, D. (1991) "Emerging Pathogens-Bacteria," *J. AWWA*, **91**, 9, 101–109.

Lee, S., Levy, D., Craun, G., Beach, M., and Calderon, R. (2002) "Surveillance for Waterborne-Disease Outbreaks—United States, 1999–2000." *Morbidity Mortality Weekly Rep.*, **45**, SS-8, 1–47.

Lew, J. F., Moe, C. L., Monroe, S. S., Allen, J. R., Harrison, B. M., Forrester, B. D., Stine, S. E., Woods, P. A., Hierholzer, J. C., Herrmann, J. E., Blacklow, N. R., Bartlett, A. V., and Glass, R. I. (1991) "Astrovirus and Adenovirus Associated with Diarrhea in Children in Day Care Settings," *J. Infect. Dis.*, **164**, 673–678.

Linnen, J., Wages, J., Jr., Zhang-Keck, Z.-Y., Fry, K. E., Krawczynski, K. Z., Alter, H., Koonin, E., Gallagher, M., Alter, M., Hadziyannis, S., Karayiannis, P., Fung, K., Nakatsuji, Y., Shih, W.-K., Young, L., Piatak, M., Jr., Hoover, C., Fernandez, J., Chen, S., Zou, J.-C., Morris, T., Hyams, K. C., Ismay, S., Lifson, J. D., Hess, G., Foung, S. K. H., Thomas, H., Bradley, D., Margolis, H., and Kim, J. P. (1996) "Molecular Cloning and Disease Association of Hepatitis G Virus: A Transfusion-Transmissible Agent," *Science*, **271**, 5248, 50–508.

Lopez, C., Jewett, E., Dortch, Q., Walton, B., and Hudnell, H. (2008) *Scientific Assessment of Freshwater Harmful Algal Blooms. Interagency Working Group on Harmful Algal Blooms*, Hypoxia, and Human Health of the Joint Subcommittee on Ocean Science and Technology, Washington, DC.

MacCallum, F. O. (1947) "Homologus Serum Jaundice," *Lancet*, **2**, 691–692.

MacDonell, M., and Hood, M. (1982) "Isolation and Characterization of Ultra-microbacteria from a Gulf Coast Estuary," *Appl. Environ. Microbiol.*, **43**,3, 566–571.

MacKenzie, W., Hoxie, N., Proctor, M., Gradus, M., Blair, K., Peterson, D., Kazmierczak, J., Addiss, D., Fox, K., Rose, J., and Davis, J. (1994) "A Massive Outbreak in Milwaukee of *Cryptosporidium* Infection Transmitted through the Public Water Supply," *N. Engl. J. Med.*, **331**, 161–167.

Madeley, C. R., and Cosgrove, B. P. (1975) "Viruses in Infantile Gastroenteritis in Infants," *Letter, Lancet*, **2**, 124.

Mason, J. (1891, Nov.) "Notes on Some Cases of Drinking WaterandDiseases," *J. Franklin Inst.*, 1–10.

Mast, E. E., and Krawczynski, K. (1996) "Hepatitis E: An Overview," *Ann. Rev. Med.* **47**, 257–266.

Melnick, J. (1947) "Poliomyelitis Virus in Urban Sewage in Epidemic and Nonepidemic Times," *Am. J. Hyg.*, **45**, 240–253.

Midthun, K., Greenberg, H. B., Kurtz, J. B., Gary, G. W., Lin, F. C., and Kapikian, A. Z. (1993) "Characterization and Seroepidemiology of a Type 5 Astrovirus Associated with an Outbreak of Gastroenteritis in Marin County, California," *J. Clin. Microbiol.*, **31**, 955–962.

MMWR (2001) "Norwalk-LikeViruses: Public Health Consequences and Outbreak Management," *J. MMWR*, **50**, RR09, 1–18.

Mosley, J. (1967) "Transmission of Viral Diseases by Drinking Water," in G. Berg (ed.), *Transmission of Viruses by the WaterRoute*, Wiley-Interscience, NewYork.

Murphy, A., Grohmann, G., and Sexton, M. (1983) "Infectious Gastroenteritis in Norfolk Island and Recovery of Viruses from Drinking Water," *J. Hyg.*, **91**, I, 139-146.

Murphy, G. S., Jr., Echeverria, P., Jackson, L. R., Arness, M. K., LeBron C., and Pitarangsi, C. (1996) "Ciprofloxacin- and Azithromycin-Resistant *Campylobacter* causing Traveler's Diarrhea in U.S. Troops Deployed to Thailand in 1994," *Clin. Infect. Dis.*, **22**, 868–869.

NAS (2004) *Indicators of Waterborne Pathogens*, National Academy of Sciences, Washington, DC.

Neefe, J., and Strokes, J. (1945) "An Epidemic of Infectious Hepatitis Apparently Due to a Waterborne Agent," *JAMA*, **128**, 1063–1071.

Noble, R., and Fuhrman, J. (1998) "Use of SYBR Green I for Rapid Epifluorescence Counts of Marine Viruses and Bacteria," *Aquat. Microbial Ecol.* **14**, 113–118.

Okhuysen, P., Chappell, C., Crabb, J., Sterling, C., and DuPont, H. (1999) "Virulence of Three Distinct *Cryptosporidium parvum* Isolates for Healthy Adults," *J. Infect. Dis.*, **180**, 5, 127–1281.

Oliver, J. (1993) Formation of viable but nonculturable cells, pp. 239–276 in *Starvation in Bacteria*, S. Kjelleberg (ed), Plenum Press, New York.

Oliver, J., Nilsson, L. and Kjelleberg, S. (1991) "Formation of nonculturable *Vibrio vulnificus* Cells and Its Relationship to the Starvation State," *Appl. Environ. Microbiol.* **57**, 9, 2640–2644.

Pacini, F. (1854) *Osservazioni microscopiche e deduzioni patologiche sul cholera asiatico (Microscopical Observations and Pathological Deductions on Cholera)*, Tipografia Bencini, Florence, Italy.

Palmer, C. M. (1959) *Algae in Water Supplies. An Illustrated Manual on the Identification, Significance and Control of Algae in Water Supplies*, Public Health Service Publication No. 657, U.S. Department of Health, Education and Welfare, Washington, DC.

Parashar, U., Bresee, J., Gentsch, J., and Glass, R. (1998) "Rotavirus," *Emerging Infect. Dis.*, **4**, 4, 561 –570.

Paul, J., Trask, J., and Gard, S. (1940) "II Poliomyelitis in Urban Sewage," *J. Exp. Med.*, **71**, 765–777.

Phillips, A.D., Rice, S.J., and Walker-Smith, J.A.(1982) "Astrovirus within Human Small Intestinal Mucosa," *Gut*, **23**, A923–924.

Rheinheimer, G. (1991) *Aquatic Microbiology*, 4th ed.,John Wiley & Sons, New York.

Rittman, B., and McCarty, P. (2001) *Environmental Biotechnology: Principles and Applications*, McGraw-Hill, New York.

Rizzetto, M. (1977) "Immunofluorescence Detection of a New Antigen-Antibody System (Delta-Antidelta) Associated with the Hepatitis B Virus in the Liver and in the Serum of HBsAg Carriers," *Gut*, **18**, 997–1003.

Rozak, D., and Colwell, R. (1987) "Metabolic Activity of Bacterial Cells Enumerated by Direct Viable Count," *Appl., Env. Microbiol.* **53**, 12, 2889–2983.

Shadix, L., and Rice, E. (1991) "Evaluation of β-Glucuronidase Assay for the Detection of *Escherichia coli* from Environmental Waters," *Can. J. Microbiol.*, **37**, 908–913.

Smith, T. (1893) A New Method for Detemining Quantitatively the Pollution of Water by Fecal Bacteria, pp. 712–722 in *Thirteenth Annual Report,* New York State Board of Health, Albany, NY.

Snow, J. (1855) *On the Mode of Communication of Cholera,* 2nd ed., J. Churchill, London.

Sobsey, M., Oglesbee, S., Wait, D. A., and Cuenca, A. I. (1985) "Detection of Hepatitis A Virus in Drinking Water," *Water Sci. Technol.*, **17**, 10, 23–38.

Tandon, B., Gandhi, B., Josh, Y., Irshad, M., and Gupta, H. (1985) "Hepatitis Virus Non-A, Non-B the Cause of a Major Public Health Problem in India," *Bull. World Health Org.*, **63**, 5, 931–934.

Todar, K. (2003) *Bacteriology 303: Procaryotic Microbiology,* University of Wisconsin. Available at: http://www.bact.wisc.edu/Bact303/Bact303mainpage.

Trask, J., and Paul, J. (1942) "Periodic Examination of Sewage for the Virus of Poliomyelitis," *J. Exp. Med.*, **73**, 1–6.

U.S. EPA (1989) "Filtration and Disinfection; Turbidity, *Giardia lamblia,* Viruses, *Legionella,* and Heterotrophic Bacteria, Final Rule," *Fed. Reg.*, **54**, 124, 27486–27541.

Von Reyn, C., Maslow, J., Barber, T., Falkinham, J., andAbreit, R. (1994) "Persistent Colonisation of Potable Water as a Source of *Mycobacterium avium* Infection in AIDS," *Lancet,* **343**, 1137–1141.

Wadell, U., Svensson, L., Olding-Stenkvist, E., Ekwall, E., and Molby, R. (1986) "Aetiology and Epidemiology of Acute Gastro-enteritis in Swedish Children," *J. Infect.*, **13**,1,73–89.

Wattie, E., and Butterfield, C. (1944) "Relative Resistance of *Escherichia coli* and *Eberthella typhosa* to Chlorine and Chloramines," *Publ. Health Rep.*, **59**, 52, 1661–1671.

Weibel, S., Dixon, R., Weidner, R., and McCabe, L. (1964) "Waterborne-Disease Outbreaks, 1946-60," *J. AWWA,* **56**, 947–958.

Wetzel, R. G. (2001) *Limnology: Lakes and River Ecosystems,* 3rd ed., Academic, San Diego, CA.

Winters, H. (2006) "Microfouling of Cartridge Filters and RO Membranes: Mechanisms and Effects," Proceedings of the IDA World Congress in Singapore.

Woese, C., and Fox, G. (1977) "Phylogenetic Structure of the Prokaryotic Domain: The Primary Kingdoms," *Proc. Natl. Acad. Sci. USA,* **74**, 5088–5090.

Yajko, D., Chin, D., Gonzalez, P., Nassos, P., Hopewell, P., Reingold, A., Horsburgh, R., Yakrus, M., Ostroff, S., and Hadley, W. (1995) "*Mycobacterium avium* Complex in Water, Food, and Soil Samples Collected from the Environment of HIV- Infected Individuals," *J. AIDS Human Retrovirol.*, **9**, 176–182.

Yokomaku, D. Nobuyasu, Y., and Nasu, M. (2000) "Improved Direct Viable Count Procedure for Quantitative Estimation of Bacterial Viability in Freshwater Environments," *Appl. Env. Microbiol.* **66**, 12, 5544–5548.

Zweifel, U., and Hagstrom, A. (1995) "Total Counts of Marine Bacteria Include a Large Fraction of Non-Nucleoid-Containing Bacteria (Ghosts)," *Appl. Env. Microbiol.*, **61**, 6, 2180–2185.

4 Water Quality Management Strategies

Stantec's Water Treatment: Principles and Design, Updated Third Edition. John C. Crittenden, R. Rhodes Trussell, David W. Hand, Kerry J. Howe, and George Tchobanoglous.

Terminology for Water Quality Management Strategies

Term	Definition
Beneficial use	Uses of water that are beneficial to society and the environment. Typically, the identification of beneficial uses is the first step in the regulatory process.
Best available technology (BAT)	Technologies defined by regulation as being suitable to meet the maximum contaminant level.
Criteria, water quality	Water quality criteria, developed by various groups, to define constituent concentrations that should not be exceeded to protect given beneficial uses.
Endocrine disruptors	Substances that interfere with the normal function of natural hormones in the human body.
Maximum contaminant level (MCL)	Enforceable standard set as close as feasible to the MCL goal, taking cost and technology into consideration.
Maximum contaminant level goal (MCLG)	Nonenforceable concentration of a drinking water contaminant, set at the level at which no known or anticipated adverse effects on human health occur and that allows an adequate safety margin. The MCLG is usually the starting point for determining the MCL.
Multiple barrier concept	Inclusion of several barriers (both activities and processes) to limit the presence of contaminants in treated drinking water. Barriers might include source protection or treatment processes.
Nanoparticles	Extremely small particles that range in size from 1 to 100 nm, used in a number of manufacturing operations and products. The implications of these particles for human health and water treatment is not well understood.
Pharmaceuticals and personal care products	Substances used for medical or cosmetic reasons that enter the wastewater system during bathing or toilet use and are now detected at low levels in many water supply sources.
Physicochemical unit processes	Treatment processes used to remove or treat contaminants using a combination of physical and chemical principles.

Term	Definition
Standards	After specific beneficial uses have been established and water quality criteria developed for those beneficial uses, standards are set to protect the beneficial uses. Typically, standards are based on (1) determining the health-based maximum contaminant level goal (MCLG) and (2) setting the maximum contaminant level (MCL).
Treatment train	Sequence of unit processes designed to achieve overall water treatment goals.
Unit process	Individual process used to remove or treat constituents from water.

Other terms and definitions are available in the U.S. EPA Terms of Environment: Glossary, Abbreviations and Acronyms. (EPA, 2011).

The previous chapters have dealt with the chemical, physical, and biological characteristics and aesthetic quality of water. In this chapter, the treatment processes used for the removal of specific constituents found in water are introduced. For many constituents, there are a variety of processes or combinations of processes that can be used to effect treatment. The selection of which process or combination of processes to utilize is dependent on several factors, including (1) the concentration of the constituent to be removed or controlled, (2) the regulatory requirements, (3) the economics of the processes, and (4) the overall integration of a treatment process in the water supply system.

The topics considered in this chapter include (1) the objectives of water treatment, (2) a review of the regulatory process for water quality, (3) water quality standards and regulations, (4) an introduction to the methods used for the treatment of water, (5) an introduction to the development of systems for water treatment, and (6) an introduction to the concept of multiple barriers. Individual treatment unit processes, their expected performance, and some of the issues related to the design of the facilities to accomplish treatment of drinking water are examined in detail in the chapters that follow.

4-1 Objectives of Water Treatment

The principal objective of water treatment, the subject of this textbook, is the production of a safe and aesthetically appealing water that is protective of public health and in compliance with current water quality standards. The primary goal of a public or private water utility or purveyor is to provide

Table 4-1
Typical constituents found in various waters that may need to be removed to meet specific water quality objectives[a]

	Typical Constituents Found In	
Class	**Groundwater**	**Surface Water**
Colloidal constituents	Microorganisms, trace organic and inorganic constituents[b]	Clay, silt, organic materials, pathogenic organisms, algae, other microorganisms
Dissolved constituents	Iron and manganese, hardness ions, inorganic salts, trace organic compounds, radionuclides	Organic compounds, tannic acids, hardness ions, inorganic salts, radionuclides
Dissolved gases	Carbon dioxide, hydrogen sulfide	—[c]
Floating and suspended materials	None	Branches, leaves, algal mats, soil particles
Immiscible liquids	—[d]	Oils and greases

[a]Specific water quality objectives may be related to drinking water standards, industrial use requirements, and effluent.
[b]Typically of anthropogenic origin.
[c]Gas supersaturation may have to be reduced if surface water is to be used in fish hatcheries.
[d]Unusual in natural groundwater aquifers.

treated water without interruption and at a reasonable cost to the consumer. Meeting these goals involves a number of separate activities, including (1) the protection and management of the watershed and the conveyance system, (2) effective water treatment, and (3) effective management of the water distribution system to ensure water quality at the point of use.

Typical constituents found in groundwater and surface waters that may need to be removed, inactivated, or modified to meet water quality standards are identified in Table 4-1. The specific levels to which the various constituents must be removed or inactivated are defined by the applicable federal, state, and local regulations. However, as the ability to measure trace quantities of contaminants in water continues to improve and our knowledge of the health effects of these compounds expands, water quality regulations are becoming increasingly complex. As a consequence, engineers in the drinking water field must be familiar with how standards are developed, the standards that are currently applicable, and what changes can be expected in the future so that treatment facilities can be designed and operated in compliance with current and future regulations and so that consumers can be assured of an acceptable quality water.

4-2 Regulatory Process for Water Quality

Water quality criteria have become an important and sometimes controversial segment of the water supply field. Concern with water quality is based on findings that associate low levels of some constituents to higher incidence of diseases such as cancer. Following the passage of the Safe Drinking Water Act (SDWA) in 1974 (Public Law 93-523), the principal responsibility for setting water quality standards shifted from state and local agencies to the federal government. Water quality standards and regulations are important to environmental engineers for a number of reasons. Standards affect (1) selection of raw-water sources, (2) choice of treatment processes and design criteria, (3) range of alternatives for modifying existing treatment plants to meet current or future standards, (4) treatment costs, and (5) residuals management.

Water quality regulation typically proceeds in the following logical stepwise fashion:

1. Beneficial uses are designated.
2. Criteria are developed.
3. Standards are promulgated.
4. Goals are set.

Although often used interchangeably, there are significant differences in the terms *criteria*, *standards*, and *goals*. However, these items all fit under the general category of water quality regulation. The interrelationships of the various regulatory process steps in determining treatment for drinking water are illustrated on Fig. 4–1.

Beneficial-Use Designation

The first step in the regulatory process is designating beneficial uses for individual water sources. Surface waters and groundwaters are typically designated by a state water pollution control agency for beneficial uses such

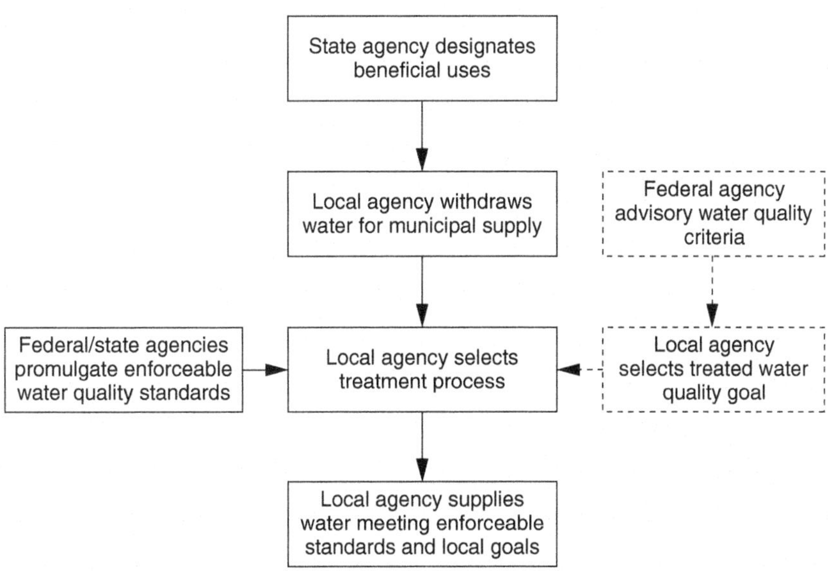

Figure 4-1
Steps in the regulatory process for setting water quality standards.

as municipal water supply, industrial water supply, recreation, agricultural irrigation, aquaculture, power and navigation, and protection or enhancement of fish and wildlife habitat. These beneficial uses are based on the quality of the water, present and future pollution sources, availability of suitable alternative sources, historical practice, and availability of treatment processes to remove undesirable constituents for a given end use.

Criteria Development

Water quality criteria have been developed by various groups to define constituent concentrations that should not be exceeded to protect given beneficial uses. Until criteria are translated into standards through rule making or adjudication, criteria are in the form of recommendations or suggestions only and do not have the force of regulation behind them. Criteria are developed for different beneficial uses solely on the basis of data and scientific judgment without consideration of technical or economic feasibility. For a single constituent, separate criteria could be set for drinking water (based on health effects or appearance), for waters used for fish and shellfish propagation (based on toxic effects), or for industry (based on curtailing interference with specific industrial processes). The primary data sources used for the development of water quality criteria are discussed below.

EARLY PUBLICATIONS DEALING WITH WATER QUALITY

Over the years, a number of publications and reports have been prepared that deal with water quality criteria for various beneficial uses, including drinking water. In 1952, the California State Water Pollution Control Board in conjunction with the California Institute of Technology published a report titled *Water Quality Criteria* in which the scientific and technical literature on water quality for various beneficial uses was summarized. The report was revised in 1963 (McKee and Wolf, 1963) and republished by the California State Water Resources Control Board (McKee and Wolf, 1971). Federal agencies have also developed water quality criteria documents in response to the federal Water Pollution Control Act and SDWA. These documents served as references for judgments concerning the suitability of water quality for designated uses, including drinking water. These references include the following:

1. *Water Quality Criteria* (U.S. EPA, 1972), National Technical Advisory Committee to the Secretary of the Interior, 1968, reprinted by the U.S. EPA.
2. *Water Quality Criteria* (NAS and NAE, 1972), prepared by the National Academy of Sciences and National Academy of Engineering for the U.S. EPA.
3. *Quality Criteria for Water* (U.S. EPA, 1976a), published by the U.S. EPA.

These three documents are often referred to as the *green book*, the *blue book*, and the *red book*, respectively.

NATIONAL ACADEMY OF SCIENCES

The NAS developed a systematic approach to establishing quantitative criteria and made a major contribution to the field of water treatment (NAS, 1977, 1980). The NAS iterated four principles for safety and risk assessment of chemical constituents in drinking water:

1. Effects in animals, properly qualified, are applicable to humans.
2. Methods do not now exist to establish a threshold for long-term effects of toxic agents.
3. The exposure of experimental animals to toxic agents in high doses is a necessary and valid method of discovering possible carcinogenic hazards in humans.
4. Material should be assessed in terms of human risk, rather than as "safe" or "unsafe."

The NAS divided criteria development into two different methodological approaches, depending on whether the compound in question was believed to be a carcinogen or a noncarcinogen. For carcinogens, the NAS used a probabilistic multistage model to estimate risk from exposure to low doses. The multistage model is equivalent to a linear model at low dosages, as illustrated on Fig. 4-2. In selecting a risk estimation model, the NAS (1980) evaluated a number of quantitative models to describe carcinogenic response at varying dose, which are described in Table 4-2. The difficulty in using any of the models summarized in Table 4-2 is the inability to determine whether predictions of risk at low dosages are accurate. It is not possible to test the large number of animals needed to statistically validate an observed response at low dosage. The effect of model selection on predicted response at low dosages for two different models is also illustrated on Fig. 4-2. On extrapolation to low doses, predicted responses will

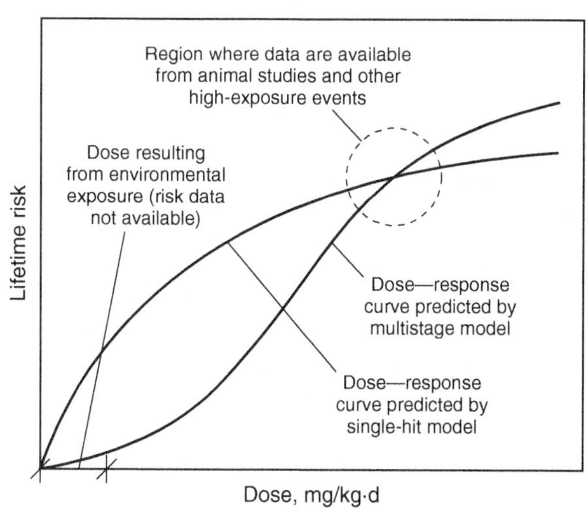

Figure 4-2
Effect of model selection on predicted response at low dosage. (Adapted from NAS, 1980).

Table 4-2
Types of quantitative models used to describe carcinogenic responses at varying doses of constituent of concern

Model Type	Description
Hitness	Based on radiation-induced carcinogenesis, in this model it is assumed that the site of action has some number of critical "targets" and that an event occurs if some number of them are "hit" by k or more radiation particles. Single- and two-hit and two-target models are used most commonly. The single-hit model is similar to the linear, no-threshold model.
Linear, no threshold	Carcinogenic risk is assumed to be directly proportional to dose.
Logistic	A logistic distribution of the logarithms of the individual tolerance is assumed along with a theoretical description of certain chemical reactions.
Probabilistic multistage	Carcinogenesis is assumed to consist of one or more stages at the cellular level beginning with a single-cell mutation, at which point cancer is initiated. The model relates doses d to the probability of response.
Time to tumor occurrence	Unlike the previous models, a latency period is assumed between exposure and carcinogenesis such that higher doses produce a shorter time to occurrence.
Tolerance distribution	Each member of the population at risk is assumed to have an individual tolerance for the toxic agent below which a dose will produce no response; higher doses will produce a response. Tolerances vary among the population according to some probability distribution F. Toxicity tests have frequently shown an approximately sigmoid relationship with the logarithm of dose, leading to development of the lognormal or log probit model. The distribution of F is normal against the logarithm of dose. Modified versions of this model have been developed.

differ significantly, with values obtained using the single-hit model being the most conservative.

Carcinogenic criteria

The NAS selected the probabilistic multistage model to estimate carcinogenic risk at low doses because (1) it was based on a plausible biological mechanism of carcinogens, a single-cell mutation, and (2) other models were empirical. For the carcinogenic compounds, the safe level could not be estimated. However, estimates were made such that concentrations of a compound in water could be correlated with an incremental lifetime cancer risk, assuming a person consumed 2 L per day of water containing the compound for 70 years. For example, a chloroform concentration of 0.29 μg/L corresponded to an incremental lifetime cancer risk of 10^{-6}. Thus, an individual's risk of cancer would increase by 1 in 1,000,000 by drinking 2 L per day of water with 0.29 μ g/L chloroform for 70 years; alternatively, in a population of 1,000,000, one person would get cancer who otherwise would not have. The NAS provided the criteria to allow correlations of contaminant levels and risks but made no judgment on an appropriate risk level. The latter decision properly falls in the sociopolitical realm of standards setting.

Noncarcinogenic criteria

For noncarcinogens, data from human or animal exposure to a toxic agent were reviewed and calculations made to determine the no-adverse-effect dosage in humans. Then, depending on the type and reliability of data, a safety factor was applied. This factor ranged from 10 (where good human chronic exposure data were available and supported by chronic oral toxicity data in other species) to 1000 (where limited chronic toxicity data were available). Based on these levels and estimates of the fraction of a substance ingested from water (compared to food, air, or other sources), the NAS method allowed calculations of acceptable daily intake and a suggested no-adverse-effect level in drinking water.

Standards

Once designation of water bodies for specific beneficial uses has been made and water quality criteria have been developed for those beneficial uses, the regulatory agency is ready to set standards. It is important to note that water quality standards, in contrast to criteria, have direct regulatory force. Quality standards in the past have been based on a number of considerations, including background levels in natural waters, analytical detection limits, technological feasibility, aesthetics, and health effects.

STANDARD PROMULGATION

The ideal method for establishing standards involves a scientific determination of health risks or benefits, a technical/engineering estimate of costs to meet various water quality levels, and a regulatory/political decision that weighs benefits and costs to set the standard.

The U.S. EPA is the governmental agency in the United States that is required to establish primary drinking water standards, which are protective of public health. Establishing standards occurs through (1) determining the health-based maximum contaminant level goal (MCLG) and (2) setting the minimum contaminant level (MCL). The MCL is the enforceable standard and is set as close as feasible to the MCLG taking costs and technology into consideration. To make the determination of where to set the MCL, the U.S. EPA gathers and assesses information on the occurrence of the contaminant, analytical methodologies and costs, and treatment technologies and costs in conjunction with the health effects information developed for the MCLG.

Outside peer review

The National Drinking Water Advisory Council (NDWAC) was created by the SDWA and consists of 15 members (appointed by the U.S. EPA administrator). The NDWA was established to provide the U.S. EPA with peer review and comment on its activities. In addition, the SDWA requires that the U.S. EPA seek review and comment from the Science Advisory Board (SAB) prior to proposing or promulgating a National Primary Drinking Water Regulation (NPDWR).

Best available technology

The SDWA requires that whenever the U.S. EPA establishes an MCL, the technology, treatment technique, or other means feasible for purposes of meeting the MCL must be listed. This approach is referred to as the best available technology (BAT). A public water system is not required to install the BAT to comply with an MCL. However, for purposes of obtaining a variance, a public water system must first install the BAT.

TREATMENT TECHNOLOGY

The SDWA provides for establishing a treatment technique instead of specifying an MCL for a given contaminant for which it is not economically or technically feasible to monitor. Examples of treatment technique regulations are the Surface Water Treatment Rule (SWTR) and the Lead and Copper Rule (LCR).

COMPLIANCE

Several factors go into the determination of whether a water system is in compliance with a drinking water regulation. For contaminants regulated by an MCL, compliance means (1) using the correct analytical method, (2) following all sample collection and preservation requirements, (3) following the required frequency and schedule for sample collection, (4) reporting sample results to the state and maintaining records on-site, and (5) maintaining measured concentration of the contaminant below the MCL.

For contaminants regulated by an MCL, compliance can be based on a single sample (e.g., when a system is monitored on an annual basis) while in other situations compliance can be based on the average of four quarterly samples.

For treatment techniques, demonstrating compliance can involve meeting operating criteria for the treatment plant (e.g., the SWTR requires water systems to meet a specific turbidity level in the effluent of the treatment plant) or taking certain steps to reduce the corrosivity of drinking water by specific deadlines (as is required under the LCR).

Reporting and record-keeping requirements

Public water systems must report compliance information to the state agency with primary enforcement responsibilities (primacy) by specified deadlines. In general, these deadlines are either 10 days after the month in which the monitoring was conducted or 10 days after the monitoring period (e.g., if conducting quarterly monitoring) in which the monitoring was conducted.

In addition to reporting compliance information to the state within specific deadlines, public water systems must maintain records onsite of monitoring results for specified periods of time. For example, under current regulations, public water systems must maintain copies of all monthly coliform reports for 5 years.

Violations, public notification, and fines
The U.S. EPA issues a notice of violation to a public water system that violates a NPDWR. As described previously, compliance includes (1) meeting the MCL or treatment technique requirements, (2) conducting monitoring at the correct frequency and at the correct locations, (3) using approved analytical methodologies, and (4) meeting all reporting and record-keeping requirements.

When a public water system violates an NPDWR, the system must provide public notification. Such notification may involve notice in a newspaper or for more acute situations could involve radio or television notice. Public notification requirements have evolved since passage of the original SDWA to better take into consideration the seriousness of the violation.

The U.S. EPA may take civil action against a water system or may issue an administrative order for a system in violation of a drinking water regulation. A public water system that is not in compliance with a drinking water regulation faces potential penalties up to $25,000 per day.

Variances and exemptions
The U.S. EPA or the state (if the state has primacy) can issue a variance or an exemption from an NPDWR, but only after the BAT has been installed in the water system and the drinking water regulation continues to be violated. The variance must include a schedule of steps to be taken by the water agency to eventually achieve compliance. A state can also grant an exemption from a drinking water regulation if, due to compelling factors, including economics, a system is unable to comply with an MCL or a treatment technique.

Goal Selection

Water quality goals represent contaminant concentrations, which an agency or water supplier attempts to achieve. Goals are typically more stringent than standards and may include constituents not covered by regulations but of particular importance to the goal-setting entity. There are two main types of water quality goals in the United States. The first type of goals is the MCLGs that are set by the U.S. EPA and the second type is set by an individual water supplier.

MAXIMUM CONTAMINANT LEVEL GOALS

The MCLG is a health-based goal for a given contaminant. These goals are nonenforceable and are set at a level at which no known or anticipated adverse effects on human health occur and that provides an adequate margin of safety. The U.S. EPA has developed different approaches for establishing MCLGs based upon whether a contaminant is considered to be a carcinogen. Typically, short- and long-term animal-feeding studies as well as available epidemiological studies are evaluated in making this determination.

INDIVIDUAL WATER SUPPLIER GOALS

Water suppliers may set operational goals that are lower than the treatment standards to ensure that the standards are always met. For example, if the turbidity standard is 0.3 NTU, a utility might choose an operating goal of 0.1 NTU to ensure meeting the standard.

Alternatively, an individual water supplier may elect to provide water quality that is better than required by the applicable standards or for constituents that are either not regulated by standards or are secondary standards. Examples include goals for turbidity or THMs in treated water lower than required by regulation or goals for unregulated parameters such as standard plate counts or secondary standards such as odor. Decisions on setting goals involve determinations of costs, benefits, and the overall philosophy or posture of a supplier.

4-3 Water Quality Standards and Regulations

The specific levels to which the various constituents must be removed are, as noted in the introduction to this chapter, defined by the applicable federal, state, and local regulations. The purpose of this section is to introduce and discuss the evolution of the current federal, state, and international drinking water standards and regulations that govern the design of water treatment plants.

Historical Development

The development of water quality criteria and standards, at least in a quantifiable sense, is a relatively recent phenomenon in the course of human history. The first standards in the United States were promulgated in 1914, but there have been numerous developments since then, particularly in the last 30 years. Key developments prior to 1900 and the actions of the U.S. PHS in establishing limits that were widely followed voluntarily are reviewed in this section along with the entry of the federal government into a standards-setting role for community water supplies.

EVENTS PRIOR TO 1900

Based on historical records, water quality standards, except for infrequent references to aesthetics, were notably absent from the time of ancient civilization through most of the nineteenth century. Typically, the sensory perceptions of taste, odor, and visual clarity were used to judge the quality of the supply. The deficiency of this system was clearly pointed out during the London Asiatic cholera epidemic of 1853 when John Snow did epidemiological investigations tracing cholera to wastewater contamination in the Broad Street Well (Snow, 1855). Even though the well was contaminated, some consumers traveled there specifically because they preferred its water, presumably on the basis of taste, appearance, or smell. From this example, it is clear that standards need to be quantifiable and related directly to measurable water quality contaminants that could have health effects and not just the appearance or aesthetics of a supply.

After the germ theory of disease, developed by Pasteur in the 1860s, was recognized, the issue of drinking water contaminated from wastewater was explored. The earliest quantitative measurements were chemical tests because bacteriological tests were not available until the end of the nineteenth century. Because it was recognized that ammonia and albumoid nitrogen from fresh wastewater were gradually oxidized in receiving water to nitrites and nitrates, these forms of nitrogen were measured in drinking water in an attempt to ensure that contamination, if present, was not recent. However, this method was an indirect measure of bacterial contamination and did not serve to curtail outbreaks of waterborne disease, particularly typhoid, in the United States. The development of a bacterial test for water supplies by Theobald Smith in 1891 (Smith, 1893) made it possible to directly analyze bacterial water quality. In 1892, the New York State Board of Health first applied the technique developed by Smith to study pollution in the Mohawk and Hudson Rivers (Clendening, 1942).

ROLE OF U.S. PUBLIC HEALTH SERVICE

The U.S. PHS, a part of the Treasury Department, has had an indirect, but nevertheless key role in setting water quality standards in the United States. In 1893 the U.S. Congress enacted the Interstate Quarantine Act authorizing the U.S. PHS to set regulations necessary to stop the spread of communicable diseases. The ability to detect bacteria, coupled with the introduction of chlorine as a disinfectant in 1902, led to the first quantitative water quality standards. In 1914, the U.S. PHS adopted the first standards for drinking water supplied to the public by any common carrier engaged in interstate commerce such as commercial trains, airplanes, and buses. Maximum permissible limits were specified for bacterial plate count and *B. coli* (a coliform bacteria).

Following the entry of the U.S. PHS into the regulatory field, standards development proceeded rapidly. Over the next 50 years, the U.S. PHS developed additional standards for minerals, metals, and radionuclides and standards for the indication of organics with revised standards issued in 1925, 1942, 1946, and 1962 (U.S. PHS, 1962). In 1969 the U.S. PHS conducted the Community Water Supply Survey (CWSS) to assess drinking water quality, water supply facilities, and bacteriological surveillance programs in the United States. The goal of the survey was to determine if drinking water in the United States met the U.S. PHS drinking water standards and to determine what kinds of surveillance programs were in place. Among other things, the results of the CWSS would play a role in the eventual enactment of the SDWA.

After the initial emphasis on controlling waterborne bacteria, new parameters were added to limit exposure to other contaminants that cause acute effects, such as arsenic, or adversely affect the aesthetic quality of the water. In 1925, a number of aesthetic parameters (color, odor, and taste) were added, along with certain minerals (chloride, copper, iron, lead, magnesium, sulfate, and zinc). Except for lead, these minerals are related to taste or aesthetics. In 1942, a number of constituents were added, including selenium, residue (dissolved solids), turbidity, fluoride, manganese, alkyl benzene sulfonate, and phenols. The latter two compounds marked the first time that specific organic constituents were covered by regulations. In 1946, standards were reissued that were similar to the 1942 standards except that a limit was set for another toxic constituent, chromium.

Following the dawning of the atomic age, the U.S. PHS standards in 1962 included ^{226}Ra, ^{90}Sr, and gross beta activity. Addition of an indicator of organics (carbon chloroform extract) plus additional toxic constituents (cadmium, cyanide, nitrate) reflected an awareness of the rapid postwar development of the chemical industry plus new data on toxicological effects. The last action of the U.S. PHS, before its standards-setting function was transferred to the newly formed U.S. EPA in 1970, was to recommend additional parameters such as pesticides, boron, and the uranyl ion be regulated.

TWO-TIERED SYSTEM

Another significant feature of the U.S. PHS standards was the development of a two-tiered system, which began in 1925. Water quality contaminants were controlled by either tolerance limits or recommended limits depending on how the effect of the contaminant was viewed. Tolerance limits were set for substances that, if present in excess of specified concentrations, constituted grounds for rejecting the supply; examples included arsenic, chromium, and lead. Alternately, recommended limits were developed for constituent concentrations that should not be exceeded if other more suitable supplies were or could be made available; examples included chloride, iron, and sulfate. This type of differentiation was the forerunner of present regulations, wherein the tolerance limits correspond to primary regulations intended for public health protection and recommended limits are analogous to secondary standards for public welfare or aesthetics.

APPLICATION OF U.S. PHS STANDARDS

The U.S. PHS standards applied only to suppliers of water engaged in interstate commerce, as the original intent was to protect the health of the traveling public. Thus, standards applied to water used on commercial trains, airplanes, buses, and similar vehicles. However, the U.S. PHS standards became recognized informally as water quality criteria and were adopted or adapted by many regulatory agencies at the state or local level as standards. Thus, prior to the entry of the U.S. EPA into the role of regulating community water supplies, many water suppliers were producing water in accordance with the levels listed in the U.S. PHS standards (U.S. PHS, 1970). A similar response occurred internationally, with agencies such as the WHO using the U.S. PHS standards as a guideline in developing their own standards (WHO, 1993, 2006). It is clear from reviewing the history of regulations, at least in the United States, that the number of regulated contaminants has continued to increase as (1) toxicological evidence has been gathered and (2) new and improved (e.g., more sensitive) analytical techniques have been developed.

Development of U.S. EPA Federal Standards and Regulations

The U.S. EPA was created through an executive reorganization plan where the goal was to consolidate federal environmental regulatory activities into one agency. On July 9, 1970, the plan to create the U.S. EPA was sent by the president to Congress and came into being on December 2, 1970.

The mandate for the U.S. EPA was to protect public health and the environment. As originally created, the U.S. EPA was headed by an administrator supported by a deputy administrator and five assistant administrators responsible for planning and management, legal enforcement, water and hazardous materials, air and waste management, and research and development. By 1974, the U.S. EPA had over 9000 employees with an operating budget of approximately $500 million and has continued to grow in size and responsibilities since then.

Online resources for laws and regulations related to water.
http://www.stantec.com/wt3r/4-1.html

SAFE DRINKING WATER ACT

The activities of the U.S. PHS related to water quality, as discussed above, were transferred to the newly formed U.S. EPA in 1970. The first major event following the transfer was the passage of the Safe Drinking Water Act (SDWA) on December 16, 1974 (Public Law 93-523). With the passage of the SDWA, the federal government, through the U.S. EPA, was given the authority to set standards for drinking water quality delivered by community (public) water suppliers. Thus, direct federal influence on water quality was authorized, as opposed to the indirect influence exerted by the U.S. PHS.

A series of steps and timetables for developing the drinking water quality regulations were outlined in the SDWA. Procedures were established for setting (1) National Interim Primary Drinking Water Regulations (NIPDWR), (2) revised National Primary Drinking Water Regulations (NPDWR), National Secondary Drinking Water Regulations (NSDWR), and (3) periodic review and update of the regulations. With each step, proposed regulations were to be developed by the U.S. EPA, published in the *Federal Register*, discussed at public hearings, commented upon by interested parties, and revised as necessary before final promulgation. A summary of major U.S. legislation and executive orders related to drinking water treatment is given in Table 4-3.

Table 4-3
Summary of major legislation and executive orders related to drinking water treatment

Law	Description
Interstate Quarantine Act, 1893	U.S. Congress authorizes the U.S. PHS to set regulations necessary to stop the spread of communicable diseases.
U.S. Environmental Protection Agency, 1970	U.S. EPA is created through an executive reorganization plan whose goal is to consolidate federal environmental regulatory activities into one agency. On July 9, 1970, the plan to create U.S. EPA is sent by the president to Congress, and the agency comes into being on December 2, 1970.
SDWA; Public Law 93-523, 1974	The SDWA requires U.S. EPA to establish drinking water regulations in two phases. (1) Establish National Interim Drinking Water Regulations (NIPDWR) within 90 days of enactment of the SDWA that specify maximum levels of drinking water contaminants and monitoring requirements that would apply to public water systems. (2) Review and revise the NIPDWRs and establish National Primary Drinking Water Regulations (NPDWR).
SDWA amendments; Public Law 99-339, 1986	Requires U.S. EPA to set standards for 83 compounds within 3 years and to establish 25 new standards every 3 years, establish criteria for filtration of surface water supplies, and establish requirements for all public water systems to provide disinfection. Requires that the MCLG and the MCL be proposed and finalized on the same schedule. Bans the use of lead pipes and solder and requires water utilities to go through a one-time public education program notifying consumers of the health effects and sources of lead in drinking water and steps that individuals can take to reduce exposure.
Lead Contamination Control Act; Public Law 100-572, SDWA amendment of 1988	Establishes a program to eliminate lead-containing drinking water coolers in schools.
SDWA amendments; Public Law 104-182, 1996	Requires the U.S. EPA to publish and seek public comment on health risk reduction and cost analyses when proposing an NPDWR that includes an MCL or a treatment technique and take into consideration the effects of contaminants upon sensitive subpopulations (i.e., infants, children, pregnant women, the elderly, and individuals with a history of serious illness) and other relevant factors. Within 5 years evaluate five contaminants from a drinking water contaminant candidate list. Establishes specific deadlines for standards for arsenic (a revised standard from the existing standard), a new standard for radon, a source water assessment and protection program, a requirement for public water systems to distribute Consumer Confidence Reports to their customers, a State Drinking Water Revolving Fund, and a program to develop operator certification requirements.

The SDWA has been amended periodically, as reported in Table 4-3. While the SDWA was amended slightly in 1977 (Public Law 95-190), 1979 (Public Law 96-63), and 1980 (Public Law 96-502), significant changes were made when the SDWA was reauthorized on June 16, 1986 (Public Law 99-339), and amended in 1996 (Public Law 104-182). The amendments of 1986 were driven by public and congressional concern over the slow process of establishing the NPDWR. The 1986 amendments also finalized the original NIPDWR and renamed the interim standards the NPDWR. The amendments enacted in 1996 emphasized the use of sound science and risk-based standard setting, increased flexibility and technical assistance for small water systems, source water assessment and protection programs, and public right to know and established a program to provide water system assistance through a multi-billion-dollar state revolving loan fund.

EVOLUTION OF NATIONAL PRIMARY DRINKING WATER REGULATIONS

A brief overview of the evolution of the key U.S. federal regulations that affect drinking water is presented in Table 4-4. As reported in Table 4-4, the current regulations for drinking water evolved from the U.S. PHS standards. As required by the SWDA, The National Interim Primary Drinking Water Regulations (NIPDWR), published on December 24, 1975, became effective June 24, 1977. The regulations contained MCLs for a number of inorganic chemicals, organic chemicals, physical parameters, radioactivity, and bacteriological factors. Maximum contaminant levels are set as concentrations that are never to be exceeded (with some minor exceptions). Perhaps the most substantial change of the NIPDWR compared to the U.S. PHS standards was the designation of turbidity as a health-related, rather than an aesthetic, parameter. The original NIPDWRs were amended several times. As noted above, on June 19, 1986 the interim standards established under the NIPDWRs were finalized and renamed the NPDWR.

NATIONAL SECONDARY DRINKING WATER REGULATIONS

The U.S. EPA has also promulgated secondary drinking water regulations (U.S. EPA, 1979a). The NSDWR pertain to those contaminants, such as taste, odor, and color, that may adversely affect the aesthetic quality of drinking water. These secondary levels represent reasonable goals for drinking water quality but are not federally enforceable; rather, they are intended as guidelines. States may establish levels as appropriate to their particular circumstances.

REGULATIONS RELATED TO CHEMICAL AND MICROBIAL AND CONTAMINANTS AND DISINFECTION BY-PRODUCTS

The regulations related to: (1) chemical contaminants and (2) microbial and disinfection by products can be found in a number of different rules and regulations. The principal rules and regulations where information can be found on microbial contaminants and disinfection by-products are

Table 4-4
Summary of key U.S. federal regulations that affect drinking water

Regulation and date[a]	Description
U.S. PHS standards, 1914 (U.S. Treasury Department, 1914)	The first drinking water standard is established in the United States. The standard establishes a maximum permissible limit for bacterial plate count and *B. coli* (a coliform bacteria) of 2 coliforms per 100 mL for water supplied to the public by any common carrier engaged in interstate commerce such as commercial trains, airplanes, and buses. These bacteriological quality standards are commonly known as the Treasury Standards.
U.S. PHS standards, revised in 1925, 1942, 1946, and 1962 (U.S. PHS, 1925, 1942, 1946, and 1962)	Bacteriological quality standards are made more restrictive, physical and chemical standards are established, and the principle of attainability is established (1925). Regulates 28 contaminants commonly found in drinking water by setting mandatory limits for health-related chemical and biological impurities and recommends limits for constituents that affect appearance, taste, and odor (1962).
National Interim Primary Drinking Water Regulations (NIPDWR); Pub. *FR* December 24, 1975; effective June 24, 1977 (U.S. EPA, 1975)	Published in December 1975, these regulations set 18 interim standards for 6 synthetic organic chemicals, 10 inorganic chemicals, turbidity, total coliform bacteria, and radionclides.
NIPDWR; Promulgation of Regulations on Radionuclides; Pub. *FR* July 9, 1976; effective June 24, 1977 (U.S. EPA 1976b)	Sets interim standards for radionuclides, gross alpha emitters, ^{226}Ra and ^{228}Ra combined, and two other classes of radionuclides. Final standard adopted December 7, 2000 (see below).
National Secondary Drinking Water Regulations (NSDWR); Pub *FR* July, 19, 1979 (U.S. EPA 1979a)	Sets nonenforceable guidelines for contaminants that may cause aesthetic problems in drinking water, including aluminum, chlorides, color, copper, corrosivity, foaming agents, iron, manganese, odor, pH, silver, sulfate, total dissolved solids, and zinc.
NIPDWR; Control of Trihalomethanes in Drinking Water; Final Rule; Pub. *FR* November 29, 1979, effective date varied depending on size of system (U.S. EPA 1979b)	Sets 0.1 mg/L as the MCL for total trihalomethanes (TTHMs).
National Primary Drinking Water Regulations (NPDWR); Pub. *FR* June 19,1986, effective June 19, 1986 (U.S. EPA 1986)	Each national interim or revised primary drinking water regulation promulgated before June 19, 1986, shall be deemed to be a national primary drinking water regulation.
NPDWR; Volatile Organic Chemicals (VOCs) Rule—Chemical Phase Rules—Phase I; July 7, 1987, effective 1989 (U.S. EPA 1987)	The chemical contaminants regulated under these rules generally pose long-term (i.e., chronic) health risks if ingested over a lifetime at levels consistently above the MCL.

Table 4-4 *(Continued)*

Regulation and date[a]	Description
NPDWR; Filtration and Disinfection; Turbidity, *Giardia lamblia*, Viruses, *Legionella*, and Heterotrophic Bacteria; Final Rule; also known as Surface Water Treatment Rule; Pub. *FR* June 29, 1989 (U.S. EPA 1989a)	Seeks to reduce the occurrence of unsafe levels of disease-causing microbes, including viruses, *Legionella* heterotrophic bacteria, and *G. lamblia*. Filtration of surface waters required. Criteria for avoiding filtration, criteria for disinfection based on *Giardia* and viruses, filtered water turbidity <0.3 NTU for 95% of time.
NPDWR; Total Coliforms, Final Rule; Pub. *FR* June 29, 1989 (U.S. EPA 1989b)	Sets an MCL with an MCLG of zero for total coliforms and changes the previous coliform MCL from a density-based standard to a presence/absence basis.
NPDWR; Synthetic Organic Chemicals (SOCs) and Inorganic Chemicals (IOCs)—Phase II; Final Rule; January 30,1991 (U.S. EPA 1991a)	The chemical contaminants regulated under these rules generally pose long-term (i.e., chronic) health risks if ingested over a lifetime at levels consistently above the MCL.
NPDWR; Lead and Copper; Final Rule; Pub. *FR* June 7, 1991 (U.S. EPA 1991b)	Sets health goals and action levels (trigger for requiring additional prevention of removal steps) for lead and copper (Pb ≤ 15 μg/L, Cu ≤ 1.3 mg/L in 90% of samples at consumer's tap).
NPDWR; Synthetic Organic Chemical and Inorganic Chemicals—Phase V; Final Rule; Pub. *FR* July 17, 1992 (U.S. EPA 1992)	The chemical contaminants regulated under these rules generally pose long-term (i.e., chronic) health risks if ingested over a lifetime at levels consistently above the MCL.
NPDWR; Monitoring Requirements for Public Drinking Water Supplies or Information Collection Rule; Final Rule; *FR* May 14, 1996 (U.S. EPA 1996)	Establishes requirements for monitoring microbial contaminants and disinfection by-products by large public water systems and requires these systems to provide operating data and descriptions of their treatment plant design, plus conducting either bench- or pilot-scale testing of advanced treatment techniques. The Information Collection Rule (ICR) is a one-time monitoring effort to gather information for future microbial and disinfection by-product regulations.
NPDWR; Stage 1 Disinfectants and Disinfection Byproducts; Final Rule; Pub. *FR* December 16,1998 (U.S. EPA 1998a)	Lowers the MCLs for disinfection by-products (DBPs) to 0.08 mg/L for THMs, 0.06 mg/L for five haloacetic acids (HAA5), 0.10 mg/L for bromate, and 1.0 mg/L for chlorite. Sets requirements for reducing total organic carbon (TOC) in surface water treatment systems based on a 3 × 3 matrix of source water TOC concentration and source water alkalinity.
NPDWR; Interim Enhanced Surface Water Treatment Rule (IESWTR); Pub. *FR* December 16, 1998 (U.S. EPA 1998b)	Lowers turbidity performance standards, requires 2 log *Cryptosporidium* removal for filtering and individual filter monitoring for turbidity, and requires disinfection profiling/benchmarking, covering of new finished water reservoirs, and sanitary surveys by the states.

(*Continued*)

Table 4-4 *(Continued)*

Regulation and date[a]	Description
NPDWR; Final Standards for Radionuclides; Final Rule; Pub. *FR* December 7, 2000 (U.S. EPA 2000)	This regulation became effective on December 8, 2003, and covers combined $^{226}Ra/^{228}Ra$ (adjusted), gross alpha, beta particle, and photon radioactivity, and uranium. This promulgation consists of revisions to the 1976 rule, as proposed in 1991.
NPDWR; Filter Backwash Recycling Rule; Final Rule; Pub. *FR* June 8, 2001 (U.S. EPA 2001a)	Any system that recycles (spent-filter backwash water, thickener supernatant, or liquids from dewatering processes) must return flows through all processes of the systems exiting conventional or direct filtration plant (or an alternate location approved by the state) by June 8, 2004, plus additional record-keeping requirements.
NPDWR; Arsenic and Clarifications to Compliance and New Source Contaminants Monitoring; Final Rule; Pub. *FR* January 22, 2001, effective February 22, 2002 (U.S. EPA 2001b)	Arsenic MCL is lowered from 50 to 10 ppb. Systems must comply by January 23, 2006.
NPDWR; Long Term 1 Enhanced Surface Water Treatment Rule (LT1ESWTR); Pub. *FR* January 14, 2002, effective February 13, 2002 (U.S. EPA 2002a)	The purposes of the LT1ESWTR are to improve control of microbial pathogens, specifically the protozoan *Cryptosporidium*, in drinking water and address risk trade-offs with disinfection by-products. The rule will require systems to meet strengthened filtration requirements as well as to calculate levels of microbial inactivation to ensure that microbial protection is not jeopardized if systems make changes to comply with disinfection requirements of the Stage 1 D/DBP Rule. The LT1ESWTR builds upon the framework established for systems serving a population of 10,000 or more in the IESWTR. Regulated entities must comply with this rule starting March 15, 2002.
NPDWR; Stage 2 Disinfectant and Disinfection Byproduct; Final Rule; proposed in 2002, Pub. *FR* January 4, 2006 (U.S. EPA 2006a)	DBP compliance method to change to be specific to each sampling location rather than systemwide and to select compliance points through an initial distribution system evaluation.
NPDWR; Long Term 2 Enhanced Surface Water Treatment Rule (LT2ESWTR); proposed in 2002 Pub. *FR* January 5, 2006 (U.S. EPA 2006b)	Sets *Cryptosporidium* removal levels based on source water concentration ranges that are established through a 24-month monitoring program and provides a toolbox of available control methods for meeting treatment requirements. Inactivation of *Cryptosporidium* is required for all unfiltered systems, disinfection profiling, and benchmarking to assure continued levels of microbial protection while systems comply with the Stage 2 D/DBP Rule and covering, treating, or implementing a risk management plan for all uncovered finished water reservoirs. The LT2ESWTR builds upon the framework established in the LT1ESWTR and the IESWTR.

Table 4-4 *(Continued)*

Regulation and date[a]	Description
NPDWR; Ground Water Rule (GWR); October 11, 2006; Pub. *FR* November 8, 2006 (U.S. EPA 2006c)	The rule establishes a risk-based approach to target ground water systems that are vulnerable to fecal contamination. The rule applies to all systems that use groundwater as a source of drinking water.

[a]The date reported is typically the date the rule or regulation was published in the *Federal Register* (*FR*). In some cases, the date the rule was proposed and or became effective is also given.
Source based on: Information in this table is taken in part from the U.S. EPA (1999), the EPA website, *Federal Register*, and Pontius and Clark (1999).

Table 4-5
Summary of U.S. EPA drinking water regulations for microbial contaminants and disinfection by-products arranged in chronological order by date enacted or most current version

Date	Regulation and/or Rule
1989	Total Coliform Rule
1989	Surface Water Treatment Rule (SWTR)
1998	Interim Enhanced Surface Water Treatment Rule (IESWTR)
1998	Stage 1 Disinfectants and Disinfection Byproducts Rule (Stage 1 DBP)
2001	Filter Backwash Recycling Rule (FBR)
2002	Long Term 1 Enhanced Surface Water Treatment Rule (LT1ESWTR)
2006	Long Term 2 Enhanced Surface Water Treatment Rule (LT2ESWTR)
2006	Stage 2 Disinfectants and Disinfection Byproducts Rule (Stage 2 DBP)
2006	Ground Water Rule (GWR)

summarized in Table 4-5. Additional specific information may be found at the following U.S. EPA website: www.epa.gov/safewater/contaminants/index.html#listsec.

UNREGULATED CONTAMINANTS

As part of its ongoing drinking water program, the U.S. EPA maintains a list of unregulated compounds. Compounds are continually added to the list as they are identified from a variety of sources. Listed unregulated compounds are (1) not scheduled for any proposed or promulgated national primary drinking water regulation (NPDWR), (2) have either been identified or are anticipated to be identified in public water systems, and (3) may ultimately need to be regulated under SDWA. Unregulated contaminants are typically grouped into the following general categories.

- Pharmaceuticals and personal care products (PPCPs)
- Endocrine disrupting chemicals (EDCs)
- Organic wastewater contaminants (OWCs)
- Persistent organic pollutants (POPs)

- Contaminants of emerging concern (CECs)
- Microconstituents
- Nanomaterials

To be current, the Drinking Water Contaminant Candidate List (CCL) website maintained by the U.S. EPA should be consulted on a periodic basis. For example, the U.S. EPA is currently examining a number of contaminants and others on the CCL list may be regulated within the next few years, including perchlorate and *N*-nitrosodimethylamine; selected endocrine disruptors, pharmaceuticals, and personal care products; and nanoparticles.

Perchlorate

Perchlorate (ClO_4^-) is a contaminant from the solid salts of ammonium, potassium, or sodium perchlorate. Ammonium perchlorate has been used as an oxygen-adding component in solid fuel propellant for rockets, missiles, and fireworks. Perchlorate is mobile in aqueous systems, and it can persist under typical groundwater and surface water conditions for decades. Beginning around 1997 (with development of a low-level detection methodology), perchlorate has been detected in various drinking water supplies throughout the United States. In January 2009, the U.S. EPA issued an Interim Health Advisory for perchlorate to assist state and local officials in addressing local contamination of perchlorate in drinking water, while the opportunity to reduce risks through a national primary drinking water standard is being evaluated.

N-Nitrosodimethylamine

N-Nitrosodimethylamine (NDMA) is a semivolatile organic chemical that is soluble in water. From the mid-1950s until 1976, it was manufactured and used as an intermediate in the production of 1,1-dimethylhydrazine, a storable liquid rocket fuel that contained approximately 0.1 percent NDMA as an impurity. NDMA has also been used as an inhibitor of nitrification in soil, a plasticizer for rubber and polymers, a solvent in the fiber and plastics industry, an antioxidant, a softener of copolymers, and an additive to lubricants. A potential link between the quaternary amines present in many consumer products including shampoos, detergents, and fabric softeners and the formation of nitrosamine in wastewater has been identified.

It has been found that NDMA, along with other nitrosamines, can cause cancer in laboratory animals. In its Integrated Risk Information System (IRIS) database, the U.S. EPA has classified a number of the nitrosamines as probable human carcinogens. Because of the presence of NDMA and other nitrosamines in drinking water, it appears likely that NDMA will be a candidate for future regulation. However, because the development of an MCL for NDMA will not be available for several years, a 10-mg/L notification level has been established by a number of states to provide information to local government agencies that may ultimately be used in the developing regulations.

Endocrine Disruptors, Pharmaceuticals, and Personal Care Products
The presence of pharmaceuticals, personal care products, and hormonally active agents in the environment is also another area of concern. One of the concerns with these products is they release chemical substances that may have possible endocrine disrupting effects in humans in the environment (Trussell, 2001). Domestic wastes are the primary sources of these personal care products and hormonally active agents in the environment. There are a broad variety of pharmaceuticals and personal care products that can be released into the environment, as listed in Table 4-6. In addition, other types of compounds are being examined as potentially being hormonally active agents. These include such compounds as pesticides, plastic additives, polychlorinated biphenyls, brominated flame retardants, dioxins, and hormones and their metabolites.

The public health impacts of exposure to low levels of these contaminants are not well defined. Potential health impacts include disruption of the male and female reproductive systems, the hypothalamus and pituitary, and the thyroid. The 1996 amendments to the SDWA required the U.S. EPA to develop a screening and testing program to determine which chemical substances have possible endocrine-disrupting effects in humans. For the development of this program the Endocrine Disruptor Screening and Testing Advisory Committee (EDSTAC) was formed. Several compounds that may turn out to be identified as hormonally active agents are already regulated in drinking water and include such contaminants as cadmium, lead, mercury, atrazine, chlordane, dichlorodiphenyl trichloroethane (DDT), endrin, lindane, methoxychlor, simazine, toxaphene, benzo[*a*]pyrene, di-(2-ethylhexyl) phthalate, dioxin, and polychlorinated biphenyls.

Nanoparticles and Nanotechnology
The manufacture and use of nanoparticles, which range from 1 to 100 nm, is a relatively new and rapidly growing field. Nanotechnology involves the

Table 4-6
Representative examples of pharmaceuticals and personal care products

Analgesics	Fragrances
Antibiotics	Hormones
Antiepileptic medicines	Hair care products
Anti-inflammatory medicines	Oral hygiene products
Bath additives	Skin care products
Blood lipid regulators	Stimulants
Cough syrups	Sunscreens
Detergents	

design, production, and application of nanoparticles in various configurations (e.g., singly, clusters, clumps, etc.) in a variety of commercial and scientific applications such as consumer products, food technology, medical products, electronics, pharmaceuticals, and drug delivery systems (SCENIHR, 2006).

Because the field of nanotechnology is so new, few research programs have been initiated that are aimed at understanding the toxicity and potential risk of nanoparticles in the environment. The potential for discharge of nanoparticles to the environment will increase as production increases, so it is important to obtain a better understanding of the health risk and environmental impact of these materials. The U.S. EPA is currently leading scientific efforts to understand the potential risks to humans, wildlife, and ecosystems from exposure to nanoparticles and nanomaterials. One nanoparticle that will likely be regulated in the near future is nanosilver because of its potential toxicity.

State Standards and Regulations

Although the U.S. EPA sets national regulations, the SDWA gives states the opportunity to obtain primary enforcement responsibility (primacy). States with primacy must develop their own drinking water standards, which must be at least as stringent as the U.S. EPA standards. Almost all states have applied for and have been granted primacy. In many instances, the state water quality standards are identical to the U.S. EPA NPDWR and amendments thereto.

International Standards and Regulations

A number of agencies outside the United States have developed drinking water regulations. These include standards for individual countries or groups of countries. The WHO has been at the forefront of developing standards. The WHO standards, known as the Guidelines for Drinking Water Quality (WHO, 1993, 2006), are meant for guidance only and are recommendations, not mandatory requirements. However, the WHO standards have been adopted in whole or in part by a number of countries as a basis of formulation for national standards. The WHO guidelines contain recommendations, health-based standards, monitoring, measurement, and removal for microbial quality and waterborne pathogens, chemical constituents, radionuclides, and aesthetic aspects.

Focus of Future Standards and Regulations

The continued process of water quality regulation is expected to produce additional standards in the future, especially as new compounds are being developed and identified continually. As the U.S. EPA continues to work toward protection of public health, it is expected that standards will be set or revised for more constituents as well as the individual processes and the distribution systems. Improved methods for risk assessment, analysis, and removal of drinking water constituents will also contribute to regulatory activity in the future. In addition, the U.S. EPA released nine white papers on potential public health risks associated with various distribution system issues in 2002 covering the following topics: (1) intrusion, (2) crossconnection control, (3) aging infrastructure and corrosion, (4) permeation and leaching, (5) nitrification, (6) biofilms/microbial growth, (7) covered storage, (8) decay in water quality over time, and (9) new and repaired water mains.

4-4 Overview of Methods Used to Treat Water

A variety of methods have been developed and new methods are being developed for the treatment of water. In most situations, a combination or sequence of methods is needed depending on the quality of the untreated water and the desired quality of the treated water. Although treating water is relatively inexpensive on a volumetric basis, there is little opportunity to modify water quality directly in most natural systems such as streams, lakes, and groundwaters because of the large volumes involved. It is common to treat the water used for public water supplies before distribution and to treat wastewater in engineered systems before it is returned to the environment. It is the purpose of this section to present an overview of the various methods and means used for the treatment of water. Topics to be considered include (1) the classification of treatment methods and (2) the application of the various methods used for the treatment of specific constituents.

Developing solutions to treat for PFAS is an evolving science.

http://www.stantec.com/wt3r/4-2.html

Classification of Treatment Methods

The constituents in water and wastewater are removed by physical, chemical, and biological means. An individual process is known throughout environmental engineering and chemical engineering literature as a *unit process,* although the phrase *unit operation* is sometimes used and the two phrases can be used interchangeably. The most common unit processes in water treatment remove constituents through a combination of physical and chemical means and are known as *physicochemical unit processes.* The unit processes used for the treatment of water are reported in Table 4-7.

Water treatment plants rarely contain a single unit process; instead, they typically have a series of processes. Multiple processes may be needed when different processes are needed for different contaminants. In addition, sometimes processes are effective only when used in concert with another; that is, two processes individually may be useless for removing a compound but together may be effective if the first process preconditions the compound so that the second process can remove it. A series of unit processes is called a *treatment train.* Although unit processes are combined into treatment trains in water treatment plants, they are usually considered separately. By considering each unit process separately, it is possible to examine the fundamental principles involved apart from their application in the treatment of water.

Table 4-7
Typical unit processes used for the treatment of water

Unit Process	Description	Typical Application in Water Treatment
Adsorption	The accumulation of a material at the interface between two phases	Removal of dissolved organics from water using granular activated carbon (GAC) or powdered activated carbon (PAC)
Advanced oxidation	Use of chemical reactions that generate highly reactive short-lived hydroxyl radicals (OH•) for purpose of oxidizing chemical compounds; typical reactions that produce these free radicals, listed from most common to least common: O_3, Peroxone (H_2O_2 and O_3), O_3 and UV radiation, and H_2O_2 and UV radiation	Oxidation of certain humic compounds, pesticides, and chlorinated organics and some taste and odor compounds such as methylisoborneol (MIB) and geosmin found in surface waters and contaminated groundwaters
Aeration	The process of contacting a liquid with air by which a gas is transferred from one phase to another: either the gas phase to the liquid phase (gas absorption) or the liquid phase to the gas phase (gas stripping)	Removal of gases from groundwater (e.g., H_2S, VOCs, CO_2, and radon); oxygenation of the water to promote oxidation of iron and manganese
Biofiltration	Rapid granular media (often activated carbon) filter operated for dual purpose of particle removal and removal of biodegradable organic matter by biological oxidation	Removal of biodegradable organic matter (BOM) following ozonation
Chemical disinfection	Addition of oxidizing chemical agents to inactivate pathogenic organisms in water	Disinfection of water with chlorine, chlorine compounds, or ozone
Chemical neutralization	Neutralization of solution though addition of chemical agents	Control of pH; optimizing operating range for other treatment processes
Chemical oxidation	Addition of oxidizing agent to bring about change in chemical composition of compound or group of compounds	Oxidation of iron and manganese for subsequent removal with other processes; control of odors; removal of ammonia
Chemical precipitation	Addition of chemicals to bring about removal of specific constituents through solid-phase precipitation	Removal of heavy metals, phosphorus
Coagulation	Process of destabilizing colloidals so that particle growth can occur during flocculation	Addition of chemicals such as ferric chloride, alum, and polymers to destabilize particles found in water

Table 4-7 *(Continued)*

Unit Process	Description	Typical Application in Water Treatment
Denitrification	Biological conversion of nitrate (NO_3^-) to nitrogen gas (N_2)	Conversion of nitrate found in some surface wastes to nitrogen gas
Disinfection	Addition of chlorine, chloramines, chlorine dioxide, ozone, or UV light followed by a specified amount of contact time	Inactivation of pathogenic organisms such as viruses, bacteria, and protozoa
Distillation	Separation of components of liquid from liquid by vaporization and condensation	Used for desalination of seawater
Filtration (granular)	The removal of particles by passing water through a bed of granular material, particles are removed by transport and attachment to the filter media	Removal of solids following coagulation, flocculation, gravity sedimentation, or flotation
Filtration (membrane)	The removal of particles by passing water through a porous membrane material; particles are removed by straining (size exclusion) because they are larger than the pores	Used to remove turbidity, viruses, bacteria, and protozoa such as *Giardia* and *Cryptosporidium*
Flocculation	Aggregation of particles that have been chemically destabilized through coagulation	Used to create larger particles that can be more readily removed by other processes such as gravity settling or filtration
Flotation, dissolved air	Removal of fine particles and flocculent particles with specific gravity less than water or very low settling velocities	Removal of particles following coagulation and flocculation for high-quality raw waters that are low in turbidity, color, and/or TOC or experience heavy algal blooms
Flow equalization	Storage basin in a process train, which can store water to equalize flow and minimize variation in water quality	Large storage tanks used to store waste washwater to permit constant return flow to head of treatment plant; clearwells used to store treated water to allow treatment plant to operate at constant rate regardless of short-term changes in system demand
Gravity separation, accelerated	Solids contact clarifiers and floc-blanket clarifiers where coagulation, flocculation, and sedimentation occur in a single basin and gravity settling occurs in an accelerated flow field	Where land area is limited, surface loading rates are typically 2.4 m/h (1 gal/ft^2•min) and higher; lime-soda softening
Ion exchange	Process in which ions of given species are displaced (exchanged) from insoluble exchange material by ions of different species in solution	Removal of hardness, nitrate, NOM, and bromide; also complete demineralization

(Continued)

Table 4-7 *(Continued)*

Unit Process	Description	Typical Application in Water Treatment
Mixing	Mixing and blending of two or more solutions through input of energy	Used to mix and blend chemicals
Nitrification	Biological conversion of ammonia (NH_3) to nitrate (NO_3^-)	Conversion of ammonia found in some surface wastes to nitrate for subsequent removal by denitrification
Reverse osmosis	A membrane process that separates dissolved solutes from water by differences in solubility or diffusivity through the membrane material; uses reverse osmosis or nanofiltration membranes	To produce potable water from ocean, sea, or brackish water; water softening; removal of specific dissolved contaminants such as pesticides and removal of NOM to control DBP formation
Screening, coarse	Passing untreated water through coarse screen to remove large particles from 20 to 150 mm and larger	Used at the intake structure to remove sticks, rags, and other large debris from untreated water by straining (i.e., interception) on screen
Screening, micro	Passing water through stainless steel or polyester media for removal of small particles from 0.025 to 1.5 mm from untreated water by straining (i.e., interception) on a screen	Used for removal of filamentous algae
Sedimentation	Separation of settleable solids by gravity	Used to remove particles greater than 0.5 mm generally following coagulation and flocculation
Stabilization	Addition of chemical to render treated water neutral with respect to formation of calcium carbonate scale	Stabilization of treated water before entry into distribution system
Ultraviolet light oxidation	Use of UV light to oxidize complex organic molecules and compounds	Used for the oxidation of *N*-nitrosodimethylamine (NDMA)

Application of Unit Processes

As discussed in Chaps. 2 and 3, a wide variety of constituents may be found in water. Representative specific physical, inorganic chemical, organic chemical, radionuclides biological, and aesthetic constituents that may have to be removed from surface and groundwater to meet specific water quality objectives are identified in Table 4-8, along with the treatment processes that can be used for their removal. For many constituents, a number of processes can be used. Final process selection will depend, in part, on what additional constituents must be removed and how complimentary are the processes being considered.

Table 4-8
Application of unit processes for the removal of specific constituents

Constituent	Process	Applicability
	Physical Constituents	
Hardness	Lime-soda softening	Applicable for moderate to extremely hard waters. Historically the most common method for removal of hardness.
	Ion exchange	Most common in small installations. Disposal of regenerate solutions can be a problem.
	Nanofiltration	Often referred to as low-pressure reverse osmosis (RO) membranes. Applicable for moderate to extremely hard waters. Disposal of concentrate may be the limiting factor in using nanofiltration.
Total dissolved solids (TDS)	Reverse osmosis, ion exchange, distillation	Used for desalination with ocean, sea, and brackish water. Reverse osmosis concentrate and ion exchange regenerate solution disposal may be the limiting factor in selecting these treatment processes.
Turbidity/particles	In-line filtration[a]	Works well in low-turbidity, low-color waters. Pilot studies should be performed to establish performance and design criteria.
	Direct filtration[b]	Applicable for low to moderate turbidity and colored waters. Pilot studies should be performed to establish performance and design criteria. Shorter filter runs than conventional treatment.
	Conventional treatment[c]	Works well in moderate- to high-turbidity waters. More operational flexibility than direct or in-line filtration options. Sedimentation basin detention time allows for NOM, taste and odor, and color removal in combination with sedimentation. Sometimes can be designed without piloting if local regulatory agency guidelines are followed.
	Membrane filtration	Effective at removing turbidity, bacteria, and protozoa-sized particles. Viruses may be removed by some types of ultrafiltration membranes. Works well on low-turbidity waters or with pretreatment for particle removal. Natural organics can foul membranes. Pilot testing required to demonstrate particle removal and potential for organic fouling. Easily automated and space requirements are much smaller than conventional plants.
	Slow sand filtration	Primary removal mechanisms are biological and physical. Works well in low-turbidity waters. When used in conjunction with granular activated carbon (GAC), effective at taste and odor removal. Surface loading rates are 50 to 100 times lower than rapid filtration so filters are very large. Most applicable to small communities, but there are very large plants in operation throughout the world.

(*Continued*)

Table 4-8 *(Continued)*

Constituent	Process	Applicability
	Inorganic Chemical Constituents	
Arsenic	Coagulation/ precipitation, activated alumina, ion exchange, reverse osmosis	Conventional coagulation with iron or aluminum salts is effective for removing greater than 90% of As(V) (with initial concentrations of roughly 0.1mg/L) at pH values of 7 or below. As with fluoride, is strongly adsorbed/exchanged by activated alumina. Arsenic(III) is difficult to remove but is rapidly converted to As(V) with chlorine (Cl_2).
Fluoride	Lime softening, coagulation/ precipitation, activated alumina	Lime softening will remove fluoride from water both by forming an insoluble precipitate and by co-precipitation with magnesium hydroxide [$Mg(OH)_2$]. Alum coagulation will reduce fluoride levels to acceptable drinking water standards but requires very large amounts of alum to do so. Contact of fluoride-containing water with activated alumina will remove fluoride.
Iron/ manganese	Oxidation, polyphosphates, ion exchange	Typically found in groundwaters or lake waters with low dissolved oxygen. Removal is most commonly through precipitation by oxidation using aeration or chemical addition (e.g., potassium permanganate or chlorine) for removal by sedimentation or filtration. Greensand filtration in which oxidation and filtration take place simultaneously is also common. The use of polyphosphate precipitation is another method that can be used for the removal of iron and manganese. Iron oxidizes much more readily than does manganese.
Nitrate	Biological denitrification, reverse osmosis, ion exchange	Biological denitrification requires the use of special organisms to reduce nitrate to nitrogen gas. Reverse osmosis will reduce nitrate levels in drinking water, but this process is used primarily for treating high TDS and salt water. Ion exchange with anionic resins is attractive when brine disposal is available.
Selenium	Coagulation/precipitation, activated alumina, ion exchange, reverse osmosis	Conventional treatment techniques using alum or ferric sulfate coagulation and lime softening have been investigated for selenium removal. Activated alumina has also been investigated for its potential to remove Se(IV) and Se(VI). Although strong-base anion exchange resins have not been thoroughly investigated for selenium removal, it appears that they could be successful, but they are not selective for selenium.
Sulfate	Reverse osmosis	Reverse osmosis is most common for removal of sulfate from seawater.
Sulfide	Oxidation	Typically found in groundwaters as H_2S and is responsible for taste and odors similar to rotten eggs. Removal is most common through aeration and chlorination.

Table 4-8 *(Continued)*

Constituent	Process	Applicability
		Organic Chemical Constituents
Disinfection by-products	Enhanced coagulation, adsorption, alternative disinfectants low-pressure reverse osmosis	Strategies for the control of disinfection by-products (DBPs) include alternative disinfectants (ozone, chlorine dioxide, chloramines, and ultraviolet light) or removal of DBP precursor material (NOM) through enhanced coagulation or adsorption on activated carbon [either GAC or powdered activated carbon (PAC)]. GAC can be used to remove bromate, a DBP formed from ozone and bromide. Reverse osmosis is very effective but expensive.
Natural organic matter	Enhanced coagulation, adsorption, ion exchange, reverse osmosis	Enhanced coagulation (low-pH coagulation) can be used to remove significant amounts of NOM as measured by TOC and is the most widely used process for NOM removal. GAC adsorption, postfiltration is also very effective in removing NOM. Ion exchange use is limited by disposal of the high-TDS regeneration brine. The high cost of RO and concentrate disposal issues limit the use of this process for NOM removal.
Volatile organics, pesticides/herbicides	Air stripping, coagulation, adsorption, advanced oxidation	For volatiles, air-stripping process is recommended. Usually, volatile organic compound (VOC) removal from the gas phase is required posttreatment. For nonvolatile components, coagulation or adsorption process can be used. Low-pH coagulation (enhanced coagulation) can be used to remove significant amounts of TOCs and some nonvolatile organic compounds.
		Radionuclides
Radium	Coagulation	The selected process depends on the level of contamination. Residuals may represent a low-level radioactive waste disposal problem.
Radon	Aeration, detention time	Simple aeration is effective. Packed tower aeration can be used for removal of very high levels. Mixing and detention time may control low-level radon contamination.
	Adsorption	Carbon used for adsorption may be a low-level radioactive disposal problem.
Uranium	Lime softening, ion exchange, reverse osmosis	The selected process depends on the level of contamination. Residuals may represent a low-level radioactive waste disposal problem.

(Continued)

Table 4-8 *(Continued)*

Constituent	Process	Applicability
		Microbial Constituents
Algae	Copper sulfate, conventional treatment, dissolved air flotation, microscreening	Copper sulfate application in raw-water storage areas (e.g., reservoirs, ponds, and lakes) has been used to control algal blooms. Moderate to severe seasonal algae bloom situations can be handled through careful control of conventional processes: sedimentation and filtration. Direct filtration and in-line filtration can experience extremely shortened filter runs during algae episodes. For persistent algae problems, dissolved air flotation processes or contact clarification devices should be considered. Microscreening may be used at the headworks of a treatment plant for filamentous algae removal.
Bacteria	Conventional treatment, membrane filtration, reverse osmosis, disinfection	Bacteria can be removed through conventional processes, including sedimentation and filtration. Membrane processes provide a positive barrier to most bacteria. Given sufficient dose and contact time, all common disinfectants (chlorine, chlorine dioxide, chloramines, UV, ozone) are effective at inactivation of bacteria.
Protozoan cysts	Conventional treatment, granular media filtration, reverse osmosis, high-pressure membranes, disinfection	Pathogenic cysts and oocysts (*Giardia* and *Cryptosporidium*) require high levels of disinfectants to inactivate. Effectiveness in decreasing order is ozone > chlorine dioxide > chlorine ≫ chloramines. Conventional treatment as well as granular media filtration is effective at removing cysts and oocysts. Membrane processes provide a positive barrier for cysts. UV irradiation is also very effective.
Viruses	Conventional treatment,[c] membrane filtration, reverse osmosis, disinfection	Viruses can be removed through conventional processes, including sedimentation and filtration. Membrane process with low-molecular-weight cutoff such as some ultrafiltration membranes can be used for virus removal. Pilot studies are required with membranes to demonstrate effective control. All common disinfectants, with the exception of chloramines, are effective at inactivation of most viruses. Chloramines require a long contact time at high doses for effective virus disinfection.
		Aesthetic Constituents
Color	Coagulation/ precipitation	High coagulation doses and low pH can be effective even for very high color levels. Bench or pilot testing is recommended.

Table 4-8 *(Continued)*

Constituent	Process	Applicability
	Adsorption with GAC or PAC	Granular activated carbon as a filter medium can be very effective for low to moderate taste and odor levels. Replacement is usually on a 3-5-year cycle. In slurry form, PAC can be added to the coagulation process for taste and odor control. PAC is especially effective in contact clarification devices.
	Oxidation with chlorine, ozone, potassium permanganate, and chlorine dioxide	Effectiveness is generally ozone > chlorine > chlorine dioxide > $KMnO_4$. pH can affect the efficiency and some colors may return after oxidation. Pilot or bench studies are recommended.
	Ozone/BAC (biologically active carbon)	Preoxidation followed by biologically active carbon treatment has proven effective for a number of colored waters.
Taste and odors[d]	Source control with copper sulfate and reservoir destratification (in situ aeration)	Many taste and odor problems are associated with algae growths and reservoir turnover. Copper sulfate applied in the source water is effective at controlling algae growth. Aeration is appropriate for use in relatively shallow raw-water storage areas where seasonal turnover of stratified water releases taste and odor compounds.
	Oxidation with chlorine, ozone, potassium permanganate, and chlorine dioxide	Chlorine may be used to control taste and odors from H_2S but is not effective at algal taste and odors and may even make these types of taste and odors worse. Chlorination of industrial chemicals such as phenols intensifies objectionable tastes. Ozone is viewed as one of the most effective oxidants for reducing taste and odors and has the additional benefit in that it can also be used for disinfection. Permanganate is effective for removal of some algae taste and odors at alkaline pH but not the most common taste and odor compounds MIB and geosmin. Additionally, overdosing results in pink water and the formation of black deposits in the distribution systems and household and industrial appurtenances. Chlorine dioxide is effective at controlling many tastes and odors but is not effective in reducing geosmin and MIB. Pilot or bench testing should be performed to determine the best oxidation approach.
	Adsorption with GAC	Granular activated carbon bed life can be short depending on the levels and empty bed contact time (EBCT).

[a]In-line filtration is comprised of coagulation followed by filtration. Also known as contact filtration in countries outside the United States.

[b]Direct filtration is comprised of coagulation followed by flocculation and filtration.

[c]Conventional treatment is comprised of coagulation, flocculation, sedimentation, and filtration.

[d]Total dissolved solids, which can also cause taste and odor problems, are considered under the heading physical parameters.

4-5 Development of Systems for Water Treatment

The design, construction, and operation of water treatment facilities follow a process that starts with the desire or need for clean potable water. The initial design is based on water quality data, regulatory requirements, water quality issues, consumer concerns, construction challenges, operational constraints, water treatment technology, and economic feasibility, all of which are combined with human creativity to develop a water treatment plant design. The design is then transformed into a permanent facility through the construction process and becomes an operational water treatment facility with the addition of raw materials and operator know-how. The general considerations involved in the development of systems for water treatment include (1) process selection, (2) the synthesis of surface water and groundwater treatment processes, (3) treatment processes for residuals management, (4) the sizing of treatment train processes and ancillary equipment, and (5) conducting pilot studies. The multibarrier concept as applied to the design of water treatment systems is considered in Sec. 4-6. Detailed analysis of the unit processes are considered in the following chapters.

General Considerations Involved in Selection of Water Treatment Processes

The water treatment process selection starts with at least three key pieces of information: source water quality and variability, required and/or desired quality of the treated water, and required plant production and operational goals. Other factors that affect treatment process selection, including the impact of BAT and the various treatment rules, are also considered (Patania Brown, 2002).

SOURCE WATER QUALITY

The first step is gathering key information about the source water quality and variability, be it surface water or groundwater (see Fig. 4-3). For surface waters, this information should include typical and event water as well as source stability. Event water quality (i.e., variability) is the water quality that coincides with specific occurrences such as spring runoff, summer algae growth, fall reservoir turnover, and minimum winter water temperatures. Impounded water sources tend to be more stable than flowing river sources, meaning the water quality in impounded sources changes much less rapidly than the water quality in flowing river sources. In general, groundwaters tend to be more stable but, depending on the pumping rate, may contain variable amounts of gases and synthetic organic compounds (SOCs).

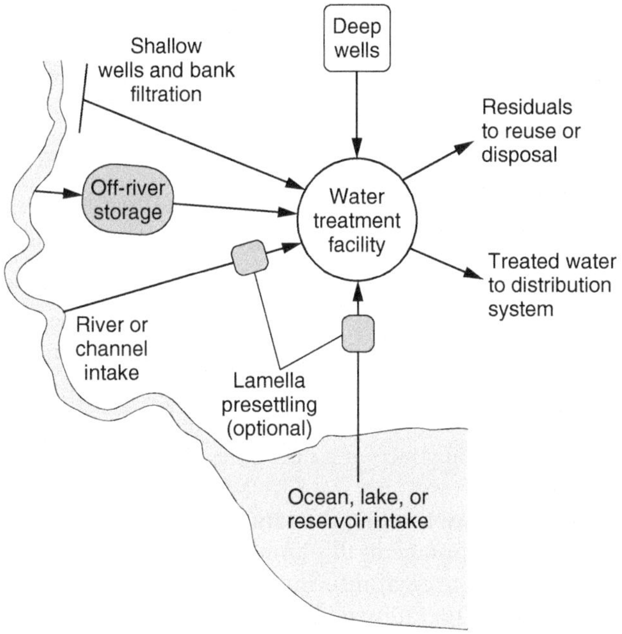

Figure 4-3
Definition sketch for water supply sources. Depending on water quality and characteristics of surface water sources, presedimentation facilities may be employed.

Many source water quality characteristics can affect treatability. Physical characteristics include turbidity and particulates. Organic characteristics include natural organic matter (NOM), disinfection by-product precursors, color, and chlorine demand. Microbial constituents include protozoa such as *Cryptosporidium* and *Giardia*, viruses, bacteria, and helminths. Chemical parameters that affect treatability include pH, alkalinity, hardness, and corrosivity. Inorganic constituents may include iron, manganese, arsenic, and bromide. Aesthetic concerns include color, taste, and odor. Constituents of regulatory concern include SOCs, VOCs, some inorganic constituents, and radionuclides plus other water quality characteristics such as algae and biological stability.

REQUIRED TREATED WATER QUALITY

The required and/or desired treated water quality also imposes constraints on the treatment process. Treated water quality is directly impacted by current and future regulatory compliance, the quality objectives of the utility, consumer expectations, and political constraints. For example, regulatory compliance with disinfection by-products may become more stringent or a water may have periodic taste and odor episodes. Achieving treated water quality goals frequently means more than one process is necessary to address a quality issue. This multiple-barrier approach is described in Sec. 4-6.

PLANT PRODUCTION AND OPERATIONAL GOALS

Plant production and operational goals can vary and sometimes can be at odds with each other. Typical productivity operational goals for a water filtration plant can include:

- Filtration efficiency, generally defined by unit filter run volume (UFRV)
- Filter run length, which may be based on operator convenience
- Terminal head loss, which is typically dictated by the hydraulic profile
- Filter maturation volume, which is the volume of water filtered to waste
- Disinfection *Ct* value [the residual disinfection concentration *C* (mg/L) multiplied by the contact time *t* (min)], which can be in the chlorine contact basin, the ozone contactors, or throughout the plant processes, depending upon where the disinfectant addition occurs

An example of specific operational goals for a new water treatment plant is given in Table 4-9.

If there are other plants that are treating the same or similar source water, it is important to take full advantage of this knowledge and experience. Operator knowledge is a huge source of information that is practical and useful in designing a new plant. Operational information that may be available could include what processes and types of chemicals are being used, how to treat changes in raw-water quality, what the operators like and do not like about their plant, and how the new plant may be more efficient, easier to operate, and more flexible for meeting future regulations.

BEST AVAILABLE TECHNOLOGY AND TREATMENT RULES

Along with the establishment of MCL values for a variety of constituents, the U.S. EPA has also developed a recommended list of BATs for treating

Table 4-9
Example operational goals for new direct-filtration water treatment plant

Parameter	Unit	Goal
Ozone disinfection	log inactivation	1.0 log *Cryptosporidium* inactivation for direct-filtration process train 0.5 log *Cryptosporidium* inactivation for clarification process train
Filter run length	h	24
Terminal head loss	m (ft)	3.0 10.0
Turbidity at breakthrough	NTU	0.1
Unit filter run volume (UFRV)	m^3/m^2 (gal/ft^2)	200–400 5000–10,000
Unit backwash volume	m^3/m^2 (gal/ft^2)	4–8 100–200

each contaminant for which EPA sets an MCL. The BAT represents the minimum acceptable level of treatment. As noted in the previous section, a water purveyor is not required to install the recommended BAT but must comply with the appropriate MCLs. It is also interesting to note that if BAT is used and the MCL values cannot be met, the water purveyor is not held responsible.

The various U.S. EPA rules that apply to water supply systems also impact the selection of treatment processes. For example, rules that deal with the removal of microbial contaminants (e.g., viruses, heterotrophic bacteria, *G. lamblia, Cryptosporidium,* etc.) were identified previously in Table 4-5. Thus, the requirements set forth in the various rules will also affect the selection, design, and operation of water treatment facilities. The challenge for the designer of water treatment facilities is to meet current regulations while at the same time trying to anticipate what changes will occur in regulations and rules over the useful life of the facility and how they might impact the design and operation of the facility in the future.

OTHER FACTORS AFFECTING PROCESS SELECTION

Other factors such as steering committee guidance, phasing options, meeting known regulations, and balancing meeting versus planning for future regulations are all important in treatment process selection. Sometimes an important driver of process selection may be a political agenda, a community issue, or an unusual health issue, but whatever the driver is, it may be the deciding factor for why a particular process is selected over another process.

Synthesis of Water Treatment Trains

As described in the previous section, many different processes are available for drinking water treatment. Some processes may be easily ruled out, as they are obviously not appropriate for the raw water being treated. Other processes may seem reasonable and warrant further investigation. Important factors that must be considered in the selection of the treatment process scheme and facility designs are reported in Table 4-10.

Experience acquired through treatment of the same or similar source waters provides an excellent guide in selecting the treatment process scheme. However, where experience is lacking or where there is desire to provide for a different degree of treatment, special studies are required. The special studies include bench-scale study in the laboratory, pilot plant testing, and plant-scale simulation testing. Examples of prototype studies used in actual treatment plant design are presented in Chap. 23.

Reference materials that provide design guidelines and standards are valuable tools to the water treatment plant design engineer. The many books published by the American Water Works Association (AWWA) contain a wealth of material gathered from a number of sources on specific topics. The AWWA handbook on *Water Quality and Treatment* (AWWA, 2011), is also useful. A popular reference for treatment plant design used primarily in the mid and eastern United States is the *Recommended Standards*

Table 4-10
Important factors that must be considered when evaluating and selecting unit processes

Factor	Comment
Adaptability	Can the process be modified to meet future treatment requirements?
Applicable flow range	The process should be matched to the expected range of flow rates. For example, slow sand filters are generally not suitable for extremely large flow rates in highly populated areas.
Applicable flow variation	Most unit processes have to be designed to operate over a wide range of flow rates. Most processes work best at a relatively constant flow rate.
Ancillary processes	What support processes are required? How do they affect the treated water quality, especially when they become inoperative? What backup provisions are necessary to ensure continued operation of vital ancillary processes, such as chemical feed?
Chemical requirements	What resources and what amounts must be committed for a long period of time for the successful operation of the unit process? What effects might the addition of chemicals have on the characteristics of the treatment residuals and the cost of treatment? Are there neighbor concerns that have to be considered both onsite and in transporting chemicals to the site?
Climatic constraints	Temperature affects the rate of reaction of most processes. Temperature may also affect the physical operation of the facilities.
Compatibility	Can the unit process be used successfully with existing facilities? Can plant expansion be accomplished easily?
Complexity	How complex is the process to operate under routine or emergency conditions? What levels of training must the operators have to operate the process?
Energy requirements	The energy requirements, as well as probable future energy cost, must be known if cost-effective treatment systems are to be designed.
Environmental constraints	Environmental factors, such as animal habitat and proximity to residential areas, may restrict or affect the use of certain processes and types of intakes. Noise and traffic may affect selection of a plant site and impact the plant deliveries.
Environmental protection	What specific steps need to be taken to protect the environment? Are chemical spills or discharges environmental threats and what ancillary processes are necessary to address these events?
Economic life-cycle analysis	Cost evaluation must consider initial capital cost and long-term operating and maintenance costs. The plant with the lowest initial capital cost may not be the most effective with respect to operating and maintenance costs. The nature of the available funding will also affect the choice of the process.
Inhibiting and unaffected constituents	What constituents are present and may be inhibitory to the treatment processes? What constituents are not affected during treatment?

Table 4-10 *(Continued)*

Factor	Comment
Land availability	Is there sufficient space to accommodate not only the facilities currently being considered but also possible future expansion? How much of a buffer zone is available to provide landscaping to minimize visual and other impacts?
Operating and maintenance requirements	What special operating or maintenance requirements will need to be provided? What spare parts will be required and what will be their availability and cost?
Other resource requirements	What, if any, additional resources must be committed to the successful implementation of the proposed treatment system using the unit process being considered?
Performance	Performance is usually measured in terms of treated water quality, which must be consistent with the governing regulations and treatment goals.
Personnel requirements	How many people and what levels of skills are needed to operate the unit process? Are these skills readily available? How much training will be required?
Process applicability	The applicability of a process is evaluated on the basis of past experience, data from full-scale plants, published data, and pilot plant studies. If new or unusual conditions are encountered, pilot plant studies are essential.
Process sizing based on mass transfer rates or process criteria	Reactor sizing is based on mass transfer coefficients. If mass transfer rates are not available, process criteria are used. Data for mass transfer coefficients and process criteria usually are derived from experience, published literature, and the results of pilot plant studies.
Process sizing based on reaction kinetics or process loading criteria	Reactor sizing is based on the governing reaction kinetics and kinetic coefficients. If kinetic expressions are not available, more general process criteria are used. Data for kinetic expressions and process loading criteria usually are derived from experience, published literature, and the results of pilot plant studies.
Process sizing based on redundancy requirements and size availability	Size processes to accommodate regular operational and maintenance activities such as when a sedimentation basin is drained for maintenance or a filter is being backwashed. Additionally, some processes (typically proprietary processes) are available only in certain sizes, which may make them either too large for some applications or so small that numerous units are required, making the process impractical.
Raw-water characteristics	The characteristics of the raw water affect the types of processes to be used and the requirements for their proper operation.
Reliability	What is the long-term reliability of the unit process being considered? Is the process easily upset? Can it stand periodic excursions in raw-water quality? If so, how do such occurrences affect the treated-water quality?

(Continued)

Table 4-10 *(Continued)*

Factor	Comment
Residuals processing	Are there any constraints that would make sludge, concentrate, or regeneration brine processing and disposal infeasible or expensive? The selection of the residuals-processing system should go hand in hand with the selection of the liquid treatment system.
Security	Secure water treatment plants are vital to protect the public from attack. Can the process be readily protected from intentional upset? What steps or facilities are required to protect the overall plant?
Treatment residuals	The types and amounts of solid, liquid, and gaseous residuals produced must be known or estimated.

Source: Adapted from Tchobanoglous et al. (2003).

for Water Works, which is also known as the *Ten State Standards* (Great Lakes Upper Mississippi River Board, 2003). The book *Integrated Design and Operation of Water Treatment Facilities* (Kawamura, 2000) is an excellent all-round reference on design and operation. The book *Pump Station Design* (Jones et al., 2008) is comprehensive with respect to the design of pumping facilities. Ancillary equipment such as chemical feed systems is generally not addressed in typical water treatment plant references but is covered in specialty references and brochures supplied by manufacturers. The U.S. EPA has published many manuals and reports, including documents for optimizing water treatment plants using the comprehensive performance evaluation/composite correction program (CPE/CCP).

TYPICAL TREATMENT PROCESSES FOR SURFACE WATER

Depending on the quality of the surface water, a number of alternative treatment trains (described on process flow diagrams) can be used. Four common treatment trains for treating surface water are identified below. Each of these treatment trains is illustrated and described in the following discussion:

- Conventional water treatment
- Direct- and in-line filtration treatment
- Membrane filtration treatment
- Reverse osmosis treatment

Conventional water treatment

Conventional treatment trains are typically used to treat surface waters with water quality issues such as high turbidity (typically >20 NTU), high color (>20 c.u.), or high TOC (>4 mg/L). A conventional process treatment train consists of coagulation, flocculation, sedimentation, granular media filtration, and disinfection, as shown on Fig. 4-4. In general, conventional plants have more operational flexibility, are hydraulically stable, and require somewhat less operator attention than other types of water treatment plants. Another consideration when selecting a conventional process train is land availability as surface loading rates are typically low, resulting in large process surface area requirements.

Direct and in-line (contact) filtration

Direct- and in-line filtration treatment trains are typically used to treat higher quality surface waters with low turbidity (typically ≤15 NTU), moderate to low color (≤20 c.u.), and low TOC (<4 mg/L). As shown on Fig. 4-5, a direct-filtration process treatment train consists of flash mixing, flocculation, granular media filtration, and disinfection. Coarse deep-bed monomedia filters (or dual-media filters consisting of a deep layer of anthracite over a shallow lower layer of coarse sand) are commonly used in direct-filtration plants as a deep coarse filter can store more suspended solids than a conventional filter. The coagulation and flocculation processes are very important and must be able to form a small, tough floc, as a large floc will readily blind the filters, resulting in shortened filter run times.

An in-line filtration process treatment train is the same as a direct-filtration process treatment train, except flocculation is incidental, meaning flocculation occurs in conveyance structures between the coagulation process and the filters and above the filter media instead of in a flocculation basin. In-line filtration treatment trains are used to treat high to excellent quality surface waters, meaning very low turbidity (typically ≤5 NTU), low color (<10 c.u.), and low TOC (<4 mg/L). Coagulation and filter designs for in-line process trains are similar to that for direct-filtration process trains. Granular filtration is considered in detail in Chap. 11.

Membrane filtration

A membrane filtration process treatment train consists of a screening system, low-pressure membranes, and disinfection, as shown on Fig. 4-6. The screening system consists of cartridge filters or microscreens and may include coarse screens at the raw-water source if needed. Membrane filtration process trains as configured on Fig. 4-6 are typically used to treat good-quality surface waters with low turbidity (typically ≤10 NTU) and moderate to low color (<10 c.u.) and TOC (<4 mg/L). However, membrane filters can be used to treat water of any quality when combined with other processes by replacing the granular filters with membrane filters in the conventional water treatment process described earlier. Filtration membranes are configured in modules, which may be stacked, and membrane plants may be readily automated, making membrane plants ideal for small, remote plants on constrained sites. Membrane filtration is considered in detail in Chap. 12.

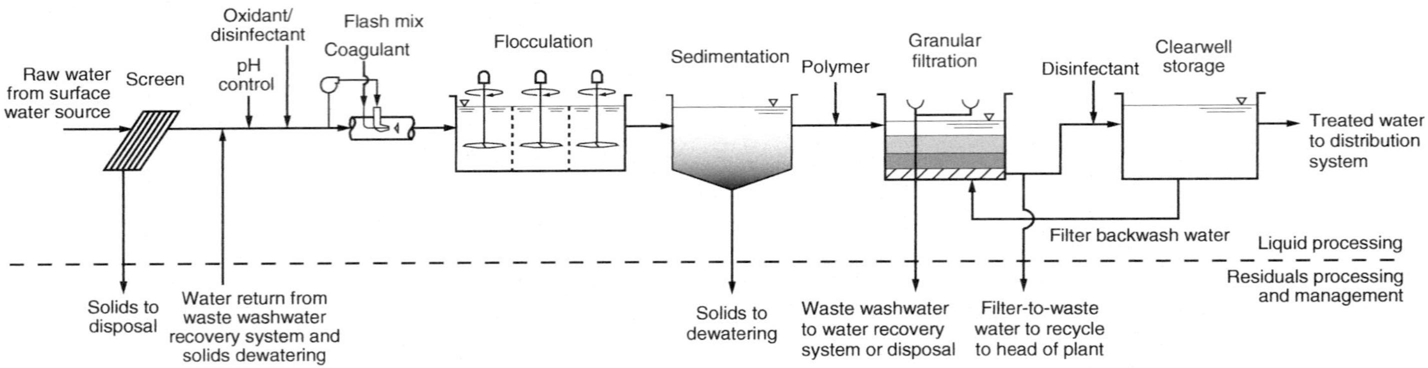

Figure 4-4
Typical process train for treatment of surface water by conventional treatment.

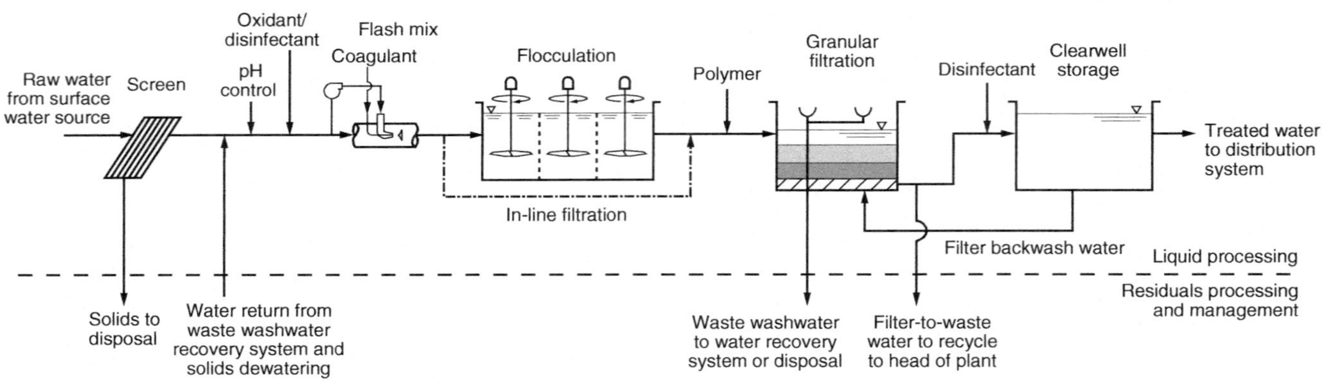

Figure 4-5
Typical process train for treatment of surface water by direct and in-line filtration.

Reverse osmosis treatment
A reverse osmosis (RO) process treatment train consists of a screening system, RO membranes, and disinfection, as shown on Fig. 4-7. The screening system typically consists of cartridge filters or microscreens and may include coarse screens at the raw-water source and other processes for pretreatment if needed. Reverse osmosis is mainly used for desalination of seawater or brackish water and may be used for specific contaminant removal such as arsenic. Reverse osmosis is considered in detail in Chap. 17.

TYPICAL TREATMENT PROCESSES FOR GROUNDWATER

As with surface water, a number of alternative treatment trains can be used depending on the quality of the groundwater. Four common treatment trains for treating groundwater are identified and described in the following discussion:

- Conventional lime-softening treatment
- Membrane water softening
- Gas removal treatment
- Iron and manganese treatment

Conventional lime softening
Conventional lime softening is used to remove hardness from groundwater. Water is generally considered hard when total hardness (the sum of carbonate and noncarbonate hardness) is greater than 150 or 180 mg/L (see discussion in Chap. 2), but many utilities do not provide softening unless the hardness is much higher or there is public and political support for water softening. When hardness is mainly from calcium and magnesium hardness is low (<40 mg/L), lime is added in a single-stage softening process, as shown on Fig. 4-8. Adding soda ash increases alkalinity, which removes noncarbonate hardness. Frequently the coagulation, flocculation, and sedimentation steps occur in one basin, called a reactor-clarifier with sludge recirculation, as shown on Fig. 4-8. A more complete discussion of the softening process may be found in Sec. 20-4.

Membrane softening
An alternative to lime softening for the removal of hardness from groundwater involves the use of membranes, as shown on Fig. 4-9. Membrane products that have been used for the softening and removal of other constituents such as color, TOC, and DBP precursors include nanofiltration (NF) and RO membranes. The difference between NF and RO is the size of ion removed, with NF removing divalent ions and RO removing monovalent ions (see Fig. 12-2). A major consideration when siting a membrane softening plant is concentrate disposal, which may be a regulatory stumbling block. Reverse osmosis membranes are configured in modules, which may be stacked, and membrane plants may be readily automated, making RO membrane plants ideal for small, remote plants on constrained sites.

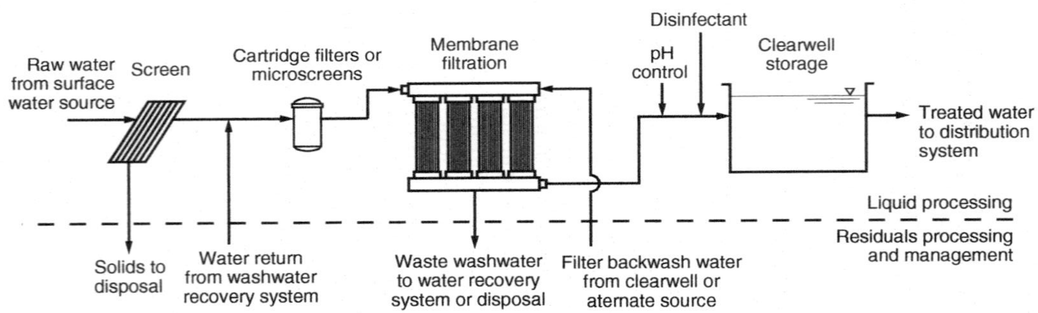

Figure 4-6
Typical process train for treatment of surface water by membrane filtration.

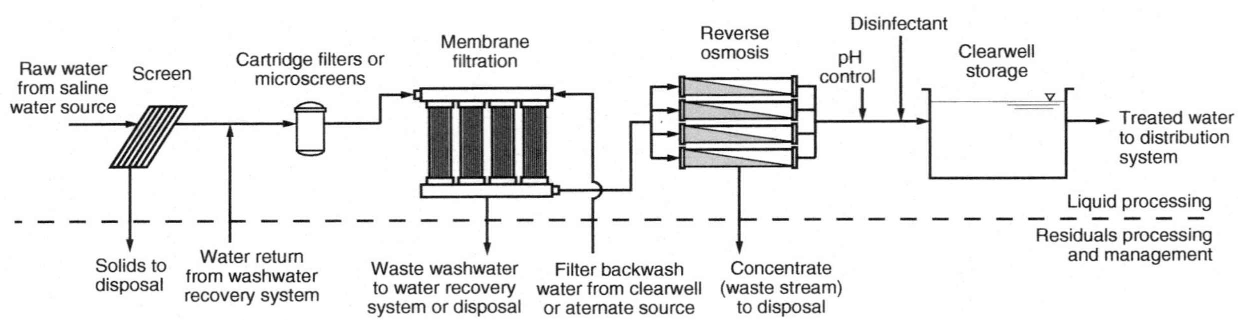

Figure 4-7
Typical process train for treatment of saline water by reverse osmosis.

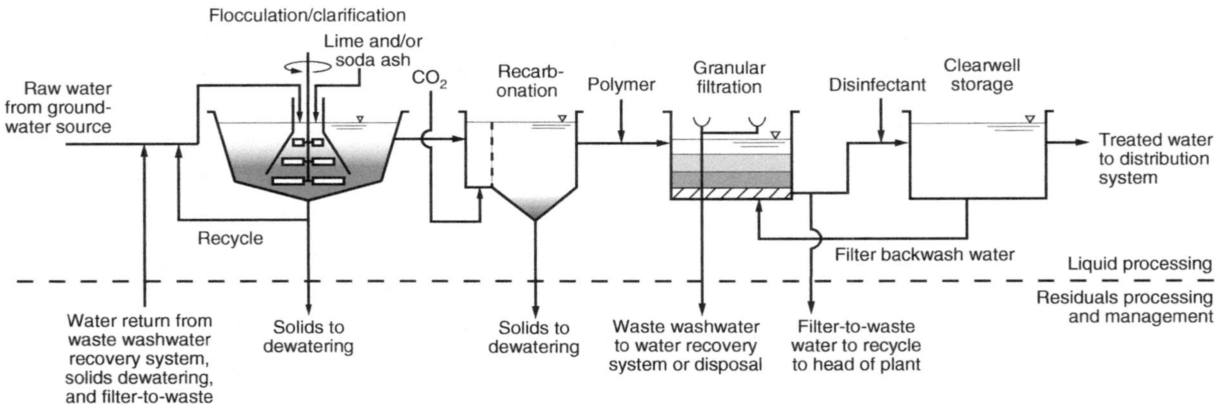

Figure 4-8
Typical process train for removal of hardness (calcium and magnesium) from groundwater with lime–soda softening process.

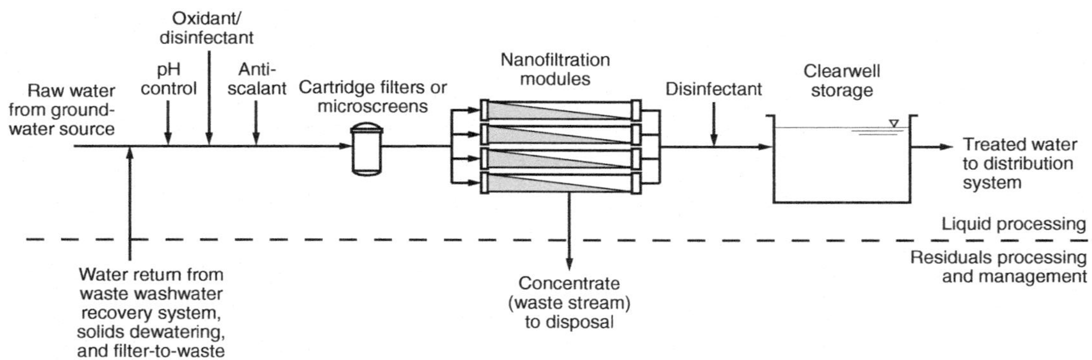

Figure 4-9
Typical process train for treatment of groundwater with nanofiltration softening process.

Air stripping

Air stripping is used to treat groundwater containing undesirable gases such as hydrogen sulfide or VOCs at levels above the MCL. An air-stripping process treatment train consists of a stripping tower followed by pressurized granular media filtration and disinfection, as shown on Fig. 4-10. In some cases the use of conditioning chemicals and/or acid addition may be necessary to prevent the formation of precipitates. A consideration when using air stripping is treating the off-gas from the stripping tower. One method for treatment of the off-gas is carbon adsorption, as shown on Fig. 4-10.

Iron and manganese treatment

Oxidation and precipitation on pressure filters is the process used most commonly for the removal of inorganic iron and manganese typically found in groundwaters with low dissolved oxygen, as shown on Fig. 4-11. Oxidants

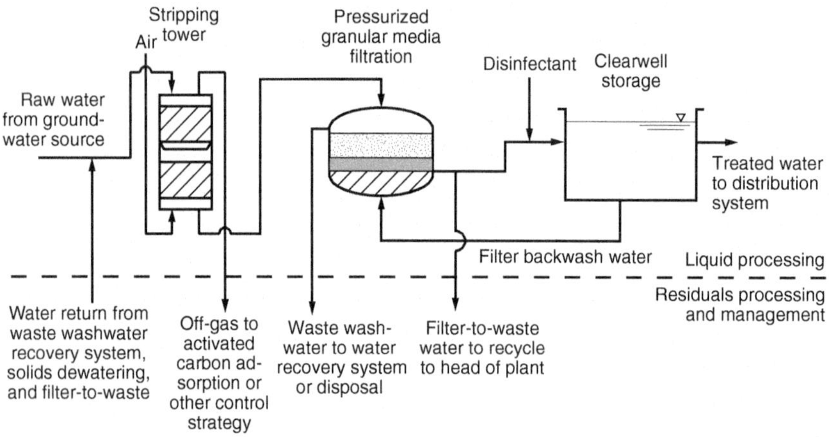

Figure 4-10
Typical process train for removal of dissolved gases and/or volatile constituents from groundwater.

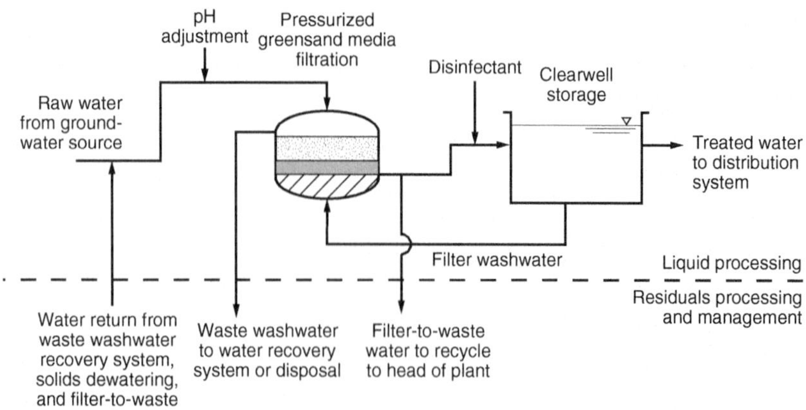

Figure 4-11
Typical process train for removal of iron and manganese from groundwater.

that are used include chlorine, potassium permanganate ($KMnO_4$), oxygen, and ozone. Chorine is used when the concentration of iron is less than 2 mg/L and little or no manganese is present. In many cases chlorine is not effective for the removal of iron because iron can be complexed with NOM. Consequently, the use of chlorine must be assessed with bench and pilot tests. In addition, the impact of chlorine on DBP formation must be considered before choosing chlorine as an oxidant. Potassium permanganate and ozone are used when both iron and manganese are present. Oxygen can be effective for oxidizing iron but is not able to break the NOM–iron complex. Pressure filters may use manganese greensand media, which is the name commonly used for sand having a high percentage of glauconite, as shown on Fig. 4-11.

Treatment Processes for Residuals Management

In the treatment processes presented on Figs. 4-4 through 4-11, the primary objective is to remove certain impurities from the water. Impurities removed during treatment, along with the added materials and transport water, are referred to as *residuals* and consist of liquid-, semisolid-, solid-, and gaseous-phase products. Typically, these residuals are comprised of the turbidity-causing materials in raw water, organic and inorganic solids, algae, bacteria, viruses, protozoa, colloids, precipitates from the raw water and those added in treatment, and dissolved salts. *Sludge* is the term used to refer to the solid, or liquid-solid, portion of some types of water treatment plant residuals, such as the underflow from sedimentation basins.

The planning, design, and operation of facilities to reuse or dispose of water treatment residuals is known as *residuals management.* The principal objective in residuals management is usually to minimize the amount of material that must ultimately be disposed by recovering recyclable materials and reducing the water content of the residuals. Typical residual management options are illustrated on Fig. 4-12. Residuals management can have an important impact on the design and operation of many water treatment plants. For existing plants, residuals management systems may limit overall plant capacity if not designed and operated properly. Frequently, residuals are stored temporarily in the process train before removal for treatment, recycle, and/or disposal. Residual removal must be optimized for the process train and coordinated with the residuals management systems to maintain water quality in the process train. Common unit processes used for residuals management are given in Table 4-11. The subject of residuals management is considered in detail in Chap. 21.

Hydraulic Sizing of Treatment Facilities and Processes

Along with selection of the treatment processes to achieve a specific treatment goal or goals, it is equally important to understand how large or of what capacity the individual treatment processes must be to meet the treated water requirements. General guidance on the hydraulic sizing of treatment processes is presented in Table 4-12. As reported in Table 4-12, the sizing of most treatment units in a water treatment facility is based on the peak-day demand at the end of the design period, with the hydraulic

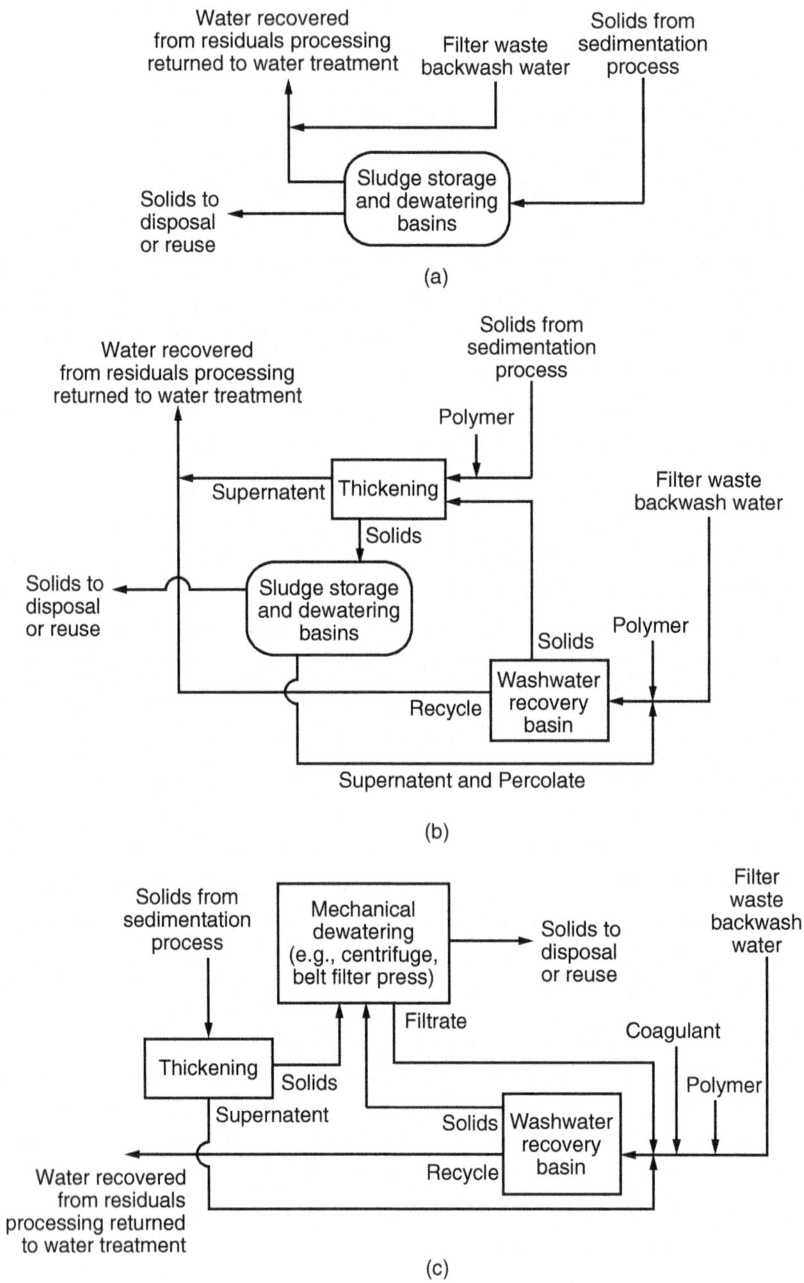

Figure 4-12
Typical residual processing options: (a) least mechanically intensive employing sludge storage and dewatering basins, typically used at small and remote treatment plants where adequate land is available; (b) more mechanically intensive employing sludge thickening, sludge storage and dewatering basins, and filter waste washwater recovery basins, typically used at intermediate size plants where adequate land is available; and (c) mechanically intensive employing sludge thickening, filter waste washwater recovery basins, and mechanical dewatering, typically used at larger water treatment plants.

Table 4-11
Typical unit processes used for residual management

Unit Process	Description	Typical Application in Water Treatment
Concentration	Reducing the volume of reject streams from reverse osmosis and ion exchange processes.	The reject stream from membrane processes is passed through additional membranes to further concentrate the reject stream and reduce the volume required for ultimate disposal.
Conditioning	Conditioning is used to improve the physical properties of the sludge so that it can be dewatered easily.	The addition of polymers or lime to alum or ferric thickened sludge from sedimentation basins and filter waste washwater clarifiers prior to mechanical dewatering.
Dewatering, mechanical	The objective is to reduce the sludge volume and produce a sludge that can be easily handled for further processing.	Thickened, conditioned sludge from sedimentation basins, waste washwater clarifiers, and lagoons is compressed using vacuum filters, filter presses, or belt filter presses.
Lime sludge pelletization	Lime sludge is formed into pellets.	Primarily used in the southeastern United States on high-calcium, warm-temperature groundwaters during the suspended-bed cold-softening water treatment process.
Thickening (sludge)	Thickening to increase the solids content of sludge involves the removal of excess water by decanting and the concentration of the solids by settling.	Sludge from sedimentation basins and clarifiers in the waste washwater recovery system is thickened using centrifuges, thickeners, lagoons, or sand drying beds.

capacity higher than the peak day to account for recycle and treated waste streams. Thus, it is very important to secure the best possible estimate of projected future population growth in the service area.

Information on the acceptable velocities in conveyance piping between unit processes is presented in Table 4-13. The information given in Table 4-13 along with the information given in Table 4-12 can be used to size the piping used to interconnect the various unit processes that comprise the treatment train. Special attention must be devoted to the piping that is used to connect flocculation facilities to downstream filters if floc breakup is to be avoided.

At the other end of the spectrum, a plant also needs to function at the low end of the anticipated flow range. Frequently, multiple basins and conveyance pipes are designed so that one or more units may be turned off when flows are low to maintain appropriate basin loading and pipe velocities to ensure treated water quality. Chemical feed pumps and storage facilities also need to be designed to cover the entire anticipated flow range and the dose range. The same approach of multiple facilities as well as pump turndown may be considered when sizing ancillary facilities such as chemical feed systems.

Table 4-12
Guidance for hydraulic sizing of treatment facilities and unit processes

Design Flow	Facility and Unit Processes	Value
Maximum hour	High-service pump station, depending on local conditions	$Q_{max\ h}$
	Water distribution reservoir in distribution system	$Q_{max\ h}$
Average day	Storage volume for one sludge lagoon	$365 \times Q_{avg\ d}$
	Unit processes with one unit out of service	$Q_{avg\ d}$ for $Q_{max\ month}/30$ d
	Bulk chemical storage	$Dose_{avg} \times 30\ d \times Q_{avg\ d}$ for $Q_{max\ yr}$
	Day tank for chemical feed	$Dose_{avg} \times 12h \times Q_{avgd}$ for $Q_{max\ yr}$
	Average capacity of chemical feeders	$Dose_{avg} \times Q_{avg\ d}$ for $Q_{max\ yr}$
Maximum day[a]	All treatment processes, including intake facilities[b]	$Q_{max\ d}$
	Plant hydraulic capacity (e.g., piping)	$1.25–1.50 \times Q_{max\ d}$
	Maximum capacity of chemical feeders	$Dose_{max} \times Q_{max\ d}$
	Sludge collection, pumping, and treatment facilities	$Q_{max\ d}$
	Clearwell capacity	$0.15–0.20 \times Q_{max\ d}$
	Low- and high-lift internal plant process pumps with largest pump out of service	$Q_{max\ d}$
	High-service pump station with largest pump out of service	$Q_{max\ d}$
	Maximum capacity of flowmeters[b]	$Q_{max\ d}$
Minimum day	Minimum capacity of chemical feeders	$Dose_{min} \times Q_{min\ d}$
	Lower capacity of flowmeters	$Q_{min\ d}$
	Minimum flow for recycle pumps	$Q_{min\ d}$

[a]The maximum day demand, $Q_{max\ d}$, is at the end of the design period.
[b]Rating the unit processes high is an alternative method for building in future plant capacity (e.g., design the flocculation basins for 15 min detention time, with the intent of operating at 20min detention time for the foreseeable future). The ancillary equipment, such as flowmeters, should be designed to match the rating for the unit processes.
Source: Adapted, in part, from Kawamura (2000).

Pilot Plant Studies

Where the applicability of a process for a given situation is unknown but the potential benefits of using the process are significant, bench- or pilot-scale tests must be conducted (see Fig. 1-3). The purpose of conducting pilot plant studies is to establish the suitability of the process in the treatment of a specific water under specific environmental conditions and to obtain the necessary data on which to base a full-scale design. Factors that should be considered in planning pilot plant studies for water treatment are presented in Table 4-14. The relative importance of the factors presented in Table 4-14 will depend on the specific application and the reasons for conducting

Table 4-13
Guidance for acceptable velocities in water treatment facility piping

	U.S. Customary Units		SI Units	
Piping Component	**Unit**	**Value**	**Unit**	**Value**
Distribution system	ft/s	4.0–10.0	m/s	1.25–3.0
Filter backwash main line	ft/s	6.0–9.0	m/s	1.8–2.75
Filter effluent line	ft/s	5.0–6.0	m/s	0.4–1.8
Line from floc basin (conventional rapid sand filter with alum floc)	ft/s	1.0–4.0	m/s	0.3–1.25
Line from floc basin (direct filtration with polymer)	ft/s	2.5–4.5	m/s	0.75–1.4
Pump discharge line	ft/s	6.0–9.0	m/s	1.8–2.75
Pump suction line	ft/s	4.0–6.0	m/s	1.2–1.8
Raw-water main	ft/s	6–8	m/s	1.8–2.5
Waste washwater line	ft/s	6.0–8.0	m/s	1.8–2.5

Source: Adapted, in part, from Kawamura (2000).

the testing program. For example, testing of UV disinfection systems is typically done to (1) verify manufacturers' performance claims, (2) quantify effects of water quality on UV performance, (3) assess the effect(s) of system and reactor hydraulics on UV performance, and (4) investigate photo reactivation and impacts.

In addition to the criteria in Table 4-14, certain general considerations are required in the design of pilot plant facilities that make these facilities useful in process evaluation and selection:

1. The equipment design is dictated by the anticipated use and objective of the pilot equipment. Permanently mounted trailer installations do offer some advantages; however, they tend to be somewhat bulky and inflexible. A modular approach provides the engineer with a reusable, flexible, and easily transportable configuration. In the module approach, each process module is self-contained; thus the experimental designer is free to vary the process train without concern for the interdependence of the various unit processes on each other.
2. At least two process trains are required to enable side-by-side comparison of various design parameters; otherwise no control is available for data evaluation comparison.
3. There should be adequate raw-water supply that mimics the supply for the full-scale application.

Table 4-14
Considerations in setting up pilot plant testing programs

Item	Consideration
Design of pilot testing program	Dependent variables including ranges Independent variables including ranges Time required Test facilities Test protocols Statistical design of data acquisition program
Nonphysical design factors	Available time, money, and labor Degree of innovation and motivation involved Quality of water or waste washwater Location of facilities Complexity of process Similar testing experience Dependent and independent variables
Pilot plant size	Bench- or laboratory-scale model Pilot-scale tests Full- (prototype) scale tests
Physical design factors	Scale-up factors Size of prototype Facilities and equipment required and setup Materials of construction
Reasons for conducting pilot testing	Test new process Simulation of another process Predict process performance Document process performance Optimize system design Satisfy regulatory agency requirements Satisfy legal requirements Verify performance claims made by manufacturer

Source: Adapted from Tchobanoglous et al. (2003).

4. Adequate bypass capabilities need to be provided around and within unit processes.
5. The effect of scale-up on system process performance should be considered (e.g., the "sidewall" effects in granular media filtration, hydrodynamics of gravity settling and mixing).
6. Flexibility in the operation should be maximized, especially with respect to multiple chemical addition points. In addition, provisions to add various types of chemicals are desirable.
7. All pumps and motors should be equipped with a variable-speed adjustment and rate control. For example, the speed of the flocculators should be adjustable to test the effect of flocculation energy input for various treatment schemes.

8. Positive flow splitting is best achieved by use of weirs or variable-speed rate-controlled pumps.
9. Accurate flowmeters are necessary.
10. If positive displacement (peristaltic) pumps are used, a pulsation dampener is needed to eliminate high and low surges created by the action of the pump.
11. Unlike full-scale facilities, the hydraulics of pilot plants should be designed to accommodate a wide range of flow conditions. For example, normal filtration rates in water treatment plants are on the order of 12 to 19 m/h (5 to 8 gpm/ft^2); however, higher filtration rate (e.g., 29 to 36 m/h or 12 to 15 gpm/ft^2) should be utilized in sizing the inlet and outlet piping of pilot equipment.
12. Provisions for the artificial injection of turbidity (or other water quality constituents) may be desirable to simulate periodic extreme raw-water quality conditions.
13. When more than one process train is utilized, an influent or transition structure may be required. Such a structure should include (1) a means of providing an even flow split to each process train, (2) a common location for the introduction of pretreatment chemicals and monitoring of influent water quality parameters, and (3) provisions for a strainer to remove gross debris such as leaves and twigs that could clog tubing.

Removal Efficiency and the Log Removal Value

The objective of treatment processes is to remove contaminants. Removal can be determined for bulk water quality measures (e.g., turbidity, total dissolved solids) or for individual constituents of interest (e.g., perchlorate, *Cryptosporidium* oocysts). The fraction of a constituent removed by a process can be calculated with the equation

$$R = 1 - \frac{C_e}{C_i} \qquad (4\text{-}1)$$

where R = removal expressed as a fraction, dimensionless
C_e = effluent concentration, mg/L
C_i = influent concentration, mg/L

In general, Eq. 4-1 is used where the removal efficiency for a given constituent is three orders of magnitude or less (i.e., 99.9%). For some constituents, such as microorganisms and trace organics, and some processes, such as membrane filtration, the concentration in the effluent is typically three or more orders of magnitude less than the influent concentration. For these situations, the removal is expressed in terms of log removal value (LRV) as given by the equation

$$\text{LRV} = \log(C_i) - \log(C_e) - \log\left(\frac{C_i}{C_e}\right) \qquad (4\text{-}2)$$

For example, if the influent and effluent concentrations of *E. coli* were 10^7 and 10^2 org/100 mL, respectively, the corresponding log removal value would be 5 [$\log(10^7/10^2) = 5$]. The log removal notation is used routinely to express the removals achieved with membrane filtration (Chap. 12) and for disinfection (Chap. 13).

4-6 Multiple-Barrier Concept

Pathogens, one of the most important targets of drinking water treatment, place special demands on the performance of the water treatment process because acute effects can result from short-term exposure. As a result, where pathogens are concerned, the reliability of the treatment train is especially important. To address this issue, public health engineers require that the water supply systems include multiple barriers to limit the presence of pathogens in the treated drinking water. Barriers might include source protection or additional treatment (Haas and Trussell, 1998). Additional security in the design and operation of the water distribution system is also helpful in protecting the treated water from further contamination. The multiple-barrier approach is not just a concept of redundancy. It can be shown that multiple barriers will increase the reliability of the system, even if the overall removal capability is not significantly different. The multiple-barrier concept is illustrated in Example 4-1.

Example 4-1 Effect of multiple barriers on reliability

A thought experiment can be used to illustrate the increased reliability associated with the use of multiple barriers. Consider two alternative treatment trains. Train 1 includes one unit process, which, when operating normally, reduces the target pathogen by six orders of magnitude (a 6 log reduction). Train 2 includes three independent unit processes in series, each of which, when operating normally, reduces the target pathogen by two orders of magnitude (a 2 log reduction in each step).

For the purpose of this analysis, assume that each of the four unit processes listed above fails to perform, at random, about 1 percent of the time and that when a unit process fails the removal it achieves is half of what it normally achieves. Use the following information estimate: (a) the overall removal for trains 1 and 2 when all the unit processes are operating normally and (b) for each train the frequency (in days per year) of various levels of removal assuming that process failures occur randomly. Present the results of the frequency in a summary table for various levels of removal.

Solution

1. Overall removal during normal operation:
   a. Train 1. Normal operation = 6 log removal.
   b. Train 2. Normal operation = 2 + 2 + 2 = 6 log removal.
2. Frequency of various levels of removal:
   a. Train 1:
      i. Provides 6 log removal 99 percent of time = 0.99 × 365 d= 361.35 d.
      ii. Provides 3 log removal 1 percent of time = 0.01 × 365 d = 3.65 d.
   b. Train 2:
      i. Provides 6 log removal when all three processes are operating normally = 0.99 × 0.99 × 0.99 × 365 d = 354 d.
      ii. Provides 5 log removal when two of the three processes are operating normally and one is in failure mode = 0.99 × 0.99 × 0.01 × 3 (failure mode combinations) × 365 d = 10.73 d.
      iii. Provides 4 log removal when one of the three processes is operating normally and two are in failure mode = 0.99 × 0.01 × 0.01 × 3 (failure mode combinations) × 365 d = 0.11 d = 2.6 h.
      iv. Provides 3 log removal when all three processes are in failure mode = 0.01 × 0.01 × 0.01 × 365 d = 0.00037 d = 32 s.
3. The results of this analysis are displayed in the following table:

Log Removal	Time of Operation During Typical Year, d	
	Train 1	Train 2
6	361.35	354.16
5		10.73
4		0.11
3	3.65	0.00037
Total	365.0	365.0

Comment

Referring to the data in the above table, it will be observed that the process train with multiple barriers (train 2) is much more robust, reducing the time in which the consumer is exposed to the poorest removal by 10,000-fold, from 3.65 d per year to 32 s per year (0.00037 d). The use of multiple barriers in treatment provides reduced exposure to the risks that are associated with process failure.

Problems and Discussion Topics

4-1 The U.S. EPA sets water treatment goals and standards. Discuss the differences between goals and standards and the differences between primary and secondary standards.

4-2 A water treatment plant design engineer is establishing the treated water quality goals for a new treatment plant. Describe the steps the engineer should take to be sure the goals that are established are in compliance with all current regulations. Identify other groups or individuals who would be valuable to have participate in the goal-setting process.

4-3 Drinking water regulations are continually evolving. Discuss how an engineer might approach the design of a new water treatment plant so that the plant is readily able to meet future regulations.

4-4 A drinking water treatment plant is located in an area that is prone to intense summer rainstorms. The source water for the plant is a river that begins high in the mountains and flows through land that is used for growing crops and grazing for cows before it reaches a lake that is the source for the plant. What constituents would be of concern in drinking water that would be expected to be in the source water and what types of unit processes would be appropriate to treat them?

4-5 Develop a plan for a pilot plant to evaluate the treatment alternatives identified in the previous problem. In the plan, include a process flow diagram, constituents to be evaluated (testing requirements), and operational information such as study duration.

4-6 Discuss how the multiple barrier concept, as it applies to the removal of pathogens, is at work in one of the process trains, to be selected by instructor, that are presented on Figs. 4-4 through 4-11 and discussed in Sec. 4-5.

4-7 Consider two alternative treatment trains. Train 1 includes two treatment processes, each of which, when operating normally, reduces the target pathogen by three orders of magnitude (a 3 log reduction). Train 2 includes three independent unit processes in series, each of which, when operating normally, reduces the target pathogen by two orders of magnitude (a 2 log reduction in each step). If each of the five unit processes listed above fails to perform, at random, about one percent of the time and if, when a unit process fails, the removal it achieves is half of what it normally achieves, estimate: (a) the overall removal for trains 1 and 2 when all the unit processes are operating normally and (b) for each train the frequency (in days per year) of various levels of removal assuming that process failures occur randomly.

4-8 Two alternative treatment trains are being considered. Train 1 includes three treatment processes, each of which, when operating normally, reduces the target pathogen by two orders of magnitude (a 2 log reduction). Train 2 includes four independent unit processes in series. Two of the processes reduce the target pathogen by two orders of magnitude (a 2 log reduction) Each of the other two processes reduce the target pathogen by one order of magnitude (a 1 log reduction in each step). If each of the seven unit processes listed above fails to perform, at random, about one percent of the time and if, when a unit process fails, the removal it achieves is half of what it normally achieves, estimate: (a) the overall removal for trains 1 and 2 when all the unit processes are operating normally and (b) for each train the frequency (in days per year) of various levels of removal assuming that process failures occur randomly.

References

AWWA (2011) *Water Quality and Treatment—A Handbook of Community Water Supplies*, 6th ed., James K. Edzwald, ed. McGraw-Hill, New York.

Clendening, L. (ed.) (1942) *Source Book of Medical History*, Paul B. Hoeber, New York.

Great Lakes Upper Mississippi River Board (2003) *Recommended Standards for Water Works (Ten State Standards)*, Health Research, Albany, NY.

Haas, C. N., and Trussell, R. R. (1998) "Frameworks for Assessing Reliability of Multiple, Independent Barriers in Potable Water Reuse," *J. Water Sci. Tech.*, **38**, 6, 1–8.

Jones, G., Sanks, R. L., Tchobanoglous, G., and Bosserman, B. (eds.) (2008) *Pumping Station Design*, Revised 3rd ed., Butterworth-Heinmann, Boston, MA.

Kawamura, S. (2000) *Integrated Design and Operation of Water Treatment Facilities*, 2nd ed., John Wiley & Sons, New York.

McKee, J., and Wolf, H. W. (1963) *Water Quality Criteria*, 2nd ed., Publication No. 3-A, California State Water Quality Control Board, Sacramento, CA.

McKee, J., and Wolf, H. W. (1971) *Water Quality Criteria*, 2nd ed., Publication No. 3-A, California State Water Quality Control Board, Sacramento, CA.

NAS (1977) *Drinking Water and Health*, National Academy of Sciences Safe Drinking Water Committee, National Academy of Sciences, Washington, DC.

NAS (1980) *Drinking Water and Health*, Vols. 2 and 3, National Academy of Sciences Safe Drinking Water Committee, National Academy Press, Washington, DC.

NAS and NAE (1972) *Water Quality Criteria*, National Academy of Sciences and National Academy of Engineering, Prepared for the U.S. Environmental Protection Agency, Washington, DC.

Patania Brown, N. (2002) Design Development, from Blank Paper to Pre-Design, presentation at the MWH Water Treatment Treasures Series, Las Vegas, NV.

Pontius, F. W., and Clark, S. W. (1999) "Drinking Water Quality Standards, Regulations, and Goals," Chap. 1, in R. D. Letterman (ed.), *Water Quality and Treatment: A Handbook of Community Water Supplies*, 5th ed., AWWA, McGraw-Hill, Inc., New York.

Public Law 95-190 (1977) *Safe Drinking Water Act Amendments of 1977.*

Public Law 96-63 (1979) *Safe Drinking Water Act Amendments of 1979.*

Public Law 96-502 (1980) *Safe Drinking Water Act Amendments of 1980.*

Public Law 99-339 (1986) *Safe Drinking Water Act Amendments of 1986.*

Public Law 100-572 (1988) *Safe Drinking Water Act Amendments of 1988.*

Public Law 104-182 (1996) *Safe Drinking Water Act Amendments of 1996.*

SCENIHR (2006) "The Appropriateness of Existing Methodologies to Assess the Potential Risks Associated with Engineered and Adventitious Products of Nanotechnologies," Scientific Committee on Emerging and Newly Identified Health Risks, European Commission, Health & Consumer Protection Directorate General, http://ec.europa.eu/health/ph_risk/committees/04_scenihr/docs/scenihr_o_003b.pdf.

Smith, T. (1893) A New Method for Determining Quantitatively the Pollution of Water by Fecal Bacteria, pp. 712–722 in *Thirteenth Annual Report,* New York State Board of Health, Albany, NY.

Snow, J. (1855) *On the Mode of Communication of Cholera,* 2nd ed., J. Churchill, London.

Tchobanoglous, G., Burton, F. L., and Stensel, H. D. (2003) *Wastewater Engineering: Treatment and Reuse,* 4th ed., Metcalf and Eddy, Inc., McGraw-Hill Book Co., New York.

Trussell, R. (2001) "Endocrine Disruptors and the Water Industry," *J. AWWA,* **93**, 2, 58–65.

U.S. EPA (1972) *Water Quality Criteria,* National Technical Advisory Committee to the Secretary of the Interior, Washington, DC.

U.S. EPA (1975) "National Interim Primary Drinking Water Regulations," *Fed. Reg.,* **40**, 248, 59566–59588.

U.S. EPA (1976a) *Quality Criteria for Water,* U.S. Environmental Protection Agency, Washington, DC.

U.S. EPA (1976b) "National Interim Primary Drinking Water Regulations; Promulgation of Regulations on Radionuclides," *Fed. Reg.,* **41**, 133, 28402–28409.

U.S. EPA (1979a) "National Secondary Drinking-Water Regulations; Final Rule," *Fed. Reg.,* **44**, 140, 42195–42202.

U.S. EPA (1979b) "National Interim Primary Drinking Water Regulations: Control of Trihalomethanes in Drinking Water; Final Rule," *Fed. Reg.,* **44**, 231, 68624–68707.

U.S.EPA (1986) "Primary and Secondary Drinking Water Regulations, Final Rule," *Fed. Reg.,* **51**, 63, 11396–11412.

U.S. EPA (1987) "Water Pollution Controls: National Primary Drinking Water Regulations; Volatile Synthetic Organic Chemicals: Para-Dichlorobenzene," *Fed. Reg.,* **52**, 12876–12883.

U.S. EPA (1989a) "National Primary Drinking Water Regulations: Filtration and Disinfection; Turbidity, *Giardia lamblia,* Viruses, *Legionella,* and Heterotrophic Bacteria; Final Rule," *Fed. Reg.,* **54**, 124, 27486–27541.

U.S. EPA (1989b) "National Primary Drinking Water Regulations: Total Coliforms; Final Rule," *Fed. Reg.,* **54**, 124, 27544–27568.

U.S. EPA (1991a) "National Primary Drinking Water Regulations: Synthetic Organic Chemicals (SOCs) and Inorganic Chemicals (IOCs) — Phase II; Final Rule," *Fed. Reg,* **56**, 30, 3526–3599.

U.S. EPA (1991b) "National Primary Drinking Water Regulations: Lead and Copper; Final Rule," *Fed. Reg.*, **56**, 110, 26460–26546.

U.S. EPA (1992) "National Primary Drinking Water Regulations: Synthetic Organic Chemicals and Inorganic Chemicals-Phase V, Final Rule," *Fed. Reg.*, **57**, 138, 31776–31000.

U.S. EPA (1996) "National Primary Drinking Water Regulations: Monitoring Requirements for Public Drinking Water Supplies or Information Collection Rule: Final Rule," *Fed. Reg.*, **61**, 94, 24354–24388.

U.S. EPA (1998a) "National Primary Drinking Water Regulations: Disinfectants and Disinfection Byproducts; Final Rule," *Fed. Reg.*, **63**, 241, 69389–69476.

U.S. EPA (1998b) "National Primary Drinking Water Regulations: Interim Enhanced Surface Water Treatment: Final Rule," *Fed. Reg.*, **63**, 241, 69478–69476.

U.S. EPA (1999) *25 Years of Safe Drinking Water Act: History and Trends*, EPA 816-R-99-007 U.S. EPA office of Water Washington, DC.

U.S. EPA (2000) "National Primary Drinking; Water Regulations: Radionuclides; Final Rule," *Fed. Reg.* **65**, 236, 76708–76753.

U.S. EPA (2001a) "National Primary Drinking Water Regulations: Filter Backwash Recycling Rule; Final Rule," *Fed. Reg*, **66**, 111, 31086–31105.

U.S. EPA (2001b) "National Primary Drinking Water Regulations: Arsenic and Clarifications to Compliance and New source Contaminants Monitoring; Final Rule," *Fed. Reg.*, **66**, 14, 6976–7066.

U.S. EPA (2002a) "National Primary Drinking Water Regulations: Long Term 1 Enhanced Surface Water Treatment Rule; Final Rule," *Fed. Reg.*, **67**, 9, 1812–1844.

U.S. EPA (2006a) "National Primary Drinking Water Regulations: Stage 2 Disinfectant and Disinfection Byproduct Rule, Final Rule," *Fed. Reg.*, **712**, 388–493.

U.S. EPA (2006b) "National Primary Drinking Water Regulations: Long Term 2 Enhanced Surface Water Treatment Rule," *Fed. Reg.*, **71**, 3, 653–702. (Note: rule has been amended a number of times since it was first published).

U.S. EPA (2006c) "National Primary Drinking Water Regulations: Ground Water Rule," *Fed. Reg.*, **71**, 216, 65574–67427.

U.S. PHS (1925) "Public Health Service Drinking Water Standards," *Pub. Health Rep.*, **40**, 693–721.

U.S. PHS (1942, published 1943) "Public Health Service Drinking Water Standards," *Pub. Health Rep.*, **58**, 3, 69–111.

U.S. PHS (1946) "Public Health Service Drinking Water Standards," *Pub. Health Rep.*, **61**, 11, 371–401,

U.S. PHS (1962) *Public Health Service Drinking Water Standards: 1962*, Publication 956, U.S. Public Health Service, U.S. Government Printing Office, Washington, DC, *Fed. Reg.*, **27**, 2152–2155.

U.S. PHS (1970) *Community Water Supply Survey: Significance of National Findings*, U.S. Govt Printing Office, Washington, D.C., available from NTIS PB 215198/BE, Springfield, VA.

U.S. Treasury Department (1914) "Bacterial Standard for Drinking Water," *Pub. Health Rep.*, **29**, 45, 2959–2966.

USEPA (2011) "U.S. EPA Terms of Environment: Glossary, Abbreviations, and Acronyms," accessed at www.epa.gov/OCEPATERMS/ on July 3, 2011.

WHO (1993) *Guidelines for Drinking-Water Quality, Vol. 1, Recommendations,* 2nd ed., World Health Organization, Geneva.

WHO (2006) *Guidelines for Drinking-Water Quality: Incorporating First Addendum, Vol. 1, Recommendations,* 3rd ed., World Health Organization, Geneva.

5 Principles of Chemical Reactions

Stantec's Water Treatment: Principles and Design, Updated Third Edition. John C. Crittenden, R. Rhodes Trussell, David W. Hand, Kerry J. Howe, and George Tchobanoglous.

Terminology for Chemical Reactions

Term	Definition
Acid	A molecule that is capable of releasing a proton.
Acid-base reactions	Reactions that involve the loss or gain of a proton. The solution becomes more acidic if the reaction produces a proton or basic if it consumes a proton. Acid/base reactions are reversible.
Activation energy	Energy barrier that reactants must exceed in order for the reaction to proceed as written.
Activity	Ability of an ion or molecule to participate in a reaction. In dilute solution, the activity is equal to the molar concentration. For ions in solution, the activity decreases as ionic strength increases.
Activity coefficient	Parameter that relates the concentration of a species to its activity.
Conjugate base	A molecule that can accept a proton and is formed when an acid releases a proton.
Conversion	Amount of a reactant that can be lost or converted to products, normally given as a moles fraction.
Catalyst	A species that Speeds up a chemical reaction, but is neither consumed nor produced by the reaction.
Complex	Species that is comprised of a metal ion and a ligand.
Elementary reaction	A chemical reaction in which products are formed directly from reactants without the formation of intermediate species.
Free energy	Thermodynamic energy in a system available to do chemical work. Associated with the potential energy of chemical reactions. Also known as the Gibbs energy.
Heterogeneous reaction	A chemical reaction in which the reactants are present in two or more phases (i.e., a liquid and a solid).
Homogeneous reaction	A chemical reaction in which all reactants are present in a single phase.

Term	Definition
Ionic strength	A measure of the total concentration of ions in solution. An increase in the ionic strength increases nonideal behavior of ions and causes activity to deviate from concentration.
Irreversible reaction	A chemical reaction that proceeds in the forward direction only, and proceeds until one of the reactants has been totally consumed.
Ligand	Anions that bind with a central metal ion to form soluble complexes. Common ligands include CN^-,OH^-, Cl^-, F^-, CO_3^{2-}, NO_3^-, SO_4^{2-}, and PO_4^{3-}.
Oxidant	A reactant that gains electrons in a oxidation/reduction reaction.
Oxidation/reduction reaction	A chemical reaction in which electrons are transferred from one molecule to another. Also known as a redox reaction. Redox reactions are irreversible.
Precipitation reaction	A chemical reaction in which dissolved species combine to form a solid. Precipitation reactions are reversible. The reverse is a dissolution reaction, in which a solid dissolved to form soluble species.
Parallel reactions	Reactions that involve the concurrent utilization of a reactant by multiple pathways.
Reaction order	The power to which concentration is raised in a reaction rate law.
Reaction rate law	Mathematical description of rate of reaction. It takes the form of a rate constant multiplied by the concentration of reactants raised to a power.
Reductant	A reactant that loses electrons in a oxidation/reduction reaction.
Reversible reaction	A chemical reaction that proceeds in either the forward or reverse direction, and reaches an equilibrium condition in which products and reactants are both present.
Selectivity	The preference of one reaction over another. Selectivity is equal to the moles of desired product divided by the moles of reactant that has reacted.
Series reactions	Individual reactions that proceed sequentially to generate products from reactants.
Stoichiometry	A quantitative relationship that defines the relative amount of each reactant consumed and each product generated during a chemical reaction.

Chemical reactions are used in water treatment to change the physical, chemical, and biological nature of water to accomplish water quality objectives. An understanding of chemical reaction pathways and stoichiometry is needed to develop mathematical expressions that can be used to describe the rate at which reactions proceed. Kinetic rate laws and reaction stoichiometry are valid regardless of the type of reactor under consideration and are used in the development of mass balances (see Chap. 6) to describe the spatial and temporal variation of reactants and products in chemical reactors. Understanding the equilibrium, kinetic, and mass transfer behavior of each unit process is necessary in developing effective treatment strategies. Equilibrium and kinetics are both introduced in this chapter, and mass transfer is discussed in Chap. 7.

Topics presented in this chapter include (1) chemical reactions and stoichiometry, (2) equilibrium reactions, (3) thermodynamics of chemical reactions, (4) reaction kinetics, (5) determination of reaction rate laws, and (6) chemical reactions used in water treatment. Water chemistry textbooks (Benefield et al., 1982; Benjamin, 2002; Pankow, 1991; Sawyer et al., 2003; Snoeyink and Jenkins, 1980; Stumm and Morgan, 1996) may be reviewed for more complete treatment of these concepts and other principles of water chemistry.

5-1 Chemical Reactions and Stoichiometry

Chemical operations used for water treatment are often described using chemical equations. These chemical equations may be used to develop the stoichiometry that expresses quantitative relationships between reactants and products participating in a given reaction. An introduction to the types of chemical reactions and reaction stoichiometry used in water treatment processes is presented below.

Types of Reactions

Chemical reactions commonly used in water treatment processes can be described in various ways. For example, the reactions of acids and bases, precipitation of solids, complexation of metals, and oxidation-reduction of water constituents are all important reactions used in water treatment. In general, reactions can be thought of as reversible and irreversible. Irreversible reactions tend to proceed to a given endpoint as reactants are consumed and products are formed until one of the reactants is totally consumed. Irreversible reactions are signified with an arrow in the chemical equation, pointing from the reactants to the products. Symbols commonly used in chemical equations are described in Table 5-1. In the following reaction, reactants A and B react to form products C and D:

$$A + B \rightarrow C + D \tag{5-1}$$

Reversible reactions tend to proceed, depending on the specific conditions, until equilibrium is attained at which point the formation of products from

Table 5-1
Symbols used in chemical equations

Symbol	Description	Comments
$\rightarrow$	Irreversible reaction	Single arrow points from the reactants to the products, e.g., $A + B \rightarrow C$
$\rightleftarrows$	Reversible reaction	Double arrows used to show that the reaction proceeds in the forward or reverse direction, depending on the solution characteristics
[ ]	Brackets	Concentration of a chemical constituent or compound in mol/L
{ }	Braces	Activity of a chemical constituent or compound
(s)	Solid phase	Used to designate chemical component present in solid phase, e.g., precipitated calcium carbonate, $CaCO_3(s)$
(l)	Liquid phase	Used to designate chemical component present in liquid phase, e.g., liquid water, $H_2O(l)$
(aq)	Aqueous (dissolved)	Used to designate chemical component dissolved in water, e.g., ammonia in water, $NH_3(aq)$
(g)	Gas	Used to designate chemical component present in gas phase, e.g., chlorine gas, $Cl_2(g)$
$\overset{x}{\rightarrow}$	Catalysis	Chemical species, represented by x, catalyzes reaction, e.g., cobalt (Co) is the catalyst in the reaction $SO_3^{2-} + \frac{1}{2}O_2 \overset{Co}{\rightarrow} SO_4^{2-}$
$\uparrow$	Volatilization	Arrow directed up following a component is used to show volatilization of given component, e.g., $CO_3^{2-} + 2H^+ \rightleftarrows CO_2(g)\uparrow + H_2O$
$\downarrow$	Precipitation	Arrow directed down following a component is used to show precipitation of given component, e.g., $Ca^{2+} + CO_3^{2-} \rightleftarrows CaCO_3(s)\downarrow$

Source: Adapted from Benefield et al., 1982.

the forward reaction is equal to the loss of products for the reverse reaction. For example, in Eq. 5-1 the reactants A and B react to form products C and D, whereas in Eq. 5-2 the reactants C and D react to form products A and B:

$$C + D \rightarrow A + B \tag{5-2}$$

The reactions presented in Eqs. 5-1 and 5-2 can be combined as follows:

$$A + B \rightleftarrows C + D \tag{5-3}$$

Theoretically, all reactions are reversible given the appropriate conditions; however, under the limited range of conditions typically experienced in water treatment processes, some reactions may be classified as irreversible for practical purposes.

HOMOGENEOUS REACTIONS

When all the reactants and products are present in the same phase, the reactions are termed *homogeneous*. For homogeneous reactions occurring in water, the reactants and products are dissolved. For example, the reactions of chlorine (liquid phase) with ammonia (liquid phase) and dissolved organic matter (liquid phase) are common homogeneous reactions.

HETEROGENEOUS REACTIONS

When reacting materials composed of two or more phases are involved, the reactions are termed *heterogeneous*. The use of ion exchange media (solid phase) for the removal of dissolved constituents (liquid phase) from water is an example of a heterogeneous reaction used in water treatment. Reactions that require the use of a solid-phase catalyst may also be considered heterogeneous.

Reaction Sequence

An understanding of the sequence of reaction steps is needed for engineering and control of reactions in water treatment reactors. Chemical reactions in water treatment can occur via a single reaction step or multiple steps in a sequential manner. In addition, reactions may occur in series or parallel or in a combination of series and parallel reactions. Due to the diverse chemistry of water originating from surface and subsurface sources, many reactions occur during water treatment processes.

SERIES REACTIONS

The conversion of a reactant to a product through a stepwise process of individual reactions is known as a series reaction. For example, reactant A forms product B, which in turn reacts to form product C:

$$A \rightarrow B \rightarrow C \tag{5-4}$$

For example, the two-step conversion of carbonic acid (H_2CO_3) to carbonate (CO_3^{2-}) takes place in water according to the following series reaction:

$$H_2CO_3 \rightleftarrows HCO_3^- + H^+ \tag{5-5}$$

$$HCO_3^- \rightleftarrows CO_3^{2-} + H^+ \tag{5-6}$$

The extent and rate of the reactions shown in Eqs. 5-5 and 5-6 are determined by the water pH, temperature, and other properties, as discussed later in this chapter.

PARALLEL REACTIONS

Reactions that involve the concurrent utilization of a reactant by multiple pathways are known as parallel reactions. Parallel reactions may be thought of as competing reactions. In the reactions shown in Eqs. 5-7 and 5-8, reactant A is simultaneously converted to products B and C:

$$\mathrm{A} \rightarrow \mathrm{B} \tag{5-7}$$

$$\mathrm{A} \rightarrow \mathrm{C} \tag{5-8}$$

When there are competing parallel reactions such as those shown in Eqs. 5-7 and 5-8, there is often a preferred reaction. The preference of one reaction over another is known as reaction selectivity. For example, if Eq. 5-7 were the preferred reaction over Eq. 5-8 due to the undesirable nature of product C, product B would be the desired product, and the selectivity would be defined as

$$S = \frac{\text{moles of desired product formed, [B]}}{\text{moles of all products formed, [B]} + \text{[C]}} \tag{5-9}$$

where S = selectivity, dimensionless

MULTIPLE REACTIONS

Many reactions in water treatment involve complex combinations of series and parallel reactions, as shown in the following reactions:

$$\mathrm{A} + \mathrm{B} \rightarrow \mathrm{C} \tag{5-10}$$

$$\mathrm{A} + \mathrm{C} \rightarrow \mathrm{D} \tag{5-11}$$

For example, the reaction of ozone (O_3) with bromide ions (Br^-)in groundwater occurs by the following three-step process:

$$\mathrm{O_3} + \mathrm{Br^-} \rightarrow \mathrm{OBr^-} \tag{5-12}$$

$$\mathrm{OBr^-} + \mathrm{O_3} \rightarrow \mathrm{BrO_3^-} \tag{5-13}$$

$$\mathrm{BrO_2^-} + \mathrm{O_3} \rightarrow \mathrm{BrO_3^-} \tag{5-14}$$

In this series of reactions, ozone converts bromide to bromate (BrO_3^-), which can be a health concern. Reactions involving ozone are discussed in more detail in Chaps. 8, 13, and 18.

Reaction Mechanisms

Many reactions proceed as a series of simple reactions between atoms, molecules, and radical species. A radical species is an atom or molecule containing an unpaired electron, giving it unusually fast reactivity. A radical species is always expressed with a dot in the formula (e.g., HO•). Intermediate products are formed during each step of a reaction leading up to the final products. An understanding of the mechanisms of a reaction may be used to improve the design and operation of water treatment processes.

ELEMENTARY REACTIONS

Reaction mechanisms involving an individual reaction step are known as elementary reactions. Elementary reactions are used to describe what is happening on a molecular scale, such as the collision of two reactants. For example, the decomposition of ozone (in organic-free, distilled water) has been described by the following four-step process (McCarthy and Smith, 1974):

$$O_3 + H_2O \rightarrow HO_3^+ + OH^- \tag{5-15}$$

$$HO_3^+ + OH^- \rightarrow 2HO_2 \tag{5-16}$$

$$O_3 + HO_2 \rightarrow HO\bullet + 2O_2 \tag{5-17}$$

$$HO\bullet + HO_2 \rightarrow H_2O + O_2 \tag{5-18}$$

In this series of elementary reactions, ozone reacts with water to form, among other compounds, HO• (hydroxyl radical) and HO_2 (superoxide), which are very reactive and sometimes used for the destruction of organic compounds.

OVERALL REACTIONS

A series of elementary reactions may be combined to yield an overall reaction. The overall reaction is determined by summing the elementary reactions and canceling out the compounds that occur on both sides of the reaction. For the elementary reactions shown in Eqs. 5-15 to 5-18, the overall reaction may be written as

$$2O_3 \rightarrow 3O_2 \tag{5-19}$$

The specific reaction mechanism and intermediate products that are formed cannot be determined from the overall reaction sequence. In many cases the elementary reaction mechanisms are not known and empirical expressions must be developed to describe the reaction kinetics.

Reaction Catalysis

A catalyst speeds up a chemical reaction, but it is neither consumed nor produced by the reaction. For a reaction between two molecules to occur, the molecules must collide with the proper orientation. However, molecules have a tendency to move in ways that make the proper orientation less likely. For example, molecules move about their axis in two directions (called a rotation and a translation) and they vibrate. Adsorption and reaction on a catalyst surface reduce this motion and increase the local concentration of reactant.

Catalysts may be homogeneous or heterogeneous in nature. Homogeneous catalysts are dissolved in solution and speed up homogeneous reactions. For example, cobalt, a homogeneous catalyst, is known to speed up the following reaction, which is used to deoxygenate water for oxygen transfer studies (Pye, 1947):

$$SO_3^{2-} + \tfrac{1}{2}O_2 \xrightarrow{Co} SO_4^{2-} \tag{5-20}$$

Example 5-1 Reactions for dissolution of carbon dioxide in water

The dissolution of carbon dioxide in water leads to the formation of several different components. Combine the following elementary reactions to determine the overall reaction with the initial product CO_2 and the final product of CO_3^{2-}:

$$
\begin{aligned}
CO_2(g) &\rightleftarrows CO_2(aq) \\
CO_2(aq) + H_2O &\rightleftarrows H_2CO_3 \\
H_2CO_3 &\rightleftarrows HCO_3^- + H^+ \\
HCO_3^- &\rightleftarrows CO_3^{2-} + H^+
\end{aligned}
$$

Solution

1. Eliminate species that occur on both sides of the elementary reaction equations:

$$
\begin{aligned}
CO_2(g) &\rightleftarrows \cancel{CO_2(aq)} \\
\cancel{CO_2(aq)} + H_2O &\rightleftarrows \cancel{H_2CO_3} \\
\cancel{H_2CO_3} &\rightleftarrows \cancel{HCO_3^-} + H^+ \\
\cancel{HCO_3^-} &\rightleftarrows CO_3^{2-} + H^+
\end{aligned}
$$

2. Determine the overall reaction by combining the remaining species from step 1:

$$CO_2(g) + H_2O \rightleftarrows 2H^+ + CO_3^{2-}$$

Heterogeneous catalysts speed up reactions at the interface of a liquid or gas with a solid phase, even if all reactants and products are in a single phase. If the products and reactants are not adsorbed too strongly, reactions at a surface can increase the rate of reaction, which demonstrates the utility of heterogeneous catalysis. Another purpose of catalysis is to improve reaction selectivity and minimize the formation of harmful by-products.

Reaction Stoichiometry

The amount of a substance entering into a reaction and the amount of a substance produced are defined by the stoichiometry of a reaction. In the general equation for a chemical reaction, as shown in Eq. 5-21, reactants A and B combine to yield products C and D:

$$aA + bB \rightleftarrows cC + dD \tag{5-21}$$

where a, b, c, d = stoichiometric coefficients, unitless

Using the stoichiometry of a reaction and the molecular weight of the chemical species, it is possible to predict the theoretical mass of reactants and products participating in a reaction. For example, calcium hydroxide [$Ca(OH)_2$] may be added to water to remove calcium bicarbonate:

$$Ca(HCO_3)_2 + Ca(OH)_2 \rightleftarrows 2CaCO_3(s)\downarrow + 2H_2O \qquad (5\text{-}22)$$

As shown in Eq. 5-22, 1 mole of $Ca(HCO_3)_2$ and 1 mole of $Ca(OH)_2$ react to form 2 moles of $CaCO_3(s)$ and 2 moles of H_2O. The molecular weights can be used to determine the theoretical mass of calcium hydroxide needed to react with a specified mass of calcium bicarbonate and the amount of calcium carbonate formed, as shown in Example 5-2.

Example 5-2 Determination of product mass using stoichiometry

For the reaction shown in Eq. 5-22, estimate the amount of $CaCO_3(s)$ that will be produced from the addition of calcium hydroxide to water containing 50 mg/L $Ca(HCO_3)_2$. Use a flow rate of 1000 m^3/d and determine the quantity of $CaCO_3(s)$ in kilograms per day. Assume that the reaction proceeds in the forward direction to completion.

Solution

1. Write the chemical equation and note the molecular weight of the reactants and products involved in the reaction. The molecular weights are written below each species in the reaction.

$$\underset{162}{Ca(HCO_3)_2} + \underset{74}{Ca(OH)_2} \rightleftarrows \underset{2\times100}{2CaCO_3(s)}\downarrow + \underset{2\times18}{2H_2O}$$

2. Determine the molar relationship for the disappearance of $Ca(HCO_3)_2$ and formation of $CaCO_3(s)$:

$$\left[\frac{2\text{ mol }CaCO_3(s)}{1\text{ mol }Ca()(HCO_3)_2}\right]\left[\frac{100\text{ g }CaCO_3(s)}{\text{mol }CaCO_3(s)}\right]\left[\frac{1\text{ mol }Ca(HCO_3)_2}{162\text{ g }Ca(HCO_3)_2}\right]$$
$$= 1.23\frac{\text{g }CaCO_3(s)}{\text{g }Ca(HCO_3)_2}$$

Therefore, for each gram of $Ca(HCO_3)_2$ removed, 1.23 g of $CaCO_3(s)$ will be produced.

3. Compute the mass of $CaCO_3(s)$ that will be produced each day.
   a. Determine the mass of $Ca(HCO_3)_2$ removed each day:

$$Ca(HCO_3)_2\text{ removed} = (0.050\text{ g/L})(1000\text{ m}^3\text{/d})(1000\text{ L/m}^3)$$
$$= 50{,}000\text{ g/d}$$

b. Estimate the amount of $CaCO_3$(s) produced each day:

$$CaCO_3(s)\text{ produced} = [50{,}000\text{ g }Ca(HCO_3)_2/d]$$
$$\times [1.23\text{ g }CaCO_3(s)]/[\text{g }Ca(HCO_3)_2](1\text{ kg}/10^3\text{ g})$$
$$= 61.5\text{ kg }CaCO_3(s)/d$$

Comment

In addition to estimating the amount of $CaCO_3$(s) produced, it is also possible to estimate the amount of calcium hydroxide that must be added to water to bring about this reaction. However, due to the nonideal nature of water treatment processing, the amount of calcium hydroxide that is required will exceed the stoichiometric amount, which is the minimum amount needed.

Reactant Conversion

As a reaction proceeds, reactants are converted into products. At any intermediate point during the reaction or when the reaction has reached equilibrium, it is possible to determine the amount (in moles) of reactants and products remaining if the stoichiometry and the amount of one of the reactants present is known. For example, consider the reaction shown in Eq. 5-21, in which *a*, *b*, *c*,and *d* are stoichiometric coefficients. For this reaction, the conversion may be determined for a reference reactant A and written per mole of A by dividing by the stoichiometric coefficient *a*:

$$A + \frac{b}{a}B \rightleftarrows \frac{c}{a}C + \frac{d}{a}D \tag{5-23}$$

For the general reaction shown in Eq. 5-23, all the reactants and products can be related to the conversion of reactant A, X_A, and the initial concentration of A, assuming there is no volume change upon reaction (which is valid for most water treatment problems):

$$X_A = \frac{\text{moles of A reacted}}{\text{moles of A present initially}} = \frac{N_{A0} - N_A}{N_{A0}} \tag{5-24}$$

where X_A = conversion of reactant A
N_{A0} = initial amount of reactant A, mol
N_A = final amount of reactant A, mol

Equation 5-24 can be written in molar concentration units by dividing each term by the volume in which the reaction is occurring. Thus, Eq. 5-24 written in concentration units is

$$X_A = \frac{C_{A0} - C_A}{C_{A0}} \tag{5-25}$$

where C_{A0} = initial concentration of reactant A, mol/L
C_A = final concentration of reactant A, mol/L

If the final amount and concentration of A, N_A, and C_A are written in terms of the conversion, the following expressions are obtained:

$$N_A = N_{A0}(1 - X_A) \tag{5-26}$$

$$C_A = C_{A0}(1 - X_A) \tag{5-27}$$

For the reaction given in Eq. 5-23, the final concentrations of B, C, and D can be computed in terms of A. The final amount of B, C, and D are determined by subtracting the product of moles of A reacted and the stoichiometric ratio of B, C, and D to A from the initial moles of B, C, and D, as shown by the following expressions. The final amount of B written in terms of moles is shown below.

$$N_B = N_{B0} - \frac{b}{a} X_A N_{A0} \tag{5-28}$$

where N_B = final amount of reactant B, mol
N_{B0} = initial amount of reactant B, mol

The final amount of B in terms of concentration is

$$\begin{aligned} C_B &= \frac{N_B}{V} = C_{B0} - \frac{b}{a} X_A C_{A0} \\ &= C_{B0} - \frac{b}{a} \frac{C_{A0} - C_A}{C_{A0}} C_{A0} = C_{B0} - \frac{b}{a}(C_{A0} - C_A) \end{aligned} \tag{5-29}$$

where C_B = final concentration of reactant B, mol/L
V = solution volume, L
C_{B0} = initial concentration of reactant B, mol/L

Similarly, for reactants C and D,

$$C_C = C_{C0} + \frac{c}{a}(C_{A0} - C_A) \tag{5-30}$$

$$C_D = C_{D0} + \frac{d}{a}(C_{A0} - C_A) \tag{5-31}$$

where C_C, C_D = final concentration of reactants C and D, mol/L
C_{C0}, C_{D0} = initial concentration of reactants C and D, mol/L

The final concentration of the various species are related to one another and to the conversion, as summarized in Table 5-2. As illustrated on Fig. 5-1, the addition of a catalyst (or other change in the reaction conditions) may improve the selectivity and the reaction conversion for a given time. The conversion from reactant to product can eventually reach the thermodynamic limit of the reaction, as discussed in the following section.

Table 5-2
Final concentration of various species related to one another and to conversion

Initial Amount Present, mol	Change in Initial Amount, mol	Final Amount Present, mol	Final Concentration, mol/L
N_{A0}	$-N_{A0}X_A$	$N_A = N_{A0}(1 - X_A)$	$C_A = C_{A0}(1 - X_A)$
N_{B0}	$-\frac{b}{a}X_A N_{A0}$	$N_B = N_{B0} - \frac{b}{a}X_A N_{A0}$	$C_B = C_{B0} - \frac{b}{a}(C_{A0} - C_A)$
N_{C0}	$\frac{c}{a}X_A N_{A0}$	$N_C = N_{C0}\frac{c}{a}X_A N_{A0}$	$C_C = C_{C0} + \frac{c}{a}(C_{A0} - C_A)$
N_{D0}	$\frac{d}{a}X_A N_{A0}$	$N_D = N_{D0} + \frac{d}{a}X_A N_{A0}$	$C_D = C_{D0} + \frac{d}{a}(C_{A0} - C_A)$

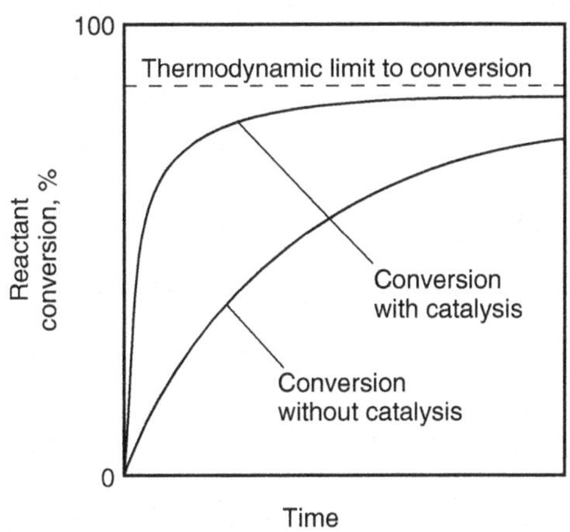

Figure 5-1
Improved reactant conversion with addition of catalyst.

5-2 Equilibrium Reactions

As discussed previously in this chapter, many of the reactions of significance in water treatment processes are reversible reactions. In other words, reactions such as that shown in Eq. 5-3 will not usually achieve complete conversion of reactants to products but instead will reach a state of dynamic equilibrium. Dynamic equilibrium is characterized by a balance between the continuous formation of products from reactants and reactants from products. If there is a change or stress to the system that affects the balance, the amount of reactants and products present will change to accommodate the stress. This concept is known as Le Chatelier's principle, which states that a reaction at equilibrium shifts in the direction that reduces a stress to the reaction. For example, in Eq. 5-21 if constituent A is removed from the system, the equilibrium will shift to form more A. In a chemical system, the difference between the actual state and the equilibrium state is the driving force used to accomplish treatment objectives.

Equilibrium Constants

When chemical reactions come to a state of equilibrium, the numerical value of the ratio of the concentration of the products over the concentration of the reactants all raised to the power of the corresponding stoichiometric coefficients is known as the equilibrium constant (K_c) and, for the reaction shown in Eq. 5-21, is written as

$$\frac{[C]^c[D]^d}{[A]^a[B]^b} = K_c \tag{5-32}$$

where K_c = equilibrium constant (subscript c used to signify equilibrium constant based on species concentration)
[ ] = concentration of species, mol/L
a, b, c, d = stoichiometric coefficients of species A, B, C, D, respectively

For example, the ionization of carbonic acid, given previously as Eq. 5-5, is shown as

$$H_2CO_3 \rightleftarrows HCO_3^- + H^+$$

The equilibrium constant at 25°C (neglecting nonidealities) for the reaction shown above may be written as

$$\frac{[H^+][HCO_3^-]}{[H_2CO_3]} = K_c = 5.0 \times 10^{-7} \tag{5-33}$$

The value of equilibrium constants and reactant and product concentrations are typically small and, therefore, are often reported in the literature using the operand "p," which is defined as

$$p[i] = -\log_{10}[i] \tag{5-34}$$

where $[i]$ = concentration of species i, mol/L

The reporting of the hydrogen ion activity as pH is a familiar example of the p-notation. Similarly, an equilibrium constant K may be reported as pK, which is defined as

$$pK = -\log_{10} K \tag{5-35}$$

Therefore, the K_c reported in Eq. 5-33 may be written as

$$pK_c = -\log_{10} K_c = -\log_{10}\left(5.0 \times 10^{-7}\right) = 6.3 \tag{5-36}$$

Ionic Strength

In dilute solutions, the ions present behave independently of each other. However, as the concentration of ions in solution increases, the activity of the ions decreases because of ionic interaction. The ionic strength may be

Example 5-3 Dependence of chemical species on pH

A drinking water contains hypochlorous acid (HOCl). Using the following relationship, determine the ratio of the hypochlorite ion (OCl^-) to HOCl at (a) pH 7.0 and (b) pH 8.0 (neglecting nonidealities):

$$HOCl \rightleftarrows OCl^- + H^+$$

The equilibrium constant K_c for the dissociation of HOCl into OCl^- and H^+ (also known as an acid dissociation constant and typically reported as K_a) is $10^{-7.5}$ ($pK_a = 7.5$).

Solution

1. Write the equilibrium relationship for the equation provided in the problem statement

$$\frac{[H^+][OCl^-]}{[HOCl]} = K_a = 10^{-7.5}$$

2. Determine the ratio of $[OCl^-]$ to [HOCl] at the given pH values.
   a. At pH 7.0, the hydrogen concentration $[H^+]$ is equal to 10^{-7} and the equilibrium relationship is written as

$$\frac{(10^{-7})[OCl^-]}{[HOCl]} = 10^{-7.5} \quad \frac{[OCl^-]}{[HOCl]} = 10^{-0.5} = 0.32$$

   b. At pH 8.0, the hydrogen concentration $[H^+]$ is equal to 10^{-8} and the equilibrium relationship is written as

$$\frac{(10^{-8})[OCl^-]}{[HOCl]} = 10^{-7.5} \quad \frac{[OCl^-]}{[HOCl]} = 10^{0.5} = 3.2$$

Comment

As shown in the calculations above, the solution pH can have a significant impact on the chemical species present. As shown in Chapter13, HOCl is a more effective disinfectant than OCl^- and is formed when chlorine is added to water. Consequently, it will be important to keep the pH 7 or less to achieve the greatest level of disinfection for a given dose of chorine.

For a given reaction, the value of the equilibrium constant, expressed in terms of concentration, will depend on the temperature and ionic strength of the solution. It should be noted that the equilibrium condition shown in Eq. 5-32 is based on the concentration of the chemical species involved in the reaction and may need to be adjusted for ionic activity, as discussed below.

determined using the equation (Lewis and Randall, 1921)

$$I = \frac{1}{2}\sum_{i} C_i Z_i^2 \tag{5-37}$$

where I = ionic strength of solution, mol/L(M)
C_i = concentration of species i, mol/L(M)
Z_i = number of replaceable hydrogen atoms or their equivalent (for oxidation–reduction reactions, Z is equal to the change in valence)

If the concentration of individual species is not known, the ionic strength may be estimated from the total dissolved solids concentration using the correlation (Stumm and Morgan, 1996)

$$I = \left(2.5 \times 10^{-5}\right)(\text{TDS}) \tag{5-38}$$

where TDS = total dissolved solids, mg/L

To account for nonideal conditions encountered due to ion–ion interactions (e.g., at high ionic strength), an effective concentration term called "activity" is used.

Activity and Activity Coefficients

The activity of a substance is defined by the standard state conditions of the substance and is based on commonly used standard conditions. The standard reference conditions for zero free energy are defined as 1 atm of pressure at a temperature of 298.15 K (25°C), elements in their lowest energy level (e.g., O_2 as a gas, carbon as graphite), and 1 *molal* hydrogen ion (1 mole of hydrogen ion per 1000 g of water). Some recent chemical references use 1 bar rather than 1 atm as the standard state, but the difference is small (1 atm = 1.01325 bar). Nonetheless, when looking up values for free energy in reference tables, note whether 1 atm or 1 bar is used for the standard state. The activity coefficient of a chemical in water may be determined as discussed below.

For ions and molecules in solution,

$$\{i\} = \gamma_i[i] \tag{5-39}$$

where $\{i\}$ = activity or effective concentration of ionic species, mol/L(M)
γ_i = activity coefficient for ionic species
$[i]$ = concentration of ionic species in solution, mol/L(M)

In general, γ_i is greater than 1.0 for nonelectrolytes and less than 1.0 for electrolytes. As the solution becomes dilute (applicable to most applications in water treatment), γ_i approaches 1 and $\{i\}$ approaches $[i]$. In the dilute aqueous solutions normally encountered in water treatment, activity coefficients are assumed to be equal to 1.

For pure solids or liquids in equilibrium with a solution $\{i\} = 1$, and for gases in equilibrium with a solution, the activity of species i is

$$\{i\} = \gamma_i P_i \tag{5-40}$$

where $\{i\}$ = activity or effective gas pressure, atm
P_i = partial pressure of i, atm

When reactions take place at atmospheric pressure (actually, much less than its critical pressure), the activity of a gas is equal to its partial pressure in atmospheres and the activity coefficient is 1.0.

For solvents or miscible liquids in a solution,

$$\{i\} = \gamma_i x_i \tag{5-41}$$

where x_i = mole fraction of species i

As the solution becomes more dilute, γ_i approaches 1. As stated above, the activity coefficient generally is assumed to be 1 for the dilute solutions, which are typical in water treatment.

When a species in water is an electrolyte, the activity should be considered but is usually ignored in routine calculations. The activity coefficient for electrolytes in solution with ionic strength less than 0.005 M may be estimated from the Debye-Hückel limiting law (Debye and Hückel, 1923):

$$\log_{10} \gamma_i = -AZ_i^2 I^{1/2} \tag{5-42}$$

where A = constant equal to 0.51 at 25°C (Stumm and Morgan, 1996)

For more concentrated solutions up to $I \leq 0.1$ M, the following modification of the Debye-Hückel equation, known as the Davies equation, can be applied with acceptable error (Davies, 1967):

$$\log_{10} \gamma_i = -AZ_i^2 \left(\frac{I^{1/2}}{1 + I^{1/2}} - 0.3I \right) \tag{5-43}$$

where A = constant (see Eq. 5-44)

The Davies equation is typically in error by 1.5 percent and 5 to 10 percent at ionic strengths between 0.1 and 0.5 M, respectively (Levine, 1988).

The constant A in Eq. 5-43 depends on temperature and can be estimated from the equation (Stumm and Morgan, 1996; Trussell, 1998)

$$A = 1.29 \times 10^6 \frac{\sqrt{2}}{(D_\varepsilon T)^{1.5}} \tag{5-44}$$

where T = absolute temperature, K (273 + °C)
D_ε = dielectric constant (see Eq. 5-45)

The dielectric constant may be determined using the equation (Harned and Owen, 1958)

$$D_\varepsilon \cong 78.54\{1-[0.004579(T-298)]+[11.9\times 10^{-6}(T-298)^2] + [28\times 10^{-9}(T-298)^3]\} \tag{5-45}$$

where T = absolute temperature, K(273 + °C)

Therefore, the constant A for water at 0, 15, and 25 °C is 0.49, 0.5, and 0.51, respectively.

The equilibrium relationship shown in Eq. 5-32 may now be expressed in terms of activities:

$$\frac{(\gamma_c[\mathrm{C}])^c(\gamma_d[\mathrm{D}])^d}{(\gamma_a[\mathrm{A}])^a(\gamma_b[\mathrm{B}])^b} = \frac{\{\mathrm{C}\}^c\{\mathrm{D}\}^d}{\{\mathrm{A}\}^a\{\mathrm{B}\}^b} = K \tag{5-46}$$

where K = equilibrium constant based on ionic activity (note absence of the subscript to signify activity basis)

The corresponding equilibrium for Eq. 5-23 is

$$\frac{\{\mathrm{C}\}^{c/a}\{\mathrm{D}\}^{d/a}}{\{\mathrm{A}\}\{\mathrm{B}\}^{b/a}} = K \tag{5-47}$$

For most water supplies, the ionic strength is less than 5 millimole/L (mM) and the activity coefficients for monovalent ions are close to one. The calculation of activity coefficients for solutions of different ionic strengths is presented in the following example.

Example 5-4 Determination of activity coefficients at different ionic strengths

Calculate the activity coefficients of Na^+, Ca^{2+}, and Al^{3+} at ionic strengths of 0.001, 0.005, and 0.01M at 25 °C.

Solution

1. Determine the activity coefficients for an ionic strength of 0.001 M at 25 °C using the Debye-Hückel limiting law (Eq. 5-42):

$$\log_{10}\gamma_{\mathrm{Na^+}} = -0.51(1)^2\sqrt{0.001} = -1.61\times 10^{-2} \qquad \therefore \gamma_{\mathrm{Na^+}} = 0.96$$

$$\log_{10}\gamma_{\mathrm{Ca^{2+}}} = -0.51(2)^2\sqrt{0.001} = -6.45\times 10^{-2} \qquad \therefore \gamma_{\mathrm{Ca^{2+}}} = 0.86$$

$$\log_{10}\gamma_{\mathrm{Al^{3+}}} = -0.51(3)^2\sqrt{0.001} = -0.14 \qquad \therefore \gamma_{\mathrm{Al^{3+}}} = 0.72$$

2. Determine the activity coefficients for an ionic strength of 0.005 M at 25 °C using the Debye-Hückel limiting law:

$$\log_{10}\gamma_{Na^+} = -0.51(1)^2\sqrt{0.005} = -3.61\times10^{-2} \qquad \therefore \gamma_{Na^+} = 0.92$$

$$\log_{10}\gamma_{Ca^{2+}} = -0.51(2)^2\sqrt{0.005} = -0.14 \qquad \therefore \gamma_{Ca^{2+}} = 0.72$$

$$\log_{10}\gamma_{Al^{3+}} = -0.51(3)^2\sqrt{0.005} = -0.32 \qquad \therefore \gamma_{Al^{3+}} = 0.48$$

3. Determine the activity coefficients for an ionic strength of 0.01 M at 25 °C using the Davies equation (Eq. 5-43):

$$\log_{10}\gamma_{Na^+} = -0.51(1)^2\left[\frac{\sqrt{0.01}}{1+\sqrt{0.01}} - 0.3(0.01)\right]$$

$$= -4.48\times10^{-2} \quad \therefore \gamma_{Na^+} = 0.90$$

$$\log_{10}\gamma_{Ca^{2+}} = -0.51(2)^2\left[\frac{\sqrt{0.01}}{1+\sqrt{0.01}} - 0.3(0.01)\right]$$

$$= -0.18 \quad \therefore \gamma_{Ca^{2+}} = 0.66$$

$$\log_{10}\gamma_{Al^{3+}} = -0.51(3)^2\left[\frac{\sqrt{0.01}}{1+\sqrt{0.01}} - 0.3(0.01)\right]$$

$$= -0.40 \quad \therefore \gamma_{Al^{3+}} = 0.40$$

Comment

The activity for all the ions decreases as the ionic strength increases. As the ionic strength of the solution increases, the impact of charge on the species has a large influence on the value of the activity coefficient. For example, as ionic strength increased from 0.001 to 0.01, the activity coefficient for Na^+ decreased by 6.3 percent as compared to Al^{3+}, which decreased by 44 percent.

5-3 Thermodynamics of Chemical Reactions

Principles from equilibrium thermodynamics provide a means for determining whether reactions are favorable and are also used in process design calculations to determine the final equilibrium state. The difference between the actual state and the equilibrium state is the driving force for many processes and reactions. Equilibrium thermodynamics can be used to determine whether the treatment process is feasible, and the reaction kinetics, described in the following sections, will provide a basis for the treatment device size.

To determine whether a reaction will proceed(i.e., is thermodynamically favorable), two fundamental thermodynamic criteria must be considered. The first thermodynamic criterion that must be satisfied is that the change in entropy of the system and its surroundings must be greater than zero for a reaction to proceed. When evaluating chemical reactions, the entropy requirement is typically satisfied, especially when heat is produced by the reaction and, therefore, is not considered further in this text. The second thermodynamic criterion necessary for a reaction to proceed is the requirement that the change in free energy (final energy state minus initial energy state) of the reaction must be less than zero.

Reference Conditions

To understand how the free energy of reaction changes as a reaction proceeds, it is useful to examine the total free energy of reaction as a function of the reaction extent, as shown on Fig. 5-2. Because the absolute free energy of reaction cannot be determined easily, it is most common to determine the change in free energy of a reaction. The free energy of the reaction curve shown on Fig. 5-2 is compared to a convenient set of standard conditions. For example, a common definition of standard conditions is as follows: (1) solids, liquids, and gases in their lowest energy state at 1 atm (or 1 bar); (2) solutes in solution referenced to a 1 molal hydrogen ion concentration; and (3) a specified temperature, usually 25°C. For most water treatment applications, the molar concentration is essentially equal to the molal concentration and a 1 M solution is 1 mole per 1000 g of solvent.

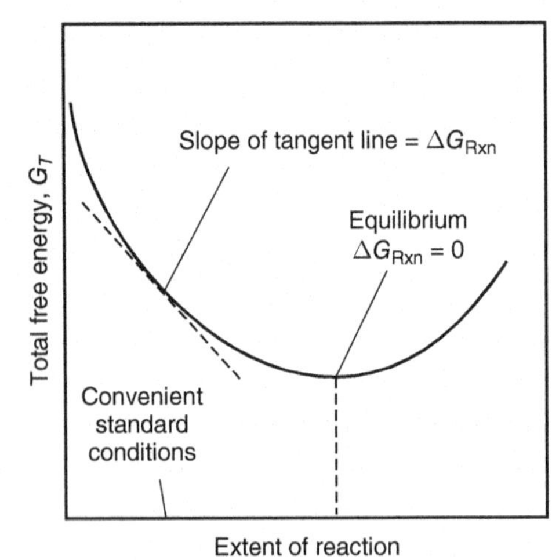

Figure 5-2
Total free energy as function of the extent of the reaction.

Free Energy of Formation

The expression for free energy was developed by J. W. Gibbs and is often referred to as the Gibbs free energy or Gibbs function G. The free-energy change of formation of a substance i is given by the expression

$$\Delta G_{F,i} = \Delta G^{\circ}_{F,i} + RT\ln\{i\} \tag{5-48}$$

where $\Delta G_{F,i}$ = free-energy change of formation of species i, kJ/mol
$\Delta G^{\circ}_{F,i}$ = free-energy change of formation per mole of i at standard conditions, kJ/mol
R = universal gas law constant, 8.314×10^{-3} kJ/mol·K
T = absolute temperature, K(273 + °C)
$\{i\}$ = activity of species i

Thermodynamic constants may be found in various reference books, including Stumm and Morgan (1996) and *Lange's Handbook* (Dean, 1992).

Free Energy of Reaction

The free energy of a reaction can be calculated using the definition of activity and the free-energy change of formation. For this purpose, consider the reaction shown in Eq. 5-21, in which a, b, c, and d are stoichiometric coefficients. For this reaction, the free-energy criterion may be determined for a reference reactant A and written per mole of A by dividing by the stoichiometric coefficient a, as shown in Eq. 5-23 and repeated here:

$$\mathrm{A} + \frac{b}{a}\mathrm{B} \rightleftarrows \frac{c}{a}\mathrm{C} + \frac{d}{a}\mathrm{D}$$

The free-energy change is defined as the final state minus the initial state (David, 2000; Dean, 1992; Poling et al., 2001). Therefore, the change in free energy of a reaction is the sum of the free-energy change of each product minus the sum of the free-energy change of the reactants, as shown in the following expression written in terms of free-energy change per mole of A:

$$\Delta G_{\mathrm{Rxn,A}} = -\Delta G_{F,\mathrm{A}} - \frac{b}{a}\Delta G_{F,\mathrm{B}} + \frac{c}{a}\Delta G_{F,\mathrm{C}} + \frac{d}{a}\Delta G_{F,\mathrm{D}} \tag{5-49}$$

where $\Delta G_{\mathrm{Rxn,A}}$ = free-energy change of reaction per mole of A, kJ/mol
$\Delta G_{F,A}$ = change in free energy of reactant A, kJ/mol
$\Delta G_{F,B}$ = change in free energy of reactant B, kJ/mol
$\Delta G_{F,C}$ = change in free energy of product C, kJ/mol
$\Delta G_{F,D}$ = change in free energy of product D, kJ/mol

The free-energy change of the formation of each species, as defined in Eq. 5-49, may be substituted into Eq. 5-49 for each reactant and product to obtain the overall free-energy change for the reaction. The resulting expression for the free-energy change of the reaction is shown in the expression

$$\begin{aligned}\Delta G_{\mathrm{Rxn,A}} = {} & \frac{c}{a}\Delta G^{\circ}_{F,\mathrm{C}} + RT\ln\{\mathrm{C}\}^{c/a} + \frac{d}{a}\Delta G^{\circ}_{F,\mathrm{D}} + RT\ln\{\mathrm{D}\}^{d/a} \\ & -\Delta G^{\circ}_{F,\mathrm{A}} - RT\ln\{\mathrm{A}\} - \frac{b}{a}\Delta G^{\circ}_{F,\mathrm{B}} - RT\ln\{\mathrm{B}\}^{b/a}\end{aligned} \tag{5-50}$$

where $\Delta G_{\text{Rxn,A}}$ = free-energy change of reaction per mole of A, kJ/mol
{A} = activity of reactant A, mol/L
{B} = activity of reactant B, mol/L
{C} = activity of product C, mol/L
{D} = activity of product D, mol/L

The free-energy change of the reaction per mole of A at standard conditions (25 °C and 1 atm pressure), $\Delta G^{\circ}_{\text{Rxn,A}}$, can be written as

$$\Delta G^{\circ}_{\text{Rxn,A}} = -\Delta G^{\circ}_{F,\text{A}} - \frac{b}{a}\Delta G^{\circ}_{F,\text{B}} + \frac{c}{a}\Delta G^{\circ}_{F,\text{C}} + \frac{d}{a}\Delta G^{\circ}_{F,\text{D}} \tag{5-51}$$

Equation 5-50 can be further simplified by substituting in the relationship shown in Eq. 5-51:

$$\Delta G_{\text{Rxn,A}} = \Delta G^{\circ}_{\text{Rxn,A}} + RT\ln\frac{\{\text{C}\}^{c/a}\{\text{D}\}^{d/a}}{\{\text{A}\}\{\text{B}\}^{b/a}} \tag{5-52}$$

The logarithmic term in Eq. 5-52 is called the reaction quotient Q:

$$Q = \frac{\{\text{C}\}^{c/a}\{\text{D}\}^{d/a}}{\{\text{A}\}\{\text{B}\}^{b/a}} \tag{5-53}$$

If the stoichiometric coefficient *a* had not been factored out of Eq. 5-21, then Eq. 5-52 would be written per *a* moles of A as

$$a\Delta G_{\text{Rxn,A}} = a\Delta G^{\circ}_{\text{Rxn,A}} + RT\ln\frac{\{\text{C}\}^{c}\{\text{D}\}^{d}}{\{\text{A}\}^{a}\{\text{B}\}^{b}} \tag{5-54}$$

When examining thermodynamic data, it is important to make certain that the free energy that is reported is per mole of A. Finally, the thermodynamic criterion that must be met for a reaction to proceed as written from the initial state toward the final state may be expressed as

$$\Delta G_{\text{Rxn,A}} \text{ must be} < 0 \tag{5-55}$$

While a reaction is thermodynamically feasible when $\Delta G_{\text{Rxn,A}} < 0$, the rate at which a reaction will proceed is not known because reactants often have to proceed through reactive intermediates that have a higher free energy than the reactants. Alternately, if $\Delta G_{\text{Rxn,A}} > 0$, the reverse reaction would be thermodynamically feasible.

Free Energy at Equilibrium

Another useful relationship, known as the equilibrium state, is obtained when $\Delta G_{\text{Rxn,A}} = 0$. When $\Delta G_{\text{Rxn,A}} = 0$ in Eq. 5-52, the reaction quotient is equal to the equilibrium constant *K* as shown below:

$$\Delta G_{\text{Rxn,A}} = \Delta G^{\circ}_{\text{Rxn,A}} + RT\ln\frac{\{\text{C}\}^{c/a}\{\text{D}\}^{d/a}}{\{\text{A}\}\{\text{B}\}^{b/a}} = 0 \tag{5-56}$$

If the relationship shown in Eq. 5-47 for the equilibrium constant is substituted for the reaction quotient, the following expression is obtained:

$$\Delta G^{\circ}_{\text{Rxn,A}} = -RT\ln\left[\frac{\{\text{C}\}^{c/a}\{\text{D}\}^{d/a}}{\{\text{A}\}\{\text{B}\}^{b/a}}\right] = -RT\ln[\text{K}] \tag{5-57}$$

Rearranging Eq. 5-57 and solving for the equilibrium constant results in the expression

$$K = e^{-\Delta G^\circ_{\text{Rxn,A}}/RT} \tag{5-58}$$

The free energy, calculated using Eq. 5-52, is actually the slope of a tangent to the total free-energy curve shown on Fig. 5-2, and equilibrium is represented by the special case where the slope is zero. This means that $\Delta G_{\text{Rxn,A}}$ is really the change in free energy that results from an infinitesimal conversion of A to products.

Calculation of Free Energy of Formation Using Henry's Constant

A difficulty often encountered when calculating the free energy of reaction is that the free-energy change of formation per mole of A in the aqueous phase, $\Delta G^\circ_{F,\text{A,aq}}$, is needed to calculate $\Delta G_{\text{Rxn,A}}$, and the free energy of formation may be reported for the gas phase, $\Delta G^\circ_{F,\text{A,gas}}$. However, the relationship shown in Eq. 5-59 can be used to develop an expression for the free energy of formation of slightly soluble gases in the aqueous phase, $\Delta G^\circ_{F,\text{A,aq}}$, based on the free energy of formation in the gas phase, $\Delta G^\circ_{F,\text{A,gas}}$:

$$\Delta G^\circ_{\text{Vol,A}} = \Delta G^\circ_{F,\text{A,gas}} - \Delta G^\circ_{F,\text{A,aq}} = -RT \ln H_{\text{PC}} \tag{5-59}$$

where $\Delta G^\circ_{\text{Vol,A}}$ = free-energy change of volatilization per mole of A at standard conditions kJ/mol

$\Delta G^\circ_{F,\text{A,gas}}$ = free-energy change of formation per mole of A in gas phase at standard conditions kJ/mol

$\Delta G^\circ_{F,\text{A,aq}}$ = free-energy change of formation per mole of A in aqueous phase at standard conditions kJ/mol

H_{PC} = Henry's law constant atm/(mol/L)

Equation 5-59 can then be rearranged to solve for the aqueous-phase concentration of A as a function of the gas-phase concentration of A:

$$\Delta G^\circ_{F,\text{A,aq}} = \Delta G^\circ_{F,\text{A,gas}} + RT \ln H_{\text{PC}} \tag{5-60}$$

Consequently, $\Delta G^\circ_{F,\text{A,aq}}$ can be calculated from $\Delta G^\circ_{F,\text{A,gas}}$ if H_{PC} is known. Henry's law is presented and discussed in detail in Chap. 14.

Temperature Dependence of Free-Energy Change

Most reactions in water treatment do not occur at 25°C because the water temperature is usually lower. The free-energy change at other temperatures can be determined from the expression

$$\left.\frac{\Delta G^\circ_{Rxn}}{T}\right|_{T=298\text{ K}}^{T} = \int_{T=298\text{ K}}^{T} -\frac{\Delta H^\circ_{\text{Rxn}}(T)}{T^2}\,dT \tag{5-61}$$

where $\Delta H^\circ_{\text{Rxn}}(T)$ = standard enthalpy of reaction that depends on temperature

The temperature-dependent standard enthalpy of reaction is defined as

$$\Delta H^\circ_{\text{Rxn}}(T) = \int_{T=298\text{ K}}^{T} \Delta C_{p,\text{Rxn}} dT + \Delta H^\circ_{\text{Rxn},298\text{ K}} \quad (5\text{-}62)$$

where $\Delta C_{p,\text{Rxn}}$ = change in heat capacity for the reaction, kJ/mol
$\Delta H^\circ_{\text{Rxn},298\text{K}}$ = standard enthalpy at 298 K

The heat capacity term may be calculated using the equation (Poling et al., 2001)

$$C_{p,i} = A + BT + CT^2 + DT^3 \quad (5\text{-}63)$$

where A, B, C, D = constants
$C_{p,i}$ = isobaric (constant-pressure) heat capacity for compound i

To calculate $\Delta C_{p,\text{Rxn}}$, the difference of each constant (A, B, C, and D) between products and reactants needs to be calculated:

$$\Delta C_{p,\text{Rxn}} = \Delta A + \Delta BT + \Delta CT^2 + \Delta DT^3 \quad (5\text{-}64)$$

For the reaction shown in Eq. 5-23, the terms in Eq. 5-64 are given by the expressions

$$\begin{aligned} \Delta A &= \frac{d}{a}A_D + \frac{c}{a}A_C - \frac{b}{a}A_B - A_A \quad \Delta B = \frac{d}{a}B_D + \frac{c}{a}B_C - \frac{b}{a}B_B - B_A \\ \Delta C &= \frac{d}{a}C_D + \frac{c}{a}C_C - \frac{b}{a}C_B - C_A \quad \Delta D = \frac{d}{a}D_D + \frac{c}{a}D_C - \frac{b}{a}D_B - D_A \end{aligned} \quad (5\text{-}65)$$

Substituting the relationships shown in Eq. 5-64 into Eq. 5-62 and subsequently into Eq. 5-61, the following expression is obtained:

$$\left.\frac{\Delta G^\circ_{Rxn}}{T}\right|_{T=298\text{ K}}^{T} = -\int_{T=298\text{ K}}^{T} \left[\frac{\Delta H^\circ_{\text{Rxn},298\text{ K}}(T)}{T^2} + \frac{\Delta A(T-298)}{T^2} + \frac{\Delta B(T^2-298^2)}{2T^2} + \frac{\Delta C(T^3-298^3)}{3T^2} + \frac{\Delta D(T^4-298^4)}{4T^2} \right] dT \quad (5\text{-}66)$$

For the case where $\Delta H^\circ_{\text{Rxn}}$ does not depend on temperature ($\Delta H^\circ_{\text{Rxn}}$ is constant), Eq. 5-66 can be simplified as

$$\left.\frac{\Delta G^\circ_{\text{Rxn}}}{T}\right|_{T=298\text{ K}}^{T} = \frac{\Delta G^\circ_{\text{Rxn},T}}{T} - \frac{\Delta G^\circ_{\text{Rxn},298\text{ K}}}{298\text{ K}} = \Delta H^\circ_{\text{Rxn},298\text{ K}}\left(\frac{1}{T} - \frac{1}{298\text{ K}}\right) \quad (5\text{-}67)$$

At equilibrium, Eq. 5-57 can be substituted into Eq. 5-61 to yield the van't Hoff relationship, which may be used to determine the equilibrium constant (K_{eq}) at different temperatures:

$$\ln K \Big|_{T=298\text{ K}}^{T} = \int_{T=298\text{ K}}^{T} \frac{\Delta H^\circ_{\text{Rxn},298\text{ K}}}{RT^2}\, dT$$

$$\ln \frac{K_T}{K_{298\text{ K}}} = \frac{\Delta H^\circ_{\text{Rxn},298\text{ K}}}{R}\left(\frac{1}{298\text{ K}} - \frac{1}{T}\right) \tag{5-68}$$

where K_T = equilibrium constant at temperature T, K (273 + °C)
$K_{298\text{ K}}$ = equilibrium constant at 298 K

Using Eq. 5-68, the linear relationship between ln K and $1/T$ can be determined by plotting the function

$$\ln K = -\frac{\Delta H^\circ_{\text{Rxn},298\text{ K}}}{RT} + \text{const} \tag{5-69}$$

For most reactions occurring in water treatment processes, $\Delta H^\circ_{\text{Rxn}}$ can be assumed to be constant because ΔH_{Rxn} does not change significantly over the temperature range encountered in water treatment (0 to 30 °C).

Example 5-5 Dependence of pH and free-energy change on temperature

For the dissociation reaction of water, the free-energy change and enthalpy change for each species in the reaction

$$H_2O \rightleftarrows H^+ + OH^-$$

are as follows:

$$\Delta G^\circ_{F,H_2O} = -237.18 \text{ kJ/mol} \quad \Delta H^\circ_{F,H_2O} = -285.83 \text{ kJ/mol}$$
$$\Delta G^\circ_{F,H^+} = 0 \text{ kJ/mol} \quad \Delta H^\circ_{F,H^+} = 0 \text{ kJ/mol}$$
$$\Delta G^\circ_{F,OH^-} = -157.29 \text{ kJ/mol} \quad \Delta H^\circ_{F,OH^-} = -230.0 \text{ kJ/mol}$$

Calculate the pH of neutrality and free-energy change of the reaction at 10 °C. Assume that $\Delta H^\circ_{\text{Rxn}}$ does not change with temperature.

Solution

1. Calculate the equilibrium constant, using Eq. 5-58, for water at 25 °C.
   a. Calculate $\Delta G^\circ_{\text{Rxn},H_2O}$ using Eq. 5-51:

$$\Delta G^\circ_{\text{Rxn},H_2O} = \Delta G^\circ_{F,OH^-} + \Delta G^\circ_{F,H^+} - \Delta G^\circ_{F,H_2O}$$
$$= -157.29 + 0 - (-237.18\text{ K}) = 79.89 \text{ kJ/mol}$$

b. Calculate the equilibrium constant at 25 °C (298 K) using Eq. 5-58:

$$K = \exp\left(\frac{-\Delta G^{\circ}_{\mathrm{Rxn,H_2O}}}{RT}\right)$$

$$= \exp\left[\frac{-79.89\ \mathrm{kJ/mol}}{\left(8.314\times10^{-3}\ \mathrm{kJ/mol\cdot K}\right)(298\ \mathrm{K})}\right]$$

$$= 9.90954\times10^{-15}\approx10^{-14}$$

The value for K for the dissociation of water at standard conditions is generally reported as K_w.

2. Calculate the equilibrium constant at 10 °C (283 K):
   a. Calculate $\Delta H^{\circ}_{\mathrm{Rxn}}$:

$$\Delta H^{\circ}_{\mathrm{Rxn,H_2O}} = \Delta H^{\circ}_{\mathrm{F,OH^-}} + \Delta H^{\circ}_{\mathrm{F,H^+}} - \Delta H^{\circ}_{\mathrm{F,H_2O}}$$
$$= -230.0 + 0 - (-285.83) = 55.83\ \mathrm{kJ/mol}$$

   b. Calculate the equilibrium constant using Eq. 5-68:

$$\ln K\Big|_{T=298\ \mathrm{K}}^{T} = \int_{T=298\ \mathrm{K}}^{T} \frac{\Delta H^{\circ}_{\mathrm{Rxn,298\ K}}}{RT^2}\,dT$$

$$\ln\frac{K_T}{K_{298\ \mathrm{K}}} = \frac{\Delta H_{\mathrm{Rxn,298\ K^{\circ}}}}{R}\left(\frac{1}{298\ \mathrm{K}} - \frac{1}{T}\right)$$

$$K_T = K_{298\ \mathrm{K}}\exp\left[\frac{\Delta H_{\mathrm{Rxn,298\ K}}}{R}\left(\frac{1}{298\ \mathrm{K}} - \frac{1}{T}\right)\right]$$

$$= 10^{-14}\exp\left[\frac{55.83\ \mathrm{kJ/mol}}{8.314\times10^{-3}\ \mathrm{kJ/mol\cdot K}}\left(\frac{1}{298\ \mathrm{K}} - \frac{1}{283\ \mathrm{K}}\right)\right]$$

$$= 3.0015\times10^{-15}$$

3. Calculate the pH at neutrality at 10 °C.
   At neutral conditions, $[H^+]$ is equal to $[OH^-]$:

$$[\mathrm{H^+}][\mathrm{OH^-}] = K_{\mathrm{eq,283}} = 3.0015\times10^{-15}$$

$$[\mathrm{H^+}] = [\mathrm{OH^-}] = \sqrt{3.0015\times10^{-15}} = 5.48\times10^{-8}$$

$$\mathrm{pH} = \mathrm{pOH} = -\log(5.48\times10^{-8}) = 7.26$$

4. Calculate ΔG°_{Rxn} at 10 °C using Eq. 5-67:

$$\left.\frac{\Delta G^\circ_{Rxn}}{T}\right|_{T=298\text{ K}}^{T} = \frac{\Delta G_{Rxn,T}}{T} - \frac{\Delta G_{Rxn,298\text{ K}}}{298\text{ K}}$$

$$= \Delta H^\circ_{Rxn,298\text{ K}}\left(\frac{1}{T} - \frac{1}{298\text{ K}}\right)$$

$$\Delta G^\circ_{Rxn,T} = T\left[\frac{\Delta G^\circ_{Rxn,298\text{ K}}}{298\text{ K}} + \Delta H^\circ_{Rxn,298\text{ K}}\left(\frac{1}{T} - \frac{1}{298\text{ K}}\right)\right]$$

$$= 283\text{ K}\left[\frac{79.89\text{ kJ/mol}}{298\text{ K}} + 55.83\text{ kJ/mol} \times \left(\frac{1}{283\text{ K}} - \frac{1}{298\text{ K}}\right)\right]$$

$$= 78.68\text{ kJ/mol}$$

Comment

According to Le Chatelier's principle, as the temperature decreases, the reaction for the dissociation of water would be less favorable because it takes energy to dissociate water; consequently, the equilibrium constant is lower at 10 °C than at 25 °C.

5-4 Reaction Kinetics

While thermodynamic calculations provide a means for estimating the likelihood and maximum possible extent of a given reaction, they cannot be used to determine the rate of the reaction. This section discusses reaction rate laws which describe how fast a reaction proceeds in the absence of mass transfer limitations.

Reaction Rate

The rate of a chemical reaction depends on the activity of the reacting species and the temperature of the system. As noted earlier for organic compounds (nonelectrolytes) or monovalent ions in water, the activity is nearly equivalent to concentration when the ionic strength is less than 0.005 M. For divalent and trivalent ions, the activity coefficients should be calculated to determine whether activity coefficients are needed.

The rate of a reaction is expressed as the change in the concentration of a constituent with time:

$$\text{Rate of reaction} = \frac{\text{change in concentration}}{\text{change in time}} \tag{5-70}$$

The reaction rate is used to describe the rate of formation of a product or the rate of decomposition of a reactant. As a reaction proceeds, the reaction rate changes; for example, as the concentration of reactants is decreased, the rate of a reaction may decrease. The reaction rate usually changes as the concentrations of the reactants and products change. The rate of change is described by an expression known as a rate law, as discussed in the following section.

Rate Law and Reaction Order

In the following discussion, the term r_A is used to represent the reaction rate. It should be noted that dC_A/dt is not the reaction rate law but is obtained from a mass balance on a completely mixed batch reactor, which has a constant volume. The subscript A is used to designate the species described by the reaction rate. The units of the reaction rate are given by the expression

$$r_A = \frac{\text{moles A lost}(-)\text{or generated}(+)\text{due to reaction}}{(\text{volume})(\text{time})} = \frac{\text{mol}}{\text{L} \cdot \text{s}} \quad (5\text{-}71)$$

A negative or positive sign for the reaction rate indicates that species A is either disappearing or appearing, respectively. The following irreversible reaction is used to develop the reaction rate.

$$a\text{A} + b\text{B} \rightarrow \text{products} \quad (5\text{-}72)$$

For an irreversible reaction, the rate law depends on the concentrations of reactants. For the reaction shown in Eq. 5-72, the rate law may take the form

$$r_A = -k[\text{A}]^m[\text{B}]^n \quad (5\text{-}73)$$

Where r_A = reaction rate, mol/L·s

k = reaction rate constant, units vary depending on reaction order as discussed later

m, n = constants, unitless

The concentration dependence of the reaction rate is accounted for in the reactant exponents m and n and is known as the reaction order. For Eq. 5-73, the reaction order is m for species A and n for species B, and the overall reaction order is $m + n$. The reaction order is typically a small positive integer; however, it may also be negative, zero, or fractional.

Relationship between Reaction Rates

For an elementary reaction with the following reaction stoichiometry, a relationship between relative rates of reaction can be determined from stoichiometry:

$$a\text{A} + b\text{B} \rightleftarrows c\text{C} + d\text{D} \quad (\text{Eq. } 5\text{-}21)$$

To simplify the expression Eq. 5-21 can be divided by the stoichiometric coefficient *a* which yields the expression

$$A+\frac{b}{a}B \rightleftarrows \frac{c}{a}C+\frac{d}{a}D \qquad \text{(Eq. 5-23)}$$

Assuming the reaction shown in Eq. 5-23 is proceeding to the right the following relationship between the reaction rates can be written using the stoichiometry from Eq. 5-23:

$$\frac{-r_A}{1}=\frac{-r_B}{b/a}=\frac{r_C}{c/a}=\frac{r_D}{d/a} \qquad (5\text{-}74)$$

The reaction rates for reactants A and B, $-r_A$ and $-r_B$ are negative because they are disappearing to form products and the reaction rates for products C and D, r_C and r_D are positive because they are being produced. If the reaction rate for A is known, the stoichiometric coefficients, as given by Eq. 5-74, can be used to determine the reaction rate for reactant B and products C and D. The use of stoichiometric coefficients is illustrated in the following example.

Example 5-6 Determination of reaction rates using stoichiometry

Given the reaction $4Fe^{2+} + O_2 + 10H_2O \rightarrow 4Fe(OH)_3 + 8H^+$, estimate the loss rates of oxygen and water and production rates of iron hydroxide and acid when the rate of loss of Fe^{2+} is 2×10^{-7} mol/L·min.

Solution

1. Write the relevant reaction using the form shown in Eq. 5-23. To cancel out the coefficient for Fe^{2+}, the reaction must be divided by 4, resulting in the expression

$$Fe^{2+}+\tfrac{1}{4}O_2+\tfrac{5}{2}H_2O \rightarrow Fe(OH)_3+2H^+$$

2. Based on Eq. 5-74, the rate expression can be written as follows:

$$-r_{Fe^{2+}}=\frac{-r_{O_2}}{\frac{1}{4}}=\frac{-r_{H_2O}}{\frac{5}{2}}=\frac{+r_{Fe(OH)_3}}{1}=\frac{+r_{H^+}}{2}$$

3. Estimate the loss rates for oxygen ($-r_{O_2}$) and water ($-r_{H_2O}$) given that the rate of loss of Fe^{2+}($-r_{Fe^{2+}}$) is 2×10^{-7} mol/L · min:

$$-r_{O_2}=\frac{1}{4}\left(-r_{Fe^{2+}}\right)=\frac{1}{4}\left(2\times10^{-7}\right)=0.5\times10^{-7}\text{mol/L}\cdot\text{min}$$

$$-r_{H_2O}=\frac{5}{2}\left(-r_{Fe^{2+}}\right)=\frac{5}{2}\left(2\times10^{-7}\right)=5\times10^{-7}\text{mol/L}\cdot\text{min}$$

4. Estimate the production rates for iron hydroxide $[+r_{Fe(OH)_3}]$ and acid $(+r_{H+})$:

$$+r_{Fe(OH)_3} = -r_{Fe^{2+}} = 2\times10^{-7}\text{mol / L}\cdot\text{min}$$
$$+r_{H+} = 2(-r_{Fe^{2+}}) = 4\times10^{-7}\text{mol / L}\cdot\text{min}$$

Comment

For reactions that are given by Eq. 5-23, the reaction rates are related to one another. If one knows the rate of reaction of any at the components, then reaction rates of the other components can be calculated using Eq. 5-74.

Rate Constants

The units of the rate constant depend on the reaction order; but it should be noted that the units of the reaction rate r are always of the form mol/ L · s. For a zero-order reaction, the rate constant would have the following units:

$$r_A = -k \quad k = \text{mol/L}\cdot\text{s} \tag{5-75}$$

For a first-order reaction, the rate constant would have the following units:

$$r_A = -k[A] \quad k = s^{-1} \tag{5-76}$$

For a second-order reaction, the rate constant would have the following units:

$$r_A = -k[A][B] \quad k = \text{L / mol}\cdot\text{s} \tag{5-77}$$

For Eq. 5-77, the rate is first order in A and B and second order overall. For a third-order reaction, the rate constant would have the following units:

$$r_A = -k[A][B][C] \quad k = \text{L}^2 / \text{mol}^2\cdot\text{s} \tag{5-78}$$

For Eq. 5-78, the rate is first order in A, B, and C and third order overall.

Factors Affecting Reaction Rate Constants

A distinctive feature of water treatment is the varying character of each water source. A wide range in values of temperature, pH, and ionic composition is encountered in water treatment practice. In addition, environmental conditions often produce wide seasonal fluctuations in these intensive variables. Quantitative estimates of the dependence of empirical rate constants on temperature, pH, ionic composition, and other factors are essential for proper control of the reactions of interest.

EFFECT OF TEMPERATURE AND CATALYSIS ON REACTION RATE CONSTANT

Reaction rate constants are known to be dependent on temperature. A relationship known as the Arrhenius equation is used to describe the temperature dependence:

$$k = Ae^{-E_a/RT} \tag{5-79}$$

where E_a = activation energy, kJ/mol
A = frequency factor, same units as k

The irreversible elementary reaction shown below may be used to illustrate the physical phenomenon causing rate constants to follow this temperature dependence:

$$A + B \rightarrow \text{products} \tag{5-80}$$

According to collision theory, only colliding pairs of molecules that have sufficient kinetic energy to overcome the activation energy E_a will react, as shown on Fig. 5-3. In addition, the molecular collisions with sufficient kinetic energy must also have proper orientation for the reaction to proceed. The rate of reaction is given by

$$-r_A = \begin{pmatrix} \text{frequency of collisions} \\ \text{that have} \\ \text{proper orientation} \end{pmatrix} \times \begin{pmatrix} \text{fraction of collisions} \\ \text{that have} \\ \text{sufficient energy} \end{pmatrix} \tag{5-81}$$

The frequency of collisions that have the proper orientation is proportional to the product of the concentrations of A and B. To estimate the fraction of collisions that have sufficient energy, Arrhenius postulated that the energy of the resulting collisions followed a Maxwell-Boltzmann distribution, which is given by the expression

$$N = Ce^{-E_a/RT} \tag{5-82}$$

where N = number of collisions with energy equal to or greater than E_a
C = constant

Accordingly, the number of collisions with energy equal to or greater than the amount of energy required to overcome E_a is proportional to $e^{-E_a/RT}$:

$$\text{Fraction of collisions that have energy} \geq E_a \propto e^{-E_a/RT} \tag{5-83}$$

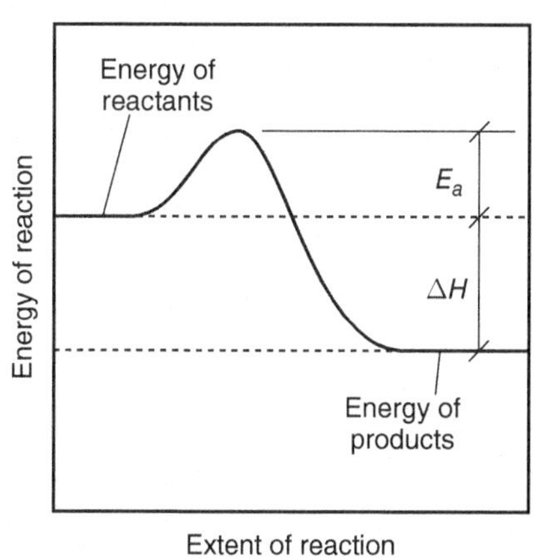

Figure 5-3
Reaction energy as function of reaction extent.

The rate expression can therefore be written as

$$-r_A = [A][B]Ae^{-E_a/RT} \tag{5-84}$$

Molecules in solution have translational, rotational and vibrational motion and this motion may make it more difficult for molecules to collide in the proper orientation and react. These effects are incorporated into the frequency factor *A*.

The constants in the Arrhenius equation can be determined by taking the natural logarithm of Eq. 5-79 which yields the expression

$$\ln(k) = \ln(A) + \left(-\frac{E_a}{R}\right) \times \left(\frac{1}{T}\right) \tag{5-85}$$

The rate constant can be determined experimentally at several temperatures and when $\ln(k)$ is plotted as a function of $1/T$, the slope is equal to $-E_a/R$ and the y intercept is equal to $\ln(A)$. The temperature dependence of rate constants is sensitive to the magnitude of the activation energy E_a. Values of E_a for reactions in solution range from 4 to 125 kJ/mol (1 to 30 kcal/mol) (Moelwyn-Hughes 1974). For example the hydrolysis of aqueous CO_2 ($CO_2 + H_2O \rightleftarrows H_2CO_3$) has an E_a of approximately 55 kJ/mol (13 kcal/mol) (Kern 1960). Reactions with high E_a values show a much greater sensitivity to temperature increases as compared to reactions with low E_a values. In general it has been observed that many rate constants double for every 10 °C increase in temperature. However as shown in the following example this rule only holds for an activation energy of approximately 50 kJ/mol (12 kcal/mol) which is in the median range of most reactions in solution (Frost and Pearson 1961).

Example 5-7 Determination of the activation energy from rate contants that are known at two temperatures

If the rate of a chemical reaction doubles for each 10 °C increase in temperature, estimate the value of the activation energy. Assume the initial temperature is 25 °C.

Solution

1. Write an expression for Eq. 5-79 to isolate the activation energy term:

$$k = Ae^{-E_a/RT} \quad \ln(k) = \ln(A) - \frac{E_a}{RT}$$

2. Write expressions for the relationship between the reaction rate k_{T_1} and k_{T_2} at temperatures T_1 and T_2, respectively.

a. At temperature T_1, the expression developed in step 1 may be written as

$$\ln\left(k_{T_1}\right) = \ln(A) - \frac{E_a}{RT_1}$$

b. A similar expression may be written for T_2:

$$\ln\left(k_{T_2}\right) = \ln(A) - \frac{E_a}{RT_2}$$

3. Solve the equations in step 2a and 2b for E_a.

a. Develop a relationship for E_a by subtracting the expression developed in step 2b from the one developed in step 2a:

$$\ln\left(\frac{k_{T_1}}{k_{T_2}}\right) = \frac{E_a}{R}\left(\frac{1}{T_2} - \frac{1}{T_1}\right)$$

$$E_a = \frac{R\ln\left(k_{T_1} / k_{T_2}\right)}{1/T_2 - 1/T_1}$$

b. Substitute known values and solve for E_a:

$$\begin{aligned}
T_1 &= \left(273 + 25^\circ\mathrm{C}\right) = 298\ \mathrm{K} \\
T_2 &= \left(273 + 25^\circ\mathrm{C} + 10^\circ\mathrm{C}\right) = 308\ \mathrm{K} \\
k_{T_2} &= 2k_{T_1} \\
E_a &= \frac{\left(8.3144 \times 10^{-3}\ \mathrm{kJ/mol \cdot K}\right)\ln\left(k_{T_1} / 2k_{T_1}\right)}{(1/308 - 1/298)\mathrm{K}^{-1}} \\
&= 52.90\ \mathrm{kJ/mol} \quad (12.63\ \mathrm{kcal/mol})
\end{aligned}$$

While increases in the reaction rate with temperature occur with chemical reactions, the reaction rate for microbial reactions do not always increase with temperature. It has been shown that Arrhenius-type behavior is exhibited up to a limiting temperature, where enzyme deactivation occurs, and then the reaction rate decreases rapidly with an additional increase in temperature.

In many cases, the addition of a catalyst may speed up a reaction by lowering the activation energy, as shown on Fig. 5-4. In heterogeneous and enzyme catalysis, the catalyst can also increase the rate because the catalyst/enzyme can help orient the reactants in the proper manner for reaction.

IMPACT OF IONIC STRENGTH ON REACTION RATE CONSTANTS

As discussed previously, the rate of chemical reactions depends on activities of the reactants and products. For an irreversible second-order elementary reaction, the rate of disappearance of either reactant would be given by the following expressions:

$$A + B \rightarrow \text{products} \tag{5-86}$$

$$r_A = r_B = -k\{A\}\{B\} = -k\gamma_a[A]\gamma_b[B] \tag{5-87}$$

The observed rate constant then becomes

$$k_{obs} = -k\gamma_a\gamma_b \tag{5-88}$$

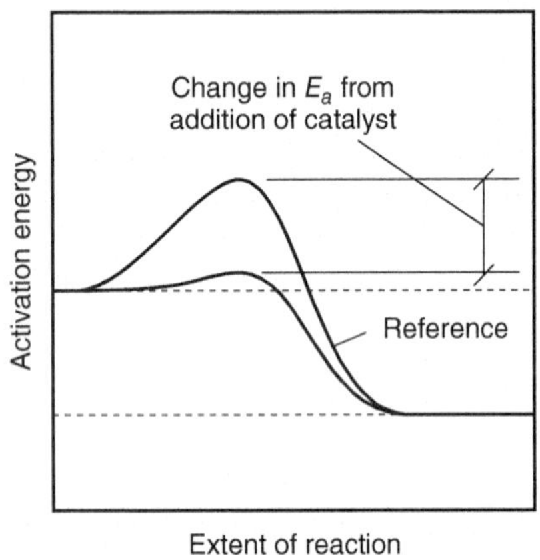

Figure 5-4
Change in activation energy when catalyst is added to reaction.

Generally, potable water sources have ionic strengths less than 10^{-2} M [approximately 400 mg/L TDS or 600 μS/cm (600 μ℧/cm)], and the effect of ionic strength on rate constants for ionic species can be estimated from the Davies equation (Eq. 5-43). The activity coefficients for monovalent and divalent ions are 0.90 and 0.66 at an ionic strength of 0.01 M, respectively. The activity coefficients for neutral species are negligible for ionic strengths less than 10^{-2} M. Weil and Morris (1949) demonstrated that the rate constant for the formation of chloramine was independent of ionic strength, supporting the reaction mechanism hypothesis that the neutral molecules NH_3 and HOCl are the principal reactants.

Accordingly, if experiments are conducted on a given water, the observed rate constant will be a function of the specific nonideality of the solution, and the observed rate constant can be used in reactor modeling as long as the ionic strength does not change. However, rate constants that were determined in low-ionic-strength waters for ionic reactions cannot be extrapolated to high-ionic-strength systems unless activity coefficients are used.

pH EFFECTS ON REACTION RATE CONSTANTS

In water treatment practice, the hydrogen ion concentration (pH) is a process variable that has a major role in the control of reaction selectivity and product distribution. The pH influences reaction rates through direct reaction pathways (e.g., precipitation of aluminum or magnesium hydroxide), determines whether reactant species are ionic, or acts as a reaction catalyst. It is only in the latter case, however, that the actual rate constant is affected; in the other cases, it is the activity (concentration) of the reactants that are affected. Often, control of the pH will permit acceleration of desired reaction pathways. The influence of pH on reaction rates is illustrated by the monochloramine formation reaction in the following example.

Example 5-8 Effect of pH on monochloramine formation

Monochloramine (NH_2Cl) is formed when ammonia (NH_3) and hypochlorous acid (HOCl) react, as shown in the following second-order reaction:

$$NH_3 + HOCl \rightarrow NH_2Cl + H_2O$$

Because of the long-lasting residual disinfecting properties of NH_2Cl, it is desirable to maximize the rate of formation of this chemical. Develop an expression for the change in the rate constant for the formation of NH_2Cl as a function of pH and present the results graphically, assuming the activity coefficients are unity. The following ionic equilibria relationships will be needed for developing the solution expression:

Dissociation of hypochlorous acid:

$$HOCl \rightleftarrows OCl^- + H^+ \quad K_{HOCl} = \frac{[H^+][OCl^-]}{[HOCl]} = 10^{-7.5}$$

Dissociation of ammonium:

$$NH_4^+ \rightleftarrows NH_3 + H^+ \qquad K_{NH_4^+} = \frac{[H^+][NH_3]}{[NH_4{}^+]} = 10^{-9.3}$$

Solution

1. Write the second-order rate expression for the formation of chloramines:

$$r_{NH_2Cl} = -r_{NH_3} = -r_{HOCl} = k[NH_3][HOCl]$$

2. Develop expressions for the concentration of hypochlorous acid [HOCl] and ammonia [NH_3], using the equilibria relationships given in the problem statement.
   a. The total concentration of HOCl, $C_{T,HOCl}$ may be written as

$$C_{T,HOCl} = [HOCl] + [OCl^-]$$

   b. The equilibrium relationship may be arranged to obtain an expression for [OCl^-]:

$$[OCl^-] = \frac{[HOCl]K_{HOCl}}{[H^+]}$$

   c. The expression from step 2b may be substituted into the $C_{T,HOCl}$ expression from step 2a and rearranged to obtain an expression for [HOCl] as a function of pH:

$$C_{T,HOCl} = [HOCl] + \frac{[HOCl]K_{HOCl}}{[H^+]}$$

$$[HOCl] = \frac{C_{T,HOCl}}{1 + K_{HOCl}/[H^+]}$$

d. The expression for ammonia is developed using a similar procedure:

$$C_{T,NH_3} = [NH_3] + \frac{[NH_3][H^+]}{K_{NH_4^+}}$$

$$[NH_3] = \frac{C_{T,NH_3}}{1+[H^+]/K_{NH_4^+}}$$

3. The rate expression can now be written in terms of the total ammonia, $C_{T,N}$, and total hypochlorous acid, $C_{T,HOCl}$, by substituting the expression developed in step 2 into the rate expression from step 1:

$$r_{NH_2Cl} = k[NH_3][HOCl] = k\frac{(C_{T,NH_3})(C_{T,HOCl})}{(1+[H^+]/K_{NH_4^+})(1+K_{HOCl}/[H^+])}$$

4. Write an expression for the observed rate constant that can be plotted as a function of pH. An observed rate constant can be expressed in terms of the actual rate constant k, hydrogen ion concentration $[H^+]$, and equilibrium constants K_{HOCl} and $K_{NH_4^+}$:

$$k_{obs} = \frac{k}{(1+[H^+]/K_{NH_4^+})(1+K_{HOCl}/[H^+])}$$

Where

$$[H^+] = 10^{-pH}.$$

5. Plot the expression from step 4:

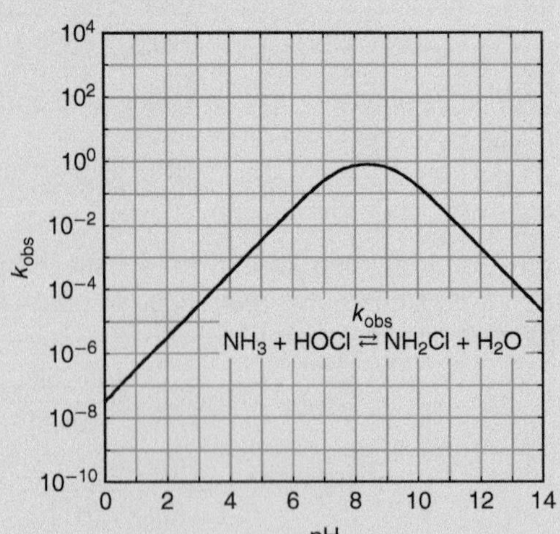

Comment

The observed rate constant depends strongly on pH, as shown on the plot from step 5, with a maximum formation rate occurring at pH 8.3, corresponding to the pH at which the product of the concentrations of the undissociated reactants is also a maximum. It is also interesting to note that the true rate constant has remained unchanged. Thus, the control of the system pH is an important factor in proper design and operation of water treatment plants.

Determination of Reaction Rate Constants

The rate at which reactions occur usually is determined experimentally by measuring the concentration of either a reactant or a product as the reaction proceeds to completion. The measured results are then compared to the corresponding results obtained from various standard rate equations by which the reaction under study is expected to proceed (see Chap. 6).

The chemical structure of a compound may also be used to predict its behavior in a chemical system, known as a quantitative structure-activity relationship (QSAR). One such method for determining rate constants is the linear free-energy relationship (LFER), which was proposed by Hammett (1935, 1938). For example, Valentine and Jafvert (1988) studied the mechanisms of monochloramine destruction, an important step involved the acid-catalyzed reaction that forms NH_3Cl^+, as shown in the reaction

$$NH_2Cl + HA_i \overset{k_{c,i}}{\rightleftarrows} NH_3Cl^+ + A^- \tag{5-89}$$

where HA_i = proton-donating species, i
$k_{c,i}$ = specific catalysis rate constant for ith proton-donating species

Valentine and Jafvert (1988) developed an LFER to relate species-specific catalysis rate constants to acid dissociation constants to predict the effect of carbonate and silicate on the rate constant. The LFER relationship can be expressed in the form

$$\log\left(\frac{k_{c,i}}{N_{P,i}}\right) = C_1\left[pK_{a,i} + \log\left(\frac{N_{P,i}}{N_{Q,i}}\right)\right] + C_2 \tag{5-90}$$

where $N_{P,i}$ = number of exchangeable protons on species HA_i
$N_{Q,i}$ = maximum number of protons with which conjugate base could combine
$K_{a,i}$ = acid dissociation constant for species HA_i
C_1, C_2 = constants determined based on experimental rate constants

Using Eq. 5-90, a linear relationship may be obtained if $\log(k_{c,i}/N_{P,i})$ is plotted versus $pK_{a,i} + \log(N_{P,i}/N_{Q,i})$, as displayed on Fig. 5-5. As shown in

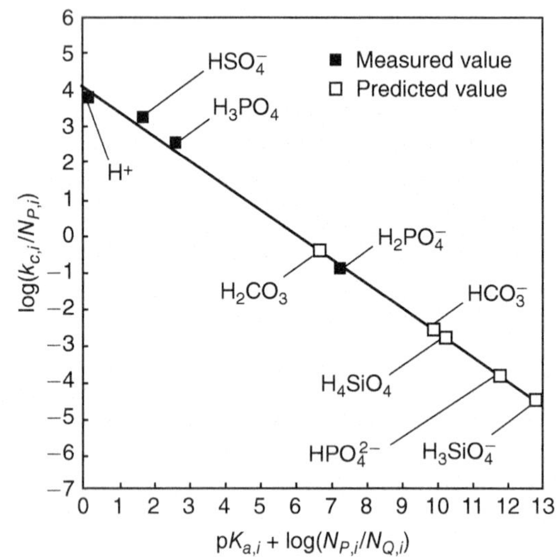

Figure 5-5
Linear free-energy relationship relating monochloroamine degradation rate constants to acid dissociation constants. (Adapted from Valentine and Jafvert, 1988).

Eq. 5-90 the activation energy is related linearly to the free energy of losing a proton and the number of protons that the acid can donate hence the name linear free-energy relationship.

As LFER methods are developed and refined they will be very useful in elucidating reaction mechanisms and predicting rate constants in water treatment. Brezonik (1990) and Schwarzenbach et al. (2003) present additional discussion on QSARs LFERs and their application to water chemistry.

5-5 Determination of Reaction Rate Laws

The relationship between the reaction mechanism and the rate expression is examined briefly in this section. This relationship can (1) explain why thermodynamics provides the necessary condition for reactions to proceed but not the sufficient condition and (2) provide insight into the functional form of the rate law. Empirical rate laws will be presented later in this chapter and are of ten used for reactor design because the reaction mechanism is not known.

Reaction Rate Laws for Individual Reaction Steps

The rate mechanism often includes numerous reaction steps and the reaction rate depends on which of these reaction steps is the slowest or rate controlling. An individual reaction rate step is called an elementary reaction and corresponds to one step in the reaction pathway. Also an elementary reaction is used to identify the number of atoms or molecules involved in the reaction mechanism. Consequently the rate law for that step depends on the concentrations of the reacting species raised to the power given by the stoichiometric coefficients. If a certain fraction of reactant A produces products, then the elementary reaction is given by the expression

$$A \rightarrow \text{products} \quad \text{(unimolecular reaction)} \tag{5-91}$$

For the reaction shown in Eq. 5-91 the rate law is first order because the reaction rate depends on the amount of reactant that is present:

$$r_A = -k[A] \tag{5-92}$$

If the elementary reaction involves the collision of two molecules, the reaction rate will depend on the product of the concentrations of both reactants because the product of the concentrations is proportional to the probability of two molecules colliding. For the collision of two molecules, A and B, the elementary reaction and the rate law are given by the following expressions, respectively:

$$A + B \rightarrow \text{products} \quad \text{(bimolecular reaction)} \tag{5-93}$$

$$r_A = -k[A][B] \tag{5-94}$$

It is also possible to have a bimolecular elementary reaction that involves the collisions of just A, and the elementary reaction and rate law are given by the following expressions, respectively:

$$A + A \rightarrow \text{products} \tag{5-95}$$

$$r_A = -k[A][A] = -k[A]^2 \tag{5-96}$$

If the elementary reaction involves the collision of A, B, and C, the rate law will depend on the product of the concentrations of A, B, and C because the product of the concentrations is proportional to the probability of the three molecules colliding. In this case, the elementary reaction and the rate law are given by the following expressions, respectively:

$$A + B + C \rightarrow \text{products} \quad \text{(trimolecular reaction)} \tag{5-97}$$

$$r_A = -k[A][B][C] \tag{5-98}$$

From a practical point of view, reactions that involve the simultaneous collision of three molecules are highly unlikely.

The reaction shown in Eq. 5-21 may be written to show the relationship between the forward- and reverse-reaction rate constants k_f and k_r when considering a reversible elementary reaction:

$$aA + bB \underset{k_r}{\overset{k_f}{\rightleftarrows}} cC + dD \tag{5-99}$$

where k_f = reaction rate constant for forward reaction
k_r = reaction rate constant for reverse reaction

The rate law for the reaction shown in Eq. 5-99 may be written as

$$r_A = -k_f[A]^a[B]^b + k_r[C]^c[D]^d \tag{5-100}$$

At equilibrium, the net rate is equal to zero; and, if the activity coefficients are equal to 1, the ratio of the product concentrations to reactant concentrations raised to the respective stoichiometric powers is equal to the equilibrium constant K_c. In addition, the equilibrium constant is equal to the ratio of the forward rate constant divided by the reverse rate constant:

$$\frac{k_f}{k_r} = \frac{[\mathrm{C}]^c[\mathrm{D}]^d}{[\mathrm{A}]^a[\mathrm{B}]^b} = K_c \quad \text{when } r_\mathrm{A} = 0 \tag{5-101}$$

Reaction Rate Expressions for Overall Reaction

The reaction mechanism is often different than what is given by the stoichiometric equation. In such cases, the rate expression is not necessarily related to the stoichiometric equation. If the details of the chemistry involved in the reaction are known, we may be able to develop the reaction pathway and rate expression from the reaction pathway. This will involve proposing a reaction pathway. For example, suppose that the stoichiometric equation for a reaction is given by the reaction

$$2\mathrm{A} + \mathrm{B} \rightarrow \mathrm{D} \tag{5-102}$$

We then must propose a pathway for this reaction to proceed. Based on the chemistry that is involved, we propose the following elementary reactions:

$$\mathrm{A} + \mathrm{B} \xrightarrow{k_1} C \tag{5-103}$$

$$\mathrm{C} \xrightarrow{k_{-1}} A + B \tag{5-104}$$

$$\mathrm{C} + \mathrm{A} \xrightarrow{k_2} D \tag{5-105}$$

When we add all the elementary reactions in Eqs. 5-103 to 5-105, the elementary reactions must add up to the stoichiometric equation, Eq. 5-102. While this is not shown here, the sum of Eqs. 5-103 to 5-105 do add up to Eq. 5-102. According to Eqs. 5-103 to 5-105, the rate of formation of A is given by this expression

$$r_\mathrm{A} = -k_1[\mathrm{A}][\mathrm{B}] + k_{-1}[\mathrm{C}] - k_2[\mathrm{C}][\mathrm{A}] \tag{5-106}$$

While Eq. 5-106 is perfectly valid, it is not particularly useful because it involves a reactive intermediate, C, which may be difficult to measure. Accordingly, the rate law that is given by Eq. 5-106 is not useful because reactor mass balances will be written on the principal reactants and products, A, B, and D. So we must develop an expression for C that we can substitute into Eq. 5-106. In this case, we assume that all the highly reactive species (e.g., radicals) achieve a pseudo-steady-state concentration. For purposes of illustration, we assume C is a highly reactive intermediate. For this example, the net rate of C formation, r_C, can be assumed to be zero (pseudo-steady-state assumption).

$$r_\mathrm{C} = 0 = k_1[\mathrm{A}][\mathrm{B}] - k_{-1}[\mathrm{C}] - k_2[\mathrm{C}][\mathrm{A}] \tag{5-107}$$

The concentration of C can be determined by solving Eq. 5-107 for C:

$$[C] = \frac{k_1[A][B]}{k_{-1} + k_2[A]} \tag{5-108}$$

According to Eq. 5-108 the concentration of C depends on A and B, and A and B are functions of time. C rapidly adjusts its concentration according to Eq. 5-107 because it is highly reactive. According to Eq. 5-107, three reactions are responsible for the creation and disappearance of C. By assuming that the net rate is zero, we assume that the reaction is at steady state. However, the concentration of C will in fact change with time because the concentrations of A and B change with time. This is why we call it pseudo–steady state. The final rate expression may be obtained by substituting Eq. 5-108 into Eq. 5-106:

$$r_A = -k_1[A][B] + \frac{k_{-1}k_1[A][B]}{k_{-1} + k_2[A]} - \frac{k_2k_1[A]^2[B]}{k_{-1} + k_2[A]} \tag{5-109}$$

After algebraic manipulation the final form of the rate law may be obtained:

$$r_A = -\frac{2k_1k_2[A]^2[B]}{k_{-1} + k_2[A]} \tag{5-110}$$

Because catalytic and/or radical species concentrations are often small compared to the products and reactants, and they are highly reactive, their concentration can change rapidly in response to solution conditions. In such situations, the formation rate of these intermediate species can be assumed to be zero and the pseudo-steady-state assumption can be invoked. By setting the rate expressions of the reactive species equal to zero, an algebraic equation for each expression is obtained and the algebraic equations maybe solved to express the concentrations of the highly reactive species in terms of the reactants or products, which are easily measured. Alternatively, in catalytic or enzyme-facilitated reactions a mass balance may be performed on the total concentration of the catalyst or enzyme, which does not change with time. The resulting equation is rearranged and solved for reactive intermediates or catalytic species in terms of the reactants, products, or constants. The rate law may then be expressed in terms of only the reactants and products.

In practice, the process of obtaining a valid rate expression includes the postulation of a rate mechanism and collecting data for a variety of reactant concentrations. The proposed rate law is compared to the data and the validity of the rate law can be examined, as discussed in Chap. 6. Finding the appropriate rate law often involves an iterative solution process. In many instances, however, the rate mechanism is complex and empirical rate laws are fit to data and used in mass balances. Empirical rate laws are discussed later in the following section.

In summary, the steps used to determine a rate law are

1. Propose the reaction pathway.
2. Write out the rate laws for each species in the pathway.
3. Invoke the pseudo-steady-state assumption for reactive intermediates and/or perform a mass balance on the catalytic species. Rearrange these expressions to obtain expressions of the reactive intermediates or catalytic species in terms of the reactants or products, which are easily measured.
4. Obtain a rate expression and eliminate the hard-to-measure species from the rate law using the results from step 3.
5. Check to see that the sum of all the elementary reactions equals the overall stoichiometric equation.
6. Collect data in various reactors and compare the concentration profiles to what would be predicted using postulated kinetic mechanisms.
7. Repeat steps 1 through 6 until good agreement is obtained.

Empirical Reaction Rate Expressions

In many cases, there is not sufficient data or resources available to determine a general expression for a rate law (especially given the unique composition of water sources); consequently, empirical forms are fit to data and used in reactor mass balances. For example, a first-order rate expression is often used to describe biochemical oxygen demand (BOD) degradation in receiving waters:

$$r_L = -k_L L \tag{5-111}$$

where r_L = rate of BOD loss
L = BOD concentration
k_L = observed first-order rate constant

Equation 5-111 represents a major simplification of all the processes that are occurring, given the variety of the types of organic compounds that are being oxidized and the enzyme pathways that are used. Nevertheless, this empirical approach appears to describe the demand for dissolved oxygen in the BOD test.

When rate laws are not available, the following rate expressions may be fit to rate data:

$$r_i = \begin{cases} -k_i & \text{(5-112)} \\ -k_i C_i & \text{(5-113)} \\ -k_i C_i^2 & \text{(5-114)} \\ -k_i C_i^{n_i} & \text{(5-115)} \end{cases}$$

where r_i = reaction rate for species i
k_i = rate constant for species i

C_i = concentration of species i (more correctly the activity should be used rather than the concentration, but the activity coefficient is often assumed to equal 1 and/or included in the rate constant)
n_i = reaction order

Another useful empirical rate law uses first-order kinetics (Eq. 5-113) and declining first-order rate constants:

$$k_{1st} = \begin{cases} \dfrac{k_{1st,0}}{(1+F_i t)^{n_i}} & \text{(5-116)} \\ \dfrac{k_{1st,0}}{(1+F_i z)^{n_i}} & \text{(5-117)} \end{cases}$$

Where k_{1st} = retarded first-order rate constant
$k_{1st,0}$ = initial first-order rate constant
F_i, n_i = empirical parameters
t = time
z = position in reactor

Equations 5-116 and 5-117 can be used to describe trends that have been observed in ozone mass transfer studies, particle removal, and biodegradation rates. When dealing with particle removal, n_i would be related to the particle size distribution. For biological reactions, the rate of reaction declines over time because the more readily degradable compounds are consumed first followed by the more recalcitrant compounds. In biological reactions, n_i is related to variability of the biodegradability of the individual constituents (soluble, colloidal, and particulate) that comprise BOD.

5-6 Reactions Used in Water Treatment

The major chemical reactions that occur in water are (a) acid-base reactions, (b) precipitation (dissolution reactions), (c) complexation reactions with ligands (metal anion reactions), and (d) redox reactions (oxidation and reduction reactions). Acid-base reactions are very fast (reaching equilibrium in less than a second) because they often involve only a proton transfer. Precipitation reactions often involve the coordination of anions around a cation and are relatively fast in the formation of amorphous solids (10 to 1000 min) and much slower (1 to 10,000 years) in the subsequent formation of crystals. Oxidation–reduction reactions follow many steps through specific single-electron transfers and, therefore, can be either very fast or very slow depending on the reaction mechanism. In general, acid–base, complexation, and precipitation reactions tend to be reversible and redox reactions are often not reversible, because a significant amount of energy is often released for each of the elementary reactions that are involved in the overall reaction.

Acid-Base Reactions

Acid–base reactions are common in water treatment, and pH has a significant effect on the chemical species present in water and on the efficiency of many treatment processes. In addition, acid-base reactions proceed faster than many other equilibrium reactions, making these reactions feasible given the time scale available for water treatment. Alkalinity, as discussed in Chap. 2, should be reviewed because of its importance to acid-base chemistry. Many acid–base reactions can be described by the loss of a proton, as shown by the expression

$$HA \rightleftarrows H^+ + A^- \tag{5-118}$$

where HA = acid species
H^+ = hydrated proton (i.e., H_3O^+)
A^- = conjugate base species

At equilibrium, the following expression can be used to relate the activities of the species in Eq. 5-118 to one another:

$$\frac{\gamma_{H^+}[H^+]\gamma_{A^-}[A^-]}{\gamma_{HA}[HA]} = \frac{\{H^+\}[A^-]}{\{HA\}} = K_a \tag{5-119}$$

where K_a = equilibrium constant for acid (HA) dissociation, used when acid donates a proton to a water molecule

Alternately, the reaction that occurs when a conjugate base accepts a proton from water may be written as

$$H^+ + A^- \rightleftarrows HA \tag{5-120}$$

The equilibrium relationship for the reaction shown in Eq. 5-120 is expressed as

$$\frac{\{HA\}}{\{H^+\}\{A^-\}} = K_b$$

where K_b = equilibrium constant for base (A^-), used when base accepts a proton from a water molecule

Another relationship that is used when analyzing acid–base reactions is a mass balance on the conjugate base:

$$C_{T,A} = [HA] + [A^-] \tag{5-121}$$

where $C_{T,A}$ = total concentration of species A

The use of the concepts shown above is presented in the following example.

The charge balance and proton condition relationships are also used to solve acid–base equilibrium problems. Charge balance is given by the expression

$$\sum \begin{pmatrix} \text{final concentration} \\ \text{of species with} \\ \text{positive charge} \end{pmatrix} = \sum \begin{pmatrix} \text{final concentration} \\ \text{of species with} \\ \text{negative charge} \end{pmatrix} \tag{5-122}$$

Example 5-9 Acid–base chemistry

For water containing an acid, HA, with $pK_a = 5$ and $C_{T,A} = 10^{-3}$ mol/L, determine the concentration and percentage of [HA] and [A^-] as a function of pH. Plot the concentration of [HA] and [A^-] as a function of pH on a logarithmic scale. Assume all activity coefficients are equal to 1.0.

Solution

1. Determine the concentration of [HA] and [A^-] as a function of pH.
   a. Solve Eq. 5-119 for [A^-]:

$$[A^-] = K_a \frac{[HA]}{[H^+]}$$

   b. Substitute the result from step 1a into Eq. 5-121:

$$C_{T,A} = [HA] + [A^-] = [HA] + K_a \frac{[HA]}{[H^+]}$$

   c. Solve the expression developed in step 1b for [HA]:

$$[HA] = \frac{C_{T,A}}{1 + K_a/[H^+]}$$

   d. The expression for [A^-] is developed following a similar procedure:

$$[A^-] = \frac{C_{T,A}}{1 + [H^+]/K_a}$$

2. Determine the [HA] and [A^-] concentrations at pH values ranging from 0 to 14. A sample calculation for pH 4 is shown below.
   a. The concentration of [HA] at pH 4, or [H^+] = 10^{-4}, is as follows:

$$[HA] = \frac{C_{T,A}}{1 + K_a/[H^+]} = \frac{10^{-3}\ \text{mol/L}}{1 + 10^{-5}/10^{-4}} = 9.09 \times 10^{-4}\ \text{mol/L}$$

   b. The concentration of [A^-] at pH 4, or [H^+] = 10^{-4}, is as follows:

$$[A^-] = \frac{C_{T,A}}{1 + [H^+]/K_a} = \frac{10^{-3}\ \text{mol/L}}{1 + 10^{-4}/10^{-5}} = 9.09 \times 10^{-5}\ \text{mol/L}$$

   c. Determine the fraction of the total concentration of [HA] and [A^-] at pH 4:

$$\frac{[A^-]}{C_{T,A}} \times 100 = \frac{9.09 \times 10^{-5}}{10^{-3}} \times 100 = 9.1\%$$

$$\frac{[HA]}{C_{T,A}} \times 100 = \frac{9.09 \times 10^{-4}}{10^{-3}} \times 100 = 90.9\%$$

3. Plot concentration profiles (log concentration) for [HA] and [A^-] versus pH:

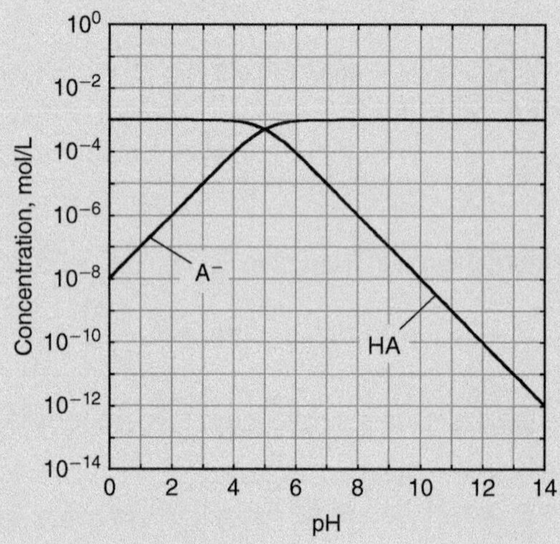

Comment

As shown in the plot constructed in step 3, the acid is undissociated at pH values less than 2 units below pK_a and nearly completely dissociated 2 units above pK_a. Further, the acid is 50 percent dissociated at a pH equal to pK_a.

The proton condition, a mass balance on protons starting with the species that are added to the solution, is given by the expression

$$\sum \begin{pmatrix} \text{concentration of species} \\ \text{that donate protons} \end{pmatrix} = \sum \begin{pmatrix} \text{concentration of species} \\ \text{that receive protons} \end{pmatrix} \quad (5\text{-}123)$$

Use of the relationships defined in the above equations is presented in the following example.

Example 5-10 Solving acid-base problems as system of equations with unknown values

The acid HA is dissolved in water. Set up the four general equations that can be used to determine [HA], [A^-], [H^+], and [OH^-].

Solution

1. Set up the general solution as a series of four equations with four unknowns ([H^+], [OH^-], [HA], [A^-]) that must be solved simultaneously to arrive at a solution.
   a. Mass balance on A:

$$C_{T,A} = [HA] + [A^-]$$

   b. Definition of equilibrium constant for water, K_w:

$$K_w = 10^{-14} = [H^+][OH^-]$$

   c. Definition of equilibrium constant for the acid, K_a:

$$K_a = \frac{[H^+][A^-]}{[HA]}$$

   d. For the proton condition the initial reactants are H_2O and HA. Consequently, the species that are formed from the loss of a proton are A^- and OH^-, and the species that is formed from the addition of a proton is H^+. In this case, the proton condition is equal to the charge balance. Thus

$$[A^-] + [OH^-] = [H^+]$$

2. The system of four equations with four unknowns can be solved using various methods; however, for complex chemical systems, chemical equilibrium models such as MINTEQA2 (U.S. EPA, 1999) or Visual MINTEQ (Gustafsson, 2011), may be required.

Comment

Note that the chemical species CO_3^{2-} also participates in the acid–base reactions (see Eqs. 5-5 and 5-6). Thus, equilibrium of the species H_2CO_3, HCO_3^-, and H^+ would have to be calculated simultaneously, which is illustrative of the complexity that can result in many of the reactions encountered in water chemistry.

Precipitation–Dissolution Reactions

The equilibrium constant for a compound in its solid phase and its ions in solution is known as the solubility product. A compound that has a low solubility in water is not likely to dissolve or, if present in excess of its equilibrium value and, given sufficient time, the compound will precipitate. A compound with a high solubility is more likely to be dissolved in water. The general solubility equilibrium equation may be written as

$$A_aB_b(s) \rightleftarrows aA^{m+} + bB^{n-} \tag{5-124}$$

The equilibrium relationship for the reaction shown in Eq. 5-124 is

$$\gamma_A^a \gamma_B^b \left[A^{m+}\right]^a \left[B^{n-}\right]^b = K_{SO} \tag{5-125}$$

Where K_{SO} = solubility equilibrium constant for reaction shown in Eq. 5-124

γ_A, γ_B = activity coefficients for species A and B, respectively

For the equilibrium relationship shown in Eq. 5-125, if the product of the reactants, $[A^{m+}]^a\,[B^{n-}]^b$, is less than the K_{SO} value (assuming $\gamma_A = \gamma_B = 1$), precipitation will not occur. However, if the product of the reactants is greater than the K_{SO} value, the solid phase will precipitate until the K_{SO} value is obtained. It is important to note that the solubility product is an equilibrium constant. As shown in Eq. 5-125, it is the ratio of the activities of the products raised to their stoichiometric power divided by the reactant activity. In this case the reactant activity is a pure solid and a pure solid has an activity of 1.0.

Example 5-11 Solubility of calcium carbonate

The pK_{s0} value for the precipitation-dissolution reaction of calcium carbonate in water at 25 °C is 8.48, and the reaction may be written as

$$CaCO_3(s) \rightleftarrows Ca^{2+} + CO_3^{2-}$$

If a sufficient amount of $CaCO_3(s)$ is added to pure water so that equilibrium is reached, determine how much $CaCO_3(s)$ is dissolved in the water. Neglect ionic strength effects and the formation of bicarbonate and carbonic acid (assume the pH is high).

Solution

1. Determine the concentration of Ca^{2+} present using the solubility constant and by noting that for every mole of calcium ion formed there is an equivalent mole of carbonate ion (i.e., $Ca^{2+} = CO_3^{2-}$):

$$K_{SO} = [Ca^{2+}][CO_3^{2-}] = [Ca^{2+}]^2$$

$$[Ca^{2+}] = (K_{S0})^{1/2} = (10^{-8.48})^{1/2} = 5.75 \times 10^{-5} \text{mol/L as } Ca^{2+}$$

2. Compute the amount of $CaCO_3$ dissolved in solution. For each mole of Ca^{2+} formed, 1 mole of $CaCO_3(s)$ is dissolved; therefore, the concentration of $CaCO_3(s)$ dissolved may be computed:

$$\begin{aligned}
[CaCO_3] &= (5.75 \times 10^{-5} \text{ mol } Ca^{2+})\left(\frac{1 \text{ mol } CaCO_3}{1 \text{ mol } Ca^{2+}}\right) \\
&= 5.75 \times 10^{-5} \text{ mol/L} \\
CaCO_3 &= (5.75 \times 10^{-5} \text{ mol/L})(100 \text{ g/mol } CaCO_3)(1000 \text{ mg/g}) \\
&= 5.75 \text{ mg/L}
\end{aligned}$$

Solubility is affected by temperature, competing ions, and solution pH. A given solid in solution may be present and in equilibrium with one or more of its dissolved species. Further, the chemical equation used to explain the precipitation–dissolution reaction may be expressed in terms of pH (see the following discussion on complexation). The concentration of the dominant species present may be plotted together for a graphical presentation of solubility, as shown in the following example. Care should be taken when selecting chemical species relevant to a particular precipitation–dissolution reaction (Morel and Hering, 1993).

Example 5-12 Solubility of aluminum hydroxide

Amorphous aluminum hydroxide $Al(OH)_3(s)$ is a form of Al(III) that is formed when alum is added to water as part of coagulation or the destabilization of particles in solution. Given the following information, calculate the total Al(III) concentration in a solution at equilibrium with $Al(OH)_3(s)$ at pH 7.5. Also develop appropriate equations for each species and plot the results to obtain the equilibrium Al(III) concentration.

$$\begin{aligned}
Al(OH)_3(s) + 3H^+ &\rightleftarrows Al^{3+} + 3H_2O & pK_{S0} &= -10.8 \\
Al(OH)_3(s) + 2H^+ &\rightleftarrows AlOH^{2+} + 2H_2O & pK_{S1} &= -5.8 \\
Al(OH)_3(s) + H^+ &\rightleftarrows Al(OH)_2^+ + H_2O & pK_{S2} &= -1.5 \\
Al(OH)_3(s) &\rightleftarrows Al(OH)_3^0 & pK_{S3} &= 4.2 \\
Al(OH)_3(s) + H_2O &\rightleftarrows Al(OH)_4^- + H^+ & pK_{S4} &= 12.2
\end{aligned}$$

Solution

1. Write the mass balance equation on Al(III):

$$Al(III) = [Al^{3+}] + [AlOH^{2+}] + [Al(OH)_2^+] + [Al(OH)_3^0] + [Al(OH)_4^-]$$

2. Replace each aluminum species with its respective *K* relationship.
   a. The relationship for Al(III) is shown below:

$$K_{S0} = 10^{10.8} = \frac{[Al^{3+}]\cancel{[H_2O]^3}}{[H^+]^3\cancel{[Al(OH)_3(s)]}}$$

   The terms that are crossed out in the above expression have an activity of 1.

$$[Al^{3+}] = 10^{10.8}[H^+]^3 = K_{S0}[H^+]^3$$

   b. Other species are derived using a similar procedure, resulting in the following expression for the total Al(III) concentration:

$$Al(III) = K_{S0}[H^+]^3 + K_{S1}[H^+]^2 + K_{S2}[H^+] + K_{S3} + K_{S4}/[H^+]$$

3. Substitute in a pH value of 7.5 and solve for Al(III):

$$[Al(III)] = (10^{10.8})(10^{-7.5})^3 + (10^{5.8})(10^{-7.5})^2 + (10^{1.5})(10^{-7.5}) + (10^{-4.2}) + (10^{-12.2})/(10^{-7.5}) = 8.4 \times 10^{-5} M$$

4. Develop appropriate equations for each species and plot the results to obtain an Al(III) concentration.
   a. Write an equation for each species as a function of pH, as shown in step 2:

$$\begin{aligned}
[Al^{3+}] &= K_{S0}[H^+]^3 = 10^{10.8}[H^+]^3 \\
[AlOH^{2+}] &= K_{S1}[H^+]^2 = 10^{5.8}[H^+]^2 \\
[Al(OH)_2^+] &= K_{S2}[H^+] = 10^{1.5}[H^+] \\
[Al(OH)_3^0] &= K_{S3} = 10^{-4.2} \\
[Al(OH)_4^-] &= K_{S4}/[H^+] = 10^{-12.2}[H^+]
\end{aligned}$$

   where $[H^+] = 10^{-pH}$

   b. Plot the equations for each species and identify the line that represents the total Al(III) concentration:

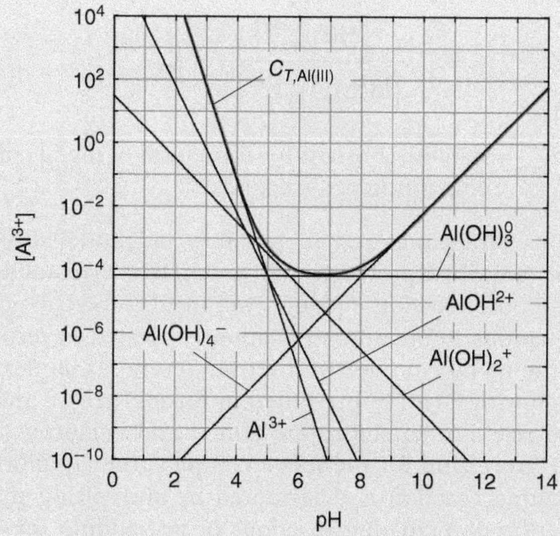

Comment

It should be noted that on a log Al species versus pH scale, the equilibrium Al species concentrations are linear functions of pH in most cases. The slope is determined by the concentration dependence of the species as a function of hydrogen ion concentration. For example, $Al(OH)_4^-$ depends on the hydrogen ion concentration raise to the −1 power and this means it will be linearly related to the pH and have slope of +1. On a pC− pH diagram it will have a slope of −1 because pC is the negative of the log of the concentration.

Complexation Reactions

Complexation reactions are important in water treatment because the reactions may be used to reduce the concentration of free-metal concentrations. In addition, complexation reactions can be used to reduce toxicity or change the adsorptive properties of metals. The formation of complexes in water generally involves the reaction between a metal ion (M) and a ligand (L). Ligands may be added individually or cumulatively, as shown in the reaction

$$M(L)_m^{x+} + nL^{y-} \rightleftarrows M(L)_{m+n}^{x-ny} \tag{5-126}$$

where M = metal ion
L = ligand
m, n = number of ligands added
x, y = valances of cationic complex and anionic ligand, respectively

The equilibrium relationship for the reaction shown in Eq. 5-126 is written as

$$\frac{\{M(L)_{m+n}^{x-ny}\}}{\{M(L)_m^{x+}\}\{L^{y-}\}^n} = K_{m+n} \tag{5-127}$$

where K_{m+n} = stability constant for formation of metal complex containing $m + n$ ligands

The reactions of ligands and metals may be modeled as a system of reactions, as described previously for precipitation–dissolution reactions. In some cases, the reaction equation will need to be balanced with other reaction equations to obtain an equation expressed in terms of the solid phase and the metal complex of interest. Several examples of Al and Cu complexation with OH^- are presented in Examples 5-12 and 5-13, respectively. When reaction equations are added and balanced to arrive at an appropriate expression for the metal complex, the equilibrium constants for the resulting reaction is determined by multiplying the equilibrium constants of the participating reactions or just adding the powers of the equilibrium constants. The sign of the power of the equilibrium constant is reversed when the chemical equation is reversed. Summing chemical equations and the powers of the equilibrium constants is illustrated in the following example.

Example 5-13 Complexation reactions for copper hydroxide

For the following dissolution and complexation reactions, determine the reactions needed to create a *pC–pH* (p*C* is the negative logarithm of the concentration) diagram for the following complexation reactions. The solution is in equilibrium with solid copper hydroxide, $Cu(OH)_2(s)$. Ignore the effects of ionic strength:

$$Cu(OH)_2(s) \rightleftarrows Cu^{2+} + 2OH^- \quad K_{SO} = 10^{-19.3}$$

$$Cu^{2+} + OH^- \rightleftarrows CuOH^+ \quad K_1 = 10^{6.3}$$

$$Cu^{2+} + 2OH^- \rightleftarrows Cu(OH)_2 \quad K_2 = 10^{11.8}$$

$$Cu^{2+} + 3OH^- \rightleftarrows Cu(OH)_3^- \quad K_3 = 10^{16.4}$$

$$H^+ + OH^- \rightleftarrows H_2O \quad K_w^{-1} = 10^{14}$$

Solution

1. Rearrange the equations given in the problem statement so that H^+ and the metal-ligand complex $Cu(OH)_x$ are the only variables:

a. For Cu^{2+},

$$\begin{array}{ll} Cu(OH)_2(s) \rightleftarrows Cu^{2+} + 2OH^- & K_{SO} = 10^{-19.3} \\ 2H^+ + 2OH^- \rightleftarrows 2H_2O & K_W^{-2} = 10^{28} \\ \hline Cu(OH)_2(s) + 2H^+ \rightleftarrows Cu^{2+} + 2H_2O & K_{SO} = 10^{8.7} \end{array}$$

b. For $CuOH^+$,

$$\begin{array}{ll} Cu(OH)_2(s) \rightleftarrows Cu^{2+} + 2OH^- & K_{SO} = 10^{-19.3} \\ Cu^{2+} + OH^- \rightleftarrows CuOH^+ & K_1 = 10^{6.3} \\ H^+ + OH^- \rightleftarrows H_2O & K_W^{-1} = 10^{14} \\ \hline Cu(OH)_2(s) + H^+ \rightleftarrows CuOH^+ + H_2O & K_{S1} = 10^{1} \end{array}$$

c. For $Cu(OH)_2$,

$$\begin{array}{ll} Cu(OH)_2(s) \rightleftarrows Cu^{2+} + 2OH^- & K_{SO} = 10^{-19.3} \\ Cu^{2+} + 2OH^- \rightleftarrows Cu(OH)_2 & K_2 = 10^{11.8} \\ \hline Cu(OH)_2(s) \rightleftarrows Cu(OH)_2 & K_{S2} = 10^{-7.5} \end{array}$$

d. For $Cu(OH)_3^-$,

$$\begin{array}{ll} Cu(OH)_2(s) \rightleftarrows Cu^{2+} + 2OH^- & K_{SO} = 10^{-19.3} \\ Cu^{2+} + 3OH^- \rightleftarrows Cu(OH)_3^- & K_3 = 10^{16.4} \\ H_2O \rightleftarrows H^+ + OH^- & K_W = 10^{-14} \\ \hline Cu(OH)_2(s) + H_2O \rightleftarrows Cu(OH)_3^- + H^+ & K_{S3} = 10^{-16.9} \end{array}$$

2. Using the equations developed in step 1, develop equations that may be plotted on a pC–pH diagram.
   a. The equation from step 1a may be rearranged for the equilibrium constant:

$$K_{SO} = \frac{[Cu^{2+}][H_2O]^2}{[H^+]^2[Cu(OH)_2(s)]}$$

   resulting in the expression

$$\log[Cu^{2+}] = \log K_{SO} - 2\text{pH}$$
$$-\log[Cu^{2+}] = -\log K_{SO} + 2\text{pH}$$
$$\text{p}[Cu^{2+}] = \text{p}K_{SO} + 2\text{pH}$$

   The corresponding expressions for the remaining equations developed in step 1 are

b. For the equation derived in step 1b, the expression may be written as

$$\log[CuOH^+] = \log K_{S1} - pH$$
$$p[CuOH^+] = pK_{S1} + pH$$

c. For the equation derived in step 1c, the expression may be written as

$$p[Cu(OH)_2] = pK_{S2}$$

d. For the equation derived in step 1d, the expression may be written as

$$\log[Cu(OH)_3^-] = \log K_{S3} + pH$$
$$p[Cu(OH)_3^-] = pK_{S3} - pH$$

Comment

Using these equations, it is possible to create a pC–pH diagram that will show the major constituents for this water at a range of pH values. The pC–pH diagrams are also useful for developing an understanding of a specific water quality system.

Ligands that are commonly involved in complexation reactions include CN^-, OH^-, Cl^-, F^-, CO_3^{2-}, NO_3^-, SO_4^{2-}, NH_3, S^{2-}, SO_3^{2-}, PO_4^{3-}, and many organic molecules with appropriate functional groups (Morel and Hering, 1993). Complexes can also form with NOM. The NOM complex that forms with aluminum ion (see Example 5-12) is thought to control the amount of alum addition in the coagulation process (see Chaps. 8 and 9 for more discussion). The NOM complex that forms with Fe(II) is very strong and makes it difficult to oxidize Fe(II) using chemical oxidation. Iron is removed by oxidizing it to Fe(III) and precipitating it as $Fe(OH)_3$ (see Chap. 8 for a more detailed discussion; see also Stumm and Morgan, 1996).

Oxidation–Reduction Reactions

Reactions that involve the transfer of electrons between two chemical species are known as oxidation–reduction, or redox, reactions. In a redox reaction, one species is reduced (gains electrons) and one species is oxidized (loses electrons). Redox reactions are typically reported as half reactions to show the number of electrons transferred. Thus, to obtain a complete oxidation–reduction reaction, an oxidation half reaction and a reduction half reaction must be combined. The general expression of a half reaction for the reduction of a species is as follows:

$$Ox_A + ne^- \rightarrow Red_A \quad (5\text{-}128)$$

where Ox_A = oxidized species A
n = number of electrons transferred
e^- = electron
Red_A = reduced species A

Although the oxidized species A is reduced during this reaction, it is called an oxidant (or electron acceptor) because the oxidized species A oxidizes another species as it is reduced. The half reaction for the oxidation of a species may be expressed as

$$Red_B \rightarrow Ox_B + ne^- \quad (5\text{-}129)$$

where Ox_B = oxidized species B
Red_B = reduced species B

Although the reduced species B is oxidized during this reaction, it is called a reductant (or electron donor) because the reduced species B reduces another species as it is oxidized.

The two half reactions may be combined to obtain the following overall oxidation–reduction reaction:

$$Ox_A + Red_B \rightarrow Ox_B + Red_A \quad (5\text{-}130)$$

Water treatment often involves oxidation–reduction reactions in a variety of processes such as disinfection and chemical oxidation. Redox reactions are discussed in detail in Chap. 8.

Problems and Discussion Topics

5-1 Using the principles of stoichiometry presented in the text, (a) balance the reaction for the coagulation of water with alum, $Al_2(SO_4)_3{\cdot}18H_2O$, shown below and (b) compute the amount of alkalinity, $Ca(HCO_3)_2$, consumed during the reaction:

$$Al_2(SO_4)_3 \cdot 18H_2O + Ca(HCO_3)_2$$

$$\rightleftarrows CaSO_4 + Al(OH)_3 + CO_2 + H_2O$$

5-2 During the process of photosynthesis, algae respiration can cause the pH and dissolved oxygen (O_2) concentration of water to increase. Photosynthesis can be described by the chemical reaction presented below. Balance the chemical reaction and calculate the milligrams of oxygen formed per milligram of carbon dioxide removed.

$$CO_2 + H_2O \rightarrow C_6H_{12}O_6 + O_2 \quad (5\text{-}131)$$

5-3 A water contains organic matter and ammonia. For disinfection, 2 mg/L of hypochlorous acid (HOCl) is added to the water, forming 5 mg/L monochloramine (desired end product) and 1 mg/L total organic chlorine (TOCl) as Cl. Determine the selectivity of the formation of monochloramine versus TOCl formation.

5-4 Lime, $Ca(OH)_2$, is added to water for the removal of calcium and magnesium. In many cases, Ca^{2+} and Mg^{2+} are associated with carbonate, as shown in the reaction below for calcium. The addition of lime results in the precipitation of $CaCO_3(s)$ and $Mg(OH)_2(s)$. For the precipitation reaction shown, write expressions for the concentration of each species after 50 percent conversion of calcium biocarbonate, $Ca(HCO_3)_2$; assuming there is no volume change upon reaction.

$$Ca(HCO_3)_2 + Ca(OH)_2 \rightleftarrows 2CaCO_3(s) + 2H_2O$$

5-5 Determine the ionic strength of a solution with the following constituents:

$$[Ca^{2+}] = 10^{-3}\ mol/L \quad [CO_3^{2-}] = 10^{-5}\ mol/L$$
$$[Mg^{2+}] = 10^{-5}\ mol/L \quad [HCO_3^-] = 5 \times 10^{-3}\ mol/L$$
$$[Na^+] = 10^{-4}\ mol/L$$

If the pH was measured at 7.0, what is the corresponding concentration of hydrogen ion?

5-6 Un-ionized ammonia (NH_3) is toxic to fish at low concentrations. The dissociation of ammonia in water has an equilibrium constant of $pK_a = 9.25$, described with the reaction

$$NH_4^+ \rightleftarrows NH_3 + H^+$$

Determine the ratio of NH_3 to NH_4^+ at pH values of 6, 7, 8, 9, and 10.

5-7 Given the following reaction and rate law for the oxidation of Fe(II), where DO = dissolved oxygen, determine the rate of production/loss of Fe^{2+}:

$$Fe^{2+} + \tfrac{1}{4}O_2 + \tfrac{5}{2}H_2O \rightleftarrows Fe(OH)_3 + 2H^+$$
$$r_{Fe^{2+}} = -k[Fe^{2+}][OH^-]^2[DO] \qquad k = 6.25 \times 10^{16}\ L^3/min \cdot mol^3$$

where $DO = 0.268$ mmol/L
$Fe^{2+} = 5.58$ mg/L
$pH = 6.0$

5-8 Using the data provided in Problem 5-7, determine the concentration and rate of production/loss of dissolved oxygen (DO), $Fe(OH)_3$, and acid (H^+)when Fe^{2+} = 0.3 mg/L at pH 6 (assume constant).

5-9 A common reaction pathway in biological systems involves the conversion of substrate (S) to product (P). The stoichiometric equation is

$$S \rightarrow P$$

The elementary reactions are given by the pathways given below that include an enzyme that is neither created nor destroyed:

$$S + E \xrightarrow{k_1} E \cdot S \quad \text{(reaction 1)}$$

$$E \cdot S \xrightarrow{k_{-1}} S + E \quad \text{(reverse reaction 1)}$$

$$ES \xrightarrow{k_2} P + E \quad \text{(reaction 2)}$$

Derive a rate law in terms ofthe total enzyme and substrate concentrations and the rate constants.

5-10 Using information obtained from the local water utility, compute the ionic strength of your drinking water. In addition, estimate the TDS concentration and electrical conductivity (EC) of the water. If available, measure the TDS and/or EC of the water and compare to the computed values.

5-11 Plot the activity coefficients of Na^+, Ca^{2+}, and Al^{3+} for ionic strengths from 0.001 M (very fresh water) to 0.5 M (seawater). Determine the ionic strength at which the activity coefficient corrections become important (activity coefficient less than 0.95) for monovalent, divalent, and trivalent ions.

5-12 The temperature dependence of the reaction rate is frequently expressed quantitatively using parameters other than E_a. For example, the following expression for the reaction rate constant for the BOD test is often used:

$$k_{T_2} = k_{20}(\theta)^{T_2 - 293}$$

where T_2 = temperature, K

k_{T_2} = rate constant at temperature T_2

a. Show that $\theta = \exp(E_a / RT_1T_2)$.

b. Determine E_a if θ = 1.047 and T_2 = 293 K.

c. If T_2 = 283 K and θ remains constant, what is the value of E_a?

5-13 In the field of biology, the Q_{10} term is frequently used to define the increase in reaction rate constant with temperature:

$$Q_{10} = \frac{k_{T+10}}{k_T}$$

Although Q_{10} does vary with E_a and temperature, if E_a is approximately a constant for certain reactions, Q_{10} values can be used as a good approximation. If the temperature is 25°C, what is E_a if $Q_{10} = 1.8$?

5-14 Prior to the design of chemical reactor systems, it is necessary to know the sensitivity of the reaction rate to temperature. It was stated by Arrhenius that the rate of most chemical reactions would double for every 10°C increase in temperature. Test the validity of this statement by calculating the temperature rise needed to double the rate of reaction for activation energies of 4, 55, and 125 kJ/mol for initial temperatures of 0, 20, 100, and 300°C using the Arrhenius law for temperature dependency of the rate constant. From the calculations, what conclusions can be drawn about the temperature sensitivity of reactions at various energies and temperatures, including conditions expected for water treatment processes.

5-15 For the reactions given below, what is the rate expression for the disappearance of A assuming that C is a highly reactive intermediate? An acceptable answer would propose a rate law that only involves the principal reactants and products as given in the following stoichiometric equation.

Stoichiometric equation:

$$2A + B \rightarrow D + F$$

Elementary reactions:

$$A + B \overset{k_1}{\rightarrow} C$$

$$C \overset{k-1}{\rightarrow} A + B$$

$$C + A \overset{k_2}{\rightarrow} D + F$$

5-16 Using a linear free-energy relationship (LFER), write an expression that could be used to estimate the reaction rate constant, $k_{c,i}$, for the following reaction:

$$NH_2Cl + HA \overset{k_{c,i}}{\rightarrow} NH_3Cl^+ + A^-$$

Outline a procedure that could be used to estimate the reaction rate constant using the expression.

5-17 Construct plots of (a) the log concentration and (b) the percent distribution of H_2CO_3, HCO_3^-, and CO_3^{2-} as a function of pH at 25 °C. Consider a pH range of 0 to 14. Use a C_{T,CO_3} value of 10^{-3} and assume the system is closed to the atmosphere and that the following reactions apply:

$$H_2CO_3 \rightleftarrows HCO_3^- + H^+ \quad pK_{a1} = 6.35$$

$$HCO_3 \rightleftarrows CO_3^{2-} + H^+ \quad pK_{a2} = 10.33$$

5-18 Using the thermodynamic data given below, calculate the equilibrium constants and free energy of reaction for the following reaction at 10, 25, and 35 °C:

$$H_2CO_3 \rightleftarrows HCO_3^- + H^+$$

Thermodynamic data:

$$\Delta H^\circ_{F,H_2CO_3(aq)} = -698.7\,\text{kJ/mol} \quad \Delta G^\circ_{F,HCO_3^-(aq)} = -587.1\,\text{kJ/mol}$$

$$\Delta H^\circ_{F,HCO_3^-(aq)} = -691.1\,\text{kJ/mol} \quad \Delta G^\circ_{F,H_2CO_3(aq)} = -623.4\,\text{kJ/mol}$$

$$\Delta H^\circ_{F,H^+(aq)} = 0\,\text{kJ/mol} \quad \Delta G^\circ_{F,H^+(aq)} = 0\,\text{kJ/mol}$$

5-19 Calculate the equilibrium constant K_w at (a) 25 °C and (b) 40 °C using the following free energy of formation values: [H^+] = 0 kJ/mol, [OH^-] =–157.29 kJ/mol, and [H_2O]=–237.18 kJ/mol. Determine if the disssociation of water is an endothermic or exothermic reaction. The enthalpy values for the various constituents are [H^+] = 0 kT/mol, [OH^-] =–230 kT/mol, and [H_2O] =–285.83 kT/mol.

5-20 Determine if HOCl is thermodynamically stable in water at 25 °C given the reaction

$$2HOCl \rightleftarrows 2Cl^- + 2H^+ + O_2$$

Assume HOCl = 5 mg/L as Cl_2, pH = 7, O_2(aq) = 9 mg/L, and Cl^- = 10^{-3} M. The free energies of formation for the compounds involved are given as

H^+ = 0 kJ/mol O_2(aq) = 16.44 kJ/mol H_2O =–237.18 kJ/mol

Cl^- = –131.29 kJ/mol HOCl = –79.91 kJ/mol

5-21 Using the reactions shown below for the solubility of $FeOH_3$(s), construct a pC – pH diagram and determine the Fe^{3+} concentration at pH values of 3, 5, 7, 9, and 11:

Reaction	Equilibrium Constant	Value
$Fe(OH)_3(s) + 3H^+ \rightleftarrows Fe^{3+} + 3H_2O$	$\log K_{S0}$	3.2
$Fe(OH)_3(s) + 2H^+ \rightleftarrows Fe(OH)^{2+} + 2H_2O$	$\log K_{S1}$	1.0
$Fe(OH)_3(s) + H^+ \rightleftarrows Fe(OH)_2^+ + H_2O$	$\log K_{S2}$	-2.5
$Fe(OH)_3(s) \rightleftarrows Fe(OH)_3^0$	$\log K_{S3}$	-12.0
$Fe(OH)_3(s) + H_2O \rightleftarrows Fe(OH)_4^- + H^+$	$\log K_{S4}$	-18.4

5-22 Manganese, Mn(II), is soluble in water and is present in many groundwaters because insoluble forms (e.g., MnO_2) that are contained in minerals are reduced to soluble forms. (The subsurface is a reducing environment because electron acceptors such as oxygen have been used up by heterotrophic bacteria in the A horizon of soil, comprised mainly of mineral material and organic detritus such as peat.) Ozone (O_3) is sometimes used to remove Mn according to the reaction

$$Mn^{2+} + O_{3(aq)} + H_2O \rightleftarrows MnO_{2(s)} + O_{2(aq)} + 2H^+$$

Compute the equilibrium constant for the reaction and plot the free energy as a function of the conversion of Mn^{2+} from 0.01 to 0.999 using the following data:

$$\Delta G^{\circ}_{Rxn} = -164.05 \text{ kJ / mol}$$

Assume that the initial reactant concentrations are DO = 10 mg/L, O_3 = 0.5 mg/L, Mn^{2+} = 2 mg/L, and MnO_2 = 0 mg/L.

References

Benefield, L. D., Judkins, J. F., and Weand, B. L. (1982) *Process Chemistry for Water and Wastewater Treatment*, Prentice-Hall, Englewood Cliffs, NJ.

Benjamin, M. M. (2002) *Water Chemistry*, McGraw-Hill, New York.

Brezonik, P. L. (1990) Principles of Linear Free-Energy and Structure-Activity Relationships and Their Applications to the Fate of Chemicals in Aquatics Systems, in W. Stumm (ed.), *Aquatic Chemical Kinetics*, John Wiley & Sons, New York.

David, R. L. (2000) *CRC Handbook of Chemistry and Physics*, 81st ed., CRC, Boca Raton, FL.

Davies, C. (1967) *Electrochemistry*, Philosophical Library, London.

Dean, J. A. (1992) *Lange's Handbook of Chemistry*, 14th ed., McGraw-Hill, New York.

Debye, V. P., and Hückel, E. (1923) "Zur Theorie der Elektrolyte," *Physik. Zeit.*, **24**, 185–206.

Frost, A. A., and Pearson, R. G. (1961) *Kinetics and Mechanism*,2nd ed., John Wiley& Sons, New York.

Gustafsson, J. P. (2011) Visual MINTEQ, Version 3.0a, KTH Royal Institute of Technology, Stockholm, Sweden.

Hammett, L. P. (1935) "Some Relations between Reaction Rates and Equilibrium Constants," *Chem. Rev.*, **17**, 125–136.

Hammett, L. P. (1938) "Linear Free Energy Relationships in Rate and Equilibria Phenomena, *Trans. Faraday Soc.*, **34**, 156–165.

Harned, H., and Owen, B. (1958) *The Physical Chemistry of Electrolyte Solution*, 3rd ed., Reinhold, New York.

Kern, D. M. (1960) "The Hydration of Carbon Dioxide," *J. Chem. Ed.*, **37**, 1, 14-23.

Levine, I. N. (1988) *Physical Chemistry*, 3rd ed., McGraw-Hill, New York.

Lewis, G. N., and Randall, M. (1921) "Activity Coefficient of Strong Electrolytes," *J. Am. Chem. Soc.*, **43**, 1111–1154.

McCarthy, J. J., and Smith, C. H. (1974) "A Review of Ozone and Its Application to Domestic Wastewater Treatment," *J. AWWA*, **66**, 718–725.

Moelwyn-Hughes, E. A. (1974) *Kinetics of Reactions in Solution*, Clarendon, Oxford.

Morel, F. M. M., and Hering, J. G. (1993) *Principles and Applications of Aquatic Chemistry*, Wiley-Interscience, New York.

Pankow, J. F. (1991) *Aquatic Chemistry Concepts*, Lewis, Chelesa, MI.

Poling, B. E., Prausnitz, J. M., and O'Connell, J.P. (2001) *The Properties of Gases and Liquids*, 5th ed., McGraw-Hill, New York.

Pye, D. J. (1947) "Chemical Fixation of Oxygen," *J. AWWA*, **39**, 11, 1121–1127.

Sawyer, C. N., McCarty, P. L., and Parkin, G. F. (2003) *Chemistry for Environmental Engineering*, 5th ed., McGraw-Hill, New York.

Schwarzenbach, R. P., Gschwend, P. M., and Imboden, D. M. (2003) *Environmental Organic Chemistry*, 2nd ed., John Wiley & Sons, Hoboken, NJ.

Snoeyink, V. L., and Jenkins, D. (1980) *Water Chemistry*, John Wiley & Sons, New York.

Stumm, W., and Morgan, J. J. (1996) *Aquatic Chemistry: Chemical Equilibria and Rates in Natural Waters*, 3rd ed., John Wiley & Sons, New York.

Trussell, R. R. (1998) "Spreadsheet Water Conditioning," *J. AWWA*, **90**, 6, 70–81.

U.S. EPA (1999) MINTEQA2, Version 4.0, U.S. Environmental Protection Agency, Washington, DC.

Valentine, R. L., and Jafvert, C. T. (1988) "General Acid Catalysis of Monochloramine Disproportionation," *Environ. Sci. Tech.*, **22**, 691–696.

Weil, I., and Morris, J. C. (1949) "The Rates of Formation of Monochloramine, *N* -chlormethylamine, and *N* -chlordimethylamine," *J. Am. Chem. Soc.*, **71**, 1664–1671.

6 Principles of Reactor Analysis and Mixing

Stantec's Water Treatment: Principles and Design, Updated Third Edition. John C. Crittenden, R. Rhodes Trussell, David W. Hand, Kerry J. Howe, and George Tchobanoglous.

Terminology for Reactor Analysis and Mixing

Term	Definition
Agitation	Motion induced in a fluid to achieve flocculation, maintain particles in suspension, or promote mass transfer.
Batch reactor	Vessel in which reactants are introduced and reactions are allowed to proceed with no additional inputs to or outputs from the reactor during the reaction period.
Blending	Process of combining two liquid streams to achieve a specified level of uniformity as defined by the COV.
Completely mixed flow reactor (CMFR)	An ideal flow reactor in which the contents are continuously mixed and completely homogenous; no variation in concentration or other condition exists from one location to another in the reactor. Called continuously-stirred tank reactor (CSTR) in some older texts.

Term	Definition
Coefficient of variation (COV)	Normalized standard deviation of the concentration in a stream used to define the uniformity (also homogeneity) of blending.
Conservative constituent	Constituent that does not react, transform, adsorb, or otherwise change as it passes through a reactor.
Control volume	System in which a mass balance analysis is performed.
Diffusion	Movement of molecules from a higher concentration to a lower concentration due to Brownian motion.
Dispersion	Mixing in which a constituent is transported from a higher concentration to a lower concentration by eddies formed by turbulent flow or shearing forces between fluid layers.
Flow reactor	Reactor that operates on a continuous basis with flow into and out of the reactor.
Hydraulic residence time	Theoretical time that fluid remains in a reactor, defined as the reactor volume divided by the flow rate.
Mass balance analysis	Application of the law of conservation of mass, to account for changes in any component due to fluid flow, mass transfer, or chemical transformations.
Mean residence time	Average time that fluid remains in a reactor, defined as the first moment of tracer curve.
Mixer	Device used to bring about motion in a fluid for the purpose of agitation or blending.
Mixing	General term used to refer to agitation and blending
Nonconservative constituent	Constituent that reacts or transforms as it passes through a reactor.
Plug flow reactor (PFR)	An ideal flow reactor in which no dispersion, diffusion, or mixing of contents occurs in the axial direction.
Reactor	Tank, basin, or other vessel used in environmental and chemical engineering as a container in which chemical or biological reactions for treatment or transformation can take place.
Residence time distribution (RTD)	Probability distribution function that describes the range of time that fluid elements remain within a reactor.
Steady-state analysis	Analysis conducted when a reactor is operated for a long enough period of time with a constant influent concentration such that the concentration profile in the reactor does not change with time.
System boundary	Border used to identify all of the material flows into and out of a control volume.
Tracer, chemical	Conservative chemical used to assess the flow conditions through a reactor.
Velocity gradient G	Measure of the power input per unit volume $(P/\mu V)^{1/2}$

In the environment, many of the contaminants in water are removed gradually by naturally occuring physical, chemical, and biological processes. In water treatment, the same processes that occur in nature are carried out in vessels or tanks, commonly known as *reactors*. Through the use of engineered reactors, the processes used to treat water can be accelerated under controlled conditions. The rate at which such processes occur depends on the constituents involved and conditions in the reactor, including temperature and hydraulic (mixing) characteristics.

Examples of Reactor Vessels.
http://www.stantec.com/wt3r/6-1.html

The topics presented in this chapter include (1) the types of reactors used in water treatment processes; (2) the mass balance analysis, which is the fundamental basis for the analysis of the physical, chemical, and biological processes used for water treatment; (3) ideal reactors used in modeling; (4) the modeling of reactions occurring in completely mixed batch reactors; (5) the modeling of reactions occurring in ideal continuous-flow reactors; (6) the use of tracer curves to characterize nonideal flow patterns; (7) the modeling of nonideal flow through reactors; (8) modeling the performance of nonideal reactors; (9) using tracer curves to model reactor performance; and (10) mixing.

6-1 Types of Reactors Used in Water Treatment

Unit operations and unit processes in water treatment can be carried out in a variety of reactors, which include large square and rectangular basins, cylindrical tanks, pipes, long channels, columns, and towers. Stoichiometric and kinetic descriptions of chemical reactions combined with knowledge of practical flow patterns provide the basis for reactor selection and design. Other factors to be considered include the quantity of material being processed and the structural requirements of the reactor selected (Froment and Bischoff, 1979; Kramer and Westerterp, 1963; Levenspiel, 1998; Green and Perry, 2007; Smith, 1981). The types of reactors and their applications and the hydraulic characteristics of reactors are introduced in this section.

Types of Reactors

The reactors used for water treatment can be categorized based on the operation pattern, hydraulic characteristics, unit operation occurring, and entrance and exit conditions. Several types of reactors are shown on Fig. 6-1.

Reactors Characterized by Operation Pattern

Batch and continuous-flow reactors are the principal types of reactors. *Batch* reactors are characterized by noncontinuous operation (see Fig. 6-1a). Reactants are mixed together, and the reaction is allowed to proceed to completion. *Continuous-flow* reactors operate on a continuous basis with flow into and out of the reactor (see Fig. 6-1b). Continuous-flow reactors may also be arranged sequentially to change the flow characteristics (see Fig. 6-1c).

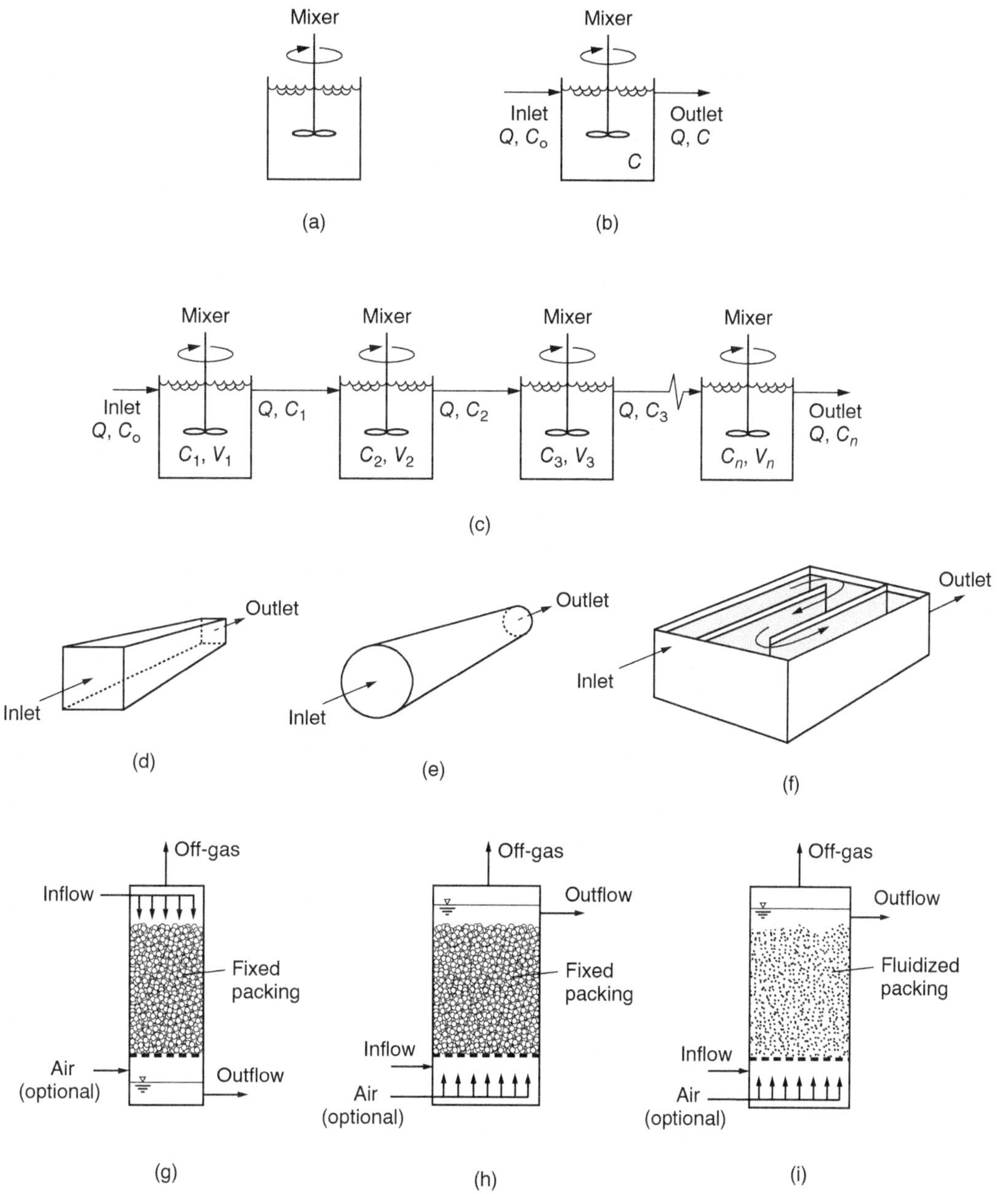

Figure 6-1
Typical reactors used in water treatment processes: (a) batch reactor; (b) continuous-flow mixed reactor; (c) continuous-flow mixed reactors in series, also known as tanks in series; (d) rectangular channel plug flow reactor; (e) circular pipe plug flow reactor; (f) serpentine configuration plug flow reactor; (g) packed-bed downflow reactor; (h) packed-bed upflow reactor; and (i) expanded-bed upflow reactor. (Adapted from Tchobanoglous et al., 2003.).

Batch reactors are used widely in the production of small-volume, specialty chemicals in the chemical processing industries. However, the use of batch reactors in water treatment applications is generally restricted to laboratory-scale investigations and chemical coagulant preparation. Continuous-flow reactors are used most commonly in full-scale water treatment plants because of the large volumes of water processed.

Reactors Characterized by Hydraulic Characteristics

Reactors used for carrying out reactions can be characterized as ideal or nonideal, according to the nature of the hydraulic and mixing conditions. In nonideal reactors, the hydraulic and mixing conditions tend to be complex (e.g., the mixing conditions in Lake Superior or a chlorine contact chamber). Ideal reactors are assumed to have uniform mixing and hydraulic conditions, depending on the specific reactor configuration. Common reactor configurations include (1) completely mixed batch reactors (CMBRs), (2) completely mixed flow reactors (CMFRs), and (3) plug flow reactors (PFRs). Reactor configurations are defined in Table 6-1 and discussed in detail in the following sections.

Table 6-1
Definition of reactors used in water treatment

Reactor	Definition
Ideal reactors	Reactors defined for purposes of modeling. Ideal assumptions, such as no dispersion or diffusion, are nearly achievable under closely controlled laboratory conditions. Definitions assume extreme fluid conditions, such as complete mixing or no mixing of reactants or products in the direction of flow.
Nonideal reactors	Mixing and/or the residence time distribution in the reactor does not meet ideal assumptions, for example, complete mixing. Often reactors that are used in practice are nonideal reactors.
Completely mixed batch reactor (CMBR)[a]	An ideal reactor in which no reactants or products flow into or out of the reactor. Complete mixing occurs instantaneously and uniformly throughout the reactor, and the reaction rate proceeds at the identical rate everywhere in the reactor.
Completely mixed flow reactor (CMFR)[b]	An ideal reactor in which reactants and products flow into and out of the reactor. Complete mixing occurs instantaneously and uniformly throughout the reactor. The reaction rate proceeds at the identical rate everywhere in the reactor, and the concentrations throughout the reactor are the same as the effluent concentration.
Plug flow reactor (PFR)	An ideal reactor in which fluid moves through the reactor as a plug and the fluid does not mix with fluid elements in front of or behind it. As a result, the reaction rate and concentrations of the reactants decrease as the fluid moves down the PFR. (Except for zero-order reactions) The composition at any travel time down the reactor is identical to the composition in the CMBR after the same period of time has passed.

[a]A CMBR may also be referred to as a batch reactor.
[b]A CMFR may also be referred to as a complete-mix reactor (CMR), continuous stirred-tank reactor (CSTR), constant-flow stirred-tank reactor (CFSTR), or backmix reactor.

Reactors Characterized by Unit Process

Reactors may be classified according to the type of water treatment process to be carried out. Many reactions of importance in water treatment are heterogeneous because they consist of reactions occurring in more than one phase, so-called multiphase reactions. For example, ozone gas is sometimes mixed with water to achieve the transformation of some undesired constituent (e.g., oxidation of manganese or inactivation of pathogens). In other cases, a solid precipitate is formed that may be removed by sedimentation or filtration.

Various types of reactors are used to carry out multiphase reactions. Some reactors are used to mix reactants and to provide high contact areas between water and gas. Others are used for reactions that occur on or within the solid phase and some are used for precipitation reactions. Various multiphase reaction processes, the type of reactors used, and specific examples of the multiphase reactions used in water treatment processes are listed in Table 6-2.

REACTORS USED FOR MIXING

In reactors used for the mixing of reactants, the mixing can be either intense or slow depending upon the reactions desired. In the mixing of

Table 6-2
Examples of reactors used in water treatment

Process	Reactor Type	Examples
Oxidation	Stirred tanks, tanks in series, diffused gas contactors, Venturi reactor	Oxidation of iron, oxidation of manganese, dechlorination by SO_2, ozone reactions
Disinfection	Tanks with serpentine baffling, long channels, diffused gas contactors, pipes	Chlorination, ozonation, chlorine dioxide, chloramination
Coagulation and flocculation	Stirred tanks in series, sludge blanket reactors	Removal of particulates and NOM using Al(III) or Fe(III), removal of As(V)
Lime softening	Stirred tanks in series, recycle reactors, sludge blanket reactors, upflow fluidized beds	Removal of hardness
Air stripping	Packed tower, diffused gas contactors	VOC removal, CO_2 removal
Adsorption	Fixed-bed reactor, stirred tanks in series	SOC removal, taste and odor control
Ion exchange	Fixed-bed reactor, stirred tanks in series	Removal of hardness, nitrate, perchlorate, barium, NOM, etc.
Filtration	Fixed bed	Particulate removal, turbidity removal, microbial removal, assimilable organic carbon (AOC) removal
Membranes	Fixed bed	Particulate removal, microbial removal

coagulants, intense mixing is desirable to disperse the reactants quickly. Flocculation, on the other hand, requires moderate agitation to increase the rate of particle collision and formation of large aggregate particles. *Venturi reactors* are in-line mixers with a section of the pipe that is restricted to a throat section, where chemicals are applied. Mixing occurs in a turbulent region following the throat section. Venturi-type reactors are often used for injection of chlorine, carbon dioxide, and other soluble gases into water. *Static mixers* and *pumped flash mixers* are useful for rapid mixing of coagulants and polymers with large volumes of water. Various types of mixing devices are discussed further in Sec. 6-10.

REACTORS USED FOR CONTACT TIME

Reactors designed for a specified detention or reaction time are commonly known as *plug flow reactors* and are not subject to backmixing. For example, the disinfection of water is typically carried out by the exposure of the water to the disinfectant of interest for a specified duration of time. Sedimentation processes also require time for particles to fall out of solution. For reactions of this nature, where holding time is important, reactors may be designed as long, narrow channels (see Fig. 6-1d), long pipe or tubular vessels (see Fig. 6-1e), or a series of long channels (see Fig. 6-1f).

REACTORS USED FOR CONTACT BETWEEN WATER AND GAS

Packed columns consist of a cylindrical column containing appropriate packing materials that provide high interfacial areas between water and a gas, usually air. Packed columns are used for stripping of undesirable gases or volatile organic compounds (VOCs) from water. Two-phase flow in packed towers is typically countercurrent, with liquid entering at the top of the reactor and air forced in the bottom of the reactor (see Fig. 6-1g). As the liquid flows over the packing, a thin liquid film is produced and volatile compounds and gases are transferred into the gas phase. Other reactors used to contact water and gases include *bubble tanks*, where a gas is bubbled into the water in tanks, and *spray towers*, where water is sprayed into the air, used primarily for removal of volatile materials. Air-liquid contactors are discussed in Chap. 14.

REACTORS USED FOR REACTIONS OCCURRING ON OR WITHIN SOLID PHASE

In adsorption (see Chap. 15) and ion exchange (see Chap. 16), the reaction occurs on or within the solid phase of the adsorbent (e.g., activated carbon) or ion exchange resin, respectively. Reactors for adsorption and ion exchange may consist of a fixed bed (i.e., packed bed; see Fig. 6-1h without airflow) or a fluidized bed (i.e., the packing media is suspended in the reactor; see Fig. 6-1i without airflow). Filters (covered in Chap. 11) and membranes (covered in Chap. 12) can be used in a reactor to retain reactants and reaction products that must be removed from the water. Combined unit processes may also be occurring in a single reactor; for example, hybrid membrane adsorption reactors are used in water treatment processes (powdered activated carbon addition before ultrafiltration).

REACTORS WITH RECYCLE USED FOR PRECIPITATION REACTIONS

Recycle reactors operate with a portion of the flow returned to the reactor inlet. Such reactors are used principally for precipitation reactions in which a portion of the precipitated solids is recycled to accelerate the rate of precipitation as in softening.

Reactors Characterized by Entrance and Exit Conditions

Open- and closed-reactor terminology is used to describe the entrance and exit conditions when dispersion (longitudinal mixing caused by fluid turbulence) and molecular diffusion (see discussion in Sec. 6-7) are important reactant/product transport mechanisms. A reactor is classified as an *open* reactor when either dispersion or diffusion contributes to solute fluxes into and out of the reactor. An example of an open reactor is groundwater with a contaminant plume that moves with the bulk groundwater flow but also by diffusion. In such situations, the contaminants may be found upstream of the groundwater flow direction. Reactors in which neither dispersion nor diffusion contributes to solute flux into or out of the reactor are classified as *closed* reactors. Most reactors used in water treatment are closed reactors because reactants and products are typically conveyed into or out of basins by pipes and weirs and no backmixing can occur at the entrance or exit.

6-2 Mass Balance Analysis

The quantitative description of a water treatment process begins with an accounting of all materials that enter, leave, accumulate in, or are transformed within the boundaries of a system. The basis for this accounting procedure, known as a *mass balance*, is the law of conservation of mass, which accounts for changes in any component due to fluid flow, mass transfer, or chemical transformations. The materials involved in a system, the scale and system chosen to write a mass balance, and the general mass balance analysis are introduced in this section.

Conservative and Nonconservative Constituents

A constituent that passes through a processing system without reacting in the reactor and remains unchanged in total mass (but not perhaps in concentration) is known as a *conservative* constituent. Constituents that undergo reaction, are transformed, or accumulate within the reactor during processing, resulting in less mass exiting with the reactor effluent than entered, are known as *nonconservative* constituents. For example, the chloride ion will pass through a water filtration plant unchanged because this inorganic ion does not adsorb on particulate matter or undergo biological transformations. Thus, the chloride ion acts as a conservative constituent. On the other hand, the ferrous ion, an ion often found in ground and surface waters, undergoes oxidation and hydrolysis in the presence of oxygen or other oxidizing agents, leading to the formation and sedimentation of an insoluble precipitate at appropriate chemical conditions. Therefore, a ferrous ion is a nonconservative constituent.

Scale and System Selection

Accounting for the fate of a constituent in a water treatment plant or in an individual treatment process can be approached from various levels. While not commonly used in engineering design, the time and spatial variations of a constituent may be predicted based on forces operating at the ionic or molecular level. At the opposite extreme, it is possible to ignore all molecular interactions as well as the internal details of the system or unit and assume no local gradients of mass or temperature exist. The loss of mechanistic insight with this approach may still be correct, depending on what design issue is being investigated. Moreover, simpler approaches have the benefit of requiring much less complex mathematical analysis. Design based on mechanistic considerations is a desirable goal, but the complexity and uncertainty of the processes involved make mechanistic design unattainable in many cases, requiring the use of simplified but often still reasonable approaches. For example, if the amount of sludge that is generated by a coagulation/flocculation process is to be determined, then an overall mass balance on the process will be adequate. On the other hand, if the flocculator is to be designed for particle agglomeration and breakup on the microscale, then fluid velocity gradients in the flocculator on which microscale mixing occurs must be considered.

General Mass Balance Analysis

Two concepts needed to write a mass balance on a reactor include correct drawing of the system boundary and the choice of a time interval over which to write the mass balance. The *system boundary* is used to identify all of the material flows into and out of the system. The guiding principle in choosing the system is to have a uniform concentration (i.e., intensive properties such as concentration, temperature, and pressure are assumed isotropic) so that the kinetic expression can be evaluated at one concentration within the system. The actual volume in which change is occurring is often referred to as the *control volume*, and the control volume should be chosen so the mass flux (in and out) across the boundaries can be easily determined. A definition sketch of a control volume for a completely mixed reactor with inflow and outflow is shown on Fig. 6-2. For any control volume, a materials balance can be expressed as

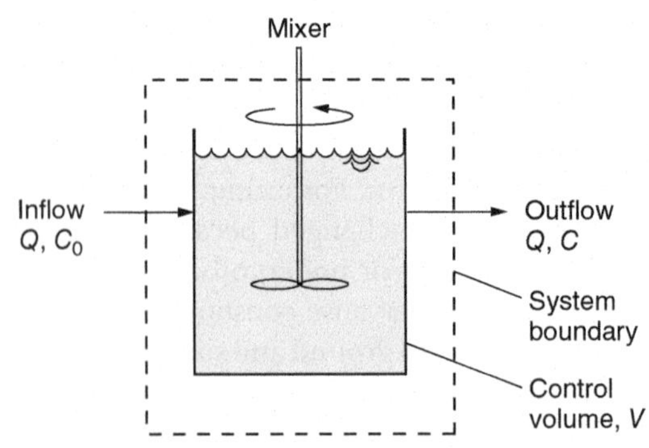

Figure 6-2
Definition sketch for a materials balance analysis of a completely mixed flow reactor (CMFR).

$$\begin{aligned}&\text{Mass of constituent entering system}\\&\quad-\text{mass of constituent leaving system}\\&\quad-\text{mass of constituent disappearing}+\text{mass of constituent}\\&\quad\ \ \text{appearing within system due to reaction}\\&\quad=\text{mass of constituent accumulation in system}\end{aligned}\tag{6-1}$$

The simplified word statement for the mass balance expression shown in Eq. 6-1 is written as

$$\text{In}-\text{out}-\text{loss}+\text{generation}=\text{accumulation}\tag{6-2}$$

Mass may be transported across system boundaries by bulk fluid flow (advection) or by molecular diffusion or turbulent mixing (dispersion). The distinction between molecular diffusion and dispersion is described in detail in Sec. 6-7. Transformations or losses may occur because of chemical reactions and mass transfer between phases within the system boundaries. The spatial and temporal variation of the concentration of conservative and nonconservative constituents, depending on the scale of its application, can be determined using Eq. 6-1.

A mass balance on a single reaction only needs to be written for one component. The concentrations and pressures of the other components can be determined from the stoichiometry of the reaction, as shown previously in Chap. 5.

6-3 Hydraulic Characteristics of Ideal Reactors

Ideal reactors provide the basic conceptual foundation upon which an understanding of real reactors can be built. On the laboratory bench, with carefully managed conditions, reactors can be made to provide nearly ideal performance. For many processes, pilot-scale systems often come close to ideal performance as well. These ideal systems can be used to provide useful estimates of the performance of even the largest scale systems. Thus, a thorough understanding of the behavior of ideal reactors, including the necessary assumptions, is essential to testing and modeling full-scale process performance.

Completely Mixed Batch Reactor

When considering an ideal CMBR, the following assumptions are made: (1) the contents of the tank are completely uniform with no density gradients or dead space, (2) the probability of a particle of water being in any one part of the tank at any time is the same, (3) the temperature is uniform throughout the reactor, and (4) any chemical added to the contents is instantly and uniformly distributed throughout the reactor. A diagram sketch of the CMBR is shown on Fig. 6-3a.

Completely Mixed Flow Reactor

The assumptions made when modeling a CMFR are similar to those made for the CMBR, namely (1) the contents of the tank are completely uniform with no density gradients or dead space, (2) the probability of a particle of

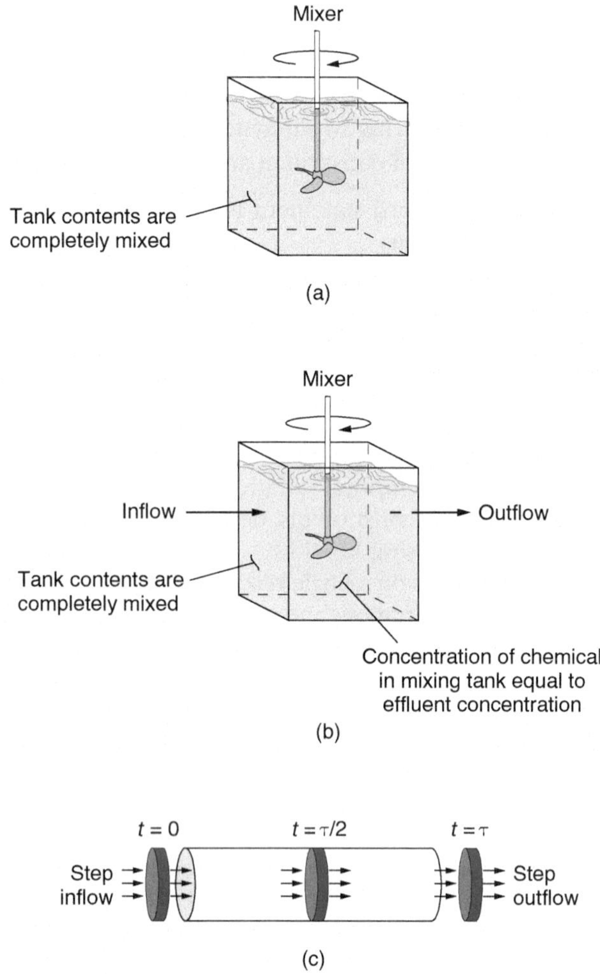

Figure 6-3
Concept diagrams for three ideal reactors: (a) completely mixed batch reactor, (b) completely mixed flow reactor, and (c) plug flow reactor.

water being in any part of the tank at any time is the same, (3) the temperature is uniform throughout the reactor, and (4) any chemical added to the contents is instantly and uniformly distributed throughout the reactor. Because of the first assumption of complete mixing, it is also assumed that the effluent of a CMFR has precisely the same composition as the contents. A diagram sketch of the CMFR is shown on Fig. 6-3b.

There are two important differences between the assumptions of the CMBR and the CMFR when reactions are occurring in the reactor: (1) In the CMBR all the reactants are in the reactor for the same residence time, whereas in the CMFR the reactants are in the reactor for a variety of residence times; and (2) in the CMBR the concentration of the reactants changes with time as the reaction takes place, whereas in the CMFR the concentration of all the reactants is the same throughout the reactor all the time (once the reactor has reached steady state).

Plug Flow Reactor

The water traversing an ideal PFR flows uniformly without mixing with the water in front of or behind it in the reactor. The plug flow concept can also be described as flow consisting of a series of plugs with the same diameter as the inside diameter of the reactor. Each time a new plug is introduced in one end of the reactor, another plug of the same size must exit the other end. If the plugs are introduced at a constant rate, then each plug is in the reactor for the same amount of time. Thus, the performance of a PFR is, by definition, precisely the same as the performance of a CMBR that has operated for the same period of time as the residence time of a PFR.

It is common for practitioners to conduct a bench-scale, batch experiment and then assume the same result will be obtained for the same conditions (e.g., chemical dosing, mixing intensity) and the same contact time in a full-scale continuous PFR facility, known as the plug flow assumption. The jar test, used to determine the optimum coagulant chemical dose, is an example of a bench-scale experiment. Reactors can be designed to operate as near-ideal PFRs; however, in many cases the performance is not ideal. Because of the desirable characteristics of the PFR for water treatment operations, it is important to understand what can be done to make a design approach plug flow and know when the plug flow assumption is erroneous. In addition, it is necessary to be able to estimate the performance of PFRs under nonideal conditions. A sketch of ideal flow in a PFR is shown on Fig. 6-3c.

Tracer Curves for Ideal Reactors

A tracer is a conservative chemical that is used to assess the flow conditions through a reactor. To compare the hydraulic performance of a continuous-flow reactor to the ideal model, a tracer is introduced into the reactor's influent, and its concentration is then observed in the reactor's effluent. Three techniques are used: (1) the instantaneous addition of a pulse or slug of tracer in the influent followed by observation of the same pulse as it exits the reactor, (2) the addition of the tracer at a steady rate followed by observation until the effluent of the reactor equals the influent concentration, or (3) the addition of a tracer at a steady rate until the effluent concentration equals the influent concentration, then a cessation of the tracer feed followed by continued observation until no tracer is found in the effluent. All three tracer addition methods yield information about the exit age distribution, the cumulative exit distribution, and the internal age distribution, respectively. Conducting tracer studies is discussed in greater depth in Sec. 6-6.

The tracer curves that occur from the addition of the same pulse input to both a PFR and a CMFR are illustrated on Fig. 6-4. A pulse that passes through the PFR has exactly the same shape it had initially and with a detention time $\tau = V/Q$ (volume/flow rate) after the tracer is added.

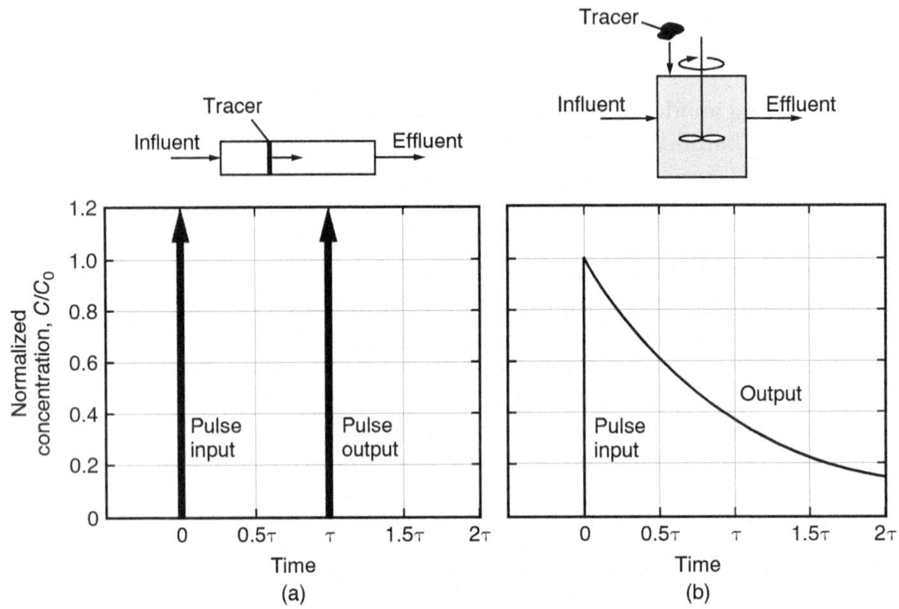

Figure 6-4
Tracer curves from ideal reactors: (a) plug flow reactor and (b) completely mixed flow reactor.

The tracer curve for a pulse input to a CMFR has a significantly different shape from that obtained for a PFR. The effluent tracer concentration from a CMFR instantly reaches a maximum as the tracer is uniformly distributed throughout the reactor and then gradually dissipates in an exponential manner as the tracer material leaves the effluent. The shape of the tracer curve is exponential because, as the tracer leaves the reactor, the concentration of the tracer is reduced, which in turn reduces the rate at which the tracer mass leaves the reactor. A mass balance analysis for a nonreactive substance (generation term is zero) can be used to determine the characteristics of a pulse tracer (in term is zero) in a CMFR:

$$\text{In} - \text{out} + \text{generation} = \text{accumulation}$$

$$0 - QC + 0 = V\frac{dC}{dt} \tag{6-3}$$

where Q = flow rate through reactor, L/s
V = reactor volume, L
C = effluent concentration of tracer at time t,mg/L
t = time since slug of tracer was added to reactor, s

Equation 6-3 can be rearranged to obtain the expression

$$-\frac{Q}{V}dt = \frac{dC}{C} \tag{6-4}$$

At $t = 0^+$ (time immediately after tracer is added), the tracer slug has entered the reactor and is uniformly dispersed within the CMFR. Consequently, Eq. 6-4 may be integrated:

$$-\int_0^t \frac{Q\,dt}{V} = \int_{C_0}^{C} \frac{dC}{C} \qquad (6\text{-}5)$$

where C_0 = initial mass of tracer added divided by volume of reactor, mg/L

The following expression is obtained after substitution of $\tau = V/Q$, the theoretical hydraulic detention time, into Eq. 6-5:

$$C = C_0 e^{-t/\tau} \qquad (6\text{-}6)$$

where $\tau = V/Q$ = hydraulic detention time, s

The tracer curve shown on Fig. 6-4b may be obtained using the expression presented in Eq. 6-6.

Completely Mixed Flow Reactors in Series

In environmental engineering, it is common to employ a series of CMFRs to improve the hydraulic performance of a reactor. The improved efficiency that may be gained from CMFRs in series is discussed in Sec. 6-5. The impact of putting a few CMFRs in series (commonly known as tanks in series) on the tracer curve is described below. A definition sketch of CMFRs in series is shown on Fig. 6-5. The development of the CMFRs in series analysis is based on a constant total volume because the tank volume is an important factor controlling capital cost, and the purpose of the exercise is to determine if dividing that volume into several smaller compartments will improve efficiency. Assuming that both the total reactor volume and the mass of tracer added remain constant, the volume of each reactor is

$$V_R = \frac{V}{n} \qquad (6\text{-}7)$$

where V_R = volume of each reactor in series, m^3
V = total volume of all reactors in series, m^3
n = number of reactors in series

If the same mass of tracer is used in all tracer studies, the initial concentration of the tracer would be equivalent to the concentration that would result if the entire mass of tracer were placed in one tank having the same volume as all of the tanks in the series:

$$C_0^* = \frac{M}{V_R} \qquad (6\text{-}8)$$

where M = mass of tracer added, g
C_0^* = initial concentration, mg/L (equivalent to g/m^3)

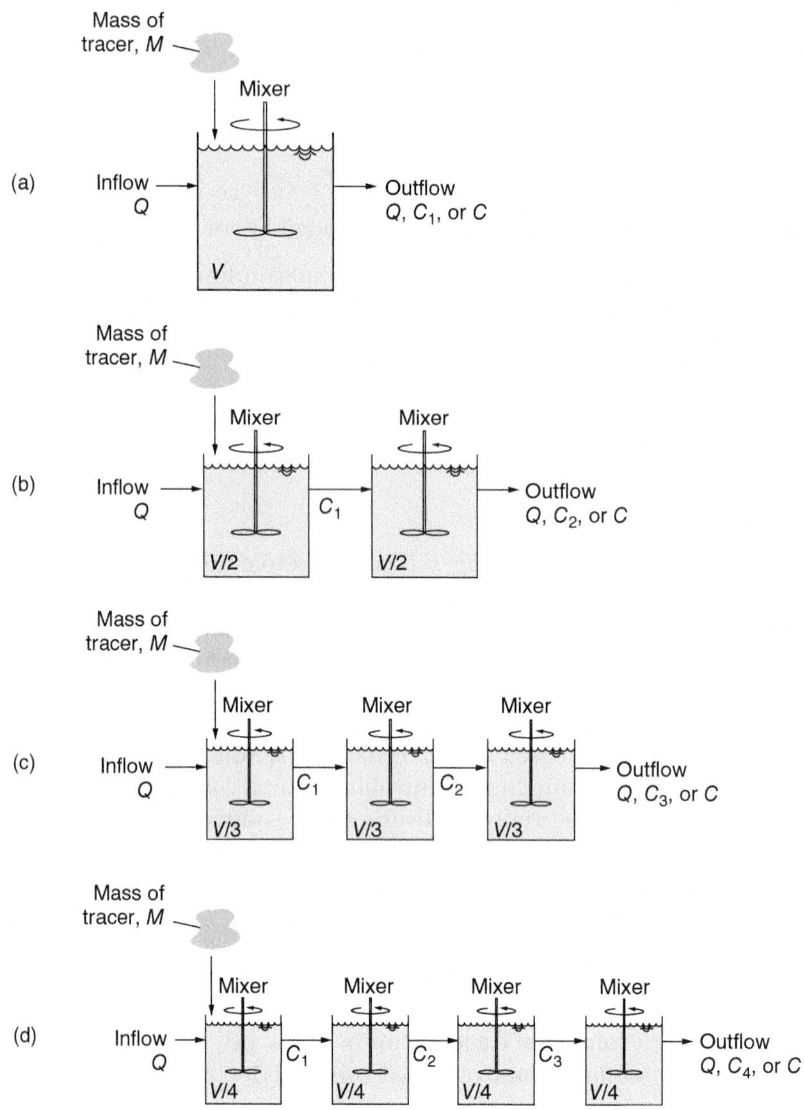

Figure 6-5
Definition sketch: comparing a tracer test in (a) one, (b) two, (c) three, or (d) four CMFRs in series. Each series receives the same mass dose of tracer, M, and each series has the same total volume, V.

If, however, all of the tracer is added to the first reactor, the initial concentration is given by the expression

$$C_0^* = \frac{nM}{V} = nC_0 \tag{6-9}$$

The mass balance (see Eq. 6-2) for a pulse input of conservative tracer in the first reactor in a series of n reactors incorporates the following assumptions: (1) the generation term can be assumed to be zero because the tracer substance is conservative (no reaction) and (2) the influent concentration of tracer is zero because the tracer is added as a pulse input (single event). These assumptions result in the mass balance expression

$$0 - QC_1 + 0 = \frac{V}{n}\frac{dC_1}{dt} \tag{6-10}$$

where C_1 = effluent concentration exiting first reactor in series, mg/L

The expression shown in Eq. 6-10 can be simplified and rearranged for integration as

$$-\int_{C_0^*}^{C_1} \frac{dC_1}{C_1} = -\int_0^t \frac{nQ}{V}dt \tag{6-11}$$

Integrating, as shown previously in Eq. 6-5, results in the expression

$$\ln\frac{C_1}{C_0^*} = -\frac{nQ}{V}t \tag{6-12}$$

Substituting nC_0 for C_0^*, the effluent concentration from the first CMFR in a series of ideal CMFRs as a function of time is written as

$$\frac{C_1}{C_0} = ne^{-(nQ/V)t} = ne^{-n\theta} \tag{6-13}$$

where θ = relative (normalized) detention time
$= Qt/V = t/\tau$, dimensionless

For the second reactor in the series, the generation term is still zero. The influent concentration changes with time and is equal to the effluent concentration from the first CMFR in the series:

$$QC_1 - QC_2 + 0 = \frac{V}{n}\frac{dC_2}{dt} \tag{6-14}$$

The expression from Eq. 6-14 can be simplified and rearranged as

$$\frac{dC_2}{dt} + \frac{nQ}{V}C_2 = \frac{nQ}{V}C_1 \tag{6-15}$$

where C_2 = effluent concentration exiting second reactor in series, mg/L

Substituting Eq. 6-13 into Eq. 6-15 yields

$$\frac{dC_2}{dt} + \frac{nQ}{V}C_2 = \frac{nQ}{V}nC_0e^{-(nQ/V)t} \tag{6-16}$$

The integrating factor method is used to solve Eq. 6-16. The method involves multiplying Eq. 6-16 by the integrating factor $e^{(nQ/V)t}$, which results in the expression

$$\frac{dC_2}{dt}e^{(nQ/V)t} + \frac{nQ}{V}e^{(nQ/V)t}C_2 = \frac{n^2Q}{V}C_0e^{-(nQ/V)t}e^{(nQ/V)t} \tag{6-17}$$

The two terms on the left-hand side of Eq. 6-17 are the derivative of the product of the functions C_2 and $\exp[(nQ/V)t]$. Combining these terms yields

$$\frac{d\left[C_2 e^{(nQ/V)t}\right]}{dt} = \frac{n^2 Q}{V} C_0 \tag{6-18}$$

Integrating Eq. 6-18 yields the following expression for the effluent concentration of tracer from the second CMFR in the series after some algebraic manipulation:

$$\frac{C_2}{C_0} = \frac{n^2 Q t}{V} e^{-(nQ/V)t} = n(n\theta)e^{-n\theta} \tag{6-19}$$

Using the same approach as shown above, the effluent concentration for any number of reactors in series can be obtained. The corresponding effluent concentration expression for the third and fourth CMFRs in series is given by Eqs. 6-20 and 6-21, respectively:

$$\frac{C_3}{C_0} = \frac{n(n\theta)^2}{2} e^{-n\theta} \tag{6-20}$$

$$\frac{C_4}{C_0} = \frac{n(n\theta)^3}{6} e^{-n\theta} \tag{6-21}$$

where C_3 = effluent concentration exiting third reactor in series, mg/L
C_4 = effluent concentration exiting fourth reactor in series, mg/L

The tracer curves from one, two, three, and four CMFRs in series, with each reactor series having an equivalent volume, are shown on Fig. 6-6. The tracer curve changes dramatically as the number of CMFRs increases. In addition, the use of CMFRs in series can also result in significant improvements in reactor performance, as discussed in the following sections.

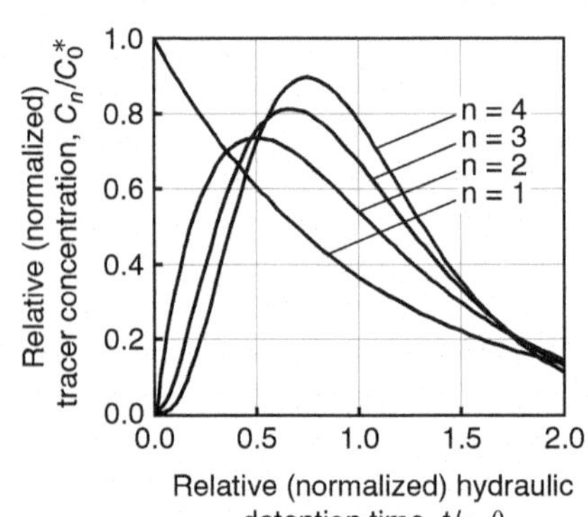

Figure 6-6
Tracer curves from one, two, three, and four CMFRs in series. Each series has the same total reactor volume and the same total mass of tracer added at the start.

6-4 Modeling Reactions in Completely Mixed Batch Reactors

The CMBR is the ideal reactor most widely used in the laboratory to gather and understand reaction data. The contents of the CMBR are mixed completely and the concentration is uniform throughout the reactor, as shown on Fig. 6-3a. The focus of this section is on modeling the reactions that occur in a CMBR.

Mass Balance Analysis

For the system shown on Fig. 6-7, a boundary can be drawn around the system (see dashed line) and a mass balance can be written that accounts for the mass entering, leaving, reacting, or accumulating within that boundary during the time period from t to $t + \Delta t$:

$$0 - 0 + \Delta t\, rV = N|_{t+\Delta t} - N|_t \tag{6-22}$$

where V = reactor volume, L
Δt = time interval, s
$N|_{t+\Delta t}$ = amount of reactant in reactor evaluated at $t + \Delta t$, mol
$N|_t$ = amount of reactant in reactor evaluated at t,mol
r = average reaction rate during interval from t to $t+\Delta t$, mol/L . s

Example of a batch reactor.
http://www.stantec.com/wt3r/6-2.html

If Eq. 6-22 is rearranged and the limit as Δt approaches 0 is taken, the following general expression is obtained for a CMBR:

$$rV = \lim_{\Delta t \to 0} \left(\frac{N|_{t+\Delta t} - N|_t}{\Delta t} \right) = \frac{dN}{dt} \tag{6-23}$$

Writing Eq. 6-23 in terms of concentrations yields

$$r = \frac{1}{V}\frac{dCV}{dt} = \frac{dC}{dt} + \frac{1}{V}\frac{C\,dV}{dt} \tag{6-24}$$

where C = concentration of reactant in CMBR, mol/L
N = amount of reactant in reactor, mol
t = time,s

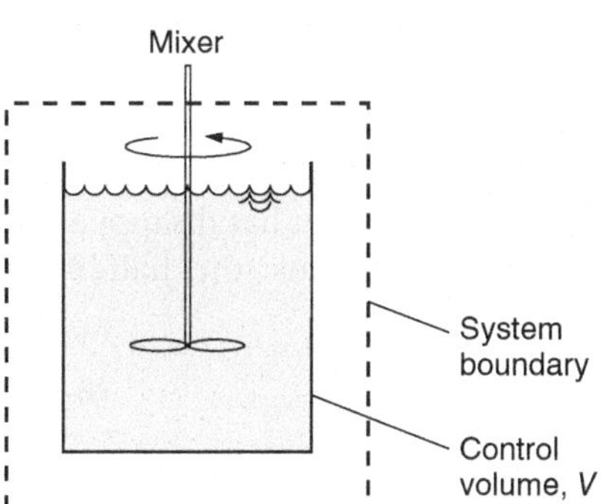

Figure 6-7
Definition sketch for mass balance analysis of batch reactor.

For a constant volume CMBR, Eq. 6-24 becomes

$$r = \frac{dC}{dt} \tag{6-25}$$

As shown in Eq. 6-25, C is a function of time. Equations 6-24 and 6-25 are valid regardless of the order of the reaction.

Reaction Rates in a Completely Mixed Batch Reactor

If the progress of a reaction is observed while it takes place in a CMBR, common kinetic rate expressions can be evaluated to determine the correlation between the data and the reaction kinetics. If the reaction is first order, Eq. 6-25 may be written as

$$r = -kC = \frac{dC}{dt} \tag{6-26}$$

where k = first-order rate constant, s^{-1}

Integrating the expression shown in Eq. 6-26 yields

$$\frac{C}{C_0} = e^{-kt} \tag{6-27}$$

where C_0 = initial concentration, mol/L

Taking the natural logarithm of both sides, the following relationship is obtained:

$$\ln(C) - \ln(C_0) = -kt \tag{6-28}$$

For a first-order reaction, a plot of $\ln(C)$ as a function of t, as shown in Eq. 6-28, will result in a linear relationship. Such a plot is illustrated on Fig. 6-8a. The slope of the line in the plot is equal to the first-order rate constant k and the intercept is equal to $\ln(C_0)$.

Similar, straightforward graphical solutions can be demonstrated for reactions of zero order (Fig. 6-8c) and second order (Fig. 6-8b). Analytical solutions for all these reaction orders as well as for nth-order reactions are shown in Table 6-3. There is no linear graphical solution for an nth-order reaction, but a spreadsheet can be used to determine which reaction order results in the best fit.

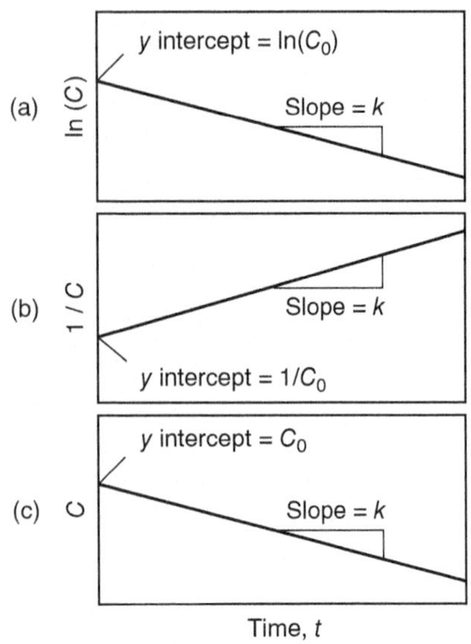

Figure 6-8
Graphical display of (a) first-, (b) second-, and (c) zero-order reactions.

Half-Life Concept for Irreversible Reactions

A widely used parameter to compare reaction rates is the half-life, the time within which half of the initial concentration of a reactant has disappeared, that is, $C/C_0 = 0.5$. For first-order, irreversible reactions, the half-life is obtained by rearrangement of Eq. 6-28 as follows:

$$\ln\left(\tfrac{1}{2}C_0\right) - \ln(C_0) = -kt_{1/2} \tag{6-29}$$

Equation 6-29 can also be written as

$$t_{1/2} = \frac{\ln(2)}{k} = \frac{0.693}{k} \tag{6-30}$$

Example 6-1 Determination of reaction rate constant for decomposition of ozone

In laboratory experiments, ozone was added to a beaker (batch) of water and the concentration of ozone remaining was measured periodically. The initial concentration of ozone, C_0, was 5 mg/L for all experiments. The fraction of ozone remaining in the water at pH values of 7.6, 8.5, and 9.2 are presented in the following table (from Stumm, 1956):

	Ozone Concentration, *C* (mg/L)		
Time, min	**pH = 7.6**	**pH = 8.5**	**pH = 9.2**
0	5	5	5
1	4.95	—	4.25
2.3	—	4.15	—
5.5	—	—	2.1
6.8	4.55	—	—
7.4	—	3.05	—
8.9	—	2.75	—
9	4.35	—	1.1
14.3	—	2	—
18	3.7	—	—

Determine the reaction order and reaction rate constant for the decomposition of ozone in water at three pH values (7.6, 8.5, and 9.2), considering zero-, first-, and second-order reactions.

Solution

1. Determine the order of the reaction by plotting various concentration quantities as a function of time.
   a. Construct a computation table for the values to be plotted.

	C			ln(*C*)			1/*C*		
Time, min	**pH = 7.6**	**pH = 8.5**	**pH = 9.2**	**pH = 7.6**	**pH = 8.5**	**pH = 9.2**	**pH = 7.6**	**pH= 8.5**	**pH = 9.2**
0	5	5	5	1.61	1.61	1.61	0.20	0.20	0.20
1	4.95	—	4.25	1.60	—	1.45	0.20	—	0.24
2.3	—	4.15	—	—	1.42	—	—	0.24	—
5.5	—	—	2.1	—	—	0.74	—	—	0.48
6.8	4.55	—	—	1.52	—	—	0.22	—	—
7.4	—	3.05	—	—	1.12	—	—	0.33	—
8.9	—	2.75	—	—	1.01	—	—	0.36	—
9	4.35	—	1.1	1.47	—	0.10	0.23	—	0.91
14.3	—	2	—	—	0.69	—	—	0.50	—
18	3.7	—	—	1.31	—	—	0.27	—	—

b. For a zero-order reaction, a plot of concentration C as a function of time t is shown in panel (a) of the figure below.
c. For a first-order reaction, a plot of the natural log of concentration, ln(C), as a function of time t is shown in panel (b) below.
d. For a second-order reaction, a plot of inverse concentration $1/C$ as a function of time t is shown in panel (c) below.

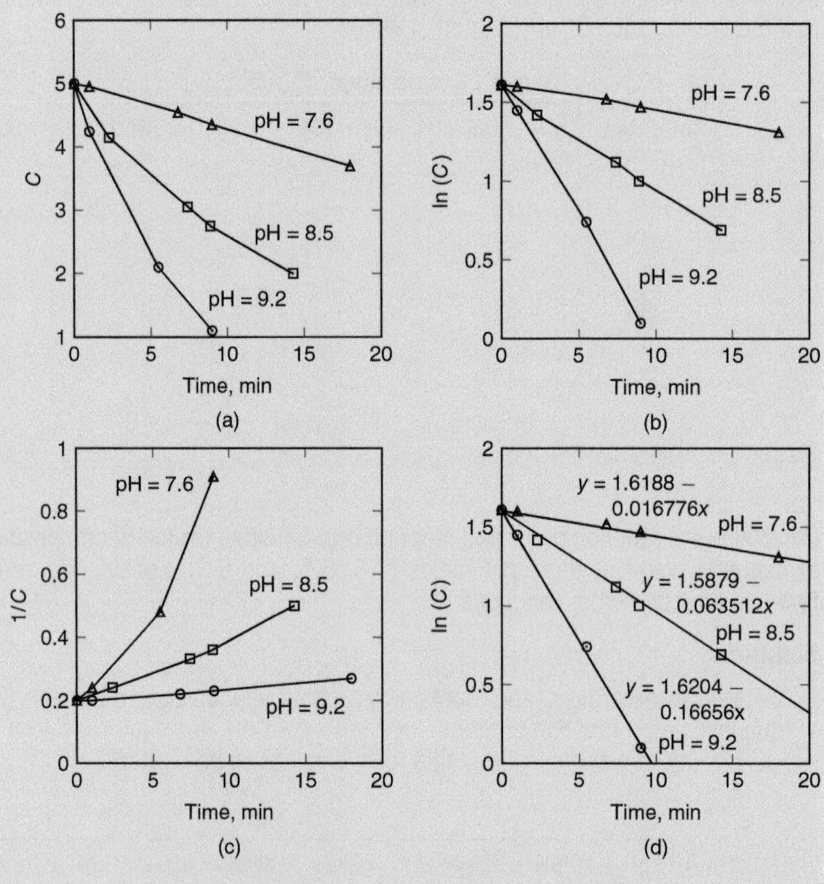

Because the plot constructed in panel (b) results in a linear relationship, ozone decomposition in water can be described using first-order kinetics.

2. Determine the reaction rate constants for each pH value. The reaction rate constant is determined by finding the slope of the best-fit line for the data. As shown in panel (d) above, the first-order reaction rate constants for the decomposition of ozone in water are 0.0168, 0.0635, and 0.167 min^{-1} for water with pH values of 7.6, 8.5, and 9.2, respectively.

Table 6-3
Solutions to reactions with different orders

Reaction Order	Rate Expression	Solution[a]	Units for Rate Constant
Zero order	$r = -k = \dfrac{dC}{dt}$	$C = C_0 - kt$	mol/L · s
First order	$r = -kC = \dfrac{dC}{dt}$	$C = C_0 e^{-kt}$	s^{-1}
Second order	$r = -kC^2 = \dfrac{dC}{dt}$	$\dfrac{1}{C} = \dfrac{1}{C_0} + kt$	L/mol · s
nth order	$r = -kC^n = \dfrac{dC}{dt}$	$C = {}_{-n+1}\sqrt{(-n+1)\left(-kt + \dfrac{C_0^{-n+1}}{-n+1}\right)}$ $(n \geq 0, n \neq 1)$	$(L/mol)^n \cdot s^{-1}$

[a]For reactions in a PFR, the reaction time t is replaced by the hydraulic detention time τ ($\tau = V/Q$) in the solution equation.

Through a similar exercise, the half-life for a second-order irreversible reaction can be determined from the second-order rate expression solution given in Table 6-3. The half-life is inversely proportional to both the rate constant and the initial reactant concentration:

$$t_{1/2} = \frac{1}{kC_0} \tag{6-31}$$

Rate constants for common first-order reactions used in water treatment applications are listed in Table 6-4. Additional reaction rate constants can be found in compendiums in the literature (e.g., Hoffmann, 1981; Hoigne and Bader, 1983; Pankow and Morgan, 1981; Stumm and Morgan, 1996). The relationship between the half-life and the rate constant for first- and second-order reactions is illustrated on Fig. 6-9. Note that, where second-order reactions are concerned, the initial reactant concentration must be considered as well as the second-order rate constant.

Table 6-4
Selected first-order rate constants for reactions common to water treatment

Reaction	Conditions of Measurement	Rate Constant, s^{-1}	Reference
$Cl_2(aq) + H_2O \rightarrow HOCl + H^+ + Cl^-$	20°C	11	Eigen and Kustin (1962)
$SO_2(aq) + H_2O \rightarrow HOSO_3^- + H^+$	20°C	3.4×10^6	Eigen et al. (1961)
$CO_2(aq) + H_2O \rightarrow HCO_3^- + H^+$	20°C	0.02	Kern (1960)
$Al(H_2O)_6^{3+} \rightarrow [Al(H_2O)_5OH]^{2+} + H^+$	25°C, pH 4	4.2×10^4	Holmes et al. (1968)

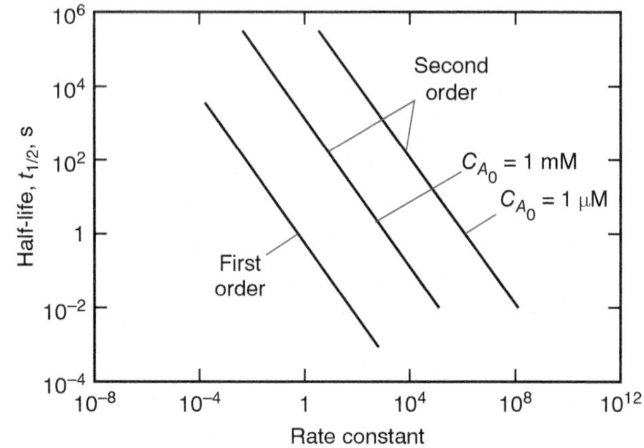

Figure 6-9
Comparison of reaction half-life for first- and second-order reactions in water treatment (C_0 = initial concentration in reactor).

First-order reactions with rate constants greater than 1 s^{-1} are characterized as fast reactions. For example, the hydrolysis of chlorine in water has a first-order rate constant of 11 s^{-1} at 20°C (Eigen and Kustin, 1962). The reaction has a $t_{1/2}$ of 0.06 s and is essentially complete (99 percent conversion) in less than 0.5 s at a pH greater than 5. When the initial concentration is 10^{-3} mol/L or greater and the rate constant is greater than 10^2 L/mol·s, second-order reactions are also considered fast.

6-5 Modeling Reactions in Ideal Continuous-Flow Reactors

Models for ideal, continuous-flow reactors are useful for gaining an understanding of the behavior of reactors that are used in full-scale systems. The models for these systems are constructed easily and often provide practical estimates of full-scale behavior. The progress of reactions in CMFRs and PFRs with and without recycle are analyzed in this section. Although CMFRs in series will also be considered, combinations of CMFRs and PFRs will not be discussed. The student may want to seek other resources (Fogler, 1999; Levenspiel, 1998) to examine these questions, which can also have practical value.

Completely Mixed Flow Reactor

Four assumptions are made when considering an ideal CMFR (Sec. 6-3), and because of the assumption that the contents are mixed completely, it follows that the effluent has precisely the same composition as the reactor contents. Therefore, the rate of reaction in a CMFR proceeds according to the effluent concentration, resulting in the need for a larger reactor volume as compared to PFRs.

CMFR MASS BALANCE

Design of a CMFR is typically based on *steady-state* conditions, meaning the effluent concentration does not change with respect to time, there is no accumulation, and reactor volume and flow rate are constant. Mathematically, this mass balance may be written as

$$Q(C_0 - C) + Vr|_C = 0 \tag{6-32}$$

where C_0 = influent concentration, which may be a function of time, mg/L
Q = flow rate, L/s
C = effluent concentration, mg/L
$r|_C$ = reaction rate in reactor at effluent concentration C, mg/L · s
V = reactor volume, L

In Eq. 6-32, the value of $r|_C$ will be less than zero if the component of interest is consumed by the reaction; the value of $r|_C$ will be greater than zero if the component of interest is produced by the reaction.

Dividing Eq. 6-32 by Q results in the expression

$$0 = C_0 - C + \frac{V}{Q} r|_C = C_0 - C + \tau r|_C \tag{6-33}$$

where τ = hydraulic detention time, which is equal to V/Q, s.

The hydraulic detention time or the volume of the reactor needed can be estimated if the chemical kinetics, treatment objective (effluent concentration), influent concentration, and flow rate are known, as follows:

$$\tau = \frac{C_0 - C}{-r|_C} \tag{6-34}$$

$$V = \frac{Q(C_0 - C)}{-r|_C} \tag{6-35}$$

The specific form that is taken by these equations depends on the rate of the reaction. The reaction rate takes different forms for zero-, first-, and second-order reactions, as shown in Table 6-3. Putting Eqs. 6-34 and 6-35 in proper form for these three reaction rates, the following expressions for determining the hydraulic detention time τ and volume V are obtained:

$$\tau = \frac{C_0 - C}{k} \qquad V = \frac{Q(C_0 - C)}{k} \qquad \text{(zero order)} \tag{6-36}$$

$$\tau = \frac{C_0 - C}{kC} \qquad V = \frac{Q(C_0 - C)}{kC} \qquad \text{(first order)} \tag{6-37}$$

$$\tau = \frac{C_0 - C}{kC^2} \qquad V = \frac{Q(C_0 - C)}{kC^2} \qquad \text{(second order)} \tag{6-38}$$

where k = reaction rate constant (see Table 6-3 for units)

EXPRESSION OF CONCENTRATION IN TERMS OF REMOVAL

In many drinking water treatment operations, the focus is on removal of contaminants, so it is often relevant to determine the amount of

Example 6-2 Effluent concentration from a CMFR

A CMFR has an influent concentration of 200 mg/L and a first-order reaction rate constant of 4 d^{-1}. Assuming steady-state conditions, calculate the effluent concentration for a hydraulic detention time of 12 h. Calculate the required hydraulic detention time for an effluent concentration of 10 mg/L.

Solution

1. Determine the effluent concentration by rearranging Eq. 6-37. For steady state, the following result is obtained:

$$C = \frac{C_0}{1+\tau k} = \frac{200\text{ mg/L}}{1+(4/\text{d})(0.5\text{ d})} = 66.6\text{ mg/L}$$

2. Determine the detention time using Eq. 6-37. The required hydraulic detention is given by the expression

$$\tau = \frac{C_0 - C}{kC} = \frac{(200-10)\text{ mg/L}}{(4/\text{d})(10\text{ mg/L})} = \frac{190\text{ mg/L}}{40\text{ mg/L}\cdot\text{d}} = 4.75\text{ d}$$

contaminant removal, that is,

$$\text{Removal} = R = 1 - \frac{C}{C_0} \tag{6-39}$$

Each of the equations for detention time given above (Eqs. 6-36, 6-37, and 6-38) can be rearranged so that (1) the amount of contaminant removal for a reactor with a given detention time may be determined and (2) the detention time necessary for a specified degree of removal may be determined. For example, rearranging Eq. 6-36, the removal for a zero-order reaction may be defined as

$$1 - \frac{C}{C_0} = R = \frac{\tau k}{C_0} \tag{6-40}$$

CMFR TANKS-IN-SERIES ANALYSIS

Treatment processes are frequently staged to meet treatment objectives (e.g., ozonation may be carried out in three or four consecutive stages). Staged treatment processes may be analyzed as tanks in series by building upon Eq. 6-32 developed for a single CMFR. The following analysis of CMFR in series is for first-order reactions, which are frequently encountered in water treatment engineering. The first step is to write Eq. 6-32 in terms of a rate law, instead of a reaction rate, and to include the number of equally sized tanks. The following expression is obtained for a first-order reaction for the first tank in a series of CMFRs.

$$0 = C_0 - C_1 - \frac{V}{nQ}kC_1 \tag{6-41}$$

where C_1 = effluent concentration of first CMFR, mg/L
n = number of CMFRs
k = first-order reaction rate constant, s^{-1}

Rearranging Eq. 6-41 yields an expression to solve for C_1:

$$C_1 = \frac{C_0}{1+(V/nQ)k} \tag{6-42}$$

Following the same steps, the expression for the effluent concentration of the second CMFR, C_2, can be obtained:

$$C_2 = \frac{C_1}{1+(V/nQ)k} \tag{6-43}$$

Substituting the equation for C_1, Eq. 6-42, into Eq. 6-43 yields the expression

$$C_2 = \frac{C_0}{[1+(V/nQ)k]^2} \tag{6-44}$$

A general expression for the effluent concentration of the nth equally sized CMFR, when the influent concentration is known, may be expressed as

$$C_n = \frac{C_0}{[1+(V/nQ)k]^n} \tag{6-45}$$

The total volume required for n CMFRs in series may be determined by rearranging Eq. 6-45 as follows:

$$V = \left[\left(\frac{C_0}{C_n}\right)^{1/n} - 1\right]\left(\frac{nQ}{k}\right) \tag{6-46}$$

UNSTEADY-STATE ANALYSIS

Reactors of concern in water treatment engineering typically operate at steady-state conditions. However, there are times that it is important to analyze reactors operating under unsteady-state conditions, such as when a reactor is first brought online. The unsteady-state analysis begins with a mass balance equation, which is written for a first-order reaction assuming constant volume and using detention time, $\tau = V/Q$, as follows:

$$\tau\frac{dC}{dt} = (C_0 - C) - \tau(kC) = C_0 - C(1+k\tau) \tag{6-47}$$

The integral form of the mass balance analysis shown in Eq. 6-47 is expressed as

$$\int_{C_0^*}^{C} \frac{dC}{C_0 - C(1+k\tau)} = \frac{1}{\tau}\int_0^t dt \tag{6-48}$$

where C_0^* = initial concentration in reactor at time zero, mg/L

Performing the integration of the unsteady-state mass balance results in the expression

$$-\left(\frac{1}{1+k\tau}\right)\ln\left[\frac{C_0 - C(1+k\tau)}{C_0 - C_0^*(1+k\tau)}\right] = \frac{t}{\tau} \tag{6-49}$$

Solving Eq. 6-49 for C results in the final form of the unsteady-state mass balance:

$$C=\left(\frac{C_0}{1+k\tau}\right)-\left[\left(\frac{C_0}{1+k\tau}\right)e^{-(1+k\tau)t/\tau}\right]+\left(C_0^*e^{-(1+k\tau)t/\tau}\right) \tag{6-50}$$

The first term in Eq. 6-50 corresponds to the steady-state condition, and as $t\rightarrow\infty$, the second and third terms in Eq. 6-50 will drop out, indicating that the system has reached steady state.

TIME REQUIRED TO ACHIEVE STEADY STATE

When a process is brought online, it may be important to determine how long it will take for the process to achieve steady state. For instance, when bringing an ozone contactor online, it is important to know how long it will take for the effluent quality to meet treatment goals. To determine the time required to achieve steady state, the case of a first-order reaction is addressed. The initial concentration in the reactor (C_0^*) is set equal to the influent concentration (C_0) in Eq. 6-48 and the expression is integrated. The final form of the unsteady-state mass balance for this special condition is

$$C=\frac{C_0}{1+k\tau}+\frac{k\tau C_0e^{-(1+k\tau)t/\tau}}{1+k\tau} \tag{6-51}$$

Equation 6-51 may now be simplified by dividing by C_0 and considering time as a fraction or multiple of the theoretical detention time, t/τ. The term t/τ, also known as normalized time, is used to allow comparison of reactors with different hydraulic detention times:

$$\frac{C}{C_0}=\frac{1}{1+k\tau}+\frac{k\tau e^{-(1+k\tau)(t/\tau)}}{1+k\tau} \tag{6-52}$$

As shown in Eq. 6-52, $k\tau$, which is the *Damköhler* number, is used to determine the dimensionless steady-state concentration and the profile of dimensionless concentration versus time.

In theory, it takes an infinite amount of time to achieve steady state; however, a reasonable steady state can generally be achieved within several detention times. To estimate an acceptable time, the following operational definition of steady state is introduced:

$$\frac{C}{C_0}-\frac{C_\infty}{C_0}\leq 0.01 \tag{6.53}$$

where C_∞ = steady-state concentration, mg/L

Substituting for $C_\infty/C_0 = 1/(1 + k\tau)$ and rearranging Eq. 6-53 yields the expression

$$\frac{C}{C_0}\leq 0.01+\frac{1}{1+k\tau} \tag{6-54}$$

The expression shown in Eq. 6-52 may be substituted for C/C_0 and Eq. 6-54 is rearranged to solve for time to steady state:

$$\frac{1}{1+k\tau}+\frac{k\tau e^{-(1+k\tau)(t/\tau)}}{1+k\tau}\leq 0.01+\frac{1}{1+k\tau} \tag{6-55}$$

The resulting expression for time to reach steady state is

$$\frac{t}{\tau}\geq\frac{1}{1+k\tau}\ln\left[\frac{100(k\tau)}{1+k\tau}\right] \tag{6-56}$$

Based on the criterion developed in Eq. 6-53, operation times greater than those estimated using Eq. 6-56 will be adequate to achieve steady state. Using the expression for first-order reactions shown in Eq. 6-37, the term $k\tau$ can be substituted into Eq. 6-56, resulting in the equation

$$\frac{t}{\tau}\geq\frac{C_\infty}{C_0}\ln\left[100\left(1-\frac{C_\infty}{C_0}\right)\right] \tag{6-57}$$

To plot all possible scenarios, all possible Damköhler numbers need to be examined, which can be accomplished by plotting the number of hydraulic detention times that are required to achieve steady state as a function of the reduced steady-state concentration. As shown on Fig. 6-10, approximately 2.5 hydraulic detention times are adequate for all of the scenarios to achieve steady state. The time required to reach steady state for zero-order reactions can be computed using the procedure shown above, and the resulting equation is

$$\frac{t}{\tau}\geq-\ln\left(\frac{0.01}{1-C_\infty/C_0}\right) \tag{6-58}$$

A plot of hydraulic detention time versus reduced steady-state concentration for zero-order reaction is shown on Fig. 6-10. Approximately five hydraulic

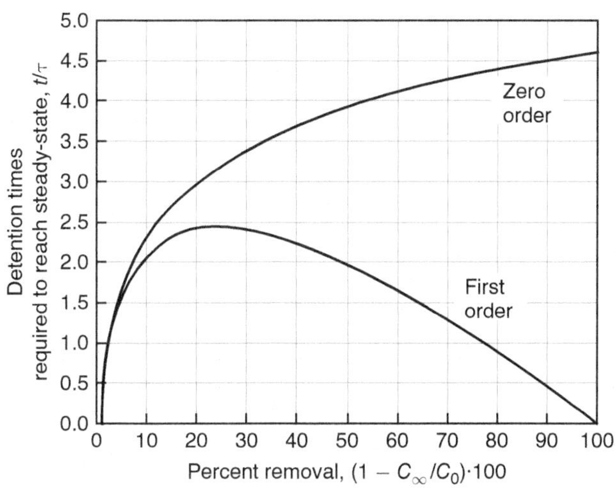

Figure 6-10
Number of detention times, τ, required to reach steady state. For purposes of this graph, steady state is defined as the condition when the effluent concentration is within 1 percent of long-term steady-state concentration.

detention times are adequate for all of the scenarios to achieve steady state, as shown on Fig. 6-10.

Plug Flow Reactor

In a PFR (see Fig. 6-3c), the fluid moves as a plug, and under perfect flow conditions there is no mixing in the axial direction, and there are neither velocity gradients nor concentration gradients in the direction perpendicular to flow. The PFRs of interest in water treatment engineering typically have a constant cross-sectional area (e.g., pipes, channels, or ditches), although a PFR can have a varying cross-sectional area, such as an irregularly shaped river (see Fig. 6-11). In general, a PFR usually requires less reactor volume than a CMFR for an equivalent degree of removal.

PFR MASS BALANCE

There are two possible points of view in which mass balances can be written for PFRs: (1) Eulerian and (2) Langrangian. For the Eulerian point of view, the observer is stationary and fluid flows through the system that the observer has chosen. For the Langrangian point of view, the observer moves with the fluid at its velocity and the system has no flow entering or leaving it.

The concentration in a PFR is dependent upon the independent variables of volume and time, and according to the mean-value theorem, there is at least one value of the dependent variable (C) that is equal to the mean value over the time and volume intervals (t, $t + \Delta t$) and (V, $V + \Delta V$), respectively. Thus, a general mass balance on the small element ΔV using the Eulerian point of view can be written as

$$\Delta t QC|_{V,t} - \Delta t QC|_{V+\Delta V,t} + r\Delta V \Delta t = \Delta V\left[C|_{t+\Delta t,V} - C|_{t,V}\right] \quad (6\text{-}59)$$

where Q = flow rate, L/s

Δt = elapsed time on interval $t \rightarrow t + \Delta t$, s

$C|_{V,t}$ = average concentration in reactor evaluated at V for interval $t \rightarrow t + \Delta t$, mg/L

$C|_{V+\Delta V,t}$ = average concentration in reactor evaluated at $V + \Delta V$ for interval $t \rightarrow t + \Delta t$, mg/L

r = average reaction rate on interval $t \rightarrow t + \Delta t$ in element ΔV, mg/L · s

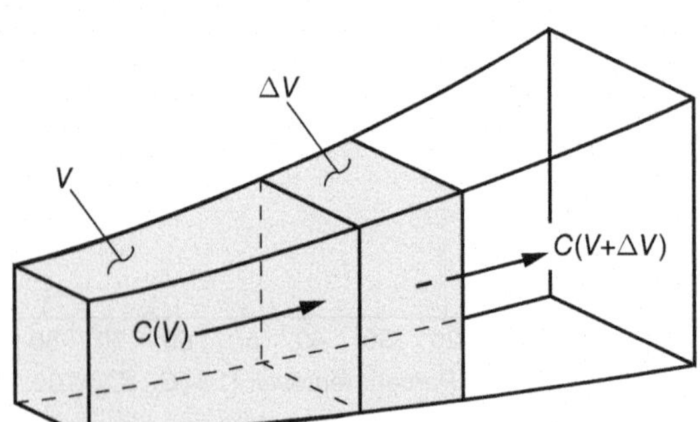

Figure 6-11
Illustration of plug flow reactor with variable cross-sectional area.

$C|_{t+\Delta t, V}$ = average concentration in element ΔV evaluated at $t + \Delta t$, mg/L

$C|_{t,V}$ = average concentration in element ΔV evaluated at t, mg/L

Dividing Eq. 6-59 by Δt and ΔV yields the equation

$$-\frac{QC|_{V+\Delta V,t} - QC|_{V,t}}{\Delta V} + r = \frac{C|_{t+\Delta t,V} - C|_{t,V}}{\Delta t} \tag{6-60}$$

Taking the limit as Δt and ΔV approach zero and assuming there is no volume change upon reaction (which is common for water treatment), the general material balance for a PFR is obtained as

$$-Q\frac{\partial C}{\partial V} + r = \frac{\partial C}{\partial t} \tag{6-61}$$

As shown in Eq. 6-61, the dependent variable C changes with the independent variables V and t. Consequently, one boundary condition and one initial condition are needed to solve this equation.

STEADY-STATE ANALYSIS

As discussed for the CMFR, when a reactor is operated for a long enough period of time with a constant influent concentration, it reaches steady state, meaning the concentration profile in the reactor does not change with time. As described in the following discussion, steady state will be established for a PFR after one hydraulic detention time. For steady-state operation, Eq. 6-61 may be written as

$$-Q\frac{\partial C}{\partial V} + r = \frac{\partial C}{\partial t} = 0 \tag{6-62}$$

The general equation for describing the fate of the component of interest in a PFR assuming steady state and constant Q is expressed as

$$\int_0^V dV = Q\int_{C_0}^{C} \frac{dC}{r} \tag{6-63}$$

The final form of the general equation is obtained by switching the order of the limits of integration (multiplying by −1):

$$V = Q\int_{C}^{C_0} \frac{dC}{-r} \tag{6-64}$$

In addition to calculating the volume of the reactor, the steady-state concentration profile in the reactor can be determined. For example, considering a first-order irreversible reaction, the following expression can be obtained for the concentration profile in the reactor by substituting the rate expression $r = -kC$ into Eq. 6-64 and performing the integration results in

$$C = C_0 e^{-k\tau} = C_0 e^{-k(V/Q)} \tag{6-65}$$

Example 6-3 Steady-state operation for PFR

For steady-state operation, calculate the required hydraulic detention time for a first-order reaction occurring in a PFR. The influent concentration of C_0 is 5 mg/L and the treatment objective is $C = 0.5$ mg/L. The reaction rate constant $k = 0.2\ \text{min}^{-1}$. Compare the detention time required for a PFR to the time obtained for a CMFR at a flow rate of 25 L/min.

Solution

1. Determine the steady-state residence time and volume for the CMFR using Eq. 6-37. At steady state for a CMFR, the required residence time is estimated as

$$\tau_{\text{CMFR}} = \frac{C_0 - C}{kC} = \frac{(5 - 0.5)\ \text{mg/L}}{\left(0.2\ \text{min}^{-1}\right)(0.5\ \text{mg/L})} = \frac{4.5\ \text{mg/L}}{0.1\ \text{mg/L}\cdot\text{min}} = 45\ \text{min}$$

The required volume is calculated as

$$V_{\text{CMFR}} = Q\tau_{\text{CMFR}} = (25\ \text{L}/\text{min})(45\ \text{min}) = 1125\ \text{L}$$

2. Determine the steady-state residence time and volume for the PFR by rearranging Eq. 6-65. At steady state the required residence time and volume are as follows:

$$\tau_{\text{PFR}} = \frac{1}{k}\ln\left(\frac{C_0}{C}\right) = \frac{1}{0.2\ \text{min}}\ln\left(\frac{5\ \text{mg}/\text{L}}{0.5\ \text{mg}/\text{L}}\right) = 11.5\ \text{min}$$

$$V_{\text{PFR}} = Q\tau_{\text{PFR}} = (25\ \text{L}/\text{min})(11.5\ \text{min}) = 288\ \text{L}$$

3. Compare the PFR with the CMFR:

$$\tau_{\text{CMFR}} > \tau_{\text{PFR}}$$
$$V_{\text{CMFR}} > V_{\text{PFR}}$$

For a PFR with constant influent concentration steady state is achieved after an elapsed time equal to the hydraulic detention time because each fluid element travels through the reactor for one detention time.

Comparison of Residence Time and Volume Required for PFRs and CMFRs

It is important to compare the residence time and volume required for PFRs to CMFRs to evaluate the efficiency of reactors. The results differ greatly for irreversible reactions, as discussed below. As defined in Chap. 5, irreversible reactions are reactions that do not proceed in the reverse direction at any measurable rate. In such reactions the concentration of reactants decreases with time, and the reaction rate decreases from its initial value.

A PFR is much more efficient as compared to a CMFR for irreversible reactions, as may be determined by comparing the rates at which the reaction proceeds in a PFR versus a CMFR. In the CMFR, the rate of reaction proceeds at a rate governed by the concentration within the reactor, which corresponds to the effluent concentration. At the inlet of a PFR, the reaction proceeds at a rate governed by the influent concentration, and the rate declines as the fluid moves through the reactor. The final reaction rate in a PFR is governed by the effluent concentration. Effluent concentrations as a function of reactor volume required for PFRs and CMFRs are compared on Fig. 6-12. For a PFR, the concentration profile illustrated on Fig. 6-12 corresponds to the concentration within the reactor, whereas the concentration profile in a CMFR is a horizontal line corresponding to the effluent concentration. For any irreversible reaction of order greater than zero, the reaction rate will proceed at its lowest rate for the CMFR and the highest rate for the PFR.

The equation to determine V or τ for nth order ($n \geq 0$, $n \neq 1$) can be obtained by substituting the appropriate rate expression into Eq. 6-64:

$$V = Q\int_{C}^{C_0} \frac{dC}{kC^n} = \frac{Q}{k}\frac{C^{-n+1}}{-n+1}\bigg|_{C}^{C_0} = \frac{Q}{k}\left(\frac{C_0^{-n+1}}{-n+1} - \frac{C^{-n+1}}{-n+1}\right) \tag{6-66}$$

$$\tau = \frac{V}{Q} = \frac{1}{k}\left(\frac{C_0^{-n+1}}{-n+1} - \frac{C^{-n+1}}{-n+1}\right) \tag{6-67}$$

where V = reactor volume, L
Q = flow rate, L/s
C_0 = influent concentration, mg/L

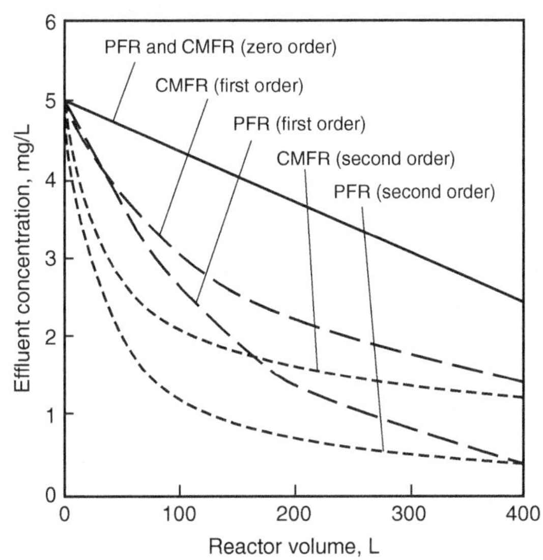

Figure 6-12
Effluent concentration as function of reactor volume for PFRs and CMFRs with zero-, first-, and second-order reactions occurring. The curves were generated using a flow Q = 26.3 L/min, rate constant k = 0.168 min^{-1}, and initial concentration C_0 = 5 mg/L.

C = effluent concentration, mg/L
k = reaction rate constant, $(\text{mg/L})^{-n+1}/\text{s}$
n = reaction order
τ = hydraulic detention time, s

The PFR and CMFR equations are now used to represent the volume of a CMFR divided by the volume of a PFR versus percent removal, as presented on Fig. 6-13. As the percent removal increases, the volume required for a CMFR compared to a PFR increases. The CMFR volume also increases relative to a PFR as the reaction order increases.

A graphical representation of the hydraulic detention time of the CMFR and PFR, as given by the following equations, is shown on Fig. 6-14a:

$$\tau_{\text{CMFR}} = \frac{V_{\text{CMFR}}}{Q} = \frac{C_0 - C}{-r} \tag{6-68}$$

$$\tau_{\text{PFR}} = \frac{V_{\text{PFR}}}{Q} = \int_C^{C_0} \frac{dC}{-r} \tag{6-69}$$

The hydraulic detention time of a CMFR can be computed as the area of a rectangle with a base equal to $C_0 - C$ and a height equal to $-1/r$. The hydraulic detention time of a PFR is equal to the area under the $-1/r$ curve from C to C_0.

It should be noted that a monotonically decreasing function for $-1/r$ versus C is obtained for reaction order n greater than zero, with the decrease being larger as n increases. Accordingly, there is a greater difference in the volumes of a CMFR and a PFR for a second-order reaction as compared to a first-order reaction. There would be no difference for a zero-order reaction; consequently, the residence times would be identical for a CMFR and a PFR.

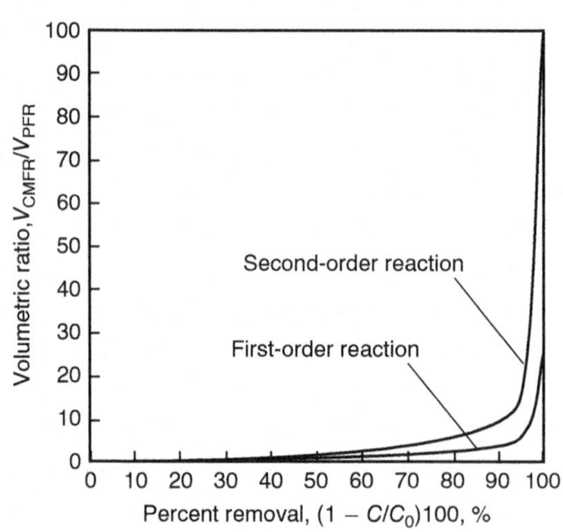

Figure 6-13
Plot of volume of CMFR divided by volume of PFR as function of percent removal.

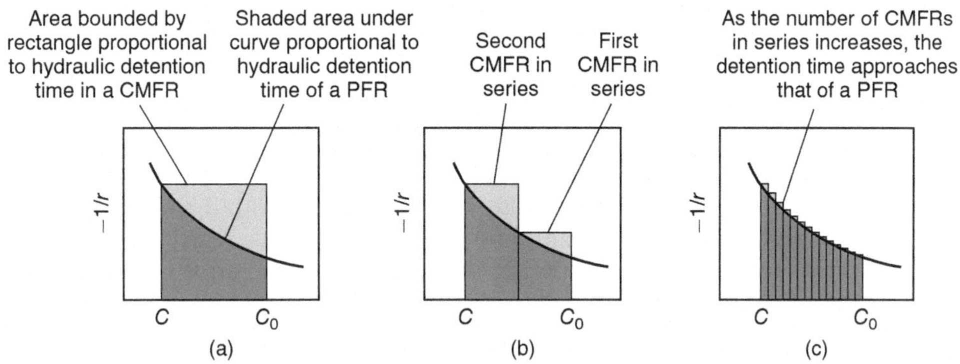

Figure 6-14
Comparison of detention time required for CMFR as compared to PFR: (a) for one CMFR, (b) for two CMFRs in series, and (c) for multiple CMFRs in series.

As shown on Fig. 6-14b, the hydraulic detention time and reactor volume are reduced for two CMFRs in series. The use of multiple CMFRs in series can approach the efficiency of a PFR, as shown on Fig. 6-14c.

Reactors with Recycle

The notation for determining chemical conversion in reactors with recycle is shown on Fig. 6-15. A portion of the throughput Q is diverted from the reactor effluent and recycled to the influent at a rate q. A portion of the solids may also be wasted from the recycle stream using a separation device such as a clarifier.

COMPLETELY MIXED FLOW REACTORS WITH RECYCLE

A steady-state materials balance around the CMFR of volume V (see Fig. 6-15) yields

$$QC_0 - \left[(Q - q_w)C + q_w C\right] - kCV = 0 \tag{6-70}$$

where Q = flow rate, L/s
C_0 = influent concentration, mg/L
q_w = waste flow rate, L/s
C = effluent concentration, mg/L
k = reaction rate constant, s^{-1}
V = reactor volume, L

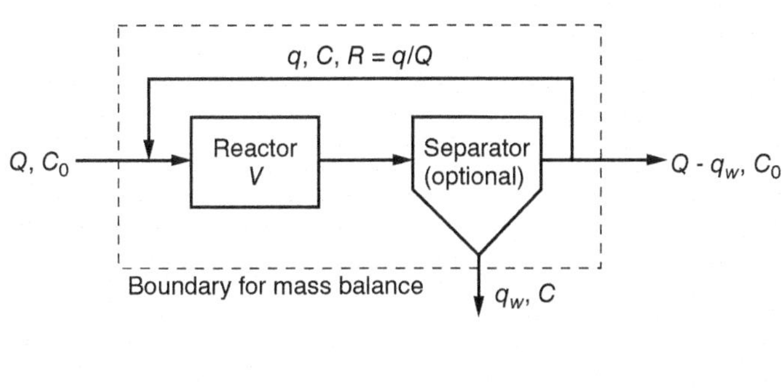

Figure 6-15
Notation used for materials balance for reactor with recycle and solids separator.

The mass balance given by Eq. 6-70 may be simplified to the expression

$$\tau_{\mathrm{CMFR}} = \frac{1}{k}\left(\frac{C_0}{C} - 1\right) \tag{6-71}$$

where τ_{CMFR} = residence time for CMFR, s
k = reaction rate constant, s^{-1}

Equation 6-71 is identical to Eq. 6-34 for a CMFR with first-order kinetics. However, the rate constant k for the heterogeneous reaction in Eq. 6-71 depends on the concentration of solids in the reactor; the reaction rate would be higher if some solids were recycled and they catalyzed the reaction.

Plug Flow Reactors with Recycle

The steady-state materials balance for the PFR with recycle around the fluid element dV may be written as

$$(Q + q)dC = r\,dV \tag{6-72}$$

where q = recycle flow rate, L/s
r = reaction rate, mg/L · s

Separating variables and integrating between the influent concentration with recycle C_I and C, the following equation for first-order kinetics is obtained:

$$\tau_{\mathrm{reactor}} = \left(\frac{1+R}{k}\right)\ln\left(\frac{C_I}{C}\right) \tag{6-73}$$

where τ_{reactor} = residence time for PFR with recycle, s
R = recycle ratio, q/Q
k = reaction rate constant, s^{-1}
C_I = influent concentration with recycle, mg/L

An expression for the influent concentration with recycle can be derived from the mass balance:

$$C_I = \frac{C_0 Q + qC}{Q + q} \tag{6-74}$$

The residence time for a PFR with recycle may be determined using the equation

$$\tau_{\mathrm{reactor}} = \frac{1+R}{k}\ln\left(\frac{C_0/C + R}{1+R}\right) \tag{6-75}$$

With no recycle ($R = 0$), the expected design equation for an ideal PFR is obtained. At the other extreme, with $R \rightarrow \infty$, the reactor approaches the CMFR model. The recycle reactor is thus a model of fluid behavior lying between the two extremes of complete mixing and no mixing. The residence time for a given reactor, τ_{reactor}, is compared to τ_{PFR} on Fig. 6-16. At the 95 percent conversion level (fraction of reactant remaining is 0.05), 100 percent recycle (PFR recycle $R = 1$) increases the required reactor volume by a factor of approximately 4 ($\tau_{\mathrm{reactor}}/\tau_{\mathrm{PFR}} \approx 4$). For the same 95 percent conversion level, a CMFR would need to be approximately

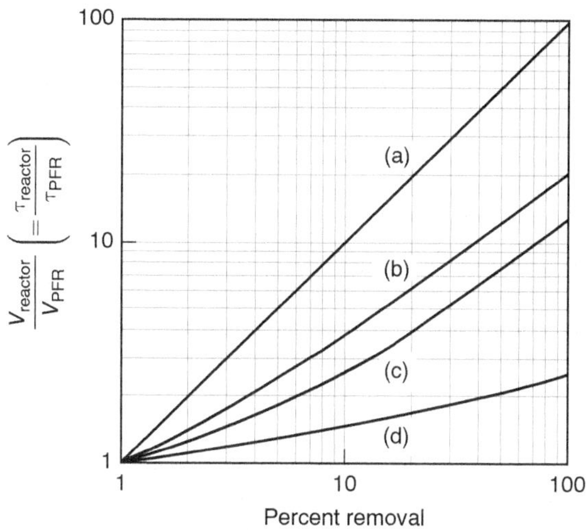

Figure 6-16
Impact of reaction order, reactor type, and required removal on reactor volume: (a) CMFR with second-order reaction, (b) CMFR with first-order reaction, (c) PFR with recycle $R = 1$, and first-order reaction, and (d) three CMFRs in series with first-order reaction.

6 times larger than a PFR for a first-order reaction and approximately 20 times larger than a PFR for a second-order reaction. As R decreases, a smaller reactor volume is required. Thus, a PFR recycle reactor should be designed with an efficient solids separator to ensure minimum recycle ratios that are consistent with the levels of solids required in the reactor. Complete process design also requires a mass balance on solids (Ferguson et al., 1973). As discussed for the CMFR with recycle, a PFR with solids recycle could have a smaller reactor volume if the recycled solids catalyzed the reaction.

6-6 Using Tracers to Characterize Hydraulic Performance of Nonideal Reactors

Water treatment facilities include the largest continuous-flow reactors in use today, and as the scale of continuous-flow reactors increases, their departure from the ideal flow behavior increases. The nonideality associated with large reactors has important implications for the treatment of drinking water because drinking water treatment plants include the largest engineered continuous-flow reactors in existence, far larger than most of the reactors of interest in chemical engineering. As a result, understanding nonideality is critical to the proper design of water treatment processes. There are two important ways in which real continuous-flow reactors behave differently than the ideal reactors discussed in the previous sections: (1) They exhibit nonideal flow behavior and tracer curves that deviate from the ideal curves shown on Fig. 6-4, and (2) the contents of nonideal reactors are not perfectly homogeneous, that is, they are not uniformly mixed to the molecular level.

Although much is known about the residence time distribution (RTD) behavior of flow through different reactor types, it is difficult to design a reactor and know the specific RTD that will result once flow passes through it at full scale. In the long term, computational fluid dynamics (CFD) promises to change this situation, but at the present time, if a reactor's application requires that its RTD be known, then tracer tests must be conducted in the full-scale reactor to measure the RTD directly. Tracer tests are also necessary to confirm the RTD of pilot-scale and demonstration-scale reactors and to develop data that can be used to predict the impact of design changes on performance. However, if the RTD is determined for a large-scale reactor, it may be possible to determine how to modify it to improve its performance by using baffles, turning vanes, and the like to make it perform closer to a PFR. Conducting and evaluating tracer tests are discussed in this section.

Methodology for Tracer Testing

The basic technique used to conduct a tracer study is to introduce the tracer at the reactor inlet and measure the response at the outlet. A tracer is a conservative element, typically a dye or salt solution. The tracer concentration may be measured using a spectrophotometer if a dye is used, a conductivity meter, or specific ion measurements (e.g., fluoride or lithium) if salts are used. Application may be all at once (slug or pulse input) or pumped continuously (step input), resulting in different tracer output curves. When the tracer is applied through a step input, a second tracer curve may be obtained after the tracer input is stopped by recording the tracer disappearance, referred to as a step-down tracer study. Effluent samples can be collected manually as grab samples at specified time intervals or by using an autosampler coupled with a detection instrument for automated sampling. Sampling should be performed frequently enough so that the tracer response may be properly characterized.

Analysis of Tracer Data

Example using a dye tracer study to analyze mixing and contact time.

http://www.stantec.com/wt3r/6-3.html

The tracer curve resulting from a step or pulse input, generally a plot of tracer concentration exiting the reactor as a function of time, is known as the *C* curve, as shown on Fig. 6-17a. To standardize the analysis of tracer curves, tracer data must be normalized in two ways: (1) with respect to the residence time and (2) with respect to the output concentration.

NORMALIZED TIME

The theoretical hydraulic residence time τ in any reactor is equal to the reactor volume divided by the bulk flow rate. For a perfect tracer in an ideal reactor, the mean residence time $\bar{t}$ is equal to τ (Fogler, 1999; Trussell et al., 1979). However, reactors are not perfect, and as a result, the measured mean residence time is always less than τ. The principal cause of this deviation is the presence of dead spaces in the reactor (spaces that do not mix well with the remainder of the contents) where the volume is not used. Thus, the effective volume of the reactor is smaller than the actual volume,

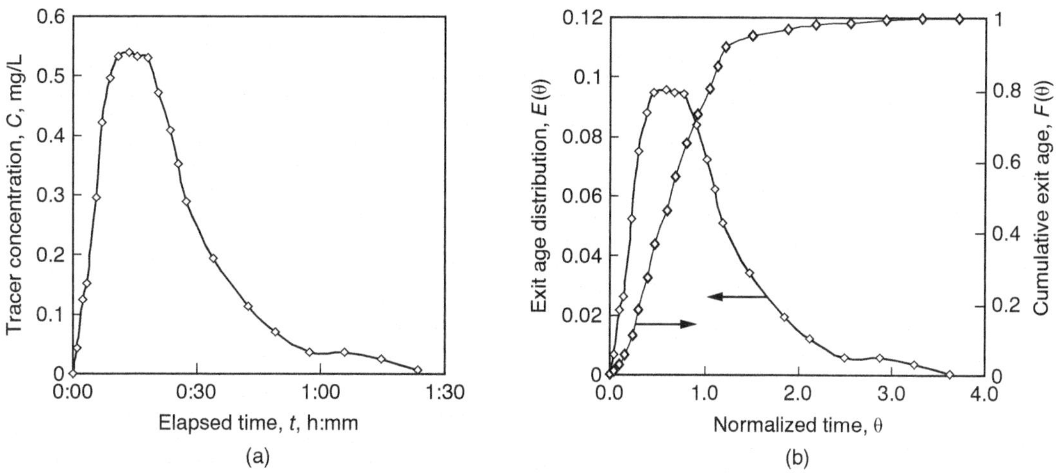

Figure 6-17
Results of tracer test from three CMFRs in series: (a) concentration *C* as function of time and (b) exit age distribution *E* and cumulative exit age distribution *F*.

and the mean residence time is smaller than τ. When choosing a tracer, care must be taken that the tracer does not adsorb on the walls of the vessel because adsorption can cause the mean residence time to be less than τ. It is important that $\bar{t}$, not τ, be used in normalizing tracer curves. The mean residence time is determined from the results of the tracer study in the following manner:

$$\bar{t} = \frac{\int_0^\infty Ct\,dt}{\int_0^\infty C\,dt} \tag{6-76}$$

where $\bar{t}$ = mean residence time of tracer in reactor, min or s
C = concentration exiting reactor at time t, mg/L
t = time since addition of tracer pulse to reactor's entrance, min or s

The denominator in Eq. 6-76 corresponds to the area under the concentration-versus-time curve. The normalized time is determined with the following expression:

$$\theta = \frac{t}{\bar{t}} \tag{6-77}$$

where θ = normalized time, dimensionless

NORMALIZATION CONCENTRATION

The measured concentration must be normalized such that the possibility for all measurements is 1.0. To normalize the concentration, the measured tracer concentration is divided by the total mass concentration.

$$C_N = \int_0^\infty C\,d(\theta) \tag{6-78}$$

$$1 = \int_0^\infty \frac{C}{C_N} d(\theta) \tag{6-79}$$

where C_N = total mass concentration of tracer recovered, mg/L

Just as τ cannot be used in place of $\bar{t}$ for normalization of the detention time, neither can the mass of tracer injected be used to normalize the concentration. The recovery of the tracer is always somewhat less than the mass injected, and the mass recovered in the measurements must be used for normalization. Nevertheless, a carefully conducted tracer study will normally account for more than 95 percent of the mass of the tracer applied, and when this is not the case, it is cause for reexamination of the entire study.

EXIT AGE DISTRIBUTION

The normalized curve for the pulse tracer study is referred to as the exit age distribution $E(\theta)$, and the normalized curve of the step input tracer study is referred to as the cumulative exit age distribution $F(\theta)$. It is also possible to do a continuous tracer study with a step down in concentration. In this case, the normalized curve is referred to as the cumulative internal age distribution $I(\theta)$. The exit age distribution is the most convenient form for performing reactor calculations and provides a distribution of the ages of the fluid elements that leave the reactor.

The exit age distribution is related to the cumulative exit age distribution and the cumulative internal age distribution by the equation

$$E(\theta) = \frac{dF(\theta)}{d\theta} = -\frac{dI(\theta)}{d\theta} \tag{6-80}$$

The exit age distribution curve has the the following important properties:

1. The integral $\int_\theta^{\theta+\Delta\theta} E(\theta)\,d\theta$ equals the fraction of material leaving the reactor having ages between θ and $\theta + \Delta\theta$.
2. The probability of having all possible ages is 1.0, that is $1.0 = \int_0^\infty E(\theta)\,d\theta$.
3. The fraction older than a specified age, θ_2, is $\int_{\theta_2}^\infty E(\theta)\,d\theta$.
4. The fraction younger than a specified age, θ_2, is $\int_0^{\theta_2} E(\theta)\,d\theta$.

COMPUTATION OF EXIT AGE DISTRIBUTION CURVE

The exit age distribution for a CMFR is given by the expression

$$E(\theta) = \frac{C}{C_N} = \frac{C_0 e^{-\theta}}{\int_0^\infty C_0 e^{-\theta} d(\theta)} = e^{-\theta} \tag{6-81}$$

The exit age distribution and cumulative exit age for the tracer curve presented on Fig. 6-17a are shown on Fig. 6-17b. In principle, the exit age distribution for a PFR tracer study should show a very high concentration spike at the hydraulic detention time of the reactor, and the area under the exit age distribution $E(\theta)$ must be 1. Accordingly, the exit age distribution for a PFR is given by the expression

$$E(\theta) = \delta(1-\theta) \tag{6-82}$$

where $\delta\,(1-\theta)$ = Dirac delta function

The Dirac delta function $\delta\,(x)$ has the following operational definition:

$$1 = \int_0^\infty \delta(1-\theta)\,d\theta \qquad \delta(1-\theta) = 0 \quad \text{for} \quad \theta \neq 1$$

and

$$\delta(1-\theta) = \delta(0) = \infty \quad \text{for} \quad \theta = 1 \tag{6-83}$$

Conceptually, the Dirac delta function has an infinite height, infinitesimally small base, and an area of 1.0. The Dirac delta function is commonly used in calculus and is defined as

$$\int_0^\infty y(x)\delta(x_0)\,d\theta = y(x_0) \tag{6-84}$$

where x, x_0 = independent variables

The term $\delta(x)$ is not evaluated except when it is used in an integral. Use of the delta function is similar to the use of imaginary numbers such as $\sqrt{-1}$ because the products of imaginary numbers are always evaluated and the mathematical representation of the numbers is needed before they are multiplied with one another.

Because not all of the tracer will be recovered, the following steps are followed to obtain the exit age distribution from a pulse tracer study:

1. Determine the mean detention time using Eq. 6-76.
2. Compare the mean detention time with the hydraulic detention time and determine if there is any short circuiting. If there is short circuiting, evaluate whether modification to the reactor can eliminate this and improve performance.
3. Determine the normalization concentration using the equation

$$C_N = \int_0^\infty C d\left(\frac{t}{\bar{t}}\right) = \frac{\int_0^\infty C\,dt}{\bar{t}} \tag{6-85}$$

4. Replot the tracer study as $E(\theta) = C/C_N$ versus $\theta = t/\bar{t}$.

COMPUTATION OF VARIANCE

The variance σ_t^2, the second moment about the mean of the data, is used to deteremine the spread of the tracer curve using the following equation:

$$\sigma_t^2 = \frac{\int_0^\infty (t-\bar{t})^2 C\,dt}{\int_0^\infty C\,dt} \tag{6-86}$$

where σ_t^2= variance with respect to t, min^2

The variance can also be calculated from the exit age distribution as follows:

$$\sigma_\theta^2 = \int_0^\infty E(\theta)\,(\theta-1)^2\,d\theta = \frac{\sigma_t^2}{\bar{t}^2} \tag{6-87}$$

where σ_θ^2 = variance with respect to θ, dimensionless

Example 6-4 Calculation of exit age distribution for pulse tracer study

A pulse study on an open-channel reactor (PFR) was conducted and the results are reported in the table below. Plot the tracer curve and normalized RTD curve and determine the variance of the tracer data.

Time, min	C, mg/L	Time, min	C, mg/L	Time, min	C, mg/L
0	0	64	34	90	31
10	0	65	38	95	21
20	1	66	46	100	15
30	2	68	58	105	10
40	4	70	62	110	6
50	9	71	63	120	3
55	15	72	63	130	1
60	24	75	64	140	0
62	28	80	58		
63	30	85	45		

Solution

1. The tracer curve is shown on panel (a) of the figure in step 2e below.
2. Construct a plot of the normalized RTD curve.
   a. Set up a computation table to determine the mean detention time (used for normalizing time data) using Eq. 6-76:

$$\bar{t} = \frac{\int_0^\infty Ct\,dt}{\int_0^\infty C\,dt} \approx \frac{\sum \overline{Ct}\Delta t}{\sum \bar{C}\Delta t}$$

The summation is shown in the table given below.

t,min	C,mg/L	$\left(\frac{c_{i-1}+c_i}{2}\right)\Delta t$	$\left(\frac{c_{i-1}t_{i-1}+c_it_i}{2}\right)\Delta t$	t,min	C,mg/L	$\left(\frac{c_{i-1}+c_i}{2}\right)\Delta t$	$\left(\frac{c_{i-1}t_{i-1}+c_it_i}{2}\right)\Delta t$
0	0	—	—	70	62	120	8,284
10	0	0	0	71	63	62.5	4,407
20	1	5	100	72	63	63	4,505
30	2	15	400	75	64	190.5	14,004
40	4	30	1,100	80	58	305	23,600
50	9	65	3,050	85	45	257.5	21,163
55	15	60	3,188	90	31	190	16,538
60	24	97.5	5,663	95	21	130	11,963
62	28	52	3,176	100	15	90	8,738
63	30	29	1,813	105	10	62.5	6,375
64	34	32	2,033	110	6	40	4,275
65	38	36	2,323	120	3	45	5,100
66	46	42	2,753	130	1	20	2,450
68	58	104	6,980	140	0	5	650
Total						2,148.5	164,627

$$\bar{t} \approx \frac{\sum \overline{Ct}\,\Delta t}{\sum \bar{C}\,\Delta t} = \frac{164{,}627}{2148.5} = 76.6\text{ min}$$

b. Determine the value to be used for normalization of the tracer concentration C_N using Eq. 6-85:

$$C_N = \int_0^\infty C d\left(\frac{t}{\bar{t}}\right) = \frac{\int_0^\infty C\,dt}{\bar{t}} \approx \frac{\sum \bar{C}\Delta t}{\bar{t}} = \frac{2148.5}{76.6} = 28\text{ mg/L}$$

c. Use $\bar{t}$ and C_N to normalize the original tracer study data:

$$\theta = \frac{t}{\bar{t}} = \frac{t}{76.6} \quad \text{and} \quad E(\theta) = \frac{C}{C_N} = \frac{C}{28}$$

d. Set up a computation table to determine the values to be used for plotting the RTD curve.

t, min	C, mg/L	θ	E(θ)	t, min	C, mg/L	θ	E(θ)
0	0	0.00	0.00	70	62	0.91	2.21
10	0	0.13	0.00	71	63	0.93	2.25
20	1	0.26	0.04	72	63	0.94	2.25
30	2	0.39	0.07	75	64	0.98	2.28
40	4	0.52	0.14	80	58	1.04	2.07
50	9	0.65	0.32	85	45	1.11	1.61
55	15	0.72	0.54	90	31	1.17	1.11
60	24	0.78	0.86	95	21	1.24	0.75
62	28	0.81	1.00	100	15	1.30	0.54
63	30	0.82	1.07	105	10	1.37	0.36
64	34	0.83	1.21	110	6	1.43	0.21
65	38	0.85	1.36	120	3	1.56	0.11
66	46	0.86	1.64	130	1	1.69	0.04
68	58	0.89	2.07	140	0	1.83	0.00

e. The final exit age distribution is shown in panel (b) below.

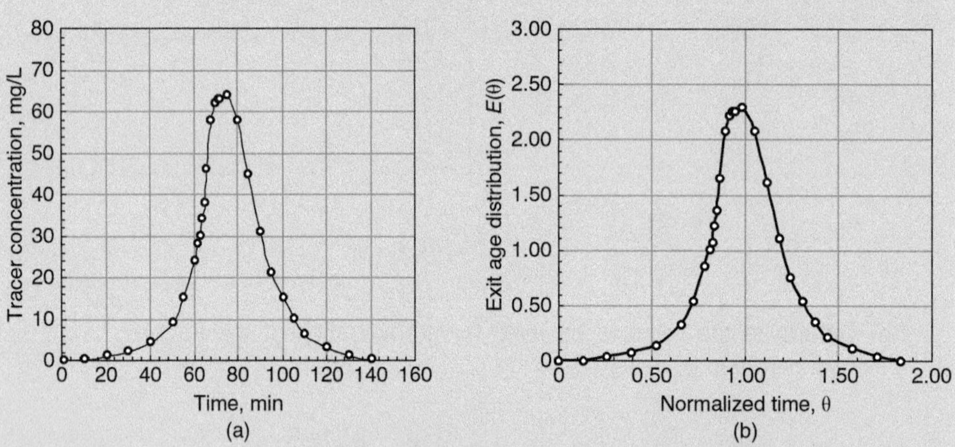

3. Compute the variance using Eqs. 6-86 and 6-87:

$$\sigma_t^2 = \frac{\int_0^\infty (t-\bar{t})^2 C\,dt}{\int_0^\infty C\,dt} \approx \frac{\sum (t-\bar{t})^2 C\,\Delta t}{\sum \bar{C}\,\Delta t}$$

$$\sigma_\theta^2 = \frac{\sigma_t^2}{\bar{t}^2}$$

Set up a computation table to compute the terms needed to determine the variance.

t, min	C, mg/L	$\bar{C}\Delta t$[a]	Term[b]	t, min	C, mg/L	$\bar{C}\Delta t$[a]	Term[b]
0	0	_	_	70	62	120	7,176.5
10	0	0	0.0	71	63	62.5	2,416.4
20	1	5	16,075.5	72	63	63	1,720.4
30	2	15	37,886.0	75	64	190.5	2,367.1
40	4	30	48,750.9	80	58	305	2,040.7
50	9	65	59,024.7	85	45	257.5	9,324.2
55	15	60	33,703.4	90	31	190	21,452.2
60	24	97.5	34,398.2	95	21	130	31,283.7
62	28	52	12,746.8	100	15	90	37,933.6
63	30	29	5,842.1	105	10	62.5	40,375.0
64	34	32	5,558.8	110	6	40	36,651.3
65	38	36	5,344.4	120	3	45	61,384.2
66	46	42	5,235.8	130	1	20	42,324.6
68	58	104	9,660.1	140	0	5	14,203.5
						2,148.5	584,880.0

[a] $\bar{C}\,\Delta t = \left(\dfrac{C_{i-1}+C_i}{2}\right)\Delta t.$

[b] $\left\{\left[(t_{i-1}-\bar{t})^2 C_{i-1} + (t_i-\bar{t})^2 C_i\right]/2\right\}\Delta t.$

$$\sigma_t^2 = \frac{\sum \overline{(t-\bar{t})^2 C\,\Delta t}}{\sum \bar{C}\,\Delta t} = \frac{584{,}880}{2{,}148.5} = 272.2\ \text{min}^2$$

$$\sigma_t = 16.50\ \text{min}$$

$$\sigma_\theta^2 = \frac{\sigma_t^2}{\bar{t}^2} = \frac{272.2\ \text{min}^2}{(76.6\ \text{min})^2} = 0.0464$$

$$\sigma_\theta = 0.221$$

Comment

The numerical integration in this example was carried out using the trapezoidal rule. It should be noted that other methods of integration, including the rectangular, Simpson, and Durand rules as well as more complex methods, may be used.

Parameters Used to Characterize Tracer Results

For some time, the environmental engineering community has understood that the RTD of a reactor, particularly a disinfection contactor, is important. As a result, tracer curves have been used to characterize these reactors and aid in understanding their performance. As this understanding has grown more sophisticated, a number of indices have been used to characterize reactors and their tracer results. Some of the most commonly used terms are summarized in Table 6-5. By far the most commonly used term to characterize reactors is the hydraulic detention time τ, and it is clear that

Table 6-5
Terms used to characterize tracer curves

Term	Definition
d	Dispersion number. Measures dispersion in reactor. For an ideal PFR, $d = 0$. For an ideal CMFR, $d = \infty$
Pe	Peclet number; Pe = $1/d$
τ	Theoretical hydraulic residence time ($\tau = V/Q$)
t_i	Time at which tracer first appears
θ	Normalized detention time $(t/\bar{t} \approx t/\tau)$
t_p, t_{modal}	Time at which peak concentration of tracer is observed (mode)
$\bar{t}$	Mean residence time, centroid of pulse tracer curve
t_{10}, t_{50}, t_{90}	Time at which 10, 50, and 90% of tracer has passed through reactor or when 10, 50, and 90% of the fluid has passed through the reactor
t_{90}/t_{10}	Morrill dispersion index (MDI)
t_i/τ	Index of short circuiting. In an ideal PFR, the ratio is 1 and approaches zero with increased short circuiting
t_p/τ	Index of modal retention time. Ratio is 1 in an ideal PFR and zero in an ideal CMFR. For values of the ratio greater than or less than 1.0, the flow distribution in the reactor is not uniform
$\bar{t}/\tau$	Index of average retention time; reflects the volume of the reactor that is not used as the fluid passes through the reactor. This volume is sometime called dead volume. If the ratio is 1.0, there is no dead volume in the reactor
t_{50}/τ	Index of mean retention time. The ratio t_{50}/τ is a measure ofthe skew of the $E(\theta)$ curve. A value of t_{50}/τ of less than 1.0 corresponds to an $E(\theta)$ curve that is skewed to the left. Similarly, for values greater than 1.0 the $E(\theta)$ curve is skewed to the right
n	Equivalent number of tanks in series in TIS model

τ has great significance. Of the remaining indices, the most important are probably the dispersion number *d*; the Peclet number Pe; the equivalent number of tanks in series (TIS), *n*; and t_{10}. The dispersion number, the Peclet number, and the equivalent number of tanks in series are important because these can be used in the single-parameter models as a comprehensive measure of dispersion, as discussed in Sec. 6-7. The time for 10 percent of the tracer to pass through the reactor, t_{10}, is important because the U.S. EPA uses that measurement for regulating the performance of disinfection reactors for drinking water.

6-7 Modeling Hydraulic Performance of Nonideal Reactors

As described in Sec. 6-6, the flow in real reactors is not ideal and RTD curves are used to characterize the nonideal flow. In nonideal reactors, the mixing conditions are complex and flow behavior often deviates substantially from the assumptions of ideal flow. The nonideal reactor analysis in this section includes (1) the factors that cause nonideal flow; (2) distinction between molecular diffusion, turbulent diffusion, and dispersion; and (3) models used for characterizing nonideal flow and predicting nonideal reactor performance.

Causes of Nonideal Flow

Mixing at two scales dramatically affects reactor efficiency. At the microscale (<~500 μm), there is concern about mixing reactants in the influent such that there is a uniform concentration of the reactants entering the reactor. At the macroscale, there is concern about how long different fluid elements (which are presumed uniform in concentration at the microscale) reside in the reactor. There are three principal types of nonideal fluid behavior at the macroscale in processing equipment: (1) inadequate initial blending, (2) short circuiting, and (3) diffusion and dispersion. Ultimately, nonideal flow can cause decreased removal efficiencies or the formation of undesirable by-products.

INADEQUATE INITIAL BLENDING

Nonideal flow in reactors may be caused by inadequate initial blending of reacting components as they enter the reactor. As blending of miscible fluids and soluble components will occur, the initial mixing issue becomes a question of whether the blending is accomplished fast enough relative to the speed of the reactions taking place. Blending is addressed in greater depth in Sec. 6-10.

SHORT CIRCUITING

Short circuiting is characterized by a segment of the fluid stream having a residence time considerably shorter than the mean hydraulic residence time. Short circuiting is a common design issue in CMFRs, rectangular basins, and packed columns or towers, and it is particularly important in processes where a high level of removal is required, such as disinfection (e.g., 99.99 percent inactivation). The impact of short circuiting on processes with low versus high removal requirements is compared in Sec. 13-4. Short circuiting may develop within the reactor due to poor fluid mechanical design of (1) internal packing material, (2) inlet and outlet structures, and (3) the aspect ratio of the reactor itself (length as compared to depth and width with a larger aspect ratio being most desirable). Short circuiting also occurs when circulation patterns develop due to wind or density differences due to temperature or the concentration of dissolved or suspended materials. For example, when the flow entering the reactor is warmer than the flow in the reactor, a portion of flow can travel to the outlet across the top of the reactor. Short circuiting also occurs in sedimentation basins due to the greater bulk density of the sludge with respect to the bulk water above it. In shallow reactors, wind can transport a portion of the incoming flow to the outlet, resulting in an observed detention time that is shorter than the theoretical residence time of the reactor. Wind can also cause backmixing. However, as discussed later, velocity gradients and short circuiting can be overcome by using baffles, increasing the aspect ratio of the reactor, and proper orientation of basins relative to the prevailing winds or covering them.

DIFFUSION AND DISPERSION

In addition to improper initial blending and short circuiting, diffusion and dispersion can contribute to the nonideal flow observed in reactors. In general, diffusion involves the movement of a constituent from a higher concentration to a lower concentration as a result of collisions with fluid molecules that move by Brownian motion. Dispersion is the mixing brought about by physical processes such as turbulence and velocity gradients.

Diffusion

Molecular diffusion occurs when dissolved constituents or very small particles move randomly within the water matrix as a result of collisions with water molecules that move randomly. This random motion is generally referred to as Brownian motion. Because molecular diffusion does not depend on any bulk movement of the water, it can occur both under laminar and turbulent flow conditions. Molecular diffusion is irreversible and is different for different constituents. Mass transfer brought about by molecular diffusion is described by Fick's law, which is considered in Chap. 7.

Dispersion

The mixing process whereby a constituent is transported from a higher concentration to a lower concentration, by eddies formed by turbulent flow or shearing forces between fluid layers is known as dispersion. Eddies can vary in size from microscale to macroscale to large circulation patterns in the ocean. Microscale mass transport is only by molecular diffusion, whereas in the macroscale mass transport can be by both molecular and dispersion, with the latter predominating. Kolmogorov (1941a,b,c), as discussed in Sec. 6-10, identified the dividing line between the microscale and macroscale and suggested a method for determining the size of the smallest eddy that could be generated as a function of the amount of energy dissipated. Dispersion is considered further in Sec. 6-10.

The constituent mixing that results from the shearing forces between fluid layers and by the random fluid motion of turbulence is also known as dispersion. The parabolic distribution of flow velocities that occurs in pipe flow is a classic example of the shearing that occurs between fluid layers and is known as Taylor dispersion. Dispersion occurs mainly in tanks, channels, pipes, or columns and is characterized by longitudinal mixing, which distorts the flat velocity profile (perpendicular to the direction of flow) that is assumed for an ideal PFR. Because dispersion coefficients are dominated by the character of turbulence at the macroscale, they are identical for all constituents, and tracer studies can be conducted using a dye or salt solution. Typical values of some observed dispersion coefficients E are illustrated on Fig. 6-18. Representative values for molecular diffusion are shown in the lower left on Fig. 6-18.

Models Used to Describe Nonideal Flow

When an engineer designs a large reactor for water treatment, nonideal flow is almost always a design issue. To ensure that the reactor will perform as intended, it is important that the design engineer consider impacts that

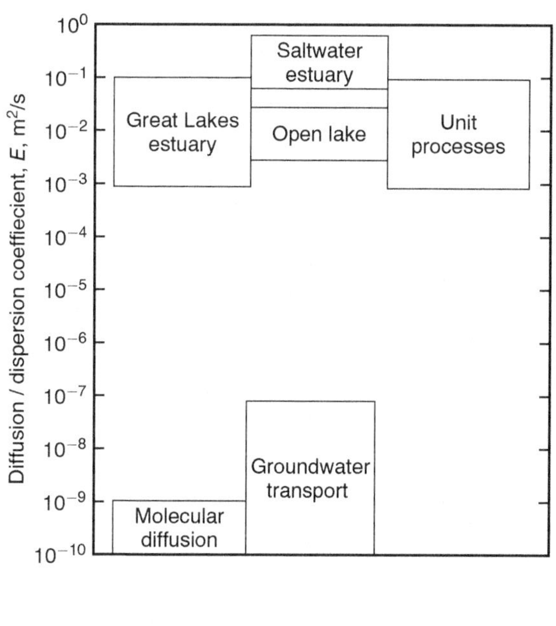

Figure 6-18
Observed coefficients of dispersion and molecular diffusion.

nonideal flow might have on reactor performance. Considering nonideal effects on reactors occurs through two important steps: (1) understanding the impact of reactor design on nonideal flow and (2) understanding the impact of nonideal flow on reactor performance. Neither of these tasks is accomplished easily, but understanding the impact of reactor design on nonideal flow presents a particular challenge. Usually, the approach taken is to (1) construct a model of the RTD in the reactor as a way of understanding its nonideal behavior and then (2) use the RTD model to predict the impact on reactor performance. Two single-parameter models are typically used to model RTD: (1) the dispersion model or dispersed-flow model (DFM) and (2) the tanks-in-series (TIS) model. Once a reactor has been built, it is common to conduct a tracer study to determine the actual RTD of the reactor and then to see how well that RTD can be fit to one of these single-parameter models. The DFM and the TIS model are introduced in the following discussion.

DISPERSED-FLOW MODEL

If longitudinal dispersion is the main cause for deviation from ideal flow in a conduit reactor (i.e., a tube, pipe, or channel), then the following mass balance may be established, assuming a constant cross-sectional area and no short circuiting in the reactor:

$$\left(QC|_{z,t} - E\cdot a\frac{\partial C}{\partial z}\bigg|_{z,t}\right)\Delta t - \left(QC|_{z+\Delta z,t} - E\cdot a\frac{\partial C}{\partial z}\bigg|_{z+\Delta z,t}\right)\Delta t \\ + r\cdot a\cdot \Delta z\cdot\Delta t = a\cdot\Delta z[C|t+\Delta t, z - C|t, z] \tag{6-88}$$

where Q = flow rate, L/s

Δt = elapsed time on interval $t \rightarrow t+\Delta t$, s

$C|_{z,t}$ = average concentration in reactor evaluated at z for interval $t \rightarrow t+\Delta t$, mg/L

$C|_{z+\Delta z,t}$ = average concentration in reactor evaluated at $z+\Delta z$ for interval $t \rightarrow t+\Delta t$, mg/L

E = dispersion coefficient, m^2/s

a = cross-sectional area of reactor, m^2

$\frac{\partial C}{\partial z}\Big|_{z,t}$ = change in concentration with position in reactor evaluated at z for the interval $t \rightarrow t+\Delta t$, mg/L · m

$\frac{\partial C}{\partial z}\Big|_{z+\Delta z,t}$ = change in concentration with position in reactor evaluated at $z+\Delta z$ for interval $t \rightarrow t+\Delta t$, mg/L · m

r = average reaction rate on interval $t \rightarrow t+\Delta t$ in element, mg/L ·s

Δz = length of element, m

$C|_{t+\Delta t,z}$ = average concentration in element evaluated at $t+\Delta t$, mg/L

$C|_{t,z}$ = average concentration in element evaluated at t, mg/L

In Eq. 6-88, there are values of the dependent variable that are equal to the mean value over the intervals (t, $t+\Delta t$) or (z, $z+\Delta z$), which can be inferred from the mean-value theorem. For this reason, the dependent values are evaluated somewhere located in the intervals (t, $t+\Delta t$) and (z, $z+\Delta z$). When taking the limit as Δt and Δz approach zero, they are evaluated at t and z because they were contained on the intervals. Dividing by $a\,\Delta z$ and Δt, multiplying by τ, taking the limit as $\Delta z \rightarrow 0$, $\Delta t \rightarrow 0$, and rearranging, the following expression is obtained:

$$-\frac{\partial C}{\partial(z/L)} + \left(\frac{E}{vL}\right)\frac{\partial^2 C}{\partial(z/L)^2} + r\tau = \frac{\partial C}{\partial(t/\tau)} \tag{6-89}$$

where L = length of reactor, m
v = average fluid velocity (Q/a), m/s
τ = hydraulic detention time, s

To evaluate the RTD, it is assumed that the tracer does not participate in any reactions (i.e., a conservative tracer). Thus, Eq. 6-89 can be rearranged to the following:

$$-\frac{\partial C}{\partial \bar{z}} + \frac{1}{\text{Pe}}\frac{\partial^2 C}{\partial \bar{z}^2} = \frac{\partial C}{\partial(t/\tau)} \tag{6-90}$$

where Pe = Peclet number = vL/E, dimensionless
$\bar{z}$ = dimensionless length ($\bar{z} = z/L$)

The Peclet number is the "single parameter" of the DFM. Specifying the Peclet number is concomitant with specifying the entire RTD. Conceptually, the Peclet number is the ratio between mass transport by advection to dispersion:

$$\text{Pe} = \frac{vL}{E} = \frac{\text{rate of transport by advection}}{\text{rate of transport by dispersion}} \tag{6-91}$$

High Peclet numbers result when advection controls mass transport within the reactor because dispersion and advection act in parallel. As the Peclet number approaches infinity, transport is only by advection; no axial dispersion occurs and the reactor performance approaches that of a PFR. Conversely, for a Peclet number approaching zero, there is no transport by advection; only axial dispersion occurs and the reactor performance approaches that of a CMFR.

Often the dispersion number d is used in place of the Peclet number. The relationship between the dispersion number and the Peclet number is as follows:

$$d = \frac{1}{\text{Pe}} \tag{6-92}$$

where d = dispersion number, dimensionless

The Peclet number and dispersion number are both used in environmental literature, and, therefore, it is important to understand the relationship between these two parameters. The dispersion number and DFM are also used in Chaps. 8 and 13. To solve Eq. 6-90, one initial condition and two boundary conditions are needed (Fogler, 1999). The boundary conditions, which apply to either steady-state or unsteady-state conditions, are discussed below.

Open and closed systems and the DFM model

There are two well-known approaches to the boundary conditions for the DFM model, the closed-system approach and the open-system approach. The differences between the open and closed models are shown on Fig. 6-19. For the closed-system model, it is assumed that plug flow occurs in and out of the reactor and dispersed flow occurs within the reactor. For the open-system model, it is assumed that dispersed-flow conditions are present throughout; that is, the reactor is essentially a segment of flow with characteristics common to the flow preceding and following it. Most unit operations used in environmental engineering are better approximated by the closed-system assumptions, although, as will be shown later, the differences between these models are not very significant when the Peclet number is greater than approximately 40 (when the dispersion number is below approximately 0.025), and most engineered PFRs exceed this performance.

Solving DFM for a closed-flow system

For a closed reactor it is assumed that plug flow conditions exist before the entrance (advection only) and dispersed-flow conditions exist after the

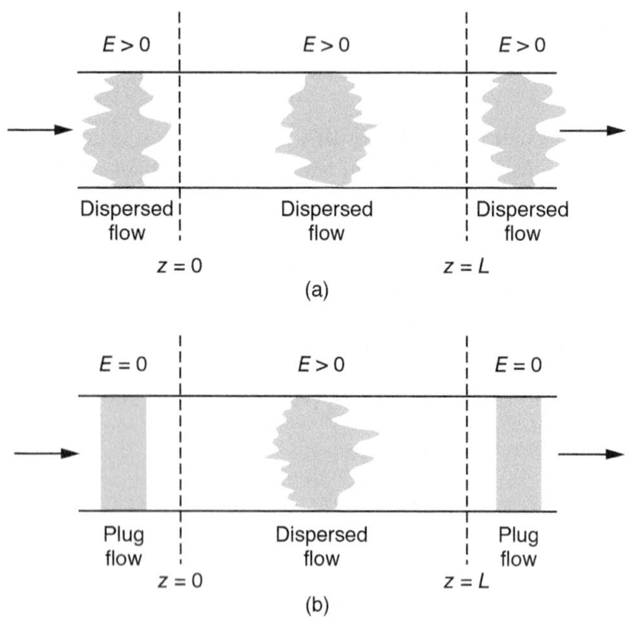

Figure 6-19
Illustration of boundary conditions for dispersion model: (a) open system and (b) closed system.

entrance (advection and dispersion). Thus, the following mass balance can be written at the entrance to the reactor:

$$QC\,|_{z=0^-} = QC\,|_{z=0^+} - E \cdot a \left.\frac{\partial C}{\partial z}\right|_{z=0^+} \tag{6-93}$$

where $QC\,|_{z=0^-}$ = mass of tracer that enters reactor, or QC_0, mg/s
$QC\,|_{z=0^+}$ = advective transport of tracer evaluated at $z = 0^+$, mg/s
$z = 0^-, 0^+$ = location just before and just after entrance of reactor

Recognizing that the velocity v is equal to flow divided by the cross-sectional area, Q/a, Eq. 6-93 may be rewritten as

$$\left.\frac{\partial C}{\partial z}\right|_{z=0^+} = \frac{v}{E}\left(C\,|_{z=0^+} - C\,|_{z=0^-}\right) \tag{6-94}$$

The final form of the boundary condition is given by the expression

$$\left.\frac{\partial C}{\partial (z/L)}\right|_{z=0^+} = \frac{vL}{E}\left(C\,|_{z=0^+} - C_0\right) = \text{Pe}\left(C\,|_{z=0^+} - C_0\right) \tag{6-95}$$

As shown in Eq. 6-95, there is a discontinuity in the concentration at the entrance of the reactor. Also for a closed system, it is assumed that dispersed-flow conditions exist before the exit (advection and dispersion) and plug flow conditions exist after the exit (advection only). Thus, a mass balance on the exit of a closed reactor yields

$$QC\,|_{z=L^-} - (E \cdot a)\left.\frac{\partial C}{\partial z}\right|_{z=L^-} = QC\,|_{z=L^+} \tag{6-96}$$

where $QC\,|_{z=L^-}$ = advective transport of tracer evaluated at $z = L^-$, mg/s
$QC\,|_{z=L^+}$ = mass of tracer that leaves reactor, mg/s
C = effluent concentration, mg/L
$z = L^-, L^+$ = locations just before and just after exit of reactor

Unlike the entrance of the reactor, there cannot be a discontinuity in concentration at the exit, and therefore the final form of the boundary condition is given by the following expressions:

$$\left.\frac{\partial C}{\partial z}\right|_{z=L} = 0 \tag{6-97}$$

$$C\,|_{L^-} = C\,|_{L^+} \tag{6-98}$$

Thomas and McKee (1944) developed a solution for a closed reactor for a conservative or nonreacting tracer, which has been modified in terms of the Peclet number:

$$\frac{C}{C_0} = 2\sum_{i=1}^{\infty} b_i \left[\frac{a \sin b_i + b_i \cos b_i}{a^2 + 2a + b_i^2}\right] \exp\left[a - \frac{(a^2 + b_i^2)\theta}{2a}\right] \tag{6-99}$$

where C = effluent concentration of nonreactive tracer, mg/L
C_0 = initial concentration of tracer in reactor, mg/L

$$b_i = \cot^{-1}\left[\frac{(b_i/a)-(a/b_i)}{2a}\right]$$

$$a = \frac{Pe}{2} = \frac{1}{2d}$$

Pe = Peclet number, dimensionless (see Eq. 6-91)
θ = normalized detention time, $t/\bar{t}$

To fit the closed-system DFM to an existing reactor, a pulse tracer study is conducted to obtain the exit age distribution. In turn, the tracer data can be fit with Eq. 6-99 to determine Pe. For a corresponding set of C/C_0 and θ values, a Pe is assumed and Eq. 6-99 is solved by successive numerical approximation. The process is repeated until a value of Pe is found such that Eq. 6-99 closely approximates the tracer curve.

Because of the difficulty in solving Eq. 6-99 and because it is easier to determine the variance of the tracer curve, the dispersion may be estimated using the following expression, which relates the Pe to σ_θ^2 (Fogler, 1999):

$$\sigma_\theta^2 = \frac{\sigma_t^2}{\bar{t}^2} = \left(\frac{2}{\text{Pe}}\right) - \left[2\left(\frac{1}{\text{Pe}}\right)^2\left(1-e^{-\text{Pe}}\right)\right] \tag{6-100}$$

where σ_θ^2 = variance with respect to θ
σ_t^2 = variance with respect to t

Consequently, a determination from variance is often used in spite of the inaccuracies that can result from determining σ_θ^2 from experimental data. For example, small concentrations far from the centroid can have a large impact on the value of σ_θ^2. Thus, for accurate estimates of reaction performance, Eq. 6-99 should be fit to the data to determine the Pe.

Solving DFM for an open-flow system

For an open system it is assumed that dispersed-flow conditions exist before and after the entrance (advection and dispersion). Thus, the following mass balance can be written at the entrance to the reactor:

$$QC\big|_{z=0^-} - E\cdot a\frac{\partial C}{\partial z}\bigg|_{z=0^-} = QC\big|_{z=0^+} - E\cdot a\frac{\partial C}{\partial z}\bigg|_{z=0^+} \tag{6-101}$$

The final form of the boundary condition at the entrance to the reactor is given by

$$C\big|_{z=0^-} - \frac{1}{\text{Pe}}\frac{\partial C}{\partial(z/L)}\bigg|_{z=0^-} = C\big|_{z=0^+} - \frac{1}{\text{Pe}}\frac{\partial C}{\partial(z/L)}\bigg|_{z=0^+} \qquad C\big|_{z=0^-} = C\big|_{z=0^+} \tag{6-102}$$

Similarly, a mass balance at the exit of an open reactor yields the final form of the boundary condition:

$$C\big|_{z=L^-} - \frac{1}{\text{Pe}}\frac{\partial C}{\partial(z/L)}\bigg|_{z=L^-} = C\big|_{z=L^+} - \frac{1}{\text{Pe}}\frac{\partial C}{\partial(z/L)}\bigg|_{z=L^+} \qquad C\big|_{z=L^-} = C\big|_{z=L^+} \tag{6-103}$$

It is important to realize that C|z=0− and C|z=0+ are not equal to the influent concentration into the reactor because dispersion transports the substance upstream and downstream.

If Eq. 6-90 is solved for a nonreactive tracer ($r = 0$) in an open reactor, the following expression for $E(\theta)$ is obtained (Levenspiel and Smith, 1957):

$$E(\theta) = \frac{1}{\sqrt{4\pi\theta(1/\text{Pe})}} e^{-\text{Pe}(1-\theta)^2/4\theta} \tag{6-104}$$

For low dispersion (high Pe), Eq. 6-104 simplifies to the following:

$$E(\theta) = \frac{1}{\sqrt{4\pi(1/\text{Pe})}} e^{-\text{Pe}(1-\theta)^2/4} \tag{6-105}$$

The mean detention time $\bar{t}$ for an open reactor is longer than the theoretical hydraulic detention time τ because some of the tracer can migrate upstream of the reactor. The mean detention time can be determined using the expression

$$\bar{t} = \left[1 + \left(\frac{2}{\text{Pe}}\right)\right]\tau \tag{6-106}$$

where $\bar{t}$ = centroid of tracer curve = $\int_0^\infty Ct\,dt \big/ \int_0^\infty C\,dt$

τ = hydraulic detention time = V/Q

It has been shown that the following relationship between the Pe and σ_θ^2 applies to open systems when the true τ is unknown (due to short circuiting or dead zones) (Fogler, 1999):

$$\sigma_\theta^2 = \frac{\sigma_t^2}{(\bar{t})^2} = \frac{2/\text{Pe} + 8(1/\text{Pe})^2}{(1+2/\text{Pe})^2} + \frac{2\text{Pe}+8}{(\text{Pe}+2)^2} \tag{6-107}$$

To fit the open-system DFM to an existing reactor, a pulse tracer study is conducted to obtain the exit age distribution. The Peclet number may be estimated in two ways: from the variance of tracer data using Eq. 6-107 or by directly fitting the tracer data to Eq. 6-104. As discussed for the closed reactor system, the latter approach usually results in a superior fit and should be used to determine the Peclet number whenever possible.

TANK-IN-SERIES MODELS

The analysis summarized in Eq. 6-21 can be used to show that the exit age distribution for a cascade of n CMFRs in series is given by the following expression (Levenspiel, 1998):

$$E(\theta)_n = \frac{n(n\theta)^{n-1}}{(n-1)!} e^{-n\theta} \tag{6-108}$$

where $E(\theta)_n$ = exit age distribution for n tanks in series

θ = relative residence time = $t/\bar{t}$, dimensionless

The variance can be determined using the expression

$$\sigma_\theta^2 = \frac{1}{n} \tag{6-109}$$

where σ_θ^2 = variance with respect to θ

To fit the TIS model to an existing reactor, a pulse tracer study is conducted to obtain the exit age distribution. The tracer data can be fit with Eq. 6-108 to determine n. Once the value of n has been estimated using this method, Eq. 6-108 can be used to construct a curve of $E(\theta)$ as a function of θ. Again, under conditions of low dispersion (high n), the results of the TIS model approach those of the DFM. Under these conditions, the following relationship may be used:

$$n = \tfrac{1}{2}\text{Pe} + 1 \approx \tfrac{1}{2}\text{Pe} \qquad (6\text{-}110)$$

The results obtained using the TIS and DFM methods are similar; however, it is much easier to use the TIS method as compared to the DFM method. Fortunately, the equivalence can be drawn between the number of tanks for the TIS model and the Pe number used in the closed-flow DFM method (see Eqs. 6-100 and 6-109), as shown in Eqs. 6-111 and 6-112 for a large Pe:

$$\sigma_\theta^2 = \frac{1}{n} = \left\{\left(\frac{2}{\text{Pe}}\right) - \left[2\left(\frac{1}{\text{Pe}}\right)^2\left(1 - e^{-\text{Pe}}\right)\right]\right\} \qquad (6\text{-}111)$$

$$n = \frac{1}{(2/\text{Pe}) - 2(1/\text{Pe})^2\left(1 - e^{-\text{Pe}}\right)} \approx \frac{1}{2}\text{Pe} \qquad (6\text{-}112)$$

The result shown in Eqs. 6-110 and 6-112 can be used to develop an appreciation for the magnitude of the Pe number. For example, a Pe of 20, 50, or 100 would correspond to 10, 25, or 50 tanks in series, respectively.

Example 6-5 Determination of single-parameter fit to RTD

Using the tracer data analysis shown in Example 6-4, estimate the Peclet number and dispersion number for the open-flow DFM by fitting the model to Eq. 6-104 and also the number of tanks in series by fitting the data to Eq. 6-108. Plot the resulting exit age curves and compare them to the original data.

Solution

1. Determine the Peclet number for the open-flow DFM by fitting Eq. 6-104 to the $E(\theta)$ curve shown in Example 6-4.
2. Determine the number of tanks in series for the tanks-in-series model by fitting Eq. 6-108 to the $E(\theta)$ curve shown in Example 6-4.

   Both fits are accomplished by setting up the data and the equation on a spreadsheet and finding the value of Pe (Eq. 6-104) or n (Eq. 6-108) that minimizes the sum of squares of the differences between the model and the data. The results of these calculations are shown in the table below.

θ	$E(\theta)$	Eq. 6-104	[col. 2 – col. 3]^2	Eq. 6-108	[col. 2 – col. 5]^2
0.013	0.00	0.00	0.0000	0.00	0.0000
0.130	0.00	0.00	0.0000	0.00	0.0000
0.261	0.04	0.00	0.0016	0.00	0.0016
0.391	0.07	0.00	0.0049	0.00	0.0049
0.522	0.14	0.00	0.0190	0.01	0.0156
0.652	0.32	0.13	0.0372	0.25	0.0045
0.717	0.54	0.42	0.0145	0.62	0.0062
0.782	0.86	0.94	0.0069	1.16	0.0925
0.808	1.00	1.20	0.0385	1.41	0.1647
0.821	1.07	1.33	0.0652	1.53	0.2075
0.834	1.21	1.45	0.0594	1.64	0.1869
0.847	1.36	1.58	0.0481	1.75	0.1554
0.860	1.64	1.70	0.0036	1.86	0.0483
0.887	2.07	1.93	0.0206	2.05	0.0003
0.913	2.21	2.10	0.0113	2.19	0.0003
0.926	2.25	2.17	0.0058	2.24	0.0000
0.939	2.25	2.23	0.0004	2.28	0.0011
0.978	2.28	2.32	0.0013	2.32	0.0015
1.043	2.07	2.20	0.0157	2.13	0.0033
1.108	1.61	1.84	0.0526	1.72	0.0130
1.173	1.11	1.39	0.0787	1.25	0.0199
1.239	0.75	0.96	0.0434	0.82	0.0045
1.304	0.54	0.62	0.0059	0.49	0.0024
1.369	0.36	0.37	0.0002	0.27	0.0076
1.434	0.21	0.21	0.0000	0.14	0.0048
1.565	0.11	0.06	0.0025	0.03	0.0063
1.695	0.04	0.01	0.0006	0.01	0.0012
1.825	0.00	0.00	0.0000	0.00	0.0000
			sum = 0.5377		sum = 0.9544
			Pe = 67.0		n = 33

3. Determine the dispersion number from the Peclet number using Eq. 6-92:

$$d = \frac{1}{\text{Pe}} = \frac{1}{67} = 0.0149$$

4. A plot comparing the exit age data with the results of the two models is illustrated below.

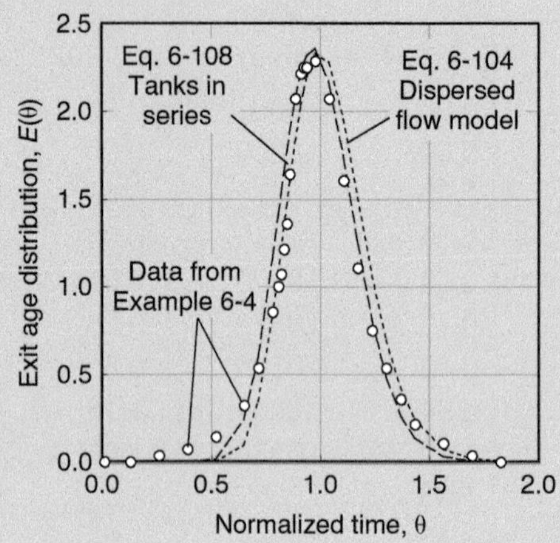

Comment

Both the TIS and DFM models fit the data well, except for the data for small exit ages. The fluid that exits the reactor with small residence times can have a large influence on reactor performance.

Application of RTDs and t_{10} Concept

The RTDs for the three models, closed-flow DFM, open-flow DFM, and TIS model, are displayed on Fig. 6-20. Under conditions where dispersion is high (high d for the DFM or low n for the TIS model), these three models produce significantly different RTDs, as shown on Fig. 6-20. Under conditions where dispersion is low (low d for the DFM or high n for the TIS model), these models produce RTDs that are nearly identical.

In subsequent sections it will be shown that the entire RTD is useful in evaluating the expected performance of a given reactor design. However, RTDs are complicated, and their use in evaluating reactor performance, while the best available technology, is also somewhat more complicated than just assuming that an actual reactor performs similar to a PFR with a certain residence time that is determined by a tracer study. As a result, regulatory authorities often regulate reactor design using certain simplified performance criteria. For example, in disinfection practice (see Chap. 13), it is assumed that the effective contact time corresponds to the length of time it takes for the first 10 percent (θ_{10}) of a tracer to pass through the reactor. This approach is conservative because credit is only received for θ_{10} and not θ_{50}, which corresponds to the actual residence time.

Each RTD model discussed was examined to assess the impact of reactor dispersion on θ_{10}, the fraction of the reactor's theoretical detention time that would be allowed as credit by the U.S. EPA rules (see example for open

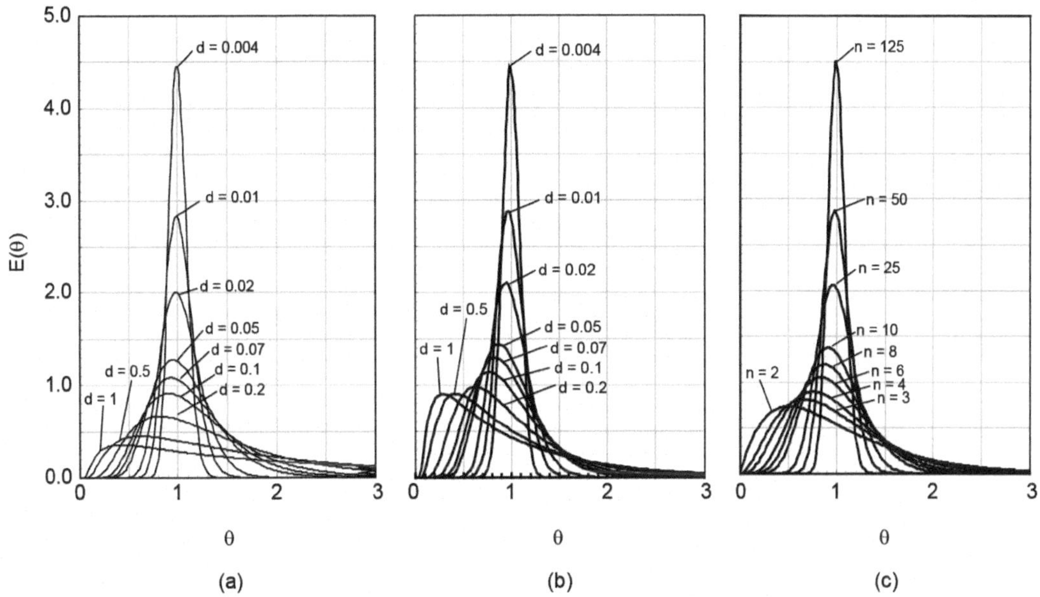

Figure 6-20
Comparison of $E(\theta)$ curves for (a) DFM open system, (b) DFM closed system, and (c) TIS model.

system DFM on Fig. 6-21a). A plot of θ_{10} $(t_{10} / \bar{t})$ as a function of increasing dispersion for all three RTD models is displayed on Fig. 6-21b. Note that the TIS model does not allow for direct application of the dispersion or Peclet numbers, so an approximate transformation represented by Eq. 6-110 was used for comparison purposes. As expected, the θ_{10} value drops consistently as dispersion increases. For example, credit would be received for more than 80 percent of the residence time of the reactor for low dispersion ($d < 0.01$, Pe > 100, $n > 50$) to as low as 25 to 50 percent of the residence time for reactors with high dispersion ($d > 0.3$, Pe < 3, $n < 2$ or 3). A credit of 80 percent or more would correspond to a well-designed reactor.

Predicting Dispersion in a Channel

Extensive studies have been conducted to determine the dispersion in open-channel flow. The dispersion in an existing reactor can be determined using a tracer study; however, when designing facilities, it is useful to be able to predict the dispersion. Davies (1972) proposed the following relationship for flow in an open channel and a high Reynolds number, which can be used to estimate the dispersion coefficient:

$$E = 1.01\nu(\text{Re})^{0.875} \tag{6-113}$$

where E = coefficient of dispersion, m^2/s
ν = kinematic viscosity, m^2/s
Re = Reynolds number = $4vR_h/\nu$, dimensionless
v = velocity in open channel, m/s
R_h = hydraulic radius = sectional area/wetted perimeter, m

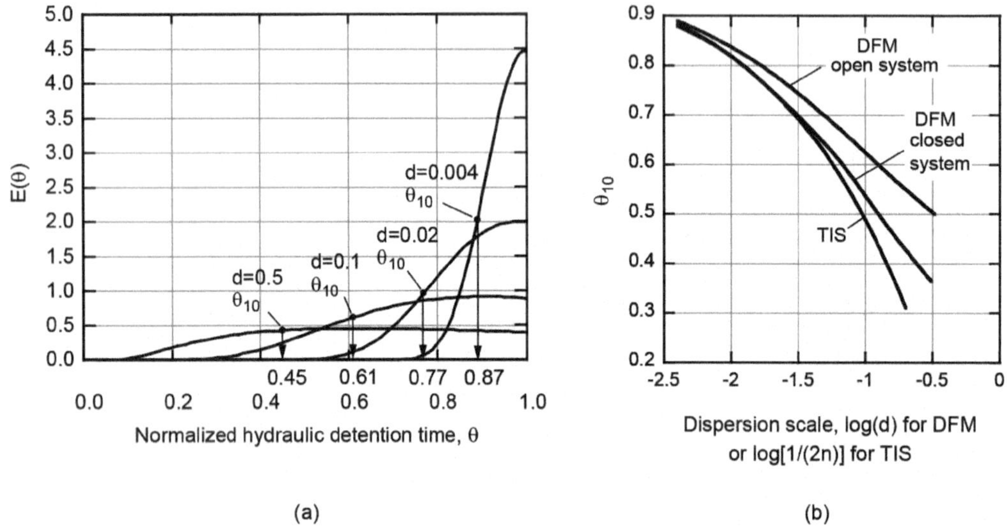

Figure 6-21
Dispersion and θ_{10}. (a) Analyzing effect of dispersion on θ_{10} using DFM model for open system. (*b*) Relationship between increasing dispersion and θ_{10} for three models: DFM closed system, DFM open system, and TIS model.

Other models developed for the dispersion coefficient are based on the Taylor equation and are discussed in Chap. 13.

Example 6-6 Estimate of dispersion in open channel

For an open-channel PFR with a water flow rate of 4320 m³/d, estimate the dispersion number and t_{10}. Assume that $\bar{t} \cong \tau$ and that the channel has a length of 40 m, a width of 3 m, and a water depth of 3 m. Use a kinematic viscosity of 1.003×10^{-6} m²/s for the water. Assume the dispersion in the channel can be estimated accurately using Eq. 6-113 and that the RTD in the channel can be predicted using the open-flow DFM model (Eq. 6-104).

Solution

1. Estimate the dispersion number.
   a. To apply Eq. 6-113, compute the Reynolds number Re:

$$\text{Average water velocity } v = \frac{(4320 \text{ m}^3/\text{d})(1 \text{ d}/86{,}400 \text{ s})}{3 \text{ m} \times 3 \text{ m}}$$

$$= 0.0056 \text{ m/s}$$

$$\text{Hydraulic radius } R_h = \frac{3 \text{ m} \times 3 \text{ m}}{3 \text{ m} + 3 \text{ m} + 3 \text{ m}} = 1 \text{ m}$$

$$\text{Reynolds number Re} = \frac{4vR_h}{\nu} = \frac{(4)(0.0056 \text{ m/s})(1 \text{ m})}{1.003 \times 10^{-6} \text{ m}^2/\text{s}}$$
$$= 22{,}156$$

b. Compute the coefficient of dispersion E:

$$E = 1.01\nu(\text{Re})^{0.875} = (1.01)\left(1.003 \times 10^{-6} \text{ m}^2/\text{s}\right)(22{,}156)^{0.875}$$
$$= 0.00643 \text{ m}^2/\text{s}$$

c. Compute the dispersion number using Eqs. 6-91 and 6-92:

$$d = \frac{E}{vL} = \frac{0.00643 \text{ m}^2/\text{s}}{(0.0056 \text{ m/s})(40 \text{ m})} = 0.0289$$

2. Estimate t_{10}.
   a. Determine the Peclet number from the dispersion number using Eq. 6-92:

$$\text{Pe} = \frac{1}{d} = \frac{1}{0.0289} = 34.5$$

   b. Set up a table to compute the cumulative exit age distribution $F(\theta)$ using Eq. 6-104 to find $E(\theta)$ from θ and Eq. 6-80 to convert $E(\theta)$ to $F(\theta)$:

θ	$E(\theta)$	$\bar{E}(\theta)$	$\bar{E}(\theta)\Delta\theta$[a]	$F(\theta)$	θ	$E(\theta)$	$\bar{E}(\theta)$	$\bar{E}(\theta)\Delta\theta$[a]	$F(\theta)$
0.0									
0.1	0.000	0.0	0.0	0.0	1.4	0.522	0.6607	0.0661	0.9008
0.2	0.000	0.0	0.0	0.0	1.5	0.320	0.4212	0.0421	0.9429
0.3	0.000	0.0	0.0	0.0	1.6	0.187	0.2539	0.0254	0.9683
0.4	0.001	0.0005	0.0001	0.0001	1.7	0.105	0.1463	0.0146	0.9829
0.5	0.031	0.0161	0.0016	0.0017	1.8	0.057	0.0811	0.0081	0.9910
0.6	0.213	0.1222	0.0122	0.0139	1.9	0.030	0.0436	0.0044	0.9954
0.7	0.652	0.4328	0.0433	0.0572	2.0	0.016	0.0228	0.0023	0.9977
0.8	1.204	0.9280	0.0928	0.1500	2.1	0.008	0.0117	0.0012	0.9989
0.9	1.589	1.3963	0.1396	0.2896	2.2	0.004	0.0059	0.0006	0.9994
1.0	1.659	1.6241	0.1624	0.4520	2.3	0.002	0.0029	0.0003	0.9997
1.1	1.462	1.5609	0.1561	0.6081	2.4	0.001	0.0014	0.0001	0.9999
1.2	1.135	1.2989	0.1299	0.7380	2.5	0.000	0.0007	0.0001	0.9999
1.3	0.800	0.9675	0.0967	0.8347	2.6	0.000	0.0003	0.0000	1

[a] $\Delta\theta = 0.1$.

   c. Plot the cumulative exit age distribution and estimate the value of θ at $F(\theta) = 0.10$. As shown in the plot below, the value of θ at $F(\theta) = 0.10$ is 0.76.

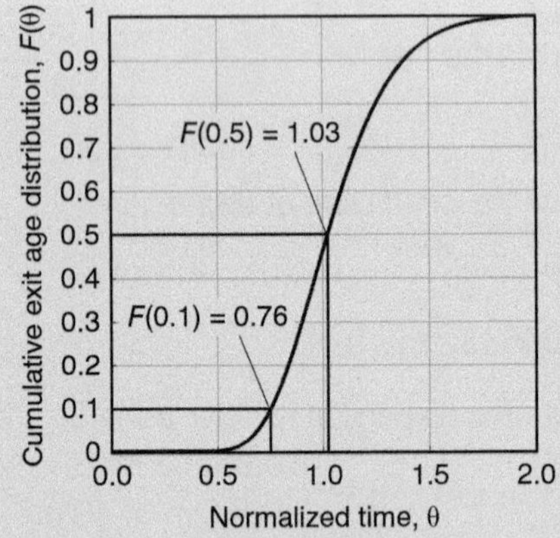

d. Determine the hydraulic detention time for the basin.

$$\tau = \frac{V}{Q} = \frac{360\ \text{m}^3}{3\ \text{m}^3/\text{min}} = 120\ \text{min}$$

e. Determine the mean residence time. The mean residence time is equal to the hydraulic detention time as given in the problem statement. Because there is no short circuiting, Eq. 6-106 should be used to estimate $\bar{t}$, but the open-flow DFM model was used to simulate a closed-flow DFM model because it is simpler to use and yields similar results at high Pe values.

$$\bar{t} = \tau \quad \text{and} \quad \bar{t} = 120\ \text{min.}$$

f. Estimate t_{10}.
From the above plot, $F(0.10) = 0.76$
Therefore, by EPA's definition $t = t_{10}$ when $\theta_{10} = 0.76$. Substituting into to Eq. 6-77 and rearranging,

$$t_{10} = \bar{t} \times \theta_{10} \quad \text{or} \quad t_{10} = 120 \times 0.76 = 91\ \text{min}$$

The influence of the aspect ratio can be determined by examining Eq. 6-113. If it is assumed that the Reynolds number is raised to the power of 1.0, the following expressions can be derived:

$$\text{Pe}_{R_h} = \frac{vR_h}{E} = 0.25 \tag{6-114}$$

$$\text{Pe}_L = \text{Pe}_{R_h} \frac{L}{R_h} \tag{6-115}$$

where Pe_{R_h} = Peclet number based on hydraulic radius

As shown in Eq. 6-115, the Pe number increases as the L/R_h increases. Thus, if the L/R_h is 400, which corresponds to a long conveyance channel, the Pe number would be 100, which corresponds to a θ_{10} of 0.8.

Improving Reactor Performance

When designing reactors for water treatment, there are generally two types of reactors: (1) a well-mixed reactor in which mixing is mechanically forced for a number of reasons (e.g., to blend chemicals, as discussed in Sec. 6-10, to strip volatile compounds, or to prevent particle settling) and (2) a reactor mixed by the flow through the process equipment (e.g., channels, ditches, or pipes).

IMPROVING CMFR PERFORMANCE

Generally, the performance of well-mixed reactors can be improved by inserting baffles in the reactor and converting the reactor from one CMFR to multiple CMFRs. The improvement in reactor performance expected after dividing a CMFR into multiple CMFRs may be estimated with the TIS model described previously. Procedures used to obtain adequate mixing are discussed in Sec. 6-10.

IMPROVING PFR PERFORMANCE

The methods used to improve the performance of a PFR are different than those described for the CMFR. If the fluid velocity is low or the inlet and outlet hydraulics are poor, as shown on Fig. 6-22a, dead volume and

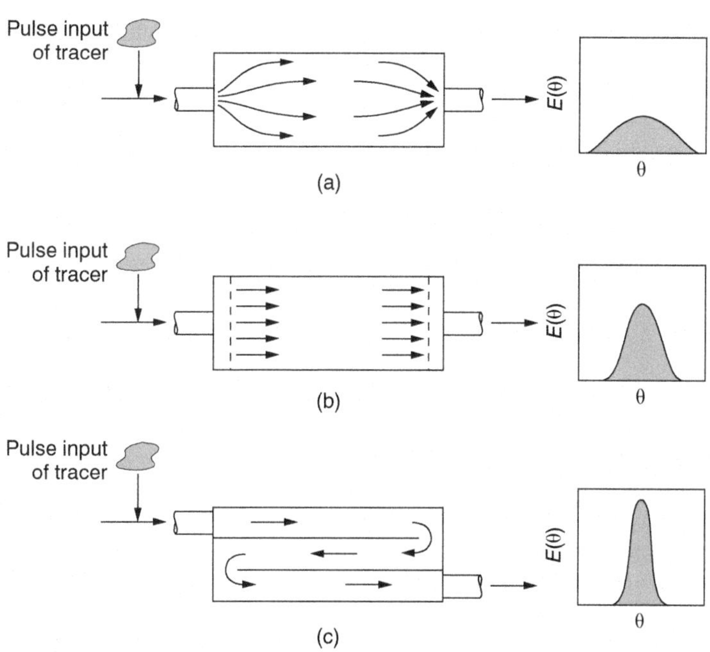

Figure 6-22
Improvement of reactor performance of a basin: (a) original basin with poor inlet hydraulics, (b) basin with improved inlet and outlet hydraulics, and (c) basin with insertion of baffles to increase aspect ratio.

short circuiting can occur. The solution to improving process performance is to distribute the flow uniformly at the inlet and outlet, such as with diffuser baffles, as shown on Fig. 6-22b. For nonsedimentation processes, reactor performance can be improved by installing baffles in the basin so that the flow serpentines through the basin, as shown on Fig. 6-22c. Increasing velocity by inserting baffles that are shown on Fig. 6-22c is not recommended for sedimentation processes, and is discussed in Chap. 10, because sedimentation processes need quiescent conditions. Tracer studies and nonideal flow models can be used to evaluate such improvements. Baffle installation as shown on Fig. 6-22c increases the reactor length-to-depth or length-to-width ratio and the impact of the aspect ratio can then be evaluated.

Turning vanes, hammer heads, filets, and submerged baffles after each turn can be used to minimize short circuiting and improve reactor performance. These features and other techniques are discussed in Chap. 13 because of the importance of minimizing dispersion in disinfection contactors. Chapter 13 also discusses approaches to design to achieve a specific level of dispersion.

6-8 Modeling Reactions in Nonideal Reactors

The modeling of chemical reactions occurring in ideal reactors was introduced in Secs. 6-4 and 6-5. However, the nonideal nature of the hydraulics of real reactors, as described in Secs. 6-6 and 6-7, affects the actual performance. Therefore, it is necessary to describe the performance of reactors in terms of the nonideal nature of reactor hydraulics. When a tracer curve is not available, the DFM and TIS model, introduced in Sec. 6-7, may be used with appropriate kinetic expressions to model reactor performance, as described below.

Dispersed-Flow Model Applied to a Reactive System

The dispersed-flow model, presented in Sec. 6-7, can be used with appropriate reaction kinetics to predict reactor performance. Steady state can only exist for a constant influent concentration, and at steady state, Eq. 6-89 becomes

$$\left(\frac{1}{\text{Pe}}\right)\frac{d^2C}{d\overline{z}^2}-\frac{dC}{d\overline{z}}-k\tau C^n=0 \qquad (6\text{-}116)$$

where Pe = Peclet number, dimensionless
C = effluent concentration, mg/L
$\overline{z}$ = dimensionless length = z/L, dimensionless
L = length of reactor, m
k = rate constant, $(\text{mg/L})^{-n+1}/\text{s}$
τ = hydraulic detention time, s
n = reaction order

If the dimensionless normalized concentration $\bar{C}$ is introduced into Eq. 6-116, the following expression is obtained:

$$\left(\frac{1}{\text{Pe}}\right)\frac{d^2\bar{C}}{d\bar{z}^2} - \frac{d\bar{C}}{d\bar{z}} + k\tau C_0^{n-1}\bar{C}^n = 0 \tag{6-117}$$

where $\bar{C}$ = normalized concentration = C/C_0, dimensionless
C_0 = initial concentration, mg/L

In Eq. 6-117, the effluent concentration is governed by three dimensionless groups: the Damköhler number $\left(k\tau C_0^{n-1}\right)$, the Peclet number (Pe), and the reaction order (n).

Equation 6-117 has been solved analytically by Danckwerts (1953) and Wehner and Wilhelm (1958) for a first-order reaction. For reactors that are either open or closed, the solution is

$$\frac{C}{C_0} = \frac{4a\exp(\text{Pe}/2)}{[(1+a)^2\exp(a\,\text{Pe}/2)] - [(1-a)^2\exp(-a\,\text{Pe}/2)]} \tag{6-118}$$

where $a = \sqrt{1 + 4k\tau(1/\text{Pe})}$

A generalized plot of Eq. 6-118 is presented on Fig. 6-23 for values of the Damköhler number ($k\tau$) and dispersion number of interest in water treatment. The design engineer can control the product $k\tau$ by adjusting the design hydraulic detention time τ. The Peclet number can also be controlled to some extent by the details of the design of the reactor itself (e.g., baffling, aspect ratio). For example, if the value of $k\tau$ is equal to 4, then the best performance that could be achieved with a PFR would be 98 percent

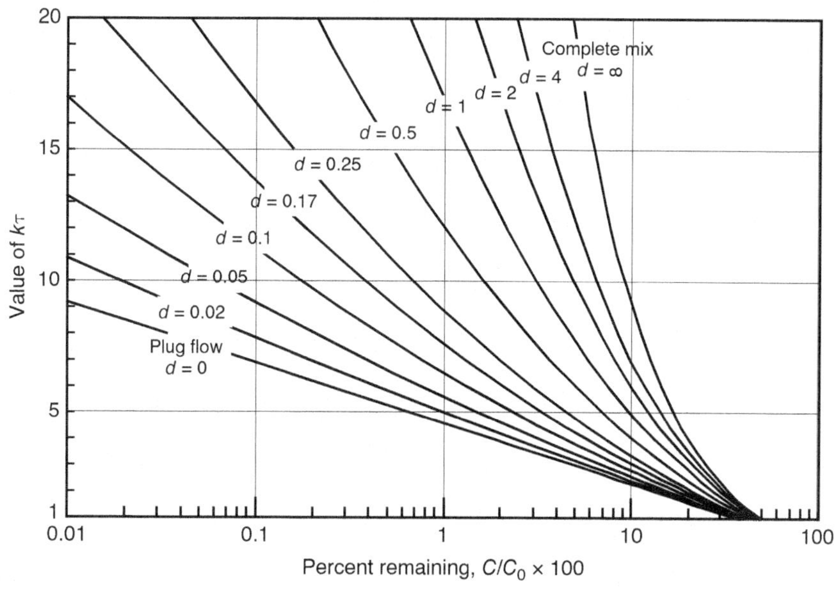

Figure 6-23
Reactor performance as function of Damköhler number and dispersion number (inverse of Peclet number) for first-order reaction.

Example 6-7 DFM calculation for first-order reaction

Using the DFM for a closed reactor and exit age distribution shown in Example 6-5, compare the reactor removal efficiency for a PFR and a CMFR. Use the following values for the calculations:

$$C_0 = 200 \text{ mg/L} \quad k = 0.0746 \text{ min}^{-1} \quad -r = kC$$

$$\bar{t} = 76.6 \text{ min (calculated in Example 6-4)}$$

Solution

1. Compute the reactor removal efficiency. For a first-order reaction, the Damköhler number is

$$k\tau \simeq k\bar{t} = (0.0746)(76.6) = 5.72$$

Use Eq. 6-118 to solve for C/C_0 (Pe = 67 from Example 6-5):

$$a = \sqrt{1 + 4k\tau\left(\frac{1}{Pe}\right)} = \sqrt{1 + 4(0.0746)(76.6)\left(\frac{1}{67}\right)} = 1.16$$

$$\frac{C}{C_0} = \frac{4a\exp(\text{Pe}/2)}{\left[(1+a)^2\exp(a\,\text{Pe}/2)\right] - \left[(1-a)^2\exp(-a\,\text{Pe}/2)\right]}$$

$$= \frac{4(1.16)\exp(67/2)}{\left[(1+1.16)^2\exp(1.16\times 67/2)\right] - \left[(1-1.16)^2\exp(-1.16\times 67/2)\right]}$$

$$= 0.0050$$

$$C = 1.0 \text{mg/L}$$

Alternatively, C/C_0 may be estimated using Fig. 6-23; however, because the dispersion number is low, the use of Fig. 6-23 is limited.

2. Compare the reactor removal efficiency with a PFR and a CMFR.
   a. The effluent concentration for a PFR with a mean detention time equal to the hydraulic detention time is determined using Eq. 6-65:

$$\frac{C}{C_0} = e^{-k\bar{t}} = e^{-(0.0746)(76.6)} = 0.00330 \text{ or } C = 0.66 \text{ mg/L}$$

   The ratio C/C_0 may also be determined using Fig. 6-23; $C/C_0 \simeq$ 0.0035 for a PFR when d approaches zero (Pe = ∞).

   b. The effluent concentration for a CMFR with a mean detention time equal to the hydraulic detention time is determined by rearranging Eq. 6-37:

$$\frac{C}{C_0}=\frac{1}{1+k\bar{t}}=0.149 \quad \text{or} \quad C=29.8 \text{ mg/L}$$

Here, C/C_0 may also be determined using Fig. 6-23; $C/C_0 \simeq 0.14$ for a CMFR when d approaches infinity (Pe = 0).

Comment

In this example, a Pe value of 67 (which corresponds to 33 reactors in series in the TIS model) yields an effluent concentration that is double that of an ideal PFR. For disinfection where reductions of 99.99 percent or greater may be required, reactors with very high Pe or low dispersion numbers must be used (see Chap. 13).

(2 percent remaining), and if the reactor Peclet number was equal to 2 (dispersion number equals 0.5), the best performance that could be achieved is 90 percent (10 percent remaining). Also, as shown on Fig. 6-23, when removal requirements are modest (less than 50 percent), the Peclet number for the reactor does not make a great deal of difference. On the other hand, when removal requirements are more stringent, maintaining a high Pe (low dispersion) is a critical design requirement.

For second-order reactions, Eq. 6-115 can only be solved numerically. The corresponding results for the DFM for a second-order reaction are displayed on Fig. 6-24.

Tanks-in-Series Model Applied to a Reactive System

The reactor performance for the TIS model can be estimated from mass balances for a number of tanks in series. For a first-order reaction, the following expression is obtained:

$$\frac{C}{C_0}=\frac{1}{(1+k\tau/n)^n} \tag{6-119}$$

For a second-order reaction, the following equation can be used for n tanks by calculating the effluent concentration from tank i from the previous tank:

$$\frac{C_i}{C_{i-1}}=\frac{-1+\sqrt{1+4k\tau C_{i-1}/n}}{2k\tau C_{i-1}/n} \tag{6-120}$$

where i = intermediate tank in series of n tanks
C_i = concentration exiting tank i, mg/L
C_{i-1} = initial concentration entering tank i, mg/L

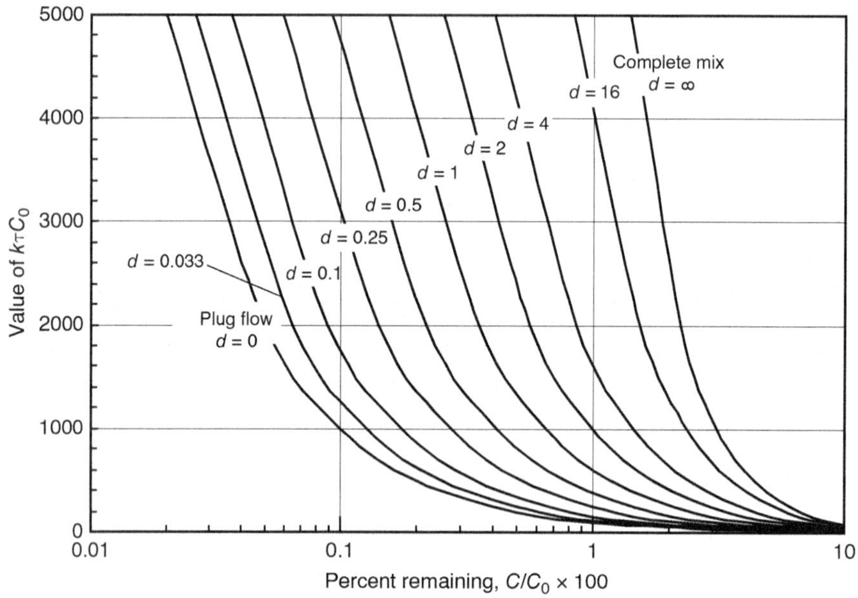

Figure 6-24
Reactor performance as function of Damköhler number and dispersion number (inverse of Peclet number) for second-order reaction. (Adapted from Levenspiel and Bischoff, 1959.).

Example 6-8 Tank-in-series model calculation

Calculate the performance for the reactor evaluated in Example 6-7 for a first-order reaction using the TIS model and compare to the results from the DFM. Estimate the reactor performance for a PFR and a CMFR with the identical hydraulic detention times.

Solution

1. Determine the reactor removal efficiency for a TIS model with $n = 33$ tanks (see Example 6-5) using Eq. 6-119 for the TIS model:

$$\frac{C}{C_0} = \frac{1}{(1 + k\tau/n)^n} = \frac{1}{(1 + 5.72 / 33)^{33}} = 0.00512$$

$$C = 1.02 \text{ mg/L}$$

2. The results from TIS and DFM analysis and the estimated performance of a reactor for a PFR and a CMFR with the identical hydraulic detention time are compared in the following table:

	TIS	DFM	PFR	CMFR
C, mg/L	1.02	1.00	0.66	29.8

Comment

As summarized in the table above, the reactor performance predicted using the TIS and DFM models is similar. In addition, the performance is closer to that of a PFR than to a CMFR because the Pe and number of tanks in series is large.

6-9 Using Tracer Curves to Model Reactions in Reactors

The DFM and TIS models presented in the previous sections are useful for generating RTD curves when actual tracer response data are not available. The DFM and TIS model can also be used for modeling reactions, where it is assumed that homogeneous conditions exist on the molecular scale (microscale). However, when a tracer curve obtained from a reactor study or CFD calculation is available, the segregated-flow model (SFM) may be used to model reactions. The SFM, limitations of the SFM, and comparison of the SFM and various nonideal flow models are presented in this section.

Segregated-Flow Model

The assumption in the SFM is that all fluid elements are segregated,meaning they do not mix or interact with each other. Consequently, the performance of a real reactor can be determined by estimating the amount of reaction that would take place for each fluid element and then mixing the elements at the end of the reactor. For an exit age distribution as shown on Fig. 6-25, the SFM is obtained by approximating the real reactor as numerous batch or plug flow reactors that have different residence times and exit age characteristics. The normalized curve for the pulse tracer study, $E(\theta)$, is used to determine the amount of fluid that has a particular age.

For development of the SFM, performance for a real reactor is estimated in terms of C/C_0. The contribution for fluid element 9 (see Fig. 6-25) is given by the expression

$$\begin{pmatrix} C/C_0 \text{ from CMBR or PFR} \\ \text{with detention time } \theta_9 \end{pmatrix} \times \begin{pmatrix} \text{fraction of exit stream} \\ \text{that has age } \theta_9 \end{pmatrix} = R(\theta_9)E_9\,\Delta\theta_9 \tag{6-121}$$

where θ_9 = exit age for fluid element 9

$R(\theta_9)$ = dimensionless effluent concentration for fluid element 9, which equals effluent concentration divided by influent concentration leaving ideal PFR or CMBR with hydraulic detention time that corresponds to θ_9

E_9 = exit age distribution for fluid element 9

$E_9\Delta\theta_9$ = fraction of exit stream that has age θ_9

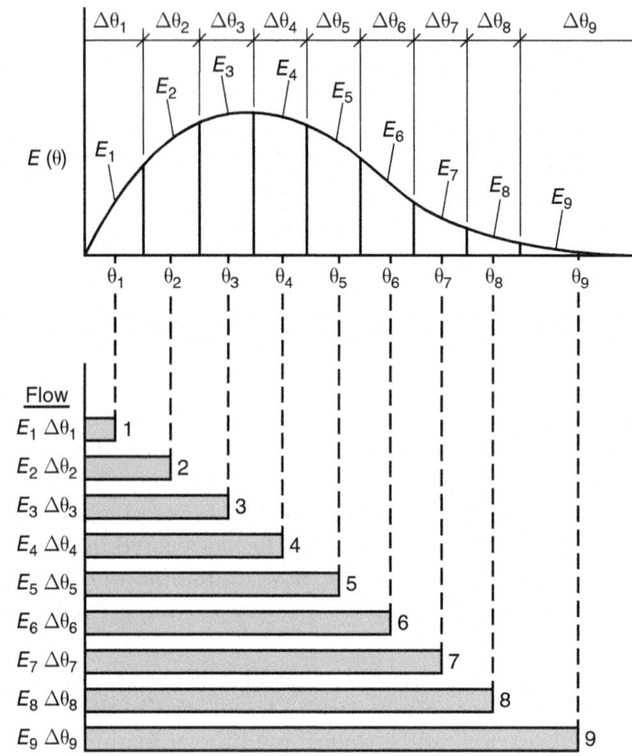

Figure 6-25
Definition sketch for SFM. Flow at each average θ corresponds to flow in ideal plug flow or batch reactor.

To determine the reactor performance, sum the contribution of each PFR as follows:

$$\frac{\bar{C}}{C_0} = R(\theta_1)E_1\,\Delta\theta_1 + R(\theta_2)E_2\,\Delta\theta_2 + \cdots + R(\theta_9)E_9\,\Delta\theta_9 \qquad (6\text{-}122)$$

where $\bar{C}$ = average effluent concentration, mg/L

In theory, an infinite number of small PFRs is needed, and taking the limit $N \to \infty$, the following expression is obtained:

$$\frac{\bar{C}}{C_0} = \sum_{i=1}^{N} R(\theta_i)E(\theta_i)\,\Delta\theta_i = \int_0^{\infty} R(\theta)E(\theta)\,d\theta \qquad (6\text{-}123)$$

The parameter $R(\theta)$ can be obtained by substitution of $\theta\bar{t}$ for τ in the following general expression for any order kinetics. The expressions given previously in Table 6-3 can also be used.

$$\tau = -\int_{C_0}^{C} \frac{dC}{-r} \qquad (6\text{-}124)$$

For example, the following expressions are obtained for first- and second-order reactions:

$$R(\theta) = \begin{cases} e^{-k\theta\bar{t}} & \text{(first order)} \quad (6\text{-}125) \\ \dfrac{1}{1 + k\bar{t}C_0\theta} & \text{(second order)} \quad (6\text{-}126) \end{cases}$$

The SFM can be used when $E(\theta)$ and $R(\theta)$ are known. For example, SFM applies when there is batch rate data because $R(\theta)$ is identical to the result that would be expected from a batch reactor. To convert C/C_0 versus time to C/C_0 versus θ, divide the elapsed time in the batch reactor study by $\bar{t}$. The SFM can also be used for sedimentation basins with $R(\theta)$ determined from settling column tests, where the settling column has a depth equal to the settling zone in the sedimentation basin.

Example 6-9 Evaluation of UV disinfection process using SFM

A UV disinfection process takes place in a pipe that contains UV lamps. The pipe is 1 m long with a diameter of 0.3 m. The process flow rate is 1000 m^3/d and the average UV intensity in the reactor is 15 mW/cm^2. The results of a tracer study conducted on the UV disinfection reactor are shown in the following table:

Time, s	Normalized Time, θ	$E(\theta)$	Time, s	Normalized Time, θ	$E(\theta)$
0.0	0.000	0.000	9.0	1.400	0.466
0.6	0.100	0.000	9.7	1.500	0.309
1.3	0.200	0.001	10.3	1.600	0.196
1.9	0.300	0.017	11.0	1.700	0.120
2.6	0.400	0.101	11.6	1.800	0.071
3.2	0.500	0.314	12.3	1.900	0.040
3.9	0.600	0.647	12.9	2.000	0.022
4.5	0.700	1.006	13.6	2.100	0.012
5.2	0.800	1.273	14.2	2.200	0.006
5.8	0.900	1.376	14.8	2.300	0.003
6.5	1.000	1.313	15.5	2.400	0.002
7.1	1.100	1.134	16.1	2.500	0.001
7.7	1.200	0.901	16.8	2.600	0.000
8.4	1.300	0.668			

The UV dose-response curve for a certain microorganism has been determined and the results are shown in the following plot:

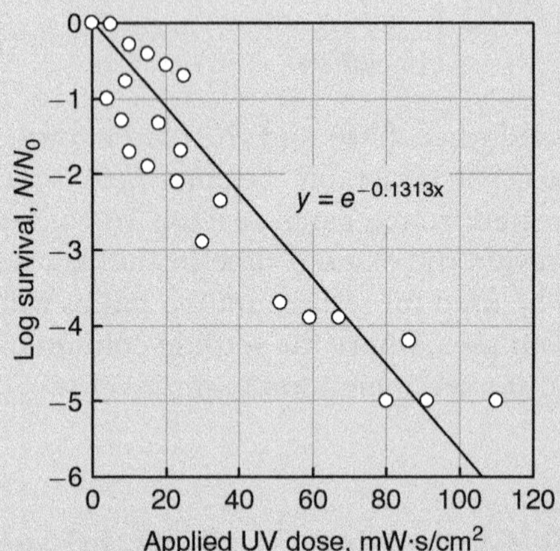

The average concentration of the microorganism in the influent to the UV reactor, N_0, is 0.1 L^{-1}. Estimate the concentration of the microorganism in the effluent from the UV disinfection process using the SFM.

Solution

1. Set up a table to compute the number of organisms surviving the disinfection process.
   a. Compute the unit exit age distribution $E(\theta)\ \Delta\theta$ with a $\Delta\theta$ value of 0.1. For the time step $t = 2.6$,

$$E(\theta)\ \Delta\theta = (0.101)(0.1) = 0.0101$$

   b. Estimate the applied dose. The applied dose is determined by multiplying the average UV intensity in the reactor by the corresponding time step. For the time step $t = 2.6$,

$$\text{Applied dose} = (15\ \text{mW/cm}^2)(2.6\ \text{s}) = 39.0\ \text{mW}\cdot\text{s/cm}^2$$

   c. Estimate the fraction of organisms surviving at the applied dose, $R(\theta)$.The fraction of organisms surviving is a function of the applied dose, and the relationship is given in the problem statement. Use the curve fit to determine the survival. For the time step $t = 2.6$,

$$\text{Estimated survival, } N/N_0 = e^{(-0.1313\times 39.0\ \text{mW}\cdot\text{s/cm}^2)} = 5.97 \times 10^{-3}$$

d. Compute the number of organisms surviving. The number of organisms surviving may be determined by multiplying the fractional survival N/N_0 at each time step by the average influent concentration and the unit exit age distribution. For the time step $t = 2.6$,

$$\text{Microorganisms surviving} = (0.0101)(0.00597)(0.1\ \text{L}^{-1})$$
$$= 6.03 \times 10^{-6}\ \text{L}^{-1}$$

e. Summarize the computations in a table.

Time, s	Normalized Time, θ	$E(\theta)$	$E(\theta)\,\Delta\theta$	Applied Dose, mW · s/cm²	Estimated Survival, $R(\theta)$, N/N_0	Effluent Microorganism Concentration, L⁻¹
0.0	0.0	0.000		0.0	1.0	0.0
0.6	0.1	0.000	0.0000	9.0	3.07×10^{-1}	0.0
1.3	0.2	0.001	0.0001	19.5	7.73×10^{-2}	7.73×10^{-7}
1.9	0.3	0.017	0.0017	28.5	2.37×10^{-2}	4.03×10^{-6}
2.6	0.4	0.101	0.0101	39.0	5.97×10^{-3}	6.03×10^{-6}
3.2	0.5	0.314	0.0314	48.0	1.83×10^{-3}	5.75×10^{-6}
3.9	0.6	0.647	0.0647	58.5	4.61×10^{-4}	2.99×10^{-6}
4.5	0.7	1.006	0.1006	67.5	1.42×10^{-4}	1.42×10^{-6}
5.2	0.8	1.273	0.1273	78.0	3.57×10^{-5}	4.54×10^{-7}
5.8	0.9	1.376	0.1376	87.0	1.09×10^{-5}	1.51×10^{-7}
6.5	1.0	1.313	0.1313	97.5	2.76×10^{-6}	3.62×10^{-8}
7.1	1.1	1.134	0.1134	106.5	8.45×10^{-7}	9.59×10^{-9}
7.7	1.2	0.901	0.0901	115.5	2.59×10^{-7}	2.34×10^{-9}
8.4	1.3	0.668	0.0668	126.0	6.53×10^{-8}	4.36×10^{-10}
9.0	1.4	0.466	0.0466	135.0	2.00×10^{-8}	9.34×10^{-11}
9.7	1.5	0.309	0.0309	145.5	5.05×10^{-9}	1.56×10^{-11}
10.3	1.6	0.196	0.0196	154.5	1.55×10^{-9}	3.04×10^{-12}
11.0	1.7	0.120	0.0120	165.0	3.90×10^{-10}	4.68×10^{-13}
11.6	1.8	0.071	0.0071	174.0	1.20×10^{-10}	8.50×10^{-14}
12.3	1.9	0.040	0.0040	184.5	3.01×10^{-11}	1.21×10^{-14}
12.9	2.0	0.022	0.0022	193.5	9.25×10^{-12}	2.03×10^{-15}
13.6	2.1	0.012	0.0012	204.0	2.33×10^{-12}	2.80×10^{-16}
14.2	2.2	0.006	0.0006	213.0	7.15×10^{-13}	4.29×10^{-17}
14.8	2.3	0.003	0.0003	222.0	2.19×10^{-13}	6.58×10^{-18}
15.5	2.4	0.002	0.0002	232.5	5.52×10^{-14}	1.10×10^{-18}
16.1	2.5	0.001	0.0001	241.5	1.69×10^{-14}	1.69×10^{-19}
16.8	2.6	0.000	0.0000	252.0	4.27×10^{-15}	0.0
			1			2.16×10^{-5}

2. Estimate the concentration of microorganisms in the process effluent. Summing the values of effluent microorganism concentration for all time steps results in a value of $2.16 \times 10^{-5}\ L^{-1}$. For a flow rate of $1000\ m^3/d$, the total number of organisms that can pass through the reactor without being inactivated in one day is

$$\text{Untreated microorganisms per day} = (2.16 \times 10^{-5}\ L^{-1})(10^6\ L/d)$$
$$= 22\ d^{-1}$$

Comment

Because of the nonideal hydraulics in the reactor, it is possible for some of the water to pass through before the detention time required for adequate treatment. Because some microorganisms can cause illness even at low dosages, reactor performance is critical to ensure safe water.

Limitations of SFM Application

The SFM is valid for any shape of reactor and has the following limitations:

1. The SFM can be used to obtain an exact solution for a first-order reaction.
2. For a zero-order reaction, the effluent concentration is identical for both PFRs and CMFRs. Mixing has no impact on the reactor performance and maximum performance can be obtained as long as $\bar{t} = \tau$.
3. For reaction orders greater than 1.0, the effluent concentration calculated using the SFM is less than the actual effluent concentration and represents the lower bound of the effluent concentration (the best possible performance).
4. For reaction orders less than 1.0, the effluent concentration calculated by the SFM is greater than the actual effluent concentration and represents the upper bound of the effluent concentration (the worst possible performance).

A comparison of the differences between the actual and calculated effluent concentration for a CMFR when using the SFM for reaction orders greater than 1 and less than 1 are presented in the following discussion.

REACTION ORDERS GREATER THAN 1

For a second-order reaction in a CMFR, the effluent concentration from a CMFR is given by the expression

$$\frac{C}{C_0} = \frac{-1 + \sqrt{1 + 4k\tau C_0}}{2k\tau C_0} \tag{6-127}$$

where k = reaction rate, L/mg · s
τ = hydraulic detention time, s

For $k\tau C_0 = 1$, the resulting value of C/C_0 is 0.618, as computed using Eq. 6-125. The terms that appear in the SFM approximation of a CMFR are

$$R(\theta) = \frac{1}{1 + k\tau C_0\theta} \tag{6-128}$$

$$E(\theta) = e^{-\theta} \tag{6-129}$$

The SFM approximation for a CMFR with $k\tau C_0 = 1.0$ results in the expression

$$\frac{\bar{C}}{C_0} = \int_0^\infty \left[\left(\frac{1}{1 + k\tau C_0\theta}\right) e^{-\theta}\right] d\theta = 0.596 \tag{6-130}$$

Thus, for a second-order reaction, the SFM method predicts a lower concentration than a mass balance on a CMFR. The difference between Eqs. 6-130 and 6-127 is shown as line (a) on Fig. 6-26. As shown, the concentration predicted by the SFM is consistently lower than the result for a CMFR. The reason the SFM predicts a lower concentration for a second- order reaction is because less backmixing occurs as the performance approaches ideal plug flow. Under ideal conditions, the results from the SFM analysis will be identical to the analytical solution for a PFR.

ORDERS LESS THAN 1

For reaction orders less than 1.0, such as $n = 0.5$, the following relationship for $R(\theta)$ is obtained from a PFR mass balance:

$$R(\theta) = \left(1 - \frac{k\tau\theta}{2C_0^{0.5}}\right)^2 \tag{6-131}$$

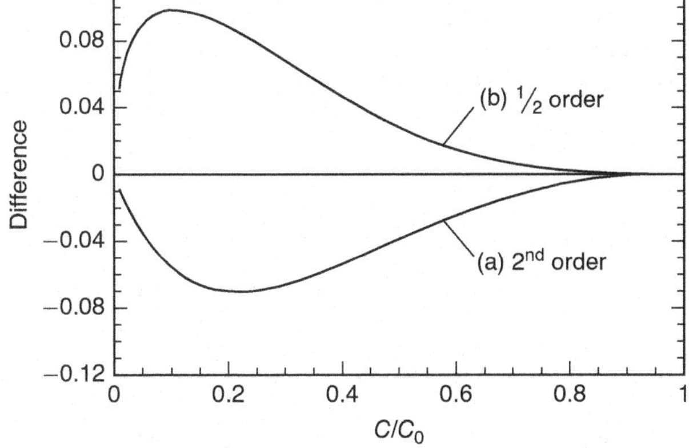

Figure 6-26
Difference between the actual effluent concentration for a CMFR and the SFM for: (a) a second-order reaction rate and (b) a half-order reaction rate.

The SFM approximation to a CMFR is given by the expression

$$\frac{\bar{C}}{C_0} = \int_0^\infty \left[\left(1 - \frac{k\tau\theta}{2C_0^{0.5}} \right)^2 e^{-\theta} \right] d\theta \qquad (6\text{-}132)$$

However, a reasonable result of Eq. 6-130 cannot be obtained using the range of integration from zero to infinity because the value of $k\tau\theta/\left(2C_0^{0.5}\right)$ needs to be between zero and unity (otherwise the effluent concentration is negative). Therefore, the SFM approximation for a CMFR with a reaction order of 0.5 is modified to

$$\frac{\bar{C}}{C_0} = \int_0^{(1/L)} \left[(1 - L\theta)^2 e^{-\theta} \right] d\theta \qquad (6\text{-}133)$$

where $L = k\tau/\left(2C_0^{0.5}\right)$

For $L = 0.5$ the resulting value of C/C_0 is 0.432, as computed using Eq. 6-133.

For a CMFR using a conventional mass balance approach, the following expressions are obtained:

$$\tau = \frac{V}{Q} = \frac{C_0 - C}{kC^{0.5}} \qquad (6\text{-}134)$$

$$\frac{C}{C_0} = \frac{1}{2} \left[2 + \frac{\tau^2 k^2}{C_0} - \sqrt{\left(2 + \frac{\tau^2 k^2}{C_0} \right)^2 - 4} \right] \qquad (6\text{-}135)$$

For $k\tau/2C_0^{0.5} = 0.5$ (equal to $\tau^2 k^2/C_0 = 1$), the resulting value of C/C_0 is 0.382, as computed using Eq. 6-135.

Thus, for a half-order reaction, the SFM method predicts a higher concentration than a mass balance on a CMFR. The difference between Eq. 6-133 and 6-135 is shown as line (b) on Fig 6-26. As shown on Fig. 6-26, the SFM consistently predicts a larger concentration than a mass balance on a CMFR. This difference is the maximum difference that can be expected using the SFM for a half-order reaction.

As shown in the analysis presented above, the differences for 0.5 and second orders represent the maximum difference between the SFM and a given reactor's actual performance, respectively. The SFM predicts an exact answer for any reaction order when used to simulate a PFR [$E(\theta) = \delta(1)$]. Accordingly, the SFM would predict reactor performance that is closer to the observed reactor performance as the reactor backmixing decreased to that of a PFR (no backmixing).

6-10 Mixing Theory and Practice

Mixing is a central part of water treatment. In some unit operations, mixing has a profound impact on the course of the reactions of interest. In other unit operations, understanding mixing is an important adjunct to process control. In still others, mixing energy actually contributes to the rate of the reaction (e.g., flocculation). In the early days, the design and implementation of mixing facilities was a haphazard process. Even today, there is a great deal of art in designing mixing facilities. Nevertheless, as discussed below, scientific tools are available that can be used to improve the design of these facilities.

Types of Mixing

Two types of mixing are applied in water treatment: (1) agitation and (2) blending. Each is considered separately in the following discussion.

AGITATION (FLUID)

Agitation is the term used to describe the motion induced in a fluid to promote processes such as flocculation, to maintain particles in suspension, and for mass transfer such as aeration. In flocculation, the water is agitated to bring about contact between particles, after the chemistry (coagulation) has been used to neutralize their natural repulsion to each other. To design mixing facilities for flocculation it is important to know (1) the particle size distribution of the particles to be flocculated and (2) the degree of agitation necessary to bring about particle contact. The subject of flocculation is considered in greater detail in Chap. 10. Agitation is also used to prevent particles from settling in equalization and related facilities (also covered in Chap. 10). Mass transfer reactions are enhanced by agitation such as occurs as water falls over a cascade of stairs or when air is bubbled through water for the addition of oxygen or to remove supersaturated gases from solution. Mass transfer reactions are discussed further in Chap. 7.

Examples of chemical mixers.
http://www.stantec.com/wt3r/6-4.html

BLENDING

The process of combining two or more liquid streams to achieve a specified level of uniformity is known as *blending*. To design mixing facilities, for the purposes of blending, it is important to (1) understand how to estimate the thoroughness of the blending required so that process sampling, analysis, and control can be accomplished (e.g., chlorination, pH control, or fluoridation) and (2) identify those situations where the speed and thoroughness of blending both have important impacts on process efficiency and effectiveness (e.g., coagulation with Al^{3+} or Fe^{3+} or disinfection of secondary effluent with chlorine). Significant differences in approach result from these different design objectives.

Some Fundamentals of Mixing

Because of their large size, virtually all water treatment processes take place in turbulent flow. As a result, to better understand mixing in water treatment, it is helpful to gain a conceptual understanding of turbulence. Thus, before discussing mixing for agitation and blending, the nature of turbulence is explored briefly below.

INTRODUCTION TO TURBULENCE

It is helpful to think of turbulent flow as consisting of a cascade of eddies—more specifically a cascade of energy from large eddies to small eddies. Kinetic energy imparted to the water through physical action (a pump, a mixer, falling over a weir, etc.) imparts momentum to large segments of water, creating the large eddies moving in a direction consistent with the motive force; however, the structure of water is such that, as these large eddies move around, their energy is immediately transferred to smaller eddies. Once the eddies become small enough, inertial forces are overcome by the viscous nature of water and they can get no smaller. This cascade of eddies and the significance of the different zones of turbulent flow are illustrated on Fig. 6-27.

The immediate turbulence resulting from the motion-inducing device is anisotropic (energy of motion is in one direction or another), but once the energy from this source has been relayed down to smaller eddies, the turbulence becomes isotropic (energy of motion is equal in all directions). The velocity gradients necessary for flocculation (see Sec. 9-6) are present under both anisotropic and isotropic turbulence, but the large-scale shear forces responsible for floc breakup are mostly found in anisotropic flow.

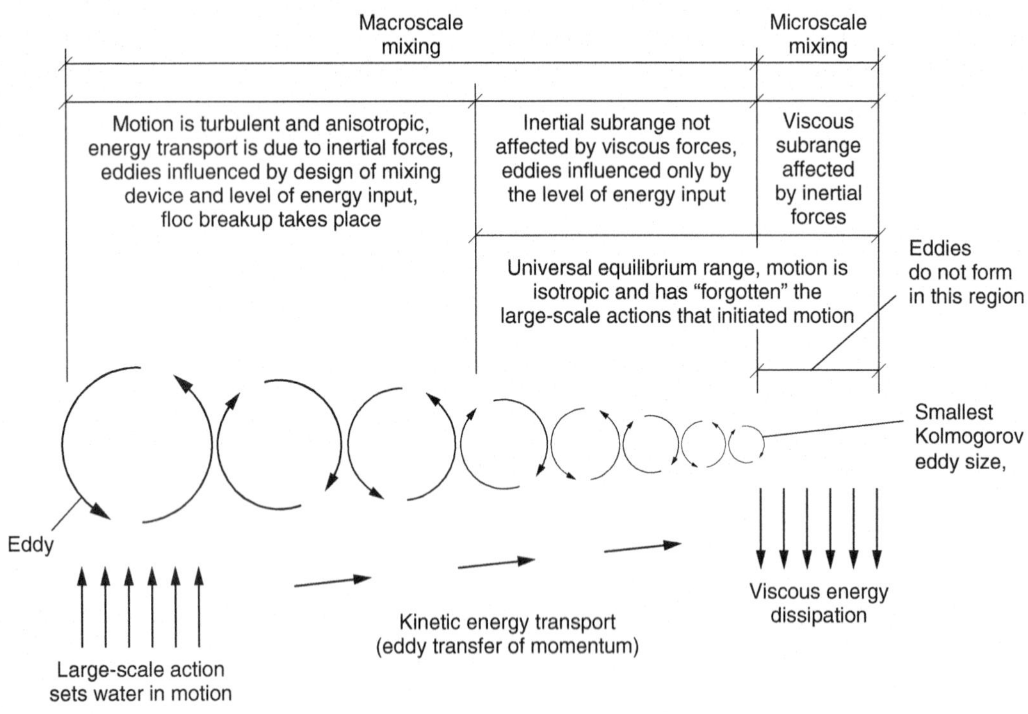

Figure 6-27
Overview of turbulence as it relates to mixing. (Adapted from Stanley and Smith, 1995.).

The design of the mixing equipment for flocculation involves a trade-off between flocculation and floc breakup.

MIXING AND SCALE OF TURBULENCE

The largest eddies start out approximately one-half to one-third the dimension of the mixing device, and the size of the smallest eddies is determined by the boundary between turbulent and viscous flow. The size of the smallest eddy is considered the dividing line between the microscale and macroscale. In the macroscale, mass transfer takes place both by turbulent diffusion and molecular diffusion, but turbulent diffusion is dominant. In the microscale, mass transfer takes place only by molecular diffusion. As more energy is input to the system, the dividing line between the microscale and the macroscale becomes increasingly smaller; that is, the smallest eddies get even smaller. A Russian mathematician named Kolmogorov recognized this dividing line and suggested that the diameter of the smallest eddy, η, could be estimated from the amount of energy being dissipated in the system (Kolmogorov 1941a,b,c):

$$\eta = \left(\frac{\nu^3}{\varepsilon}\right)^{1/4} \tag{6-136}$$

where η = diameter of smallest eddy, m
ν = kinematic viscosity, m^2/s
ε = energy dissipation rate at point of interest, J/kg · s

The energy dissipation rate in a mixing vessel, ε, is not uniform throughout the vessel, but because energy must be dissipated at the same rate at which it is input to the system, the overall average rate of energy dissipation, $\overline{\varepsilon}$, is equal to the power input:

$$\overline{\varepsilon} = \frac{P}{M} \tag{6-137}$$

where $\overline{\varepsilon}$ = average energy dissipation per unit mass for vessel, J/kg · s
P = power of mixing input to entire mixing vessel, J/s
M = mass of water in mixing vessel, kg

CAMP-STEIN ROOT-MEAN-SQUARE VELOCITY GRADIENT

About the same time as Kolmogorov did his work, Camp and Stein (1943) proposed a similar parameter, the root-mean-square (RMS) velocity gradient $\overline{G}$. Camp and Stein proposed that $\overline{G}$ could be used as a design parameter for flocculation facilities and that the speed of flocculation is directly proportional to the velocity gradient. In subsequent studies it was demonstrated that the direct proportionality that Camp and Stein hypothesized occurred with both metal ion coagulants (Harris et al., 1966) and polymers (Birkner and Morgan, 1968) at both bench and pilot scale (Harris et al., 1966).

Camp and Stein (1943) developed a simple equation for $\overline{G}$ by equating the velocity gradient to the power dissipated per unit volume (P/V).

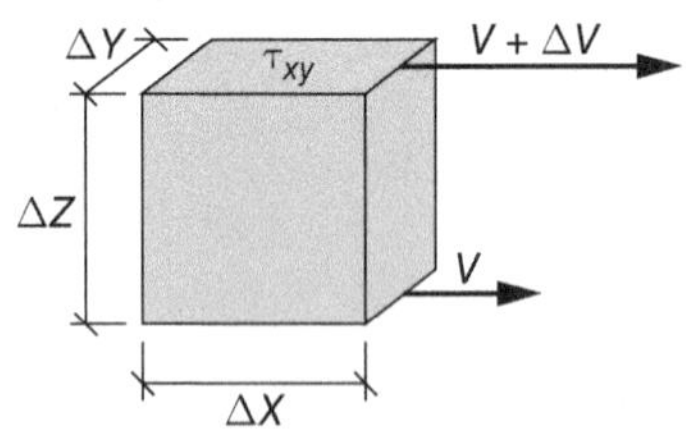

Figure 6-28
Schematic of forces acting on fluid element in flocculator.

Consider the fluid element illustrated on Fig. 6-28 and the forces acting on it. The shear stress in the x–y plane, τ_{xy}, is due to the velocity gradient in the z direction, and the force exerted on it is given by the expression

$$\text{Force} = \tau_{xy}\Delta x\ \Delta y = \mu\frac{dv}{dz}\Delta x\ \Delta y \tag{6-138}$$

where μ = dynamic viscosity of water, $N \cdot s/m^2$ and Δx, Δy, Δz are fluid element dimensions on Fig. 6-28. The product of force and velocity is power, so, using the velocity increment due to the shear stress in the fluid element, the power per unit volume can be written as

$$\frac{P}{V} = \frac{\text{Force} \times \text{velocity}}{\Delta x \Delta y \Delta z} = \frac{[\mu(dv/dz)\Delta x\ \Delta y][(dv/dz)\ \Delta z]}{\Delta x \Delta y \Delta z} = \mu\left(\frac{dv}{dz}\right)^2 \tag{6-139}$$

where P/V = power dissipated in selected fluid element, $J/m^3 \cdot s$

Under turbulent-flow conditions, the velocity gradient is not well defined and varies both in time and space throughout the flocculation vessel. Camp and Stein proposed that the velocity gradient, averaged over the volume of the entire vessel, could be used as a design parameter for flocculation. They named the parameter the RMS velocity gradient. Rearranging Eq. 6-139 and defining the RMS velocity gradient dv/dz as $\overline{G}$,

$$\overline{G} = \sqrt{\frac{P}{\mu V}} \tag{6-140}$$

where $\overline{G}$ = global RMS velocity gradient (energy input rate), s^{-1}
P = power of mixing input to vessel, J/s (same as P in Eq. 6-137)
V = volume of mixing vessel, m^3

The Camp -Stein RMS velocity gradient $\overline{G}$ has since become a widely adopted standard used by engineers for assessing energy input in all kinds of mixing processes, particularly flocculation, and $\overline{G}$ is the parameter that will be used to characterize mixing energy throughout this book.

Example 6-10 Determination of relationship between smallest eddy size η and $\overline{G}$ for water at 10°C

Find the relationship between η and $\overline{G}$ and then produce a semilog plot of this relationship for water at 10°C and for $\overline{G}$ values between 1 and 1000 s^{-1}. At 10°C the kinematic viscosity of water is 1.31×10^{-6} m^2/s. Assume that all the energy is dissipated uniformly throughout the vessel (e.g., $\varepsilon \cong \overline{\varepsilon}$).

Solution

1. Develop a relationship between η and $\bar{G}$.
   a. Solving Eq. 6-137 and Eq. 6-140 for P and setting them equal to each other yields

$$M\bar{\varepsilon} = \bar{G}^2 \mu V$$

   b. Rearranging gives

$$\bar{\varepsilon} = \nu \bar{G}^2$$

   where $$\nu = \frac{\mu}{\rho} = \frac{\mu V}{M}$$

   c. Rearrange Eq. 6-136 to yield

$$\varepsilon = \frac{\nu^3}{\eta^4}$$

   d. Assuming $\varepsilon \cong \bar{\varepsilon}$, equating the equations from steps b and c yields

$$\eta = \left(\frac{\nu}{\bar{G}}\right)^{1/2}$$

$$= \left[\frac{(1.31 \times 10^{-6}\ \text{m}^2/\text{s})}{\bar{G},\ \text{s}^{-1}}\right]^{1/2}$$

2. Substitute values for $\bar{G}$ between 1 and 1000 s^{-1} in the expression developed above and plot the results. The required plot is given below.

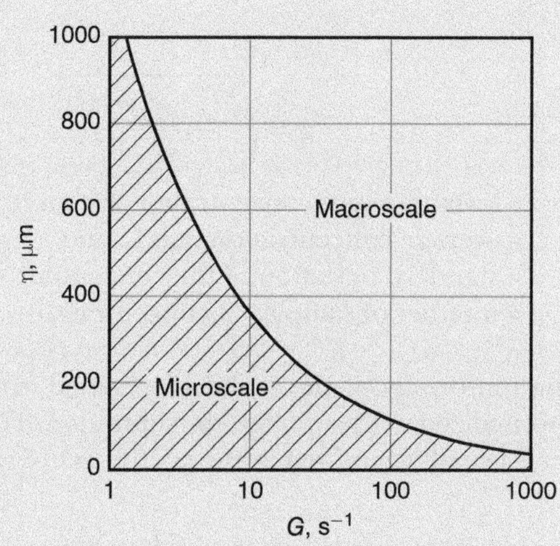

Uniformity and Time Scales in Mixing

One of the principal objectives of mixing, as noted previously, is the blending of two flows, usually the blending of a small flow of chemical solution with the much larger flow of the water to be treated. The greater the difference in the flow of the water to be treated and the flow of the chemical being blended, the more difficult the blending is to accomplish. There are two ways that the degree of blending can be assessed: (1) the uniformity of the concentration of components in solution with time (both in time and in space) and (2) the time it takes to accomplish a specified level of uniformity. The uniformity of blending with time and/or space is generally addressed by specifying an objective for the variation of the concentration of the chemical to be blended.

In water treatment there are two principal circumstances when blending is important: (1) blending must be achieved before samples can be taken for the analysis that confirms that the quality of the blend meets the goals (e.g., before chlorine residual analysis) and (2) rapid blending must be achieved with water treatment chemicals that involve irreversible competitive consecutive reactions (e.g., coagulation with alum or ferric chloride or the breakpoint reactions between chlorine and ammonia described in Chap. 13).

UNIFORMITY OF BLENDING

When a chemical is to be added to a water stream and blending is required, there are two tasks that must be addressed: (a) specifying the uniformity of the blend produced and (b) specifying the magnitude of the task that the mixer must accomplish. Assessing the uniformity of the blend can be determined with respect to variations in time and/or space, but it is most often determined in the context of variation in time. Important variations in the concentration with time are identified on Fig. 6-29. The standard deviation is

$$\sigma = \sqrt{\frac{\sum_{t=1}^{n}\left(C_t - \bar{C}\right)^2}{n-1}} \tag{6-141}$$

where C_t = instantaneous concentration at time t, mg/L
$\bar{C}$ = average concentration mg/L, and
σ = standard deviation of the concentration in the stream
n = number of samples of concentration

When the uniformity of the blend is specified, the standard deviation is usually normalized to the average concentration. This normalized standard deviation is usually identified as the coefficient of variation (COV):

$$\text{COV} = \left(\frac{\sigma}{\bar{C}}\right) \times 100\% \tag{6-142}$$

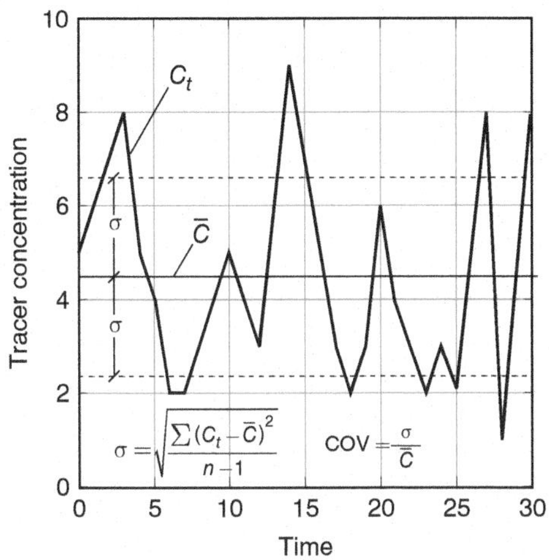

Figure 6-29
Illustration of typical concentration variation immediately downstream of tracer addition point.

where COV = coefficient of variation with time
σ = standard deviation of concentration, mg/L
$\bar{C}$ = average concentration over time, mg/L

An analogous exercise can be used to estimate the quality of a blend in space, for example, samples taken across the cross section of flow a specified distance downstream of the point of chemical addition.

The magnitude of the blending that must be accomplished is determined by the COV that is sought in the design and the segregation of the unmixed streams. Danckwerts (1952) defined the term *intensity of segregation* I_s as given by Eq. 6-143 to characterize the state of blending:

$$I_s = \left(\frac{\sigma_b}{\sigma_u}\right)^2 \tag{6-143}$$

where I_s = Danckwerts' intensity of segregation
σ_b = standard deviation of concentration in blended stream
σ_u = standard deviation between two streams in unblended condition

Danckwerts' intensity of segregation I_s is a description of the degree to which the two streams have been blended. The general description of how completely the streams are blended can be put in the following way:

When $I_s = 0$, the two streams are blended completely.
When $I_s = 1$, the two streams are completely unblended.

The term I_s can also be used to specify what a mixing device must accomplish. The value of I_s for a mixer can be determined from Eq. 6-143 if σ_u is known and a decision is made on the σ_b to be specified.

The standard deviation between the two unblended streams, σ_u,canbe estimated on the basis of their relative flow rates. When chemical A is being added to a stream of water, the volume fraction of the chemical solution being added to the original water stream can be estimated as follows:

$$\bar{X}_A = \frac{Q_A}{Q_w + Q_A} \tag{6-144}$$

By definition,

$$\bar{X}_w = 1 - \bar{X}_A \tag{6-145}$$

where $\bar{X}_A$ = volume fraction of stream containing chemical A in unblended condition
Q_A = flow rate of solution stream of chemical A, m^3/s
Q_w = flow rate of water stream being treated, m^3/s
$\bar{X}_w$ = volume fraction of water in unblended condition

and if it is assumed that a large number of random samples are taken from the two streams, it can be shown that

$$\sigma_{u(\text{vol})} \cong \sqrt{\bar{X}_A\left(1 - \bar{X}_A\right)} \tag{6-146}$$

where $\sigma_{u(\text{vol})}$ = standard deviation of concentrations before blending (expressed as volume fraction)

When designing systems to dose chemicals in a water treatment plant, concentration data are often more readily available than flow rate data. A mass balance can be used to relate the flows and concentrations:

$$Q_A C_A = Q_w C_{\text{dose}} \tag{6-147}$$

or

$$\frac{Q_A}{Q_w} = \frac{C_{\text{dose}}}{C_A} \tag{6-148}$$

where Q_A = flow rate of feed stream for chemical A, m^3/s
Q_w = flow rate of water stream being treated, m^3/s
C_A = concentration of chemical A in feed stream, kg/m^3
C_{dose} = dose of chemical A to be applied to water stream, kg/m^3

The volume fraction can then be found by substituting Eq. 6-148 into Eq. 6-144:

$$\bar{X}_A = \frac{C_{\text{dose}}}{C_A + C_{\text{dose}}} \tag{6-149}$$

The application of the above equations is illustrated in Example 6-11.

TIME REQUIRED TO ACCOMPLISH BLENDING

The uniformity of the blend is important, but the time required to accomplish blending is equally important. There are two circumstances where the time required to meet a blending goal, t_b, is important: (1) when blending must be complete for purposes of analysis and control and (2) when blending must be completed rapidly to prevent adverse outcomes.

Example 6-11 Estimating value of I_s

A water treatment plant must dose the water with 30 mg/L of alum so that the coefficient of variation of the blend is 5 percent or less (COV ≤ 5 percent). Estimate the I_s that characterizes the magnitude of the blending job (the mixer's specification). It may be assumed that the alum solution has a strength of 651 g/L. *Hint*: Both σ_u and σ_b must be on the same basis. Because σ_u is estimated as a volume fraction ($\sigma_{u(vol)}$), σ_b must also be expressed as a volume fraction ($\sigma_{b(vol)}$) for I_s to be properly determined.

Solution

1. Determine the volume fraction using Eq. 6-149:

$$\overline{X}_A = \frac{C_{dose}}{C_A + C_{dose}}$$

As $C_A \gg C_{dose}$,

$$\overline{X}_A \cong \frac{C_{dose}}{C_A} = \frac{30 \text{ mg/L}}{651{,}000 \text{ mg/L}} = 4.61 \times 10^{-5}$$

2. Estimate the standard deviation of the concentration using Eq. 6-142: (Eq. 6-142 is equally valid whether concentration or volume fraction is used).

$$\text{COV} = 5\% = \frac{\sigma_{b(vol)}}{X_A} \times 100\%$$

$$\sigma_{b(vol)} = 0.05 \times 4.61 \times 10^{-5} = 2.30 \times 10^{-6}$$

3. Determine the uniformity of the unblended streams. Substituting 4.61 × 10^{-5} into Eq. 6-146 results in the unblended uniformity on a volume fraction basis ($\sigma_{u(vol)}$):

$$\sigma_{u(vol)} = \sqrt{4.61 \times 10^{-5}\left(1 - 4.61 \times 10^{-5}\right)} = 0.00679$$

4. Determine the intensity of segregation using Eq. 6-143:

$$I_S = \left(\frac{2.30 \times 10^{-6}}{0.00679}\right)^2 = \left(3.39 \times 10^{-4}\right)^2 = 1.15 \times 10^{-7}$$

Blending time and control

A situation where blending must be complete for purposes of analysis and control is illustrated on Fig. 6-30. As shown, a simple point discharge is introduced into a conduit, and the variation of concentration is observed

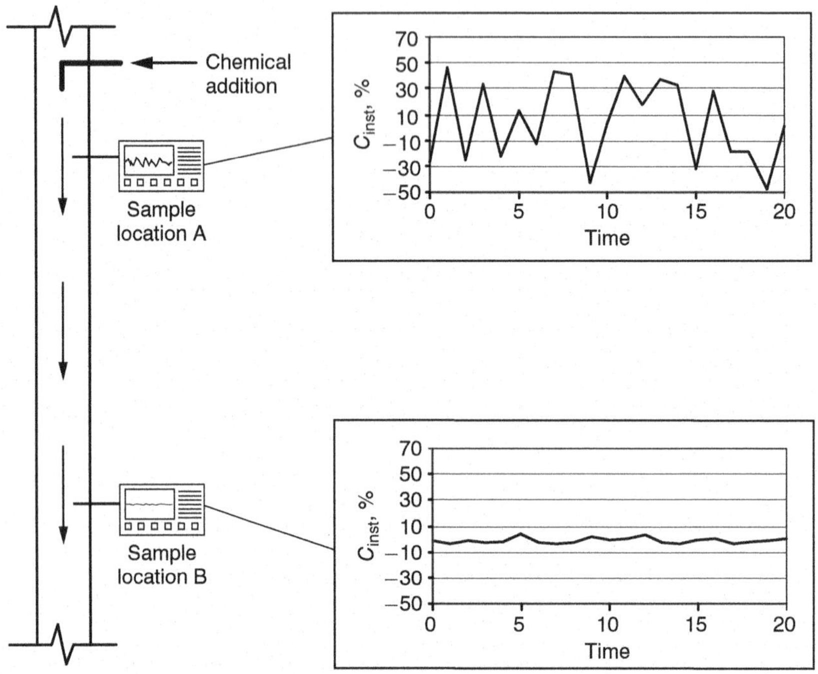

Figure 6-30
To achieve reliable monitoring for control, the downstream sampling point must be beyond the point where uniform blending is achieved.

at two sampling points downstream. Sampling at point A, the signal that is generated would make control very difficult. Sampling at point B solves this problem, but at the expense of time. Sampling at point B introduces another control problem, the travel time from point A to point B. When a chemical feed is simply discharged into turbulent flow in a pipe or channel, a travel distance of as much as 100 conduit diameters in length may be required before acceptable blending is achieved. When such travel distances are not acceptable, a special mixing or blending device is required. The design engineer needs to know where the blend will be sufficiently uniform and what may be done if uniformity is not achieved soon enough.

Rapid blending

When irreversible competitive consecutive reactions are involved, it is often important to accomplish blending rapidly to avoid adverse outcomes. It is also important that blending be accomplished in a manner that prevents backmixing (recirculation) so that components already formed do not gain access to the chemical being added a second time. Put another way, the two components must be rapidly blended across the flow cross section. A relatively simple model for characterizing situations where rapid mixing is important was proposed by Toor (1969). Toor proposed that reactions be broken into three classes:

$$\frac{t_k}{t_b} \gg 1 \text{ (slow reaction)} \tag{6-150}$$

$$\frac{t_k}{t_b} \approx 1 \text{ (moderate reaction)} \tag{6-151}$$

$$\frac{t_k}{t_b} \ll 1 \text{ (fast reaction)} \tag{6-152}$$

where t_k = time characteristic of reaction of interest, e.g., reaction half-life, s
t_b = time characteristic of blending, e.g., time required to achieve COV < 5 percent, s

For slow reactions ($t_k \gg t_b$), blending is generally accomplished at the molecular level before the reaction of interest makes significant progress. With moderate reactions ($t_k \sim t_b$), the reaction occurs at the same pace as blending, and with fast reactions ($t_k \ll t_b$) the state of the reaction at any moment in time is limited by the remaining segregation, that is, the degree to which the blending is not yet complete. To make deliberate decisions about rapid blending, information must be available on both the time characteristic of the reaction t_k and the time characteristic of mixing t_b.

Toor recognized that comparing t_k and t_b alone is not enough to make sound decisions about the importance of rapid blending. For example, when a simple, reversible reaction with no competitive side reactions is being considered, the outcome will be the same no matter how fast is the blending. The outcome of fast, competitive, consecutive, poorly reversible reactions can be markedly influenced by the time of blending. Consider the following competitive, consecutive model reactions (Toor, 1969):

$$A + B \xrightarrow{k_1} R \tag{6-153}$$

$$A + B \xrightarrow{k_2} S \tag{6-154}$$

Consider the case where a concentrate of chemical A is being added to a dilute solution of chemical B and the objective is to produce product R, that is, S is considered an undesirable by-product. When mixing is slow ($t_b \gg t_k$), B is soon depleted in the A-rich zones of the mixture and the formation of S is favored. When blending is fast ($t_b \ll t_k$), the formation of R is favored unless too much A is added.

The time characteristic of blending t_b increases with the scale of the water stream being treated. For example, in a 0.007- to 0.03-ML/d pilot plant, t_b ranges from 0.1 to 1 s. Whereas in treatment plants with capacities of 10, 100, and 1000 ML/d, with efficient blending, t_b is on the order of 3, 10, and 30 s, respectively. When rapid blending is expected to be important, large-scale testing is important.

Table 6-6
Blending assessment of some typical reactions

Example Reactions	Blending Time, s	Is Rapid Initial blending Important?	Comments
Coagulation with Fe^{3+} or Al^{3+}	<0.3	Yes	Fast, poorly reversible, competitive, consecutive
pH adjustment	≪1	No	Fast, easily reversible
$CaCO_3$ nucleation	~20	Perhaps	Somewhat fast, poorly reversible
HOCl/NOM to DBPs	~90,000	No	Very slow, not reversible, competitive, consecutive
Chlorine hydrolysis	0.06	Yes	Fast, easily reversible, Cl_2 has low solublity
Chlorine/ammonia (high Cl_2/N)	<0.1–2000	Yes	Fast in early stages, poorly reversible, competitive, consecutive
Chlorine/ammonia (low Cl_2/N)	10,000–1,000,000	No	Extremely slow, poorly reversible, competitive, consecutive
HOCl/coliform	~10	No	Fast, persistent residual kill
HOCl/*Giardia*	~200	No	Slow, persistent residual kill
HOCl/*Cryptosporidium*	~90,000	No	Slow, persistent residual kill

Several reactions commonly encountered in the treatment of drinking water are summarized in Table 6-6. Three reactions are discussed further below: (1) the addition of sulfuric acid to reduce the pH, (2) the reaction of chlorine with NOM, and (3) the reaction of chlorine with ammonia to form chloramines:

1. *pH adjustment with H_2SO_4* This reaction is extremely fast ($t_k \ll t_b$). The reaction is also reversible. The time of blending will control the apparent rate of the reaction, but the outcome will be the same for any reasonable mixing time. Thus, it is only important that blending be complete before the water is sampled and analyzed for pH or before another reaction is introduced, which depends on the pH goal being sought.
2. *Chlorination of NOM to form DBPs.* Though hardly simple, the reaction of chlorine with NOM is generally regarded as slow. The time for completion is generally 10 to 20 h or more. The time of blending is rarely more than a few minutes. Thus, $t_k \gg t_b$, and this is a slow reaction where the speed of initial blending would not be of great importance.
3. *Reaction of chlorine with ammonia to form chloramines.* These reactions are discussed further in Chaps. 8 and 13. They consist of a series of

competitive, consecutive reactions some of which are very fast and many of which are poorly reversible (Saunier and Selleck, 1979; Wei and Morris, 1974). Consequently, blending can and does have a great impact on the outcomes achieved.

Blending below Microscale

All the discussion so far has been concerned with managing the dispersion created by the turbulence at the macroscale (turbulent diffusion) to accomplish a specified level of blending. When rapid blending is being pursued for the purposes of facilitating chemical reactions, blending must be accomplished to the molecular level. Consequently, diffusion must be responsible for transport of the treatment chemical within the microscale.

The time scale that is required for this transport into the smallest eddies can be estimated by using the following expression (Crank, 1979):

$$t_d = \frac{3R^2}{4D_l} \tag{6-155}$$

where t_d = time for molecules to diffuse in or out of eddy, s
R = radius of eddy, m
D_l = liquid diffusivity of chemical molecule ($\sim 10^{-9}$ m²/s)

Earlier in this section it was shown that eddy size is influenced by the energy input and that the eddy diameter is equal to the Kolmogorov microscale η, introduced previously in the earlier subsection on mixing and the scale of turbulence (Logan, 1999). Hence the radius of the smallest eddy is equal to half of η:

$$R_{\text{avg}} = \tfrac{1}{2}\eta \tag{6-156}$$

where R_{avg} = radius of smallest eddy, m

Expressing energy input in terms of the Camp-Stein $\overline{G}$ and combining Eqs. 6-136, 6-137, 6-140, and 6-155, it can be demonstrated that the product $\overline{G}t_d$ that must be sustained to ensure that mixing occurs throughout the microscale can be calculated as shown below for 10°C:

$$\overline{G}t_d = \frac{3\nu}{16D_l} \cong \frac{3\times 1.31\times 10^{-6}\ \text{m}^2/\text{s}}{16\times 10^{-9}\ \text{m}^2/\text{s}} = 246 \tag{6-157}$$

Mixing Devices Used for Blending

Obtaining the engineering data necessary for the design of a mixing device using intensity of segregation I_s is not very easy for the following reasons: (1) most of the work on specification of blending has been done in the chemical engineering field, (2) even in that field the important blending problems are in laminar flow in liquids and turbulent-flow studies are largely limited to combustion, and (3) the mixing devices used in chemical engineering are designed to handle much smaller flows.

As a result, environmental engineers more commonly rely on mixing devices that are designed to achieve a certain intensity of mixing ($\overline{G}$) rather than mixing devices designed to produce a specified quality of mix (COV < 5 percent) (Kawamura, 2000). Several of the devices that are commonly used are described in Table 6-7 and illustrated on Fig. 6-31. The advantages and disadvantages of the various devices are also noted in Table 6-7.

Nevertheless, design data for the intensity of segregation approach are available for some devices. In the remainder of this chapter, the limited design data available are used to highlight some key design issues.

Design of Mixers to Achieve a Specified Blend

The design of mixers to achieve a specified blend is discussed briefly below. The design is based on the principles described earlier and published statistics on four mixing devices that are useful for blending for control or rapid mixing in small facilities. Unfortunately, design information is not available for devices used for the rapid mixing of flow streams, which are the size of those found in the larger full-scale water treatment plants. For these designs, empirical rules are generally used, although there is

Table 6-7
Rapid mixing devices that avoid or minimize backmixing

Mixing Device	Advantages	Disadvantages	Upper Flow Limit, ML/d[a]
Centerline diffuser	Simple, reliable, inexpensive, data for Eq. 6-158 available	Turbulence is essential	5
Venturi injector	Simple, reliable	Data for Eq. 6-158 not available, subject to clogging	30
Static mixer	Simple, reliable, data for Eq. 6-158 available	Expensive, subject to clogging	100
Axial pumped jets	Can be simple, reliable	Data for Eq. 6-158 not available, subject to clogging, some backmixing	100
Lateral pumped jets	Can be simple, reliable	Data for Eq. 6-158 not available, subject to clogging	150
Conventional stirred tanks	Familiar, can be effective at blending	Expensive to maintain, high energy costs, extensive backmixing	150

[a]Approximate upper limit for efficient mixing in reasonable time. Above these limits, multiple units in parallel are recommended.

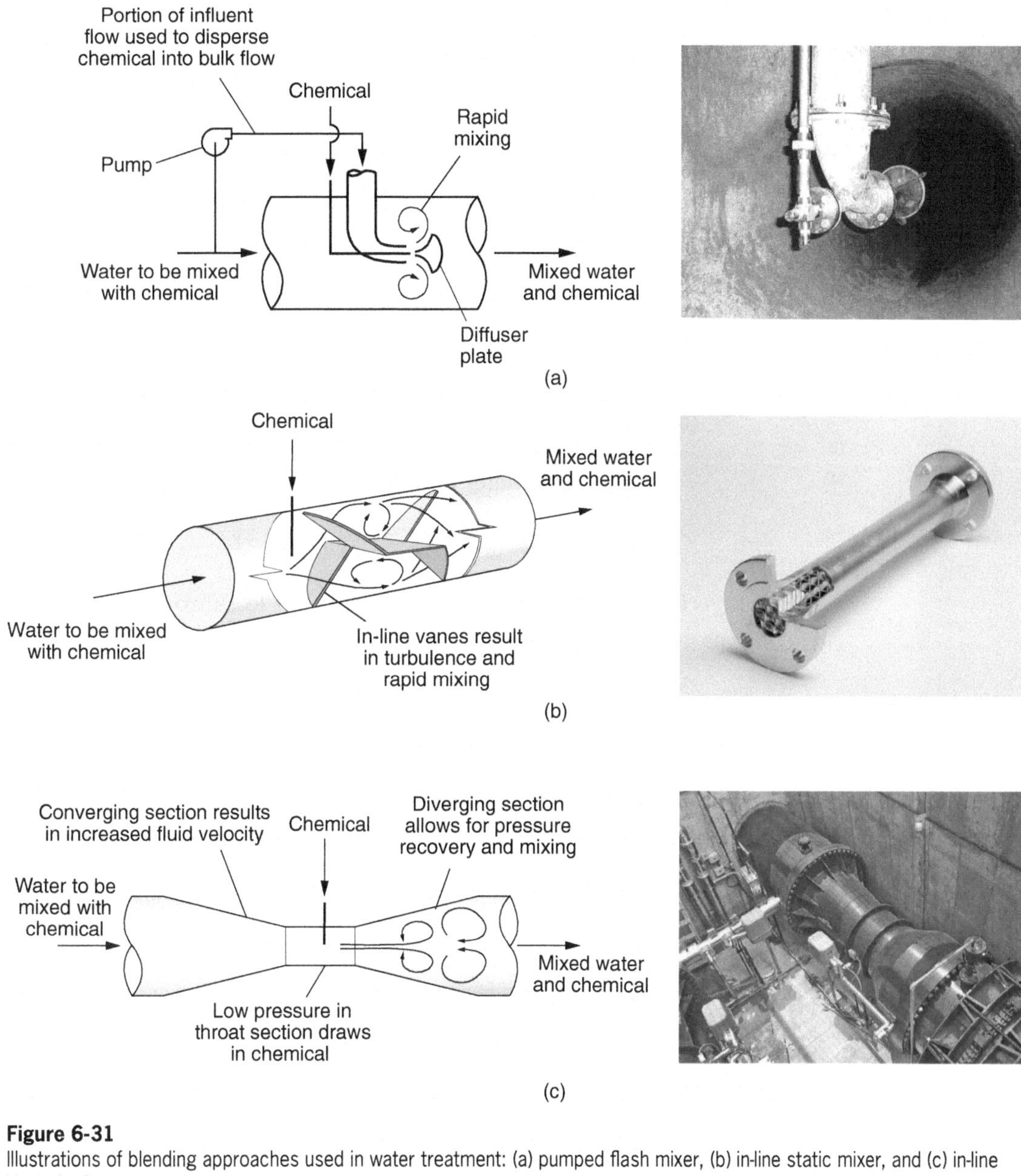

Figure 6-31
Illustrations of blending approaches used in water treatment: (a) pumped flash mixer, (b) in-line static mixer, and (c) in-line Venturi mixer.

increasing interest in applying computational fluid dynamics (DuCoste and Oritz, 2003). Despite these limitations, the illustrative calculations are given here because they demonstrate important features regarding the scalability of rapid mixing.

Godfrey (1985) proposed the following related rules for modeling a variety of mixers that operate in the centerline of pipe flow: (1) a center-line discharge, (2) a Kenics static mixer, (3) a Koch static mixer, and (4) a pipe with trapezoidal baffles:

$$\left(\frac{\sigma_b}{\sigma_u}\right)_{\mathrm{AR}=L/D} = \left(\frac{\sigma_b}{\sigma_u}\right)_{\mathrm{AR}=1}^{L/D} \tag{6-158}$$

where $(\sigma_b/\sigma_u)_{\mathrm{AR}=L/D}$ = σ_b/σ_u achieved by mixer with length-to-width ratio (aspect ratio, AR) equal to L/D

$(\sigma_b/\sigma_u)_{\mathrm{AR}=1}$ = σ_b/σ_u achieved by mixer with aspect ratio of unity

L = length of mixer, m

D = diameter of mixer, m

$$(N_{\mathrm{VH}})_{\mathrm{AR}=L/D} = \frac{L}{D}(N_{\mathrm{VH}})\,\mathrm{AR}=1 \tag{6-159}$$

where $(N_{\mathrm{VH}})_{\mathrm{AR}=L/D}$ = number of velocity heads lost through mixer with aspect ratio equal to L/D

$(N_{\mathrm{VH}})_{\mathrm{AR}=1}$ = number of velocity heads lost through mixer with aspect ratio of unity

Godfrey (1985) reported values for $(\sigma_b/\sigma_u)_{\mathrm{AR}=1}$ and $(N_{\mathrm{VH}})_{\mathrm{AR}=1}$ for four mixing devices as shown in Table 6-8.

Equation 6-159 can be used to estimate the L/D of the mixing device required to meet a specified level of uniformity, $(\sigma_b\sigma_u)_{\mathrm{AR}=L/D}$, and Eq. 6-158 can be used to estimate the head loss that will occur through the device. The mixing time required to meet the specified level of uniformity, t_b, is the time it takes for the water to pass through the mixer and is determined from the velocity of flow and the length of the mixer.

Table 6-8
Performance and design parameters for mixing devices

Mixing Device	$(\sigma_b/\sigma_u)_{\mathrm{AR}=1}$	$(N_{\mathrm{VH}})_{\mathrm{AR}=1}$
Pipeline	0.95	0.02
Koch static mixer	0.21	3.8
Kenics static mixer	0.63	2.0
Trapezoidal baffles	0.38	6.6

Example 6-12 Design of static mixer to achieve specified degree of blending

A pilot plant is to be designed to add 30 mg/L of alum to a design flow of 5 L/min. The blend specification is that COV ≤ 5 percent. Estimate the length of a static mixer required for this degree of blending. Use a design velocity of ≤2 m/s. Assume that the mixing device inserts come in units 1.5 diameters in length and are available in diameters of 12.5 and 20 mm. Use a whole number of inserts to achieve a reasonable safety factor. Finally, estimate the length of time required for the water to achieve the required degree of blending and the head loss through the mixer.

Solution

1. Determine the mixer diameter using a design velocity ≤2 m/s:
   a. The area of the pipe is

$$A_{\text{pipe}} = \frac{Q}{V}$$

$$= \frac{(5\text{ L/min})\left(1\text{ m}^3/1000\text{ L}\right)}{(2\text{ m/s})(60\text{ s/min})} = 4.17 \times 10^{-5}\text{ m}^2$$

   b. The corresponding pipe diameter is

$$d_{\text{pipe}} = \sqrt{\frac{4A_{\text{pipe}}}{\pi}} = \sqrt{\frac{4 \times 4.17 \times 10^{-5}\text{ m}^2}{\pi}} = 0.0073\text{ m} = 7.3\text{ mm}$$

   A pipe diameter of 12.5 mm (0.5 in.) is about the minimum practical size to use in a pilot plant. Choose D_{mixer} = 12.5 mm.

2. Determine the number of mixer inserts:
   a. From Example 6-11, the value of the intensity of segregation $[I_s]_{\text{mixer}}$ is 1.15 × 10^{-7}. Thus, from Eq. 6-143.

$$\left(\frac{\sigma_b}{\sigma_U}\right)_{\text{mixer}} = \sqrt{1.15 \times 10^{-7}} = 3.39 \times 10^{-4}$$

   Substituting into Eq. 6-158, yields

$$3.39 \times 10^{-4} = (0.21)^{L/D} \quad \text{and} \quad \frac{L}{D} = 5.12$$

b. According to the problem statement, these mixing devices come with standard length inserts of $L/D = 1.5$. One mixer insert will provide $L/D = 1.5$. Four will provide $L/D = 6$. Use four inserts.

3. Estimate blending time t_m to meet the blend requirement:
   a. The total length of the mixer is $4 \times 1.5 \times 0.0125 = 0.075$ m. The velocity v in the mixer will be

$$v = \frac{Q}{A} = \frac{(5\ \text{L/min})(1\ \text{m}^3/1000\ \text{L})}{\left[\pi \times (0.0125\ \text{m})^2/4\right](60\ \text{s/min})} = 0.679\ \text{m/s}$$

   b. The blending time is

$$t_m = \frac{0.075\ \text{m}}{(0.679\ \text{m/s})} = 0.11\ \text{s}$$

4. Determine the total head loss:
   a. The velocity head is

$$\text{Velocity head} = \frac{v^2}{2g} = \frac{(0.679\ \text{m/s})^2}{2\left(9.8\ \text{m/s}^2\right)} = 0.0235\ \text{m}$$

   b. From Gray's data,

$$\left[N_{VH}\right]_{AR=1} = 3.8$$

   The total head loss is

$$\Delta H_L = 0.0235 \times 3.8 \times 6 = 0.54\ \text{m}$$

Comment

It seems practical to meet this blending specification with $t_b \sim 0.1$ s at the small scale of this pilot plant. Further analysis will show that using the same static mixer takes much more time and head loss at full scale.

Design of Blending for Process Control

Equation 6-158 along with the $(\sigma_b/\sigma_u)_{AR=1}$ value for flow in a straight pipe or channel can be used to estimate the blending requirements for process control. However, the value of $(\sigma_b/\sigma_u)_{AR=1}$ given above is an approximation, and field measurement can often provide a more accurate estimate. Likewise, field measurements of the blending being achieved in a channel can be used to evaluate alternatives for remediating control problems.

Example 6-13 Blending to achieve process control

In a large water treatment plant with a flow of 200 ML/d, hydrofluosilicic acid is added to the product water. The acid is added, with mixing, at the beginning of a 100-m (328-ft) channel and the fluoride residual is monitored 23m (75 ft) downstream of the point of addition. Unfortunately, the COV of the fluoride signal at the monitoring point is unacceptably high for control purposes (25 percent). Estimate how far down the channel the sample point would have to be moved to meet a COV criterion of 5 percent. Also estimate the water travel time between the point of fluoride addition and the point of fluoride analysis. The hydrofluosilicic acid solution has a fluoride concentration of 220 g/L. *Hint*: It is possible to estimate $[I_s]_{\text{mixer}}$ by assuming that segregation will decline as the water flows down the channel as it would in a pipe. The channel has a depth of 3 m. The depth should be used as D in Eqs. 6-157 and 6-158. The width of the channel is also 3 m.

Solution

1. Using Eq. 6-142, the COV downstream of the old monitor ($\text{COV}_{\text{old monitor}}$) and the COV downstream of the new monitor ($\text{COV}_{\text{new monitor}}$) can be used to relate the standard deviation downstream of the old monitor ($\sigma_{\text{old monitor}}$) to the standard deviation downstream of the new monitor ($\sigma_{\text{new monitor}}$):

$$\frac{\text{COV}_{\text{new monitor}}}{\text{COV}_{\text{old monitor}}} = \frac{\left(\sigma_{\text{new monitor}} / \bar{C}\right) \times 100\%}{\left(\sigma_{\text{old monitor}} / \bar{C}\right) \times 100\%} = \frac{\sigma_{\text{new monitor}}}{\sigma_{\text{old monitor}}}$$

Thus

$$\frac{\sigma_{\text{new monitor}}}{\sigma_{\text{old monitor}}} = \frac{\text{COV}_{\text{new monitor}}}{\text{COV}_{\text{old monitor}}} = \frac{5\%}{25\%} = 0.2$$

2. Estimate the required length L:
   a. Estimate the L/D value using Eq. 6-158:

$$\left(\frac{\sigma_b}{\sigma_u}\right)_{AR=L/D} = \left(\frac{\sigma_b}{\sigma_u}\right)_{AR=1}^{L/D}$$

$$\left(\frac{\sigma_b}{\sigma_u}\right)_{AR=L/D} = \frac{\sigma_{\text{new monitor}}}{\sigma_{\text{old monitor}}} = 0.2$$

$$0.2 = \left(\frac{\sigma_b}{\sigma_u}\right)_{AR=1} = 0.95^{L/D}$$

Taking the logarithm of both sides and rearranging,

$$\frac{L}{D}=\frac{-0.699}{-0.0223}=31.4$$

b. The required length L is

$$L=31.4\times 3=94\ \text{m}$$

Thus, the sampling point will have to be moved 71 m (94 – 23) downstream from the current monitoring point.

3. Estimate the water travel time, t_b:

a. Estimate the velocity in the channel:

$$v=\frac{Q}{A}=\frac{(200\ \text{ML/d})(1\ \text{m}^3/1000\ \text{L})(10^6\ \text{L}/1\ \text{ML})}{3\ \text{m}\times 3\ \text{m}}$$

$$=2.2\times 10^4\ \text{m/d}=15.4\ \text{m/min}$$

b. The blending time is

$$t_b=\frac{L}{\nu}=\frac{0.94\ \text{m}}{15.4\ \text{m/min}}=6.1\ \text{min}$$

Problems and Discussion Topics

6-1 Derive the solution to mass balance equations for the second- and nth-order reactions occurring in a CMBR.

6-2 A second-order irreversible elementary reaction 2A → products is carried out in a batch reactor. For a certain set of conditions, it is found that it took 20 min for a reaction to reach 60 percent completion. What would be the time required to reach the same degree of completion if (a) the initial concentration of A were doubled and (b) the reaction rate constant were doubled?

6-3 A first-order reaction A → products is to be carried out in a CMFR. The reaction rate constant was determined to be 1.0 h^{-1}. What is the residence time required for 90 percent conversion of the reactant provided there are no changes in temperature and density of the system?

6-4 A given reactant decomposes in water with a second-order rate constant of k = 0.43 L/mol · s. If a feed stream with an initial concentration of 1.0 mol/L is passed through a CMFR at the rate of 0.1 m^3/s, what will be the reactor volume required for 90 percent decomposition of the reactant?

6-5 If two CMFRs in series each having one-half the volume calculated in Problem 6-4 were used, what would be the resulting percent decomposition of the reactant?

6-6 Derive and solve the differential equation that can be used to describe the effluent concentration leaving a CMFR for a first-order reaction. Assume $C_0^* = C_0$.

6-7 Rearrange the solution obtained in Problem 6-6 and solve for the time (expressed as multiples of the hydraulic detention time) required after startup such that the effluent concentration is within 1 percent of the steady-state value C_∞:

$$\frac{C - C_\infty}{C_0} \leq 0.01$$

Plot the time required to achieve the above criteria versus the ratio of the steady-state concentration to the influent concentration. Is there a maximum time?

6-8 A reaction follows the rate expression of $r = -kC^{1.5}$. The feed concentration is 1 mol/L and the reaction rate constant k was determined to be 1.0 $(\text{L/mol})^{1/2}/\text{h}$. For a flow rate of 100 L/h, compute the volume requirement for a CMBR at 95 percent conversion if the down time for the CMBR is 1 h between batches? What is the volume requirement for a CMFR at 95 percent conversion? What is the volume requirement for a PFR at 95 percent conversion?

6-9 If A → products follows a second-order reaction mechanism, what is the volume requirement for a CMFR as compared to a PFR for 95 percent conversion of A? How could the volume efficiency be improved?

6-10 Because of the ease of construction and operation of a CMFR, its use is sometimes preferred over a PFR. For the following conditions, indicate the preferred reactor (PFR or CMFR), where an additional CMFR volume requirement of 25 percent is acceptable. Justify each answer by computing the volume requirements for the CMFR and PFR under each condition presented in the table below.

Reaction Order	Degree of Completion of Reaction, %	PFR	CMFR
2	99		
1	99		
0	99		
0	10		
1	10		
2	10		

Reaction Order	Degree of Completion of Reaction, %	PFR	CMFR
0	40		
1	40		
2	40		
$\frac{1}{3}$	40		

6-11 Explain graphically using a plot of $-1/r$ versus concentration of reactants as a function of time why the PFR is more efficient in terms of volume requirements than the CMFR when the order of the reaction is greater than zero. Also show graphically why a large number of equal-volume CMFRs connected in series can achieve the same volume efficiency as a PFR.

6-12 The following kinetic data were obtained to determine the order of reaction with respect to one reactant. Determine the reaction order that yields the best fit and estimate the rate constant for the reaction.

Time, min	Concentration, mg/L
0	40.00
1	31.50
2	21.50
3	17.85
4	12.16
5	10.08
6	6.84
7	5.25
8	4.30
9	2.95
10	2.42

6-13 The reaction A → B is autocatalytic (i.e., the product itself is also participating in the reaction), yielding a rate expression of the form

$$r_A(\text{mol}/\text{L}\cdot\text{min}) = -kC_AC_B \quad k = 0.05\ \text{L}/\text{mol}\cdot\text{min}$$

a. Calculate the residence time required in a CMFR at steady state to achieve 80 percent conversion of A when the feed concentration of A and B are 2 mol/L and 0.05 mol/L.

b. Because the reaction follows second-order kinetics, a PFR should in theory yield a lower residence time and thus require a smaller reactor volume than a CMFR. Is a PFR preferred over a CMFR for the case of an autocatalytic reaction? State clearly the reasons for either recommending or not recommending a PFR, based on a plot of $-1/r_A$ versus concentration. What combination of reactors will result in the smallest volume?

6-14 Derive the exit age distribution for n reactors (tanks) in series.

6-15 Calculate the effluent concentration C using the DFM for a second-order reaction in the case of a closed reactor. Assume Pe = 0.4, τ = 10 min, k = 0.05 L/mg · min, and C_0 = 2000 mg/L.

6-16 Consider a PFR in which the conversion (extent of reaction) is 99.5 percent for a first-order reaction.

a. What would be the conversion in a completely mixed flow reactor of the same volume?

b. What would be the conversion in that reactor if the dispersion number $E/vL = 1$?

c. What would the conversion be if the length was doubled but the residence time remained the same? (Assume E is the same as the original.)

d. What would the conversion be if the length was halved and the residence time remained the same? (Assume E is the same as the original reactor.)

e. Suppose it is desired to obtain 95 percent conversion in this non-ideal reactor system with $E/vL = 1$. How many times larger than the volume of a PFR would be needed in the real reactor?

6-17 Rework Problem 6-16 for second-order kinetics.

6-18 Derive the expression to obtain the effluent concentration for n reactors (tanks) in series for a second-order reaction.

6-19 The following concentration data expressed in mg/L were obtained from tracer studies conducted on five different reactors. For a given reactor (to be selected by the instructor), plot the tracer curve, the normalized RTD curve, and the cumulative RTD curve.

Time, min	Reactor 1	Reactor 2	Reactor 3	Reactor 4	Reactor 5
0	0	0	0	0	0
10	0	2	0	0	0
20	1	5.4	0	0	0
30	2	8.4	0.1	0	0
40	5.1	11. 4	0.2	0	0
50	8.9	13	0.5	0	2
60	11.2	12. 1	6.3	0	6.2
70	10. 5	9.3	15.2	4.5	13
80	9.2	7.2	18.1	9	10. 4
90	8	5.2	8.5	14.1	5.1
100	6.5	3.6	3.2	15.6	2.8
110	5	2.5	1. 8	12.9	1. 1
120	3.5	1. 4	1. 2	9.2	0.5
130	2	0.9	0.8	5.3	0.4
140	1.4	0.4	0.6	2.3	0.1

Time, min	Reactor 1	Reactor 2	Reactor 3	Reactor 4	Reactor 5
150	0.8	0.1	0.3	1.1	0
160	0.4	0	0.2	0.8	0
170	0.2	0	0.2	0.5	0
180	0	0	0.1	0.2	0
190	0	0	0	0.1	0
200	0	0	0	0	0

6-20 Using the data provided in Problem 6-19, compute the dispersion number, Peclet number, and equivalent number of tanks in series for the selected reactor (to be selected by the instructor). Plot the results of the single-parameter model and compare to the actual data.

6-21 Using the tracer data (to be selected by the instructor) provided in Problem 6-19, calculate the expected effluent concentration using the TIS model assuming a first-order reaction rate constant $k = 0.2\ \text{min}^{-1}$.

6-22 Using the tracer study results given in Problem 6-19 (reactor data to be selected by the instructor), determine the expected effluent concentration using the SFM. Assume a first-order reaction with the rate constant $k = 0.2\ \text{min}^{-1}$.

6-23 For a reaction A → P, $C_0 = 100$ mg/L:

a. Calculate the rate constants for first- and second-order kinetics assuming a PFR with the same detention time calculated in Problem 6-19 and an effluent concentration $C = 0.5$ mg/L.

b. Using figures developed for the DFM, calculate the expected effluent concentrations for first- and second-order kinetics. Use the rate constants determined in (a) above and the Pe number determined in Problem 6-20.

c. Calculate the expected effluent concentration for first- and second-order kinetics using the TIS model. Use the rate constants determined in (a) above and the tank number determined in Problem 6-20.

d. Calculate the expected effluent concentration at steady state for the SFM using second-order kinetics and the rate constants in (a) above.

6-24 Consider an open channel with the following characteristics: (a) depth = width = 2 m, (b) depth = width = 2.5 m, and (c) depth = width = 3 m. The flow rate is 4 m^3/s and the detention time is 60 s. If a first-order irreversible reaction is occurring with a rate constant equal to 0.1 s^{-1}, determine the expected effluent concentration C/C_0 for the various channel widths. Comment on the impact of increasing the aspect ratio on reactor performance. Use Eq. 6-113 to estimate the dispersion factor. Assume the kinematic viscosity of water is 1.003×10^{-6} m^2/s.

6-25 For a reactor with a length of 25 m, depth of 3.5 m, flow rate of 2.5 m^3/ min, and width of 2, 3, 4, or 5 m (to be selected by the instructor), estimate the dispersion number, hydraulic detention time, and θ_{10} value. Assume the kinematic viscosity of water is $1.003 \times 10{-}6$ m^2/s.

6-26 Using the following dose-response data for a particular microorganism found in a water supply, apply the SFM to the reactor from Problem 6-19 (to be selected by the instructor) to estimate the effluent concentration:

Exposure Time, min	Number of Organisms Remaining, No./100 mL
0	100,000
10	10,000
20	1,000
30	100
40	10
50	1
60	0.1
70	0.01
80	0.001
90	0.0001

6-27 A treatment plant has been designed with the capability to add hydrofluosilicic acid. The mixer installed at the point of chemical addition is warranted to achieve a COV of ≤5 percent across the cross section of flow at the point where the automatic analyzer draws samples for analysis. In a performance test during plant startup (commissioning) samples were taken at nine representative points across the cross section and the results are reported below. Determine if the mixer is meeting its specification.

Sample Point	Fluoride, mg/L
1	1.02
2	1.20
3	0.95
4	0.90
5	1.10
6	0.85
7	0.95
8	0.98
9	0.95

6-28 A water treatment plant is to be designed to add 30 mg/L of alum that has a strength of 651 g/L to a design flow of 50 ML/min. The blend specification is for a COV value of ≤ 5 percent. Estimate the length of a Koch static mixer required for this application. Use a design velocity of 2 m/s. Assume that the mixer inserts come in units 1.5 diameters in length and are available in diameters of 12.5 and 20 mm. Use a whole number of inserts to achieve a reasonable safety factor. Finally, estimate the length of time required for the water to achieve the mixing and the head loss through the mixer.

6-29 Derive Eq. 6-146 for σ_u, the standard deviation of the volume fraction of two unmixed streams. *Hint 1*: Assume that 1000 samples are taken at random from the two streams and calculate the standard deviation of each. *Hint 2*: Samples drawn from Q_A will have a volume fraction equal to 1, whereas samples drawn from Q_w will have a volume fraction of Q_A equal to 0.

References

Birkner, F., and Morgan, J. (1968) "Polymer Flocculation Kinetics of Dilute Colloidal Suspensions," *J. AWWA*, **60**, 2, 175–191.

Camp, T. R., and Stein, P. C. (1943) "Velocity Gradients and Hydraulic Work in Fluid Motion," *J. Boston Soc. Civil Eng.*, **30**, 203–221.

Crank, J. (1979) *Mathematics of Diffusion*, Oxford University Press, Oxford.

Danckwerts, P. (1952) "The Definition and Measurement of Some Characteristics of Mixtures," *App. Sci. Res.*, **A3**, 11, 279–296.

Danckwerts, P. V. (1953) "Continuous Flow Systems: Distribution of Residence Times," *Chem. Eng. Sci.*, **2**, 1–13.

Davies, J. T. (1972) *Turbulence Phenomena*, Academic Press, New York.

Ducoste, J., and Oritz, V. (2003) Characterization of Drinking Water Treatment Chemical Mixing Performance Using CFD, paper presented at the ASCE Conference, Toronto, Canada.

Eigen, M., and Kustin, K. (1962) "The Kinetics of Halogen Hydrolysis," *J. Am. Chem. Soc.*, **84**, 1355–1361.

Eigen, M., Kustin, K., and Maas, G. (1961) "Die Geschwindigkeit der Hydratation von SO_2 in wässriger Löisung," *Phys. Chem.*, **30**, 130–136.

Ferguson, J. F., Jenkins, D., and Eastman, J. (1973) "Calcium Phosphate Precipitation at Slightly Alkaline pH Values," *J. WPCF*, **45**, 4, 620–631.

Fogler, H. S. (1999) *Elements of Chemical Reaction Engineering*, 3rd ed., Prentice-Hall, Upper Saddle River, NJ.

Froment, G. F., and Bischoff, K. B. (1979) *Chemical Reactor Analysis and Design*, John Wiley & Sons, New York.

Godfrey, J. (1985) Static Mixers, Chap. 13, in N. Harnby, M. Edwards, and A. Nienow (eds.), *Mixing in the Process Industries*, Butterworths, London.

Green, D., and Perry, R. H. (2007) *Chemical Engineers' Handbook,* 8th ed., McGraw-Hill, New York.

Harris, H. S., Kaufman, W. F., and Krone, R. B. (1966) "Orthokinetic Flocculation in Water Purification," *J. Div. Sanit. Eng. Proc. ASCE,* **92**, 95–111.

Hoffmann, M. R. (1981) "Thermodynamic, Kinetic, and Extrathermodynamic Considerations in the Development of Equilibrium Models for Aquatic Systems," *Environ. Sci. Technol.*, **15**, 3, 345–353.

Hoigne, J., and Bader, H. (1983) "Rate Constants of Reactions of Ozone with Organic and Inorganic Compounds in Water. I. Nondissociating Organic Compounds," *Water. Res.*, **17**, 173–185.

Holmes, L. P., Cole, D. L., and Eyring, E. M. (1968) "Kinetics of Aluminum Ion Hydrolysis in Dilute Solutions," *J. Phys. Chem.*, **72**, 301–304.

Kawamura, S. (2000) *Integrated Design and Operation of Water Treatment Facilities,* 2nd ed., Wiley-Interscience, New York.

Kern, D. M. (1960) "The Hydration of Carbon Dioxide," *J. Chem. Educ.*, **37**, 14–23.

Kolmogorov, A. (1941a) "The Local Structure of Turbulence in Incompressible Viscous Fluid for Very Large Reynolds Numbers," *Dokl. Akad. Nauk. SSSR.*, **30**, 299–303.

Kolmogorov, A. (1941b) "The Local Structure of Turbulence in Incompressible Viscous Liquid," *Dokl. Akad. Nauk. SSSR.*, **31**, 538–541.

Kolmogorov, A. (1941c) "Dissipation of Energy in Locally Isotropic Turbulence," *Dokl. Akad. Nauk. SSSR.*, **32**, 19–21.

Kramer, H., and Westerterp, K. P. (1963) *Elements of Chemical Reactor Design and Operation,* Academic Press, New York.

Levenspiel, O. (1998) *Chemical Reaction Engineering,* 3rd ed., John Wiley & Sons, New York.

Levenspiel, O., and Bischoff, K. B. (1959) "Backmixing in the Design of Chemical Reactor," *Ind. Eng. Chem.*, **51**, 1431–1434.

Levenspiel, O., and Smith, W. K. (1957) "Notes on the Diffusion-Type Model for the Longitudinal Mixing of Fluids in Flow,'" *Chem. Eng. Sci.*, **6**, 227–235.

Logan, B. E. (1999) *Environmental Transport Processes,* Wiley-Interscience, New York.

Pankow, J. F., and Morgan, J. J. (1981) "Kinetics for the Aquatic Environment," *Environ. Sci. Technol.*, **15**, 11, 1306–1313.

Saunier, B., and Selleck, R. (1979) "The Kinetics of Breakpoint Chlorination in Continuous Flow Systems," *J. AWWA,* **71**, 3, 164–172.

Smith, J. M. (1981) *Chemical Engineering Kinetics,* 3rd ed., McGraw-Hill, New York.

Stanley, S. J., and Smith, D. W. (1995) "Measurement of Turbulent Flow in Standard Jar Test Apparatus," *J. Environ. Eng.*, **121**, 12, 902–910.

Stumm, W. (1956) "Chemical Aspects of Water Ozonation," *Schw. Z. Hydrol.*, **18**, 201.

Stumm, W., and Morgan, J. J. (1996) *Aquatic Chemistry,* 3rd ed., John Wiley & Sons, New York.

Tchobanoglous, G., Burton, F. L., and Stensel, H. D. (2003) *Wastewater Engineering: Treatment and Reuse,* 4th ed., McGraw-Hill, Boston, MA.

Thomas, H. A., Jr., and Mckee, J. E. (1944) "Longitudinal Mixing in Aeration Tanks," *Sewage Works,* **16**, 42–55.

Toor, H. (1969) "Turbulent Mixing with and without Chemical Reactions," *Ind. Eng. Chem., Fundam.*, **8**, 655–659.

Trussell, R. R., Selleck, R. E., and Chao, J. L. (1979) "Discussion of Hydraulic Analysis of Model Treatment Units by F. Hart and S. Gupta," *ASCE*, **105**, EE4, 796–798.

Wehner, J. F., and Wilhelm, R. F. (1958) "Boundary Conditions of Flow Reactor," *Chem. Eng. Sci.*, **6**, 89–93.

Wei, I., and Morris, J. (1974) Dynamics of Breakpoint Chlorination, Chap. 1 in A. Rubin (ed.), *Chemistry of Water Supply Treatment and Distribution*, Ann Arbor Publishers, Ann Arbor, MI.

7 Principles of Mass Transfer

Stantec's Water Treatment: Principles and Design, Updated Third Edition. John C. Crittenden, R. Rhodes Trussell, David W. Hand, Kerry J. Howe, and George Tchobanoglous.

Terminology for Mass Transfer

Term	Definition
Absorption	Process in which a solute is transferred from one bulk phase and is homogeneously spread throughout another bulk phase (as opposed to collecting at the interface between phases. See Adsorption).
Adsorbent	Solid phase onto which a solute accumulates during the process of adsorption.
Adsorption	Process in which a solute is transferred from one bulk phase and accumulates at the surface of another phase (such as a solid surface), resulting in an increased concentration of molecules in the immediate vicinity of the surface (in contrast to absorption).
Air stripping	Transfer of volatile components from water to air.
Batch system	System with no flow in or out during the mass transfer operation. Typically, two phases are brought together, mass transfer is allowed to proceed until nearly at equilibrium, and then the phases are separated.
Brownian motion	Random motion of solute molecules or particles caused by collisions with solvent molecules.
Co-current flow	Process in which two phases (e.g., liquid and gas, water and powdered activated carbon) contact each other with their mass flow in the same direction.

Term	Definition
Continuous contact operation	A process in which two phases are in continuous contact with each other from the inlet to the outlet of the system, with continuously changing concentrations in each phase as a function of position (e.g., a column filled with adsorbent, a countercurrent packed tower, etc.).
Countercurrent flow	Process in which two phases (e.g., water and air) contact each other with flow in opposite directions, with contact either in stages or continuously.
Cross flow	Process in which two phases (e.g., water and air) contact each other with flows perpendicular to each other.
Desorption	Mass transfer process involving the removal of substances from an adsorbent surface.
Diffusion	Mass transfer process in which solute molecules or small particles are transported from a region of high concentration to a region of lower concentration as a result of Brownian motion.
Diffusion coefficient	Parameter that relates proportionality of the flux of a solute in a solvent to the concentration gradient. Frequently used synonymously with diffusivity.
Diffusivity	Used as a synonym for diffusion coefficient.
Extracting phase	Phase to which compounds are transferred in water treatment (e.g., gas phase for stripping, liquid phase for absorption, solid phase for adsorption or ion exchange).
Flow-through system	System in which one or both phases flow continuously through it during the mass transfer operation.
Fluid–fluid process	Mass transfer process in which fluid is in contact with another fluid (e.g., air and water in a stripping tower).
Fluid–solid process	Mass transfer process in which fluid is in contact with a solid (e.g., packing), operated as fixed or fluidized beds.
Mass transfer	Transport of components (molecules, particles, etc.) from one location to another (typically from one phase to another).

Term	Definition
Staged operation	Mass transfer system in which two phases contact each other in discrete steps. Typically, the two phases are completely mixed with each other in each stage, and then separated before being sent to the next stage, where they are remixed (often with the two contacting phases traveling in different directions).
Solute	Dissolved substance.
Solvent	Liquid in which other compounds (solutes) are dissolved.
Sorption	General term for the many phenomena commonly included under the terms adsorption and absorption when the nature of the phenomenon involved is unknown or indefinite.
Stripping	Removal of a component from one phase by transfer to another (such as air stripping, see above)

Several water treatment processes involve the transfer of material from one phase to another (i.e., from liquid to gas, or liquid to solid). Aeration and air stripping (Chap. 14), adsorption, (Chap. 15), ion exchange (Chap. 16), and reverse osmosis (Chap. 17) are all processes that involve mass transfer between phases. In these processes, the contaminant removal efficiency, the rate of separation, and/or the size of the equipment can be governed by the rate of mass transfer.

Mass transfer, in the broadest possible definition, is the movement of matter from one location to another, and the rate at which this occurs can be the governing factor in treatment processes. Consider a contaminant removal process that relies on an instantaneous reaction at a surface. Since the reaction is instantaneous, the rate at which the contaminant is degraded is controlled not by the rate of the reaction but by the rate at which the reactants can be transported to the surface. Such a process is called *mass transfer limited.*

Mass transfer is a complex topic. Books have been written about the topic and the chemical engineering curriculum at many universities includes an entire course in mass transfer. This chapter focuses on key principles that are relevant to environmental engineering and water treatment processes. Topics discussed in this chapter include an introduction to mass transfer, molecular diffusion and diffusion coefficients, models and correlations for mass transfer coefficients, operating diagrams, and mass transfer across a gas–liquid interface with and without chemical reactions.

7-1 Introduction to Mass Transfer

To introduce the subject of mass transfer, the concept of flux and the fundamental equation for mass transfer are explained in this section.

Concept of Flux

In mass transfer operations, the movement of matter is measured as flux. Mass flux is defined as the amount of material that flows through a unit area per unit time:

$$J_A = \frac{m}{At} \tag{7-1}$$

where J_A = mass flux of solute A across an interface, mg/m²·s
m = mass of solute A, mg
A = area perpendicular to the direction of flow, m²
t = time, s

Because flux is defined per unit area, it is an intensive property (intensive properties, like concentration or temperature, do not depend on the size of the system). Thus, for two systems with the same mass flux, the system with the larger amount of area will have more mass transfer. Mass flow is the product of the flux and the area:

$$M_A = J_A A \tag{7-2}$$

where M_A = mass flow of solute A, mg/s

As will be seen later in this chapter, increasing the surface area is a key method for increasing the rate of mass transfer (and hence, increasing the efficiency of a separation process that relies on mass transfer).

In some cases (principally membrane processes), the material moving across the interface is measured in units of volume instead of mass, and the corresponding flux is called a volumetric flux instead of a mass flux. An example of units for a volumetric flux is L/m²·s. Other situations are best described with molar units, where the units of molar flux are mol/m²·s. Molar fluxes can be converted to mass fluxes by multiplying by the molecular weight.

Fundamental Equation for Mass Transfer

Mass transfer occurs in response to a driving force. Forces that can move matter include gravity, magnetism, electrical potential, pressure, and others. In each case, the flux of material is proportional to the driving force.

In environmental engineering, the driving force of interest is a concentration gradient or, in more general terms, a gradient in chemical potential, or Gibbs energy. When a concentration gradient is present between two locations within a single phase, matter will flow from the region of higher concentration to the region of lower concentration at a rate that is proportional to the difference between the two concentrations, as given by the following equation:

$$J_A = k_f(\Delta C_A) \tag{7-3}$$

where J_A = mass flux of component A, g/m²·s
k_f = mass transfer coefficient, m/s
ΔC_A = difference in concentration of component A, mg/L

Equation 7-3 has only two components (the mass transfer coefficient and the concentration gradient), and while this equation seems simple, it has profound implications for many treatment processes. The bulk of the rest of this chapter is devoted to the examination of variations of this equation. The next four sections are devoted to development of the mass transfer coefficient and models that describe mass transfer. Following that, Sect. 7-6 will explore how operating diagrams can be used to describe the concentration gradient, and the last two sections describe mass transfer across a gas–liquid interface.

7-2 Molecular Diffusion

In the previous section, it was noted that mass flux is the product of a mass transfer coefficient and a driving force (see Eq. 7-3). A special case of mass transfer is *molecular diffusion,* in which solute molecules or particles flow from a region of higher concentration to a region of lower concentration solely due to kinetic energy of the solution molecules, that is, when no external forces are present to cause fluid movement. Molecular diffusion is a fundamental concept in many mass transfer problems. Although mass transfer coefficients are often determined using empirical correlations, the correlations are based on models of mass transfer that in turn depend on molecular diffusion at some level, and the diffusion coefficient will be a required parameter. As a result, an understanding of molecular diffusion is a necessary part of an understanding of mass transfer. Several important concepts related to molecular diffusion, including Brownian motion, Fick's first and second laws, and the Stokes–Einstein equation, are described in this section.

Brownian Motion

Brownian motion is the random motion of a particle or solute molecule due to the internal energy of the molecules in the fluid. As a result of this internal thermal energy, all molecules are in constant motion. A solute molecule or small particle suspended in a gas or liquid phase will be bombarded on all sides by the movement of the surrounding gas or liquid molecules. The random collisions cause unequal forces that cause the

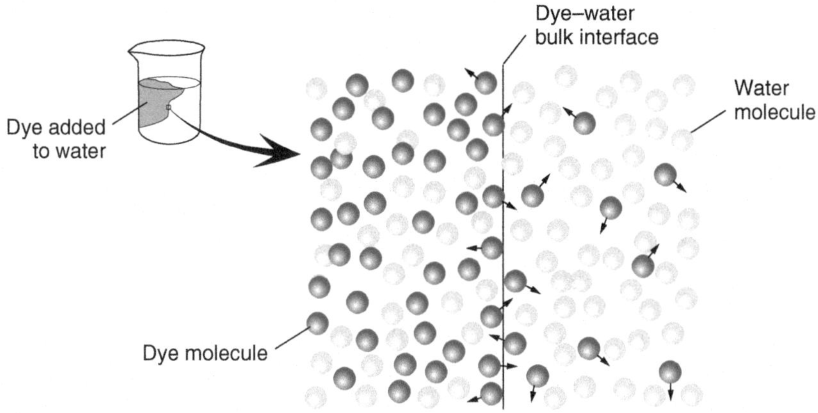

Figure 7-1
Mechanism by which Brownian motion leads to diffusion. The left side has about 4 times as many dye molecules, consequently about 4 times as many pass the interface from left to right compared to the number passing in the other direction.

solute molecule to move in random directions. The random motion caused by these collisions is called Brownian motion after Robert Brown, who described it (Brown, 1827).

In a completely quiescent fluid, molecular diffusion by Brownian motion will cause matter to flow from regions of high concentration to regions of low concentration. If Brownian motion is strictly random, how does it result in the movement of matter in a specific direction defined by the concentration gradient? That question can be answered by considering the probability associated with the movement of groups of molecules. Consider a beaker containing water in which one drop of a blue dye has been placed. Molecules, both water molecules and dye molecules, are randomly moving in all directions. An imaginary boundary in the solution, as shown on Fig. 7-1, has a greater concentration of dye molecules on one side than the other. In response to completely random movement, the rate at which dye molecules cross the boundary in each direction is proportional to the number of dye molecules on each side; that is, the more dye molecules present, the more that can randomly cross the boundary from that direction. The net result is a bulk movement from concentrated regions to dilute ones. Net movement of dye molecules across any particular interface ceases when the concentration is the same on both sides. In this way, molecular diffusion stops (although Brownian motion continues) when the dye is uniformly distributed throughout the beaker; that is, the concentration is the same everywhere. When the concentration is the same everywhere, the solution in the beaker has reached equilibrium.

Fick's First Law

With Brownian motion as a foundation, molecular diffusion can be described by Fick's first law (Fick, 1855):

$$J_A = -D_{AB}\frac{dC_A}{dz} \tag{7-4}$$

where J_A = mass flux of component A due to diffusion, $mg/m^2 \cdot s$
D_{AB} = diffusion coefficient of component A in solvent B, m^2/s
C_A = concentration of component A, mg/L
z = distance in direction of concentration gradient, m

The term dC_A/dz is the concentration gradient, that is, the change in concentration per unit change in distance. The negative sign in Fick's first law arises because material flows from regions of high concentration to low concentration; thus, positive flux is in the direction of a negative concentration gradient.

The diffusion coefficient describes the proportionality between a measured concentration gradient and the measured flux of material. Typical values of diffusion coefficients for solutes in gases and liquids are as follows:

$$\text{Liquids:} \sim 10^{-10} \text{ to } 10^{-9}\ \text{m}^2/\text{s}\ (10^{-6} \text{ to } 10^{-5}\ \text{cm}^2/\text{s}).$$

$$\text{Gases:} \sim 10^{-6} \text{ to } 10^{-5}\ \text{m}^2/\text{s}\ (10^{-2} \text{ to } 10^{-1}\ \text{cm}^2/\text{s}).$$

Diffusion in the Presence of Fluid Flow

Strictly speaking, Fick's first law describes the flux of component A with respect to the centroid of the diffusing mass of solute. In other words, Fick's first law describes the rate of diffusion from a relative point of view; if the fluid is moving, the mass transfer due to diffusion is superimposed on top of, or in addition to, mass transfer due to the movement of the fluid.

The mass flow of component A due strictly to advection (in the absence of diffusion) may be written as

$$M_A = QC_A \tag{7-5}$$

where M_A = mass flow of solute A due to advection, mg/s
Q = flow rate of fluid, m^3/s

In terms of flux, the mass flow is divided by the perpendicular area:

$$J_A = \frac{QC_A}{A} = v(C_A) \tag{7-6}$$

where J_a = mass flux of component A due to advection, $mg/m^2 \cdot s$
A = cross-sectional area perpendicular to direction of flow, m^2
v = fluid velocity in direction of concentration gradient, where $v = Q/A$

Consequently, when matter is being transported by both fluid flow and diffusion, Eqs. 7-2, 7-4, 7-5, and 7-6 can be combined to define the net mass flow and mass flux as follows:

$$M_A = QC_A - D_{AB}\frac{dC_A}{dz}A \tag{7-7}$$

and

$$J_A = v(C_A) - D_{AB}\frac{dC_A}{dz} \tag{7-8}$$

Diffusion in Fixed and Relative Frames of Reference

The governing equations for unit processes are often developed by writing mass balance expressions around a control volume using a fixed point of view (stationary frame of reference). It is useful, therefore, to examine the difference between the mathematical expression for diffusion that is defined from a relative reference frame (flux = J) and from a stationary reference frame (flux = N). The expression for the molar flux of component A in solvent B can be written as a fraction of total molar flux, where the total molar flux is the sum of the fluxes of components A and B:

$$N_A = x_A N_{TOT} = x_A(N_A + N_B) \tag{7-9}$$

where N_A = molar flux of component A relative to stationary frame of reference, mol/m^2·s
N_B = molar flux of solvent B relative to stationary frame ofreference, mol/m^2·s
N_{TOT} = total molar flux ($N_A + N_B$), mol/m^2 · s
x_A = mole fraction of A in solution, mol/mol

When matter is transported by both fluid flow and diffusion, the overall flux from a stationary reference frame is the sum of fluxes described in Eqs. 7-4 and 7-9:

$$N_A = x_A(N_A + N_B) - D_{AB}\frac{dC_A}{dz} \tag{7-10}$$

where D_{AB} = diffusion coefficient of component A in solvent B, m^2/s
C_A = molar concentration of component A, mol/L
z = position in direction of flow and diffusion flux (or in direction of concentration gradient), m

In Eq. 7-10, the first term on the right side describes the molar flux of A due to the movement of the fluid, and the second term describes the molar flux of A due to diffusion, superimposed on the movement of the fluid.

For the case of no advective flow of the solvent ($N_B = 0$), Eq. 7-10 can be algebraically rearranged to yield the expression

$$N_A = \frac{1}{1 - x_A}\left(-D_{AB}\frac{dC_A}{dz}\right) \tag{7-11}$$

where N_A = molar flux of component A relative to stationary frame of reference, mol/m^2·s

An example of a situation where there is advective flow of the solution but no advective flow of the solvent is when a solute evaporates from a surface.

The solute evaporates and moves away from the surface, and because the solute is moving and the solute is a component of the solution, the solution can be seen as moving. However, the solvent (in this case, the air) is not moving toward the surface.

For many environmental applications, particularly in aqueous solutions, x_A is very small. For example, the aqueous solubilities of chloroform and oxygen are about 9.3 g/L and 9.3 mg/L at 20°C, respectively; consequently, the largest mole fractions that can be found in water are 0.0014 and 5.23×10^{-6}, respectively. In these cases, the $1/(1 - x_A)$ is negligible and $J_A = N_A$, that is, molar flux due to diffusion is the same regardless of whether a stationary or relative frame of reference is used. Fick's first law (Eq. 7-4) is also valid when the sum of the fluxes N_A and N_B are equal to zero, as in a case where the diffusion of species A is countered by the diffusion of B (equal molar counterdiffusion). For highly miscible solvents in water or VOCs in gases, however, it is advisable to examine whether the $1/(1 - x_A)$ factor is important. These cases are rare, and for most applications throughout the remainder of the book, Fick's first law (Eq. 7-4) is applied directly even though a stationary frame of reference is being used.

Fick's Second Law

Fick's first law describes diffusion when the concentration gradient is constant. Fick's second law describes the rate of change of concentration when the diffusion into a control volume is different from the diffusion leaving a control volume. Fick's second law can be derived by a mass balance on a differential element with volume $a\Delta z$, in which the only mass transport is due to diffusion:

$$[\text{accum}] = [\text{mass in}] - [\text{mass out}] \tag{7-12}$$

$$V\frac{dC}{dt} = J_{A,z}A - J_{A,z+\Delta z}A \tag{7-13}$$

where V = volume, m^3
C_A = concentration of component A, mg/L
t = time, s
$J_{A,z}$ = flux of component A entering the control volume, $mg/m^2.s$
$J_{A,z+\Delta z}$ = flux of component A leaving the control volume, $mg/m^2.s$
A = cross-sectional area of control volume, m^2

Substituting Eq. 7-4 and replacing the volume of the control volume with the differential element volume $A\Delta z$ results in

$$A\Delta z\frac{\partial C_A}{\partial t} = -D_{AB}\frac{\partial C_{A,z}}{\partial z}A + D_{AB}\frac{\partial C_{A,z+\Delta z}}{\partial z}A \tag{7-14}$$

where Δz = length of differential element, m^3
D_{AB} = diffusion coefficient of component A in solvent B, m^2/s

The partial derivative arises because Eq. 7-14 contains derivatives in both time and distance. Dividing all terms by the area and rearranging yields

$$\frac{\partial C_A}{\partial t} = D_{AB}\left(\frac{\dfrac{\partial C_{A,z+\Delta z}}{\partial z} - \dfrac{\partial C_{A,z}}{\partial z}}{\Delta z}\right) \tag{7-15}$$

Taking the limit of the term in parentheses as $\Delta z \to 0$ results in Fick's second law:

$$\frac{\partial C_A}{\partial t} = D_{AB}\frac{\partial^2 C_A}{\partial z^2} \tag{7-16}$$

Stokes–Einstein Equation

Based on the principle that diffusion is caused by Brownian motion, and Brownian motion is caused by collisions with the solvent molecules, it ought to be possible to derive a theoretical value for the diffusion coefficient from the kinetic theory of matter. Albert Einstein derived this relationship in papers published in 1905 and 1908, and the derivation is explained in Laidler and Meiser (1999). As noted earlier, movement of a solute molecule or particle by Brownian motion is random in all directions, so net average distance over a period of time would be zero (movement in the x direction would be balanced by movement in the negative x direction). However, evaluating distance traveled by the molecule as the mean square displacement, $\overline{x^2}$, solves this problem. Relating the diffusion coefficient to kinetic energy involves two components; first, relating the mean square distance traveled by a molecule (or particle) during diffusion to the diffusion coefficient defined by Fick's laws, and second, determining the mean square distance traveled by a solute molecule as a result of collisions with solvent molecules. Equating the two relationships for mean square distance traveled provides a relationship between kinetic energy and the diffusion coefficient.

Einstein solved the first part of this relationship using arguments from statistical mechanics. The derivation is beyond the scope of this text, but the relationship established is

$$\overline{x^2} = 2Dt \tag{7-17}$$

where $\overline{x^2}$ = mean square displacement of solute molecule in x direction, m²
D = diffusion coefficient, m²/s
t = time, s

The second portion, relating distance traveled to kinetic energy, can be derived as follows. As shown on Fig. 7-2, for a perfect elastic collision, the solvent molecule hits the solute molecule and moves in the opposite

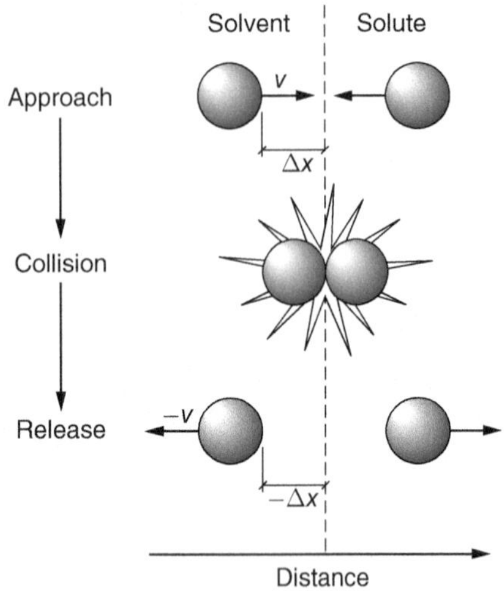

Figure 7-2
Schematic of solvent molecule collision with solute molecule.

direction after the collision. For one dimension (in the direction of the concentration gradient), the average collision force exerted due to momentum change can be calculated using the following equation:

$$F_c = m\frac{v-(-v)}{t} = m\frac{2v}{t} \tag{7-18}$$

where F_c = average collision force exerted due to momentum change, N
m = mass of solute molecule, kg
v = average velocity of solvent molecule in one dimension, m/s
t = time interval between collisions, s

The kinetic theory of gases states that the kinetic energy of gas molecules is

$$\text{KE} = \tfrac{3}{2}kT \tag{7-19}$$

where KE = average kinetic energy, N·m
k = Boltzmann's constant, 1.381×10^{-23} J/K
T = absolute temperature, K (273 + °C)

Kinetic energy is isotropic and can be partitioned equally in the three coordinate directions (i.e., kinetic energy in the x, y, and z directions are each $\frac{1}{2}kT$), so the kinetic energy in one direction is given by

$$\text{KE} = \tfrac{1}{2}mv^2 = \tfrac{1}{2}kT \tag{7-20}$$

Noting that $v = x/t$, Eq. 7-20 can be rearranged algebraically and substituted into Eq. 7-18 as follows:

$$mv = \frac{kT}{v} = \frac{kTt}{x} \tag{7-21}$$

$$F_c = \frac{2mv}{t} = \frac{2kT}{x} \tag{7-22}$$

After a collision, the solute molecule briefly accelerates due to unbalanced forces in the direction of the collision, but quickly reaches a steady-state velocity due to drag forces, which are a function of velocity. In 1851, Stokes deduced that the drag force of a solute molecule moving through a continuous fluid would be subjected to a drag force equal to

$$F_d = 3\pi\mu_l d_m v = 3\pi\mu_l d_m \frac{x}{t} \tag{7-23}$$

where F_d = drag force, N
μ_l = viscosity of liquid, kg/m·s

d_m = solute molecule diameter, m
x = average distance traveled by particle in time t

(It should be noted that Eq. 7-23 is exactly the same as Eq.10-4 when the drag coefficient $Cd = 24/\text{Re}$). The velocity of a diffusing solute molecule or particle is constant when the collision forces equal the drag forces, and the average distance a molecule moves due to Brownian motion during a period of time can be determined by equating these equations. Equating the collision force (Eq. 7-22) and drag force (Eq. 7-23) and solving for the mean square distance traveled yields

$$\overline{x^2} = \frac{2kT}{3\pi\mu_l d_m} t \tag{7-24}$$

The mean square distance has now been related to the kinetic energy of molecules and the diffusion coefficient as defined in Fick's laws. Equating these two relationships (Eqs. 7-17 and 7-24) yields the Stokes-Einstein equation:

$$D_l = \frac{\overline{x^2}}{2t} = \frac{kT}{3\pi\mu_l d_m} \tag{7-25}$$

where D_l = liquid-phase diffusion coefficient, m^2/s

Equation 7-25 was derived from the kinetic theory of gases and does not strictly apply to liquids. Nonetheless, Eq. 7-25 can be used to obtain a good prediction of the liquid diffusion coefficient for large molecules and particles. Furthermore, it predicts that diffusion increases with temperature and decreases with viscosity and molecular size, which have been observed experimentally.

For gases, the viscous force is not the same as it is for liquids because the fluid does not appear continuous for small particles or solutes. The size of a molecule or atom is on the order of 0.1 to 1 nm, and the average distance that a solute molecule travels before it collides with another molecule is on the order of 100 nm in a gas at 1 atm and room temperature. This effect is considered using the Cunningham correction factor, as shown in the following equation (Seinfeld and Pandis, 1998):

$$D_g = \frac{kTC_c}{3\pi\mu_g d_m} \tag{7-26}$$

where D_g = gas-phase diffusion coefficient, m^2/s
C_c = Cunningham correction factor
μ_g = viscosity of gas, kg/m·s

The Cunningham correction factor is determined from the expression

$$C_c = 1 + \frac{\lambda}{r}\left[1.257 + 0.4\exp\left(-\frac{r}{\lambda}\right)\right] \tag{7-27}$$

where λ = mean free path, m
r = particle radius, m

When λ/r is large, the following expression is obtained:

$$D_g = \frac{kT}{3\pi\mu_g d_m}\left(1+\frac{\lambda}{r}\right) \tag{7-28}$$

The Cunningham correction factor gives rise to the collision function that appears in correlations for gas-phase diffusion coefficients.

The Brownian velocity is given by the following expression:

$$v_b = \frac{2D_l}{\bar{x}} \tag{7-29}$$

where v_b = Brownian velocity, m/s

The value of $\bar{x}$ that can be used to calculate the Brownian velocity is approximately 0.5 nm for liquids and 100 nm for gases at room temperature and pressure. Equation 7-29 can be used to compare the distance a particle moves due to Brownian motion in comparison to other forces, such as settling due to gravity.

7-3 Sources for Diffusion Coefficients

The diffusion coefficient is an essential parameter for calculating the rate of mass transfer in a wide variety of situations. Diffusion coefficients can be obtained from:

1. Laboratory measurements
2. Reference books or published literature on previous laboratory measurements
3. Models and empirical correlations

Diffusion coefficients can be determined experimentally in the laboratory and procedures for doing so are available in the literature (Robinson and Stokes, 1959; Malik and Hayduk, 1968). When diffusion coefficients are needed for designing treatment systems or predicting process performance, however, it is more practical to obtain existing values or use correlations than to measure them. Measured diffusion coefficients of some common solutes found in water treatment are presented in Table 7-1. Diffusion coefficients for other constituents are available in the literature and reference books, such as Robinson and Stokes (1959), Marrero and Mason (1972), Poling et al. (2001), and CRC (2003).

For many compounds of interest, however, measured values of diffusion coefficients are not readily available. In addition, diffusion varies with temperature, and coefficients in reference books are often not at the temperature desired for the process application. In these cases, it is possible to estimate the diffusion coefficient based on chemical properties and

Table 7-1
Measured values of molecular diffusion coefficients in water (at 25°C, unless noted otherwise)

Constituent	D_l, m^2/s	Constituent	D_l, m^2/s
Neutral species		Strong electrolytes (0.001 M)	
Acetic acid	1.29×10^{-9}	$BaCl_2$	1.32×10^{-9}
Acetone	1.28×10^{-9}	$CaCl_2$	1.25×10^{-9}
Benzene (20°C)	1.02×10^{-9}	KCl	1.96×10^{-9}
Carbon dioxide	2.00×10^{-9}	KNO_3	1.90×10^{-9}
Ethanol	1.24×10^{-9}	NaCl	1.58×10^{-9}
Ethylbenzene (20°C)	0.81×10^{-9}	Na_2SO_4	1.18×10^{-9}
Glycerol	1.06×10^{-9}	$MgCl_2$	1.19×10^{-9}
Methane	1.49×10^{-9}	$MgSO_4$	0.77×10^{-9}
Phenol (20°C)	0.89×10^{-9}	$SrCl_2$	1.27×10^{-9}
Propylene	1.44×10^{-9}		
Sucrose	0.52×10^{-9}		
Toluene (20°C)	0.85×10^{-9}		
Vinyl chloride	1.34×10^{-9}		

Source: Robinson and Stokes (1959), Poling et al. (2001), CRC (2003).

structure using various models and empirical correlations. For each class of compound, a variety of calculation methods are available (Lyman et al., 1990; Poling et al., 2001). Some of the most common correlations for diffusion coefficients are summarized in Table 7-2 and presented in more detail in the following sections. Because of the difficulty in finding measured values at the appropriate temperature, use of these correlations is the most common way of estimating diffusion coefficients for many applications.

Liquid-Phase Diffusion Coefficients for Large Molecules and Particles

The Stokes-Einstein equation (Eq. 7-25) relates the kinetic energy of molecules and the drag force of moving through a fluid to the diffusion coefficient. This equation can be used to calculate diffusion coefficients for large spherical molecules [molecular weight (MW) > 1000 Da] or particles in liquids, although values calculated with the Stokes–Einstein equation are slightly larger than measured values. Conversely, the Stokes–Einstein equation underestimates the diffusivity for small molecules with MW on the order of 100 Da. Substituting the parameters for water (at 20°C) into the Stokes–Einstein equation results in the expression

$$D_l = \frac{kT}{3\pi\mu_l d_m} = \frac{\left(1.381\times10^{-23}\,\mathrm{N\cdot m/K}\right)(293\ \mathrm{K})}{3\pi\left(1.00\times10^{-3}\,\mathrm{N\cdot s/m^2}\right)d_m} = \frac{4.29\times10^{-19}\,\mathrm{m^3/s}}{d_m} \tag{7-30}$$

Table 7-2
Models used for estimating molecular diffusion coefficients in liquids

Application	Model	Comments
Large spherical molecules or particles	Stokes–Einstein equation: $D_l = \dfrac{kT}{3\pi\mu_l d_m}$	Diffusion coefficients are slightly larger than measured values for large spherical molecules (MW > 1000 Da) or particles in liquids. Diffusivity for small molecules with MW on order of 100 Da is underestimated.
For nonelectrolytes and small molecules	Hayduk–Laudie correlation: $D_l = \dfrac{13.26 \times 10^{-9}}{(\mu_l)^{1.14}(V_b)^{0.589}}$	Hayduk and Laudie (1974); convert units for μl to cP and use Table 7-3 for V_b (cm³/mol), D_l is expressed in units of m²/s.
For electrolytes in absence of electric field	Nernst–Haskell equation: $D^{\circ}_{AB} = \dfrac{RT}{F^2}\left[\dfrac{1/n^{+} + 1/n^{-}}{1/\lambda^{\circ}_{+} + 1/\lambda^{0}_{-}}\right]$	Poling et al. (2001); use Table 7-4 for λ values.

where D_l = liquid-phase diffusion coefficient at 20°C, m²/s
k = Boltzmann constant, 1.381×10^{-23} J/K (N·m/K)
T = absolute temperature, K (273 + °C)
μ_l = liquid viscosity at 20°C, 1.00×10^{-3} N·s/m²
d_m = molecular diameter, m

Equation 7-30 is valid only for spherical molecules.

The validity of the Stokes–Einstein equation can be evaluated by comparing it against measured values of diffusion coefficients for proteins. Polson (1950) correlated experimental data for globular proteins and obtained the empirical expression:

$$D_l = 2.74 \times 10^{-9}(\text{MW})^{-1/3} \tag{7-31}$$

where D_l = liquid-phase diffusion coefficient, m²/s
MW = molecular weight, Da or g/mol

The molecular density of a globular protein is about 1.35 g/cm³, so a protein molecule with a MW of 50,000 Da has a diameter of about 4.9 nm (i.e., MW = number of molecules per mole × density of a molecule × volume of a molecule). Substituting the protein diameter into Eq. 7-30 results in a diffusion coefficient of 8.8×10^{-11} m²/s, compared to 7.4×10^{-11} m²/s calculated by Eq. 7-31. Thus, the value calculated with the Stokes–Einstein equation is about 15 percent higher than the value calculated by Eq. 7-31. The similarity of the two values is noteworthy considering that the Stokes–Einste equation is derived from first principles and Eq. 7-31 is an empirical expression.

Liquid-Phase Diffusion Coefficients for Small Neutral Molecules

The diffusivities of small uncharged molecules (such as synthetic organic chemicals) in water can be calculated using the Hayduk–Laudie correlation (Hayduk and Laudie, 1974), which is a revised version of a correlation developed by Othmer and Thakar (1953). The Hayduk–Laudie correlation is an empirical equation given by

$$D_l = \frac{13.26 \times 10^{-9}}{(\mu_l)^{1.14} (V_b)^{0.589}} \tag{7-32}$$

where D_l = liquid-phase diffusion coefficient of solute, m^2/s
μ_l = viscosity of water, cP (1 cP = 10^{-3} kg/m·s)
V_b = molar volume of solute at normal boiling point, cm^3/mol

Because the Hayduk–Laudie correlation was developed as a regression of experimental data and is not dimensionally consistent, it is important to use the units given for the equation. The molar volume at the normal boiling point, V_b, can be estimated using the LeBas method (LeBas, 1915). The atomic volumes for different elements, mixtures, and functional groups for use in calculation of molar volume at the normal boiling point via the LeBas method are presented in Table 7-3. Contributions of the various functional groups are added together along with deductions for certain ring structures. Calculation of the diffusion coefficient of a small neutral molecule using the Hayduk–Laudie correlation is illustrated in Example 7-1.

Liquid-Phase Diffusion Coefficients for Electrolytes

Electroneutrality requires that positive and negative ions migrate together, so diffusion coefficients are calculated for electrolytes (solutions of charged ions) instead of being calculated for each ion individually. As an example, the values of diffusion coefficients in Table 7-1 demonstrate that sodium and magnesium each diffuse faster when the counterion is chloride than when it is sulfate. In the absence of an electric field, diffusion of ions will generate an electric current in a solution. Conversely, the current through a unit area that results from applying an electric field for a given electrolyte concentration is known as the equivalent conductance. Thus, liquid-phase diffusion coefficients of electrolytes in the absence of an electric field are related to the equivalent conductance and can be calculated using the Nernst–Haskell equation:

$$\overset{\circ}{D_l} = \frac{RT}{F^2}\left(\frac{1/n^+ + 1/n^-}{1/\overset{\circ}{\lambda}_+ + 1/\overset{\circ}{\lambda}_-}\right) \tag{7-33}$$

where $\overset{\circ}{D_l}$ = liquid-phase diffusion coefficient at infinite dilution, cm^2/s
R = universal gas constant, 8.314J/mol·K
T = absolute temperature, K (273 + °C)

Table 7-3
Atomic volumes for use in computing molar volumes at normal boiling point with LeBas method

Element, Mixture, or Functional group	Atomic Volume, cm^3/mol	Circumstance	Element, Mixture, or Functional group	Atomic Volume, cm^3/mol	Circumstance
Air	29.9		Oxygen	7.4	Doubly bond, as carbonyl oxygen
Antimony	34.2			7.4	In aldehydes or ketones
Arsenic	30.5			9.1	In methyl esters
Bismuth	48.0			9.9	In methyl ethers
Bromine	27.0			11.0	In higher ethers and esters
Carbon	14.8			12.0	In acids
Chlorine	21.6	Terminal as in R—Cl		8.3	In union with S, P, or N
		Medial as in R—CHCl—R	Phosphorus	27.0	
Chromium	27.4		Silicon	32.0	
Fluorine	8.7		Sulfur	25.6	
Germanium	34.5		Tin	42.3	
Hydrogen	3.7	In organic compound	Titanium	35.7	
	7.15	In hydrogen molecule	Vanadium	32.0	
Iodine	37.0		Water	18.8	
Lead	46.5–50.1		Zinc	20.4	
Mercury	19.0		Ring deductions	6.0	Three-membered ring
Nitrogen	15.6			8.5	Four-membered ring
	10.5	In primary amines		11.5	Five-membered ring
	12.0	In secondary amines		15	Six-membered ring
				30	Naphthalene ring
				47.5	Anthracene ring

Source: Adapted from LeBas (1915).

n^+ = cation valence
n^- = anion valence
F = Faraday's constant, 96,500 C/eq
$\overset{\circ}{\lambda}_+$ = limiting positive ionic conductance, $cm^2 \cdot S/eq$
$\overset{\circ}{\lambda}_-$ = limiting negative ionic conductance, $cm^2 \cdot S/eq$

Values for limiting positive ionic conductance at 25°C are tabulated in Table 7-4. Values at other temperatures are available in reference books such as Robinson and Stokes (1959). The limiting ionic conductance is related to electric current and electric field strength in an infinitely dilute

Example 7-1 Estimating the diffusion coefficient for small neutral molecules in water

Estimate the diffusion coefficients of the following contaminants found in a groundwater: (1) vinyl chloride at 25°C and (2) benzene at 20°C. Use Table 7-3 to find the contributions of the various functional groups to the molar volume. Density and viscosity of water are available in App. C [convert the given units of viscosity to g/cm·s, i.e., centipoise (cP)]. Compare the values to the measured values reported in Table 7-1.

Solution

1. Calculate the liquid-phase diffusion coefficient for vinyl chloride.
   a. Estimate the molar volume at the boiling point using the information in Table 7-3. The chemical formula for vinyl chloride is C_2H_3Cl. The contribution of each atom to the molar volume is

$$2C = 2\,(14.8) = 29.6\ \text{cm}^3/\text{mol}$$
$$3H = 3\,(3.7) = 11.1\ \text{cm}^3/\text{mol}$$
$$Cl = (21.6) = 21.6\ \text{cm}^3/\text{mol}$$

   The molar volume is determined by adding the contributions of each atom:

$$V_b = 29.6 + 11.1 + 21.6 = 62.3\ \text{cm}^3/\text{mol}$$

   b. Calculate the diffusion coefficient using Eq. 7-32. The viscosity of water at 25°C is 0.89×10^{-3}kg/m·s = 0.89 cP.

$$D_l = \frac{13.26\times10^{-9}}{(0.89\text{cP})^{1.14}(62.3\ \text{cm}^3/\text{mol})^{0.589}} = 1.33\times10^{-9}\text{m}^2/\text{s}$$

   c. Compare the calculated value to the measured value in Table 7-1.

$$\frac{1.34\times10^{-9} - 1.33\times10^{-9}}{1.34\times10^{-9}}\times100 = 1\%\ \text{error}$$

2. Calculate the liquid-phase diffusion coefficient for benzene.
   a. Benzene is an aromatic compound (6 carbon ring) with the chemical formula C_6H_6. The contribution of each atom to the molar volume using the information in Table 7-3 is

   $6C = 6\,(14.8) = 88.8\ \text{cm}^3/\text{mol}$ $6H = 6\,(3.7) = 22.2\ \text{cm}^3/\text{mol}$

   Six-member ring = $-15\ \text{cm}^3/\text{mol}$

   The molar volume is determined by adding the contributions of each atom:

$$V_b = 88.8 + 22.2 - 15 = 96\ \text{cm}^3/\text{mol}$$

b. Calculate the diffusion coefficient using Eq. 7-32. The viscosity of water at 20°C is 1.00×10^{-3} kg/m · s = 1.00 cP.

$$D_l = \frac{13.26\times10^{-9}}{(1.00\ \text{cP})^{1.14}(96\ \text{cm}^3/\text{mol})^{0.589}} = 0.90\times10^{-9}\ \text{m}^2/\text{s}$$

c. Compare the calculated value to the measured value in Table 7-1.

$$\frac{1.02\times10^{-9} - 0.90\times10^{-9}}{1.02\times10^{-9}}\times100 = 11\%\ \text{error}$$

Comment

The value estimated with the Hayduk–Laudie correlation is within 1 percent of the measured value for vinyl chloride and within 11 percent of the measured value for benzene. These results are typical; the Hayduk–Laudie correlation is often within 10 percent of measured values for many compounds (it should be noted that measured values by different researchers with different methods also vary). As a result of this level of accuracy, it is common to estimate liquid-phase diffusion coefficients with the Hayduk–Laudie correlation rather than obtaining measured values for the species of interest.

Table 7-4
Limiting ionic conductances in water at 25°C [$cm^2 \cdot S/eq$ or $(cm^2 \cdot C^2)/(J \cdot s \cdot eq)$]

Cation	Formula	λ_+°	Anion	Formula	λ_-°
Hydrogen	H^+	349.8	Hydroxide	OH^-	199.1
Lithium	Li^+	38.6	Fluoride	F^-	55.4
Sodium	Na^+	50.1	Chloride	Cl^-	76.4
Potassium	K^+	73.5	Bromide	Br^-	78.1
Rubidium	Rb^+	77.8	Iodide	I^-	76.8
Cesium	Cs^+	77.2	Bicarbonate	$HCO3^-$	44.5
Ammonium	$NH4^+$	73.5	Nitrate	$NO3^-$	71.5
Silver	Ag^+	61.9	Perchlorate	$ClO4^-$	67.3
Magnesium	Mg^{2+}	53.0	Bromate	BrO_3^-	55.7
Calcium	Ca^{2+}	59.5	Formate	$HCOO^-$	54.5
Strontium	Sr^{2+}	59.4	Acetate	$CH3COO^-$	40.9
Barium	Ba^{2+}	63.6	Chloroacetate	$ClCH2COO^-$	42.2
Copper	Cu^{2+}	53.6	Propionate	$CH3CH2COO^-$	35.8
Zinc	Zn^{2+}	52.8	Benzoate	$C6H5COO^-$	32.3
Lead	Pb^{2+}	69.5	Carbonate	CO_3^{2-}	69.3
Lanthanum	La^{3+}	69.7	Sulfate	SO_4^{2-}	80.0

Ref: Robinson and Stokes (1959).

solution. Conductance is measured in siemens (S, where 1 S = 1 A/V); current is measured in amperes (A, where 1 A = 1 C/s); and electric field strength is measured in volts/cm (V/cm, where 1 V = 1 J/C). Specifically, the equivalent conductance relates the current flow through an area (A/cm^2) to the electric field strength and the concentration ofions in solution (measured in equivalents, or "mole of charge"), as given by the equation

$$\text{Current flow} = \lambda_i \times \text{electric field strength} \times C_i \qquad (7\text{-}34)$$

where λ_i = equivalent conductance of electrolyte i, cm^2·S/eq
C_i = concentration of electrolyte i,eq/cm^3

The charge of 1 electron is 1.60×10^{-19} C so the charge of a mole of electrons is the Faraday constant, 96,500 C/eq. As shown in Table 7-4, small ions have a higher equivalent conductance because they migrate through water more rapidly in response to an imposed electric field. Consequently, their diffusion coefficient is higher than that of large ions. Calculation of the diffusion coefficient of electrolytes with the Nernst–Haskell equation is shown in Example 7-2.

Example 7-2 Estimating diffusion coefficients for electrolytes in water

Estimate the diffusion coefficient of $MgCl_2$ in a dilute aqueous solution at 25°C and compare it to the measured value in Table7-1.Use the information in Table 7-4 to find the limiting ionic conductances for the ions.

Solution

1. From Table 7-4, the limiting ionic conductances are 53.0 (cm^2·C^2)/(J·s·eq) for Mg^{2+} and 76.4 (cm^2·C^2)/(J·s·eq) for Cl^-.
2. Calculate the diffusion coefficients at infinite dilution using Eq. 7-33:

$$D_i^\circ = \frac{(8.314 \text{ J/mol·K})(298 \text{ K})}{(96{,}500 \text{ C/eq})^2} \times \left\{ \frac{\left(\frac{1}{2}+\frac{1}{1}\right)\frac{\text{mol}}{\text{eq}}}{\left(\frac{1}{53.0}+\frac{1}{76.4}\right)\frac{\text{J}\cdot\text{s.eq}}{\text{cm}^2\cdot\text{c}^2}} \right\}$$

$$= 1.25 \times 10^{-5} \text{ cm}^2/\text{s}$$

3. Compare the calculated diffusion coefficient to the measured value reported in Table 7-1.

$$\frac{1.25 \times 10^{-5} - 1.19 \times 10^{-5}}{1.19 \times 10^{-5}} \times 100 = 5\% \text{ error}$$

Comment

The value calculated with the Nernst–Haskell equation is the diffusion coefficient in an infinitely dilute solution and the measured value in Table 7-1 is for a 0.001 M solution, but the values are within 5 percent of each other.

In a process such as ion exchange, the movement of the ions by diffusion will generate an electric field that can exert an additional force that influences mass transfer. As a result, the mass transfer rate can be many times greater than would be calculated from Fick's law. The Nernst–Planck equation (not covered in this book) can be used to calculate the flux due to the combined forces of a concentration gradient and an electrical field.

Liquid-Phase Diffusion Coefficient for Oxygen

The liquid-phase diffusion coefficient of oxygen in water can be determined from a correlation that was obtained from a best fit of literature values (Holmén and Liss, 1984):

$$D_{l,O_2} = 10^{(A+B/T)}\left(1.0 \times 10^{-9}\right) \tag{7-35}$$

where D_{l,O_2} = liquid-phase diffusion coefficient of oxygen, m^2/s
A = fitting parameter, 3.15, unitless
B = fitting parameter, −831.0, unitless
T = absolute temperature, K (273 + °C)

Gas-Phase Diffusion Coefficients

The diffusion coefficient of an organic compound in the gas phase can be calculated using a variety of correlations (Lyman, et al. 1990). Consider the Wilke–Lee correlation (Wilke and Lee, 1955), which is a modification of the Hirschfelder–Bird–Spotz correlation (Hirschfelder et al., 1949):

$$D_g = \frac{\left(1.084 - 0.249\sqrt{1/M_A + 1/M_B}\right)\left(T^{1.5}\right)\sqrt{1/M_A + 1/M_B}}{P_l\left(r_{AB}\right)^2 f\left(kT/\varepsilon_{AB}\right)} \tag{7-36}$$

where D_g = gas-phase diffusion coefficient of organic compound A in stagnant gas B, cm^2/s
T = absolute temperature, K (273 + °C)
M_A, M_B = molecular weights of A and B, respectively, Da or g/mol
P_1 = absolute pressure, N/m^2
r_{AB} = molecular separation at collision, equal to $(r_A + r_B)/2$, nm
r_A = molecular separation at collision for component A, nm
r_B = molecular separation at collision for stagnant gas B, nm
ε_{AB} = energy of molecular attraction, equal to $\sqrt{\varepsilon_A \varepsilon_B}$, erg (1 erg = 10^{-7} J)

ε_A = energy of molecular attraction for component A, erg
ε_B = energy of molecular attraction for stagnant gas B, erg
k = Boltzmann constant, $1.381 \times 10^{-16}\ \text{g} \cdot \text{cm}^2/\text{s}^2 \cdot \text{K}$
$f(kT/\varepsilon_{AB})$ = collision function

The collision function originates from the Cunningham correction factor, discussed in Sec. 7-2. The values of r_A and ε_A can be estimated for each component from the following equations:

$$r_A = 1.18 V_{b,A}^{1/3} \quad (\text{in nm for } V_{b,A} \text{ in L/mol}) \tag{7-37}$$

$$\frac{\varepsilon_A}{k} = 1.21 T_{b,A} \tag{7-38}$$

where $V_{b,A}$ = molar volume of component A at normal boiling point, L/mol
$T_{b,A}$ = normal boiling point of component A, K

The diffusion coefficient of a substance when the stagnant gas B is air can be calculated by assuming that air behaves like a single substance with respect to molecular collisions. The required parameters for air are

$$r_B = 0.3711 \text{ nm} \tag{7-39}$$

$$\frac{\varepsilon_B}{k} = 78.6 \tag{7-40}$$

$$\frac{\varepsilon_{AB}}{k} = \sqrt{\frac{\varepsilon_A}{k} \times \frac{\varepsilon_B}{k}} \tag{7-41}$$

$$f\left(\frac{kT}{\varepsilon_{AB}}\right) = 10^{\xi} \tag{7-42}$$

$$\xi = \left(\begin{array}{l} -0.14329 - 0.48343(\text{ee}) + 0.1939(\text{ee})^2 + 0.13612(\text{ee})^3 \\ -0.20578(\text{ee})^4 + 0.083899(\text{ee})^5 - 0.011491(\text{ee})^6 \end{array}\right) \tag{7-43}$$

$$\text{ee} = \log_{10}\left(\frac{kT}{\varepsilon_{AB}}\right) \tag{7-44}$$

Calculation of gas-phase diffusion coefficients with the Wilke-Lee correlation is demonstrated in the following example.

Example 7-3 Estimating gas-phase diffusion coefficients

Calculate the gas-phase diffusion coefficient of trichloroethene (TCE) in air at 20°C at 1 atm. Given:

A = TCE B= air

M_A = 131.39 g/mol M_B = 29 g/mol
$V_{b,A}$ = 98.1 cm^3/mol = 0.0981 L/mol
$T_{b,A}$ = 87°C = (87 + 273) K = 360 k

Solution

1. Determine pressure and temperature in proper units:

$$T = (273 + 20°C)\ K = 293\ K$$
$$P_l = 1\ atm = 101{,}325\ N/m^2$$

2. Calculate r_{AB} using Eq. 7-37 for r_A and Eq. 7-39 for r_B.

$$r_A = 1.18\left(V_{b,A}\right)^{1/3} = 1.18(0.0981)^{1/3} = 0.544\ \text{nm}$$
$$r_B = 0.3711\ \text{nm}$$
$$r_{AB} = \tfrac{1}{2}\left(r_A + r_B\right) = \tfrac{1}{2}(0.544\ \text{nm} + 0.3711\ \text{nm}) = 0.458\ \text{nm}$$

3. Determine ε_{AB}/k by calculating ε_A/k with Eq. 7-38, ε_B/k with Eq. 7-40, and ε_{AB}/k with Eq. 7-41.

$$\frac{\varepsilon_A}{k} = 1.21 T_{b,A} = 1.21(360K) = 435.6$$
$$\frac{\varepsilon_B}{k} = 78.6$$
$$\frac{\varepsilon_{AB}}{k} = \sqrt{\frac{\varepsilon_A}{k} \times \frac{\varepsilon_B}{k}} = \sqrt{(435.6)(78.6)} = 185$$

4. Determine the collision function $f(k\tau/\varepsilon_{AB})$.

   a. Calculate $k\tau/\varepsilon_{AB}$:

$$\frac{kT}{\varepsilon_{AB}} = \frac{T}{\varepsilon_{AB}/k} = \frac{293}{185} = 1.58$$

   b. Calculate ee using Eq. 7-44:

$$ee = \log_{10}\left(\frac{kT}{\varepsilon_{AB}}\right) = \log_{10}(1.58) = 0.200$$

   c. Calculate ξ using Eq. 7-43:

$$\xi = \left\{\begin{array}{l} -0.14329 - [0.48343(0.200)] + \left[0.1939(0.200)^2\right] + \left[0.13612(0.200)^3\right] \\ -\left[0.20578(0.200)^4\right] + \left[0.083899(0.200)^5\right] - \left[0.011491(0.200)^6\right] \end{array}\right\}$$
$$= -0.231$$

d. Calculate $f(kT/\varepsilon_{AB})$ using Eq. 7-42:

$$f\left(\frac{kT}{\varepsilon_{AB}}\right) = 10^{\xi} = 10^{-0.231} = 0.587$$

5. Determine the gas-phase diffusion coefficient of TCE in air using Eq. 7-36:

$$D_g = \frac{\left(1.084 - 0.249\sqrt{1/M_A + 1/M_B}\right)\left(T^{1.5}\right)\sqrt{1/M_A + 1/M_B}}{P_t\left(r_{AB}\right)^2 f\left(kT/\varepsilon_{AB}\right)}$$

$$= \frac{(1.084 - 0.249\sqrt{1/131.39 + 1/29})(293)^{1.5}\sqrt{1/131.39 + 1/29}}{(101{,}325)(0.458)^2(0.587)}$$

$$= 8.52 \times 10^{-2}\,\text{cm}^2/\text{s}$$

7-4 Models for Mass Transfer at an Interface

In many common treatment processes, such as air stripping, adsorption, ion exchange, and reverse osmosis, mass transfer occurs at an interface. The interface is the phase boundary between the phase containing the solute or contaminant (typically the water) and the extracting phase (e.g., air or activated carbon). An understanding of mass transfer at an interface is essential to understanding the principles of these processes. This section describes three common models for how mass transfer occurs at an interface.

The mass that is transferred from one phase to another per unit time depends on the mass transfer coefficient, the driving force, and the surface area available, as was introduced in Sec. 7-1. The driving force is caused by contacting the contaminated phase with an extracting phase that does not contain the contaminant. When mass transfer occurs at an interface, the concentration gradient is given by the concentrations in the bulk solution and at the interface, as shown on Fig. 7-3 and in following expression:

$$J_A = k_f(C_b - C_s) \tag{7-45}$$

where J_a = mass flux of solute A to interface, mg/m²·s
k_f = mass transfer coefficient, m/s
C_s = concentration of solute A at interface, mg/L
C_b = concentration of solute A in bulk solution, mg/L

The mass transfer coefficient depends on the diffusion coefficient and the mass transfer boundary layer thickness δ, as shown on Fig. 7-3. As shown on

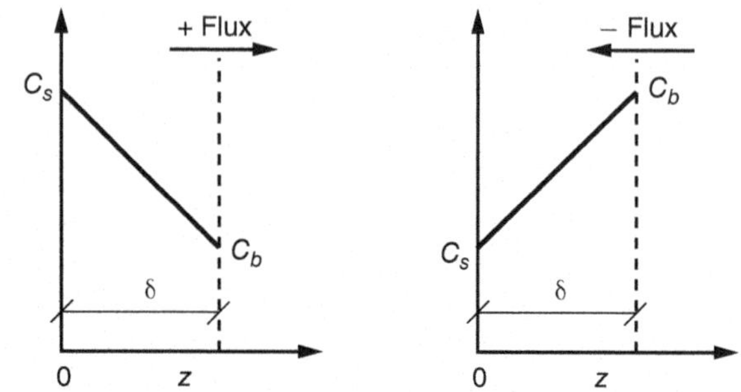

Figure 7-3
Hypothetical fluxes at interface at steady state.

Fig. 7-3, the direction of flux depends on the direction of the concentration gradient. Common models used to predict the mass transfer coefficient include (1) the film model, (2) the surface renewal or penetration model, and (3) the boundary layer model. Many investigators have taken the theoretical forms of the mass transfer correlations and modified them to fit data that were collected for a variety of geometries (e.g., sphere, cylinder, plate) and flow regimes (e.g., laminar, transition, and turbulent). Several of the common correlations used to determine mass transfer coefficients are given in Sec. 7-5.

Surface Area Available for Mass Transfer

To calculate the mass flow rate, the flux must be multiplied by the surface area (see Eq. 7-2). It is common to express the area of the interface between phases as a function of the contactor volume (e.g., the surface area of carbon grains is expressed as a function of the volume of the carbon bed). Thus, the mass flow rate is given by the expression

$$M_A = k_f a(C_b - C_s)V \tag{7-46}$$

where M_A = mass flow of solute A, mg/s
a = specific area, or surface area, available for mass transfer per unit volume of the contactor, m^2/m^3
V = contactor volume, m^3

The specific area is an important concept. For a given contactor volume, the mass transfer rate can increase linearly with an increase in specific area. Thus, designing a mass transfer device with a high specific area can result in a high rate of mass transfer in a small contactor. Mass transfer devices are often designed to have the highest possible specific area within the limitations imposed by hydraulic considerations. Increases in specific area often come at the expense of higher headloss. For example, in a packed bed of activated carbon it would be advantageous to use small carbon granules to increase the specific area, but the pressure drop would become too large and the cost of pumping water through the contactor would be high. In addition, the contactor would have to withstand the increased pressure.

Example 7-4 Calculating area available for mass transfer

Determine the specific area for the transport of a solute to granular activated carbon (GAC) particles in a carbon adsorber. The porosity (ε, fraction of void volume) of the carbon bed is 0.45 and the GAC particle diameter is d_p = 1 mm. Assume the surface of the GAC is like that of a smooth sphere.

Solution

$$a = \left(\frac{\text{surface area of particle}}{\text{volume of particle}}\right)\left(\frac{\text{volume of particles}}{\text{volume of contactor}}\right)$$

$$= \left(\frac{\pi d_p^2}{\left(\frac{1}{6}\right)\pi d_p^3}\right)(1-\varepsilon) = \frac{6(1-\varepsilon)}{d_p} = \frac{6(1-0.45)}{0.001\text{ m}} = 3300\text{ m}^2/\text{m}^3$$

Comment

The grain diameter is in the denominator, so decreasing the size will increase the specific area for the same amount of GAC in the contactor (decreasing the grain size to 0.1 mm would increase the specific area to 33,000 m^2/m^3, which would increase the rate of mass transfer by a factor of 10 for the same size contactor if diffusion from the bulk solution to the particle surface is the limiting rate). This action, however, would increase the headloss and make it more difficult to pass water through the contactor.

The relationship between grain size and specific area is demonstrated in Example 7-4.

Film Model

The film model is the most straightforward of the models that explain mass transfer at an interface. The system is considered to be composed of a well-mixed bulk solution (either gas or liquid), a stagnant film layer, and an interface to another phase (e.g., a solid surface), as shown on Fig. 7-3. As a result of the solution being well mixed, solutes are transported continually to the edge of the stagnant film layer, and no concentration gradients exist in the bulk solution. Mass transfer to the interface occurs when the concentration at the interface to the other phase is different than the concentration in the bulk solution, causing a concentration gradient across the film layer. Because this layer is quiescent, the sole mechanism for transport across this layer is molecular diffusion. Processes that occur at the actual interface (such as a chemical reaction or adsorption to the surface) are assumed to occur much faster than the rate of diffusion and, as a result, the rate of mass transfer is described by Fick's first law for diffusion across the film layer:

$$J_A = -D_f \frac{dC}{dz} = -\frac{D_f}{\delta}(C_s - C_b) = k_f (C_b - C_s) \tag{7-47}$$

where J_A = mass flux, mg/m^2· s
D_f = fluid-phase diffusion coefficient of solute A, m^2/s
k_f = fluid-phase mass transfer coefficient of solute A, m/s
δ = film thickness, as shown on Fig. 7-3, m
C = concentration, mg/L
z = distance in direction of mass transfer (or in direction of concentration gradient), m

In the film model, the mass transfer coefficient is explicitly related to the film thickness, as shown in the expression

$$k_f = \frac{D_f}{\delta} \tag{7-48}$$

The theoretical stagnant film thickness will vary from 10 to 100 μm for liquids and from 0.1 to 1 cm for stagnant gases. Unfortunately, there is no way to calculate the film thickness based on fluid mixing; consequently, the film model cannot be used to calculate the local mass transfer coefficient. Nevertheless, the film model is used frequently to develop a conceptual view of mass transfer across an interface and to illustrate the importance of diffusion in controlling the rate of mass transfer.

Penetration and Surface Renewal Models

According to the penetration model (Higbie, 1935) and the surface renewal model (Danckwerts, 1951, 1955), packets of water move up to the gas–water interface and transfer solute either from the gas to the water or from the water to the gas, and then the packets of fluid return to the bulk solution. The transport of fluid packets to and from the surface is shown on Fig. 7-4a and 7-4b. When a packet of water moves to the interface, the concentration of dissolved gases increases during aeration as shown on the bottom of Fig. 7-4c.

The essential difference between the penetration and surface renewal models is that for a penetration model a fixed residence time at the surface is assumed. For a surface renewal model, it is assumed that there is an equal probability for the fluid elements to move to and from the surface up to a certain residence time. It is likely that a surface renewal model is more plausible because fluid turbulence is thought to be responsible for transporting the fluid element to and from the surface. Consequently, the time that fluid packets remain on the surface is random and all residence times are equally plausible. The surface renewal model can be used to derive the theoretical basis for predicting mass transfer coefficients.

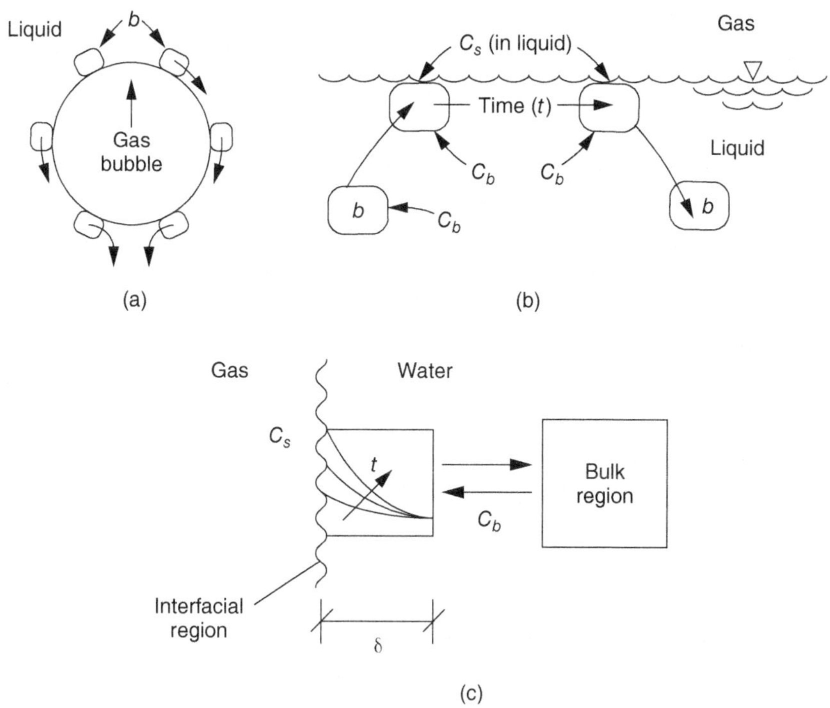

Figure 7-4
Schematic of penetration and surface renewal models for gas transfer: (a) gas bubble rising through liquid in which liquid packets *b* are in contact with the bubble for a time period *t* equal to the time it takes for the bubble to rise one diameter, (b) liquid packets *b* in turbulent eddy rising to liquid surface, as in an open channel, and remaining in contact with gas for a time period *t*, and (c) increase in concentration of a dissolved gas at liquid surface interface with time for both cases (a) and (b). (Adapted from Treybal, 1980).

Boundary Layer Models

When fluid passes a solid flat plate, a velocity gradient forms because the fluid velocity is assumed to be zero at the surface (no slip condition). Simultaneously, the solutes are transported to or from the surface, resulting in a concentration gradient between the concentration at the surface and the concentration in the bulk solution. The limit of the concentration gradient is not necessarily the same as the velocity gradient. The rate of mass transfer and relationship between the concentration and velocity gradients is related to conditions of the fluid flow. The concentration gradient shown on Fig. 7-5 is for the case where solute is transported from the surface of the plate to the bulk solution and laminar flow conditions exist. For laminar flow past a flat plate, the following theoretical mass transfer correlation can be derived:

$$\frac{k_{f(\text{avg})} L}{D_f} = 0.664\,\text{Re}^{1/2}\,\text{Sc}^{1/3} \tag{7-49}$$

where $k_{f(\text{avg})}$ = average fluid-phase mass transfer coefficient, m/s
L = length of channel, m
D_f = fluid-phase diffusion coefficient, m^2/s
Re = Reynolds number, dimensionless
Sc = Schmidt number, dimensionless

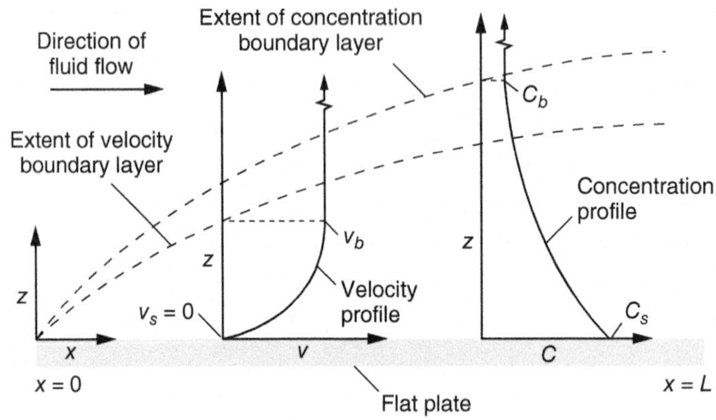

Figure 7-5
Boundary layer model diagram showing velocity and concentration profiles for laminar flow across flat plate.

The Reynolds and Schmidt numbers for flow past a flat plate are defined as follows:

$$\mathrm{Re} = \frac{\rho_f v L}{\mu_f} \tag{7-50}$$

$$\mathrm{Sc} = \frac{\mu_f}{\rho_f D_f} \tag{7-51}$$

where ρ_f = fluid-phase density, kg/m^3
v = velocity above boundary layer, m/s
μ_f = fluid-phase viscosity, kg/m · s

The parameter group on the left side of Eq. 7-49 is known as the Sherwood number:

$$\mathrm{Sh} = \frac{k_f L_c}{D_f} \tag{7-52}$$

where Sh = Sherwood number, dimensionless
L_c = characteristic length, m

The Sherwood number describes the relationship between the mass transfer coefficient and the diffusion coefficient. For instance, in the film model presented earlier the characteristic length scale is the stagnant film layer thickness (δ), mass transfer occurs only by molecular diffusion, and Sh = 1. Equation 7-49 expresses that the relationship between the mass transfer coefficient and the diffusion coefficient depends on the values of the Reynolds and Schmidt numbers; for laminar flow past a flat plate, the mass transfer coefficient will increase relative to the diffusion coefficient as the Reynolds and Schmidt numbers increase.

Numerous theoretical analyses have demonstrated that equations similar to Eq. 7-49 can be derived for many situations. A more general form of this equation is

$$\text{Sh} = A\text{G}r^c + B\,\text{Re}^a\,\text{Sc}^b \tag{7-53}$$

$$= \frac{k_f L_c}{D_f} = f(\text{Gr}, \text{Sc}, \text{Re}) \tag{7-54}$$

$$\text{Sc} = \frac{\mu_f}{\rho_f D_f} = \frac{\nu_f}{D_f} \tag{7-55}$$

$$\text{Re} = \frac{\rho_f v L_c}{\mu_f} = \frac{v L_c}{\nu_f} \tag{7-56}$$

$$\text{Gr} = \frac{g L_c^3 (\rho_p - \rho_f)}{\rho_f \nu_f^2} \tag{7-57}$$

where Sh = Sherwood number, dimensionless
Sc = Schmidt number, equal to ratio of momentum diffusivity to mass diffusivity, dimensionless
Gr = Grashof number, equal to ratio of buoyant forces to viscous forces, dimensionless
Re = Reynolds number, equal to ratio of inertial forces to viscous forces, dimensionless

A, B, a, b, c = constants, unitless
k_f = mass transfer coefficient, m/s
L_c = characteristic length, m
D_f = diffusion coefficient, m^2/s
μ_f = fluid-phase viscosity, kg/m · s
ρ_f = fluid-phase density, kg/m^3
ρ_p = particle density, kg/m^3
v = velocity, m/s
ν_f = fluid-phase kinematic viscosity, equal to $\mu f/\rho f$, m^2/s
g = acceleration due to gravity, 9.81 m/s^2

The coefficients in Eq. 7-53 (i.e., A, B, a, b, c) depend on the geometry (e.g., particles, bubbles, packed bed) and flow regime (i.e., laminar, transition, or turbulent).

The first term on the right side of Eq. 7-53, $A\text{Gr}^c$, accounts for molecular diffusion conditions under which there is no advection. The Grashof number accounts for mass transfer due to natural advection, which can be caused by density differences at the interface. A familiar example from heat transfer is a space heater in which air flows past the heater as a result of lighter, heated air that rises. A similar phenomenon is observed for salts dissolved in water because the solution at the salt interface is usually denser and fluid flow will occur at the interface without forced advection. However, in many cases, the buoyant force is not important in environmental problems because the solutions are very dilute. In such a case, the first term is simply a constant and Eq. 7-53 becomes

$$\mathrm{Sh} = A + B\,\mathrm{Re}^a\mathrm{Sc}^b \quad (7\text{-}58)$$

The second term on the right side of Eqs. 7-53 and 7-58, $B\,\mathrm{Re}^a\,\mathrm{Sc}^b$, accounts for mass transfer that is enhanced by advection. Many investigators have developed the theoretical bases for Eq. 7-53 for various geometries and flow regimes and have also developed mass transfer correlations by fitting data to Eq. 7-58 for various geometries and flow regimes. The forms of mass transfer correlations for different geometry and flow conditions are discussed briefly in the next subsection. In spite of the plethora of papers on mass transfer (there are at least 100,000 articles in the archival literature), mass transfer correlations for complex geometry and flow conditions do not exist for every situation. For such cases, it is possible to estimate the mass transfer coefficient for component A if the mass transfer for another component, B, is known. This relationship is discussed at the end of this section.

The characteristic length in Eq. 7-50 was the length of the flat plate. A common geometry in environmental applications is spheres (such as where the extracting phase is a bubble or air or a particle or activated carbon). Re is defined differently for a sphere than it is for flat surfaces:

$$\mathrm{Re} = \frac{\rho_f v d_p}{\mu_f} \quad (7\text{-}59)$$

where d_p = diameter of sphere or particle, m

7-5 Correlations for Mass Transfer Coefficients at an Interface

Based on the models in the previous section, numerous mass transfer correlations have been developed to estimate mass transfer coefficients for various geometries and flow regimes. This section describes common forms of mass transfer correlations, and the relationship between mass transfer coefficients and diffusing species.

Common Mass Transfer Correlations

Mass transfer coefficient correlations for a variety of transfer situations are shown in Table 7-5. Some of these correlations are obtained by analogy to heat transfer, others by measurement or theoretical approximation. Correlations presented in Table 7-5 include (1) Gnielinski correlation for packed beds, which can be used for calculating mass transfer coefficients in gas- and liquid-phase adsorption (see Chap. 15); (2) Onda correlation for absorption and air stripping, which can be used to determine the mass transfer coefficient in packed-tower air stripping (see Chap. 14); and (3) the Gilliland correlation for gases and liquids in pipes. Many other correlations for other geometries and flow regimes are available in the literature. The procedure for calculating a mass transfer coefficient using the Gnielinski correlation is presented in Example 7-5.

Table 7-5
Common mass transfer correlations

Description/Application	Empirical Correlation	Nomenclature
Gnielinski correlation[a] for packed beds; used for calculating external liquid-phase mass transfer coefficient in packed beds; application demonstrated in Chap. 15	$k_f = \frac{[1+1.5(1-\varepsilon)]D_f}{d_p}\left(2+0.644\,\mathrm{Re}^{1/2}\,\mathrm{Sc}^{1/3}\right)$ $\mathrm{Re} = \frac{\rho_f \Phi d_p V_l}{\varepsilon \mu_f}$ $\mathrm{Sc} = \frac{\mu_f}{\rho_f D_f}$ *Constraints:* $Pe = \mathrm{Re} \cdot \mathrm{Sc} > 500$ $0.7 < \mathrm{Sc} < 10^4, \mathrm{Re} < 2 \times 10^4$ $0.26 < \varepsilon < 0.935$	k_f = fluid-phase mass transfer coefficient, m/s D_f = fluid-phase diffusion coefficient, m^2/s ε = bed void fraction, dimensionless d_p = diameter of particle, m ρ_f = fluid-phase density, kg/m^3 Φ = sphericity, equal to ratio of surface area of equivalent-volume sphere to actual surface area of particle, dimensionless v_l = superficial liquid velocity, m/s μ_f = fluid-phase viscosity, kg/m · s Re = Reynolds number, dimensionless Sc = Schmidt number, dimensionless[b] Pe = Peclet number, dimensionless

(Continued)

Table 7-5 *(Continued)*

Description/Application	Empirical Correlation	Nomenclature
Onda correlation[c] for absorption or stripping in packed tower; used for calculating liquid- and gas-phase mass transfer coefficient in packed air absorber or stripper; application demonstrated in Chap. 14	$k_l = 0.0051\left(\frac{L_m}{a_w \mu_l}\right)^{2/3}\left(\frac{\mu_l}{\rho_l D_l}\right)^{-0.5}(a_t d_p)^{0.4}\left(\frac{\rho_l}{\mu_l g}\right)^{-1/3}$ $k_g = 5.23(a_t D_g)\left(\frac{G_m}{a_t \mu_g}\right)^{0.7}\left(\frac{\mu_g}{\rho_f D_g}\right)^{1/3}(a_t d_p)^{-2}$ $a_w = a_t\left\{1 - \exp\left[-1.45\left(\frac{\sigma_c}{\sigma}\right)^{0.75}(\text{Re})^{0.1}(\text{Fr})^{-0.05}(\text{We})^{0.2}\right]\right\}$ $\text{Re} = \frac{L_m}{a_t \mu_l}$ $\text{Fr} = \frac{(L_m)^2 a_t}{(\rho_l)^2 g}$ $\text{We} = \frac{(L_m)^2}{\rho_l a_t \sigma}$ *Constraints:* $d_p < 0.0508$ m (2 in.) $0.8 < L_m < 43$ kg/m$^2 \cdot$ s ($1.1 < L_m < 63$ gpm/ft^2) $0.014 < G_m < 1.7$kg/m$^2 \cdot$s ($2.21 < G_m < 268$ cfm/ft^2)	k_l = liquid-phase mass transfer rate coefficient, m/s L_m = liquid mass loading rate, kg/m$^2 \cdot$ s a_w = wetted surface area of packing, m^{-1} a_t = specific surface area of packing, m^{-1} D_l = liquid-phase diffusion coefficient, m^2/s d_p = nominal packing diameter, m ρ_l = liquid-phase density, kg/m^3 μ_l = liquid-phase viscosity, kg/m $\cdot$ s g = acceleration of gravity, 9.81 m/s^2 k_g = gas-phase mass transfer coefficient, m/s μ_g = gas-phase viscosity, kg/m $\cdot$ s D_g = gas-phase diffusion coefficient, m^2/s ρ_g = gas-phase density, kg/m^3 G_m = gas mass loading rate, kg/m$^2 \cdot$ s σ_c = critical surface tension of packing, kg/s^2 σ = surface tension of water, kg/s^2 Re = Reynolds number, dimensionless Fr = Froude number, dimensionless We = Weber number, dimensionless

Gilliland Correlation[d] for gases or liquids in pipes; used for calculating gas- or liquid-phase mass transfer coefficient in pipes	$\mathrm{Sh} = \dfrac{k_f D}{D_g\,(\text{or } D_l)} = 0.023\mathrm{Re}^{0.83}\ \mathrm{Sc}^{0.33}$ $\mathrm{Re} = \dfrac{\rho_g D v_g}{\mu_g}\left(\text{or } \dfrac{\rho_l D v_l}{\mu_l}\right)$ $\mathrm{Sc} = \dfrac{\mu_g}{\rho_g D_g}\left(\text{or } \dfrac{\mu_l}{\rho_l D_l}\right)$ For turbulent flow: Re >2100 $0.6 < \mathrm{Sc} < 3000$	Sh = Sherwood number, dimensionless k_f = fluid-phase mass transfer coefficient, m/s D = diameter of pipe, m D_l = liquid-phase diffusion coefficient, m^2/s D_g = gas-phase diffusion coefficient, m^2/s *Re* = Reynolds number, dimensionless *Sc* = Schmidt number, dimensionless ρ_g = gas density, kg/m^3 μ_g = gas viscosity, kg/m · s ρ_l = liquid density, kg/m^3 μ_l = liquid viscosity, kg/m · s V_g = superficial gas velocity, m/s v_l = superficial liquid velocity, m/s

[a]Adapted from Gnielinski (1978–1981).
[b]Typical values for the Schmidt number for gas and liquid phases are 0.7 and 1000, respectively.
[c]Adapted from Onda et al. (1968).
[d]Adapted from Gilliland and Sherwood (1934) and Linton and Sherwood (1950).

Example 7-5 Application of a correlation to determine a mass transfer coefficient

A resort in the mountains has a good water source; however, the water is extremely soft (no hardness) and acidic, which makes cleaning and bathing difficult. One solution is to pass the low-pH water through a packed bed containing crushed limestone ($CaCO_3$). Determine the film transfer coefficient for limestone media. Given: The media diameter d_p is 1.0 cm, the bed porosity ε is 0.43, the particle sphericity Φ is 0.8, the temperature is 20°C, and the superficial velocity v_l through the bed is 12 m/h.

Solution

Determine the mass transfer coefficient k_f for limestone particles using the Gnielinski correlation in Table 7-5.

1. Calculate the diffusion coefficient for aqueous calcium carbonate using the Nernst–Haskell equation (see Example 7-2). From Table 7-4, the limiting conductances are 59.5 $(cm^2 \cdot C^2)/(J \cdot s \cdot eq)$ for Ca^{2+} and 69.3 $(cm^2 \cdot C^2)/(J \cdot s \cdot eq)$ for CO_3^{2-}.

$$D_l = \frac{(8.314 \text{ J/mol} \cdot \text{K})\ (298 \text{ K})}{(96{,}500 \text{ C/eq})^2} \left[\frac{\left(\frac{1}{2}+\frac{1}{2}\right)\frac{\text{mol}}{\text{eq}}}{\left(\frac{1}{59.5}+\frac{1}{69.3}\right)\left(\frac{\text{J} \cdot \text{s} \cdot \text{q}}{\text{cm}^2 \cdot \text{c}^2}\right)} \right]$$

$$= 8.52 \times 10^{-6} \text{ cm}^2/\text{s} = 8.52 \times 10^{-10} \text{ m}^2/\text{s}$$

2. Calculate Re from the equation in Table 7-5. From App. C, ρ_l = 998.2 kg/m^3 and $\mu_l = 1.002 \times 10^{-3}$ kg/m·s at 20°C.

$$\text{Re} = \frac{\rho \Phi d_p v_l}{\varepsilon \mu} = \frac{\rho_l \Phi d_p v_l}{\varepsilon \mu_l}$$

$$= \frac{\left(998.2 \text{ kg/m}^3\right)(0.8)\ [1.0 \text{ cm} \times (1 \text{ m}/100 \text{ cm})]\ [12 \text{ m/h} \times (1 \text{ h}/3600 \text{ s})]}{(0.43)\ (1.002 \times 10^{-3} \text{ kg/m} \cdot \text{s})}$$

$$= 61.8$$

3. Calculate Sc using Eq. 7-55:

$$\text{Sc} = \frac{\mu_f}{\rho_f D_f} = \frac{\mu_l}{\rho_l D_l} = \frac{1.002 \times 10^{-3} \text{ kg/m} \cdot \text{s}}{\left(998.26 \text{ kg/m}^3\right)\left(8.52 \times 10^{-10} \text{ m}^2/\text{s}\right)}$$

$$= 1180$$

4. Calculate k_f using the empirical correlation given in Table 7-5:

$$\begin{aligned} k_f &= \frac{[1+1.5\,(1-\varepsilon)]D_f}{d_p}\left(2+0.644\,\mathrm{Re}^{1/2}\,\mathrm{Sc}^{1/3}\right) \\ &= \frac{[1+1.5\,(1-0.43)]\left(8.52\times10^{-10}\ \mathrm{m^2/s}\right)}{1\ \mathrm{cm}\times(1\ \mathrm{m}/100\ \mathrm{cm})}\left[2+0.644(61.8)^{1/2}(1180)^{1/3}\right] \\ &= 8.76\times10^{-6}\ \mathrm{m/s} \end{aligned}$$

Relationship between Mass Transfer Coefficients and Diffusing Species

Despite the availability of mass transfer correlations for diverse situations, there are many cases for which mass transfer correlations do not exist. Under such circumstances, the mass transfer coefficient of one solute can be used to estimate the mass transfer coefficient of another. For example, if the mass transfer coefficient for oxygen is known, the mass transfer coefficient for other compounds can be calculated. The procedure is convenient because the mass transfer coefficient for dissolved oxygen is relatively easy to measure and many correlations are available for oxygen transfer.

Mass transfer correlations depend on the exponent of the Schmidt number (i.e., the b that appears in Eqs. 7-53 and 7-58, Sc^b), and this dependency allows for estimation of the mass transfer coefficient of one compound from the mass transfer coefficient of another compound. Situations that arise include (1) mass transfer occurs only by molecular diffusion, (2) the fluid moves freely to the surface and surface renewal occurs, and (3) the fluid cannot move freely to the surface because the interface is a solid and a boundary layer forms. If mass transfer occurs only by molecular diffusion, the relationship between the mass transfer coefficient and diffusion coefficient is

$$\mathrm{Sh} = \frac{k_f L_c}{D_f} = 1 \tag{7-60}$$

where k_f = fluid-phase mass transfer coefficient, m/s
L_c = characteristic length, m
D_f = fluid-phase diffusion coefficient, $\mathrm{m^2/s}$

For pure molecular diffusion, k_f is directly proportional to D_f, that is, $k_f \propto D_f$. Thus, if $k_{f,A}$ for one compound is known, then $k_{f,B}$ can be determined from the relationship

$$\frac{k_{f,\mathrm{B}}}{k_{f,\mathrm{A}}} = \frac{D_{f,\mathrm{B}}}{D_{f,\mathrm{A}}} \tag{7-61}$$

Typical packed column used for mass transfer operations.
http://www.stantec.com/wt3r/7-1.html

where $k_{f,B}$ = fluid-phase mass transfer coefficient for B, m/s
$k_{f,A}$ = fluid-phase mass transfer coefficient for A, m/s
$D_{f,B}$ = fluid-phase diffusion coefficient of B, m^2/s
$D_{f,A}$ = fluid-phase diffusion coefficient of A, m^2/s

The surface renewal model presented in Sec. 7-4 yields a dependence on the Sherwood number as shown here:

$$\mathrm{Sh} = \frac{k_f L_c}{D_f} \propto \mathrm{Sc}^{0.5} = \left(\frac{\mu_f}{\rho_f D_f}\right)^{0.5} \tag{7-62}$$

Thus, from Eq. 7-61, it can be seen that the surface renewal model predicts that k_f depends on the diffusion coefficient according to the expression

$$k_f \propto D_f^{0.5} \tag{7-63}$$

Thus, if mass transfer occurs according to the surface renewal model and if $k_{f,A}$ is known, then $k_{f,B}$ can be determined from the relationship

$$\frac{k_{f,\mathrm{B}}}{k_{f,\mathrm{A}}} = \left(\frac{D_{f,\mathrm{B}}}{D_{f,\mathrm{A}}}\right)^{0.5} \tag{7-64}$$

A third situation is described by the penetration model or boundary layer model. As discussed in Sec. 7-4, the boundary layer model yields a dependence of Sh as shown here:

$$\mathrm{Sh} = \frac{k_f L}{D_f} \propto \mathrm{Sc}^{1/3} = \left(\frac{\mu_f}{\rho_f D_f}\right)^{1/3} \tag{7-65}$$

From Eq. 7-65, k_f depends on the diffusion coefficient according to the expression

$$k_f \propto D_f^{2/3} \tag{7-66}$$

If $k_{f,A}$ is known, then $k_{f',B}$ can be determined from the following expression:

$$\frac{k_{f,\mathrm{B}}}{k_{f,\mathrm{A}}} = \left(\frac{D_{f,\mathrm{B}}}{D_{f,\mathrm{A}}}\right)^{2/3} \tag{7-67}$$

The dependency of the ratio of the mass transfer coefficients for constituents A and B on the ratio of the diffusivities of A and B for some common situations that are encountered in water treatment are given in Table 7-6. The procedure to calculate one mass transfer coefficient from another is demonstrated in Example 7-6.

Table 7-6
Dependency of ratio of mass transfer coefficients for compounds A and B on ratio of diffusivities of A and B for some common situations in water treatment

Situation	$\frac{k_{f,B}}{k_{f,A}} = \left(\frac{D_{f,B}}{D_{f,A}}\right)^n$	Comment
Transport from fluid to solid or from solid to liquid	$\frac{k_{f,B}}{k_{f,A}} = \left(\frac{D_{f,B}}{D_{f,A}}\right)^{2/3}$	A boundary layer forms because the velocity at the solid–fluid interface is zero.
Mass transfer resistance in water at air–water interface	$\frac{k_{l,B}}{k_{l,A}} = \left(\frac{D_{l,B}}{D_{l,A}}\right)^{0.5}$	For an air–water interface, no velocity gradient exists in the water at the boundary because the viscosity of air is 50 times lower than that for water at 1 atm (the air offers no resistance against which a liquid velocity gradient would form).
Mass transfer resistance in air at air–water interface	$\frac{k_{g,B}}{k_{g,A}} = \left(\frac{D_{g,B}}{D_{g,A}}\right)^{2/3}$	At the air–water interface, a boundary layer forms in the air because the higher viscosity of the water causes the water to essentially act as a solid surface. At very high air Reynolds numbers the air velocity gradient can decay and the dependency would tend toward $\frac{1}{2}$.

where k_f, k_l, and k_g = fluid-, liquid-, and gas-phase mass transfer coefficients for solutes A and B, m/s
D_f, D_l, and D_g, = fluid-, liquid-, and gas-phase diffusion coefficients for solutes A and B, m²/s
n = exponent used to describe the relationship between the ratio of the mass transfer coefficients of compounds A and B to the ratio of diffusion coefficients of compounds A and B

Example 7-6 Determination of mass transfer coefficient by relating mass transfer coefficients and diffusion coefficients

The mass transfer coefficient of oxygen in a mass transfer device measured at 20°C is $k_{l,O_2} = 0.0045\ \text{m/s}$. Estimate the mass transfer coefficient of benzene at 20°C on the water side of an air–water interface in the device by relating mass transfer coefficients and diffusivities of benzene to those of oxygen. From Table 7-1, the liquid diffusion coefficient for benzene at 20°C is $D_{l,\text{benzene}} = 1.02 \times 10^{-9}\ \text{m}^2/\text{s}$.

Solution

Determine the mass transfer coefficient for benzene, $k_{l,\text{benzene}}$, using the relationship for mass transfer resistance in the water at the air–water interface in Table 7-6.

1. Determine the diffusion coefficient of oxygen at 20°C using Eq. 7-35:

$$D_{l,O_2} = 10^{(A+B/T)}\left(1.0\times10^{-9}\right)$$
$$= 10^{(3.15-(831/(273+20)))}.\left(1.0\times10^{-9}\right)$$
$$= 2.06\times10^{-9}\,\text{m}^2/\text{s}$$

2. Determine $k_{l,\text{benzene}}$ using Eq. 7-64:

$$\frac{k_{l,B}}{K_{l,A}} = \left(\frac{D_{l,B}}{D_{l,A}}\right)^{0.5} \Rightarrow \frac{k_{l,\text{benzene}}}{k_{l,O_2}} = \left(\frac{D_{l,\text{benzene}}}{D_{l,O_2}}\right)^{0.5} \Rightarrow$$

$$k_{l,\text{ benzene}} = k_{l,O_2}\left(\frac{D_{l,\text{ benzene}}}{D_{l,O_2}}\right)^{0.5} = (0.0045\text{ m/s})\left(\frac{1.02\times10^{-9}\text{m}^2/\text{s}}{2.06\times10^{-9}\text{m}^2/\text{s}}\right)^{0.5}$$
$$= 0.0032\text{ m/s}$$

7-6 Design of Treatment Systems Controlled by Mass Transfer

When treatment devices are controlled by the rate of mass transfer, the concepts presented in this chapter can be used for design. The relationship between mass transfer and process design will be developed more fully in Chaps. 14 (Air Stripping and Aeration), 15 (Adsorption), and 17 (Reverse Osmosis), but an illustration of how mass transfer concepts can be used to design a treatment system is presented in the following example.

Example 7-7 Design of a packed column treatment system controlled by mass transfer

An acid waste stream is to be neutralized by continuous (steady state) flow through a column packed with a rapidly dissolving calcium carbonate media. Determine the hydraulic residence time and bed depth necessary so that the column effluent is 99 percent saturated with calcium carbonate, assuming that the dissolution is limited only by the rate of diffusion through the boundary layer. The characteristics of the limestone media are the same as in Example 7-5; the media diameter d_p is 1.0 cm, the bed porosity ε is 0.43, the particle sphericity Φ is 0.8, and the superficial velocity v_l through the bed is 12 m/h.

Solution

1. A schematic of the media bed is shown below. Set up a mass balance (see Chap. 6) using the liquid in a differential unit of depth in the bed as the control volume. Since the system is at steady state, the accumulation term is zero. The differential element has two mass input terms, one from advective flow of water containing calcium carbonate from the previous differential element (QC_x), and the other from the dissolution of the calcium carbonate. The mass flow due to diffusion through the boundary layer is described by Eq. 7-46, except that the concentration gradient is ($C_s - C_x$), where C_s is the calcium carbonate concentration at the surface of the media (which is at the saturation concentration) and C_x is the concentration in the bulk solution. The mass output term is $QC_{x+\Delta x}$. The volume of the differential element is $V = A\Delta x$, where A is the cross-sectional area of the column. The mass balance can be set up and algebraically rearranged as follows:

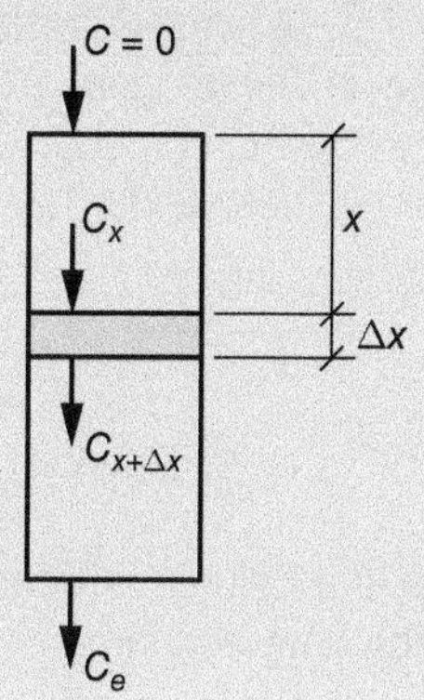

$$0 = [\text{mass in}] - [\text{mass out}] \quad (1)$$

$$0 = QC_x + k_f a\,(C_s - C_x)\,V - QC_{x+\Delta x} \quad (2)$$

$$\frac{C_{x+\Delta x} - C_x}{\Delta x} = \frac{k_f a}{Q}(C_s - C_x)A \quad (3)$$

2. The term on the left is the derivative dC/dx after taking the limit as $\Delta x \rightarrow 0$. The equation is algebraically rearranged again so that both sides of the equation can be integrated across the depth of the column:

$$\frac{dC}{dx} = \frac{k_f a}{Q}(C_s - C)A \quad (4)$$

$$\int_0^{C_e} \frac{dC}{C_s - C} = \frac{k_f a A}{Q} \int_0^L dx \tag{5}$$

3. Integrating Eq. 5 and recognizing that $A \times L = V$ and $V/Q = \tau$ (the hydraulic residence time) yields

$$-\ln\left(\frac{C_s - C_e}{C_s}\right) = \frac{k_f a A}{Q} L = k_f a \frac{V}{Q} = k_f a \tau \tag{6}$$

4. The specific area is determined as in Example 7-4, except that the sphericity increases the surface area to be greater than that of a sphere. Substituting the values from the problem statement yields

$$a = \frac{6(1-\varepsilon)}{\Phi d_p} = \frac{6(1-0.43)}{(0.8)\,(0.01\text{ m})} = 428\text{ m}^{-1} \tag{7}$$

5. In Example 7-5, k_f was determined to be 8.76×10^{-6} m/s. The concentration in the effluent of the column is $0.99C_s$. Then $(C_s - C_e)/C_s = (C_s - 0.99C_s)/C_s = 0.01$. Rearranging Eq. 6 and plugging in the necessary values yields

$$\tau = \frac{-\ln\left[(C_s - C_e)/C_s\right]}{k_f a} = \frac{-\ln(0.01)}{\left(8.76\times10^{-6}\text{ m/s}\right)\left(428\text{ m}^{-1}\right)}$$

$$= 1230\text{ s} = 20.5\text{ min} = 0.34\text{ h}$$

6. The depth of the column is determined by multiplying the superficial velocity by the detention time:

$$L = v\tau = (12\text{ m/h})(0.34\text{ h}) = 4.1\text{ m}$$

Comment

When dissolution is rapid as in this example, the size of the column will be controlled by the rate of mass transfer. Examining steps 4 through 6 indicates that reducing the size of the media would increase the specific area and thereby decrease the required depth of the column. For instance, reducing the diameter of the media to 1 mm would increase the specific area to 4280 m^{-1}, increase the mass transfer coefficient to 2.99×10^{-5} m/s, and decrease the depth of the column to 0.12 m. These calculations demonstrate the importance of specific area in designing mass transfer equipment.

In many situations, dissolution kinetics are slower and the process is controlled by the rate of dissolution and not by the rate of mass transfer (i.e., calcium carbonate diffuses through the boundary layer faster than it dissolves, so that the concentration at the surface (C_s) is not the saturated concentration). The difference between reaction-limited and mass-transfer-limited processes is an additional complication in designing treatment equipment.

7-7 Evaluating the Concentration Gradient with Operating Diagrams

The last sections have dealt with development of theory and correlations needed to determine mass transfer coefficients. This section explores the other half of the primary mass transfer equation (Eq. 7-3), the concentration gradient. The concentration gradient and the impact it has on mass transfer can be evaluated graphically. Graphical analysis of concentration gradients depends on the type of contacting equipment. The major types of contacting equipment are described next, followed by a discussion of operating diagrams, also known as McCabe–Thiele diagrams.

Contact Modes

Two major methods are used for bringing two phases into contact: batch operation and continuous operation. Continuous systems may be operated with or without discrete stages.

BATCH OPERATION

A batch operation is a contained system with no flow in or out. Typically, the two phases are brought together, allowed to approach equilibrium, and then separated. An example of a batch operation is an adsorption equilibrium isotherm conducted by adding powdered activated carbon (PAC) to a bottle.

CONTINUOUS OPERATION

A continuous operation involves flow within the system. Three flow patterns are possible with continuous operations: (1) co-current, (2) countercurrent, and (3) cross flow. A co-current operation consists of a system in which the two contacting phases flow in the same direction. In water treatment, there are natural draft co-current stripping devices in which the falling water creates a natural draft of air to flow through the device.

A countercurrent operation consists of a system in which the two contacting phases flow in opposite directions. Countercurrent operation represents the preferred mode of operation for many air-stripping processes, such as packed towers, as will be discussed in Chap. 14.

In a cross-flow operation, the two contacting phases flow perpendicular to one another. Sedimentation is an example of a cross-flow process and is widely used in water treatment, but cross-flow operations in general are less common for mass-transferred-controlled processes in water treatment.

In a staged operation, the process is operated as a series of stages. The two phases are mixed with each other in each phase, and the concentrations in both phases are uniform within each stage. The flow pattern in a staged operation can be co-current, cross flow, or countercurrent. An example of a countercurrent staged operation is a low-profile air stripper, which will be examined in Chap. 14.

Development of Operating Diagrams

The impact of the concentration gradient on the rate of mass transfer between two phases can be evaluated graphically using a concept called operating diagrams, or McCabe–Thiele diagrams (McCabe and Thiele, 1925). Operating diagrams are drawn by plotting the solute concentration in the extracting phase (e.g., air for gas transfer, activated carbon for adsorption) as a function of the solute concentration in the aqueous phase. The operating diagram consists of two lines: (1) an equilibrium line and (2) an operating line. Operating diagrams can be used to determine the minimum amount of the extracting phase needed for treatment and to examine graphically the trade-off between the size of the mass transfer contacting device and the quantity of extracting phase needed (e.g., air-water ratio for stripping or PAC required for adsorption).

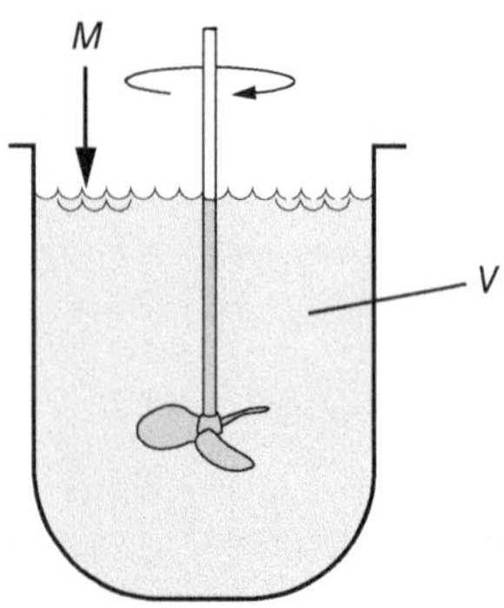

Figure 7-6
Batch contactor for powdered activated carbon.

EQUILIBRIUM LINE

The equilibrium line is derived from two-phase equilibrium relationships and gives the solute concentration in the extracting phase that exists when the extracting and aqueous phases are in equilibrium with each other. Examples of two-phase equilibrium relationships are Henry's law for air stripping and the Freundlich isotherm for adsorption. Equilibrium relationships were introduced in Chap. 5, and additional details on Henry's law and Freundlich isotherms will be provided in Chaps. 14 and 15, respectively.

OPERATING LINE

The operating line is derived from a mass balance on the contacting device, relating the solute concentration in each phase initially to the solute concentration in each phase after contact has begun. An example using a batch reactor, in which PAC is added to a vessel containing a solution of water and an organic solute, is shown on Fig. 7-6. Initially, there is no solute adsorbed onto the PAC. The mass balance for this system is as follows:

$$\begin{pmatrix}\text{Mass present}\\ \text{initially in solution}\end{pmatrix} = \begin{pmatrix}\text{mass}\\ \text{adsorbed}\end{pmatrix} + \begin{pmatrix}\text{mass remaining}\\ \text{in solution after}\\ \text{adsorption}\end{pmatrix} \tag{7-68}$$

$$VC_0 = Mq + VC \tag{7-69}$$

where V = volume of liquid in vessel, L
C_0 = initial concentration of solute in vessel, mg/L
M = mass of carbon, g

q = concentration of solute adsorbed to the activated carbon at any time, mg/g
C = concentration of the solute in the water after adsorption, mg/L

Equation 7-69 can be rearranged as follows:

$$q = \frac{V}{M}(C_0 - C) \tag{7-70}$$

The operating line, which is the solute concentration in the extracting phase as a function of the concentration in the aqueous phase at any point in time after contact has started, is defined by Eq. 7-70. When the PAC is first added to the vessel, there is no solute on the PAC. As time proceeds, the solute becomes adsorbed onto the PAC, and q and C at a particular time are related to one another by the operating line. It should be noted that although adsorption in a batch reactor proceeds toward equilibrium over the passage of time, the operating line does not identify the time progression of the process but only relates the dependent variables q and C.

The operating diagram for the relationship described in Eq. 7-70 is shown on Fig. 7-7. Equation 7-70 is the equation of a straight line with a slope of $-V/M$, and several operating lines with different values for V/M have been shown. The equilibrium line is shown on Fig. 7-7 as a dashed line.

DRIVING FORCE

The driving force for mass transfer, as shown on Fig. 7-7, is the difference between the actual solute concentration in solution and the concentration in solution that would be in equilibrium with the extracting phase. Initially, the solute is entirely in the aqueous phase, and the solute is transferred rapidly to the PAC. As time progresses, the concentration on the PAC increases and the concentration in the aqueous phase decreases, which slows

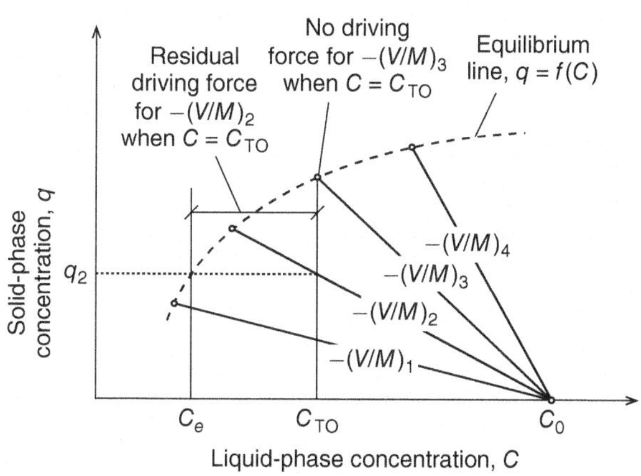

Figure 7-7
Operating lines for a constant initial concentration C_0 and different adsorbent doses, V/M (equilibrium line is also plotted for reference).

the rate of mass transfer. After a very long time, the solute concentration in the water is in equilibrium with the concentration on the PAC, and bulk mass transfer ceases. Thus, the concentration gradient, or driving force, is defined as the difference between the actual and equilibrium concentration C_e in the aqueous phase.

Because the equilibrium concentration is identified by the equilibrium line and the actual concentration (determined by mass balance) is identified by the operating line, the horizontal distance between these lines describes the concentration gradient. Equilibrium occurs and mass transfer ceases when the operating line and equilibrium line intersect.

Analysis Using Operating Diagrams

The operating diagram can be used to determine the minimum amount of extracting phase required for treatment, which is an initial indicator of the feasibility of a process. For example, if millions of tons of activated carbon are required to treat a given water, then adsorption with activated carbon is not a feasible treatment option and no further analysis is necessary. If a separation process appears to be feasible based on the amount of extracting phase, then more detailed design and economic calculations are warranted.

An operating line analysis for an adsorption process is shown on Fig. 7-7. For a given volume of water, the quantity of PAC required can be defined by the V/M ratio, with greater values of V/M (greater slope of the operating line) corresponding to smaller amounts of PAC. If the treatment objective is the concentration shown as C_{TO} on Fig. 7-7, the minimum amount of PAC required can be determined from the operating line with the slope of $(V/M)_3$, which is the operating line that intersects the equilibrium line at the value of C_{TO}. Operating lines with greater slope, shown as $(V/M)_4$, intersect the equilibrium line at a concentration higher than C_{TO} and therefore would be unable to meet the treatment objective.

The operating diagram also qualitatively demonstrates the trade-off between the quantity of the extracting phase and the size of the contacting device. For the operating line identified as $(V/M)_3$, the driving force (horizontal distance between the equilibrium and operating lines) becomes infinitesimally small as equilibrium is approached. The small driving force results in a slow rate of mass transfer, requiring an exceedingly long time to reach the treatment objective. In a flow-through system treating a specified water flow rate, a long time corresponds to a long detention time and hence a very large contactor. The operating lines labeled as $(V/M)_1$ and $(V/M)_2$ have lower slopes, which correspond to greater quantities of carbon, but have larger concentration gradients when the actual concentration (operating line) reaches the treatment objective, resulting in shorter contact times. Thus, for the operating lines shown, the line labeled $(V/M)_1$ would use the most carbon but have the smallest contactor, the line labeled $(V/M)_2$ would have an intermediate carbon usage rate and contactor size, the line labeled $(V/M)_3$ would use the minimum amount of carbon but have a large (theoretically, infinitely large) contactor, and the line labeled $(V/M)_4$ would be unable to meet the treatment objective.

Equation 7-70 is for a batch operation, but a similar relationship can be derived for cocurrent continuous plug flow operation. If a quantity of PAC per time, M_r, is added to water with a flow rate, Q, the mass balance is the same as presented previously in Eq. 7-68.

$$\left(\begin{array}{c}\text{Mass present}\\ \text{initially in solution}\end{array}\right)=\left(\begin{array}{c}\text{mass}\\ \text{adsorbed}\end{array}\right)+\left(\begin{array}{c}\text{mass remaining}\\ \text{in solution after}\\ \text{adsorption}\end{array}\right)$$

$$QC_0 = M_r q + QC \tag{7-71}$$

$$q = \frac{Q}{M_r}(C_0 - C) \tag{7-72}$$

where Q = flow rate, L/s
C_0 = initial concentration of solute in the solution, mg/L
M_r = PAC feed rate, mass added per time, g/s
C = concentration of the solute in the water at any time, mg/L
q = concentration of solute adsorbed to the activated carbon at any time, mg/g

The PAC dosage in the plug flow system, M_r/Q, is identical to the PAC dosage in the batch reactor, M/V, and Eqs. 7-70 and 7-72 are essentially identical.

An example calculation of the minimum amount of extracting phase required for treatment is presented for PAC in Example 7-8.

Example 7-8 Minimum amount of PAC required to achieve given level of treatment

Many adsorption equilibrium lines, as discussed in Chap. 15, can be described by the Freundlich isotherm:

$$q_e = KC_e^{1/n}$$

where q_e = equilibrium concentration of solute in solid phase,mg/g
K = Freundlich capacity factor, (mg/g)[L/mg]$^{1/n}$
C_e = equilibrium concentration of solute in aqueous phase, mg/L
$1/n$ = Freundlich intensity factor, dimensionless

Calculate the minimum dose of PAC that is required for the removal of geosmin, an odor-producing compound. The initial concentration is 50 ng/L, and the treatment objective C_{To} is 5 ng/L. The K and $1/n$ values for geosmin are 200 (mg/g)[L/mg]$^{1/n}$ and 0.39, respectively. A reasonable PAC dose would be less than 10 to 20 mg/L. Is the process feasible and should more detailed studies be conducted?

Solution

1. The lowest PAC dose occurs when the PAC is used to capacity, which is when the concentration on the PAC would be in equilibrium with C_{To}. The concentration on the PAC at equilibrium is calculated with the equilibrium relationship:

$$q_e = KC_{\text{TO}}^{1/n}$$

2. The minimum PAC dose occurs when the operating line (Eq. 7-70) intersects the equilibrium line at the treatment objective:

$$q_e = \frac{V}{M}(C_0 - C_{\text{TO}})$$

3. The intersection of the equilibrium and operating lines is determined by equating the two equations given above and solving for the minimum dose:

$$\left(\frac{M}{V}\right)_{\text{min}} = \frac{C_0 - C_{\text{TO}}}{KC_{\text{TO}}^{1/n}} = \frac{\left(50\times10^{-6} - 5\times10^{-6}\right)\text{mg/L}}{200\times\left(5\times10^{-6}\right)^{0.39}\ \text{mg/g}}$$

$$= 2.63\times10^{-5}\,\text{g/L} = 0.0263\,\text{mg/L}$$

Comment

A dosage of 0.0263 mg/L is within the acceptable range, and additional tests that simulate water plant conditions (jar tests) can be planned. The tests would be needed because the presence of natural organic matter (NOM) will reduce the adsorption capacity. Further, the computed value is the minimum dose of PAC, which yields an exceedingly small driving force as equilibrium is approached, resulting in an extremely low rate of mass transfer and an unreasonably large PAC contactor. In practice, the required dosage to remove geosmin is at least 100 times greater than 0.0263 mg/L because of the impact of NOM and mass transfer.

7-8 Mass Transfer across a Gas–Liquid Interface

The various models and correlations that have been developed to describe the transport across a single interface have been introduced and discussed in Secs. 7-4 and 7-5. In this section, the methods for describing the transport

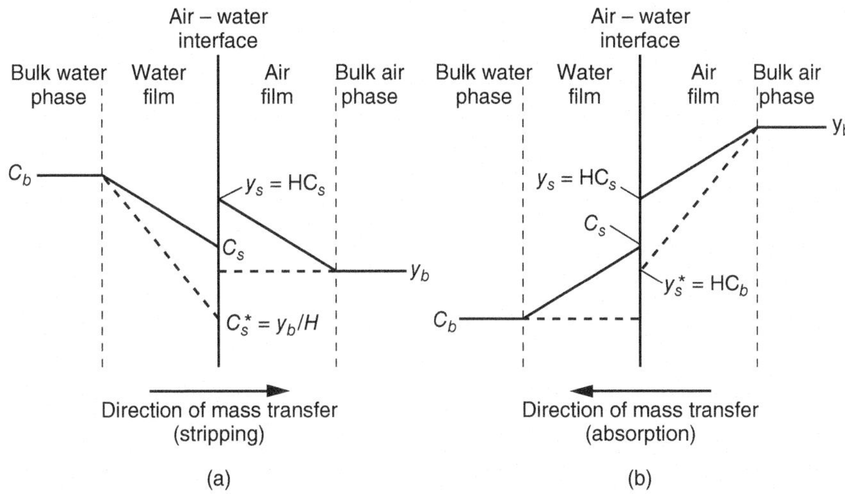

Figure 7-8
Two-film model: mass transfer driving gradients that occur for (a) stripping and (b) absorption.

of solute across a gas-liquid interface are considered. Because boundary layers can form on both the liquid and gas sides of the interface, the two-film model can be used to describe mass transfer. Chemical reactions can increase the rate of mass transfer and are considered in the next section.

The driving force for mass transfer between one phase and another results from the displacement of the system from equilibrium. The two-film model describes the interaction of two films (one gas and one liquid) at the gas-liquid interface. The two situations where mass transfer occurs between air and water at steady state are shown on Fig. 7-8. The situation for stripping where mass is transferred from the water to the air is shown on Fig. 7-8a, and the situation for absorption in which mass is transferred from the air to the water is shown on Fig. 7-8b. A detailed explanation is only provided for stripping because the mechanisms and assumptions for mass transfer are essentially identical for both cases, and the only difference is that mass is transferred in the opposite direction.

Conditions in Bulk Solution

The two-film model is used to describe the mass transfer rate for (a) the air stripping of VOCs such as methane, trichloroethane, and tetrachloroethane and other gases such as hydrogen sulfide and (b) the absorption of gases such as oxygen, carbon dioxide, nitrogen, or ozone. The following discussion will address the stripping of a volatile component A from water. As shown on Fig. 7-8a, the concentration of A in the bulk water is larger than the concentration of A at the air-water interface. Consequently, A diffuses from the bulk solution, where its liquid concentration is C_b, to the air-water interface, where its liquid concentration is C_s. The difference between C_b

and C_s is the driving force for stripping in the liquid phase. There is a discontinuity in the concentration at the air-water interface because A partitions in air at a different concentration based on equilibrium or Henry's law (see Chap. 14). Similarly, the concentration of A in the air at the air-water interface, y_s, is larger than the concentration of A in the bulk air, y_b, and it diffuses from the air-water interface to the bulk air. The difference between y_s and y_b is the driving force for stripping in the gas phase.

Conditions at Interface

Local equilibrium occurs at the air-water interface because random molecular movement (on a local scale of nanometers in water and thousands of nanometers on the air side) causes constituent A to dissolve in the aqueous phase and volatilize into the air at a rate more rapidly than diffusion to or away from the air -water interface. Accordingly, local equilibrium may be assumed and Henry's law can be used to relate y_s to C_s (Lewis and Whitman, 1924):

$$y_s = HC_s \tag{7-73}$$

where y_s = gas-phase concentration of A at air-water interface, mg/L
H = Henry's law constant, L of water/L of air, dimensionless
C_s = liquid-phase concentration of A at air-water interface, mg/L

For a dilute solution where no accumulation occurs at the surface, the flux of A through the gas-phase film must be equal to the flux through the liquid-phase film. Thus

$$J_A = k_l\left(C_b - C_s\right) = k_g\left(y_s - y_b\right) \tag{7-74}$$

where J_A = flux of A across air-water interface, mg/m^2·s
k_l = liquid-phase mass transfer coefficient for rate at which contaminant A is transferred from bulk aqueous phase to air-water interface, m/s
C_b = liquid-phase concentration of A in bulk solution, mg/L
C_s = liquid-phase concentration of A at air-water interface, mg/L
k_g = gas-phase mass transfer coefficient for rate at which contaminant A is transferred from air–water interface to bulk gas phase, m/s
y_s = gas-phase concentration of A at air-water interface, mg/L
y_b = gas-phase concentration of A in bulk solution, mg/L

Both k_l and k_g are sometimes referred to as local mass transfer coefficients for the liquid and gas phases because they depend upon the conditions at or near the air-water interface in their particular phase. The flux cannot be determined directly from Eq. 7-74 because the interfacial concentrations y_s and C_s are not known and cannot be measured easily. Consequently, it is necessary to define another flux equation in terms of hypothetical concentrations that are easy to determine. If it is hypothesized that all the resistance to mass transfer is on the liquid side, then there is no concentration gradient on the gas side and a hypothetical concentration, C_s^*, can be defined as shown on Fig. 7-8a:

$$y_b = HC_s^* \tag{7-75}$$

where C_s^* = liquid-phase concentration of A that is in equilibrium with bulk air concentration, mg/L

Alternatively, it can be hypothesized that all the resistance to mass transfer is on the gas side, in which case there is no concentration gradient on the liquid side and a hypothetical concentration y_s^* can be defined as shown on Fig. 7-8b:

$$y_s^* = HC_b \tag{7-76}$$

where y_s^* = gas-phase concentration of A that is in equilibrium with bulk water concentration, mg/L

Overall Mass Transfer Relationship

For stripping operations, mass balances are normally written on the liquid side, and it is convenient to calculate the mass transfer rate using the hypothetical concentration C_s^* and an overall mass transfer coefficient K_L, as shown in the equation

$$J_A = K_L\left(C_b - C_s^*\right) \tag{7-77}$$

where J_a = mass flux of A across air-water interface, mg/m^2·s
K_L = overall mass transfer coefficient, m/s
C_b = liquid-phase concentration of A in bulk solution, mg/L
C_s^* = liquid-phase concentration of A at air-water interface assuming no concentration gradient in air phase, mg/L

Since no mass accumulates at the interface, the hypothetical, gas-side, and liquid-side mass fluxes given in Eqs. 7-74 and 7-77 must all be equal to one another:

$$J_A = k_l\left(C_b - C_s\right) = k_g\left(y_s - y_b\right) = K_L\left(C_b - C_s^*\right) \tag{7-78}$$

Equation 7-78 relates K_L to k_l and k_g and accounts for mass transfer resistances on both the gas and liquid sides of the interface, which is known as the two-film model. The individual expressions in Eq. 7-78 can be rearranged as follows:

$$C_b - C_s = \frac{J_A}{k_l} \tag{7-79}$$

$$y_s - y_b = \frac{J_A}{k_g} \tag{7-80}$$

$$C_b - C_s^* = \frac{J_A}{K_L} \tag{7-81}$$

The overall mass transfer coefficient can be related to the local mass transfer coefficients starting with the relationship

$$C_b - C_s^* = (C_b - C_s) + (C_s - C_s^*) \tag{7-82}$$

Substituting Eqs. 7-73 and 7-75 into Eq. 7-80, and then substituting Eqs. 7-79 to 7-81 into Eq. 7-82 yields

$$\frac{J_A}{K_L} = \frac{J_A}{k_l} + \frac{J_A}{Hk_g} \tag{7-83}$$

or

$$\frac{1}{K_L} = \frac{1}{k_l} + \frac{1}{Hk_g} \tag{7-84}$$

Thus, according to the two-film model, the mass flux across the interface can be calculated using the expression

$$J_A = K_L\left(C_b - \frac{y_b}{H}\right) \tag{7-85}$$

Equation 7-85 is convenient to use because the driving force for stripping $(C_b - y_b/H)$ involves concentrations that are easy to measure. The overall mass transfer coefficient can be estimated from the local mass transfer coefficients, and the local mass transfer coefficients can be determined from correlations.

Determining the Phase That Controls Mass Transfer

Evaluating which phase controls the mass transfer rate is important in optimizing the design and operation of aeration and air-stripping processes. For example, when the liquid-phase resistance controls the mass transfer rate, increasing the mixing of the air will have little impact on the removal efficiency. From Eq. 7-84, the overall resistance to mass transfer is equal to the sum of the resistance in the liquid and gas phases and can be rewritten as

$$R_T = R_L + R_G / H \tag{7-86}$$

where R_T = overall resistance to mass transfer, equal to $1/K_L$, s/m
R_L = liquid-phase resistance to mass transfer, equal to $1/k_l$, s/m
R_G = gas-phase resistance to mass transfer, equal to $1/k_g$, s/m

To evaluate which phase controls the rate of mass transfer, Eq. 7-86 can be rearranged as follows to evaluate the liquid resistance as a fraction of total resistance:

$$\frac{R_L}{R_T} = \frac{1/k_l}{1/k_l + 1/Hk_g} = \frac{H}{H + k_l/k_g} \tag{7-87}$$

Based on Eq. 7-87, the fraction of total resistance contributed by the liquid resistance depends on the value of H relative to k_l/k_g. Reported values of the ratio k_g/k_l range are (1) 40 to 200 (Munz and Roberts, 1989) for surface aerators, (2) 5 to 50 for packed towers, and (3) 2.2 to 3.6 for diffused bubble aeration (Hsieh et al., 1993). Inverting these values to obtain k_l/k_g yields 0.005 to 0.025 for surface aerators, 0.02 to 0.2 for packed towers, and 0.28 to 0.45 for diffused bubble aeration. Assuming $k_l/k_g = 0.01$, the liquid phase controls the rate of mass transfer for compounds with H values greater than about 0.05. The gas phase controls mass transfer of compounds with H values less than 0.002. For compounds with H values between 0.002 and 0.05, the liquid and the gas phase both control the rate of mass transfer. A higher value of H indicates that the solute will have a greater concentration in the gas phase for a given concentration in the liquid phase, so the general trend expressed by Eq. 7-87 is that the phase that is less preferred by the solute is the phase that controls the mass transfer rate.

Application of the Two-Film Model

When designing aeration and stripping processes, the rate of mass transfer is often expressed on a volumetric basis rather than an interfacial area basis. The flux term is converted to a volumetric basis by multiplying by the surface area available for mass transfer per contactor vessel volume, a, as defined in Sec. 7-4. Equation 7-84 can be expressed in terms of a volumetric mass transfer rate by dividing by the area a:

$$\frac{1}{K_L a} = \frac{1}{k_l a} + \frac{1}{H k_g a} \qquad (7\text{-}88)$$

where K_L = overall liquid mass transfer coefficient, m/s
a = specific surface area m^2/m^3
k_l = liquid-phase mass transfer coefficient, m/s
k_g = gas-phase mass transfer coefficient, m/s

The combined coefficient $K_L a$ can then be incorporated into equations for mass transfer across a gas-liquid interface, using Eqs. 7-46 and 7-73:

$$M_A = K_L a\left(C_b - \frac{y_b}{H}\right)V \qquad (7\text{-}89)$$

where M_A = mass flow of A, mg/s
$K_L a$ = overall liquid-side mass transfer coefficient, s^{-1}
V = volume of contactor, m^3

Relationship between Overall Mass Transfer Coefficients and Diffusing Species

To relate the mass transfer coefficients of one compound to another when mass transfer resistances exist on both the water and gas side of the interface, Eq. 7-84 or 7-88 is combined in several ways depending on which mass transfer coefficients are known. In one approach the overall mass transfer coefficient of B can be determined from the gas- and liquid-side mass transfer coefficients for compound A. The following equation can be derived by combining Eq. 7-88 with Eqs. 7-64 and 7-67:

$$\frac{1}{K_{L,\mathrm{B}}a} = \frac{1}{k_{l,\mathrm{A}}a\left(\dfrac{D_{l,\mathrm{B}}}{D_{l,\mathrm{A}}}\right)^{n}} + \frac{1}{H_{\mathrm{B}}k_{g,\mathrm{A}}a\left(\dfrac{D_{g,\mathrm{B}}}{D_{g,\mathrm{A}}}\right)^{m}} \tag{7-90}$$

where $K_{L,\mathrm{B}}$ = overall mass transfer rate of B, s^{-1}
a = specific surface area m^2/m^3
$k_{l,\mathrm{A}}$ = liquid-phase mass transfer coefficient of compound A, m/s
$D_{l,\mathrm{B}}$ = liquid-phase diffusion coefficient of compound B, m^2/s
$D_{l,\mathrm{A}}$ = liquid-phase diffusion coefficient of compound A, m^2/s
n = empirical exponent
H_{B} = Henry's constant for compound B, L of water/L of air, dimensionless
$k_{g,\mathrm{A}}$ = gas-phase mass transfer coefficient of compound A, m/s
$D_{g,\mathrm{A}}$ = gas-phase diffusion coefficient of compound A, m^2/s
$D_{g,\mathrm{B}}$ = gas-phase diffusion coefficient of compound B, m^2/s
m = empirical exponent

As discussed in Sec. 7-5 and shown in Table 7-6, n and m have values between $\frac{1}{2}$ for no boundary layer (no velocity gradient at the interface) and $\frac{2}{3}$ for a boundary layer (velocity gradient at the interface). Normally, the gas side has a velocity gradient and the water side has no velocity gradient, so that n and m are $\frac{1}{2}$ and $\frac{2}{3}$, respectively.

In another approach, the ratio of k_g/k_l is assumed to be a constant for all compounds. This ratio is relatively constant for a given device and does not depend on the compound (Hsieh et al., 1993; Munz and Roberts, 1989). Therefore, the following simplification can be made:

$$\frac{k_{g,i}a}{k_{l,i}a} = \frac{k_g}{k_l} \tag{7-91}$$

where $k_{g,i}$ = gas-phase mass transfer coefficient of compound i, m/s
$k_{l,i}$ = liquid-phase mass transfer coefficient of compound i, m/s
a = specific surface area m^2/m^3
k_g/k_l = ratio of gas-phase mass transfer coefficient to liquid-phase mass transfer coefficient, which tends to be constant for a given separation device

Values of the inverse of the ratio k_g/k_l were presented in the earlier discussion on determination of the phase that controls the mass transfer rate. Rewriting Eq. 7-88 in terms of a compound c yields

$$\frac{1}{K_{L,i}a} = \frac{1}{k_{l,i}a} + \frac{1}{H_i k_{g,i}a} \tag{7-92}$$

Multiplying both sides of Eq. 7-92 by $k_{l,i}a$ and solve for $K_{L,i}a$ yields

$$K_{L,i}a = k_{l,i}a\left[1+\frac{1}{H_i\left(k_{g,i}a / k_{l,i}a\right)}\right]^{-1} \tag{7-93}$$

where $K_{L,i}$ = overall mass transfer coefficient for compound i, m/s
H_i = Henry's constant for compound i, L of water/L of air, dimensionless

Substituting Eq. 7-91 into Eq. 7-93 results in the expression

$$K_{L,i}a = k_{l,i}a\left[1+\frac{1}{H_i\left(k_g / k_l\right)}\right]^{-1} \tag{7-94}$$

It is convenient to choose a reference compound that is easy to measure and has all resistance to mass transfer on the liquid side, such as oxygen, because the overall mass transfer coefficient is equal to the liquid-phase mass transfer coefficient for such a compound:

$$\frac{1}{K_{L,O_2}a} = \frac{1}{k_{l,O_2}a} + \frac{1}{H_{O_2}k_{g,O_2}a} \cong \frac{1}{k_{l,O_2}a} \tag{7-95}$$

$$K_{L,O_2}a = k_{l,O_2}a \tag{7-96}$$

where K_{L,O_2} = overall mass transfer coefficient for oxygen, s^{-1}
k_{l,O_2} = liquid-phase mass transfer rate constant for oxygen, s^{-1}
a = specific surface area m^2/m^3
H_{O_2} = Henry's law constant for oxygen, L of water / L of air, dimensionless
k_{g,O_2} = gas-phase mass transfer coefficient for oxygen, m/s

The mass transfer coefficient of a given compound, i, can be related to the mass transfer coefficient for oxygen by dividing Eq. 7-94 by Eq. 7-96:

$$\frac{K_{L,i}a}{K_{L,O_2}a} = \frac{k_{l,i}a}{k_{l,O_2}a}\left[1+\frac{1}{H_i\left(k_g / k_l\right)}\right]^{-1} \tag{7-97}$$

where $K_{L,i}$ = overall mass transfer coefficient for component i,s^{-1}
$k_{l,i}$ = liquid-phase mass transfer coefficient for component i,s^{-1}
H_i = Henry's law constant for component i, L of water/L of air, dimensionless
a = specific surface area m^2/m^3
k_l = liquid-phase mass transfer coefficient, m/s

k_g = gas-phase mass transfer coefficient, m/s
k_g/k_l = ratio of gas-phase to liquid-phase mass transfer coefficients, which tends to be relatively constant for a given device and does not depend on the compound

The ratio of the mass transfer coefficients $k_{l,i}a / k_{l,O_2}a$ in Eq. 7-97 can be determined from Table 7-6 for the case where the mass transfer resistance is in the water at the air-water interface:

$$\frac{k_{l,i}a}{k_{l,O_2}a} = \frac{k_{l,i}}{k_{l,O_2}} = \left(\frac{D_{l,i}}{D_{l,O_2}}\right)^{1/2} \tag{7.98}$$

where $D_{l,i}$ = liquid-phase diffusion coefficient of compound i,m²/s
D_{l,O_2} = liquid-phase diffusion coefficient of oxygen, m²/s

The final expression for determining the mass transfer coefficient of any compound i using oxygen as a reference compound results from substituting Eq. 7-98 into Eq. 7-97:

$$K_{L,i}a = K_{L,O_2}a\left(\frac{D_{l,i}}{D_{l,O_2}}\right)^{1/2}\left[1+\frac{1}{H_i\left(k_g/k_l\right)}\right]^{-1} \tag{7-99}$$

The procedure for determining an overall mass transfer coefficient from diffusion coefficients is demonstrated in Example 7-9.

Example 7-9 Determining the overall mass transfer coefficient using oxygen as reference compound

Calculate the overall mass transfer coefficient K_La of benzene at 20°C using oxygen as a reference compound for a mechanical surface aerator. Given: The mass transfer rate constant of oxygen $(K_{L,O_2}a)$ for the device has been measured at 0.0015 s^{-1}. From Table 7-1, the liquid diffusion coefficient of benzene at 20°C is $D_{l,b}$ = 1.02 × 10^{-9} m²/s. From Example 7-6, the liquid diffusion coefficient of oxygen at 20°C is $D_{l,O_2} = 2.06\times10^{-9}\,m^2/s$. The Henry's law constant of benzene at 20°C is H = 0.188 (see Chap. 14).

Solution

1. Determine k_g/k_l. As discussed in the section on determination of the phase that controls mass transfer, for mechanical surface aeration devices, k_g/k_l values vary from 40 to 200. Choose a value of k_g/k_l = 40.

2. Determine K_La for benzene from Eq. 7-99:

$$K_La_{(\text{benzene})} = K_{L,O_2}a\left(\frac{D_{(\text{benzene})}}{D_{l,O_2}}\right)^{1/2}\left[1+\frac{1}{H(k_g/k_l)}\right]^{-1}$$

$$= (0.0015\,\text{s}^{-1})\left(\frac{1.02\times10^{-9}}{2.06\times10^{-9}}\right)^{1/2}\left[1+\frac{1}{(0.188)\,(40)}\right]^{-1}$$

$$= 0.00093\,\text{s}^{-1}$$

Comment

Choosing a value of k_g/k_l= 40 represents a conservative estimate of the mass transfer rate constant. The mass transfer rate constant predicted by Eq. 7-99 will be at its lowest value for the range of k_g/k_l values applicable to mechanical surface aeration ($40 \le k_g/k_l \le 200$) when k_g/k_l = 40.

7-9 Enhancement of Mass Transfer across an Interface by Chemical Reactions

If a chemical reaction occurs after a solute enters the water at the air-water interface, the mass transfer rate of absorption may be faster than the rate of transfer by diffusion alone. The reason for the increase is that the reaction occurs within the mass transfer boundary layer. The reaction causes a much sharper concentration gradient; and, as predicted by Fick's first law, the larger concentration gradient causes a faster mass transfer rate. There are two possible situations: (a) as shown on Fig. 7-9a, some solute is left after the mass transfer boundary layer and (b) as shown on Fig. 7-9b, there is no solute left after the mass transfer boundary layer because the chemical reaction proceeds rapidly.

Several gases commonly used in water treatment applications, notably chlorine, sulfur dioxide, carbon dioxide, and ozone, undergo hydrolysis reactions or rapid chemical reactions with other solutes and water. The increase in rate of mass transfer must be considered in the process design of absorption equipment. The magnitude of this increase depends on the type of chemical reaction (e.g., reversible, irreversible, series), reaction order, reaction rate constants, concentration of reactants, and the solute diffusion coefficients (Danckwerts, 1970).

While innumerable reaction combinations are possible, chemical reactions for gases in water used in water treatment fall mainly in two categories: first-order, irreversible reactions and rapid or instantaneous reversible reactions.

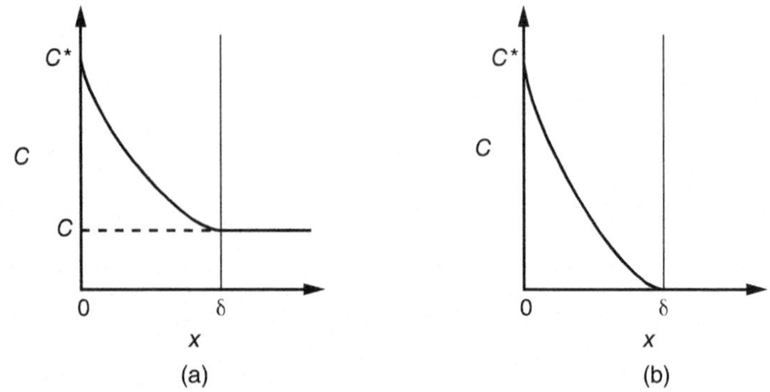

Figure 7-9
Concentration profiles in the mass transfer boundary layer for two situations where a chemical reaction is occurring: (a) some solute is left after the mass transfer boundary layer and (b) no solute is left after the mass transfer boundary layer.

Consider a first-order, irreversible reaction in the water phase with rate constant k_r, as illustrated by the nearly irreversible reaction of SO_2 in water:

$$SO_2(g) + H_2O \xrightarrow{k_r} HSO_3^- + H^+ \tag{7-100}$$

where k_r = first-order reaction rate constant, s^{-1}

Assuming that mass transfer resistances in the gas phase are negligible, it has been shown that the gas transfer flux at steady state is given by the expression (Danckwerts, 1970)

$$J_A = k_l \left[C^* - \frac{C}{\cosh\,(\mathrm{Ha})} \right] \frac{\mathrm{Ha}}{\tanh\,(\mathrm{Ha})} \tag{7-101}$$

where J_A = gas transfer flux, mg/m²·s
k_l = liquid-phase mass transfer coefficient, m/s
C^* = liquid-phase concentration in equilibrium with bulk gas concentration, mg/L
C = concentration in bulk water, mg/L
Ha = Hatta number, dimensionless
cosh = hyperbolic cosine
tanh = hyperbolic tangent

The Hatta number is given by the expression

$$\mathrm{Ha} = \frac{\sqrt{D_l k_r}}{k_l} \tag{7-102}$$

where D_l = diffusion coefficient of solute in water, m^2/s
k_r = first-order reaction rate constant, s^{-1}

When a single gas such as pure chlorine or carbon dioxide is absorbed, mass transfer resistances in the gas phase are negligible, and Eq. 7-101 can be used to predict the enhancement of the mass transfer rate due to reaction. The mass transfer coefficient k_l shown in Eq. 7-101 can be estimated using a mass transfer correlation, and the Hatta number can be calculated from the diffusion coefficient and first-order rate constant of the species.

The second term that appears on the right-hand side of Eq. 7-101 defines the driving force. If Ha > 5, the reaction occurs entirely in the water near the interface, and the bulk water concentration of the dissolving solute is zero, as shown on Fig. 7-9b. In this case, Eq. 7-101 simplifies to the following expression [note that tanh(5) = 1]:

$$J_A = k_l Ha\, C^* \tag{7-103}$$

The flux in the absence of reaction for a zero bulk solution concentration would be

$$J_A = k_l C^* \tag{7-104}$$

By comparing Eqs. 7-103 and 7-104, it is seen that the rate of mass transfer is enhanced by a factor that is equal to the Hatta number, Ha:

$$E_{flux} = \frac{\text{flux with reaction}}{\text{flux without reaction}} = \frac{(\text{Ha})\, k_l}{k_l} = \text{Ha} \tag{7-105}$$

where E_{flux} = enhancement in mass transfer due to reaction

Enhancement factors for many other chemical reaction types are summarized by several authors (Danckwerts, 1970; Levenspiel, 1998; Sherwood et al., 1975). Enhancements of Ha = 1000 are possible for fast radical reactions. The impact of hydrolysis on the rate of mass transfer is demonstrated in Example 7-10.

Example 7-10 Impact of rapid hydrolysis reactions on mass transfer

Many gases added to water during water treatment that undergo rapid hydrolysis reactions can accelerate the rate of absorption. Calculate the enhancement of initial mass transfer rate for the absorption of pure gaseous chlorine into water. The initial chlorine concentration in the bulk solution is equal to zero. Given: The temperature is 20°C. Tests were conducted using oxygen, and the oxygen liquid-phase mass transfer coefficient k_{l,O_2} was determined to be 10^{-4} m/s. Using Eq. 7-32, the liquid-phase diffusion coefficient for chlorine, D_{l,Cl_2}, based on a molar volume of 43.2 cm^3/mol (2 × 21.6, see Table 7-3), is 1.41×10^{-9} m^2/s. Using Eq. 7-35, the liquid-phase diffusion coefficient for oxygen, D_{l,O_2}, is 2.06×10^{-9} m^2/s. The rate constant k_r for chlorine is 11 s^{-1} at 20°C (Eigen and Kustin, 1962). Chlorine reacts with water as follows:

$$Cl_2 + H_2O \rightleftarrows HOCl + HCl \tag{a}$$

$$HOCl \rightleftarrows H^+ + OCl^- \tag{b}$$

The reaction to form HOCl is nearly complete and can be considered a pseudo-first-order, irreversible reaction.

$$Cl_2 \xrightarrow{k_r} \text{products}$$

Solution

The enhancement is equal to the relative initial mass transfer rate, which is equal to the ratio of the mass transfer flux for chlorine enhanced by reaction to the mass transfer flux for chlorine in the absence of reaction.

1. The initial mass transfer flux for chlorine without considering the hydrolysis reaction is given by Eq. 7-45, for which the initial chlorine concentration in the bulk solution is equal to 0:

$$J_{Cl_2} = k_{l,Cl_2} C^*_{Cl_2}$$

2. The enhanced gas transfer flux due to reaction is described by the following expression (Eq. 7-101) when the bulk concentration is zero:

$$J_{Cl_2} = k_{l,Cl_2} C^*_{Cl_2} \frac{\text{Ha}}{\tanh(\text{Ha})}$$

3. Thus, the enhancement of the initial mass transfer rate for the absorption of pure gaseous chlorine is given by the ratio of the mass transfer with (step 2) and without (step 1) the hydrolysis reaction.

$$\text{Relative initial mass transfer rate} = \frac{k_{l,Cl_2}[\text{Ha}/\tanh(\text{Ha})]}{k_{l,Cl_2}} = \frac{\text{Ha}}{\tanh(\text{Ha})}$$

a. Determine k_{l,Cl_2} using the relationship for mass transfer resistance in the water at the air-water interface in Table 7-6. Rearranging the expression in Table 7-6 to solve for k_{l,Cl_2} yields

$$k_{l,Cl_2} = k_{l,O_2}\left(\frac{D_{l,Cl_2}}{D_{l,O_2}}\right)^{1/2} = (10^{-4}\,\text{m/s})\left(\frac{1.41\times10^{-9}}{2.06\times10^{-9}}\right)^{1/2}$$
$$= 8.26\times10^{-5}\,\text{m/s}$$

b. Determine Ha using Eq. 7-102:

$$\text{Ha} = \frac{\sqrt{D_l k_r}}{k_l} = \frac{\sqrt{D_{l,Cl_2} k_r}}{k_{l,Cl_2}} = \frac{\left[(1.41\times10^{-9}\,\text{m}^2/\text{s})(11\,\text{s}^{-1})\right]^{1/2}}{(8.26\times10^{-5}\,\text{m/s})}$$
$$= 1.51$$

c. Determine tanh(Ha):

$$\tanh(\text{Ha}) = \tanh(1.51) = 0.907$$

d. Determine the relative rate:

$$\text{Relative rate} = \frac{\text{Ha}}{\tanh(\text{Ha})} = \frac{1.51}{0.907} = 1.66$$

Comment

The initial mass transfer rate for chlorine is 66 percent faster because of the hydrolysis reaction. The value of cosh(Ha) for chlorine is 2.37;consequently, the second term in the mass transfer rate expression (see Eq. 7-101) is lower by a factor of 2.37 because of the hydrolysis of chlorine, and the mass transfer rate for chlorine would always be significantly enhanced by chemical reaction. Therefore, the mass transfer device can be smaller than expected by considering mass transfer alone.

Problems and Discussion Topics

7-1 Explain which method or method(s) of estimating diffusion coefficients are best suited for the following cases:

a. Methane gas diffusing in water

b. A sugar cube diffusing in a cup of iced tea

c. A globular protein diffusing in blood

d. NaCl diffusing in water

7-2 Using the Stokes–Einstein equation, derive an expression for diffusion coefficient D_l of a large molecule diffusing through water as a function of the water temperature T, water viscosity μ_p, molecular weight of the molecule (MW), and molecular density of the molecule (ρ_M). *Hint*: Use the definition of density to relate ρ_M to MW and molecular diameter d_M. Be sure to specify units of measure.

7-3 Estimate the diffusion coefficient for humic acid in water at 25°C using (a) the Stokes–Einstein equation and (b) the Polson equation. The MW of humic acid is 3000 and molecular diameter is 0.8 nm.

7-4 Use (a) the Hayduk–Laudie correlation and (b) the Nernst–Haskell equation to estimate the diffusion coefficient of acetic acid (CH_3COOH) in water at 25°C assuming no electrical field is present in the solution. Compare the results from these two different methods to the measured value cited in Table 7-1. Assume CH_3COOH is fully protonated for the Hayduk–Laudie correlation and fully dissociated for the Nernst–Haskell equation. *Hint*: One of the oxygen atoms in CH_3COOH has a double bond to carbon while the other oxygen has a single bond to hydrogen and a single bond to carbon.

7-5 Using (a) the Hayduk–Laudie correlation and (b) the Nernst–Haskell equation, estimate the diffusion coefficient in water of one of the following compounds (compound to be selected by the instructor): chloroacetate ($ClCH_2COOH$), cyanoacetate ($CNCH_2COOH$), or benzoic acid (C_6H_5COOH). Calculate the diffusion coefficient at the following temperatures: 10, 15, 20, 25, 30, and 40°C. Compare the two methods (Hayduk–Laudie correlation and Nernst-Haskell equation) by plotting the diffusion coefficient as a function of temperature.

7-6 Estimate the diffusion coefficient of the dissolved gas CO_2 in water at 25°C. Compare the result to the value given in Table 7-1.

7-7 Estimate the diffusion coefficient of one of the following dissolved gases (compound to be selected by the instructor) in water: O_2, CO_2, Cl_2, or NH_3. Calculate the diffusion coefficient at the following temperatures: 10, 15, 20, 25, 30, and 40°C and plot as a function of temperature.

7-8 Determine the specific surface area (the area available for mass transfer per volume) of the particle, for the following particles: sphere (diameter d), square (side length L), and cylinder (length L and diameter d).

7-9 Determine the specific surface area (the area available for oxygen transfer per volume) for a concrete sewer pipe that has a circular cross section of radius r filled to a channel height h and having a length L for the following cases: (a) $h \leq r$ and (b) $h > r$.

7-10 Raw water that has an influent pH of 2.8 is to be fed to a packed bed of crushed limestone to raise the pH and add hardness (as Ca^{2+}). The temperature is 25°C, the bed porosity is 0.5, and the particle sphericity is 0.75. Calculate the film transfer coefficient for limestone media for 0.5-, 1.5-, 2-, or 3-cm limestone particles (particle size to be specified by the instructor). The flow rate is 800 L/min and the superficial velocity is 10 m/h. Use the Nernst–Haskell equation to estimate the diffusion coefficient.

7-11 Determine the mass transfer coefficient of tetrachloroethene (PCE) on the water side of an air-water interface by relating mass transfer coefficients and diffusivities of PCE to those of oxygen. Given: The temperature is 15°C. Calculate the liquid diffusion coefficient of PCE at 15°C from the Hayduk–Laudie correlation. The mass transfer coefficient of oxygen in the mass transfer device at 15°C is $k_{l,O_2} = 0.0045$ m/s.

7-12 During an experiment, various amounts of PAC are added to separate 500-mL bottles filled with water containing 25 mg/L of an organic contaminant. The contents of the sealed bottles were mixed and allowed to equilibrate for 2 weeks (this is adequate time to achieve equilibrium). Analysis of the water revealed that the aqueous-phase organic contaminant decreased to the concentrations specified in the table below. Plot the operating lines (McCabe–Thiele diagram) and specify the *V/M* ratios. Draw the phase equilibrium line.

	Bottle				
Item	**A**	**B**	**C**	**D**	**E**
PAC dosage, mg	6	10	20	40	200
Aqueous equilibrium concentration, mg/L	16.2	11.7	6.1	3.1	2.1

7-13 Derive the following expression based on the two-film model:

$$\frac{1}{K_G a} = \frac{H}{k_l a} + \frac{1}{k_g a}$$

7-14 Many gases added to water during water treatment undergo rapid hydrolysis reactions that accelerate the rate of absorption. Calculate the enhancement of initial mass transfer rate for the absorption of carbon dioxide into water. The temperature is 20°C. Tests were conducted using oxygen, and the oxygen liquid-phase mass transfer coefficient was determined to be 10^{-4} m/s. The liquid diffusion coefficient for oxygen, D_{l,O_2}, is 2.067×10^{-9} m^2/s.
The rate constant k_1 for CO_2 is 0.02 s^{-1}. The CO_2 reacts with water as follows:

$$CO_2(g) + H_2O \rightleftarrows H_2CO_3$$

$$H_2CO_3 \rightleftarrows HCO_3^- + H^+$$

$$HCO_3^- \rightleftarrows CO_3^{2-} + H^+$$

At neutral pH, the absorption of carbon dioxide can be thought of as the following irreversible reaction:

$$CO_2 \xrightarrow{k_1} \text{products}$$

7-15 Estimate the enhancement of the mass transfer rate for So_2 absorption. For So_2, the pertinent reactions are

$$SO_2(g) + H_2O \rightleftarrows H_2SO_3$$

$$H_2SO_3 \rightleftarrows HSO_3^- + H^+$$

$$HSO_3 \rightleftarrows SO_3^- + H^+$$

At neutral pH, H_2So_3 rapidly dissociates, and the So_2 reaction with water becomes pseudo-first order, given as

$$SO_2 \xrightarrow{k_1} \text{products}$$

For SO_2, the rapid hydrolysis reaction dramatically enhances the rate of absorption. The first-order rate constant k_1 at 20°C is 3.4× 10^6 s^{-1} (Eigen et al., 1961).

7-16 Calculate the mass transfer rate constant of tetrachloroethene (PCE), K_La, at 13°C using oxygen as a reference compound for a mechanical surface aerator. Given: The mass transfer rate constant of oxygen is $K_{L,O_2}a = 0.0015$ s^{-1}. Calculate the liquid diffusion coefficient of PCE at 13°C using the Hayduk–Laudie correlation. The Henry's law constant of PCE at 13°C is $H = 0.50$.

References

Brown, R. (1827) "A Brief Account of Microscopical Observations on the Particles Contained in the Pollen of Plants and the General Existence of Active Molecules in Organic and Inorganic Bodies," unpublished work available at <http://sciweb.nybg.org/science2/pdfs/dws/Brownian.pdf>; accessed on Dec. 1, 2010.

CRC (2003) *CRC Handbook of Chemistry and Physics*, 84th ed., CRC Press, Boca Raton, FL.

Danckwerts, P. V. (1951) "Significance of Liquid-Film Coefficients in Gas Absorption," *Ind. Eng. Chem.*, **43**, 6, 1460–1467.

Danckwerts, P. V. (1955) "Gas Absorption Accompanied by Chemical Reaction," *AIChE J.*, **1**, 4, 456–463.

Danckwerts, P. V. (1970) *Gas-Liquid Reactions*, McGraw-Hill, NewYork.

Eigen, M., and Kustin, K. (1962) "The Kinetics of Halogen Hydrolysis," *J. Am. Chem. Soc.*, **84**, 8, 1355–1361.

Eigen, M., Kustin, K., and Mass, G. (1961) "Die Geschwindigkeit der Hydratation von SO_2 in wäßäriger Lösung," *Z. Phys. Chem. Neue Folge*, **30**, 130–136.

Fick, A. (1855) "On Liquid Diffusion," *Phil. Mag. Series 4*, **10**, 63, 30–39.

Gilliland, E.R., and Sherwood, T.K. (1934) "Diffusion of Vapors into Air Streams," *Ind. Eng. Chem.*, **26**, 5, 516-523.

Gnielinski, V. (1978) "Gleichungen Zur Berechnung Des Warme- Und Stoffaustausches in Durchstromten Ruhenden Kugelschüttungen Bei Mittleren Und Grossen Pecletzahlen," *Verf. Tech.*, **12**, 6, 363–366.

Gnielinski, V. (1981) "Equations for the Calculation of Heat and Mass Transfer During Flow through Stationary Spherical Packings at Moderate and High Peclet Numbers," *Int. Chem. Eng.*, **21**, 3, 378–383.

Hayduk, W., and Laudie, H. (1974) "Prediction of Diffusion Coefficients for Nonelectrolytes in Dilute Aqueous Solutions," *AIChE J.*, **20**, 3, 611–615.

Higbie, R. (1935) "The Rate ofAbsorption of a Pure Gas into a Still Liquid During Short Periods of Exposure," *Trans. Am. Inst. Chem. Eng.*, **31**, 365–389.

Hirschfelder, J. O., Bird, R. B., and Spotz, E. L. (1949) "The Transport Properties of Gases and Gaseous Mixtures. II," *Chem. Rev.*, **44**, 1, 205–231.

Holmén, K., and Liss, P. (1984) "Models for Air-Water Gas Transfer: An Experimental Investigation," *Tellus*, **36B**, 92–100.

Hsieh, C., Ro, K. S., and Stenstrom, M. (1993) "Estimating Emissions of 20 VOCs: I. Surface Aeration," *J. Environ. Eng.*, **119**, 6, 1077–1098.

Laidler, K. J., and Meiser, J. H. (1999) *Physical Chemistry*, Houghton Mifflin, Boston.

LeBas (1915) *The Molecular Volumes of Liquid Chemical Compounds*, Longmans, London.

Levenspiel, O. (1998) *Chemical Reaction Engineering*, 3rd ed., John Wiley & Sons, New York.

Lewis, W. K., and Whitman, K. L. (1924) "Principles of Gas Absorption," *Ind. Eng. Chem.*, **16**, 12, 1215–1220.

Linton, W. H. J., and Sherwood, T. K. (1950) "Mass Transfer from Solid Spheres to Water in Streamline and Turbulent Flow," *Chem. Eng. Progr.*, **46**, 258–264.

Lyman, W. J., Reehl, W. F., and Rosenblatt, D. H. (1990) *Handbook of Chemical Property Estimation Methods: Environmental Behavior of Organic Compounds*, American Chemical Society, Washington, DC.

Malik, V. K., and Hayduk, W. (1968) "A Steady-State Capillary Cell Method for Measuring Gas-Liquid Diffusion Coefficients," *Canadian J. Chem. Eng.*, **46**, 6, 462–466.

Marrero, T. R., and Mason, E. A. (1972) "Gaseous Diffusion Coefficients," *J. Phys. Chem. Ref. Data*, **1**, 1, 3–118.

McCabe, W. L., and Thiele, E. W. (1925) "Graphical Design of Fractionating Columns," *Ind. Eng. Chem.*, **17**, 6, 605–611.

Munz, C., and Roberts, P. V. (1989) "Gas-and Liquid-Phase Mass Transfer Resistance of Organic Compounds During Mechanical Surface Aeration," *Water Res.*, **23**,5, 589–601.

Onda, K., Takeuchi, H., and okumoto, Y. (1968) "Mass Transfer Coefficients between Gas and Liquid Phases in Packed Columns," *J. Chem. Eng. Jpn.*, **1**, 1, 56–62.

Othmer, D. F., and Thakar, M. S. (1953) "Correlating Diffusion Coefficients in Liquids," *Ind. Eng. Chem.*, **45**, 3, 589–593.

Poling, B. E., Prausnitz, J. M., and O'Connell, J. P. (2001) *The Properties of Liquids and Gases*, 5th ed., McGraw-Hill, New York.

Polson, A. (1950) "Some Aspects of Diffusion in Solution and a Definition of a Colloidal Particle," *J. Phys. Colloid Chem.*, **54**, 649–652.

Robinson, R. A., and Stokes, R. H. (1959) *Electrolyte Solutions: The Measurement and Interpretation of Conductance, Chemical Potential and Diffusion in Solutions of Simple Electrolytes*, 2nd ed., Butterworths, London.

Seinfeld, J. H., and Pandis, S. N. (1998) "Dynamics of Single Aerosol Particles," Chap. 8, *Atmospheric Chemistry and Physics*, John Wiley & Sons, New York.

Sherwood, T. K., Pigford, R. L., and Wilke, C. R. (1975) *Mass Transfer*, McGraw-Hill, New York.

Treybal, R. E. (1980) *Mass-Transfer Operations*, 3rd ed., McGraw-Hill, New York.

Wilke, C. R., and Lee, C. Y. (1955) "Estimation of Diffusion Coefficients for Gases and Vapors," *Ind. Eng. Chem.*, **47**, 6, 1253–1257.

8 Chemical Oxidation and Reduction

Stantec's Water Treatment: Principles and Design, Updated Third Edition. John C. Crittenden, R. Rhodes Trussell, David W. Hand, Kerry J. Howe, and George Tchobanoglous.

Terminology for Chemical Oxidation and Reduction

Term	Definition
Advanced oxidation processes	Processes that generate hydroxyl radical at room temperature and pressure.
Anode	Electrode in a electrochemical cell where oxidation takes place.
Cathode	Electrode in a electrochemical cell where reduction takes place.
Chromophores	Functional groups or bonds on chemical compounds responsible for the absorption of light.
Conventional oxidation processes	Oxidation processes that achieve oxidation without the generation of hydroxyl radicals,
Electron acceptor	Reactant that gains electrons in a redox reaction; an oxidant.
Electron donor	Reactant that loses electrons in a redox reaction; a reductant.
Oxidant	Reactant that causes the oxidation of a reduced species in a redox reaction. Oxidants are electron acceptors.
Oxidation reaction	Chemical half-reaction in which a reactant loses electrons.
Reductant	Reactant that causes the reduction of an oxidized species in a redox reaction. Reductants are electron donors.
Redox reaction	Abbreviated name for oxidation–reduction reaction
Reduction reaction	Chemical half-reaction in which a reactant gains electrons.

In water treatment, chemical oxidation and reduction processes are used for the treatment of specific inorganic or organic species found in water. For organic compounds, the purpose is to convert compounds into harmless or nonobjectionable forms. For example, it is desirable to oxidize toxic organic compounds into carbon dioxide and mineral acids (e.g., HCl) or taste and odor compounds into nonodorous compounds. Inorganic metal species (e.g., iron or manganese) are oxidized to insoluble forms and are removed by precipitation. Other inorganic species such as hydrogen sulfide, an odorous gas, is oxidized to nonodorous sulfate.

Because many types of oxidation processes have been developed and are used in various applications, it is useful to note some important differences between (1) conventional oxidation processes, (2) oxidation processes carried out at elevated temperatures and/or pressure, and (3) advanced oxidation processes. Conventional chemical oxidation processes employing such oxidants as chlorine, chlorine dioxide, or potassium permanganate do not produce highly reactive species, such as the hydroxyl radical (HO•), which are produced in the other two types of oxidation processes (the dot placed after the hydroxyl and other radical species indicates that there is an unpaired electron in the outer orbital). Hydroxyl radicals are reactive electrophiles that readily react with most organic compounds by undergoing addition reactions with double bonds or extracting hydrogen atoms from organic compounds. Reaction with conventional oxidants are more specific with regard to the types of organic molecules that can be oxidized, and the reaction rates for conventional oxidants are slower than the reaction rates involving HO•. Nevertheless, conventional oxidation processes can be effective in oxidizing certain organic and inorganic compounds. Wet oxidation, supercritical oxidation, gas-phase combustion, and catalytic oxidation processes are also known to oxidize organic matter. These processes require elevated temperatures and/or high pressures and are carried out by free-radical reactions involving HO•. In advanced oxidation processes (AOPs), HO• radicals are generated at ambient temperature and atmospheric pressure.

The purpose of this chapter is to introduce the general subject of conventional oxidation. AOPs will be described in Chap. 18. Topics to be considered include (1) an introduction to the use of chemical oxidation in water treatment, (2) the fundamentals of chemical oxidation and reduction, (3) discussion of the common chemical oxidants used in water treatment, and (4) photolysis theory and applications. The chemistry, storage, and production of oxidants used for disinfection are considered in Chap. 13. Ultraviolet disinfection is also discussed in Chap. 13. By-products formed during disinfection are discussed in Chap. 19. The process engineering aspect of iron and manganese oxidation and removal and arsenic [As(III)] oxidation are addressed in Chap. 20.

8-1 Introduction to Use of Oxidation Processes in Water Treatment

Historically, the term "oxidation" was used to describe the combining of an element with oxygen to form an oxide and "reduction" was used to describe the removal of an oxygen from an oxide to yield the element (McMurry and Fay, 2003). Today, the terms oxidation and reduction have new and more inclusive definitions. *Oxidation* involves the loss of one or more electrons and *reduction* involves the gain of one or more electrons. Taken together, oxidation and reduction reactions are referred to as *redox* reactions. Before discussing the details of conventional oxidation, it is important to introduce the oxidants used in water treatment and their principal applications.

Water treatment can employ either oxidation or reduction as a treatment process, although oxidation is the most common of the two. Oxidation is used to destroy chemical constituents that are in a reduced state, such as toxic organic or odorous compounds and inorganic compounds such as iron, manganese, or hydrogen sulfide. Reduction is used for denitrification and quenching of residual oxidants.

Commonly Used Oxidants

The principal oxidants used in water treatment and their corresponding applications are summarized in Table 8-1. With the exception of the hydroxyl radical, which is involved in AOPs, the other oxidants are often termed *conventional* in that they are in common use. Oxidants that are frequently used in water treatment are (1) chlorine, (2) ozone, (3) chlorine dioxide, (4) permanganate, and (5) hydrogen peroxide. The oxidants are usually added at the beginning (e.g., preoxidation) or end (e.g., disinfection) of the water treatment process; however, oxidants are also added at a variety of intermediate points depending on the treatment objectives.

Application of Conventional Oxidants in Water Treatment

The principal applications of chemical oxidation are for

1. Taste and odor control
2. Hydrogen sulfide removal
3. Color removal
4. Iron and manganese removal
5. Disinfection

Table 8-1
Oxidants and their applications in water treatment

Purpose	Oxidants	Applications
Oxidation of reduced inorganic species	Chlorine, hydrogen peroxide, permanganate, chlorine dioxide	Convert soluble metals such as Fe(II) and Mn(II) to insoluble forms; oxidize odorous sulfide; destroy metal organic complexes
Oxidation of organics	Ozone, AOPs, ultraviolet light, permanganate, chlorine dioxide	Destroy taste- and odor-causing compounds; destroy toxic organics [e.g., pesticides, benzene, trichloroethene, methyl tertiary-butyl ether (MTBE)]; eliminate color; reduce natural organic matter and disinfection by-product precursors
Coagulation aids	Ozone	Reduce amount of coagulant and/or improve coagulation process
Biocidal agents	Ozone, chlorine, iodine, ultraviolet light	Control nuisance growths such as algae in pretreatment basins or reservoirs; as primary disinfectants to meet *Ct*[a] regulations (discussed in Chap. 13)

[a]*Ct* = product of oxidant residual concentration (mg/L) and contact time (min).

Each of the above applications, with the exception of disinfection, is introduced in the following discussion. Because of the importance of disinfection in water treatment, a separate chapter (Chap. 13) is devoted to this subject.

TASTE AND ODOR CONTROL

Because of the various combinations of inorganic and organic compounds that cause tastes and odors in water supplies, a wide variety of treatment processes are employed to treat taste and odors. Because most known taste and odor compounds are present in a reduced form, some form of oxidation is usually effective. Generally, no simple treatment process is cost effective for all taste and odor issues, and a case-by-case analysis is recommended. The use of granular and powdered activated carbon (GAC and PAC) for the control of taste and odor is discussed in Chap. 15.

Both surface waters and groundwaters can be contaminated with anthropogenic chemicals that impart taste and odor. Taste and odor in surface waters and groundwaters are discussed separately in the following sections.

Taste and odor in surface waters

Both inorganic and organic compounds can cause taste and odor problems. The most significant taste and odor problem in surface waters is from naturally occurring organic compounds that are produced by algal blooms and bacteria. Taste and odor outbreaks are seasonal, and according to a recent survey in North America, outbreaks usually occur between June and October (Graham et al., 2000). The three principal organoleptic compounds, geosmin, 2-methylisoborneol (MIB), and cyclocitral are thought to be produced and released into the water by actinomycetes and cyanobacte- ria. Reported threshold odor concentrations for geosmin and MIB are very low, 4 and 9 ng/L, respectively (McGuire et al., 1981). Geosmin and MIB concentrations above 7 and 12 ng/L have resulted in consumer complaints (Simpson and MacLeod, 1991). Accordingly, the treatment objective for these compounds must be in the low-nanogram-per-liter concentrations. Achieving exceptionally low values (below 5 ng/L) can be a challenge because during peak summer months the concentrations of geosmin and MIB in the raw water can reach 17 and 70 μg/L, respectively (Bruce et al., 2002). A comparison of a number of taste and odor control methods for geosmin and MIB is presented in Table 8-2. No single technology or oxidant does an excellent job with the exception of ozone and ozone/hydrogen peroxide.

Taste and odor in groundwaters

The most important taste and odor problems for groundwater are from naturally occurring inorganic compounds and mercaptans (organic sulfides), which are caused by the reducing environment found in groundwater.

Table 8-2
Removal of geosmin and methylisoborneol (MIB) that were spiked into filtered water at initial concentration of 100 ng/L

Chemical	Chemical Feed Rate, mg/L	Removal, %	
		Geosmin	MIB
Powdered activated carbon	10	40	62
	25	52	65
Potassium permanganate	0.8	42	28
Chlorine	2	45	33
Hydrogen peroxide	1	50	72
Ozone	2.5	94	77
Ozone and hydrogen peroxide	2.5, 0.5	97	95

Source: Adapted from Kawamura (2000).

The most important components that cause taste and/or odor are iron, manganese, and hydrogen sulfide. Manganese and hydrogen sulfide removal are discussed in this section following the discussion of the removal of organic taste- and odor-causing compounds.

Commonly used oxidants for taste and odor control
The oxidants most commonly employed for the destruction of chemicals that cause tastes and odors are hydrogen peroxide, chlorine, permanganate, ozone, and chlorine dioxide. AOPs (discussed in Chap. 18) are also effective at destroying geosmin and MIB (Glaze et al., 1990). However, it is unlikely that AOPs would be used for taste and odor control because ozone alone appears to be effective in eliminating geosmin and MIB (ozone can be an AOP because it generates HO• when it reacts with natural organic matter). Other benefits of ozone are (1) it is the only effective oxidant that does not increase total dissolved solids (TDS) and (2) water purveyors can receive disinfection credit for its use.

Chlorine often increases odor problems, especially when used to destroy odors of industrial or algal origin due to (1) formation of volatile products or (2) lyses of algae cells and release of odorants (Burttschell et al., 1959). For example, when low dosages of chlorine are added to water that contains phenols, chlorophenol compounds are formed and impart an objectionable medicinal taste to the water. The taste-producing intensity of the water increases up to a maximum after which increasing chlorine doses reduces and finally eliminates chlorophenolic tastes (Ettinger and Ruchhoft, 1951; Riddick, 1951). However, application of large doses of chlorine are not recommended because of the formation of chlorination by-products; consequently, measures must be taken to remove the phenol before chlorination.

HYDROGEN SULFIDE REMOVAL

Hydrogen sulfide (H_2S) is occasionally present in groundwaters. Hydrogen sulfide has an objectionable and readily identifiable "rotten-egg" odor, so it must be removed from drinking water to make the water aesthetically acceptable. Hydrogen sulfide also increases the corrosiveness of some waters to metal and concrete, and sulfides promote the growth of various filamentous sulfur bacteria, leading to a general degradation of water quality. Oxidants that have been used for the removal of hydrogen sulfide include (1) chlorine, (2) hydrogen peroxide, (3) potassium permanganate, and (4) ozone.

The biggest problem associated with hydrogen sulfide removal using oxidation is the formation of polysulfides (usually S_8) and turbidity. The formation of polysulfides is unavoidable if the hydrogen sulfide concentration is greater than about 1 mg/L, and oxidant dosages in excess of the stoichiometric requirement and pH values greater than 8 are required to assure conversion to sulfate. At pH values above 9, it appears that polysulfides do not form, which may be the reason that alkaline groundwaters that contain sulfides and have been lime softened do not exhibit threshold odors after chlorination.

There are significant problems with polysulfides that include (1) removal difficulty, (2) unique taste and odor problems, and (3) the ability to complex with metals in distribution systems, leading to the formation of black water. In most cases, pilot studies are required to determine the most suitable treatment methods to avoid the formation of polysulfides and/or to evaluate liquid–solid separation methods to remove polysulfides once formed.

COLOR REMOVAL

Color, primarily imparted to water by the degradation of dead plant matter (also known as natural organic matter, or NOM), is characteristically yellow in color and is often associated with double bonds in polyaromatic hydrocarbons. The soluble organic carbon that is formed from the degradation of dead plant matter includes humic acids and other substances that are generally referred to as humic substances. As discussed in Chap. 2, color can be expressed in platinum–cobalt units or light absorption at a specified wavelength. The double bonds that absorb visible light also absorb UV light, and color and UV light absorption at 254 nm correlate with one another. Furthermore, UV light absorption and disinfection by-product (DBP) formation are related; consequently, color, UV light absorption at 254 nm, and DBP formation are all related to one another. The reason that these parameters are related to one another is that the reaction centers correspond to the chromophores on NOM (e.g., double bonds and metal humic complexation sites) (Benjamin et al., 1997).

Chlorine has been used to remove color, but it is no longer considered a viable option because chlorine reacts with NOM to form chlorinated by-products. Chlorine dioxide is effective at color removal, but the production of the by-product chlorite has to be considered. Ozone is also effective at color removal, depending on the ozone-to-dissolved organic carbon (DOC) dosage ratio. However, ozone produces biodegradable compounds such as aldehydes and ketones, and these may stimulate biofilm growth in the distribution system. Thus, when ozone is used as an oxidant, the production of biodegradable organic matter has to be considered and biological treatment downstream of the coagulation process may be required. As discussed in Chap. 11, biologically active filtration can be used to remove the biodegradable organic matter. Because of the relationship between color and DBP formation, processes that are used to reduce the concentration of NOM for DBP formation control, presented in Sec. 19-2, can also be used for color removal.

OXIDATION AS A COAGULATION AID

Oxidants may aid the coagulation and flocculation process in several ways. First, it appears that particles adsorb negatively charged NOM that imparts a negative charge on the particles, causing particle repulsion and stability. Oxidant addition is thought to react with the adsorbed negatively charged NOM and make it more polar, which causes some of the NOM to desorb, leading to particles with a lower net negative surface charge. In addition, the oxidant may react with the adsorbed organics and make them bind more readily with Al(III) and Fe(III). As a result, the particles lose some of their negative charge or are destabilized more easily using metal salts and therefore flocculate more readily. Second, oxidants can react with NOM in the bulk solution and produce carboxylic acid groups that bind calcium ions, and this binding can cause direct precipitation of NOM.

Depending on the oxidant and dosage, chemical oxidants are also thought to destroy the functional groups that are responsible for metal complexation. The amount of metal coagulant that must be added for particle destabilization will be reduced by either NOM precipitation or reduction in NOM metal complexation sites (Reckhow et al., 1986).

IRON AND MANGANESE REMOVAL

Some of the chemistry of iron and manganese removal is reviewed here, but process engineering details are discussed in Chap. 20. Oxidants that have been used to oxidize and precipitate iron and manganese include (1) oxygen, (2) chlorine, (3) chlorine dioxide, (4) hydrogen peroxide, (5) ozone, and (6) potassium permanganate. However, because iron forms a strong complex with NOM, it has been found that oxygen, permanganate, chlorine dioxide, and free chlorine are unable to oxidize iron in many waters (Knocke et al., 1991). As a result, the feasibility of using chemical oxidation for iron removal has to be evaluated on a case-by-case basis using batch or pilot testing. Chemical oxidation may have to be combined with processes such as coagulation and adsorption, which are used to remove NOM.

OXIDATION OF SELECTED TRACE ORGANIC CONSTITUENTS

Another important role of chemical oxidation is the destruction of anthropogenic or synthetic toxic organics. Conventional oxidants that have been used for this purpose include (1) hydrogen peroxide, (2) ozone, (3) chlorine, (4) chlorine dioxide, and (5) potassium permanganate. As noted previously, the use of HO• for the oxidation of trace constituents is considered in Chap. 18.

8-2 Fundamentals of Chemical Oxidation and Reduction

The fundamental concepts involved in oxidation and reduction reactions are introduced and discussed in this section. These concepts include (1) fundamentals of redox reactions, (2) standard electrode potentials and redox equilibrium reactions, (3) E_H–pH predominance area diagrams, and (4) rate of oxidation–reduction processes.

Introduction to Redox Reactions

Redox reactions, as discussed in the previous section, are processes in which electrons are exchanged between reacting constituents (atoms, molecules, or ions). The driving force for the exchange of electrons between constituents is a decrease in the electrical potential, which is analogous to what happens when a live electrical wire is grounded and electrons flow from wire to ground (McMurry and Fay, 2003).

HALF REACTIONS

When an oxidant is added to water and a redox reaction takes place, electrons are transferred from the reductant to the oxidant. The constituent that gains electrons is reduced and is sometimes called the *oxidant,* whereas the constituent that loses electrons is oxidized and is called the *reductant.* For example, consider the oxidation reduction reaction:

$$Mn^{2+} + O_3(aq) + H_2O \rightarrow MnO_2(s) + O_2(aq) + 2H^+ \qquad (8\text{-}1)$$

In the above reaction, the manganese (Mn^{2+}) ion is oxidized with ozone to produce manganese oxide precipitate, while ozone is reduced to aqueous oxygen. Manganese loses two electrons while ozone gains two electrons.

Because electrons are exchanged in the reaction, the redox reaction in Eq. 8-1 can be separated into the following two half reactions:

$$Mn^{2+} + 2H_2O \rightarrow MnO_2(s) + 4H^+ + 2e^- \quad \text{(oxidation)} \qquad (8\text{-}2)$$

$$O_3(aq) + 2H^+ + 2e^- \rightarrow O_2(aq) + H_2O \quad \text{(reduction)} \qquad (8\text{-}3)$$

Equation 8-2 is referred to as the oxidation half-reaction because the manganese ion loses two electrons, and Eq. 8-3 is referred to as the reduction half reaction because the ozone ion gains two electrons. Ozone is an oxidant because it causes manganese to be oxidized, and ozone itself is reduced. Manganese is a reductant because it causes ozone to be reduced, and manganese is oxidized. Sometimes to reduce confusion the terms *electron acceptor* or *electron donor* are used. In this reaction, ozone is the electron acceptor and manganese is the electron donor.

BALANCING REDOX REACTIONS

Characterizing redox reactions requires that the reactions be balanced. Oxidation–reduction reactions are balanced most commonly using either (1) the half-reaction method or (2) the oxidation number method. The half-reaction method for balancing oxidation–reduction reactions is as follows:

1. Write down all principal reactants and products for one of the half reactions, except for the hydrogen and oxygen atoms.
2. Balance all atoms, except for the hydrogen and oxygen atoms, with probable forms that may be found in solution (e.g., for Cl in an oxidation reaction, it would be Cl^-).
3. Balance the oxygen atoms with the oxygen in water (H_2O).
4. Balance the hydrogen atoms with H^+.
5. Balance the charge with electrons. If the reactants generate electrons, then the half reaction is an oxidation reaction (loss of electrons is oxidation—LEO). If the reactants consume electrons, then the half reaction is a reduction reaction (gain of electrons is reduction—GER).
6. Write down all principal reactants and products for the other half reaction except for the hydrogen and oxygen atoms.
7. Balance all atoms except for the hydrogen and oxygen atoms with probable forms that may be found in solution (e.g., for C in an oxidation reaction, it may be CO_2).
8. Balance the oxygen atoms with the oxygen in water (H_2O).
9. Balance the hydrogen atoms with H^+.
10. Balance the charge with electrons.
11. The final step is to obtain a balanced reaction. For this step, both half reactions are added together so that electrons are eliminated from the equation.

The following example illustrates the application of the half-reaction method for balancing redox reactions.

Example 8-1 Balancing redox reactions

Balance the oxidation–reduction reaction for hydrogen peroxide (H_2O_2) oxidation of 1,1-dichloroethene ($C_2H_2Cl_2$).

Solution

1. Write the unbalanced reaction involving H_2O_2 and $C_2H_2Cl_2$:

$$C_2H_2Cl_2 + H_2O_2 \rightarrow HCl + CO_2$$

2. Balance the reaction, starting with the half reaction for hydrogen peroxide:

$$H_2O_2 \rightarrow ?$$

   a. First, the expected reactants and products other than oxygen and hydrogen should be balanced as shown. In this case, there are no atoms other than hydrogen or oxygen. Consequently, the next step is to balance the oxygen on the left side of the expression with the oxygen in water on the right side:

$$H_2O_2 \rightarrow 2H_2O$$

   b. Next, balance hydrogen by placing $2H^+$ on the left-hand side of the expression:

$$H_2O_2 + 2H^+ \rightarrow 2H_2O$$

   c. Finally, balance the charge by placing two electrons on the left side of the expression:

$$H_2O_2 + 2H^+ + 2e^- \rightarrow 2H_2O$$

   The above expression represents the reduction half reaction because the reactant gains electrons; however, hydrogen peroxide is the oxidant (or electron acceptor) because it accepts electrons from the oxidation half reaction.

3. Evaluate the oxidation half reaction in which the reactant loses electrons.

   a. Balance the expected reactants and products other than hydrogen and oxygen:

$$C_2H_2Cl_2 \rightarrow ?$$

$$C_2H_2Cl_2 \rightarrow CO_2 + 2Cl^-$$

$$C_2H_2Cl_2 \rightarrow 2CO_2 + 2Cl^-$$

   b. Balance the reaction for oxygen using the oxygen in water:

$$C_2H_2Cl_2 + 4H_2O \rightarrow 2CO_2 + 2Cl^-$$

c. Balance the reaction for hydrogen with H^+.

$$C_2H_2Cl_2 + 4H_2O \rightarrow 2CO_2 + 2Cl^- + 10H^+$$

d. Balance the charge with electrons:

$$C_2H_2Cl_2 + 4H_2O \rightarrow 2CO_2 + 2Cl^- + 10H^+ + 8e^-$$

The above expression represents the oxidation half reaction because electrons are lost by the reactant. Dichloroethene is called the reductant (or electron donor) in this case because it causes the reduction of the oxidant.

4. Add the reduction and oxidation half reactions and eliminate electrons from the reaction. By multiplying the reduction half reaction by a factor of 4 and adding it to the oxidation half reaction, electrons are eliminated from the reaction and the final form of the equation is obtained.
   a. Multiplying the reduction half reaction by a factor of 4 yields the expression

$$4H_2O_2 + 8H^+ + 8e^- \rightarrow 8H_2O$$

   b. Adding the reduction half reaction from step 4a to the oxidation half reaction from step 3d yields

$$\begin{array}{r} 4H_2O_2 + 8H^+ + 8e^- \rightarrow 8H_2O \\ C_2H_2Cl_2 + 4H_2O \rightarrow 2CO_2 + 2Cl^- + 10H^+ + 8e^- \\ \hline 4H_2O_2 + 8H^+ + 8e^- + C_2H_2Cl_2 + 4H_2O \rightarrow 8H_2O + 2CO_2 + 2Cl^- + 10H^+ + 8e^- \end{array}$$

5. Obtain the final redox reaction by eliminating molecules that have stoichiometric coefficients on both sides of the equation from step 4b. The final expression is obtained by eliminating molecules that have stoichiometric coefficients appearing on both sides of the equation (e.g., eight electrons). Subtracting the number of molecules on one side of the equation or the other such that the molecule no longer appears on both sides of the equation (e.g., subtract $4H_2O$ from both sides of the equation), the final redox reaction is

$$4H_2O_2 + C_2H_2Cl_2 \rightarrow 2CO_2 + 2Cl^- + 2H^+ + 4H_2O$$

Comment

While hydrogen peroxide is capable of oxidizing dichloroethene, the reaction is generally not practical in full-scale treatment systems because the reaction rate is too slow. Removal of dichloroethene requires require advanced oxidation (Chap. 18).

Standard Electrode Potentials and Redox Equilibrium Reactions

The gain or loss of electrons from redox reactions can be characterized from the standard electrode potentials for oxidation and reduction half reactions. Every oxidation or reduction half reaction can be characterized by the electrical potential, or electromotive force (emf). This potential is called the standard electrode potential and is measured in volts. The standard electrode potentials for many of the reactions that occur in water treatment are provided in the electronic Table E1 at the website listed in App. E.

Using the International Union of Pure and Applied Chemists (IUPAC) convention:

1. Half reactions are written as reduction reactions.
2. To obtain the oxidation reaction, the direction of the reduction reaction is reversed and the reduction potential is multiplied by a factor of –1.
3. The reported standard electrode potential values are given with respect to a reference standard hydrogen electrode [sometimes referred to as a standard hydrogen electrode (SHE) or normal hydrogen electrode (NHE)].

The value of the redox potential can be illustrated using oxygen. The value corresponds to the following two half reactions:

$$O_2(aq) + 4H^+ + 4e^- \rightleftarrows 2H_2O \qquad \text{(reduction)} \quad E^\circ_{red} = 1.27 \text{ V} \tag{8-4}$$

$$H_2 \rightleftarrows 2H^+ + 2e^- \quad \text{(oxidation)} \quad E^\circ_{ox} = 0 \text{ V} \tag{8-5}$$

The overall redox reaction can be obtained by multiplying Eq. 8-5 by 2, adding Eqs. 8-4 and 8-5, and eliminating electrons and H^+ from both sides of the equation:

$$O_2 + 2H_2 \rightleftarrows 2H_2O \qquad \text{(overall)} \quad E^\circ_{Rxn} = ? \tag{8-6}$$

where E°_{Rxn} = standard electrode potential for overall redox reaction, V

The value of E°_{Rxn} can be determined by simply adding the reduction and oxidation potentials together, noting the sign convention, because the numbers of electrons transferred in the reaction are identical for reduction and oxidation reactions. The value of E°_{Rxn} is obtained using the equation

$$\begin{aligned} E^\circ_{Rxn} &= E^\circ_{red} + E^\circ_{ox} \\ &= 1.27 + 0 = 1.27 \text{ V} \end{aligned} \tag{8-7}$$

A positive value of E°_{Rxn} can be taken as a general indication that a reaction will proceed as written. However, as will be demonstrated later, both

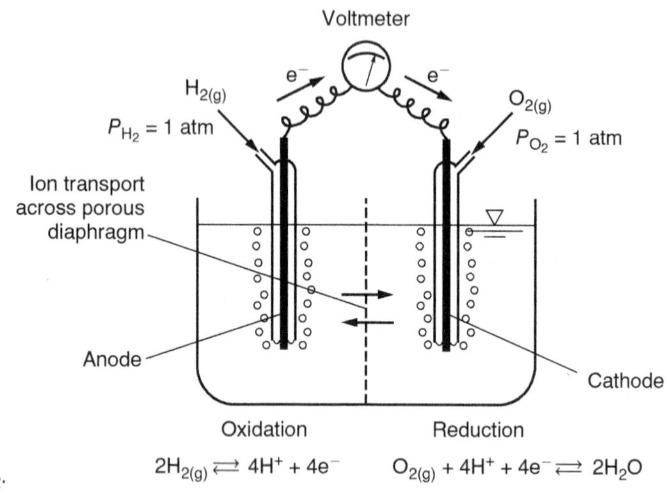

Figure 8-1
Cell potential for reaction between H_2 and O_2.

the electrical potential and free energy of a given reaction must be evaluated at the expected concentrations in solution.

MECHANISTIC DESCRIPTION OF ELECTRODE POTENTIALS WITH AN ELECTROCHEMICAL CELL

An electrochemical cell is set up as shown on Fig. 8-1, and 1.27 V (ignoring the resistance of the wires and assuming that suitable electrodes are present) is measured for unit activities of all species in the cell under the following conditions: 1 atm of hydrogen and oxygen in equilibrium with the water phase and 1 M concentration of H^+. Oxidation takes place at the *anode*, where hydrogen gas is oxidized, and reduction takes place at the *cathode*, where oxygen is reduced. Electrons flow from the anode to the cathode and ions in solution migrate either to the cathode or anode depending on their charge to ensure that electroneutrality is maintained. Cations migrate toward the anode, and anions migrate toward the cathode. The anode is negatively charged because electrons are produced at this electrode, and the cathode is positively charged because electrons are used at this electrode.

ASSESSING REACTION FEASIBILITY

Every reaction is written with respect to the hydrogen electrode, so it is easy to assess whether a reaction is possible by examining the respective electrode potentials of each half reaction because the hydrogen reaction can be eliminated from the overall reaction when the two half reactions are added together.

Returning to the manganese/ozone example (Eqs. 8-1 to 8-3), the standard reduction potential for $MnO_2(s)$ is +1.230 V, corresponding to the combination of the following two half reactions:

$$MnO_2(s) + 4H^+ + 2e^- \rightleftarrows Mn^{2+} + 2H_2O \quad \text{(reduction)} \quad E^o_{red} = +1.230V \tag{8-8}$$

$$H_2 \rightleftarrows 2H^+ + 2e^- \quad \text{(oxidation)} \quad E^o_{ox} = 0V \tag{8-9}$$

$$MnO_2(s) + H_2 + 2H^+ \rightleftarrows Mn^{2+} + 2H_2O \quad \text{(overall)} \quad E^o_{Rxn} = +1.230 \tag{8-10}$$

The standard reduction reaction for ozone corresponds to a combination of the following two half reactions:

$$O_3(aq) + 2H^+ + 2e^- \rightleftarrows O_2(aq) + H_2O \quad \text{(reduction)} \quad E^o_{red} = +2.08V \tag{8.11}$$

$$H_2 \rightleftarrows 2H^+ + 2e^- \quad \text{(oxidation)} \quad E^o_{ox} = 0V \tag{8-12}$$

$$O_3(aq) + H_2 \rightleftarrows O_2 + H_2O \quad \text{(overall)} \quad E^o_{Rxn} = +2.08V \tag{8-13}$$

Ozone is a much more powerful electron acceptor (oxidant) than manganese oxide because E°_{Rxn} for ozone is +2.08 V, as compared to +1.230 V for manganese oxide. Consequently, ozone can bring about the oxidation of the manganese ion when the activity of the products and reactants is equal to one.

DETERMINING WHETHER A REACTION WILL PROCEED

Thus far, only the potential for a redox reaction has been evaluated under standard conditions (unit activity is covered in Chap. 5). However, to consider whether a reaction proceeds as written, the free energy and electrical potential of a given reaction must be evaluated at the expected concentrations in solution. The methodology used to determine if a reaction will occur is presented below. Consider the generic redox reaction

$$A + \frac{b}{a}B \rightarrow \frac{c}{a}C + \frac{d}{a}D \tag{8-14}$$

where a, b, c, d = stoichiometric coefficients, unitless

Relating free-energy change to electrical potential

The free-energy change and cell potential for the reaction in Eq. 8-14 are given by the following expressions (refer to Chap. 5):

$$\Delta G_{Rxn} = \Delta G^{\circ}_{Rxn} + RT\ln\left(\frac{\{C\}^{c/a}\{D\}^{d/a}}{\{A\}\{B\}^{b/a}}\right) \tag{8-15}$$

$$E_{Rxn} = \frac{\Delta G_{Rxn}}{-nF} \tag{8-16}$$

$$E_{Rxn} = E^{\circ}_{Rxn} - \frac{RT}{nF} \ln\left(\frac{\{C\}^{c/a}\{D\}^{d/a}}{\{A\}\{B\}^{b/a}}\right)$$

$$\text{or} \quad E^{\circ}_{Rxn} - \frac{2.303RT}{nF} \log\left(\frac{\{C\}^{c/a}\{D\}^{d/a}}{\{A\}\{B\}^{b/a}}\right) \tag{8-17}$$

where ΔG_{Rxn} = free-energy change of reaction, J/mol
ΔG°_{Rxn} = free-energy change of reaction under standard conditions, J/mol
E_{Rxn} = electrode potential of reaction = $E_{ox} + E_{red}$, V
E°_{Rxn} = standard electrode potential of reaction, V
n = number of electrons transferred in reaction, eq/mol
F = Faraday's constant, 96,500 C/eq
R = universal gas constant, 8.314 J/mol · K
T = absolute temperature, K
$\{i\}$ = activity of constituent i (A, B, C, or D), mol/L

Equation 8-17 is known as the Nernst equation. The base-10 form of the equation is useful when dealing with reactions involving the hydrogen ion because, as shown later, the hydrogen ion concentration can be expressed as pH.

The electrode potential E°_{Rxn} may be calculated from ΔG°_{Rxn} by adding all the reactions that are involved using the $-nFE^{\circ}$ values because the ΔG°_{Rxn} and $-nFE^{\circ}$ values are additive. For example, E°_{Rxn} may be calculated from the oxidation and reduction reaction as shown in the following equations:

$$\Delta G^{\circ}_{Rxn} = -n_{Rxn}FE^{\circ}_{Rxn} = \Delta G^{\circ}_{ox} + \Delta G^{\circ}_{red} = -n_{ox}FE^{\circ}_{ox} - n_{red}FE^{\circ}_{red} \tag{8-18}$$

where n_{Rxn} = number of electrons transferred in overall reaction, eq/mol
E°_{Rxn} = standard electrode potential of redox reaction, V
ΔG°_{ox} = standard free-energy change of oxidation half reaction, J/mol
ΔG°_{red} = standard free-energy change of reduction half reaction, J/mol
E°_{ox} = standard electrode potential of oxidation half reaction, V
E°_{red} = standard electrode potential of reduction half reaction, V
n_{ox} = number of electrons produced in oxidation half reaction, eq/mol
n_{red} = number of electrons obtained in reduction half reaction, eq/mol

Example 8-2 Oxidation power of oxygen and hydrogen peroxide

Investigate whether oxygen (O_2) or hydrogen peroxide (H_2O_2) is the more powerful oxidant from a consideration of free energy.

Solution

1. Write the reduction and oxidation half reactions of H_2O_2 as well as the overall reaction. From Table E1 at the website listed in App. E, the half reaction of H_2O_2 can be written with respect to the hydrogen electrode as follows:

$$H_2O_2 + 2H^+ + 2e^- \rightleftarrows 2H_2O \qquad \text{(reduction)} \qquad E^o_{red} = 1.780\text{ V}$$
$$H_2 \rightleftarrows 2H^+ + 2e^- \qquad \text{(oxidation)} \qquad E^o_{ox} = 0\text{ V}$$
$$H_2O_2 + H_2 \rightleftarrows 2H_2O \qquad \text{(overall)} \qquad E^o_{Rxn,\,H_2O_2} = 1.780\text{ V}$$

2. Calculate $\Delta G^\circ_{Rxn,\,H_2O_2}$ using Eq. 8-18 (note J/C = V):

$$\Delta G^\circ_{Rxn,\,H_2O_2} = -nFE^\circ_{Rxn,\,H_2O_2} = -(2\text{ eq/mol})(96{,}500\text{ C/eq})(1.780\text{ J/C})$$
$$= -343{,}540\text{ J/mol} \quad (-343.5\text{ kJ/mol})$$

3. Write the reduction and oxidation half reactions of O_2 as well as the overall reaction: The half reaction of O_2 can be written with respect to the hydrogen electrode as follows:

$$O_2(aq) + 4H^+ + 4e^- \rightleftarrows 2H_2O \qquad \text{(reduction)} \qquad E^o_{red} = 1.27\text{ V}$$
$$2\left(H_2 \rightleftarrows 2H^+ + 2e^-\right) \qquad \text{(oxidation)} \qquad E^o_{ox} = 0\text{ V}$$
$$O_2 + 2H_2 \rightleftarrows 2H_2O \qquad \text{(overall)} \qquad E^o_{Rxn,O_2} = 1.27\text{ V}$$

4. Calculate $\Delta G^\circ_{Rxn,O_2}$ by rearranging Eq. 8-18:

$$\Delta G^\circ_{Rxn,O_2} = -nFE^\circ_{Rxn,O_2} = -(4\text{ eq/mol})(96{,}500\text{ C/eq})(1.27\text{ J/C})$$
$$= -490{,}220\text{ J/mol} \quad (-490\text{ kJ/mol})$$

Comment

The free-energy change per mole of oxidant is greater for oxygen than it is for peroxide. (*Note*: Only two electrons are transferred in the hydrogen peroxide reaction versus four in the oxygen reaction.) Consequently, it is impossible to tell which is the more powerful oxidant by examining the free energy. Using the standard electrode potential, it can be seen that H_2O_2 is a more powerful oxidant than O_2 by comparing E°_{Rxn,H_2O_2} to E°_{Rxn,O_2} because the redox potential of H_2O_2 is greater than O_2. In addition, the power of H_2O_2 as compared to O_2 can be illustrated by the fact that the following reaction is feasible from a thermodynamic point of view (at unit activity):

$$2H_2O_2 \rightleftarrows 2H_2O + O_2$$

The free energy describes the overall energy transferred in a chemical reaction, and therefore cannot be used to tell which oxidant is more powerful. The important factor in oxidant strength is the amount of energy in each electron; thus, the standard electrode potential, which is the amount of energy released per coulomb of electron and has units of volts (J/C), can be used to compare oxidant strength.

Potential of combined reactions

When the potential of an oxidation (or reduction) reaction is a combination of several reactions (e.g., reactions 1 and 2), the following expression must be used:

$$-(n_{ox,1} + n_{ox,2})FE^\circ_{ox,sum} = -n_{ox,1}FE^\circ_{ox,1} - n_{ox,2}FE^\circ_{ox,2} \tag{8-19}$$

$$E^\circ_{ox,sum} = \frac{n_{ox,1}E^0_{ox,1} + n_{ox,2}E^\circ_{ox,2}}{n_{ox,1} + n_{ox,2}} \tag{8-20}$$

$$E^\circ_{red,sum} = \frac{n_{red,1}E^\circ_{red,1} + n_{red,2}E^\circ_{red,2}}{n_{red,1} + n_{red,2}} \tag{8-21}$$

where $E^\circ_{ox,sum}$ = standard electrode potential of combined oxidation half reaction, V
$E^\circ_{red,sum}$ = standard electrode potential of combined reduction half reaction, V
$E^\circ_{ox,i}$ = standard electrode potential of ith oxidation half reaction, V
$E^\circ_{red,i}$ = standard electrode potential of ith reduction half reaction, V
$n_{ox,i}$ = number of electrons produced in ith oxidation half reaction, eq/mol
$n_{red,i}$ = number of electrons obtained in ith reduction half reaction, eq/mol

The application of Eq. 8-20 is illustrated in the following example.

Example 8-3 Oxidation of bromide to bromate

Determine the oxidation potential for converting bromide (Br^-) to bromate (BrO_3^-) from the hypobromous/bromide acid reaction and bromate/hypobromous reaction.

Solution

1. From Table E1 at the website listed in in App. E, determine the two half reactions, their potentials, and the overall reaction:

$$Br^- + H_2O \rightleftarrows HOBr + H^+ + 2e^- \quad \text{(oxidation)} \quad E^\circ_{ox} = -1.33\text{ V}$$

$$HOBr + 2H_2O \rightleftarrows BrO_3^- + 5H^+ + 4e^- \quad \text{(oxidation)} \quad E^\circ_{ox} = -1.49\text{ V}$$

$$Br^- + 3H_2O \rightleftarrows BrO_3^- + 6H^+ + 6e^- \quad \text{(oxidation)} \quad E^\circ_{ox,sum} = ?$$

2. Determine the potential for the oxidation reaction using Eq. 8-20 and the values from the online table of standard redox potentials

$$E^\circ_{ox,sum} = \frac{2(-1.33\text{ V}) + 4(1.49\text{ V})}{2+4} = -1.437\text{ V}$$

Comment

It is important to recognize that the potential is the sum of the potentials of the oxidation and reduction components (see Eq. 8-21) only when the overall redox reaction is being considered.

Determining equilibrium constant from electrical potential

In the Nernst equation (Eq. 8-15), equilibrium is achieved when either the free-energy change or electrical potential is zero. Thus, the equilibrium constant can be calculated from the free energy or electrical potential as shown below:

$$0 = E^\circ_{Rxn} - \frac{RT}{nF} \ln K_{eq} \tag{8-22}$$

$$K_{eq} = e^{nFE^\circ_{Rxn}/RT} = e^{-\Delta G^\circ_{Rxn}/RT} \tag{8-23}$$

where K_{eq} = equilibrium constant

The utility of the equilibrium constant in redox reactions is illustrated in the following example.

Example 8-4 Oxidation of Fe(s) with dissolved oxygen

Calculate the equilibrium constant for the corrosion of Fe(s) to Fe^{2+} by dissolved oxygen. Determine the oxygen concentrations that are needed for pH values equal to 5.5, 7.0, and 8.5 such that the reaction is thermodynamically favorable. *Given*: $[Fe^{2+}] = 10^{-6}$ M and temperature is 298 K.

Solution

1. Write the reduction and oxidation half reactions for corrosion of Fe(s) to Fe^{2+} by reacting with oxygen as well as the overall reaction: From Table E1 at the website listed in App. E, the corrosion of Fe(s) to Fe^{2+} by reacting with oxygen can be written as

$$2\left[Fe(s) \rightleftarrows Fe^{2+} + 2e^-\right] \quad \text{(oxidation)} \quad E^\circ_{ox} = 0.44\ V$$

$$\frac{O_2(aq) + 4H^+ + 4e^- \rightleftarrows 2H_2O \quad \text{(reduction)} \quad E^\circ_{red} = 1.27\ V}{2Fe(s) + O_2(aq) + 4H^+ \rightleftarrows 2Fe^{2+} + 2H_2O \quad \text{(overall)} \quad E^\circ_{Rxn} = 1.71\ V}$$

2. Calculate the equilibrium constant K_{eq} using Eq. 8-23:

$$K_{eq} = \exp\left(\frac{nFE^\circ_{Rxn}}{RT}\right) = \exp\left[\frac{(4\ \text{eq/mol})\ (96{,}500\ \text{C/eq})\ (1.71\ \text{J/C})}{(8.314\ \text{J/mol}\cdot\text{K})\ (298\ \text{K})}\right]$$

$$= 4.97 \times 10^{115}$$

3. Determine the equilibrium oxygen concentrations at pH values equal to 5.5, 7.0, and 8.5 using the expression developed in step 1 [$2Fe(s) + O_2(aq) + 4H^+ \rightleftarrows 2Fe^{2+} + 2H_2O$] and the equilibrium value developed in step 2:

$$K_{eq} = \frac{\{Fe^{2+}\}^2\{H_2O\}^2}{\{Fe(s)\}^2\{O_2(aq)\}\{H^+\}^4}$$

Neglecting activity coefficient corrections and assuming the activity of solids and liquids is equal to 1,

$$[O_2(aq)] = \frac{[Fe^{2+}]^2}{[H^+]^4 K_{eq}} = \frac{(10^{-6})^2}{[H^+]^4 (4.97 \times 10^{115})}$$

Therefore,

$$[O_2(aq)] = \begin{cases} \dfrac{[Fe^{2+}]^2}{[H^+]^4 K_{eq}} = \dfrac{(10^{-6})^2}{(10^{-8.5})^4(4.97\times10^{115})} = 2.01\times10^{-94}\text{ M at pH 8.5} \\ \dfrac{[Fe^{2+}]^2}{[H^+]^4 K_{eq}} = \dfrac{(10^{-6})^2}{(10^{-7})^4(4.97\times10^{115})} = 2.01\times10^{-100}\text{ M at pH 7} \\ \dfrac{[Fe^{2+}]^2}{[H^+]^4 K_{eq}} = \dfrac{(10^{-6})^2}{(10^{-5.5})^4(4.97\times10^{115})} = 2.01\times10^{-106}\text{ M at pH 5.5} \end{cases}$$

Note: When the reaction quotient $Q = K_{eq,}$ $\Delta G = 0$. If $\Delta G < 0$, the reaction is thermodynamically favorable. Thus, if the actual oxygen concentrations are greater than the equilibrium values computed above as a function of pH, the reaction will proceed as written in step 1.

Comment

For all cases, the reactions are thermodynamically favorable for small oxygen concentrations. The oxygen concentrations that are calculated here are extremely small. In fact, for a pH of 5.5, there would be one molecule of oxygen per 8.26×10^{81} L of water, which may, in fact, be more than all the water in the entire universe. Many redox reactions have high equilibrium constants, and as a result they tend to be irreversible reactions when the rate of reaction is fast.

Impact of pH on reduction potential

Reaction conditions, especially pH, can have an important impact on reduction potential. For example, pH can have a large influence on the standard potential, and if 1 mole of hydrogen ion appears on the left-hand side of the equation (as a reactant), then the potential drops according to the following equation for a unit increase in pH:

$$\Delta E^\circ_{red} = -\left(\frac{2.303RT}{nF}\right)\log\left(\frac{1}{[H^+]}\right)$$

$$= -\left[\frac{(2.303)(8.314\text{ J/mol}\cdot\text{K})(298\text{ K})}{(96{,}500\text{ C/eq})(n\text{ eq/mol})}\right]\text{pH}$$

$$= -\left(\frac{0.0591}{n}\right)\times\text{pH} \tag{8-24}$$

where ΔE°_{red} = change in potential, V
R = universal gas constant, 8.314 J/mol · K
T = absolute temperature, K
n = number of electrons transferred, eq/mol
F = Faraday's constant, 96,500 C/eq
$[H^+]$ = concentration of hydrogen ion, mol/L
pH = $\log([H^+]) = \log(1/[H^+])$, unitless

Consequently, the electrode potential for HO• is 0.413 V lower at pH 7 than for unit activity assuming that all other species are at unit activity or it would be 2.18 V instead of 2.59 V as reported in the electronic Table E1 at the website listed in App. E. If there are 2 moles of hydrogen ion on the left-hand side of the reduction reaction, then the reduction potential would be 0.826 V lower at pH 7 for unit activity of all other species if the number of electrons accepted is 1. For ozone and oxygen, the electrode potential would be 0.413 V less than is reported in the table of standard redox potentials at pH 7 because the number of hydrogen ions on the left side is the same as the number of electrons.

Evaluating free-energy change and electrical potential over a concentration range
To determine whether a redox reaction can proceed from the initial concentrations in solution to the treatment objective, ΔG°_{Rxn} and E°_{Rxn} must be evaluated over that concentration range. Because the concentrations of all the reactants and products change over that concentration range, as described in Chap. 5, the concentrations of the constituents can be related to one another and to the conversion using stoichiometry. The following relationships from Chap. 5, repeated here for convenience, are valid if there are no competing reactions and no volume change upon reaction:

$$C_A = C_{A0}(1 - X_A) \tag{8-25}$$

$$C_B = C_{B0} - \frac{b}{a} C_{A0} X_A \tag{8-26}$$

$$C_C = C_{C0} + \frac{c}{a} C_{A0} X_A \tag{8-27}$$

$$C_D = C_{D0} + \frac{d}{a} C_{A0} X_A \tag{8-28}$$

$$C_B = C_{B0} - \frac{b}{a}(C_{A0} - C_A) \tag{8-29}$$

$$C_C = C_{C0} + \frac{c}{a}(C_{A0} - C_A) \tag{8-30}$$

$$C_D = C_{D0} + \frac{d}{a}(C_{A0} - C_A) \tag{8-31}$$

where C_i = concentration of constituent i (A, B, C, D), mol/L
C_{i0} = initial concentration of constituent i (A,B,C,D), mol/L
X_A = conversion of constituent A, dimensionless
a, b, c, d = stoichiometry coefficients

Equations 8-25 to 8-28 or Eqs. 8-29 to 8-31 may be substituted into Eq. 8-15 to determine whether ΔG is negative as a function of X_A or C_A. Similarly, Eqs. 8-25 to 8-28 or Eqs. 8-29 to 8-31 may be substituted into Eq. 8-17 to determine whether E_{Rxn} is positive as a function of X_A or C_A. Equations 8-25 to 8-28 or Eqs. 8-29 to 8-31 can also be used to determine if one of the reactants will be exhausted before the desired conversion is achieved. The following example is used to illustrate the application of these types of thermodynamic calculations in water treatment applications.

Example 8-5 Oxidation of manganese with ozone

Manganese [Mn(II)] is soluble in water and is present in many groundwaters because insoluble forms (e.g., MnO_2) that are contained in minerals are reduced to soluble forms. The subsurface is a reducing environment because electron acceptors such as oxygen have been used up by heterotrophic bacteria in the top organic-rich layer of soil. Ozone (O_3) is sometimes used to remove Mn^{2+}. Assume that the ozone and dissolved oxygen (DO) do not react with anything but Mn^{2+} (a simplifying assumption because ozone will react with many other constituents in a real water) and the pH is constant. Then:

1. Balance the overall redox reaction for the oxidation of Mn^{2+} to MnO_2(s) with O_3.
2. Calculate the equilibrium constant.
3. Calculate the equilibrium Mn^{2+} concentration when the pH is 7, DO is 5 mg/L, and the ozone concentration is 0.5 mg/L.
4. Obtain expressions for DO, ozone, and MnO_2(s) concentrations in terms of Mn^{2+}(s) concentration. The initial reactant concentrations are [DO] = 10 mg/L, ozone concentration = 3 mg/L, and Mn^{2+} = 2 mg/L and pH is 7.
5. Plot the free energy as a function of the conversion of Mn^{2+} from 0.01 to 0.999.

Use the stoichiometric table to determine all reacting species as a function of X_A and then eliminate X_A by using the final concentration C_A. After substituting C_A back into the expressions, DO, ozone, and MnO_2 concentrations can be obtained in terms of the final concentration of Mn^{2+}.

Solution

1. Balance the overall redox reaction:
   a. Identify the oxidation and reduction reactions and determine the standard electrode potential:

Oxidation reaction: $Mn^{2+} + 2H_2O \rightleftarrows MnO_2\,(s) + 4H^+ + 2e^- \quad E° = -1.23\ V$

Reduction reaction: $O_3\,(aq) + 2H^+ + 2e^- \rightleftarrows O_2\,(aq) + H_2O \quad E° = +2.08\ V$

   b. Balance the overall redox reaction by adding the two half reactions:

$$
\begin{array}{lll}
Mn^{2+} + 2H_2O \rightleftarrows MnO_2(s) + 4H^+ + 2e^- & \text{(oxidation)} & E^o_{ox} = -1.23V \\
O_3(aq) + 2H^+ + 2e^- \rightleftarrows O_2(aq) + H_2O & \text{(reduction)} & E^o_{red} = +2.08V \\
\hline
Mn^{2+} + O_3(aq) + H_2O \rightleftarrows MnO_2(s) + O_2(aq) + 2H^+ & \text{(overall)} & E^o_{Rxn} = 0.850V
\end{array}
$$

$$E^{\circ}_{Rxn} = 2.08 + (-1.23) = 0.850\ V$$

2. Calculate the equilibrium constant:
   a. Calculate ΔG°_{Rxn} using Eq. 8-16:

$$\Delta G^{\circ}_{Rxn} = -nFE^{\circ}_{Rxn} = -(2\ \text{eq/mol})\ (96{,}500\ \text{C/eq})\ (0.850\ \text{J/C})$$
$$= -164{,}050\ \text{J/mol} = -164.05\ \text{kJ/mol}$$

   b. Calculate the equilibrium constant K_{eq} using Eq. 8-23:

$$K_{eq} = e^{-\Delta G^{\circ}_{Rxn}/RT} = e^{-(-164.05)/(8.314\times10^{-3}\times298)} = 5.71\times10^{28}$$

3. Calculate the equilibrium Mn^{2+} concentration when the pH is 7, DO is 5 mg/L, and the ozone concentration is 0.5 mg/L:

$$[H^+] = 10^{-7}\ M$$
$$[O_3(aq)] = 0.5\ \text{mg/L} = 1.04\times10^{-5}\ M$$
$$[O_2(aq)] = 5\ \text{mg/L} = 1.56\times10^{-4}\ M$$

$$K_{eq} = \frac{[O_2(aq)][H^+]^2}{[Mn^{2+}][O_3(aq)]}$$

$$[Mn^{2+}] = \frac{[O_2(aq)][H^+]^2}{K_{eq}[O_3(aq)]} = \frac{(1.56\times10^{-4})(10^{-7})^2}{(5.71\times10^{28})(1.04\times10^{-5})}$$
$$= 2.63\times10^{-42}\ M = 1.44\times10^{-37}\ \text{mg/L}$$

4. Obtain an expression for DO, ozone, and MnO_2 in terms of Mn^{2+}: Initial concentrations:

$$[Mn^{2+}]_0 = 2 \text{ mg/L} = 3.64 \times 10^{-5} \text{ M}$$
$$[O_3(aq)]_0 = 3 \text{ mg/L} = 6.25 \times 10^{-5} \text{ M}$$
$$[O_2(aq)]_0 = 10 \text{ mg/L} = 3.13 \times 10^{-4} \text{ M}$$

According to the overall reaction and Eqs. 8-29 to 8-31, the molar concentrations of $O_3(aq)$ and $O_2(aq)$ can be expressed in terms of $[Mn^{2+}]$ as

$$\begin{aligned}[O_3(aq)] &= [O_3(aq)]_0 - ([Mn^{2+}]_0 - [Mn^{2+}]) \\ &= 6.25 \times 10^{-5} - (3.64 \times 10^{-5} - [Mn^{2+}]) \\ &= 2.61 \times 10^{-5} + [Mn^{2+}]\end{aligned}$$

$$\begin{aligned}[O_2(aq)] &= [O_2(aq)]_0 + ([Mn^{2+}]_0 - [Mn^{2+}]) \\ &= 3.13 \times 10^{-4} + (3.64 \times 10^{-5} - [Mn^{2+}]) \\ &= 3.49 \times 10^{-4} - [Mn^{2+}]\end{aligned}$$

Because $MnO_2(s)$ is a solid in water, its activity is unity.

5. Plot the free energy as a function of the conversion of Mn^{2+} from 0.01 to 0.999:
   a. Obtain an expression of free energy as a function of the conversion of Mn^{2+}:
   Assume the conversion of Mn^{2+} is $X_{Mn\ 2+}$. According to the overall reaction and Eqs. 8-25 to 8-28, the molar concentrations of Mn^{2+}, H^+, $O_3(aq)$, and $O_2(aq)$ can be expressed as

$$\begin{aligned}[Mn^{2+}] &= [Mn^{2+}]_0(1 - X_{Mn^{2+}}) \\ &= 3.64 \times 10^{-5}(1 - X_{Mn^{2+}})\end{aligned}$$

$$\begin{aligned}[H^+] &= [H^+]_0 + 2[Mn^{2+}]_0 X_{Mn^{2+}} \\ &= 10^{-7} + 7.28 \times 10^{-5} X_{Mn^{2+}}\end{aligned}$$

$$[O_3(aq)] = [O_3(aq)]_0 - [Mn^{2+}]_0 X_{Mn^{2+}}$$
$$= 6.25\times10^{-5} - 3.64\times10^{-5} X_{Mn^{2+}}$$
$$[O_2(aq)] = [O_2(aq)]_0 - [Mn^{2+}]_0 X_{Mn^{2+}}$$
$$= 3.13\times10^{-4} + 3.64\times10^{-5} X_{Mn^{2+}}$$

The free-energy expression (Eq. 8-15) can be written as given below by substituting the quotient Q for the logarithmic term in Eq. 8-15. The quotient is described in greater detail in Chap. 5.

$$\Delta G_{Rxn} = \Delta G^{\circ}_{Rxn} + RT\ln(Q) = \Delta G^{\circ}_{Rxn} + RT\ln\left(\frac{[O_2(aq)][H^+]^2}{[Mn^{2+}][O_3(aq)]}\right)$$

$$= -164.05 + (8.314\times10^{-3})(298)$$
$$\times\ln\left\{\frac{(3.13\times10^{-4} + 3.64\times10^{-5}X_{Mn^{2+}})(10^{-7} + 7.28\times10^{-5}X_{Mn^{2+}})^2}{\left[3.64\times10^{-5}(1 - X_{Mn^{2+}})\right](6.25\times10^{-5} - 3.64\times10^{-5}X_{Mn^{2+}})}\right\}$$

b. Plot the free energy as a function of the conversion of Mn^{2+} from 0.01 to 0.999:

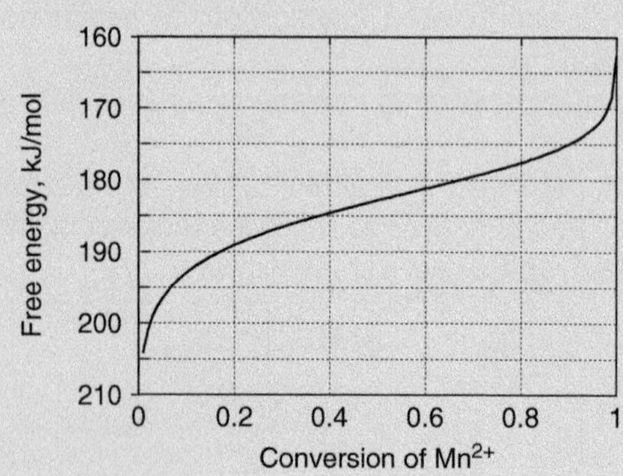

E_H-pH Predominance Area Diagrams: Definition and Example for Chlorine

The E_H–pH (or $p\varepsilon$–pH) diagram is a visual tool used for determining predominant chemical species at various pH values and is useful when analyzing redox equilibria. Because most redox reactions depend on pH and the electrical potential, the thermodynamically preferred species can be shown on a two-dimensional diagram in which pH and the electrical potential are the axes. Acid–base, complexation, and precipitation reactions can also be displayed on these diagrams because oxidants and reductants are involved in these types of reactions. These types of diagrams are called predominance area diagrams and are constructed based on the following rules:

1. Boundaries between major species are drawn for a given set of conditions (e.g., total chlorine concentration).
2. The boundary lines are drawn where the concentrations of the two species involved in the redox reaction are equal; consequently, one species predominates in concentration on one side of the line.

The construction of a predominance area diagram will be illustrated for chlorine; however, first the stability domain for water/oxygen will be identified.

OXYGEN

For the reduction of gaseous oxygen (from the electronic Table E1 at the website listed in App. E) the following half reaction may be written:

$$O_2(g) + 4H^+ + 4e^- \rightleftarrows 2H_2O \quad E^\circ_{red} = 1.23 \text{ V} \tag{8-32}$$

$$E_H = E^\circ - \frac{0.059}{n}\log\frac{1}{[H^+]^4 P_{O_2}} \tag{8-33}$$

$$= E^\circ - \frac{0.059}{4}(4pH - \log P_{O_2}) \tag{8-34}$$

where E_H = electrode potential as function of pH, V
E = standard electrode potential, V
n = number of electrons transferred, eq/mol
P_{O_2} = partial pressure of oxygen, atm

For gas-phase concentrations the equilibrium expression can be written in terms of the partial pressure (McMurry and Fay, 2003). It should be noted that the partial pressure is used because it can be measured easily. Assuming P_{O_2} = 0.21 atm (see Table B-2, App. B), Eq. 8-34 results in the expression

$$E_H = 1.24 - 0.059pH \tag{8-35}$$

Equation 8-35 can be plotted as a straight line (see Fig. 8-2). For a given pH, with E_H values above the line, water would be reduced and O_2 would be formed. For E_H values below the line, water is stable and the preferred species.

HYDROGEN

For the reduction of hydrogen, the following half reaction may be written:

$$2H_2O + 2e^- \rightleftarrows H_2(g) + 2OH^- E^\circ = -0.828 \text{ V} \tag{8-36}$$

$$E_H = E^\circ - \frac{0.059}{2}(-2pOH + \log P_{H_2}) \tag{8-37}$$

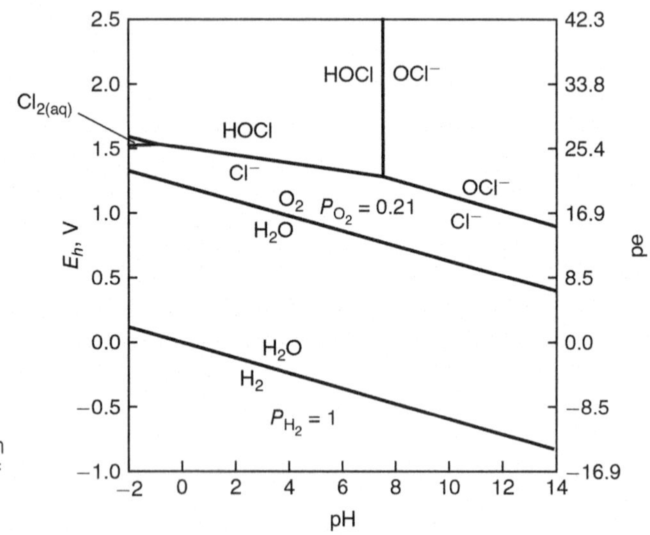

Figure 8-2
Predominance area diagram for chlorine system with total chlorine species concentration $C_{T,Cl}$ of 10^{-4} mol/L ($C_{T,Cl} = 10^{-4}$ mol/L = $2[Cl_2(aq)]$ + [HOCl] + $[OCl^-]$ + $[Cl^-]$).

Assuming P_{H_2} = 1 atm, Eq. 8-37 results in the expression

$$E_H = -0.059\,\text{pH} \tag{8-38}$$

Equation 8-38 can also be plotted as a straight line as shown on Fig. 8-2. For a given pH, E_H values above the line water would be stable. For E_H values below the line water would be reduced and hydrogen would be formed. The lines for oxidation and reduction of water are plotted on Fig. 8-2 and the region where water is stable is identified on the figure.

CHLORINE SPECIES

The species that are formed when chlorine is added to water are $Cl_2(aq)$, HOCl, OCl^-, and Cl^- and the predominance of these species can be plotted if the total mass of chlorine species is fixed. A concentration of 10^{-4} M is used to illustrate the development of a predominance diagram for chlorine. The total concentration of chlorine is given by the relationship

$$C_{T,\text{Cl}} = 10^{-4}\ \text{M} = 2[\text{Cl}_2(\text{aq})] + [\text{HOCl}] + [\text{OCl}^-] + [\text{Cl}^-] \tag{8-39}$$

When chlorine is added to water, it reacts with the water to form hypochlorous acid:

$$\text{Cl}_2\,(\text{aq}) + \text{H}_2\text{O} \rightleftarrows \text{Cl}^- + \text{HOCl} + \text{H}^+ \tag{8-40}$$

The equilibrium constant for the above reaction is 4×10^{-4} at 25°C. The equilibrium constant for 0, 15, 35, and 45°C has been reported as approximately 1.5×10^{-4}, 3×10^{-4}, 5×10^{-4}, and 6×10^{-4}, respectively (Faust and

Table 8-3
Values of dissociation constant for hypochlorous acid (HOCl) at different temperatures[a]

Temperature, °C	$K_a \times 10^8$, mol/L
0	1.5
5	1.7
10	2.0
15	2.3
20	2.6
25	2.9

[a]Computed using equation from Morris (1966).

Aly, 1998). Hypochlorous acid is a weak acid that can dissociate at or near neutral pH:

$$HOCl \rightleftarrows H^+ + OCl^- \tag{8-41}$$

The equilibrium constant for the reaction is $10^{-7.5}$ ($pK_a = 7.5$ at 25°C). The dissociation constants for hypochlorous acid at several temperatures are provided in Table 8-3. The reduction reaction and potential for hypochlorous acid are

$$HOCl + H^+ + 2e^- \rightarrow H_2O + Cl^-(aq) \qquad E^\circ = 1.5\ V \tag{8-42}$$

where E°_{SHE} = standard hydrogen electrode potential, V

The following equation can be used to estimate the pK_a of hypochlorous acid as a function of temperature (Morris, 1966):

$$pK_a = \frac{3000.0}{T} - 10.0686 + 0.0253\ T \tag{8-43}$$

where T = absolute temperature, K (273 + °C)

Calculation of the equilibrium concentrations of the chlorine species formed when chlorine is added to water is illustrated in Example 8-6.

Example 8-6 Hydrolysis of chlorine

Calculate the equilibrium concentrations of HOCl and $Cl_2(aq)$ in solution for a chlorine addition of 2 mg/L at 25°C. Assume that the pH is 5 and does not change and that at pH 5 the amount of HOCl that disassociates into H^+ and OCl^- is insignificant. Express the chlorine concentrations in terms of milligrams per liter of $Cl_2(aq)$.

Solution

1. Calculate the initial $Cl_2(aq)$ concentration in moles per liter:

$$[Cl_2(aq)]_0 = \frac{2\text{ mg/L}}{(35.45\text{ g/mol})\times(2\text{ mol})\times(10^3\text{ mg/g})}$$
$$= 2.82\times10^{-5}\text{ mol/L}$$

2. Calculate the percentage of $Cl_2(aq)$ that is hydrolyzed: The percentage of $Cl_2(aq)$ that is converted into HOCl is $X_{Cl_2(aq)}$, and the molar concentrations of Cl^-, HOCl, H^+, and $Cl_2(aq)$ can be expressed as

$$[Cl_2(aq)] = [Cl_2(aq)]_0[1 - X_{Cl_2(aq)}] = 2.82\times10^{-5}[1 - X_{Cl_2(aq)}]$$
$$[Cl^-] = [HOCl] = [Cl_2(aq)]_0 X_{Cl_2(aq)} = 2.82\times10^{-5} X_{Cl_2(aq)}$$

The equilibrium constant for Eq. 8-40 can be written as

$$K_{eq} = \frac{[Cl^-][HOCl][H^+]}{[Cl_2(aq)]} = \frac{(2.82\times10^{-5} X_{Cl_2(aq)})^2(10^{-5})}{2.82\times10^{-5}(1 - X_{Cl_2(aq)})}$$
$$= 4\times10^{-4}$$
$$X_{Cl_2(aq)} = 0.999999$$

3. Calculate the equilibrium concentrations of HOCl and $Cl_{2,aq}$:
   a. In units of moles per liter:

$$[Cl_2(aq)] = (2.82\times10^{-5}\text{ mol/L})(1 - 0.999999)$$
$$= 2.82\times10^{-11}\text{ mol/L}$$
$$[HOCl] = [Cl^-] = (2.82\times10^{-5}\text{ mol/L})(0.999999)$$
$$= 2.82\times10^{-5}\text{ mol/L}$$

   b. In units of milligrams per liter as $Cl_2(aq)$:

$$C_{Cl_2(aq)} = (2.82\times10^{-11}\text{ mol/L})(70.9\text{ g/mol})(1000\text{ mg/g})$$
$$= 2\times10^{-6}\text{ mg/L as }Cl_2$$
$$C_{HOCl} = \frac{(2.82\times10^{-5}\text{ mol HOCl/L})(70.9\text{ g/mol }Cl_2)(1000\text{ mg/g})}{1\text{ mol HOCl/mol }Cl_2}$$
$$= 1.9998\text{ mg/L as }Cl_2$$

Comment

Chlorine is hydrolyzed almost completely by the reaction with water at 25°C. For the temperatures that are normally encountered in water treatment, the hydrolysis equilibrium constant is large even at low temperatures. Regarding the influence of pH, if the pH were higher, then HOCl would dissociate, resulting in even less $Cl_2(aq)$.

The species that are initially formed when chlorine is added to water further react with chlorine. The reactions of hypochlorous acid (HOCl), hypochlorite (OCl^-), and chloride (Cl^-) are necessary to develop a predominance diagram for chlorine and are discussed in the following sections.

Hypochlorous acid

The reactions of hypochlorous acid must be written as reduction reactions. The reduction reaction involving hypochlorous acid and chlorine (from Table E1 at the website listed in App. E) is given by the following:

$$2HOCl + 2H^+ + 2e^- \rightleftarrows Cl_2(aq) + 2H_2O \qquad E^\circ = 1.61\ \text{V} \tag{8-44}$$

$$E_H = E^\circ - \frac{0.059}{2}\log\left\{\frac{[Cl_{2(aq)}]}{[H^+]^2[HOCl]^2}\right\} \tag{8-45}$$

At the boundary, [HOCl] equals [$Cl_2(aq)$], and when HOCl and $Cl_2(aq)$ predominate in the solution, $C_{T,Cl} \approx 2[Cl_2(aq)] + [HOCl] = 10^{-4}$ M and $[HOCl] = [Cl_2(aq)] = 3.33 \times 10^{-5}$ M, and Eq. 8-45 can be rewritten as

$$E_H = 1.61 - \frac{0.059}{2}\left\{2\text{pH} + \log\left[\frac{3.33\times10^{-5}}{(3.33\times10^{-5})^2}\right]\right\} \tag{8-46}$$

$$= 1.47 - 0.059\ \text{pH} \tag{8-47}$$

The reduction reaction involving chlorine and the chloride ion (from Table E1 at the website listed in App. E) is given by the following:

$$Cl_2(aq) + 2e^- \rightleftarrows 2Cl^- \qquad E^\circ = 1.396\ \text{V} \tag{8-48}$$

$$E_H = E^\circ - \frac{0.059}{2}\log\left(\frac{[Cl^-]^2}{[Cl_2(aq)]}\right) \tag{8-49}$$

Similarly, at the boundary [Cl^-]equals [$Cl_2(aq)$], and when Cl^- and Cl_2 (aq) predominate in the solution, $C_{T,Cl} \approx 2[Cl_2(aq)] + [Cl^-] = 10^{-4}$ M and $[Cl^-] = [Cl_2(aq)] = 3.33 \times 10^{-5}$ M, and Eq. 8-49 can be rewritten as

$$E_H = 1.52\ \text{V} \tag{8-50}$$

The acid–base equilibria for hypochlorous acid is given by the following:

$$HOCl \rightleftarrows OCl^- + H^+ pK_a = 7.5 \tag{8-51}$$

$$K_a = \frac{[OCl^-][H^+]}{[HOCl]} \tag{8-52}$$

$$pK_a = pH - \log\left(\frac{[OCl^-]}{[HOCl]}\right) \tag{8-53}$$

At the boundary, $[OCl^-]$ equals [HOCl] and the pH equals the pK_a:

$$pH = pK_a = 7.5 \tag{8-54}$$

The reduction reaction involving hypochlorous acid and the chloride ion is given by the expression

$$HOCl + H^+ + 2e^- \rightleftarrows Cl^- + H_2O \tag{8-55}$$

The standard cell potential must be determined using the combined reactions shown below:

$$\begin{array}{ll} 2HOCl + 2H^+ + 2e^- \rightleftarrows Cl_2(aq) + 2H_2O & E^o_{red} = 1.61\ V \\ Cl_2(aq) + 2e^- \rightleftarrows 2Cl^- & E^o_{red} = 1.396\ V \\ \hline HOCl + H^+ + 2e^- \rightleftarrows Cl^- + H_2O & E^o_{red,sum} = ? \end{array}$$

The value of $E^\circ_{red,sum}$ can be computed using Eq. 8-21:

$$E^\circ_{red,sum} = \frac{n_{red,1}E^\circ_{red,1} + n_{red,2}E^\circ_{red,2}}{n_{red,1} + n_{red,2}} = \frac{2(1.61\ V) + 2(1.396\ V)}{2+2} = 1.50\ V$$

The E_H can be determined using the expression

$$E_H = E^\circ - \frac{0.059}{2}\log\left(\frac{[Cl^-]}{[HOCl][H^+]}\right) \tag{8-56}$$

At the boundary, $[Cl^-]$ equals [HOCl] and Eq. 8-56 may be rewritten as

$$E_H = 1.50 - \frac{0.059}{2} pH \tag{8-57}$$

Hypochlorite ion

The reduction reaction involving hypochlorite and the chloride ion is given by the expression

$$OCl^- + 2H^+ + 2e^- \rightleftarrows Cl^- + H_2O \tag{8-58}$$

The two reactions involved are shown below. Because the reaction involving the hypochlorite does not entail the gain or loss of an electron, the standard cell potential must be determined using the free energy as follows:

$$\begin{array}{ll} HOCl + H^+ + 2e^- \rightleftarrows Cl^- + H_2 & \Delta G^o_{Rxn,1} = ? \\ OCl^- + H^+ \rightleftarrows HOCl & \Delta G^o_{Rxn,2} = ? \\ \hline OCl^- + 2H^+ + 2e^- \rightleftarrows Cl^- + H_2O & \Delta G^o_{Rxn,sum} = ? \qquad E^o_{red,sum} = ?\,V \end{array}$$

The value of $\Delta G^\circ_{Rxn,1}$ for the reaction involving hypochlorous acid can be determined using Eq. 8-18, rewritten as follows:

$$\begin{aligned} \Delta G^\circ_{Rxn,HOCl} &= -nFE^\circ_{Rxn,1} \quad E^\circ_{Rxn,1} = 1.5\ V = 1.5\ J/C \\ &= -(2\ eq/mol)(96{,}500\ C/eq)(1.5\ J/C) \\ &= -289{,}000\ J/mol \end{aligned}$$

The value of $\Delta G^\circ_{Rxn,OCl^-}$ for the reaction involving the hypochlorite ion can be determined using Eq. 8-23, rewritten as follows:

$$\begin{aligned} \Delta G^\circ_{Rxn,OCl^-} &= -RT\ln K_{eq} \quad \text{where } K_{eq} = 10^{7.5} \\ &= -(8.314\ J/mol\cdot K)(298\ K)\ln\left(10^{7.5}\right) \\ &= -42{,}800\ J/mol \end{aligned}$$

The value of $\Delta G^\circ_{Rxn,sum}$ can be obtained using Eq. 8-18 as follows:

$$\begin{aligned} \Delta G^\circ_{Rxn,sum} &= \Delta G^\circ_{Rxn,HOCl} + \Delta G^\circ_{Rxn,OCl^-} \\ &= (-289{,}000\,J/mol) + (-42{,}800\,J/mol) \\ &= -331{,}800\,J/mol \end{aligned}$$

The value of $E^\circ_{red,sum}$ can be obtained by rearranging Eq. 8-18:

$$\begin{aligned} E^\circ_{red,sum} &= \frac{\Delta G^\circ_{Rxn,sum}}{-n_{Rxn,sum}F} = \frac{-331{,}800\,J/mol}{-(2\ eq/mol)(96{,}500\ C/eq)} \times \frac{1V}{1J/C} \\ &= 1.72\ V \end{aligned}$$

The E_H can be determined from the equation

$$E_H = E^\circ - \frac{0.059}{2}\log\left(\frac{\left[Cl^-\right]}{\left[OCl^-\right]\left[H^+\right]^2}\right) \tag{8-59}$$

At the boundary, $[Cl^-]$ equals [HOCl] and the following expression may be obtained:

$$E_H = 1.72 - 0.059\ pH \tag{8-60}$$

The predominance area diagram can then be constructed by noting that the reduction reaction is favored and the products predominate when E_H values are higher than the line, as shown on Fig. 8-2. Based on this diagram, the following conclusions may be drawn: (1) $Cl_2(aq)$ predominates at low pH, (2) $Cl_2(aq)$ disproportionates into HOCl and Cl^- at higher pH values, (3) Cl^- predominates for typical E_H–pH values in natural waters, and (4) HOCl, OCl^-,and $Cl_2(aq)$ are more powerful oxidants than oxygen from a thermodynamic point of view.

Rate of Oxidation–Reduction Processes

The reaction between an organic compound R (reductant) and an oxidant proceeds as shown in the following elementary reaction; and, based on numerous laboratory and full-scale studies, it has been found that second-order rate constants may be used to assess the rate of reaction:

$$\mathrm{R} + \text{oxidant}\left(\text{electron acceptor, } O_2,\ O_3,\ HO\bullet\right) \rightarrow \text{intermediate by-products} \quad (8\text{-}61)$$

$$\text{intermediate by-products} + \text{oxidant} \rightarrow CO_2 + H_2O + \text{mineral acids}\left(\text{e.g., HCl}\right) \quad (8\text{-}62)$$

$$r_{\mathrm{R}} = -k_{\mathrm{ox}} C_{\mathrm{ox}} C_{\mathrm{R}} \quad (8\text{-}63)$$

where r_{R} = rate of disappearance of organic compound R, mol/L · s
k_{ox} = second-order rate constant for oxidation reaction, L · mol/s
C_{ox} = concentration of oxidant, mol/L
C_{R} = concentration of organic compound R, mol/L

For an oxidant dosage of 0.1 mM, the half lives of compounds with second-order rate constants of 10, 100, 1000, and 10000 L/mol · s are 11.5 min, 1.15 min, 6.93 s, and 0.693 s, respectively. These half-lives are for the oxidation of the parent compound only, and destruction of by-products needs to be considered if the by-products are toxic.

8-3 Conventional Chemical Oxidants

Common chemical oxidants used in water treatment are (1) oxygen, (2) chlorine, (3) chlorine dioxide, (4) hydrogen peroxide, (5) ozone, and (6) permanganate. The forms of these oxidants and the method of application are summarized in Table 8-4. With respect to oxidation rate, the following general trend is typically observed; however, there will be exceptions depending on the type of compound that is oxidized:

$$HO\bullet > O_3 > H_2O_2 > HOCl > ClO_2 > KMnO_4 > Cl_2 > O_2 \quad (8\text{-}64)$$

Table 8-4
Common oxidants, forms, and application methods

Oxidant	Forms	Application Methods
Chlorine, free	Chlorine gas, sodium hypochlorite (NaOCl) solution	Gas eductors and spray jets
Chlorine dioxide	Chlorine dioxide gas produced onsite using 25% sodium chlorite solution; sodium chlorite solution reacted with following constituents to form chlorine dioxide [$ClO_2(g)$]: gaseous chlorine (Cl_2), aqueous chlorine (HOCl), or acid (usually hydrochloric acid, HCl)	Gas eductors
Hydrogen peroxide	Liquid solution	Concentrated solution mixed with water to be treated
Oxygen	Gas and liquid	Pure oxygen or oxygen in air is applied as gas
Ozone	A gas generated onsite by passing compressed air or pure oxygen across an electrode	Applied to water as a gas; mass transfer is an important issue; ozone contactors are usually bubble columns to ensure high transfer efficiency
Permanganate	Available in bulk as granules	Added as dry chemical using feeder or as concentrated solution (no more than 5% by weight due to its limited solubility)

The behavior of the hydroxyl radical, HO•, is discussed in Chap. 18. The purpose of this section is to present information on the conventional oxidants used in water treatment, including (1) the physical and chemical characteristics of the oxidants and (2) their application as oxidants in water treatment operations.

Oxygen (O_2)

Although the oxygen in the atmosphere has always been with us, Joseph Priestly is credited with the discovery of oxygen in 1775. However, it was Lavoisier who later explained correctly that oxygen was an active constituent of air. He called the gas *oxygen*, which means "acid former," because he incorrectly assumed that all acids contained it. Because oxygen is so readily available, it has been used to oxidize a variety of constituents and compounds found in water. However, as will be demonstrated, the oxidation kinetics are usually too slow to be of practical use in water treatment.

PHYSICAL AND CHEMICAL CHARACTERISTICS

Oxygen is a colorless, odorless, and tasteless gas. Under standard conditions (0° C and 1 atm) about 5 volumes of oxygen will dissolve in 100 volumes of water. Air is comprised of 21 and 23 percent oxygen by volume and weight, respectively (see App. B).

Oxidation potential

The reduction reaction for oxygen at 25°C is given by the expression

$$O_2(aq) + 4H^+ + 4e^- \rightarrow 2H_2O \quad E^\circ_{red} = 1.27 \text{ V} \tag{8-65}$$

Predominance area diagram for oxygen

The predominance area diagram for oxygen is shown on Fig. 8-3. The lines that are plotted show the point at which the concentrations of the species are equal; consequently, the species that are indicated on the opposite side of the line are preferred for the particular *EH* and pH value. For example, for an oxygen partial pressure of 0.21 atm, oxygen is the preferred species above the line and water is the preferred species below the line.

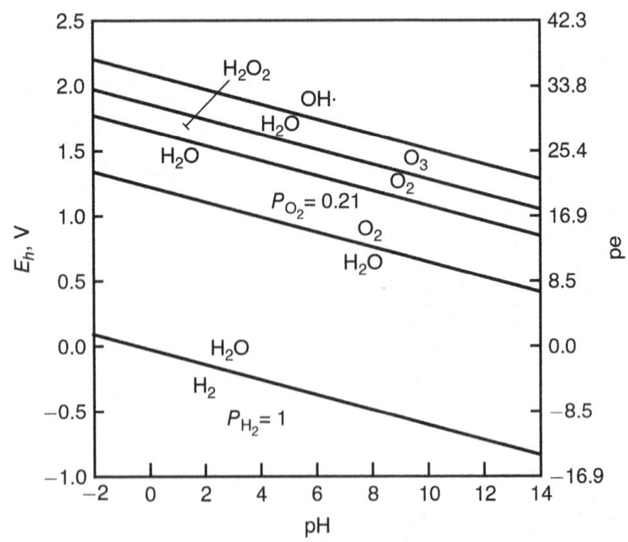

Figure 8-3
Predominance area diagram for oxygen system for oxygen concentration of 0.21 atm.

APPLICATIONS OF OXYGEN AS AN OXIDANT

In surface water bodies such as lakes, rivers, and oceans, oxygen is the oxidant that dominates conditions, determining, for example, the forms that iron, manganese, and sulfur will take. But the rate at which oxygen acts on these species is generally too slow to be useful as a unit process in water treatment plants. The oxidation of ferrous iron Fe(II) is the one notable exception to this observation. Ferrous iron can be oxidized quickly by oxygen under controlled conditions, particularly at alkaline pH. However, Fe(II) is occasionally complexed by NOM to the degree that oxygen is unable to oxidize it in reasonable time.

Oxidation of iron

Oxidation of Fe(II) with oxygen can be described by the reaction

$$4Fe^{2+} + O_2 + 4H^+ \rightarrow 4Fe^{3+} + 2H_2O \quad (8\text{-}66)$$

The amount of oxygen required for the oxidation of iron is 0.14 mg O_2/mg Fe^{2+}.

Since the pioneering work of Stumm and Lee (1961), more than 300 papers have been written on the rates and mechanisms of ferrous iron oxidation (King, 1998). Based on this body of work, and particularly the work of Millero (Millero and Izaguirre, 1989; Millero and Sotolongo, 1989; Millero, 1989, 1990a,b; Millero et al., 1987) and Wehrli (1990), it has been demonstrated that the overall oxidation rate of ferrous iron in a completely mixed batch reactor (CMBR) can be described as a pseudo-second-order reaction:

$$-\frac{d[\text{Fe(II)}]}{dt} = -r_{\text{Fe(II)}} = P_{O_2}[\text{Fe(II)}]k_{\text{app}} \quad (8\text{-}67)$$

where $[\text{Fe(II)}]$ = concentration of ferrous iron, mol/L

$r_{\text{Fe(II)}}$ = overall oxidation rate of ferrous iron, mol/L · min

t = time, min

P_{O_2} = partial pressure of oxygen, atm

k_{app} = apparent rate constant, 1/min · atm

It has also been demonstrated that Fe(II) forms a number of ligand complexes in solution. The following are among the more common inorganic ligand complexes: $FeCO_3^0$, $Fe(CO_3)_2^{2-}$, $Fe(CO_3)(OH)^-$, $FeCl^+$, $FeSO_4^0$, $Fe(OH)^+$, and $Fe(OH)_2^0$. As mentioned earlier, complexes with NOM have also been shown to have important influence on the rate of oxidation (Theis and Singer, 1974). In the presence of these ligand complexes, the apparent rate constant k_{app} can be computed as the sum

$$k_{\text{app}} = 4\left(k_1\alpha_{Fe^{2+}} + k_2\alpha_{FeOH^+} + \cdots + k_n\alpha_n\right) \quad (8\text{-}68)$$

where $k_1, k_2, \ldots, k_n$ = first-order rate constant for 1st, 2nd, … nth Fe(II) species

$\alpha_{Fe^{2+}}, \alpha_{FeOH^+}, \ldots, \alpha_n$ = fraction of total Fe(II) in solution present as species Fe^{2+}, $FeOH^+$, ..., species n

and the factor 4 comes from reaction stoichiometry (King, 1998).

Although thermodynamic constants and rate constants are available for Fe^{2+} and for the inorganic ligand complexes listed above, it is usually not practical to calculate k_{app} from first principles. The fact that the rate of oxidation takes the form of a pseudo-first-order reaction when P_{O_2} is constant in Eq. 8-65 means that bench-scale testing can be used to develop a value of k_{app} characteristic of a particular water quality. Such testing should always be conducted to assess the feasibility and gain design criteria for facilities used to oxidize Fe(II) using oxygen.

Example 8-7 Oxidation of Fe(II) in presence of oxygen

Bench-scale tests have been conducted to examine the rate of oxidation of Fe(II) in a particular well water in the presence of oxygen. The results are shown below:

Time, min	Fe(II), mg/L
0	5
8.3	2.0
16.7	0.67
24.8	0.32
41.7	0.05

Assuming that the partial pressure of oxygen is maintained at 0.21 atm during the tests, determine k_{app}.

Solution

1. Set up a spreadsheet with the following columns:
   Column 1 —time, min
   Column 2—concentration of Fe(II), mg/L
   Column 3—C/C_0
   Column 4—$\ln(C/C_0)$
   The spreadsheet values are given below:

Time, min	Fe(II), mg/L	Fe(II), C/C_o	Fe(II), $\ln(C/C_o)$
0	5	1	0.00
8.3	2.0	0.4	−0.92
16.7	0.67	0.134	−2.01
24.8	0.32	0.064	−2.75
41.7	0.05	0.01	−4.61

2. Plot time (column 1) versus $\ln(C/C_0)$ (column 4), and plot the best-fit regression line through the intercept at $t = 0$ and $\ln(C/C_0) = 0$.

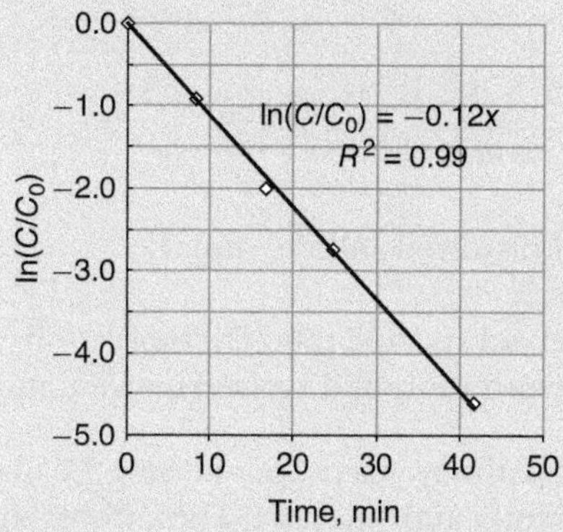

The slope of the line is $-P_{O_2}k_{app}$:

$$P_{O_2}k_{app} = 0.12\ \text{min}^{-1}$$

$$k_{app} = \left(\frac{0.12/\text{min}}{0.11\ \text{atm}}\right) \simeq 1.091/\text{min}\cdot\text{atm}$$

The oxidation of iron is relatively straightforward in normal, low-TOC groundwaters of low mineral content; however, high levels of TOC have been shown to complex with iron, which prevents its expeditious oxidation (Theis and Singer, 1974). As a practical matter, iron is the only constituent material that oxygen can be relied upon to accomplish removal by direct oxidation.

Oxidation of manganese

Aeration can also be used for the oxidation of Mn(II), but it has been found that the direct oxidation of Mn(II) is slow, even at elevated pH (Chen 1974; Morgan, 1967). As a result, phenomena other than direct oxidation probably occur when aeration is successful (e.g., biologically mediated oxidation), and unless these phenomena are well understood or there are successful projects of the same scale in the same vicinity with the same water quality, the engineer should avoid relying on these designs. The stoichiometric oxygen dose for the oxidation of Mn(II) is 0.29 mg O_2/mg Mn^{2+}.

Oxidation of sulfide

Aeration can be used to oxidize hydrogen sulfide, but it has been found that the direct oxidation of hydrogen sulfide with oxygen is very slow, even at elevated pH (Chen 1974; Morgan, 1967). Chen and Morris (1972) did a study of the rate of sulfide oxidation by dissolved oxygen and found that the rate in a CMBR could be described by the empirical rate equation

$$-\left[\frac{d\left(\sum S^{2-}\right)}{dt}\right]_{t=0} = k\left(\sum S^{2-}\right)_0^{1.34} [O_2]^{0.56} \tag{8-69}$$

where (ΣS^{2-}) = total dissolved sulfide, mol/L
t = time, h
k = observed specific rate, $(L/mol)^{0.9} \cdot h$
$[O_2]$ = concentration of dissolved oxygen, mol/L

A complex reaction pathway was proposed, and the observed specific rate was shown to change significantly with pH. The authors showed that the reaction also requires an induction time. During the induction time, the concentrations of oxygen and sulfide remain unchanged. As a practical matter, this research and practical experience indicate that, unless the reaction between oxygen and sulfide is catalyzed or biologically mediated, its half-life is on the order of several days. Thus, oxidation of sulfide by oxygen alone is not practical in the typical treatment process time frame in a water treatment facility.

Chlorine (Cl_2)

Chlorine is the most commonly used compound for the disinfection of water in the United States. The use of chlorine for disinfection is examined in detail in Chap. 13. In the following discussion, the focus is on the use of chlorine as an oxidant.

PHYSICAL AND CHEMICAL CHARACTERISTICS

Chlorine is a heavy greenish-yellow gas with a characteristic penetrating odor, corrosive and intensely irritating to the respiratory organs of all animals. Chlorine gas is easily liquefied. The physical characteristics of common oxidizing agents including chlorine are summarized in Table 8-5.

Oxidation potential

The reduction half reaction for chlorine at 25°C and its dissociation products hypochlorous acid and hypochlorite ion are given by the reactions

$$Cl_2(aq) + 2e^- \rightleftarrows 2Cl^- \qquad E^\circ_{red} = 1.396\ V \tag{8-70}$$

$$2HOCl + 2H^+ + 2e^- \rightleftarrows Cl_2(aq) + 2H_2O \qquad E^\circ_{red} = 1.61\ V \tag{8-71}$$

$$HOCl + H^+ + 2e^- \rightleftarrows Cl^- + H_2O \qquad E^\circ_{red} = 1.50\ V \tag{8-72}$$

$$OCl^- + H_2O + 2e^- \rightleftarrows Cl^- + 2OH^- \qquad E^\circ_{red} = 0.90\ V \tag{8-73}$$

Table 8-5
Properties of common oxidizing agents: chlorine, chlorine dioxide, ozone, hydrogen peroxide, and potassium permanganate

Property	Unit	Chlorine, Cl_2	Chlorine dioxide, ClO_2	Ozone, O_3	Hydrogen peroxide, H_2O_2	Potassium permanganate, $KMnO_4$
Molecular weight	g/mol	70.91	67.45	48.0	34.02	158.04
Boiling point (liquid)	°C	−33.97	11	−111.9 ± 0.3	114	
Melting point	°C	−100.98	−59	−192.5 ± 0.4	−50	150
Latent heat of vaporization at 0°C	kJ/kg	253.6	27.28	14.90		
Liquid density at 15.5°C	kg/m^3	1422.4	1640[a]	1574	1460	
Solubility in water at 15.5°C	g/L	7.0	70.0[a]	2.154		70
Specific gravity of liquid at 0°C (water = 1)	unitless	1.468		12.07	1.48	2.70
Vapor density at 0°C and 1 atm	kg/m^3	3.213	2.4	11	1.15	
Vapor density compared to dry air at 0°C and 1 atm	Unitless	2.486	1.856	1.666		
Specific volume of vapor at 0°C and 1 atm	m^3/kg	0.3112	0.417	0.464		
Critical temperature	°C	143.9	153	−12.1		
Critical pressure	kPa	7811.8		5532.3		

Source: Adapted in part from U.S. EPA (1986) and White (1999). [a]At 20°C.

Predominance diagram for chlorine

The chlorine predominance diagram for a total chlorine concentration of 10^{-4} M is shown on Fig. 8-2. The details involved in the preparation of the diagram are given in Sec. 8-2.

APPLICATION OF CHLORINE AS AN OXIDANT

Chlorine can be an effective oxidant for some species, particularly sulfide, Fe(II), and Mn(II). Chlorine has also been used for the removal of taste and odor compounds. Occasionally, chlorine is used to remove ammonia.

Oxidation of sulfide

Under alkaline conditions, sulfide reacts rapidly with chlorine, and the following two reactions compete with each other:

$$Cl_2 + H_2S \rightarrow 2HCl + S_0 \tag{8-74}$$

$$4Cl_2 + H_2S + 4H_2O \rightarrow 8HCl + H_2SO_4 \tag{8-75}$$

Although research has been conducted on the oxidation of sulfide by chlorine (Black and Goodson, 1952; Chen, 1974; Powell and Lossberg, 1948), it has not been done in a way that allows more than a qualitative understanding of the chemistry involved. By the first reaction (Eq. 8-74), 2.1 mg Cl_2 is required per milligram of H_2S removed. By the second reaction (Eq. 8-75), 8.4 mg Cl_2 is required per milligram of H_2S removed. Both the kinetics and the stoichiometry of the reaction are influenced by the pH. Above pH 8, the chlorine requirement corresponds to Eq. 8-74. As the pH decreases from this point, the chlorine requirement increases until it approaches Eq. 8-75. Generally, the reaction of chlorine with sulfide is rapid, reaching completion in a few minutes.

Oxidation of iron

Oxidation of Fe(II) with chlorine is normally quite rapid following the stoichiometry shown below (approximately):

$$2Fe^{2+} + Cl_2 \rightarrow 2Fe^{3+} + 2Cl^- \tag{8-76}$$

The stoichiometric chlorine dose for the oxidation of iron is 0.63 mg Cl_2/mg Fe^{2+}. The reaction is relatively rapid. Knocke (1990) has found that the reaction is generally completed in less than 15 min, even at low pH. Like the activity of oxygen, the oxidation of ferrous iron is accelerated by high pH and can be decelerated substantially by the presence of high levels of organic matter (Knocke, et al., 1992).

Oxidation of manganese

The oxidation of Mn(II) with free chlorine is more difficult than the oxidation of Fe(II). In solution, the reaction between chlorine and Mn(II) is too slow to be useful in water treatment unless the pH is elevated above approximately 9. When chlorinated water is passed through a filter containing media coated with MnO_2, removal will occur by adsorption to the media, and the adsorbed Mn(II) will gradually be oxidized to MnO_2 on the filter media surface. This behavior is analogous to the oxidation of Mn(II) by oxygen on MnO_2 surfaces as demonstrated by Morgan (1967). Coffey et al. (1993) proposed the following steps in the oxidation process:

Step 1: Adsorption of Mn(II) on the MnO_2 surface:

$$Mn^{2+} + MnO(OH)_2 \underset{k_2}{\overset{k_1}{\rightleftarrows}} MnO_2MnO + 2H^+ \qquad (8\text{-}77a)$$

Step 2: Oxidation of the adsorbed species by chlorine:

$$MnO_2MnO + Cl_2 + H_2O \underset{k_4}{\overset{k_3}{\rightleftarrows}} 2MnO_2 + 2HCl \qquad (8\text{-}77b)$$

Thus, chlorine is commonly used for manganese removal when filtration is available. Some time (on the order of months) is required for the filter media to become coated with the necessary MnO_2 surface, but once the media is coated, manganese removal is fairly efficient and complete. During the acclimation period, the pH needs to be elevated to accelerate Mn(II) oxidation. Little work has been done on the kinetics of adsorption, but it is likely that the rate of adsorption may be the controlling factor when low effluent concentrations of Mn(II) are required.

Oxidation of tastes and odors

Given sufficient reaction time, free chlorine is also effective in the control of a wide variety of tastes and odors associated with drinking water, with sulfide odors being the most common, but also many fishy, grassy, and swampy odors. The earthy musty odors associated with geosmin and MIB are not removed with chlorine.

Reactions of chlorine with ammonia

The removal of ammonia with chlorine is called breakpoint chlorination. Breakpoint chlorination reactions are presented and discussed in Chap. 13. More common is the addition of ammonia to an existing free chlorine residual to stabilize the disinfectant residual and to arrest the formation of undesirable by-products.

Chlorine Dioxide (ClO_2)

One of the principal appeals of chlorine dioxide is that it can oxidize a variety of constituents without producing the trihalomethanes (THMs) and haloacetic acids (HAAs) associated with free chlorine. At the same time, both chlorite and chlorate ions are by-products of oxidation with chlorine dioxide, and regulations on these ions (see Chaps. 4 and 19) sometimes limit the dose of chlorine dioxide to concentration levels that cannot be used for oxidation.

PHYSICAL AND CHEMICAL CHARACTERISTICS

Chlorine dioxide must be produced onsite because it is unstable at high concentrations. The physical characteristics of chlorine dioxide are summarized in Table 8-5; production methods are discussed in Chap. 13. The following issues must be considered with regard to handling, use, and storage: (1) chlorine dioxide is volatile (dimensionless Henry's constant is 0.0409 at 25°C, which is much larger than Henry's constant for chlorine, which is 4.42×10^{-5} at 25°C) and it can be stripped from aqueous solution if precautions against volatilization are not taken; (2) after it is generated and dissolved in water, chlorine dioxide is stable when it is not exposed to light or high temperatures; and (3) at high pH, chlorine dioxide disproportionates to form both chlorite (ClO_2^-) and chlorate (ClO_3^-), which are regulated by-products (see Chaps. 4 and 19).

Oxidation potential

The reduction half reaction for chlorine dioxide at 25°C is given by the reaction

$$ClO_2\ (g) + 2H_2O + 5e^- \rightarrow Cl^- + 4OH^- \quad E^\circ_{red} = 0.799\ V \qquad (8\text{-}78)$$

Formation of chlorite and chlorate

The formation of chlorite (ClO_2^-) and chlorate (ClO_3^-) is given by the reaction

$$2ClO_2 + 2OH^- \rightarrow ClO_2^- + ClO_3^- + H_2O \qquad (8\text{-}79)$$

Predominance diagram for chlorine dioxide

The chlorine dioxide predominance diagram for a total concentration of 10^{-4} M is shown on Fig. 8-4. The lines plotted show the point at which the concentrations of the species are equal; consequently, the species that are indicated on the opposite side of the line are preferred for the particular E_H and pH value. For example, for pH values greater than 2, chlorine dioxide is unstable and forms either ClO_2^- or ClO_3^- depending on the E_H and pH value. The line for oxygen and water is not shown because it coincides with the chlorate line.

APPLICATIONS OF CHLORINE DIOXIDE AS AN OXIDANT

Relatively little has been published on the effectiveness of chlorine dioxide in oxidizing Fe(II) in drinking water, but it is likely that it can be effective when Fe(II) is not strongly complexed with NOM. Based on work with chlorine dioxide and Mn(II), it appears that the reaction is relatively rapid with the formation of colloidal particles of MnO_2 (Knocke et al., 1988). The typical chlorine dioxide dose that has been reported for the

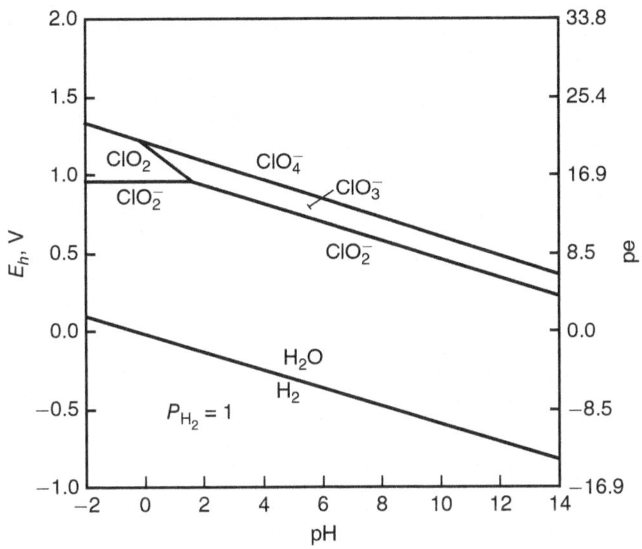

Figure 8-4
Predominance area diagram for chlorine dioxide system for total species concentration of 10^{-4} mol/L.

oxidation of iron is 1.2 mg ClO_2/mg Fe^{2+}; the corresponding dose for manganese is 2.5 mg ClO_2/mg Mn^{2+}. Little information is available on the oxidation of hydrogen sulfide via chlorine dioxide, although, once again, rapid oxidation seems likely. Chlorine dioxide will also remove many of the swampy, grassy, and fishy odors associated with drinking water, but like chlorine, it is of little use against the earthy musty odors associated with MIB and geosmin.

Hydrogen Peroxide (H_2O_2)

Hydrogen peroxide is one of the strongest oxidizing agents available. A metabolite of many microorganisms, hydrogen peroxide is decomposed by the same organisms to oxygen and water. Hydrogen peroxide is also produced by the action of sunlight on water. The use of hydrogen peroxide in the water field can be traced back to the early 1990s.

PROPERTIES AND CHEMICAL CHARACTERISTICS

Hydrogen peroxide is a nearly colorless liquid, with blue tinge, which may be solidified. It is soluble in water in all proportions. The physical properties of hydrogen peroxide are summarized in Table 8-5. Commercial solutions typically contain about 35, 50, or 75 percent hydrogen peroxide. In the absence of sunlight and exposure to foreign particles, concentrated solutions can be held for brief periods of time. Hydrogen peroxide becomes quite unstable, particularly in the presence of foreign particles such as dust. The concentrated solution can explode violently at temperatures above 100°C. Hydrogen peroxide is added to water as a dilute solution.

Oxidation potential
The half reaction for hydrogen peroxide at 25°C is

$$H_2O_2 + 2H^+ + 2e^- \rightarrow 2H_2O \qquad E^\circ_{red} = 1.780\ V \tag{8-80}$$

Predominance diagram for hydrogen peroxide
The predominance area diagram for a total hydrogen peroxide species concentration of 10^{-4} mol/L is shown on Fig. 8-3. Hydrogen peroxide is preferred above the line and water is preferred below the line.

APPLICATIONS OF HYDROGEN PEROXIDE AS AN OXIDANT
One conventional application in water treatment for which hydrogen peroxide has been used successfully is in the oxidation of sulfides. Hydrogen peroxide has also been used for the removal of iron.

Oxidation of sulfide
The redox reaction for the oxidation of sulfide is

$$H_2S + H_2O_2 \rightarrow S_0 + 2H_2O \tag{8-81}$$

The stoichiometric hydrogen peroxide dose for the oxidation of sulfide is 1.0 mg H_2O_2/mg H_2S.

Oxidation of iron
The redox reaction for the oxidation of iron is

$$2Fe^{2+} + H_2O_2 + 2H^+ \rightarrow 2Fe^{3+} + 2H_2O \tag{8-82}$$

The stoichiometric hydrogen peroxide dose for the oxidation of iron is 0.30 mg H_2O_2/mg Fe^{2+}.

Other oxidation applications
Little information is available on the effectiveness of peroxide in oxidizing manganese. Based on the limited information that is available, it appears that hydrogen peroxide alone is not particularly effective at controlling tastes and odors. In advanced oxidation, hydrogen peroxide is an important precursor in processes that involve UV light and ozone, as discussed in Chap. 18.

Ozone (O_3)

The use of ozone as a conventional oxidant is discussed below. However, as discussed in Chap. 18, ozone will react with NOM and produce the hydroxyl radical and thus it can be argued that ozonation is an advanced oxidation process. Advanced oxidation processes involving ozone (including the production of the hydroxyl radical via reactions with NOM) are discussed in Chap. 18.

PHYSICAL AND CHEMICAL CHARACTERISTICS

Ozone is applied to water as a gas, which is generated onsite by passing dry compressed air or pure oxygen across an electrode (see Chap. 13). The physical properties of ozone are reported in Table 8-5.

Oxidation potential

The reduction half reaction for ozone at 25°C is

$$O_3(g) + 2H^+ + 2e^- \rightarrow O_2 + H_2O \quad E^\circ_{red} = 2.08\ V \tag{8-83}$$

Predominance area diagram for ozone

The predominance area diagram for a total ozone concentration of 10^{-4} mol/L is shown on Fig. 8-3. The point at which the concentrations of the species are equal corresponds to the lines that are plotted. Consequently, the species that are indicated on the opposite side of the line are preferred for the particular E_H and pH value. The line that separates the reactants ozone and oxygen clearly indicates that the hydroxyl radical is a much more powerful oxidant than ozone.

APPLICATION OF OZONE AS AN OXIDANT

Ozone is used in water treatment in a variety of applications, including (1) disinfection, (2) oxidation of iron and manganese, (3) oxidation of sulfides, (4) oxidation of taste and odor compounds, (5) oxidation of micropollutants, (6) removal of color, primarily through oxidation, (7) control of DBP precursors, and (8) the reduction of chlorine demand through oxidation. Because ozone is considered in Chap. 18, the following discussion is limited to the oxidation of iron, manganese, and sulfide; a brief mention of the oxidation of taste and odor compounds and NOM; and rate constants for ozone oxidation.

Oxidation of iron and manganese

The redox reactions for the oxidation of iron and manganese with ozone are as follows:

For iron

$$2Fe^{2+} + O_3(aq) + 5H_2O \rightarrow 2Fe(OH)_3(s) + O_2 + 4H^+ \tag{8-84}$$

For manganese

$$Mn^{2+} + O_3(aq) + H_2O \rightarrow MnO_2(s) + O_2 + 2H^+ \tag{8-85}$$

The stoichiometric ozone dose for the oxidation of iron is 0.43 mg O_3/mg Fe^{2+}; the corresponding stoichiometric dose for manganese is 0.88 mg O_3/mg Mn^{2+}. In both of the above reactions, alkalinity is consumed as a result of acid production. The amount of alkalinity consumed is 1.79 and 1.82 mg/L as $CaCO_3$ per milligrams per liter of Fe^{2+} and Mn^{2+} oxidized,

Example of an ozone generator

http://www.stantec.com/wt3r/8-1.html

respectively. Alkalinity is an important issue, especially where alum coagulation is involved, because both consume alkalinity. Ozone should be used with caution for the removal of Mn(II) because it converts Mn(II) to MnO_2 so rapidly that MnO_2 tends to clog ozone diffusers and the MnO_2 that forms in solution is of an extremely fine colloidal nature and can be difficult to remove in filtration.

Oxidation of sulfide

The redox reaction for the oxidation of sulfide with ozone is

$$H_2S + O_3(aq) \rightarrow S_0 + O_2(aq) + H_2O \tag{8-86}$$

Based on the above reaction, the required ozone dose for the oxidation of sulfide is 1.41 mg O_3/mg H_2S. In practice, it has been found that the required ozone dose will vary from 2.0 to 4.0 mg O_3/mg H_2S. Ozone dosages greater than the stoichiometric requirement occur because of the presence of other oxidizable constituents and because a portion of the sulfide is often converted to sulfate (SO_4^{2-}) rather than elemental sulfur.

Oxidation of taste and odor compounds

Ozone has been found to be effective for the oxidation of taste and odor compounds in water. Typical ozone doses are in the range of 1 to 3 mg/L with a minimum contact time of 10 to 15 min. However, because the compounds that contribute to taste and odor are site specific, bench- and pilot-scale testing is usually required to establish the appropriate dose and the points of application. As discussed in Chap. 18, ozone can oxidize the taste and odor compounds geosmin and methyl isoborneol through the production of the hydroxyl radical.

Use of Ozone in Advanced Oxidation Processes

http://www.stantec.com/wt3r/8-2.html

Oxidation of NOM

In addition to the above uses, ozone reacts with NOM to form lower molecular weight polar compounds (e.g., aldehydes, organic acids, and ketones). Many polar compounds are biodegradable and are not believed to be harmful, but they can cause biofouling problems in the water distribution system. As a result, ozonated water can be followed by a biologically active filtration process, which sometimes includes GAC, to remove the biodegradable fraction.

Rate constants for ozone oxidation

The second-order rate constants for ozone, provided in the electronic Table E2 at the website listed in App. E, are useful in assessing possible reactions and their kinetics. However, NOM can also initiate the production of hydroxyl radicals, which is more important for the degradation of most compounds (Elovitz and von Gunten, 1999; Westerhoff et al., 1999), as discussed in Chap. 18. The second-order rate constants for organics are highly dependent on the type of organic being oxidized. The reaction rate is high for the hydroxyl- or amine-substituted benzenes and low for aliphatics without nucleophilic sites. Most of the rate constants are too low to allow for the use of ozone in water treatment, unless initiators (e.g., NOM, organic compounds, UV, or hydrogen peroxide) are used to produce hydroxyl radicals, which react rapidly with organic compounds. Reactions that involve hydroxyl radicals are referred to as advanced oxidation processes (see Chap. 18).

Permanganate (MnO_4^-)

Permanganate was first used for water treatment in 1910 in London but did not begin to grow in use until the 1960s, when it was applied successfully for taste and odor control. Since then, potassium permanganate has been accepted by the water industry as one of the most versatile oxidants available.

PHYSICAL AND CHEMICAL CHARACTERISTICS

As an oxidant, potassium permanganate is typically more expensive than chlorine and ozone, but for iron and manganese removal, it has been reported to be as efficient and may require considerably less equipment and capital investment. The physical properties of potassium permanganate are summarized in Table 8-5. Potassium permanganate can be purchased in bulk as granules and be added using a dry chemical feeder or as a concentrated solution.

Oxidation potential

Potassium permanganate will oxidize a wide variety of inorganic and organic compounds. Under acidic conditions the principal reduction half reactions are

$$MnO_4^- + 4H^+ + 3e^- \rightleftarrows MnO_2(s) + 2H_2O \quad E^\circ_{red} = 1.68\,V \qquad (8\text{-}87)$$

$$MnO_4^- + 8H^+ + 5e^- \rightleftarrows Mn^{2+} + 4H_2O \quad E^\circ_{red} = 1.510 \qquad (8\text{-}88)$$

Under alkaline conditions the corresponding reduction half reaction is

$$MnO_4^- + 2H_2O + 3e^- \rightleftarrows MnO_2(s) + 4OH^- \quad E^\circ_{red} = 0.590\,V \qquad (8\text{-}89)$$

Predominance diagram for permanganate

The predominance area diagram for a total permanganate species concentration of 10^{-4} M is shown on Fig. 8-5. The lines that are plotted show the point at which the concentrations of the species are equal; consequently, the species shown on the opposite side of the line are preferred for the particular E_H and pH value. For example, the order required for increasing E_H potential to be thermodynamically favored is Mn^{2+}, MnO_2, and MnO_4^-.

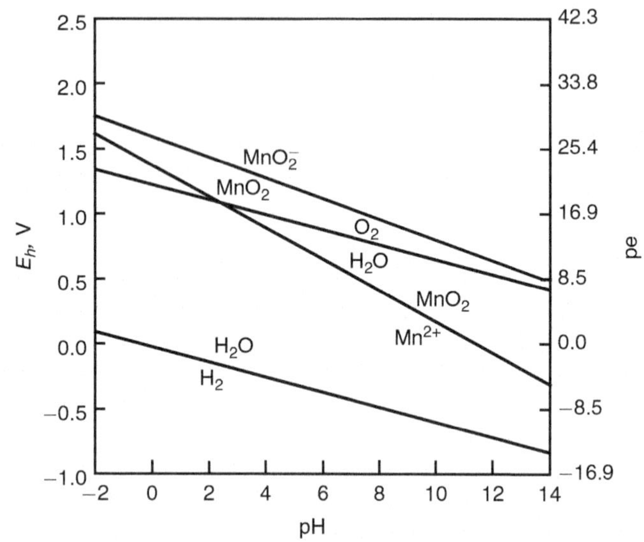

Figure 8-5
Predominance area diagram for permanganate system for total species concentration of 10^{-4} mol/L.

APPLICATION OF POTASSIUM PERMANGANATE AS AN OXIDANT

Potassium permanganate is used in drinking water treatment for a variety of purposes. The principal applications involving its use as an oxidant include (1) oxidation of iron and manganese, (2) oxidation of taste and odor compounds, (3) control of nuisance organisms, and (4) control of the formation of THMs and other DBPs by oxidizing precursors and reducing the demand for other disinfectants.

Oxidation of iron and manganese

The principal use of potassium permanganate in water treatment is the oxidation of iron and manganese. The corresponding redox reactions are as follows:

For iron

$$3Fe^{2+} + KMnO_4 + 7H_2O \rightarrow 3Fe(OH)_3(s) + MnO_2(s) + K^+ + 5H^+ \qquad (8\text{-}90)$$

For manganese

$$3Mn^{2+} + 2KMnO_4 + 2H_2O \rightarrow 5MnO_2(s) + 2K^+ + 4H^+ \qquad (8\text{-}91)$$

The stoichiometric potassium permanganate dose for the oxidation of iron is 0.94 mg $KMnO_4$/mg Fe^{2+}, and the corresponding dose for manganese is 1.92 mg $KMnO_4$/mg Mn^{2+}. In both of the above reactions alkalinity is consumed as a result of acid production. The amount of alkalinity consumed is 1.49 and 1.21 mg/L as $CaCO_3$ per milligram per liter of Fe^{2+} and Mn^{2+}, respectively. Alkalinity is an important issue, especially where alum coagulation is involved because both consume alkalinity. The oxidation of manganese is considered in the following example.

Example 8-8 Oxidation of Mn(II) with permanganate

Determine how much Mn(II) remains after 30 s of oxidation with permanganate in a CMBR. The initial concentrations of permanganate and Mn(II) are 1.5 times the stoichiometric requirement and 1 mg/L, respectively. Assume that the second-order rate constant is 10^5 L/mol · s.

Solution

1. Calculate the initial concentration of permanganate using Eq. 8-91. Assume the initial concentrations of Mn^{2+} and MnO_4^{2-} are C_{A0} and C_{B0}, respectively:

$$3Mn^{2+} + 2MnO_4^- + 2H_2O \rightarrow 5MnO_2 + 4H^+$$

$$C_{A0} = [Mn^{2+}]_0 = \frac{1\times 10^{-3}\,\text{g/L}}{54.938\ \text{g/mol}} = 1.82\times 10^{-5}\ \text{mol/L}$$

$$C_{B0} = [MnO_4^-]_0 = 1.5\times\frac{2}{3}\times[Mn^{2+}]_0$$

$$= 1.5\times\frac{2}{3}\times(1.82\times 10^{-5}\ \text{mol/L})$$

$$= 1.82\times 10^{-5}\ \text{mol/L}$$

2. Calculate the remaining concentrations of Mn(II):
   a. Write a mass balance for a CMBR, and develop the rate expression:

$$\frac{dC_A}{dt} = r_A = -kC_AC_B = -kC_A\left[C_{B0} - \frac{2}{3}(C_{A0} - C_A)\right]$$

   b. Integrate the rate expression:

$$\frac{1}{C_{B0} - \frac{2}{3}C_{A0}}\int_{C_{A0}}^{C_A}\left[\frac{dC_A}{C_A} - \frac{d\left[C_{B0} - \frac{2}{3}(C_{A0} - C_A)\right]}{C_{B0} - \frac{2}{3}(C_{A0} - C_A)}\right] = -\int_0^t kdt$$

$$\frac{1}{C_{B0} - \frac{2}{3}C_{A0}}\ln\left(\frac{C_A}{C_{B0} - \frac{2}{3}(C_{A0} - C_A)}\times\frac{C_{B0}}{C_{A0}}\right) = -kt$$

$$\frac{C_A}{C_{B0} - \frac{2}{3}(C_{A0} - C_A)} = \frac{C_{A0}}{C_{B0}}\exp\left[-\left(C_{B0} - \tfrac{2}{3}C_{A0}\right)kt\right]$$

c. Solve for Mn(II) concentration at t = 30 s:

$$\frac{C_A}{1.82\times10^{-5} - \frac{2}{3}(1.82\times10^{-5} - C_A)}$$
$$= \exp\,(-0.333\times1.82\times10^{-5}\times10^{5}\times30)$$
$$C_A = [\mathrm{Mn}^{2+}] = 7.56\times10^{-14}\ \mathrm{M}$$

Oxidation of taste and odor compounds
The application of potassium permanganate is also quite common for the oxidation of the odorous compounds occasionally found in water. Typical dosages of potassium permanganate, which are site specific, vary from 0.25 to 20 mg/L. Potassium permanganate also seems to work fairly well for sulfide oxidation and for the removal of fishy/grassy odors that are produced by methyl sulfides. However, it does a poor job of removing geosmin and MIB.

Importance of dosage control with permanganate
Permanganate gives an easily detected pink color to water with overdoses in the 0.05-mg/L range. Due to this fact, the dose range is critical in avoiding consumer complaints. It is also important to remove unreacted permanganate because it will form black deposits in distribution systems and on plumbing fixtures. In addition, the manganese concentration in the final treated water may exceed the levels prescribed in the secondary regulations. A typical permanganate dose is on the order of 1 to 3 mg/L, and a contact time of at least 1 to 2 h is usually provided for the oxidation reaction to be complete. As exceptions, doses as high as 10 mg/L have been used without adverse effects and contact times of 10 to 15 min are sometimes adequate. To obtain longer contact times than would be available in their treatment plant basins, many utilities add $KMnO_4$ to raw-water pipelines at the source. If excessive permanganate is used in the oxidation process, it will pass through the downstream filters and enter the distribution system. However, if the excess dose appears to be necessary to remove a target compound, raising the pH prior to solids separation will accelerate the kinetics of permanganate oxidation.

Settling out the excess permanganate in the sedimentation basin generally controls overdosing, which corresponds to the disappearance of permanganate's characteristic pink color. However, for plants without flocculation or sedimentation steps, such as in-line or direct-filtration plants, special monitoring equipment must be used to prevent permanganate from passing through the filters. Excess permanganate residual can sometimes be removed by adding PAC after the oxidation step to avoid reducing the permanganate's efficiency.

8-4 Photolysis

Photolysis is a process by which photons are absorbed by compounds, and the energy released is used to drive light-induced oxidation or reduction processes. The rate at which a compound is photolyzed can be estimated from the rate at which the compound absorbs light and the quantum yield (photonic efficiency of the reaction). Estimating these rates is discussed in this section.

Energy Required for Photolysis and Wavelength of Light

Usually, photons in the UV range (200 to 400 nm) are capable of providing enough energy to drive photolytic reactions. There are three major options for UV lamps: (1) low-pressure, low-intensity, (2) low-pressure, high-intensity (sometimes referred to as pulsed UV lamps), and (3) mediumpressure, high-intensity lamps. Low-pressure lamps emit all their energy at a wavelength of 254 nm. The medium-pressure lamps emit energy at several wavelengths. Additional discussion of UV lamp technologies is provided in Chap. 13.

In photolysis, the photons from a light source supply the energy required for a reaction to proceed. For a given photolytic reaction, a reaction occurs when an electron in the outer orbital absorbs a photon and forms an unstable compound that undergoes reaction or splits apart. The photonic energy that is required for such a reaction to proceed depends on the specific electron structure of a given compound, but basic thermodynamics can be used to estimate the minimum photonic energy that is required. The minimum energy for a given reaction may be calculated from the reaction potential. The free energy of the reaction and its electrochemical potential are related to each other as given by Faraday's law:

$$\Delta G^{\circ}_{\text{Rxn}} = -nFE^{\circ}_{\text{Rxn}} \tag{8-92}$$

where F = Faraday's constant, 96,500 C/eq
n = number of photons, eq/mol
$\Delta G^{\circ}_{\text{Rxn}}$ = free energy of reaction, J/mol
E°_{Rxn} = reaction potential, V or J/C

The reaction potential can also be obtained from summary tables, such as the one in the electronic Table E1 at the website listed in App. E. Often, it is necessary to know the wavelength of light that is necessary for the photolysis reaction to occur. The energy required for the reaction to proceed can be calculated from the reaction potential as follows:

$$E = ne^{-}E^{\circ}_{\text{rxn}} \tag{8-93}$$

where E = energy for photolysis reaction, J
e^- = charge of an electron = 1.602×10^{-19} C

The frequency of light is related to the energy of a photon by Planck's constant, as shown in the expression

$$\nu = \frac{E}{h} \tag{8-94}$$

where ν = frequency of light, s^{-1}
h = Planck's constant, 6.62×10^{-34} J · s

The wavelength of light is inversely related to the frequency, as shown in the expression

$$\lambda = \frac{c}{\nu} \tag{8-95}$$

where c = speed of light, 3.00×10^8 m/s
λ = wavelength of light, m

Because photonic energy is inversely related to the wavelength, the wavelength of light that is expressed in Eq. 8-95 represents the longest wavelength required to power the photolytic reaction.

Example 8-9 Determining the longest wavelength of light required for photolysis of hydrogen peroxide

Advanced oxidation processes take advantage of the extreme reactivity of the hydroxyl radical. The hydroxyl radical and its behavior are discussed in Sec. 8-5. In one important advanced oxidation process, hydroxyl radicals are produced by causing UV light to act on hydrogen peroxide. The overall reaction is

$$H_2O_2 \xrightarrow{hv} 2HO\bullet$$

The two half reactions for this overall equation can be combined to find the potential of this reaction (see the electronic Table E1 at the website listed in App. E, for half reactions):

$$H_2O \rightleftarrows HO\bullet + H^+ + e^- \qquad E^\circ = -2.59\,V$$
$$H_2O_2 + 2H^+ + 2e^- \rightleftarrows 2H_2O \qquad E^\circ = 1.763\,V$$

Find the potential of the overall reaction. From the overall reaction potential, estimate the frequency and wavelength of the light that will best promote it. Note 1 V = 1 J/C.

Solution

1. Calculate the potential of the overall reaction: Using the procedure for balancing redox reactions (see Sec. 8-2), the potential of the overall reaction can be determined from the potentials of the two half reactions:

$$E^\circ_{\text{Rxn}} = E^\circ_{\text{ox}} + E^\circ_{\text{red}} = -2.59 + 1.763 = -0.827 \text{ V}$$

2. Because ΔG°_{Rxn} is positive and E°_{Rxn} is negative, the reaction requires energy in order to proceed. Calculate the energy required for the reaction using Eq. 8-93:

$$\begin{aligned} E &= ne^- E^\circ_{Rxn} = (2)\,(1.602 \times 10^{-19} \text{C})\,(0.827 \text{ J/C}) \\ &= 2.65 \times 10^{-19} \text{ J} \end{aligned}$$

3. Calculate the frequency of the light using Eq. 8-94:

$$\nu = \frac{E}{h} = \frac{2.65 \times 10^{-19} \text{J}}{6.62 \times 10^{-34} \text{J} \cdot \text{s}} = 4.00 \times 10^{14} \text{ s}^{-1}$$

4. Calculate the wavelength of the light using Eq. 8-95:

$$\lambda = \frac{c}{\nu} = \frac{\left(3.00 \times 10^{8} \text{ m/s}\right)}{4.00 \times 10^{14} \text{ s}^{-1}} = 7.50 \times 10^{-7} \text{ m} = 750 \text{ nm}$$

Comment

This is the longest wavelength of light that can split hydrogen peroxide based on thermodynamics. Shorter wavelengths have higher frequency and higher energy as shown in Eqs. 8-94 and 8-95. However, application of light with shorter wavelengths does not guarantee that the reaction will proceed because the wavelength of light that will split hydrogen peroxide also depends on the electronic structure of the molecule's orbital. Based on experimental evidence, wavelengths much shorter than 750nm are required to split hydrogen peroxide.

Estimating Photolysis for Single Absorbing Solute

Photolysis generally takes place in circumstances where multiple solutes absorb light; however, to introduce the concepts involved in photolysis, the photolysis of a single absorbing solute will be examined first. As previously stated, photolysis occurs when an electron in the outer orbital absorbs a photon and forms an unstable compound that undergoes a chemical reaction. The photonic absorption efficiency may be calculated from the Beer–Lambert law, which is discussed below. The instability of the compound that absorbed a photon is determined from the quantum yield, as discussed below.

Example of UV oxidation reactors

http://www.stantec.com/wt3r/8-3.html

ABSORPTION OF UV LIGHT BY A COMPOUND IN AQUEOUS SOLUTION

Lambert's law and Beer's law are two empirical laws used to describe light absorption in aqueous solution. When these laws are combined, the Beer–Lambert law (repeated here from Chap. 2) is used to relate the light intensity emerging from solution to the incident light intensity for a one-dimensional light source (e.g., Cartesian coordinates and a plate light source):

$$\log\left(\frac{I}{I_0}\right) = -\varepsilon(\lambda)\,Cx = -k(\lambda)\,x = -A(\lambda) \tag{8-96}$$

where
- I = light intensity after passing through solution containing constituents of interest at wavelength λ, einstein/cm$^2\cdot$s
- I_0 = initial detector reading for blank (i.e., distilled water) after passing through solution of known depth (typically 1.0 cm) at wavelength λ, einstein/cm$^2\cdot$s
- $\varepsilon(\lambda)$ = base-10 extinction coefficient or molar absorptivity of light-absorbing solute at wavelength λ, L/mol $\cdot$ cm
- λ = wavelength, nm
- C = concentration of light-absorbing solute, mol/L
- x = length of light path, cm
- $k(\lambda)$ = absorptivity (base 10), = $\varepsilon(\lambda)\,C$, cm^{-1}
- $A(\lambda)$ = absorbance, = $k(\lambda)\,x$, dimensionless

The terms in Eq. 8-96 can be confusing because they may not be familiar and may be used differently in the literature than in this text. For instance, the term "absorbance" is often used to refer to both the absorbance $A(\lambda)$ and the absorptivity $k(\lambda)$. While the two terms are not equivalent, their values are equal when the path length of absorption, x, is 1 cm. The unit of einstein is in recognition of the work Albert Einstein did to establish that light was comprised of particles now called photons. One mole of photons is referred to as one einstein. The extinction coefficient is the representation of the general phenomenon that, as the wavelength decreases, more energetic photons are absorbed; thus, the molar absorptivity of a light absorbing solute increases. Values of the extinction coefficients at various wavelengths, $\varepsilon(\lambda)$, for several common compounds are summarized in the electronic Table E3 at the website listed in App. E.

Example 8-10 Absorption of UV_{254} by NDMA

N-Nitrosodimethylamine (NDMA), an undesirable compound sometimes found in drinking water, can be removed by photolysis. Estimate the absorptivity of NDMA at a wavelength of 254 nm for both base e and base 10, assuming NDMA is present at a concentration of 20 ng/L. The extinction coefficient of NDMA is 1974 L/mol · cm and the molecular weight is 74.09 g/mol.

Solution

1. Calculate the concentration C in moles per liter:

$$C = \frac{20\ \text{ng/L}}{(74.09\ \text{g/mol})\,(10^9\text{ng/g})} = 2.7 \times 10^{-10}\ \text{mol/L}$$

2. Convert $\varepsilon(\lambda)$ (base 10) to $\varepsilon'(\lambda)$ (base e):

$$\varepsilon'(254) = \ln(10) \times \varepsilon(254) = 2.303 \times 1974 = 4546\ \text{L/mol} \cdot \text{cm}$$

3. Calculate base 10 absorptivity:

$$k(\lambda) = \varepsilon(254)\,C = (1974\ \text{L/mol} \cdot \text{cm})\,(2.7 \times 10^{-10}\ \text{mol/L}) = 5.33 \times 10^{-7}\ \text{cm}^{-1}$$

4. Calculate base-e absorptivity:

$$k'(\lambda) = \varepsilon'(254)\,C = (4546\ \text{L/mol} \cdot \text{cm})\,(2.7 \times 10^{-10}\ \text{mol/L}) = 1.23 \times 10^{-6}\ \text{cm}^{-1}$$

Comment

The absorptivity of NDMA is very low because NDMA is present at very low concentrations.

RATE OF PHOTON ABSORPTION AND QUANTUM YIELD

As described earlier, the rate at which a compound is photolyzed depends on the rate of photon absorption and the fraction of adsorbed photons that results in a reaction. For a single compound, the rate at which photons are absorbed can be determined by differentiating the intensity of light over the distance the light travels, as shown in the following derivation. First, Eq. 8-96 is converted to base *e*:

$$\ln\left(\frac{I}{I_0}\right) = -\varepsilon'(\lambda)\,Cx \qquad (8\text{-}97)$$

where $\varepsilon'(\lambda)$ = base-*e* extinction coefficient or molar absorptivity of light-absorbing solute at wavelength, $\lambda = 2.303\varepsilon(\lambda)$, L/mol · cm

Raising both sides of the equation to the *e* power and solving for I yields:

$$I = I_0 e^{-\varepsilon'(\lambda)Cx} \qquad (8\text{-}98)$$

Differentiating with respect to distance yields the volumetric photon absorption rate I_a:

$$I_a = -\frac{dI}{dx} = \varepsilon'(\lambda)\, C I_0 e^{-\varepsilon'(\lambda)Cx} \tag{8-99}$$

where I_a = number of photons absorbed per volume of solution at particular point, einstein/cm$^3 \cdot$ s

The fraction of adsorbed photons that result in a photolysis reaction must be known to estimate the rate of photolysis of a particular compound. This fraction is called the quantum yield and depends on the type of compound and the wavelength. The quantum yield $\phi(\lambda)$is defined as the rate of photolysis divided by photon absorption rate as follows:

$$\phi(\lambda) = \frac{-r_R}{I_a} = \frac{\text{photolysis reaction rate}}{\text{photon absorption rate}} \tag{8-100}$$

where $\phi(\lambda)$ = quantum yield at wavelength λ, mol/einstein
r_R = reaction rate, mol/cm$^3 \cdot$ s

As a general rule, the quantum yield $\phi(\lambda)$ increases as wavelength decreases (increasing photonic energy). Selected quantum yields at wavelength 254 nm are summarized in Table 8-6.

RATE OF PHOTOLYSIS IN A COMPLETELY MIXED FLOW REACTOR

For modeling a UV reactor in Cartesian coordinates, it is convenient to assume the light source to be a flat plate and that the photonic flux is in the *x* direction. The resulting light intensity is constant in the *y–z* plane for a given *x* value and is illustrated on Fig. 8-6. Rearranging Eq. 8-100 and substituting Eq. 8-99 yields an expression that describes the reaction rate at a local point in the reactor

$$r_x = -\phi(\lambda)\, I_a = -\phi(\lambda)\varepsilon'(\lambda)\, C I_0 e^{-\varepsilon'(\lambda)Cx} \tag{8-101}$$

where r_x = photolysis rate at a point in reactor, mol/cm$^3 \cdot$ s
x = distance from light source, cm

Because the contents of the reactor are mixed completely in the *y–z* plane, the overall average photolysis rate in the reactor can be determined by integrating Eq. 8-100 over the path length of the light, as shown below:

$$r_{avg} = \frac{1}{b}\int_0^b r_x dx = -\frac{1}{b}\int_0^b \phi(\lambda)\, \varepsilon'(\lambda)\, C I_0 e^{-\varepsilon'(\lambda)Cx}\, dx \tag{8-102}$$

$$r_{avg} = -\frac{\phi(\lambda) I_0}{b}\left(1 - e^{-\varepsilon'(\lambda)Cb}\right) = -\phi(\lambda)\, P_{U-V}\left(1 - e^{-\varepsilon'(\lambda)Cb}\right) \tag{8-103}$$

where r_{avg} = average photolysis rate, mol/cm$^3 \cdot$ s
b = effective light path length, cm
$P_{U-V} = I_0/b$ = photonic intensity per volume, einstein/cm$^3 \cdot$ s

Table 8-6
Selected quantum yields

Compound	Primary Quantum Yield in Aqueous Phase, mol/einstein	Extinction Coefficient at 253.7 nm, (Base 10) L/mol · cm
NO_3^-	—	3.8
HOCl	0.23[a]	15[a,c]
OCl^-	0.23[b]	190[a,c]
HOCl	—	53.4[c]
OCl^-	0.52[d]	155[c]
O_3	0.5 (0.48 ± 0.6)[e,f]	3300[f]
ClO_2	0.44[b,g]	108[h]
Sodium chlorite	0.72[g]	—
TCE	0.54[i]	9 (8)[i]
PCE	0.29 (0.31 ± 0.08)[i,j]	205[i]
NDMA	0.3[k]	1974[l]
Water	—	6.1×10^{-6}

[a]330 nm.
[b]Independent of wavelength.
[c]Nowell and Hoigne (1992a).
[d]Nowell and Hoigne (1992b).
[e]Reisz et al. (2003).
[f]Gurol and Akata (1996).
[g]Cosson and Ernst (1994).
[h]Zika et al. (1984).
[i]Taku and Tanaka (2000).
[j]Mertens and von Sonntag (1995).
[k]Sharpless and Linden (2003).
[l]Ho et al. (1996).

Photoreactors are designed in such a way that all the light remains within the reactor and is absorbed (e.g., using reflective surfaces). The effective path length is then much longer than the physical dimensions of the reactor, as the light bounces back and forth between reflective surfaces until all the light is adsorbed. When the path length b is sufficiently large, the exponent term in Eq. 8-103 approaches zero.

Accordingly, the average rate of reaction is given by the zero-order rate expression

$$r_{avg} = -\frac{\phi(\lambda) I_0}{b} = -\phi(\lambda) P_{U-V} \qquad (8\text{-}104)$$

Thus, the reaction rate in a completely mixed, reflective reactor is independent of the concentration of compound being photolyzed and depends only on two parameters: (1) the photonic energy per unit volume and (2) the quantum yield. This relationship is true regardless of the geometry of the light source and reactor as long as the solution is mixed completely. Most UV lamps are cylinders, but if all the light is absorbed, the photolysis rate is still given by Eq. 8-104.

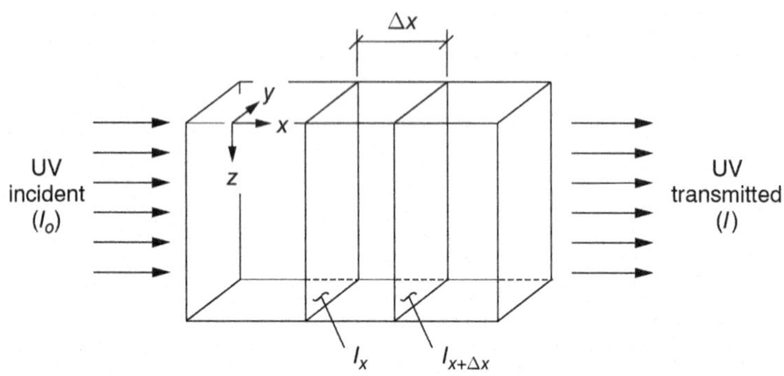

Figure 8-6
Definition sketch for light flux in one dimension.

DETERMINING QUANTUM YIELD IN A BENCH-SCALE COLLIMATED BEAM APPARATUS

Bench-scale experiments can be conducted in a collimated beam apparatus. A collimated beam apparatus is a device in which coherent light is produced that irradiates a Petri dish. The objective is to achieve uniform irradiation of a sample by using a one-dimensional light source and a shallow reactor.

If the reactor is sufficiently shallow, the distance from light source (x in Eq. 8-98) is small and the exponent term is approximately equal to one. Under these circumstances, Eq. 8-98 can be simplified to

$$I_a = \varepsilon'(\lambda) C I_0 \tag{8-105}$$

Multiplying the photon adsorption rate by the quantum yield (Eq. 8-100) yields the following expression for the average photolysis rate:

$$r_{avg} = -\phi\,(\lambda)\,\varepsilon'\,(\lambda)\,C I_0 \tag{8-106}$$

Applying a mass balance for a completely mixed batch reactor (the Petri dish) and solving for the concentration results in the following expression:

$$\frac{C}{C_0} = e^{-\phi(\lambda) I_0 \varepsilon'(\lambda) t} \tag{8-107}$$

The use of Eq. 8-107 to estimate the quantum yield from laboratory data is illustrated in Example 8-11.

Example 8-11 Estimation of the quantum yield from collimated beam experiments

Calculate the quantum yield of NDMA from the collimated beam data obtained from a bench-scale experiment using a low-pressure UV lamp. The molar absorption coefficient for NDMA at a wavelength of 253.7 nm is 1974 L/mol · cm (see Table 8-6).

Time, s	Time, min	UV dose, mJ/cm^2	[NDMA], μM
0	0.0	0	1.00
180	3.0	200	0.67
350	5.8	400	0.45
420	7.0	580	0.29
600	10.0	770	0.17
800	13.3	920	0.11
950	15.8	1120	0.07
1100	18.3	1300	0.05
1250	20.8	1550	0.02

Solution

1. Plot the logarithm (base *e*) of the ratio of the concentration to the initial concentration of NDMA as a function of UV dose.

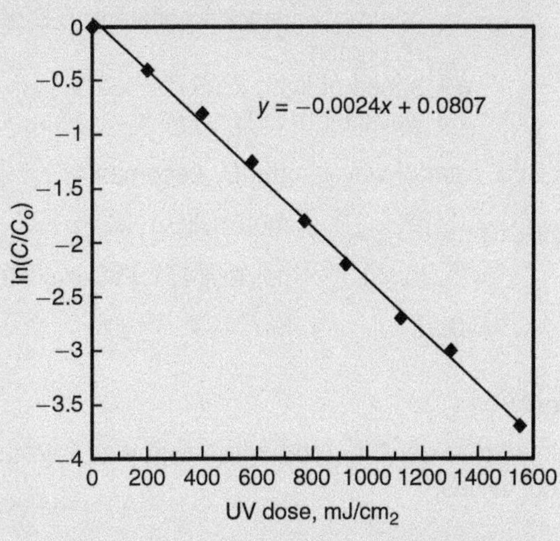

2. Convert Eq. 8-107 to a linear form by taking the natural log of both sides of the equation:

$$\ln\left[\frac{C}{C_0}\right] = -\phi_i(\lambda) \in_i (\lambda)\ln(10)I_0 t$$

3. Rearrange the equation given in step 2 to solve for $\phi_i(\lambda)$:

$$\phi_i(\lambda) = -\left[\frac{\ln(C/C_0)}{I_0 t}\right]\left[\frac{1}{\varepsilon_i(\lambda)\ln(10)}\right]$$

It should be noted that the first term on the right side of the equations corresponds to the slope of the line in the plot created in step 1 above.

4. Substitute the slope of the line and the value of the extinction coefficient to determine the value for $\phi_i(\lambda)$:

$$\phi_i(\lambda) = -(-0.0024\ \text{cm}^2/\text{mJ})\left[\frac{1}{(1974\ \text{L/mol} \cdot \text{cm}) \times 2.303}\right]\left[\frac{\text{L}}{10^3\ \text{cm}^3}\right]$$

$$= 5.28 \times 10^{-10}\ \text{mol/mJ}$$

5. Convert $\phi_i(\lambda)$ to units of mol/einstien using the following expression

$$U_\lambda = \frac{A_V hc}{\lambda}$$

where U_λ = energy per einstein for photons of wavelength λ, J/einstein
A_V = Avagadro's number, 6.02214×10^{23} photons/einstein h = Planck's constant, 6.62607×10^{-34} J · s/photon
c = speed of light, 2.99792×10^8 m/s
λ = wavelength of light, m

At 253.7 nm of wavelength, U_λ becomes 471,155 J/einstein

$$\phi_i(\lambda) = (5.28 \times 10^{-10}\ \text{mol/mJ})(U_\lambda)$$

$$= (5.28 \times 10^{-10}\ \text{mol/mJ})(471{,}155\ \text{J/einstein})(1000\ \text{mJ/J})$$

$$= 0.25$$

Comment

The estimated value is in good agreement with literature-reported experimental values.

Photolysis in Presence of Multiple Absorbing Compounds

The principles of UV absorbance and photolysis have been examined for a single absorbing compound, and now these principles will be extended to cover the more common circumstance where a number of absorbing compounds are present in solution. An analogous extension of the same concepts can be employed to examine the performance of UV lamps emitting multiple wavelengths. Although this approach will be outlined, the details of that analysis are beyond the scope of this discussion. The discussion of photolysis in the presence of multiple absorbing compounds will be on the basis of a single incident wavelength, which is the circumstance that occurs when low-pressure, low-intensity and low-pressure, high-intensity UV lamps are used.

ABSORPTION BY MULTIPLE COMPOUNDS

Functional groups or bonds on chemical compounds responsible for the absorption of light are referred to as *chromophores.* When more than one chromophore is present in a solution, their impact on the absorption of light is additive. As a result, the absorption of light as it passes through a solution containing several different compounds may be determined by summing the absorption that would result from each individual compound, as shown in the expression

$$\ln\left(\frac{I}{I_0}\right) = -\left[\sum \varepsilon'(\lambda)_i C_i\right]x \tag{8-108}$$

where $\varepsilon'(\lambda)_i$ = extinction coefficient of compound i at wavelength λ (base e), L/mol · cm
C_i = concentration of compound i, mol/L

ABSORPTION OF UV BY NOM

The extinction coefficient for NOM varies over a wide range and is typically site specific. Because the specific UV absorption ratio at 254 nm (SUVA, L/mg · m) is used frequently in estimating the potential for formation of disinfection by-products (see Chap. 19), SUVA data are widely available. The SUVA is the extinction coefficient expressed in L/mg · m and is related to absorptivity $k(\lambda)$ and extinction coefficient $\varepsilon(254)$, as shown in the equations

$$\text{SUVA}\left(\frac{\text{L}}{\text{mg C} \cdot \text{m}}\right) = \frac{k\,(254)\ \text{cm}^{-1}}{(\text{DOC mg C/L})} \times \frac{100\ \text{cm}}{\text{m}} \tag{8-109}$$

$$\varepsilon\,(254)\frac{\text{L}}{\text{mg} \cdot \text{cm}} = \frac{k\,(254)\ \text{cm}^{-1}}{\text{DOC mg C/L}} \tag{8-110}$$

$$\text{SUVA} = \varepsilon\,(254)\frac{\text{L}}{\text{mg} \cdot \text{cm}} \times 100\frac{\text{cm}}{\text{m}} \tag{8-111}$$

The range of values and the average value for extinction coefficients for NOM at 254 nm that have been reported (Westerhoff et al., 1999) are 0.013 to 0.107 L/mg C · cm, which correspond to SUVA values between 1.3 and 10.7.

Example 8-12 Estimating absorptivity $k(\lambda)$ of UV_{254} of water sample

A potential raw-water source for drinking water is analyzed and found to contain the following constituents:

Constituent	Unit	Value
DOC	mg/L as C	3.0
Fe(II)	mg/L as Fe	0.3
Nitrate	mg/L as NO_3^-	5.5
SUVA	L/mg · m	2

Estimate the absorptivity (both base 10 and base *e*) of the water at a wavelength of 254 nm.

Solution

1. Convert the constituent concentrations to moles per liter:

		Concentration	
Constituent	MW	mg/L	mol/L
Nitrate	62	5.5	8.87×10^{-5}
DOC	—	3.0	—
Fe(II)	56	0.3	0.54×10^{-5}
Water	18	10^6	55.6

2. Calculate the extinction coefficient for DOC using Eq. 8-111:

$$\varepsilon(254) = \frac{\text{SUVA}}{100} = 0.02\,\text{L/mg} \cdot \text{cm}$$

3. Estimate the absorptivity.
   a. Find the extinction coefficients for the other constituents using the data given in the electronic Table E3 at the website listed in App. E, and estimate the absorptivity of each constituent:

	Concentration,		
Constituent	C (mol/L)	ε(254)	ε(254)C
Nitrate	8.87×10^{-5}	3.8	0.0003
DOC	3 (mg/L)	0.02	0.0600
Fe(II)	0.54×10^{-5}	465	0.0025
Water	55.6	6.1×10^{-6}	0.0003
			$\sum = 0.0631$

b. Sum the absorptivity of each component:

$$k\,(254) = \sum \varepsilon\,(254)\,C = 0.0631\ \text{cm}^{-1}$$

c. Convert to absorptivity base 10 to base e:

$$k'\,(254) = 2.303k\,(254) = 0.145\ \text{cm}^{-1}$$

Comment

The absorptivity is dominated by the DOC, and the absorptivity of both the nitrate and the water itself is insignificant.

RATE OF PHOTON ABSORPTION WITH MULTIPLE COMPOUNDS PRESENT

To determine the rate at which the target compound absorbs photons when multiple compounds are present, the photon absorption rate for the other compounds must also be determined. The absorption rate of all the species in a solution with n compounds present can be obtained by rearranging and differentiating Eq. 8-108 as follows:

$$I_a = -\frac{dI}{dx} = I_0\left[\sum_1^n \varepsilon'(\lambda)_i C_i\right]\exp\left[-x\sum_1^n \varepsilon'(\lambda)_i C_i\right] \tag{8-112}$$

where $\varepsilon'(\lambda)_i$ = extinction coefficient of compound i (base e), L/mol · cm
C_i = concentration of compound i, mol/L
n = number of compounds, unitless
x = length of light path, cm
I_a = combined rate at which all n compounds are absorbing photons at wavelength λ, einstein/cm^3 · s
I_0 = irradiance entering the reactor, einstein/cm^2 · s

Thus, the relative rate at which each particular compound absorbs photons can be determined using the expression

$$I_{aj} = -\frac{dI_j}{dx} = I_0\varepsilon'(\lambda)_j\ C_j\ \exp\left[-x\sum_1^n \varepsilon'(\lambda)_i\ C_i\right] \tag{8-113}$$

where I_{aj} = rate at which compound j is absorbing photons, einstein/cm^3 · s
I_j = light intensity after passing through solution containing compound j at wavelength λ, einstein/cm^2 · s
$\varepsilon'(\lambda)_j$ = extinction coefficient of compound j (base e), L/mol · cm
C_j = concentration of compound j, mol/L

MODELING REACTOR PERFORMANCE

Equation 8-113 can be combined with Eq. 8-101 to estimate the rate of photolysis for compound j at a point that is a distance x from the light source:

$$r_{x,j} = -\phi(\lambda)_j \, I_0 \varepsilon'(\lambda)_j \, C_j \exp\left[-x \sum_1^n \varepsilon'(\lambda)_i \, C_i\right] \tag{8-114}$$

where r_{xj} = photolysis rate of compound j at a point x in reactor, $\text{mol/cm}^3 \cdot \text{s}$

x = distance from light source, cm

$\phi(\lambda)_j$ = quantum yield of compound j at wavelength λ, mol/einstein

The average rate of photolysis of compound j in a photoreactor with an optical path length b can be determined by substituting Eq. 8-114 in Eq. 8-102 and integrating to obtain the expression

$$\begin{aligned} r_{\text{avg},j} &= -\phi(\lambda)_j f_j \left(\frac{I_0}{b}\right)\left\{1 - \exp\left[-b \sum_1^n \varepsilon'(\lambda)_i \, C_i\right]\right\} \\ &= -\phi(\lambda)_j f_j P_{\text{U-V}} \left\{1 - \exp\left[-b \sum_1^n \varepsilon'(\lambda)_i \, C_i\right]\right\} \end{aligned} \tag{8-115}$$

where $r_{\text{avg},j}$ = overall average photolysis rate of compound j in reactor, $\text{mol/cm}^3 \cdot \text{s}$

f_j = fraction of light absorbed by component j, dimensionless and calculated by the expression

$$f_j = \left[\frac{\varepsilon'(\lambda)_j \, C_j}{\sum_{i=1}^n \varepsilon'(\lambda)_i \, C_i}\right] \tag{8-116}$$

Equation 8-115 becomes first order in component j if j absorbs only a fraction of the light. Many photoreactors are designed with reflective surfaces so that all the light emitted by the lamps is retained in the reactor. In these circumstances, Eq. 8-115 reduces to the form

$$r_{\text{avg},j} = -\phi(\lambda)_j \, f_j P_{\text{U-V}} \tag{8-117}$$

Addressing Multiple Wavelengths

When medium-pressure lamps are used in the photooxidation process, the spectral distribution of the lamp must be considered. Usually, the incident UV light intensity is measured at specific wavelength intervals (e.g., 5 nm) within the effective UV radiation range. The UV light intensity can be assumed to be monochromatic within small wavelength increments. The quantum yield $\phi(\lambda)$ and the extinction coefficient $\varepsilon(\lambda)$ of an absorbing compound are dependent upon wavelength. Knowing the UV light intensities I_{01}, I_{02}, I_{03}, Λ, I_{0k} at every kth-wavelength band (represented as λ_1, λ_2,

$\lambda_3, \ldots, \lambda_k$) as well as the quantum yields $\phi(\lambda_1), \phi(\lambda_2), \phi(\lambda_3), \ldots, \phi(\lambda_k)$ and extinction coefficients $\varepsilon'(\lambda_1), \varepsilon'(\lambda_2), \varepsilon'(\lambda_3), \ldots, \varepsilon'(\lambda_k)$, the following formula can be developed to estimate the photolysis rate:

$$r_{\text{avg}} = -\sum_{j=1}^{k}\left[\phi(\lambda_j)P_{\text{U-V},j}f(\lambda_j)\right] \tag{8-118}$$

where r_{avg} = overall average photolysis rate of an absorbing compound in a reactor with multiple wavelengths, $\text{mol/cm}^3 \cdot \text{s}$

$\varphi(\lambda_j)$ = quantum yield at wavelength λ_j, mol/einstein

$P_{\text{U-V},j}$ = photonic intensity per volume for wavelength j, = I_{0j}/b $\text{einstein/cm}^3 \cdot \text{s}$

$f(\lambda_j)$ = fraction of light absorbed at wavelength λ_j, dimensionless

k = number of wavelength, unitless

b = optical path length, cm

The extinction coefficients for common inorganic species in water as a function of wavelength are reported in the electronic Table E3 at the website listed in App. E. The absorbances of several natural water resources are illustrated as a function of wavelength on Fig. 8-7.

Application of Photolysis in Water Treatment

Photolysis may be used to remove some organic compounds, notably NDMA and several oxidants and disinfectants, such as chlorine, chlorine dioxide, and combined chlorine and ozone. However, for the destruction of most organic compounds, photolysis is often more efficient when used in combination with hydrogen peroxide so that hydroxyl radicals can be produced. The use of photolysis with hydrogen peroxide oxidation is a form of advanced oxidation and is discussed in greater detail in Chap. 18.

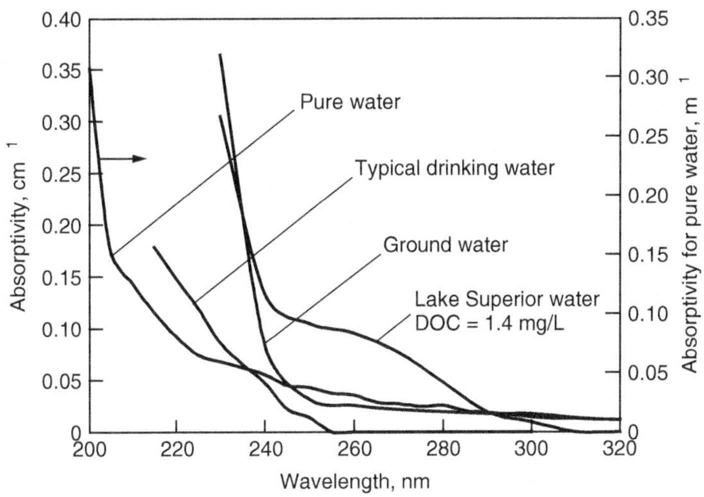

Figure 8-7
The UV absorbance of typical natural water and pure water (Lake Superior sample taken 9 km from the shore of Eagle Harbor, MI).

CONVERTING LAMP POWER EINSTEIN UNITS

In theory, the photons entering the reactor per unit volume of the reactor and time, P_{U-V} (einstein/L · s), can be determined using the expression

$$P_{U-V} = \frac{P \times \eta}{N_{av} \times V \times hv} \tag{8-119}$$

where P = lamp power, W
h = Planck's constant, 6.62×10^{-34} J · s
η = electrical efficiency (as a fraction), dimensionless
N_{av} = Avogadro's number, 6.023×10^{23} molecules/mol
V = reactor volume, L
ν = frequency of light, = c/λ, s^{-1}
c = speed of light, 3.00×10^8 m/s
λ = wavelength of light, m

The following assumption was made to develop Eq. 8-119: There are no photon losses upon reflection off the reactor wall or through the exterior surface of sleeves that cover the lamps (precipitate builds up on the exterior surface because of the elevated temperature). Consequently, Eq. 8-119 represents the highest possible photonic intensity per volume. Performance of an actual photoreactor is much lower than would be expected by Eq. 8-119, partly due to light being absorbed and blocked by the precipitate that forms on the lamp sleeve and reactor walls.

PHOTOREACTOR DESIGN

A typical UV photolysis reactor is a stainless steel column that contains UV lights in a criss-crossing pattern, as shown in the schematic on Fig. 8-8a and the photo on Fig. 8-8b. The approximate locations of the lamps are shown on the left, and the perpendicular arrangement is shown on the right of Fig. 8-8a. Most lamps are cylindrical in shape, and, in this design, they are arranged perpendicular to the direction of flow. The UV lamps are covered with quartz-insulating sleeves to allow the lamps to operate at the appropriate temperature. The elevated temperature of the lamp sleeves causes inorganic precipitates to form, which are removed on a routine basis by collars that move back and forth across the sleeves. Other options for photoreactors, including those that are used for disinfection, are discussed in Chap. 13.

The rate of destruction (for a single wavelength λ) of a compound by photolysis may be described by Eq. 8-117. Equation 8-117 is expanded further in the expression

$$r_{avg,j} = -\phi(\lambda)_j f_j P_{U-V} = -\phi(\lambda)_j P_{U-V} \frac{\varepsilon'(\lambda)_j C_j}{\sum_{i=1}^{n} \varepsilon'(\lambda)_i C_i} \tag{8-120}$$

where $r_{avg,j}$ = overall average photolysis rate of compound j in reactor, mol/L · s

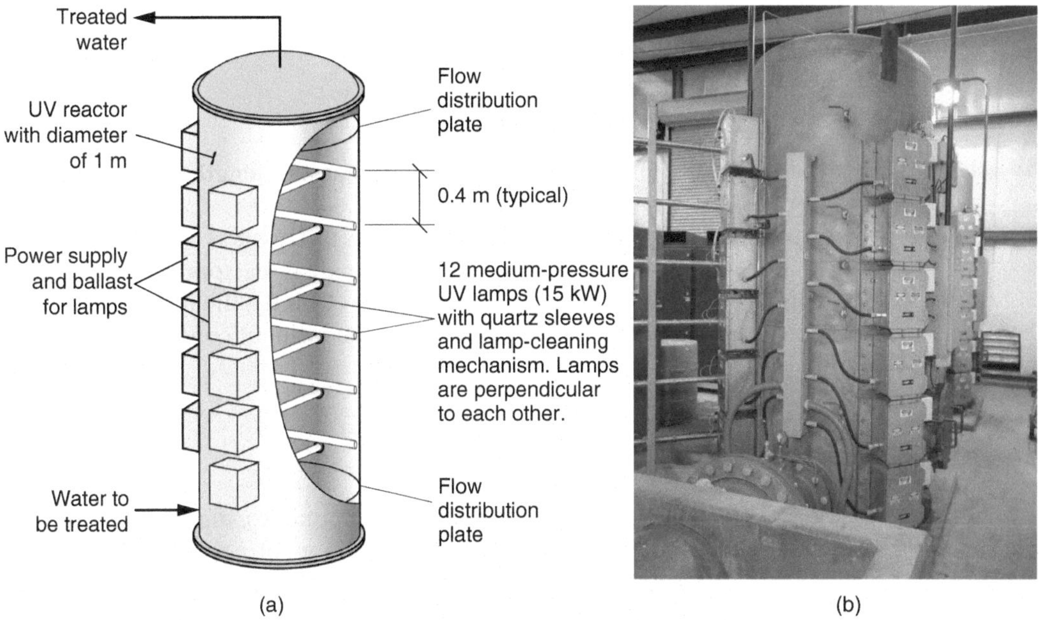

Figure 8-8
UV reactor used for advanced oxidation: (a) schematic and (b) photograph.

$\phi(\lambda)_j$ = quantum yield of compound j at wavelength λ, mol/einstein
f_j = fraction of light absorbed by compound j, dimensionless
$P_{U\text{-}V}$ = photonic intensity per unit volume, einstein/cm$^3 \cdot$ s
$\varepsilon'(\lambda)_i$ = extinction coefficient of compound i (base e), L/mol · cm
$\varepsilon'(\lambda)_j$ = extinction coefficient of compound j (base e), L/mol · cm
C_i = concentration of compound i, mol/L
C_j = concentration of compound j, mol/L

In most cases, the light absorption by the component that is targeted for removal is minor as compared to the light absorption by the background water matrix. For this situation, the rate law becomes pseudo–first order, as shown in the following expression:

$$r_{\text{avg},j} = k_j C_j \tag{8-121}$$

$$k_j = \phi(\lambda)_j P_{U-V} \frac{\varepsilon'(\lambda)_j}{\sum_{i=1}^{n} \varepsilon'(\lambda)_i C_i} \tag{8-122}$$

where k_j = pseudo-first-order rate coefficient, s^{-1}

If all the chromophores in the water have been measured, then the pseudo-first-order rate coefficient can be estimated from that analysis. However, it is more accurate and easier to measure the absorptivity of the water matrix. The pseudo-first-order rate coefficient of the water matrix can be determined by simplifying Eq. 8-122 as follows:

$$k_j = \phi(\lambda)_j P_{\text{U-V}} \frac{\varepsilon'(\lambda)_j}{k'(\lambda)} \tag{8-123}$$

where $k'(\lambda)$ = absorptivity of water matrix at wavelength (base e) λ, cm^{-1}

PERFORMANCE OF A COMPLETELY MIXED FLOW REACTOR

The mass balances for various ideal and nonideal reactors are discussed in Chap. 6. For a complete mixed flow reactor (CMFR), the following expression relates the effluent concentration to the influent concentration:

$$\frac{C_{j,e}}{C_{j,o}} = \frac{1}{1 + \text{k}_j\tau} \tag{8-124}$$

where $C_{j,e}$ = effluent concentration of component j, mg/L
$C_{j,o}$ = influent concentration of component j, mg/L
k_j = pseudo-first-order rate constant, s^{-1}
τ = hydraulic detention time of photoreactor, = V/Q, s
V = reactor volume, L
Q = volumetric flow rate, L/s

PERFORMANCE OF PLUG FLOW REACTOR

For a plug flow reactor (PFR), the following expression relates the effluent concentration to the influent concentration:

$$\frac{C_{j,e}}{C_{j,o}} = e^{-k_j\tau} \tag{8-125}$$

where the terms are as defined above.

PERFORMANCE OF NONIDEAL REACTOR

Most reactors are not PFRs or CMFRs, and, as discussed in Chap. 6, tracer studies can be conducted on the reactors to determine the degree of nonideal mixing that occurs. Models that describe nonideal mixing may be fit to the tracer curve and then used to describe reactor performance. The TIS model and the DFM are two models that were discussed in Chap. 6 and are repeated here. The SFM may also be used to estimate reactor performance when a tracer curve is available, as discussed in Chap. 6.

Tanks in series model

The TIS model is used to describe nonideal mixing in a photoreactor by varying the number of tanks in series. The following expression relates the effluent concentration to the influent concentration for the TIS model:

$$\frac{C_{j,e}}{C_{j,o}} = \frac{1}{(1 + k_j\tau / n)^n} \tag{8-126}$$

where n = number of tanks, unitless

Other terms are as defined previously.

Dispersed flow model
The nonideal mixing in a photoreactor may be described by the DFM using the Peclet number. The following expression relates the effluent concentration to the influent concentration for the DFM model:

$$\frac{C_{j,e}}{C_{j,o}} = \frac{4a\exp(\text{Pe}/2)}{(1+a)^2\exp(a\text{Pe}/2) - (1-a^2)\exp(-a\text{Pe}/2)} \tag{8-127}$$

$$a = \sqrt{1 + \frac{4k_j\tau}{\text{Pe}}} \tag{8-128}$$

where Pe = vL/E, also known as the Peclet number, dimensionless
v = average velocity of fluid, m/s
L = reactor length, m
E = dispersion coefficient, m^2/s

Other terms as defined previously.

Example 8-13 Removal of NDMA by photolysis

The source water for a drinking water treatment plant in California contains 20 ng/L NDMA. The treatment objective for NDMA has been set at 2 ng/L. After considering various treatment options, UV photolysis has been chosen for NDMA removal. The commercial reactors being evaluated are 1 m in diameter and 3 m in height (water volume 2300 L). Each reactor has 12 lamps that use 15 kW per lamp. The lamps are 20 percent efficient; that is, 20 percent of the energy consumed by the lamp is produced as UV light at the relevant wavelength (15 kW × 20% = 3 kW). From a previous tracer study it has been found that the reactor can be treated as four tanks in series. Estimate the flow rate that can be treated in this reactor. The extinction coefficient $\varepsilon(254)$ and quantum yield ϕ (254) of NDMA can be found in Table 8-6. For calculation simplicity, assume there are no losses on the reactor walls, the lamp sleeves do not block any light, the UV light intensity is monochromatic at 254 nm, and the water has exactly the same quality as the water used in Example 8-12: $k'(\lambda) = 0.145\ \text{cm}^{-1}$.

Solution

1. Calculate the rate constant for NDMA:
   a. Calculate the photonic intensity per volume, $P_{U\text{-}V}$:
      i. Calculate the total lamp power:
      $$P = (12 \text{ lamps} \times 15 \text{ kW/lamp}) = 180 \text{ kW}$$
      ii. Calculate the UV photonic intensity $P_{U\text{-}V}$ using Eq. 8-115:

$$P_{U-V} = \frac{(180\,\text{kW})(10^3\,\text{W/kW})[(1\,\text{J/s})/1\,\text{W}](254\times10^{-9}\,\text{m})(0.2)}{(6.023\times10^{23}\text{ photon/einstein})(6.62\times10^{-34}\,\text{J}\cdot\text{s})(3.0\times10^{8}\,\text{m/s})(2300\,\text{L})}$$
$$= 3.32\times10^{-5}\text{ einstein/L}\cdot\text{s}$$

   b. Calculate the rate constant for NDMA, k_{NDMA}:
      i. From Table 8-6, the extinction coefficient of NDMA at 254 nm, $\varepsilon(254)$, is 1974 L/mol · cm. Determine $\varepsilon'(254)$:
      $$\varepsilon'(254) = 2.303\varepsilon(254) = 2.303 \times 1974 = 4546 \text{ L/mol} \cdot \text{cm}$$
      ii. From Table 8-6, the quantum yield for NDMA, $\phi(\lambda)_{NDMA}$, is equal to 0.3 mol/einstein. Determine k_{NDMA} according to Eq. 8-119:

$$k_{NDMA} = \phi(\lambda)_{NDMA} P_{U-V} \frac{\varepsilon'(\lambda)_{NDMA}}{k'\lambda}$$
$$= (0.3\text{ mol/einstein})\,(3.32\times10^{-5}\text{einstein/L}\cdot\text{s})$$
$$\times\left(\frac{4546\text{ L/mol}\cdot\text{cm}}{0.1451/\text{cm}}\right)$$
$$= 0.313\,\text{s}^{-1}$$

2. Calculate the flow rate that can be treated:
   a. Calculate hydraulic detention time by rearranging Eq. 8-126 to solve for τ. Rewriting Eq. 8-126 for four reactors in series whose contents are completely mixed with a first-order reaction yields

$$\frac{C_{NDMA,e}}{C_{NDMA,O}} = \frac{1}{(1+k_{NOMA}\tau/4)^4}$$

Rearranging and solving for τ give

$$\tau = \frac{4\left[\left(C_{\mathrm{NDMA},0}/C_{\mathrm{NDMA},e}\right)^{1/4} - 1\right]}{k_{\mathrm{NDMA}}} = \frac{4\left[\left(\frac{20}{2}\right)^{1/4} - 1\right]}{0.313\ \mathrm{s}^{-1}}$$
$$= 9.95\ \mathrm{s}$$

b. Calculate the flow rate:

$$Q = \frac{V}{\tau} = \frac{2300\ \mathrm{L}}{9.95\ \mathrm{s}} = 231\ \mathrm{L/s}$$

ELECTRICAL EFFICIENCY PER ORDER OF CONTAMINANT DESTRUCTION

Photolytic reactions require a significant amount of electrical energy and the associated costs are significant. Consequently, it is important to compare process efficiency on the basis of electrical usage per amount of compound destruction. One such measure is the electrical efficiency per log order (EE/O) of compound destruction (Bolton and Carter, 1994). The definition of EE/O is the electrical energy (in kWh) required to reduce the concentration of a pollutant by one order of magnitude for 1000 U.S. gallons (3785 L) of water, and it may be calculated using the following equations for batch and flowing systems, respectively:

$$\mathrm{EE/O} = \frac{P \times t}{V \times \log(C_i/C_f)} \tag{8-129}$$

$$\mathrm{EE/O} = \frac{P}{Q \times \log(C_i/C_f)} \tag{8-130}$$

where EE/O = electrical efficiency per log order reduction, $\mathrm{kWh/m^3} = 3.785\ \mathrm{kWh/10^3\ gal}$
P = lamp power output, kW
t = irradiation time, h
V = reactor volume, $\mathrm{m^3}$
C_i = initial concentration, mg/L
C_f = final concentration, mg/L
Q = water flow rate, $\mathrm{m^3/h}$

For a flow-through system, the power input can be divided by the EE/O to obtain an estimate of the flow rate that can be treated in a given reactor and achieve one order-of-magnitude reduction in concentration.

A theoretical basis for EE/O for a flowing system may be derived from the photolysis model for a PFR and a pseudo-first-order reaction. However, it must be recognized that using a photolysis model will yield the most optimistic (i.e., smallest) value of EE/O because photons are absorbed by the reactor walls and blocked by precipitate that builds up on the lamp sleeves, which is not accounted for in the models. The theoretical value for EE/O is referred to as EE/O_{min}. Substituting Eq. 8-123 into Eq. 8-125 yields the expression

$$\frac{C_{j,e}}{C_{j,o}} = \exp\left[-\phi(\lambda)_j P_{U-V} \frac{\varepsilon'(\lambda)_j}{k'(\lambda)} \tau\right] \tag{8-131}$$

where $C_{j,e}$ = effluent concentration of compound j,mg/L
$C_{j,o}$ = influent concentration of compound j,mg/L
$\phi(\lambda)_j$ = quantum yield of compound j at wavelength λ, mol/einstein
P_{U-V} = photonic intensity per unit volume, einstein/$cm^3 \cdot s$
$\varepsilon'(\lambda)_j$ = extinction coefficient of compound j (base e), L/mol · cm
$k(\lambda)$ = absorptivity of water matrix at wavelength (base e) λ, cm^{-1}
τ = hydraulic detention time, s

Equation 8-131 can be rearranged after substituting in the definition of the hydraulic detention time to the form

$$\frac{k'(\lambda)}{2.303\varepsilon'(\lambda)_j\phi(\lambda)_j} = \frac{VP_{U-V}}{2.303Q\ln(C_{j,o}/C_{j,e})} = \frac{VP_{U-V}}{\log(C_{j,o}/C_{j,e})} \tag{8–132}$$

The left side of Eq. 8-132 is equal to the moles of photons per volume (einstein/L) that are required to reduce the concentration of the contaminant by one order of magnitude. Finally, it can be shown that the EE/O_{min} is related to the quantum yield and the fraction of the light that is absorbed by the targeted component as follows:

$$EE/O_{min} = \frac{N_{av}h\nu k'(\lambda)}{2.303\eta\varepsilon'(\lambda)_j\phi(\lambda)_j} \tag{8-133}$$

where EE/O_{min} = minimum electrical efficiency per order,
J/L = 0.00105 kWh/10^3 gal
N_{av} = Avagodro's number, 6.023×10^{23} molecules/mol
h = Planck's constant, 6.62×10^{-34} J · s
ν = frequency of light, s^{-1}
η = electrical efficiency of lamps, dimensionless

Other terms are as defined above.

Based on Eq. 8-133, the EE/O_{min} is independent of the light intensity and the concentration of the contaminant and inversely depends upon the extinction coefficient and quantum yield of the targeted compound. In real-world applications the actual EE/O is a multiple of EE/O_{min}, but it is independent of light intensity and the concentration of the target component. Consequently, EE/O is a convenient measure because it can be used to quickly estimate the energy that is required to reduce the contaminant concentration by one order of magnitude.

Example 8-14 Calculation of EE/O

Estimate the minimum EE/O for NDMA for a lamp efficiency of 0.2 in a PFR. Compare this value to the EE/O that would be calculated for Example 8-13. Calculate EE/O values in units of kWh/m^3 and $kWh/10^3$ gal.

Solution

1. Calculate the EE/O for Example 8-13 (4 CMFRs in series):
   a. In units of kWh/m^3:

$$\begin{aligned}EE/O &= \frac{P}{Q\log(C_i/C_f)} \\ &= \frac{180\text{ kW}\times(10^3\text{ L/m}^3)}{(231\text{ L/s})[\log(20\text{ ng/L})/(2\text{ ng/L})](3600\text{s/h})} \\ &= 0.216\text{ kWh/m}^3\end{aligned}$$

   b. In units of kWh/1000 gal:

$$\begin{aligned}EE/O &= \frac{0.216\text{ kWh/m}^3}{264.2\text{ gal/m}^3}\times\frac{10^3\text{gal}}{10^3\text{gal}} \\ &= 0.818\text{ kWh/}10^3\text{gal}\end{aligned}$$

2. Calculate the minimum EE/O for a PFR using Eq. 8-95:
   a. Calculate the frequency of light:

$$\nu = \frac{c}{\lambda} = \frac{3\times10^8\text{ m/s}}{254\times10^{-9}\text{ m}} = 1.18\times10^{15}\text{s}^{-1}$$

   b. From Example 8-13:

$$k'(\lambda) = 0.145\text{ cm}^{-1}$$
$$\varepsilon'(\lambda)_i = 4546\text{ L/mol}\cdot\text{cm}$$
$$\phi(\lambda)_i = 0.3\text{ molecules/photon}$$

c. Calculate minimum EE/O using Eq. 8-133:

$$
\begin{aligned}
EE/O_{min} &= \frac{N_{av} h v k'(\lambda)}{\eta \varepsilon'(\lambda)_j \phi(\lambda)_j \times 2.303} \\
&= \frac{(6.023 \times 10^{23}\ \text{molecules/mol})(6.62 \times 10^{-34}\ \text{J} \cdot \text{s / photon})(1.18 \times 10^{15}\ \text{s}^{-1})(0.145\ \text{cm}^{-1})}{0.2 \times (4546\text{L/mol} \cdot \text{cm})(0.3\ \text{molecules/photon}) \times 2.303} \\
&= 109\ \text{J / L}
\end{aligned}
$$

i. In units of kWh/m^3:

$$EE/O_{min} = \frac{(109\ \text{J/L})[1\ \text{W/(J/s)}](10^3\ \text{L/m}^3)}{(3600\ \text{s/h})10^3\text{W/kW}(10^3\ \text{W/kW})} = 0.0303\ \text{kWh/m}^3$$

ii. In units of kWh/103 gal:

$$
\begin{aligned}
EE/O_{min} &= \frac{0.0303\ \text{kWh/m}^3}{264.2\ \text{gal/m}^3} \times \frac{10^3\text{gal}}{10^3\text{gal}} \\
&= 0.115\ \text{kWh/}10^3\ \text{gal}
\end{aligned}
$$

Comment

The EE/O for the existing reactor is reasonable for reducing the NDMA concentration from 20 to 2 ng/L, considering the energy required. The EE/O value of 109 J/L (0.115 kWh/10^3 gal) is the EE/O$_{min}$ for a PFR, which is much lower than 0.818 kWh/10^3 gal, the value obtained for actual mixing conditions in the reactor (represented by four tanks in series). For a PFR, an effluent concentration of 0.2 ng/L can be obtained by doubling the energy input from 0.115 kWh/10^3 gal to 0.230 kWh/10^3 gal because photolysis is a pseudo-first-order reaction in this case. The value of the EE/O concept is demonstrated by the ease with which the energy required for a lower effluent concentration can be determined.

Problems and Discussion Topics

8-1 What oxidants are used most frequently in water treatment? What are the principal applications of these oxidants? What oxidants are employed most commonly for taste and odor control?

8-2 One problem associated with H_2S removal using oxidation is the formation of polysulfides when H_2S concentration is higher than

1 mg/L. What measures can be taken to avoid the formation of polysulfides?

8-3 Discuss the reason that permanganate, chlorine dioxide, free chlorine, ozone, and hydrogen peroxide are unable to oxidize iron in many waters and how to assess the feasibility of using chemical oxidation for iron removal?

8-4 Balance the oxidation-reduction reaction for the oxidation of benzene (C_6H_6) using (a) hydrogen peroxide and (b) ozone.

8-5 Balance the oxidation-reduction reaction for the oxidation of *tert*-butyl methyl ether (MTBE) [$(CH_3)_3COCH_3$] using (a) hydrogen peroxide and (b) ozone.

8-6 Determine whether chlorine or ozone is the more powerful oxidant from a consideration of free energy and reduction potential.

8-7 Determine the oxidation potential for converting chloride (Cl^-) to chlorate (ClO_3^-) from (a) the chlorine dioxide/chloride reaction, (b) the chlorine dioxide/chlorite reaction, and (c) the chlorate/chlorite reaction.

8-8 Calculate the equilibrium constant and ΔG°_{Rxn} for the oxidation of Mn(II) to manganese oxide by dissolved oxygen. What oxygen concentrations are needed for pH values of 6.0, 7.0, and 8.0 such that the reaction is thermodynamically favorable? *Given*: $[Mn^{2+}] = 10^{-6}$ M and the temperature is 298 K.

8-9 Manganese [Mn(II)] is soluble in water and is present in many groundwaters because insoluble forms (e.g., MnO_2) that are contained in minerals are reduced to soluble forms. The initial reactant concentrations are as follows: potassium permanganate ($KMnO_4$), 8 mg/L; Mn^{2+}, 2 mg/L. Permanganate is sometimes used to remove Mn^{2+} and the half reactions are

$$Mn^{2+} \rightarrow MnO_2 \qquad E^\circ = -1.21 \text{ V}$$

$$MnO_4^- \rightarrow MnO_2 \qquad E^\circ = 0.590 \text{ V}$$

a. Balance the overall redox reaction. Which reaction is the oxidation reaction? Which is the reduction reaction? Identify the electron acceptor and donor as well as the reductant and the oxidant.
b. Calculate the equilibrium constant.
c. Calculate the equilibrium Mn^{2+} concentration when the pH is 7 and the concentration of potassium permanganate is 1 mg/L.
d. Obtain expressions for permanganate and MnO_2 concentrations in terms of Mn^{2+} concentration.
e. Plot the free energy as a function of the conversion of Mn^{2+} from 0.01 to 0.999.

Hint: Use the stoichiometric table to determine all reacting species as a function of X_a and then eliminate X_a by using the final concentration, C_a. After substituting C_a back into the expressions, the permanganate and MnO_2 concentrations can be obtained in terms of the final concentration of Mn^{2+}.

8-10 Can hydrogen sulfide be oxidized using hydrogen peroxide for the following conditions: $[H_2S] = 10^{-2}$ M, $[H_2O_2] = 10^{-12}$ M, $[Cl^-] = 1$ M, $P_{CO_2} = 1$ atm, and pH 7?

8-11 Can nitrate (NO_3^-) be reduced to nitrogen gas (N_2) under aerobic conditions? For this problem, assume the following conditions are applicable for aerobic fresh water: $[NO_3^-] = 10^{-2}$ M, $P_{N_2} = 1$ atm, $[H^+] = 10^{-7}$ M, and $[O_2(aq)] = 8.24$ mg/L (2.58×10^{-4} M) at 25°C.

8-12 Is it thermodynamically possible to form bromate (BrO_3^-)from bromide (Br^-) using hydroxyl radicals ($HO^\bullet$) for the following conditions? Also, determine whether it is possible to form bromate concentrations in excess of 10 μg/L, which is its maximum contaminant level.

$$[HO\bullet] = 10^{-11}\text{ M} \quad Br^- = 0.3\text{ mg/L} \quad BrO_3^- = 10\ \mu\text{g/L}$$

8-13 Calculate the equilibrium concentrations of HOCl and $Cl_2(aq)$ in solution for a chlorine addition of 4 mg/L at 25°C. Assume that the pH is 6.0 and does not change and that the HOCl does not dissasociate into H^+ and OCl^-. Express the chlorine concentrations in terms of milligrams per liter of Cl_2.

8-14 Chlorine has a unitless Henry's law constant of 0.480 at 25°C and a reduction potential of 1.390 V when it is a gaseous reactant. Calculate the reduction potential when it is present as an aqueous reactant as shown in the reaction

$$Cl_2(aq) + 2e^- \rightarrow 2Cl^-$$

8-15 Rank the following oxidants according to redox potential: oxygen, chlorine, chlorine dioxide, potassium permanganate, ozone, hydroxyl radical, and hydrogen peroxide. What is the general trend typically observed with respect to oxidation rate for these oxidants?

8-16 Bench-scale tests have been used to develop a value of the apparent rate constant (K_{app}) for a particular well water in the presence of oxygen. The results are shown below. Assuming that the partial pressure of oxygen is maintained at 0.1 atm during the tests, determine K_{app}.

Time, min	Fe(II), mg/L
0	5
5.1	3
10.2	1.8
15.3	1
25.8	0.4
40.6	0.08

8-17 How much Mn(II) remains after 5, 10, 20, and 30 s of oxidation with chlorine dioxide? The initial concentrations of chlorine dioxide and Mn(II) are 1.5 times the stoichiometric requirement and 1.5 mg/L, respectively. Assume that the second-order rate constant is 5 × 104 L/ mol · s.

$$Mn^{2+} + 2ClO_2(aq) + 2H_2O \rightarrow MnO_2 + 2ClO_2^- + 4H^+$$

8-18 How much Fe(II) remains after 10 and 30 s of oxidation with permanganate? The initial concentration of permanganate and Fe(II) are 1.5 times the stoichiometric requirement and 1 mg/L, respectively. Assume that the second-order rate constant is $10^5\ M^{-1}\ s^{-1}$.

8-19 Estimate the absorptivity of ozone at a wavelength of 254 nm for both base e and base 10, assuming ozone is present at a concentration of 0.50 mg/L. The extinction coefficient of ozone is 3300 L/mol · cm and the molecular weight of ozone is 48.0 g/mol.

8-20 A potential raw-water source for drinking water is analyzed and found to contain the constituents given below. Using the given values, estimate the absorptivity (both base 10 and base *e*) of the water at a wavelength of 254 nm.

Constituent	Unit	Value	ε(254) L/mol · cm
TOC	mg/L as C	2.0	
water	mol/L		6.1×10^{-6}
Fe(II)	mg/L as Fe	1.3	466
Nitrate	mg/L as NO_3^-	3.0	3.4
SUVA	L/mg · m	3.0	

8-21 The source water for a drinking water treatment plant in California contains 50 ng/L NDMA. The treatment objective for NDMA has been set at 2 ng/L. After considering various treatment options, UV photolysis has been chosen for NDMA removal. The commercial reactors being evaluated are 1 m in diameter and 3 m in height (water volume 2300 L). Each reactor has 15 lamps that use 15 kW per lamp. The lamps are 30 percent efficient; that is, 30 percent of the energy consumed by the lamp is produced as UV light at the relevant wavelength (15 kW × 0.30 = 4.5 kW). A previous dye study has shown that the reactor can be modeled as three tanks in series. Estimate the flow rate that can be treated in this reactor. The extinction coefficient $\varepsilon(254)$ and quantum yield $\phi(254)$ for NDMA can be found in Table 8-6. For calculation simplicity, assume there are no losses on the reactor walls, the lamp sleeves do not block any light, the UV light intensity is monochromatic at 254 nm, and the water has exactly the same quality as the water used in Example 8-12: $k'(\lambda) = 0.145\ \text{cm}^{-1}$.

8-22 Calculate the EE/O for NDMA in Problem 8-21. Compare this value to the EE/O estimated in a PFR with the same lamp efficiency as Problem 8-21. Calculate the EE/O values in units of kWh/m^3 and $\text{kWh}/10^3$ gal.

8-23 Calculate the half-life and time to convert 95 percent of the hypochlorous acid to hypobromous acid. The initial concentrations of HOCl and Br^- are 2 mg/L as chlorine (2.82×10^{-5} mol/L) and 0.3 g/L, respectively. Assume that the second-order rate constant is 2.95×10^3 L/mol · s.

References

Benjamin, M. M., Korshin, G. V., and Li, C. W. (1997) "The Decrease of UV Absorbance as an Indicator of TOX Formation," *Water Res.*, **31**, 4, 946–949.

Black, A., and Goodson, J. (1952) "The Oxidation of Sulfides by Chlorine in Dilute Aqueous Solution," *J. AWWA*, **44**, 4, 309–316.

Bolton, J. R.,and Carter, S. R. (1994) Homogeneous Photodegradation of Pollutants in Contaminated Water: An Introduction, Chap. 33, in G. R. Helz, R. G. Zepp, and D. G. Crosby (eds.), *Aquatic and Surface Photochemistry*, CRC Press, Boca Raton, FL.

Bruce, D., Westerhoff, P., and Brawley-Chesworth, A. (2002) "Removal of 2-Methylisoborneol and Geosmin in Surface Water Treatment Plant in Arizona," *J. Water Supply: Res. Technol.—Aqua*, **51**, 4, 183–197.

Burttsechell, H., Rosen, A., Middleton, F., and Ettinger, M. (1959) "Chlorine Deviations of Phenol Causing Taste and Odor," *J. AWWA*, **51**, 2, 205–214.

Chen, K. (1974) Chemistry of Sulfur Species and Their Removal from Water Supply, Chap. 6 in A. J. Rubin (ed.), *Chemistry of Water Supply, Treatment, and Distribution*, Ann Arbor Science, Ann Arbor, MI.

Chen, K. Y., and Morris, J. C. (1972) "Kinetics of Oxidation of Aqueous Sulfide by O_2," *Environ. Sci. Technol.*, **6**, 6, 529–537.

Coffey, B. M., Gallagher, D. L., and Knocke, W. R. (1993) "Modeling Soluble Manganese Removal by Oxide-Coated Filter Media," *J. Environ. Eng.*, **119**, 4, 679–694.

Cosson, H., and Ernst, W. R. (1994) "Photodecomposition of Chlorine Dioxide and Sodium Chlorite in Aqueous Solution by Irradiation with Ultraviolet Light," *Ind. Eng. Chem. Res.*, **33**, 1468–1475.

Elovitz, M. S., and von Gunten, U. (1999) "Hydroxyl Radical/Ozone Ratios During Ozonation Processes," *Ozone: Sci. Eng.*, **21**, 3, 239–260.

Ettinger, M., and Ruchhoft, C. (1951) "Stepwise Chlorination and Taste and Odor Producing Intensity of Some Phenolic Compounds," *J. AWWA*, **43**, 5, 561–569.

Faust, S. D., and Aly, O. M. (1998) *Chemistry of Water Treatment*, 2nd ed., Ann Arbor Press, Chelsea, MI.

Glaze, W. H., Schep, R., Chauncey, W., Ruth, E. C., Zarnoch, J. J., Aieta, E. M., Tate, C. H., and McGuire, M. J. (1990) "Evaluating Oxidants for the Removal of Model Taste and Odor Compounds from a Municipal Water Supply," *J. AWWA*, **82**, 5, 79–84.

Graham, M., Najm, I., Simpson, M., MacLeod, B., Summers, S., and Cummings, L. (2000) *Optimization of Powdered Activated Carbon: Application for Geosmin and MIB Removal*, American Water Works Association Research Foundation, Denver, CO.

Gurol, M. D., and Akata, A. (1996) "Kinetics of Ozone Photolysis in Aqueous Solution," *AlChE J.*, **42**, 3283–3292.

Ho, T. L., Bolton, J. R., and Lipzynska-Kochany, E. (1996) "Quantum Yield for the Photodegradation of Pollutants in Dilute Aqueous Solution: Phenol, 4-Chlorophenol and *N*-Nitrosodimethylamine", *J. Adv. Oxidation Technol.*, **1**, 2, 170–178.

Kawamura, S. (2000) *Integrated Design and Operation of Water Treatment Facilities*, 2nd ed., Wiley-Interscience, New York.

King, D. W. (1998) "Role of Carbonate Speciation on the Oxidation Rate of Fe(II) in Aquatic Systems," *Environ. Sci. Technol*, **32**, 19, 2997–3003.

Knocke, W. (1990) *Alternative Oxidants for the Removal of Soluble Iron and Manganese*, American Water Works Association Research Foundation, Denver, CO.

Knocke, W. R., Conley, L., and Van Benschoten, J. E. (1992) "Impact of Dissolved Organic Carbon on the Removal of Iron During Surface Water Treatment," *Water Res.*, **26**, 11, 1515–1522.

Knocke, W., Hamon, J., and Thompson, C. (1988) "Soluble Manganese Removal on Oxide-Coated Filter Media," *J. AWWA*, **80**, 12, 65–70.

Knocke, W. R., Van Benschoten, J. E., Kearney, M. J., Soborski, A. W., and Reckhow, D. A. (1991) "Kinetics of Manganese and Iron Oxidation by Potassium Permanganate and Chlorine Dioxide," *J. AWWA*, **83**, 6, 80–87.

McGuire, M. J., Krasner, S. W., Hwang, C. J., and Lzaguirre, G. (1981) "Closed-Loop Stripping Analysis as a Tool for Solving Taste and Odor Problems," *J. AWWA*, **73**, 530–537.

McMurry, J., and Fay, R. C. (2003) *Chemistry*, 4th ed., Prentice-Hall, New York.

Mertens, R., and von Sonntag, C. (1995) "Photolysis (L = 254 nm) of Tetrachloroethene in Aqueous Solutions," *J. Photochem. Photobiol., A: Chem.*, **85**, 1–9.

Millero, F. J. (1989) "Effect of Ionic Interactions on the Oxidation of Fe II and Cu I in Natural Waters," *Marine Chem.*, **28**, 1–3, 1–18.

Millero, F. J. (1990a) "Marine Solution Chemistry and Ionic Interactions," *Marine Chem.*, **30**, 1–3, 205–229.

Millero, F. J. (1990b) "Effect of Ionic Interactions on the Oxidation Rates of Metals in Natural Waters," Chap. 34, *Chemical Modeling of Aqueous Systems*, Vol. **2**, American Chemical Society, Washington, DC.

Millero, F. J., and Izaguirre, M. (1989) "Effect of Ionic Strength and Ionic Interactions on the Oxidation of Fe(II)," *J. Solution Chem.*, **18**, 6, 585–599.

Millero, F. J., and Sotolongo, S. (1989) "The Oxidation of Fe(II) with H_2O_2 in Seawater," *Geochim. Cosmochim. Acta*, **53**, 8, 1867–1873.

Millero, F. J., Sotolongo, S., and Izaguirre, M. (1987) "The Oxidation Kinetics of Fe(II) in Seawater," *Geochim. Cosmochim. Acta*, **51**, 793–801.

Morgan, J. (1967) Chemical Equilibria and Kinetic Properties of Manganese in Natural Waters, 561–624, in S. Faust and J. Hunter (eds.), *Principles and Applications of Water Chemistry*, John Wiley & Sons, New York.

Morris, J. C. (1966) "The Acid Ionization Constant of HOCl from 5° C to 35°C," *J. Phys. Chem.*, **70**, 12, 3798–3805.

Nowell, L. H., and Hoigne, J. (1992a) "Photolysis of Aqueous Chlorine at Sunlight and Ultraviolet Wavelengths—II. Hydroxyl Radical Production," *Water Res.*, **26**, 5, 599–605.

Nowell, L. H., and Hoigne, J. (1992b) "Photolysis of Aqueous Chlorine at Sunlight and Ultraviolet Wavelengths—I. Degradation Rates," *Water Res.*, **26**, 5, 593–598.

Powell, S., and Lossberg, L. (1948) "Removal of Hydrogen Sulfide from Well Water," *J. AWWA* **40**, 12, 1277–1290.

Reckhow, D. A., Singer, P. C., and Trussell, R. R. (1986) *Ozone as a Coagulant Aid, AWWA Seminar Proceedings—Ozonation: Recent Advances and Research Needs*, American Water Works Association, Denver, CO.

Reisz, E., Schmidt, W., Schuchmann, H.-P., and Von Sonntag, C. (2003) "Photolysis of Ozone in Aqueous Solutions in the Presence of Tertiary Butanol," *Environ. Sci. Technol.*, **37**, 9, 1941–1948.

Riddick, T. M. (1951) "Controlling Taste, Odor, and Color with Free Residual Chlorination," *J. AWWA*, **43**, 545–552.

Sharpless, C. M., and Linden, K. G. (2003) "Experimental and Model Comparisons of Low- and Medium-Pressure Hg Lamps for the Direct and H_2O_2 Assisted UV Photodegradation of N-nitrosodimethylamine in Simulated Drinking Water," *Environ. Sci. Technol.*, **37**, 9, 1933–1940.

Simpson, M. R., and MacLeod, B. W. (1991) Using Closed Loop Stripping and Jar Tests to Determine Powdered Activated Carbon Dose Needed for Removal of Geosmin: Manatee County's Experience, paper presented at the American Water Works Association Water Quality Technology Conference, Orlando, FL.

Stumm, W., and Lee, G. (1961) "Oxygenation of Ferrous Iron," *Ind. Eng. Chem.*, **53**, 143–146.

Taku, K., and Tanaka, S. (2000) "Photodegradation and Reaction Rate Analysis of TCE and PCE," *Yosui to Haisui*, **42**, 228–234.

Theis, T., and Singer, P. (1974) "Complexation of Iron (II) by Organic Matter and Its Effecton Iron (II) Oxygenation," *Environ. Sci. Technol.*, **8**, 6, 569–573.

u.S. EPA (1986) *Municipal Wastewater Disinfection Design Manual*, EPA 625/1-86/021, u.S. Environmental Protection Agency, Cincinnati, OH.

Wehrli, B. (1990) Redox Reactions of Metal Ions at Mineral Surfaces, in W. Stumm (ed.) *Aquatic Chemical Kinetics*, Wiley Interscience, New York.

Westerhoff, P., Aiken, G., Amy, G., and Debroux, J. (1999) "Relationships between the Structure of Natural Organic Matter and Its Reactivity Towards Molecular Ozone and Hydroxyl Radicals," *Water Res.*, **33**, 10, 2265–2276.

White, G. C. (1999) *Handbook of Chlorination and Alternative Disinfectants*, 4th ed., John Wiley & Sons, New York.

Zika, R. G., Moore, C. A., Gidel, L. T., and Cooper, W. J. (1984) Sunlight-Induced Photodecomposition of Chlorine Dioxide, Chap. 82, in R. L. Jolley, R. J. Bull, W. P. Davis, S. Katz, M. H. Roberts, Jr., and V. A. Jacobs (eds.), *Water Chlorination: Chemistry Environmental Impact and Health Effects*, Vol. **5**, Lewis Publishers, Boca Raton, FL.

9 Coagulation and Flocculation

Stantec's Water Treatment: Principles and Design, Updated Third Edition. John C. Crittenden, R. Rhodes Trussell, David W. Hand, Kerry J. Howe, and George Tchobanoglous.

Terminology for Coagulation and Flocculation

Term	Definition
Coagulation	Addition of a chemical to water with the objective of destabilizing particles so they aggregate or forming a precipitate that will sweep particles from solution or adsorb dissolved constituents.
Coagulant aid	Chemicals (typically synthetic polymers) added to water to enhance the coagulation process.
Counterions	Ions of opposite charge to the surface charge of particles.
Critical coagulation concentration (CCC)	Concentration of coagulant that reduces the electric double layer to the point where flocculation can occur.
Destabilization	Process of eliminating the surface charge on a particle so that flocculation can occur.
Electric double layer (EDL)	Electrostatic potential surrounding a charged particle in solution, consisting of a layer of counterions adsorbed directly to the surface and a diffuse layer of ions forming a cloud of charge around the particle.
Enhanced coagulation	Coagulation process with the objective of removing natural organic matter, typically for minimizing the formation of disinfection by-products (see Sec 9-5).
Enmeshment or sweep floc	Entrapment or capture of particles by amorphous precipitates that form when a coagulant is added to water.
Flocculation	Aggregation of destabilized particles into larger masses that are easier to remove from water than the original particles.

Term	Definition
Flocculant aid	Organic polymers used to enhance settleability and filterability of floc particles.
Inorganic metal coagulant	Metal salts such as aluminum sulfate and ferric chloride that will hydrolyze, forming mononuclear and polynuclear species of varying charge. When added in excess, metal coagulants form chemical precipitates.
Jar test	Procedure to study effect of coagulant addition to water; used to determine required doses and operating conditions for effective coagulation and flocculation.
Stable particle suspension	Suspension of particles that will stay in solution indefinitely; stable particles have a surface charge that causes them to repel each other and prevent aggregation into larger particles that would settle on their own.
Synthetic organic coagulant	High-molecular-weight (typically 10^4 to 10^7 g/mol) organic molecules that can carry positive (cationic), negative (anionic), or neutral (nonionic) charge.
Zeta potential	Measurement of the charge at the shear plane of particles, used as a relative measure of particle surface charge.

Natural surface waters contain inorganic and organic particles. Inorganic particulate constituents, including clay, silt, and mineral oxides, typically enter surface water by natural erosion processes. Organic particles may include viruses, bacteria, algae, protozoan cysts and oocysts, as well as detritus litter that have fallen into the water source. In addition, surface waters will contain very fine colloidal and dissolved organic constituents such as humic acids, a product of decay and leaching of organic debris. Particulate and dissolved organic matter is often identified as natural organic matter (NOM).

Removal of particles is required because they can (1) reduce the clarity of water to unacceptable levels (i.e., cause turbidity) as well as impart color to water (aesthetic reasons), (2) be infectious agents (e.g., viruses, bacteria, and protozoa), and (3) have toxic compounds adsorbed to their external surfaces. The removal of dissolved NOM is of importance because many of the constituents that comprise dissolved NOM are precursors to the formation of disinfection by-products (see Chap. 19) when chlorine is used for disinfection. NOM can also impart color to the water.

The most common method used to remove particulate matter and a portion of the dissolved NOM from surface waters is by sedimentation and/or filtration following the conditioning of the water by coagulation and flocculation, the subject of this chapter. Thus, the purpose of this chapter is to present the chemical and physical basis for the phenomena occurring in the coagulation and flocculation processes. Specific topics include (1) the role of coagulation and flocculation processes in water treatment, (2) stability of particles in water, (3) coagulation theory, (4) coagulation practice, (5) coagulation of dissolved and organic constituents, (6) flocculation theory, and (7) flocculation practice.

9-1 Role of Coagulation and Flocculation Processes in Water Treatment

The importance of the coagulation and flocculation processes in water treatment can be appreciated by reviewing the process flow diagrams illustrated on Fig. 9-1. As used in this book, *coagulation* involves the addition of a chemical coagulant or coagulants for the purpose of conditioning the suspended, colloidal, and dissolved matter for subsequent processing by flocculation or to create conditions that will allow for the subsequent removal of particulate and dissolved matter. *Flocculation* is the aggregation of destabilized particles (particles from which the electrical surface charge has been reduced) and precipitation products formed by the addition of coagulants into larger particles known as flocculant particles or, more commonly, "floc." The aggregated floc can then be removed by gravity sedimentation and/or filtration. Coagulation and flocculation can also be differentiated on the basis of the time required for each of the processes. Coagulation typically occurs in less than 10 s, whereas flocculation occurs over a period of 20 to 45 min. An overview of the coagulation and flocculation processes is provided below.

Coagulation Process

The objective of the coagulation process depends on the source of the water and the nature of the suspended, colloidal, and dissolved organic

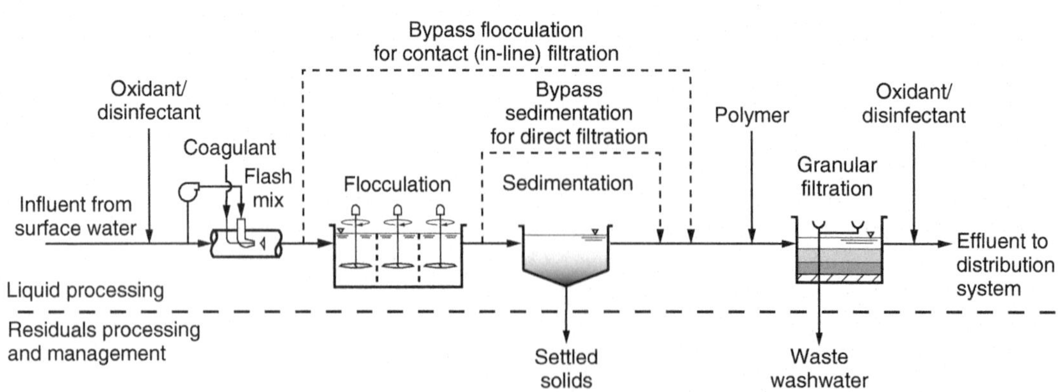

Figure 9-1
Typical water treatment process flow diagram employing coagulation (chemical mixing) with conventional treatment, direct filtration, or contact filtration.

constituents. Coagulation by the addition of the hydrolyzing chemicals such as alum and iron salts and/or organic polymers can involve

1. Destabilization of small suspended and colloidal particulate matter
2. Adsorption and/or reaction of portions of the colloidal and dissolved NOM to particles
3. Creation of flocculant particles that will sweep through the water to be treated, enmeshing small suspended, colloidal, and dissolved material as they settle

Coagulants such as alum, ferric chloride, and ferric sulfate hydrolyze rapidly when mixed with the water to be treated. As these chemicals hydrolyze, they form insoluble precipitates that destabilize particles by adsorbing to the surface of the particles and neutralizing the charge (thus reducing the repulsive forces) and/or forming bridges between them. Natural or synthetic organic polyelectrolytes (polymers with multiple charge-conferring functional groups) are also used for particle destabilization. Because of the many competing reactions, the theory of chemical coagulation is complex. Thus, the simplified reactions presented in this and other textbooks to describe the various coagulation processes can only be considered approximations, as the reactions may not necessarily proceed as indicated (Letterman et al., 1999).

Flocculation Process

The purpose of flocculation is to produce particles, by means of aggregation, that can be removed by subsequent particle separation procedures such as gravity sedimentation and/or filtration. Two general types of flocculation can be identified: (1) microflocculation (also known as perikinetic flocculation) in which particle aggregation is brought about by the random thermal motion of fluid molecules (known as Brownian motion) and (2) macroflocculation (also known as orthokinetic flocculation) in which particle aggregation is brought about by inducing velocity gradients and mixing in the fluid containing the particles to be flocculated. Another form of macroflocculation is brought about by differential settling in which large particles overtake small particles to form larger particles.

Practical Design Issues

When it comes to the practical design of coagulation and flocculation facilities, designers must consider four process issues: (1) the type and concentration of coagulants and flocculant aids, (2) the mixing intensity and the method used to disperse chemicals into the water for destabilization, (3) the mixing intensity and time for flocculation, and (4) the selection of the liquid–solid separation process (e.g., sedimentation, flotation, and granular filtration). With the exception of sedimentation and flotation (considered in Chap. 10) and filtration (considered in Chaps. 11 and 12), these subjects are addressed in the subsequent sections of this chapter.

9-2 Stability of Particles in Water

The particles in water may, for practical purposes, be classified as suspended and colloidal, according to particle size. Because small suspended and colloidal particles and dissolved constituents will not settle in a reasonable period of time, chemicals must be used to help remove these particles. The physical characteristics of particles found in water including particle size, number, distribution, and shape have been discussed previously in Chap. 2, Sec 2-3.

To appreciate the role of chemical coagulants and flocculant aids, it is important to understand particle solvent interactions and the electrical properties of the colloidal particles found in water. These subjects along with the nature of particle stability and the compression of the electrical double layer are considered in this section.

Particle–Solvent Interactions

Particles in natural water can be classified as hydrophobic (water repelling) and hydrophilic (water attracting). Hydrophobic particulates have a well-defined interface between the water and solid phases and have a low affinity for water molecules. In addition, hydrophobic particles are thermodynamically unstable and will aggregate irreversibly over time.

Hydrophilic particles such as clays, metal oxides, proteins, or humic acids have polar or ionized surface functional groups. Many inorganic particulates in natural waters, including hydrated metal oxides (iron or aluminum oxides), silica (SiO_2), and asbestos fibers, are hydrophilic because water molecules will bind to the polar or ionized surface functional groups (Stumm and Morgan, 1996). Many organic particulates are also hydrophilic and include a wide diversity of biocolloids (humic acids, viruses) and suspended living or dead microorganisms (bacteria, protozoa, algae). Because biocolloids can adsorb on the surfaces of inorganic particulates, the particles in natural waters often exhibit heterogeneous surface properties. Some particulate suspensions such as humic or fulvic acids can be reversibly aggregated because of their hydrogen bonding tendencies.

Electrical Properties of Particles

The principal electrical property of fine particulate matter suspended in water is surface charge, which contributes to relative stability, causing particles to remain in suspension without aggregating for long periods of time. The particulate suspensions are thermodynamically unstable and, given sufficient time, colloids and fine particles will flocculate and settle. However, this process is not economically feasible because it is very slow. A review of the causes of particulate stability will provide an understanding of the techniques that can be used to destabilize particulates, which are discussed in the following section.

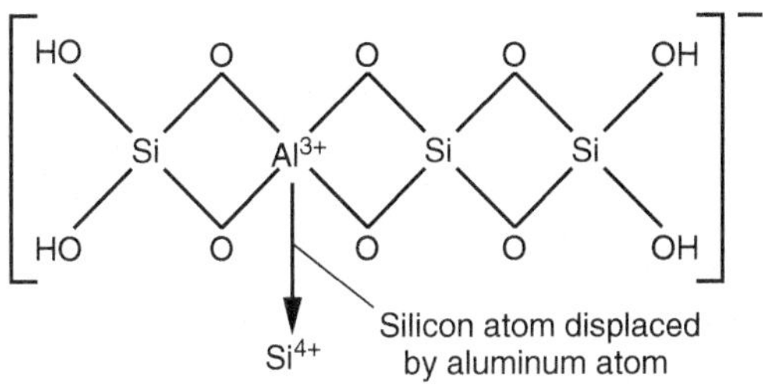

Figure 9-2
Charge acquisition through isomorphous substitution of Al for Si.

ORIGIN OF PARTICLE SURFACE CHARGE

Most particulates have complex surface chemistry and surface charges may arise from several sources. Surface charge arises in four principal ways, as discussed below (Stumm and Morgan, 1996).

Isomorphous replacement (crystal imperfections)

Under geological conditions, metals in metal oxide minerals can be replaced by metal atoms with lower valence, and this will impart a negative charge to the crystal material. An example where an aluminum atom replaced a silicon atom in a clay particle is shown on Fig. 9-2. This process, known as isomorphous replacement, produces negative charges on the surface of clay particles (van Olphen, 1963).

Structural imperfections

In clay and similar mineral particles, imperfections that occur in the formation of the crystal and broken bonds on the crystal edge can lead to the development of surface charges.

Preferential adsorption of specific ions

Particles adsorb NOM (e.g., fulvic acid), and these large macromolecules typically have a negative charge because they contain carboxylic acid groups:

$$R-COOH \rightleftarrows R-COO^- + H^+ \quad (pK_a = 4 \text{ to } 5) \tag{9-1}$$

Consequently, particle surfaces that have adsorbed NOM will be negatively charged for pH values greater than ~5.

Ionization of inorganic groups on particulate surfaces

Many mineral surfaces contain surface functional groups (e.g., hydroxyl) and their charge depends on pH. For example, silica has hydroxyl groups on its exterior surface, and these can accept or donate protons as shown here:

$$\begin{array}{ccccc} Si-OH_2^+ & \rightleftarrows & Si-OH + H^+ & \rightleftarrows & Si-O^- + 2H^+ \\ pH \ll 2 & & pH = 2 & & pH \gg 2 \end{array} \tag{9-2}$$

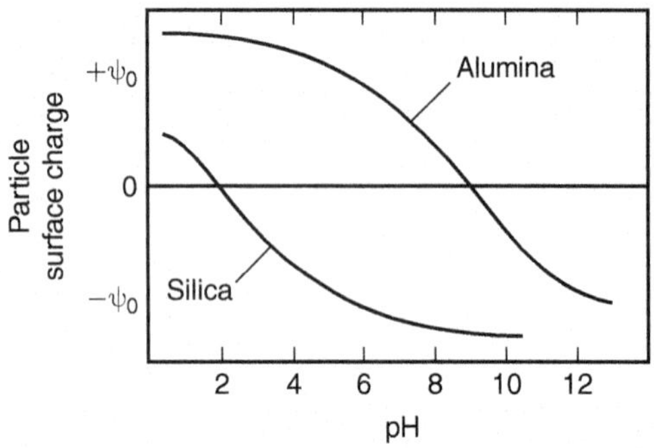

Figure 9-3
Variation in particle charge with pH.

The zero point of charge, as shown on Fig. 9-3, for silica is at pH 2, whereas the zero point of charge for alumina is about pH 9. The pH corresponding to a surface charge of zero is defined as the *zero point of charge* (ZPC). Above the ZPC the surface charge will be negative (anionic), and below the ZPC the charge will be positive (cationic). The ZPC for other particles that commonly occur in water are listed in Table 9-1. When examining Table 9-1, it is important to realize that many of the measurements that are reported are in low-ionic-strength waters (i.e., distilled water); consequently, the reported pH_{zpc} values are higher than is observed in natural waters.

Table 9-1
Surface characteristics of inorganic and organic particulates commonly found in natural waters

Type of Particle	Zero Point of Charge, pH_{zpc}
Inorganic	
$Al(OH)_3$ (amorphous)	7.5–8.5
Al_2O_3	9.1
CuO_3	9.5
$Fe(OH)_3$ (amorphous)	8.5
MgO	12.4
MnO_2	2–4.5
SiO_2	2–3.5
Clays	
Kaolinite	3.3–4.6
Montmorillonite	2.5
Asbestos	
Chrysotile	10–12
Crocidolite	5–6
$CaCO_3$	8–9
$Ca_5(PO_4)_3OH$	6–7
$FePO_4$	3
$AlPO_4$	4
Organic	
Algae	3–5
Bacteria	2–4
Humic acid	3
Oil droplets	2–5

Source: From Parks (1967) and Stumm and Morgan (1981).

ELECTRICAL DOUBLE LAYER

In natural waters, negatively charged particulates accumulate positive counterions on and near the particle's surface to satisfy electroneutrality. As shown on Fig. 9-4, a layer of cations will bind tightly to the surface of a negatively charged particle to form a fixed adsorption layer. This adsorbed layer of cations, bound to the particle surface by electrostatic and adsorption forces, is about 5 Å thick and is known as the *Helmholtz layer* (also known as the *Stern layer* after Stern, who proposed the model shown on Fig. 9-4). Beyond the Helmholtz layer, a net negative charge and electric field is present that attracts an excess of cations (over the bulk solution concentration) and repels anions, neither of which are in a fixed position. These cations and anions move about under the influence of diffusion (caused by collisions with solvent molecules), and the excess concentration of cations extends out into solution until all the surface charge and electric potential is eliminated and electroneutrality is satisfied.

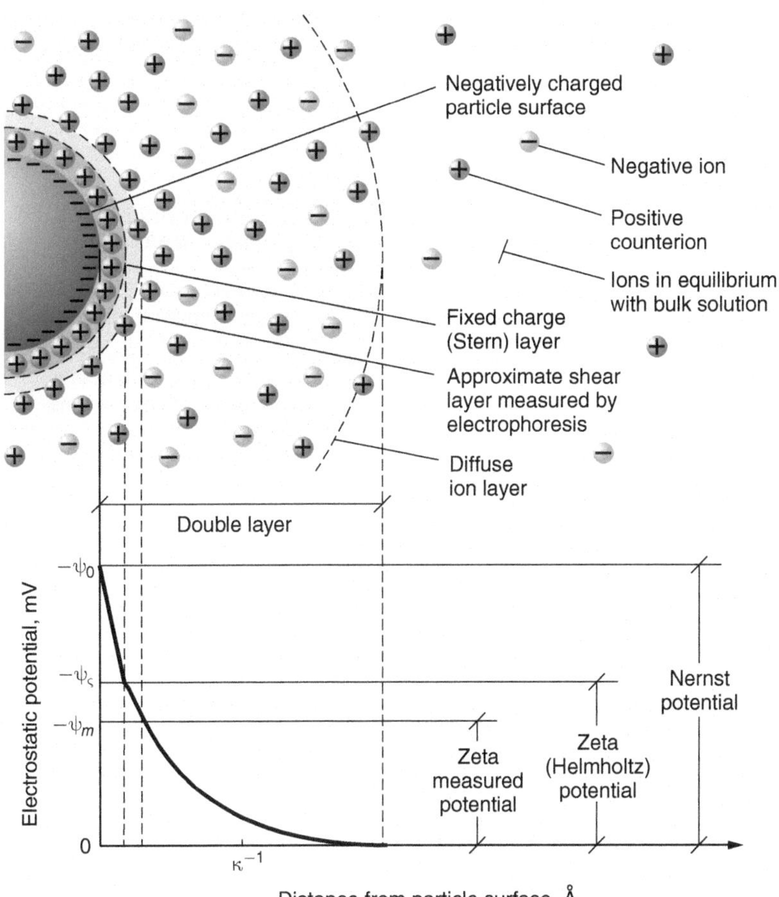

Figure 9-4
Structure of the electrical double layer. The potential measured at the shear plane is known as the zeta potential. The shear plane typically occurs in diffuse layer.

The layer of cations and anions that extends from the Helmholtz layer to the bulk solution where the charge is zero and electroneutrality is satisfied is known as the diffuse layer. Taken together the adsorbed (Helmholtz) and diffuse layer are known as the *electric double layer* (EDL). Depending on the solution characteristics, the EDL can extend up to 300 Å into the solution (Kruyt, 1952). It is interesting to note that the double-layer model proposed by Stern (see Fig. 9-4) is a combination of the earlier models proposed by Helmholtz–Perrin and Gouy–Chapman. In fact, the diffuse layer is often identified as the *Gouy–Chapman diffuse layer* (Voyutsky, 1978).

MEASUREMENT OF SURFACE CHARGE

The electrical properties of highly dispersed particle systems having a solid dispersed phase and a liquid dispersion medium can be defined in terms of four phenomena: (1) *electrophoresis*, (2) *electroosmosis*, (3) *sedimentation potential* (also known as the *Dorn effect*), and (4) *streaming potential.* Collectively these four phenomena, described in Table 9-2, are known as *electrokinetic phenomena* because they involve the movement of particles (or a liquid) when a potential gradient is applied or the formation of the potential

Table 9-2
Description and application of electrochemical phenomena

Phenomena	Description	Application in Water Treatment
Electrophoresis, discovered by R. Reuss, circa 1808	Refers to the movement of charged particles relative to a stationary liquid subject to an applied electrical field. The particles move along the lines of the electrical field.	Used to assess the destabilization of particles subject to the addition of coagulant chemicals. Also used in laboratory studies to isolate new proteins and other organic molecules.
Electroosmosis, discovered by R. Reuss, circa 1808	Refers to the movement of liquid relative to a stationary charged surface (e.g., a porous plug) subject to an applied electrical field.	
Streaming potential, discovered by G. Quincke, circa 1859	Refers to the creation of a potential gradient when liquid is made to flow along a stationary charged surface (e.g., when forced through a porous plug). The charges from the particles are carried along with the fluid.	Used to assess the destabilization of particles subject to the addition of coagulant chemicals. Online instruments are now available that can be used to control chemical addition in water treatment.
Sedimentation potential, discovered by Dorn, circa 1878	Refers to the creation of a potential gradient when charged particles move (e.g., settling) relative to a stationary liquid medium. Sedimentation potential is the opposite of electrophoresis.	

Source: Adapted from Voyutsky (1978) and Shaw (1966).

gradient when particles (or liquid) move. It should be noted that these aforementioned electrical phenomena are caused by the opposite charge of the particle (solid) and liquid. Although there is no direct measure of the electrical field surrounding a particle or method to determine when particles have been destabilized from the addition of coagulants, the surface charge on a particle can be measured indirectly using one of the four electrokinetic phenomena (Voyutsky, 1978).

ZETA POTENTIAL

When a charged particle is subjected to an electric field between two electrodes, a negatively charged particle will migrate toward the positive electrode, as shown on Fig. 9-5, and vice versa. This movement is termed *electrophoresis.* It should be noted that when a particle moves in an electrical field some portion of the water near the surface of the particle moves with it, which gives rise to the shear plane, as shown on Fig. 9-4. Typically, as shown on Fig. 9-4, the actual shear plane lies in the diffuse layer to the right of the theoretical fixed shear plane defined by the Helmholtz layer. The electrical potential between the actual shear plane and the bulk solution is what is measured by electrophoretic measurements. This potential is called the *zeta potential* or the electrical potential and is given by the expression

$$Z = \frac{\upsilon^0 k_z \mu}{\varepsilon \varepsilon_0} \tag{9-3}$$

where Z = zeta potential, V

υ^0 = electrophoretic mobility, (μm/s)/(V/cm)

$= \nu_E / E$

ν_E = electrophoretic velocity of migrating particle, μm/s (also reported as nm/s and mm/s)

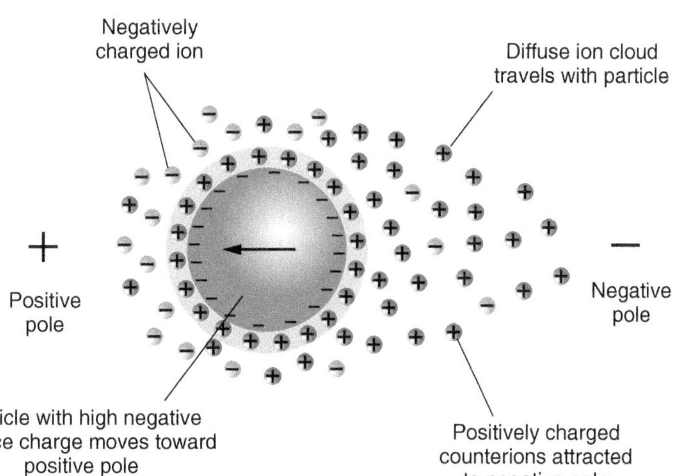

Figure 9-5
Schematic illustration of electrophoresis in which a charged particle moves in an electrical field, dragging with it a cloud of ions.

E = electrical field at particle, V/cm
k_z = constant that is 4π or 6π
μ = dynamic viscosity of water, $N \cdot s/m^2$
ε = permitivity relative to a vacuum (ε for water is 78.54, unitless)
ε_0 = permitivity in a vacuum, $8.854188 \times 10^{-12}\ C^2/J \cdot m$ (note that $C^2/J \cdot m$ is equivalent to N/V^2)

Typical values for the electrophoretic mobility for particles in natural waters vary from about − 2 to + 2 (μm/s)/(V/cm). The constant k_z is used to account for the shape of the particle. The value of 4π appears in the derivation put forth by Smoluchowski and applies if the extent of the diffuse layer is small relative to the curvature of the particle. The value of 6π is used when the particle is much smaller than the thickness of the double layer (Kissa, 1999).

For example, if the value of the constant is 4π and the electrical mobility is 0.5 (μm/s)/(V/cm), the value of the zeta potential at 25°C is 80.4 mV as given below:

$$Z = \frac{(0.5\ \mu m \cdot cm/s \cdot V)(4\pi)(0.890 \times 10^{-3}\ N \cdot s/m^2)(1\ m/10^6\ \mu m)(1\ m/10^2\ cm)}{(78.54)(8.854188 \times 10^{-12}\ C^2/J \cdot m)}$$

$$= 0.0804\ V = 80.4\ mV$$

Empirically, when the absolute value of the zeta potential is reduced below approximately 20 mV, rapid flocculation occurs (Kruyt, 1952). The zeta potential will vary with the size and shape of the particle, with the number of charges on the particle, with the strength of the electric field, and with the nature of the ions in the diffuse layer.

Particle Stability

The stability of particles in natural waters depends on a balance between (1) the repulsive electrostatic force and (2) the attractive force known as the van der Waals force.

REPULSIVE ELECTROSTATIC FORCES

The principal mechanism controlling the stability of hydrophobic and hydrophilic particles is electrostatic repulsion. Electrostatic repulsion occurs, as discussed above, because particles in water have a net negative surface charge. The magnitude of the electrostatic force will depend on the charge of the particle and the solution characteristics.

VAN DER WAALS ATTRACTIVE FORCE

Van der Waals forces originate from magnetic and electronic resonance that occurs when two particles approach one another. This resonance is caused by electrons in atoms on the particle surface, which develop a strong attractive force between the particles when these electrons orient themselves in such a way as to induce synergistic electric and magnetic fields. Van der Waals forces are proportional to the polarizability of the particle surfaces. Van der Waals attractive forces (<~20 kJ/mol) are strong enough to overcome electrostatic repulsion, but they are unable to do so because electrostatic repulsive forces and the EDL extend further into solution than do the van der Waals forces. As a result, an energy barrier is formed that must be overcome for flocculation to occur and coagulants are added to reduce the repulsive force, which allows for rapid flocculation.

PARTICLE–PARTICLE INTERACTIONS

Particle–particle interactions are extremely important in bringing about aggregation by means of Brownian motion. The theory of particle–particle interaction is based on the interaction of the repulsive and attractive forces on two charged particles as they are brought closer and closer together. The theory, first worked out by Derjaguin, later improved upon together with Landau, and subsequently extended by Verwey and Overbeek, is known as the DLVO theory after the scientists who developed it (Derjaguin and Landau, 1941; Verwey and Overbeek, 1948).

A conceptual diagram of the DLVO model is provided on Fig. 9-6, in which the interaction between two particles represented by flat plates with similar charge is illustrated. As shown on Fig. 9-6, the two principal forces involved are the forces of repulsion due to the electrical properties of the charged plates and the van der Waals forces of attraction. Two cases are illustrated on Fig. 9-6 with respect to the forces of repulsion. In the first case, the repulsive force extends far into solution. In the second case, the extent of the repulsive force is reduced considerably. The net total energy shown by the solid lines on Fig. 9-6 is the difference between the forces of repulsion and attraction. For case 1, the forces of attraction will predominate at very short and long distances. The net energy curve for condition 1 contains a repulsive maximum that must be overcome if the particles are to be held together by the van der Waals force of attraction.

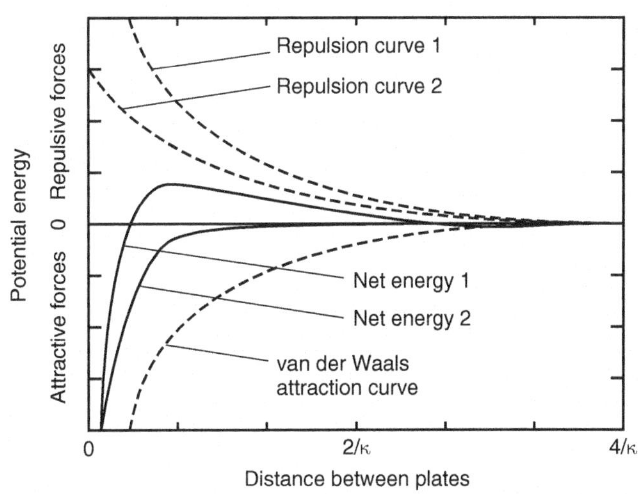

Figure 9-6
Attractive and repulsive potential energy that result when two particles are brought together.

Although floc particles can form at long distances as shown by the net energy curve for case 1, the net force holding these particles together is weak and the floc particles that are formed can be ruptured easily. In case 2, there is no energy barrier to overcome. Clearly, if colloidal particles are to be flocculated by microflocculation, the repulsive force must be reduced as shown in case 2. With the addition of a coagulant, which reduces the extent of the electrical double layer, rapid flocculation can occur.

Compression of the Electrical Double Layer

It has been observed that, if the electrical double layer is compressed, particles in water will come together as a result of Brownian motion and remain attached due to van der Waals forces of attraction, as discussed above. As the ionic strength of a solution is increased, the extent of the double layer decreases, which in turn reduces the zeta potential. The thickness of the double layer and the effects of ionic strength and electrolyte addition on the compression of the double layer are described below.

DOUBLE-LAYER THICKNESS

The thickness of the electrical diffuse layer as a function of the ionic strength and electrolyte is given in Table 9-3. The thickness of the diffuse layer may be calculated using the following equation (Gouy, 1910):

$$\kappa^{-1} = 10^{10}\left[\frac{(2)(1000)\,e^2 N_A I}{\varepsilon\varepsilon_0\, kT}\right]^{-1/2} \tag{9-4}$$

where κ^{-1} = double-layer thickness, Å
10^{10} = length conversion, Å /m
1000 = volume conversion, L/m^3
e = electron charge, 1.60219×10^{-19} C

Table 9-3
Thickness of electrical double layer (EDL) as function of ionic strength and valence at 25°C

Molarity	$z^+ : z^-$	I, mol/L	κ, cm^{-1}	$1/\kappa$, Å
0.001	1:1	0.001	1.04×10^6	96.2
	2:2	0.004	2.08×10^6	48.1
	3:3	0.009	3.12×10^6	32.1
0.01	1:1	0.01	3.29×10^6	30.4
	2:2	0.04	6.57×10^6	15.2
	3:3	0.09	9.86×10^6	10.1
0.1	1:1	0.1	1.04×10^7	9.6
	2:2	0.4	2.08×10^7	4.8
	3:3	0.9	3.12×10^7	3.2

N_A = Avagadro's number, 6.02205×10^{23}/mol
I = ionic strength, $\frac{1}{2}\sum z^2 M$, mol/L
z = magnitude of positive or negative charge on ion
M = molar concentration of cationic or anionic species, mol/L
ε = permittivity relative to a vacuum (ε for water is 78.54, unitless)
ε_0 = permittivity in a vacuum, 8.854188×10^{-12} $C^2/J \cdot m$
k = Boltzmann constant, 1.38066×10^{-23} J/K
T = absolute temperature, K (273 + °C)

The relationship given in Eq. 9-4 is not actually the double-layer thickness but is related to how far out into the solution the repulsive force will reach. It is approximately equal to the distance at which the electrical potential is 37 percent of the value at the particle surface. However, it is still important to know the EDL thickness because it provides insight into the particle stability and the coagulation process.

EFFECT OF IONIC STRENGTH

Of the many factors that affect double-layer thickness, ionic strength is perhaps the most important. As reported in Table 9-3, the EDL thickness shrinks dramatically with increasing ionic strength and valance. According to the DLVO theory, van der Waals forces extend out into solution about 10 Å; consequently, if the double layer is smaller than this, a rapidly flocculating suspension is formed. While it is possible to reduce the thickness of the EDL by increasing the ionic strength, this is not a practical method for destabilizing particles in drinking water treatment because the required ionic strengths are greater than are considered acceptable in potable water. It is interesting to note that ionic strength can be used to explain why particles are stable in freshwater (low ionic strength but high electrical repulsive forces) and flocculate rapidly in salt water (high ionic strength but low electrical repulsive forces). Determination of the thickness of the double layer as function of the ionic strength is illustrated in Example 9-1.

EFFECT OF COUNTERIONS

If the charge on the counterions in solution is altered, the thickness of the EDL will be reduced, as illustrated in Table 9-3. The ionic concentration that results in the reduction of the EDL to the point where flocculation occurs is defined as the critical coagulation concentration (CCC) and will depend on the type of particulate as well as the dissolved ions. According to the DLVO theory, the CCC is inversely proportional to the sixth power of the charge on the ion. Thus, the CCC values for mono-, di-, and trivalent ions are in the ratio of $1 : \frac{1}{2}^6 : \frac{1}{3}^6$, or 100: 1.6: 0.14 percent, assuming that the electrolytes do not adsorb or precipitate. The above relationship is known as the Schultz–Hardy rule, which was originally observed in the 1880s

Example 9-1 Determination of thickness of electrical double layer

Verify that the values in Table 9-3 are correct for 0.001 M solutions of monovalent and divalent ions using Eq. 9-4.

Solution

1. Determine the ionic strength I for a molarity of 0.001 for chemical constituents with a charge of 1 and 2.
   a. Determine the ionic strength for $Z = +1$ and -1:

$$1 = \frac{1}{2}\sum Z^2 M = \frac{1}{2}\sum (+1)^2\ (0.001) + (-1)^2\ (0.001) = 0.001\ \text{mol/L}$$

   b. Determine the ionic strength for $Z = +2$ and -2:

$$1 = \frac{1}{2}\sum Z^2 M = \frac{1}{2}\sum (+2)^2\ (0.001) + (-2)^2\ (0.001) = 0.004\ \text{mol/L}$$

2. Substitute known terms in Eq. 9-4 and solve for $1/\kappa$:
   a. For $M = 0.001$, $Z = +1$, -1, and $I = 0.001$ mol/L,

$$\kappa^{-1} = \left(10^{10}\ \text{Å/m}\right) \times \left[\frac{(2)\left(1000\ \text{L/m}^3\right)\left(1.60219 \times 10^{-19}\ \text{C}\right)^2\left(6.02205 \times 10^{23}\ \text{mol}^{-1}\right)(0.001\ \text{mol/L})}{(78.54)\left(8.854188 \times 10^{-12}\ \text{C}^2/\text{J}\cdot\text{m}\right)\left(1.38066 \times 10^{-23}\ \text{J/K}\right)(273 + 25\ \text{K})}\right]^{-1/2}$$

$$= 96.2\ \text{Å}$$

   b. For $M = 0.001$, $Z = +2$, -2, and $I = 0.004$ mol/L,

$$\kappa^{-1} = \left(10^{10}\ \text{Å/m}\right) \times \left[\frac{(2)\left(1000\ \text{L/m}^3\right)\left(1.60219 \times 10^{-19}\ \text{C}\right)^2\left(6.02205 \times 10^{23}\ \text{mol}^{-1}\right)(0.004\ \text{mol/L})}{(78.54)\left(8.854188 \times 10^{-12}\ \text{C}^2/\text{J}\cdot\text{m}\right)\left(1.38066 \times 10^{-23}\ \text{J/K}\right)(273 + 25\ \text{K})}\right]^{-1/2}$$

$$= 48.1\ \text{Å}$$

Comment

The above computation illustrates the importance of the charge of the ionic species, as reported in Table 9-3.

Kruyt, 1952). Thus, if 3000 mg/L of NaCl will produce rapid flocculation of hydrophobic particulates, then 47 mg/L of $CaCl_2$ will achieve similar results. It should also be noted that if multivalent ions comprise the fixed layer next to a negatively charged particle, the EDL will be reduced significantly and the CCC value would be much lower than predicted by the theory (for the Schultz–Hardy rule).

9-3 Coagulation Theory

The electrical properties of particles were considered in the previous section. Coagulation, as described in Sec. 9-1, is the process used to destabilize the particles found in waters so that they may be removed by subsequent separation processes. The purpose of this section is to introduce the principal coagulation mechanisms responsible for particle destabilization and removal. Coagulation practice including the principal chemicals used for coagulation in water treatment and jar testing is presented and discussed in Sec. 9-4.

Mechanisms that can be exploited to achieve particulate destabilization include (1) compression of the electrical double layer, (2) adsorption and charge neutralization, (3) adsorption and interparticle bridging, and (4) enmeshment in a precipitate, or "sweep floc." While these mechanisms are discussed separately here, it will become apparent that each one has certain pitfalls, and this is the reason that destabilization strategies exploit several mechanisms simultaneously. It should also be noted that compression of the electrical double layer, discussed in the previous section, is also considered a coagulation mechanism but is not discussed here because increasing the ionic strength is not practiced in water treatment.

Adsorption and Charge Neutralization

Particulates can be destabilized by adsorption of oppositely charged ions or polymers. Most particulates in natural waters are negatively charged (clays, humic acids, bacteria) in the neutral pH range (pH 6 to 8); consequently, hydrolyzed metal salts, prehydrolyzed metal salts, and cationic organic polymers can be used to destabilize particles through charge neutralization. Cationic organic polymers can be used as primary coagulants, but they are most often used in conjunction with inorganic coagulants to form particle bridges, as discussed below. Generally, the optimum coagulant dose occurs when the particle surface is only partially covered (less than 50 percent). Polymers of high charge density and low to moderate molecular weights (10,000 to 100,000) are believed to be adsorbed on negatively charged particles as a patch on the surface and do not extend much from the surface. The optimum dose appears to increase in proportion to the surface area concentration of the particulates.

When the proper amount of polymer has adsorbed, the charge is neutralized and the particle will flocculate. When too much polymer has been

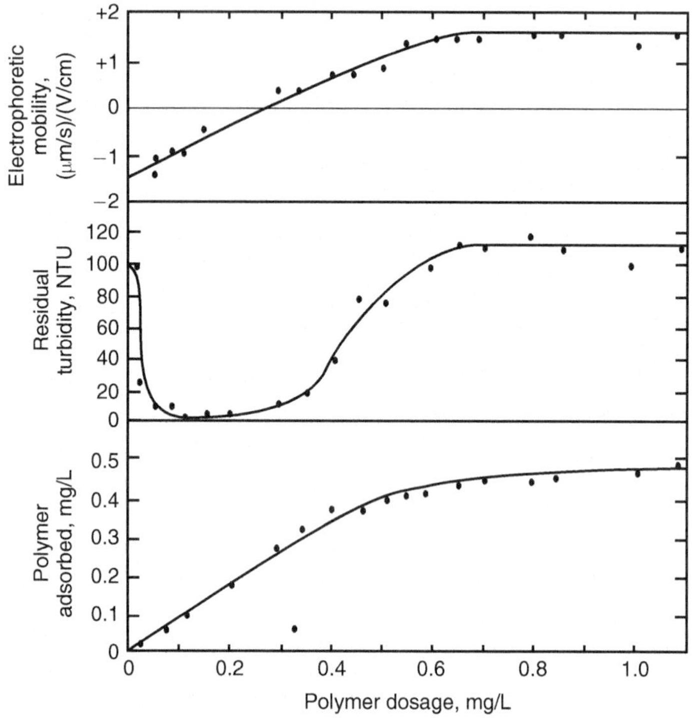

Figure 9-7
Destabilization of a kaolinite clay suspension with cationic polymer No. 4. Initial clay concentration = 73.2 mg/L. (Adapted from Black et al., 1966.)

added, the particles will attain a positive charge and become stable once again. This phenomenon is demonstrated by the classical experiments of Black et al. (1966), which are shown on Fig. 9-7. For polymer dosages up to 0.7 mg/L, the electrophoretic mobility becomes more positive and the amount adsorbed increases. Higher dosages cause charge reversal, particle stability, and a higher residual turbidity. Cationic polymers and polyaluminum chloride (PACl) are said to exhibit stoichiometry because a certain amount of charge exists on the particle suspension surface, and when the precise amount of coagulant is added, a rapidly flocculating suspension is created.

Adsorption and Interparticle Bridging

Polymer bridging is complex and has not been adequately described analytically. Schematically, polymer chains adsorb on particle surfaces at one or more sites along the polymer chain as a result of (1) coulombic (charge–charge) interactions, (2) dipole interaction, (3) hydrogen bonding, and (4) van der Waals forces of attraction (Hunter, 2001). The rest of the polymer may remain extended into the solution and adsorb on available surface sites of other particles, thus creating a "bridge" between particle surfaces. If the extended polymer cannot find vacant sites on the surface of other particulates, no bridging will occur. Thus, there is an optimum degree of coverage or extent of polymer adsorption at which the rate of aggregation will be a maximum. Polymer bridging is an adsorption phenomenon; consequently, the optimum coagulant dose will generally be proportional to the concentration of particulates present. Adsorption and interparticle bridging occur with nonionic polymers and high-molecular-weight (MW 10^5 to 10^7), low-surface-charge polymers. High-molecular-weight cationic polymers have a high charge density to neutralize surface charge.

REACTION MECHANISMS FOR POLYMERS

A schematic of the reaction mechanisms for polymers is shown on Fig. 9-8. At the optimum dosage of polymer shown in reaction (a), the particles are destabilized and can subsequently flocculate, as shown in reaction (b). If the particle concentration is very low or if adequate mixing does not allow flocculation, then nonadsorbed ends of the polymers will eventually adsorb on the destabilized particle, causing it to restabilize, as shown in reaction (c). If too much polymer is added, all adsorption sites will be taken up and the particle will not flocculate, as shown in reaction (d). If the particles are mixed for too long or too intensively, they will break up, as shown in reaction (e).

POLYMER SELECTION

Because polymer–solution interactions are complex, polymer selection is based on empirical testing. In general, though, anionic polymers have been shown to be effective flocculant aids, while nonionic polymers have been effective as filter aids. Polymer selection for sludge conditioning is dependent on sludge properties, polymer properties, and the mixing environment (O'Brien and Novak, 1977). Polymer bridging is the dominant mechanism in sludge conditioning, and thus polymer molecular weight is the dominant property of interest. For each system, the optimum polymer dose, mixing conditions, and pH must be determined empirically.

Precipitation and Enmeshment

When high enough dosages are used, aluminum and iron form insoluble precipitates and particulate matter becomes entrapped in the amorphous precipitates. This type of destabilization has been described as *precipitation and enmeshment* or *sweep floc* (Packham, 1965; Stumm and O'Melia, 1968). Although the molecular events leading to sweep floc have not been defined clearly, the steps for iron and aluminum salts at lower coagulant dosages are as follows: (1) hydrolysis and polymerization of metal ions, (2) adsorption of hydrolysis products at the interface, and (3) charge neutralization. At high dosages, it is likely that nucleation of the precipitate occurs on the surface of particulates, leading to the growth of an amorphous precipitate with the entrapment of particles in this amorphous structure. This mechanism predominates in water treatment applications where pH values are generally maintained between pH 6 and 8, and aluminum or iron salts are

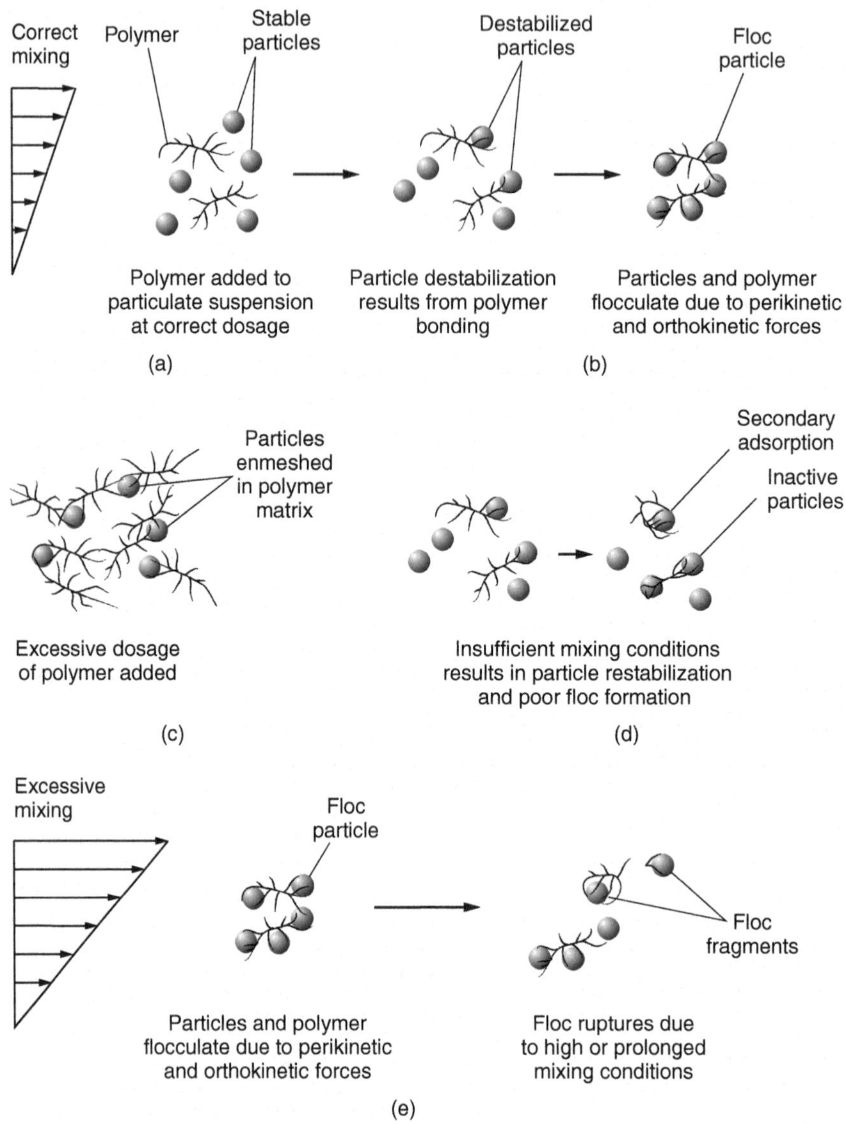

Figure 9-8
Schematic representation of the bridging model for the destabilization of particles by polymers. (Adapted from O'Melia, 1972.)

used at concentrations exceeding saturation with respect to the amorphous metal hydroxide solid that is formed.

One interesting finding regarding sweep floc is that, in general, the sweep floc mechanism does not depend on the type of particle, and, thus, the same dosage of coagulant is required for sweep floc formation regardless of the type of particles that may be present (in the absence of

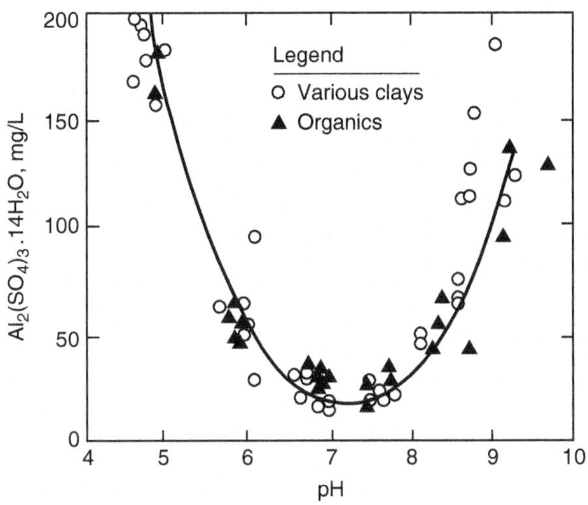

Figure 9-9
Coagulation of various types of clays and organics, which supports hypothesis that sweep floc is not influenced by type of particles present: (○) clays and (▲) organics. (Adapted from Packman, 1962.)

NOM). The dosage of alum required to reduce the turbidity of a variety of particles is displayed on Fig. 9-9. Although the dosage does not depend on the type of particles, it does depend on the pH, as expected. However, a caveat that should be mentioned is that the coagulant demand exerted by NOM is not reflected on Fig. 9-9. Thus, the concentration of hydrolyzing metal salts that is required for sweep floc will depend on the type and concentration of NOM, which unfortunately is site specific. The effects of NOM on coagulation practice are considered in Sec. 9-5.

9-4 Coagulation Practice

Selection of the type and dose of coagulant depends on the characteristics of the coagulant, the concentration and type of particulates, concentration and characteristics of NOM, water temperature, and water quality. At present, the interdependence of these five elements is only understood qualitatively, and prediction of the optimum coagulant combination from characteristics of the particulates and the water quality is not yet possible. The purpose of this section is to introduce coagulation practice, including the types of inorganic and organic coagulants and coagulant aids used, and alternative techniques used to reduce coagulant dosages.

Inorganic Metallic Coagulants

Inorganic coagulants, coagulant aids, and the chemicals used for alkalinity and pH adjustment are summarized in Table 9-4. Of the chemicals listed in Table 9-4, the principal inorganic coagulants used in water treatment are salts of aluminum and ferric ions and prehydrolyzed salts of these metals. These hydrolyzable metal cations are readily available as sulfate or

Table 9-4
Common inorganic coagulants, coagulant aids, and pH and alkalinity adjusting chemicals used in water treatment

Classification	Chemical Formula	Molecular Weight, g/mol	Application
Coagulants			
Aluminum sulfate	$Al_2(SO_4)_3 \cdot 14H_2O$	594.4	Primary coagulant
Sodium aluminate	$Na_2Al_2O_4$	163.9	Used with alum; provides alkalinity and pH control
Aluminum chloride	$AlCl_3$	160.3	Used in blends with organic polymers
Polyaluminum chloride (PACl)[a]	$Al_a(OH)_b(Cl)_c(SO_4)_d$	Variable	Primary coagulant
Polyaluminum sulfate (PAS)[b]	$Al_a(OH)_b(Cl)_c(SO_4)_d$	Variable	Primary coagulant, produced onsite
Polyiron chloride[c]	$Fe_a(OH)_b(Cl)_c(SO_4)_d$	Variable	Primary coagulant, produced onsite
Ferric chloride	$FeCl_3$	162.2	Primary coagulant
Ferric sulfate	$Fe_2(SO_4)_3$	400.0	Primary coagulant
Coagulant aids Activated silica	SiO_2	60.0	Coagulant aid used with alum during cold winter months
Sodium silicate	$Na_2O(SiO_2)_{3-25}$	242–1562	Coagulant aid, produced onsite
Bentonite	$Al_2Si_2O_5(OH)_4$	258	Used to provide nucleation sites for enhanced coagulation
Alkalinity and pH adjustment Calcium hydroxide	$Ca(OH)_2$	56.1 as CaO	Used to provide alkalinity and adjust pH
Sodium hydroxide	NaOH	40.0	Used to provide alkalinity and adjust pH
Soda ash	Na_2CO_3	106.0	Used to provide alkalinity and adjust pH

[a]Prehydrolyzed metal salts made from aluminum chloride.
[b]Prehydrolyzed metal salts made from aluminum sulfate.
[c]Prehydrolyzed metal salts made from iron chloride.

chloride salts in both liquid and solid (dry) forms. In the United States, the predominant water treatment coagulant is aluminum sulfate, or "alum," sold in a hydrated form as $Al_2(SO_4)_3 \cdot xH_2O$, where x is usually 14 because it is the least expensive coagulant. The action, solubility, and application of these coagulants are considered in the following discussion.

ACTION OF ALUM AND IRON SALTS

When ferric or aluminum ions are added to water, a number of parallel and sequential reactions occur. Initially, when a salt of Al(III) and Fe(III) is added to water, it will dissociate to yield trivalent Al^{3+} and Fe^{3+} ions, as given below:

$$Al_2(SO_4)_3 \rightleftarrows 2Al^{3+} + 3SO_4^{2-} \tag{9-5}$$

$$FeCl_3 \rightleftarrows Fe^{3+} + 3Cl^- \tag{9-6}$$

The trivalent ions of Al^{3+} and Fe^{3+} then hydrate to form the aquometal complexes $Al(H_2O)_6{}^{3+}$ and $Fe(H_2O)_6{}^{3+}$, as shown on the left-hand side of Eq. 9-7. As shown, the metal ion (aluminum in this case) has a coordination number of 6 and six water molecules orient themselves around the metal ion:

$$\begin{bmatrix} H_2O & & OH_2 \\ H_2O & -Al- & OH_2 \\ H_2O & & OH_2 \end{bmatrix}^{3+} \rightleftarrows \begin{bmatrix} H_2O & & OH \\ H_2O & -Al- & OH_2 \\ H_2O & & OH_2 \end{bmatrix}^{2+} + H^+ \tag{9-7}$$

These aquometal complexes then pass through a series of hydrolytic reactions, as illustrated on the right-hand side of Eq. 9-8, which give rise to the formation of a variety of soluble mononuclear (one aluminum ion) and polynuclear (several aluminum ions) species, as illustrated on Fig. 9-10. The mononuclear species—$Al(H_2O)_5(OH)^{2+}$ [or just $Al(OH)^{2+}$] and $Al(H_2O)_4(OH)_2{}^+$ [or just $Al(OH)_2{}^+$]—are among the many species formed. Similarly, iron forms a variety of soluble species, including mononuclear species (one iron ion) such as $Fe(H_2O)_5(OH)^{2+}$ [or just $Fe(OH)^{2+}$] and $Fe(H_2O)_4(OH)_2{}^+$ [or just $Fe(OH)_2{}^+$].

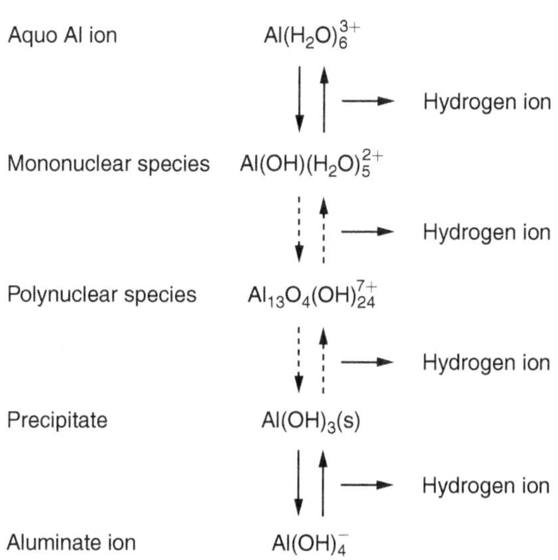

Figure 9-10
Aluminum hydrolysis products. The dashed lines are used to denote an unknown sequence of reactions. (Adapted from Letterman, 1981)

Polynuclear species such as $Al_{18}(OH)_{20}{}^{4+}$ form via hydroxyl bridges. For example, a hydroxyl bridge for two aluminum atoms is shown below:

$$2\left[Al(H_2O)_5(OH)\right]^{2+} \rightleftarrows \left[(H_2O)_4Al \begin{array}{c} OH \\ \diagup \quad \diagdown \\ \diagdown \quad \diagup \\ OH \end{array} Al(H_2O)_4\right]^{4+} + 2H_2O \quad (9\text{-}8)$$

It should be noted that all of these mononuclear and polynuclear species can interact with the particles in water, depending on the characteristics of the water and the number of particles. Unfortunately, it is difficult to control and know which mononuclear and polynuclear species are operative. As will be discussed later, this uncertainty gave rise to the development of prehydrolyzed metal salt coagulants.

SOLUBILITY OF METAL SALTS

The solubility of the various alum [Al(III)] and iron [Fe(III)] species are illustrated on Figs. 9-11a and 9-11b, respectively, in which the log molar concentrations have been plotted versus pH. The equilibrium diagrams shown on Figs. 9-11a and 9-11b were created using equilibrium constants for the major hydrolysis reactions that have been estimated after approximately 1 h of reaction time (upper limit of coagulation/flocculation detention times). Accordingly, the composition of aluminum and iron species in contact with the freshly precipitated hydroxide (amorphous) is illustrated on Figs. 9-11a and 9-11b. In preparing these diagrams, only the mononuclear species for alum

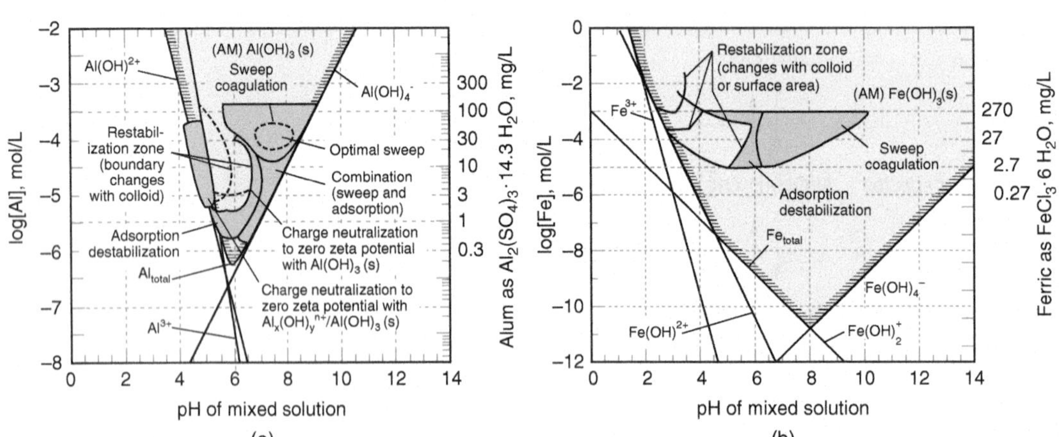

Figure 9-11
Solubility diagram for (a) Al(III) and (b) Fe(III) at 25°C. Only the mononuclear species have been plotted. The metal species are assumed to be in equilibrium with the amorphous precipitated solid phase. Typical operating ranges for coagulants: (a) alum and (b) iron. (Adapted from Amirtharajah and Mills, 1982)

Table 9-5
Reactions and associated equilibrium constants for aluminum and iron species in equilibrium with amorphous aluminum hydroxide and ferric hydroxide

Reaction	Acid Equilibrium Constants		
	Equilibrium Constant	Range, log *K*	Used for Fig. 9-11
Aluminum, Al(III)			
$Al(OH)_3(s) + 3H^+ \rightarrow Al^{3+} + 3H_2O$	K_{s0}	9.0–10.8	10.8
$Al(OH)_3(s) + 2H^+ \rightarrow Al(OH)^{2+} + 2H_2O$	K_{s1}	4.0–5.8	5.8
$Al(OH)_3(s) + H^+ \rightarrow Al(OH)_2{}^+ + H_2O$	K_{s2}	0.7–1.5	0.7
$Al(OH)_3(s) \rightarrow Al(OH)_3^0$	K_{s3}	- 4.2 to -6.1	- 6.1
$Al(OH)_3(s) + H_2O \rightarrow Al(OH)_4{}^- + H^+$	K_{s4}	- 7.7 to -12.5	- 11.9
Species not considered: $Al_2(OH)_2{}^{4+}$, $Al_8(OH)_{20}{}^{4+}$, $Al_{13}O_4(OH)_{24}{}^{7+}$, $Al_{14}(OH)_{32}{}^{10+}$			
Iron, Fe(III)			
$Fe(OH)_3(s) + 3H^+ \rightarrow Fe^{3+} + 3H_2O$	K_{s0}	3.2–4.891	3.2
$Fe(OH)_3(s) + 2H^+ \rightarrow Fe(OH)^{2+} + 2H_2O$	K_{s1}	0.91–2.701	1.0
$Fe(OH)_3(s) + H^+ \rightarrow Fe(OH)_2{}^+ + H_2O$	K_{s2}	- 0.779 to -2.5	- 2.5
$Fe(OH)_3(s) \rightarrow Fe(OH)_3{}^0$	K_{s3}	- 8.709 to -12.0	- 12.0
$Fe(OH)_3(s) + H_2O$ ® $Fe(OH)_4{}^- + H^+$	K_{s4}	- 16.709 to -19	- 18.4
Species not considered: $Fe_2(OH)_2{}^{4+}$, $Fe_3(OH)_4{}^{5+}$			

Source: Benefield et al. (1982), McMurry and Fay (2003), Morel and Hering (1993), Nordstrom and May (1989a, b), Pankow (1991), Snoeyink and Jenkins (1980), Sawyer et al. (2002), and Stumm and Morgan (1981).

and iron have been plotted. The various mononuclear species for alum and iron are given in Table 9-5, along with the corresponding range of acid solubility products reported in the literature and the values used to prepare Figs. 9-11a and 9-11b. The approximate total concentration of residual soluble alum (see Fig. 9-11a) and iron (see Fig. 9-11b) in solution after precipitation is identified by the solid line. Aluminum hydroxide and ferric hydroxide are precipitated within the shaded areas, and polynuclear and polymeric species are formed outside of the shaded areas at higher and lower pH values. It should also be noted that the structure of the precipitated iron is far more compact and inert as compared to the amorphous nature of precipitated aluminum.

In most water treatment applications for removal of turbidity, disinfection by-product precursors (NOM), and color, the pH during coagulation ranges between 6 and 8. The lower limit is imposed to prevent accelerated corrosion rates that occur at pH values below pH 6. The regions shown on Figs. 9-11a and 9-11b correspond to the operating pH and dosage ranges that are normally used in water treatment when alum and iron are used in the sweep floc mode of operation. The operating region for aluminum hydroxide precipitation is in a pH range of 5.5 to about 7.7, with minimum solubility occurring at a pH of about 6.2 at 25°C, and from about 5 to 8.5 for iron precipitation, with minimum solubility occurring at a pH of 8.0. The importance of pH in controlling the concentration of soluble metal species that will pass through the treatment process is illustrated on Figs. 9-11a and 9-11b. The effect of temperature on the solubility products for aluminum is also illustrated on Fig. 9-11a. As shown, the point of minimum solubility for alum shifts with temperature, which has a significant impact on the operation of water treatment plants where alum is used as the coagulant. Comparing the solubility of alum and ferric species, the ferric species are more insoluble than aluminum species and are also insoluble over a wider pH range. Thus, ferric ion is often the coagulant of choice to aid destabilization in the lime-softening process, which is carried out at higher pH values (pH 9).

STOICHIOMETRY OF METAL ION COAGULANTS

The overall stoichiometric reactions for aluminum and ferric ion in the formation of hydroxide precipitates are given by Eqs. 9-9 and 9-10. As shown, each mole of trivalent ion will produce 1 mole of the metal hydroxide and 3 moles of hydrogen ions:

$$Al^{3+} + 3H_2O \rightleftarrows Al(OH)_{3,\,am}\downarrow + 3H^+ \tag{9-9}$$

$$Fe^{3+} + 3H_2O \rightleftarrows Fe(OH)_{3,\,am}\downarrow + 3H^+ \tag{9-10}$$

The "am" subscripts in Eqs. 9-9 and 9-10 refer to amorphous solids (hours old), which have a much higher solubility product than crystalline precipitates.

When alum is added to water and aluminum hydroxide precipitates, the overall reaction is

$$Al_2(SO_4)_3 \cdot 14H_2O \rightarrow 2Al(OH)_{3\downarrow} + 6H^+ + 3SO_4^{\,2-} + 8H_2O \tag{9-11}$$

Similarly, the overall reactions for ferric chloride and ferric sulfate are as follows:

Ferric chloride:

$$FeCl_3 + 3H_2O \rightarrow Fe(OH)_{3\downarrow} + 3H^+ + 3Cl^- \tag{9-12}$$

Ferric sulfate:

$$Fe_2(SO_4)_3 \cdot 9H_2O \rightarrow 2Fe(OH)_{3\downarrow} + 6H^+ + 3SO_4^{\,2-} + 3H_2O \tag{9-13}$$

After $Al(OH)_3$ or $Fe(OH)_3$ precipitates, the species remaining in water are the same as if H_2SO_4 or HCl had been added to the water. Thus, adding alum or ferric is like adding a strong acid. A strong acid will lower the pH and consume alkalinity. Alkalinity is the acid-neutralizing capacity of water and is consumed on an equivalent basis; that is, 1 meq/L of alum or ferric will consume 1 meq/L of alkalinity. Since alkalinity buffers water against changes in pH, the change in pH following coagulant addition depends on the initial alkalinity. If the natural alkalinity of the water is not sufficient to buffer the pH, it may be necessary to add alkalinity to the water to keep the pH from dropping too low. Alkalinity can be added in the form of caustic soda (NaOH), lime [$Ca(OH)_2$], or soda ash (Na_2CO_3). In many water plants, caustic soda is often used because it is easy to handle and the required dosage is relatively small. The reaction for alum with caustic soda is

$$Al_2(SO_4)_3 \cdot 14H_2O + 6NaOH \rightarrow 2Al(OH)_{3,am}\downarrow + 3Na_2SO_4 + 14H_2O \quad (9\text{-}14)$$

The corresponding reaction for lime is given by the expression

$$Al_2(SO_4)_3 \cdot 14H_2O + 3Ca(OH)_2 \rightarrow 2Al(OH)_{3,am}\downarrow + 3CaSO_4 + 14H_2O \quad (9\text{-}15)$$

Coagulants are typically purchased in a concentrated liquid form. Calculating coagulant doses can be confusing because the stock chemical concentration is often reported in percent by weight and the density of the stock solution will be significantly heavier than water. In addition, the extent of hydration of the alum or ferric will vary or be unknown in the stock solution, which affects the formula weight of the chemical. To get around this issue, chemical manufacturers will sometimes report the concentration of the coagulant as a different formula entirely, for example, stock alum concentration is often reported as percent as Al_2O_3, even though the chemical present is $Al_2(SO_4)_3 \cdot xH_2O$. Ferric chloride may be reported with or without waters of hydration (i.e., $FeCl_3 \cdot 6H_2O$ or $FeCl_3$) or as soluble iron (Fe^{3+}). To calculate doses accurately, the density and chemical formula used by the chemical manufacturer to report the concentration must be known. The application of these principles and the above equations is illustrated in Example 9-2. Note that the sludge produced during coagulation consists of both the precipitate formed in the reactions shown above and the solids that were present in the source water. Example 21-2 in Chap. 21 demonstrates the procedure for calculating the amount of sludge produced considering both components.

Stantec Water Webinar-Coagulation 101

http://www.stantec.com/wt3r/9-1.html

Example 9-2 Calculation of coagulant doses, alkalinity consumption, and precipitate formation

A chemical supplier reports the concentration of stock alum chemical as 8.37 percent as Al_2O_3 with a specific gravity of 1.32. For the stock chemical, calculate (a) the molarity of Al^{3+} and (b) the alum concentration if reported as g/L $Al_2(SO_4)_3 \cdot 14H_2O$. Also, for a 30-mg/L alum dose applied to a treatment plant with a capacity of 43,200 m^3/d (0.5 m^3/s), calculate (c) the chemical feed rate in L/min, (d) the alkalinity consumed (expressed as mg/L as $CaCO_3$), (e) the amount of precipitate produced in mg/L and kg/day, and (f) the amount of NaOH that would need to be added to counteract the consumption of alkalinity by alum.

Solution

1. Calculate the formula weights (FW) for Al_2O_3, $Al_2(SO_4)_3 \cdot 14H_2O$, $Al(OH)_3$, and NaOH, given molecular weights: Al = 27, O = 16, H = 1, S = 32, and Na = 23 g/mol.

   FW: $Al_2O_3 = 2(27) + 3(16) = 102$ g/mol

   FW: $Al_2(SO_4)_3{\cdot}14H_2O = 2(27) + 3(32) + 26(16) + 28(1) = 594$ g/mol

   FW: $Al(OH)_3 = 27 + 3(16) + 3(1) = 78$ g/mol

   FW: $NaOH = 23 + 16 + 1 = 40$ g/mol

2. Calculate the molar concentration of Al^{3+} in the stock alum chemical.
   a. Calculate the density of stock chemical:

   $$\rho_{stock} = 1.32\,(1\text{ kg/L}) = 1.32\text{ kg/L}$$

   b. Calculate the concentration of alum in the stock chemical as mg/L Al_2O_3:

   $$C_{stock} = 0.0837\,(1.32\text{ kg/L})\,(10^3\text{ g/kg}) = 110.5\text{ g/L }Al_2O_3$$

   c. Calculate the molar concentration of Al^{3+} in the stock alum chemical:

   $$\left[Al^{3+}\right] = 110.5\text{ g/L}\,Al_2O_3\left(\frac{\text{mol }Al_2O_3}{102\text{ g }Al_2O_3}\right)\left(\frac{2\text{ mol }Al^{3+}}{\text{mol }Al_2O_3}\right) = 2.17\text{ mol/L}$$

3. Calculate the stock alum concentration if reported as g/L $Al_2(SO_4)_3 \cdot 14H_2O$.

   $$C_{stock} = 110.5\text{ g/L }Al_2O_3\left(\frac{594\text{ g/mol alum}}{102\text{ g/mol }Al_2O_3}\right) = 643.5\text{ g/L alum}$$

4. Calculate the chemical feed rate.
   By mass balance:

   $$C_{stock}\,Q_{feed} = C_{process}\,Q_{process}$$

$$Q_{feed} = \frac{C_{process}Q_{process}}{C_{stock}} = \frac{(30\text{ mg/L})\left(43{,}200\text{m}^3\text{/d}\right)\left(10^3\text{ L/m}^3\right)}{643.5\text{ g/L}\left(10^3\text{ mg/g}\right)(1440\text{ min/d})} = 1.40\text{ L/min}$$

5. Calculate the alkalinity consumed using Eq. 9-11:

$$\text{Alk} = [30\ \text{mg/L alum}]\left(\frac{1\ \text{mmol alum}}{594\ \text{mg alum}}\right)\left(\frac{3\ \text{mmol}\ SO_4^{\,2-}}{\text{mmol alum}}\right)\left(\frac{2\ \text{meq}\ SO_4^{\,2-}}{\text{mmol}\ SO_4^{\,2-}}\right)$$

$$\times\left(\frac{1\ \text{meq alk}}{\text{meq}\ SO_4^{\,2-}}\right)\left(\frac{50\ \text{mg}\ CaCO_3}{\text{meq alk}}\right) = 15\ \text{mg/L as}\ CaCO_3$$

6. Calculate the precipitate formed using Eq. 9-11:

$$\left[Al(OH)_3\right] = [30\ \text{mg/L alum}]\left(\frac{1\ \text{mmol alum}}{594\ \text{mg alum}}\right)\left[\frac{2\ \text{mmol}\ Al(OH)_3}{\text{mmol alum}}\right]\left[\frac{78\ \text{mg}\ Al(OH)_3}{\text{mmol}\ Al(OH)_3}\right]$$

$$= 7.88\ \text{mg / L}\ Al(OH)_3$$

$$\left[Al(OH)_3\right] = \frac{(7.88\ \text{mg/L})\left(43{,}200\ \text{m}^3/\text{d}\right)\left(10^3\ \text{L/m}^3\right)}{\left(10^6\ \text{mg/kg}\right)} = 340\ \text{kg/d}$$

7. Calculate the NaOH dose required to counteract the alkalinity consumption using Eq. 9-14:

$$[NaOH] = [30\ \text{mg/L alum}]\left(\frac{1\ \text{mmol alum}}{594\ \text{mg alum}}\right)\left(\frac{6\ \text{mmol NaOH}}{\text{mmol alum}}\right)\left(\frac{40\ \text{mg NaOH}}{\text{mmol NaOH}}\right)$$

$$= 12.1\ \text{mg/L NaOH}$$

Comment

The sludge produced by coagulation has two components the precipitate formed by the reactions shown above and the particles from the raw water. Calculation of the total amount of sludge produced during coagulation considering both components is illustrated in Example 21-2.

APPLICATION OF METAL SALTS IN WATER TREATMENT

Because of the sequence of reactions that follow the addition of alum or iron salts, as discussed above and illustrated on Fig. 9-10, it is not possible to predict a priori the performance of a coagulation process. Consequently, jar testing is typically used for coagulant/coagulant aid screening, and these results must be evaluated in the full-scale operation. Nevertheless, it is useful to review some general aspects of coagulation practice, including (1) the operating regions for the alum and iron, (2) interactions with other constituents in water, (3) typical dosages, and (4) the importance of initial blending when using metal salts. As noted in Chap. 6, blending is a mixing process to combine two liquid streams to achieve a specified level of uniformity. Guidance on the use of coagulants is provided in Table 9-6. Additional effects of NOM on the coagulation process are considered in Sec. 9-5.

Table 9-6
Application guidance for Al(III) and Fe(III) as coagulants and prehydrolyzed metal coagulants used in water treatment

Water Quality Parameter	Coagulant		
	Alum(III)	**Fe(III)**	**PACl**
Turbidity	For low-turbidity waters (i.e., low particle concentration), sweep floc will be required.	For low-turbidity waters (i.e., low particle concentration), sweep floc will be required.	For low-turbidity waters (i.e., low particle concentration), sweep floc will be required. Medium-basicity PACls (40 – 50%) are suitable for cold waters with low turbidity.
Alkalinity	High alkalinity values make pH adjustment for optimum coagulation more difficult. If sufficient alkalinity is not present, soluble aluminum is formed, which can result in postflocculation in downstream processes. Supplemental alkalinity should be added before coagulant.	Although high alkalinity values make pH adjustment for optimum coagulation more difficult, its impact on coagulation using Fe is less than Al.	
pH	The optimum pH range is between 5.5 and 7.7 but will fluctuate seasonally (see Fig. 9-11). Typically, the optimum pH will be nearer 6 in the summer and 7 in the colder winter months. Higher pH levels often correspond to periods of algal growth, which in turn will affect the coagulant dose.	The optimum pH range is from 5 to 8.5 or more (see Fig. 9-11).	PACls are less sensitive to pH. Can be used over the pH range of 4.5–9.5.

NOM	The removal of NOM will normally control the coagulant dose. Removal of NOM tends to increase as pH is reduced. Removal of up to 70% has been achieved.	The removal of NOM will normally control the coagulant dose. Removal of NOM tends to increase as pH is reduced. Removal of up to 80% has been achieved.	The removal of NOM will normally control the coagulant dose. Removal of NOM tends to increase as the pH is reduced. Removals of up to 70% have been achieved. Low-basicity PACls (up to 20%) are suitable for waters high in color and total organic carbon.
Temperature	Temperature affects solubility products. Floc formed in colder water tends to be weaker.	Floc formed in colder water tends to be weaker.	
Mixing	Hydrolysis reactions are very fast. Mixing times should be less than 1 s and preferably less than 0.5 s.	Hydrolysis reactions are very fast. Mixing times should be less than 1 s and preferably less than 0.5 s.	Because the PACl is prehydrolyzed, the initial blending time is somewhat less critical.

OPERATING REGIONS FOR METAL SALTS

Because the chemistry of the various reactions discussed above is so complex, there is no complete theory to explain the action of hydrolyzed metal ions. To qualitatively describe the application of alum as a function of pH, taking into account the action of alum as described above, Amirtharajah and Mills (1982) developed the diagrams shown on Fig. 9-11. It is important to note that the generalized information represented on Fig. 9-11 does not reflect the effects of NOM on the dosages of coagulant required. The approximate regions in which the different phenomena associated with particle removal in conventional sedimentation and filtration applications are plotted as a function of the alum dose and the pH of the treated effluent after alum has been added. For example, optimum particle removal by sweep floc occurs in the pH range of 7 to 8 with an alum dose of 20 to 60 mg/L. With proper pH control it is possible to operate with extremely low alum dosages.

Interactions with other constituents in water

As with all cations in water, hydrolysis products of aluminum and iron react with various ligands (e.g., SO_4^{2-}, NOM, F^-, PO_4^{3-}) forming both soluble and insoluble products that will influence the quantity or dose of the coagulant required to achieve a desired level of particle destabilization. Thus, the optimum dose of a coagulant depends strongly on the particular water chemistry and the types of particles.

Typical dosages

A typical dosage of alum ranges from 10 to 150 mg/L, depending on raw-water quality and turbidity. Typical dosages of ferric sulfate $[Fe_2(SO_4)_3 \cdot 9H_2O]$ and ferric chloride $(FeCl_3 \cdot 6H_2O)$ range from 10 to 250 mg/L and 5 to 150 mg/L, respectively, depending on raw-water quality and turbidity. Ferric chloride is more commonly used than ferric sulfate and comes as a liquid.

Importance of initial mixing with metal salts

The rapid initial mixing (known as blending) of the metal salts in water treatment is extremely important. The sequence of reactions shown on Fig. 9-10 occurs rather rapidly (Rubin and Kovac, 1974). For example, at a pH of 4, half of the Al^{3+} hydrolyzes to $Al(OH)^{2+}$ within 10^{-5} s (Base and Mesmer, 1976). Hudson and Wolfner (1967) noted that "coagulants hydrolyze and begin to polymerize in a fraction of second after being added to water." Hahn and Stumm (1968), studying the coagulation of silica dispersions with Al(III), reported that the time required for the formation of mono- and polynuclear hydroxide species was on the order of 10^{-3} s, and the time of formation for the polymer species was on the order of 10^{-2} s. The importance of initial and rapid mixing is also discussed by Amirtharajah and Mills (1982) and Vrale and Jorden (1971).

Clearly, based on the literature and actual field evaluations, the instantaneous rapid and intense mixing of metal salts is of critical importance, especially where the metal salts are to be used as coagulants to lower the surface charge of the colloidal particles. It should be noted that, although achieving extremely low blending times in large treatment plants is often difficult, low blending times can be achieved by using multiple mixers. Typical blending times for various chemicals are reported in Table 6-10 in Sec. 6-10, where the subject of mixing is considered in detail.

Prehydrolyzed Metal Salts

From the previous discussion of the use of alum and iron salts, it is clear that it is difficult to control the metal species formed, especially at low dosages. The unpredictability associated with alum and iron salts led to the development of prehydrolyzed metal salts. Prehydrolyzed metal salts are prepared by reacting alum or ferric with various salts (e.g., chloride, sulfate) and water and hydroxide under controlled mixing conditions. Several advantages of preformed aluminum metal salts include the following: (1) lower dosages may be required for effective coagulation (on the basis of Al^{3+}) for cases where NOM does not dictate the coagulant dosage at neutral or slightly acidic conditions, (2) flocs tend to be tougher and denser (although flocculation aids are still necessary in many cases), and (3) the performance of prehydrolyzed alum salts is less temperature dependent as compared to unmodified alum salts. General guidance on the application of prehydrolyzed metal salts is given in Table 9-6.

CHEMICAL COMPOSITION

The commercial prehydrolyzed alum salts, commonly known as PACl, have the following overall formula: $Al_a(OH)_b(Cl)_c(SO_4)_d$. Although many formulations do not contain any sulfate; the presence of sulfate ions helps to stabilize the aluminum polymers and keep them from precipitating. These polymers can be more effective than those formed by simply adding aluminum salts to solution because the larger cationic polymers can be formed by increasing the hydroxide-to-aluminum ratio ($R = OH/Al$, see following basicity discussion), which can lead to enhanced charge neutralization. Another benefit is that, as the polymer becomes larger, it becomes more crystalline, compact, and dense. However, as the value of R increases, the polymers become less stable and may begin to precipitate, which can cause a problem in the storage of PACl.

BASICITY

As given by Eqs. 9-9 and 9-10, when metal salts such as alum and iron hydrolyze, hydrogen ions are released, which will react with the alkalinity of the water. In the formulation of PACl coagulants, some of the acid that would have been released is neutralized with base (OH^-) when the coagulant is manufactured. The degree to which the hydrogen ions that would be released by hydrolysis are preneutralized is known as the basicity of the product and is given by the following relationship for prehydrolyzed metal salts that do not contain oxygen:

$$\text{Basicity}, \% = \frac{[\text{OH}]}{[\text{M}]\, Z_{\text{M}}} \times 100 \tag{9-16}$$

where $[OH]/[M]$ = molar ratio of hydroxide bound to metal ion
Z_M = charge on metal species

For example, the basicity of the PACl $Al_2(OH)_4Cl_2$ is 66.7 percent {[4/(3 × 2)] × 100}. It should be noted that, if oxygen is included in the formulation, the basicity of the compound will increase by the effect of the oxygen. For example, the basicity for the compound $Al_{13}O_4\ (OH)_{24}$ is 82.1 percent {[24 + (4 × 2)]/(13 × 3)] × 100}. In effect, each mole of oxygen will neutralize 2 moles of hydrogen. Most prehydrolyzed alum products have an OH/Al ratio of 0.45 to 2.5, which corresponds to basicity values of 15 [(0.45/3) × 100] and 83.3 [(2.5/3) × 100] percent.

Organic Polymers

Organic polymers are long-chain molecules consisting of repeating chemical units with a structure designed to provide distinctive physicochemical properties. The chemical units usually have an ionic functional group that imparts an electrical charge to the polymer chain. Hence, organic polymers are often termed *polyelectrolytes*. Organic polymers have two principal uses in water treatment: (1) as a coagulant for the destabilization of particles and (2) as a filter aid to promote the formation of larger and more shear-resistant flocs. While destabilization occurs primarily through charge neutralization, nonionic and anionic polymers can be used to form a bridge between particles. Organic polymers are not generally used as primary coagulants and are often used after the particles have been destabilized to some degree with metal coagulants. Polymers are broadly classified as being natural or synthetic in origin. Because of their greater use in water treatment, the synthetic polymers are discussed first.

SYNTHETIC ORGANIC POLYMERS

Generally, synthetic organic polymers are much cheaper than those made from natural sources and consequently are used more often in the United States than natural organic polymers. The principal synthetic organic polymers used for water treatment are summarized in Table 9-7. Synthetic organic polymers are made either by homopolymerization of the monomer or by copolymerization of two monomers. Polymer synthesis can be manipulated to produce polymers of varying size (molecular weight), charge groups, number of charge groups per polymer chain (charge density), and varying structure (linear or branched). A typical example is the production of polyacrylamide in which the monomer, acrylamide, homopolymerizes under appropriate conditions to form the polymer. Polyacrylamide carries no ionic charge and is referred to as a *nonionic* polymer. Subsequent

Table 9-7
Typical organic coagulants used in water treatment

Type	Charge	Molecular Weight, g/mole	Common Applications	Typical Examples[a]	Other Examples
Anionic	Negative	$10^4 - 10^7$	Coagulant aid, filter aid, flocculant aid, sludge conditioning	Hydrolyzed polyacrylamides 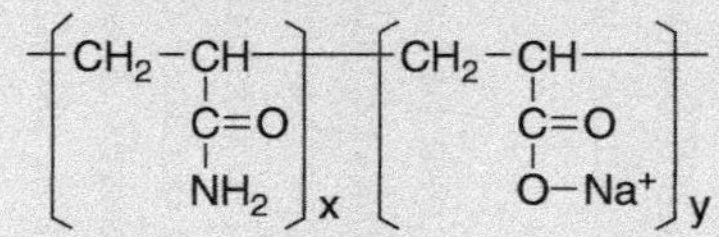	Hydrolyzed polyacrylamides, polyacrylates, polyacrylic acid, polystyrene sulfonate
Cationic	Positive	$10^4 - 10^6$	Primary coagulant, turbidity and color removal	Epichlorohydrin dimethylamine (epi-DMA) 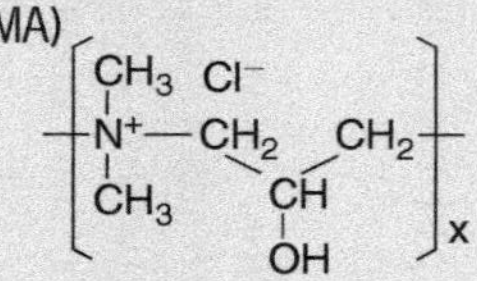	Aminomethyl polyacrylamide, polyalkylene, polyamines, Polyethylenimine
			Sludge conditioning	Polydiallyldimethyl ammonium chloride (poly-DADMAC) 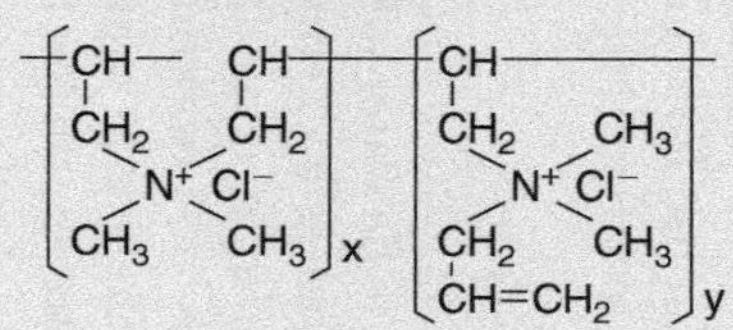	Polydimethyl aminomethyl polyacrylamide, polyvinylbenzyl, trimethyl ammonium chloride
Nonionic	Neutral	$10^5 - 10^7$	Coagulant aid, filter aid, filter conditioning	Polyacrylamides 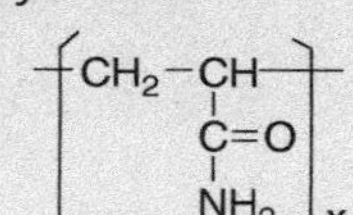	Polyacrylamides, polyethylene oxide
Others	Variable	Variable	—	Sodium alginate	Alginic acid, dextran, guar gum, starch derivatives

[a]Number of monomer molecules in polymer designated by *x* and *y*.

hydrolysis of polyacrylamide under basic conditions produces a polymer with *anionic* charges. Thus, the number of anionic groups, in this case a carboxyl group, can be controlled, providing anionic polymers of different molecular weights and charge density. The third type of polymer has a *cationic* or positive charge group incorporated in the polymer chain, usually by a copolymerization process.

APPLICATION OF POLYMERS

Since their introduction in the early 1950s, the use of organic polyelectrolytes such as poly diallyl-dimethyl ammonium chloride (poly-DADMAC) and epichorohydrin dimethylamine (epi-DMA) (see Table 9-7) has gained widespread use for water treatment in the United States. The MW ranges from 10^4 to 10^5 and the basic polymer units are shown in Table 9-7.

Cationic polymers

In water treatment applications, cationic organic polymers are generally designed to be water soluble, to adsorb on or react rapidly with particulates, and to possess a chemical structure suitable for the intended use. When used as primary coagulants, cationic polymers, in contrast to aluminum or ferric ions, do not produce large floc volumes because organic coagulants can be effective at much lower coagulant dosages than inorganic coagulants. However, sludge from organic coagulants is usually denser and stickier than sludge from inorganic coagulants. Consequently, cationic organic coagulants are not suitable for every type of separation process.

It should be noted that, because organic coagulants do not always produce the same water quality as is obtained with metallic ion coagulants, cationic organic polymers are rarely used alone except for direct filtration. Furthermore, if cationic organic polymers are used alone, they are ineffective in removing dissolved substances (e.g., NOM, As, F). It is common to use cationic organic polymers and metallic ion coagulants together. The main advantage of the combined usage is that the dosage of metallic ion coagulants can be reduced by 40 to 80 percent. The lower metallic ion coagulant dosage in turn reduces sludge and alkalinity consumption. With lower alkalinity consumption, the pH will not be depressed as much, which can improve metallic ion coagulation.

Polymer dosages

Because of the complex interactions between polymers and particulates and the uncertain influence of water quality on these interactions, polymer selection is empirical. The typical dosage rates for sedimentation are on the order of 1 to 10 mg/L for DADMAC and epi-DMA. Low dosages of high-molecular-weight nonionic polymers (0.005 to 0.05 mg/L) are often added before granular filtration and to the backwash water to improve filter performance. Incorrect dosing can cause mudball formation in the filters, which are not usually broken apart during normal backwashing operations.

Impact of solution parameters

Solution parameters will also impact polymer dose. If the polymer charge density depends on pH, as with nonquaternized polyamines (see Table 9-7), then the optimum polymer dose will vary with pH, generally decreasing as the pH decreases [the charge on secondary and tertiary amines depends on pH because the amine group will tend to protonate at lower pH values (less than 6) and will remain uncharged at neutral pH]. The charge density of quaternized polymers such as poly-DADMAC are only slightly affected by pH. Changes in ionic strength and composition do not appear to affect polymer dose strongly over typical ranges encountered in water treatment (TDS between 50 and 500 mg/L).

MIXING OF POLYMERS

Most polymers are available in liquid form and can be used without a preparation stage, but they must be injected directly following in-line dilution. Successful use of polymers in water treatment requires adequate dispersion of the polymer to promote more uniform polymer adsorption. Jar testing, as described in the previous section, may be used to assess the effect of mixing.

NATURAL POLYMERS

Sodium alginate is a natural organic polymer extracted from brown seaweed. The polymeric structure of sodium alginate is comprised of mannuronic acid and glucuronic acid and the molecular weight is on the order of 10^4 to 2×10^5 (Degrémont, 2007). Sodium alginate is particularly effective as a flocculant aid with ferric salts, and good results have also been obtained with aluminum salts, with typical dosages ranging from 0.5 and 2 mg/L. Chitosan, another natural organic, is obtained from chitin shells (crab, lobster, etc.). Natural starches are also classified as natural polymers. Starches can be obtained from a number of sources, including potatoes, tapioca, or plant seed extracts. Starches are branched and nonlinear glucopyranose polymers, which are sometimes partially broken down with OH^- or derivatized to form carboxy-ethyl-dextrose. Starches are used in concentrations of 1 to 10 mg/L, preferably together with aluminum salts.

Coagulant and Flocculant Aids

A variety of chemicals and additives known as *coagulant aids* and *flocculant aids,* used to enhance coagulation and flocculation processes, are described below. Some of the commonly used inorganic coagulant aids are given in Table 9-4. Polymers used as flocculant aids are given in Table 9-7.

COAGULANT AIDS

Coagulant aids, typically insoluble particulate materials, are added to enhance the coagulation process. Clay (bentonite, kaolinite), sodium silicate, pure precipitated calcium carbonate, diatomite, powdered activated carbon (used as an adsorbent), and fine sand have all been used as coagulation aids. Coagulant aids are often added to waters that contain low concentrations of particles to form nucleating sites for the formation of larger flocs. Coagulant aids are used in conjunction with inorganic coagulants, organic polyelectolytes, or both. Because the density of these particles is higher than that of most floc particles, the settling velocities of flocculated particles is increased.

FLOCCULANT AIDS

Uncharged and negatively charged organic polymers that were discussed in the previous section are used as flocculant aids as opposed to primary coagulants. As previously discussed, the main advantage of using flocculant aids is that a stronger floc is formed. Flocculant aids are added after the coagulants are added and the particles are already destabilized. The time required for destabilization depends on water temperature and the type of particles; consequently, jar tests have to be conducted. The important factors that need to be evaluated in jars and full-scale implementation are floc strength, size, and settling rate. It should be noted that improper dosing of flocculant aids can cause mudballs to form in gravity filters that are not easily eliminated by backwashing.

Activated silica is an important inorganic flocculant aid that is used in combination with alum and can be effective in cold water. It is usually stored as sodium silicate, which is soluble at high pH. Usually, the concentrated sodium silicate is partially neutralized (usually with sulfuric acid) prior to use and then added immediately to the water. In some instances, aluminosilicate is used where alum is the primary coagulant because the acidity of the alum can be used to activate the silica, and this will produce aluminum hydroxide floc and the SiO_2 flocculant aid. A typical dosage is from 0.5 to 4 mg/L as SiO_2.

Jar Testing for Coagulant Evaluation

Because of the many competing reactions and mechanisms that are operative in the coagulation process, the selection of coagulants and dosage is typically determined empirically using bench-scale and pilot-scale studies. The standard bench-scale testing procedure for determining coagulant doses and types is the "jar test" procedure. Developed originally by Langelier (1921) and refined over the years (see Black et al., 1957; Tekippe and Ham, 1970), jar testing permits rapid evaluation of a range of coagulant types and doses. A modern jar test apparatus is shown on Fig. 9-12. As shown on Fig. 9-12, the apparatus consists of six batch reactors, each equipped with a paddle mixer. Square-shaped jars are used to avoid vortex flow, which can occur if circular beakers are used.

JAR TEST PROCEDURE

The purpose of the jar test is to simulate, to the extent possible, the expected or desired conditions in the coagulation–flocculation facilities.

Figure 9-12
Jar test apparatus. Note use of square containers to limit the formation of vortex flow in which the particles rotate in the same position relative to each other.

Standard jar testing procedures are available in Kawamura (2000), ASTM (2008), and AWWA (2011). Generally, the test consists of a rapid-mix phase (blending) with simple batch addition of the coagulant or coagulants followed by a slow-mix period to simulate flocculation. Flocs are allowed to settle and samples are taken from the supernatant. These parameters should be measured as part of the jar test routine: (1) turbidity or suspended solids removal; (2) NOM removal as measured by dissolved organic carbon (DOC) or a surrogate measure of dissolved NOM, such as UV at 254 nm; (3) residual dissolved coagulant concentrations of Fe or Al coagulants; and (4) sludge volume that is produced. If direct filtration is to be used, the filterability should be evaluated using a filterability test. The filterability is evaluated by filtering the mixed suspension through a 5- or 8-μm laboratory filter to simulate a granular medium filter.

Benchtop testing establish water quality parameters required for coagulation and flocculation system design.

http://www.stantec.com/wt3r/9-2.html

The results of a series of jar tests to determine the optimum alum dose and pH for turbidity removal for given water are summarized on Fig. 9-13. As shown on Fig. 9-13, the optimum alum dose and pH would be approximately 8 mg/L and 7, respectively, because the turbidity is minimized under these conditions. However, it must be emphasized that the raw-water particle concentration and NOM vary with water quality, and thus the optimum coagulant dosage also changes as the water quality changes.

ANALYSIS OF COAGULATION PROCESS USING JAR TEST PROCEDURE

A conceptual diagram in which the residual turbidity from jar tests conducted for waters with different particle concentrations is illustrated on Fig. 9-14. The diagram on Fig. 9-14 applies to a limited pH range for the two coagulants: for alum, it is pH ~5.5 and for Fe it is pH ~5. If the pH is

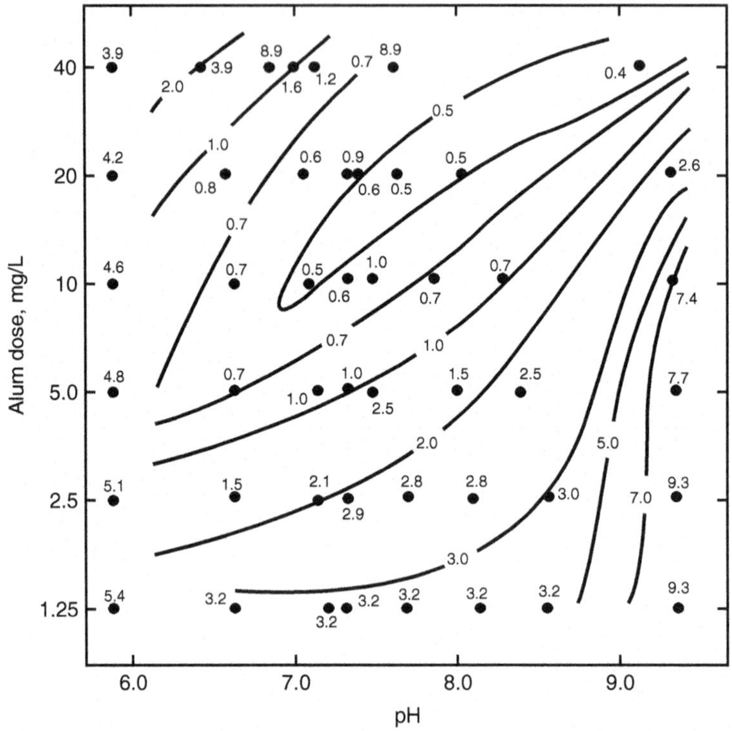

Figure 9-13
Turbidity topogram as function of pH and alum dosage. (Adapted from Trussell, 1978.) Points shown on the plot represent turbidity values and the isopleths represent constant turbidity at the value denoted on the isopleth.

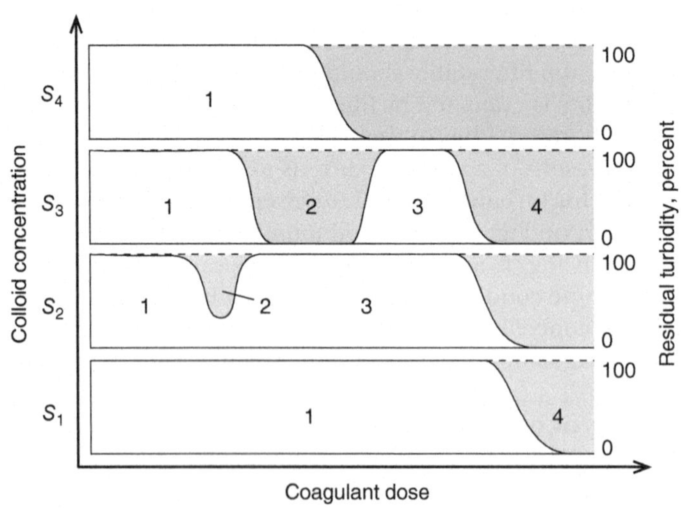

Figure 9-14
Simulated residual turbidity remaining after jar tests as a function of particle and coagulant concentration for Al(III) or Fe(III) salts at constant pH. (Adapted from O'Melia, 1972.)

higher, then no cationic metal ion species would be formed for charge neutralization. Furthermore, the impact of NOM on the coagulant dose has not been considered.

Water with low concentration of particles
For the lowest particle concentration, S_1, particles are removed by sweep flocculation (precipitation) because the concentration is too low for them to flocculate and settle. These particles are also destabilized by the hydrolysis products of the coagulant. The particle concentration S_1 may be a good candidate for direct filtration or dissolved air flotation for liquid–solid separation.

Water with moderate concentration of particles
At a slightly higher particle concentration, S_2, some flocculation and settling occurs in zone 2, in which adsorption and charge neutralization occur. However, the concentration is too low for effective flocculation and a high degree of turbidity removal is not possible. Further, if more coagulant is added, the particles stabilize with a positive charge and the turbidity increases, as shown in zone 3. However, as the coagulant dosage is increased even further, sweep flocculation is again observed but at a lower coagulant concentration than for a particle concentration S_1 because the particle concentration is higher, which allows for more effective flocculation and settling at a lower coagulant dosage.

Water with high concentration of particles
At a higher particle concentration, S_3, all four zones can clearly be found. Nearly complete removal by charge neutralization occurs in zone 2 and the particles stabilize at higher coagulant dosages. Sweep flocculation occurs in zone 4, and it occurs at lower coagulant dosages than for particle concentration S_2 or S_1. Sweep flocculation occurs because a substantial portion of the floc volume comes from the particles, and this allows for effective flocculation at lower coagulant dosages.

Water with very high concentration of particles
For the highest particle concentration, S_4, the sweep floc and charge neutralization regions merge because the coagulant concentration required to neutralize the charges on the particles coincides with the onset of precipitation. Another noteworthy effect is that zone 2 starts at successively higher coagulant dosages as the particle concentration increases. As a result, the charge neutralization region is said to exhibit stoichiometry.

Analysis of results
By examining these results, it can be said that the addition of clays, such as bentonite, may be an effective coagulant aid for particle concentration S_2 or

S_1, if sedimentation is used for liquid–solid separation, because they would increase the floc volume, reduce the coagulant dosage, and increase the settling velocity. Jar testing and cost analysis would have to be conducted to examine the economic benefits that may result from such a strategy.

Alternative Techniques to Reduce Coagulant Dose

In addition to the use of cationic polymers, several techniques have been evaluated to reduce the coagulant dose, including (1) preozonation and (2) the use of selective ion exchange resins, as discussed below.

COAGULANT REDUCTION THROUGH USE OF OZONATION

It has been reported that preozonation can be used to improve NOM removal in some instances; however, the results have proven to be quite variable.

For high DOC concentrations, the optimum coagulant dosage is dictated by the required DOC removal, and ozone dosages in the range of 0.1 to 2.0 mg/L do not appear to improve DOC or turbidity removal. In fact, ozonation at higher dosages (i.e., >2 mg/L) can be deleterious to DOC and turbidity removal for high DOC concentrations when using alum coagulation. It is likely that ozonation leads to the formation of more polar and reactive functional groups (e.g., carboxylic acid groups) in the DOC, and these react with aluminum hydroxide surfaces, increasing the coagulant demand.

At low DOC concentrations, there is some evidence that preozonation can reduce the required alum dosage. In this case, preozonation appears to affect adsorption of the DOC onto the particles in a beneficial way without increasing the reactive functional groups that in turn increase the coagulant demand. Although preozonation may be beneficial for low- DOC waters (especially when using direct filtration), it is likely that these situations will rarely occur, and it is better to coagulate and remove NOM before ozonating because this will reduce the ozone dose used for other purposes, such as disinfection.

The impact of ozonation on coagulant dosage using organic polymers is also quite variable. At high DOC levels and low to moderate turbidity, low dosages of ozone appear to improve DOC or turbidity removal and lower the coagulant dosage. Unfortunately, the DOC reduction decreases as ozone dose increases. It is likely that ozonation degrades the DOC into smaller polar compounds that cannot interact with most organic polymers and cannot be removed. Also, there are fewer large DOC molecules remaining to interact with the organic polymer. Because of the variable effects that have been reported, if preozonation is to be used, bench-scale and/or pilot plant testing will be required.

Additional information on ozonation, including equipment for ozone generation and ozone contactors, is presented in Chap. 13.

COAGULANT REDUCTION THROUGH USE OF ION EXCHANGE RESINS

Another approach that has been developed to reduce the coagulant dose involves the use of an ion exchange resin to remove the DOC (i.e., NOM) before the coagulant is added. The principal resin used for DOC removal, known as the MIEX DOC resin, was developed in Australia for use in water treatment. The specially developed resin beads, about 180 μm in diameter, contain a magnetized component within their structure such that the resin beads act as weak individual magnets. Thus, in a sedimentation tank the magnetized resin beads readily aggregate and settle rapidly. The MIEX process is described in more detail in Chap. 16.

9-5 Coagulation of Dissolved Constituents

While the original objective of coagulation was to remove suspended particles from water, it can also be useful in removing natural organic matter and some dissolved inorganic constituents. This section discusses the impact of natural organic matter on the coagulation process and the removal of dissolved constituents by coagulation, including the process known as enhanced coagulation.

Effects of NOM on Coagulation for Turbidity Removal

Natural organic matter (NOM), as described in Chap. 2, is the term used to describe the complex matrix of organics originating from natural sources that are present in all water bodies. Hydrophilic in nature, the constituents that comprise NOM (e.g., low-molecular-weight acids, amino acids, proteins and polysaccharides, fulvic and humic acids) have a wide range of molecular weights. In the literature, the concentration of NOM in water has been measured as total organic carbon (TOC), DOC, and UV_{254} absorbance. Dissolved organic carbon is the fraction of TOC remaining in solution after filtering the water through a 0.45-μm filter. The particulate fraction of NOM is easily removed from water following coagulation because particulate NOM is destabilized in the same way that inorganic particles are destabilized. The dissolved fraction of NOM, however, also interacts with coagulants and can complicate efforts to determine the correct coagulant dose for turbidity removal.

It has been observed that dissolved NOM reacts or binds with metal ion coagulants, and some evidence suggests that the coagulant dosages at many, if not most, operating plants are determined by the dissolved NOM–metal ion interactions and not particle–metal ion interactions (O'Melia et al., 1999). No quantitative relationships about coagulant dosages for turbidity removal have been developed because solution conditions that affect dosage and effectiveness of coagulants—such as pH, hardness, and temperature—also affect the speciation of NOM. However, qualitatively, as pH increases, NOM becomes more ionized because the carboxyl groups lose protons, and the positive charge on metal coagulants will decrease.

Consequently, higher coagulant dosages will be required at higher pH values. At neutral pH, the amount of positively charged coagulant (Al or Fe) species decreases with increasing temperature and a higher coagulant dosage may be required.

Enhanced Coagulation

As discussed in Chap. 19, disinfection by-products (DBPs) are formed as a result of chemical reactions between chlorine and NOM. While trihalomethanes (THMs) and haloacetic acids (HAAs) are the primary DBPs that form during chlorination, the DBP regulations in the United States recognize that MCLs for specific DBPs may not address the total risk associated with adding chlorine to water containing NOM. Consequently, the regulations include a treatment technique that requires the removal of NOM prior to disinfection under certain conditions. The process of performing coagulation for the purpose of achieving specified removal of DBP precursors (NOM) is known as *enhanced coagulation*. The treatment technique uses a TOC removal requirement because TOC is a practical measure for the amount of NOM in water. The TOC removal requirements range from 15 to 50 percent removal depending on the raw water TOC and alkalinity at the specific site. Utilities can meet the treatment technique without practicing enhanced coagulation by meeting one of several alternate compliance criteria, which depend on factors such as raw or finished water TOC concentrations, specific UV absorbance (SUVA) values, disinfectant usage, and other factors. Specific requirements associated with enhanced coagulation are described in the Stage 1 D/DBP Rule (U.S. EPA, 1998) and the Enhanced Coagulation Guidance Manual (U.S. EPA, 1999).

Coagulation tends to preferentially remove the higher-MW, more hydrophobic fractions of NOM (White, et al. 1997). Fortunately, the portion of NOM preferentially removed by enhanced coagulation tends to correspond to the fraction that preferentially forms DBPs; the hydrophobic fraction of NOM typically forms more DBPs than the hydrophilic fraction (Kavanaugh, 1978).

The dose required to achieve enhanced coagulation is typically higher than the dose for turbidity removal. Typical results from flocculation and sedimentation jar tests are shown on Fig. 9-15. As shown on Fig. 9-15, when turbidity and DOC removals are plotted as a function of coagulant dose, the DOC coagulant demand and the required degree of DOC removal for enhanced coagulation, not turbidity, will usually dictate the coagulant dosage. Of the metal salts and prehydrolyzed metal salts, the most effective for the removal of NOM is typically iron, followed by alum and PACl (see Table 9-6).

The previous section noted that the solubility of coagulants is dependent on pH; the minimum solubility of alum precipitate is around a pH of 6.3 at 25° C. As a result, the optimum NOM removal with alum is at a pH ranging from 5.5 to 6.5, depending on the water temperature and total dissolved solids (TDS) concentration. Removal of NOM with alum can also occur at

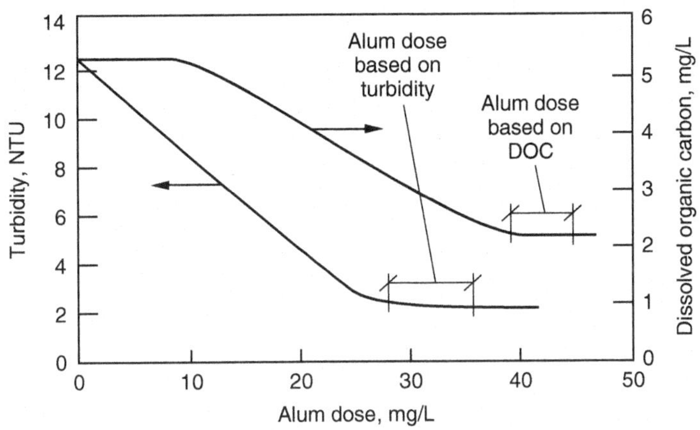

Figure 9-15
Residual turbidity and dissolved organic matter as function of alum dose.

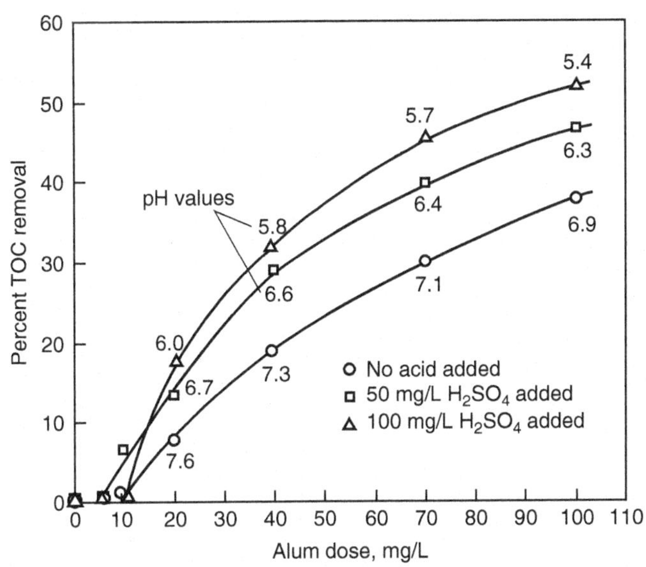

Figure 9-16
Removal of NOM with alum in bench-scale jar tests (data from of the Elsinore Valley Municipal Water District). (Untreated water quality: Temperature 20° C, TOC 9 mg/L, alkalinity 160 mg/L as $CaCO_3$, turbidity 3.8 NTU.)

higher pH values, but higher alum doses are required to meet the same NOM removal that can be achieved at optimum pH. In instances of high-pH conditions at the point of coagulation, acid addition to lower the pH can help improve NOM removal.

The impact of pH on NOM removal is shown on Fig. 9-16. In this study, three scenarios of NOM removal with alum from a natural water sample were investigated in bench-scale jar tests. The three scenarios were (1) without preaddition of sulfuric acid, (2) with preaddition of 50 mg/L sulfuric acid, and (3) with preaddition of 100 mg/L sulfuric acid. Without acid addition to this water, an alum dose of about 90 mg/L was required to achieve 35 percent reduction in the TOC concentration (resulting in a settled-water pH of about 7.0). With the addition of 50 mg/L sulfuric acid, the alum dose required to achieve the same TOC reduction was about 60 mg/L (with a settled-water pH of about 6.5), a 33 percent reduction in coagulant usage.

Using pH adjustment for NOM removal has a number of consequences that need to be considered before pH adjustment is adopted by a water treatment plant. These consequences include:

- A lower alum dose, which will reduce the amount of sludge produced at the plant.
- A lower settled-water pH, which will require a substantially higher dose of an alkaline chemical (such as lime or sodium hydroxide) to raise the pH of the finished water to acceptable levels (in the range of 8 to 8.5).
- The high doses of acid and caustic will increase the TDS concentration in the finished water.
- Costs associated with pH adjustment.

One potential problem associated with the use of a high alum dosage, sulfuric acid, and the resulting low pH is the residual aluminum ion in the settled water. Accordingly, the residual aluminum ion must be monitored to ensure that it does not exceed the MCL.

Determination of Coagulant Dose for DOC Removal

Edwards (1997) developed an empirical model to estimate DOC removal during enhanced coagulation. The model was based on 21 water sources coagulated with ferric salts (250 jar tests) and 39 water sources coagulated with alum (608 jar tests). The model assumes that DOC is composed of absorbable and nonabsorbable fractions. A portion of the adsorbable DOC adsorbs to the floc and the rest remains in solution after coagulation in accordance with an adsorption isotherm. Accordingly, the DOC remaining in solution after coagulation is the sum of the nonadsorbable DOC and the adsorbable DOC that is not adsorbed and is given by the expression

$$\mathrm{DOC}_f = \mathrm{DOC}_{\mathrm{na}} + \mathrm{DOC}_{a,f} \tag{9-17}$$

where DOC_f = final DOC concentration, mg/L
$\mathrm{DOC}_{\mathrm{na}}$ = nonadsorbable DOC concentration, mg/L
$\mathrm{DOC}_{a,f}$ = adsorbable DOC remaining in solution after coagulation, mg/L

Edwards found a linear relationship between the nonadsorbable fraction of DOC and the SUVA of the influent water, which is UV_{254} absorbance of the water divided by the DOC:

$$(\mathrm{SUVA})_i = (100)\left(\frac{\mathrm{UV}_{254,i}}{\mathrm{DOC}_i}\right) \tag{9-18}$$

where $(\text{SUVA})_i$ = specific UV absorbance of influent water, L/mg·m
DOC_i = influent DOC concentration, mg/L
$\text{UV}_{254,\text{i}}$ = influent UV_{254} absorbance, cm^{-1}

The nonadsorbable DOC is determined from the expression

$$\text{DOC}_{\text{na}} = \text{DOC}_i \times [K_1 \, (\text{SUVA})_i + K_2] \tag{9-19}$$

where K_1, K_2 = empirical constants from Table 9-8

The equilibrium between the amounts of adsorbable DOC that adsorb and remain in solution is described by a Langmuir isotherm:

$$q = \frac{Q_M b\left(\text{DOC}_{a,f}\right)}{1 + b\left(\text{DOC}_{a,f}\right)} \tag{9-20}$$

$$Q_M = x_3 (pH)^3 + x_2 (pH)^2 + x_1 (pH) \tag{9-21}$$

where q = DOC adsorbed at equilibrium, mg DOC/mmol of Al^{3+} or Fe^{3+} added (the adsorbent is the floc that forms after the coagulant is added)
Q_M = total adsorbent capacity at monolayer coverage, mg DOC/mmol of Al^{3+} or Fe^{3+} added
x_1, x_2, x_3 = empirical constants from Table 9-8, unitless
pH = coagulation pH
b = Langmuir equilibrium constant from Table 9-8, L/mg DOC

A mass balance is used to relate the fate of each portion of the initial DOC (i.e., at equilibrium, the initial DOC is divided between the nonadsorbable

Table 9-8
Summary of best-fit model coefficients for DOC removal with iron and aluminum

	DOC Model Coefficients	
Parameter	**Iron**	**Aluminum**
Standard error, mg/L	0.47	0.4
Standard error, %	9.3	9.5
90% confidence, %	± 21	± 21
x_3	4.96	4.91
x_2	−73.9	−74.2
x_1	280	284
K_1, mg · m/L	−0.028	−0.075
K_2	0.23	0.56
b, L/mg	0.068	0.147

Source: Adapted from Edwards (1997).

fraction, the adsorbable fraction that is not adsorbed, and the adsorbable fraction that is adsorbed):

$$\text{DOC}_i = \text{DOC}_{\text{na}} + \text{DOC}_{a,f} + q(M) \tag{9-22}$$

where $M = Al^{3+}$ or Fe^{3+} added as coagulant, mmol/L (mM)

Substituting Eqs. 9-19 and 9-20 into Eq. 9-22 and rearranging algebraically reveals a quadratic equation that can be solved for the $\text{DOC}_{a,f}$ concentration:

$$(\text{DOC}_{a,f})^2 + B\,(\text{DOC}_{a,f}) + C = 0 \tag{9-23}$$

where B and C are defined as

$$B = Q_M M + \frac{1}{b} - \text{DOC}_i + \text{DOC}_{\text{na}} \tag{9-24}$$

$$C = \frac{\text{DOC}_{\text{na}} - \text{DOC}_i}{b} \tag{9-25}$$

The adsorbable DOC remaining in solution is then found as the positive root (because concentration cannot be negative) of the quadratic equation

$$\text{DOC}_{a,f} = \frac{-B + \sqrt{B^2 - 4C}}{2} \tag{9-26}$$

The total DOC remaining after coagulation is calculated with Eq. 9-17. The DOC removal model can be used to plan jar tests for a water supply and assess how much DOC may be removed using coagulation. The trihalomethane formation potential (THMFP) reductions, which typically have to be determined from testing, can be estimated from the DOC remaining in solution. The model parameters and associated statistics shown in Table 9-8 were determined for iron and aluminum coagulation. The accuracy of the model can be improved by calibrating it to a specific site by determining actual nonadsorbable DOC or other parameters (Edwards, 1997).

This model is appropriate for preliminary evaluation to determine the proper coagulant dose and pH for enhanced coagulation. Jar testing can be used to provide more site-specific information prior to design, and more detailed investigation (i.e., pilot testing) may also be appropriate depending on the size of the facility and the NOM removal requirements. The use of this model to determine DOC remaining after enhanced coagulation is demonstrated in Example 9-3.

Example 9-3 Removal of DOC by enhanced coagulation

Predict the DOC removal using an alum dose of 30 mg/L for the following conditions: initial DOC = 4.0 mg/L, initial UV_{254} absorbance = 0.1 cm^{-1}, and pH 7.

Solution

1. Calculate the molar concentration of Al^{3+} added with 30 mg/L alum [$Al_2(SO_4)_3 \cdot 14H_2O$, formula weight 594 g/mol; see also Example 9-2].

$$\left[Al^{3+}\right] = 30\ \text{mg/L alum}\left(\frac{2\ \text{mmol Al}^{3+}}{594\ \text{mg alum}}\right) = 0.10\ \text{mmol/L Al}^{3+}$$

2. Calculate the specific UV absorbance using Eq. 9-18.

$$(\text{SUVA})_i = \left(10^2\ \text{cm/m}\right)\left(\frac{0.1\ \text{cm}^{-1}}{4.0\ \text{mg/L}}\right) = 2.5\ \text{L/mg}\cdot\text{m}$$

3. Calculate the nonadsorbable DOC using Eq. 9-19 and model parameters from Table 9-8.

$$\text{DOC}_{na} = 4.0\ \text{mg/L}\ [-0.075\ (2.5\ \text{L/mg}\cdot\text{m}) + 0.56] = 1.49\ \text{mg/L}$$

4. Calculate the total adsorbent capacity using Eq. 9-21 and model parameters from Table 9-8.

$$Q_M = 4.91\ (7)^3 - 74.2\ (7)^2 + 284\ (7) = 36.33\ \text{mg DOC/mmol Al}^{3+}$$

5. Calculate the quadratic coefficients B and C using Eqs. 9-24 and 9-25.

$$B = (36.33\ \text{mg/mmol})\ (0.1\ \text{mmol/L}) + \frac{1}{0.147\ \text{L/mg}} - 4.0\ \text{mg/L}$$
$$+ 1.49\ \text{mg/L} = 7.926\ \text{mg/L}$$

$$C = \frac{1.49\ \text{mg/L} - 4.0\ \text{mg/L}}{0.147\ \text{L/mg}} = -17.07\ \text{mg}^2/\text{L}^2$$

6. Calculate the adsorbable DOC remaining in solution using Eq. 9-26.

$$\text{DOC}_{a,f} = \frac{-7.926\ \text{mg/L} + \sqrt{(7.926\ \text{mg/L})^2 - 4\left(-17.07\ \text{mg}^2/\text{L}^2\right)}}{2}$$
$$= 1.76\ \text{mg/L}$$

7. Calculate the total DOC remaining in solution using Eq. 9-17.

$$\text{DOC}_f = 1.49 + 1.76 = 3.25\ \text{mg/L}$$

8. Calculate the DOC removal and percent removal.

$$\text{DOC removal} = \text{DOC}_i - \text{DOC}_f = 4.0 - 3.25 = 0.75 \text{ mg/L}$$

$$\text{Percent DOC removal} = \frac{0.75 \text{ mg/L}}{4.0 \text{ mg/L}} \times 100 = 19\%$$

Removal of Dissolved Inorganics

The coagulation process can sometimes effectively be used to remove dissolved constituents such as arsenic, lead, iron, manganese, and uranium (see Table 20-1). For example, coagulants such as alum, ferric chloride, and ferric sulfate have been used to remove arsenic. A detailed discussion of the coagulation process applied to the removal of selected dissolved constituents is discussed in Chap. 20.

9-6 Flocculation Theory

Flocculation theories have evolved from the following observations: (1) small particles undergo random Brownian motion due to collisions with fluid molecules resulting in particle–particle collisions (Smoluchowski, 1917) and (2) stirring water containing particles creates velocity gradients that bring about particle collisions (Langelier, 1921). An understanding of the theory of flocculation may be used to provide insight into process design and operation, which are discussed in the next section. For example, the theory can be used to assess the importance of mixing and what particle sizes would flocculate as a result of mixing. The three prevailing models used to describe the flocculation process are (1) spherical particles in a linear flow field, (2) spherical particles in a nonlinear flow field, and (3) fractal-based models. The main differences between the flocculation models are the rate and manner in which particles are predicted to grow in size as a result of flocculation.

Mechanisms of Flocculation

The action of flocculation depends on the characteristics of the particles as well as the fluid-mixing conditions. The addition of a coagulant to water containing small particles causes the particulates to become destabilized and begin flocculating. The mechanisms of particle flocculation are described below. A schematic of the processes controlling the rate of particulate aggregation during coagulation and flocculation is shown on Fig. 9-17.

MICROSCALE (PERIKINETIC) FLOCCULATION

The rate of flocculation of small particles is relative to the rate at which particles diffuse toward one another (Smoluchowski, 1917). Thus, for small particles (less than 0.1 μm), the primary mechanism of aggregation

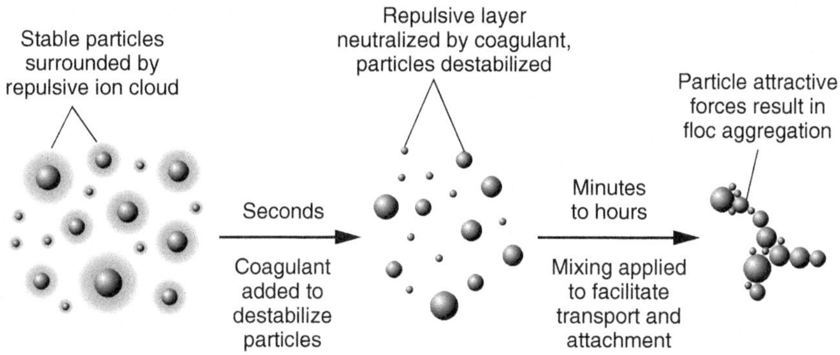

Figure 9-17 Schematic illustrating the progression of the particle aggregation process.

is Brownian motion, or microscale flocculation. Microscale flocculation is also known as perikinetic flocculation. As aggregation of small particles proceeds, larger particles are formed. After a short time (seconds), microflocs ranging in size from 1 to about 100 μm are formed (Argaman and Kaufman, 1970).

MACROSCALE (ORTHOKINETIC) FLOCCULATION

The major mechanism for flocculation in water treatment for particles larger than about 1 μm is the gentle mixing of the water, typically with mechanical mixers. The mixing causes velocity gradients that causes collisions between suspended particles (Langelier, 1921), known as macroscale flocculation or orthokinetic flocculation. During the mixing processes in macroscale flocculation, however, the particle flocs are subjected to shear forces, resulting in erosion and disruption of some of the floc aggregates, or floc breakup. After some period of mixing, a steady-state floc size distribution is reached, and the formation and breakup of floc particles is roughly equal (Parker et al., 1972). The rate at which the steady-state size distribution is achieved, as well as the form of the size distribution, will depend upon the hydrodynamics of the system and the chemistry of the coagulant–particulate interactions.

DIFFERENTIAL SETTLING

Aggregation and growth of flocs can result from particles settling at different velocities. As large particles are formed, they begin to settle due to gravitational forces. The velocity of particles of similar densities settling in a water column is proportional to the square of their size. Differences in settling velocities cause particles with size and/or density differences to collide and flocculate. Thus, differential particle settling occurs in heterogeneous suspensions (differing particle sizes) during sedimentation, providing an additional mechanism for promoting flocculation. For suspensions containing a wide range of particle sizes, differential sedimentation can be a significant flocculation mechanism (O'Melia, 1978). Flocculation by differential settling will not have an impact on direct filtration, dissolved air flotation (DAF), or high-rate sedimentation processes such as inclined plates because settling distances or times are too short.

Particle Collisions

The fundamental problem in mathematical modeling of the flocculation process is predicting the change of the particle size distribution as a function of time for a given set of chemical and hydrodynamic conditions. Any general kinetic model must account for changes in the number of particles found in all size classes. Particles of size d_i collide with particles of size d_j, forming particles of size d_k when collisions are successful. At the same time, aggregates of size d_k may break up into smaller aggregates due to hydrodynamic shearing forces.

RATE OF PARTICLE COLLISION

The overall particle collision rate is a function of the rate of macroscale flocculation (r_M), rate of microscale flocculation (r_μ), and rate of differential settling flocculation (r_{DS}) between particles i and j.

The rate of particle attachments r_{ij} is a function of the particle concentrations and a collision frequency function β_{ij}:

$$r_{ij} = \alpha\beta_{ij} n_i n_j \tag{9-27}$$

where r_{ij} = rate of attachment between i and j particles
α = collision efficiency factor (attachments per collision)
β_{ij} = collision frequency function for particles of i and j size classes (rate constant for collisions between particles)
n_i = concentration of i particles
n_j = concentration of j particles

The collision efficiency factor α, defined as the ratio of collisions that result in attachment to total collisions, has a range of values between $0 \le \alpha \le 1$. The collision efficiency factor depends on the effectiveness of destabilization; for example, if particles have been destabilized completely, then $\alpha = 1$. Solution of mass balances on flocculation reactors that use Eq. 9-27 require the use of appropriate values of β to predict the change in the size distribution of the suspension as aggregation occurs (Lawler et al., 1980).

COLLISION FREQUENCY FUNCTION

The collision frequency function β_{ij} depends on the size of the particles, the flocculation transport mechanism, and the efficiency of particulate collisions. The overall collision frequency function is a function of the three individual mechanisms of flocculation as follows:

$$\beta_{ij} = \beta_M + \beta_\mu + \beta_{DS} \tag{9-28}$$

where β_{ij} = overall collision frequency between particles i and j
β_M = macroscale collision frequency, $= r_M/\alpha n_i n_j$

β_μ = microscale collision frequency, $= r_\mu/\alpha n_i n_j$
β_{DS} = differential settling collision frequency, $= r_{DS}/\alpha n_i n_j$
r_M = rate of attachment due to macroscale collisions
r_μ = rate of attachment due to microscale collisions
r_{DS} = rate of attachment due to differential setting

The development of the equations used to model the collision frequency factor for the various flocculation mechanisms is presented in the following discussion for spherical particles in a linear flow field.

OVERALL RATE OF PARTICLE COLLISION

The formation rate of aggregates in size class d_k is the sum of all collisions between i and j particles minus the subsequent disappearance of aggregates from the k size class due to collisions with other (e.g., i and j) particles. The general model for aggregation, assuming no particle breakup, is given as follows (Swift and Friedlander, 1964):

$$r_k = \frac{1}{2}\alpha \sum_{j=1;i+j=k}^{j=k-i} \beta_{ij} n_i n_j - n_k \alpha \sum_{i=1}^{N} \beta_{ik} n_i \tag{9-29}$$

where r_k = net formation rate of k-sized particles
n_k = concentration of k-sized particles
β_{ik} = collision frequency function for particles of size classes i and k
N = total number of n_i particles

Mathematical expressions of the collision functions are derived below by considering the various flocculation mechanisms. An empirical model for particle breakup is discussed after particle formation theories.

Flocculation of Spherical Particles

In the linear flow field model, it is assumed that particles agglomerate as spheres and that the total floc volume fraction does not change with time. An analogy is agglomeration of small drops of oil into larger droplets in which the total volume of oil does not change. Linear flow field models for flocculation of spherical particles by macroscale, microscale, and differential sedimentation mechanisms are discussed below.

MACROSCALE FLOCCULATION OF SPHERICAL PARTICLES IN LINEAR FLOW FIELD

Consider particles i and j with diameters d_i and d_j, respectively, suspended in and moving in fluid streamlines in the x direction with water subjected to a velocity gradient dv_x/dz, as shown on Fig. 9-18. When the distance between the centers of the particles, R_{ij}, becomes equal to $(d_i + d_j)/2$, a collision will occur.

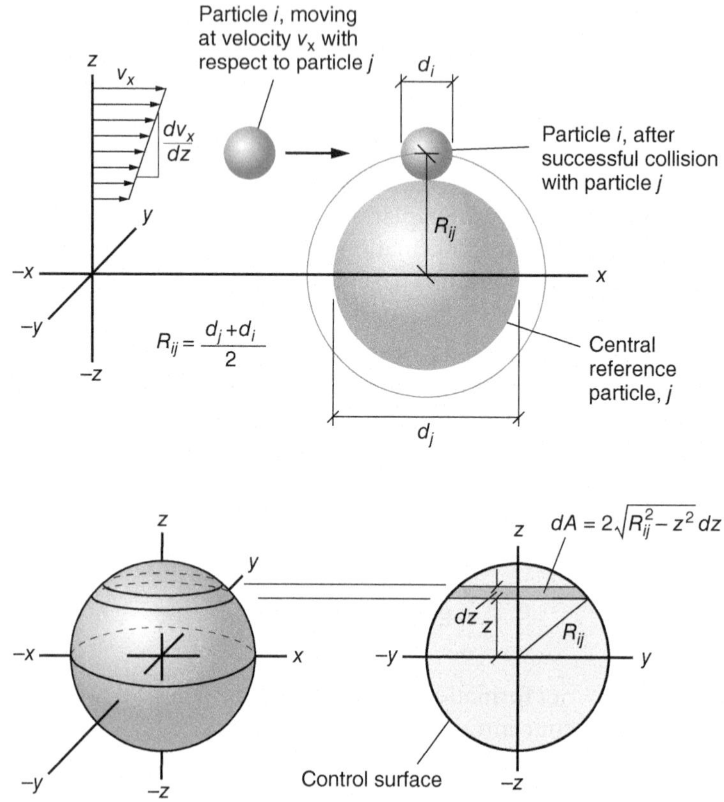

Figure 9-18
Definition sketch for analysis of the flocculation process.

Velocity gradients
When fluid flow is laminar and steady, the velocity gradients are well defined, as shown on Fig. 9-18. The velocity gradient on Fig. 9-18 is proportional to the shear stress on the fluid elements because it is a Newtonian fluid. Given a uniform velocity gradient, the rate of flocculation can be determined from geometric considerations, as illustrated below.

Rate of flocculation of heterodisperse particles
The rate of macroscale flocculation in a system of unequal size (heterodisperse) particles subjected to uniform mixing may be derived using the relationships shown on Fig. 9-18. The flow rate of fluid into an area element dA of the control surface is given by the following expression (Swift and Friedlander, 1964; Smoluchowski, 1917):

$$dq = (\text{ velocity})\ (\text{differential area}) = \left(z\frac{dv_x}{dz}\right)\left(2\sqrt{R_{ij}^2 - z^2}\,dz\right) \quad (9\text{-}30)$$

where dq = differential flow of fluid through area element dA, m^3/s
q = fluid flow rate through particle area projected onto $y - z$ plane, m^3/s
z = vertical direction, m
dv_x/dz = velocity gradient in x direction, s^{-1}
R_{ij} = distance between centers of particles i and j, m

In a heterogeneous solution, the flow rate of particles through the control area may be expressed as the product of the i and j particle concentrations (n_i and n_j, respectively) and the differential flow of fluid through the control surface. Assuming that the velocity gradient is constant,

$$\text{Particle flow through control surface} = 2n_i n_j \int_{z=0}^{z=R_{ij}} dq = 4n_i n_j \left(\frac{dv_x}{dz}\right) \int_0^{R_{ij}} z\sqrt{R_{ij}^2 - z^2}\,dz \tag{9-31}$$

Recalling from calculus that

$$\int x\sqrt{a^2 - x^2}\,dx = -\tfrac{1}{3}\left(a^2 - x^2\right)^{3/2} + c,$$

the integrated form of Eq. 9-31 is given by the expression

$$\text{Particle flow} = 4n_i n_j \left(\frac{dv_x}{dz}\right)\left[-\tfrac{1}{3}\left(R_{ij}^2 - z^2\right)^{3/2}\right]_0^{R_{ij}} = \frac{4}{3}\left(\frac{dv_x}{dz}\right) R_{ij}^3 n_i n_j \tag{9-32}$$

The rate of flocculation is equal to the flow rate of particles times the collision efficiency α (i.e., fraction of collisions that result in attachment):

$$r_{ij} = \frac{4}{3}\left(\frac{dv_x}{dz}\right) R_{ij}^3 n_i n_j \alpha \tag{9-33}$$

where r_{ij} = rate of collision between i and j particles (rate of flocculation)

Substituting the term $(d_i + d_j)/2$ for R_{ij} (see Fig. 9-18) results in the following expression for the rate of flocculation, by macroscale mechanisms, between i- and j-sized particles:

$$r_M = \frac{1}{6}\left(\frac{dv_x}{dz}\right)\left(d_i + d_j\right)^3 n_i n_j \alpha \tag{9-34}$$

where r_M = rate of macroscale flocculation (sometimes referred to as J_m)

Under turbulent-flow conditions, the velocity gradient is not well defined and varies both in time and space throughout the flocculation basin. When averaged over the entire basin, the velocity gradient is known as the root-mean-square (RMS) velocity gradient and is given the symbol $\overline{G}$ (see Sec. 6-10 and Eqs. 6-139 and 6-140).

Thus, for unequal-sized (heterodisperse) particles the collision frequency function for the macroscale flocculation rate β_M can be computed using Eq. 9-34 and the relationship given in the nomenclature for Eq. 9-28, resulting in

$$\beta_M = \tfrac{1}{6}\overline{G}\left(d_i + d_j\right)^3 \tag{9-35}$$

where $\overline{G}$ = RMS velocity gradient, s^{-1}

Rate of flocculation of monodisperse particles

When the suspension is heterodispersed, with a wide size distribution, the rate of aggregation is increased (Swift and Friedlander, 1964). Thus, the kinetic model for monodispersed suspensions is relatively conservative. The monodisperse system, initially composed of only particles with size d_i at concentration n_i, may be considered as a special case where each particle can serve as a central particle. Therefore, the net rate of aggregation is obtained by substituting $n_i^2/2$ for $n_i n_j$ in Eq. 9-33, as given by Eq. 9-36. The n_i^2 term is divided by 2 to reflect the fact that one-half of the particles serve as central particles:

$$r_M = \frac{4}{3}\left(\frac{dv_x}{dz}\right)R_{ij}^3 n_i n_j \alpha = \frac{4}{3}\left(\frac{dv_x}{dz}\right)d_i^3\frac{n_i^2}{2}\alpha = \frac{2}{3}\overline{G}d_i^3 n_i^2 \alpha \tag{9-36}$$

If the particles agglomerate with no void volume (e.g., Euclidean objects, coalescing oil droplets), the floc volume Ω is time invariant because it only depends on the total number of particles initially present. Consequently, Eq. 9-36 is actually pseudo–first order (i.e., depends only on the number concentration to the first power) because the term $n_i d_i^3$ is related to Ω, a fixed quantity, as described below. The floc volume is given by

$$\Omega = \frac{\pi d_i^3 n_i}{6} \tag{9-37}$$

where Ω = floc per unit of solution volume, cm^3/L ($cm^3/10^3\ cm^3$)
d = particle diameter, cm
n_i = number concentration of particles, L^{-1}

Rearranging Eq. 9-37 for substitution into Eq. 9-36,

$$n_i d^3 = \frac{6\Omega}{\pi} \tag{9-38}$$

Substituting Eq. 9-38 into Eq. 9-36 results in the following equation for the rate of flocculation of a monodisperse system:

$$r_M = \frac{4\Omega \bar{G} \alpha n_i}{\pi} \tag{9-39}$$

Thus, the macroscale flocculation rate for a monodispersed suspension is a first-order rate expression with respect to n_i, and the rate constant is directly proportional to the velocity gradient and the floc volume fraction.

Rate correction for turbulent conditions

In practice, fluid flow in most flocculation units is turbulent. Under turbulent conditions, the velocity gradient is not well defined and will vary locally in the flocculation unit. As discussed in Sec. 6-10, Camp and Stein (1943) developed an expression relating $\bar{G}$ to mixing power (see Eq. 6-140) by equating the velocity gradients to the power dissipated per unit volume (P/V) for uniform shear flow:

$$\bar{G} = \sqrt{\frac{P}{\mu V}} \tag{9-40}$$

where $\bar{G}$ = RMS velocity gradient, s^{-1}
P = power of mixing input to flocculation basin, W (note 1 W = 1 $kg{\cdot}m^2/s^3$)
V = volume of flocculation basin, m^3
μ = dynamic viscosity of water, $kg/m \cdot s$

In turbulent flow, the rate of aggregation for particles smaller than the Kolmogorov eddy size (see Eq. 6-136) is approximately the same as it is for laminar flow because flow within eddies is laminar. Using this argument, the flocculation rate for turbulent shear should be similar to the rate for laminar flow. While this may be true in theory, it has been found that the rate is much higher due to interactions between eddies (Logan, 1999). Consequently, the rate becomes proportional to $\bar{G}$ (Harris et al., 1966), and an empirical rate constant must be employed, shown below for the monodisperse system:

$$r_{M,T} = K_A \bar{G} \Omega n_i \tag{9-41}$$

where $r_{M,T}$ = rate of flocculation for turbulent flow
K_A = empirical aggregation constant

The aggregation constant depends on system chemistry, the heterogeneity of the suspension, and variations in the scale and intensity of turbulence, which are not incorporated in the velocity gradient. Because of the different flow patterns and distributions of velocity gradients promoted by various mixing devices, K_A must be determined experimentally.

Example 9-4 Time needed for macroscale flocculation

Calculate the time required to reduce the number of particles by 50 percent under laminar conditions for macroscale flocculation, assuming first-order kinetics, for 10-μm particles. Assume the initial particle concentration is 10,000/mL, $\bar{G} = 60\ s^{-1}$, and $\alpha = 1.0$.

Solution

1. Determine the volume fraction of particles using Eq. 9-37. Note $10\ \mu m = 10^{-3}$ cm.

$$\Omega = \frac{\pi d_i^3 n_i}{6} = \frac{\pi \left(10^{-3}\ \text{cm}\right)^3 \left(10^4/\text{mL}\right)}{6} = 5.2 \times 10^{-6}$$

2. For first-order kinetics, Eq. 9-38 may be written as

$$\frac{dn_i}{dt} = -r_M = -kn_i \qquad \text{where} \quad k = \frac{4\Omega\bar{G}\alpha}{\pi}$$

3. Integrating the above expression yields

$$n_i = n_0\ e^{-kt}$$

where n_0 = initial particle concentration

4. Determine the time needed to achieve 50 percent particle reduction using the equation for half-life:

$$t_{1/2} = \frac{\ln\left(0.5n_0/n_0\right)}{k} = \frac{-\pi \ln(0.5)}{4\alpha\Omega\bar{G}}$$

$$= \frac{-\pi \ln(0.5)\ (1\ \text{min}/60\ \text{s})}{4\ (1)\left(5.2\times10^{-6}\right)(60/\text{s})} = 28.9\ \text{min}$$

MICROSCALE FLOCCULATION OF SPHERICAL PARTICLES

The flux of *j*-size particles to the surface of a single *i*-size particle by diffusion is given by the expression

$$J_A = -D_{lj}\left(\frac{\partial n_j}{\partial r}\right)_{r=d_i/2} = \frac{-2D_{lj}n_j}{d_i} \tag{9-42}$$

where J_A = flux of particles, m · number of particle/s.

D_{lj} = liquid-phase diffusion coefficient for particle j to particle i, m^2/s

Thus, the flocculation rate $r_{\mu,j}$ is given by the expression

$$r_{\mu,j} = \text{sphere surface area} \times \text{flux} = \left(\pi d_i^2\right)\left(\frac{2D_{lj}n_j}{d_i}\right) = 2\pi d_i D_{lj} n_j \quad (9\text{-}43)$$

Rate of microscale flocculation of heterodisperse particles

Substituting the Stokes–Einstein equation $D_{lj} = kT/3\pi\mu d_j$ (see Sec. 7-2, Eq. 7-25) into Eq. 9-43 and incorporating the collision efficiency factor α and the number of particles, n_i, an expression for the rate of flocculation, $r_{\mu,ji}$ of all j-size particles diffusing to the surface of all i-size particles can be obtained:

$$r_{\mu,ji} = 2\pi d_i D_{lj} n_j n_i \alpha = 2\pi d_i \left(\frac{kT}{3\pi\mu d_j}\right)\alpha n_i n_j = \frac{2}{3}\alpha\left(\frac{kT}{\mu}\right)\left(\frac{d_i}{d_j}\right) n_i n_j \quad (9\text{-}44)$$

where k = Boltzmann constant, 1.3807×10^{-23} J/K
T = absolute temperature, K (273 + °C)
μ = dynamic viscosity of water, N · s/m^2

Generalizing to all possible combinations of i and j to form a particle of size k, the overall rate of r_μ is given by

$$r_\mu = \underbrace{\frac{2}{3}\alpha\left(\frac{kT}{\mu}\right)\left(\frac{d_i}{d_j}\right)n_i n_j}_{\substack{j \text{ diffusing to } i \\ \text{(different sizes)}}} + \underbrace{\frac{2}{3}\alpha\left(\frac{kT}{\mu}\right)\left(\frac{d_i}{d_j}\right)n_i n_j}_{\substack{i \text{ diffusing to } j \\ \text{(different sizes)}}} + \underbrace{\frac{2}{3}\alpha\left(\frac{kT}{\mu}\right)\left(\frac{d_i}{d_j}\right)n_i\left(2n_j\right)}_{\substack{i,j \text{ diffusing toward} \\ \text{each other (equal size)}}} \quad (9\text{-}45)$$

Grouping terms and simplifying the rate expression in Eq. 9-45 result in the expression

$$r_\mu = \frac{2}{3}\alpha\left(\frac{kT}{\mu}\right)n_i n_j\left(\frac{1}{d_i} + \frac{1}{d_j}\right)\left(d_i + d_j\right) \quad (9\text{-}46)$$

The collision frequency function for microscale flocculation of heterodisperse particles can now be written as

$$\beta_\mu = \left(\frac{2kT}{3\mu}\right)\left(\frac{1}{d_i} + \frac{1}{d_j}\right)\left(d_i + d_j\right) \quad (9\text{-}47)$$

Brownian motion affects the movement of colloidal particles but has only a minor influence on transport of particles larger than about 1 μm (Smoluchowski, 1917).

Rate of microscale flocculation of monodisperse particles

The relationship shown in Eq. 9-47 can be simplified further for a system of uniform particle size. The collision frequency function for Brownian transport for a suspension of monodisperse particles is given by the expression

$$\beta_\mu = \frac{8}{3}\frac{kT}{\mu} \tag{9-48}$$

If considering the flocculation of only one size of particles, the first term in Eq. 9-29 represents the formation (+) of doublets and the second term represents the loss (–) of singlets. Combining Eqs. 9-29 and 9-48, the instantaneous loss of singlets due to Brownian or microscale flocculation is

$$r_u = -\frac{4}{3}\alpha\frac{kT}{\mu}n_i^2 \tag{9-49}$$

where units are as defined previously.

The second-order rate constant of $\frac{4}{3}\alpha(kT/\mu)$ is 5.4×10^{-12} L/s·particle at 20°C, assuming $\alpha = 1$. The term $\frac{4}{3}\alpha(kT/\mu)$ is the largest second-order rate constant for a chemical reaction because it describes the rate at which two molecules collide by molecular diffusion. Multiplying the term $\frac{4}{3}\alpha(kT/\mu)$ by Avogadro's number yields a second-order rate constant of 3.25×10^{12} L/s·mol. Accordingly, microscale flocculation can be a relatively fast process if the concentration of small particles (<0.1 μm) is high.

Example 9-5 Collosion Frequency Function for microscale flocculation

A suspension contains small colloids and 10-μm coagulant floc particles. Estimate the collosion frequency function for the transport of the colloids to the floc particles by microscale flocculation if the colloids are 0.01μm (the size of a virus). The water temperature is 15°C. Assume there is no floc breakup and $\alpha = 1.0$.

Solution

1. Determine the collosion frequency function for 0.01-μm particles using Eq. 9-47. Note that 1 J = 1 N·m: The viscosity of water from Table C-1 in App. C is 1.139×10^{-3} N · s/m²

$$\beta_\mu = \left[\frac{(2)\left(1.3805\times10^{-23}\,\text{N}\cdot\text{m/K}\right)(288\,\text{K})}{(3)\left(1.139\times10^{-3}\,\text{N}\cdot\text{s/m}^2\right)}\right]$$

$$\times\left(\frac{1}{10^{-5}\,\text{m}}+\frac{1}{10^{-8}\,\text{m}}\right)\left(10^{-5}\,\text{m}+10^{-8}\,\text{m}\right)$$

$$= 2.33\times10^{-15}\,\text{m}^3/\text{s}$$

FLOCCULATION OF SPHERICAL PARTICLES BY DIFFERENTIAL SETTLING

Differential particle settling is a special case of macroscale flocculation where collisions occur because each particle settles at a specific terminal setting velocity that depends on the square of the particle diameter for laminar conditions ($\mathrm{Re}_i < 1$):

$$v_{s,i} = \frac{(\rho_p - \rho_l) g d_i^2}{18\mu} \tag{9-50}$$

$$\mathrm{Re}_i = \frac{\rho_l v_{s,i} d_i}{\mu} \tag{9-51}$$

where $v_{s,i}$ = settling velocity for particle i
$\rho_p - \rho_l$ = difference in density between particle and fluid
g = acceleration due to gravity, 9.81 m/s^2
Re_i = Reynolds number for particle i

Differential flow of particles through given unit area

$$= \pi (R_{ij})^2 (v_{s,i} - v_{s,j}) = \pi \left(\frac{d_i}{2} + \frac{d_j}{2} \right)^2 (v_{s,i} - v_{s,j}) \tag{9-52}$$

where $v_{s,i} - v_{s,j}$ = velocity difference between particles i and j
R_{ij} = distance between centers of particles i and j

The velocity difference allows either particle i to overtake particle j or vice versa. As with the macroscale case, the flow rate of fluid into a unit area, defined by R_{ij} (see Fig 9-18), results in the following expression for the collision rate. The resulting final form of the rate of flocculation due to differential settling r_{DS} is given as

$$r_{\mathrm{DS}} = -\frac{\pi (\rho_p - \rho_l) g}{72\mu} (d_i + d_j)^2 (d_i^2 - d_j^2) n_i n_j \alpha \tag{9-53}$$

where r_{DS} = rate of flocculation due to differential setting

After simplification, the rate of flocculation by differential settling shown in Eq. 9-53 may be expressed as

$$r_{\mathrm{DS}} = -\frac{\pi (\rho_p - \rho_l) g}{72\mu} (d_i + d_j)^3 (d_i - d_j) n_i n_j \alpha \tag{9-54}$$

By inspection, the collision frequency function (Friedlander, 2000) is

$$\beta_{\mathrm{DS}} = \frac{\pi (\rho_p - \rho_l) g}{72\mu} \left[(d_i + d_j)^3 (d_i - d_j) \right] \tag{9-55}$$

Collision by differential sedimentation will not occur in a monodisperse system of particles of the same size and density because $d_i - d_j = 0$.

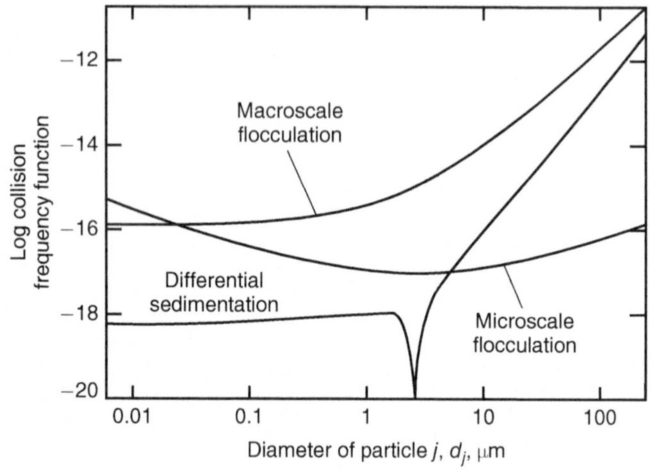

Figure 9-19
Collision frequency functions for macroscale (orthokinetic) flocculation, microscale (perikinetic) flocculation, and differential settling.

COMPARISON OF COLLISION FREQUENCY FUNCTIONS

The collision frequency functions for macroscale flocculation, microscale flocculation, and flocculation due to differential settling are given by Eqs. 9-35, 9-47, and 9-55, respectively. The collision frequency function may be plotted for a given system to assess the relative effect of each type of flocculation mechanism. A plot of the collision frequency functions is presented on Fig 9-19 for a system containing particles d_i of size 2.0 μm and particles d_j with sizes ranging from 0.01 to 50 μm. The curves shown on Fig. 9-19 are for a $\bar{G}$ value of 100 s^{-1}, water temperature of 15°C, and particle density of 1.1 g/cm^3. As shown on Fig 9-19, the dominant flocculation mechanism is microscale flocculation for particles smaller than about 0.035 μm and macroscale flocculation for larger particles at these conditions.

FLOCCULATION OF SPHERICAL PARTICLES IN NONLINEAR FLOW FIELD

The linear flow field model tends to overemphasize the importance of macroscale flocculation by not accounting for the collision efficiency associated with the hydrodynamics of particle–particle interactions and short-range attractive forces (van der Waals forces). Han and Lawler (1992) solved the appropriate equations for the nonlinear flow model, which more accurately considers the hydrodynamics and short-range forces. A further improvement in flocculation theory, in which flocculated particles are represented by fractal configurations that have a much larger size than is predicted by spherical particle models, is considered in the next section.

Fractal Flocculation Models

Spherical particles are considered in the linear and nonlinear flow field models described above. However, in these models the growth of floc particles is oversimplified and not considered properly because it is assumed that the floc volume does not change with extent of flocculation. Flocs do, however, form large amorphous flakes and the size and floc volume do increase with the degree of flocculation. As a result, the flocculation rate in practice is typically faster than would be estimated by the spherical particle models.

FRACTAL THEORY OF PARTICLE FORMATION

The growth rate of large particles from small particles depends on the shape and number of the small particles. For example, if large floc particles are spherical Euclidean objects, then the diameter of the larger particles, d_k, is proportional to the number of small particles raised to the third power, n_i^3. However, when rapid flocculation of small particles occurs, dendrites (or snowflake-like structures) form, and this forms particles that are much larger than would be predicted from the number of small particles using Euclidean geometry, as shown on Fig. 9-20. These snowflake-shaped particles are known as fractals.

FRACTAL PARTICLE SHAPE AND SIZE

In the spherical particle models, it is assumed that the particles are spherical and that spherical particles are formed as a result of flocculation. However, as shown on Fig. 9-21, many shapes other than spherical are formed and aggregate size for a given number of flocculated small particles varies. Fractals may be used to describe the size of a floc particle that is constructed of small particles. The smaller particles that form the floc particle are referred to as fractal generators. For a three-dimensional object, a sphere that contains closely packed particles with a diameter d_p would scale with aggregate diameter d_a as given by the equation

$$\text{Volume} \propto \left(\frac{d_a}{d_p}\right)^3 \propto n_p^3 \tag{9-56}$$

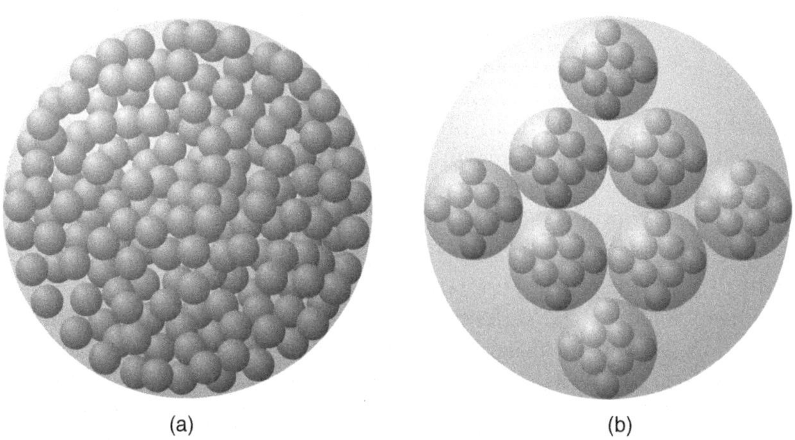

Figure 9-20
Particles made up of flocculated smaller particles: (a) Euclidean geometry and (b) fractal geometry.

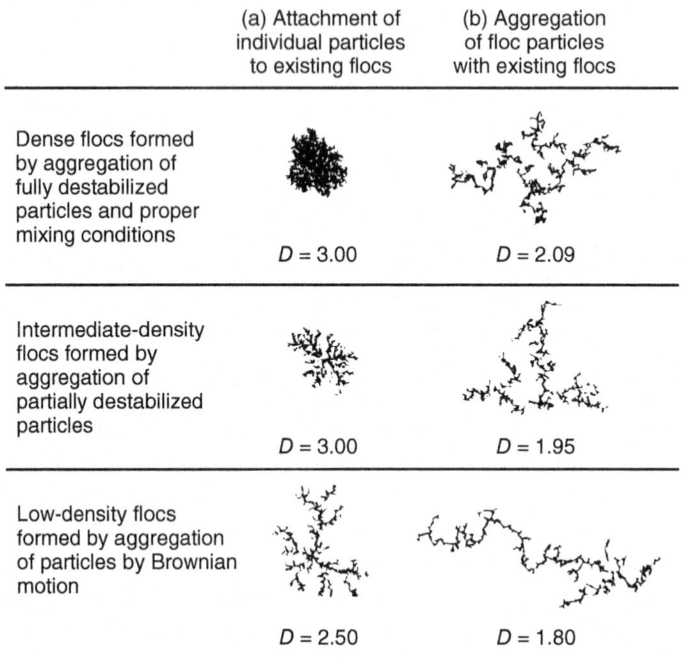

Figure 9-21
Aggregate shapes formed by flocculation: (a) that arise from adding monomers to preformed clusters (monomer particles are added after the formation of some clusters), which is called monomer–cluster, and (b) that arise from adding all the particles at one time, which is called cluster–cluster. The fractal dimension D is in three dimensions. Aggregates that are formed from reaction-limited (slow coagulation), diffusion-limited (rapid coagulation), or ballistic (particles added on a straight-line trajectory) conditions form different aggregate shapes and have different D values. (Adapted from Schaefer, 1989).

where n_p = number of particles

The object described in Eq. 9-56 is a Euclidean object (see Fig. 9-20a) because the volume depends on the diameter of the aggregates raised to the third power.

FRACTAL DIMENSION

The fractal dimension D is used to describe the fractal volume by accounting for differences in particle shape:

$$\text{Volume} \propto n_p^D \tag{9-57}$$

A three-dimensional Euclidean object has a fractal dimension of 3. If a straight row of particles forms the floc particle, then D would have a numerical value of 1. If a flat circular disk of particles forms (somewhat like our galaxy), D would have a numerical value of 2. Consequently, as shown on Fig. 9-21, the size of a sphere that contains the floc particle increases dramatically with decreasing fractal number.

As shown on Fig. 9-21, different fractal aggregates and fractal dimensions are formed during flocculation depending on the suspension conditions. As presented on Fig. 9-21, (1) reaction-limited flocs are formed from particles that are fairly stable and will flocculate occasionally when there is sufficient energy to overcome the repulsive forces, (2) ballistic flocs are formed from macroscale flocculation, and (3) diffusion-limited flocs are formed from microscale flocculation. The fractal dimension provides insight into the general shape of the aggregate that is formed.

Example 9-6 Calculating the size of a fractal particle

Calculate the size of a spherical aggregate floc that contains 10,000 particles with a size of 0.1 μm, assuming that the floc porosity is 0.4. Recalculate the size assuming that 100 particles of diameter 0.1 μm form clumps and 100 of these clumps come together to form the aggregate with a porosity of 0.4. An example of the differences in packing arrangement is shown on Fig. 9-20.

Solution

1. Determine the volume of a spherical aggregate that contains 10,000 particles with a diameter of 0.1 μm.
   a. The volume of a spherical aggregate is given by the following equation and depicted on Fig. 9-20a:

$$\text{Volume of aggregate} = \frac{4}{3}\pi\left(0.05\times10^{-6}\right)^3 \times \frac{10{,}000}{0.4}$$
$$= 1.309\times10^{-17}\ \text{m}^3$$

   b. The diameter of this Euclidean object is

$$d_a = 2\left(\frac{3}{4\pi}1.309\times10^{-17}\text{m}^3\right)^{1/3} = 2.924\times10^{-6}\ \text{m} = 2.92\ \mu\text{m}$$

2. Determine the size of aggregate that contains 100 particles.
   a. The volume of a spherical aggregate is given by the equation

$$\text{Volume of aggregate} = \frac{4}{3}\pi\left(0.05\times10^{-6}\right)^3 \times \frac{100}{0.4}$$
$$= 1.309\times10^{-19}\ \text{m}^3$$

   b. The diameter of this Euclidean object is

$$d_a = 2\left(\frac{3}{4\pi}1.309\times10^{-19}\ \text{m}^3\right)^{\frac{1}{3}} = 6.30\times10^{-7}\ \text{m} = 0.63\ \mu\text{m}$$

3. Compute the volume of a spherical aggregate containing 100 small clumps each comprised of a hundred particles.

a. The volume is given by the following equation and depicted on Fig. 9-20b:

$$\text{Volume of aggregate} = \frac{4}{3}\pi\left(0.315\times10^{-6}\right)^3\times\frac{100}{0.4}$$
$$= 3.273\times10^{-17}\,\text{m}^3$$

b. The diameter of a spherical aggregate containing 100 aggregates of 100 particle aggregates is given by the equation

$$d_a = 2\left(\frac{3}{4\pi}3.273\times10^{-17}\,\text{m}^3\right)^{\frac{1}{3}} = 3.97\times10^{-6}\,\text{m} = 3.97\,\mu\text{m}$$

Comment

The sphere that is made of clumps of 100 flocculated particles is much larger than a sphere made of the small floc particles.

The conditions under which the aggregates are formed provide insight into the three-dimensional appearance of the fractal. If a slowly flocculating suspension of small particles is completely destabilized, particles that come in contact with an aggregate will form branches on the surface of the aggregate. Further, as more branches are formed, the complexity and intricate nature of the branches increase. Consequently, completely destabilized suspensions have smaller fractal dimensions, as shown on Fig. 9-21. However, for particles that are not destabilized completely, attachment will not occur after every collision, and more forceful collisions are more likely to overcome the repulsive energy barrier. Thus, floc particles that form are typically dense and compact aggregates with fractal dimensions that approach the value used for Euclidean objects.

FRACTAL COLLISION FREQUENCY

The collision frequency for the spherical particle models (Euclidean objects) is smaller than what is observed for fractal particles because fractal particles are much larger (including numerous branches) and have a greater porosity than Euclidean particles. The greater porosity of fractals must be considered when particle sizes are measured because, according to Logan (1999), certain particle measurement devices that measure particles according to solid volume, such as a Coulter counter, will report a size that is significantly smaller than the actual aggregate size. Other types of instruments (e.g., HIAC Royco) that use light blockage may be more appropriate for the determination of the true floc size. The collision efficiencies

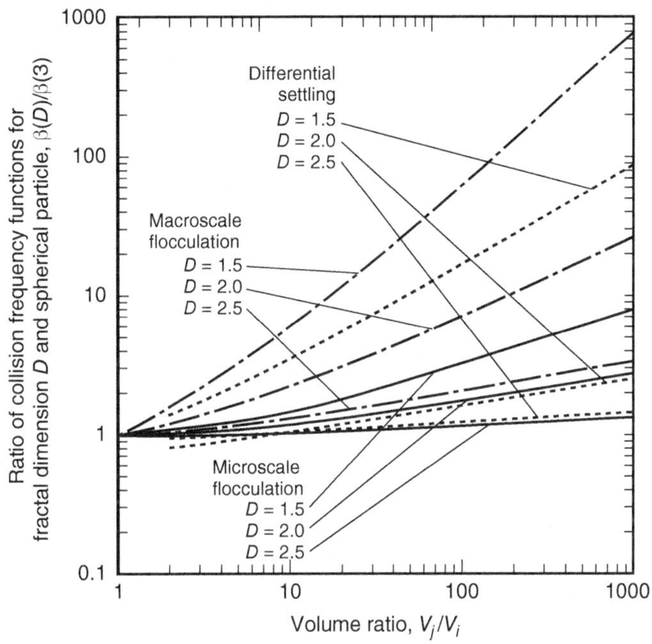

Figure 9-22
Ratio of collision functions for fractal and Euclidean geometry as function of fractal dimension and volume ratio.

(assuming complete destabilization) for the fractal model are compared to those of the spherical particle model on Fig. 9-22 for the three flocculation mechanisms. The collision function increases as the size difference of the particles increases. The impact is greater for smaller fractal dimensions because the fractals grow more branches and have a much larger size as a result of flocculation. It is interesting to note that the collision functions predicted from fractal geometry are up to 1000 times higher than from Euclidean geometry. Accordingly, the flocculation rate is much faster than is predicted from spherical particle models.

Floc Breakup

When flow conditions are turbulent, floc breakup cannot be neglected. Small particles are sheared from larger aggregates when the local shear stress exceeds the internal binding forces of the aggregate. The principal mechanisms of aggregate or floc breakup are surface erosion (Argaman and Kaufman, 1970) and floc splitting (Thomas, 1964).

As discussed in Chap. 6, microscopic eddies form due to turbulent mixing, and it is likely that floc particles cannot grow much larger than 2η (η is the Kolmogorov scale, discussed in Sec. 6-10) because the turbulent eddies would break up the particles. Logan (1999) compared the shear rate and eddy sizes to floc particle sizes for various mixing intensities and coagulants. The following conclusions were drawn: (1) polymer strengthens the floc and allows it to grow larger, (2) the mean floc size decreases with increasing $\overline{G}$ and decreasing eddy size, and (3) clay particles appear to strengthen the floc.

Based on a surface erosion model, it has been shown (Argaman and Kaufman, 1970; Parker et al., 1972) that the formation rate of particle fragments due to breakup is dependent on the velocity gradient. If only primary particles are considered, the rate expression is

$$r_B = K_B \overline{G}^{\delta} \tag{9-58}$$

where r_B = rate of change of number of primary particles
K_B = floc breakup constant
δ = turbulence constant

The floc breakup constant is dependent on the internal binding forces or floc strength of the aggregate. The turbulence constant varies between 2 and 4 depending on the hydraulic regime of the turbulence (Parker et al., 1972). The net rate of disappearance of primary particles under turbulent mixing conditions and the spherical particle model for macroscale flocculation may be written by combining Eqs. 9-41 and 9-58 as follows:

$$r_N = -K_A \overline{G} \Omega n_i + K_B \overline{G}^{\delta} \tag{9-59}$$

where r_N = net rate of floc disappearance
K_A = aggregation constant

The aggregation constant K_A and the breakup constant K_B can be determined empirically in laboratory or pilot-scale tests (Argaman, 1971; Bratby et al., 1977; Odegaard, 1979; Parker et al., 1972). The ranges of reported values for the aggregation and breakup constants are shown in Table 9-9.

Table 9-9
Reported kinetic parameters for flocculation kinetics[1]

System	Kinetic Parameters		Reference
	Aggregation, K_A	Breakup, K_B, s	
Kaolin–alum	4.5×10^{-5}	1×10^{-7}	Argaman and Kaufman 1970
	2.5×10^{-4}	4.5×10^{-7}	Bratby et al. (1977)
Natural particulates–alum	1.8×10^{-5}	0.8×10^{-7}	Argaman 1971
Alum–phosphate precipitate	2.8×10^{-4}	3.4×10^{-7}	
Alum–phosphate plus polymer	2.7×10^{-4}	1×10^{-7}	Odegaard 1979
Lime–phosphate, pH 11	5.6×10^{-4}	2.4×10^{-7}	

[1]The parameters K_A and K_B are defined differently throughout the flocculation literature, including different nomenclature and parameters with different units. When using values from the literature, caution should be exercised to ensure that parameter values apply to the selected equation.

Use of Spherical Particle Models for Reactor Design

The simplified kinetic models of particle aggregation provide a basic understanding of the design issues of coagulation/flocculation systems, including selection of the flocculation configuration (number of tanks), type and intensity of mixing, and flocculation residence times to achieve the desired removal efficiency. From a design perspective, flocculation can be considered to be a pseudo-first-order reaction with respect to total particle number because flocculators are designed to gently mix the water and cause macroscale flocculation. Moreover, Argaman (1971) was able to use Eq. 9-59 to describe residual turbidity data.

The greatest efficiency in terms of both volume and mixing energy is obtained using a plug flow reactor. However, flocculation reactors must be mixed, and the best way to achieve an efficiency approaching that of a plug flow reactor is with a number of completely mixed flow reactors (CMFRs) in series. For equal-volume CMFRs in series, neglecting floc breakup, the flocculator performance equation for flocculation of primary particles is expressed as follows (see Eq. 6-119 in Chap. 6 for the tanks-in-series model for first-order reactions):

$$\frac{N}{N_0} = \frac{1}{\left(1 + K_A\bar{G}\Omega\tau / m\right)^m} \tag{9-60}$$

where τ = hydraulic residence time of flocculator (V/Q)
m = number of tanks in series, unitless
N = number of particles in effluent per unit volume
N_0 = initial number of particles per unit volume

The effect on flocculation performance of increasing the number of reactors in series is illustrated on Fig. 9-23a at various $\bar{G}$ values, as determined by changes in the turbidity (concentration of primary particles). As m increases, the optimum $\bar{G}$ value decreases. If floc breakup is included (Eq. 9-59), the flocculation performance equation becomes (Argaman and Kaufman, 1970, Argaman, 1971)

$$\frac{N_0}{N} = \frac{\left[1 + K_A\bar{G}(\tau/m)\right]^m}{1 + K_B\bar{G}^{\delta}(\tau/m)\sum_{i=0}^{m-1}[1 + K_A\bar{G}(\tau/m)]^i} \tag{9-61}$$

The use of Eq. 9-61 for process design of flocculation basins is illustrated on Fig. 9-23b. As shown on Fig. 9-23b, for a given $\bar{G}$ and desired performance, there is a minimum required hydraulic residence time. It should be noted that, for each configuration and performance goal, an optimum value of $\bar{G}$ can be determined. Finally, and of particular importance, several CMFRs in series decrease the required residence time and mixing power needed to achieve a given performance goal, as would be predicted by reactor design principles presented in Chap. 6.

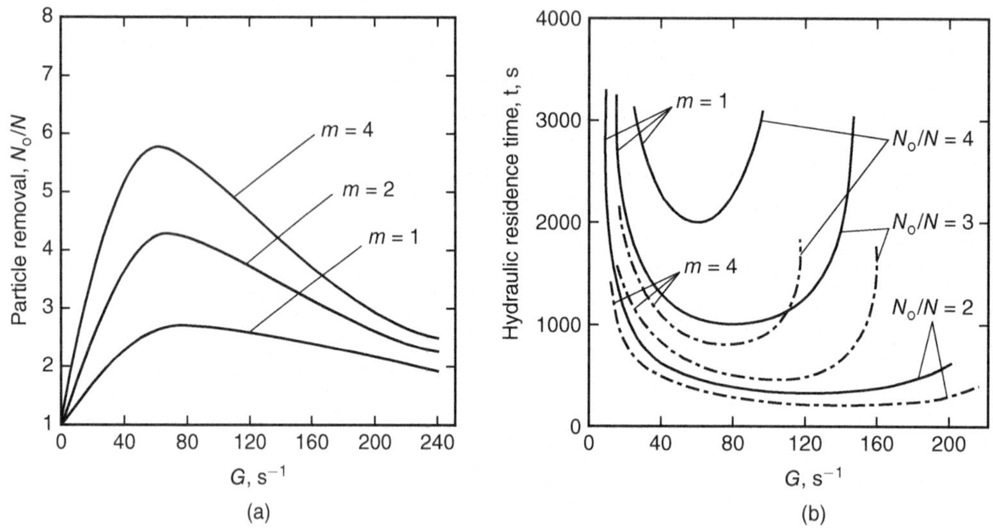

Figure 9-23
(a) Performance of multicompartment systems, $\tau/m = 8$ min, and (b) performance of flocculator as related to $\bar{G}$, τ, and m (number of tanks). (Adapted from Argaman, 1971.)

9-7 Flocculation Practice

The principal idea behind flocculation practice dates back to work done by Hyde and Langelier in 1921 while designing a new water treatment plant for Sacramento, California (Langelier, 1921). The mixing required for flocculation is provided by horizontal and vertical mechanical devices as well as arrangements that promote turbulence by hydraulic means alone. It has also been shown that flocculation benefits from compartmentalization so that the process operates as a series of CMFRs.

Alternative Methods of Flocculation

In the 1920s, the first flocculators were large, flat, vertical blades rotating in cylindrical tanks made to emulate large jars. Since that time many innovative designs have come forth, some more successful than others. Today's flocculation installations can be divided into two groups: mechanical and hydraulic. In mechanical flocculation horizontal paddles and vertical turbines have become the most common configurations for the prime mover, although new innovations continue to be developed. No particular arrangement dominates in hydraulic flocculation. Occasionally designers have used agitation with air or pumped water jets to create the velocity gradients for flocculation, but these efforts have met with limited success.

Some views of these three most common approaches to flocculation are given on Fig. 9-24. Information on how these approaches compare to each

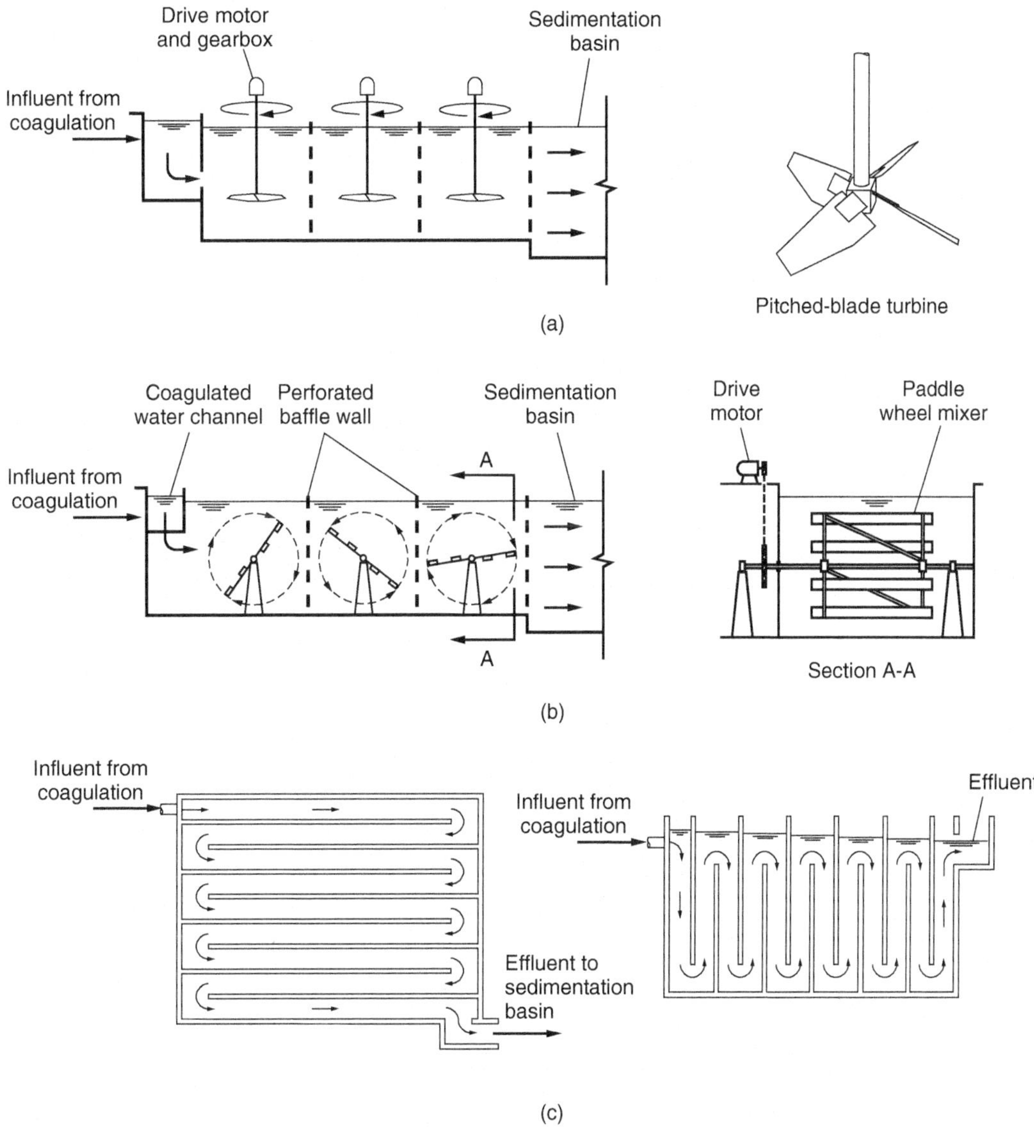

Figure 9-24
Common types of flocculation systems: (a) vertical-shaft turbine flocculation system, (b) horizontal paddle wheel flocculation system, and (c) hydraulic flocculation systems.

other with respect to a number of design and operational issues is presented in Table 9-10. All three of these approaches have been used successfully in numerous operations, and design details for a number of variations of each of them can be found in other sources (e.g., AWWA/ASCE, 2004; Kawamura, 2000).

Table 9-10
Comparison of basic approaches to flocculation

Process Issue	Horizontal Shaft with Paddles	Vertical-Shaft Turbines	Hydraulic Flocculation
Type of floc produced	Large and fluffy	Small to medium, dense	Very large and fluffy
Head loss	None	None	0.05–0.15 m
Operational flexibility	Good, limited to low $\overline{G}$	Excellent	Moderate to poor
Capital cost	Moderate to high	Moderate	Low to moderate
Construction difficulty	Moderate	Easy to moderate	Easy to difficult
Maintenance effort	Moderate	Low to moderate	Low to moderate
Compartmentalization	Moderate compartmentalization	Excellent compartmentalization	Excellent compartmentalization, some designs nearly plug flow
Advantages	❑ Generally produces large floc ❑ Reliable ❑ No head loss ❑ One shaft for several mixers	❑ Flocculators can be maintained or replaced without basin shutdown ❑ No head loss ❑ Very flexible, reliable ❑ Highest energy input potential	❑ Simple and effective ❑ Easy, low-cost maintenance ❑ No moving parts ❑ Can produce very large flocs
Disadvantages	❑ Compartmentalization more difficult ❑ Replacement and some maintenance requires shutdown of basin ❑ Shaft breakage on startup because of high initial torque	❑ Difficult to specify proper impellers and reliable gear drives in competitive bidding process	❑ Little flexibility

APPLICATION OF ALTERNATIVE METHODS

The choice among these three alternatives is usually driven by personal preference, by downstream processes, and by the level of operational expertise available. Horizontal-shaft paddles are more common in conventional treatment (includes sedimentation), although vertical turbines have been used successfully. Vertical turbines tend to dominate in direct filtration (no sedimentation) where horizontal-shaft paddles are rarely used. Hydraulic flocculation is usually employed with conventional treatment, although it has also been successfully used for direct filtration. Hydraulic flocculation is particularly popular in locations with poor access to resources and trained personnel for maintenance and operation, but it also plays an important role in some developed countries, particularly Japan (Kawamura and Trussell, 1991). In recent years, vertical turbine flocculators have gained in popularity as impeller designs have improved and as design engineers learn how to specify them properly. One special attraction of vertical turbines is that these flocculators can be replaced or maintained while the process is operating.

FLOC CHARACTERISTICS

Provided there is sufficient flocculation time, the flocs produced by hydraulic flocculation are virtually always of settleable size. With either type of mechanical mixer, large flocs suitable for sedimentation can be attained by tapering down the power input in subsequent flocculators. However, when sedimentation is the goal, mesh-type impellers appear to have an advantage in the last stage of the flocculation process where floc breakup is particularly important (Sajjad and Cleasby, 1995). To promote growth of very large flocs in this last stage, the power input must be tuned after construction is complete and, sometimes, from one season to the next. As a result, variable-speed drives are usually provided. Less expensive two-speed drives may perform satisfactorily, particularly in the earlier stages of flocculation. Often two- or three-speed drives, judiciously chosen, are all that are necessary.

DESIGN APPROACH

The basic design criteria for mechanical flocculators are the Camp–Stein RMS velocity gradient $\overline{G}$ and the hydraulic detention time t. Requirements of hydraulic detention time depend more on the downstream process than on the means of flocculation. Somewhat shorter flocculation times are often used for direct filtration (10 to 20 min) than for conventional treatment (20 to 30 min). Longer flocculation times are also required in colder climates. Representative design parameters for horizontal-shaft paddles and vertical turbines are shown in Table 9-11.

Corrosion of submerged metal components of the flocculator assemblies can be a serious maintenance problem. Specifying Type 316 stainless steel for submerged portions of the flocculator assembly and a cathodic protection system for structural steel are common solutions to this problem.

Vertical Turbine Flocculators

Vertical-shaft turbine flocculators are impellers attached to a vertical shaft that is rotated by an electric motor through a speed reducer. The impellers used for mixing can be placed in two broad classifications: (1) radial flow impellers and (2) axial flow impellers. Examples of the two types of impellers and the differences between their performance are illustrated on Fig. 9-25. The radial impeller directs flow outward from the impeller blades in a horizontal direction, through centrifugal force, with a velocity profile that peaks at the center of the blades. The axial impeller directs the flow parallel to the vertical shaft. The circulation pattern in the mixing tank is also substantially different for these two types of impellers. Two circulation

Table 9-11
Typical design criteria for horizontal-shaft paddles and vertical-shaft turbines

Design Parameter	Unit	Horizontal Shaft with Paddles	Vertical-Shaft Turbines
Velocity gradient, $\bar{G}$	s^{-1}	5–40	10–80
Tip speed, maximum	m/s	<0.5	1–3
Rotational speed	rev/min	0.5–3	5–20
Compartment [a] dimensions (plan)			
Width	m	3–25	3–8
Length	m	3–8	3–8
Number of stages	No.	2–6	2–4
Variable-speed drives	—	Common	Common

[a]The compartment is the region influenced by an individual flocculator. Horizontal-shaft flocculators often have multiple paddle wheel assemblies on a single flocculator shaft. Vertical turbine flocculators may or may not have baffle walls between the compartments in a single stage.

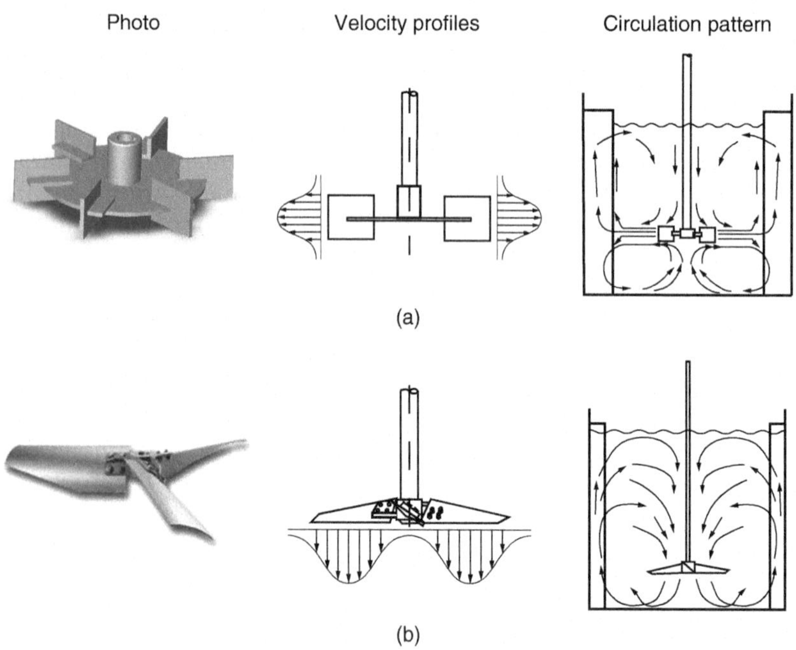

Figure 9-25
Comparison of (a) radial and (b) axial flow mixers with respect to shape, velocity profiles, and circulation patterns. (Adapted from Oldshue and Trussell, 1991.)

loops are generated from radial flow mixers: one above the impeller and one below. Axial flow impellers, on the other hand, create one circulation pattern from the bottom of the tank to the top and back through the impeller again.

Axial flow impellers can be configured in two ways: to pump downward or to pump upward. Down pumping is usually employed in flocculation because it helps keep the particles in the tank in suspension. The motors that drive the impellers are usually designed to rotate in a clockwise direction when viewed in plan view, as if looking down into the water. The axial impeller shown on Fig. 9-25 is arranged to pump downward.

IMPELLER DESIGN CRITERIA

Important design considerations for vertical turbine impellers are the displacement capacity (the rate at which the impeller pumps water), the power consumption, and the pumping head. Together, these determine much about the nature of the flow in the impeller's operating environment.

To evaluate the impeller's performance, it is important to know the nature of the flow in the mixing tank, specifically if the flow is laminar or turbulent as determined by the Reynolds number. Virtually all flocculation impellers operate in the turbulent-flow regime. The Reynolds number for a vertical turbine flocculator is given by the expression

$$\mathrm{Re} = \frac{D^2 N \rho}{\mu} \tag{9-62}$$

where Re = Reynolds number, dimensionless
D = diameter of impeller, m
N = impeller's rotational speed, s^{-1}
ρ = density of water, kg/m^3
μ = dynamic viscosity of water, $N \cdot s/m^2$

For the vertical turbines used in flocculation, full turbulence is developed at Reynolds numbers of 10,000 and greater.

Example 9-7 Estimating Reynolds number of vertical turbine flocculator

A vertical turbine 1.6 m in diameter is used to mix the contents of a flocculation tank 4 m in diameter. The turbine rotates at a speed of 20 rev/min. The absolute viscosity of the water is 1.31×10^{-3} kg/m · s. Determine if turbulent conditions are present.

Solution

1. Determine the Reynolds number using Eq. 9-62:

$$\mathrm{Re} = \frac{D^2 N_\rho}{\mu} = \frac{(1.6^2\ \mathrm{m}^2)(20\ \mathrm{min}^{-1})(998\ \mathrm{kg/m}^3)}{(60\ \mathrm{s/min})(1.31 \times 10^{-3}\ \mathrm{kg/m \cdot s})} = 6.5 \times 10^5$$

2. Because the computed value of R is greater than 10^4, the flow regime is turbulent.

Three parameters that are important to the design of mixing devices are the power number, the pumping number, and the head number. These have the following form:

$$N_p = \frac{P}{\rho N^3 D^5} \tag{9-63}$$

$$N_Q = \frac{Q}{ND^3} \tag{9-64}$$

$$N_H = \frac{\Delta H g}{(ND)^2} \tag{9-65}$$

where P = power requirement, J/s (W)
N_p = power number, dimensionless
D = diameter of impeller, m
N_Q = pumping number, dimensionless
N_H = head number, dimensionless
ρ = fluid density, kg/m^3
N = rotational speed, rev/s
Q = flow rate imparted by impeller, m^3/s
ΔH = head impeller imparts to impeller flow, m
g = acceleration due to gravity, 9.81 m/s^2

Power number is the most straightforward of these numbers to determine. All that is required is a torque meter on the shaft of the mixer and a tachometer to measure its rate of rotation. As a consequence, power numbers are available for most commercial impellers. The availability of power numbers is convenient because it is the power number and the rotational speed that determine the nominal Camp–Stein RMS velocity gradient $\bar{G}$ for the basin.

In general, as the pumping number increases, the circulation pattern becomes prevalent. As the head number increases for a given pumping number, more turbulence occurs. In addition, if the pumping number and head number are available, they can be used to determine whether a particular impeller mixer is suitable for the mixing tank. For example, the circulation time is related to the pumping rate and mixing time required to achieve completely mixed conditions.

VARIATIONS IN POWER NUMBER

The power number changes with the flow conditions in the basin being mixed. Significant factors include the depth and shape of the basin, the submergence of the impeller, the baffling provided, the type of impeller, and the Reynolds number. Fortunately, once the basin reaches full turbulent flow, power numbers for a variety of impellers are relatively constant. The relationship of power number and Reynolds number is compared for the two classic types of impellers on Fig. 9-26. Note that, for both impellers, N_p is constant above a Reynolds number of 10^4. Pumping numbers and head numbers are substantially more difficult to measure. As a result, these are not as readily available as the power number.

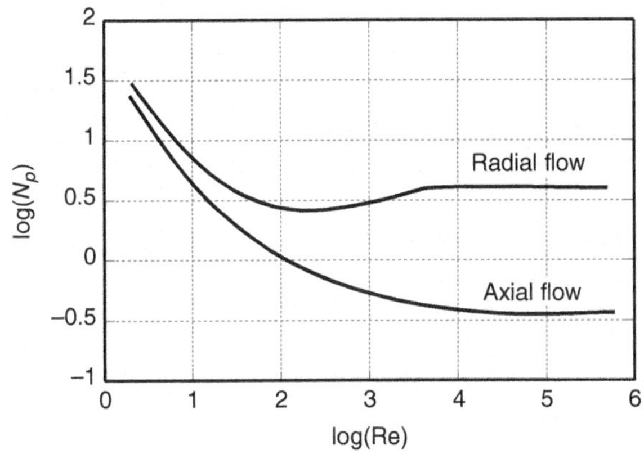

Figure 9-26
Change in power number as function of Reynolds number in baffled tank. (Adapted from Oldshue and Trussell, 1991.)

IMPACT OF IMPELLER SHAPE

Several types of impellers used in water treatment along with their typical uses are displayed in Table 9-12. When impellers on vertical shafts were first used for flocculation, some radial flow turbines were used, particularly Rushton turbines and flat-bladed turbines. But it has been determined that, as these impellers move through the water, they create substantial trailing vortices (Van't Reit et al., 1976). Vortices represent anisotropic turbulence that contributes significantly to floc breakup. Long pitch blade turbines subsequently became more popular, but, as illustrated on Fig. 9-27, even these produce substantial trailing vortices (Shäfer et al., 1998). Today hydrofoils, or pitched-blade turbines with cambered blades (blades with an upper surface shaped like an airplane wing), are the impellers of choice. Properly designed, flocculators using these devices can form large floc similar to that formed by more traditional horizontal paddle flocculators, but Sajjad and Cleasby (1995) demonstrated that these devices are still less effective in flocculation than an ideal wire mesh impeller.

OTHER DESIGN CONSIDERATIONS

In addition to the choice of the impeller itself, the following design parameters should be carefully scrutinized: (1) the ratio of the blade diameter to equivalent tank diameter should be greater than 0.35, preferably between 0.4 and 0.5, and (2) the velocity profile caused by the mixing blade should have a maximum of 2.5 m/s (8 ft/s) in the first stage and less than 0.6 m/s (2 ft/s) in the last stage of the flocculator. Design criteria are summarized in Table 9-13.

Baffling

Another issue related to the circulation rate and head is providing enough baffling to prevent vortexing around the impeller shaft. Vortexing occurs from the centrifugal forces created when the entire contents of the mixing chamber are brought into rotation around the impeller. These circumstances are not optimum for creating the velocity gradients that promote

Table 9-12
Power and pumping numbers for common impellers

Impeller Type	Photograph	Power Number	Pumping Number	Application
Flat-bladed turbine (FBT)		3.6	0.9	Blending, maintaining suspensions, flocculation
Pitched-blade turbine (45° PBT)		1.26	0.75	Blending, maintaining suspensions, flocculation
Pitched-blade turbine with camber (hydrofoil, 3 blades)		0.2–0.3	0.45–0.55	Blending, maintaining suspensions, flocculation
Cast foil with proplets		0.23	0.59	Blending viscous liquids
Rushton turbine (6 blades)		4.5–5.5	0.72	Gas–liquid dispersion, solids suspension, flocculation
Propeller (pitch of 1:1)		0.32–0.36	0.4	Blending viscous liquids

flocculation. Though circular tanks are rarely used, when they are, four baffles evenly placed around the outside of the tank are essential. The baffles should be about 10 percent of the tank diameter. The standard in the United States is $\frac{1}{12}$th of the tank diameter and the standard elsewhere is $\frac{1}{10}$th the tank diameter. Similar baffles can also be important in rectangular tanks. The appropriate placement of baffling in some alternative tank shapes is shown on Fig. 9-28.

Depth and shape of flocculation chamber
In addition to baffles, the depth and shape of the flocculation chamber can be important. Most mixing tests are conducted in square tanks with the impeller held at two-thirds of the depth of the tank. The more closely

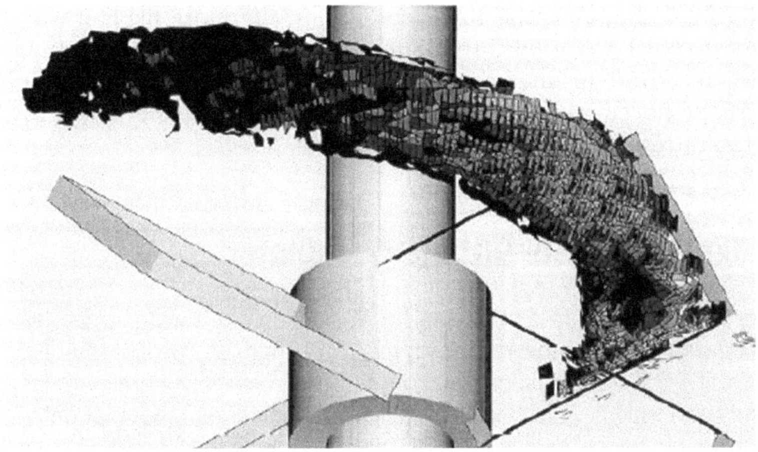

Figure 9-27
Trailing vortex behind 45° pitched-blade turbine in turbulent flow. (From Shäfer et al., 1998.)

Table 9-13
Key design criteria for vertical-turbine flocculator

Parameter	Range	Definition Sketch
Impeller	Hydrofoil or 45° pitched-blade turbine (PBT), hydrofoil preferred	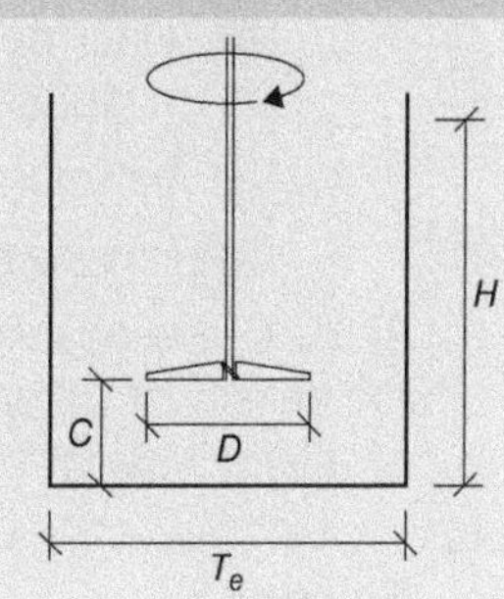
D/T_e[a]	0.3–0.6, 0.4–0.5 preferred	
H/T_e	0.9–1.1	
C/H	0.5–0.33	
N	10–30 rev/min	
Tip speed	2–3 m/s	

[a] $T_e = \sqrt{4A_{\text{plan}} / \pi}$.

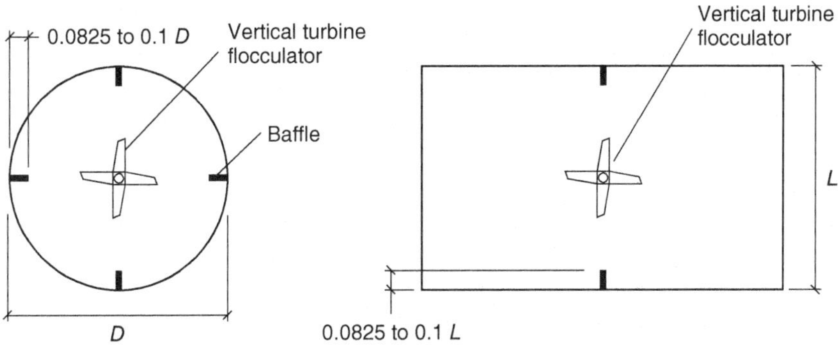

Figure 9-28
Baffle placement in flocculation tanks using vertical-turbine impellers: (a) circular tank and (b) rectangular tank.

the full-scale design emulates those conditions, the more likely it is that the full-scale performance will replicate the manufacturer's test data. As a result, when vertical turbine impellers are used, it is wise to stick to a nearly cubical shape flocculation chamber and to locate the impeller at approximately two thirds of the chamber's water depth.

Example 9-8 Design of vertical turbine flocculator

Vertical turbines are to be used for flocculation in a water treatment plant with a design flow rate of 75 ML/d (20 mgd) and design temperature of 10°C. Flocculation is to be designed with four parallel trains, and each train is to be made of four stages in series. The total detention time in flocculation is to be 20 min. Determine the following design features for the first stage in each flocculation train:

1. The dimensions of the stage
2. The diameter of the impeller (assume a turbine having three pitched blades with camber, a foil)
3. The water power required to achieve a $\bar{G}$ of 80 s^{-1} (the power that must be input to the water through the impeller shaft)
4. The maximum rotational speed
5. The pumping capacity of the impeller and circulation time in the tank

At 10°C, the absolute viscosity of water is 1.31×10^{-3} kg/m · s and the density of water is 999.7 kg/m^3. The circulation time is the volume of the flocculation chamber divided by the impeller pumping rate.

Solution

1. Determine the dimensions of the compartment:

$$\text{Volume} = \frac{(75\ \text{ML/d})\left(1000\ \text{m}^3/\text{ML}\right)(20\ \text{min})}{(1440\ \text{min/d})(4\ \text{trains})(4\ \text{stages/train})} = 65.1\ \text{m}^3$$

Assume a perfect cube of length L. The size is in the range for a single flocculator compartment in each stage.

$$L = \sqrt[3]{65.1\ \text{m}^3} = 4.0\ \text{m}\ \ (13.2\ \text{ft})$$

2. Determine the diameter of the impeller. Based on Table 9-13, choose an impeller diameter of 0.45Te:

$$T_e = \sqrt{\frac{4 \times A_{\text{plan}}}{\pi}}$$

Assume $A_{plan} = 4.0 \text{ m} \times 4.0 \text{ m} = 16 \text{ m}^2$:

$$T_e = 4.51 \text{ m}$$
$$D = 0.45 \times 4.51 \text{ m} = 2.03 \text{ m}$$

Choose $D = 2$ m.

3. Determine the power input to the water: The water power is determined by the requirement for $\bar{G} = 80 \text{ s}^{-1}$. Rearranging Eq. 9-40,

$$P = \bar{G}^2 \mu V$$
$$= \left(80 \text{ s}^{-1}\right)^2 \left(1.31 \times 10^{-3} \text{ kg/m} \cdot \text{s}\right)\left(65.1 \text{ m}^3\right) = 546 \text{ kg} \cdot \text{m}^2/\text{s}^3$$
$$= 546 \text{ W}$$

4. Determine the maximum rotational speed: From Table 9-12, for a three-bladed foil, N_p values of 0.2 to 0.3, use 0.25. Rearranging Eq. 9-63,

$$N = \sqrt[3]{\frac{P}{N_p \rho D^5}}$$
$$= \sqrt[3]{\frac{546 \text{ kg} \cdot \text{m}^2/\text{s}^3}{(0.25)\left(999.7 \text{ kg/m}^3\right)(2 \text{ m})^5}} = 0.409 \text{ s}^{-1}$$
$$= \left(0.409 \text{ s}^{-1}\right)(60 \text{ s/min}) = 24.5 \text{ min}^{-1} \text{ (rev/min)}$$

Note: N is within the operating range of 10 to 30 rev/min recommended in Table 9-13.

5. Determine the pumping capacity and circulation time:
   a. Pumping capacity: From Table 9-12, $N_Q \sim 0.5$. Rearranging Eq. 9-64,

$$Q = N_Q N D^3$$
$$= (0.5)(0.409 \text{ s}^{-1})(2 \text{ m})^3 = 1.64 \text{ m}^3/\text{s}$$

   b. Circulation time:

$$t_c = \frac{V}{Q} = \frac{65.1 \text{ m}^3}{1.64 \text{ m}^3/\text{s}} = 39.8 \text{ s}$$

The circulation time is a little less than 1 min.

Horizontal Paddle Wheel Flocculators

Horizontal-shaft paddle wheel flocculators are often employed if conventional treatment is used and a high degree of solids removal by sedimentation is required (see Fig. 9-29). However, they require more maintenance and expense, mainly because bearings and packings are typically submerged.

(a) (b) (c)

Figure 9-29
Views of paddle flocculators: (a) horizontal paddle wheel arrangement and (b) and (c) vertical paddle arrangements. (Courtesy AMWELL A Division of McNish Corp.)

By comparison, high-energy, vertical-shaft turbine flocculators are the unit of choice for liquid–solid separation using high-rate filtration systems and dissolved air flotation. Another advantage of horizontal-shaft flocculators is that one shaft flocculates a larger basin volume, but with that advantage comes the liability that a significant amount of the mixing capacity is lost when one drive is out of commission. When these units first start rotating, a tremendous torque is suddenly applied. Consequently, most failures occur during startup, especially if the unit is started at maximum rotational speed. Consequently, these mixers should be started at the lowest speed possible to minimize the initial torque.

The power input to the water by horizontal paddles may be estimated from the expression

$$P = \frac{C_D A_P \rho v_R^3}{2} \tag{9-66}$$

where C_D = drag coefficient on paddle (for turbulent flow), unitless
A_p = projected area of paddle, m^2
ρ = fluid density, kg/m^3
v_R = velocity of paddle relative to fluid, m/s

Here, v_R is usually assumed to be 70 to 80 percent of the paddle speed without tank baffles. With tank baffles, 100 percent of the paddle speed is

Table 9-14
Design criteria for paddle wheel flocculator

Parameter	Unit	Value
Diameter of wheel	m	3–4
Paddle board section	mm	100 × 150
Paddle board length	m	2–3.5
A_P/tank section area	%	<20
C_D (for use in Eq. 9-66)	$L/W = 1$	$C_D = 1.16$
	$L/W = 5$	$C_D = 1.20$
	$L/W = 20$	$C_D = 1.5$
	$L/W \gg 20$	$C_D = 1.90$
Paddle tip speed	m/s	Strong floc, 4
	m/s	Weak floc, 2
Spacing between paddle wheels on same shaft	m	1
Clearance from basin walls	m	0.7
Minimum basin depth	m	1 m greater than diameter of paddle wheel
Minimum clearance between stages	m	1

approached. The Reynolds number for a paddle flocculator is

$$\text{Re} = \frac{D_{\text{pw}}^2 N \rho}{\mu} \tag{9-67}$$

where D_{pw} = diameter of paddle wheel

For Reynolds numbers greater than 1000 (computed using Eq. 9-67), the drag coefficients for flat paddles are $C_D = 1.16, 1.20, 1.50$, and 1.90 for length-to-width ratios of 1.0, 5.0, 20.0, and infinity, respectively. Criteria that are useful for the design of paddle wheel flocculators are summarized in Table 9-14. Two things can be done to increase or decrease the $\bar{G}$ that is produced by a paddle wheel: (1) change the number of paddle boards or (2) change the rotational speed. It is difficult to achieve 50 to 60 s^{-1} with paddle wheel flocculators. Typical values of $\bar{G}$ for paddle wheel flocculators are 20 to 50 s^{-1}.

Example 9-9 Design of horizontal paddle wheel flocculator

Horizontal-shaft paddle wheel flocculators are to be used for flocculation in a water plant with a design flow rate of 150 ML/d (40 mgd) and water temperatue of 10°C. Flocculation is to be designed with two parallel trains and each train is to be made of five stages of flocculation in series. The total detention time for flocculation is to be 20 min. The paddle wheel flocculators to be used will have the design shown below:

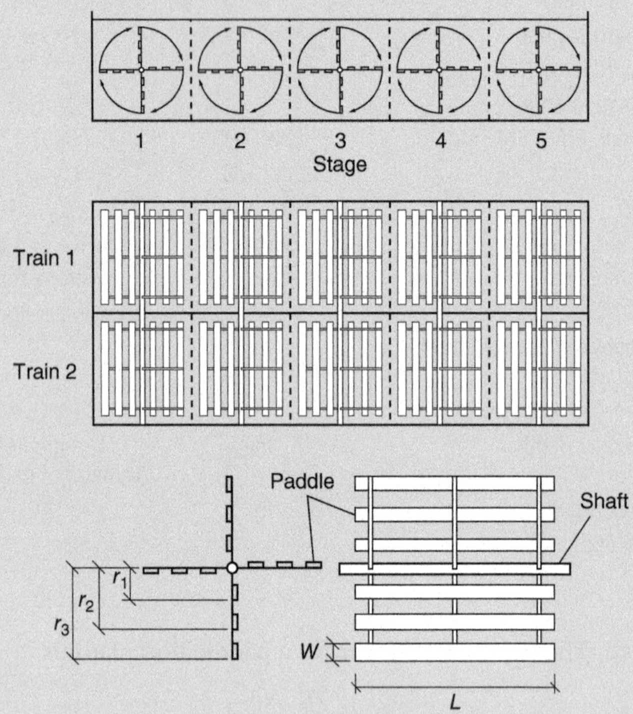

Two paddle wheels will be on the shaft in each stage. The paddle wheel design should include three paddle boards per arm with leading edges located at 0.67, 1.33, and 2.0 m from the shaft centerline. The width of the paddle boards is 0.15 m. Determine the following design features for the second stage in each flocculation train:

1. Dimensions of the compartment in the stage (including the number of paddle wheels and their length)
2. Water power input required to achieve a $\bar{G}$ value of 40 s^{-1}
3. Rotational speed of the paddle shaft

Solution

1. Determine the physical features of the flocculation basins.

a. The dimensions of the compartment are as follows:

$$\text{Basin depth} = (2\text{ m})(2) + 1\text{ m} = 5\text{ m}$$

$$\text{Volume} = \frac{(150\text{ ML/d})\left(1000\text{ m}^3/\text{ML}\right)(20\text{ min})}{(1440\text{ min/d})(2\text{ trains})(5\text{ stages/train})}$$

$$= 208.3\text{ m}^3/\text{stage}$$

$$\text{Basin area (plan)} = \frac{208.3\text{ m}^3}{5\text{ m}} = 41.7\text{ m}^2$$

$$\text{Minimum length of stage} = 4\text{ m} + 2(0.5\text{ m}) = 5\text{ m}$$

$$\text{Nominal width} = \frac{41.7\text{ m}^2}{5\text{ m}}$$

$$= 8.33\text{ m (perpendicular to flow)}$$

b. Determine paddle configuration: Two paddle wheel assemblies are needed. Clearance is needed at each end of each paddle and between the paddles.

$$\text{Required clearance} = 2(0.7\text{ m}) + 1\text{ m} = 2.4\text{ m}$$

$$\text{Length of both paddles} = 8.33\text{ m} - 2.4\text{ m} = 5.93\text{ m}$$

$$\text{Length of each paddle} = \frac{5.93\text{ m}}{2} = 2.97\text{ m}$$

c. Summary:

Compartment:

Depth = 5 m

Length = 5 m

Width = 8.33 m

Paddle wheel assemblies:

Number = 2

Length of paddles = 2.97 m

2. Determine the water power input required to achieve a $\bar{G}$ value of 40 s^{-1} using Eq. 9-40:

$$P = \bar{G}^2 \mu V$$

$$= \left(40\text{ s}^{-1}\right)^2 \left(1.31\times10^{-3}\text{ kg/m}\cdot\text{s}\right)\left(208.3\text{ m}^3\right) = 436.7\text{ kg}\cdot\text{m}^2/\text{s}^3$$

$$= 436.7\text{ J/s}$$

3. Determine the power required by the paddles by rearranging Eq. 9-66 and noting that the areas and shapes of the first, second, and third

boards are the same; therefore

$$P = \frac{\rho C_D A_p}{2}\left(v^3_{\text{R inside paddles}} + v^3_{\text{R middle paddles}} + v^3_{\text{R outside paddles}}\right)$$

a. Determine the areas of the boards at each position (inside, middle, and outside):

$$A_p = (2\text{ wheels})(4\text{ boards/wheel})(0.15\text{ m})(2.97\text{ m}) = 3.56\text{ m}^2$$

b. Check the length-to-width ratio and select the drag coefficient C_D:

$$\text{Paddle } L/W = 2.97/0.15 = 19.8$$

$$C_D \sim 1.5 \text{ (from Table 9-14)}$$

c. Develop parameters needed to determine the paddle power requirements:

$$\text{Velocity of paddles} = v_R = \frac{r2\pi N(0.75)}{60\text{s / min}}$$

where r = distance to centerline of paddle from center of rotation

N = shaft rotational speed, rev/min

0.75 = relative velocity of paddle with respect to fluid

$r_{\text{inside}} = r_1 = 0.67 - 0.15/2 = 0.595\text{ m}$

$r_{\text{middle}} = r_2 = 1.33 - 0.15/2 = 1.255\text{ m}$

$r_{\text{outside}} = r_3 = 2.0 - 0.15/2 = 1.925\text{ m}$

d. Substitute known values in the paddle power equation:

$$P = \frac{\rho C_D A_p}{2}\left(v^3_{\text{R inside paddles}} + v^3_{\text{R middle paddles}} + v^3_{\text{R outside paddles}}\right)$$

$$= \frac{\rho C_D A_p}{2}\left[\frac{2\pi N(0.75)}{60\text{ s/min}}\right]^3\left(r_1^3 + r_2^3 + r_3^3\right)$$

$$= \left(\frac{\left(999.7\text{ kg/m}^3\right)(1.5)\left(3.56\text{ m}^3\right)}{2}\right)\left[\frac{2\pi N(0.75)}{60\text{ s/min}}\right]^3$$

$$\times\left[(0.595)^3 + (1.255)^3 + (1.925)^3\right]$$

$$= (2664.7)\left(4.85\times10^{-4}\text{N}^3\right)(9.321)$$

4. Equate the required power determined in step 2 to meet the $\bar{G}$ value to the power required by the paddles as determined in step 3 above and

solve for N:

$$436.7 = (2664.7)\left(4.85\times10^{-4}\ N^3\right)(9.321)$$

$$N = \sqrt[3]{\frac{436.7}{(2664.7)\left(4.85\times10^{-4}\right)(9.321)}} = 3.31\ \text{rev/min}$$

Hydraulic Flocculation

There are a number of approaches to hydraulic flocculation. Monk and Trussell (1991) divided hydraulic flocculators into three groups. With some minor modifications those groups are (1) baffled channels, (2) hydraulic-jet flocculators, and (3) coarse-media flocculators. Examples of the first two types are illustrated on Fig. 9-30. Baffled channels are probably the most common application. Although most hydraulic flocculators have some disadvantages, such as inflexible mixing and a large head loss across the basin, most designs produce good floc, often without much short circuiting. Most hydraulic flocculators work best if the plant flow rate is fairly constant.

The main design issue for hydraulic flocculators is whether there is head available in the plant profile to provide the required power input. Hydraulic flocculators often operate well at low-flow conditions (even if $\overline{G}$ is as low as 10 s^{-1}), because the longer detention time provides for an adequate $\overline{G}t$. Helicoidal or tangential flow baffled channels perform as well as traditional designs and have lower head loss (Kawamura, 2000). Around-the-end baffled channels are preferred over the under-and-over baffled flocculators because they have fewer problems with scum and silt/grit buildup on the upstream side of each baffle. Several rules of thumb that prove useful in the design of hydraulic flocculators are summarized in Table 9-15.

BAFFLED CHANNELS

Baffled channels are the most common form of hydraulic flocculators. In these flocculators energy dissipation is achieved by changing the direction of flow of the water, either by over–under or around-the-end baffles. In principle, these are plug flow devices, but, as a result of the flow separation that occurs at each turn, their operation is actually closer to a series of CFMRs. Nevertheless, these devices have excellent compartmentalization and virtually no short circuiting, two of their greater strengths. A variety of different approaches to around-the-end flocculators are illustrated schematically on Fig. 9-30 and photographically on Fig. 9-31.

The proper input by a hydraulic system is the product of the pressure and flow rate. Because pressure can be related to head loss by ($\Delta P = \rho g \Delta H$), the RMS velocity gradient in a baffled basin can be determined from the

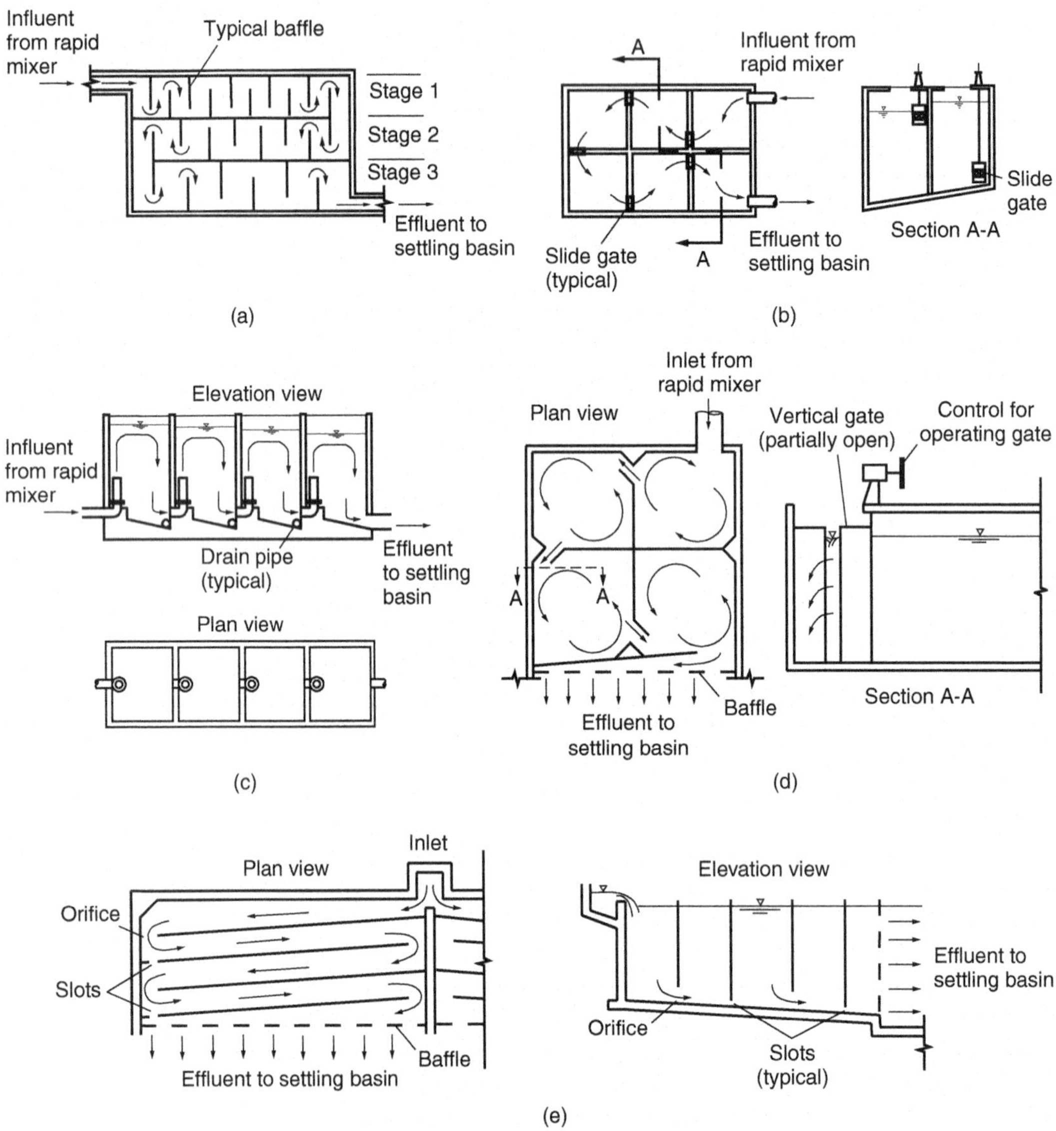

Figure 9-30
Examples of some alternative designs of hydraulic flocculators: (a) tapered horizontal baffled hydraulic flocculator, (b) helicoidal flow flocculator, (c) Alabama-type hydraulic flocculator, (d) variable-gate hydraulic flocculator, and (e) orifice and channel over and under flocculator. (Adapted from Kawamura, 2000.)

Table 9-15
Design criteria for hydraulic flocculation

Parameter	Unit	Value
Average $\bar{G}$	s^{-1}	30–40
Channel velocities	m/s	0.15–0.45
Minimum residence time	min	20
Head loss coefficient at 180° turn	—	3.2–3.5
Head loss coefficient through slit or port	—	1.5
Minimum distance between baffles[a]	m	0.75
Minimum water depth	m	1

[a]To facilitate cleaning.

(a) (b)

Figure 9-31
Views of tapered horizontal, baffled channel hydraulic flocculators.

following expression (Monk and Trussell, 1991):

$$\bar{G} = \left(\frac{P}{\mu V}\right)^{1/2} = \left(\frac{\rho g h Q}{\mu V}\right)^{1/2} = \left(\frac{\rho g h}{\mu \tau}\right)^{1/2} \tag{9-68}$$

where ρ = density of water, kg/m^3
g = acceleration due to gravity, 9.81 m/s^2
h = head loss through basin, m
μ = dynamic viscosity of water, $N \cdot s/m^2$ ($kg/m \cdot s$)
τ = detention time, s

In the most common horizontal-flow baffled hydraulic flocculator, the head loss in a baffled mixing channel from turbulence and friction on the sides of the channel can be calculated using the expression

$$h = \frac{Lv^2}{C^2 R_h} \tag{9-69}$$

where L = length of channel, m
v = velocity of flow in channel, m/s

$C =$ Chezy coefficient, unitless
$R_h =$ hydraulic radius of channel, m

Head loss resulting from each 180° turn can be estimated as follows (Monk and Trussell, 1991):

$$h = k\left(\frac{v^2}{2g}\right) \tag{9-70}$$

where $k =$ head loss coefficient (typically 2.5 to 4), unitless

HYDRAULIC-JET FLOCCULATORS

In this flocculator design energy dissipation is achieved by using the energy of a hydraulic jet created as the flow enters each flocculation compartment. There are three types in use today: (1) the helicoidal flow flocculator, (2) the Alabama flocculator, and (3) the variable-gate flocculator. Each is shown on Fig. 9-30. The helicoidal flow flocculator organizes the flow from one sector of the flocculator compartment to the next so that flow travels in a helical pattern. Turbulence is created by the discharge jet as the flow enters each compartment. The Alabama flocculator uses a simple up/down flow arrangement with the jet being created by the entrance pipe. The variable-gate flocculator is a more complex, but flexible, design that enables hydraulic flocculation to achieve constant mixing at variable flow. The head loss through each type of device that is used to create a jet is slightly different.

COARSE-MEDIA FLOCCULATORS

In this flocculator design energy dissipation is achieved by turbulent flow through a coarse media. Coarse-media flocculators are also called roughing filters or adsorption clarifiers because the coarse media used for flocculation also have excellent properties for storing coagulated solids. Roughing filters are described in more detailed in Chap. 10. The coarse-media flocculation process has been successful in small package plants throughout much of the United States. One particularly successful application is the Siemens Microfloc Tri-Mite, which uses a buoyant plastic media that is easily cleaned (Monk and Trussell, 1991).

Combining hydraulic and mechanical flocculation sometimes allows the water utility to capitalize on the strengths of both approaches. Using such a combination, the number of mechanical flocculators is reduced, reducing the capital and maintenance costs and increasing the reliability. In such combinations, Kawamura (2000) recommends that mechanical flocculators be located at the end of the process to keep the floc in suspension during low-flow conditions. The Houston East (Houston, Texas) plant [150 mgd (6.5 m^3/s)] and the Mohawk (Tulsa, Oklahoma) plant [100 mgd (4.4 m^3/s)] both utilize this design and achieve excellent settled water turbidity and operate effectively during low-flow conditions by isolating some of the treatment trains (Kawamura, 2000).

Important Design Features in Flocculation

The size and shape of a flocculation basin are generally determined by the type of flocculator selected and the type of sedimentation process employed downstream. If mechanical flocculators are paired with rectangular, horizontal-flow sedimentation basins, the width and depth of the flocculation basins should match the width and depth of the sedimentation basins. Similar dimensions enhance constructability and reduce overall project costs.

SIZE OF FLOCCULATION BASIN

The size of the flocculation basin and the flocculation time are determined by the downstream liquid–solid separation technology used. Typically, flocculation times range from 20 to 45 min for plants that use conventional settling and plate-and-tube settlers, depending on the characteristics of the raw water, water temperature, and type of coagulant used. For low-turbidity raw water in cold regions, the flocculation time should be at least 30 min. A flocculation time of 15 min is typical for direct filtration and a flocculation of time of 5 to 10 min is typical for dissolved air flotation (DAF). Jar or bench tests, conducted in conjunction with pilot studies, will aid in producing accurate design criteria. However, it is important that such tests be conducted at a representative temperature.

Although no mechanistic principle has been developed to define the relationship between basin area and water depth, Kawamura (2000) notes that basins with depths in excess of 5 m (16.5 ft) sometimes display unstable flow patterns and poor flocculation.

INLET AND OUTLET ARRANGEMENTS

Another important consideration in flocculation basin design is uniform hydraulic loading to each basin. There are three basic types of basin inlet structures: a simple pipe connection to the basin, a weir inlet, and a submerged orifice inlet. The plant layout (especially the symmetry of the basin layout to inlet line) and maintaining an appropriate flow velocity in the distribution pipe or channel will greatly minimize uneven flow distribution to each basin regardless of the inlet type selected (Chao and Trussell, 1980).

DIFFUSER WALLS

Diffuser walls are often used to divide flocculation basins into separate compartments (see Fig. 9-32), to place a hydraulic division between flocculation and sedimentation basins, as well as in other situations where an even velocity profile is required and backmixing is undesirable.

Separating flocculation and sedimentation

As water leaves the flocculation tank after treatment, the flocs that have been formed should not be broken before they enter the liquid–solid

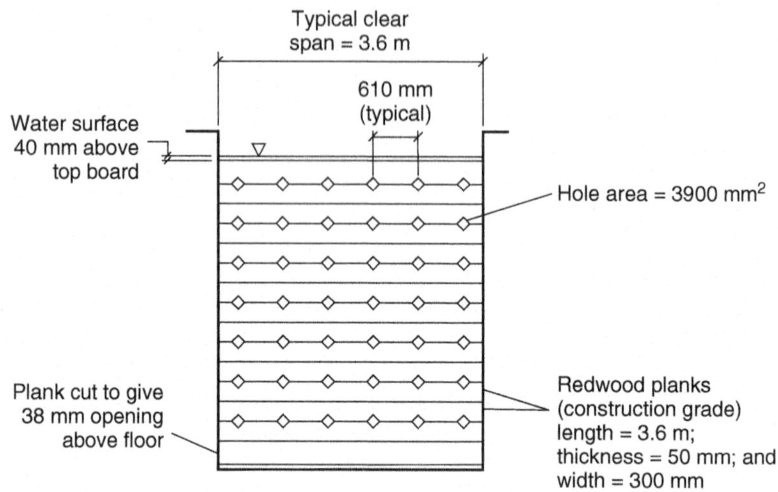

Figure 9-32
Typical design of diffuser wall (see also Table 9-16 for additional design details).

separation process. The problem with this sensible requirement is that it is in direct conflict with the need to establish a uniform velocity profile across the entrance of the sedimentation tank, a requirement that is usually met through the dissipation of head loss. When conventional settling is used, a simple approach is to build the flocculation tanks as an integral part of the settling tank and provide a diffuser wall between the two tanks. The diffuser wall must have enough head loss to establish a uniform discharge profile, and yet it must not create turbulence that will shear floc. Based on operating experience, it has been found that a permeable baffle wall between the flocculation and sedimentation basins with a head loss of approximately 3 to 4 mm can be effective.

Separating flocculation compartments

As discussed earlier in the chapter, the performance of flocculation may be improved through compartmentalization to minimize short circuiting of flow. Diffuser walls are typically used for this purpose, as shown on Fig. 9-33. Baffles are placed after each flocculation stage, perpendicular to the flow path. Baffles can be specified in various shapes and arrangements, but the baffle opening must be sized correctly. The top of the baffle should be slightly submerged (30 to 40 mm) so scum does not accumulate behind the baffle, and the bottom of the baffle should also be 30 to 40 mm above the floor of the flocculator to facilitate drainage and sludge removal. If the flocculation tank is designed as an integral part of the sedimentation basin, a diffuser wall should be provided at the end of the flocculation tank to assure uniform flow distribution into the sedimentation tank.

(a)

(b)

Figure 9-33
Views of vertical-turbine flocculators in basin separated by diffuser walls. Typical design details for a diffuser wall are presented in Fig. 9-32 and Table 9-16.

Table 9-16
Diffuser wall design guidelines for flocculation basins

Parameter	Unit	Guideline
Opening area	Percent of flow cross section	2–5
Velocity through orifice		
Dividing first and second floc basins	m/s	0.55
Dividing floc and sedimentation basins	m/s	0.35
Head loss across baffle		
Dividing first and second floc basins	mm	7–9
Dividing floc and sedimentation basins	mm	3–4
Submergence of highest port	mm	15
Clearance below baffle for sludge	mm	25

Source: Adapted in part from Kawamura (2000).

Design of diffuser walls

Design criteria for diffuser walls are summarized in Table 9-16. Generally such diffusion walls are constructed of lumber, redwood when it is available, but there is no requirement other than that a durable, water-resistant material be employed. The diffuser wall shown on Fig. 9-33 is illustrative of a simple design that is easily constructed, requiring no special sophistication on the part of the contractor.

Problems and Discussion Topics

9-1 If the electrophoretic velocity of a migrating particle is 12.5 μm/s and the electrical field is 100 V/cm, estimate the zeta potential at 25°C if the value of the constant K_2 is 4π. The viscosity of water at 25°C is 0.89×10^{-3} N · s/m^2, and the relative permittivity for water is 78.54.

9-2 Particles are present in a water with the following chemical characteristics. Estimate the thickness of the particle electrical double layer.

Cation	Concentration, mg/L	Anion	Concentration, mg/L
Ca^{2+}	40.0	HCO_3^-	91.5
Mg^{2+}	12.2	SO_4^{2-}	72
Na^+	15.1	Cl^-	22.9
K^+	5.1	NO_3^-	5.0

9-3 Determine the thickness of the particle electrical double layer if the ionic strength is 0.0025 mol/L.

9-4 Prepare a plot of the thickness of the electrical double layer versus the ionic strength.

9-5 Calculate the amount of $Fe(OH)_3$ precipitate formed and amount of alkalinity consumed (expressed as $CaCO_3$) by a 25 mg/L dose of ferric sulfate [$Fe_2(SO_4)_3$].

9-6 A natural water with a flow of 3800 m^3/d is to be treated with an alum dose of 60 mg/L. Determine the chemical feed rate for the alum, the amount of alkalinity consumed by the reaction, and the amount of precipitate produced in mg/L and kg/day.

9-7 Determine the amount of $Fe(OH)_3$ produced and the amount of alkalinity consumed when 60 mg/L of ferric chloride [$FeCl_3 \cdot 6H_2O$] is added to a natural water. Assume enough alkalinity is present.

9-8 Determine the basicity of the following PACl compounds: (a) $Al_4(OH)_6(Cl_2)_2$, (b) $Al_2(OH)_2Cl_2SO_4$, and (c) $Al_{15}O_6(OH)_{24}SO_4$.

9-9 The following data were obtained from a bench-scale coagulation flocculation test. Using these data, estimate the alum dosage for turbidity removal and for NOM removal.

Alum Dose, mg/L	DOC, mg/L	Turbidity, NTU
0.00	5.00	2.50
10.0	5.10	2.05
20.0	5.25	1.65
30.0	5.00	1.25
40.0	4.50	0.800
50.0	4.00	0.500
60.0	3.40	0.480
70.0	2.80	0.470
80.0	2.45	0.450
100	2.00	0.440
120	1.95	
140	1.90	

9-10 Estimate the DOC removal using alum as a function of dose for a concentration range varying from 10 to 90 mg/L for the following conditions: initial DOC = 5 mg/L, initial UV_{254} absorbance = 0.138 cm^{-1}, and pH = 7.

9-11 Estimate the DOC removal using ferric chloride as a function of dose varying from 5 to 50 mg/L for the following conditions: initial DOC = 5 mg/L, initial UV_{254} absorbance = 0.138 cm^{-1}, and pH = 7.

9-12 Using the collision frequency functions for macroscale flocculation, microscale flocculation, and flocculation due to differential settling given by Eqs. 9-35, 9-47, and 9-55, respectively, demonstrate the correctness of Fig. 9-19. Use a particle size d_i of 2.0 μm and particles d_j ranging in size from 100 Å to 50 μm, a $\overline{G}$ value of 100 s^{-1}, and a water temperature of 15° C.

9-13 Based on your reading of this chapter, provide a brief answer to the following questions: (a) Who first came up with the principal idea behind flocculation theory? When? (b) Who were the first people to put flocculation into practice? When? (c) Who proposed the concept of RMS velocity gradient? When?

9-14 Based on your reading of this chapter, provide a brief answer to the following questions: (a) Flocculation installations can be divided into what two groups? (b) Which type of flocculator produces very large and fluffy floc? (c) What are two advantages of vertical turbines versus horizontal-shaft paddles for flocculation? (d) What are the two principal impeller types used for flocculation? (e) What is the principal disadvantage of hydraulic flocculators?

9-15 Which impeller is better at keeping solids suspended in a tank?

9-16 A first-stage flocculator uses a six-bladed Rushton turbine 2 m in diameter rotating at 25 rev/min. What is the Reynolds number? How much power must be applied to the shaft to rotate it? What flow does the impeller pump?

9-17 The impeller in Problem 9-16 is in a tank 4 m square and 4 m deep. Calculate the tank turnover time and $\overline{G}$ values.

9-18 What is the largest paddle wheel that meets the design criteria in Table 9-14? How many paddle boards may be used on such a wheel?

9-19 Design a flocculation compartment for a horizontal-shaft flocculator with two paddles like that in Example 9-9. How fast must the paddle wheel rotate in that compartment to generate a $\overline{G}$ value 30 s^{-1}?

9-20 What is the minimum water depth for a hydraulic flocculator?

9-21 Describe, in your own words, the principal advantages of an Alabama flocculator versus a variable gate flocculator.

References

Amirtharajah, A., and Mills, K. M. (1982) "Rapid-Mix Design for Mechanisms of Alum Coagulation," *J. AWWA*, **74**, 4, 210–216.

Argaman, Y. A. (1971) "Pilot-Plant Studies of Flocculation." *J. AWWA*, **63**, 12, 775–777.

Argaman, Y. A., and Kaufman, W. J. (1970) "Turbulence and Flocculation." *J. Div. Sanit. Eng. Proc. Am. Soc. Civil Eng.*, **96**, 223–241.

ASTM (2008) Standard Recommended Practice for Coagulation–Flocculation Jar Test of Water, in *ASTM 2008 Annual Book of Standards*, ASTM D2035-08, Philadelphia, PA.

AWWA (2011) *Manual M37 — Operational Control of Coagulation and Filtration Processes*, American Water Works Assocciation, Denver, CO.

AWWA/ASCE (2004) *Water Treatment Plant Design*, 4th ed., McGraw-Hill, New York.

Base, C. F., and Mesmer, R. E. (1976) *The Hydrolysis of Cations*, Wiley-Interscience, New York.

Benefield, L. D., Judkins, J. F., and Weand, B. L. (1982) *Process Chemistry for Water and Wastewater Treatment*, Prentice-Hall, Englewood Cliffs, NJ.

Black, A. P., Birkner, F. B., and Morgan, J. J. (1966) "The Effect of Polymer Adsorption on the Electrokinetic Stability of Dilute Clay Suspensions," *J. Colloid Interface Sci.*, **21**, 626–648.

Black, A. P., Buswell, A. M., and Eidniess, F. A. (1957) "Review of the Jar Test," *J. AWWA*, **49**, 1414–1424.

Bratby, J., Miller, M. W., and Marais, G. V. R. (1977) "Design of Flocculation Systems from Batch Test Data," *Water S. Afr.*, **3**, 173–182.

Camp, T. R., and Stein, P. C. (1943) "Velocity Gradients and Hydraulic Work in Fluid Motion," *J. Boston Soc. Civil Eng.*, **30**, 203–221.

Chao, J., and Trussell, R. R. (1980) "Hydraulic Design of Flow Distribution Channels," *J.*

Degrémont (2007) *Water Treatment Handbook*, Vols. 1 and 2, 7th ed., Lavoisier, Paris.

Derjaguin, B. V., and Landau, L. D. (1941) "Theory of Stability of Strongly Charged Lyophobic Soles and Coalesance of Strongly Charged Particles in Solutions of Electrolytes," *Acta Physicochim. URSS*, **14**, 733–762.

Edwards, M. (1997) "Predicting Doc Removal during Enhanced Coagulation," *J. AWWA*, **89**, 5, 78–89.

Friedlander, S. K. (2000) *Smoke, Dust and Haze*, 2nd ed. Wiley-Interscience, New York.

Gouy, G. (1910) "Sur la Constitution de la Charge Electrique a la Surface, d'un Electrolyte," *J. Phys. Chem.*, **9**, 457–467.

Hahn, H. H., and Stumm W., (1968) "Kinetics of Coagulation with Hydrolyzed Al(III)," *J. Colloidal Interface Sci.*, **28**, 1, 134–144.

Han, M., and Lawler, D. (1992) "The (Relative) Insignificance of G in Flocculation," *J. AWWA*, **84**, 10, 79–91.

Harris, H. S., Kaufman, W. F., and Krone, R. B. (1966) "Orthokinetic Flocculation in Water Purification," *J. Div. Sanit. Eng. Proc. ASCE*, **92**, 95–111.

Hudson, H. E., and Wolfner J. P., (1967) "Design of Mixing and Flocculating Basins," *J. AWWA*, **59**, 10, 1257–1267.

Hunter, R. J. (2001) *Foundations of Colloid Science*, Vos. 1 and 2, Oxford University Press, Oxford, UK.

Kavanaugh, M. C. (1978) "Modified Coagulation for Improved Removal of Trihalomethane Precursors," *J. AWWA*, **70**, 11, 613–620.

Kawamura, S. (2000) *Integrated Design and Operation of Water Treatment Facilities*, 2nd ed., Wiley-Interscience, New York.

Kawamura, S., and Trussell, R. (1991) "Main Features of Large Water Treatment Plants in Japan," *J. AWWA*, **83**, 6, 56–62.

Kissa, E. (1999) *Dispersions: Characterization, Testing, and Measurement*, Marcel Dekker, New York.

Kruyt, H. R. (1952) *Colloid Science*, Elsevier, New York.

Langelier, W. F. (1921) "Coagulation of Water with Alum by Prolonged Agitation," *Eng. News-Record*, **86**, 924–928.

Lawler, D. F., O'Melia, C. R., and Tobiason, J. E. (1980) Integral Water Treatment Plant Design; from Particle Size to Plant Performance, in M. C. Kavanaugh and J. E. Leckie (eds.), *Particulates in Water, Advances in Chemistry Series*, No. 189, American Chemical Society, Washington, DC.

Letterman, R. D., Amirtharajah, A., and O'Melia C. R. (1999) Coagulation and Flocculation, Chap. 6, in R. D. Letterman (ed.), *Water Quality and Treatment: A Handbook of Community Water Supplies*, 5th ed., American Water Works Association, McGraw-Hill, New York.

Letterman, R. D. (1981) Theoretical Principles of Flocculation, paper presented at Seminar Proceedings, AWWA Sunday Seminar Series, AWWA Annual Conference, St. Louis, June.

Logan, B. E. (1999) *Environmental Transport Processes*, Wiley-Interscience, New York.

McMurry, J., and Fay, R. C. (2003) *Chemistry*, 4nd ed., Prentice-Hall, Upper Saddle River, NJ.

Monk, R., and Trussell, R. (1991) Design of Mixers for Water Treatment Plants: Rapid Mixing and Flocculators, in *Mixing in Coagulation and Flocculation*, pp. 380–419, in A. Amirtharajah, M. Clark, and R. Trussell (eds.), American Water Works Association Research Foundation, Denver, CO.

Morel, F. M. M., and Hering, J. G. (1993) *Principles and Applications of Aquatic Chemistry*, Wiley-Interscience, New York.

Nordstrom, D., and May, H. (1989a) Aqueous Equilibrium Data for Mononuclear aluminum Species, in G. Sposito (ed.), *Chemical Modeling in Aqueous Systems*, Vol. **2**, CRC Press, Boca Raton, FL.

Nordstrom, D. K., and May, H. M. (1989b) Aqueous Equilibrium Data for Mononuclear Species, pp. 29–53, in G. Sposito (ed.), *The Environmental Chemistry of Aluminum*, CRC Press, Boca Raton, FL.

O'Brien, J. H., and Novak, J. T. (1977) "Effects of Ph and Mixing Polymer Conditioning of Chemical Sludges," *J. AWWA*, **69**, 11, 600–605.

Odegaard, H. (1979) "Orthokinetic Flocculation of Phosphate Precipitates in a Multicomponent Reactor with Non-Ideal Flow," *Progr. Water Tech.*, **11**, 61–88.

Oldshue, J. Y., and Trussell, R. R. (1991) Design of Impellers for Mixing, in A. Amirtharajah, M. M. Clark, and R. R. Trussell (eds.), *Mixing in Coagulation and Flocculation*, AWWARF, Denver CO.

O'Melia, C. R. (1972) Coagulation and Flocculation, in W. J. Weber, Jr. (ed.), *Physicochemical Processes for Water Quality Control*, Wiley-Interscience, New York.

O'Melia, C. R. (1978) Coagulation in Wastewater Treatment, in K. J. Ives (ed.), *Scientific Basis of Flocculation*, Noordhoff International, Leyden, The Netherlands.

O'Melia, C. R., Becker, W. C., and Au, K. K. (1999) "Removal of Humic Substances by Coagulation," *Water Sci. Technol.*, **40**, 47–54.

Packham, R. F. (1962) "The Coagulation Process," *J. Appl. Chem.*, **12**, 556–568.

Packham, R. F. (1965) "Some Studies of the Coagulation of Dispersed Clays with Hydrolyzing Salts," *J. Coll. Science*, **20**, 81–92.

Pankow, J. F. (1991) *Aquatic Chemistry Concepts*, Lewis, Chelsea, MI.

Parker, D. S., Kaufmann, W. J., and Jenkins, D. (1972) "Floc Breakup in Turbulent Flocculation Processes," *J. Sanit. Eng. Div.*, **98**, 79–99.

Parks, G. A. (1967) Aqueous Surface Chemistry of Oxides and Complex Oxide Minerals; Isolectric Point and Zero Point of Charge, in *Equilibrium Concepts in Natural Water Systems, Advances in Chemistry Series*, No. 67, American Chemical Society, Washington, DC.

Rubin, A. J., and Kovac, T. W. (1974) Effect of Aluminum III Hydrolysis on Alum Coagulation, in A. J. Rubin (ed.), *Chemistry of Water Supply, Treatment and Distribution*, Ann Arbor Science, Ann Arbor, MI.

Sajjad, M., and Cleasby, J. (1995) "The Effect of Impeller Geometry and Various Mixing Patterns on Flocculation Kinetics of Kaolin Using Ferric Salt," *Proc. AWWA Annual Conf.*, 265–305.

Sawyer, C. N., McCarty, P. L., and Parkin, G. F. (2002) *Chemistry for Environmental Engineering*, 5th ed., McGraw-Hill, New York.

Schaefer, D. V. (1989) "Polymers, Fractals, and Ceramic Materials," *Science*, **243**, 1023–1027.

Shäfer, M., Yianneskis, M., Wächter, P., and Durst, F. (1998) "Trailing Vortices around a 45° Pitched Blade Impeller," *AIChE J.*, **44**, 1233–1246.

Shaw, D. J (1966) *Introduction to Colloid and Surface Chemistry*, Butterworth, London.

Smoluchowski, M. (1917) "Versuch einer mathematischen Theorie der Koagulationskinetic Kolloider Losunger," *Zeit. Phys. Chemie*, **92**, 129–168.

Snoeyink, V. L., and Jenkins, D. (1980) *Water Chemistry*, 2nd ed., John Wiley & Sons, New York.

Stumm, W., and Morgan, J. J. (1981) *Aquatic Chemistry*, 2nd ed., Wiley-Interscience, New York.

Stumm, W., and Morgan, J. J. (1996) *Aquatic Chemistry*, 3rd ed., Wiley, New York.

Stumm, W., and O'Melia, C. R. (1968) "Stoichiometry of Coagulation," *J. AWWA*, **60**, 514–539.

Swift, D. L., and Friedlander, S. K. (1964) "The Coagulation of Hydrosols by Brownian Motion and Laminar Shear Flow," *J. Colloid Sci.*, **19**, 621–647.

Tekippe, R. J., and Ham, R. K. (1970) "Coagulation Testing: A Comparison of Techniques," *J. AWWA*, **62**, 9, 594–628.

Thomas, D. G. (1964) "Turbulent Disruption of Flocs in Small Particle Size Suspensions," *AIChE J.*, **10**, 517–523.

Trussell, R. R. (1978) Chapter 3, Predesign Studies, in *Water Treatment Plant Design*, R. L. Sanks (ed.), Ann Arbor Science, Ann Arbor, MI.

U.S. EPA (1998) "National Primary Drinking Water Regulations: Disinfectants and Disinfection Byproducts; Final Rule," *Fed. Reg.*, **63**, 241, 69390.

U.S. EPA (1999) *Enhanced Coagulation and Enhanced Precipitative Softening Guidance Manual*, U.S. EPA, Washington, DC.

van Olphen, H. (1963) *An Introduction to Clay Colloid Chemistry*, Wiley-Interscience, New York.

Van't Reit, K., Bruijn, W., and Smith, J. (1976) "Real and Pseudo Turbulence in the Discharge Stream from a Rushton Turbine," *Chem. Engrg. Sci.*, **31**, 407–412.

Verwey, E. J. W., and Overbeek, J. T. G. (1948) *Theory of the Stability of Lyophobic Colloids*, Elsevier, Amsterdam.

Vrale, L., and Jorden, R. M. (1971) "Rapid Mixing in Water Treatment," *J. AWWA*, **63**, 1, 52–58.

Voyutsky, S. (1978) *Colloid Chemistry*, MIR Publishers, Moscow, Russia.

White, M.C., Thompson, J.D., Harrington, G.W., and Singer, P.C. (1997) "Evaluating Criteria for Enhanced Coagulation Compliance," *J. AWWA*, **89**, 5, 64–77

10 Gravity Separation

Stantec's Water Treatment: Principles and Design, Updated Third Edition. John C. Crittenden, R. Rhodes Trussell, David W. Hand, Kerry J. Howe, and George Tchobanoglous.

Terminology for Gravity Separation

Term	Definition
Colloidal particles	Very small particles that do not settle out of water for a long time because of their small size and electrical charge.
Compression settling (Type IV)	Particle blanket containing a high concentration of particles that compresses by allowing water to move up through the voids within the blanket due to the weight of the particles in the blanket.
Density currents	Nonideal flow patterns that occur in sedimentation basins caused by differences in fluid density (e.g., temperature, solids concentrations) within the tank and outside forces (e.g.,wind) acting on the fluid in the tank.
Discrete particles	Particles that are completely surrounded by water and have no interactions with other particles in the water.
Discrete settling (Type I)	Settling of discrete particles in water by gravity with no interactions with other particles in the water.
Dissolved air flotation	Gravity separation process in which dissolved air forms tiny bubbles in the water at the bottom of a basin. As the bubbles float upward, they attach to particles, causing the particles to float to the surface where they can be removed by skimming.
Flocculent settling (Type II)	Process in which initially discrete or flocculated particles interact with one another during the settling and form larger particles with higher settling velocities.

Term	Definition
Gravity separation	Process of removing particles suspended in water by the force of gravity.
Hindered settling (Type III)	Process in which particles interact to retard the settling of nearby particles, resulting in the formation of a layer or blanket that settles and also overtakes other particles in its path.
Particle settling velocity	Rate at which a particle settles in water.
Sedimentation basins	Tanks or chambers designed to separate and remove settleable particles from water.
Settleable particles	Suspended particles in water that can be removed by the process of gravity separation.
Stokes' law	Mathematical expression used to predict the terminal settling velocity of a discrete particle falling in a viscous fluid under the force of gravity.
Suspended particles	Small organic or inorganic material or solids found in water.

Gravity separation of suspended material from water is the oldest and most widely used process in water treatment. Gravity separation historically has meant sedimentation, where water is introduced into a large quiescent basin for a long enough period of time so that the majority of the particles in the water settle to the bottom of the basin. Most raw surface waters will contain mineral particles and organic particles. Mineral particles usually have densities ranging from 2000 to 3000 kg/m^3 and will settle out readily by gravity, whereas organic particles with densities of 1010 to 1100 kg/m^3 may require a long time to settle by gravity. In a conventional treatment train, sedimentation follows coagulation and flocculation, which are used to destabilize particles and form large aggregates that will settle out of suspension in a reasonable time period. Over time, sedimentation practice has undergone a number of changes. Sedimentation tanks have been modified to accelerate the separation process. More recently, air bubbles have been used to float particles to the water surface for removal. Adding air bubbles to the water is known as dissolved air flotation (DAF), which is discussed in this chapter along with more traditional gravity separation processes.

The removal of suspended matter from water at low cost and low energy consumption is conceptually simple but often involves complications that render proper sedimentation basin design a challenge for many engineers. The performance of a sedimentation basin for a given raw-water quality can be understood with the help of particle-settling theories. When supplemented with the understanding of the practical aspects of sedimentation basin design, sedimentation basins can be designed to perform reliably and consistently.

The topics discussed in this chapter include (1) the classification of particles for settling, (2) principles of discrete particle settling, (3) discrete particle settling in sedimentation basins, (4) principles of flocculant settling, (5) principles of hindered settling, (6) conventional sedimentation basin design, (7) alternative sedimentation processes, (8) physical factors affecting sedimentation, and (9) dissolved air flotation.

10-1 Classification of Particles for Settling

Particles are separated into four classifications based on their concentration and morphology, as shown on Fig. 10-1. Type I particles are discrete and do

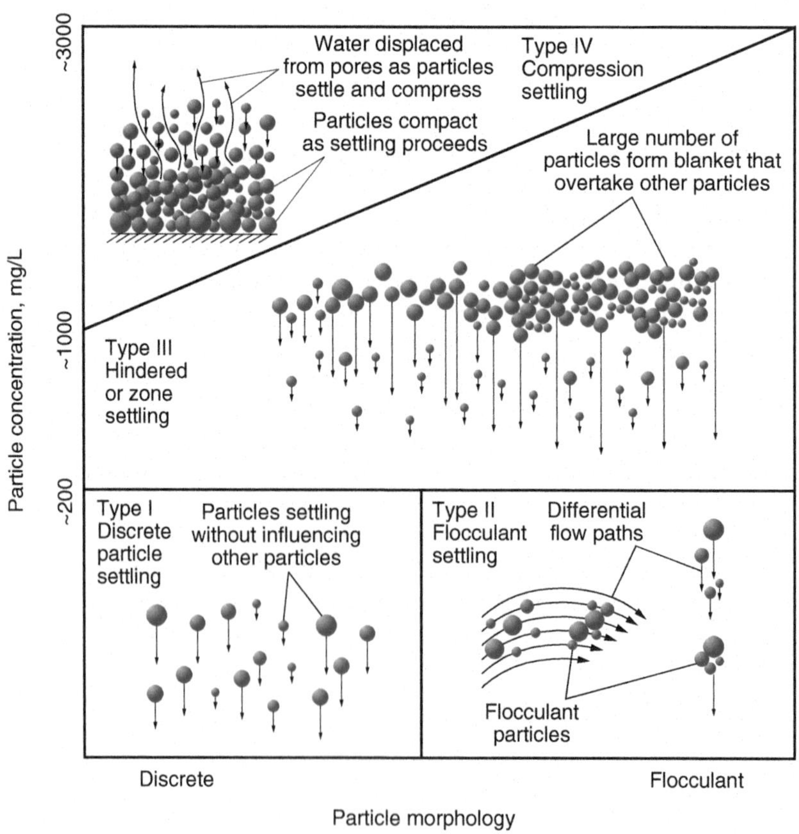

Figure 10-1
Relationship between settling type, concentration, and flocculent nature of particles.

not interfere with one another during settling because the concentration is low and they do not flocculate. Type I suspensions are found in grit chambers, presedimentation basins for sand removal prior to coagulation, and settling of sand particles during backwashing of rapid sand filters. Type II suspensions consist of particles that can adhere to each other if they bump into each other (i.e., they are capable of flocculating). As particles aggregate and grow in size, they can settle faster. Type II suspensions are found when settling occurs following iron and alum coagulation and in most conventional sedimentation basins.

At concentrations higher than Type I and II suspensions, hindered, or Type III, settling occurs. In hindered settling, a blanket of particles is formed. The blanket traps particles below it as it settles; consequently, a clear interface is found above the blanket. The settling velocity of the blanket depends on the suspended solids concentration, with the blanket velocity decreasing with increasing concentration. Type III suspensions are found in thickeners (sludge disposal) and the bottom of some sedimentation basins (e.g., lime-softening sedimentation).

At much higher concentrations than are found in Type III settling, the suspension begins to consolidate slowly. This type of settling or consolidation is known as Type IV settling or compression settling. For Type IV suspensions, the particles may not really settle, and a more correct visualization of what is occurring is that water flows or drains out of a mat of particles very slowly. Type IV suspensions are found in dewatering operations, and once they are dewatered, the suspension may become a paste or cake.

10-2 Principles of Discrete (Type I) Particle Settling

In a dilute suspension, individual particles settle based on their size and density and do not interact with each other. Settling only occurs if the vertical movement overcomes the random movement of particles. This section develops the equations that describe settling velocity and then compares the settling velocity to the velocity due to Brownian motion.

Settling Velocity of Discrete Particles

A particle moving vertically through a fluid is subjected to gravitational and drag forces. The vertical forces acting on the particle are shown in the free-body diagram on Fig. 10-2 and the force balance is given by the expression

$$\sum F = F_g - F_b - F_d \tag{10-1}$$

where F_g = gravitational force, N
F_b = buoyant force, N
F_d = drag force, N

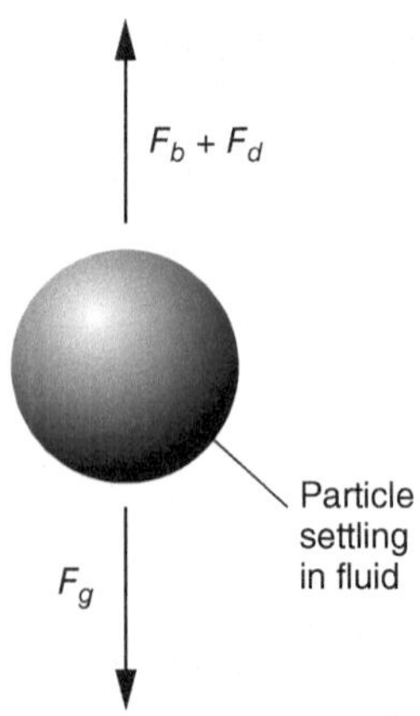

Figure 10-2
Forces acting on settling particle.

The force balance is written so that the direction of gravitational force is positive. Therefore, a positive settling velocity means that the particle settles and a negative settling velocity means the particle rises. The gravitational and buoyant forces are given by $F = ma$, as follows:

$$F_g = ma = \rho_p V_p g \tag{10-2}$$

$$F_b = ma = \rho_w V_p g \tag{10-3}$$

where m = mass, kg
a = acceleration, m/s^2
ρ_p = density of particle, kg/m^3
ρ_w = density of water, kg/m^3
V_p = volume of particle, m^3
g = acceleration due to gravity, 9.81 m/s^2

In 1647, Issac Newton proposed that the drag force could be described by the expression

$$F_d = \frac{1}{2} C_d \rho_w A_p v_s^2 \tag{10-4}$$

where C_d = drag coefficient, unitless
A_p = projected area of the particle in direction of flow, m^2
v_s = settling velocity of the particle, m/s

If the particles are spherical, the volume and projected area are given by the following expressions:

$$V_p = \frac{\pi}{6} d_p^3 \tag{10-5}$$

$$A_p = \frac{\pi}{4} d_p^2 \tag{10-6}$$

where d_p = particle diameter, m

If a particle starts at rest, it will accelerate due to an imbalance in forces. As the particle velocity increases, the drag force increases until the vertical forces are balanced (i.e., $\Sigma F = 0$). At that time, the particle reaches a constant velocity known as its terminal settling velocity. For conditions typical in water treatment, the period of initial acceleration is extremely short and not relevant in sedimentation basin design. By substituting Eqs. 10-2 to 10-6 into Eq. 10-1 and setting $\Sigma F = 0$, the following expression for terminal settling velocity is obtained:

$$v_s = \sqrt{\frac{4g(\rho_p - \rho_w) d_p}{3 C_d \rho_w}} \tag{10-7}$$

The drag coefficient as defined in Eq. 10-4 generally cannot be predicted theoretically. Drag coefficients are determined experimentally by measuring settling velocity in laboratory experiments and then calculating the drag coefficient using Eq. 10-7. Analysis of experimental data reveals that the drag coefficient depends on the Reynolds number, where the Reynolds number is defined as

$$\mathrm{Re} = \frac{\rho_w v_s d_p}{\mu} = \frac{v_s d_p}{\nu} \tag{10-8}$$

where Re = Reynolds number, dimensionless
μ = dynamic viscosity, $\mathrm{N \cdot s/m^2}$ or $\mathrm{kg/m \cdot s}$
ν = kinematic viscosity, $\mathrm{m^2/s}$

The drag coefficient for spheres as a function of Reynolds number is shown on Fig. 10-3. At low Reynolds numbers (laminar region), viscous forces control the drag force and the drag coefficient is larger because momentum is transferred farther into the fluid. As the Reynolds number increases, inertial forces become more significant. In the turbulent regime, inertial forces of displaced fluid control the drag force (the particle basically punches a hole in the fluid equal to the size of the projected area) and the drag coefficient becomes a constant.

Over the years, many researchers have collected experimental settling velocity data and developed various empirical correlations for the drag coefficient, some easier to use than others. Brown and Lawler (2003) conducted a rigorous reevaluation of much of the existing settling velocity data

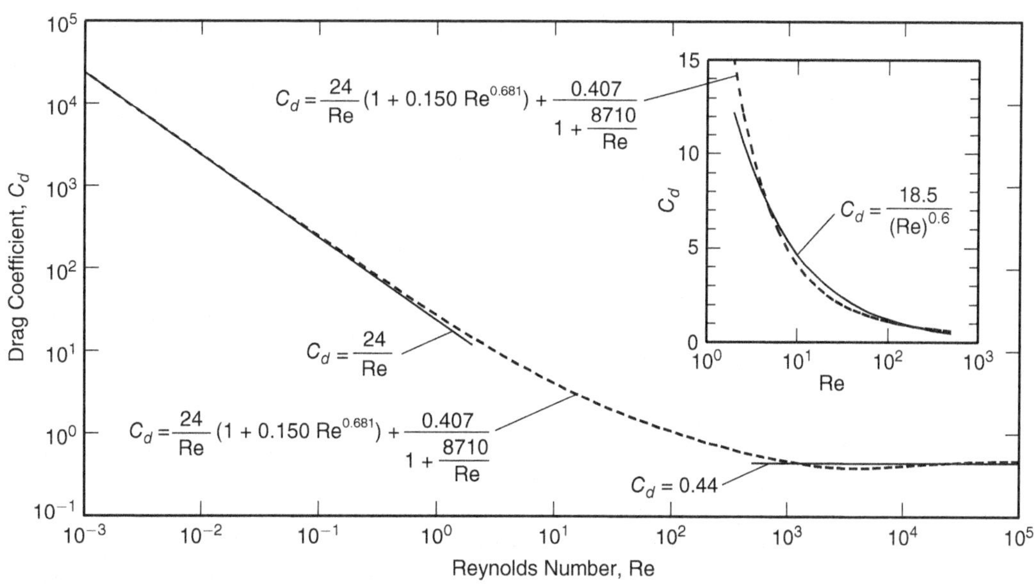

Figure 10-3
Newton's coefficient of drag for varying magnitudes of Reynolds numbers.

and proposed an empirical correlation for the drag coefficient that fit all experimental data less than Re = 2×10^5. The proposed correlation is

$$C_d = \frac{24}{\text{Re}}\left(1 + 0.150\,\text{Re}^{0.681}\right) + \frac{0.407}{1 + 8710/\text{Re}} \tag{10-9}$$

While this correlation results in a single equation for drag coefficient that covers a wide range of Reynolds numbers, it cannot be substituted into Eq. 10-7 and easily manipulated to produce an equation for settling velocity. Thus, it is useful to develop simpler correlations that are reasonably accurate over smaller ranges of Reynolds numbers. For spherical particles, the drag coefficient C_d can be approximated by the following expressions, depending on the magnitude of the Reynolds number (Clark, 1996):

$$C_d = \frac{24}{\text{Re}} \quad \text{for Re} < 2 \qquad \text{(laminar flow)} \tag{10-10}$$

$$= \frac{18.5}{\text{Re}^{0.6}} \quad \text{for } 2 \leq \text{Re} \leq 500 \qquad \text{(transition flow)} \tag{10-11}$$

$$= 0.44 \quad \text{for } 500 < \text{Re} \leq 2 \times 10^5 \qquad \text{(turbulent flow)} \tag{10-12}$$

For comparison purposes, drag coefficients calculated using Eqs. 10-9 to 10-12 are shown in Fig. 10-3. Equation 10-9 should be considered for rigorous laboratory studies, but in full-scale systems, confounding factors such as heterogeneities in particle size and geometry and currents in fluid flow reduce the usefulness of a highly accurate equation for drag coefficient. In those circumstances, the simpler Eqs. 10-10 to 10-12 are sufficient.

The equations for drag coefficients can be substituted into Eq. 10-7 to develop equations for settling velocity as a function of flow regime. In water treatment, particle settling generally occurs in the laminar and transition flow regimes. For laminar and transition flow, respectively, the equation becomes

$$v_s = \frac{g\left(\rho_p - \rho_w\right) d_p^2}{18\mu} \qquad \text{(laminar flow)} \tag{10-13}$$

$$= \left[\frac{g\left(\rho_p - \rho_w\right) d_p^{1.6}}{13.9\,\rho_w^{0.4}\,\mu^{0.6}}\right]^{1/1.4} \qquad \text{(transition flow)} \tag{10-14}$$

Equation 10-13, for spherical particles and laminar flow, is commonly referred to as Stokes' law.

Equation 10-13 or 10-14 is used to calculate the settling velocity depending on whether the particle is in laminar or transition flow. However, the flow regime depends on the settling velocity, so it is not possible to predict a priori which equation applies. It is necessary to calculate the settling velocity using one of the equations, then calculate the flow regime based on the resultant settling velocity, and recalculate the settling velocity with the other equation if necessary. For sand particles (density = 2650 kg/m^3) and a temperature of 20°C, Stokes' law is valid for particles up to 0.13 mm in diameter, and Eq. 10-14 is valid for particles up to 1.7 mm in diameter. Calculating terminal settling velocity is demonstrated in Example 10-1.

Example 10-1 Calculating terminal settling velocity

Calculate the terminal settling velocity for sand in water at 10°C having particle diameters of 75 and 180 μm and a density of 2650 kg/m^3. The density and viscosity of water at 10°C are available in Table C-1 in App. C.

Solution

1. Calculate the settling velocity and Reynolds number for the 75-μm sand particles.
   a. Since the settling velocity is unknown, the Reynolds number is also unknown. First, calculate settling velocity using Eq. 10-13 (Stokes' law). From Table C-1 in App. C, μ = 1.307 N · s/m^2 (or kg/m · s) and ρ_w = 999.7 kg/m^3.

$$V_s = \frac{(9.81\ \text{m/s}^2)(2650 - 999.7\ \text{kg/m}^3)(75\times10^{-6}\ \text{m})^2}{18(1.307\times10^{-3}\ \text{kg/m·s})}$$
$$= 0.00387\ \text{m/s}$$

   b. Check the Reynolds number using Eq. 10-8:

$$\text{Re} = \frac{\rho_w V_s d_p}{\mu}$$
$$= \frac{(999.7\ \text{kg/m}^3)(75\times10^{-6}\ \text{m})(0.00387\ \text{m/s})}{1.307\times10^{-3}\ \text{kg/m·s}}$$
$$= 0.22$$

   Because Re < 2, laminar flow exists and Stokes' law is valid. The settling velocity of a 75-μm sand particle in water is 0.00387 m/s (13.9 m/h).

2. Calculate the settling velocity and Reynolds number for the 180-μm sand particles.
   a. Calculate the settling velocity using Eq. 10-13:

$$v_s = \frac{(9.81\text{m/s}^2)(2650 - 999.7\ \text{kg/m}^3)(1.80\times10^{-4}\ \text{m})^2}{18(1.307\times10^{-3}\ \text{kg/m·s})}$$
$$= 0.0223\ \text{m/s}$$

b. Check the Reynolds number using Eq. 10-8:

$$Re = \frac{\left(999.7\,\text{kg}/\text{m}^3\right)\left(1.80\times10^{-4}\ \text{m}\right)(0.0223\,\text{m}/\text{s})}{1.307\times10^{-3}\ \text{kg}/\text{m}\cdot\text{s}}$$

$$= 3.07$$

Since Re > 2.0, Stokes' law is not valid.

c. Calculate the settling velocity using Eq. 10-14:

$$v_s = \left[\frac{\left(9.81\,\text{m}/\text{s}^2\right)\left(2650-999.7\,\text{kg}/\text{m}^3\right)\left(1.80\times10^{-4}\ \text{m}\right)^{1.6}}{13.9\left(999.7\,\text{kg}/\text{m}^3\right)^{0.4}\left(1.307\times10^{-3}\ \text{kg}/\text{m}\cdot\text{s}\right)^{0.6}}\right]^{1/1.4}$$

$$= 0.0195\,\text{m}/\text{s}$$

d. Check the Reynolds number using Eq. 10-8:

$$Re = \frac{\left(999.7\ \text{kg}/\text{m}^3\right)\left(1.80\times10^{-4}\ \text{m}\right)(0.0195\ \text{m}/\text{s})}{1.307\times10^{-3}\ \text{kg}/\text{m}\cdot\text{s}}$$

$$= 2.68$$

Because Re > 2, transition flow exists and Eq. 10-14 is valid. The settling velocity of a 180-μm sand particle in water is 0.0195 m/s (70.1 m/h).

If the particles are hard spheres, the settling velocity as a function of particle size does follow Eq. 10-13 or 10-14. However, flocculated particles have fractal morphology and are composed of many flocculated small particles. Consequently, fractal particles do not settle as rapidly as would be estimated using a hard-sphere model. Fractal particles are discussed in Chap. 9 and a detailed discussion of fractals can be found in Logan (1999).

Brownian Motion

Particles in natural waters can be so small that they do not settle because random movement caused by collisions with fluid molecules, known as Brownian motion, can overwhelm the vertical movement due to gravity. As presented in Chap. 7, Einstein equated the drag force (Eq. 10-4) to the force of the collisions between fluid molecules and particles resulting from the kinetic energy of the fluid molecules. For spherical particles and perfect elastic collisions, the mean square distance traveled by a particle due to Brownian motion can be described by the following relationship:

$$\overline{x^2} = \frac{2kT}{3\pi\mu d_p}t \qquad (10\text{-}15)$$

where x = distance traveled due to Brownian motion, m

k = Boltzmann's constant, 1.38×10^{-23} J/K

T = absolute temperature, K (273 + °C)
t = time, s
μ = dynamic viscosity, N · s/m^2
d_p = particle diameter, m

When the movement due to Brownian motion is large relative to the movement due to settling, particles will not settle out of solution because the motion of the particle will be governed by collisions with water molecules. A comparison of Brownian motion and settling velocity is demonstrated in Example 10-2.

Example 10-2 Comparison between distance traveled by Brownian motion and Stokes' settling

Estimate the size of a sand particle (sg_p = 2.65) that would move the same distance in 1 s due to Brownian motion as it would settle in water at 20°C based on Stokes' law.

Solution

1. Rearrange Stokes' law to solve for the distance settled in 1 s, noting that $v = x/t$:

$$x_s = \frac{g(\rho_p - \rho_w)d_p^2}{18\mu}t$$

2. Set the equation from step 1 equal to the distance traveled by a particle by Brownian motion, and solve for particle size:

$$\left(\frac{2kT}{3\pi\mu d_p}t\right)^{0.5} = \frac{g(\rho_p - \rho_w)d_p^2}{18\mu}t$$

$$\frac{2kT}{3\pi\mu d_p}t = \left[\frac{g(\rho_p - \rho_w)t}{18\mu}\right]^2 d_p^4$$

$$d_p^5 = \frac{2kT(18\mu)^2}{3\pi\mu g^2(\rho_p - \rho_w)^2 t}$$

$$d_p = \left[\frac{216kT\mu}{\pi g^2(\rho_p - \rho_w)^2 t}\right]^{1/5}$$

3. Substitute the values from the problem statement and solve for the particle size. The density and viscosity of water at 20°C is available

in Table C-1 in App. C. At 20°C, $\rho = 998.2\ \text{kg/m}^3$ and $\mu = 1.002 \times 10^{-3}\ \text{N} \cdot \text{s/m}^2$. Note that the units for joules can be converted 1 J = 1 kg · m²/s².

$$d_p = \left[\frac{(216)(1.38\times10^{-23}\ \text{kg}\cdot\text{m}^2/\text{s}^2\cdot\text{K})(293\ \text{K})(1.002\times10^{-3}\ \text{kg/m}\cdot\text{s})}{\pi(9.81\ \text{m/s}^2)^2(2650-998.2\ \text{kg/m}^3)^2(1\ \text{s})}\right]^{1/5}$$

$$= 1.01\times10^{-6}\ \text{m} = 1.01\ \mu\text{m}$$

Comment

Based on this estimate, a 1-μm sand particle with a specific gravity of 2.65 will move about the same distance by Brownian motion and it would by settling in 1 s. Thus, particles smaller than this are unlikely to settle in a reasonable period of time because Brownian motion would keep them suspended. Clearly, the relationship between Brownian motion and settling velocity is not exact (i.e., particles will settle more slowly as the Brownian motion increases relative to the settling velocity, but there is not an exact point below which particles suddenly become nonsettleable). Nevertheless, the observation that particles smaller than about 1 μm exhibit poor settling characteristics is correct. Coagulation and flocculation (Chap. 9) can be used to coalesce these particles into a form that settles more readily.

10-3 Discrete Settling in Ideal Sedimentation Basins

Particle settling is dependent on the nature of the particle and geometry of the sedimentation process. As introduced in Sec. 10-1, there are four main types of particle settling (see Fig. 10-1). The analysis of discrete particle settling in sedimentation basins, based on the principles presented in Sec. 10-2, is introduced in this section.

The theory for ideal settling was originally put forth by Hazen (1904). Camp (1936) later developed a rational theory for the removal of discrete particles in an ideal sedimentation basin. Camp divided a settling tank into four zones, as illustrated on Fig. 10-4. The inlet, sludge, and outlet zones were considered special tank areas that permit ideal settling in the settling zone but do not achieve particulate removal. In addition, the following assumptions were made by Camp to develop a theoretical basis for the removal of discrete particles: (1) plug flow conditions exist in the settling zone, (2) there is uniform horizontal velocity in the settling zone, (3) there

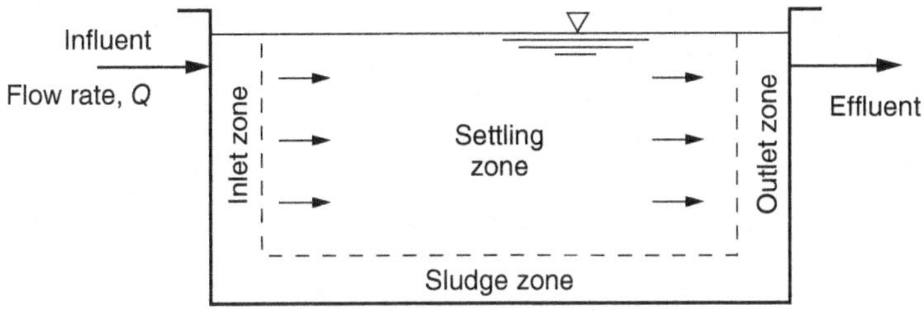

Figure 10-4
Functional regions within rectangular clarifier.

is uniform concentration of all size particles across a vertical plane at the inlet end of the settling zone, (4) particles are removed once they reach the bottom of the settling zone, and (5) particles settle discretely without interference from other particles at any depth.

Rectangular Sedimentation Basins

Particle trajectories have two components in the settling zone: the settling velocity v_s and the fluid velocity v_f, as shown on Fig. 10-5. For a rectangular sedimentation basin, the fluid velocity is constant. The settling velocity for discrete particles is also constant because the particles do not flocculate or interfere with one another. Since both horizontal and vertical components of the velocity are constant, the particle trajectories are linear. As noted above, every particle that enters the sludge zone is removed. A particle from the inlet zone that enters at the top of the basin and settles in the sludge zone just before the outlet is assigned a settling velocity of v_c, or a critical settling velocity (particle 2 in Fig. 10-5). The critical particle settling velocity is given by the equation

$$v_c = \frac{h_o}{\tau} \tag{10-16}$$

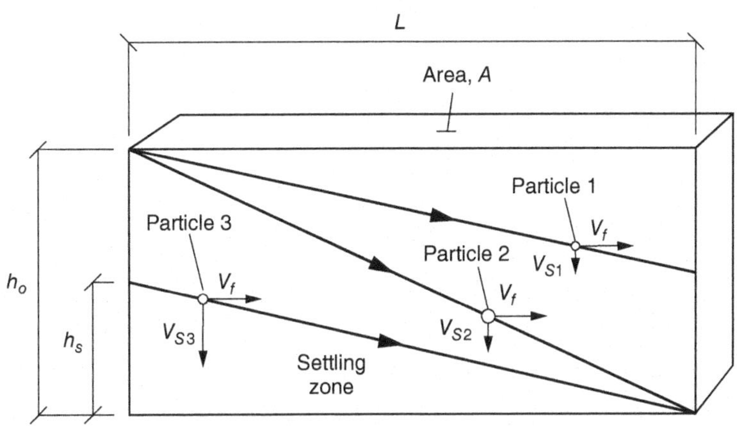

Figure 10-5
Discrete particle trajectories in settling zone of a rectangular clarifier.

where v_c = particle settling velocity such that particle at surface of inlet is removed in sludge zone just before outlet, m/h
ho = depth of sedimentation basin, m
τ = hydraulic detention time of sedimentation basin, h

The critical settling velocity may be defined as the overflow rate using the relationships

$$v_c = \frac{h_o}{\tau} = \frac{h_o Q}{h_o A} = \frac{Q}{A} = \text{OR} \quad (10\text{-}17)$$

where OR = overflow rate, $m^3/m^2 \cdot h$ (equal to v_c)
A = area of top of basin settling zone (see Fig. 10-5), m^2
Q = process flow rate, m^3/h

The inlet zone is assumed to be homogenous; therefore, particles may enter the settling zone at any height h_s. Any particles in the inlet zone with a settling velocity v_s greater than or equal to the critical settling velocity v_c will be removed regardless of the starting position because their trajectories will take them into the sludge zone before they exit the basin.

Particles with a settling velocity less than v_c may also be removed, depending on their position at the inlet, as shown on Fig. 10-5. Particles at the top of the basin will pass through the settling zone and exit in the outlet zone and will not be removed. However, particles starting at position h_s and lower will enter the sludge zone before exiting the basin and will be removed. The fraction of particles that will be removed is given by the expression

$$\text{Fraction of particles removed} = \frac{h_s}{h_o} = \frac{h_s/\tau}{h_o/\tau} = \frac{v_s}{v_c}(v_s < v_c) \quad (10\text{-}18)$$

where h_s = height of particle from bottom of tank at position entering settling zone, m
v_s = particle settling velocity smaller than v_c, m/h

Other terms are as defined above. Removal of particles as a function of size is demonstrated in Example 10-3.

Example 10-3 Particle removal in sedimentation basin

Calculate the particle removal efficiency in a rectangular sedimentation basin with a depth of 4.5 m, width of 6 m, length of 35 m, and process flow rate of 525 m^3/h. Compute the required sedimentation basin design parameters and plot the influent and effluent particle concentrations as a function of particle size using a histogram. Assume the following influent particle-settling characteristics (adapted from Tchobanoglous et al., 2003):

Settling Velocity, m/h	Number of Particles, #/mL
0–0.4	511
0.4–0.8	657
0.8–1.2	876
1.2–1.6	1168
1.6–2.0	1460
2.0–2.4	1314
2.4–2.8	657
2.8–3.2	438
3.2–3.6	292
3.6–4.0	292
Total	7665

Solution

1. Compute the sedimentation basin overflow rate and critical settling velocity using Eq. 10-17:

$$\text{OR} = v_c = \frac{Q}{A} = \frac{525\ \text{m}^3/\text{h}}{(6\ \text{m})(35\ \text{m})} = 2.5\ \text{m}^3/\text{m}^2\cdot\text{h}$$

2. Compute the percent removal of particles in each size range using a data table.
   a. Compute the average settling velocity for each particle size range; see column 2 in the table below.
   b. Compute the fraction of particles removed using Eq. 10-18. For particles with an average settling velocity of 1.0 m/h, the fraction of particles removed is (1.0 m/h)/(2.5 $\text{m}^3/\text{m}^2\cdot\text{h}$) = 0.4; see column 4. Note that for particle-settling ranges with a fraction removed greater than 1, a value of 1 should be used.
   c. Estimate the number of particles that will be removed and remaining in each size range. The number of particles removed is determined by multiplying the influent particle concentration for a given settling velocity range by the corresponding percent removal, (876)(0.4) = 350; see column 5. The number of remaining particles is determined by subtracting the removed particles from the influent particles for each size range, 876 – 350 = 526, for the range 0.8 to 1.2 m/h; see column 6.

d. The remaining values are summarized in the following table:

Settling Velocity, m/h (1)	Average Settling Velocity, m/h (2)	Number of Influent Particles, #/mL (3)	Fraction of Particles Removed (4)	Number of Particles Removed, #/mL (5)	Number of Particles in Effluent, #/mL (6)
0–0.4	0.2	511	0.08	41	470
0.4–0.8	0.6	657	0.24	158	499
0.8–1.2	1.0	876	0.40	350	526
1.2–1.6	1.4	1168	0.56	654	514
1.6–2.0	1.8	1460	0.72	1051	409
2.0–2.4	2.2	1314	0.88	1156	158
2.4–2.8	2.6	657	1	657	0
2.8–3.2	3.0	438	1	438	0
3.2–3.6	3.4	292	1	292	0
3.6–4.0	3.8	292	1	292	0
Total		7665		5090	2575

3. Compute the overall particle removal efficiency:

$$\text{Removal efficiency} = \frac{5090}{7665} = 0.664 = 66.4\%$$

4. Plot the influent and effluent particle concentrations for each settling velocity range using a histogram.

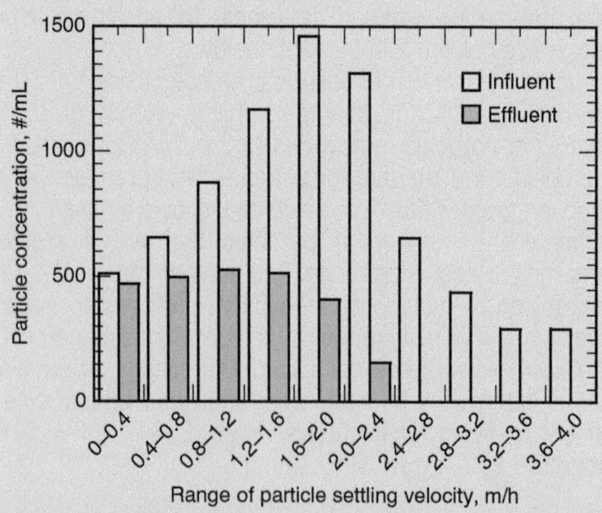

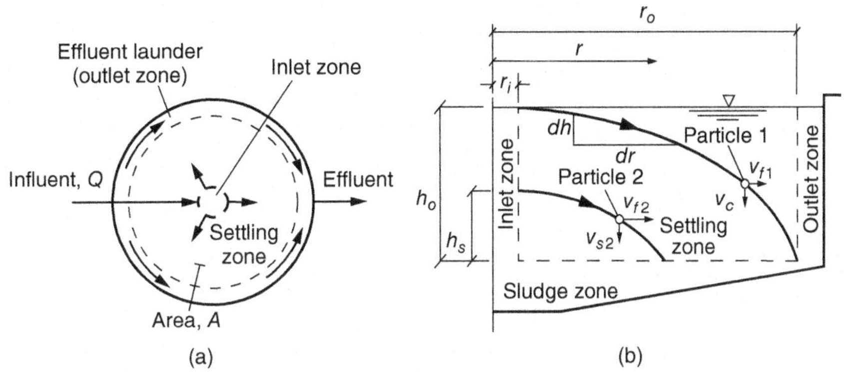

Figure 10-6
Analysis of particle settling in circular clarifier: (a) plan view of circular clarifier and (b) particle trajectory of discrete particles in settling zone of circular clarifier.

Circular Sedimentation Basins

The removal of particles in a circular sedimentation tank can also be described using Eqs. 10-17 and 10-18, as shown on Fig. 10-6. As shown in Fig. 10-6, the settling zone in the circular sedimentation basin extends from radius r_i to r_o. As the fluid moves from the center of the tank (inlet zone) through the settling zone, the fluid velocity changes according to the equation

$$v_f = \frac{Q}{2\pi r h_o} \tag{10-19}$$

where v_f = fluid velocity, m/h
Q = flow rate, m³/h
r = distance measured from center of clarifier, m
h_o = depth of settling zone, m

The trajectory of a particle that starts at the top of the inlet zone and enters the sludge zone just before the outlet zone is shown for particle 1 on Fig. 10-6b. For a given settling time t, the particle moves a horizontal distance given by $v_f\ \Delta r$ and a vertical distance given by $v_c\ \Delta h$. Equating these and integrating, the distance that particle 1 has settled as a function of r is given by the equation

$$h = tv_c = \frac{\pi\left(r^2 - r_i^2\right)h_o}{Q}v_c \tag{10-20}$$

where r_i = radius of inlet zone, m
h = distance from water surface for particle 1 (see Fig. 10-6b), m
t = settling time, h
v_c = critical particle settling velocity, m/h

Accordingly, discrete particles have a parabolic trajectory in an ideal circular sedimentation basin. Particles with a settling velocity greater than or equal to v_c are all removed. The settling velocity v_c can be related to the overflow rate according to the equation

$$v_c = \frac{h_o}{\tau} = \frac{h_o Q}{h_o \pi\left(r_o^2 - r_i^2\right)} = \frac{Q}{\pi\left(r_o^2 - r_i^2\right)} = \frac{Q}{A} = \text{OR} \qquad (10\text{-}21)$$

where τ = hydraulic detention time of basin, h
r_o = radius of outer edge of the settling zone, m
r_i = radius of inlet zone, m
OR = overflow rate, $m^3/m^2 \cdot h$
A = area of top of basin in settling zone, m^2

The result shown in Eq. 10-21 is identical to the result given in Eq. 10-17. Consequently, the critical design parameter (overflow rate) for rectangular and circular sedimentation basins is identical.

10-4 Principles of Flocculant (Type II) Settling

Type II settling typically occurs in conventional sedimentation basins following coagulation. There are two principal mechanisms of flocculation during sedimentation: (1) differences in the settling velocities of particles whereby faster settling particles overtake those that settle more slowly and coalesce with them and (2) velocity gradients within the liquid that cause particles in a region of a higher velocity to overtake those in adjacent stream paths moving at slower velocities.

Advantages of Flocculant Settling

Flocculation within a sedimentation basin is considered beneficial for two principal reasons. First, the combination of smaller particles to form larger particle aggregates results in faster settling particles because of the increase in size. Second, flocculation tends to have a sweeping effect in which large particles settling at a velocity faster than smaller particles tend to sweep some of the smaller particles from suspension. Consequently, many tiny particles and particles that settle slowly are removed. The net effect of flocculation during settling is a reduction in the size of the sedimentation basin necessary for effective clarification or improved water quality exiting the sedimentation basin.

Analysis of Flocculant Settling

Design equations for Type II suspensions using the flocculation equations have proven to be impractical for sedimentation and flotation basin design. Design of sedimentation basins is usually based on overflow rates and detention times that have been reported in design manuals as guidelines or by regulatory agencies. For waters with unusual settling characteristics, a number of investigators have developed design equations based on column experiments. In a technique developed by O'Connor and Ecken-felder (1958), measured solids concentrations taken at regular intervals throughout the depth of a quiescent settling column, slightly deeper than the proposed sedimentation basin, are related to the overall percent removal at a particular basin residence time. The water to be treated is placed in the column and allowed to settle for the detention time of the basin. The effluent concentration is equal to the average concentration in the column. The average concentration can be obtained by draining off the settled solids and then mixing the particles remaining in the column (typically with air) and then sampling the mixed liquid. The concept behind this approach is that the column represents a fluid element that travels as a plug through the basin and has a settling time equal to the basin residence time.

Several fundamentals of sedimentation basin design that are different from design principles arrived at through discrete particle settling have been established. The depth of the basin is important because flocculent particles tend to grow in size during their downward movement through the basin. A greater depth facilitates floc growth and allows for sweep flocculation at high solids concentrations at the bottom of the basin. In general, more flocculent particles are removed in deeper basins.

10-5 Principles of Hindered (Type III) Settling

Type III settling, also known as zone settling, occurs when the settling velocities of particles are affected by the presence of other particles. When particles are dispersed in solution, the movement of the fluid that is displaced by the particle motion has little impact on the drag force. However, when particle concentrations are high enough to restrict the fluid velocity fields around individual particles, a settling particle experiences increased frictional forces. In water treatment, hindered settling typically occurs in the lower regions of the sedimentation basin, where the concentration of suspended particles is highest. When Type III settling occurs, particle aggregates form a blanket of particles with a distinct interface with the clarified liquid in the basin. Zone settling is of primary importance in water treatment in sludge thickening and dewatering operations, as discussed in Chap. 21.

Solids Flux Analysis

The solids flux in a sedimentation basin or solids thickener (see Fig. 10-7) is comprised of the downward movement of particles due to gravity settling and the downward movement of particles due to fluid flow toward the underdrain, as shown in the expression

$$J_T = J_s + J_u \tag{10-22}$$

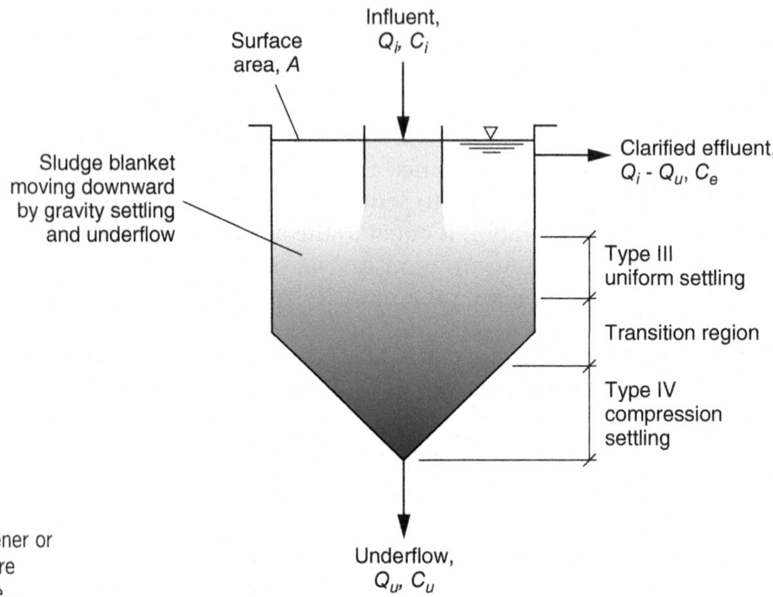

Figure 10-7
Diagram of sludge thickener or sedimentation basin where thickening is taking place.

where J_T = total solids flux toward the bottom of the basin, $kg/m^2 \cdot h$
J_s = solids flux due to particle settling, $kg/m^2 \cdot h$
J_u = solids flux due to fluid flow from the underflow, $kg/m^2 \cdot h$

To determine the solids flux from gravity settling J_s, the depth of the blanket interface is measured as a function of time in a column that is initially uniformly mixed with a specified solids concentration C. Data from a settling column test is shown on Fig. 10-8. The settling velocity is determined from

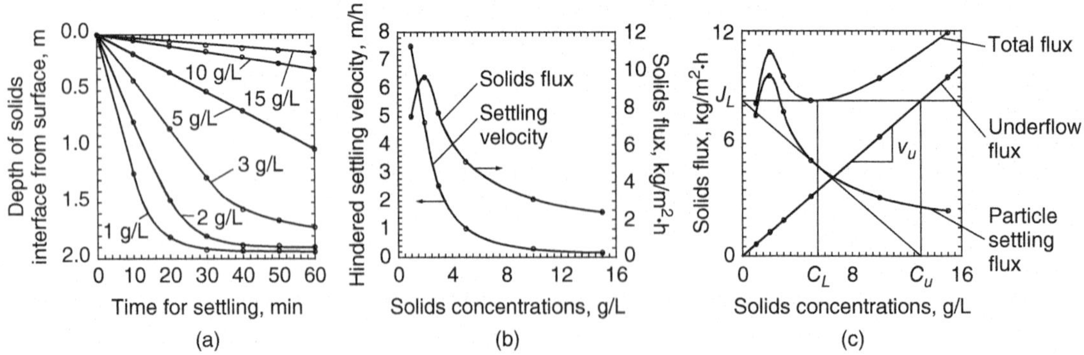

Figure 10-8
Analysis of zone-settling (Type III) data shown in Table 10-1: (a) interfacial settling velocity as function of concentration, (b) initial settling velocity and solids flux due to settling as function of concentration, and (c) limiting solids flux analysis for Type III settling. (Adapted from Tchobanoglous et al., 2003.)

the initial slopes of the concentration curves shown on Fig. 10-8a. The solids flux values due to particle settling is determined by multiplying the concentrations of particles by their respective initial settling velocities, as shown in the equation

$$J_s = v_s C \tag{10-23}$$

where v_s = settling velocity for particle concentration C, m/h
C = suspended solids concentration, kg/m^3

The resultant settling velocity and solids flux values are reported in Table 10-1 for the data presented on Fig. 10-8a. The initial settling velocities and the values for solids flux as a function of solids concentration are presented graphically on Fig. 10-8b.

The solids flux due to the fluid flow to the underdrain, J_u, is defined as

$$J_u = \frac{Q_u C}{A} = v_u C \tag{10-24}$$

where Q_u = flow rate leaving the bottom of basin/thickener, m^3/h
A = cross-sectional area of basin, m^2
v_u = bulk downward fluid velocity, m/h

The total flux at a suspended solids concentration C can be written in terms of bulk fluid velocity and sludge blanket settling velocity by substituting Eqs. 10-23 and 10-24 into Eq. 10-22, resulting in the equation

$$J_T = (v_s + v_u)C \tag{10-25}$$

where terms are as defined previously. The use of the solids flux equations to size solids thickening basins is discussed below.

Limiting Flux Rate

The solids loading for a basin can be determined from an analysis of the limiting flux rate. If solids loading exceeds the limiting flux rate,

Table 10-1
Settling velocity and solids flux values

Solids Concentration, C, g/L	Initial Settling Velocities, v_i		Solids Flux, J_s, kg/m² · h
	m/min	m/h	
1	0.125	7.50	7.5
2	0.080	4.80	9.6
3	0.043	2.55	7.7
5	0.017	1.02	5.1
10	0.005	0.31	3.1
15	0.003	0.16	2.4

solids will accumulate and eventually overflow. To determine the limiting flux rate, an underdrain solids concentration C_u must be selected. On Fig. 10-8c an underdrain concentration of 13 g/L is shown. The underdrain solids concentration is typically determined based on the requirements of downstream residuals processing operations. The limiting solids flux J_L for a given C_u can be determined by drawing a line from the desired underdrain concentration on the x axis and through the tangent to the particle settling flux curve. The intersect of the tangent line with the y axis is the value of the limiting solids flux J_L for the given particle settling flux curve and selected underdrain concentration C_u. For the case shown on Fig. 10-8c, the limiting particle concentration, C_L, is about 5.5 g/L and the limiting solids flux shown is 8.25 kg/m² · h. The downward velocity of the bulk fluid may be determined using the relationship

$$v_u = \frac{J_L}{C_u} \tag{10-26}$$

where v_u = downward velocity of bulk fluid, m/h
J_L = limiting solids flux, kg/m² · h
C_u = concentration of solids in underflow, kg/m³

Area Required for Solids Thickening The flow rate through the underdrain can be estimated using the following mass balance analysis. For the solids thickener shown on Fig. 10-7, a solids mass balance is given by the expression

Suspended solids entering thickener
= suspended solids leaving thickener in effluent (10-27)
+ settled solids leaving thickener in underflow

$$Q_i C_i = (Q_i - Q_u)C_e + Q_u C_u \tag{10-28}$$

where Q_i = influent flow rate to basin/thickener, m³/h
C_i = influent suspended solids concentration, mg/L
Q_u = flow rate leaving the bottom of basin/thickener, m³/h
C_u = solids concentration leaving bottom of basin/thickener, mg/L
C_e = effluent solids concentration, mg/L

If it is assumed that $C_e \ll C_u$ and $C_e \ll C_i$, C_e may be considered negligible and the following expression is obtained for the flow rate through the underdrain:

$$Q_u = \frac{Q_i C_i}{C_u} \tag{10-29}$$

Once the flow rate of the underflow is determined, the area required for the basin can be determined using Eq. 10-26, substituting Q_u/A for v_u, and solving for A, as shown below:

$$A = \frac{Q_u C_u}{J_L} = \frac{Q_i C_i}{J_L} \tag{10-30}$$

where A = area required for thickening, m^2

Other terms are as defined previously. Sizing of a thickener is demonstrated in Example 10-4.

Example 10-4 Area required for thickening

Determine the area required for thickening for a basin that receives 600 mg/L of solids and a flow rate of 4000 m^3/h for an underdrain concentration of 15,000 mg/L. Assume the settling velocity of the sludge blanket follows the relationship plotted on Fig. 10-8b. Also determine J_L, C_L, and Q_u.

Solution

1. Determine J_L and C_L. From the data plotted on Fig. 10-8b and an underflow solids concentration of C_u = 15,000 mg/L, the gravity flux is determined by drawing a line from the x axis at a solids concentration of 15,000 mg/L to the y axis such that it is tangent to the solids flux curve and intersects the y axis. The point at which the line intersects the y axis is the limiting gravity flux and is equal to 7.45 kg/m^2 · h. The value for C_L can also be determined by drawing a vertical line from the tangent point to the x axis and is equal to 6500 mg/L.

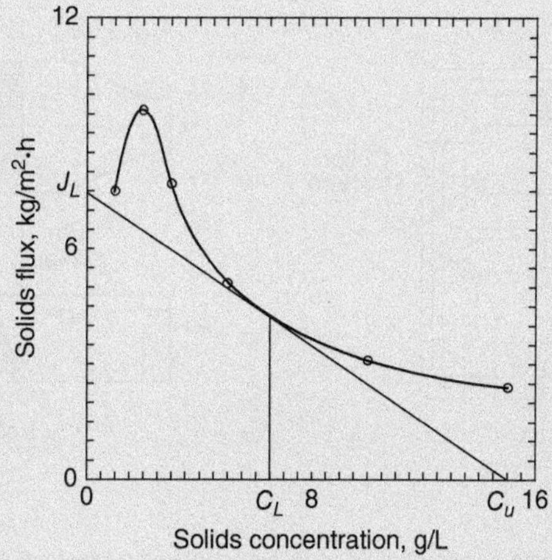

2. Determine Q_u using Eq. 10-29:

$$Q_u = \frac{Q_i C_i}{C_u} = \frac{\left(4000\text{m}^3/\text{h}\right)(600\text{mg/L})}{15{,}000\text{mg/L}} = 160\text{m}^3/\text{h}$$

3. Determine the area for thickening, A, using Eq. 10-30:

$$A = \frac{Q_i C_i}{J_L} = \frac{\left(4000\ \text{m}^3/\text{h}\right)\left(600\ \text{g/m}^3\right)\left(1\ \text{kg}/10^3\ \text{g}\right)}{7.45\ \text{kg/m}^2\cdot\text{h}} = 322\ \text{m}^2$$

4. Summary:

$J_L = 7.45\ \text{kg/m}^2\cdot\text{h}$ $\quad C_L = 6500\ \text{mg/L}$

$Q_u = 120\ \text{m}^3/\text{h}$ $\quad A = 322\ \text{m}^2$

10-6 Conventional Sedimentation Basin Design

Sedimentation basin design is based on applied theoretical principles and practical considerations, including basin location in the overall process treatment train, basin size, and basin geometry. Topics discussed in this section include design considerations for presedimentation basins and conventional sedimentation processes utilizing rectangular, circular, and square basin configurations. Alternative sedimentation processes for improved performance are described in Sec. 10-7.

Presedimentation Facilities Presedimentation facilities (see Fig. 10-9) are used to remove easily settleable sand and silt, often present in surface water supplies, especially rivers,

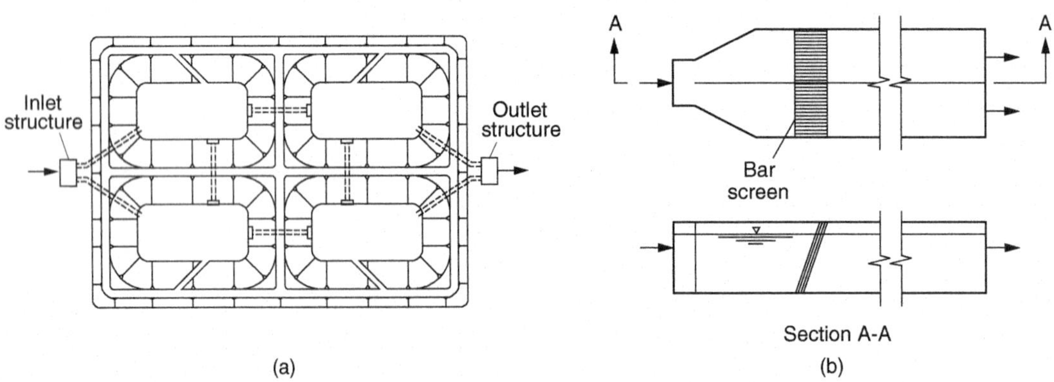

Figure 10-9
Typical presedimentation facilities: (a) earthen basins (both lined and unlined) and (b) rectangular tank shown without continuous mechanical sediment removal facilities.

to avoid silting in treatment plant inlet piping. In general, presedimentation basins and tanks should be located upstream of any raw-water pumping facility (low-lift pumps) and as close as possible to the intake structure to avoid silting problems in the plant intake pipeline. A minimum of two basins or tanks, either a divided single tank or two separate tanks, is required so that one can be emptied for routine maintenance and repairs without having to take the entire plant off-line. Where sand carryover is not a major problem, a single tank with a bypass pipeline may be satisfactory. Rectangular presedimentation tanks can be designed with hopper bottoms or be equipped with continuous sediment removal facilities. Typical design criteria for presedimentation tanks are listed in Table 10-2.

Assuming ideal settling in a rectangular basin as presented in Sec. 10-2, the required length of a presedimentation tank can be estimated by the equation

$$L = K\left(\frac{h_o}{v_s}\right)v_f \tag{10-31}$$

where L = length, m
K = safety factor (typically 1.5 to 2), unitless
h_o = effective water depth, m
v_s = settling velocity of particle to be removed, m/s
v_f = mean water velocity at maximum day flow rate, m/s

The settling velocities of various sizes of fine sand particles are listed in Table 10-3. An example of presedimentation tank design is presented in Example 10-5.

Table 10-2
Typical presedimentation tank design criteria

Parameter	Units	Value
Type	–	Horizontal flow, rectangular tank
Minimum number of tanks	Dimensionless	2
Depth (without automated sediment removal)	m (ft)	3.5–5 (11.5–16)
Depth (with automated sediment removal)	m (ft)	3–4 (10–13)
Minimum length-to-depth ratio	Dimensionless	6:1
Length-to-width ratio	Dimensionless	4:1 – 8:1
Surface loading rate[a]	$m^3/m^2 \cdot d$ (gpm/ft^2)	200–400 (3.3–6.6)
Horizontal mean flow velocity (at maximum daily flow)	m/s (ft/s)	0.05–0.07 (0.16–0.23)
Detention time[b]	min	6–15
Minimum size of particle to be removed	mm	0.1
Bottom slope	m/m	Minimum 1:100 longitudinal slope

Source: Adapted from Kawamura (2000).
[a]The surface loading rate is also known as the overflow rate.
[b]Detention time in earthen basins is typically on the order of 2 to 3h or more, depending on available space.

Table 10-3
Settling velocities of various sized discrete particles[a]

Particle diameter, mm	1	0.6	0.4	0.2	0.15	0.1	0.08	0.06
Settling velocity, m/s	0.138	0.077	0.048	0.022	0.015	0.0069	0.0044	0.0025

[a]With a specific gravity of 2.65, in still water, and a water temperature of 10°C (50°F) calculated using Eq. 10-13 or 10-14, as appropriate.

Example 10-5 Presedimentation tank design

A grit chamber (one concrete structure divided into two tanks) is designed to remove sand of 0.1 mm and larger for an average flow of 1.0 m^3/s (22.8 mgd). The maximum flow rate is to be 1.5 times the average flow and the water temperature is 10°C. Assuming a typical water depth of 3.0 m and a factor of safety of 1.75, use the information in Table 10-2 to determine the length and width of each tank and check that the surface loading (overflow) rate and the detention time are within the recommended design criteria ranges. Particle settling velocities are listed in Table 10-3.

Solution

1. Determine the cross-sectional area for each tank. From Table 10-2, the horizontal-flow velocity at maximum flow is 0.05 m/s. The cross-sectional area for each tank can be calculated as

$$A = \frac{(1.5)(1.0\ \text{m}^3/\text{s})}{(2\ \text{tanks})(0.05\ \text{m/s})} = 15\ \text{m}^2$$

For a water depth of 3.0 m, the width will be 5.0 m (15 m^2/3.0 m).

2. Calculate the tank length. The length of each tank can be determined using Eq. 10-31 as shown below:

$$L = K\left(\frac{h_o}{v_s}\right)v_f \quad \text{where } K = 1.75$$

From Table 10-3, the settling velocity v_s for a 0.1-mm sand particle is 0.0069 m/s. The horizontal velocity at maximum daily flow rate from Table 10-2 was

$$v_f = 0.05\ \text{m/s}$$

Substituting into Eq. 10-31,

$$L = (1.75)\times\left(\frac{3.0\ \text{m}}{0.0069\ \text{m/s}}\right)\times 0.05\ \text{m/s} = 38.0\ \text{m}$$

3. Verify the length-to-depth (L/d) and length-to-width (L/w) ratios.
   a. From Table 10-2, the minimum length-to-depth ratio is 6:1:

$$\frac{L}{d}=\frac{38}{3.0}=\frac{12.7}{1}>\frac{6}{1} \quad \text{OK}$$

   b. From Table 10-2, the minimum length-to-width ratio is 4:1:

$$\frac{L}{w}=\frac{38}{5}=\frac{7.6}{1}>\frac{4}{1} \quad \text{OK}$$

4. Verify the detention time and surface loading rates.
   a. Determine the detention time.

$$\tau=\frac{V}{Q}=\frac{(38.0\text{ m})(5\text{ m})(3\text{ m})}{\left(\dfrac{1.0\text{ m}^3/\text{s}}{2\text{ tanks}}\right)(60\text{ s/min})}=19\text{ min}$$

   The calculated value is higher than the typical range of detention time given in Table 10-2 (6 to 15 min) for average flow conditions, but more detention time is acceptable. The detention time will be 12.7 min at maximum flow conditions.

   b. Determine the surface loading rate using Eq. 10-17.

$$\text{OR}=\frac{Q}{A}=\frac{\left(1.0\text{ m}^3/\text{s}\right)(3600\text{ s/h})(24\text{ h/d})}{(38\text{ m})(5\text{ m})(2\text{ basins})}=227\text{ m}^3/\text{m}^2\cdot\text{d}$$

   The surface loading rate range recommended in Table 10-2 is 200 to 400 $\text{m}^3/\text{m}^2\cdot\text{d}$. Thus, the computed value is within the acceptable range.

Although coarse screens are used with river intakes, a fine debris screen with approximately 20-mm openings is often provided at the front end of the presedimentation tank. Because the screen also acts as an effective diffuser wall, it should be installed close to the tank inlet. If a separate diffuser wall is specified, the total area of openings at the wall should be about 15 percent of the tank cross-sectional area.

Rectangular Sedimentation Basins

Many sedimentation basins are rectangular with horizontal flow, as shown on Fig. 10-10. A minimum of two basins should be provided so that one may be taken off-line for inspection, repair, and periodic cleaning while the other basin(s) remain in operation. Basins arranged longitudinally side by side, sharing a common wall, have proven to be a cost-effective approach. In addition, a flocculation process may be incorporated into the head end of the sedimentation basin, minimizing piping, improving flow distribution to sedimentation basins, and potentially reducing floc damage during transfer between the flocculation stage and the sedimentation stage. Providing an

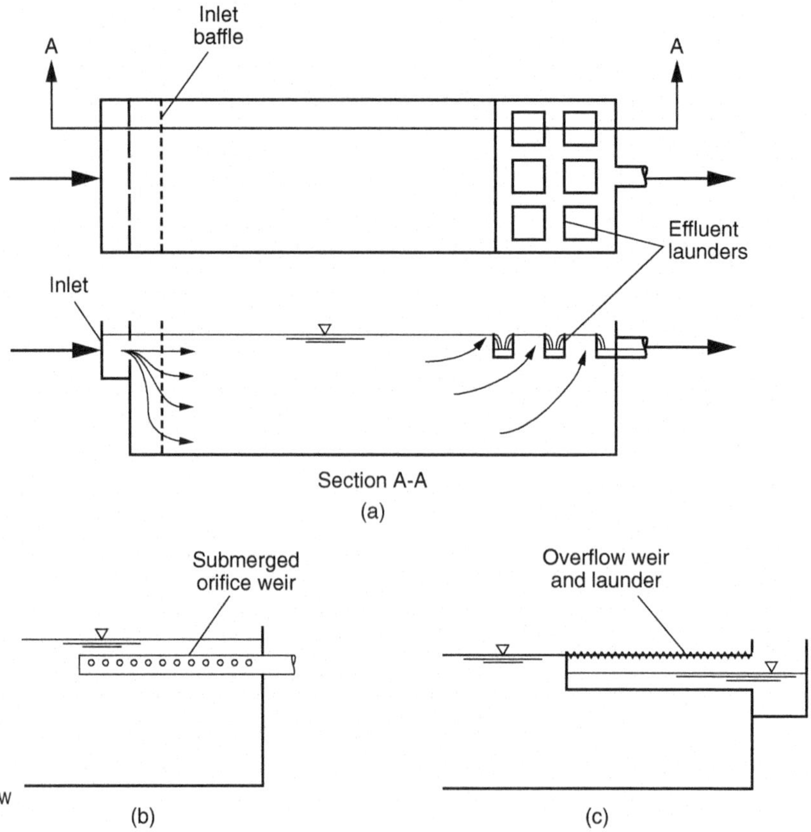

Figure 10-10
Rectangular, horizontal-flow sedimentation basin with various outlet types: (a) inboard effluent launders, (b) submerged orifice withdrawal (see also Figure. 10-14), and (c) overflow weir and launder.

access road around the basins will aid in operation and maintenance work and may facilitate future plant additions.

Sedimentation basins at a variety of water treatment plants.
http://www.stantec.com/wt3r/10-1.html

INLET STRUCTURE

The inlet to a rectangular sedimentation basin should be designed to distribute the flocculated water uniformly over the entire cross section of the basin at low velocity. The flow pattern in the basin is strongly controlled by inertial currents, density flows (e.g., temperature gradients), and wind direction.

A well-designed inlet permits water from the flocculation basin to enter directly into the sedimentation basin without channels or pipelines. Flow velocity in a pipe or flume can be either too slow or too fast depending upon the daily and seasonal plant flow variations and may cause floc settling or breakage to occur in the pipe or flume. The permissible flow velocity to maintain floc suspension generally ranges from 0.15 to 0.60 m/s (0.5 to 2 ft/s). A diffuser wall is one of the most effective and practical flow distribution methods used at the basin inlet when the flocculation basin is directly attached to the sedimentation basin. The openings should be small holes (100 to 200 mm diameter circular or equivalent) of identical size, evenly distributed on the wall (see discussion of diffuser walls, Sec. 9-7, Chap. 9).

When sedimentation basins are fed from a common channel, the basin inlet structure may consist of weirs or submerged ports, with a permeable baffle about 2 m (6.5 ft) downstream in the sedimentation basin. Uniform or equal distribution of flow to each sedimentation basin is also essential.

SETTLING ZONE

The basic design criteria to be considered for the horizontal-flow settling zone are (1) surface loading rate, (2) effective water depth, (3) detention time, (4) horizontal-flow velocity, and (5) minimum length-to-width ratio. Typical design parameters used for rectangular sedimentation facilities are summarized in Table 10-4 and discussed below.

Surface loading rate and settling velocity

The relationship between surface loading and the settling velocity of discrete particles was developed by Hazen (1904) and discussed previously in Sec. 10-3. Hazen stated that the efficiency of an idealized, horizontal-flow settling tank is solely a function of the settling velocity of discrete particles and of the surface loading rate (the flow rate of the basin divided

Table 10-4
Typical design criteria for horizontal-flow rectangular tanks

Parameter	Units	Value
Type	—	Horizontal-flow rectangular tank
Minimum number of tanks	Unitless	2
Water depth	m (ft)	3–5 (10–16)
Length-to-depth ratio, minimum	Dimensionless	15:1
Width-to-depth ratio	Dimensionless	3:1–6:1
Length-to-width ratio, minimum	Dimensionless	4:1–5:1
Surface loading rate (overflow rate)	m/h (gpm/ft^2)	1.25–2.5 (0.5–1.0)
Horizontal mean-flow velocity (at maximum daily flow)	m/min (ft/min)	0.3–1.1 (1–3.5)
Detention time	h	1.5–4
Launder weir loading	m^3/m · h (gpm/ft)	9–13 (12–18)[a]
Reynolds number	Dimensionless	<20,000
Froude number	Dimensionless	>10^{-5}
Bottom slope for manual sludge removal systems	m/m	1:300
Bottom slope for mechanical sludge scraper equipment	m/m	1:600
Sludge collector speed for collection path	m/min (ft/min)	0.3–0.9 (1–3)
Sludge collector speed for the return path	m/min (ft/min)	1.5–3 (5–10)

Source: Adapted from Kawamura (2000).
[a]Can be higher, depending upon characteristics of floc.

Table 10-5
Settling velocity of selected floc types

	Setting Velocity at 15°C	
Floc Type	**m/h**	**ft/min**
Small fragile alum floc	2–4.5	0.12–0.24
Medium-sized alum floc	3–5	0.18–0.28
Large alum floc	4.0–5.5	0.22–0.30
Heavy lime floc (lime softening)	4.5–6.5	0.25–0.35
Fe floc	2–4	0.12–0.22
PACl floc	2–4	0.12–0.22

by the surface area) and is independent of the basin depth and detention time. However, most settling basins treat flocculated suspended matter (not discrete particles) and do not have idealized flow patterns. Furthermore, flocculent particles may increase in size while in the basin and settle faster than predicted for a discrete particle. The settling velocities of selected floc particles are presented in Table 10-5.

Effective depth
Sedimentation basins can be made shallow with a large surface area, but there is a practical minimum basin depth necessary (2.5 to 3 m minimum effective water depth) for mechanical sludge removal equipment. Also, other factors such as flow velocity, effect of wind and sun, and required basin area make shallow basins less practical. Effective water depth is even more important for a basin without mechanical sludge removal facilities since the basin must provide adequate volume for sludge deposit. With an efficient flocculation process, about 70 percent of the floc will settle within the first one-third of the basin length at average flow. Estimated sludge height for well-flocculated water under normal conditions and without a mechanical sludge removal mechanism may be 2 to 3 m (6.5 to10 ft) at the influent end of the basin but only 0.3 m (1 ft) in the last half of the basin.

Horizontal-flow velocity
Settling characteristics and surface loading are generally the main basis of design, with Reynolds and Froude numbers being used as a check on turbulence and backmixing. The Reynolds number is determined as

$$\mathrm{Re} = \frac{v_f R_h}{\nu} \tag{10-32}$$

where Re = Reynolds number based on hydraulic radius, dimensionless
v_f = average horizontal fluid velocity in tank, m/s

R_h = hydraulic radius, A_x/P_w, m
A_x = cross-sectional area, m^2
P_w = wetted perimeter, m
v = kinematic viscosity, m^2/s

The Froude number may be determined using the equation

$$\text{Fr} = \frac{v_f^2}{gR_h} \tag{10-33}$$

where Fr = Froude number, dimensionless
g = acceleration due to gravity, 9.81 m/s^2

Recommended values for settling zone design determined using Eqs. 10-32 and 10-33 are Re < 20,000 and Fr > 10^{-5} (Kawamura, 2000). These dimensionless numbers are useful for general design guidelines because a large Reynolds number indicates a high degree of turbulence and a low Froude number implies that the water flow is not dominated by horizontal flow, and backmixing may occur. The criteria for Re and Fe are of less significance and may be exceeded for conservatively designed basins; a basin with an appropriate length:width ratio, low overflow rate, and detention time of 3 to 4 h will often achieve satisfactory performance even if the Re and Fr criteria are not met. It is more important to check these criteria for high rate rectangular basins with detention times of 2 h or less.

Placing longitudinal baffles (in the direction of flow) can help alleviate poor sedimentation basin performance. Adding longitudinal baffles produces a number of parallel narrow channels and reduces the Reynolds number and increases the Froude number. For example, if one parallel baffle is placed into the tank and the tank is divided into two parallel tanks, then the Reynolds number is decreased by 50 percent and the Froude number is increased by a factor of 2. To allow for sludge removal equipment, the baffles should be separated by at least 3 m (10 ft) and can be made of wooden planks or concrete. Baffles should never be placed in sedimentation basins where they would cause serpentine flow (180° turns) to occur because the turbulence that is caused by abrupt turns will significantly reduce particle settling.

Length-to-width ratio

The proportions of rectangular horizontal-flow sedimentation tanks can be determined from design criteria that are listed in Table 10-4. In general, long, narrow, and relatively deep (5 m) basins are preferred to minimize short circuiting. To promote plug flow in rectangular sedimentation basins, a minimum length-to-width ratio of 4:1 to 5:1 should be maintained. Approximately 0.5 m of tank freeboard should be provided to act as a wind barrier. This will also have the additional benefit of preventing waves that are produced by wind from splashing onto walkways (Kawamura, 2000).

OUTLET STRUCTURE

Outlet structures for rectangular tanks are generally composed of launders running parallel to the length of the tank, shown on Fig. 10-10, or a simple weir at the end of the tank. Cross baffles may be added in the vicinity of the effluent launders to prevent the return of surface currents from the end of the basin back toward the inlet. Water leaving the sedimentation basin should be collected uniformly across the width of the basin. Inadequate weir length may result in solids being carried over the effluent weir due to excessive approach velocity. Long weirs have at least three advantages for rectangular sedimentation tanks: (1) a gradual reduction of flow velocity toward the end of the tank, (2) minimization of wave action from wind, and (3) collection of clarified water located in the middle of the tank when a distinct density flow occurs in the basin. Some disadvantages of long effluent launders are that they are expensive and the support columns for them must be designed so they do not interfere with sludge collection devices. With proper sedimentation basin design, long effluent launders may provide only a marginal improvement in effluent turbidity and a simple weir at the end of the tank may provide a satisfactory result.

The water level in the sedimentation basin is controlled by the end wall or overflow weirs. V-notch weirs are commonly attached to launders and broad-crested weirs are attached to the end wall. Submerged orifices or weirs have sometimes been used on the outlet structure when discharging clarified water to a rapid sand filtration system to avoid breakup of fragile alum floc and turbidity breakthrough in rapid sand filters. For high-rate filter designs (dual and monomedia), there is little concern over floc breakage because high-rate filters require a small, strong floc, and filter aids are added prior to filtration primarily for improved particle attachment in the filter. The optimal weir-loading rate will depend on the individual design of the facility and a general rule does not exist. For example, lowering the weir loading rate by 50 percent may not result in a significant improvement of sedimentation efficiency, partly due to density currents.

In the past, installation of a permeable baffle at the tank outlet was a popular design, but the effect of the outlet baffle was often not beneficial and, in fact, may have an adverse effect on basin performance due to floc carryover, as shown on Fig. 10-11.

SLUDGE ZONE

Sludge collects in the bottom of the sedimentation basin, and in a rectangular basin, more sludge settles near the inlet than the outlet end of the basin.

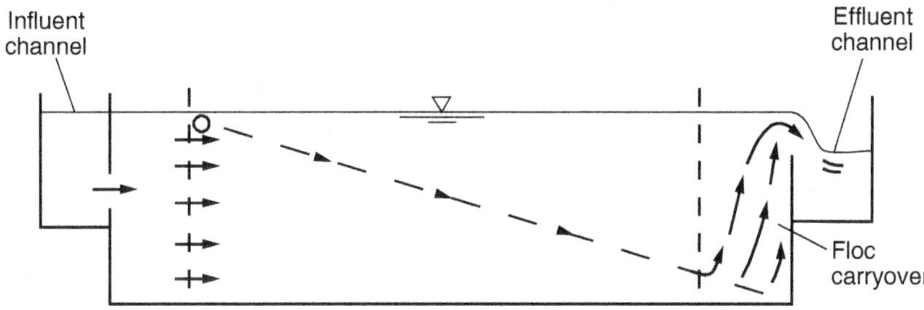

Figure 10-11
Floc carryover effect resulting from presence of effluent permeable baffle.

To facilitate sludge removal, the bottom of the basin is typically sloped toward a sludge hopper. For manual sludge removal systems, water is drained from the basin and pressurized water is used for solids flushing; the bottom slope should have a slope of at least 1:300 to ensure gravity movement of sludge. If mechanical sludge scraper equipment is used, the bottom slope should be at least 1:600. The basin bottom may be level when mechanical equipment uses a vacuum to remove sludge. If local labor is inexpensive or if funds for investment are limited, sedimentation tanks may be designed without mechanical sludge removal. However, provisions should be made for possible future installation of mechanical sludge removal equipment.

Manufacturers produce several types of mechanical collectors for rectangular sedimentation basins. The major types of mechanical collectors for rectangular basins are (1) chain-and-flight (plastic material) collectors (see Fig. 10-12), (2) a traveling bridge with sludge-scraping squeegees and a mechanical cross collector at the influent end of the tank, (3) a traveling bridge with sludge suction headers and pumps, and (4) sludge suction headers supported by floats and pulled by wires.

Figure 10-12
Chain-and-flight-type sludge collector.

The standard maximum width of the chain-and-flight sludge collector is 6 m (20 ft), and the operation and maintenance cost usually increases for the chain-and-flight collectors if the length of the basin exceeds 60 m. When mechanical scraper units are used, the velocity of the scraper should be kept below 18.0 m/h to prevent resuspending the settled sludge. For suction sludge removal units, the velocity can be 60 m/h because the principal concern is not the resuspension of settled sludge but the disruption of the settling process.

Traveling bridges can span up to 30 m (100 ft) with widths 12 to 30 m (40 to 100 ft) usually being the most cost effective. Because the width of sedimentation basins is often less than 15 m (50 ft), using one bridge to span two or three tanks can significantly reduce the capital investment for sludge removal equipment. Both the drain and sludge draw-off pipelines should have a minimum diameter of 150 mm (6 in.) to prevent clogging problems. Additionally, traveling bridges are susceptible to high winds, and in cold-weather climates, the pumps and piping need cold-weather protection as they are exposed above the water. Sedimentation basin design is demonstrated in Example 10-6.

Example 10-6 Sedimentation basin design

A water treatment plant with a maximum daily flow of 3 m^3/s (1.5 times the average flow of 2 m^3/s) is treating surface water. The water is coagulated with alum and the alum floc was measured to have a settling velocity of 2.2 m/h at 10°C (50°F). The dynamic viscosity of water at 10°C is 0.00131 kg/m · s and the density is 999.7 kg/m^3. Design a horizontal flow rectangular sedimentation basin with a chain and flight sludge removal system including the number of basins and the basin dimensions. The design is based on the maximum flow rate.

Solution

1. Determine the number of basins. Two basins would satisfy the minimum requirement for maintenance purposes. However, if one basin were off-line, the entire plant flow would be directed through the remaining basin, possibly resulting in overloading of the basin. To minimize the risk of basin overloading, three basins will be selected.
2. Determine the size of each basin.
   a. Select the basin width and depth. The basin width will be governed by the standard size of sludge removal equipment. The standard maximum width of the chain-and-flight sludge collector is 6 m, so basin widths in increments of 6m will be considered, starting with 18 m. Water depths from 3 to 5 m are appropriate, according to

the design criteria listed in Table 10-4. As previously mentioned, deeper basins are recommended over shallower basins, so a depth of 4m will be selected.

b. Determine the basin area. The settling velocity such that the particle is removed in the sludge zone just before the outlet, v_c, is given as 2.2 m/h at 10°C. (This value is also equal to the overflow rate.) Use Eq. 10-17 to determine the basin surface area:

$$A = \frac{Q}{v_c} = \frac{3\ \text{m}^3/\text{s}}{(2.2\ \text{m/h})(1\ \text{h}/3600\ \text{s})} = 4909\ \text{m}^2$$

c. Determine the length using the design guidelines in Table 10-4 for length-to-width ratios. For three tanks that are 18 m (60 ft) wide, the tank length and length-to-width ratio can be estimated:

$$L = \frac{4909\ \text{m}^2}{3\ \text{basins} \times 18\ \text{m}} = 90.9\ \text{m} \qquad \frac{L}{W} = \frac{90.9}{18} = \frac{5.05}{1}$$

The length-to-width ratio is greater than the minimum recommendation of 4:1 to 5:1.

3. Check the various design parameters listed in Table 10-4.

a. Check the detention times at Q_{max} and Q_{ave}:

$$\begin{aligned}\text{Detention time for } Q_{max} &= \frac{(18 \times 90.9 \times 4)\ \text{m}^3 \times 3\ \text{basins}}{\left(3\ \text{m}^3/\text{s}\right)(3600\ \text{s/h})} \\ &= 1.82\ \text{h}\end{aligned}$$

Detention time for Q_{ave} = 1.5 × 1.82 h = 2.73 h

These detention times are within the acceptable range of 1.5 to 4 h.

b. Check the length-to-depth ratio:

$$\frac{L}{D} = \frac{90.9}{4} = \frac{22.7}{1}$$

The basin length-to-depth ratio is 22.7:1, which is greater than the minimum recommendation of 15:1.

c. Check the horizontal-flow velocity. The mean velocity is given by the expression

$$v_f = \frac{Q}{A} = \frac{\left(3\ \text{m}^3/\text{s}\right)(60\ \text{s/m})}{18\ \text{m} \times 4\ \text{m} \times 3\ \text{basins}} = 0.833\ \text{m/min}$$

The mean velocity is greater than 0.3 m/min and less than 1.1 m/min.

d. Check the Reynolds and Froude numbers using Eqs. 10-32 and 10-33:

$$\text{Re} = \frac{\rho v_f R_h}{\mu}$$

$$R_h = \frac{A_x}{P_w} = \frac{4\text{ m} \times 18\text{ m}}{18\text{ m} + 2(4\text{ m})} = 2.77\text{ m}$$

$$v_f = \frac{0.833\text{ m/min}}{60\text{ s/min}} = 0.014\text{ m/s}$$

$$\text{Re} = \frac{\left(999.7\text{ kg/m}^3\right)(0.014\text{ m/s})(2.77\text{ m})}{0.00131\text{ kg/m}\cdot\text{s}} = 29{,}594$$

The Reynolds number of 29,594 is higher than the recommended value of 20,000 for a horizontal sedimentation basin. The Froude number is given by Eq. 10-33:

$$\text{Fr} = \frac{v^2}{gR_h} = \frac{(0.014)^2\text{ m}^2/\text{s}^2}{\left(9.81\text{ m/s}^2\right)(2.77\text{ m})} = 7.2\times10^{-6}$$

The Froude number is lower than the recommended value for sedimentation tanks, so the tank design must be modified.

4. Consider the addition of two longitudinal baffles per basin and recompute the Reynolds and Froude numbers.

$$R_h = \frac{A_x}{P_w} = \frac{4\text{ m} \times 6\text{ m}}{6\text{ m} + 2(4\text{ m})} = 1.71\text{ m}$$

$$\text{Re} = \frac{(999.7\text{ kg/m}^3)(0.014\text{ m/s})(1.71\text{ m})}{0.00131\text{ kg/m}\cdot\text{s}} = 18{,}162 < 20{,}000 \quad \text{OK}$$

$$\text{Fr} = \frac{(0.014)^2\text{ m}^2/\text{s}^2}{(9.81\text{ m/s}^2)(1.71\text{ m})} = 1.17\times10^{-5} > 10^{-5} \quad \text{OK}$$

Comment

The values of the Reynolds and Froude numbers after the addition of longitudinal baffles are within the acceptable range; however, they are evaluated at the maximum daily flow. As water demand changes with the season, the number of basins that are online needs to be selected to keep the basins operating within the Reynolds and Froude number guidelines. Note that the Re and Fr criteria are not as significant for conservatively designed basins.

Circular Sedimentation Basins and Upflow Clarifiers

Circular sedimentation tanks, also known as upflow clarifiers, have been used in many cases because they provide an opportunity to use relatively trouble-free circular sludge removal mechanisms and, for small plants, can be constructed at a lower capital cost per unit surface area. However, circular tanks tend to need more piping for water and sludge conveyance to and from the tanks than a rectangular basin configuration.

Circular tank diameters are calculated on the basis of overflow rates using approximately the same criteria that are used for rectangular basin design (see Table 10-4). Circular tanks, as shown on Fig. 10-13, may have center feed or peripheral feed. A circular sedimentation basin with center feed and peripheral collection using radial submerged orifice weirs is shown on Fig. 10-14. The inlet structure used for center-feed configurations is a circular weir around the influent vertical rise pipe. For peripheral-feed tanks, the inlet weir is located around the perimeter of the tank. Inlet weirs provide energy dissipation and direct the flow downward into the depths of the settling tank where particles are removed. Particles settle as the water rises to the outlet structure. Baffles near the outlet and surface-skimming devices are not used unless the influent water has problems with debris and floatable material.

Circular clarifiers used at water treatment plants.

http://www.stantec.com/wt3r/10-2.html

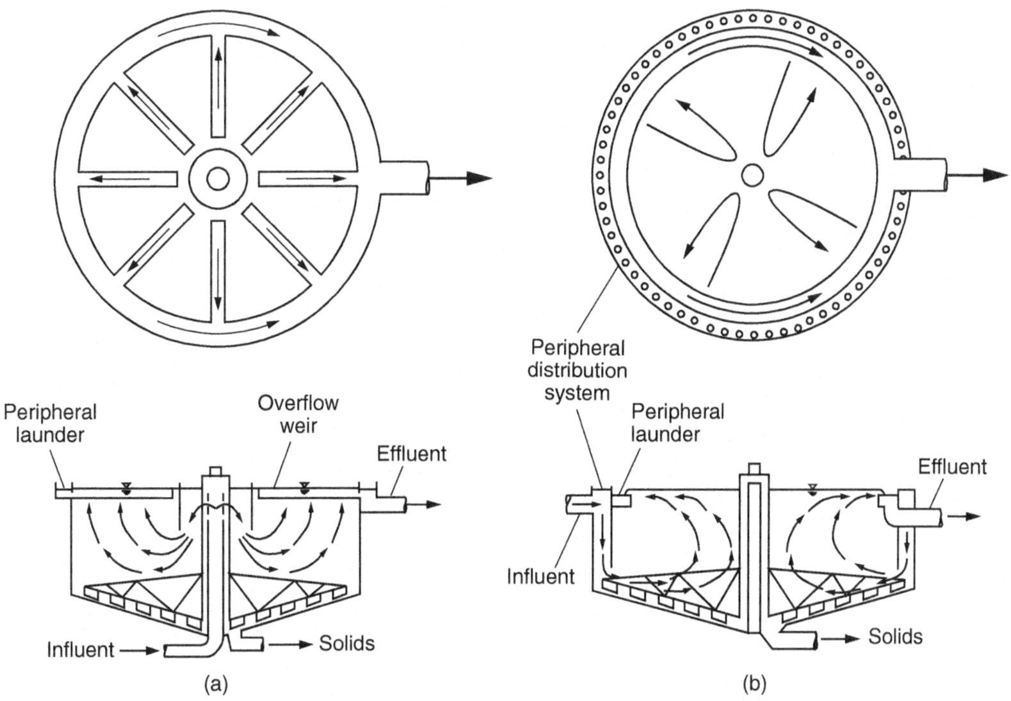

Figure 10-13
Circular sedimentation basins: (a) center feed with radial collection and (b) peripheral feed with peripheral collection.

Figure 10-14
View of circular sedimentation basin with radial collection troughs with submerged orifices.

The design of circular clarifiers requires careful consideration of factors such as surface loading, uniform flow distribution into the settling zone, minimization of flow short circuiting from hydraulic and density currents, uniform withdrawal of clarified water, and sludge withdrawal without disturbing settling efficiency. For upflow clarifiers, the vertical-flow rise rate becomes an additional criterion; at any selected level, the flow rise must be less than the respective floc-settling rate. Refer to Table 10-5 for settling velocities of various flocs.

The most significant potential problem of center-feed circular clarifiers is short circuiting of the upward flow of water. Hydraulic short circuiting can be particularly significant when the peripheral collection channel is not equipped with radial weirs or when the influent contains a high solids concentration that flows along the tank bottom. Circular tanks may experience density currents along the bottom when the turbidity of the raw water exceeds about 50 NTU or when there is a temperature difference (as little as 0.3°C or 0.5°F) between the inflow and the ambient water. Placing the peripheral launder trough two-thirds to three-fourths of the radial distance from the center minimizes the density currents and produces better quality water. Because of the potential problems with hydraulic short circuiting, the best use of upflow clarifiers is for clarification of waters with heavy, non-colloidal solids loading such as filter-to-waste washwater.

To address problems with short circuiting in center-feed clarifiers, a peripheral-feed clarifier was developed that introduces flow between the tank wall and an annular skirt. The peripheral-feed design allows the inflow to enter the settling zone near the tank bottom. To ensure uniform flow distribution and additional loading capacity, the orifices in the annular inlet channel should be designed so that the head loss across each orifice inlet is approximately 10 to 15 mm. Peripheral-feed designs have two to three times the loading permissible with center-feed clarifiers. Peripheral-feed,

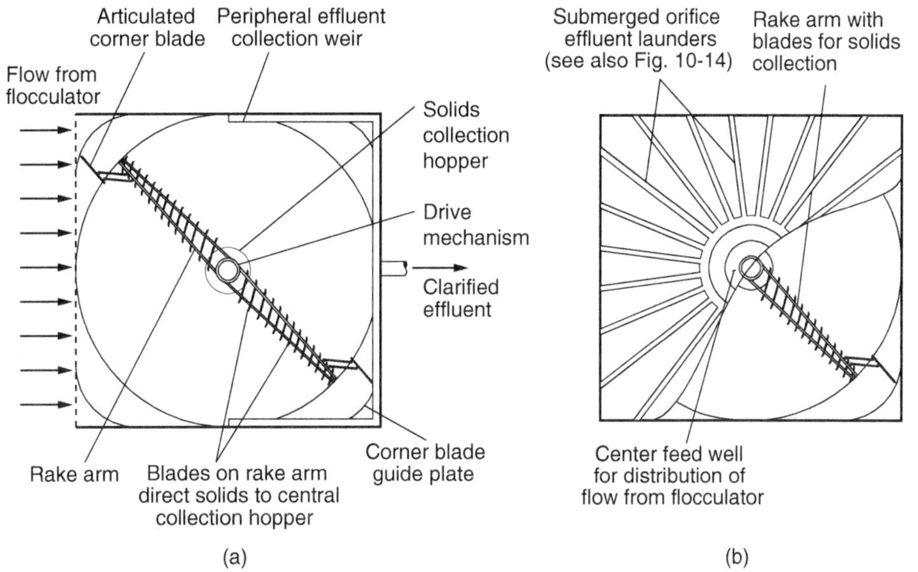

Figure 10-15
Square sedimentation basins: (a) square basin close coupled with flocculation facilities and (b) square basin with center feed.

peripheral-collection clarifiers (e.g., Kraus-Fall peripheral-feed sedimentation tank) have proven more efficient for high solids influent water.

Square Sedimentation Basins

Square sedimentation tanks, shown on Fig. 10-15, were developed in an effort to combine the advantages of common-wall construction of rectangular basins with the simplicity of circular sludge collectors. However, several features of square basins have presented difficulties for sedimentation processes. For example, the effluent launders are constructed along the perimeter of the basins, resulting in the corners having more weir length per degree of radial arc. Thus, flows are not distributed equally and solids preferentially accumulate in the corners of the basin. Corner sweeps, added to the circular sludge collector mechanisms to remove sludge settling in the corners, have been a source of mechanical difficulty. While the corners can be steeply sloped so that the sludge may flow by gravity to the circular sludge collectors, there are relatively few square basins constructed for water treatment.

10-7 High-Rate Sedimentation Processes

The use of large quiescent basins such as those described above to settle particles out of water has been established as an appropriate method for particle removal. However, these basins require large land areas, which are not always available, and plant upgrades to accommodate increasing water demand may be constrained by the available site area. Increasing the overflow rate in sedimentation basins and achieving the same or better water quality would allow new water treatment plants to fit on smaller sites and existing water treatment plants to expand without having to use additional land area. For example, a high-rate tube settler module, as described below, can be installed under the long launders, significantly increasing the tank loading rate without adding basin volume. Alternative approaches to sedimentation, such as high-rate clarification using parallel-plate or tube settlers, upflow clarifiers, sludge blanket clarifiers, and ballasted sedimentation, are discussed in this section.

Tube and Lamella Plate Clarifiers

Increasing particle size or decreasing the distance a particle must fall prior to removal can accelerate sedimentation of aqueous suspensions. Particle size increase is achieved by coagulation and flocculation prior to sedimentation. Reducing the settling distance can be achieved by making the entire basin shallower, but practical aspects of sludge storage, equipment movement, and wind effects on the surface limit this approach.

To decrease the distance a particle must fall, the clarification process must be separated from the process of sludge withdrawal and surface current effects. One approach is to provide parallel plates or tubes in the sedimentation basin, permitting solids to reach a surface after a short settling distance. If these settling surfaces (plates or tubes) were oriented in a horizontal direction, they would eventually fill with solids, which would increase the head loss and eventually increase velocities to a point that the suspended materials would be scoured back into suspension. Inclining the surfaces to a degree where the solids can slide from the plate or tube surface results in the settled particles depositing in the sludge zone. Inclined plate settlers are illustrated on Fig. 10-16. Some design aspects and process selection criteria for high-rate settlers are discussed below.

Gravity separation systems can be enhanced by incorporating tube settlers and plate settlers.

http://www.stantec.com/wt3r/10-3.html

SETTLING CHARACTERISTICS AND SURFACE LOADING RATE

The settling characteristics of the suspended particles to be removed and the portion of the total tank surface area that is covered by the settler modules primarily control the surface loading for high-rate settlers. Design criteria for Lamella settlers in rectangular sedimentation basins are provided in Table 10-6. The surface loadings presented in Table 10-6 are based on the footprint area and not the top area of the plates or projected area. In cold regions where alum floc is to be removed, the maximum surface loading should be limited to $150\ m^3/m^2 \cdot d$ ($2.6\ gpm/ft^2$). Pilot testing may help establish design criteria, but criteria used for design should be more conservative than pilot test results to allow for poor inlet conditions, density flow, inappropriate coagulant dosage,

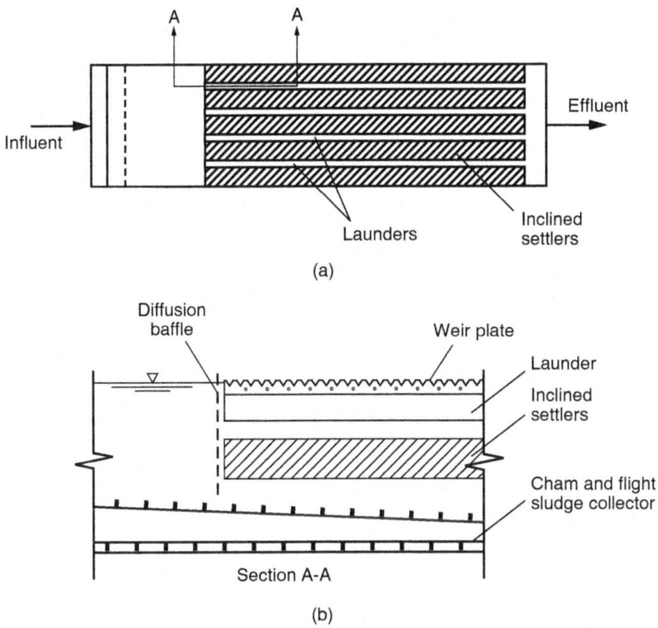

Figure 10-16
Rectangular sedimentation basin with inclined plate settlers: (a) plan view of basin and (b) section through inclined plates. (Adapted from Kawamura, 2000.)

Table 10-6
Typical design criteria for horizontal-flow rectangular tanks with tube settlers

Parameter	Units	Value
Type	—	Horizontal-flow rectangular tank
Minimum number of tanks	Unitless	2
Depth	m (ft)	3–5 (10–16)
Surface loading for plate or tube settlers: alum floc[a]	m/h (gpm/ft^2)	2.5–6.25 (1–2.5)
Surface loading for plate or tube settlers: heavy floc[a]	m/h (gpm/ft^2)	3.8–7.5 (1.5–3.0)
Typical hydraulic diameter	mm	50 – 80
Maximum-flow velocity in plate or tube settlers	m/min (ft/min)	0.15 (5)
Detention time in tube settlers	min	6–10
Detention time in plate settlers	min	15–25
Fraction of basin covered by plate or tube settlers	%	<75
Launder weir loading[b]	m^3/m · h (gpm/ft)	3.75–15 (5–20)
Flow direction	—	Normally countercurrent upflow
Plate or tube angle	deg	60°
Mean horizontal velocity	m/min (ft/min)	0.05 –0.13 (0.15–0.5)
Reynolds number	Dimensionless	<20,000
Froude number	Dimensionless	>10^{-5}

Source: Adapted from Kawamura (2000).
[a]Before the plate or tube settlers are installed.
[b]Can be higher depending on the characteristics of the floc and the type of plate or tube settlers used.

or other unforeseen negative factors. While Reynolds and Froude numbers discussed in Sec. 10-6 have only limited use in conventional sedimentation basin design, they are good design guides for high-rate settlers.

DETENTION TIME IN PLATE AND TUBE SETTLERS

The discussion on the theoretical performance of a sedimentation basin demonstrated that the removal of Type I particles depended on the overflow rate. For a basin depth h_0 and a theoretical detention time τ, particles with a settling velocity v_s would be removed if $v_s \geq h_0/\tau$ (which is equal to the overflow rate). Consequently, if plates or tubes are inserted into a sedimentation basin, then greater particle removal is expected because the detention time remains the same, but the distance that particles must settle before they are removed is greatly reduced. Both parallel-plate settlers and tube settlers typically have detention times less than 20 min, but they still have a settling efficiency comparable to that of a rectangular settling tank with a minimum 2 h detention time.

SOLIDS REMOVAL

Usually settled floc or sludge does not adhere to the plates or tubes but rather slides down to the tank bottom for subsequent mechanical sludge removal. Settlers may be designed with a scouring device, such as water jets or compressed air, to remove flocs such as biological floc that may adhere to the settler surface. For certain applications, downflow settlers can be used to enhance the self-cleaning action. A tank equipped with high-rate settler modules must provide continuous sludge removal because of the high sludge accumulation rate produced by the high-rate settler modules. A chain-and-flight sludge collector is one of the most suitable types of sludge removal mechanisms when used with high-rate settler modules, although traveling bridge collectors may also be used if channels for the sludge header passage are included.

PROCESS CONFIGURATION

To allow for better inlet flow conditions, the front one-quarter length of a rectangular horizontal-flow sedimentation basin is free from settler modules, as shown on Fig. 10-16a. Following the settling modules, a launder system is used to collect clarified water uniformly from the area covered by the settler modules. The modules can be hung from the launders, thereby eliminating an elaborate support system. Launders are usually spaced 3 to 4 m on center.

There are three alternatives for placement of tubes or plates in a sedimentation basin: (a) countercurrent, (b) cocurrent, and (c) cross flow, as shown on Fig. 10-17. The following settling analysis is presented for plate settlers; however, tube settlers may be modeled using a similar technique.

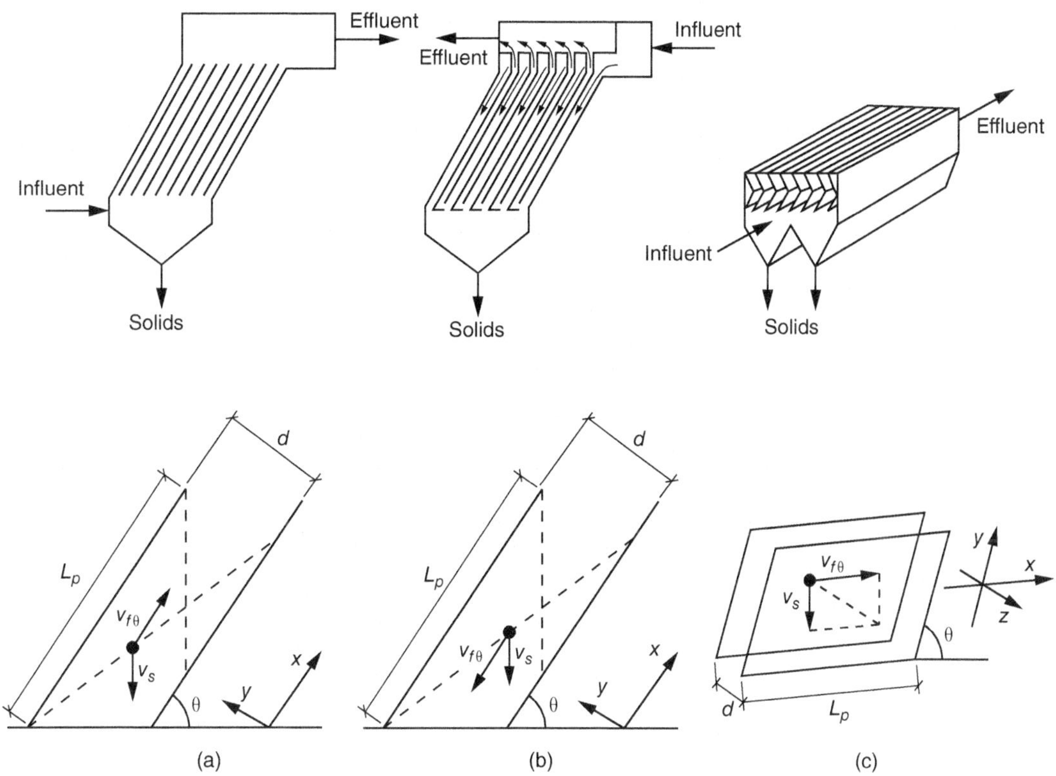

Figure 10-17
Flow patterns for inclined settling systems: (a) countercurrent, (b) cocurrent, and (c) cross flow. (Courtesy of Infilco Degrémont, Inc.)

Countercurrent settlers

The settling time for a particle to move between countercurrent parallel plates is given by the expression

$$t = \frac{d}{v_s \cos\theta} \tag{10-34}$$

where t = settling time, s
d = distance between two parallel plates (perpendicular to plates), m
v_s = particle settling velocity, m/s
θ = inclination angle of plates from horizon, deg

If a uniform velocity is assumed, then the particle travel time spent in the plates is given by the expression

$$t_p = \frac{L_p}{v_{f\theta} - v_s \sin\theta} \tag{10-35}$$

where t_p = particle travel time spent in plates, s
L_p = length of plate, m
$v_{f\theta}$ = fluid velocity in channel, m/s

If the trajectory of a particle that is shown on Fig. 10-17 is considered, then all of the particles with settling velocity v_s are removed and t_p is equal to the settling time t (equating Eq. 10-34 to Eq. 10-35). Further, those particles with a larger settling velocity are also removed, as shown by the expression

$$v_s \geq \frac{v_{f\theta} d}{L_p \cos\theta + d \sin\theta} \quad (10\text{-}36)$$

The fluid velocity $v_{f\theta}$ may be determined from the number of channels:

$$v_{f\theta} = \frac{Q}{N d w} \quad (10\text{-}37)$$

where Q = flow rate, m^3/s
N = number of channels, dimensionless
d = distance between two parallel plates (perpendicular to plates), m
w = width of channel, m

The fluid velocity $V_{f\theta}$ is also related to the overflow rate of the basin assuming that the surface area of the basin is comprised of plates and the area occupied by the plates is ignored:

$$v_{f\theta} = \frac{Q}{N d w} = \frac{Q}{A \sin\theta} \quad (10\text{-}38)$$

where A = top area of basin, m^2

Other terms are as defined above.

Particles with settling velocities less than v_s may also be removed, depending on where they enter the plate.

Co-current settlers

For co-current settling, the settling time for a particle to move between two parallel plates is given by Eq. 10-34. The time that particles moving with the fluid spend in the plates is given by the expression

$$t_p = \frac{L_p}{v_{f\theta} + v_s \sin\theta} \quad (10\text{-}39)$$

where t_p = particle travel time spent in plates, s
L_p = length of plate, m
$v_{f\theta}$ = fluid velocity in channel, m/s
v_s = particle settling velocity, m/s

If t_p is equal to settling time t (equating Eq. 10-34 to Eq. 10-39), then the particles with settling velocity v_s are removed. Further, those particles with a larger settling velocity are also removed, as shown by the expression

$$v_s \geq \frac{v_{f\theta} d}{L_p \cos\theta - d \sin\theta} \tag{10-40}$$

Cross-current settlers

For cross-current settling, the settling time for a particle to move between two parallel plates is also given by Eq. 10-34. The time that particles moving with the fluid spend in the plates is given by the expression

$$t_p = \frac{L_p}{v_{f\theta}} \tag{10-41}$$

where t_p = particle travel time spent in plates, s
L_p = length of plate, m
$v_{f\theta}$ = fluid velocity in channel, m/s

If t_p is equal to settling time t (equating Eq. 10-34 to Eq. 10-41), then the particles with settling velocity v_s are removed. Further, those particles with a larger settling velocity are also removed, as shown by the expression

$$v_s \geq \frac{v_{f\theta} d}{L_p \cos\theta} \tag{10-42}$$

The design of plate settlers is demonstrated in Example 10-7.

Example 10-7 Design of sedimentation process with plate settlers

A sedimentation basin has been retrofitted with 2.0-m (6.6-ft) square inclined plates spaced 50 mm (2.0 in.) apart. The angle of inclination of the plates can be altered from 0° to 80°. Assuming that the sedimentation basin can be used for countercurrent, co-current, or cross-flow sedimentation, determine which flow pattern is the most efficient for particle removal, ignoring any hydraulic problems that may arise as a result of flow distribution and sludge resuspension (adapted from Gregory et al.,1999).

Solution

1. Develop equations that can be used to compare the ratio of settling velocity to the fluid velocity for the three flow types from Eqs. 10-36, 10-40,and 10-42. The following equations for the ratio of the settling

velocity to the fluid velocity can be used to evaluate which configuration can capture the solids with the lowest settling velocity. Laminar flow or a constant velocity across the plate is assumed:

$$\text{Countercurrent flow}: \frac{v_s}{v_{f\theta}} = \frac{d}{L_p \cos\theta + d \sin\theta}$$

$$= \frac{0.05\,\text{m}}{(2.0\,\text{m} \times \cos\theta) + (0.05\,\text{m} \times \sin\theta)}$$

$$\text{Co-current flow}: \frac{v_s}{v_{f\theta}} = \frac{d}{L_p \cos\theta - d \sin\theta}$$

$$= \frac{0.05\,\text{m}}{(2.0\,\text{m} \times \cos\theta) - (0.05\,\text{m} \times \sin\theta)}$$

$$\text{Cross flow}: \frac{v_s}{v_{f\theta}} = \frac{d}{L_p \cos\theta} = \frac{0.05\,\text{m}}{2.0\,\text{m} \times \cos\theta}$$

2. Calculate the settling velocity-fluid velocity ratio for various plate angles for all flow types. The following table was prepared using the equations developed in step 1:

	$v_s/v_{f\theta}$ for Given Angle of Inclination							
Flow Pattern	**0°**	**10°**	**20°**	**30°**	**40°**	**60°**	**75°**	**80°**
Countercurrent flow	0.025	0.025	0.026	0.028	0.032	0.048	0.088	0.126
Co-current flow	0.025	0.026	0.027	0.029	0.033	0.052	0.106	0.168
Cross flow	0.025	0.025	0.027	0.029	0.033	0.050	0.096	0.144

3. Compare the settling velocity-fluid velocity ratios calculated in step 2. For angles less than 60°, there is very little difference in the calculated ratios between the various flow arrangements. However, above 60° countercurrent flow provides the most efficient operation as particles with the smallest settling velocity are removed.

Comment

Based on the computations shown in this example, there is little difference between the various flow arrangements for angles less than 60°. Countercurrent flow provides the most efficient operation above 60° as it allows settlement of particles with the smallest settling velocity.

PROCESS SELECTION

While the theoretical design and operation of high-rate settlers appear to provide an excellent alternative to conventional settlers, the performance of high-rate settlers can be greatly reduced due to the following conditions (Kawamura, 2000): (1) poor flocculation, (2) poor inlet flow distribution, (3) scaling (e.g., $CaCO_3$), and (4) algal growth. In addition to these performance issues, the following aspects of high-rate clarifiers should be considered.

Angle of inclination

Early development studies were conducted using flat parallel plates, shallow trays, and circular pipes or tubes. The feed rates were set so as to maintain laminar flow at all times. Tests indicated that for alum-coagulated sludge and countercurrent flow solids would remain deposited in the tubes until the angle of inclination was increased to 60° or more from horizontal. However, for co-current flow, the angle of inclination could be reduced to approximately 30°.

Flow pattern

In theory, the three flow patterns described above have little difference in performance when the angle of inclination is great enough for sludge to move toward the sludge zone when under the influence of gravity. However, in practice, co-current or cross-flow designs do not perform as well as countercurrent designs. Co-current designs have the problem of keeping the effluent water from resuspending the sludge. For cross-flow designs, it is difficult to obtain good flow distribution because water flow through the sludge zone has a smaller resistance than through the plates. Consequently, countercurrent designs are the most common.

Tube shape

Based on an analysis of tube shapes, there appears to be little difference between the efficiency of the various tubes, but the hexagon tube or chevron (see Fig.10-18) may have some advantages because sludge can collect in the notch of these channels. While parallel plates are efficient, their installation and manufacture can be difficult to maintain even spacing and ultimately uniform flow.

Solids Contact Clarifiers

Solids contact clarifiers have been used to achieve suspended solids removal in less space than conventional sedimentation basins. Solids contact units are usually found in industrial and municipal applications, where lime softening or softening clarification is the major treatment process and uniform flow rates and constant water quality prevail. Solids contact clarifiers can

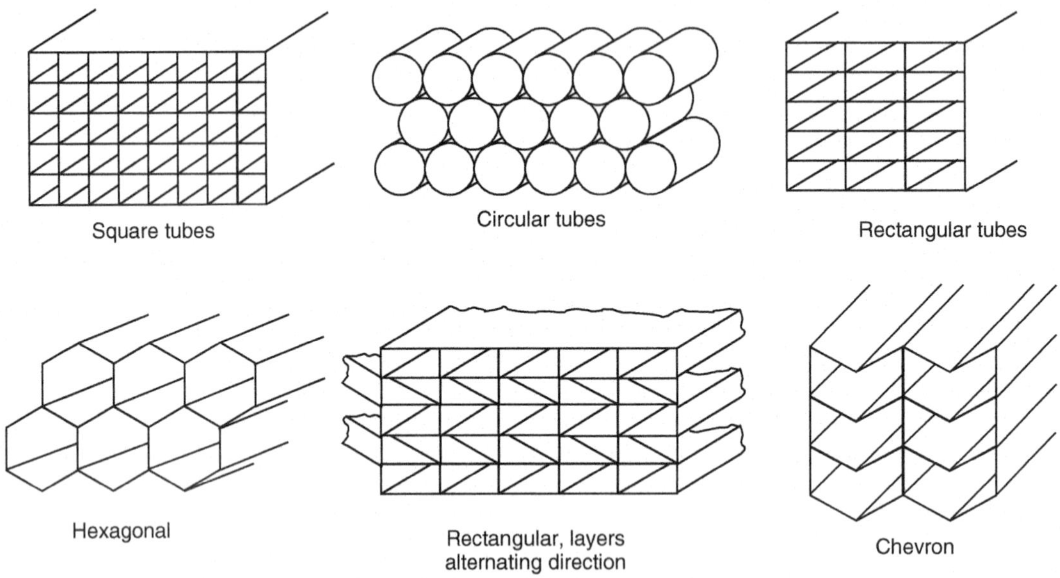

Figure 10-18
Typical configurations for tube settlers. (Adapted from Gregory et al., 1999.)

be categorized as reactor clarifiers, sludge blanket reactors, and adsorption reactors. Design criteria and other data for the solids contact units are summarized and compared with conventional rectangular clarifiers in Table 10-7.

REACTOR CLARIFIERS

In conventional treatment process design, the unit operations of rapid mixing, flocculation, and sedimentation are in individual unit processes in series. In a reactor clarifier, all these processes are combined in one unit. This combined process has significant advantages, such as reduced cost and more effective use of the sludge blanket. On the other hand, because all the unit operations are in one unit, usually stacked vertically, sludge blanket or reactor clarifiers reduce, somewhat, the ability of both the designer and the operator to refine the design and operating criteria for each of these operations. Most of these devices are preengineered, packaged proprietary devices that trade reduced flexibility to achieve greater optimization of a particular process option. In some circumstances these products are an excellent choice.

There are several proprietary reactor clarifier designs. Generally, these devices are used in smaller plants; however, some of them see regular use in large water treatment facilities. Most are preengineered units that involve a

Table 10-7
Typical design criteria for sedimentation processes and their principal applications

Typical Applications	Design Criteria	Advantages	Disadvantages
	Rectangular Basin (Horizontal Flow)		
• Many municipal and industrial water works • Particularly suited to large-capacity plants	• Surface loading: 30–60 $m^3/m^2 \cdot d$ (0.3–1.0 gpm/ft^2) • Water depth: 3–5 m (10–16 ft) • Detention time: 1.5–4 h • Minimum length-to-width ratio 4:1 to 5:1 • Weir loading <9–13 m^2/h (12–18 gpm/ft)	• More tolerance to shock loads • Predictable performance under most conditions • Easy operation and low maintenance costs • Easily adapted for high-rate settler modules	• Subject to density flow creation in basin • Requires careful design of inlet and outlet structures • Usually requires separate flocculation facilities
	Upflow (Radial Flow)		
• Small to midsize municipal and industrial water treatment plants • Best suited where rate of flow and raw-water quality are constant	• Surface loading: 30 – 45 $m^3/m^2 \cdot d$ (0.5 – 0.75 gpm/ft^2) • Water depth: 3 – 5 m (10 – 16 ft) • Settling time: 1 – 3 h • Weir loading: 170 $m^3/m \cdot d$ (13,700 gpd/ft)	• Economical compact geometry • Easy sludge removal • High clarification efficiency	• Problems of flow short circuiting • Less tolerance to shock loads • Need for more careful operation • Limitation on practical size unit • May require separate flocculation facilities
	Solids Contact Clarifiers[a]		
• Water softening • Flocculation–sedimentation treatment of raw water that has constant quality and rate of flow • Plants treating a raw water with low solids concentration	• Flocculation time: ~20 min • Settling time: 1 – 2 h • Surface loading: 50 – 75 $m^3/m^2 \cdot d$ (0.85 – 1.28 gpm/ft^2) • Weir loading: 175 – 350 $m^3/m \cdot d$ (14,000 – 28,000 gpd/ft) • Upflow velocity: <10 mm/min (2 in./min) • Higher maintenance costs and need for greater operator skill • Slurry circulation rate: up to 3 – 5 times raw-water inflow rate	• Good softening and turbidity removal • Flocculation and clarification in one unit • Compact and economical design	• Sensitive to shock loads and changes in flow rate • Sensitive to temperature change • Two to 3 days required to build up necessary sludge blanket • Plant operation dependent on single mixing motor

Source: Adapted from Kawamura (2000).
[a]Reactor clarifiers and sludge blanket clarifiers are often considered as one category, solids contact clarifiers.

trade-off between flexibility and experience with refinement of a particular design. Three of the more common design types—reactor clarifiers, sludge blanket clarifiers, and adsorption clarifiers—are illustrated on Fig. 10-19 and are described below.

Reactor clarifiers are center-feed clarifiers with a flocculation zone built into the central compartment (see Fig. 10-19a). Generally, these units contain a single motor-driven mixer, with recirculation of the sludge slurry (sometimes optional), followed by a settling zone in a separate outer annular compartment. When slurry recirculation is featured, these are often called solids contact clarifiers and generally include an impeller that provides considerable recirculation. The concentration in the unit is controlled by an adjustable timer on a sludge blow-off line. When using alum, it is common practice to maintain the slurry concentration in the mixing zone at 5 to 20 percent of the sludge volume after 10 min of settling. The slurry concentration is somewhat higher in softening. Reactor clarifiers work well in both clarification and softening, but, in the case of clarification using aluminum or iron salts, sludge recirculation improves performance at the expense of a significantly increased chemical requirement. In the case of lime softening, sludge recirculation improves both performance and chemical consumption.

SLUDGE BLANKET CLARIFIERS

The sludge blanket clarifiers are solids contact clarifiers with a distinct solids layer that is maintained as a suspended filter through which flow passes (see Fig. 10-19b). The sludge blanket unit contains a central mixing zone for partial flocculation and a fluidized sludge blanket in the lower portion of the settling zone. Partially flocculated water flows through the sludge blanket where flocculation is completed and solids are retained by adsorption and filtration. The sludge level is normally 1.5 to 2 m (4.5 to 6 ft) below the water surface, and clarified water is collected in launder troughs along the top of the unit. Sludge blanket clarifiers are made with or without sludge recirculation mechanisms. Sludge blanket clarifiers are compared with other processes in Table 10-7.

Design criteria for sludge blanket clarifiers are essentially the same as for sludge recirculation reactor clarifiers. However, the launder troughs should be sufficiently spaced to avoid sludge spillover. The launder spacing should be less than twice the distance between the water level and the top of the fluidized sludge blanket, approximately 4.5 m (15 ft). The troughs are typically arranged radially or in parallel.

Generally, sludge blanket clarifiers should be used only where the raw water characteristics and flow rates are relatively uniform. Given these parameters, the most effective applications are for lime softening and clari-

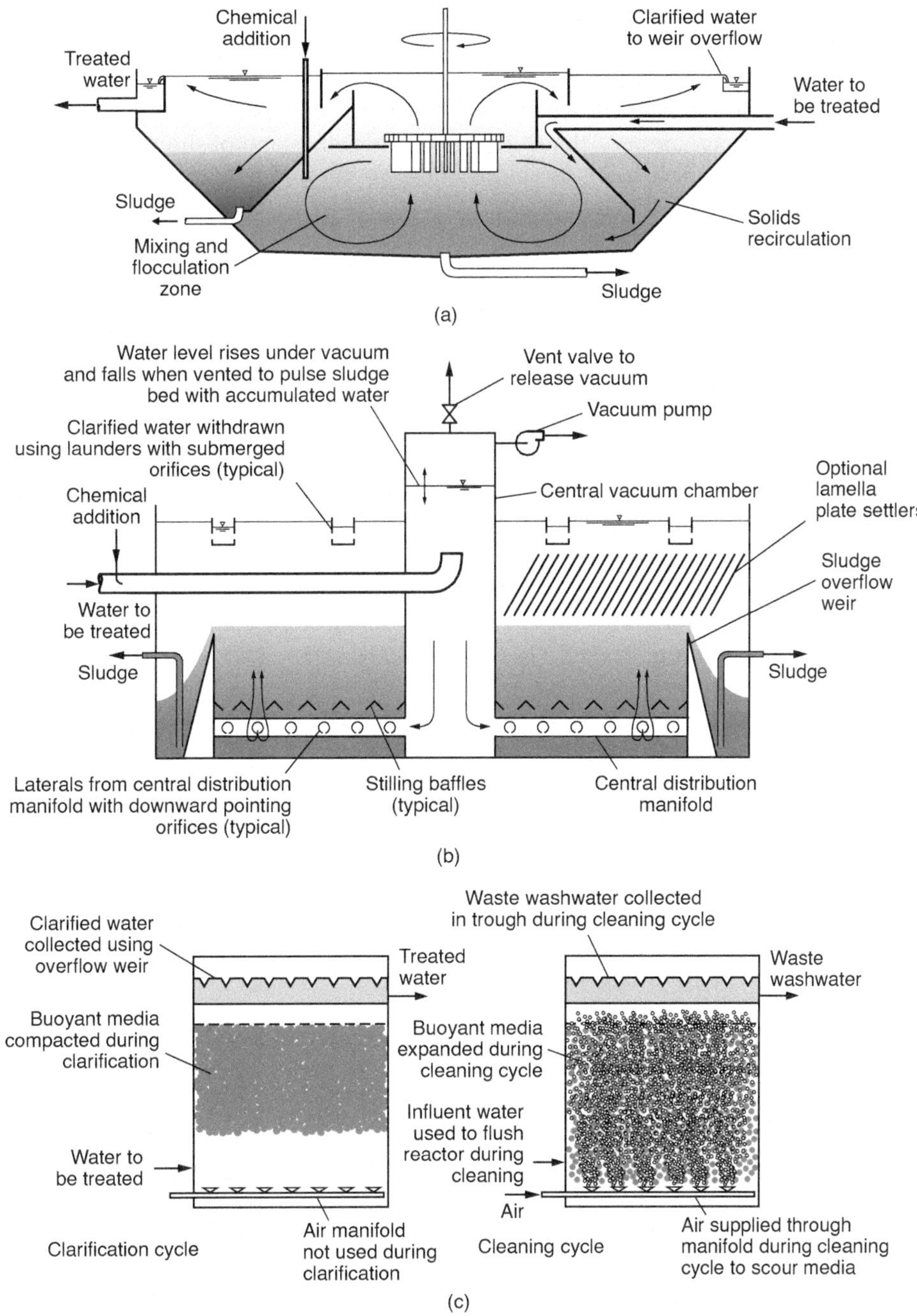

Figure 10-19
Three common types of proprietary reactor clarifiers: (a) reactor clarifier (Accelator), (b) sludge blanket clarifier (Pulsator, when optional lamella plates are added, the unit is known as the Super Pulsator), and (c) absorption clarifier. (Courtesy of Infilco Degrémont, Inc., now a subsidiary of Suez Water Technologies & Solutions)

fication of low-turbidity water. These units may also be used for clarification of highly turbid water (exceeding 500 NTU) if a sludge-scraping mechanism is provided.

One of the more difficult problems in operating sludge blanket clarifiers is the management of the sludge blanket itself. Some of the more popular designs accomplish this by simply allowing the sludge blanket to fall over a submerged weir that is kept a significant distance below the free surface. The Pulsator and its progeny the Super Pulsator are widely used examples of this principle. The Pulsator is shown on Fig 10-19b. Operationally, a portion of the flow is brought in the central vacuum chamber and allowed to rise above the operating water level in the clarifier by pulling a vacuum. When the water level in the vacuum chamber is about 0.5 to 1.0 m above the operating level in the clarifier, the vacuum is released by opening a valve to the atmosphere, allowing the water in the chamber to flow as a pulse through the influent distribution system located at the bottom of the tank below the sludge blanket. The pulse of water is used to contact the incoming water with the sludge blanket and to suspend and redistribute the sludge blanket. The depth of the sludge blanket is controlled by the overflow weir. The sludge blanket is typically pulsed once every 60 s (40 s to fill the vacuum chamber and 5 to 20 s to drain into the clarifier). The Super Pulsator is similar to the Pulsator but employs Lamellae settling.

ROUGHING FILTERS AND ADSORPTION CLARIFIERS

Roughing filters are used to create a zone of laminar flow during clarification. Similar in objective to the plate and tube settlers, a bed of granular material is used to establish a zone of laminar flow. The media in the bed may be heavy material such as gravel or buoyant plastic media. Suspended material deposits on the media as the water flows through the channels in the media bed. To remove the sludge, the media must be agitated to loosen the particles, which in turn fall to the bottom of the tank or are flushed from the tank with backwash water.

Most of the experimental work on gravel bed roughing filters has been done in Europe and South America. Gravel roughing filters are most frequently used ahead of slow sand filters when the raw-water source contains high turbidity or is subject to frequent runoff events. To date, gravel roughing filters have not been used on a full-scale basis in the United States, and they are not discussed further in this text. For additional information about roughing filters, refer to Collins et al. (1994).

The adsorption clarifier uses buoyant plastic media as a roughing filter. As the coagulated water travels upward through the media, flocculation takes place as the tortuous path of the water causes mixing and collisions between particles. Collisions between particles and the media causes particles to stick to the media, and most of the flocculated solids can be collected in the media. The media is then occasionally washed by introducing air from below, which reduces the bulk specific gravity of the water surrounding the media and allows it to expand and be cleaned. Both the adsorption (clarification mode) and the cleansing (flush cycle) are shown on Fig. 10-19c.

Ballasted Sedimentation

As the process name suggests, ballasted sedimentation involves the addition of ballast (usually microsand) that increases the settling velocity of the floc particles by increasing their density (providing ballast). The concept of ballasted sedimentation was first applied to water treatment in the 1970s. Currently, there are a number of proprietary sedimentation processes that employ the ballasted flocculation principle. Two well-known processes are the Actiflo® process and the Densadeg® process. These processes have been used in water treatment for both the production of potable water and the treatment of filter-to-waste washwater.

A schematic of the Actiflo® process is shown on Fig. 10-20. The Actiflo process involves adding an inorganic coagulant (alum or ferric) to the raw water and allowing floc to form in the coagulation stage. Subsequently, a high-molecular-weight cationic polymer and microsand particles (20 to 200 μm) are added to the injection stage, and the microsand particles flocculate with the preformed floc particles in the maturation stage. After flocculation, the ballasted floc is settled and the sludge containing the microsand is sent through a hydrocyclone (not shown) where the microsand is recovered and reused and the sludge is sent on for further treatment. The microsand is fed at a rate that is approximately 0.15 to 0.4 percent of the influent flow rate and the sludge ultimately contains 10 to 12 percent sand by weight.

The surface loading rate for an Actiflo sedimentation tank ranges from 35 to 62 m/h, which can be up to 50 times greater than the surface loading rate for a conventional rectangular sedimentation basin. The small

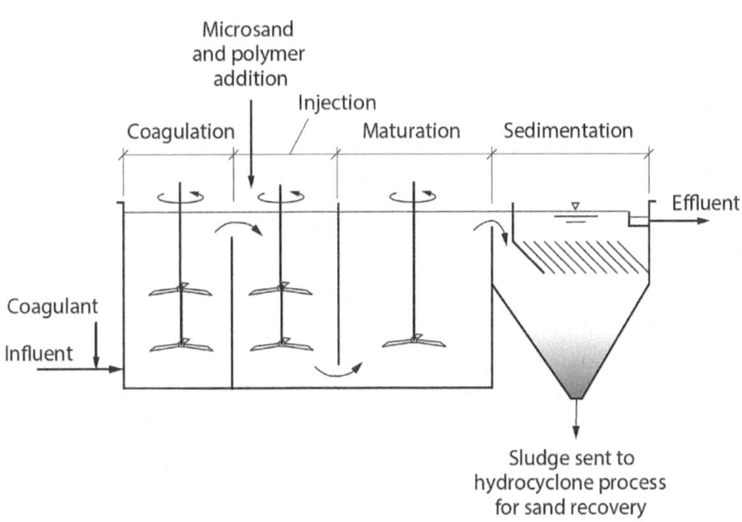

Figure 10-20
Schematic of Actiflo® process.

size of the Actiflo unit can be attributed to the use of high mixing energy ($\bar{G}$ values ranging from 150 to 400 s^{-1}), shorter detention times for flocculation (between 9 and 10 min), floc settling velocities 20 to 60 times greater than conventional flocculation and sedimentation, and the use of lamella plate settler modules to accelerate particle removal.

The advantages of the high-rate settling processes include (1) a small footprint requirement at water treatment plants with site constraints; (2) turbidity removal down to the 0.5-NTU level, but treating to 2.0 NTU is more common to reduce polymer usage and potential polymer carryover into the filters; (3) a quick process startup, about 15 min; (4) robust process that is not easily upset by changes in raw-water quality; and (5) potential savings in capital costs based on the small footprint. The disadvantages are (1) a heavy dependence on mechanical equipment and a short processing time; (2) the entire process must be shut down when there is a power outage lasting more than 10 min; (3) a higher coagulant dosage is required than for conventional processes with a high proportion of polymers, which may cause problems in downstream processes such as filter blinding and reduced filter run time; (4) potential for sand carryover (e.g., Actiflo process) into downstream processes; and (5) proprietary processes, which may limit competitive bidding.

10-8 Physical Factors Affecting Sedimentation

Accurate prediction of settling tank performance by mathematical and experimental methods is a challenge to even the best design engineers. Model testing using tracers and settling columns are limited by scale-up, which cannot be expressed adequately by principles of similitude, primarily because solid particles are not easily scaled down. In addition, many of the simplifying assumptions of modeling do not hold true in prototype units. Factors such as temperature gradients, wind effects, inlet energy dissipation, outlet currents, and equipment movement affect tank performance but are not easily modeled. Density currents, inlet energy dissipation, outlet currents, and equipment movement are presented and discussed in this section. Most of the information presented below on the physical factors related to sedimentation is directed toward conventional sedimentation basins and less toward innovative designs.

Density Currents

When feed water is entering the sedimentation basin, it can form a surface or a bottom density current, as illustrated on Figs. 10-21a and 10-21b, respectively, depending on the relative densities of the feed water and water in the basin. Under these flow conditions, actual flow-through velocity will depart from the theoretical, idealized average basin velocity. The theoretical velocity is equal to the total incoming flow divided by the total

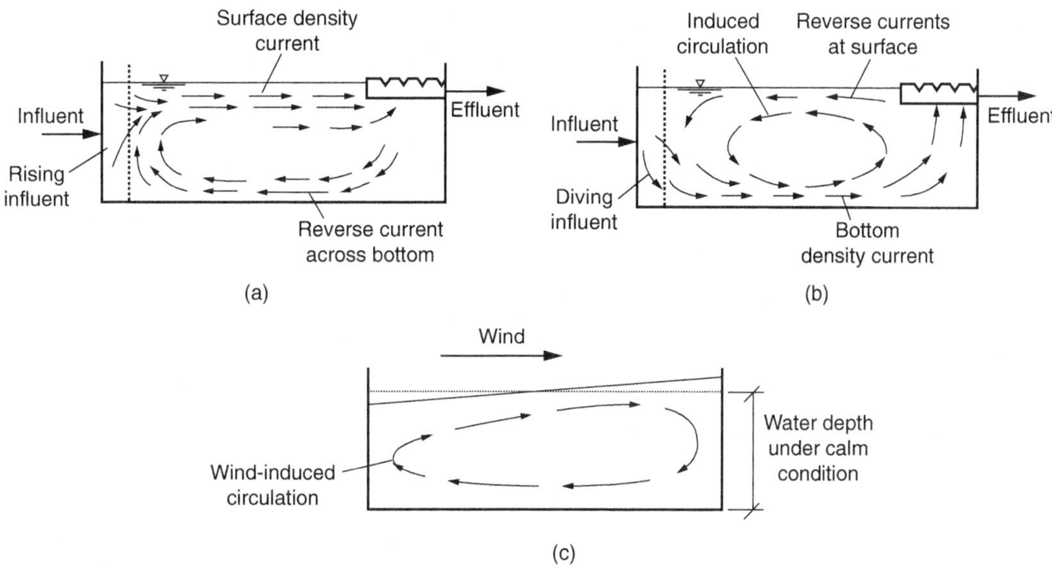

Figure 10-21
Nonideal flow in conventional sedimentation basin: (a) surface density currents, (b) bottom density currents, and (c) wind-induced currents.

cross-sectional area in the basin. Short circuiting caused by density currents has been observed in many water treatment plants (Camp, 1946; Harleman, 1961; Kao, 1977). Methods used to minimize the effects of density currents have been reported by a number of investigators (Camp, 1946; Fitch and Lutz, 1960; Harleman, 1961; Hudson, 1972; Kao, 1977; Sank, 1978). Kawamura (2000) reported density flow velocities of 0.8 to 1.8 m/min (2.6 to 6 ft/min) in the bottom of sedimentation tanks as compared to design velocity of 0.4 m/min (1.3 ft/min). These density flow velocities were observed during the day when the surface temperature was 0.2 to 0.5° C warmer than the influent raw water.

The velocity of the density current can be evaluated by the following expression (Harleman, 1961):

$$v_{fd} = \left\{8g\frac{\Delta\rho}{\rho}\left[\frac{hs}{f(1+\alpha)}\right]\right\}^{0.5} \qquad (10\text{-}43)$$

where v_{fd} = velocity of density flow, m/s
g = acceleration due to gravity, 9.81 m/s^2
$\Delta\rho$ = density difference between two liquids, kg/m^3
ρ = density of influent, kg/m^3
h = depth of density current flow, m

s = slope of channel bottom, m/m
f = Darcy-Weisbach friction factor, unitless
α = correction factor for kinetic energy ranging from 0 to 1 (0.43 for turbulent flow), unitless

Basin inlet and outlet arrangements should be designed to provide some degree of control in minimizing the effects of density currents. At the inlet, the following techniques have been used: (1) feed flow is distributed uniformly through the basin cross section in the plane perpendicular to the flow by employing diffuser walls and (2) devices that will break up the feed stream and dissipate the energy by turbulence.

Improvements can be made in the basin to control the density currents. These improvements include tube settlers, redistribution baffles, or intermediate diffuser walls. Launders extending into sedimentation basins have been used to control the effluent flow distribution, which is more effective for controlling bottom density currents than surface density currents.

TEMPERATURE DIFFERENTIALS

The addition of warm influent water to a sedimentation basin containing cooler water can lead to a short-circuiting phenomenon in which the warm water rises to the surface and reaches the effluent launders in a fraction of the nominal detention time. Conversely, cold water added to a basin containing warm water tends to force the incoming water to dive to the bottom of the basin, flow along the bottom, and rise at the basin outlet.

Studies were conducted by Kawamura (1981) using a 1:25 scale model with dynamic similitude following the Froude law. Temperatures in the tank and the feed flow were kept constant with the feed flow about 0.3° C lower than the water in the tank. Without diffuser walls in the tank, the density current flowed along the bottom of the tank at a relatively shallow depth and took less time to reach the outlet. Under these conditions, the top two-thirds of the tank depth was not used effectively. To improve the hydraulic efficiency, diffuser walls with approximately 7 percent net opening were added. With the modified diffuser walls, head losses created at the diffuser walls forced the retardation and mixing of the density current with the ambient water, improving the flow distribution in the tank and the efficiency of hydraulic performance.

When conducting tracer tests in pilot plant investigations, Tekippe and Cleasby (1968) found that minor temperature differences greatly reduced the reproducibility of experiments and overshadowed minor differences in inlet and outlet design variables. The performance of sedimentation tanks that are constructed with metallic walls and exposed to sunlight may also be unpredictable. The heat transmitted through the wall on the sunny side of the basin tends to warm the water, making it less dense than water on the shaded side. The warm water in turn rises, forming a density current that, if sufficiently severe, can invert the contents of the basin.

SOLIDS CONCENTRATION EFFECTS

Density current problems similar to those discussed above may also be caused by changes in influent solids concentrations resulting from flash floods or strong winds on lake water surfaces. A rapid increase in turbidity increases the density of the influent and causes it to plunge as it enters the sedimentation basin. In a center-feed circular tank, when the influent is denser than the water in the tank, it falls to the bottom and flows outward from the center in a radial direction; rising toward the effluent launders near the perimeter. Water not leaving the basin through the effluent launders returns along the surface of the basin. These currents tend to leave the center of the basin in a relatively stagnant position, thereby reducing the effective volume and detention time for settling. The effect of solids concentration on the mass of water in a basin is presented in Example 10-8.

Example 10-8 Effect of solids concentration on mass of water

Calculate the difference in mass of water in a sedimentation basin 20 m wide by 100 m long by 5 m deep (10,000 m^3 of water) containing 20 mg/L of solids and 2000 mg/L of solids. Assume the solids have a density of 2.5 g/mL and the water has a density of 0.99823 g/mL. The volume of solids and water in a 1-L solution can be determined from the following expressions:

$$V_{\text{solids}} = \frac{\text{mass of solids (g)}}{\text{density (g/mL)}}$$

$$V_{\text{water}} = 1000\text{ mL} - V_{\text{solids}}\text{ (mL)}$$

Solution

1. Calculate the densities of the two solutions.
   a. Determine the volume of solids and water in a 1-L solution with solids concentrations of 20 and 2000 mg/L. For solids concentration of 20 mg/L,

$$V_{\text{solids}} = \frac{\text{mass of solids (g)}}{\text{density (g/mL)}} = \frac{0.02\text{ g}}{2.5\text{ g/mL}} = 0.008\text{ mL}$$

$$V_{\text{water}} = 1000\text{ mL} - V_{\text{solids}}\text{ (mL)} = 1000\text{ mL} - 0.008\text{ mL}$$
$$= 999.992\text{ mL}$$

For solids concentration of 2000 mg/L,

$$V_{\text{solids}} = \frac{\text{mass of solids(g)}}{\text{density(g/mL)}} = \frac{2\text{ g}}{2.5\text{ g/mL}} = 0.8\text{ mL}$$

$$V_{\text{water}} = 1000\text{ mL} - V_{\text{solids}}(\text{mL}) = 1000\text{ mL} - 0.8\text{ mL}$$
$$= 999.2\text{ mL}$$

b. Determine the density of a 1-L solution with solids concentrations of 20 and 2000 mg/L. For solids concentration of 20 mg/L,

$$\rho_{\text{solution}} = \frac{V_{\text{solids}}\,\rho_{\text{solids}} + V_{\text{water}}\,\rho_{\text{water}}}{V_{\text{solids}} + V_{\text{water}}}$$
$$= \frac{(0.008\text{ mL})(2.5\text{ g/mL}) + (999.992\text{ mL})(0.99823\text{ g/mL})}{0.002 + 999.992\text{ mL}}$$
$$= 0.998242\text{ g/mL}$$

For solids concentration of 2000 mg/L,

$$\rho_{\text{solution}} = \frac{(0.8\text{ mL})(2.5\text{ g/mL}) + (999.2\text{ mL})(0.99823\text{ g/mL})}{0.8 + 999.2\text{ mL}}$$
$$= 0.998431\text{ g/mL}$$

2. Calculate the mass of the two solutions.
   a. The mass of 10,000 m^3 (10^7 L or 10^{10} mL) of water with a solids concentration of 20 mg/L is given as

$$\text{Mass of solution} = (0.998242\text{ g/mL})(10^{10}\text{ mL})$$
$$= 9.98242 \times 10^9\text{ g}$$

   b. The weight of 10,000 m^3 of a 2000-mg/L solution is given as

$$\text{Mass of solution} = (0.998431\text{ g/mL})(10^{10}\text{ mL})$$
$$= 9.98431 \times 10^9\text{ g}$$

   c. The weight difference between a solution containing 20 and 2000 mg/L of solids in a sedimentation basin is 1890 kg.

Comment

The increased mass of the water with solids concentration of 2000 mg/L would result in density flow and poor clarifier performance if appropriate measures were not taken.

The remediation of problems of varying influent turbidity are similar to those for incoming temperature differences and include diffuser walls in the basin. Additionally, the source of water should be carefully selected and the method of removing water from the source should minimize quality variations. It should be noted that changes in water density resulting from variable dissolved solids (salinity) concentration may also lead to density flow and short circuiting.

Wind Effects

Wind can have a pronounced effect on the performance of large, open gravity settling basins. High wind velocity tends to push the water to the downwind side of a basin and produces a surface current moving in the direction of the wind. An underflow current in the opposite direction is also created, which moves along the bottom of the tank. The resulting circulating current, shown on Fig. 10-21c, can lead to short circuiting of the influent to the effluent weir and scouring of settled particles from the sludge zone. For open sedimentation basins with length or diameter greater than 30 m, wind effects can be significant and result in reduced effluent quality. Wind-induced currents were studied by Baines and Knapp (1965) and Hidy and Plate (1966) in laboratory wind-water tunnels. Liu and Perez (1971) have also modeled wind-induced currents by numerically solving the governing equations of motion with assumptions of the kinematic eddy viscosity of water.

When long and shallow rectangular settling basins are used, orienting the basin with the local prevailing wind direction should be considered. In areas with strong predictable winds, sedimentation basins should be positioned so that the water flow parallels the wind, and wave breakers (launders or baffles) should be placed at approximately 20- to 30-m (65- to 100-ft) intervals. Changes in water surface elevation are minimized when the wind blows across the length of the rectangular settling basin, as opposed to across the width, and the effects of wind currents on sedimentation basin performance are minimized. Wind-induced wave patterns on the surface of the water may cause the formation of circulation patterns in the basins, which in turn affect settling rates. Where sandstorms or other wind-induced particle deposition is expected, a windbreak or cover over the basins may also be necessary. Trees may not be the best windbreaks because leaves and small branches can drop into the basin and clog the sludge withdrawal system.

Inlet Energy Dissipation

Sedimentation basin performance is strongly influenced by inlet energy dissipation. When the flocculation basin is not directly attached to the sedimentation basin, water may be conveyed to the basin through a pipe at high enough velocities to keep solids in suspension. The influent flow must be slowed down and distributed over a broad area to begin the sedimentation process. In rectangular basins, flow is often distributed with a channel across the inlet end of the basin. Baffles are then used to distribute the flow across the tank cross section in a horizontal and vertical direction, simultaneously. If the energy dissipation is not engineered carefully, density and eddy currents and excessive velocity vectors will be created.

Outlet Currents

Outlet currents in a sedimentation basin are often related to design details of effluent weirs and launders. Initially, these weirs were simply flat plates across the end of a rectangular basin. The width of the basin established the length of the weir. When tanks were designed in a long, narrow configuration, the weir length was relatively short and was believed to contribute to formation of outlet currents that, if severe, could sweep settleable particles into the tank effluent. The problem of currents was compounded in early designs because the flat weir plates were sometimes not level. Concern for this led to the development of V-notch weirs, which provide better lateral distribution of outlet flow when leveling is imperfect.

For upflow clarifiers, such as solids contact basins, launders carefully spaced across the surface are considered of vital importance to good performance. The launders, which are often arranged in a radial pattern, serve an important role in directing the vertical flow through the solids contact zone. As solids contact tanks become larger, strategic location of radial weirs becomes more critical.

In general, for most water treatment sedimentation basins, performance is primarily a function of density currents and inlet energy dissipation rather than outlet currents. Careful design of effluent weirs will not solve problems associated with density currents created by other design deficiencies.

Equipment Movement

Another potential effect on sedimentation basin performance is the movement of equipment within the basin. Sludge collection mechanisms, normally consisting of chain-and-flight scrapers, bridge-mounted scrapers, or hydraulic vacuum units, must move through the liquid contents of the tank to remove settled sludge. If equipment movement is excessive, currents may be introduced that upset the sedimentation process. Most equipment moves at a rate of 15 to 30 m/h and has a minimal effect. However, equipment movement in the vicinity of the effluent launders is important because of the potential for disturbed settled solids to be caught in the effluent currents and carried over the effluent weir.

10-9 Dissolved Air Flotation

Dissolved air flotation (DAF) is a unit operation that involves the use of fine air bubbles for the separation of solid and semisolid (floc) particles from a liquid. Fine air bubbles are introduced near the bottom of the basin containing the water to be treated. As the bubbles move upward through the water, they become attached to particulate matter and floc particles, and the buoyant force of the combined particle and air bubbles causes the aggregated particles to rise to the surface. Thus, particles that are heavier than the liquid can be made to float. Particles that rise to the surface are removed for further processing as residuals, and the clarified liquid is filtered to remove any residual particulate matter. A summary overview of the DAF process, including a process description, DAF applications, theoretical factors affecting DAF, and design considerations for DAF systems, is presented in this section. Additional details may be found in Zabel and Hyde (1977), Zabel (1985), Gregory et al. (1999), Haarhoff and Edzwald (2004), Haarhoff (2008), Edzwald (2010), and Gregory and Edzwald (2010).

Process Description

The DAF process dates back to the 1920s when a vacuum system was used. Two such plants were still in operation in Sweden in the 1970s. The pressurized process in use today was developed in Finland and Sweden in the 1960s. Since then, the use of pressurized DAF technology has spread throughout the world. The DAF was first used in the United States in the early 1980s (Haarhoff, 2008). It is now estimated that there are more than 150 DAF plants in the United States and Canada, and the popularity of the process is continuing to increase.

Like other gravity separation processes, raw water is coagulated and flocculated prior to entering the DAF basin. The water is introduced into the contact zone of the flotation basin near the floor, as shown on Fig. 10-22a. A baffle wall separates the contact zone from the separation (clarification) zone and limits short circuiting. In the contact zone (see Fig. 10-22b), a cloud of air bubbles called white water, typically 10 to 100 μm in diameter, adheres to floc particles in the influent creating floc–bubble aggregates with a net specific gravity below that of surrounding liquid.

In the separation zone (see Fig. 10-22b), floc particles rise to the surface and form a discrete layer of solids known as float (as the float thickens before removal it is termed sludge). The float layer collects at the effluent end of the basin and is removed into a float collection trough. Float removal is accomplished either with a mechanical skimming device or hydraulically by solids overflow into the collection trough. The hydraulic removal of float is achieved by temporarily prohibiting water from leaving the basin, which causes the water level in the basin to rise and float to overflow into the float collection trough. Clarified water is removed through a perforated pipe lateral system, a false-floor system with a plenum, or an underflow baffle wall. An effluent weir is the ultimate point of discharge of the clarified water, as shown on Fig. 10-22a.

Dissolved Air Flotation Applications

In water treatment plants, the DAF process is used for the removal of various forms of particulate matter from water including:

1. Low-density particulate matter such as algae
2. Dissolved organic matter (natural color)

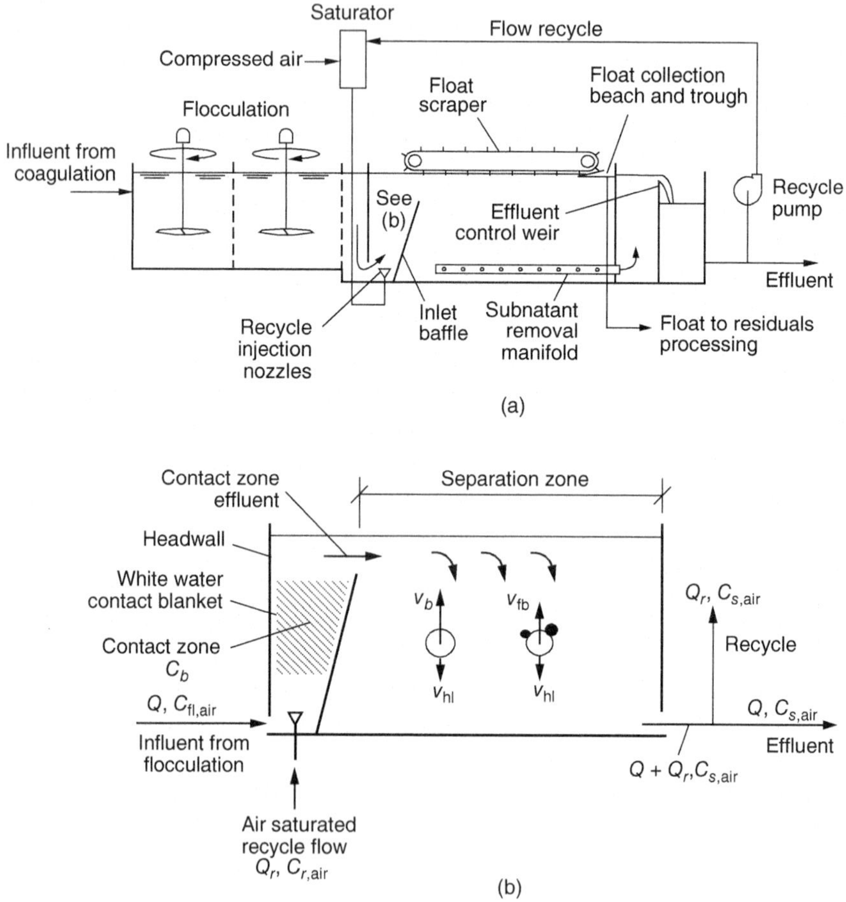

Figure 10-22
Dissolved air flotation process schematic.

3. Particulate matter from low- to moderate-turbidity waters
4. Algae, dissolved organic matter, and turbidity from low-temperature waters
5. Suspended material from filter waste washwater

Because gravity separation processes rely on density differences between particulate matter and water, separation of light particles or separation at low temperatures can be challenging. Dissolved air flotation takes advantage of these characteristics by widening the density difference by making the floc–bubble aggregate buoyant. The advantages and disadvantages of the DAF process compared to other clarification processes are presented in Table 10-8.

Table 10-8
Advantages and disadvantages of DAF in comparison with other clarification technologies

Advantages	Disadvantages
• High loading rate: Typically 10–20 m/h. New process variants have operated successfully up to 40–45 m/h. • Very thick float (sludge) product: Typically 2–3% total solids float can be achieved using hydraulic or mechanical skimming devices. Float can be dewatered without intermediate thickening. • Often, no polymer is required, as DAF does not require a large, dense floc. Coagulant dosages may also be reduced in some circumstances. • Shorter flocculation times, as compared to gravity separation, are possible, because a smaller floc particle size is required. • Rapid startup, typically < 30–60 min to reach steady state, depending on size. • Excellent algae removal efficiencies. • Excellent *Giardia* and *Cryptosporidium* removal efficiencies (~2–2.5 log), depending on temperature. • Smaller footprint required as compared to conventional flocculation and gravity sedimentation	• Requires a cover or housing to protect the float layer from wind and precipitation. • Mechanically more complex than conventional gravity clarifiers. • More power intensive as compared to conventional flocculation and sedimentation (2.5 – 3 to 0.75 – 1 kWh/ 10^3 $m^3 \cdot d$). • Generally not well suited for clarification of high-turbidity silt-laden waters. • Because DAF is more mechanically intensive, may not be suitable for locations where equipment maintenance is likely to be neglected.

Factors Affecting DAF Performance

Proper coagulation is, perhaps, the most important factor affecting DAF performance. If destabilized floc particles (low particle charge and fairly hydrophobic) are not produced, then floc particle attachment to air bubbles will be poor (Edzwald, 2010). Other factors that are important in the DAF process include (1) floc characteristics, (2) bubble size and rise velocity, (3) air loading, (4) floc–bubble attachment, and (5) the solubility of gases at elevated pressure.

FLOC CHARACTERISTICS

A large, readily settleable floc is appropriate for sedimentation, while a small, low-density floc is appropriate for DAF. A low-density floc contributes to the density of the floc–bubble aggregate being significantly less than the density of water, allowing the floc–bubble aggregate to float. In addition, the higher inherent turbulence of the DAF process, particularly in the contact zone of the basin, is more suitable for a small floc that is able to withstand high shear forces without disintegration.

The requirements for a small, low-density floc result in some significant differences in pre-DAF flocculation system design as compared to presedimentation flocculation as follows:

- ❑ Flocculation times can typically be reduced. Flocculation times as short as 5 min have been used successfully in DAF flocculation design.

In full-scale systems, 10 to 15 min is typical. In small-package plants, detention times are on the order of 20 min.

- Higher flocculation energy is usually beneficial to develop a small, tough floc. Typical design flocculation $\bar{G}$ values range from 50 to 100 s^{-1}.
- Polymers used as flocculation aids are rarely needed or used.
- Reductions in coagulant dosages may be possible as a result of improved performance with smaller floc (Edzwald, 1995).
- Design of the baffling between the last stage of the flocculator and the DAF inlet should be an over–under baffle arrangement, so that flocculated water flows downward and into the DAF basin at floor level, encouraging interaction between the flocculated water and the recycle stream (see Fig. 10-22).

BUBBLE SIZE AND RISE VELOCITY

Generation of microbubbles is achieved by supersaturating a recycled side stream with air at an elevated pressure. Microbubbles are formed when the pressure of the saturated side stream is released. The side stream may be taken from the clarified water when the solids content is low enough not to cause saturator or injection nozzle fouling problems. Filtered water is used when the solids content is too high or the DAF and filtration processes are in the same concrete tank with flotation over the filters, typical in most small plants and in large plants where land (or space) is limited.

Generation of bubbles between 10 and 100 μm in size is important for adequate floc–bubble attachment and flotation. Because floc–bubble attachment is a surface phenomenon, it is important that the bubbles be small enough that the total surface area for adherence is large. However, the bubbles must not be larger than 130 μm to maintain laminar flow conditions (Gregory et al., 1999).

The bubble rise velocity depends upon the size of the bubble and the water temperature. Optimum conditions for maximum collision and attachment between floc particles and bubbles occur when the bubbles rise in the laminar flow regime, minimizing the impact of particles detaching from the bubbles due to the fluid shear forces. In the laminar flow region, Stokes' law (Eq. 10-13) can be used to estimate bubble rise velocity.

FLOC-BUBBLE AGGREGATE RISE VELOCITY

An important consideration is the rise velocity of the aggregates that form when bubbles attach to the floc. Floc–bubble aggregates are not spherical particles so the drag coefficients tend to vary somewhat from the drag coefficients for spherical particles presented in Sec. 10-2. Furthermore, the density and size of the floc–bubble aggregates must be determine from the density and size of the bubbles and particles. The density of the floc–bubble aggregate is calculated from the expression (Edzwald, 2010)

$$\rho_{pb} = \frac{\rho_p d_p^3 + N_b \rho_b d_b^3}{d_p^3 + N_b d_b^3} \tag{10-44}$$

where ρ_{pb} = particle–bubble aggregate density, kg/m^3
ρ_p = particle density, kg/m^3
ρ_b = air bubble density, kg/m^3
d_p = particle diameter, m
d_b = mean bubble diameter, m
N_b = number of bubbles attached to floc particle

The equivalent spherical diameter of the floc–bubble aggregate can be determined from the expression

$$d_{pb} = \left(d_p^3 + N_b d_b^3\right)^{1/3} \tag{10-45}$$

where d_{pb} = equivalent spherical diameter of floc–bubble aggregate, m

For laminar flow, the drag coefficient varies gradually from $C_d = 24/\text{Re}$ for flocs of 40 μm and smaller to $C_d = 45/\text{Re}$ for flocs of 170 μm (Edzwald, 2010). Thus, Stokes' law is appropriate for floc of 40 μm (using the floc diameter and density from Eqs. 10-44 and 10-45) but the coefficient in the denominator should be adjusted for larger floc rising in laminar flow conditions. When flow is transitional ($1 < \text{Re} < 50$), the drag coefficient has been reported as (Haarhoff and Edzwald, 2004)

$$C_d = \frac{45}{\text{Re}^{0.75}} \tag{10-46}$$

Substituting Eq. 10-45 into Eq. 10-7 and solving for the rise velocity as was done in Sec. 10-2 yields

$$v_{pb} = \left[\frac{g\left(\rho_w - \rho_{pb}\right) d_p^{1.75}}{33.75 \rho_w^{0.25} \mu^{0.75}}\right]^{1/1.25} \tag{10-47}$$

where v_{pb} = rise velocity of floc particle–bubble aggregate, m/h
g = acceleration due to gravity, 9.81 m/s^2
ρ_w = water density, kg/m^3
d_{pb} = equivalent spherical diameter of floc–bubble aggregate, m
μ = dynamic viscosity of water, N · s/m^2

AIR LOADING

Recycle systems are designed to provide a target air loading per unit volume of raw water treated to ensure a dense bubble cloud forms within the contact zone. As noted previously, this dense bubble cloud, which to the naked eye appears as "milky" water, has been termed the "white water blanket (WWB)." The WWB contact zone provides an opportunity to capture the relatively low density of solids.

The effects of air loading on the DAF process are illustrated on Fig. 10-23a. Effluent turbidity declines with increasing air loading until a break point is reached where the application of additional air provides no corresponding increase in process performance.

The mass concentration of air released can be calculated from the following expression (Gregory et al., 1999):

$$C_b = \frac{e(C_r - C_s)r - k}{1 + r} \tag{10-48}$$

where C_b = mass concentration of air released, mg/L
e = efficiency factor, dimensionless
C_r = mass concentration of air in recycle flow, mg/L
C_s = mass concentration of air at saturation for atmospheric pressure, mg/L
r = recycle ratio, dimensionless
k = factor to account for air deficit in incoming flocculated water

The efficiency factor is used to account for minor pressure losses between the saturator and the point of air release. Henry's law is used to calculate the mass concentration of air in the recycle flow, as will be illustrated

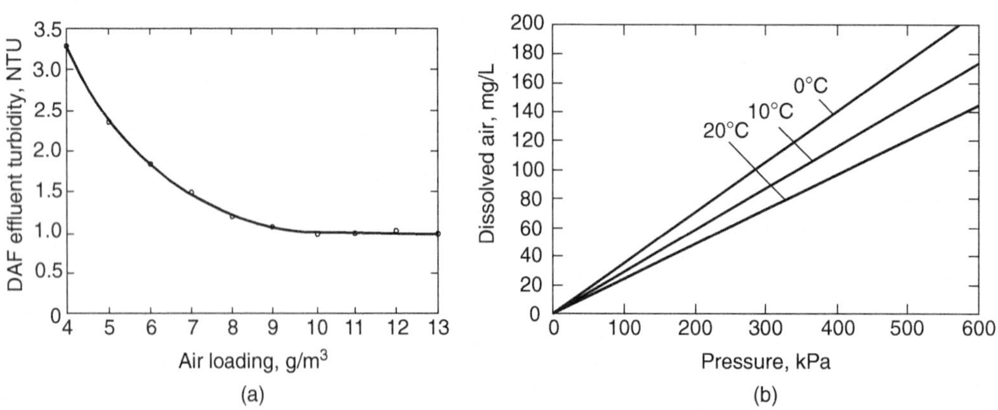

Figure 10-23
Effects of (a) air loading on DAF effluent turbidity (adapted from Zabel, 1985) and (b) pressure and temperature on air solubility.

in Example 10-9. The air concentration in the saturator can be expressed either as the bubble number concentration (N_b) or the bubble volume concentration (ϕ_b). These values can be determined from the expressions

$$\phi_b = \frac{C_b}{\rho_{\text{air}}} \tag{10-49}$$

$$N_b = \frac{10^{12} \times 6\phi_b}{\pi d_b^3} \tag{10-50}$$

where ϕ_b = bubble volume concentration, L/L
C_b = mass concentration of air released, mg/L
ρ_{air} = density of air saturated with water vapor, mg/L
N_b = bubble number concentration, no./mL
10^{12} = conversion factor, μm^3/mL
d_b = mean bubble diameter, μm

It has been found that flotation performance increases as N_b increases because there are more collision and attachment opportunities between the bubbles and particles. Attached air bubbles provide lower floc particle density and larger volume, producing floc particle–bubble aggregates that have high upward velocities (Gregory et al., 1999).

MINIMUM VOLUME OF GAS NEEDED

The minimum volume of gas needed to achieve flotation can also be approximated with the following expression:

$$\frac{\phi_g}{\phi_p} = \frac{\rho_p - \rho_w}{\rho_w - \rho_g} \tag{10-51}$$

where ϕ_g = minimum volume of gas required for flotation, mL/L, ppm
ϕ_p = volume of particle, mL/L, ppm
ρ_p = density of particle, kg/m^3
ρ_w = density of water, kg/m^3
ρ_b = density of air, kg/m^3

Because the density of air relative to water is small Eq. 10-51 can be written as follows:

$$\frac{\phi_g}{\phi_p} \simeq \rho_p - 1 \tag{10-52}$$

In general, the amount of air relative the amount of floc must be significantly greater.

SOLUBILITY OF AIR AT ELEVATED PRESSURE

The increased solubility of gases in water at elevated pressure is the fundamental principle that allows the formation of microbubbles. The solubility of air in water increases in an essentially linear fashion, in the temperature range of concern in water treatment, with increasing pressure according to Henry's law, as presented and discussed in Chap. 14 and repeated here for convenience. The application of Henry's law is illustrated in Example 10-9.

$$Y = H_{YC}C \tag{14-25}$$

where Y = gas-phase concentration, mg/L
H_{YC} = dimensionless Henry's constant
C = liquid-phase concentration in equilibrium with gas-phase concentration y, mg/L

Example 10-9 Estimating the saturation concentration of air in water as function of temperature and pressure

Determine the mass concentration of air in solution at 0 and 20°C if a DAF saturator is to operate at a pressure of 600 kPa (500 gauge).

Solution

1. Determine the mass concentration of air in solution at 0°C and 600 kPa.
   a. From App. B, the mass concentration of air at 0°C and 1 atm, which corresponds to the density of air, is 1.2928 g/L.
   b. Using the temperature coefficients given in Table 14-4 and Eq. 14-11, determine the value of the dimensionless Henry's constant, H_{YC}, at 0°C and 1 atm:

$$H_{YC} = K_C \exp\left(-\frac{\Delta H^\circ_{dis}}{RT}\right)$$

$$K_c = 3{,}368$$

$$\Delta H^\circ_{dis} = 10.28 \times 10^3 \text{ J/mol}$$

$$H_{YC} = K_C \exp\left(-\frac{\Delta H^\circ_{dis}}{RT}\right)$$

$$= 3368 \exp\left\{-\frac{10.28 \times 10^3 \text{ J/mol}}{(8.314 \text{ J/mol}\cdot\text{K})[(273+0)\text{ K}]}\right\}$$

$$= 36.3$$

c. Rearrange Eq. 14-25 and solve for the mass of air in solution:

$$C = \frac{Y}{H_{YC}} = \frac{1292.8 \text{ mg/L}}{36.3} = 35.6 \text{ mg/L}$$

d. Determine the mass of air in solution at 600 kPa. Note: 1 atm = 101.325 kPa

$$C = 35.6 \text{ mg/L}\left(\frac{600}{101.325}\right) = 210.8 \text{ mg/L}$$

2. Determine the mass concentration of air in solution at 20°C and 600 kPa.
   a. From equation in App. B, the density of air at 20°C and 1 atm is 1.204 g/L.
   b. From Table 14-4 the value of the dimensionless Henry's constant, H_{YC}, at 20°C and 1 atm is 49.58.
   c. Solve for the mass of air in solution:

$$C = \frac{Y}{H_{YC}} = \frac{1204 \text{ mg/L}}{49.58} = 24.28 \text{ mg/L}$$

   d. Determine the mass of air in solution at 600 kPa:

$$C = 24.28 \text{ mg/L}\left(\frac{600}{101.325}\right) = 143.8 \text{ mg/L}$$

Comment

The approximate saturation concentration of air can also be determined by considering oxygen and nitrogen separately and then combining the two values. The relationship is not exact because argon is neglected.

Using the computational procedures given in Example 10-9 a plot of the saturation concentration of air versus pressure for various temperatures is presented in Fig. 10-23b. As shown in Fig. 10-23b, the solubility of air in water is reduced at elevated temperatures. In DAF designs in tropical areas, or anywhere where raw-water temperatures routinely exceed 25° C, it is sometimes necessary to adopt higher than normal design recycle rates to ensure sufficient air can be delivered to the process at warmer temperatures.

FLOC-BUBBLE ATTACHMENT

The mechanisms for floc–bubble attachment are not well understood and may be any of the following four mechanisms individually or in combination, also shown on Fig. 10-24: (1) fine air bubbles adhere to floc in WWB due to electrostatic attraction or other mechanisms (Fig. 10-24a); (2) fine air bubbles become physically entrapped within preformed floc in

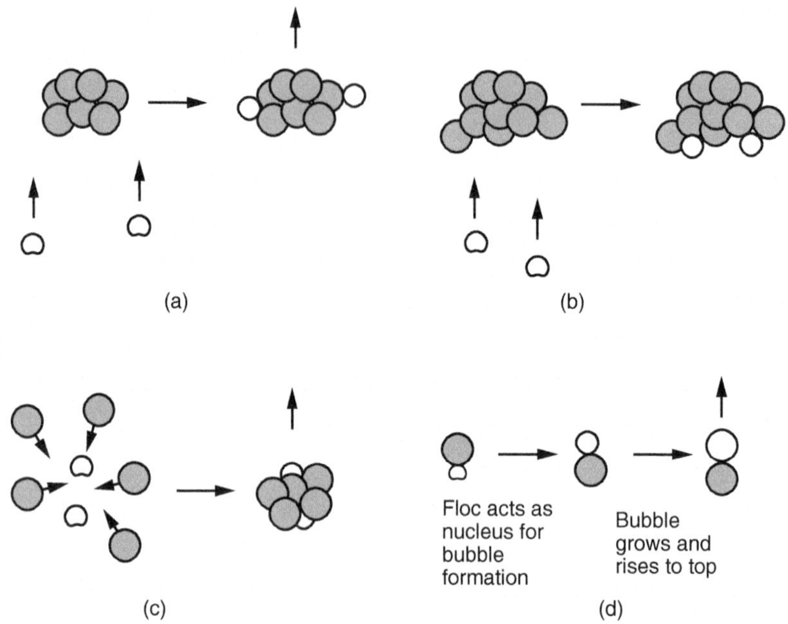

Figure 10-24
Plausible mechanisms for floc – bubble aggregation: (a) bubbles adhere to preformed floc, (b) bubbles trapped in preformed floc, (c) bubbles trapped as floc forms, and (d) floc acts as nucleus for bubble formation.

the WWB (Fig. 10-24b); (3) fine air bubbles become entrapped in floc particles as they aggregate further within the DAF basin (Fig. 10-24c); and (4) floc particles act as nuclei for bubble formation (Fig. 10-24d).

Because any or all of the above mechanisms may be operative in floc–bubble attachment, it is important to perform at least bench-scale, and ideally pilot-scale, trials to confirm the viability of DAF early in process design. Use of design equations for DAF design is presented in Example 10-10. Additional details on floc–bubble attachment including proposed models can be found in Edzwald (2010) and Gregory and Edzwald (2010).

Example 10-10 DAF design

A DAF plant is operating at 10 percent recycle with a saturator pressure of 600 kPa (500 kPa gage). Flocculated water enters the contact zone with a floc particle concentration (N_p) of 2500 particles/mL and a floc volume concentration (ϕ_p) of 2.5×10^{-6} L/L (2.5 ppm). Calculate the air mass concentration (C_b), bubble zone volume concentration (ϕ_b), and bubble number concentration (N_b) in the WWB contact zone of the DAF tank and compare the concentrations of bubbles to floc particles. Assume the water temperature is 20°C (ρ_{air} = 1.204 kg/m^3 = 1204 mg/L, see App. B), the flocculated water has an oxygen deficit of 2 mg/L, the transfer efficiency between the saturator and the point of air release is 90 percent, and the mean bubble diameter is 40 μm. (Adapted from Gregory et al., 1999.)

Solution

1. Calculate the mass concentration of air in the DAF tank. The dissolved air in the water through the saturator is calculated first. From Example 10-9, the mass of air in the recycle water at 0°C and 600 kPa is about 143.5 mg/L. Based on the air mass balance in the WWB contact zone of the DAF tank, the concentration of air released can be calculated using Eq. 10-48:

$$\begin{aligned} C_b &= \frac{e(C_r - C_{fl})r - k}{1 + r} \\ &= \frac{0.9(143.9\ \text{mg/L} - 24.2\ \text{mg/L})0.10 - 2\ \text{mg/L}}{1 + 0.10} \\ &= 7.98\ \text{mg/L} = 7.98\ \text{g/m}^3 \end{aligned}$$

2. Calculate the bubble volume concentration ϕ_b using Eq. 10-49:

$$\phi_b = \frac{C_b}{\rho_{air}} = \frac{11.36\ \text{mg/L}}{1200\ \text{mg/L}} = 9464 \times 10^{-6}\ \text{L/L}(9464\ \text{ppm})$$

3. Calculate the bubble number concentration N_b using Eq. 10-50:

$$\begin{aligned} N_b &= \frac{10^{12} \times 6\phi_b}{\pi d_b^3} = \frac{\left(10^{12}\ \mu\text{m}^3/\text{mL}\right)(6)(0.009464\ \text{L/L})}{\pi(40\ \mu\text{m})^3} \\ &= 2.8 \times 10^5\ \text{bubbles/mL} \end{aligned}$$

4. Compare the concentrations of bubbles to floc particles. The ratio of the concentration of bubbles to floc particles is calculated for the bubble number concentration and the bubble volume concentration:
   a. Bubble number concentration:

$$\frac{N_b}{N_p} = \frac{2.8 \times 10^5\ \text{bubbles/mL}}{2500\ \text{particles/mL}} = 112$$

   Because the ratio of bubbles to particles is high, there is a lot of opportunity for particle collision and attachment:

   b. Bubble volume concentration

$$\frac{\phi_b}{\phi_p} = \frac{9464\ \text{ppm}}{2.5\ \text{ppm}} = 3786$$

Because the ratio of bubble volume to particle volume is high, the floc–bubble density is low, resulting in high rise velocities of the particle–bubble aggregate.

Comment

The calculation may be a best-case scenario because it was assumed that there is no loss of air between the saturator and the DAF contact tank.

FLOC – BUBBLE SEPARATION (FLOTATION) ZONE

In the floc–bubble separation zone the removal of free air bubbles and floc–bubble aggregates is based on Hazen's sedimentation theory (Edzwald, 2010). The flow through the separation zone is assumed to be plug flow in the vertical direction, neglecting the horizontal flow that occurs over the end of the baffle and below the float layer. Under ideal conditions, referring to Fig. 10-22b, the following relationships must apply (Edzwald, 2010):

$$v_b \geq v_{hl} = \frac{Q + Q_r}{A_s} \qquad (10\text{-}53)$$

$$v_{fb} \geq v_{hl} = \frac{Q + Q_r}{A_s} \qquad (10\text{-}54)$$

where v_b = rise rate of air bubble, m/h
v_{hl} = downward vertical velocity based on hydraulic loading rate, m/h
Q = flow rate to be treated, m^3/h
Q_r = recycle flow rate, m^3/h
A_s = cross-sectional area of separation zone perpendicular to flow, m^2
v_{fb} = rise rate of floc–bubble aggregate, m/h

Clearly, the rise rate for air bubbles and floc–bubble aggregates must be greater than the downward velocity due to the hydraulic loading rate. A number of theories have been advanced to explain the removals achieved under less than ideal conditions that often exist within the separation zone including the effects of stratified flow that can occur below the float (sludge) layer. The flow characteristics within the separation zone have also been modeled using computational fluid dynamics (Edzwald, 2010).

Design Considerations for DAF Systems

Important design considerations for DAF systems, such as shown on Fig. 10-25, include the basin layout and geometry, recycle systems, subnatant removal systems, and float removal systems. Each of these design considerations is discussed below.

(a) (b)

Figure 10-25
Views of typical dissolved air flotation units: (a) rectangular type with a chain and flight system for float removal and (b) circular type (used for small systems) with a skimming arms for float removal.

BASIN LAYOUT AND GEOMETRY

The design hydraulic loading rate fixes the total surface area available for flotation in the basin. Basin geometry is a balance between the most costeffective geometry of the individual basins and the total number of basins in the plant. A summary of design parameters and recommended values is given in Table 10-9. Some alternative DAF systems are also reviewed in this section.

Basin length

Basin length is restricted typically to less than 11 m from the headwall to the float beach (see Fig. 10-22) to control the density of the bubble blanket. The bubble blanket formed in the contact zone is typically thick but becomes thinner as it flows through the contact zone. Restricting the length of the overall basin prevents the blanket from becoming too thin, which would cause the buoyancy of the float layer to be reduced or even lost, potentially resulting in resuspension of float into the subnatant.

Basin width

Once the basin length is fixed, the width of the basin is selected based on the design loading rate, while considering mechanical equipment and backmixing. Structurally spanning large widths may not be practical, and standard mechanical equipment may not be available to fit large widths. Narrow basins are also preferred to reduce backmixing by promoting plug flow through the basin. A length-to-width ratio (L/W) slightly greater

Table 10-9
Typical design criteria for a DAF clarification system

Design Parameter	Unit	Value
Flocculation Process		
Number of basins	Minimum number	2
Number of stages	Number	2
Water depth	m	3.5–4.5
Basin length-to-width ratio	Ratio	1
Detention time, large plants	min	10–15[a]
Detention time, small package plants	min	20
Energy input range, *G*	s^{-1}	50–100
Basin Design for Rectangular Configuration		
Number of basins	Minimum number	2
Surface hydraulic loading rate, conventional, based on separation zone	m/h	10–20
Surface hydraulic loading rate, high rate, based on separation zone	m/h	40–45
Basin length, from headwall to float beach (separation zone)	m	<11
Basin length-to-width ratio	—	1–1.25
Surface area	m^2	90–110
Maximum hydraulic capacity for single basin	m^3/s	0.25–0.5
Basin cross-flow velocity	m/h	18–100
Basin depth	m	2.5–3
Contact zone detention time	s	60–240
Contact zone hydraulic loading rate	m/h	35–100
Baffle clearance velocity	m/h	55
Contact zone baffle angle	deg to horizontal	60–90
Recycle System		
Recycle ratio	% of influent flow	6–12
Recycle system pressure	kPa (gauge)	400–700
Nozzle spacing	m	0.2–0.3
Saturator hydraulic loading rate	m/h	60–80
Saturator packing depth	m	1.0 –1.5
Injection nozzles	Fixed-orifice nozzles recommended	
Air loading	g air/m^3 raw water	6–10
Air bubble size	μm	10–100
Bubble concentration	bubbles/mL	$1.0–2.0 \times 10^5$
Raw-water bubble number-particle number ratio	Dimensionless	10:1–200:1
Bubble volume concentration	ppm	3500–8000

Table 10-9 *(Continued)*

Design Parameter	Unit	Value
Flocculated water bubble volume ratio volume to particle	Dimensionless	350:1–8000:1
Recycle pump	Centrifugal	Provides one stand-by pump per recycle system
Air compressor	Oil-lubricated rotary screw type, with good posttreatment for particulate and oil removal; oil-free compressors not recommended due to maintenance requirements	
	Float Removal	
System type	Mechanical or hydraulic. Choice of type is site specific, but should include cost implications on residuals-handling systems. For mechanical skimming, reciprocating scrapers are recommended.	
	Subnatant Removal	
Type of removal system	Perforated pipe laterals or underflow end wall. Special underdrain systems consisting of a false bottom and plenum	

Source: Adapted in part from Longhurst and Graham (1987), Adkins et al. (1994), Breese (1995), Haarhoff and van Vuuren (1995), Gregory et al. (1999), Kawamura (2000), Degrémont (2007), Edzwald (2010), and Gregory and Edzwald (2010).
[a]Flocculation times as low as 5 min with high surface loading rates have been demonstrated (Edzwald, 2010; Gregory and Edzwald, 2010).

than 1.0 is recommended for full-scale DAF basins. In some new high-rate DAF systems, as described at the end of this section, the length is less than the width.

Basin depth

Although in theory basin depth should not impact solids removal, in practice it is important to control cross-flow velocity to limit scouring of float from underneath. Recommended basin depths and cross-flow velocities are listed in Table 10-9.

Baffle placement

The inlet baffle placement should create a contact zone volume large enough to provide good floc–bubble contact time (see Table 10-9), but not so large as to encourage short circuiting. Key parameters governing placement of the baffle are the distance from the headwall to the baffle at floor level, the angle of inclination of the baffle, and the height of the baffle, which defines the clearance between the top of the baffle and the water surface. Criteria ranges for baffle placement are listed in Table 10-9. The lower range of the criteria and baffle angles of at least 75° are recommended for basins with higher hydraulic loading rates to ensure that the bubble cloud will entirely fill the contact zone. If an excessive contact zone length (or a low baffle angle) is selected, the potential exists for raw-water short circuiting within the zone, whereby water flows along the bottom of the contact zone and up the face of the baffle wall. Circulatory patterns in the contact zone may result in less desirable floc–bubble interaction.

RECYCLE SYSTEMS

The recycle system is comprised of the saturator, the injection system manifold with orifice nozzles, the recycle flow pressurization pump, an air compressor, and associated piping. An effective DAF recycle system provides the required quantity of air while using as little recycle water as possible, minimizing unnecessary turbulence in the contact zone and keeping pumping costs down. Typical operating pressures optimal for generating microbubbles while minimizing capital cost for saturator vessels are listed in Table 10-9. Additional details on recycle system design are presented below.

Saturator vessel

The saturator is typically packed with an inert mass transfer media (see Fig. 10-26a) to enhance saturation efficiency (Rees et al., 1979) and is designed based upon surface hydraulic loading (recycle flow divided by superficial area). Higher loading rates tend to cause flooding of the packed bed and significant loss of efficiency. Recycle water is drawn from the DAF effluent collector and pumped into the saturator near the top, where a flow distribution header sprays the water over the entire superficial area of the vessel.

The recycle water trickles downward, forming a thin film on the media and maximizing surface area for transfer of air into the water. A separate air connection at the top of the vessel maintains the saturator at a relatively

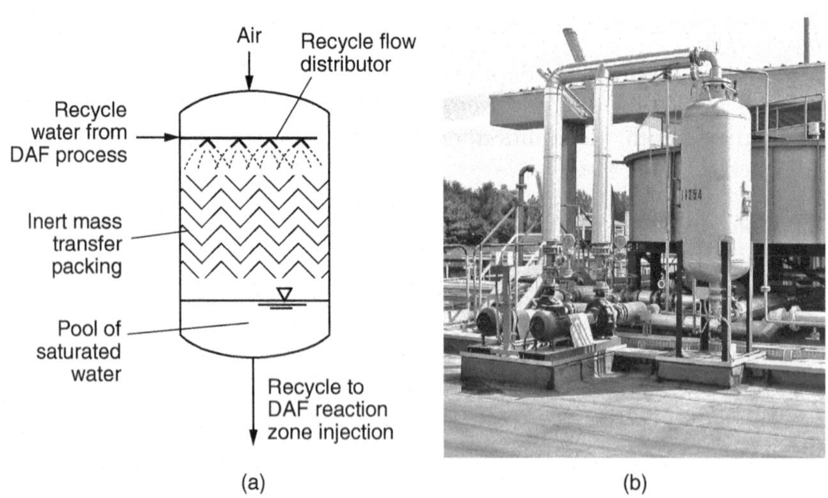

Figure 10-26
Typical DAF saturator: (a) schematic cross section and (b) photograph.

constant operating pressure. The packed bed rests on a perforated plate, below which saturated water collects in a clearwell at the base of the vessel. With the vessel discharge open to the recycle injection manifolds in the DAF basin, the pressure in the saturator provides the force necessary to drive injection of the saturated fluid back into the basin.

Injection manifolds
One or more injection manifolds are mounted within the basin contact zone either at floor level or just above the point where the raw water is introduced. Each manifold is fitted with fixed-orifice nozzles spaced at 8 to 12 even intervals across the basin width to release the pressureforming microbubbles across the superficial area of the contact zone.

Fixed-orifice nozzles
Several manufacturers have devised proprietary fixed-orifice nozzles for generating bubbles in the ideal size range. These nozzles usually share two common principles (Haarhoff and Rykaart, 1995):

- The nozzles utilize some means for rapidly reducing the pressure from the saturator pressure to atmospheric pressure. The rapid pressure drop causes immediate formation of air bubbles, potentially in an uncontrolled fashion.
- Immediately downstream of the pressure release, some form of impingement or rapid change in direction of flow shears the newly formed bubbles and breaks them down into the desired size range.

Needle valves may be used for generating bubbles; however, they are substantially more difficult to adjust to ensure an even balance of recycle flow across the basin and are therefore not recommended. Fixed-orifice nozzles provide a passive means to balance this flow and are generally not prone to fouling due to solids once the plant is operating.

Head loss in manifold systems
Once saturation of the recycle stream has been achieved, water is delivered to the recycle injection manifolds with as little head loss as possible, as any loss of head will result in the premature formation of bubbles, which leads to bubble coalescence. From a design standpoint, several approaches to limit head loss are recommended:

- The saturator vessel should be kept elevated so the water level in the saturator clearwell is above the water level in the basin. If the saturator is placed too low, a reduction in static head as the recycle travels toward the injection point will result in air precipitation.
- The saturator should be placed as close as possible to the injection point, and interconnecting piping downstream of the saturator should be designed to keep velocities low and with a minimum of fittings.

SUBNATANT REMOVAL

Clarified water, or subnatant, from the separation zone of the DAF basin is removed at a submerged location near the basin floor. An important design consideration for the subnatant removal system is the variation of the hydraulic profile through the basin at different flow rates. A discussion of the two most common approaches to separating the subnatant from the separation zone follows.

Underflow baffle

An underflow baffle wall may be used at the end of the basin, allowing subnatant to flow under the baffle across the full width of the basin. Systems with an underflow baffle typically have significant downward velocities at the end of the basin. The downward-flow velocities occur in an area of the basin where float concentrations are highest and bubble cloud densities are lowest, so the potential exists for resuspension of float at the end of the basin.

Perforated-pipe laterals

A series of parallel, perforated-pipe laterals (false floor) across the floor of the basin are used to evenly withdraw subnatant. In theory, the perforated-pipe or false-floor approach offers an even withdrawal of subnatant; however, in practice, there is about 0.3 m of head loss with a perforated-pipe lateral system. Usually the laterals can be designed to achieve a reasonably even withdrawal of flow, and operating an individual basin across a fairly narrow range of flows limits water surface level variation.

FLOAT REMOVAL

Float may be removed mechanically or hydraulically from a DAF basin. Because float removal occurs at the surface, varying the basin water level may impact the density of sludge removed, especially if the water level varies more than 0.05 to 0.08 m (2 to 3 in.) across the full range of basin operating flows.

Mechanical float removal

A surface skimmer is used to scrape the floated solids over a dewatering bench (float beach) and into a hopper from where they can be pumped to residual-handling facilities (see Fig. 10-22a). There are three types of mechanical float-skimming systems: (1) a chain and flight system where flights are mounted across the basin and attached at either side to a continuous chain, as shown on Fig. 10-25a; (2) a reciprocating-type system, where a series of blades are mounted on a carriage and the entire carriage is moved forward when in the scraping mode; and (3) a beach scraper, where a drum with a number of blades is fixed at the far end of the DAF tank at the beach. As the drum rotates, the blades sweep the float over the beach plate.

Hydraulic float removal

When the float is removed hydraulically, the effluent flow is closed off, but the basin is kept on-line so that the water level in the basin rises, causing float to flow over a sludge weir and into the float collection trough. For a hydraulic sludge system, the basin outlet needs to include a convenient point where the effluent flow can be temporarily shut off, using a gate (sluice or slide) or a valve. A hydraulic float removal system is substantially less complex than a mechanical system, typically requiring only a single moving component (gate or valve) on the basin outlet, making this type of system less expensive to install and easier to operate and maintain.

Comparison of mechanical and hydraulic float removal systems

The most significant difference between mechanical and hydraulic systems is the consistency of the sludge product, which impacts the residualhandling system. Because the float separates out as a discrete layer on the surface of the basin, a well-designed mechanical skimmer system can physically remove the float while minimizing carryover of additional liquid. The result is a very thick sludge product that can be removed from the basin using a mechanical skimming system, typically on the order of 2 to 3 percent total solids (percent total solids by weight). By comparison, hydraulic float removal systems allow a significant carryover of water into the float collection trough, producing a sludge that is typically 0.5 percent total solids. The end result is a total sludge volume substantially larger than that produced by mechanical systems.

Both mechanical and hydraulic float removal systems commonly make use of spraying systems to help move sludge as required. In mechanical systems, a spray header is typically provided along the length of the float collection trough to assist sludge flow, if required. Hydraulic systems usually do not require this feature; however, spray headers are commonly supplied along both sidewalls of the basin and are used to cut the sludge blanket away from the wall during the desludging cycle.

OTHER TYPES OF DAF SYSTEMS

Over the years since becoming an effective and accepted process for the treatment of water, a number of process variations and proprietary DAF processes have been developed. Two proprietary units are described below: (1) the AquaDAF and (2) the countercurrent dissolved air flotation/ filtration (CoCoDAFF) process (Degrémont, 2007).

AquaDAF

The AquaDAF unit was developed in Finland and is licensed by SUEZ Water Technologies & Solutions. The unit involves the use of an improved proprietary air release manifold system across the entire bottom of the unit to produce a deeper WWB zone (1 to 2 m) in the contact chamber to enhance the coalescence of the air bubbles and aggregation of the coalescence air

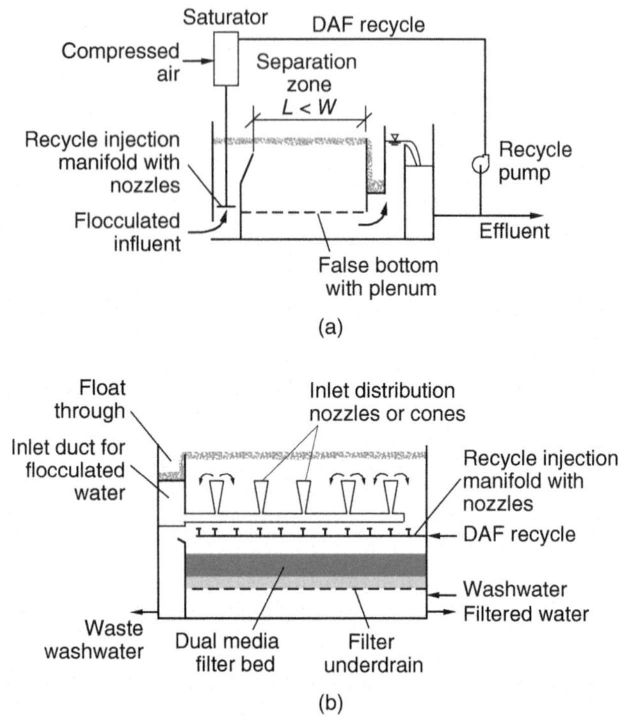

Figure 10-27
Alternative dissolved air flotation units: (a) AquaDAF with proprietary discharge nozzle system and shorter length to width ratio than conventional DAF unit and (b) CoCoDAFF, which combines a flotation unit with a granular media filter in a single basin.

bubbles with the floc particles. The discharge baffle is curved, which when coupled with a false bottom in the flotation part of the reactor creates a circulation pattern within the unit (see Fig. 10-27). Because of the presence of a deeper WWB zone and the circulation pattern, the rise velocity the resulting air/floc particles is significantly greater than that encountered in conventional DAF units. Because of the increased rise rate, the length of the unit is less than the width of the unit.

CoCoDAFF

The CoCoDAFF involves placing the dissolved air flotation cell directly above a granular media filter as shown on Fig. 10-27b. This process was developed in the early 1990s by Thames Water to overcome filter operational problems resulting from seasonal algal blooms. Both mono- and dual- media filter beds have been used. As shown in Fig 10-27b, flocculated water is discharged below the surface through a series of special high- flow-rate conical nozzles connected to an inlet manifold. The pressurized recycle water is introduced below the inlet manifold. The air released from the pressurized recycle flow creates a bubble blanket, which moves countercurrent to the water. Flocculent material is removed from the water and floated to the surface as the water moves downward through the bubble blanket. In combined systems, the filter hydraulic loading rate, being lower than that of the DAF unit, will control the design of the combined flotation–filter unit. When algae are not present or their concentration is low, the DAF portion of the combined system does not need to be operated.

Problems and Discussion Topics

10-1 Calculate the terminal settling velocity and the Reynolds number of the particle given (to be selected by instructor).

Parameter	A	B	C	D	E
Particle diameter, μm	50	500	300	150	210
Particle density, kg/m^3	2650	1050	1050	2600	1700
Water temperature, °C	10	15	5	20	15

10-2 Estimate the size of sand particle with density of 2650 kg/m^3 that would move the same distance in 1 h due to Brownian motion as it would settle in water at 10°C based on Stokes' law.

10-3 Consider the particle shown below with the values in the table (to be selected by instructor). Calculate the overflow rate that corresponds to the settling velocity of the particle on the trajectory shown (report your answers in m/h and gpm/ft^2). If it is desired to achieve complete removal of particles of this size, what adjustment in the length of the basin would be required?

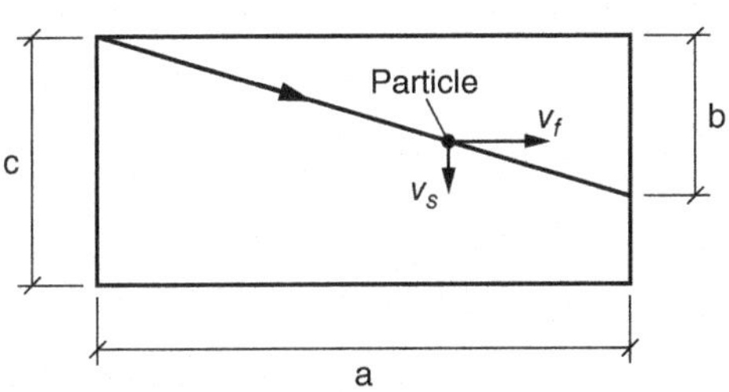

Parameter	A	B	C	D	E
Fluid velocity, cm/s	20	1.4	0.5	1	0.28
Dimension a, m	4	100	72	80	30
Dimension b, m	0.6	3.5	1.7	0.85	3.6
Dimension c, m	0.9	4.2	3.5	1	4.2

10-4 For the particle-settling data shown in Example 10-3, plot the removal efficiency as a function of overflow rate for overflow rates ranging from 0.5 to 4 m/h. Determine the overflow rate required to achieve 75 percent removal. If the depth of the basin is 4 m, what is the corresponding detention time?

10-5 For the rectangular horizontal-flow sedimentation basin and influent particle-settling characteristics given (to be selected by instructor), calculate the particle removal efficiency and plot the influent and effluent particle concentrations as a function of settling velocity.

Parameter	A	B	C	D	E
Flow rate, m^3/d	7,570	19,000	19,000	56,800	56,800
Length, m	30	72	60	100	80
Width, m	5	12	8	18	12

	Number of Particles, #/mL				
Settling Velocity, m/h	A	B	C	D	E
0–0.4	511	511	460	560	255
0.4–0.8	657	657	578	720	314
0.8–1.2	876	876	891	880	454
1.2–1.6	1168	1168	1285	1110	584
1.6–2.0	1460	1460	1748	1320	761
2.0–2.4	1314	1314	1577	1110	639
2.4–2.8	657	657	719	620	321
2.8–3.2	438	438	436	440	219
3.2–3.6	292	292	263	320	141
3.6–4.0	292	292	241	160	116
Total	7665	7665	8198	7240	3804

10-6 What are the principal causes of flocculation during sedimentation? Why is flocculation beneficial in a sedimentation tank?

10-7 Determine the area of a clarifier required for solids thickening for the parameters given below (to be selected by instructor). The settling velocity of the sludge blanket follows the data given in Table 10-1 and plotted on Fig. 10-8. Also determine J_L, C_L, and Q_u.

Parameter	A	B	C	D	E
Influent flow rate, m^3/h	3,000	1,500	2,500	3,300	4,500
Influent solids conc., mg/L	500	800	400	500	800
Underflow solids conc., mg/L	10,000	12,000	14,000	14,000	15,000

10-8 A grit chamber is designed to remove sand of 0.08 mm and larger for a design flow of 2.2 m^3/s (50 mgd). The maximum flow rate is approximately 1.5 times the average flow and the water temperature is 15° C. Assume a typical water depth of 3.5 m for each tank and a factor of safety of 1.5 is to be used. Determine the number and shape of tanks required and their size. Determine the configuration of the diffuser wall.

10-9 A water treatment plant is to be designed to treat water with the maximum daily flow and design given below (to be selected by instructor). For the given overflow rate, design a horizontal rectangular sedimentation basin.

Parameter	A	B	C	D	E
Influent flow rate, m^3/d	15,000	380,000	90,000	220,000	220,000
Overflow rate, m/h	1.10	2.15	2.6	1.65	2.0
Water temperature, °C	10	15	20	20	10

10-10 A continuous horizontal flow setting basin is designed to treat a flow rate of 1.1 m^3/s (25 mgd) and remove floc particles with an average settling velocity of 3.6 m/h. This settling velocity was measured in a laboratory in a quiescent condition at 15° C. Determine the area of the basins and the basin area that needs to be covered with tube settlers. The tube settlers consist of a series of 50-mm (2-in.) square honeycomb cells tilted at a 60° angle. The vertical height of the tube settlers is 0.6 m. Assume a countercurrent flow pattern is used.

10-11 Calculate the theoretical settling efficiency (v_s/v_θ) of parallel square plates, square tubes, hexagons, and circular tubes. The hydraulic diameter, tube length, and angle of the tubes to the horizon are 80 mm, 1.5 m, and 60°. Assume a consistent hydraulic diameter is used because it guarantees that the Reynolds number is the same for the same velocity. Assume a countercurrent flow pattern is used and that any hydraulic problems associated with poor design such as flow distribution and sludge resuspension can be ignored. Note: The hydraulic diameter is equal to two times the hydraulic radius.

10-12 Evaluate and compare the settling velocity and rise velocity of a 15-μm floc particle that has a density of 1250 kg/m^3 for summer (25°C) and winter (4°C) water temperatures. For the rise velocity calculations assume that the floc attaches to one air bubble that is 30 μm in size.

10-13 Using the values for the Henry's law constants for oxygen and nitrogen, as given in Table 14-4, verify the values for the amount of air dissolved at 0°C and 101.325 kPa (1 atm) and at 20° C and 600 kPa as determined in Example 10-9.

10-14 A DAF plant is operating at 8 percent recycle with a saturator pressure of 650 kPa. Flocculated water enters the contact zone with a floc particle concentration (N_p) of 1000 particles/mL and a floc volume concentration (ϕ_p) of 1.0 ppm. Calculate the air mass concentration (c_b), bubble zone volume concentration (ϕ_b), and bubble number concentration (N_b) in the contact zone of the DAF tank and compare the concentrations of bubbles to floc particles.

Assume the water temperature is 15° C (ρ_{air} = 1.2250 kg/m^3), the flocculated water has no oxygen deficit, and the mean bubble diameter is 30 μm.

References

Adkins, M. F., Harris, D. I., and Pfeifer, B. (1994) Dissolved Air Flotation Pilot Trials at Castor and Lac La Biche Water Treatment Plants, Alberta, Canada, Proceedings of the 46th Annual Conference—Western Canada Water and Wastewater Association, pp. 146–170.

Baines, W. D., and Knapp, D. J. (1965) "Wind Driven Water Currents," *J. Hydr. Div. ASCE,* **91**, 2, 205–221.

Breese, S. M. (1995) Discussion Paper Design Parameters for the DAF Treatment Trains for the Greenville Water System, Greenville, S.C., Reid Crowther & Partners, Greenville, SC.

Brown, P. P., and Lawler, D. F. (2003) "Sphere Drag and Settling Velocity Revisited," *J. Eng. Div. ASCE,* **129**, 222–231.

Camp, T. R. (1936) "A Study of the Rational Design of Settling Tanks," *Sewage Works J.*, **8**, 9, 742–758.

Camp, T. R. (1946) "Sedimentation and the Design of Settling Tanks, paper no. 2285" *ASCE Trans.*, **3**, 895–936.

Clark, M. M., (1996) *Transport Modeling for Environmental Engineers and Scientists,* John Wiley & Sons, New York, NY.

Collins, M. R., Westersund, C. M., Cole, J. O., and Roccaro, J. (1994) *Evaluation of Roughing Filtration Design Variables*, America Water Works Association Research Foundation, Denver, CO.

Degrémont (2007) *Water Treatment Handbook,* Vols. 1 and 2, 7th ed., Lavoisier, Paris.

Edzwald, J. K. (1995) "Principles and Applications of Dissolved Air Flotation," *Water Sci. Technol.*, **31**, 3/4, 1–23.

Edzwald, J. K. (2010) "Dissolved Air Flotation and Me," *Water Res.*, **44**, 7, 2077–2106.

Fitch, E. B., and Lutz, W. A. (1960) "Feedwell for Density Stabilization," *J. WPCF,* **32**, 2, 147–156.

Gregory, R., and Edzwald, J.K. (2010) Sedimentation and Flotation, Chap. 9, in J. K. Edzwald (ed.), *Water Quality and Treatment,* 6th ed., McGraw-Hill, New York.

Gregory, R., Zabel, T. F., and Edzwald, J. K. (1999) Sedimentation and Flotation, Chap. 7, in R. D. Le Herman (ed.), *Water Quality and Treatment: A Handbook of Community Water Supplies,* American Water Works Association, McGraw-Hill, New York.

Haarhoff, J. (2008) "Dissolved Air Flotation: Progress and Prospects for Drinking Water Treatment," *J. Water Supply: Res. and Technol.—Aqua,* **57**, 8, 555–567.

Haarhoff, J., and Edzwald, J. K. (2004) "Dissolved Air Flotation Modelling: Insights and Shortcomings," *J. Water Supply: Res. Technol. (Aqua),* **53**, 3, 127 – 150.

Haarhoff, J., and Rykaart, E. M. (1995) "Rational Design of Packed Saturators," *Water Sci. Technol.,* **31**, 3–4, 179–190.

Haarhoff, J., and van Vuuren, L. R. J. (1995) "Design Parameters for Dissolved Air Flotation in South Africa," *Water Sci. Technol.*, **31**, 3–4, 203–212.

Harleman, D. F. (1961) Stratified Flow, in V. S. (ed.), *Handbook of Fluid Mechanics*, McGraw-Hill, New York.

Hazen, A. (1904) "On Sedimentation," *Trans. ASCE*, **53**, 45–71.

Hidy, G. M., and Plate, E. J. (1966) "Wind Action on Water Standing in a Laboratory Channel," *J. Fluid Mech.*, **26**, part 4, 651–688.

Hudson, E. R. (1972) "Density Considerations in Sedimentation," *J. AWWA*, **64**, 6, 382–386.

Kao, T. W. (1977) "Density Currents and Their Applications," *J. Hydrol. Div. ASCE*, **103**, 5, 543–555.

Kawamura, S. (1981) "Hydraulic Scale-Model Simulation of the Sedimentation Process," *J. AWWA*, **73**, 7, 372–379.

Kawamura, S. (2000) *Integrated Design and Operation of Water Treatment Facilities*, 2nd ed., Wiley-Interscience, New York.

Liu, H., and Perez, H. J. (1971), "Wind-Induced Circulation in Shallow Water," *J. Hydraul. Div. ASCE*, **97**, 7, 923–935.

Logan, B. E. (1999) *Environmental Transport Processes*, Wiley-Interscience, New York.

Longhurst, S. J., and Graham, N. J. D. (1987) "Dissolved Air Flotation for Potable Water Treatment: A Survey of Operational Units in Great Britain," *Public Health Eng.*, **14**, 71–76.

O'Connor, D. J., and Eckenfelder, W. W., Jr. (1958) Evaluation of Laboratory Settling Data for Process Design, in W. W. Eckenfelder and B. J. McCabe (eds.), *Biologic Treatment of Sewage and Industrial Wastes*, Reinhold, New York.

Rees, A. J., Rodman, D. J., and Zabel, T. F. (1979) *Evaluation of Dissolved Air Flotation Saturator Performance*, Technical Report TR143, Water Research Centre. Medmenham, UK.

Sank, R. L. (1978) *Water Treatment Plant Design*, Ann Arbor Science, Ann Harbor, MI.

Tchobanoglous, G., Burton, F. L., and Stensel, H. D. (2003) *Wastewater Engineering: Treatment, and Reuse*, 4th ed., McGraw-Hill, New York.

Tekippe, R. J., and Cleasby, J. L. (1968) "Model Studies of Peripheral Feed Settling Tank," *J. ASCE*, **94**, No. 541, 85–102.

Zabel, T. F. (1985) "The Advantages of Dissolved Air Flotation for Water Treatment" *J. AWWA*, **77**, 5, 42–46.

Zabel, T. F., and Hyde, R. A. (1977) Factors Influencing Dissolved Air Floatation as Applied to Water Clarification, in J. D. Melbourne and T. F. Zabel (eds.), *Papers and Proceedings of Water Research Centre Conference on Flotation for Water and Waste Treatment*, Water Research Center, Medmenham, UK.

11 Granular Filtration

Stantec's Water Treatment: Principles and Design, Updated Third Edition. John C. Crittenden, R. Rhodes Trussell, David W. Hand, Kerry J. Howe, and George Tchobanoglous.

Terminology for Granular Filtration

Term	Definition
Air scour	Optional feature during backwash in which air is introduced into filter underdrains along with backwash water; the vigorous scouring action helps clean deep-bed filters.
Backwash	Process for removing accumulated solids from a filter bed by reversing the water flow.
Bag and cartridge filtration	Pressure driven separation processes that remove particles larger than 1 μm using an engineered porous filtration media consisting of fabric or self-supporting filter elements.
Conventional treatment	Process train consisting of coagulation, flocculation, sedimentation, and filtration.
Contact filtration	Process train consisting of coagulation and filtration.
Depth filtration	Filtration mechanism in which particles accumulate throughout the depth of a granular filter bed by colliding with and adhering to the media. Captured particles can be many times smaller than the pore spaces in the bed.

Term	Definition
Diatomaceous earth	Granular material of nearly pure silica, mined from natural deposits of fossilized diatoms that is used as a filtration media in precoat filtration.
Direct filtration	Process train consisting of coagulation, flocculation, and filtration.
Effective size (ES)	Measure of the size of granular media; the size at which 10 percent of the media has a smaller diameter (d_{10}) as determined by a sieve analysis.
Filtration	Removal of particles (solids) from a suspension (two-phase system containing particles and liquid) by passage of the suspension through a porous medium. In granular filtration, the porous medium is a bed of granular material.
Filtration rate	Key process variable; the superficial water velocity through the filter bed, calculated as the flow rate divided by the cross-sectional area of the bed.
In-line filtration	Contact filtration.
Precoat filtration	Granular filtration process in which a fine granular material is introduced into the filter module and collects as a thin cake against a support septum; filtration occurs by straining at the surface of this cake layer.
Rapid filtration	Granular filtration process engineered to achieve filtration rates about 100 times greater than slow sand filtration. Key requirements include coagulation pretreatment, granular media sieved for greater uniformity, and backwashing to remove accumulated particles.
Ripening	Process of granular media conditioning at the beginning of a filter run during which clean media captures particles and becomes more efficient at capturing additional particles. During ripening filter effluent water may not meet quality requirements and must be wasted; typically it is recycled to the head of the plant.
Schmutzdecke	Layer of particles and microorganisms that forms in the top few centimeters of a slow sand filter.
Slow sand filtration	Granular filtration process during which water passes slowly down through a bed of sand. Filtration occurs primarily by straining at the surface of the Schmutzdecke located at the top of the bed.
Specific deposit	Mass of accumulated particles in a filter per unit of filter volume

Term	Definition
Straining	Filtration mechanism in which particles are captured at the surface of a filter because they are too large to fit through the pore spaces in the filter.
Underdrain	Components installed at the base of a filter bed. Underdrains must support the media and evenly collect filter effluent and distribute backwash water (and air) to avoid channeling in the filter bed.
Uniformity coefficient (UC)	Measure of the uniformity of granular media; the ratio of the 60th percentile (d_{60}) to the 10th percentile (d_{10}) media sizes as determined by a sieve analysis.
Unit filter run volume (UFRV)	Quantity of water that passes through a filter over the course of an entire filter run.

Filtration is widely used for removing particles from water. Filtration can be defined as any process for the removal of solid particles from a suspension (a two-phase system containing particles in a fluid) by passage of the suspension through a porous medium. In granular filtration, the porous medium is a thick bed of granular material such as sand. The most common granular filtration technology in water treatment is *rapid filtration.* The term is used to distinguish it from *slow sand filtration,* an older filtration technology with a filtration rate 50 to 100 times lower than rapid filtration. Key features of rapid filtration include granular media sieved for greater uniformity, coagulation pretreatment, backwashing to remove accumulated particles, and a reliance on *depth filtration* as the primary particle removal mechanism. In depth filtration, particles accumulate throughout the depth of the filter bed by colliding with and adhering to the media. Captured particles can be many times smaller than the pore spaces in the bed.

Nearly all surface water treatment facilities and some groundwater treatment facilities use filtration. Most surface waters contain algae, sediment, clay, and other organic or inorganic particles. Filtration improves the clarity of water by removing these particles. All surface waters also contain microorganisms that can cause illness, and filtration is nearly always required in conjunction with chemical disinfection to assure that water is free of these pathogens. Groundwater is often free of significant concentrations of microorganisms or particles, but may require filtration when other treatment processes (such as oxidation or softening) generate particles that must be removed.

This chapter presents a brief history of granular filtration, a description of the rapid filtration process, properties of filter media, hydraulics of flow through granular media, particulate removal in rapid filtration, and design of rapid filters. A variety of other filtration options and technologies are used in water treatment, including pressure filtration, slow sand filtration, greensand filtration, biologically active filtration, diatomaceous earth filtration, and cartridge or bag filtration. These technologies are introduced briefly at the end of this chapter. Membrane filtration is another common filtration technology used in water treatment but will be discussed in a different chapter (Chap. 12) because of the substantial differences between granular and membrane filtration technologies. Granular media filters are still the most common type of filters in use today.

11-1 Brief History of Filtration

Filters have been used to clarify water for thousands of years. Medical lore written in India, dating to perhaps 2000 BC, mentions filtration through sand and gravel as a method of purifying water. Hippocrates advocated filtration through cloth bags in the fourth century BC. The Romans dug channels parallel to lakes to take advantage of natural filtration through soil when using lakes for water supplies. Venice, Italy, stored rainwater in cisterns but drew the fresh water from wells in sand that surrounded the cisterns (Baker, 1948).

The commercialization and patenting of filtration technologies started in France around 1750, using various filter media such as sponges, charcoal, wool, sand, crushed sandstone, or gravel. The practice of filtering surface water through engineered systems and distributing it on a municipal scale began in England and Scotland around 1800. Various filtration concepts were tested, including flow direction (downflow, upflow, and horizontal flow), sand and gravel media graded from smaller to larger sizes, and backwashing by reverse flow. The first modern slow sand filter, designed by James Simpson for the Chelsea Water Works Company in London in 1829, incorporated an underdrain system, graded gravel and sand media, a filtration rate of about 0.12 m/h (0.05 gpm/ft^2), and cleaning by scraping (Baker, 1948). These design features are still used today.

The first regulation mandating filtration, passed in 1852, required all river water supplied by the Metropolitan District of London to be filtered. The regulation was prompted by rampant pollution in the Thames River and suspicions that cholera was transmitted by water (Fuller, 1933), a suspicion confirmed by Dr. John Snow in his famous investigation of a cholera outbreak in London just 2 years later.

Interest in filtration grew as people realized that it prevented waterborne disease. In 1892, the city of Altona, Germany, largely escaped a cholera epidemic that ravaged neighboring Hamburg. Both cities used the Elbe River as a water supply, but Altona was protected by its slow sand filters even though its water was withdrawn downstream of Hamburg and was contaminated with Hamburg's waste (Hamburg had no filtration system). Similarly, a dramatic reduction in typhoid cases resulted when filters were installed in Lawrence, Massachusetts. Many communities in the United States first began filtering their water supplies during the first couple of decades of the twentieth century.

Examples of granular filters used in water treatment plants.

http://www.stantec.com/wt3r/11-1.html

Rapid filtration originated in the United States during the 1880s. The first municipal plant employing coagulation and other critical elements of rapid filtration was in Somerville, New Jersey, in 1885 (Fuller, 1933). Both slow sand and rapid filters were common in early filter installations (Fuller, 1933), but by the middle of the twentieth century, rapid filters were commonplace and slow sand filters were rarely used.

By the latter part of the twentieth century, most surface waters were filtered before municipal distribution, with rapid filters used in almost all cases (99 percent). Nevertheless, the Surface Water Treatment Rule (SWTR), passed in 1989, was the first regulation in the United States requiring widespread (but not universal) mandatory filtration of municipal water (U.S. EPA, 1989), with the recognition that chemical disinfection alone was insufficient for protozoa such as *Giardia lamblia* and *Cryptosporidium parvum.* Surface water treatment regulations have continued to get more stringent, particularly the remaining utilities with unfiltered surface water supplies that have been under increasing pressure to install filtration. In short, filtration is and will continue to be a central feature in surface water treatment plants.

11-2 Principal Features of Rapid Filtration

Rapid filtration has several features that allow it to operate at rates up to 100 times greater than slow sand filtration. The most important of these features are (1) a filter bed of granular material that has been processed to a more uniform size than typically found in nature, (2) the use of a coagulant to precondition the water, and (3) mechanical and hydraulic systems to efficiently remove collected solids from the bed.

Uniformity of Filter Media

The filter material in rapid filters is processed to a fairly uniform size. Media uniformity allows the filters to operate at a higher hydraulic loading rate with lower head loss but results in a filter bed with void spaces significantly larger than the particles being filtered. As a result, straining is not the dominant removal mechanism. Instead, particles are removed when they adhere to the filter grains or previously deposited particles. Particles are removed throughout the entire depth of the filter bed by a process called depth filtration, which gives the filter a high capacity for solids retention without clogging rapidly.

Coagulation Pretreatment

Coagulation pretreatment is required ahead of rapid filtration. If particles are not properly destabilized, the natural negative surface charge on the particles and filter media grains cause repulsive electrostatic forces that prevent contact between particles and media. The origin of surface charge on particles in nature and the proper use of coagulants for destabilizing

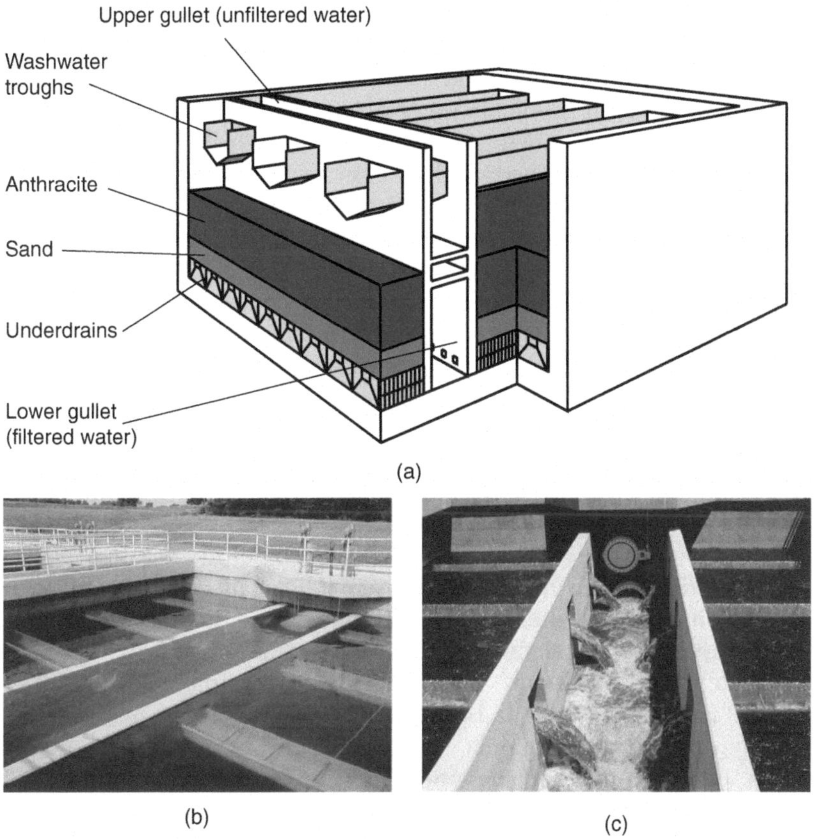

Figure 11-1
Typical dual-media rapid filter. (a) Schematic representation of dual-media filter. (b) View of an operating rapid filter. Washwater troughs are visible below the water surface. Influent water enters through the central channel, flows through the wall openings for the washwater troughs, and then down through the filter media, which is below the water surface. (c) Rapid filter during the backwash cycle. Washwater flows up through the media, pours over into the troughs, and then runs into the central channel.

particles were discussed in detail in Chap. 9. Properly designed and operated rapid filters can fail quickly if the coagulant feed breaks down or the rawwater quality changes and the coagulant dose is not adjusted accordingly.

Basic Process Description

A typical configuration for rapid filters is illustrated on Fig. 11-1. The filter bed is contained in a deep structure that is typically constructed of reinforced concrete and open to the atmosphere. The rapid filtration cycle consists of two stages: (1) a filtration stage, during which particles accumulate, and (2) a backwash stage, during which the accumulated material is flushed from the system. During the filtration stage, water flows downward through the filter bed and particles collect within the bed. The filtration stage typically lasts from 1 to 4 days.

During the backwash stage, water flows in the direction opposite to remove the particles that have collected in the filter bed. Efficient removal of collected solids is a key component of rapid filtration systems, so while the backwashing stage is very short compared to the filtration stage, it is a very important part of the filtration cycle.

The physical steps that occur during the backwashing stage include the following: (1) the filter influent and effluent lines are isolated with valves and the backwash supply and waste washwater valves are opened; (2) backwash water, which is potable water produced by the plant, is directed upward through the filter bed; (3) the upward flow flushes captured particles up and away from the bed; and (4) after backwash, the valve positions are reversed and the filter is placed back in service.

Most filters also contain supplemental systems to assist the backwashing process. One option is the surface wash system, which is a fixed grid or rotating system of nozzles that blast the surface of the filter bed to break up any mat of solids that may have formed. Another option is to introduce pressurized air underneath the media with the backwash water. Often the air is introduced while the backwash water flows at a low rate, and the consequent pulsing efficiently scours retained solids from the media. For deep filter beds, both air and surface wash are often provided. The backwashing step typically takes 15 to 30 min.

Filtration Effectiveness During the Filtration Stage

The efficiency of particle capture as reflected by effluent turbidity and head loss varies during the filtration stage (also called a filter run), as illustrated on Fig. 11-2. Filter effluent turbidity during the filter run follows a characteristic pattern with three distinct segments. During the first segment (immediately after backwash), the filter effluent turbidity rises to a peak and then falls. This segment is called filter ripening (or maturation). Ripening is the process of media conditioning and occurs as clean media captures particles and becomes more efficient at collecting additional particles. Some studies indicate that 90 percent of the particles that pass through a well-operating filter do so during the initial stage of filtration (Amirtharajah, 1988). The ripening curve sometimes contains two distinct peaks, the first corresponding to residual backwash water being flushed from the media and the second corresponding to particles from the water column above the backwashed filter. Ripening periods of 15 min to 2 h are possible. The magnitude and duration of the ripening peak can be sizable but can be substantially reduced by proper backwashing procedures, such as minimizing the duration of the backwash stage or using filter aid polymers in the backwash water. Modern filtration plants are designed with a filter-to-waste line, and the water produced during ripening is discharged to waste or recycled to the head of the plant.

The particles captured during ripening improve the overall efficiency of the filter by providing a better collector surface than uncoated media grains. After ripening, effluent turbidity typically can be maintained at a steady-state value below 0.1 NTU. Even though effluent turbidity is essentially constant after ripening, head loss through the filter continuously increases because of the collection of particles in the filter bed. After the period of effective filtration, the filter can experience breakthrough. During breakthrough, the filter contains so many particles than it can no longer filter effectively and the effluent turbidity increases.

Several events can trigger the end of the filter run and lead to backwash. First, if the filter reaches breakthrough, it must be backwashed to prevent high-turbidity water from entering the distribution system. Second, the head loss can increase beyond the available head through the process. Rapid filters typically operate by gravity and are designed with 1.8 to 3 m (6 to 10 ft) of available head. When head loss exceeds this available head (also called the limiting head), the filter must be backwashed even if it has not reached breakthrough. Some filters do not reach breakthrough or the limiting head within several days. In these cases, utilities backwash filters after a set period to maintain a convenient schedule for plant operators, even though the filter has additional usable capacity.

On Fig. 11-2, the filter reaches breakthrough before reaching the available head, but these events can occur in either order depending on the filter design and raw-water quality. A filter design is optimized when both events occur simultaneously.

Classifications of Rapid Filtration Systems

Rapid filtration is classified by the level of pretreatment, as presented on Fig. 11-3. The most important factors that determine the required level of pretreatment are the raw-water quality and the preference and resources of the operating utility. Rapid filters are also classified by the number of layers of filter material, as shown in Table 11-1. The common filter materials are sand, anthracite, granular activated carbon (GAC), garnet, and ilmenite. Some are used alone, and others are used only in combination with other media. Additional information on the use and characteristics of these media materials is presented later in the chapter.

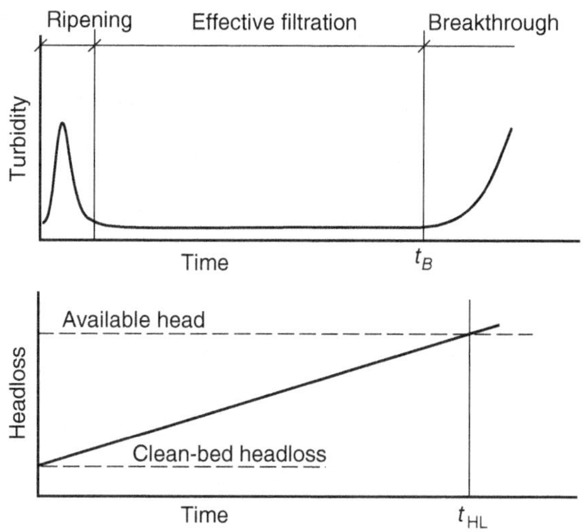

Figure 11-2
Operation of a rapid filter: (a) effluent turbidity versus time and (b) head loss development versus time.

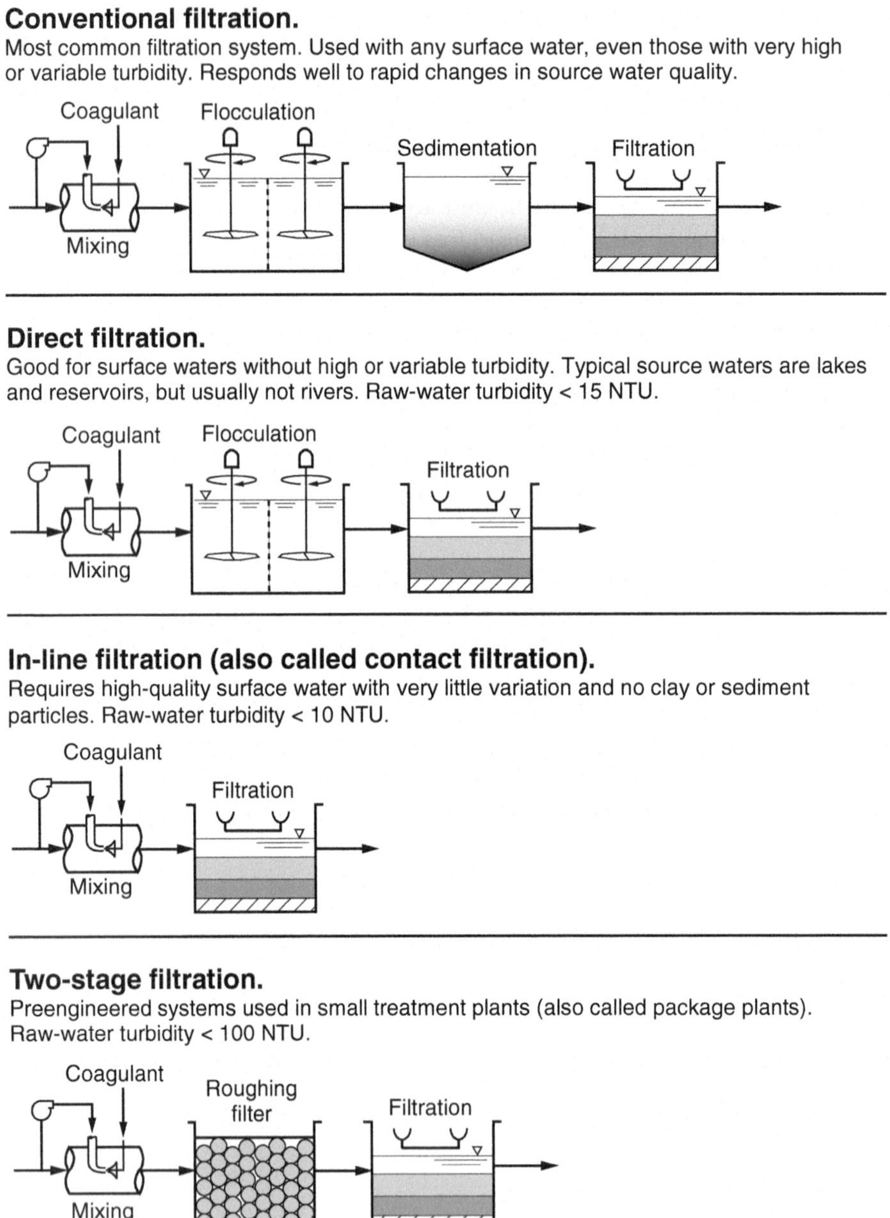

Figure 11-3
Classification of rapid filtration by pretreatment level.

Table 11-1
Classification of rapid filtration by media type

Filter Classification	Description
Monomedia	One layer of filter material, usually sand. Sand monomedia filters are typically about 0.6–0.76 m (24–30 in.) deep. Sand monomedia filters are an older design and have been largely superseded by other designs.
Deep-bed monomedia	One layer of filter material, usually anthracite or granular activated carbon. Deep-bed monomedia filters are typically 1.5–1.8 m (5–6 ft) deep. They are used to provide greater filtration capacity (longer run time) when feed water of consistent quality can be provided.
Dual media	Two layers of filter media. The older design is 0.45–0.6 m (18–24 in.) of anthracite over 0.23–0.3 m (9–12 in.) of sand, with filtration in the downflow direction. Deep-bed dual-media filters using 1.5–1.8 m (5–6 ft) of anthracite in the top layer are now common. GAC is sometimes used instead of anthracite in the top layer. Dual-media filters are more robust than monomedia filters.
Trimedia or mixed media	Three media, typically anthracite as the top layer, sand as the middle layer, and garnet or ilmenite as the bottom layer. The anthracite layer is typically 0.45–0.6 m (18–24 in.) deep, the sand layer is typically 0.23–0.3 m (9–12 in.) deep, and the garnet or ilmenite layer is 0.1–0.15 m (4–6 in.) deep. These have sometimes been called mixed-media filters when the media properties were selected to promote intermixing rather than the formation of distinct layers.

11-3 Properties of Granular Filter Media

Granular media filtration is affected by properties of the filter media and the filter bed, including grain size and size distribution, density, shape, hardness, bed porosity, and specific surface area. The types of media used in water filtration and their properties are addressed below.

Materials Used for Rapid Filtration Media

Naturally occurring granular minerals are mined and processed specifically for use as filter media. The common materials are sand, anthracite coal, garnet, and ilmenite. Anthracite is harder and contains less volatile material than other types of coal. Garnet and ilmenite are heavier than sand and are used as the bottom layer in trimedia filters. Garnet is comprised of a group of minerals containing a variety of elements, often appearing reddish or pinkish, and ilmenite is an oxide of iron and titanium. In addition to these four minerals, GAC is sometimes used as a filter material when adsorption and filtration are combined in a single unit process. Standard requirements for filtering materials are described in *ANSI/AWWA B100-01 Standard for Filtering Material* (AWWA, 2001a). Additional information on GAC is provided in Chap. 15.

Effective Size and Uniformity Coefficient

Filter materials are found in a granular state in nature or must be crushed to the desired size. Naturally occurring granular materials have nearly a lognormal size distribution, meaning that the distribution plots roughly as a straight line on lognormal graph paper. The size distribution is determined by sieve analysis (ASTM, 2001a), in which a sample of material is sifted through a stack of calibrated sieves (ASTM, 2001b), the weight of material retained on each sieve is measured, and the cumulative weight retained is plotted as a function of sieve size. The results of sieve analyses for naturally occuring sand and processed filter media are illustrated on Fig. 11-4. The size distribution of naturally occurring material is broader than desirable for rapid filter media. Thus, rapid filter media is processed to remove the largest (by sieving) and smallest (by washing) grain sizes, producing a narrower size distribution.

In North America, the standard method for characterizing the media size distribution is by effective size and uniformity coefficient. The *effective size* (ES or d_{10}) is the media grain diameter at which 10 percent of the media by weight is smaller, as determined by a sieve analysis. The *uniformity coefficient* (UC) is the ratio of the 60th percentile media grain diameter (the diameter at which 60 percent of the media by weight is smaller) to the

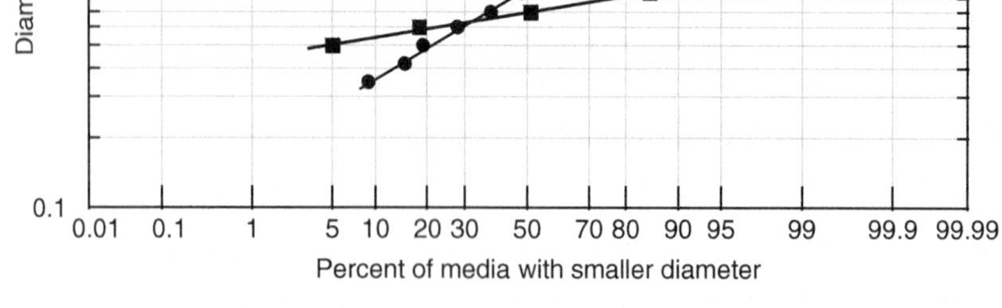

Figure 11-4
Size distribution of typical naturally occurring and processed filter sand.

effective size, as shown in the equation

$$UC = \frac{d_{60}}{d_{10}} \tag{11-1}$$

where UC = uniformity coefficient, dimensionless
d_{10} = 10th percentile media grain diameter, mm
d_{60} = 60th percentile media grain diameter, mm

The concept of the effective size was proposed by Allen Hazen in the 1890s because the hydraulic resistance of an unstratified granular bed tends to be unaffected by size variation as long as the effective size remains constant (Fair et al., 1971). Filter media tends to stratify during backwash. Fine grains collect at the top of the filter bed, where they cause excessive head loss and reduce overall effectiveness of the filter bed. Large grains settle to the bottom of the bed and are difficult to fluidize during backwash. A low UC can minimize these effects and is an important factor in the design of rapid filters. The ES and UC of typical filtration materials are provided in Table 11-2, along with typical values of other material properties. Granular activated carbon is typically specified by the maximum and minimum sieve sizes. For instance, an 8 × 20 mesh GAC refers to media that passes through a No. 8 sieve but is retained by a No. 20 sieve. Determination of the ES and UC from sieve data is demonstrated in Example 11-1.

Grain Shape

Mathematical models often assume that particles and filter grains are spherical for simplicity, but actual filter grains are not spherical, as shown on Fig. 11-5. The shape of individual grains affects filter design and performance in at least three ways. First, shape affects the size determined by sieve analysis. For spheres, the sieve opening will correspond to the diameter, but for nonspherical media, the sieve opening theoretically corresponds to the largest dimension of the smallest particle cross section (visualize rods going through a sieve lengthwise; the sieve opening corresponds to the largest rod diameter). The grain diameter determined by sieve

Table 11-2
Typical properties of filter media used in rapid filters[a]

Property	Unit	Garnet	Ilmenite	Sand	Anthracite	GAC
Effective size, ES	mm	0.2–0.4	0.2–0.4	0.4–0.8	0.8–2.0	0.8–2.0
Uniformity coefficient, UC	UC	1.3–1.7	1.3–1.7	1.3–1.7	1.3–1.7	1.3–2.4
Density, ρ_p	g/mL	3.6–4.2	4.5–5.0	2.65	1.4–1.8	1.3–1.7
Porosity, ϵ	%	45–58	N/A	40–43	47–52	N/A
Hardness	Moh	6.5–7.5	5–6	7	2–3	Low

[a]N/A = not available.

Example 11-1 Determination of effective size and uniformity coefficient

Determine the effective size and uniformity coefficient of the processed filter sand shown on Fig. 11-4.

Solution

1. Find the 10th percentile line on the x axis and follow it up to the intersection of the line for the processed filter sand. The corresponding value on the y axis is 0.54 mm.
2. The size (y axis) corresponding to the 60th percentile (x axis) for the processed filter sand is 0.74 mm.
3. The effective size is $ES = d_{10} = 0.54$ mm. The uniformity coefficient is $UC = d_{60}/d_{10} = 0.74/0.54 = 1.37$.

Comment

Probability paper is not required to determine the effective size and uniformity coefficient. Either an arithmetic scale or a probability scale can be used on the x axis. As long as a smooth curve can be drawn through the data, the d_{10} and d_{60} values can be determined. In addition, some spreadsheet software has functions for determining standard deviations and probability functions that can assist with the process of determining ES and UC.

analysis is typically smaller than the diameter of an "equivalent-volume" sphere. Cleasby and Woods (1975) compared size determinations from sieve analysis to equivalent-volume spheres and found that the equivalent-volume sphere diameter was 5 to 10 percent larger than the sieve size for sand and anthracite and 2 percent larger for garnet. Second, shape affects

(a) (b) (c)

Figure 11-5
Typical filter media: (a) anthracite coal, (b) sand, and (c) garnet. The sand shown is a worn river sand; suppliers may provide worn or crushed sand, depending on the source, which would change the shape factor.

how filter grains pack together in a bed. The porosity (defined below) of a randomly packed bed of spherical beads is typically about 38 percent, but the porosity of beds of filter grains typically ranges from 40 to 60 percent. Third, the hydraulics of flow through a bed of grains with sharp, angular surfaces is different from that through a bed of spherical beads.

Although grain shape has important implications in filter design, there is no easy way to account for it. Throughout filtration literature, grain shape is often characterized by either sphericity (ψ) or shape factor (ξ), which are interrelated as follows:

$$\psi = \frac{\text{surface area of equivalent-volume sphere}}{\text{actual surface area of grain}} \tag{11-2}$$

$$\xi = \frac{6}{\psi} \tag{11-3}$$

where ψ = sphericity, dimensionless
ξ = shape factor, dimensionless

For spherical grains, $\psi = 1$ and $\xi = 6$. Because a sphere has the minimum surface area of any geometric shape with the same volume, other shapes will have $\psi < 1$ and $\xi > 6$ based on the definitions in Eqs. 11-2 and 11-3.

Unfortunately, ψ and ξ have limited value in actual practice for several reasons. First, filter media is routinely measured and specified using the diameter determined by sieve analysis, not by equivalent-volume diameter. The equivalent-volume diameter can be determined by counting and weighing a representative number of media grains and calculating volume from the weight and density (Cleasby and Woods, 1975), a tedious procedure that is not done for commercially available filter media. Second, sphericity and shape factors are difficult to measure directly, and indirect means are normally used. The literature values for sphericity of common filter materials, such as the values available in Carman (1937), Cleasby and Fan (1981), and Dharmarajah and Cleasby (1986), were calculated from head loss experiments with the implicit assumption that the head loss equation coefficients (discussed in Sec. 11-4, see Eqs. 11-11 and 11-13) are independent of grain shape characteristics. As a result, many of the sphericity values for filter media available in the literature are really just empirical fitting parameters for head loss rather than true independent measurements of shape. Finally, other variables such as porosity have more impact on design, and arbitrary selection of a value for sphericity does little to improve the accuracy of design equations.

Material Density

The fluidization and settling velocities of filter media during and after backwash are influenced by material density. Backwash flow requirements are higher for denser materials of equal diameter. In addition, multimedia filters are constructed in a reverse-graded fashion (larger filter grains are near the top of the bed after backwashing) by using materials of different density. In a dual-media filter, anthracite is above sand because of differences in density. In a trimedia filter, the media are arranged from top to bottom as anthracite, sand, and garnet or ilmenite.

Material Hardness

Hardness affects the abrasion and breakdown of filter material during the backwash cycle. Hardness is ranked on the Moh table, a relative ranking of mineral hardness (talc = 1, diamond = 10). Sand, garnet, and ilmenite are hard enough to be unaffected by abrasion, but anthracite and GAC are friable, and design specifications must identify minimum hardness values to avoid excessive abrasion. A minimum Moh hardness of 2.7 is often specified for anthracite. The hardness of GAC is evaluated with procedures (the stir-ring abrasion test or the Ro-Tap abrasion test) outlined in *ANSI/AWWA B604-96 Standard for Granular Activated Carbon* (AWWA, 1996).

Granular Bed Porosity

The filter bed porosity (not porosity of the individual grains) has a strong influence on the head loss and filtration effectiveness in a filter bed. Porosity, or fraction of free space, is the ratio of void space volume to total bed volume and is calculated using the expression

$$\varepsilon = \frac{V_V}{V_T} = \frac{V_T - V_M}{V_T} \tag{11-4}$$

where ε = porosity, dimensionless
V_V = void volume in media bed, m^3
V_T = total volume of media bed, m^3
V_M = volume of media, m^3

Filter bed porosity ranges from 40 to 60 percent, depending on the type and shape of the media and how loosely it is placed in the filter bed.

Granular Bed Specific Surface Area

The specific surface area of a granular bed is defined as the total surface area of the filter material divided by the bed volume and is described by the expression

$$S = \frac{(\text{number of grains})\,(\text{surface area of each grain})}{\text{bulk volume of filter bed}} \tag{11-5}$$

where S = specific surface area, m^{-1}

For a uniform bed of monodisperse spheres, the specific surface area is given by the expression

$$S = \frac{6(1-\varepsilon)}{d} \tag{11-6}$$

where d = diameter of sphere, m

For nonspherical media, Eq. 11-6 is written as

$$S = \frac{6(1-\varepsilon)}{\psi d} = \frac{\xi(1-\varepsilon)}{d} \tag{11-7}$$

where d = equivalent-volume diameter, m

Equation 11-7 is useful only when the equivalent-volume diameter is known.

11-4 Hydraulics of Flow through Granular Media

The head loss through a clean filter bed and the flow rate needed to fluidize the filter bed during backwashing are discussed in this section. The increase in head loss as particles are captured during filtration is discussed in Sec. 11-5.

An important aspect of hydraulic behavior is the flow regime. The flow regime in granular media is identified by the Reynolds number for flow around spheres, which uses the media grain diameter for the length scale:

$$\mathrm{Re} = \frac{\rho_w v d}{\mu} \tag{11-8}$$

where Re = Reynolds number for flow around a sphere, dimensionless
ρ_w = fluid density, kg/m^3
v = filtration rate (superficial velocity), m/s
d = media grain diameter, m
μ = dynamic viscosity of fluid, kg/m·s

Flow in granular media does not experience a rapid transition from laminar to turbulent, as observed in pipes, but can be divided into four flow regimes (Trussell and Chang, 1999). The low end, called Darcy flow or creeping flow, occurs at Reynolds numbers less than about 1 and is characterized entirely by viscous flow behavior. The next regime, called Forchheimer flow after the first investigator to describe it, occurs at Reynolds numbers between about 1 and 100. Both Darcy flow and Forchheimer flow can be described as steady laminar flow because dye studies demonstrate that the fluid follows distinct streamlines. Forchheimer flow, however, is influenced by both viscous and inertial forces. In purely viscous flow, momentum is transferred between streamlines solely via molecular interactions. In twisting, irregular voids of a granular media bed, however, the fluid must accelerate and decelerate as void spaces turn, contract, and expand. The complex fluid motion through passageways of varying dimensions complicates the momentum transfer between streamlines, leading to an additional component of head loss that can be ascribed to inertial forces. Head loss due to viscous forces is proportional to v and head loss due to inertial forces is proportional to v^2. The third regime, a transition zone, has an upper limit Reynolds number between 600 and 800, and full turbulence occurs at higher Reynolds numbers.

Typical rapid filters have Reynolds numbers ranging from 0.5 to 5, straddling the transition between the Darcy and Forchheimer flow regimes. High-rate rapid filters have been designed with filtration rates as high as 33 m/h (13.5 gpm/ft^2), resulting in a Reynolds number of about 18. Backwashing of rapid filters occurs between Reynolds numbers of 3 and 25, completely in the Forchheimer flow regime.

Head Loss through Clean Granular Filters

The head loss through a filter increases as particles are retained. The net head available for particle retention is the difference between the available head and the clean-bed head loss.

FILTRATION RATE

The filtration rate is the flow rate through the filter divided by the area of the surface of the filter bed. The filtration rate has units of volumetric flux (reported as m/h in SI units, gpm/ft^2 in U.S. customary units) and is sometimes referred to as the superficial velocity because it is the velocity the water would have in an empty filter box (actual average velocity within the bed is higher due to the volume taken up by the filter grains).

DARCY FLOW REGIME

In 1856, Henry Darcy published a report stating the relationship between velocity, head loss, and bed depth in granular media under creeping-flow conditions (Darcy, 1856):

$$v = k_p \frac{h_L}{L} \tag{11-9}$$

where v = superficial velocity (filtration rate), m/s
k_p = coefficient, known as hydraulic permeability, m/s
h_L = head loss across media bed, m
L = depth of granular media, m

Darcy's law contains no mathematical descriptors of the porous material and therefore has no predictive value for filter system design. In 1927, Kozeny (1927a, b) developed an equation to relate granular media hydraulics to properties of the media by postulating an analogy between a bed of granular media and a system of parallel cylindrical channels. Laminar flow through cylindrical tubes is described by Poiseuille's law (Poiseuille, 1841), which can be written as

$$\frac{h_L}{L} = \frac{32\mu v}{\rho_w g d^2} \tag{11-10}$$

where g = acceleration due to gravity, 9.81 m/s^2

By equating the bed void volume to total internal channel volume and the media surface area to internal channel surface area, the Kozeny equation can be developed:

$$\frac{h_L}{L} = \frac{\kappa_k \mu S^2 v}{\rho_w g \varepsilon^3} \tag{11-11}$$

where κ_k = Kozeny coefficient, unitless
S = specific surface area from Eq. 11-5, m^{-1}
ε = porosity from Eq. 11-4, dimensionless

The Kozeny coefficient is an empirical coefficient introduced to fit the model results to experimental data. Other experimenters determined the value of κ_k to be about 5 (Carman, 1937; Fair and Hatch, 1933) for spherical media. Carman (1937) and Fair and Hatch (1933) proposed that the value of κ_k was independent of media properties and introduced a correction factor to account for the nonspherical nature of filter grains, Carman using sphericity and Fair and Hatch using the shape factor. Carman's correction factors were calculated from head loss data, which suggests that they might not be independent of κ_k. The origin of Fair and Hatch's shape factors was not clear, and their factors were based on diameter determined by sieve analysis rather than equivalent volume.

FORCHHEIMER FLOW REGIME

Subsequent studies demonstrated that head loss in granular media was greater than predicted by Eq. 11-11 when the Reynolds number was greater than 1. Forchheimer (1901) proposed a nonlinear equation that more accurately described head loss with higher velocity or larger media:

$$\frac{h_L}{L} = \kappa_1 v + \kappa_2 v^2 \tag{11-12}$$

where κ_1 = permeability coefficient for linear term, s/m
κ_2 = permeability coefficient for square term, s^2/m^2

Ahmed and Sunada (1969) showed that an equation with the form of Eq. 11-12 could be derived from the Navier–Stokes equation with only two assumptions: (1) the medium and the fluid are homogeneous and isotropic and (2) thermodynamic and chemical effects are small.

Ergun (1952) developed an equation with this form to describe head loss through granular media under Forchheimer flow conditions. The result was

$$h_L = \kappa_V \frac{(1-\varepsilon)^2}{\varepsilon^3} \frac{\mu L v}{\rho_W g d^2} + \kappa_I \frac{1-\varepsilon}{\varepsilon^3} \frac{L v^2}{g d} \tag{11-13}$$

where κ_V = head loss coefficient due to viscous forces, unitless
κ_I = head loss coefficient due to inertial forces, unitless

Equation 11-13 is known as the Ergun equation. Ergun (1952) compiled data from 640 experiments covering a range of Reynolds numbers between about 1 and 2000 when the diameter d was an effective diameter based on specific surface, and proposed values of $\kappa_V = 150$ and $\kappa_I = 1.75$ to fit the experimental data. The first term of the Ergun equation is identical to Eq. 11-11 (with substitution of Eq. 11-6) with the exception of the numerical value of the coefficient. Ergun proposed that the first term in Eq. 11-13 represented viscous energy losses and the second term represented kinetic energy losses, which is consistent with the mathematical construct of the equation. The dependence on μ, L, v, ρ_W, g, and d in the first term is consistent with the Poiseuille equation (i.e., laminar flow), while the dependence on these six variables in the second term is consistent flow under turbulent conditions, where kinetic energy losses predominate (Streeter and Wylie, 1979).

PRACTICAL CONSIDERATIONS FOR CALCULATION OF CLEAN-BED HEAD LOSS

Although some filters may operate in the Darcy flow regime, the transition between Darcy flow and Forchheimer flow is gradual. Based on past experience, it has been found that equations based on the Forchheimer flow regime can be used to determine the clean-bed head loss over the full range of values of interest in rapid filtration; therefore, Eq. 11-13 is the recommended equation for clean-bed head loss in rapid filters.

An important issue is the selection of values for each parameter in the clean-bed head loss equation. The coefficients proposed by Ergun are based on an effective diameter that is not easily measured. A more recent study has reexamined head loss through granular media (Chang et al., 1999; Trussell and Chang, 1999; Trussell et al., 1999). For spherical glass beads, the values proposed by Ergun were found to be reasonable. Different values are proposed for sand and anthracite, as shown in Table 11-3 (Trussell and Chang, 1999). The values for the head loss coefficients and porosity in Table 11-3 are based on the use of the effective size as determined by sieve analysis for the diameter (e.g., d = ES) and take media shape into account so that a separate shape factor is not needed. In the absence of pilot data or other site-specific information, the midpoint values in Table 11-3 are recommended for model use.

The sensitivity of clean-bed head loss to filtration rate, porosity, and media diameter is illustrated on Fig. 11-6. The significant impact of filtration rate is evident. In addition, clean-bed head loss doubles as the effective size of anthracite decreases from 1.2 to 0.8 mm, and increases by about 65 percent as porosity declines from 0.52 to 0.47. Head loss is also dependent on temperature because fluid viscosity increases as temperature decreases. The clean-bed head loss at 5°C is 60 to 70 percent higher than at 25°C.

Table 11-3
Recommended parameters for use with Eq. 11-13[a]

Medium	κ_V	κ_I	ε_l,
Sand	110–115	2.0–2.5	0.40–0.43
Anthracite	210–245	3.5–5.3	0.47–0.52

[a]When effective size as determined by sieve analysis is used for the diameter.

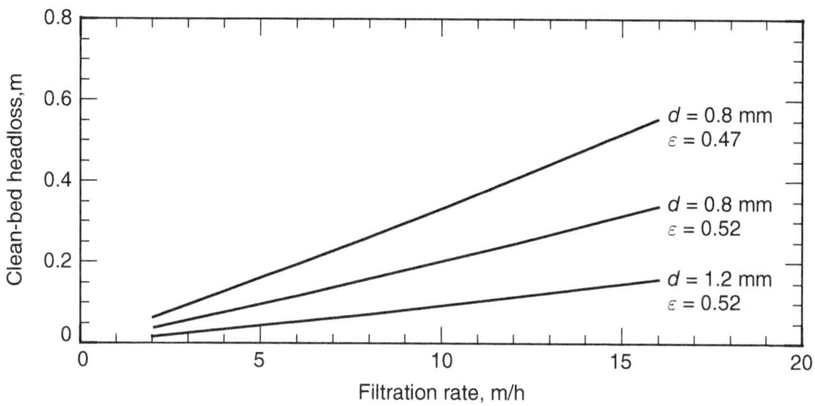

Figure 11-6
Effect of media size, bed porosity, and filtration rate on head loss through a clean granular filter bed. Calculated using Eq. 11–13 for anthracite ($L = 1$ m, $T = 15°C$, $\kappa_V = 228$, $\kappa_I = 4.4$).

Fortunately, it is common for water treatment plants to operate at a lower capacity during the winter than during the summer, and the reduction in filtration rate typically counteracts the increase in viscosity. Calculation of clean-bed head loss is demonstrated in Example 11-2.

Example 11-2 Clean-bed head loss through rapid filter

Calculate the clean-bed head loss through a deep-bed anthracite filter with 1.8 m of ES = 0.95 mm media at a filtration rate of 15 m/h and a temperature of 15°C.

Solution

The head loss through anthracite is calculated first using Eq. 11-13.

1. No pilot or site-specific information is given, so midpoint values are selected from Table 11-3; $\kappa_V = 228$, $\kappa_I = 4.4$, and $\varepsilon = 0.50$. Values of ρ_W and μ are available in Table C-1 in App. C ($\rho_W = 999$ kg/m^3 and $\mu = 1.14 \times 10^{-3}$ kg/m · s).
2. Calculate the first term in Eq. 11-13:

$$\frac{(228)(1-0.50)^2(1.14\times10^{-3}\ \text{kg/m}\cdot\text{s})(1.8\ \text{m})(15\ \text{m/h})}{(0.50)^3(999\ \text{kg/m}^3)(9.81\ \text{m/s}^2)(0.95\ \text{mm})^2(10^{-3}\ \text{m/mm})^2(3600\ \text{s/h})}$$
$$= 0.44\ \text{m}$$

3. Calculate the second term in Eq. 11-13:

$$\frac{(4.4)(1-0.50)(1.8\ \text{m})(15\ \text{m/h})^2}{(0.50)^3(9.81\ \text{m/s}^2)(0.95\ \text{mm})(10^{-3}\ \text{m/mm})(3600\ \text{s/h})^2} = 0.06\ \text{m}$$

4. Add the two terms together:

$$h_L = 0.44 \text{ m} + 0.06 \text{ m} = 0.50 \text{ m} \quad (1.6 \text{ ft})$$

Comments

A relatively small contribution to head loss comes from the inertial term. The inertial term becomes more important for the larger media and higher velocities used in high-rate rapid filters. If the filter is designed with 2.5 m (8.2 ft) of available head, the clean-bed head loss consumes about 20 percent of the available head. Note that if multiple layers of media are present, the head loss through each layer is additive.

Backwash Hydraulics

At the end of a filter run, rapid filters are backwashed by filtered water flowing upward through the filter bed. The backwash flow rate must be great enough to flush captured material from the bed, but not so high that the media is flushed out of the filter box. To prevent loss of media, it is important to determine the bed expansion that occurs as the filter media is fluidized, which is a function of the backwash flow rate and can be calculated using head loss equations for fixed beds.

FORCES ON PARTICLES

The forces on an individual particle (either a particle from the influent or a media grain) in upward-flowing water are exactly the same as were developed for the terminal settling velocity in Chap. 10 (see the free-body diagram on Fig 10-2). The particle will settle (or fail to fluidize) when downward forces predominate, be washed away when upward forces predominate, and remain suspended (fluidized) when the forces are balanced. The downward force is equal to the buoyant weight of the media and the upward force is the drag caused by the backwash flow. As noted in Chap. 10, the sum of forces on a particle is given by the expression

$$\sum F = F_g - F_b - F_d \tag{10-1}$$

where F_g = gravitational force on a particle, N
F_b = buoyant force on a particle, N
F_d = drag force on a particle, N

Combining Eqs. 10-1 through 10-4 from Chap. 10 yields the equation

$$\sum F = \rho_P V_p g - \rho_W V_p g - C_d \rho_W A_p \frac{v_s^2}{2} \tag{11-14}$$

where ρ_P = particle density, kg/m^3
ρ_W = water density, kg/m^3
V_p = volume of particle, m^3
A_p = projected area of particle in direction of flow, m^3
C_D = drag coefficient, unitless
g = acceleration due to gravity, 9.81 m/s^2
v_s = settling velocity of the particle, m/s

The drag coefficient is dependent on the flow regime (Clark, 1996). As noted in Chap. 10, the drag coefficient is described by the following expressions:

$$C_d = \frac{24}{\text{Re}} \quad \text{for Re} < 2 \quad \text{(laminar flow)} \tag{10-10}$$

$$C_d = \frac{18.5}{\text{Re}^{0.6}} \quad \text{for } 2 \le \text{Re} \le 500 \quad \text{(transition flow)} \tag{10-11}$$

Video fo filter backwash cycle

http://www.stantec.com/wt3r/11-2.html

The fluid velocity required to keep an individual particle suspended can be determined by substituting Eq. 10-10 or Eq. 10-11 into Eq. 11-14 and solving for velocity. As was shown in Chap. 10, the fluid velocity is given as Stokes' law (Eq. 10-13) for laminar flow, and the following expression for transition flow:

$$v_s = \frac{g(\rho_P - \rho_W)d_p^2}{18\mu} \quad \text{(laminar flow)} \tag{10-13}$$

$$v_s = \left[\frac{g(\rho_P - \rho_W)d_p^{1.6}}{13.9\rho_W^{0.4}\mu^{0.6}}\right]^{1/1.4} \quad \text{(transition flow)} \tag{10-14}$$

The velocity required to suspend an isolated particle in a uniform flow field (i.e., above the filter bed, away from the influences of the bed) may be determined using Eqs. 10-13 or 10-14, as appropriate. Within a filter bed, velocities (and therefore drag forces) are higher due to the volume taken up by the media. The balance of forces on an individual particle is demonstrated in Example 11-3.

Example 11-3 Forces on suspended particle

A filter is backwashed at 50 m/h at 15°C. Determine whether a 0.1-mm diameter particle of sand will be washed from the filter.

Solution

1. Calculate the gravitational force on the particle using the F_g term from Eq. 11-14. The value for ρ_P is available in Table 11-2:

$$F_g = \rho_P V_p g = (2650\,\text{kg/m}^3)\left(\frac{\pi}{6}\right)\left(\frac{0.1\ \text{mm}}{10^3\ \text{mm/m}}\right)^3 (9.81\ \text{m/s}^2)$$

$$= 1.36 \times 10^{-8}\ \text{kg} \cdot \text{m/s}^2 = 1.36 \times 10^{-8}\ \text{N}$$

2. Calculate the buoyant force on the particle using the F_b term from Eq. 11-14. The value for ρ_W is available in Table C-1 in App. C:

$$F_b = \rho_W V_p g = (999\ \text{kg/m}^3)\left(\frac{\pi}{6}\right)\left(\frac{0.1\ \text{mm}}{10^3\ \text{mm/m}}\right)^3 (9.81\ \text{m/s}^2)$$

$$= 5.13 \times 10^{-9}\ \text{kg} \cdot \text{m/s}^2 = 5.13 \times 10^{-9}\ \text{N}$$

3. Calculate the Reynolds number using Eq. 11-8 to determine in what flow regime the particle is:

$$\text{Re} = \frac{\rho_W v d}{\mu} = \frac{(999\ \text{kg/m}^3)(50\ \text{m/h})(0.1\ \text{mm})}{(1.139 \times 10^{-3}\ \text{kg/m} \cdot \text{s})(3600\ \text{s/h})(10^3\ \text{mm/m})} = 1.22$$

4. The Reynolds number is less than 2, so Eq. 10-10 can be used to calculate drag forces:

$$C_d = \frac{24}{\text{Re}} = \frac{24}{1.22} = 19.7$$

$$F_d = C_d \rho_W A_p \frac{v_s^2}{2} = \frac{19.7(999\ \text{kg/m}^3)}{2}\left(\frac{\pi}{4}\right)\left(\frac{0.1\ \text{mm}}{10^3\ \text{mm/m}}\right)^2 \left(\frac{50\ \text{m/h}}{3600\ \text{s/h}}\right)^2$$

$$= 1.49 \times 10^{-8}\ \text{N}$$

5. Calculate the sum of the forces:

$$\sum F = F_g - F_b - F_d = 1.36 \times 10^{-8}\ \text{N} - 5.13 \times 10^{-9}\ \text{N} - 1.49 \times 10^{-8}\ \text{N}$$

$$= -6.43 \times 10^{-9}\ \text{N}$$

The net force is negative (upward), so the particle will be flushed away with the backwash water.

BED EXPANSION AND POROSITY

A state of equilibrium between gravitational and drag forces is established in the filter bed. During backwash, the velocity in a filter bed is higher than for an isolated particle due to the presence of the media, causing higher drag forces that lift the media. As the media rises, increasing porosity reduces the velocity until the drag force is balanced by the gravitational force. The relationship between bed expansion and porosity is described in the following equation and on Fig. 11-7:

$$\frac{L_E}{L_F} = \frac{1-\varepsilon_F}{1-\varepsilon_E} \tag{11-15}$$

where L_E = depth of expanded bed, m
L_F = depth of bed at rest (fixed bed), m
ε_E = porosity of expanded bed, dimensionless
ε_F = porosity of bed at rest (fixed bed), dimensionless

The drag force on the media exerts an equal and opposite force on the water, which is manifested as head loss. Head loss through a fluidized bed is calculated as the gravitational force (fluidized weight) of the entire bed, as shown in the expression

$$F_g = mg = (\rho_p - \rho_w)(1-\varepsilon)aLg \tag{11-16}$$

where F_g = weight of the entire filter bed, N
a = cross-sectional area of filter bed, m^2

The weight of the bed must be divided by the filter area to convert the weight of the bed to units of pressure (i.e., convert N to N/m^2) and divided by $\rho_w g$ to convert units of pressure (N/m^2) to units of head (m) as follows:

$$h_L = \frac{F_g}{a\rho_w g} = \frac{(\rho_p - \rho_w)(1-\varepsilon)L}{\rho_w} \tag{11-17}$$

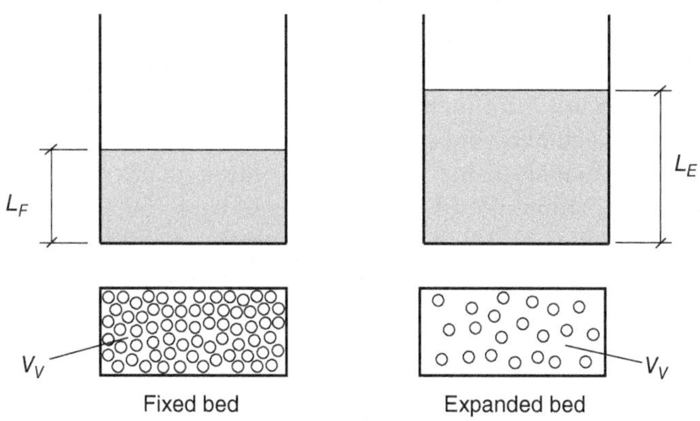

Figure 11-7
Fixed and expanded beds during backwashing of rapid filters. During filtration, the media grains are touching each other, but when media are fluidized during backwashing, the void volume increases, causing an overall expansion of the bed.

Akgiray and Saatçi (2001) demonstrated that the Eq. 11-13 is equally valid for fixed and expanded beds. Thus, in a fluidized bed, the head loss due to the weight of the media is equal to the head loss calculated from Eq. 11-13. Equating Eqs. 11-13 and 11-17 yields the expression

$$\kappa_V \frac{(1-\varepsilon)^2}{\varepsilon^3} \frac{\mu L v}{\rho_w g d^2} + \kappa_I \frac{1-\varepsilon}{\varepsilon^3} \frac{L v^2}{g d} = \frac{(\rho_p - \rho_w)(1-\varepsilon) L}{\rho_w} \tag{11-18}$$

An analytical solution for Eq. 11-18 in terms of v would allow the backwash velocity to be calculated directly for any set of filter conditions. Equation 11-18 can be seen to be a quadratic equation in v with a multitude of other terms, but it can be solved directly by making use of the Reynolds number. Equation 11-18 can be rearranged as follows after incorporating Eq. 11-8:

$$\kappa_I \,/\, \text{Re}^2 + \kappa_V (1-\varepsilon)\text{Re} - \beta = 0 \tag{11-19}$$

$$\beta = \frac{g\rho_w (\rho_p - \rho_w) d^3 \varepsilon^3}{\mu^2} \tag{11-20}$$

where β = backwash calculation factor, dimensionless

Equation 11-19 is a quadratic equation in terms of Re. One root of Eq. 11-19 is necessarily negative because both κ_I and κ_V are positive. The remaining meaningful solution of the quadratic equation is

$$\text{Re} = \frac{-\kappa_V (1-\varepsilon) + \sqrt{\kappa_V^2 (1-\varepsilon)^2 + 4\kappa_I \beta}}{2\kappa_I} \tag{11-21}$$

Once the Reynolds number is obtained from Eq. 11-21, the velocity that will maintain the bed in an expanded state corresponding to a specific porosity value can be determined from Eq. 11-8. The minimum fluidization flow rate can be calculated by determining the velocity that produces head loss equal to the buoyant weight of the media at the fixed-bed porosity. The minimum fluidization velocity is a function of grain size, with smaller particles fluidizing at lower velocity. The backwash rate must be above the minimum fluidization velocity of the largest media, typically taken as the d_{90} diameter (Cleasby and Logsdon, 1999). After fluidization, head loss may decrease slightly because the media grains are no longer in contact and extremely small or dead-end void spaces disappear. Akgiray and Saatçi (2001) performed an analysis using equivalent-volume diameters and sphericity factors and recommended Ergun's values of $\kappa_I = 150$ and $\kappa_V = 1.75$ for fixed beds but that $\kappa_V = 1.0$ fit the data better for expanded beds. The problems associated with equivalent-volume diameters and sphericity factors have been discussed previously. Thus, the values of κ_I and κ_V from Table 11-3 are recommended for backwash expansion calculations. Calculation of the backwash flow rate to achieve a certain level of bed expansion is illustrated in Example 11-4.

Example 11-4 Backwash flow rate for bed expansion

Find the backwash flow rate that will expand an anthracite bed by 30 percent given the following information: $L_F = 2$ m, $d = 1.3$ mm, $\rho_P = 1700$ kg/m^3, $\varepsilon = 0.52$, and $T = 15$°C.

Solution

1. Calculate L_E that corresponds to a 30 percent expansion:

$$L_E = L_F + 0.3\,L_F = 2\text{ m} + 0.3\,(2\text{ m}) = 2.6\text{ m}$$

2. Calculate ε_E using Eq. 11-15:

$$\varepsilon_E = 1 - \left[\frac{L_F}{L_E}(1-\varepsilon_F)\right] = 1 - \left[\left(\frac{2\text{ m}}{2.6\text{ m}}\right)(1-0.52)\right] = 0.63$$

3. Calculate β using Eq. 11-20. Values of ρ_W and μ are available in Table C-1 in App. C.

$$\begin{aligned}\beta &= \frac{g\rho_w(\rho_p-\rho_w)d^3\varepsilon^3}{\mu^2}\\ &= \frac{(9.81\text{ m/s}^2)(999\text{ kg/m}^3)(1700-999\text{ kg/m}^3)(0.0013\text{ m})^3(0.63)^3}{(1.139\times10^{-3}\text{ kg/m}\cdot\text{s})^2}\\ &= 2910\end{aligned}$$

4. Calculate Re using Eq. 11-21. Because no pilot or site-specific data are given, use values of κ_V and κ_I from midpoint values from Table 11-3 (e. g., $\kappa_V = 228$ and $\kappa_I = 4.4$):

$$\begin{aligned}\text{Re} &= \frac{-\kappa_V(1-\varepsilon)+\sqrt{\kappa_V^2(1-\varepsilon)^2+4\kappa_I\beta}}{2\kappa_I}\\ &= \frac{-228(1-0.63)+\sqrt{(228)^2(1-0.63)^2+4(4.4)(2910)}}{2(4.4)} = 17.9\end{aligned}$$

5. Calculate v using Eq. 11-8:

$$\begin{aligned}v &= \frac{\mu\,\text{Re}}{\rho_W d} = \frac{(1.139\times10^{-3}\text{ kg/m}\cdot\text{s})(17.9)(3600\text{ s/h})}{(999\text{ kg/m}^3)(0.0013\text{ m})}\\ &= 56.5\text{ m/h}\quad(22.6\text{ gpm/ft}^2)\end{aligned}$$

Alternatively, it is frequently necessary to determine the bed expansion that occurs for a specific backwash rate. Equation 11-18 is a cubic equation in porosity, which was analytically solved by Akgiray and Saatçi (2001). Akgiray and Saatçi showed that two roots of the cubic equation are complex numbers, leaving only one meaningful solution as follows:

$$\varepsilon = \sqrt[3]{X + (X^2 + Y^3)^{1/2}} + \sqrt[3]{X - (X^2 + Y^3)^{1/2}} \qquad (11\text{-}22)$$

where X = backwash calculation factor, dimensionless
Y = backwash calculation factor, dimensionless

The factors X and Y are defined as

$$X = \frac{\mu v}{2g(\rho_p - \rho_w)d^2}\left(\kappa_V + \frac{\kappa_I \rho_w v d}{\mu}\right) \qquad (11\text{-}23)$$

$$Y = \frac{\kappa_V \mu v}{3g(\rho_p - \rho_w)d^2} \qquad (11\text{-}24)$$

The targeted expansion rate is about 25 percent for anthracite and about 37 percent for sand (Kawamura, 2000). The procedure for calculating the expansion of media during backwashing is demonstrated in Example 11-5.

Example 11-5 Filter bed expansion during backwash

Find the expanded bed depth of a sand filter at a backwash rate of 40 m/h given the following information: $L = 0.9$ m, $d = 0.5$ mm, $\rho_P = 2650$ kg/m^3, and $T = 15°$C.

Solution

1. Calculate X using Eq. 11-23. Values of ρ_W and μ are available in Table C-1 in App. C. Because no pilot or site-specific data are given, use values of κ_V and κ_I from midpoint values in Table 11-3 (e. g., $\kappa_V = 112$ and κ_I 2. 25):

$$X = \frac{\mu v}{2g(\rho_p - \rho_w)d^2}\left(\kappa_V + \frac{\kappa_I \rho_w v d}{\mu}\right)$$

$$= \frac{(1.14 \times 10^{-3}\,\text{kg/m·s})[(40\text{ m/h})/(3600\text{ s/h})]}{2(9.81\text{ m/s}^2)(2650 - 999\text{ kg/m}^3)[0.5\text{ mm}/(10^3\text{ mm/m})]^2}$$

$$\times\left[112 + \frac{(2.25)(999\text{ kg/m}^3)[(40\text{ m/h})/(3600\text{ s/h})][0.5\text{ mm}/(10^3\text{ mm/m})]}{1.14 \times 10^{-3}\text{ kg/m·s}}\right]$$

$$= 0.1921$$

2. Calculate Y using Eq. 11-24:

$$Y = \frac{\kappa_V \mu v}{3g(\rho_p - \rho_w)d^2}$$

$$= \frac{(112)(1.14 \times 10^{-3}\ \text{kg/m}\cdot\text{s})(40\ \text{m/h})(10^3\ \text{mm/m})^2}{3(9.81\ \text{m/s}^2)(2650 - 999\ \text{kg/m}^3)(0.5\ \text{mm})^2(3600\ \text{s/h})} = 0.1168$$

3. Calculate porosity using Eq. 11-22:

$$\varepsilon_E = \sqrt[3]{X + (X^2 + Y^3)^{1/2}} + \sqrt[3]{X - (X^2 + Y^3)^{1/2}}$$

$$= \sqrt[3]{0.1921 + [(0.1921)^2 + (0.1168)^3]^{1/2}}$$

$$+ \sqrt[3]{0.1921 - [(0.1921)^2 + (0.1168)^3]^{1/2}} = 0.57$$

4. Calculate the expanded bed depth using Eq. 11-15. Because no site-specific porosity value is given, the fixed-bed porosity is taken from Table 11-3 and is assumed to be $\varepsilon_F = 0.42$.

$$L_E = L_F \frac{1 - \varepsilon_F}{1 - \varepsilon_E} = 0.9\ \text{m}\left(\frac{1 - 0.42}{1 - 0.57}\right) = 1.21\ \text{m}$$

5. Calculate the percent expansion of the bed:

$$\left(\frac{L_E}{L_F} - 1\right) \times 100 = \left(\frac{1.21\ \text{m}}{0.9\ \text{m} - 1}\right) \times 100 = 34\%$$

Comment

The bed expansion under the example conditions is 34 percent, which is about equal to the desired expansion rate of 37 percent for sand.

Backwash hydraulics depends on the viscosity of water, which varies with temperature. To achieve the same expansion, it is necessary to use a higher backwash rate in summer, when the water is warmer, than the backwash rate used when the water is cold.

Several aspects of rapid filter design and operation result directly from requirements for effective backwashing. These include selection of a low uniformity coefficient to minimize stratification, skimming to remove fines, and selecting media for dual- and multimedia filters.

STRATIFICATION

Stratification is an important side effect of backwashing of rapid media filters. As shown in Eq. 10-14, the settling velocity of individual grains of filter media depends on diameter, with larger grains requiring a larger fluidization velocity. When a graded media filter bed (of constant grain density) is backwashed at a uniform rate, the smallest particles fluidize most and rise to the top of the filter bed, while the largest particles collect near the bottom of the bed.

Stratification has several adverse effects on filter performance. First, the accumulation of small grains near the top of the bed causes excessive head loss in the first few centimeters of bed depth (because head loss is a function of grain size). Second, the ability of a filter to remove particles is also a function of grain size (as will be presented subsequently), so small grains at the top of a bed cause all particles to be filtered in the first few centimeters of bed depth, which means the entire bed depth is not being used effectively.

The method for minimizing stratification is proper selection of filter media. The uniformity coefficient determines the stratification of the media. Low values of the uniformity coefficient are recommended specifically to minimize stratification of the filter bed during backwashing. A uniformity coefficient less than 1.4 is recommended for all rapid filter media, and uniformity coefficient values less than 1.3 are becoming common.

REMOVAL OF FINES

Stratification is particularly problematic if the media has excessive fines (particles considerably smaller than the effective size), even if a low uniformity coefficient has been specified. Fines are normally removed by backwashing and skimming immediately after the installation of new media. After media installation, backwashing normally proceeds at a low rate, just above the minimum fluidization velocity to bring as many fines as possible to the top of the bed, which are then skimmed with a flat-bladed shovel after the filter is drained. It is usually necessary to repeat the backwashing and skimming several times to remove the fines. Multimedia filters should be backwashed and skimmed after each layer of media is installed.

MULTIMEDIA FILTERS

Backwash hydraulics have important implications for the selection of media in dual- and trimedia filters. The media in multimedia filters must be matched so that all media fluidize at the same backwash rate. Otherwise, one media may be washed out of the filter during attempts to fluidize the other media, or alternatively, one media may fail to fluidize. Fluidization of media can be balanced by selecting a ratio of grain sizes that is matched to the ratio of grain densities. During backwash, the settling velocity is in the transition flow regime. Equating an equivalent fluidization velocity for two media in Eq. 10-14 and solving for the ratio of particle sizes yield the expression

$$\frac{d_1}{d_2} = \left(\frac{\rho_2 - \rho_w}{\rho_1 - \rho_w}\right)^{0.625} \tag{11-25}$$

where d_1 = grain diameter of one filter medium, m
d_2 = grain diameter of a second filter medium, m
ρ_1 = density of medium with diameter d_1, kg/m^3
ρ_2 = density of medium with diameter d_2, kg/m^3

If the two media have approximately the same uniformity coefficient, the effective size can be used in Eq. 11-25.

INTERMIXING

Proper selection of media and proper backwashing procedures result in the layers of media staying segregated, with only a few centimeters of intermixing. Two layers of media tend to stay segregated when the bulk densities of the two layers are different. Bulk density is a function of the grain density, water density, and bed porosity (Cleasby and Woods, 1975), as shown in the expression

$$\rho_B = \rho_p(1-\varepsilon) + \rho_w \varepsilon \tag{11-26}$$

where ρ_B = bulk density of bed, kg/m^3

The vigorous agitation of media during backwashing can cause intermixing. Segregation of media types is maintained by reducing the backwash rate gradually at the end of the backwash cycle, which allows the media to segregate before the backwash cycle is terminated.

The size and uniformity coefficient of media for trimedia filters are sometimes selected to encourage intermixing rather than segregation. Filters with media that is intermixed are called *mixed-media filters* and are thought to have a better distribution of media, from coarse grains on the top of the bed to fine grains at the bottom, which would minimize the porosity of the filter bed, sacrificing head loss but improving removal.

11-5 Particle Removal in Rapid Filtration

Filters can remove particles from water by several mechanisms. When particles are larger than the void spaces in the filter, they are removed by straining. When particles are smaller than the voids, they can be removed only if they contact and stick to the grains of the media. Transport to the media surface occurs by interception, sedimentation, and diffusion, and attachment occurs by attractive close-range molecular forces such as van der Waals forces.

Straining causes a cake to form at the surface of the filter bed, which can improve particle removal but also increases head loss across the filter. Rapid filters quickly build head loss to unacceptable levels if a significant cake layer forms. In addition, filtration at the surface leaves the bulk of the rapid filter bed unused. Consequently, rapid filters are designed to minimize straining and encourage depth filtration.

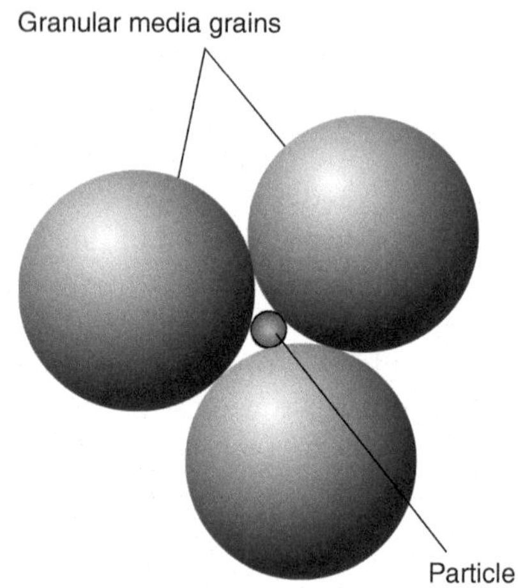

Figure 11-8
Capture of spherical particle by spherical media grains. If the ratio of particle diameter to media diameter is greater than 0.15, the particle will be strained by the media. If it is smaller, straining is not possible and particle capture must occur by other means. For typical rapid filtration, straining is limited to particles about 80 μm and larger.

Straining

A bed of granular media can strain particles smaller than the grain size. For spherical media, a close-packed arrangement will cause straining when the ratio of particle diameter to grain diameter is greater than 0.15; smaller particles can pass through the media, as shown on Fig. 11-8. The effective size of the smallest media specified in rapid filters is typically around 0.5 mm, although some trimedia filters use garnet or ilmenite with an effective size as small as 0.2 mm. With the use of engineered media that minimizes the quantity of finer grains, straining becomes insignificant for particles smaller than about 30 to 80 μm, depending, of course, on the shape and variability of the media and how it packs together. The vast majority of particles in the influent to rapid filters are smaller, particularly when sedimentation is used ahead of filtration. For example, viruses can be more than 1000 times smaller than particles that would be strained in a conventional filter, and clearly would not be removed without transport and attachment mechanisms.

Depth Filtration

In depth filtration, particles are removed continuously throughout the filter through a process of transport and attachment to the filter grains. Particle removal within a filter is dependent on the concentration of particles, similar to a first-order rate equation (Iwasaki, 1937), as described by

$$\frac{\partial C}{\partial z} = -\lambda C \tag{11-27}$$

where λ = filtration coefficient, m^{-1}
C = mass or number concentration of particles, mg/L or L^{-1}
z = depth in filter, m

If the filtration coefficient was known, it would be possible to calculate the effluent particle concentration from a filter. Unfortunately, filtration is a complex process, and the filtration coefficient can vary in both time and depth in the filter and depend on properties of the filter bed (grain shape and size distribution, porosity, depth), influent suspension (turbidity, particle concentration, particle size distribution, particle and water density, water viscosity, temperature, level of pretreatment), and operating conditions (filtration rate).

Two types of models have evolved to explain rapid filter behavior. Fundamental (or microscopic) models examine the importance of actual transport and attachment mechanisms. Phenomenological (or macroscopic) models attempt to explain the physical progression of the filtration cycle, through ripening, effective filtration, and breakthrough, though they do so with empirical parameters obtained from site-specific pilot studies rather than fundamental mechanisms. Phenomenological models are useful for evaluating pilot data and can be used to predict filter performance for conditions that were not specifically addressed within a pilot study. Because of the complexity of filtration mechanisms and the wide variation in source water properties, neither type of model can predict filter performance without site-specific pilot studies; nevertheless, they provide insight and understanding into the filtration process.

Fundamental Depth Filtration Theory

Fundamental filtration models examine the relative importance of mechanisms that cause particles to contact media grains. They can explain how particles are removed during depth filtration and the importance of various design and operating parameters under time-invariant conditions. For instance, fundamental filtration models are used later in this section to demonstrate the advantages of dual-media over monomedia filters and the importance of a low uniformity coefficient. Fundamental filtration models can also be used to examine the relative impact of varying other parameters on filter performance, such as porosity, filtration rate, or temperature. For these reasons, fundamental filtration models are valuable to a student acquiring a conceptual understanding of the filtration process.

Although they assist with conceptual understanding, fundamental filtration models are not very effective at quantitatively predicting the effluent turbidity in actual full-scale filters for the following reasons: (1) the models are based on an idealized system in which spherical particles collide with spherical filter grains; (2) the hydrodynamic variability and effect on streamlines introduced by the use of angular media are not addressed; (3) the models predict a single value for the filtration coefficient, which does not change as a function of either time or depth, whereas in real filters the filtration coefficient changes with both time and depth as solids collect on the media; and (4) the models assume no change in grain dimensions or bed porosity as particles accumulate. For these reasons, fundamental depth filtration models are often called clean-bed filtration models, and experimental validation generally focuses on the initial performance of laboratory filters (with spherical particles and media grains).

Yao Filtration Model

The basic model for water treatment applications was presented by Yao et al. (1971). Yao et al.'s theory is based on the accumulation of particles on a single filter grain (termed a "collector"), which is then incorporated into a mass balance on a differential slice through a filter. The accumulation on a single collector is defined as the rate at which particles enter the region of influence of the collector multiplied by a transport efficiency factor and an attachment efficiency factor. The transport efficiency η and the attachment efficiency α are ratios describing the fraction of particles contacting and adhering to the media grain, respectively, as described by the equations

$$\eta = \frac{\text{particles contacting collector}}{\text{particles approaching collector}} \tag{11-28}$$

$$\alpha = \frac{\text{particles adhering to collector}}{\text{particles contacting collector}} \tag{11-29}$$

where η = transport efficiency, dimensionless
α = attachment efficiency, dimensionless

The mass flow of particles approaching the collector is the mass flux through the cross-sectional area of the collector:

$$\text{Mass flow to one collector} = vC\frac{\pi}{4}d_c^2 \tag{11-30}$$

where v = superficial velocity, m/s
C = concentration of particles, mg/L
d_c = diameter of collector (media grain), m

The model development was based on an isolated single collector in a uniform-flow field, so the velocity in Eq. 11-30 is the filtration rate.

The accumulation of particles on a single collector is applied to a mass balance in a filter using a differential element of depth as the control volume, as shown on Fig. 11-9. The number of collectors in the control volume must be determined, which is the total volume of media within the control volume divided by the volume of a single collector:

$$\text{Number of collectors} = \frac{(1-\varepsilon)a\Delta z}{(\pi/6)d_c^3} \tag{11-31}$$

where ε = porosity
a = cross-sectional area of filter bed, m^2
Δz = incremental unit of depth in filter, m

The total accumulation of particles within the control volume is the product of the number of collectors and the accumulation on a single isolated collector. These terms can then be applied to a mass balance on the differential element:

$$[\text{accum}] = [\text{mass in}] - [\text{mass out}] \pm [\text{rxn}] \tag{11-32}$$

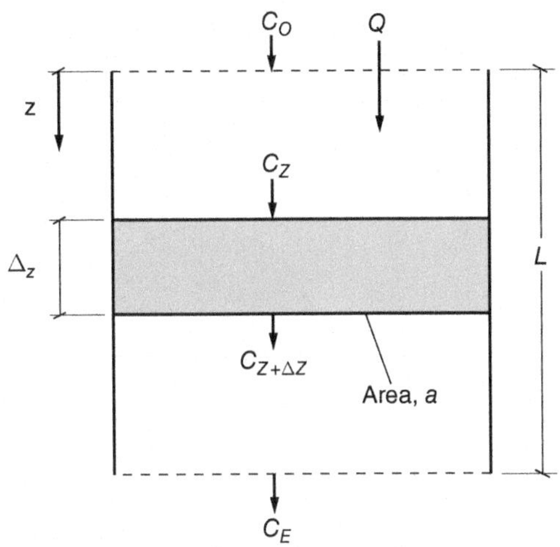

Figure 11-9
Differential element of filter bed for filtration models.

Generation or loss of particles due to reactions (i.e., production of biomass or consumption of particles via chemical or biological activity) is not included in the model. Diffusion is negligible compared to convective flux, so the mass balance can be written as

$$\left(\upsilon C\frac{\pi}{4}d_c^2\eta\alpha\right)\left[\frac{(1-\varepsilon)a\Delta z}{(\pi/6)d_c^3}\right]=QC_Z-QC_{Z+\Delta Z}=-\upsilon a(C_{Z+\Delta Z}-C_Z) \tag{11-33}$$

where Q = flow through filter, m^3/s

Taking the limit as Δz goes to zero, Eq. 11-33 can be rearranged as

$$\frac{dC}{dz}=\frac{-3(1-\varepsilon)\,\eta\alpha C}{2d_c} \tag{11-34}$$

Equation 11-34 has the same form as Eq. 11-27 and defines the filter coefficient as

$$\lambda=\frac{3(1-\varepsilon)\eta\alpha}{2d_c} \tag{11-35}$$

If the parameters in Eq. 11-35 (ε, η, α, and d_c) are constant with respect to depth in the filter, Eq. 11-34 can be integrated to yield the expression

$$C=C_O\exp\left[\frac{-3(1-\varepsilon)\eta\alpha L}{2d_c}\right] \tag{11-36}$$

where C_O = particle concentration in filter influent, mg/L
L = depth of filter, m

The next step in the development of the Yao model is to evaluate the mechanisms that influence the transport of particles to the media surface and the forces that influence attachment to the media.

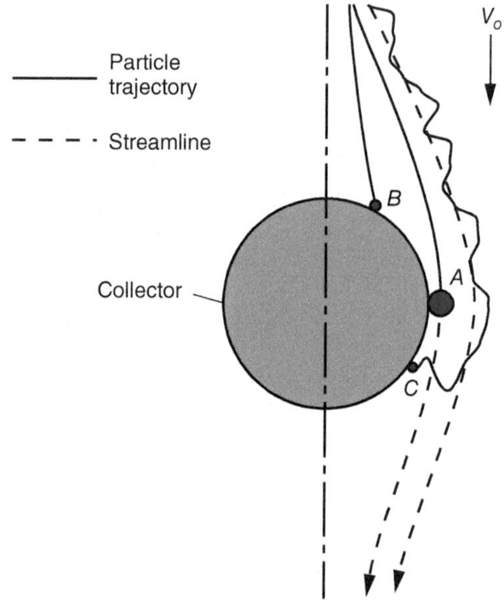

Figure 11-10
Particle transport mechanisms in fundamental filtration theory: (a) interception, particle A follows streamline but collides with the collector because of the proximity between the streamline and the collector; (b) sedimentation, particle B deviates from the streamline and collides with the collector because of gravitational forces; (c) diffusion, particle C collides with collector due to random Brownian motion.

Transport Mechanisms

The mechanisms for transporting particles to media grains are shown on Fig. 11-10. Water approaching a spherical collector in a uniform-flow field under laminar flow conditions follows streamlines to either side of the collector. Some particles will contact the collector because they follow a fluid streamline that passes close to the grain, while others must deviate from their fluid streamline to reach the collector surface. Details for each transport mechanism are as follows.

INTERCEPTION

Particles remaining centered on fluid streamlines that pass the collector surface by a distance of half the particle diameter or less will be intercepted. For laminar flow, spherical particles, and spherical collectors, particle transport by interception is given by the following expression (Yao et al., 1971):

$$\eta_I = \frac{3}{2}\left(\frac{d_p}{d_c}\right)^2 \tag{11-37}$$

where η_I = transport efficiency due to interception, dimensionless
d_p = diameter of particle, m

As shown in Eq. 11-37, interception increases as the ratio of particulate size to collector size increases. For 10-μm particles passing through a filter with 0.5-mm sand, $\eta_I < 10^{-3}$. In other words, only about one of a thousand possible collisions with a single collector due to interception will actually occur. However, a particle will pass thousands of collectors during its passage through a filter bed, increasing the chance of being removed somewhere in the filter bed.

SEDIMENTATION

Particles with a density significantly greater than water tend to deviate from fluid streamlines due to gravitational forces. The collector efficiency due to gravity has been shown to be the ratio of the Stokes settling velocity (see Chap. 10) to the superficial velocity (Yao et al., 1971), as shown in the expression

$$\eta_G = \frac{v_S}{v_F} = \frac{g\left(\rho_p - \rho_w\right)d_p^2}{18\mu v_F} \tag{11-38}$$

where η_G = transport efficiency due to gravity, dimensionless
v_S = Stokes' settling velocity, m/s
v_F = filtration rate (superficial velocity), m/s

DIFFUSION

Particles move by Brownian motion and will deviate from the fluid streamlines due to diffusion. The transport efficiency due to diffusion is given by the following expression (Levich, 1962):

$$\eta_D = 4\ \mathrm{Pe}^{-2/3} \tag{11-39}$$

$$\mathrm{Pe} = \frac{3\pi\mu d_p d_C v_F}{k_B T} \tag{11-40}$$

where η_D = transport efficiency due to diffusion, dimensionless
Pe = Peclet number, dimensionless
k_B = Boltzmann constant, 1.381×10^{-23} J/K
T = absolute temperature, K (273 + °C)

The Peclet number is a dimensionless parameter describing the relative significance of advection and dispersion in mass transport and is discussed further in Chap. 6. For physically similar systems, a lower value of the Peclet number implies greater significance of diffusion. The formulation of the Peclet number in Eq. 11-40 uses the Stokes–Einstein equation (see Chap.7) to relate the diffusion coefficient to the diameter of a spherical particle. In rapid filtration, diffusion is most significant for particles less than about 1 μm in diameter.

TOTAL TRANSPORT EFFICIENCY

The relative importance of these various mechanisms for transporting the particle to the surface depends on the physical properties of the filtration system. The Yao model assumes that the transport mechanisms are additive:

$$\eta = \eta_I + \eta_G + \eta_D \tag{11-41}$$

where η = total transport efficiency, dimensionless

The importance of each mechanism can be evaluated as a function of system properties. The effect of particle diameter on the importance of each

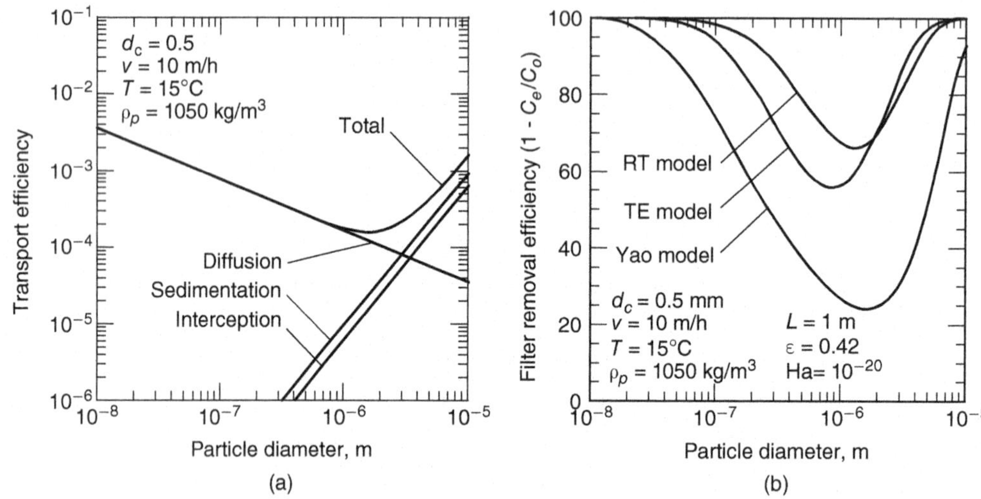

Figure 11-11
Predictions of fundamental filtration models: (a) importance of each transport mechanism on particles of different size as predicted by the Yao model and (b) comparison of predictions by each model for removal efficiency.

mechanism is shown on Fig. 11-11a. Small particles are efficiently removed by diffusion, whereas larger particles are removed mainly by sedimentation and interception. The Yao model predicts that the lowest removal efficiency occurs for particles of about 1 to 2 μm in size, which has been verified experimentally (Yao et al., 1971).

Advanced Fundamental Filtration Models

The Yao filtration model frequently underpredicts the number of collisions between particles and collectors when compared to experimental data. Several groups of researchers have tried to refine the Yao model by using a different flow regime or incorporating addition transport mechanisms. Rajagopalan and Tien (1976) developed a fundamental depth filtration model (the RT model) that (1) used a sphere-in-cell model of granular media, (2) accounted for the attraction between the collectors and particles caused by van der Waals forces (for interception and sedimentation only), and (3) accounted for reduced collisions due to viscous resistance of the water between the particle and collector. Following Rajagopalan and Tien's work, Tufenkji and Elimelech (2004) expanded the correlation further (the TE model) and more fully integrated van der Waal forces and hydrodynamic interactions into all transport mechanisms. The RT and TE models are semiempirical expressions that were correlated with the results of a numerical simulation model. The equations for the individual transport mechanisms in the Yao, RT, and TE models, along with the underlying expressions, are summarized in Table 11-4. In each case, the total transport efficiency is expressed by Eq. 11-41. The filter removal

Table 11-4
Transport mechanism equations in the Yao, RT[a], and TE fundamental filtration models

Mechanism	Yao, Habibian, and O'Melia	Rajagopalan and Tien[a]	Tufenkji and Elimelech
Interception	$\eta_I = \frac{3}{2} N_R^2$ (11-37)	$\eta_I = A_S \left(\frac{4}{3} N_A\right)^{1/8} N_R^{15/8}$ (11-43)	$\eta_I = 0.55\, A_S N_A^{1/8} N_R^{1.675}$ (11-46)
Sedimentation	$\eta_G = N_G$ (11-42)	$\eta_G = 0.00338 A_S N_R^{-0.4} N_G^{1.2}$ (11-44)	$\eta_G = 0.22 N_R^{-0.24} N_{vdW}^{0.053} N_G^{1.11}$ (11-47)
Diffusion	$\eta_D = 4\mathrm{Pe}^{-2/3}$ (11-39)	$\eta_D = 4 A_S^{1/3} \mathrm{Pe}^{-2/3}$ (11-45)	$\eta_D = 2.4 A_S^{1/3} N_R^{-0.081} N_{vdW}^{0.052} \mathrm{Pe}^{-0.715}$ (11-48)
$N_R = \dfrac{d_p}{d_c}$	(11-48)	N_r = relative size group, dimensionless N_g = gravity number, dimensionless N_a = attraction number, dimensionless N_{vdw} = van der Walls number, dimensionless	
$N_G = \dfrac{v_S}{v_F} = \dfrac{g(\rho_p - \rho_w) d_p^2}{18 \mu v_F}$	(11-50)	Pe = Peclet number (see Chap. 6, Eq. 6-91), dimensionless D_L = diffusion coefficient (Chap. 7, Stokes–Einstein equation, Eq. 7-25), m^2/s Ha = Hamaker constant, J	

(Continued)

Table 11-4 *(Continued)*

$\mathrm{Pe} = \dfrac{v_F d_C}{D_L} = \dfrac{3\pi\mu d_p d_C v_F}{k_B T}$	(11-40)	A_s = porosity function, dimensionless d_p = particle diameter, m d_c = collector diameter, m
$N_A = \dfrac{N_{vdW}}{N_R \mathrm{Pe}} = \dfrac{\mathrm{Ha}}{3\pi\mu d_p^2 V_F}$	(11-51)	k_B = Boltzmann constant, 1.381 × 10^{-23} J/K T = absolute temperature, K 273 + °C v_F = filtration rate, m/s
$N_{vdW} = \dfrac{\mathrm{Ha}}{k_B T}$	(11-52)	v_s = Stokes' settling velocity (see Chap. 10, Eq. 10-13), m/s ε = bed porosity, dimensionless
$\gamma = (1-\varepsilon)^{1/3}$	(11-53)	γ = porosity coefficient, dimensionless ρ_p = particle density, kg/m^3 ρ_w = liquid density, kg/m^3 μ = liquid viscosity, kg/m-s
$A_S = \dfrac{2(1-\gamma^5)}{2-3\gamma+3\gamma^5-2\gamma^6}$	(11-54)	

[a]The RT model does not consider η_I and η_G independent; they are shown here separately for convenience of comparison to the other models.

efficiency predicted by each model under comparable conditions is shown on Fig. 11-11b. The Hamaker constant is a parameter used in describing van der Waals forces. The theory necessary to calculate a value for the Hamaker constant is beyond the scope of this text, but the value ranges from 10^{-19} to 10^{-20} J (Hiemenz and Rajagopalan, 1997).

Fundamental filtration models can be used to examine the effect of important variables on filter performance, as shown in Example 11-6.

Example 11-6 Application of the TE Model

Use the TE model to examine the effect of media diameter (ranging from 0.4 to 2mm in diameter) on the removal of 0.1-µm particles in a filter bed of monodisperse media under the following conditions: porosity $\varepsilon = 0.50$, attachment efficiency $\alpha = 1.0$, temperature $T = 20°C$ (293.15 K), particle density $\rho_p = 1050$ kg/m^3, filtration rate $v = 15$ m/h, bed depth $L = 1.0$ m, Hamaker constant Ha $= 10^{-20}$ J (10^{-20} kg · m^2/s^2), and Boltzmann constant $k_B = 1.381 \times 10^{-23}$ J/K (1.381×10^{-23} kg · m^2/s^2 · K).

Solution

1. Calculate N_R for a media diameter of 0.4 mm using Eq. 11-49:

$$N_R = \frac{d_p}{d_c} = \frac{1\times10^{-7}}{4\times10^{-4}} = 2.5\times10^{-4}$$

2. Calculate N_G using Eq. 11-50. The values of ρ_w and μ are available in Table C-1 in App. C:

$$\begin{aligned} N_G &= \frac{g(\rho_p - \rho_w)d_p^2}{18\mu v_F} \\ &= \frac{(1050 - 998\ \text{kg/m}^3)(9.81\ \text{m/s}^2)(1\times10^{-7}\ \text{m})^2(3600\ \text{s/h})}{18(1\times10^{-3}\ \text{kg/m}\cdot\text{s})(15\ \text{m/h})} \\ &= 6.76\times10^{-8} \end{aligned}$$

3. Calculate Pe for a media diameter of 0.4 mm using Eq. 11-40:

$$\begin{aligned} \text{Pe} &= \frac{3\pi\mu d_p d_c v_F}{k_B T} \\ &= \frac{3\pi(1\times10^{-3}\ \text{kg/m}\cdot\text{s})(1\times10^{-7}\ \text{m})(4\times10^{-4}\ \text{m})(15\ \text{m/h})}{(1.381\times10^{-23}\ \text{kg}\cdot\text{m}^2/\text{s}^2\ \text{K})(293.15\ \text{K})(3600\ \text{s/h})} \\ &= 3.89\times10^5 \end{aligned}$$

4. Calculate N_A using Eq. 11-51:

$$N_A = \frac{\text{Ha}}{3\pi\mu\, d_p^2 v_F} = \frac{(1\times10^{-20}\ \text{kg}\cdot\text{m}^2/\text{s}^2)(3600\ \text{s/h})}{3\pi(1\times10^{-3}\ \text{kg/m}\cdot\text{s})(1\times10^{-7}\ \text{m})^2(15\ \text{m/h})}$$
$$= 2.54\times10^{-2}$$

5. Calculate N_{vdW} using Eq. 11-52:

$$N_{vdW} = \frac{\text{Ha}}{k_B T} = \frac{1\times10^{-20}\ \text{kg}\cdot\text{m}^2/\text{s}^2}{(1.381\times10^{-23}\ \text{kg}\cdot\text{m}^2/\text{s}^2)(293.15\ \text{K})} = 2.47$$

6. Calculate γ using Eq. 11-53:

$$\gamma = (1-\varepsilon)^{1/3} = (1-0.50)^{1/3} = 0.7937$$

7. Calculate A_S using Eq. 11-54:

$$A_S = \frac{2(1-\gamma^5)}{2-3\gamma+3\gamma^5-2\gamma^6}$$
$$= \frac{2\left[1-(0.7937)^5\right]}{2-3(0.7937)+3(0.7937)^5-2(0.7937)^6} = 21.46$$

8. Calculate η_I using Eq. 11-46:

$$\eta_I = (0.55)(21.46)(2.54\times10^{-2})^{1/8}(2.5\times10^{-4})^{1.675}$$
$$= 6.91\times10^{-6}$$

9. Calculate η_G using Eq. 11-47:

$$\eta_G = (0.22)(2.5\times10^{-4})^{-0.24}(2.47)^{0.053}(6.76\times10^{-8})^{1.11}$$
$$= 1.86\times10^{-8}$$

10. Calculate η_D using Eq. 11-48:

$$\eta_D = (2.4)(21.46)^{1/3}(2.50\times10^{-4})^{-0.081}$$
$$\times(2.47)^{0.052}(3.89\times10^{5})^{-0.715} = 1.38\times10^{-3}$$

11. Calculate η using Eq. 11-41:

$$\eta = 6.91\times10^{-6} + 1.86\times10^{-8} + 1.38\times10^{-3} = 1.39\times10^{-3}$$

12. Calculate C/C_0 using Eq. 11-36:

$$\frac{C}{C_0} = \exp\left[\frac{-3(1-0.50)(1.39\times10^{-3})(1.0)(1\text{ m})}{2(4\times10^{-4}\text{ m})}\right] = 0.074$$

13. Set up a computation table to determine particle removal for other diameters. Repeat steps 1 through 12 for additional media sizes between 0.4 and 2.0 mm. These calculations are best done with a spreadsheet. The results are as follows:

Media Diameter (mm)	C/C_0	Log Removal
0.4	0.074	1.13
0.6	0.262	0.58
0.8	0.434	0.36
1.0	0.560	0.25
1.2	0.650	0.19
1.4	0.716	0.15
1.6	0.764	0.12
1.8	0.801	0.10
2.0	0.830	0.08

Comment

The initial removal of small particles is highly sensitive to media size. While these particles are removed relatively efficiently by 0.4-mm-diameter media, removal drops dramatically as the media size increases.

Attachment Efficiency

As particles approach the surface of the media, short-range surface forces begin to influence particle dynamics. The attachment efficiency varies from a value of zero (no particles adhere) to a value of 1.0 (every collision between a particle and collector results in attachment). The attachment efficiency is affected by London–van der Waals forces, surface chemical interactions, electrostatic forces, hydration, hydrophobic interactions, or steric interactions (Tobiason and O'Melia, 1988; O'Melia, 1985; O'Melia and Stumm, 1967). Laboratory studies have found values of attachment efficiency ranging from about 0.002 to 1.0 (Chang and Chan, 2008; Elimelech and O'Melia, 1990; Tobiason and O'Melia, 1988). A number of correlations have been developed to relate the attachment efficiency value to properties of the collectors, particles, and solution (Chang and Chan, 2008; Bai and Tien, 1999; Elimelech, 1992). A key property that makes attachment unfavorable is the presence of repulsive electrostatic forces.

In water treatment, the interest is not so much the ability to predict the value of the attachment efficiency when attachment conditions are unfavorable, but to modify the system so that attachment is as favorable as possible, that is, an attachment efficiency value very nearly 1.0. The most important factor in achieving high attachment efficiency is eliminating the repulsive electrostatic forces; that is, proper destabilization of particles by coagulation. The need for a high attachment efficiency is exactly why coagulation is a critical part of rapid filtration. Particle stability and destabilization by coagulation was discussed in Chap. 9.

Predicting Filter Performance

The RT model has been used to demonstrate the value of dual-media filter beds when poorly pretreated water enters the filter (O'Melia and Shin, 2001). A similar analysis is shown on Fig. 11-12a. When water is properly conditioned ($\alpha = 1.0$), both the monomedia and dual-media filters perform well. When the water is not properly conditioned ($\alpha = 0.25$), both filters perform worse, but the degradation of quality is much more dramatic with the monomedia filter, suggesting that dual-media filters are more robust during periods of inadequate chemical pretreatment.

Similarly, the TE model can be used to demonstrate the effect of specifying filter media with a low uniformity coefficient. The particle concentrations through three filter beds with different UC values, but the same filter effluent quality are shown on Fig. 11-12b. The monodisperse media (i.e., UC = 1.0) has a constant filter grain size through the depth of the bed, but the two polydisperse media have been stratified by backwashing, sending the smallest grains to the top and the largest grains to the

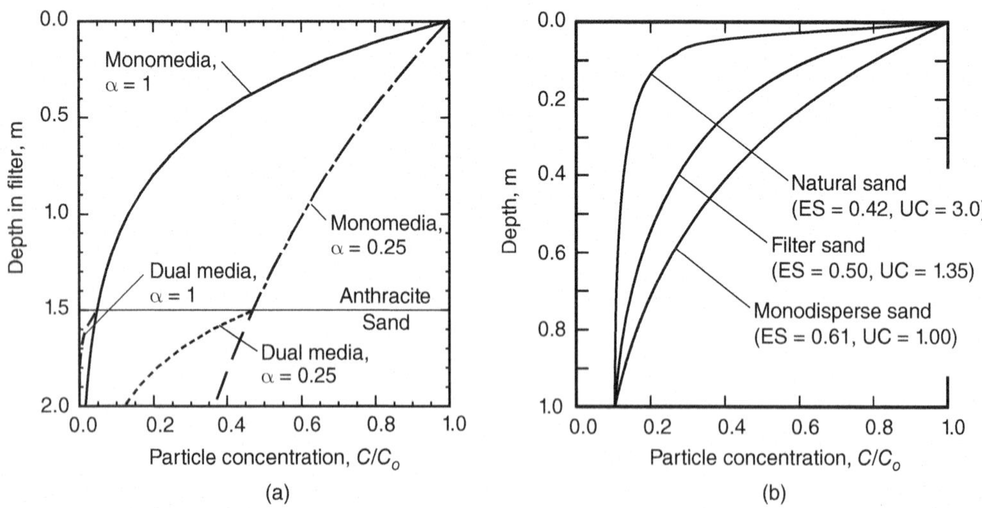

Figure 11-12
(a) Effect of attachment efficiency on effluent from monomedia and dual-media filters, as predicted by the RT model and (b) effect of media uniformity on solids penetration in filter bed, as predicted by the TE model.

bottom. Stratification increases the removal efficiency near the top of the bed. As a result, nearly all of the particles are removed near the top of the bed with the natural sand (UC = 3.0), whereas the particles are distributed throughout the monodisperse bed. In addition, the rapid collection of particles at the top of the natural sand bed will lead to clogging and onset of straining as an additional removal mechanism, leading to rapid head loss buildup and short filter runs. Thus, a low uniformity coefficient leads to a more effective use of the entire depth of the filter bed, less straining and cake formation at the top of the bed, and longer filter runs.

As filtration progresses, the media bed physically changes due to the accumulation of particles. Thus, filtration efficiency changes with time, a phenomenon not addressed in the fundamental filtration models. More sophisticated models have been developed to incorporate time-dependent phenomena, such as ripening (Darby et al., 1992), but current models are unable to predict changes in particle removal or head loss when design or operating conditions change. To examine the change in filter performance as solids collect within the filter bed, phenomenological filter models have been developed.

Phenomenological Depth Filtration Models

The primary function of phenomenological filtration models is to explore the progression of a filter run and the change in performance as solids collect within the filter. Performance variables of interest include (1) the duration of ripening and water quality during ripening, (2) the water quality during the effective filtration cycle, (3) the time to breakthrough, and (4) the time to limiting head.

PHENOMENOLOGICAL MODEL DEVELOPMENT

The mathematical formulation of phenomenological filtration models is based on the same overall mass balance through the filter bed that was used for the fundamental filtration models, using Eq. 11-32 to describe the accumulation of solids in a differential element of depth in the filter bed, as shown on Fig. 11-9. Phenomenological models do not focus on the accumulation of particles on a single collector but instead consider the increase of mass within the differential element. The basic mass balance equation for phenomenological models is developed with the following simplifying assumptions: (1) although particles are present in the interstitial fluid and at the surface of the media, the accumulation of particles in the interstitial fluid is negligible compared to the accumulation of particles on the media; (2) the number of particles entering and exiting the element by diffusion is negligible; and (3) the generation or loss of particles due to reaction is ignored. Thus, the mass balance for a differential element is described by the expression

$$\frac{\partial \sigma}{\partial t} = -v_F \frac{\partial C}{\partial z} \tag{11-55}$$

where σ = specific deposit, mass of accumulated particles per filter bed volume, mg/L
t = time, s

Phenomenological models are empirical and based on experimental data using the specific deposit as a master variable. By combining Eqs. 11-27 and 11-55, the basic form of the phenomenological model can be developed as

$$\frac{\partial \sigma}{\partial t} = \lambda v_F C \tag{11-56}$$

As noted earlier, filter performance changes as a function of time as solids collect in the filter bed. Thus, the filtration coefficient is normally expressed as a function of the specific deposit. For instance, filter efficiency improves as a filter ripens, and the filtration capabilities of the clean media are quickly superseded by the filtration capabilities of the retained particles. Ripening can be viewed as a condition that increases the value of the filtration coefficient as solids are collected in the filter bed. Thus, the filtration coefficient could be expressed as follows (Iwasaki, 1937):

$$\lambda = \lambda_0 + k\sigma \tag{11-57}$$

where λ_0 = initial filtration coefficient, m^{-1}
k = filtration rate constant, L/mg · m

As solids accumulate in the filter, the value of the filtration coefficient increases, leading to greater capture of solids and a lower concentration of particles in the filter effluent. Removal efficiency increases with depth during ripening and is dependent on the size and charge of particles in the water and other factors (Kim and Lawler, 2008; Kim et al., 2008). Consequently, the value of the filtration coefficient must be calculated as a function of depth because solids do not collect uniformly throughout the entire depth of the filter.

Breakthrough is a decrease of the filtration coefficient that causes an increase in effluent turbidity, which is opposite to what occurs when particles are being captured in the filter. A filtration coefficient that initially increases (ripening) and eventually decreases to zero (breakthrough) can be expressed in several forms, such as (Tien, 1989)

$$\lambda = \lambda_0 + k\sigma - \frac{k_T \sigma^2}{\varepsilon_0 - \sigma/\rho_P} \tag{11-58}$$

where k_T = breakthrough rate constant, $L^2/mg^2 \cdot m$
ε_0 = initial porosity, dimensionless

Filtration models must account for ripening and breakthrough and must also consider other processes such as detachment of previously attached particles. A number of filtration models have been proposed over the years and have been summarized elsewhere in the filtration literature (Tien, 1989; Tien and Payatakes, 1979).

A solution to a phenomenological model allows the particle concentration at any depth in the filter bed as well as the effluent concentration to be calculated at any point in time. Unfortunately, phenomenological model equations are complex and not easily solved. The rate of particle capture at any point in the filter bed is dependent on the quantity of previously captured solids, which in turn varies with bed depth (more solids collect at the top of the bed because the concentration of solids is higher near the top of the bed, in accordance with Eq. 11-56) and filtration time. Development of a phenomenological model involves the simultaneous solution of Eqs. 11-27 and 11-55 under conditions where λ is a function of the specific deposit, which in turn varies in both space and time. The filter rate coefficient λ is also site specific because of variations in local water quality, characteristics of the particles, characteristics of the media, stratification, and operating parameters. Because determining the filter rate coefficient is complex, phenomenological models are frequently solved numerically, although analytical solutions are possible depending on the complexity of the equation for λ.

STEADY-STATE PHENOMENOLOGICAL MODEL

A simplified phenomenological model can be developed to allow easier analysis of pilot data. The basic assumptions of a simplified phenomenological model are (1) the specific deposit is averaged over the entire filter bed, (2) solids accumulate at a steady rate over the entire filter run (the reduced accumulation of solids during ripening is ignored, under the legitimate assumption that the relatively small quantity of solids retained in the bed during ripening has little impact on the specific deposit over the entire filter run), and (3) head loss increases at a constant rate. With these assumptions, the specific deposit can be determined by performing a mass balance over the entire bed:

$$\sigma_t V = C_O Qt - C_E Qt \tag{11-59}$$

Where σ_t = specific deposit at time t, mg/L
V = bed volume, m^3
C_O = influent concentration, mg/L
C_E = effluent concentration, mg/L

Dividing by the filter bed area and rearranging yields an expression for the specific deposit as a function of time:

$$\sigma_t = \frac{v_F (C_O - C_E) t}{L} \tag{11-60}$$

where L = filter bed depth, m

The specific deposit increases at a steady rate as solids accumulate in the filter bed. Pilot filters can be operated until breakthrough occurs, and the value of the specific deposit at breakthrough can be related to the time to breakthrough by the expression

$$\sigma_B = \frac{v_F (C_O - C_E) t_B}{L} \tag{11-61}$$

where σ_B = specific deposit at breakthrough, mg/L
t_B = time to breakthrough, h

Equation 11-61 can be rearranged and expressed as a function of t_B:

$$t_B = \frac{\sigma_B L}{v_F (C_O - C_E)} \tag{11-62}$$

Specific deposit depends on process parameters (influent water quality, filtration rate, bed depth, media diameter, etc.). The value of the specific deposit at breakthrough can be recorded for a number of pilot filter runs in which these process parameters are varied. A regression analysis of the data can determine the dependence of the specific deposit at breakthrough, σ_B, on the process parameters (Kavanaugh et al., 1977). The dependence of the specific deposit varies because of site-specific conditions and can be determined only by analyzing pilot data.

Similarly, the rate of head loss buildup has been observed to depend on the rate of solids deposition in a filter. If $h_{L,O}$ is the clean-bed head loss determined from the Ergun equation (Eq. 11-13) and head loss increases at a constant rate, then the head loss at any time during the filtration run can be determined using the expression (Ives, 1967)

$$h_{L,t} = h_{L,O} + k_{HL}\sigma_t \tag{11-63}$$

where $h_{L,t}$ = filter head loss at time t, m
$h_{L,O}$ = initial head loss, m
k_{HL} = head loss increase rate constant, L · m/mg

Like the specific deposit, the head loss increase rate constant depends on site-specific conditions and process parameters. Some evidence suggests that filtration rate is an important factor in determining the type of deposit that forms. Higher filtration rates tend to cause particle to penetrate more deeply into the bed, spreading the deposit over a larger area of the bed. In addition, higher filtration rates lead to more compact deposits whereas lower filtration rates tend to form more open, porous deposits (Veerapanani and Weisner, 1997; Weisner, 1999). Compact deposits can be characterized by a higher fractal dimension (spheres have a fractal dimension of 3 whereas lines have a fractal dimension of 1). The rate of head loss increase as the deposit accumulates (i.e., the head loss increase rate constant k_{HL}) appears to be higher for deposits with a low fractal dimension; thus, low filtration rates cause a faster increase in head loss than high filtration rates for the same specific deposit. Quantitative models relating the rate of head loss increase to the characteristics of the deposit have not been successfully developed, and the head loss increase rate constant is best determined through a pilot study. Incorporating Eq. 11-60 and rearranging yields an expression for the rate constant:

$$k_{\mathrm{HL}} = \frac{(h_{L,t} - h_{L,O})\mathrm{L}}{v_F (C_O - C_E)t} \tag{11-64}$$

Once the rate constant for head loss buildup is determined, it can be used to determine the specific deposit that can be accumulated before reaching the limiting head as follows:

$$t_{\mathrm{HL}} = \frac{(H_T - h_{L,O})\mathrm{L}}{k_{\mathrm{HL}} v_F (C_O - C_E)} \tag{11-65}$$

where t_{HL} = time to limiting head, h
H_T = total available head, m

Once dependence of the specific deposit at breakthrough and the rate of head loss buildup on process parameters are determined, the phenomenological model can be used to determine the duration of filter runs and whether filter runs are limited by breakthrough or limiting head. Use of the simplified phenomenological model to analyze pilot data is shown in Example 11-7.

Example 11-7 Determination of optimum media size from pilot data

Four pilot filters with different effective sizes of anthracite (UC < 1.4, ρ = 1700 kg/m^3) were operated over multiple runs. The results are summarized in the table below. The media depth in each filter was 1.8 m, the filtration rate was 15 m/h, and the temperature was relatively constant at 20°C. Based on turbidity, you can assume the solids concentration was constant at 2.2 mg/L in the influent and negligible in the effluent.

Media ES (mm)	No. of Runs	Ave. Eff. Turbidity (NTU)	Ave. Time to Breakthrough (h)	Ave. Initial Head (m)	Ave. Final Head (m)
0.73	7	0.08	55	0.77	3.35
0.88	6	0.07	49	0.56	2.38
1.02	9	0.08	41	0.44	1.87
1.23	8	0.13	38	0.29	1.44

Determine: (a) the relationship between specific deposit at breakthrough (σ_B) and the media ES, (b) the relationship between the head loss rate constant (k_{HL}) and the media ES, and (c) the required available head and optimal media size if the full-scale system is to have a design run length of at least 48 h.

Solution

1. Any type of equation that relates media ES to σ_B and k_{HL} and results in a linear graph can be used. Often the relationships between the media ES and σ_B or k_{HL} can be described by a power function, and that type of equation is used in this example. Thus:

$$\sigma_B = b_1(d)^{m1} \quad \text{and} \quad k_{HL} = b_2(d)^{m2}$$

To find the value of the coefficients b and m, take the log of each equation and plot $\log(\sigma_B)$ and $\log(k_{HL})$ as a function of $\log(d)$. The slope of the straight line is m and the intercept is $\log(b)$:

$$\log(\sigma_B) = \log(b_1) + m_1 \log(d)$$

and

$$\log(k_{HL}) = \log(b_2) + m_2 \log(d)$$

2. Calculate the necessary values for the first effective size
   a. Calculate $\log(d)$

$$\log(d) = \log(0.73 \text{ mm}) = -0.137$$

   b. Calculate σ_B using Eq. 11-61

$$\sigma_B = \frac{v_F(C_0 - C_E)t_B}{L} = \frac{(15 \text{ m/h})(2.2 - 0 \text{ mg/L})(55 \text{ h})}{1.8 \text{ m}}$$
$$= 1008 \text{ mg/L}$$

   c. Calculate $\log(\sigma_B)$

$$\log(\sigma_B) = \log(1008) = 3.00$$

   d. Calculate k_{HL} using Eq. 11-64

$$k_{HL} = \frac{(3.35 - 0.77 \text{ m})(1.8 \text{ m})}{15 \text{ m/h}(2.2 - 0 \text{ mg/L})(55 \text{ h})} = 0.00256 \text{ L}\cdot\text{m/mg}$$

   e. Calculate $\log(k_{HL})$

$$\log(k_{HL}) = \log(0.00256) = -2.59$$

3. Repeat step 2 for the remaining effective sizes. The results are summarized in the following table:

ES	log(d)	σ_B	log (σ_B)	k_{HL}	log (k_{HL})
0.73	−0.137	1008	3.00	0.00256	−2.59
0.88	−0.056	898	2.95	0.00203	−2.69
1.02	0.0086	752	2.88	0.00190	−2.72
1.23	0.090	697	2.84	0.00165	−2.78

4. Plot $\log(\sigma_B)$ against $\log(d)$.

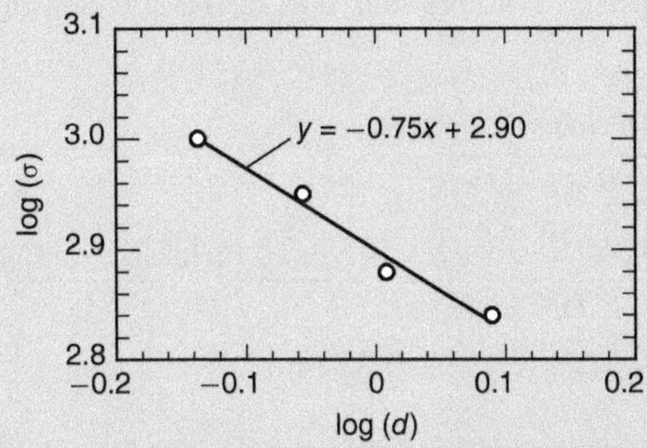

5. Perform a linear regression of the data (shown in the graph in step 4 using the Excel trendline function) and determine the slope and intercept of the regression line. From the graph in step 4, $m_1 = -0.75$ and $\log(b_1) = 2.90$. Therefore, $b_1 = 794$. The relationship between σ_B and d is

$$\sigma_B = 794(d)^{-0.75} \tag{1}$$

when the units of σ_B are mg/L and the units of d are mm.

6. Plot $\log(k_{HL})$ against $\log(d)$.

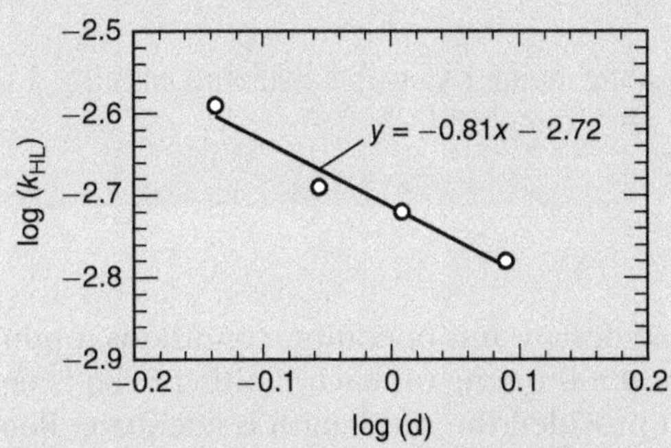

7. Perform a linear regression of the data (shown in the graph in step 6 using the Excel trendline function) and determine the slope and intercept of the regression line. From the graph in step 6, $m_2 = -0.81$ and $\log(b_2) = -2.72$. Therefore, $b_2 = 0.00191$. Thus, the relationship between k_{HL} and d is

$$k_{HL} = 0.00191(d)^{-0.81} \tag{2}$$

when the units of k_{HL} are L · m /mg and the units of d are mm.

8. Calculate the required size to reach 48 h before breakthrough by substituting Eq. 1 above into Eq. 11-62 and solving for the media size.

$$t_B = \frac{794\,(d)^{-0.75}\,L}{v_F\,(C_0 - C_E)}$$

$$(d)^{-0.75} = \frac{t_B v_F\,(C_0 - C_E)}{794\,L} = \frac{(48\text{ h})(15\text{ m/h})(2.2 - 0\text{ mg/L})}{794(1.8\text{ m})} = 1.108$$

$$d = (1.108)^{1/-0.75} = 0.87\text{ mm}$$

9. Calculate the required head to reach 48 h before reaching the limiting head by substituting Eq. 2 above into Eq. 11-65 and solving for the available head.

$$t_{HL} = \frac{(H_T - h_{L,0})L}{0.00181(d)^{-0.81}v_F(C_0 - C_E)}$$

$$H_T = \frac{t_{HL}0.00181(d)^{-0.81}v_F(C_0 - C_E)}{L} + h_{L,0}$$

$$= \frac{(48\text{ h})(0.00181)(0.87\text{ mm})^{-0.81}(15\text{ m/h})(2.2 - 0\text{ mg/L})}{1.8\text{ m}} + 0.53\text{ m}$$

$$= 2.3\text{ m}$$

where the initial head loss was calculated with Eq. 11-13 (the Ergun equation; see Example 11-2).

OPTIMIZATION

For a given set of design and operating conditions, optimum water production occurs when the time to reach limiting head is equal to the time to breakthrough provided the run length is adequate. Phenomenological filtration models can be used to optimize filtration design by allowing the engineer to vary design parameters until the time to breakthrough and limiting head are equal. Optimization of the filter design presented in Example 11-7 for two conditions of available head with respect to media size is shown on Fig. 11-13. Increasing the media depth will tend to increase the time to reach breakthrough (t_B) but decrease the time to reach the limiting head (t_{HL}). For 2.5 m of available head, the optimum design is achieved at a media size of 1.0 mm. An increase in available head to 3.0 m would have no effect on the run length if the media stayed the same size but would increase the run length by about 5 h if the media effective size were decreased to 0.90 mm.

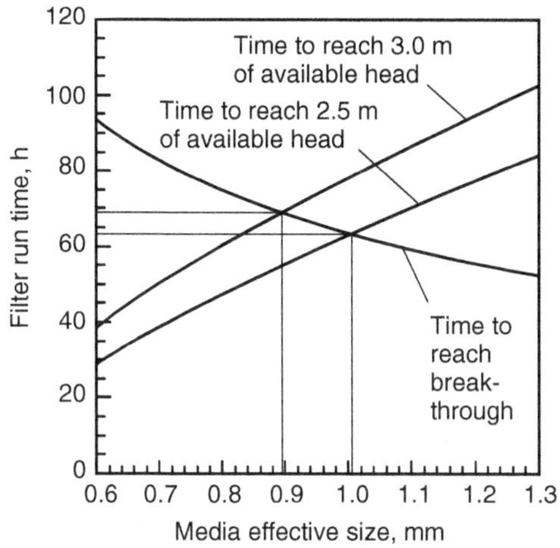

Figure 11-13
Optimization of media size with respect to time to breakthrough and time to limiting head.

The effect of significant design parameters on t_B and t_{HL} is summarized in Table 11-5. The effects summarized in Table 11-5 can be predicted from the theory presented earlier in the chapter and have generally been observed in actual filter operation. These design variables will influence filter performance, and thus t_B and t_{HL}, based on the rate of particulate capture and the rate of head loss increase. Some of the design variables, such as media size, media depth, and flow rate, are subject to designer selection. The limits of media size should be chosen to minimize interlayer mixing, which tends to decrease porosity and thereby lead to a rapid increase in head loss (decrease in t_{HL}) (Cleasby and Woods, 1975). Other variables, such as influent solids concentration, will depend upon the location of the filter in the process scheme. Variables such as floc strength and deposit

Table 11-5
Effect of design parameters on time to breakthrough and limiting head loss

	Effect of Parameter Increase on	
Parameter	**Time to Breakthrough, t_B**	**Time to Limiting Head Loss, t_{HL}**
Effective size	Decrease	Increase
Media depth	Increase	Decrease
Filtration rate	Decrease	Decrease
Influent particle Concentration	Decrease	Decrease
Floc strength	Increase	Decrease
Deposit density	Decrease	Decrease
Porosity	Decrease	Increase

density are difficult to control, but the use of polymers can be employed to improve floc strength.

Particle Detachment

Particle removal in granular filters is not an irreversible process, and detachment of particles may occur during the filtration cycle. Several studies have examined the breakthrough phenomenon and noted that it may be due to a decrease in particle capture or an increase in detachment, with evidence suggesting that the latter may be the dominant cause (Moran et al., 1993). Detachment occurs when there are perturbations to the system (Bergendahl and Grasso, 2003) but may occur at a low rate during steadystate filtration in the absence of perturbations. Perturbations may include changes in hydraulic forces or changes in water quality, including ionic strength and pH (Amirtharajah and Raveendran, 1993; Raveendran and Amirtharajah, 1995).

Detachment occurs when the forces shearing the particles away from the media grain are greater than the adhesive forces holding the particle. The primary forces between particles and media grains include van der Waals forces, electric double-layer interactions, Born repulsion, and hydration and hydrophobic forces (Raveendran and Amirtharajah, 1995). Under constant-flow (shear) conditions, detachment increases as pH increases or ionic strength decreases. Under constant chemical conditions, detachment increases as hydraulic shear increases. Models suggest that the shear stress from hydraulic perturbations has a greater effect on large particles than on smaller colloids (Bergendahl and Grasso, 2003). In addition, experimental evidence suggests that solids aggregate on the filter media (in a manner analogous to the aggregation of solids during flocculation) and that the aggregated flocs can then detach from the media (Darby and Lawler, 1990; Kau and Lawler, 1995; Moran et al., 1993).

Experimental research has not produced a quantitative model describing particle attachment and detachment mathematically. It is evident, however, that changes in filtration rate or influent water quality can have negative effects on filter performance and lead to the detachment of previously retained particles as well as changes in the ability to retain new particles, resulting in higher effluent turbidity. For these reasons, granular filters perform best when operated under constant conditions and when any changes are made gradually.

11-6 Rapid Filter Design

Preliminary design of rapid filters consists of the following:

1. Setting performance criteria, such as effluent turbidity, filter run length, recovery, and unit filter run volume (UFRV)
2. Selecting process design criteria, such as required level of pretreatment; filter media type, size, and depth; filtration rate; number of filters; and available head

3. Selecting a method for flow distribution and control
4. Selecting major process components, including backwashing systems, underdrains, wash troughs, and process piping

These topics are considered in the following sections.

Performance Criteria

The primary performance criteria for rapid filter design are the effluent water quality, the length of the filter run, and the recovery (i.e., the ratio of net to total water filtered). Filter design must also consider nonperformance criteria, such as minimizing capital and operation and maintenance (O&M) costs, reliability, and ease of maintenance.

EFFLUENT WATER QUALITY

Filter performance is primarily monitored by measuring effluent turbidity. To be in compliance with current U.S. regulations, the turbidity must be measured in the combined filter effluent at least every 4 h and at least 95 percent of the measurements must be below 0.3 NTU (maximum 1 NTU) (U.S. EPA, 2006). Most utilities set a design turbidity goal below the regulated limits, with a typical goal being 0.1 NTU. Additional information about the measurement and interpretation of turbidity values is presented in Chap. 2. Facilities can get additional credit for *Cryptosporidium* removal by achieving lower turbidity levels or by achieving low levels in the effluent from each individual filter.

A second method of measuring effluent water quality is particle counts. Particle counters provide both the number and size distribution of particles in water. Some utilities have installed particle counters and set water quality goals for particles, but there are no standard methods for the measurement of particles or regulatory requirements for the number of particles in filtered water. Additional information about particle counters is presented in Chap. 2.

FILTER RUN LENGTH

The length of the filter run dictates how often backwashes must be performed and has an impact on recovery. Since operators either perform backwashes manually or supervise automated backwash procedures, the frequency of backwashing has a direct impact on the amount of labor involved in filter operation. Typically, the minimum desirable filter run length is about 1 day, with filter designs that produce a filter run length between 1 and 4 days being common. Design parameters that can affect the length of a filter run are presented in Table 11-5.

RECOVERY

Recovery is the ratio between the net and total quantity of water filtered. Portions of the filtered water are used for backwashing and discharged as filter-to-waste volume, so the net water production is lower than the total volume of water processed through the filter. Recovery is evaluated using the concepts of unit filter run volume (UFRV) and unit backwash volume (UBWV) (Trussell et al., 1980). The UFRV is the volume of water that passes through the filter during a run, and the UBWV is the volume required to backwash the filter, defined as

$$\text{UFRV} = \frac{V_F}{a} = v_F t_F \qquad (11\text{-}66)$$

$$\text{UBWV} = \frac{V_{\text{BW}}}{a} = v_{\text{BW}} t_{\text{BW}} \qquad (11\text{-}67)$$

$$\text{UFWV} = \frac{V_{\text{FTW}}}{a} = v_F t_{\text{FTW}} \qquad (11\text{-}68)$$

where UFRV = unit filter run volume, m^3/m^2
UBWV = unit backwash volume, m^3/m^2
UFWV = unit filter-to-waste volume, m^3/m^2
V_F = volume of water filtered during one filter run, m^3
V_{BW} = volume of water required to backwash one filter, m^3
V_{FTW} = volume of water discharged as filter-to-waste, m^3
v_F = filtration rate (superficial velocity), m/h
v_{BW} = backwash rate, m/h
t_F = duration of filter run, h
t_{BW} = duration of backwash cycle, h
t_{FTW} = duration of filter-to-waste period, h
a = filter cross-sectional area, m^2

The ratio of net to total water filtered is the recovery:

$$r = \frac{V_F - V_{\text{BW}} - V_{\text{FTW}}}{V_F} = \frac{\text{UFRV - UBWV - UFWV}}{\text{UFRV}} \qquad (11\text{-}69)$$

where r = recovery, expressed as a fraction

Filters should be designed for a recovery of at least 95 percent. Typical wash water quantities are about 8 m^3/m^2 (200 gal/ft^2). Thus, to achieve a recovery greater than 95 percent, a UFRV of at least 200 m^3/m^2 (5000 gal/ft^2) is required. Calculation of the parameters for net water production is demonstrated in Example 11-8.

Utility operators sometimes use excessive backwash rates or time in the belief that cleaning the media thoroughly will result in longer filter runs or lower effluent turbidity. Excessive backwashing, however, is counterproductive because it lowers the recovery and can result in longer ripening periods, which further reduce recovery.

Process Design Criteria

Design criteria are established to meet the performance requirements, given the source water quality and site-specific constraints.

Example 11-8 Calculation of parameters for net water production

A filter is operated at a rate of 12.5 m/h for 72 h, of which 30 min was discharged as filter-to-waste volume. After filtration, it is backwashed at a rate of 40 m/h for 15 min. Calculate the UFRV, UBWV, UFWV, and recovery.

Solution

1. Calculate UFRV using Eq. 11-66:

$$\text{UFRV} = (12.5\ \text{m/h})(72\ \text{h}) = 900\ \text{m} = 900\ \text{m}^3/\text{m}^2$$

2. Calculate UBWV using Eq. 11-67:

$$\text{UBWV} = (40\ \text{m/h})(0.25\ \text{h}) = 10\ \text{m} = 10\ \text{m}^3/\text{m}^2$$

3. Calculate UFWV using Eq. 11-68:

$$\text{UFWV} = (12.5\ \text{m/h})(0.5\ \text{h}) = 6.25\ \text{m} = 6.25\ \text{m}^3/\text{m}^2$$

4. Calculate recovery using Eq. 11-69:

$$r = \frac{(900 - 10 - 6.25)\ \text{m}^3/\text{m}^2}{900\ \text{m}^3/\text{m}^2} = 0.982 = 98.2\%$$

FILTER TYPE

Classifications of rapid filters were presented on Fig. 11-3 (by level of pretreatment) and in Table 11-1 (by type of media). The level of pretreatment is typically based on raw-water quality. In borderline cases, pilot studies can verify whether a lower level of pretreatment is effective (i.e., direct filtration instead of conventional treatment).

The selection of monomedia versus dual-media filters is also based on raw-water quality and pilot study data. Deep-bed monomedia anthracite filters have a low rate of head loss accumulation, a high capacity for solids retention, and long filter runs. Dual-media filters, however, can provide a more robust design when filters are subjected to influent water that has not been properly conditioned, as was shown with the RT model on Fig. 11-12a. Thus, monomedia filters may be appropriate in situations when the raw-water quality is fairly constant and predictable, and dual-media filters may be more appropriate for variable water supplies. Deep-bed dual-media filters might be an attractive option when robustness and long filter runs are needed with variable water supplies.

FILTRATION RATE

Filtration rate influences the required area of the filter beds, clean-bed head loss, rate of head loss accumulation, distribution of solids collection in the bed, effluent water quality, and run length. Filters are typically designed to treat the maximum plant capacity at the design filtration rate with at least one filter out of service for backwashing. A low filtration rate increases the capital cost because it increases the required area of the filter beds, whereas a high filtration rate can increase the clean-bed head loss and decrease the length of the filter runs. Typically, the highest filtration rate that yields good filter performance is recommended. In filtering floc resulting from alum or ferric coagulation with polymeric coagulant aids, reasonable filter run lengths with no degradation of effluent quality can generally be achieved up to 25 m/h (10 gpm/ft^2). Filter effluent quality tends to degrade at filtration rates above 12.5 m/h (5 gpm/ft^2) with weak chemical floc such as alum floc without polymer or poorly flocculated biological floc. Higher filtration rates tend to increase solids penetration (if media size is properly selected), and the rate of head loss accumulation may be slower at higher rates because of more efficient use of the filter bed.

Filtration rates are often subject to regulatory limits. For instance, West Coast states restrict the filtration rate to 15 m/h (6 gpm/ft^2) or less, unless pilot testing demonstrates a higher rate is justified (Kawamura, 1999). Most rapid-filtration plants have design filtration rates between 5 and 15 m/h (2 and 6 gpm/ft^2), although some high-rate rapid filters have been constructed with filtration rates as high as 33 m/h (13.5 gpm/ft^2).

NUMBER AND DIMENSIONS OF FILTERS

The number of filters is influenced by the overall capacity of the plant, the maximum dimensions of a single filter, the effect of filtration rate changes during backwashing, and economic considerations. Most water treatment plants have a minimum of four filters, although small plants may have as few as two.

A small number of filters can reduce cost by minimizing the number of components, but a large number of filters minimizes the filtration rate change on the remaining filters when one is taken out of service for backwash. With only two filters, the filtration rate in one filter would double when the other was backwashed. As noted in Sec. 11-5, significant changes in the filtration rate can have adverse effects on filter performance by causing particle detachment and increasing effluent turbidity.

The maximum dimensions of a single filter are generally determined by the economic sizing of the filter backwash facility and possible difficulties in providing uniform distribution of backwash water over the entire filter bed. The practical maximum size of a typical high-rate gravity filter is about 100 m^2 (1100 ft^2).

AVAILABLE HEAD

The head for filtration is the difference between the clean-bed head loss and the available head in the structure. Because rapid filters typically operate by gravity, the available head is dependent on the elevation of the filter building relative to upstream and downstream structures (sedimentation basins and clearwells). Selection of the available head involves a trade-off between longer filter runs (greater available head) and economics (smaller available head). Due to construction costs, filter designs rarely provide more than 2 to 3 m (6.5 to 10 ft) of available head through the filter bed.

FILTER MEDIA

The selection of filter media is critical to meeting the performance criteria established for the treatment plant. Selection of filter media involves a trade-off between filtration efficiency (smaller media captures particles better) and head loss (larger media minimizes head loss).

The primary design criteria for filter media are the ES and UC. As noted earlier, a low UC is important for effective utilization of the entire filter bed. The ES must be selected in concert with other design parameters, such as filtration rate and filter depth. Factors such as clean-bed head loss as calculated in Sec. 11-4 must be considered along with effluent quality.

Filter depth and media size are interrelated. Some design engineers recommend a rule-of-thumb relationship, the ratio of depth to effective size (L/d ratio), to be between 1000 and 2000. As described in Secs. 11-4 and 11-5, head loss and particle removal are not simple inverse relationships of media diameter. Thus, the L/d ratio can provide some general guidance on the adequacy of a particular design, but it cannot be used to predict that two filters with the same L/d ratio will perform identically, particularly if they are composed of different media.

Pilot Testing

As a result of many years of successful operation of rapid filters, it is frequently possible to design an effective rapid-filtration system using the principles presented in this chapter without the use of pilot testing. In situations where higher filtration rates are warranted, pilot testing may be necessary to verify acceptable performance or satisfy regulatory agencies.

A typical rapid-filtration pilot plant is shown on Fig. 11-14. The pilot equipment typically consists of a feed pump, acrylic columns with an inside diameter of 0.1 to 0.15 m (4 to 6 in), and associated piping, valves, and instrumentation. In general, the diameter of the column should be at least 50 times the diameter of the media (Lang et al., 1993). The most common variables to be considered in a filtration pilot study are media size, media depth, and filtration rate. In addition, mono- and dual-media filters are often compared in pilot studies. For each pilot experiment, it is important to collect data on (1) clean-bed head loss, (2) filtration rate, (3) duration and magnitude of ripening, (4) influent and effluent turbidity, (5) water temperature, (6) time to breakthrough, and (7) head loss at the end of the filter run. It is normally desirable to collect some of these data continuously, particularly the turbidity, filtration rate, and head loss data.

Examples of pilot scale filter columns
http://www.stantec.com/wt3r/11-3.html

Figure 11-14 Typical rapid granular filter pilot plant and associated facilities. The filters are the tall Plexiglas columns near the front of the photograph.

Flow Control

Flow control is an important part of any filter system. Filter flow control can be accomplished in a variety of ways, and texts and design manuals describe four or five different control strategies (Cleasby and Logsdon, 1999; Kawamura, 2000). All systems for flow control have advantages and disadvantages but must accomplish three objectives: (1) control the filtration rate of individual filters, (2) distribute flow among individual filters, and (3) accommodate increasing head loss. Figure 11-15 shows several options for filter control systems.

Three basic methods are used for controlling filtration rates and distributing the flow to individual filters: (1) modulating control valve, (2) influent weir flow splitting, and (3) declining-rate filtration (no active flow control or distribution). The features of these methods are described in Table 11-6.

As noted earlier, the total available head in a gravity rapid filter system is fixed by the water elevation in the upstream and downstream structures (i.e., sedimentation basins and clearwells). The head loss through the filter bed increases as the filter collects solids, so provisions must be made to accommodate the variation in head loss. Three basic strategies are used: (1) maintain constant head above the filter (e.g., constant water level) and vary the head in the filter effluent by modulating a control valve;

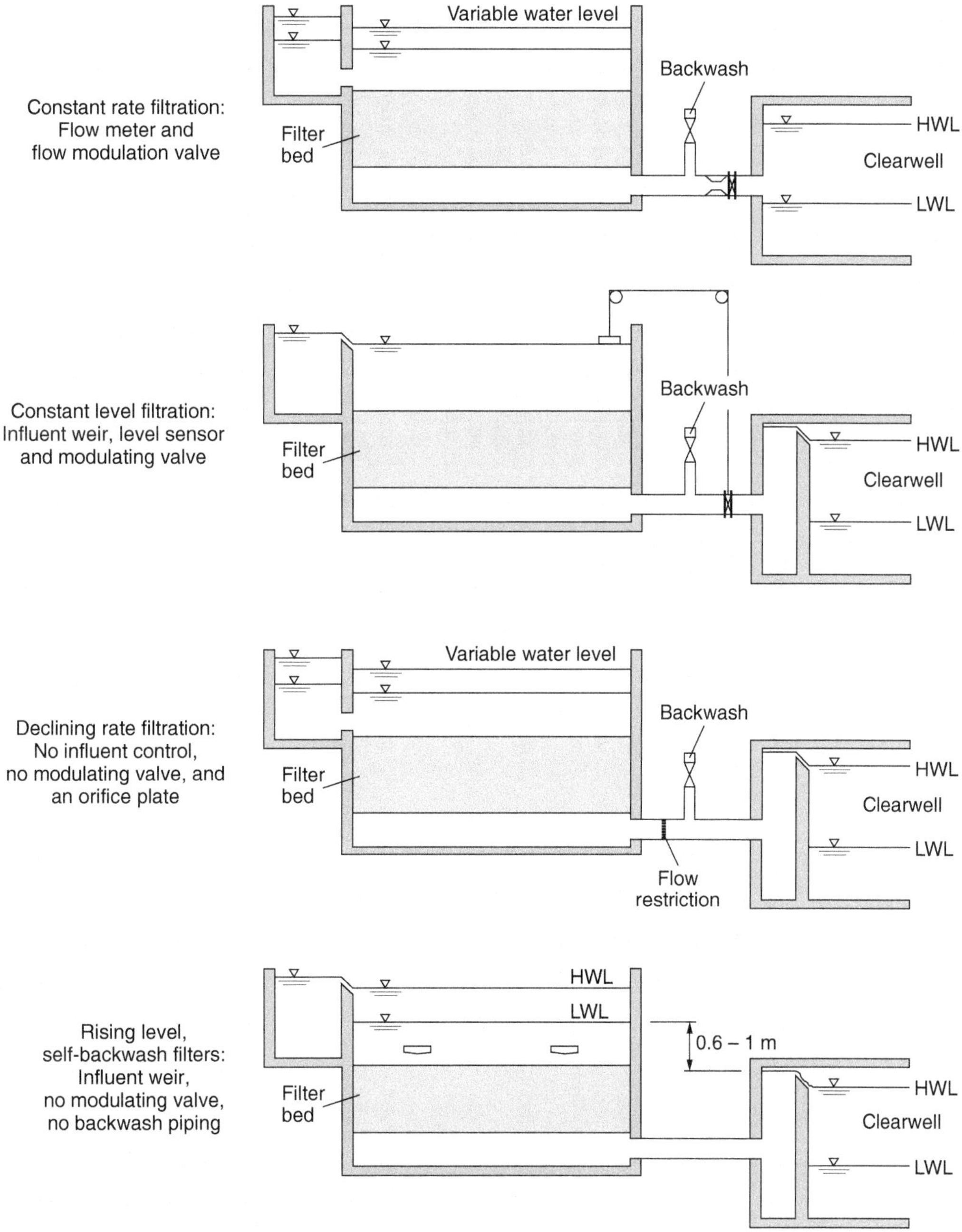

Figure 11-15
Rapid granular filter flow control strategies.

Table 11-6
Filter control options for distributing flow to filters and controlling filtration rate

	Modulating Control Valve	Influent Weir Flow Splitting	Declining Rate Filtration
Description	Effluent piping from each filter has a flowmeter and modulating control valve. The control system calculates a flow setpoint for each filter (from total flow and number of filters in service, or an influent level indicator) and automatically adjusts the control valve until the signal from the flowmeter matches the setpoint.	Water from a common influent channel flows to each filter over a weir. Water splits equally to all operating filters. When a filter is removed from service for backwashing, the flow is automatically distributed to the remaining filters.	Water from a common influent channel flows to each filter without any devices to measure or control the flow. All filters operate at the same head loss but different filtration rates. The flow is greatest in the cleanest filters and declines as solids accumulate. The common water level rises as all filters accumulate solids and then falls after the dirtiest (i.e., longest operating) filter is backwashed. After backwash, the filter that had been producing the lowest flow now produces the highest flow.
Advantages	Maximum flexibility and control. No need for influent weirs or orifices that can cause turbulence that can break fragile floc and degrade filter performance.	Simple and effective. No sophisticated instrumentation required. Total flow always divided evenly with each filter having the same filtration rate. Filtration rate changes gradually as water level rises or drops.	Simple. No instrumentation required. May be appropriate in developing countries or remote locations.
Disadvantages	Greatest complexity. Without proper maintenance, components can break down and lead to poor filter control. Valve movement in discrete steps can repeatedly overshoot target position, resulting in the valve "hunting" for the right position and causing large swings in the filtration rate, which can cause particle detachment and an increase in effluent turbidity (see Sec. 11-5).	Floc may break up and degrade filter performance if floc is fragile or if drop over weir is large (drop can be minimized by proper design and floc can be strengthened by using polymers during coagulation or flocculation).	No indication of the flow rate in any individual filter is provided. No way to control the filtration rate. The filtration rate at the beginning of a filter run may greatly exceed the design filtration rate. The high filtration rate during the filter-to-waste (FTW) period can waste water. Flow restrictors in FTW piping can reduce water loss but may cause particle detachment during increase in filtration rate with the FTW period is over.

(2) maintaining constant head in the filter effluent and vary the head upstream of the filter (allowing the water level to rise); and (3) maintaining nearly constant head loss and allowing the filtration rate to decline as solids accumulate in the bed (declining-rate filtration). Declining rate filtration was described in Table 11-6, and the features of the other two methods are described in Table 11-7.

It should be evident from Tables 11-6 and 11-7 that all flow control methods have advantages and disadvantages. No method is clearly superior

Table 11-7
Filter control options for accommodating the increase in head loss in a rapid granular filter during filtration

	Constant Level	Rising Water Level
Description	The water column above the filter is maintained near the maximum level. When head loss in the filter bed is low (immediately after backwash), an effluent valve is maintained in a nearly closed position. As head loss builds up in the filter, the valve gradually opens so that the total head loss across the filter bed and effluent valve stays constant. The effluent valve is controlled by an effluent flowmeter or a level transmitter. Head loss is monitored with a differential-pressure transmitter.	Filter effluent piping is configured to keep the filter bed submerged, typically with a filter effluent control weir with an elevation just above the top of the filter bed. When head loss in the filter bed is low, the water level is just above the top of the filter bed. As head loss builds up in the filter, the water level in the filter box gradually increases.
Used with (see Table 11-6)	Influent weir flow splitting. Modulating control valve.	Influent weir flow splitting.
Advantages	The filter box can be shorter (minimizing construction costs) than for rising water level systems because no head is lost over an effluent control weir, and even more height can be saved if there is no influent flow splitting weir. Minimizes cascade over the influent weir, when one is present.	Simple. No individual flow meters or modulating control valves on filters. Changes in filtration rate occur gradually because it takes time for the water level in the filter box to rise or fall.
Disadvantages	With no effluent control weir, poor design can lead to potential for negative pressure in the filter bed (see section on negative pressure in filter beds).	The influent weir must be above the maximum water level in the filter boxes to feed all filters equally. If the influent water cascades into the filter box from a large height, the top of the filter bed may be disturbed or the floc may be broken up by turbulence.

to the others. Selection is typically made on the basis of designer and owner preferences. Cost, complexity, and reliability are important issues. Whichever method is used, proper design is important because poor flow control can have a significant negative impact on filter performance.

Backwashing Systems

Backwashing is an indispensable part of rapid filtration. Improper or inadequate backwashing is one of the most frequent causes of problems in filters. Backwash criteria are established based on the flow rates necessary to fluidize the media and carry away deposited solids. The design equations for calculating these backwash rates were presented in Sec. 11-4. Alternatives for backwashing and backwash water delivery systems are discussed below.

ALTERNATIVES FOR BACKWASHING

Backwashing consists of upflowing water and a supplemental scouring system. The typical options for supplemental scouring systems are (1) fixed nozzle surface wash, (2) rotating-arm surface wash, and (3) air scour. Supplemental scouring causes vigorous agitation of the bed and causes collisions and abrasion between media grains that break deposited solids loose from the media grains. Once the solids are separated from the media grains, the upflowing wash water can flush the solids from the filter. Design criteria for water-jet-type surface wash systems and air scour washing systems are shown in Table 11-8.

SURFACE WASH SYSTEMS

Surface wash systems typically have water nozzles on a rotating header or on a fixed pipe grid located just above the surface of the bed. Subsurface

Table 11-8
Typical design criteria for supplemental backwash systems

Criteria	Units	Fixed-Nozzle Surface Wash	Rotating-Arm Surface Wash	Air Scour
Surface wash water flow rate	m/h gpm/ft^2	7–10 2.8–4	1.2–1.8 0.5–0.7	— —
Air flow rate	m^3/m^2 · h scfm/ft^2	— —	— —	36–72 2–4
Pressure at discharge point	bar psi	0.5–0.8 7.2–11.6	5–7 73–100	0. 3–0.5 4. 3–7.3
Duration of washing	min	4–8	4–8	8–15
Backwash water flow rate	m/h gpm/ft^2	30–60 12–24	30–60 12–24	15–45 6–18

agitators or dual-arm agitators (with one set of nozzles above the bed and a second set located near the interface in a dual- or multimedia bed) are also available but are not common. Surface wash systems typically start operation 1 to 2 min before the backwash water starts flowing and continue for 5 to 10 min after the bed is fluidized. As the media fluidizes, it rises above the level of the nozzles, so the surface wash system is able to provide vigorous agitation of the fluidized media. Surface wash systems are effective for cleaning traditional filters with depths of 0.6 to 0.9 m (2 to 3 ft) but are less effective for cleaning deep-bed filters.

AIR SCOUR SYSTEMS

Air scour systems are necessary for cleaning deep-bed filters. Air and water are introduced simultaneously at the bottom of the filter bed for a portion of the backwash cycle followed by a water-only wash for the remainder of the cycle. The method of air introduction depends on the type of underdrain and support system used. Most modern underdrain systems can introduce air through the underdrains along with the water. When the filter design includes support gravel, however, air must be introduced through a piping system located just above the support gravel to prevent dislodging the gravel.

The most effective air scouring occurs when the water is flowing between 25 and 50 percent of the minimum fluidization velocity (Amirtharajah, 1993). At this water flow rate, the air forms cavities within the media that subsequently collapse (a phenomenon that has been called *collapse pulsing*), causing a substantial amount of agitation of the bed. With no water flow, air moves through the media as bubbles or channels with little movement of the media. At water flow rates above 75 percent of the minimum fluidization velocity, the air moves with the water as bubbles.

Air scour provides such vigorous agitation that media can be lost if air is flowing while waste wash water is being discharged from the filter. Thus, the common procedure for using air scour is to (1) drain the water to a level about 150 mm (6 in.) above the top of the media, (2) start the water air at appropriate rates for collapse pulsing, (3) continue the air scour while the water level gradually rises in the filter box, (4) terminate the air flow rate just before the water level reaches the lip of the wash water troughs, (5) increase the backwash water rate to a fluidization velocity and continue to wash the filter for several more minutes to flush solids from the bed, and (6) terminate backwash water slowly to allow dual-media filters to restratify. This procedure typically allows several minutes for air scour, which is sufficient for cleaning the bed. Additional details of the air scour process are available in Amirtharajah (1993).

BACKWASH WATER DELIVERY SYSTEMS

Backwashing requires a large volume of water to flow through the filter in a short time. Backwash rates typically range from 30 to 60 m/h (12 to

24 gpm/ft^2) for 10 to 20 min. Backwash water can be delivered to the filter through one of three methods: (1) backwash pumps, (2) an elevated backwash water tank, or (3) a head difference between the effluent channel and filter box. These systems are described in Table 11-9. Most filters require between 2 and 4 m (6.6 and 13 ft) of static head at the filter bottom, although many backwash systems are designed to provide up to 10m (33 ft) of head at the pump or elevated tank, with the remaining head being dissipated by delivery piping, a throttling valve, and a flow controller to ensure a relatively constant backwash rate.

Table 11-9
Backwash water delivery systems

Water Supply	Backwash Pumps	Elevated Tank	Effluent Channel (Self-Backwashing Filter)
Description	Pumps, sized to provide the entire backwash flow, withdraw water from the filter effluent channel or finished water clearwell and provide it directly to the filter bottom.	Small pumps withdraw water from the filter effluent channel or finished water clearwell and send it to an elevated tank. The minimum water level in the tank is typically 9–12 m (30–40 ft) above the filter media. During backwash, water flows from the tank to the filters by gravity.	Filter effluent flows to a common effluent channel whose water elevation is controlled by a weir set several feet above the top of the media. During backwash, the water level in the filter box drops so that the head in the effluent channel is sufficient to provide the necessary backwash flow. This type of filter is often called a self-backwashing filter.
Advantages	Provides the maximum amount of control over backwash flow rates.	Smaller pumps are required because the volume of water required for backwash can be pumped to the tank over a period of hours.	Simplicity of design and operation; no pumps required.
Disadvantages	Large pumps are required.	Backwash flow rate can decline as water level in the elevated tank declines.	Less control over backwash flow rates. Deep filter box is required to provide sufficient head for filtering (maximum water level in filter to water level in effluent channel) and sufficient head for backwashing (water level in effluent channel to minimum water level in filter).

Filter System Components

Filters can be designed with a wide variety of configurations and various alternatives for the positioning of influent and effluent channels and piping. Detailed design of the structural aspects of filters is beyond the scope of this text, and students are referred to several references for detailed design, such as Kawamura (1975a, b, c, 1999, 2000). The primary components of a filter, other than the media, control system, and backwashing systems, are the underdrains and wash troughs.

FILTER SUPPORT MEDIA AND UNDERDRAINS

The function of filter underdrains is to support the filter media, collect and convey filtered water away from the filter system, and distribute backwash water and air. The underdrains must capture and distribute water uniformly to avoid localized variations in filtration rate or backwash rate that would jeopardize the effectiveness of the filter. Historically, a common design was a grid of perforated pipes overlain by several layers of gravel. The perforated pipe collects filtered water and distributes backwash water, and the gravel prevents the filter media from entering the perforations and provides additional distribution of backwash water. The gravel is installed in several layers, each 75 to 150 mm (3 to 6 in.) thick. The gravel on the bottom is typically 40 to 60 mm (1.5 to 2.5 in.) in diameter, and each overlying layer has stones of smaller diameter, with the top layer having an effective size of 0.8 to 2.0 mm (known as torpedo sand). Each layer is able to physically retain the overlying layer (recall from Fig. 11-8 that smaller grains can be physically retained by larger grains if the ratio of sizes is larger than about 0.15). A wide variety of other systems have been used, including false bottoms with strainers, underdrain blocks, precast concrete underdrains, teepee-type underdrains, and porous plates. Examples of underdrains are shown on Fig. 11-16. Uniform backwash flow distribution, durability, and cost are the three most important factors in selecting filter underdrains.

(a) (b) (c)

Figure 11-16
Typical filter underdrains: (a) Type S and SL with and without integral media support (IMS) cap (courtesy F. B. Leopold Company, Inc.), (b) direct retention underdrain system (courtesy Johnson Screens), and (c) nozzle-type underdrain system (courtesy of Ondeo-Degremont).

Modern underdrains typically have porous plates or fine mesh screens that can retain the filter media directly without the layers of gravel. Eliminating the gravel reduces the height of the filter box by about 0.5 m (1.6 ft). In addition, gravel can be dislodged by surges in backwash water or air scour.

To achieve an even distribution of backwash flow, (1) the orifice openings should be small enough to introduce a controlling loss of head and (2) the flow velocity in the pipe or channel in the underdrain system should be reasonably low and uniform throughout the entire filter area. Head loss in the underdrain system during backwash ranges from 0.1 to 3 m (0.3 to 10 ft) depending on the type of underdrain and backwash rates. For a false-bottom-type underdrain system, the required head loss is low (0.1 m for some systems) because the pressure is constant throughout the plenum if the inlet is properly designed. Perforated pipe grid systems use small orifices to create the necessary head loss to provide even distribution of backwash water.

WASH TROUGHS

Wash troughs provide a channel to collect the waste washwater so that dislodged suspended matter will be carried away without losing filter media. Backwash troughs are nearly universal in filtration plants in the United States but are typically not used in Europe, where waste washwater flows over a single overflow weir or side-channel weirs.

Wash troughs are generally of two basic types: (1) troughs with a shallow but wide cross section and a slight V-shaped bottom and (2) deeper troughs with a U-shaped cross section. The bottom of the wash trough should not be flat because froth and suspended matter are often trapped under the trough bottom and may never be washed out. Typical wash troughs are shown on Fig. 11-17. If the troughs are located close to the top of the filter bed, it may be difficult to provide sufficient depth to fluidize the media without losing media. In these cases, troughs can be modified to retain media while allowing suspended solids to flow to the wash troughs (Kawamura, 2000; Kawamura et al., 1997).

The troughs should have enough capacity to carry the maximum expected wash rate without flooding. A uniform flow over the lip of the trough can be guaranteed if the trough lips behave as free-flowing weirs along their entire length. Weirs should also provide a free-fall to the main collection outlet gullet. The bottom of the trough may be either horizontal or sloping. Spacing and dimensions of troughs can be obtained from standard design texts (Kawamura, 2000) or from manufacturer's literature.

Negative Pressure in Filter Beds

During filtration, the hydraulic gradient (head loss per unit depth) can be greater near the top of the bed because of the greater collection of solids near the top of the bed (Darby and Lawler, 1990). If the hydraulic gradient is greater than the static head gradient, low or even negative pressure

(a) (b)

Figure 11-17
Typical filter washwater troughs: (a) plastic adjustable type and (b) cast-in-place concrete type (see also Figure 1-1e in Chap. 1).

(below atmospheric pressure) can develop in the filter bed. The pressure within a filter bed and the potential for negative pressure development are depicted on Fig. 11-18. Negative pressure can cause bubbles to form as dissolved gases (oxygen and nitrogen) and come out of solution. Bubbles can be trapped by the media and cause a dramatic increase in head loss, a phenomenon called air binding.

Air binding can be avoided with proper filter design. A weir in the effluent channel that maintains sufficient water depth over the media can prevent the problem, but with proper understanding of the hydraulics of the filtration process, it is not necessary to provide an excessive water depth (Monk, 1984).

Residual Management

Filters produce two waste streams. The filter-to-waste water is water that has been fully treated through the water treatment plant but does not meet effluent requirements such as turbidity. Because filter-to-waste water is fully treated, it may be recycled to the head of the plant for retreatment instead of being sent to the waste washwater recovery system. The second waste stream is the waste washwater produced from filter backwashing, which contains the accumulated solids from a filter run and can have significant concentrations of microorganisms such as *Giardia* and *Cryptosporidium*. Waste washwater is regulated by the Filter Backwash Recycling Rule (U.S. EPA, 2001) and should typically be treated before being recycled to the head of the plant or discharged to a receiving stream. Additional details on the treatment and ultimate disposition of filter waste streams are discussed in Chap. 21.

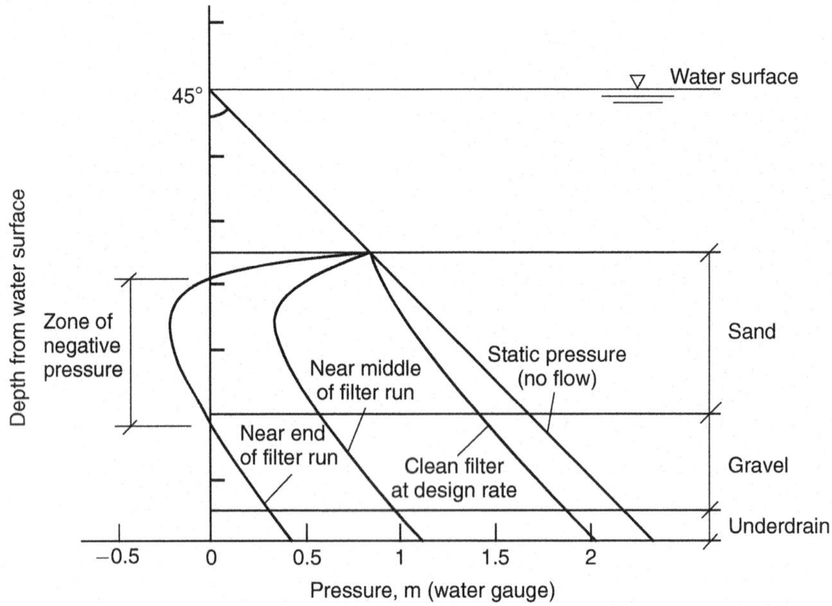

Figure 11-18
Pressure development within filter bed during filtration. (Adapted from Kawamura, 2000).

11-7 Rapid Filter Design Example

A new water treatment plant is to be built to treat water from a river in Arkansas. The average turbidity in the river is typically about 40 NTU but will frequently spike to 150 NTU during storm events. The plant is to have a capacity of 230,000 m^3/d (60 mgd). The owners and engineers have set the target effluent turbidity to be 0.1 NTU. Other treatment plants upstream of the proposed site have been successful at treating the water with dual media sand/anthracite filters. Pilot testing similar to that presented in Example 11-7 has been conducted.

The process design criteria that need to be determined are:

1. Type of filtration process (conventional, direct, or contact filtration)
2. Type of filter bed (monomedia or dual media)
3. Type of flow control
4. Media size (ES and UC)
5. Media depth
6. Filtration rates with all filters in service and with one out of service
7. Available head
8. Number and size of filters
9. Backwash rate
10. UFRV and recovery

Solution

Pretreatment requirements for the filtration process can be influenced by raw water quality, regulatory agency requirements, and the experience and preference of the operators. In this case, the turbidity spikes to 150 NTU dictate that the appropriate rapid granular filtration process is conventional filtration (see Fig 11-3). Based on successful pilot testing, deep-bed monomedia filters containing 1.8m (6 ft) of anthracite are selected.

As noted earlier, several options are available for flow control in filters and no one method has clear advantages over the other methods. Selection is based on engineer and owner preferences with respect to cost, complexity, and reliability. For this design, the engineers have chosen the influent weir split, constant level flow control system.

The principles presented in this chapter demonstrate that media effective size, media depth, filtration rate, and available head are interrelated. None of these design parameters can be set without considering the impact of the others. The goal for selecting these parameters is to minimize capital and operating costs while achieving the effluent turbidity goal. The regulatory limit on turbidity is 0.3 NTU but a typical goal is to keep effluent turbidity below 0.1 NTU. Many combinations of size, depth, and rate can achieve this turbidity goal. Capital costs are reduced by minimizing the required filter area, which is accomplished by operating at the highest possible filtration rate. The cost of media does not change significantly as the size changes. The depth of the bed is only a portion of the overall structure depth, so an increase in media depth causes only a moderate increase in capital cost. However, capital costs increase almost linearly as the filter area increases. Operating costs are minimized by decreasing the frequency of back- washing; thus, long filter runs are desirable.

Increasing the filtration rate will moderately reduce the time to breakthrough but significantly reduce the time to reach the available head. The reduced time to reach the available head at a higher rate is almost entirely due to the higher clean-bed head loss. Clean-bed head loss is also sensitive to media diameter. Thus, increasing the media diameter can compensate for the reduced run length resulting from a higher filtration rate. As noted earlier, smaller media captures particles better, so an increase in media effective size might reduce effluent quality. Equation 11-36 demonstrates that filter effluent turbidity improves as the media depth increases, so an increase in media effective size can be compensated for by increasing the media depth.

Using data from the pilot plant shown in Example 11-7 and the optimization graph shown on Fig 11-13, the engineers design the system with 2.5 m (8.2 ft) of available head. To get the longest filter runs at this condition, the optimal media effective size is about 1.0 mm. From the other pilot information, the filtration rate is 15 m/h (6.1 gpm/ft^2) and the media depth is 1.8 m (6 ft). With this filter bed design, filter runs of about 62 h are expected. The uniformity coefficient is specified to be below 1.4 to minimize stratification after backwashing.

With the media specifications and filtration rate set, the number of filters and filtration area can be determined. The required filter area is based on the capacity and filtration rate:

$$A = \frac{Q}{v_F} = \frac{230{,}000\,\text{m}^3/\text{d}}{(15\,\text{m/h})\,(24\,\text{h/d})} = 640\,\text{m}^2\left(6900\,\text{ft}^2\right)$$

Capital costs are minimized by using the lowest possible number of filters (reducing valves, piping connections, etc). Because the largest practical size of a filter is about 100 m^2 (1100 ft^2), the required filter area could be met with seven filters. It is necessary to operate at full capacity while one filter is backwashing or out of service for maintenance, so eight filters are required. The area of each filter is

$$A_F = \frac{230{,}000\ \text{m}^3/\text{d}}{(15\ \text{m/h})\ (24\ \text{h/d})\ (7)} = 91\ \text{m}^2\left(980\ \text{ft}^2\right)$$

The dimensions of individual filters are determined by options for components such as underdrains and wash troughs. For this plant, filters were configured to have two cells separated by a central gullet (the channel that provides water flow), with each cell being 10 m × 4.55 m (32.8 ft × 15 ft).

The effective filtration rate when all filters are in service is

$$v_F = \frac{230{,}000\ \text{m}^3/\text{d}}{(91\ \text{m}^2)\,(24\ \text{h/d})(8)} = 13.2\ \text{m/h}\left(5.4\ \text{gpm/ft}^2\right)$$

The slightly lower filtration rate when all filters are in service will have a positive effect on the length of filter runs.

Backwash flow requirements are determined using principles shown in Example 1 1-4. A 1. 8-m (6-ft) bed of 1.0 mm anthracite with a density of 1700 kg/m^3 and porosity of 0. 50 can be expanded by 25 percent at a temperature of 20°C using a normal backwash flow rate of 38.6 m/h (15.8 gpm/ft^2). The required backwash flow rate is

$$Q_B = v_{\text{BW}}A_F = (38.6\ \text{m/h})\left(91\ \text{m}^2\right) = 3500\ \text{m}^3/\text{h}(15{,}400\ \text{gpm})$$

Backwash pumps are frequently specified with additional capacity to accommodate occasional as more vigorous backwashing or changes in temperature. Twenty-five percent additional capacity is specified in this plant.

The unit filter run volume at design capacity is determined with Eq. 11-66:

$$\text{UFRV} = v_F t_F = (13.2\ \text{m/h})(62\ \text{h}) = 818\ \text{m}^3/\text{m}^2\left(20{,}000\ \text{gal/ft}^2\right)$$

Because a deep-bed monomedia filter is specified, collapse-pulse backwashing will be necessary. The expected volume of backwash water based on a backwash flow rate of 12 m/h for 5 min followed by a flow of 38.6 m/h for 10 min is 676 m^3 (23, 900 ft^3). Including a filter-to-waste period of 30 min (found during the pilot study), the net water recovery is determined from Eqs. 11-66 to 11-69:

$$\text{UFWV} = v_F t_{FTW} = (13.2\ \text{m/h})(0.5\ \text{h}) = 6.6\ \text{m}^3/\text{m}^2\left(162\ \text{gal/ft}^2\right)$$

$$\text{UBWV} = \frac{V_{\text{BW}}}{a} = \frac{676\ \text{m}^3}{91\ \text{m}^2} = 7.4\ \text{m}^3/\text{m}^2\left(182\ \text{gal/ft}^2\right)$$

$$r = \frac{818 - 7.4 - 6.6}{818} \times 100 = 98.3\%$$

The UFRV and recovery meet typical goals for granular media filters. The design parameters for this filter system are summarized in Table 11-10.

11-8 Other Filtration Technologies and Options

This chapter has focused on gravity-driven rapid filtration because that is the most common granular filtration technology used in water treatment. Several other filtration options, however, are used in specific applications. The following filtration options are briefly introduced in this section: pressure filtration, biologically active filtration, slow sand filtration, greensand

Table 11-10
Summary of design criteria for rapid filter design example

Parameter	Units	Value
Filter type	—	Conventional, deep-bed monomedia
Flow control	—	Influent weir split, constant level
Number	—	8
Inside dimensions	m · m	10× 4.55× 2 cells
Media surface area (each filter)	m^2	91
Media surface area (total)	m^2	728
Maximum available head	m	2.5
Filtration rate (at plant design flow rate)		
One filter off-line	m/h	15
All filters in service	m/h	13.2
Filter media		
Type	—	Anthracite
Depth	m	1.8
Effective size	mm	1.0
Uniformity coefficient	—	<1.4
Density	—	1700
Backwash criteria		
Maximum rate	m/h	48.2
Normal rate	m/h	38.6
Duration	min	15

filtration, diatomaceous earth filtration, and bag or cartridge filtration. Membrane filtration is discussed in Chap. 12.

Pressure Filtration

Pressure filtration is largely similar to gravity-driven rapid filtration with the exception that the filter is housed in a pressure vessel. Filter media used, pretreatment requirements, mechanisms for filtration, backwashing requirements, and other features of gravity-driven rapid filtration are applicable to pressure filters. Design equations and procedures are similar. An example of a pressure filter system is shown on Fig. 11-19. The primary advantage of a pressure filtration system is that the water remains under continuous pressure; that is, any excess pressure in the influent water (beyond that needed to overcome the head loss in the filter bed and piping) is available in the filtered water. Because of this, the filter effluent can be delivered to the point of use without additional pumping facilities.

A key disadvantage of pressure filters is that it is not easy to observe the filtration process, backwash process, or condition of the filter bed. Sudden changes in pressure can disturb the media and lead to channeling and poor filtration. Because poor filter conditions might be difficult to detect and compromise filtration effectiveness, some state regulatory agencies do not allow pressure filters to be used for surface water treatment, where prevention of waterborne illness is a primary concern (GLUMRB, 2007). Pressure filters can be suitable for groundwater applications. An example of an appropriate use for pressure filters is for iron and manganese removal from well water (see Chap. 20). The well pump can supply the head for the filter and deliver the water to the distribution system. Pressure filters are also used in tertiary wastewater treatment, swimming pools, and industrial applications.

Figure 11-19
Typical pressure filters. The vessels in the left foreground are pressure filters, right foreground are cartridge filters, and background are GAC contactors.

Biologically Active Filtration

Biologically active filtration, or biofiltration, incorporates biological activity into the granular filtration process. Heterotrophic bacteria colonize the surface of the media, forming a biofilm that is able to degrade some organic compounds and micropollutants, such as phenol, trichlorobenzene, ozonation by-products, ammonia, and odor-causing compounds such as geosmin and MIB. Biofiltration can improve water quality by decreasing the potential for bacterial regrowth in distribution systems, decreasing DBP formation during final disinfection, decreasing chlorine demand, and decreasing chlorine potential (Urfer et al, 1997).

Biofiltration is frequently used after ozonation because ozone can break down large recalcitrant humic acid molecules to smaller, more biologically degradable compounds. Without biofiltration, this increase in biodegradable organic matter (BOM) may encourage regrowth in the distribution system. With biofiltration, DOC removal of 35 to 40 percent may be possible, as shown on Fig 11-20. Thus, ozonation and biofiltration might be considered as a coupled process.

In Europe, dedicated biofilters using GAC media are sometimes installed following a conventional filtration process. In the United States, granular media filters are typically configured to accomplish both biodegradation and particle filtration in a single process. A thorough review of the use of biofiltration in water treatment is provided in Urfer et al. (1997).

Several factors affect the removal of BOM with biological filtration. These include (1) BOM type and concentration, (2) filter media type (i.e., GAC, anthracite, and/or sand), (3) water temperature, and (4) and empty bed contact time (EBCT) through the filter. Depending on these factors, steadystate biological performance will be reached within a maximum period of 1 to 2 months.

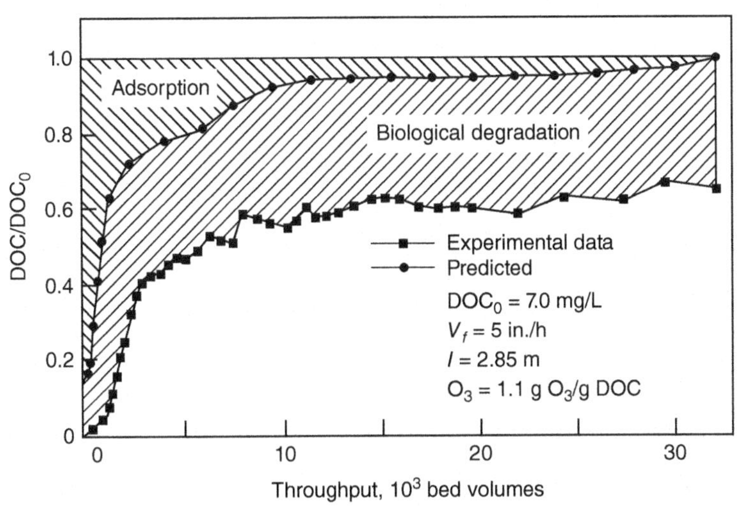

Figure 11-20
DOC removal by adsorption and biodegradation during biofiltration of an ozonated humic acid solution (Adapted from Sontheimer and Hubele, 1987).

BIOFILTRATION MEDIA

Sand, anthracite, and GAC have all been used as biofilter media and some studies suggest that they all perform similarly. Other studies suggest that GAC can be a more effective biofiltration media in some situations. The advantages of GAC appear to be that its irregular surface may provide a better attachment surface for bacteria, it may adsorb potentially inhibitory chemicals and thereby protect the biofilm, it may adsorb slowly biodegradable compounds that may then be degraded by the bacteria, and it can degrade oxidant residuals in the top few centimeters of the bed, thereby protecting the remaining biofilm from inadvertent exposure to oxidants such as chlorine or ozone. Because of these advantages, GAC may be able to establish a biofilm more rapidly, work better at colder temperatures, and be more robust following upsets or inadvertent oxidant exposure.

Results from studies comparing GAC and anthracite as biofilter media are shown on Fig. 11-21. Data shown on Fig. 11-21a were gathered under relatively warm temperature conditions of 10 to 15°C. The anthracite–sand filter performed as well as the GAC–sand filter for removal of oxalate, a common by-product of ozonation, regardless of EBCT. Similar performance in warm conditions was reported by Price et al. (1993) and Krasner et al. (1993). On the other hand, under cold temperature conditions as shown on Fig. 11-21b, the GAC–sand biofilter was still capable of removing a fraction of the oxalate (albeit with high EBCT values), while no removal was achieved with the anthracite–sand biofilter. The removal of total

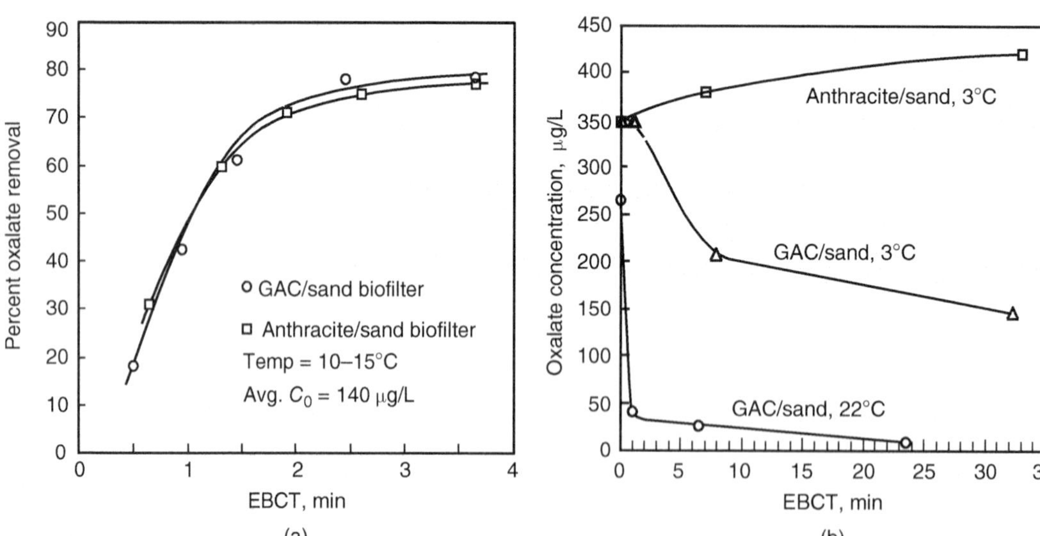

Figure 11-21
Impact of media type and EBCT on the removal of oxalate with biofiltration (a): under warm temperature conditions (Adapted from Coffey et al., 1997) and (b) under cold temperature conditions (Adapted from Emelko et al., 1997).

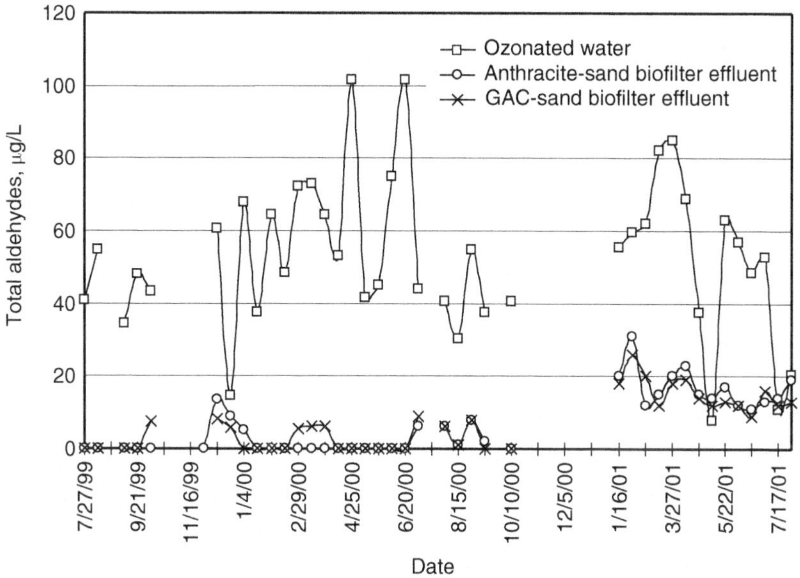

Figure 11-22
Removal of aldehydes with biological filtration at a full-scale water treatment plant (temperature = 12 to 24°C) (Data Courtesy of the Alameda County Water District, Fremont, California).

aldehydes (another common by-product of ozonation) with biofiltration in a full-scale water treatment plant in Fremont, California, is shown on Fig. 11-22. The 3-year operational data confirm that under relatively warm temperature conditions, anthracite–sand biofilters perform as well as GAC–sand biofilters.

BIOFILM MAINTENANCE AND BACKWASHING

A biofilm will establish itself in granular media as long as conditions (substrate, oxygen, temperature, etc.) are favorable. A key requirement is to prevent exposure to inhibitory substances; thus, the primary factor in operating a biofilter is to ensure that the filter influent has no chlorine or ozone residual.

Biofilters develop head loss similar to conventional filters, requiring backwashing on a regular basis. Numerous studies have indicated that biofilms adhere more strongly to filter media than nonbiological particles and that no major losses of biomass occur with proper backwashing procedures for particle removal. There are two types of concerns about backwashing of biological filters: (1) whether backwashing (especially with air scour) causes excessive detachment of the biomass off the filter media and (2) whether biofilters must be backwashed exclusively with nonchlorinated water to maintain their viability. Evaluations of these issues has concluded that rigorous backwashing does not adversely impact biofiltration performance (Miltner et al., 1995; Ahmad and Amirtharajah, 1995; Ahmad et al., 1998). While AOC removal was impaired immediately after backwashing with chlorinated water, its removal was back to normal levels within a few hours of filter run. Amirtharajah (1993) concluded that air scour helped control the long-term buildup of head loss in a biofilter, which was confirmed by Teefy (2001) who noted that occasional backwashing with chlorinated water was necessary at a full-scale water treatment plant to prevent excessive biological growth, which otherwise results in shorter filter runs. The biofiltration performance shown on Fig. 11-22 was achieved with filters backwashed with chlorinated water approximately every third wash.

FILTRATION PERFORMANCE

Biological filters will generally achieve good turbidity removal, and many studies have observed no difference in turbidity removal between biological and prechlorinated filters. In other cases, differences in particle counts were observed. Bacterial counts in filter effluent may be higher with biofiltration; but postdisinfection will be provided in either case. The rate of head loss buildup may be higher in biologically active filters, leading to somewhat shorter filter runs if the length of the filter run is limited by the available head. Temperature is also an important factor with respect to biological activity and will vary through the year depending upon the climate. For example, biological activity is usually high in summer months but then drops in the winter months (Servais and Joret, 1999).

DESIGN OF BIOFILTERS

When used for both biological activity and particle removal, design of the filtration aspects of a biofilter is essentially identical to that presented earlier in this chapter for rapid filters. Contact time, expressed as empty bed contact time (EBCT), is a primary design variable for biofiltration. Removal of BOM generally increases as EBCT increases. Removal of ozone by-products can generally be accomplished with an EBCT of 2 to 4 min, and most biofilters in the United States have EBCT between 1 and 15 min. An appropriate value for the EBCT can be determined during pilot testing.

Slow Sand Filtration

Slow sand filtration is substantially different from rapid filtration. Some of the most significant differences are that slow sand filters use sand media that does not need to be as uniform in size as in rapid filters, do not need coagulation pretreatment, and do not need backwashing. While these differences suggest that slow sand filters are simpler to build and operate than rapid filters, another key difference is that the filtration rate is 50 to 100 times lower, which means that slow sand filters require that much more land to treat the same amount of water. Other differences between rapid filtration and slow sand filtration are summarized in Table 11-11.

In slow sand filtration, the low filtration rate and the use of smaller, less uniform sand causes particles to be removed in the top few centimeters of the bed. The surface of the bed forms a mat of material, called a *schmutzdecke*. The schmutzdecke forms an additional filtration layer, physically straining smaller particles from the influent water. In addition, the schmutzdecke

Table 11-11
Comparison between rapid and slow sand granular filtration design criteria[a]

Process Characteristic	Slow Sand Filtration	Rapid Filtration
Filtration rate	0.08–0.25 m/h (0.03–0.10 gpm/ft^2)	5–15 m/h (2–6 gpm/ft^2)
Media effective size	0.15–0.30 mm	0.50–1.2 mm
Media uniformity coefficient	<2.5	<1.4
Bed depth	0.9–1.5 m (3–5 ft)	0.6–1.8 m (2–6 ft)
Required head	0.9–1.8 m (3–6 ft)	1.8–3.0 m (6–10 ft)
Run length	1–6 months	1–4 days
Ripening period	Several days	15 min–2 h
Pretreatment	None required	Coagulation
Dominant filtration mechanism	Straining, biological activity	Depth filtration
Regeneration method	Scraping	Backwashing
Maximum raw-water turbidity	10 NTU	Unlimited with proper pretreatment

[a]Values represent typical ranges. Some filters are designed and operated outside of these ranges.

forms a complex biological community that degrades some organic matter. Because particles are physically strained at the surface of the filter bed, destabilization by coagulation pretreatment is not necessary.

A typical configuration for a slow sand filter is illustrated on Fig. 11-23. Filter effluent passes through a support layer of graded gravel about 0.3 to 0.6m (1 to 2 ft) deep and is collected in an underdrain system constructed of perforated pipes or concrete blocks. The water level in the structure is several feet above the top of the media, with the maximum water level dictating the available head. The filtration rate is controlled by valves in either the influent or effluent piping.

OPERATING CYCLE FOR SLOW SAND FILTRATION

As with rapid filtration, slow sand filtration operates over a cycle with two stages, consisting of a filtration stage and a regeneration stage. Head loss builds slowly during a filter run that lasts weeks or months. Head loss builds slowly because of the low filtration rate and because the microorganisms degrade some of the accumulated particles. Slow sand filters typically never reach breakthrough and are always terminated when the head loss reaches the available head in the system, typically 0.9 to 1.8m (3 to 6 ft). Instead of being backwashed when the available head is reached, the filter is drained and the top 1 to 2 cm (0.4 to 0.8 in.) of media is scraped off, hydraulically

Figure 11-23
Typical slow sand filter.

cleaned, and stockpiled onsite. The filter is then placed back in service. The operation and scraping cycle can be repeated many times, often over a period of several years, before the sand must be replenished. When the sand reaches a minimum depth of 0.5 m (20 in.), the stockpiled sand is replaced in the filter to restore the original depth.

A filter with new media typically has a ripening period that can last several days, during which the schmutzdecke forms and the effluent quality improves. Filter-to-waste piping is provided to allow filtered water to be returned to the source during the ripening period. After several filter runs and scrapings, the microbial community can become established deeper in the bed and the ripening period can be shorter or nonexistent.

ADVANTAGES AND USE OF SLOW SAND FILTRATION

The primary advantage of slow sand filtration is that the filters are simple to operate and can run without constant supervision. Operators do not need to have knowledge of coagulation chemistry. Simple operational requirements are particularly attractive to small utilities that do not have the resources for full-time, highly trained treatment plant operators. The simplicity may make slow sand filtration appropriate in developing countries.

With no pretreatment, slow sand filters cannot adequately treat poor-quality surface waters. Slow sand filtration should be used only when source waters have turbidity less than 10 NTU, color less than 15 color units, and no colloidal clay (GLUMRB, 2007). While most slow sand filter plants treat water with less than 10 NTU of turbidity (Slezak and Sims, 1984), some research suggests that the upper limit should be 5 NTU (Cleasby et al., 1984).

For small facilities that have high-quality source water but want to avoid the use of coagulants, membrane filtration (see Chap. 12) or cartridge or bag filters should also be considered.

Slow sand filtration continues to be used successfully in Europe, including facilities supplying large communities such as London and Amsterdam (Joslin, 1997). However, it has been largely superseded by rapid filtration in the United States. A survey conducted in the early 1980s (Slezak and Sims, 1984) identified fewer than 50 operating slow sand filtration plants in the United States [for comparison, there are more than 50,000 community water systems in the United States (U.S. EPA, 2001)]. Most of the facilities listed in Slezak and Sims's (1984) survey were more than 50 years old and served populations of less than 10,000.

DESIGN OF SLOW SAND FILTERS

Slow sand filters can be housed in steel, fiberglass, or reinforced concrete structures. Details of slow sand filter design can be found in the literature, such as Hendricks et al. (1991), Visscher (1990), and Seelaus et al. (1986).

Greensand Filtration

Greensand filtration combines oxidation and filtration in a single granular media filtration process. The filtration media is coated with a layer of manganese dioxide, which oxidizes soluble iron, manganese, hydrogen sulfide, and other reduced species. The underlying media can be sand, anthracite, or naturally occurring glauconite mineral. After oxidation, iron and manganese precipitate and can be removed by the filter. The manganese dioxide coating must periodically be regenerated by feeding an oxidant to the filter; typically, potassium permanganate is used. Continuous regeneration can be practiced by feeding potassium permanganate continuously.

The design of the filtration process using greensand media is essentially identical to rapid filtration. The media is typically specified with an effective size of about 0.3, uniformity coefficient < 1.6, and depth similar to rapid filtration. Filtration rates typically range from 7.5 to 12.5 m/h (3 to 5 gpm/ft^2). Greensand filters can be designed as either gravity or pressure rapid filters. Additional information about the use of greensand filtration for iron and manganese removal is presented in Chap. 20.

Diatomaceous Earth Filtration

Diatomaceous earth (DE) filtration, also known as precoat filtration, uses a thin cake (2 to 5 mm) of fine granular material as a filter medium. Particle removal occurs primarily at the surface of the cake, with straining as the predominant removal mechanism.

Diatomaceous earth (also known as Fuller's earth) is the microscopic remnants of the siliceous shells of diatoms, occurs in natural deposits, and is almost pure silica. DE used for filter media typically has a diameter between 4 and 30 μm (Baumann, 1978). Smaller sizes produce higher quality effluent but at the expense of more head loss. The properties and characteristics of precoat media are covered in *ANSI/AWWA B101-01 Standard for Precoat Filtering Media* (AWWA, 2001b).

The DE filtration cycle has three stages: precoat, filtration, and backwash (AWWA, 1988). In the first stage, a slurry of DE is fed into the filter vessel and deposited on the septum, which is a porous plate or screen designed to support the precoat material. After precoat is complete, raw water enters the filtration vessel and filtration occurs across the precoat layer. Additional DE (called body feed) is added to the influent water during filtration. The body feed reduces the rate at which head loss builds up by maintaining the porosity of the cake and extends the length of the filter run. Run lengths range from 10 min to 30 days (Baumann, 1957).

Backwash starts after the pressure drop reaches the limiting head, typically 2 to 3 bars (29 to 44 psi). During backwash, water is pumped through the septum in reverse and the filter cake and all accumulated solids slough off. In some instances, a surface wash or agitation system is used to break up an encrusted filter cake, particularly in high-pressure filters. After backwash is complete, the entire cycle begins again.

DE filtration is not used extensively in drinking water treatment (AWWA, 1988), but some small utilities have used it for compliance with the SWTR and subsequent surface water treatment rules. DE filters strain particles larger than about 1 μm, so they can achieve high removal of *Giardia* and *Cryptosporidium* without coagulation. Under the SWTR and LT1ESWTR, precoat filters receive filtration credit for 2 log removal of *Giardia* and *Cryptosporidium* if the filtered water turbidity is equal to or less than 1 NTU (U.S. EPA, 1989, 2002). DE filters are less effective for particles smaller than 1 μm and should only be used for high-quality source waters (turbidity of 10 NTU or less). Additional information on precoat filtration is available in the published literature (AWWA, 1988; Baumann, 1965).

Bag and Cartridge Filtration

Bag and cartridge filtration are not granular filtration processes and therefore do not really belong in this chapter. They are, however, considered as alternatives for compliance for surface water treatment regulations and are discussed briefly here.

Bag and cartridge filters are filter elements installed inside a pressure vessel. Cartridge filters are typically a self-supporting filter element with either a pleated-fabric or string-wound construction. The flow is from the outside of the cartridge to the inside. Bag filters are a nonrigid, fabric filter with the flow from the inside of the bag to the outside. Although some cartridge filters are backwashable, both cartridge and bag filters are typically considered to be disposable media and are discarded when the head loss exceeds the available head. Thus, they are only suitable for small systems treating relatively high-quality water. Cartridge filter vessels are shown in Fig. 11-19.

Problems and Discussion Topics

11-1 Samples of filter media was sifted through a stack of sieves and the weight retained on each sieve is recorded below. For a given sample (A, B, C, D, or E, to be selected by instructor), determine the effective size and uniformity coefficient for the media.

Sieve Designation	Sieve Opening, mm	Weight of Retained Media, g				
		A	B	C	D	E
8	2.36	0				4
10	2.00	35	0			11
12	1.70	178	11		0	60
14	1.40	216	315		4	227
16	1.18	242	242		16	343
18	1.00	51	116	0	33	216
20	0.85	12	55	23	75	40
25	0.71	5	26	217	285	16
30	0.60	3	14	325	270	3
35	0.50	0	2	151	121	1
40	0.425		0	71	21	0
45	0.355			49	8	
50	0.300			4	3	
Pan	—			20	4	

11-2 Explain why a low uniformity coefficient is important in rapid filtration.

11-3 Explain (a) the process of ripening, (b) how ripening affects recovery, or net water production, and (c) how to minimize the duration of ripening.

11-4 A filter is designed with the following specifications. The anthracite and sand have density of 1700 and 2650 kg/m^3, respectively, and the design temperature is 10°C. For a given sample (A, B, C, D, or E; to be selected by instructor), calculate the clean-bed head loss.

Item	A	B	C	D	E
Bed type	Mono-media	Mono-media	Dual media	Dual media	Dual media
Filtration rate (m/h)	8	15	15	10	10
Anthracite specifications:					
Effective size (mm)		1.0	1.0	1.2	1.6
Depth (m)		1.8	1.5	1.4	1.2
Sand specifications:					
Effective size (mm)	0.55		0.5	0.55	0.55
Depth (m)	0.75		0.3	0.4	0.7

11-5 For the media specification given in Problem 7-4 (C, D, or E to be selected by instructor), determine if the two media layers are matched to each other.

11-6 A filter contains 0.55-mm sand that has a density of 2650 kg/m^3. Calculate the effective size of 1550 kg/m^3 anthracite that would be matched to this sand.

11-7 In dual-media filter containing sand and anthracite, which material will be the top layer? Why?

11-8 Using the sieve analysis from Problem 11-1, determine the clean-bed head loss through a stratified filter bed by calculating the head loss contribution from each layer of media. Assume that the media stratifies into layers based on grain size, that the depth of each layer is proportional to the mass of media retained on each sieve pan, that the grain diameter of the media in each layer is equal to the arithmetic average of two adjacent sieve pans (i.e., the layer formed by the 178 g of media in sample A that passed through the 2.0-mm pan and was retained on the 1.70-mm pan has an average size of 1.85 mm), and that the total head loss is the sum of the head loss from each layer. In addition, assume that the sand that passed through the smallest pan has an average grain diameter of 0.1 mm. The total bed depth is 0.9 m, the filtration rate is 10 m/h, and the temperature is 15°C.

a. Calculate the total clean-bed head loss using the entire sand sample in Problem 11-1, including the material smaller than the smallest sieve pan.

b. Calculate the total clean-bed head loss assuming that the top 5 percent of the filter bed has been scraped to remove fines.

c. Discuss the importance of scraping the surface of rapid-filter beds after media installation and the impact it has on clean-bed head loss.

11-9 Compare the clean-bed head loss at 15°C through a rapid filter with a filtration rate of 15 m/h to that through a slow sand filter with a filtration rate of 0.15 m/h for media with the following specifications: effective size 0.5 mm, density 2650 kg/m^3, depth 1 m, and porosity 0.42. What implications do these calculations have on the significance of clean-bed head loss in the design of rapid and slow sand filters?

11-10 Given a backwash flow rate of 45 m/h and temperature of 20°C, calculate the largest (a) sand particle (density = 2650 kg/m^3) and (b) floc solid particle (density = 1050 kg/m^3) that can be washed from a filter bed.

11-11 Calculate and plot the size of particles that will be washed from a filter as a function of backwash velocity ranging from 10 to

100 m/h at 20°C for (a) sand particles (density = 2650 kg/m^3), (b) anthracite particles (density = 1650 kg/m^3), and (c) floc solid particles (density = 1050 kg/m^3). In addition, calculate the minimum fluidization velocities for 0.5-mm sand (porosity 0.40) and 1.0 mm anthracite (porosity 0.50) and indicate these velocities on your graph. What is an appropriate range for the backwash velocity for dual-media filters? Assuming the backwash troughs are placed high enough, will any media be lost with backwash velocities in this range?

11-12 A monomedia anthracite filter is designed with the following specifications: effective size 1.1 mm, uniformity coefficient 1.4, and density 1650 kg/m^3.

a. Calculate backwash rate to get a 25 percent expansion at the design summer temperature of 22°C.

b. Calculate the expansion that occurs at the backwash rate determined in part (a) at the minimum winter temperature of 3°C.

c. Discuss the implications of these results on backwash operations for plants that experience a large seasonal variation in water temperature.

11-13 Using the Yao filtration model, examine the effect of filtration rate on filter performance for particles with diameters of 0.1, 1.0, and 10 μm. Assume a monodisperse media of 0.5 mm diameter, porosity 0.42, particle density 1020 kg/m^3, filtration rate 10 m/h, filter depth 1 m, temperature 20°C, and attachment efficiency 1.0. Plot the results as C / C_O as a function of filtration rate over a range from 1 to 25 m/h. Comment on the effect of filtration rate and particle size on filter performance.

11-14 Using the Tufenkji and Elimelech filtration model, examine the effect of water temperature on filter efficiency for particles with diameters of 0.1, 1.0, and 10 μm. Assume a monodisperse media of 0.5 mm diameter, porosity 0.42, particle density 1020 kg/m^3, filtration rate 10 m/h, filter depth 1 m, and attachment efficiency 1.0. Plot the results as log (C / C_O) as a function of temperature over a temperature range of 1 to 25°C. What implications do these calculations have on filtration in cold climates? Is temperature more important for filtration of certain particle sizes?

11-15 Using the Rajagopalan and Tien filtration model, calculate and plot the concentration profile of 4 μm particles (i.e., the size of *Cryptosporidium* oocysts), through monodisperse filter with 0.5-mm-diameter media under filtration conditions typical of rapid filtration (v = 10 m/h) and slow sand filtration (v = 0.1 m/h). Assume porosity 0.40, particle density 2650 kg/m^3, filter depth 1 m, and temperature 20°C. Assume an attachment efficiency of 1.0

for the rapid filter and 0.05 for the slow sand filter. Explain why rapid and slow sand filtration should be modeled with different values for the attachment efficiency. Plot the results as C/C_O as a function of depth. Using these results, comment on the methods used to restore the filtration capacity of slow sand and rapid filters (i.e., scraping vs. backwashing).

11-16 The results of pilot filter experiments are summarized in the tables below. For each set of experiments, the independent variable was either the media effective size or the media depth, as given in the second column below. For a given set of experiments (A, B, C, or D, to be selected by instructor), determine equations for how the specific deposit at breakthrough (σ_B) and the head loss rate constant (k_{HL}) each depend on the independent variable. Also determine the optimal value of the independent variable and the corresponding filter run duration. For all problems, assume $C_O = 2.0$ mg/L and $C_E = 0$ mg/L.

a. Design conditions: $v_F = 15$ m/h, media = anthracite, depth = 1.75 m, max available head = 2.8 m.

A Filter	Media ES, mm	Time to Breakthrough, h	Initial Head Loss, m	Head Loss When Breakthrough Occurred, m
1	0.8	112	0.65	4.6
2	1.0	85	0.39	2.9
3	1.1	72	0.33	2.4
4	1.2	71	0.30	2.0
5	1.4	58	0.24	1.5

b. Design conditions: $v_F = 15$ m/h, media = GAC, depth = 2.0 m, max available head = 3.0 m.

B Filter	Media ES, mm	Time to Breakthrough, h	Initial Head Loss, m	Head Loss When Breakthrough Occurred, m
1	0.83	54	0.65	6.1
2	1.05	43	0.40	4.3
3	1.25	38	0.33	2.9
4	1.54	32	0.22	2.0

c. Design conditions: $v_F = 33.8$ m/h, media = anthracite, ES = 1.55 mm, max available head = 3 m (adapted from pilot results for the LADWP Aqueduct Filtration Plant).

C Filter	Media ES, mm	Time to Breakthrough, h	Initial Head Loss, m	Head Loss When Breakthrough Occurred, m
1	0.6	4.0	0.16	1.0
2	1.0	6.7	0.30	1.7
3	1.8	11.9	0.50	3.2
4	2.0	13.4	0.58	3.6
5	2.2	14.5	0.65	4.1

d. Design conditions: $v_F = 25$ m/h, media = anthracite, ES = 1.50 mm, max available head = 3 m (adapted from pilot results for the Bull Run water supply)

D Filter	Media ES, mm	Time to Breakthrough, h	Initial Head Loss, m	Head Loss When Breakthrough Occurred, m
1	2.0	41	0.43	1.8
2	2.3	49	0.51	2.0
3	2.5	55	0.51	2.5
4	3.0	65	0.63	2.9

11-17 For the filter design selected in Problem 11-16, what is the UFRV at the design condition? If the filters are designed to be backwashed at 40 m/h for 15 min, what are the UBWV and recovery, assuming there is no filter-to-waste period?

11-18 A filter has been designed to have a run length of 48 h while operating at a filtration rate of 12 m/h. The design backwash rate is 40 m/h, backwash duration is 15 min, and filter-to-waste duration is 10 min. The plant operator decides to clean the filters more thoroughly and backwashes at a rate of 55 m/h for 25 min. As a result, ripening takes longer and the filter-to-waste duration is 50 min. Calculate the UFRV, UBWV, UFWV, and recovery (a) as designed and (b) as operated. What is the percent increase in the volume of treated water lost as waste backwash water and filter-to-waste water?

11-19 Explain (a) the importance of flow control in proper filtration operation and (b) the main types of flow control systems.

11-20 Discuss (a) the impact of rapid variations in filtration rate on filter performance, (b) causes of rapid variations in filtration rate, and (c) design features and operational methods for preventing rapid variations in filtration rate.

11-21 Discuss factors that influence the selection of the number of filters in a treatment plant design.

11-22 Discuss the benefits and salient features of air scour.

References

Ahmad, R., and Amirtharajah, A. (1995) "Detachment of Biological and Nonbiological Particles from Biological Filters During Backwashing," pp. 1057–1085, in *Proceedings of the AWWA Annual Conference*, Anaheim, CA, Volume on Water Research, American Water Works Association, Denver, CO.

Ahmad, R., Amirtharajah, A., Al-Shawwa, A., and Huck, P. M. (1998) "Effects of Backwashing on Biological Filters," *J. AWWA*, **90**, 12, 62–73.

Ahmed, N., and Sunada, D. (1969) "Nonlinear Flow in Porous Media," *J. Hydr. Div. ASCE*, **95**, 6, 1847–1857.

Akgiray, Ö., and Saatçi, A. M. (2001) "A New Look at Filter Backwash Hydraulics," *Water Sci. Technol. Water Supply*, **1**, 2, 65–72.

Amirtharajah, A. (1988) "Some Theoretical and Conceptual Views of Filtration," *J. AWWA*, **80**, 12, 36–46.

Amirtharajah, A. (1993) "Optimum Backwashing of Filters with Air Scour: A Review," *Water Sci. Technol.*, **27**, 10, 195–211.

Amirtharajah, A., and Raveendran, P. (1993) "Detachment of Colloids from Sediments and Sand Grains," *Colloids Surfaces A: Physicochem. Eng. Aspects*, **73**, 211–227.

ASTM (2001a) *C136-01 Standard Test Method for Sieve Analysis of Fine and Coarse Aggregates*, American Society for Testing and Materials, Philadelphia, PA.

ASTM (2001b) *E11-01 Standard Specification for Wire Cloth and Sieves for Testing Purposes*, American Society for Testing and Materials, Philadelphia, PA.

AWWA (1988) *Precoat Filtration*, AWWA Manual M30, American Water Works Association, Denver, CO.

AWWA (1996) *ANSI/AWWA B604-96 Standard for Granular Activated Carbon*, American Water Works Association, Denver, CO.

AWWA (2001a) *ANSI/AWWA B100-01 Standard for Filtering Material*, American Water Works Association, Denver, CO.

AWWA (2001b) *ANSI/AWWA B101-01 Standard for Precoat Filtering Media*, American Water Works Association, Denver, CO.

Bai, R. B., and Tien, C. (1999) "Particle Deposition under Unfavorable Surface Interactions," *J. Colloid Interface Sci.*, **218**, 2, 488–499.

Baker, M. N. (1948) *The Quest for Pure Water; The History of Water Purification from the Earliest Records to the Twentieth Century*, American Water Works Association, New York.

Baumann, E. R. (1957) "Diatomite Filters for Municipal Installations," *J. AWWA*, **49**, 2, 174–186.

Baumann, E. R. (1965) "Diatomite Filters for Municipal Use," *J. AWWA*, **57**, 2, 157–180.

Baumann, E. R. (1978) Precoat Filtration, p. 845, in R. L. Sanks (ed.), *Water Treatment Plant Design for the Practicing Engineer*, Ann Arbor Science, Ann Arbor, MI.

Bergendahl, J. A., and Grasso, D. (2003) "Mechanistic Basis for Particle Detachment from Granular Media," *Environ. Sci. Technol.*, **37**, 10, 2317–2322.

Carman, P. C. (1937) "Fluid Flow through Granular Beds," *Trans. Inst. Chem. Eng.*, **15**, 150–165.

Chang, Y-I., and Chan, H-C. (2008) "Correlation Equation for Predicting Filter Coefficient under Unfavorable Deposition Conditions," *J. AICHE*, **54**, 5, 1235–1253.

Chang, M., Trussell, R. R., Guzman, V., Martinez, J., and Delaney, C. K. (1999) "Laboratory Studies on the Clean Bed Headloss of Filter Media," *Aqua (Oxford)*, **48**, 4, 137–145.

Clark, M. M. (1996) *Transport Modeling for Environmental Engineers and Scientists*, Wiley-Interscience, New York.

Cleasby, J. L., and Fan, K.-S. (1981) "Predicting Fluidization and Expansion of Filter Media," *J. Environ. Eng.*, **107**, EE3, 455–471.

Cleasby, J. L., Hilmoe, D. J., and Dimitracopoulos, C. J. (1984) "Slow Sand and Direct In-line Filtration of a Surface Water," *J. AWWA*, **76**, 12, 44–55.

Cleasby, J. L., and Logsdon, G. S. (1999) Granular Bed and Precoat Filtration, Chap. 8, in R. D. Letterman (ed.), *Water Quality and Treatment: A Handbook of Community Water Supplies*, McGraw-Hill, New York.

Cleasby, J. L., and Woods, C. F. (1975) "Intermixing of Dual Media and Multi-Media Granular Filters," *J. AWWA*, **67**, 4, 197–203.

Coffey, B. M., Huck, P. M., Bouwer, E. J., Hozalski, R. M., Pett, B., and Smith E. F. (1997) The Effect of BOM and Temperature on Biological Filtration: An Integrated Comparison at Two Treatment Plants, paper presented at the American Water Works Association Annual Conference, Denver, CO.

Darby, J. L., Attanasio, R. E., and Lawler, D. F. (1992) "Filtration of Heterodisperse Suspensions. Modeling of Particle Removal and Head Loss," *Water Res.*, **26**, 6, 711–726.

Darby, J. L., and Lawler, D. F. (1990) "Ripening in Depth Filtration. Effect of Particle Size on Removal and Head Loss," *Environ. Sci. Technol.*, **24**, 7, 1069–1079.

Darcy, H. (1856) *Les fontaines publiques de la vile de Dijon* [in French], Victor Dalmont, Paris.

Dharmarajah, A. H., and Cleasby, J. L. (1986) "Predicting the Expansion Behavior of Filter Media," *J. AWWA*, **78**, 12, 66–76.

Elimelech, M. (1992) "Predicting Collision Efficiencies of Colloidal Particles in Porous Media," *Water Res.*, **26**, 1, 1–8.

Elimelech, M., and O'Melia, C. R. (1990) "Kinetics of Deposition of Colloidal Particles in Porous Media," *Environ. Sci. Technol.*, **24**, 10, 1528–1536.

Emelko, M. B., Huck, P. M., and Smith, E. F. (1997) Full-Scale Evaluation of Backwashing Strategies for Biological Filtration, paper presented at the American Water Works Association Annual Conference, Atlanta, GA.

Ergun, S. (1952) "Fluid Flow through Packed Columns," *Chem. Eng. Prog.*, **48**, 2, 89–94.

Fair, G. M., Geyer, J. C., and Okun, D. A. (1971) *Elements of Water Supply and Wastewater Disposal*, Wiley, New York.

Fair, G. M., and Hatch, L. P. (1933) "Fundamental Factors Governing the Streamline Flow of Water through Sand," *J. AWWA*, **25**, 11, 1551–1565.

Forchheimer, P. (1901) "Wasserbewegung durch Boden" [in German], *Forschtlft ver. D. Ing.*, **45**, 1782–1788.

Fuller, G. W. (1933) "Progress in Water Purification," *J. AWWA*, **25**, 11, 1566–1576.

GLUMRB (Great Lakes Upper Mississippi River Board). (2007) *Recommended Standards for Water Works (Ten State Standards)*, Health Research Inc., Albany, NY.

Hendricks, D. W., Barrett, J. M., and AWWA Research Foundation (1991) *Manual of Design for Slow Sand Filtration*, American Water Works Association Research Foundation, Denver, CO.

Hiemenz, P. C., and Rajagopalan, R. (1997) *Principles of Colloid and Surface Chemistry*, Marcel Dekker, New York.

Ives, K. J. (1967) "Deep Filters," *Filtration Separation*, **4**, 3/4, 125–135.

Iwasaki, T. (1937) "Some Notes on Sand Filtration," *J. AWWA*, **29**, 10, 1592–1602.

Joslin, W. R. (1997) "Slow Sand Filtration: A Case Study in the Adoption and Diffusion of a New Technology," *J. N. Engl. Water Works Assoc.*, **111**, 3, 294–303.

Kau, S. M., and Lawler, D. F. (1995) "Dynamics of Deep-Bed Filtration: Velocity, Depth, and Media," *J. Environ. Eng.*, **121**, 12, 850–859.

Kavanaugh, M., Evgster, J., Weber, A., and Boller, M. (1977) "Contact Filtration for Phosphorus Removal," *J. WPCF*, **49**, 10, 2157–2171.

Kawamura, S. (1975a) "Design and Operation of High-Rate Filters—Part 1," *J. AWWA*, **67**, 10, 535–544.

Kawamura, S. (1975b) "Design and Operation of High-Rate Filters—Part 2," *J. AWWA*, **67**, 11, 653–662.

Kawamura, S. (1975c) "Design and Operation of High-Rate Filters—Part 3," *J. AWWA*, **67**, 12, 705–708.

Kawamura, S. (1999) "Design and Operation of High-Rate Filters," *J. AWWA*, **91**, 12, 77–90.

Kawamura, S. (2000) *Integrated Design and Operation of Water Treatment Facilities*, Wiley, New York.

Kawamura, S., Najm, I. N., and Gramith, K. (1997) "Modifying a Backwash Trough to Reduce Media Loss," *J. AWWA*, **89**, 12, 47–59.

Kim, J., and Lawler, D. F. (2008) "Influence of particle characteristics on filter ripening," *Separation Sci. and Technol.*, **43**, 7, 1583–1594.

Kim, J., Nason, J. A., and Lawler, D. F. (2008) "Influence of Surface Charge Distributions and Particle Size Distributions on Particle Attachment in Granular Media Filtration," *ES&T*, **42**, 7, 2557–2562.

Kozeny, J. (1927a) Ueger Kapillare Leutung des Wassers im Boden (On Cappillary Conduction of Water in Soil), *Sitzungsbericht Akad. Wiss.* **136**, 271–306, Wein, Austria.

Kozeny, J. (1927b) "Ueger Kapillare Leutung des Wassers im Boden (On Cappillary Conduction of Water in Soil)," *Wasserkraft Wasserwirtschaft*, **22**, 67–78.

Krasner, S. W., Sclimenti, M. J., and Coffey, B. M. (1993) "Testing Biologically Active Filters for Removing Aldehydes Formed During Ozonation," *J. AWWA*, **85**, 5, 62–71.

Lang, J. S., Giron, J. J., Hansen, A. T., Trussell, R. R., and Hodges, W. E. J. (1993) "Investigating Filter Performance as a Function of the Ratio of Filter Size to Media Size," *J. AWWA*, **85**, 10, 122–130.

Levich, V. G. (1962) *Physicochemical Hydrodynamics*, Prentice-Hall, Englewood Cliffs, NJ.

Miltner, R. J., Summers, R. S., and Wang, J. Z. (1995) "Biofiltration Performance: Part 2, Effect of Backwashing," *J. AWWA*, **87**, 12, 64–70.

Monk, R. D. G. (1984) "Improved Methods of Designing Filter Boxes," *J. AWWA*, **76**, 8, 54–59.

Moran, M. C., Moran, D. C., Cushing, R. S., and Lawler, D. F. (1993) "Particle Behavior in Deep-Bed Filtration: Part 2—Particle Detachment," *J. AWWA*, **85**, 12, 82–93.

O'Melia, C. R. (1985) "Particles, Pretreatment, and Performance in Water Filtration," *J. Environ. Eng.*, **111**, 6, 874–890.

O'Melia, C. R., and Shin, J. Y. (2001) "Removal of Particles Using Dual Media Filtration: Modeling and Experimental Studies," *Water Sci. Technol.: Water Supply*, **1**, 4, 73–79.

O'Melia, C. R., and Stumm, W. (1967) "Theory of Water Filtration," *J. AWWA*, **59**, 11, 1393–1411.

Poiseuille, J. (1841) *Reserches Experimentales sur le Mouvement des Liquides dans les Tubes de tres Petits Dimetres* [in french], Comptes Rendus de l'Academie des Sciences, Paris, France.

Price, M. L., Enos, A. K., Bailey, R., Hermanowicz, S. W., and Jolis, D. (1993) "Evaluation of Ozone-Biological Treatment for Reduction of Disinfection ByProducts and Production of Biologically Stable Water," *Ozone Sci. Eng.*, **15**, 2, 95–130.

Rajagopalan, R., and Tien, C. (1976) "Trajectory Analysis of Deep-Bed Filtration with the Sphere-in-Cell Porous Media Model," *AIChEJ.*, **22**, 3, 523–533.

Raveendran, P., and Amirtharajah, A. (1995) "Role of Short-Range Forces in Particle Detachment During Filter Backwashing," *J. Environ. Eng.*, **121**, 12, 860–868.

Seelaus, T. J., Hendricks, D. W., and Janonis, B. A. (1986) "Design and Operation of a Slow Sand Filter," *J. AWWA*, **78**, 12, 35–41.

Servais, P., and Joret, J. C. (1999) BOM in Water Treatment, in M. Prevost (ed.), *Biodegradable Organic Matter in Drinking Water*, American Water Works Association Research Foundation, Lewis Publisher, Boca Raton, FL.

Slezak, L. A., and Sims, R. C. (1984) "The Application and Effectiveness of Slow Sand Filtration in the United States," *J. AWWA*, **76**, 12, 38–43.

Streeter, V. L., and Wylie, E. B. (1979) *Fluid Mechanics*, McGraw-Hill, New York.

Sontheimer, H., and Hubele, C. (1987) The Use of Ozone and Granular Activated Carbon in Drinking Water Treatment, pp. 7-8, in P. M. Huck and P. Toft (eds.), *Treatment of Drinking Water Organic Contaminants*, Pergamon Press, Oxford, UK.

Teefy, S. (2001) *Personal communication*, Alameda County Water District, Fremont, CA.

Tien, C. (1989) *Granular Filtration of Aerosols and Hydrosols*, Butterworths, Boston.

Tien, C., and Payatakes, A. C. (1979) "Advances in Deep Bed Filtration," *AIChE J.*, **25**, 5, 737–759.

Tobiason, J. E., and O'Melia, C. R. (1988) "Physicochemical Aspects of Particle Removal in Depth Filtration," *J. AWWA*, **80**, 12, 54–64.

Trussell, R. R., and Chang, M. (1999) "Review of Flow through Porous Media as Applied to Head Loss in Water Filters," *J. Environ. Eng.*, **125**, 11, 998–1006.

Trussell, R. R., Chang, M. M., Lang, J. S., and Hodges, W. E., Jr. (1999) "Estimating the Porosity of a Full-Scale Anthracite Filter Bed," *J. AWWA*, **91**, 12, 54–63.

Trussell, R. R., Trussell, A., Lang, J. S., and Tate, C. (1980) "Recent Developments in Filtration System Design," *J. AWWA*, **73**, 12, 705–710.

Tufenkji, N., and Elimelech, M. (2004) "Correlation Equation for Predicting Single-Collector Efficiency in Physicochemical Filtration in Saturated Porous Media," *Environ. Sci. Technol.*, **38**, 2, 529–536.

Urfer, D., Huck, P. M., Booth, S. D. J., and Coffey, B. M., (1997) "Biological filtration for BOM and Particle Removal: A critical Review," *J. AWWA*, **89**, 12, 83–98.

U. S. EPA (1989) "National Primary Drinking Water Regulations: Filtration and Disinfection; Turbidity, *Giardia lamblia*, Viruses, *Legionella*, and Heterotrophic Bacteria; Final Rule," *Fed. Reg.*, **54**, 124, 27486.

U. S. EPA (2001) *Factoids: Drinking Water and Ground Water Statistics for 2000*, U. S. Environmental Protection Agency, Office of Water, Washington, DC.

U. S. EPA (2002) "National Primary Drinking Water Regulations: Long Term 1 Enhanced Surface Water Treatment Rule; Final Rule," *Fed. Reg.*, **67**, 9, 1812.

U. S. EPA (2006) "National Primary Drinking Water Regulations: Long Term 2 Enhanced Surface Water Treatment Rule; Final Rule," *Fed. Reg.*, **71**, 3, 654–786.

Veerapaneni, S. and Wiesner, M.R., (1997) "Deposit Morphology and Head Loss Development in Porous Media," *Environ. Sci. Technol.*, **31**, 10, 2738–2744.

Visscher, J. T. (1990) "Slow Sand Filtration: Design, Operation, and Maintenance," *J. AWWA*, **82**, 6, 67–71.

Wiesner, M.R., (1999) "Morphology of Particle Deposits," *J. Environ. Eng.*, **125**, 12, 1124–1132.

Yao, K.-M., Habibian, M. T., and O'Melia, C. R. (1971) "Water and Waste Water Filtration: Concepts and Applications," *Environ. Sci. Technol.*, **5**, 11, 1105–1112.

12 Membrane Filtration

Stantec's Water Treatment: Principles and Design, Updated Third Edition. John C. Crittenden, R. Rhodes Trussell, David W. Hand, Kerry J. Howe, and George Tchobanoglous.

Terminology for Membrane Filtration

Term	Definition
Asymmetric membrane	Membrane whose morphology (structure) varies significantly across the thickness of the membrane.
Cross-flow filtration	Filtration technique in which the feed stream is pumped at high velocity parallel to the membrane surface to reduce the collection of retained species at the membrane surface.
Dalton	Unit for molecular weight, equal to one-twelfth of the mass of a carbon-12 atom. Also equal to the molar mass in units of grams per mole. Equivalent to atomic mass units (amu).
Dead-end filtration	Filtration technique in which the feed stream is directed toward and perpendicular to the membrane surface.
Fouling	Process resulting in loss of performance of a membrane due to the deposition of suspended or dissolved substances on its external surfaces, at its pore openings, or within its pores.
Homogeneous membrane	Membrane with consistent morphology and transport properties throughout its thickness.
Lumen	Bore, or cavity, in the center of a hollow fiber membrane.
Molecular weight cutoff	See *Retention rating*.
Packing density	Membrane area per unit volume in a membrane module.
Permeability	Specific flux of clean, deionized water through a new membrane.

Term	Definition
Permeate	Water and permeable components that pass through a membrane.
Retentate	Solution containing water and impermeable components retained on the feed side of a semipermeable membrane.
Retention rating	Designation for the size of materials retained by a membrane. The retention rating is called the pore size in micrometers for microfiltration (MF) membranes and the molecular weight cutoff (MWCO) in daltons for ultrafiltration (UF) membranes.
Semipermeable membrane	Membrane that is permeable to some components in a feed solution and impermeable to other components.
Specific flux	Flux divided by transmembrane pressure.
Straining	Process in which particles are retained because they are physically larger than the void spaces in the filter medium (often called sieving).
Transmembrane pressure	Differential pressure between the feed and permeate sides of a membrane.

Note: Additional membrane nomenclature is available in Koros et al. (1996) and ASTM (2001b).

Membrane processes are modern physicochemical separation techniques that use differences in permeability (of water constituents) as a separation mechanism. During membrane treatment, water is pumped against the surface of a membrane, resulting in the production of product and waste streams, as shown on Fig. 12-1. The membrane, typically a synthetic material less than 1 mm thick, is *semipermeable* —meaning that it is highly permeable to some components in the feed stream and less permeable (or impermeable) to others. During operation, permeable components pass through the membrane and impermeable components are retained

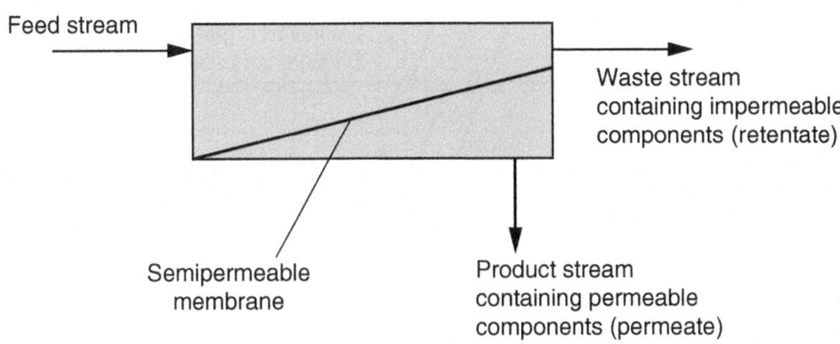

Figure 12-1
Schematic of separation process through semipermeable membrane.

on the feed side. As a result, the product stream is relatively free of impermeable constituents and the waste stream is concentrated in impermeable constituents.

12-1 Classification of Membrane Processes

Four types of pressure-driven membranes are currently used in municipal water treatment: microfiltration (MF), ultrafiltration (UF), nanofiltration (NF), and reverse osmosis (RO) membranes. The hierarchy of membrane processes is shown on Fig. 12-2. The distinction between the types of membranes is somewhat arbitrary and subject to differing interpretations, but the membranes are loosely identified by the types of materials rejected, operating pressures, and nominal pore dimensions (which are identified on an order-of-magnitude basis on Fig. 12-2). A "loose" NF membrane marketed by one manufacturer might be substantially similar to a "tight" UF membrane marketed by another manufacturer. As used in water treatment, these membranes can be classified into two distinct physicochemical processes: (1) membrane filtration and (2) reverse osmosis.

Membrane Filtration

Membrane filtration is the focus of this chapter and encompasses the use of MF and UF membranes. Filtration can be broadly defined as a process that separates suspended particles (a dispersed solid phase) from a liquid phase by passage of the suspension through a porous medium (either membranes or granular media). In membrane filtration, the feed stream is a suspension,

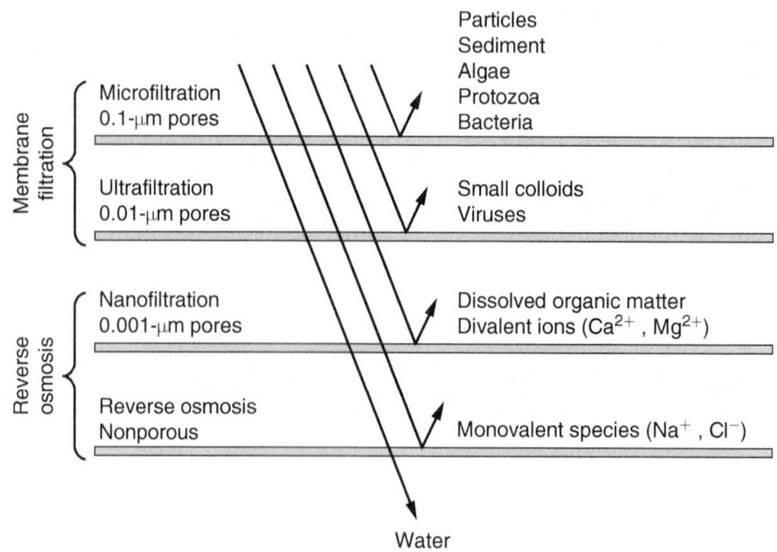

Figure 12-2
Hierarchy of pressure-driven membrane processes.

or two-phase system, in which the dispersed solid phase to be separated may include sediment, algae, bacteria, protozoa, viruses, or colloids. The primary goal of membrane filtration is to produce a product stream (water) from which the targeted solids have been completely removed, which is similar to the goal of granular filtration. While used for similar purposes, MF and UF membranes have important differences that will be described later in this chapter.

Reverse Osmosis

The other fundamental physicochemical membrane process is *reverse osmosis*. Reverse osmosis is the focus of Chap. 17 and encompasses the use of NF and RO membranes. *Osmosis* is the preferential diffusion of water through a semipermeable membrane in response to a concentration gradient. In reverse osmosis, the feed stream is a solution, or single-phase system, in which the constituents targeted for removal are truly dissolved solutes (ions such as sodium, chloride, calcium, or magnesium, and dissolved NOM). The primary goal of reverse osmosis is to reduce the concentration of these solutes in the product water. Reverse osmosis membranes are used to produce potable water from ocean or brackish water and to remove specific dissolved contaminants (e.g., pesticides, arsenic, nitrate, radionuclides). Nanofiltration membranes are used to soften hard waters (remove calcium and magnesium ions), freshen brackish waters, and reduce the concentration of NOM to control disinfection by-product (DBP) formation.

Differences between Membrane Processes

The differences between membrane filtration and reverse osmosis are substantial. The predominant removal mechanism in membrane filtration is straining, or size exclusion, so the process can theoretically achieve perfect exclusion of particles regardless of operational parameters such as influent concentration and pressure. Mass transfer in reverse osmosis, however, involves a diffusive mechanism so that separation efficiency is dependent on influent solute concentration, pressure, and water flux rate. Differences between membrane filtration and reverse osmosis are evident in the materials used for the membranes, the configuration of the membrane elements, the equipment used, the flow regimes, and the operating modes and procedures. Additional comparisons between membrane filtration and reverse osmosis are detailed in Table 12-1. It should be noted that membranes are used for many purposes in a wide variety of fields and industries, and the distinction between membrane types as used in water treatment may not be appropriate in other industries. For instance, UF membranes are used in food-processing and pharmaceutical industries for purifying, concentrating, and fractionating concentrated solutions of macromolecules such as proteins and polysaccharides; UF membrane use in those industries involves phenomena (such as concentration polarization) described in Chap. 17.

Table 12-1
Comparison between membrane filtration and reverse osmosis

Process Characteristic	Membrane Filtration	Reverse Osmosis
Objectives	Particle removal, microorganism removal	Seawater desalination, brackish water desalination, softening, NOM removal for DBP control, specific contaminant removal
Target contaminants	Particles	Dissolved solutes
Membranes types	Microfiltration, ultrafiltration	Nanofiltration, reverse osmosis
Typical source water	Fresh surface water (TDS < 1000 mg/L)	Ocean or seawater, brackish groundwater (TDS = 1000–20,000 mg/L), colored groundwater (TOC > 10 mg/L)
Membrane structure	Homogeneous or asymmetric	Asymmetric or thin-film composite
Most common membrane configuration	Hollow fiber	Spiral wound
Dominant exclusion mechanism	Straining	Differences in solubility or diffusivity
Removal efficiency of targeted impurities	Frequently 99.9999% or greater	Typically 50–99%, depending on objectives
Most common flow pattern	Dead end	Tangential
Operation includes backwash cycle	Yes	No
Influenced by osmotic pressure	No	Yes
Influenced by concentration polarization	No	Yes
Noteworthy regulatory issues	Challenge testing and integrity monitoring	Concentrate disposal
Typical transmembrane pressure	0.2–1 bar (3–15 psi)	5–85 bar (73 –1200 psi)
Typical permeate flux	30 –170 $L/m^2 \cdot h$ (18 –100 $gal/ft^2 \cdot d$)	1–50 $L/m^2 \cdot h$ (0.6–30 $gal/ft^2 \cdot d$)
Typical recovery	>95%	50% (for seawater) to 90% (for colored groundwater)
Competing processes	Granular filtration	Carbon adsorption, ion exchange, precipitative softening, distillation

Table 12-2
Non-pressure-driven membrane processes

Membrane Process	Driving Force
Dialysis	Concentration gradient
Electrodialysis	Electrical potential
Electrodialysis reversal	Electrical potential
Pervaporation	Pressure gradient
Forward osmosis	Osmosis
Membrane distillation	Vapor pressure
Thermoosmosis	Temperature gradient

It should be noted that membrane filtration and reverse osmosis are both pressure-driven membrane processes. Driving forces other than pressure are used in other membrane processes, including some that are occasionally used in water treatment, such as electrodialysis. Other membrane processes (not covered in this text because of their limited applicability in water treatment) and their driving forces are identified in Table 12-2.

12-2 History of Membrane Filtration in Water Treatment

Microporous membranes were first patented in the 1920s (Belfort et al., 1994) and were limited primarily to laboratory use until the 1950s. They were used primarily for enumerating bacteria, removing microorganisms and particles from liquid and gas streams, fractionating and sizing macromolecules such as proteins, and diffusion studies. The U.S. Public Health Service (U.S. PHS) adopted membrane filtration as a method for identifying coliform bacteria in 1957.

In the 1950s, industrial users started applying membrane filtration to larger scale industrial use, with one common use being sterilization of liquid pharmaceuticals and intravenous solutions. Membrane filtration was used in food-processing industries for clarifying, concentrating, purifying, or sterilizing various products such as fruit juices, dairy products, vegetable oils, and alcoholic beverages. Membrane filtration also began to be used for industrial process and waste treatment—such as oily wastewater treatment, caustic acid, and brine recovery—and treatment or recovery of various other industrial waste streams.

Application to Drinking Water Treatment

The first interest in membrane filtration for potable water production began in the 1980s as utilities and regulators became increasingly concerned about microbiological contamination. Advances in industrial equipment design and operation, including the introduction of dead-end flow regimes, outside-in hollow-fiber flow configurations, and backwashing systems, made the production of drinking water by membrane filtration an economically realistic possibility. Very small utilities, in particular, began to consider membrane filtration. Rapid granular filter equipment was expensive and required a level of operator attention and sophistication that was sometimes unaffordable for small communities, and membrane filtration offered an attractive, highly automated, operationally simple alternative. The first membrane filtration plant used for drinking water production in the United States was a 225-m^3/d (0.06-mgd) plant at Keystone Resorts in Colorado in 1987 (U.S. EPA, 2001). Similar developments occurred in Europe, and a 250-m^3/d (0.07-mgd) UF plant was installed in France in 1988 (Anselme et al., 1999).

The passage of the SWTR (U.S. EPA, 1989) in 1989 provided utilities with another reason to consider membrane filtration. Regulatory agencies were focusing greater attention on microorganisms in water supplies and lower turbidity levels were required. Membrane filtration offered the potential of higher quality treatment than granular filtration. Still, the use of membrane filtration grew slowly, and by 1993 there were only eight systems installed in the United States, all considerably smaller than 3800 m^3/d (1 mgd) (U.S. EPA, 2001).

Effectiveness of Membrane Filtration for Removing Protozoa

As noted in Chap. 1, an outbreak of cryptosporidiosis in Milwaukee, Wisconsin, in 1993 caused over 400,000 illnesses and 50 deaths (Craun et al., 1998; U.S. EPA, 1998). During the incident, *Cryptosporidium* oocysts had passed through the conventional water treatment plant, including the rapid granular filters. The outbreak underscored the fact that the effluent water quality from rapid granular filters is dependent on proper chemical conditioning of the feed water, which is ultimately dependent on operators' judgment, experience, and knowledge of water chemistry. In contrast, membrane filtration removes particles by straining so that complete removal of protozoa is virtually guaranteed as long as the membranes are intact.

Afterward, the view that membranes provided superior filtration helped fuel a rapid increase in the installation of membrane filtration plants, with growth rates in installed capacity of 50 to 100 percent per year over the next several years. Costs of membrane filtration facilities dropped dramatically during this period as a result of advances in technology, mass production, and the entry of additional manufacturers into the market. A survey of equipment manufacturers revealed over 700 membrane filtration facilities in operation worldwide by the end of 2003. In North America, 213 plants [with capacity greater than 379 m^3/d (0.1 mgd)] had a total installed capacity of 2.3 million m^3/d (620 mgd) by the end of 2003 (Adham et al., 2005). Interest in membrane filtration has continued unabated since then. Membrane filtration is now considered a viable option for any surface water treatment facility of any size. The decision to use granular filtration or membrane filtration in any particular facility depends on site-specific circumstances.

12-3 Principal Features of Membrane Filtration Equipment and Operation

Membrane filtration occurs when water is forced through a thin wall of porous material. The filter medium is not woven or fibrous like cloth but is a continuous mass with tortuous interconnecting voids, as shown in the scanning electron microscope (SEM) images on Fig. 12-3. Nearly all membrane filtration systems installed in the United States use polymeric membranes. Polymeric membranes are almost always configured as hollow fibers, as shown on Figs. 12-4a and 12-4b. The fibers have an outside diameter ranging from about 0.65 to 2 mm and a wall thickness (i.e.,membrane thickness) ranging from about 0.1 to 0.6 mm. Although the hollow fiber configuration is the most common used in water treatment, other configurations exist and are in widespread use in other industries. Membrane filtration is a rapidly evolving field, and other configurations might be used in the future. Ceramic membranes are used in some systems in Japan, and the first large ceramic membrane system in the United States was designed for Parker, Colorado, and is expected to be operational in 2012. Ceramic membranes have a tubular configuration with many parallel channels in a rigid monolithic element, as shown on Fig. 12-4c. The configuration has a strong effect on the *packing density*, or membrane area per unit of volume of equipment module, which can be

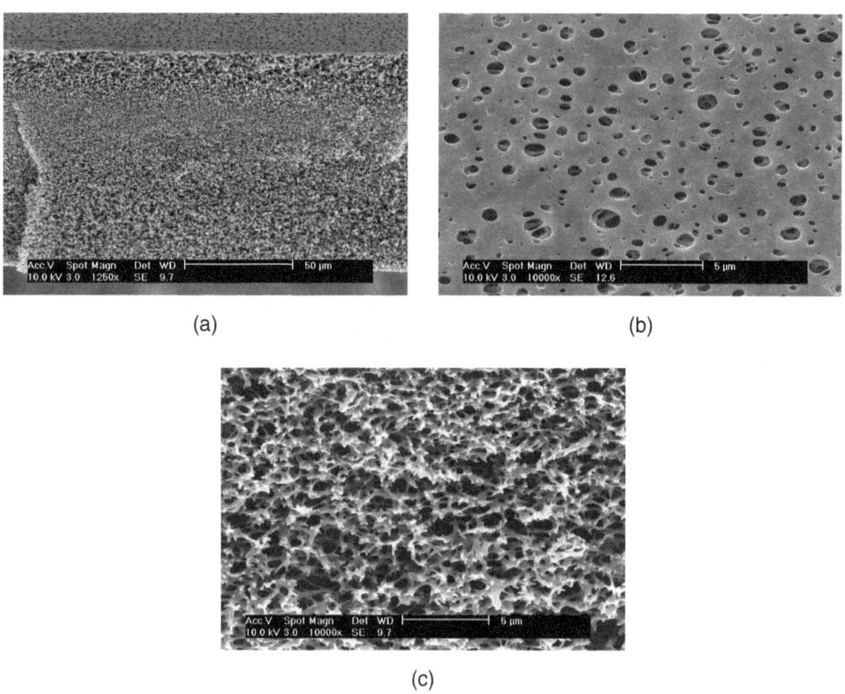

Figure 12-3
Scanning electron microscope images of a 0.2-μm polyethersulfone microfiltration membrane: (a) cross section of the entire membrane, (b) high magnification of the membrane surface, and (c) high magnification of the membrane internal structure.

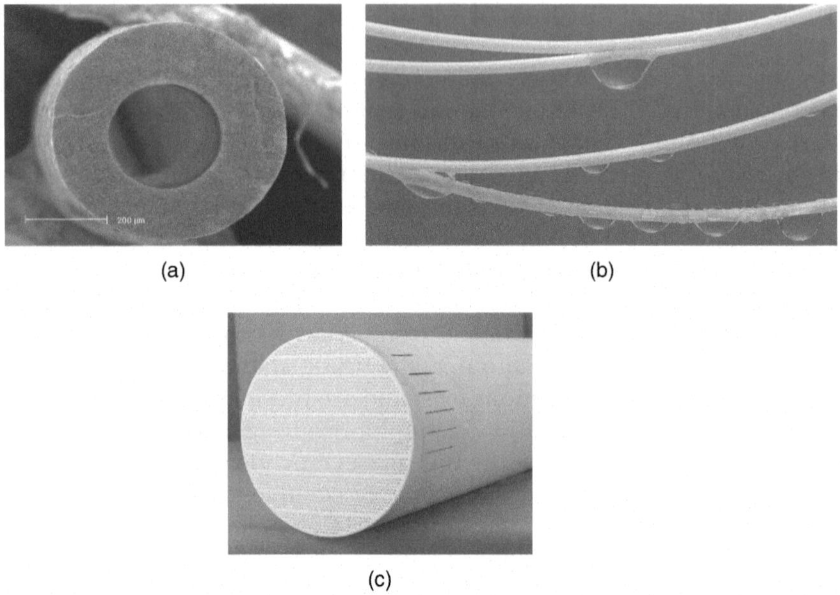

Figure 12-4
(a) Scanning electron microscope image of end view of a hollow-fiber membrane (courtesy of US Filter Memcor Products), (b) water permeating hollow-fiber membranes (courtesy of Suez Environnement), and (c) end view of a ceramic tubular membrane (courtesy of NKG).

an important consideration in the cost effectiveness of membrane plants. Other membrane configurations are described in Table 12-3.

The water passing through the membrane is called permeate, and water remaining on the feed side is called retentate. As solids accumulate against the filter medium, the head across the membrane required to maintain constant flux increases. The difference in pressure between the feed and permeate is known as the transmembrane pressure. The transmembrane pressure is between 0.2 and 1 bar (3 and 15 psi) for most membrane filtration systems. Keeping pressure below 1 bar (15 psi) helps minimize membrane fouling.

Membrane filters operate over a cycle consisting of two stages, just like granular filters: (1) a filtration stage, during which particles accumulate, and (2) a backwash stage, during which the accumulated material is flushed from the system. During the backwash cycle, air and/or water is used to remove accumulated solids. Typical permeate flux, operating pressure, and duration of filter and backwash cycles, along with a comparison to rapid granular filtration, are presented in Table 12-4. Although the backwash removes accumulated solids, a gradual but continuous loss of performance is observed over a period of days or weeks, as shown on Fig. 12-5. The loss of performance, or fouling, is due to slow adsorption or clogging of material that cannot be removed during backwash. Fouling affects the cost effectiveness of membrane filtration and will be discussed in detail later in this chapter. Fouling is minimized by periodically adding chemicals to the backwash cycle, known as chemically enhanced backwash (CEB), and periodic

Table 12-3
Membrane configurations

Configuration	Description
Hollow fiber	Membranes are cast as hollow tubes and filtration occurs as water passes through the wall of the fibers (see Fig. 12-4b). The module packing density (specific surface area) is 750–1700 m^2/m^3.
Tubular	Membranes are constructed as a monolithic structure with one or more channels through the structure (see Fig. 12-4c). Ceramic membranes are typically tubular membranes. These membranes can be operated at a high cross-flow velocity, which is ideal for applications where the particle concentration is high. The module packing density is up to 400–800 m^2/m^3.
Flat sheet	Membranes are cast as a sheet and used as a single layer or as a stack of sheets. Common in laboratory separations but not as common at an industrial scale. Packing density depends on spacing of the sheets.
Spiral wound	Flat-sheet membranes, stacked in layers separated by permeate and retentate spacers, then rolled around a central tube so that the permeate travels in a spiral flow path toward the central collection tube. Common in NF and RO membranes but not in wide use for membrane filtration due to clogging of flow paths with particulate matter and problems with backwashing effectively. See Chap. 17 for additional details on the construction of spiral-wound elements. The packing density is 700–1000 m^2/m^3.
Hollow fine fiber	Membranes cast as hollow tubes with an outside diameter of 0.085 mm (about the thickness of human hair). Hollow fine fibers are used only as RO membranes; see Chap. 17 for additional details. The packing density is 5600–7400 m^2/m^3.
Track etched	Flat-sheet membranes that are cast as a dense sheet of polymer material and exposed to a radioactive beam, which damages the material along "tracks," or straight pathways through the material. The material is then immersed in an etching bath that dissolves the material along the pathways, widening the tracks to form pores of uniform cylindrical dimensions. The result is a flat-sheet membrane with a narrow, controllable, and extremely uniform pore size distribution, which is advantageous in laboratory separations. Track-etched membranes are not currently used in industrial-scale applications.

Table 12-4
Operating characteristics of membrane and rapid granular filters

Criteria	Membrane Filtration	Rapid Granular Filtration
Filtration rate (permeate flux)	0.03–0.17 m/h[a] (0.01–0.07 gpm/ft^2)	5–15 m/h[a] (2–6 gpm/ft^2)
Operating pressure	0.2–1 bar (7–34 ft)	0.18–0.3 bar (6–10 ft)
Filtration cycle duration	30–90 min	1–4 d
Backwash cycle duration	1–3 min	10–15 min
Ripening period	None	15–120 min
Recovery	>95 %	>95 %
Filtration mechanism	Straining	Depth filtration

[a]Conventional units for membrane permeate flux are L/m^2 · h and gal/ft^2 · d. The conversions to the units shown in this table are 1 L/m^2 · h = 10^{-3} m/h and 1440 gal/ft^2 · d = 1 gpm/ft^2.

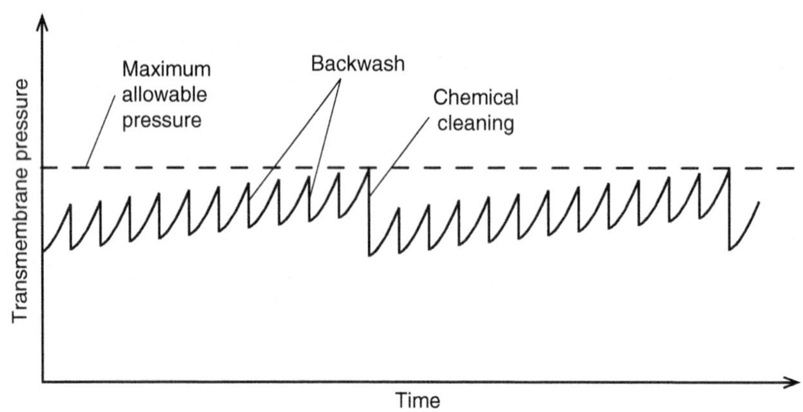

Figure 12-5
Transmembrane pressure development during membrane filtration.

chemical cleaning, known as the clean-in-place (CIP) cycle. CIP typically involves soaking the membranes for several hours in one or more warm solutions containing surfactants, acids, or bases. The cleaning frequency may range from a few days to several months, depending on the membrane material, operating conditions, and raw-water quality. The membranes degrade over a longer period of time, and replacement may be necessary after a period of 5 to 10 years.

The increase in transmembrane pressure when filters are operated in a constant-flux, rising-pressure mode is shown on Fig. 12-5. Full-scale facilities are operated in this mode because of production capacity requirements. In contrast, laboratory studies are sometimes performed in a constant-pressure, declining-flux mode to accommodate equipment capabilities and data analysis procedures.

Module Configuration

As shown in Table 12-4, the flux through a membrane filter is typically about two orders of magnitude lower than the flux through a rapid granular filter; consequently, a membrane filtration plant needs 100 times the filter area of a rapid granular filtration plant to produce the same quantity of water. One characteristic of membrane filtration plants, however, is that they are frequently more compact than granular filtration plants. This apparent contradiction is possible because membrane plants are constructed by packing thousands of hollow fibers into modules; thus, 1 m^2 of floor space at a membrane plant may contain more than 100 m^2 of membrane area. Membrane modules are available in two basic configurations: pressure-vessel systems or submerged systems.

Pressure vessel membrane filtration system.

http://www.stantec.com/wt3r/12-1.html

PRESSURE-VESSEL CONFIGURATION

Pressure-vessel modules are generally 100 to 300 mm (4 to 12 in.) in diameter, 0.9 to 5.5 m (3 to 18 ft) long, and arranged in racks or skids. Typical pressure-vessel membrane elements are shown on Fig. 12-6. A single module has thousands of fibers and typically contains between 40 and 80 m^2 (430 and 860 ft^2) of filter area. The rack or skid is the basic production unit, and all modules within one rack are operated in parallel simultaneously (see Fig. 12-7). Racks can contain between 2 and 100 modules, depending on capacity requirements. Feed pumps typically deliver water

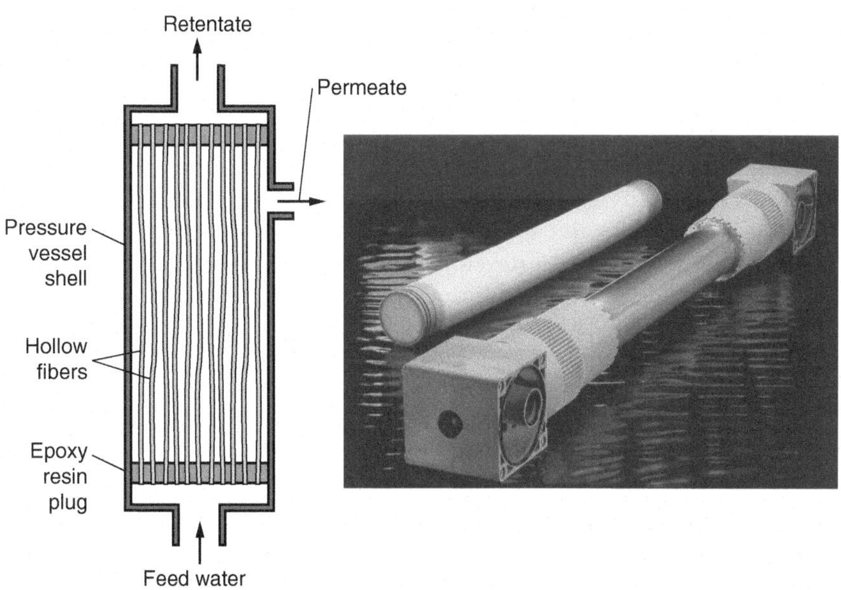

Figure 12-6 Pressure-vessel configuration for membrane filtration: (a) schematic of a single cross-flow membrane module and (b) photograph (courtesy of US Filter Memcor Products).

Figure 12-7
Full-scale membrane filtration facility using the pressure-vessel configuration.

to a common manifold that supplies each rack. Each module must be piped individually for feed and permeate water, so large racks involve a substantial number of piping connections. Transmembrane pressure is developed by a feed pump that increases the feed water pressure, while the permeate stays at near-atmospheric pressure. Pressure-vessel systems typically operate at transmembrane pressures between about 0.4 and 1 bar (6 and 15 psi).

SUBMERGED CONFIGURATION

Submerged systems (also called immersed membranes) are modules of membranes suspended in basins containing feed water, as shown on Fig. 12-8. The basins are open to the atmosphere, so pressure on the influent side is limited to the static pressure provided by the water column. Transmembrane pressure is developed by a pump that develops suction on the permeate side of the membranes; thus submerged systems are sometimes called suction- or vacuum-based systems. Net positive suction head (NPSH) limitations on the permeate pump restrict submerged membranes to a maximum transmembrane pressure of about 0.5 bar (7.4 psi), and they typically operate at a transmembrane pressure of 0.2 to 0.4 bar (3 to 6 psi). Submerged systems are configured with multiple basins so that individual basins can be isolated for cleaning or maintenance without shutting down the entire plant. Each basin typically has its own permeate pump.

Because clean water is extracted from the feed basin through the membranes and solids are returned directly to the feed tank during the backwash

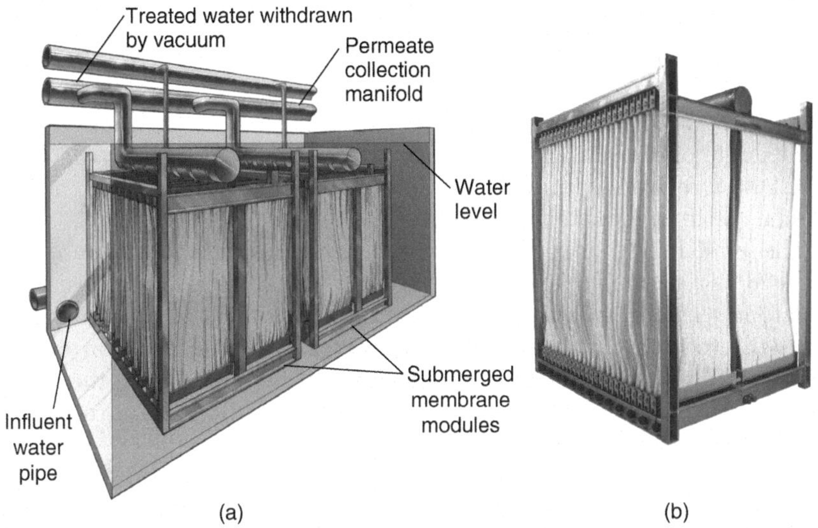

Figure 12-8
Submerged configurations for membrane filtration: (a) schematic of a submerged membrane module and (b) photograph of a single module. (© 2011 General Electric Company. All rights reserved. Reprinted with permission.)

Photographs of submerged MF systems:

http://www.stantec.com/wt3r/12-2.html

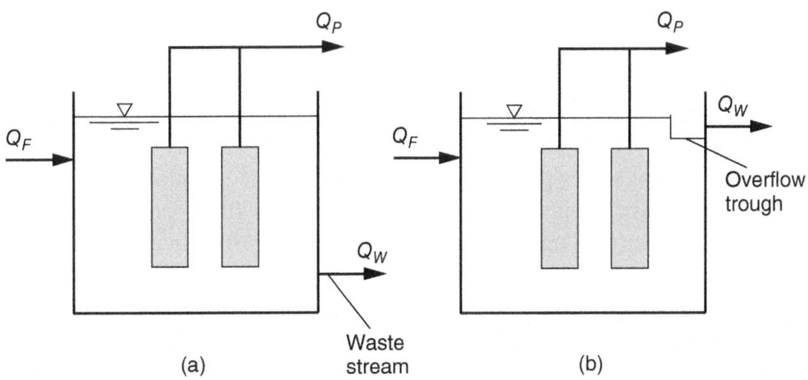

Figure 12-9
Feed-and-bleed and semibatch modes of operation. In feed-and-bleed, Q_P and Q_W are both continuous, the sum of the two flows equals Q_F. In semibatch, *QP* is continuous and equal to *QF*, *QW* only flows when solids are being wasted.

cycle, the solids concentration in the feed tank can be significantly higher than in the raw water. A high solids concentration can be advantageous when using treatment additives (i.e., coagulants or PAC) to remove dissolved contaminants but can have an adverse impact on the solids loading on the membrane during filtration. Two basic strategies are used to maintain the proper solids concentration in the feed tank, as shown on Fig. 12-9: (1) the feed-and-bleed strategy and (2) the semibatch strategy. In the feed-and-bleed strategy, a small waste stream is continuously drawn from the feed tank. The average solids concentration in the tank will be a function of the size of the waste stream:

$$C_w = \left(\frac{Q_f}{Q_w}\right) C_f \qquad (12\text{-}1)$$

where C_f = solids concentration in influent, mg/L
C_w = solids concentration in tank and waste stream, mg/L
Q_f = influent flow rate, m^3/h
Q_w = waste flow rate, m^3/h

Some guide books, such as the *Membrane Filtration Guidance Manual* (U.S. EPA, 2005), refer to the ratio C_w/C_f, and therefore the ratio Q_f/Q_w, as the volume concentration factor (VCF).

The semibatch strategy operates without a continuous waste stream, and the feed and permeate flows are at the same rate. As a result, solids accumulate in the feed tank during the filtration cycle. During the backwash cycle, the volume of water in the tank increases due to addition of the backwash flow (raw water continues to flow to the tank during the backwash cycle), and the excess water (and solids) exits the basin through an overflow trough or port.

In currently available equipment, submerged systems tend to accommodate larger modules than pressure-vessel systems. Furthermore, submerged systems have substantially fewer valves and piping connections. As larger membrane plants are designed and built, membrane manufacturers have tried to improve the economy of scale by developing larger modules to reduce the number of individual modules and piping connections necessary in large facilities, and these trends are expected to continue to lead to the development of larger modules.

Flow Direction through Hollow Fibers

Filtration occurs as water passes through the wall of the hollow fiber. Some manufacturers have designed membrane systems to filter from outside to inside (the feed water is against the shell, or outside the fiber, and the permeate is in the lumen, or inside the fiber), and other manufacturers have designed systems to filter in the opposite direction (inside out). The advantages and disadvantages of each flow configuration are described in Table 12-5. Pressure vessels use either outside-in or inside-out membranes, while submerged systems use only outside-in membranes. The difference in flow that can be achieved with outside-in versus inside-out systems is demonstrated in Example 12-1.

Cross-Flow and Dead-End Flow Regimes

Permeate flux and fouling are affected by the flow regime of the feed water near the membrane surface. Two filtration strategies, cross-flow filtration and dead-end filtration, have been developed to influence this flow regime and are shown schematically on Fig. 12-10.

CROSS-FLOW FILTRATION

In cross-flow filtration, the feed water is pumped at a high rate through the lumen of inside-out membrane fibers. The cross-flow velocity, typically

Table 12-5
Comparison of hollow-fiber membrane configurations

Configuration	Advantages	Disadvantages
Outside in 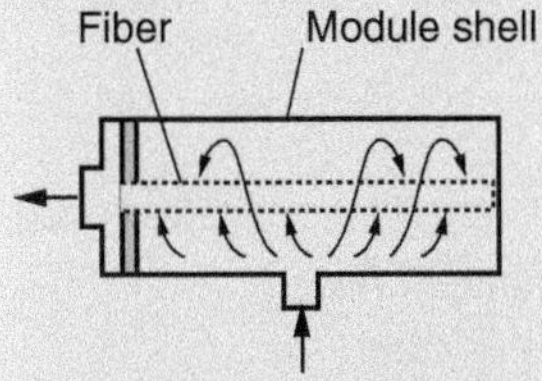	❑ Can treat more water at same flux because outside of fiber has more surface area ❑ Less sensitive to presence of large solids in the feed water	❑ Cannot be operated in cross-flow mode
Inside out (dead-end mode) 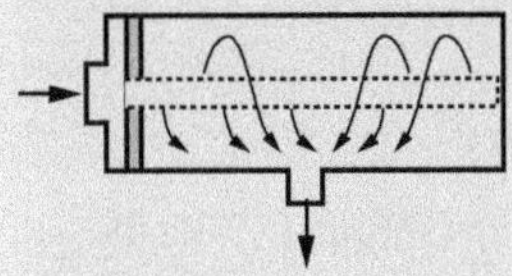	❑ Less expensive to operate than inside out in cross-flow mode	❑ Large solids in feed water can clog lumen ❑ Can treat less water at same flux because inside of fiber has less surface area
Inside out (cross-flow mode) 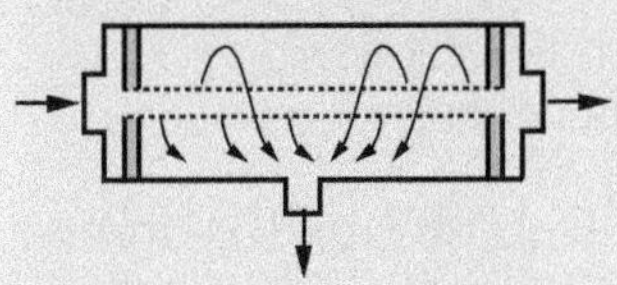	❑ Can be operated at higher flux with high-turbidity feed water because cross-flow velocity flushes away solids and reduces impact of particles forming cake at membrane surface	❑ Large solids in feed water can clog lumen ❑ Can treat less water at same flux because inside of fiber has less surface area ❑ Pumping costs associated with recirculating feed water through lumen can be expensive

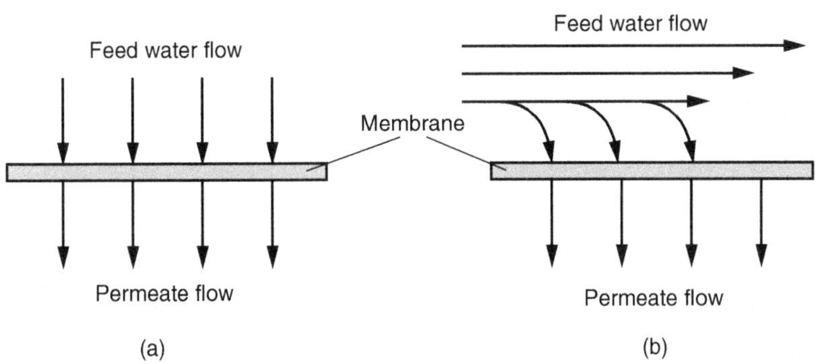

Figure 12-10
Flow regimes in membranes: (a) dead-end filtration and (b) cross-flow filtration.

Example 12-1 Comparison of outside-in and inside-out filtration

A Dow Filmtec SFX-2860 membrane module contains 5760 fibers. The fibers are 1.87 m long with an outside diameter of 1.3 mm and inside diameter of 0.7 mm. Calculate the water production from one module if the flux is 75 $L/m^2 \cdot h$ and the flow direction is (1) outside in and (2) inside out. Compare the two answers.

Solution

1. Compute the product water flow for outside-in flow.
   a. Determine the outside surface area per fiber:

$$a \text{ (per fiber)} = \pi dL = \pi(1.3 \text{ mm})(1.87 \text{ m})(10^{-3} \text{ m/mm})$$
$$= 7.64 \times 10^{-3} \text{ m}^2\text{/fiber}$$

   b. Compute the product water flow:

$$Q = Ja = (75 \text{ L/m}^2 \cdot \text{h})(7.64 \times 10^{-3} \text{ m}^2\text{/fiber})(5760 \text{ fibers})$$
$$= 3300 \text{ L/h}$$

2. Compute the product water flow for inside-out flow.
   a. Determine the inside surface area per fiber:

$$a \text{ (per fiber)} = \pi dL = \pi(0.7 \text{ mm})(1.87 \text{ m})(10^{-3} \text{ m/mm})$$
$$= 4.11 \times 10^{-3} \text{ m}^2\text{/fiber}$$

   b. Compute the product water flow:

$$Q = Ja = (75 \text{ L/m}^2 \cdot \text{h})(4.11 \times 10^{-3} \text{ m}^2\text{/fiber})(5760 \text{ fibers})$$
$$= 1780 \text{ L/h}$$

3. Compare the outside-in and inside-out flow configurations:

$$\text{Ratio} = (3300/1780) \times 100\% = 186\%$$

Comment

Operating at the same flux, the outside-in system produces nearly double the product water flow (86 percent more) as the inside-out system. Based on the results presented in this example, membrane systems cannot be compared or specified on the basis of flux if the flow configuration is different (the total flow per module and cost per module would be more important indicators than flux).

0.5 to 1 m/s (1.6 to 3.3 ft/s), is parallel to the membrane surface and about four orders of magnitude greater than the superficial velocity of water toward the membrane surface. The velocity parallel to the membrane surface creates a shear force that reduces the development of a surface cake (Wiesner and Chellam, 1992). Because many solids are carried away with the retentate instead of accumulating on the membrane surface, the system can be operated at a higher flux or with longer intervals between backwashes. The retentate is recirculated to the feed water, so cross-flow filtration requires a substantial recirculation flow—the permeate flow is typically less than 25 percent of the feed flow. The recirculation requirements can be prohibitive in a large facility—a 50,000-m^3/d (13.2-mgd) membrane filtration plant must recirculate 200,000 to 250,000 m^3/d (53 to 66 mgd) to maintain sufficient cross-flow velocity.

The retentate can be returned directly to the feed line to the membrane modules or to a mixing basin upstream of the modules. In either case, the solids content of the feed water will increase due to the recirculation. Either the feed-and-bleed or the semibatch procedure can be used to control the solids content in the recycle line.

DEAD-END FILTRATION

Dead-end filtration operates without a cross-flow component to the feed stream. The bulk feed water flow is transverse (perpendicular) to and toward the membrane surface during dead-end filtration, so all solids accumulate on the membrane during the filtration cycle and are removed during the backwash cycle. The greater solids accumulation during the filter run may result in lower average flux values than those achieved with cross-flow filtration.

PRACTICAL CONSIDERATIONS IN WATER TREATMENT

The dead-end flow regime is most common in membrane filtration for water treatment, in contrast to many industrial applications of microfiltration and ultrafiltration. Many industrial feed streams have high solids concentrations (e.g., the solids concentration in many food-processing operations can be 1 to 30 percent), and cross-flow operation is critical for achieving reasonable flux and filter run length. Surface waters are fairly dilute (many membrane plants operate with feed water turbidity of 100 NTU or less, which corresponds to a solids concentration on the order of 0.01 percent) so the advantages of cross-flow filtration are less significant. In addition, the piping and pumping costs of recirculating a large fraction of the feed water become prohibitive as the facility size gets larger, and water treatment facilities are built with considerably higher capacity than most industrial applications. The electrical costs of cross-flow pumping can triple the operating costs (Glucina et al., 1998) over dead-end operation. Some cross-flow systems are designed to operate in a dead-end mode by closing a valve in the retentate line when raw-water quality conditions permit (turbidity is low) and switch to a cross-flow mode only when necessary to maintain flux.

Comparison to Rapid Granular Filtration

A comparison between membrane filtration and rapid granular filtration has been presented in Table 12-4. Membrane filtration has several advantages over granular filtration. Effective rapid filtration with granular media, as noted in Chap. 11, depends on properly destabilizing particles with a coagulant to facilitate the attachment process. The void spaces in a membrane filter are much smaller; particles are literally strained from the water so destabilization is not necessary. As a result, membrane filtration plants do not require coagulation, flocculation, and sedimentation facilities for effective particle removal. These differences can reduce requirements for chemical storage and handling and residual-handling facilities and allow membrane plants to be more compact and automated. Furthermore, the more compact installation can result in considerable cost savings in densely populated areas or other areas where land costs are high.

The most significant advantage, however, is that the filtered water turbidity from membrane filters is independent of the concentration of particulate matter in the feed. The effluent from rapid granular filters is not independent of influent quality. Changes in raw-water chemistry without changes in pretreatment (i.e., adjustment of the coagulant dose) can cause the rapid granular filtration process to fail. Rapid granular filtration is sensitive to fluctuations in raw-water quality and the experience of the plant operators.

12-4 Properties of Membrane Materials

An understanding of the mechanisms that control membrane filtration begins with an understanding of the filtration medium. Important material properties, membrane chemistry, and physical structure are discussed in this section. Although MF and UF membranes are used for similar purposes in water treatment, some of the properties are different.

Material Properties

Membrane performance is affected strongly by the physical and chemical properties of the material. The ideal membrane material is one that can produce a high flux without clogging or fouling and is physically durable, chemically stable, nonbiodegradable, chemically resistant, and inexpensive. Important characteristics of membrane materials, methods of determination, and effects on membrane performance are described in Table 12-6.

One of the important characteristics in Table 12-6 with respect to membrane fouling is hydrophobicity. Hydrophilic materials, which like contact

Table 12-6
Important properties of membrane materials[a]

Property	Method of Determination	Impact on Membrane Performance
Retention rating (pore size or molecular weight cut-off)	Bubble point, challenge tests	Controls the size of material retained by the membrane, making it one of the most significant parameters in membrane filtration. Also affects head loss.
Hydrophobicity	Contact angle	Reflects the interfacial tension between water and the membrane material. Hydrophobic materials "dislike" water; thus, constituents from the water accumulate at the liquid–solid interface to minimize the interfacial tension between the water and membrane. In general, hydrophobic materials will be more susceptible to fouling than hydrophilic materials.
Surface or pore charge	Streaming potential	Reflects the electrostatic charge at the membrane surface. Repulsive forces between negatively charged species in solution and negatively charged membrane surfaces can reduce fouling by minimizing contact between the membrane and fouling species. In UF, electrostatic repulsion can reduce the passage of like-charged solutes. Membranes fabricated of uncharged polymers typically acquire some negative charge while in operation.
Surface roughness	Atomic force microscopy	Affects membrane fouling; some studies have shown rough materials will foul more than smooth materials.
Porosity (surface and bulk)	Thickness/weight measurements	Affects the head loss through the membrane; higher porosity results in lower head loss.
Thickness	Thickness gauge, electron microscopy	Affects the head loss through the membrane; thinner membranes have lower head loss.
Surface chemistry	ATR/FTIR, SIMS, XPS	Affects fouling and cleaning by influencing chemical interactions between the membrane surfaces and constituents in the feed water.
Chemical and thermal stability	Exposure to chemicals and temperature extremes	Affects the longevity of the membrane; greater chemical and temperature tolerance allows more aggressive cleaning regimes with less degradation of the material.
Biological stability	Exposure to organisms	Affects the longevity of the membrane; low biological stability can result in the colonization and physical degradation of the membrane material by microorganisms.

(Continued)

Table 12-6 *(Continued)*

Property	Method of Determination	Impact on Membrane Performance
Chlorine/oxidant tolerance	Exposure to chlorine/ oxidants	Affects the ability to disinfect the membrane equipment. Routine disinfection prevents microbial growth on the permeate side of membrane surfaces and prevents biological degradation of membrane materials (increasing the longevity of the membrane).
Mechanical durability	Mechanical tests	Affects the ability of the material to withstand surges due to operation of valves and pumps.
Internal physical structure, tortuosity	Electron microscopy	Affects the hydrodynamics of flow and particle capture. There are no standard procedures for quantifying the tortuosity or internal structure of membranes.
Cost	Material cost	Affects the cost of the membrane system.

[a]Abbreviations: ATR/FTIR = attenuated total reflectance/Fourier transform infrared spectrometry, SIMS = secondary ion mass spectrometry, XPS = X-ray photoelectron spectrometry.

with water, tend to have low fouling tendencies, whereas hydrophobic materials may foul extensively. Hydrophobicity is quantified by contact angle measurements, in which a droplet of water or bubble of air is placed against a membrane surface, and the angle between the surface and water or air is measured. Hydrophobic surfaces have a high contact angle (the water beads like on a freshly waxed car), whereas hydrophilic surfaces have a low contact angle (the water droplets spread out). Techniques for measuring contact angle are demonstrated on Fig. 12-11. Contact angle measurements vary widely because of differences in measurement techniques and variables such as surface roughness but typically range from about 40° to 50° for cellulose acetate to about 110° for polypropylene (Cheryan, 1998).

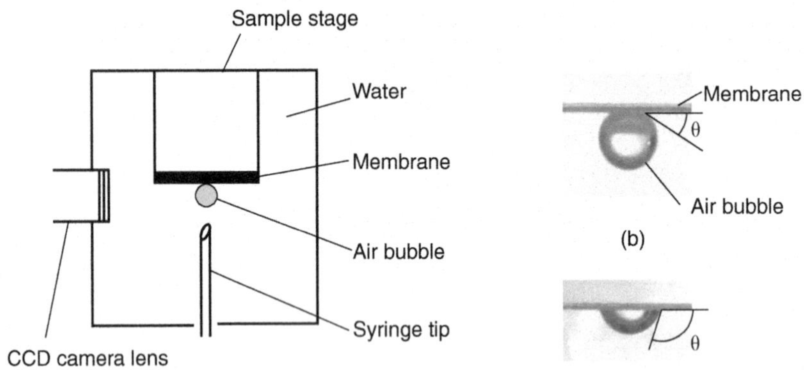

(a) (c)

Figure 12-11
Captive bubble contact angle measurements for determination of hydrophobicity: (a) contact angle measurement apparatus, (b) hydrophilic surface (low contact angle), and (c) hydrophobic surface (high contact angle).

Hydrophobicity is affected strongly by the chemical composition of the polymer comprising the material. Polymers that have ionized functional groups, polar groups (water is very polar), or oxygen-containing and hydroxyl groups (for hydrogen bonding) tend to be very hydrophilic. Unfortunately, chemical properties that improve hydrophilicity tend to reduce the chemical, mechanical, and thermal stability because water molecules act as plasticizers for hydrophilic materials (Anselme and Jacobs, 1996).

Material Chemistry

Lacking the existence of a perfect material, a variety of materials have been used. The two most common materials in early commercial membrane filtration systems were cellulose acetate (CA) and polypropylene (PP), but their use has been declining. The most common polymeric materials currently used in water treatment are polyvinylidene fluoride (PVDF), polysulfone (PS), and polyethersulfone (PES). Ceramic membrane may also be gaining in popularity. Some membrane manufacturers consider the composition of their membranes to be proprietary and do not release information on their material chemistry. The chemical structure of these polymers is shown on Fig. 12-12 and important properties are given in Table 12-7.

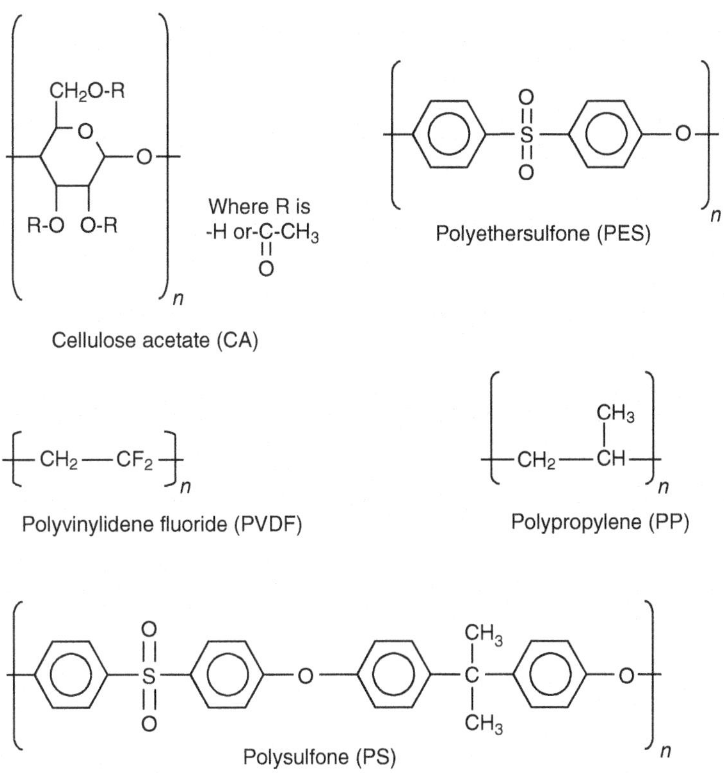

Figure 12-12
Chemical structure of common polymeric MF and UF membrane materials.

Table 12-7
Characteristics of common membrane materials

Membrane Material	Characteristics
Cellulose acetate (CA)	CA is the most hydrophilic of common industrial-grade membrane materials, which helps to minimize fouling and maintain high flux values. The material is easy to manufacture, inexpensive, and available in a wide range of pore sizes. Has been losing favor for membrane filters because of higher susceptibility to biological degradation, lack of tolerance to continuous exposure or high concentrations of free chlorine, gradual decline in the flux over its lifetime due to compaction, and lack of tolerance to aggressive cleaning chemicals or temperatures above 30°C.
Polysulfone (PS)/ polyethersulfone (PES)	PS and PES are moderately hydrophobic and have excellent chemical tolerance and biological resistance. They can withstand free chlorine contact to 200 mg/L for short periods of time for cleaning, pH values between 1 and 13, and temperatures to 75°C. Aggressive cleaning and disinfecting is possible.
Polyvinylidene fluoride (PVDF)	PVDF is moderately hydrophobic and has excellent durability, chemical tolerance, and biological resistance. It can withstand continuous free chlorine contact to any concentration, pH values between 2 and 10, and temperatures to 75°C. Aggressive cleaning and disinfecting is possible.
Polypropylene (PP)	PP is the most hydrophobic of common industrial-grade membrane materials. Only MF membranes are available in PP; the material is too hydrophobic to allow water to pass through the small pore spaces in UF membranes. It is durable, chemically and biologically resistant, and tolerant of moderately high temperatures and pH values between 1 and 13, which allows aggressive cleaning. It has been losing favor for membrane filters because it is not tolerant to chlorine, which hinders the ability to control biological growth.
Ceramic	Ceramic membranes are configured as rigid monolithic elements. The material is hydrophilic, rough, and can withstand high operating pressure and temperature. They have excellent chemical and pH tolerance. Aggressive cleaning and disinfecting is possible.

Membrane Structure

The structure, porosity, and transport properties of most MF membranes are relatively constant throughout their depth (this structure is called homogeneous). Theoretically, homogeneous membranes perform identically regardless of which direction filtration is proceeding. In contrast, UF membranes have an asymmetric (also called anisotropic or "skinned") structure, which means that the morphology varies significantly across the depth of the membrane. A homogeneous membrane was shown on Fig. 12-4a, and the structure of an asymmetric membrane, consisting of an active layer and a support layer, is shown on Fig. 12-13. The active and support layers have separate functions.

Filtration occurs at the active layer in asymmetric membranes, which is a thin skin with low porosity and very small void spaces. The low porosity and small pores generate significant resistance to flow, which must be minimized by making the active layer as thin as possible. The active layer is so thin that it has no mechanical durability. Thus, the remainder of the membrane is a highly porous layer that provides support but produces very little hydraulic resistance. The support layer accounts for the majority of the membrane thickness. Asymmetric membranes allow the filtration and mechanical properties to be designed separately. Filtration through an asymmetric membrane is not the same in both directions. Filtration in the "wrong" direction would cause the voids in the support layer to become clogged and may cause the active layer to separate from the rest of the membrane. To prevent clogging, some commercial asymmetric membranes have active layers on both surfaces of the membrane with a support layer sandwiched between the two active layers.

Microfiltration and UF membranes have different porosities. The porosity of MF membranes varies widely, and values ranging from 30 to 90 percent have been reported. Theoretically, the porosity of homogeneous membranes should be constant throughout the depth of the membrane.

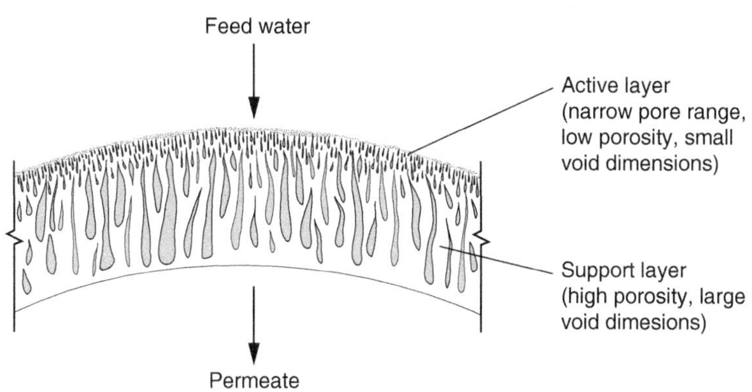

Figure 12-13
Structure of an asymmetric UF membrane.

The porosity of the active layer of UF membranes is low and ranges from 0.5 to 10 percent, whereas the porosity in the support structure is considerably higher (50 to 90 percent).

12-5 Particle Capture in Membrane Filtration

For regulatory purposes in the United States, membrane filtration is defined as "a pressure or vacuum driven separation process in which particulate matter larger than 1 μm is rejected by an engineered barrier primarily through a size exclusion mechanism and which has a measurable removal efficiency of a target organism that can be verified through the application of a direct integrity test" (U.S. EPA, 2006, p. 702). The principles by which membranes are rated, particles are captured, and performance is demonstrated is discussed in this section.

Retention Rating

One of the most significant parameters in membrane filtration is the size of material retained. Microfiltration and UF membranes are currently rated with different systems, making them difficult to compare. The retention rating of MF membranes is based on the diameter of material that is retained by the membrane; for example, a 0.2-μm MF membrane should hypothetically retain 100 percent of 0.2-μm-diameter particles. The MF retention rating is frequently called the nominal pore diameter or pore size throughout the membrane industry, but those terms are misnomers. As was shown on Fig. 12-3, the "pores" in MF membranes are tortuous voids with a wide size distribution, not cylindrical holes of a particular diameter. It is generally accepted that the average void space dimension is somewhat larger than the membrane retention rating (Cheryan, 1998). The pore size value is a nominal rating, so some particles smaller than the pore size can be retained and some particles larger may be able to penetrate the membrane. The retention rating for MF membranes used in water treatment is typically between 0.1 and 1 μm. Membranes with retention ratings in this range will completely retain bacteria and protozoan structures such as *Giardia lamblia* cysts and *Cryptosporidium parvum* oocysts. Because of their small size, viruses may not be completely retained by MF membranes.

Membrane manufacturers use two approaches for defining the retention rating of UF membranes. Some manufacturers use a pore size rating similar to MF membranes, with pore sizes of 0.01 to 0.04 μm being common. For others, the retention rating for UF membranes is based on the molecular weight of material retained by the membrane and is called the molecular weight cutoff (MWCO) or nominal molecular weight limit (NMWL). The first applications of UF membranes were for fractionating macromolecules, so the original focus for classifying UF membranes was on molecular weight rather than size. Membrane filtration for water treatment, however, is principally concerned with retaining materials of a particular size, such as viruses, which can be as small as 0.025 μm. Unfortunately, the diameter of solids retained by a UF membrane is only loosely related to the MWCO value and depends on various physical and chemical properties (shape, electrostatic charge, etc.) of the solid. The standard procedure for determining the MWCO value of a UF membrane involves filtration of dextran solutions with varying average molecular weights (ASTM, 2001c). Dextran is a branched polysaccharide that might be expected to have substantially different physical and chemical properties from viruses or other particles. In addition, the MWCO of UF membranes is based on the molecular weight at 90 percent rejection. The difference between the rating of MF and UF membranes is shown on Fig. 12-14.

The hydrodynamic diameter of molecules can be roughly estimated from the molecular weight. For instance, data in Ioan et al. (2000) suggest the following empirical relationship between the hydrodynamic diameter and molecular weight of dextran:

$$d_H = 0.11\,(\text{MW})^{0.46} \qquad (12\text{-}2)$$

where d_H = hydrodynamic diameter of dextran molecule, nm
MW = molecular weight, g/mol

Researchers have attempted to use relationships such as Eq. 12-2 to estimate the pore size of UF membranes. In addition, analytical techniques such as electron microscopy, atomic force microscopy, thermoporometry, and biliquid permporometry have been used to estimate the pore size of UF membranes (Kim et al., 1994).

The MWCO for UF membranes range from about 1000 daltons (Da) to about 500,000 Da. These MWCO values correspond to an ability to retain particles ranging from about 1 to 30 nm in diameter (Cheryan, 1998). Comparing these values to the size of viruses, it is clear that some UF membranes can completely retain viruses but others may not, depending

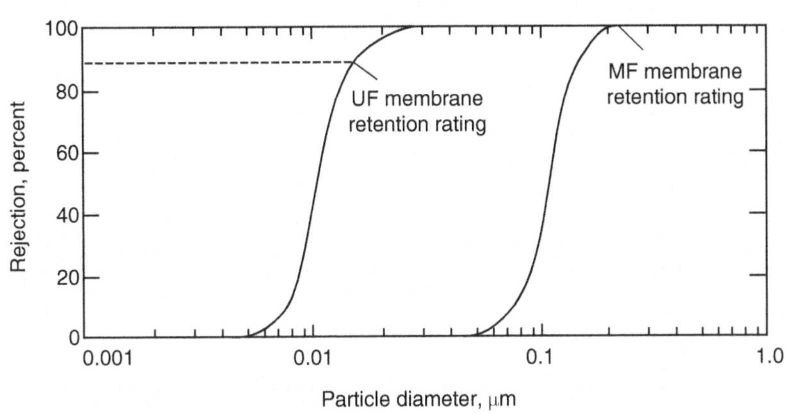

Figure 12-14
Determination of retention ratings for MF and UF membranes.

on the MWCO. Studies have shown, for instance, greater than 7-log removal of MS2 bacteriophage (a model virus) with a 100,000-Da UF membrane but less than 1-log removal with a 500,000-Da UF membrane (Jacangelo et al., 1995).

It should be noted that there is overlap between the size of pores of MF and UF membranes, and there are no standard specifications that classifies a particular product as one or the other. Often, the classification of a particular product as an MF or UF membrane depends on the marketing strategy of the manufacturer and whether the retention rating was measured as a pore size or MWCO. Furthermore, classification as an MF or UF membrane does not guarantee a particular level of removal efficiency for specific pathogen organisms. For that, challenge testing is required, as described later in this section.

Rejection and Log Removal

The fraction of material removed from the permeate stream is called rejection:

$$R = 1 - \frac{C_p}{C_f} \tag{12-3}$$

where R = rejection, dimensionless
C_p = permeate concentration, mol/L or mg/L
C_f = feed water concentration, mol/L or mg/L

Rejection can be calculated for bulk measures of particulate matter (e.g., turbidity, particle counts) or individual components of interest (e.g., *Cryptosporidium* oocysts). In membrane filtration, the concentration of some components in the permeate can be several orders of magnitude lower than in the feed. Many significant figures must be retained to quantify rejection if Eq. 12-3 is used. In these cases, the log removal value (LRV) (see Sec. 4-5) is used:

$$\text{LRV} = \log(C_f) - \log(C_p) = \log\left(\frac{C_f}{C_p}\right) \tag{12-4}$$

A comparison of the calculation of rejection and log removal value is demonstrated in Example 12-2.

Filtration Mechanisms

The primary mechanism for removing particles from solution in membrane filtration is straining, but removal is also affected by adsorption and cake formation. These removal mechanisms are depicted on Fig. 12-15.

STRAINING

Straining (also called sieving or steric exclusion) is the dominant filtration mechanism in membrane filtration. Nominally, particles larger than the retention rating of the membrane collect at the surface while water and

Example 12-2 Calculation of rejection and log removal value

During testing of a prototype membrane filter, bacteriophage concentrations of 10^7 mL^{-1} and 13 mL^{-1} were measured in the influent and effluent, respectively. Calculate the rejection and log removal value.

Solution

1. Calculate rejection using Eq. 12-3:

$$R = 1 - \frac{C_p}{C_f} = 1 - \frac{13\ \text{mL}^{-1}}{10^7\ \text{mL}^{-1}} = 0.9999987$$

2. Calculate log removal value using Eq. 12-4:

$$\text{LRV} = \log\left(\frac{C_f}{C_p}\right) = \log\left(\frac{10^7\ \text{mL}^{-1}}{13\ \text{mL}^{-1}}\right) = 5.89$$

Comment

Note that seven significant digits are necessary to express rejection adequately in arithmetic units, but only three significant digits are necessary to express log removal value for this example. Also note that LRV = 5 corresponds to 99.999 percent and LRV = 6 corresponds to 99.9999 percent rejection (i.e., the log removal value equals the "number of 9's").

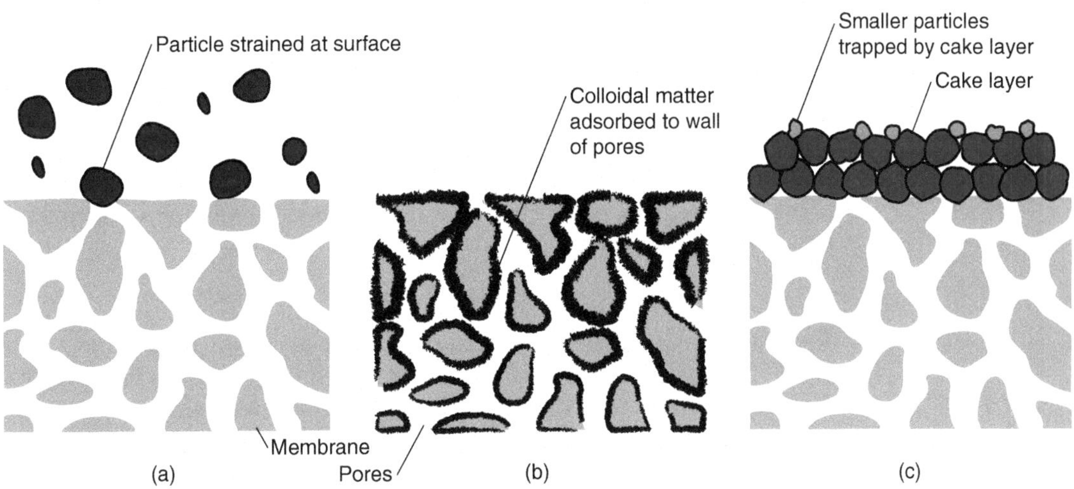

Figure 12-15
Mechanisms for rejection in membrane filtration. (a) Straining occurs when particles are physically retained because they are larger than the pores. (b) Adsorption occurs when material small enough to enter pores adsorbs to the walls of the pores. (c) Cake filtration occurs when particles that are small enough to pass through the membrane are retained by a cake of larger material that collects at the membrane surface.

smaller particles pass through. This view, however, suggests that the relationship between particle size and retention rating is a step function, with $R = 100\%$ for all particles larger than the retention rating and $R = 0\%$ for all smaller particles. As shown on Fig. 12-14, particle removal is not a simple step function at the retention rating. This nonideal performance occurs when the particle size is close to the membrane retention rating and is caused by the variability of pore size dimensions, nonspherical shape of the particles, and other interactions such as electrostatic repulsion.

As is evident from Fig. 12-3c, the tortuous interconnecting voids in membrane filters have a distribution of sizes, including some larger than the retention rating. Thus, particles smaller than the retention rating may be trapped in smaller passageways, and larger particles may pass through the membrane in other areas. The existence of larger void spaces is particularly important when high rejection values are required.

A second source of nonideal rejection arises from particle dimensions. Particles in natural systems can have shape characteristics significantly different from the materials used to determine the retention rating. Rodshaped bacteria and linear macromolecules may be very long in one dimension and considerably smaller in others and may not be adequately described by an average diameter. Rod-shaped bacteria near the retention rating of the membrane may or may not be rejected depending on orientation when they approach the membrane. Some small particles, particularly large proteins or other macromolecules, may be flexible, assuming a spherical shape when in solution but becoming more linear when forced through a membrane under pressure. Thus, particles that appear to be slightly larger than the retention rating may pass through the membrane.

Interactions between particles and the membrane can affect rejection when the particle size is close to the membrane retention rating. Typically, both particles and membrane surfaces are negatively charged. Electrostatic interactions may prevent the particles from entering the pores even if the physical size would permit passage. Ferry (1936) reported nonideal behavior even under near-ideal filtration conditions (filtration of a monodisperse dispersion through a track-etched membrane with uniform cylindrical pores).

ADSORPTION

Natural organic matter adsorbs to membrane surfaces (Jucker and Clark, 1994). Thus, soluble materials may be rejected even though their physical dimensions are orders of magnitude smaller than the membrane retention rating. Adsorption may be an important rejection mechanism during the early stages of filtration with a clean membrane. The adsorption capacity is quickly exhausted, however, and adsorption is not an effective mechanism in long-term operation of membrane filters. Nevertheless, adsorption has a profound impact on membrane operations. Adsorbed material can reduce the size of voids throughout the membrane, increasing the ability of the membrane to retain smaller material by straining. In addition, adsorption has been implicated as a prime cause of membrane fouling by NOM.

CAKE FORMATION

During filtration, a clean membrane will quickly accumulate a cake of solids at the surface due to straining. This surface cake acts as a filtration medium, providing another mechanism for rejection. The surface cake is often called a "dynamic" membrane, because its filtering capability varies with time, growing in thickness during filtration but being partially or wholly removed during backwashing. Mathematical modeling of cake filtration is similar to granular filtration, and equations such as the Kozeny equation (see Chap. 11) are used to calculate the additional resistance to flow resulting from the formation of surface cakes.

Removal of Microorganisms

The principal microorganisms of concern in water treatment are (1) protozoa and helminths, (2) bacteria, and (3) viruses. The removal of each of these is considered below.

REMOVAL OF PROTOZOA AND HELMINTHS

Giardia lamblia cysts and *C. parvum* oocysts, which are highly resistant to chemical disinfectants, are at least 10 times larger than the retention ratings of MF and UF membranes. Rejection of greater than 7 log has been observed for both MF and UF membranes (Jacangelo et al., 1991, 1995), as would be expected due to straining. Indeed, reported rejection values are limited only by the influent concentration of organisms. For instance, if 10^6 cysts/L are measured in the feed and the detection limit in the permeate is 1 cyst/L, the calculated log removal value is LRV $\geq$ 6.

REMOVAL OF BACTERIA

Bacteria range in size from 0.1 to 100 μm (Madigan et al., 1997). Being considerably larger than the retention rating, complete rejection via straining is expected for UF membranes. In addition, most species of bacteria should be rejected completely by MF membranes through straining. In many studies, bacteria are removed to below the detection limit; in some cases, rejection was higher than 8 log (U.S. EPA, 2001). In other studies, reported rejection was lower because of contamination and regrowth interferences. Some bacterial species are near the retention ratings of MF membranes. In these cases, lower rejection may be possible due to the factors mentioned earlier.

REMOVAL OF VIRUSES

The smallest viruses have a diameter of about 25 nm. At this size, viruses are considerably smaller than the retention rating of MF membranes and are similar to that of UF membranes. For MF membranes, straining, adsorption, and cake filtration all contribute to rejection, and virus rejection can vary from LRV < 1 to LRV > 4. Madaeni et al. (1995) studied poliovirus rejection by 0.2-μm MF membranes and found complete rejection (LRV ≥ 4.5) during the first 15 min of filtration, but rejection quickly dropped to LRV = 1.31 after 30 min of filtration and gradually rose to LRV = 1.71 during the next 5 h. Adsorption appears to be significant in initial rejection, but the adsorption capacity was quickly exhausted. After 30 min, rejection may have been due to straining through small void spaces or continued adsorption. The gradual increase in rejection over time suggests that straining was increasing as the pores became more restricted with previously deposited virus particles. Jacangelo et al. (1995) also observed a gradual increase in virus rejection that correlated with a decrease in specific flux over a period of 45 days. Fouling caused a reduction in the pore diameter or formation of a surface cake that was able to increase virus rejection. Cake filtration also plays an important role in virus rejection by MF membranes. Preloading an MF membrane with kaolinite clay (Jacangelo et al., 1995) and increasing the turbidity of the feed water (Madaeni et al., 1995) result in higher rejection of viruses. Increased rejection of viruses with higher turbidity may also be due to attachment of the viruses to the particle prior to filtration.

Ultrafiltration membranes with low MWCO ratings may be able to achieve complete rejection of viruses, but UF membranes with higher MWCO ratings might not. It was noted earlier that the pore size of UF membranes may range from 1 to 30 nm depending on the MWCO. Jacangelo et al. (1995) found complete rejection (LRV > 7.2) of MS2 bacteriophage, a model virus with a diameter of about 25 nm, with a 100,000-Da UF membrane but LRV < 1 with a 500,000-Da UF membrane.

The overall implication for water treatment is that straining is only effective for particles significantly larger than the retention rating of the membrane. Cake filtration and adsorption may provide added rejection of smaller material, but are not considered effective removal mechanisms from a regulatory point of view. Thus, MF membranes should not be relied upon for complete removal of viruses (although some removal credit may be warranted), and poor selection of UF membranes may also provide insufficient removal for viruses. To validate the ability of MF and UF membranes to remove specific microorganisms, challenge testing is performed.

Challenge Testing

Challenge testing is a process in which the ability of a membrane product to remove specific target organisms is determined in carefully controlled tests. Specific requirements in the Long Term 2 Enhanced Surface Water Treatment Rule (LT2ESWTR) (U.S. EPA, 2005) are focused on challenge testing for *Cryptosporidium* removal efficiency, but the general methods could be used for other microbial contaminants such as viruses, bacteria, or other protozoa. Challenge tests establish the maximum removal credit allowed for a particular product. Once performance has been verified, site-specific testing by utilities is not required. The LT2ESWTR requires that the modules tested be similar in design, materials, and construction to full-scale modules, although the test modules do not necessarily need to be full size. The number of modules testing should be determined on a scientifically defensible basis to provide results that take manufacturing variability into account. Operating conditions representative of full-scale operation, such as maximum flux and recovery are necessary. The target microorganisms or particles should be measured directly and not inferred from a surrogate like turbidity. The concentration of particles in the feed water should be 6.5 log units higher than the detection limit in the permeate water to ensure that high removal efficiency can be demonstrated. Finally, nondestructive performance tests (NDPT) (such as bubble point or pressure decay tests, discussed later) should be performed at the same time, so that the NDPT can be performed on modules after manufacture and related to the results of the challenge testing.

12-6 Hydraulics of Flow through Membrane Filters

The flow of water through MF and UF membranes follows the fundamental law for flow through porous media known as Darcy's law:

$$v = k_P \frac{h_L}{L} \tag{12-5}$$

where v = superficial fluid velocity, m/s
k_P = hydraulic permeability coefficient, m/s
h_L = head loss across porous media, m
L = thickness of porous media, m

The hydraulic permeability coefficient in Darcy's law is an empirical parameter that is used to describe the proportionality between head loss and fluid velocity and is dependent on media characteristics such as porosity and specific surface area. Although flow through membranes follows Darcy's law, the standard equation for membrane flow is written in a substantially different form. Flow is expressed in terms of volumetric flux J rather than superficial velocity, the driving force is expressed as transmembrane pressure ΔP rather than head loss (which are related by $\Delta P = \rho_w g h_L$), and media characteristics are expressed as a resistance coefficient (the inverse of a permeability coefficient). In addition, the membrane flow equation includes the fluid viscosity explicitly (Darcy's law buries it in the permeability coefficient) because viscosity has a significant impact on flux and is easy to determine (via temperature). Finally, the membrane flux equation incorporates the membrane thickness into the resistance coefficient. The equation for membrane flux is

$$J = \frac{Q}{a} = \frac{\Delta P}{\mu \kappa_m} \tag{12-6}$$

where J = volumetric water flux through membrane, $L/m^2 \cdot h$ or m/s
Q = flow rate, L/h
a = membrane area, m^2
ΔP = differential pressure across membrane, bar
μ = dynamic viscosity of water, $kg/m \cdot s$
κ_m = membrane resistance coefficient, m^{-1}

The membrane resistance coefficient can be calculated from laboratory experiments so that flux through a new membrane can be determined for other pressure or temperature conditions.

The linear relationship between flux and pressure in Eq. 12-6 suggests that the flux can be maximized by operating at the highest possible transmembrane pressure. While this relationship may be true for deionized water, high-pressure operation may not be preferable during filtration of natural waters. Fouling can be exacerbated by high-pressure operation (Cheryan, 1998), so a balance must be struck between flux and fouling. Fouling will be presented in detail later in this chapter.

Ideally, it would be desirable to calculate flux from measurable parameters that describe the internal structure of MF and UF membranes, such as porosity, nominal pore diameter, specific surface area, and membrane thickness, as is done for clean-bed head loss in granular filtration. These parameters, however, are difficult to measure, and the amorphous internal structure of MF and UF membranes (refer to Fig. 12-3c) cannot be described mathematically with any great accuracy. In addition, it will be shown later in this chapter that the volumetric flux through a full-scale membrane filter is influenced more by fouling than by the intrinsic clean-membrane resistance. As a result, currently no reliable models allow flux to be calculated from easily measurable fundamental parameters. To account for membrane fouling, Eq. 12-6 can be extended by adding additional resistance terms in the denominator, as will be presented later in the chapter. Calculation of the membrane resistance coefficient is demonstrated in Example 12-3.

Temperature and Pressure Dependence

During operation, changes in permeate flux due to fouling are monitored to determine when cleaning is necessary. Because flux is dependent on pressure and water viscosity (see Eq. 12-6), determination of the extent of fouling is confounded by simultaneous changes in pressure and temperature (which changes viscosity). In temperate climates, water temperatures can vary by more than 20°C between summer and winter. Due to the

Example 12-3 Calculation of membrane resistance coefficient

An MF membrane is tested in a laboratory by filtering clean, deionized water and the flux is found to be 850 $L/m^2 \cdot h$ at 20°C and 0.9 bar. Calculate the membrane resistance coefficient.

Solution

Rearrange Eq. 12-6 to solve for the membrane resistance coefficient. The dynamic viscosity of water at 20°C, from Table C-1 in App. C, is 1.00×10^{-3} $kg/m \cdot s$. Also recall that 1 bar = 100 kPa = 10^5 N/m^2 = 10^5 $kg/s^2 \cdot m$.

$$\begin{aligned}\kappa_m &= \frac{\Delta P}{\mu J} = \frac{(0.9 \times 10^5 \text{ kg/s}^2 \cdot \text{m})(3600 \text{ s/h})(10^3 \text{ L/m}^3)}{(1.00 \times 10^{-3} \text{ kg/m} \cdot \text{s})(850 \text{ L/m}^2 \cdot \text{h})} \\ &= 3.81 \times 10^{11} \text{ m}^{-1}\end{aligned}$$

temperature dependence of water viscosity, the flux through a membrane can be 70 percent higher in the summer than in the winter. Temperature variations are usually accommodated by calculating the equivalent flux at a standard temperature:

$$J_s = J_m \left(\frac{\mu_m}{\mu_s} \right) \tag{12-7}$$

where J_m = flux at measured temperature, $L/m^2 \cdot h$
J_s = flux at standard temperature (typically 20°C), $L/m^2 \cdot h$
μ_m = dynamic viscosity of water at measured temperature, $kg/m \cdot s$
μ_s = dynamic viscosity of water at standard temperature, $kg/m \cdot s$

The dynamic viscosity can be obtained from tabular data or calculated from one of a variety of expressions that relate the viscosity of water to temperature. A relationship often used in membrane operations is (ASTM, 2001a)

$$J_s = J_m (1.03)^{T_s - T_m} \tag{12-8}$$

where T_m = measured temperature, °C
T_s = standard temperature, °C

When using a standard temperature of 20°C, Eq. 12-8 is accurate to within 5 percent over a temperature range of 1 to 28°C, which covers most natural waters. At temperatures above 28°C, Eq. 12-8 becomes increasingly inaccurate and a more rigorous expression should be used.

The factor in Eq. 12-8 accounts only for the effect of water viscosity. Temperature may also have an effect on the membrane material, such as swelling of the material at higher temperature. Some manufacturers have developed temperature correction formulas that account for changes in both the water viscosity and material properties. Manufacturer or sitespecific correction equations should be used, if available.

Flux is normalized for pressure by calculating specific flux, which is the flux at a standard temperature divided by the transmembrane pressure:

$$J_{\text{sp}} = \frac{J_s}{\Delta P} \tag{12-9}$$

where J_{sp} = specific flux at standard temperature, $\text{L/m}^2 \cdot \text{h} \cdot \text{bar}$

The specific flux is called membrane permeability when clean water (reagent-grade water, typically deionized to a resistivity of >10 MΩ· cm and DOC <0.2 mg/L) is being filtered through a new, unused membrane. Membrane permeability is measured in laboratory experiments but rarely or never determined in full-scale installations because a large supply of deionized water is not available. Specific flux and membrane permeability are typically reported in units of $\text{L/m}^2 \cdot \text{h} \cdot \text{bar}$ or $\text{gal/ft}^2 \cdot \text{d} \cdot \text{atm}$.

When flux has been normalized to account for temperature and pressure variations, the effect of fouling can be determined, as illustrated in Example 12-4.

12-7 Membrane Fouling

Fouling is widely perceived to be the most significant issue affecting the design and operation of membrane filtration facilities (AWWA, 1992, 1998, 2005a). The results of a laboratory experiment of natural water filtration through a membrane are shown on Fig. 12-16. In this experiment, the membrane lost about half of its flow capacity in just a few hours. The individual steep trend lines represent individual 30-min filter runs. Backwash after each filter run recovers most of the lost flux as solids are flushed from the membrane surface; however, not all of the flux is recovered and a longer-term decline in performance is also evident. Fouling can be more dramatic in laboratory tests conducted at constant pressure (as this experiment was), but the overall trends shown in Fig. 12-16 are routinely observed in fullscale operating membrane filtration facilities. Fouling is characterized by the mechanism (pore blockage, pore constriction, and cake formation), by whether it can be removed (i.e., reversible or irreversible), and by the material causing it (particles, biofouling, and natural organic matter). This section examines each of these views of membrane fouling, followed by an introduction of several ways to model fouling.

Example 12-4 Calculation of specific flux

A membrane plant has a measured flux in March of 80 L/m^2 · h at 0.67 bar and 7°C. Four months later, in July, the measured flux is 85 L/m^2 · h at 0.52 bar and 19°C. Has a change in specific flux occurred? What is the change in percent? Has fouling occurred?

Solution

1. Calculate the specific flux in March.
   a. Calculate the flux in March at a standard temperature of 20°C using Eq. 12-8:

$$J_S = J_m(1.03)^{T_s - T_m} = (80\ \text{L/m}^2.\text{h})(1.03)^{20^\circ\text{C} - 7^\circ\text{C}}$$
$$= 117\ \text{L/m}^2 \cdot \text{h}$$

   b. Calculate the specific flux in March using Eq. 12-9:

$$J_{sp} = \frac{J_S}{\Delta P} = \frac{117\ \text{L/m}^2 \cdot \text{h}}{0.67\ \text{bar}} = 175\ \text{L/m}^2 \cdot \text{h} \cdot \text{bar}$$

2. Calculate the specific flux in July.
   a. Calculate the flux in July at a standard temperature of 20°C using Eq. 12-8:

$$J_s = J_m(1.03)^{Ts - Tm} = (85\ \text{L/m}^2 \cdot \text{h})(1.03)^{20^\circ\text{C} - 19^\circ\text{C}}$$
$$= 87.6\ \text{L/m}^2 \cdot \text{h}$$

   b. Calculate the specific flux in July using Eq. 12-9:

$$J_{sp} = \frac{J_s}{\Delta P} = \frac{87.6\ \text{L/m}^2 \cdot \text{h}}{0.52\ \text{bar}} = 168\ \text{L/m}^2 \cdot \text{h} \cdot \text{bar}$$

3. Calculate the percent loss of performance due to fouling:

$$\frac{175\ \text{L/m}^2 \cdot \text{h} \cdot \text{bar} - 168\ \text{L/m}^2 \cdot \text{h} \cdot \text{bar}}{175\ \text{L/m}^2 \cdot \text{h} \cdot \text{bar}} \times 100$$
$$= 4\%\ \text{loss of flux due to fouling}$$

Comment

The specific flux at 20°C has declined from 175 to 168 L/m^2 · h · bar. Thus, although the plant is operating at a higher flux with a lower pressure in July than it was in March, there has been a 4 percent loss of performance due to fouling.

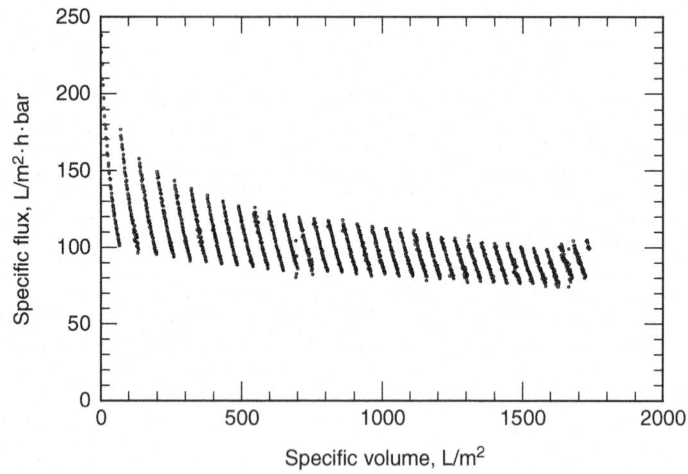

Figure 12-16
Fouling of a membrane filter during filtration of natural water.

Mechanisms of Fouling

Membrane fouling is traditionally visualized as occurring through three mechanisms—pore blocking, pore constriction, and cake formation—as shown on Fig. 12-17. The similarity between these mechanisms and the mechanisms for particle retention on Fig 12-15 should be evident. Pore blocking occurs when the entrance to a pore is completely sealed by a particle. For this mechanism, the membrane is viewed as a plate with orifices in it, and hydraulic resistance to flow is proportional to the net area of open pores. As a portion of the pore area is blocked at the surface, flow declines by a commensurate amount. While this mechanism may be

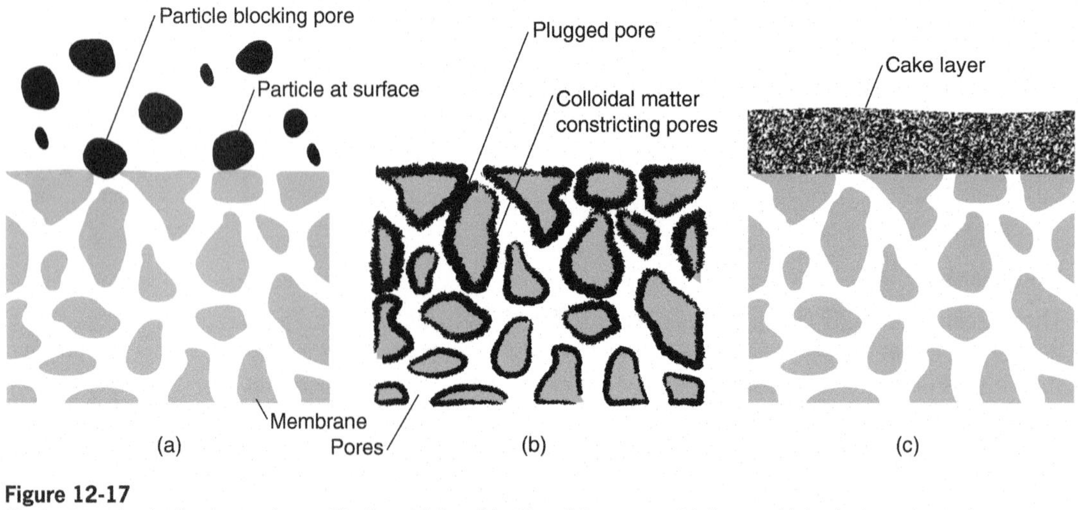

Figure 12-17
Mechanisms for fouling in membrane filtration: (a) Pore blocking, (b) pore constriction, and (c) cake layer formation.

relevant on a track-etched membrane (see Table 12-3), it might have minimal significance in the fouling of commercial membranes for water treatment. As shown on Fig. 12-3, commercial membranes have a depth of tortuous interconnecting voids with a flow path much longer than the pore width. Since typical membrane thickness ranges from 0.1 to 0.6 mm and the pore width is 0.01 to 0.1 μm, the pores are 1000 to 10,000 times longer than they are wide. Hydraulic resistance occurs throughout this depth, and the interconnectedness of the pores would allow flow to redistribute within the membrane matrix if a portion of the pores were plugged at the surface.

Pore constriction is the reduction of the void volume within a membrane due to adsorption of materials within the pores. Several essential elements must take place for pore constriction to occur. First, the materials must be smaller than the pore size of the membrane so they can penetrate into the membrane matrix instead of being sieved at the surface. Second, they must be transported to the pore walls either by diffusion or hydrodynamic conditions. Although the residence time through the membrane is short (less than 0.5 s), the characteristic diffusion time of a colloid to a pore wall is about 4 orders of magnitude shorter, so sufficient opportunity for diffusion exists (Howe, 2001). Third, materials must have an affinity for attaching to the pore walls, without which they would pass right through the membrane. Research has demonstrated that hydrophobic membranes foul more than hydrophilic ones, and hydrophobic materials in the feed water can cause greater fouling. Concepts of particle stability presented in Chap. 9 are also relevant here. Finally, the attached material must be sufficiently large to constrict the pore dimensions. Research has shown that high-MW and colloidal organics cause more fouling than low-MW dissolved materials. Low-MW dissolved materials would not have as large of an impact on pore dimensions as colloidal materials.

Particles that are too large to enter the pores collect on the membrane surface in a porous mat called a filter cake. Sediment larger than 0.2 mm is typically prefiltered with cartridge filters to protect the membrane, so the cake is initially composed of material between 0.2 mm and the membrane retention rating. The cake layer acts as a "dynamic" filter and can retain additional smaller material, but also generates hydraulic resistance to flow as it does so. The cake layer can prevent particles smaller than the retention rating from reaching the membrane, improving filtration effectiveness and possibly minimizing fouling from pore constriction.

Reversibility of Fouling

Fouling can be characterized as irreversible or reversible. An idealized graph of specific flux over time is shown on Fig. 12-18. Full-scale facilities generally operate in a constant-flux mode; therefore, a decline in specific flux is caused by an increase in transmembrane pressure while flux remains constant. A decline in specific flux occurs during initial operation, and a portion of this flux loss cannot be recovered during backwashing and

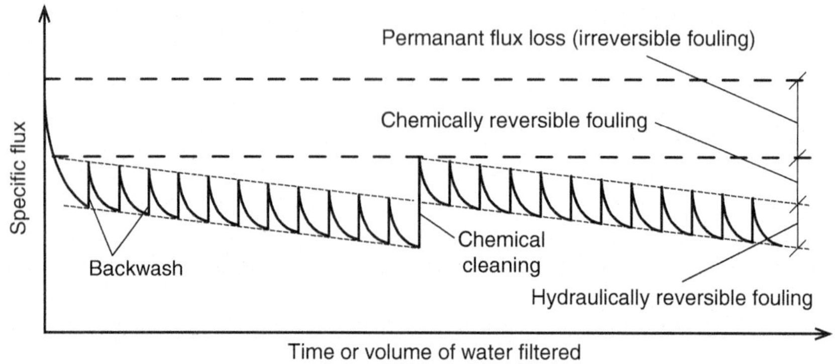

Figure 12-18 Variation in specific flux during filtration of natural waters. The loss of specific flux from the initial clean membrane permeability, which cannot be recovered by backwashing or cleaning, is called irreversible fouling; that which can be recovered is called reversible fouling.

cleaning operations. Permanent flux loss is called irreversible fouling and depends on the source water quality as well as the type of membrane used. The specific flux declines further during each filter run (normally recorded as an increase in transmembrane pressure) but a significant portion can be recovered during backwashing. This loss of flux that can be recovered during backwashing is called hydraulically reversible fouling. Fouling due to cake formation is largely reversible during backwash. The longer-term, slower decline in specific flux over multiple filter runs is due to the slow adsorption and clogging of materials within the membrane matrix (pore constriction), which can be dissolved and removed during chemical cleaning. The loss of flux that can be recovered during cleaning is called chemically reversible fouling.

Resistance-in-Series Model

As noted in the previous sections, several factors may contribute to resistance to flow. The resistance-in-series model applies a resistance value to each component of membrane fouling, assuming that each contributes to hydraulic resistance and that they act independently from one another. Typical forms of the resistance-in-series model are

$$J = \frac{\Delta P}{\mu\left(\kappa_m + \kappa_{ir} + \kappa_{hr} + \kappa_{cr}\right)} \tag{12-10}$$

$$= \frac{\Delta P}{\mu\left(\kappa_m + \kappa_c + \kappa_p\right)} \tag{12-11}$$

where κ_m = membrane resistance coefficient, m^{-1}
κ_{ir} = irreversible fouling resistance coefficient, m^{-1}
κ_{hr} = hydraulically reversible fouling resistance coefficient, m^{-1}
κ_{cr} = chemically reversible fouling resistance coefficient, m^{-1}
κ_c = cake layer resistance coefficient, m^{-1}
κ_p = pore constriction resistance coefficient, m^{-1}

The resistance-in-series equation can be applied to any number of individual resistances, which may be due to irreversible and reversible components, specific fouling materials (organic fouling resistance, biological fouling resistance, etc.), or fouling mechanisms (cake fouling resistance, pore constriction fouling resistance, etc.).

Individual resistance coefficients can be calculated by selecting operating conditions in which individual forms of fouling can be isolated. The procedure for calculating individual resistance coefficients is demonstrated in Example 12-5.

Example 12-5 Calculation of resistance coefficients

The MF membrane in Example 12-3 is used under full-scale conditions in a water treatment facility, producing a flux of 84 L/m^2 · h at 1.1 bar just before cleaning and 106 L/m^2 · h at 0.52 bar immediately after cleaning, both at a standard temperature of 20°C. Calculate values for the membrane resistance coefficient, irreversible fouling resistance coefficient, and chemically reversible fouling resistance coefficient.

Solution

1. The membrane resistance coefficient was calculated in Example 12-3 (under conditions when $\kappa_{ir} = 0$ and $\kappa_{cr} = 0$) and found to be 3.81×10^{11} m^{-1}.
2. Determine the irreversible fouling resistance coefficient.
   a. The viscosity of water at 20°C from Table C-1 in App. C is 1.00×10^{-3} kg/m · s. Also recall that 1 bar = 100 kPa = 10^5 N/m^2 = 10^5 kg/s^2 · m.
   b. The reversible component of fouling is removed by cleaning, so immediately after cleaning the chemically reversible fouling resistance coefficient is zero ($\kappa_{cr} = 0$). The only factors that cause resistance to flow are the membrane resistance and the irreversible fouling resistance, so the resistance-in-series equation can be written

$$J = \frac{\Delta P}{\mu\left(\kappa_m + \kappa_{ir}\right)} \tag{1}$$

   c. Rearrange Eq. 1 to solve for κ_{ir}:

$$\kappa_{ir} = \frac{\Delta P}{\mu J} - \kappa_m = \frac{(0.52 \times 10^5 \text{ kg/s}^2 \cdot \text{m})(3600 \text{ s/h})(1 \times 10^3 \text{ L/m}^3)}{(1.00 \times 10^{-3} \text{ kg/m} \cdot \text{s})(106 \text{ L/m}^2 \cdot \text{h})}$$
$$-3.81 \times 10^{11} \text{ m}^{-1}$$
$$= 1.39 \times 10^{12} \text{ m}^{-1}$$

3. Determine the chemically reversible fouling resistance coefficient.
   a. Prior to cleaning, three components of resistance are present:

$$J = \frac{\Delta P}{\mu\left(\kappa_m + \kappa_{ir} + \kappa_{cr}\right)}$$

   b. Rearrange the above equation to solve for κ_{cr}:

$$\begin{aligned}\kappa_{cr} &= \frac{\Delta P}{\mu J} - \kappa_m - \kappa_{ir}\\ &= \frac{(1.1\times10^5\ \text{kg/s}^2\cdot\text{m})(3600\ \text{s/h})(1\times10^3\ \text{L/m}^3)}{(1.00\times10^{-3}\ \text{kg/m}\cdot\text{s})(84\ \text{L/m}^2\cdot\text{h})}\\ &\quad -3.81\times10^{11}\ \text{m}^{-1} - 1.39\times10^{12}\ \text{m}^{-1}\\ &= 2.94\times10^{12}\ \text{m}^{-1}\end{aligned}$$

Comment

In this example, the chemically reversible fouling resistance coefficient is the largest resistance and nearly an order of magnitude larger than the membrane resistance coefficient, which demonstrates the importance of fouling on overall membrane performance.

Fouling by Particles

Large particles form a cake on the membrane surface. Fouling due to cake formation is often modeled using the resistance-in-series model. In laboratory studies using well-defined synthetic suspensions such as monodisperse spherical latex particles, the cake layer resistance coefficient in Eq. 12-11 can be calculated using the Kozeny equation for flow through a granular medium (see Chap. 11):

$$\kappa_C = \frac{36\kappa_K(1-\varepsilon)^2\delta_C}{\varepsilon^3 d_P^2} \tag{12-12}$$

where κ_C = cake layer resistance coefficient, m^{-1}
κ_K = Kozeny coefficient, unitless (typically 5)
ε = cake porosity, dimensionless
δ_C = thickness of cake layer, m
d_P = diameter of retained particles, m

In dead-end filtration, the thickness of the cake layer as a function of time can be calculated from the mass flux of particles toward the membrane surface, assuming that the backward migration of particles due to diffusion can be neglected:

$$\delta_C(t) = \frac{CV}{\rho_P a(1-\varepsilon)} \tag{12-13}$$

where $\delta_C(t)$ = thickness of cake layer at time t, m
C = concentration of particles, mg/L
V = volume of feed water filtered, m^3
ρ_P = density of particles, kg/m^3
a = membrane area, m^2

In natural systems, the cake layer is an amorphous mat of polydisperse solids, so the cake layer resistance cannot be related easily to parameters such as particle diameter, porosity, and cake thickness. In this case, the cake resistance coefficient can be defined in terms of a specific cake resistance, or resistance per unit of mass loading. The mass loading is the mass of dry solids retained by the membrane per unit of membrane area. Because the influent solids concentration, volume filtered, and membrane area can be readily measured, the cake resistance as a function of time (via volume filtered) can be determined:

$$\kappa_C = \alpha_C \frac{CV}{a} \tag{12-14}$$

where α_C = specific cake resistance, m/g

During cross-flow filtration, cake layer formation is more complex because additional phenomena contribute to the transport of particles away from the membrane surface. The convective flow of particles toward the membrane is opposed by at least three mechanisms (Kim and DiGiano, 2009; Cheryan, 1998). First, the flow of water parallel to the membrane surface causes surface shear forces that drag particles downstream and minimize the formation of the cake. Second, the cross-flow velocity decreases the thickness of the concentration boundary layer, which increases the concentration gradient and enhances diffusion of particles away from the membrane surface. Finally, a velocity gradient exists close to the membrane surface such that particles in this velocity field are exposed to a higher velocity on the side opposite the membrane.

The differential velocity between the near and far sides of particles leads to inertial lift that draws them away from the membrane surface. The random movement of particles in this velocity field also enhances their migration toward streamlines moving at a higher velocity, a phenomenon that has been called shear-enhanced diffusion (Zydney and Colton, 1986). Shear-enhanced diffusion coefficients can be more than two orders of magnitude greater than Brownian diffusion coefficients (Belfort et al., 1994). Additional phenomena, such as electrostatic interactions (McDonogh et al.,

1988; Welsch et al., 1995), boundary layer separation (Treybal, 1980), gravitational forces, or van der Waals attractions (Wiesner and Chellam, 1992), may contribute to the movement of particles toward or away from the membrane surface.

A substantial number of models have been developed to predict the flux of membranes during cross-flow filtration of particle suspensions (Kim and DiGiano, 2009; Bacchin et al., 1995; Cheryan, 1998; Field et al., 1995; Koltuniewicz, 1992; Lee and Clark, 1997; Wiesner and Chellam, 1992). The trajectory of particles is evaluated in these models by considering all forces on individual particles. An important prediction of many of these models is that under certain conditions the net forces pushing particles away from the membrane are greater than the convective forces drawing particles to the membrane surface. As a result, no particle deposition and no fouling should occur. Wiesner et al. (1989) suggested that particles above a certain size should not deposit on the membrane, and others (Field et al., 1995) extended this to define the concept of a "critical flux," the flux at which no particles deposit on the membrane surface and no fouling occurs. In addition, many of the models suggest that particles between about 0.1 and 1 μm in diameter exhibit the least backmigration from the membrane surface and will cause the greatest fouling during cross-flow filtration.

Biofouling

Biofouling is the loss of system performance due to the formation of a biofilm (Ridgway and Flemming, 1996). During filtration, microorganisms are transported to the membrane surface, where they can attach with sufficient force as to prevent removal during backwashing. Once attached, they can excrete extracellular material that causes additional fouling. Biofouling is particularly relevant in membrane filtration systems operated in wastewater applications, such as membrane bioreactors. In water treatment, biofouling is addressed by the use of chemical disinfectants such as chlorine in the feed water, backwash water, or both. The recent trend by manufacturers to use membrane materials with good chlorine resistance has helped reduce the importance of biofouling.

Natural Organic Matter Fouling

Fouling by particles can be managed by proper backwashing (as long as the particles are significantly larger than the membrane pores) and biofouling can be managed with proper disinfection. The most problematic and least controllable membrane fouling is due to the adsorption of natural organic matter (NOM) to the membrane surface. Fouling by NOM (or the dissolved fraction, DOM) has been confirmed with laboratory experiments with solutions of commercially available dissolved organic matter. The ability of DOM to adsorb to membranes has been demonstrated with traditional adsorption isotherms (Crozes et al., 1993; Jucker and Clark, 1994). Filtration of commercially available humic, fulvic, and tannic acid solutions has been shown to lead to rapid membrane fouling (Crozes et al., 1993; Lahoussine-Turcaud et al., 1990; Yuan and Zydney, 1999, 2000). Surface cake formation and pore constriction have both been proposed as mechanisms for fouling (Combe et al., 1999; Kim et al., 1992; Yuan and Zydney, 1999; Yuan et al., 2002). The relationship between DOM adsorption and flux has not been successfully described mathematically, and there are currently no models that can predict the extent of DOM fouling as a function of water quality measurements. Fouling depends on characteristics of the DOM, the membrane material, and the solution properties, although the size and stability of the DOM appear to be the most important factors. The effects of several important factors on DOM fouling are identified in Table 12-8.

Some studies have found that removing a large fraction of DOM (up to 85 percent) by carbon adsorption (Laîné et al., 1990; Carroll et al., 2000; Lin et al., 2000) can have little or no reduction of fouling or, alternatively, removing a small fraction of DOM (less than 10 percent) by prefiltration through a 3-kD membrane (Howe, 2001) can completely eliminate fouling. Collectively, this research suggests that only a fraction of DOM causes the majority of fouling in membrane filtration, and this and other research suggests that the high-MW and colloidal fractions are the necessary components because they have the necessary dimensions to constrict membrane pores. Chemical properties and particle stability are also important (Huang et al., 2008a; Huang and O'Melia, 2008) because fouling will not occur unless the colloids have an affinity for attachment to the membrane pore walls.

Significant fouling by a small amount of DOM is possible when the relationship between transmembrane pressure and pore diameter under laminar flow conditions is considered. Poiseuille's law for laminar flow (see Eq. 11-10 in Chap. 11) indicates that pressure drop varies as the inverse fourth power of diameter under constant flow conditions (i.e., transmembrane pressure could double following a mere 16 percent decrease in pore diameter).

Strategies to minimize fouling by NOM, such as coagulation pretreatment, are discussed in Sec. 12-8.

Blocking Filtration Laws for Membrane Fouling

Models that simulate fouling mechanisms have been developed for filtration under specific laboratory operating conditions and have come to be known as the blocking laws. The analytical solution for each fouling mechanism is shown in Table 12-9. The equations predict a different rate of flux decline for each type of fouling.

Hermia (1982) demonstrated that the equations in Table 12-9 can be written in a consistent format. By multiplying flux by membrane area and integrating with respect to time, the total volume of permeate produced can be determined. Rearranging the equations with time as the dependent

Table 12-8
Factors contributing to membrane fouling by dissolved organic matter (DOM)

Factor	Observed Effects
Hydrophobicity	Hydrophobic membranes adsorb more DOM and therefore foul more rapidly than hydrophilic membranes (Matthiasson, 1983; Laîné et al., 1989; Cheryan, 1998). Hydrophobic fractions of DOM and hydrophobic sources of DOM are expected to cause greater fouling, which has been observed in some research (Crozes et al., 1993; Yuan and Zydney, 1999; Schäfer et al., 2000). However, researchers have also reported that hydrophilic fractions of DOM may be implicated in greater fouling (Amy and Cho, 1999; Carroll et al., 2000; Lin et al., 2000).
Electrostatic charge	Most DOM is negatively charged, and many MF and UF membranes acquire a slight negative charge during operation. Conditions that increase electrostatic repulsion might reduce fouling. The magnitude of the negative charge on membrane (Causserand et al., 1994; Nyström et al., 1994; Combe et al., 1999) and the negative charge on DOC both tend to increase at higher pH. As expected, low-pH conditions increase the adsorption of DOM to membranes (Jucker and Clark, 1994; Combe et al., 1999) and the fouling due to DOM adsorption (Kulovaara et al., 1999).
Size/molecular weight	Size may be an essential factor in determining which components of DOM cause fouling. Several studies suggest that high-MW and colloidal materials cause greater fouling (Lin et al., 1999, 2000; Yuan and Zydney, 1999, 2000; Habarou et al., 2001; Howe and Clark, 2002). Fouling by this colloidal fraction is consistent with the ability of larger material to constrict pores more efficiently than dissolved materials.
Colloidal stability	Since colloids must be smaller than the pore size to enter the membrane matrix, an additional mechanism must explain their attachment to the pore walls. A model developed by Huang et al. (2008a) and supported by experimental results indicated that colloids with low particle–membrane stability and high particle–particle stability caused the greatest fouling.
Ionic strength	High ionic strength reduces electrostatic repulsion (and particle stability) by compressing the double layer, and irreversible fouling has been shown to increase at high ionic strength (e.g., seawater) (Kulovaara et al., 1999).
Calcium concentration	Calcium ions may act as a positively charged bridge between DOM and membrane surfaces. Calcium has been shown to neutralize the negative charge on DOM and increase the adsorption of NOM on membranes (Jucker and Clark, 1994) and contribute to greater flux decline (Schäfer et al., 2000).

Table 12-9
Blocking filtration laws

Flux Equation	Equation Number	Filtration Coefficient, k	Filtration Exponent, n	Major Features and Assumptions
Complete Blocking Filtration Law (Pore Sealing)				
$J_t = J_0 \exp\left(-1.5\dfrac{CJ_0t}{\rho_P d_P}\right)$	(12-16)	$\dfrac{1.5CJ_0}{\rho_P d_p}$	2	❑ Models blockage of the entrance to pores by particles retained at the membrane surface. ❑ Each retained particle blocks an area of the membrane surface equal to the particle's cross-sectional area. ❑ Flux declines in proportion to the membrane area that has been covered. ❑ No superposition of particles occurs. Each particle lands on the membrane surface and not on other particles, so flux reaches zero when a monolayer of particles has been retained.
Standard Blocking Filtration Law (Internal Pore Constriction)				
$J_t = \dfrac{J_0}{\left(1+\dfrac{CJ_0t}{L\rho_P}\right)^2}$	(12-17)	$\dfrac{2C}{L\rho_P}\left(\dfrac{J_0}{a}\right)^{0.5}$	1.5	❑ Models the reduction of the void volume within the membrane. ❑ Assumes the membrane is composed of cylindrical pores of constant and uniform diameter. ❑ Particles deposit uniformly on the pore walls; pore volume decreases proportionally to the volume of particles deposited. ❑ L = membrane thickness, m

(Continued)

Table 12-9 *(Continued)*

Flux Equation	Equation Number	Filtration Coefficient, k	Filtration Exponent, n	Major Features and Assumptions
Intermediate Blocking Filtration Law (Pore Sealing with Superposition)				
$J_t = \dfrac{J_0}{\left(1+1.5\dfrac{CJ_0t}{\rho_P d_P}\right)}$	(12-18)	$\dfrac{1.5C}{\rho_P d_P a}$	1	❑ Models blockage of the entrance to pores by particles retained at the membrane surface. ❑ Extension of the complete blocking filtration law. ❑ Relaxes the "monolayer" assumption in the complete blocking filtration law by allowing particles to land on previously retained particles or on the membrane surface by evaluating the probability that a particle will block a pore.
Cake Filtration Law				
$J_t = \dfrac{J_0}{\left(1+2\dfrac{\alpha_C CJ_0t}{\kappa_M}\right)^{0.5}}$	(12-19)	$\dfrac{\alpha_C C}{\kappa_M J_0 a^2}$	0	❑ Models the formation of a cake on the surface of a membrane using the resistance-in-series model. ❑ The retained particles have no impact on the membrane itself, i.e., no pore blocking or pore constriction.

variable and taking the first and second derivatives with respect to volume, all four filtration laws can be written in the form

$$\frac{d^2t}{dV^2} = k\left(\frac{dt}{dV}\right)^n \tag{12-15}$$

where t = time, s
V = volume, L
k = blocking law filtration coefficient, units vary depending on n
n = blocking law filtration exponent, unitless

The parameters k and n are defined for each filtration law in Table 12-9. The form of the filtration laws shown in Eq. 12-15 can be used to analyze experimental data to determine which law most closely approximates the data. Permeate volume is measured at constant time intervals and tabulated in a spreadsheet. Using the spreadsheet, the derivatives d^2t/dV^2 and dt/dV can be determined. The slope of the line is the coefficient n from Table 12-9.

Many researchers have fit laboratory flux data to these equations in an attempt to isolate which mechanism causes fouling under specific conditions. Despite their use in research studies, however, the equations have failed to have much predictive value to relate bench studies to full-scale applications for several reasons. First, membranes are modeled as perfectly flat and uniformly porous with vertical cylindrical pores, which does not match the membranes used in commercial applications. Second, filtration is modeled as being under constant-pressure, declining flux conditions, compared to the constant-flux, rising pressure situation normally used during water treatment Third, the original models envision flux as declining due to a single mechanism and the shape of the flux decline curve reveals the fouling mechanism, whereas in full-scale operation several fouling mechanisms probably occur simultaneously. Finally, the original model equations have few or no parameters related to membrane properties so it is not possible to analyze, for instance, the impact of pore size on which mechanism might be most important for fouling by a particular size particle. More sophisticated models have been developed recently to address some of these limitations (Ho and Zydney, 2000; Chellam and Xu, 2006; Cogan and Chellam, 2009), but current models do not yet have predictive value for full-scale applications.

The filtration blocking laws apply only to constant-pressure, dead-end filtration. Models for flux decline have been developed for other operating conditions with varying levels of complexity (Belfort et al., 1994; Chang and Benjamin, 2003; Fane and Fell, 1987; Yuan et al., 2002; Kim and DiGiano, 2009). Theory regarding fouling of MF and UF membranes is evolving, and the current technical literature should be consulted if an in-depth understanding of flux modeling is required.

Membrane Fouling Index

In the absence of fundamental models that can predict full-scale performance, it is useful to have empirical models that can compare fouling under different conditions, such as with different source waters, different membrane products, or at different scales. A fouling index can be derived using the resistance-in-series model with two resistance terms: one for clean membrane resistance and another for fouling resistance:

$$J = \frac{\Delta P}{\mu\left(\kappa_m + \kappa_f\right)} \tag{12-20}$$

where κ_f = resistance due to all forms of fouling, m^{-1}

With the basic assumption that fouling resistance is directly proportional to the mass of foulants that have been transported to the membrane surface with the feed water, the fouling resistance can be related to the amount of water filtered per unit of membrane area, that is,

$$\kappa_f = kV_{sp} \tag{12-21}$$

where k = rate of increase in resistance, m^{-2}
V_{sp} = specific throughput, volume of water filtered per membrane area, m^3/m^2

After conversion to flux at a standard temperature using Eq. 12-8 and substituting Eq. 12-21, Eq. 12-20 can be written in terms of specific flux:

$$J_{sp} = \frac{J_s}{\Delta P} = \frac{1}{\mu\left(\kappa_m + kV_{sp}\right)} \tag{12-22}$$

For a new membrane, $V_{sp} = 0$ so $\kappa_f = 0$, so

$$J_{sp0} = \frac{1}{\mu\kappa_m} \tag{12-23}$$

Membrane filtration performance is typically evaluated by comparing the flux over time to the initial flux through the membrane when it was new. Clean membrane permeability can vary from one membrane sample to another due to slight variations in membrane pore dimensions, thickness, or porosity because of manufacturing variability. Normalizing against new membrane performance eliminates membrane sample variability when comparing experiments. Dividing by clean membrane specific flux yields

$$J'_{sp} = \frac{J_{sp}}{J_{sp0}} = \frac{1/\left[\mu\left(\kappa_m + kV_{sp}\right)\right]}{1/\left(\mu\kappa_m\right)} = \frac{\kappa_m}{\kappa_m + kV_{sp}} \tag{12-24}$$

A fouling index can be defined as the slope of the line when the inverse of J'_{sp} is plotted as a function of the specific throughput:

$$\frac{1}{J'_{sp}} = 1 + (\text{MFI})V_{sp} \tag{12-25}$$

where $\text{MFI} = k/\kappa_m$ = membrane fouling index, m^{-1}

The MFI is an empirical fouling index that can be used to compare the rate of fouling between experiments or between bench and pilot-scale results. The MFI is valid for any form of fouling as long as the fouling resistance is directly proportional to the specific throughput. The use of specific flux allows filter runs at either constant pressure or constant flux to be compared, because J_{sp} declines as membranes foul regardless of whether filtration occurs at constant pressure (J declines at constant ΔP) or constant flux (ΔP increases at constant J). Specific throughput allows runs of different duration or systems with different membrane area (i.e., different scale) to be compared. Huang et al. (2008b) demonstrated that Eq. 12-25 could be derived as an approximation for standard blocking, intermediate blocking, or cake filtration mechanisms under either constant-flux or constant-pressure filtration scenarios. The MFI has been used to compare fouling between different membrane products and source waters, and studies have shown reasonably good agreement between MFI values using bench-scale and pilot-scale data with the same membrane and source water (Huang et al., 2008b, 2009b).

The MFI can be calculated using either a linear regression of flux data or as the slope of the line between two points, depending on the data available. Calculation of the MFI using both methods is demonstrated in Example 12-6.

Evaluating Fouling with Bench-Scale Studies

Early bench-scale studies to explore membrane filtration and understand fouling were often done with flat-sheet membranes in a constant-pressure operating mode over a single filter run. A typical flat-sheet membrane experimental setup of this type is shown on Fig. 12-19a. While the research

(a) (b)

Figure 12-19
Equipment for bench-scale testing of membrane filtration: (a) flat-sheet membrane cell configuration and (b) backwashable hollow-fiber configuration.

Example 12-6 Calculation of the membrane fouling index

A laboratory membrane experiment using a backwashable single-fiber membrane module was carried out to collect the data in Fig. 12-16. The membrane had a total area of 23.0 cm^2 and the initial permeability of the new membrane was 225.0 $L/m^2 \cdot h \cdot bar$. The test was run at a constant pressure of 1.023 bar and temperature of 22°C. The membrane was backwashed every 30 min. Time and volume filtered were recorded at 2-min intervals and the data from filter run 6 is shown in the first two columns of Table 1 below. The flux at the beginning of each of the first 10 filter runs is also shown in Table 2 below. Calculate the fouling index during filter run 6 and the hydraulically irreversible fouling index (fouling that corresponds to the flux that could not be recovered by backwashing).

Solution

1. Divide the volume filtered by the membrane area to determine the specific throughput. Results are in column (3) in Table 1. For the second row,

$$V_{sp} = \frac{(743.92 \text{ mL})(10^4 \text{ cm}^2/\text{m}^2)}{(23.0 \text{ cm}^2)(10^3 \text{ mL/L})} = 323.4 \text{ L}/\text{m}^2$$

2. Calculate the volume filtered in each time increment by subtracting the previous volume. Results are in column (4) in Table 1. For the second row,

$$\Delta V = 743.92 \text{ mL} - 732.63 \text{ mL} = 11.29 \text{ mL}$$

3. Divide the volume filtered in each increment by membrane area and time to determine flux. Then correct for temperature and pressure using Eqs. 12-8 and 12-9 to determine specific flux. Results are in column (5) in Table 1. For the second row,

$$J_m = \frac{(11.29 \text{ mL})(10^4 \text{ cm}^2/\text{m}^2)(60 \text{ min/h})}{(23.0 \text{ cm}^2)(2 \text{ min})(10^3 \text{ mL/L})} = 147.3 \text{ L/m}^2 \cdot \text{h}$$

$$J_{sp} = \frac{J_m(1.03)^{T_s - T_m}}{\Delta P} = \frac{147.3 \text{ L/m}^2 \cdot \text{h}(1.03)^{20-22}}{1.023 \text{ bar}}$$

$$= 135.7 \text{ L/m}^2 \cdot \text{h} \cdot \text{bar}$$

4. Divide the specific flux (J_{sp}) by the initial specific flux (J_{sp0}). Results are in column (6) in Table 1. For the second row,

$$J'_{sp} = \frac{135.7}{225.0} = 0.60$$

5. Invert the normalized flux from column 6. Results are in column (7) in Table 1.

Example 12-6 Table 1

(1) Filtration Time, min	(2) Volume Filtered, mL	(3) Specific throughput, L/m^2	(4) Delta volume, mL	(5) Specific flux, L/m^2 · h	(6) Normalized specific flux, J'_{sp}	(7) Inverse normalized specific flux, $1/J'_{sp}$
0	732.63	—	—	—	—	—
2	743.92	323.4	11.29	135.7	0.60	1.66
4	754.79	328.2	10.87	130.6	0.58	1.72
6	765.26	332.7	10.47	125.8	0.56	1.79
8	775.40	337.1	10.14	121.9	0.54	1.85
10	785.17	341.4	9.77	118.4	0.53	1.90
12	794.63	345.5	9.46	113.7	0.51	1.98
14	803.79	349.5	9.16	110.1	0.49	2.04
16	812.70	353.3	8.91	107.1	0.48	2.10
18	821.34	357.1	8.64	103.8	0.46	2.17
20	829.73	360.8	8.39	100.8	0.45	2.23
22	837.88	364.3	8.15	97.9	0.44	2.30
24	845.85	367.8	7.97	95.8	0.43	2.35
26	853.62	371.1	7.77	93.4	0.42	2.41
28	861.22	374.4	7.60	91.3	0.41	2.46

6. Plot the inverse of the normalized specific flux ($1/J'_{sp}$) as a function of the specific throughput (V_{sp}), as shown in the following figure:

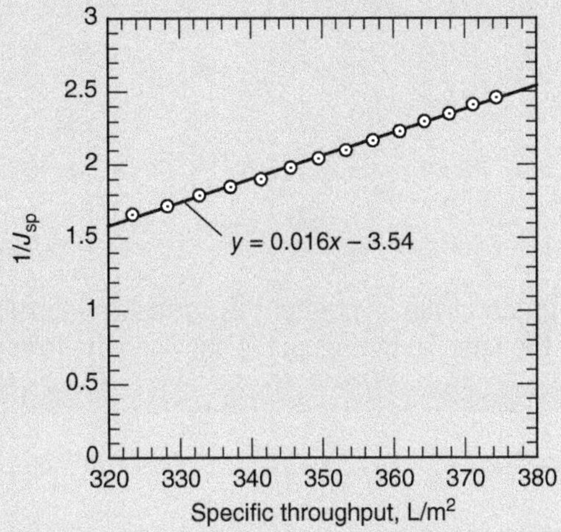

The slope of the line is the membrane fouling index for filter run 6, $MFI_6 = 0.016\ m^2/L = 16\ m^{-1}$. Note that the intercept of the graph is not 1.0 as is suggested by Eq. 12-25. This result is because backwashes remove foulants and reset membrane performance to a higher flux, whereas the specific volume progresses continuously. For an initial filter run (i.e., before any backwashes or cleanings), the intercept is very close to 1.0.

7. Determine the hydraulically irreversible membrane fouling index (MFI_{hi}). The MFI_{hi} represents the flux that cannot be recovered by backwashing and can be evaluated by considering the net reduction in flux at the beginning of each filter run (immediately after backwashing). Data from the first 10 filter runs of the experiment shown in Fig. 12-16 is shown in Table 2 below. Column (1) is the filter run number, Column (2) is the specific throughput at the beginning of each filter run, and Column (3) is the average specific flux over the first 30 of each filter run.

Example 12-6 Table 2

(1) Filter Run	(2) Specific throughput, L/m^2	(3) Specific flux, $L/m^2 \cdot h$	(4) Normalized specific flux, J'_{sp}	(5) Inverse normalized specific flux, $1/J'_{sp}$
1	2.2	238.0	1.06	0.95
2	71.3	176.9	0.79	1.27
3	137.6	157.7	0.70	1.43
4	200.0	149.0	0.66	1.51
5	260.5	143.3	0.64	1.57
6	319.0	138.0	0.61	1.63
7	376.4	133.6	0.59	1.68
8	432.6	129.3	0.57	1.74
9	487.9	125.5	0.56	1.79
10	542.4	121.6	0.54	1.85

8. A graph of the inverse of the normalized flux ($1/J'_{sp}$) as a function of the specific throughput is shown in the following figure:

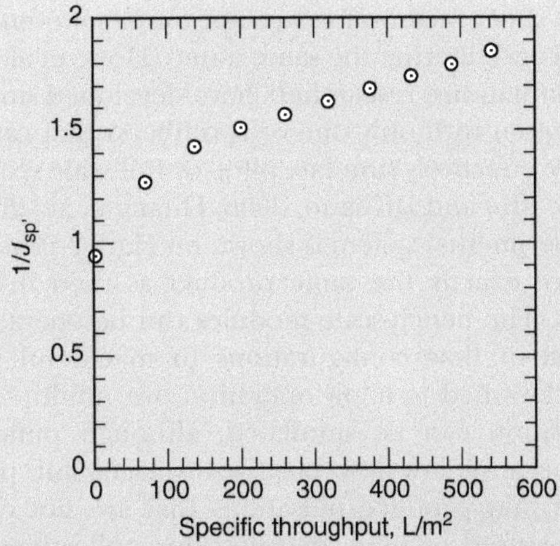

The graph indicates more rapid fouling during the first two filter runs (i.e., the first two runs are not linear with the rest of the data), and a linear regression through all of the data would not reflect the longer-term fouling index. The long-term hydraulically irreversible membrane fouling index can be calculated as a straight line between runs 3 and 10:

$$MFI_{hi} = \frac{(1/J'_{sp})_{10} - (1/J'_{sp})_3}{(V_{sp})_{10} - (V_{sp})_3} = \frac{1.850 - 1.427}{542.4\ L/m^2 - 137.6\ L/m^2}$$

$$= 0.00104\ m^2/L$$

$$MFI_{hi} = (0.00104\ m^2/L)(10^3\ L/m^3) = 1.04\ m^{-1}$$

has helped improved the industry's understanding of membrane filtration, researchers have gradually realized that performance with this system did not effectively simulate full-scale commercial membrane filtration systems. New membranes tend to experience an initial period of irreversible fouling that might mask longer-term fouling trends. Constant-pressure filtration through a flat-sheet membrane can cause very high initial flux that leads to rapid fouling in the first moments of operation. Studies have shown different fouling with constant-pressure operation than with constant-flux operation (Tarabara et al., 2002). Commercial membrane manufacturers often have proprietary chemical formulations or fabrication methods to optimize performance, and use of a laboratory-grade flat-sheet membrane may not adequately represent the commercial product, even if nominally of the same material. Studies have demonstrated that rates of fouling of hollow-fiber and flat-sheet membranes are not the same even when made of the same material and filtering the same water (Howe et al., 2007).

To rectify these limitations, researchers have developed small-scale hollow-fiber modules (often with only one or two fibers) that can be used at bench scale but more effectively simulate pilot- or full-scale systems (Chang and Benjamin, 1996; Kim and DiGiano, 2006; Huang et al., 2009b). A typical hollow-fiber experimental system is shown on Fig. 12-19b. The hollow-fiber membranes are exactly the same product as used in commercial membrane modules. The bench-scale modules can be operated in either inside-out or outside-in flow configurations to match full-scale operation and can be backwashed to allow operation over multiple filter runs. Constant flux operation can be simulated, although pumps must be selected carefully because low-flow positive-displacement pumps often used in laboratories may produce pulsations that are not characteristic of centrifugal pumps used in pilot- or full-scale applications. While the hydraulics of small-scale systems are not identical to commercial modules, they are more representative of larger systems than earlier flat-sheet bench-scale systems were. The membrane fouling index presented in the previous section offers the potential to relate bench-scale membrane performance to pilot- or full-scale performance, although more experimental validation of this approach is needed.

12-8 Process Design

Membrane filtration design concepts are changing rapidly, so the application of this technology presents unique challenges for the design engineer. Design based on previous projects or "tried-and-true" rules of thumb may fail to capitalize on recent technological advancements. On the other hand, undue reliance on manufacturers' claims about unproven technologies may lead to failure. A critical role for the design engineer is to stay abreast of technical advancements and provide appropriate guidance to facility owners. An understanding of the fundamentals of membrane materials, modules, fouling, and performance is necessary to evaluate new technologies with the objective of allowing the design to capitalize on valuable technological advancements while avoiding unproven alternatives that have an unreasonable chance of failure. Detailed guidance and design manuals for membrane filtration systems have recently been published by EPA and AWWA (U.S. EPA, 2005; AWWA, 2005b, 2008).

The primary tasks for the design engineer during preliminary design of a membrane filtration installation include:

1. Set performance criteria, such as retention criteria, plant capacity, and recovery, based on raw-water resources, finished-water quality goals, and projected water demand.

2. Evaluate the need for supplementary unit processes such as pretreatment or disinfection in the overall treatment facility, based on raw- and finished-water quality and the capabilities of membrane filtration.
3. Evaluate process alternatives, including new developments and technological advancements. Frequently, this evaluation will include pilot studies and evaluation of the construction and O&M costs of the appropriate alternatives.
4. Establish process design criteria and develop specifications for major system components. Pilot studies contribute to the basis of design.
5. Predesign ancillary and support facilities, such as transfer piping, pumping facilities, chemical storage facilities, laboratory space, buildings, and electrical systems. Evaluate hydraulic-grade line and waste washwater disposal options.

Performance Criteria

Important performance criteria include retention capabilities, capacity requirements, and recovery. Performance criteria also provide a basis for determining whether membrane filtration must be coupled with other treatment technologies.

RETENTION CAPABILITIES

As noted in Sec. 12-5, MF membranes achieve excellent rejection of most microorganisms but are not a complete barrier to viruses. Some UF products do provide excellent virus rejection, suggesting that there might be an advantage to specifying UF membrane systems exclusively. However, viruses are readily inactivated by free chlorine, and most regulatory agencies, if not all, will require primary disinfection as part of a multibarrier approach to water treatment. Thus, while some UF membranes may provide a more robust treatment train with respect to virus removal, as a practical matter there is generally no regulatory or cost advantage to selecting UF membranes over MF membranes when specifying acceptable membrane technologies.

Because membrane filtration is still a relatively new water treatment process, the regulatory system in the United States does not have uniform criteria for the rejection capabilities of membrane filters. Criteria are applied at the state level and vary between states. In the United States, most states with guidelines for membrane filtration grant between 2 and 4 log rejection credits for *Giardia* and *Cryptosporidium* and between 0 and 4 log rejection credits for viruses. Numerous states grant the same long rejection credit regardless of whether the system is marketed as an MF or UF system (Herschell, 2007). The *Cryptosporidium* challenge testing requirements in the LT2ESWTR (U.S. EPA, 2005) are becoming the de facto standard for specifying rejection requirements for membrane systems.

Effluent turbidity is an important performance criterion in the design of granular filtration facilities. For membrane filtration design, turbidity provides little guidance because all properly operating membrane filters will reduce turbidity to very low levels.

PLANT CAPACITY AND RECOVERY

Plant capacity is governed by the anticipated water demand at the end of the design life. Summer and winter demand must be considered separately because of the effect of temperature on permeate flux. In most locales, summer water demand is higher than winter demand, which fortunately corresponds to the seasonal variation in water temperatures. For each season, required plant size should be determined for the peak-day demand and minimum water temperature, which are worst-case conditions.

Recovery is the ratio of net water production to gross water production over a filter run:

$$r = \frac{Q_p}{Q_f} = \frac{V_f - V_{bw}}{V_f} \tag{12-26}$$

where r = recovery
Q_p = permeate flow rate, m^3/d
Q_f = feed flow rate, m^3/d
V_f = volume of water fed to membrane over filter run, m^3
V_{bw} = volume of water used during backwash, m^3

The calculation for recovery is identical to that for granular filters, except that common terminology in granular filtration (UFRV and UBWV) currently is not used in membrane filtration. Recovery in membrane filtration is typically 95 to 98 percent, which is comparable to rapid granular filters. If waste washwater is recovered, processed, and recycled to the feed stream, even higher recovery (greater than 99 percent) can be achieved.

Integration with Other Treatment Processes

Application of membrane filtration for water treatment started with small facilities with straightforward treatment requirements, so the membranes were a stand-alone process with no significant pretreatment or integration with other processes (except postmembrane disinfection). Now membrane filtration is an alternative to granular filtration regardless of source water quality. Integration of membrane filters as part of a larger process train is now the norm (Adham et al., 2005). Membrane filtration is used as pretreatment for other processes, such as granular activated carbon (GAC) adsorption or reverse osmosis. Processes that precede membrane filtration include coagulation and flocculation, sedimentation, granular filtration, adsorption, oxidation, and softening. A substantial amount of technical literature about pretreatment for membrane filtration is available, including reviews by Farahbakhsh et al. (2004) and Huang et al. (2009a).

Pretreatment is employed for two major reasons. First, additional processes can remove contaminants that are not removed by membrane filtration alone, such as dissolved constituents. For example, coagulation is used for removal of arsenic; GAC is used for removal of taste, odor, and synthetic organic chemicals; oxidants are used for the removal of iron and manganese; and lime is used for the removal of hardness. Other technologies are being explored for other treatment objectives.

Second, pretreatment can enhance membrane performance by reducing the solids loading on the membrane process or by reducing membrane fouling. High turbidity sources may require excessive backwashing unless coagulation, flocculation, and sedimentation are used to remove a significant fraction of the solids, a strategy identical to conventional treatment using granular filtration. Processes can also remove membrane foulants, allowing the membrane system to operate at higher flux or with decreased backwashing and cleaning requirements. If the flux enhancement is significant, the cost of pretreatment may be offset by decreased membrane system costs.

Regardless of the reason for applying pretreatment, the impact on the membrane process must be considered. While some pretreatment can substantially improve membrane performance, other pretreatment can foul or even damage the membranes. Bench or pilot testing is recommended to evaluate the impact of chemicals on membranes. The two most common pretreatment strategies, coagulation and adsorption, are considered in more detail in the following sections.

COAGULATION PRETREATMENT

Coagulation and flocculation, sometimes with settling, is the most common pretreatment for membrane filtration. Coagulation can remove 15 to 50 percent of the NOM in natural waters and may be appropriate for utilities that need some NOM removal to meet DBP regulations. Coagulation with settling allows membrane filtration to be applied to source waters with higher and more variable turbidity.

Coagulation frequently reduces membrane fouling, and many studies and full-scale facilities have reported higher flux or reduced cleaning requirements when coagulation is used (Huang et al., 2009a). As noted earlier, high-MW and colloidal constituents may be a primary factor in fouling by NOM, and coagulation is effective at removing the higher-MW and more hydrophobic components of NOM. In some cases, however, coagulation has made fouling worse (Adham et al., 2006; Schäfer et al., 2001; Shorney et al., 2001; Shrive et al., 1999). The effect of coagulation on membrane fouling is site specific and is due to specific interactions between coagulants, feed water components, and membrane materials. One factor in the differences between the studies reported above is the coagulant dose: When the dose is sufficient to remove a significant fraction of the NOM, membrane performance may be improved due to the reduction of fouling (Howe and Clark, 2006). Until further research clearly identifies the interactions between coagulants, foulants, and membranes, pilot studies are necessary to assess the effect of coagulation on membrane performance.

Experience with polymeric coagulant aids is mixed. Some studies have reported improved membrane performance when coagulant aids were used during coagulation, but others have reported that the polymers irreversibly fouled the membranes. Some manufacturers discourage the use of cationic polymers as pretreatment to their membrane filter systems.

POWDERED ACTIVATED CARBON–MEMBRANE REACTORS

Powdered activated carbon has been applied in the raw water of conventional treatment facilities for the removal of taste and odors as well as a variety of synthetic organic compounds. In conventional facilities PAC use can be complicated by competition from other treatment chemicals, limited contact time due to settling of the PAC in sedimentation basins, or breakthrough of the PAC through the filters and into the distribution system. The coupling of PAC treatment with membrane filtration provides a unique process without these disadvantages. Additional details of adsorption by PAC are provided in Chap. 15.

In the typical PAC–membrane reactor, PAC is added to the feed water upstream of the membranes. By using pressure-vessel membranes in a cross-flow mode or by using submerged membranes, the PAC is recycled to the feed water, increasing contact time and maximizing carbon utilization. Powdered activated carbon is considerably larger than the membrane pores, so passage of PAC into the distribution system is avoided. A recirculating or submerged membrane filtration system provides an ideal reactor for PAC adsorption, and this process is in use at full-scale water treatment plants (Anselme et al., 1999; Petry et al., 2001).

Powdered activated carbon adsorption is affected by the operating mode. The PAC can be added and withdrawn continuously during filtration in a feed-and-bleed configuration or the PAC can be added to an upstream reactor and allowed to accumulate over a filtration cycle. Each operating mode provides a different contact time distribution between the PAC and the contaminants. Campos et al. (2000a, b) determined that dosing carbon at the beginning (i.e., pulse input) of the filtration run rather than continuously resulted in lower average concentrations of contaminants.

While PAC is generally effective at removing synthetic organics, its ability to improve membrane performance is less evident. Although some studies have concluded that PAC can improve membrane performance, others have shown only minor improvement or worse performance. The primary reason PAC is less effective than coagulation at improving membrane performance is that activated carbon tends to remove the lower-MW components of NOM. The high-MW components of NOM that are implicated in membrane fouling tend to be excluded from the small pores of activated carbon.

System Components

Currently, membrane filtration systems are available as modular systems from several manufacturers. Current vendors of membrane filtration equipment are listed in AWWA (2005b). Other suppliers are expected to enter the market as the technology evolves, including suppliers experienced in the reverse osmosis (see Chap. 17) market. Systems are often offered as a package containing all necessary filtration and ancillary equipment.

Integral parts of membrane filtration facilities are the pretreatment processes, backwashing facilities, and cleaning facilities. These components are discussed in the following sections.

PRETREATMENT

When processes for nonparticulate treatment goals are not present, the pretreatment requirements for membrane filtration are minimal. Pretreatment is necessary to protect the filter fibers from damage or clogging of the lumen (in the case of inside-out membranes). Microscreening or prefiltration to remove coarse sediment larger in diameter than 0.1 to 0.5 mm, depending on the manufacturer, is required. Prefiltration is accomplished with self-cleaning screens, cartridge filters, or bag filters.

Because the primary removal mechanism is straining, chemical conditioning to destabilize particles is not required, although coagulation is now often applied as pretreatment for other reasons. The lack of a requirement for particle destabilization can be an advantage over granular filtration because the elimination of coagulation and flocculation facilities reduces chemical handling and storage facilities and residual management requirements.

BACKWASHING

The objective of backwashing is to remove the surface cake that develops during the filtration cycle. Backwashing occurs automatically at timed intervals ranging from 30 to 90 min. The increase in transmembrane pressure during the filtration cycle is typically 0.01 to 0.07 bar (0.2 to 1 psi). Most systems also initiate the backwash cycle early if the increase in transmembrane pressure during the filter run exceeds a preset limit. The backwash cycle lasts 1 to 3 min, and the sequence is run entirely by the control system. All modules in a rack are backwashed simultaneously. Backwashing of MF membranes involves forcing either air or permeate water through the fiber wall in the reverse direction at a pressure higher than the normal filtration pressure. Ultrafiltration membranes are backwashed with permeate water because the air pressure required to force water from the small pores in UF membranes can be excessive. In some pressure-vessel systems, the backwash flow is supplemented by a high-velocity flush in the feed channels to assist with removing the surface cake, and the wastewater is piped to a washwater handling facility. The backwash water in submerged systems flows directly into the feed tank.

Many membrane systems periodically add chemicals to backwash water to improve the backwash process, a sequence called chemically enhanced backwash (CEB). CEB chemicals can include hypochlorite or other cleaning chemicals. Chemically enhanced backwash reduces the rate of membrane fouling and can decrease the required frequency for the more extensive clean-in-place (CIP) cycle, which is discussed below.

Many membrane systems use a single backwash system to service multiple racks. The system is sized to backwash a single rack at a time, so backwashing of multiple racks must be staggered. The automated backwash sequence is a complex operation that involves the sequencing of numerous valves, which may take several minutes. The design engineer must consider the time requirements to complete one backwash cycle and ensure that there is sufficient time for all units to be washed within the allowable time between backwashes, including a factor of safety. If sufficient time is not available, more than one backwash system may be necessary.

CLEANING

Despite frequent backwashing, membrane filters gradually lose filtration capacity due to clogging or adsorption of material that cannot be removed during the backwash cycle. When the transmembrane pressure increases to a preset maximum limit or when a preset time interval has elapsed, the membranes are chemically cleaned. Chemical cleaning frequency ranges from a few days to several months depending on the membrane system characteristics and source water quality. The cleaning procedure typically takes several hours and involves circulating cleaning solutions that have been heated to 30 to 40°C. Cleaning solutions are proprietary mixtures provided by membrane manufacturers but are often high-pH solutions containing detergents or surfactants, which are effective for removing organic foulants. Low-pH solutions such as citric acid can be used for removing inorganic foulants. For some membrane materials (e.g., cellulose acetate), the pH of the cleaning solution is limited by the pH compatibility of the material. The membranes in both pressure-vessel and submerged systems are typically cleaned without removing the membranes from the modules, so the process is typically called the clean-in-place cycle.

POSTTREATMENT

The membrane filtration process has no inherent posttreatment requirements. Fluoridation or pH adjustment may be added after membrane filtration to fulfill other treatment objectives. Although membrane filtration is capable of completely removing microorganisms, disinfection is normally practiced after filtration as part of the multibarrier concept and to provide a disinfectant residual in the distribution system. Most state regulatory agencies have specific regulations for chemical disinfection following filtration.

Integrity Testing and Monitoring

Membrane integrity monitoring involves procedures to verify that membrane filters are meeting treatment objectives. Integrity monitoring is important because of the physical characteristics of the filtration barrier. In a granular filtration plant, water is cleaned gradually as it flows through a series of processes, ending with a thick bed of filter media; clean water and dirty water are separated in both time and space. In a membrane filtration plant, water is cleaned nearly instantaneously as it flows through a thin membrane; clean water and dirty water are separated by a distance less than 1 mm and time less than 1 s. In addition, broken fibers or leaking O-ring connectors may compromise the filtration system. The regulatory framework for integrity monitoring is provided as part of the LT2ESWTR (U.S. EPA, 2006).

Integrity monitoring for membrane filtration has both direct and indirect components. As defined in the LT2ESWTR, the direct integrity test is a physical test that is sensitive enough to detect a 3-μm breach in the membrane system, is conducted at least one per day, and can verify the log removal value awarded to the membrane process. Continuous indirect integrity monitoring is the measurement of a water quality parameter that is indicative of particle removal, such as turbidity or particle counts.

INDIRECT INTEGRITY MONITORING

Indirect integrity monitoring is the continuous (at least every 15 min) monitoring of a suitable effluent water quality parameter, such as turbidity or particle levels. Indirect integrity monitoring is not as sensitive as direct integrity testing, but it has the advantages that it can be applied continuously and uses commercially available equipment that can be used with any membrane system (whereas most direct integrity testing equipment is proprietary to the specific membrane system manufacturer). Therefore, it is complementary to direct integrity testing in an overall integrity verification program.

Membrane filter effluent may be monitored for turbidity or particle concentrations. Particle counters, in which particles are classified according to size, are in common use in water treatment facilities (see Chap. 2). Particle monitors are a less expensive alternative to particle counters and provide only a relative measurement of total particulate matter in water. Particle counting is generally more sensitive than turbidity monitoring.

The sensitivity of effluent monitors is dependent on the filtration area being monitored. A study performed by Adham et al. (1995) showed that turbidity monitoring was able to detect a single pinhole in 5 m^2 of membrane area, whereas a particle monitor and a batch particle counter were able to detect similar breaches in 12 and 720 m^2 of membrane area, respectively. However, individual membrane modules may have 40 to 80 m^2 of membrane area and racks may contain 2 to 100 modules. The limitation of turbidity for integrity monitoring, which drives the need for direct integrity testing, is illustrated in Example 12-7.

Example 12-7 Indirect integrity monitoring with turbidity

A membrane filtration plant treating a feed water with 10^6 microorganisms/L and turbidity of 10 NTU reduces both to below detection limits. For this example, assume the detection limits are 1 org/L for microorganisms and 0.03 NTU for turbidity. Using this information, calculate (1) the log rejection value for microorganisms under normal operation and (2) the effluent turbidity and microorganism concentration and the rejection of microorganisms assuming the membrane develops a hole that allows 0.01 percent of the water to pass through untreated.

Solution

1. Calculate the log rejection value under normal operation. Use the detection limit for the effluent concentration and Eq. 12-4:

$$\text{LRV} = \log\left(\frac{C_f}{C_p}\right) = \log\left(\frac{10^6 \text{ org / L}}{1 \text{ org / L}}\right) = 6.0 \text{ for microorganisms}$$

Because the effluent concentration is below the detection limit, the log rejection is greater than calculated, that is, LRV ≥ 6.0 for microorganisms.

2. Calculate log rejection under compromised operation.
   a. Draw a mass balance diagram for membrane breach:

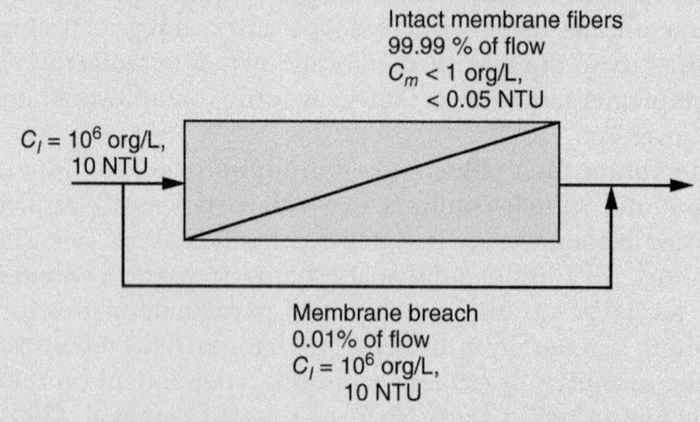

   b. Write the mass balance equation for microorganisms and solve:

$$C_e = \frac{Q_m C_m + Q_b C_b}{Q_e} = \frac{(1 \text{ org /L})(0.9999) + (10^6 \text{ org/L})(0.0001)}{1}$$
$$= 101 \text{ org / L}$$

c. Write the mass balance equation for turbidity and solve:

$$C_e = \frac{Q_m C_m + Q_b C_b}{Q_e} = \frac{(0.03\ \text{NTU})(0.9999) + (10\ \text{NTU})(0.0001)}{1}$$
$$= 0.031\ \text{NTU}$$

d. Calculate the log rejection value under compromised operation using Eq. 12-4:

$$\text{LRV} = \log\left(\frac{C_p}{C_f}\right) = \log\left(\frac{10^6\ \text{org/L}}{101\ \text{org/L}}\right) = 4.0 \text{ for microorganisms}$$

Comment

As demonstrated in this example, effluent turbidity with a 0.01 percent breach would not be distinguishable from the turbidity with an intact membrane (turbidity = 0.031 NTU versus 0.03 NTU). For microorganisms, however, C_E = 101 org/L and LRV = 4.0 compared to $C_E < 1$ org/L and LRV > 6.0 with no breach—a dramatic reduction in rejection capabilities. In other words, a small breach has a substantial effect on microorganism log rejection value but is undetectable by turbidity measurements.

Most membrane filtration systems produce effluent water turbidity between 0.03 and 0.07 NTU, which is near the limit of detection. Indirect integrity monitoring involves establishing a control limit, typically between 0.10 and 0.15 NTU. Readings greater than the control limit for more than 15 min may indicate an integrity problem and would trigger the need to perform a direct integrity test.

DIRECT INTEGRITY TESTS

Direct integrity tests can be either pressure-based or marker-based tests. Pressure-based tests involve applying pressurizing one side of the membrane with air and monitoring the change in air pressure, flow of air, or volume of displaced water, based on the concept that passage of air through a hole in the membrane is much faster than the diffusion of air through the water-filled pores in the membrane. Marker-based tests involve spiking the influent with particles or molecular markers and measuring the concentration of the marker in the effluent, which is similar to the challenge test that manufacturers have performed to get systems approved for use under the LT2ESWTR. Pressure-based integrity tests are most common for commercial membrane filtration systems, and the equipment, instrumentation, and procedures for conducting the test are built into the skid and implemented automatically.

The general principle of pressure-based integrity testing is to isolate a group of modules being tested (typically, one rack), drain the water from one side of the membrane, and pressurize the system with air. In the pressure-hold test, the rate of decay in the pressure is monitored. In a membrane with no breaches, air will diffuse through the water in the membrane pores, and pressure will decay slowly. Air can flow more rapidly through pinholes. The pressure required must be high enough to detect a 3-μm breach but below the bubble point of the membrane material. This pressure range depends on membrane properties and the procedure to determine the appropriate pressure is defined in the *Membrane Filtration Guidance Manual* (U.S. EPA, 2005). The *Guidance Manual* also describes procedures necessary to calculate the sensitivity of the test, which is related to the log removal value that the system would achieve if a portion of the water was flowing through a hole in the membrane untreated and the rest of the water was flowing through membrane pores and achieving complete removal of microorganisms. Acceptable rates of pressure decay vary with the system being monitored according to calculations in the *Membrane Filtration Guidance Manual* . Decay rates of 0.007 to 0.03 bar/min (0.1 to 0.5 psi/min) are typical limits (U.S. EPA, 2001).

Pressure-hold tests have been reported to be capable of detecting one broken fiber in a module containing 20,000 fibers. Entire racks of membranes can be monitored simultaneously, but these tests are sensitive to the size of the system being monitored because the air diffusion through the pore water may exceed the airflow through a breach if the filter area being monitored is too large. Breakage of several fibers in a large pressure-vessel rack, containing 90 modules with 20,000 fibers each, cannot be detected with pressure-hold tests (Landsness, 2001).

The pressure-hold test is the most common method and most manufacturers include all necessary equipment as part of the skid. Less common variations of this test involve measuring the rate of airflow through the membrane at constant pressure or measuring the volume of water displaced by the flowing air.

Current direct integrity testing methods require that the plant be taken out of service, thus reducing the available time for water production. The required frequency of pressure-hold test requirements in state regulations vary from once every 4 h to weekly (Allgeier, 2001).

SONIC TESTING

Sonic testing is a method of identifying leaks in individual pressure vessels. The test involves manually placing an acoustic sensor against a module and listening for changes in the sounds emanating from within the module. During a pressure-hold test, air bubbling through a damaged fiber will make enough noise to be detected by the sensor. The sonic test is highly sensitive when performed by a skilled operator but is subject to background equipment noise. A single broken fiber in a module containing 20,000 fibers can be detected (Landsness, 2001). The sonic test is typically used in conjunction with other integrity monitoring techniques. Effluent monitoring or pressure-hold tests are typically used to identify possible problems in a group of modules, and sonic testing can then be used to confirm the presence of a damaged fiber and identify the individual module containing the breach.

REPAIR OF MODULES

Modules can be repaired by isolating damaged fibers in a process called pinning. The module is typically removed from the rack and placed in a special test casing. The module can be pressurized with air while the ends of the fibers are exposed and covered with water. In this setting, bubbles will emit from the end of the broken fiber, so that the damaged fiber can be identified. A pin is glued into the end of the fiber, effectively taking the individual fiber out of service. The repaired module can then be placed back in service.

Design Criteria Development Based on Pilot Testing

Typical operating criteria for membrane filtration facilities are given in Table 12-10. As demonstrated previously, steady-state membrane performance is controlled not by intrinsic membrane properties but by the fouled state of the membrane after it has been in contact with natural water. Fouling depends on interactions between the source water and membrane material, but current understanding of fouling is not sufficient for predicting basic design criteria from measurable water quality parameters and membrane properties. Thus, pilot testing is typically part of the process evaluation procedure. AWWA's *Manual M53* (AWWA, 2005b) describes pilot testing efforts for two different water supplies where the same two membrane products were tested. For one water supply, one membrane performed well and the other fouled rapidly; for the other water supply, the results were exactly reversed. Such information is critical for selecting the most appropriate membrane product and operating conditions for a particular water supply. In additional, nearly all states require pilot testing as part of the permitting process (Herschell, 2007).

PILOT TESTING

When done properly, pilot testing can be used to demonstrate the effectiveness of innovative technologies or to provide a basis for comparing alternative systems. Pilot testing should incorporate all pretreatment processes that are being considered for the full-scale facility. For most MF and UF membrane studies the following parameters should be studied:

1. Feed and permeate water quality, including pH, turbidity, particle counts, TOC or DOC, UV_{254} absorbance, and other parameters relevant to the specific site
2. Feed water temperature

3. Feed water and permeate flow rates
4. Transmembrane pressure
5. Backwash requirements (frequency, duration, flow rate, pressure)
6. Cleaning requirements (frequency, duration, chemical dosages, procedures)
7. Other constituents of concern in specific applications, such as NOM

PILOT TESTING PERIOD

Pilot testing should be performed over an extended time, ranging from several months to a year, depending on seasonal variations in water quality and temperature. For instance, spring runoff or lake turnover may lead to water quality conditions that cause considerably greater fouling than other times of the year. When complete, pilot testing will provide data on each of the design criteria shown in Table 12-10, with the exception of membrane life. For comparisons between alternatives to be meaningful, data for permeate flux, pressure, and temperature must be combined to determine specific

Table 12-10
Typical operating characteristics of membrane filtration facilities

Parameter	Units	Range of Typical Values
Permeate flux		
Pressurized systems	$L/m^2 \cdot h$	30–170
	$gal/ft^2 \cdot d$	18–100
Submerged systems	$L/m^2 \cdot h$	25–75
	$gal/ft^2 \cdot d$	15–45
Normal transmembrane pressure		
Pressurized systems	bar	0.4–1
	psi	6–15
Submerged systems	bar	0.2–0.4
	psi	3–6
Maximum transmembrane pressure		
Pressurized systems	bar	2
	psi	30
Submerged systems	bar	0.5
	psi	7.4
Recovery	%	>95
Filter run duration	min	30–90
Backwash duration	min	1–3
Time between chemical cleaning	d	5–180
Duration of chemical cleaning	h	1–6
Membrane life	yr	5–10

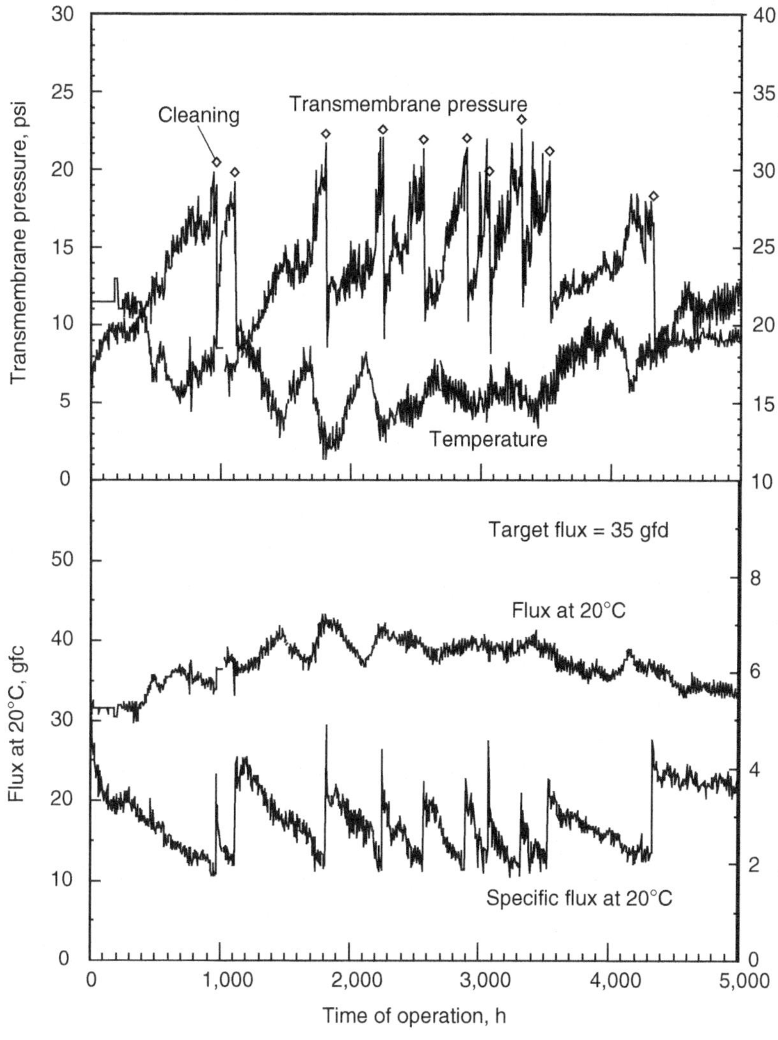

Figure 12-20
Variation in flux, pressure, temperature, and specific flux during pilot testing.

flux at a standard temperature. Variations in flux, pressure, temperature, and specific flux from a 7-month pilot study are shown on Fig. 12-20. The effect of chemical cleaning (manifested as lower pressure or higher specific flux) can be observed on Fig. 12-20.

OPERATING PARAMETERS

A basic operating parameter that should be varied during pilot testing is permeate flux. Pilot facilities and full-scale facilities normally operate with constant-flux, rising-pressure conditions. Flux and recovery determine the filtration area required to provide the required capacity, which is a significant factor in determining the capital cost of a facility. Often, pilot testing demonstrates the existence of a critical limit to permeate flux, below which long-term operation is successful and above which pressure rise, backwash frequency, and cleaning frequency are unacceptable. Backwash and cleaning strategies should also be evaluated during the pilot study.

PILOT PLANT UNITS

Most membrane manufacturers provide self-contained pilot plant units for use in evaluating performance. A typical skid-mounted membrane filtration pilot plant is shown on Fig. 12-21. Manufacturer-provided pilot plants typically contain all necessary equipment for their membrane system, including membrane modules, a feed tank, a feed pump, a backwash system with either an air compressor or liquid feed backwash pump, a clean-in-place system, permeate storage tank, all piping, valves, and instrumentation, and a programmable logic controller (PLC). The membrane modules are standard full-size modules identical to what would be provided on a fullscale system; the only difference is that a pilot unit typically will contain only 1 to 6 modules whereas a full-scale system may have 50 to 100 modules. Since the modules are identical and are tested with operating conditions that are identical to full-scale operation, the performance and fouling of the membranes can be expected to be very similar to that which would occur at full scale. Pilot plants may be designed with more instrumentation and operational flexibility than full-scale units to allow a range of testing conditions. Manufacturers typically supply specifications for pilot plant systems so they can be operated properly.

PILOT TESTING EXPECTATIONS

Pilot testing establishes the minimum performance requirements that can be accomplished by specific systems. For instance, the flux observed in pilot testing should be achievable in a full-scale facility by the same manufacturer, and the pilot testing can be used to set the minimum performance requirements for each manufacturer that will submit a bid for the project. Pilot testing provides a basis for comparing the effectiveness of alternative systems or new technologies. Individual design parameters, however, should not be compared directly when evaluating alternative systems. For instance, it would be inappropriate to use pilot testing to establish a minimum flux value as a requirement for all systems. A system operating at a low flux may be more cost effective if it operates at a lower pressure with less frequent backwash and clean sequences and has a lower cost per unit of filter area. Physical dimensions, capacity, and filtration area of individual modules, permeate flux, operating pressure, and backwashing and cleaning requirements, taken individually, generally are not a basis for comparing systems. Many parameters are interrelated and can only be compared on the basis of total system performance and cost.

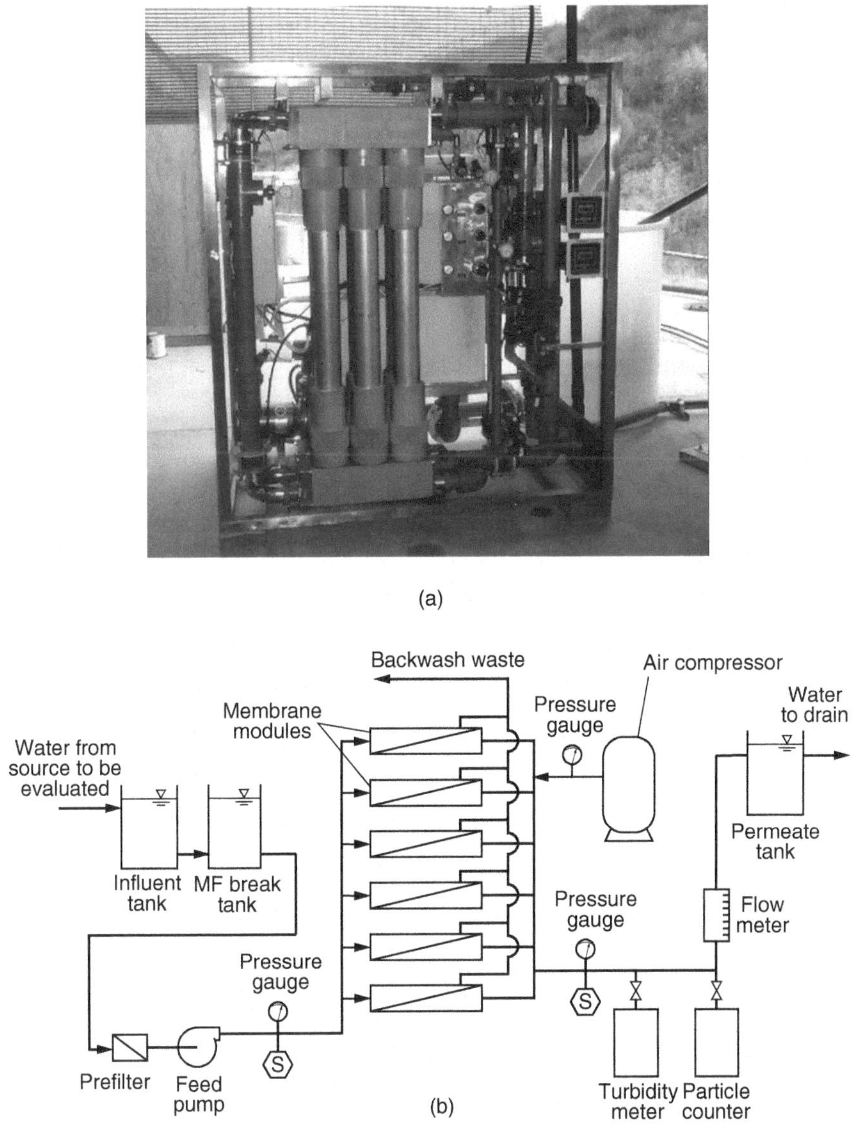

Figure 12-21
Typical skid-mounted pressure-vessel configuration membrane filtration pilot plant: (a) photograph and (b) schematic of the pilot plant.

Photographs of MF pilot systems:

http://www.stantec.com/wt3r/12-3.html

SYSTEM DESIGN FROM PILOT DATA

The data generated during pilot testing can be used to design the full-scale facility. Membrane systems are routinely taken off-line for backwashing, integrity testing, and cleaning, which reduces the time available for permeate production. The percent of time that permeate is produced, or online production factor (AWWA, 2005b), is expressed as

$$\eta = \frac{1440 - t_{\text{bw}} - t_{\text{dit}} - t_{\text{cip}}}{1440} \tag{12-27}$$

where η = online production factor
t_{bw} = time per day for backwashing, min
t_{dit} = time per day for direct integrity testing, min
t_{cip} = time for cleaning, prorated per day, min

The water produced during each filter run can be determined from the flux, membrane area, and run duration:

$$V_f = Jat_f \tag{12-28}$$

where V_f = volume of water filtered per filter run, L
t_f = duration of filter run (excluding backwash time), min

The water consumed during backwashing should be recorded during the pilot testing. With that information and the volume of water filtered from Eq. 12-28, the recovery and the required feed flow rate can be calculated with Eq. 12-26. The amount of time that the system is not producing permeate and the quantity of water that must be used for backwashing both increase the required total membrane area:

$$a_t = \frac{Q_p}{J\eta r} = \frac{Q_f}{J\eta} \tag{12-29}$$

where a_t = total membrane area, m^2
r = recovery, unitless

Once the total membrane area is determined, the number of racks and modules per rack can be determined by relating the total required membrane area to the capabilities of the system. An example of the sizing of a full-scale membrane system from pilot data is demonstrated in Example 12-8.

Residual-Handling Requirements

Residual handling from membrane filters is similar in many respects to residual handling from granular filters. However, the reduced or eliminated use of coagulants reduces the generation of sludge and simplifies sludge disposal in some cases. Some utilities discharge the waste washwater to the wastewater collection system and allow the sludge to be handled at the wastewater treatment plant rather than have separate sludge-handling facilities at the water treatment plant. Waste washwater can be clarified and returned to the plant influent or the source water, depending on regulatory constraints. The sludge can be thickened and dewatered similar to sludge from granular filters, and when coagulants are not used, the sludge is generally easier to thicken and dewater. Residual management is discussed further in Chap. 21.

Example 12-8 Determining system size from pilot data

A treatment plant is to be designed to produce 75,700 m^3/d (20 mgd) of treated water at 20°C. Pilot testing demonstrates that it can operate effectively at a flux of 65 L/m^2 · h at 20°C with a 2-min backwash cycle every 45 min and cleaning once per month. The membrane modules have 50 m^2 of membrane area. The pilot unit contained 3 membrane modules and the full-scale racks can contain up to 100 modules. Backwashes for the pilot unit consumed 300 L of treated water. Cleaning takes 4 h. Regulations require direct integrity testing, which takes 10 min, once per day.

Determine the following: (a) the online production factor, (b) system recovery, (c) feed flow rate, (d) total membrane area, (e) number of racks, and (f) number of modules per rack.

Solution

1. Determine the fraction of time the system is producing permeate.

$$t_{bw} = (2\text{ min})\left(\frac{1440\text{ min/d}}{45\text{ min}}\right) = 64\text{ min / d}$$

$$t_{dit} = 10\text{ min/d}$$

$$t_{cip} = \frac{(4\text{ h})(60\text{ min/h})}{30\text{ d}} = 8\text{ min/d}$$

$$\eta = \frac{1440 - t_{bw} - t_{dit} - t_{cip}}{1440} = \frac{1440 - 64 - 10 - 8\text{ min/d}}{1440\text{ min/d}} = 0.943$$

2. Determine the system recovery. The system recovery is the same for one element as for all elements and can be calculated using Eq. 12-26. For one element that filters for 43 min per cycle (2 min out of every cycle is backwash),

$$V_f = Jat_f = \frac{(65\text{ L/m}^2\cdot\text{h})(50.0\text{ m}^2)(43\text{ min})}{60\text{ min/h}} = 2330\text{ L}$$

$$V_{bw} = \frac{300\text{ L}}{3\text{ modules}} = 100\text{ L}$$

$$r = \frac{V_f - V_{bw}}{V_f} = \frac{2330\text{ L} - 100\text{ L}}{2330\text{ L}} = 0.957$$

3. Calculate required feed flow:

$$Q_f = \frac{Q_p}{r} = \frac{75{,}700\text{ m}^3\text{/d}}{0.957} = 79{,}100\text{ m}^3\text{/d}$$

4. Calculate the total membrane area required:

$$a = \frac{Q_f}{J\eta} = \frac{(79{,}100\ \text{m}^3/\text{d})(10^3\ \text{L/m}^3)}{(65\ \text{L/m}^2\cdot\text{h})(24\ \text{h/d})(0.943)} = 53{,}800\ \text{m}^2$$

5. Calculate the total number of modules required:

$$N_{\text{MOD}} = \frac{53{,}800\ \text{m}^2}{50\ \text{m}^2} = 1076$$

6. Determine the number of racks and modules/rack. Since the racks can accommodate up to 100 modules, at least 11 racks will be required. Dividing the required modules evenly among racks is preferred. In addition, leaving space in the racks is recommended as an inexpensive way to provide flexibility to reduce flux or increase capacity by adding additional modules in the future. Twelve racks are chosen in this example.

$$N_{\text{Racks}} = 12$$

$$N_{\text{MOD/Rack}} = \frac{1076}{12} = 90$$

The system will have 12 racks that each have 90 modules.

Problems and Discussion Topics

12-1 Discuss the differences between MF and UF membranes. What impact do these differences have on their use in water treatment?

12-2 Discuss the similarities, differences, advantages, and disadvantages between membrane filtration and rapid granular filtration. This is an essay question.

12-3 How effective do you think membrane filtration is for each of the following treatment issues? Explain your reasoning.

a. Arsenic

b. Anthrax spores

c. Hardness

d. Taste and odor

12-4 Calculate rejection and log removal value for the following filtration process (to be selected by instructor). Use the number of significant figures necessary to correctly illustrate the removal being obtained.

	A	B	C	D	E
Influent concentration (#/mL)	10^6	6.85×10^5	7.1×10^5	1.65×10^7	2.8×10^6
Effluent concentration (#/mL)	10	136	0.16	65	96

12-5 An inside-out hollow-fiber membrane system is operated with a cross-flow configuration. Each module contains 10,200 fibers that have an inside diameter of 0.9 mm and a length of 1.75 m. Calculate the following for one module:

a. Feed flow necessary to achieve a cross-flow velocity of 1 m/s at the entrance to the module.

b. Permeate flow rate if the system maintains an average permeate flux of 80 $L/m^2 \cdot h$.

c. Cross-flow velocity at the exit to the module.

d. Ratio of the cross-flow velocity at the entrance of the module to the flow velocity toward the membrane surface. Given the magnitude of this ratio, what effect would you expect cross-flow velocity to have on fouling in crossflow versus dead-end filtration?

e. Ratio of permeate flow rate to feed flow rate (known as the single-pass recovery). What impact does this ratio have on operational costs in cross-flow versus dead-end filtration?

12-6 Hollow-fiber membranes with a membrane area of 23.3 cm^2 were tested in a laboratory and found to have the clean-water flow shown in the table below, at the given temperature and pressure.

	A	B	C	D	E
Flow (mL/min)	4.47	4.22	2.87	6.05	1.22
Temperature (°C)	16	22	23	25	22
Pressure (bar)	0.67	0.80	0.71	1.25	0.21

For the data set selected by your professor,

a. Calculate the specific flux at 20°C.

b. Calculate the membrane resistance coefficient.

c. Does membrane resistance coefficient depend on the pressure and temperature used for the tests? Why or why not?

12-7 The 0.2 μm polyethersulfone microfiltration membrane shown on Fig. 12-3 was tested in the laboratory and found to have a cleanwater flux of 6500 $L/m^2 \cdot h$ at 23°C and 0.69 bar. Assume that the flow through a microfiltration membrane can be modeled using the Kozeny equation for flow through porous media (Eq. 11-11 in Chap. 11).

a. Calculate the specific surface area of the membrane assuming a porosity of 0.6, thickness of 0.10 mm, and Kozeny coefficient of 5.0.

b. What would the theoretical grain diameter be if the membrane were composed of spherical granular media with the same specific surface area (see Eq. 11-6)?

c. How does the theoretical grain diameter compare to the retention rating for the membrane? Using concepts of particle retention through granular media from Chap. 11, what does this comparison suggest about the mechanisms for particle removal in microfiltration?

d. Using the theoretical grain diameter as the characteristic dimension, calculate the Reynolds number for flow through a microfiltration membrane. Is the flow laminar or turbulent?

12-8 A membrane plant is operated at a volumetric flux of 75 $L/m^2 \cdot h$ at 17°C and 0.85 bar. Calculate the specific flux at 20°C.

12-9 Feed water pressure and temperature and permeate flux at a membrane filtration plant are reported on two dates below. For the plant selected by your instructor, calculate the specific flux on each date, and indicate whether fouling has occurred between the first and second dates.

	A	B	C	D	E
Day 1					
Flux ($L/m^2 \cdot h$)	72	26	31	86	112
Temperature (°C)	21	17	17	22	19
Pressure (bar)	0.62	0.24	0.24	0.72	0.66
Day 2					
Flux ($L/m^2 \cdot h$)	56	26	27	90	120
Temperature (°C)	4	15	10	25	11
Pressure (bar)	0.80	0.29	0.26	0.77	1.05

12-10 A new membrane plant is being designed. Pilot testing indicates that the membrane will be able to operate at a specific flux of 120 $L/m^2 \cdot h \cdot bar$ at 20°C. The full-scale plant will operate at 0.8 bar, online production factor of 95 percent, and recovery of 97 percent. Water demand projections predict a summer peak-day demand of 90,000 m^3/d and a winter peak-day demand of 60,000 m^3/d. Historical records indicate that the source water has a minimum temperature of 3°C in winter and 18°C in summer.

a. Which season will govern the size (membrane area) of the plant?

b. What is the required membrane area?

12-11 An ultrafiltration membrane with a membrane resistance coefficient of 2.7×10^{12} m^{-1} is used to filter a 150-mg/L suspension of 0.5-μm latex particles in a laboratory unstirred dead-end filtration cell. The experiment is operated at a constant flux of 120 $L/m^2 \cdot h$ and temperature of 20°C, and the membrane has an area of 28.2 cm^2. Assume that fouling is due to cake formation, the particle density is 1050 kg/m^3, the cake porosity is 0.38, and the Kozeny coefficient is 5. Neglecting the backmigration of particles due to diffusion, calculate and plot the transmembrane pressure and specific flux over the first 90 min of the filter run.

12-12 Show how the cake layer resistance coefficient (Eq. 12-12) can be derived from the Kozeny equation (Eq. 11-11 in Chap. 11) when the membrane feed water is a suspension of monodisperse, well-characterized particles.

12-13 A membrane plant containing 1200 m^2 of membrane area operates at a constant permeate flux of 45 $L/m^2 \cdot h$ at a temperature of 15°C and pressure of 0.25 bar immediately after backwash. The feed water contains 12 mg/L of suspended solids. After 40 min of operation, the pressure rises to 0.30 bar.

a. Assuming that pressure rise between backwashes is due to formation of a cake layer, calculate the specific cake resistance.

b. If permeate flux is increased to 50 $L/m^2 \cdot h$, calculate the pressure immediately after backwash and the pressure after 40 min of operation.

12-14 Calculate the membrane fouling index for the following data, for the data set specified by your instructor.

a. Experimental flat-sheet laboratory filter, membrane area = 30 cm^2, initial flux = 3560 $L/m^2 \cdot h \cdot bar$, test pressure = 0.69 bar, test temperature = 23.9°C.

Time, min	Permeate Volume, mL	Time, min	Permeate Volume, mL
0	0	6	458.3
1	108.8	7	506.8
2	199.8	8	552.1
3	277.4	9	594.1
4	345.0	10	634.1
5	404.2	11	670.8

b. Full-scale plant operating at constant permeate flow of 15,000 m^3/day, temperature = 20°C, 5800 m^2 of membrane area, pressure each day as shown below. Use day 0 as the initial flux.

Time, Day	Transmemb. Pressure, Bar	Time, Day	Transmemb. Pressure, Bar	Time, Day	Transmemb. Pressure, Bar
0	0.704				
2	0.712	12	0.747	22	0.786
4	0.721	14	0.754	24	0.794
6	0.726	16	0.765	26	0.801
8	0.735	18	0.770	28	0.812
10	0.740	20	0.777	30	0.812

c. Data from a 30-min filter run in the middle of a day of laboratory testing of coagulated feed water, membrane area = 23 cm^2, initial flux = 238 $L/m^2 \cdot h \cdot bar$, test pressure = 2.07 bar, test temperature = 21.5°C.

Time, min	Permeate Volume, mL	Time, min	Permeate Volume, mL	Time, min	Permeate Volume, mL
0	2276.64				
2	2292.62	12	2370.17	22	2444.76
4	2308.41	14	2385.31	24	2459.35
6	2324.05	16	2400.33	26	2473.88
8	2339.53	18	2415.24	28	2488.26
10	2354.92	20	2430.04		

12-15 A membrane filtration plant is to be designed using results from a pilot study. Treatment plant requirements and pilot results are given in the table below. For the selected system (to be specified by the instructor), determine (a) the online production factor, (b) system recovery, (c) feed flow rate, (d) total membrane area, (e) number of skids, and (f) number of modules per skid. The pilot system contained two membrane elements that had 45 m^2 of membrane area each. In the full-scale plant, integrity testing will be required by regulations once per day and will take 15 min. Chemical cleaning (CIP) will take 4 h.

	A	B	C	D	E
Design capacity (m^3/d)	56,000	115,000	38,000	76,000	227,000
Memb. area in full-scale modules (m^2)	45	55	45	45	80
Max. modules in skid	80	90	80	80	100

	A	B	C	D	E
Pilot results					
Flux (L/m^2 · h)	80	125	40	80	110
Backwash frequency (min)	30	25	25	22	30
Backwash duration (min)	1.5	0.5	1	2	1
Backwash volume (L)	270	100	200	240	240
Cleaning frequency (day)	45	30	60	30	30

References

Adham, S., Chiu, K., Gramith, K., and Oppenheimer, J. (2005) *Development of a Microfiltration and Ultrafiltration Knowledge Base,* American Water Works Association Research Foundation, Denver, CO.

Adham, S., Chiu, K., Lehman, G., Howe, K., Marwah, A., Mysore, C., Clouet, J., Do-Quang, Z., and Cagnard, O. (2006) *Optimization of Membrane Treatment for Direct and Clarified Water Filtration,* American Water Works Association Research Foundation, Denver, CO.

Adham, S. S., Jacangelo, J. G., and Laîné, J.-M. (1995) "Low-Pressure Membranes: Assessing Integrity," *J. AWWA,* **87**, 3, 62–76.

Allgeier, S. C. (2001) Overview of Regulatory Issues Facing Microfiltration and Ultrafiltration, paper presented at the American Water Works Association Membrane Technology Conference, San Antonio, TX.

Amy, G. L., and Cho, J. (1999) "Interactions between Natural Organic Matter (NOM) and Membranes: Rejection and Fouling," *Water Sci. Technol.*, **40**, 9, 131–139.

Anselme, C., Baudin, I., and Chevalier, M. R. (1999) Drinking Water Production by Ultrafiltration and PAC Adsorption, First Year of Operation for a Large Capacity Plant, paper presented at the American Water Works Association Membrane Technology Conference, Long Beach, CA.

Anselme, C., and Jacobs, E. P. (1996) Ultrafiltration, Chap. 10, in J. Mallevialle, P. E. Odendaal, and M. R. Wiesner (eds.), *Water Treatment Membrane Processes,* McGraw-Hill, New York.

ASTM (2001a) D5090-90 Standard Practice for Standardizing Ultrafiltration Permeate Flow Performance Data, in *Annual Book of Standards,* Vol. 11.01, American Society for Testing and Materials, Philadelphia, PA.

ASTM (2001b) D6161-98 Standard Terminology Used for Crossflow Microfiltration, Ultrafiltration, Nanofiltration and Reverse Osmosis Membrane Processes, in *Annual Book of Standards,* Vol. 11.02, American Society for Testing and Materials, Philadelphia, PA.

ASTM (2001c) E1343-90 Standard Test Method for Molecular Weight Cutoff Evaluation of Flat Sheet Ultrafiltration Membranes, in *Annual Book of Standards,* Vol. 11.01, American Society for Testing and Materials, Philadelphia, PA.

AWWA (1992) "Membrane Processes in Potable Water Treatment, AWWA Membrane Technology Research Committee Report," *J. AWWA,* **84**, 1, 59–67.

AWWA (1998) "Membrane Processes, AWWA Membrane Technology Research Committee Report," *J. AWWA*, **90**, 6, 91–105.

AWWA (2005a) "Committee Report: Recent Advances and Research Needs in Membrane Fouling," *J. AWWA*, **97**, 8, 79–89.

AWWA (2005b) *Microfiltration and Ultrafiltration Membranes for Drinking Water: Manual of Water Supply Practices M53*, AWWA, Denver, CO.

AWWA (2008) "Microfiltration and Ultrafiltration Membranes for Drinking Water," *J. AWWA*, **100**, 12, 84–97.

Bacchin, P., Aimar, P., and Sanchez, V. (1995) "Model for Colloidal Fouling of Membranes," *AIChE J.*, **41**, 2, 368–376.

Belfort, G., Davis, R. H., and Zydney, A. L. (1994) "Behavior of Suspensions and Macromolecular Solutions in Crossflow Microfiltration," *J. Memb. Sci.*, **96**, 1/2, 1–58.

Campos, C., Mariñas, B. J., Snoeyink, V. L., Baudin, I., and Laîné, J.-M. (2000a) "PAC–Membrane Filtration Process I: Model Development," *J. Environ. Eng.*, **126**, 2, 97–103.

Campos, C., Mariñas, B. J., Snoeyink, V. L., Baudin, I., and Laîné, J.-M. (2000b) "PAC–Membrane Filtration Process II: Model Application," *J. Environ. Eng.*, **126**, 2, 104–111.

Carroll, T., King, S., Gray, S. R., Bolto, B. A., and Booker, N. A. (2000) "Fouling of Microfiltration Membranes by NOM after Coagulation Treatment," *Water Res.*, **34**, 11, 2861–2868.

Causserand, C., Nyström, M., and Aimar, P. (1994) "Study of Streaming Potentials of Clean and Fouled Ultrafiltration Membranes," *J. Memb. Sci.*, **88**, 2/3, 211–222.

Chang, Y.-J., and Benjamin, M. M. (1996) "Iron Oxide Adsorption and UF to Remove NOM and Control Fouling," *J. AWWA*, **88**, 12, 74–88.

Chang, Y.-J., and Benjamin, M. M. (2003) "Modeling Formation of Natural Organic Matter Fouling Layers on Ultrafiltration Membranes," *J. Environ. Eng.*, **129**, 1, 25–32.

Chellam, S., and Xu, W. (2006) "Blocking Laws Analysis of Dead-End Constant Flux Microfiltration of Compressible Cakes," *J. Colloid Interface Sci*, **301**, 1, 248–257.

Cheryan, M. (1998) *Ultrafiltration and Microfiltration Handbook*, Technomic, Lancaster, PA.

Cogan, N. G., and Chellam, S. (2009) "Incorporating Pore Blocking, Cake Filtration, and EPS Production in a Model for Constant Pressure Bacterial Fouling During Dead-End Filtration," *J. Memb. Sci.*, **345**, 1–2, 81–89.

Combe, C., Molis, E., Lucas, P., Riley, R., and Clark, M. M. (1999) "The Effect of CA Membrane Properties on Adsorptive Fouling by Humic Acid," *J. Memb. Sci.*, **154**, 1, 73–87.

Craun, G. F., Hubbs, S. A., Frost, F., Calderon, R. L., and Via, S. H. (1998) "Waterborne Outbreaks of Cryptosporidiosis," *J. AWWA*, **90**, 9, 81–91.

Crozes, G., Anselme, C., and Mallevialle, J. (1993) "Effect of Adsorption of Organic Matter on Fouling of Ultrafiltration Membranes," *J. Memb. Sci.*, **84**, 1/2, 61–77.

Fane, A. G., and Fell, C. J. D. (1987) "A Review of Fouling and Fouling Control in Ultrafiltration," *Desalination*, **62**, 117–136.

Farahbakhsh, K., Svrcek, C., Guest, R. K., and Smith, D. W. (2004) "A Review of the Impact of Chemical Pretreatment on Low-Pressure Water Treatment Membranes," *J. Env. Eng. Sci*, **3**, 4, 237–253.

Ferry, J. D. (1936) "Statistical Evaluation of Sieve Constants in Ultrafiltration," *J. Gen. Physiol.*, **20**, 95–104.

Field, R. W., Wu, D., Howell, J. A., and Gupta, B. B. (1995) "Critical Flux Concept for Microfiltration Fouling," *J. Memb. Sci.*, **100**, 3, 259–272.

Glucina, K., Laîné, J.-M., and Durand-Bourlier, L. (1998) "Assessment of Filtration Mode for the Ultrafiltration Membrane Process," *Desalination*, **118**, 1/3, 205–211.

Habarou, H., Makdissy, G., Croue, J.-P., Amy, G. L., Buisson, H., and Machinal, C. (2001) Toward an Understanding of NOM Fouling of UF Membranes, paper presented at the American Water Works Association Membrane Technology Conference, San Antonio, TX.

Hermia, J. (1982) "Constant Pressure Blocking Filtration Laws—Application to Power-Law Non-Newtonian Fluids," *Trans. IChemE*, **60**, 183–187.

Herschell, J. A. (2007) Survey of State Regulatory Approaches for Approval of Low Pressure Membrane Systems, presentation at the 2007 AWWA Annual Conference, Toronto, ON.

Ho, C. C., and Zydney, A. L. (2000) "A Combined Pore Blockage and Cake Filtration Model for Protein Fouling During Microfiltration," *Journal of Colloid and Interface Science*, **232**, 2, 389–399.

Howe, K. J. (2001) Effect of Coagulation Pretreatment on Membrane Filtration Performance, Ph.D. Thesis, University of Illinois at Urbana-Champaign, Urbana, IL.

Howe, K. J., and Clark, M. M. (2002) "Fouling of Microfiltration and Ultrafiltration Membranes by Natural Waters," *Environ. Sci. Technol.*, **36**, 16, 3571–3576.

Howe, K. J., and Clark, M. M. (2006) "Effect of Coagulation Pretreatment on Membrane Filtration Performance," *J. AWWA*, **98**, 4, 133–146.

Howe, K. J., Marwah, A., Chiu, K. P., and Adham, S. S. (2007) "Effect of Membrane Configuration on Bench-Scale MF and UF Fouling Experiments," *Water Res.*, **41**, 17, 3842–3849.

Huang, H., and O'Melia, C. R. (2008) "Direct-Flow Microfiltration of Aquasols II. On the Role of Colloidal Natural Organic Matter," *J. Memb. Sci.*, **325**, 2, 903–913.

Huang, H., Spinette, R., and O'Melia, C. R. (2008a) "Direct-Flow Microfiltration of Aquasols I. Impacts of Particle Stabilities and Size," *J. Memb. Sci.*, **314**, 1–2, 90–100.

Huang, H., Young, T. A., Jacangelo, J. G. (2008b) "Unified Membrane Fouling Index for Low Pressure Membrane Filtration of Natural Waters: Principles and Methodology," *Environ. Sci. Technol.*, **42**, 3, 714–720.

Huang, H., Schwab, K., Jacangelo, J. G. (2009a) "Pretreatment for Low Pressure Membranes in Water Treatment: A Review," *Environ. Sci. Technol.*, **43**, 9, 3011–3019.

Huang, H., Young, T. A., Jacangelo, J. G. (2009b) "Novel Approach for the Analysis of Bench-Scale, Low Pressure Membrane Fouling in Water Treatment," *J. Memb. Sci.*, **334**, 1–2, 1–8.

Ioan, C. E., Aberle, T., and Burchard, W. (2000) "Structure Properties of Dextran 2. Dilute Solution," *Macromolecules*, **33**, 15, 5730–5739.

Jacangelo, J. G., Adham, S. S., and Laîné, J.-M. (1995) "Mechanism of *Cryptosporidium, Giardia*, and MS2 Virus Removal by MF and UF," *J. AWWA*, **87**, 9, 107–121.

Jacangelo, J. G., Laîné, J.-M., Carns, K. E., Cummings, E. W., and Mallevialle, J. (1991) "Low-Pressure Membrane Filtration for Removing *Giardia* and Microbial Indicators," *J. AWWA*, **83**, 9, 97–106.

Jucker, C., and Clark, M. M. (1994) "Adsorption of Aquatic Humic Substances on Hydrophobic Ultrafiltration Membranes," *J. Memb. Sci.*, **97**, 37–52.

Kim, J. and DiGiano, F.A. (2006) "A Two-Fiber, Bench-Scale Test of Ultrafiltration (UF) for Investigation of Fouling Rates and Characteristics," *J. Memb. Sci.*, **271**, 1–2, 196–204.

Kim, J. and DiGiano, F. A. (2009) "Fouling Models for Low-Pressure Membrane Systems," *Separation Purification Technol.*, **68**, 3, 293–304.

Kim, K. J., Fane, A. G., Ben Aim, R., Liu, M. G., Jonsson, G., Tessaro, I. C., Broek, A. P., and Bargeman, D. (1994) "Comparative Study of Techniques Used for Porous Membrane Characterization: Pore Characterization," *J. Memb. Sci.*, **87**, 1/2, 35–46.

Kim, K. J., Fane, A. G., Fell, C. J. D., and Joy, D. C. (1992) "Fouling Mechanisms of Membranes During Protein Ultrafiltration," *J. Memb. Sci.*, **68**, 1/2, 79–91.

Koltuniewicz, A. (1992) "Predicting Permeate Flux in Ultrafiltration on the Basis of Surface Renewal Concept," *J. Memb. Sci.*, **68**, 1/2, 107–118.

Koros, W.J., Ma, Y. H., and Shimidzu, T. (1996) "Terminology for Membranes and Membrane Processes," *J. Memb. Sci.*, **120**, 2, 149–159.

Kulovaara, M., Metsämuuronen, S., and Nyströim, M. (1999) "Effects of Aquatic Humic Substances on a Hydrophobic Ultrafiltration Membrane," *Chemosphere*, **38**, 15, 3485–3496.

Lahoussine-Turcaud, V., Wiesner, M. R., and Bottero, J.-Y. (1990) "Fouling in Tangential-Flow Ultrafiltration: The Effect of Colloid Size and Coagulation Pretreatment," *J. Memb. Sci.*, **52**, 2, 173–190.

Laîné, J.-M., Clark, M. M., and Mallevialle, J. (1990) "Ultrafiltration of Lake Water: Effect of Pretreatment on the Partitioning of Organics, THMFP, and Flux," *J. AWWA*, **82**, 12, 82–87.

Laîné, J.-M., Hagstrom, J. P., Clark, M. M., and Mallevialle, J. (1989) "Effects of Ultrafiltration Membrane Composition," *J.AWWA*, **81**, 11, 61–67.

Landsness, L. B. (2001) Accepting MF/UF Technology, Making the Final Cut, paper presented at the American Water Works Association Membrane Technology Conference, San Antonio, TX.

Lee, Y., and Clark, M. M. (1997) "Numerical Model of Steady-State Permeate Flux During Cross-Flow Ultrafiltration," *Desalination*, **109**, 3, 241–251.

Lin, C.-F., Huang, Y.-J., and Hao, O.J. (1999) "Ultrafiltration Processes for Removing Humic Substances: Effect of Molecular Weight Fractions and PAC Treatment," *Water Res.*, **33**, 5, 1252–1264.

Lin, C.-F., Lin, T.-Y., and Hao, O. J. (2000) "Effects of Humic Substance Characteristics on UF Performance," *Water Res.*, **34**, 4, 1097–1106.

Madaeni, S. S., Fane, A. G., and Grohmann, G. S. (1995) "Virus Removal from Water and Wastewater Using Membranes," *J. Memb. Sci.*, **102**, 65–75.

Madigan, M. T., Martinko, J. M., and Parker, J. (1997) *Brock Biology of Microorganisms*, Prentice-Hall, Upper Saddle River, NJ.

Matthiasson, E. (1983) "The Role of Macromolecular Adsorption in Fouling of Ultrafiltration Membranes," *J. Memb. Sci.*, **16**, 23–26.

McDonogh, R. M., Welsch, K., Fane, A. G., and Fell, C. J. D. (1988) "Flux and Rejection in the Ultrafiltration of Colloids," *Desalination*, **70**, 1/3, 251–264.

Nyströim, M., Pihlajamaki, A., and Ehsani, N. (1994) "Characterization of Ultrafiltration Membranes by Simultaneous Streaming Potential and Flux Measurements," *J. Memb. Sci.*, **87**, 3, 245–256.

Petry, M., Thonney, D., Roux, J. P., Moulart, P., and Bonnelye, V. (2001) Lausanne: Specific Design for a Challenging Project, paper presented at the American Water Works Association Membrane Technology Conference, San Antonio, TX.

Ridgway, H. F., and Flemming, H.-C. (1996) Membrane Biofouling, Chap. 6, in J. Mallevialle, P. E. Odendaal, and M. R. Wiesner (eds.), *Water Treatment Membrane Processes*, McGraw-Hill, New York.

Schäfer, A. I., Fane, A. G., and Waite, T. D. (2001) "Cost Factors and Chemical Pretreatment Effects in the Membrane Filtration of Waters Containing Natural Organic Matter," *Water Res.*, **35**, 6, 1509–1517.

Schäfer, A. I., Schwicker, U., Fischer, M. M., Fane, A. G., and Waite, T. D. (2000) "Microfiltration of Colloids and Natural Organic Matter," *J. Memb. Sci.*, **171**, 2, 151–172.

Shorney, H. L., Vernon, W. A., Clune, J., and Bond, R. G. (2001) Performance of MF/UF Membranes with In-Line Ferric-Salt Coagulation for Removal of Arsenic from a Southwest Surface Water, paper presented at the American Water Works Association Membrane Technology Conference, San Antonio, TX.

Shrive, C. A., DeMarco, J., Metz, D. H., Braghetta, A., and Jacangelo, J. G. (1999) Assessment of Microfiltration for Integration into a Granular Activated Carbon Facility, paper presented at the American Water Works Association Membrane Technology Conference, Long Beach, CA.

Tarabara, V. V., Hovinga, R. M., and Wiesner, M. R. (2002) "Constant Transmembrane Pressure vs. Constant Permeate Flux: Effect of Particle Size on Crossflow Membrane Filtration," *Environ. Eng. Sci.*, **19**, 2, 343–355.

Tchobanoglous, G., Burton, F. L. and Stensel, H. D. (2003) *Wastewater Engineering: Treatment and Reuse*, 4th ed., Metcalf and Eddy, McGraw-Hill, New York.

Treybal, R. E. (1980) *Mass-Transfer Operations*, McGraw-Hill, New York.

U.S. EPA (1989) "National Primary Drinking Water Regulations: Filtration and Disinfection; Turbidity, *Giardia lamblia*, Viruses, *Legionella*, and Heterotrophic Bacteria; Final Rule," *Fed. Reg.*, **54**, 124, 27486.

U.S. EPA (1998) "National Primary Drinking Water Regulations: Interim Enhanced Surface Water Treatment: Final Rule," *Fed. Reg.*, **63**, 241, 69478–69521.

U.S. EPA (2001) *Low-Pressure Membrane Filtration for Pathogen Removal: Application, Implementation, and Regulatory Issues*, U.S. Environmental Protection Agency, Cincinnati, OH.

U.S. EPA (2005) *Membrane Filtration Guidance Manual*, EPA 815-R-06-009, U.S. Environmental Protection Agency, Cincinnati, OH.

U.S. EPA (2006) "National Primary Drinking Water Regulations: Long Term 2 Enhanced Surface Water Treatment; Final Rule," *Fed. Reg.*, **71**, 3, 654–786.

Welsch, K., McDonogh, R. M., Fane, A. G., and Fell, C. J. D. (1995) "Calculation of Limiting Fluxes in the Ultrafiltration of Colloids and Fine Particulates," *J. Memb. Sci.*, **99**, 3, 229–239.

Wiesner, M. R., and Chellam, S. (1992) "Mass Transport Considerations for Pressure-Driven Membrane Processes," *J. AWWA*, **84**, 1, 88–95.

Wiesner, M. R., Clark, M. M., and Mallevialle, J. (1989) "Membrane Filtration of Coagulated Suspensions," *J. Environ. Eng.*, **115**, 1, 20–40.

Yuan, W., Kocic, A., and Zydney, A. L. (2002) "Analysis of Humic Acid Fouling During Microfiltration Using a Pore Blockage-Cake Filtration Model," *J. Memb. Sci.*, **198**, 1, 51–62.

Yuan, W., and Zydney, A. L. (1999) "Humic Acid Fouling During Microfiltration," *J. Memb. Sci.*, **157**, 1, 1–12.

Yuan, W., and Zydney, A. L. (2000) "Humic Acid Fouling During Ultrafiltration," *Environ. Sci. Technol.*, **34**, 23, 5043–5050.

Zydney, A. L., and Colton, C. K. (1986) "A Concentration Polarization Model for the Filtration Flux in Cross-Flow Microfiltration of Particulate Suspensions," *Chem. Eng. Commun.*, **47**, 1/3, 1–21.

13 Disinfection

Stantec's Water Treatment: Principles and Design, Updated Third Edition. John C. Crittenden, R. Rhodes Trussell, David W. Hand, Kerry J. Howe, and George Tchobanoglous.

Terminology for Disinfection

Term	Definition
Absorbance	Amount of light of a specified wavelength absorbed by the constituents in water.
Biodosimetry	Determination of the dose of a disinfectant to inactivate a specific biological test organism.
Breakpoint chlorination	Process in which chlorine is added to react with all oxidizable substances in water so that if additional chlorine is added it will remain as free chlorine (see below, $HOCl + OCl^-$).
Combined chlorine residual	Concentration of chlorine species resulting from the reaction of chlorine and ammonia, specifically the sum of monochloramine (NH_2Cl), dichloramine ($NHCl_2$), and trichloramine (NCl_3), expressed as mg/L as Cl_2.
Ct	Product of chlorine residual expressed in mg/L and contact time expressed in min. The term *Ct* is used to assess the effectiveness of the disinfection process for regulatory purposes.
Disinfection	Partial destruction and inactivation of disease-causing organisms from exposure to chemical agents (e.g., chlorine) or physical processes (e.g., UV irradiation).

Term	Definition
Decay rate	Rate at which the concentration of a disinfectant decreases over time.
Disinfection by-products (DBPs)	Undesirable products of reactions between disinfectants and other species in the feed water. DBPs of concern are those that are carcinogenic or have other negative health effects.
Dose–response curve	Relationship between the degree of microorganism inactivation and the dose of a disinfectant.
Free chlorine residual	Sum of the hypochlorous acid (HOCl) and hypochlorite ion (OCl^-) in solution, expressed as mg/L as Cl_2.
Inactivation	Rendering microorganisms incapable of reproducing and thus limiting their ability to cause disease.
Pathogens	Microorganisms capable of causing disease.
Photoreactivation and dark repair	Methods used by microorganisms to repair the damage caused by exposure to UV irradiation.
Reactivation	Process by which organisms repair the damage caused by exposure to a disinfectant.
Sterilization	Total destruction of disease-causing and other organisms.
Transmittance	Ability of water to transmit light. Transmittance is related to absorbance.
Total chlorine residual	Sum of the concentrations of free and combined chlorine.
UV light	Portion of the electromagnetic spectrum between 100 and 400 nm.

The threat of microbiological contaminants in drinking water is eliminated by three complementary strategies: (1) preventing their access to the water source, (2) employing water treatment to reduce their concentration in the water, and (3) maximizing the integrity of the distribution system for finished water. Early in the history of public drinking water systems, the emphasis was almost entirely on gaining access to a protected source. In recent years, greater emphasis has been directed toward providing effective water treatment to reduce microbiological contaminants. Today, there is increasing emphasis on employing both source protection and treatment to ensure that safe water is produced and on improving distribution system integrity to ensure that contamination does not occur during transport from the treatment plant to the consumer's tap.

In the water treatment process, *reducing* microbiological contaminants is accomplished by two basic strategies, *removing* them from the water or *inactivating* them. Inactivated microorganisms, although still present in the water, are no longer able to cause disease in the consumer. Processes that use inactivation as their strategy are traditionally referred to as disinfection, the focus of this chapter.

In water works practice, the term *disinfection* is used to refer to two activities: (1) primary disinfection—the inactivation of microorganisms in the water—and (2) secondary disinfection—maintaining a disinfectant residual in the treated-water distribution system. The characteristics that make a disinfectant the best choice for each of these purposes are not the same.

Primary disinfection is discussed in this chapter, along with the role disinfection plays in protecting the public, the strengths and weaknesses of inactivation versus removal, the kinetics of the disinfection process, and some specific details about the design of disinfection facilities. Disinfection by-products are discussed in Chap. 19.

13-1 Historical Perspective

Beginning a decade before the work of Dr. John Snow (1849 and 1853, see Chap. 3) and continuing for five decades after, two principal means were employed to control waterborne disease: (1) using water supplies not exposed to fecal contamination and (2) filtration through sand. At first, slow sand filtration was the dominant water treatment process; however, it was not always effective. The first efforts in rapid sand filtration were even less effective. Eventually George W. Fuller (1897) demonstrated that it is essential that complete coagulation precede the filtration step. Even with proper coagulation, however, filtration alone was not consistently successful in reducing the microorganisms to safe levels (Johnson, 1911; Whipple, 1906).

In 1881, not long before Fuller did his work on coagulation and filtration, Koch, the German scientist who demonstrated the role bacteria play in waterborne disease, also demonstrated that chlorine could inactivate pathogenic bacteria. The first continuous use of chlorination for disinfection of drinking water occurred in Middelkerke, Belgium, in 1902. The first continuous application to drinking water in the United States was at the Boonton Reservoir for the water works of Jersey City, New Jersey, in late 1908. In these first applications, disinfection was accomplished by feeding solid calcium hypochlorite. Soon after, liquid chlorine gas became available, making large-scale continuous chlorination more feasible. The first water treatment facility to use liquid chlorine gas on a permanent basis was in Philadelphia in 1913. Most of these early installations were used to address serious contamination or to avoid filtration, but in the three decades following the installation in Philadelphia, the practice of chlorination was expanded rapidly to include most surface water supplies, even those that were filtered. By 1941, 85 percent of the drinking water supplies in the United States were chlorinated (U.S. PHS, 1943). Also, by the 1940s, disinfection with chlorine had become a world water treatment standard and, even today, many water supplies are treated with chlorination alone.

The presence of a free chlorine residual in water at the tap was generally taken as a guarantee of microbiological safety by health officials and the public. Disinfection thus became established as the most important water treatment process. A more detailed discussion of the use of chlorine can be found in Baker (1948) and White (1999).

From the beginning, the use of chlorine has been contentious with many of its opponents arguing for the use of protected supplies in place of disinfection (Drown, 1893/1894). Equally important, a significant portion of the population has always had an aversion to the use of chlorine, complaining about its impact on the water's aesthetic qualities and wishing to avoid exposure to a chemical with such toxic properties, even at low concentrations. Largely for this second reason, ozone became the preferred primary disinfectant in much of mainland Europe in the late 1960s and 1970s.

In the mid-1970s, events took place that stimulated a reevaluation of disinfection practice. In Holland and the United States, researchers demonstrated that free chlorine reacts with natural organic matter (NOM) in water to produce chlorinated organics, specifically the trihalomethanes (THMs) (Bellar and Lichtenberg, 1974; Rook, 1974). Not long thereafter, limits were set on the allowable THM concentrations in potable water (U.S. EPA, 1979; WHO, 1994). Since then, more by-products have been identified resulting from chlorination and the use of other disinfectants (Bull et al., 1990). Limits have also been established for many of these by-products (U.S. EPA, 1998). It is likely that chemical by-products are formed any time an oxidant is employed in water treatment and that some of these by-products will be regulated in the future (Trussell, 1992, 1993).

During the last two decades of the twentieth century, events occurred that have also resulted in the questioning of the effectiveness of chlorination in controlling waterborne disease. In the 1980s, the protozoa *Giardia lamblia* was identified as an important waterborne pathogen. Because *G. lamblia* is more resistant to chlorine than other targets of disinfection, more stringent standards for reduction of pathogens were established (U.S. EPA, 1989). More recently, another protozoa, *Cryptosporidium parvum*, has also been identified as an important source of waterborne disease and is even more resistant to chlorine than *G. lamblia*. In fact, chlorination is ineffective for *C. parvum*.

The discovery of chlorination by-products and chlorine-resistant organisms is causing a reevaluation of the use of chlorine as the primary disinfectant and a reevaluation of the role of inactivation itself in the control of pathogens. For example, because methods are not available to determine if *C.parvum* oocysts found in water supplies will cause disease if ingested by a consumer, the Drinking Water Inspectorate in the United Kingdom recognizes only *removal*, not *inactivation*, as a viable strategy for addressing the control of this pathogen (U.K. Department of the Environment, 1999a,b).

New treatment processes have also come to the fore that show promise for the removal or inactivation of chlorine-resistant organisms and others as well. Membrane filtration processes, developed originally in the mid-1950s and later employed for sterilizing laboratory solutions, juices, and eventually brewed beverages, have now reached a stage in their development where they are commercially viable at large scale. Membranes are capable of removing pathogens much more effectively than traditional physical treatment processes such as coagulation and granular media filtration. In fact, the removals that have been demonstrated using membranes are on the same order of magnitude of inactivation of bacteria customarily achieved by chlorine (Jacangelo et al., 1989). Disinfection with UV light is also effective for inactivating *Giardia* (Stolarik et al., 2001) and *Cryptosporidium* (Craik et al., 2001). While chlorine remains the dominant drinking water disinfectant and disinfection (inactivation) remains the cornerstone of water treatment, this situation may change in the future.

13-2 Methods of Disinfection Commonly Used in Water Treatment

Five disinfection agents are commonly used in drinking water treatment today: (1) free chlorine, (2) combined chlorine (chlorine combined with ammonia, also known as chloramines), (3) chlorine dioxide, (4) ozone, and (5) UV light. The first four are chemical oxidants, whereas UV light involves the use of electromagnetic radiation. Of the five, by far the most common in the United States is free chlorine. As shown on Fig. 13-1, surveys of disinfectant use by the American Water Works Association Disinfection Systems Committee in 1978, 1989, 1998, and 2007 found that nearly all water utilities in the United States use free chlorine, although the method of application has been changing over time (AWWA, 2008). In 1978, 91 percent of utilities used chlorine gas to apply free chlorine to the water and 7 percent used sodium hypochlorite (i.e., bleach). By 2007, however, only 63 percent of utilities were using chlorine gas and nearly 40 percent were using either bulk liquid or onsite generation of sodium hypochlorite. The transition from chlorine gas to hypochlorite is primarily because of safety and security reasons because chlorine gas is highly toxic.

As shown on Fig. 13-1, the number of utilities using chloramines for disinfection has increased to 30 percent by 2007. Its use, however, is often limited to residual maintenance, and typically a different disinfectant is used for primary disinfection when chloramine is used.

Ozone is the strongest of the four oxidants and its use has increased from less than 1 percent of utilities in 1989 to 9 percent in 2007. The increasing use is in part because of its stronger disinfecting properties and in part because it controls taste and odor compounds, specifically geosmin

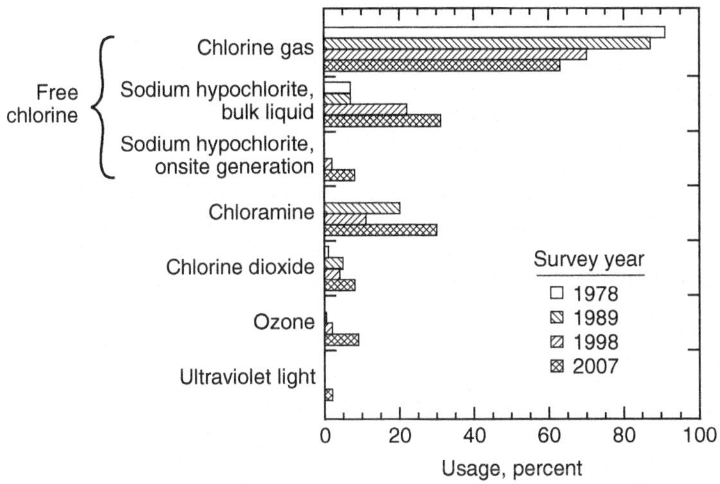

Figure 13-1
Disinfectant use in municipal drinking water treatment in the United States. (Adapted from AWWA 2008.)

and methyl is oborneol. UV light is not frequently used for disinfecting in drinking water applications, with only 2 percent of utilities reporting to use it in 2007. Its use may increase in the future, however, because of its lack of by-product generation and its effectiveness against protozoa. Information on each of these common disinfectants is summarized in Table 13-1.

Historically, chlorine was added to the raw water at a treatment plant and disinfection occurred during contact over the residence time of the entire plant. This practice has become obsolete and disinfection is now best applied as a separate unit process. The chemical disinfectants are most often applied in baffled, serpentine contact chambers or long pipelines when these are available. Both types of contactors can be designed to be highly efficient, closely approaching ideal plug flow. Additionally, ozone can be introduced in over–under baffled contactors. Over–under baffled contactors, however, have bigger problems with short circuiting, so pipeline and serpentine basin contactors have become more common for ozone disinfection. Design of contactors for chemical disinfectants is discussed in Sec. 13-8 in this chapter.

Ultraviolet light disinfection is often applied in proprietary reactors. Short circuiting is a special concern for UV reactors, particularly the proprietary reactors because their contact times are so short. Proprietary pressure vessels are particularly common where medium-pressure UV lamps are used because the high intensity of the UV lamps enables the delivery of a high UV dosage in a small space. Standards to address these issues exist in Europe (DVGW, 1997) and are being developed in the United States (NWRI, 2003; U.S. EPA, 2006).

Table 13-1
Characteristics of five most common disinfectants

	Disinfectant				
Issue	**Free Chlorine**	**Combined Chlorine**	**Chlorine Dioxide**	**Ozone**	**Ultraviolet Light**
Effectiveness in disinfection					
Bacteria	Excellent	Good	Excellent	Excellent	Good
Viruses	Excellent	Fair	Excellent	Excellent	Fair
Protozoa	Fair to poor	Poor	Good	Good	Excellent
Endospores	Good to poor	Poor	Fair	Excellent	Fair
Regulatory limit on residuals	4 mg/L	4 mg/L	0.8 mg/L	—	—
Formation of chemical by-products					
Regulated by-products	Forms 4 THMs[a] and 5 HAAs[b]	Traces of THMs and HAAs	Chlorite	Bromate	None
By-products that may be regulated in future	Several	Cyanogen halides, NDMA[c]	Chlorate	Biodegradable organic carbon	None known
Typical application					

Dose, mg/L (kg/ML)	1–6	2–6	0.2–1.5	1–5	20–100 mJ/cm^2
Dose, lb/MG	8–50	17–50	2–13	8–42	—
Chemical source	Delivered: as liquid gas in tank cars, 1 tonne and 68-kg (150-lb) cylinders, or as liquid bleach. Onsite generation from salt and water using electrolysis.Calcium hypochlorite powder is used for very small applications.	Same sources for chlorine. Ammonia is delivered as aqua ammonia solution, liquid gas in cylinders, or solid ammonium sulfate. Chlorine and ammonia are mixed in treatment process.	ClO_2 is manufactured with an onsite generator from chlorine and chlorite. Same sources for chlorine. Chlorite as powder or stabilized liquid solution.	Manufactured onsite using a corona discharge in dry air or pure oxygen. Oxygen is usually delivered as a liquid. Oxygen can also be manufactured onsite.	Uses (1) low-pressure, low-intensity UV (2) low-pressure, high-intensity UV (254-nm), or (3) medium pressure UV (several wavelengths) lamps in the contactor itself.

[a]THMs = trihalomethanes.
[b]HAAs = haloacetic acids.
[c]NDMA = N: nitrosodimethy lamine.

13-3 Disinfection Kinetics

For chemical disinfectants, the specific mechanisms of microorganism inactivation are not well understood. Inactivation depends on the properties of each microorganism, the disinfectant, and the water. As will be shown later, the reaction rates that have been observed can vary by as much as six orders of magnitude from one organism to the next, even for one disinfectant. Even for disinfection reactions where the reaction mechanism is well understood, for example, UV light, reaction rates vary by one and one-half orders of magnitude.

Nevertheless, there is one simple kinetic model that is widely used, and there is enough commonality in the behavior of all these reactions to allow the development of some phenomenological laws that are useful in modeling all of these reactions. As these disinfection processes are physiochemical processes, they are also subject to the rules of analysis discussed in Chaps. 6 and 7. In the following discussion, the form of disinfection data resulting from laboratory experiments is examined by considering the shape of classical disinfection kinetic plots. Following this discussion, useful phenomenological kinetic models are discussed along with the merits of each.

Classical Disinfection Kinetics—Chick–Watson

Near the beginning of the twentieth century, Dr. Harriet Chick, a research assistant at the Lister Institute of Preventive Medicine in Chelsea, England, proposed that disinfection could be modeled as a first-order reaction with respect to the concentration of the organisms. Chick demonstrated her concept by plotting the concentration of viable organisms versus time on a semilog graph for disinfection data for a broad variety of disinfectants and organisms (Chick, 1908). Chick worked with disinfectants such as phenol, mercuric chloride, and silver nitrate and organisms such as *Salmonella typhi, Salmonella paratyphi, Escherischia coli, Staphylococcus aureus, Yersinia pestis,* and *Bacillus anthracis.*Over the subsequent years "Chick's law" has been shown to be broadly applicable to disinfection data. Chick's law takes the form

$$r = -k_c N \tag{13-1}$$

where r = reaction rate for the decrease in viable organisms with time, org/L·min
k_c = Chick's law rate constant, min^{-1}
N = concentration of organisms, org/L

Application of Chick's concept met with immediate success, and that success has continued through the years and across all the disciplines interested in disinfection.

While Chick's law has broad applicability, one important effect not addressed in the model is the effect of the concentration of the disinfectant. Frequently, different concentrations of disinfectant will lead to different

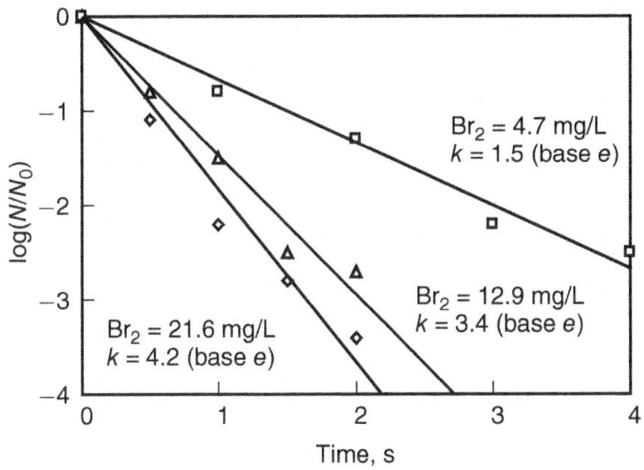

Figure 13-2
Inactivation of poliovirus type I with three concentrations of bromine in a batch reactor. (Adapted from Floyd et al., 1978.)

rates in the decrease in viable organisms, as illustrated on Fig. 13-2. Note that there is a different slope for each concentration of bromine and, using Eq. 13-1, the reaction has a different rate constant for each concentration. Thus, while Chick's first-order concept is consistent with the data, a better means for accounting for disinfectant concentration is necessary.

In the same year that Chick proposed her model, Herbert Watson proposed that the time needed to reach a specific level of disinfection was related to the disinfectant concentration by the equation (Watson, 1908)

$$C^n t = \text{constant} \tag{13-2}$$

where C = concentration of disinfectant, mg/L
n = empirical constant related to concentration, unitless
t = time required to achieve a constant percentage of inactivation (e.g., 99%)
constant = value for given percentage of inactivation, dimensionless

Watson demonstrated the concept by plotting data showing equal inactivation on a plot of log(C) versus log(t). The slope of the log–log plot, n, is often called the coefficient of dilution, which reflects the effect of diluting the disinfectant (Morris, 1975). Such plots are still used today, and an example is shown on Fig. 13-3. As a matter of convention, Watson plots are generally constructed with data corresponding to a removal of 99 percent. In such plots, the dilution coefficient is generally found to be approximately 1, and given the inaccuracies involved in collecting disinfection data, there is little evidence for a dilution coefficient other than unity. A dilution coefficient equal to 1 suggests that disinfection concentration and time are of equal importance for inactivating microorganisms.

With the knowledge that disinfection concentration and time are of equal importance, Chick's law and the Watson equation can be combined

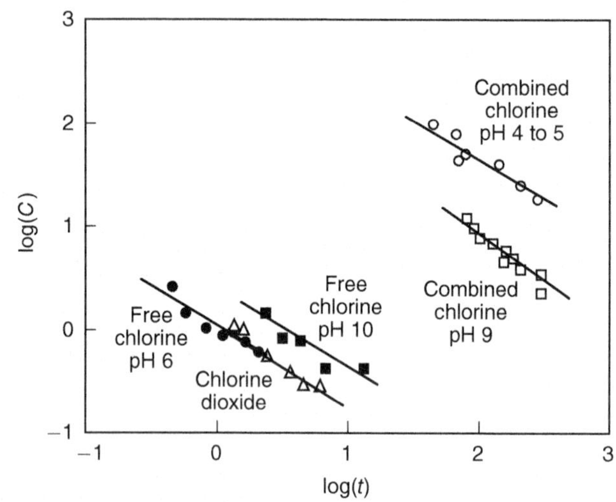

Figure 13-3
Watson plot of requirements for 99 percent inactivation of poliovirus type I. (Adapted from Scarpino et al., 1977.)

and are often referred to as the "Chick–Watson model" (Haas and Karra, 1984):

$$r = -\Lambda_{CW}\, CN \tag{13-3}$$

where Λ_{CW} = coefficient of specific lethality (disinfection rate constant), L/mg·min
C = concentration of disinfectant, mg/L

Most laboratory disinfection studies are conducted using completely mixed batch reactors (CMBR). Using concepts presented in Chap. 6, a mass balance on a batch reactor can be written and integrated, leading to

$$\ln\left(\frac{N}{N_0}\right) = -\Lambda_{CW}\, Ct \tag{13-4}$$

where N_0 = concentration of organisms at time = 0, org/L
t = time, min

It is important to note that even though laboratory disinfection studies typically use batch reactors, the rate equation (Eq. 13-3) can be applied to other reactors using the concepts presented in Chap. 6.

When Chick did her work, she plotted the organism concentration directly against time on a semilog graph [log(N) vs. t]. Now that Eq. 13-4 has received broad recognition, it is more common to plot the log or natural log of the survival ratio, where $S = N/N_0$, versus time [$\ln(N/N_0)$ or $\log(N/N_0)$ vs. t]. In disinfection studies, however, it is typically difficult to get an accurate measurement of the initial concentration of organisms, N_0, even with several replicates of the tests. As a result, a line fit through the data may not pass through zero [i.e., $\ln(N/N_0)_{t=0} \neq 0$]. Although it is not consistent with the definition of N_0[at $t = 0$, $\ln(N/N_0) \equiv 0$], it is often best to find the coefficient of specific lethality without forcing the regression line to pass through zero.

Equation 13-4 was derived using calculus so the term on the left is a natural logarithm (i.e., base e). However, disinfection effectiveness is typically expressed using the log removal value (LRV), which uses base 10 logarithms as described in Sec. 4-5. Thus, it is necessary to convert between natural and base 10 logarithms when evaluating disinfection data. The use of Eq. 13-4 to determine the coefficient of specific lethality for a disinfection reaction is demonstrated in Example 13-1.

Example 13-1 Application of the Chick–Watson model

Plot the data shown on Fig. 13-2, as given below, according to Eq. 13-4. Determine the coefficient of specific lethality and the coefficient of determination (r^2). The data for the inactivation of poliovirus type I with bromine (Floyd et al.,1978) are provided in the following table:

C, mg/L	Time, s	log(N/N$_0$)	C, mg/L	Time, s	log(N/N$_0$)
21.6	0	0	12.9	1.5	−2.5
21.6	0.5	−1.1	12.9	2	−2.7
21.6	1	−2.2	4.7	1	−0.8
21.6	1.5	−2.8	4.7	2	−1.3
21.6	2	−3.4	4.7	3	−2.2
12.9	0.5	−0.8	4.7	4	−2.5
12.9	1	−1.5			

Solution

1. Determine the values of Ct and $\ln(N/N_0)$ for each organism survival value.
   a. Ct is calculated simply by multiplying C by t.
   b. To convert from base 10 to base e logarithms, recall the logarithmic identity $\log_b(x) = \log_a(x)/\log_a(b)$, thus:

$$\ln\left(N/N_0\right) = \frac{\log\left(N/N_0\right)}{\log\left(e\right)} = 2.303 \log\left(\frac{N}{N_0}\right)$$

c. The required data table is shown below:

Time, s	C, mg/L	Ct, mg · s/L	ln(N/N_0)	Time, s	C, mg/L	Ct, mg · s/L	ln(N/N_0)
0.5	21.6	10.8	−2.53	1.5	12.9	19.4	−5.76
1	21.6	21.6	−5.07	2	12.9	25.8	−6.22
1.5	21.6	32.4	−6.45	1	4.7	4.7	−1.84
2	21.6	43.2	−7.83	2	4.7	9.4	−2.99
0.5	12.9	6.5	−1.84	3	4.7	14.1	−5.07
1	12.9	12.9	−3.45	4	4.7	18.8	−5.76

2. Prepare a plot of ln(N/N_0) as a function of Ct and fit a linear trendline through the data. Select trendline options to display the equation and r^2 value.
3. The required plot is shown below.

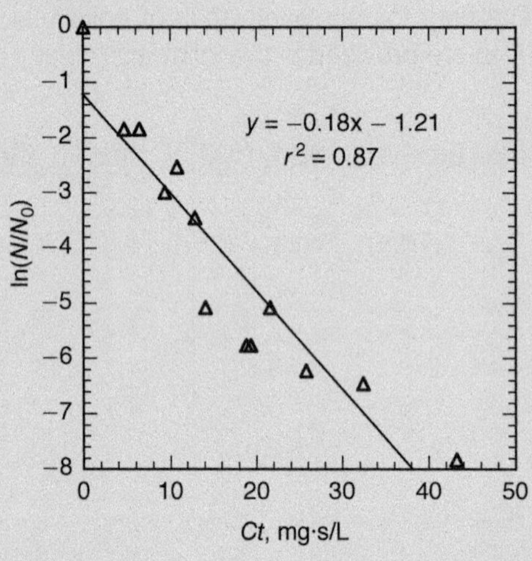

4. The slope of the line in the above plot corresponds to the coefficient of specific lethality, Λ_{CW}. From the plot $\Lambda_{CW} = 0.18$ and $r^2 = 0.87$.

Disinfection data do not always conform to Chick's linear semilog plot. Two anomalies, accelerating rate and decelerating rate, as illustrated on Fig. 13-4, sometimes occur. Reasons often cited in the literature for these particular curve shapes and the circumstances (organism, disinfectant, and magnitude of disinfection) under which each type of curve is sometimes found are also given. Contemporary kinetic models that describe these alternate forms of disinfection data are described in the next section.

Shape of semilog plot of disinfection data	Reasons for shape	Examples
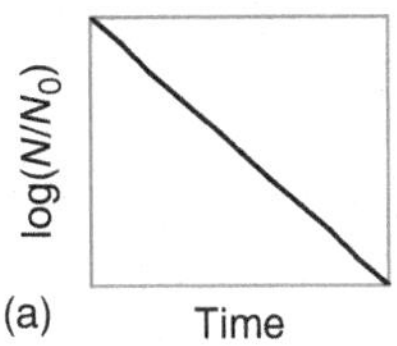 (a)	**Pseudo-first order** The most common form of disinfection data. 1. Data fit Chick's law.	**Free chlorine**: *E. coli,* poliovirus **Ozone**: Poliovirus, *E. coli, G. Lamblia,* and *C. parvum* **UV**: *C. parvum*, MS2 (<4 log), and *G. lamblia* (<3 log)
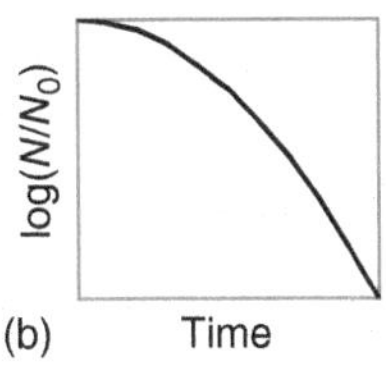 (b)	**Accelerating rate** Often observed at low disinfectant doses. Possible reasons include: 1. Disinfectant must react with more than one critical site in organism (Rahn, 1973; et al., 1975). 2. Disinfectant must take time to diffuse to critical site (Collins and Selleck, 1971). 3. Natural heterogeniety in resistance among organisms (Kim et al., 2002a).	**Combined chlorine**: Most organisms at low inactivation **Any disinfectant**: Suspension of aggregated virus particles of multicellular organisms **Chlorine dioxide**: *C. parvum*, endospores
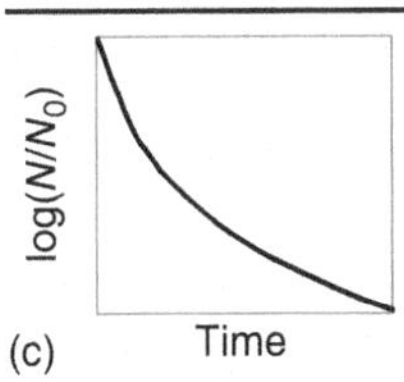 (c)	**Decelerating rate** Often observed after several logs of inactivations. Possible reasons include: 1. Decrease in germicidal properties of the disinfecting agent with time (Gard, 1957; Collins and Selleck, 1971). 2. Resistance to the disinfectant increases with increasing exposure (Gard, 1957; Collins and Selleck, 1971). 3. Natural heterogeniety in resistance among organisms (Hess, 1953). 4. Interference of particles with disinfection (Severin, 1980; Qualls et al, 1983; Parker and Darby, 1995). 5. Organisms are in clumps that test as one unit but must be inactivated individually (Hunt and Mariñas, 1997).	**Combined chlorine**: Most any organism at high removals **UV**: Total coliform in secondary effluent, *G. lamblia* above 3 log removal

Figure 13-4
Graphical forms of disinfection data.

Contemporary Kinetic Models

As discussed earlier, only a limited understanding of the specific mechanisms for the various disinfection reactions is now available. Substantially different kinetics mechanisms may control the rate of inactivation of different microorganisms with the same disinfectant or the same microorganisms with different disinfectants. There is extensive literature on disinfection modeling; two of these models are presented in the following discussion because they are useful in modeling many common disinfection reactions. The models discussed below may be used to model disinfection data for reactions with accelerating and/or decelerating rates on a semilog plot (Figs. 13-4b,c).

RENNECKER–MARIÑAS MODEL (ACCELERATING RATE)

Some organisms do not exhibit significant inactivation until a certain Ct value has been exceeded. This inactivation response is observed, for example, when chemical disinfectants are applied to oocysts and endospores. The Mariñas group at the University of Illinois has recently addressed this situation by proposing the use of a lag coefficient, b for Eq. 13-4 (Kim et al., 1999; Rennecker et al., 1997, 2001). The Rennecker–Mariñas model can be summarized as follows:

$$\ln\left(\frac{N}{N_0}\right)=\begin{cases}0 & \text{for } Ct<b \quad (13\text{-}5)\\ -\Lambda_{\text{CW}}(Ct-b) & \text{for } Ct\geq b \quad (13\text{-}6)\end{cases}$$

where b = lag coefficient, mg · min/L

The lag coefficient b is the maximum value of Ct at which $\ln(N/N_0)$ = $\ln(S_o)$ = 0 (i.e., no inactivation has occurred). When b is zero, Eq. 13-6 corresponds to Eq. 13-4. It should be noted that the presentation of the mathematics used in the analysis of Eqs. 13-5 and 13-6 is consistent with but not identical to the approach used by Rennecker et al. (1997). The Rennecker–Mariñas model is demonstrated in Example 13-2.

Example 13-2 Application of the Rennecker–Mariñas model

Apply the Rennecker–Mariñas model to evaluate the coefficient of specific lethality and the lag coefficient for the inactivation of *C. parvum* using chlorine dioxide (ClO_2) based on the data given below. As shown in the data table, inactivation was measured at three concentrations of ClO_2 and at several time intervals. In analyzing the data, do not assume that it was possible to measure N_0 accurately (i.e., $N/N_0 \neq 0$ for $Ct < b$; instead, require N/N_0 = constant for $Ct < b$). Analyze the data by developing a spreadsheet solution and use the Solver function in Excel to determine the model parameters. Also calculate the coefficient of determination (r^2). Data for the inactivation of *C. parvum* by ClO_2 (Corona-Vasquez et al., 2002) are provided in the following table:

C, mg/L	t, min	log(N/N_0)	C, mg/L	t, min	log(N/N_0)
0.96	0.0	−0.21	0.48	122.0	-1.08
0.96	15.5	−0.25	0.48	152.0	-1.68
0.96	30.8	−0.38	4.64	0.0	−0.15
0.96	46.1	−0.55	4.64	2.1	0.02

C, mg/L	t, min	$\log(N/N_0)$	C, mg/L	t, min	$\log(N/N_0)$
0.96	61.2	−1.04	4.64	4.2	−0.11
0.96	76.2	−1.66	4.64	6.2	−0.19
0.96	91.1	−2.03	4.64	8.2	−0.29
0.48	0.0	−0.17	4.64	10.0	−0.56
0.48	32.0	−0.12	4.64	12.0	−0.79
0.48	61.6	−0.31	4.64	13.9	−1.19
0.48	92.0	−0.60	4.64	15.8	−1.47

Solution

Construct a spreadsheet, as shown below, for analysis of the data and determination of the model parameters. Because the Solver function will be used, some of the calculations will need to be automated using advanced features of Excel, including the IF function, as described below.

1. Compute the value of *Ct* for each experiment and enter the corresponding inactivation value into the spreadsheet.
2. The measured log survival ratios are entered into the column labeled Data in the table below.
3. The value of the model parameters can be determined using an IF statement of the form "IF [$Ct < b$,$\log(S_o)$, $\log(S_o) + \text{slope}(b - Ct)$]."
4. Solver is used to minimize the sum of the [Data–model]^2 column by varying *b*, slope, and $\log(S_o)$. The results are displayed in the following table and figure:

Spreadsheet setup for model evaluation

Ct, mg · min/L	$\log(N/N_0)$			
	Data	Model	[Data-Model]^2	[Data-Data$_{avg}$]^2
0.0	−0.21	−0.2	0.001	0.216
14.9	−0.25	−0.2	0.004	0.182
29.6	−0.38	−0.2	0.036	0.089
44.3	−0.55	−0.5	0.001	0.015
58.8	−1.04	−1.0	0.000	0.138
73.2	−1.66	−1.6	0.010	0.972
87.5	−2.03	−2.1	0.001	1.852
0.0	−0.17	−0.2	0.000	0.259
15.4	−0.12	−0.2	0.005	0.310
29.6	−0.31	−0.2	0.016	0.130
44.2	−0.60	−0.5	0.007	0.005

(*Continued*)

(Continued)

Ct, mg · min/L	log(N/N_0) Data	Model	[Data-Model]^2	[Data-Data$_{avg}$]^2
58.6	−1.08	−1.0	0.002	0.167
73.0	−1.68	−1.6	0.017	1.016
0.0	−0.15	−0.2	0.001	0.277
9.7	0.02	−0.2	0.043	0.483
19.3	−0.11	−0.2	0.006	0.319
28.8	−0.19	−0.2	0.000	0.230
38.2	−0.29	−0.3	0.000	0.147
46.6	−0.56	−0.6	0.002	0.012
55.6	−0.79	−0.9	0.020	0.013
64.6	−1.19	−1.3	0.003	0.271
73.5	−1.47	−1.6	0.009	0.642
Average:	−0.67	Sum:	0.184	7.745

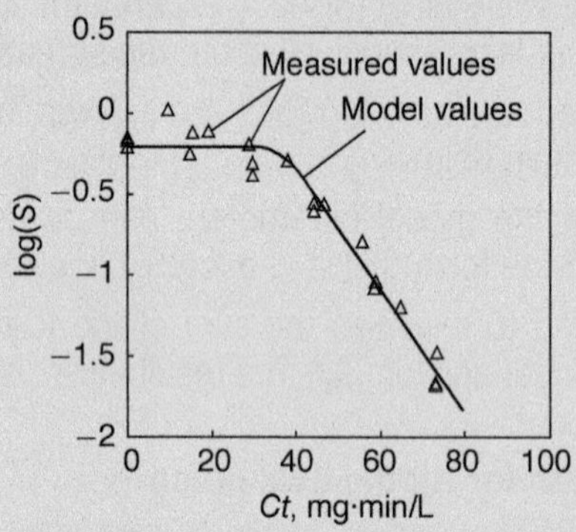

5. Solver minimizes the value of the sum of the [Data–model]^2 when the following values are used:

$$b = 34.9\,\text{mg}\cdot\text{min/L}$$
$$\text{Slope} = 0.036$$
$$\log(N/N_0) = -0.19$$

6. Since the data was plotted on a log (base 10) scale, the coefficient of specific lethality is calculated by dividing the slope by log(e):

$$\Lambda_{cw} = (\text{slope base } e)/\log(e) = 0.036(2.303) = 0.083\ \text{L/mg}\cdot\text{min}$$

7. The coefficient of determination (r^2) is calculated using the data in the spreadsheet as follows:

$$r^2 = 1 - \frac{\sum[\text{Data} - \text{model}]^2}{\sum[\text{Data} - \text{data}_{ave}]^2} = 1 - \frac{0.184}{7.745} = 0.98$$

COLLINS–SELLECK MODEL (DECELERATING RATE)

The Collins–Selleck model was developed specifically to address the inactivation of coliform organisms in domestic wastewater using free and combined chlorine (Collins and Selleck, 1971; Selleck and Saunier, 1978; Selleck, et al., 1970), but it has proven valuable for modeling the behavior of a number of other disinfection alternatives as well. The Collins–Selleck model is particularly useful when a declining rate of disinfection is observed (convex curve on Fig. 13-4c). The form of the Collins–Selleck model first published in the literature did not include a lag effect (Selleck et al., 1970). Although the final formulation was published shortly after that (Collins and Selleck, 1971), it did not appear in the peer-reviewed literature until sometime later (Selleck and Saunier 1978), and there was some discussion about its form even at that time (Haas, 1979; Selleck et al., 1980). Collins and Selleck began with a simple formulation that describes a declining rate of inactivation with time as proposed by Gard (1957) and adapted it to the lag in inactivation often observed in real systems. For batch reactors, the model has the form

$$\ln\left(\frac{N}{N_0}\right)=\begin{cases} 0 & \text{for } Ct < b \qquad (13\text{-}7) \\ -\Lambda_{CS}[\ln(Ct)-\ln(b)] & \text{for } Ct \geq b \qquad (13\text{-}8) \end{cases}$$

where Λ_{CS} = Collins–Selleck coefficient of specific lethality, unitless
b = lag coefficient, mg · min/L

An anomaly in the Collins–Selleck model is that the initial conditions are undefined because $\ln(Ct)$ at $t = 0$ is indeterminate. Nevertheless, there is a close parallel between the Collins–Selleck and Rennecker–Mariñas models. They have a similar form, but in the Rennecker–Mariñas model, the observed data are fit with a straight line when log survival is plotted versus Ct, whereas in the Collins–Selleck model, the data can be fit with a straight line when log survival is plotted versus $\log(Ct)$. Each model has only two parameters, Λ_{CW} or Λ_{CS} and b. Based on a large number of tests, it has been found that most disinfection data can be fit to one of these two models.

Example 13-3 Comparison of Chick–Watson and Collins–Selleck models

The Collins–Selleck model is particularly useful for modeling a declining rate of disinfection such as observed in the disinfection of coliform in wastewater above a Ct value of approximately 100 mg · min/L. Data from Selleck and Saunier (1978) for the disinfection of total coliform in wastewater with chloramines are presented below. Fit the Chick–Watson model to the data for Ct values less than 100 mg · min/L and construct a plot of the results.

Also plot the remainder of the data for *Ct* values greater than 100 mg · min/L on the same plot. Prepare a separate plot to fit the data using the Collins–Selleck model.

Ct, mg · min/L	log(N/N$_0$)	*Ct*, mg · min/L	log(N/N$_0$)	*Ct*, mg · min/L	log(N/N$_0$)
5	−0.79	48	−3.21	244	−5.39
6	−0.61	54	−3.14	244	−5.96
6	−1.36	58	−3.57	291	−5.39
10	−1.79	64	−3.86	300	−5.54
15	−1.18	106	−4.64	319	−5.64
19	−1.86	130	−4.68	390	−5.68
26	−2.21	130	−5.16	417	−6
39	−3.14	175	−4.89	476	−5.74
				602	−5.75

Solution

1. Application of the Chick–Watson model:
   a. The 12 data points with *Ct* values less than 100 mg · min/L are put in an Excel spreadsheet and log(N/N_0)is plotted as a function of *Ct*.
   b. Use the trendline function to fit a line through the data. Select options to fit the line through the origin and to display the equation. As shown in the plot, the slope of the linear trendline (base 10) is 0.067 L/mg · min. Converting to base *e* by dividing by log(*e*), the following value is obtained:

$$\Lambda_{CW}\,(\text{base } e) = 0.067(2.303) = 0.15\ \text{L/mg}\cdot\text{min}$$

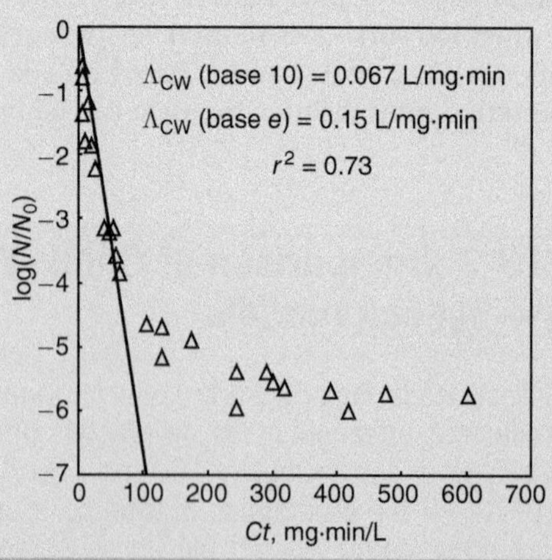

2. Application of the Collins–Selleck model:
   a. All 25 data points are put in an Excel spreadsheet and $\log(N/N_0)$ is plotted as a function of $\log(Ct)$.
   b. The linear trendline is used to determine the slope of the best-fit line.
   c. The slope of the trendline that corresponds to Λ_{CS} (base 10) is equal to 2.76. The value of the intercept b (base 10) is 1.32 mg · min/L. The corresponding base e values are the same. These results are displayed below:

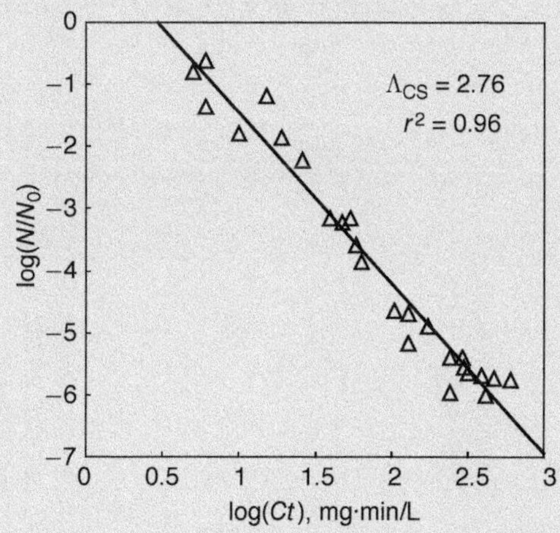

Comment

As shown in this example, the Collins–Selleck model was used effectively to model declining rate disinfection. Based on numerous studies, it has been found that the Collins–Selleck model can be used for a variety of organism–disinfectant combinations where both a declining rate of disinfection and a lag in disinfection are important.

Comparison of Disinfection Models

Some important characteristics of the disinfection models presented in this section are summarized in Table 13-2. The Chick–Watson model (Eq. 13-4) is not shown because it is a special case of the Rennecker–Mariñas model when $b = 0$. A selection of kinetic constants gathered from the literature are reported in Table 13-3. Constants are offered for the disinfection of total coliform from wastewater because this has long been a target organism in effluent reuse. Constants are presented for *E. coli* and poliovirus because these organisms have long been the targets of classical disinfection regulations. Constants are presented for *Giardia* and *Cryptosporidium* because the difficulty in inactivating these organisms is having profound effects on

Table 13-2
Comparison of disinfection models

Disinfection Model	Form of Data	Plots as Straight Line On	Number of Coefficients	Comment
Chick[a]: $\ln\left(\frac{N}{N_0}\right) = -k_c t$ where k_C = Chick's law rate constant, time^{-1}	Pseudo–first order	Semilog graph: $\log(N/N_0)$ vs. t	1	Widely used in microbiology. Approximates a lot of disinfection data.
Rennecker–Mariñas: For $Ct < b$ $\ln\left(\frac{N}{N_0}\right) = 0$; For $Ct > b$ $\ln\left(\frac{N}{N_0}\right) = -\Lambda_{CW}(Ct - b)$ where Λ_{CW} = coefficient of specific lethality, 1/(mg · min/L); b = lag coefficient, mg·min/L	Pseudo–first order with lag	Semilog graph: $\log(N/N_0)$ vs. Ct	2	Equation is consistent with "*Ct*" concept. Can approximate most disinfection data. Performs poorly only if the disinfection reaction truly shows an accelerating rate from the start or if a decelerating rate of reaction is observed. When $b = 0$, this equation simplifies to the Chick–Watson equation.
Collins–Selleck: For $Ct < b$ $\ln\left(\frac{N}{N_0}\right) = 0$; For $Ct > b$ $\ln\left(\frac{N}{N_0}\right) = -\Lambda_{cs}[\ln(Ct) - \ln(b)]$ where Λ_{CS} = log-based coefficient of specific lethality; b = lag coefficient, mg · min/L	Decelerating rate with lag	Log–log graph: $\log(N/N_0)$ vs. $\log(Ct)$	2	Equation is also consistent with "*Ct*" concept. Can approximate most disinfection data. Performs poorly if only the accelerating phase is of interest or if several logs of first-order behavior are observed.

[a]The Chick–Watson model is not shown because it is a special case of the Rennecker–Mariñas model when $b = 0$.

Table 13-3
Selected kinetic parameters (base *e*) based on data in the literature[a]

Organism	Disinfectant	Chick–Watson and Rennecker–Mariñas: Λ_{cw}, L/mg·min or m^2/J	*b*, mg·min/L or J/m^2	Collins–Selleck: Λ_{cs}	Source of constant or data used to develop constant
E. coli	Cl_2, pH 8.5, T =2–5°C	3.75	0.2	—	Butterfield et al. (1943)
	NH_2Cl	0.0375	10	—	Butterfield and Wattie (1946)
	NH_2Cl	0.0327	—	—	Butterfield and Wattie (1946)
	ClO_2	3.3	0.33	—	Scarpino et al. (1977)
	O_3	8330	—	—	Hunt and Mariñas (1999)
	UV	0.83	—	—	Harris et al. (1987)
Total coliform (wastewater or wastewater seed)	HOCl	—	0.005	1.2	Selleck and Saunier (1978)
	OCl^-	—	0.1	1.9	Selleck and Saunier (1978)
	NH_2Cl	—	3.0	2.8	Selleck and Saunier (1978)
	ClO_2	—	0.9	2.2	Roberts et al. (1980)
	UV	—	4	26	Tchobanoglous et al. (2003)
Poliovirus	HOCl	0.2	—	—	Floyd and Sharp (1979)
	NH_2Cl	—	—	—	—
	ClO_2	0.47	28	—	Scarpino et al. (1977)
	O_3	0.85	—	—	Katzenelson et al. (1974)
	UV	3	—	—	Cooper et al. 2001 (sewage)
MS-2	Cl_2	3.4	—	—	Haas et al. (1996)
	NH_2Cl	0.005	—	—	Cooper et al. (2001) (buffer)
	ClO_2	—	—	—	—
	O_3	—	—	—	—
	UV	0.96	—	—	Oppenheimer et al. (2001)

(*Continued*)

Table 13-3 *(Continued)*

Organism	Disinfectant	Chick–Watson and Rennecker– Mariñas: Λ_{cw}, L/mg·min or m^2/J	*b*, mg·min/L or J/m^2	Collins–Selleck: Λ_{cs}	Source of constant or data used to develop constant
Giardia	Cl_2, pH 7	—	68	3.8	Haas and Heller (1990)
	NH_2Cl	—	300	5	JMM (1991) (G. *muris)*
	ClO_2	0.21	—	—	Wallis et al. (1989)
	O_3	—	0.02	1.77	JMM (1991) (G. *muris)*
	O_3	1.9	—	—	Wallis et al. (1989)
	UV	38	—	—	Oppenheimer et al. (2001) (G. *muris)*
C. parvum	Cl_2, pH 6	0.0013	375	—	Driedger et al. (2000)
	NH_2Cl	0.00077	5500	—	Rennecker et al. (2001)
	ClO_2	0.083	35	—	Corona-Vasquez et al. (2002)
	O_3	1.7	0.22	—	Driedger et al. (2001)
	O_3	0.83	—	—	Oppenheimer et al. (2000)
	UV	25	—	—	Oppenheimer et al. (2001)
B. subtilis	Cl_2, pH 6	0.0006	—	—	Brazis et al. (1958) (B. *anthracis)*
	NH_2Cl	0.00054	4560	—	Larson and Mariñas (2003)
	ClO_2	0.13	—	—	Radziminski et al. (2002)
	O_3	2.12	4.91	—	Larson and Mariñas (2003)
	UV	0.004	170	—	Knudson (1986) (B. *anthracis)*

[a]Unless otherwise noted all kinetic parameters are given for 25∘C.

water treatment regulations. Constants are presented for *Bacillus subtilis* because its behavior in disinfection is thought to be similar to *B. anthracis*, a possible organism that may be used by terrorists.

Declining Concentration of Chemical Disinfectant

Chick's experiments were conducted with constant disinfectant concentrations because excess disinfectant was present. In the laboratory, researchers generally attempt to maintain a constant disinfectant concentration so that disinfection rates can be measured with maximum precision. Given the complexities that exist in the microbiological world that can influence the outcome of such experiments, it is important to minimize variations in chemistry and physical conditions. A constant residual of combined chlorine can usually be achieved in full-scale contactors as well. With free chlorine and chlorine dioxide, a constant residual concentration can be maintained for short contact times. For these same disinfectants at longer contact times or for ozone at any contact time, once must account for residual decay.

Accounting for varying disinfection concentration can be addressed by dividing the problem into two parts: (1) modeling the decay of the disinfectant and (2) integrating that work into the model of the disinfection reaction itself. For all the common oxidizing disinfectants (chlorine, combined chlorine, chlorine dioxide, and ozone), it is often assumed that disinfectant decay can be modeled as first order, that is,

$$r_d = -k_d\, C \tag{13-9}$$

where r_d = reaction rate for the decline in disinfectant concentration with time, mg/L·s or mol/L·s
k_d = first-order decay rate, s^{-1}
C = disinfectant concentration, mg/L or mol/L

The decay of these disinfectants is often characterized by two phases, an early phase of rapid decay followed by a later phase with slower decay. When two-phase decay occurs, a second-order model with a fast reaction step and a slow reaction step has been used successfully (Kim et al., 2002a; Lev and Regli, 1992), but this model is rather difficult to use because it cannot be solved analytically. Another alternative is the parallel first-order decay model proposed by Haas and Karra (1984b), in which it is assumed that decay may proceed through two mechanisms, each first order but involving a different component of the chlorine residual:

$$r_d = -xk_{d1}\, C - (1-x)\, k_{d2}\, C \tag{13-10}$$

where x = fraction of disinfectant decaying by the first mechanism, unitless
C = concentration of disinfectant, mg/L or mol/L
k_{d1}, kd_2 = decay coefficient for two different mechanisms, s^{-1}

The first component, with an initial concentration of xC_0, is subject to first-order decay with a faster rate constant, k_{d1}, and the second component, with an initial concentration of $(1-x)\,C_0$, is subject to first-order decay with a slower rate constant, k_{d2}. As noted above, the value of x, by definition, is between 0 and 1. When $x = 0$, the parallel first-order model becomes the simple first-order model; the same is true when $x = 1$.

Finding an analytical solution in which the decay reaction and the disinfection reaction are integrated together adds to the complexity of the mathematics used to describe the disinfection process. Where analytical solutions are not available, it is possible to use computer models to simulate the two processes in parallel. Haas and Joffe (1994) have developed an analytical solution for the Chick–Watson model.

Influence of Temperature on Disinfection Kinetics

The effect of temperature on the rate of a chemical reaction is described by the Arrhenius equation, as discussed in Chap. 5, and is used here to describe the influence of temperature on the Pseudo-first-order disinfection rate constant:

$$\ln(k_r) = \ln(A) + \left(-\frac{E_a}{R}\right)\left(\frac{1}{T}\right) \tag{5-85}$$

where k_r = appropriate reaction rate constant, k_c, Λ_{CW}, Λ_{CS}, or k_d
E_a = activation energy, J/mol
R = universal gas constant, 8.314 J/(mol · K)
T = reaction temperature, K (273 + °C)
A = collision frequency parameter

Once the rate is known at one temperature, the rate at another temperature can be determined if the activation energy E_a is known. In the disinfection literature, an empirical approach used is to specify θ in the following equation:

$$\frac{k_{r,T_1}}{k_{r,T_2}} = \theta^{T_1 - T_2} \tag{13-11}$$

where k_{r,T_1} = reaction rate constant at temperature 1
k_{r,T_2} = reaction rate constant at temperature 2
θ = empirical constant, dimensionless
T_1 = temperature corresponding to known rate constant k_{r,T_1}, K (273 + °C)
T_2 = temperature corresponding to known rate constant k_{r,T_2}, K (273 + °C)

Combining Eqs. 5-85 and 13-11 and solving for θ, the following expression is obtained:

$$\theta = e^{E_a/RT_1T_2} \tag{13-12}$$

Because the product T_1T_2 is somewhat insensitive to changes in temperature, it is reasonable to assume θ is constant in empirical approach. Values of E_a from the literature are summarized in Table 13-4.

Table 13-4
Activation energies for a variety of disinfection reactions

Microorganism	Disinfectant	E_a, kJ/mol	$K_{25°C}/K_{5°C}$	Reference
C. parvum	HOCl	71.9		Rennecker et al. (2001)
C. parvum	HOCl	64.7		Corona-Vasquez et al. (2002)
		72[a]	6.4	
C. parvum	ClO_2	67.5		Corona-Vasquez et al. (2002)
C. parvum	ClO_2	86.3		Ruffell et al. (2000)
		77[a]	8.0	
C. parvum	NH_2Cl	75.6		Driedger et al. (2001)
C. parvum	NH_2Cl	78.7		Rennecker et al. (2001)
C. parvum	NH_2Cl	59.2[b]		Corona-Vasquez et al. (2002)
		77[a]	8.0	
C. parvum	O_3	102		Oppenheimer et al. (2001)
C. Parvum	O_3	75.7		Driedger et al. (2001)
C. Parvum	O_3	81.2		Rennecker et al. (1999)
C. Parvum	O_3	47.6		Finch et al. (2001)
		76[a]	7.8	
C. muris	O_3	92.8	12	Kim et al. (2002b)
E. coli	O_3	37.1	2.7	Hunt and Mariñas (1997)
G. lamblia	O_3	39.2	2.9	Wickramanyake et al. (1984b)
G. muris	O_3	70	6.6	Wickramanyake et al. (1984a)
N. gruberi	O_3	31.4	2.3	Wickramanyake et al. (1984a)
B. subtilis	O_3	46.8	3.6	Larson and Mariñas (2003)
B. subtilis	NH_2Cl	79.6	8.7	Larson and Mariñas (2003)

[a]Recommended value.
[b]Old oocysts.

Approaches to Relating Disinfection Kinetics to Disinfection Effectiveness

The true, detailed kinetics of most chemical disinfectants are exceedingly complex, and they are influenced by the chemistry of the disinfectant as well as the nature of the susceptibility in the organism. Moreover, measuring disinfection effectiveness is difficult to do with great precision, partly because of the complexity of the chemical conditions but also due to the imprecise nature of most microbiological measurements. As a result, it is probably best to employ the simplest approach possible to describe the results of disinfection experiments. In order of increasing complexity, the following alternatives might be considered:

1. *Ct tables.* Numerical *Ct* (concentration × time) values are established to achieve a given degree of inactivation of a specific organism using a defined disinfectant under controlled conditions. This approach is

consistent with all the models presented in this section. Furthermore, the U.S. EPA uses this approach in regulating disinfection of drinking water. When required, different tables can be offered for a range of concentrations, as the U.S. EPA did for the inactivation of *G. lamblia* with free chlorine.

2. *Semilog plots of survival versus Ct values.* The use of semilog plots of log survival as a function of *Ct* is consistent with the Chick–Watson model and the Rennecker–Mariñas model. In this approach, it is assumed that the log survival values will plot as a linear function of time or the product *Ct* on a semilog plot and only one or two constants, Λ_{CW} and *b*, are required for application of the model. This approach is often successful when a modest degree of disinfection is required, a reduction of approximately 3 log inactivation, for example.
3. *Log-log plots of survival versus Ct values.* The use of log–log plots is consistent with the Collins–Selleck model. This approach is useful for situations where a lag time is present (complex organisms, slow disinfectants, etc.) or where a declining rate of disinfection with time is observed. This approach is also useful when disinfection requirements are substantial, for example, 4 log reduction or more. In the Collins–Selleck model, it is assumed that the log survival will plot as a linear function against log(*Ct*) and that two constants, Λ_{cs} and *b*, are required for application of the model.

Generally, as disinfection models become more complex, the precision with which they can be used to describe the results of a given disinfection experiment improves. However, comparing the constants of the simpler models provides better perspective on the performance of different disinfectants and on the resistance of different organisms. The ability to compare results is one of the reasons that Chick's law and the Chick–Watson equation continue to be popular.

The *Ct* Approach to Disinfection

In each of the approaches discussed in the previous section, disinfection effectiveness was related to the product *Ct*. In fact, the product *Ct* has long been used as a basis for disinfection requirements. It is equally practical when the Collins–Selleck and Rennecker–Mariñas models are used. The *Ct* product required for achieving a given level of disinfection for a specific microorganism under defined conditions is a useful way of comparing alternate disinfectants and for comparing the resistance of a variety of pathogens. Indeed, the product *Ct* can be thought of as the dose of disinfectant.

The dose concept, analogous to *Ct*, is also applicable when UV light is used for disinfection. The product of the UV light intensity (mW/cm^2) and the time of exposure is used to compute the dose ($mW/cm^2 \times s = mJ/cm^2$). This product is often referred to as *It* (intensity × time). Modeling disinfection with UV light using *It* in place of *Ct* in Eq. 13-4 has been successful. There is probably greater justification for this equation for UV light because the mechanism of inactivation is not so much a function of light intensity but a function of exposure of the organism to a quantity of potentially damaging photons.

The *Ct* concept also allows for the development of a broad overview of the relative effectiveness of different disinfectants and the resistance of different organisms. The *Ct* required to produce a 99 percent (2 log) inactivation of several microorganisms using the five disinfection techniques most often used in water treatment is illustrated on Fig. 13-5. Because of the difference in the behavior from one organism and one disinfectant combination to the next, *Ct* and *It* products range over seven orders of magnitude. For example, the *Ct* product required to inactivate *C. parvum* must be three orders of magnitude higher with combined chlorine than with ozone. Comparing

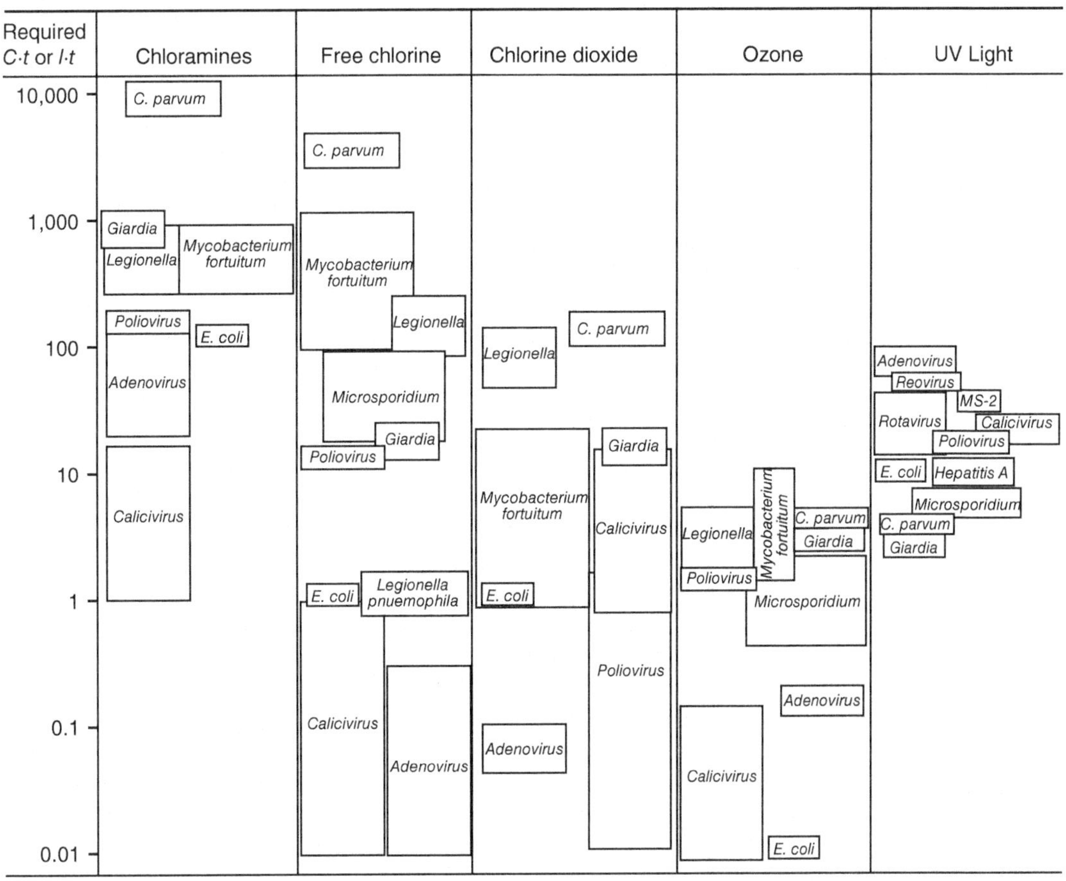

Figure 13-5
Overview of disinfection requirements for 99 percent inactivation. (Adapted from Jacangelo et al., 1997.)

UV disinfection to disinfection with chemical oxidants, little similarity exists between the *It* values and *Ct* values for a single organism. While the required UV doses vary over a range of two orders of magnitude, their variation is much less than that for other disinfectants. The reduced variation may be a result of the fact that UV disinfection of all microorganisms results from a similar protein dimerization mechanism.

The U.S. EPA began the practice of specifying *Ct* products that must be met as a way of regulating the control of pathogens in water treatment with the promulgation of the Surface Water Treatment Rule (U.S. EPA, 1989). Tables of *Ct* and *It* values required to meet the primary disinfection requirements are available in the *Surface Water Treatment Rule Guidance Manual* (U.S. EPA, 1991) available on the EPA website.

A limitation of the *Ct* approach is that the microorganisms in a real disinfection contactor are exposed to a distribution of contact times according to the contactor's residence time distribution (RTD), rather than all microorganisms being exposed to the disinfectant for the same amount of time. The RTD has a significant impact on disinfection effectiveness, as discussed in detail in the next section.

13-4 Disinfection Kinetics in Nonideal Flow-Through Reactors

The disinfection kinetics described in Sec. 13-3 were based on studies conducted in completely mixed batch reactors (CMBRs). While the insight obtained from batch reactors is useful, full-scale continuous-flow systems exhibit more complex nonideal behavior. Of particular importance is the impact of dispersion on the progress of the reaction.

Three approaches to modeling the performance of real (nonideal) reactors are introduced in Chap. 6: (1) the tanks-in-series (TIS) model, (2) the dispersed-flow model (DFM), and (3) the segregated-flow model (SFM). The TIS model simulates the effects of dispersion on the RTD curve by an analogy between a real reactor and a series of completely mixed flow reactors (CMFRs). The parameter that describes dispersion in the TIS model is the number of reactors in series, n. A high value of n corresponds to low dispersion.

The DFM simulates the effects of dispersion on the RTD by including mass transport by axial dispersion in addition to advection into the mass balance of a plug flow reactor (PFR) In the DFM, dispersion is described using the Peclet number (Pe) or the dispersion number (d,Pe $= 1/d$). A high value of Pe or a low value of d corresponds to low dispersion.

The SFM, presented in Sec. 6-9, simulates the effects of nonideal mixing by an analogy between a real reactor and a series of parallel PFRs having detention times that, in sum, match the RTD of the real reactor. While the TIS model and the DFM incorporate assumptions about the nature of the RTD curve, an RTD curve must be provided to use the SFM.

In the TIS model, it is assumed that all the reactants are mixed completely throughout each reactor at all times. In the DFM, it is assumed that all reactants are mixed completely in the lateral direction but axial transport occurs by advection and dispersion. When dispersion is low, the TIS and DFM models produce similar results. In the SFM, it is assumed that the reactants are never completely blended in the reactor; rather the target reactant travels through the reactor in small cells or discrete elements that react with the bulk solution.

Application of the SFM Model to Disinfection

Disinfection processes are an ideal application of the SFM because microorganisms actually do travel through the reactor as particles, separate from the fluid, but react with disinfectants in their environment as they pass through (Trussell and Chao, 1977). If disinfection conditions are uniform throughout the reactor (e.g., the reactant residual or the intensity of inactivating irradiance is uniform throughout), the inactivation of each individual microorganism is the same as it would be in a batch reactor after the same residence time. The RTD of a conservative tracer can reasonably be used to describe the RTD of the microorganisms themselves. Thus, the disinfection process can be modeled by the SFM (see Sec. 6-9):

$$\frac{N}{N_0} = \sum_{i=1}^{n} R(\theta_i)E(\theta_i)\Delta\theta_i \tag{6-123}$$

where

N = number of microorganisms in the effluent from the real reactor, org/mL

N_0 = number of microorganisms in the influent to the real reactor, org/mL

$R(\theta_i) = N_i/N_0$ = inactivation of microorganisms achieved in CMBR (or PFR) after reaction time equal to θ_i

θ_i = normalized time (time divided by mean residence time, $t_i/\bar{t}$), dimensionless

$E(\theta_i)$ = exit age distribution at time θ_i (see Chap. 6)

$\Delta\theta_i$ = differential normalized time step

i = time step in RTD

n = total number of time steps in RTD

Selleck first introduced the approach outlined above to modeling in the early 1970s (Selleck et al., 1970). Trussell and Chao (1977) then employed this approach to demonstrate the influence of dispersion on chlorine contactor performance. Both authors worked on disinfection of coliform bacteria using combined chlorine and, in both studies, the disinfectant residual was assumed to be constant and uniform throughout the reactor. Scheible (1987) introduced a similar approach to the modeling of UV disinfection in the U.S. EPA disinfection design manual (U.S. EPA, 1986). The approach is appropriate for UV disinfection if it is assumed that turbulent flow exists, no short circuiting occurs, and each organism takes a path through the contactor such that its average exposure to UV light is equal to the average intensity of UV light in the reactor.

The promulgation of the U.S. EPA's Surface Water Treatment Rule substantially increased disinfection requirements for drinking water in the United States (U.S. EPA, 1989) and, as a result, has stimulated further interest in methods of refining the rule's approach to specifying disinfection. Lawler and Singer (1993) reintroduced the concept again and later Haas demonstrated its application (Haas et al., 1995). Subsequently, the SFM concept was incorporated in the *integrated disinfection design framework*, an effort to further optimize the design and operation of water disinfection systems (Bellamy et al., 1998; Ducoste et al., 2001). An example of the application of the SFM to disinfection is demonstrated in Example 13-4.

Example 13-4 Application of SFM to estimate disinfection efficiency

Use the disinfection data from Example 13-2 to determine the hydraulic detention time of a contactor designed for the inactivation of *C. parvum* using chlorine dioxide. The contact chamber is to be designed with a hydraulic detention time to provide a C_T value equal to the Ct value that achieves 4 log inactivation in the batch tests. The target chlorine dioxide residual in the full-scale contactor is 0.8 mg/L.

After the full-size contactor was built, tracer tests were conducted to evaluate the hydraulic characteristics of the contactor. Using the procedures outlined in Chap. 6, the tracer curve has been analyzed to produce the exit age distribution. The mean residence time was found to be 178 min and the results of the tracer study are given in the following table:

θ_i	$E(\theta_i)$	θ_i	$E(\theta_i)$	θ_i	$E(\theta_i)$
0.15	0	0.91	1.995	1.67	0.128
0.31	0	1.06	1.541	1.82	0.067
0.46	0.017	1.21	0.928	1.98	0.046
0.61	0.279	1.36	0.446	2.13	0.036
0.76	0.895	1.52	0.251	2.28	0.015

Use the tracer study data and contactor design information: (a) plot the exit age distribution $E(\theta_i)$ versus θ_i; (b) use the SFM to estimate the level of inactivation, $\log(N/N_0)$, that will actually occur in the full-scale reactor with dispersion.

Solution

1. Determine the hydraulic detention time for the full-scale contact chamber using the batch data.

a. The values for the disinfection parameters found in Example 13-2 were $\Lambda_{CW} = 0.083$ L/mg·min and $b = 34.9$ mg·min/L. Thus, inactivation as expressed by the Rennecker–Mariñas model is

$$\ln\left(\frac{N}{N_0}\right) = \begin{cases} 0 & \text{for } Ct < 34.9 \\ -0.083(Ct - 34.9) & \text{for } Ct \geq 34.9 \end{cases} \qquad (1)$$

b. Find the value of Ct that corresponds to 4 log inactivation. For 4 log inactivation, $\log(N_0/N) = 4$, so $N/N_0 = 10^{-4} = 0.0001$. Rearranging Eq. 1 to solve for Ct yields

$\ln(0.0001) = -0.083 \text{ L/mg} \cdot \text{min} \, (Ct - 34.9 \text{ mg} \cdot \text{min/L})$

$$Ct = \frac{-\ln(0.0001)}{0.083 \text{ L/mg} \cdot \text{min}} + 34.9 \text{ mg} \cdot \text{min/L} = 145.5 \text{ mg} \cdot \text{min/L}$$

c. Find the hydraulic detention time that provides a Ct value of 146 mg · min/L when the chlorine dioxide residual is 0.8 mg/L:

$$\tau = t = \frac{Ct}{C} = \frac{145.5 \text{ mg} \cdot \text{min/L}}{0.8 \text{ mg/L}} = 182 \text{ min}$$

2. Plot the exit age distribution using the data provided in the problem statement. The exit age distribution is plotted below:

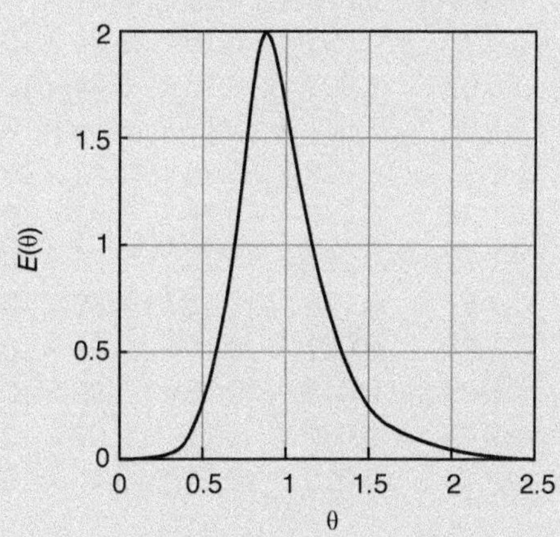

3. Determine the degree of inactivation achieved with the contactor using the SFM. Using the data given in the problem statement, a spreadsheet is developed using the principles of the SFM shown in Chap. 6. The resulting spreadsheet is shown below. As an example, calculations for the fifth row of the spreadsheet are as follows:

a. Columns 1 and 2 contain values of θ_i and $E(\theta_i)$ given in the problem statement.

b. Column 3 ($\Delta\theta_i$):

$$\Delta\theta_i = \theta_i - \theta_{i-1} = 0.61 - 0.46 = 0.15$$

c. Column 4 [$R(\theta_i)$] is developed using an IF statement because the value changes depending on whether Ct_i is less or greater than b: If $Ct_i < b$, then $N_i/N_0 = e^0 = 1$.
If $Ct \geq b$, then $N_i/N_0 = \exp[-\Lambda_{CW}(Ct_i - b)]$, where $t_i = \theta_i \bar{t}$.

$$\frac{N_0}{N_i} = \exp\{-0.083\ \text{L/mg}\cdot\text{min}\ [(0.8\ \text{mg/L})(0.61)(178\ \text{min}) - 34.5\ \text{mg}\cdot\text{min/L}]\} = 0.013$$

d. Column 5 [$R(\theta_i)E(\theta_i)\Delta\theta_i$]:

$$R(\theta_i)\,E(\theta_i)\,\Delta\theta_i = (0.0130)\,(0.279)\,(0.15) = 5.42\times10^{-4}$$

θ_i	$E(\theta_i)$	$\Delta\theta_i$	$R(\theta_i)$	$R(\theta_i)E(\theta_i)\Delta\theta_i$
0				
0.15	0	0.15	1	0
0.31	0	0.16	0.449	0
0.46	0.017	0.15	0.0763	1.95×10^{-4}
0.61	0.279	0.15	0.0130	5.42×10^{-4}
0.76	0.895	0.15	2.20×10^{-3}	2.95×10^{-4}
0.91	1.995	0.15	3.74×10^{-4}	1.12×10^{-4}
1.06	1.541	0.15	6.35×10^{-5}	1.47×10^{-5}
1.21	0.928	0.15	1.08×10^{-5}	1.50×10^{-6}
1.36	0.446	0.15	1.83×10^{-6}	1.22×10^{-7}
1.52	0.251	0.16	2.76×10^{-7}	1.11×10^{-8}
1.67	0.128	0.15	4.69×10^{-8}	9.01×10^{-10}
1.82	0.067	0.15	7.97×10^{-9}	8.01×10^{-11}
1.98	0.046	0.16	1.20×10^{-9}	8.85×10^{-12}
2.13	0.036	0.15	2.04×10^{-10}	1.10×10^{-12}
2.28	0.015	0.15	3.47×10^{-11}	7.81×10^{-14}
				$\Sigma = 0.00116$

e. The degree of inactivation is the sum of column 5 [$R(\theta_i)E(\theta_i)\Delta\theta_i$]. Thus, $N/N_0 = 0.00116$ and the log removal value (LRV) is

$$LRV = \log\left(\frac{N}{N_0}\right) = -\log(0.00116) = 2.94$$

Comment

Due to dispersion, the full-scale contactor achieves less than 3 log of inactivation when the hydraulic residence time was set equal to the time necessary to achieve 4 log of inactivation in laboratory batch tests (see Sec. 4-5 for additional discussion of expressing removal in terms of log removal) Clearly, a different approach must be used to determine the hydraulic residence time of a full-scale reactor when dispersion is important.

When Dispersion Is Important in Disinfection

Minimizing dispersion and short circuiting in disinfection contactors is widely accepted. The U.S. EPA limits the credit for disinfection contact time to the time it takes for the first 10 percent of a tracer to pass through a disinfection contactor (t_{10}), that is, the value of t in Ct is t_{10} instead of τ. California requires the minimum time to the peak concentration on the tracer curve (t_{modal}) to be 90 min and a minimum length-to-width ratio of 40 : 1 for baffled chlorine contactors in its regulations for reclaiming wastewater for nonrestricted reuse (Cal DHS, 1999).

As a general rule, reducing dispersion is more important when disinfection goals are substantial. For example, dispersion is more important in the design of a contactor that must achieve 4 log of inactivation than in the design of a contactor that must achieve 1 log of reduction. This effect is true regardless of the organism under consideration or its specific reaction kinetics.

A thought experiment can be used to illustrate this effect. Assume a disinfection process is designed to achieve a 4 log reduction of a particular virus and a 1 log reduction of a certain protozoa. Further assume the reactor operates as designed and achieves exactly those objectives. A small bypass pipe is installed and 1 percent of the flow coming into the reactor is diverted so that it flows around the reactor and blends, without disinfection, with the treated water from the reactor. The result of the experiment is illustrated on Fig. 13-6. As illustrated, the small diversion has almost no impact on the removal of protozoa (only 9 percent increase in effluent concentration) but severely compromises the removal of the virus, exposing the consumer to virus levels over 100 times higher than the goal that was being sought.

Using the reactor dispersion models presented in Chap. 6, it is possible to compare the performance of a real reactor with dispersion with an ideal plug flow reactor. A model may be prepared to estimate the amount of dispersion that could be allowed without compromising plug flow performance more than 5 percent (in other words, without reducing the log removal more than 5 percent). As shown on Fig. 13-7, which was developed for a first-order reaction and with a removal goal that spans several orders of magnitude, the requirements for controlling dispersion are modest when the required removal is modest. As the removal requirements increase to 3 log or more,

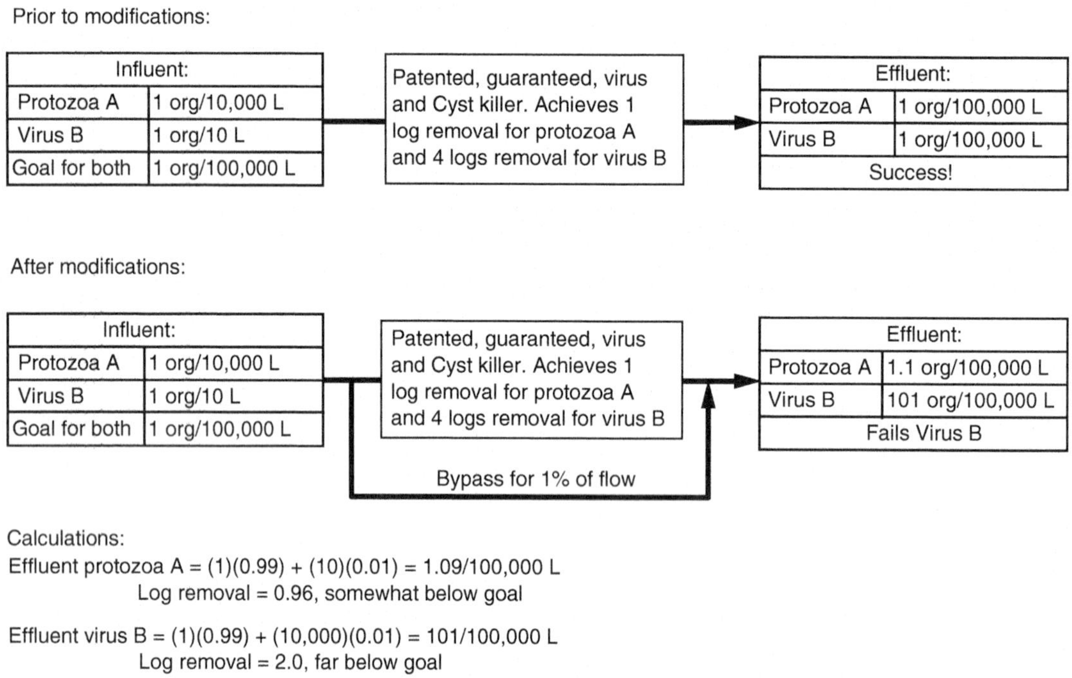

Figure 13-6
Thought experiment: Dispersion and short circuiting are more important when removal goals are high.

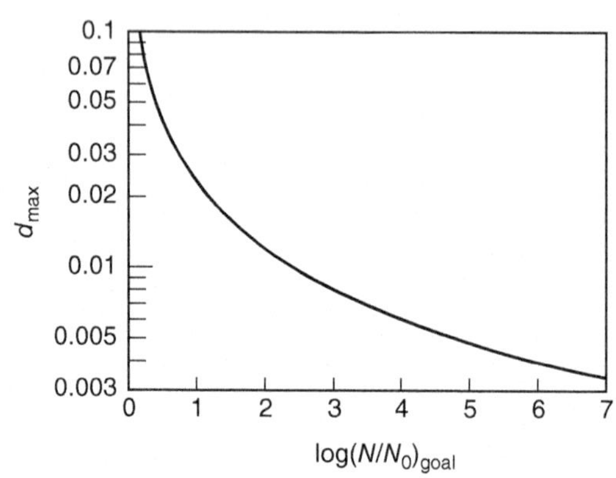

Figure 13-7
Allowable dispersion for contactor versus inactivation goals. At d_{max}, performance is short of goal by 5 percent.

it becomes difficult to prevent the dispersion from being greater than the allowable amount. Removals higher than 3 log generally can only be accomplished by constructing reactors with a significantly greater hydraulic detention time (and greater *Ct* value) than predicted by removal measured in a batch reactor. For instance, the reactor in Example 13-4 could meet 4 log inactivation requirements if the mean residence time were 30 percent greater than the time predicted by the batch tests.

Assessing Dispersion with the t_{10} Concept

A number of indices have been used to assess performance of full-scale disinfection contactors. Some of the more common indices are the dispersion number d, t_{10}/τ, t_{10}/t_{90}, and t_{modal}. A reasonable simulation of the original RTD of a reactor can be produced using the dispersion number and the DFM for a closed system (see Chap. 6). The RTD curve generated by the DFM for a given dispersion number can be used as a substitute for actual tracer data to estimate reactor performance using the SFM (Trussell and Chao, 1977) as demonstrated in Example 13-4. As a result, the dispersion number is perhaps the best measure of the suitability of a reactor for accomplishing disinfection. Nevertheless, regulators tend to prefer parameters such as t_{10} (US EPA, 1986) or t_{modal} (Cal DHS, 1999) as these values are easier to determine and more readily understood by operating personnel. Because of the U.S. EPA's regulations, t_{10} deserves special attention where water treatment is concerned.

To assess whether using the t_{10} value provides the same level of protection as controlling dispersion, Fig. 13-8 was constructed using Eq. 6-123 and a reaction that would achieve 4 log of removal in a plug flow reactor. The performance estimated by the SFM for the reactor with dispersion (middle curve) is compared to the performance credit that would be allowed for the reactor based on the batch equation and the product Ct_{10} (bottom curve). The inactivations estimated by the SFM and by the product Ct_{10} both improve as dispersion is reduced. From the presentation on Fig. 13-8 it can be concluded that the U.S. EPA's t_{10} approach is effective, but conservative.

The design of disinfection contact chambers that exhibit low dispersion is presented later in this chapter, after sections that describe each of the chemical disinfectants.

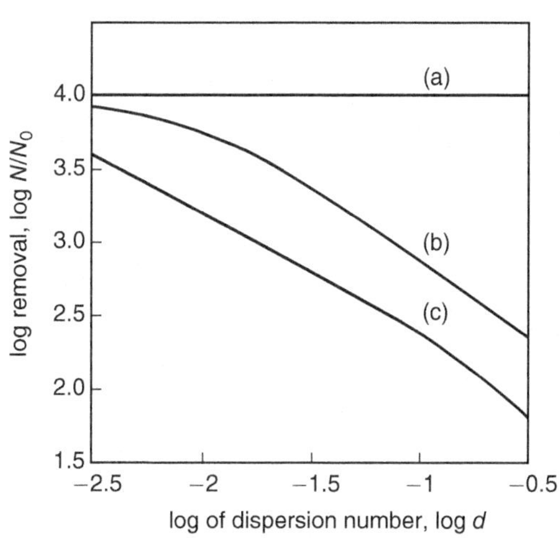

Figure 13-8
Reactor disinfection performance predictions: (a) ideal plug flow; (b) segregated-flow model (SFM) with dispersion, derived from residence time distribution (RTD) curve produced using closed-system dispersion flow model (DFM); and (c) predicted using t_{10} values derived from $E(\theta)$ curves based on closed-system DFM.

13-5 Disinfection with Free and Combined Chlorine

Until approximately World War II, free and combined chlorine (chlorine combined with ammonia, also known as chloramines) were both commonly used and viewed as effective disinfectants. In 1943, the U.S. PHS demonstrated that free chlorine exhibits more rapid kinetics in the disinfection of several bacteria (Wattie and Butterfield, 1944). As a result, the use of combined chlorine declined between 1943 and the mid-1970s. In the mid-1970s, it became widely recognized that free chlorine formed chemical by-products (Bellar and Lichtenberg, 1974; Rook, 1974) and that combined chlorine did so to a much lesser degree (Stevens and Symons, 1977). Since that realization, the use of combined chlorine has increased, particularly the addition of ammonia to convert a free-chlorine residual to a combined chlorine residual once primary disinfection has been accomplished. By 2004, about one in four utilities in the United States were using combined chlorine (U.S. EPA, 2004).

Chemistry of Free Chlorine

When chlorine gas is injected into water, it dissolves according to Henry's law and then rapidly reacts with the water to form hydrochloric acid and hypochlorous acid:

$$Cl_2(g) + H_2O \rightarrow HCl + HOCl \tag{13-13}$$

Photographs of various chlorine system components

http://www.stantec.com/wt3r/13-1.html

Hydrochloric acid is a strong acid that dissociates completely, causing a reduction in alkalinity and pH:

$$HCl \rightarrow H^+ + Cl^- \tag{13-14}$$

Hypochlorous acid is a weak acid and the extent of dissociation depends on pH (see Chap. 5 for discussion of weak acids):

$$HOCl \rightleftarrows H^+ + OCl^- \tag{13-15}$$

$$K_a = \frac{[H^+][OCl^-]}{[HOCl]} \tag{13-16}$$

The pK_a for HOCl is 7.6 at 20°C; hypochlorous acid (HOCl) is the predominant form below this pH value and hypochlorite ion (OCl^-)is the predominant form above it. The distribution between HOCl and OCl^- is illustrated on Fig. 13-9 as a function of pH and temperature. Hypochlorous acid (HOCl) exhibits faster disinfection kinetics than does hypochlorite ion (OCl^-) (see Table 13-3). Consequently, a pH of 7 or less is desirable where disinfection alone is concerned. As can be seen from Fig. 13-9, temperature has a small effect, warmer waters causing hypochlorous acid to dissociate at somewhat lower pH.

Chlorine is relatively stable in pure water but reacts slowly with the organic matter naturally present in drinking waters and rapidly with sunlight. Where sunlight is concerned, photons react with hypochlorite ion to produce

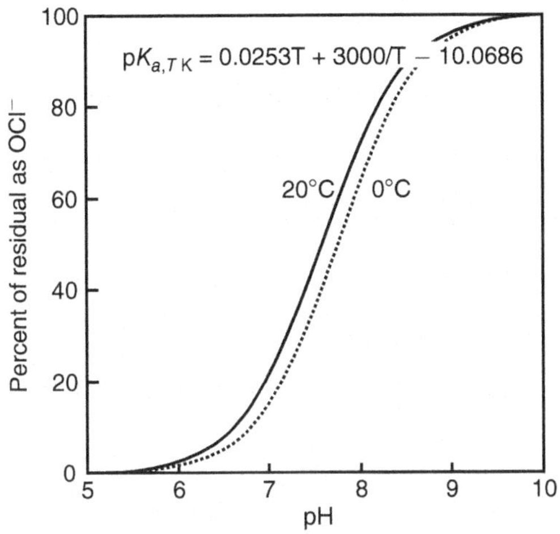

Figure 13-9
Effect of temperature and pH on fraction of free chlorine present as hypochlorous acid. (Adapted from Morris, 1966.)

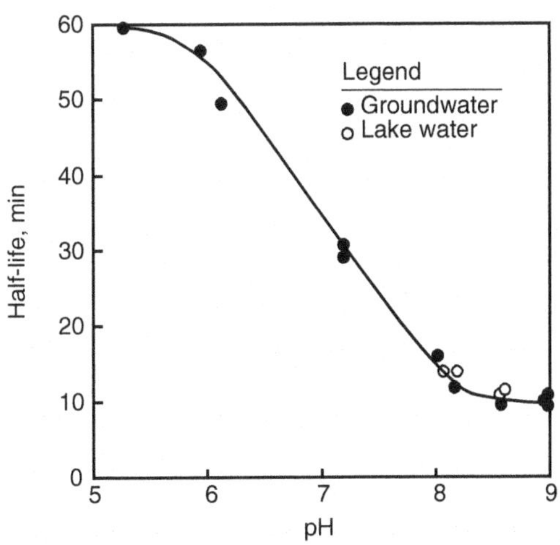

Figure 13-10
Half-life of free chlorine residual in sunlight.

oxygen, chlorite ion, and chloride ion (Buxton and Subhani, 1971). The sensitivity of the reaction rate to pH, a consequence of the fact that the photolytic reaction is with hypochlorite ion and not hypochlorous acid, is illustrated on Fig. 13-10, constructed using the data of Nowell and Hoigné (1992a, b).

Although significant research has investigated the decay of chlorine residuals in the presence of natural organic matter, no universal relationships have evolved. Rather, the decay of free chlorine is often best modeled with the simple first-order reaction depicted in Eq. 13-9. Sometimes the process is modeled as two reactions operating in parallel, a fast reaction with rapidly reducible substances and a slower first-order reaction (Eq. 13-10). Studying data from multiple sources, Powell et al. (2000) concluded that the activation energy for the chlorine decay reaction is on the order of 62 kJ/mol. Modeling chlorine decay as a first-order reaction is illustrated in Example 13-5.

Example 13-5 Evaluating chlorine residual decay data

One milligram of chlorine was added to 1 L of water. The water was stored in the dark at a constant temperature of 10°C and the chlorine residual was measured periodically. The results of the chlorine decay experiment are given below. Assuming a simple first-order decay reaction, estimate the constant for first-order decay, k_d, of chlorine. Assuming that the activation energy E_a, is 62 kJ/mol, what would k_d be at 25°C? What would the residual have been at the end of the same decay test at 25°C?

Time, h	Concentration, mg/L
0	0.97
1	0.80
2	0.69
3	0.63
5	0.54
8	0.45
9	0.39
12	0.30

Solution

1. Determine the first-order decay rate constant for 10°C.
   a. To find the rate, $\ln(C/C_0)$ is plotted as a function of time and a linear best fit is forced through zero as shown below:

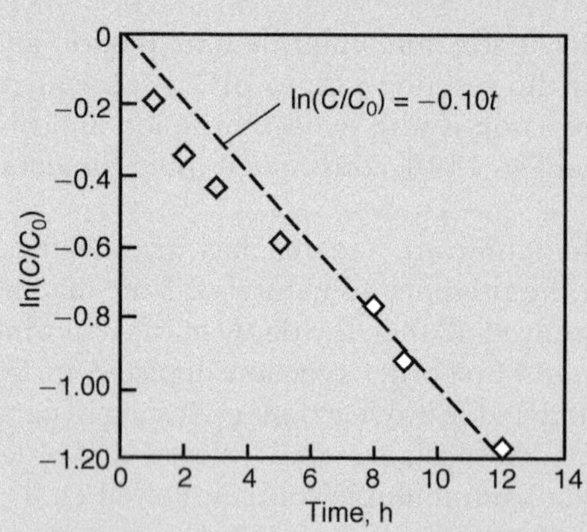

b. From the plot, k_d at 10°C is estimated to be approximately 0.10 h^{-1}.

c. Determine the value of k_d at 25°C. The value of k_d at 25°C can be computed using Eq. 5-85 (see Chap. 5):

$$\ln(k) = \ln(A) + \left(-\frac{E_a}{R}\right) \times \left(\frac{1}{T}\right)$$

The Arrhenius factor A needed for computing the k_d at 25°C is determined using the k_d value for 10°C:

$$\ln(A) = \ln(0.10) + \frac{62{,}000\,\text{J/mol}}{(8.314\,\text{J/mol}\cdot\text{K})(273+10)\,\text{K}}$$

$$\ln(A) = 24.05$$

The k_d at 25°C is given below:

$$\ln(k_d) = 24.05 - \frac{62{,}000\ \text{J/mol}}{(8.314\,\text{J/mol}\cdot\text{K})(273+25)\,\text{K}} = -0.976$$

$$k_d = e^{-0.976} = 0.377$$

2. Determine the residual concentration of chlorine:

$$C_{t=12} = C_{t=0}\ e^{-0.377t} = C_{t=0}\ e^{(-0.377)(12)}$$
$$= (0.97)(0.0109) = 0.0106\ \text{mg/L}$$

Chemistry of Combined Chlorine

When ammonia is present in water, chlorine reacts to form species that combine chlorine and ammonia, known as chloramines. In general, chlorine reacts successively with ammonia to form the three chloramine species as more chlorine is added.

$$NH_3 + HOCl \rightarrow NH_2Cl + H_2O \quad \text{(monochloramine formation)} \qquad (13\text{-}17)$$

$$NH_2Cl + HOCl \rightarrow NHCl_2 + H_2O \quad \text{(dichloramine formation)} \qquad (13\text{-}18)$$

$$NHCl_2 + HOCl \rightarrow NCl_3 + H_2O \quad \text{(trichloramine formation)} \qquad (13\text{-}19)$$

The sum of these three reaction products is called combined chlorine. The total chlorine residual is the sum of the combined residual and any free-chlorine residual. A summary of these definitions is provided below:

$$\text{Free chlorine} = HOCl + OCl^- \qquad (13\text{-}20)$$

$$\text{Combined chlorine} = NH_2Cl + NHCl_2 + NCl_3 \qquad (13\text{-}21)$$

$$\text{Total chlorine} = \text{free chlorine} + \text{combined chlorine} \qquad (13\text{-}22)$$

All chlorine species are expressed as milligrams per liter as Cl_2 and the ammonia concentration is expressed as mg/L as nitrogen (i.e., mg/L NH_3–N). When small amounts of chlorine are added to water, the reactions are much like the simple model above. However, as the amount of chlorine added increases, the reactions become more complex. These reactions and their behavior are partially illustrated by the three zones on Fig. 13-11. At first, as depicted in zone A, the total chlorine residual increases by approximately the amount of chlorine added until the molar ratio of chlorine to ammonia approaches 1 (a weight ratio of 5.07 as Cl_2 to NH_3–N), assuming no other species that consume chlorine are present.

Beyond a molar ratio of 1, the addition of more chlorine decreases, rather than increases, the total chlorine residual (zone B) because the chlorine is oxidizing some of the chloramine species. Eventually, essentially all of the chloramines species are oxidized. The point at which the

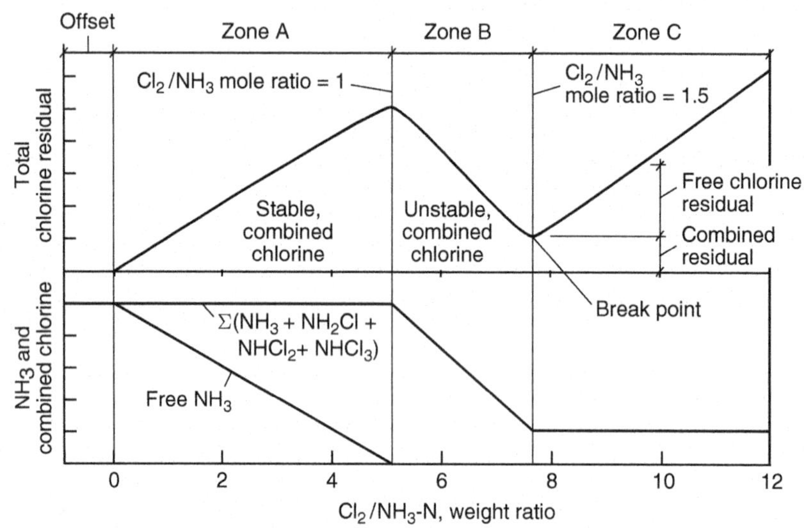

Parameter	Offset	Zone A	Zone B	Zone C
Time to metastable equilibrium	Fraction of a second	Seconds to a few minutes	10 to 60 min	10 to 60 min
Composition of metastable residual.	Reduction of readily oxidizable substances such as Fe(II), Mn(II), and H_2S.	Mostly monochloramine, some dichloramine and traces of trichloramine at neutral or acid pH or at high Cl_2/NH_3 ratios.	A mixture of monochloramine and dichloramine, some free chlorine and traces of trichloramine at low pH.	Mostly free chlorine, trichloramine can be significant (aesthetically, but not as fraction of residual) at neutral pH, but especially in acid region.

Figure 13-11
Overview of chlorine break-point stoichiometry.

oxidation of chloramine species is complete is called the break point and is the beginning of zone C. The exact locations of maximum residual and breakpoint (minimum residual) are influenced by the presence of dissolved organic matter, organic nitrogen, and reduced substances [e.g., S^{2-}, Fe(II), Mn(II)]. The presence of any of these will shift all three zones to the right. The degree to which they shift the point of maximum residual depends on how easily they are oxidized. The shift in the breakpoint corresponds to their stoichiometric chlorine demand. After the breakpoint is reached, the free-chlorine residual increases in proportion to the amount of additional chlorine added. Prior to concerns about disinfection by-products, "break-point" chlorination was often used as a simple means of ammonia removal.

In zone A, monochloramine forms rapidly and with little interference. Nevertheless, the species present in zone A are influenced by pH. At low pH values, dichloramine can form via the following reactions:

$$NH_2Cl + H^+ \leftrightarrows NH_3Cl^+ \tag{13-23}$$

$$NH_3Cl^+ + NH_2Cl \leftrightarrows NHCl_2 + NH_4{}^+ \tag{13-24}$$

Monochloramine is the only chloramine present in zone A at pH 8, but significant amounts of dichloramine can be present at pH 6 (Palin, 1975). In zone B, which is richer in chlorine, some dichloramine will be present even at pH 8 (Palin, 1975). In zone B, hypochlorous acid can oxidize ammonia to nitrogen gas and nitrate ion, resulting in the complete loss of chlorine residual. Between these, the conversion to nitrogen gas is the dominant reaction commonly observed (Saunier and Selleck, 1979):

$$3HOCl + 2NH_3 \rightarrow N_2(g) + 3H_2O + 3HCl \quad \text{(ammonia to nitrogen gas)} \tag{13-25}$$

$$4HOCl + NH_3 \rightarrow H^+ + NO_3{}^- + H_2O + 4HCl \quad \text{(ammonia to nitrate ion)} \tag{13-26}$$

Although break-point chlorine can be described with equilibrium reactions, the behavior of the $Cl_2 - NH_3$ system is actually quite dynamic, and the break-point curve shown on Fig. 13-11 should be considered more of a metastable than an equilibrium state. As a result, laboratory studies to construct a breakpoint curve require precise timing to be reproducible, especially for Cl_2/NH_3 mole ratios above 1. Above this ratio the reaction proceeds rapidly until the metastable state is reached. Anywhere along the curve, the rate at which the reaction progresses is strongly influenced by the pH (Fig. 13-12), particularly in the vicinity of the break point itself. Near the break point, the reaction is at its maximum rate at a pH between 7 and 8. The rate decreases rapidly at pH values outside that range. Facilities engineered to accomplish ammonia removal via the break-point reaction should be designed to accommodate the time for this reaction. Even in the optimum range, the reaction time can be significant.

Example 13-6 Estimating break-point chlorine requirements

Ammonia is added to pure water in the laboratory to reach a concentration of 1 mg N/L. Estimate the chlorine dose needed to reach break point for the following conditions: (1) all the ammonia is converted to nitrogen gas and (2) all the ammonia is converted to nitrate ion. When using breakpoint chlorination to remove ammonia, which reaction requires less chlorine?

Solution

1. Determine the chlorine dose needed to convert ammonia to nitrogen gas. From Eq. 13-25, 3 mol of HOCl is needed for every 2 mol of NH_3:

$$\text{Weight ratio} = (1.5\ \text{mol/mol})\frac{71\ \text{g}\ Cl_2}{14\ \text{g N}} = 7.61\ \text{mg}\ Cl_2/\text{mg N}$$

$$\text{Required dose} = 7.61\ \text{mg}\ Cl_2/\text{mg N} \times 1\ \text{mg N/L} = 7.61\ \text{mg}\ Cl_2/\text{L}$$

2. Determine the chlorine dose to convert ammonia to nitrate. From Eq. 13-26, 4 moles of HOCl is needed for each mole of NH_3:

$$\text{Weight ratio} = (4\ \text{mol/mol})\frac{71\ \text{g}\ Cl_2}{14\ \text{g N}} = 20.2$$

$$\text{Required dose} = 20.2\ \text{mg}\ Cl_2/\text{mg N} \times 1\ \text{mg N/L} = 20.2\ \text{mg}\ Cl_2/\text{L}$$

3. The reaction to nitrogen gas uses less chlorine.

Forms of Chlorine (Liquid, Gas, Hypochlorite, etc.)

The forms of chlorine most often used in the treatment of drinking water are chlorine gas and sodium hypochlorite solution. Calcium hypochlorite powder is also used in some smaller systems. In the United States, chlorine gas can be purchased in 68-kg (150-lb) cylinders, in 908-kg (1-ton) cylinders (in Europe 1000-kg cylinders are used), by tank truck, or in railroad tank cars of between 14.5 and 49.9 metric tons in capacity (16 and 55 American tons). Generally the cost of chlorine is lower when it is shipped in larger volumes, the cost delivered in 1-ton cylinders being approximately half the cost delivered in 68-kg cylinders but nearly twice that when delivered by rail. As a result, some very large utilities purchase liquid chlorine by rail and repackage it for use at various sites.

Liquid Chlorine

The elements of a chlorination facility address each of the following:

1. Storage of liquid chlorine gas
2. Conduits for the transport of liquid chlorine
3. Evaporation of liquid chlorine

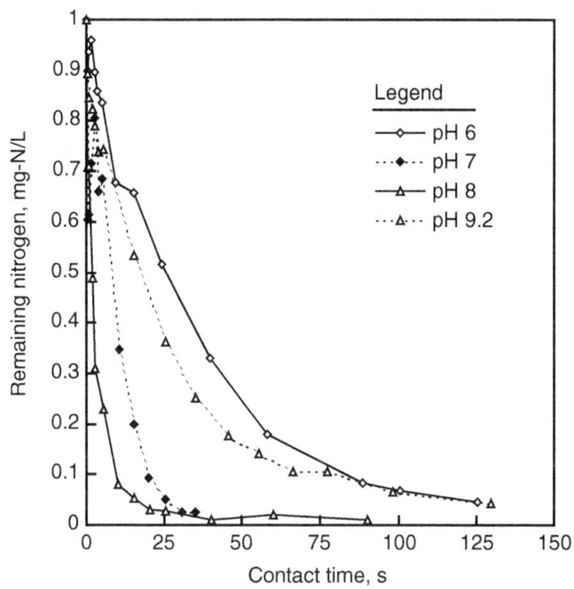

Figure 13-12
Effect of pH on break-point chlorination. (Data from Saunier and Selleck, 1979, Temp. 15 to 18.5°C, $[NH_3]_0 = 1$ mg/L, and $[Cl_2/NH_3]_0 \sim 10$.)

4. Conduits for the transport of chlorine gas under pressure
5. Regulation of the chlorine feed rate
6. Conduits for the transport of chlorine gas under vacuum
7. Chlorine-to-water mass transfer
8. Mixing of chlorine water with the process flow
9. The chlorine contact facility
10. Chlorine sampling and analysis
11. Chlorination control system

In small systems many of these elements are found in one device and other elements, such as the control system, are very rudimentary. In large chlorination systems, each of these elements can sometimes present a separate, specific design challenge. Each of these elements requires different materials and different design considerations apply to each.

DESIGN ISSUES WITH LIQUID CHLORINE

The details of the design of systems for handling the delivery, storage, and dosing of liquid chlorine are beyond the scope of this book. An overview of a variety of the more important issues is provided in Table 13-5. Chlorine is truly a hazardous material so it is important that care be taken in the design of these facilities. White's (1999) handbook is an excellent source for design details.

Table 13-5
Overview of key design issues for chlorination systems

Item	Description
Delivery	In cylinders 68 kg (150 lb) and 908 kg (1 ton); in tank trucks 13,600–18,200 kg (15–20 tons); and in rail cars 14,500–49,900 kg (16–55 tons).
Storage	In cylinders; in tank trucks; in rail cars or in custom tanks.
Conduits for liquid chlorine	Schedule 80 stainless steel (SS), schedule 80 carbon steel, or cast iron (DO NOT USE PVC). Should be seamlessly welded. Use cast-iron valves. Use pipe sizes recommended by White (1999) to avoid "flashing."
Evaporation of liquid	Can use vapor pressure of container to feed up to 19 kg/d (40 lb/d) with 68-kg cylinder and up to 150 kg/d (330 lb/d) with 908-kg cylinders. Multiple cylinders are often used with automatic switchover. At feed rates above 680 kg/d (1.5 tons/d) a separate evaporator is recommended to convert liquid chlorine to chlorine gas.
Conduits for chlorine gas under pressure	Use schedule 80 SS, schedule 80 carbon steel-or cast iron (DO NOT USE PVC). Should be seamlessly welded. Use cast-iron valves.
Regulation of chlorine gas feed rate	Accomplished by the chlorinator. Most chlorinators include four principal elements: (a) a pressure-reducing valve, (b) a rotometer, (c) a control valve, and (d) a vacuum regulator.
Conduits for transport of chlorine gas under vacuum	Often constructed of schedule 80 PVC or reinforced fiberglass pipe. Piping should be carefully sized to maintain pressure drop below 50–60 mm Hg (see White, 1999).
Chlorine-to-water mass transfer	Chlorine is highly soluble and reacts vigorously with water to form hypochlorous and hydrochloric acids. Chlorine-to-water mass transfer is normally accomplished via chlorine injector, a venturi-type device. The maximum solution strength downstream of the injector is approximately 3.5 g/L. The injector is also used to create the vacuum in the system.
Blending of chlorine water into process flow	Under normal conditions, blending must be accomplished before the chlorine residual monitoring point. With normal turbulent flow in a conduit, this requires travel down the conduit an axial distance of 40–200 times the hydraulic radius. Blending can be expedited with devices normally used for rapid mixing or via flow structures that dissipate energy (e.g., a hydraulic jump or a fall over a sharp-crested weir). When ammonia is present, it is important that chlorine be rapidly blended with the bulk flow. If not, both chlorine and ammonia are lost in localized breakpoint reactions and disinfection is compromised. In this case rapid mixing is required.
Chlorine contact facility	Historically, the contact time in existing facilities (e.g., sedimentation basins, clearwells) has been used. Modern treatment plants use specially designed chlorine contact tanks. The most efficient designs, from the standpoint of dispersion, are long, straight pipelines and carefully designed, serpentine contact chambers. Most contact chambers are of concrete.
Chlorine sampling and analysis	Reliable equipment for the sampling and analysis of free and total chlorine has been available for some time. Several devices are available on the market.
Chlorination control system	Historically control systems were manual, feed-forward, feedback, and compound loop in design. Today control systems and methods of sampling and analysis have improved so complex control is possible.

Control of Gas Chlorination

Four methods have traditionally been used for controlling the feed rate of chlorine gas when it is used for residual control in drinking water systems. Each is displayed on Fig. 13-13: (1) manual control, (2) feedforward control, (3) residual feedback control, and (4) compound loop control. Through the middle of the twentieth century, manual control was the most common. Significant operator attention was required to ensure that a suitable residual was reliably provided, especially when the flow rate through the plant was adjusted. By the mid-1950s flow measurement and chlorine metering techniques improved until feed-forward control systems began to appear. This important advance allowed automatic adjustment for flow but still required the operator to adjust for any changes in the water quality (chlorine demand) or any drifts in monitoring and feed rates.

By the mid-1960s direct residual control began to appear. In principle, the feedback method of control is more robust than feed-forward control, but residual measurement did not approach suitable levels of reliability and precision for two more decades. As a result, compound loop control evolved as a compromise. With this method, changes in flow could be

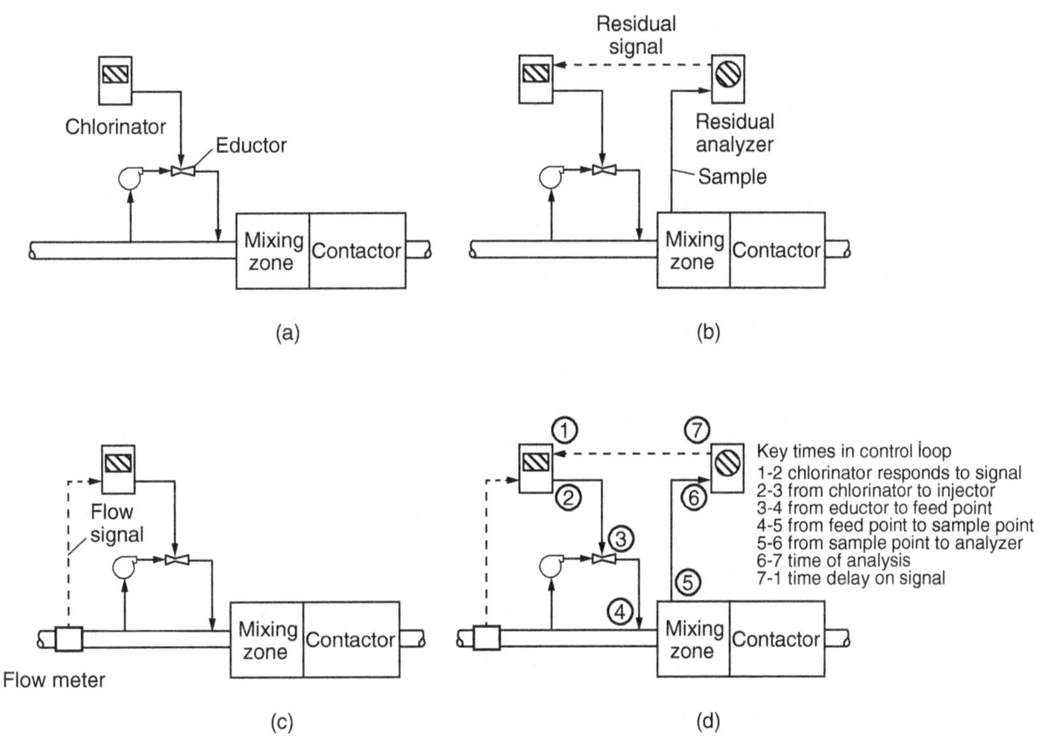

Figure 13-13
Control of chlorine gas feed rate: (a) manual control, (b) feedback or residual control, (c) feed forward control, and (d) compound loop control.

accommodated via the flow signal and an additional control increment could be added via residual control for minor water quality changes. Properly maintained, compound loop control was the first system to provide reliable, continuous residual control.

During the last decade of the twentieth century, computerized supervisory control systems had evolved to the point where these same inputs (flow and residual) could be combined with other measurements to provide improved reliability. None of these control systems, however, is an adequate substitute for vigilant attention from the operator.

Residual control system

The sequence of events in the residual control system must be carefully designed and controlled. All the elements shown and labeled in the diagram as "compound loop control" on Fig. 13-13 must be considered in designing the system and envisioning its method of control. The time between the instant when a change in chloride feed rate is made and when the change in residual is detected by the control system has a significant effect on the effectiveness of the control system. The instructions of the supervisory control and data acquisition (SCADA) system must be designed with a full understanding of the effect of timing delays in each element of the system.

The time required for the chlorinator to completely respond to an instruction from the SCADA system is normally not too significant. The time required for the change in feed rate established by the chlorinator to be recognized at the eductor must be considered. This time to change feed rates is normally not too long either, but it can be too long when the chlorinator is located a long way from the injector and when the chlorine feed rate is very low. White (1999) suggests that this time be estimated by assuming that the change in pressure will travel in a wave about three times as fast as the gas flow in the line.

Next, the time for the water in the chlorine water line to travel from the eductor to the application point must be considered. This time is a function of the distance between the eductor and the application point and the flow rate (velocity) in the chlorine water line. Again, designs with long distances between the eductor and the application point can cause trouble for control. Ideally the chlorine is stored near the application point so that both the time in the vacuum line and the time in the chlorine line are minimized. When nearby storage is not possible, it is usually best to lengthen the vacuum line, not the chlorine water line, as a signal ordinarily travels much faster down a vacuum line.

Sampling point

Another important constraint is the time between the chlorine application point and sampling point. There is an inherent design conflict in the distance between these locations. Putting these points too close together can result in poor blending before the treated water reaches the sampling point. When this happens, the control system constantly "searches" for control but can never quite find it. Putting them too far apart can result in too much delay between when a change in dose is made and when it is detected by the control system. To avoid control problems, the residual for sample control must be taken after mixing is satisfactory. Depending on the method of chlorine introduction and the criteria used for mixing, the distance downstream to accomplish satisfactory blending is between 40 and 200 times the hydraulic radius of the water conduit. Problems associated with this distance are aggravated in larger applications because of the larger conduit diameters that are used.

The time required for the sample to travel from the sample point to the analyzer is also important. Sample travel time can be a significant complication if the analyzer is located in some central location far from the sampling site. The time for the analyzer to assay the sample (normally between 15 and 20 s) can also be important in some applications. In designing such a control system, it is important to analyze all these times and the sequence in which they operate at both high- and low-flow conditions, both early and late in the life of the design, to ensure that problems do not occur after the installation is complete.

Example 13-7 Establishing time between chlorine application and residual control sampling

Consider two treatment plants A and B. In plant A the filtered water line is 305 mm (12 in.) in diameter. In plant B, the filtered water line is 3050 mm (120 in.) in diameter. Both pipelines are designed for a velocity of 1.5 m/s (5 ft/s). Assume that both have equivalent mixing at the point of chlorine injection and that the flow in both pipelines will be suitably blended for sampling at a point 50 pipe diameters downstream (100 hydraulic radii).

Estimate how far down the pipeline the sample point must be and how long it will take for the water to travel from the point of chlorine injection to the sampling point in each case.

Solution

1. Estimate the travel time from application point to sampling point (4 to 5 on Fig 13-13d):
   a. For plant A, pipe diameter is 305 mm, 50 pipe diameters equal 15 m, and at a velocity of 1.5 m/s the travel time is $\sim$10 s.
   b. For plant B, pipe diameter is 3050 mm, 50 pipe diameters equal 150 m, and at a velocity of 1.5 m/s the travel time is $\sim$100 s.

Sodium Hypochlorite

When chlorine was first used for disinfection, it was often applied in the form of hypochlorite. Sodium hypochlorite (NaOCl), or liquid bleach, came into use near the beginning of the Great Depression in the late 1920s. Later, chlorination using liquid chlorine became predominant because of its lower cost, but now hypochlorite is again becoming more common because of the hazardousness of liquid chlorine.

Sodium hypochlorite is the most widely used form of hypochlorite today. It is widely used not only in disinfection of water but also for a myriad of other household and industrial uses. Calcium hypochlorite [$Ca(OCl)_2$] is used by some small utilities.

Whereas chlorine gas is prepared by an electrolytic process that breaks sodium chloride solution into chlorine gas and sodium hydroxide, ironically, sodium hypochlorite is generally prepared by mixing sodium hydroxide and chlorine gas together:

$$2NaOH + Cl_2 \rightarrow NaOCl + NaCl + H_2O \tag{13-27}$$

On a weight basis, 1.128 kg of NaOH reacts with 1 kg of chlorine to produce 1.05 kg of NaOCl and 0.83 kg of NaCl. The process is complicated by the fact that the reaction generates a significant amount of heat. It is common practice to add an excess of NaOH because, as will be shown, hypochlorite is more stable at higher pH values. As a result, the density of one hypochlorite solution is not necessarily the same as another, even if both have the same strength (percent Cl_2). This density difference occurs because the final density depends on the amount of excess NaOH added during manufacture. Liquid bleach usually has a pH between 11 and 13. Hypochlorite can also be manufactured via onsite generation; this process is becoming more common.

STABILITY OF HYPOCHLORITE

Under some conditions, the strength of hypochlorite can decline significantly in just a few days. In fact, stability is one of the major issues that must be addressed in both designing and operating a hypochlorite facility. A utility should not consider using hypochlorite unless it is prepared to dedicate time and energy to a regular program of monitoring and controlling its decay. Of considerable significance is the fact that, when hypochlorite does decay, chlorate ion is one of the principal by-products of the reaction. The stability of hypochlorite is affected by the strength of the solution, the storage temperature, the pH, and the contamination of heavy metals, which can catalyze its decay. Light is also a problem. As a general rule, the rate of decay is accelerated by (1) higher concentration, (2) higher temperature, (3) lower pH, (4) exposure to sunlight, and (5) the presence of certain heavy metals, notably copper and nickel.

Under basic conditions, the decomposition of hypochlorite ion to chlorate ion follows a disproportionation reaction, which exhibits second-order reaction kinetics and the following overall stoichiometry (Gordon et al., 1995a, b):

$$OCl^- + OCl^- \rightarrow ClO_2^- + Cl^- \quad (13\text{-}28)$$

$$OCl^- + ClO_2^- \rightarrow ClO_3^- + Cl^- \quad (13\text{-}29)$$

$$3OCl^- \rightarrow ClO_3^- + 2Cl^- \quad (13\text{-}30)$$

The second reaction, as given by Eq. 13-29, is the faster of the two. As a result, the first reaction is the rate-limiting step in the consumption of hypochlorite ion. Bleach also decays via a slow reaction that forms oxygen and an acid-forming reaction that also forms chlorate ion, as shown in the following reactions:

$$OCl^- + OCl^- \rightarrow O_2 + 2Cl^- \quad (11\text{-}31)$$

$$2HOCl + OCl^- \rightarrow ClO_3^- + 2H^+ + 2Cl^- \quad (13\text{-}32)$$

Gordon and colleagues (1995a, b) have shown that copper and nickel catalyze oxygen formation (see Eq. 13-31) and research at the Swiss Federal Institute of Aquatic Science and Technology (EAWAG) has shown that a similar reaction occurs via proteolysis (Nowell and Hoigné, 1992a,b). The relationships between the three principal reactions that result in hypochlorite decay are displayed on Fig. 13-14.

The pH at which a sodium hypochlorite solution is stored has important impacts on its rate of decay, as shown on Fig. 13-15a (Gordon et al. 1995a, b). The rate of decay is low at pH 11 and above but increases rapidly below pH 10. Some evidence suggests that a decay minimum occurs between pH 12 and 13. As liquid bleach is normally delivered at pH 12 or above, low-pH decay is normally not a problem with the undiluted product. Often

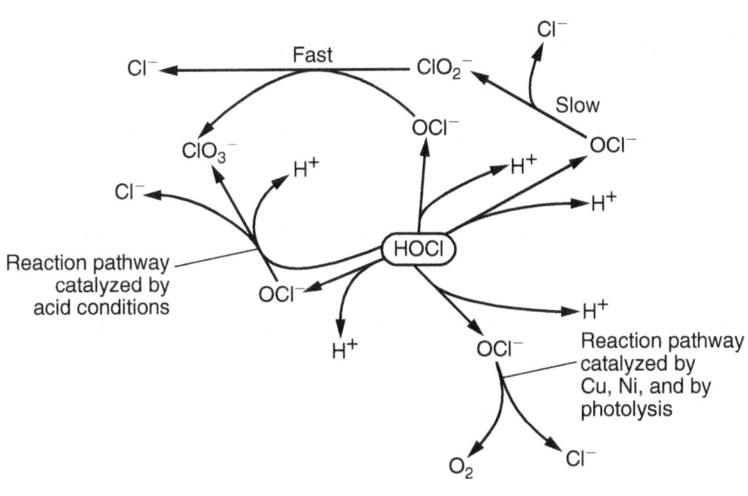

Figure 13-14
Decay reactions of hypochlorite.

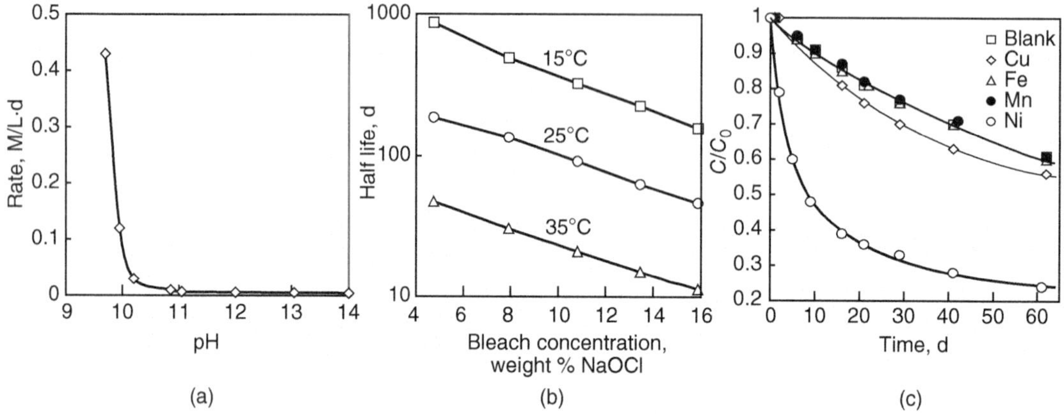

Figure 13-15
Factors that influence decay of sodium hypochlorite: (a) effect of pH on rate of decay of hypochlorite; (b) effect of temperature and concentration on decay of hypochlorite; and (c) effect of trace metals on rate of decay of hypochlorite. (Data from Gordon et al., 1993, 1995a, b.)

it is delivered with enough excess hydroxide to allow a 50 percent dilution without increasing the rate of decay. Nevertheless, pH monitoring and control are important in a hypochlorite management program.

As mentioned earlier, the rate of the dominant decay reaction in liquid bleach (Eq. 13-30) is a second-order reaction (Gordon et al., 1995a, b). As a result, a stronger bleach solution will decay faster. This effect can be illustrated by the solution of the second-order rate equation for a completely mixed batch reactor:

$$\frac{C}{C_0} = \frac{1}{1 - k_d C_0 t} \tag{13-33}$$

where C = bleach concentration after time t, mol/L
C_0 = bleach concentration at time 0, mol/L
k_d = second-order decay coefficient, L/mol · s
t = time, s

The effects of bleach strength and temperature are illustrated on Fig. 13-15b. Based on the data in this figure, diluting bleach delivered at a concentration of 15 percent to a concentration of 7.5 percent will increase its half-life from about 50 to about 140 days (at 25°C). If the 7.5 percent bleach is also stored at 15°C instead of 25°C, the combined effect of dilution and temperature control will increase its half-life to more than 500 days.

Finally, since the work of Lister (1952, 1956), bleach technologists have understood that certain metals can catalyze the decomposition of bleach. In the mid-1950s rhodium, iridium, cobalt, copper, manganese, iron, and nickel were issues. Today the principal concerns are copper and nickel, and manganese has also been shown to exacerbate the destructive effect of nickel. Gordon et al. (1995a, b) conducted tests to examine the effect of a concentration of 1 mg/L of copper, iron, manganese, and nickel, individually, on the decay of a 13.5 percent bleach. These are illustrated on Fig. 13-15c. The authors recommended that copper and nickel be kept below 0.1 mg/L. Bleaches are also filtered in an attempt to reduce metals contamination, but, with one exception, additional filtering of modern commercial bleaches showed only small improvements (Gordon et al., 1995a, b). It appears that many modern bleaches are produced in such a condition that additional filtering is of little benefit.

FORMATION OF CHLORITE AND CHLORATE ION

In 1992, the U.S. EPA discovered that hypochlorite solutions containing significant concentrations of chlorate ion were responsible for introducing chlorate ions into drinking water (Bolyard et al., 1992). Of special significance in this regard is the fact that the principal bleach decay reaction results in the production of chlorite (ClO_2^-) and chlorate (ClO_3^-) ions (see Eqs. 13-28 and 13-29).Chlorite is regulated by the U.S. EPA. Chlorate is regulated in some jurisdictions; for example, the State of California has set an action limit of 0.8 mg/L (Cal DHS, 2002). As a result, it seems prudent to limit the production of chlorate as well.

As noted earlier, the disassociation of chlorite to chlorate and chloride (Eq. 13-29) is much faster than the disproportionation of hypochlorite ion to chlorite and chloride (Eq. 13-28), and this minimizes the formation of chlorite ion. As a result, it is estimated that chlorite normally stays below 0.5 percent of the hypochlorite concentration (Gordon et al., 1997). Thus, a chlorine dose of 1 mg/L delivers less than 0.005 mg/L of chlorite ion into solution (Gordon et al., 1997), considerably less than the MCL of 0.8 mg/L. Thus, even though chlorite generally does not pose a problem in hypochlorite solutions; the same is not true for chlorate.

If hypochlorite decomposition were only the result of Eq. 13-30, the chlorate generated would be about 33 percent of the hypochlorite decomposed (molar basis). But other pathways for hypochlorite decay (decomposition catalyzed by light and metals) normally produces oxygen and not chlorate (Eq. 13-31). Gordon et al. (1995b) examined chlorate production in 12 tests with commercial bleaches and found that the actual production of chlorate was slightly less, about 31 percent (Fig. 13-16a). As a rule of thumb, it is conservative to assume that one-third of the bleach lost to decomposition ends up as chlorate ion.

Two surveys were also conducted to evaluate the contribution of chlorate ion to water systems using sodium hypochlorite for disinfection (Bolyard et al., 1993; Gordon et al., 1993). Both authors looked at the ratio of chlorate ion and hypochlorite ion in the bleaches being used as well as the concentration of chlorate ion in the drinking water system itself. A probability plot of the chlorate/hypochlorite ratio in the bleaches from both surveys is presented on Fig. 13-16b. In both cases, the median was slightly less than 0.1 mol[ClO_3^-]/mol[OCl^-]. On the other hand, levels greater than

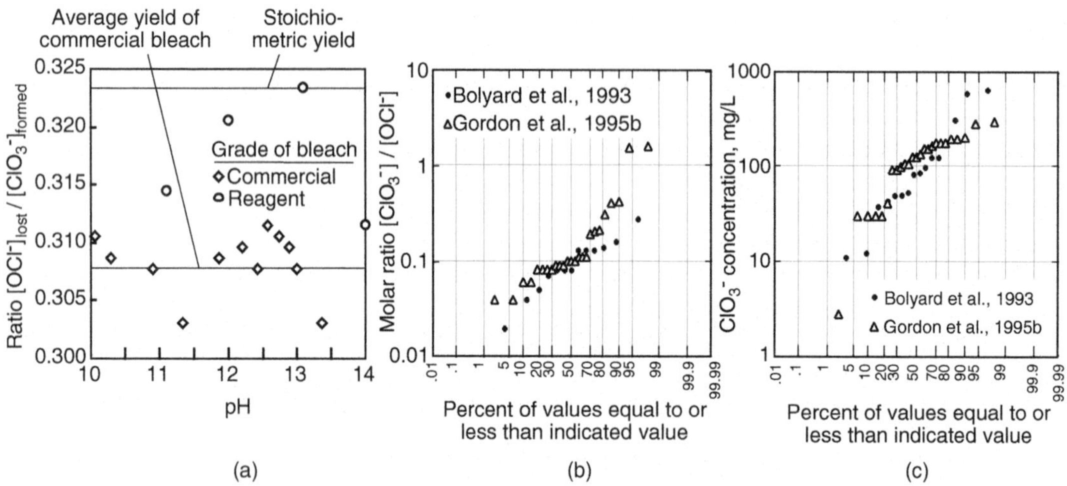

Figure 13-16
Formation of and impacts of chlorate in hypochlorite feedstock: (a) chlorate formation during decomposition of reagent and commercial hypochlorite; (b) surveys of chlorate content in bleach; and (c) surveys of chlorate content in systems using bleach. (Data from Bolyard et al., 1993; Gordon et al., 1993, 1995a, b, 1997.)

1 mol $[ClO_3^-]/mol[OCl^-]$ and as low as 0.02 mol $[ClO_3^-]/mol[OCl^-]$ were observed, indicating that bleach manufacturing and storages practice can result in substantial differences. At a ratio of 0.1, a chlorine dose of 3 mg/L would cause chlorate concentrations of approximately 0.1 mg/L. Thus the chlorate that is found in bleach under typical conditions of use should not be a significant issue. Many of the considerations that affect the stability of bleach are also important in limiting its chlorate content. Nevertheless, surveys of chlorate in systems using hypochlorite did sometimes show the presence of significant chlorate (Fig. 13-16c), suggesting that utilities using hypochlorite should occasionally monitor for chlorate and consider modifying their practice if significant amounts are observed.

STORAGE AND FEEDING OF SODIUM HYPOCHLORITE

Experience with materials for the construction of large hypochlorite tanks has not been uniformly good. Early projects in Chicago had unsatisfactory experience with filament-wound fiberglass tanks and with underground concrete tanks with fiberglass lining. These tanks were replaced with hand lay-up fabricated fiberglass tanks using a vinyl resin binder and with plastic, continuous-weld, full-weight carbon steel tanks lined with a fiberglass-reinforced polyester material at a thickness of 0.9 mm (35 mil). The latter gave acceptable performance (White, 1999). Properly fabricated fiberglass tanks or steel tanks with a rubber or polyvinyl chloride (PVC) lining give satisfactory service as well.

Hypochlorite is an extremely aggressive chemical, and no equipment used to store or feed it can be expected to last indefinitely. Some particularly robust diaphragm and solenoid metering pumps have been successfully used, and this is the approach found in most plants (White, 1999). In very large plants (>380 ML/d or 100 mgd), White recommends metering the chemical by gravity from the storage tank through a Teflon-lined magnetic flowmeter and rate-modulating valve to the point of application.

Hypochlorite can be transported in schedule 80 PVC piping; except where exposed to sunlight, chlorinated polyvinyl chloride (CPVC) should be used. Ball valves and plug valves made of steel lined with PVC or propylene should be avoided. In general, precautions should also be taken for the potential for precipitation of calcium carbonate whenever the hypochlorite is mixed with carrying water or at the application point with the water being treated.

The high specific gravity of hypochlorite solution must be overcome to accomplish effective mixing at the point of application. This can be accomplished by using a diffuser and carrying water (be cautious about the potential formation of $CaCO_3$) or by the use of a pumped jet mixer like that often used for coagulants. Mixing can also be accomplished by introducing the hypochlorite at a point of significant turbulence.

Ammonia

Ammonia can be supplied for water treatment applications in three forms: as a pure liquid (anhydrous ammonia), dissolved in water (aqueous ammonia), or as a dry ammonium salt, usually ammonium sulfate. Ammonia is not expensive, but the relative cost of these alternative forms of ammonia varies from one location and one application to another. For reasons of convenience, aqua ammonia is the form most commonly used. Exposure to high concentrations of ammonia vapor can be fatal. At concentrations of several hundred parts per million by volume (ppm_v), it causes throat and eye irritation, and at higher concentrations it can cause convulsions or even rapid asphyxia. While not addressed in this discussion, appropriate precautions should be taken both in design and operation of ammonia facilities.

STORAGE AND FEEDING OF ANHYDROUS AMMONIA

At normal temperatures and pressures, anhydrous ammonia (>99 percent NH_3) is a gas. However, it can be easily liquefied and is commonly stored and transported in liquid form in pressurized containers of the same size and same design as those used for chlorine (they are usually a different color). At atmospheric pressure, liquid anhydrous ammonia has a density of 680 kg/m^3 (42.6 lb/ft^3 or 5.7 lb/gal), approximately two-thirds that of water. Anhydrous ammonia containers comply with International Code Council (ICC) regulations, which require a minimum working pressure of 1760 kPa (256 psig) with safety valves set to release at that pressure. Valves and fittings used for anhydrous ammonia are normally rated at 2070 kPa (300 psig).

In the United States, bulk shipments of anhydrous ammonia are normally made in 23- and 73-metric-ton (25- and 80-U.S.-ton) rail tank cars, in 18-tonne (20-U.S.-ton) tank trailers, and cylinders the same size and design as those used to deliver 908 and 68 kg of liquid chlorine. It is common for vendors to provide storage tanks. Permanent (stationary) storage tanks for anhydrous ammonia can also be custom fabricated to any desired size. Such tanks must meet the same pressure restrictions as the shipping containers and are usually made of carbon steel. No copper, bronze, or brass fittings should be used because ammonia attacks copperbased alloys. Storage tanks should be sheltered from the sun to prevent excessive pressure buildup. The vapor pressure of anhydrous ammonia at 10°C is slightly more than 611 kPa (89 psig). At 30°C it nearly doubles to 1183 kPa (172 psig). The formula below may be used to estimate the vapor pressure at temperatures between 0 and 40°C:

$$P_{v,\mathrm{NH_3}} = 434.9 + 13.96T + 0.3645T^2 \tag{13-34}$$

where $P_{v,\mathrm{NH_3}}$ = vapor pressure of anhydrous ammonia, kPa
T = temperature, °C

Anhydrous ammonia can be fed by two methods: direct feed and solution feed. In direct gas feed, the ammonia gas is directly introduced into the water to be treated. Unless the plant is very small, this method often suffers from poor distribution at the application point because of the low flow rate of ammonia gas. The solution feed method is analogous to the technology used to feed chlorine, except the vapor pressure of ammonia is higher. Precipitation of $CaCO_3$ is often a problem in the vicinity of the application point.

Direct gas feed

Direct gas ammonia feeders are commercially available and differ only with respect to minor material changes from chlorinators in that they include an ammonia pressure-regulating valve, pressure gauges, a pressure relief valve, rotameters, and a control valve with back-pressure regulator, all in a modular cabinet. The ammoniator meters gaseous ammonia into the process stream under positive pressure. The high pressure in the storage tank is normally reduced to approximately 276 kPa (40 psi) using the pressure regulator. At this reduced pressure the ammonia flows through the rotameter where the gas flow can be read directly in mass/time units (In the United States the units are usually pounds per hour or pounds per day). Finally, the gas flows through the back-pressure valve, which maintains a constant back pressure on the system. This pressure is limited to a range of 101 to 122 kPa (15 to 18 psig). The backpressure valve is used to keep the feed rate constant with changes in the pressure at the application site. Ammoniators should be housed separately from chlorination equipment. A direct-feed ammonia application is illustrated on Fig. 13-17a. For completeness, an evaporator is

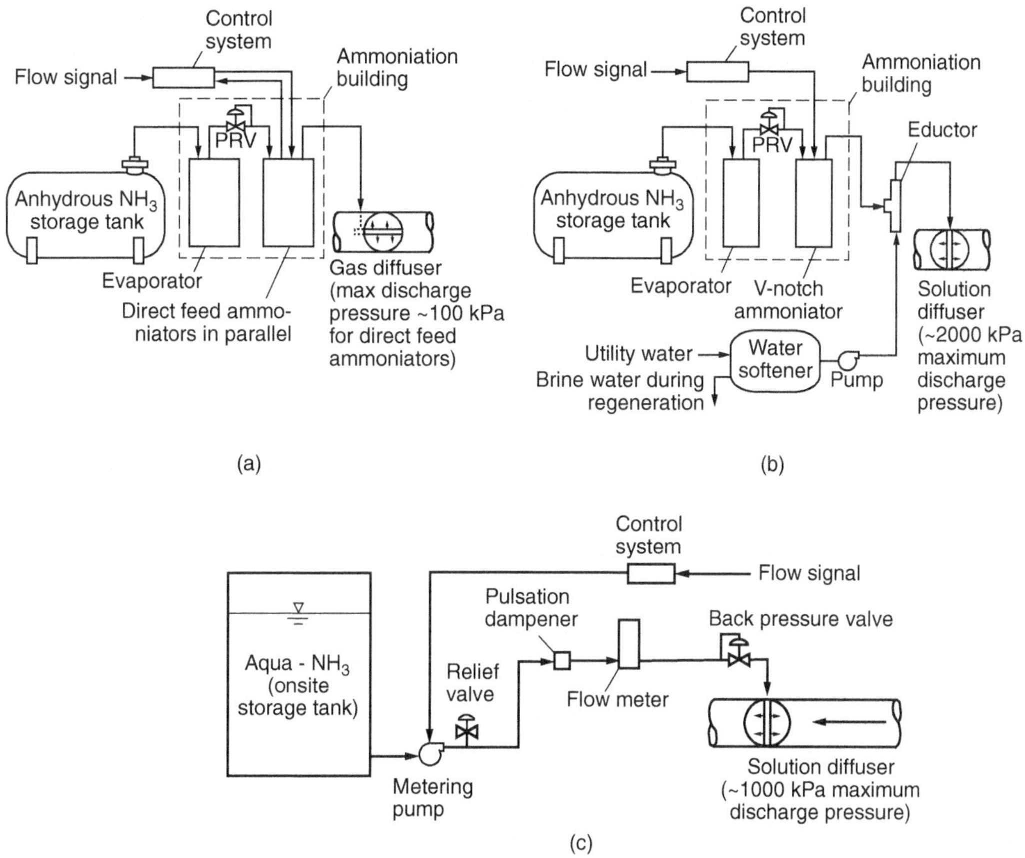

Figure 13-17
Schematics of alternate ammonia feed systems: (a) direct feed of anhydrous ammonia; (b) solution feed of anhydrous ammonia; and (c) aqua ammonia feed system.

shown, although these are not always required. If ammonia feed rates are high enough, the anhydrous liquid would be withdrawn from the bottom of the storage tank and converted to gas in the evaporator prior to entering the ammoniator. The largest direct-feed ammoniators have a maximum feed capacity (determined by the rotameter rating) of 455 kg/d (1000 lb/d).

Solution feed of anhydrous ammonia
The design of these systems closely parallels the design of modern gas chlorination systems. An ammoniator and a gas diffuser are often used to feed the anhydrous ammonia solution (see Fig. 13-17a). A solutionfeed ammoniator (see Fig. 13-17b) is typically recommended when higher feed rates or greater discharge pressures prohibit the use of direct-feed ammoniators. (It is important to remember that direct-feed ammoniators

are limited by their back-pressure valve to a pressure of approximately 100 kPa.) An important difference between ammoniation and chlorination systems is that the utility water for a solution ammoniation system must be softened to prevent the deposition of $CaCO_3$ in the system.

STORAGE AND FEEDING OF AQUA AMMONIA

Ammonia is very soluble in water. As an example, 1 volume of water will dissolve 1150 volumes of anhydrous ammonia at a temperature of 0°C and atmospheric pressure. As a consequence, ammonia is commercially available as an aqueous solution of between 20 and 30 percent strength "aqua ammonia." It is usually dissolved in deionized or softened water and stored in low-pressure tanks. The vapor pressure of 30 percent aqua ammonia at 37.8°C (100°F), a temperature common in many parts of the world, is greater than 1 atm. To prevent off-gassing of ammonia in these locations, a slightly pressurized tank should be used. In contrast, the vapor pressure of 20 percent aqua ammonia is less than 1 atm, permitting storage in a nonpressure tank with a minimum of off-gassing. Aqua ammonia is not commonly shipped long distances; hence the largest transport vessel in the United States is a 28,300-L (7500-gal) tank trailer. There seems to be less standardization for onsite aqua ammonia storage tanks, probably because low-pressure tanks are acceptable. Steel and fiberglass tanks are both used in water treatment applications.

Depending on the concentration of aqua ammonia, excessive temperatures can cause ammonia gas to come out of solution. Off-gassing should be considered in design, and a slightly pressurized storage tank with a relief valve vented to a water trap or ammonia scrubber may be necessary to keep vapors from escaping to the atmosphere.

Aqua ammonia can sometimes be fed directly to the process stream using a metering pump. Suitable metering pumps are commercially available. Progressive cavity pumps have also been successfully used. The storage tank is a permanent onsite facility and should have enough storage for at least 10 days of maximum usage. The tank should have a liquid-level monitor to allow monitoring of the inventory in the tank. The flow metering pump should be located in the proximity of the tank and below its hydraulic grade to minimize chances of ammonia vaporization in the piping. If necessary, the metering pumps can be sheltered in a building; however, the pumps themselves do not necessarily require shelter as do the anhydrous ammonia feed equipment mentioned earlier. An aqua ammonia feed system is illustrated on Fig. 13-17c.

STORAGE AND FEEDING OF AMMONIUM SULFATE

The most common salt of ammonia used in water treatment is ammonium sulfate, $(NH_4)_2SO_4$. This form of ammonia has the advantage that it does not raise the pH as much as the others do. As a result, it is easier to combine it with dilution water to obtain proper mixing. Mixing can be an important consideration when adding ammonia to water containing free chlorine to arrest the formation of DBPs.

MIXING

Adding chlorine to water that already has ammonia in it can result in undesirable reactions while mixing takes place. To prevent free ammonia and thus minimize nitrification, it is common for water systems today to add ammonia at a total dose that is at the peak of the breakpoint curve (a 1:1 molar ratio). By definition, the ratio of chlorine to ammonia in the entire mixing zone is on the left side of the breakpoint curve. This condition necessitates that the mixing be rapid relative to the time for the irreversible oxidation of ammonia. That is,

$$t_{\text{mix}} \ll t_{\text{rx}} \tag{13-35}$$

where t_{mix} = time required to obtain mixing to microscale, s
t_{rx} = half-life of breakpoint reaction, s

Although this circumstance is easily described in a qualitative way, it is quite difficult to characterize quantitatively because t_{rx} is a function of not only the pH but also the local Cl_2/NH_3 ratio (affected by the degree of mixing). When ammonia is added to a chlorinated water to arrest the formation of disinfection by-products, very good mixing is required to ensure that chemicals are efficiently used (see Chap. 6).

MANAGING COMBINED CHLORINE (CHLORAMINE) RESIDUALS

Maintaining a combined chlorine residual involves some considerations that are not important when a free-chlorine residual is used. Chloramines have the advantage that their odor threshold is lower (Krasner and Barrett, 1984), that they are more effective in controlling microbial growth on pipe surfaces (Le Chevallier et al., 1988), and that they are generally much more stable (Trussell and Kreft, 1984). It should be noted that combined chlorine residuals are subject to destruction by biological nitrification, especially if temperatures are warm and if ammonia is used in excess. Also there is recent evidence that the use of combined chlorine can result in the formation of low levels of NDMA, a suspected human carcinogen (Najm and Trussell, 2000, 2001). Some of the conditions that aggravate NDMA formation, namely a high chlorine-to-ammonia ratio, are the same things that discourage nitrification.

13-6 Disinfection with Chlorine Dioxide

When the regulation of the chlorination by-products began, chlorine dioxide (along with ozone) was a fairly high-profile disinfection alternative (U.S. EPA, 1979). Chlorine dioxide is widely used in continental Europe, particularly Germany, Switzerland, and France, and produces almost no identifiable organic by-products, except low levels of a few aldehydes and ketones (Bull et al., 1990). Chlorine dioxide was known to produce two inorganic by-products, chlorite and chlorate ion. As a result, most applications of chlorine dioxide were on low-TOC waters that did not require a high dose to overcome oxidant demand. Late in the 1980s, concern about the toxicity of chlorite ion and chlorine dioxide itself reached a peak. Also, based on field experience, it was found that the use of chlorine dioxide was sometimes responsible for a very undesirable "cat urine" odor (Hoehn et al., 1990). As a precautionary measure, the State of California banned the use of chlorine dioxide for the disinfection of drinking water and several other states followed.

Eventually, when the disinfectant by-product rule was promulgated (U.S. EPA, 1998), an MCL of 0.8 mg/L was set for chlorite ion and a maximum disinfectant residual limit (MDRL) of 1 mg/L was set for chlorine dioxide. No MCL was placed on chlorate ion, but utilities were encouraged to be cautious about the production of chlorate and, again as a precautionary measure, the State of California has set an action level of 0.8 mg/L. Methods for reducing the concentration of chlorite ion downstream of the use of chlorine dioxide have been demonstrated (Griese et al., 1991; Iatrou and Knocke, 1992), and it has been established that the cat urine odor only occurs when chlorite ion is exposed to a free chlorine residual. As a result, it appears that chlorine dioxide may indeed play an important role in DBP control, particularly in systems using combined chlorine for residual maintenance and looking for a small boost in primary disinfection.

Generation of Chlorine Dioxide

The principal reactions that occur in most chlorine dioxide generators have been known for a long time. In industry, large-scale chlorine dioxide generators use chlorate as a feedstock, but for potable water applications chlorine dioxide is usually generated onsite using a 25 percent sodium chlorite solution. Although a sodium chlorite feedstock is a common starting point, a number of different approaches are used to convert the chlorite to chlorine dioxide. These include reactions with gaseous chlorine (Cl_2), aqueous chlorine (HOCl), or acid (usually hydrochloric acid, HCl). The reactions are

$$2NaClO_2 + Cl_{2(g)} \rightarrow 2ClO_{2(g)} + 2Na^+ + 2Cl^- \tag{13-36}$$

$$2NaClO_2 + HOCl \rightarrow 2ClO_{2(g)} + 2Na^+ + Cl^- + OH^- \tag{13-37}$$

$$5NaClO_2 + 4HCl \rightarrow 4ClO_{2(g)} + 5Na^+ + 5Cl^- + 2H_2O \tag{13-38}$$

The stoichiometry of Eq. 13-36 requires 0.5 kg of chlorine and 1.34 kg of sodium chlorite to produce 1 kg of chlorine dioxide. Several of the alternative approaches used for the generation of chlorine dioxide are summarized in Table 13-6.

Table 13-6
Chlorine dioxide generation alternatives

Generator Type	Main Reactions, Reactants, By-products, Key Reactions, and Chemistry Notes	Special Attributes
Acid–chlorite: (direct acid system)	$5NaClO_2 + 4HCl \rightarrow 4ClO_2(g) + 5NaCl + 2H_2O$ ❑ Low pH ❑ ClO_3^- also possible ❑ Slow reaction rates	Chemical feed pump interlocks required; production limit ~10–15 kg/d (25–30 lb/d); maximum yield is ~80% of stoichiometric yield.
Aqueous chlorine–chlorite: (Cl_2 gas ejectors with chemical pumps for liquids or booster pump for ejector water)	$Cl_2 + H_2O \rightarrow HOCl + HCl$ $HOCl + 2NaClO_2 \rightarrow ClO_2(g) + NaCl + NaOH$ ❑ Low pH ❑ ClO_3^- also possible ❑ Relatively slow reaction rates ❑ Excess Cl_2 or acid to neutralize NaOH	Production rates limited to ~450 kg/d (1000 lb/d); high conversion but yield only 80–92%; more corrosive effluent due to low pH (~2.8–3.5); three chemical systems pump HCl, hypochlorite, chlorite, and dilution water to reaction chamber
Recycled aqueous chlorine–chlorite: (saturated Cl_2 solution via a recycling loop prior to mixing with chlorite solution)	$2HOCl + 2NaClO_2 \rightarrow 2ClO_2 + Cl_2 + 2NaOH$ ❑ Excess Cl_2 or HCl needed due to NaOH formed ❑ Concentration of ~3 g/L required for maximum efficiency	Production rate limited to ~450 kg/d (1000 lb/d); yield of 92–98% with ~10% excess Cl_2 reported; highly corrosive to pumps; drawdown; calibration needed; maturation tank required after mixing
Gaseous chlorine–chlorite: (gaseous Cl_2 and 25% solution of sodium chlorite; pulled by ejector into the reaction column)	$Cl_2(g) + 2NaClO_2 \rightarrow 2ClO_2(g) + 2NaCl$ ❑ Neutral pH ❑ Rapid reaction ❑ Potential scaling in reactor under vacuum due to hardness of feedstock	Production rates 2300–55,000 kg/d (5000–120,000 lb/d); ejector based, with no pumps; motive water is dilution water; near-neutral pH effluent; no excess Cl_2; turndown rated at 5–10X with yield of 95–99%; less than 2% excess Cl_2; highly calibrated flowmeters with minimum line pressure ~275 kPa (40 psig) needed

Source: Adapted in part from U.S. EPA (1999).

The differences between Eqs. 13-36, 13-37, and 13-38 help to explain how generators can differ even though the same feedstock chemicals are used and why some should be pH controlled and others are less dependent on pH. In most generators, more than one reaction may be taking place.

Chlorine dioxide generators are relatively simple mixing chambers. The reactors are frequently filled with media (Teflon chips, ceramic, or Raschig rings) to generate hydraulic turbulence for mixing. A sample petcock valve on the discharge side of the generator is desirable to allow for monitoring of the generation process. An excellent source for additional information on chlorine dioxide generation may be found in Masschelein (1992).

Sodium Chlorite

Sodium chlorite is used as a solution, normally with a concentration of approximately 25 percent sodium chlorite or less. It is commercially available as a 25 or 38 percent solution. The major safety concern for solutions of sodium chlorite is the unintentional and uncontrollable release of high levels of chlorine dioxide gas. Levels that approach an explosive mix can sometimes occur if the sodium chlorite is exposed to acid.

Another concern to be addressed with sodium chlorite is crystallization. Like most salts, sodium chlorite solutions are prone to crystallization at low temperatures and/or higher concentrations. When crystallization occurs, it may obstruct flow in pipelines, valves, and other equipment.

Sodium chlorite is not stable as a powder. If dried, it is a fire hazard and can ignite when in contact with combustible materials. A sodium chlorite explosion may occur if too much water and inappropriate firefighting techniques are used to quench such a fire. Burning sodium chlorite will quickly generate enough heat to turn water to steam. At high temperatures, the breakdown products of sodium chlorite include oxygen. As a result, highly trained firefighters are required to extinguish closed containers or dry material that has been ignited.

Stratification in holding tanks for sodium chlorite solutions may also occur and, when it does, will adversely influence the chlorine dioxide yield in the generator. As stratification develops, the sodium chlorite solution being fed gradually changes from low to high density as the generator operates. In stratified tanks, excess chlorite will be fed to the generator as the bottom of the tank will have denser material, and this material will have more chlorite than required. Similarly, the bulk tank will later yield chlorite that is too dilute. Although infrequent, such stratification is not readily apparent and may likely remain unnoticed by an operator unless the generator performance is evaluated frequently. Operators should be aware of the possibility of stratification and crystallization during delivery conditions.

13-7 Disinfection with Ozone

Ozone is the strongest of the chemical disinfectants and its use is becoming increasingly common. Ozone (O_3) is an allotrope of oxygen with three oxygen atoms. The word *ozone* comes from the Greek word *ozein,* which means "to smell." In air, ozone has a pungent odor that is noticeable to most persons at levels above 0.1 ppm_v. Ozone is generated at the treatment plant site as a gas and is then injected into water.

Once dissolved in water, ozone begins a process of decay that results in the formation of the hydroxyl radical (HO•). Ozone reacts in two ways with contaminants and microbes: (1) by direct oxidation and (2) through the action of hydroxyl radicals generated during its decomposition. The consensus is that action of ozone as a disinfectant is primarily dependent on its direct reactions; hence it is the residual of the ozone itself that is important.

Ozone Demand and Ozone Decay

The ozone demand is the ozone dose that must be added before any ozone residual is measured in the water. It corresponds to the amount of ozone consumed during rapid reactions with readily degradable compounds. Ozone decay is the rate at which the residual ozone concentration decreases over time when the ozone dose is greater than the ozone demand. The overall rate of ozone decay in water is generally consistent with first-order kinetics. Like chlorine, it can be modeled successfully using a parallel first-order decay model, as shown in Eq. 13-9. Although simple reactions serve as good phenomenological models for ozone decay, it is unlikely that they correctly characterize the actual mechanism of decay. From work done in this area (Grasso and Weber, 1989; Gurol and Singer, 1982; Hermanowicz et al., 1999; Staehelin and Hoigné, 1982, 1985; Tomiyasu et al., 1985), it appears more likely that ozone decay consists of a large number of *n*th-order reactions operating in parallel that, in sum, appear to be simple first order.

Photographs of ozone generators and other ozone system components

http://www.stantec.com/wt3r/13-2.html

An introduction to ozone decay based on the models developed by Staehelin and Hoigné (1982, 1985) is provided on Fig. 13-18. The cyclic nature of the ozone decay process in pure water is illustrated on Fig. 13-18a. The process must be initiated by a reaction between ozone and the hydroxide ion to form superoxide radicals (O_2^-) and peroxide ions (HO_2^-), a slow process. As a result, decay is accelerated at higher pH. Once completed, the decay reactions enter a cyclic process represented in the figure by a circle. The cyclic reactions are promoted by ozone itself. If the concentration of ozone is increased, the cycle is accelerated.

In natural waters, other "initiators" besides hydroxide ion can be present as shown on Fig. 13-18b. Prominent among them are the ferrous ion and hydrogen peroxide. In natural waters certain natural organic materials have also been shown to promote the cycle, accelerating decay. Finally, the continuation of the cycle depends on the action of the hydroxyl radical on the ozone residual. As a result, scavengers that react with the hydroxyl radical, removing it from the process, also slow the rate of decay. The carbonate and bicarbonate ions are important examples of such inhibitors. The data of Reckhow and co-workers (Reckhow et al., 1986), are shown on Fig. 13-18c to illustrate the action of fulvic acids as initiators and promoters and carbonate and bicarbonate ions as HO· traps or inhibitors. The factors that influence the stability of ozone residuals are summarized in Table 13-7.

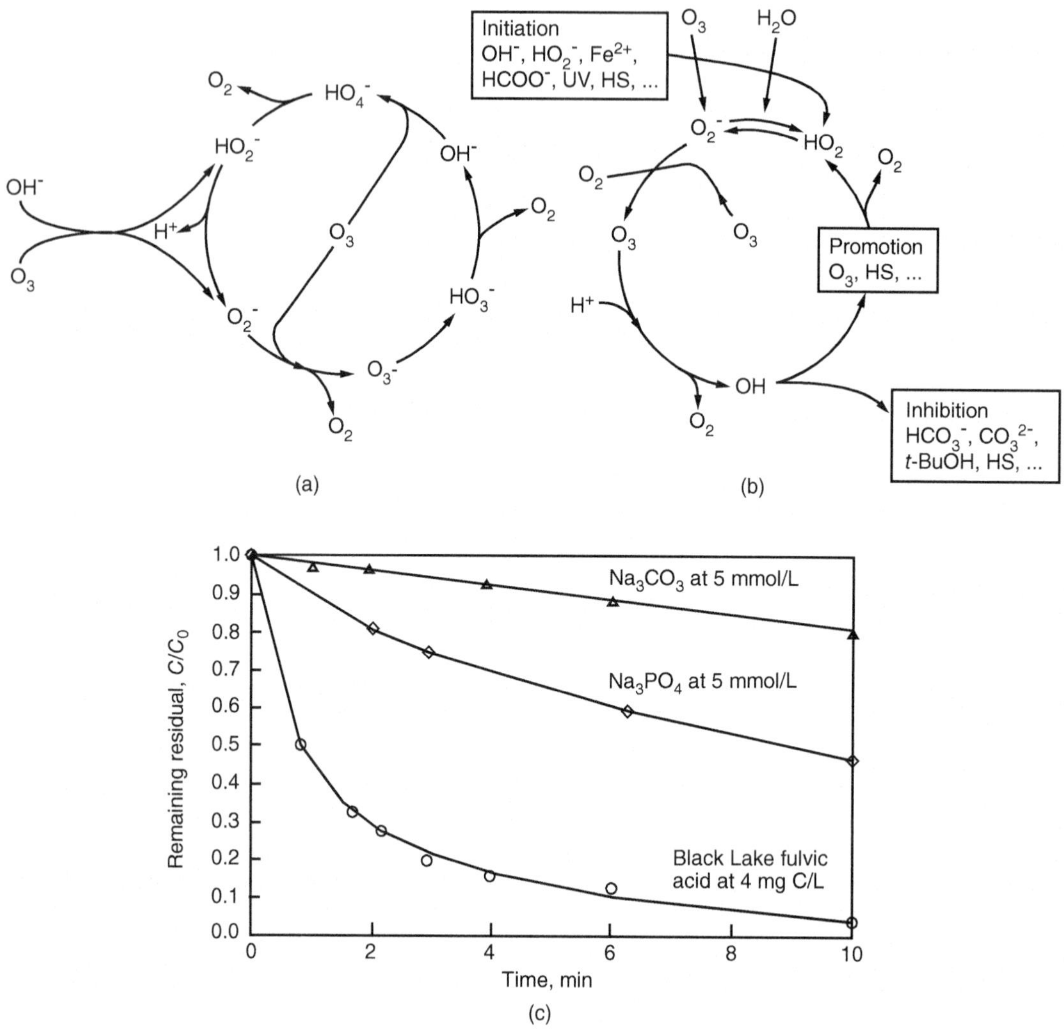

Figure 13-18
Understanding ozone reaction pathways and decay of residual ozone in natural waters: (a) the ozone decay wheel—reaction pathways in pure water (adapted from Hoigné and Bader, 1976); (b) influence of initiators, promotors, and inhibitors (adapted from Hoigné and Bader, 1976); (c) effect of fulvic acid and carbonate on ozone decay—all tests conducted at 20°C with GAC filtered, deionized tap water adjusted to pH 7, and $C_0 \sim 8$ mg/L. (Adapted from Reckhow et al., 1986).

Bench Testing for Determinin Ozone Disinfection Kinetics

The conceptual design of any ozonation system requires a means for estimating mass transfer of ozone into the water, an understanding of the kinetics of ozone decay, and an understanding of the disinfection kinetics. These components are often investigated using bench and pilot testing. Both batch and flow-through reactors have been used for bench testing, as described in the following sections.

Table 13-7
Factors that influence stability of aqueous ozone residuals

Increases Stability	Reduces Stability
Low pH	High pH
High alkalinity	Low alkalinity
Low TOC	High TOC
Low temperature	High temperature

ANALYSIS USING BATCH REACTORS

Batch testing is often conducted by bubbling ozone directly into a gas wash bottle containing the sample of interest. The ozone concentration is measured in the gas entering and exiting the bottle, and the difference constitutes the ozone added to or consumed by the sample. For a number of reasons, the preferred technique is to prepare the ozonated water first and then add that to the sample of interest. In this case, the batch reactor might be a 1- or 2-L jar or a 0.5- to 1-L Teflon bag containing the water of interest. The concentrated ozone solution is prepared in a separate container by continuously bubbling ozone gas into a small volume of distilled–deionized (DI) water. At ambient temperature, the maximum ozone solution concentration may be about 15 mg/L. To prepare a more concentrated solution, the DI water can be chilled in an ice bath. At temperatures close to 1°C, the concentration of ozone in the stock solution can be as high as 40 mg/L. Aliquots of the ozone stock solution are then drawn and injected into the test water sample. The volume of each aliquot is calculated to deliver a predetermined ozone dose to the test water sample. Water samples are then drawn from the test water at various times after the ozone is added and analyzed for ozone residual concentration. This test is repeated using various ozone doses.

The profile of ozone residual concentration versus time can then be plotted. Two example ozone decay profiles in two waters dosed with 1.0 mg/L ozone are shown on Fig. 13-19a. Both waters have relatively high ozone demand, particularly water B. The profile of ozone decay in water A is typical of most moderate TOC, well-oxygenated surface waters. The curve fit through the data points is that of a Pseudo-first-order decay equation with a decay coefficient of 0.3 min^{-1}. The decay of ozone in some waters does not always follow this uniform first-order decay model. Water B is an example of common ozone decay profiles where the ozone experiences an initial period of a high decay rate followed by a second period of slower decay. The curve fit through the data points for water B was accomplished using Eq. 13-9: Although this equation is based on the progress of two parallel first-order reactions, it should be viewed as a phenomenological model, not a mechanistic one. Based on experimental evidence, ozone, and particularly

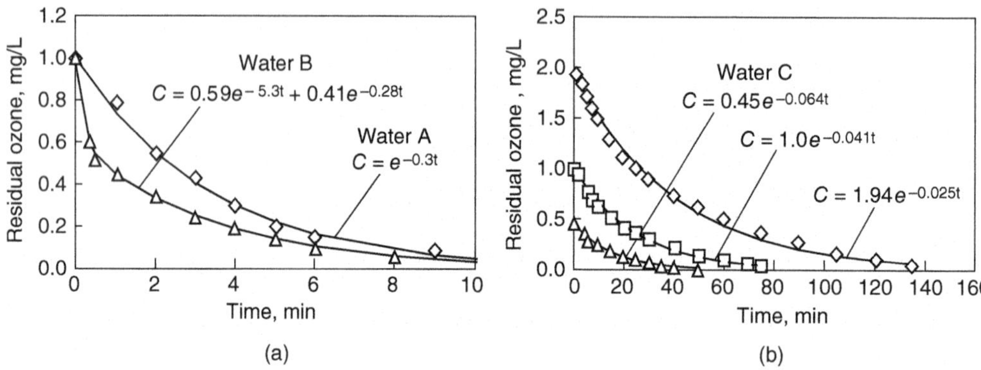

Figure 13-19
Typical batch ozone decay curves for three different waters: (a) waters A and B and (b) water C.

the hydroxyl radical that is produced when it decays, participate simultaneously in many reactions of different orders at the same time.

Another result of this complexity is that with ozone, as with chlorine, the rate of decay observed in a batch test is also influenced by the ozone residual at the beginning of the decay period, as illustrated on Fig. 13-19b using the data from a pure mountain water supply. In general, these curves exhibit a low rate of decay; nevertheless they also show a rate of decay that decreases as the residual at the beginning of the decay process (C_0) increases. As with chlorine, the change in decay rate is approximately inversely proportional to C_0. As a result of these complexities, if only batch testing is conducted to determine the basis of design, a wide variety of test conditions must be evaluated to get an adequate database for design. Even with such data, a number of assumptions and approximations must be made during the process of design.

ANALYSIS USING FLOW-THROUGH REACTORS

Continuous-flow reactors (see Fig. 13-20) are better than batch reactors for gathering information for design of ozonation facilities, especially for an over–under ozone contactor with ozone addition via diffusers. In fullscale designs of this type, the ozone is generally added in the first few compartments of the design, and then the residual is allowed to decay as the water travels throughout the rest of the reactor. This approach to design and operation can be simulated by operating the small-scale continuous-flow unit so that it has the same detention time as the ozone addition compartments will have in the full-scale design. Once the reactor has reached steady-state operation, both the flow of water and the ozone dosing can be stopped and the decay of ozone residual can be observed as a function of time. The continuous operation simulates the ozone addition

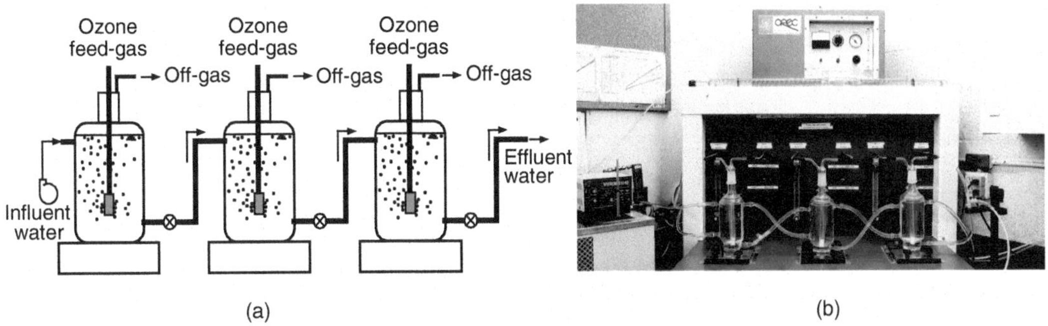

Figure 13-20
Bench-scale continuous-flow ozonation test system: (a) schematic and (b) photograph.

compartments and the decay curve can be used to estimate the residual in downstream compartments.

A continuous-flow setup requires measuring the ozone gas flow rate, water flow rate, ozone concentration in the feed gas, and ozone concentration in the off-gas. The ozone dose to the reactor is then calculated with a mass balance as

$$\text{Ozone dose, mg/L} = \frac{Q_g}{Q_l} \times \left(C_{g,\text{in}} - C_{g,\text{out}}\right) \tag{13-39}$$

where Q_g = gas flow rate, L/min
Q_1 = water flow rate, L/min
$C_{g,\text{in}}$ = concentration of ozone in feed gas, mg/L
$C_{g,\text{out}}$ = concentration of ozone in off-gas, mg/L

For each ozone dose, the operating conditions are kept constant until steady-state conditions are reached. This stabilization period can be between three and five times the hydraulic residence time of the reactor. It is essential that the continuous reactors be operated with approximately the same detention time as the ozone addition compartments in the full-scale design. An RTD curve similar to the full-scale reactor is also highly desirable. Unfortunately, tall, narrow pilot columns with long aspect ratios are often used because they achieve more efficient ozone transfer. The use of tall columns is not a particularly good choice because they much more closely approach plug flow than full-scale designs. This test must also be conducted at various doses because it is important to understand the relationship between the ozone dose and the ozone residual in the water exiting the ozone addition section of the reactor. The ozone decay rate downstream of these compartments will vary with this residual as well.

Example 13-8 Analysis of bench-scale ozone data

A municipality wishes to build a treatment plant incorporating an ozonation reactor and using water from a particular lake as a raw-water source. The lake water was studied using a bench-scale continuous-flow test unit (see Fig. 13-20), which included a three-compartment ozonation system, providing a total of 3.8 min of contact time (all three compartments). The system was operated at four different ozone doses. Samples were collected using two methods: (a) continuous-flow tests and (b) batch decay tests. In the continuous-flow test, the effluent from the third compartment was sampled for ozone residual after 15 min operation at each dose. For the batch test the system was shut down and the residual in the final compartment was sampled with time. The summary results of the testing program are given below. Using these data, estimate the ozone demand and the decay rate constant at each initial ozone residual.

For the sizing of the full-scale ozonation system, estimate the *Ct* product that can be achieved if the system is designed for an ozone dose of 3 mg/L at a temperature of 27°C. Assume the following conditions apply: (1) all ozone is added in the first compartment, which has a residence time of 3.8 min, (2) no *Ct* credit is allowed for the first compartment, and (3) the total residence time of the remaining compartments is 15 min.

Results from Continuous-Flow Tests		Results from Batch Decay Tests			
Ozone Dose, mg/L	O_3 Residual, mg/L	Time, min	O_3 Residual, mg/L	Time, min	O_3 Residual, mg/L
		0.0	1.23	0.0	0.60
2.10	0	1.0	0.98	1.0	0.42
2.72	0.36	2.0	0.85	2.0	0.29
3.00	0.60	3.0	0.71	3.0	0.23
3.80	1.05	4.0	0.59	4.0	0.17
4.01	1.23	5.0	0.53		
		7.0	0.42		
		9.0	0.31		
		11.0	0.23		
		13	0.16		
		15	0.14		

Solution

1. Analysis of continuous-flow data: The continuous-flow data for $\tau = 3.8$ min and $T = 27°C$ are plotted below. The best-fit line can be described using the equation

$$C_{Residual} = a\,(C_{dose} - C_{demand})$$

Using the form of the equation shown above, the best-fit line parameters from the plot are $a = 0.64$ and $C_{\text{demand}} = 2.1$ mg/L. The above equation can be used to estimate the dose required to achieve a specified residual exiting the ozone dosing compartment in the reactor.

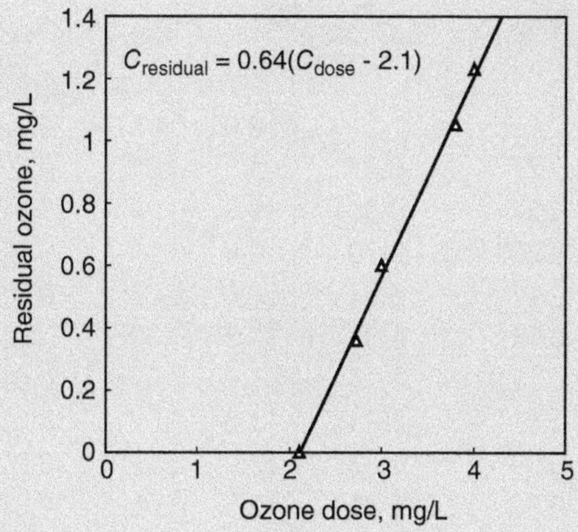

2. Analysis of the batch decay data: The batch decay data are plotted below. The best-fit parameters are obtained using an exponential curve fit. The corresponding equations are

$$\text{For } C_0 = 1.23 \text{ mg/L(dose} = 4 \text{ mg/L)}: \quad C_{\text{residual}} = 1.13e^{-1.14t}$$

$$\text{For } C_0 = 0.60 \text{ mg/L(dose} = 3 \text{ mg/L)}: \quad C_{\text{residual}} = 0.58e^{-0.31t}$$

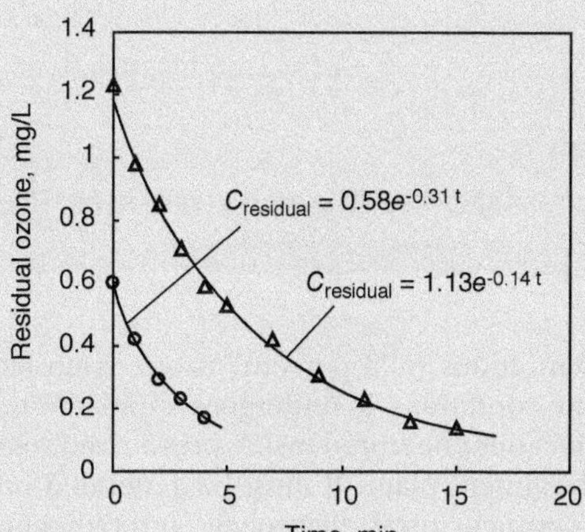

3. Determine the maximum *Ct* credit for the full-scale system assuming an ozone dose of 3 mg/L. Use the results from the batch decay data for the ozone dose of 3 mg/L.
   a. The maximum *Ct* credit can be estimated by numerical integration of the equation developed in step 2.

t, min	C, mg/L	C Δt, mg · min/L	Σ C Δt, mg · min/L	t, min	C, mg/L	C Δt, mg · min/L	Σ C Δt, mg · min/L
0.0	0.58	—	—	8.0	0.05	0.03	1.72
0.5	0.50	0.27	0.27	8.5	0.04	0.02	1.74
1.0	0.43	0.23	0.50	9.0	0.04	0.02	1.76
1.5	0.36	0.20	0.70	9.5	0.03	0.02	1.78
2.0	0.31	0.17	0.87	10.0	0.03	0.01	1.79
2.5	0.27	0.14	1.01	10.5	0.02	0.01	1.80
3.0	0.23	0.12	1.14	11.0	0.02	0.01	1.81
3.5	0.20	0.11	1.24	11.5	0.02	0.01	1.82
4.0	0.17	0.09	1.33	12.0	0.01	0.01	1.83
4.5	0.14	0.08	1.41	12.5	0.01	0.01	1.84
5.0	0.12	0.07	1.48	13.0	0.01	0.01	1.84
5.5	0.11	0.06	1.53	13.5	0.01	0.00	1.85
6.0	0.09	0.05	1.58	14.0	0.01	0.00	1.85
6.5	0.08	0.04	1.62	14.5	0.01	0.00	1.85
7.0	0.07	0.04	1.66	15.0	0.01	0.00	1.86
7.5	0.06	0.03	1.69				

   b. The maximum *Ct* credit is

$$\sum Cdt = 1.86 \text{ mg} \cdot \text{min/L}$$

Comment

Dispersion and short circuiting are not considered in the above calculations.

Generation of Ozone

At high concentrations (>23 percent) ozone is unstable (explosive) and under ambient conditions it undergoes rapid decay. Therefore, unlike chlorine gas, it cannot be stored inside pressurized vessels and transported to the water treatment plant. It must be generated onsite. Ozone can be generated by photochemical, electrolytic, and radiochemical methods, but the corona discharge method is the most commonly used in water treatment. In this method, oxygen is passed through an electric field that is generated by applying a high-voltage potential across two electrodes separated by a dielectric material (see Fig. 13-21). The dielectric material

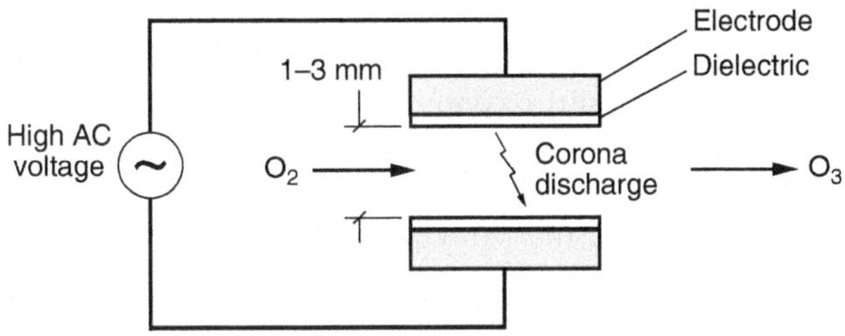

Figure 13-21
Ozone generation by corona discharge.

Table 13-8
Influence of increasing four key design factors on generator performance

Design Factor	Effect on Ozone Production
Frequency of applied current	Increases ozone production
Voltage of applied current	Increases ozone production
Gap between generator electrodes	Decreases ozone production
Dielectric constant of dielectric separating electrodes	Decreases ozone production

prevents arcing and spreads the electric field across the entire surface of the electrode. As the oxygen molecules pass through the electric field, they are broken down to highly reactive oxygen singlets (O·), which then react with other oxygen molecules to form ozone. The thickness of the gap through which the oxygen-rich gas stream passes is 1 to 3 mm wide. Because most of the energy used in ozone generation is lost as heat, cooling of the ozone generator is necessary to avoid overheating and subsequent decomposition of the ozone generated. Cooling is normally accomplished by passing a continuous stream of cooling water next to the ground electrode. Some of the key design factors that influence ozone generator performance are summarized in Table 13-8.

The equation below, while not intended to be quantitative, provides a general idea of the significance of a number of the variables of importance to the design of a corona discharge ozone generator:

$$Q_{O_3} \propto \left(f \frac{V^2 A}{d\varepsilon} \right) Q_{O_2} \qquad (13\text{-}40)$$

where Q_{O_3} = ozone generation, kg/s
f = frequency of applied emf
V = emf across electrodes, V
A = surface area of electrodes, m^2
d = distance between electrodes, m
ε = dielectric constant
Q_{O_2} = oxygen flow rate, kg/s

Oxygen Source

Ozone can be generated directly from the oxygen in air or from pure oxygen. Pure oxygen is generated onsite from ambient air at larger plants or provided through the use of liquid oxygen (commonly referred to as LOX), which is generated offsite and transported to the plant. The most suitable method for providing oxygen for ozone generation in a particular plant depends on economic factors, the principal ones being the scale of the facility and the availability of industrial sources of liquid oxygen.

USE OF PREPARED, AMBIENT AIR

The most accessible oxygen source is ambient air, which contains about 21 percent oxygen by volume. Ambient air used to be the most common source of oxygen for ozone systems, but it has largely been replaced by liquid oxygen except for small, remote systems. Ambient air contains significant levels of particulates and water vapor, which must be removed. Water vapor is detrimental to corona discharge ozone generators for two reasons: (1) the presence of water vapor significantly reduces the ozone generation efficiency and (2) trace levels of water can react with the nitrogen present in the air and the generated ozone to form nitric acid, which attacks the ozone generator itself:

$$O_3 + N_2 + O_2 + H_2O \xrightarrow{h\nu} 2HNO_3 \qquad (13\text{-}41)$$

The moisture content of a gas is often defined by its dew point, which is the temperature to which the gas needs to be cooled to reach 100 percent saturation. The lower the dew point of a gas, the lower is its moisture content. For example, air with a dew point of 30°C contains about 28,000 ppm_v of water, whereas air with a dew point of 5°C contains about 5000 ppm_v of water. The dew point specified for many ozone generators is as low as −80°C, which corresponds to a moisture content of less than 1.5 ppm_v. Drying ambient air to this level is usually accomplished by a three-step process of compression, refrigeration, and desiccant drying. Compression and refrigeration help because the water vapor capacity of air decreases with increased pressure and decreased temperature, reducing the load on the desiccant system. Desiccant drying, however, is required to achieve the specifications for ozone generation. A schematic of all the components of such a system is shown on Fig. 13-22.

LIQUID OXYGEN DELIVERY

Liquid oxygen is widely available as a commercial, industrial-grade chemical and is the most common source of oxygen for ozone systems. Water treatment plants can purchase commercially available LOX, store it at the plant, and use it as the oxygen source for ozone generation. Liquid oxygen is delivered in trucks and stored in insulated pressurized tanks. Liquid oxygen is then drawn from the tank and piped to a vaporizer that warms and converts the oxygen to the gaseous form. Commercially available

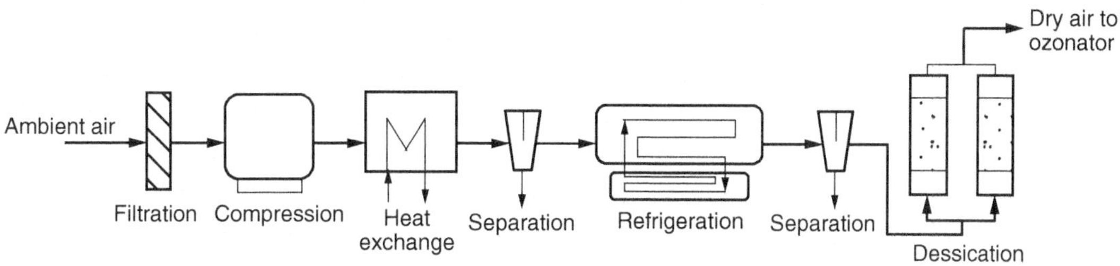

Figure 13-22
Preparation system for ozone generation from ambient air.

Figure 13-23
Liquid oxygen (LOX) storage container tanks at a large water treatment plant.

LOX is inherently low in contaminants and water vapor as a result of the manufacturing process. Therefore, minimal additional processing of the oxygen stream is required before it is introduced to the ozone generator. A LOX storage system at a large water treatment plant is shown on Fig. 13-23.

The use of LOX for ozone generation has several advantages over the use of ambient air, including (1) simpler operation and maintenance because fewer processes are required, (2) a smaller facility with lower capital cost, and (3) a smaller number of ozone generators. The disadvantages of LOX include (1) increased truck traffic caused by the need for regular LOX deliveries and (2) susceptibility to market pricing. Safety concerns associated with the storage of a large volume of concentrated oxygen must also be addressed. However, the advantages are significant and LOX has largely displaced the use of ambient air as the most common source of oxygen for ozone systems.

ONSITE OXYGEN GENERATION

Two types of onsite oxygen separation and concentration processes are used in water treatment plants that require oxygen: (1) pressure swing or vacuum swing adsorption (PSA or VSA) processes and (2) cryogenic oxygen generation processes. Generally, the economics of these processes improve as oxygen requirements increase. None are more economical than LOX feed systems in small applications. VSA systems are viable for systems requiring as much as 100 tonnes/d (110 tons/d) (Lotepro, 2002), and PSA systems can be used for smaller systems needing onsite oxygen generation. Cryogenic oxygen generation was installed at a few very large ozone systems in the past but are generally not economically competitive today for drinking water applications.

The PSA and VSA processes take advantage of the effect of gas pressure on the differences in the adsorption characteristics of the various constituents of ambient air on specialty adsorption resins. For the generation of oxygen, the affinity of the resin for nitrogen, water, and carbon dioxide is higher than that for oxygen and increases with increased pressure. Therefore, the PSA or VSA system cycles between "high" and "low" pressures. During the high-pressure period, water moisture, carbon dioxide, nitrogen, and any hydrocarbons present preferentially adsorb onto the resin while oxygen, now constituting about 90 to 95 percent of the remaining gas, passes through. Once the resin is saturated with the constituents removed, the system cycles to the low pressure, resulting in the desorption of the adsorbed material, which is then exhausted to the atmosphere before the cycle is repeated. For a PSA system, the high-pressure setting ranges from 200 to 400 kPa (30 to 60 psig), while the low setting is atmospheric pressure. In a VSA system, the high-pressure setting is at 20 to 70 kPa (3 to 10 psig), while the low setting is achieved using a vacuum pump. VSA systems are favored over PSA for large systems because they utilize less energy. However, a VSA system requires additional equipment compared to a PSA system in the form of a vacuum pump as well as a downstream compressor to boost the pressure of the oxygen stream to the level required by the ozone generator. The need for an extra pump translates into higher capital cost and higher maintenance cost. The schematic layout of a typical PSA system is illustrated on Fig. 13-24.

Regardless of whether air or pure oxygen is used for ozone generation, the efficiency of ozone generators is relatively low. When ambient air is used as the feed gas, ozone content in the generator outlet is typical between 1 and 4 percent by weight. With pure oxygen, typical generators produce about 6 to 16 percent ozone by weight.

Ozone Injection Systems

The addition of ozonation in a water treatment plant requires two components in the process treatment train: (1) a device for injecting the ozone

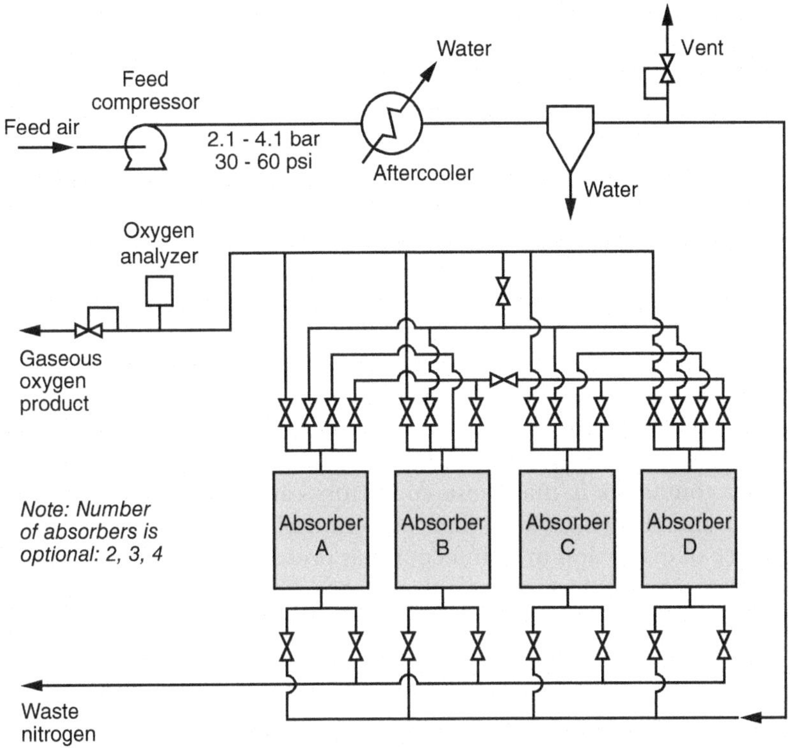

Figure 13-24
Schematic of pressure swing adsorption system for producing pure oxygen. (Adapted from Lotepro, Inc.)

into the water and (2) a contact chamber in which the disinfection reaction takes place. For several decades, the most common approach to ozonation has been to combine these components by introducing the ozone into the water in large, deep basins using porous diffusers. More recently, the injection and contact systems are designed separately. For injection systems, side-stream injection using venturi injectors with or without side-stream degassing has become more common than fine bubble diffusers. Ozone contactors can be pipeline contactors, serpentine basins, or over–under baffled contactors and are described in Sec. 13-8. Details of the design of side-stream ozone injection systems can be found in Rakness (2005) and are described briefly below.

In side-stream injection, a portion of the process flow is withdrawn from the main process line and pumped through a venturi injector. Low pressure in the throat of the injector draws ozone gas in from the ozone generator. After dissolution of the ozone gas, the side stream is injected back into the process stream through nozzles that provide good blending of the ozonated side stream into the main flow. In some systems, the side stream passes through a degassing tower before being injected into the process stream. After the ozone is injected, the process water flows to a pipeline or serpentine basin contactor. Design of contactors is presented in the next section.

The purpose of the degasser in the side stream is to allow undissolved and supersaturated gases to separate from the water prior to injection to the main process flow and to minimize bubbles in the ozone contactor. Since the carrier gas for the ozone is typically pure oxygen, the process flow can become supersaturated with oxygen, which can lead to problems with downstream processes such as air binding in rapid granular filters. If the side stream does not contain a degas vessel, a mechanism for stripping supersaturated oxygen, such as by diffusing air after the ozone contactor, should be provided.

An advantage of side-stream injection coupled with pipeline or serpentine basin contactors is that these contactors can be designed with less dispersion and short-circuiting than over–under baffled contactors. The importance of dispersion in disinfection was presented in Sec. 13-4. In the case of ozonation, dispersion not only reduces the effectiveness of the disinfection reaction but also increases the formation of bromate.

Off-Gas Treatment

Because ozone is a strong oxidant, extended exposure to ozone-containing air is harmful. Even with the most efficient ozone contactor designs, off-gas ozone concentrations substantially exceed acceptable levels and, as a result, off-gas treatment is required. In the United States, the Occupational Safety and Health Administration (OSHA) sets an 8-h workday ozone exposure limit of 0.1 ppm_v by volume at standard temperature and pressure (STP), which is equivalent to 0.0002 mg/L in air (*Federal Register*, 1993). In general, the concentration in the ozone gas entering the contactor can range anywhere from 5000 to 160,000 ppm_v; so ozone contactors would have to achieve removals in excess of 99.998 percent to meet these standards directly. The efficiencies actually achieved in these reactors range from 90 to 99 percent, rarely higher. Therefore, the off-gas cannot be vented to the atmosphere before the residual ozone is destroyed.

Ozone in the off-gas stream can be destroyed thermally with or without the use of solid catalysts. When a catalyst is not used, ozone destruction is accomplished by heating the off-gas to a temperature between 300 and 350°C. At this temperature, the required contact time through the destruction unit is less than 5 s. Newer destruction units combine the use of specialty metal catalysts with moderate heating to achieve ozone destruction. Depending on the type of catalyst used, the off-gas temperature need only be raised to somewhere between 30 and 70°C (AWWARF, 1991).

Example 13-9 Estimating ozone concentration in contactor off-gas

An ozonation system produces ozone from air at a concentration of approximately 12 percent by volume. Assume the ozone contactor achieves a transfer efficiency of 99.5 percent. Estimate the concentration of ozone in the contactor off-gas from the contact chamber and compare it to OSHA standards.

Solution

1. Determine the downstream ozone concentration.
   a. Convert 12 percent by volume to ppm as follows:

$$12\% = \frac{12}{100}\left(\frac{10{,}000}{10{,}000}\right) = \frac{120{,}000}{1{,}000{,}000} = 120{,}000\ \text{ppm}_\text{v}$$

   b. Downstream of the contactor, the concentration is

$$C_{\text{off-gas}} = 120{,}000\,\text{ppm}_\text{v} \times (1 - 0.995) = 600\ \text{ppm}_\text{v}$$

2. How does the off-gas concentration compare to OSHA standards? To reduce the ozone concentration from 600 to 0.1 ppm, greater than 99.9 percent additional removal is required.

13-8 Design of Disinfection Contactors with Low Dispersion

Throughout much of the twentieth century, the design of specialized disinfectant contactors was not a particular concern. Chlorine was added early in the treatment process and the chlorine residual carried throughout the plant. Following the THM rule in 1980, many utilities moved the point of chlorine addition to the end of the treatment process. Later, when the first Surface Water Treatment Rule came about, many utilities struggled to find a way to get more credit for contact time in their existing facilities, often by baffling them to increase t_{10}. Because dispersion is so important in disinfection effectiveness (see Sec. 13-4), disinfectant contactors are now typically designed as a separate unit process. Engineered disinfectant contactors are typically of three types: (1) pipelines, (2) serpentine basins, and (3) over–under baffled contactors. Chlorine, combined chlorine, and chlorine dioxide contactors are typically pipelines or serpentine basins. Ozone contactors can be any of the three common types, and additionally deep U-tube contactors have also been used. Additional detail on dispersion and the design of reactors is presented in Chap. 6.

Design of Pipeline Contactors

A long channel or pipeline with plug flow characteristics can be an ideal disinfectant contactor. Occasionally, a long pipeline leaving the plant has sufficient contact time to make it an attractive alternative for chlorine or chloramines disinfection. Axial (longitudinal) dispersion in pipeline flow is the most straightforward case that will be considered. Taylor (1954) demonstrated that the longitudinal dispersion coefficient (D_L) can be described as

$$D_L = 5.05 D v^* \tag{13-42}$$

where D_L = longitudinal dispersion coefficient, m^2/s
D = diameter of conduit, m
v^* = shear velocity, m/s

In the above formula the shear velocity or friction velocity (v^*) may be defined in terms of the velocity of flow and the friction factor:

$$v^* = \sqrt{\frac{fv^2}{8}} \tag{13-43}$$

where f = Darcy–Weisbach friction factor, unitless
v = velocity of flow in pipe, m/s

The dispersion number is defined in terms of the longitudinal dispersion coefficient, the velocity of flow, and a characteristic length, in this case, the length of the pipe:

$$d = \frac{D_L}{vL} \tag{13-44}$$

where d = dispersion number, dimensionless

Combining Eqs. 13-42 through 13-44 results in a formula that can be used to describe the dispersion of flow in a pipe:

$$d = 5.05\left(\frac{D}{L}\right)\sqrt{\frac{f}{8}} \tag{13-45}$$

Available data from laboratory experiments confirm Taylor's theory within a factor of 2. Generally, more dispersion is found in field-scale measurements than is predicted from the theory. For this reason, Sjenitzer (1958) gathered a great number of measurements, both in the laboratory and in the field, and correlated them to produce the empirical expression

$$d = 89{,}500 f^{3.6}\left(\frac{D}{L}\right)^{0.859} \tag{13-46}$$

Using Sjenitzer's data, Trussell and Chao (1977) demonstrated that Eq. 13-46 provides a significantly better fit of the data than Eq. 13-45. Even Sjenitzer's equation, however, is only accurate for a long pipeline without bends, restrictions, or other disturbances to flow. Generally, the flow in a pipeline with 30 min of contact time, a flow rate greater than 3785 m^3/d (1 mgd), and a velocity greater than 0.6 m/s (2 ft/s) will be nearly ideal plug flow in behavior.

Example 13-10 Dispersion in pipelines

A treatment plant with a capacity of 25,000 m^3/d (6.6 mgd) is planning to use a 1-km treated-water pipeline as a chlorine contactor. Determine the diameter of the pipeline needed for a hydraulic residence time (τ) of 30 min and the resulting dispersion number of the flow in the pipeline. Using Fig. 13-7, determine whether dispersion will have a significant impact on achieving 4 log of inactivation with this pipeline. The Darcy–Weisbach friction factor is 0.018.

Solution

1. Determine the diameter D of the pipeline.

$$\tau = \frac{V}{Q} = \frac{AL}{Q} = \frac{\left[\pi / 4D^2\right]L}{Q}$$

Rearranging and solving for D yields

$$D = \sqrt{\frac{4Q\tau}{\pi L}} = \sqrt{\frac{(4)\left(25{,}000\ \text{m}^3/\text{d}\right)(30\ \text{min})}{(\pi)(1000\ \text{m})(1440\ \text{min/d})}} = 0.81\ \text{m}$$

2. Estimate the dispersion factor using Eq. 13-46:

$$d = 89{,}500(0.018)^{3.6}\left(\frac{0.81\ \text{m}}{1000\ \text{m}}\right)^{0.859} = 0.000104$$

3. Assess whether dispersion will have a significant impact on achieving 4 log of inactivation in the pipeline.

Comment

To achieve 4 log of inactivation with less than 5 percent deviation from the inactivation goal, the dispersion number must be less than 0.006 (see Fig. 13-7). Since the calculated dispersion number is less than that, the impact of dispersion on this contactor will be minimal. It should also be noted that pipe must be purchased in standard sizes, and the actual inside diameter of the pipeline would likely be larger than the calculated value, leading to an increase in τ, which would provide additional inactivation.

Design of Serpentine Basin Contactors

A pipeline is convenient if it is already necessary for some other purpose, but long, baffled, serpentine basins are generally more cost-effective means of achieving low dispersion. Serpentine basins are capable of achieving dispersion numbers less than 0.01 (Markse and Boyle 1973; Sepp, 1981; Trussell and Chao, 1977) and t_{10}/τ of 0.8 (Crozes et al., 1999). An optimal basin would be long and narrow, similar to the contactor discussed in the previous section. In the following discussion, the design of serpentine basins to achieve a specified level of dispersion is addressed first and then, because of U.S. regulatory requirements, designing these same facilities to meet a specified t_{10} will also be discussed. Computational fluid dynamics can be used to optimize the design of any large disinfection contactor (DuCoste, 2001; Hannoun et al., 1999).

DESIGNING FOR A SPECIFIED DISPERSION NUMBER

To develop a better understanding of design criteria, it is useful to start with a more general form of Eq. 13-42 (the Taylor equation):

$$D_L = CR_h v^* \tag{13-47}$$

where D_L = longitudinal dispersion coefficient, m^2/s
C = coefficient, unitless
R_h = hydraulic radius of channel, m
v^* = shear velocity, m/s

The coefficient C is a function of channel geometry and the Reynolds number. Elder (1959) applied Taylor's concept of dispersion to a logarithmic velocity profile and suggested that the coefficient C should have a value of approximately 5.9 for the circumstances in most chlorine contact chambers. For uniform flow in an open channel, the shear velocity can be defined as follows:

$$v^* = \frac{3.82\ nv}{R_h^{1/6}} \tag{13-48}$$

where v = velocity of flow in channel, m/s
n = Manning coefficient, unitless

Combining Eqs. 13-48, 13-47, and 13-44, the following approximate formula for dispersion coefficient in a long open channel is obtained:

$$d = \frac{22.7\ nR_h^{5/6}}{L} \tag{13-49}$$

Equation 13-49 may be rewritten to describe dispersion using the channel volume and height and length aspect ratios (Trussell and Chao, 1977):

$$d = 22.7\frac{n}{\beta_L}\left(\frac{\beta_H}{2\beta_H + 1}\right)^{5/6}\left(\frac{\beta_H\beta_L}{V_{ch}}\right)^{1/18} \tag{13-50}$$

where β_H = height aspect ratio or H/W (channel height/channel width)
β_L = length aspect ratio or L/W (channel length/channel width)
V_{ch} = channel volume, m^3

The dispersion values computed using Eq. 13-50 are not sensitive to the range of β_H values typical for concrete contact chambers (1 to 3). As a result, the following abbreviated form of Eq. 13-50 can be used satisfactorily (Trussell and Chao, 1977):

$$d = \frac{0.14}{\beta_L} \tag{13-51}$$

A plot of dispersion coefficients from field-scale tracer studies conducted on 17 different field-scale basins is illustrated on Fig. 13-25. Because the field tests were conducted in baffled, serpentine contactors, not long straight channels, none of the studies resulted in the performance predicted using Eq. 13-51. These basins include entrance effects, exit effects, 180° turns, and other nonidealities that would be expected to increase dispersion. Nevertheless, the results shown on Fig. 13-25 are encouraging for two reasons: (1) confirmation of the implication of Eq. 13-51 that dispersion is inversely proportional to the length aspect ratio and (2) the basins fall short of ideal performance, as expected. Recognizing this situation, a coefficient of ideality C_i was proposed (Trussell and Chao, 1977) such that

$$d = \frac{0.14C_i}{\beta_L} \tag{13-52}$$

where C_i = coefficient of ideality

Lines corresponding to C_i values between 3 and 15 are also displayed on Fig. 13-25 and all the data lie on or between them. Based on the data

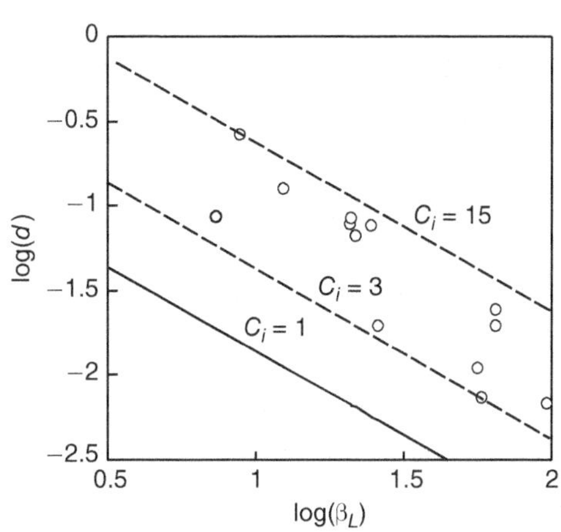

Figure 13-25
Impact of contactor aspect ratio on dispersion. (Adapted from Trussell and Pollock, 1983).

presented on Fig. 13-25, it appears that a good design should be able to equal or exceed the performance estimated by Eq. 13-52 with a C_i value of 3. A best-fit line corresponding to a C_i of 7.1 approximates the performance of a typical older reactor design.

DESIGNING FOR A SPECIFIED t_{10}/τ

Although the dispersion number is probably the most suitable means of assuring disinfection performance, a means of estimating t_{10}/τ must be used to be sure that the design will meet regulations (U.S. EPA, 1989). The impact of baffling rectangular contact tanks to improve hydraulic performance was evaluated by Crozes et al. (1999). A pilot contactor was baffled with nine different configurations having length aspect ratios ranging from 4.8 to 98. In addition, tracer tests were conducted on a full-scale, 34 ML/d (9-mgd) contactor before ($\beta_L = 6.1$) and after ($\beta_L = 52$) modifications. Finally, an empirical correlation between t_{10}/τ and β_L was developed and confirmed (Ducoste et al., 2001):

$$\left(\frac{t_{10}}{\tau}\right) = 0.198 \ln(\beta_L) - 0.002 \tag{13-53}$$

The data and correlation from the study are shown on Fig. 13-26. Note the results from full-scale tests lie close to model predictions.

Although the design of an effective disinfection contact basin requires attention to the length aspect ratio, other design details are also important. Any design detail that causes disturbances in flow is undesirable. Unnecessary gates, ports, or objects that constrict the flow lines are examples. In addition to minimizing the presence of these features, however, special attention should be given to three elements of design in every contactor:

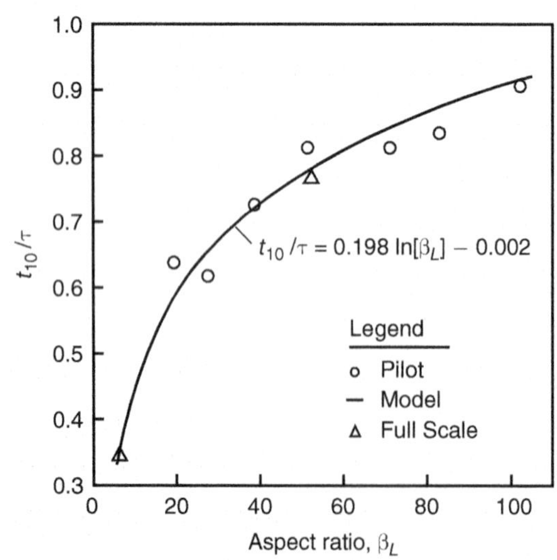

Figure 13-26
Impact of contactor aspect ratio on t_{10}. (Data from Crozes et al., 1999, and DuCoste et al., 2001.)

(1) inlet configuration, (2) outlet configuration, and (3) turns. Without proper attention, each of these is a likely cause of poor basin performance.

BASIN INLETS AND OUTLETS

Basin inlets are designed ordinarily as flow over a weir, through a pipe, or through a gate or gate valve into the basin. The momentum of the incoming water can cause significant dispersion in the first pass. When the entrance is a pipe, it is best for the water to exit through a tee so that the flow is not directed down the basin. With any of these inlet configurations (including a pipe with a tee) it is desirable to install a diffuser wall between the inlet of the basin and the first pass. Basin outlets are similar to inlets and have similar problems, although outlet effects are not quite as significant because outlets do not impart momentum to the basin flow. Often a diffuser wall is the best way to manage flow to outlets.

180° TURNS

To build a compact basin with the best possible length aspect ratio, rectangular basins are baffled in a serpentine fashion. However, the impact of baffling is not entirely benign. While increasing the tank's length–width ratio, the baffles also introduce flow separations at the 180° turns (Graber, 1972). Computational fluid dynamics (CFD) can be used to evaluate the flow in a chlorine contactor design and produce an estimate of the resulting tracer curve as illustrated on Fig. 13-27. A more complete discussion of CFD modeling may be found in Hannoun et al. (1999). Note that although the overall t_{10}/τ of the design shown on Fig. 13-27 is quite good, the CFD images illustrate the adverse impact of 180° turns on basin flow patterns. Flow separations can be observed at each turn, and these impact the character of the flow for some distance down each pass. Based on some estimates, as much as 40 percent of the volume in a baffled tank behaves as a dead zone (Louie and Fohrman, 1968). The increased dispersion decreases the effective contact time (early tracer appearance and a great deal of tailing in the tracer curve). Most of the nonideality in the tracer curve on Fig. 13-27 results from the 180° turns.

The primary way to minimize this dispersion is to keep the width of the flow path constant around a turn. A number of methods have been devised for controlling the problem, and some of them, illustrated on Fig. 13-28, are hammerheads and fillets (Louie and Fohrman, 1968; White, 1999), turning vanes (Crozes et al., 1999; Graber, 1972; Louie and Fohrman, 1968), and diffuser walls (Crozes et al., 1999; Hart, 1979). Turning vanes, hammerheads, and fillets are used to reduce or eliminate the flow separation. Diffuser walls, in contrast, redistribute the flow across the channel after the turn is complete. As a result, turning vanes, hammerheads, and fillets have the potential to actually *reduce* the head loss due to the turn as well as to reduce the flow nonideality introduced by the turn.

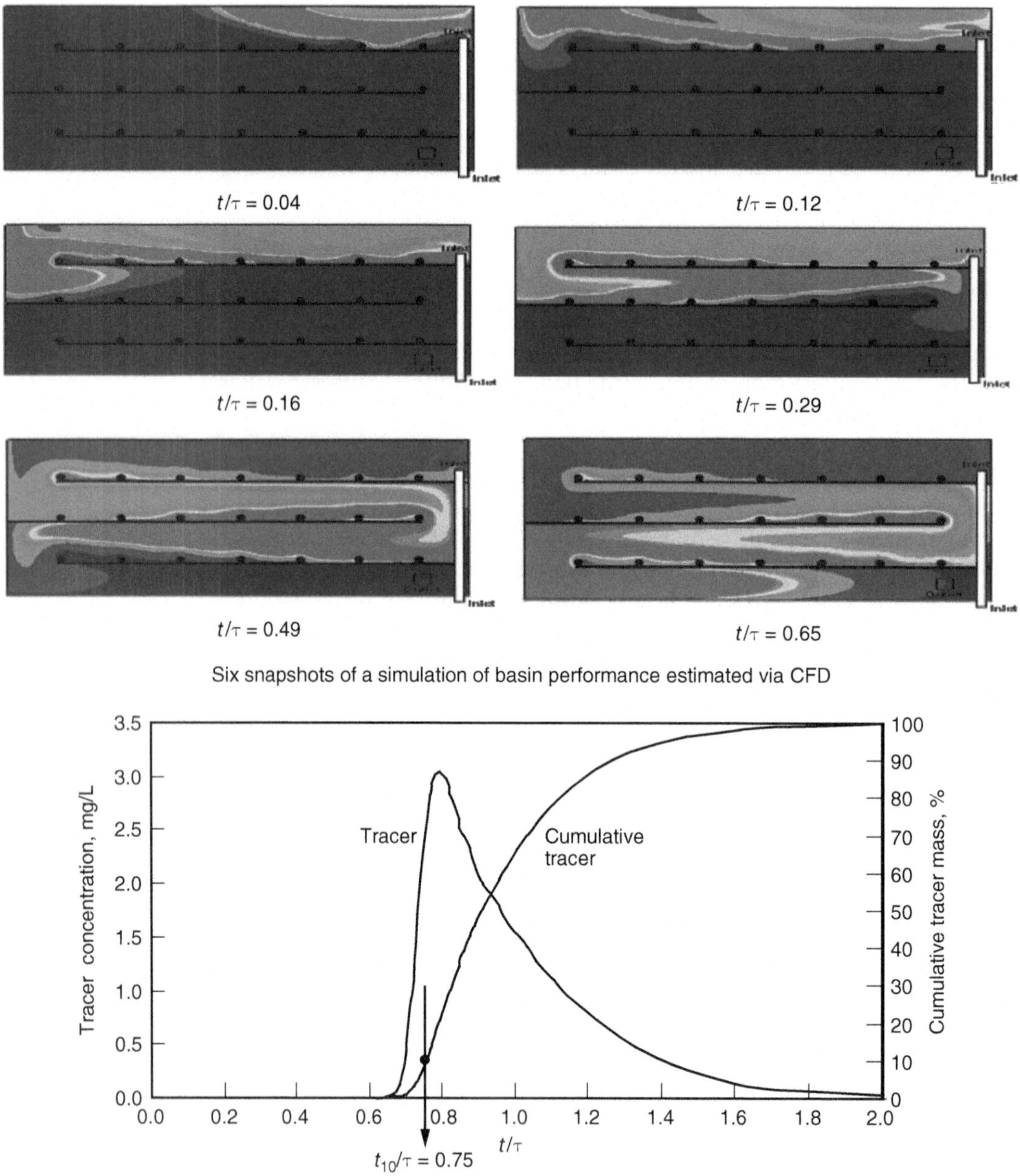

Figure 13-27
Using computational fluid dynamics (CFD) to evaluate RTD of disinfection contactor (CFD by Flow Science for an optimized design for the Weber Basin Water Conservancy District in Utah; $\tau = 110$ min, $t_{10} = 83$ min).

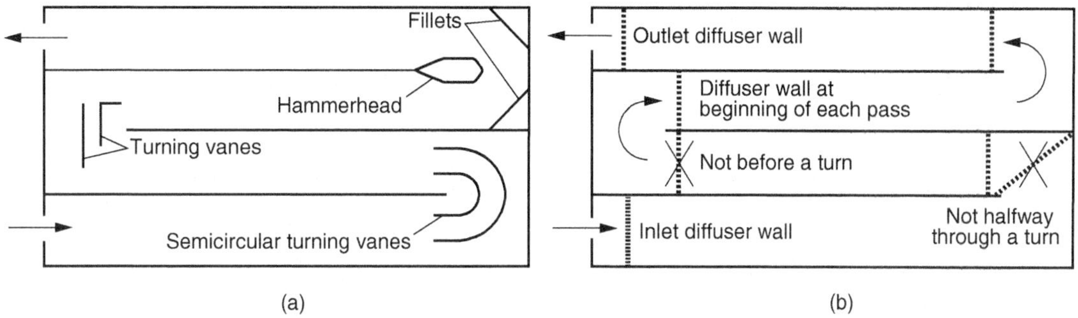

Figure 13-28
Controlling flow separation in serpentine basins using various devices: (a) fillets, hammerhead, and turning vanes (adapted from Louie and Fohrman, 1968) and semicircular turning vanes (adapted from Graber, 1972) and (b) diffusion walls (adapted from Trussell and Chao, 1977; Kawamura, 2000).

Diffuser walls always increase head loss because they depend on head loss to redistribute the flow.

Kawamura (2000) presented some useful criteria for designing diffuser walls between flocculation basins and sedimentation basins. These criteria are also useful for disinfection contact basins:

- Port openings should be uniformly distributed across the baffle wall.
- A maximum number of ports should be provided so that flow is evenly distributed.
- The size of the ports should be uniform in diameter.
- Ports should be 75 mm or larger to avoid clogging.
- Ports should be spaced with consideration to the structural integrity of the baffle. For wood baffles, this leads to 250- to 500-mm spacing.
- Ports should be designed to cause a head loss of 0.3 to 0.9 mm.

While diffuser walls have the advantage that some design criteria are available and they improve flow, they have the disadvantage that they increase the head loss. In fact, head loss and construction are the two major limitations on designing baffled, serpentine basins. Many baffle and channel designs become so narrow that construction is difficult. Moreover the head loss from the 180° turn can become significant. Nevertheless, baffled contactors with length aspect ratios as high as 100 and dispersion numbers below 0.01 are common.

Design of Over–Under Baffled Contactors

As noted in Sec. 13-7, over–under baffled contactors were the most common type of ozone contactor for many years but are less common now because of increased use of pipe contactors or serpentine basins for ozone contact systems. Pipeline and serpentine basins have better hydraulic characteristics that improve disinfection and minimize bromate formation.

Multichamber over–under baffled contactors often have several chambers where the water alternately flow up over a baffle and down under the next baffle (Rakness, 2005). Schematics of such a contactor are shown on Fig. 13-29. Ozone is typically added to the first one or two chambers via porous stone diffusers situated at the bottom of the chambers. Water enters the first chamber from the top and exists from the bottom. This countercurrent flow configuration (between the water and the air) helps increase the overall ozone transfer efficiency. The water depth in the contactor is typically between 4.6 and 6 m (15 and 20 ft) to achieve high transfer efficiency of the added ozone.

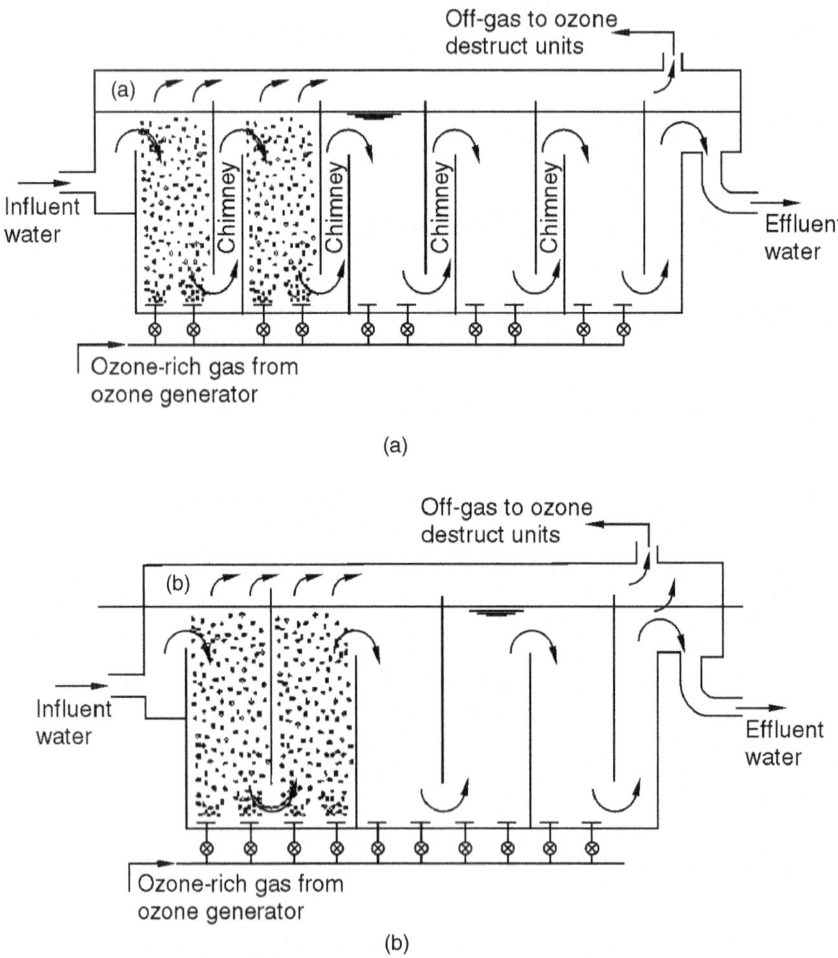

Figure 13-29
Schematics cross-sectional views of two alternate designs for five-chamber, over–under ozone contact chamber: (a) with chimneys and (b) without chimneys.

To achieve countercurrent flow in subsequent chambers, the contactor is also designed with segments that return the flow back to the top. A design is shown on Fig. 13-29a, where the water exiting the bottom of the first chamber rises to the surface through a narrow chamber, commonly called a *chimney*, before it enters the top of the second chamber. The chimney design achieves countercurrent flow in all chambers where ozone is added. A design with no chimneys is shown on Fig. 13-29b. In this design, the flow configuration alternates from countercurrent to co-current as the water moves from one chamber to the next. While lower transfer efficiency may take place in the co-current chambers, experience has shown that the impact is minimal. The passage of the water through the narrow chimneys of the alternate design causes a significant flow separation as the water enters and exits each down-flow contact chamber, resulting in high dispersion. On Fig. 13-30, schematic renderings of possible hydraulic flow patterns are shown in a multichamber contactor where the water is forced through a narrow pathway. Chimneys between chambers are indicated on Fig. 13-30a. The design shown on Fig. 13-30b no longer has chimneys but still exhibits significant flow separation at the turns.

The problem with the contactor design on Fig. 13-30b is that the openings through which the water flows between chambers are too narrow. The same principle that applies in the design of the serpentine basin contactors discussed previously applies here: the width of the flow path must be

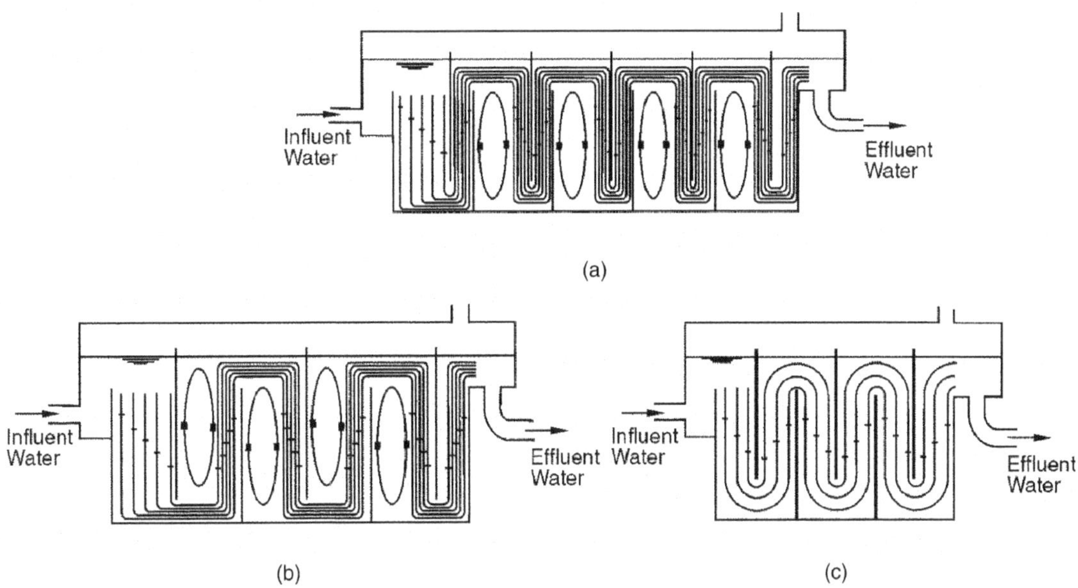

Figure 13-30
Conceptual impact of ozone contactor design flow hydrodynamics: (a) with chimneys, (b) without chimneys, and (c) with uniform flow path. (Adapted From Henry and Freeman, 1996.)

maintained. The flow path can be maintained by ensuring that the opening between two consecutive chambers is approximately the same width as the downstream chamber. The hydraulic flow pattern in a contactor designed with these considerations in mind is illustrated on Fig. 13-30c. It is noted that the hydraulic flow lines shown on Fig. 13-30 are only conceptual. A more accurate determination of the true hydraulic behavior can be determined using computational fluid dynamic (CFD) modeling of the contactor. Henry and Freeman (1996) conducted such modeling on various ozone contactor designs and determined that the contactor-baffling ratio (defined as the ratio of t_{10}/τ) is greatly impacted by the internal geometry of the contactor. The impact of the H/L ratio on the baffling ratio, where H is the water depth and L is the longitudinal width of the chamber, is shown on Fig. 13-31a (Henry and Freeman, 1996). Increasing the H/L ratio from 2 to 4 increases the t_{10}/τ ratio from 0.55 to 0.65. The impact of the G/L ratio, where G is the depth of the flow path under the baffle, on the baffling ratio is illustrated on Fig. 13-31b. Increasing the G/L ratio from 0.2 to 1.0 increases the t_{10}/τ ratio from 0.45 to 0.65. Based on this work, a maximum t_{10}/τ ratio can be achieved with an H/L ratio of 4 : 1 and a G/L ratio of 1:1.

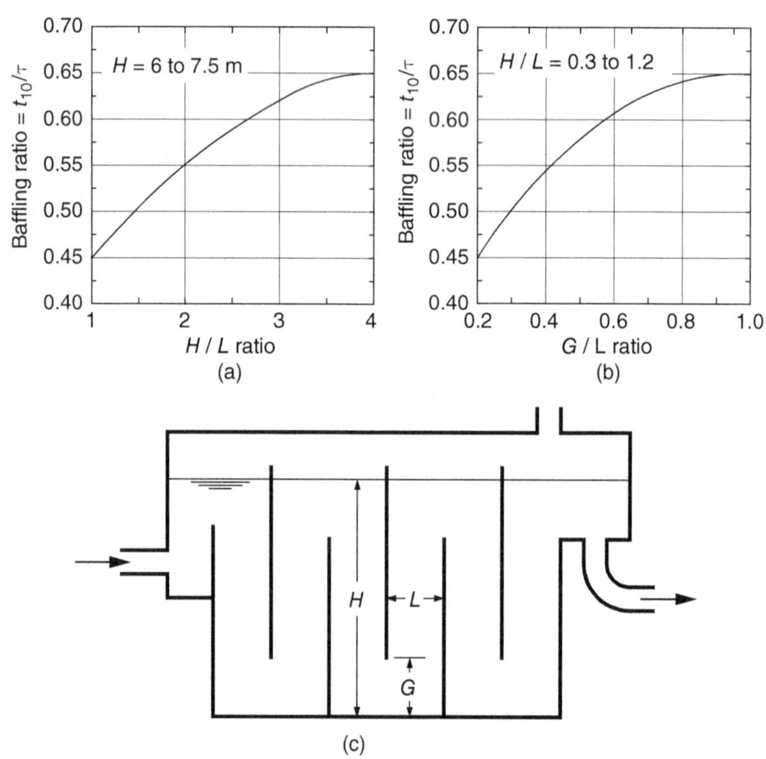

Figure 13-31
Impact of internal contactor design on its baffling ratio: (a) impact of H/L ratio; (b) impact of G/L ratio; and (c) contactor schematic. Dimensions H, G, and L are defined in contactor schematic. (Adapted from Henry and Freeman, 1996.)

Porous stone diffusers are used in ozone contactors to produce fine bubbles, which greatly increases the overall ozone transfer efficiency from the gas phase to the water, especially when compared to the use of a perforated-pipe diffuser. While both types of diffusers are used, experience has shown that perforated-pipe diffusers produce an excessively large bubble size. The cause of this problem is attributed to the way air exits the diffuser. When the diffuser is positioned horizontally, the air that exits on the underside of the diffuser seems to creep along the circumference of the diffuser before it rises into the water. As this creep occurs, the initial fine bubbles pick up more air and grow to large bubbles by the time they rise into the water column. Dome diffusers do not have this problem as the bubbles rise into the water column immediately after they exit the diffuser. Due to head loss limitations, a commercially available diffuser typically has a maximum gas flow rating that should not be exceeded.

13-9 Disinfection with Ultraviolet Light

All of the disinfectants discussed previously in this chapter are oxidizing chemicals. Disinfection can also be accomplished by other means, heat and electromagnetic radiation among them. Heat is used to disinfect, or "pasteurize," beverages and even to disinfect water through boiling. Electromagnetic radiation, specifically gamma radiation and UV radiation, is also used for disinfection: gamma radiation in the case of food products and UV radiation in the case of air, water, and some medical surfaces. Of these, only UV radiation has so far found a place in the routine disinfection of drinking water.

Photographs of UV reactors

http://www.stantec.com/wt3r/13-3.html

Ultraviolet disinfection is not common for drinking water disinfection in the United States, as was shown in Fig. 13-1. It is used more commonly in other countries, however, and its use is growing in the United States. The purpose of this section is to provide a basis for understanding the use of UV radiation for the inactivation of microorganisms. In practice, the design and implementation of UV radiation for water treatment is governed by U.S. EPA (2006) and state guidelines, also discussed in this section.

What Is Ultraviolet Light?

Ultraviolet light is the name used to describe electromagnetic radiation having a wavelength between 100 and 400 nm. As illustrated on Fig. 13-32, electromagnetic radiation of slightly shorter wavelength has been named "x-rays" and electromagnetic radiation of slightly longer wavelength, visible to the human eye, is referred to as "visible light." Radiation just long enough to be outside the visible range is referred to as infrared radiation. Light in the UV spectrum is often further subdivided into four segments, vacuum UV, short-wave UV (UV-C), middle-wave UV (UV-B), and long-wave UV (UV-A). These classifications can also be described as follows:

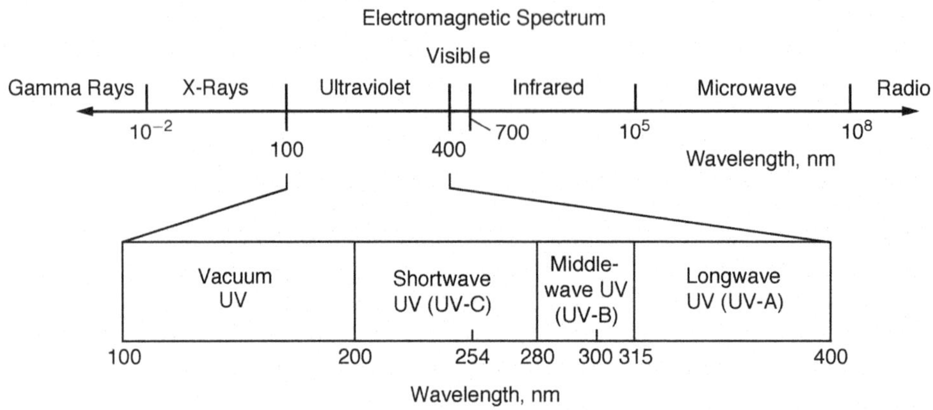

Figure 13-32
Location of the ultraviolet light region within the electromagnetic spectrum.

1. Both UV-A and UV-B activate the melanocytes in the skin to produce melanin ("a tan").
2. UV-B radiation also causes "sunburn."
3. UV-C radiation is absorbed by the DNA and is the most likely of the three to cause skin cancer.

If electromagnetic radiation is thought of as photons, then the energy associated with each photon is related to the wavelength of the radiation (Einstein, 1905):

$$E = \frac{hc}{\lambda} \tag{13-54}$$

where E = energy in each photon, J
h = Planck's constant (6.6×10^{-34} J · s)
c = speed of light, m/s
λ = wavelength of radiation, m

As a general rule, the more energy associated with each photon in electromagnetic radiation, the more dangerous it is for living organisms. Thus, visible and infrared light have relatively little affect on organisms, whereas both x-rays and gamma rays can be quite dangerous. Beyond these broad considerations, there are other factors that determine the fraction of the UV spectrum that is effective in disinfection. The portion of the UV spectrum that is more effective in disinfection is called the "germicidal range." On the lower end, the germicidal range is limited by the absorption of UV radiation by water. As wavelengths decrease, water becomes an increasingly efficient barrier for UV. For practical purposes, vacuum UV, the fraction of UV with a wavelength below 200 nm, cannot penetrate water. So radiation having a wavelength of 200 nm or less is not considered germicidal. It is also

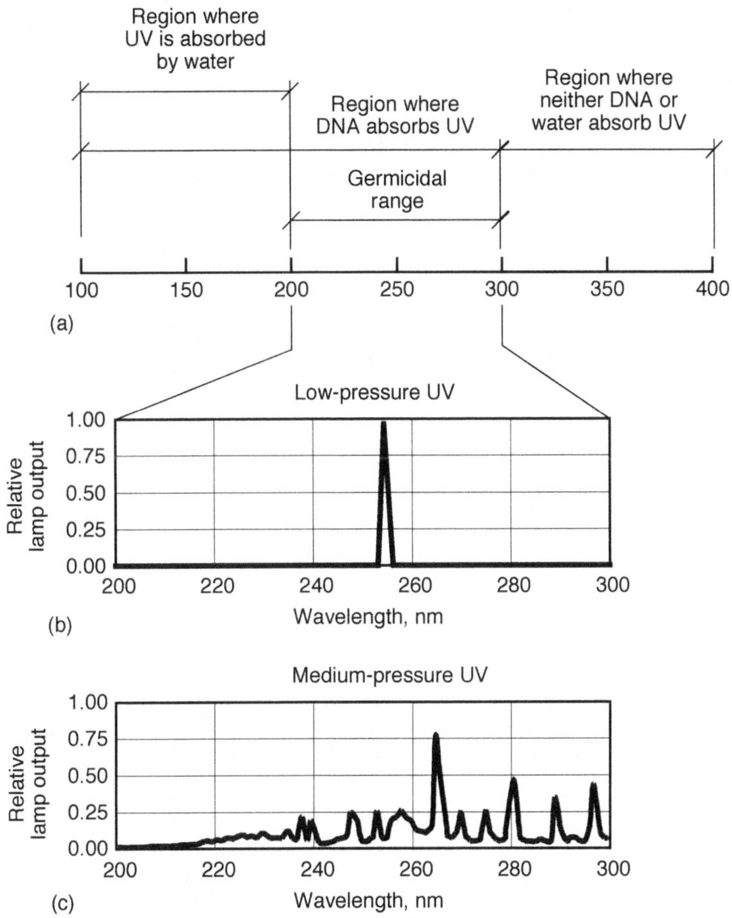

Figure 13-33
Ultraviolet sources and germicidal range: (a) ultraviolet portion of electromagnetic spectrum, (b) output from low-pressure UV lamp, and (c) output from medium-pressure UV lamp.

well established that UV inactivates microorganisms by transforming their DNA. This transformation cannot happen unless the UV is at a wavelength at which DNA will absorb it, and this absorption does not occur above wavelengths of approximately 300 nm. Therefore the germicidal range for UV is between approximately 200 and 300 nm (Fig. 13-33a).

Sources of Ultraviolet Light

The UV disinfection units used most commonly in the water industry employ three different types of UV lamps: (1) low-pressure low-intensity lamps, (2) low-pressure high-intensity lamps (also called low-pressure high-output lamps), and (3) medium-pressure high-intensity lamps. The design of these lamps closely approximates that of the common fluorescent light bulb. Low- and medium-pressure, high-intensity lamps are able to achieve a higher UV output in an equivalent space. Of the three technologies, medium-pressure UV has the greatest output. The spectrum of the UV light output by both types of low-pressure lamps is essentially the same, a very small amount of the light energy emanating at a wavelength of 188 nm and the vast majority of it emanating at a wavelength of 254 nm. The spectrum of the UV light output by medium-pressure lamps includes a number of wavelengths. These spectra are illustrated and compared with the germicidal range on Fig. 13-33b and 13-33c.

Several important characteristics of each of these UV lamps are compared in Table 13-9; however, it must be noted that UV lamp technology is evolving continuously. One of the design engineer's more important challenges is to evaluate the technologies available at the time a design is prepared and to write specifications that will enable new technologies while protecting the owner against innovative, but unproven, alternatives where the prospect for failure can be significant. New UV technologies under development and testing include pulsed UV, narrowband excimer UV (Naunovic et al., 2008), and deep UV (DUV) semiconductor light-emitting diodes (LEDs). The pulsed UV lamp produces polychromatic light at very high intensity, the narrowband excimer lamp produces nearly monochromatic light at

Table 13-9
Characteristics of three types of UV lamps

Item	Unit	Type of lamp		
		Low pressure Low intensity	Low pressure High Intensity	Medium Pressure
Power consumption	W	40–100	200–500[a]	1,000–10,000
Lamp current	mA	350–550	Variable	Variable
Lamp voltage	V	220	Variable	Variable
Germicidal output/input	%	30–40	25–35	10–15[b]
Lamp output at 254 nm	W	25–27	60–400	Variable
Lamp operating temperature	°C	35–45	60–100	600–900
Partial pressure of Hg vapor	kPa	0.00093	0.0018–0.10	40–4000
Lamp length	m	0.75–1.5	Variable	Variable
Lamp diameter	mm	15–20	Variable	Variable
Sleeve life	yr	4–6	4–6	1–3
Ballast life	yr	10–15	10–15	1–3
Estimated lamp life	h	8,000–10,000	8,000–12,000	4,000–8,000
Decrease in lamp output at estimated lamp life	%	20–25	25–30	20–25

[a]Up to 1200 W in very high output lamp.
[b]Output in the most effective germicidal range (~255–265 μm).

wavelengths of 172, 222, and 308 nm, and UV LED lamps emit light at 280 to 285 nm.

Equipment Configurations

Before discussing the fundamentals of UV disinfection, it will be useful to consider the types of reactors used for UV disinfection, as many of the factors that affect the effectiveness of UV disinfection are related to the reactor configuration. The components of a UV disinfection system consists of (1) the UV lamps; (2) transparent quartz sleeves that surround the UV lamps, protecting them from the water to be disinfected; (3) the structure that supports the lamps and sleeves and holds them in place; (4) the power supply for the UV lamps and cleaning system; (5) online UV dose monitoring sensors and associated equipment, and (6) the cleaning system used to maintain the transparency of the quartz sleeves. By themselves, UV lamps, which use an electrical arc, are not electrically stable because their electrical resistance decreases as their current increases. As a consequence, the electrical system must be ballasted to limit the current to the lamp. Cleaning systems are necessary for low-pressure high-intensity and mediumpressure UV lamps because they operate at such high temperatures (see Table 13-9) that salts with inverse solubility can precipitate, fouling the outer surface of the quartz sleeve and reducing the net UV output. These UV system components are installed in closed-vessel pressurized systems or as open-channel gravity flow systems, as shown on Fig. 13-34. Closed-vessel systems are used most commonly for the disinfection of drinking water, whereas open-channel systems are more common in wastewater disinfection.

CLOSED-VESSEL SYSTEMS

Whereas most low-pressure systems are designed with open-channel flow, most low-pressure high-intensity and medium-pressure systems for drinking water are designed using closed vessels. These closed-vessel systems have the advantage that they can (and usually do) operate under pressure, and this feature makes them particularly attractive in upgrades and retrofits because it is not necessary to "break head" to use them. The placement of UV lamps in closed systems can be either perpendicular to the flow (see Fig. 13-34a) or parallel to the flow (see Fig. 13-34b). Because low-pressure high-intensity and the medium-pressure systems, operate with a limited number of lamps, more care is required to ensure that short circuiting does not occur. Biodosimetry methods, as discussed subsequently, have evolved that can be used to assess whether a UV reactor will perform as specified.

Of critical importance in the application of UV radiation for the inactivation of microorganisms is the ability to monitor the UV reactor online to be assured that the required UV dose is being delivered. The method used to monitor the UV dose is of importance both in the validation of the of

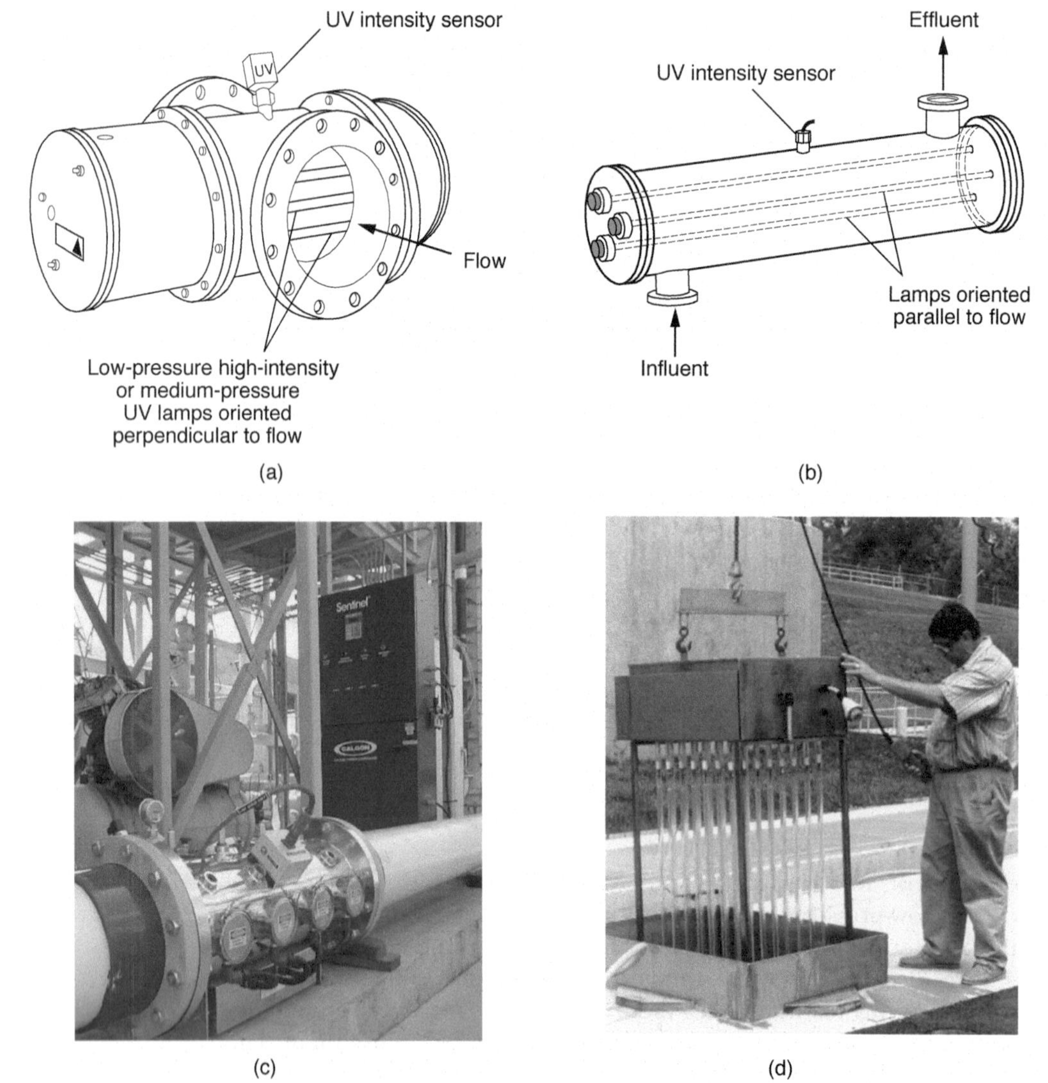

Figure 13-34
Common UV configurations: (a) medium pressure lamps placed perpendicular to the flow in a closed reactor, (b) low-pressure high-intensity lamps placed parallel to flow, (c) view of medium-pressure closed reactor, and (d) view of vertical low-pressure lamp arrangement in open reactor.

UV reactors as well as for monitoring the long-term performance of the UV reactor. The most common methods are:

1. ***UV Intensity Set Point*:** The reactor UV dose is monitored based on UV intensity, flow rate, and lamp status.

2. ***UV Transmittance and UV Intensity Set Point***: The reactor UV dose is monitored based on UV intensity, UV transmittance, flow rate, and lamp status.
3. ***Calculated Dose***: The UV dose received by a microorganism is calculated continuously, using a predetermined algorithm, based on the UV transmittance, flow rate, and lamp status including the effects of aging and lamp fouling.

OPEN-CHANNEL SYSTEMS

Open-channel designs are available for all types of UV systems. Typically, the UV lamps are retained in modules or racks that are placed in the flow channel (see Fig. 13-34d). Designs are available with lamps placed horizontally parallel to the flow and with lamps placed vertically perpendicular to the flow. Conventional low-pressure low-intensity systems are typically designed so that they can be removed and cleaned easily. Most low-pressure high-intensity and all medium-pressure systems are provided with mechanical or mechanical/chemical self-cleaning systems.

Mechanism of Inactivation

More is known about the specific mechanisms of disinfection by UV than for any other disinfectant used in water treatment. The photons in UV light react directly with the nucleic acids in the target organism, damaging them. The genetic code that guides the development of every living organism is made up of nucleic acids. These nucleic acids are either in the form of deoxyribonucleic acid (DNA) or ribonucleic acid (RNA). The DNA serves as the databank of life while the RNA directs the metabolic processes in the cell. Ordinarily DNA is a double-stranded helical structure that includes the nucleotides adenine, guanine, thymine, and cytosine. Ordinarily RNA is a single-stranded structure with the nucleotides adenine, guanine, uracil, and cytosine (refer to Chap. 3).

Ultraviolet light damages DNA by dimerizing adjacent thymine molecules, inhibiting further transcription of the cell's genetic code (see Fig. 13-35). While not usually fatal to the organism, such dimerization will

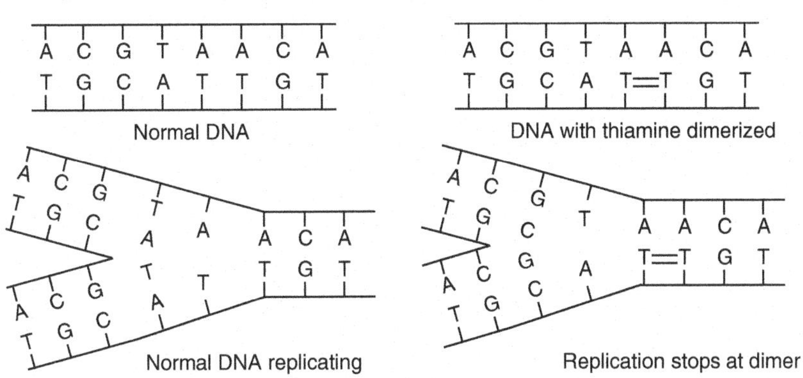

Figure 13-35
Formation of thiamine dimers by UV light interferes with normal replication of microorganisms.

prevent its successful reproduction (Setlow, 1967). Ultraviolet light also forms cytosine–cytosine and cytosine–thymine dimers, but these reactions have a lower quantum yield (they occur less frequently). As a result, organisms rich in thymine tend to be more sensitive to UV irradiation. For example, *C. parvum* and *G. lamblia* both contain DNA and both are inactivated by UV at relatively low doses (see Table 13-3). Most viruses of significance in drinking water have only RNA (which contains uracil instead of thymine) and, thus, are less sensitive to UV radiation. Among the most resistant organisms are viruses such as rotavirus and adenovirus, which incorporate a special double-stranded RNA. Other factors also influence the rate of inactivation, and some are not as well understood. Ultraviolet radiation can also cause damage of a more severe kind, breaking chains, crosslinking DNA with itself, crosslinking DNA with other proteins, and forming other by-products. These effects have an even lower quantum yield, and they are usually observed only at high doses of irradiation.

Reactivation

Reactivation is a more important consideration in UV disinfection than it is with disinfection by other methods. It is important to note that most forms of life evolved with some exposure to the sun and that sunlight includes significant amounts of UV irradiation. As a result, the process of evolution has addressed UV-induced damage by generating mechanisms for repairing the damage it causes. These mechanisms fall into two basic classes: (1) photoreactivation and (2) dark repair. Photoreactivation only takes place in the presence of light, whereas dark repair has no such requirement. Organisms capable of dark repair generally show much greater UV resistance; however, understanding the importance of photoreactivation requires that special tests be conducted, evaluating samples with and without light exposure to understand its effects.

Certainly when water is being disinfected for discharge into the environment, only the net inactivation after photoreactivation is important. Even in the case of drinking water systems, where light exposure is often more limited, the most conservative approach is to consider photoreactivation as well. Eventually, it may be possible to determine if an organism is capable of photorepair by using its genetic fingerprint to map its position on the evolutionary tree. In general, it is not safe to assume that any organism is incapable of photorepair, unless through testing it has been demonstrated to be the case. Even some viruses have been shown to be capable of photorepair, apparently taking advantage of enzymes in the host organism following infection.

Concept of Action Spectrum

Until recent years, low-pressure low-intensity lamps were the only source of ultraviolet light available for disinfection of drinking water. The principal light output of these lamps is at only one wavelength, 254 nm. Medium-pressure lamps, on the other hand, emit light at a variety of wavelengths (see Fig. 13-33c). There is no reason to expect that light will have the same disinfecting power at each wavelength. Earlier, the boundaries of the *germicidal range* of wavelengths were broadly established, the lower boundary (200 nm) being defined by the absorption of light by water and the upper boundary (300 nm) being defined by the lack of absorption of light by DNA. To compare the effectiveness of medium- and low-pressure low- and high-intensity lamps for disinfection, a better understanding is required of possible significance of UV radiation at different wavelengths. A number of researchers have looked at this issue and the results of their research are generally expressed in the form of an action spectrum. To generate the action spectrum, a modification of Eq. 13-3 for UV light of a particular wavelength λ can be used:

$$r_{N_\lambda} = -N\Lambda_\lambda I_\lambda \tag{13-55}$$

where r_{N_λ} = rate of change in number of organisms exposed to light of wavelength λ

N = number of organisms exposed to light, organisms/100 mL

Λ_λ = coefficient of specific lethality for light of wavelength λ, m^2/J

I_λ = intensity of light at wavelength λ, W/m^2

The action spectrum is a representation of Λ_λ over a range of wavelengths. Often it is displayed as a plot of the ratio $\Lambda_\lambda/\Lambda_{254\text{ nm}}$ versus wavelength. The action spectrums for *C. parvum* (Linden et al., 2001) and MS2 (Rauth, 1965) are compared with the absorption spectrum for DNA on Fig. 13-36. A close correlation between Λ_λ and DNA absorption is observed. The action spectra of a number of organisms have been determined and are similar to the results shown on Fig. 13-36. As a result, many scientists believe that the germicidal efficiency determined for one species of microorganism to medium-pressure UV may be used to represent the relative response of other microorganisms as well (Giese and Darby, 2000).

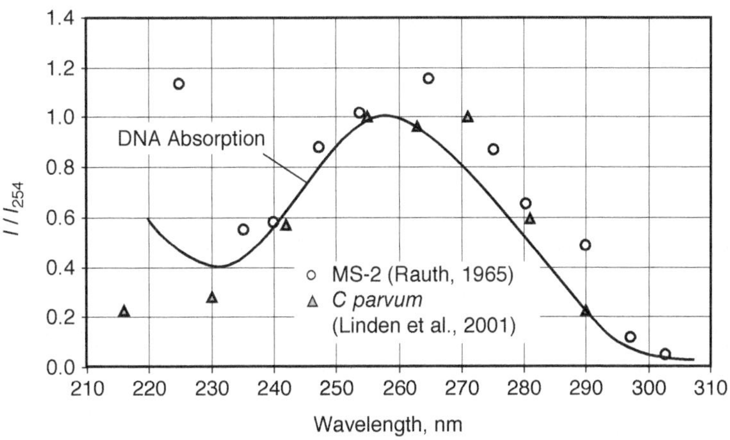

Figure 13-36
Comparing action spectra for *C. parvum* and MS-2 coliphage with absorption spectrum for DNA.

Ultraviolet Light Dose

The effectiveness of UV disinfection is based on the UV dose to which the microorganisms are exposed. The UV dose D is defined as

$$D = I_{avg}t \tag{13-56}$$

where D = *UV* dose, mJ/cm^2 (note $mJ/cm^2 = mW \cdot s/cm^2$)
I_{avg} = average UV intensity, mW/cm^2
t = exposure time, s

Note that the UV dose term is analogous to the dose term used for chemical disinfectants (i.e., *Ct*). As given by Eq. 13-56, the UV dose can be varied by changing either the average UV intensity or the exposure time. Determination of the average UV intensity, as a function of the distance from the light source, was illustrated previously in Example 2-2 in Chap. 2. The impact of dissolved and suspended substances on average UV intensity, and ultimately dose, are discussed below (Linden and Rosenfeldt, 2011; U.S. EPA, 2006).

Influence of Water Quality

The quality of the water being treated can have an important influence on the performance of UV disinfection systems. The two most important impacts stem from the action of dissolved and suspended substances.

DISSOLVED SUBSTANCES

Pure water absorbs light in the lower UV wavelengths. A number of dissolved substances also have important influence on the absorption of UV radiation as it passes through the water on its way to the target organism. Among the more significant are iron, nitrate, and natural organic matter. Chlorine, hydrogen peroxide, and ozone can also have important effects.

The absorption of light in aqueous solution by dissolved substances is described by the Beer–Lambert law. This relationship, discussed in Chaps. 2 and 8, takes the form

$$\log\left(\frac{I}{I_0}\right) = -\varepsilon(\lambda)\, Cx \tag{13-57}$$

where I = light intensity at distance x from light source, mW/cm^2
I_0 = light intensity at light source, mW/cm^2
C = concentration of light-absorbing solute, mol/L
x = light path length, cm
$\varepsilon(\lambda)$ = molar absorptivity of light-absorbing solute at wavelength λ, $L/mol \cdot cm$

The term on the right-hand side of Eq. 13-57 is defined as the absorbance A, which is unitless. As discussed in Chap. 2, the absorptivity is the absorbance corresponding to a path length of 1 cm, or

$$k(\lambda) = \varepsilon(\lambda)C = \frac{A}{x} \tag{13-58}$$

where $k(\lambda)$ = absorptivity, cm^{-1}

The absorptivity of the water is an important aspect of UV reactor design. Waters with higher absorptivity absorb more UV light and need a higher energy input for an equivalent level of disinfection. Absorbance is measured using a spectrophotometer typically using a fixed sample path length of 1.0 cm. The absorbance of water is typically measured at a wavelength of 254 nm.

In the application of UV radiation for microorganism inactivation, transmittance, which reflects the amount of UV radiation that can pass through a specified length at a particular wavelength, is the water quality parameter used in the design and monitoring of UV systems. The transmittance of a solution is defined as

$$\text{Transmittance, } T, \% = \left(\frac{I}{I_0}\right) \times 100 \tag{13-59}$$

The transmittance at a given wavelength can also be derived from absorbance measurements using the following relationship:

$$T = 10^{-A(\lambda)} \tag{13-60}$$

Thus, for a perfectly transparent solution $A(\lambda) = 0$, $T = 1$ and for a perfectly opaque solution $A(\lambda) \rightarrow \infty$, $T = 0$. At a UV radiation wavelength of 254 nm, Eq. 13-60 is written as follows:

$$\text{UVT}_{254} = 10^{-A_{254}} \tag{13-61}$$

The term percent transmittance, commonly used in the literature is

$$\text{UVT}_{254,}\% = 10^{-A_{254}} \times 100 \tag{13-62}$$

Typical absorbance and transmittance values for various waters are presented in Table 13-10.

PARTICULATE MATTER

Particulate matter can also interfere with the transmission of UV light. Particulates are an aspect of water quality that can be particularly important where UV disinfection is concerned. Two mechanisms of particular importance are shading and encasement, as shown on Fig. 13-37. Interference of this kind has been studied at great depth for the case of coliform organisms in secondary wastewater effluents, and models have been developed that do an excellent job of characterizing the situation (Loge et al., 2001). The effect of shading can be integrated into models for the absorption of

Table 13-10
Typical absorbance and transmittance values for various waters

Type of Water	UV_{254} Absorbance, cm^{-1}	Transmittance UVT_{254},%
Groundwater	0.0706–0.0088	85–98
Surface water, untreated	0.3010–0.0269	50–94
Surface water, after coagulation, flocculation, and sedimentation	0.0969–0.0132	80–97
Surface water, after coagulation, flocculation, sedimentation, and filtration	0.0706–0.0088	85–98
Surface water after microfiltration	0.0706–0.0088	85–98
Surface water after reverse osmosis	0.0458–0.0044	90–99

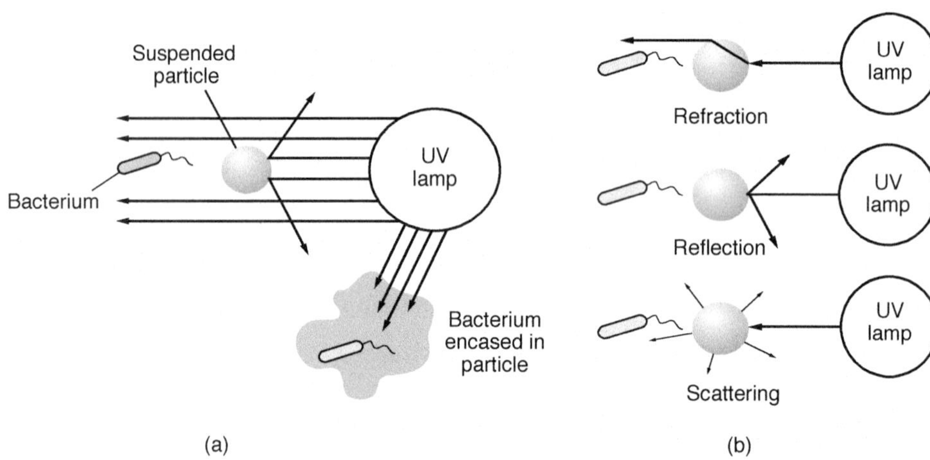

Figure 13-37
Illustration of mechanisms for interference in disinfection by particles: (a) overview of mechanisms for interference and (b) mechanisms of "shading."

light. Beyond that, the number of organisms is dominated by the effect of organisms associated with particles. Particles can "shade" target organisms from UV light via three mechanisms: refraction, reflection, and scattering. Where filtration is used, these effects are not very important, but in the treatment of unfiltered water supplies and unfiltered wastewater effluents, these effects can be quite significant.

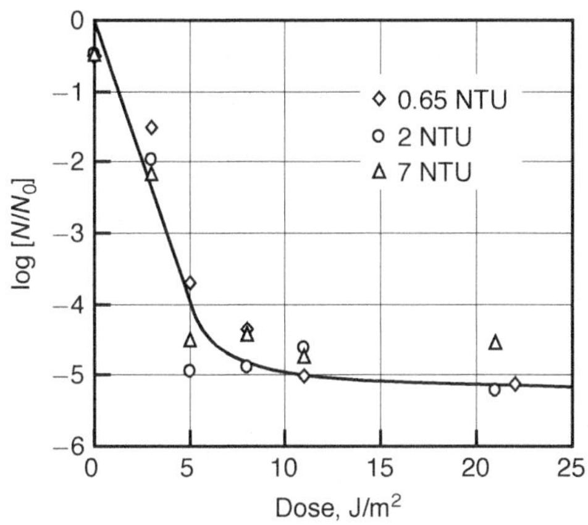

Figure 13-38
Impact of low levels of turbidity on inactivation of *G. muris* with UV radiation. (Adapted from Oppenheimer et al., 2001).

The effects of particle shading are not particularly significant at low turbidities, as illustrated by the work of Oppenheimer et al. (2002), who examined the inactivation of *G. muris* added to waters with turbidities ranging from 0.65 to 7 NTU (see Fig. 13-38). A collimated beam apparatus (see Fig. 13-41) was used to study the inactivation of *G. muris* with waters at three different turbidity levels ranging from 0.65 to 7 NTU. After the UV dose was corrected for apparent absorbance (absorbance including the effects of particle shading), turbidities at these levels seemed to have little significance.

Influence of UV Reactor Hydraulics

Ultraviolet disinfection systems, particularly medium-pressure systems, are characterized by overall residence times that are much shorter than other kinds of disinfection systems. In these systems short circuiting and dispersion are difficult design issues. Designing these systems to achieve good performance requires a greater appreciation of the factors that influence dispersion and short circuiting than is required for the design of most other disinfection systems. The issues are the same as those discussed earlier with contactors for disinfection with chlorine, chloramines, chlorine dioxide, and ozone; however, with UV disinfection contactors, the time spent in transition zones becomes much more important.

In chlorine contactors, for example, inlet conditions can have a big influence on performance. If the contactor is designed with a sufficiently long aspect ratio, good performance can be achieved in spite of nonideal inlet conditions. In many UV reactors, the zones of flow transition can dominate most of the contact time. Also because for the short contact time it is extremely difficult to conduct a meaningful tracer study. The outcome of a tracer study often depends on the UV reactor configuration and precisely where the tracer is introduced. A further complication in UV reactors is that the UV light intensity varies throughout the reactor. As a result, the UV dose that an organism receives is not only a function of the length of time the organism spends in the reactor and the amount of light being emitted by the UV lamps but also of the specific path the organism takes as it makes its way through the reactor. Thus, the issue is not just the contact time the organism receives, but its cumulative exposure to UV.

Because there are so many complications in determining the performance of a given full-scale UV reactor, it is increasingly common for regulators to require full-scale tests of each reactor design to establish, by actual disinfection measurements, how much of a UV dose a given reactor design will be credited with delivering. The use of a test microorganism to determine the performance of a UV reactor is known as *biodosimetry*. The principal limitation with biodosimetry, in light of the above discussion, is that it cannot be used to measure the dose distribution. Computational fluid dynamics (CFD) modeling and chemical actinometry, employing dyed microspheres, are also being used in conjunction with biodosimetry to assess the performance of UV reactors including the UV dose distribution. CFD modeling and chemical actinometry and are discussed briefly below. Biodosimetry is considered subsequently in greater detail because it is the method now used most commonly for the assessment of UV reactor performance.

COMPUTATIONAL FLUID DYNAMICS

Because of the expense of conducting biodosimetry testing, CFD modeling is now used routinely to simulate mathematically the movement of particles (e.g., microorganisms) through a UV reactor. One of the earliest simulations of the movement of microorganisms through a hypothetical UV reactor was conducted be Chiu et al. 1999. Examples of their model simulation results are illustrated on Fig. 13-39. As shown on Fig. 13-39b, the dose a microorganism depends not only on the intensity of the lamps and the time the organism spends in the reactor but also on the specific path the organism takes through the reactor. The early CFD modeling studies have been extended by a number of researchers, including Lyn and Blatchley (2005) and Ducoste et al. (2005). Because so many different operating conditions can be modeled quickly, CFD modeling is now used essentially by all UV reactor manufacturers to develop new UV reactor configurations. When CFD modeling is coupled with chemical actinometry, and biodosimetry, the performance of UV reactors can be predicted with a greater degree of reliability as compared to the use of a single method.

CHEMICAL ACTINOMETRY

Determination of UV intensity from the measurement of the quantum yield of a chemical reaction induced by UV radiation is known as *chemical actinometry*. The quantum yield of a reaction, as given by Eq. 8-100, is a measure of the number of photolysis reactions (e.g., fluorescence) divided

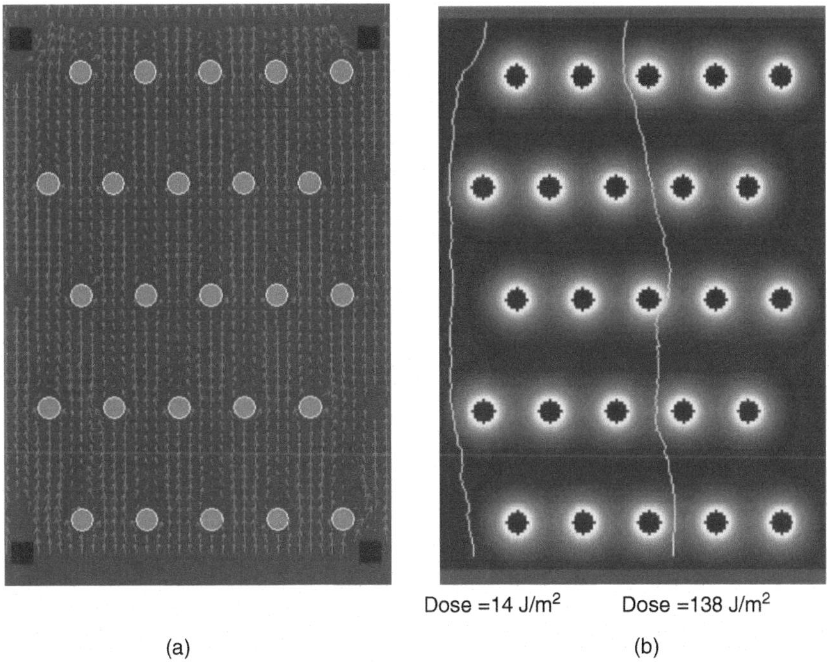

Figure 13-39 Performance of UV reactor: (a) flow pattern and (b) UV dose based on two alternative microorganism travel tracks. The microorganism on the left was exposed to a UV dose of 14 J/m^2 whereas the microorganism on the right was exposed to a UV dose of 138 J/m^2. (Adapted from Chiu et al., 1999.)

by the number of photons adsorbed. Ideally, chemical actinometry involves the use of a chemical that is easy to measure and has a known quantum yield. Microsphere chemical actinometry involves coating, imbedding, or attaching a chemical to polystyrene microspheres (specific gravity 1.05, mean diameter 5.6 μm) that will fluoresce when exposed to UV light (Bohrerova et al., 2005; Blatchley et al., 2008; Shen et al., 2009). If the fluorescence of the individual microsphere particles is measured, the increase in fluoresce intensity can be related to the UV dose received by an individual microsphere. If a sufficient number of microspheres are measured, the UV dose distribution can be assessed. This method has been demonstrated at full scale and the results have been compared with CFD modeling and biodosimetry results (Blatchley et al., 2008; Shen et al., 2009). When all three techniques are used together to evaluate the performance of new UV reactor designs, a high degree of predictability can be achieved.

BIODOSIMETRY

Biodosimetry, as illustrated on Fig. 13-40, involves conducting both bench-scale laboratory and field-scale tests with the same biological test organism. The laboratory study is conducted to establish the relationship between UV dose and the inactivation of a test organism. The field-scale test is conducted at design flow and under conditions designed to represent a conservative

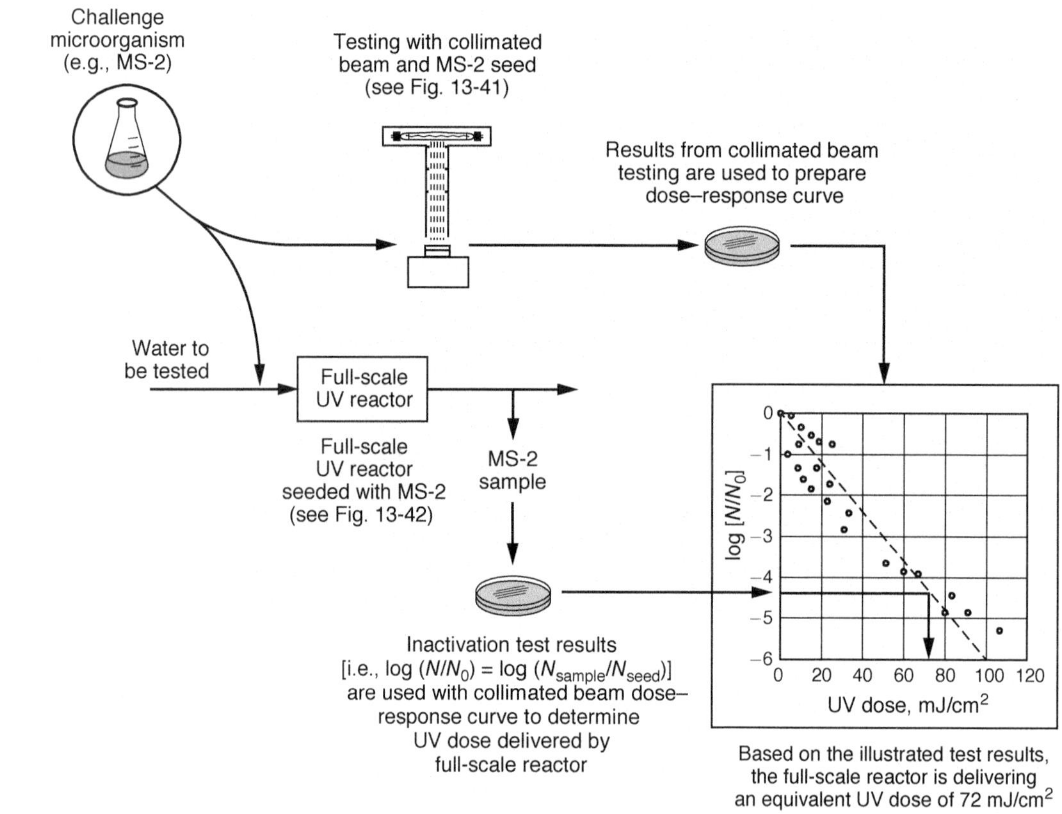

Figure 13-40
Schematic illustration of the application of biodosimetry as used to determine the performance of a full-scale UV reactor.

simulation of full-scale operation. The specifics of the conduct of this test are outlined in the appropriate guidelines (see subsequent section). The disinfection dose that a UV reactor is credited with is determined by the dose that accomplishes the same level of inactivation under laboratory conditions. Biodosimetry is most effective when it is conducted with an organism that shows approximately the same resistance to UV radiation as the target organism. The principal limitation of biodosimetry, as discussed previously is that the test cannot be used to assess the UV dose distribution within the reactor. The elements of biodosimetry are examined in what follows.

Determination of UV Dose Using Collimated Beam

The most common procedure for determining the required UV dose for the inactivation of challenge microorganism involves the exposure of well-mixed water sample in a small batch reactor (i.e., a Petri dish) to collimated beam of UV light of known UV intensity for a specified period of time, as

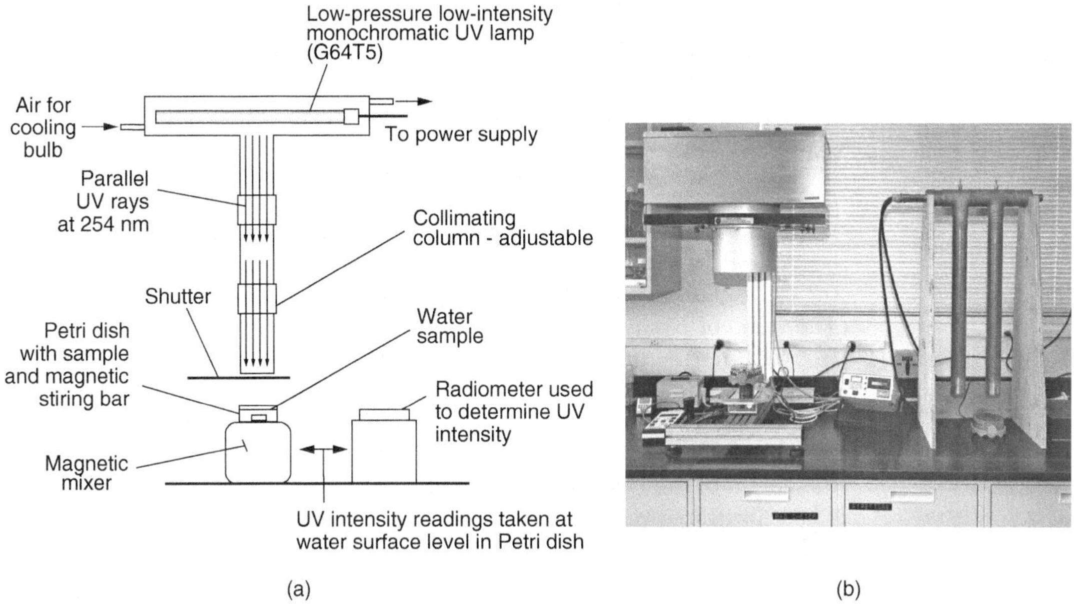

Figure 13-41
Collimated beam devices used to develop dose–response curves for UV disinfection: (a) schematic of the key elements of a collimated beam setup and (b) view of two different types of collimated beam devices. The collimated beam on the left is of European design; the collimated beam on the right is of the type shown in the schematic on the left.

illustrated on Fig. 13-41. Use of a monochromatic low-pressure low-intensity lamp in the collimated beam apparatus allows for accurate characterization of the applied UV intensity. Use of a batch reactor allows for accurate determination of exposure time. The applied UV dose, as defined by Eq. 13-56, can be controlled either by varying the UV intensity or the exposure time. Because the geometry is fixed, the depth-average UV intensity within the Petri dish sample (i.e., the batch reactor) can be computed using the following relationship, which also takes into account other operational variables that may affect the UV dose:

$$D_{CB} = E_s t(1-R)P_f \left[\frac{1-10^{-k_{254}d}}{2.303(k_{254}d)}\right]\left(\frac{L}{L+d}\right) \tag{13-63}$$

$$D_{CB} = E_s t(1-R)P_f \left[\frac{1-e^{-2.303k_{254}d}}{2.303(k_{254}d)}\right]\left(\frac{L}{L+d}\right) \tag{13-64}$$

where D_{CB} = average collimated beam UV dose, mW/cm^2
E_S = incident UV intensity at the center of the surface of the sample, before and after sample exposure, mW/cm^2

t = exposure time, s
R = reflectance at the air–water interface at 254 nm
P_f = Petri dish factor
k_{254} = absorptivity, cm^{-1} (base 10)
d = depth of sample, cm
L = distance from lamp centerline to liquid surface, cm

Without the other correction factors, the basic form of Eqs. 13-63 and 13-64 is the same as that derived in Example 2-2 in Chap. 2. The term (1 – R) on the right-hand side accounts for the reflectance at the air–water interface. The value of R is typically about 2.5 percent. The term P_f accounts for the fact that the UV intensity may not be uniform over the entire area of the Petri dish. The value of P_f is typically greater than 0.9. The term within the brackets is the depth averaged UV intensity within the Petri dish and is based on the Beer–Lambert law. The final term is a correction factor for the height of the UV light source above the sample. The application of Eqs. 13-63 illustrated in Example 13-11. The uncertainty of the computed UV dose at a given UV intensity can be estimated using the sum of the variances as given by either of the following expressions:

Maximum uncertainty:

$$U_D = \sum_{n=1}^{N} \left| U_{V_n} \frac{\partial D}{\partial V_n} \right| \tag{13-65}$$

Best estimate of uncertainty

$$U_D = \left[\sum_{n=1}^{N} \left(U_{V_n} \frac{\partial D}{\partial V_n} \right)^2 \right]^{1/2} \tag{13-66}$$

where U_D = uncertainty of UV dose value, %
U_{Vn} = uncertainty or error in variable n
V_n = variable n
$\partial D/\partial V_n$ = partial derivative of the expression with respect to the variable V_n
N = number of variables

The maximum estimate of uncertainty as given by Eq. 13-65 represents the condition where every error will be a maximum value. The best estimate of uncertainty, as given by Eq. 13-66, is used most commonly because it is unlikely that every error will be a maximum at the same time and the fact that some errors may cancel each other. The application of Eq. 13-66 is illustrated in Example 13-11.

Example 13-11 Estimation of UV dose using collimated beam

A collimated beam, with the following characteristics, is to be used for biodosimetry testing. Using these data estimate the average UV dose delivered to the sample and best estimate of the uncertainty associated with the measurement.

$E_S = 15 \pm 0.75$ mW/cm^2 (accuracy of meter $\pm 5\%$), $t = 10 \pm 0.2$ s, $R = 0.025$ (assumed to be the correct value), $P_f = 0.94 \pm 0.02$, $k_{254} = 0.065 \pm 0.005$ cm^{-1}, d = 1 $\pm$ 0.05 cm, L = 40 $\pm$ 0.5 cm.

Solution

1. Using Eq. 13-63 estimate the delivered dose:

$$D_{CB} = E_s t(1-R)P_f\left[\frac{1-10^{-k_{254}d}}{2.303(k_{254}d)}\right]\left(\frac{L}{L+d}\right)$$

$$D_{CB} = (15)\times(10)(1-0.025)(0.94)\left[\frac{1-10^{-(0.065\times 1)}}{(2.303)(0.065)\times(1)}\right]\left(\frac{40}{40+1}\right)$$

$$D_{CB} = (150)(0.975)(0.94)(0.928)(0.976) = 124.6 \text{ mJ/cm}^2$$

2. Determine the best estimate of uncertainty for the computed UV dose. The uncertainty of the computed dose can be estimated using Eq. 13-66. The procedure is illustrated for one of the variables and summarized for the remaining variables.
   a. Consider the variability in the measured time, t. The partial derivative of the expression used in step 1 with respect to t is

$$U_t = U_{t_e}\frac{\partial D}{\partial t_n} = t_e E_S(1-R)P_f\left[\frac{1-10^{-k_{254}d}}{2.303(k_{254}d)}\right]\left(\frac{L}{L+d}\right)$$

where t_e is the uncertainity of the measured value of (0.25). Substituting known values and solving for u_t the uncertainity with respect to t, yields

$$U_t = (0.2)(15)(1-0.025)(0.94)\left[\frac{1-10^{-(0.065)\times 1}}{(2.303)(0.065)\times(1)}\right]\left(\frac{40}{40+1}\right)$$

$$U_t = 2.49 \text{ mJ/cm}^2$$

Percent = 100 U_t/D = (100) × (2.49)/124.6 = 2.0%

b. Similarly for the remaining variables, the corresponding uncertainity values are given below:

$$U_{ES} = 6.23\ \text{mJ/cm}^2 \text{ and } 5.0\%$$
$$U_{Pf} = 2.65\ \text{mJ/cm}^2 \text{ and } 2.13\%$$
$$U_a = -0.7\ \text{mJ/cm}^2 \text{ and } -0.56\%$$
$$U_d = -0.61\ \text{mJ/cm}^2 \text{ and } -0.49\%$$
$$U_L = 0.038\ \text{mJ/cm}^2 \text{ and } 0.03\%$$

c. The best estimate of uncertainty using Eq. 13-66 is

$$U_D = \left[(2.49)^2 + (6.23)^2 + (2.65)^2 + (-0.7)^2 + (-0.61)^2 + (0.038)^2\right]^{1/2}$$
$$U_D = 7.27\ \text{mJ/cm}^2$$
$$\text{Percent} = (100) \times (7.27)/124.6 = 5.84\%$$

3. Based on the above uncertainty computation the most likely UV dose is 124.6 ± 7.27 mJ/cm²

Comment

Thus, the most conservative estimate of the UV dose that can be delivered consistently is 117.3 mJ/cm² (124.6 – 7.27). If a similar analysis is carried for each of the UV doses evaluated, a curve of the most likely UV dose can be drawn as a function of the microorganism inactivation achieved with each UV dose, as discussed below.

DEVELOPMENT OF UV DOSE RESPONSE CURVE USING COLLIMATED BEAM

To assess the degree of inactivation that can be achieved at a given UV dose, the concentration of microorganism is determined before and after exposure in a collimate beam apparatus (see Fig. 13-41). Microorganism inactivation is measured using an most probable number (MPN) procedure for bacteria, a plaque count procedure for viruses, or an animal infectivity procedure for protozoa. To verify the accuracy of the laboratory collimated beam dose–response test data, the collimated beam test must be repeated to obtain statistical significance. To be assured that stock solution of the challenge microorganisms is monodispersed, the laboratory inactivation test data must fall within an accepted set of quality control limits. Quality control limits proposed by the National Water Research Institute (NWRI, 2003) and the U.S. EPA (2006) for bacteriophage MS2 spores are as follows:

NWRI:

$$\text{Upper bound: } -\log_{10}(N/N_0) = 0.040 \times D + 0.64 \qquad (13\text{-}67)$$

$$\text{Lower bound: } -\log_{10}(N/N_0) = 0.033 \times D + 0.20 \qquad (13\text{-}68)$$

U.S. EPA:

$$\text{Upper bound: } -\log_{10}(N/N_0) = -1.4\times10^{-4}\times D^2 + 7.6\times10^{-2}\times D \qquad (13\text{-}69)$$

$$\text{Lower bound: } -\log_{10}(N/N_0) = -9.6\times10^{-5}\times D^2 + 4.5\times10^{-2}\times D \qquad (13\text{-}70)$$

where D = UV dose, mJ/cm^2

As illustrated in Example 13-12, the bounds proposed by the U.S. EPA are more lenient as compared to those used by NWRI. Similar bounding curves have been proposed for *B. subtilus* (U.S. EPA, 2006; AWWARF and NYSERDA, 2007). The NWRI guidelines are used for water reuse applications in California.

Example 13-12 Develop dose response curve for bacteriophage MS2 using a collimated beam.

Bacteriophage MS2 (ATCC 15597) is to be used to validate the performance of a full-scale UV reactor. The following collimated beam test results were obtained for MS2 in a phosphate buffer solution with a UVT_{254} in the range from 95 to 99 percent (Data courtesy B. Cooper, BioVir Labs). Verify that the laboratory test results are acceptable and develop the dose–response curve for use in the full-scale validation. Also, estimate the UV dose required to achieve 2 log of inactivation.

Dose, mJ/cm^2	Surviving Concentration, phage/mL	Log Survival,Log (phage/mL)	Log Inactivation
0	5.00×10^6	6.70	
20	4.00×10^5	5.60	1.10[a]
40	4.30×10^4	4.63	2.07
60	6.31×10^3	3.80	2.9
80	8.70×10^2	2.94	3.76
100	1.20×10^2	2.08	4.62

[a]Sample calculation: log inactivation = 6.70 – 5.60 = 1.10.

Solution

1. Plot the collimated beam test results and compare to the quality control range expressions provided in the NWRI (Eqs. 13-67 and 13-68) and (Eqs. 13-69 and 13-70) U.S. EPA UV Guidelines. The results are plotted in the figure given below.

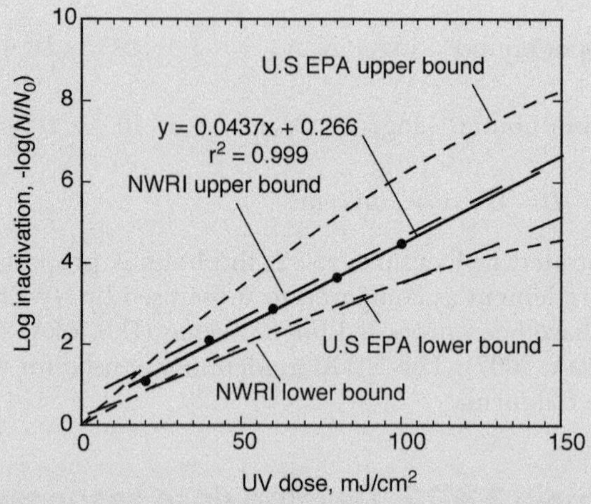

2. As shown in the above plot, all of the data points fall within the acceptable range.
3. Dose–response curve for bacteriophage MS2. The slope of the line, based on a linear fit, is

$$y = 0.0437 \times + 0.266$$

which corresponds to

$$-\log(N/N_0) = 0.266 + (0.0437\ \text{cm}^2/\text{mJ})(\text{UV dose,mJ/cm}^2)$$

4. UV dose required for 2 log of inactivation of MS2. Using the equation from step 3, the required UV dose is

$$\text{UVdose} = -\frac{-\log N/N_0(N/N_0) - 0.266}{0.0437\ \text{cm}^2/\text{mJ}} = \frac{2 - 0.266}{0.0437\ \text{cm}^2/\text{mJ}}$$

$$= 39.7\ \text{mJ/cm}^2$$

Comment

As shown in the above plot, there is a considerable difference in the upper quality control limit between the NWRI and the U.S. EPA UV guidelines (U.S. EPA, 2006). Also note that the U.S. EPA guidelines are curvilinear, whereas the NWRI guidelines are linear. Clearly, the NWRI guidelines are more restrictive.

UV DOSE REQUIRED FOR INACTIVATION OF *CRYPTOSPORIDIUM, GIARDIA,* AND VIRUSES

Using the biodosimetry approach, outlined above, the U.S. EPA has developed minimum UV dose requirements for various levels of inactivation for *Cryptosporidium, Giardia,* and virus (U.S. EPA, 2006). Adenovirus was utilized as the test virus because it is considered the most difficult to inactivate by UV radiation. It is important to note that the UV values reported in Table 13-11 are based on tests conducted using the specific organisms and take into account the uncertainty associated with dose–response relationships. Other sources of uncertainty associated with the full-scale installation such as the design of the UV reactors, the system hydraulics, the measured UV intensity, and monitoring approach are not included but are considered during the validation testing of UV reactors.

When a surrogate microorganism, such as MS2, is used, the values reported in Table 13-11 must be adjusted to reflect the differences in resistance between the target organism and the surrogate (see discussion under Validation of UV Reactors). The ideal surrogate should be

- Nonpathogenic
- Easy to culture at high titers (on the order of 10^{11} to 10^{12} org./mL)
- Stable over long periods
- Easy to enumerate

In the United States, the organism of choice is MS2 bacteriophage, whereas in Europe *B. subtilis* is the microorganism of choice. Other organisms such as the T1 and Q beta phage that more closely mirror the response of *Cryptosporidium* are also under investigation. Also, it is important to note that the host organism used for the culture of MS2 or other phage organisms must be specified if comparable results are to be obtained. Additional information on the types of microorganisms that have been examined may

Table 13-11
UV dose required for inactivation of *Cryptosporidium, Giardia,* and virus

Log Inactivation Credit	UV Dose (mJ/cm^2)		
	Cryptosporidium	Giardia	Virus[a]
0.5	1.6	1.5	39
1.0	2.5	2.1	58
1.5	3.9	3.0	79
2.0	5.8	5.2	100
2.5	8.5	7.7	121
3.0	12	11	143
3.5	15	15	163
4.0	22	22	186

[a]UV dose for virus based on adenovirus.
Source: Adapted from *Fed. Reg.*, Vol. 68, No. 154, August 11, 2003.

be found in an extensive report prepared by AWWARF and NYSERDA (2007).

Validation Testing of UV Reactors

At the present time there are a number of UV manufacturers that produce UV reactors suitable for the inactivation of microorganisms. Unfortunately, the performance of the various UV reactors varies from unit to unit and manufacturer to manufacturer. Because of the interest in utilizing UV by the water industry to obtain partial inactivation credit for *Cryptosporidium*, *Giardia*, and viruses (in some cases) and the need to protect public health, the United States and many other countries have established regulations and guidelines for the use of UV radiation for water and wastewater treatment. The regulations typically involve validation testing of the UV reactors to verify minimum levels of performance (i.e., specifically the delivered UV dose) under varying the conditions of operation including:

1. High and low water transmittance
2. Varying flow rate
3. Varying power levels
4. Simulated lamp aging

Testing is also used to determine a set of operating conditions that can be monitored on a continuing basis to be assured that the UV dose needed for the inactivation credit is delivered at all times. Operationally, the method of controlling the UV dose, as discussed previously, is of critical importance. A number of prevalidated UV reactors, varying in size from 40 L/min (10 gal/min) to 225 ML/d (60 mgd), are available from a number of manufacturers.

In general, validation testing must be done and certified by an independent third party. Typically, as illustrated in Fig. 13-40, validation testing involves:

1. Generation of a UV dose response curve for the challenge microorganism.
2. Determination of the inactivation achieved with the full-scale reactor, at the actual installation location or at an approved test site (see Fig. 13-42), using the challenge microorganism.
3. Determination of the UV dose corresponding to the measured inactivation achieved with the full-scale reactor using the dose–response curve developed with the collimated beam. The computed UV dose delivered by the reactor is known as the reduced equivalent dose (RED).
4. Determination of a validated UV dose by dividing the RED value by a validation factor VF. The VF is used to account for the fact that a challenge microorganism was used instead of the target organism and for the experimental uncertainty associated with the testing program.

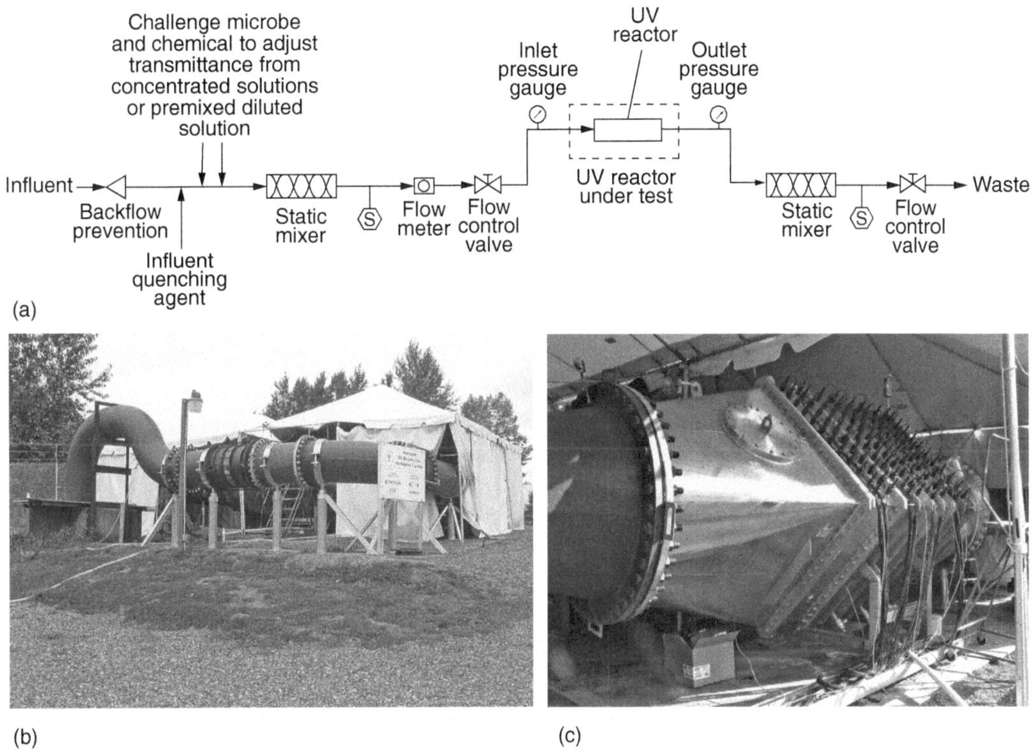

Figure 13-42
Experimental setup for validation of UV reactors under controlled conditions: (a) schematic of setup requirements for testing full-scale UV reactor, (b) view of test facility at Portland, OR, and (3) large UV reactor instrumented for UV dose validation by dosimetry.

For most drinking water applications the target RED value is 40 mJ/cm^2. The principal validation guidelines now used for the validation of various UV reactors are summarized in Table 13-12. Validation test centers in the United States are located in Johnstown, New Yark, and Portland, Oregen. While the approach of using a prevalidated UV reactor is favored by most Public Water Systems because of simplicity, it tends to be more conservative as compared to the onsite validation.

U.S. EPA UV Disinfection Guidance Manual Validation Process

As discussed previously in Chap. 4, the U.S. EPA developed the Long Term 2 Enhanced Surface Water Treatment Rule (LT2) to protect public health by further reducing the microbial contamination of drinking water. Based on the source water *Cryptosporidium* concentrations and current treatment practices, additional treatment may be required for some public water systems (PWS). Public water systems utilizing surface that must provide additional treatment under the LT2 rule can utilize UV radiation as one of the many different treatment options to meet the treatment requirements.

Table 13-12
UV reactor validation protocols used in the United Stated and Europe

Test Protocol	Discussion
German DVGW W294-3 (GAGW, 2003)	Use of reference sensor with multiple set points and a minimum number of monitoring ports. Although the protocol has a 10-year history, many feel the protocol is too prescriptive. UV validation based on a dose of 40 mJ/cm^2. Test results guarantee a UV dose of 40 mJ/cm^2 or more.
Austrian ONORM M5873-1 Low pressure, and M5873-2 Medium pressure (Onorm, 2001, 2003)	Use of reference sensor with multiple set points. UV validation based on a dose of 40 mJ/cm^2. Test results guarantee a UV dose of 40 mJ/cm^2 or more.
U.S. EPA UV Disinfection Guidance Manual (UVDGM) (U.S. EPA, 2006)	Greater flexibility as compared to German and Austrian guidelines, but more complex to understand. With proper testing, potential to reduce cost. Used for validation of community scale UV systems
ANSI/NSF Standard 55 (ANSI/NSF, 2004)	Unit must produce UV dose of 40 mJ/cm^2 at the alarm set point. A UV sensor to measure UV intensity continuously at 254 nm, a flow control device, and other related appurtenances are required. The challenge microorganism is MS2. Protocol is applied to residential point of use devices primarily.
National Water Research Institute (NWRI, 2003)	Developed primarily for wastewater reuse applications. Discussion of water applications is limited

Recognizing the desire of PWSs to use UV radiation to meet drinking water disinfection standards established under the Safe Drinking Water Act (SDWA), the U.S. EPA developed the UV Disinfection Guidance Manual (UVDGM) (U.S. EPA, 2006) to (1) delineate the design, operation, and maintenance needs for UV disinfection systems, which are quite different from those traditionally used in drinking water applications, (2) clarify the requirements for UV disinfection in the LT2 rule, and (3) familiarize states and PWSs with these distinctions, as well as associated regulatory requirements contained in the LT2 rules.

Two validation protocols are set forth in the UVDGM, the details of which are beyond the scope of this book. The two approaches are as follows:

1. PWS purchases a prevalidated UV reactor(s).

a. If the UV reactor(s) are installed in accordance with specified hydraulic constraints, onsite validation is not necessary.
b. Onsite validation may be necessary if the full UVT range was not tested in the offsite validation, if the hydraulic constraints cannot be met, and/or if more information is needed to match current operation conditions.

2. PWS purchases a UV reactor that has not been prevalidated.
   a. In this case, the PWS can develop a plan for offsite validation and has the flexibility of using any hydraulic installation option.
   b. PWS develops a validation plan and conducts an onsite validation, as outlined in the UVDGM.

Problems and Discussion Topics

13-1 Based on your reading of this chapter, provide brief responses to the following questions:

a. In waterworks practice, what two activities are described with the term disinfection?
b. What were the two principal means of controlling waterborne disease for the first five decades after John Snow did his work with cholera in the 1850s?
c. Why did chlorination encounter difficulties from the start?
d. What was the main discovery that caused concern about disinfection by-products?
e. What organism caused more stringent standards to be established for pathogen reduction?
f. What other organism was found to be so chlorine resistant that it began to raise questions about inactivation as a strategy for pathogen reduction?

13-2 Based on your reading of this chapter, discuss briefly two different ways in which the effect of concentration on the disinfection process can be handled. What are the advantages and disadvantages of each approach?

13-3 Describe how Watson proposed that the effect of concentration be handled in modeling disinfection.

13-4 Given below are some data from Wattie and Butterfield (1944) on the inactivation of *E. coli* with free chlorine at 2 °C and pH 8. Fit the data to the Chick–Watson, Rennecker–Mariñas, and Collins–Selleck models and comment on the results.

C, mg/L	T, min	log(N/N$_0$)
0.05	1.0	-0.02
0.05	3.0	-0.09
0.05	4.9	-0.15
0.05	9.6	-0.68
0.05	18	-2.52
0.07	1.0	-0.06
0.07	3.0	-0.22
0.07	4.9	-0.58
0.07	9.7	-2.28
0.14	1.0	-0.24
0.14	2.8	-0.95
0.14	4.5	-2.15

13-5 Fit the Rennecker–Mariñas model to the following disinfection data and determine the coefficient of leathality and the lag coefficient *b*:

C, mg/L	T, min	log(N/N$_0$)
1.0	5	0.0
1.1	10	0.0
1.05	25	-1.0
1.03	30	-1.5
1.05	35	-2.1
2.05	20	-2.55
2.0	23	-3.1
2.03	25	-3.45
5.02	11	-4.1

13-6 Using data from Table 13-3, estimate the *Ct* required for a 3 log reduction of *C. parvum* and *B. subtilus* using combined chlorine and chlorine dioxide. Is either one practical? What about UV?

13-7 From an examination of Fig. 13-5, which organism varies the least in *Ct* or *It* between one disinfectant and the next? Which varies the most? Which disinfectant shows the smallest range of *Ct* or *It* values required for all organisms?

13-8 A treatment plant has been designed to achieve 99 percent inactivation of *C. parvum* using ozonation. The engineer used data on ozonation of *C. parvum* at 20 °C for the design, but the plant operates in a northern climate and current estimates are that the low water temperature in some winters will be 0.5 °C. Estimate how much inactivation the plant will actually achieve when the water is at that temperature. You may assume that the inactivation of *C.*

parvum follows the Chick–Watson relationship (Eq. 13-4). Use the E_α value of Rennecker et al. (1999) as reported in Table 13-4.

13-9 Use the Segregated Flow Model (SFM) to redo the dispersion estimate in Example 13-4 two times. In the first estimate, assume the contactor is operated so that the product $C\tau$ is adequate to accomplish an 8 log reduction in the target organism, and in the second estimate assume the contactor is operated so that the product $C\tau$ is adequate to accomplish a 0.5 log reduction in the target. Discuss the implications of the results.

13-10 A treatment plant with a design capacity of 80 ML/d has a pipeline between the plant and the clearwell that the operators would like to use as a contactor for disinfection. The pipeline was built with future expansions in mind and is 4 m in diameter and 80 m in length. What would be the dispersion in this pipeline when the plant is operating at design flow? Assume the Darcy–Weisbach friction factor is 0.02.

13-11 A water treatment plant with a capacity of 80 ML/d is being constructed. The plant includes a baffled chlorine contact chamber that has a length-to-width ratio of 40 : 1. Estimate both the dispersion and the t_{10}/τ ratio for the chamber. Assume the coefficient of nonideality for the design, C_i, is 5. Other than baffling, what sort of provisions might have been made to improve the basin's performance? What might the design engineer have done to confirm this performance before going to construction?

13-12 Given below are data on the decay of ozone gathered by Gurol and Singer (1982). Fit the data to the first-order decay model and to the parallel first-order decay model and discuss.

T,min	C, mg/L
0.0	8.15
0.3	6.95
0.6	5.80
1.0	5.05
1.4	4.95
1.6	4.07
2.0	3.95
2.4	3.60

13-13 A treatment plant doing color removal by coagulation is using combined chlorine as a means of residual control. There have been complaints about chlorinous odors. The plant is operating with a chlorine-to-ammonia molar ratio of 5 : 1 and at a pH of 7. What precautions might be taken to reduce the odor complaints?

13-14 A water plant has influent ammonia levels of about 0.5 mg/L as N. The utility plans on installing a basin to remove the ammonia by breakpoint chlorination prior to using free chlorine for disinfection. What should the hydraulic detention time of that basin be to ensure that the ammonia is completely removed? The water is highly buffered at a pH of approximately 7.

13-15 What can the second plant in Example 13-7 do to improve performance of its residual control system?

13-16 A utility in south Florida has converted its plant to sodium hypochlorite because of community complaints about the safety of using chlorine gas. The hypochlorite is delivered at a concentration of 7 percent by weight and stored in a new fiberglass tank that was installed behind the maintenance building. The plant delivers an average of 8 ML/d of water with a chlorine dose of 4 mg/L. The storage tank is just large enough for one delivery, about 40,000 L. Recently, the local health department sampled the system and found high levels of chlorate ion. Also, periodically, especially during the summer, the utility finds that the strength of its bleach has dropped substantially. What precautions might be considered to improve the situation?

13-17 A gas chlorine system is being designed for residual control in the discharge line of a water treatment plant. The maximum and minimum design flows are 19 and 1.90 ML/d, respectively. The treated water discharge pipe is 0.36 m in diameter. The velocity of the chlorine gas in the vacuum line from the chlorinator to the injectors is 2 m/s, and the line is 20 m (66 ft) in length. The pipe from the injector to the application point is 152 m long, and the design velocity in the pipe is 1.5 m/s. The chlorine application point and the residual sampling point on the discharge line are 150 m apart. The sample runs for 100 m in a 6.35-mm sample line. The sample pump is designed for a flow of 200 mL/min. Sample analysis takes 20 s, and signal response times are assumed to be instantaneous. Prepare a sketch of the control loop similar to Fig. 13-13. Prepare a table analyzing the loop time and comment on the strengths and weaknesses of this design.

13-18 A continuous-flow pilot ozonation system was used to ozonate surface water at several different doses. The results are tabulated below. Assuming the pilot system successfully emulated the ozone dosing stage of the full-scale design, plot a curve of the ozone residual versus ozone dose and estimate the ozone demand and the ozone dose required to achieve a residual of 1 mg/L entering the disinfection section downstream.

Ozone Applied, mg/L	Residual, mg/L
1.30	0.04
2.45	0.28
2.74	0.43
3.05	0.56
3.39	0.56
4.01	0.90
4.49	1.12
6.01	1.50
6.05	1.74

13-19 The data below by Hermanowicz et al. (1999) show the decay of ozone residual in treated water from the upper Hackensack River. Estimate the *Ct* that can be achieved after 20 min of contact time.

T, min	*C*, mg/L	*T*, min	*C*, mg/L
0	0.97	12	0.155
1	1.02	13	0.135
2	0.85	14	0.12
3	0.71	15	0.115
4	0.58	16	0.11
5	0.49	17	0.105
6	0.41	18	0.1
7	0.35	19	0.1
8	0.295	20	0.095
9	0.25	21	0.09
10	0.22	22	0.09
11	0.185		

13-20 A full-scale UV reactor was tested with MS 2 bacteriophage and was rated to have an effective UV dose of 25 mJ/cm^2. Using an analogy to the thought experiment shown on Fig. 13-1, how much flow could have been bypassed around the reactor during the test without changing $\log(N/N_0)$ for MS 2 by more than 10 percent? Assuming no short circuiting, how many logs of reduction should the reactor achieve with *C. parvum*? How many logs reduction in *C. parvum* would the reactor achieve if the bypass discussed earlier were to occur? Discuss the significance of these results.

13-21 Given the following UV disinfection data (courtesy B. Cooper, BoiVir Labs)determine for water sample number (to be selected by instructor) whether the results are consistent with the NWRI and U.S. EPA quality control limits and the expected log inactivation as a UV dose of 50 mJ/cm^2.

UV dose, mJ/cm²	Titer, Pfu/mL Water sample number				
	1	2	3	4	5
0.00	5.30E+05	1.60E+05	2.80E+05	5.00E+06	2.60E+06
20.00	3.10E+04	1.30E+04	2.30E+04	4.00E+05	1.50E+05
40.00	5.30E+03	1.70E+03	1.90E+03	4.30E+04	1.70E+04
80.00	1.20E+02	6.00E+01	6.70E+01	8.70E+02	3.60E+02
100.00	2.20E+01	1.40E+01	1.30E+01	1.20E+02	7.00E+01

13-22 In Example 13-12, a linear relationship was used to define the UV dose response for MS2. What difference will it make with respect to the required UV dose if the linear relationship is replaced with a polynomial of the following form.

$$\log(N/N_0) = a + b(\text{UV dose}) + \text{c}(\text{UV dose})^2$$

where *a*, *b*,and *c* are empirical constants.

13-23 Verify the results given in Example 13-11 for the error of the following variables are correct.

$$U_{\text{ES}} = 6.23\ \text{mJ/cm}^2 \text{ and } 5.0\%$$
$$U_{\text{Pf}} = 2.65\ \text{mJ/cm}^2 \text{ and } 2.13\%$$
$$U_a = -0.7\ \text{mJ/cm}^2 \text{ and } -0.56\%$$
$$U_d = -0.61\ \text{mJ/cm}^2 \text{ and } -0.49\%$$
$$U_L = 0.038\ \text{mJ/cm}^2 \text{ and } 0.03\%$$

13-24 Review the current literature on the use of light emitting diode (LED) UV lamps for disinfection and prepare a brief assessment of their feasibility. A minimum of three articles, dating no further back than the year 2000, should be cited in your assessment.

References

ANSI/NSF 2004. Standard Number 55, *Ultraviolet Microbiological Water Treatment Systems* American National Standards Institute/National Sanitation Foundation, Ann Arbor, MI.

AWWA Disinfection Systems Committee (2008) "Committee Report: Disinfection Survey, Part 1—Recent Changes, Current Practices, and Water Quality," *J. AWWA*, **100**, 10, 76–90.

AwwaRF (1991) *Ozone in Water Treatment: Applicatjon and Engineering*, Cooperative Research Report, B. Langlais, D. Reckhow, and D. Brink (eds.), American Water Works Association, Research Foundation, Denver, CO, and Lewis Publishers, Chelsea, MI.

AwwaRF and NYSERDA (2007) *Optimizing UV Disinfection*, American Water Works Association. Research Foundation and New York State Energy Research and Development Authority, Denver, CO.

Baker, M. (1948) *The Quest for Pure Water*, Vol. *I*, 2nd ed., American Water Works Association, Denver, CO.

Bellamy, W., Haas, C., and Finch, G. (1998) *Integrated Disinfection Design Framework*, American Waterworks Research Foundation, Denver, CO.

Bellar, T. A., and Lichtenberg, J. J. (1974) "Determining Volatile Organics at Microgram-per-Litre Levels by Gas Chromatography," *J. AWWA*, **66**, 12, 739–744.

Blatchley, E. R., Shen, C., Scheible, O.K., Robinson, J.P., Ragheb, K., Bergstrom, D.E., and Rokjer, D. (2008) "Validation of Large-Scale, Monochromatic UV Disinfection Systems for Drinking Water using Dyed Microspheres," *Water Res.*, **42**, 3, 677–688.

Bohrerova, Z., Bohrer, G., Mohanraj, S., Ducoste, J., and Linden, K.G. (2005) "Experimental Measurements of Fluence Distribution in a UV Reactor Using Fluorescent Dyed Microspheres," *Environ. Sci. Technol.*, **29**, 22, 8925–8930.

Bolyard, M., Fair, P., and Hautman, D. (1992) "Occurrence of Chlorate in Hypochlorite Solutions Used for Drinking Water Disinfection," *Environ. Sci. Tech.*, **26**, 8, 1663–1665.

Bolyard, M., Fair, P., and Hautman, P. (1993) "Sources of Chlorate Ion in US Drinking Water," *J. AWWA*, **85**, 9, 81–88.

Brazis, A., Leslie, J., Kabler, P., and Woodward, R. (1958) "The Inactivation of Spores of *Bacillus globiglii* and *Bacillus anthracis* by Free Available Chlorine", *Public Health Reports*, **6**, 338–342.

Bull, R., Gerba, R., and Trussell, R. (1990) "Evaluation of Health Risks Associated with Disinfection," *Crit. Rev. Environ. Control*, **20**, 2, 77–114.

Butterfield, C., and Wattie, E. (1946) "Influence of pH and Temperature on the Survival of Coliforms and Enteric Pathogens When Exposed to Chloramine," *Public Health Reports*, **61**, 6, 157–193.

Butterfield, C., Wattie, E., Megregian, S., and Chambers, C. (1943). "Influence of pH and Temperature on the Survival of Coliforms and Enteric Pathogens When Exposed to Free Chlorine," *Public Health Reports*, **58**, 51, 1837–1866.

Buxton, G., and Subhani, M. (1971) "Radiation Chemistry and Photochemistry of Oxychlorine Ions," *Faraday Trans.*, **68**, 5, 958–971.

Cal DHS (1999) *Proposed Regulations: Water Recycling Criteria*, California Department of Health Services, Drinking Water Technical Programs Branch, Sacramento, CA.

Cal DHS (2002) *Drinking Water Action Levels: Contaminants of Current Interest*, California Department of Health Services, Sacramento, CA; also available at http://www.dhs.ca.gov/ps/ddwem/chemicals/AL/actionlevels.htm.

Chick, H. (1908) "An Investigation of the Laws of Disinfection," *J. Hygiene*, **8**, 92–158.

Chiu, K., Lyn, D., Savoye, P., and Blatchley, E. (1999) "Integrated UV Disinfection Model Based on Particle Tracking," *J. Environ. Engr., ASCE*, **125**, 1, 7–15.

Collins, H., and Selleck, R. (1971) "Problems in Obtaining Adequate Sewage Disinfection," *JSAE, ASCE*, SA5, **97**, 10, 549–562.

Cooper, R., Salveson, A., Sakaji, R., Tchobanoglous, G., Requa, D., and Whitley, R. (2001) Comparison of the Resistance of MS-2 and Poliovirus to UV and Chlorine Disinfection, paper presented at the Proceedings WateReuse 2000, Napa Valley, CA, Fountain Valley, CA.

Corona-Vasquez, B., Rennecker, J., Driedger, A., and Mariñias, B. (2002) "Sequential Inactivation of *Cryptosporidium parvum* Oocysts with Chlorine Dioxide Followed by Free Chlorine or Monochloramine," *Water Res.*, **36**, 1, 178–188.

Craik, S., Weldon, D., Finch, G., Bolton, J., and Belosevic, M. (2001) "Inactivation of *Cryptosporidium parvum* Oocysts Using Medium Pressure and Low Pressure Ultraviolet Light," *Water Res.*, **35**, 6, 1387–1398.

Crozes, G., Hagstrom, J., Clark, M., Ducoste, J., and Burns, C. (1999) *Improving Clearwell Design for Ct Compliance*, American Water Works Association Research Foundation, Denver, CO.

Driedger, A., Rennecker, J., and Mariñas, B. (2000) "Sequential Inactivation of *Cryptosporidium parvum* Oocysts with Ozone and Free Chlorine," *Water Res.*, **34**, 14, 3591–3597.

Driedger, A., Rennecker, J., and Mariñas, B. (2001) "Inactivation of *Cryptosporidium parvum* Oocysts with Ozone and Monochloramine at Low Temperature," *Water Res.*, **35**, 1, 41–48.

Drown, T. (1893/1894) "Electrical Purification of Water," *J. NEWWA*, **8**, 183–186.

Ducoste, J., Carlson, K., and Bellamy, W. (2001) "The Integrated Disinfection Design Framework Approach to Reactor Hydraulics Characterization," *J. Water Supply Res. Technol.-Aqua*, **50**, 44, 245–261.

Ducoste, J., Liu, D., and Linden, K.G. (2005) "Alternative Approaches To Modeling Dose Distribution and Microbial Inactivation in Ultraviolet Reactors: Lagrangian vs Eulerian," *J. of Environ. Engr., ASCE*, **131**, 10, 1393–1403.

DVGW (1997) *DVGW-W294 UV Disinfection Devices for Drinking Water Supply— Requirements and Testing*, German Gas and Water Management Union, Bonn, Germany.

Einstein, A. (1905) "Über einen die Erzeugung und Verwandlung des Lichtes betreffenden heuristischen Gesichtspunkt," *Ann. Physik*, **17**, 3, 131–148.

Elder, J. (1959) "The Dispersion of Marked Fluid in Turbulent Shear Flow," *J. Fluid Mech.*, **5**, 544–560.

Federal Register (1993) *Code of Federal Regulations* (29 CFR), Part 1915, **58**, FR 35514.

Finch, G., Haas, C., Openheimer, J., Gordon, G., and Trussell, R. (2001) "Design Criteria for Inactivation of *Cryptospordium* by Ozone in Drinking Water," *Ozone: Sci Eng.*, **23**, 4, 259–284.

Floyd, R., and Sharp, D. (1979) "Inactivation by Chlorine of Single Poliovirus Particles in Water," *Environ. Sci. Tech.*, **13**, 4, 138–442.

Floyd, R., Sharp, D., and Johnson, J. (1978) "Inactivation of Single Poliovirus Particulates in Water by Hypobromite Ion, Molecular Bromine, Dibromamine and Tribromamide," *Environ. Sci. Technol.*, **16**, 7, 377–383.

Fuller, G. W. (1897) *Report on the Investigations into the Purification of Ohio River Water at Louisville, KY,* Van Nostrand, New York.

Galasso, G., and Sharp, D. (1965) "Effect of Particle Aggregation on Survival of Irradiated Viruses," *J. Bacteriol.*, **90**, 4, 1138–1142.

Gard, S. (1957) "Chemical Inactivation of Viruses," pp. 123–146 in *CIBA Foundation Symposium on the Nature of Viruses,* Little Brown, Boston, MA.

GAGW (2003) *Technical Standard DVGW 294, UV Systems for German Association on Gas and Water,* 2nd ed., German Association on Gas and Water, Bonn, Germany.

Giese, N., and Darby, J. (2000) "Sensitivity of Microorganisms to Different Wavelengths of UV Light—Implications on Modeling of Medium Pressure UV Systems," *Water Res.*, **34**, 16, 4007–4013.

Gordon, G., Adam, L., and Bubnis, B. (1995a) *Minimizing Chlorate Formation in Drinking Water When Hypochlorite Ion Is the Chlorinating Agent,* AWWA American Water Works Assocation Research Foundation, Denver, CO.

Gordon, G., Adam, L., and Bubnis, B. (1995b) "Minimizing Chlorate Ion Formation," *J. AWWA,* **87**, 6, 97–106.

Gordon, G., Adam, L., Bubnis, B., Hoyt, B., Gillette, S., and Wilczek, A. (1993) "Controlling the Formation of Chlorate Ion in Hypochlorite Feedstocks, *J. AWWA,* **85**, 9BI, 89–97.

Gordon, G., Adam, L., Bubnis, B., Kuo, C., Cushing, R., and Sakaji, R. (1997) "Predicting Liquid Bleach Decomposition," *J. AWWA,* **89**, 4, 142–149.

Graber, D. (1972) "Discussion/Communication on: 'Hydraulic Model Studies of Chlorine Contact Tanks,"' *J. WPCF,* **44**, 10, 2029–2035.

Grasso, D., and Weber, W. (1989) "Mathematical Interpretation of Aqueous-Phase Ozone Decomposition Rates," *J. Environ. Engr. ASCE,* **115**, 541–559.

Griese, M., Hauser, K., Berkemeier, M., and Gordon, G. (1991) "Using Reducing Agents to Eliminate Chlorine Dioxide and Chlorite Ion Residuals in Drinking Water," *J. AWWA,* **85**, 5, 56–61.

Gurol, M., and Singer, P. (1982) "Kinetics of Ozone Decomposition: A Dynamic Approach," *Environ. Sci. Technol.*, **16**, 7, 377–383.

Haas, C. (1979) "Discussion of Kinetics of Bacterial Deactivation with Chlorine," *J. Environ. Eng. ASCE,* **105**, 1198–1199.

Haas, C., and Heller, B. (1990) "Kinetics of Inactivation of *Giardia lamblia* by Free Chlorine," *Water Res.*, **24**, 2, 233–238.

Haas, C., and Joffe, J. (1994) "Disinfection under Dynamic Conditions: Modification of Hom's Model for Decay," *Environ. Sci. Technol.*, **28**, 7, 1367–1369.

Haas, C., Joffe, J., Anmangandla, U., Hornberger, J., Heath, M., Jacangelo, J., and Glicker, J. (1995) *Development and Validation of Rational Design Methods of Disinfection,* American Water Works Association Research Foundation, Denver, CO.

Haas, C., Joffe, J., Anmangandla, U., Jacangelo, J., and Heath, M. (1996) "The Effect of Water Quality on Disinfection Kinetics," *J. AWWA,* **88**, 95–103.

Haas, C., and Karra, S. (1984) "Kinetics of Microbial Inactivation by Chlorine—I. Review of Result in Demand-Free Systems," *Water Res.*, **18**, 11, 1443–1449.

Haas, C., and Karra, S. (1984) "Kinetics of Waste water Chlorine Demand Exertion," *J. WPCF,* **56**, 170–182.

Hannoun, I., Boulos, P., and List, J. (1999) "Using Hydraulic Modeling for CT Compliance," *J. AWWA*, **90**, 8, 77–87.

Harris, G., Adam, V., Sorenson, D. L., and Curtis, M. S. (1987) "Ultraviolet Inactivation of Selected Bacteria and Viruses," *Water Res.*, **6**, 687–692.

Hart, F. (1979) "Improved Hydraulic Performance of Chlorine Contact Chambers," *J. WPCF*, **51**, 12, 2868–2875.

Henry, D. J., and Freeman, E. M. (1996) "Finite Element Analysis and T10 Optimization of Ozone Contactors," *Ozone Sci. Eng.*, **17**, 587–606.

Hermanowicz, S., Bellamy, W., and Fung, L. (1999) "Variability of Ozone Reaction Kinetics in Batch and Continuous Flow Reactors," *Water Res.*, **33**, 2130–2138.

Hess, S., Diachishin, A., and De Falco, Jr., P. (1953) "Bactericidal Effects of Sewage Chlorination, Theoretical Aspects," *Sewage Ind. Wastes*, **25**, 909–917.

Hoehn, R. C., Dietrich, A. M., Farmer, W. S., Orr, M. P., Lee, R. G., Aieta, M., Wood, D. W. III, and Gordon, G. (1990) "Household Odors Associated with the Use of Chlorine Dioxide," *J. AWWA*, **81**, 4, 166–172.

Hoigné, J., and Bader, H. (1976) "Role of Hydroxyl Radical Reactions in Ozonation Processes in Aqueous Solutions," *Water Res.*, **10**, 377–386.

Hunt, N., and Mariñas, B. (1997) "*Escherichia coli* Inactivation with Ozone" *Water Res.*, **31**, 1355–1267.

Hunt, N., and Mariñas, B. (1999) "Inactivation of *Escherischia coli* with Ozone: Chemical and Inactivation Kinetics," *Water Res.*, **33**, 11, 2633–2641.

Iatrou, A., and Knocke, W. (1992) "Removing Chlorite by the Addition of Ferrous Iron," *J. AWWA*, **86**, 11, 63–68.

Jacangelo, J. G., Laîné, J. M., Carns, K. E., Cummings, E. W., and Mallevialle, J. (1989) "Low-Pressure Membrane Filtration for Removing *Giardia* and Microbial Indicators," *J. AWWA*, **83**, 9, 97–106.

Jacangelo, J., Patania, N., Haas, C., Gerba, C., and Trussell, R. (1997) *Inactivation of Waterborne Emerging Pathogens by Selected Disinfectants*, Report No. 442, American Water Works Research Foundation, Denver, CO.

JMM (1991) *Disinfection Report for the Water Treatment Pilot Study*, The City of Portland Bureau of Water Works, Portland, OR.

Johnson, G. A. (1911) "Hypochlorite Treatment of Public Water Supplies," *Am. J. Public Health*, 562–565.

Katzenelson, E., Kletter, B., Schechter, H., and Shuval, H. (1974) Inactivation of Viruses and Bacteria by Ozone, in A. Rubin (ed.), *Chemistry of Water Supply, Treatment, and Distribution*, Ann Arbor Science, Ann Arbor, MI.

Kawamura, S. (2000) *Integrated Design and Operation of Water Treatment Facilities*, 2nd ed., Wiley-Interscience, New York.

Kimball, A. (1953) "The Fitting of Mulit-Hit Survival Curves," *Biometrics*, **9**, 6, 201–211.

Kim, J., Tomiak, R., and Mariñas, B. (2002a) "Inactivation of *Cryptosporidium* Oocysts in a Pilot-Scale Ozone Bubble-Diffuser Contactor. I: Model Development," *J. Environ. Eng. ASCE*, **128**, 6, 514–521.

Kim, J., Tomiak, R., Rennecker, J., Mariñas, B., Miltner, R., and Owens, J. (2002b) "Inactivation of *Cryptosporidium* in a Pilot-Scale Ozone Bubble-Diffuser Contactor. Part II: Model Verification and Application." *J. Environ. Eng.*, **128**, 6, 522–532.

Kim, J., Urban, M., Echigo, S., Minear, R., and Mariñas, B. (1999) Integrated Optimization of Bromate Formation and *Cryptosporidum parvum* Oocyst Control in Batch and Flow-Through Ozone Contactors, Proc. 1999 American Water Works Association Water Quality Technology Conference, on CD, Tampa, FL.

Knudson, G. (1986) "Photoreactivation of Ultraviolet-Irradiated, Plasmid-Bearing and Plasmid-Free Strains of *Bacillus anthracis,*" *Appl. Environ. Microbiol.*, **52**, 3, 444–449.

Krasner, S., and Barrett, S. (1984) Aroma and Flavor Characteristics of Free Chlorine and Chloramines, pp. 381–389. *Proc. AWWA WQTC*, American Water Works Association, Denver, CO.

Larson, M., and Mariñas, B. (2003) "Inactivation of *Bacillus subtilis* Spores with Ozone and Monochloramine," *Water Res.*, **37**, 4, 833–844.

Lawler, D., and Singer, P. (1993) "Analyzing Disinfection Kinetics and Reactor Design: A Conceptual Approach versus the SWTR," *J. AWWA*, **97**, 11, 67–76.

Le Chevallier, M., Cawthon, C., and Lee, R. (1988) "Factors Promoting Survival of Bacteria in Chlorinated Water Supplies," *Appl. Environ. Microbiol.*, **54**, 2492–2499.

Lev, O., and Regli, S. (1992) "Evaluation of Ozone Disinfection Systems: Characteristic Time *T*," *J. Environ. Eng. ASCE*, **118**, 268.

Linden, K., Shin, G., and Sobsey, M. (2001) "Comparative Effectiveness of UV Wavelengths for the Inactivation of *Cryptosporidium parvum* oocysts in water," *Water Sci. Technol.*, **43**, 12, 171–174.

Linden, K.G. and Rosenfeldt, E.J. (2011) "Ultraviolet Light Processes," Chap. 18, in J.K. Edzwald (ed.) *Water Quality And Treatment: A Handbook Drinking Water*, 6th ed., American Water Works Association, Denver CO.

Lister, M. (1952) "Decomposition of Sodium Hypochlorite," *Can. J. Chem*, **30**, 879.

Lister, M. (1956) "Uncatalyzed and Catalyzed Decomposition of Sodium Hypochlorite," *Can. J. Chem.*, **34**, 6, 465–478.

Loge, F., Bourgeous, K., Emerick, R., and Darby, J. (2001) "Variations in Wastewater Quality Parameters Influencing UV Disinfection Performance: Relative Impact of Filtration," *J. Environ. Eng. ASCE*, **127**, 9, 832–837.

Lotepro (2002) Technical Bulletin. Available at: www.loteproesg.com/DownLoads/OXGEN3.pdf.

Louie, D., and Fohrman, M. (1968) "Hydraulic Model Studies of Chlorine Mixing and Contact Chambers," *J. WPCF*, **40**, 174.

Lyn, D.A. and Blatchley, E.R. (2005) "Numerical Computational Fluid Dynamics Based Models of Ultraviolet Disinfection Channels," *Journal of Environmental Engineering, ASCE*, **131**, 6, 838–849.

Marske, D., and Boyle, J. (1973) "Chlorine Contact Chamber Design—a Field Evaluation," *Water Sewage Works*, **120**, 1, 70–76.

Masschelein, W. J. (1992) *Unit Processes in Drinking Water Treatment,* Marcel Decker, New York.

Morris, C. (1975) Aspects of the Quantitative Assessment of Germicidal Efficiency, pp. 1–10, in D. Johnson (ed.), *Disinfection: Water and Wastewater,* Ann Arbor Science, Ann Arbor, MI.

Morris, J.C. (1966) "The Acid Ionization Constant of HOCl from 5 to 35°," *J. Phys. Chem.*, **70**, 12, 3798–3806.

Najm, I., and Trussell, R. (2000) NDMA Formation in Water and Wastewater, in Proceedings American Water Works Association Water Quality Technology Conference, on CD, Denver, CO.

Najm, I., and Trussell, R. (2001) "NDMA Formation in Water and Wastewater," *J. AWWA*, **93**, 2, 92–99.

Naunovic, Z., Lim, S., and Blatchley, E.R. (2008) "Investigation of Microbial Inactivation Efficiency of a UV Disinfection System Employing an Excimer Lamp," *Water Res.*, **42**, 4838–4846.

Nowell, L., and Hoigné, J. (1992a) "Photolysis of Aqueous Chlorine at Sunlight and Ultraviolet Wavelengths—I. Degradation Rates," *Water Res.*, **26**, 5, 593–598.

Nowell, L., and Hoigné, J. (1992b) "Photolysis of Aqueous Chlorine at Sunlight and Ultraviolet Wavelengths—II. Hydroxyl Radical Production," *Water Res.*, **26**, 5, 599–605.

NWRI (2003) *Ultraviolet Disinfection Guidelines for Drinking Water and Water Reuse,* 2nd ed., National Water Research Institute, Fountain Valley, CA.

NWRI (2003) *Ultraviolet Disinfection Guidelines for Drinking Water and Water Reuse,* 2nd ed., National Water Research Institute, Fountain Valley, CA, in collaboration with American Water Works Association Research Foundation.

ÖNORM. 2001. Plants for the Disinfection of Water Using Ultraviolet Radiation— Requirements and Testing—Part 1: Low Pressure Mercury Lamp Plants. ÖNORM M 5873-1. Osterreichisches Normungsinstitut, Vienna, Austria.

ÖNORM. 2003. Plants for the Disinfection of Water Using Ultraviolet Radiation— Requirements and Testing—Part 2: Medium Pressure Mercury Lamp Plants. ÖNORM M 5873-2. Osterreichisches Normungsinstitut, Vienna, Austria.

Oppenheimer, J. A., Aieta, E. M., Trussell, R. R., Jacangelo, J. G., and Najm, I. N. (2000) *Evaluation of Cryptosporidium Inactivation in Natural Waters,* American Water Works Association Research Foundation, Denver, CO.

Oppenheimer, J., Gillogly, T., and Trussell, R. (2001) *Technical Memorandum to the Los Angeles Department of Water and Power,* Los Angeles, CA.

Oppenheimer, J., Gillogly, T., Stolarik, G., and Ward, G. (2002) Comparing the Efficiency of Low and Medium Pressure UV Light for Inactivating *Giardia muris* and *Cryptosporidium parvum* in Waters with Low and High Levels of Turbidity, in *Proc. 2002 Annual AWWA Conference and Exhibition,* New Orleans, LA. American Water Works Association, Denver, CO.

Palin, A. (1975) Water Disinfection—Chemical Aspects and Analytical Control, pp. 71–93 in J. Johnson (ed.) *Disinfection— Water and Wastewater,* Ann Arbor Science, Ann Arbor, MI.

Parker, J. A., and Darby, J. L. (1995) "Particle-Associated Coliform in Secondary Effluents: Shielding from Ultraviolet Light Disinfection," *Water Environ. Res.*, **67**, 1065.

Powell, J., Hallam, N., West, J., Forster, C., and Simms, J. (2000) "Factors Which Affect Bulk Chlorine Decay Rates," *Water Res.*, **34**, 1, 117–126.

Qualls, R., Flynn, M., and Johnson, J. (1983) "The Role of Suspended Particles in Ultraviolet Disinfection," *J. WPCF*, **55**, 1280–1285.

Radziminski, C., Ballantyne, L., Hodson, J., Creason, R., Andrews, R., and Chauret, C. (2002) "Disinfection of *Bacillus subtilis* Spores with Chlorine Dioxide: A Bench-Scale and Pilot Scale Study," *Water Res.*, **36**, 1629–1639.

Rahn, O. (1973) *Physiology of Bacteria*, Blankston's and Son, Philadelphia, PA.

Rakness, K. L. (2005) *Ozone in Drinking Water Treatment: Process Design, Operation, and Optimization*, American Water Works Association, Denver, CO.

Rauth, A. (1965) The Physical State of Viral Nucleic Acid and the Sensitivity of Viruses to Ultraviolet Light, *Biophys. J.*, **5**, 257–273.

Reckhow, D., Legube, B., and Singer, P. (1986) "The Ozonation of Organic Halide Precursors: Effect of Bicarbonate," *Water Res.*, **20**, 8, 987–998.

Rennecker, J., Kim, J., Corona-Vasquez, B., and Mariñas, B. (2001) "Role of Disinfectant Concentration and pH in the Inactivation Kinetics of *Cryptosporidium parvum* Oocysts with Ozone and Monochloramine," *Environ. Sci. Tech.*, **35**, 13, 2752–2757.

Rennecker, J., Mariñas, B., Owens, J., and Rice, E. (1999) "Inactivation of *Cryptosporidium parvum* oocysts with ozone," *Water Res.*, **33**, 11, 2481–2488.

Rennecker, J., Mariñas, B., Rice, E., and Owns, J. (1997) Kinetics of *Cryptosporidium parvum* Oocyst Inactivation with Ozone, pp. 299–316, in *Proc. 1997 Annual AWWA Conference, Water Research, Vol. C.*

Roberts, P., Aieta, E., Berg, J., and Chow, B. (1980) *Chlorine Dioxide for Wastewater Disinfection: A Feasibility Evaluation*, Tech. Rep., No. 251, Stanford University, Palto Alto, CA.

Rook, J. J. (1974) Formation of Haloforms During the Chlorination of Natural Water, *Water Treatment Exam.*, **23**, 234–243.

Ruffell, K., Rennecker, J., and Mariñas, B. (2000) "Inactivation of *Cryptosporidium parvum* Oocysts with Chlorine Dioxide," *WaterRes.*, **34**, 3, 868–876.

Saunier, B., and Selleck, R. (1979) "The Kinetics of Breakpoint Chlorination in Continuous Flow Systems," *J. AWWA*, **71**, 3, 164–172.

Scarpino, P., Cronier, S., Zink, M., and Brigano, F. (1977) "Effect of Particulates on Disinfection of Enteroviruses and Coliform Bacteria in Water by Chlorine Dioxide,"paper 2B-3 Proceedings of AWWA Water Quality Technology Conference, Denver, CO.

Schieble, O. (1987) "Development of a Rationally Based Design Protocol for the Ultraviolet Light Disinfection Process," *J. WPCF*, **59**, 1, 25–31.

Selleck, R., Collins, H., and White, G. (1970) Kinetics of Wastewater Chlorination in a Continuous Flow Process, paper presented at the International Water Pollution Research Conference, San Francisco, CA.

Selleck, R., and Saunier, B. (1978) "Kinetics ofBacterial Deactivation with Chlorine," *J. Environ. Eng. ASCE*, **104**, 1197–1212.

Selleck, R., Saunier, B., and Collins, H. (1980) "Closure to Discussion of Kinetics of Bacterial Deactivation with Chlorine," *J. Environ. Eng. ASCE*, **106**, 1000–1002.

Sepp, E. (1981) "Optimization of Chlorination Disinfection Efficiency," *ASCE JEED*, **107**, EE1, 139–153.

Setlow, J. (1967) "The Effects of Ultraviolet Radiation and Photoreactivation," *Comprehensive Biochem.*, **27**, 157–209.

Severin, B. (1980) "Disinfection of Municipal Wastewater Effluents with Ultraviolet Light," *J. WPCF*, **52**, 7, 2007–2018.

Shen, C., Scheible, O. K., Chan, P., Mofidi, A., Yun, T.I., Lee, C.C., and Blatchley, E. R. (2009) "Validation of Medium-Pressure UV Disinfection Reactors by Lagrangian Actinometry using Dyed Microspheres," *Water Res.*, **43**, 1370–1380.

Sjenitzer, F. (1958) "How Much Do Products Mixin a Pipeline?" *Pipeline Eng.*, **12**, D-31–34.

Staehelin, J., and Hoigné, J. (1982) "Decomposition of Ozone in Water: Rate of Initiation by Hydroxide Ions and Hydrogen Peroxide," *Environ. Sci. Tech.*, **16**, 10, 676–681.

Staehelin, J., and Hoigné, J. (1985) "Decomposition of Ozone in Water in the Presence of Organic Solutes Acting as Promoters and Inhibitors of Radical Chain Reactions," *Environ. Sci. Tech.*, **19**, 12, 1206–1213.

Stevens, A. and Symons, J. (1977) "Measurement of Trihalomethanes and Precursor Concentration Changes," *J. AWWA* **69**, 10, 546–554.

Stolarik, G. F., Christie, D., Prendergast, R., Gillogly, T. E. T., and Oppenheimer, J. A. (2001) Long-Term Performance and Reliability of a Demonstration-Scale UV Reactor, in *Proc. of the First International Congress on Ultraviolet Technologies*, Washington, DC. International Ultraviolet Association, Ontario, Canada.

Taylor, G. (1954) "The Dispersion of Matter in Turbulent Flow through a Pipe," *Proc. Royal Soc.*, **A223**, 446–484.

Tchobanoglous, G., Burton, F., and Stensel, H. (2003) *Wastewater Engineering*, 4th ed., Metcalf and Eddy, McGraw-Hill, New York.

Tomiyasu, H., Fukutomi, H., and Gordon, G. (1985) "Kinetics and Ozone Decomposition in Basic Aqueous Solution," *Inorg. Chem.* **24**, 2962.

Trussell, R. R. (1992) Control Strategy I: Alternate Oxidants and Disinfectants and Disinfectant Residuals, pp. 43–95, in *Seminaron Control of Disinfectant By-products*, Proceedings 1992 Annual AWWA Conference, Philadelphia, PA. American Water Works Association, Denver, CO.

Trussell, R. R. (1993) Treatment for the Control of Disinfectant Byproducts and Disinfectant Residuals, in G. F. Craun (ed.), *Safety of Water Disinfection: Balancing Chemical and Microbial Risks*, International Life Science Institute (ILSI), Washington, DC.

Trussell, R., and Chao, J. (1977) "Rational Design of Chlorine Contact Tanks," *J. WPCF*, **49**, 4, 659–667.

Trussell, R. R., and Kreft, P. (1984) Engineering Considerations of Chloramine Application, pp. 47–73 in *Chlorination for THM Control: Principles and Practices*, AWWA Special Workshop, Dallas, TX.

Trussell, R. R., and Pollock, T. (1983) Design of Chlorination Facilities for Wastewater Disinfection, "paper presented at Wastewater Disinfection Alternatives—Design, Operation, Effectiveness," Preconference Workshop for 56th WPCF Conference, Atlanta, GA.

U.K. Department of the Environment (1999a) Transport and the Regions. Water Supply (Water Quality) (Amendment) *Regulations 1999.* Statutory Instruments 1999 No. 1524. *Cryptosporidium* in Water Supplies.

U.K. Department of the Environment (1999b) Transport and the Regions. Standard Operating Protocols for the Monitoring of *Cryptosporidium* Oocysts in Treated Water Supplies to Satisfy Water Supply (Water Quality) Amendment Regulations 1999 Statutory Instruments No. 1524.

U.S. EPA (1979) "Control of Trihalomethanes in Drinking Water. Final Rule," *Fed. Reg.*, **44**, 231, Nov. 29, 68624.

U.S. EPA (1986) *Design Manual: Municipal Wastewater Disinfection*, EPA/625/1-86/021, U.S. Environmental Protection Agency, Washington, DC.

U.S. EPA (1989) "Filtration and Disinfection; Turbidity, *Giardia lamblia*, Viruses, *Legionella*, and Heterotrophic Plate Count Bacteria. Final Rule," *Fed. Reg.* **54**, 124, June 29, 27486–27541.

U.S. EPA (1991) *Guidance Manual for Compliance with the Filtration and Disinfection Requirements for Public Water Systems Using Surface Water Sources*, U.S. Environmental Protection Agency, Washington, D.C.

U.S. EPA (1998) "Disinfectants and Disinfection By-Products Rule: Final Rule," *Fed. Reg.*, **63**, 241, Dec. 16, 69390.

U.S. EPA (1999) *Alternative Disinfectants and Oxidants Guidance Manual*, 815-R-99-014, U.S. Environmental Protection Agency, Washington, DC.

U.S. EPA (2006) *Ultraviolet Disinfection Guidance Manual, for the Final Long Term 2 Enhanced Surfance Water Treatment Ru EPA 815-R-06–00*, U.S. Environmental Protection Agency, Washington, DC.

U.S. EPA (2004) "National Primary Drinking Water Regulations: Stage 2 Disinfectants and Disinfection Byproducts Rule; National Primary and Secondary Drinking Water Regulations; Approval of Analytical Methods for Chemical Contaminants; Proposed Rule", *Fed. Reg.*, **68**, 159, 49548–49681.

U.S. PHS (1943) "National Census of Water Treatment Plants of the United States," *Water Works Eng.*, **96**, 63–117.

Wallis, P., van Roodselaar, A., Neurwirth, M., Roach, P., Buchanan-Mappin, J., and Mack, H. (1989) Inactivation of *Giardia* Cysts in a Pilot Plant Using Chlorine Dioxide and Ozone, pp. 695–708, in *Proceedings AWWA WQTC*, Philadelphia, PA. American Water Works Association, Denver, CO.

Watson, H. (1908) "A Note on the Variation of the Rate of Disinfection with Change in Concentration of the Disinfectant," *J. Hygiene*, **8**, 536.

Wattie, E., and Butterfield, C. (1944) "Relative Resistance of *Escherichia coli and Eberthella typhosa* to Chlorine and Chloramines," *Public Health Reports*, **59**, 52, 1661–1671.

Whipple, G. C. (1906) Disinfection as a means of water purification, pp. 266–288, in *Proc. AWWA*. American Water Works Association, New York.

White, G. C. (1999) *Handbook of Chlorination and Alternative Disinfectants*, 4th ed., Wiley-Interscience, New York.

WHO (1994) *Guidelines for Drinking Water Quality*, World Health Organization, Geneva, Switzerland.

Wickramanyake, G., Rubin, A., and Sproul, O. (1984a) "Inactivation of *Nagleria* and *Giardia* Cysts in Water by Ozonation," *J. WPCF*, **56**, 983–988.

Wickramanyake, G., Rubin, A., and Sproul, O. (1984b) "Inactivation of *Giardia lamblia* Cysts with Ozone," *Appl. Environ. Microbiol.*, **48**, 3, 671–672.

14 Air Stripping and Aeration

Stantec's Water Treatment: Principles and Design, Updated Third Edition. John C. Crittenden, R. Rhodes Trussell, David W. Hand, Kerry J. Howe, and George Tchobanoglous.

Terminology for Air Stripping and Aeration

Term	Definition
Absorption	Transfer of volatile substances from air to water.
Air Stripping	Process of removing or desorbing volatile and gaseous constituents from water into air by contacting fresh air with the contaminated water.
Aeration	Process of adding or absorbing gases (e.g., oxygen, ozone) from air into water by contacting the gaseous-laden air with the water.
Aspirator contactors	Devices that force pressurized water through a constriction, changing the pressure head of the water to velocity head, creating a low-pressure zone for atmospheric air or gas to be drawn into the water (e.g., Venturi tube).
Countercurrent packed tower	Tower in which water enters at the top and flows downward over a packing material while air is blown up from the bottom of the column and flows up through the voids of the column.
Desorption	Process for transfer of volatile substances from water to air.
Diffusion contactors	Devices that force compressed air into the water, forming bubbles and creating the surface area available for transfer of a constituent from one phase to another.
Droplet air–water contactors	Devices creating small water drops that are dispersed into fresh air for a given contact time to allow transfer of a constituent from one phase to the other depending on objective (e.g., absorption or desorption).

Term	Definition
Height of a transfer unit (HTU)	Parameter that is a measure of the effectiveness of a particular air–water contact device.
Mechanical contactors	Devices that agitate the water, creating renewed air- water surface for transfer of a constituent from one phase to the other.
Number of transfer units (NTUs)	Dimensionless number that is a measure of the effectiveness of stripping a constituent from water to air. As NTU increases, the maximum possible removal efficiency increases.
Packing factor	Parameter for a random packing material that is used in conjunction with the Eckert pressure drop correlation to estimate the gas pressure drop through the packing.
Random packing	Small geometrically designed irregularly-shaped pieces (typically plastic), randomly packed to a specified height in the tower to provide a high surface area and efficient air–water contact within the tower.
Stripping factor (*S*)	Dimensionless number defined as the slope of the equilibrium line to the slope of the operating line for countercurrent packed towers and that can be used to assess the ability of a constituent to be removed from water.
Thin-film contactors	Devices that allow water to flow over surfaces, creating a thin water film that is exposed to flowing air and allowing the constituent to be transferred from one phase to the other.

Air stripping and aeration are two water treatment unit processes that utilize the principles of mass transfer to accomplish specific water treatment objectives. Both of these water treatment unit processes bring air and water into intimate contact to transfer volatile substances from the water (e.g., hydrogen sulfide, carbon dioxide, volatile organic compounds) into the air or from the air (e.g., carbon dioxide, oxygen) into the water. The mass transfer process involving the removal of volatile substances from water into the air is known as *desorption.* Air stripping is one of the most common desorption processes used in water treatment. The addition of gases from air into water is the mass transfer process known as *absorption.* Aeration involving the addition of oxygen to water is a commonly used absorption process.

An understanding of the principles of the underlying mass transfer processes, including how to calculate diffusion coefficients and the basis for mass transfer correlations (discussed previously in Chap. 7), is necessary to design air strippers and aerators effectively. In this chapter, the focus is on the application of the aforementioned mass transfer principles to water treatment unit processes. Specific topics considered in this chapter include (1) an introduction to air stripping and aeration, (2) gas-liquid equilibria (Henry's law), (3) the classification of air stripping and aeration systems, (4) the fundamentals of packed tower air stripping, (5) analysis and design for packed tower air stripping, (6) an analysis of low-profile air strippers, (7) an analysis of spray aerators, and (8) other air stripping and aeration processes. Other gas-liquid contacting systems are presented in other chapters, such as Chap. 13, where ozone contactors are discussed.

14-1 Introduction to Air Stripping and Aeration

Water treatment objectives that can be achieved through the gas–liquid mass transfer process are summarized in Table 14-1. In both air stripping and aeration, air–water contactors are used to increase the contact opportunities between the gas and liquid phases. By increasing the air–water contact opportunities, the desorption or absorption mass transfer process is accelerated above the rate that would occur naturally, meaning volatile

Table 14-1
Applications of air–water mass transfer in water treatment

Examples	Water Treatment Objectives
	Absorption
O_2	Oxidation of Fe^{2+}, Mn^{2+}, S^{2-}; lake destratification
O_3	Disinfection, color removal, oxidation of selected organic compounds
Cl_2	Disinfection; oxidation of Fe^{2+}, Mn^{2-}, H_2S
ClO_2	Disinfection
CO_2	pH control
SO_2	Dechlorination
NH_3	Chloramine formation for disinfection
	Desorption
CO_2	Corrosion control
O_2	Corrosion control
H_2S	Odor control
NH_3	Nutrient removal
Volatile organics (e.g., $CHCl_3$)	Taste and odor control, removal of potential carcinogens

substances move more rapidly from the water into the air or gases that are not as soluble move more rapidly from the air into the water. The increase in contact opportunities between the two phases occurs through increasing the air–water interface in the air–water contactor by increasing the air–water ratio.

Bringing about Air–Water Contact

Over the years, a number of methods have been developed to bring about effective air–water contact. Packed towers or slat countercurrent flow towers, known as gas-phase contactors, have a continuous gas phase and a discontinuous water phase and are typically used to remove (or strip) gases or volatile chemicals from water. Air–water contactors such as basins with diffused aeration, also called bubble columns, are known as flooded contactors. In flooded contactors the water phase is continuous and the gas phase is discontinuous because the air is present as discontinuous bubbles. Flooded contactors are typically used to add gases (e.g., O_2, CO_2, O_3) into water.

One confusing concept with air stripping and aeration is that aerators can be used to accomplish air–water contact in both air-stripping and aeration processes. In general, aerators are a relatively simple method for increasing the air–water ratio by (1) spraying water into the air or (2) introducing air into the water through surface turbines or submerged nozzles and diffusors (bubble columns). Thus, aerators allow both of the mass transfer processes, desorption and absorption, to occur in a relatively cost-effective manner. However, because backmixing can occur in aeration systems, a high degree of removal may be difficult to achieve.

Air Stripping

Two major types of air–water contactors are used for air stripping: (1) towers and (2) aerators. The two principal factors that control the selection of the type of air–water contactor for stripping are (1) the desired degree of removal of the compound and (2) the Henry's law constant of the compound. Towers are used when either a high degree of removal is desired or the compound has a high affinity for water (is not very volatile so it has a low Henry's law constant), as shown on Fig. 14-1. Aerators are used when either the desired degree of removal is not very high or the gas has a low affinity for water (is very volatile so it has a high Henry's law constant). When removals less than 90 percent are required, both spray and diffused aeration systems, including mechanical aeration, may be economically attractive. More information on the various types of air-stripping and aeration systems used in water treatment is presented in Sec. 14-3.

Aeration

Aeration is used to increase the oxygen content in the water by adding air into water through (1) diffusors in a pipe, channel, or process basin; (2) cascading water over stacked trays; or (3) surface turbines and wheels that mix air into water at the top of basins. Oxygenation can also be

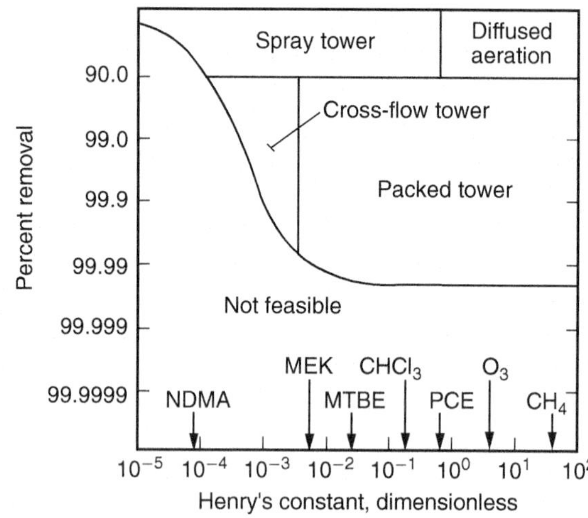

Figure 14-1
Schematic diagram for selection of feasible aeration process for control of volatile compounds (Adapted from Kavanaugh and Trussell, 1981).

accomplished using pure oxygen. The various types of aeration systems are presented in Sec. 14-3 and the specific details of spray aerators are covered in Sec. 14-7.

14-2 Gas–Liquid Equilibrium

When gas-free water is exposed to air, compounds such as oxygen and nitrogen will diffuse from the air into the water until the concentration of these gases in the water reaches equilibrium with the gases in the air. Conversely, if water in deep wells is brought to the ground surface, dissolved gases such as methane or carbon dioxide will be released to the air because their concentrations in groundwater typically exceed equilibrium conditions with air. The eruption of a carbonated beverage after it is opened is a more familiar example of carbon dioxide release after a pressure change. In each case, the driving force for mass transfer is the difference between the existing and equilibrium concentrations in the two phases, as discussed in Chap. 7.

Vapor Pressure and Raoult's Law

Consider water poured into a closed container that contains some headspace as shown on Fig. 14-2a. Some water molecules will have enough energy to overcome the attractive forces among the liquid water molecules and escape into the headspace above the liquid water, which is called evaporation. At the same time, some water molecules that have escaped into the gas phase above the liquid water may lose energy and move back into the liquid water, which is called condensation. When the rates of evaporation and condensation are equal, the system is at equilibrium. The partial pressure exerted by the water vapor above the liquid water in the

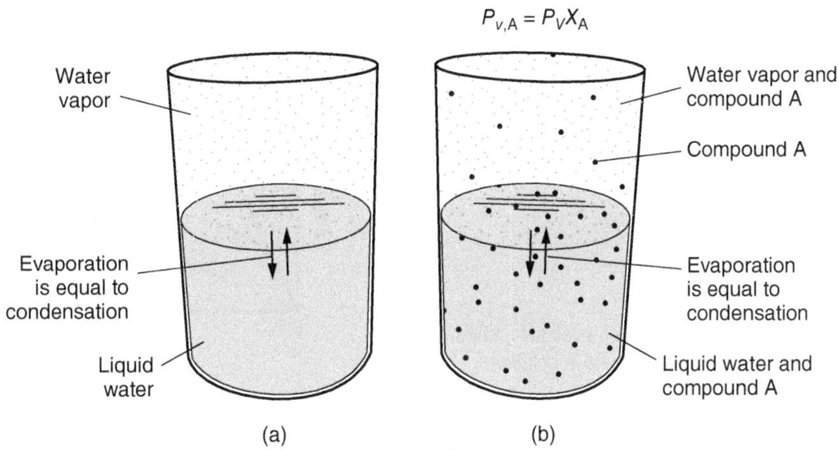

Figure 14-2
Schematic diagram for solution equilibrium description of vapor pressure with (a) vapor pressure of water and (b) partial pressure of compound A in the presence of water.

container at equilibrium is called the vapor pressure. Vapor pressure is dependent on temperature and increases with increasing temperature. For example, the vapor pressure of water is 1.23 kPa at 10°C and 3.17 kPa at 25°C. Other volatile liquids (e.g., acetone, benzene) behave the same way and also have a vapor pressure.

If a volatile compound (A) is placed in the same closed container containing water as shown on Fig. 14-2b, it too would come to equilibrium between the liquid and gas phases and exert a partial pressure above the liquid water. If the solution is assumed to behave ideally in which the molecular forces between the solute (A) and the solvent (water) are identical to the solvent-solvent forces, and the solute (or solvent) molecule behaves identically regardless of whether it is surrounded by solute or solvent molecules, then the partial pressure of the solute would be a function of its vapor pressure and the mole fraction of the solute. The partial pressure of solute A can be calculated from the following expression, known as Raoult's law:

$$P_A = P_{v,A} X_A \tag{14-1}$$

where P_A = partial pressure of solute A, Pa
$P_{v,A}$ = vapor pressure of pure liquid A, Pa
X_A = mole fraction of solute A in water, dimensionless

The mole fraction of A is defined as

$$X_A = \frac{n_A}{\sum_i n_i} = \frac{n_A}{n_A + n_{H_2O}} \tag{14-2}$$

where n = amount of A (solute) and water (solvent), mol

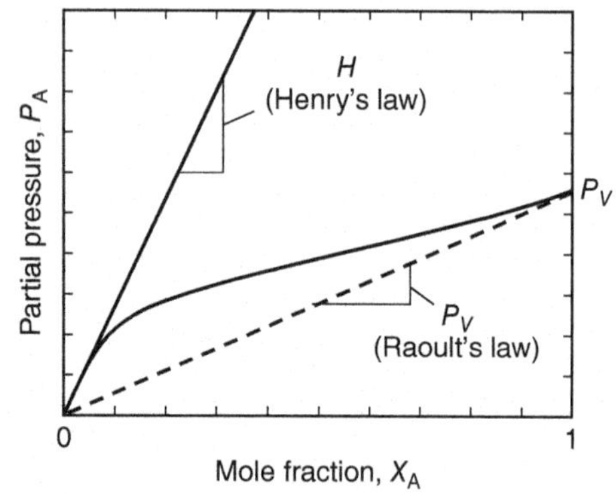

Figure 14-3
Relationship between partial pressure of a volatile compound and the mole fraction of the volatile compound in solution.

The relationship between partial pressure and mole fraction for solute A is illustrated on Fig. 14-3, and ideal solutions follow Raoult's law and the slope is $P_{v,A}$. For nonideal solutions the molecular forces between the solute and solvent are not identical to the solvent-solvent forces because the molecular forces between water molecules are very strong, so the solute–solvent attractions are smaller than the solvent-solvent attractions. Since there are smaller attractive forces holding the solute in solution, it is pushed out of solution and into the gas phase. Consequently, as shown, the partial pressure of the solute is higher than predicted by Raoult's law (a positive deviation from Raoult's law).

Henry's Law

For very dilute solutions most often found in environmental applications, the molecular interactions do not change significantly as additional solute is added, so partial pressure is proportional to mole fraction as shown in Figure 14-3, and this relationship is known as Henry's law. The equilibrium partitioning of a chemical solute between a liquid and gas phase is governed by Henry's law when the solute is very dilute in the mixture. Henry's law in equation form is

$$P_A = H_{PX} X_A \tag{14-3}$$

where $H_{PX,A}$ = Henry's law constant for solute A in solvent (water), atm

Henry's law is valid and constant up to mole fractions of 0.01 and has been shown to be valid for concentrations up to 0.1 mol/L (Rogers, 1994). Solvent–solvent forces are unaffected by small amounts of solute and the solvent follows Raoult's law for dilute solutions. Henry's law constants are valid for binary systems (e.g., component A in water). For systems where there are several solutes in a solvent (water) and the solution is still considered dilute, Henry's law will be valid for each solute. The presence of air does not affect the Henry's law constant for volatile organic chemicals (VOCs) or gases because the constituents of interest have low concentrations in air.

Other Units for Henry's Law Constants

The units of Henry's law constant, H_{PX}, in Eq. 14-3 are in atm because the standard conditions for pressure in the gas phase and concentration in the liquid phase were given in atmospheres and mole fractions, respectively. Henry's law constants can also be expressed in terms of concentration or partial pressure of A for the gas phase and mole fraction or concentration for the liquid or water phase. The gas-phase concentration expressed as either partial pressure (atm) or concentration in mol/L is related through the ideal gas law as shown below:

$$P_A V = n_A RT \quad \text{or} \quad Y_A = \frac{n_A}{V} = \frac{P_A}{RT} \tag{14-4}$$

where R = universal gas constant, 0.082057 atm · L/mol · K
$Y_A = n_A/V$ = gas-phase concentration (mol/L)
V = volume of the gas phase, m^3

The liquid-phase concentration can be expressed as either mole fraction (mol/mol) or concentration (mol/L) as

$$X_A = \frac{n_A}{n_A + n_W} \approx \frac{n_A}{n_W} = \frac{C_A}{C_W} \tag{14-5}$$

where n_W = amount of water in solution, mol

$$C_W = \frac{\text{density of water}}{\text{molecular weight of water}} = \frac{1000\ \text{g/L}}{18\ \text{g/mol}} = 55.56\ \text{mol/L}$$

$$C_A = X_A C_W = X_A(55.56\ \text{mol/L})$$

Applying these relationships results in three common forms of expressing Henry's law, which are summarized in Table 14-2. A particularly useful set of units is when the solute is expressed as concentration (either mass

Table 14-2
Unit conversions for Henry's law constants

$$H_{YC}\left(\frac{L_{H_2O}}{L_{Air}}\right) = \frac{H_{PC}[\text{atm/(mol/L)}]}{RT}$$

$$= \frac{H_{PX}(\text{atm})}{RT \times (55.6\ \text{mol}\ H_2O/\text{L}\ H_2O)}$$

$$H_{PC}[\text{atm/(mol/L)}] = \frac{H_{PX}(\text{atm})}{55.6\ \text{mol}\ H_2O/\text{L}\ H_2O}$$

$$H_{PX}(\text{atm}) = H_{YC}\left(\frac{L_{H_2O}}{L_{Air}}\right) \times RT \times (55.6\ \text{mol}\ H_2O/\text{L}\ H_2O)$$

Note: subscripts on H correspond to units as follows:
Y = gas phase concentration, C = liquid-phase concentration, P = partial pressure, X = mole fraction

or molar) in both the gas and liquid phases. These units are known as "dimensionless" forms of Henry's law and are widely used in environmental engineering. Use of the relationships displayed in Table 14-2 is illustrated in Example 14-1.

Example 14-1 Henry's law constants

What is the dimensionless Henry's law constant for a compound that has a value of 250 atm? What is the Henry's law constant in atmospheres and atm/(mol/L) for a compound that has a dimensionless Henry's law constant of 0.0545? Assume the temperature is 25°C.

Solution

1. Determine the dimensionless Henry's law constant. Inserting Henry's law constant of 250 atm into the relationship shown in Table 14-2 for converting H_{PC} to H_{YC} results in the expression

$$H_{YC} = \frac{H_{PX}}{RT(55.6)}$$
$$= \frac{250 \text{ atm}}{[0.082057 \text{ atm} \cdot \text{L/mol} \cdot \text{K}][(273 + 25)\text{K}](55.6 \text{ mol/L})}$$
$$= 0.183$$

2. Determine Henry's law constant in atmospheres. Rearranging the expression for H_{PX} in terms of H_{YC} and solving for H_{PX} for an H_{YC} of 0.0545, the following result is obtained:

$$H_{PX} = RT \times 55.6 H_{YC}$$
$$= [0.082057 \text{ atm} \cdot \text{L/mol} \cdot \text{K}][(273 + 25)\text{K}](55.6 \text{ mol/L})(0.0545)$$
$$= 74.1 \text{ atm}$$

3. Determine Henry's law constant in atm/(mol/L) for an H_{YC} equal to 0.0545. Inserting Henry's law constant of 0.0545 dimensionless into the relationship shown in Table 14-2 for converting H_{PC} to H_{YC} and solving for H_{PC} results in the expression

$$H_{PC} = RTH_{YC} = [0.082057 \text{ atm} \cdot \text{L/mol} \cdot \text{K}[(273 + 25)\text{K}](0.0545)$$
$$= 1.33 \text{ atm/(mol/L)}$$

Sources of Henry's Law Constants

Methods have been developed to determine Henry's law constants for volatile compounds. In the early 1980s methods that included measuring the compound's vapor pressure and solubility, direct measurement of a compound's vapor pressure and aqueous concentrations in an equilibrium system, and using batch air stripping techniques to determine Henry's law constants were evaluated and compared (Mackay and Shiu, 1981). However, these techniques can be unreliable, and a more suitable method was developed called the equilibrium partitioning in closed systems (EPICS) technique (Lincoff and Gossett, 1984). This method consists of the addition of equal masses of a volatile solute to two sealed serum bottles that are identical in all respects except they possess different water volumes. The gas-phase concentrations are measured, and the following equation is used to determine Henry's law constant (Gossett, 1987):

$$H_{YC} = \frac{V_{W2} - \left[\left(C_{g1}/M_1\right)/\left(C_{g2}/M_2\right)\right]V_{W1}}{\left[\left(C_{g1}/M_1\right)/\left(C_{g2}/M_2\right)\right]V_{g1} - V_{g2}} \tag{14-6}$$

where V_{W1} = volume of water in bottle 1, L
V_{W2} = volume of water in bottle 2, L
V_{g1} = volume of headspace in bottle 1, L
V_{g2} = volume of headspace in bottle 2, L
M_1 = total mass of of solute added to bottle 1, mol
M_2 = total mass of of solute added to bottle 2, mol
C_{g1} = concentration of solute in the gas in bottle 1, mol/L
C_{g2} = concentration of solute in the gas in bottle 2, mol/L

In the evaluation of H_{YC} using Eq. 14-6 the actual masses M_1 and M_2 do not need to be known but only their ratio. This means that if the same stock solution of a solute is used and injected into the two serum bottles, the actual concentration of the stock required need not be known because a gravimetric measure of the relative quantity of the stock added to the bottles is all that is needed (Gossett, 1987). The reported precision or coefficient of variation of this method is within 2 to 5 percent. This technique has become widely used to experimentally determine Henry's constants for VOCs (Gossett, 1987; Ashworth et al., 1988; Robbins et al., 1993; Dewulf et al., 1995; Heron et al., 1998; Ayuttya et al., 2001).

Henry's law constants are readily available in the published literature (Yaws et al., 1976; Mckay et al., 1979; Nicholson et al., 1984; U.S. EPA, 1986; Ashworth et al., 1988; Sander, 1999). Values can also be found in Internet databases, including at sites maintained by NIST (2011) and SRC (2011). Table 14-3 displays Henry's constants for several VOCs encountered in water supplies. Note their values change with temperature, and a discussion of the impact of temperature on Henry's constants is presented below. Most Henry's constants reported in the literature are performed using organic-free laboratory water. Natural waters used for drinking supply may contain concentrations of dissolved solids (50 to 600 mg/L TDS) and

Table 14-3
Dimensionless Henry's law constants for selected volatile organic chemicals

	Henry's Law Constants, H				
Component	**10°C**	**15°C**	**20°C**	**25°C**	**30°C**
Benzene	0.142	0.164	0.188	0.216	0.290
Carbon tetrachloride	0.637	0.808	0.96	1.210	1.520
Chloroform	0.0741	0.0968	0.1380	0.1720	0.2230
cis-1,2-Dichloroethylene	0.116	0.138	0.150	0.186	0.231
Dibromochloromethane	0.0164	0.0190	0.0428	0.0483	0.0611
1,2-Dichlorobenzene	0.0702	0.0605	0.0699	0.0642	0.0953
1,3-Dichlorobenzene	0.0952	0.0978	0.1220	0.1170	0.1700
1,2-Dichloropropane	0.0525	0.0533	0.0790	0.1460	0.1150
Ethylbenzene	0.140	0.191	0.250	0.322	0.422
Methyl ethyl ketone	0.01210	0.01650	0.00790	0.00532	0.00443
Methyl *t*-butyl ether (MTBE)	0.0117	0.0177	0.0224	0.0292	0.0387
m-Xylene	0.177	0.210	0.249	0.304	0.357
n-Hexane	10.3	17.5	36.7	31.4	62.7
o-Xylene	0.123	0.153	0.197	0.199	0.252
1,1,2,2-Tetrachloroethane	0.01420	0.00846	0.03040	0.01020	0.02820
Tetrachloroethylene	0.364	0.467	0.587	0.699	0.985
Toluene	0.164	0.210	0.231	0.263	0.325
Trichloroethylene	0.237	0.282	0.350	0.417	0.515

Source: Adapted from Ashworth et al. (1988).

natural organic matter (0.5 to 15 mg/L as DOC). The value of Henry's constants is not impacted by the range of these dissolved constituents in natural waters (Nicholson et al., 1984). The impact of high dissolved constituent concentrations on Henry's constants is discussed below. For water supplies that contain multiple VOCs in low concentrations, their Henry's constant values are not impacted by the other VOCs present in the water.

When experimental values of Henry's constant are not available, they can be estimated using software that incorporates molecular group or bond contribution calculations or from solubility and vapor pressure data as discussed in the following section.

Estimation of Henry's Constant Using Molecular Techniques or Using Vapor Pressure and Solubility

ESTIMATION OF HENRY'S CONSTANTS USING MOLECULAR TECHNIQUES

Molecular methods have been developed to estimate Henry's constants for solutes that lack reliable experimental data. Methods involving group contributions are widely used to estimate Henry's constants for solutes in water. Group contribution methods separate a molecular structure of a molecule into smaller parts, known as functional groups or fragments such that the molecule property is obtained from summing its functional groups. The small functional groups or segments are determined by fitting to a set of experimental data containing many solutes, while assuming the interaction parameter values for a functional group or fragment are independent of the molecules in which they make up and can be used to estimate Henry's constants of more complex solutes (Lin and Sandler, 2002). Group and bond contribution methods to predict Henry's constants of solutes in water have been developed and presented in the literature (Hine and Mookerjee, 1975; Nirmalakhandan and Speece, 1988; Meylan and Howard, 1991; Suzuki et al., 1992; Meylan, 1999; Lin and Sandler, 2000, 2002). Many of these methods are incorporated into user-friendly programs that can be used to estimate Henry's constants for solutes in water. The most widely known software program is HENRYWIN, which is a part of EPI (Estimation Programs Interface) Suite. EPI Suite is a Windows-based suite of physical/chemical property and environmental fate estimation programs developed by the EPA's Office of Pollution Prevention Toxics and Syracuse Research Corporation (Meylan, 1999). EPI Suite is available for free download on the EPA's website. Another commonly used software package is the System to Estimate Physical Properties (StEPP). StEPP contains a database of over 600 compounds and their physical properties including experimentally determined values of Henry constants as well as estimation methods using a group contribution method (Hokanson, 1996).

ESTIMATION OF HENRY'S CONSTANTS FROM VAPOR PRESSURE AND SOLUBILITY DATA

Compounds with strong repulsive interactions with water molecules have low aqueous solubilities. These compounds usually have large deviations between Henry's constant and vapor pressure. The ratio of vapor pressure to Henry's constant is approximately equal to the aqueous solubility of the compound as shown in the expression

$$C_{S,\,A} \approx \frac{P_{V,A}}{H_{PC,A}} \tag{14-7}$$

where $C_{S,A}$ = aqueous solubility of compound A, mol/L
$P_{V,A}$ = vapor pressure of compound A, atm

Consequently, Henry's constant for a slightly soluble compound A can be estimated from vapor pressure and aqueous solubility as

$$H_{PC,A} \approx \frac{P_{V,A}}{C_{S,A}} \tag{14-8}$$

For completely miscible compounds, the Henry law constant approaches the value of the vapor pressure:

$$H_{PX,A} \approx P_{V,A} \tag{14-9}$$

Using this approach, the estimated values are typically within ±50 to 100 percent of the experimentally reported values and should, therefore, only be used when measured values of the constants are not available. This approach is valid for compounds that are liquid at standard conditions.

Factors Influencing Henry's Constant

Temperature, pressure, ionic strength, surfactants, and solution pH (for ionizable species such as NH_3 and CO_2) can influence the equilibrium partitioning between air and water. The impact of total system pressure on H_{YC} is negligible because other components in air have limited solubility in water.

EFFECT OF TEMPERATURE

Henry's constants for several compounds (at different temperatures) and gases (at 20°C) are listed in Table 14-4. Assuming that the standard enthalpy change (ΔH°_{dis}) for the dissolution of a component in water is constant over the temperature range of interest, the change in H_{YC} with temperature can be estimated using the van't Hoff equation (see Chap. 5):

$$H_{YC,T_2} = H_{YC,T_1} \exp\left[\frac{-\Delta H^{\circ}_{vol}}{R}\left(\frac{1}{T_2} - \frac{1}{T_1}\right)\right] \quad (14\text{-}10)$$

where H_{YC,T_2} = dimensionless Henry's law constant at temperature T_2
H_{YC,T_1}, = dimensionless Henry's law constant at temperature T_1
ΔH°_{vol} = standard enthalpy change of dissolution in water, J/mol
R = universal gas constant, 8.314 J/mol · K
T_1, T_2 = absolute temperature, K (273 + °C)

Equation 14-10 can be simplified to the following expression, and K_C and ΔH°_{vol} values for selected compounds are reported in Table 14-4:

$$H_{YC} = K_C \exp\left(-\frac{\Delta H^{\circ}_{vol}}{RT}\right) \quad (14\text{-}11)$$

Table 14-4
Dimensionless Henry's law constants at 20°C and temperature dependence for gases in water

Compound	ΔH°_{vol}[a]	K_C	H_{YC}
Air	10.28	3,368	49.58
Ammonia	36.12	1,526	0.0006
Carbon dioxide	19.97	4,013	1.1
Chlorine	16.80	420	0.43
Chlorine dioxide	28.26	4,300	0.04
Hydrogen sulfide	17.84	567	0.38
Methane	14.86	12,402	28.41
Nitrogen	7.94	1,563	60.01
Oxygen	13.40	7,537	30.75
Ozone	24.28	83,848	3.92
Sulfur dioxide	23.15	358	0.03

[a]ΔH°_{vol} in units of kJ/mol.

The application of Eq. 14-11 to calculate H_{YC} as a function of temperature is illustrated in Example 14-2. Another common method of expressing the temperature dependence of H_{YC} is to define K_C and $\Delta H^\circ_{\text{vol}}/R$ as fitting parameters K_A and K_B, respectively, using the equation

$$H_{YC} = \exp\left(K_A - \frac{K_B}{T}\right) \tag{14-12}$$

Values of K_A and K_B for several compounds valid for temperatures ranging from 283 to 303 K are presented in Ashworth et al. (1988).

The relationship between H_{YC} and temperature over a wide temperature range is displayed in Fig. 14-4 for benzene and hexane. For most

Example 14-2 Henry's law constant and temperature effect

Calculate Henry's law constant at 5 and 20°C for ozone using Eq. 14-11 and Table 14-4.

Solution

1. At 5°C: Using Eq. 14-11 and the constants provided in Table 14-4, Henry's law constant for ozone at 20°C can be estimated as follows:

$$K_C = 83{,}848$$

$$\Delta H^\circ_{\text{vol}} = 24.28 \times 10^3 \text{ J/mol}$$

$$H_{YC} = K_C \exp\left(-\frac{\Delta H^\circ_{\text{vol}}}{RT}\right)$$

$$= 83{,}848 \exp\left\{-\frac{24.28 \times 10^3 \text{ J/mol}}{(8.314 \text{ J/mol·K})[(273+5)\text{ K}]}\right\}$$

$$= 2.3$$

2. At 20°C:

$$H_{YC} = K_C \exp\left(-\frac{\Delta H^\circ_{\text{vol}}}{RT}\right)$$

$$= 83{,}848 \exp\left\{-\frac{24.28 \times 10^3 \text{ J/mol}}{(8.314 \text{ J/mol·K})[(273+20)\text{K}]}\right\}$$

$$= 3.93$$

Comment

From the above computations, it is clear that temperature has an effect on the dimensionless form of Henry's law.

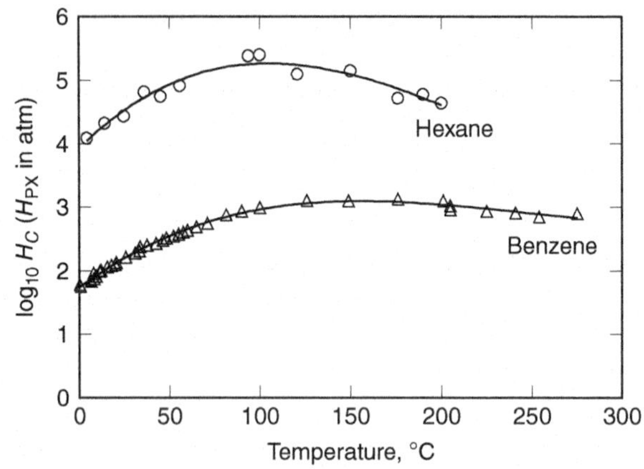

Figure 14-4
Relationship between Henry's law constant and temperature for benzene and heptane. (Adapted from Lan, 2009).

environmental applications where the temperature is less than 30°C, the relationship is nearly linear and can be approximated by the van't Hoff and related equations (Eqs. 14-11 and 14-12). Depending on the volatility of the compound, for temperatures greater than 30°C, H_{YC} increases nonlinearly with temperature reaching a maximum and then decreasing as temperature continues to increase (Miller and Hawthorne, 2000; Lau 2009). For benzene, H_{YC} reaches a maximum at 100°C followed by a decrease as temperature continues to increase. While this may not be important for air stripping at ambient temperatures, it may be important for other applications such as steam stripping of VOCs from industrial wastes.

IONIC STRENGTH

Gases or synthetic organic chemicals (SOCs) have a higher apparent Henry's law constant ($H_{YC,\text{app}}$) when the dissolved solids are high because equilibrium depends on activity, not concentration. Thus, the apparent Henry's law constant takes activity into account:

$$Y_A = H_{YC}\{A\} = H_{YC}\gamma_A C_A = H_{YC,\text{app}}\, C_A \tag{14-13}$$

where
$H_{YC,\text{app}} = H_{YC}\,\gamma_A$ = apparent Henry's law constant, dimensionless
Y_A = gas-phase concentration, mol/L
γ_A = activity coefficient of A
C_A = concentration of A, mol/L
H_{YC} = dimensionless Henry's constant

The activity coefficient γ_A is a function of ionic strength and can be calculated using the following empirical equation for neutral species:

$$\log \gamma_A = K_s \times I \tag{14-14}$$

where K_s = Setschenow, or "salting-out," constant, L/mol
I = ionic strength of water, mol/L

Values of K_s need to be determined experimentally because there is no general theory for predicting them. Salting-out coefficients for several compounds in seawater are available in the literature (Gossett, 1987; Schwarzenbach et al., 1993). Significant increases in volatility and the apparent Henry constant are observed only for high-ionic-strength waters such as seawater.

EFFECT OF SURFACTANTS

Surfactants can impact the volatility of compounds. In most natural waters, surfactant concentrations are relatively low; consequently, surfactants do not affect the design of most aeration devices. However, when surfactants are present in relatively high concentrations, the volatility of other compounds may decrease by several mechanisms. The dominant mechanism is collection of surfactants at the air–water interface, decreasing the mole fraction of the volatile compound at the interfacial area, thereby lowering the apparent Henry law constant. For example, the solubility of oxygen in water can decrease by 30 to 50 percent due to the presence of surfactants.

Another surfactant effect for hydrophobic organic compounds is the incorporation of dissolved organic compounds into micelles in solution. Above the critical micelle concentration, the formation of additional micelles will decrease the concentration of the organic compound at the air–water interface and decrease the compound's volatility.

IMPACT OF PH

The pH does not affect the Henry's constant directly, but it does affect the distribution of species between ionized and un-ionized forms, which influences the overall gas–liquid distribution of the compound because only the un-ionized species are volatile.

Uncharged weak acids such as H_2CO_3, HCN, or H_2S cannot be stripped at pH values significantly above their pK_a value. For example, if hydrogen sulfide is a weak acid with the following equilibrium (note, the second ionization constant can be neglected because p$K_{a2} > 14$):

$$H_2S \rightleftarrows HS^- + H^+ \tag{14-15}$$

Ionization constants for weak acids were described in Chap. 5. The equilibrium constant (K_a) for the reaction in Eq. 14-15 is 7.94×10^{-8} (pK_a = 7.1). Since only the un-ionized species is volatile, Henry's law can be written as

$$Y_A = H_{YC}[H_2S] = H_{YC}\alpha_0 C_{T,S} = H_{YC,\text{app}} C_{T,S} \tag{14-16}$$

$$\alpha_0 = \frac{[H^+]}{[H^+] + K_a} \tag{14-17}$$

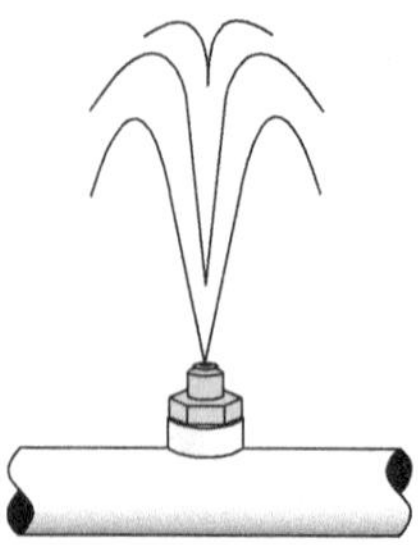

Figure 14-5
Schematic of fountain spray aerator.

where
$[H_2S]$ = hydrogen sulfide concentration, mol/L
$C_{T,S}$ = total sulfide concentration, mol/L
α_0 = fraction of total sulfide present as hydrogen sulfide
$H_{YC,app}$ = apparent Henry's law constant, dimensionless
K_a = acid dissociation constant

If the pH is equal to 5.1, two pH units lower than the pK_a, then sulfide is only 1 percent ionized and the apparent Henry's constant is essentially the same as the H value. If the pH is two units higher than the pK_a value, then sulfide is 99 percent ionized and the apparent Henry's constant is 1 percent of the H value.

14-3 Classification of Air-Stripping and Aeration Systems

Gas–liquid contactors are classified as either gas phase contactors or flooded contactors and then further classified into four subgroups based on the method that is used to either remove gas from water or add gas to water. The four basic types of air–water contact systems that are discussed in this section are (1) droplet or thin–film air–water contactors, (2) diffusion or bubble aerators, (3) aspirator–type aerators, and (4) mechanical aerators. Characteristics and typical applications of air–water contact systems that fall into one of these four groups are summarized in Table 14-5. Some of these systems may be used to contact water with gases other than air, and while the seuses are listed in Table 14-5, they are discussed in other chapters such as Chap. 13.

Droplet or Thin-Film Air–water Contactors

Droplet or thin-film air–water contactors are gas-phase contactors designed to produce small droplets of water or thin films, which promote rapid mass transfer.

Photographs of aeration and air stripping systems

http://www.stantec.com/wt3r/14-1.html

DROPLET AIR–WATER CONTACTORS

Contactors that use droplets are spray devices, such as towers and the fountain spray aerator shown on Fig. 14-5. Spray devices are designed to provide the desired droplet size for the desired contact time with the gas phase, which is typically air. Spray aerators are an efficient method of gas transfer; however, for efficient operation spray aerators should be placed in large basins or reservoirs in favorable climatic conditions.

THIN-FILM AIR–WATER CONTACTORS

Thin films of water are created in cascade and multiple-tray aerators and packed columns and towers. Air–water contact occurs when water flows by gravity over the surfaces of packing materials that are placed in

Table 14-5
Characteristics of gas-liquid contacting systems

Type of Contacting Device	Process Description	Method of Gas Introduction	Typical Applications	Oxygen Transfer Rate, kg O_2/kW · h	Number of Transfer Units (NTU)	Hydraulic Head Required, m (ft)	Loading Factor
Spray aerator	Water to be treated is sprayed through nozzles to form disperse droplets; typically, a fountain configuration. Nozzle diameters usually range from 2.5 to 4cm (1-1.6 in.) to minimize clogging.	Natural aeration through convection	H_2S, CO_2, and marginal VOC removal; taste and odor control, oxygenation		0.5–0.7	1.5–7.6 (5-25)	Surface area of 0.10–0.30 $m^2 \cdot s/L$
Spray tower	Water to be treated is sprayed downward through nozzles to form disperse droplets in a tower configuration; air–water ratio is controlled; typically countercurrent flow.	Forced-draft aeration	H_2S, CO_2, and VOC removal; taste and odor control		1–1.5	1.5–7.6 (5-25)	Surface area of 0.10–0.30 $m^2 \cdot s/L$
Packed tower	Water to be treated is sprayed onto high-surface-area packing to produce a thin-film flow; process configuration typically countercurrent.	Forced-draft aeration	H_2S, CO_2, and VOC removal; taste and odor control		1-4	3–12 (10-40)	

(Continued)

Table 14-5 *(Continued)*

Type of Contacting Device	Process Description	Method of Gas Introduction	Typical Applications	Oxygen Transfer Rate, kg O_2/kW · h	Number of Transfer Units (NTU)	Hydraulic Head Required, m (ft)	Loading Factor
Cascade	Water to be treated flows over the side of sequential pans, creating a waterfall effect to promote droplet-type aeration.	Aeration primarily by natural convection	CO_2 removal, taste and odor control, aesthetic value, oxygenation	—	0.5–0.7	0.9–3 (3–10)	
Multiple tray	Water to be treated trickles by gravity through trays containing media [layers 0.1–0.15m (4–6 in.) deep] to produce thin-film flow. Typical media used include coarse stone or coke [50–150 mm (2–6 in.) in diameter] or wood slats.	Natural or forced-draft aeration	H_2S, CO_2 removal, taste and odor control	—	<1	1.5–3 (5–10)	0.007–0.014 m/s(10-20 gpm/ft^2)
Low profile (sieve tray)	Water flows from entry at the top of the tower horizontally across series of perforated trays. Large air flow rates are used, causing frothing upon air–water contact, which provides large surface area for mass transfer. Units are typically less than 3m (10 ft) high.	Air introduced at bottom of tower	VOC removal	—		—	Water flow rates less than 0.065 m^3/s (1000 gpm)

Diffuser	Fine bubbles are supplied through porous diffusers submerged in the water to be treated; tank depth is typically restricted to 4.5 m (15 ft).	Compressed air or ozone	Fe and Mn removal, CO_2 removal, taste and odor control, oxygenation, ozonation	0.5	0.5–1.5	—	0.1–1 L air/L water
Dispersed air	Compressed air is created through a stationary sparger orifice-type dispersion apparatus located directly below a submerged high-speed turbine.	Compressed air or ozone	Ozonation, especially when high concentrations of Fe and Mn are present. Used due to clogging of porous diffusers	1.5	1-2		
Hydraulic aspirator	A gas stream is educted into the liquid stream with a venturi-type device.	Compressed ozone, CO_2, Cl_2	Ozonation, CO_2 addition, Cl_2 disinfection	1.5–3.5	—	3–6 (10-20)	
Mechanical aspirator	A hollow-blade impeller rotates at a speed sufficient to aspirate and discharge a gas stream into the water.	Compressed air or ozone	Ozonation, CO_2 addition	0.7			
Mechanical aerator	Surface aerators (brush or turbine types) sweep air into the water and fling water into the air. Aeration pumps pull air into the water.	Mechanical agitation of water into surrounding air	O_2 absorption, VOC removal when <90% removal required	1.5–4.5 (turbine) 2.5 (brush)			

trays, columns, or towers and between trays. The thin liquid film that forms as water flows downward is disrupted continuously by the irregular surfaces of the packing material, maximizing the exposure of the water to the atmosphere and encouraging air–water mixing.

Cascade aerators

Cascade aerators are commonly used for treating groundwater and may be located at the groundwater source or reservoir. Cascade aerators are also called step aerators as water flows downward in a thin film over a series of steps or baffles, sometimes constructed of concrete. Cascade aerators are generally less efficient than other types of thin-film aerators because water flow is less turbulent, resulting in less air–water mixing.

Multiple-tray aerators

There are several types of multiple-tray aerators, all based on the same design concept of stacked trays, where water is distributed over the top tray with a spray nozzle or special distribution trough and then flows from the upper tray over the tray sides into lower trays. Tray-type aerators may be either natural draft type such as coke tray aerators or forced draft type such as wood slat aerators.

Of all the types of multiple-tray aerators, wood slat towers are the most efficient. The slat towers are either forced or induced draft and are enclosed in a wood, fiberglass, or metal shell, as shown on Fig. 14-6. The slats are generally stacked and centered vertically and the horizontal spaces between the slats are staggered so that the water trickling down one tray strikes the middle of the slat of the next tray.

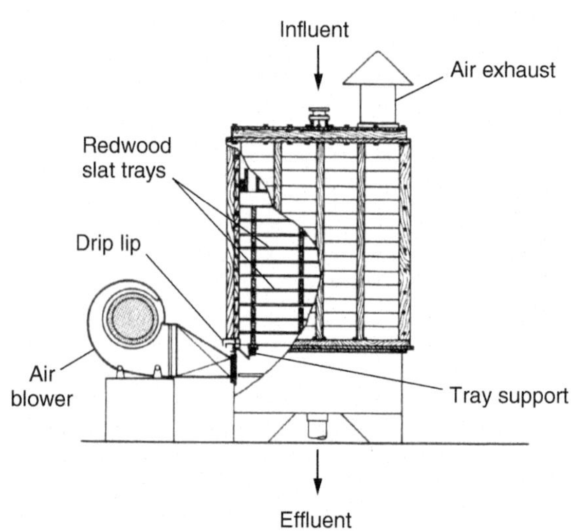

Figure 14-6
Forced-draft multiple-slat tower cascade aerator.

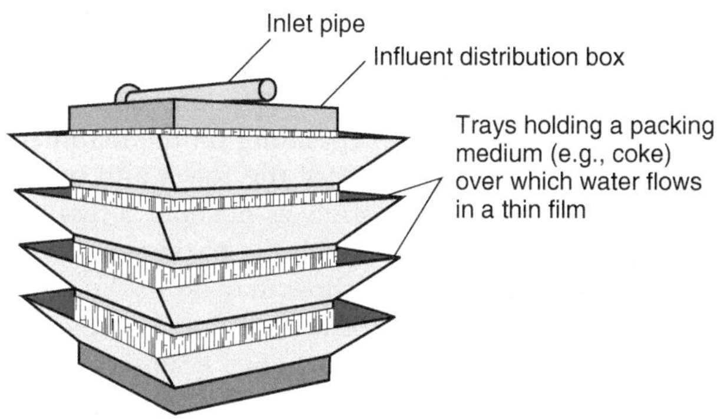

Figure 14-7
Natural draft coke tray aerator.

Tray aerators are typically designed for natural draft, as shown on Fig. 14-7, and typically the tray is filled with packing material such as coke. Coke tray aerators provide somewhat more turbulence in air–water contact because the large surface area of the coke provides a large air–water contact area. Tray aerators are built with splash skirts to reduce the water loss and icing of the protective retaining screens.

A type of multiple-tray aerator that has recently grown in use for removal of VOCs from contaminated waters is the low-profile air stripper, also called the sieve tray column, as shown on Fig. 14-8a. Because the water flows horizontally across each tray, the desired removal efficiency can be obtained by increasing the length or width of the trays instead of the height.

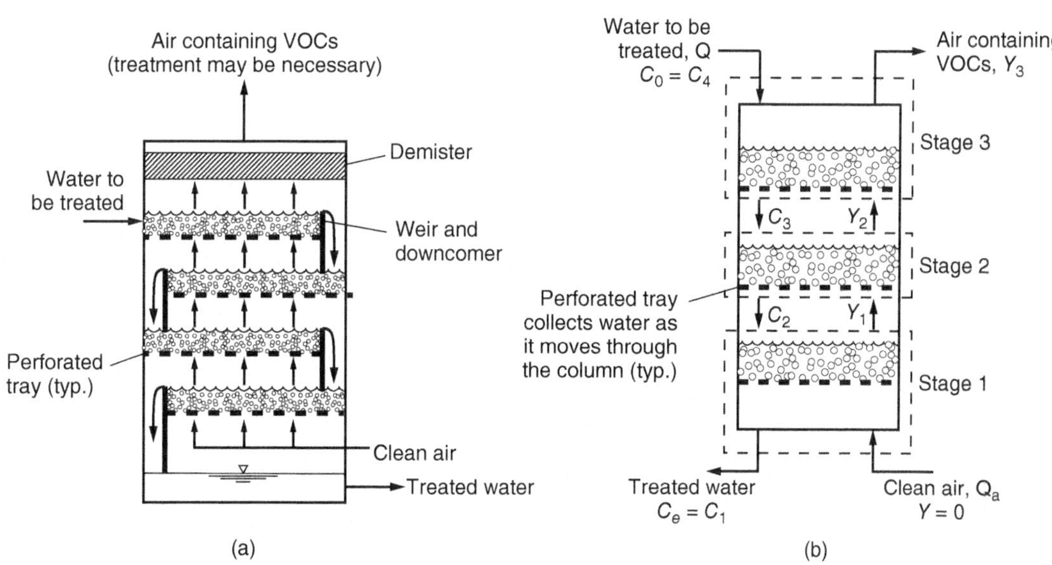

Figure 14-8
Low-profile air stripping: (a) schematic (adapted from U.S. ACE, 2001) and (b) diagram of low-profile air stripping as multistage, countercurrent process.

Packed towers

Packed towers are circular or square towers that are filled with an irregular shaped inert packing material, as shown on Fig. 14-9. Packing material is available in a wide variety of sizes and shapes depending on the manufacturer. Operationally, water is pumped to the top of the tower and into a liquid distributor where it is dispersed as uniformly as possible across the packing surface, and then it flows by gravity through the packing material and is collected at the bottom of the tower. Airflow may be in the same direction as the water (co-current), in the opposite direction as the water (countercurrent), or across the water (cross flow). For countercurrent operation, a blower is used to introduce fresh air into the bottom of the tower and the air flows countercurrent to the water up through the void spaces between the wetted packing material, as shown on Fig. 14-9.

An important part of the packed tower system that is not shown on Fig 14-9 is a demistor, which eliminates entrained water drops and aerosols in the off-gas of a packed tower. Aerosols must be eliminated because they can be displeasing to local communities from an aesthetic point of view and they can freeze in cold climates, causing icing problems.

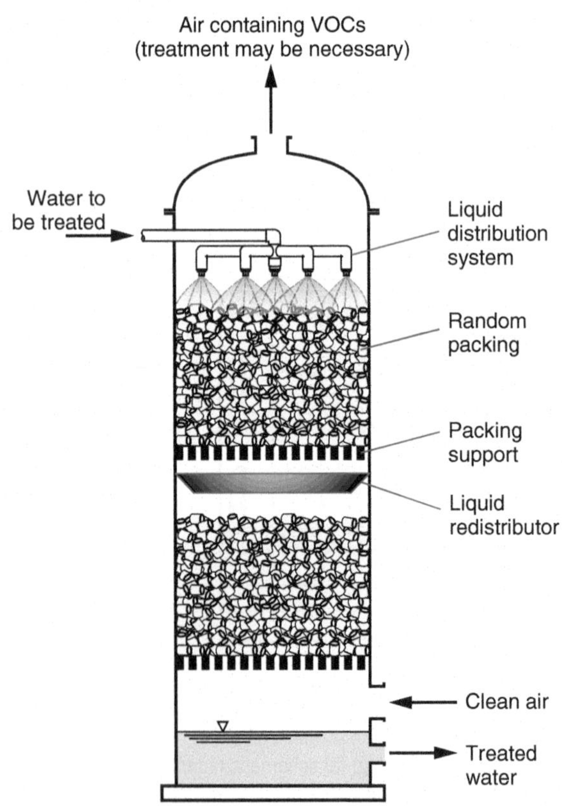

Figure 14-9
Schematic of a counter current packed tower.

Diffusion and Dispersion Aerators

Diffusion-type aeration systems (either diffused-air or dispersed-air systems) force air into the water using compressed air. Blowers used for aerating systems are either multiple-stage or single-stage centrifugal or rotary positive displacement. Rotary blowers are often used for small installations, where water depth varies significantly.

DIFFUSED AIR

Compressed air is generally introduced through porous membranes, porous plates or tubes, or wound fiber or metallic filaments at the bottom of a basin or tank, as illustrated by the bubble column on Fig. 14-10a. Diffused-air systems generally require filters to screen out particulates in the air because the air is forced to flow through very fine pores, which can easily plug.

DISPERSED AIR

Mechanical mixers and a stationary orifice-type sparger air dispersion system are used to force air into water in dispersed-air systems, as shown on Fig. 14-10b. The mechanical mixer in the contactor aids in the air–water mixing and, therefore, the gas transfer efficiencies are generally much better than in simple diffused-air systems. The air dispersion outlet is generally located a small distance above the tank bottom to reduce the pressure requirements of the air compressor. Dispersed-air systems usually do not require air filtration as the orifice-type spargers do not readily plug.

Aspirators

Air aspiration is commonly accomplished with either hydraulic aspirators or mechanical devices. A typical hydraulic aspirator is a type of hydraulic eductor or injector in which pressurized water flows through a throat similar to a venturi tube to create a low-pressure condition that in turn draws atmospheric air or gas into the water. A second type of aspirator is a mechanical aspirator that consists of a hollow-blade impeller that rotates at a speed sufficient to aspirate and discharge atmospheric air into the

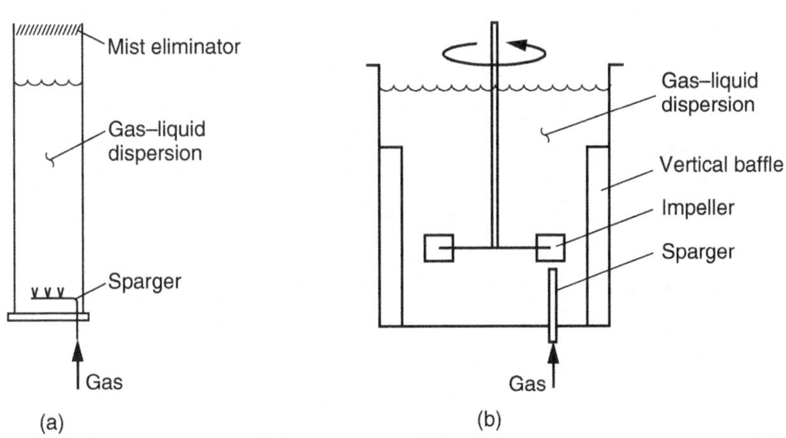

Figure 14-10
Common diffused aeration systems: (a) bubble column and (b) agitated vessel.

water. Hydraulic aspirators are mostly small in size, but a single unit under favorable conditions and operating with atmospheric air produces twice the oxygen transfer rate of ordinary mechanical aspirators.

Mechanical Aerators

Surface aerators and aeration pumps are the two basic types of mechanical aerators. Surface aerators may be of the turbine type or the brush type, as shown on Fig. 14-11. Aeration pumps consist of a turbine mixer with a draft tube. A number of patented mechanical aerators can be purchased from manufacturers.

Advantages and Disadvantages of Various Air-Stripping and Aeration Systems

Each type of air-stripping and aeration system offers process and cost advantages for a specific mass transfer problem. Low-profile air strippers versus packed towers and diffusion-type aerators versus droplet and thin-film aerators are compared in the following discussion.

LOW-PROFILE AIR STRIPPERS VERSUS PACKED TOWERS

The advantages of using a low-profile air stripper over a conventional packed tower stripper are that (1) the low-profile air stripper is smaller and more compact and (2) periodic maintenance is easier to perform on the low-profile air stripper. The disadvantage is that for a given removal the low-profile air stripper requires a significantly higher airflow than a conventional countercurrent packed tower. Consequently, the operational costs can be greater for the low-profile air stripper. In addition, low-profile air strippers are limited to lower water flow rates than countercurrent packed towers.

DIFFUSION-TYPE AERATORS VERSUS DROPLET AND THIN-FILM AERATORS

The advantages of diffusion-type aerators include (1) negligible head loss for the process water system and (2) less space requirements than for

(a) (b)

Figure 14-11
Typical examples of surface aeration devices: (a) brush type and (b) turbine type.

the droplet and thin-film aerators. Diffused-air systems may be extremely effective for reservoir management. Many successful diffused-air applications for reservoir destratification have been reported in the literature (AWWA, 1978a,b; Biederman and Fulton, 1971; Garton, 1978; Laverty and Nielsen, 1970; Steichen et al., 1979; Symons et al., 1970).

Selection of Appropriate Equipment

Selection of the appropriate equipment is based on relative transfer efficiencies, available hydraulic head, ease of maintenance, and cost considerations (capital and operating costs). Transfer efficiency and some cost considerations are considered below.

TRANSFER EFFICIENCY

An important consideration in selecting a process for a specific application is the upper limit of transfer efficiency that can be economically achieved by the process. Many of the processes described in Table 14-5 have limited transfer efficiency. Commercially available cascade aerators cannot achieve greater than 50 or 60 percent removal of chloroform, a relatively volatile organic contaminant. In contrast, packed towers can achieve >99% removal of chloroform. Cascade and multiple-tray aerators encounter corrosion and algae and slime growth problems, particularly if the process water contains hydrogen sulfide. Chlorination and copper sulfate treatment of the process water may help control these problems, but that is an additional operational issue with which to contend.

COST CONSIDERATIONS

For many applications of air–water mass transfer in water treatment, such as stripping of volatile contaminants or addition of a reactant gas, one or more of the four types of air–water contacting systems described above may be used. When more than one system may be appropriate to address the air–water mass transfer issue, capital and operating costs generally are the deciding factors.

Capital cost

The capital cost of each of the processes discussed is closely related to the mass transfer efficiency of the process. In general, the lower the transfer efficiency, the larger the facility required for achieving a certain removal.

Operating cost

Operating cost is primarily a function of the hydraulics of the process and the method of gas dispersion. Equipment complexity or heavy maintenance is typically not a major consideration because most mass transfer equipment is relatively simple, although it should be noted that fouling by chemical precipitation and/or biological growth must be controlled.

When adding a reactive gas such as oxygen, the operating cost must also include chemical costs. The principal cost may be the chemical itself and/or generation of the compound and to a lesser degree the feeding equipment and gas transfer contact tank (if any).

14-4 Fundamentals of Packed-Tower Air Stripping

Packed towers are either cylindrical columns or rectangular towers containing packing that disrupts the flow of liquid, thus producing and renewing the air–water interface. A schematic of a countercurrent cylindrical packed tower is shown on Fig 14-9, and the operation was described in Sec. 14-3. Packed towers have high liquid interfacial areas and void volumes greater than 90 percent, which minimizes air pressure drop through the tower.

The random packing material is important to the efficient transfer of volatile contaminants from the water to the air because it provides a large air–water interfacial area. Various types of packing shapes, sizes, and their physical properties are available commercially, as shown on Fig. 14-12. The packing can be structured packing or individual pieces that are randomly placed in the tower.

Mass Balance Analysis for a Countercurrent Packed Tower

The model for countercurrent packed towers requires the relationship that relates the bulk water-phase concentration to the bulk air-phase concentration. To obtain a relationship between the bulk air and water concentrations that are shown on Fig. 14-13, a mass balance is written around the lower half of the tower as follows:

1. The general statement of mass balance in words is

   Mass of chemical entering in liquid phase per unit time
   + mass of chemical entering in gas phase per unit time
   = mass of chemical exiting in liquid phase per unit time
   + mass of chemical exiting in gas phase per unit time (14-18)

2. The mass balance representation using symbols on Fig. 14-13 is

$$QC_b(z) + Q_a Y_0 = QC_e + Q_a Y_b(z) \tag{14-19}$$

where
Q = liquid flow rate, m^3/s
$C_b(z)$ = bulk liquid-phase concentration at axial position z along tower, mg/L
Q_a = gas flow rate, m^3/s
Y_0 = gas-phase concentration entering tower, mg/L
C_e = effluent liquid-phase concentration, mg/L
$Y_b(z)$ = bulk gas-phase concentration at axial position z along tower, mg/L

Intalox Saddles

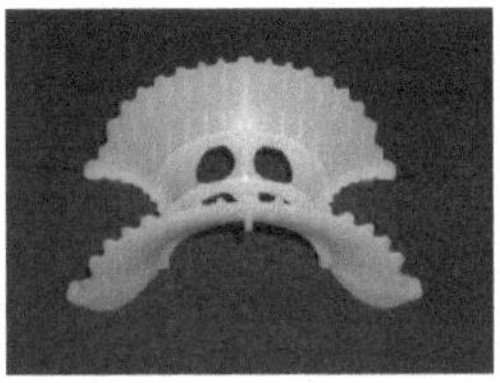

d_p	5.08 cm (2.0 in.)	7.62 cm (3.0 in.)
C_f	69 m^{-1} (21 ft^{-1})	52 m^{-1} (16 ft^{-1})
a_t	98 m^{-1} (30 ft^{-1})	66 m^{-1} (20 ft^{-1})

Jaeger Tri-Packs

d_p	5.08 cm (2.3 in.)	8.89 cm (3.5 in.)
C_f	52 m^{-1} (16 ft^{-1})	39 m^{-1} (12 ft^{-1})
a_t	157 m^{-1} (48 ft^{-1})	125 m^{-1} (38 ft^{-1})

Tellerettes (Type K)

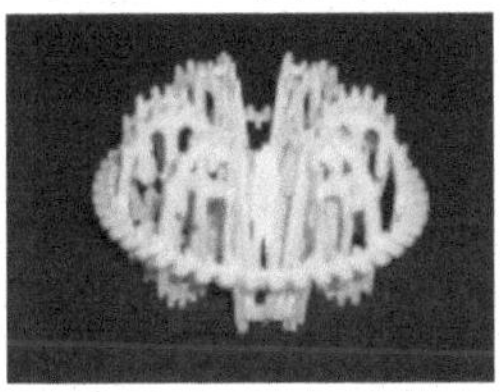

	No. 2	No. 3
d_p	5.08 cm (2.0 in.)	7.62 cm (3.0 in.)
C_f	36 m^{-1} (11 ft^{-1})	30 m^{-1} (9 ft^{-1})
a_t	52 m^{-1} (28 ft^{-1})	72 m^{-1} (22 ft^{-1})

Lantec Q-PAC

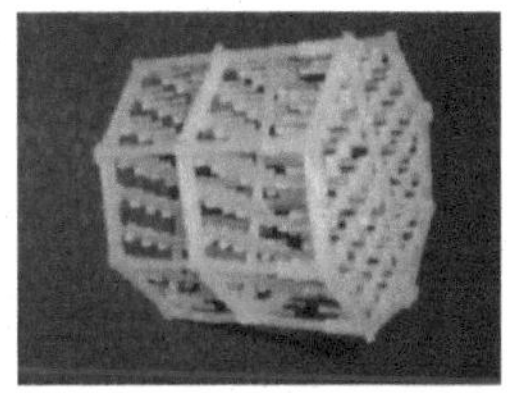

d_p	8.89 cm (3.5 in.)
C_f	23 m^{-1} (7 ft^{-1})
a_t	98 m^{-1} (30 ft^{-1})

Lantec NUPAC

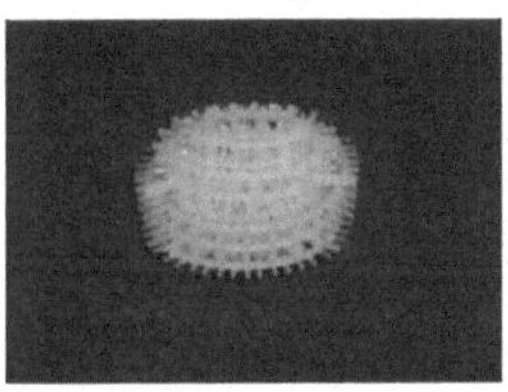

d_p	6.32 cm (2.5 in.)	11.4 cm (4.5 in.)
C_f	53 m^{-1} (16 ft^{-1})	26 m^{-1} (8 ft^{-1})
a_t	180 m^{-1} (55 ft^{-1})	131 m^{-1} (40 ft^{-1})

Lantec LANPAC

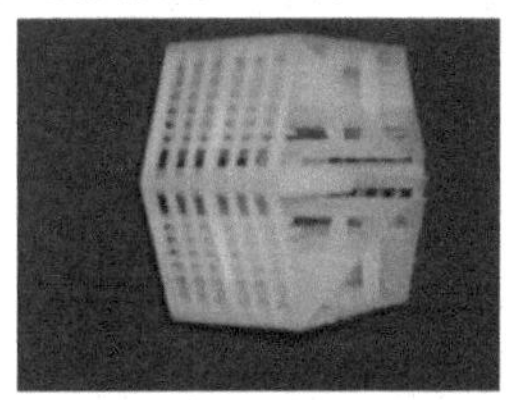

d_p	5.84 cm (2.3 in.)	8.89 cm (3.5 in.)
C_f	69 m^{-1} (21 ft^{-1})	46 m^{-1} (14 ft^{-1})
a_t	223 m^{-1} (68 ft^{-1})	144 m-1 (44 ft-1)

Lantec LANPAC-XL

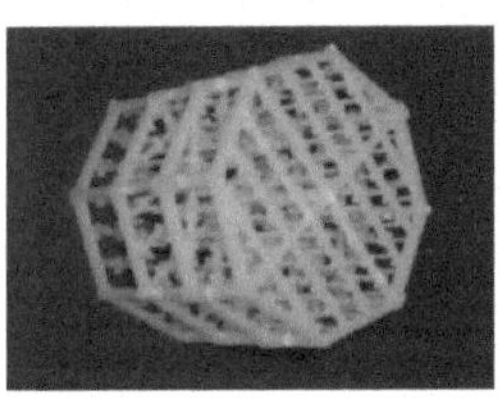

d_p	8.89 cm (3.5 in.)
C_f	33 m^{-1} (10 ft^{-1})
a_t	242 m^{-1} (74 ft^{-1})

Plastic rings

d_p	5.08 cm (2.0 in.)	7.62 cm (3.0 in.)
C_f	82 m^{-1} (25 ft^{-1})	30 m^{-1} (9 ft^{-1})
a_t	108 m^{-1} (33 ft^{-1})	72 m-1 (22 ft-1)

Critical surface tension depends on material, for polypropylene, σ_c = 0.033 N/m

Figure 14-12
Typical examples of polyethylene packing materials used in air-stripping towers and their physical characteristics.

Combining terms, Eq. 14-19 can be written as

$$[Y_b(z) - Y_0] = \left(\frac{Q}{Q_a}\right)[C_b(z) - C_e] \tag{14-20}$$

A mass balance on the overall tower shown in Fig. 14-13 is

$$QC_0 + Q_aY_0 = QC_e + Q_aY_e \tag{14-21}$$

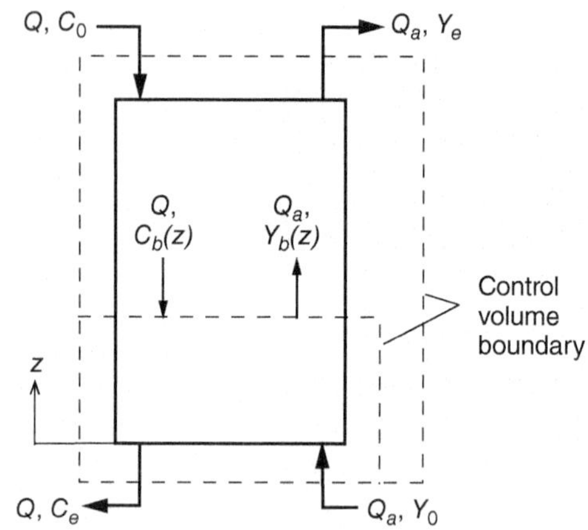

Figure 14-13
Definition drawing for mass balances on a packed tower.

Combining terms, Eq. 14-21 can be written as

$$Y_e - Y_0 = \left(\frac{Q}{Q_a}\right)(C_0 - C_e) \tag{14-22}$$

where C_0 = influent liquid-phase concentration to tower, mg/L
Y_e = effluent gas-phase concentration from tower, mg/L

Assuming clean air entering the bottom of the tower ($Y_0 = 0$), Eq. 14-20 becomes

$$Y_b(z) = \left(\frac{Q}{Q_a}\right)[C_b(z) - C_e] \tag{14-23}$$

Under the same assumption, Eq. 14-22 can be written as

$$Y_e = \left(\frac{Q}{Q_a}\right)(C_0 - C_e) \tag{14-24}$$

Equations 14-23 and 14-24 represent the operating line equation for packed-tower aeration (see Chap. 7 for an introduction to operating lines). The operating and equilibrium lines for packed-tower aeration are presented on Fig. 14-14, which is known as a McCabe-Thiele diagram (McCabe and Thiele, 1925). The operating line is labeled 1 on Fig. 14-14, and the equilibrium line is labeled 2. Equilibrium is described by a straight line known as Henry's law (see Sec. 14-2):

$$Y = H_{YC}C \tag{14-25}$$

where Y = gas-phase concentration, mg/L
C = liquid-phase concentration in equilibrium with gas-phase concentration Y, mg/L

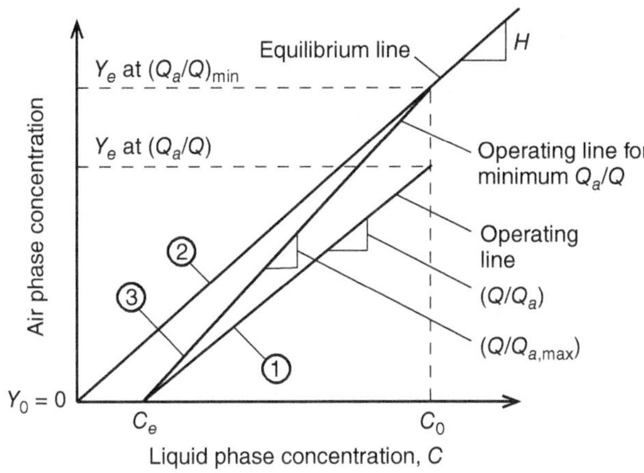

Figure 14-14
Operating line diagram for packed tower.

As discussed in Chap. 7, the operating and equilibrium lines are an important concept in separation processes, such as air stripping, because they can be used to determine the minimum amount of extracting phase (e.g., air in packed-tower aeration), in terms of mass or volume required to remove a component (e.g., from water in packed-tower aeration) to a desired removal efficiency.

STRIPPING FACTOR

A parameter commonly used in the evaluation of packed towers is the stripping factor (S); S is defined as the ratio of the slope of the equilibrium line to the operating line slope. As shown in Fig. 14-14, the equilibrium line divided by the operating line yields the following expression for the S:

$$S = \frac{\text{slope of equilibrium line}}{\text{slope of operating line}} = \frac{H_{YC}}{Q/Q_a} = \left(\frac{Q_a}{Q}\right) H_{YC} \tag{14-26}$$

where S = stripping factor, dimensionless
Q_a/Q = operating air-to-water ratio of tower
Q_a = air flow rate, m^3/s
Q = water flow rate, m^3/s

When $S = 1$, the slopes of the equilibrium and operating lines are parallel to one another and removal to the treatment objective is possible but will require a very long or infinite tower length. When $S < 1$, the slope of the operating line is greater than the slope of the equilibrium line, the desired removal will be equilibrium limited and the treatment objective will not be obtained if a very low effluent concentration is needed. When $S > 1$, the slope of the operating line is less than the slope of the equilibrium line and the treatment objective can be met using stripping.

MINIMUM AIR-TO-WATER RATIO

A special case of the operating line shown in Fig. 14-14 is line number 3. This line intersects the equilibrium line where the influent concentration, C_0, is in equilibrium with the exiting gas-phase concentration (i.e., $Y_e = H_{YC}C_0$). The slope of this line represents the inverse of the minimum air-to-water ratio that can meet the treatment objective if the packed-tower length is infinite. If it is assumed the influent gas-phase concentration, Y_0, is equal to zero, and the influent liquid-phase concentration is in equilibrium with the exiting air according to Eq. 14-25, then Eq. 14-22 can be rearranged to yield the following expression for the minimum air-to-water ratio:

$$\left(\frac{Q_a}{Q}\right)_{\text{min}} = \frac{C_0 - C_e}{H_{YC}C_0} \tag{14-27}$$

where $(Qa/Q)_{\text{min}}$ = minimum air-to-water ratio, dimensionless
C_0 = influent liquid-phase concentration, mg/L
C_e = treatment objective, mg/L

The minimum air-to-water ratio $(Q_a/Q)_{\text{min}}$ represents the minimum air-to- water ratio that can be applied for a packed tower to meet its treatment objective C_e. If the air-to-water ratio applied is less than the minimum air-to-water ratio, it will not be possible to design a packed tower capable of meeting the treatment objective because equilibrium will be established in the tower before the treatment objective is reached.

With respect to the selection of the optimum air-to-water ratio, it has been demonstrated that minimum tower volume and power requirements are achieved using approximately 3.5 times the minimum air-to-water ratio for contaminants with Henry's law constants greater than 0.05 for high percentage removals, corresponding to a stripping factor of 3.5 (Hand et al., 1986).

RELATIONSHIP BETWEEN S AND $(Q_a/Q)_{\text{min}}$

The stripping factor can be related to the minimum air-to-water ratio when the treatment efficiency is very high, and Eq. 14-27 can be approximated as

$$\left(\frac{Q_a}{Q}\right)_{\text{min}} = \frac{C_0 - C_e}{H_{YC}C_0} \approx \frac{1}{H_{YC}} \quad (C_e \ll C_0) \tag{14-28}$$

Substitution of Eq. 14-28 into Eq. 14-26 yields a relationship for stripping factor in terms of minimum air to water ratio

$$S = \frac{Q_a/Q}{(Q_a/Q)_{\text{min}}} \tag{14-29}$$

When $C_e \ll C_0$, the stripping factor is approximately equal to the ratio of the actual air flow rate to the minimum air flow rate for treating a given flow of water.

Example 14-3 Calculation of minimum air-to-water ratio and stripping factor

Calculate the minimum air-to-water ratio and operating air-to-water ratio for 1,2-dichloropropane (DCP) and tetrachloroethylene (PCE) with 90 percent removal at 10°C for a countercurrent packed tower. Calculate the operating air-to-water ratio for a packed tower that minimizes the tower volume and and power requirements, and the stripping factor for each compound at the operating air-to-water ratio.

Solution

1. Determine H_{PC} for each compound using data in Table 14-3.
   a. DCP:
   $$H_{YC,DCP} = 0.0525 \text{ (dimensionless)}$$
   b. PCE:
   $$H_{YC,PCE} = 0.364 \text{ (dimensionless)}$$
2. Calculate the minimum air-to-water ratio for each compound using Eq. 14-28.
   a. DCP:
   $$\left(\frac{Q_a}{Q}\right)_{\text{min,DCP}} = \frac{C_0 - C_e}{H_{YC,DCP}C_0} = \frac{C_0 - 0.1C_0}{0.0525C_0} = 17.14$$
   b. PCE:
   $$\left(\frac{Q_a}{Q}\right)_{\text{min,PCE}} = \frac{C_0 - C_e}{H_{YC,PCE}C_0} = \frac{C_0 - 0.1C_0}{0.364C_0} = 2.47$$
3. To calculate the operating air-to-water ratio that minimizes tower volume and power consumption, multiply the minimum air-to-water ratio by 3.5.
   a. DCP:
   $$\left(\frac{Q_a}{Q}\right)_{\text{DCP}} = 3.5\left(\frac{Q_a}{Q}\right)_{\text{DCP,min}} = (3.5)(17.14) = 60$$
   b. PCE:
   $$\left(\frac{Q_a}{Q}\right)_{\text{PCE}} = 3.5\left(\frac{Q_a}{Q}\right)_{\text{PCE,min}} = (3.5)(2.47) = 8.65$$

4. Calculate the stripping factor for each compound using Eq. 14-26. Note that the largest air-to-water ratio from step 3 applies to all compounds being stripped in the tower.
   a. DCP:

$$S_{DCP} = \left(\frac{Q_a}{Q}\right) H_{YC,DCP} = (60)\,(0.0525) = 3.15$$

   b. PCE:

$$S_{PCE} = \left(\frac{Q_a}{Q}\right) H_{YC,PCE} = (60)\,(0.364) = 21.8$$

Comment

The compound with the lower Henry's law constant (DCP) requires much larger minimum and operating air-to-water ratios to achieve the desired removal. This is expected because a smaller Henry's constant indicates lower volatility, i. that is, a greater preference of the compound for the water phase and a lower tendency for stripping from the water phase to the air phase.

Mass Balance for Multistage Stripping Tower

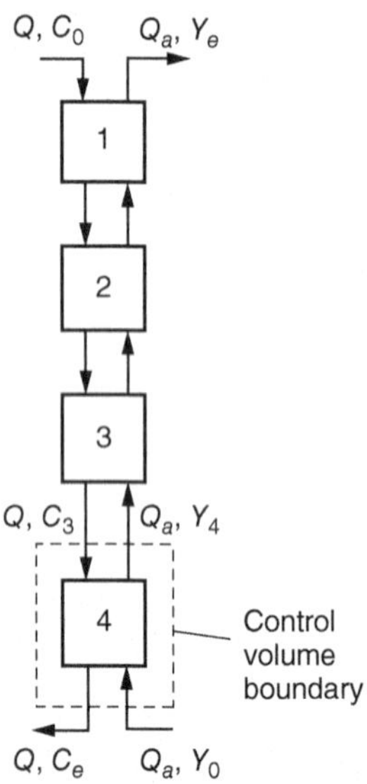

Figure 14-15
Definition drawing for mass balances on multistage stripping tower.

In the design of air-stripping towers, the number of equilibrium stages required for stripping is often determined. Determining the number of equilibrium stages is analogous to representing plug flow as a series of completely mixed flow reactors (CMFRs), as described in Chap. 6. It is assumed that equilibrium conditions prevail within each stage.

A mass balance on the lower section of a staged countercurrent tower shown in Fig.14-15 is

$$QC_3 + Q_aY_0 = QC_e + Q_aY_4 \tag{14-30}$$

Assuming clean air enters the tower ($Y_0 = 0$) and rearranging Eq. 14-30,

$$Y_4 = \left(\frac{Q}{Q_a}\right)(C_3 - C_e) \tag{14-31}$$

The McCabe-Thiele (1925) graphical method for determining the number of equilibrium stages is demonstrated on Fig. 14-16 for a four-stage stripping tower. Both the operating line and the equilibrium line are shown on Fig. 14-16. The method for constructing the McCabe-Thiele diagram for finding the number of equilibrium stages or number of transfer units is described as follows:

1. The point (C_0, Y_e) represents the influent water-phase concentration and exiting air-phase concentration of the contaminant of interest at the top of the tower. Draw a horizontal line from the point (C_0, Y_e) to the point (C_1, Y_e), which represents the location

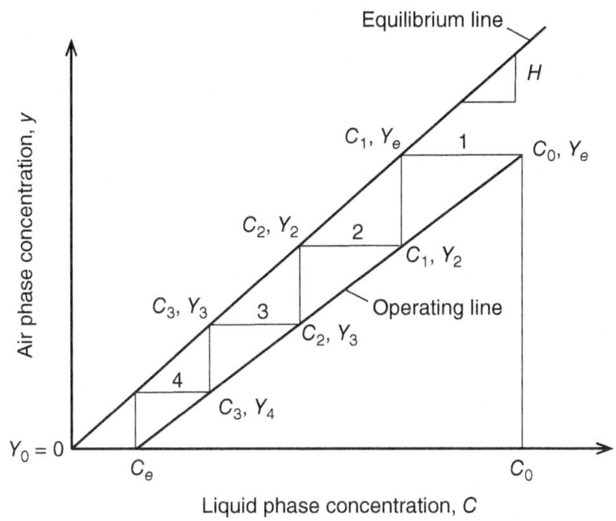

Figure 14-16
Operating line diagram for multistage stripping tower.

where the water-phase concentration of the constituent is in equilibrium with the exiting air, as shown on Fig. 14-16.

2. Draw a vertical line from the point (C_1, Y_e) to the point (C_1, Y_2) on the operating line. The numerical value of Y_2 can be determined based on a mass balance around stage 1 shown on Fig. 14-16:

$$Y_2 = \left(\frac{Q}{Q_a}\right)C_1 + Y_e - \left(\frac{Q}{Q_a}\right)C_0 \tag{14-32}$$

3. Draw a horizontal line from the point (C_1, Y_2) on the operating line to the point (C_2, Y_2) on the equilibrium line.
4. Draw a vertical line from the point (C_2, Y_2) on the equilibrium line to the point (C_2, Y_3) on the operating line.
5. Draw a horizontal line from the point (C_2, Y_3) on the operating line to the point (C_3, Y_3) on the equilibrium line.
6. Draw a vertical line from the point (C_3, Y_3) on the equilibrium line to the point (C_3, Y_4) on the operating line.
7. Continue until point (C_n, Y_{n+1}) is reached and the effluent concentration C_e is surpassed.

If the final stage does not intersect the effluent concentration, then the minimum number of stages would include the stage that overshoots the effluent concentration. This procedure is used to estimate the minimum number of stages or transfer units, because equilibrium is not usually attained and additional stages are required beyond the minimum number.

The McCabe–Thiele method described above can also be used to determine the minimum number of trays required in low-profile air stripping, which is a countercurrent, staged process and is discussed in Sec. 14-6.

Design Equation for Determining Packed-Tower Height

Predicting the required height of a packed tower to meet a given airstripping treatment objective is one of the goals of packed-tower design. The design equation for tower height can be derived using these assumptions: (1) steady-state conditions prevail in the tower, (2) air flow rate and water flow rate are constant through the column, (3) no chemical reactions occur, and (4) plug flow conditions prevail for both the air and water.

LIQUID-PHASE MASS BALANCE AROUND A DIFFERENTIAL ELEMENT

A liquid-phase mass balance around the differential element surrounded by a dashed box on Fig. 14-17a serves as the basis for the design equation. A schematic of the differential element applicable to the case of a liquid-side mass balance is presented on Fig. 14-17b.

The liquid-phase mass balance around the differential element is written in words as

Mass of organic entering per unit time
− mass of organic leaving per unit time
+ mass of organic generated per unit time
= mass of organic accumulated per unit time (14-33)

Equation 14-33 can be written symbolically as

$$[QC_b(z+\Delta z)]-[QC_b(z)]+0-[J_A(a\Delta V)]=0 \qquad (14\text{-}34)$$

where
Q = water flow rate, m^3/s
C_b = bulk liquid-phase concentration, mg/L
z = axial position along tower, m
Δz = height of differential element, m
J_A = flux across air–water interface, $mg/m^2 \cdot s$

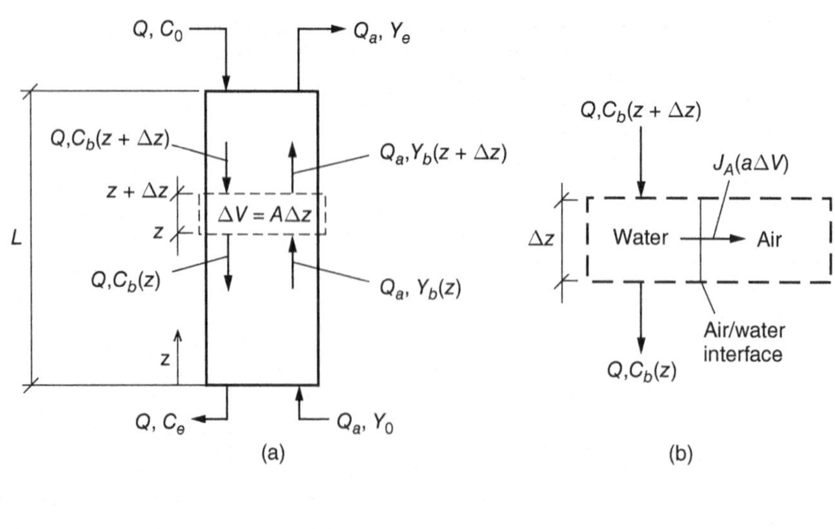

Figure 14-17 Packed-tower design equation definition drawing: (a) schematic of packed tower showing differential element and (b) schematic of differential element used in liquid-side mass balance.

a = area available for mass transfer divided by vessel volume, m^2/m^3
ΔV = volume of differential element, m^3

As shown in Chap. 7, the term J_A in Eq. 14-34 is obtained from the two-film theory:

$$J_A = K_L[C_b(z) - C_S^*(z)] \tag{14-35}$$

where K_L = overall liquid-phase mass transfer coefficient, m/s
$C_S^*(z)$ = liquid-phase concentration at air–water interface in equilibrium with the bulk gas-phase concentration, mg/L

Inserting Eq. 14-35 and $\Delta V = A\,\Delta z$ into Eq. 14-34 yields

$$[QC_b(z+\Delta z)] - [QC_b(z)] - \left\{K_L[C_b(z) - C_S^*(z)](aA\Delta z)\right\} = 0 \tag{14-36}$$

where A = cross-sectional area of packed tower, m^2

Rearranging Eq. 14-36 and dividing by $A\,\Delta z$ yields the equation

$$\frac{Q}{AK_La}\left[\frac{C_b(z+\Delta z) - C_b(z)}{\Delta z}\right] = C_b(z) - C_s^*(z) \tag{14-37}$$

where K_La = overall mass transfer rate constant, s^{-1}

Taking the limit as Δz approaches zero results in

$$\frac{Q}{AK_La}\lim_{\Delta z\to 0}\left[\frac{C_b(z+\Delta z) - C_b(z)}{\Delta z}\right] = \frac{Q}{AK_La}\frac{dC_b}{dz} = \left[C_b(z) - C_s^*(z)\right] \tag{14-38}$$

Separating variables in Eq. 14-39 results in

$$\frac{Q}{AK_La}\int_{C_e}^{C_0}\frac{dC_b}{C_b - C_s^*} = \int_0^L dz = L \tag{14-39}$$

where L = height of packed tower, m
C_0 = influent liquid-phase concentration, mg/L
C_e = treatment objective, mg/L

RELATING CONCENTRATION AT AIR–WATER INTERFACE TO CONCENTRATION IN BULK LIQUID

To solve Eq. 14-39, C^*_s needs to be expressed in terms of C_b. The following relationship, obtained by using the definition of C^*_s and the operating line, can be used:

$$C_s^* = \frac{Y_b}{H_{YC}} = \frac{Y_b(z)}{H_{YC}} \tag{14-40}$$

where Y_b = bulk gas-phase concentration, $Y_b(z)$, mg/L

Substituting Eq. 14-23 into Eq. 14-40 yields the desired result:

$$C_s^* = \frac{Y_b}{H_{YC}} = \frac{(Q/Q_a)(C_b - C_e)}{H_{YC}} \tag{14-41}$$

where Q_a = air flow rate, m^3/s

DETERMINATION OF TOWER HEIGHT

Substituting Eq. 14-41 into Eq. 14-39 results in the following:

$$L = \frac{Q}{AK_La} \int_{C_e}^{C_0} \frac{dC_b}{C_b\left[1-(Q/Q_a)/H_{YC}\right] + C_e(Q/Q_a)/H_{YC}} \tag{14-42}$$

$$= \frac{Q}{AK_La}\left[\frac{1}{1-(Q/Q_a)/H_{YC}}\right]\ln\left[C_b\left(1-\frac{Q/Q_a}{H_{YC}}\right)+C_e\left(\frac{Q}{Q_a}\right)/H_{YC}\right]\Bigg|_{C_e}^{C_0} \tag{14-43}$$

$$= \frac{Q}{AK_La}\left[\frac{1}{1-(Q/Q_a)/H_{YC}}\right]\ln\left[\frac{C_0+(C_e-C_0)(Q/Q_a)/H_{YC}}{C_e}\right] \tag{14-44}$$

Additional details on the development of the design equation for countercurrent packed-tower aeration may be found in the literature (Ball and Edwards, 1992; Ball et al., 1984; Cummins and Westrick, 1983; Dzombak et al., 1993; Gross and TerMaath, 1985; Hand et al., 1986; Kavanaugh and Trussell, 1980, 1981; McKinnon and Dyksen, 1984; Roberts and Levy, 1985; Roberts et al., 1985; Sherwood and Hollaway, 1940; Singley et al., 1980, 1981; Treybal, 1980; Umphres et al., 1983).

Expressing tower height in terms of stripping factor

Equation 14-44 can be further simplified using the definition of the stripping factor, $S = (Q_a/Q)H_{YC}$ (see Eq. 14-26):

$$L = \frac{Q}{AK_La}\left[\frac{1}{1-(1/S)}\right]\ln\left[\frac{(1/S)(C_e-C_0)+(C_0)}{C_e}\right] \tag{14-45}$$

$$= \frac{Q}{AK_La}\left(\frac{S}{S-1}\right)\ln\left[\frac{(C_e-C_0)+S(C_0)}{SC_e}\right] \tag{14-46}$$

The design equation for packed-tower aeration is given by

$$L = \frac{Q}{AK_La}\left[\frac{S}{(S-1)}\right]\ln\left[\frac{1+(C_0/C_e)(S-1)}{S}\right] \tag{14-47}$$

where L = packed-tower height, m
A = cross-sectional area of packed tower, m^2
K_La = overall liquid-phase mass transfer rate constant, 1/s
S = stripping factor, dimensionless

C_0 = influent liquid-phase concentration, mg/L
C_e = treatment objective, mg/L

Expressing tower height in terms of transfer units
In packed-tower aeration, the tower length is often defined as

$$L = \text{HTU} \times \text{NTU} \tag{14-48}$$

where HTU = height of transfer unit, m
NTU = number of transfer units or number of equilibrium stages

With additional algebraic rearranging, the HTU and NTU are defined as

$$\text{HTU} = \frac{Q}{AK_L a} \tag{14-49}$$

$$\text{NTU} = \frac{S}{S-1} \ln \left[\frac{1 + (C_0/C_e)(S-1)}{S} \right] \tag{14-50}$$

The height of a transfer unit is determined by the superficial velocity (Q/A) divided by the overall mass transfer rate constant. For packed towers, the height of a transfer unit is a measure of the stripping effectiveness of particular packings for a given stripping process. Packing that is typically smaller in size has higher specific surface area causing more efficient transfer of solute from one phase to another, thereby increasing $K_L a$ and decreasing the HTU. The HTU and tower length will decrease as the superficial velocity decreases or the rate of mass transfer increases.

The number of transfer units can be thought of as a measure of the difficulty of stripping a solute from the liquid to the gas phase. The more difficult it is to strip the solute, the more NTUs needed to achieve a given removal efficiency. The number of transfer units in a packed column can be determined from Fig. 14-18, which is a plot of numerous solutions of Eq. 14-50. For a given S, the removal efficiency increases with increasing

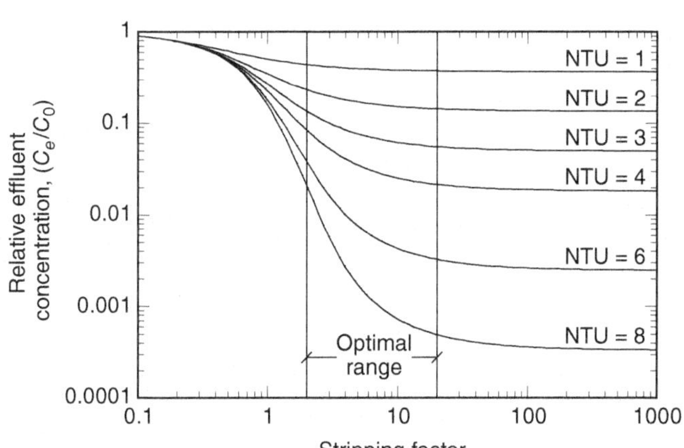

Figure 14-18
Dependence of relative effluent concentration on NTU and stripping factor.

NTU (or number of hypothetical completely mixed tanks). In addition, for a given removal efficiency, increasing S or the air-to-water ratio will decrease the NTU required. As shown on Fig. 14-18, the optimal range for the stripping factor might be considered between 2 and 20 because high removal efficiency is not possible at S less than 1, and no additional improvements in removal occurs at values of S greater than about 20. The best efficiency point for minimum power requirements and tower volume tends to occur at an air-to-water ratio of 3.5 times the minimum air-to-water ratio required for stripping, which would correspond to a low value of the stripping factor (Hand et al., 1986).

The NTU concept bridges the gap between continuous fluid contact and stage operation (see discussion in Chap. 7). For the hypothetical driving force shown on Fig. 14-19a, four equilibrium countercurrent stages would be required to meet an effluent concentration of 2.5 mg/L for an influent concentration of 12.5 mg/L. The stripping factor and Henry's constant for this situation are 1.0 and 0.5, respectively.

The operating and equilibrium lines are parallel, and $C_b - C_s^*$ differs by exactly 2.5 mg/L, resulting in exactly four equilibrium stages. The NTU for continuous contact can be determined from this equation and the integration of $1/(C_b - C_s^*)$, which is shown on Fig. 14-19b:

$$\text{NTU} = \int_{C_e}^{C_0} \frac{dC_b}{C_b - C_s^*} \tag{14-51}$$

where $C_s^* = C_s^*(z)$ represents the liquid-phase concentration at the air–water interface in equilibrium with $Y_b(z)$ (see Eq. 14-41) in milligrams per liter and $C_b = C_b(z)$, which is defined above.

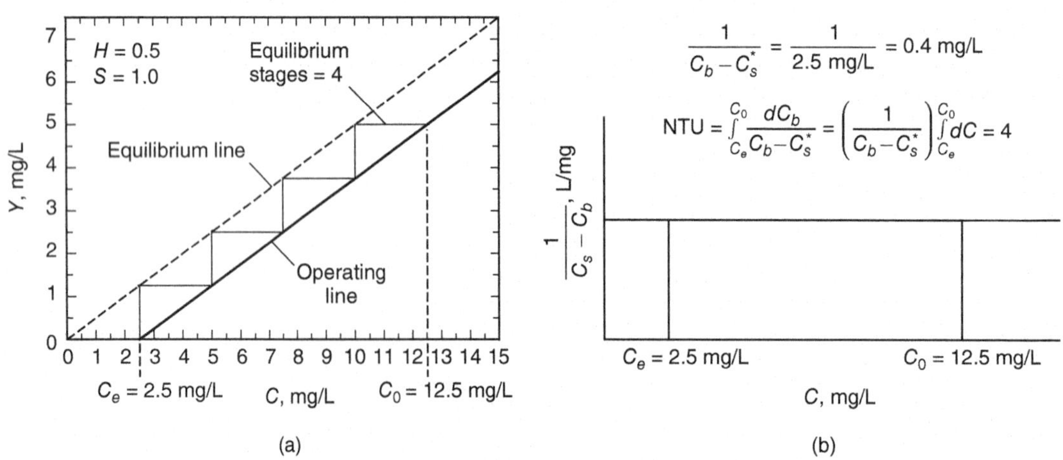

Figure 14-19
(a) Determination of the number of equilibrium stages (NTU) using an operating line diagram for a packed tower.
(b) Sample integration to determine the number of transfer units (NTU).

In this case, the NTU for equilibrium stages and continuous contact are equal. In general, this will not be true, but this does establish the relationship between NTU and the number of equilibrium tanks in series. The HTU is the height of one equivalent stage at equilibrium.

14-5 Analysis and Design of Packed-Tower Air Stripping

The two main design categories for packed-tower air stripping are (1) modifications to existing towers (rating analysis) and (2) designing new towers (design analysis). Modifications are made to existing towers to either treat greater volumes of water or modify constituent removal (e.g., lower levels, different constituents). Process efficiency may be improved by increasing the air flow rate, decreasing the water flow rate, replacing the packing with a more efficient packing type, or increasing the packed-tower height. Designing new towers includes the selection of packing type, air-to-water ratio, gas pressure drop, and operational flexibility.

Because packed-tower analysis involves repetitive calculations and the opportunity for introducing errors, commercially available software is commonly used to evaluate the impact of process variables on process performance. Software design tools typically contain the design equations, Henry's constants, mass transfer correlations, databases for many commercially available packing types, and physical properties of many VOCs that have been encountered in water supplies. In addition, graphical user interfaces make the software user friendly (Dzombak et al., 1993; Hokanson, 1996).

Packed-tower air stripping is analyzed in this section including (1) determination of properties required to calculate packed-tower height, (2) description of process variables, and (3) representation of the equations applicable to design versus rating analysis of packed towers. The following design considerations for packed-tower air stripping are also discussed: (1) design variables, (2) design approach, and (3) factors influencing packed-tower performance.

Properties Needed to Determine Packed-Tower Height

To determine the packed-tower height as described above, the following properties are needed: (1) gas pressure drop, (2) cross-sectional area of the tower, and (3) mass transfer rate constant. Determination of the properties required to calculate packed-tower height is discussed below.

GAS PRESSURE DROP

The gas pressure drop in packed columns is an important design and operational parameter because the electrical costs of the blower account for a significant fraction of the operational costs. Consequently, it is important to operate at a low gas pressure drop to minimize the blower costs. Methods used to determine the gas pressure drop through the packing includes: pilot and full-scale data collection, manufacturer's pressure drop specifications, and generalized Eckert pressure drop curves. Pressure drop data obtained from pilot and full-scale testing is the best way to determine the operating gas pressure drop. However, in many instances engineers use software programs and spreadsheets to design these systems because much of the design information is known or can be easily calculated.

Manufacturers provide gas pressure drop information on most all of their packing materials, which can be used to determine the pressure drop through the packing. The pressure drop of gas rising countercurrent to liquid flowing through a packed tower typically follows the pattern illustrated on Fig. 14-20. This pressure drop data is for Jaeger 3.5-in. nominal diameter Tripack plastic packing. The pressure drop per unit depth of packing is typically plotted in terms of a C factor, which is defined by the following equation:

$$\text{C Factor} = V_S\left[(\rho_g)/(\rho_L - \rho_g)\right]^{0.5} \tag{14-52}$$

where V_s = superficial gas velocity, m/s
ρ_g = gas density, kg/m^3
ρ_L = liquid density, kg/m^3

C factor is the density-corrected superficial gas velocity through the column packing and describes the balance between the gas momentum force, which acts to entrain bundles of liquid droplets, and the gravity force, which resists the upward entrainment of water (Kister et al., 2007). Given

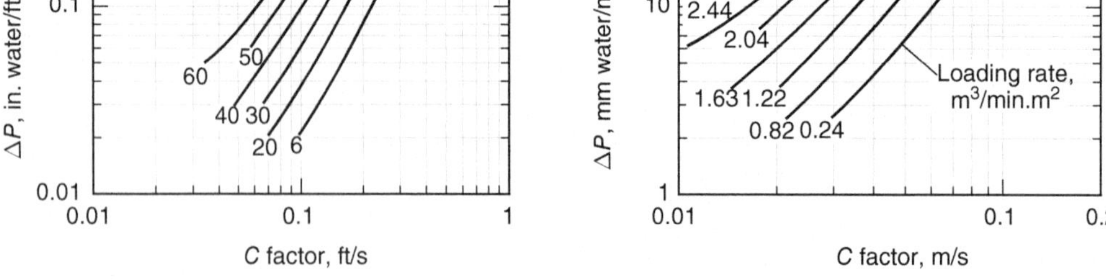

Figure 14-20
Gas pressure drop curves as a function of C factor for 3.5-in. nominal diameter plastic tripacks (Adapted from Jaeger Products, Brochure 600).

the temperature, packed-tower area, packing height, gas and liquid loading rates, Fig. 14-20 can be used to determine the pressure drop across the packing material. Manufacturers typically supply pressure drop information on their various packing materials.

When manufacturers data is not available, a common method of estimating the gas pressure drop through random packing in towers is the use of the generalized Eckert pressure drop correlation (see Fig. 14-21). The Eckert correlation relates the gas pressure drop to the capacity parameter on the ordinate (y axis) as a function of the flow parameter on the abscissa (x axis). For high gas loading rates, entrainment of the liquid by the rising gas can occur, characterized by a sudden rapid increase in the gas pressure drop, and eventually the column will become a flooded contactor because of the back pressure caused by the rising gas. However, as discussed above, most all air-stripping applications operate at low gas pressure drops to minimize energy costs associated with the blower operation and flooding is never a problem.

The Eckert correlation shown in Fig. 14-21 was developed based on data for packings such as small intalox saddles, rashig, and pall rings. Incorporated in the capacity parameter on the ordinate scale is an empirical parameter characteristic of the shape, size, and material property of the packing type and is called the packing factor (C_f). C_f has units of inverse length and is used to relate the packing type to the relative gas pressure drop through the packing in the tower. Figure 14-12 displays C_f values for several commonly used plastic packing types. Since C_f is incorporated in the numerator of the capacity parameter on the ordinate scale, packing

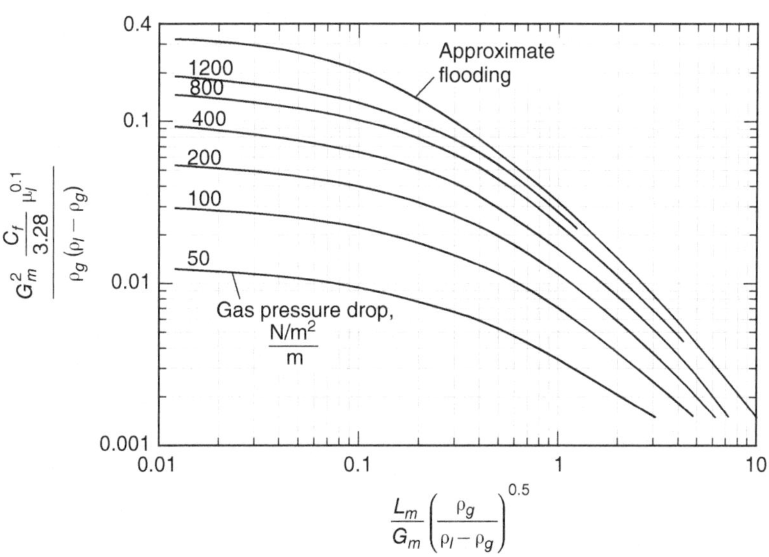

Figure 14-21
Generalized Eckert gas pressure drop and liquid and gas loading correlation in SI units for random packed tower. The coefficient 3.28 is a conversion factor when the packing factor in SI units (m^{-1}) is used because the Eckert diagram was originally drawn in English units. (Adapted from Eckert, 1961; Treybal, 1980).

materials with a higher C_f value will have a higher gas pressure drop than packing materials with a lower C_f value. In general, the gas pressure drop will increase with increasing packing factor.

The practical operating range for packed towers is between abscissa (x axis) values of 0.02 and 4.0 on the generalized Eckert pressure drop curves (see Fig. 14-21). For abscissa values greater than 4, large water loading rates can reduce the water–air contact area provided by the packing surface and inhibit proper airflow through the column, causing a decrease in removal efficiency. Similarly, high air flow rates (abscissa values less than about 0.02) can cause entrained water in the tower as well as channeling of the air through the tower. For situations where high air flow rates are required for high removal efficiencies (>95 percent), it is important to provide an even air inlet distribution at the bottom of the tower (Thom and Byers, 1993). Towers operating in the regions described above may require additional packing depth to compensate for efficiency reductions.

For an air–water system at 20°C and 1 atm and S value of 3, the dimensionless Henry's constants can only range from about 0.0021 to 0.42, corresponding to an abscissa between 0.02 and 4 on the Eckert curves. Thus, use of Fig. 14-21 for stripping tower design is restricted, and pilot studies are recommended for tower design when the abscissa values are greater than 4, unless manufacturer's data on gas pressure drop applies in that higher range. Nonvolatile compounds with Henry's constants below 0.0021 should not be used as the basis for a packed-tower aeration design given the difficulty of their removal in the air-stripping process.

In summary, the best method for evaluating gas pressure drop for a given design is from pilot or full-scale studies. If these studies are not available, the next best way would be to use manufacturer's data for gas pressure drop, followed by the use of the Eckert curves. The Eckert curves could be applied initially to estimate the tower diameter followed by the use of the manufacturers data to determine the actual pressure drop for the given tower diameter.

CROSS-SECTIONAL AREA

The cross-sectional area of a packed tower can be estimated from the generalized Eckert pressure drop curves shown on Fig. 14-21 (see above discussion of gas pressure drop). The gas loading rate, liquid loading rate, and tower area may be determined from Fig. 14-21 using the following procedure:

1. Specify the following design parameters:
   a. Packing factor for the media (see Fig. 14-12)
   b. Air-to-water ratio [typically 3.5 times $(Q_a/Q)_{\min}$]
   c. Gas pressure drop (typically 50 $N/m^2/m$)

2. Determine the value on the x axis on the Eckert curve shown on Fig. 14-21:

$$x = \left(\frac{1}{G_m / L_m}\right)\left(\frac{\rho_g}{\rho_l - \rho_g}\right)^{0.5} \quad (14\text{-}53)$$

where x = value on x axis on Eckert curve
G_m = air mass loading rate, $kg/m^2 \cdot s$
L_m = water mass loading rate, $kg/m^2 \cdot s$
ρ_g = air density, kg/m^3
ρ_l = water density, kg/m^3

The value of G_m/L_m can be determined knowing the air-to-water ratio, water density, and air density:

$$\frac{G_m}{L_m} = \left(\frac{Q_a}{Q}\right)\left(\frac{\rho_g}{\rho_l}\right) \quad (14\text{-}54)$$

3. Graphically determine the numerical value y on the y axis on the Eckert curve shown on Fig. 14-21 knowing the gas pressure drop and x.
4. Determine the gas loading rate based on the following relationship for the y-axis value on the Eckert curve shown on Fig. 14-21:

$$y = \frac{G_m^2 (C_f / 3.28)\mu_l^{0.1}}{\rho_g(\rho_l - \rho_g)} \quad (14\text{-}55)$$

Rearrange Eq. 14-55 and solve for G_m:

$$G_m = \left[\frac{y\rho_g(\rho_l - \rho_g)}{(C_f / 3.28)\mu_l^{0.1}}\right]^{0.5} \quad (14\text{-}56)$$

where y = numerical value on y axis of Eckert curve determined in step 3
C_f = packing factor, m^{-1}
μ_l = dynamic viscosity of water, $kg/m \cdot s$

5. Determine the water mass loading rate based on the following relationship:

$$L_m = \frac{G_m}{(Q_a/Q)(\rho_g/\rho_l)} \quad (14\text{-}57)$$

6. Determine the cross-sectional area of the packed tower based on the following relationship:

$$A = \frac{Q\rho_l}{L_m} \quad (14\text{-}58)$$

where A = cross-sectional area of packed tower, m^2
Q = water flow rate, m^3/s

Correlations describing the Eckert pressure drop curves to predict gas loading rate and tower area were fit by Cummins and Westrick (1983). The Eckert pressure drop correlations are useful for performing packed-tower aeration design calculations using spreadsheets or computer programs, but the correlations are beyond the scope of this book.

Example 14-4 Diameter, area, and pressure drop of a packed tower

Determine the cross-sectional area and tower diameter for a packed-tower design based on 1,2-dichloropropane (DCP) at 10°C for a water flow rate Q of 0.1 m^3/s (1585 gal/min).The basis for design is given by the operating air-to-water ratio of 60 (see Example 14-3), gas pressure drop $\Delta P/L = 50\ N/m^2 \cdot m$, and the 8.9-cm (3.5-in.) plastic tripacks. The physical properties of air and water at 10°C are as follows: water density $\rho_l = 999.7\ kg/m^3$, water viscosity $\mu_l = 1.307 \times 10^{-3}\ kg/m \cdot s$, and air density, $\rho_g = 1.247\ kg/m^3$ (see Apps. B and C). The packing factor from Fig. 14-12, C_f, for 8.9-cm (3.5-in.) plastic tripacks is 39.0 m^{-1}. The dimensionless Henry's law constant of DCP is $H_{yc,dcp} = 0.0525$ (see Table 14-3).

Solution

1. Specify the packing factor, air-to-water ratio, and gas pressure drop.
   a. *Packing factor*: Given in problem statement:

$$C_f = 39\ m^{-1}$$

   b. *Air-to-water ratio*: Determined in Example 14-3 for 90 percent removal:

$$\left(\frac{Q_a}{Q}\right)_{DCP} = 60$$

   c. *Gas pressure drop*: Given in problem statement:

$$\frac{\Delta P}{L} = 50\ N/m^2 \cdot m$$

2. Determine the value on the x axis on the Eckert curve shown on Fig. 14-21:
   a. Determine G_m/L_m using Eq. 14-54:

$$\frac{G_m}{L_m} = \left(\frac{Q_a}{Q}\right)\left(\frac{\rho_g}{\rho_l}\right) = 60\left(\frac{1.247}{999.7}\right)$$
$$= 0.075\ kg\ air/kg\ water$$

b. Determine x using Eq. 14-53:

$$X = \left[\frac{1}{(G_m/L_m)}\right]\left(\frac{\rho_g}{\rho_l - \rho_g}\right)^{0.5} = \left(\frac{1}{0.075}\right)\left(\frac{1.247}{999.7 - 1.247}\right)^{0.5}$$
$$= 0.47$$

3. Graphically determine the numerical value y on the y axis on the Eckert curve shown on Fig. 14-21 knowing the gas pressure drop and x. At the location on Fig. 14-21 where x = 0.47 and $\Delta P/L = 50\ \text{N/m}^2 \cdot \text{m}$,

$$y = 0.005$$

4. Determine the gas loading rate based on the relationship for the y-axis value on the Eckert curve shown on Fig. 14-21. Solve for G_m using Eq. 14-56:

$$G_m = \left[\frac{y\rho_g(\rho_l - \rho_g)}{(C_f/3.28)\mu_l^{0.1}}\right]^{0.5} = \left[\frac{0.005(1.247)(999.7 - 1.247)}{(39.0/3.28)(1.307 \times 10^{-3})^{0.1}}\right]^{0.5}$$
$$= 1.01\ \text{kg/m}^2 \cdot \text{s}$$

5. Determine the water mass loading rate using Eq. 14-57:

$$L_m = \frac{G_m}{(Q_a/Q)(\rho_g/\rho_l)} = \frac{1.01\ \text{kg/m}^2 \cdot \text{s}}{(60)(1.247\ \text{kg/m}^3/999.7\ \text{kg/m}^3)}$$
$$= 13.5\ \text{kg/m}^2 \cdot \text{s}$$

6. Determine the cross-sectional area of the packed tower using Eq. 14-58:

$$A = \frac{Q\rho_l}{L_m} = \frac{(0.1\ \text{m}^3/\text{s})(999.7\ \text{kg/m}^3)}{13.5\ \text{kg/m}^2 \cdot \text{s}}$$
$$= 7.4\ \text{m}^2$$

7. Determine the tower diameter assuming a circular tower area.

$$D = \sqrt{\frac{4A}{\pi}} = \sqrt{\frac{4(7.4\ \text{m}^2)}{\pi}} = 3.07\text{m}$$

Standard tower sizes of 1.22m (4 ft), 1.83m (6ft), 2.44m (8ft), 3.048 m (10 ft), 3.66m (12 ft), and sometimes 4.27m (14 ft) in diameter are usually the norm for most packed-tower equipment manufacturers. For this case we will use a 3.048-m (10-ft) diameter tower. For a tower diameter of 3.048 m, the operating values of G_m and L_m are $1.02\ \text{kg/m}^2 \cdot \text{s}$ and $13.7\ \text{kg/m}^2 \cdot \text{s}$, respectively.

8. Determine the tower gas pressure drop based on the manufacturer's data presented in Fig. 14-20.
   a. Calculate the C-factor form Eq. 14-52,

$$\begin{aligned} \text{C factor} &= V_S((\rho_g)/(\rho_L - \rho_g))^{0.5} \\ &= \left(\frac{1.02\ \text{kg/m}^2\cdot\text{s}}{1.247\ \text{kg/m}^3}\right)\left(\frac{1.247}{999.7-1.247}\right)^{0.5} \\ &= 0.03\ \text{m/s} \end{aligned}$$

   b. Calculate the liquid loading rate in $L/m^2 \cdot min$ for a tower diameter of 3.048 m (10 ft):

$$\frac{Q}{A} = \frac{0.1\ \text{m}^3/\text{s}}{\pi(3.048\ \text{m})^2/4}\ \frac{1000\ \text{L}}{\text{m}^3}\ \frac{60\ \text{s}}{\text{min}} = 822\frac{\text{L}}{\text{m}^2\cdot\text{min}}$$

   c. Determine the relative head loss through the packing using Fig. 14-20. For a C factor of 0.03 m/s and a volumetric liquid loading of 822 $L/m^2 \cdot$ min, the head loss per unit length of packing is about 4.5 mm water per m of packing height or 44 $N/m^2 \cdot m$. In this case the gas pressure drop determined from the Eckert correlation is in close agreement with the manufacturer's gas pressure drop data.

Comment

If multiple compounds are to be removed, the compound with the lower Henry's law constant in the water to be treated is used as the basis for determining the cross-sectional area of the tower, because it will require the highest air-to-water ratio to have a stripping factor in the optimal range.

MASS TRANSFER RATE CONSTANT

The general equation for calculating the overall mass transfer rate constant $K_L a$ in aeration processes was derived earlier based on the two-film theory of mass transfer in Chap. 7:

$$\frac{1}{K_L a} = \frac{1}{k_l a} + \frac{1}{H_{YC} k_g a} \qquad (7\text{-}88)$$

where
$K_L a$ = overall mass transfer rate constant, s^{-1}
k_l = liquid-phase mass transfer coefficient, m/s
k_g = gas-phase mass transfer coefficient, m/s
a = area available for mass transfer divided by vessel volume, m2/m^3

The $K_L a$ values for packed towers can be determined by performing pilot plant studies or from packing manufacturers and previously reported field studies. They can also be estimated from mass transfer correlations.

Determination by pilot plant studies

A pilot plant study is the preferred way to determine $K_L a$ for a given VOC in water, but, as discussed below, fairly accurate estimates can be made from the correlations. Pilot-scale packed towers range in size from 2 to 6m (6.5 to 20 ft) in height and 0.3 to 0.6m (1 to 2 ft) in diameter. The column diameter used will depend upon the desired packing size. It is generally recommended that ratios of column diameter to nominal packing diameter be greater than 10:1 (>15:1 desired) to minimize error caused by channeling of the water down the walls of the column (Treybal, 1980). It is also recommended that the packed-column height-to-diameter ratio be greater than 1:1 to provide for proper liquid distribution (Tryebal, 1980; Roberts et al., 1985).

Equation 14-48 is used in conjunction with pilot plant data to determine $K_L a$ for a given VOC. The value of $K_L a$ is based on VOC removal due to the packed height portion of the tower. However, VOC removal also occurs as the water contacts the air above the packing at the top of the tower and at the bottom as the water falls into the clearwell below the packing. This incidental additional removal is sometimes referred to as "end effects" (Umphres et al., 1983). To determine the $K_L a$, $\text{NTU}_{\text{measured}}$ is plotted versus the packing height and the NTU value corresponding to zero packing depth is referred to as $\text{NTU}_{\text{end effects}}$ (dimensionless):

$$\text{NTU}_{\text{measured}} = \left(\frac{1}{\text{HTU}}\right) \times Z + \text{NTU}_{\text{end effects}} \tag{14-59}$$

where Z = distance from top of packing to sample port location along packed portion of tower, L

For a given water and air loading rate, aqueous-phase concentration measurements are evaluated at the influent, effluent, and various sample port locations along the packed column and $\text{NTU}_{\text{measured}}$ is calculated from Eq. 14-50. The plot of $\text{NTU}_{\text{measured}}$ versus Z should result in a straight line (Eq. 14-59), and $K_L a$ is determined from the slope (1/HTU). Experimentally determined $K_L a$ values can be correlated as a function of water loading rate for several air-to-water ratios that would be expected during operation of the full-scale column. The full-scale packed-tower height can be determined from $K_L a$ and the design equations.

A full-scale packed-tower height calculated using a $K_L a$ value determined from a pilot study is generally conservative. For a given packing size, $K_L a$ values generally increase as the tower diameter increases (Roberts et al., 1985). The increase is caused by minimizing channeling of the water down the inside of the column walls (wall effects), which occurs to a greater degree in small columns. The VOC removal rate is lower along the walls of

Table 14-6
Packed-tower air-stripping pilot plant studies that determined $K_L a$ values for several VOCs

Water Matrix	VOCs	Reference
Sacramento–San Joaquin Delta water in northern California	Chloroform, dibromochloromethane, bromodichloromethane, bromoform	Umphres et al. (1983)
Potomac tidal fresh estuary water mixed with nitrified effluent wastewater	Carbon tetrachloride, tetrachloroethene, trichloroethene, chloroform, bromoform	Ball et al. (1984)
City of Tacoma, WA, groundwater	1,1,2,2-Tetrachloroethane, trans-1,2-dichloroethene, trichloroethene, tetrachloroethene	Byers and Morton (1985)
Laboratory-grade organic free water	Oxygen, tetrachloroethene, Freon-12, 1,1,1-trichloroethane, trichloroethene, carbon tetrachloride	Roberts et al. (1985)
North Miami Beach, FL, groundwater and City of Gainsville, FL, groundwater	Chloroform	Bilello and Singley (1986)
Village of Brewster, NY, groundwater	*cis*-1,2-Dichloroethene, trichloroethene, tetrachloroethene	Wallman and Cummins (1986)
Gloucester, Ottawa, Ontario, groundwater	Chloroform, toluene, 1,2-dichloroethane, 1,1-dichloroethane, trichloroethene, diethyl ether	Lamarche and Droste (1989)
Miami, FL, tap water	Bromoform, bromodichloromethane, chloroform, dibromochloromethane, carbon tetrachloride, tetrachloroethene, trichloroethene, 1,1,1-trichloroethane, chlorobenzene, *m*-dichlorobenzene, *m*-xylene, toluene	Narbaitz et al. (2002)

the column than within the packing because the air/water contact time, surface area, and mixing are smaller. As the tower diameter increases, the percentage of flow down the walls of the column decreases, minimizing the wall effects.

Based on the results from several packed-tower field studies (see Table 14-6), experimentally determined $K_L a$ values for several VOCs and various contaminated water sources can be obtained. The $K_L a$ values reported in these studies can be used to design towers if the operating conditions (temperature, water and air loading rate) and packing type and size are identical.

Estimation with empirical correlations

From an evaluation of available mass transfer models for packed-tower aeration (Onda et al., 1968; Sherwood and Hollaway, 1940; Shulman et al., 1955), it has been found that the Onda model or a modification of it provides the best predictions of mass transfer coefficients (Lamarche and Droste, 1989; Djebbar and Narbaitz, 1995, 1998). For several VOCs, it has been demonstrated that $K_L a$ values obtained from the Onda correlations compare favorably to pilot plant data using smaller packing sizes (Cummins and Westrick, 1983; Roberts et al., 1985). The Onda correlations for determination of the liquid-phase mass transfer coefficient (k_l), gas-phase mass transfer coefficient (k_g), and specific interfacial area (a) are presented in Chap. 7. The packing properties needed for the Onda correlations are shown in Fig. 14-12. The Onda correlations were developed for nominal packing sizes up to 5.1cm (2 in.).

Studies have found that the Onda correlations for mass transfer coefficients overestimate the K_La values for larger packing sizes [greater than 2.5 cm (1 in.) nominal diameter] (Djebbar and Narbaitz, 1995; Lenzo et al., 1990; Thom and Byers, 1993; Dvorak et al., 1996). Djebbar and Narbaitz (1998) modified the Onda model in an effort to improve its predictive capabilities. The modified Onda model included recalibration to a new extensive set of mass transfer data that included adjustments to the constants and exponents in the model, incorporation of an additional dimensionless parameter (L/d_p) into the liquid-phase mass transfer correlation k_l, and incorporation of the gas-phase Reynolds number (RE_g) and the packing efficiency number ($a_t\ d_p$) into the interfacial area equation. As compared to the Onda model, the modified Onda model reduced the average absolute error to 21 percent as compared to 30 percent for the Onda model, which is about a 30 percent reduction in the error. The modified Onda model requires a trial-and-error method for design because an initial guess of the tower length is needed to calculate the k_l. More recently Dejebbar and Narbaitz (2002) used neural network nonparametric approach to analyze gas and liquid mass transfer data from packed tower technology to predict K_La values. They were able to improve upon the predictions for K_La with an average absolute error of less than 19 percent, which is perhaps the best prediction to date. Unfortunately, it would be too time consuming for design engineers to use this technique to obtain K_La values.

At present, there is no correlation that can be used to predict K_La within ±10 percent for larger packing sizes. Thus, based on the literature cited above, it is recommended that a safety factor of 0.70 (design K_La/Onda K_La) be applied for packing diameters greater than 2.5 cm (1 in.) as a conservative estimate of packing height required.

Example 14-5 Mass transfer coefficients in packed-tower aeration

Determine the mass transfer coefficients for DCP and PCE at 10°C in packed-tower aeration for the air mass loading rate and water mass loading rate determined in Example 14-4 using the Onda correlations and a safety factor of 0.70 (design K_La/Onda K_La) for 8.9cm (3.5 in.) plastic tripacks. The water flow rate, Q, is 0.1 m^3/s (1585 gal/min). The physical properties of air and water from Apps. B and C at 10°C are water density ρ_l= 999.7

kg/m³, the dynamic viscosity of water $\mu_l = 1.307 \times 10^{-3}$ kg/m · s, water surface tension $\sigma = 0.0742$ N/m, air density $\rho_g = 1.247$ kg/m³, and air viscosity $\mu_g = 1.79 \times 10^{-5}$ kg/m · s. The properties of the packing material from Fig. 14-12 are nominal diameter of packing $d_p = 0.0889$ m, packing factor $C_f = 39.0\text{m}^{-1}$, specific surface area of packing $a_t = 125.0$ m²/m³, and critical surface tension of packing $\sigma_c = 0.033$ N/m. The liquid diffusivity D_l and gas diffusivity D_g for DCP and PCE were determined from the Hayduk–Laudie correlation and the Wilke–Lee modification of the Hirschfelder–Bird–Spotz method, respectively, to be equal to $D_{l,\text{DCP}} = 6.08 \times 10^{-10}$ m²/s, $D_{l,\text{PCE}} = 5.86 \times 10^{-10}$ m²/s, $D_{g,\text{DCP}} = 7.65 \times 10^{-6}$ m²/s, and $D_{g,\text{PCE}} = 7.13 \times 10^{-6}$ m²/s (see Chap. 7). From Example 14-4, the air loading rate G_m and water loading rate L_m are equal to 1.02 and 13.7 kg/m² · s, respectively. As obtained in Table 14-3 at 10°C, the dimensionless Henry's law constants of DCP and PCE are $H_{YC,\text{DCP}} = 0.0525$ and $H_{YC,\text{DCE}} = 0.364$.

Solution

1. Calculate the specific surface area available for mass transfer a_w, which is determined from the Onda correlations (see Table 7–5):

$$a_w = a_t\left\{1-\exp\left[-1.45\left(\frac{\sigma_c}{\sigma}\right)^{0.75}\left(\frac{L_m}{a_t\mu_l}\right)^{0.1}\left(\frac{L_m^2 a_t}{\rho_l^2 g}\right)^{-0.05}\left(\frac{L_m^2}{\rho_l a_t \sigma}\right)^{0.2}\right]\right\}$$

$$= 125\left\{1-\exp\left[\begin{array}{l}-1.45\left(\dfrac{0.0330}{0.0742}\right)^{0.75}\left(\dfrac{13.7}{125.0\times1.307\times10^{-3}}\right)^{0.1}\\ \times\left(\dfrac{13.7^2\times125.0}{999.7^2\times9.81}\right)^{-0.05}\\ \times\left(\dfrac{13.7^2}{999.7\times125.0\times0.0742}\right)^{0.2}\end{array}\right]\right\}$$

$$= 67\text{m}^2/\text{m}^3$$

2. Calculate the liquid-phase mass transfer coefficient k_l using the Onda correlations (see Table 7-5).

   a. DCP:

$$k_\ell = 0.0051\left(\frac{L_m}{a_w\mu_l}\right)^{2/3}\left(\frac{\mu_l}{\rho_l D_l}\right)^{-0.5}(a_t d_p)^{0.4}\left(\frac{\rho_l}{\mu_l g}\right)^{-1/3}$$

$$= 0.0051\left\{\begin{array}{l}\left[\dfrac{13.7}{67\times(1.307\times10^{-3})}\right]^{2/3}\left[\dfrac{1.307\times10^{-3}}{999.7\times(6.08\times10^{-10})}\right]^{-0.5}\\ \times(125.0\times0.0889)^{0.4}\left[\dfrac{999.7}{(1.307\times10^{-3})\times9.81}\right]^{-1/3}\end{array}\right\}$$

$$= 1.95\times10^{-4}\ \text{m/s}$$

b. PCE:

$$k_\ell = 0.0051\left(\frac{L_m}{a_w\mu_l}\right)^{2/3}\left(\frac{\mu_l}{\rho_l D_l}\right)^{-0.5}(a_t d_p)^{0.4}\left(\frac{\rho_l}{\mu_l g}\right)^{-1/3}$$

$$= 0.0051\left\{\begin{array}{l}\left[\dfrac{13.7}{67\times(1.307\times10^{-3})}\right]^{2/3}\left[\dfrac{1.307\times10^{-3}}{999.7\times(5.87\times10^{-10})}\right]^{-0.5}\\ \times(125.0\times0.0889)^{0.4}\left[\dfrac{999.7}{(1.307\times10^{-3})\times9.81}\right]^{-1/3}\end{array}\right\}$$

$$= 1.92\times10^{-4}\ \text{m / s}$$

3. Calculate the gas-phase mass transfer coefficient k_g using the Onda correlations (see Table 7-5).

a. DCP:

$$k_g = 5.23(a_t D_g)\left(\frac{G_m}{a_t\mu_g}\right)^{0.7}\left(\frac{\mu_g}{\rho_g D_g}\right)^{1/3}(a_t d_p)^{-2}$$

$$= 5.23\left\{\begin{array}{l}\left[125.0\times(7.65\times10^{-6})\right]\left[\dfrac{1.02}{125.0\times(1.79\times10^{-5})}\right]^{0.7}\\ \times\left[\dfrac{1.79\times10^{-5}}{1.247\times(7.65\times10^{-6})}\right]^{1/3}(125.0\times0.0889\ \text{m})^{-2}\end{array}\right\}$$

$$= 3.63\times10^{-3}\ \text{m/s}$$

b. PCE:

$$k_g = 5.23(a_t D_g)\left(\frac{G_m}{a_t\mu_g}\right)^{0.7}\left(\frac{\mu_g}{\rho_g D_g}\right)^{1/3}(a_t d_p)^{-2}$$

$$= 5.23\left\{\begin{array}{l}\left[125.0\times(7.13\times10^{-6})\right]\left[\dfrac{1.02}{125.0\times(1.79\times10^{-5})}\right]^{0.7}\\ \times\left[\dfrac{1.79\times10^{-5}}{1.247\times(7.13\times10^{-6})}\right]^{1/3}(125.0\times0.0889\ \text{m})^{-2}\end{array}\right\}$$

$$= 3.46\times10^{-3}\ \text{m/s}$$

4. Calculate the overall mass transfer rate constant K_La based on a_w, k_l, and k_g from the Onda correlations using Eq. 7–88.
   a. DCP:

$$\frac{1}{K_La} = \frac{1}{k_1a_w} + \frac{1}{k_ga_wH_{YC}}$$

$$= \frac{1}{(1.95\times10^{-4})\times67} + \frac{1}{(3.63\times10^{-3})\times67\times0.0525}$$

$$\Rightarrow K_La = 0.00645\ \text{s}^{-1} \quad \text{(based on Onda correlations)}$$

   b. PCE:

$$\frac{1}{K_La} = \frac{1}{k_la_w} + \frac{1}{k_ga_wH_{YC}}$$

$$= \frac{1}{(1.92\times10^{-4})\times67} + \frac{1}{(3.46\times10^{-3})\times67\times0.364}$$

$$\Rightarrow K_La = 0.011\ \text{s}^{-1} \quad \text{(based on Onda correlations)}$$

5. Calculate actual K_La applying a safety factor (SF) of 0.70 on the Onda K_La.
   a. DCP:

$$K_La = K_La\,(\text{Onda})\times(\text{SF}) = (0.00645\ \text{s}^{-1})\times0.70$$
$$= 0.00452\ \text{s}^{-1}$$

   b. PCE:

$$K_La = K_La\,(\text{Onda})\times(\text{SF})K_La = (0.011\ \text{s}^{-1})\times0.70$$
$$= 0.0077\ \text{s}^{-1}$$

Power Requirements

The total operating power for a single air-stripping packed-tower system is the sum of the blower and pump brake power requirements. The blower brake power P_{blower} can be determined from the following relationship (Tchobanoglous et al., 2003):

$$P_{\text{blower}} = \left(\frac{G_{\text{me}}RT_{\text{air}}}{\text{MW}n_a\ \text{Eff}_b}\right)\left[\left(\frac{P_{\text{in}}}{P_{\text{out}}}\right)^{0.283} - 1\right] \tag{14-60}$$

where P_{blower} = blower brake power, kW
Eff_b = blower net efficiency, expressed as decimal fraction, which accounts for both fan and motor on blower

G_{me} = mass flow rate of air, kg/s
n_a = constant used in determining blower brake power,
= 0.283 for air
P_{in} = inlet air pressure in packed tower (bottom of tower), atm or N/m^2
P_{out} = outlet air pressure in packed tower (top of the tower), usually equal to ambient pressure, atm or N/m^2
R = universal gas constant, 8.314 J/mol · K
T_{air} = absolute air temperature, typically assumed equal to T, K (273 + °C)
T = absolute water temperature, K (273 + °C)
MW = molecular weight of air, 28.97 g/mol

The term P_{in} refers to the pressure at the bottom of the tower, which is the inlet for the airstream:

$$P_{in} = P_{ambient} + [(\Delta P/L) \times L] + \Delta P_{losses} \tag{14-61}$$

where $P_{ambient}$ = ambient pressure, atm or N/m^2
ΔP = pressure drop caused by packing media, atm or N/m^2
L = packing height, m
ΔP_{losses} = pressure drop by demister, packing support plate, duct work, inlet and outlet of tower, atm or N/m^2

The pressure drop ΔP_{losses} may be estimated by using the empirical constant k_p, which was determined by fitting full-scale tower data (Hand et al., 1986):

$$\Delta P_{losses} = \left(\frac{Q_a}{A}\right)^2 k_p \tag{14-62}$$

where Q_a = volumetric air flow rate, m^3/s
A = tower cross-sectional area, m^2
k_p = empirical constant, 275 N · s^2/m^4

The air pressure drop through the demister, packing support plate, duct work, and inlet and outlet of the tower is accounted for in Eq. 14-62. It is assumed that turbulent-flow conditions prevail and most of the losses occur in the tower (i.e., in the packing support plate and the demister).

The pump power requirement can be determined from the equation

$$P_{pump} = \frac{\rho_l Q H g}{\text{Eff}_p} \tag{14-63}$$

where P_{pump} = power required to pump water to top of tower, W
ρ_l = water density, kg/m^3

Q = water flow rate, m^3/s
H = vertical distance from pump to liquid distributor at top of tower, m
g = acceleration due to gravity, 9.81 m/s^2
Eff_p = pump efficiency, expressed as fraction

Equation 14-63 only accounts for the additional head required to pump the water to the top of the tower.

Example 14-6 Power requirements for packed-tower aeration

Determine the total power requirement (blower and pump brake power) and specific energy per unit volume of water treated for a packed tower aeration design removing DCP at 10°C and 1 atm (101,325 N/m^2) for a water flow rate of 0.1 m^3/s, a stripping factor of 3.5, and a gas pressure drop of 50 $(N/m^2)/m$. Assume the blower efficiency is 35 percent ($Eff_b = 0.35$) and the pump efficiency is 80 percent ($Eff_p = 0.80$).

From Example 14-4, the air-to-water ratio $Q_a/Q = 60$ and the tower area $A = 7.3\ m^2$ (based on a 3.048-m (10-ft) tower diameter). From Example 14-7, the tower length $L = 7.8$ m. The water density and air density at 10°C are $\rho_l = 999.7\ kg/m^3$ and $\rho_g = 1.247\ kg/m^3$.

Solution

1. Calculate blower power requirements.
   a. Calculate the air mass flow rate from the volumetric air flow rate.
      i. Calculate volumetric air flow rate Qa:

$$Q_a = \left(\frac{Q_a}{Q}\right)Q = (60)\left(0.1\ m^3/s\right)$$
$$= 6.0\ m^3/s$$

      ii. Calculate the air mass flow rate G_{me}:

$$G_{me} = Q_a\rho_g = \left(6.0\ m^3/s\right)\left(1.247\ kg/m^3\right)$$
$$= 7.48\ kg/s$$

   b. Calculate the pressure drop through the demister, the packing support plate, duct work, and inlet and outlet (ΔP_{losses}) using Eq. 14-62:

$$\Delta P_{losses} = \left(\frac{Q_a}{A}\right)^2 k_p = \left(\frac{6.0\ m^3/s}{7.3\ m^2}\right)^2 \left(275\ N\cdot s^2/m^4\right)$$
$$= 186\ N/m^2$$

c. Calculate the inlet pressure to the packed tower, P_{in}, using Eq. 14-61:

$$P_{in} = P_{ambient} + \left(\frac{\Delta P}{L}\right)L + \Delta P_{losses} = 101{,}325 \text{ N/m}^2$$
$$+ \left\{\left[\left(50 \text{ N/m}^2\right)/\text{m}\right] \times 7.8 \text{ m}\right\} + 186 \text{ N/m}^2$$
$$= 101{,}901 \text{ N/m}^2$$

d. Calculate the blower brake power P_{blower} using Eq. 14-60:

$$P_{blower} = \left(\frac{G_{me}RT_{air}}{\text{MW } n_a \text{ Eff}_b}\right)\left[\left(\frac{P_{in}}{P_{out}}\right)^{n_a} - 1\right]$$
$$= \left[\frac{(7.48)(8.314) \times (273+10)}{(28.97)(0.283)(0.35)}\right]\left[\left(\frac{101{,}901}{101{,}325}\right)^{0.283} - 1\right]$$
$$= 9.85 \text{ kW}$$

2. Calculate pump power requirements P_{pump} to move the water to the top of the tower using Eq. 14-63:

$$P_{pump} = \frac{\rho_l QLg}{\text{Eff}_p}$$
$$= \left[\frac{(999.7 \text{ kg/m}^3)(0.1 \text{ m}^3/\text{s})(7.8 \text{ m})(9.81 \text{ m/s}^2)}{0.80}\right]\left(\frac{1 \text{ W}}{1 \text{ kg} \cdot \text{m}^2/\text{s}^3}\right)$$
$$= 9{,}561 \text{ W} = 9.56 \text{ kW}$$

3. Calculate total power requirements P_{total}:

$$P_{total} = P_{blower} + P_{pump} = 9.56 \text{ kW} + 9.85 \text{ kW} = 19.4 \text{ kW}$$

4. Calculate the specific energy:

$$E = \frac{19.4 \text{ kW}}{(0.1 \text{ m}^3/\text{s})(3600 \text{ s/h})} = 0.0539 \text{ kWh/m}^3$$

Design versus Rating Analysis of Packed Towers

There are two types of analyses commonly performed for packed-tower air stripping, termed *design analysis* and *rating analysis*. In a design analysis, it is desired to *size a new packed tower* to exactly meet the treatment objective C_{TO}. Substituting $C_{TO} = Ce$ into Eq. 14-47 results in the design equation for packed tower aeration:

$$L = \frac{Q}{AK_La}\left(\frac{S}{S-1}\right)\ln\left[\frac{1+(C_0/C_{TO})(S-1)}{S}\right] \qquad (14\text{-}64)$$

where L = packed tower height, m
Q = water flow rate, m^3/s
A = cross-sectional area of packed tower, m^2
K_La = overall liquid-phase mass transfer rate constant, s^{-1}
S = stripping factor, dimensionless
C_0 = influent liquid-phase concentration, mg/L
C_{TO} = treatment objective, mg/L

Estimation of mass transfer rate constant K_La and cross-sectional area A for packed-tower air stripping is described above.

In a rating analysis, the effluent concentrations of various compounds *for an existing tower* can be determined. The following variables are known in a rating analysis: (1) tower height, (2) tower diameter, (3) type of packing, (4) water flow rate, (5) air flow rate, (6) pressure, (7) temperature, (8) influent concentration, and (9) mass transfer coefficient. Knowing these variables, it is possible to determine effluent concentration and gas pressure drop for the tower. The effluent concentration is found by rearranging Eq. 14-47 and solving for effluent concentration C_e:

$$C_e = \frac{C_0(S-1)}{S\exp\left[LK_La(S-1)/(Q/A)S\right]-1} \tag{14-65}$$

where C_e = effluent liquid-phase concentration, mg/L

In a rating analysis, the pressure drop for the tower can be calculated using an iterative method based on correlations for the Eckert curves.

Design Variables

Design variables for packed-tower air stripping include (1) the air-to-water ratio, (2) the gas pressure drop, and (3) the type of packing material. Once the physical properties of the contaminant(s) of interest, the influent concentration(s), treatment objective(s), water, and air properties are known, design parameters can be selected to obtain the lowest capital and operation and maintenance costs.

AIR-TO-WATER RATIO

It has been shown that air-to-water ratios of approximately 3.5 times the minimum air-to-water ratio provide the minimum tower volume and power requirement, which corresponds to a stripping factor of about 3.5 for a range of Henry's law constants from 0.003 to 0.3 (Hand et al., 1986).

GAS PRESSURE DROP

A low gas pressure drop should be chosen to minimize the blower power consumption. Packed towers are usually designed to operate with a gas pressure drop well below flooding conditions. Many stripping towers are designed for gas pressure drops of 200 to 400 $(N/m^2)/m$ of packing depth (0.25 to 0.5 in. H_2O/ft of packing) (Treybal, 1980). Based on detailed cost analyses, it has been found that using a lower gas pressure drop between 50 and 100 $(N/m^2)/m$ and a stripping factor of 3.5 yields the lowest total annual treatment cost for removal of volatile compounds with dimensionless H_{YC} values greater than 0.05 (Cummins and Westrick, 1983; Dzombak et al., 1993; Hand et al., 1986). Towers have been constructed with gas pressure drops as low as 30 $(N/m^2)/m$, but gas pressure drops that are too low may result in very low liquid loading and poor water distribution across the packing, which will reduce the area available for mass transfer and tower performance.

An additional advantage of operating at a low gas pressure drop is that, if the blower is sized conservatively, the air flow rate can be increased to improve removal efficiency without major changes in the process operation. However, the required tower height may sometimes be too large for a particular application (mostly for aesthetic reasons, the local community may object to tall towers). To obtain a smaller tower height for a given removal, the air-to-water ratio can be increased.

TYPE AND SIZE OF PACKING MATERIAL

The competing requirements of a low gas pressure drop and high surface area available for mass transfer per vessel volume must be weighed when selecting packing because the preferred packing characteristics work against each other as high surface area per volume causes higher gas pressure drop. The surface area per volume and packing factor for commonly available packing materials are reported on Fig. 14-12. For a given type of packing, the packing factor and surface area increase as the size of packing decreases. However, different types of packing can have lower packing factors for the same surface area per volume. For example, the 75-mm (3-in.) saddles and 50-mm (2-in.) tripacks have packing factors of 16 and 15, respectively, and yet the tripacks have 76 percent more surface area per volume because of their unique shape, which is shown on Fig. 14-12. A packing material with a low packing factor and a high specific surface area is desired for optimal tower performance.

Tower volume and power requirements have been compared for a number of packing types and sizes reported in Fig. 14-12. The comparisons show that the type of packing media does not have a large impact on the tower volume or the total operating power requirements. However, it has been shown for the same type of packing (e.g., plastic tripacks), smaller nominal diameters result in lower tower volume and power requirements (Hand et al., 1986).

A major concern with respect to choosing the type and size of packing is the possibility of calcium, iron, and manganese precipitates forming on the packing during extended periods of operation and causing reduced removals and higher gas pressure drops, which is discussed in this section. To alleviate precipitation problems, larger packing sizes, which have smaller specific surface area, may be preferable because there would be less surface area upon which precipitate can form as well as larger spaces for airflow.

The criteria for choosing the type and size of packing will depend upon the water flow rate and the desired degree of operational flexibility of the design. For small water flow rates, it is recommended that nominal diameter packing of 50 mm (2 in.) or less be used to minimize channeling or short circuiting of the water down the wall of the tower. Minimizing the impact of channeling requires that the ratio of tower diameter to nominal packing diameter be greater than 10:1 (> 15:1 is desired).

Design Approach

In most situations in water treatment, multiple contaminants are present in the water, and the packed tower must be designed to remove all the contaminants to some specified treatment level. At the design stage, the limiting contaminant that controls the design must first be determined. In general, the contaminant with the lowest Henry's constant is used to determine the required air-to-water ratio and the contaminant with the highest removal efficiency is used to determine the required packing height.

Once the influent concentration of the organic contaminant, treatment objective, flow rate, and design temperature are known, the following steps are followed for design:

1. Select an efficient packing material that is expected to give good mass transfer at low gas head loss. For the selected packing, determine head loss and mass transfer characteristics from commercially available data. Based on the data provided in Table 14-5, tripacks and lantec packing are among the best packing material.
2. Select a gas-phase pressure drop per unit tower length. A value of 50 $N/m^2/m$ usually provides an economical choice and the largest flexibility.
3. Select an operating air-to-water ratio. For most situations, an operating air-to-water ratio that is 3.5 times the minimum air-to-water ratio required for stripping provides the most economical design. For multicomponent systems, the compound with the lowest Henry's constant is used to calculate the operating air-to-water ratio.
4. Given the packing type, stripping factor, and gas pressure drop, the gas loading rate, liquid loading rate, and tower diameter can be determined based on the Eckert curve.
5. Compare the liquid loading rate to allowable liquid loading rates in commercially available equipment. If the liquid loading rate exceeds recommended values, reduce the gas pressure drop and repeat the computation. If the liquid loading rate is less than recommended values, increase the gas pressure drop and repeat the computation.
6. Determine the K_La from the Onda correlation using a safety factor of 0.70.

7. Determine the HTU from Eq. 14-49.
8. Determine the NTU from Eq. 14-50. For multicomponent systems, the contaminant with the highest degree of removal is used to determine the NTU.
9. Determine the height of the tower from Eq. 14-48. Typical packed-tower heights usually do not exceed about 9.0 m (30 ft). Should the calculated tower length exceed 9.0 m, the air-to-water ratio could be increased by increasing the air flow rate to achieve the same treatment objective but with a smaller tower height.
10. Repeat for various values of the stripping factor and the gas pressure drop and determine the optimum or least-cost design. The optimal design will usually be obtained with an operating air-to-water ratio equal to 3.5 times the minimum air-to-water ratio required for stripping and a gas pressure drop of 50 $N/m^2/m$.
11. At this point, preliminary design is complete and a pilot test should be conducted to be certain that the mass transfer correlations are correct.
12. Once the mass transfer parameters are confirmed to be correct, the design can be finalized by examining the operational flexibility of the system.

Several of these steps have been demonstrated in previous examples. Calculation of HTU, NTU, and height of the tower is demonstrated in Example 14.7.

Example 14-7 Height of a packed tower

Determine the packed-tower height required to remove DCP and PCE at 10°C for a water flow rate Q of 0.1 m^3/s (1585 gal/min). The basis for design is DCP removal, gas pressure drop $\Delta P/L = 50$ $N/m^2 \cdot m$, and 8.9-cm (3.5-in.) plastic tripacks. From Table 14-3, the dimensionless Henry's constants of DCP and PCE at 10°C are $H_{YC,DCP} = 0.0525$ and $H_{YC,PCE} = 0.364$. As shown in Example 14-4, the air-to-water ratio is determined based on the contaminant with lower Henry constant, DCP, and a factor of 3.5 $(Q_a/Q)_{min,DCP} = 60$. From Example 14-4, the tower area based on DCP (the compound with the lower Henry constant), for the conditions described above, is 7.3 m^2 [based on a tower diameter of 3.048 m (10 ft)]. From Example 14-5, the actual liquid-phase mass transfer rate constants after applying a safety factor of 0.70 on the Onda K_La values for DCP and PCE at 10°C for the conditions described above are $K_La_{DCP} = 0.0045$ s^{-1} and $K_La_{PCE} = 0.0077$ s^{-1}. The influent concentrations of DCP and PCE are $C_{0,dcp} = 40$ μg/L and $C_{0,pce} = 35$ μg/L. Both DCP and PCE have a treatment objective C_e equal to 5 μg/L.

Solution

1. The required tower depth will be different for each compound. It is necessary to calculate the packing depth for each compound and use the largest value. In this example, DCP requires the largest depth. The calculations for DCP are:
   a. Calculate the stripping factor, S, from Eq. 14-26:

$$S = \frac{Q_a}{Q} H_{YC} = (60)(0.0525) = 3.15$$

   b. Calculate the height of a transfer unit, HTU, using Eq. 14-49:

$$\text{HTU} = \frac{Q}{AK_L a_{DCP}} = \frac{0.1 \text{ m}^3/\text{s}}{(7.3 \text{ m}^2)(0.0045 \text{ s}^{-1})} = 3.04 \text{ m}$$

   c. Calculate the number of transfer units, NTU, using Eq. 14-50:

$$\begin{aligned}\text{NTU} &= \frac{S}{S-1} \ln\left[\frac{1+(C_0/C_e)(S-1)}{S}\right] \\ &= \frac{3.15}{3.15-1} \ln\left[\frac{1+(40/5)(3.15-1)}{3.15}\right] \\ &= 2.57\end{aligned}$$

   d. Calculate the packed-tower height L using Eq. 14-48:

$$L = (\text{HTU})\,(\text{NTU}) = (2.57 \text{ m})\,(3.04) = 7.8 \text{ m}$$

2. Determine the effluent concentration of PCE for the given design tower area and height to make sure it meets its treatment objective.
   a. Calculate the stripping factor for PCE given the air-to-water ratio and Henry's constant using Eq. 14-26:

$$S_{PCE} = \left(\frac{Q_a}{Q}\right) H_{YC,PCE} = (60)(0.364) = 21.8$$

   b. Calculate the effluent concentration C_e of PCE using Eq.14-65:

$$\begin{aligned}C_e &= \frac{C_{0,PCE}(S_{PCE}-1)}{S_{PCE}\exp\left[LK_L a_{PCE}(S_{PCE}-1)/(Q \times S_{PCE}/A)\right]-1} \\ &= \frac{(35\ \mu\text{g/L})(21.8-1)}{21.8\exp\left\{(7.8\text{ m})\left(0.0077\text{ s}^{-1}\right)(21.8-1)/\left[\left(0.1\text{ m}^3/\text{s}\right)(21.8)/\left(7.3\text{ m}^2\right)\right]\right\}-1} \\ &= 0.51\ \mu\text{g}/\text{L}\end{aligned}$$

Comments

The design based on DCP for this example resulted in both components meeting their treatment objectives. While calculation of tower height based on the compound with the highest removal efficiency is suggested as a guideline, there are cases where the guideline will break down because tower height depends on more than just removal efficiency (see Eqs. 14-49 to 14-51) and a design based on the compound with the higher removal efficiency may not allow the treatment objectives of the other compounds to be met. The situation described above is particularly likely to occur if the compound with the highest removal efficiency has a dimensionless Henry's law constant much higher than one or more of the other compounds. The examples in this chapter demonstrate that the design of countercurrent packed towers is a computationally-intensive process. The spreadsheet identified as Resource E10 at the website listed in App. E can be used to perform the calculations.

Factors Influencing Packed-Tower Performance

Packed-tower performance may be impacted by environmental conditions such as water temperature and water quality such as dissolved solids.

TEMPERATURE

Temperature influences both the rate of mass transfer and Henry's constant and thus impacts equipment size, as well as the removal efficiency, in an existing packed tower. A packed tower that is designed to meet treatment objectives at one temperature may not be able to achieve the same treatment objectives at a lower temperature, as shown in Table 14-7. For example, if the temperature decreases from 15 to 5°C, the effluent concentration increases threefold. The information in Table 14-7 is based on a packed tower designed with the following specifications:

1. Trichloroethylene removal with an influent concentration of 100 μg/L = 95 percent. The 95 percent removal value is used to determine the NTU, as shown in Eq. 14-50
2. Design temperature 15°C
3. $H_{YC} = 0.282$
4. $Q_a/Q = 12$

Table 14-7
Effect of temperature on packed-tower operation

Temperature T, °C	$C_{E,T}/C_{E,15°C}$
0	5.2
5	3.3
10	2.0
15	1
20	0.45

5. Packing: Plastic tripacks = 0.089 m (3.5in.)
6. Flow rate 0.095 m^3/s (1500 gal/min)
7. Pressure drop 50 $N/m^2/m$

DISSOLVED SOLIDS

During operation of a packed tower, dissolved inorganic chemicals such as calcium, iron, and manganese may precipitate onto packing media, which can cause a pressure drop increase and a void volume decrease in the tower. The main methods for controlling the negative effects of chemical precipitates are cleaning the precipitate off the packing and controlling precipitate formation.

Precipitate potential

The potential for fouling of packing material by precipitates is especially great in waters containing appreciable amounts of carbon dioxide. Groundwater often contains 30 to 50 mg/L of carbon dioxide. Carbon dioxide can be removed in an air-stripping tower, particularly at high air-to-water ratios, but removal of carbon dioxide tends to raise the pH of the water. As pH increases, bicarbonate is converted to carbonate. In natural waters containing significant quantities of calcium ion, calcium carbonate will precipitate when the carbonate ion concentration is high enough that the solubility product of calcium carbonate is surpassed.

Based on a dimensionless Henry's constant for carbon dioxide of 1.3 at 25°C (calculated using data given in Table 14-4) and the fact that air contains about 0.035 percent by volume of carbon dioxide, the aqueous concentration of carbon dioxide in equilibrium with air can be determined as 0.48 mg/L. The concentration of free carbon dioxide can be reduced to its equilibrium concentration with air via air stripping. The amount of carbonate in the tower effluent depends on both the final carbon dioxide concentration and the pH.

Since carbon dioxide (carbonic acid when in solution) is a weak acid, the rate of stripping depends on the apparent Henry's law constant and pH as presented earlier in this chapter. As carbon dioxide is stripped, the pH will rise and the rate of total carbonate stripping will decrease as water flows through the packing. Acid–base chemistry can be incorporated into the design equations presented earlier in this chapter to develop equations that predict the rate of carbonate stripping and pH of the tower effluent (Howe and Lawler, 1989). Once the tower effluent pH is known, the maximum amount and rate of precipitation that will result in fouling of the tower can be estimated using theoretical precipitation calculations [e.g., using a chemical equilibrium model such as MINTEQA2 (U.S. EPA, 1999) or Visual MINTEQ (Gustafsson, 2002)]. Because the free carbon dioxide concentration is most often reduced to a level greater than the concentration in equilibrium with air, the time taken to foul the tower will be much longer than predicted by the theoretical maximum precipitation calculations. Pilot plant testing is the only method available for determining the actual precipitation rate.

Cleaning
Plastic packing can be removed periodically and put into a tumbler so that the precipitate can be broken off. Acid treatment dramatically deteriorates the plastic packing (making it very brittle) over time and is not recommended. In some instances, conditioning chemicals may be necessary to add to the cleaning process because precipitates can form within weeks in hard water.

Controlling precipitate
Larger packing size, which has smaller specific surface area, may be preferable because there is less surface area upon which precipitate can form as well as larger spaces for airflow. Controlling precipitation with scale inhibitors represents a significant cost in certain situations; therefore, the potential for precipitation must be carefully analyzed.

14-6 Analysis of Low-Profile Air Strippers

Over the past 20 years, low-profile air stripping has become increasingly common. Unit compactness is a key advantage of low-profile air strippers compared to packed towers. Design guidelines for low-profile air strippers, including a comparison with countercurrent packed-tower air strippers, are presented in the following discussion.

Description

A schematic of a low-profile air stripper, which consists of a stack of sieve trays with contaminated water entering the top of the unit and exiting the bottom as treated water and clean air entering the bottom of the unit and exiting at the top containing VOCs, is shown on Fig. 14-8. The operation of a low-profile air stripper has been described by Treybal (1980), and several other researchers have expanded upon that seminal work to explain sieve tray (low-profile) air-stripping columns in detail (LaBranche and Collins, 1996; Mead and Leibbert, 1998; Notthakun et al., 1994; U.S. ACE, 2001). Low-profile air strippers operate as a countercurrent process with water entering at the top of the unit and flowing across each sieve tray, as shown on Fig. 14-8a. Inlet and outlet channels or downcomers are placed at the ends of each tray to allow the water to flow from tray to tray. Fresh air flows upward from a blower positioned beneath the bottom tray through perforated holes into a water layer on each tray. Large air flow rates are typically used, causing very small bubbles or frothing to form upon air contact with the water. The frothing provides a high air–water surface area for mass transfer to occur. Low-profile air stripping can be described conceptually as a countercurrent, staged operation, as demonstrated on Fig. 14-8b.

Both packed towers and low-profile air strippers are capable of providing greater than 99 percent removal of most VOCs. There are numerous advantages and disadvantages of low-profile air strippers when compared to packed towers. Advantages include the following:

- *Unit Compactness.* Because the water flows horizontally across each tray, augmenting the length or width of the trays, instead of the height of the unit, will increase the removal efficiency. A typical low-profile air stripper is less than 3m (10 ft) tall, whereas packed towers are often on the order of 8m (26 ft) in height. There are many situations when architectural or height restrictions require use of a compact, low-profile air stripper even when cost analysis favors a packed tower.
- *Fouling.* Low-profile air strippers are less susceptible to fouling by inorganics because there is no packing. Low-profile air strippers are also much easier to disassemble and clean, compared to packed towers, as the trays are stackable and can be easily removed for cleaning.

Disadvantages include:

- *Narrow Range of Air Flow Rates.* A low-profile air stripper must operate under a narrow range of air flow rates. If the air flow rate is too high, a flooding condition results. If the air flow rate is too low, water will flow through the holes in the sieve trays, a condition known as weeping. Because of the importance of operating the low-profile air stripper under proper conditions, it is necessary that the manufacturer design the sieve tray columns to assure proper performance. Because the air flow rate is finely tuned by the manufacturer, it is not possible to adjust the air flow rate downward should the amount of water treated decrease. In contrast, the air flow rate for a packed tower can be more readily adjusted should a shift in water flow rate occur.
- *Higher Air-to-Water Ratios.* The air-to-water ratio required for a low-profile air stripper is on the order of 100 to 900 (LaBranche and Collins, 1996), compared to a typical air-to-water ratio of 30 for a packed tower. The higher air flow rate for low-profile air strippers is an important consideration, especially when off-gas treatment is required. The higher air-to-water ratio for low-profile air strippers will result in higher costs to operate the blower due to a higher pressure drop and higher costs to treat the off-gas.
- *Foaming.* If the water has a tendency to foam, then packed-tower aeration must be used.

Design Approach Design equations for low-profile air stripping are not currently available in the literature. The diffused aeration approach is not applicable because of the frothing that occurs in low-profile air stripping. An empirical Fickian approach to mass transfer was applied to low-profile air stripping, and it was shown that the mass transfer rate constants for low-profile air stripping of

TCE and PCE compare favorably to mass transfer rate constants of VOCs in packed-tower aeration and mechanical aeration (LaBranche and Collins, 1996).

The following methods are available for determining the size of a low-profile air stripper:

- Analytical equations: Treybal (1980)
- Manufacturer-supplied software: Carbonair Environmental Systems (2003); North East Environmental Products (2003).
- McCabe–Thiele graphical method: See Sec. 14-4

A description of the recommended method for preliminary sizing of a low-profile air stripper from the U.S. ACE (2001) design manual follows:

1. Determine the minimum and maximum volume of water to be treated, the minimum temperature of the water, and the maximum concentration of VOCs in the untreated water.
2. Determine the desired concentration (percent removed) of the VOCs in the treated water.
3. Calculate the theoretical number of sieve trays needed to remove the VOCs to the desired concentration.
4. Estimate the tray efficiency and the number of actual trays needed.
5. Estimate the size (cross-sectional area) of the perforated plate section of each tray.
6. Estimate the pressure drop through the air stripper.
7. Estimate the size of the air blower motor (in kilowatts).

The McCabe–Thiele graphical approach for determining the number of equilibrium stages (theoretical trays) for low-profile air stripping was discussed in Sec. 14-4. It is also possible to determine the number of theoretical trays from the following relationship (Li and Hsiao, 1990; Treybal, 1980):

$$N_{\text{th}} = \frac{\ln\left[1 + (C_0/C_{\text{TO}})(S-1)\right]}{\ln(S)} - 1 \tag{14-66}$$

where N_{th} = number of theoretical trays
S = stripping factor, dimensionless
C_0 = influent liquid-phase concentration, mg/L
C_{TO} = treatment objective, mg/L

Once the number of theoretical trays is known, the number of actual trays can be calculated based on the tray efficiency:

$$N_{\text{act}} = \frac{N_{\text{th}}}{\text{Eff}_{\text{tray}}} \tag{14-67}$$

where N_{act} = number of actual trays
Eff_{tray} = tray efficiency, expressed as decimal fraction

Example 14-8 Low-profile air stripping

Determine the actual number of trays needed for a low-profile air stripper compared to the theoretical number of trays for the following conditions. The influent concentration of PCE is 15 mg/L and the treatment objective is 0.005 mg/L. The water flow rate is 0.003 m^3/s (48 gpm) and the water temperature is 10°C. The air flow rate is 0.7 m^3/s (1500 cfm). The dimensionless Henry's constant of PCE at 10°C is equal to 0.364.

Solution

1. Calculate the theoretical number of sieve trays required to remove the compound to the desired concentration.
   a. Determine air-to-water ratio and the stripping factor:

$$\frac{Q_a}{Q} = \frac{0.7\,\text{m}^3/\text{s}}{0.003\,\text{m}^3/\text{s}} = 233 \qquad S = \frac{Q_a}{Q} \times H_{YC,PCE} = 233 \times 0.364 = 85$$

   b. Use Eq.14-66 to determine the number of theoretical trays:

$$N_{th} = \frac{\ln\left[1 + (C_0/C_{TO})(S-1)\right]}{\ln(S)} - 1$$
$$= \frac{\ln[1 + (15/0.005)(85-1)]}{\ln(85)} - 1 = 1.8$$

   c. The appropriate number of theoretical trays is thus equal to 2.
2. Determine the number of actual trays using Eq. 14-67. Use a tray efficiency of 0.5, which is within the appropriate range of 0.4 to 0.6:

$$N_{act} = \frac{N_{th}}{\text{Eff}_{tray}} = \frac{2}{0.5} = 4$$

Comment

The actual number of trays needed for low-profile air stripping is greater than the theoretical number of trays by a factor of approximately 2 based on manufacturer's data.

Based on manufacturers' data, an appropriate Eff_{tray} value appears to be in the range of 0.4 to 0.6 (U.S. ACE, 2001).

14-7 Analysis of Spray Aerators

Spray towers and spray fountains are the two main types of spray aerators. A fixed grid of nozzles is used to either spray water in towers (spray towers) or spray water vertically into the air from the water surface (spray fountains), as shown on Fig. 14-5. The primary type of spray aerator used in water treatment is a fountain spray aerator, which is popular because existing reservoirs and large basins may be readily retrofit with them. When used in reservoirs and large basins, spray aerators are used to strip taste- and odor-causing compounds from raw water stored in reservoirs, oxygenate groundwater to remove iron and manganese, and strip VOCs. Spray fountain aerators are considered in this section; spray towers are discussed in Sec. 14-8.

Description

Air–water contact occurs by spraying fine water droplets from pressurized nozzles into the air, which creates a large air–water surface for mass transfer. Three types of pressurized spray nozzles are typically used in water treatment applications: (1) hollow cone, (2) full cone, and (3) fan spray (see Fig. 14-22). Full-cone nozzles create a uniform pattern of droplets over the entire angle of spray, while hollow-cone nozzles create a circular pattern of droplets, primarily around the circumference of the angle of spray. Although hollow-cone nozzles do not distribute droplets as well as full-cone nozzles and have a larger pressure drop requirement, hollow cones are generally preferred over full cones because they create smaller diameter drops and have a larger nozzle orifice. Hollow-cone spray nozzles are also prone to plugging and may require strainers upstream of the nozzle to discourage nozzle plugging.

Design Approach

Contaminant removal occurs during the time the water droplet is in contact with the air, so the basis for spray aeration design equations is the mass transfer from the droplet across the air–water interface. A mass balance on water droplets of equal size and equal air exposure is given in words in Eq. 14-68 and mathematically in Eq. 14-69 (Hand et al., 1999):

$$\frac{\text{Mass lost from water}}{\text{dropperunittime}} = \frac{\text{Mass transferred across air–water}}{\text{interface of water drop per unit time}} \tag{14-68}$$

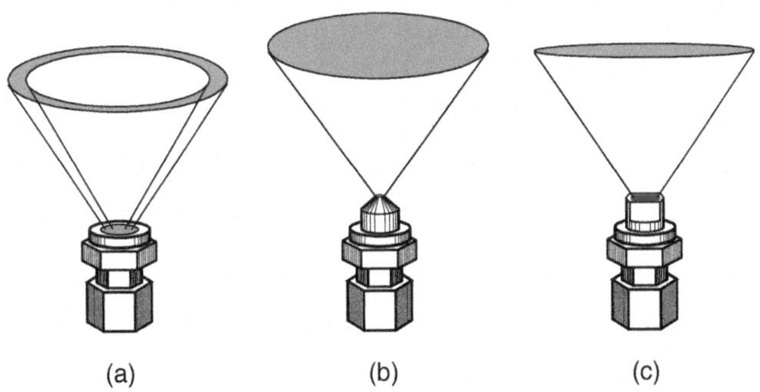

Figure 14-22
Common spray nozzles: (a) hollow cone, (b) full cone, and (c) fan spray.

$$V_d \frac{dC}{dt} = K_L a[C(t) - C_s(t)]V_d \tag{14-69}$$

where V_d = volume of drop, m^3
K_L = overall mass transfer coefficient, m/s
a = interfacial surface area available for mass transfer for water drop, m^2/m^3
$C(t)$ = concentration in water drop at time t, mg/L
$C_s(t)$ = concentration at air–water interface at time t, mg/L

The gas-phase contaminant concentration in the open air is assumed to be zero, and the concentration at the air–water interface, C_s, is assumed to be in equilibrium with the air, so it is also zero. The final contaminant concentration of the water drop after exposure to air may be determined by rearranging Eq. 14-69 and integrating over the time the drop is exposed to the air:

$$C_e = C_0 e^{-K_L a \theta} \tag{14-70}$$

where C_e = final contaminant concentration of water drop after being exposed to air, mg/L
C_0 = initial contaminant liquid-phase concentration of drop before being exposed to air, mg/L
θ = time of contact between water drop and air, s

The time of contact between the water drop and the air, θ, is dependent upon the exiting velocity and trajectory and can be estimated from the equation

$$\theta = \frac{2v_d \sin\alpha}{g} \tag{14-71}$$

where α = angle of spray measured from horizontal, deg
g = acceleration due to gravity, 9.81 m/s^2
v_d = exit velocity of water drop from nozzle, m/s

The exit velocity v_d can be determined using the orifice equation

$$v_d = C_v\sqrt{2gh} \tag{14-72}$$

where C_v = coefficient of velocity for orifice, unitless
h = total head of nozzle, m

The coefficient of velocity C_v is provided by the nozzle manufacturer and typically varies from 0.4 to 0.65. The area of flow is typically less than the nozzle area so the flow rate Q may be calculated by using a revised form of Eq. 14-72, as shown in the equation

$$Q_n = C_d A_n\sqrt{2gh} \tag{14-73}$$

where Q_n = flow rate through nozzle, m^3/s
A_n = area of nozzle opening, m^2
C_d = coefficient of discharge from nozzle, which is supplied by nozzle manufacturer, unitless

The overall mass transfer coefficient may be computed from either Eq. 14-74 or Eq. 14-75, depending on the value of the dimensionless quantity $[2(D_l\theta)^{0.5}/d_d]$ (Calvert et al., 1972; Higbie, 1935; Jury, 1967):

$$K_L = \begin{cases} 2\left(\dfrac{D_l}{\pi\theta}\right)^{1/2} & \text{for } \dfrac{2(D_l\theta)^{1/2}}{d_d} < 0.22 \quad (14\text{-}74) \\ \dfrac{10D_l}{d_d} & \text{for } \dfrac{2(D_l\theta)^{1/2}}{d_d} > 0.22 \quad (14\text{-}75) \end{cases}$$

where K_L = overall mass transfer coefficient, m/s
d_d = Sauter mean diameter (SMD) of water drop, equal to total volume of spray divided by total surface area, m
D_l = contaminant liquid diffusivity, m^2/s
θ = contact time of water drop with air, s

The Sauter mean diameter is a design parameter provided by the nozzle manufacturers. The area on the water droplets that is available for mass transfer can be calculated as

$$a = \frac{6}{d_d} \quad (14\text{-}76)$$

where a = interfacial surface area available for mass transfer, m^2/m^3

Example 14-9 Spray aeration

It is necessary to strip carbon dioxide, CO_2, from a groundwater. Determine the number of nozzles required and the expected carbon dioxide removal efficiency for treating the water with a spray aeration system. The groundwater has a temperature of 25°C and a dissolved CO_2 concentration of 100 mg/L. The water is pumped from the well at a flow rate of 0.050 m^3/s, and the pump has the capacity to deliver an additional 30m of head. The nozzle manufacturer has supplied the following data: SMD = 0.0010 m, $\alpha = 90°$, $C_v = 0.45$, $C_d = 0.25$, nozzle diameter = 0.0125 m. Refer to Table 7-1 in Chap. 7 for diffusion coefficients.

Solution

1. Determine the number of nozzles required.
   a. Calculate the area of one nozzle:

$$A_n = \left(\frac{1}{4}\pi\right)(0.0125\ \text{m})^2 = 1.2 \times 10^{-4}\ \text{m}^2$$

b. Calculate the flow rate through one nozzle, Q_n, using Eq. 14-73:

$$Q_n = C_d A_n \sqrt{2gh} = 0.25(1.2 \times 10^{-4}\ \text{m}^2)\sqrt{2(9.81\ \text{m/s}^2)(30\ \text{m})}$$
$$= 7.3 \times 10^{-4}\ \text{m}^3/\text{s}$$

c. The number of nozzles can be calculated by dividing the total flow by the flow through each nozzle:

$$\text{Required number of nozzles} = \frac{Q}{Q_n} = \frac{0.050\ \text{m}^3/\text{s}}{7.3 \times 10^{-4}\ \text{m}^3/\text{s}} = 68$$

2. Determine the CO_2 removal efficiency.
   a. Calculate the velocity of the water exiting the nozzle, v_d, using Eq. 14-72:

$$v_d = C_v \sqrt{2gh} = 0.45\sqrt{2(9.81\ \text{m/s}^2)(30\ \text{m})} = 11\ \text{m/s}$$

   b. Determine the contact time of the water drop with the air, t, using Eq. 14-71:

$$\theta = \frac{2v_d \sin\alpha}{g} = \frac{2(11\ \text{m/s})\sin(90^\circ)}{9.81\ \text{m/s}^2} = 2.2\ \text{s}$$

   c. Calculate the overall liquid-phase mass transfer coefficient K_L. From Table 7-1, the liquid-phase diffusion coefficient of CO_2 is $2.0 \times 10^{-9}\ \text{m}^2/\text{s}$.
      i. Calculate the dimensionless quantity $2(D_l\theta)^{1/2}/d_d$:

$$\frac{2(D_l\theta)^{1/2}}{d_d} = \frac{2\left[(2.0 \times 10^{-9}\ \text{m}^2/\text{s})(2.2\ \text{s})\right]^{1/2}}{0.0010\ \text{m}} = 0.13 < 0.22$$

      ii. Because the dimensionless quantity determined in the previous step is less than 0.22, calculate the overall liquid-phase mass transfer coefficient K_L using Eq. 14-74:

$$K_L = 2\left(\frac{D_l}{\pi\theta}\right)^{0.5} = 2\left(\frac{2.0 \times 10^{-9}\ \text{m}^2/\text{s}}{\pi \times 2.2\ \text{s}}\right)^{0.5}$$
$$= 3.4 \times 10^{-5}\ \text{m/s}$$

   d. Calculate the interfacial area for mass transfer, a, using Eq. 14-76:

$$a = \frac{6}{d_d} = \frac{6}{0.0010\ \text{m}} = 6.0 \times 10^3/\text{m}$$

   e. Calculate $K_L a$:

$$K_L a = (3.4 \times 10^{-5}\ \text{m/s})(6.0 \times 10^3/\text{m})$$
$$= 0.20\ \text{s}^{-1}$$

f. Calculate the effluent liquid-phase CO2 concentration after stripping using Eq. 14-70:

$$C_e = C_0 e^{-K_L a \times \theta} = (100\ \text{mg/L}) e^{-0.20\ \text{s}^{-1} \times 2.2\ \text{s}}$$
$$= 64\ \text{mg/L}$$

g. Calculate the carbon dioxide removal efficiency due to stripping:

$$\text{Carbon dioxide removal efficiency} = \frac{C_0 - C_e}{C_e} \times 100$$
$$= \frac{100 - 64}{100} = 36\%$$

14-8 Other Air-Stripping and Aeration Processes

Other types of air-stripping and aeration processes, such as spray towers, diffused aerators, and mechanical aerators, are introduced and discussed briefly in this section.

Spray Towers

There are a variety of configurations in which spraying can be used. Some configurations are analogous to packed-tower designs, and some are more complex designs, which are typically used for air pollution control such as cyclone scrubbers and Venturi scrubbers. Historically, spray systems have been used in water treatment for aeration, degasification of well water, and odor removal.

Only a few studies of spray towers have been conducted on mass transfer in clean-water systems. Based on these studies, it has been found that spray systems are limited with respect to the removals that can be achieved, and a substantial portion of removal in a spray system may occur at the nozzle. Typically, one to three transfer units are reported as a maximum limit that can be achieved in spraying systems. The apparent limitation in percent removal is the product of backmixing of air and spray disturbance due to wall or adjacent spray contact (Davies and Ip, 1981; Ip and Raper, 1977; Pigford and Pyle, 1951).

The NTU as a function of the height of the spray tower is shown on Fig. 14-23. The residual NTU at zero height, in this case between 0.1 and 0.2 transfer units, is the result of the removal occurring at the nozzle. Some process designs may take advantage of the removal occurring at the nozzles by recycling flow or by incorporating several banks of nozzles. The aforementioned nonideal effects have hindered development of a general empirical design model. With the data presently available, a spray tower cannot be designed for a precise removal. Rather, the design

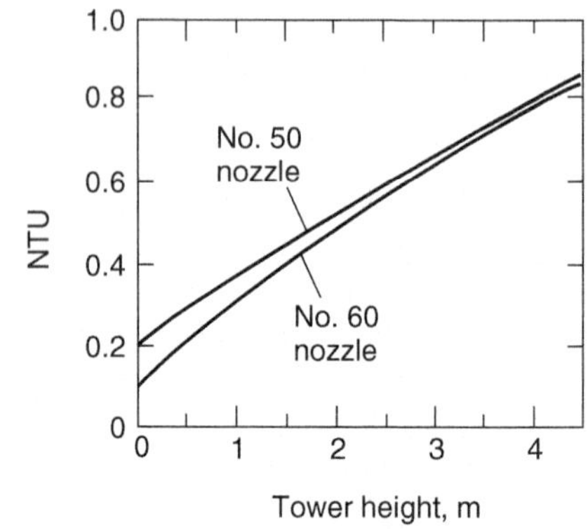

Figure 14-23
Number of transfer units in pilot spray tower as function of tower height (Adapted from Davies and Ip, 1981).

approach serves merely as a basis to estimate the approximate removal efficiency of a spray system.

Diffused Aeration

The diffused, or bubble, aeration process consists of contacting water with gas bubbles for the purposes of transferring gas to the water (e.g., O_3, CO_2, O_2) or removing VOCs from the water by stripping. The process can be carried out in a clearwell or special rectangular concrete tanks (contactors) typically 2.7 to 4.6 m (9 to 15 ft) in depth.

A typical diffused-air aeration system is shown on Fig. 14-24. The most commonly used diffuser system consists of a matrix of perforated tubes (or membranes) or porous plates arranged near the bottom of the tank to provide maximum gas-to-water contact. Various types of diffusers and diffuser system layouts are presented in the U.S. EPA's (1989) technology transfer design manual on fine pore aeration systems. Jet aerators, which consist of jets that discharge fine gas bubbles and provide enhanced mixing

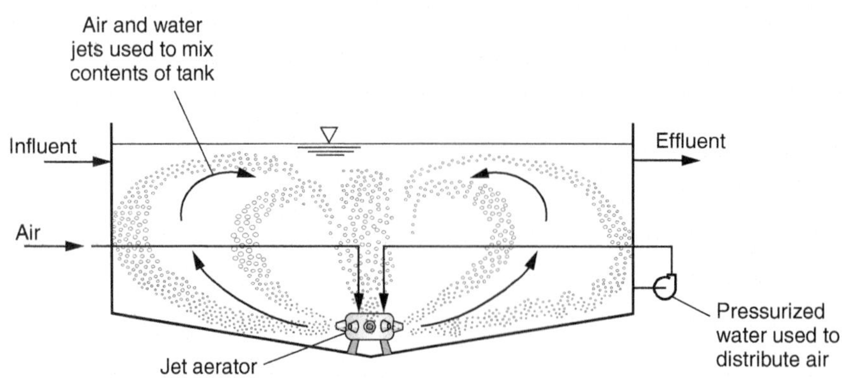

Figure 14-24
Typical example of a diffused-air aeration system.

for increased absorption efficiency, are used to provide good air-to-water contact (Mandt and Bathija, 1977). Model development for bubble aeration has been described in the literature (Matter-Müller et al., 1981; Munz and Roberts, 1982; Roberts et al., 1984).

Mechanical Aerators

Mechanical aerators typically used in water treatment are surface or submerged turbines or brushes, as shown on Fig. 14-11. Surface aerators can be used in water treatment as an alternative to diffused aeration systems for stripping of volatile contaminants. Surface aeration has been primarily used for oxygen absorption and the stripping of gases and volatile contaminants when the required removals are less than about 90 percent.

The brush-type aerator consists of several blades attached to a rotary drum that is half submerged in water in the center of the tank, as shown on Fig. 14-11. As the drum rotates, it disperses the water into the surrounding air, thus providing interfacial contact between the air and water for mass transfer to take place.

The turbine-type aerator consists of a submerged propeller system located in the center of the tank and surrounded by a draft tube. As the submerged propeller rotates, it draws water from outside the draft tube through the inner section and into the air, creating contact between the air and water.

Model development for surface aeration as applied to water treatment has been described in the literature (Matter-Müller et al., 1981; Munz and Roberts, 1989; Roberts and Dandliker, 1983; Roberts et al., 1984, 1985).

Problems and Discussion Topics

Note: Several of these problems pertain to the design of countercurrent packed towers, which is a computationally-intensive process. The spreadsheet identified as Resource E10 at the website listed in App. E can be used to perform the calculations.

14-1 What is the dimensionless Henry's law constant for a compound that has a value of 400 atm? What is the Henry's law constant in atmospheres and atm/(mol/L) for a compound that has a dimensionless Henry's law constant of 0.2? Assume the temperature is 15°C.

14-2 What is the Henry's law constant in dimensionless form and atmospheres for a compound that has a Henry's law constant of 2.0 atm/ (mol/L)? What is the dimensionless Henry's law constant for a compound that has a value of 200 atm? Assume the temperature is 10°C.

14-3 Calculate the dimensionless Henry's law constant at 10 and 25°C for benzene using $\Delta H^0_{\text{vol}} = 35.44$ kJ/mol and $K_c = 357{,}678$.

14-4 Calculate the dimensionless Henry's law constant at 5 and 15°C for chloroform using $\Delta H^0_{vol} = 38.53$ kJ/mol and $K_c = 940{,}789$.

14-5 Calculate the apparent Henry's constant of hydrogen sulfide at pH 6.3 at 25°C.

14-6 Calculate the minimum air-to-water ratio for chloroform and benzene with 95 percent removal at 10°C for a countercurrent packed tower.

14-7 Using the McCabe–Thiele graphical method, determine the number of equilibrium stages required to strip chloroform from an influent concentration of 200 μg/L to its treatment objective of 5 μg/L in a countercurrent, packed tower at 15°C. Assume clean air enters the tower and $S = 3.5$.

14-8 Determine the cross-sectional area and diameter for a packed-tower design based on chloroform at 20°C for a water flow rate Q of 0.15 m^3/s (2400 gal/min). The basis for design is stripping factor $S = 3.5$, gas pressure drop $\Delta P/L = 50\ N/m^2 \cdot m$, and 5.1-cm (2-in.) plastic saddles.

14-9 Determine the cross-sectional area and diameter for a packed–tower design based on benzene at 10°C for a water flow rate Q of 0.05 m^3/s (800 gal/min). The basis for design is stripping factor $S = 3.5$, gas pressure drop $\Delta P/L = 50\ N/m^2 \cdot m$, and 5.8-cm (2.3-in.) LANPAC packing material.

14-10 Determine the mass transfer coefficients for benzene and chloroform at 20°C in packed-tower aeration for the air mass loading rate and water mass loading rate determined in Problem 14-8 using the Onda correlations and a safety factor of 0.75 (actual K_La/Onda K_La) for 5.1-cm (2-in.) plastic saddles. The water flow rate Q is 0.15 m^3/s (2400 gal/min). Determine the liquid diffusivity D_l and gas diffusivity D_g for benzene and chloroform using the Hayduk–Laudie correlation and the Wilke–Lee modification of the Hirschfelder-Bird–Spotz method, respectively. The viscosity of air, μ_g, at 20°C is 1.837×10^{-5} kg/m · s.

14-11 Determine the mass transfer coefficients for benzene and chloroform at 10°C in packed-tower aeration for the air mass loading rate and water 14-9 using the Onda correlations and a safety factor of 0.75 (actual K_La/Onda K_La) for 5.8-cm (2.3-in.) LANPAC packing material. The water flow rate Q is 0.05 m^3/s (800 gal/min). Determine the liquid diffusivity D_l and gas diffusivity D_g for benzene and chloroform using the Hayduk–Laudie correlation and the Wilke–Lee modification of the Hirschfelder–Bird–Spotz method, respectively. The viscosity of air, μ_g, of 10°C is 1.787×10^{-5} kg/m · s.

14-12 Determine the packed-tower height required to remove chloroform at 20°C for a water flow rate Q of 0.15 m^3/s (2400 gal/min). Use the tower area determined in Problem 14-8 and the mass transfer coefficients determined in Problem 14-10 in the solution of the problem. The influent concentration for benzene is 50 μg/L and its treatment objective is 5 μg/L. The influent concentration for chloroform is 100 μg/L and its treatment objective is 5 μg/L.

14-13 Determine the packed-tower height required to remove benzene at 20°C for a water flow rate Q of 0.15 m^3/s (2400 gal/min). Use the tower area determined in Problem 14-8 and the mass transfer coefficients determined in Problem 14-10 in the solution of the problem. The influent concentration for benzene is 50 μg/L and its treatment objective is 5 μg/L. The influent concentration for chloroform is 100 μg/L and its treatment objective is 5 μg/L.

14-14 Determine the packed-tower height required to remove chloroform at 10°C for a water flow rate Q of 0.05 m^3/s (800 gal/min). Use the tower area determined in Problem 14-9 and the mass transfer coefficients determined in Problem 14-11 in the solution of the problem. The influent concentration for benzene is 75 μg/L and its treatment objective is 5 μg/L. The influent concentration for chloroform is 100 μg/L and its treatment objective is 5 μg/L.

14-15 Determine the packed-tower height required to remove benzene at 10°C for a water flow rate Q of 0.05 m^3/s (800 gal/min). Use the tower area determined in Problem 14-9 and the mass transfer coefficients determined in Problem 14-11 in the solution of the problem. The influent concentration for benzene is 75 μg/L and its treatment objective is 5 μg/L. The influent concentration for chloroform is 75 μg/L and its treatment objective is 5 μg/L.

14-16 Using the packed-tower height and conditions in Problem 14-12, perform a rating analysis to determine if benzene will be removed to meet its treatment objective.

14-17 Using the packed-tower height and conditions in Problem 14-13, perform a rating analysis to determine if chloroform will be removed to meet its treatment objective.

14-18 Using the packed-tower height and conditions in Problem 14-14, perform a rating analysis to determine if benzene will be removed to meet its treatment objective.

14-19 Using the packed-tower height and conditions in Problem 14-15, perform a rating analysis to determine if chloroform will be removed to meet its treatment objective.

14-20 Determine the total power requirement (blower and pump brake power) for a packed-tower aeration design removing chloroform at 20°C and 1 atm (101,325 N/m^2) for a water flow rate of 0.15 m^3/s, a stripping factor of 3.5, and a gas pressure drop of 50 $N/m^2/m$. Assume the blower efficiency is 35 percent ($Eff_b = 0.35$) and the pump efficiency is 80 percent ($Eff_p = 0.80$). Use the air-to-water ratio and tower area determined in Problem 14-8. Use the tower length determined in Problem 14-12.

14-21 Determine the total power requirement (blower and pump brake power) for a packed-tower aeration design removing benzene at 10°C and 1 atm (101,325 N/m^2) for a water flow rate of 0.05 m^3/s, a stripping factor of 3.5, and a gas pressure drop of 50 $N/m^2/m$. Assume the blower efficiency is 35 percent ($Eff_b = 0.35$) and the pump efficiency is 80 percent $Eff_p = 0.80$. Use the air-to-water ratio and tower area determined in Problem 14-9. Use the tower length determined in Problem 14-13.

14-22 Design a packed-tower aeration system to treat 0.075 m^3/s of water at 15°C and remove benzene (influent concentration that equals 40 µg/L.), chloroform (influent concentration equals 60 µg/L), and carbon tetrachloride (influent concentration equals 30 µg/L) to a treatment objective of total VOC concentration that equals 5 µg/L. Determine an appropriate tower diameter and tower length. Use 0.0508-m (2-in.) plastic tripacks as the packing material. The viscosity of air at 15°C is 1.75×10^{-5} kg/m · s.

14-23 Determine the actual number of trays needed for a low-profile air stripper for the following conditions. The influent concentration of trichloroethylene (TCE) is 30 mg/L and the treatment objective is 0.005 mg/L. The water flow rate is 0.00630 m^3/s (100 gpm) and the water temperature is 10°C. The air flow rate is 1.42 m^3/s (3000 cfm). The Henry's constant of TCE at 10°C is 0.230.

14-24 A well water contains 1.0 mg/L of methane with temperature of 10°C. The well pumps 0.0600 m^3/s (950 gpm) and the pump has the capacity to deliver an additional 28 m (40 psi) of head. Determine the number of nozzles required and the expected methane removal efficiency. The following information was obtained from the nozzle manufacturer: SMD = 0.10 cm, $\alpha = 90°$, $C_v = 0.45$, $C_d = 0.25$, nozzle diameter = 1.25 cm. Let D_l for methane be 1.1×10^{-5} cm^2/s at 10°C.

References

Ashworth, R. A., Howe, G. B., Mullins, M. E., and Rogers, T. N. (1988) "Air–water Partitioning Coefficients of Organics in Dilute Aqueous Solutions," *J. Hazardous Mat*, **18**, 1, 25–36.

AWWA (1978a) "Water Treatment Plant Sludges—An Update of the State of the Art, Part 1, American Water Works Association Sludge Disposal Committee Report", *J. AWWA*, **70**, 9, 498–503.

AWWA (1978b) "Water Treatment Plant Sludges—An Update of the State of the Art, Part 2, American Water Works Association Sludge Disposal Committee Report", *J. AWWA*, **70**, 10, 548–554.

Ayuttaya, P. C. N., Rogers, T. N., Mullins, M. E., and Kline, A. A., (2001) "Henry's Law Constants Derived from Equilibrium Static Cell Measurements for Dilute Organic-Water Mixtures," *Fiuid Phase Equiiibria*, **185**, 359–377.

Ball, B. R., and Edwards, M. D. (1992) "Air Stripping VOCs from Groundwater: Process Design Considerations," *Environ. Progr.*, **11**,1,39–48.

Ball, W. P., Jones, M. D., and Kavanaugh, M. C. (1984) "Mass Transfer of Volatile Organic Compounds in Packed Tower Aeration," *J. WPCF*, **56**, 2, 127–135.

Biederman, W. J., and Fulton, E. E. (1971) "Destratification Using Air," *J. AWWA*, **63**, 7, 462–466.

Bilello, L. J., and Singley, J. E. (1986) "Removing Trihalomethanes by Packed-Column and Diffused Aeration," *J. AWWA*, **78**, 2, 62–71.

Byers, W. D., and Morton, C. M. (1985) "Removing VOC from Groundwater; Pilot, Scale-up, and Operating Experience," *Environ. Progr.*, **4**, 2, 112–118.

Calvert, S., Lundgren, D., and Mehta, D. S. (1972) "Venturi Scrubber Performance," *J. Air Poiiut. Controi Assoc.*, **22**, 7, 529–532.

Carbonair Environmental Systems (2003) 2731 Nevada Avenue North, New Hope, MN.

Cummins, M. D., and Westrick, J. J. (1983) "Trichlorethylene Removal by Packed Column Air Stripping: Field Verified Design Procedure," in Proceedings, American Society of Civil Engineers Environmental Engineering Conference, Boulder, CO, pp. 442–449.

Davies, T. H., and Ip, S. Y. (1981) "Droplet Size and Height Effects in Ammonia Removal in a Spray Tower," *Water Res.*, **15**, 5, 525–533.

Dewulf, J., Drijvers, D., and Langenhove, H. V., (1995) "Measurement of Henry's Law Constant as Function of Temperature and Salinity for the Low Temperature Range," *Atmosph. Environ.*, **29**, 4, 323–331.

Djebbar,Y., and Narbaitz, R.M. (1995) "MassTransfer Correlations for Air Stripping Towers," *Environ. Progr.*, **14**, 3, 137–145.

Djebbar, Y., and Narbaitz, R. M. (1998) "Improved Correlations in Packed Towers," *Water Sci. Technoi.*, **38**, 6, 295–302.

Djebbar,Y., and Narbaitz, R.M. (2002) "Neural Network Prediction of Air Stripping KLa," *J. Environ. Eng.*, ASCE, **128**, 5, 451–460.

Dvorak, B. I., Lawler, D. F., Fair, J. R., and Handler, N. E. (1996) "Evaluation of the Onda Correlations for Mass Transfer with Large Random Packings," *Environ. Sci. Technoi.*, **30**, 2, 945–953.

Dzombak, D. A., Roy, S. B., and Fang, H.-J. (1993) "Air Stripper Design and Costing Computer Program," *J. AWWA*, **85**, 10, 63–72.

Eckert, J. S. (1961) "Design Techniques for Sizing Packed Towers," *Chem. Eng. Progr.*, **57**, 9, 54–58.

Garton, J. E. (1978) "Improve Water Quality through Lake Destratification," *Water Wastes Eng.*, **15**, 5, 42–44.

Gossett, J. M. (1987) "Measurement of Henry's Law Constants for C1 and C2 Chlorinated Hydrocarbons," *Environ. Sci. Technoi.*, **21**, 2, 202–208.

Gross, R. L., and TerMaath, S. G. (1985) "Packed Tower Aeration Strips Trichloroethylene from Groundwater," *Environ. Progr.*, **4**, 2, 119–124.

Gustafsson, J. P. (2002) *Visual MINTEQ, Version 2.12a*, KTH Royal Institute of Technology, Stockholm, Sweden.

Hand, D. W., Crittenden, J. C., Gehin, J. L., and Lykins, B. W., Jr. (1986) "Design and Evaluation of an Air Stripping Tower for Removing VOCs from Groundwater," *J. AWWA*, **78**, 9, 87–97.

Hand, D. W., Hokanson, D. R., and Crittenden, J. C. (1999) Air Stripping and Aeration, Chap. 5 in R. D. Letterman (ed.), *Water Quality and Treatment: A Handbook of Community Water Supplies*, 5th ed., American Water Works Association, McGraw-Hill, New York.

Heron, G., Christensen, T. H., and Enfield, C. G. (1998) "Henry's Law Constant for Trichloroethylene between 10 and 95 C," *Environ. Sci. Technol.*, **32**, 10, 1433–1437.

Higbie, R. (1935) "The Rate of Absorption of a Pure Gas into a Still Liquid During Short Periods of Exposure," *Trans. Am. Inst. Chem. Eng.*, **31**, 365–389.

Hine, J., and Mookerjee, P. K. (1975) "The Intrinsic Hydrophobic Character of Organic Compounds, Correlations in Terms of Structural Contributions," *J. Org. Chem.*, **40**, 3, 292–298.

Hokanson, D. R. (1996) Development of Software Design Tools for Physical Property Estimation, Aeration, and Adsorption, M.S. Thesis, Michigan Technological University, Houghton, MI (http://cpas.mtu.edu/etdot).

Howe, K. J., and Lawler, D. L. (1989) "Acid-Base Reactions in Gas Transfer: A Mathematical Approach," *J. AWWA*, **81**, 1, 61–66.

Ip, S. Y., and Raper, W. (1977) "Ammonia Stripping with Spray Towers," *Progr. Water Technol.*, **10**, 587–605.

Jury, S. H. (1967) "An Improved Version of the Rate Equation for Molecular Diffusion in a Dispersed Phase," *AIChE J.*, **13**, 6, 1124–1126.

Kavanaugh, M. C., and Trussell, R. R. (1980) "Design of Aeration Towers to Strip Volatile Contaminants from Drinking Water," *J. AWWA*, **72**, 12, 684–692.

Kavanaugh, M.C., and Trussell, R.R. (1981) Air Stripping as a Treatment Process, in Proceedings of AWWA Symposium on Organic Contaminants in Groundwater, St. Louis, MO. American Water Works Association, Denver, CO. Paper S2-6, pp. 83–106.

Kister, H. Z., Scherffius, J., Afshar, K., and Abkar, E. (2007) "Realistically Predict Capacity and Pressure Drop for Packed Columns," *Chem. Eng. Prog.*, **103**, 7, 28–38.

LaBranche, D. F., and Collins, M. R. (1996) "Stripping Volatile Organic Compounds and Petroleum Hydrocarbons from Water," *Water Environ. Res*, **68**, 3, 348–358.

Lamarche, P., and Droste, R. L. (1989) "Air Stripping Mass Transfer Correlations for Volatile Organics," *J. AWWA*, **81**, 1, 78–89.

Lau, K. A. (2009) Theoretical and Experimental Studies of the Temperature Dependence of the Henry's Law Constant of Organic Solutes in Water, Ph.D. Dissertation, Michigan Technological University, Houghton, MI.

Laverty, G. L., and Nielsen, H. L. (1970) "Quality Improvements by Reservoir Aeration," *J. AWWA*, **62**, 11, 711–714.

Lenzo, F. C., Frielinghaus, T. J., and Zienkiewicz, A. W. (1990) The Application of the Onda Correlation to Packed Column Air Stripper Design: Theory Versus Reality, pp. 1301–1321, in Proceedings American Water Works Association Annual Conference, Cincinnati, OH.

Li, K. Y., and Hsiao, K. J. (1990) "VOC Strippers: How Many Trays?" *Hydrocarbon Process.*, **69**, 2, 79–81.

Lin, S. T., and Sandler, S.I, (2000) "Multipole Corrections to Account for Structure and Proximity Effects in Group Contribution Methods: Octonal-water Partition Coefficients," *J. Phys. Chem.*, **104**, 30, 7099–7105.

Lin, S. T., and Sandler, S. I, (2002) "Henry's Law Constant of Organic Compounds in Water from Group Contribution Model with Multipole Corrections," *Chem. Eng. Sci.*, **57**, 2727–2733.

Lincoff, A. H., and Gossett, J. M. (1984) "The Determination of Henry's Law Constant for Volatile Organics by Equilibrium Partitioning in Closed Systems," pp. 17–25 in *Gas Transfer at Water Surfaces*, W. Brutsaert and G. H. Jurka (eds.) Reidel, Germany.

Mackay, D., and Shiu, W. Y. (1981) "A Critical Review of Henry's Law Constants for Chemicals of Environmental Interest," *J. Phys. Chem. Ref. Data*, **10**, 1175–1199.

Mackay, D., Shiu, W. Y., and Sutherland, R. P. (1979) "Determination of Air–water Henry's Law Constant for Hydrophobic Pollutants," *Environ. Sci. Technoi.*, **13**, 3, 333–337.

Mandt, M. G., and Bathija, P. R. (1977) "Jet Fluid Gas/Liquid Contacting and Mixing," *AIChE Symp. Ser.*, **73**, 167, 15–22.

Matter-Müller, C., Gujer, W., and Giger, W. (1981) "Transfer of Volatile Substances from Water to the Atmosphere," *WaterRes.*, **15**, 11, 1271–1279.

McCabe, W. L., and Thiele, E. W. (1925) "Graphical Design of Fractionating Columns," *Ind. Eng. Chem.*, **17**, 6, 605–611.

McKinnon, R. J., and Dyksen, J. E. (1984) "Removal of Organics from Groundwater through Aeration Plus GAC," *J. AWWA*, **76**, 5, 42–47.

Mead, E., and Leibbert, J. (1998) A Comparison of Packed-Column and Low- Profile Sieve Tray Air Strippers, in Proceedings of the 1998 Conference on Hazardous Waste Research, The Great Plains/Rocky Mountain Hazardous Substance Research Center, Snowbird, UT, pp. 328–334.

Meylan, W. M. (1999) HENRYWIN v 3.05, Syracuse Research Corporation, Syracuse, NY.

Meylan, W. M., and Howard, P. H. (1991) "Bond Contribution Method for Estimating Henry's Law Constants," *Environ. Toxicoi. Chem.*, **10**, 10, 1283–1293.

Miller, D. J., and Hawthorne, S. B. (2000) "Solubility of Liquid Organic Flavor and Fragrance Compounds in Subcritical (Hot/Liquid) Water from 298 K to 473 K," *J. Chem. Eng. Data*, **45**, 2, 315–318.

Munz, C., and Roberts, P. V. (1982) Mass Transfer and Phase Equilibria in a Bubble Column, paper presented at the American Water Works Association Annual Conference, Miami, FL.

Munz, C., and Roberts, P. V. (1989) "Gas and Liquid-Phase Mass Transfer Resistance of Organic Compounds During Mechanical Surface Aeration," *Water Res.*, **23**, 5, 589–601.

Narbaitz, R. M., Mayorga, W. J., Torres, F. D., Greenfield, J. H., Amy, G. L., Minear, R. A., (2002) "Evaluating Aeration Stripping Media on the Pilot Scale," *J. AWWA*, **94**, 9, 97–111.

Nicholson, B. C., Maguire, B. P., and Bursell, D. B. (1984) "Henry's Law for the Trihalomethanes: Effects of Water Composition and Temperature," *Environ. Sci. Technol.*, **18**, 7, 518–521.

Nirmalakhandan, N. N., and Speece, R. E. (1988) "QSAR Model for Predicting Henry's Constant," *Environ. Sci. Technol.*, **22**, 11, 1349–1357.

NIST (2011) Accessed at http://webbook.nist.gov/chemistry on Jan. 3, 2011.

North East Environmental Products (2003) 17 Technology Drive, West Lebanon, NH.

Notthakun, S., Bros, D. E., and Riddle, C. S. (1994) *Sieve Tray Air Strippers*, Carbonair Environmental Systems, Minneapolis, MN.

Onda, K., Takeuchi, H., and Okumoto, Y. (1968) "Mass Transfer Coefficients between Gas and Liquid Phases in Packed Columns," *J. Chem. Eng. Jpn.*, **1**, 1, 56–62.

Pigford, R. L., and Pyle, C. (1951) "Performance Characteristics of Spray-Type Absorption Equipment," *Ind. Eng. Chem. Process Des. Devel.*, **43**, 1649–1662.

Roberts, P. V., and Dandliker, P. G. (1983) "Mass Transfer of Volatile Organic Contaminants from Aqueous Solution to the Atmosphere During Surface Aeration," *Environ. Sci. Technol.*, **17**, 8, 484–489.

Roberts, P. V., Hopkins, G. D., Munz, C., and Riojas, A. H. (1985) "Evaluating Two-Resistance Models for Air Stripping of Volatile Organic Contaminants in a Countercurrent, Packed Column," *Environ. Sci. Technol.*, **19**, 2, 164–173.

Roberts, P. V., and Levy, J. A. (1985) "Energy Requirements for Air Stripping Trihalomethanes," *J. AWWA*, **77**, 4, 138–146.

Roberts, P. V., Munz, C., and Dandliker, P. (1984) "Modeling Volatile Organic Solute Removal by Surface and Bubble Aeration," *J. WPCF*, **56**, 2, 157–163.

Robbins, G. A., Wang, S., and Stuart, J. D. (1993) "Using Static Headspace Method to Determine Henry's Law Constants," *Anal. Chem.*, **65**, 21, 3113–3118.

Rogers, T. N. (1994) Predicting Environmental Physical Properties from Chemical Structure Using a Modified Unifac Model, Ph.D. Dissertation, Michigan Technological University, Houghton, MI.

Sander, R. (1999) "Compilation of Henry's Law Constants for Inorganic and Organic Species of Potential Importance in Environmental Chemistry (Version 3)." Accessed at http://www.henrys-law.org on Jan. 3, 2011.

Schwarzenbach, R. P., Gschwend, P. M., and Imboden, D. M. (1993) *Environmental Organic Chemistry*, Wiley, New York.

Sherwood, T. K., and Hollaway, F. A. (1940) "Performance of Packed Towers—Liquid Film Data for Several Packings," *Trans. Am. Inst. Chem. Eng.*, **36**, 39–70.

Shulman, H. L., Ullrich, C. F., and Wells, N. (1955) "Performance of Packed Columns I. Total, Static, and Operating Holdups," *AIChE J.*, **1**, 2, 247.

Singley, J. E., Ervin, A. L., and Mangone, M. A. (1980) Trace Organics Removal by Air Stripping, report to AWWA Research Foundation, Denver, CO.

Singley, J. E., Ervin, A. L., and Mangone, M. A. (1981) Trace Organics Removal by Air Stripping, supplemental report to AWWA Research Foundation, Denver, CO.

SRC (2011). Accessed at http://www.syrres.com/what-we-do/databaseforms.aspx?id=386 on Jan. 3, 2011.

Steichen, J. M., Garton, J. E., and Rice, C. E. (1979) "Effect of Lake Destratification on Water Quality," *J. AWWA*, **71**, 4, 219–225.

Suzuki, T., Ohtaguchi, K., and Koide, K. (1992) "Application of Principle Components Analysis to Calculate Henry's Constant from Molecular Structure," *Computer Chem.*, **16**, 1, 41–52.

Symons, J. M., Carswell, J. K., and Robeck, G. G. (1970) "Mixing of Water Supply Reservoirs for Quality Control," *J. AWWA*, **62**, 5, 322–334.

Tchobanoglous, G., Burton, F. L., and Stensel, H. D. (2003) *Wastewater Engineering: Treatment and Reuse*, 4th ed., McGraw-Hill, New York.

Thom, J. E., and Byers, W. D. (1993) "Limitations and Practical Use of a Mass Transfer Model for Predicting Air Stripper Performance," *Environ. Progr.*, **12**, 1, 61–66.

Treybal, R. E. (1980) *Mass-Transfer Operations*, 3rd ed., McGraw-Hill, New York.

Umphres, M. D., Tate, C. H., Kavanaugh, M. C., and Trussell, R. R. (1983) "Tri- halomethane Removal by Packed Tower Aeration," *J. AWWA*, **75**, 8, 414–418.

U.S. ACE (2001) *Engineering and Design: Air Stripping*, No. 1110-1-3, U.S. Army Corps of Engineers, Department of the Army, Washington, DC.

U.S. EPA (1986) *Superfund Pubiic Heaith Evaiuation Manuai*, EPA/540/1-86/060 U.S. Environmental Protection Agency, Office of Emergency and Remedial Response, Washington, D.C.

U.S. EPA (1989) *Design Manuai: Fine Pore Aeration Systems*, EPA 625/1-89/023, U.S. Environmental Protection Agency, Washington, DC.

U.S. EPA (1999) *MINTEQA2*, Version 4.0, U.S. Environmental Protection Agency, Washington, DC.

Wallman, H., and Cummins, M. D. (1986) "Design Scale-up Suitability for Air Stripping Columns," *Pubiic Works*, **117**, 10, 74–78.

Yaws, C. L., Miller, J. W., Shah, P. N., Schorr, G. R., and Patel. P. M. (1976) "Correlation Constants for Chemical Compounds," *Chem. Eng.*, **83**, 25, 153–162.

15 Adsorption

Stantec's Water Treatment: Principles and Design, Updated Third Edition. John C. Crittenden, R. Rhodes Trussell, David W. Hand, Kerry J. Howe, and George Tchobanoglous.

Terminology for Adsorption

Term	Definition
Particle properties	
Adsorbent	Solid media on which adsorption occurs.
Adsorbent particle density in an adsorber, ρ_s	Weight of the dry (and fresh) adsorbent particles divided by the solid volume. The density of activated carbon is approximately equal to the density of graphite ($\approx$2 g/mL).
Apparent particle density, ρ_a	Weight of the dry (and fresh) adsorbent particles divided by the total volume of the adsorbent particle. The total volume includes the solid and pore volume.
Particle porosity, ε_p	Ratio of the pore volume to the total volume of an adsorbent particle. This parameter characterizes the fraction of the adsorbent volume that is not occupied by the carbon material. $\varepsilon_p = 1 - (\rho_a/\rho_s)$.
Sphericity, ϕ	External surface area of a particle divided by the surface area of a sphere that would have the same volume. Describes the increase in surface area due to a particle having an irregular shape.
Specific surface area	External surface area per weight of a dry particle. Because most adsorbent particles have an irregular shape, the external surface area per unit mass is defined as $3/R\phi\rho_a$ where R is equal to particle radius.
Adsorber properties	
Bed porosity, ε	Void volume in the contactor divided by the total volume that is occupied by the adsorbent particles. This parameter characterizes the fraction of the bed volume in which the fluid moves. $\varepsilon = 1 - (\rho_f/\rho_a)$.

Term	Definition
Contactor or adsorber density, ρ_f	Weight of the dry (and fresh) adsorbent particles divided by the total volume of the packed bed, including the bed pore volume.
Performance properties	
Adsorbate	Molecule that accumulates or adsorbs onto the adsorbent material.
Breakthrough profile	Relationship between the adsorbate concentration leaving the adsorber as a function of the adsorber run time.
Carbon use rate	Mass of adsorbent used per volume of water treated to a given treatment objective.
Equilibrium isotherm	Equilibrium partitioning relationship between the bulk aqueous-phase adsorbate concentration and the solid-phase adsorbate concentration at a constant temperature.
Specific throughput	Volume of water treated per mass of adsorbent used at a given treatment objective.
Treatment objective	Aqueous-phase adsorbate concentration that determines the bed life of a GAC adsorber or maximum value leaving a PAC contactor.
Empty-bed contact time (EBCT)	Volume of the bed occupied by the GAC (including voids) divided by the flow rate to the column.

Adsorption is a mass transfer operation in which substances present in a liquid phase are adsorbed or accumulated on a solid phase and thus removed from the liquid. Adsorption processes are used in drinking water treatment for the removal of taste- and odor-causing compounds, synthetic organic chemicals (SOCs), color-forming organics, and disinfection by-product (DBP) precursors. Inorganic constituents, including some that represent a health hazard, such as perchlorate, arsenic, and some heavy metals, are also removed by adsorption. Reactions with granular activated carbon (GAC), a common adsorbent, can also be used to dechlorinate drinking water.

The primary adsorbent materials used in the adsorption process for drinking water treatment are powdered activated carbon (PAC) and GAC. Powdered activated carbon is added directly to water and can be applied at various locations within a water treatment plant and is usually removed by sedimentation or filtration. Granular activated carbon is usually employed after filtration just prior to postdisinfection and is operated in a fixedbed mode. Granular activated carbon is also used in the upper layer of dual- or multimedium filters or as a substitute for conventional granular filter media.

The discussion that is presented in the following sections is intended to provide an introduction to adsorption processes and methods used for the design of PAC and GAC systems. The topics discussed include (1) development of the adsorption phenomena; (2) manufacture, regeneration, and reactivation; (3) fundamentals of adsorption; (4) development of isotherms and equations used to describe adsorption equilibrium; (5) applications using PAC; and (6) applications using GAC.

15-1 Introduction to Adsorption Phenomena

To provide a perspective for the material to be presented in this chapter, the historical development of adsorption processes and present applications of adsorption materials in water treatment is discussed in this section.

Adsorption Phenomena

The constituent that undergoes adsorption onto a surface is referred to as the *adsorbate*, and the solid onto which the constituent is adsorbed is referred to as the *adsorbent*. During the adsorption process, dissolved species are transported into the porous solid adsorbent granule by diffusion and are then adsorbed onto the extensive inner surface of the adsorbent. Dissolved species are concentrated on the solid surface by chemical reaction (chemisorption) or physical attraction (physical adsorption) to the surface. Physical adsorption and chemisorption mechanisms are listed in Table 15-1. Physical adsorption is a rapid process caused by nonspecific

Table 15-1
Comparison of adsorption mechanisms between physical adsorption and chemisorption

Parameter	Physical Adsorption	Chemisorption
Use for water treatment	Most common type of adsorption mechanism	Rare in water treatment
Process speed	Limited by mass transfer	Variable
Type of bonding	Nonspecific binding mechanisms such as van der Waals forces, vapor condensation	Specific exchange of electrons, chemical bond at surface
Type of reaction	Reversible, exothermic	Typically nonreversible, exothermic
Heat of adsorption	4–40 kJ/mol	>200 kJ/mol

binding mechanisms such as van der Waals forces and is similar to vapor condensation or liquid precipitation. Physical adsorption is reversible, that is, the adsorbate desorbs in response to a decrease in solution concentration. Physical adsorption is the most common mechanism by which adsorbates are removed in water treatment.

The physical adsorption process is exothermic with a heat of adsorption that is typically 4 to 40 kJ/mol (about two times greater than the heat of vaporization or dissolution for gases and liquids, respectively). Chemisorption is more specific because a chemical reaction occurs that entails the transfer of electrons between adsorbent and adsorbate, and a chemical bond with the surface can occur. The heat of adsorption for chemisorption is typically above 200 kJ/mol. Chemisorption is usually not reversible, and desorption, if it occurs, is accompanied by a chemical change in the adsorbate. What is commonly referred to as "irreversible adsorption" is chemisorption because the adsorbate is chemically bonded to the surface. While physical adsorption and chemisorption can be distinguished easily at their extremes, some cases fall between the two, as a highly unequal sharing of electrons may not be distinguishable from the high degree of distortion of an electron cloud that occurs with physical adsorption (Adamson, 1982; Kipling, 1965; Satterfield, 1980). Because most water treatment applications involve physical adsorption, physical adsorption mechanisms are discussed in greater detail in this chapter.

Historical Development

Modern purification of water supplies by adsorption has a short history as compared to other processes, although the use of adsorption has been reported in a 4000-year-old Sanskrit text (Sontheimer et al., 1988). Adsorption was first observed in solution by Lowitz in 1785 and was soon applied as a process for removal of color from sugar during refining (Hassler, 1974). In the latter half of the nineteenth century, charcoal adsorbers (charcoal is not activated and contains underdeveloped pores) were used in U.S. water treatment plants (Croes, 1883). The first GAC units for treatment of water supplies were constructed in Hamm, Germany, in 1929 and Bay City, Michigan, in 1930. In the 1920s, Chicago meat packers used PAC to remove taste and odor in water supplies that were contaminated by chlorophenols (Baylis, 1929). Powdered activated carbon was first used in municipal water treatment in New Milford, New Jersey, in 1930 and its use became widespread in the next few decades, primarily for taste and odor control.

During the mid-1970s, interest in adsorption as a process for removal of organics from drinking water was heightened because the public became increasingly concerned about water sources that were contaminated by industrial wastes, agricultural chemicals, and municipal discharges. Another major concern was the formation of DBPs during chlorination of water containing background organic matter (referred to as DBP precursors).

It has been found that activated carbon can be effective in removing some of the DBP precursors.

Applications of Adsorption Materials

Three types of commercially available adsorbents merit consideration in water treatment: zeolites, synthetic polymeric adsorbents, and activated carbon. Most activated carbons have a wide range of pore sizes and can accommodate large organic molecules such as natural organic matter (NOM) and synthetic organic compounds (SOCs) such as pesticides, solvents, and fuels. Synthetic polymeric adsorbents usually have only micropores, which prevents them from adsorbing NOM. Zeolites (aluminosilicates with varying Alto-Si ratios) tend to have very small pores, which will exclude some synthetic organic compounds. Granular ferric hydroxide and iron-impregnated GACs have been developed to remove arsenic. Ammonia-treated GAC has been use to increase the adsorption capacity of GAC for bromated and perchlorate, and it is likely that this would increase GAC adsorption capacity for other anionic species; however, there are no commercially available GACs. Properties of several commercially available adsorbents are reported in Table 15-2.

Porous adsorbents can have a large internal surface area (400 to 1500 m^2/g) and pore volume (0.1 to 0.8 mL/g) and as a result can have an adsorption capacity as high as 0.2 g of adsorbate per gram of adsorbent,

Table 15-2
Properties of several commercially available adsorbents

Adsorbent	Manufacturer	Type	Surface Area, m^2/g (BET)[a]	Packed Bed Density, g/cm^3	Pore Volume, cm^3/g
Filtrasorb 300 (8×30)	Calgon	GAC	950–1050	0.48	0.851
Filtrasorb 400	Calgon	GAC	1075	0.4	1.071
CC-602	US Filter/Wastates	Coconut-shell-based GAC	1150–1250	0.47–0.52	0.564
Aqua Nuchar	MWV	PAC	1400–1800	0.21–0.37	1.3–1.5
Dowex Optipore L493	Dow	Polymeric	>1100	0.62	1.16
Lewatit VP OC 1066	Bayer	Synthetic polymer	700	0.5	0.65–0.8

[a]BET is the Brunauer, Emmett, and Teller method for measuring surface area based on gas (usually nitrogen) adsorption.
Source: Adapted from Sontheimer et al. (1988), Crittenden (1976), Lee et al. (1981), Munakata et al. (2003), and Sigama_Aldrich Online Catalog (2004).

depending on the adsorbate concentration and type. Synthetic polymeric resins, zeolites, and activated alumina have been used in water treatment applications, but activated carbon is the most commonly used adsorbent because it is much less expensive than the alternatives. Activated carbon is manufactured from natural, carbonaceous materials such as coal, peat, and coconuts by several inexpensive processes (e.g., high temperatures ~800°C and steam). Consequently, most of the discussion in this chapter centers on the use of activated carbon; where appropriate, alternative adsorbents are discussed. Activated carbon is available in essentially two particle size ranges: PAC (mean particle size 20 to 50 μm) and GAC (mean particle size 0.5 to 3 mm). The principal uses, advantages, and disadvantages of using PAC versus GAC are reported in Table 15-3.

At present, the applications of adsorption in water treatment in the United States are predominantly for taste and odor control. In a 1984 survey, 29 percent of the water utilities used PAC (AWWA, 1986), and in a 1989 survey it was reported that 63 percent of the water plants used PAC and 7 percent used GAC for taste and odor control (Suffet et al., 1996). Currently, it is thought that about 90 percent of the surface water treatment plants worldwide use PAC on a seasonable basis (Hansen, 1975; Sontheimer, 1976). In 1996, there were 300 GAC surface water plants and

Table 15-3
Principal uses, advantages, and disadvantages of granular and powdered activated carbon

Parameter	Granular Activated Carbon (GAC)	Powdered Activated Carbon (PAC)
Principal uses	❑ Control of toxic organic compounds that are present in groundwater ❑ Barrier to occasional spikes of toxic organics in surface waters and control of taste and odor compounds ❑ Control of disinfection by-product precursors or DOC	❑ Seasonal control of taste and odor compounds and strongly adsorbed pesticides and herbicides at low concentration (<10 μg/L)
Advantages	❑ Easily reactivated ❑ Lower carbon usage rate per volume of water treated as compared to PAC	❑ Easily added to existing water intakes or coagulation facilities for occasional control of organics
Disadvantages	❑ Need contactors and yard piping to distribute flow and replace exhausted carbon ❑ Previously adsorbed compounds can desorb and in some cases appear in the effluent at concentrations higher than present in influent	❑ Hard to reactivate and impractical to recover from sludge from coagulation facilities ❑ Much higher carbon usage rate per volume of water treated as compared to GAC

several hundred groundwater plants (Snoeyink, 2001). European water treatment plants have had longer experience using GAC to remove SOCs in water from polluted rivers (Sontheimer et al., 1988). In the future, it is expected that the removal of low concentrations of toxic or carcinogenic compounds using adsorption technology will increase.

For continuous removal of SOCs (e.g., pesticides, herbicides, trichloroethene, tetrachloroethene, benzene), GAC is preferred because much less GAC is needed than if PAC was used (details provided later in this chapter). In most cases, SOCs are present at concentrations of 1 to 500 μg/L, and the background organic matter concentration ranges from 0.5 to 20 mg/L of dissolved organic carbon (DOC), with typical values of 1 to 3 mg/L DOC. Because the background DOC concentration is so high relative to SOCs, SOCs are sometimes referred to as micropollutants. There are two principal application scenarios for the removal of SOCs by GAC: (1) removal of organics from contaminated groundwaters and (2) as a barrier against spikes of organics that occur in susceptible water supplies such as the Ohio and Mississippi Rivers. Contaminated groundwaters typically have a consistent (10 to 300 μg/L) and most often declining concentration of SOCs. Surface waters typically have very low concentrations of SOCs (most often below the maximum contaminant levels) with occasional spikes of SOCs due to spills or the seasonal application of pesticides and herbicides.

15-2 Manufacture, Regeneration, and Reactivation of Activated Carbon

The production of activated carbon consists of the pyrolytic carbonization of the raw material and the subsequent or parallel activation. During the carbonization step, volatile components are released and graphite is formed. Further, the carbon realigns to form a pore structure that is developed during the activation process. In the activation step, carbon is removed selectively from an opening of closed porosity and the average size of the micropores is increased. A generalized flow diagram for the production of both GAC and PAC is shown on Fig. 15-1.

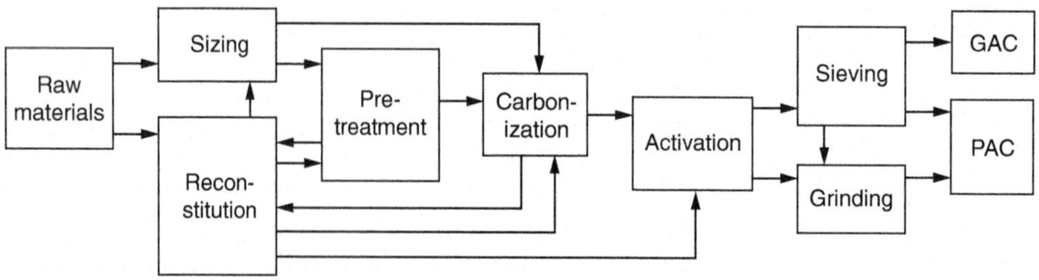

Figure 15-1
General flow scheme for production of activated carbon.

Manufacture from Raw Materials

The most direct production method involves sizing of the raw materials, carbonization, activation, and sieving of the product. The direct production method can be applied to coconut shells, relatively hard coals, and materials to be used as powdered carbons. Reconstitution and pretreatment are normally required for peat, lignite, petrol coke, and bituminous coals. For bituminous coals, pretreatment is necessary to control the loss of microporosity during carbonization due to swelling and softening of the coal.

In practice, activated carbon is produced by either chemical or physical activation. Chemical activation is normally utilized for raw materials that contain cellulose and combines the carbonization and activation steps. Dehydrating chemicals, such as zinc chloride or phosphoric acid, are added to the raw materials at elevated temperatures. The resulting product is then heated pyrolytically (causing a degradation of the cellulose) and cooled, and the activating agent is then extracted. Carbons produced using this method are of low density and, without special treatment, have a low proportion of micropores, which makes them less suitable for use in the removal of micropollutants and odor-causing substances.

Carbons that are produced for water treatment utilize an endothermic thermal activation process that involves the contacting of a gaseous activating agent, typically steam, with the char at elevated temperatures, typically 850 to 1000°C. Thermal activation causes a slight reduction in the size of the adsorbent grain due to external oxidation as the oxidizing gas diffuses into the unactivated internal domain of the carbon. While the rudimentary pore structure is determined by the raw material or occasionally by pretreatment, the type of activating agent, and the length and temperature of activation can have a major influence on the adsorbent properties, as shown on Fig. 15-2. The decrease in particle density and particle size for increasing mass burnoff is shown on Fig. 15-2a. The effect of increasing mass burnoff on adsorbent surface area per weight of original char can be seen on Fig. 15-2b for activation with steam at three temperatures and with CO_2 at one temperature.

For most thermally activated carbons, a maximum surface area per weight of original char is found at about 40 to 50 percent mass burnoff. Activation up to this point opens closed pores and enlarges existing pores, resulting in a net increase in surface area. Continued activation beyond this point results in a net surface area decrease as most closed pores are now open and pore walls are burned away (Jüntgen, 1968, 1976; Walker, 1986). As will be shown in Example 15-1, for a given pore volume the surface area decreases with increasing pore size and the concomitant development of more pore volume and large pore sizes gives rise to the trends displayed on Fig. 15-2b.

The types of base materials can influence the distribution of the pores. For the purpose of classifying pore sizes (diameter D_p), the International

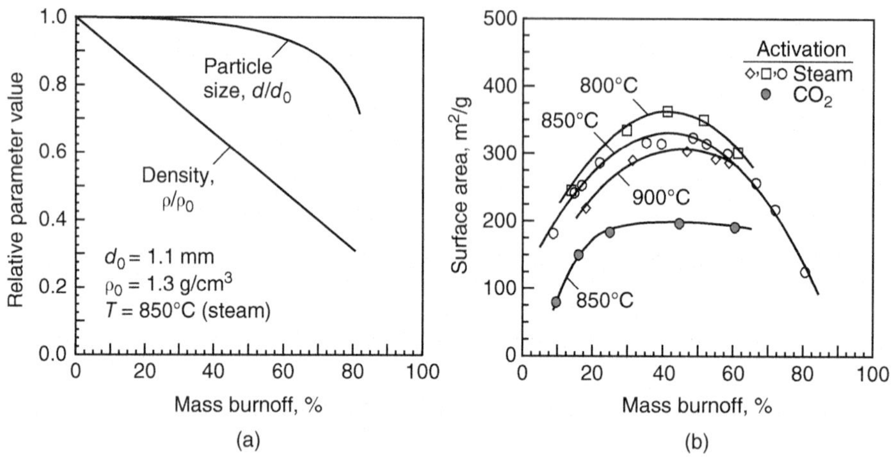

Figure 15-2
Impact of mass burnoff of coal-based carbon on (a) apparent particle density and particle diameter and (b) surface area per weight of original char at three temperatures and with two gases. (Adapted from Hashimoto et al., 1979.)

Union of Pure and Applied Chemistry (IUPAC) uses the following convention:

Micropores: $d_p < 2$ nm.
Mesopores: 2 nm $< d_p < 50$ nm.
Macropores: 50 nm $< d_p$.

Coconut shell carbons are considered a microporous carbon because the majority of their total void volume is micropores, as shown on Fig. 15-3. Wood-based carbons have a more even distribution of micro-, meso-, and macropores.

The macropore structure of a lignite-based carbon with increasing magnification is displayed on Fig. 15-4. In the scanning electron micrograph

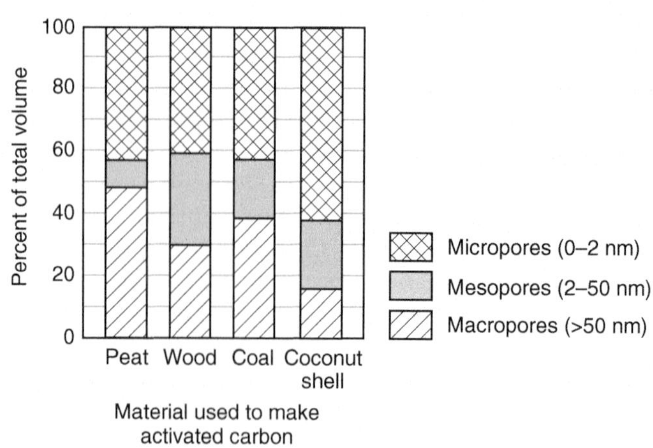

Figure 15-3
Pore size distributions for activated carbons with different starting materials.

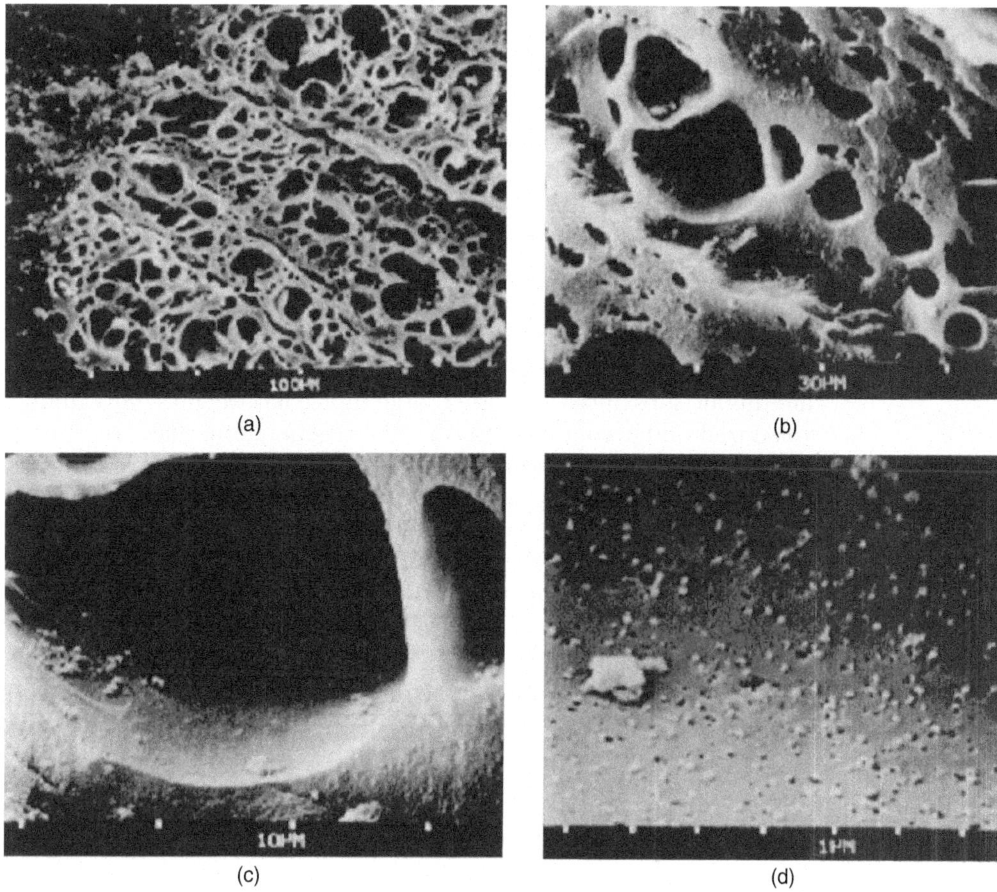

Figure 15-4
Scanning electron micrographs for lignite-based activated carbon. The scale line for all the SEMs is the interval between the white marks (clockwise starting in the upper left the scales are 100, 30, 10, and 1 μm).

(SEM) shown on Fig. 15-4a, both the external surface and the largest macropores with sizes ranging from about 30 nm to 0.3 μm are visible. In the SEM image shown on Fig. 15-4d, pores with a size of 50 nm are branching off of the large pore, which is also shown under different magnifications on Figs. 15-4b and 15-4c. At all magnification levels pores of different sizes are visible, and even at the greatest magnification small pores can be seen branching off from the large pores.

Regeneration and Reactivation of Spent GAC

When the adsorption capacity of the activated carbon has been exhausted, it must be removed from the contactor and replaced with fresh or reactivated carbon. *Regeneration* occurs when adsorbed solute molecules are removed from the carbon surface through desorption in their original or a modified state with no change in the carbon surface. Methods of regeneration that have been proposed include thermal, physicochemical, and biologically induced regeneration. *Reactivation* of GAC involves restoration of the adsorption capacity through partial desorption of the solute molecules and then the burnoff of carbonaceous residual on the carbon surface. Reactivation conditions are similar to those in the manufacturing of activated carbon by thermal activation, where part of the carbon surface can be burned off during the process. A summary of regeneration and reactivation methods is presented in Table 15-4.

Regeneration of water treatment carbons is seldom practiced because complete restoration of the adsorption capacity cannot be achieved. In water treatment, the concentrations of even volatile solutes are very low, and humic substances and other large molecular weight compounds, not volatilized under conditions of thermal regeneration, make regeneration of water treatment carbons ineffective. Further, because spent carbon that is reactivated at a central facility is typically commingled with other spent carbons, reactivation is seldom used unless large quantities of carbon are involved. Onsite reactivation facilities only make economic sense if carbon usage is greater than 150,000 kg/yr (Sontheimer et al., 1988). Because regeneration and reactivation are not used in water treatment practice, these subjects are not considered further in this chapter. Detailed information on regeneration and reactivation may be found in Sontheimer et al. (1988).

15-3 Fundamentals of Adsorption

Knowledge of the fundamental phenomena and factors involved in the adsorption process will provide a basis for understanding the PAC and GAC processes and the process design considerations. The adsorption process on a molecular level and the interactions between the adsorbing compound and the adsorbent and how these interactions are impacted by physical and chemical forces within and surrounding the adsorbing compound and the adsorbent are discussed in this section.

Interfacial Equilibria for Adsorption and Other Solute Surface Phenomena

In aqueous solution, three interactions compete when considering physical adsorption: (1) adsorbate–water interactions, (2) adsorbate–surface interactions, and (3) water–surface interactions. The extent of adsorption is determined by the strength of adsorbate–surface interactions as compared to the adsorbate–water and water–surface interactions. Adsorbate–surface interactions are determined by surface chemistry, and adsorbate–water are related to the solubility of the adsorbate. Water–surface interactions are determined by the surface chemistry, for example, a graphitic surface is hydrophobic and oxygen containing functional groups are hydrophilic. For

Table 15-4
Summary of regeneration and reactivation methods

Processes	Subprocesses	Advantages/Disadvantages
Regeneration		
Thermal	Steam	❑ Used for high concentrations in industrial vapor solvent recovery systems ❑ Not used for drinking water treatment applications because large amounts of condensed water vapor containing SOCs require further treatment and nonvolatile SOCs and natural organic matter not removed leading to loss in capacity over time. Has met with some success when removing VOCs and SOCs from expensive synthetic resins
	Hot air	❑ Can successfully desorb and oxidize VOCs, but nonvolatile SOCs and natural organic matter are not removed leading to loss in capacity over time
Physicochemical	Aqueous solution extraction	❑ Use of acid/base solutions to desorb some ionizing organic compounds (e.g., phenol) from GAC ❑ Practical only if acid/base solution can be recycled ❑ Nonvolatile SOCs and natural organic matter not removed leading to a loss in capacity over time ❑ Liquid carbon dioxide is an excellent solvent because it can volatilize off after extraction, but VOCs may be lost during carbon dioxide evaporation
	Supercritical carbon dioxide extraction	❑ Does not remove very strongly adsorbed SOCs and some natural organic matter ❑ Requires special facilities to handle liquid carbon dioxide
	Organic solvent extraction	❑ Easy process to apply ❑ Natural organic matter difficult to extract from adsorbent resulting in loss in capacity over time ❑ Requires disposal of spent solvent and solvent-laden water ❑ Solvent can desorb into finished drinking water

(Continued)

Table 15-4 *(Continued)*

Processes	Subprocesses	Advantages/Disadvantages
Biological	—	❑ Can reduce loading of carbon through desorption of compound in response to decrease in liquid-phase concentration of biodegradable compounds; may be promising for reduction in DBP precursor material ❑ Has been shown to work for high concentrations of biodegradable SOCs ❑ Does not achieve high regeneration efficiencies ❑ Concerns about use of biological processes in drinking water treatment ❑ For taste and odor (T&O) removal, GAC appears to last for several years due to biological degradation of T&O compounds
Reactivation		
Multiple hearth furnace	—	❑ Most commonly used reactivation process; has long residence time without back mixing, good mass transfer, low energy requirements, low carbon losses (3–5%) and adequate burner control ❑ Has long startup time and not recommended for intermittent use
Rotary kiln furnace	—	❑ Has low energy and equipment costs, low GAC losses (5–8%) ❑ Has poor mass transfer characteristics, residence time distributions, and control of reaction environment
Fluidized-bed reactor	—	❑ Has relatively good mass transfer characteristics that lead to low energy costs and few moving parts that contribute to low maintenance; provides good flexibility in terms of reaction conditions and GAC throughput ❑ A major disadvantage is that it has backmixing that causes a wide residence time distribution that leads to some overreactivated and some underreactivated GAC particles. Carbon losses can be as high as 12%

chemisorption, the primary factor controlling the extent of reaction is the type of reaction that occurs on the surface. In either case, it is important to provide enough surface area for adsorption. The volumetric filling of small pores is also important.

Important Factors Involved in Adsorption

The surface area and pore size are important factors that determine the number of adsorption sites and the accessibility of the sites for adsorbates. Generally, there is an inverse relationship between the pore size and surface area: the smaller the pores for a given pore volume, the greater the surface area that is available for adsorption. In addition, the size of the adsorbate that can enter a pore is limited by the pore size of the adsorbent, and is referred to as steric effects. The relationship between pore size and surface area is shown in Example 15-1.

The porosity of adsorbents generally does not exceed 50 percent, partly due to the manufacturing process and the skeletal strength of the adsorbent. If adsorbents become very porous, they become brittle and break apart when transported into and out of adsorption vessels, which can result in significant adsorbent losses and expense.

Surface Chemistry and Forces Involved in Adsorption

There are three interfaces involved in adsorption: adsorbate–adsorbent, adsorbate–water, and water–adsorbent. The forces active at each of these interfaces are summarized in Table 15-5. Some of the forces that occur between the adsorbent surface and adsorbates are illustrated on Fig. 15-5.

CHEMICAL ADSORPTION

Chemical adsorption, or chemisorption, occurs when the adsorbate reacts with the surface to form a covalent bond or an ionic bond. In chemisorption,

Table 15-5
Summary of forces that are active at the three interfaces involved in adsorption

Force	Approximate Energy of Interaction, kJ/mol	Interface		
		Adsorbate/ Adsorbent	Adsorbate/ Water	Water/ Adsorbent
Coulombic repulsion	>42	Yes	No	No
Coulombic attraction	>42	Yes	No	No
Ionic species–neutral species attraction		Yes	No	No
Covalent bonding	>42	Yes	No	No
Ionic species–dipole attraction	<8	Yes	Yes	Yes
Dipole–dipole attraction	<8	Yes	Yes	Yes
Dipole–induced dipole attraction	<8	Yes	Yes	Yes
Hydrogen bonding	8–42	Yes	Yes	Yes
van der Waal's attraction	8–42	Yes	Yes	Yes

Source: Stumm and Morgan (1981).

Example 15-1 Determination of surface area

Determine the surface area of an adsorbent that has a bulk porosity of 50 percent, particle density of 1 g/cm^3, and pore sizes (diameters) of 1 and 5 nm. Assume the pore shape is cylindrical in the quantification of surface area and pore volume.

Solution

1. Develop a relationship for the ratio of surface area to pore volume for the adsorbent.
   a. The volume of cylindrical pores in an adsorbent, V_{ad} (m^3/g), can be computed based on the number of pores n (no./g), the pore radius R (m), and the pore height H (m):

$$V_{ad} = n\pi R^2 H$$

   b. The surface area of pores in an adsorbent, A_{ad} (m^2/g), is also determined assuming a cylindrical pore shape:

$$A_{ad} = 2n\pi RH$$

   c. The surface area–pore volume ratio for the adsorbent, A_{ad}/V_{ad}, can now be written by combining the expressions developed in steps 1a and 1b:

$$\frac{A_{ad}}{V_{ad}} = \frac{2}{R}$$

2. Determine the surface area for adsorbents with pore sizes of 1 and 5 nm.
   a. Compute the adsorbent volume using the porosity and adsorbent density provided in the problem statement. By definition, porosity = pore volume/total volume, so 1 g of adsorbent with a porosity of 0.5 would have a total volume of 1 cm^3 and a pore volume of 0.5 cm^3. Therefore $V_{ad} = 0.5$ cm^3/g.
   b. For a pore diameter $d_p = 1.0$ nm $= 10 \times 10^{-8}$ cm,

$$A_{ad} = V_{ad}\frac{2}{d_p/2} = \left(0.5\ \text{cm}^3/\text{g}\right)\frac{2}{\left(10 \times 10^{-8}\text{cm}\right)/2}$$

$$= 20{,}000{,}000\ \text{cm}^2/\text{g}\left(2000\ \text{m}^2/\text{g}\right)$$

   c. For a pore diameter $d_p = 5.0$ nm $= 50 \times 10^{-8}$ cm,

$$A_{ad} = V_{ad}\frac{2}{d_p/2} = \left(0.5\ \text{cm}^3/\text{g}\right)\frac{2}{\left(50 \times 10^{-8}\text{cm}\right)/2}$$

$$= 4{,}000{,}000\ \text{cm}^2/\text{g}\left(400\ \text{m}^2/\text{g}\right)$$

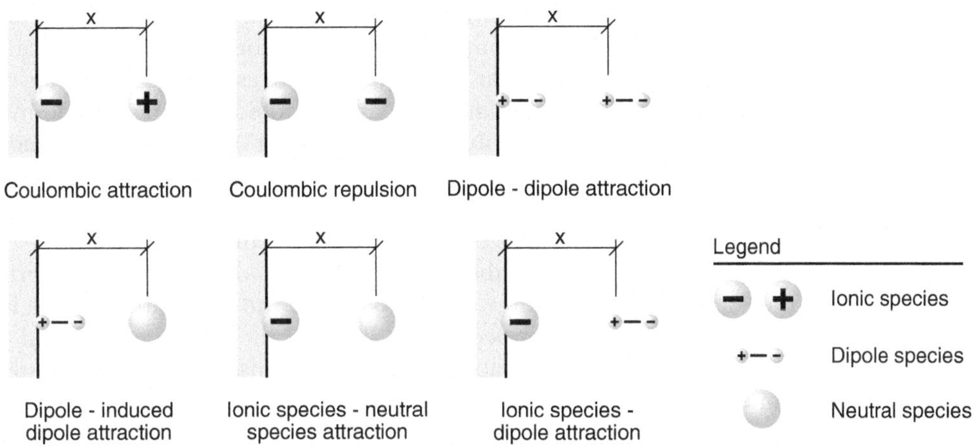

Figure 15-5
Surface functional groups and forces of attraction.

the attraction between adsorbent and adsorbate approaches that of a covalent or electrostatic chemical bond between atoms, with shorter bond length and higher bond energy. Adsorbates bound by chemisorption to a surface generally cannot accumulate at more than one molecular layer because of the specificity of the bond between adsorbate and surface. The bond may also be specific to particular sites or functional groups on the surface of the adsorbent. The charged surface groups attract the opposite charges and repel like charges according to Coulomb's law, as discussed in Chap. 16.

PHYSICAL ADSORPTION

Adsorbates are said to undergo physical adsorption if the forces of attraction include only physical forces that exclude covalent bonding with the surface and coulombic attraction of unlike charges. In some cases, the difference between physical adsorption and chemisorption may not be that distinct. Physical adsorption is less specific for which compounds sorb to surface sites, has weaker forces and energies of bonding, operates over longer distances (multiple layers), and is more reversible.

In water treatment there is often interest in the adsorption of organic adsorbates from water (polar solvent) onto a nonpolar adsorbent (activated carbon). Because activated carbon has crosslinked graphitic crystallite planes that form micropores, the major attractive force between organics and the adsorbent is van der Waals forces that exist between organic compounds and the graphitic carbon basal planes.

In general, attraction between an adsorbate and polar solvent is weaker for adsorbates that are less polar or have lower solubility. The attraction between an adsorbate and activated carbon surface increases with increasing polarizability and size, which are directly related to van der Waals forces. More nonpolar and larger compounds tend to adsorb more strongly to nonpolar adsorbents such as activated carbon. This form of adsorption is also known as hydrophobic bonding (Nemethy and Scheraga, 1962); hydrophobic ("disliking water") compounds will adsorb on carbon more strongly.

ADSORBABILITY OF VARIOUS CLASSES OF COMPOUNDS

Applying what is known about the adsorption of organics to determine their adsorbability requires consideration of the summation of the interactions and forces described above. Because these interactions and forces are not readily measurable, in a general sense they can be related to some properties of the adsorbate and solvent. For example, solubility is a direct indication of adsorption strength or magnitude of the adsorption force. The lower the solubility of an adsorbate in the solvent, the higher the adsorption strength. Adsorption strength is inversely proportional to solubility. Unfortunately, all other factors are different for different classes of organics (e.g., aliphatic, aromatic, or polar compounds); consequently, solubility alone is not the only indicator of adsorbability. For example, water–adsorbent interactions are only important for adsorption onto polar or ionized surface functional groups, as discussed below. To make more specific statements regarding adsorbability, polar, neutral, and charged compounds must be considered separately.

Polar species

Polar organics and adsorbates with ionic functional groups will not be removed from water because water–adsorbate forces will be strong and polar functional groups on the adsorbent surface will attract water to the surface. The adsorption of acids and bases on nonpolar adsorbents such as activated carbon depends strongly on pH. The adsorption of neutral forms is generally much stronger (Getzen and Ward, 1969), and the pH of maximum adsorbability depends on the particular dissociation constant of the acid or base. Furthermore, pH affects the charge on activated carbon, which generally tends to be negative at neutral pH and neutral in the pH range of 4 to 5. From a practical point of view, most ionic organics are negatively charged and adsorbents are negatively charged at neutral pH. Thus, lowering the pH to less than 4 to 5 increases adsorbability but is not practical from an operational perspective.

Neutral species

Neutral organics are strongly held to nonpolar surfaces such as the graphitic surface of activated carbon. The adsorbate–water interaction force is related to the solubility. The adsorbate–adsorbent interaction force is related to the polarizability, which is related directly to the size of the organic compound. The adsorbent–water interactions are related to how many water molecules must be pulled away from the surface of the adsorbent (the attractive force in this case is a dipole–neutral species interaction) and, in turn, how many water molecules must be removed from the surface of the adsorbent to make room for the adsorbate. Adsorbability is related to the size of the organic compound. Consequently, the adsorbability of neutral organics increases with increasing polarizability and molecule size and decreases with increasing solubility.

Ionic species

With respect to adsorption of inorganics onto inorganic adsorbents, one class of chemical bonding to specific surface sites is the acid–base reaction at a functional group. An example is the reaction of hydrated metal ions from solution with hydroxide sites on metal oxides (Parks, 1975):

$$\text{MeOH(aq)} + \text{SOH}_{\text{surface}} \rightarrow \text{SOMe}_{\text{surface}} + \text{H}_2\text{O} \tag{15-1}$$

where MeOH = metal ion adsorbate
SOH = hydroxide site on metal oxide adsorbent

Studies comparing theory and experimental evidence for this specific chemical bonding that have relevance to water treatment for removing heavy metals by adsorption onto silicon and aluminum-oxide-based clays and sands are available (James and Healy, 1972; Parks, 1967).

For adsorption of ionic species to surfaces, the most important mechanism is electrostatic attraction, which is highly dependent on pH and ionic strength. This mechanism is described in Chap. 9 for forces controlling coagulation and in Chap. 16 for the process of ion exchange. Adsorption of electrolytes onto metal oxide adsorbents can be used to control heavy metals, fluoride, and a few other minerals.

15-4 Development of Isotherms and Equations Used to Describe Adsorption Equilibrium

The affinity of the adsorbate for an adsorbent is quantified using adsorption isotherms, which are used to describe the amount of adsorbate that can be adsorbed onto an adsorbent at equilibrium and at a constant temperature. For most applications in water treatment, the amount of adsorbate adsorbed is usually a function of the aqueous-phase concentration and this relationship is commonly called an isotherm. Several researchers have presented procedures, protocols, and problems associated with performing adsorption equilibrium isotherms (Crittenden et al., 1987b; Luft, 1984; Randtke and Snoeyink, 1983; Summers, 1986).

Equilibrium Isotherm

Adsorption isotherms are performed by exposing a known quantity of adsorbate in a fixed volume of liquid to various dosages of adsorbent. To prevent the loss of adsorbate in situations where the adsorbate is volatile, adsorbs onto the container, and is light sensitive, amber glass bottles (250 to 1000 L) with Teflon screw caps are used for aqueous-phase isotherms. If the adsorbent is granular, it is powdered (<200 mesh or <0.074 mm), washed, and dried to a moisture-free constant weight and stored in a sealed container in a desiccator before using. Approximately 12 headspace free bottles (no air voids in bottle) are used with various dosages of adsorbent and allowed to equilibrate in a rotating tumbler at 25 rev/min at a constant temperature for a period of no less than 6 days. At the end of the equilibration period, the aqueous-phase concentration of the adsorbate is measured and the adsorption equilibrium capacity is calculated for each bottle using the mass balance expression

$$q_e = \frac{V}{M}(C_0 - C_e) \tag{15-2}$$

where q_e = equilibrium adsorbent-phase concentration of adsorbate, mg adsorbate/g adsorbent

C_0 = initial aqueous-phase concentration of adsorbate, mg/L

C_e = equilibrium aqueous-phase concentration of adsorbate, mg/L

V = volume of aqueous phase added to bottle, L

M = mass of adsorbent, g

Equations developed by Lamgmuir, Freundlich, and Brunauer, Emmet, and Teller (BET isotherm) are used to describe the equilibrium capacity of adsorbents. A discussion of these isotherm expressions is presented following Example 15-2.

Example 15-2 Determination of adsorption isotherm

A trichloroethene (TCE) isotherm was performed on Calgon F400 GAC. A total of 25 isotherm points were determined using 250-mL amber bottles with Teflon-lined screw caps. The dosage of GAC varied in each bottle. The GAC used was powdered from virgin stock GAC, washed, and dried to a constant weight before use. Pure TCE was added to a solution containing organic free laboratory water to yield a TCE initial concentration of about 10,000 μg/L. The weight of the bottles and the caps were recorded prior to filling the bottles with the GAC dosage and the TCE solution. The bottles were filled headspace free to prevent any TCE from volatizing out of solution. A total of eight extra empty bottles were filled and allowed to equilibrate. The extra bottles were used as blanks to measure the initial concentration used in the isotherm. All the bottles were placed on a rotating device and rotated at 25 rev/min for a period of 14 days. The bottles were then removed from the tumbler and the carbon was allowed to settle for a few hours, and a sample was drawn from each bottle and the TCE concentration was analyzed using a gas chromatograph.

Based on the raw data given below, calculate the average initial liquidphase concentration from the equilibrated blanks and the equilibrium adsorbent-phase concentration. Plot the corresponding values of q_e and C_e on arithmetic and log–log paper to determine the nature of the distribution. A summary of the GAC dosages, solution volume, and equilibrated blanks is provided below.

Experimental data:

- ☐ Carbon type: F-400
- ☐ Chemical: trichloroethene
- ☐ Carbon size: 200 × 400
- ☐ Temperature: 13°C
- ☐ pH: 6.8

Sample No.	Dosage M, g	Volume V, mL	TCE Liquid-Phase Concentration C_e, μg/L
1	0.44254	247.1	3
2	0.39002	251.2	4.5
3	0.34427	252.5	4.1
4	0.26784	252.4	8.1
5	0.20674	253.6	15.5
6	0.18305	251.1	18.9
7	0.16521	251.4	24.5
8	0.14041	252.1	74.3
9	0.12416	252.1	57.0
10	0.10836	249.6	109.0
11	0.09418	254.7	162.5
12	0.08320	253.0	213.6
13	0.07332	251.0	144.9
14	0.05380	251.2	643.1
15	0.04752	255.1	872.6
16	0.03956	252.3	1109.1
17	0.03315	251.5	1476.9
18	0.02696	255.1	2699.8
19	0.02189	254.6	3271.9
20	0.01609	253.0	4858.4
21	0.01072	251.7	6263.2
22	0.00544	251.5	8427.3
23	0.00343	252.3	10009.8
24	0.00164	252.9	9875.5
25	0.06273	253.0	352.6

- Equilibrated blank data:

Sample No.	Equilibrated Blank C_0, μg/L
1	10,486
2	8,401
3	11,355
4	10,205
5	10,415
6	12,912
7	12,025
8	11,123

Solution

1. Calculate the average initial TCE aqueous-phase concentration in micrograms per liter:

$$C_0 = \frac{10,486 + 8,401 + 11,355 + 10,205 + 10,415 + 12,912 + 12,025 + 11,123}{8}$$
$$= 10,865\ \mu\text{g/L}$$

2. Calculate the equilibrium adsorbent-phase concentration in micrograms per gram. Equation 15-2 can be used to calculate the adsorbent-phase TCE concentration. The required computations for sample 1 is shown below:

$$q_e = \frac{V}{M}(C_0 - C_e) = \frac{V}{M}\left[(10,865 - C_e)\ \mu\text{g/L}\right]$$
$$q_e(1) = \frac{V(1)}{M(1)}\left[10,865 - C_e(1)\right] = \frac{0.2471\ \text{L}}{0.44254\ \text{g}}[(10,865 - 3)\ \mu\text{g/L}]$$
$$= 6065\ \mu\text{g/g}$$

The q_e values are summarized in the following table:

Sample No.	TCE Aqueous-Phase Concentration C_e, μg/L	TCE Adsorbent-Phase Concentration q_e, μg/g
1	3	6065.0
2	4.5	6995.0
3	4.1	7966.0
4	8.1	10231.0
5	15.5	13309.0

Sample No.	TCE Aqueous-Phase Concentration C_e, μg/L	TCE Adsorbent-Phase Concentration q_e, μg/g
6	18.9	14877.8
7	24.5	16496.4
8	74.3	19374.7
9	57.0	21945.6
10	109.0	24776.4
11	162.5	28944.6
12	213.6	32390.3
13	144.9	36699.6
14	643.1	47728.9
15	872.6	53643.6
16	1,109.1	62221.6
17	1,476.9	71227.2
18	2,699.8	77263.3
19	3,271.9	88317.9
20	4,858.4	94452.8
21	6,263.2	108054.9
22	8,427.3	112712.7
25	352.6	42339.3

3. Plot the TCE isotherm data on arithmetic and log–log paper.

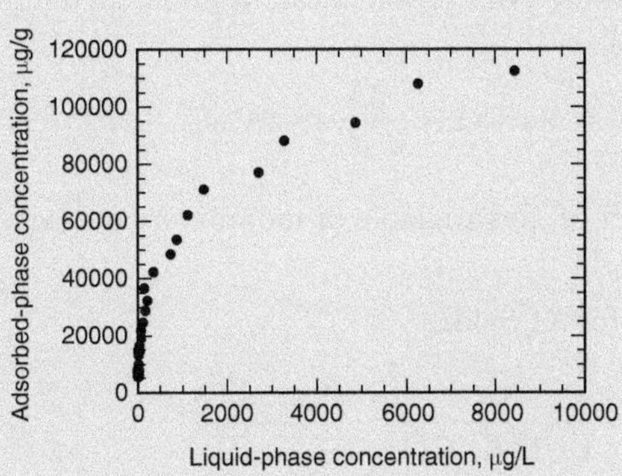

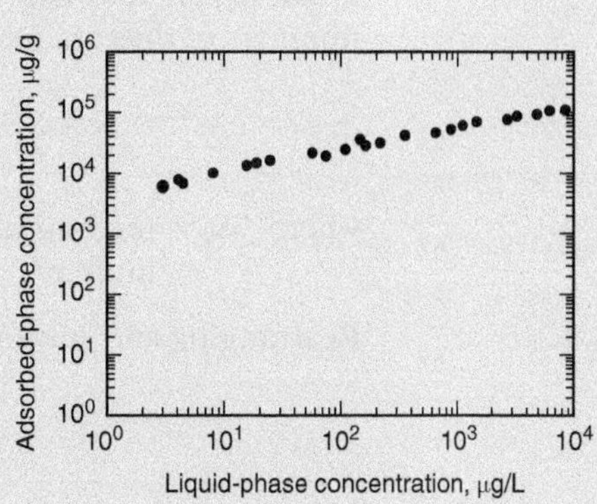

4. Based on the above plots, the isotherm is linear on a log–log plot.

Langmuir Isotherm Equation

The Langmuir adsorption isotherm is used to describe the equilibrium between surface and solution as a reversible chemical equilibrium between species (Langmuir, 1918). The adsorbent surface is made up of fixed individual sites where molecules of adsorbate may be chemically bound.

The following reaction describes the relationship between vacant surface sites and adsorbate species and adsorbate species bound to surface sites:

$$S_V + A \rightleftarrows SA \tag{15-3}$$

where S_V = vacant surface sites, mmol/m^2
A = adsorbate species A in solution, mmol
SA = adsorbate species bound to surface sites, mmol/m^2

In the Langmuir expression it is assumed that the reaction has a constant free-energy change ($\Delta G^\circ_{\text{ads}}$) for all sites (see Chap. 5). Further, each site is assumed to be capable of binding at most one molecule of adsorbate; that is, the Langmuir model allows accumulation only up to a monolayer. Accordingly, the equilibrium condition may be written as

$$K_{\text{ad}} = \frac{SA}{S_V C_{\text{A}}} = e^{-\Delta G^\circ_{\text{ads}}/RT} \tag{15-4}$$

Where K_{ad} = Langmuir adsorption equilibrium constant, L/mg
C_{A} = equilibrium concentration of adsorbate A in solution, mg/L
$\Delta G^\circ_{\text{ads}}$ = free-energy change for adsorption, J/mol
R = universal gas constant, 8.314 J/mol · K
T = absolute temperature, K (273 + °C)

The expression shown in Eq. 15-4 is not a convenient way to express the amount adsorbed as a function of concentration because there are two unknowns, S_V and C_{A}. However, this problem can be eliminated if the total numbers of sites are fixed:

$$S_T = S_V + SA = \frac{SA}{K_{\text{ad}} C_{\text{A}}} + SA \tag{15-5}$$

where S_T = total number of sites available or monolayer coverage, mol/m^2

Rearranging and solving for SA yields

$$SA = \frac{S_T}{1 + 1/(K_{\text{ad}} C_{\text{A}})} = \frac{K_{\text{ad}} C_{\text{A}} S_T}{1 + K_{\text{ad}} C_{\text{A}}} \tag{15-6}$$

The concentration of occupied sites that are expressed as mmol/m^2 is not particularly useful in mass balances; mass loading per mass of adsorbent is much more useful. Multiplying both sides of Eq. 15-6 by the surface area per gram and molecular weight, Eq. 15-6 can be expressed in terms of a mass loading q:

$$q_A = (SA)(A_{ad})(\mathrm{MW}) = \frac{K_{ad}C_A S_T A_{ad}\mathrm{MW}}{1 + K_{ad}C_A} = \frac{Q_M K_{ad} C_A}{1 + K_{ad}C_A} \quad (15\text{-}7)$$

$$= \frac{Q_M b_A C_A}{1 + b_A C_A} \quad (15\text{-}8)$$

where q_A = equilibrium adsorbent-phase concentration of adsorbate A, mg adsorbate/g adsorbent (see Eq. 15-2)
A_{ad} = surface area per gram of adsorbent, m^2/g
MW = molecular weight of adsorbate, g/mol
C_A = equilibrium concentration of adsorbate A in solution, mg/L
Q_M = maximum adsorbent-phase concentration of adsorbate when surface sites are saturated with adsorbate, $S_T A_{ad}$ MW, mg adsorbate/g adsorbent
b_A = Langmuir adsorption constant of adsorbate A, K_{ad}, L/mg

It is convenient to rearrange Eq. 15-8 to a linear form:

$$\frac{C_A}{q_A} = \frac{1}{b_A Q_M} + \frac{C_A}{Q_M} \quad (15\text{-}9)$$

A plot of C_A/q_A versus C_A using Eq. 15-9 results in a straight line with slope of $1/Q_M$ and intercept $1/b_A Q_M$.

Freundlich Isotherm Equation

The Freundlich adsorption isotherm (Freundlich, 1906), originally proposed as an empirical equation, is used to describe the data for heterogeneous adsorbents such as activated carbon:

$$q_A = K_A C_A^{1/n} \quad (15\text{-}10)$$

where K_A = Freundlich adsorption capacity parameter, $(\mathrm{mg/g})(\mathrm{L/mg})^{1/n}$
$1/n$ = Freundlich adsorption intensity parameter, unitless

The linear form of Eq. 15-10 is

$$\log(q_A) = \log(K_A) + \left(\frac{1}{n}\right)\log(C_A) \quad (15\text{-}11)$$

A log–log plot of q_A versus C_A using the form shown in Eq. 15-11 will result in a straight line, as shown on Fig. 15-6 for tetrachloroethene, TCE, and 1,1,1-trichloroethane. While it is possible to graph the data on a log–log plot and determine the Freundlich parameters, nonlinear regression should be used and all data should be weighted according to their precision (Sontheimer et al., 1988). If the logs of q_A and C_A are taken and linear regression is

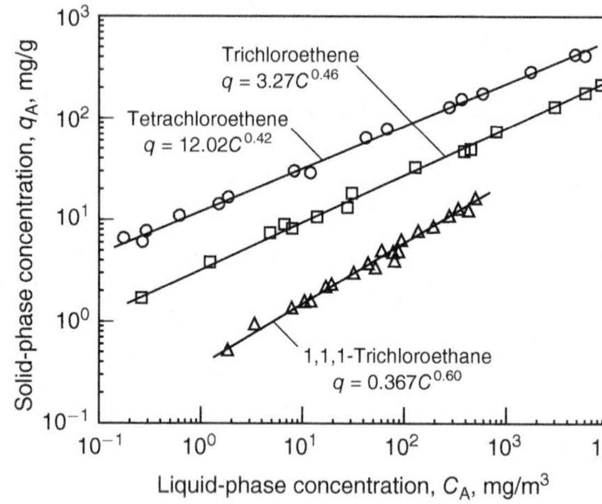

Figure 15-6
Single-solute isotherms for tetrachloroethene, trichloroethene, and 1,1,1-trichlorethane over a wide concentration range (Zimmer et al., 1988).

applied, more consideration is given to all data, but it does not weight all the data equally.

The Freundlich equation is consistent with the thermodynamics of heterogeneous adsorption (Halsey and Taylor, 1947). The Freundlich equation can be derived using the Langmuir equation to describe the adsorption onto sites of a given free energy and considering the following two assumptions: (1) the site energies for adsorption follow a Boltzmann distribution and the mean site energy is $\Delta \overset{\circ}{H}_M$, and (2) the change in site entropy increases linearly with increasing site enthalpy $-\Delta \overset{\circ}{H}_{\text{ad}}$ and the proportionality constant is r. Based on this development, $1/n$ will depend on temperature, as shown by the expression

$$n = \frac{\Delta \overset{\circ}{H}_M}{RT} - \frac{r\Delta \overset{\circ}{H}_{\text{ad}}}{R} \tag{15-12}$$

where $\Delta \overset{\circ}{H}_M$ = mean site energy, J/mol
R = universal gas constant, 8.314 J/mol · K
$\Delta \overset{\circ}{H}_{\text{ad}}$ = change in site enthalpy, J/mol
T = absolute temperature, K (273 + °C)
r = proportionality constant

The Freundlich isotherm equation always provides a better fit to the isotherm data for GAC than the Langmuir equation because many layers of adsorbates can adsorb to the GAC and there is a distribution of sites with different adsorption energies. Examples of Freundlich isotherm parameters are shown in Table 15-6, and the procedure for determining Langmuir and Freundlich isotherm parameters is demonstrated in Example 15-3.

Example 15-3 Determination of Freundlich and Langmuir isotherm parameters

For the experimental isotherm data given below, determine the Freundlich and Langmuir isotherm parameters. Apply linear regression to determine the isotherm parameters. A spreadsheet can be used for this purpose.

Experimental data:

- ☐ Carbon type: F-400
- ☐ Carbon size: 200 × 400
- ☐ Chemical: Trichloroethene
- ☐ Temperature: 13°C
- ☐ pH: 7.5–8
- ☐ Equilibrium time: 31 days

Sample number	TCE Liquid-Phase Concentration C_A, μmol/L	TCE Adsorbent-Phase Concentration q_A, μmol/g
1	23.6	737
2	6.67	450
3	3.26	318
4	0.322	121
5	0.169	85.2
6	0.114	75.8

Solution

1. Determine Langmuir isotherm parameters.
   a. For the Langmuir equation, the plot of C_A/q_A versus C_A along with the Langmuir isotherm fit obtained from a spreadsheet using linear regression is shown in the following figure:

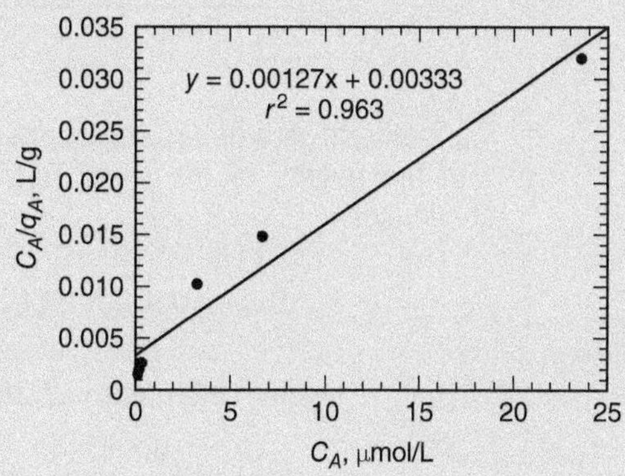

b. The Langmuir parameters are obtained by comparing Eq. 15-9 with the results of the linear regression as shown in the above plot.

$$\text{slope} = \frac{1}{Q_M} = 0.00127\ \text{g}/\mu\text{mol}$$

$$Q_M = 787.4\ \mu\text{mol/g}$$

$$Q_M = (787.4\ \mu\text{mol/g})(131.39\ \mu\text{g}/\mu\text{mol}) = 1.03 \times 10^5\ \mu\text{g/g}$$

$$\text{intercept} = \frac{1}{b_A Q_M} = 0.00333\ \text{g/L}$$

$$b_A = \frac{1}{(0.00333\ \text{g/L})(787.4\ \mu\text{mol/g})} = 0.381\ \text{L}/\mu\text{mol}$$

$$b_A = \frac{0.381\ \text{L}/\mu\text{mol}}{131.39\ \mu\text{g}/\mu\text{mol}} = 2.90 \times 10^{-3}\ \text{L}/\mu\text{g}$$

2. Determine Freundlich isotherm parameters.
   a. The Freundlich isotherm plot using linear regression is shown in the following figure:

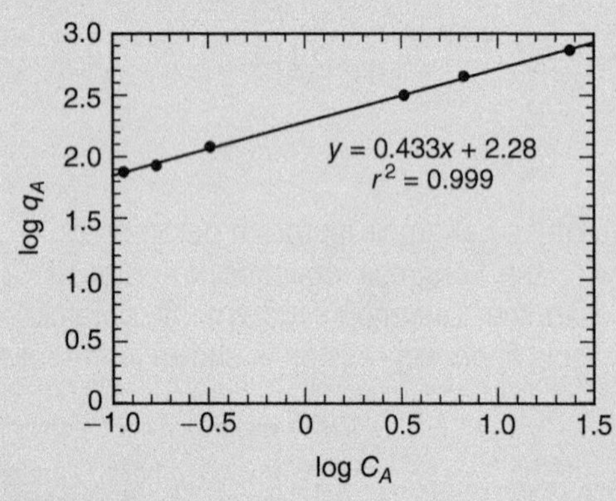

   b. The isotherm parameters are obtained by comparing Eq. 15-10 with the results of the linear regression as shown in the above plot.

$$\frac{1}{n} = \text{slope} = 0.43$$

$$\log K = \text{intercept} = 2.28$$

$$K = 190 \frac{\mu\text{mol}}{\text{g}} \left(\frac{\text{L}}{\mu\text{mol}} \right)^{0.43}$$

$$= 190 \frac{\mu\text{mol}}{\text{g}} \times \frac{131.39\ \mu\text{g}}{\mu\text{mol}} \times \frac{1\ \text{mg}}{1000\ \mu\text{g}}$$

$$\times \left(\frac{\text{L}}{\mu\text{mol}} \times \frac{\mu\text{mol}}{131.39\ \mu\text{g}} \times \frac{1000\ \mu\text{g}}{1\ \text{mg}} \right)^{0.43}$$

$$= 59.92 \frac{\text{mg}}{\text{g}} \left(\frac{\text{L}}{\text{mg}} \right)^{0.43}$$

Comment

The Freundlich isotherm equation provides a better fit of the data than the Lanqmiur model.

Table 15-6
Aqueous-phase Freundlich isotherm parameters K and $1/n$ for selected organic adsorbates[a]

Compound	K[b]	$1/n$	pH	Tmin (°C)	Name of Carbon[c]	Reference
Atrazine	182	0.18	7.1	20	F 100	Haist-Gulde (1991)
Benzoic acid	0.7	1.8	7	20	F 300	Dobbs and Cohen (1980)
Chlorodibromomethane	45	0.517	6	11	F 400	Crittenden et al. (1985)
Chloroform	15	0.47	7.1	20	F 100	Haist-Gulde (1991)
Cyclohexanone	6.2	0.75	7.3	20	F 300	Dobbs and Cohen (1980)
Cytosine	1.1	1.6	7	20	F 300	Dobbs and Cohen (1980)
1,2-Dichlorobenzene	242.2	0.4	7.1	20	F 100	Haist-Gulde (1991)
1,3-Dichlorobenzene	458.8	0.63	7.9	24	F 400	Speth and Miltner (1998)
1,2-*trans*-Dichloroethene	3.1	0.51	6.7	20	F 300	Dobbs and Cohen (1980)
2,4-Dichlorophenol	141	0.29	9	20	F 300	Dobbs and Cohen (1980)
Ethylbenzene	53	0.79	7.4	20	F 300	Dobbs and Cohen (1980)
Methyl ethyl ketone	19.4	0.295	8	24	F 400	Speth and Miltner (1998)
N-Nitrosodimethylamine	0	0	7.5	20	F 300	Dobbs and Cohen (1980)
Pentachlorophenol	150	0.42	7	20	F 300	Dobbs and Cohen (1980)
Tetrachloroethene	218.2	0.42	7.1	20	F 100	Zimmer et al. (1988)
Trichloroethene	55.9	0.48			F 400	Speth and Miltner (1998)
1,1,1-Trichloroethane	23.2	0.6	7.1	20	F 100	Zimmer et al. (1988)

[a]Additional Freundlich isotherm parameters are available in the electronic resource E5 at the website listed in App. E.
[b]Units of K are $(\text{mg/g})(\text{L/mg})^{1/n}$.
[c]Calgon Carbon Corporation.

Brunauer–Emmett–Teller Isotherm Equation

The BET adsorption isotherm (Brunauer et al., 1938) extends the Langmuir model from a monolayer to several molecular layers. Above the monolayer, each additional layer of adsorbate molecules is assumed to equilibrate with the layer below it, and layers of different thickness are allowed to coexist. To develop the BET equation, the adsorption of the first layer is described using Eq. 15-8. Equation 15-8 is also used to describe the adsorption in subsequent layers by assuming that the free-energy change for layers 2 and higher is equal to the free energy of precipitation and not equal to $\Delta G^{\circ}_{\text{ads}}$. The basic assumption is that the first layer adsorbs according to forces between the adsorbent and adsorbate and subsequent layers adsorb as if they were forming precipitating layers $\Delta G^{\circ}_{\text{prec}}$. The resulting equation is

$$\frac{q_{\text{A}}}{Q_M} = \frac{B_{\text{A}} C_{\text{A}}}{(C_{S,\text{A}} - C_{\text{A}})[1 + (B_{\text{A}} - 1)(C_{\text{A}}/C_{S,\text{A}})} \tag{15-13}$$

$$B_{\text{A}} = \frac{K_{1,\text{ad}}}{K_{i,\text{ad}}} = \frac{e^{-\Delta G^{\circ}_{\text{ads}}}}{e^{-\Delta G^{\circ}_{\text{prec}}}} \tag{15-14}$$

where
- q_{A} = equilibrium adsorbent-phase concentration of adsorbate A, mg adsorbate/g adsorbent
- Q_M = maximum adsorbent-phase concentration of adsorbate when surface sites are saturated with adsorbate, mg adsorbate/g adsorbent
- $K_{1,\text{ad}}$ = equilibrium constant for first layer, L/mg
- $K_{i,\text{ad}}$ = equilibrium constant for subsequent layers, L/mg
- B_{A} = ratio of $K_{1,\text{ad}}$ and $K_{i,\text{ad}}$
- C_{A} = equilibrium concentration of adsorbate A in solution, mg/L
- $C_{S,\text{A}}$ = saturated solution concentration of A, mg/L
- $\Delta G^{\circ}_{\text{ads}}$ = free energy of adsorption, J/mol
- $\Delta G^{0}_{\text{prec}}$ = free energy of precipitation, J/mol

Here, B_A is greater than 1 because $-\Delta G^{\circ}_{\text{ads}}$ is greater than $-\Delta G^{\circ}_{\text{prec}}$. The BET isotherm has the general form shown on Fig. 15-7, with surface concentration reaching a plateau as the monolayer is filled, then increasing again with increasing C_{A}.

In the Langmuir model, it is assumed that the site energy for adsorption is the same for all surface sites and does not depend on degree of coverage and that the largest capacity corresponds to only one monolayer. These assumptions are not valid for most adsorbents because, for example, activated carbon has a wide range of pore sizes that continue to adsorb organics as the concentration increases. While the BET isotherm does allow for multiple layers, it is assumed in the BET equation that site energy is the same for the first layer and equal to the free energy of precipitation for

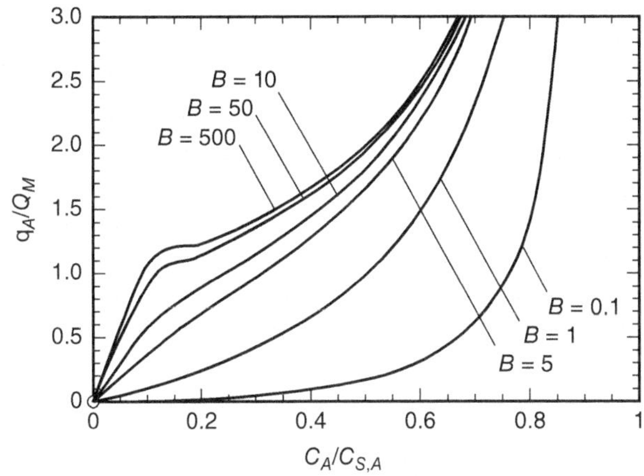

Figure 15-7
The BET isotherm for different values of B (capacity/monolayer capacity, q/Q, as function of degree of saturation).

subsequent layers. In reality, the site energy of adsorption varies widely for most adsorbents because adsorbents, such as activated carbon, are very heterogeneous and the site energy varies considerably with surface coverage. The single-solute data that are displayed on Fig. 15-6 cannot be described with either the BET or Langmuir equations. The Freundlich equation is used to describe isotherm data for heterogeneous adsorbents (varying site energies) and does a much better job describing the data than the Langmuir or BET equations.

Polanyi Correlation for Liquids

Because there are many organic compounds of health concern in drinking water treatment and there are also many different commercially activated carbon adsorbents to choose from, the chances that isotherms are available for all compounds on a given adsorbent are small. Engineers are then faced with either performing the isotherm experiments on the compound and adsorbent of interest or using some existing correlations to estimate the isotherm parameters. In preliminary design phases, engineers may use correlations to estimate isotherm parameters to evaluate the feasibility of using adsorption. A method for estimating isotherms and isotherm parameters for organic compounds onto activated carbon adsorbents based on the Polanyi potential theory is presented in the following discussion.

POLANYI POTENTIAL THEORY

Polanyi potential theory (Polanyi, 1916) has been used to correlate adsorption isotherms of volatile organic compounds in air (Grant and Manes, 1964, 1966; Reucroft et al., 1971; Tang, 1986). In addition, it has also been used to correlate aqueous-phase adsorption of a wide variety of compounds (Crittenden et al., 1999; Greenbank and Manes, 1981, 1982, 1984). While complicated forms of this theory exist, only the simplest form is presented here because of the many adjustable parameters required for various compound classes (Sontheimer et al., 1988; Crittenden et al., 1999).

The following assumptions are made in Polanyi potential theory: (1) a fixed pore volume exists that is close to the adsorbent surface where adsorption occurs; (2) the adsorptive forces originate from London–van der Waals interactions; and (3) adsorbing molecules will concentrate at high-energy sites on the adsorbent surfaces and undergo enhanced precipitation within the pores of the adsorbent. Polanyi defines the adsorption potential ε as the work or free energy required for any molecule to move from the bulk solution to the adsorption space assuming the adsorbed state is a saturated solution. If the displacement of the adsorbed fluid (water) is ignored, the following equation for ε may be obtained (Polanyi, 1916):

$$\varepsilon = RT \ln\left(\frac{C_s}{C}\right) \tag{15-15}$$

where ε = adsorption potential, J/mol
R = universal gas constant, 8.314 J/mol · K
T = absolute temperature, K (273 + °C)
C_s = aqueous solubility of adsorbate, mg/L
C = concentration of adsorbate in bulk solution, mg/L

Equation 15-15 is used to calculate the minimum energy required to extract a dissolved species in water. As applied, ε is assumed to vary from a maximum value close to the adsorbent surface with low bulk solution concentrations to zero when the pores are filled and the solubility limit is reached in the bulk solution (see Fig. 15-8).

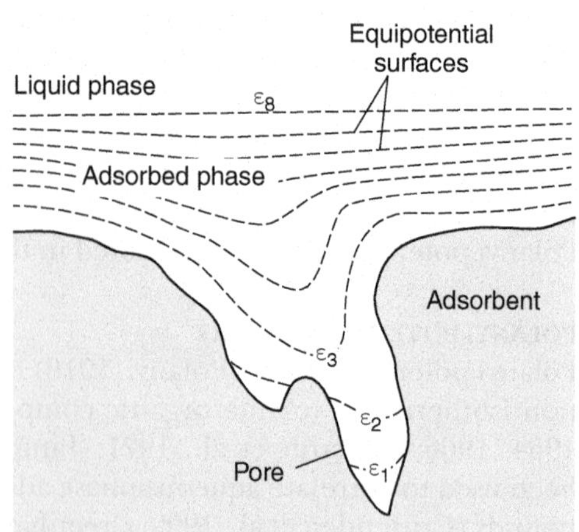

Figure 15-8
Schematic model of aqueous adsorption with equipotential surfaces.

ADSORPTION PARAMETERS DERIVED FROM POLANYI THEORY

A correlation can be developed to obtain adsorption parameters by plotting the volume adsorbed versus the adsorption potential. The adsorption potential needs to be divided by a normalizing physical property accounting for the major cause for adsorption. Previous researchers have used a number of normalizing physical properties such as molar volume, polarizability, and parachlor. For application of Polanyi potential theory to a wide variety of compounds in which more adjustable parameters are used to improve the correlation, it is necessary to consult the literature (Crittenden et al., 1999; Greenbank and Manes, 1981, 1982, 1984).

For several liquid aromatic and chlorinated alkanes and alkenes, Crittenden et al. (1987b) found that molar volume V_m was the best normalizing physical property. Accordingly, the correlation using V_m can be described by the equation

$$q = \rho_l W = \rho_l W_0 \exp\left[-\beta\left(\frac{\varepsilon}{V_m}\right)^{\sigma}\right] \tag{15-16}$$

$$\ln q = \ln(\rho_l W) = -\beta\left(\frac{\varepsilon}{V_m}\right)^{\sigma} + \ln(\rho_l W_0) \tag{15-17}$$

where q = adsorbent-phase concentration of adsorbate, mg adsorbate/g adsorbent
ρ_l = liquid density of adsorbate, g/L
W = volume of adsorbate adsorbed on adsorbent, mL adsorbate/g adsorbent
W_0 = maximum volume of adsorbate adsorbed on adsobent, mL adsorbate/g adsorbent
β = Polanyi constant determined for particular adsorbent, $(\text{L/J})^{\sigma}$
ε = adsorption potential, J/mol
V_m = molar volume of adsorbate, L/mol
σ = Polanyi constant determined for particular adsorbent, unitless

For liquids, it has been found that ρ_l could be taken as the liquid density (Crittenden et al., 1987b). To obtain a correlation for solids and liquids, correlations were proposed for ρ_l, and it has been demonstrated that data for a wide variety of compounds could be correlated (Greenbank and Manes, 1981, 1982, 1984).

Isotherms were conducted with solutes that are listed in Table 15-7 on Calgon's Filtrasorb 400 carbon. The concentration range, equilibration time, and 95 percent confidence limits for the isotherms have been reported (Crittenden et al., 1985). Isotherms with F400 carbon plotted as volume adsorbed versus adsorption potential divided by the molar volume are shown on Fig. 15-9. The data essentially fall on a straight line with a small curvature at low adsorption potentials.

Table 15-7
Experimental and predicted Freundlich isotherm parameters for F400 carbon[a]

	Freundlich K Value [(mmol/kg) (m^3/mmol)^n]			1/n Value		
Compound	Experimental	Predicted	Percent Error	Experimental	Predicted	Percent Error
Chloroform	30.4	40.9	34.70	0.5325	0.5756	−8.09
cis-1,2-Dichloroethene	46.9	48.4	3.20	0.5562	0.6047	8.72
1,2-Dibromoethane	118.4	152.0	28.40	0.4808	0.5064	5.32
Bromoform	160.5	135.8	−15.40	0.5629	0.5148	8.55
Trichloroethene	191.9	186.1	2.92	0.4327	0.4857	12.20
Toluene	475.1	376.9	−20.70	0.3282	0.3901	18.90
Tetrachloroethene	435.2	452.2	3.90	0.3847	0.4069	5.77
Ethylbenzene	714.4	706.2	−1.15	0.2953	0.3211	8.74
o,p-Xylene	894.6	862.1	−3.63	0.2587	0.3026	17.00
n-Xylene	1044.0	865.2	−17.10	0.2458	0.3013	22.60

[a]Temperature = 10.0–13.8°C.
Source: Speth (1986).

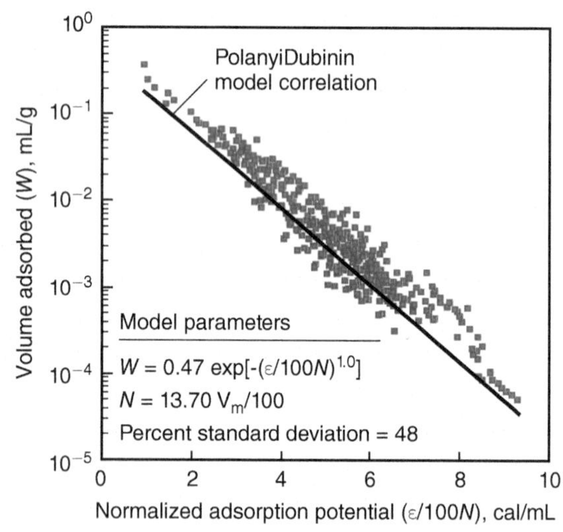

Figure 15-9
Correlation of aqueous adsorption isotherm data, using the Polanyi potential theory for isotherms, for halogenated aliphatic organic compounds on F-400 using molar volume as a normalizing factor. The best-fit line was obtained by weighting all of the data equally (adapted from Crittenden et al., 1999). The individual compounds studied are shown on Fig. 2 of the original paper, along with a discussion of appropriate methods of data analysis.

To demonstrate the utility of Eq. 15-16, the Freundlich K and $1/n$ values were calculated from the correlation curve and the liquid density and solubility of the solute. Based on the results presented in Table 15-7, if the data are fitted by Eq. 15-16, a reasonable approximation of the Freundlich parameters can be obtained from the correlation. When using the correlation, it is important to recognize that the Freundlich K and

$1/n$ values depend on the liquid-phase concentration; consequently, the appropriate concentration range must be used to estimate K and $1/n$. As expected, $1/n$ increases with decreasing concentration, and this is considered in Eq. 15-15.

For hydrophobic compounds, plotting the volume adsorbed versus adsorption potential divided by the molar volume may correlate isotherm data, and the error associated with estimating isotherm capacity may be generally low. In the case of other classes of compounds such as ketones and alcohols, the error may be larger and more testing may be required to obtain accurate predictions. The correlation shown on Fig. 15-9 is valid only for F-400 GAC. A similar approach has been used for other adsorbents, and some of the Polanyi parameters are summarized in Table 15-8.

DETERMINATION OF FREUNDLICH PARAMETERS USING POLANYI THEORY

For the correlation shown in Eq. 15-17, when the value of σ is 1.0, the following equations can be used to directly calculate the Freundlich single-solute K and $1/n$ parameters:

$$\frac{1}{n} = \frac{\beta RT\rho_s}{\text{MW}} \tag{15-18}$$

$$K = \frac{W_0\rho_s}{(C_s)^{1/n}} \tag{15-19}$$

where ρ_s = liquid density of adsorbate, g/L

MW = molecular weight of adsorbate, g/mol

R = universal gas constant, 8.314 J/mol · K

T = absolute temperature, K (273 + °C)

C_s = aqueous-phase solubility of adsorbate, mg/L

Table 15-8

Summary of various observed Polanyi potential parameters for selected adsorbents

Adsorbent	W_0, cm^3/g	β, mL/J$^\sigma$	σ
Ambersorb 563, Rohm and Haas	0.3815	0.001974	1.366
XAD-7, Rohm and Haas	0.3988	0.06580	0.8943
XAD-4, Rohm and Haas	1.48	0.02978	1.016
580–26, Barneby Suttcliffe	1.944	0.01468	1.117
Filtrasorb 300 (8 × 30), Calgon Carbon	0.172	0.01772	1.00
Filtrasorb 400 (12 × 40), Calgon Carbon	0.63	0.00490	1.208
APA, Calgon Carbon	1.53	0.01020	1.169
WV-G (12 × 40), Westvaco	2.994	0.02261	1.00

If the value of σ is not 1.0, then the following procedure must be used to determine the Freundlich K and $1/n$ parameters. For the aqueous-phase concentration range of interest, Eq. 15-16 is used to calculate the volume adsorbed for several liquid phase values in that range. The q values are determined by multiplying each value of W by the liquid density (make certain the mass units are consistent). Constructing a plot in which the values of C and q are fit using the Freundlich isotherm equation results in the parameters K and $1/n$. The procedure used to determine the isotherm parameters is demonstrated in Example 15-4.

Example 15-4 Estimate isotherm parameters using the Polanyi potential theory

Using Calgon Filtrasorb 400 GAC and a water temperature of 10°C, calculate the Freundlich isotherm parameters for TCE using Polanyi potential theory.

Solution

1. Summarize the available information. From Table 15-8, the Polanyi parameters for Calgon Filtrasorb 400 (12 × 40) are $W_0 = 0.63\ \text{cm}^3/\text{g}$, $\beta = 0.0049\ (\text{mL/J})^\sigma$, $\sigma = 1.208$. From a handbook of physical and chemical properties, the following properties for TCE were obtained: $V_m = 88.6$ mL/mol, $\rho_l = 1480\ \text{kg/m}^3$, and $C_s = 821$ mg/L.
2. Choose several TCE liquid-phase concentrations that span the desired range of liquid-phase concentrations that will be observed in the fixed-bed adsorber. Typically, a spread of two orders of magnitude on either side of the expected average influent is desirable. If the expected influent concentration is 1 mg/L, the following values for C are selected:

$$C = 0.01, 0.1, 1.0, 10, 100\ \text{mg/L}$$

3. Calculate the adsorption potential of the liquid-phase concentrations using Eq. 15-15. The required calculations for $C = 1$ mg/L are shown below:

$$\begin{aligned}\varepsilon &= RT \ln\left(\frac{C_s}{C}\right)\\ &= (8.314\ \text{J/mol}\cdot\text{K})(283\ \text{K}) \ln\left(\frac{821\ \text{mg/L}}{1.0\ \text{mg/L}}\right)\\ &= 15{,}789\ \text{J/mol}\end{aligned}$$

4. Calculate q using the V_m correlation shown in Eq. 15-16.
   a. Calculate ε/V_m for each adsorption potential. The required calculations for $C = 1$ mg/L are shown below:

$$\frac{\varepsilon}{V_m} = \frac{15{,}789 \text{ J/mol}}{88.6 \text{ mL/mol}} = 178.2 \text{ J/mL}$$

   b. Calculate the volume adsorbed for each liquid-phase concentration. The required calculations for $C = 1$ mg/L are shown below:

$$W = W_0 \exp\left[-\beta\left(\frac{\varepsilon}{V_m}\right)^{\sigma}\right]$$

$$= (0.63 \text{ cm}^3/\text{g}) \exp\left[-0.00490(\text{mL/J})^{1.208}(178.2 \text{ J/mL})^{1.208}\right]$$

$$= 0.0484 \text{ cm}^3/\text{g}$$

   c. Calculate q for each liquid-phase concentration. The required calculations for $C = 1$ mg/L are shown below:

$$q = W \times \rho_l$$

$$= (0.0484 \text{ cm}^3/\text{g})(1480 \text{kg/m}^3)\left(\frac{1\text{m}^3}{10^6 \text{cm}^3}\right)(10^6 \text{mg/kg})$$

$$= 71.6 \text{ mg TCE/g adsorbent}$$

5. Calculate the log of q and C. The required calculations for $C = 1$ mg/L are shown below:

$$\log(q) = \log(71.6) = 1.86 \qquad \log(C) = \log(1.00) = 0.0$$

6. Tabulate log q and log C. Make a table of several log q and log C values spanning at least two orders of magnitude on either side of the expected average liquid-phase influent concentration.

C, mg/L	**ε, J/mol**	**ε/V_m, J/mL**	**W, cm^3/g**	**q, mg/g**	**log q**	**log C**
0.01	26,624	300.5	0.0051	7.5	0.87	−2
0.1	21,207	239.4	0.0161	23.9	1.38	−1
1	15,789	178.2	0.0484	71.6	1.86	0
10	10, 371	117.1	0.1344	199.0	2.30	1
100	4,954	55.9	0.3346	495.3	2.69	2

7. Construct a plot of log q as a function of log C. The required plot is shown below:

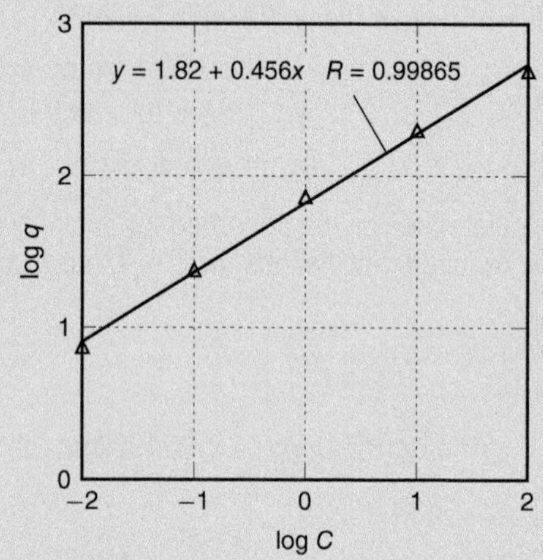

8. Determine the Freundlich isotherm parameters by linear regression:

$$\text{Slope} = \frac{1}{n} = 0.46 \quad Y \text{ intercept} = 1.82 = \log K$$

$$K = 66.1\ \text{mg/g (L/mg)}^{1/n}$$

Comment

The TCE isotherm values calculated using the Polanyi potential theory [$1/n = 0.46$, $K = 66.1$ mg/g (L/mg)$^{1/n}$] compare favorably with the values determined from the isotherm data in Example 15-3 [$1/n = 0.43$, $K = 60.7$ mg/g (L/mg)$^{1/n}$].

Multicomponent Equilibrium

In water treatment, the ideal case of one adsorbate being removed onto an adsorbent is seldom encountered, and the objective of adsorption in most real systems is to remove several adsorbates. This complicates both the theoretical picture of adsorbates in equilibrium with an adsorbent and the ability of the engineer to apply the theory to practice. The following theory and discussion can elucidate the phenomena of multicomponent adsorption.

Ideal adsorbed solution theory (IAST) has been used successfully to describe the competitive interactions of adsorbates on the surface of adsorbents. To begin, consider the fact that adsorption at an interface causes a reduction in surface tension. (Surface tension has the units of surface energy per area.) According to Gibbs (1906), this change in surface tension, which represents the two-dimensional work of adsorption, can be related to the amount adsorbed.

$$-A\,d\sigma_i = A\,d\pi_i = q_i RT\,d\left(\ln C_i\right) \tag{15-20}$$

where A = adsorption area per mass unit of adsorbent
$d\sigma_i$ = change in surface tension due to adsorption
$d\pi_i$ = change in the spreading pressure due to adsorption
$d\ln(C_i)$ = change in the natural log of concentration

The spreading pressure is defined as the difference between the surface tension, σ_i, that exists between the interfaces of solid with the pure solvent and with the solution containing the adsorbate (termed sorptive solution below). In the case of an aqueous solution,

$$d\pi_i = -\,d\sigma_i = \sigma_{\text{Pure Water/Adsorbent}} - \sigma_{\text{Sorptive Solution/Adsorbent}} \tag{15-21}$$

In a sense, the spreading pressure is the tendency of the adsorbate to spread out on the adsorbent surface and lower the surface tension (energy per area) or increase the spreading pressure. Unfortunately, Eq. 15-21 cannot be used to develop single-solute isotherms because it cannot quantify the change in the surface tension at the solution/adsorbent interface. (However, Eq. 15-21 can be used to estimate how much surfactant will concentrate at the air–water interface because we can measure the change in surface tension due to the addition of a surfactant.) The Gibbs isotherm can be used to extend single-solute isotherms to describe multicomponent isotherms using IAST because one can estimate the change in spreading pressure from single-solute isotherms. The IAST makes the following assumptions: (1) single-component concentrations, C_i°, are in equilibrium with the spreading pressure of the mixture and (2) the single-component concentration, C_i, that is in equilibrium with the mixture is equal to the product of the mole fraction of component i in the mixture and $\overset{\circ}{C_i}$. This is a surface—solution version of Raoult's law. The five basic equations for IAST are:

$$q_T = \sum_{i=1}^{N} q_i \tag{15-22}$$

$$z_i = \frac{q_i}{q_T} \tag{15-23}$$

$$C_i = z_i C_i^\circ \tag{15-24}$$

$$\frac{1}{q_T} = \sum_{i=1}^{N} \frac{z_i}{\overset{\circ}{q_i}} \tag{15-25}$$

$$\frac{\pi_m A}{RT} = \frac{\pi_i A}{RT} = \int_0^{q_i^\circ} \frac{d \ln C_i^\circ}{d \ln q_i^\circ} dq_i^\circ \tag{15-26}$$

where q_T = total surface loading
q_i = single component solid phase loading
z_i = mole fraction of component i on the adsorbent surface
C_i = concentration of solute i in multicomponent system, mg/g
C_i° = concentration of solute i in single solute system, mg/g
A = adsorption area per mass unit of adsorbent, m^2/g
π_i = spreading pressure of component i, Pa

Equation 15-24 is analogous to Raoult's law because the mixture concentration is equal to the surface mole fraction times the single component concentration, which is identical to the spreading pressure of the mixture; q_i° is the single component solid-phase loading that corresponds to the spreading pressure of the mixture. Equation 15-25 states that there is no area change per mole upon mixing in the mixture as compared to the single-solute isotherms, which are evaluated at the spreading pressure of the mixture. Equation 15-26 equates the spreading pressures of the pure component systems to the spreading pressure of the mixture.

If the Freundlich isotherm equation is used to represent single-solute behavior in Eq. 15-26, then the following expression can be obtained:

$$n_i q_i = n_j q_j \quad j = 2 \text{ to } N \tag{15-27}$$

By means of algebraic manipulation, the IAST for a Freundlich isotherm single-solute isotherm is obtained:

$$C_i = \frac{q_i}{\sum_{j=1}^{N} q_j} \left[\frac{\sum_{j=1}^{N} n_j q_j}{n_i K_i} \right]^{n_i} \tag{15-28}$$

The following expression is another useful form of IAST, which may be derived from Eq. 15-28 by assuming all the $1/n$ values are identical:

$$q_i = K_i^n C_i \left(\sum_{j=1}^{N} K_j^n C_j \right)^{1/n-1} \tag{15-29}$$

Crittenden et al. (1985) have compared Eq. 15-28, combined with the following mass balance equation, to data for mixtures of two to six chlorinated aliphatic compounds:

$$q_i = \frac{V}{M}(C_{i,0} - C_i) \tag{15-30}$$

To visualize the difference between the predictions and data, the worst and best predictions are displayed on Figs. 15-10a and 15-10b for trichloroethene and chloroform, respectively.

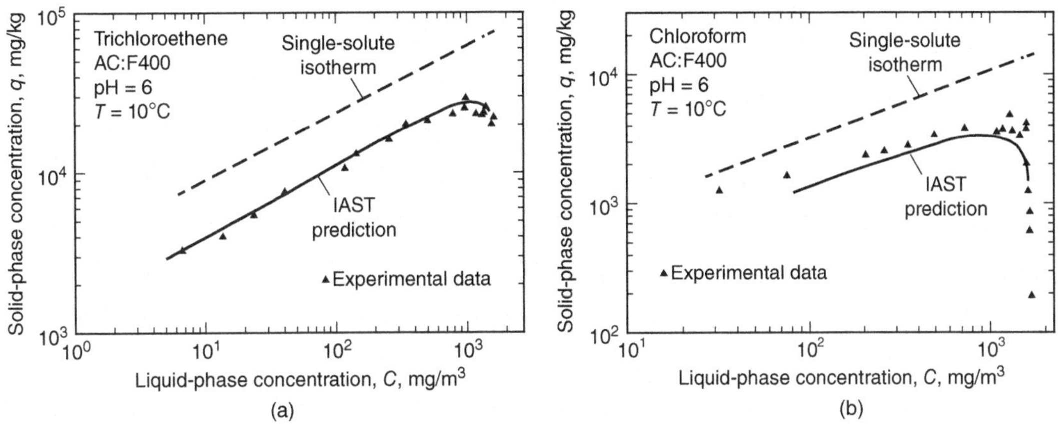

Figure 15-10
Trichloroethene and chloroform single-solute isotherms, isotherms in a six-component mixture, and IAST predictions.

Comparing the dashed lines, which represent the single-solute isotherms to the multicomponent data and predictions, shows the effect of adsorption competition on the adsorption capacity of a particular solute. This displacement is the greatest at high liquid-phase concentrations, that is, at small carbon dosages, because high solid-phase concentrations of the strongly adsorbing components cause more competitive interactions.

Dubinin–Radushkevich Correlation for Air Stripping Off-Gases

One technology that may be used for control of off-gases from stripping operations is gas-phase GAC. The Dubinin–Radushkevich (DR) equation, derived from the Polanyi potential theory, has been shown to describe adsorption isotherms of volatile organic compounds for GAC. Many researchers have shown that this form of the DR equation may be used for correlating data of VOCs:

$$W = W_0 \exp\left(\frac{-B\varepsilon^2}{\mu^2}\right) \tag{15-31}$$

$$\varepsilon = RT \ln\left(\frac{P_s}{P}\right) \tag{15-32}$$

where W = volume of adsorbate adsorbed on adsorbent, mL adsorbate/g adsorbent
W_0 = maximum volume of adsorbate adsorbed on adsobent, mL adsorbate/g adsorbent
B = microporosity constant, m^6/J^2
μ = polarizability, $C \cdot m^2/V$
ε = adsorption potential, equal to minimum energy to remove adsorbate from air and form a liquid or solid precipitate, J/mol

T = absolute temperature, K(273+°C)
P_s = vapor pressure of adsorbate at T, Pa
P = partial pressure of adsorbate in gas, Pa

Polarizability appears to be the best normalizing constant for gas-phase adsorption because VOC adsorption appears to be governed primarily by van der Waals force. Unlike adsorption from the aqueous phase, the solvent does not have to be displaced from the adsorbent surface; consequently, polarizability, not molecular size, appears to be the most important factor.

For nonpolar compounds, polarizability can be estimated from the refractive index using the Lorenz–Lorentz equation:

$$\mu = \frac{(\eta^2 - 1)\,\text{MW}}{(\eta^2 + 2)\,\rho_l} \qquad (15\text{-}33)$$

where η = refractive index, unitless
MW = molecular weight, g/mol
ρ_l = liquid density of VOC, g/L

Because W_0 and B in the DR equation are dependent only on the nature of the adsorbent, the relationship between W and ε/μ is unique for a given adsorbent. The plot of W versus ε/μ is called the "characteristic curve" and is defined by constant values of W_0 and B for a given adsorbent. Once the values of W_0 and B are determined experimentally for an adsorbent, the DR equation 15-31 may be used to predict the adsorption capacity or isotherm for other compounds for that adsorbent. The DR parameters for some widely used gas-phase adsorbents are summarized in Table 15-9.

There are some limitations to the DR equation (Reucroft et al., 1971): (1) It cannot be used to describe equilibrium at relative pressures P/P_s greater than 0.2 because capillary condensation occurs; (2) it only works for molecules with permanent dipole moments less than 2 debyes (D) because, if the molecule has a larger dipole moment, then dipole induction may be the principal adsorption force rather than van der Waals forces; and (3) it is only valid for relative humidity (RH) values less than when water vapor begins to adsorb, which corresponds to an RF of 40 to 50 percent.

Water vapor can have a large influence on the adsorption capacity of adsorbents such as GAC. When the concentration of the water in the gas

Table 15-9
W_0 and B values for various adsorbents

Adsorbent Type	W_0, mL/g	B, cm^6/J^2
Calgon BPL (4 × 6 mesh)	0.515	1.9898×10^{-6}
Calgon BPL (6 × 16 mesh)	0.460	1.8478×10^{-6}
CECA GAC-410G	0.503	1.300×10^{-6}

stream is high, a phenomenon called capillary condensation takes place where the water vapor will begin to condense in the micropores of the adsorbent. The polar oxygen-containing functional groups, which were discussed above, are responsible for attracting water by dipole–dipole interactions. The condensed water will then reduce the adsorption capacity for VOCs because water that undergoes dipole induction with the carbon surface must be displaced from the surface. For this reason aqueous-phase capacities are about a factor of 10 lower than that of the gas phase. If the off-gas from an air stripper has a relative humidity of 100 percent and is fed to a GAC adsorber, then the pores will fill completely with water and the capacity of the adsorbent will be no greater than that observed for aqueous-phase GAC systems. Furthermore, the rate of adsorption will be reduced by a factor of 10^5 because it will proceed by diffusion in water-filled pores rather than by gas diffusion. The relative humidity of air stripping off-gases is usually 100 percent and must be reduced to less than 50 percent to obtain reasonable capacities by heating the off-gas.

Photographs of GAC adsorbers

http://www.stantec.com/wt3r/15-1.html

15-5 Powdered Activated Carbon

With a small particle size (mean size about 24 μm) PAC can be added to water at various locations in the water treatment process to provide time for adsorption to take place and then remove the PAC by sedimentation and/or filtration. The topics considered in this section include the uses of PAC in water treatment, the points of PAC application, and the determination of dosage and how it is related to percent removal.

Uses of PAC in Water Treatment

Powdered activated carbon is primarily used in the treatment of taste and odor compounds and the treatment of low concentrations of pesticides and other organic micropollutants. The convenience of PAC is that it can be employed periodically (when needed) in a conventional water treatment plant with minimum capital costs. For example, PAC can be used during summer months for surface water sources containing taste and odor compounds resulting from algal blooms. It can also be employed to remove chemical pollution (pesticides and herbicides) carried in spring runoff. Discussion of the point of PAC addition in a water treatment train is presented later in this section.

Experimental Methods for Determining PAC Dosages

Standard jar testing can be used to evaluate PAC addition in conventional water treatment facilities. Standard jar testing procedures, as discussed in Chap. 9, can be used to determine PAC dosages for use in conventional water treatment plants. A site-specific bench-scale protocol has been developed to predict full-scale PAC performance for geosmin and 2-methyl-isoborneol (MIB) removal (Graham et al., 2000). The steps involved in the protocol employ a standard jar test procedure. This protocol enables site-specific details of the water treatment plant to be incorporated into the testing procedure. Raw water from the plant is used to perform the jar tests and may be spiked if necessary. The water to be tested is poured into the jars, and the mixing velocities, timing of chemical addition, retention times, and dosages need to be closely mimicked in the jar test. Retention times are inversely related to plant flows and will be the smallest for the largest rates. In other words, PAC will have less contact time with odorous compounds, which will lead to less adsorption. The settling and filtration follow the plant-specific jar testing procedure.

The above protocol is used to develop dose–response curves to evaluate PAC dosages needed for taste and odor episodes. An example of dose–response curves obtained for five different PAC types removing 40 ng/L of odorants is presented on Fig. 15-11. The dose–response curves are given in terms of odorant percent removal as a function of PAC dosage. Five PAC dosages were used for each PAC to develop their dose–response curves. If the treatment objective is 80 percent removal of geosmin (8 ng/L), PAC type B provides the best result with a 34-mg/L PAC dosage.

Comparison of Carbon Usage Rates for PAC and GAC

To begin the discussion on PAC, it is instructive to compare the theoretical carbon usage rates of PAC and GAC. The following equation is a mass balance on the PAC and water shown on Fig. 15-12:

$$QC_{\text{inf}} = q\dot{M}_{\text{PAC}} + QC_{\text{eff}} \tag{15-34}$$

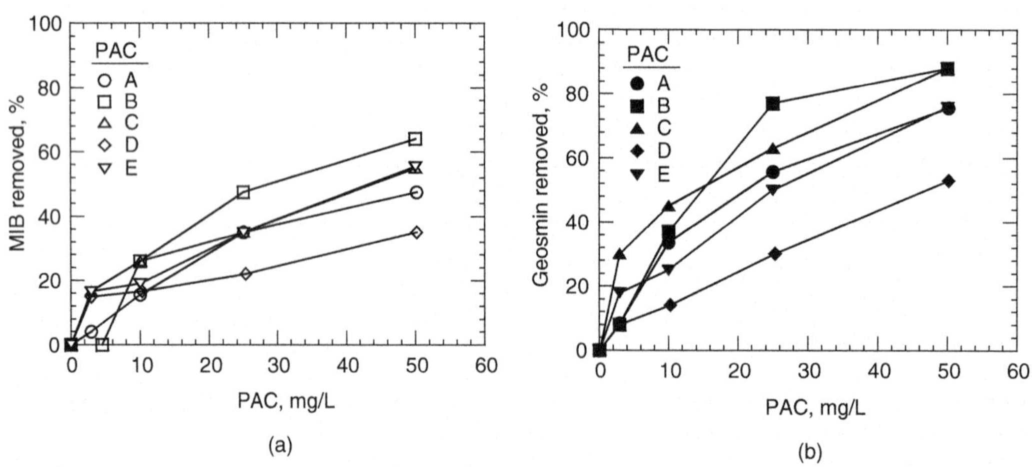

Figure 15-11
Percent removal of MIB and geosmin using Manatee Lake water and testing protocol and 40-ng/L initial contaminant concentrations. Letters A through E correspond to carbons shown in Table 15-10. (Adapted from Graham et al., 2000).

where Q = water flow rate, L/d
C_{inf} = influent liquid-phase concentration of adsorbate, mg/L
C_{eff} = effluent liquid-phase concentration of adsorbate (should meet the treatment objective, C_{to}), mg/L
q = adsorbent-phase concentration, mg adsorbate/g PAC
$\dot{M}_{PAC}$ = mass of PAC added per unit time, g/d

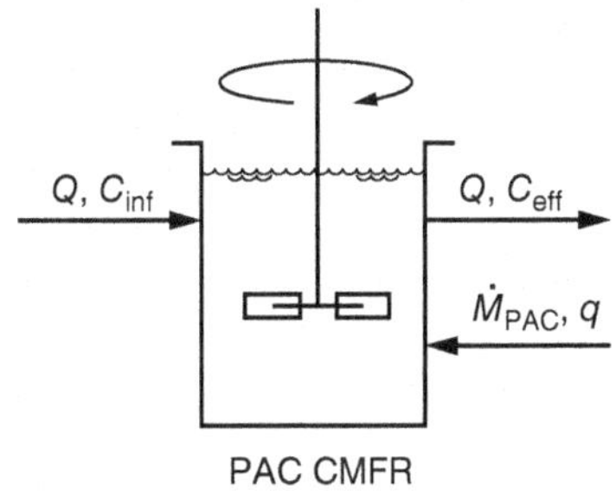

Figure 15-12
Sketch of CMFR PAC reactor.

If the carbon leaving the process is in equilibrium with the exiting water, then the following expression is obtained:

$$QC_{inf} = q_e|_{C_{eff}}\, \dot{M}_{PAC} + QC_{eff} \tag{15-35}$$

where $q_e|_{C_{eff}}$ = adsorbent-phase concentration of adsorbate in equilibrium with C_{eff}, mg adsorbate/g adsorbent

The required PAC dosage is given by the expression

$$D_{PAC} = \frac{\dot{M}_{PAC}}{Q} = \frac{C_{inf} - C_{eff}}{q_e|_{C_{eff}}} \tag{15-36}$$

where D_{PAC} = powered activated carbon dosage, g/L

The concentration profile for a single adsorbate in a GAC bed where mass transfer is present is shown on Fig. 15-13. If the mass transfer zone is very

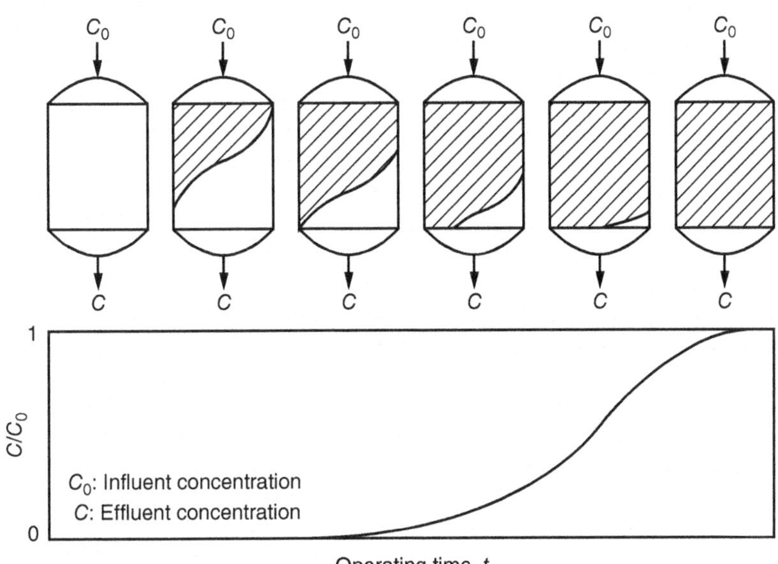

Figure 15-13
Concentration profiles and breakthrough curves for granular activated carbon columns (Based on Vermeulen, 1958).

small, then no adsorbate will appear in the effluent prior to complete exhaustion of the bed and the GAC will be at equilibrium with the influent concentration. A mass balance for this condition is given by the expression

$$t_{ex} Q C_{inf} = q_e|_{C_{inf}} M_{GAC} \tag{15-37}$$

The GAC carbon dosage is then given by

$$D_{GAC} = \frac{M_{GAC}}{t_{ex} Q} = \frac{C_{inf}}{q_e|_{C_{inf}}} \tag{15-38}$$

where D_{GAC} = granular activated carbon dosage, g/L
M_{GAC} = mass of GAC, g
t_{ex} = time to GAC exhaustion, d
$q_e|_{C_{inf}}$ = adsorbent-phase concentration of adsorbate in equilibrium with influent concentration, mg adsorbate/g adsorbent

If adsorption equilibrium can be described by the Freundlich equation, then the ratio of the PAC dosage to GAC dosage can be compared and depends only on $1/n$:

$$q_e = K C_e^{1/n} \tag{15-39}$$

$$\frac{D_{PAC}}{D_{GAC}} = \frac{1-(C_{eff}/C_{inf})}{q_e|_{C_{eff}}/q_e|_{C_{inf}}} = \frac{1-(C_{eff}/C_{inf})}{(C_{eff}/C_{inf})^{1/n}} \tag{15-40}$$

where C_e = equilibrium liquid-phase concentration, mg/L

The ratio of the PAC to GAC dramatically increases for a higher percentage removal, but the increase decreases as the value of $1/n$ decreases, as shown on Fig. 15-14, because the PAC is in equilibrium with the effluent concentration and the GAC is in equilibrium with the influent concentration. The difference in capacity is much less as $1/n$ becomes smaller. Further, GAC and PAC are countercurrent and co-current processes, respectively (see Chap. 7). As a point of reference, most $1/n$ values are around 0.5 to 0.7 for the compounds that are considered for removal using adsorption. Thus, if removals of less than 95 percent are required and the problem is seasonal, PAC may be the most economical solution. It should be noted that the curves shown on Fig. 15-14 apply to organic-free water and that the presence of NOM in natural waters can reduce the adsorption capacity of GAC and PAC significantly in a number of different ways.

Factors That Influence PAC Performance

As stated previously, one of the most common uses of activated carbon is the removal of taste and odor compounds. Taste and odor outbreaks are seasonal, and, according to a recent survey in North America, outbreaks

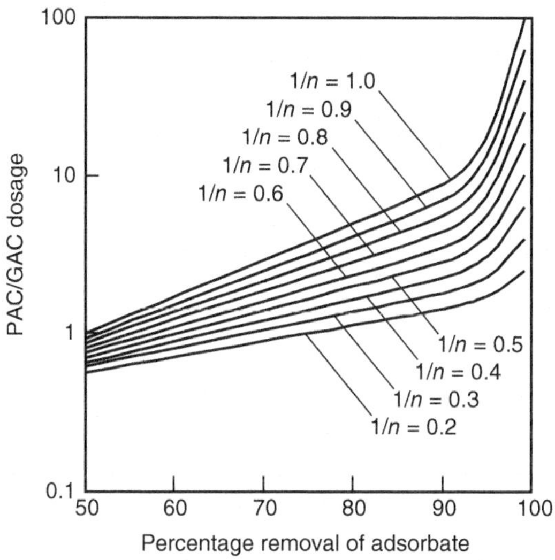

Figure 15-14
Comparison of adsorption capacity for PAC and GAC.

usually occur between June and October (Graham et al., 2000). The two principal odor-causing compounds that are not removed by chlorine are geosmin and MIB. Cyanobacteria are thought to produce and release these compounds into the water. Reported odor threshold concentrations for geosmin and MIB are 4 and 9 ng/L (McGuire et al., 1981). Accordingly, the treatment objective for these compounds must be below these threshold concentrations.

Powdered activated carbon is added to water as a suspension using adsorbent doses in the range of 5 to 25 mg/L. According to a 1994 survey of U.S. water utilities, 90 percent of the plants surveyed used a dosage between 0.5 and 18 mg/L and the average dosage was 5.1 mg/L (Graham et al., 2000). The efficacy of the PAC process is dependent upon these variables: type of PAC, PAC dosage, location of PAC addition, contact time, and presence of competing compounds and oxidants. These variables and their impact on PAC performance are discussed below.

TYPE OF PAC

The significant properties of five commercially available brands of PAC are reported in Table 15-10. There is considerable variability in the physical and chemical parameters. Studies have shown that these properties do not correlate well with the removal of odorants such as geosmin and MIB (Graham et al., 2000). However, these parameters are useful for quality assurance and control during PAC production and are used to control the activation and carbonization steps during manufacture.

The results of a study on the removal of geosmin and MIB using the brands of PAC in Table 15-10 was shown on Fig. 15-11. At a PAC dose of 50 mg/L, the percent removal for MIB ranged from about 35 percent to

Table 15-10
Reported values for five commercially available powdered activated carbons

PAC Type and Code	PAC-20B Atochem, A	Nuchar SA-20 Westvaco, B	HydroDarco B American Norit, C	WaterCarb Acticarb, D	WPL Calgon, E
Carbon source	Bituminous coal	Wood	Lignite	Wood/bark/ flyash primarily soft pine	Bituminous coal
Activation method	Steam	Phosphoric acid and steam	Steam –1000°C, rotary kiln	Steam 1600°C	Steam 800–1050°C, ground F300
Iodine number, mg/g	848	1040	547	604	897
Tannin value	750	30	281	1115	952
Molasses number	204	1076	286	113	179
Molasses decolor index	4.72	25.17	16.15	0.70	4.32
Phenol value	2.4	5.1	3.5	2.0	2.4
Percent ash, %	11.2	4.5	28.7	5.6	6.6
Pore volume, mL/g	0.494	1.258	0.555	0.280	0.214
Mean particle size, μm	28.9	46.27	23.44	48.54	21.36
Median particle size, μm	23.47	38.72	19.63	30.8	16.77
Modal particle size, μm	35.52	56.00	32.43	32.43	32.43

Source: Adapted from Graham et al. (2000).

over 60 percent, and the removal of geosmin ranged from about 50 percent to nearly 90 percent. Within experimental error, PAC B performed the best for this source water and these adsorbates. Due to the variability in performance of different adsorbents, an assortment of adsorbents should be evaluated for a specific application.

LOCATION OF PAC ADDITION

The most promising locations for the addition of PAC are (1) at the raw-water intake, (2) in the rapid-mix tank, and (3) in a slurry contactor (specially designed for PAC). The advantages and disadvantages of the common points of PAC addition are summarized in Table 15-11. With respect to the point of addition of PAC in a water plant, jar test studies optimizing PAC performance for taste and odor removal of MIB and geosmin show that PAC should be added before coagulation (termed precoagulation time).

DISINFECTANTS AND OXIDANTS

For the removal of MIB and geosmin, oxidants such as chlorine and potassium permanganate have a negative impact on PAC removal of taste and odor compounds. The impact is the greatest when the oxidant is added simultaneously with the PAC. For example, when the oxidant is

Table 15-11
Advantages and disadvantages of different points of addition of PAC

Point of Addition	Advantages	Disadvantages
Intake	Long contact time, good mixing	Interferes with preoxidation process (Cl_2 or $KMnO_4$). Some substances may be adsorbed that would otherwise probably be removed by coagulation, thus increasing carbon usage rate (this still needs to be demonstrated).
Rapid mix	Good mixing during rapid mix and flocculation, reasonable contact time	Interferes with preoxidation process (Cl_2 or $KMnO_4$). Possible reduction in rate of adsorption because of interference by coagulants; contact time may be too short for equilibrium to be reached for some contaminants; some competition may occur from molecules that would otherwise be removed by coagulation.
Filter inlet	Efficient use of PAC	Filter breakthrough, compromising finished water quality and making it difficult to meet turbidity requirements.
Slurry contactor preceding rapid mix	Excellent mixing for design contact time, no interference by coagulants, additional contact time possible during flocculation and sedimentation	A new basin and mixer may have to be installed; some competition may occur from the molecules that may otherwise be removed by coagulation.

Source: Adapted from Graham et al. (2000).

added with the PAC, removal efficiencies for MIB decrease by as much as 50 to 75 percent (Graham et al., 2000). Removal efficiencies for geosmin can decrease by as much as 20 to 40 percent (Graham et al., 2000). Consequently, it is recommended that PAC be added prior to the addition of oxidants or disinfectants.

NATURAL ORGANIC MATTER (NOM)

Most waters contain NOM and other organic compounds of anthropogenic origin, as described in Chap. 2. The NOM in water is comprised of thousands of different compounds. Higher-MW compounds will not compete with micropollutants for adsorption sites in smaller pores but can block the entrance to pores. However, micropollutants and smaller competing organics diffuse much faster than larger molecules and can diffuse into the smaller pores and compete for adsorption sites. By either pore blockage or competing for adsorption sites, NOM can reduce the adsorption capacity of micropollutants in PAC. Consequently, single-solute isotherms performed

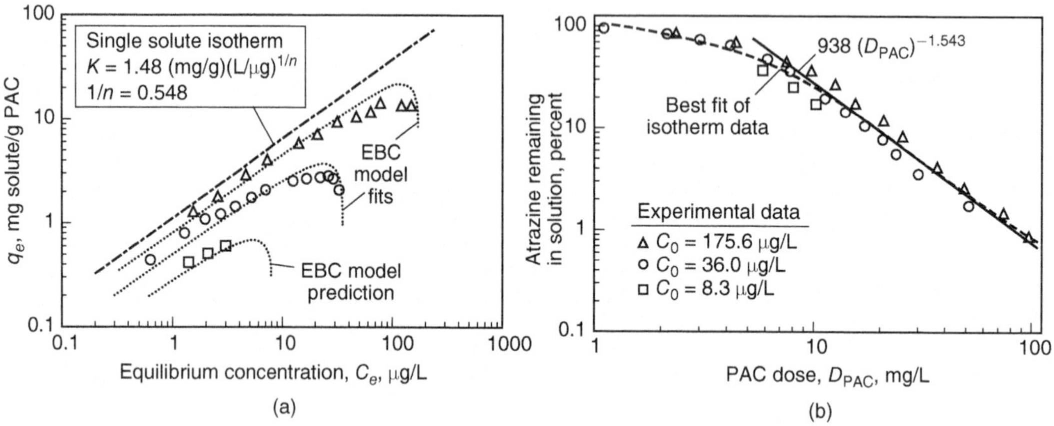

Figure 15-15
Atrazine isotherms for PICA B PAC conducted in organic-free water and Illinois groundwater (adapted from Knappe et al., 1998): (a) atrazine isotherm data and (b) atrazine isotherm data plotted as percentage remaining versus PAC dose.

in organic-free water will predict a higher capacity than would be observed for PAC dosages in natural waters containing NOM. The single-solute isotherm for atrazine is displayed on Fig. 15-15a with the isotherm for atrazine in the presence of a groundwater from a well located in Urbana, Illinois, for different initial atrazine concentrations. The DOC and pH of this water were 3 mg/L and 7.3, respectively. As the initial concentration of atrazine decreases, the impact of NOM is greater and the isotherm capacity for atrazine is less as compared to the single-solute capacity. This is the result of competitive interactions between the NOM and the micropollutant at low carbon dosages. At these low dosages, the strongly adsorbed species from the NOM are adsorbed first and are present at high surface concentrations. As a result, NOM has a greater impact on the micropollutant. As the dosage of PAC increases beyond this point, no additional strongly adsorbed species from the NOM are adsorbed. Consequently, the surface concentration of the strongly adsorbed species decreases, and its competitive impact decreases up to a point where the micropollutant isotherm begins to decrease in a manner that is similar to the single solute isotherm. The unusual feature of the isotherms is that the capacity is lower at high concentrations near the initial concentration and this can be described by IAST.

When the data on Fig. 15-15a are plotted in terms of percent atrazine remaining in solution as a function of PAC dosage, all the data correlate to the same line as shown on Fig. 15-15b. The percent atrazine remaining is only a function of PAC dosage and is independent of initial concentration. This result was demonstrated for several PACs, adsorbates, and natural waters containing NOM (Campos et al., 2000c; Gillogly et al., 1998; Knappe et al., 1998). Consequently, if an equilibrium isotherm test is conducted for a given initial concentration and the percentage removals for various PAC dosages are determined, then percentage removal for other initial concentrations may be determined from this result. The constant percent reduction that is observed in Fig. 15-15b can be derived from ideal adsorbed solution theory, as shown below.

As discussed above, multicomponent interactions can be predicted using the ideal adsorbed solution theory (IAST), assuming that the entire adsorbent surface is equally available to all solutes. If the Freundlich equation describes the single-solute isotherm, then multicomponent equilibrium interactions can be described using the following equation, which was derived from IAST:

$$C_{i,e} = \frac{q_{i,e}}{\sum_{j=1}^{N} q_{j,e}} \left(\frac{\sum_{j=1}^{N} n_j q_{j,e}}{n_i K_i} \right)^{n_i} \tag{15-41}$$

where $C_{i,e}$ = liquid-phase equilibrium concentration of component i
$q_{i,e}$ = solid-phase equilibrium concentration of component i
n_i = inverse of slope of the single-solute isotherm data on a $\log(q_{i,e})$ versus $\log(C_{i,e})$ graph
K_i = Freundlich single-solute capacity term for component i.
N = number of components in solution

A mass balance that was written on the PAC process is rewritten here in terms of the nomenclature that is used here:

$$QC_{i,0} = q_{i,e}\dot{M} + QC_{i,e} \tag{15-42}$$

$$C_{i,0} - q_{i,e}D_0 - C_{i,e} = 0 \tag{15-43}$$

where $C_{i,0}$ = influent liquid-phase equilibrium concentration of component i;
Q = flowrate of water, L/s
$\dot{M}$ = PAC feed rate, mg/s
$D_0 = \dot{M}/Q$ = dose of PAC, mg/L

Equation 15-43 can be combined with Eq. 15-41 to yield Eqs. 15-44 and 15-45 which can be used to predict the equilibrium solid-phase concentrations as a function of carbon dosage. Once $q_{i,e}$ is known, $C_{i,e}$ can be predicted from Eq. 15-43; $q_{m,e}$ and $q_{\text{EBC},e}$ are the the solid-phase concentrations of the micropollutant and equivalent background concentration (EBC) at equilibrium, respectively; $C_{m,0}$ and $C_{\text{EBC},0}$ are the liquid-phase concentrations of the micropollutant and EBC at equilibrium, respectively; K_{EBC} and $1/n_{\text{EBC}}$ are the Freundlich single-solute isotherm parameters for the EBC; and K_m and $1/n_m$ are the Freundlich single-solute isotherm parameters for the micropollutant, $q_{m,e}$ and $q_{\text{EBC},e}$. (This assumes that $K_{\text{EBC}}, 1/n_{\text{EBC}}, C_{\text{EBC},0}, C_{m,0}$ and D_0 are specified, and this is the case when K_{EBC}, $1/n_{\text{EBC}}$, and $C_{\text{EBC},0}$ are being determined.) Once $q_{m,e}$ and $q_{\text{EBC},e}$ are known, the equilibrium liquid-phase micropollutant concentration, $C_{m,e}$ and equilibrium liquidphase EBC concentration, $C_{\text{EBC},0}$, can be determined from these equations

$$C_{m,e} = \frac{q_{m,e}}{q_{m,e} + q_{\text{EBC},e}} \left(\frac{n_m q_{m,e} + n_{\text{EBC}} q_{\text{EBC},e}}{n_m K_m} \right)^{n_m} \tag{15-44}$$

$$C_{\text{EBC},e} = \frac{q_{\text{EBC},e}}{q_{m,e} + q_{\text{EBC},e}} \left(\frac{n_m q_{m,e} + n_{\text{EBC}} q_{\text{EBC},e}}{n_{\text{EBC}} K_{\text{EBC}}} \right)^{n_{\text{EBC}}} \tag{15-45}$$

As shown on Fig. 15-15, Knappe et al. (1998) fit the data initial concentrations of 175 and 36 μg/L and determined EBC values for initial concentration and Freundlich parameters. The EBC parameters were then used to predict the isotherm for an initial concentration of 8.3 μg/L. Accordingly, once the EBC properties are determined, they may be used to predict the isotherms at other concentrations.

It must be emphasized that the EBC properties only describe the impact of NOM on the absorbability of micropollutants, and they are not related to the absorbability of NOM. For example, Graham et al. (2000) demonstrated that DOC adsorption isotherms for NOM were not related to the impact of NOM on micropollutant removal. It is likely that the adsorbent surface is not available to all size fractions of the NOM due to its molecular weight distribution and the adsorbent pore size distribution.

Another limitation of the EBC method is that it is dependent on the target compound. For example, Speth and Adams (1993) used the EBC method to describe the adsorption isotherms for 22 compounds in Ohio River water, and they found that EBC parameters were very different depending on the compound. Accordingly, the EBC must be thought of as just a fitting exercise that can only describe the initial concentration impact of NOM on a given target compound.

The EBC is not really needed to describe the impact of NOM on the initial concentration because Eq. 15-44 can be simplified if we make the following assumptions: (1) $q_{m,e}$ is much less than $q_{\text{EBC},e}$; and (2) $1/n_m$ and $1/n_{\text{EBC}}$ are similar and between 0.1 and 1 (Knappe et al., 1998):

$$q_{m,e} = \frac{C_{m,0}}{D_0 + \dfrac{1}{q_{\text{EBC},e}} \left(\dfrac{n_{\text{EBC}} q_{\text{EBC},e}}{n_m K_m} \right)^{n_m}} \tag{15-46}$$

in which, $q_{m,e}$ = solid phase concentrations of the micropollutant
$q_{\text{EBC},e}$ = equivalent background concentration (EBC) at equilibrium.

$C_{m,0}$ = liquid phase concentrations of the micropollutant
K_m = Freundlich single solute isotherm capacity factor
n_m = Freundlich single solute isotherm intensity factor

Multicomponent isotherms can be predicted using Equation 15-46, if the initial concentrations and single solute isotherm parameters of all components are known. Natural organic matter is comprised of thousands of different compounds and some can block pores and not compete with micropollutants in smaller pores. However, micropollutants and smaller competing organics (from NOM) diffuse much faster than larger molecules and if this is ignored, then IAST has been used to describe competitive interactions between NOM and micropollutants. However, when it comes to the removal of micropollutants from water such as pesticides and taste and odor compounds, the concentrations and the Freundlich isotherm parameters are not known for the natural organic matter.

If we combine Eq. 15-45 with Eq. 15-42, this equation can be obtained:

$$\frac{C_{m,e}}{C_{m,0}} = \frac{\dfrac{1}{q_{\text{EBC},e}}\left(\dfrac{n_{\text{EBC}}q_{\text{EBC},e}}{n_m K_m}\right)^{n_m}}{D_0 + \dfrac{1}{q_{\text{EBC},e}}\left(\dfrac{n_{\text{EBC}}q_{\text{EBC},e}}{n_m K_m}\right)^{n_m}} \tag{15-47}$$

Knappe et al. (1998) demonstrated that for high PAC dosage that $C_{\text{EBC},e}$ is negligible and $q_{\text{EBC},e}$ is given by this expression:

$$q_{\text{EBC},e} = \frac{C_{\text{EBC},0} - C_{\text{EBC},e}}{D_0} \approx \frac{C_{\text{EBC},0}}{D_0} \tag{15-48}$$

Substituting Eq. 15-48 into Eq. 15-47 yields the final form of the equation that relates the percent removal as a function of PAC dosage:

$$\frac{C_{m,e}}{C_{m,0}} = \frac{\dfrac{1}{C_{\text{EBC},0}}\left(\dfrac{n_{\text{EBC}}C_{\text{EBC},0}}{n_m K_m}\right)^{n_n}}{D_0^{n_m} + \dfrac{1}{C_{\text{EBC},0}}\left(\dfrac{n_{\text{EBC}}C_{\text{EBC},0}}{n_m K_m}\right)^{n_m}} \tag{15-49}$$

Based on EBC values for atrazine that were reported by Campos et al. (2000c), the second term in the denominator of Eq. 15-49 is less than 1% of the first term and can be ignored:

$$\frac{C_{m,e}}{C_{m,0}} = \frac{\dfrac{1}{C_{\text{EBC},0}}\left(\dfrac{n_{\text{EBC}}C_{\text{EBC},0}}{n_m K_m}\right)^{n_m}}{D_0^{n_m}} \tag{15-50}$$

As predicted from Eq. 15-50, the percent reduction in the micropollutant is the same for a given PAC dosage, and it does not depend on the initial concentration of the micropollutant. This is what is seen in Fig. 15-15b. Snoeyink and co-workers (Campos et al., 2000c; Knappe et al., 1998; Gillogly et al., 1998) demonstrated this remarkably simple result is valid for several PACs, adsorbates, and natural waters containing NOM. Consequently, if an equilibrium isotherm test is conducted for a given initial concentration and the percentage removal for various PAC dosages are determined, then percentage removal for other initial concentrations may be determined from this result, as demonstrated in Example 15-5.

Seasonal variation in NOM and pH are two additional issues that can impact the removal of micropollutants. Atrazine isotherms that were conducted in distilled deionized water also did not show an impact for these pH values. Consequently, it would appear that pH values that are typical of finished water would not have an impact on the removal of micropollutants. It appears that the season does not change the competitive impact of NOM for Missouri River water even during spring flush. Much of the modeling work that has been presented was for bench-scale tests, and this begs the question as to whether bench-scale PAC tests can be used to predict full-scale performance. Based on the results presented earlier, Graham et al. (2000) developed the optimum scheme for PAC application given the unique features of the plants that were examined. These schemes were developed using bench tests, and the bench test agree with full-scale testing. In total, Graham et al. (2000) compared 140 full-scale data and similar agreement was found for all the data. It is surprising how close the bench tests are to the full-scale data because the bench test protocol used jar tests that simulated a plug flow PAC contactor. Clearly, bench-scale protocols can be used to predict full-scale performance.

CONTACT TIME

Typically, PAC added in a conventional plant has contact times between 0.5 and 2 h, which is not sufficient to utilize fully the capacity of the PAC for micropollutants. For example, it was reported that for 90 percent removal of atrazine the contact time could be decreased from 4 h to 30 min if the PAC dosage was increased from 23 to 32 mg/L (Gillogly et al., 1998). Other studies have shown similar results (Knappe et al., 1998).

The impact of MIB removal as a function of PAC dosage for various contact times is plotted on Fig. 15-16. As the contact time increases for a given removal efficiency, the PAC dosage decreases. For example, given an MIB removal efficiency of 90 percent (or 10 percent remaining) the PAC dosage for 7.5 min contact time is about 65 mg/L as compared to only about 25 mg/L for a contact time of 4 h. Jar tests can be used to simulate various contact times for various dosages for a given initial micropollutant concentration, if PAC contactor is a plug flow reactor.

Example 15-5 Adsorption of atrazine on PAC in natural waters

Isotherm experiments were conducted in bottles with three different initial concentrations to measure the adsorption isotherm of atrazine on PAC in a groundwater and the following data were obtained. Plot the percentage of atrazine remaining in the solution as a function of PAC dosage, and determine the PAC dosage corresponding to 90 percent removal of atrazine in a batch reactor for an initial concentration of 50 μg/L.

C_0, μg/L	PAC Dosage, mg/L	C_e, μg/L
8.0	1.0	4.72
	2.0	1.92
	3.0	1.04
35.0	0.3	34.5
	1.0	20.8
	3.0	4.50
	5.0	2.10
	10.0	0.76
	20.0	0.35
	50.0	0.09
100.0	2.0	23.5
	5.0	5.95
	10.0	2.25
	15.0	1.40

Solution

1. Calculate the percentage of atrazine remaining in the solution under various experimental conditions:

C_0, μg/L	PAC Dosage, mg/L	C_e, μg/L	$(C_e/C_0) \times 100\%$
8.0	1.0	4.72	59.0
	2.0	1.92	24.0
	3.0	1.04	13.0
35.0	0.3	34.5	98.6
	1.0	20.8	59.4
	3.0	4.50	12.9
	5.0	2.10	6.00
	10.0	0.76	2.17
	20.0	0.35	1.00
	50.0	0.09	0.25
100.0	2.0	23.5	23.5
	5.0	5.95	5.95
	10.0	2.25	2.25
	15.0	1.40	1.40

2. Plot the percentage of atrazine remaining in the solution as a function of PAC dosage on a log–log scale:

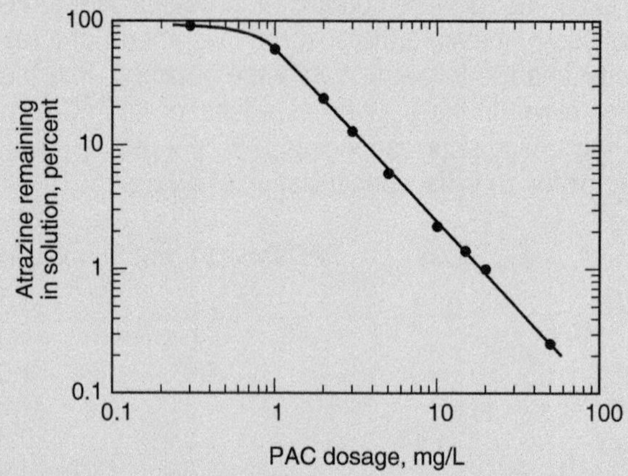

3. Determine the PAC dosage corresponding to 90 percent removal of atrazine in a batch reactor for an initial concentration of 50 μg/L. The micropollutant removal percentage at a given PAC dosage is independent of its initial concentration. Ninety percent atrazine removal corresponds to 10 percent remaining in the solution. Using the plot developed in step 2, 10 percent atrazine remaining in the solution requires a PAC dosage of 3.8 mg/L.

Use of PAC in Unit Operations

In addition to the application of PAC for control of seasonal water quality problems, PAC may be combined with specific unit processes for improved performance. More efficient methods for usage of PAC include the direct addition to floc blanket reactors or the use of PAC in conjunction with membrane processes.

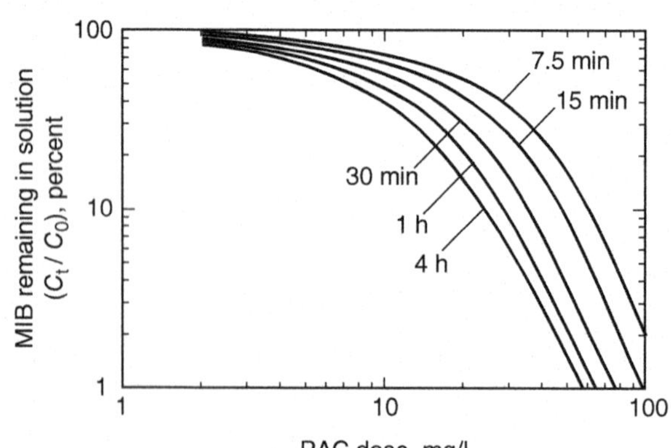

Figure 15-16
MIB remaining in solution as function of PAC dose and contact time.

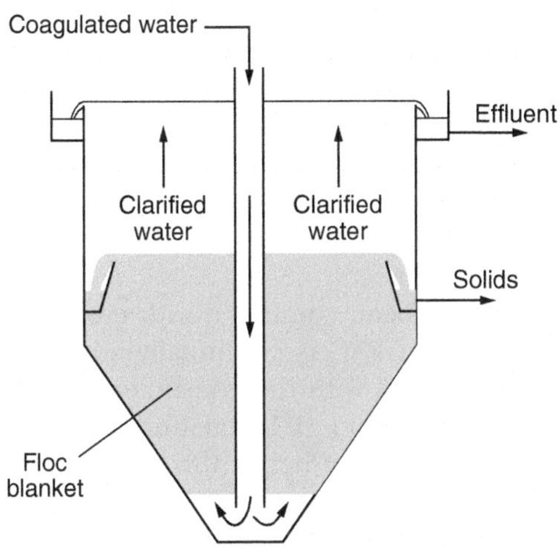

Figure 15-17
Floc blanket reactor for application in PAC systems (Adapted from AWWA, 1999).

PAC APPLICATIONS WITH FLOC BLANKET REACTORS

Floc blanket reactor (FBR) systems may be incorporated into conventional systems to provide for more efficient solid–liquid contact. A schematic of an upflow FBR is displayed on Fig. 15-17. The water containing PAC and coagulated particles is fed down the center of the clarifier and distributed in an upflow manner in the clarifier. Near the center of the clarifier, the sludge is trapped and withdrawn at a specified design rate while the clarifier-treated water continues to travel upward in the clarifier and is removed in the weir.

The detention time of PAC in FBRs may be determined from the following equation:

$$\text{CRT} = \frac{\rho_{\text{bl}} V_{\text{bl}}}{Q D_0} = \frac{\rho_{\text{bl}} H_{\text{bl}}}{v_u D_0} \tag{15-51}$$

where CRT = carbon residence time, h
ρ_{bl} = carbon density in blanket or blanket concentration, mg/L
V_{bl} = volume of floc blanket, L
v_u = upflow velocity or hydraulic loading in solids contactor, m/h
H_{bl} = depth of floc blanket, m
Q = water flow rate, L/d
D_0 = carbon dosage, mg/L

The PAC dosage, blanket depth, and v_u all impact the CRT. In full-scale operations, the typical blanket depth is 2 to 3m and the hydraulic loading ranges from 1.3 to 3 m/h (see Chap. 10). If the hydraulic loading is 1.5 m/h, the depth is 3 m, and PAC dosage is 10 mg/L, a blanket concentration of

120 mg/L would be required for a 24-h CRT based on Eq. 15-51. If the blanket were 2 m deep and the hydraulic loading were 4 m/h, a blanket concentration of 480 mg/L would be required and this may be infeasible. In practice, operators will not usually be able to adjust the hydraulic loading and depth, and floc blanket clarifiers require a fairly consistent water quality and flow rate (see Chap. 10).

PAC APPLICATIONS WITH MEMBRANE REACTORS

Powdered activated carbon treatment has been combined with cross-flow micro- or ultrafiltration membranes. Fresh PAC is continually added to the raw water, mixed to contact the particles with the constituents to be removed, and sent to a cross-flow ultrafiltration (UF) membrane where PAC particles are concentrated as the water is filtered through the UF membrane. The concentrated PAC solution is sent to waste and recycled back to the mixing basin. A number of full-scale plants operating in France have used this process (Anselme et al., 1997).

The PAC/UF technologies appear to have a great deal of potential for the reduction of DOC and DBPs. For example, the Vigneux-sur-Seine plant uses the PAC/UF process and, as shown in Table 15-12, significant removal of TOC, BDOC, and DBP was achieved after installation in 1998. The design parameters are reported in Table 15-13. The PAC/UF Vigneux-sur-Seine plant includes preozonation to break down organic matter and increase its adsorbability. Given the reliability, performance, and increasingly stringent treatment objectives, the PAC/UF process was considered to be the more viable alternative (Anselme et al., 1999).

Homogeneous Surface Diffusion Model

Adsorption kinetics for PAC systems can be quantified using the pore surface diffusion model (PSDM). As shown on Fig. 15-18, an adsorbate can diffuse from the bulk solution to the exterior surface of the PAC, which is

Table 15-12
Impact of PAC/UF process on Vigneux-sur-Seine finished water

Parameter	Unit	1997[a]	1998[b]	Reduction, %
TOC	mg/L	2.6	0.8	69
BDOC	mg/L	0.7	0.2	70
UV	OD/m[c]	2.4	0.8	67
THM	μg/L	73	8	89
Turbidity	NTU	0.1	0.1	0
HPC[d]	CFU/mL[e]	5	0	100

[a]Before PAC/UF installation.
[b]After PAC/UF installation.
[c]OD/m = optical density per meter at a wavelength of 254 nm.
[d]HPC = heterotrophic plate count.
[e]CFU/mL = colony-forming units per milliliter.
Source: Anselme et al. (1999).

Table 15-13
Key UF membrane properties and design parameters for the Vigneux-sur-Seine plant

Item	Unit	Value
General Design Specifications		
Water source, clarified Seine water		
Flow rate	m^3/d (mgd)	55,000 (14.5)
Average PAC dose	mg/L	8
Final chlorination dose (network residual)	mg/L	0.2
Treated-water turbidity	NTU	<0.1
Membrane Parameters		
Manufacturer, Aquasource Rueil, France		
Molecular weight cutoff	Da[a]	100,000
Membrane material, cellulosic derivative		
Internal fiber diameter	mm	0.93
Maximum recommended operating temperature	°C	30
pH range	Unitless	4–8.5
Recommended free chlorine during backwash	mg/L	3–5
Membrane configuration, inner skinned hollow fiber (inside out)		
Maximum recommended transmembrane pressure	bar (psi)	2 (29)
Maximum recommended backwash pressure	bar (psi)	2.5 (36)
Average clean-water flux	L/h · m^2 · bar (gfd/psi)	250 (10)
Number of racks	no.	8
Number of membrane modules per rack	no.	28
Total number of membrane modules	no.	224
Membrane surface area per module	m^2 (ft^2)	55 (590)
Production flux at 20°C	L/h · m^2	200
Backwash frequency	min	60
Feed water recovery	%	>95
Estimated chemical cleaning frequency	times/yr	4–6

[a]Da = Dalton.
Source: Adapted from Anselme et al. (1999).

called film diffusion. The adsorbate can then diffuse into the PAC particle by diffusing in the liquid in the pores, which is called pore diffusion, or the adsorbate can adsorb to the surface and then diffuse along the surface, which is called surface diffusion. The following expression can be used to determine the intraparticle flux:

$$J = -D_s \rho_a \frac{\partial q}{\partial r} - \frac{D_l \varepsilon_p}{\tau_p} \frac{\partial C_p}{\partial r} \tag{15-52}$$

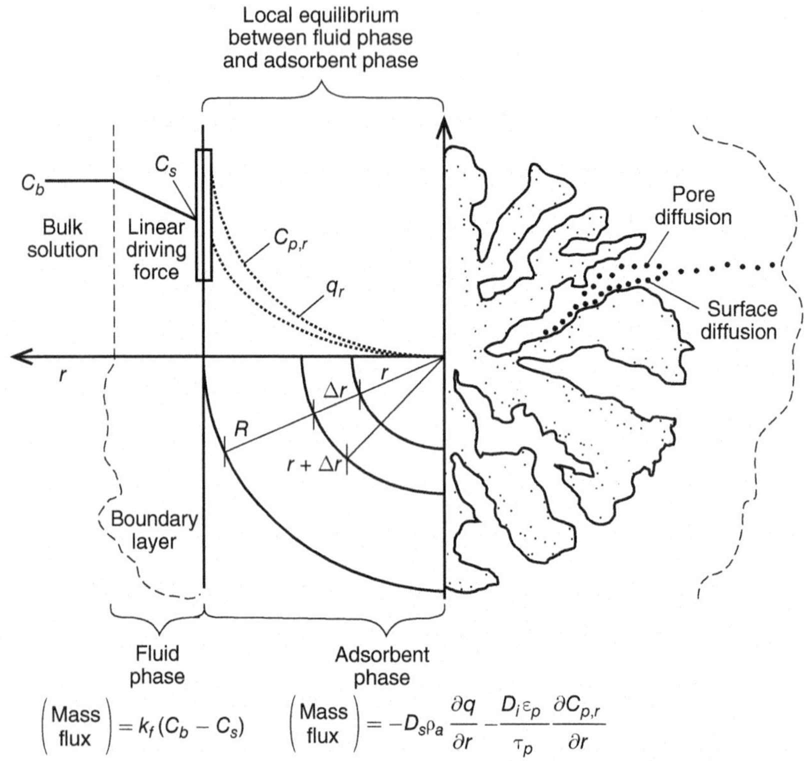

Figure 15-18 Mechanisms involved in adsorption kinetics.

where J = intraparticle flux, mg/m$^2 \cdot$ s
D_s = surface diffusion coefficient, m^2/s
D_p = pore diffusion coefficient, $D_l\varepsilon_p/\tau_p$, m^2 /s
ρ_a = adsorbent particle density (carbon mass divided by the total particle volume including pore volume), kg/m^3
D_l = liquid phase diffusion coefficient, m^2/s
ε_p = porosity of the particle, dimensionless
τ_p = tortuosity of the path that the adsorbate must take as compared to the radius, dimensionless
C_p = liquid-phase concentration of the adsorbate in the PAC pores, mg/L
q = adsorbent-phase concentration, mg adsorbate/g PAC
r = radial coordinate, m

Pore diffusion can be ignored in most cases with a single component because the pore concentration is small compared to the adsorbate surface concentration. When the model includes only surface diffusion, this model is called the homogeneous surface diffusion model (HSDM). The reason that the model is called the HSDM is that the surface diffusion coefficient and particle porosity are assumed to be isotropic throughout the particle. This assumption is clearly not completely correct because activated carbon has a distribution of pore sizes and various locations within PAC. For the HSDM, pore diffusion is ignored, and, if a mass balance is written on the spherical shell ignoring the second term in the right-hand side of Eq. 15-52, then the following expression may be written:

$$\text{In} - \text{out} = \text{accumulation} \tag{15-53}$$

$$-D_s\,\rho_a \left.\frac{\partial q}{\partial r}\right|_r 4\pi r^2 - \left[-D_s\,\rho_a \left.\frac{\partial q}{\partial r}\right|_{r+\Delta r} 4\pi(r+\Delta r)^2\right] = \rho_a 4\pi r^2 \Delta r \frac{\partial q}{\partial t} \tag{15-54}$$

Dividing Eq. 15-54 by $\rho_a 4\pi r^2\,\Delta r$ yields the following expression:

$$\frac{-D_s \left.\frac{\partial q}{\partial r}\right|_r 4\pi r^2 - \left[-D_s \left.\frac{\partial q}{\partial r}\right|_{r+\Delta r} 4\pi(r+\Delta r)^2\right]}{4\pi r^2 \Delta r} = \frac{\partial q}{\partial t} \tag{15-55}$$

If the limit as $\Delta r \rightarrow 0$ is taken, then the following expression is obtained.

$$\frac{D_s}{r^2}\frac{\partial}{\partial r}\left(r^2 \frac{\partial q}{\partial r}\right) = \frac{\partial q}{\partial t} \tag{15-56}$$

The model must be made dimensionless in order to provide general answers. The model is based on the following dimensionless variables:

$$\bar{r} = \frac{r}{R} \tag{15-57}$$

$$\bar{t} = \frac{tD_s}{R^2} \tag{15-58}$$

$$\bar{q} = \frac{q}{q_e} \tag{15-59}$$

where R = adsorbent particle radius, m
q_e = solid-phase concentration in equilibrium with C_e
t = elapsed time, min
$\bar{r}$ = dimensionless radial coordinate
$\bar{t}$ = dimensionless elapsed time
$\bar{q}$ = dimensionless adsorbent-phase concentration

Substitution of Eqs. 15-57, 15-58, and 15-59 into 15-56 yields the following dimensionless expression, which is known as Fick's second law in spherical coordinates:

$$\frac{1}{\bar{r}^2}\frac{\partial}{\partial \bar{r}}\left\{\bar{r}^2 \frac{\partial \bar{q}}{\partial \bar{r}}\right\} = \frac{\partial \bar{q}}{\partial \bar{t}} \tag{15-60}$$

There are two boundary conditions and one initial condition for Eq. 15-60. Because of symmetry, the first derivative of the solid-phase loading is equal to zero for dimensionless time values greater than or equal to zero:

$$\frac{\partial \bar{q}}{\partial \bar{r}} = 0 \tag{15-61}$$

As shown in Fig. 15-18, the other boundary conditions come from equating the exterior mass flux to the intraparticle mass flux:

$$k_f (C - C_s) = D_s \, \rho_a \frac{\partial q}{\partial r} \tag{15-62}$$

where k_f = external mass transfer coefficient, m/s
C = liquid-phase concentration of the adsorbate, mg/L
C_s = liquid-phase concentration of the adsorbate at the particle surface, mg/L

Converting Eq. 15-62 to dimensionless by substituting Eqs. 15-57, 15-58, and 15-59 into Eq. 15-62 and noting that $\bar{C}_s = C_s/C_0$ yields the second dimensionless boundary condition.

$$\text{Bi}_s \left(\bar{C} - \bar{C}_s \right) = \frac{\partial \bar{q}}{\partial \bar{r}} \tag{15-63}$$

$$\text{Bi}_s = \frac{k_f C_0 R}{D_s \rho_a q_e} = \frac{\text{external mass transfer rate}}{\text{surface diffusion intraparticle mass transfer rate}} \tag{15-64}$$

where Bi_s = Biot number
C_0 = initial liquid-phase concentration in the reactor at time zero, mg/L

The Biot number is a good indicator of which phase controls the rate of mass transfer. For low Bi_s numbers ($\text{Bi}_s < 1.0$), external mass transfer controls the adsorption rate. For large Bi numbers ($\text{Bi}_s > 30$) surface diffusion controls the adsorption rate. For Bi numbers between 1 and 30 both external and intraparticle mass transfer rates contribute to the adsorption rate.

The initial condition for Eq. 15-60 states that there is no adsorbate within the adsorbent:

$$\bar{q}(\bar{t} = 0, 0 \leq \bar{r} \leq 1) = 0 \tag{15-65}$$

Mixing conditions affect the liquid-phase mass balance. If it is assumed that the PAC moves along with the fluid and plug flow conditions prevail, then the liquid-phase mass balance is identical to that which is obtained for a completely mixed batch reactor, and time corresponds to the contact time of PAC with the water. A mass balance on the liquid phase that moves at the fluid velocity may be written as follows:

~~In~~ − ~~out~~ + ~~generation~~ − loss = accumulation (15-66)

$$-k_f a(C - C_s)V_p = V_p \frac{dC}{dt} \tag{15-67}$$

$$a = \frac{3\phi}{R}\frac{D_0}{\rho_a} = \frac{\text{area available for mass transfer}}{\text{solution volume}} \tag{15-68}$$

$$-\frac{3\phi k_f(1-\varepsilon_p)}{R}(C - C_s) = \frac{dC}{dt} \tag{15-69}$$

where a = area available for mass transfer per volume of solution, m^2/m^3
ϕ = particle sphericity, dimensionless
V_p = solution volume per particle, L/particle
D_0 = PAC dosage = $\rho_a(1 - \varepsilon_p)$

Applying the dimensionless variables to Eq. 15-69 yields the following dimensionless form of the equation:

$$\text{Sh}_s(\overline{C} - \overline{C}_s) = \frac{d\overline{C}}{d\overline{t}} \tag{15-70}$$

$$\text{Sh}_s = \frac{3\phi k_f(1-\varepsilon_p)R}{D_s} \tag{15-71}$$

where Sh_s = Sherwood number based on the surface diffusivity, dimensionless.

The initial condition for Eq. 15-70 is

$$\overline{C}(\overline{t} = 0) = 0 \tag{15-72}$$

The concentration on the adsorbed phase is represented by q and the liquid-phase concentration is represented by C. Equations 15-60 and 15-70 can be coupled to the Freundlich equation. When both phases are at local equilibrium, the Freundlich expression can be written as follows (by noting that the adsorbed-phase concentration at the exterior or the adsorbent is in equilibrium with the liquid-phase adsorbate concentration at the exterior of the adsorbent):

$$q(r = R,\ t) = KC_s^{1/n} \tag{15-73}$$

$$q_e = KC_e^{1/n} \tag{15-74}$$

$$\overline{q}(\overline{r} = 1, \overline{t}) = (\overline{C}_e)^{-1/n}\,\overline{C}_s^{1/n} \tag{15-75}$$

where $\overline{C}_s = C_s/C_0$
$\overline{C}_e = C_e/C_0$

For most PAC applications, neglecting the impact of external mass transfer in Eq. 15-67 can provide a good model/data comparison (Najm, 1996; Campos et al., 2000a). If the overall mass balance does not include external mass transfer, the following expression is obtained:

$$0 = \text{accumulation} = \text{final mass} - \text{initial mass} \tag{15-76}$$

$$0 = q_{\text{ave}} M_p + V_p C - V_p C_0 \tag{15-77}$$

where M_p = mass of PAC, g
q_{ave} = average adsorbent-phase concentration, mg/g

The average adsorbent-phase concentration can be calculated using the following equation:

$$q_{\text{ave}} = \frac{3}{4\pi R^3} \int_0^R q 4\pi r^2 dr \tag{15-78}$$

The final form of the overall mass balance in dimensionless form is developed by substituting Eq. 15-78 into Eq. 15-77, taking the derivative with respect to time, and inserting the dimensionless variables from Eqs. 15-57 to 15-59:

$$0 = \frac{3D_0}{4\pi R^3} \int_0^R q 4\pi r^2 dr + C - C_0 \tag{15-79}$$

$$\frac{3D_0 q_e}{C_0} \int_0^1 \frac{\partial \bar{q}}{\partial \bar{t}} \bar{r}^2 d\bar{r} = -\frac{d\bar{C}}{d\bar{t}} \tag{15-80}$$

Equation 15-79 may be simplified further by substituting the following equation into Eq. 15-79:

$$q_e = \frac{C_0 - C_e}{D_0} \tag{15-81}$$

$$\frac{d\bar{C}}{d\bar{t}} = -3\left(1 - \bar{C}_e\right) \int_0^1 \frac{\partial \bar{q}}{\partial \bar{t}} \bar{r}^2 d\bar{r} \tag{15-82}$$

The final set of dimensionless equations that ignore external mass transfer resistance and pore diffusion can be presented by noting that $\bar{C}^{1/n} = \bar{C}_s^{1/n}$ because there is assumed to be no concentration gradient in the liquid phase:

$$\frac{d\bar{C}}{d\bar{t}} = -3\left(1 - \bar{C}_e\right) \int_0^1 \frac{\partial \bar{q}}{\partial \bar{t}} \bar{r}^2 d\bar{r} \tag{15-83}$$

$$\frac{1}{\bar{r}^2} \frac{\partial}{\partial \bar{r}} \left\{ \bar{r}^2 \frac{\partial \bar{q}}{\partial \bar{r}} \right\} = \frac{\partial \bar{q}}{\partial \bar{t}} \tag{15-84}$$

$$\bar{q}(\bar{r} = 1, \bar{t}) = \left(\bar{C}_e\right)^{-1/n} \bar{C}^{1/n} \tag{15-85}$$

$$\frac{\partial \bar{q}}{\partial \bar{r}} = 0 \tag{15-86}$$

$$\bar{q}(\bar{t}=0, 0 \le \bar{r} \le 1) = 0 \tag{15-87}$$

$$\bar{C}(\bar{t}=0) = 0 \tag{15-88}$$

Inspection of Eq. 15-83 to 15-88 implies that only two parameters determine $\bar{C} = C/C_0$ at any $\bar{t} = D_s t/R^2$ (Hand et al., 1983). Consequently, all possible solutions can be presented here by varying $\bar{C}_e$ for several $1/n$ values. Figure 15-19 displays $(C - C_e)/(C_0 - C_e)$ as a function of time for $1/n$ values from 0.1 to 0.9 and a C_e/C_0 value of 0.1. $(C - C_e)/(C_0 - C_e)$ was used instead of C/C_0 because it tends to bring the curves together and make them easier to display and interpolate. The following equation was fit to the HSDM for various $1/n$ values, and may be used to analyze contaminant removal by PAC batch reactors and plug flow reactors with a given residence time.

$$\frac{C - C_e}{C_0 - C_e} = A_0 + A_1(\ln \bar{t}) + A_2(\ln \bar{t})^2 + A_3(\ln \bar{t})^3 \tag{15-89}$$

An example of the constants in Eq. 15-89 for a $1/n$ value of 0.2 are shown in Table 15-14 [constants for additional $1/n$ values are provided in Zhang et al. (2009) and the electronic Table E-6 at the website listed in App. E].

Steady state can be achieved in a completely mixed flow PAC contactor in approximately three hydraulic contact times or $\bar{t} = 3$, whichever is larger. Under steady-state conditions, the concentration of the micropollutant in water is a constant and the appropriate mass balance for the liquid phase is as follows:

$$\text{in} - \text{out} + \cancel{\text{generation}} - \cancel{\text{loss}} = 0 = \cancel{\text{accumulation}} \qquad \text{in} = \text{out} \tag{15-90}$$

$$QC_0 = QC(t) + q_{\text{ave}}\dot{M} \tag{15-91}$$

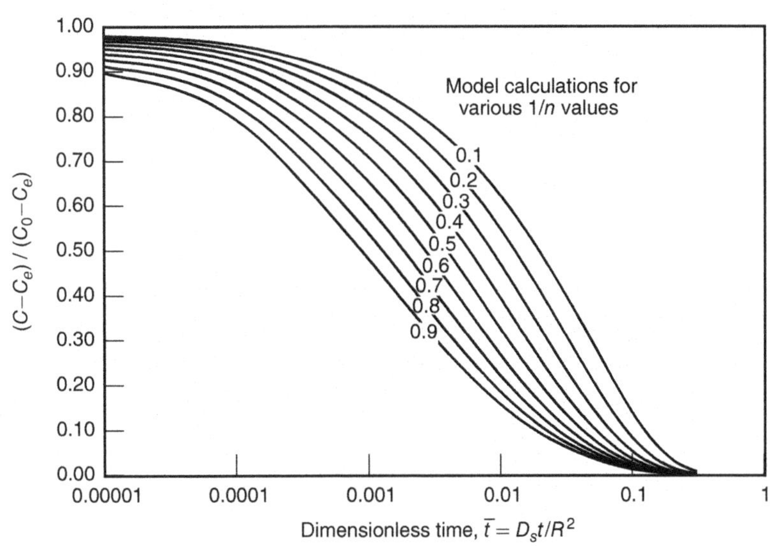

Figure 15-19
HSDM calculations for either a batch reactor or plug flow reactor with a similar detention time.

Table 15-14
Parameters used for the empirical equation that describe solutions to the HSDM for a batch reactor for Freundlich adsorption intensity parameter $(1/n) = 0.2$[a]

	Equation Coefficients				Equations Valid for the Following $\bar{t}$ Range	
C_e/C_0	A_0	A_1	A_2	A_3	Lower	Upper
0.001	1.17237	7.85118×10^{-1}	1.58099×10^{-1}	8.23844×10^{-3}	5.05×10^{-4}	3.50×10^{-2}
0.005	6.45253×10^{-1}	5.50680×10^{-1}	1.39005×10^{-1}	8.36855×10^{-3}	2.05×10^{-4}	8.50×10^{-2}
0.01	5.13173×10^{-1}	4.92388×10^{-1}	1.39344×10^{-1}	9.08314×10^{-3}	3.65×10^{-4}	1.10×10^{-1}
0.05	2.24322×10^{-1}	3.05970×10^{-1}	1.21216×10^{-1}	9.26458×10^{-3}	7.05×10^{-4}	2.00×10^{-1}
0.1	1.22475×10^{-1}	2.12696×10^{-1}	1.05034×10^{-1}	8.51555×10^{-3}	8.50×10^{-4}	3.00×10^{-1}
0.2	8.97853×10^{-2}	1.84347×10^{-1}	1.08293×10^{-1}	9.48046×10^{-3}	9.90×10^{-4}	3.00×10^{-1}
0.3	5.48862×10^{-2}	1.44294×10^{-1}	1.01682×10^{-1}	9.26875×10^{-3}	9.90×10^{-4}	3.00×10^{-1}
0.4	2.47976×10^{-2}	1.07206×10^{-1}	9.33323×10^{-2}	8.72285×10^{-3}	9.90×10^{-4}	3.00×10^{-1}
0.5	1.84338×10^{-3}	7.83684×10^{-2}	8.66479×10^{-2}	8.27250×10^{-3}	9.90×10^{-4}	3.00×10^{-1}
0.6	-1.62317×10^{-2}	5.53300×10^{-2}	8.12304×10^{-2}	7.90238×10^{-3}	9.90×10^{-4}	3.00×10^{-1}
0.7	-3.07867×10^{-2}	3.68066×10^{-2}	7.68789×10^{-2}	7.60532×10^{-3}	9.90×10^{-4}	3.00×10^{-1}
0.8	-4.28149×10^{-2}	2.13154×10^{-2}	7.32217×10^{-2}	7.35463×10^{-3}	9.90×10^{-4}	3.00×10^{-1}
0.9	-5.29236×10^{-2}	8.23859×10^{-3}	7.01365×10^{-2}	7.14335×10^{-3}	9.90×10^{-4}	3.00×10^{-1}

[a]Parameters for other values of 1/n are available in the electronic Table E-6 at the website listed in App. E.

where $\dot{M}$ = mass flow rate of PAC, g/min

$$\bar{C}(t) = 1 - \frac{q_{\text{ave}}}{C_0}\frac{\dot{M}}{Q} = 1 - \frac{q_e|_{C(t)}}{C_0}\bar{q}_{\text{ave}} D_0 \qquad (15\text{-}92)$$

where $\bar{q}_{\text{ave}}$ = dimensionless average adsorbed-phase concentration q_{ave}/q_e

q_e = solid-phase concentration in equilibrium with $C(t)$ and equals $[KC(t)^{1/n}]$, mg/g

Substituting $q_e = KC(t)^{1/n} = K\left[C_0\bar{C}(t)\right]^{1/n}$ and $\bar{t}$ into Eq. 15-92 yields a nonlinear equation in $\bar{C}(\bar{t})$ for the liquid-phase mass balance:

$$C_0[1 - \bar{C}(\bar{t})] - K\left[C_0\bar{C}(\bar{t})\right]^{1/n} \bar{q}_{\text{ave}} D_0 = 0 \qquad (15\text{-}93)$$

The appropriate mass balance for the adsorbent phase is Eq. 15-60:

$$\frac{1}{\bar{r}^2}\frac{\partial}{\partial \bar{r}}\left\{\bar{r}^2\frac{\partial \bar{q}}{\partial \bar{r}}\right\} = \frac{\partial \bar{q}}{\partial \bar{t}} \qquad (15\text{-}94)$$

The first boundary condition utilizes the fact that the solution concentration is equal to effluent concentration of the contactor:

$$\bar{q}(\bar{r} = 1, \bar{t}) = [\bar{C}(t)]^{1/n} = 1 \qquad (15\text{-}95)$$

The second boundary condition and initial condition for Eq. 15-93 are as follows:

$$\frac{\partial \bar{q}}{\partial \bar{r}} = 0 \qquad (15\text{-}96)$$

$$\bar{q}(\bar{t} = 0, 0 \le \bar{r} \le 1) = 0 \qquad (15\text{-}97)$$

Crank (1964) has provided a solution to this problem for the average solid-phase loading:

$$\bar{q}_{\text{ave}} = 1 - \frac{6}{\pi^2}\sum_{n=1}^{\infty}\frac{1}{n^2}\exp\left(n^2\pi^2\bar{t}\right) \qquad (15\text{-}98)$$

Traegner et al. (1996) noted that the PAC is completely mixed and the PAC has an exponential exit age distribution corresponding that of to a CMFR. The exit age distribution for a CMFR is discussed in Chap. 7. Using the CMFR exit age distribution and Eq. 15-98, Traegner et al. (1996) developed the following closed-form analytical solution:

$$\bar{q}_{\text{ave}} = 3\bar{t}\left[\frac{1}{\sqrt{\bar{t}}}\coth\left(\frac{1}{\sqrt{\bar{t}}}\right) - 1\right] \qquad (15\text{-}99)$$

To use these equations, first select a $\bar{t}$ and calculate $\bar{q}_{\text{ave}}$ using Eq. 15-99. Then use Eq. 15-93 to calculate $\bar{C}(\bar{t})$ for a given PAC dose.

Although the HSDM assumes that the surface diffusion coefficient and particle porosity are isotropic throughout the particle, the HSDM can predict the surface diffusion coefficient that is consistent with the experimental data. Example 15-6 demonstrates the HSDM user-oriented solutions with the batch rate data. Examples 15-7 and 15-8 demonstrate the use of PAC kinetic models using different reactors.

Example 15-6 Estimation D_s of using the HSDM

Given the following batch rate data for atrazine, initial atrazine concentration of 175.1 ng/L, PAC particle radius $R = 5\ \mu m$, and PAC dose of 11.5 mg/L, calculate D_s using the HSDM user-oriented solutions. The Freundlich isotherm parameters are $1/n = 0.216$ and $K = 10^{0.402} = 2.52$ (ng/mg)(L/ng)$^{1/n}$.

t (min)	C_t/C_0	t (min)	C_t/C_0
0	1.012	30	0.730
0	1.007	45	0.662
1.42	0.924	59.5	0.691
4.08	0.856	90	0.635
7.08	0.836	120	0.600
13.52	0.753	186	0.576
16.37	0.774	240	0.569

Solution

Estimation of D_s involves the following steps:

1. Calculate the equilibrium concentration C_e by equating Eq. 15-74 with 15-81:

$$q_e = \frac{C_0 - C_e}{D_0} = KC_e^{1/n}$$

$$\frac{175.1 - C_e}{11.5} = 2.52C_e^{0.216}$$

$$C_e = 97.2163$$

$$\frac{C_e}{C_0} = 0.5552$$

2. Calculate the dimensionless time $\bar{t}$ and dimensionless concentration $\bar{C}$ for the batch rate data given in the above table:

$$\bar{t} = \frac{tD_s}{R^2}$$

$$\bar{C}_{\text{data}} = \frac{C_t - C_e}{C_0 - C_e} = \frac{\dfrac{C_t}{C_0} - \dfrac{C_e}{C_0}}{1 - \dfrac{C_e}{C_0}}$$

3. Pick the appropriate empirical equation from Table 15-14 for $1/n = 0.216$ and $C_e/C_0 = 0.5552$. The appropriate equation for $1/n = 0.2$, $C_e/C_0 = 0.6$ in Table 15-14 was used to calculate the dimensionless concentration for the given time using Eq. 15-89:

$$\bar{C}_{\text{model}} = \frac{C(\bar{t}) - C_e}{C_0 - C_e} = A_0 + A_1(\ln \bar{t}) + A_2(\ln \bar{t})^2 + A_3(\ln \bar{t})^3$$

$C_e/C_0 = 0.6$:

$$A_0 = -1.6232 \times 10^{-2}$$

$$A_1 = 5.5330 \times 10^{-2}$$

$$A_2 = 8.1230 \times 10^{-2}$$

$$A_3 = 7.9024 \times 10^{-3}$$

4. Use Excel Solver to find the optimum D_s/R^2 by minimizing the objective function (OF) value, which can be calculated using the following equation:

$$\text{OF} = \sqrt{\frac{\sum_{i=1}^{n} \left(\frac{\bar{C}_{\text{data},i} - \bar{C}_{\text{model},i}}{\bar{C}_{\text{data},i}} \right)^2}{n-1}}$$

Excel table of HSDM user-oriented model fit and PAC data

t (min)	C_t/C_0 (data)	$\bar{t}$	$\bar{C}$ (data)	$\bar{C}$ (model)	C_t/C_0 (model)
0	1.012	0	—	—	1
0	1.007	0	—	—	1
1.42	0.925	0.00176	0.829315	8.84×10^{-1}	9.49×10^{-1}
4.10	0.856	0.00507	0.676721	7.94×10^{-1}	9.08×10^{-1}
7.10	0.8357	0.00880	0.630528	7.04×10^{-1}	8.68×10^{-1}
13.5	0.753	0.01680	0.444397	5.75×10^{-1}	8.11×10^{-1}
16.4	0.774	0.02930	0.491019	5.34×10^{-1}	7.93×10^{-1}
30	0.730	0.03730	0.393238	3.99×10^{-1}	7.33×10^{-1}
45	0.662	0.05590	0.240476	3.10×10^{-1}	6.93×10^{-1}
59.5	0.691	0.07390	0.304178	2.51×10^{-1}	6.67×10^{-1}
90	0.634	0.11200	0.178435	1.69×10^{-1}	6.31×10^{-1}
120	0.600	0.14900	0.101117	1.18×10^{-1}	6.08×10^{-1}
186	0.576	0.23100	0.046059	5.21×10^{-2}	5.78×10^{-1}
240	0.568	0.2982	0.03058	2.18×10^{-2}	5.65×10^{-1}

The optimum $D_s/R^2 = 1.2424 \times 10^{-3}\ \text{min}^{-1}$.

The final result for $D_s = 1.2424 \times 10^{-3}\ \text{min}^{-1} \times (0.0005\ \text{cm})^2/(60\ \text{s/min}) = 5.18 \times 10^{-12}\ (\text{cm}^2/\text{s})$ and a plot of the experimental data and model fit is shown in the following graph:

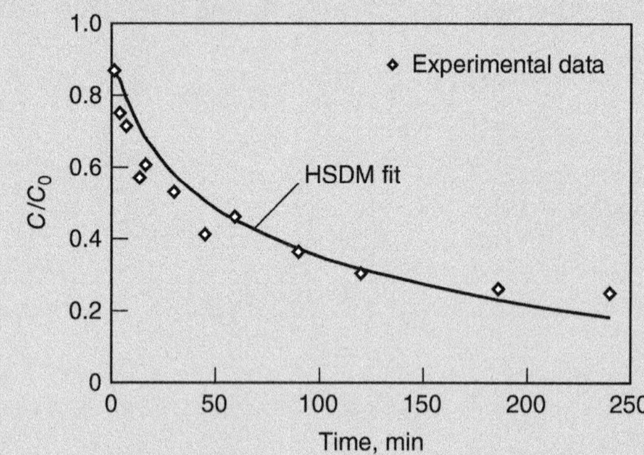

Example 15-7 Generate a plot of $C(t)/C_0$ versus PAC dosage for plug flow reactor times using the parameters for atrazine

Using the data presented in Example 15-6 (i.e., $C_0 = 174.5$ ng/L, $R = 0.0005$ cm, Freundlich isotherm parameters K and $1/n$, and D_s), generate a plot of reduced concentration as a function of contact times for 7.5, 15, 30, 60, and 240 min. Assume the PAC process follows a plug flow reactor.

Solution

To generate the plot, $C(t)/C_0$ needs to be calculated for different PAC dosage using the following steps:

1. Calculate PAC dosage for each C_e/C_0 using Eq. 15-36:

$$D_0 = \frac{C_0 - C_e}{q_e} = \frac{C_0 - C_e}{KC_e^{1/n}}$$

2. Calculate dimensionless time $\bar{t}$ for $t = 7.5$ min, 15 min, 30 min, 1 h, and 4 h using Eq. 15-58.

$$\bar{t} = \frac{(t\ \text{min})(5.18 \times 10^{-12}\ \text{cm}^2/\text{s})(60\ \text{s/min})}{(0.0005\ \text{cm})^2}$$

3. Calculate equilibrium adsorbed-phase concentrations, PAC dosages, and the HSDM user-oriented solutions associated with the C_e/C_0 values and $1/n = 0.2$ from Table 15-14.

C_e/C_0	0.005	0.01	0.05	0.1
C_e (ng/L)	0.8725	1.745	8.725	17.45
q_e	2.446842	2.842028	4.023517	4.67335
D_0 (mg/L)	70.95984	60.78582	41.20151	33.60544
A_0	0.645253	0.513173	0.224322	0.122475
A_1	0.55068	0.492388	0.30597	0.212696
A_2	0.139005	0.139344	0.121216	0.105034
A_3	0.008369	0.009083	0.009265	0.008516

Using these parameters with Eq. 15-89 calculate the $\bar{C}$ values for each C_e/C_0 and $\bar{t}$ values. Calculate a C_t/C_0 value from each $\bar{C}$ for each C_e/C_0 and $\bar{t}$ value. An Excel spreadsheet can be used to perform the calculations. The following table summarizes the C_t/C_0 values for each retention time for C_e/C_0 values of 0.005, 0.01, 0.05, and 0.1.

		C_t/C_0			
t (min)	$\bar{t}$	$C_e/C_0 = 0.005$	$C_e/C_0 = 0.01$	$C_e/C_0 = 0.05$	$C_e/C_0 = 0.1$
7.5	9.33×10^{-3}	0.257445	0.335248	0.521619	0.59817
15	1.87×10^{-2}	0.132426	0.196458	0.375738	0.462845
30	3.73×10^{-2}	0.044742	0.087018	0.239495	0.330394
60	7.46×10^{-2}	0.011033	0.024898	0.130477	0.216131
240	2.99×10^{-1}	0.005	0.01	0.05	0.103422

4. Plot C_t/C_0 versus PAC dosage for different times.

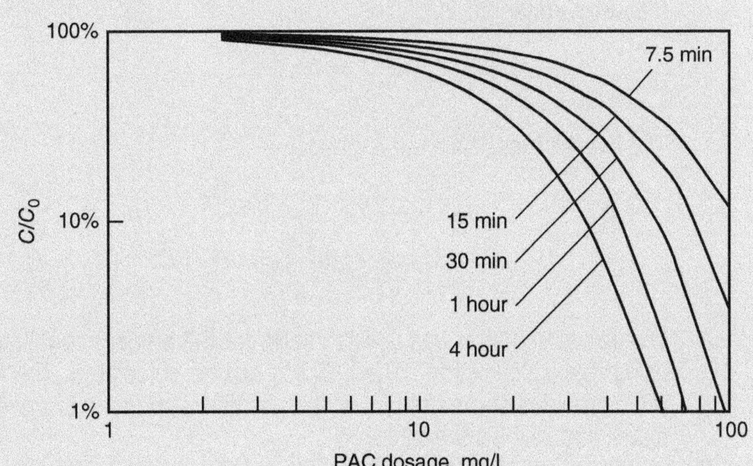

Example 15-8 Compare PFR performance to CMFR performance for the same PAC dosage

Using an initial atrazine concentration of 174.5 ng/L, compare the performance of a PFR with a $C_e/C_0 = 0.1$ to a CMFR. Use the equilibrium (K and $1/n$) and kinetic (D_s) parameters from Example 15-7.

Solution

The results for $1/n = 0.2$ and a $C_e/C_0 = 0.1$ were determined in Example 15-7. The dosage of PAC for the CMFR will be the same as that for the PFR, and the PAC dosage for the PFR is determined from this mass balance on the PFR.

$$C_0 - C_e - q_e \times D_0 = 0$$

$$C_0 - C_e - KC_e^{1/n} \times D_0 = 0$$

$$D_0 = \frac{C_0 - C_e}{KC_e^{1/n}} = \frac{(174.5 - 17.4)\ \text{ng/L}}{2.52\ (\text{ng/mg})\ (\text{L/ng})^{1/n}\ 17.4^{1/n}} = 33.6\ \text{mg/L}$$

To solve for CMFR results, the overall mass balance equation that includes both the liquid and PAC phases for a completely mixed flow reactor is written as follows. First, the average loading on the PAC is given by Eq. 15-99 for a CMFR:

$$\overline{q}_{ave} = 3\overline{t}\left[\frac{1}{\sqrt{\overline{t}}}\coth\left(\frac{1}{\sqrt{\overline{t}}}\right) - 1\right]$$

For a CMFR, the equilibrium solid-phase concentration is that which is in equilibrium with the steady-state liquid concentration as shown in the following equation:

$$q_e = KC_{\overline{t}}^{1/n}$$

$$q_{ave} = \overline{q}_{ave} q_e$$

The overall mass balance is given by

$$C_0 = C_t + q_{ave} D_0$$

$$C_0 - C_t - 3\overline{t}\left(\frac{1}{\sqrt{\overline{t}}}\coth\left(\frac{1}{\sqrt{\overline{t}}}\right) - 1\right) \times KC_t^{1/n} \times D_0 = 0$$

The above equation was used to solve for the effluent concentration C_t using Excel Solver tool. Then, C_t/C_0 can be calculated. The following table provides some of the calculations that were used to generate the figure that follows the table.

t (min)	$\bar{t}$	PFR C_t/C_0	CMFR $\bar{q}_{ave}$	CMFR C_t/C_0
0	0	1	0	1
5	6.22×10^{-3}	0.672545	2.18×10^{-1}	0.701265
10.00	1.24×10^{-2}	0.542638	2.97×10^{-1}	0.605253
15.00	1.87×10^{-2}	0.462845	3.54×10^{-1}	0.541367
20.00	2.49×10^{-2}	0.406647	3.99×10^{-1}	0.493583
25.00	3.11×10^{-2}	0.364095	4.36×10^{-1}	0.455765
30.00	3.73×10^{-2}	0.330394	4.68×10^{-1}	0.424787

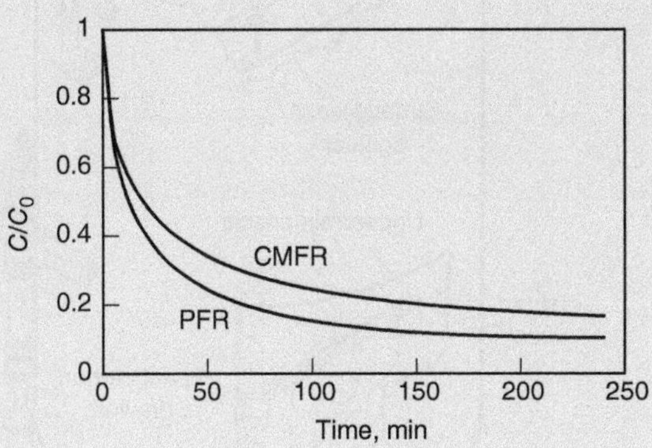

15-6 Granular Activated Carbon

Granular activated carbon operations can be divided into the following two categories: (1) trace contaminant removal and (2) DOC removal. In water treatment, physical adsorption is typically the mechanism responsible for the removal of organics; surface reactions, complexation, and ion exchange with surface functional groups are responsible for the removal of inorganic constituents. Biological activity on the carbon surface can also play a role in extending GAC bed life by using adsorbed molecules for electron donors or acceptors. Three GAC contactor options are (1) gravity feed contactors, (2) pressure contactors, and (3) upflow and/or fluidized-bed contactors. Granular activated carbon can also be used both as a filter and an adsorber in sand replacement filtration operations. Gravity feed contactors have the same features as granular media filters, which are described in Chap. 11, but can be deeper than conventional granular filters. A typical schematic of a pressure GAC contactor is illustrated on Fig. 15-20.

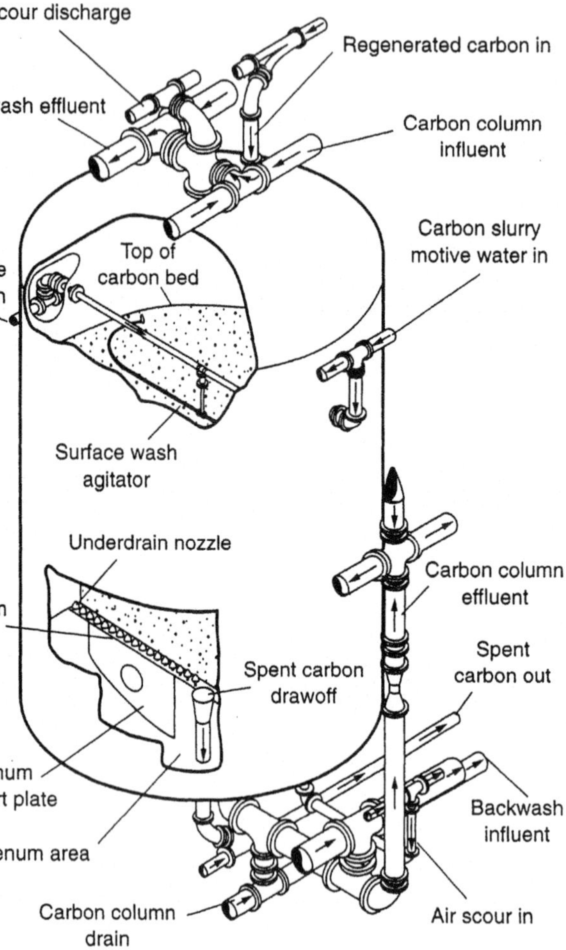

Figure 15-20
Schematic of GAC pressure adsorber (from Tchobanoglous et al., 2003).

Terms Used in GAC Application

Before continuing the discussion on GAC contactors, it is useful to provide a few definitions of important terms. The definitions for some of the terms defined at the beginning of the chapter have been expanded upon here.

MASS TRANSFER ZONE

The concentration–history profile for a GAC contactor was shown previously on Fig. 15-13. As time proceeds, the adsorbate slowly saturates the GAC in the contactor near the inlet, and a concentration profile known as the mass transfer zone develops and moves through the bed. The mass transfer zone (MTZ) is the length of bed needed for the adsorbate to be transferred from the fluid into the adsorbent. Eventually, the adsorbate at the front of the MTZ appears in the effluent, and the time when the concentration exceeds the treatment objective in the effluent is called breakthrough. The time when the effluent concentration essentially equals the influent is called the point of exhaustion because the bed is no longer able to remove the adsorbate.

For a single component and constant influent concentration, the following expression can be derived by writing a mass balance on the mass transfer zone that moves with the mass transfer zone:

$$\frac{C(z)}{C_{\text{inf}}} = \frac{q(z)}{q_e|_{C_{\text{inf}}}} \tag{15-100}$$

Where
$C(z)$ = liquid-phase concentration of adsorbate at location z within the mass transfer zone, mg/L
C_{inf} = influent liquid-phase concentration of the adsorbate, mg/L
$C(z)/C_{\text{inf}}$ = normalized liquid-phase concentration at location z within the mass transfer zone, dimensionless
$q(z)$ = adsorbent-phase concentration of the adsorbate at location z, mg adsorbate/g adsorbent
$q_e|_{C_{\text{inf}}}$ = adsorbent-phase concentration of the adsorbate in equilibrium with the influent concentration, mg adsorbate/g adsorbent
$q(z)/q_e|_{C_{\text{inf}}}$ = degree of saturation at location z within the mass transfer zone, dimensionless

The relationship given in Eq. 15-100 is useful because the ratio of the liquid concentration as compared to the influent concentration equals the degree of saturation at any point in the fixed bed.

EMPTY-BED CONTACT TIME

The empty-bed contact time (EBCT) equals the volume of the bed occupied by the adsorbent divided by the flow rate:

$$\text{EBCT} = \frac{V_F}{Q} = \frac{A_F L}{v A_F} = \frac{L}{v} \tag{15-101}$$

where
EBCT = empty-bed contact time, h
V_F = volume occupied by adsorber media including porosity volume, m^3
Q = flow rate to adsorber, m^3/h
A_F = adsorber area available for flow, m^2
L = adsorber or media depth, m
v = superficial flow velocity (Q/A_F), m/h

The range of EBCTs in fixed-bed adsorption processes can vary from 5 to 60 min for GAC. For removal of SOCs from water, EBCTs in the range of 5 to 30 min are common. The superficial flow velocity is equal to the flow rate divided by the cross-sectional area perpendicular to the flow. Typical adsorber velocities (approach velocity) range from 5 to 15 m/h (2 to 7 gpm/ft^2).

SPECIFIC THROUGHPUT

Specific throughput is used to quantify the performance of a GAC adsorber and is defined as the volume fed to the adsorber divided by the mass of GAC in the adsorber:

$$\text{Specific throughput} = \frac{Qt_{\text{bk}}}{M_{\text{GAC}}} = \frac{V_F\ t_{\text{bk}}}{\text{EBCT}\ M_{\text{GAC}}} = \frac{V_F\ t_{\text{bk}}}{\text{EBCT}\rho_F\ V_F} = \frac{t_{\text{bk}}}{\text{EBCT}\rho_F} \tag{15-102}$$

where specific throughput = volume fed to adsorber divided by mass of GAC in adsorber, m^3/kg

M_{GAC} = mass of GAC, kg

t_{bk} = time to breakthrough at the treatment objective, d

ρ_F = adsorber density or filter bed density, g/L (or kg/m^3)

ρ_F is defined using the following equation:

$$\rho_F = \frac{M_{\text{GAC}}}{V_F} \tag{15-103}$$

The range of adsorber densities for GAC is 350 to 550 kg/m^3 (22 to 34 lb/ft^3).

CARBON USAGE RATE

A more common way to quantify the performance of a GAC adsorber is in terms of carbon usage rate (CUR):

$$\text{CUR} = \frac{M_{\text{GAC}}}{Qt_{\text{bk}}} = \frac{1}{\text{specific throughout}} \tag{15-104}$$

PARTICLE SIZE

For GAC, two of the most important physical properties are hardness and particle size. Much of the operating cost of GAC results from losses by attrition during handling and reactivation. It is important to have large sweeping turns in GAC transport lines to the contactors to reduce clogging and GAC attrition. Losses are smaller for harder carbons. Similarly, the friability of the carbon used in adsorber beds controls the rate with which particles are broken down in size, leading to short adsorber runs (because high head loss) and loss during backwashing (which also mixes up the bed and results in poor performance). Particle size also influences head loss across a bed of GAC; if very small particles are used, higher head loss and crushing of the bed may result. The typical GAC particle diameters are 0.6 to 2.36 mm (8 × 30 U.S. mesh) and 0.425 to 1.70mm (12 × 40 U.S. mesh).

BED POROSITY

The other important bed and particle properties, including bed porosity, were reported in the nomenclature table at the beginning of the chapter. The bed porosity of a GAC contactor fixed bed is complicated by the fact that GAC is porous and the activated carbon itself can have a porosity of 0.2 to 0.7, and this needs to be considered when bed porosity is estimated. In this regard, it is important to discuss how one could obtain the various carbon densities. The apparent density, ρ_a, is the density of the GAC per volume of GAC particle. The solid density, ρ_s, is that of graphite, which is about 2.0 to 2.2 g/cm^3. The filter or bulk density, ρ_F, is the density of the GAC per volume of bed and is about 0.35 to 0.5 g/cm^3. These relationships can be established for the porosity of a particle ε_p and the porosity of void fraction of the bed ε:

$$\varepsilon_p = 1 - \frac{\rho_a}{\rho_s} \qquad (15\text{-}105)$$

$$\varepsilon = 1 - \frac{\rho_F}{\rho_a} \qquad (15\text{-}106)$$

Determination of Specific Throughput and Carbon Usage Rate

If the MTZ is short, the GAC column will be completely saturated at the point when the adsorbate reaches the end of the column, which corresponds to the largest specific throughput or smallest carbon usage rate that can be achieved. In effect, all the adsorbate fed is adsorbed in the column, and the adsorption capacity is in equilibrium with the influent concentration. Relating the total quantity of adsorbate fed to the column to the ultimate capacity of the GAC in the column, expressions for the maximum specific throughput and minimum carbon usage rate can be derived as follows:

$$QC_{\text{inf}}\ t_{\text{bk}} = M_{\text{GAC}}\ q_e|_{C_{\text{inf}}} \qquad (15\text{-}107)$$

The maximum specific throughput and minimum carbon usage rate are then given by the expressions

$$\text{Maximum specific throughout} = \frac{Qt_{\text{bk}}}{M_{\text{GAC}}} = \frac{q_e|_{\text{inf}}}{C_{\text{inf}}} \qquad (15\text{-}108)$$

$$\text{Minimum CUR} = \frac{M_{\text{GAC}}}{Qt_{\text{bk}}} = \frac{C_{\text{inf}}}{q_e|_{C_{\text{inf}}}} \qquad (15\text{-}109)$$

The fraction of utilized capacity increases for a GAC column as the length of column is increased, as shown on Fig. 15-21. The increase in capacity results from a mass transfer zone that has a constant shape and size, which occurs for compounds that have favorable isotherms (Freundlich $1/n <$ 1.0). The use of Eqs. 15-108 and 15-109 is demonstrated in Example 15-9.

If a treatment objective of $C_{\text{eff}}/C_{\text{inf}}$ or $\overline{C}_{\text{to}}$ is chosen, then the specific throughput and carbon usage rate can be calculated using the following equations:

$$QC_{\text{inf}}\ t_{\text{bk}} = Q\int_0^{t_{\text{bk}}} C_{\text{eff}}\ dt + q_c M_{\text{GAC}} \qquad (15\text{-}110)$$

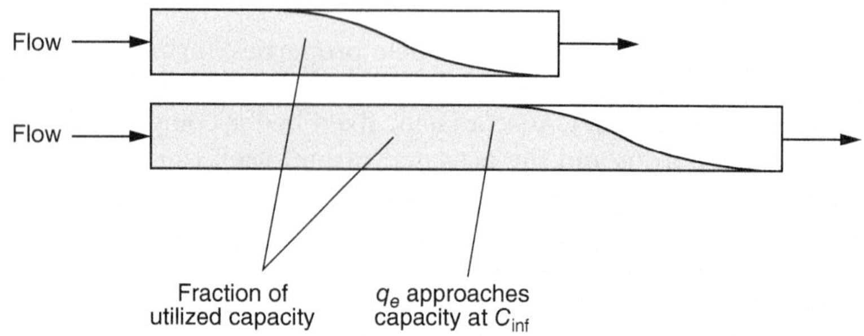

Figure 15-21
Utilized capacity for two GAC column lengths.

Example 15-9 GAC column analysis

Calculate the bed life, volume of water treated, minimum CUR, and maximum specific throughput for trichloroethene (TCE) given the following design specifications:

Influent concentration = 1 mg/L

Carbon type: Calgon Filtrasorb 400 (12 × 40 mesh); $\rho_b = 450$ g/L

Freundlich parameters for TCE can be obtained from Table 15-6.

- Freundlich capacity parameter $K = 55.9\ (\text{mg/g})(\text{L/mg})^{1/n}$
- Freundlich intensity parameter $1/n = 0.48$

Treatment objective = 0.005 mg/L TCE

Flow rate = 378.5 L/min (100 gal/min)

EBCT = 10 min

Solution

1. Calculate the minimum GAC usage rate and maximum specific throughput. The best performance for the GAC is given by assuming that only TCE is in the influent and the MTZ is small compared to the column length. The GAC usage rate and specific throughput are given by Eqs. 15-109 and 15-108, respectively:

$$\text{CUR} = \frac{M_{\text{GAC}}}{Qt_{\text{bk}}} = \frac{C_{\text{inf}}}{q_e} = \frac{C_{\text{inf}}}{K(C_{\text{inf}})^{1/n}}$$

$$= \frac{1.0\ \text{mg/L}}{\left[55.9\ (\text{mg/g})(\text{L/mg})^{0.48}\right](1.0\ \text{mg/L})^{0.48}}$$

$$= 0.018\ \text{g GAC/L H}_2\text{O treated}$$

$$\text{Specific throughput} = \frac{1}{\text{GAC usage rate}} = \frac{1}{0.018\ \text{g/L}}$$
$$= 55.9\ \text{L } H_2O\ \text{treated/g GAC}$$

2. Determine the volume of water treated. The volume of water treated can be calculated in the following manner. First, the mass of carbon in the vessel is calculated:

$$\begin{Bmatrix}\text{Mass of GAC in}\\ \text{10-min EBCT bed}\end{Bmatrix} = V_F\ \rho_F = (\text{EBCT})(Q)(\rho_F)$$
$$= (10\ \text{min})(378.5\ \text{L/min})(450\ \text{g/L}) = 1.7\times10^6\,\text{g}$$

The volume treated can be determined from the definition of the GAC usage rate:

$$\begin{Bmatrix}\text{Volume of } H_2O \text{ treated}\\ \text{for 10-min EBCT bed}\end{Bmatrix} = \begin{Bmatrix}\dfrac{\text{mass of GAC in 10-min EBCT bed}}{\text{GAC usage rate}}\end{Bmatrix}$$
$$= \frac{1.7\times10^6\ \text{g}}{0.018\ \text{g/L } H_2O} = 9.4\times10^7\ \text{L } H_2O$$

3. Calculate the bed life. The bed life can be determined from the volume of water treated and the flow rate:

$$\text{Bed life} = \frac{\text{volume of } H_2O \text{ treated for 10-min EBCT bed}}{Q}$$
$$= \frac{9.4\times10^7\ \text{L}}{(378.5\ \text{L/min})(1440\ \text{min/d})}$$
$$= 172\ \text{d}$$

Comment

The presence of NOM will reduce bed life significantly. The bed life determined in this example represents the maximum expected value, as the influence of NOM is ignored and the MTZ is assumed to be very short.

where q_c = average adsorbent-phase concentration of adsorbate in GAC column, mg adsorbate/g adsorbent
C_{eff} = effluent liquid-phase concentration at t, mg/L

The adsorbent-phase concentration is given as

$$q_c = \frac{\int_0^{t_{bk}} Q(C_{\text{inf}} - C_{\text{eff}})\,dt}{M_{\text{GAC}}} \tag{15-111}$$

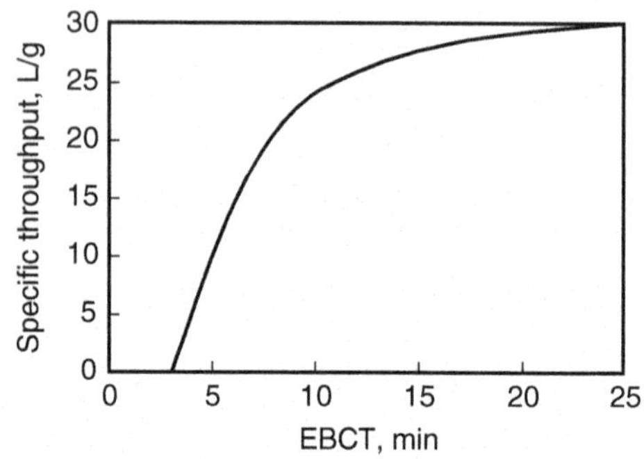

Figure 15-22
Specific throughput versus EBCT for a single GAC column.

The specific throughput can then be determined by rearranging Eq. 15-109:

$$\frac{Qt_{\text{bk}}}{M_{\text{GAC}}} = \frac{Q\int_0^{t_{bk}} C_{\text{eff}}\ dt}{C_{\text{inf}}\ M_{\text{GAC}}} + \frac{q_c}{C_{\text{inf}}} \tag{15-112}$$

If $\overline{C}_{\text{eff}}/\overline{C}_{\text{inf}}$ or $\overline{C}_{\text{to}}$ is small (e.g., < 0.05), then the first term on the right-hand side is negligible and the specific throughput is equal to the expression:

$$\frac{Qt_{\text{bk}}}{M_{\text{GAC}}} = \frac{q_c}{C_{\text{inf}}} \tag{15-113}$$

Consequently, the specific throughput increases as the EBCT increases as shown on Fig. 15-22 and demonstrated in Example 15-10. The specific throughput approaches the maximum value given by Eq. 15-108 when the mass transfer zone is much smaller than the length determined by the EBCT and q_c approaches q_e. As shown in Fig. 15-22, when the bed length equals the MTZ, the specific throughput approaches zero because the effluent concentration of the adsorbate appears in the effluent very quickly.

As shown on Fig. 15-22, the specific throughput is zero up to a minimum EBCT because the column must be longer than the MTZ or the effluent concentration immediately exceeds the treatment objective. From a cost perspective, it is important to realize that as the specific throughput increases (by increasing the EBCT) the operation and maintenance costs decrease, but it comes at the expense of increasing capital cost because the contactor size needed is larger.

Example 15-10 Analysis of pilot plant adsorption data

A GAC pilot plant study was performed on a groundwater containing *cis*-1,2-dichloroethene (DCE). The impact of EBCT on GAC performance was evaluated by conducting column experiments for EBCTs of 3, 5, 10, 21, and 32 min. The DCE effluent concentration for each EBCT was plotted in terms of specific throughput (liters of water treated per gram GAC) using Eq. 15-102 and is displayed below. Using the column data, plot the specific throughput for a treatment objective of 5 μg/L as a function of EBCT and determine a reasonable EBCT for DCE in this groundwater.

Solution

1. Construct a plot of the specific throughput for a treatment objective of 5 μg/L as a function of EBCT.
   a. On the *y* axis locate the 5-μg/L treatment objective and draw a line parallel to the *x* axis so it intersects the effluent profile.
   b. Where the 5-μg/L line intersects effluent profiles, draw a line down to the *x* axis to obtain the specific throughput for each EBCT as shown. For EBCTs of 3, 5, 10, 21, and 32 the specific throughputs are 6.5, 16.0, 22.0, 27.5, and 29.0 m^3 water treated per gram of GAC, respectively.
   c. Plot the specific throughput as a function of EBCT.

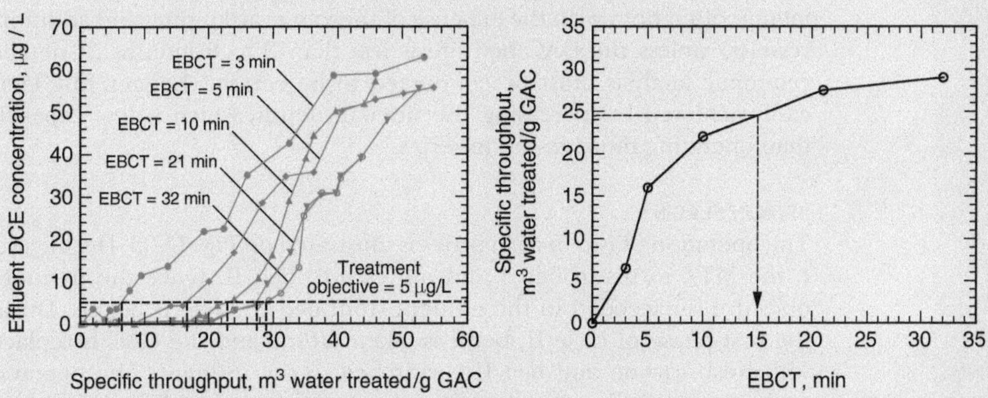

2. Determine a reasonable EBCT for a single adsorber for DCE in this groundwater. From the plot constructed in step 1c, it is clear that the specific throughput reaches a point of diminishing returns at about 15 min of EBCT.

Comments

The concepts that were introduced above for evaluating the specific throughput as a function of EBCT are useful for fixed beds that are used to treat contaminated groundwaters containing mixtures of organics.

However, the pilot data that was presented in this example took one year to collect. Not only is this time in most cases unacceptably long, but a pilot test like this is very costly. Accordingly, rapid small-scale column tests may be useful in determine the carbon usage rate. These tests are presented later in this chapter.

GAC Operation

Countercurrent operation is the most efficient operation when high degrees of removal are needed ($C_{to}/C_{inf} < 0.05$) because, in principle, the GAC can be saturated with respect to the influent concentration. Furthermore, the effluent concentration will be less than the treatment objective if the column is longer than the MTZ. However, GAC is friable and cannot withstand the movement in a countercurrent operation without suffering significant losses (fines generation). Although countercurrent GAC columns have been commercially available, it has been found that the increase in specific throughput did not warrant the extra cost. Consequently, GAC operations may be operated as two beds (or perhaps more beds) in series to achieve similar efficiencies. Series operation will reduce the amount of GAC needed by as much as 25 to 50 percent. The series arrangement is often not worth the expense of the extra yard piping and additional vessel(s) unless the GAC bed life is less than 3 to 6 months. A detailed economic analysis must be conducted to be certain. Longer EBCTs are easily achieved by increasing the sidewall depth, which is less expensive than operating more vessels in series.

BEDS IN SERIES

The operation of two beds in series is illustrated on Fig. 15-23. During cycle I, the MTZ forms in bed I and moves into bed II. Once the treatment objective is exceeded in the effluent from bed II, cycle II begins. During the first phase of cycle II, bed I is taken offline and the GAC is replaced with fresh carbon and bed II is switched to the influent. The operation continues until the mass transfer zone moves from bed II into bed I and the effluent from bed I exceeds the treatment objective. At this point, cycle III begins and bed I receives the influent, and bed II is recharged with fresh carbon and put into operation just as shown in cycle I. If the length

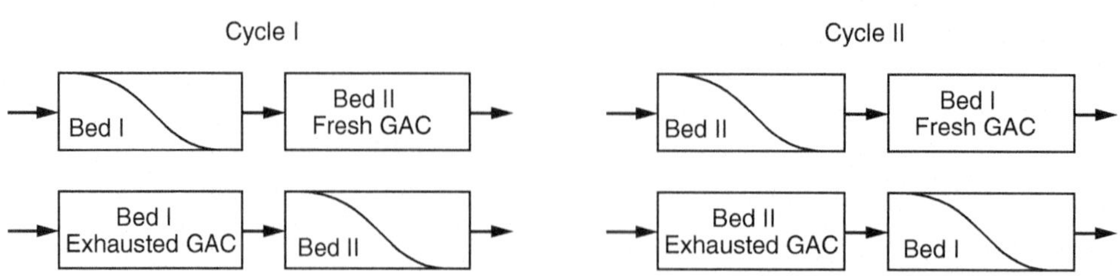

Figure 15-23
Operation of two beds in series.

of beds I and II are greater or equal to the length of the MTZ, then the GAC will be saturated fully and the GAC usage rate can be calculated using Eq. 15-109. The CUR and specific throughput for this type of operation can be determined from a pilot study that is conducted in this manner using the following expressions:

$$\text{Specific throughput} = \frac{Qt_c}{M_1} \tag{15-114}$$

$$\text{CUR} = \frac{M_1}{Qt_c} \tag{15-115}$$

where M_1 = mass of carbon in first bed that is removed after a number of cycles, g
Q = water flow rate, L/s
t_c = cycle time, d

The specific volume for two beds in series operation can be determined from pilot study data, as shown in the previous example, by considering what happens to the mass of adsorbate during a cyclic operation after several cycles and how that is related to a two-beds-in-series pilot column operation. To use the data from the example, which is a noncyclic pilot plant, the loading on the first and second columns will be assumed to be the same as for a cyclic operation. This condition will be met in most instances for the following two reasons. First, the two columns in a cyclic pilot study will have a similar history to a noncyclic pilot study if the influent to the second column in a noncyclic pilot study is close to the influent concentration when the effluent concentration exceeds the treatment objective. Second, similar loadings will also occur if the total EBCT is long enough to reestablish a MTZ that is similar to that observed in pilot plant data, even if the lag column sees an increase in influent concentration after being switched from the lag position to the lead position. Consequently, the specific throughput can be determined by ignoring the adsorbate in the second column because it would be equal to the mass that was in the first column at startup, and this would move to the second column at the end of the cycle and balance out. With these arguments in mind, the mass fed to a cyclic operation can be determined from a noncyclic pilot study:

$$\begin{matrix}\text{Mass fed for a} \\ \text{cyclic operation}\end{matrix} = \begin{matrix}\text{mass retained} \\ \text{on first column}\end{matrix} + \begin{matrix}\text{mass in effluent of} \\ \text{the second column}\end{matrix} \tag{15-116}$$

$$QC_{\text{inf}}\ t_c = q_1 M_1 + Q\int_0^{t_o} C_2\ dt \tag{15-117}$$

$$q_1 = \frac{Q\left[C_{\text{inf}}\ t_c - \int_0^{t_o} (C_1)\, dt\right]}{M_1} \tag{15-118}$$

where C_{inf} = influent liquid-phase concentration of adsorbate, mg/L
q_1 = loading observed in column 1 in noncyclic pilot study, mg adsorbate/g adsorbent
C_1 = effluent concentration from column 1 in noncyclic pilot study, mg/L
C_2 = effluent concentration from column 2 in noncyclic pilot study, mg/L
t_o = time of operation of noncyclic pilot study up to time when effluent exceeds treatment objective, d

The specific throughput and CUR can then be determined from the following equations:

$$\text{Specific throughput} = \frac{Qt_c}{M_1} = \frac{q_1}{C_{inf}} + \frac{Q\int_0^{t_o} \overline{C}_2\, dt}{M_1} \tag{15-119}$$

$$\text{CUR} = \frac{M_1}{Qt_c} = \frac{1}{q_1/C_{inf} + Q\int_0^{t_o} \overline{C}_2\, dt/M_1} \tag{15-120}$$

where $\overline{C}_2$ = dimensionless effluent concentration from column 2 in a noncyclic pilot study, $= C_2/C_{inf}$

Example 15-11 Analysis of carbon beds in series

For the GAC pilot plant data presented in Example 15-10, determine the specific throughput for DCE for two beds in series with each bed having an EBCT of 5.0 min. The flow rate is 232 L/d, the adsorber density is 0.457 g/cm^3, the average DCE influent concentration is 80 μg/L, and the DCE treatment objective is 5 μg/L. The concentration–history profiles for DCE as a function of elapsed time in days are given below.

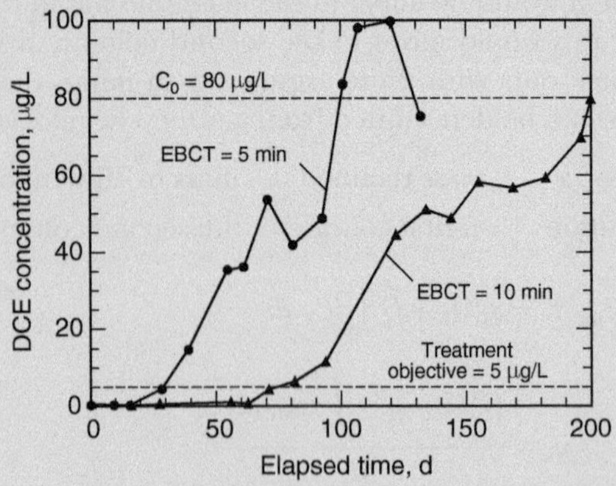

Column data:

EBCT, min	*M*, g	*t*, d	*Q* · *t*, L	L/g
5.0	395.5	27.3	6,334	16.0
10.0	791.1	75	17,400	22.0

Solution

1. Calculate the specific throughput. The specific throughput can be calculated using Eq. 15-119. The contribution from both the first and second columns is given by:

$$\text{Specific throughput} = \frac{Qt_c}{M_1} = \underbrace{\frac{q_1}{C_0}}_{\substack{\text{contribution}\\ \text{from}\\ \text{frst column}}} + \underbrace{\frac{Q\int_0^{t_o}\bar{C}_2 dt}{M_1}}_{\substack{\text{contribution}\\ \text{from}\\ \text{second column}}}$$

   a. Calculate the contribution from the amount of adsorbate in the effluent from the second column using the expression shown in Eq. 15-119 for the second column. For two beds in series with 5.0-min EBCT, the treatment objective from the second column is exceeded after 75 days and the average influent concentration is 80 μg/L. The contribution from the amount of adsorbate in the effluent from the pilot study in cycle is given by the expressions:

$$\frac{Q}{M_1 C_{\text{inf}}}\int_0^{t_o} C_2 dt = \frac{Q}{V_{F_p F} C_{\text{inf}}}\int_0^{t_o} C_2\, dt = \frac{1}{\text{EBCT}_1 \rho_F C_{\text{inf}}}\int_0^{t_o} C_2\, dt$$

$$\frac{1}{\text{EBCT}_1 \rho_F C_{\text{inf}}}\int_0^{t_o} C_2\, dt \approx \left(\frac{1}{5.0\text{ min}}\right)\left(\frac{\text{L}}{457\text{ g}}\right)\left(\frac{\text{L}}{80\ \mu\text{g}}\right)\frac{1}{2}(75-50)\text{d}$$
$$\times (1440\text{ min/d})(5\ \mu\text{g/L})$$
$$= 0.49\text{ L/g}$$

   b. Calculate the contribution to the specific throughput due to adsorption onto the first column. The loading on the first column, when the effluent is 5 μg/L from the second column, is computed using Eq. 15-118 as follows:

$$q_1 = \frac{Q\left(C_{\text{inf}}\, t_0 - \int_0^{t_o} C_1\, dt\right)}{M_1} = \frac{1}{\text{EBCT}_1 \rho_F}\left(C_{\text{inf}}\, t_0 - \int_0^{t_o} C_1\, dt\right)$$

$$\int_0^{t_o} C_1\,dt \approx \frac{1}{2}(75-20)\text{d}\times(48\ \mu\text{g/L})(1440\ \text{min/d})$$

$$= 1.90\times10^6\ \mu\text{g}\cdot\text{min/L}$$

$$C_{inf}\,t_0 = (80\ \mu\text{g/L})(75\ \text{d})(1440\ \text{min/d}) = 8.64\times10^6\ \mu\text{g}\cdot\text{min/L}$$

$$q_1 = \frac{(8.64\times10^6 - 1.90\times10^6)\mu\text{g}\cdot\text{min/L}}{5.0\ \text{min}\times(457\ \text{g/L})} = 2950\ \mu\text{g/g}$$

The contribution to the specific throughput due to adsorption onto the first column is given by the following expression, which is the expression shown in Eq. 15-119 for the first column:

$$\frac{q_1}{C_0} = \frac{2950\ \mu\text{g/g}}{80\ \mu\text{g/L}} = 37\ \text{L/g}$$

2. Calculate the specific throughput using Eq. 15-119 and the values computed in step 1:

$$\text{Specific throughput} = \frac{Qt_c}{M_1} = \frac{q_1}{C_0} + \frac{Q\int_0^{t_0}\bar{C}_2\,dt}{M_1} = 37 + 0.49 = 37.5\ \text{L/g}$$

Comment

If one compares the specific throughput for two 5-min-EBCT beds in series (37.5 L/g) with a single adsorber with 10-min EBCT (22 L/g), the beds-in-series operation can treat about 70 percent more water.

BEDS IN PARALLEL

Dissolved organic carbon can also be removed using GAC, but a high degree of DOC removal cannot be achieved using reasonable specific throughputs and EBCTs. Typically, 30 to 70 percent of the DOC can be removed using GAC. Using beds that are operated in parallel can significantly increase specific throughput and can reduce the amount of GAC that is required. Backwashing of the columns is not as much of an issue for DOC removal because high degrees of removal are not achieved and performance, therefore, will not be significantly impacted by backwashing. Further, the mass transfer rate for DOC is slow; consequently, the concentration profiles within particles at different depths may not be significantly different.

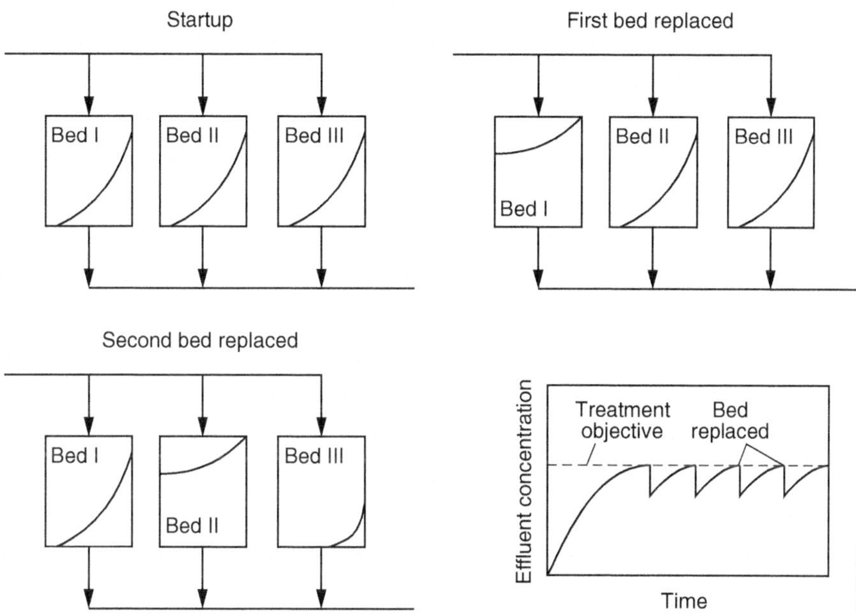

Figure 15-24
Operation of three beds in parallel.

The blending of effluent from three GAC adsorbers operating in parallel after startup and after several cycles is shown on Fig. 15-24. At startup, all three adsorbers have similar bed profiles; once the treatment objective is exceeded, the first adsorber is replaced with fresh GAC. After replacement, the treatment objective can be met with blended effluent from the adsorbers. Operation continues until the treatment objective cannot be met and then the second bed is replaced. At this point, there are three adsorbers with different degrees of saturation, and the treatment objective is still being met because effluent from nearly exhausted adsorbers is blended with effluent from fresh adsorbers. After the treatment objective is exceeded, the third bed is replaced and the cycle begins again by replacing the first column, which will be the column that has been online for the longest period of time. A pilot study can be used to determine the activated carbon usage rate for GAC beds in parallel. The following simple approach can be used to calculate the concentration of organic compound remaining in the effluent of parallel adsorbers just prior to the replacement of one of the beds operated in parallel (Roberts and Summers, 1982):

$$\bar{f} = \frac{1}{n}\sum_{i=1} f_i \tag{15-121}$$

where $\bar{f}$ = concentration of organic compound remaining in effluent, mg/L

f_i = concentration of organic compound remaining in effluent from *i*th adsorber, which is determined by dividing the effluent profile from a single adsorber into a number of intervals, mg/L
n = number of equal-capacity adsorbers in parallel

The value of f_i can be determined from a single breakthrough curve. Assume that replacement of GAC in each adsorber will take place at equal intervals. If θ_n is the number of bed volumes processed through each adsorber in the parallel system at the time of replacement, the abscissa of the breakthrough curve for the individual contactor from 0 to θ_n is divided into θ_n/n equal increments. For a given value of θ_n, Eq. 15-121 can be used with a single breakthrough profile to calculate the concentration of the blended water and the specific throughput. To determine θ_n, one needs to choose the number of bed volumes from a single adsorber such that, if you divide the bed volumes by n and add the effluent concentrations of the beds in parallel, the sum will equal the treatment objective. The idea is that these beds would be operating parallel, and one of them would be replaced when the sum of the effluent concentrations from the adsorbers equals the treatment objective. The starting concentration for the next cycle would shift to the left by θ_n and then would be equal to the sum of all the adsorbers operating in parallel. The effluent profile of all the adsorbers would then operate for another θ_n before the adsorber that has been online for the longest period is replaced. An example of this approach may be found in Snoeyink and Summers (1999). It must be stated that this method assumes the same flow for each adsorber and that if biological activity does occur, it does not change with time.

MULTIPLE BEDS IN PARALLEL

The utility of using multiple beds in parallel will be illustrated by discussing a pilot study that was conducted by the Metropolitan Water District of Southern California, MWH Global, and Michigan Technological University (Crittenden et al., 1993; McGuire et al., 1991; McGuire, 1989). The purpose of the study was to determine the cost associated with removing TOC and the associated trihalomethane formation potential (THMFP) from two Southern California water supplies. During the study, mixtures of California State Project water (SPW) and Colorado River water (CRW) were used, and the mean TOCs for SPW and CRW were 2.64 and 2.52 mg/L, respectively, and were very consistent. The concentration of TOC as a function of operation time and EBCT is shown on Fig. 15-25a. The concentration of TOC as a function of specific throughput and EBCT is shown on Fig. 15-25b. The column with 60-min EBCT was backwashed at 38 and 112 days of operation, and the column with 30-min EBCT was backwashed at 108 days of operation. Backwashing had essentially no impact on performance, as shown on Fig. 15-25a.

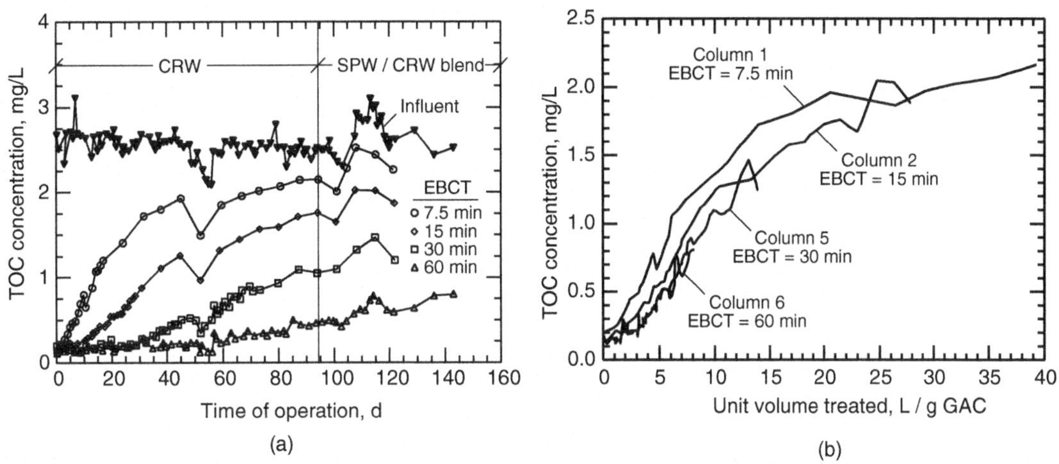

Figure 15-25
TOC concentration history profiles for EBCTs of 7.5, 15, 30, and 60 min: (a) TOC concentration versus time of operation and (b) TOC concentration versus specific throughput.

The following equations were used to describe the relationship between TOC and THMFP for the two waters (McGuire, 1989):

$$\text{THMFP}(\mu\text{g/L}) = \text{TOC}\,(\text{mg/L}) \times 59 - 2.3 \qquad (15\text{-}122)$$

$$\text{THMFP}\,(\mu\text{g/L}) = \text{TOC}\,(\text{mg/L}) \times 97.41 - 6.36 \qquad (15\text{-}123)$$

Assuming a treatment objective for THMFP of 50 μg/L, the effluent TOC concentration cannot exceed 1 mg/L for CRW, which corresponds to a $\bar{C}_{to}$ value of 0.4. The specific throughput as a function of EBCT for a treatment object of 1 mg/L is shown on Fig. 15-26. The specific throughput

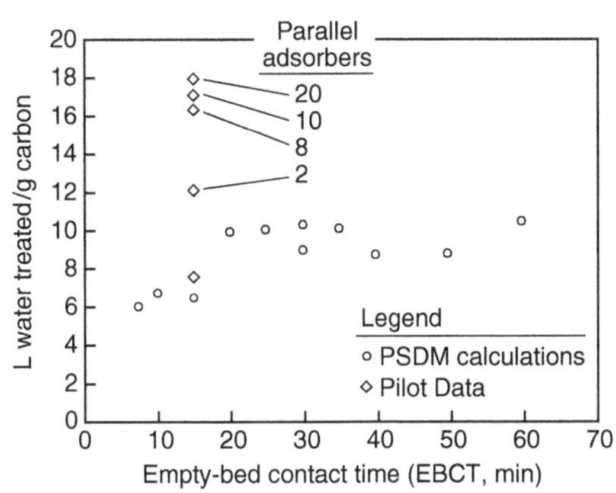

Figure 15-26
Liters of water treated per gram of GAC for a DOC treatment objective of 1 mg/L as function of EBCT. The specific throughput values at some EBCTs were calculated using the pore surface diffusion model (PSDM).

for this DOC treatment objective using a single GAC contactor reaches diminishing returns after about 15 to 25 min of EBCT. The variability in the specific throughput is due to the time–variable influent concentration for the pilot studies and modeling predictions of the time–variable influent concentration. However, as shown on Fig. 15-26, no significant change in the specific throughput is observed after more than eight adsorbers are operated in parallel and about 2.4 times more water can be treated as compared to a single adsorber with 15 min of EBCT. This type of information has been used to optimize the GAC process in terms of costs (Lee et al., 1983; McGuire, 1989; McGuire et al., 1991). A number of researchers have shown that UV absorption can be used to assess the DOC and DBP precursor removal and for process control (Sontheimer et al., 1988).

SUMMARY OF THE BENEFITS OF SERIES AND PARALLEL BED OPERATION

The largest specific throughput is obtained for EBCTs around 10 to 20 min for the removal of SOCs (MW < ~300) with stringent treatment objectives. Two beds that are operated in series may increase the specific throughput by 20 to 50 percent. For TOC removal [disinfection by-product formation potential (DBPFP)], beyond EBCTs of 15 to 30 min the specific throughput does not increase. In addition, it may not be reasonable to achieve more than 70 percent removal of TOC using GAC. In situations where only 30 to 80 percent removal of organics is required, single beds that are operated in parallel may be the least expensive option because the flow from exhausted columns can be blended with the flow from fresh columns. Furthermore, there will always be a GAC barrier that can removal spikes of organic contaminants.

Modeling GAC Performance

To describe the migration of an adsorbate through a fixed bed, mass balances on the solid (immobile phase) and liquid (mobile phase) phases are written and the following assumptions are made: (1) liquid-phase concentration gradients in a fixed-bed adsorber exist only in the axial direction; that is, concentration gradients only exist in the flow direction; (2) the liquid-phase concentration gradient in the axial direction is small enough such that the concentration difference across any single adsorbent particle is negligible (this implies that the bulk solution concentration surrounding any single-adsorbent particle is identical); (3) the adsorbent is in a fixed position in the adsorber (backwashing that may disturb the mass transfer zone is not considered); (4) the adsorbate contained in the liquid phase (not adsorbed onto the surface) in the carbon pores can be neglected; and (4) the hydraulic loading is constant. The liquid-phase mass balance on a thin or differential element of the bed, which is shown in

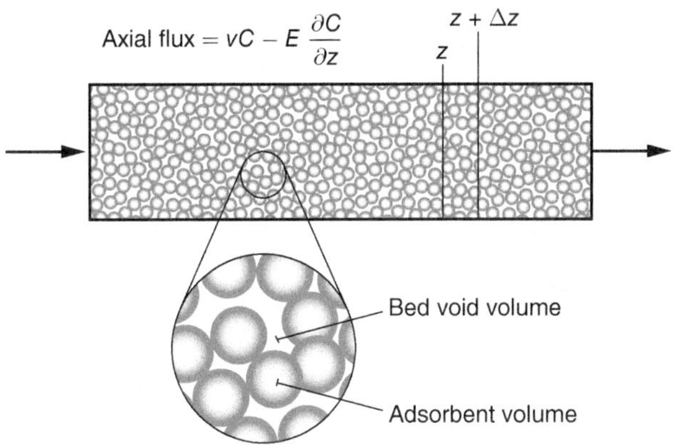

Figure 15-27
Axial transport mechanisms in a fixed-bed adsorber.

Fig. 15-27, is given by the following word equation:

$$\begin{Bmatrix} \text{Mass of adsorbate } i \\ \text{entering by advection} \\ \text{and dispersion} \\ \text{in liquid phase at } z \end{Bmatrix} - \begin{Bmatrix} \text{Mass of adsorbate } i \\ \text{exiting by advection} \\ \text{and dispersion} \\ \text{in liquid phase at } z+\Delta z \end{Bmatrix}$$
$$= \begin{Bmatrix} \text{Mass of adsorbate } i \\ \text{accumulating in} \\ \text{the liquid phase} \\ \text{between } z \text{ and } z+\Delta z \end{Bmatrix} + \begin{Bmatrix} \text{Mass of adsorbate } i \\ \text{accumulating in} \\ \text{the solid phase} \\ \text{between } z \text{ and } z+\Delta z \end{Bmatrix} \quad (15\text{-}124)$$

The four terms that appear in Eq. 15-124 are given by these expressions:

$$\begin{Bmatrix} \text{Mass of adsorbate } i \\ \text{entering by advection} \\ \text{and dispersion in} \\ \text{liquid phase at } z \end{Bmatrix} = v\varepsilon AC\big|_z - E\varepsilon A \frac{\partial C}{\partial z}\bigg|_z \quad (15\text{-}125)$$

$$\begin{Bmatrix} \text{Mass of adsorbate } i \\ \text{exiting by advection} \\ \text{and dispersion in} \\ \text{liquid phase at } z+\Delta z \end{Bmatrix} = v\varepsilon AC\big|_{z+\Delta z} - E\varepsilon A \frac{\partial C}{\partial z}\bigg|_{z+\Delta z} \quad (15\text{-}126)$$

$$\begin{Bmatrix} \text{Mass of adsorbate } i \\ \text{accumulating in} \\ \text{the liquid phase} \\ \text{between } z \text{ and } z+\Delta z \end{Bmatrix} = \varepsilon A \; \Delta z \frac{\partial C}{\partial t} \quad (15\text{-}127)$$

$$\begin{Bmatrix} \text{Mass of adsorbate } i \\ \text{accumulating in} \\ \text{the solid phase} \\ \text{between } z \text{ and } z+\Delta z \end{Bmatrix} = A_p k_f \left(C - C_s\right) \quad (15\text{-}128)$$

where ε = bed void fraction, dimensionless
A = fixed-bed cross-sectional area, m^2
E = dispersion coefficient, m^2/s
v = interstitial velocity, m/s
A_p = total external surface area of adsorbent particle available for mass transfer, m^2
z = axial position in bed, m
t = elapsed time, s
k_f = film transfer coefficient, m/s
C = adsorbate liquid-phase concentration, mg/L
C_s = adsorbate liquid-phase concentration at adsorbent exterior surface, mg/L

Combining Eq. 15-124 to 15-128 and dividing by Δz results in the expression:

$$-v\varepsilon A\frac{C|_{z+\Delta z}-C|_z}{\Delta z}+\varepsilon EA\frac{\left.\frac{\partial C}{\partial z}\right|_{z+\Delta z}-\left.\frac{\partial C}{\partial z}\right|_z}{\Delta z}=\varepsilon A\frac{\partial C}{\partial t}+\frac{A_p k_f\,(C-C_s)}{\Delta z} \tag{15-129}$$

Dividing Eq. 15-129 by εA and taking the limit as Δz approaches zero results in

$$-v\frac{\partial C}{\partial z}+E\frac{\partial^2 C}{\partial z^2}=\frac{\partial C}{\partial t}+\frac{A_p k_f\,(C-C_s)}{A\varepsilon\Delta z} \tag{15-130}$$

The total adsorbent surface area per volume of bed, A_p, may be expressed as

$$A_p=\left(\frac{\text{Adsorbent area}}{\text{Adsorbent volume}}\right)\left(\frac{\text{Adsorbent volume}}{\text{Bed volume}}\right) \tag{15-131}$$

The adsorbent area per adsorbent volume is independent of the number of particles (as long as a representative sample is used).

$$\frac{\text{Adsorbent area}}{\text{Adsorbent volume}}=\frac{4\pi R^2 3}{\phi 4\pi R^3}=\frac{3}{\phi R} \tag{15-132}$$

where R = radius of adsorbent particle, m
ϕ = adsorbent particle sphericity, dimensionless

$$\phi=\frac{\dfrac{\text{Sphere area}}{\text{Sphere volume}}}{\dfrac{\text{Adsorbent area}}{\text{Adsorbent volume}}} \tag{15-133}$$

$$\frac{\text{Adsorbent volume}}{\text{Bed volume}}=1-\varepsilon \tag{15-134}$$

Substituting Eqs. 15-132 and 15-134 into Eq. 15-131 yields:

$$A_p=\frac{3}{\phi R}(1-\varepsilon) \tag{15-135}$$

The sphericity value, ϕ, depends on how the radius is determined. A wide variety of ϕ values based on volume average diameters have been reported in the literature for as-received carbons and a few are reported here: F-300, 1.41 ± 0.10; ROW 0.8S, 1.40 ± 0.07; SNK12, 1.37 ± 0.08; SLSS, 1.24 ± 0.09 (Sontheimer et al., 1988). These values were obtained using image analysis. If sieve analysis is used to determine the particle diameter, sphericity values of 1.0 yield adequate results because the GAC particles usually have a cylindrical shape and tend to pass through smaller sieve openings. For example, Sontheimer et al. (1988) reported a 30 percent larger A_p value for F-300, if the diameter from sieve analysis is used as compared to the volume average diameter. This is very similar to the sphericity value that has been estimated by image analysis.

Substituting Eq. 15-135 into Eq. 15-130 yields the final form of the liquid-phase mass balance for adsorbate i in the fixed bed:

$$-v\frac{\partial C}{\partial z} + E\frac{\partial^2 C}{\partial z^2} = \frac{\partial C}{\partial t} + \frac{3k_f(1-\varepsilon)}{\phi R\varepsilon}(C - C_s) \tag{15-136}$$

To solve Eq. 15-136, one initial condition and two boundary conditions are needed. The initial condition is given by

$$C(0 \le z \le L, \quad t = 0) = 0 \tag{15-137}$$

where L = length of the bed occupied by the GAC, m

The first boundary condition used to solve Eq. 15-136 can be derived by writing a mass balance over the entire fixed bed and may be considered a dynamic version of the Dankwerts boundary condition (Dankwerts, 1953). In words, this may be expressed as

$$\left\{\begin{matrix}\text{Mass of adsorbate } i\\ \text{entering the fixed bed}\end{matrix}\right\} - \left\{\begin{matrix}\text{Mass of adsorbate } i\\ \text{leaving the fixed bed}\end{matrix}\right\} = \left\{\begin{matrix}\text{Mass of adsorbate } i\\ \text{accumulating in}\\ \text{the liquid phase}\\ \text{in the fixed bed}\end{matrix}\right\} + \left\{\begin{matrix}\text{Mass of adsorbate } i\\ \text{accumulating in}\\ \text{the solid phase}\\ \text{in the fixed bed}\end{matrix}\right\} \tag{15-138}$$

The four terms appearing in Eq. 15-138 are show below:

$$\left\{\begin{matrix}\text{Mass of adsorbate } i\\ \text{entering the fixed}\end{matrix}\right\} = vA\varepsilon C_0(t) \tag{15-139}$$

$$\left\{\begin{matrix}\text{Mass of adsorbate } i\\ \text{leaving the fixed}\end{matrix}\right\} = vA\varepsilon C(z = L,\ t) \tag{15-140}$$

$$\left\{\begin{matrix}\text{Mass of adsorbate } i\\ \text{accumulating in}\\ \text{the liquid phase}\\ \text{in the fixed bed}\end{matrix}\right\} = A\varepsilon\int_0^L \frac{\partial C}{\partial t}\, dz \tag{15-141}$$

$$\left\{\begin{array}{c}\text{Mass of adsorbate } i\\ \text{accumulating in}\\ \text{the solid phase}\\ \text{in the fixed bed}\end{array}\right\} = \rho_a(1-\varepsilon)\, A \int_0^L \frac{\partial q_{\text{ave}}}{\partial t}\, dz \tag{15-142}$$

The final form of the first boundary condition is given by

$$vC_0(t) - vC(z=L,\ t) = \int_0^L \frac{\partial C}{\partial t}\, dz + \frac{\rho_a(1-\varepsilon)}{\varepsilon} \int_0^L \frac{\partial q_{\text{ave}}}{\partial t}\, dz \tag{15-143}$$

where C_0 = adsorbate influent concentration entering fixed bed, mg/L
q_{ave} = adsorbate average solid-phase concentration, mg/g

The average solid-phase loading is given by

$$q_{\text{ave}} = \frac{3}{4\pi R^3} \int_0^R q 4\pi r^2\, dr \tag{15-144}$$

where q = adsorbate solid-phase concentration at radial coordinate in adsorbent and axial coordinate within fixed bed, mg/g
r = radial coordinate within adsorbent particle, m

The second boundary condition for the liquid-phase mass balance is the Dankwerts condition at the exit boundary, $z = L^-$.

$$\frac{\partial^2 C\left(z=L^-, t\right)}{\partial t\, \partial z} = 0 \tag{15-145}$$

$$\frac{\partial C(z, t=0)}{\partial z} = 0 \tag{15-146}$$

The solution to Eq. 15-136 requires the liquid-phase concentration at the adsorbent surface C_s. To obtain C_s, a mass balance on the adsorbent phase has to be derived. Equation 15-56, which was developed for PAC applications, is also valid for GAC, but the solid-phase concentration depends on the axial position in the bed:

$$\frac{D_s}{r^2}\frac{\partial}{\partial r}\left(r^2 \frac{\partial q}{\partial r}\right) = \frac{\partial q}{\partial t} \tag{15-147}$$

The boundary and initial conditions for Eq. 15-147 are given by these expressions:

$$k_f\left(C - C_s\right) = D_s \rho_a \frac{\partial q}{\partial r} \tag{15-148}$$

$$\frac{\partial q(r=0, 0 \le z \le L,\ t \ge 0)}{\partial r} = 0 \tag{15-149}$$

$$q(0 \le r \le R, 0 \le z \le L,\ t=0) = 0 \tag{15-150}$$

The model must be made dimensionless to provide general answers and using dimensionless time, and positions, $\overline{z} = z/L$ and $\overline{r} = r/R$ yields the desired equation. Dimensionless time is defined by the following word equation and mathematical equivalent for a single component:

$$\begin{pmatrix} \text{Mass} \\ \text{throughput} \end{pmatrix} = T = \frac{\text{mass fed}}{\text{mass adsorbed at equilibrium}} \tag{15-151}$$

$$T = \frac{QC_0 t}{Mq_e} = \frac{QC_0 t}{V\rho_a(1-\varepsilon)q_e} \tag{15-152}$$

$$\tau = \frac{L}{v} = \frac{L}{Q/\varepsilon A} = \frac{\varepsilon V}{Q} \tag{15-153}$$

$$T = \frac{\varepsilon C_0 t}{\tau\rho_a(1-\varepsilon)q_e} \tag{15-154}$$

where T = mass throughput, dimensionless
Q = liquid flow rate, m^3/s
M = mass of adsorbent, g
q_e = solid-phase concentration in equilibrium with influent concentration C_0, mg/g
τ = EBCT × ε = fluid residence time in bed, min
EBCT = V/Q, min
V = bed volume, m^3
v = fluid velocity in bed pore space, m/s

A dimensionless group, the surface solute distribution parameter is defined as

$$D_g = \frac{\rho_a q_e(1-\varepsilon)}{\varepsilon C_0} = \frac{\text{adsorbate on the adsorbent}}{\text{adsorbate in bed voids}}\bigg|_{\text{equilibrium}} \tag{15-155}$$

Substituting Eq. 15-153 into Eq. 15-152 results in the final expression of dimensionless time:

$$T = \frac{t}{\tau D_g} = \frac{t}{\varepsilon\,(\text{EBCT})\,D_g} \approx \frac{C_0 t}{\text{EBCT}\rho_a q_e\,(1-\varepsilon)} \tag{15-156}$$

The throughput is a valuable way to express time because the area above the effluent profile must be equal to 1.0 when the effluent concentration is plotted as C/C_0. Substitution of Eqs. 15-57, 15-59, and 15-156 into Eq. 15-136 and rearrangement yields the following dimensionless liquid-phase mass balance:

$$-\frac{\partial \bar{C}}{\partial \bar{z}} + \frac{1}{\text{Pe}}\frac{\partial^2 \bar{C}}{\partial \bar{z}^2} = \frac{1}{D_g}\frac{\partial \bar{C}}{\partial T} + 3\text{St}\left(\bar{C} - \bar{C}_s\right) \tag{15-157}$$

Two dimensionless groups arise in the process of conversion and they are defined as

$$\text{Pe} = \frac{Lv}{E} \tag{15-158}$$

$$\text{St} = \frac{k_f \tau(1-\varepsilon)}{\varepsilon R} \tag{15-159}$$

Table 15-15
Dimensionless groups used in the modeling of adsorption processes

Dimensionless Group	Equation	Definition
D_g	$\frac{\rho_a q_e (1-\varepsilon)}{\varepsilon C_0}$	$\left.\frac{\text{Mass of solute in solid phase}}{\text{Mass of solute in liquid phase}}\right\vert_{\text{equilibrium}}$
Pe	$\frac{Lv}{E}$	$\frac{\text{Solute transfer rate by advection}}{\text{Solute transfer rate by axial dispersion}}$
St	$\frac{k_f \tau (1-\varepsilon)}{\varepsilon R}$	$\frac{\text{Solute liquid-phase mass transfer rate}}{\text{Solute transfer rate by advection}}$
Bi	$\frac{k_f R (1-\varepsilon)}{D_S D_g \varepsilon}$	$\frac{\text{Solute liquid-phase mass transfer rate}}{\text{Solute intraparticle mass transfer rate}}$
Ed_s	$\frac{D_S D_g \tau}{R^2}$	$\frac{\text{Solute transfer rate by intraparticle surface diffusion}}{\text{Solute transfer rate by advection}}$
Ed_p^a	$\frac{D_p \tau (1-\varepsilon)\varepsilon_p}{R^2 \varepsilon}$	$\frac{\text{Solute transfer rate by intraparticle pore diffusion}}{\text{Solute transfer rate by advection}}$

[a] Ed_p is a dimensionless intraparticle mass transfer group that characterizes pore diffusion contribution for the pore surface diffusion model (PSDM).

The Peclet number, Pe, is a measure of the amount of dispersion present in the fixed bed and the Stanton number, St, is a measure of the film mass transfer rate as compared to the rate by advection. Insight into the physical meaning of these and other dimensionless groups is summarized in Table 15-15.

The initial condition for Eq. 15-157 was given by Eq. 15-137. Converting to dimensionless terms yields

$$\overline{C}(0 \le \overline{z} \le 1, T = 0) = 0 \tag{15-160}$$

Substituting the dimensionless positions and Eqs. 15-153, 15-155, 15-156, and 15-157 into Eq. 15-143 results in the final form of the first boundary condition:

$$1 - \overline{C}(\overline{z} = 1, T) = \frac{1}{D_g} \frac{\partial}{\partial T} \left[\int_0^1 \left(\overline{C} + 3D_g \int_0^1 \overline{q}\overline{r}^2 \, d\overline{r} \right) d\overline{z} \right] \tag{15-161}$$

The second boundary condition was given by the system of Eq. 15-145 and 15-146. Converting to dimensionless form yields

$$\frac{\partial^2 \overline{C}(\overline{z} = 1, T)}{\partial T \, \partial \overline{z}} = 0 \tag{15-162}$$

$$\frac{\partial \overline{C}(\overline{z}, T = 0)}{\partial \overline{z}} = 0 \tag{15-163}$$

The solid-phase mass balance was given in Eq. 15-147. Converting to dimensionless form yields this expression and one dimensionless group, Ed_s.

$$\frac{\mathrm{Ed}_s}{\bar{r}^2}\frac{\partial}{\partial \bar{r}}\left(\bar{r}^2\frac{\partial \bar{q}}{\partial \bar{r}}\right)=\frac{\partial \bar{q}}{\partial T} \tag{15-164}$$

The dimensionless group, the surface diffusion modulus, Ed_s, is defined by (see Table 15-15)

$$\mathrm{Ed}_s=\frac{D_s D_g \tau}{R^2} \tag{15-165}$$

The initial condition for Eq. 15-164 was given in Eq. 15-148. Converting to dimensionless terms yields

$$\bar{q}(0\le\bar{r}\le1, 0\le\bar{z}\le1,\ T=0)=0 \tag{15-166}$$

The first boundary condition for Eq. 15-164 was given in Eq. 15-148. Substituting in dimensionless groups in this expression, and the Biot number (see Table 15-15) yields

$$\mathrm{Bi}\left(\bar{C}-\bar{C}_s\right)=\frac{\partial \bar{q}}{\partial \bar{r}} \tag{15-167}$$

The Biot number based on surface diffusivity, Bi, is defined by

$$\mathrm{Bi}=\frac{k_f R(1-\varepsilon)}{D_s D_g \varepsilon}=\frac{\dfrac{k_f \tau(1-\varepsilon)}{\varepsilon R}}{\dfrac{D_s D_g \tau}{R^2}}=\frac{\mathrm{St}}{\mathrm{Ed}_s} \tag{15-168}$$

The second boundary condition for Eq. 15-164 was given in Eq. 15-149. Converting to dimensionless variables yields

$$\frac{\partial \bar{q}_i(\bar{r}=0,\ 0\le\bar{z}\le1,\ T\ge0)}{\partial \bar{r}}=0 \tag{15-169}$$

For convenience the final form of the dimensionless mass balances are reported in Table 15-16. The conversion of the equations into dimensionless form yields several important dimensionless parameters. The dimensionless group that would appear if pore diffusion were included are also shown in Table 15-15 (Crittenden et al., 1986).

For high Pe numbers, plug flow conditions exist and dispersion is negligible. For plug conditions, the Danckwerts boundary conditions are no longer necessary, and the boundary simply becomes the concentration at the entrance, which is equal to the influent concentration.

The asymptotic solutions for long fixed-bed adsorbers and linear adsorption can be used to develop relationships between the dimensionless groups and the controlling transport mechanism, as well as between the dimensionless groups and the relative size of the mass transfer zone due to each mass transfer mechanism. To produce the same size mass transfer zone, the following relationships between the Pe number, Ed_s, Ed_p, and St can be derived from the analytical solutions to the plug flow pore and surface

Table 15-16
Dimensionless form of equations for dispersed-flow homogenous surface diffusion model (DFHSDM)

Phase	Equation	Initial condition	Boundary conditions
Liquid	$-\frac{\partial \bar{C}}{\partial \bar{z}} + \frac{1}{\text{Pe}} \frac{\partial^2 \bar{C}}{\partial \bar{z}^2} = \frac{1}{D_g} \frac{\partial \bar{C}}{\partial T} + 3\text{St}\left(\bar{C} - \bar{C}_s\right)$	$\bar{C}\begin{pmatrix} 0 \leq \bar{z} \leq 1 \\ T = 0 \end{pmatrix} = 0$	$1 - \bar{C}(\bar{z} = 1, T) = \frac{1}{D_g} \frac{\partial}{\partial T}\left[\int_0^1 \left(\bar{C} + 3D_g \int_0^1 \bar{q}\bar{r}^2 \, d\bar{r}\right) d\bar{z}\right]$ $\frac{\partial^2 \bar{C}(\bar{z} = 1, T)}{\partial T \partial \bar{z}} = 0$ $\frac{\partial \bar{C}(\bar{z}, T = 0)}{\partial \bar{z}} = 0$ $\text{Bi}\left(\bar{C} - \bar{C}_S\right) = \frac{\partial \bar{q}}{\partial \bar{r}}$
Solid	$\frac{\text{Ed}_s}{\bar{r}^2} \frac{\partial}{\partial \bar{r}}\left(\bar{r}^2 \frac{\partial \bar{q}}{\partial \bar{r}}\right) = \frac{\partial \bar{q}}{\partial T}$	$\bar{q}\begin{pmatrix} 0 \leq \bar{r} \leq 1 \\ 0 \leq \bar{z} \leq 1 \\ T = 0 \end{pmatrix} = 0$	$\frac{\partial \bar{q}_i \begin{pmatrix} \bar{r} = 0, \\ 0 \leq \bar{z} \leq 1 \\ T \geq 0 \end{pmatrix}}{\partial \bar{r}} = 0$

diffusion model by Rosen (1954), and to the dispersed flow model which was proposed by Dankwerts (1953) and includes only axial dispersion (Crittenden et al., 1986). The value of the dimensionless groups can be related such that equivalent amounts of spreading occurs in the mass transfer zone due to each of the various mass transport mechanisms:

$$\text{Pe} = 3 \cdot \text{St} = 15 \cdot \text{Ed}_s = 15 \cdot \text{Ed}_p \tag{15-170}$$

Accordingly, the mass transfer zone for a linear isotherm is similar in size if only dispersion, film transfer, or surface diffusion are the primary cause for spreading in the mass transfer zone, and the values of Pe, St, Ed_s, and Ed_p follow Eq. 15-170. For example, Pe, St, Ed_s, and Ed_p values equal to 45, 15, 3, and 3, respectively, would result in equal spreading of the mass transfer zone when the respective mechanism is controlling.

If surface diffusion controls the mass transfer rate, it is interesting to determine the controlling mechanism from the relative magnitudes of St and Ed_s. As shown in Eq. 15-158, the Biot number compares the film transfer rate to the surface diffusion rate. For linear isotherms, as shown in Eq. 15-170, a Biot number of 5.0 indicates that both film transfer and surface diffusion have an equal impact on the breakthrough curve for a $1/n$ of 1.0. As far as when external mass transfer or intraparticle resistance is concerned, external mass transfer controls for Bi numbers less than 1 and intraparticle mass transfer controls for Bi greater than 20.

However, the Freundlich exponent, $1/n$, affects the shape of the breakthrough curve and influences the relative importance of the two mass transfer mechanisms with regard to control of the adsorption rate. For cases where external mass transfer had an equal importance, Hand et al. (1984) showed that Bi numbers of 3, 3.2, and 4 for $1/n$ values of 0.2, 0.6 and 0.8, respectively, were required. For cases where intraparticle mass transfer controlled, Hand et al. (1984) showed that Bi numbers greater than 8, 16, and 20 for $1/n$ values of 0.2, 0.6, and 0.8, respectively, were required. For cases where external mass transfer controlled, (Hand et al., 1984) Bi numbers less than 1.0 were required for all the $1/n$ values 0 to 1.

When an adsorbate enters a fresh GAC column, it establishes a mass transfer zone that spreads out to a constant shape. This shape is called a constant pattern. Hand et al. (1984) presented constant pattern solutions to the plug flow homogeneous surface diffusion model (PFHSDM), which can be used in many cases to obtain the identical results as given by the numerical solution of the PFHSDM. In this method, the solutions to the PFHSDM were fit by polynomials. The following four dimensionless groups and parameters characterize the constant pattern solutions to the PFHSDM: Solute distribution parameter, D_g; Stanton number, St; Biot number, Bi; and Freundlich isotherm intensity constant, $1/n$. With these characteristic parameters, the breakthrough curve can be expressed in the generalized form:

$$\bar{C} = f\left(T,\ D_g,\ \text{St},\ \text{Bi},\ \frac{1}{n}\right) \tag{15-171}$$

According to Eq. 15-154, the throughput ratio, T, is linked to the D_g.Thus, the number of independent parameters in Eq. 15-171 is reduced to four. However, the multitude of possible solutions is nevertheless very great if all four parameters are considered. Therefore, two essential assumptions were made by Hand et al. (1984): (1) The isotherm can be described by a Freundlich equation over the whole concentration range, and the Freundlich $1/n$ value is between 0 and 1; and (2) the constant pattern of the mass transfer zone is completely developed. Under these conditions, Hand et al. (1984) determined the minimum packed-bed residence time, $\tau_{\min}$, which is necessary to achieve constant pattern.

The minimum $\tau_{\min}$ values are expressed as Stanton numbers, $\text{St}_{\min}$, on Fig. 15-28a for different Freundlich exponents and Bi numbers between 0.5 and 100 ($0.5 \le \text{Bi} \le 100$). In the case of $\text{Bi} < 0.5$, the mass transfer rate is controlled by film diffusion, which allows the use of the analytical solution, which was proposed by Fleck et al. (1973). In the case of Bi>100, surface diffusion controls the adsorption rate and identical constant pattern breakthrough curves are obtained for all Bi numbers >100 for a given $1/n$ when they are expressed in terms of T for $\text{St}_{\min}$.

Using Fig. 15-28a, the minimum Stanton number can be determined in a simple and quick manner, if the Biot number is known. The equations corresponding to the curves in Fig. 15-28a are given in Table 15-17

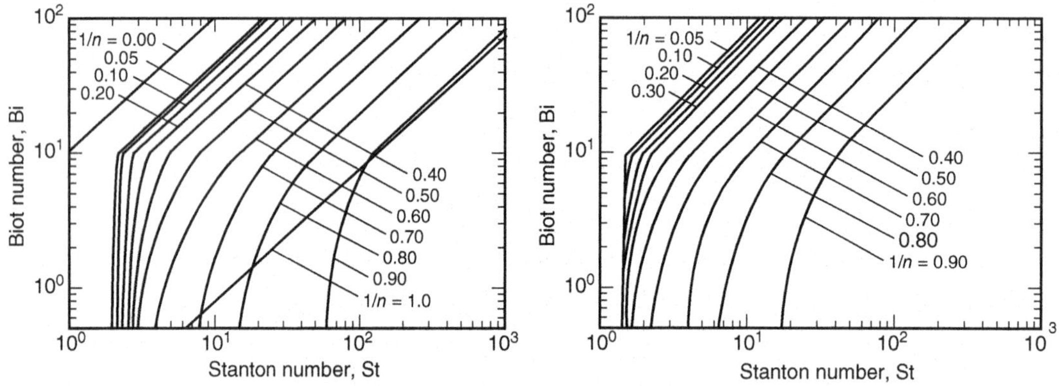

Figure 15-28
(a) Minimum Stanton number necessary to reach a constant pattern for various Bi numbers and $1/n$ values and (b) minimum Stanton number necessary to reach within 10 percent constant pattern for various Bi values (Hand et al., 1984).

Table 15-17
Minimum Stanton number required to achieve constant pattern conditions as function of Bi for various $1/n$ values

Freundlich Isotherm Exponent Parameter $1/n$	Empirical Equation $St_{min} = A_0 (Bi) + A_1$			
	$0.5 \le Bi \le 10$		$10 \le Bi \le \infty$	
	A_0	A_1	A_0	A_1
0.05	2.10526×10^{-2}	1.98947	0.22	—
0.10	2.10526×10^{-2}	2.18947	0.24	—
0.20	4.21053×10^{-2}	2.37895	0.28	—
0.30	1.05263×10^{-1}	2.54737	0.36	—
0.40	2.31579×10^{-1}	2.68421	0.50	—
0.50	5.26316×10^{-1}	2.73684	0.80	—
0.60	1.15789×10^{0}	3.42105	1.50	—
0.70	1.78947×10^{0}	7.10526	2.50	—
0.80	3.68421×10^{0}	13.1579	5.00	—
0.90	6.31579×10^{0}	56.8421	12.00	—

Adapted from Hand et al. (1984).

(Hand et al., 1984). If $St > St_{min}$, the shape of the breakthrough curve is no longer dependent on the residence time or on the adsorber length. Consequently, Eq. 15-171 reduces to this expression as only the solution at St_{min} and all others may be derived from it:

$$\bar{C} = f\left(T_{min}, St_{min}, Bi, 1/n\right) \tag{15-172}$$

where $T_{\min}$ is equal to the ratio of the adsorption time required to reach constant pattern; $t_{\min}$ is the minimum packed-bed residence time in the fixed bed, $\tau_{\min}$; $\tau_{\min}$ is the packed-bed contact time required for constant pattern and is directly related to a measure of the minimum length of the fixed bed, $L_{\min}$, required to achieve constant pattern; and the corresponding St value is equal to $\mathrm{St}_{\min}$, which is the minimum Stanton number required to achieve constant pattern.

According to Eq. 15-172, Bi and $1/n$ are the only two parameters necessary for all possible solutions to the PFHSDM under constant pattern conditions. Accordingly, Hand et al. (1984) has provided all solutions to Eq. 15-172 for Bi ≥ 0.5 and Freundlich exponents $0.05 \le 1/n \ge 0.90$ in the form of the following empirical equation:

$$T_{\min}\left(\mathrm{St}_{\min}, \mathrm{Bi}, \frac{1}{n}\right) = A_0 + A_1\left(\frac{C}{C_0}\right)^{A_2} + \frac{A_3}{1.01 - (C/C_0)^{A_4}} \qquad (15\text{-}173)$$

The coefficients, A_0 through A_4, and the validity range of Eq. 15-173 for a $1/n$ value of 0.5 are given in Table 15-18 [constants for additional values of $1/n$ are provided in Hand et al. (1984) and the electronic Table E-7 at the website listed in App. E]. Thus, for each given effluent concentration, C/C_0, the respective throughput, $T_{\min}$, can be calculated.

The breakthrough curve, which is given by Eq. 15-173, can be used to calculate breakthroughs for any residence time, $\tau > \tau_{\min}$, because it remains constant in shape and travels at a constant velocity. Since the breakthrough curves are parallel, the operation time t for other residence times can be calculated from $T_{\min}$ and $\tau_{\min}$ according to

$$t = \tau_{\min} D_g T_{\min} + (\tau - \tau_{\min}) D_g \qquad (15\text{-}174)$$

Substituting the expression $T = t/\tau D_g$ into Eq. 15-174 yields

$$T = 1 + (T_{\min} - 1) \cdot \frac{\tau_{\min}}{\tau} \qquad (15\text{-}175)$$

Table 15-18
Parameter values used in Eq. 15-173 for constant pattern solutions to the plug flow homogeneous surface diffusion model for $1/n = 0.5$

Bi	A_0	A_1	A_2	A_3	A_4	$\left(\frac{C}{C_0}\right)_{\min}$	$\left(\frac{C}{C_0}\right)_{\max}$
0.5	−0.040800	1.099652	0.158995	0.005467	0.139116	0.01	0.99
4.0	−0.040800	0.982757	0.111618	0.008072	0.111404	0.01	0.99
10.0	0.094602	0.754878	0.092069	0.009877	0.090763	0.01	0.99
14.0	0.023000	0.802068	0.057545	0.009662	0.084532	0.01	0.99
25.0	0.023000	0.793673	0.039324	0.009326	0.082751	0.01	0.99
≥100.0	0.529213	0.291801	0.082428	0.008317	0.075461	0.01	0.99

Adapted from Hand et al. (1984). Parameters for other values of $1/n$ are available in the electronic Table E-7 at the website listed in App. E.

St_{min} is calculated using the equation in Table 15-17 (Hand et al., 1984) and τ_{min} and L_{min} can then be calculated from St_{min} using the following equation:

$$\tau_{min} = \frac{St_{min}\ \varepsilon R}{k_f\ (1-\varepsilon)} \tag{15-176}$$

$$L_{min} = v \times \frac{\tau_{min}}{\varepsilon} = v \times EBCT_{min} \tag{15-177}$$

Finally, Eq. 15-173 and 15-176 can be substituted into Eq. 15-175 and an expression for T can be derived for specified Bi and $1/n$ values:

$$T = 1 + \left[A_0 + A_1\left(\frac{C}{C_0}\right)^{A_2} + \frac{A_3}{1.01-(C/C_0)^{A_4}} - 1\right] \cdot \frac{St_{min}\varepsilon R}{k_f(1-\varepsilon)\tau} \tag{15-178}$$

Equation 15-178 can be used to calculate the value of T for each given value of C/C_0 if the Biot number and $1/n$ value are known. The parameters A_0 through A_4 for every Bi and $1/n$ combination are reported by Hand et al. (1984) and the electronic Table E-7 at the website listed in App. E. It is recommended to use the A_0 through A_4 parameters for a larger Biot number and/or $1/n$ values because this will give the largest mass transfer zone and result in the most conservative design. For example, if $1/n$ is 0.245, use the solution for a $1/n$ value of 0.3. For the parameters A_0 through A_4 interpolation is not possible because significant errors would result with respect to determining St_{min}. The equations for higher and lower $1/n$ values could also be used and the two values for St_{min} should be interpolated.

Equation 15-178 can also be used for adsorbers with a length smaller than the length required to establish constant pattern, L_{min}. When the adsorbate first enters the bed, a very steep mass transfer zone is established. As it migrates into the bed, the mass transfer zone expands until it reaches the constant pattern shape. Consequently, the breakthrough profile expands as bed length increases. Accordingly, if breakthrough curves are desired for residence times less than τ_{min}, the constant pattern solution, given by Eq. 15-178 is conservative because the constant pattern breakthrough profile and mass transfer zone would be broader than the actual profile. If a 10 percent error in T can be tolerated using the constant pattern solution, then the smallest St that can be used is given in Fig. 15-28b. To determine St for a 10 percent error in T, the equations are given in Hand et al. (1984) or Table E-8 at the website listed in App. E can be considered.

To complete the presentation and provide all solutions to the PFHSDM, the analytical solution to the PFHSDM for liquid-phase mass transfer controls the adsorption rate (Bi < 1.0) and $1/n$ is less than 1.0 is presented. Fleck et al. (1973) has provided the following analytical solution to the PFHSDM for this situation:

$$T = \frac{1}{3St}\left\{1 + \ln\left(\frac{C}{C_0}\right) - \frac{1}{n-1}\ln\left[1-\left(\frac{C}{C_0}\right)^{n-1}\right] + \gamma\right\} + 1 \tag{15-179}$$

$$\gamma = \frac{1/n}{(1/n-1)} \sum_{k=1}^{\infty} \frac{1/n}{k[k(1-1/n)+1/n]} \tag{15-180}$$

The series given by Eq. 15-180 does not converge very rapidly so the computed values of γ can be obtained for a given $1/n$ from the electronic Table E-9 at the website listed in App. E and Hand et al. (1984). Fleck et al. (1973) assumed constant pattern conditions exist, and this can be guaranteed for the various $1/n$ values by examining Fig. 15-28a.

Intraparticle mass transfer will control the adsorption rate for $1/n$ equal to 0.0 (irreversible adsorption). Wicke (1939) provided the following PFHSDM solution for the effluent concentration in the case of irreversible adsorption and constant pattern conditions:

$$\bar{C}(\bar{z}=1,\ T) = 1 - \frac{6}{\pi^2} \sum_{k=1}^{\infty} \frac{1}{k^2} \exp\left\{-k^2\left[\pi^2 \mathrm{Ed}_s\left(\frac{TD_g - 1}{D_g} - 1\right) + 0.64\right]\right\} \tag{15-181}$$

Constant pattern conditions require that Ed_s must be greater than 0.101, which is located to the right of the line drawn on Fig. 15-28a for $n = 0.0$. By examining Eq. 15-181, it can be demonstrated that as Ed_s increases, the mass transfer zone occupies a smaller fraction of the bed. To obtain a convergent solution, T must be greater than the following:

$$T \geq \frac{1}{D_g}\left[1 + D_g\left(1 - \frac{0.64}{\pi^2 \mathrm{Ed}_s}\right)\right] \tag{15-182}$$

Usually, no more than three to six terms in the infinite series are needed to obtain an accurate solution for Eq. 15-181. The only exceptions are when the adsorbate first begins to appear in the effluent and when the exponential argument in the series does not vanish very rapidly with increasing k.

To decide how many terms in Eq. 15-181 are needed, the following equation can be used to evaluate the error associated with ignoring higher order terms, Err. $(k < N_0)$:

$$\mathrm{Err.}(k < N_0) \leq \left(\frac{1}{N_0^2} + \frac{1}{N_0}\right) \exp\left\{-N_0^2\left[\pi^2 \mathrm{Ed}_s\left(\frac{TD_g - 1}{D_g} - 1\right) + 0.64\right]\right\} \tag{15-183}$$

Rosen (1954) has provided the following solution to PFHSDM for linear adsorption isotherm ($1/n = 1.0$), which expresses the effluent concentration as a function of time:

$$\bar{C}(\bar{z}=1,\ T) = \frac{1}{2}\left\{1 + \mathrm{erf}\left[\frac{(TD_g - 1)/D_g - 1}{2\sqrt{(1+5\,\mathrm{Bi})/(15\,\mathrm{Ed}_s)}}\right]\right\} \tag{15-184}$$

This solution requires Ed_s to be greater than 13.33, and this region is located to the right of the line for $1/n = 1.0$ in Fig. 15-28a.

The surface diffusion coefficient can be obtained by relating the surface diffusion flux to the pore diffusion flux. This results in following correlation, which may be used to calculate the surface diffusion coefficient (Crittenden et al., 1987a):

$$D_s = (\text{SPDFR})(\text{PDFC}) \tag{15-185}$$

$$\text{PDFC} = \left(\frac{\varepsilon_P \; C_0 \; D}{\rho_A \; q_0 \; \tau_P}\right) \tag{15-186}$$

where SPDFR = surface-to-pore diffusion flux ratio, dimensionless
PDFC = pore diffusion flux, m^2/s

The SPDFR is the correlating parameter for determining D_s, and, for single solutes, Crittenden et al. (1987a) has found that it is between 4 and 9 using the maximum PDFC ($\tau_p = 1.0$). To be conservative, a value of 4 may be used as long as there is no impact of background DOC on the breakthrough of the single component. (Single-solute SOC concentration > 5 times DOC concentration.)

However, it is rare to encounter situations where the background DOC does not have an impact on SOC removal with the possible exception of industrial waste treatment. Most often the presence of DOC has a tremendous impact on SOC removal in drinking water applications, and it usually blocks surface diffusion and intraparticle transport occurs only by pore diffusion. Experience has shown that one can use SPDFR values of 0.4 to 1 with the maximum PDFC in order to perform hand calculations using the CPHSDM (Hand et al., 1989). Crittenden et al. (1987a) and Sontheimer et al. (1988) have shown that good comparisons with data can be obtained using this approach. More complex protocols have been developed when using the pore surface diffusion model, and this protocol and the pore surface diffusion model have been built into AdDesignS software, which is currently commercially available.

Procedure for Application of CPHSDM Solutions

1. Calculate D_g and Bi from the following equations:

$$D_g = \frac{\rho_a q_e (1-\varepsilon)}{\varepsilon C_0} \tag{15-187}$$

where $q_e = KC_0^{1/n}$

$$\text{Bi} = \frac{k_f R(1-\varepsilon)}{D_s D_g \varepsilon} \tag{15-188}$$

2. Using the appropriate equation relating St_{min} to Bi from Table 15-17 (Hand et al., 1984), calculate St_{min} for the observed Bi and $1/n$, and

then calculate $\text{EBCT}_{\min}$ or $\tau_{\min}$:

$$\text{St}_{\min} = A_0(\text{Bi}) + A_1 \tag{15-189}$$

$$\text{EBCT}_{\min} = \frac{\tau_{\min}}{\varepsilon} = \frac{\text{St}_{\min} R}{k_f(1-\varepsilon)} \tag{15-190}$$

3. Obtain the constant pattern solution in terms of C/C_0 versus T using the parameters that are given in Table 15-18 for a $1/n$ value of 0.5 (or the electronic Table E-7 at the website listed in App. E for other values of $1/n$).

$$T\left(\text{St}_{\min}, \text{Bi}, \frac{1}{n}\right) = A_0 + A_1\left(\frac{C}{C_0}\right)^{A_2} + \frac{A_3}{1.01 - (C/C_0)^{A_4}} \tag{15-191}$$

4. Convert the T values obtained for constant pattern solution to elapsed time using the following equation:

$$t_{\min} = \tau_{\min}\left(D_g + 1\right)T \tag{15-192}$$

This is the constant pattern solution that corresponds to an adsorber with $\text{EBCT}_{\min}$.

5. To convert elapsed time corresponding to the $\text{EBCT}_{\min}$ to the desired EBCT, the travel time of the wave is added or subtracted according to the following equation:

$$t = t_{\min} + (\tau - \tau_{\min})\left(D_g + 1\right) \tag{15-193}$$

6. Convert the time values to usage rates.

$$\left\{\begin{matrix}\text{Adsorbent} \\ \text{usage} \\ \text{rate}\end{matrix}\right\} = \frac{M_{\text{Adsorbent}}}{Qt} \tag{15-194}$$

7. The predicted breakthrough profiles can be use for GAC beds with shorter lengths than $L_{\min}$. The length that corresponds to an error of 10 percent of the breakthrough time can be estimate using the parameters that are given in the electronic Table E-8 at the website listed in App. E and Hand et al. (1984).

8. The EBCT of the mass transfer zone can be estimated using the following equation:

$$\text{EBCT}_{\text{MTZ}} = \left[T\left(\frac{c}{c_0} = 0.95\right) - T\left(\frac{c}{c_0} = \text{treatment objective}\right)\right]\text{EBCT}_{\min} \tag{15-195}$$

This length corresponds to a breakthrough that corresponds to the treatment objective, for example, 5 percent of the influent and a saturation of 95 percent because as stated above it can be shown that $C/C_0 = q/q_e$ so the upstream end of the MTZ for this calculation is 95 percent saturated.

Example 15-12 Using the constant pattern HSDM

GAC is being used to treat a groundwater containing 500 μg/L of trichloroethene (TCE). The design flow is 0.89 m^3/min and the treatment objective is 5 μg/L. Calculate the size of the adsorber, $EBCT_{min}$, constant pattern solution, GAC usage rate, and $EBCT_{MTZ}$ using the CPHSDM. The properties of the GAC and water are provided below.

GAC Properties:

Calgon Filtrasorb F-400 (12 × 40 mesh), $\rho_F = 0.45$ g/cm^3, $\rho_a = 0.8034$ g/cm^3

$d_P = 0.1026$ cm particle porosity $\varepsilon_P = 0.641$, EBCT = 10 min, $\varepsilon = 0.44$

Single-solute Freundlich K = 2030 (μg/gm)(L/μg)$^{1/n}$, Freundlich $1/n$ = 0.48

Assume the TCE Freundlich K is reduced from 2030 (μg/gm)(L/μg)$^{1/n}$ to 1062 (μg/g)(L/μg)$^{1/n}$ due to background organic compounds in the groundwater.

Water Properties: $T = 10°C$, $\rho_W = 999.7$ kg/m^3, $\mu = 1.307 \times 10^{-3}$ N · s/m^2.

1. Calculate Bi using Eq. 15-168:

$$\text{Bi} = \frac{k_f\ R\ (1-\varepsilon)}{D_s D_g \varepsilon}$$

The film transfer rate is estimated using Gnielinski correlation from Table 7-5:

$$k_f = \frac{[1 + 1.5\ (1-\varepsilon)]\ D_1}{d_p}\left[2 + 0.644\ \text{Re}^{1/2}\ \text{Sc}^{1/3}\right]$$

See Table 7-5 to calculate k_f, Re, and Sc, and Table 7-2 to calculate D_l. Typical superficial fluid velocities in GAC fixed beds are from 5.0 to 10 m/h. For this problem assume 5.0 m/h:

$V_s = 5$ m/h, the interstitial velocity $v_i = \dfrac{v_s}{\varepsilon} = \dfrac{5\text{ m/h}}{0.44} = 11.36$ m/h

$$\begin{aligned}\text{Re} &= \frac{\rho_W d_p v_i}{\mu}\\ &= \frac{(999.7\text{ kg/m}^3)\ (0.001026\text{ m})\ [(11.36\text{ m/h})(1\text{ h}/3600\text{ s})]}{(0.44)(1.3097 \times 10^{-3}\text{N} \cdot \text{s/m})}\\ &= 5.63\end{aligned}$$

For TCE, molal volume V_b is calculated using the values in Table 7-3:

$$V_b = 2\,(14.8) + 3\,(21.6) + 3.7 = 98.1\,\frac{\text{cm}^3}{\text{mol}}$$

$$D_1 = \frac{13.26\times10^{-9}}{(\mu_w)^{1.14}(V_b)^{0.589}} = \frac{13.26\times10^{-9}}{(1.3097)^{1.14}\,(98.1)^{0.589}}$$

$$= 6.54\times10^{-10}\,\frac{\text{m}^2}{\text{s}}$$

$$\text{Sc} = \frac{\mu}{\rho D_1} = \frac{1.3097\times10^{-3}\,\text{N}\cdot\text{s/m}}{(999.7\,\text{kg/m}^3)(6.54\times10^{-10}\,\text{m}^2/\text{s})} = 2000$$

$$\text{k}_f = \frac{[1+1.5(1-0.44)]\,(6.54\times10^{-10}\,\text{m}^2/\text{s})}{0.001026\text{m}}$$

$$\times\left[2+0.644(5.63)^{1/2}(2000)^{1/3}\right] = 2.49\times10^{-5}\,\text{m/s}$$

Apply shape correction factor to k_f:

$$k_f = \text{SCF}\times k_f = (1.5)(2.49\times10^{-5}\text{ m/s}) = 3.73\times10^{-5}\text{ m/s}$$

Use Eqs. 15-185 and 15-186 to calculate D_s and PDFC, respectively:

$$D_s = \text{PDFC}\times\text{SPDFR}$$

$$\text{PDFC} = \frac{D_{l\varepsilon P}\,C_0}{\tau_P K C_0^{1/n}\rho_a} \qquad \text{where } \tau_P = 1.0$$

$$\text{PDFC} = \frac{(6.54\times10^{-10}\,\text{m}^2/\text{s})(0.641)(500\,\mu\text{g/L})}{(1.0)(1062\,\mu\text{g/g}(\text{L}/\mu\text{g})^{0.48})(500\,\mu\text{g/L})^{0.48}(803.4\text{ g/L})}$$

$$= 1.24\times10^{-14}\,\text{m}^2/\text{s}$$

As we will see later in the chapter, the background organic matter can reduce surface diffusion, and in this case we will assume that the SPDFR is 1.0:

$$D_s = (1.24\times10^{-14}\text{ m}^2/\text{s})\,(1.0) = 1.24\times10^{-14}\text{ m}^2/\text{s}$$

$$\mathrm{Bi} = \frac{(3.73\times10^{-5}\,\mathrm{m/s})\left(\dfrac{0.001026\ \mathrm{m}}{2}\right)(1-0.44)}{(1.24\times10^{-14}\mathrm{m^2/s})(42{,}870)(0.44)(1.0)} = 45.7$$

2. Calculate $\mathrm{St_{min}}$ from Eq. 15-189:
From Table 15-17, the coefficient A_0 for $1/n = 0.48$ and $\mathrm{Bi} = 45.7$ can be interpolated between $A_0 = 0.5$ ($1/n = 0.4$) and $A_0 = 0.8$ ($1/n = 0.5$). Since $0.48 \cong 0.5$, use $A_0 = 0.8$ and calculate $\mathrm{St_{min}}$ using Eq. 15-189.

$$\mathrm{St_{min}} = A_0\,(\mathrm{Bi}) = 0.8\,(45.7) = 36.5$$

Calculate $\mathrm{EBCT_{min}}$ required for constant pattern using Eq. 15-190:

$$\mathrm{EBCT_{min}} = \frac{\mathrm{St_{min}}R}{k_f(1-\varepsilon)} = \frac{(36.5)\left(\dfrac{0.001026\ \mathrm{m}}{2}\right)}{(3.73\times10^{-5}\,\mathrm{m/s})(1-0.44)} = 897\ \mathrm{s}$$

$$= 14.95\ \mathrm{min}$$

$$\tau_{min} = (0.44)(14.95\ \mathrm{min}) = 6.58\ \mathrm{min}$$

3. Calculation of constant pattern solution using Eq. 15-191:

$$T\left(\mathrm{Bi_s}, \frac{1}{n}, \mathrm{St_{min}}\right) = A_0 + A_1\left(\frac{C}{C_0}\right)^{A_2} + \frac{A_3}{1.01-(C/C_0)^{A_4}}$$

In Table 15-18, the closest values to $1/n = 0.48$ and $\mathrm{Bi} = 38.8$ are used. These are the values for $1/n = 0.5$ and $\mathrm{Bi} = 25$: $A_0 = 0.023000$, $A_1 = 0.793673$, $A_2 = 0.039324$, $A_3 = 0.009326$, and $A_4 = 0.08275$. Calculate T, t_{min}, and t for C/C_0 values from 0.01 to 0.95 as shown in the table below using Eqs. 15-191, 15-192, and 15-193; respectvely.

$$T = 0.023 + 0.793673\left(\frac{C}{C_0}\right)^{0.039324} + \frac{0.009326}{1.01-(C/C_0)^{0.08275}}$$

$$t_{min}(\mathrm{d}) = \tau_{min}(D_g+1)\,T = \frac{(6.58\ \mathrm{min})(42{,}870+1)(T)}{1440\ \mathrm{min}/d} = 196\times T$$

$$t(\mathrm{d}) = t_{min} + (\tau - \tau_{min})(D_g+1)$$

$$= t_{min} + \frac{(4.4-6.58)(\mathrm{min})(42{,}870+1)}{1440\ \mathrm{min/d}} = t_{min} - 65$$

HSDM solution using constant pattern

C/C_0	T	t_{min} (days)	t (days)
0.01	0.71	140	75
0.05	0.77	151	86
0.10	0.80	156	92
0.20	0.84	164	99
0.30	0.87	170	105
0.40	0.90	176	112
0.50	0.94	184	119
0.60	0.98	192	128
0.70	1.04	205	140
0.80	1.14	223	158
0.90	1.31	257	192
0.95	1.47	288	223

4. Calculate GAC usage rate using Eq. 15-194:

   For treatment objective $C/C_0 = 5\ \mu\text{g/L}/500\ \mu\text{g/L} = 0.01$:

   $$T = 0.71,\ t_{min} = 140 \text{ days},\ t = 75 \text{ days}$$

   After 75 days of operation the effluent will reach or exceed the MCL of 5 μg/L:

   $$\begin{Bmatrix}\text{Bed}\\ \text{volumes}\\ \text{treated}\end{Bmatrix} = \text{BVT} = \frac{t\varepsilon}{\tau} = \frac{(75\text{ d})\,(0.44)}{\dfrac{4.4\text{ min}}{1440\text{ min/d}}} = 10{,}800$$

   $$\begin{Bmatrix}\text{Specific}\\ \text{throughput}\\ \text{m}^3/\text{kg}\end{Bmatrix} = \frac{\text{BVT}}{\rho_F} = \frac{10{,}800}{450\text{ kg/m}^3} = 24\,\frac{\text{m}^3\text{ water treated}}{\text{kg GAC}}$$

5. Calculate the EBCT_{MTZ} using Eq. 15-195:

   $$\begin{aligned}\text{EBCT}_{MTZ} &= [T\,(C/C_0 = 0.95) - T\,(C/C_0 = 0.01)] \times \text{EBCT}_{min}\\ &= [1.47 - 0.71] \times 14.95\text{ min} = 11.4\text{ min}\end{aligned}$$

   The MTZ for TCE should (almost) be contained in the adsorber with an adsorber EBCT of 10 min.

6. Calculate the size of the adsorber. The diameter of the adsorber can be calculated by dividing the flow rate by the superficial velocity:

   $$A_{\text{Adsorber}} = \frac{Q}{V} = \frac{0.89\text{ m}^3/\text{min}}{(5.0\text{ m/h})(\text{h}/60\text{ min})} = 10.68\text{ m}^2$$

   $$D_{\text{Adsorber}} = \sqrt{\left(\frac{4}{\pi}\right)(10.68\text{ m}^2)} = 3.68\text{ m} = 12\text{ ft}$$

Comment

The adsorption capacity for TCE is reduced from 2030 $(\mu g/g)(L/\mu g)^{1/n}$ to 1062 $(\mu g/g)(L/\mu g)^{1/n}$ due to background organic compounds. This corresponds to the worst case where GAC is preloaded with background organic compounds.

Evaluating the Impact of Natural Organic Matter on GAC Performance

One of the most crucial applications of GAC in drinking water treatment is the removal of micropollutants. Most waters have concentrations of micropollutants that are only 0.5 to 5 percent of the concentration of the NOM with which they compete for the adsorption sites on the carbon surface. In addition to the NOM, which is comprised mainly of humic substances, there can be many unidentified synthetic organic chemicals that compete for adsorption sites with micropollutants.

Methods have yet to be determined to predict the competitive interactions between the organic background and micropollutants, even when kinetic and equilibrium data for the unknown background and the micropollutants are available. Moreover, the competitive interactions between the organic background and micropollutants are not completely understood, and this section reviews the empirical evidence of the phenomena.

Pilot plant studies have been the only reliable method of obtaining design data for GAC adsorbers. But these studies are very time consuming and expensive. To reduce the time for the column tests, columns of small particle sizes known as rapid small-scale column tests have been utilized (see discussion later in this section). Very often, a correct simulation of large adsorbers using mathematical models has not been possible for the removal of micropollutants, unless they have been calibrated with field experience.

This section presents calibrated models that can describe micropollutant removal in the presence of unknown and adsorbing organics. The model draws upon many years of experience of using GAC columns treating polluted waters. This experience includes observations from these full-scale columns, and when combined with some specific laboratory studies, it has clearly shown that the presence of NOM decreases the adsorption capacity and kinetics of micropollutants in a GAC column (Jarvi et al. 2005).

An example of the reduction in the adsorption capacity and kinetics in a GAC column in the presence of NOM is shown in Fig. 15-29 (Baldauf, 1986). The breakthrough curve of groundwater that is spiked with trichloroethene (open circles) yields a capacity at complete breakthrough of about 35 percent of the single-solute expected isotherm value.

In a study based on much data from full-scale and pilot columns, Baldauf and Zimmer (1985) compared the adsorption capacities in GAC columns of different waterworks, utilizing different groundwater sources

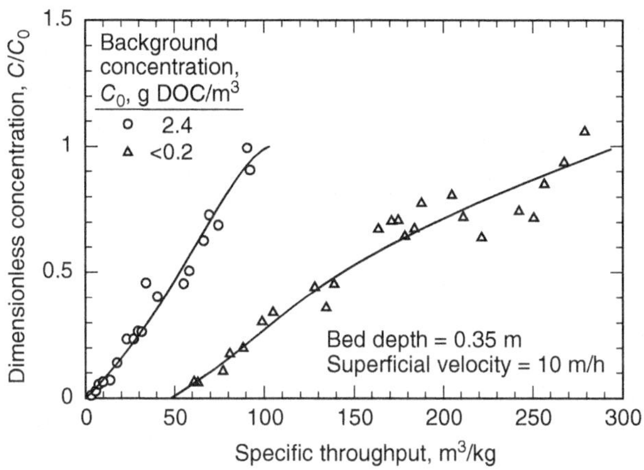

Figure 15-29
Breakthrough curves of trichloroethene in the presence of DOC (Adapted from Baldauf, 1986).

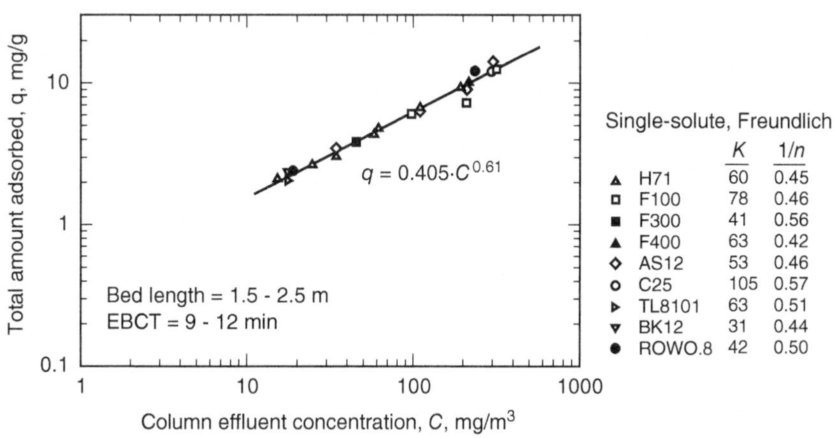

Figure 15-30
Solid-phase concentration, at complete breakthrough, as a function of influent concentration for several groundwater sources and nine activated carbon (Adapted from Baldauf and Zimmer, 1985).

and activated carbons. The solid-phase concentration was calculated by integrating the complete breakthrough curve to determine the column capacity at exhaustion. The column capacities found for trichloroethene are shown in Fig. 15-30 as a function of the influent concentration and EBCT values between 9 and 12 min.

Figure 15-30, termed an adsorber correlation curve, yielded an unexpected result. Despite the different single-solute capacities of the diverse activated carbon types and different concentrations and adsorbability of the groundwater organic matter, a single adsorption relationship for trichloroethene was found adequate for all activated carbon and groundwater sources examined. This filter correlation provides a connection between the influent concentration and the maximum solid-phase concentration of the carbon at total breakthrough for GAC columns with EBCTs of 9 to 12 min.

Filter correlations with major reductions in capacity were also found for tetrachloroethene and 1,1,1-trichloroethane (Baldauf and Zimmer, 1985). The column adsorption capacity of the strongly adsorbing tetrachloroethene was effected the most by the presence of NOM with a 90 percent reduction in its single-solute capacity, while the capacity of the weakly adsorbing 1,1,1-trichloroethane was reduced by about 50 percent. The additional intriguing result is that higher influent concentrations have less impact due to the organic matter.

The differences in the diffusivity of humic substances versus micropollutants cause very different breakthrough behavior of micropollutants and humic substances. Within a fixed bed, micropollutants build up a well-defined mass transfer zone, which migrates slowly through the column with increasing elapsed time. The large humic molecules have slow adsorption kinetics, which leads to a long mass transfer zone. This, in turn, yields a faster breakthrough of the NOM. Thus, deep in the bed only NOM is present and adsorbing, which is termed preadsorption.

Consequently, humics have a greater preadsorption time for micropollutants that have lower concentrations or are more strongly adsorbed because low concentrations or strongly adsorbing components take much longer to move through the bed. As a result, the bed has a greater exposure to NOM before the micropollutant arrives at a given bed depth. The impact of the preadsorption of NOM on the adsorption behavior of micropollutants was shown by preadsorbing an activated carbon in fixed-bed columns with organic matter from Karlsruhe (West Germany) tap water. The results for the different preloading times are displayed in Fig. 15-31. This groundwater

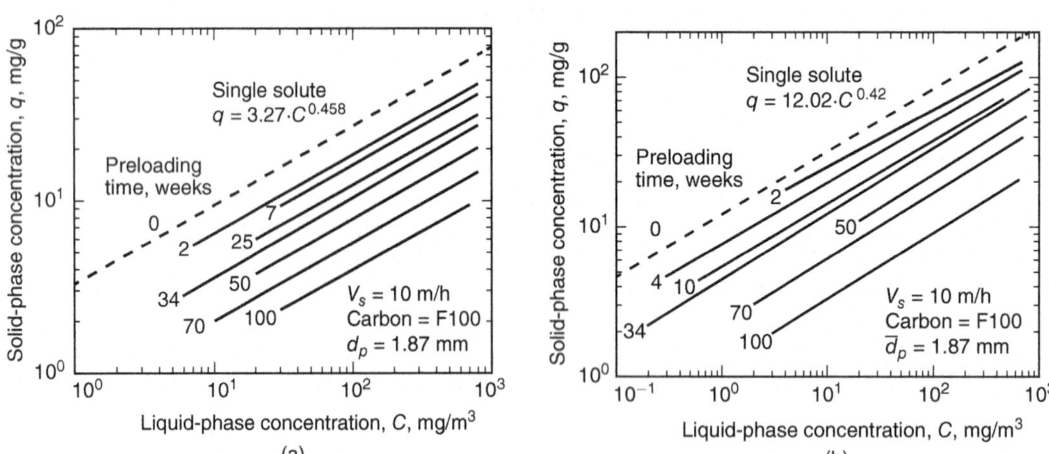

Figure 15-31
Influence of preloading time with NOM on adsorption isotherms for (a) trichloroethene and (b) tetrachloroethene (Adapted from Zimmer et al., 1987).

had an average DOC value of 0.8 mg/L and no detectable concentrations of chlorinated hydrocarbons as measured by total organic halides (TOX). At time intervals over the course of 2 years, carbon samples from the top of the column were taken and isotherm measurements for the chlorinated hydrocarbons were conducted.

For trichloroethene and tetrachloroethene, the isotherms show a parallel shift with time, compared to the isotherm of organic-free water (dashed line). After a great initial decrease in the first weeks, a further steady reduction is still observed after 2 years exposure to the tap water organic matter. In addition, the capacity of trichloroethene and tetrachloroethene, despite their different adsorbability in humic-free water, is reduced by 70 percent after 50 weeks of preloading and by 80 to 85 percent after 100 weeks preloading as compared to the single-solute isotherm value. Similar results were found for other organic substances. Consequently, the impact of preloading time on adsorption equilibrium can be expressed as a reduction in the Freundlich *K* as a function of time.

Figure 15-32 displays the reduction in the adsorption capacity as measured by the Freundlich capacity parameter *K* for other organic substances. Comparisons of the adsorption capacities with time can be made with the Freundlich capacity parameter *K* because all isotherms that were preloaded with tap water had a constant Freundlich $1/n$ value. Thus, the adsorption capacity at any concentration is equally reduced. According to Eq. 15-33, the capacity for the chlorinated aliphatic hydrocarbons is significantly reduced after a few weeks preloading as compared to aromatic compounds, which have lower $1/n$ values. After the initial rapid decrease, all substances have about the same linear reduction in capacity.

The results in Fig. 15-32 demonstrate that carbon fouling is greater for substances that have the lowest adsorbability, that is, low *K* and high *n*

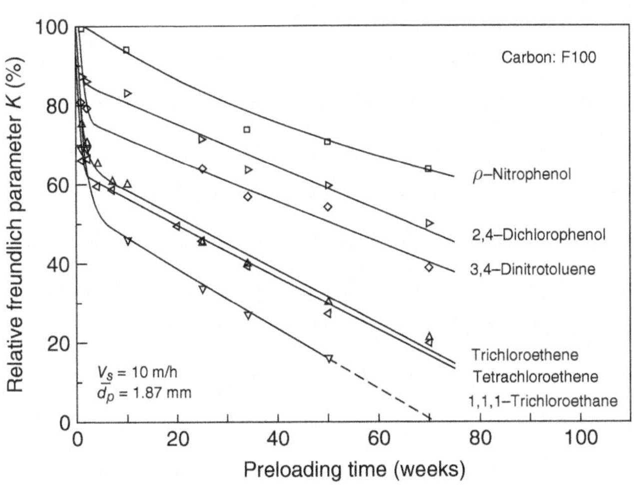

Figure 15-32
Relative Freundlich *K* parameter in percentof the value (Adapted from Sontheimer et al., 1988).

values. However, it must also be remembered that weakly adsorbable substances such as 1,1,1-trichloroethane have relatively short breakthrough times, and shorter breakthrough times result in shorter preloading times and less fouling than is observed for strongly adsorbing substances.

In order to describe the changing Freundlich K values with time, a function, which has an exponential portion for the initial rapid decrease and a linear portion for the long-term slow decrease, was used. The coefficients for that function, which follow, have been determined for various raw waters and trichloroethene and are reported in Table 15-19.

$$\frac{K(t)}{K} = 0.01\left[A_1 - A_2 t + A_3 e^{-A_4 t}\right] \tag{15-196}$$

In order to estimate the reduction in the Freundlich K values for other compound, the correction factors given in Table 15-20 are used. The reduction in Freundlich K values that are reported in Tables 15-19 and 15-20 have been incorporated into commercially available software that has been compared to numerous data sets, and so it can give preliminary design information (Hand et al., 1997).

NOM or background organic matter has an impact on the adsorption kinetics. It was found that surface diffusion is eliminated. The CPHSDM can still be used to simulate the impact of NOM by reducing the surface diffusion coefficient to a point that only pore diffusion is included.

Table 15-19
Empirical kinetic constants describing reduction in Freundlich isotherm capacity parameter for TCE in presence of various background water matrices

	Empirical Kinetic Constants			
Background Water Matrix	A_1 (—)	A_2 (d^{-1})	A_3 (—)	A_4 (d^{-1})
Surface water with significant anthropogenic input (Rhine River, Germany)[a]	35.0	8.86×10^{-4}	65.0	1.29×10^{-1}
Surface water with a small amount of anthropogenic input (Portage Lake, Michigan)[b]	51.0	1.33×10^{-1}	49.0	4.03×10^{-2}
Groundwater in Germany that caused reduction in capacity similar to six other German groundwaters (Karlsruhe, Germany)[a]	65.0	9.66×10^{-2}	35.0	1.44×10^{-1}
Rural Midwestern groundwater (Wausau, Wisconsin)[b]	83.0	1.31×10^{-1}	17.0	3.82×10^{-1}
Rural northern groundwater (Houghton, Michigan)[b]	66.0	2.23×10^{-2}	34.0	1.05×10^{-1}

[a]Calgon F100 GAC.
[b]Calgon F400 GAC.

Table 15-20
Correction factors for reduction in Freundlich isotherm capacity parameter for different classes of compounds

Class	Group	Surrogate Compound	Equation Relative to the Reference Compound—TCE
Purgeables	Halogenated alkanes	1,1,1-Trichloroethane	$K(t)/K = 1.2[K(t)/K]_{TCE} - 0.2$
	Halogenated alkenes	Trichloroethene	$K(t)/K = [K(t)/K]_{TCE}$
	Trihalomethanes	Chloroform	$K(t)/K = [K(t)/K]_{TCE}$
	Aromatics	Toluene	$K(t)/K = 0.9[K(t)/K]_{TCE} + 0.1$
Base Neutrals	Nitro compounds	3,4-Dinitrotoluene	$K(t)/K = 0.75[K(t)/K]_{TCE} + 0.25$
	Chlorinated Hydrocarbons	1,4-Dichlorobenzene	$K(t)/K = 0.59[K(t)/K]_{TCE} + 0.41$
Acids	Phenols	2,4-Dichlorophenol	$K(t)/K = 0.65[K(t)/K]_{TCE} + 0.35$
Polynuclear aromatics (PNAs)		Methylene blue	$K(t)/K = 0.32[K(t)/K]_{TCE} + 0.68$
Pesticides		Atrazine	$K(t)/K = 0.05$

Example 15-13 Using the constant pattern HSDM to simulate the impact of NOM

Evaluate the efficacy of using GAC to treat a groundwater containing 50 μg/L of tetrachloroethylene (PCE)to a treatment objective of 5 μg/L. Assume the groundwater is similar to Karlsruhe groundwater. Given a design EBCT of 10 min and superficial water velocity of 5.0 m/h, and GAC properties and water properties below, calculate $EBCT_{min}$, constant pattern solution, GAC usage rate, and $EBCT_{MTZ}$ using the CPHSDM.

GAC Properties:

Calgon Filtrasorb F-400 (12 × 40 mesh), $\rho_f = 0.45$ g/cm^3, $\rho_a = 0.8034$ g/cm^3

$d_P = 0.1026$ cm particle porosity $\varepsilon_P = 0.641$, EBCT = 10 min, $\varepsilon = 0.44$, $\phi = 1.5$, single-solute Freundlich $K = 200$ (mg/g)(L/mg)$^{1/n}$, Freundlich $1/n = 0.50$

Water properties: $T = 10°C$, $\rho_w = 999.7$ kg/m^3, $\mu_w = 1.307 \times 10^{-3}$ kg/m · s

Solution:

1. Determine the reduced Freundlich K due to the background and the solute distribution parameter D_g.

   The correction factor for the reduction in the Freundlich K is obtained from Table 15-20 for halogenated alkenes because PCE belongs to this group.

$$\left(\frac{K}{K_0}\right)_{\text{PCE}} = \left(\frac{K}{K_0}\right)_{\text{TCE}}$$

   For the groundwater similar to Karlsruhe groundwater using Eq. 15-196 and the values from Table 15-19,
   Convert the units on the coefficients for Karlsruhe groundwater as follows:

$$A1 = 65.0$$

$$A2 = 9.66\times 10^{-2}\ d^{-1} = 6.71\times 10^{-5}\ \text{min}^{-1}$$

$$A3 = 35.0$$

$$A4 = 1.44\times 10^{-1}\ d^{-1} = 1.0\times 10^{-4}\ \text{min}^{-1}$$

$$\left(\frac{K}{K_0}\right)_{\text{PCE}} = \left(\frac{K}{K_0}\right)_{\text{TCE}} = 0.01\left\{65 - 6.71\times 10^{-5}(t) + 35\exp\left[-1.0\times 10^{-4}(t)\right]\right\}$$

   Assume $T = 1$, and negative values for K and D_g will occur using trial-and-error method. Substituting in

$$t = T\tau(D_g + 1) = \tau(D_g + 1) \quad \text{and} \quad D_g = \frac{\rho_a K C_0^{0.5}(1-\varepsilon)}{\varepsilon C_0}$$

   yields a nonlinear equation:

$$\left(\frac{K}{K_0}\right)_{\text{PCE}} = 0.01\left(\begin{array}{l} 65 - 6.71\times 10^{-5}\times T\tau\left(\dfrac{\rho_a K C_0^{0.5}(1-\varepsilon)}{\varepsilon C_0} + 1\right) \\ +35\exp\left\{-1.0\times 10^{-4}\times T\tau\left[\dfrac{\rho_a K C_0^{0.5}(1-\varepsilon)}{\varepsilon C_0} + 1\right]\right\} \end{array}\right)$$

   where τ = (EBCT) (ε) = (10 min) (0.44) = 4.4 min
   Use $K = 200\ (\text{mg/g})(\text{L/mg})^{0.5} = 6325\ (\mu\text{g/g})(\text{L}/\mu\text{g})^{0.5}$ to obtain the initial guess for K. MathCAD is used to solve the above equation, $K = 1111(\mu\text{g/g})(\text{L}/\mu\text{g})^{0.5}$.

$$D_g = \frac{(803.4\ \text{g/L})\left[1111\ \mu\text{g/g(L}/\mu\text{g)}^{0.5}\right](50\ \mu\text{g/L})^{0.5}(1-0.44)}{(0.44)(50\ \mu\text{g/L})}$$
$$= 160{,}662$$

2. Calculate Bi using Eq. 15-168 as

$$\text{Bi} = \frac{k_f R(1-\varepsilon)}{D_s D_g \varepsilon}$$

The film transfer rate is estimated using Gnielinski correlation (see Table 7-5):

$$k_f = \frac{[1+1.5(1-\varepsilon)]D_l}{d_p}\left(2+0.644\ \text{Re}^{1/2}\ \text{Sc}^{1/3}\right)$$

To calculate k_f, Re, liquid diffusivity D_l and Sc has to be determined first.

$V_s = 5$ m/h, the interstitial velocity $v_i = \frac{V_s}{\varepsilon} = \frac{5\ \text{m/h}}{0.44} = 11.36\ \text{m/h}$

$$\text{Re} = \frac{\rho_w d_p v_i}{\mu}$$
$$= \frac{(999.7\ \text{kg/m}^3)(0.1026\ \text{cm})(11.36\ \text{m/h})}{(0.44)(1.307\times10^{-3}\ \text{kg/m}\cdot\text{s})(3600\ \text{s/h})(100\ \text{cm/m})}$$
$$= 5.61$$

For PCE, Table 7-3 is used to calculate the molar volume V_b as

$$V_b = 2\ (14.8) + 4\ (21.6) = 116\ \text{cm}^3/\text{mol}$$

The Hayduk–Laudie correlation shown in Table 7-2 can be used to calculate the liquid-phase diffusion coefficient for PCE as

$$D_l = \frac{13.26\times10^{-9}}{(\mu_w)^{1.14}(V_b)^{0.589}} = \frac{13.26\times10^{-9}}{(1.307)^{1.14}(116)^{0.589}} = 5.93\times10^{-10}\text{m}^2/\text{s}$$

$$\text{Sc} = \frac{\mu}{\rho D_l} = \frac{1.307\times10^{-3}\ \text{kg/m}\cdot\text{s}}{(999.7\ \text{kg/m}^3)(5.93\times10^{-10}\ \text{m}^2/\text{s})} = 2209$$

$$k_f = \frac{\left[1+1.5\ (1-0.44)\right](5.93\times10^{-10}\ \text{m}^2/\text{s})}{(0.1026\ \text{cm})(0.01\ \text{m/cm})}$$
$$\times\left[2+0.644\ (5.61)^{1/2}(2209)^{1/3}\right] = 2.33\times10^{-5}\text{m/s}$$

Apply sphericity correction factor to k_f:

$$k_f = (\phi)(k_f) = (1.5)(2.33 \times 10^{-5}\ \text{m/s}) = 3.49 \times 10^{-5}\ \text{m/s}$$

Equations 15-185 and 15-186 can be used to calculate D_s and PDFC, respectively, as

$$D_s = (\text{PDFC})(\text{SPDFR})$$

$$\text{PDFC} = \frac{D_l \varepsilon_P C_0}{\tau_P K C_0^{1/n} \rho_a} \qquad \text{where } \tau_P = 1.0$$

$$\text{PDFC} = \frac{(5.93\times10^{-10}\text{m}^2/\text{s})(0.641)(50\ \mu\text{g/L})}{(1.0)\left(1111\frac{\mu\text{g}}{\text{g}}\left(\frac{\text{L}}{\mu\text{g}}\right)^{0.5}\right)(50\ \mu\text{g/L})^{0.5}(803.4\ \text{g/L})}$$

$$= 3.01\times10^{-15}\text{m}^2/\text{s}$$

Assume an SPDFR = 1.0 for NOM fouling:

$$D_s = (3.01\times10^{-15}\text{m}^2/\text{s})(1.0) = 3.01\times10^{-15}\ \text{m}^2/\text{s}$$

$$\text{Bi} = \frac{(2.33\times10^{-5}\ \text{m/s})\left(\frac{0.001026\ \text{m}}{2}\right)(1-0.44)}{(3.01\times10^{-15}\ \text{m}^2/\text{s})(160{,}662)(0.44)(1.0)} = 31.4$$

3. Calculate St_{min} from Eq. 15-189:

   From Table 15-17, for $1/n = 0.5$ the coefficient $A_0 = 0.8(1/n = 0.5)$:

$$\text{St}_{\text{min}} = A_0(\text{Bi}) = 0.8\,(31.4) = 25.2$$

   Calculate EBCT_{min} required for constant pattern using Eq. 15-190:

$$\text{EBCT}_{\text{min}} = \frac{\text{St}_{\text{min}}R}{k_f(1-\varepsilon)} = \frac{(25.2)\left(\frac{0.001026\ \text{m}}{2}\right)}{(2.33\times10^{-5}\text{m/s})(1-0.44)}$$

$$= 991\ \text{s} = 16.52\ \text{min}$$

$$\tau_{\text{min}} = (0.44)(16.52\text{min}) = 7.27\ \text{min}$$

4. Calculation of constant pattern solution using Eq. 15-191:

$$T\left(\text{Bi}_s, \frac{1}{n}, \text{St}_{\text{min}}\right) = A_0 + A_1\left(\frac{C}{C_0}\right)^{A_2} + \frac{A_3}{1.01-(C/C_0)^{A_4}}$$

   From Table 15-18, use the values closest to $1/n = 0.5$ and $\text{Bi} = 31.5$, the appropriate value are for $1/n = 0.5$ and $\text{Bi} = 25$: $A_0 = 0.023000$, $A_1 = 0.793673$, $A_2 = 0.039324$, $A_3 = 0.009326$, $A_4 = 0.08275$.

Calculate T, t_{min}, and t for various values of reduced concentration as shown in the table below:

$$T = 0.023 + 0.793673\left(\frac{C}{C_0}\right)^{0.039324} + \frac{0.009326}{1.01-(C/C_0)^{0.08275}}$$

$$t_{min}\ (\mathrm{d}) = \tau_{min}(D_g+1)\,T = \frac{(7.27\ \mathrm{min})\,(160{,}662+1)\,(T)}{1440\ \mathrm{min/d}} = 811\times T$$

$$t\ (d) = t_{min} + (\tau - \tau_{min})\,(D_g+1)$$

$$= t_{min} + \frac{(4.4-7.27)\ (\mathrm{min})\,(160{,}662+1)}{1400\ \mathrm{min/d}} = t_{min} - 320$$

HSDM solution using constant pattern

C/C_0	T	t_{min} (d)	t (d)
0.01	0.71	579	259
0.05	0.77	624	304
0.10	0.80	648	328
0.20	0.84	679	359
0.30	0.87	705	385
0.40	0.90	731	411
0.50	0.94	760	440
0.60	0.98	797	477
0.70	1.04	847	527
0.80	1.14	924	604
0.90	1.31	1065	744
0.95	1.47	1193	872

5. Calculate GAC usage rate using Eq. 15-194:
   For treatment objective $C/C_0 = 5\ \mu g/L / 50\ \mu g/L = 0.1$, $T = 0.80$, $t_{min} = 648$ days, $t = 328$ days.
   After around 328 days of column operation, the effluent will reach or exceed the MCL of 5 μg/L.

$$\begin{Bmatrix}\text{Bed}\\ \text{volumes}\\ \text{treated}\end{Bmatrix} = \mathrm{BVT} = \frac{t\varepsilon}{\tau} = \frac{(328\ \mathrm{d})(0.44)}{\left(\dfrac{4.4\ \mathrm{min}}{1440\ \mathrm{min/d}}\right)} = 47{,}232$$

$$\begin{Bmatrix}\text{Usage}\\ \text{rate}\\ \mathrm{m^3/kg}\end{Bmatrix} = \frac{\mathrm{BVT}}{\rho_F} = \frac{47{,}232}{450\ \mathrm{kg/m^3}} = 105\frac{\mathrm{m^3\ water\ treated}}{\mathrm{kg\ GAC}}$$

6. Calculate the $EBCT_{MTZ}$ using Eq. 15-195:
   The MTZ for PCE will be contained in the adsorber with an EBCT of 10 min:

$$EBCT_{MTZ} = [T\ (C/C_0 = 0.95 - T)\ (C/C_0 = 0.1)] \times EBCT_{min}$$

$$EBCT_{MTZ} = [1.47 - 0.80] \times 16.52 \text{ min} = 11.1 \text{ min}$$

   The MTZ for PCE should almost be contained in the adsorber with an EBCT of 10 min.

Comment

The adsorption capacity parameter K for PCE is revduced from 6325 (μg/g)(L/μg)$^{1/n}$ to 1111 (μg/g)(L/μg)$^{1/n}$ due to background organic compounds. The reduction of the adsorption capacity is approximately 82 percent compared with 50 percent for TCE for a similar calculation. Therefore, the NOM has much greater influence on PCE removal by adsorption than TCE removal. Again, this corresponds to the worst case where GAC is preloaded with background organic compounds. The CPHSDM was used for a length shorter than the minimum value required to establish constant pattern. This would provide a conservative approach because this would be the longest the MTZ would be. We could calculate the length that compares to the 10 percent breakthough error criteria to get an idea of the error that is committed using this approach.

Rapid Small-Scale Column Tests

Design of GAC systems using single, parallel, and series beds was illustrated earlier using pilot plant studies. Rapid small-scale column tests (RSSCTs) may be used to determine GAC performance, and these are discussed in this section. The advantages and disadvantages of the various methods for predicting GAC performance are reported in Table 15-21.

Mathematical models of the GAC process are not completely accurate because the organic matter present in water has an impact on the intraparticle diffusion and adsorption capacity that is not completely understood. However, a smaller, scaled-down fixed bed that utilizes the actual raw water can be used to predict the performance of full-scale adsorbers if the transport processes scale according to the dimensionless groups that appear in the fixed-bed models. Such tests are called RSSCTs. The three primary advantages of using RSSCTs to predict performance are (1) the RSSCT may be conducted in a fraction of the time required to conduct pilot studies; (2) unlike predictive mathematical models, extensive isotherm or kinetic studies are not required; and (3) a small volume of water is required to conduct the RSSCT, which can be transported to a central laboratory for

Table 15-21
Methods for estimating full-scale GAC performance

Method	Reliability	Advantages	Disadvantages
Pilot studies	Excellent	1. Can predict full-scale GAC performance very accurately.	1. Can take a very long time to obtain results. 2. Expensive and must be conducted onsite.
RSSCTs	Good if scaling factor is known	1. Can predict full-scale GAC performance accurately. 2. Small volume of water is required for the test, which can be transported to a central laboratory for evaluation. 3. Can be conducted in a fraction of the time and cost required to conduct pilot studies.	1. Cannot predict GAC performance for different concentrations. 2. Biological degradation that may prolong GAC bed life is not considered. 3. The impact of NOM on micropollutant removal is less than is observed in full-scale plants.
Models	Good if calibrated; fair if not calibrated	1. Once calibrated, models can be used to predict impact of EBCT and changes in influent concentration. 2. Can predict breakthrough of SOCs with 20–50% error.	1. Cannot predict TOC breakthrough and must be used in conjunction with pilot or RSSCT data. 2. Accurate prediction of SOC removal requires calibration with pilot or RSSCT data.

evaluation. Consequently, replacing a pilot study with an RSSCT significantly reduces the time and cost of a full-scale design. However, the results from a RSSCT are site specific and only valid for the raw-water conditions that are tested. Unfortunately, even RSSCTs seem to show less impact of TOC than is observed in pilot-scale plants (Corwin and Summers, 2010).

SCALING DOWN A FULL-SCALE ADSORBER TO RSSCT

In the RSSCT method, mathematical models are used to scale down the full-scale adsorber to an RSSCT and maintain perfect similarity between the RSSCT and full-scale performance. Perfect similarity is obtained by setting the dimensionless groups that describe adsorbate transport in a small-scale RSSCT adsorber (SC) equal to those for a large-scale column (LC). In principle, if perfect similarity is maintained, the RSSCT, which uses a smaller adsorbent particle size than the full-scale adsorber, will have identical breakthrough profiles as the full-scale process. Accordingly, a number of RSSCTs could be used to evaluate important design variables such as GAC selection, EBCT, or bed operations such as beds in series or in parallel.

The scaling procedure was developed using the dispersed-flow pore and surface diffusion model (DFPSDM) (Crittenden et al., 1986, 1987a, 1991). To derive the scaling equations, one only equates the dimensionless groups that characterize the large column to those that characterize the small column. The independent dimensionless groups that characterize the DFPSDM are defined in Table 15-15. Three independent dimensionless groups, used to describe adsorbate transport, appear in the dispersed- flow homogeneous pore surface diffusion model: (1) Peclet number, Pe; (2) surface diffusion modulus, Ed_s, pore diffusion modulus, Ed_p, and (3) the Stanton number, St.

SCALING EBCT

The relationship between the empty-bed contact time of the full-scale column (EBCT_{LC}) and the empty-bed contact time of the rapid small-scale column (EBCT_{SC}) is obtained by equating the surface and pore diffusion modulus of the full-scale column and the RSSCT:

$$\text{Ed}_{s,\text{SC}} = \text{Ed}_{s,\text{LC}} \tag{15-197}$$

$$\frac{D_{s,\text{SC}}\ \tau_{\text{SC}} D_g}{R_{\text{SC}}^2} = \frac{D_{s,\text{LC}}\ \tau_{\text{LC}}\ D_g}{R_{\text{LC}}^2} \tag{15-198}$$

where $D_{s,\text{SC}}$ = effective intraparticle surface diffusion coefficient in small-scale column, m^2/s
τ_{SC} = small-scale column packed-bed contact time ($\text{EBCT}_{\text{LC}} \cdot \varepsilon$, ε is bed porosity), s
R_{SC} = particle radius of adsorbent in small-scale column, mm
D_g = solute distribution parameter defined in Table 15-15, dimensionless
$D_{s,\text{LC}}$ = effective intraparticle surface diffusion coefficient in full-scale column, m^2/s
τ_{LC} = full-scale column packed-bed contact time ($\text{EBCT}_{\text{LC}} \cdot \varepsilon$), s
R_{LC} = particle radius of adsorbent in full-scale column, m

Solving for the ratio $\tau_{\text{SC}}/\tau_{\text{LC}}$ yields

$$\frac{\tau_{\text{SC}}}{\tau_{\text{LC}}} = \left(\frac{R_{\text{SC}}}{R_{\text{LC}}}\right)^2 \left(\frac{D_{s,\text{LC}}}{D_{s,\text{SC}}}\right) = \left(\frac{d_{\text{SC}}}{d_{\text{LC}}}\right)^2 \left(\frac{D_{\text{s,LC}}}{D_{\text{s,SC}}}\right) \tag{15-199}$$

where d_{SC} = particle diameter of adsorbent in small-scale column, mm
d_{LC} = particle diameter of adsorbent in full-scale column, mm

The same result can be obtained by equating Ed_p for both the small-scale and full-scale columns as shown:

$$\frac{\tau_{\text{SC}}}{\tau_{\text{LC}}} = \left(\frac{R_{\text{SC}}}{R_{\text{LC}}}\right)^2 \left(\frac{D_{p,\text{LC}}}{D_{p,\text{SC}}}\right) = \left(\frac{d_{\text{SC}}}{d_{\text{LC}}}\right)^2 \left(\frac{D_{p,\text{LC}}}{D_{p,\text{SC}}}\right) \tag{15-200}$$

where $D_{p,\text{SC}}$ = pore diffusion coefficient in small-scale column, m^2/s
$D_{p,\text{LC}}$ = pore diffusion coefficient in full-scale column, m^2/s

In Eqs. 15-199 and 15-200 it is assumed that the adsorption capacity and physical properties of the adsorbents and bed do not depend on particle size. If the intraparticle pore and surface diffusivities of the small and large GAC are identical, then the following expression may be obtained:

$$\frac{\text{EBCT}_{\text{SC}}}{\text{EBCT}_{\text{LC}}} = \frac{\tau_{\text{SC}}/\varepsilon}{\tau_{\text{LC}}/\varepsilon} = \left(\frac{d_{\text{SC}}}{d_{\text{LC}}}\right)^2 = \left(\frac{R_{\text{SC}}}{R_{\text{LC}}}\right)^2 \tag{15-201}$$

However, the diffusivity has been observed to depend on particle size as shown in this equation:

$$\frac{D_{p,\text{SC}}}{D_{p,\text{LC}}} = \left(\frac{d_{\text{SC}}}{d_{\text{LC}}}\right)^x \quad \text{or} \quad \frac{D_{p,\text{SC}}}{D_{p,\text{LC}}} = \left(\frac{R_{\text{SC}}}{R_{\text{LC}}}\right)^x \tag{15-202}$$

where x = power dependency of the diffusivity

If the controlling intraparticle diffusivity is dependent on particle size, then the ratio of the EBCTs are given by this equation:

$$\frac{\text{EBCT}_{\text{SC}}}{\text{EBCT}_{\text{LC}}} = \left(\frac{d_{\text{SC}}}{d_{\text{LC}}}\right)^{2-x} \tag{15-203}$$

If the diffusivity's dependence on particle size is linear (i.e., $x = 1$), then the ratio of the EBCTs are given by this equation:

$$\frac{\text{EBCT}_{\text{SC}}}{\text{EBCT}_{\text{LC}}} = \left(\frac{d_{\text{SC}}}{d_{\text{LC}}}\right) \tag{15-204}$$

To minimize the impact of bulk density and void fraction differences between the pilot and RSSCT columns, the following equation should be used to calculate the mass of adsorbent in the RSSCT:

$$M_{\text{SC}} = \text{EBCT}_{\text{LC}}\left[\frac{R_{\text{SC}}}{R_{\text{LC}}}\right]^{2-x} Q_{\text{SC}}\rho_{F,\text{SC}} \tag{15-205}$$

where M_{SC} = mass of adsorbent in small-scale column, kg
Q_{SC} = water flow rate in small-scale column, L/s
$\rho_{F,\text{SC}}$ = bulk density of small-scale column, g/mL

SCALING OPERATION TIME

The duration of the RSSCT as compared to a full-scale column test is determined by noting that the mass throughput (Eq. 15-152) of the RSSCT must be equal to that of the full-scale column:

$$T_{\text{SC}} = \frac{C_0 t_{\text{SC}}}{\text{EBCT}_{\text{SC}}\ (1-\varepsilon)\ q_e\rho_a} = T_{\text{LC}} = \frac{C_0 t_{\text{LC}}}{\text{EBCT}_{\text{LC}}\ (1-\varepsilon)\ q_e\rho_a} \tag{15-206}$$

$$\frac{t_{\text{SC}}}{t_{\text{LC}}} = \frac{\text{EBCT}_{\text{SC}}}{\text{EBCT}_{\text{LC}}} = \left[\frac{d_{\text{SC}}}{d_{\text{LC}}}\right]^2 \tag{15-207}$$

where T_{SC} = mass throughput for small-scale column, dimensionless
t_{SC} = small-scale column operation time, d
T_{LC} = mass throughput for full-scale column, dimensionless
t_{LC} = full-scale column operation time, d
C_0 = average influent to GAC column, mg/L
q_e = column adsorbent capacity evaluated at C_0, mg/g

Or it is given by this equation if the diffusivity depends on particle size:

$$\frac{t_{SC}}{t_{LC}} = \frac{\text{EBCT}_{SC}}{\text{EBCT}_{LC}} = \left[\frac{d_{SC}}{d_{LC}}\right]^{2-x} \tag{15-208}$$

SCALING HYDRAULIC LOADING

The relationship between the hydraulic loading of the full-scale column and the hydraulic loading of the RSSCT is obtained by equating the Stanton numbers of the full-scale column and the RSSCT:

$$\text{St}_{SC} = \text{St}_{LC} \tag{15-209}$$

$$\frac{k_{f,SC}\ \tau_{SC}\ (1-\varepsilon)}{\varepsilon R_{SC}} = \frac{k_{f,LC}\ \tau_{LC}\ (1-\varepsilon)}{\varepsilon R_{LC}} \tag{15-210}$$

$$\frac{k_{f,SC}\ \tau_{SC}/\varepsilon}{R_{SC}} = \frac{k_{f,LC}\ \tau_{LC}/\varepsilon}{R_{LC}} \tag{15-211}$$

$$\frac{k_{f,SC}\text{EBCT}_{SC}}{R_{SC}} = \frac{k_{f,LC}\text{EBCT}_{LC}}{R_{LC}} \tag{15-212}$$

where St_{SC} = Stanton number of small-scale column, dimensionless
$k_{f,SC}$ = film transfer coefficient of small-scale column, m/s
St_{LC} = Stanton number of full-scale column, dimensionless
$k_{f,LC}$ = film transfer coefficient of full-scale column, m/s
ε = adsorber bed void fraction, dimensionless

Substituting Eq. 15-201 into 15-212, the following expression is obtained:

$$k_{f,SC}R_{SC} = k_{f,LC}R_{LC} \tag{15-213}$$

As given by Eq. 15-213, the Sherwood numbers for the small-scale column (Sh_{SC}) and full-scale column (Sh_{LC}) (see Chap. 7) are identical:

$$\frac{k_{f,SC}d_{SC}}{D_l} = \frac{k_{f,LC}d_{LC}}{D_l} \tag{15-214}$$

where D_l = liquid diffusivity of adsorbate, m²/s

$$\text{Sh}_{SC} = \text{Sh}_{LC} \tag{15-215}$$

The Sherwood number depends on the Reynolds number (Re) and Schmidt number (Sc) as shown in the following:

$$\text{Sh} = f(\text{Re}, \text{Sc}) \tag{15-216}$$

$$\text{Sc} = \frac{\mu_l}{\rho_l D_l} \tag{15-217}$$

where ρ_l = liquid density of adsorbate, g/L
μ_l = dynamic viscosity of liquid, g/m · s

Accordingly, if the Reynolds numbers of the small and large columns are identical, then the Sherwood numbers of the large and small columns are identical. Consequently, the interstitial and approach velocity of the RSSCT can be determined from the full-scale column by equating the Reynolds numbers:

$$\text{Re}_{\text{SC}} = \text{Re}_{\text{LC}} \tag{15-218}$$

$$\frac{\rho_l v_{i,\text{SC}} d_{\text{SC}}}{\mu_l} = \frac{\rho_l v_{i,\text{LC}} d_{\text{LC}}}{\mu_l} \tag{15-219}$$

$$\frac{v_{i,\text{SC}}}{v_{i,\text{LC}}} = \frac{v_{\text{SC}}/\varepsilon}{v_{\text{LC}}/\varepsilon} = \frac{v_{\text{SC}}}{v_{\text{LC}}} = \frac{d_{\text{LC}}}{d_{\text{SC}}} \tag{15-220}$$

where $v_{i,\text{SC}}$ = interstitial velocity of small-scale column, m/h
$v_{i,\text{LC}}$ = interstitial velocity of full-scale column, m/h
v_{SC} = superficial velocity (hydraulic loading) of small-scale column, m/h
v_{LC} = superficial velocity (hydraulic loading) of full-scale column, m/h

Because Pe depends upon Re and Sc, Eq. 15-220 will also guarantee that $\text{Pe}_{\text{SC}} = \text{Pe}_{\text{LC}}$.

Granular activated carbon particle size distributions have been shown to be a lognormal distribution, and the log mean of the diameters should be used for scaling. In general, the scaling relationships should be verified by comparing RSSCTs to pilot- or full-scale tests or by conducting batch tests to determine how the intraparticle rate is influenced by particle size (Crittenden et al., 1986, 1987a, 1991).

Sometimes using Eq. 15-220 leads to a design with a high head loss, which increases dramatically with operation time, as the GAC is crushed due to a large pressure drop across the RSSCT. In general, intraparticle diffusion causes most of the spreading in the MTZ, and the main factors that need to remain the same are the surface and pore diffusion modulus and the Re for the RSSCT, and large columns do not have to be the same.

Consequently, the high head loss may be avoided by lowering the superficial velocity as long as dispersion does not become the dominant transport mechanism and intraparticle mass transfer is limiting the adsorption rate.

It has been reported that the Peclet number based on diameter can be estimated from the following equation (Fried, 1975):

$$\mathrm{Pe}_d = 0.334 \text{ for } 160 \leq \mathrm{Re} \times \mathrm{Sc} \leq 40{,}000 \tag{15-221}$$

When the velocity is reduced below what is given in Eq. 15-221, axial dispersion, which is caused by molecular diffusion, can be more important in the RSSCT than in the full-scale process. Consequently, Eq. 15-221 can be used to check whether dispersion becomes important as the velocity of the RSSCT is reduced in an effort to reduce the head loss. A typical Sc value for SOCs is ~2000; consequently, Re for the RSSCT must be kept greater than ~0.1 and the $\mathrm{Pe}_{\mathrm{MTZ}}$ must be kept above 50 for the length of the mass transfer zone:

$$\mathrm{Pe}_{\mathrm{MTZ}} = \frac{L_{\mathrm{MTZ}}}{d_p}\mathrm{Pe}_d \tag{15-222}$$

$$\mathrm{Re}_{\mathrm{SC,min}} = 0.1 = \frac{\rho_l v_i 2R_{\mathrm{sc}}}{\mu} \tag{15-223}$$

$$v_{\mathrm{sc}} = v_i\varepsilon = 0.1\frac{\mu_l}{\rho_l 2R_{\mathrm{sc}}}\varepsilon \tag{15-224}$$

where $\mathrm{Pe}_{\mathrm{MTZ}}$ = Peclet number based on mass transfer zone length, dimensionless
L_{MTZ} = length of the mass transfer zone, m
d_p = diameter of the adsorbent particle, m
Pe_d = Peclet number based on particle diameter, dimensionless
$\mathrm{Re}_{\mathrm{SC,min}}$ = minimum Re number for RSSCT, dimensionless
v_i = interstitial velocity, m/h

A larger v_{sc} can be chosen if the column is too short and it can be increased until the head loss is too large.

Because the EBCT of the MTZ is typically between 5 and 10 min, the Peclet number, $\mathrm{Pe}_{\mathrm{MTZ}}$, is greater than 50, and the Pe_d of the RSSCT is equal to $0.334(L_{\mathrm{MTZ}}/d_p)$. For example, if the MTZ of the LC is 5 min (which would be short), then the MTZ would be 4150 and the $\mathrm{Pe}_{\mathrm{MTZ}}$ would be 1390. Obviously, axial dispersion can be ignored in nearly all cases.

CONSTANT-DIFFUSIVITY RSSCT DESIGN

The final set of design equations for a constant-diffusivity RSSCT design is given as

$$\frac{\mathrm{EBCT}_{\mathrm{SC}}}{\mathrm{EBCT}_{\mathrm{LC}}} = \frac{d_{\mathrm{SC}}^2}{d_{\mathrm{LC}}^2} = \frac{t_{\mathrm{SC}}}{t_{\mathrm{LC}}} \tag{15-225}$$

$$\frac{v_{\mathrm{SC}}}{v_{\mathrm{LC}}} = \frac{d_{\mathrm{LC}}}{d_{\mathrm{SC}}} \tag{15-226}$$

Or for a reduced head loss the superficial velocity would be given by this equation as long as the $Pe_{MTZ} \gg 50$, which can be calculated from the experimental results. (This criterion also ensures that intraparticle diffusion resistance is the most significant transport process.)

$$v_{sc} = 0.1 \frac{\mu_l}{\rho_l 2R_{sc}} \varepsilon \tag{15-227}$$

NONCONSTANT-DIFFUSIVITY RSSCT DESIGN

The final set of design equations for a nonconstant-diffusivity RSSCT design is given as

$$\frac{t_{SC}}{t_{LC}} = \frac{EBCT_{SC}}{EBCT_{LC}} = \left(\frac{d_{SC}}{d_{LC}}\right)^{2-x} \tag{15-228}$$

The St, Pe, and Re cannot be matched unless $x = 0$, and we can use this equation to determine the v_{sc}:

$$v_{sc} = 0.1 \frac{\mu_l}{\rho_l 2R_{sc}} \varepsilon \tag{15-229}$$

A larger v_{sc} can be chosen if the column is too short and it can be increase until the head loss is too large.

CARBON PREPARATION FOR USE IN RSSCT

Preparation of the GAC for the RSSCT is important because a representative sample is required for good results. Mixing and splitting the GAC_{LC} must first be performed on a bag of GAC that is used in the full-scale columns to obtain a representative sample. The smaller GAC used in RSSCT studies, GAC_{SC}, is obtained by crushing a representative sample of GAC_{LC}. The crushed carbon is then sieved, retaining the desired sieve fraction. The crushing of the GAC_{LC} should be done carefully so that a lot of carbon fines are not generated, which will increase the yield of GAC_{SC}. The crushing and sieving must be continued until all the crushed carbon from the GAC_{LC} sample passes through the largest sieve used to obtain GAC_{SC}. If care is used in grinding the GAC_{LC}, then the yield of GAC_{SC} can be more than 40 percent of the GAC_{LC} by weight.

The direct scaling of the full-scale column to a small column is demonstrated in Example 15-14. However, as discussed above, sometimes the RSSCT design does not work well because of excessive pressure drop caused by the small particle size coupled with a high velocity. Using a smaller RSSCT column length and a lower velocity will produce a manageable pressure drop, and still provide RSSCT predictive capabilities.

A study summarizing 22 case studies in which RSSCTs are compared to pilot columns that involved 12 SOCs, including weakly adsorbing trihalomethanes and strongly adsorbing pesticides, was performed (Crittenden et al., 1991). The background water matrices included water that had been

Example 15-14 Development of the design and operating parameters of an RSSCT

Calculate the design and operating parameters of an RSSCT that has a particle diameter of 0.21mm compared to a full-scale unit that has a particle diameter of 1.0 mm. The RSSCT is to be designed assuming that the intraparticle diffusivities do not change with particle size. The RSSCT column diameter is 1.10 cm. Typical operating conditions for pilot-scale columns are given in the following table:

Design Parameters	Unit	Pilot Scale
Particle diameter	mm	1.0 (12 × 40)
Bulk density	g/mL	0.49 (F-400)
EBCT	min	10.0
Loading rate	m/h	5.0
Flow rate	mL/min	170.1
Column diameter	cm	5.1
Column length	cm	83.3
Mass of adsorbent	g	833.8
Time of operation	d	100.0
Water volume required	L	24,501

Solution

1. Calculate the $EBCT_{SC}$ using Eq. 15-225:

$$EBCT_{SC} = EBCT_{LC}\frac{d_{SC}^2}{d_{LC}^2} = 10\left(\frac{0.21^2}{1.0^2}\right) = 0.44 \text{ min}$$

2. Calculate the hydraulic loading rate using Eq. 15-226:

$$v_{SC} = v_{LC}\frac{d_{LC}}{d_{SC}} = 5.0\left(\frac{1.0}{0.21}\right) = 23.8 \text{ m/h}$$

3. Calculate the run time using Eq. 15-225:

$$t_{SC} = t_{LC}\frac{d_{SC}^2}{d_{LC}^2} = 100\left(\frac{0.21}{1.0}\right)^2 = 4.4 \text{ d}$$

4. Calculate the bed length, flow rate, and mass of carbon using the RSSCT column diameter, hydraulic loading rate and EBCT:

$$L_{SC} = v_{SC}\,EBCT_{SC} = \frac{(23.8 \text{ m/h})(100 \text{ cm/m})(0.44 \text{ min})}{60 \text{ min/h}} = 17.4 \text{ cm}$$

$$Q_{SC} = v_{SC}A_{SC} = v_{SC}\left(\frac{\pi D_{SC}^2}{4}\right) = \frac{(23.8\ \text{m/h})\pi(1.10\ \text{cm})^2(100\ \text{cm/m})}{(4)(60\ \text{min/h})}$$

$$= 37.7\ \text{cm}^3/\text{min}\ (37.7\ \text{mL/min})$$

$$M_{sc} = Q_{sc}\ \text{EBCT}_{scPsc} = (37.7\ \text{mL/min})(0.44\ \text{min})(0.49\ \text{g/mL}) = 8.1\ \text{g}$$

5. Calculate the volume of water required to run the RSSCT

$$V_w = Q_{SC}t_{SC} = \frac{(37.7\ \text{mL/min})\ (4.4\ \text{d})\ (1440\ \text{min/d})}{10^3\ \text{mL/L}} = 239\ \text{L}$$

The design parameters for the RSSCT are:

$D = 1.1$ cm	$v = 23.4$ m/h
$L = 17.4$ cm	$Q = 37.7$ mL/min
$d = 0.21$ mm	$M = 8.1$ g
EBCT $= 0.44$ min	$V = 239$ L
$t = 4.4$ d	

Comment

The quantity of water required to simulate 100 days of pilot column operation is 239 L, which can be transported to an off-site laboratory to conduct the test.

Example 15-15 Using a RSSCT to predict full-scale performance

A proportional diffusivity RSSCT was performed to evaluate the removal of 1,2-dichloropropane from a groundwater supply using GAC adsorption. The RSSCT was designed to mimic a full-scale adsorber with an EBCT of 4.9 min and a superficial velocity of 8.75 m/h (0.243 cm/s). Based on the RSSCT results shown in the following table, scale up the data and plot the expected full-scale performance in terms of GAC specific throughput. From the predicted full-scale performance,determine the GAC specific throughput for a treatment objective of 5 μg/L and the annual GAC usage (kg/yr).

Table of bed parameters

Item	Unit	Full-Scale Column	RSSCT
Apparent particle density	g/cm^3	0.759	0.759
Bulk or bed density	g/cm^3	0.44	0.5067
Bed porosity	—	0.42	0.33
Particle radius	mm	0.81	0.106
Column diameter	mm		11
Column length	mm	714	94.1
Bed volume	cm^3	92,618	8.94
GAC mass (dry)	g		4.53
Flow rate	L/min	6,940	0.1043
Superficial velocity	cm/s	0.243	1.83
EBCT	s	294	5.16
Operating temperature	°C	13.0±2	13.0±2
Operating time	d	288	5.05

Table of RSSCT column results

Elapsed Time, min	Concentration, μg/L Influent	Effluent
0.0	19.0	0.0
79.0	19.0	0.0
128.0	18.0	0.0
1023.0	19.0	5.0
1174.0	18.0	6.0
1409.0	19.0	8.0
1551.0	19.0	8.0
1826.0	19.0	10.0
1876.0	19.0	10.0
2474.0	19.0	12.0
2612.0	19.0	10.0
2851.0	18.0	13.0
3007.0	18.0	11.0
3287.0	18.0	13.0

Solution

1. Determine the full-scale time based on RSSCT data. The RSSCT effluent time scale can be converted to the full-scale time using Eq. 15-225:

$$t_{LC} = t_{SC}\left(\frac{\text{EBCT}_{LC}}{\text{EBCT}_{SC}}\right) = t_{SC}\left(\frac{294\text{ s}}{5.16\text{ s}}\right) = \frac{57.0}{1440\text{ min/d}}t_{SC}$$
$$= (0.0396\text{ d/min})t_{SC}\text{ (min)}$$

2. Predict full-scale performance. Multiplying the RSSCT time scale by 0.0396 will scale up the RSSCT to predict the full-scale performance as shown in the following table. In addition, the following table shows the specific throughput for each time as determined using Eq. 15-102:

$$\text{Specific throughput} = \frac{t_{LC}}{\text{EBCT}_{LC}\rho_{F_{LC}}} = \frac{t_{LC}(1440\ \text{min/d})}{(4.90\ \text{min})(0.44\ \text{g/cm}^3)(1000\ \text{cm}^3/\text{L})}$$
$$= (0.67\ \text{L/g}\cdot\text{d})(t_{LC})$$

Full-scale column prediction based on the RSSCT data

RSSCT Elapsed Time, t_{SC} (min)	Predicted Full-Scale Elapsed Time, t_{LC} (d)	Predicted Specific Throughput (L/g)	Effluent Concentration (μg/L)
0.0	0.0	0.0	0.0
79.0	3.12	2.1	0.0
128.0	5.07	3.4	0.0
1023.0	40.5	27.1	5.0
1174.0	46.5	27.1	6.0
1409.0	55.8	37.4	8.0
1551.0	61.4	41.1	8.0
1826.0	72.3	48.4	10.0
1876.0	74.2	49.7	10.0
2474.0	98.0	65.7	12.0
2612.0	103.4	69.3	10.0
2851.0	112.9	75.6	13.0
3007.0	119.0	79.7	11.0
3287.0	130.1	87.2	13.0

3. Plot the effluent concentration versus specific throughput from the data in the table above.

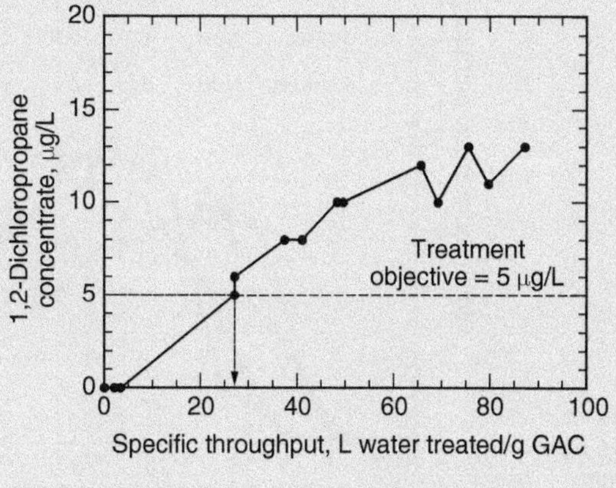

Based on a treatment objective of 5 μg/L, as shown in the figure, the specific throughput is 27.1 L of water treated per gram of GAC. If the flow rate is 10 ML/d, the annual GAC consumption is calculated as

$$\text{Annual GAC consumption} = \frac{(10 \times 10^6\ \text{L/d})(365\ \text{d/yr})}{(27.1\ \text{L/g})(1000\ \text{g/kg})}$$

$$= 134{,}700\ \text{kg GAC/yr}$$

distilled, deionized, and GAC filtered, four groundwaters, lake water, and river water. Three scenarios were represented: (1) a low concentration of background organic matter and a relatively high concentration of the SOC > 1.0 mg/L, (2) adsorbable background organic matter and a relatively high concentration of the SOC, and (3) adsorbable background organic matter and a relatively low concentration of SOCs < 0.38 mg/L.

The pilot and RSSCTs data for organic-free water and surface water containing approximately 4.0 mg DOC/L are compared on Fig. 15-33. The RSSCTs were designed using Eqs. 15-225 and 15-226 [the analysis is called a constant-diffusivity (CD) design]. The influent concentrations and operational parameters for this study are reported in Table 15-22. The source of organic matter was drainage from a swamp in Lake Superior basin (Houghton, Michigan). Based on the data presented on Fig. 15-33, the

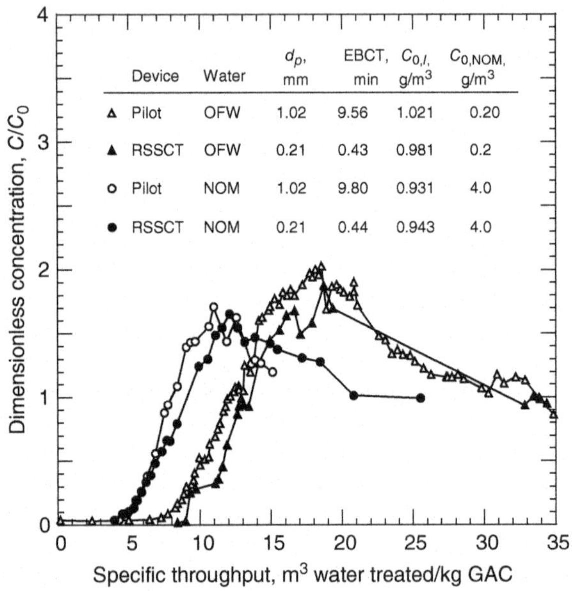

Figure 15-33
Comparison of RSSCTs and pilot column data for chloroform that were collected in organic-free water (OFW) and surface water (NOM). (Adapted from Sontheimer et al., 1988).

Table 15-22
Results of pilot-scale and RSSCT testing for removal of synthetic organic chemicals

Target Compounds Removed[a]	Influent Concentration, mg/L	DOC Concentration, mg/L	Pilot Particle, Size, cm	EBCT, min	Loading Rate, m/h	Column Capacity, mg/g	Influent Concentration, mg/L	DOC Concentration, mg/L	Particle Size, cm	RSSCT EBCT, s	Loading Rate, m/h	125 Column Capacity, mg/g
Chloroform	1.021	0.2	0.1026	4.9	5.1	8.75	0.981	0.2	0.0212	12.8	24.9	10.24
DBCM	1.775					34.81	1.839					41.33
EDB	1.577					34.70	1.692					33.95
Bromoform	2.111					69.95	2.191					73.74
TCE	1.062					35.27	1.201					37.63[b]
PCE	1.139					79.70[b]	1.345					98.21[c]
Chloroform	1.021	0.2	0.1026	9.6	5.1	10.51	0.981	0.2	0.0212	25.6	24.9	9.30
DBCM	1.775					35.21	1.839					44.22
EDB	1.577					34.44	1.692					35.27
Bromoform	2.111					76.69[b]	2.191					78.08[b]
TCE	1.062					32.92[c]	1.201					40.83[c]
PCE	1.139					78.46[c]	1.345					146.02[c]
Chloroform	0.931	4.0	0.1026	4.8	5.2	5.66	0.943	4.0	0.0212	12.8	23.9	5.72
DBCM	1.615					21.45	1.714					22.31
EDB	1.409					21.07	1.549					24.04
Bromoform	1.821					36.35	1.948					43.10[a,b]
TCE	0.875					20.32[b]	1.022					27.39[a,b]
PCE	0.995					28.27[c]	1.372					41.71[b,c]
Chloroform	0.931	4.0	0.1026	9.8	5.2	5.93	0.943	4.0	0.0212	26.1	23.9	6.23
DBCM	1.615					20.07	1.714					24.42[a,b]
EDB	1.409					18.96[b]	1.549					23.30[c]
Bromoform	1.821					26.54[c]	1.948					29.81[c]
TCE	0.875					13.09[c]	1.022					15.76[c]
PCE	0.995					14.99[c]	1.372					21.10[c]

[a]DBCM = dibromochloromethane, EDB = ethylene dibromide, TCE = trichloroethylene, PCE = tetrachloroethylene.
[b]The column capacity was determined by extrapolating the effluent concentration profiles.
[c]The column capacity was determined by extrapolating the effluent concentration profiles; however, the profiles were too short to extrapolate with precision.
Source: Adapted from Crittenden et al. (1991).

pilot and RSCCT data for the low-DOC water have higher capacity than the data for a high-DOC background. The RSSCT data for organic-free water agrees well with the pilot column data. The comparisons between RSSCTs and pilot data for the SOCs in the presence of high DOC are very good, but the RSSCTs exhibit slightly higher capacity.

The results obtained by assuming that the intraparticle diffusivity is constant for SOCs (MW < 300) are reasonable for preliminary design. Typically, using this type of design will give a column capacity that is 20 to 40 percent larger than those observed for pilot columns. Greater precision will require comparisons between RSSCT and pilot- or full-scale data. Alternatively, batch isotherm and rate data conducted on carbon exposed to the water matrix could be used to properly select the best RSSCT design (Sontheimer et al., 1988). Recently, ASTM (2000) accepted the RSSCT method as a standard test using a constant diffusivity design. Corwin and Summers (2010) examined the precision of RSSCTs and made recommendations for designing RSSCTs (x values) for several SOCs. Their findings show that proportional diffusivity may yield better results. It appears that using small GAC in the RSSCT reduces the impact of NOM on GAC capacity because there are more pore openings on the outside of the GAC that is used in RSSCTs. Accordingly, the GAC that is used in full-scale plants has fewer pore openings and NOM can plug or foul the GAC in full-scale plants more easily than GAC that is used in RSSCTs.

Studies have evaluated using RSSCTs to predict DOC removal (Crittenden et al., 1991). The studies included the following water sources: (1) Colorado River water (CRW), (2) California State Project water (SPW), from Northern California, (3) Ohio River water, (4) Mississippi River water, and (5) Delaware River water. It was determined that the intraparticle diffusivity was proportional to particle size for CRW and SPW; consequently, the RSSCTs with proportional diffusivity (PD)designs were compared to RSSCTs with CD designs. In all cases, a good comparison was reported between PD RSSCTs and pilot column results. The results for CRW and a PD design are presented on Fig. 15-34. Breakthrough time is expressed as the equivalent operation time in the pilot column, as given by Eq. 15-225; the RSSCT (using a 60 × 80-mesh GAC) can be conducted in 20 percent of the time of the pilot test. Comparisons of the breakthrough of the RSSCTs and pilot columns for 30 and 60 min EBCT show that the RSSCT breakthroughs appeared slightly after the pilot breakthrough profiles. However, the PD RSSCT design yielded good comparisons between the results of the RSSCT and the pilot columns for the other sites. However, if biodegradation of the DOC occurs in the full-scale process, it will not be reflected in the RSSCT predictions.

Factors That Impact Adsorber Performance

The three main factors that impact adsorber performance, as discussed below, are particle size, backwashing, and hydraulic loading.

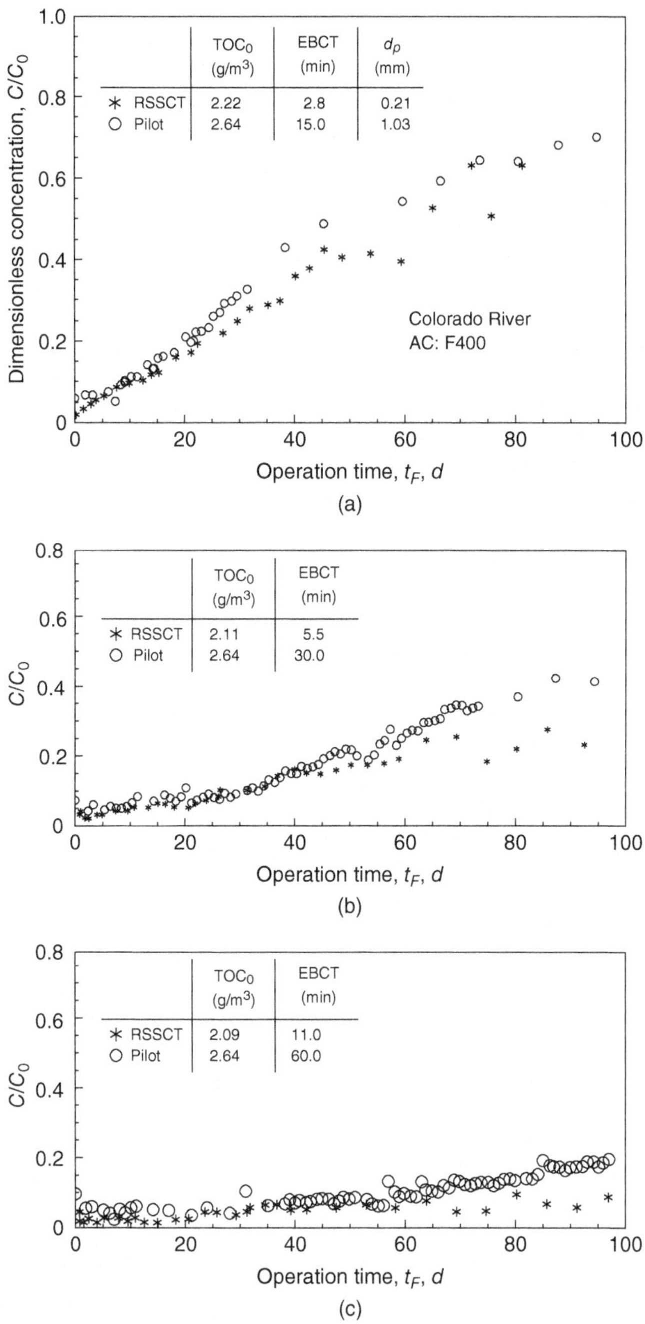

Figure 15-34
Comparison of TOC breakthrough curves for pilot columns and RSSCTs designed based on proportional diffusivity for full-scale column EBCTs of (a) 15 min, (b) 30 min, and (c) 60 min. (Adapted from Sontheimer et al., 1988).

PARTICLE SIZE

Particle size influences the rate of adsorption and head loss in GAC columns. As particle size decreases, the length of the MTZ decreases. The head loss across a GAC bed will vary with particle size. For deeper beds and longer absorber runs, the particle size is typically 0.6 to 2.36 mm (U.S. sieve sizes of 8 × 30). For lower hydraulic loading rates, particle size will typically vary from 0.425 to 1.7 mm (U.S. sieve sizes of 12 × 40).

BACKWASHING

To obtain the best performance for SOCs, GAC contactors should be operated in the postfiltration mode or receive low-turbidity water because backwashing will greatly reduce their performance. The mass transfer zone will be disrupted due to backwashing, which in turn causes premature breakthrough of contaminants. Backwashing decreases adsorber performance, as shown on Fig. 15-35. The profile for the 7.4-min EBCT decreases because during backwashing exhausted GAC is mixed up into the bed and less exhausted carbon is mixed in this section of the bed. Backwashing is usually not needed for treatment of groundwater from deep wells as long as there is no potential for precipitation of calcium carbonate or metals. Care must be taken not to introduce oxygen or other gases that may cause precipitation or significant biological growth. In cases where there is precipitation potential, dissolved species that may precipitate must be removed prior to the GAC process. When treating turbid surface waters, turbidity must be removed prior to the GAC process, otherwise backwashing will be required and the GAC cannot achieve a high degree of removal of SOCs. Based on operating experience, it has been found that backwashing does not appear to affect DOC removal because high degrees of removal cannot be achieved with reasonable EBCTs.

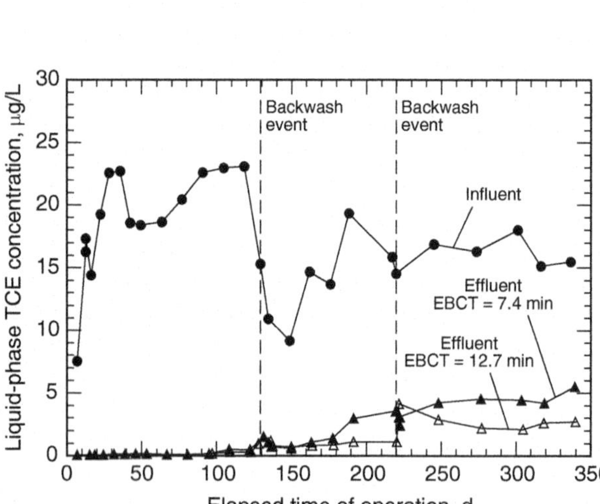

Figure 15-35
Impact of backwashing on full-scale pressure GAC contactor.

HYDRAULIC LOADING

The review of Cover and Pieroni (1969) reported that hydraulic loading does not influence the performance of adsorbers with the same EBCT. However, increasing hydraulic loading will increase head loss. One common mistake for pressure GAC contactors is to use too few GAC beds, operated in parallel, in an effort to reduce capital expense. The larger hydraulic loading can cause a high pressure drop, which can increase dramatically over time. Because GAC can be crushed by the high pressure, the bed void fraction will be reduced with a concomitant increase in pressure. The problem can be exacerbated to a point where the bed must be backwashed.

Problems and Discussion Topics

15-1 Compare chemisorption and physical adsorption.

15-2 List the types of commercially available adsorbents in water treatment. Which type is the most commonly used and why?

15-3 Describe the production method of activated carbon and list the methods of regeneration and reactivation of spent GAC.

15-4 List the forces that may be operative during adsorption. Discuss the origin of each force, and the properties of the adsorbate and adsorbent that influence the force.

15-5 Derive the Langmuir equation from the elementary reaction rate steps. List the assumptions that are required to derive the Langmuir equation and the Freundlich isotherm equations.

15-6 Determine the Freundlich and Langmuir parameters for the data given below. You may use linear regression, and plot C_e/q_e versus C_e for the Langmuir equation and log q_e versus log C_e for the Freundlich equation.

Adsorption isotherm data: Carbon type, F-400; chemical, tetrachloroethene; temperature, 13.8°C.

Isotherm Data:

C_e, μmol/L	q_e, μmol/g
15.7	1,246
1.27	489
0.396	298
0.225	250
0.161	213

15-7 Determine the Freundlich isotherm parameters for tetrachloroethene (PCE) using Polanyi potential theory and compare the parameters with those determined in Problem 15-6. Use Calgon

F-400 GAC and a water treatment temperature of 13.8°. For PCE, the following properties at 13.8°C are given: V_m = 102.4 mL/mol, ρ_l = 1620 kg/m^3, and C_s = 347.0 mg/L.

15-8 Compare the GAC usage rates for TCE concentrations of 100, 50, and 25 μg/L in water and compare the gas phase usage rate to the usage rate in water. Assume the GAC is completely saturated at the influent concentration. Use the Freundlich parameters in Table 15-6. From Chapter 14, assume a stripper with an air-to-water ratio of 3.5/H is used to strip out all TCE and transfer it to the air. T = 10°C and the following properties at 10°C for TCE are given: solubility = 821 mg/L, vapor pressure = 36.7 mm Hg, refractive index = 1.4773, ρ_l = 1620 kg/m^3. Use Calgon BPL (4 × 6) for gas-phase adsorption. Hint: ideal gas law can be used to calculate partial pressure of TCE from the TCE concentration.

15-9 Derive the expression of adsorption potential (ε) in the Polanyi and DR equations for water and air. The adsorption in water can be described using the following reaction.

$$A_{\text{aq}} \rightarrow A_{\text{ad}}$$

Assume adsorbed state is a saturated solution. Hint: $\varepsilon = \Delta G$.

15-10 Using TOC data for Colorado River water in Fig. 15-25 and a treatment objective of 1.0 mg/L of TOC (trihalomethane formation potential is 50 μg/L), calculate the volume of water treated per gram GAC for the following: (a) 2 × 7.5-min EBCT columns in series, (b) 2 columns in parallel with EBCT = 15 min, (c) single contactor with 7.5- and 15-min EBCT. The filter density is 0.457 g/mL, the average TOC influent concentration is 2.52 mg/L.

15-11 Calculate the dosage of activated carbon to reduce an influent concentration of 300 μg/L of chloroform to 100 μg/L (treatment objective) using powdered (PAC) and granular activated carbon (GAC). Assume for the GAC and PAC process that the carbons are saturated at the influent concentration and treatment objective, respectively. *Given:* Q = 10 mgd.

$$K = 159 \frac{\mu\text{g of chloroform}}{\text{g of activated carbon}} \left(\frac{L}{\mu\text{g}}\right)^{0.625}$$

How long will the GAC last if the filter density ρ_f = 0.37 g/cm^3 and EBCT = 15 min?

15-12 Derive the scaling equations (Eqs. 15-225 and 15-226) needed to simulate a full-scale adsorber by a constant diffusivity RSSCT.

15-13 Design a RSSCT from the pilot plant data for the removal of methyl *tert*-butyl ether (MTBE) from a raw-water source obtained from a reservoir based on constant diffusivity design. The design should include the column length, EBCT, time of operation, hydraulic loading rate, flow rate, mass of carbon, and volume of water needed.

Pilot Data	RSSCT
$d_p = 1.026$ mm	$d_p = 0.1643$ mm
$\rho_a = 0.803$ g/cm^3	$\rho_a = 0.803$ g/cm^3
$\rho_F = 0.480$ g/cm^3	$\rho_F = 0.480$ g/cm^3
$\varepsilon = 0.40$	$\varepsilon = 0.40$
Column diameter = 5.1 cm	Column diameter = 1.1 cm
EBCT = 10.0 min	
$V_S = 10$ m/h	
t (time of operation) = 10 wk	

15-14 Design an RSSCT from the pilot plant data for the removal of DOC (molecular weight = 10,000) from a raw-water source obtained from a reservoir based on D_s varying linearly with d_p: The design should include the column length, EBCT, time of operation, hydraulic loading rate, flow rate, mass of carbon, and volume of water needed.

Pilot Data	RSSCT
$d_p = 1.026$ mm	$d_p = 0.1643$ mm
$\rho_a = 0.803$ g/cm^3	$\rho_a = 0.803$ gm/cm^3
$\rho_F = 0.480$ g/cm^3	$\rho_F = 0.480$ gm/cm^3
$\varepsilon = 0.40$	$\varepsilon = 0.40$
Column diameter = 5.1 cm	Column diameter = 1.1 cm
EBCT = 10.0 min	
$V_S = 10$ m/h	
t (time of operation) = 10 wk	

15-15 Derive the expression comparing PAC/GAC usage rates.

15-16 For the GAC pilot plant data plotted in Example 15-10, calculate the specific volume for two beds in series with the first bed having a 10-min EBCT and the second bed a 22-min EBCT. The flow rate is 161 mL/min, and $\rho_f = 0.457$ g/mL. The treatment objective is 5 μg/L. The average DCE influent concentration is 80 μg/L. The effluent from the first bed is 64 μg/L when the treatment objective from the second column is exceeded.
Column data:

EBCT, min	M, g	T, d	Q_t,L	L/g
10	791.1	75	17400	22.0
32	2373.3	290	67280	29.0

15-17 Calculate removal in a floc blanket reactor (FBR) for 25, 50, 100, and 500 ng/L MIB and PAC dosages of 5, 10, 25, 50, and 75 mg/L for a CMFR that considers the influence of NOM. Redo this for a CMFR that does not consider the presence of NOM. Given the following single-solute adsorption isotherm parameters: $K = 9.56$ (ng/mg) (L/ng)$^{1/n}$, $1/n = 0.492$. Laboratory studies determined that the adsorption capacity for MIB was reduced by 25 percent due to NOM adsorption. For simplification, assume that the adsorption of MIB reaches equilibrium in the CMFR. [Comments: Adsorption equilibrium is rarely reached in real practice. A longer carbon retention time (CRT) can cause the adsorption closer to equilibrium. See Example 15-6.]

15-18 Isotherm experiments were conducted in bottles with two different initial concentrations to measure the adsorption isotherm of MIB on PAC in a natural water and the following data were obtained (Gillogly et al., 1998). Plot the percentage of MIB remaining in the solution as a function of PAC dosage, and determine the PAC dosage corresponding to 90 percent removal of MIB in a batch reactor for an initial concentration of 200 ng/L.

C_0, ng/L	PAC Dosage, mg/L	C_e, ng/L
150	2.2	137.7
	4.1	122.7
	9.9	81.6
	32.4	16.2
	45.7	5.85
1245	2.1	1088.13
	4	949.94
	14.6	329.68
	40.2	51.04
	60.3	14.94

15-19 A municipality wants to treat 2.7 ML/d of a groundwater that contains 85 μg/L of 1,1-dichloroethylene (DCE) using granular activated carbon (GAC) adsorption. It is recommended that a 3.66-m diameter pressure vessel containing 9000 kg of Calgon F- 400 GAC be used to treat the DCE from the water. Using the constant pattern solutions, calculate the time it will take to reach the treatment objective of 5 μg/L assuming continuous pumping, the specific throughput in m^3 water treated per kg of GAC, and the mass transfer zone length. Assume no NOM fouling is important and SPDFR = 4. The properties of the GAC and water are as follows: GAC properties: Calgon Filtrasorb F-400 (12 × 40 mesh), $\rho_f = 0.45$ g/cm^3, $\rho_a = 0.8034$ g/cm^3; $d_P = 0.1026$ cm, particle porosity $\varepsilon_P = 0.641$, EBCT = 10 min, $\varepsilon = 0.44$, temperature = 14°C; single-solute Freundlich parameters are $K = 470$ (μg/g) (L/μg)$^{1/n}$, $1/n = 0.515$.

15-20 Redo Problem 15-19 assuming NOM fouling of the GAC using Karlsruhe groundwater correlation and compare your answer to the case of no NOM fouling. Assume SPDFR = 1 when NOM is present.

15-21 Redo Problem 15-19 assuming there is 30 μg/L of methyl-*tert*-butyl ether (MTBE) and a treatment objective of 5 μg/L. Apply each NOM fouling correlations listed in Table 15-20 and compare the results. Assume MTBE behaves like a halogenated alkene and SPDFR = 1 when NOM is present.

15-22 If packed-tower air stripping is used to treat DCE in Problem 15-19, design a gas-phase GAC contactor to treat the off-gas from the packed tower. Assume the optimum air-to-water ratio is equal to 3.5 times the minimum air-to-water ratio require for stripping (you do not need to design the air stripper). Assume a typical superficial gas velocity of 0.8 m/s, EBCT of 40 s, and a treatment objective of <1 μg/L. Determine the dimensions of the fixed bed, mass of GAC required, GAC usage rate in m^3/kg, and time to breakthrough in days. Use Calgon BPL GAC with the following properties: $\rho_F = 0.525$ g/cm^3; $\rho_a = 0.525$ g/cm^3; $\varepsilon_p = 0.595$; $D_P = 0.3715$ cm, Freundlich $K = 1111$ $(\mu g/g)(L/\mu g)^{1/n}$, and $1/n = 0.838$. Assume SPDFR = 16.0 for gas-phase operation.

References

Adamson, A. W. (1982) *Physical Chemistry of Surfaces*, 4th ed., Wiley, New York.

Anselme, C., Baudin, I., and Chevalier, M. R. (1999) Drinking Water Production by Ultrafiltration and PAC Adsorption. First Year of Operation for a Large Capacity Plant, pp. 138–146 in Proceedings AWWA Membrane Technology Conference, Long Beach, CA, American Water Works Association, Denver, CO.

Anselme, C., Laîné, J. M., and Baudin, I. (1997) Drinking Water Production by UF and PAC Adsorption: First Months of Operation for a Large Capacity Plant, pp. 783–803, in Proceedings, Membrane Technology Conference, New Orleans, LA. American Water Works Association, Denver, CO.

ASTM (2000) *Standard Practice for the Prediction of Contaminant Adsorption on GAC in Aqueous Systems Using Rapid Small-Scale Column Tests*, American Society for Testing and Materials, West Conshohocken, PA.

AWWA (1986) *1984 Utility Operating Data*, American Water Works Association, Denver, CO.

AWWA (1999) *Water Quality and Treatment: A Handbook of Community Water Supplies*, 5th ed., R. D. Letterman (ed.), American Water Works Association, McGraw-Hill, New York.

Baldauf, G. G. (1986) "Adsorptive Entfernung leichtfluchtiger Halogenkohlenwasserstoffe bei der Wasseraufbereitung," *Vom Wasser.* **66**, 21–31.

Baldauf, G. G., and Zimmer, G. (1985) "Removal of Volatile Chlorinated Hydrocarbons by Stripping and/or Activated Carbon Adsorption," *Water Supply*, **3**, 187–196.

Baylis, J. R. (1929) "The Activated Carbons and Their Use in Removing Objectionable Tastes and Odors from Water." *J. AWWA*, **21**, 787–814.

Brunauer, S., Emmett, P. H., and Teller, E. (1938) "Adsorption of Gases in Multimolecular Layers," *J. Am. Chem. Soc.*, **60**, 309–319.

Campos, C., Marias, B. J., Snoeyink, V. L., Baudin, I., and Laine, J. M. (2000a) "PAC-Membrane Filtration Process. I: Model Development," *J. Environ. Eng.*, **126**, 2, 97–103.

Campos, C., Snoeyink, V. L., Marinas, B., Baudin, I., and Laine, J. M. (2000c) "Atrazine Removal by Powdered Activated Carbon in Floc Blanket Reactors," *Water Res.*, **34**, 4070–4080.

Corwin, J., and Summers, R.S. (2010) "Scaling Trace Organic Contaminant Adsorption Capacity by Granular Activated Carbon," *J. Environ. Sci. Technol*, **44**, 14, 5403–5408.

Cover, A. E., and Pieroni, L. J. (1969) *Evaluation of the Literature on the Use of GAC for Tertiary Waste Treatment*, TWRC-11, U. S. Department of the Interior, Washington, DC.

Crank, J. (1964) *The Mathematics of Diffusion*, Oxford University Press, London.

Crittenden, J. C. (1976) Mathematic Modeling of Fixed Bed Adsorber Dynamics—Single Component and Multicomponent, Dissertation, University of Michigan, Ann Arbor.

Crittenden, J. C., Berrigan, J. K., and Hand, D. W. (1986) "Design of Rapid Small Scale Adsorption Tests for a Constant Diffusivity," *J. WPCF*, **58**, 4, 312–319.

Crittenden, J. C., Berrigan, J. K., Hand, D. W., and Lykins, B. (1987a) "Design of Rapid Fixed Bed Adsorption Tests for Non-Constant Diffusivities," *J. Environ. Eng.*, **113**, 2, 243–259.

Crittenden, J. C., Hand, D. W., Arora, H., and Lykins, Jr., B. W. (1987b) "Design Considerations for GAC Treatment of Organic Chemicals," *J. AWWA*, **79**, 1, 74–82.

Crittenden, J. C., Luft, P. J., Hand, D. W., Oravitz, J., Loper, S., and Ari, M. (1985) "Prediction of Multicomponent Adsorption Equilibria Using Ideal Adsorbed Solution Theory," *Environ. Sci. Technol.*, **19**, 11, 1037–1043.

Crittenden, J. C., Reddy, P. S., Arora, H., Trynoski, J., Hand, D. W., Perram, D. L., and Summers, R.S. (1991) "Predicting GAC Performance with Rapid Small-Scale Column Tests," *J. AWWA*, **83**, 77–87.

Crittenden, J. C., Sanongraj, S., Bulloch, J. L., Hand, D. W., Rogers, T. N., Speth, T. F., and Ulmer, M. (1999) "Correlation of Aqueous Phase Adsorption Isotherms," *Environ. Sci. Technol.*, **33**, 17, 2926–2933.

Crittenden, J. C., Vaitheeswaran, K., Hand, D. W., Howe, E. W., Aieta, E. M., Tate, C. H., McGuire, M. J., and Davis, M. K. (1993) "Removal of Dissolved Organic Carbon Using Granular Activated Carbon," *Water Res.*, **27**, 4, 715–721.

Croes, J. J. R. (1883) "The Filtration of Public Water Suppliers in America," *Eng. News Am. Control J.*, **10**, 277–281.

Dankwerts, P.V. (1953) "Continuous Flow Systems," *Chem. Eng. Sci.*, **2**, 1–13.

Dobbs, R. A., and Cohen, J. M. (1980) *Carbon adsorption isotherms for toxic organics*, EPA-600/8-80-023, U.S. Environ. Protection Agency, Cincinnati, Ohio.

Fleck, R. D., Jr., Kirwan, D. J., and Hall, K. R. (1973) "Mixed-Resistance Diffusion Kinetics in Fixed-Beds under Constant Pattern Conditions," *Ind. Eng. Chem. J.*, **12**, 95–110.

Freundlich, H. (1906) "Über die Adsorption in Lösungen," *Z. Phys. Chem. A*, **57**, 385–470.

Fried, J. J. (1975) *Groundwater Pollution*, Elsevier Scientific, Amsterdam.

Getzen, F. W. and Ward, T. M. (1969) "Model for the Adsorption of Weak Electrolytes of Solids as a Function of Ph I. Carboxylic Acid-Charcoal Systems," *J. Colloid Interface Sci.*, **31**, 441–453.

Gibbs, J. W. (1906) *Scientific Papers*, **1**, 219 (Longmans, Green and Co., London).

Gillogly, T. E. T., Snoeyink, V. L., Elarde, J. R., Wilson, C. M., and Royal, E. P. (1998) "Kinetic and Equilibrium Studies of ^{14}C-MIB Adsorption on PAC in Natural Water," *J. AWWA*, **90**, 98–108.

Graham, M., Najm, I., Simpson, M., Macleod, B., Summers, S., and Cummings, L. (2000) *Optimization of Powdered Activated Carbon Application for Geosmin and MIB Removal*, American Water Works Association Research Foundation, Denver, CO.

Grant, R. J., and Manes, M. (1964) "Correlation of Some Gas Adsorption Data Extending to Low Pressures and Supercritical Temperatures," *Ind. Eng. Chem. Fund.*, **3**, 3, 221–224.

Grant, R.J., and Manes, M. (1966) "Adsorption of Binary Hydrocarbon Gas Mixtures on Activated Carbon," *Ind. Eng. Chem. Fund.*, **5**, 4, 490–498.

Greenbank, M., and Manes, M. (1981) "Application of the Polanyi Adsorption Potential Theory to Adsorption from Solution on Activated Carbon," *J. Phys. Chem.*, **85**, 20, 3050–3059.

Greenbank, M., and Manes, M. (1982) "Application of the Polanyi Adsorption Potential Theory to Adsorption from Solution on Activated Carbon. 12. Adsorption of Organic Liquids from Binary Liquid-Solid Mixtures in Water," *J. Phys. Chem.*, **86**, 21, 4216–4221.

Greenbank, M., and Manes, M. (1984) "Application of the Polanyi Adsorption Potential Theory to Adsorption from Solution on Activated Carbon. 14. Adsorption of Organic Solids from Binary Liquid-Solid Mixtures in Water," *J. Phys. Chem.*, **88**, 20, 4684–4688.

Haist-Gulde, B. (1991) "Zur Adsorption von Spurenverunreinigungen aus Oberflächenwässern," Ph.D. Dissertation, University of Karlsruhe, Germany.

Halsey, G. D., and Taylor, H. S. (1947) "Adsorption of Hydrogen on Tungsten Powders," *J. Chem. Phys.*, **15**, 624–630.

Hand, D. W., Crittenden, J. C. Arora, H., Miller, J. and Lykins, B. W. Jr. (1989) "Design of Fixed-Beds to Remove Multicomponent Mixtures of Volatile and Synthetic Organic Chemicals," *J. AWWA*. **81**, 1.

Hand, D. W., Crittenden, J.C. Hokanson, D.R. and Bulloch, J. (1997) "Predicting the Performance of Fixed-Bed Granular Activated Carbon Adsorbers," *Water Sc. Technol.* **35** (7), 235–241.

Hand, D. W., Crittenden, J.C. and Thacker, W.E. (1983) "User Oriented Batch Reactor Solutions to the Homogeneous Surface Diffusion Model," *J. Env. Eng. Div.,Proce. ASCE*, **109** (EE1), 82.

Hand, D. W., Crittenden, J.C. and Thacker, W.E. (1984) "Simplified Models for Design of Fixed-Bed Adsorbers," *J. Env. Eng. Div., Proc. ASCE,* **110** (EE2).

Hansen, R. E. (1975) The Costs of Meeting the New Water Quality Standards for Total Organics and Pesticides, in Proceedings, AWWA Annual Conference, Minneapolis, MN, American Water Works Association, Denver, CO.

Hashimoto, K., Miura, K., Yoshikawa, F., and Imai, I. (1979) "Change in Pore Structure of Carbonaceous Materials During Activation and Adsorption Performance of Activated Carbon," *Ind. Eng. Chem. Process Des. Devel.*, **18**, 1, 72–80.

Hassler, J. W. (1974) *Activated Carbon,* Chemical Publishing, New York.

James, R. O., and Healy, T. W. (1972) "Adsorption of Hydrolyzable Metal Ions at the Oxide Water Interface," *J. Colloid Interface Sci.*, **40**, 42–81.

Jarvie, M., Hand, D. W., Bhuvendralingam, S., Crittenden, J. C., and Hokanson, D. R. (2005) "Simulating the Performance of Fixed-Bed Granular Activated Carbon Adsorbers: Removal of Synthetic Organic Chemicals in the Presence of Background Organic Matter," *Water Res.* **39**, 2407–2421.

Jüntgen, H. (1968) "Entstehung Des Porensystems Bei Der Teilvergasung Von Koksen Mit Wasserdampf," *Carbon,* **6**, 297–308.

Jüntgen, H. (1976) Manufacture and Properties of Activated Carbon, in *Translation of Reports on Special Problems of Water Technology,* EPA-600/9-76-030, U.S. Environmental Protection Agency, Washington, DC.

Kipling, J. J. (1965) *Adsorption from Solutions of Non-Electrolytes,* Academic, New York.

Knappe, D. U., Matsui, Y., Snoeyink, V. L., Roche, P., Prados, M. J., and Bourbigot, M. M. (1998) "Predicting the Capacity of Powdered Activated Carbon for Trace Organic Compounds in Natural Waters," *Environ. Sci. Technol.*, **32**, 11, 1694–1698.

Langmuir, I. (1918) "The Adsorption of Gases on Plane Surfaces of Glass, Mica, and Platinum," *J. Am. Chem. Soc.*, **40**, 1361–1402.

Lee, M. C., Crittenden, J.C., Ari, M., and Snoeyink, V.L. (1983) "Design of Carbon Beds to Remove Humic Substances," *J. Environ. Eng.*, **109**, 3, 631-645.

Lee, M. C., Snoeyink, V. L., and Crittenden, J. C. (1981) "Activated Carbon Adsorption of Humic Substances," *J. AWWA,* **73**, 8, 440–446.

Luft, P. J. (1984) Modeling of Multicomponent Adsorption onto Granular Carbon in Mixtures of Known and Unknown Composition, M.S. Thesis, Michigan Technological University, Houghton, MI.

McGuire, M. J. (1989) *Optimization and Economic Evaluation of Granular Activated Carbon for Organic Removal,* American Water Works Association Research Foundation, Denver, CO.

McGuire, M. J., Davis, M. K., Tate, C. H., Aieta, E. M., Howe, E. W., and Crittenden, J. C. (1991) "Evaluating GAC for Trihalomethane Control," *J. AWWA,* **83**, 1, 38–48.

McGuire, M. J., Krasner, S. W., Hwang, C. J., and Lzaguirre, G. (1981) "Closed-Loop Stripping Analysis as a Tool for Solving Taste and Odor Problems," *J. AWWA,* **73**, 10, 530–537.

Munakata, K., Kanjo, S., Yamatsuki, S., Koga, A., and Lanovski, D. (2003) "Adsorption of Noble Gases on Silver-Mordenite," *J. Nucl. Sci. Tech.*, **40**, 9, 695–697.

Najm, I.N., (1996) "Mathematical Modelling PAC Adsorption Processes," *J. AWWA,* **88**, 10, 79–89.

Nemethy, G., and Scheraga, H. A. (1962) "Structure of Water and Hydrophobic Bonding in Proteins. I. A Model for the Thermodynamic Properties of Liquid Water," *J. Chem. Phys.*, **36**, 3382–3401.

Parks, G. A. (1967) Aqueous Surface Chemistry of Oxides and Complex Oxide Minerals, Isoelectric Point and Zero Point of Charge, in W. Stumm (ed.), *Equilibrium Concepts in Natural Water Systems*, Advances in Chemistry Series, Vol. **67**, American Chemical Society, Washington, DC.

Parks, G. A. (1975) Adsorption in the Marine Environment, pp. 241–308, in J. P. Riley and G. Skirrow (eds.) *Chemical Oceanography*, 2nd ed., Academic, New York.

Polanyi, M. (1916) "Adsorption Von Gasen (Dämpfen) Durch Ein Festes Nichtflüchtigers Adsorbens," *Ber. Deutsche Phys. Ges.*, **18**, 55–80.

Randtke, S. J., and Snoeyink, V. L. (1983) "Evaluating GAC Adsorptive Capacity," *J. AWWA*, **75**, 8, 406–413.

Reucroft, P. J., Simpson, W. H., and Jonas, L. A. (1971) "Sorption Properties of Activated Carbon," *J. Phys. Chem.*, **75**, 23, 3526–3531.

Roberts, P. V., and Summers, R. S. (1982) "Performance of Granular Activated Carbon for Total Organic Carbon Removal," *J. AWWA*, **74**, 2, 113–118.

Rosen, J.B. (1954) "General Numerical Solution for Solid Diffusion in Fixed-Beds," *Ind. Eng. Chem. Jo.*, **46**, 1950–1955.

Satterfield, C. N. (1980) *Heterogeneous Catalysis in Practice*, McGraw-Hill, New York.

Sigama_Aldrich Online Catalog (2004, June) http://www.sigmaaldrich.com/Brands/Supelco_Home/Datanodes.html?Fcat_path=982049,1005395,1005413&supel-co_name=Liquid Chromatography&id=D1005413.

Snoeyink, V. L. (2001) Adsorption of Trace Organic Compounds from Water Supplies. Association of Environmental Engineering and Science Professors Distinguished Lecture, presented at Michigan Technological University, Houghton, MI.

Snoeyink, V. L., and R. S. Summers (1999) Adsorption of Organic Compounds, Chap. 13, in R. D. Letterman (ed.), *Water Quality and Treatment: A Handbook of Community Water Supplies*, 5th ed., American Water Works Association, McGraw-Hill, New York.

Sontheimer, H. (1976) *The Use of Powdered Activated Carbon. Translation of Reports on Special Problems of Water Technology*, Vol. **9**, *Adsorption*, EPA 600/9-76-030, U.S. Environmental Protection Agency, Washington, DC.

Sontheimer, H., Crittenden, J. C., and Summers, R. S. (1988) *Activated Carbon for Water Treatment*, 2nd ed., DVGW-Forschungsstelle, University of Karlsruhe, Karlsruhe, Germany. Distributed in the U.S. by the American Water Works Association.

Speth, T. F. (1986) Predicting Equilibria for Single Solute and Multicomponent Aqueous Phase Adsorption onto Activated Carbon, Master's Thesis, Michigan Technological University, Houghton, MI.

Speth, T. F., and Adams, J. Q. (1993) "GAC and Air Stripping Design Support for the Safe Drinking Water Act," (in R. Clark and R. S. Summers ed.), *Strategies and Technologies for Meeting SDWA Requirements*. Technomic Publishing, Lancaster, PA.

Speth, T. F., and Miltner, R. J. (1998). "Adsorption Capacity of GAC for Synthetic Organics," *J. AWWA*, **90**, 4, 171–174.

Stumm, W., and Morgan, J. J. (1981) *Aquatic Chemistry*, 2nd ed., Wiley, New York.

Suffet, I. H., Corado, A., Chou, D., McGuire, M. J., and Butterworth, S. (1996) "AWWA Taste and Odor Survey." *J. AWWA*, **88**, 4, 168–180.

Summers, R. S. (1986) Activated Carbon Adsorption of Humic Substances: Effect of Molecular Size and Heterodispersity, Ph.D. Dissertation, Stanford University, Stanford, CA.

Tang, S. R. (1986) Predicting Equilibria for Gas-Phase Adsorption of Volatile Organic Compounds: The Impact of Relative Humidity and Multicomponent Interactions, M.S. Thesis, Michigan Technological University, Houghton, MI.

Tchobanoglous, G., Burton, F. L., and Stensel, H. D. (2003) *Wastewater Engineering: Treatment and Reuse*, 4th ed., Metcalf and Eddy, McGraw-Hill, New York.

Traegner, U. K., Suidan, M. T., and Kim, B. R., (1996) "Considering Age and Size Distributions of Activated-carbon Particles in a Completely-mixed Adsorber at Steady State," *Water Res.*, **30**, 1495–1501.

Vermeulen, T. (1958) "Separation by Adsorption Methods," *Adv. Chem. Eng.*, **2**, 147–203.

Walker, Jr., P. L. (1986) "Coal Derived Carbons," *Carbon*, **24**, 379–386.

Wicke, E., (1939) "Empirische und theoretische Untersuchugen der Sorptions- geschivindigkeit von Gasen an porosen Stoffen I and II," *Kolliod-Z.*, **86**, 295–305.

Zhang, Q.; Crittenden, J. C., Hristovski, K., Hand, D. W., and Westerhoff, P. (2009) "User-oriented Batch Reactor Solutions to the Homogeneous Surface Diffusion Model for Different Activated Carbon Dosages," *Water Res.*, **43** (7), 1859–1866.

Zimmer, G., Crittenden, J. C., and Sontheimer, H. (1988) Design Considerations for Fixed-Beds Adsorbers That Remove Synthetic Organic Chemicals in the Presence of Natural Organic Matter, paper presented at the American Water Works Association Annual Conference, Orlando, FL.

Zimmer, G., Haist, B., and Sontheimer, H. (1987) "The Influence of Organic Matter on Adsorption Behavior of Chlorinated Hydrocarbons," Proceedings of the 1987 AWWA Conference, Kansas City, Mi.

16 Ion Exchange

Stantec's Water Treatment: Principles and Design, Updated Third Edition. John C. Crittenden, R. Rhodes Trussell, David W. Hand, Kerry J. Howe, and George Tchobanoglous.

Terminology for Ion Exchange

Term	Definition
Counterion	Ion in solution that can exchange with an ion attached to a stationary functional group.
Empty-bed contact time	Volume of the resin in the bed including pore volume divided by the volumetric flow rate to the fixed bed.
Film diffusion	Effective rate at which ions migrate across the stagnant film surrounding the resin particles in the fixed bed.
Gel-type resin	Translucent polymeric resin with low degree of crosslinking and a high water content with an open matrix.
Helfferich number	Ratio of the rate of mass transfer by film diffusion to the rate of mass transfer by intraparticle diffusion.
Intraparticle diffusion	Effective rate at which ions migrate inside the resin particles.
Ion exchange	Process in which ions attached to a stationary functional group exchange for ions in a solution. Ions are exchanged on an equivalence basis.
Macroreticular resin	Opaque polymeric resin having a high degree of crosslinking and low water content with a closed matrix.
Presaturant ion	Ion that comes attached to the virgin resin or is exchanged onto the resin during the regeneration process.
Regeneration curves	Breakthrough curves obtained from a fixed-bed ion exchange operation during the regeneration cycle.
Resin swelling	Enlarging of an ion exchange resin due to the exchange of a larger preferred ion over a smaller less preferred ion.
Saturation loading curves	Breakthrough curves obtained from a fixed-bed ion exchange operation during the loading cycle.

Term	Definition
Selectivity	Preference of one ion over another for exchange onto an ion exchange site on a resin.
Separation factor	Quantitative description of the preference of one ion over another for a given ion exchange resin.
Service flow rate	Volumetric flow rate to the ion exchange column divided by the volume of the resin in the bed including pore volume.
Strong acid cation resin	Ion exchange resin that will readily give up a proton over a wide pH range.
Strong base anion resin	Ion exchange resin that will readily give up a hydroxide ion if the pH is less 13.
Synthetic resins	Spherical beads that contain a network of crosslinked polymers containing functional groups.
Total ion exchange capacity	Total amount of chemical equivalents available for exchange per unit weight or unit volume of resin.

Ion exchange is a process used to remove dissolved ionic constituents that can cause aesthetic and health issues. The ion exchange process for water treatment is considered to be a nonconventional process because it is not widely used in large-scale plants. The types of ion exchange materials used in water treatment, basic mechanisms involved in the ion exchange process, process design considerations, and example problems that apply ion exchange fundamentals to system design and operation are discussed in this chapter.

16-1 Evolution of Ion Exchange Technology

In drinking water treatment applications, ion exchange is primarily used for water softening and demineralization (e.g., removal of Ca^{2+}, Mg^{2+}, Na^{+}, Cl^{-}, SO_4^{2-}, NO_3^{-}). The vast majority of ion exchange installations in the United States are small, point-of-use devices at individual households. The application of ion exchange to municipal systems has been limited. Applications include the removal of hardness (softening), nitrate, barium, radium, arsenic, perchlorate, and chromate. There have been several full-scale systems designed for industrial applications, such as the demineralization of water for prevention of scale formation in power plant boilers, removal of calcium and magnesium in car-washing facilities, and production of ultrapure water for making pharmaceuticals and semiconductor materials.

With increased concern for the health effects of other contaminant ions such as barium, radium, fluoride, nitrate, arsenate, perchlorate, and uranium, the use of ion exchange and inorganic adsorbents for full-scale applications in water treatment has increased. While much attention has been placed on the use of conventional synthetic ion exchange resins, research is ongoing to develop specialty resins that are selective for some contaminant ions. Natural and synthetic inorganic materials (e.g., zeolites) that have adsorption properties and exhibit favorable capacities for contaminant ions are also being developed and evaluated.

Natural Exchange Materials

In water treatment applications, ion exchange involves the exchange of an ion in the aqueous phase for an ion in the solid phase. The solid phase or ion exchanger is insoluble and can be of natural origin such as kaolinite and montmorillonite minerals or a synthetic material such as a polymeric resin. These exchangers have fixed charged functional groups located on their external and/or internal surface, and associated with these groups are ions of opposite charge called "counterions" (see Fig. 16-1a). These mobile counterions are associated by electrostatic attraction to each of the charged functional groups to satisfy the criterion that electroneutrality is maintained at all times within the exchange material as well as in the bulk aqueous solution. Depending on the charge of the functional group on the exchanger, the counterion can either be a cation if the functional group is negative or an anion if the functional group is positive and can exchange with another counterion in the aqueous phase.

The application of natural ion exchange materials for water treatment may have been used as far back as biblical times when Moses sweetened the waters of Mariah (Exodus 15:23–25). In approximately 320 BC, Aristotle

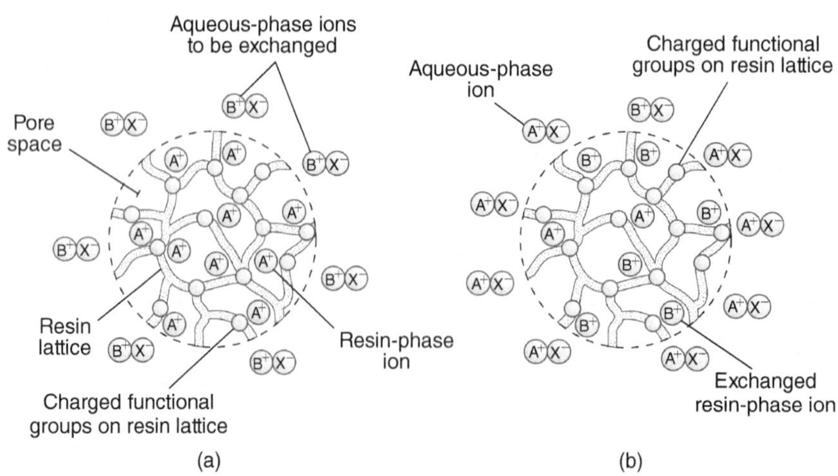

Figure 16-1
Schematic framework of functional cation exchange resin: (a) resin initially immersed in an aqueous solution containing B^+ cations and X^- anions and (b) cation exchange resin in equilibrium with aqueous solution of B^+ cations and X^- anions.

used 20 earthen containers containing a material with ion exchange properties to produce freshwater from seawater (Wachinski and Etzel, 1997). It was not until 1854 that the first reported systematic study of the ion exchange phenomenon was reported by Thompson and Way (Kunin and Myers, 1950). They observed ammonium ions adsorbing onto soils, releasing calcium ions in equivalent amounts, and the aluminum silicates present in the soils were responsible for the exchange. In 1876, Lemberg observed that the mineral leucite ($K_2O \cdot Al_2O_3 \cdot 4SiO_2$) could be transformed into analcite ($Na_2O \cdot Al_2O_3 \cdot 4SiO_2 \cdot 2H_2O$) by leaching the mineral with a solution of sodium chloride (Kunin and Myers, 1950).

In the early twentieth century, many contributions were made in the understanding of the ion exchange phenomenon in clays, peat, charred bone, soils, and other silicates. In fact, one of the first known synthetic mineral ion exchange materials was developed and composed of processed natural greensand, which is referred to as a natural zeolite material. Although these zeolite materials exhibited low capacities and poor abrasion characteristics, they were the first exchangers used in large quantities for water treatment application.

Synthetic Exchange Materials

In 1935, development of sulfonated coal exchangers by Leibknecht in Germany and synthetic phenol–formaldehyde exchangers by Adams and Holmes in England led to the discovery of the first synthetic resin materials that were stable and had high anion exchange capacities (Kunin and Meyers, 1950). Unlike aluminosilicate zeolites, these materials were highly resistant to regenerant solutions of mineral acids. The discoveries of these researchers led D'Alelio to develop and patent sulfonated, crosslinked polystyrene resins in 1945 (Kunin and Meyers, 1950). The work of Leibknecht, Adams, Holmes, and D'Alelio forms the basis of modern-day synthetic organic ion exchangers. The first large domestic ion exchange facility was a softening plant (756 ML/d or 20 mgd) built by the Metropolitan Water District of Southern California (Streicher et al., 1947). The plant was first built in 1946 using silica-based synthetic media but was later converted to modern polystyrene divinyl benzene resin. It was operated until the early 1970s.

Exchange Mechanisms for Synthetic Resins

For most ion exchange applications in water treatment, synthetic organic resins are utilized because of their relatively large available exchange capacities and ease of regeneration. The schematic framework of a synthetic organic cation ion exchange resin initially immersed in an aqueous solution containing cation B^+ with its coion X^- is illustrated on Fig. 16-1a. The resins are spherical beads consisting of a network of crosslinked polymers containing functional groups with fixed coions, which are negative $(-)$ charges located on each functional group along the polymer matrix. Associated with these fixed coions are A^+ cations, which are mobile and free to move in the pores of the polymer matrix. Cation A^+ is referred to as the presaturant ion.

When the cation exchange resin (saturated with A^+) is immersed in solution, there is a tendency for A^+ to diffuse into the bulk solution and X^- into the resin because of the concentration differences between the solution and resin phases. If these ions carried no charge, then their concentration differences would equalize by diffusion, and equilibrium would be obtained. However, because the ions are charged and electroneutrality is maintained, initially there is a small migration of A^+ into the bulk solution and X^- into the resin, causing a net positive charge in the bulk solution and a net negative charge in the ion exchanger. The first few A^+ ions diffusing into the bulk solution and X^- ion diffusing into the resin establish an electric potential difference between the bulk solution and the resin phase. This potential difference, called the *Donnan potential,* causes a small shortage of anions in the bulk solution and cations within the resin phase, which acts to maintain electroneutrality. The Donnan theory was developed to explain the equilibrium behavior of ion distributions across membranes (Helfferich, 1995; Weber, 1972). Consequently, counterion B^+ can diffuse into the negatively charged resin phase and replace A^+ stoichiometrically while X^- anions are repelled back into the positively charged bulk solution due to this Donnan exclusion potential. As shown on Fig. 16-1b, equilibrium is established eventually where the concentration differences of the ions are balanced by the electric field. In other words, the electrical potentials are equal in both phases because the voltage difference is balanced by the concentration difference. At equilibrium, the resin phase will still contain a higher concentration of counterions A^+ and B^+ than in the bulk solution, and electroneutrality will be maintained between the bulk solution and the resin phase. A similar explanation can be made for a strong-base anion exchange resin with the exception being the charges are reversed.

16-2 Synthetic Ion Exchange Media

Synthetic ion exchange resins are almost always used in water treatment practices. Synthetic polymeric resins are very durable and their properties can be modified to selectively remove both anions and cations. The resins can be regenerated using various salt or acid solutions depending upon the particular application. A discussion of the structure, manufacturing process, and various types of synthetic resins is presented below.

Resin Structure

Ion exchange polymeric resin is composed of a three-dimensional, crosslinked polymer matrix that contains covalently bonded functional groups with fixed ionic charges. Vinyl polymers (typically, polystyrene and polyacrylic) are used for the resin matrix backbone. Divinylbenzene (DVB) is used to crosslink the polymer backbone. The overall steps involved in making both polyacrylic and polystyrene cation exchange resins are displayed on Fig. 16-2.

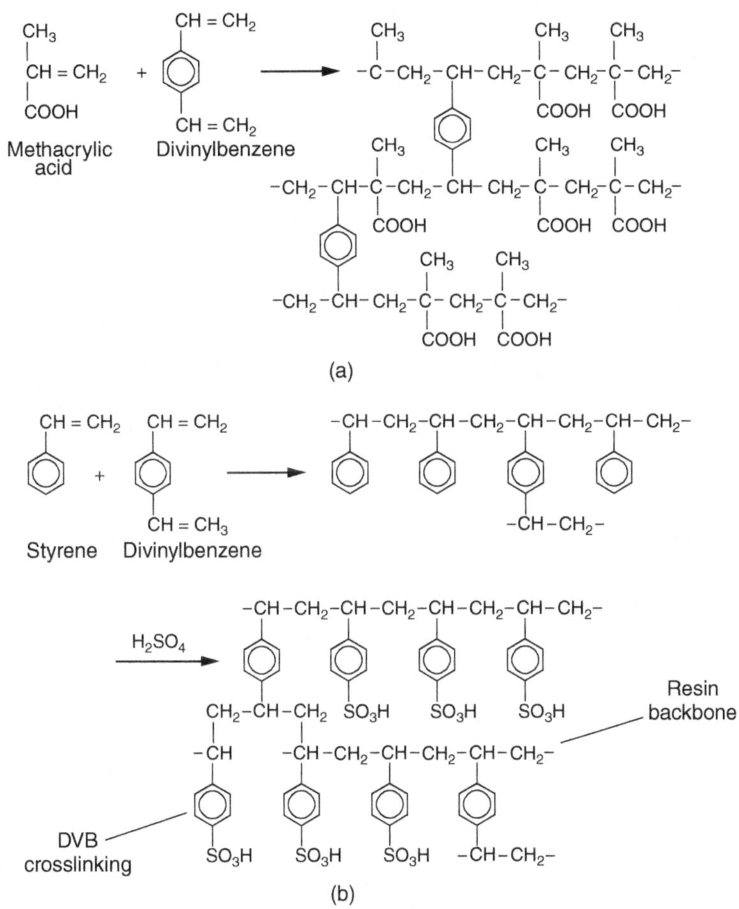

Figure 16-2
Major steps involved in synthesis of cation ion exchange resin by polymerization of (a) methacrylic acid with divinylbenzene crosslinking and (b) styrene with divinylbenzene crosslinking followed by sulfonation. (Adapted from Weber, 1972)

An important distinction for resins, with respect to their polymeric backbone, is whether the resin is a microreticular (gel) or macroreticular resin, which depends upon the degree of crosslinking within the resin's polymer backbone. Gel-type resins have about 4 to 10 percent DVB crosslinking with a typical value of 8 percent. The pore structure of a gel-type resin is determined by the distances between the polymer chains and crosslinking member that vary with (1) the exchanging ions, (2) ionic strength, (3) temperature, (4) the number of fixed ionic charges, and (5) the degree of crosslinking. Gel-type resins are translucent with high water content and can exhibit a significant amount of swelling or shrinking depending on the presaturant ion. Because gel-type resins lose their pore structure upon drying, they have very low Brunauer–Emmett–Teller (BET) nitrogen surface areas (see Chap. 15 for discussion of BET surface area), typically less than 2 m^2/g. The ion exchange rate is very fast for a gel-type resin due to its rather open matrix (Kunin, 1979; Rohm and Haas, 1975).

Macroreticular resins have approximately 20 to 25 percent DVB cross-linking and are opaque. Macroreticular resins are made up of microspheres linked together to form a resin bead (Kunin, 1979; Rohm and Haas, 1975). Macroreticular resins retain their structural integrity when dried, and as a result, they have BET surface areas of about 7 to 600 m^2/g and particle porosities of 20 to 60 percent. In addition, macroreticular resins do not swell or shrink significantly due to ion exchange reactions.

Classification of Resins by Functional Groups

Based on the functional groups bonded to the resin backbone, the four general types of exchange resins are (1) strong-acid cation (SAC), (2) weak-acid cation (WAC), (3) strong-base anion (SBA), and (4) weak-base anion (WBA). The distinctions are based on the pK values of the functional groups as summarized in Table 16-1. These resin types are discussed in more detail in the following sections.

STRONG-ACID EXCHANGERS

In SAC exchange resins, a charged sulfonate group typically acts as the exchange site. The term "strong" in SAC has nothing to do with the physical strength of the resin but is derived from Arrhenius theory of electrolyte strength in which the functional group on the resin is dissociated completely in its ionic form at any pH. In other words, the resin's low pK_a (<0) implies SAC exchangers will readily give up a proton over a wide pH range (1 to 14). The general exchange and regeneration reactions for the hydrogen form can be written as

$$n\left[\overline{RSO_3^-}\right]H^+ + M^{n+} \rightleftarrows \left[\overline{nRSO_3^-}\right]M^{n+} + nH^+ \quad \text{(exchange reaction)} \tag{16-1}$$

$$\left[\overline{nRSO_3^-}\right]M^{n+} + nHCl \rightleftarrows n\left[\overline{RSO_3^-}\right]H^+ + MCl_n \quad \text{(regeneration reaction)} \tag{16-2}$$

In Eq. 16-1, the overbar refers to the immobile resin phase, H^+ is the presaturant ion associated with the resin before exchange, M^{n+} is the counterion in solution being exchanged out of solution, and n is the charge on the counterion in solution. For the reaction shown in Eq. 16-1, based on the pK_a of SAC resins and the large hydrated radius of hydrogen, SAC resins have little affinity for hydrogen ion and will readily exchange it for another cation. Because the hydrated radius of the H^+ ion in a SAC exchanger is much larger than other cations, the resin will typically shrink upon exchange (≈7 percent for a gel-type resin, 3 to 5 percent for macroreticular type resin). The sodium form of a SAC will also behave in a similar manner, although the shrinkage will be less than observed for the hydrogen form. Equation 16-2 represents the regeneration expression for a strong-acid cation exchange resin using HCl as the regenerant solution.

Table 16-1
Characteristics of ion exchange resins used in water treatment processes

Resin Type	Acronym	Fundamental Reaction[a]	Regenerant Ions (X)	pK	Exchange Capacity, meq/mL	Constituents Removed
Strong-acid cation	SAC	$n\,[RSO_3^-]X^+ + M^{n+} \rightleftarrows [nRSO_3^-]M^{n+} + nX^+$	H^+ or Na^+	<0	1.7–2.1	H^+ form: any cation; Na^+ form: divalent cations
Weak-acid cation	WAC	$n\,[RCOO^-]X^+ + M^{n+} \rightleftarrows [nRCOO^-]M^{n+} + nX^+$	H^+	4–5	4–4.5	Divalent cations first, then monovalent cations until alkalinity is consumed
Strong-base anion (type 1)	SBA-1[b]	$n\,[R(CH_3)_3\,N^+]X^- + A^{n-} \rightleftarrows [nR(CH_3)_3\,N^+]A^{n-} + nX^-$	OH^- or Cl^-	>13	1–1.4	OH^- form: any anion; Cl^- form: sulfate, nitrate, perchlorate, etc.
Stong-base anion (type 2)	SBA-2[c]	$n\,[R(CH_3)_2(CH_3CH_2OH)N^+]X^- + A^{n-} \rightleftarrows [nR(CH_3)_2(CH_3CH_2OH)N^+]A^{n-} + nX^-$	OH^- or Cl^-	>13	2–2.5	OH^- form: any anion; Cl^- form: sulfate, nitrate, perchlorate, etc.
Weak-base anion	WBA	$[R(CH_3)_2N]HX + HA \rightleftarrows [R(CH_3)_2N]HA + HX$	OH^-	5.7–7.3	2–3	Divalent anions first, then monovalent anions until strong acid is consumed

[a]Term within brackets represents the solid phase of the resin.
[b]Greater regeneration efficiency and capacity than SBA-2.
[c]Greater chemical stability than SBA-1.

WEAK-ACID EXCHANGERS

In WAC exchange resins the functional group on the resin is usually a carboxylate and the exchange reaction can be written as

$$n\left[\overline{\mathrm{RCOO^-}}\right]\mathrm{H^+} + \mathrm{M}^{n+} \rightleftarrows \left[\overline{n\mathrm{RCOO^-}}\right]\mathrm{M}^{n+} + n\mathrm{H^+} \text{ (exchange reaction)} \tag{16-3}$$

$$\left[\overline{n\mathrm{RCOO^-}}\right]\mathrm{M}^{n+} + n\mathrm{HCl} \rightleftarrows n\left[\overline{\mathrm{RCOO^-}}\right]\mathrm{H^+} + \mathrm{MCl}_n \quad \text{(regeneration reaction)} \tag{16-4}$$

Weak-acid cation resins have pK_a values in the range of 4 to 5 and thus will not readily give up a proton unless the pH is greater than 6. At a pH less than 6, WAC resins have a great affinity for hydrogen and will not exchange it for another cation; therefore, the apparent cation exchange capacity of a WAC exchanger is a function of pH. As the pH increases, the apparent capacity increases to a maximum total capacity between pH values of 10 and 11.

Weak-acid exchangers usually require alkaline species in the water to react with the more tightly bound hydrogen ions. Consider the following reaction between a weak-acid exchanger and alkalinity:

$$2\left[\overline{\mathrm{RCOO^-}}\right]\mathrm{H^+} + \mathrm{Ca(HCO_3)_2} \rightleftharpoons \left[\overline{2\mathrm{RCOO^-}}\right]\mathrm{Ca^{2+}} + 2(\mathrm{H_2CO_3}) \tag{16-5}$$

The exchange is essentially neutralization with bicarbonate alkalinity neutralizing the H^+ on the resin. Weak acids will dissociate alkaline salts such as $NaHCO_3$ but not nonalkaline salts like NaCl or $NaSO_4$. Because weak-acid exchangers exhibit a higher affinity for H^+ than strong-acid exchangers do, they exhibit higher regeneration efficiencies. Weak-acid resins do not require as high a concentration of regenerant as required for regenerating strong-acid resins to the hydrogen form. The carboxylic functional groups will utilize up to 90 percent of the acid (HCl or H_2SO_4) regenerant, even with low acid concentrations. By comparison, strong-acid resin regeneration requires a large excess of regenerant solution to provide the driving force for exchange to take place.

Weak-acid exchangers have been used in water treatment to remove cations in high-alkaline (e.g., high CO_3^{2-}, OH^-, and HCO_3^- concentrations) waters with low dissolved carbon dioxide and sodium. Simultaneous softening and dealkalization can be accomplished with weak-acid exchangers. It has also been reported that sometimes WBA resins are used in conjunction with strong-acid exchangers to reduce regenerant requirements and produce treated water with the same quality as just using strong-acid exchangers alone.

STRONG-BASE EXCHANGERS

Strong-base anion exchange resins typically have a quaternary amine group as the fixed positive charge. For a type 1 quaternary group, the exchange reaction for a resin in the hydroxide form can be written as

$$n\left[\overline{R(CH_3)_3N^+}\right]OH^- + A^{n-} \rightleftarrows \left[\overline{nR(CH_3)_3N^+}\right]A^{n-} + nOH^- \quad \text{(exchange reaction)} \quad (16\text{-}6)$$

$$\left[\overline{nR(CH_3)_3N^+}\right]A^{n-} + nNaOH \rightleftarrows n\left[\overline{R(CH_3)_3N^+}\right]OH^- + (Na^+)_nA^{n-} \quad \text{(regeneration reaction)} \quad (16\text{-}7)$$

For a type 2 quaternary group, the exchange reaction for a resin in the hydroxide form can be written as

$$n\left[\overline{R(CH_3)_2(CH_3CH_2OH)N^+}\right]OH^- + A^{n-} \rightleftarrows \left[\overline{nR(CH_3)_2(CH_3CH_2OH)N^+}\right]A^{n-} + nOH^- \quad (16\text{-}8)$$

As shown in Eqs. 16-6 and 16-8, the main difference between type 1 and type 2 resins is the ethanol group in the type 2 quaternary amine. The purpose of the ethanol group is to reduce the resin's affinity for hydroxide ions. Strong-base anion resins have pK_b values of 0 to 1, implying that they will readily give up a hydroxide ion if the pH value is less than 13. The operational pH of SBA resins (pH < 13) makes the apparent anionic exchange capacity independent of pH. Strong-base anion exchangers in the hydroxide form will shrink upon exchange due to other anions typically having hydrated radii smaller than hydroxide. Type 1 has a slightly greater chemical stability, while type 2 has a slightly greater regeneration efficiency and capacity.

Strong-base anion resins are less stable than strong-acid resins and are characterized by the fishy odor of amines even at room temperature. Strong-base anion exchange resins will degrade to release both the tertiary amine and methanol at 60°C (Bolto and Pawlowski, 1987).

Strong-base exchangers traditionally have been used for many years to demineralize water. However, more recently SBA exchangers are increasingly being used to treat waters contaminated with nitrate, arsenic, and perchlorate ions and are usually operated in the chloride cycle, where the resin is regenerated with NaCl (Clifford and Weber, 1978; Clifford et al., 1987; Ghurye et al., 1999; Najm et al., 1999).

WEAK-BASE EXCHANGERS

In WBA exchange resins the exchange site is a tertiary amine group, which does not have a permanent fixed positive charge. Weak-base anion exchange resins are available in either chloride or freebase forms. The freebase designation indicates that the tertiary amine group is not ionized but has a water molecule (HOH) associated with it. The tertiary amine groups will adsorb ions without the exchange of an ion (Helfferich, 1995).

The exchange reaction for a WBA resin in the freebase form can be written as

$$\left[\overline{R(CH_3)_2N}\right]HOH + H^+ + A^- \rightleftarrows \left[\overline{R(CH_3)N}\right]HA + HOH$$
$$\text{(exchange reaction)} \qquad (16\text{-}9)$$

$$\left[\overline{R(CH_3)_2N}\right]HA + NaOH \rightleftarrows \left[\overline{R(CH_3)N}\right]HOH + NaA$$
$$\text{(regeneration reaction)} \qquad (16\text{-}10)$$

The reaction in Eq. 16-9 can be viewed as the ionization of the resin by hydrogen and the consequent uptake of the anion.

Weak-base anion resin behavior can also be described as the adsorption of a weak acid with the tertiary amine group acting as a Lewis base and the release of HOH as shown:

$$\left[\overline{R(CH_3)_2N}\right]HOH + HA \rightleftarrows \left[\overline{R(CH_3)_2N}\right]HA + HOH \qquad (16\text{-}11)$$

The weak-base designation is derived from the WBA resin's pK_b values of 5.7 to 7.3. Weak-base anion resins will not readily give up hydroxide ion unless the pOH is greater than the pK_b of the resin (pH values less than 8.3 to 6.7 at 25°C). In many respects, the weak-base exchangers behave much like weak-acid exchangers. The weak-base resins remove free mineral acidity such as HCl or H_2SO_4 but will not remove weakly ionized acids such as silicic and carbonic, which is why these resins are sometimes called "acid adsorbers."

Weak-base resins can be regenerated using NaOH, NH_4OH, or Na_2CO_3. The regeneration efficiencies of these resins are much greater than for strong-base resins. Weak-base exchangers are used in conjunction with strong-base exchangers in demineralizing systems to reduce regenerant costs and to attract organics that might otherwise foul the strong-base resins. Where silica removal is not critical, weak-base resins may be used alone or in conjunction with an air stripper to remove CO_2.

16-3 Properties of Ion Exchange Media

Two types of properties are important for ion exchange: (1) engineering properties and (2) physical properties. The engineering properties consist of the exchange capacity and selectivity of the resin. Engineers use exchange capacity and selectivity relationships to determine the performance of the resin under specific operating conditions. The physical properties consist of particle size, stability, swelling, moisture, and density of the resin. Physical properties are important in the selection of resins for specific water treatment applications. For example, the resin particle size must be large enough to minimize column pressure drop while in operation but small enough to enable fast mass transfer of the ions for exchange. The resins must also be durable enough to undergo swelling and shrinking of the resin during regeneration and loading. All the properties described above and the forces that impact them are discussed below.

Engineering Properties of Resins

The exchange capacity and selectivity are two important engineering parameters when considering the column design and operation of the column. The exchange capacity allows the engineer to determine how much the ionic constituent can be retained by the resin for a given resin volume. The selectivity provides the engineer with information on which ionic constituents in the water are preferred by the resin.

EXCHANGE CAPACITY

An important property of an ion exchange resin is the quantity of counterions that can be exchanged onto the resin. This exchange capacity can be expressed in terms of total (theoretical) capacity or effective capacity. The effective capacity is that part of the total capacity that can be utilized in a column operation, which is dependent on operating conditions such as service flow rate (SFR), regeneration level, regeneration flow rate, and water quality. Because the effective capacity is site specific, the total capacity is discussed here.

Total exchange capacity

The total capacity is dependent upon the quantity of functional groups in the copolymer bead. For SAC exchange resins, one sulfonate group, on average, can be attached to each benzene ring in the matrix. Hence, the dry-weight capacity of the resin can be determined and is usually expressed in terms of milliequivalents per gram of dry resin (meq/g). For example, if a gel-type SAC in the hydrogen form has a functional monomer with an empirical formula of $C_8H_7 \cdot SO_3{}^{-}H$ (molecular weight 184) and 1 eq of exchangeable hydrogen ion, the theoretical capacity would be 1 eq per 184 g of dry resin or 5.4 meq/g dry resin (Harland and Prud'homme, 1992). Reported values are actually lower than the calculated values because some of the benzene rings are occupied by the DVB crosslinking and the materials are not homogeneous. Measurement of dry-weight capacities can be determined from direct titration of a known volume of resin. For sulfonated styrene–DVB resins, the reported dry-weight capacity is typically 5.0 ± 0.1 meq/g (Anderson, 1979). For SBA exchange resins, more or less than one functional group can be attached to the benzene ring. Consequently, the dry-weight capacity is more variable than with strong-acid resins and can range from 2 to 5 meq/g.

Expressions for exchange capacity

In most ion exchange literature, the capacity is expressed in terms of a wet-volume capacity. The wet-volume capacity depends upon the moisture

Table 16-2
Properties of styrene–divinylbenzyl, gel-type strong-acid cation and strong-base anion resins

Parameter	Unit	Strong-Acid Cation Resin	Type I, Strong-Base Anion Resin
Screen size, U.S. mesh	—	16 × 50	16 × 50
Shipping weight	kg/m^3	850	700
	(lb/ft^3)	(53)	(44)
Moisture content	%	45–48	43–49
pH range	—	0–14	0–14
Maximum operating Temperature	°C	140	OH^- form 60, Cl^- form 100
Turbidity tolerance	NTU	5	5
Iron tolerance	mg/L as Fe	5	0.1
Chlorine tolerance	mg/L Cl_2	1.0	0.1
Backwash rate	M/h	12–20	4.9–7.4
	$(gal/min \cdot ft^2)$	(5–8)	(2–3)
Backwash period	min	5–15	5–20
Expansion volume	%	50	50–75
Regenerant and concentration[a]	%	NaCl, 3.0–14	NaCl, 1.5–14
Regenerant dose	kg $NaCl/m^3$ resin	80–320	80–320
	(lb/ft^3)	(5–20)	(5–20)
Regenerant rate	BV/min	0.067	0.067
	$(gal/min\ ft^3)$	(0.5)	(0.5)
Rinse volume	BV	2–5	2–10
	(gal/ft^3)	(15–35)	(15–75)
Exchange capacity	meq/mL as $CaCO_3$,	1.8–2.0	1–1.3
	$(kgr/ft^3$ as $CaCO_3)$[b]	(39–41)	(22–28)
Operating capacity[c]	meq/mL as $CaCO_3$,	0.9–1.4	0.4–0.8
	$(kgr/ft^3$ as $CaCO_3)$[b]	(20–30)	(12–16)
Service flow rate	BV/h	8–40	8–40
	$(gal/min \cdot ft^3)$	(1–5)	(1–5)

[a]Other regenerants such as H_2SO_4, HCl, and $CaCl_2$ can also be used for SAC resins while NaOH, KOH, and $CaCl_2$ can be used for SBA regeneration.
[b]Kilograins $CaCO_3/ft^3$ are the units commonly reported in resin manufacturer literature. To convert kgr $CaCO_3/ft^3$ to meq/mL, multiply by 0.0458.
[c]Operating capacity is based on Amberlite IR-120 SAC resin. Operating capacities depend on method of regeneration and amount of regenerant applied. Manufacturers should provide regeneration data in conjunction with operating capacities for their resins.
Source: Adapted from Clifford et al., (2011).

content of the resin, which is dependent upon the functional form of the resin and will vary for a given type of resin. The wet-volume capacity is commonly expressed in milliequivalents per milliliter of resin (meq/mL), although it may also be expressed in terms of kilograins as $CaCO_3$ per cubic foot (kgr/ft^3) of resin. There are 21.8 meq/mL in 1 kgr/ft^3. As shown in Table 16-2, typical SAC exchange capacities are 1.8 to 2.0 meq/mL in the sodium form and SBA exchange capacities are 1.0 to 1.3 meq/mL in the chloride form. Weak-acid cation exchange capacities are about 4.0 meq/mL in the H^+ form and WBA exchange capacities are around 1.0 to 1.8 meq/mL in the freebase form, although WAC and WBA resin capacities are variable due to their partially ionized conditions and because exchange capacity is also a function of pH.

Given two different ion forms of the same resin, the capacity on a volume basis will be different due to differences in water content. The same is true for the same resin but with different degrees of crosslinking. The volume capacity is inversely proportional to the swelling of the resin. The resin volume is determined in a column in the presence of excess water after tapping the column to settle the resin. This resin volume includes the volume of the water within the interstices between resin particles. In a backwashed and settled bed, this void volume is usually 35 to 40 percent of the total bed volume.

SELECTIVITY

Ion exchange resins have a certain affinity or preference for ions in aqueous solution. This affinity or preference for a given resin is called selectivity. The direction, forward or reverse, of the ion exchange reactions shown in Eqs. 16-1 to 16-11 will depend upon the resin selectivity for a particular ion system. Take, for example, the exchange reaction shown in Eq. 16-6 for an SBA. If a dilute aqueous solution containing NO_3^- and Cl^- ions are being treated with a type I SBA resin in the OH^- form, both NO_3^- and Cl^- ions will be exchanged over the presaturant ion OH^- because they are preferred by the resin. In this case the reaction proceeds in the forward direction. Type I SBA resins also have a higher selectivity for NO_3^- ions over Cl^- ions so NO_3^- will occupy more exchange sites in a dilute solution.

Basis for selectivity

Resin selectivity depends upon the physical and chemical characteristics of the exchanging ion and resins. Chemical properties of the ions that impact selectivity are the magnitude of the valence and the atomic number of the ion. The physical properties of the resins that influence selectivity include pore size distribution and the type of functional groups on the polymer chains. The following discussion provides insight into these properties.

Example 16-1 Estimate resin requirements

A small public water system is considering removing nitrate from its water using ion exchange. The major ions contained in the well water are listed in the following table. The average daily flow rate is about 2000 m^3/d. If an SBA exchange resin is used, estimate the minimum daily volume of resin that would be required assuming that nitrate is removed completely and is the only anion exchanging on the resin. Use the information provided in Table 16-2.

Cation	meq/L	Anion	meq/L
Ca^{2+}	1.8	Cl^-	4.3
Mg^{2+}	1.0	NO_3^-	0.5
Na^+	2.0		
Total	4.8	Total	4.8

Solution

1. Determine the minimum volume of resin required per day.
   a. From Table 16-2 the typical exchange capacity for a type I SBA resin is about 1.2 meq/mL resin (1 to 1.3 meq/mL).
   b. The required volume is

$$\begin{Bmatrix} \text{Minimum} \\ \text{resin volume} \end{Bmatrix} = (0.5 \text{ meq NO}_3-\text{N/L})(2.0\times10^6 \text{ L/d})$$
$$(\text{mL resin}/1.2 \text{ meq})$$
$$= 8.32\times10^5 \text{ mL/d} = 0.832 \text{ m}^3/\text{d}$$

Comment

This type of "back-of-the-envelope" calculation is valuable when a first estimate of the resin requirements is needed for a preliminary calculation. In most cases, the resin requirement will be higher due to the presence of other anions in the water (e.g., sulfate) that will compete with nitrate ions for exchange sites, as discussed in the section on ion exchange equilibrium.

For dilute aqueous-phase concentrations at temperatures encountered in water treatment, ion exchange resins prefer the counterion of higher valence, as shown below:

Cations: $Th^{4+} > Al^{3+} > Ca^{2+} > Na^+$

Anions: $PO_4^{3-} > SO_4^{2-} > Cl^-$

In the preference shown above, it is assumed that the spacing of the functional groups allow for the exchange of multivalent ions. Counterion preference increases with dilution of solution and is strongest with ion exchangers of high internal molality (Helfferich, 1995). Although empirical, this rule of thumb can be explained using the Donnan potential theory. As explained earlier for a cation exchanger, when a resin comes in contact with a dilute aqueous solution, large concentration gradients exist between the ions in the aqueous phase and the resin phase. The tendency is for the aqueous cations and anions to migrate into the resin phase and cations in the resin phase to migrate into the aqueous phase. However, the initial migration of ions establishes the Donnan potential, which repels any further anions from entering the resin that would cause any significant deviation from electroneutrality.

The potential attracts aqueous-phase cations into the resin to balance the diffusion of the resin-phase cations entering the aqueous solution and approaches zero when equilibrium is established. The force exerted by the Donnan potential on an ion is proportional to the ionic charge of the ion (Helfferich, 1995). A counterion with a higher charge is attracted more strongly and is preferred by the resin phase. The Donnan potential increases as the aqueous-phase concentration becomes more dilute and the molality of the fixed ionogenic groups on the resin increases. For large aqueous-phase ion concentrations, the exchange potentials of ions of different charge become negligible and ions of lower valence can sometimes be preferred over ions of higher valence.

There are some exceptions to the above general rule. For example, divalent CrO_4^{2-} has a lower preference than monovalent I^- and NO_3^- ions, as shown in the following series:

$$SO_4^{2-} > I^- > NO_3^- > CrO_4^{2-} > Br^-$$

Effect of physical properties on selectivity

Resin selectivity can also be influenced by the degree of swelling or pressure within the resin bead. In an aqueous solution, both resin-phase ions and ions in aqueous solution have water molecules that surround them. The group of water molecules surrounding each ion is called the radius of hydration and is different for different ions. Typically, the radius of hydration becomes larger as the size of the ion decreases (see Table 16-3). When these ions diffuse in solution, the water molecules associated with them move as well. The crosslinking bonds that hold the resin matrix together oppose the osmotic forces exerted by these exchanged ions. These opposing forces cause the swelling pressure. In a dilute aqueous phase containing ion exchange resins, the ions with a smaller hydrated radius are preferred because they reduce the swelling pressure of the resin and are more tightly bound to the resin. As shown in Table 16-3 for a series of ions of equal charge, the hydrated radius is inversely proportional to the unhydrated

Table 16-3
Comparison of ionic, hydrated radii, molecular weight, and atomic number for a number of cations

Ion	Ionic Radii,[a] Å	Hydrated Radii,[b] Å	Molecular Weight	Atomic Number
Li^+	0.60	10.0	6.941	3
Na^+	0.95	7.9	22.98977	11
K^+	1.33	5.3	39.0983	19
Rb^+	1.48	5.09	85.4678	37
Cs^+	1.69	5.05	132.9054	55
Mg^{2+}	0.65	10.8	24.305	12
Ca^{2+}	0.99	9.6	40.08	20
Sr^{2+}	1.13	9.6	87.62	38
Ba^{2+}	1.35	8.8	137.33	56

[a]From Mortimer (1975).
[b]From Kunin and Myers (1950).

ionic radius (Weber, 1972). For some alkali metals the order of preference for exchange is

$$Cs^+ > Rb^+ > K^+ > Na^+ > Li^+$$

For alkaline earth metals the preference for exchange is

$$Ba^{2+} > Sr^{2+} > Ca^{2+} > Mg^{2+} > Be^{2+}$$

For a given series, anion exchange follows the same selectivity relationship with respect to ionic and hydrated radii as cations:

$$ClO_4^- > I^- > NO_3^- > Br^- > Cl^- > HCO_3^- > OH^-$$

Consequently, for a given series of ions, the resin selectivity for ions increases with increasing atomic number, increasing ionic radius, and decreasing hydrated radius.

With the exception of specialty resins, WAC resins with carboxylic functional groups behave similar in preference to SAC resins with the exception that hydrogen is the most preferred ion. In a similar manner, the preference of anions for WBA resins is the same as for SBA resins with the exception that the hydroxide ion is the most preferred ion.

The above general rules for order of selectivity apply to ions in waters that have total dissolved solids (TDS) values less than approximately 1000 mg/L. The preference for divalent ions over monovalent ions diminishes as the ionic strength of a solution increases. For example, consider a sulfonic cation exchange resin operating on the sodium cycle. In dilute concentrations, calcium ion is much preferred over sodium; hence calcium will replace sodium on the resin structure. However, at high salt concentrations (≈100,000 mg/L TDS), the preference reverses and this enhances

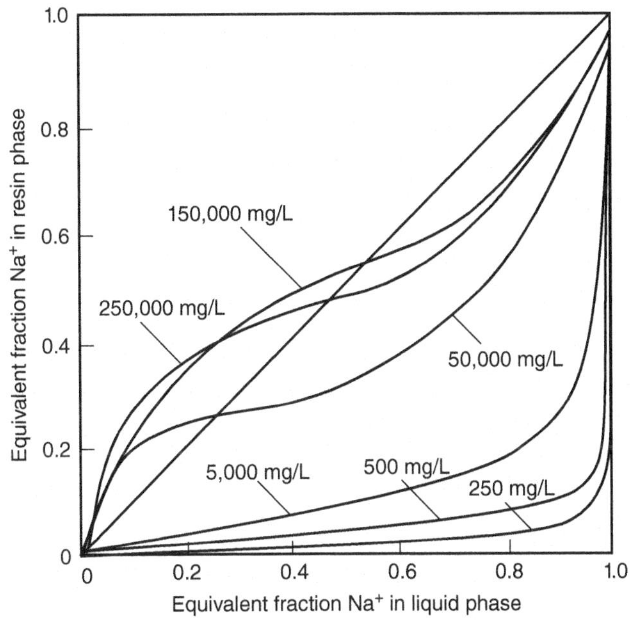

Figure 16-3
The Na^{+} –Ca^{2+} equilibria for sulfonic acid cation exchange resin. (Courtesy of Rohm and Haas.)

regeneration efficiency. Equilibrium isotherms for Na^{+} –Ca^{2+} exchange are shown on Fig. 16-3. As the TDS concentration increases, a higher concentration of sodium can be found in the resin phase. This is because as the salt concentration increases, the sodium concentration increases, and the activity coefficient for calcium decreases such that sodium is preferred over calcium.

Another important factor to consider in determining selectivity is the size of organic ions or inorganic complexes. A resin will exclude some of these ions by screening or sieving. Resins that exhibit this phenomenon are called molecular sieves. Ions too large to penetrate the resin matrix can be specifically excluded by proper selection of the resin properties. Increasing the crosslinking in the resin will produce a greater screening effect.

Physical Properties of Resins

Most synthetic ion exchange resin materials have physical properties that make them ideal for water treatment applications such as softening, demineralization, and removal of potentially toxic ionic contaminants. For example, polystyrene and polyacrylic resins are highly stable, are very durable, and retain their predictable capacities over many years of operation. In fact, some of these resins have been in use for over 15 years. Physical properties such as mechanical, chemical, and thermal stability, water content or swelling potential, total and apparent capacities, ion exchange equilibrium, and kinetics are influenced by the resin polymeric structure.

SWELLING, MOISTURE CONTENT, AND DENSITY

When resins exchange ions, the volume of the resin beads can change to reflect the differing magnitude of resin–counterion interactions, degree of resin crosslinking, and hydration. The swelling, moisture content, and density for several SAC, WAC, SBA, and WBA resins are summarized in Table 16-4. For all the resin types, the percentage of swelling decreases as the degree of crosslinking increases. Swelling of a resin due to exchange of ions can be reversible or irreversible. Reversible swelling is when the resin beads undergo a reversible volume change between one ionic form and another. The internal osmotic pressure of the resin bead increases when the resin swells and decreases when the beads shrink. Over time, the osmotic pressure swings can cause the beads to fracture, which is more likely to occur on macroporous resins than on gels. Swelling should be considered in the design of the ion exchange columns.

Table 16-4
Physical properties of several ion exchange resins

Resin	Bulk Wet Density, kg/m^3	Moisture Content Drained, % by Weight	Swelling Due to Exchange, %
SAC resins—sulfonated polystyrene			
2% crosslinked	720	72–82	12–15
5–6% crosslinked	769–849	58–65	5–10
12% crosslinked	753–929	37–49	4–9
Sulfonated phenolic	640–961	44–68	7
WAC resins—acrylic or methacrylic			
Weakly crosslinked	800	51–75	60–90
Medium crosslinked	721–800	46–62	10–90
Macroporous	688–800	56	5–10
SBA resins— polystyrene matrix, trimethyl benzyl ammonium			
2% crosslinked	705	70–78	20
4% crosslinked	673	60	14
8% crosslinked	720	34–56	15–20
WBA resins			
Aminopolystyrene	640–720	25–45	5–30
Aminated acrylic polymers	240–304	1–5	3
Epoxy-polyamine	689	50–62	6–25

Source: Adapted from Perry and Chilton (1973).

Irreversible swelling is observed with acrylic SBA resins where during the first few regeneration cycles the resins can irreversibly swell 7 to 10 percent over and above the reversible volume changes (Harland, 1994).

Because the water content of a resin can vary, the resin densities of different forms of the resin can also vary. The density will depend upon the quantity of water and the ionic form of the resin. The specific gravity of wet SAC resins will vary from 1.10 to 1.35, while the wet specific gravity of SBA resins vary from 1.05 to 1.15. The bulk or shipping weight of most wet strong-acid and strong-base resins vary from 675 to 900 kg/m^3 (42 to 56 lb/ft^3). Several physical properties of both strong-acid and strong-base resins are shown in Table 16-2.

PARTICLE SIZE

Ion exchange resin beads are spherical in shape and are commercially available in particle diameter sizes of 0.04 to 1.0 mm. In the United States, the particle sizes are listed according to standard screen sizes, or "mesh" values. A comparison of metric mesh sizes is given in Table 16-5. The most common size ranges used in large-scale applications are 16 to 50 and 50 to 100 mesh size.

Manufacturers usually provide three parameters related to particle size: (1) particle size range, (2) effective size (ES), and (3) uniformity coefficient (UC). The size range provides the minimum and maximum particle sizes for a given manufactured lot of resin beads. The ES is the mesh size in millimeters that passes 10 percent of a sieved sample, d_{10}. The UC is defined as the ratio of the d_{60} to the d_{10} resin sizes. For ion exchange resins UC values are usually in the range of 1.4 to 1.6; however, it is possible to obtain resins with smaller UCs required by kinetic or hydraulic restrictions. The ES and UC of resins can be obtained from the resin manufacturer.

Particle size has two major influences on ion exchange applications. First, the rate of ion exchange decreases with increasing particle size. An increase in resin particle size for the same mass of resin will decrease the film diffusion rate and increase the intraparticle diffusion path length. Second, the head loss through the bed increases with decreasing particle

Table 16-5
Particle size in U.S. mesh and millimeters

U.S. Standard Screen Size	Particle Diameter, mm	Geometric Mean Size, mm
16–20	1.2–0.85	1.01[a]
20–50	0.85–0.30	0.50
50–100	0.30–0.15	0.21
100–200	0.15–0.08	0.11
200–400	0.08–0.04	0.056

[a]Calculated as $\sqrt{0.85 \times 1.2} = 1.01$ mm.

size, subjecting the beads to situations that could cause breakage. In many ion exchange applications, the design is based on hydraulic requirements of the resin beads and the vessel rather than on ion exchange kinetics.

STABILITY

The stability of an ion exchange resin can be an important process design consideration under certain physical, chemical, and/or radioactive conditions. Chemical reactions between the resin matrix and dissolved constituents in water, physical impairment of resin performance due to fouling by organic and inorganic constituents, and some process operating conditions can significantly affect the resin performance and cost. Consequently, it is important to understand these interactions and take steps in the design stages to identify and prevent conditions that would negatively alter the resin's performance and the possible release of material from the resin to the finished water.

Effect of physical factors

As stated above, the stability of the resin may be impacted by either chemical or physical means. Physical stresses such as excessive swelling and shrinking, mechanical compression due to large hydraulic pressure drops across the resin bed, and abrasion due to excessive backwashing can significantly reduce the structural integrity of the resin bead and thereby shorten the resin operating life. With respect to swelling and shrinking, the acrylic resin is particularly durable due to its more elastic properties compared to the more rigid polystyrene matrix. However, in column operations with high operating pressures, the elasticity of acrylic resins can cause the beads to compress and result in inadequate liquid distribution and reduced flow.

Effect of chemical factors

Strong-acid cation exchange resins, especially the sulfonated polystyrene–DVB type, can be susceptible to oxidation. For example, oxidation can occur from free-chlorine attack of the DVB crosslinking, causing increased moisture retention of the resin, weakening the resin structurally, leading to compression of the beads, and affecting the service cycle. In addition, the resin can also lose capacity and must eventually be replaced. If an oxidant in the process water is unavoidable, it may be beneficial to use a resin with higher DVB crosslinking. Strong-acid cations with higher crosslinking (10 to 15 percent) will last longer than a typical SAC with 8 percent crosslinking. Chloride-regenerated SBA (type II) resins used to treat groundwater have lasted for more than 8 years and have still maintained over 90 percent of their capacity. Strong-base anions resins are also susceptible to releasing amines, which can lead to the formation of *N*-nitrosodimethylamine (NDMA), which can be a health concern (Kimoto et al., 1979; Najm and Trussell, 2001).

Precipitates such as calcium sulfate and ferric hydroxide can foul SACs. If the resin contains a large quantity of exchanged calcium and is regenerated with sulfuric acid, calcium sulfate precipitate will form inside the resin particles and reduce its capacity. In addition, excessive quantities of iron and manganese, if oxidized, can form precipitates and foul the resin. Weak-acid cation exchange resins can be fouled by calcium sulfate, but are not as susceptible to oxidation as SACs.

Effect of fouling

Two major types of fouling can occur with SBA resins, silica fouling and organic fouling. When used in the hydroxide form in a demineralization process, silicic acid is concentrated at the exchange front within the bed. Silicic acid will polymerize into an inorganic solid that will not behave as an exchangable anion. The silica can accumulate in the SBA resin until silica-free water cannot be produced.

Natural organic matter composed of humic and fulvic acids is negatively charged and can irreversibly exchange/adsorb onto SBA resins. Consequently, it usually requires large volumes of regenerant and rinses to bring the fouled resin back to its original capacity.

Weak-base resins are also subject to oxidation and fouling, depending on the type of resin. Special care should be taken with these resins prior to their use to ensure their stability will not be adversely affected. Manufacturers will normally provide a user guide to proper selection and use of these types of resins to prevent degradation.

16-4 Ion Exchange Equilibrium

As shown in Eqs. 16-1 through 16-11, the general equilibrium expressions for ion exchange are most often reversible. The reversibility implies that equilibrium is independent of the direction from which the equilibrium state is approached. Based on the previous discussion of selectivity (i.e., the inherent preference of a resin for one ion over another), the ratios of concentrations of various ions in solution will be different from the concentration ratios in the resin phase at equilibrium. In this section, methods for calculating ion exchange performance based on equilibrium expressions is presented for single (or binary) and multiple ions present in water.

Two methods have been used to develop ion exchange equilibrium expressions. One method treats ion exchange as a chemical reaction and applies the laws of mass action to obtain an equilibrium description. In the second method, the same equilibrium description can also be developed using the principles of Donnan exclusion theory. As mentioned above, the Donnan theory is used to describe the behavior of ions based on the unequal distribution of ions across a membrane when an electrolyte solution on one side of the membrane contains ionic species that cannot diffuse through the membrane. The Donnan theory provides a more rigorous thermodynamic basis for ion exchange equilibrium description, which the mass action laws do not include. A more rigorous thermodynamic approach may be found in Helfferich (1995). Because both methods result in the same equilibrium expression, the equilibrium description based on the mass action laws will be used to develop an expression for the apparent equilibrium constant.

Ion Exchange Selectivity

If it is assumed that ion exchange is a simple stochiometric reaction, then the mass action laws can be applied to obtain an equilibrium expression. For simplification, a generalized form of the stochiometric reaction for Eqs. 16-1 through 16-10 can be written as

$$n\left[\overline{R^{\pm}}\right]A^{\pm} + B^{n\pm} \rightleftarrows \left[n\overline{R^{\pm}}\right]B^{n\pm} + nA^{\pm} \qquad (16\text{-}12)$$

where $\overline{R^{\pm}}$ is the ionic group attached to an ion exchange resin, A and B are exchanging ions, and n is the valence of the exchanging ion. In water treatment, ion exchange applications most often involve dilute ionic solutions where the ions behave independently of one another and are treated as ideal solutions (i.e., activity coefficients are assumed to be unity). In the resin phase, the ion concentrations can be much larger (5 to 6 M, 10 percent DVB SAC; Weber, 1972) and the activity is not unity and will be a function of ionic strength. With respect to Eq.16-12, all binary exchange reactions can be expressed as

$$K_{A}^{B} = \frac{\left[A^{\pm}\right]^{n}\left\{\overline{R^{\pm}B^{n\pm}}\right\}}{\left\{\overline{R^{\pm}A^{\pm}}\right\}^{n}\left[B^{n\pm}\right]} \qquad (16\text{-}13)$$

where

$K_{A}{}^{B}$ = selectivity coefficient or apparent equilibrium constant for A exchanging with ion B onto resin

$[A^{\pm}]$ = aqueous-phase concentration of presaturant ion, mol/L

$[B^{\pm}]$ = aqueous-phase concentration of counterion, mol/L

$\left\{\overline{R^{\pm}A^{\pm}}\right\}, \left\{\overline{R^{\pm}B^{n\pm}}\right\}$ = activities of resin-phase presaturant ion and counterion, respectively

Because concentrations are measured more easily than activities, the resin phase can be expressed in terms of concentrations, and Eq. 16-12 can be written in general terms as

$$K_{j}^{i} = \frac{C_{j}^{n} q_{i}}{q_{j}^{n} C_{i}} \qquad (16\text{-}14)$$

where C_j = aqueous-phase concentration of presaturant ion, mol/L
q_i = resin-phase concentration of counterion, mol/L
q_j = resin-phase concentration of presaturant ion, mol/L
C_i = aqueous-phase concentration of counterion, mol/L

Ion exchange resin manufacturers provide equilibrium data as selectivity coefficients. The selectivity coefficient can depend on the valence, the type of resin and its saturation, and the nature and concentration of the ion in the raw water.

Selectivity coefficients for SAC and SBA resins are presented in Table 16-6. For both SAC and SBA resins, the ion preference for the resin increases as the value of the selectivity coefficient increases. Generally, selectivity increases with increasing valence of both SAC and SBA resins.

Example 16-2 Determination of selectivity expression

Write a selectivity expression for the exchange of calcium onto a SAC resin in the sodium form and for the exchange of nitrate onto an SBA resin in the chloride form.

Solution to Part A

1. For the exchange of calcium, Ca^{2+}, onto an SAC resin in the sodium, Na^+, form, the following stochiometric expression can be written:

$$2\left[\overline{R^-}\right]Na^+ + Ca^{2+} \rightleftharpoons \left[\overline{2R^-}\right]Ca^{2+} + 2Na^+$$

2. Using the general form of Eq. 16-14, the following selectivity expression can be written:

$$K_{Na^+}^{Ca^{2+}} = \frac{q_{Ca^{2+}} C_{Na^+}^2}{C_{Ca^{2+}} q_{Na^+}^2}$$

Solution to Part B

1. For the exchange of nitrate, NO_3^-, onto an SBA resin in the chloride, Cl^-, form, the following stochiometric expression can be written:

$$\left[\overline{R^+}\right]Cl^- + NO_3^- \leftrightarrows \left[\overline{R^+}\right]NO_3^- + Cl^-$$

2. Using the general form of Eq. 16-14, the following selectivity expression can be written:

$$K_{Cl^-}^{NO_3^-} = \frac{q_{NO_3^-} C_{Cl^-}}{C_{NO_3^-} q_{Cl^-}}$$

Table 16-6
Selectivity coefficients for SAC and SBA resins

Cation	Selectivity, $K^i_{Li^+}$	Anion	Selectivity, $K^i_{Cl^-}$
Li^+	1.0	HPO_4^{2-}	0.01
H^+	1.3	CO_3^{2-}	0.03
Na^+	2.0	OH^- (type I)	0.06
UO^{2+}	2.5	F^-	0.1
NH_4^+	2.6	SO_4^{2-}	0.15
K^+	2.9	CH_3COO^-	0.2
Rb^+	3.2	HCO_3^-	0.4
Cs^+	3.3	OH^- (type II)	0.65
Mg^{2+}	3.3	BrO_3^-	1.0
Zn^{2+}	3.5	Cl^-	1.0
Co^{2+}	3.7	CN^-	1.3
Cu^{2+}	3.8	NO^-	1.3
Cd^{2+}	3.9	HSO_4^-	1.6
Ni^{2+}	3.9	Br^-	3
Mn^{2+}	4.1	NO_3^-	4
Pb^{2+}	5.0	I^-	8
Ca^{2+}	5.2	SeO_4^{2-}	17
Sr^{2+}	6.5	CrO_4^{2-}	100
Ag^{2+}	8.5		
Ba^{2+}	11.5		
Ra^{2+}	13.0		

Source: Adapted, in part, from Weber (1972).

Ion exchange reactions with inorganic ions have a relatively constant free-energy change; thus equilibrium constants do not vary significantly with solution conditions. However, it has been reported that the equilibrium constants for organic ions do change with resin loading (Semmens, 1975). The concept of selectivity coefficients or apparent equilibrium constants is used primarily in the theoretical treatment of ion exchange equilibrium and in qualitatively assessing the ion exchange preference. For quantitative analysis or process design evaluation, separation factors are used more commonly than selectivity coefficients.

Separation Factors

Equilibrium can be expressed in terms of equivalent fractions instead of concentration because equivalent charges are exchanged. The binary separation factor α^i_j is a measure of the preference for one ion over another during ion exchange and can be expressed as

$$\alpha^i_j = \frac{Y_i X_j}{X_i Y_j} \tag{16-15}$$

where X_j = equivalent fraction or mole fraction of presaturant ion in aqueous phase
X_i = equivalent fraction or mole fraction of counterion in aqueous phase
Y_j = resin-phase equivalent fraction or mole fraction of presaturant ion
Y_i = resin-phase equivalent fraction or mole fraction of counterions

The equivalent fraction in the aqueous phase is calculated from the following:

$$X_i = \frac{C_i}{C_T} \quad X_j = \frac{C_j}{C_T} \tag{16-16}$$

where C_T = total aqueous ion concentration, eq/L
C_i = aqueous-phase concentration of counterion, eq/L
C_j = aqueous-phase concentration of presaturant ion, eq/L

The equivalent fraction in the resin phase is expressed as

$$Y_i = \frac{q_i}{q_T} \quad Y_j = \frac{q_j}{q_T} \tag{16-17}$$

where q_T = total exchange capacity of resin, eq/L

For process design calculations, binary separation factors are primarily used in ion exchange calculations because they are experimentally determined and account for the solution concentration and the total ion exchange capacity.

Substituting Eqs. 16-16 and 16-17 into Eq. 16-15 yields

$$\alpha_j^i = \frac{q_i C_j}{C_i q_j} \tag{16-18}$$

where α^i_j = separation factor of ion *i* with respect to ion *j*, unitless (concentrations are in eq/L)

For the special case of monovalent ion exchange with a monovalent presaturant ion (all ions are 1 eq/mol), the separation factor is constant and equal to the selectivity coefficient. For multivalent ion (*i*) exchange with a resin having a monovalent presaturant ion (*j*), the separation factor and selectivity coefficient are related by the ratio of presaturatant ion concentrations in the liquid and resin phases (Harland, 1994):

$$K_j^i = \alpha_j^i \left(\frac{C_j}{q_j} \right)^{|\pm Z|-1} \tag{16-19}$$

where Z = charge on ion, unitless (concentrations are in mol/L)

As shown in Eq. 16-19, for low multivalent ion concentrations the separation factor for multivalent/monovalent exchange is inversely proportional to the equivalent aqueous-phase concentration of ion *j* raised to the power $|\pm Z| - 1$. It is common to have sodium or chloride as the presaturant ion for cationic or anionic resins, respectively. The separation factor is inversely proportional to the aqueous-phase sodium or chloride concentration for low concentrations of exchanging multivalent ions because the amount of resin in the sodium or chloride forms will not change significantly for small cation or anion concentrations. Accordingly, ion exchange resins are very efficient for scavenging low concentrations of multivalent cations and anions.

It is important to note that the separation factor may not be a constant but rather is influenced by various factors: exchangeable ions (size and charge), properties of the resins, including particle size, degree of cross-linking, capacity, and type of functional groups occupying the exchange sites; water matrix, which includes total concentration, type, and quantity of organic compounds present in solution; reaction period; and temperature. Because separation factors can be influenced by several factors, they are usually determined by performing an equilibrium experiment called a binary isotherm. A binary isotherm involves performing a batch equilibrium experiment for a binary system. Both binary component systems and isotherms are discussed in the following sections.

Binary Component Systems

A binary component system involves the exchange of a presaturant ion with only one other component ion present in solution. For the binary system, the total aqueous-phase equivalent concentration can be expressed as

$$C_T = C_i + C_j \tag{16-20}$$

where C_T = total aqueous ion concentration, eq/L
C_i = counterion concentration, eq/L
C_j = presaturant ion concentration, eq/L

Total resin-phase equivalent concentration can be expressed as

$$q_T = q_i + q_j \tag{16-21}$$

where q_T = total resin-phase ion concentration, eq/L resin
q_i = counterion concentration, eq/L resin
q_j = presaturant ion concentration, eq/L resin

Consequently, substitution of the expression for q_T into Eq. 16-18 yields the following expression for calculating the resin-phase concentration of the counterion of interest:

$$q_i = \frac{C_i q_T}{C_i + C_j \alpha_i^j} \tag{16-22}$$

Note that $\alpha^j{}_i = 1/\alpha^i{}_j$ in the above expression. For a given counterion concentration, Eq. 16-22 can be used to estimate the resin-phase concentration provided the binary separation factor and the total resin capacity are known. Similarly, the following equation can be used to calculate the aqueous-phase concentration of the counterion given the total aqueous-phase concentration, binary separation factor, and the resin-phase concentrations of the counterion and presaturant ion:

$$C_i = \frac{q_i C_T}{q_i + \alpha^i_j q_j} \tag{16-23}$$

Separation factors for commercially available SAC and SBA exchange resins are given in Table 16-7. Based on the definition of Eq. 16-15, a separation factor greater than 1 means that ion *i* is preferred over ion *j*. For example, if $\alpha^{NO_3^-}_{Cl^-} = 2.3$, expressed in equivalents, at equal aqueous-phase concentrations, NO_3^- is preferred over chloride by 2.3 to 1.0. The magnitude of the separation factors is different for WAC and WBA resins from those shown in Table 16-7 for SAC and SBA resins. When separation factors for a given resin are unknown, they may be determined experimentally using binary isotherms. Clifford (1999) provides a detailed experimental procedure and example for determining separation factors.

Table 16-7
Separation factors for several commercially available cation and anion exchange resins[a]

Strong-Acid Cation Resins[b]		Strong-base Anion Resins[c]	
Cation	$\alpha^i_{Na^+}$	**Anion**	$\alpha^i_{Cl^-}$
Ra^{2+}	13.0	$UO_2(CO_3)_3^{4-}$	3200
Ba^{2+}	5.8	ClO_4^-[d]	150
Pb^{2+}	5.0	CrO_4^{2-}	100
Sr^{2+}	4.8	SeO_4^{2-}	17
Cu^{2+}	2.6	SO_4^{2-}	9.1
Ca^{2+}	1.9	$HAsO_4^{2-}$	4.5
Zn^{2+}	1.8	HSO_4^-	4.1
Fe^{2+}	1.7	NO_3^-	3.2
Mg^{2+}	1.7	Br^-	2.3
K^+	1.7	SeO_3^{2-}	1.3
Mn^{2+}	1.6	HSO_3^-	1.2
NH_4^+	1.3	NO_2^-	1.1
Na^+	1.0	Cl^-	1.0
H^+	0.67	BrO_3^-	0.9
		HCO_3^-	0.27
		CH_3COO^-	0.14
		F^-	0.07

[a]Values are approximate separation factors for 0.005–0.010 N solutions (TDS = 250–500 mg/L as $CaCO_3$).
[b]SAC resin is polystyrene divinylbenzene matrix with sulfonate functional groups.
[c]SBA resin is polystyrene divinylbenzene matrix with $-N^+(CH_3)_3$ functional groups (i.e.,a type 1 resin).
[d]ClO_4^-/Cl^- separation factor is for polystyrene SBA resins; on polyacrylic SBA resins, the ClO_4^-/Cl^- separation factor is approximately 5.0.
Source: Adapted From Clifford et al. (2011).

Example 16-3 Binary exchange calculation

Nitrate is to be removed from water containing high chloride concentration. The chemical composition of the water is given below. The water contains some bicarbonate and sulfate, but for this calculation it is assumed they are negligible. Using an SBA resin with a total capacity of 1.4 eq/L, estimate the maximum volume of water that can be treated per liter of resin.

Cation	meq/L	Anion	meq/L
Ca^{2+}	0.9	Cl^-	2.5
Mg^{2+}	0.8	SO_4^{2-}	0.0
Na^+	2.6	HCO_3^-	0.0
		NO_3^-	1.8
Total	4.3	Total	4.3

Solution

1. Estimate the maximum useful capacity of nitrate on the SBA resin in the chloride form using Eq. 16-22.
   a. The separation factor for nitrate over chloride can be obtained from Table 16-7:

$$\alpha_j^i = \alpha_{Cl^-}^{NO_3-} = 3.2$$

$$\alpha_i^j = \alpha_{NO_3^-}^{Cl^-} = \frac{1}{\alpha_j^i} = \frac{1}{\alpha_{Cl^-}^{NO_3^-}} = \frac{1}{3.2} = 0.3125$$

   b. The maximum useful capacity of the resin for nitrate using Eq. 16-22 is

$$q_{NO_3-} = \frac{C_{NO_3-} q_T}{C_{NO_3-} + C_{Cl^-} \alpha_{NO_3-}^{Cl-}}$$

$$= \frac{(1.8\,\text{meq}\,NO_3- / \text{L}\,H_2O)(1.4\,\text{eq/L resin})(1000\,\text{meq/eq})}{(1.8\,\text{meq}\,NO_3- / \text{L}\,H_2O) + \left(2.5\,\text{meq}\,Cl^- / \text{L}\,H_2O\right)(0.3125)}$$

$$= 976\,\text{meq}\,NO_3- / \text{L resin}$$

2. The volume of water that can be treated per volume of resin per cycle is calculated by dividing the nitrate capacity by the influent nitrate concentration:

$$V = \frac{q_{NO_3-}}{C_{NO_3-}} = \frac{976\,\text{meq}\,NO_3 - /\text{L resin}}{1.8\,\text{meq}\,NO_3 - /\text{L}H_2O} = 542\,\text{L}\,H_2O / \text{L resin}$$

Multicomponent Systems

In water treatment, the application of ion exchange involves treatment of groundwaters containing multiple cations and anions (e.g., Na^+, Ca^{2+}, Mg^{2+}, Cl^-, HCO_3^-, SO_4^{2-}). Some waters may also contain ions of more significant health threat, such as Ba^{2+}, Ra^{2+}, Pb^{2+}, Cu^{2+}, NO_3^-, $HAsO_4^-$, F^-, and ClO_4^-. Consequently, a multicomponent expression is needed to describe the competitive interactions between the ions for the resin site at equilibrium. In a multicomponent system, the total capacity of the resin and the total concentration of exchanging ions in solution can be expressed as

$$q_T = q_i + q_j + \cdots + q_n \tag{16-24}$$

$$C_T = C_i + C_j + \cdots + C_n \tag{16-25}$$

where q_T = total resin-phase ion concentration, eq/L resin
q_i = resin-phase concentration of counterion i, eq/L resin
q_j = resin-phase concentration of presaturant j, eq/L resin
q_n = resin-phase concentration of counterion n, eq/L resin
C_T = total aqueous-phase ion concentration, eq/L
C_i = aqueous-phase concentration of counterion i,eq/L
C_j = aqueous-phase concentration of presaturant j,eq/L
C_n = aqueous-phase concentration of counterion n,eq/L

Applying Eq. 16-24 to Eq. 16-22 yields the following expression for q_i in terms of n exchanging ions:

$$q_i = \frac{q_T C_i}{\sum_{k=1}^{n} \alpha_i^k C_k} \tag{16-26}$$

where C_k = aqueous-phase concentration for ion k (presaturant ion when $k = j$), eq/L resin
α_i^k = separation factor for counterion i with respect to ion k

Note that α_i^k assumes the separation factors are known with respect to the ion concentrations being sought on the resin phase for ion i. Since the separation factors are reported in terms of the prestaurant ion, Eq. 16-26 would be easier to use if the separation factors were with respect to the presaturant instead of the resin phase ion. If the subscript j is set equal to p where p is equal to the presaturant ion, the following expression for the separation factor in Eq. 16-26 can be obtained:

$$\alpha_i^k = \alpha_i^p \alpha_p^k = \frac{\alpha_p^k}{\alpha_p^i} \tag{16-27}$$

Substitution of Eq. 16-27 into Eq. 16-26 yields the following expression:

$$q_i = \frac{q_T C_i}{\sum_{k=1}^{N} \left(\frac{\alpha_p^k}{\alpha_p^i} C_k \right)} = \frac{q_T C_i}{\frac{1}{\alpha_p^i} \sum_{k=1}^{N} \left(\alpha_p^k C_k \right)} = \frac{q_T \alpha_p^i C_i}{\sum_{k=1}^{N} \left(\alpha_p^k C_k \right)} \tag{16-28}$$

If all the liquid-phase ion concentrations and the total resin capacity are known, the resin-phase concentrations can be calculated using the separation factors referenced to the presaturant ion as reported in Table 16-7. Similarly, Eq. 16-28 can be solved for C_i in terms of n exchanging ions:

$$C_i = \frac{C_T q_i}{\alpha^i_p \sum_{k=1}^{n} q_k / \alpha^k_p} \tag{16-29}$$

where q_k = resin-phase concentration of ion k (presaturant ion when $k = j$), eq/L resin
α^i_k = separation factor for ion i with respect to ion k
α^k_p = separation factor for ion k with respect to presaturant ion p

Example 16-4 Multicomponent equilibrium calculation

Consider the removal of nitrate from well water using an SBA exchange resin in the chloride form. The major ions contained in the well water are given below. Assuming nitrate is removed completely from solution, calculate the maximum volume of water that can be treated per liter of resin assuming equilibrium conditions. Assume total resin capacity of the SBA is 1.4 eq/L.

Cation	meq/L	Anion	meq/L
Ca^{2+}	0.9	Cl^-	1.0
Mg^{2+}	0.8	SO_4^{2-}	1.5
Na^+	2.6	NO_3^-	1.8
Total	4.3	Total	4.3

Solution

1. Applying Eq. 16-28 with the use of the separation factors provided in Table 16-7,the summation term in the denominator can be calculated.

$$\sum_{k=1}^{N}\left(\alpha^k_p C_k\right) = (1.0)(1\text{ meq/L}) + (9.1)(1.5\text{ meq/L}) + (3.2)(1.8\text{ meq/L}) = 20.41\text{ meq/L}$$

2. Calculate q_i for each ion.

$$q_{Cl} = \frac{(1.4\text{ eq/L})(1.0)(1\text{ meq/L})}{20.41\text{ meq/L}} = 0.069\text{ eq/L}$$

$$q_{SO_4^{2-}} = \frac{(1.4\ \text{eq/L})\,(9.1)\,(1.5\ \text{meq/L})}{20.41\ \text{meq/L}} = 0.936\ \text{eq/L}$$

$$q_{NO_3-} = \frac{(1.4\ \text{eq/L})\,(3.2)\,(1.8\ \text{meq/L})}{20.41\ \text{meq/L}} = 0.395\ \text{eq/L}$$

Check: $0.069 + 0.936 + 0.395 = 1.4$ eq/L total capacity.

Note that because the sulfate concentration is more preferred over nitrate ($9.1 \gg 3.2$), the equilibrium capacity of nitrate is low. In other words, nitrate will occupy only about 28 percent (0.395/1.4) of the exchange sites on the resin.

3. Calculate the maximum quantity of water that can be treated per cycle before nitrate breakthrough occurs.

$$\begin{array}{l}\text{Maximum}\\ \text{volume}\\ \text{treated}\end{array} = \frac{(0.395\ \text{eq/L resin})\left(10^3\ \text{meq/eq}\right)}{1.8\ \text{meq/L water}} = 219\ \text{L water/ L resin}$$

Comment

When comparing the maximum bed volumes treated in Examples 16-3 and 16-4, with sulfate present, the capacity of nitrate is reduced by 60 percent. The impact of divalent anions on exchange capacity is significant. Note that this example applies to equilibrium applied in a batch reactor. In a column system, only the portion of the resin that is exhausted will be in equilibrium with the feed water. In the mass transfer zone, the resin will be in local equilibrium with the concentrations in the water in that region of the bed.

16-5 Ion Exchange Kinetics

The transport mechanisms for fixed-bed ion exchange processes are similar to those for fixed-bed adsorbers as discussed in Chap. 15 where the combined effects of liquid- and solid-phase transport is coupled with equilibrium thermodynamics. In the ion exchange process, these effects may include diffusion and convection coupled with the process exchange rate, electrochemical effects, and sometimes chemical reaction. Since ions diffuse at different rates, charge separation can arise inducing an electric field causing ionic migration to satisfy electroneutrality within the resin particle as discussed in Sec. 16-1. For example, as cation A diffuses into the resin particle, it is transferring charge to the resin, and this charge must be offset by an equivalent charge by another ion (e.g., presaturant ion) or ions diffusing out of the resin particle into solution to satisfy the local electrical balance. As explained by Helfferich and Hwang (1991) if the ions diffusing out of the resin particle carried a weaker charge, a larger flux of the faster ion would result. For the charged ions, a net transfer of electric charge would result and violate the requirement of electrical neutrality. A small deviation from electrical neutrality causes an electric field that produces a force that enables all the charged ions in the electric field to move with a certain velocity (e.g., electrophoresis) called drift velocity. The direction of the drift of ions is that of diffusion of the slower ion. Consequently, drift velocity of the ion increases the flux of the slow diffusing ion and decreases the flux of the faster one, equalizing the net fluxes and so preventing any further buildup of the net charge.

For ion exchange processes, the ion flux (J_i) contains both the diffusive and electrical flux terms as given by the Nernst–Plank (NP) equation and can be written for both the aqueous phase ($J_{l,i}$) and resin phase ($J_{s,i}$) as

$$J_{l,i} = -D_{l,i}\left[\frac{\partial C_i(z,t)}{\partial r} + \frac{Z_i F C_i(z,t)}{RT}\frac{\partial \phi(z,t)}{\partial r}\right] \tag{16-30}$$

$$J_{s,i} = -D_{s,i}\left[\frac{\partial C_{p,i}(r,z,t)}{\partial r} + \frac{Z_i F C_{p,i}(r,z,t)}{RT}\frac{\partial \phi(r,z,t)}{\partial r}\right] \tag{16-31}$$

where $J_{l,i}$ = flux of ion i into the resin particles, eq/m$^2 \cdot$ s
$J_{s,i}$ = flux of ion i inside the resin particles, eq/m$^2 \cdot$ s
$D_{l,i}$ = aqueous-phase diffusion coefficient of ion i, m^2/s
$D_{s,i}$ = solid- or resin-phase diffusion coefficient of ion i, m^2/s
C_i = aqueous-phase concentration of diffusing ion i, eq/L
$C_{p,i}$ = aqueous-phase concentration of diffusing ion i in the resin pores, eq/L
ϕ = electrical potential caused by migration of ion in solution, mV
F = Faraday constant 96,484 C/mol
R = universal gas constant, 0.08205 L · atm/mol · K
T = temperature, K
Z_i = charge of the diffusing ion, (−)
r = dependant parameter in radial direction of the particle, m
z = dependent parameter in axial direction of fixed bed, m
t = dependent parameter of time of operation, d

For most ion exchange problems encountered in water treatment, the flux terms in Eqs. 16-30 and 16-31 can be simplified by assuming electroneutrality exists within aqueous film surrounding the resin phase and within the resin phase, and the flux of the nonexchanging coion across the resin–liquid interface is negligible (Hokanson, 2004).

$$\sum C_i(z,t) = 0 \quad \text{(aqueous phase)} \tag{16-32}$$

$$\sum C_{p,i}(r,z,t) = 0 \quad \text{(resin phase)} \tag{16-33}$$

$$J_i = (\text{ nonexchanging coion}) = 0 \tag{16-34}$$

Based on these assumptions the following condition of no net current flow can be derived:

$$\sum J_i = 0 \quad \text{(no net current flow)} \tag{16-35}$$

Equations 16-30 and 16-31 reduce to a form similar to Fick's law, and the flux is equal to the product of an "effective" diffusion coefficient and a concentration gradient. The effective diffusion coefficient includes electrical effects and is not constant and depends upon concentrations, diffusion coefficients, and charges of all the individual exchanging ions. Equation 16-30 and 16-31 can be rewritten in terms of an effective diffusion or mass transfer coefficient as

$$J_{l,i} = k_{f,i}\left[C_{b,i}(z,t) - C_{s,i}(z,t)\right] \quad \text{(aqueous phase)} \tag{16-36}$$

$$J_{s,i} = -\frac{D_{l,i}\varepsilon_p}{\tau_p}\frac{\partial C_{p,i}(r,z,t)}{\partial r} \quad \text{(resin phase)} \tag{16-37}$$

where $k_{f,i}$ = film diffusion or mass transfer coefficient, m/s
$C_{b,i}$ = bulk aqueous-phase concentration of ion i, g/m^3
$C_{s,i}$ = aqueous-phase concentration of ion i at the external surface of the resin particle, g/m^3
ε_p = void fraction of the resin particle, dimensionless
τ_p = resin particle tortuosity, dimensionless

Equations 16-36 and 16-37 are incorporated into fixed-bed and intraparticle mass balances, respectively; to provide a set of equations or model that can describe the fixed-bed ion exchange process (Hokanson, 2004; Wagner and Dranoff, 1967; Graham and Dranoff, 1972, Wildhagen et al., 1985; Haub and Foutch, 1986). The model mechanisms consist of advective transport of exchanging ions through the fixed-bed exchanger, diffusion of exchanging ions through the film surrounding the resin particles, and intraparticle diffusion of the exchanging ions within the resin particles. It is typically assumed that the rate of ion exchange on the resin is fast as compared to the mass transfer rates in the fluid and solid phases.

Rate-Controlling Step in Fixed-Bed Ion Exchange Process

Most ion exchange applications in water treatment involve two rate controlling steps in-series, liquid-phase (film), and effective intraparticle mass transfer. Determination of the rate-controlling step must consider the mass flux. According to Eqs. 16-36 and 16-37, the flux terms contain the product of the mass transfer rate and the driving force for mass transfer. The slower flux will determine the rate-controlling step. Intuitively, one may think that since the stagnant film thickness surrounding the resin particles is small as compared to the diffusion path length (surface of the resin particle to the center of the resin particle) within the resin particle, that the intraparticle mass transfer rate controls the overall rate of mass transfer. Film transfer coefficients are typically on the order of 10^{-5} to 10^{-6} m/s versus intraparticle diffusion coefficients on the order of 10^{-9} to 10^{-10} m^2/s. However, the driving force for each phase must also be considered. For Eq. 16-36 the driving force between the bulk solution and the surface of the resin particle cannot be greater than the bulk solution concentration. For Eq. 16-37, the driving force can be as high as the concentration of the fixed charges on the resin particle, which can be very large. As a result, the rate-controlling step can be difficult to determine.

As pointed out by Helfferich and Hwang (1991), liquid-phase mass transfer rate usually controls the ion exchange process when: (1) liquid-phase concentration is low, causing a small driving force in the liquid phase; (2) the resin exchange capacity is high, causing a large driving force in the resin phase; (3) the resin particle size is small, causing a small mass transfer length in the resin; (4) there is a low degree of crosslinking in the resin particle, causing an open resin matrix; and (5) the advective flow in the fixed bed is slow, causing the thickness of the stagnant film surrounding the resin particles to be large. In addition, the selectivity of the resin may also play a small role in impacting the driving force. The following expression was developed for predicting which phase would control the mass transfer rate (Helfferich, 1995):

$$\text{He} \equiv \frac{q_T D_p \delta}{C D_l r_0}\left(5 + 2\alpha_j^i\right) \tag{16-38}$$

where He = Helfferich number, dimensionless
δ = stagnant film thickness between the bulk solution and the resin particle external surface, m
r_0 = resin particle radius, m

When He ≪ 1, intraparticle diffusion will control the rate of mass transfer for the ion exchange process, and when He ≫ 1 liquid-phase diffusion will control the mass transfer rate in the ion exchange process. For He values near unity both rates will contribute in some degree to the control of the mass transfer rate.

A number of variables can influence the mass transfer rate and include particle size, flow rate, resin particle pore structure, and solution concentration. The resin particle size has a significant impact on process kinetics. As the resin particle size decreases, mass transfer rates increase in both the liquid and resin phases. The liquid-phase mass transfer rate is inversely proportional to resin particle size and as the particle size decreases the mass transfer rate increases. Similarly, the intraparticle mass transfer rate increases as the inverse of particle size raised to a higher power.

There is a trade-off between process kinetics and head loss in the bed. As the particle size decreases, the process kinetics increase, which will provide for a smaller mass transfer zone and higher capacity utilization in a given fixed bed, but at the expense of a higher head loss in the bed. The benefits of increased exchange rates with smaller resin particles must be weighed against the increased head loss within the fixed bed.

Flow rate can influence both the length of the mass transfer zone and fixed bed. If the film diffusion is controlling the rate of mass transfer, increasing the flow rate may decrease the film thickness for diffusion, decrease the length of the mass transfer zone and increase the capacity of the fixed bed. This will occur as long as the length of the mass transfer zone is shorter than the length of the fixed bed. If intraparticle diffusion is controlling the rate of mass transfer, and the flow rate is increased, the mass transfer zone will not be impacted, but the empty-be contact time (EBCT) will be less and the bed usage rate will be lower. If the flow rate was decreased, the mass transfer zone length could increase; depending upon the magnitude of the He number, and the EBCT would increase causing the usage rate to increase as long as the mass transfer zone length is less than the bed length.

As discussed above, for gel-type resins with an open matrix (e.g., microreticular resin), film transfer is usually controlling the overall rate of mass transfer in the fixed bed. However, if the degree of resin crosslinking is high for a given resin (e.g., macroreticular resin), the resin matrix becomes very tortuous and intraparticle diffusion may control the rate of ion exchange in the fixed bed.

With respect to initial ion concentration, film diffusion is likely to be controlling for low initial concentrations. At high initial ion concentrations, intraparticle diffusion is more likely to control the rate of mass transfer. This usually occurs during regeneration when using high concentrations of regenerate solutions.

16-6 Ion Exchange Process Configurations

The ion exchange process is conducted in a fixed bed of resin with the water passing through the resin until a certain treatment objective is reached. The resin is then taken offline and regenerated, while another column is used to supply continuous treatment (if needed). A typical full-scale ion exchange plant operating in the down-flow mode is shown on Fig. 16-4.

Regeneration Methods

The regeneration steps of an ion exchange resin are important to the overall efficiency of the process. There are two methods for regenerating an ion exchange resin: (1) co-current, where the regenerant is passed through the resin in the same flow direction as the solution being treated, and (2) countercurrent, where the regenerant is passed through the resin

Figure 16-4
Full-scale ion exchange plant operating in downward mode.

in the opposite direction as the solution being treated. Co-current and countercurrent regenerations are considered in the following discussion.

Photographs of Ion Exchange columns and components
http://www.stantec.com/wt3r/16-1.html

CO-CURRENT OPERATION

Co-current operation consists of the regeneration step being conducted in the same flow direction as the solution being treated. The direction of both flows is usually downward. When small concentrations of the unwanted ion(s) can be tolerated in the effluent (referred to as leakage) and the exchange in the regeneration step is favorable, co-current operation is chosen. However, in recent studies it has been found that co-current operation can reduce leakage of some unwanted ions more effectively than countercurrent systems (Clifford et al., 1987; Ghurye et al., 1999). For nitrate and arsenate ions it was found that the co-current process produced less leakage than the countercurrent process because the exchanged mass of these ions is located near the outlet of the resin bed. Consequently, flushing these ions back through the column with the countercurrent process produces more leakage. The location of these ions within the bed will depend upon the ions in the water matrix and their separation factors for a given resin. For example, for many SBA resins, sulfate has a higher affinity than either nitrate or arsenate. Consequently, the sulfate will push most of the exchanged arsenate and nitrate toward the end of the column. Upon regeneration, the preference for sulfate over chloride is reversed at high chloride concentrations and sulfate is easily removed from the resin.

COUNTERCURRENT OPERATION

In most cases, countercurrent operation will result in lower leakage levels and higher chemical efficiencies than co-current operation. In situations where (1) high-purity water is necessary, (2) chemical consumption must be reduced to a minimum, or (3) the least waste volume is produced, the countercurrent method of operation is used.

Countercurrent operation with the service flow operated in the upward direction will only be effective if the resin can be prevented from fluidizing. Any resin movement during the upflow cycle will destroy the ionic interface (exchange front) that ensures good exchange. A number of methods have been devised to prevent resin particle movement during upflow operation. Some of the more commonly used methods are presented in Table 16-8.

Advances in the use of fixed beds make certain processes more economical or provide better product purity. Two types of advances in fixed-bed design are the use of mixed beds of strong-acid and strong-base resins and the use of layered beds of a weakly ionized resin above a strong ionized resin, as discussed below.

MIXED BEDS

For the production of deionized water, a column containing intimately mixed strong-acid (in the H^+ form) and strong-base resins (in the OH^- form) provides better water quality than the individual resins segregated in series. With mixed resins, the effluent from the contactor will be deionized water. The reactions from salt to base to water or from salt to acid

Table 16-8
Methods for preventing particle movement during upflow operation

Type of Method	Description
Completely filled column	The ion exchange column is completely filled with resin and service and regeneration steps are run counterflow. A reservoir tank above the column provides space for occasional backwashing.
Use of inert granules to fill headspace	Compressible inert granules are used to fill the column's headspace during the service cycle, and they prevent the upward movement of the particles during upflow regeneration. A small reservoir is used periodically to withdraw the inert granules to backwash the resin.
Use of air or water blocking	Air or water can be introduced at the top of the column during upflow regeneration to block movement of particles. Blocking will result in increased waste volumes and has been virtually abandoned. However, only moderate air pressure is required and has been used successfully in some designs.

to water will occur so rapidly that there are virtually no back reactions. Unfortunately, mixed beds cannot be regenerated in place and regeneration will require the separation of the two resin types into layers. Properly selecting the densities and particle sizes of the resins is the best method. Strong-base anion resins are normally lighter than SAC resins; hence backwashing prior to regeneration will place the SBA resin above the SAC resin. During regeneration, the regenerant solution is introduced at the top (e.g., NaOH) and bottom (e.g., HCl) of the respective beds and is simultaneously withdrawn at the interface. After regeneration, resins must be well mixed by an air-scouring operation. Probably the most important factor in achieving good product water is how well the resins are mixed, especially close to the exit of the bed.

FIXED-BED OPERATION

In the production of deionized water, the use of fixed beds has been the traditional approach to ion exchange. There are many important design features that must be addressed when considering ion exchange columns. Proper distribution and collection of flow is critical to good operation. For traditional downflow systems, the influent can be distributed with either a water hold-down system, where the entire vessel is kept completely filled with liquid at all times, or an air hold-down system, where the liquid level is kept several millimeters above the resin level when liquid is being introduced into the column. A water hold-down design is easier to install, operate, and control and is used most often in normal applications.

A schematic of an air or water hold-down system is provided on Fig. 16-5. In hold-down systems relatively large volumes of air or water are needed to maintain the packed resin bed in place during upflow regeneration. Although hold-down systems provide good water quality and regeneration

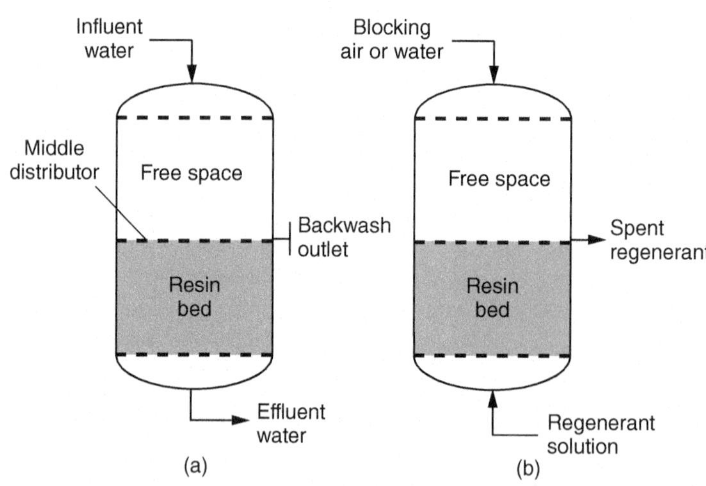

Figure 16-5
Air and water ion exchange hold-down system: (a) loading cycle and (b) regeneration cycle.

efficiency, they are equipment intensive. The issues associated with hold-down systems include the following: (1) the middle distributor can undergo mechanical damage due to resin swelling and shrinking, (2) there is high water and air consumption, (3) the regeneration process is time consuming, (4) the process is labor intensive, and (5) large vessels are required.

Types of Ion Exchange Processes

Recently, new design applications have made it possible to operate an ion exchange system continuously with ion removal and regeneration occurring simultaneously in different portions of a moving resin bed. Three ion exchange continuous operation systems that utilize the Dow UPCORE, the Bayer-Lewatit, and the Calgon ISEP systems discussed in this section. In addition, the MIEX process, which operates in a completely mixed reactor, was recently developed for removing dissolved organic carbon is also discussed.

UPCORE SYSTEM

The UPCORE system, a countercurrent system in which the service flow is in the downward direction and the regeneration flow is in the upward direction, is illustrated on Fig. 16-6. If needed, the system can be converted to a co-current process. The top collector/distributor is surrounded by a small layer of floating inert material that enables the service water, spent regenerant, rinse water, and fine particles to pass while retaining the resin particles. The process is insensitive to fluctuations in service flow rates. During upflow regeneration, the resin bed is lifted in the compacted form moving up against the inert material at the top of the bed to maintain the packed state. During regeneration, fine particles (dirt, fine resin particles, etc.), which are trapped during the service cycle, are washed out with the regenerant and rinse waters. Usually such systems are used on groundwaters or similar supplies where backwashing is rarely required. To provide backwashing, a separate backwashing vessel is normally provided and the

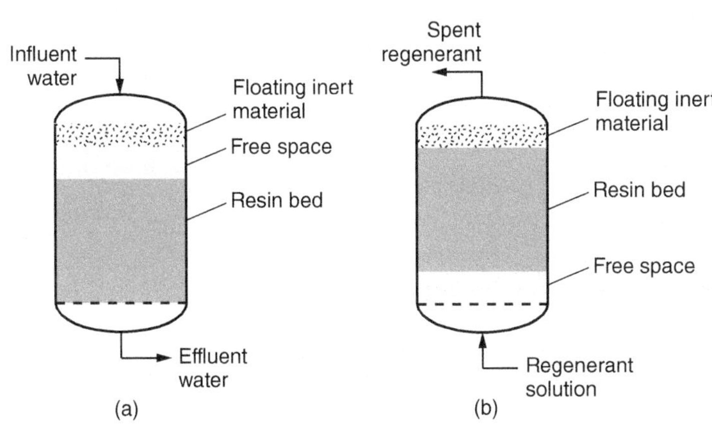

Figure 16-6
UPCORE countercurrent ion exchange systems: (a) loading cycle and (b) regeneration cycle.

media is moved hydraulically from its normal vessel to the backwashing vessel, washed, and then moved back. Obviously, this arrangement is not attractive unless backwashing is rarely required.

BAYER–LEWATIT UPFLOW FLUIDIZED SYSTEM

The Bayer–Lewatit upflow fluidized system, a countercurrent system in which the service flow is in the upward direction and the regeneration flow is in the downward direction, is shown on Fig. 16-7. Nozzle plate distributors are located at the top and bottom of the resin bed to ensure the resin is evenly distributed and held in place. A fine polishing resin layer is placed in the upper layer and is the first resin to be regenerated to reduce leakage. A small amount of inert floatable material is placed between the resin and the upper nozzle plate. The purpose of the floatable material is to prevent small beads or particles of resin from clogging the upper nozzle plate, and it also provides a more even flow distribution of the regenerant through the resin bed. Enough freeboard is provided to allow for expansion of the resin. The use of the upflow fluidized process is supposed to minimize the formation of clumps of resin and mechanical stress on the resins, causing swelling breakdown and attrition. The downflow rinse step minimizes the quantity of rinse water because the density of the rinse water is lower than the regenerant solution.

CALGON ISEP SYSTEM

The Calgon ISEP process, a countercurrent process, is illustrated schematically on Fig. 16-8 and photographically on Fig. 16-9. The process consists of about 20 to 30 small ion exchanger columns on a rotating platform. As the platform rotates at any given time, most of the columns are in the treatment mode while the others are in various phases of the regeneration cycle. The configuration provides for a flexible operation. A large valve is

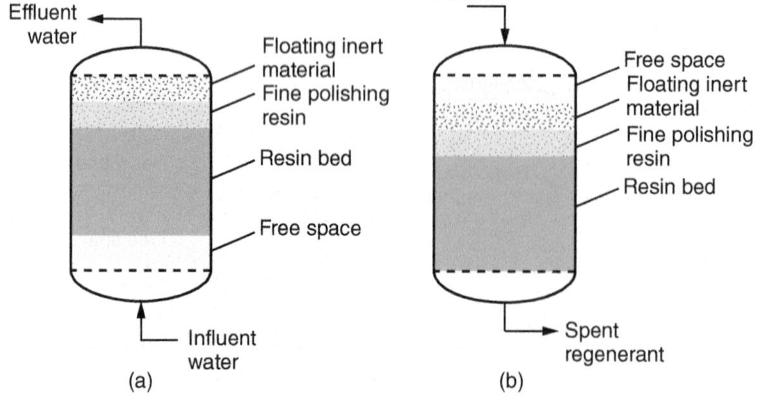

Figure 16-7
Bayer–Lewatit upflow fluidized ion exchange system: (a) loading cycle and (b) regeneration cycle.

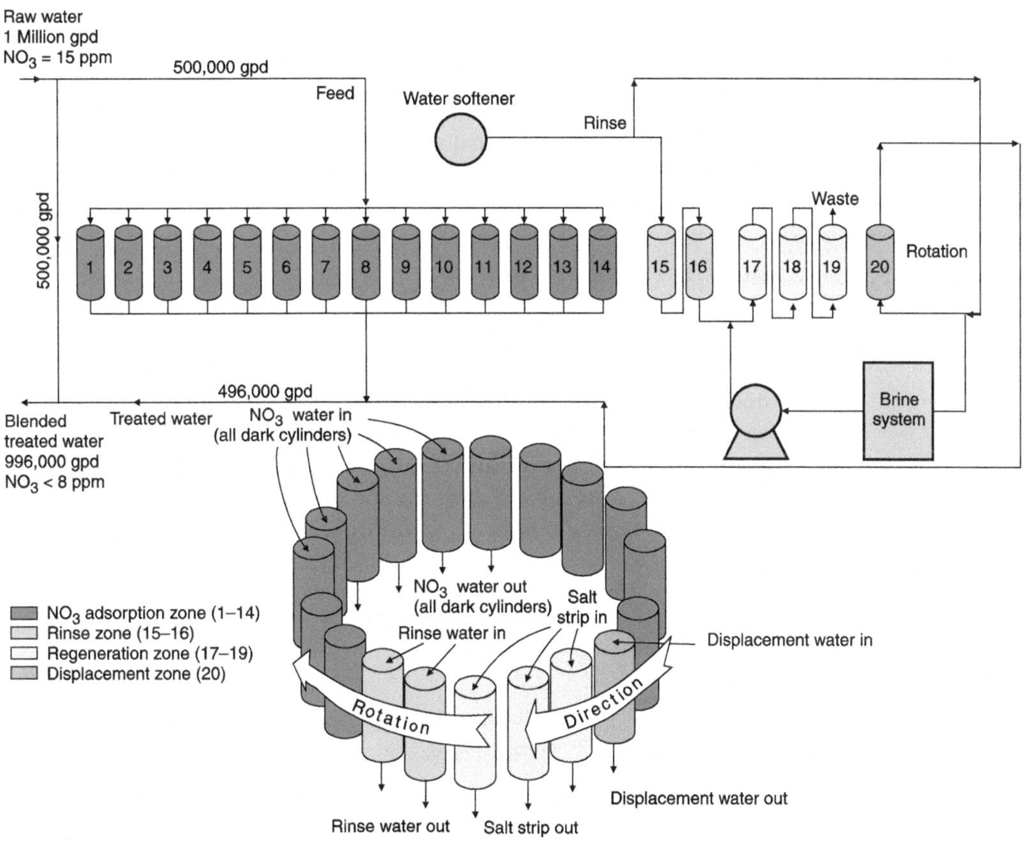

Figure 16-8
Schematic of Calgon ISEP ion exchange process.

Figure 16-9
Photograph of ion exchange cannisters on rotating platform in Calgon ISEP ion exchange process (see Fig. 16-8).

used to control the various types of flows (service, regenerant, and rinse) to the columns. The process provides continuous treatment and is fully automated. The service cycles are short as compared to other conventional ion exchange processes, which enable ISEP to have a relatively low resin inventory. The ISEP process typically produces less brine waste as compared to a conventional system and produces low leakage by providing better control over the mass transfer zone.

MIEX MAGNETIC ION EXCHANGE RESIN

The Orica Limited Company of Australia developed the MIEX process for removal of dissolved organic carbon (DOC) from drinking water supplies. The process consists of a SBA exchange resin, usually in the chloride form, with a magnetic component built into it. The resin beads, which are smaller than the conventional resin beads (i.e., diameter ≈ 180 μm), are contacted with the water in a completely mixed reactor. A typical process flow diagram employing the MIEX resin is shown on Fig. 16-10. The negatively charged DOC molecules exchange with presaturant chloride ion on the resin and are removed from the water. The resin and water are then separated in an upflow settler as the resin beads will agglomerate due to their magnetic properties and rapidly settle out of the water. The settling rate can be as

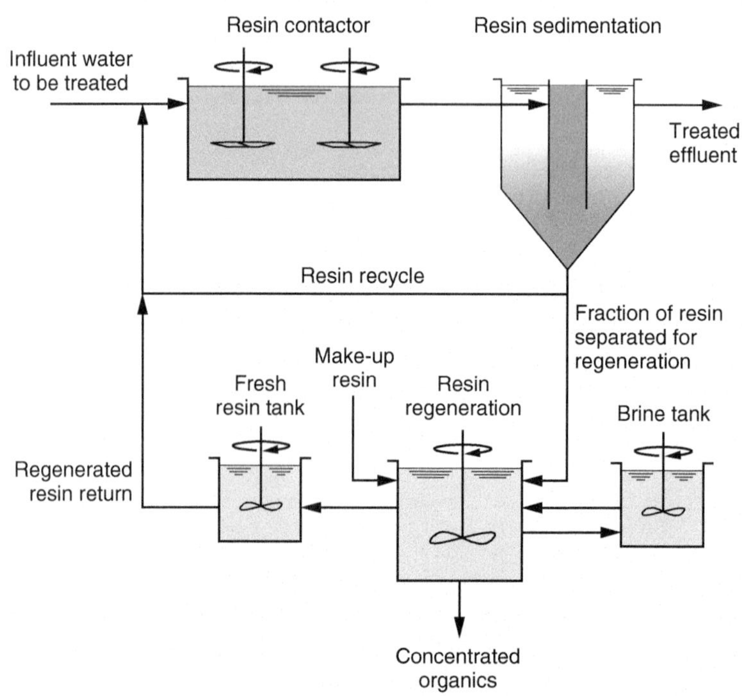

Figure 16-10
Schematic process flow diagram for use of MIEX ion exchange resin for pretreatment of surface water to reduce concentration of natural organic matter (NOM) before addition of coagulating chemical.

high as 15 m/h. The treated water goes on to further treatment. The settled resins are recovered and recycled to the front of the process. A portion of the recovered resin beads (5 to 10 percent) is removed for regeneration. The resin is regenerated with about 10 percent by weight NaCl for 30 min. The regenerated resin beads are stored and reintroduced into the process as needed. An important advantage of the MIEX DOC resin, compared to other ion exchange resins, is its apparent abrasion-resistant properties.

Because the DOC removal remains consistent in the contactor, the DOC leakage is controlled at a predetermined level. Also, because the resin has a high selectivity for DOC, the only inorganic anion that is exchanged is SO_4^{2-}.

Based on preliminary test results, it appears that the removal of DOC on the resin is a surface phenomenon. While other ion exchange resins may be suitable, the time it takes for the DOC to diffuse into the resin may limit their applicability. The performance of MIEX depends on the resin dose, the concentration and nature of the DOC, and the contact time. Reported DOC removal values have been as high as 80 percent, but site-specific testing is required. A pilot study for the City of West Palm Beach, Florida, achieved 67 percent TOC removal with MIEX, compared to 57 percent TOC removal with enhanced coagulation (MWH, 2010). Use of MIEX also reduced coagulant use and sludge production by about 80 percent compared to enhanced coagulation alone.

MIEX is a relatively new technology; as of the end of 2010, about 15 MIEX systems had been installed at treatment plants greater than 3785 m^3/d (1 mgd) in North America.

16-7 Engineering Considerations in Ion Exchange Process Design

Important engineering considerations in developing an ion exchange treatment process include (1) problem definition, (2) establishment of treatment goals, (3) preliminary process analysis, (4) bench- and pilot-scale studies, and (5) development of process design criteria. These considerations are summarized in Table 16-9, discussed below briefly, and illustrated in a case study in the following section.

Problem Definition

Initial consideration of an ion exchange process requires definition of the problem. Defining the problem will require characterization of the composition of the water to be treated (see Table 16-9). The presence of oxidants or reductants in the stream should also be evaluated. Depending upon the specific conditions, the most likely location to apply treatment should be determined so that possible design constraints such as process size, geography, and utility services (sewers, brine waste lines) can be considered in the initial phases of the design.

Table 16-9
Summary of engineering considerations for analysis of ion exchange process

Item	Key Elements/Objectives
Problem definition	1. Characterize water to be treated, including (a) quantitative analysis of the ionic and nonionic constituents and their concentrations, (b) water temperature, (c) pH, (d) turbidity, and (d) density. 2. Evaluate the presence of oxidants and/or reductants in the process stream. 3. Determine the location of the treatment process that minimizes the impact of brine disposal and provides utility services (i.e., sewers, brine waste lines, roadways for salt transport).
Define treatment goals/ design criteria and constraints	1. Required purity of the water. 2. Maximum allowable waste volume. 3. Design constraints (availability of chemicals, space requirements, regulatory permitting, and cost considerations).
Preliminary process analysis	1. Literature survey of previous studies to determine process capabilities and limitations. 2. Select several resins for preliminary assessment and followup bench and pilot plant studies. 3. Using published and manufacturers' data for the resins selected, conduct equilibrium and/or mass transfer model calculations to evaluate process capabilities and limitations.
Bench-scale studies	1. Assess performance of ion exchange resin types. 2. Develop preliminary operating parameters and characteristics. Operating parameters may include (a) saturation and elution curves to assess ion exchange performance, (b) hydraulic considerations (flow rate, head loss, backwashing rate), (c) regeneration requirements (i.e., salt requirements, backwash cycle time, rinse requirements, column requirements), and (d) scaleup requirements.
Pilot plant studies	1. Pilot-scale tests to validate bench-scale test results. 2. Develop long-term operational information, including information on fouling.
Develop design critera for full-scale plant	1. Based on the results of the bench-scale and pilot plant studies, develop design criteria for full-scale design, including (a) scaleup considerations; (b) column design details, including volume of resin, surface area of columns, number of columns, sidewall height, pressure drop, and inlet and outlet arrangements; (c) overall cycle time; and (d) regeneration requirements, including volume, salt quantity and concentration, rinse water, and regeneration cycle time.

Treatment Goals and Objectives

The next step is outlining the actual goals of the process such as required purity of the treated water and maximum waste volumes allowable (see Table 16-9). This step should include identifying possible design constraints such as the availability of chemicals, space requirements, regulatory permitting requirements and/or guidelines, and cost limitations.

Preliminary Process Analysis

Preliminary studies start with selection of promising ion exchange resins for bench-scale testing. Ions that can be removed by each type of ion exchange resin and the regenerant typically used for water treatment applications are summarized in Table 16-10. Preliminary calculations and a literature review combined with resin manufacturer's performance specifications can be used to assess and choose promising resins for bench-scale testing. When choosing a resin, the capacity, selectivity, and ease of regeneration need to be considered. Typical operating conditions for SBA and SAC resins in removing a number of common contaminants are summarized in Table 16-2. Ion exchange modeling software that includes column equilibrium and mass transfer models has been developed to describe the ion exchange process (Clifford and Majano, 1993; Guter, 1998; Hokanson et al., 1995; Liang et al., 1999; Snoeyink et al., 1987).

Bench- and Pilot-Scale Studies

Bench-scale studies are used to identify ion exchange resins and operating parameters that will provide the best possible performance and cost effectiveness over the design life period. For a specific application, the main criteria to be developed in a bench-scale study are length of removal run, service flow rate, regenerant dose, backwash flow rate, and regenerant concentration. Other variables such as resin stability under cyclic operation must be monitored over long periods of time and will require pilot-scale

Table 16-10
Types and characteristics of ion exchange resins

Resin Type	Functional Group	Ions Removed	Regenerant	Operating pH Range
Strong-acid cationic (SAC) resin	Sulfonate, SO_3^-	Ca^{2+}, Mg^{2+}, Ra^{2+}, Ba^{2+}, Pb^{2+}	HCl or NaCl	1–14
Weak-acid cationic (WAC) resin	Carboxylate, $RCOO^-$	Ca^{2+}, Mg^{2+}, Ra^{2+}, Ba^{2+}, Pb^{2+}	HCl	>7
Strong-base anionic (SBA) resin	Quaternary amine, $RN(CH_3)_3^+$	NO_3^-, SO_4^{2-}, ClO_4^-, $HAsO_3^{2-}$, SeO_3^{2-}	NaOH or NaCl	1–13
Weak-base anionic (WBA) resin	Tertiary amine, $RN(CH_3)_2H^+$	NO_3^-, SO_4^{2-}, ClO_4^-, $HAsO_3^{2-}$, SeO_3^{2-}	NaOH or $Ca(OH)_2$	<6

Source: Adapted from Najm and Trussell (1999).

testing. Before considering these variables, it will be useful to discuss the use of small laboratory columns.

USE OF SMALL-DIAMETER COLUMNS

Small-diameter columns can be used to develop meaningful process data if operated properly. Column studies are used primarily to evaluate and compare resin performance in terms of capacity and ease of regeneration. For example, an automated small-column system used to perform laboratory studies for the removal of perchlorate from a groundwater is shown on Fig. 16-11 and schematically on Fig. 16-12. Operational parameters that correspond to full-scale values are summarized in Table 16-2 for SBA resins. Because the main issues of concern are mass transfer and operating exchange capacity, small (1.0- to 5.0-cm-inside-diameter) columns can be scaled directly to full-scale design if the loading rate and empty-bed contact time are the same. Because resin particles are small and the ratio of column diameter to particle diameter is large (>25), the error due to channeling of the water down the walls of the column is minimized.

The hydraulics of full-scale operation cannot be modeled completely by small-scale columns because deviations in flow patterns can exist and should be evaluated at the pilot scale (see Fig. 16-13). However, if full-scale depth is not possible to match in the preliminary studies, a minimum packed-bed depth of 0.6 to 0.9 m (2 to 3 ft) should be adequate to properly design a

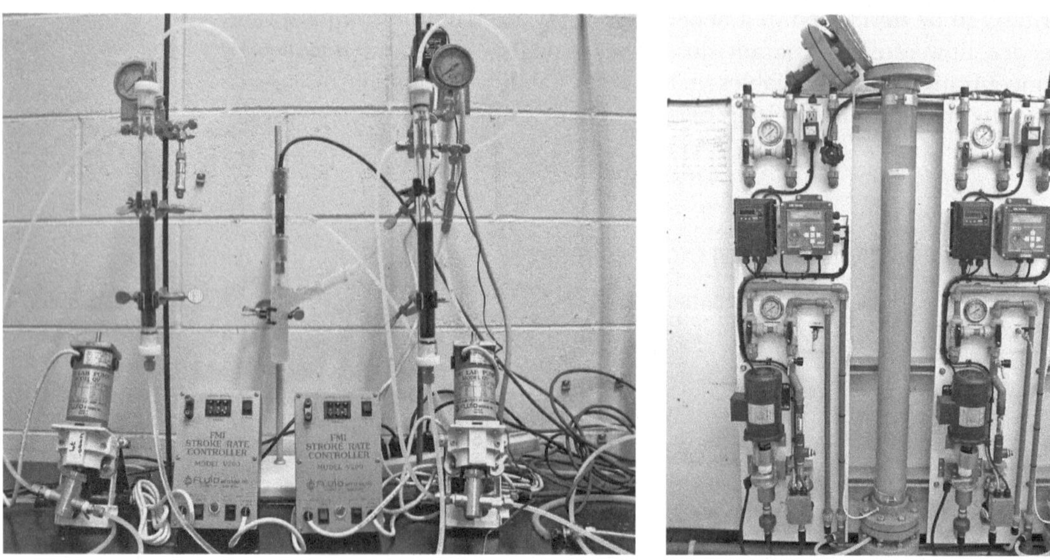

(a) (b)

Figure 16-11
Ion exchange system, used to perform preliminary experiments: (a) small-scale laboratory columns and (b) larger laboratory-type ion exchange column.

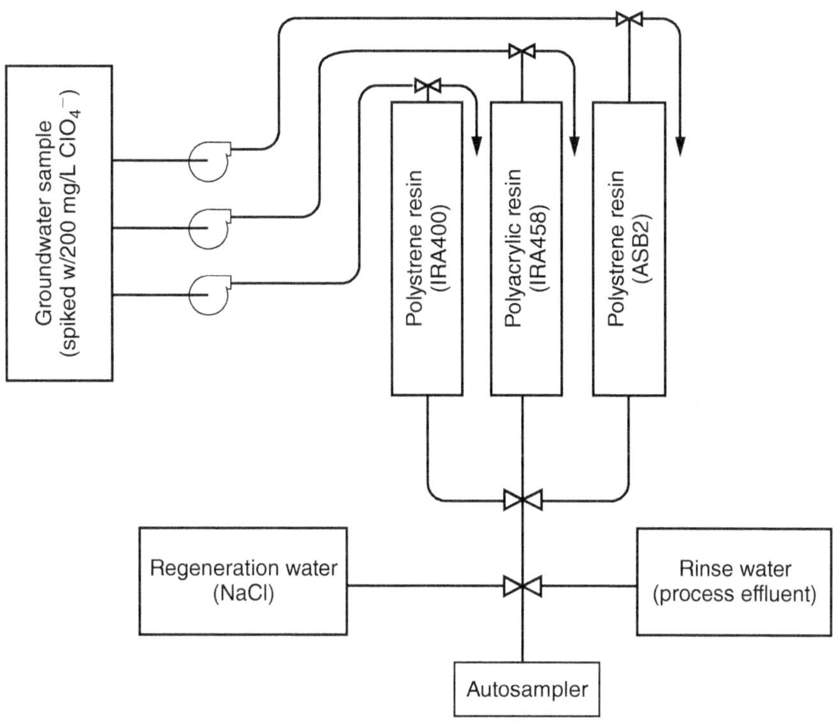

Figure 16-12
Schematic of small-scale laboratory ion exchange system shown in Fig. 16-11.

Figure 16-13
Pilot-scale ion exchange column used to verify bench-scale column tests and obtain data on fouling.

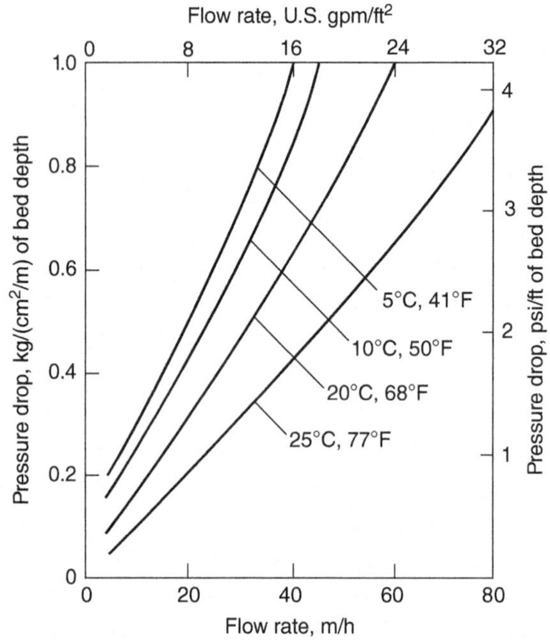

Figure 16-14
Pressure drop curves at various water temperature as function of filtration flow rate for strong-base type I acrylic anion exchange resin (A-850, Purolite).

laboratory or pilot exchange column. For most commercially available resins pressure drop curves versus flow rate and temperature and bed expansion (for backwash) versus flow rate and temperature can be obtained from the manufacturer. For example, performance curves for both the pressure drop and bed expansion as a function of flow rate are given on Figs. 16-14 and 16-15, respectively. During operation, the maximum pressure drop in an ion exchange column should not exceed 172 kPa (25 psi).

SATURATION LOADING AND ELUTION CURVES

The two main types of data collected from small-scale column testing are saturation loading curves and elution curves. Data developed from these curves form the basis for the pilot plant studies and for the development of full-scale designs.

Saturation loading curves

The saturation loading curve is obtained by passing the process stream or a simulated stream of the same chemical composition through a fully regenerated column of resin. A sample of the actual process stream should be fed through the column for a couple of runs. During the runs, samples of the effluent are collected and analyzed until the effluent concentration of the contaminant of interest equals the influent concentration. The effluent concentration is plotted in terms of percent breakthrough, equivalents per liter, or normality as a function of the number of bed volumes of process

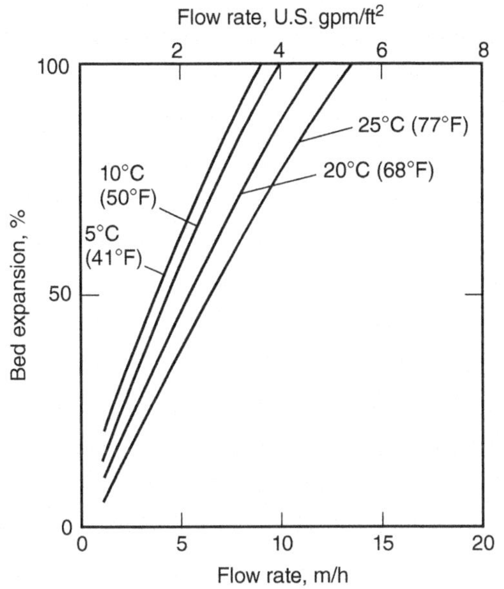

Figure 16-15
Filter bed expansion as function of backwash flow rate at various water temperatures for strong-base type I acrylic anion exchange resin (A-850, Purolite).

stream treated to develop a saturation loading curve. Bed volumes are defined as the average flow rate through the ion exchange column divided by the volume of the resin in the column, including the void fraction.

Generalized saturation loading curves for water containing three ions (A, B, and C) that were treated through an exchange column are presented on Fig. 16-16. As shown on Fig. 16-16, each anion has an effluent profile with the less preferred ions (i.e., A and B) appearing first in the effluent followed by the preferred anion (i.e., C). The observed chromatographic effect shown on Fig. 16-16 depends upon the equilibrium and mass transfer conditions within the column. Percentage concentrations greater than 100 are possible because of the competitive effects among the competing ions, which force previously exchanged ions off the resin. For example, the highest observed effluent concentration for ion B is about 120 percent, or

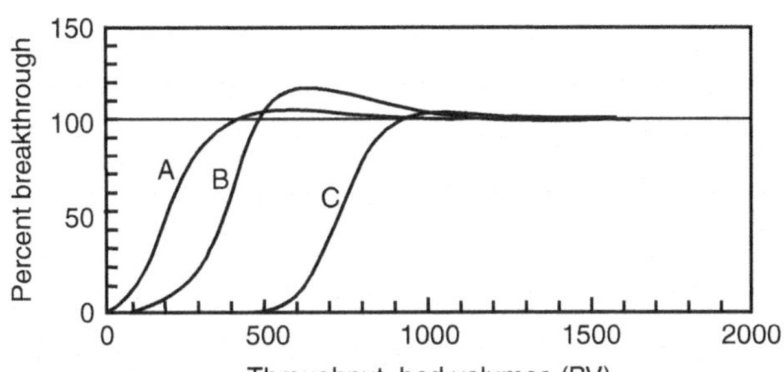

Figure 16-16
Generalized saturation loading curves for compounds A, B, and C.

1.2 times its average influent concentration. In the previous sections, both binary and multicomponent equilibria were discussed and mathematical descriptions were developed. The chromatographic effect within a column can be described when these equilibrium descriptions are incorporated into mass balance expressions. Saturation loading curves similar to the ones shown on Fig. 16-16 but over several loading and regeneration cycles provide the performance data necessary for design engineers to size the columns and determine the operational aspects of the column design.

Regeneration curves

After completing each saturation loading curve, the resin must be eluted with an excess of regenerant to fully convert it back to its presaturant form. A regeneration curve is obtained, similar to a breakthrough curve, by collecting sample volumes of regenerant after it has passed through the bed and determining the concentrations of the ions of interest in each sample volume. The bed volumes of regenerant used can be converted in terms of a salt loading rate by multiplying it by the salt concentration used and dividing by the volume of the resin bed. These data can be used to choose a regeneration level that will be optimum with respect to operating capacity (resin conversion) and regenerant efficiency.

Generalized regeneration curves for ions A, B, and C for the regeneration of a resin are presented on Fig. 16-17. Notice that with a salt loading of about 240 kg/m^3 all of ion A elutes rapidly and is replaced by chloride ions if the resin is an SBA form and sodium if the resin is an SAC form. Ion B requires a little longer to be removed and requires about 350 kg/m^3. Ion C requires about 850 kg/m^3 to ensure that a significant fraction is removed. From equilibrium theory it is known that divalent ions (i.e., ion

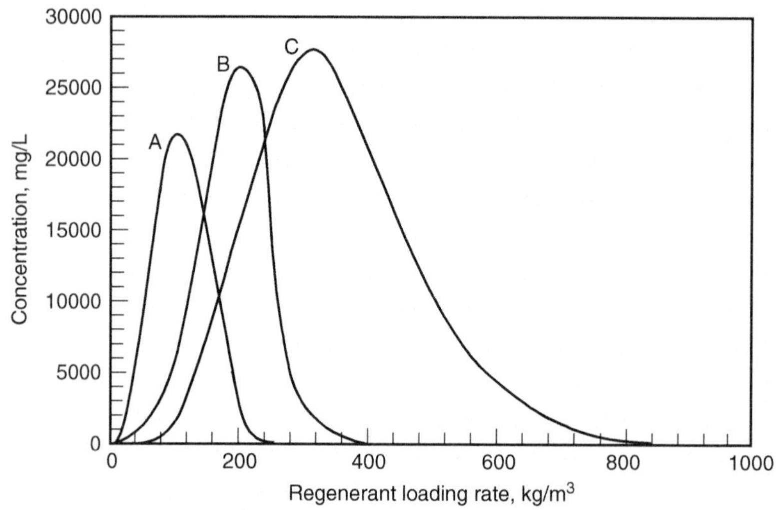

Figure 16-17
Generalized regeneration curves for regeneration of a resin loaded with compounds A, B, and C.

A on Fig. 16-17) will not be preferred in concentrated solutions and hence are easily replaced by sodium or chloride ions.

SERVICE FLOW RATE (SFR) ASSESSMENT

There are two types of flow rates of interest in ion exchange applications: (1) the volumetric flow rate and (2) the surface area loading rate. The volumetric flow rate, usually expressed in L/L · h (gpm/ft^3) or bed volumes per hour (BV/h), is inversely related to the contact time between the solution and the resin and thus the kinetics of exchange. The surface area loading rate, expressed in m/h (gpm/ft^2), is a measure of the superficial flow velocity through the resin bed. The superficial flow velocity must be considered in the scaleup to ensure that excessive flow rates that could damage the resin do not occur.

To determine the optimum SFR, the rate must be varied during the saturation loading tests over a range of choices to see if any noticeable maximum in breakthrough capacity is achieved at a specific flow rate. Typically, the volumetric flow rate is the criterion used because it is directly related to the film mass transfer rate. The main goal in determining the optimum SFR is to reduce the capital cost of equipment. The optimum SFR will minimize the impact of the film mass transfer resistance and consequently shorten the length of the mass transfer zone. The higher the acceptable flow rate, the smaller the contactor can be for a given treatment flow because the mass transfer zone length can be contained in a smaller column. Typical service flow rates range from 8 to 40 BV/h (1 to 5 gpm/ft^3).

Example 16-5 Calculation of BV/h

An ion exchange column has a column loading time of 56 h at a service flow of 6.0 ML/d. The column has a diameter of 3.66 m and a resin depth of 1.1 m. Calculate the service flow in BV/h.

Solution

1. Calculate the volume of the bed occupied by the resin:

$$\text{BV} = \text{area} \times \text{depth} = \tfrac{1}{4}\pi(3.66\ \text{m})^2 \times 1.1\ \text{m} = 11.67\ \text{m}^3$$

2. Calculate the service flow rate in BV/h:

$$\begin{aligned}\text{BV/h} &= Q \times \left(1\ \text{BV}/11.67\ \text{m}^3\right)\\ &= \left(6.0 \times 10^6\ \text{L/d}\right)(\text{d}/24\ \text{h})\left(\text{m}^3/1000\ \text{L}\right)\left(1\ \text{BV}/11.67\ \text{m}^3\right)\\ &= 21.6\ \text{BV/h}\end{aligned}$$

REGENERATION REQUIREMENTS

The three variables of concern during regeneration are (1) concentration of the regenerant, (2) regenerant flow rate, and (3) regenerant dosage.

Concentration and flow rate

A typical scheme to determine optimum conditions would be to choose a fairly slow (2 to 5 BV/h or less) rate and an excess of regenerant, then vary the concentration of regenerant and develop elution curves for each concentration. An optimum concentration would be one that elutes the resin as rapidly as possible. Next, the optimum rate can be determined by keeping the optimum concentration and excess regenerant dose constant while varying the flow rate. Normally, a slower flow rate will allow for a more complete attainment of equilibria conditions but may not be as important a factor when the separation factor favors the ion already on the resin. In many ion exchange applications the only way to fully convert a resin is to use an excess of regenerant.

Regenerant dose

Using the experimentally determined values of concentration and flow rate, the optimum dose of regenerant can be determined. The dose is usually expressed in grams of regenerant per liter of resin (pounds of regenerant per cubic foot of resin). An elution curve should be developed for the optimum set of conditions. Using this curve, it is possible to determine regeneration efficiency and column utilization curves as a function of regenerant dosage.

During regeneration, if $\alpha < 1$, an excess of regenerant must be used to convert the resin to 100 percent regenerant form. Instead of converting the resin completely to this form, the amount of regenerant is chosen so that the column will be converted to a degree that will give the required quality of effluent for a reasonable run length. A plot of regeneration efficiency and column utilization versus regeneration level for a strong-acid exchanger being used for softening is given on Fig. 16-18.

Regeneration efficiency is the actual hardness (or other species of interest) removed by the given amount of salt (or other regenerant) divided by the theoretical hardness that could be removed by that amount of salt assuming 100 percent conversion. Column utilization represents the actual hardness removed by the regenerant divided by the total available exchange capacity of the resin in the column. The product of the two percentages for a given dosage is plotted on Fig. 16-18. The curve peaks roughly where the two curves intersect, and this peak usually indicates the optimum conditions.

OPERATION TO BREAKTHROUGH

Once the above parameters have been established, it is necessary to operate the column to an allowable breakthrough point and leakage level for a

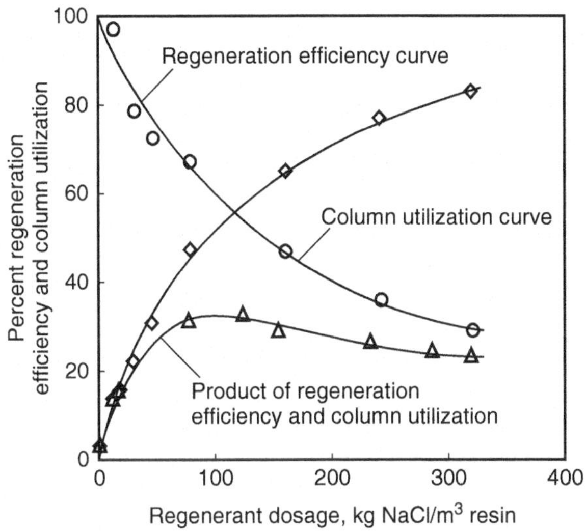

Figure 16-18
Efficiency and column utilization as function of regeneration level for strong-acid resin used for softening. Conditions: influent hardness, 500 mg/L as $CaCO_3$; service flow rate, 267 L/min · m^3; regeneration rate, 67 L/min · m^3; 10% NaCl.

number of cycles (3 to 5) to stabilize the system. The break point is defined as the point where the target ions first appear in the column effluent. Leakage is defined as the appearance of a low concentration of undesired influent ions in the column effluent during the initial part of the exhaustion. Leakage is caused by residual ions in the resin at the bottom of the column due to incomplete regeneration that are displaced by other ionic species coming down the column. A cycle can then be run to get a good indication of what can be expected in the full-scale column.

Development of Column Design Criteria

Once the results of the bench-scale studies are available, the findings are used to develop design criteria for the full-scale installation. Scale-up considerations, the selection of service flow rates, regeneration requirements, and inlet and outlet considerations are considered below.

SCALEUP CONSIDERATIONS

Data derived from small-column experiments can be scaled up directly to any diameter column should the height of the bed remain constant. If the small-column experiments were done at a reasonable height (0.6 to 1 m), then increasing the height in a full-scale design usually will not change the shape of the breakthrough curve when plotted as concentration versus time, but will extend service time. In exchanges where the separation factor is greater than 1 for the ion to be removed, the mass transfer zone length or exchange zone will be relatively small with respect to the column height. Increasing the column depth for the same flow rate will not increase the breakthrough capacity with respect to bed volumes.

COLUMN DESIGN DETAILS

Maintaining the same volumetric flow rate as determined in the small-scale experiments will produce similar cycle times and effluent concentration profiles. If the height of the column is kept constant, then the superficial velocity will also remain equal. If the column is deepened and the volumetric flow rate is kept the same, the superficial velocity will be increased by the proportion that the height has increased. The increased flow should not be a problem, unless a critical range of flow velocities is reached. Typical superficial velocities are in the range of 10 to 36 m/h (4 to 15 gpm/ft^2). Excessive velocities will increase the pressure drop through the column and could adversely affect the stability of the resin beads. Once the optimum service flow rate is known, the design details of the full-scale columns, including volume of resin, surface area of columns, number of columns, side wall height, and pressure drop, can be determined. For example, the amount of resin volume needed to treat a given flow of water will be

$$\text{Required resin volume, m}^3 = \frac{\text{treated-water flow rate(L/min)}}{\text{service flow rate}\left(\text{L/min}\cdot\text{m}^3\right)} = \frac{Q}{\text{SFR}} \tag{16-39}$$

Based on this volume and the desired depth of the resin, the diameter of a single column can be determined. Should the required diameter be much larger than 4 m (12 ft), two or more columns should be used. Typical bed depths used in the industry range from 0.75 to 3 m (2.5 to 10 ft). Determination of the column design details is illustrated in the following section.

One of the major reasons for poor ion exchange performance is the poor design of the feed distribution and outlet effluent collection facilities in contactors. The feed must be distributed uniformly over the resin surface and collected uniformly from the bottom of the column to prevent channeling, maldistribution of flow, and density currents. If the ion exchange columns are not properly designed, premature breakthrough and excessive leakage can result.

REGENERATION REQUIREMENTS

Unless the treated-water flow demand is intermittent, to prevent interruption of the service cycle for regeneration, two or more columns or a treated-water storage reservoir are required. If the exhaustion cycle is long (16 to 24 h), a reservoir can provide sufficient water during regeneration time, normally 1 to 2 h. Based on manufacturer's design data or laboratory studies, the regeneration requirements can be calculated. For most ion exchange applications, a typical regeneration cycle is as follows:

1. End of service run
2. Backwash
3. Regeneration

4. Slow displacement rinse
5. Fast rinse
6. Stand-by [optional if extra column(s) in service and the regeneration time is breakthrough volume/(number of columns −1)]
7. Beginning of service cycle

Backwashing

Backwashing is typically done to reclassify the resin so that there will be a gradual increase in particle size from top to bottom and to help prevent channeling. Ion exchange media will act as good filter media; hence backwashing will remove trapped particulate matter from the resin. Fifty to 75 percent bed expansion is normal, and proper freeboard should be allowed for in-column design. Backwashing will typically last 5 to 15 min. Every ion exchange system should be designed so it can be backwashed, but backwashing is often only necessary at infrequent intervals. This is particularly true when treating groundwater, which is relatively free of particles.

Backwashing can have important impacts on leakage. When regeneration is in the co-current mode and leakage is an issue, backwashing after each regeneration cycle is performed to thoroughly mix the resin and dramatically reduce leakage. When regeneration is in the countercurrent mode, backwashing is best avoided altogether, but if required, it should be done before regeneration so leakage is minimized.

Regenerant consumption

Regenerant consumption per cycle based on design criteria must be determined. The rinses following regeneration are normally operated in the co-current mode: the slow rinse for one to two bed volumes at the regeneration flow rate to displace most of the regenerant from the bed and the fast rinse at the rate of service flow rate for 10 to 30 min. The rinse can be monitored using an online conductivity meter at the effluent of the column to determine when the cycle is complete. An inventory of used brine and rinse volumes must be calculated to adequately prepare for disposal. The disposal of brine is typically a costly part of operation and maintenance cost along with regenerant chemical costs. The disposal of concentration brines may be the critical factor in many potential applications.

16-8 Ion Exchange Process Design Case Study

The purpose of the case study presented in this section is to illustrate the steps required in developing design criteria for an ion exchange plant. Although the approach is developed for the removal of perchlorate ions from a groundwater, the same steps would be required for other ionic constituents. The laboratory and pilot plant information for this design approach case study was taken from Najm et al. (1999).

Problem Definition

A groundwater that is being considered for use as a municipal drinking water source was found to have a perchlorate concentration of 90 μg/L. At the time of this discovery, the regulatory agency was considering a maximum contaminant level of 4 μg/L. The municipality is proposing to pump and treat about 0.16 m^3/s (2500 gpm) using an ion exchange process. Water quality parameters for the groundwater are presented in Table 16-11.

Treatment Goals/Design Criteria and Constraints

A full-scale ion exchange system including a regeneration facility to treat the groundwater to the above regulatory requirement is to be designed so that an assessment of the cost of treatment can be made. The design criteria that need to be determined are

1. Column requirements (number of columns, column dimensions)
2. Maximum SFR and head loss requirements
3. Cycle times (regeneration time, rinse time)
4. Regeneration and rinse requirements (quantities)
5. Type of resin

Preliminary Process Analysis

To determine the important design and operational parameters required for effective treatment of perchlorate from this groundwater, laboratory and pilot plant studies are needed. Based on a review of the literature and past experience, three SBA ion exchange resins (two polystyrene and one polyacrylic resin) were selected for bench and pilot plant studies.

Laboratory and Pilot Plant Studies

The design sequence begins with performing laboratory and pilot plant studies to determine the most efficient resin in terms of operational or

Table 16-11
Water quality parameters for groundwater for ion exchange process design case study

Parameter	Unit	Value
Alkalinity	mg/L as $CaCO_3$	122
Hardness	mg/L	163
pH	Unitless	7.8
Nitrate	mg/L as N	6.6
Sulfate	mg/L	53
Perchlorate	μg/L	90
TOC	mg/L	0.9
Temperature	°C	15

working capacity, regeneration requirements including regenerant salt concentration and loading, rate and regeneration, and rinse volume requirements. Once these parameters are established, pilot plant testing is performed to evaluate long-term performance of the most promising resin(s). The results of the pilot plant studies are used to establish the full-scale design criteria listed above.

LABORATORY STUDIES

It should be noted that all of the laboratory and pilot plant data developed in this case study are not presented. Only the information pertinent to determining the most efficient and least cost design is presented (see Najm et al., 1999). Laboratory studies were performed on the groundwater to evaluate the three ion exchange resins selected for study, characterize the perchlorate breakthrough, evaluate the regeneration efficiency, and identify the conditions for pilot testing. Column breakthrough was defined as exceeding the perchlorate minimum reporting limit of 4 μg/L so as to minimize perchlorate leakage. Typical saturation and elution curves for an SBA resin are shown on Fig. 16-19.

Countercurrent regeneration was employed using salt loading rates of 240, 480, and 720 kg NaCl/m^3 (15, 30, and 45 lb NaCl/ft^3). Rinse volumes were determined when the conductivity of the effluent rinse decreased to less than 700 S/cm. From the results of the laboratory study it was found that SBA resins are effective for the removal of perchlorate from groundwater, but the process requires optimization. Polystyrene resins had a higher affinity for perchlorate but are difficult to regenerate, whereas the polyacrylic resins have a moderate affinity for perchlorate and can be regenerated effectively. Perchlorate leakage occurred at a salt loading of 240 kg NaCl/m^3 (15 lb NaCl/ft^3), and further testing is necessary to identify long-term working capacity and a salt loading of 480 kg NaCl/m^3 (30 lb NaCl/ft^3) or greater for effective regeneration to eliminate perchlorate leakage.

PILOT PLANT STUDIES

Based on laboratory studies, pilot plant studies were conducted to demonstrate the performance of the three resins operated under full-scale operating conditions for several regeneration cycles and to validate the laboratory results. The pilot plant design and operational parameters presented in Table 16-12 were based on the typical design values summarized in Table 16-2 for SBA resins. The important parameters of the pilot plant study are displayed below.

Based on the results of the pilot study, the polyacrylic resin was found to provide the best working capacity while minimizing the salt quantity requirements. A pilot plant result for the most promising SBA resin operated for 31 loading cycles is displayed on Fig. 16-20. The resin was regenerated using 16 BV of 480 kg/m^3 (30 lb/ft^3) salt regeneration, salt strength of

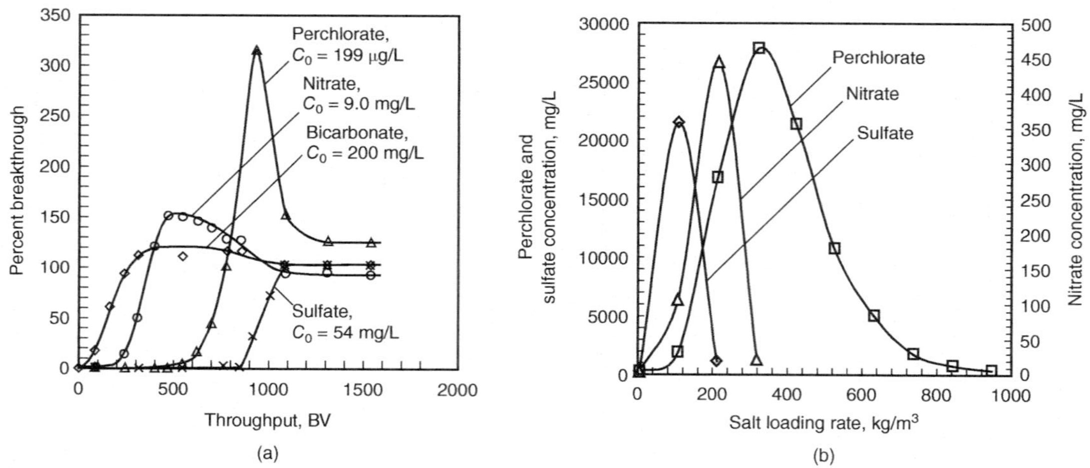

Figure 16-19
Typical saturation and regeneration curves: (a) saturation loading curves for perchlorate contaminated groundwater on SBA resin and (b) regeneration curves for regeneration of strong-base resin loaded with perchlorate, sulfate, and nitrate (Adapted from Najm and Trussell, 2001).

Table 16-12
Operational parameters utilized for pilot plant studies to assess resin performance

Parameter	SI Units	Value	U.S. Customary Units	Value
Operational mode	—	Countercurrent	—	Countercurrent
EBCT	min	1.5	min	1.5
Column diameter	m	0.0509	ft	0.167
Resin depth	m	0.862	ft	2.83
Service flow rate	BV/h	40	gpm/ft^3	5.0
Flow rate per column	m^3/h	0.0681	gpm	0.3
Column resin volume	m^3	0.00176	ft^3	0.062
Regenerant type	—	NaCl	—	NaCl
Regenerant strength	%	3	%	3
	mg/L	30,000	—	—
Salt loading rate	kg^3/m^3	480	lb/ft^3	30
Regeneration and rinse flow rate	m^3/h	0.0363	gpm	0.16
Regeneration volume	BV	16	BV	16
Rinse volume	BV	2–6[a]	BV	2–6[a]
Backwash rate	m/h	6.0	gpm/ft^2	2.5

[a]Column rinsed until effluent conductivity decreased to less than 700 S/cm.

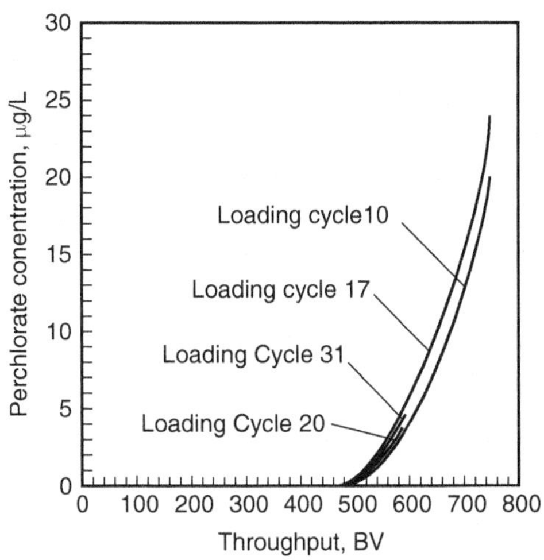

Figure 16-20
Pilot plant effluent profiles for perchlorate on SBA resin operated for total of 31 cycles at 480 kg/m^3 (30 lb/ft^3) salt regeneration rate (Adapted from Najm and Trussell, 2001).

10 percent (specific gravity 1.07), and between 2 and 6 BV of rinse water. The initial breakthrough occurred at 560 BV and consistently produced regenerated column runs with a design leakage of less than 4 μg/L. Consequently, the columns could be loaded up to 560 BV prior to regeneration without exceeding the allowable design leakage point. Because perchlorate

breakthrough (at 4 μg/L) consistently occurred at 560 BV for 31 cycles, full regeneration of the resin was accomplished.

Development of Full-Scale Design Criteria

Design criteria for a full-scale ion exchange treatment plant were developed based on the results of the bench-scale and pilot plant study. The plant is sized for a maximum finished water capacity of 0.160 m^3/s (2500 gpm). The plant is sized such that one column is in the regeneration mode on standby while the others are in the operational mode. The service flow rate used in the pilot plant study was 40 BV/h, which is on the high end of service flow rates. The high rate was used to decrease the time required to perform the pilot studies. Consequently, a lower SFR of 28 BV/h (3.5 gpm/ft^3) was used for the full-scale design (see summary design criteria in Table 16-13).

ION EXCHANGE COLUMN DESIGN

Design of the ion exchange columns involves the determination of the volume of resin, the surface area of resin required, the number of columns, the sidewall height, and the pressure drop.

Volume of resin

The number of columns can be found by first calculating the total volume of resin required assuming a typical SFR of 28 BV/h (3.5 gpm/ft^3):

$$\text{Total required resin volume} = \frac{Q}{\text{SFR}} = \frac{0.160\ \text{m}^3/\text{s}}{(28\ \text{BV/h})\ (1\ \text{h}/3600\ \text{s})}$$

$$= 20.6\ \text{m}^3\left(727\ \text{ft}^3\right)$$

Surface area of resin required

As discussed above, the EBCT of the pilot plant should be about the same as the EBCT used in the full-scale design. Because a resin depth of 0.863 m (2.83 ft) was used in the pilot plant study, a similar full-scale design with a depth of 1.0 m (3 ft) will be used. Consequently, the total ion exchange surface area required is determined to be

$$\text{Total required surface area} = \frac{\text{resin volume}}{\text{resin depth}} = \frac{20.6\ \text{m}^3}{1.0\ \text{m}} = 20.6\ \text{m}^2\left(223\ \text{ft}^2\right)$$

Ion exchange columns come in standard sizes from the manufacturer. Typically, they may have column diameters of 1.0 m (4 ft), 2.0 m (6 ft), 3.0 m (10 ft), 4.0 m (13 ft), and 5.0 m (16 ft). If a 3-m column diameter is chosen for the design, the column would provide 7.1 m^2 (76.4 ft^2)of service area and the volume occupied by the resin would be 7.1 m^3.

Table 16-13
Summary of design criteria for perchlorate removal case study

Parameter	SI Units	Value	U.S. Customary Units	Value
Design product water capacity	m^3/s	0.160	gpm	2,536
Minimum water temperature	°C	15	°F	59
Resin type	—	SBA, polyacrylic, type I	—	SBA, polyacrylic, type I
Effective resin size	mm	0.6	in.	0.024
SFR	BV/h	28	gpm/ft^3	3.6
EBCT	min	2.14	min	2.14
Resin depth	m	1.0	ft	3.0
Total minimum sidewall depth	m	3.15	ft	10.3
Required resin volume	m^3	20.6	ft^3	728
Column diameter	m	3.0	ft	10
Number of columns	—	4	—	4
BVs to perchlorate breakthrough (single column)	BV	560	BV	560
Salt loading rate (NaCl)	kg/m^3	480	lb/ft^3	30
Salt strength	%	10	%	10
Rinse volume	BV	6	BV	6
Clean-water head loss rate	kPa/m	60.8	psi/ft	2.7
Clean-water head loss	kPa	60.8	psi	8.8
Regeneration volume per column	BV	4.5	BV	4.5
Number of regenerations for each column per year	—	438	—	438
Spent regeneration solution volume per column	m^3	32	gal	8,454
Annual regeneration solution volume per column	m^3/yr	14,016	Mgal/yr	3.7
Salt quantity required per column	kg	3,408	lb	7,513
Annual salt quantity required per column	kg/yr	1.5×10^6	lb/yr	3.3×10^6
Rinse volume required per column	m^3	43	gal	11,360
Annual rinse volume per column	m^3/yr	18,834	Mgal/yr	5.0
Total annual salt requirements	kg/yr	4.50×10^6	lb/yr	9.9×10^6
Total annual regeneration solution volume	m^3/yr	42,048	Mgal/yr	11.1
Total annual rinse requirements	m^3/yr	56,502	Mgal/yr	15.0
Total regeneration cycle time	min	32.4	min	32.4

Number of columns
If the total column area is divided by the area of one column, the number of columns required can be calculated as

$$\text{Required number of columns} = \frac{\text{total column area}}{\text{area of one column}} \frac{20.6 \text{ m}^2}{7.1 \text{ m}^2} = 2.9 \approx 3$$

With one column in the regeneration or standby mode a total of four 3.0-m-(10-ft) diameter columns are required.

Sidewall height

The total sidewall height of the column must include the depth of the resin, height for inlet distributor, height for resin support, and height for backwashing. As discussed above, the depth of the resin was chosen as 1.0 m. The heights for the inlet distributor and resin support underdrain can be obtained from the manufacturer, which in this case were 1.0 and 0.5 m, respectively. The sidewall height of the ion exchange columns must include room for bed expansion due to backwashing. Bed expansion can be calculated using manufacturer's performance curves such as those shown on Fig. 16-15.

If the backwash superficial velocity from the pilot study is 6.0 m/h (2.43 gpm/ft^2) and the water temperature is 15°C, the percent bed expansion required during backwashing and extra height required for backwashing can be determined. From Fig. 16-14, the percent expansion required is about 65 percent. The expanded height required for backwash is

$$\text{Expanded bed height} = 0.65 \times 1.0\text{ m} = 0.65\text{ m}$$

The total sidewall height should be a minimum of 3.15 m (1.0 + 0.65 + 1.0 + 0.5) (10.33 ft).

Pressure drop

Before continuing the design calculations, the column pressure drop needs to be checked. As noted previously, the maximum pressure drop for the ion exchange resin bed should not exceed 172 kPa (25 psi). Manufacturers provide pressure drop curves for commercially available resins such as shown previously on Fig. 16-14. The superficial velocity for this system is 28 m/h, the initial pressure drop through the resin is 0.62 kg/cm^2/m of bed depth, as shown on Fig. 16-14. For 1.0 m of resin depth, the clean-water pressure drop is 0.62 kg/cm^2, or 60.8 kPa (8.8 lb/in.^{2}). In this case, the clean-water pressure drop column design is well below the maximum allowable pressure drop (60.8 kPa $\ll$ 172 kPa). If these curves are not available, the column head loss can be calculated (see Chap. 11, Eq. 11-13). Typically, the pressure drop can be determined in the pilot plant studies if the loading rate and EBCT used in the pilot columns are the same as those in the full-scale design.

OVERALL CYCLE TIME

As discussed above, perchlorate breakthrough in the pilot plant study consistently occurred at 560 BV for 31 cycles, at which time full regeneration of the resin was accomplished. The time for each column loading cycle can be calculated by dividing 560 BV by the SFR as shown:

$$\frac{\text{Time}}{\text{Loading cycle}} = \frac{\text{bed volumes/loading cycle}}{\text{SFR}} = \frac{560\text{ BV}}{28\text{ BV/h}} = 20\text{ h}$$

If the columns are staggered or started at different times, then each column will be regenerated every 20 h and the blended effluent will not exceed 4 μg/L perchlorate concentration.

REGENERATION REQUIREMENTS

Based on the results of the pilot plant studies, it was found that the perchlorate-loaded columns could be regenerated fully using 480 kg $NaCl/m^3$ (30 lb $NaCl/ft^3$) of resin with a salt strength of 10 percent. For the full-scale design, 480 kg $NaCl/m^3$ (30 lb/ft^3) salt loading rate with a 10 percent salt strength (specific gravity 1.07)will be used. The salt solution can be calculated from the specific gravity of the salt and the salt strength as

$$\begin{aligned} 10\% \text{ salt solution} &= (0.1 \text{ kg NaCl/kg soln})(1070 \text{ kg soln/m}^3 \text{ soln}) \\ &= 107 \text{ kg NaCl/m}^3 \text{ soln} \end{aligned}$$

Regeneration volume

The regeneration volume can be calculated by dividing the salt requirements per volume of resin by the salt solution concentration:

$$\begin{aligned} \text{Required regeneration volume per bed volume} &= \frac{480 \text{ kg NaCl/m}^3 \text{ resin}}{107 \text{ kg NaCl/m}^3 \text{soln}} \\ &= 4.5 \text{ L m}^3 \text{ soln/m}^3 \text{ resin} \end{aligned}$$

Salt quantity

The total quantity of salt required on an annual basis can be calculated by multiplying the number of regenerations in a year by the quantity of salt required per regeneration. The number of regenerations can be calculated by dividing the number of hours in a year by the loading cycle time per column:

$$\begin{aligned} \text{Number of regenerations for each column per year} &= \frac{(365 \text{ d/yr})(24 \text{ h/d})}{20 \text{ h/regen}} \\ &= 438/\text{yr} \end{aligned}$$

The quantity of salt per regeneration per column is calculated as

$$\begin{aligned} &\text{Salt quantity per column regeneration} \\ &= (7.1 \text{ m}^3 \text{ resin/regen})(480 \text{ kg NaCl/m}^3 \text{ resin}) \\ &= 3408 \text{ kg NaCl (7531 lb)} \end{aligned}$$

The annual salt consumption requirement per column is given as

$$\begin{aligned} &\text{Annual salt quantity required per column} \\ &= (438 \text{ regen/yr})\ (3408 \text{ kg NaCl/regen}) \\ &= (1.5\times10^6 \text{ kg NaCl/yr})\ (3.3\times10^6 \text{ lb/yr}) \end{aligned}$$

The volume of spent regeneration solution per column regeneration is given as

$$\begin{aligned} &\text{Spent regeneration solution per column} \\ &= (7.1 \text{ m}^3 \text{ resin/BV})\times 4.5 \text{ BV} \\ &= 32 \text{ m}^3\text{/column} \quad \text{or} \quad 1130 \text{ gal/column} \end{aligned}$$

The total annual volume of spent regeneration solution per column is calculated as

$$\begin{aligned}\text{Total annual spent regeneration solution per column}\\ &= (32\ \text{m}^3/\text{column})\,(438\ \text{columns/yr})\\ &= 14{,}016\ \text{m}^3/\text{yr} \quad \text{or} \quad 3.7\ \text{Mgal/yr}\end{aligned}$$

The total annual quantity of salt required and regeneration solution generated for the whole plant will be three times the above quantities because within every 20-h period each of the three columns in service will be regenerated. The total plant quantity values are shown in Table 16-11.

Rinse water requirement

The quantity of rinse water can be determined based on the rinse quantity used in the pilot plant study. In the pilot plant study, 2 to 6 BV were used to reduce the conductivity of the rinse water below 700 S/cm. To be conservative, 6 BV will be used for the full-scale design. The quantity of rinse volume per regeneration is calculated as

$$\begin{aligned}\text{Rinse volume per column} &= (7.1\,\text{m}^3\ \text{resin/BV})\ (6\ \text{BV})\\ &= 43\ \text{m}^3/\text{column} \quad \text{or} \quad 11{,}360\ \text{gal/column}\end{aligned}$$

The total annual rinse volume is given as

$$\begin{aligned}\text{Annual rinse volume per column} &= (43\ \text{m}^3/\text{column})\,(438\ \text{columns/yr})\\ &= 18{,}834\ \text{m}^3/\text{yr} \quad \text{or} \quad 5.0\ \text{Mgal/yr}\end{aligned}$$

Regeneration cycle time

The cycle time for the salt regeneration is calculated by multiplying the EBCT by the number of bed volumes of regeneration solution per column. The EBCT is first calculated by dividing the resin depth in the column by the superficial velocity as shown:

$$\text{EBCT} = \frac{1\,\text{m}}{28\,\text{m/h}}(60\ \text{min/h}) = 2.14\ \text{min}$$

$$\begin{aligned}\text{Regeneration time per column} &= \text{EBCT}\left(\frac{\text{BV}}{\text{regen}}\right)\\ &= (2.14\ \text{min/BV})\,(4.5\ \text{BV}) = 9.6\ \text{min}\end{aligned}$$

Similarly, the cycle time for the rinse step is calculated as

$$\begin{aligned}\text{Rinse time per column} = \text{EBCT}\left(\frac{\text{BV}}{\text{regen}}\right) &= (2.14\ \text{min/BV})\,(6\ \text{BV})\\ &= 12.8\ \text{min}\end{aligned}$$

Typical backwash times range from 5 to 20 min, so choosing a backwash time of 10 min, the total time a column will be out of service for the regeneration cycle, can be estimated to be

Total regeneration cycle time per column

$$= \text{regeneration time per column} + \text{rinse time per column} + \text{backwash time per column}$$
$$= 9.6\ \text{min} + 12.8\ \text{min} + 10\ \text{min} \qquad (16\text{-}40)$$
$$= 32.4\ \text{min}$$

Case Study Design Summary

The design parameters developed for the full-scale process design for perchlorate removal are summarized in Table 16-13.

Problems and Discussion Topics

16-1 A SAC exchanger is employed to remove calcium hardness from water. The capacity of the resin is 2.0 meq/mL in the sodium form. If calcium concentrations in the influent and effluent are 44 and 0.44 mg/L, determine the maximum volume of water that can be treated per cycle given the following:

Cations	meq/L	Anions	meq/L
Ca^{2+}	2.2	HCO_3^-	2.9
Mg^{2+}	1.0	Cl^-	3.1
Na^+	3.0	SO_4^{2-}	0.2
Total	6.2	Total	6.2

16-2 Consider the removal of perchlorate from well water using an SBA exchange resin. The following table lists the major anions contained in the well water. Assuming perchlorate is completely removed from solution, calculate the maximum volume of water that can be treated per liter of resin assuming equilibrium conditions. Assume total resin capacity of the SBA is 1.4 eq/L.

Parameter	Unit	Value
Alkalinity	mg/L as $CaCO_3$	200
Perchlorate	mg/L	200
Nitrate	mg/L	9.0
Chloride	mg/L	15
Sulfate	mg/L	55
pH	Unitless	8.0

16-3 A small public water system is considering removing calcium from its water using ion exchange. The average daily flow rate is about 2 ML/d and the influent calcium concentration is 200 mg/L as $CaCO_3$. If a SAC exchange resin in the sodium form is to be used, estimate the minimum daily volume of resin that would be required assuming that calcium is completely removed and is the only cation exchanging on the resin. Assume the total resin capacity of the SAC resin is 2.0 eq/L in the sodium form.

16-4 Describe the differences between SAC exchanger resins and SBA exchanger resins.

16-5 Explain the differences between type I and type II exchanger resins.

16-6 Describe and explain the operational advantages of using cocurrent regeneration versus countercurrent regeneration.

16-7 A small public water system is considering removing barium from its well water using ion exchange. The average daily flow rate is about 1.5 ML/d (400,000 gpd) and the influent barium concentration is 11.3 mg/L. If an SBC exchange resin is to be used, estimate the minimum daily volume of resin that would be required assuming that barium is completely removed and is the only cation exchanging on the resin.

16-8 An SBA exchanger resin is used to remove nitrate ions from well water that contains high chloride concentration. Usually bicarbonate and sulfate are present in the water (assume they are negligible). The total resin capacity is 1.5 eq/L. Calculate the maximum volume of water that can be treated per liter of resin. The water has the following composition:

Cations	meq/L	Anions	meq/L
Ca^{2+}	1.4	SO_4^{2-}	0.0
Mg^{2+}	0.8	Cl^-	3.0
Na^+	2.6	NO_3^-	1.8
Total	4.8	Total	4.8

16-9 A small municipal water supply treats a maximum daily flow of 5.0 ML/d, maximum weekly flow of 25 ML/wk, and a maximum nitrate concentration of 18 mg/L. The plant treats 5 ML of water and operates only 7 h per day and 5 days per week, and there is sufficient storage capacity for the weekend demand. The treatment objective for the ion exchange process is 0.6 mg/L NO_3–N and will be blended with untreated water at 18 mg/L NO_3–N to produce a final product water of 8 mg/L or less NO_3–N. With a standard of 10 mg/L as NO_3–N, determine the flow rate of the ion exchanger and blending rate.

16-10 A groundwater contains the following anion concentration exchanger (NO_3–N = 18 mg/L, SO_4^{2-} = 50 mg/L, Cl^- = 35 mg/L, and HCO_3^- = 85 mg/L). Assuming nitrate is removed completely from solution, calculate the equilibrium exchange capacity for each ion, and the maximum volume of water that can be treated per liter of resin, assuming equilibrium conditions. Assume total resin capacity of the SBA is equal to 1.4 eq/L.

16-11 Given the information in Problems 16-8 and 16-9, design a 5-ML/d ion exchanger for nitrate removal. Determine the number of columns required assuming 4-m diameter columns and a minimum bed depth of 0.762 m and the regenerant requirements including salt used, brine production, total volume of brine storage tank, and regeneration cycle time for the ion exchanger. Based on pilot studies, it was found that adequate regeneration can be obtained with a salt dose of 320 kg/m^3 resin, a salt concentration of 14 percent, and the specific weight of the salt is 2.165. The capacity of the brine storage tank must be sufficient to handle 10 resin regenerations. The water temperature is 10°C. Assume the working capacity for the SBA resin is the same as the maximum volume treated determined in Problem 16-8.

16-12 Perchlorate at a concentration of 100 μg/L was discovered recently in a groundwater that is being considered for use as a drinking water source. Because the action level for perchlorate is 4 μg/L, the regulatory agency is requiring remediation of the groundwater. The municipality is proposing to pump and treat about 0.158 m^3/s (2500 gpm) using ion exchange process. To obtain information on the treatment of the water that can be used for the design of a fullscale treatment plant, pilot plant ion exchange studies have been performed. Using the information given below on the water quality and the pilot plant study parameters, design an ion exchange system including a regeneration facility to treat the groundwater to the above regulatory requirement. Determine the following full-scale design criteria: plant size (number of columns, column dimensions) maximum service loading rate, single-column service time, single-column regeneration and rinse volume requirements and regeneration time, head loss requirements, and spent-brine disposal. Assume the inlet distributor and resin support underdrain require 1.0 and 0.5 m of column height, respectively.

Water quality parameters

Parameter	Unit	Value
Alkalinity	mg/L as $CaCO_3$	150
Hardness	mg/L	140
pH	Unitless	7.8
Nitrate	mg/L as N	4.0
Sulfate	mg/L	50
Perchlorate	μg/L	85
TOC	mg/L	1.5
Temperature	°C	10

Summary of the pilot plant operational parameters

Parameter	SI Units	Value
Operational mode	—	Countercurrent
EBCT	min	1.5
Column diameter	m	0.0509
Resin depth	m	0.862
Service flow rate	BV/h	30
Flow rate per column	m^3/h	0.0681
Backwash rate	m/h	6.0
Column resin volume	m^3	0.00176
Regenerant type	—	NaCl
Regenerant strength	%	10
	mg/L	100,000
Salt loading rate	kg/m^3	480
Regeneration and rinse flow rate	m^3/h	0.0363
Regeneration volume	BV	10
Rinse volume	BV	6
Initial BV to breakthrough[a]	BV	550

[a]Full regeneration and no leakage occurred for 31 cycles.

References

Anderson, R. A. (1979) *Ion Exchange Separations*, McGraw-Hill, New York.

Bolto, B. A., and Pawlowski, L. (1987) *Wastewater Treatment by Ion-Exchange*, E. & F. N. Spon, London.

Clifford, D. A., Sorg, T.J., and Ghurye, G.L. (2011) Ion Exchange and Adsorption of Inorganic Contaminants, Chap. 12, in J. E. Edzwald (ed.) *Water Quality and Treatment: A Handbook on Drinking Water*, 6th ed., American Water Works Association, McGraw-Hill, New York.

Clifford, D. A., Lin, C. C., and Al., E. (1987) *Nitrate Removal from Drinking Water in Glendale, Arizona*, U.S. Environmental Protection Agency, Washington, DC.

Clifford, D., and Majano, R. E. (1993) "Computer Prediction of Ion Exchange," *J. AWWA*, **85**, 4, 20.

Clifford, D. A., and Weber, W. J. (1978) "Multicomponent Ion-Exchange: Nitrate Removal Process with Land Disposal Regenerant," *Ind. Water. Eng.*, **15**, 18–26.

Ghurye, G. L., Clifford, D. A and Tripp, A (1999) "Combined Arsenic and Nitrate Removal by Ion Exchange," *J. AWWA*, **91**, 10, 85–96.

Graham, E. E., and Dranoff, J. S. (1972) "Kinetics of Anion Exchange Accompanied by Fast Irreversible Reaction," *J. AICHE*, **18**, 3, 606–613

Guter, G. A. (1998) *IX Windows Pro*, Cathedral Peak Software, Bakersfield, CA.

Harland, C. E. (1994) *Ion Exchange: Theory and Practice*, 2nd ed., Royal Society of Chemistry, Cambridge, UK.

Harland, R. S., and Prud'homme, R. K. (1992) *Polyelectrolyte Gels: Properties, Preparation and Applications*, American Chemical Society, Washington, DC.

Haub, C. E., and Foutch, G. L., (1986) "Mixed-Bed Ion Exchange at Concentrations Approaching the Dissociation of Water. 1. Model Development," *Ind. Eng. Chem. Fundam.*, **25**, 3, 373–381.

Helfferich, F. (1995) *Ion Exchange*, Dover, New York.

Helfferich, F., and Hwang, Y. L. (1991) "Ion Exchange Kinetics," Chap. 6, in K. Dorfner (ed.), *Ion Exchangers*, Walter de Gruyter, Berlin, New York, pp. 1471.

Hokanson, D. R. (2004) Development of Ion Exchange Models for Water Treatment and Application to the International Space Station Water Processor, PhD Dissertation, Michigan Technological University, Houghton, MI.

Hokanson, D. R., Clancey, B. L., Hand, D. W., Crittenden, J. C., Carter, D. L., and Ii, J. D. G. (1995) "Ion Exchange Model Development for the International Space Station Water Processor," *SAE Trans.: J. Aerospace*, **104**, 977–987.

Kimoto, W. I., Dooley, C. J., Carre, J., and Fiddler, W. (1979) "Role of Strong Ion Exchange Resins in Nitrosamine Formation in Water," *Water Res.*, **14**, 869–876.

Kunin, R. (1979) "Amber-Hi-Lites: Acrylic-Based Ion Exchange Resins," *Rohm and Haas Product Bulletin*, Philadelphia, PA, 1 –10.

Kunin, R., and Myers, R.J. (1950) *Ion Exchange Resins*, Wiley, New York.

Liang, S., Mann, M. A., Guter, G. A., Kim, P., and Harden, D. L. (1999) "Nitrate Removal from Contaminated Groundwater," *J. AWWA*, **91**, 2, 79–91.

Mortimer, C. E. (1975) *Chemistry: A Conceptual Approach*, 4th ed., Van Norstrand, New York.

MWH (2010) Work Authorization No. 3, Task 4, Pilot Plant Report, Phase One Operations, Final Report submitted to City of West Palm Beach. January, 2010.

Najm, I., and Trussell, R. R. (1999) "New and Emerging Drinking Water Treatment Technologies," Chap. 11, in W. R. Muir, R. R. Trussell, F. J. Bove, and L. J. Fischer (eds.), *Identifying Future Drinking Water Contaminants*, National Research Council, National Academics Press, Washington, DC.

Najm, I., and Trussell, R. R. (2001) "NDMA Formation in Water and Wastewater," *J. AWWA*, **93**, 2, 92–99.

Najm, I., Trussell, R. R., Boulos, L., Gallagher, B., Bowcock, R., and Clifford, D. (1999) Application of Ion-Exchange Technology for Perchlorate Removal from Drinking Water, in Proceedings of 1999 AWWA Annual Conference, Chicago, IL, American Water Works Association, Denver, CO.

Perry, R. H., and Chilton, C. H. (1973) *Chemical Engineers' Handbook*, 5th ed., McGraw-Hill, New York.

Rohm and Haas (1975) Summary Bulletin: Amberlite Polymeric Adsorbents, Technical Bulletin Fluid Process Chemicals Department, Philadelphia, PA.

Semmens, M. J. (1975) "A Review of Factors Influencing the Selectivity of Ion Exchange Resins for Organic Ions," *AIChE Symp. Series No. 152*, **71**, 214–223.

Snoeyink, V. L., Cairns-Chambers, C., and Pfeffer, J. L. (1987) "Strong-Acid Ion E for Removing Barium, Radium and Hardness," *J. AWWA*, **79**, 8, 66–78.

Streicher, L., Pearson, H., and Bowers, G. (1947) "Operating Characteristics of Synthetic Siliceous Zeolite," *J. AWWA*, **39**, 11, 1133–1151.

Wachinski, A. M., and Etzel, J. E. (1997) *Environmental Ion Exchange*, CRC, Boca Raton, FL.

Wagner, J. D., and Dranoff, J. S. (1967) "The Kinetics of Ion Exchange Accompanied by Irreversible Reaction. III. Film Diffusion Controlled Neutralization of a Strong Acid Exchanger by a Weak Base," *J. Phys. Chem.*, **71**, 13, 4551–4553.

Weber, W. J., Jr. (1972) *Physicochemical Processes for Water Quality Control,* Wiley-Interscience, New York.

Wildhagen, G. R. S., Qassim, R. Y., and Rajagopal, K. (1985) "Effective Liquid-Phase Diffusivity in Ion Exchange," *Ind. Eng. Chem. Fundam.*, **24**, 4, 423–432.

17 Reverse Osmosis

Stantec's Water Treatment: Principles and Design, Updated Third Edition. John C. Crittenden, R. Rhodes Trussell, David W. Hand, Kerry J. Howe, and George Tchobanoglous.

Terminology for Reverse Osmosis

Term	Definition
Active layer	Layer of membrane that provides the separation capabilities.
Asymmetric structure	Membrane formed of a single material but with multiple layers that are structurally different and have different functions.
Array	Full unit of water production in a reverse osmosis system, which may include multiple stages.
Concentrate	Portion of feed water that has not passed through the membrane. Constituents removed from the permeate are concentrated in the concentrate. Also known as brine.
Concentration polarization	Accumulation of solutes near a membrane surface due to boundary layer effects and the rejection of solutes by the membrane as water passes through the membrane.
Dense membrane	Material that is permeable to certain constituents, such as water, even though it does not have pores.
Limiting salt	Salt that reaches its saturation concentration first as water is concentrated in a reverse osmosis system.
Membrane element	Smallest distinct unit of production capacity in a reverse osmosis system; several membrane elements are arranged in series in a pressure vessel.
Nanofiltration membrane	Reverse osmosis membrane product engineered for selective removal of divalent ions or natural organic matter while allowing passage of smaller monovalent ions.
Osmosis	Flow of solvent through a semipermeable membrane from a dilute solution into a concentrated one.

Term	Definition
Osmotic pressure	Pressure required to balance the difference in chemical potential between two solutions separated by a semipermeable membrane.
Permeate	Portion of feed water that has passed through the membrane. Solutes have been largely removed from this stream so that it is usable for potable purposes. Also known as product water.
Reverse osmosis	Physicochemical separation process in which water flows through a semipermeable membrane due to the application of an external pressure in excess of the osmotic pressure.
Semipermeable membrane	Material that is permeable to some components in a solution but not others; e.g., a material permeable to water but not permeable to salts.
Spiral wound element	Most common type of reverse osmosis membrane element, in which envelopes of membrane material are wrapped around a permeate tube and treated water flows spirally through the envelope to the tube.
Stage	Group of pressure vessels operated in parallel as a single component of water production.
Thin-film composite	Reverse osmosis membranes composed of two or more materials cast on top of one another, where one material is the active layer and other materials form the support layers.

Reverse osmosis (RO) is a membrane treatment process used to separate dissolved solutes from water. It includes any pressure-driven membrane that uses preferential diffusion for separation. A typical RO membrane is made of synthetic *semipermeable* material, which is defined as a material that is permeable, to some components in the feed stream and impermeable to other components and has an overall thickness of less than 1 mm. Water is pumped at high pressure across the surface of the membrane, causing a portion of the water to pass through the membrane, as shown schematically on Fig. 17-1. Water passing through the membrane, called *permeate*, is relatively free of targeted dissolved solutes, while the remaining water, called *concentrate* (also commonly called retentate, reject water, or brine), exits at the far end of the pressure vessel. The delineation of membrane processes, applications for RO, a historical perspective, a process description, process fundamentals, and process design are presented in this chapter.

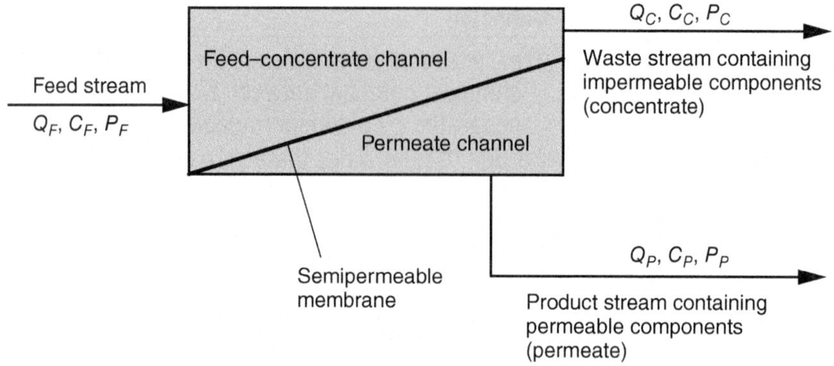

Figure 17-1
Schematic of separation process through reverse osmosis membrane.

17-1 Classification of Membrane Processes

Membrane processes were introduced in Chap. 12, where it was noted that the membranes used in municipal water treatment include microfiltration (MF), ultrafiltration (UF), nanofiltration (NF), and reverse osmosis (RO) membranes. From a physicochemical perspective, these four types of membranes are used in two distinct processes in water treatment (1) membrane filtration and (2) reverse osmosis. They are differentiated by the types of materials rejected, characteristic pore dimensions, and operating pressures. Membrane filtration is used primarily for the removal of particulate matter, whereas RO accomplishes a variety of treatment objectives involving the separation of dissolved solutes from water.

Membrane filtration is covered in Chap. 12, in which a hierarchy of membranes used in water treatment is described (Fig. 12-2), and additional details are provided on the delineation between membrane filtration and RO (Sec. 12-1) including a table of significant differences between these processes (Table 12-1). Common membrane nomenclature is included in Chap. 12 as well as here.

Nanofiltration membranes were designed by FilmTec Corporation around 1983 to remove divalent anions from seawater for applications in the oil industry. The word *nanofiltration* was coined because the separation cutoff size was about 1 nm, and the membranes were designed for removal of specific ionic species, whereas other RO membranes of that era were indiscriminate with respect to the ionic species removed. The ability of NF membranes to simultaneously remove divalent cations (hardness) and natural organic matter while achieving only low monovalent ion removal made them ideal for certain water treatment applications. While NF membranes were a unique product in the 1980s, membrane manufacturers have since engineered a variety of RO membranes with different formulations, permeation capabilities, and rejection characteristics. These products provide a full range of different capabilities, and some new RO membranes have characteristics similar to the original NF membranes. A variety of names have been applied to these new products, including "loose" RO membranes, softening membranes, and low-pressure RO membranes. Manufacturers will continue to develop new RO membranes to achieve specific goals, and NF membranes are just one in a succession of many innovative developments in the field of RO.

17-2 Applications for Reverse Osmosis

Uses for RO in water treatment as well as alternative processes are listed in Table 17-1. These objectives encompass the desalination of ocean or brackish water, advanced treatment for water reuse, softening, natural organic matter (NOM) removal for controlling disinfection by-product (DBP) formation, and specific contaminant removal.

Desalination of Ocean Water or Seawater

The scarcity of freshwater sources may mean a strong future for the use of RO for desalination of ocean water or seawater. About 97.5 percent of the earth's water is in the oceans, and about 75 percent of the world's population live in coastal areas (Bindra and Abosh, 2001). The salinity of the ocean ranges from about 34,000 to 38,000 mg/L as total dissolved solids (TDS) (Stumm and Morgan 1996), nearly two orders of magnitude higher than that of potable water [the World Health Organization's

Table 17-1
Reverse osmosis objectives and alternative processes

Process Objective	Membrane Process Name	Alternative Processes
Ocean or seawater desalination	High-pressure RO, seawater RO	Multistage flash (MSF) distillation, multieffect distillation (MED), vapor compression distillation (VCD)
Brackish water Desalination	RO, low-pressure RO, NF	Multistage flash distillation,[a] multieffect distillation,[a] vapor compression,[a] electrodialysis, electrodialysis reversal
Softening	Membrane softening, NF	Lime softening, ion exchange
NOM removal for DBP control	NF	Enhanced coagulation/softening, GAC
Specific contaminant removal[b]	RO	Ion exchange, activated alumina, coagulation, lime softening, electrodialysis, electrodialysis reversal
Water reuse	RO	Advanced oxidation
High-purity process water	RO	Ion exchange, distillation

[a]MSF, MED, and VCD are rarely competitive economically for brackish water desalination.
[b]Applicability of alternative processes depends on the specific contaminants to be removed and their concentration.

Table 17-2
Typical concentration of important solutes in seawater

Salt	Concentration, mg/L
Cations	
Sodium, Na^+	10,800
Magnesium, Mg^{2+}	1,290
Calcium, Ca^{2+}	412
Potassium, K^+	399
Strontium, Sr^{2+}	7.9
Barium, Ba^{2+}	0.02
Anions	
Chloride, Cl^-	19,400
Sulfate, SO_4^{2-}	2,700
Total carbonate, CO_3^{2-}	142
Bromide, Br^-	67
Fluoride, F^-	1.3
Phosphate, HPO_4^{2-}	0.5
Total	35,200

Source: Stumm and Morgan (1996).

(WHO's guidance level for TDS is 1000 mg/L and the United States has a secondary standard of 500 mg/L)]. The concentration of important ions in seawater is shown in Table 17-2. Seawater also contains several important neutral species, including 3 mg/L of silicon (present as H_4SiO_4) and 4.6 mg/L of boron (present as H_3BO_3). Boron is a concern because neutral species are poorly removed by conventional RO membranes, as will be presented later, and California has a notification limit of 1 mg/L for boron in drinking water.

Desalination costs are dropping, and the process is becoming more competitive with other treatment options in areas where freshwater is scarce, although desalination of ocean water is an energy-intensive process.

The Middle East is currently the most prominent market for desalination of ocean water. Virtually 100 percent of the drinking water in Kuwait and Qatar and 40 to 60 percent of the drinking water in Bahrain, Saudi Arabia, and Malta is produced by desalination (Bremere et al., 2001). Thermal processes such as multistage flash (MSF) distillation and multi-effect distillation (MED) are common in the Middle East, which has vast energy resources but little freshwater. Worldwide, 43 percent of desalination is done with thermal processes and 56 percent is done with membrane processes (NRC, 2008).

Interest in the oceans as a source water is growing in other areas, including coastal areas of the United States. Tampa, Florida, commissioned a 95,000-m^3/d (25-mgd) RO plant in 2003, and a number of communities in California are considering the Pacific Ocean as a source of municipal water.

Desalination of Brackish Groundwater

Interest in desalination of brackish groundwater has increased in areas short on freshwater, such as the southwest region of the United States. Communities in that area are rapidly growing beyond the availability of local freshwater supplies. Brackish groundwater with low to moderate salinity (1000 to 5000 mg/L TDS) are relatively common, and use of these resources has become reasonable as desalination costs have dropped and costs to obtain additional freshwater resources has increased. The difference in feed water quality between brackish water and seawater can lead to differences in design and operation, including differences in pretreatment, feed pressure, configuration of stages, water recovery, fouling prevention, and waste disposal (Greenlee et al., 2009). Since energy consumption is directly related to feed water TDS, brackish water desalination is not nearly as energy intensive as seawater desalination. However, disposal of the concentrate is a significant challenge.

Water Reuse

Along with brackish groundwater as an alternative source of water, many communities in water-scarce areas are considering the increased use of recycled treated wastewater. Water reuse for nonpotable uses such as irrigation of municipal greenscapes (parks, golf courses, road medians, etc.) and industrial process water is practiced in some areas, but treating wastewater to sufficient quality for potable reuse would increase flexibility for using the resource and eliminate the need for community dual-pipe systems. A concern in potable reuse applications, however, is the presence of pharmaceuticals, personal care products, endocrine disrupting compounds, unregulated contaminants, and other contaminants of emerging concern. RO's ability to remove virtually all contaminants in water, including many synthetic organic chemicals, has increased the interest in incorporating RO into wastewater treatment process trains as an advanced treatment process.

Softening and NOM Removal

Nanofiltration or softening membranes are capable of removing 80 to 95 percent of divalent ions such as calcium and magnesium with low removal of low-molecular-weight (MW) monovalent ions such as sodium and chloride. By allowing passage of sodium and chloride, the osmotic pressure differential is minimized. Nanofiltration membranes can soften water without the voluminous sludge production of lime softening, although concentrate disposal can be a significant regulatory obstacle. Nanofiltration membranes that effectively remove hardness are also effective at removing NOM, making them an excellent treatment option for color removal and DBP formation control because removing NOM and color from water before disinfection with free chlorine typically reduces the formation of DBPs. Nanofiltration membranes have widespread use in Florida, where the groundwater is either brackish or very hard, highly colored freshwater.

Specific Contaminant Removal

An additional use for RO is specific contaminant removal. The EPA has designated RO as a best available technology (BAT) for removal of numerous inorganic contaminants, including antimony, arsenic, barium, fluoride, nitrate, nitrite, and selenium, and radionuclides, including beta-particle and photon emitters, alpha emitters, and radium-226. Reverse osmosis has also been demonstrated to be effective for removing larger MW synthetic organics such as pesticides (Baier et al., 1987). Use of RO for specific contaminants, however, is less common because alternative technologies are frequently more cost effective and the disposal of the concentrate stream may present challenges.

17-3 History of Reverse Osmosis in Water Treatment

The process of osmosis through semipermeable membranes was first observed in 1748 by Jean Antoine Nollet (Laidler and Meiser, 1999). The feasibility of desalinating seawater with semipermeable membranes was first seriously investigated in 1949 at the University of California at Los Angeles (UCLA) and in about 1955 at the University of Florida, with funding provided by the newly formed U.S. Department of Interior Office of Saline Water (Glater, 1998). Researchers at both UCLA and the University of Florida successfully produced freshwater from seawater in the mid-1950s, but the flux was too low to be commercially viable. Research focused on reducing the membrane thickness, and in 1959, Loeb and Sourirajan of UCLA succeeded in producing the first asymmetric RO membrane (Lonsdale, 1982). *Asymmetric* membranes are formed from a single material that develops into active and support layers during the casting process (in other words, the membranes are chemically homogeneous but physically heterogeneous). Due to the thinness of the *active layer*, which provides separation capabilities, the asymmetric membrane was a major breakthrough. That advancement, along with the development of the spiral wound element to increase packing density and thin-film composite membranes, led to the commercial viability of membrane desalination.

In June of 1965, the first commercial membrane desalination plant began providing potable water to the City of Coalinga, California. The plant, with combined experimental and production capabilities, produced 19 m^3/d (0.005 mgd) of potable water from 2500 mg/L TDS feed water by operating at 41 bar (600 psi) pressure, 34 $L/m^2 \cdot h$ (20 $gal/ft^2 \cdot d$) flux, and 50 percent recovery (Stevens and Loeb, 1967). Other plants soon followed. The construction of Water Factory 21 in California helped the industry standardize on specific configurations, such as the 8-in. spiral wound element. In the mid-1970s, RO applications were extended from desalting to the softening applications mentioned earlier. The first membrane softening plant was built in Pelican Bay, Florida, in 1977 (AWWA, 2007). The use of membranes to remove NOM paralleled the development of membrane softening (Taylor et al., 1987) because many groundwater supplies in Florida are both hard and colored, and NOM and hardness can be removed simultaneously by membranes.

By the end of 2008, the total installed capacity of desalination plants was 42×10^6 m^3/d (11 billion gallons per day) worldwide. Over 1100 RO plants are operating in the United States with a total capacity of around 5.7×10^6 m^3/day (1500 mgd) (NRC, 2008), which represents about 3 percent of water withdrawn by public water systems. Reverse osmosis plants have been built in every state in the United States.

The future of RO is promising. Growth in the world population, the urbanization of coastal and arid areas, the scarcity of freshwater supplies, the increasing contamination of freshwater supplies, greater reliance on oceans and poorer quality supplies (brackish groundwater, treated wastewater), and improvements in membrane technology suggest continued rapid growth of reverse osmosis installations. The installation of desalination facilities is expected to double between 2005 and 2015 (Wang et al., 2010).

17-4 Reverse Osmosis Process Description

Reverse osmosis relies on differences between the physical and chemical properties of the solutes and water to achieve separation. A high-pressure feed stream is directed across the surface of a semipermeable material, and due to a pressure differential between the feed and permeate sides of the membrane, a portion of the feed stream passes through the membrane. As water passes through the membrane, solutes are rejected and the feed stream becomes more concentrated. The permeate stream exits at nearly atmospheric pressure, while the concentrate remains at nearly the feed pressure. Reverse osmosis is a continuous separation process; that is, there is no periodic backwash cycle.

Reverse Osmosis is used in Florida to reduce TDS in brackish source water.
http://www.stantec.com/wt3r/17-1.html

A typical RO facility is shown on Fig. 17-2. The smallest unit of production capacity in a membrane plant is called a *membrane element.* The membrane elements are enclosed in pressure vessels mounted on skids, which have piping connections for feed, permeate, and concentrate streams. A group of pressure vessels operated in parallel is called a *stage.* The concentrate from one stage can be fed to a subsequent stage to increase water recovery (a multistage system, sometimes called a brine-staged system) or the permeate from one stage can be fed to a second stage to increase solute removal (a two-pass system, also sometimes called a permeate-staged system). In multistaged systems, the number of pressure vessels decreases in each succeeding stage to maintain sufficient velocity in the feed channel

Figure 17-2
Typical reverse osmosis facility.

as permeate is extracted from the feed water stream. A unit of production capacity, which may contain one or more stages, is called an *array*. Schematics of various arrays are shown on Fig. 17-3. The ratio of permeate to feed water flow (recovery) ranges from about 50 percent for seawater RO systems to about 90 percent for low-pressure RO systems. Several factors limit recovery, most notably osmotic pressure, concentration polarization, and the solubility of sparingly soluble salts.

Pretreatment and Posttreatment

A schematic of an RO system with typical pretreatment and posttreatment processes is shown on Fig. 17-4 and described below.

PRETREATMENT

Feed water pretreatment is required in virtually all RO systems. When sparingly soluble salts are present, one purpose of pretreatment is to prevent

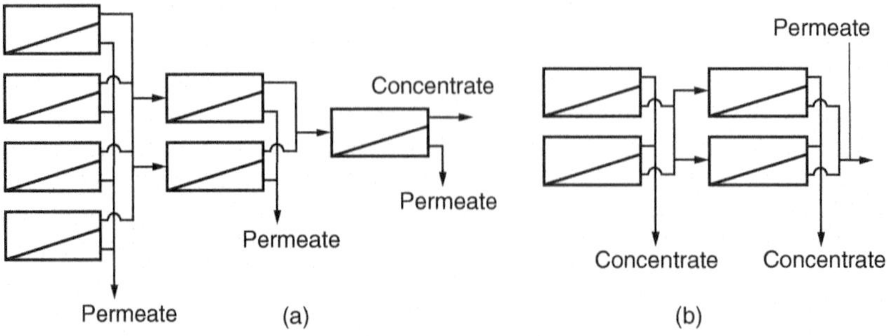

Figure 17-3
Array configurations of reverse osmosis facilities: (a) 4 × 2 × 1 concentrate-staged array, (b) two-pass system.

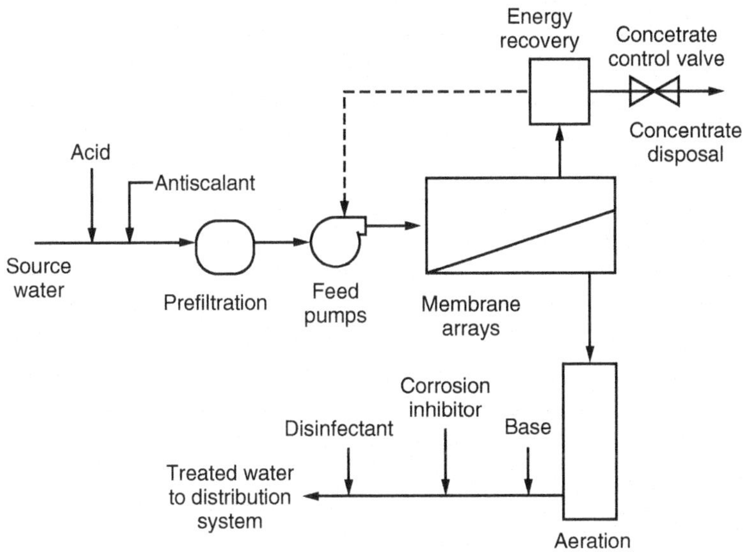

Figure 17-4
Schematic of typical reverse osmosis facility.

scaling. Solutes are concentrated as water is removed from the feed stream, and the resulting concentration can be higher than the solubility product of various salts. Without pretreatment, these salts can precipitate onto the membrane surface and irreversibly damage the membrane. Scale control consists of pH adjustment and/or antiscalant addition. Adjusting the pH changes the solubility of precipitates and antiscalants interfere with crystal formation or slow the rate of precipitate formation.

The second pretreatment process is filtration to remove particles. Without a backwash cycle, particles can clog the feed channels or accumulate on the membrane surface unless the concentration is low. As a minimum, cartridge filtration with a 5-μm strainer opening is used, although granular filtration or membrane filtration pretreatment is often necessary for surface water sources. Disinfection is another typical pretreatment step and is used to prevent biofouling. Some membrane materials are incompatible with disinfectants, so the disinfectant can only be applied in specific situations and must be matched to the specific membrane type.

After pretreatment, the feed water is pressurized with feed pumps. The feed water pressure ranges from 5 to 10 bar (73 to 145 psi) for NF membranes, from 10 to 30 bar (145 to 430 psi) for low-pressure and brackish water RO, and from 55 to 85 bar (800 to 1200 psi) for seawater RO.

POSTTREATMENT

Permeate typically requires posttreatment, which consists of removal of dissolved gases and alkalinity and pH adjustment. Membranes do not efficiently remove small, uncharged molecules, in particular dissolved gases.

If hydrogen sulfide is present in the source groundwater, it must be stripped prior to distribution to consumers. If sulfides are removed in the stripping process, provisions must be made to scrub the sulfides from the stripping tower off-gas to prevent odor and corrosion problems. The stripping of carbon dioxide raises pH and reduces the amount of base needed to stabilize the water. Permeate is typically low in hardness and alkalinity and frequently has been adjusted to an acidic pH value to control scaling. Consequently, the permeate is corrosive to downstream equipment and piping. Alkalinity and pH adjustments are accomplished with various bases, and corrosion inhibitors are used to control corrosion.

Concentrate Stream

The concentrate stream is under high pressure when it exits the final membrane element. This pressure is dissipated through the concentrate control valve, which can be a significant waste of energy. Seawater RO systems utilize energy recovery equipment on the concentrate line, and some brackish water RO systems are starting to use energy recovery as well. Unlike crossflow membrane filtration, the concentrate stream is not recycled to the head of the plant but is a waste stream that must be discarded. Concentrate disposal can be a significant issue in the design of RO facilities and the concentrate may require treatment before disposal. Methods for concentrate disposal are discussed in Chap. 21 and include ocean, brackish river, or estuary discharge; discharge to a municipal sewer; and deep-well injection. Other concentrate disposal options, including evaporation ponds, infiltration basins, and irrigation, are used by a small number of facilities.

Membrane Element Configuration

Reverse osmosis membrane elements are fabricated in either a spiral-wound configuration or a hollow-fine-fiber (HFF) configuration.

SPIRAL-WOUND MODULES

Spiral-wound modules are constructed of several elements in series. The basic construction of a spiral-wound element is shown on Fig. 17-5, and a photograph of typical elements is shown on Fig. 17-6. An envelope is formed by sealing two sheets of flat-sheet membrane material along three sides, with the active membrane layer facing out. A permeate carrier spacer material inside the envelope prevents the inside surfaces from touching each other and provides a flow path for the permeate inside the envelope. The open ends of several envelopes are attached to a perforated central tube known as a permeate collection tube. Feed-side mesh spacers are placed between the envelopes to provide a flow path and create turbulence in the feed water. By rolling the membrane envelopes around the permeate collection tube, the exterior spacer forms a spirally shaped feed channel. This channel, exposed to element feed water at one end and concentrate at the other end, is known as the feed–concentrate channel. Membrane feed water passes through this channel and is exposed to the membrane surface.

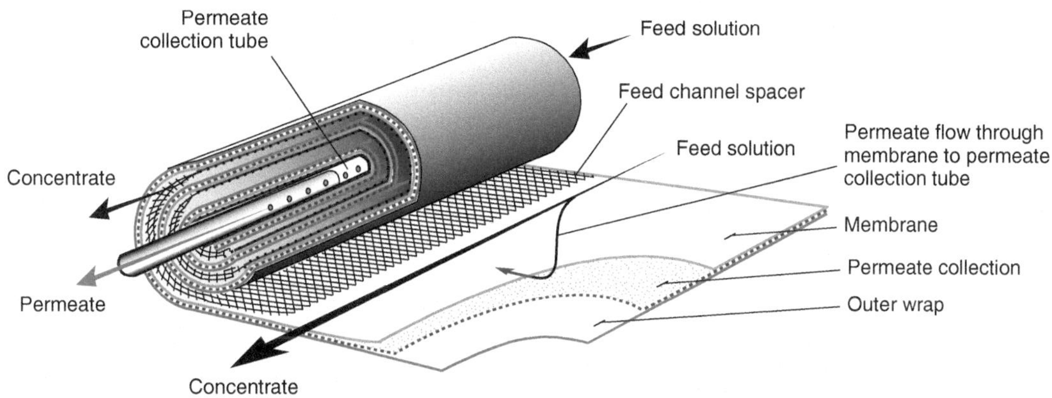

Figure 17-5
Construction of spiral-wound membrane element.

Figure 17-6
Photograph of spiral-wound membrane elements. (Courtesy GE Infrastructure Water Technologies.)

Spiral-wound elements are typically 1 m (40 in.) to 1.5 m (60 in.) long and 0.1 m (4 in.) to 0.46 m (18 in.) in diameter, although 0.2 m (8 in.) diameter elements are most common. Four to seven elements are arranged in series in a pressure vessel, with the permeate collection tubes of the spiral-wound elements coupled together.

During operation, pressurized feed water enters one side of the pressure vessel and encounters the first membrane element. As the water flows tangentially across the membrane surface, a portion of the water passes through the membrane surface and into the membrane envelope and flows spirally toward the permeate collection tube. The remaining feed water, now concentrated, flows to the next element in series, and the process is repeated until the concentrate exits the pressure vessel. Individual spiral-wound membrane elements have a permeate recovery of 5 to 15 percent per element. Head loss develops as feed water flows through the feed channels and spacers, which reduces the driving force for flow through the membrane surface. This feed-side head loss across a membrane element is low, typically less than 0.5 bar (7psi) per element.

HOLLOW-FINE-FIBER MODULES

The HFF configuration is similar to the hollow fibers used in membrane filtration. Feed water passes over the outside of the fiber and is forced through the wall of the fiber, and the permeate is collected in the lumen (or inner annulus) of the fiber. The original manufacturer of HFF membranes was DuPont, which manufactured fiber with an outside diameter (OD) of 0.085 mm (about the thickness of human hair) and inside diameter of 0.042 mm, considerably thinner than the hollow fibers used in membrane filtration, which have an OD of 1 to 2 mm (about the thickness of pencil lead) (Lonsdale, 1982). The active surface of the membrane is on the outside surface of the fiber and is 0.1 to 1 μm thick. DuPont HFFs are still in widespread use but are no longer commercially available. The only current manufacturer of hollow-fiber RO membranes is Toyobo in Japan. In a typical HFF module, the feed enters one end of the module and the concentrated brine exits from the opposite end. The fibers are folded and suspended lengthwise in the module, with the open ends of a set of fibers exposed at each end of the module. The fiber bundles are wound helically around a center tube. A single module can contain several hundred thousand fibers and have surface area up to 10 times that of spiral wound elements. Product water recovery per element is 30 percent.

17-5 Reverse Osmosis Fundamentals

The fundamentals of RO include the membrane material properties, the phenomenon of osmotic pressure, the mechanisms for water and solute permeation, the equations used to predict water and solute flux, and the phenomenon of concentration polarization. These topics are addressed in this section.

Membrane Structure, Material Chemistry, and Rejection Capabilities

An understanding of the mechanisms that control RO begins with an understanding of the membrane. Important properties include the physical structure, chemistry, and rejection capabilities of the membranes.

MEMBRANE STRUCTURE

The resistance to flow through a membrane is inversely proportional to thickness. To achieve any appreciable water flux, the active membrane layer must be extremely thin, which in RO and NF membranes ranges from about 0.1 to 2 μm. Material this thin lacks structural integrity, so these membranes are comprised of several layers, with a thin active layer providing separation capabilities and thicker, more porous layers providing structural integrity. Multilayer membranes are fabricated in two ways. As previously mentioned, asymmetric membranes are formed from a single material that develops into active and support layers during the casting process (in other words, the membranes are chemically homogeneous but physically heterogeneous). *Thin-film composite* membranes are composed of two or more materials cast on top of one another. An advantage of thin-film membranes is that separation and structural properties can be optimized independently using appropriate materials for each function. A cross section of an RO membrane is shown on Fig. 17-7.

The active layer of RO membranes must selectively allow water to pass through the material while rejecting dissolved solutes that may have

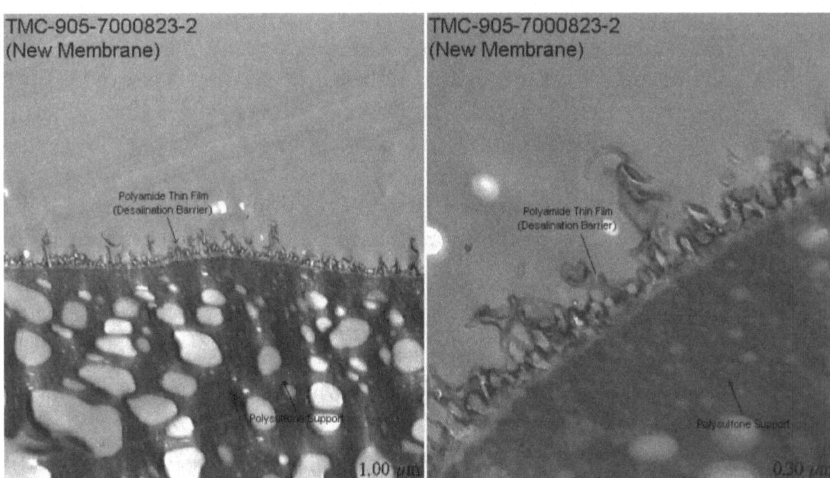

Figure 17-7
Microphotographs of asymmetric reverse osmosis membrane. (TEM images courtesy Bob Riley.)

dimensions similar to water molecules. Separation of small ions cannot be accomplished if they are convectively carried with liquid water. Thus, RO membranes are fabricated of a *dense* material, meaning a permeable but not porous material with no void spaces through which liquid water travels. Water and solutes dissolve into the solid membrane material, diffuse through the solid, and reliquefy on the permeate side of the membrane. The mechanics of permeation through a dense material will be discussed in detail later in this chapter. Low-pressure RO or NF membranes may have void spaces large enough for the convective flow of liquid water through the membrane.

MEMBRANE MATERIAL

Membrane performance is strongly affected by the physical and chemical properties of the material. The ideal membrane material is one that can produce a high flux without clogging or fouling and is physically durable, chemically stable, nonbiodegradable, chemically resistant, and inexpensive. Important characteristics of membrane materials, methods of determination, and effects on membrane performance were discussed in Chap. 12 and shown in Table 12-7. The materials most widely used in RO are cellulosic derivatives and polyamide derivatives.

Cellulose acetate membranes

The original RO membrane developed by Loeb and Sourirajan in 1960 was fabricated of cellulose acetate (CA), and RO membranes using this material are still commercially available. Membranes composed of CA are typically of asymmetric construction. Cellulose acetate is hydrophilic, which helps to maintain high flux values and to minimize fouling. The structural properties of CA are not ideal, however, and the material is not tolerant of temperatures above 30°C, tends to hydrolyze when the pH value is below 3 or above 8, is susceptible to biological degradation, and degrades with free-chlorine concentrations above 1 mg/L, depending on the concentration and duration of contact. In addition, membrane compaction due to the high operating pressure and asymmetric construction causes a reduction of flux over time.

Polyamide membranes

Polyamide (PA) membranes are chemically and physically more stable than CA membranes, generally immune to bacterial degradation, stable over a pH range of 3 to 11, and do not hydrolyze in water. Under similar pressure and temperature conditions, PA membranes can produce higher water flux and higher salt rejection than CA membranes. However, PA membranes are more hydrophobic and susceptible to fouling than CA membranes and are not tolerant of free chlorine in any concentration. Any residual oxidant such as chlorine in the feed will cause rapid deterioration of the membrane. For most applications, dechlorination is required if the feed water is chlorinated and can be done with sodium bisulfite, sulfur dioxide, or activated carbon. Sensors and instrumentation must be provided to monitor the feed water for oxidants that may damage the material and shut down the system if any are detected. Some PA membranes have a rougher surface than CA membranes, which can increase susceptibility to biological and particulate fouling. Polyamide membranes are typically of thin-film construction. The PAs are used for the active layer, and durable materials such as polyethersulfone are used for the support material. The support layer is essentially a standard UF membrane and provides little resistance to flow.

REJECTION CAPABILITIES

The rejection capabilities of RO and NF membranes are designated with either a percent salt rejection or a molecular weight cutoff (MWCO) value. Salt rejection is typically used for RO membranes:

$$\text{Rej} = 1 - \frac{C_P}{C_F} \tag{17-1}$$

where Rej = rejection, dimensionless (expressed as a fraction)
C_P = concentration in permeate, mol/L
C_F = concentration in feed water, mol/L

Rejection can be calculated for bulk parameters such as TDS or conductivity. For membrane rating, however, rejection of specific salts is specified. Sodium chloride rejection is normally specified for high-pressure RO membranes, whereas $MgSO_4$ rejection is often specified for NF or low- pressure RO membranes.

Nanofiltration membranes can also be characterized by MWCO. The MWCO of NF membranes is typically determined by passage of solutes such as sodium chloride and magnesium sulfate. The MWCO of NF membranes is typically 1000 Daltons (Da), also known as atomic mass units (amu), or less.

Osmotic Pressure

Osmosis is the flow of solvent through a semipermeable membrane, from a dilute solution into a concentrated one. Osmosis reduces the flux through an RO membrane by inducing a driving force for flow in the opposite direction.

The physicochemical foundation for osmosis is rooted in the thermodynamics of diffusion, as described in this section.

DIFFUSION AND OSMOSIS

Consider a vessel with a removable partition that is filled with two solutions to exactly the same level, as shown on Fig. 17-8a. The left side is filled with

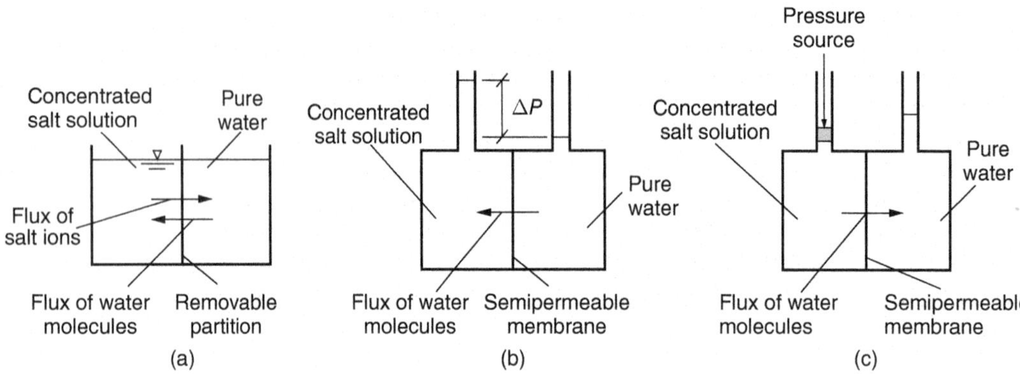

Figure 17-8
Diffusion sketch for reverse osmosis: (a) diffusion, (b) osmosis, and (c) reverse osmosis.

a concentrated salt solution, the right with pure water, and the partition is gently removed without disturbing the solutions. Initially, the contents are in a nonequilibrium state and the salt will eventually diffuse through the water until the concentration is the same throughout the vessel. With salt ions diffusing from left to right across the plane originally occupied by the partition, conservation of mass requires a flux of water molecules in the opposite direction. Without a flux of water molecules from right to left, mass accumulates on the right side of the container, which is unthinkable with a continuous water surface. Equilibrium requires mass transport in both directions.

On Fig. 17-8b, the top of the vessel has been closed and fitted with manometer tubes and the removable partition has been replaced with a semipermeable membrane. The semipermeable membrane allows the flow of water but prevents the flow of salt. Filling the chambers with salt solution and pure water again creates a thermodynamically unstable system, which must be equilibrated by diffusion. Because the membrane prevents the flux of salt, however, mass accumulates in the left chamber, causing the water level in the left manometer to rise and in the right manometer to drop. This flow of water from the pure side to the salt solution is osmosis. Water flux occurs despite the difference in hydrostatic pressure that develops due to the difference in manometer levels.

OSMOTIC PRESSURE

The driving force for diffusion is typically described as a concentration gradient, although a more rigorous thermodynamic explanation is a gradient in Gibbs energy (Laidler and Meiser, 1999). The concept of Gibbs energy (G) and its relationship to concentration were introduced in Chap. 5. When the vessels on Fig. 17-8 were filled with water and salt solutions, the two sides had different values of Gibbs energy due to differences in salt concentration. Equilibrium is defined thermodynamically when $\Delta G = 0$, so the gradient in Gibbs energy across the first vessel caused the simultaneous diffusion of salt ions and water molecules, and the system was driven toward an equilibrium condition in which the Gibbs energy (and concentration and water level) was equal throughout the system. In the second vessel, water stops flowing from right to left when the vessel reaches thermodynamic equilibrium but both pressure and concentration are unequal between the chambers. Although Gibbs energy is constant throughout the second vessel at equilibrium, the Gibbs energy includes components to account for both the pressure and concentration differences.

The discussion of Gibbs energy in Chap. 5 was done under conditions of constant temperature and pressure. To describe osmosis, a more general description of Gibbs energy is needed. The general form of the Gibbs function is

$$\partial G = V\,\partial P - S\,\partial T + \sum_i \overset{\circ}{\mu}_i\,\partial n_i \tag{17-2}$$

where G = Gibbs energy, J
V = volume, m^3
P = pressure, Pa
S = entropy, J/K
T = absolute temperature, K (273 + °C)
$\overset{\circ}{\mu}_i$ = chemical potential of solute i, J/mol
n_i = amount of solute i in solution, mol

Chemical potential is defined as the change in Gibbs energy resulting from a change in the amount of component i when temperature and pressure are held constant:

$$\overset{\circ}{\mu}_i = \left.\frac{\partial G}{\partial n_i}\right|_{P,T} \tag{17-3}$$

Therefore, the last term in Eq. 17-2 ($\overset{\circ}{\mu}_i\ \partial n_i$) describes the difference in Gibbs energy resulting from the difference in the amount of solute between the chambers (when volume is constant, the difference in amount equals the difference in concentration). Under constant-temperature conditions (i.e., $\partial T = 0$), Eq. 17-2 says equilibrium ($\partial G = 0$) will be achieved when the sum of the Gibbs energy gradient due to chemical potential is offset by the pressure gradient between the two chambers:

$$\partial G = 0 \quad \text{when} \quad V\,\partial P = -\sum_i \overset{\circ}{\mu}_i\,\partial n_i \tag{17-4}$$

The pressure required to balance the difference in chemical potential of a solute is called the *osmotic pressure* and is given the symbol π. When the vessel in the second experiment reaches equilibrium, the difference in hydrostatic pressure between the manometers is equal and opposite to the difference in osmotic pressure between the two chambers. An equation for osmotic pressure can be derived thermodynamically using assumptions of incompressible and ideal solution behavior:

$$\pi = \frac{-RT}{V_b} \ln x_W \tag{17-5}$$

where π = osmotic pressure, bar
V_b = molar volume of pure water, L/mol
x_W = mole fraction of water, mol/mol
R = universal gas constant, 0.083145 L · bar/mol · K

In dilute solution (i.e., $x_W \cong 1$), Eq. 17-5 can be approximated by the van't Hoff equation for osmotic pressure (Eq. 17-6), which is identical in form to the ideal gas law ($PV = nRT$):

$$\pi = \frac{n_S}{V} RT \quad \text{or} \quad \pi = CRT \tag{17-6}$$

where n_S = total amount of all solutes in solution, mol
C = concentration of all solutes, mol/L
V = volume of solution, L

Equation 17-6 was derived assuming infinitely dilute solutions, which is often not the case in RO systems. To account for the assumption of diluteness, the nonideal behavior of concentrated solutions, and the compressibility of liquid at high pressure, a nonideality coefficient (osmotic coefficient ϕ must be incorporated into the equation:

$$\pi = \phi CRT \tag{17-7}$$

where ϕ = osmotic coefficient, unitless

It should be noted that the thermodynamic equation for osmotic pressure (Eq. 17-5) contains no terms identifying the solute. Osmotic pressure is strictly a function of the concentration, or mole fraction, of water in the system. Solutes reduce the mole fraction of water, and the effect of multiple solutes is additive because they cumulatively reduce the mole fraction of water. Solutes that dissociate also have an additive effect on the mole fraction of water (e.g., addition of 1 mol of NaCl produces 2 mol of ions in solution, doubling the osmotic pressure compared to a solute that does not dissociate). If multiple solutes are added on an equal-mass basis, the solute with the lowest molecular weight produces the greatest osmotic pressure. The use of Eq. 17-7 is demonstrated in Example 17-1.

Example 17-1 Osmotic pressure calculations

Calculate the osmotic pressure of 1000-mg/L solutions of the following solutes at a temperature of 20°C assuming an osmotic coefficient of 0.95: (1) NaCl, (2) $SrSO_4$, and (3) glucose ($C_6H_{12}O_6$). Note that NaCl and $SrSO_4$ both dissociate into 2 ions when dissolved into water.

Solution

1. Determine the osmotic pressure for NaCl, first by calculating the molar concentration of ions and then using Eq. 17-7:

$$C = \frac{(2\ \text{mol ion/mol NaCl})(1000\ \text{mg/L})}{(10^3\ \text{mg/g})(58.4\ \text{g/mol})} = 0.0342\ \text{mol/L}$$

$$\pi = \phi CRT = (0.95)(0.0342\ \text{mol/L})(0.083145\ \text{L} \cdot \text{bar/K} \cdot \text{mol})(293\ \text{K})$$
$$= 0.79\ \text{bar}$$

2. Determine the osmotic pressure for $SrSO_4$:

$$C = \frac{(2\ \text{mol ion/mol SrSO}_4)\ (1000\ \text{mg/L})}{(10^3\ \text{mg/g})\ (183.6\ \text{g/mol})} = 0.0109\ \text{mol/L}$$

$$\pi = (0.95)(0.0109\ \text{mol/L})\ (0.083145\ \text{L} \cdot \text{bar/K} \cdot \text{mol})\ (293\ \text{K})$$
$$= 0.25\ \text{bar}$$

3. Determine the osmotic pressure for glucose (no dissociation):

$$C = \frac{1000\,\text{mg/L}}{(10^3\,\text{mg/g})\ (180\,\text{g/mol})} = 0.0056\,\text{mol/L}$$

$$\pi = (0.95)(0.000556\,\text{mol/L})(0.0083145\,\text{L} \cdot \text{bar/K} \cdot \text{mol})(293\text{K})$$
$$= 0.13\text{bar}$$

Comment

Each solution contains the same mass of solute. Because NaCl and SrSO4 dissociate into two ions, the molar ion concentration is double the molar concentration of added salt. The NaCl has a higher osmotic pressure because it has a lower molecular weight. Even though $SrSO_4$ and glucose have nearly the same molecular weight, the osmotic pressure of $SrSO_4$ is nearly double that of glucose because it dissociates.

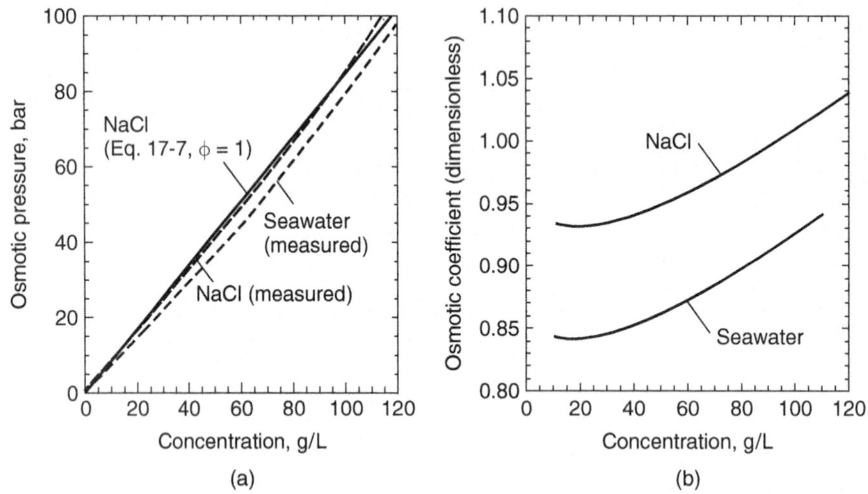

Figure 17-9
(a) Osmotic pressure of aqueous solutions of sodium chloride. (b) Osmotic coefficients for sodium chloride and seawater (osmotic coefficient for seawater with the van't Hoff equation is based on a concentration of NaCl equal to the TDS of the seawater).

The osmotic pressure of a solution of sodium chloride, calculated with Eq. 17-7 and $\phi = 1$, is shown on Fig. 17-9a along with experimentally determined values. Over the range of salt concentrations of interest in seawater desalination, the osmotic coefficient for sodium chloride ranges from 0.93 to 1.03 and is shown as a function of solution concentration on Fig. 17-9b. Osmotic coefficients for other electrolytes are available in Robinson and Stokes (1959). The deviation between measured and calculated values of osmotic pressure can be significantly greater for other solutes and higher concentrations, as shown for sucrose solutions on Fig. 17-10.

Reported values for the osmotic pressure of seawater (Sourirajan, 1970) are about 10 percent below measured values for sodium chloride, as shown on Fig. 17-9a, due to the presence of compounds with a higher molecular weight than sodium chloride. The osmotic pressure for seawater can be calculated with Eq. 17-7 and an equivalent concentration of sodium chloride by using the osmotic coefficient for seawater shown on Fig. 17-9b.

Two opposing forces contribute to the rate of water flow through the semipermeable membrane on Fig. 17-8b: (1) the concentration gradient and (2) the pressure gradient. These opposing forces are exploited in RO. Consider a new experiment using the apparatus on Fig. 17-8, modified so that it is possible to exert an external force on the left side, as shown on Fig. 17-8c. Applying a force equivalent to the osmotic pressure places the system in thermodynamic equilibrium, and no water flows. Applying a force in excess of the osmotic pressure places the system in nonequilibrium, with a

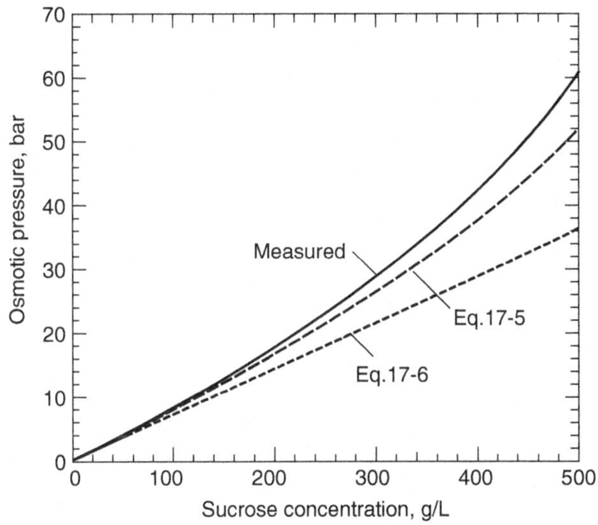

Figure 17-10
Osmotic pressure of aqueous solutions of sucrose.

pressure gradient exceeding the chemical potential gradient. Liquid would flow from left to right, that is, from the concentrated solution to the dilute solution. The process of causing water to flow from a concentrated solution to a dilute solution across a semipermeable membrane by the application of an external pressure in excess of the osmotic pressure is called reverse osmosis.

Models for Water and Solute Transport through RO Membranes

Models have been developed to describe the flux of water and solutes through RO membranes using two basic approaches. The first approach relies on fundamental thermodynamics and does not depend on a physical description of the membrane. The other approach uses physical and chemical descriptions of the membrane and feed solution, such as membrane thickness and porosity. Mathematical development of the models that include descriptions of the membrane and feed solution is beyond the scope of this text but can be found in the published literature (Cheryan and Nichols, 1992; Lonsdale, 1972; Lonsdale et al., 1965; Merten, 1966; Noordman and Wesselingh, 2002; Reid, 1972; Spiegler and Kedem, 1966; Wiesner and Aptel, 1996). For a student learning about RO, the important issue is to develop a conceptual understanding of how water and solutes pass through RO membranes. To promote this understanding, a basic qualitative description of the solution–diffusion, pore flow, and preferential sorption–capillary flow models are presented in the following sections.

SOLUTION–DIFFUSION MODEL

The solution–diffusion model (Lonsdale et al., 1965) describes permeation through a dense membrane where the active layer is permeable but does not have pores. Water and solutes dissolve into the solid membrane material, diffuse through the solid, and reliquefy on the permeate side of the membrane. Dissolution of water and solutes into a solid material occurs if the solid is loose enough to allow individual water and solute molecules to travel along the interstices between polymer molecules of the membrane. Liquids behave as liquids because of attractive interactions with surrounding liquid molecules. Thus, even if water molecules travel along a defined path (which hypothetically could be called a pore), they are surrounded by polymer molecules and not other water molecules and therefore are dissolved in the solid, not present as a liquid phase. Diffusion occurs by movement of the water and solute molecules in the direction of the Gibbs energy gradient. Separation occurs when the flux of the water is different from the flux of the solutes.

Equation 17-7 describes a proportionality between osmotic pressure and concentration. Therefore, the driving force (Gibbs energy gradient) for any component can be written equivalently in terms of either pressure or concentration provided the mass transfer coefficient has the proper units. For water, the driving force is expressed in terms of the net pressure gradient, that is, the applied pressure in excess of the osmotic pressure. Solute transport is expressed in terms of the concentration gradient, and most models neglect the effect of applied pressure on solute transport. Flux through the membrane is determined by both solubility and diffusivity. Components of low solubility have a low driving force, and components of low diffusivity have a low diffusion coefficient. The solution–diffusion model predicts that separation occurs because the solubility, diffusivity, or both of the solutes are much lower than those of water, resulting in a lower solute concentration in the permeate than in the feed.

PORE FLOW MODELS

The solution–diffusion model does not consider convective flow through the membrane. Other models consider RO membranes to have void spaces (pores) through which liquid water travels. The pore flow models consider water and solute fluxes to be coupled, meaning the solutes are convected through the membrane with the water. Thus, rejection occurs through mechanisms similar to those described in Chap. 12 for membrane filtration, meaning the solute molecules are "strained" at the entrance to the pores. Because solute and water molecules are similar in size, the rejection mechanism is not a physical sieving and must consider chemical effects such as electrostatic repulsion between the ions and membrane material.

PREFERENTIAL SORPTION–CAPILLARY FLOW MODEL

A third description of water and salt permeation through membranes is provided by the preferential sorption–capillary flow model, which assumes that the membrane has pores. Separation occurs when one component of the feed solution (either the solute or the water) is preferentially adsorbed to the pore walls and is transported through the membrane by surface diffusion. Membrane materials with a low dielectric constant, such as cellulose acetate, repel ions and preferentially adsorb water, forming a sorbed layer with a reduced concentration of salts. The sorbed layer moves through the membrane by surface diffusion, leaving behind solution components that are repelled from the membrane surface. Separation is a function of the surface chemistry of the membrane and solutes, rather than pore dimensions, although the maximum pore dimension to effect good removal of solutes is two times the thickness of the adsorbed layer, as shown on Fig. 17-11.

COUPLING

Other models consider a combination of permeation mechanisms. The solution–diffusion–imperfection model (Sherwood et al., 1967) assumes that water and solute permeate the membrane by both solution–diffusion and pore flow. The permeation by solution–diffusion is uncoupled but the pore flow is completely coupled. The flux of water by solution–diffusion is proportional to the net applied pressure ($\Delta P - \Delta\pi$), the diffusion of solutes is proportional to the concentration gradient (ΔC), and pore flow is proportional to the applied pressure gradient (ΔP). To achieve high rejection, the pore flow must be a small fraction of the total flow.

In addition to coupling between water and solutes, coupling between solutes must be considered. Electroneutrality must be maintained in both the permeate and the concentrate streams. Thus, preferential transport of ions of one charge can influence the transport of ions of the opposite charge. For instance, negative rejection of hydrogen ions (the concentration of hydrogen ions in the permeate is higher than in the feed solution, manifested as a lower pH in the permeate) is typically observed in RO operations. This occurs because of higher flux of negatively charged ions, such as chloride, than the salt's coion, sodium. Because hydrogen ions are more mobile than sodium ions, the flux of hydrogen ions increases to maintain electroneutrality in the permeate.

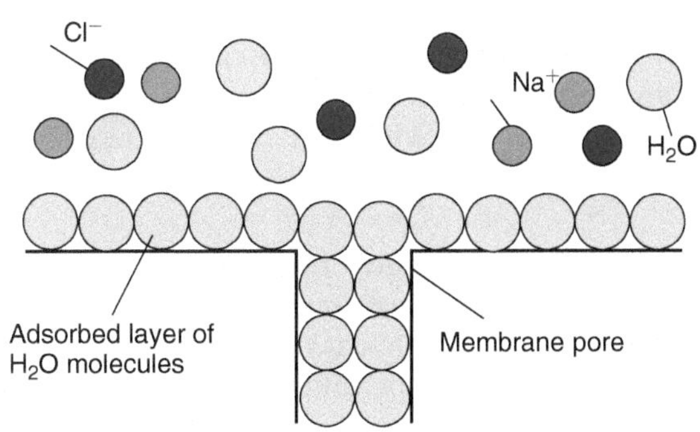

Figure 17-11
Preferential-sorption capillary flow model. Ions are repelled from the membrane surface, resulting in an adsorbed layer of water. The adsorbed water flows through capillary pores in the membrane surface, and the repelled species are left in the feed solution. Good separation can be obtained if the pore diameter is less than 2 times the adsorbed layer thickness.

Mechanisms of Solute Rejection

The membrane permeation models suggest various mechanisms for rejection. The basic mechanisms of rejection are electrostatic repulsion at the membrane surface, solubility and diffusivity through the membrane material due to chemical effects, or straining due to the size and other chemical properties of molecules.

Reverse osmosis and NF membranes are often negatively charged in operation because of the presence of ionized functional groups, such as carboxylates, in the membrane material. Negatively charged ions may be rejected at the membrane surface due to electrostatic repulsion, and positively charged ions may be rejected to maintain electroneutrality in the feed and permeate solutions. The presence of polar and hydrogen-bondable functional groups in the membrane increases the solubility of polar compounds such as water over nonpolar compounds, providing a mechanism for greater flux of water through the membrane. Large molecules would be expected to have lower diffusivity through the membrane material or be unable to pass through the membrane at all.

Experimental observations are consistent with these rejection mechanisms. Small polar molecules such as water generally have the highest flux. Dissolved gases such as H_2S and CO_2, which are small, uncharged, and polar, also permeate RO membranes well. Monovalent ions such as Na^+ and Cl^- permeate better than divalent ions (Ca^{2+} , Mg^{2+}) because the divalent ions have greater electrostatic repulsion. Acids and bases (HCl, NaOH) permeate better than their salts (Na^+,Cl^-) because of decreased electrostatic repulsion.

Silica is present in water as uncharged silicic acid (H_4SiO_4) below the pK_a of 9.84 and is poorly rejected by RO membranes. Similarly, boron is present in water as uncharged boric acid (H_3BO_3) below the pK_a of 9.24 and also permeates well. The poor removal of boron, coupled with a 1 mg/L notification level in California, often requires specific design considerations for seawater RO systems in that state, such as design of two-pass systems. Increasing the pH to above the pK_a values results in good removal for both silica and boron.

Within a homologous series, permeation increases with decreasing molecular weight. High-molecular-weight organic materials do not permeate RO membranes at all. Reverse osmosis membranes are capable of rejecting up to 99 percent of monovalent ions. Nanofiltration membranes reject between 80 and 99 percent of divalent ions while achieving low rejection of monovalent ions.

Equations for Water and Solute Flux

Based on the models presented above, a variety of equations have been developed for the rate of water and solute mass transfer through an RO membrane. Ultimately, these models express flux as the product of a mass transfer coefficient and a driving force. The driving force for water flux through RO membranes is the net pressure differential, or the difference between the applied and osmotic pressure differentials:

$$\Delta P_{\mathrm{NET}} = \Delta P - \Delta\pi = (P_F - P_P) - (\pi_F - \pi_p) \quad (17\text{-}8)$$

where ΔP_{NET} = net transmembrane pressure, bar

Subscripts F and P refer to the feed and permeate, respectively.

The water flux through RO membranes is described by the expression

$$J_W = k_W(\Delta P - \Delta\pi) \quad (17\text{-}9)$$

where J_W = volumetric flux of water, L/m^2. h
k_W = mass transfer coefficient for water flux, L/m^2 · h · bar

Water flux is normally reported as a volumetric flux (L/m^2 · h or gal/ft^2 · d) and the mass transfer coefficient is typically reported with units of L/ m^2 · h. bar or gal/ft^2 · d · atm. Equation 17-9 is valid at any point on the membrane surface between the feed water entrance and concentrate discharge in a membrane element, but it should be noted that both applied and osmotic pressures change continuously along the length of a spiral-wound element due to head loss and the changing solute concentration. As a result, overall flux must be determined by integrating Eq. 17-9 across the length of the membrane element, as will be demonstrated in the design section of this chapter.

The driving force for solute flux is the concentration gradient, and the flux of solutes through RO membranes is expressed as

$$J_S = k_S(\Delta C) \quad (17\text{-}10)$$

where J_S = mass flux of solute, mg/m^2 · h
k_S = mass transfer coefficient for solute flux, L/m^2 · h or m/h
ΔC = concentration gradient across membrane, mg/L

Solute flux is normally reported as a mass flux with units of mg/m^2 · h or lb/ft^2 · d. Values of k_W and k_S are determined experimentally by membrane manufacturers. The solute concentration in the permeate is the ratio of the fluxes of solutes and water, as shown by

$$C_P = \frac{J_S}{J_W} \quad (17\text{-}11)$$

Thus, the lower the flux of solutes or the higher the flux of water,the better removal of solutes is achieved and the permeate will have a lower solute concentration. The ratio of permeate flow to feed water flow, or recovery, is calculated as

$$r = \frac{Q_P}{Q_F} \quad (17\text{-}12)$$

where Q = flow, m^3/s
r = recovery, dimensionless

Using flow and mass balance principles, the solute concentration in the concentrate stream can be calculated from the recovery and solute rejection. The pertinent flow and mass balances using flow and concentration terminology as shown on Fig. 17-1 are

$$\text{Flow balance:} \qquad Q_F = Q_P + Q_C \tag{17-13}$$

$$\text{Mass balance:} \qquad C_F Q_F = C_P Q_P + C_C Q_C \tag{17-14}$$

where C = concentration, mol/L or mg/L

Combining the mass and flow balances with Eq. 17-1 (rejection) and Eq. 17-12 (recovery) yields the following expression for the solute concentration in the concentrate stream:

$$C_C = C_F \left[\frac{1-(1-\text{Rej})r}{1-r} \right] \tag{17-15}$$

where Rej = rejection (dimensionless, expressed as a fraction)

Rejection is frequently close to 100 percent, in which case Eq. 17-15 can be simplified as follows:

$$C_C = C_F \left(\frac{1}{1-r} \right) \tag{17-16}$$

As shown in Eqs. 17-9 and 17-10, water flux depends on the pressure gradient and solute flux depends on the concentration gradient. As feed water solute concentration increases at constant pressure, the water flux decreases (because of higher $\Delta\pi$) and the solute flux increases (because of higher ΔC), which reduces rejection and causes a deterioration of permeate quality. As the feed water pressure increases, water flux increases but the solute flux is essentially constant. Therefore, as the water flux increases, the permeate solute concentration decreases, and the rejection increases. These relationships are illustrated on Fig. 17-12.

Temperature and Pressure Dependence

Membrane performance declines (water flux decreases, solute flux increases) due to fouling and membrane aging. However, fluxes of water and solute also vary because of changes in feed water temperature, pressure, velocity, and concentration. To evaluate the true decline in system performance due to fouling and aging, permeate flow rate and salt passage must be compared at standard conditions. Reverse osmosis design manuals present equations for normalizing RO membrane performance in slightly different ways; the equations presented here are adapted from ASTM (2001e) and AWWA (2007). These procedures incorporate the

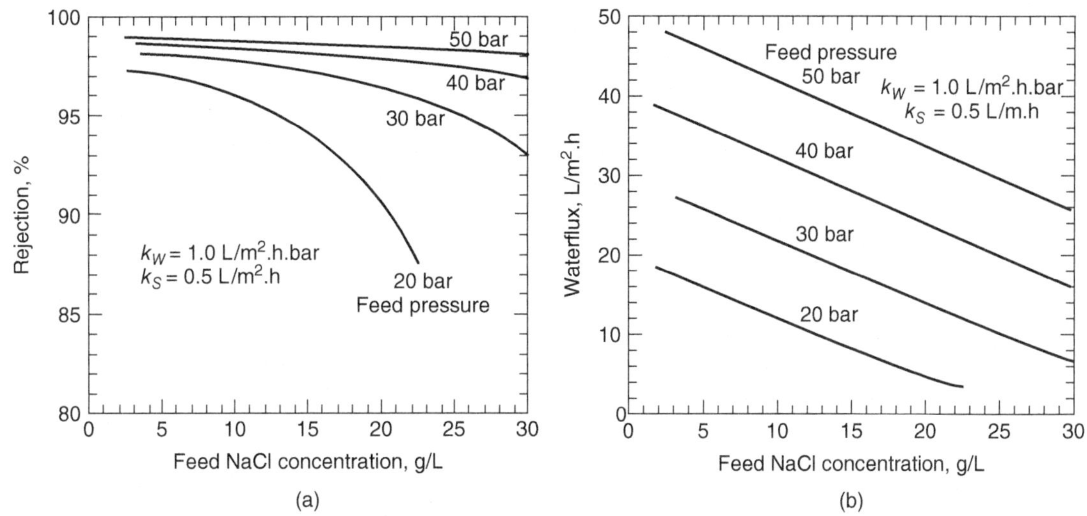

Figure 17-12
Effect of feed water concentration and pressure on (a) percent solute rejection and (b) water flux.

use of temperature and pressure correction factors, evaluated at standard (subscript S) and measured (subscript M) conditions:

$$J_{W,S} = J_{W,M}(\text{TCF})\frac{\text{NDP}_S}{\text{NDP}_M} \tag{17-17}$$

or

$$Q_{P,S} = Q_{P,M}(\text{TCF})\frac{\text{NDP}_S}{\text{NDP}_M} \tag{17-18}$$

where TCF = temperature correction factor (defined below), dimensionless
NDP = net driving pressure (defined below), bar

Temperature affects the fluid viscosity and the membrane material. The relationship between membrane material, temperature, and flux is specific to individual products, so TCF values should normally be obtained from membrane manufacturers, who determine values experimentally. If manufacturer TCF values are unavailable, the relationship between flux and fluid viscosity can be approximated by the following expression, which may be appropriate for membranes containing pores:

$$\text{TCF} = (1.03)^{T_S - T_M} \tag{17-19}$$

where T = temperature, °C

The standard temperature is typically taken to be 25°C for reverse osmosis operation.

The net driving pressure accounts for changes in feed and permeate pressures, feed channel head loss, and osmotic pressure. In spiral-wound elements, the applied pressure decreases and osmotic pressure increases continuously along the length of the feed–concentrate channel as permeate flows through the membrane. Thus, the net driving pressure must take average conditions into account, as shown in

$$\text{NDP} = \Delta P - \Delta\pi = (P_{\text{FC,ave}} - P_p) - (\pi_{\text{FC,ave}} - \pi_P) \tag{17-20}$$

where $P_{\text{FC,ave}}$ = average pressure in the feed–concentrate channel, bar
$= \frac{1}{2}(P_F + P_C)$
P_P = permeate pressure, bar
$\pi_{\text{FC,ave}}$ = average feed–concentrate osmotic pressure (see below), bar
π_p = permeate osmotic pressure, bar

Feed, concentrate, and permeate pressures are easily measured using system instrumentation. Osmotic pressure must be calculated from solute concentration using Eq. 17-7. Although osmotic pressure increases continuously along the length of a spiral-wound element, solute concentration normally can only be measured in the feed and concentrate streams. Manufacturers use various procedures for determining the average concentration in the feed–concentrate channel and must be contacted for procedures for calculating the average concentration in the feed–concentrate channel. The two most common approaches for determining the average concentration in the feed channel are (1) an arithmetic average (Eq. 17-21) and (2) the log mean average (Eq. 17-22):

$$C_{\text{FC,ave}} = \frac{1}{2}(C_F + C_C) \tag{17-21}$$

$$C_{\text{FC, ave}} = \frac{C_F}{r}\ln\left(\frac{1}{1-r}\right) \tag{17-22}$$

Because head loss is a function of feed flow and osmotic pressure is a function of solute concentration, the system design must establish standard conditions for these parameters in addition to applied pressure.

Solute flux across the membrane is affected by temperature and solute concentration, so it is standardized by multiplying the measured flux by the TCF and ratio of concentration at standard and measured conditions, as follows:

$$J_{S,S} = J_{S,M}(\text{TCF})\frac{C_{\text{FC,S}}}{C_{\text{FC},M}} \tag{17-23}$$

Membrane performance, however, is usually evaluated in terms of salt passage rather than solute flux. Salt passage is defined as the ratio of permeate concentration to feed concentration:

$$\mathrm{SP} = \frac{C_P}{C_F} = 1 - \mathrm{Rej} \tag{17-24}$$

where SP = salt passage

By rearranging and substituting Eqs. 17-11, 17-17, and 17-24 into Eq. 17-23, standard membrane performance in terms of salt passage is obtained (ASTM, 2001e) as follows:

$$\mathrm{SP}_S = \mathrm{SP}_M \left(\frac{\mathrm{NDP}_M}{\mathrm{NDP}_S}\right)\left(\frac{C_{\mathrm{FC},S}}{C_{\mathrm{FC},M}}\right)\left(\frac{C_{F,M}}{C_{F,S}}\right) \tag{17-25}$$

Rearranging Eq. 17-25 in terms of rejection yields the expression

$$\mathrm{Rej}_S = 1 - (1 - \mathrm{Rej}_M)\left(\frac{\mathrm{NDP}_M}{\mathrm{NDP}_S}\right)\left(\frac{C_{F,M}}{C_{F,S}}\right)\left(\frac{C_{\mathrm{FC},S}}{C_{\mathrm{FC},M}}\right) \tag{17-26}$$

In multistage systems, it is necessary to standardize the water flux and recovery for each stage independently. The procedures for standardizing RO performance data are shown in Example 17-2.

Example 17-2 Standardization of RO operating data

An RO system uses a shallow brackish groundwater that averages around 4500 mg/L TDS composed primarily of sodium chloride. Permeate flow is maintained constant, but temperature, pressure, and feed concentration change over time as shown in the table below. The operators need to determine whether fouling has occurred between January and May.

Parameter	Unit	January 1	May 31
Permeate flow	m^3/d	7500	7500
Feed pressure	bar	34.5	32.1
Concentrate pressure	bar	31.4	29.1
Permeate pressure	bar	0.25	0.25
Feed TDS concentration	mg/L	4612	4735
Permeate TDS concentration	mg/L	212	230
Recovery	%	0.69	0.72
Water temperature	°C	11	18

The pressure vessels contain seven membrane elements. The manufacturer has stated that performance data for this membrane element were developed using the following standard conditions:

Parameter	Unit	Standard
Temperature	°C	25
Feed pressure	bar	30
Permeate pressure	bar	0
Head loss per element	bar	0.4
Feed TDS concentration	mg/L	2000
Permeate TDS concentration	mg/L	100
Recovery	%	80

Determine the change in system performance (permeate flow and salt passage) that has occurred between January 1 and May 31. Assume $\phi = 1.0$.

Solution

1. Calculate the TCF for the January operating condition:

$$\text{TCF}_{\text{Jan}} = (1.03)^{T_S - T_M} = (1.03)^{25-11} = 1.512$$

2. Calculate the NDP for the January operating condition.
   a. Calculate the average molar solute concentration in the feed–concentrate channel using Eq. 17-22:

$$C_{\text{CF,Jan}} = \frac{C_F}{r}\ln\left(\frac{1}{1-r}\right) = \frac{4612 \text{ mg/L}}{0.69}\ln\left(\frac{1}{1-0.69}\right)$$
$$= 7828 \text{ mg/L}$$

$$C_{\text{CF,Jan}} = \frac{(7828 \text{ mg/L})(2 \text{ mol ions/mol NaCl})}{(10^3 \text{mg/g})(58.4 \text{ g/mol})}$$
$$= 0.268 \text{ mol/L}$$

   b. Calculate the osmotic pressure in the feed–concentrate channel using Eq. 17-7:

$$\pi_{\text{CF,Jan}} = \phi CRT$$
$$= (0.268 \text{ mol/L})(0.083145 \text{ L}\cdot\text{bar/K}\cdot\text{mol})(284 \text{ K})$$
$$= 6.33 \text{ bar}$$

   c. Calculate the molar concentration and osmotic pressure in the permeate:

$$C_{P,\text{ Jan}} = \frac{(212 \text{ mg/L})(2 \text{ mol ions/mol NaCl})}{(10^3 \text{ mg/g})(58.4 \text{ g/mol})} = 0.0073 \text{ mol/L}$$

$$\pi_{P,\text{Jan}} = (0.0073 \text{ mol/L})(0.083145 \text{ L}\cdot\text{bar/K}\cdot\text{mol})(284 \text{ K})$$
$$= 0.17 \text{ bar}$$

d. Calculate the NDP for the January operating condition using Eq. 17-20:

$$P_{FC,\,ave} = \frac{1}{2}(P_F + P_C) = \frac{1}{2}(34.5 + 31.4) = 32.95 \text{ bar}$$

$$\text{NDP} = (32.95 \text{ bar} - 0.25 \text{ bar}) - (6.33 \text{ bar} - 0.17 \text{ bar})$$
$$= 26.5 \text{ bar}$$

3. Repeat the calculations in steps 1 and 2 for the standard condition and the May operating condition. The concentrate pressure is not given for the standard operating condition, but can be calculated from the given head loss information:

$$h_L = (0.4 \text{ bar/element})(7 \text{ elements}) = 2.8 \text{ bar}$$

$$P_C = 30 \text{ bar} - 2.8 \text{ bar} = 27.2 \text{ bar}$$

The remaining calculations are summarized in the table below:

Parameter	Unit	Standard Conditions	January 4 Conditions	May 23 Conditions
TCF		1.0	1.51	1.23
$C_{CF,ave}$	mg/L	4024	7828	8372
π_{CF}	bar	3.36	6.33	6.94
π_p	bar	0.08	0.17	0.19
$P_{CF,ave}$	bar	28.6	32.95	30.6
NDP	bar	25.3	26.5	23.6

4. Calculate the standard permeate flow for each date using Eq. 17-17:

$$Q_{W,S(Jan)} = 7500 \text{ m}^3/\text{d}\ (1.51)\left(\frac{25.3 \text{ bar}}{26.5 \text{ bar}}\right)$$
$$= 10{,}800 \text{ m}^3/\text{d}$$

$$Q_{W,S(May)} = 7500 \text{ m}^3/\text{d}\ (1.23)\left(\frac{25.3 \text{ bar}}{23.6 \text{ bar}}\right)$$
$$= 9900 \text{ m}^3/\text{d}$$

5. Calculate the actual salt passage for each date using Eq. 17-24:

$$SP_{M,\text{Jan}} = \frac{212 \text{ mg/L}}{4612 \text{ mg/L}} = 0.046$$

$$SP_{M,\text{May}} = \frac{230 \text{ mg/L}}{4735 \text{ mg/L}} = 0.049$$

6. Calculate the standard salt passage for each date using Eq. 17-25:

$$SP_{S(\text{Jan})} = (0.046)\left(\frac{26.5\text{ bar}}{25.3\text{ bar}}\right)\left(\frac{4612\text{ mg/L}}{2000\text{ mg/L}}\right)\left(\frac{4024\text{ mg/L}}{7828\text{ mg/L}}\right)$$

$$= 0.057$$

$$SP_{S(\text{May})} = (0.049)\left(\frac{23.6\text{ bar}}{25.3\text{ bar}}\right)\left(\frac{4735\text{ mg/L}}{2000\text{ mg/L}}\right)\left(\frac{4024\text{ mg/L}}{8372\text{ mg/L}}\right)$$

$$= 0.052$$

Comment

Even though the membrane system is producing the same permeate flow with less pressure in May than in January, there has been an 8 percent loss of system performance because the standard permeate flow has declined from 10800 to 9900 m^3/d. The standard salt passage also decreased between January and May, even though a higher permeate concentration was observed.

Concentration Polarization

Concentration polarization (CP) is the accumulation of solutes near the membrane surface and has adverse effects on membrane performance. The flux of water through the membrane brings feed water (containing water and solute) to the membrane surface, and as clean water flows through the membrane, the solutes accumulate near the membrane surface. In membrane filtration, particles contact the membrane and form a cake layer. Because the rejection mechanisms for reverse osmosis are different, solutes stay in solution and form a boundary layer of higher concentration at the membrane surface. Thus, the concentration in the feed solution becomes polarized, with the concentration at the membrane surface higher than the concentration in the bulk feed water in the feed channel.

Concentration polarization has several negative impacts on RO performance:

1. Water flux is lower because the osmotic pressure gradient is higher due to the higher concentration of solutes at the membrane surface.
2. Rejection is lower due to an increase in solute transport across the membrane from an increase in the concentration gradient and a decrease in the water flux.
3. Solubility limits of solutes may be exceeded, leading to precipitation and scaling.

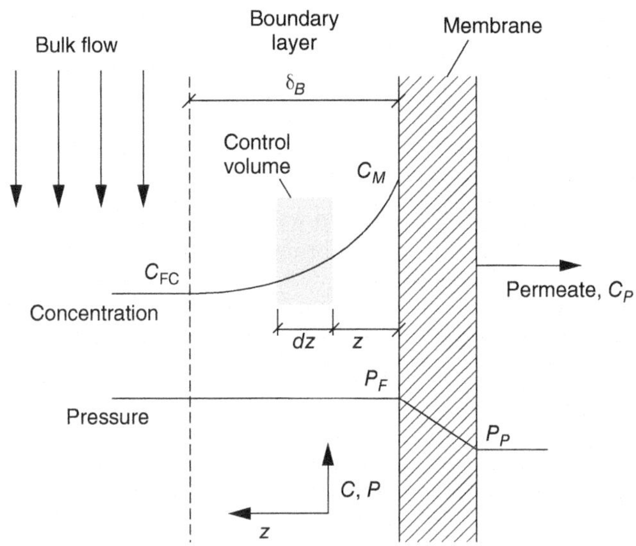

Figure 17-13
Schematic of concentration polarization.

Equations for concentration polarization can be derived from film theory (see Chap. 7) and mass balances. In the membrane schematic shown on Fig. 17-13, feed water is traveling vertically on the left side of the membrane and water is passing through the membrane to the right. According to film theory, a boundary layer forms at the surface of the membrane. Water and solutes move through the boundary layer toward the membrane surface. As water passes through the membrane, the solute concentration at the membrane surface increases. The concentration gradient in the boundary layer leads to diffusion of solutes back toward the bulk feed water. During continuous operation, a steady-state condition is reached in which the solute concentration at the membrane surface is constant with respect to time because the convective flow of solutes toward the membrane is balanced by the diffusive flow of solutes away from the surface. The solute flux toward the membrane surface due to the convective flow of water is described by the expression

$$J_S = J_W C \tag{17-27}$$

A mass balance can be developed at the membrane surface as follows:

$$\text{Mass accumulation} = \text{mass in} - \text{mass out} \tag{17-28}$$

With no accumulation of mass at steady state, the solute flux toward the membrane surface must be balanced by fluxes of solute flowing away from the membrane (due to diffusion) and through the membrane (into the permeate) as follows:

$$\frac{dM}{dt} = 0 = J_W\, Ca - D_L \frac{dC}{dz} a - J_W\, C_P a \tag{17-29}$$

where M = mass of solute, g
t = time, s
D_L = diffusion coefficient for solute in water, m^2/s
z = distance perpendicular to membrane surface, m
a = surface area of membrane, m^2

Equation 17-29 applies not only at the membrane surface but also at any plane in the boundary layer because the net solute flux must be constant throughout the boundary layer to prevent the accumulation of solute anywhere within that layer (the last term in Eq. 17-29 represents the solute that must pass through the boundary layer and the membrane to end up in the permeate). Rearranging and integrating Eq. 17-29 across the thickness of the boundary layer with the boundary conditions $C(0) = C_M$ and $C(\delta_B) = C_{FC}$, where C_{FC} is the concentration in the feed–concentrate channel and C_M is the concentration at the membrane surface, are done in the following equations:

$$D_L \int_{C_M}^{C_{FC}} \frac{dC}{C - C_P} = -J_W \int_0^{\delta_B} dz \tag{17-30}$$

Integrating yields

$$\ln\left(\frac{C_M - C_P}{C_{FC} - C_P}\right) = \frac{J_W \delta_B}{D_L} \tag{17-31}$$

$$\frac{C_M - C_P}{C_{FC} - C_P} = e^{(J_W \delta_B)/D_L} = e^{J_W / k_{CP}} \tag{17-32}$$

where $k_{CP} = D_L/\Delta_B$ concentration polarization mass transfer coefficient, m/s

The concentration polarization mass transfer coefficient describes the diffusion of solutes away from the membrane surface. Concentration polarization is expressed as the ratio of the membrane and feed–concentrate channel solute concentrations as follows:

$$\beta = \frac{C_M}{C_{FC}} \tag{17-33}$$

where β = concentration polarization factor, dimensionless

Combining Eq. 17-33 with Eqs. 17-1 and 17-32 results in the following expression:

$$\beta = (1 - \text{Rej}) + \text{Rej}\,(e^{J_W / k_{CP}}) \tag{17-34}$$

If rejection is high (greater than 99 percent), Eq. 17-34 can be reasonably simplified as follows:

$$\beta = e^{J_W / k_{CP}} \tag{17-35}$$

To predict the extent of concentration polarization, the value of the concentration polarization mass transfer coefficient is needed. As demonstrated in Chap. 7, mass transfer coefficients are often calculated using a correlation between Sherwood (Sh), Reynolds (Re), and Schmidt (Sc) numbers. Correlations for mass transfer coefficients depend on physical characteristics of the system and the flow conditions (e.g., laminar or turbulent). To promote turbulent conditions and minimize concentration polarization in RO membrane elements, spiral-wound elements contain mesh feed channel spacers and maintain a high velocity flow parallel to the membrane surface. The feed channel spacer complicates the flow patterns and promotes turbulence. The superficial velocity (assuming an empty channel) in a spiral-wound element typically ranges from 0.02 to 0.2 m/s, but the actual velocity is higher because of the space taken up by the spacer.

In the spacer-filled feed channel of a spiral-wound element, Schock and Miquel (1987) found that the concentration polarization mass transfer coefficient could be predicted by the following equation, when calculations for the velocity in the channel and the hydraulic diameter took the presence of the spacer into account:

$$k_{CP} = 0.023 \frac{D_L}{d_H} (\text{Re})^{0.875} (\text{Sc})^{0.25} \tag{17-36}$$

$$\text{Re} = \frac{\rho v d_H}{\mu} \tag{17-37}$$

$$\text{Sc} = \frac{\mu}{\rho D_L} \tag{17-38}$$

where Re = Reynolds number, dimensionless
Sc = Schmidt number, dimensionless
v = velocity in feed channel, m/s
ρ = feed water density, kg/m^3
μ = feed water dynamic viscosity, kg/m · s
d_H = hydraulic diameter, m

The hydraulic diameter is defined as

$$d_H = \frac{4\ (\text{volume of flow channel})}{\text{wetted surface}} \tag{17-39}$$

For hollow-fiber membranes (circular cross section), the hydraulic diameter is equal to the inside fiber diameter. Spiral-wound membranes can be approximated by flow through a slit, where the width is much larger than the feed channel height ($w \gg h$). In an empty channel (i.e., the spacer is neglected), the hydraulic diameter is twice the feed channel height, as shown in the equation

$$d_H = \frac{4wh}{2w + 2h} \approx 2h \tag{17-40}$$

where $h =$ feed channel height, m
$w =$ channel width, m

The feed channel height in typical spiral-wound elements ranges from about 0.4 to 1.2 mm and is governed by the thickness of the spacer.

Because the mesh spacer affects mass transfer in the feed channel and many feed spacer configurations have been developed, numerous other correlations have been developed for the mass transfer coefficient. Mariñas and Urama (1996) developed a correlation using the channel height and the superficial velocity, which eliminates the task of determining the parameters of the spacer. Their correlation is

$$k_{CP} = \lambda \frac{D_L}{d_H} (\text{Re})^{0.50} (\text{Sc})^{1/3} \tag{17-41}$$

where λ ranged from 0.40 to 0.54 for different elements. Many spacer configurations have been evaluated in small flat-sheet membrane cells instead of spiral-wound elements, and in those cases, the mass transfer correlation often has an additional term for the ratio of the channel height (d_H) to channel length (L). For instance, the correlation presented by Shakaib et al. (2009) for spacers with axial and transverse filaments is

$$k_{CP} = 0.664 \frac{D_L}{d_H} (\text{Re})^{0.5} (\text{Sc})^{0.33} \left(\frac{d_H}{L} \right)^{0.5} \tag{17-42}$$

Concentration polarization varies along the length of a membrane element; the parameters that change most significantly are the velocity in the feed channel (v) and the permeate flux (J_W). Variation in the concentration polarization factor as a function of these parameters is shown on Fig. 17-14. As might be expected, concentration polarization increases as the permeate flux increases and as the velocity in the feed channel decreases. The

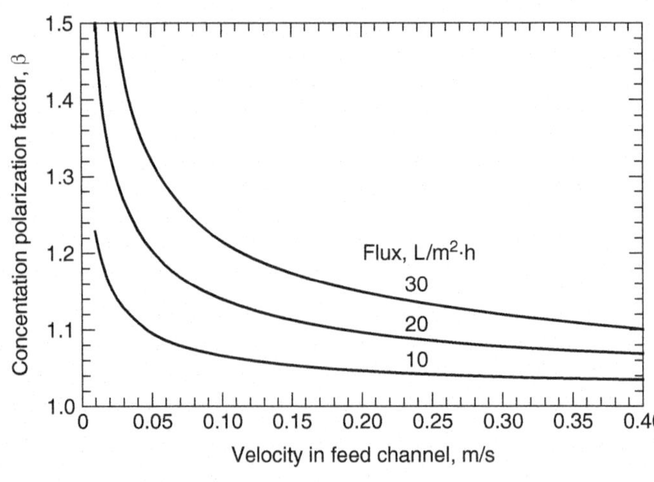

Figure 17-14
Concentration polarization factors as function of feed channel velocity and permeate flux.

maximum concentration polarization allowed for membrane elements is specified by manufacturers; $\beta = 1.2$ is a typical value. The importance of maintaining a high velocity in the feed–concentrate channel, particularly for membranes that achieve higher permeate flux, is clearly demonstrated on Fig. 17-14. Calculation of the concentration polarization factor is illustrated in Example 17-3.

Example 17-3 Concentration polarization

For a spiral-wound element, calculate the concentration polarization factor and the concentration of sodium at the membrane surface given the following information: water temperature 20°C, feed channel velocity 0.15 m/s, feed channel height 0.86 mm, permeate flux 25 $L/m^2 \cdot h$, sodium concentration 6000 mg/L, and diffusivity of sodium in water 1.35×10^{-9} m^2/s. Use the correlation in Eq. 17-41 and a value of 0.47 for the coefficient. Assume that the rejection is high enough that the impact of sodium flux through the membrane can be ignored. Water density and viscosity at 20°C can be found in Table C-1 in App. C.

Solution

1. Calculate the Reynolds and Schmidt numbers using Eqs. 17-37 and 17-38. Because the feed channel height is 0.86 mm, the hydraulic diameter is 1.72 mm:

$$\text{Re} = \frac{\rho v d_H}{\mu} = \frac{(998\ \text{kg/m}^3)(0.15\ \text{m/s})(1.72\ \text{mm})}{(1.0\times10^{-3}\ \text{kg/m}\cdot\text{s})(10^3\ \text{mm/m})} = 257$$

$$\text{Sc} = \frac{\mu}{\rho D_L} = \frac{1.0\times10^{-3}\ \text{kg/m}\cdot\text{s}}{(998\ \text{kg/m}^3)(1.35\times10^{-9}\ \text{m}^2/\text{s})} = 742$$

2. Calculate k_{CP} using Eq. 17-41:

$$k_{CP} = \frac{(0.47)(1.35\times10^{-9}\ \text{m}^2/\text{s})(257)^{0.5}(742)^{1/3}}{(1.72\ \text{mm})(10^{-3}\ \text{m/mm})} = 5.36\times10^{-5}\ \text{m/s}$$

3. Because the rejection is high, β can be calculated using Eq. 17-35 (otherwise, Eq. 17-34 must be used):

$$\beta = \exp\left(\frac{J_W}{k_{CP}}\right) = \exp\left[\frac{(25\ \text{L/m}^2\cdot\text{h})(10^{-3}\ \text{m}^3/\text{L})}{(5.36\times10^{-5}\ \text{m/s})(3600\ \text{s/h})}\right] = 1.14$$

4. Calculate the sodium concentration at the membrane surface using Eq. 17-33:

$$C_M = (1.14)(6000\ \text{mg/L}) = 6840\ \text{mg/L}$$

17-6 Fouling and Scaling

Nanofiltration and RO membranes are susceptible to fouling via a variety of mechanisms. The primary sources of fouling and scaling are particulate matter, precipitation of insoluble inorganic salts, oxidation of soluble metals, and biological matter.

Particulate Fouling

Particulate fouling is a concern in RO because the operational cycle does not include a backwashing step to remove accumulated solids (in fact, backwashing might cause the active layer of thin-film membranes to separate from the support layers). Virtually all RO systems require pretreatment to minimize particulate fouling. Fouling by residual particulate matter affects the cleaning frequency.

PLUGGING AND CAKE FORMATION

Both inorganic and organic materials, including microbial constituents and biological debris, can cause particulate fouling, which includes plugging and cake formation. Plugging is the entrapment of large particles in the feed channels and piping. Hollow-fine-fiber membranes are reported to have more significant plugging problems because the high packing density of the fibers inside the pressure vessel results in very small spaces between the fibers. The mesh spacers in spiral-wound elements are sized to minimize plugging, but an excessive load of particulate matter may cause plugging anyway. Plugging is minimized by prefiltration of the feed water, and RO membrane manufacturers recommend prefiltration through 5-μm cartridge filters as a minimum prefiltration step for protection of the membrane elements.

Particulate matter forming a cake on the membrane surface adds resistance to flow and affects system performance. Source waters with excessive potential for particulate fouling require advanced pretreatment to lower the particulate concentration to an acceptable level. Coagulation and filtration (using sand, carbon, or other filter media) are sometimes used for pretreatment as well as MF and UF.

ASSESSMENT OF PARTICLE FOULING

It is important to assess the fouling tendency prior to design and construction of an RO facility and to monitor the fouling tendency during operation. Empirical tests have been developed to assess particulate fouling, including the silt density index (SDI) and the modified fouling index (MFI). The SDI (ASTM, 2001b) is a timed filtration test using three time intervals through a gridded membrane filter with a mean pore size of 0.45 ± 0.02 μm and a diameter of 47 mm at a constant applied pressure of 2.07 bar (30 psi). The first interval is the duration necessary to collect 500 mL of permeate. Filtration continues through the second interval without recording volume until 15 min has elapsed (including the first time interval). Occasionally, a duration shorter than 15 min is used for waters with high fouling tendency. At the end of 15 min, the third interval is started, during which an additional 500-mL aliquot of water is filtered through the now-dirty membrane, and the time is recorded. The SDI is calculated from these time intervals:

$$\text{SDI} = \frac{100\,(1 - t_I/t_F)}{t_T} \qquad (17\text{-}43)$$

where SDI = silt density index, min^{-1}
t_I = time to collect first 500-mL sample, min
t_F = time to collect final 500-mL sample, min
t_T = duration of first two test intervals (15 min)

The MFI (Schippers and Verdouw, 1980) uses identical test equipment but different procedures from the SDI. The volume filtered is recorded at 30-s intervals during the MFI test. The flow rate is determined from volume and time data, and the inverse of the flow rate is plotted as a function of volume filtered. An example of the plotted data is shown on Fig. 17-15. A portion of the graph is generally linear, and the MFI is the slope of the graph in this region, that is,

$$\frac{\Delta t}{\Delta V} = \frac{1}{Q} = (\text{MFI})V + b \qquad (17\text{-}44)$$

where MFI = modified fouling index, s/L^2
V = volume of permeate, L
b = intercept of linear portion of graph

The SDI and MFI have been criticized as being too simplistic to accurately predict RO membrane fouling. They operate in a dead-end, constantpressure filtration mode, whereas full-scale RO systems operate with a significant cross flow and constant flux. They use a 0.45-μm filter so they only nominally measure fouling by material larger than that size. Research suggests that colloidal matter smaller than 0.45 μm may cause significant fouling of RO membranes. As a result, a revised MFI test that uses a 13-kDa UF membrane has also been developed (Boerlage et al., 2002, 2003).

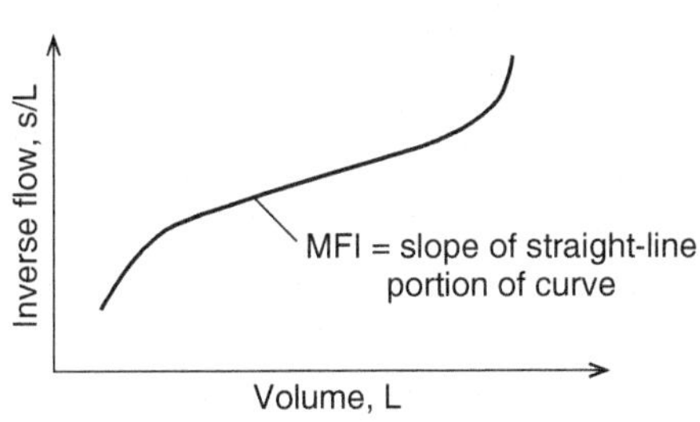

Figure 17-15
Determination of modified fouling index (MFI).

The SDI and MFI might best be considered as screening tests that can indicate unacceptable feed water quality. A high value is a good indicator of fouling problems in RO systems, but a low value does not necessarily mean the source water has a low fouling tendency. RO manufacturers typically specify a maximum SDI value of 4 to 5 min^{-1}. High SDI or MFI values indicate pretreatment is required to remove particulate matter. When lower SDI or MFI values are measured, pilot tests are often necessary to determine the appropriate level of pretreatment to minimize fouling.

Precipitation of Inorganic Salts and Scaling

Inorganic scaling occurs when salts in solution are concentrated beyond their solubility limits and form precipitates. Common sparingly soluble salts are listed in Table 17-3. If the ions that comprise these salts are concentrated past the solubility product, precipitation occurs. Precipitation reactions and solubility calculations were introduced in Chap. 5. The precipitation reaction for a typical salt is as follows:

$$CaSO_4\,(s) \rightleftarrows Ca^{2+} + SO_4{}^{2-} \tag{17-45}$$

The solubility product is written as

$$K_{SP} = \left\{Ca^{2+}\right\}\left\{SO_4{}^{2-}\right\} = \gamma_{Ca}[Ca^{2+}]\gamma_{SO_4}[SO_4{}^{2-}] \tag{17-46}$$

where
K_{SP} = solubility product
$\{Ca^{2+}\}$ = calcium activity
$\{SO_4{}^{2-}\}$ = sulfate activity
γ_{Ca} = activity coefficient for calcium
γ_{SO_4} = activity coefficient for sulfate

Table 17-3
Typical limiting salts and their solubility products

Salt	Equation	Solubility Product (pK_{sp} at 25°C)
Calcium carbonate (aragonite)	$CaCO_3(s) \rightleftarrows Ca^{2+} + CO_3{}^{2-}$	8.2
Calcium fluoride	$CaF_2(s) \rightleftarrows Ca^{2+} + 2F-$	10.3
Calcium orthophosphate	$CaHPO_4(s) \rightleftarrows Ca^{2+} + HPO_4{}^{2-}$	6.6
Calcium sulfate (gypsum)	$CaSO_4(s) \rightleftarrows Ca^{2+} + SO_4{}^{2-}$	4.6
Strontium sulfate	$SrSO_4(s) \rightleftarrows Sr^{2+} + SO_4{}^{2-}$	6.2
Barium sulfate	$BaSO_4(s) \rightleftarrows Ba^{2+} + SO_4{}^{2-}$	9.7
Silica, amorphous	$SiO_2(s) + 2H_20 \rightleftarrows Si(OH)_4(aq)$	2.7

[a]From Stumm and Morgan (1996).

$[Ca^{2+}]$ = calcium concentration, mol/L
$[SO_4^{2-}]$ = sulfate concentration, mol/L

The ionic strength of feed solutions for RO is sufficiently high that ion products must be calculated using activity, rather than the common practice of assuming that activity is equal to concentration. Several factors in RO operation affect how much ions are concentrated. The system recovery is the most important factor because the concentration of the rejected solutes increases as more clean water is withdrawn from solution. In fact, precipitation is one of the important factors that limit recovery in RO systems (osmotic pressure being the other). The rate of ion or salt rejection is also important, as an ion with 99 percent rejection will be concentrated more than one with 10 percent rejection. Finally, the degree of concentration polarization is important because precipitation occurs in the more concentrated zone near the membrane surface. The inorganic scale that forms on the membrane surface can reduce water permeability or permanently damage the membrane.

In the absence of pretreatment, precipitation must be avoided by minimizing concentration polarization, limiting salt rejection, or limiting recovery.Concentration polarization is minimized by promoting turbulence in the feed channels and maintaining minimum velocity conditions specified by equipment manufacturers. Limiting rejection is impractical because it conflicts with process objectives. Limiting recovery, however, is often necessary to prevent precipitation. The highest recovery possible before any salts precipitate is the *allowable recovery*, and the salt that precipitates at this condition is the *limiting salt*. The most common scales encountered in water treatment applications are calcium carbonate ($CaCO_3$) and calcium sulfate ($CaSO_4$).

The allowable recovery without pretreatment that can be achieved in RO is determined by performing solubility calculations for each of the possible limiting salts. The highest solute concentrations occur in the final membrane element immediately prior to the feed water exiting the system as the concentrate stream, so concentrate stream concentrations are used to evaluate solubility limits. In addition, the concentration in the concentrate steam must be adjusted for the level of concentration polarization that is occurring. Incorporating the concentration polarization factor defined in Eq. 17-40 with the expression for the solute concentration in the concentrate stream defined by Eq. 17-15 yields

$$C_M = \beta C_F \left[\frac{1-(1-\text{Rej})r}{1-r}\right] \tag{17-47}$$

Allowable recovery is determined by substituting the activities at the membrane into a solubility product calculation (from Chap. 5) and solving for the recovery, as demonstrated in Example 17-4.

Example 17-4 Allowable recovery from limiting salt calculations

Determine the limiting salt and allowable recovery for a brackish water RO system containing the following solutes: calcium 74 mg/L, barium 0.008 mg/L, and sulfate 68 mg/L. Assume 100 percent rejection of all solutes and a polarization factor of 1.15 and ignore activity coefficients (i.e., activity = concentration).

Solution

1. Calculate the molar concentration for each component:

$$\left[Ca^{2+}\right] = \frac{74\ \text{mg/L}}{(40\ \text{g/mol})(10^3\text{mg/g})} = 1.85 \times 10^{-3}\ \text{mol/L}$$

$$\left[Ba^{2+}\right] = \frac{0.008\ \text{mg/L}}{(137.3\ \text{g/mol})(10^3\text{mg/g})} = 5.83 \times 10^{-8}\ \text{mol/L}$$

$$\left[SO_4^{\ 2-}\right] = \frac{68\ \text{mg/L}}{(96\ \text{g/mol})(10^3\text{mg/g})} = 7.08 \times 10^{-4}\ \text{mol/L}$$

2. Simplify the expression for concentration at the membrane. Let $y = 1 - r$. Because Rej = 1, Eq. 17-47 becomes

$$C_M = \frac{\beta C_F}{y}$$

3. Substitute the concentrations at the membrane surface into the equation for solubility products and calculate recovery. Solubility product constants are available in Table 17-3.
   a. For calcium sulfate,

$$K_{SP} = 10^{-4.6} = [Ca^{2+}]_M[SO_4^{\ 2-}]_M = \left(\frac{\beta[Ca^{2+}]_F}{y}\right)\left(\frac{\beta[SO_4^{\ 2-}]_F}{y}\right)$$

$$= \frac{\beta^2}{y^2}[Ca^{2+}]_F[SO_4^{\ 2-}]_F$$

$$y = \left(\frac{\beta^2}{K_{SP}}[Ca^{2+}]_F[SO_4^{\ 2-}]_F\right)^{1/2}$$

$$= \left[\frac{(1.15)^2}{10^{-4.6}}(1.85 \times 10^{-3}\ \text{mol/L})\ (7.08 \times 10^{-4}\ \text{mol/L})\right]^{1/2}$$

$$= 0.26$$

$$r = 1 - y = 1 - 0.26 = 0.74$$

b. For barium sulfate,

$$y = \left[\frac{(1.15)^2}{10^{-9.7}}(5.83 \times 10^{-8}\ \text{mol/L})(7.08 \times 10^{-4}\ \text{mol/L})\right]^{1/2}$$
$$= 0.52$$
$$r = 1 - y = 1 - 0.52 = 0.48$$

Comments

1. The allowable recovery before barium sulfate precipitates is 48 percent, compared to 74 percent before calcium sulfate precipitates. Therefore, barium sulfate is the limiting salt and the allowable recovery is 48 percent.
2. Activity coefficients affect solubility calculations and, therefore, recovery. The ionic strength of the feed solution can be calculated from feed ion concentrations. However, the activity coefficients must be calculated from the ionic strength of the concentrate at the allowable recovery, so a simultateous solution procedure must be used.

The complexity of limiting salt calculations is greatly oversimplified in Example 17-4. As noted above, activity coefficients cannot be ignored. The ionic strength is dependent on recovery and rejection, however, so the activity coefficients cannot be calculated until the recovery is determined. Ignoring ionic strength may yield a significantly lower value for allowable recovery than could actually be achieved. The assumption of 100 percent rejection is often justified because divalent ions typically have rejection near 100 percent. An assumption of 100 percent rejection yields a slightly conservative value for allowable recovery because lower rejection will produce concentrate stream concentrations that are actually slightly lower. For NF and low-pressure RO systems that have divalent ion rejection significantly below 100 percent, however, this assumption would be inappropriate.

Another complicating factor is the formation of ion complexes. For instance, calcium and sulfate form a neutral $CaSO_4^0$ species that increases the solubility of $CaSO_4(s)$. The solubility of calcium sulfate in distilled water would be calculated as 680 mg/L as $CaSO_4$ using Eq. 17-48 if ionic strength and complexation were ignored. With ionic strength and complexation, the solubility of calcium sulfate in distilled water is 2170 mg/L, an error of over 200 percent.

Several models are available to calculate activity coefficients, and the applicability of each model depends on the ionic strength. Seawater has an ionic strength of about 0.7 M. Assuming 50 percent recovery, the ionic strength of the concentrate from a seawater RO plant would be about 1.4 M. This ionic strength is significantly above the range of applicability of the extended Debye–Huckel or Davies equations. The specific interaction model or Pitzer model are suitable for calculating activity coefficients when the ionic strength is above 1 M (Pitzer, 1975).

Another complicating factor is that carbonate and phosphate concentrations are dependent on pH. As can be imagined, accounting for ionic strength, recovery, complexation, and pH in the calculations in Example 17-4, and then calculating activity coefficients with the Pitzer equations, would result in equations that cannot be easily manipulated algebraically.

Furthermore, the calculations must be repeated for each limiting salt in Table 17-3. Example 17-4 demonstrates that barium was a limiting solute even though its concentration in the feed water was very low. When alternative systems with different rejection capabilities are being evaluated, the calculations must be repeated for each rejection scenario. Temperature and supersaturation considerations further complicate the calculations. Clearly, the computational requirements of limiting salt calculations can be daunting and are rarely done manually. Membrane manufacturers provide computer programs to perform these calculations. These programs account for the concentration polarization factor and rejection capabilities of specific products, temperature and pH effects, and the degree of supersaturation that can be accommodated with various pretreatment strategies. Use of an equilibrium speciation program (Visual MINTEQ) to solve Example 17-4 reveals that the barium sulfate reaches saturation at 84 percent recovery instead of 48 percent recovery.

ACID ADDITION AND ANTISCALANTS TO PREVENT SCALING

Pretreatment is necessary in virtually all RO systems to prevent scaling due to precipitation of sparingly soluble salts. Calcium carbonate precipitation is common, and most systems require pretreatment for this compound. In addition to the limiting salt calculations presented in the above example, calcium carbonate solubility can also be expressed in terms of the Langelier saturation index (LSI) and Stiff and Davis stability index (ASTM, 2001a, 2001f), and manufacturers' solubility programs often report these values. Calcium carbonate precipitation can be prevented by adjusting the pH of the feed stream with acid to convert carbonate to bicarbonate and carbon dioxide. Sulfuric or hydrochloric acids are normally used, but using sulfuric acid can increase the sulfate concentration enough to cause precipitation of sulfate compounds. The pH of most RO feed waters is adjusted to a pH value of 5.5 to 6.0. At this pH, most carbonate is in the form of carbon dioxide and passes through the membrane.

Scaling of other limiting salts is commonly prevented with the addition of antiscalant chemicals. Antiscalants allow supersaturation without precipitation occurring by preventing crystal formation and growth. At one time, sodium hexametaphosphate (SHMP) was commonly used as an antiscalant, but it is rarely used anymore because it has limited ability to extend the supersaturation range and adds phosphate compounds to the concentrate, which causes disposal problems. SHMP has been largely replaced with polymeric antiscalants. The degree of supersaturation allowed because of antiscalant addition depends on properties of the antiscalant, which are often proprietary, and characteristics of specific equipment configurations. It is appropriate to rely on the recommendations of equipment and antiscalant manufacturers when determining appropriate antiscalant selection and doses necessary for a specific feed water analysis and design recovery.

In addition to acid and antiscalant addition, newer installations are incorporating a variety of strategies to minimize scaling with the goal of reducing the quantity of waste concentrate that must be disposed and increasing the recovery of water. These strategies are discussed in more detail in Sec. 17-7 under the heading Concentrate Management.

SILICA SCALING

Silica scaling can be particularly problematic because silica chemistry is complex and silica can occur in several forms in groundwater, including monomeric, polymeric, and colloidal forms. Many brackish groundwater sources in the Southwestern United States have sufficiently high silica concentrations such that silica is the species that limits recovery. Silica precipitates in an amorphous rather than crystalline form; thus, antiscalants that prevent crystal growth are ineffective for preventing silica precipitation. The presence of metals can increase silica precipitation and change its form (Sahachaiyunta et al., 2002; Sheikholeslami and Bright, 2002), complicating the presence of silica in RO feed water.Recent advances and new antiscalant formulations are now available for both minimizing silica precipitation and cleaning silica from membranes, but these proprietary compounds have had varying degrees of success. When high silica concentrations are present, high-pH softening (resulting in co-precipitation with magnesium hydroxide) may be necessary to remove silica from the feed water to prevent precipitation on the membrane.

A cost trade-off exists between methods of preventing scaling: operating at a lower recovery or the use of pretreatment processes and chemicals. In some cases, it may be more cost effective to operate at a lower recovery to minimize pretreatment costs. Pretreatment and membrane equipment costs must be considered simultaneously and the design recovery set at the point that minimizes overall system costs.

Metal Oxide Fouling

Groundwater used as the source water for RO and NF systems is often anaerobic. Iron and manganese, soluble compounds in their reduced states, can oxidize, precipitate, and foul membranes if oxidants enter the feed water system. Iron fouling is more prevalent and can occur rapidly if any air enters the feed system. Fouling may be avoided by preventing oxidation or removing the iron or manganese after oxidation. If iron concentrations are low, precautions to prevent air from entering the feed system may be sufficient;

antiscalants often include additives to minimize fouling by low concentrations of iron. Pretreatment to control iron might include oxidation with oxygen or chlorine followed by adequate mixing and hydraulic detention time and granular media or membrane filtration or greensand filtration in which oxidation and filtration take place simultaneously. When oxidants are used, precautions must be made to prevent them from reaching the membranes, particularly for polyamide membranes or other materials that are not oxidant resistant. Iron-fouling deposits are usually removable from RO membrane surfaces by commercially available cleaning agents and procedures.

An additional constituent present in many anaerobic groundwaters is hydrogen sulfide. If air enters the feed water system, hydrogen sulfide can oxidize to colloidal sulfur, which can foul membranes. As with iron oxidation, precautions to prevent air from entering the feed system are important to prevent colloidal sulfur fouling. Sulfur deposits on membrane surfaces are typically irreversible.

Biological Fouling

Biological fouling refers to the attachment or growth of microorganisms or extracellular soluble material on the membrane surface or in the membrane element feed channels. Biological fouling is common in many RO systems and can have a variety of negative effects on performance, including loss of flux, reduced solute rejection, increased head loss through the membrane modules, contamination of the permeate, degradation of the membrane material, and reduced membrane life (Ridgway and Flemming, 1996). An example of biological fouling is shown on Fig. 17-16. The primary source of microbial contamination is the feed water. Biological fouling is a significant problem in many RO systems.

Biological fouling is prevented by maintaining proper operating conditions, applying biocides, and flushing membrane elements properly when not in use. Many RO and NF feed waters (groundwater in many cases) have low microbial populations. When operated properly, the shear in the feed channels helps to keep bacteria from accumulating or growing to unacceptable levels. When membrane trains are out of service, however, bacteria can quickly multiply. To avoid this problem, membranes should be flushed with permeate periodically or filled with an approved biocide if out of service for any significant period. Chlorine solutions can be used as a biocide for cellulose acetate membranes within recommended limits, but other chemicals such as sodium bisulfite must be used with polyamide membranes because of their susceptibility to degradation by chlorine. An excellent review of the issues involved in biological fouling of membranes is provided in Ridgway and Flemming (1996).

The feed water to cellulose acetate membranes can be continuously chlorinated within limited concentrations to prevent biological growth, if necessary. Ultraviolet radiation, chloramination, or chlorination followed by dechlorination can sometimes be used for polyamide membranes.

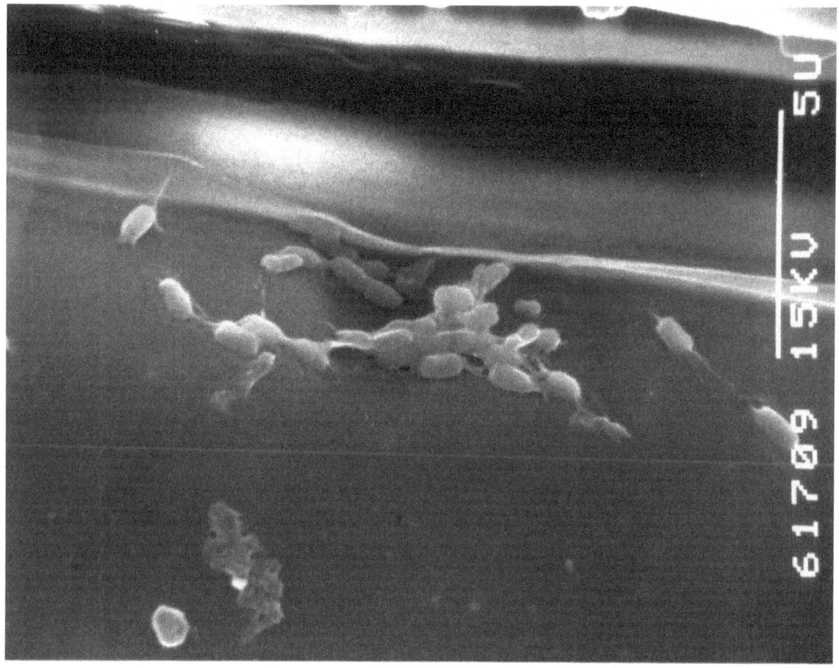

Figure 17-16
Scanning electron micrograph (SEM) image of biological fouling of membrane. (Courtesy Orange County Water District.)

17-7 Reverse Osmosis Process Design

During preliminary design of an RO system, the design engineer must perform the following activities:

1. Select the basic performance criteria: capacity, recovery, rejection, and permeate solute concentrations.
2. Evaluate alternatives for membrane equipment and operation, select the type of membrane element, and determine the array configuration (number of stages, number of passes, number of elements in a pressure vessel, number of vessels in each stage, feed pressure).
3. Select feed water pretreatment requirements (methods to control fouling).
4. Select permeate posttreatment requirements.
5. Select concentrate management and disposal requirements.
6. Select ancillary membrane system features such as permeate backpressure control and interstage booster pumps.
7. Select equipment and procedures for process monitoring.

These elements of design are not independent of one another. For instance, recovery is often constrained by the solubility of limiting salts. As a result, selection of pretreatment requirements, recovery, and array design must be done simultaneously and iteratively to determine the most economical design.

The basis for design information typically includes characteristics of the feed water (solute concentrations, turbidity, SDI and MFI values) from laboratory or historical data, required treated-water quality (established by the client or regulatory limits), and required treated-water capacity (established by demand requirements). The process design criteria for a hypothetical brackish water RO facility are shown in Table 17-4. Frequently, pilot testing is part of the design process.

The following discussion focuses primarily on the design of the membrane components of an RO system. Design of additional components, such as intakes and pretreatment systems, are available in design manuals such as AWWA (2007).

Element Selection and Membrane Array Design

Membrane array design involves determination of the quantity and quality of water produced by each membrane element in an array. This involves calculation of the flow, velocity, applied pressure, osmotic pressure, water flux, and solute flux in each element, which leads to the determination of the number of stages, number of passes, number of elements in each pressure vessel, and number of vessels in each stage. Membrane array design is a complex and iterative process using a large number of interrelated design parameters. Several important design parameters such as mass transfer coefficients are specific to individual products and available only from membrane manufacturers. Because of the complexity of the calculations and dependence on manufacturer information, array design is often done with design software provided by membrane manufacturers. Nevertheless, an understanding of the mechanics of the design procedure as described in the following paragraphs is important to interpreting the results from manufacturer design software.

DESIGN CALCULATIONS

The most common type of membrane element in use is the spiral-wound element. As described earlier, feed water enters one end of the pressure vessel and flows through several spiral-wound elements in series. As the water passes through each element, some water passes through the membrane into the permeate carrier channel, resulting in continuously changing conditions along the length of the membrane element. The net transmembrane pressure declines continuously across the length of a membrane element because of changes in both applied pressure (due to head loss in the feed channels) and osmotic pressure (due to concentration of salts). As a result, fluxes of both water and solute are dependent on the position within a spiral-wound element, and the design procedure must integrate along the length of the membrane element.

Table 17-4
Design criteria for a hypothetical reverse osmosis facility

Operating Parameter	Units	Value
Feedwater pretreatment		
Capacity	m^3/d	37,900
Strainers		
Number	Number	5
Nominal particle size rating	μm	5
Capacity, each	m^3/d	9,480
Chemicals		
Sulfuric acid, max. dose	mg/L	200
Scale inhibitor, max. dose	mg/L	2
Feed pumps		
Number	Number	5
Capacity, each	m^3/d	9,480
Pressure	bar	40
Membrane system		
Feed water flow rate	m^3/d	37,900
Permeate flow rate	m^3/d	30,300
Concentrate flow rate	m^3/d	7,580
Recovery	%	80
Number of arrays	Number	4
Capacity per array	m^3/d	9,480
Array design criteria		
Membrane area per element	m^2	32.5
Elements per pressure vessel	Number	7
Number of stages per array	Number	2
Number of pressure vessels (stage 1)	Number	40
Stage 1 avg. permeate flux	$L/m^2 \cdot h$	21
Number of pressure vessels (stage 2)	Number	20
Stage 2 avg. permeate flux	$L/m^2 \cdot h$	17
Posttreatment[a]		
Caustic soda, max. dose	mg/L	10
Corrosion inhibitor, max. dose	mg/L	1
Chlorine, max. dose	mg/L	2
Fluoride, max. dose	mg/L	1
Concentrate disposal	Deep-well injection	

[a]Posttreatment may also include a countercurrent packed tower for hydrogen sulfide or carbon dioxide removal. See Chap. 14 for details of packed-tower design.

A differential slice of a membrane element is shown on Fig. 17-17. In this figure, the center plane represents the membrane surface, with the feed–concentrate channel above the membrane and the permeate channel

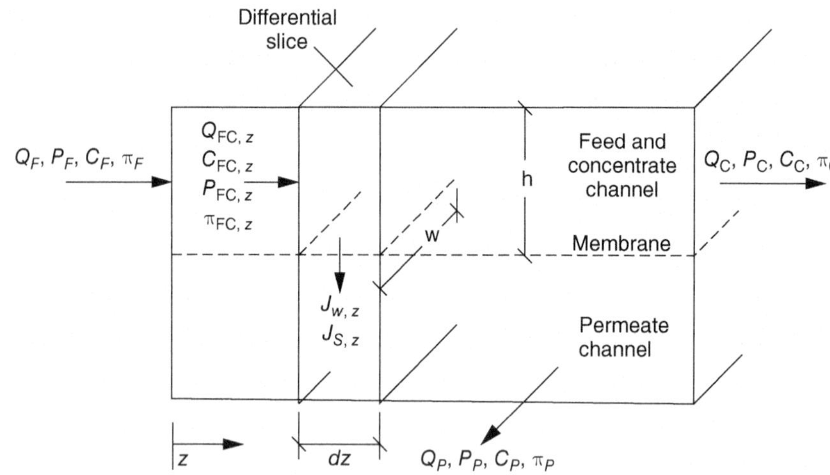

Figure 17-17
Differential slice of spiral-wound membrane element. Because the feed flows axially along the pressure vessel and the permeate flows spirally toward the center of the vessel, the feed and permeate flows are perpendicular to each other.

below the membrane. The fluxes of water and solute are described by Eqs. 17-9 and 17-10, but the applied pressure differential, osmotic pressure differential, and concentration differential depend on the location within the pressure vessel:

$$J_{W,Z} = k_W\left(\Delta P_Z - \Delta\pi_Z\right) = k_W\left[\left(P_{FC,Z} - P_{P,Z}\right) - \left(\pi_{M,Z} - \pi_{P,Z}\right)\right] \quad (17\text{-}48)$$

$$J_{S,Z} = k_S\left(\Delta C_Z\right) = k_S\left(C_{M,Z} - C_{P,Z}\right) \quad (17\text{-}49)$$

where $C_{M,Z}$ = concentration at the membrane surface, $C_{M,Z} = \beta_Z C_{FC,Z}$, mol/L

$\pi_{M,Z}$ = osmotic pressure at the membrane surface, bar

Other terms are defined on Fig. 17-17

The water and solute mass transfer coefficients (k_W and k_S) are dependent on the properties and configurations of specific membrane elements and cannot be generalized. These values are embedded in the manufacturer's design software and are typically not publicized but can be generated from pilot data if they cannot be obtained from the manufacturer.

Solute flux calculations are complicated by the presence of multiple solutes, which may have different value for the mass transfer coefficient. For instance, a low-pressure NF membrane has low rejection of monovalent ions but high rejection of divalent ions, and the mass transfer coefficients would reflect this difference in rejection.

The permeate flow and mass solute flow through the membrane are equal to the flux times the membrane area in the differential element, and the cumulative transfer of water and solute across the membrane is determined by integrating the flow between the feed end and the position z within the pressure vessel, as shown in the following:

$$Q_{P,Z} = \int_0^z J_{W,Z}\, w\, dz \tag{17-50}$$

$$M_{S,Z} = \int_0^z J_{S,Z}\, w\, dz \tag{17-51}$$

where w = effective width of feed–concentrate flow channel, m
$M_{S,Z}$ = mass of solute transferred, mg/s

The water flow rate (and velocity) in the feed–concentrate channel declines as permeate is produced, and the flow rate at any point in the channel can be determined by subtracting the net permeate production up to that point from the feed water flow rate as follows:

$$Q_{\text{FC},Z} = Q_F - Q_{P,Z} \tag{17-52}$$

Similarly, the solute concentration in the feed–concentrate channel can be determined by performing a mass balance on the solute as follows:

$$C_{\text{FC},Z} = \frac{Q_F C_F - M_{S,Z}}{Q_{\text{FC},Z}} \tag{17-53}$$

Water and solute flux are affected by concentration polarization and the concentration of solute at the membrane surface. Some manufacturers have developed relationships describing concentration polarization for specific element designs, and these relationships should be used if available. If no manufacturer information is available, the correlations presented earlier in this chapter can be used to estimate the concentration polarization factor. Because both flux and velocity are changing, β must be calculated using Eq. 17-41, but as a function of position, as shown in the equation

$$\beta_Z = \text{Rej}(e^{J_{W,Z}/k_{\text{CP},Z}}) + (1 - \text{Rej}) \tag{17-54}$$

The mass transfer coefficient k_{CP} depends on velocity in the feed–concentrate channel, which can be calculated from the expression

$$v_Z = \frac{Q_{\text{FC},Z}}{hw} \tag{17-55}$$

where h = height of feed–concentrate channel, m

The solute concentration at the membrane surface is defined by Eq. 17-40, using concentrations as a function of position.

$$C_{M,Z} = \beta_Z C_{\text{FC},Z} \tag{17-56}$$

Pressure in the feed channel drops due to head loss, but head loss is not constant across the length of the membrane element. Turbulent conditions are maintained, so head loss in the channel is proportional to the square of the velocity and the first power of length (consistent with the Darcy-Weisbach equation) as given by the expression

$$h_L = \delta_{\text{HL}} v^2 L \tag{17-57}$$

where h_L = head loss in feed–concentrate channel, bar
δHL = head loss coefficient, bar · s^2/m^3
v = water velocity in feed–concentrate channel, m/s
L = channel length, m

Finally, the permeate solute concentration can be calculated from the ratio of the solute and water fluxes per Eq. 17-11:

$$C_{P,Z} = \frac{J_{S,Z}}{J_{W,Z}} \tag{17-58}$$

Additional design calculations, such as the calculation of osmotic pressure from concentration, have been presented earlier in this chapter. The use of these equations in system array design is demonstrated in Example 17-5.

Example 17-5 Calculation of permeate flux and concentration

Calculate the quantity and quality of water produced by a single membrane element (permeate concentration, rejection, and recovery) given the following information:

Parameter	Unit	Value
Membrane properties		
Element length	m	1
Element membrane area	m^2	32.5
Effective feed channel height	mm	0.125
Water mass transfer coefficient (k_W)	L/m^2 · h · bar	2.87
Solute mass transfer coefficient (k_S)	m/h	6.14×10^{-4}
Element head loss (at design velocity of 0.5 m/s)	bar	0.2
Operating conditions		
Feed flow (Q_F)	m^3/d	270
Feed pressure (P_F)	bar	14.2
Feed concentration (C_F)	mg/L NaCl	2000
Feed temperature (T_F)	°C	20
Permeate pressure (P_P)	bar	0.3
Osmotic coefficient (ϕ)		1.0

Assume $D_{\text{NaCl}} = 1.35 \times 10^{-9}$ m^2/s, $\phi = 1$, and $\text{MW}_{\text{NaCl}} = 58.4$.

Solution

The basic solution strategy is to (1) divide the membrane element into a number of increments; (2) determine P, v, C, and π on both sides of the membrane in the first increment; (3) calculate the water and solute flux across the membrane in the first increment; (4) determine Q, P, C, v,and π on both sides of the membrane in the next increment; (5) calculate the water and solute flux across the membrane in the next increment; and (6) repeat steps 4 and 5 for all remaining increments.

Part 1

Divide the element into 10 increments 0.1m length each. Determine v, P, C, and π on both sides of the membrane in the first increment. The subscript FC is used to designate the feed–concentrate side of the membrane, and the subscript P designates the permeate side of the membrane.

1. The following values are given in the problem statement:

$$Q_{FC,Z} = Q_F = 270 \text{ m}^3/\text{d}$$
$$P_{FC,Z} = P_F = 14.2 \text{ bar}$$
$$P_{P,Z} = 0.3 \text{ bar}$$

2. The feed channel velocity is determined by dividing the feed flow by the channel cross-sectional area. The effective channel height is given as 0.125 mm, but the width is not given. The width can be determined by dividing the membrane area by the element length, both of which are readily available information:

$$w = \frac{a}{L} = \frac{32.5 \text{ m}^2}{1 \text{ m}} = 32.5 \text{ m}$$

It should be noted that the element is not 32.5m wide. Spiral-wound elements are typically 0.2 to 0.3 m in diameter, and 32.5m is the unit width of the membrane surface (which includes multiple feed channels because multiple envelopes are used, see Sec. 17-4) as wrapped around the permeate tube. Then,

$$Q_{FC,Z} = \frac{270 \text{ m}^3/\text{d}}{86{,}400 \text{ s/d}} = 3.125 \times 10^{-3} \text{ m}^3/\text{s}$$

$$V_Z = \frac{Q_{FC,Z}}{hw} = \frac{(3.125 \times 10^{-3} \text{ m}^3/\text{s})\,(10^3 \text{ mm/m})}{(0.125 \text{ mm})\,(32.5 \text{ m})}$$
$$= 0.769 \text{ m/s}$$

3. Calculate the osmotic pressure in the feed channel using Eq. 17-7:

$$\pi_{FC,Z} = \frac{(2\text{ mol ion/mol NaCl})\,(1.0)\,(2000\text{ mg/L})\,(0.0831451\text{ L}\cdot\text{bar/K}\cdot\text{mol})\,(293\text{ K})}{(10^3\text{ mg/g})\,(58.4\text{ g/mol})}$$
$$= 1.67\text{ bar}$$

4. The water and solute fluxes depend on the concentration and osmotic pressure in the permeate, which of course depend on the water and solute fluxes. Although a simultaneous numerical solution procedure could be used, it is acceptable to assume C_P and π_P are zero in the first increment for this example. Values calculated in the first increment will be used as an approximation of the values in the next increment.

Part 2

Calculate the water and solute flux and flow rate across the membrane in the first increment.

1. The concentration and osmotic pressure at the membrane wall are higher than in the feed channel because of concentration polarization. However, the concentration polarization factor is dependent on permeate flux, so values for the concentration polarization factor and permeate flux must be determined concurrently by simultaneously solving Eqs. 17-48 and 17-54.

   a. Calculate the Reynolds number, Schmidt number, and k_{CP} using Eqs. 17-37, 17-38, and 17-36. The hydraulic diameter is $2h = 2 \times (0.125\text{ mm}) = 0.25\text{ mm}$. Water density and viscosity at 20°C are $\rho_W = 998\text{ kg/m}^3$ and $\mu_W = 10^{-3}\text{ kg/m . s}$ (Table C-1, App. C):

$$\text{Re} = \frac{\rho v d_H}{\mu} = \frac{(998\text{ kg/m}^3)(0.769\text{ m/s})(0.25\text{ mm})}{(1.0\times10^{-3}\text{ kg/m}\cdot\text{s})(10^3\text{ mm/m})} = 192$$

$$\text{Sc} = \frac{\mu}{\rho D_L} = \frac{1.0\times10^{-3}\text{ kg/m}\cdot\text{s}}{(998\text{ kg/m}^3)(1.35\times10^{-9}\text{ m}^2/\text{s})} = 742$$

$$k_{CP} = \frac{(0.023)(1.35\times10^{-9}\text{ m}^2/\text{s})(192)^{0.83}(742)^{0.33}}{(0.25\text{ mm})(10^{-3}\text{ m/mm})}$$
$$= 8.64\times10^{-5}\text{ m/s}$$

   b. The parameter β can be calculated using Eq. 17-34. Rej is not yet known and is assumed to be 1.0 in the first increment.

In subsequent increments, Rej will be taken as equal to the value calculated in the previous increment:

$$\beta_Z = \exp\left(\frac{J_{W,Z}}{k_{CP}}\right)\text{Rej} + (1 - \text{Rej})$$

$$= \exp\left[\frac{(J_{W,Z}\ \text{L/m}^2\cdot\text{h})(10^{-3}\ \text{m}^3/\text{L})}{(8.64\times10^{-5}\ \text{m/s})(3600\ \text{s/h})}\right] \quad \text{(a)}$$

c. The osmotic pressures in the feed water and at the membrane surface are related by β_Z:

$$C_{M,Z} = \beta_Z C_{FC,Z}$$

Therefore

$$\pi_{M,Z} = \beta_Z \pi_{FC,Z} \quad \text{(b)}$$

d. Substituting Eq. (b) into Eq. 17-48 yields

$$J_{W,Z} = K_W[(P_{FC,Z} - P_{P,Z}) - (\beta_Z \pi_{CF,Z} - \pi_{P,Z})] \quad \text{(c)}$$

e. Solving Eqs. (a) and (c) simultaneously using values given in the problem statement yields $\beta_Z = 1.12$ and $J_{W,Z} = 35.1\ \text{L/m}^2\cdot\text{h}$.

2. The permeate flow rate is calculated by multiplying the flux by the area of the increment:

$$Q_{P,Z} = J_{W,Z}(w)(dz) = \frac{(35.1\ \text{L/m}^2\cdot\text{h})\ (32.5\ \text{m})\ (0.1\ \text{m})}{(10^3\text{L/m}^3)\ (3600\text{s/h})}$$

$$= 3.17\times10^{-5}\ \text{m}^3/\text{s}$$

3. The solute flux can be calculated using Eq. 17-49 after substituting in Eq. (b) (see step 1c above):

$$J_{S,Z} = k_S(\beta_Z C_{FC,Z} - C_{P,Z})$$

$$J_{S,Z} = (6.14\times10^{-4}\text{m/h})[(1.12)(2000\ \text{mg/L}) - 0\ \text{mg/L}](10^3\text{L/m}^3)$$

$$J_{S,Z} = 1375\ \text{mg/m}^2\cdot\text{h}$$

4. Calculate the solute transport across the membrane:

$$M_{S,Z} = J_{S,Z}(w)(dz) = \frac{(1375\ \text{mg/m}^2\cdot\text{h})(32.5\ \text{m})\ (0.1\ \text{m})}{3600\ \text{s/h}} = 1.24\ \text{mg/s}$$

Part 3

Determine P, C, and π on both sides of the membrane in the next increment along with v in the feed channel.

1. The flow in the feed channel is equal to the influent flow minus any permeate production and is calculated using Eq. 17-52:

$$Q_{FC,Z} = Q_F - Q_{P,Z} = 3.125 \times 10^{-3}\,\text{m}^3/\text{s} - 3.17 \times 10^{-5}\,\text{m}^3/\text{s}$$
$$= 3.09 \times 10^{-3}\text{m}^3/\text{s}$$

2. Calculate feed channel velocity:

$$V_Z = \frac{Q_{FC,Z}}{hw} = \frac{(3.09 \times 10^{-3}\ \text{m}^3/\text{s})\ (10^3\ \text{mm/m})}{(0.125\ \text{mm})(32.5\ \text{m})} = 0.761\ \text{m/s}$$

3. The solute concentration in the feed channel of the next increment can be calculated using Eq. 17-53:

$$C_{FC,Z} = \frac{Q_F C_F - M_{S,Z}}{Q_{FC}, Z}$$
$$= \frac{[(3.125 \times 10^{-3}\ \text{m}^3/\text{s})(2000\ \text{mg/L})\ (10^3\ \text{L/m}^3)]\ (-1.24\ \text{mg/s})}{(3.09 \times 10^{-3}\ \text{m}^3/\text{s})\ (10^3\ \text{L/m}^3)}$$
$$= 2020\ \text{mg/L}$$

4. The solute concentration in the permeate of the next increment can be calculated from the water and solute fluxes in the first increment using Eq. 17-58:

$$C_{P,Z} = \frac{J_{S,Z}}{J_{W,Z}} = \frac{1371\ \text{mg/m}^2 \cdot \text{h}}{35.1\ \text{L/m}^2 \cdot \text{h}} = 39.2\ \text{mg/L}$$

5. Calculate the feed channel and permeate osmotic pressures using Eq. 17-7:

$$\pi_{FC,Z} = \frac{(2\ \text{mol ion/mol NaCl})\ (1.0)\ (2020\text{mg/L})\ (0.0831451\ \text{L}\cdot\text{bar/K}\cdot\text{mol})\ (293\ \text{K})}{(10^3\ \text{mg/g})(58.4\ \text{g/mol})}$$
$$= 1.68\text{bar}$$

$$\pi_{P,Z} = \frac{(2\ \text{mol ion/mol NaCl})\ (1.0)\ (39.2\text{mg/L})\ (0.0831451\ \text{L}\cdot\text{bar/K}\cdot\text{mol})\ (293\ \text{K})}{(10^3\text{mg/g})\ (58.4\ \text{g/mol})}$$
$$= 0.03\text{bar}$$

6. The pressure in the feed channel drops due to head loss through the channel, and the head loss is a function of the feed velocity. The head loss in the first increment and pressure in the next increment can be calculated:
   a. The head loss in an incremental length of the membrane element as a function of velocity must be determined from the given head

loss information using Eq. 17-57 rearranged as follows:

$$\delta_{HL} = \frac{h_L}{v^2 L} = \frac{0.2\text{bar}}{(0.5\text{m/s})^2(1\text{ m})} = 0.8\text{bar}\cdot\text{s}^2/\text{m}^3$$

b. Determine the head loss in the increment using Eq. 17-57:

$$h_{L,Z} = \delta_{HL} v_Z^2 dz = (0.8\text{bar}\cdot\text{s}^2/\text{m}^3)(0.769\text{ m/s})^2(0.1\text{m}) = 0.047\text{bar}$$

c. Determine pressure in the next increment:

$$P_{FC,Z} = 14.2\text{ bar} - 0.047\text{ bar} = 14.15\text{ bar}$$

Part 4

Repeat Parts 2 and 3 for the second and subsequent increments. The results are shown in the table below:

Increment (z)	Unit	1	2	3	4	5	...	10
$Q_{FC,Z}$	m³/s	3.125×10^{-3}	3.093×10^{-3}	3.062×10^{-3}	3.030×10^{-3}	2.999×10^{-3}		2.845×10^{-3}
V_Z	m/s	0.7692	0.7614	0.7536	0.7459	0.7382		0.7003
$P_{FC,Z}$	bar	14.20	14.15	14.11	14.06	14.02		13.81
$h_{L,Z}$	bar	0.047	0.046	0.045	0.045	0.044		0.039
$C_{FC,Z}$	mg/L	2000	2020	2041	2062	2084		2196
$\pi_{FC,Z}$	bar	1.67	1.68	1.70	1.72	1.74		1.83
$Q_{P,Z}$	m³/s	3.17×10^{-5}	3.16×10^{-5}	3.14×10^{-5}	3.13×10^{-5}	3.11×10^{-5}		3.03×10^{-5}
$P_{P,Z}$	bar	0.3	0.3	0.3	0.3	0.3		0.3
$C_{P,Z}$	mg/L	0	39.2	39.0	39.6	40.3		43.6
$\pi_{P,Z}$	bar	0	0.03	0.03	0.03	0.03		0.04
$K_{CP,Z}$	m/s	8.64×10^{-5}	8.56×10^{-5}	8.49×10^{-5}	8.42×10^{-5}	8.35×10^{-5}		$7.99\text{X}10^{-5}$
β_Z		1.120	1.120	1.121	1.121	1.122		1.124
$J_{W,Z}$	L/m²h	35.1	35.0	34.8	34.7	34.5		33.6
$J_{S,Z}$	mg/m²h	1374.77	1365.46	1380.43	1395.16	1410.15		1489.01
M_Z	mg/s	1.24	1.23	1.25	1.26	1.27		1.34
RejZ		0.980	0.981	0.981	0.981	0.981		0.981

Part 5

After calculating Part 4 for all increments in the element, the overall performance can be determined.

1. Permeate production from the element is the sum of the permeate produced in each increment:

$$Q_P = \sum_{Z=1}^{10} Q_{P,Z} = 3.1\times10^{-4}\text{ m}^3/\text{s}$$

2. Salt transfer from the element is the sum of the salt transferred in each increment:

$$M_S = \sum_{Z=1}^{10} M_{S,z} = 12.8 \text{ mg/s}$$

3. Permeate concentration:

$$C_P = \frac{M_S}{Q_P} = \frac{12.8 \text{ mg/s}}{(3.10\times10^{-4} \text{ m}^3/\text{s})(10^3 \text{ L/m}^3)} = 41.3 \text{ mg/L}$$

4. Rejection (Eq. 17-1):

$$\text{Rej} = 1 - \frac{C_P}{C_F} = 1 = \frac{41.3 \text{ mg/L}}{2000 \text{ mg/L}} = 0.98$$

5. Recovery (Eq. 17-12):

$$r = \frac{Q_P}{Q_F} = \frac{3.1\times10^{-4} \text{ m}^3/\text{s}}{3.12\times10^{-3} \text{ m}^3/\text{s}} = 0.099$$

Comment

In this example, the performance of a single membrane element has been determined. The concentrate from this element becomes the feed to the next element in series; that is, $Q_{C,1}$, $P_{C,1}$, and $C_{C,1}$ are $Q_{F,2}$, $P_{F,2}$, and $C_{F,2}$. The system permeate flow rate is the sum of the permeate flow from each element. The system permeate concentration is the flow-averaged permeate concentration from each element.

MANUFACTURER SOFTWARE

In Example 17-5 pressure was used as an input variable and a value for recovery was generated. Normally, the desired recovery is determined from limiting salt calculations (taking acid and antiscalant addition into account), and design calculations generate the feed pressure required for a particular membrane element. Using these equations, an iterative solution would be necessary. The design calculations are also repeated with varying membrane elements and array configurations. In addition, other process parameters, such as permeate backpressure and interstage booster pumps, can affect system design and performance. Thus, design is an iterative process and typically takes place with the cooperation of several membrane system manufacturers. Manufacturers provide design software to perform these calculations, which are based on the principles presented in this chapter, and incorporate issues such as osmotic pressure, limiting salt solubility, mass transfer rates, concentration polarization, and permeate water quality. As such, manufacturers' software is reliable for predicting effluent water quality from a specific membrane system design and a given set of operating conditions. An example of the output from a vendor-supplied RO design program is shown in Table 17-5.

FUNCTIONAL SPECIFICATIONS

Because design criteria cannot be developed independently of manufacturer data, procurement of RO systems is often accomplished by means of a functional specification. By this method, an engineer develops the system requirements, designs the pretreatment processes, designs the RO system support facilities, and defines the basic requirements of the RO system. The functional specifications outline the operating requirements of the system, physical constraints of the system, and warranty agreements between the manufacturer and the owner. Bid proposals are returned by the interested manufacturers that outline the particulars of the system being supplied, estimates of system product quality as a function of time, system capital costs, and system operating costs as a function of time. The proposals are typically reviewed by the engineer to determine the optimum life-cycle cost.

Pilot Testing

An important aspect of long-term RO operation is loss of performance due to compaction, fouling, or degradation of the membrane. Limiting salt calculations can be a good predictor of the recovery that can be achieved without causing scaling. Antiscalants can allow supersaturation (i.e., higher recovery) without scaling, but their effectiveness might be dependent on other water quality parameters. SDI and MFI tests can indicate when feed water quality is unacceptable, but low values do not assure that fouling will be minimal. Therefore, it is necessary to perform pilot testing for nearly all RO installations. Pilot testing is guided by membrane system selection and operating conditions developed during array design and serves to verify the array design criteria and identify pretreatment requirements to prevent excessive fouling.

COMMERCIAL RO PILOT PLANTS

Reverse osmosis pilot plant systems are typically available from membrane manufacturers or consulting engineering firms. A typical commercially available skid-mount system is shown on Fig. 17-18. This skid unit contains six pressure vessels, each containing spiral-wound membrane elements in series. The pressure vessels can be operated as two independent systems, with each system containing three pressure vessels that can be piped as a 2×1 array, which allows membranes from two manufacturers to be tested simultaneously. The pilot plant system is operated with a programmable logic controller (PLC). Chemicals are added to the feed water to prevent fouling of the membrane. Manufacturer-supplied specifications for pilot plant systems are usually provided so that the pilot unit can be properly operated. These specifications are usually obtained from the manufacturer

Table 17-5
Example output from vendor-supplied RO design program[a]

Hydranautics Membrane System Design Software, v. 8.00 © 2002 3/11/03
RO program licensed to: K Howe
Calculation created by: K Howe
Project name: MWH Example

HP pump flow:	4666.7 gpm	Permeate flow:	3500.0 gpm
Recommended pump press:	204.4 psi	Raw-water flow:	4666.7 gpm
Feed pressure:	175.4 psi	Booster pump pressure:	10.0 psi
Feed water temperature:	15.0°C (59°F)	Permeate recovery ratio:	75.0%
Raw water pH:	8.00	Element age:	5.0 years
Acid dosage, ppm (100%):	131.1 H_2SO_4	Flux decline % per year:	7.0
Acidified feed CO_2:	127.3	Salt passage increase, %/yr:	10.0
Average flux rate:	15.8 gfd	Feed type:	Well water

Stage	Perm. Flow, gpm	Flow/Vessel Feed, gpm	Flow/Vessel Conc, gpm	Flux, gfd	Beta	Concentration and Throt. Pressures psi	Concentration and Throt. Pressures psi	Element Type	Element No.	Array
1-1	2623.6	53.0	23.2	17-9	1.16	149.5	0.0	ESPA3	528	88 × 6
1-2	876.4	45.4	25.9	11.7	1.08	133.1	0.0	ESPA3	270	45 × 6

	Raw water		Feed water		Permeate		Concentrate	
Ion	mg/L	$CaCO_3$	mg/L	$CaCO_3$	mg/L	$CaCO_3$	mg/L	$CaCO_3$
Ca	8.0	20.0	8.0	20.0	0.27	0.7	31.2	77.7
Mg	2.0	8.2	2.0	8.2	0.07	0.3	7.8	32.1
Na	734.3	1596.3	734.3	1596.3	115.11	250.2	2591.9	5634.5
K	8.0	10.3	8.0	10.3	1.52	2.0	27.4	35.2
NH_4	0.0	0.0	0.0	0.0	0.00	0.0	0.0	0.0
Ba	0.004	0.0	0.004	0.0	0.000	0.0	0.016	0.0
Sr	2.000	2.3	2.000	2.3	0.069	0.1	7.794	8.9
CO_3	3.0	5.0	0.2	0.4	0.00	0.0	0.8	1.4
HCO_3	631.0	517-2	473.5	388.1	174.26	142.8	1371.3	1124.0
SO_4	79.0	82.3	207.5	216.1	7.41	7.7	807.7	841.3
Cl	730.0	1029.6	730.0	1029.6	72.28	101.9	2703.2	3812.6
F	1.1	2.9	1.1	2.9	0.28	0.7	3.6	9.4
NO_3	0.0	0.0	0.0	0.0	0.00	0.0	0.0	0.0
SiO_2	24.0			24.0	5.83		78.5	
TDS	2222.4		2190.6		377.1		7631.2	
pH	8.0		6.8		6.4		7.3	

	Raw Water	Feed Water	Concentrate
$CaSO_4/K_{sp} \times 100$:	0%	0%	2%
$SrSO_4/K_{sp} \times 100$:	2%	5%	29%
$BaSO_4/K_{sp} \times 100$:	7%	17%	97%
SiO_2 saturation:	20%	20%	65%
Langelier saturation index (LSI)	−0.14	−1.47	0.04
Stiff–Davis saturation index	−0.20	−1.53	−0.24
Ionic strength	0.03	0.04	0.13
Osmotic pressure	22.2 psi	21.3 psi	74.2 psi

Table 17-5 *(Continued)*

Stage	Element No.	Feed Pressure, psi	Pressure Drop, psi	Permeate Flow, gpm	Permeate Flux, gfd	Beta	Permeate TDS	Concentrate Osmotic Pressure	Concentrate Saturation Level, %				
									$CaSO_4$	$SrSO_4$	$BaSO_4$	SiO_2	LSI
1-1	1	175.4	6.5	5.7	20.5	1.11	116.6	23.8	1	6	20	22	−0.9
1-1	2	168.9	5.5	5.4	19.4	1.12	126.5	26.7	1	7	23	25	−0.7
1-1	3	163.4	4.6	5.1	18.3	1.12	137.8	30.2	1	8	27	28	−0.6
1-1	4	158.8	3.8	4.8	17-2	1.13	151.0	34.4	2	9	32	32	−0.4
1-1	5	155.0	3.1	4.5	16.1	1.15	166.2	39.6	2	11	38	36	−0.3
1-1	6	151.8	2.5	4.1	14.9	1.16	203.0	45.9	2	14	47	42	−0.1
1-2	1	156.3	5.4	4.1	14.6	1.09	225.4	49.8	3	16	52	45	0.0
1-2	2	150.9	4.7	3.7	13.4	1.09	251.4	54.0	3	18	59	49	0.1
1-2	3	146.3	4.1	3.4	12.2	1.09	279.6	58.5	3	20	66	53	0.1
1-2	4	142.1	3.6	3.1	11.1	1.09	309.1	63.2	4	22	74	56	0.2
1-2	5	138.5	3.2	2.8	10.0	1.09	341.4	68.2	4	25	84	60	0.3
1-2	6	135.4	2.8	2.5	8.9	1.08	374.9	73.3	5	28	94	64	0.3

[a]These calculations are based on nominal element performance when operated on a feed water of acceptable quality. No guarantee of system performance is expressed or implied unless provided in writing by Hydranautics.

Figure 17-18
Typical reverse osmosis pilot plant.

and provide useful guidelines when planning and operating the pilot plant units.

PILOT TEST PARAMETERS

For most RO pilot studies, the following parameters should be recorded:

1. Date and time of sample analysis
2. Flow rates (feed, concentrate, and permeate)
3. Pressure (feed, concentrate, and permeate)
4. Feed water temperature
5. Conductivity (online reading recommended)
6. Power consumption
7. Chemical usage
8. pH (feed, concentrate, and permeate)

Additional reporting and recording requirements are available elsewhere (ASTM, 2001c, 2001d).

Pretreatment

Pretreatment is necessary to prevent scaling and fouling. The common pretreatment strategies include the injection of acids and antiscalants to prevent the precipitation of sparingly soluble salts and filtration to prevent plugging by particulate matter. Very clean source water (such as groundwater) often can operate with only cartridge filtration prior to the membrane units, but more advanced filtration methods, including coagulation, flocculation, sedimentation, and granular filtration, or membrane filtration, are commonly required with surface water intake facilities. Pretreatment must be selected and designed in concert with the array design because the membrane element performance is dependent on the level of pretreatment. Additional details on the design of pretreatment systems is available in design manuals such as AWWA (2007).

Posttreatment

The permeate from an RO facility typically requires additional treatment. Feed water pH adjustment prior to RO, along with extensive removal of divalent ions by the RO process, produces treated water with low pH, low alkalinity, and low hardness, which are conditions that cause water to be corrosive. Anaerobic groundwater frequently contains hydrogen sulfide, which passes through the membrane and causes odor problems in the treated water. Finally, residual disinfection is always required for municipal water distribution.

PERMEATE STABILITY

A number of strategies can be used to increase the stability (reduce the corrosivity) of the water. When the feed water is acidified for scale control, carbonate alkalinity in the raw water is converted to carbonic acid, which passes through the membrane. Thus, addition of a base such as caustic soda can restore both pH and alkalinity to acceptable levels. Without additional measures, however, such water will still be corrosive. Stability can be improved by adding hardness ions to the water, so base addition with chemicals containing calcium is sometimes preferred over caustic soda. Lime and soda ash are common chemicals for increasing the stability of RO permeate. Small systems sometimes can add an acceptable amount of hardness by passing the permeate through a bed of calcareous media such as dolomite or calcite. In lieu of adding hardness to the water, corrosion inhibitors may be effective. Another strategy for producing a stable finished water is to blend the permeate with a by pass stream of raw water that meets all other water treatment requirements (such as filtration if a surface water source is used). Proper blending of raw and permeate water may produce a finished water with the desired pH, alkalinity, and hardness. However, DBP precursor concentration in the raw water and the potential for DBP formation need to be evaluated when considering blending options. The importance of finished-water stability is discussed in additional detail in Chap. 22.

HYDROGEN SULFIDE

Anaerobic groundwater can contain hydrogen sulfide, a highly odorous compound that is not removed during RO. Hydrogen sulfide can be removed by oxidation or aeration. Oxidation to sulfate can be accomplished with oxidants such as chlorine, but large doses are needed (the stoichiometric chlorine requirement is about 9 times the hydrogen sulfide concentration on a mass basis and insufficient amounts can oxidize sulfide to elemental sulfur, which is equally undesirable). Thus, hydrogen sulfide is commonly removed after the membrane process in an air-stripping process using countercurrent packed towers, which are discussed in Chap. 14. Since hydrogen sulfide is a weak acid, the pH of the water will have a significant impact on its removal efficiency (Howe and Lawler, 1989). Odor control can be a significant issue when stripping water that contains sulfide.

It is necessary to consider all posttreatment goals simultaneously and select treatment options that achieve all objectives. For instance, air stripping to remove sulfide before base addition will strip carbon dioxide and increase the permeate pH; subsequent pH adjustment with caustic soda will not restore alkalinity because the carbonate will be gone. Alternatively, pH adjustment before stripping can prevent effective stripping because sulfide is present as ionic hydrogen bisulfide rather than gaseous hydrogen sulfide.

DISINFECTION

Chlorine is commonly used for disinfection and is discussed in Chap. 13. The RO process is effective at removing DBP precursors; thus, free chlorination can typically be practiced without forming significant quantities of DBPs. However, care must be used if the RO permeate is to be blended with either raw water (for stability, see above) or a fresh water supply. Blending may increase DBP formation when using free chlorine. Cases have been observed when the blending of desalinated seawater into freshwater can increase the DBP formation of the freshwater, even though the desalinated water has a very low DBP formation potential on its own. Desalinated seawater can have a higher bromide concentration than freshwater sources, so that interactions between bromide from the desalinated seawater and NOM from the freshwater can increase overall DBP formation after chlorination to above what it would be with either water source individually. Thus, bromide removal can be one of the factors that controls the design of RO facilities.

Concentrate Management

A significant concern in the design and operation of inland brackish water RO facilities is the low product water recovery compared to other water treatment processes. Recovery is limited by osmotic pressure in seawater systems and by scaling from sparingly soluble salts in inland brackish water systems. For inland systems, the low recovery has two negative consequences. First, brackish water desalination is typically considered because of a lack of adequate freshwater resources, and inability to recover a high fraction of the feed water is simply a poor use of scarce natural resources. Second, the unrecovered water becomes the concentrate stream and must be disposed of. The increase salinity of the concentrate stream greatly limits available disposal options because of the potential for contaminating the scarce freshwater resources. Thus, there is significant interest in increasing recovery of product water and decreasing the volume of concentrate that must be disposed of.

Increasing recovery from inland brackish water RO facilities involves preventing the precipitation of sparingly soluble salts. As noted earlier, scale inhibitors are used to prevent precipitation and increase recovery up to a point. However, scale inhibitors are limited in their effectiveness, and more aggressive strategies typically must be employed to achieve recovery of greater than 90 percent.

One strategy is to provide an intermediate treatment process between two stages of RO membranes. Since calcium is often the limiting cation, lime softening can be an effective intermediate strategy. Softening can also be effective at removing other scale-causing constituents. Gabelich et al. (2007) found that increasing pH to between 10.5 and 11.5 with NaOH was able to remove 88 to 98 percent of Ca^{2+}, Ba^{2+}, Sr^{2+}, and 67 percent of silica. However, the high alkalinity and hardness present after a first stage of RO can lead to high doses of lime or NaOH; doses in excess of 1000 mg/L have been reported in experimental studies. Similarly high doses of acid can be necessary to reduce the pH after softening. The high doses also lead to a large amount of waste production. Seeding with calcite or gypsum crystals has also been explored as a way of improving the effectiveness of the intermediate precipitation process (Rahardianto et al., 2007). Fluidized bed crystallization using sand as a seed material has also proved effective in bench-scale testing (Sethi et al., 2008). Ion exchange is another possibility for interstage treatment for the removal of scale-causing constituents that may result in less waste production (Howe et al., 2010).

Several patented or proprietary processes have been developed to increase recovery from brackish RO systems. The patented high-efficiency reverse osmosis (HERO) process involves pretreatment to reducing scaling, followed by pH adjustment and additional stages of reverse osmosis. Hardness is typically removed using a cation exchange column that removes calcium and magnesium, and carbonate is removed by stripping carbon dioxide in a countercurrent packed column (see Chap. 14). The pH is then increased using caustic soda, typically above pH = 10. Since calcium and carbonate have been removed, calcium carbonate scaling at high pH is no longer a concern and the concentrate is fed into another stage of reverse osmosis. At pH above 10, silica and borate are transformed from neutral to ionic species, the solubility of silica is increased and scaling potential is reduced, the rejection of silica and borate is increased, the potential for organic fouling or biofouling is decreased, and cleaning costs are reduced. Recovery of 90 to 98 percent has been achieved.

Another proprietary system is the SAL-PROC system developed by Geo-Processors, Inc. This process uses are variety of treatment steps, including chemical addition, heating, cooling,and sequential concentration steps that may include more RO or evaporation. The SAL-PROC system is potentially capable of producing usable and possibly sellable salt products and slurries from the RO concentrate.

Another option that has been explored in research to prevent scaling and potentially increase recovery include the vibratory shear-enhanced process (VSEP) in which a membrane system is vibrated to prevent scale from forming on the membrane surface (Chang, 2008). Researchers have also explored other electrodialysis reversal, membrane distillation, or other desalination processes as a second-state desalination system after an intermediate-scale reduction process (Sethi et al., 2008).

Brine concentrators and crystallizers are additional technologies to reduce the volume of concentrate, and can lead to zero liquid discharge (ZLD), in which the only residuals from the facility are solids, which are then easier to dispose of (Mickley, 2006). While brine concentrators and crystallizers are used in some industrial processes such as the power generation industry, they are expensive, energy intensive, and have not yet been used in municipal water treatment industry. Brine concentrators and crystallizers are discussed in more detail in Chap. 21.

Disposal of Residuals

Disposal of the concentrate stream is frequently a challenge in RO plant design. The factors that contribute to this problem are identified in Table 17-6. In addition to the concentrate stream, RO plants must also dispose of spent cleaning solutions. Both of these residuals are discussed in this section.

CONCENTRATE

Several surveys of concentrate disposal methods are available (Kenna and Zander, 2001; Mickley et al., 1993; Truesdall et al., 1995). The most common concentrate disposal options in the United States are (1) discharge to a brackish surface water (include oceans, brackish rivers, or

Table 17-6
Factors affecting concentrate disposal

Issue	Description
Volume	The waste stream volume from many water treatment processes is less than 5% of the feed stream volume. In RO, the waste stream volume ranges from 15 to 50% of the feed stream volume.
Salinity/toxicity	The high salinity of the concentrate stream makes it toxic to many plants and animals, limiting options for land application or surface water discharge and rendering it unusable for recycling or reuse. Many concentrate streams are anaerobic, which can be toxic to fish without sufficient dilution. In addition, RO processes used for specific contaminant removal (i.e., arsenic, radium) may produce concentrate streams that can be classified as a hazardous material.
Regulations	Concentrate is classified as an industrial waste by the U.S. EPA. Concentrate disposal is regulated under several different federal, state, and local laws, and the interaction between these regulatory requirements can be complex (Kimes, 1995; Pontius et al., 1996). Regulatory considerations are often as important as cost and technical considerations for determining viable concentrate disposal options.

estuaries), (2) discharge to a municipal sewer, and (3) deep-well injection. In the United States, about half of all plants discharge concentrate to a surface water, a third discharge to a municipal sewer, and about 10 percent discharge to a deep well. Deep-well injection is most common in Florida. Evaporation ponds are used by a small number of facilities. Concentrate disposal is an integral part of the design of RO facilities and disposal options are discussed in more detail in Chap. 21 of this text.

An alternative to disposal of concentrate is to identify beneficial uses for the concentrate or its constituent salts and minerals. Possible beneficial uses that have been explored in various research projects include (1) land application or irrigation of salt-tolerant crops, (2) saline aquaculture, farming of brine shrimp or other saltwater species, (3) restoration of brackish waterways or development of saltwater marshes, wetlands, or habitats, (4) energy generation using solar gradient ponds, (5) industrial uses as feedstock or process stream, (6) production of marketable salts or mineral commodities (Ahuja and Howe, 2005; Everest and Murphree, 1995). At the current time, however, beneficial uses for the concentrate have not been identified at most facilities.

CLEANING SOLUTIONS

Spent cleaning solutions from RO plants are frequently acidic or basic solutions and contain detergents or surfactants. In many cases, the cleaning solution volume is small compared to the concentrate stream and can be diluted into and disposed of with the concentrate. In some cases, treatment of the cleaning solution may be required prior to disposal, but treatment may consist only of pH neutralization. Detergents and surfactants should be selected with disposal issues in mind.

Energy Recovery

Reverse osmosis is an energy-intensive process. The theoretical thermodynamic minimum energy requirement for desalinating seawater, based solely on the pressure required to overcome the osmotic pressure, is 0.70 kWh/m^3. This value is significantly higher than the typical energy required for the treatment of freshwater. A significant component of operating costs is electrical power for the feed pumps because of the high pressure necessary to operate RO membranes. Although pressure drops significantly as permeate passes through the membrane, the head loss through the feed channels is relatively small, and the concentrate exits the final membrane element at 80 to 90 percent of the feed pressure, with backpressure maintained by a concentrate control valve. If concentrate is discharged to a deep well, a portion of this pressure can be used to drive the disposal process. If, however, the concentrate is discharged to a surface water, this pressure must be dissipated prior to discharge. Pressure in the concentrate stream dissipated across the concentrate control valve is wasted energy because it performs no useful work in the treatment system. Because the concentrate steam is both high energy and relatively high volume, the amount of wasted energy is substantial.

Energy recovery devices are being used more frequently to reclaim the wasted energy in the concentrate stream. Several types of devices are available, including reverse-running turbines, Pelton wheels, pressure exchangers, and electric motor drives (Geisler et al., 1999; Harris, 1999; Oklejas and Pergande, 2000; Tomkins and Nemeth, 2001). Typically, recovered energy from the residual pressure of the concentrate stream is used to pressurize the feed stream. In some systems, the concentrate stream spins a rotor, losing energy in the process, and exits the energy recovery device at a significantly lower pressure. In the reverse-running turbine and pressure exchanger, the energy recovery device is in contact with both the feed and concentrate streams, with a single rotor transferring pressure from the concentrate to the feed stream. Pressure exchanges allow direct contact between the feed and concentrate streams via a rotating rotor, and are thus able to transfer the pressure from the concentrate stream directly to the feed stream. Pelton wheel devices use a rotor connected directly to the feed pump via an extended shaft, and the energy recovered from the concentrate stream provides hydraulic assistance to the operation of the feed pumps. The main moving part is the Pelton wheel and shaft. Electric motor drives are more complex, utilizing a hydraulic drive system connected to the pump motor.

More than 90 percent of the energy expended to pressurize the concentrate stream can be recovered. Depending on the price for electricity, capital costs of energy recovery equipment may be recouped within 3 to 5 years. Energy recovery devices were first utilized on seawater RO systems because they operate at high pressure and low recovery, compounding the energy loss. Recent trends and improvements in energy recovery equipment and rising electricity prices suggest that energy recovery will be applied in more and more low-pressure systems.

In addition to providing pressure to the feed stream, another application is to use the energy recovery system to add pressure between stages (Duranceau et al., 1999). In normal operation, the second or later stages produce less permeate because of lower applied pressure (due to pressure drop in the first stage) and higher osmotic pressure (due to concentration of the feed stream in the first stage). The lower permeate flow and higher feed concentration also increase salt passage and degrade permeate quality. These effects are sometimes counteracted by installing booster pumps between stages, so that a higher feed pressure is available to offset the higher osmotic pressure. By using energy recovery devices to boost pressure between stages, the booster pumps can be eliminated, which offsets a portion of the capital cost of the energy control device.

Problems and Discussion Topics

17-1 Discuss key similarities and differences between membrane filtration and RO.

17-2 Explain why dissolved gases such as CO_2 and H_2S are poorly rejected by RO membranes.

17-3 Calculate the total osmotic pressure of seawater at a temperature of 20°C using the ion concentrations shown in Table 17-2 and $\varphi = 1$. Calculate the osmotic pressure of a solution containing an equivalent concentration of sodium chloride (i.e., 35,200 mg/L NaCl) also using $\varphi = 1$. Explain and discuss the difference between the two results and discuss Fig. 17-9 in the context of these results.

17-4 The following solutions are representative of common applications of reverse osmosis. Calculate the osmotic pressure of each at 20°C. Discuss the importance of osmotic pressure and how it affects the applied pressure for these applications.

a. NaCl = 35,000 mg/L (representative of seawater RO).

b. NaCl = 8000 mg/L (representative of brackish water RO).

c. Hardness = 400 mg/L as CaCO3 (representative of softening NF).

d. Natural Organic Matter (NOM) = 25 mg/L (representative of using NF to control DBP precursors. Assume the NOM has an average MW of 1000 g/mol.).

17-5 Seawater RO facilities are restricted to a maximum applied pressure of about 85 bar (1200 psi) because of equipment limitations. Using the seawater composition shown in Table 17-2, calculate the maximum recovery that can be achieved before the osmotic pressure at the membrane surface (at the exit from a membrane module) is equal to the applied pressure. Assume 100 percent rejection, a temperature of 15°C, and a concentration polarization factor of 1.12. Discuss how the results of this calculation compare to the typical recovery achieved by seawater RO facilities. Does osmotic pressure lead to any practical limitations on the size of the waste stream from a seawater RO facility?

17-6 Operating data for a low-pressure RO system on two different days are shown in the table below:

	Unit	Day 1	Day 2
Water temperature	°C	13	22
Water flux	L/m^2 · h	17-5	18.8
Feed pressure	bar	41.9	38.7
Concentrate pressure	bar	39.0	35.8
Permeate pressure	bar	0.25	0.25
Feed TDS concentration	mg/L	10,500	10,200
Permeate TDS concentration	mg/L	120	120
Recovery	%	66	68

Performance data for this membrane element were developed using the following standard conditions:

	Unit	Standard
Temperature	°C	20
Feed pressure	bar	40
Permeate pressure	bar	0
Head loss per element	bar	0.4
Number of elements	no.	7
Feed TDS concentration	mg/L	10,000
Permeate TDS concentration	mg/L	100
Recovery	%	70

Determine the difference in system performance (water flux and rejection) between the two days using the temperature correction formula in this text and an arithmetic average for the solute concentration in the feed–concentrate channel. Assume the salts in the feed water are sodium chloride for the purpose of calculating osmotic pressures.

17-7 In Eq. 17-10 the solute flux is dependent on the concentration gradient and independent of pressure; also it was noted that solute flux is dependent on temperature. However, Eq. 17-26 includes a correction factor for pressure and not temperature, from which it appears that rejection is dependent on pressure and independent of temperature. Show mathematically and explain (1) how rejection can be dependent on pressure when solute flux is independent of pressure and (2) why there is no temperature correction factor for rejection when there is a temperature correction factor for water flux.

17-8 Examine the importance of the diffusion coefficient on concentration polarization by plotting β as a function of the diffusion coefficient for diffusion coefficient values between 10^{-10} m^2/s (typical of NOM with a diameter of 5 nm) and 1.35 x 10^{-9} m^2/s

(sodium chloride). Use feed channel velocity 0.65 m/s, permeate flux 25 L/m^2 · h, hydraulic diameter 0.5 mm, and temperature 20°C. Discuss the implications that this graph has on the accumulation of material at the membrane surface.

17-9 Examine the importance of temperature on concentration polarization by plotting β as a function of temperature for values between 1 and 30°C. Use feed channel velocity 0.65 m/s, permeate flux 25 L/m^2 · h, hydraulic diameter 0.5 mm, and calculate the diffusion coefficient from the Nernst–Haskell equation given in Chap. 7 (Eq. 7-36) for sodium chloride. Discuss how temperature will impact water and solute flux across the membrane from the perspective of concentration polarization.

17-10 An SDI test was performed to evaluate the fouling tendency of potential RO source water. The time to collect 500mL of water was measured as 24 s. Filtration continued for a total of 15 min, and then a second 500 mL was collected. The time necessary to collect the second 500-mL sample was 32 s. Calculate the SDI.

17-11 Calculate the MFI from the following experimental data:

Time, min	Volume Filtered, L	Time, min	Volume Filtered, L	Time, min	Volume Filtered, L
0	0	5.5	5.37	11.0	9.86
0.5	0.63	6.0	5.80	11.5	10.24
1.0	1.17	6.5	6.23	12.0	10.61
1.5	1.68	7.0	6.65	12.5	10.98
2.0	2.16	7.5	7.07	13.0	11.35
2.5	2.64	8.0	7.48	13.5	11.71
3.0	3.11	8.5	7.89	14.0	12.06
3.5	3.58	9.0	8.29	14.5	12.41
4.0	4.03	9.5	8.69	15.0	12.75
4.5	4.48	10.0	9.08		
5.0	4.93	10.5	9.47		

17-12 An RO facility is being designed to treat groundwater containing the ions given below. Calculate the allowable recovery before scaling occurs and identify the limiting salt. Assume 100 percent rejection, a concentration polarization factor of 1.08, and $T = 25$°C, and ignore the impact of ionic strength. The water contains calcium = 105 mg/L, strontium = 2.5 mg/L, barium = 0.0018 mg/L, sulfate = 128 mg/L, fluoride = 1.3 mg/L, and silica = 9.1 mg/L as Si.

17-13 A groundwater has a calcium concentration of 125 mg/L, alkalinity of 180 mg/L as $CaCO_3$, and pH of 7.1. Calculate the degree of supersaturation of calcium carbonate (ratio of actual concentration to the saturated concentration for each ion) at 60 percent recovery. Calculate the adjusted pH value and acid (HCl) dose necessary to prevent calcium carbonate precipitation at this recovery. Assume 100 percent rejection, $\beta = 1.12$, and $T = 25°C$, and ignore ionic strength.

17-14 Feed water to a proposed low-pressure RO facility has a barium concentration of 0.2 μg/L and a sulfate concentration of 420 mg/L. The planned recovery is 80 percent. Calculate the concentration polarization allowable before the solubility of barium sulfate is exceeded. Assume 100 percent rejection and $T = 25°C$, and ignore the impact of ionic strength.

17-15 Reverse osmosis facilities can be designed with multiple stages (concentrate from one stage is fed to the next stage) or multiple passes (permeate from one stage is fed to the next stage). Explain the difference in permeate quantity and quality expected from these systems.

17-16 Concentrate-staged membrane arrays can be designed with a booster pump in the concentrate line between stages. Explain the benefits of this interstage booster pump and the impact it has on permeate quantity and quality.

17-17 Design criteria for an RO system are given in the following table:

Item	Unit	Value
Membrane properties		
Element length	m	1
Element membrane area	m^2	32.5
Feed channel height (spacer thickness)	mm	0.125
Water mass transfer coefficient (k_W)	$L/m^2 \cdot h \cdot bar$	1.25
Solute mass transfer coefficient (k_S)	m/h	3.29×10^{-4}
Element head loss (at design velocity of 0.5 m/s)	bar	0.1
Operating conditions		
Feed flow (Q_F)	m^3/d	19,000
Feed pressure (P_F)	bar	34
Feed concentration (C_F)	mg/L NaCl	8500
Feed temperature (T_F)	°C	20
Permeate pressure (P_p)	bar	0.3

The system is to be designed as a 2 × 1 array with 80 pressure vessels in the first stage and 40 pressure vessels in the second stage, and with 7 membrane elements in each pressure vessel.

a. Using a spreadsheet or computer program, calculate and graph (1) the feed flow rate entering each element, (2) the feed concentration entering each element, (3) the concentration polarization factor β at each element, (4) the permeate flow rate produced by each element, and (5) permeate salt concentration produced by each element. For the purposes of this problem, assume that the operating conditions are constant across the length of each individual element. Assume that the feed water salinity is due entirely to NaCl, $\varphi = 0.94$, and $D_{NaCl} = 1.58 \text{ x } 10^{-9}$ m^2/s (from Table 7-1 in Chap. 7).

b. Calculate the average permeate flow rate and concentration for each stage and for the whole array.

c. Calculate overall recovery, rejection, and average water flux.

d. Discuss any observations about the quantity and quality of water produced by the first element compared to the last element, and explain the observed trend in β.

17-18 Calculate and plot water flux and salt rejection as a function of recovery, for recovery ranging from 50 to 85 percent, given $C_F = 10{,}000$ mg/L NaCl, $\Delta P = 50$ bar, $k_W = 2.2\ L/m^2 \cdot h \cdot bar$ and $k_S = 0.75\ L/m^2 \cdot h$, $\varphi = 1$, and $T = 20°C$. Comment on the effect of recovery on RO performance.

17-19 A new brackish water RO system is being proposed. The water quality is as shown in the table below. Using RO manufacturer design software (provided by the instructor or obtained from a membrane manufacturer website), develop the process design criteria for the plant. The required water demand is 38,000 m^3/d and the finished-water TDS should be 500 mg/L or lower.

Constituent	Concentration, mg/L	Constituent	Concentration, mg/L
Ammonia	1.3	Bicarbonate	680
Barium	0.04	Chloride	890
Calcium	20	Fluoride	0.7
Iron	0.5	Orthophosphate	0.7
Magnesium	2.5	Sulfate	105
Manganese	0.02	Silica	21.5
Potassium	17	Nitrate	1.2
Sodium	875	Hydrogen sulfide	0.3
Strontium	2.17		
pH	7.8	Turbidity	0.3 NTU
SDI	$<1\ min^{-1}$	Temperature	15°C

17-20 A new seawater RO system is being proposed. The water quality is as shown in the table below. Using RO manufacturer design software (provided by the instructor or obtained from a membrane manufacturer website), develop the process design criteria for the plant. The required water demand is 4000 m^3/d and the finished-water TDS should be 500 mg/L or lower.

Constituent	Concentration, mg/L	Constituent	Concentration, mg/L
Aluminum	0.15	Strontium	6.6
Ammonia	0.092	Bromide	51
Barium	0.00	Bicarbonate	112
Boron	4.3	Chloride	18,900
Calcium	439	Fluoride	0.61
Iron	0.1	Phosphate	0.12
Magnesium	1,240	Sulfate	2380
Potassium	425	Silica	0.86
Sodium	10,100	Hydrogen sulfide	0.0
Strontium	6.6		
pH	8.0	Turbidity	3.3 NTU
SDI	<1 min^{-1}	UV_{254}	0.03/cm
Temperature	15°C		

17-21 A new membrane softening system is being proposed. The water quality is as shown in the table below. Using RO manufacturer design software (provided by the instructor or obtained from a membrane manufacturer website), develop the process design criteria for the plant. The required water demand is 14,200 m^3/d and the finished-water hardness should be between 50 and 75 mg/L as CaCO3.

Constituent	Concentration, mg/L	Constituent	Concentration, mg/L
Ammonia	1.5	Bicarbonate	135.1
Barium	0.0	Bromide	0.0
Calcium	100	Carbonate	0.11
Magnesium	10	Chloride	98.8
Manganese	0.002	Fluoride	0.5
Sodium	60	Phosphate	0.5
Strontium	1.0	Sulfate	167.6
		Silica	15.0
pH	7.0	Temperature	20°C
SDI	<1min^{-1}		

References

Ahuja, N., and Howe K. J. (2005) "Beneficial Use of Concentrate from Reverse Osmosis Facilities," paper presented at the American Water Works Association Annual Conference, San Francisco, CA.

ASTM (2001a) D3739-94 Standard Practice for Calculation and Adjustment of the Langelier Saturation Index for Reverse Osmosis, in *Annual Book of Standards*, Vol. 11.02, American Society for Testing and Materials, Philadelphia, PA.

ASTM (2001b) *D4189-95 Standard Test Method for Silt Density Index (SDI) of Water*, American Society for Testing and Materials, Philadelphia, PA.

ASTM (2001c) D4195-88 Standard Guide for Water Analysis for Reverse Osmosis Application, in *Annual Book of Standards*, Vol. 11.02, American Society for Testing and Materials, Philadelphia, PA.

ASTM (2001d) D4472-89 Standard Guide for Recordkeeping for Reverse Osmosis Systems, in *Annual Book of Standards*, Vol. 11.02, American Society for Testing and Materials, Philadelphia, PA.

ASTM (2001e) D4516-00 Standard Practice for Standardizing Reverse Osmosis Performance Data, in *Annual Book of Standards*, Vol. 11.02, American Society for Testing and Materials, Philadelphia, PA.

ASTM (2001f) D4582-91 Standard Practice for Calculation and Adjustment of the Stiff and Davis Stability Index for Reverse Osmosis, in *Annual Book of Standards*, Vol. 11.02, American Society for Testing and Materials, Philadelphia, PA.

AWWA (2007) *Reverse Osmosis and Nanofiltration*, AWWA Manual M46, American Water Works Association, Denver, CO.

Baier, J. H., Lykins, B. W., Jr., Fronk, C. A., and Kramer, S. J. (1987) "Using Reverse Osmosis to Remove Agricultural Chemicals from Groundwater," *J. AWWA*, **79**, 8, 55–60.

Bindra, S. P., and Abosh, W. (2001) "Recent Developments in Water Desalination," *Desalination*, **136**, 1–3, 49–56.

Boerlage, S. F. E., Kennedy, M. D., Aniye, M. P., Abogrean, E., Tarawneh, Z. S., and Schippers, J. C. (2003) "The MFI-UF as a Water Quality Test and Monitor," *J. Memb. Sci.*, **211**, 2, 271–289.

Boerlage, S. F. E., Kennedy, M. D., Dickson, M. R., El-Hodali, D. E. Y., and Schippers, J. C. (2002) "The Modified Fouling Index Using Ultrafiltration Membranes (MFI-UF): Characterization; Filtration Mechanisms and Proposed Reference Membrane," *J. Memb. Sci.*, **197**, 1–2, 1–21.

Bremere, I., Kennedy, M., Stikker, A., and Schippers, J. (2001) "How Water Scarcity Will Effect the Growth in the Desalination Market in the Coming 25 Years," *Desalination*, **138**, 1–3, 7–15.

Chang, Y. (2008) Brine Minimization for Brackish Water Desalination Using Vibratory Shear Enhanced Process (VSEP): An Engineering Assessment, Paper presented at the American Water Works Association Annual Conference, Atlanta, GA.

Cheryan, M., and Nichols, D. J. (1992) Modelling of Membrane Processes, pp. 49–98 in S. Thorne (ed.), *Mathematical Modelling of Food Processing Operations*, Elsevier Applied Science, London.

Duranceau, S. J., Foster, J., Losch, H. J., Weis, R. E., Harn, J. A., and Nemeth, J. E. (1999) Interstage Turbines as Self-Regulating Osmotic Pressure Control Valves in a Desalting Facility, paper presented at the American Water Works Association Membrane Technology Conference, Long Beach, CA.

Everest, W. R., and Murphree, T. (1995) "Desalting Residuals: A Problem or a Beneficial Resource?" *Desalination*, **102**, 1–3, 107–117.

Gabelich, C. J., Williams, M. D., Rahardianto, A., Franklin, J. C., and Cohen, Y. (2007) "High-Recovery Reverse Osmosis Desalination Using Intermediate Chemical Demineralization," *J. Memb. Sci.*, **301**, 1–2, 131–141.

Geisler, P., Hahnenstein, F. U., Krumm, W., and Peters, T. (1999) "Pressure Exchange System for Energy Recovery in Reverse Osmosis Plants," *Desalination*, **122**, 2–3, 151–156.

Glater, J. (1998) "The Early History of Reverse Osmosis Membrane Development," *Desalination*, **117**, 1–3, 297–309.

Greenlee, L. F., Lawler, D. F., Freeman, B. D., Marrot, B., and Moulin, P. (2009) "Reverse osmosis Desalination: Water Sources, Technology, and Today's Challenge," *Water Res.* **43**, 9, 2317–2348.

Harris, C. (1999) "Energy Recovery for Membrane Desalination," *Desalination*, **125**, 1-3, 173-180.

Howe, K. J., and Lawler, D. F. (1989) "Acid-Base Reactions in Gas Transfer: A Mathematical Approach." *J. AWWA*, **81**, 1, 61–66.

Howe K. J., Goldman, J., and Thomson, B. M. (2010) Selective Recovery of Desalination Concentrate Salts Using Interstage Ion Exchange, paper presented at the American Water Works Association Annual Conference, Chicago, IL.

Kenna, E., and Zander, A. K. (2001) Survey of Membrane Concentrate Reuse and Disposal, Chap. 29, in S. J. Duranceau (ed.), *Membrane Practices for Water Treatment*, American Water Works Association, Denver, CO.

Kimes, J. K. (1995) "The Regulation of Concentrate Disposal in Florida," *Desalination*, **102**, 1–3, 87–92.

Laidler, K. J., and Meiser,J. H. (1999) *Physical Chemistry*, Houghton Mifflin, Boston.

Lonsdale, H. K. (1972) Theory and Practice of Reverse Osmosis and Ultrafiltration, in R. E. Lacey and S. Loeb (eds.), *Industrial Processing with Membranes*, Wiley- Interscience, New York.

Lonsdale, H. K. (1982) "The Growth of Membrane Technology," *J. Memb. Sci.*, **10**, 2–3, 81–181.

Lonsdale, H. K., Merten, U., and Riley, R. (1965) "Transport Properties of Cellulose Acetate Osmotic Membranes," *J. Appl. Polym. Sci.*, **9**, 1341–1362.

Mariñas, B. J., and Urama, R. I. (1996) "Modeling Concentration-Polarization in Reverse Osmosis Spiral-Wound Elements," *J. Env. Engr — ASCE*, **122**, 4, 292–298.

Merten, U. (1966) *Desalination by Reverse Osmosis, MIT Press*, Cambridge, MA.

Mickley, M. (2006) *Membrane Concentrate Disposal: Practices and Regulation*, U.S. Bureau of Reclamation, Denver, CO.

Mickley, M., Hamilton, R., Gallegos, L., and Truesdall, J. (1993) *MembraneConcentrate Disposal*, American Water Works Association Research Foundation, Denver, CO.

Noordman, T. R., and Wesselingh, J. A. (2002) "Transport of Large Molecules through Membranes with Narrow Pores: The Maxwell-Stefan Description Combined with Hydrodynamic Theory," *J. Memb. Sci.*, **210**, 2, 227–243.

NRC (National Research Council) (2008) *Desalination: A National Perspective* National Academies Press, Washington, DC.

Oklejas, E., and Pergande, W. F. (2000) "Integration of Advanced High-Pressure Pumps and Energy Recovery Equipment Yields Reduced Capital and Operating Costs of Sea water RO Systems," *Desalination*, **127**, 2, 181–188.

Pitzer, K. S. (1975) "Thermodynamics of Electrolytes: 5. Effects of Higher-Order Electrostatic Terms," *J. Solution Chem.* **4**, 3, 249–265.

Pontius, F. W., Kawczynski, E., and Koorse, S. J. (1996) "Regulations Governing Membrane Concentrate Disposal," *J. AWWA*, **88**, 5, 44–52.

Rahardianto, A., Gao, J., Gabelich, C. J., Williams, M. D., and Cohen, Y. (2007) "High-Recovery Membrane Desalination of Low-Salinity Brackish Water: Integration of Accelerated Precipitation Softening with Membrane RO," *J. Memb. Sci.*, **289**, 123–137.

Reid, C. E. (1972) Principles of Reverse Osmosis, in R. E. Lacey and S. Loeb (eds.), *Industrial Processing with Membranes*, Wiley-Interscience, New York.

Ridgway, H. F., and Flemming, H.-C. (1996) Membrane Biofouling, Chap. 9, in J. Mallevialle, P. E. Odendaal, and M. R. Wiesner (eds.), *Water Treatment Membrane Processes*, McGraw-Hill, New York.

Robinson, R. A., and Stokes, R. H. (1959) *Electrolyte Solutions; the Measurement and Interpretation of Conductance, Chemical Potential, and Diffusion in Solutions of Simple Electrolytes*, Butterworths, London.

Sahachaiyunta, P., Koo, T., and Sheikholeslami, R. (2002) "Effect of Several Inorganic Species on Silica Fouling in RO Membranes," *Desalination*, **144**, 1–3/SISI, 373–378.

Schippers, J.C., and Verdouw, J. (1980) "The Modified Fouling Index, a Method for Determining the Fouling Characterstics of Water," *Desalination*, **32**, 137–148.

Schock, G., and Miguel, A. (1987) "Mass Transfer and Pressure Loss in Spiral Wound Modules," *Desalination*, **64**, 339–352.

Sethi, S., Walker, S., Drewes, J., and Xu, P. (2008) Comparison of an Innovative Approach for Concentrate Minimization with Traditional Approaches for Zero Liquid Dicharge, paper presented at the American Water Works Association Annual Conference, Atlanta, GA.

Shakaib, M., Hasani, S. M. F., and Mahmood, M. (2009) "CFD Modeling for Flow and Mass Transfer in Spacer–Obstructed Membrane Feed Channels," *J. Memb. Sci.*, **326**, 2, 270–284.

Sheikholeslami, R., and Bright, J. (2002) "Silica and Metals Removal by Pretreatment to Prevent Fouling of Reverse Osmosis Membranes," *Desalination*, **143**, 3, 255–267.

Sherwood, T. K., Brian, P. L. T., and Fisher, R. E. (1967) "Desalination by Reverse Osmosis," *Ind. Eng. Chem. Fund.*, **6**, 2–12.

Sourirajan, S. (1970) *Reverse Osmosis*, Academic, New York.

Spiegler, K.S., and Kedem, O. (1966) "Thermodynamics of Hyperfiltration (Reverse Osmosis): Criteria for Efficient Membranes," *Desalination*, **1**, 4, 311–326.

Stevens, D., and Loeb, S. (1967) "Reverse Osmosis Desalination Costs Derived from the Coalinga Pilot Plant Operation," *Desalination*, **2**, 1, 56–74.

Stumm, W., and Morgan, J. J. (1996) *Aquatic Chemistry: An Introduction Emphasizing Chemical Equilibria in Natural Waters*, 3rd ed., Wiley, New York.

Taylor, J. S., Thompson, D. M., and Carswell, J. K. (1987) "Applying Membrane Processes to Groundwater Sources for Trihalomethane Precursor Control," *J. AWWA*, **79**, 8, 72–82.

Tomkins, B., and Nemeth, J. E. (2001) Energy Recovery Devices: Comparisons and Case Studies, paper presented at the American Water Works Association Membrane Technology Conference, San Antonio, TX.

Truesdall, J., Mickley, M., and Hamilton, R. (1995) "Survey of Membrane Drinking Water Plant Disposal Methods," *Desalination*, **102**, 1–3, 93–105.

Wang, S., Veerapaneni, V., and Ozekin, K. (2010) Desalination Facility Design and Operation for Maximum Energy Efficiency paper presented at the American Water Works Association Sustainable Water Management Conference, Albuquerque, NM.

Wiesner, M. R., and Aptel, P. (1996) Mass Transport and Permeate Flux and Fouling in Pressure-Driven Processes, Chap. 4, in M. R. Wiesner (ed.), *Water Treatment Membrane Processes*, McGraw-Hill, New York.

18 Advanced Oxidation

Stantec's Water Treatment: Principles and Design, Updated Third Edition. John C. Crittenden, R. Rhodes Trussell, David W. Hand, Kerry J. Howe, and George Tchobanoglous.

Terminology for Advanced Oxidation

Term	Definition
Advanced oxidation process	Oxidation process that produces hydroxyl radicals at room temperature and pressure.
Electrical efficiency per log order reduction (EE/O)	Electrical energy (in kWh/m^3) required to reduce the concentration of a pollutant by one order of magnitude.
Extinction coefficient	Measure of how strongly a substance absorbs light at a given wavelength per mass unit.
Hydroxyl radical	Chemical species containing hydrogen and oxygen, differentiated from hydroxide ion (OH^-) because it has an unpaired electron in its outer shell. The unpaired electron makes it a powerful, unselective electrophile that is able to oxidize a wide range of organic compounds.
Photocatalysis	Acceleration of a photolysis reaction due to the presence of a catalyst.
Photolysis	Process in which compounds absorb photons and release energy that is able to initiate advanced oxidation.
Photon	Small packet of energy that carries electromagnetic radiation.
Quantum yield	Ratio of the number of molecules degraded in a photolysis reaction to the number of adsorbed photons.
Relative quenching rate (Q_R)	Ratio of the rate of target compound destruction by hydroxyl radical reaction to the rate of all hydroxyl radical reactions in solution.
Second-order hydroxyl radical rate constant	Indicator of the speed of an advanced oxidation process (AOP) reaction. AOP reactions tend to be quite rapid with second-order hydroxyl radical rate constants on the order of 10^8 to 10^{10} L/mol · s.
Sonolysis	Formation of radicals using ultrasound to break apart chemical bonds of organic compounds, causing oxidation.
Ultraviolet light (UV)	Electromagnetic radiation with wavelengths between 100 and 400 nm.

Three types of oxidation processes are used in water treatment. They include (1) conventional chemical oxidation processes, (2) oxidation processes carried out at elevated temperatures and/or pressure, and (3) advanced oxidation processes. Conventional chemical oxidation processes employ oxidants such as chlorine, chlorine dioxide, and potassium permanganate and do not produce highly reactive species such as the hydroxyl radical (HO•), which are produced in the other two types of oxidation processes. The "dot" written as part of the hydroxyl and other radical species indicates that the outer electron orbital has an unpaired electron. Hydroxyl radicals are reactive electrophiles that readily react with most organic compounds by undergoing addition reactions with double bonds or extracting hydrogen atoms from organic compounds. The reaction rates for conventional oxidants are much slower than the reaction rates involving HO•, and conventional chemical oxidants are more selective in terms of the types of organic molecules that they oxidize. Oxidation processes carried out at elevated temperatures and/or pressures such as wet oxidation, supercritical oxidation, gas-phase combustion, and catalytic oxidation processes can oxidize organic matter by free-radical reactions involving HO•. In advanced oxidation processes (AOPs), HO• radicals are generated at ambient temperature and atmospheric pressure.

The purpose of this chapter is to introduce the general subject of advanced oxidation. Topics to be considered include (1) an introduction to advanced oxidation processes, (2) ozonation as an advanced oxidation process, (3) the hydrogen peroxide/ozone process for potable water, and (4) other advanced oxidation processes.

18-1 Introduction to Advanced Oxidation

Some water supplies may contain toxic synthetic organic compounds (SOCs) that must be removed or destroyed to protect public health. These chemicals include agricultural pesticides and herbicides, fuels, solvents, human and veterinary drugs, and other potential endocrine disruptors. Given the uncertainty of the toxicity of the by-products of chemical oxidation, any oxidation process that is used to remove these compounds must oxidize the SOCs completely into carbon dioxide, water, and mineral acids (e.g., HCl). Conventional oxidants, such as chlorine, are selective as to which compounds they can degrade, whereas AOPs are able to completely convert organic compounds into carbon dioxide, water and mineral acids, making AOPs a viable option to destroy SOCs. Advanced oxidation processes also have several inherent advantages over other processes, such as adsorption onto activated carbon or air stripping: (1) the contaminants can be destroyed completely, (2) contaminants that are not adsorbable or volatile can be destroyed, and (3) mass transfer processes such as adsorption or stripping only transfer the contaminant to another phase, which becomes a residual and may require additional treatment.

Advanced oxidation processes are feasible for full-scale use to destroy organic compounds because they generate hydroxyl radicals at ambient temperature and atmospheric pressure (Glaze et al., 1987), whereas other processes that generate hydroxyl radicals (wet oxidation, supercritical oxidation, gas-phase combustion, and catalytic oxidation processes) require elevated temperatures and/or high pressures and are mediated by freeradical chain reactions involving hydroxyl radicals (HO•). As noted in the introduction to the chapter, a dot is often given after each radical species indicating that there is an unpaired electron in the outer orbital. This notation is an abbreviated version of the Lewis structure, which normally is used to denote whether all eight electrons in the outer orbitals of carbon, oxygen, and nitrogen are filled. In the case of HO•, there are three sets of paired electrons that are not shown and one unpaired electron, which is shown as the dot.

Hydroxyl radicals are effective in destroying organic chemicals because they are reactive electrophiles (electron preferring) that react rapidly and nonselectively with nearly all electron-rich organic compounds. Moreover, the second-order hydroxyl radical rate constants for most organic pollutants in water are on the order of 10^8 to 10^9 L/mol · s (Buxton and Greenstock, 1988), which is about the magnitude of diffusion-limited acid–base reactions ($\sim 10^9$; Stumm and Morgan, 1972). Acid–base reactions are considered to be the fastest aqueous-phase chemical reactions because they only involve the transfer of a hydrated proton. These rate constants are three to four orders of magnitude greater than any of the second-order rate constants reported for other oxidants.

Types of Advanced Oxidation Processes

The major advantages and disadvantages of various AOPs are provided in Table 18-1. Of the processes listed only the commercially available AOPs can be considered for full-scale water treatment, which are (1) ozone and hydrogen peroxide, (2) UV light and ozone, (3) UV light and hydrogen peroxide, (4) UV light and titanium dioxide, and (5) combinations of the aforementioned technologies.

Estimating Performance of AOPs

One of the most important considerations in advanced oxidation is the quantity of oxidants that are required to destroy the organics targeted for destruction and that are scavenged (HO• radicals) by background organic and inorganic matter. The influence of background matter on AOP performance is discussed later, but insight into the type of compounds that may be degraded in a reasonable time can be evaluated by using typical HO• concentrations and reported rate constants.

Advanced oxidation processes that operate in the field generate HO• concentrations between 10^{-11} and 10^{-9} mol/L (Glaze and Kang, 1988; Glaze et al., 1987). The second-order hydroxyl radical rate constants for several commonly encountered water pollutants are provided in Table 18-2. A more comprehensive list is provided in the electronic Table E-4 at the website listed in App. E. The reaction mechanism and the rate law for HO• that

Table 18-1
Advantages and disadvantages of various oxidation processes that produce hydroxyl radicals in decreasing order of preference

Advanced Oxidation Process	Advantages	Disadvantages
Hydrogen peroxide/UV light	❑ H_2O_2 is quite stable and can be stored onsite for long periods prior to use.	❑ H_2O_2 has poor UV absorption characteristics, and if the water matrix absorbs a lot of UV light energy, then most of the light input to the reactor will be wasted. ❑ Special reactors designed for UV illumination are required. ❑ Residual H_2O_2 must be addressed.
Hydrogen peroxide/ozone	❑ Waters with poor UV light transmission may be treated. ❑ Special reactors designed for UV illumination are not required.	❑ Volatile organics will be stripped from the ozone contactor. ❑ Production of O_3 can be an expensive and inefficient process. ❑ Gaseous ozone present in the off-gas of the ozone contactor must be removed. ❑ Maintaining and determining the proper O_3/H_2O_2 dosages may be difficult. ❑ Low pH is detrimental to the process.
Titanium dioxide/UV	❑ Activated with near-UV light; consequently greater light transmission is achievable.	❑ Fouling of the catalyst may occur. ❑ When used as slurry, the TiO_2 must be recovered.
Ozone/UV	❑ No need to maintain precise dosages of O_3/H_2O_2. ❑ Residual oxidant will degrade rapidly (typical half-life of O_3 is 7 min). ❑ Ozone absorbs more UV light than an equivalent dosage of hydrogen peroxide (~200 times more at 254 nm).	❑ Must use O_3 and UV light to produce H_2O_2, which is the primary means of producing HO• and using O_3 to produce H_2O_2 is very inefficient as compared to just adding H_2O_2. ❑ Special reactors designed for UV illumination are required. ❑ Ozone in the off-gas must be removed. ❑ Volatile compounds will be stripped from the process.

(Continued)

Table 18-1 *(Continued)*

Advanced Oxidation Process	Advantages	Disadvantages
Ozone/UV/H_2O_2	❑ There are commercially available processes that utilize the technology. ❑ H_2O_2 promotes ozone mass transfer.	❑ Special reactors designed for UV illumination are required. ❑ Ozone in the off-gas must be removed. ❑ Volatile compounds will be stripped from the process.
Ozone at elevated pH (8 to >10)	❑ Process would not require the addition of UV light or hydrogen peroxide.	❑ Ozone in the off-gas must be removed. ❑ pH adjustment is not practical. ❑ There are no commercial applications. ❑ Process is inefficient in removing SOCs for the reasons provided in Sec. 18-2.
Fenton's reactions (Fe/hydrogen peroxide, photo-Fenton's or Fe/ozone)	❑ Some groundwaters may contain sufficient Fe to drive Fenton's reaction. ❑ Commercial processes are available that utilize the technology.	❑ Process requires low pH.
Sonolysis	❑ Process may be used for waters that have low light penetration.	❑ Technology is not available commercially and requires too much energy input.
Ozone/sonolysis	❑ Process may have faster destruction rates as compared to sonolysis alone.	❑ There are no commercial applications. ❑ Requires specialized reactor and significant amount of energy.
Supercritical water oxidation	❑ Complete mineralization can be obtained for complex hazardous mixtures.	❑ Requires specialized reactor that operates at high pressure and significant amount of energy (e.g., to heat the water to the critical point). ❑ Corrosion can be significant if high concentration of chloride is present. ❑ Expensive process and normally designed for small flows (<~50 L/min).

Table 18-1 *(Continued)*

Advanced Oxidation Process	Advantages	Disadvantages
Ozone/TiO_2	❑ Process may have faster destruction rates as compared to TiO_2 alone.	❑ There are no commercial applications.
Ozone/TiO_2/H_2O_2	❑ Process may have faster destruction rates as compared to TiO_2 alone.	❑ There are no commercial applications.
Pulsed corona discharges/ nonthermal plasma	❑ Process does not require elevated temperatures.	❑ Works only in the gas phase.
Catalytic oxidation	❑ For gas-phase applications, process does not require temperatures that are as high as thermal incineration. ❑ For aqueous-phase applications, process does not require temperatures that are as high as supercritical water oxidation.	❑ Only certain compounds in the aqueous phase may be degraded. ❑ For gas-phase applications, the production of dioxins and furans must be avoided.
Electron beam irradiation	❑ Process does not require elevated temperatures.	❑ There are no commercial applications. ❑ Requires specialized reactor and significant amount of energy.
Ozone/electron beam irradiation	❑ Process does not require elevated temperatures. ❑ Process may have faster destruction rates as compared to electron process alone.	❑ Same as electron beam.
Gamma radiation	❑ Process does not require elevated temperatures.	❑ Same as electron beam.
Electrohydraulic cavitation	❑ Process does not require elevated temperatures.	❑ Same as electron beam.

reacts with an organic compound is given by the expressions

$$HO\bullet + R \rightarrow \text{by-products} \tag{18-1}$$

$$r_R = -k_R C_{HO\bullet} C_R \tag{18-2}$$

where r_R = destruction rate of R with HO• radicals, mol/L · s
k_R = second-order rate constant for destruction of R with HO• radicals, L/mol · s

Table 18-2
Reaction rate constants and half lives for degradation of selected inorganic and organic species by hydroxyl radicals[a]

Compound	HO·Rate Constant, L/mol · s	Half-Life, min		
		$[HO\cdot] = 10^{-9}$ M	$[HO\cdot] = 10^{-10}$ M	$[HO\cdot] = 10^{-11}$ M
Inorganics				
Ammonia	9.0×10^{7}	0.13	1.3	13
Bicarbonate	8.5×10^{6}	1.4	14	140
Bromide	1.1×10^{10}	0.001	0.01	0.1
Carbonate	3.9×10^{8}	0.03	0.3	3
Chloride	4.3×10^{9}	0.003	0.03	0.3
Iron(II)	3.2×10^{8}	0.04	0.4	4
Hydrogen peroxide	2.7×10^{7}	0.43	4.3	43
Manganese(II)	3.0×10^{7}	0.39	3.9	39
Ozone	1.1×10^{8}	0.11	1	11
Organics				
Acetate ion	7.0×10^{7}	0.2	2	17
Acetone	1.1×10^{8}	0.11	1.1	11
Atrazine	2.6×10^{9}	0.004	0.04	0.44
Benzene	7.8×10^{9}	0.001	0.01	0.1
Chloroacetic acid	4.3×10^{7}	0.3	2.7	27
Chlorobenzene	4.5×10^{9}	0.003	0.03	0.3
Chloroform	5.0×10^{6}	2	23	231
2-Chlorophenol	1.2×10^{10}	0.001	0.01	0.1
Formate ion	2.8×10^{9}	0.004	0.0	0
Geosmin	$(1.4 \pm 0.3) \times 10^{10}$	0.00083	0.0083	0.083
Methyl ethyl ketone	9.0×10^{8}	0.01	0.1	1
Methyl *tert*-butyl ether	1.6×10^{9}	0.01	0.1	1
MIB	$(8.2 \pm 0.4) \times 10^{9}$	0.0014	0.014	0.14
Natural organic matter	1.4 to 4.5×10^{8}	0.03	0.3	3.0
Oxalic acid	1.4×10^{6}	8	83	825
Oxalic ion	1.0×10^{7}	1	12	116
p-Dioxane	2.8×10^{9}	0.004	0.04	0.4
Phenol	6.6×10^{9}	0.002	0.02	0.2
Tetrachloroethylene	2.6×10^{9}	0.004	0.04	0.4
1,1,1-Trichloroethane	4.0×10^{7}	0.3	3	29
1,1,2-Trichloroethane	1.1×10^{8}	0.11	1	11
Trichloroethylene	4.2×10^{9}	0.003	0.03	0.3
Trichloromethane	5.0×10^{6}	2	23	231
Urea	7.9×10^{5}	15	146.2	1462
Vinyl chloride	1.2×10^{10}	0.001	0.01	0.1

[a]Additional values are available in the electronic Table E-4 at the website listed in App. E.
Source: Adapted from Buxton and Greenstock (1988), Lal et al. (1988), and Mao et al. (1991).

$C_{HO\bullet}$ = concentration of hydroxyl radical, mol/L
C_R = concentration of target organic R, mol/L

The half-life of the target organic compounds may be calculated assuming that the concentration of HO• is constant and equal to a typical value that is encountered in the field. The expression for the half-life of an organic compound is obtained by substituting the rate expression into a mass balance on a batch reactor whose contents are mixed completely and solving and rearranging the result, as follows:

$$\frac{dC_R}{dt} = -k_R C_{HO\bullet} C_R \tag{18-3}$$

$$t_{1/2} = \frac{\ln(2)}{k_R C_{HO\bullet}} \tag{18-4}$$

where $t_{1/2}$ = half-life of organic compound R, s

The half-lives of the target organic compounds for HO• concentrations of 10^{-9}, 10^{-10}, and 10^{-11} mol/L are also provided in Table 18-2. Based on the reported half-life, it is possible to mineralize many organic compounds completely within a matter of minutes. However, it is clear that if background matter reduces the HO• concentration to 10^{-11} mol/L, then AOPs may not be effective. The influences of NOM, carbonate, bicarbonate, and pH on AOPs are considered later in this chapter.

Example 18-1 Half life and detention time in a PFR for advanced oxidation of MTBE

Methyl *tert*-butyl ether (MTBE) was used as an octane enhancer that has been found in groundwater underneath a gasoline station at a concentration of 100 μg/L. From Table 18-2, the second-order rate constant of HO• for MTBE is 1.6×10^9 L/mol · s. Calculate the half-life and the time it would take to lower the concentration of MTBE to 5 μg/L for a HO• concentration of10^{-11} mol/L in a completely mixed bateh reactor (CMBR). Determine the detention time for an ideal plug flow reactor (PFR) to achieve a treatment objective of 5 μg/L.

Solution

1. Obtain an expression of the concentration of MTBE as a function of time in a CMBR. The following expression can be written for a CMBR, where CR represents the concentration of MTBE:

$$\frac{dC_R}{dt} = r_R = -k_R C_{HO\bullet} C_R$$

Let $k = k_R C_{HO\bullet} = (1.6 \times 10^9\ \text{L/mol} \cdot \text{s})(10^{-11}\ \text{mol/L}) = 1.6 \times 10^{-2}\ \text{s}^{-1}$. The rate expression can be rewritten as

$$\frac{dC_R}{dt} = -kC_R$$

Integrating the above expression yields

$$\int_{C_{RO}}^{C_R} \frac{dC_R}{C_R} = -\int_0^t kt \quad C_R = C_{RO}e^{-kt}$$

2. Calculate the half-life of MTBE: Substituting $C_R = C_{RO}/2$ into the equation $C_R = C_{RO}e^{-kt}$ and defining t for $C_R = C_{RO}/2$ as the half-life $t_{1/2}$ yields

$$\tfrac{1}{2} = e^{-kt_{1/2}}$$

$$t_{1/2} = \frac{\ln(2)}{k} = \frac{0.693}{\left(1.6 \times 10^{-2}\text{s}^{-1}\right)} = 43.3\ \text{s}$$

3. Calculate the time it would take to achieve a concentration of 5 μg/L. Rearranging the equation $C_R = C_{RO}e^{-kt}$ and solving for t,

$$t = \frac{1}{k}\ln\frac{C_{RO}}{C_R}$$

$$t = \frac{1}{1.6 \times 10^{-2}}\ln\left(\frac{100}{5}\right) = 187\ \text{s} \qquad (3.12\ \text{min})$$

4. The residence time for an ideal PFR would also be 3.12 min because the elapsed time in a CMBR is equivalent to residence time in an ideal PFR.

Comment

Many AOPs have much shorter residence times than 3 min. Consequently, the hydroxyl radical concentration must be much higher than 10^{-11} M for AOPs to be feasible.

Two common reactions of HO• with organic compounds are addition reactions with double bonds and extraction of hydrogen atoms. Doublebond addition is much more rapid than hydrogen abstraction. For example, TCE reacts much more rapidly than 1,1,2-trichloroethane, as shown in Table 18-2. Some toxic organic compounds of interest, such as halogenated organic contaminants that do not contain double bonds, are more difficult to degrade. For example, as reported in Table 18-2, chloroform, 1,1,2- trichloroethane, and 1,1,1-trichloroethane react more slowly with HO•, and these compounds will require longer reaction times and/or high concentrations of HO•.

By-products of AOPs

Both hydrogen abstraction and double-bond addition produce reactive organic radicals that rapidly undergo subsequent oxidation and most often combine with dissolved oxygen to form peroxy organic radicals (ROO•). These peroxy organic radicals undergo radical chain reactions that produce a variety of oxygenated by-products. The following general pattern of oxidation is observed for AOPs (Bolton and Carter, 1994):

$$\text{Organic pollutant} \rightarrow \text{aldehydes} \rightarrow \text{carboxylic acids} \rightarrow \text{carbon dioxide and mineral acids} \quad (18\text{-}5)$$

Some of the significant by-products and the highest yields observed are listed in Table 18-3. The most significant observed by-products are the carboxylic acids, due to the fact that the second-order rate constants for these compounds are much lower than those for most other organics. However, if adequate reaction time is provided, all by-products (>99 percent as measured by a TOC mass balance) are destroyed (Stefan and Bolton, 1998, 1999, 2002; Stefan et al., 2000). Other by-products of concern are the halogenated acetic acids, formed from the oxidation of halogenated alkenes such as trichloroethylene. The rate constant and half-life for chloroacetic acid is

Table 18-3
By-products observed following advanced oxidation for four selected organic compounds

Target Compound	Observed By-products	Approximate Yield: Mole By-product per Mole Compound, %
Acetone[a]	Acetic, pyruvic, and oxalic acids, pyruvaldehyde	10–30
	Formic and glyoxylic acids, hydroxyacetone, formaldehyde	2–5
Methyl *tert*-butyl ether[b]	Acetone, acetic acid, formaldehyde, *tert*-butyl formate (TBF), pyruvic acid, *tert*-butyl alcohol (TBA), 2-methoxy-2-methyl propionaldehyde (MMP), formic, methyl acetate	10–30
	Hydroxy-*iso*-butyraldehyde, hydroxyacetone, pyruvaldehyde and hydroxy-*iso*-butyric, oxalic acid	2–5
Dioxane[c]	1,2-Ethanediol diformate, formic acid, oxalic acid, glycolic acid, acetic acid, formaldehyde,1,2-ethanediol monoformate	10–30
	Methoxyacetic acid glyoxal	2–5
	Acetaldehyde	<1
Trichloroethylene[d]	Formic acid	10–40
	Oxalic acid	2–5
	1,1-Dichloroacetic acid, 1-monochloro acetic acid	<1

[a]Stefan and Bolton (1999).
[b]Stefan et al. (2000).
[c]Stefan and Bolton (1998).
[d]Stefan and Bolton (2002).

reported in Table 18-2, and longer retention time and/or higher HO• radical concentrations are needed to destroy this compound. For example, it has been demonstrated that it is possible to completely mineralize TCE using an AOP that utilizes TiO_2, O_3, and UV light in a few minutes of reaction time (Zhang et al., 1994).

Another problem with advanced oxidation processes (and processes that use ozone) is the production of brominated by-products and bromate in waters containing bromide ion. A complete discussion of ozone by-products is contained in Chap. 19.

Major Factors Affecting AOPs

The following chemical and physical properties of the water matrix have a major impact on AOPs because they either scavenge HO• radicals or absorb UV light that is used to produce HO• from hydrogen peroxide or ozone: (1) carbonate species (HCO_3^- and CO_3^{2-}), (2) pH, (3) NOM, (4) reduced metal ions (iron and manganese), (5) reactivity of the parent component with hydroxyl radical, and (6) UV light transmission of the water matrix. The first five of these factors are considered briefly in the following discussion. The effect of UV light transmission has been discussed previously in Sec. 8-4.

IMPACT OF CARBONATE SPECIES

Bicarbonate and carbonate ions are known scavengers of HO• radicals and significantly reduce the rate of organics destruction. As shown in Table 18-2, the rate constants between HO• and HCO_3^- and CO_3^{2-} are much lower than for many organic compounds. Unfortunately, the concentrations of HCO_3^- and CO_3^{2-} are often three orders of magnitude higher than the organic pollutants targeted for destruction. The degree of quenching of the AOP oxidation rate can be estimated from the expression

$$Q_R = \frac{k_R C_R}{k_R C_R + k_{HCO_3^-} - C_{HCO_3^-} + k_{CO_3^{2-}} C_{CO_3^{2-}}} \tag{18-6}$$

where Q_R = target organics (R) reaction rate with hydroxyl radical divided by total reaction rate of hydroxyl radical with both R and alkalinity ($1/Q_R$ equals the reduction in reaction due to presence of carbonate species), dimensionless

k_R = second-order rate constant for destruction of R with HO• radicals, L/mol · s

$k_{HCO_3^-}$ = second-order hydroxyl radical rate constants for HCO_3^-, L/mol · s

$k_{CO_3^{2-}}$ = second-order hydroxyl radical rate constants for CO_3^{2-}, L/mol · s

C_R = concentration of target organic R, mol/L

Table 18-4
Relative rates of destruction of TCE (Q_{TCE} at concentration of 0.100 mg/L) for various pH values and alkalinities

Relative rate Q_{TCE}, %	pH	C_{T,CO_3}, mM	HCO_3^-, mM	CO_3^{2-}, mM
10.9	7.0	1	0.997	0.003
5.78	7.0	2	1.994	0.006
2.98	7.0	4	3.988	0.012
2.00	7.0	6	5.982	0.018
1.51	7.0	8	7.976	0.024
5.55	8.0	2	1.990	0.010
3.04	9.0	2	1.904	0.096
0.754	10.0	2	1.333	0.667

Source: Adapted from Glaze and Kang (1988).

$C_{HCO_3^-}$ = concentration of bicarbonate, mol/L
$C_{CO_3^{2-}}$ = concentration of carbonate, mol/L

The relative rates of destruction of TCE for various pH values and total carbonate concentration are provided in Table 18-4. If only the influence of alkalinity is considered, it appears that even low alkalinities (50 mg/L) reduce the rate of TCE destruction by a factor of 10 at pH 7. However, at high pH, the alkalinity is more detrimental because the second-order rate constant with CO_3^{2-} is much larger than HCO_3^-. In summary, water matrices with high pH and alkalinity are more difficult to treat using AOPs, and AOPs are more effective if pretreatment processes such as softening are used to remove the alkalinity. However, NOM as discussed later has a much greater influence on quenching the reaction rate than alkalinity.

IMPACT OF PH

The performance of AOPs is affected by pH in three ways: (1) pH affects the concentration of HCO_3^- and CO_3^{2-} as discussed above; (2) the concentration of HO_2^- (H_2O_2 has a pK_a of 11.6),which is important in the UV/H_2O_2, UV/O_3, and H_2O_2/O_3 advanced oxidation processes, is a function of pH; and (3) pH affects the charge on the organic compounds if they are weak acids or bases.

For the O_3/UV and H_2O_2/O_3 processes, O_3 reacts with HO_2^- to form HO•, which is the rate-limiting step, especially for the H_2O_2/O_3 AOP. Accordingly, low pH (<5.0) greatly reduces the rate of production of HO• and the AOP reactions. High pH (11) is also thought to catalyze the formation of HO• radicals directly from O_3, but significant rates of organics destruction have not been observed with O3 at high pH (Hoigné and Bader, 1976).

The reactivity and light absorption properties of the compound can be affected by its charge. For example, in the H_2O_2/UV process, HO_2^- has about 10 times the UV molar absorptivity at 254 nm (228 L/mol · cm) than does H_2O_2; consequently, H_2O_2/UV may be more effective at higher pH, especially if the background water matrix absorbs a lot of UV light (this would only be practical if the pH was raised for other purposes and carbonate was removed, such as softening).

IMPACT OF NOM

Natural organic matter reacts with hydroxyl radicals and quenches the reaction. The quenching of the reaction rate can be estimated using the expression

$$Q_R = \frac{k_R C_R}{k_{NOM} C_{DOC} + k_R C_R} \tag{18-7}$$

where Q_R = target organics (R) reaction rate with hydroxyl radical divided by total reaction rate of hydroxyl radical with both R and NOM ($1/Q_R$ equals the reduction in reaction due to presence of NOM), dimensionless

k_R = second-order rate constant for destruction of R with HO• radicals, L/mol · s

k_{NOM} = second-order hydroxyl radical rate constants for NOM, L/mol NOM C · s

C_R = concentration of target organic R, mol/L

C_{DOC} = concentration of NOM, mol/L

In a survey of the reactivity of NOM with hydroxyl radicals, it was found that most values for NOM were between 1.4 and 4.5×10^8 L/s · mol NOM carbon (Westerhoff et al., 2007). The average value for 17 water sources was 3.9×10^8 and the standard deviation was 1.2×10^8 (Westerhoff et al., 1999).

Example 18-2 Quenching of the AOP reaction rate due to NOM

Examine the difference in the rate of oxidation of trichloroethylene (TCE) and chloroacetic acid (CAA) in distilled deionized water versus the rate in a typical tap water. The rate constants are (1) $k_{NOM} = 3 \times 10^8$ L/mol NOM C · s, $k_{TCE} = 4.2 \times 10^9$ L/mol · s, and $k_{CAA} = 4.3 \times 10^7$ L/mol · s. The NOM concentrations in distilled water and tap water are 0.05 and 3 mg/L, respectively. The initial concentrations of TCE and CAA are 35 and 48 μg/L. Estimate Q_R for each compound in distilled water and then in tap water. Can rates observed in distilled water be used to estimate the rate of oxidation that will be observed in the field?

Solution

1. Determine Q_R for TCE and chloroacetic acid in distilled water using Eq. 18-7
   a. Convert concentration of NOM in mg/L to mol/L:

$$C_{DOC} = \frac{0.05 \text{ mg C/L}}{(12 \text{ g/mol})(1000 \text{ mg/g})} = 4.167 \times 10^{-6} \text{mol C/L}$$

$$k_{NOM} = 3.0 \times 10^{8} \text{ L/mol C} \cdot \text{s}$$

   b. Q_R for TCE:

$$C_R = \frac{35 \ \mu/\text{gL}}{(131.39 \text{ g/mol})(10^6 \ \mu\text{g/g})} = 2.66 \times 10^{-7} \text{ mol/L}$$

$$Q_R = \frac{(4.2 \times 10^9 \text{ L/mol} \cdot \text{s})(2.66 \times 10^{-7} \text{ mol/L})}{(3.0 \times 10^8 \text{ L/mol} \cdot \text{s})(4.167 \times 10^{-6} \text{ mol/L}) + (4.2 \times 10^9 \text{ L/mol} \cdot \text{s})(2.66 \times 10^{-7} \text{ mol/L})}$$

$$= 0.472$$

   c. Q_R for chloroacetic acid:

$$C_R = \frac{48 \ \mu\text{g/L}}{(81.5 \text{ g/mol})(10^6 \ \mu\text{g/g})} = 5.89 \times 10^{-7} \text{mol/L}$$

$$k_R = 4.3 \times 10^7 \text{L / mol} \cdot \text{s}$$

$$Q_R = \frac{(4.3 \times 10^7 \text{ L/mol} \cdot \text{s})(5.89 \times 10^{-7} \text{ mol/L})}{(3.0 \times 10^8 \text{ L/mol} \cdot \text{s})(4.167 \times 10^{-6} \text{ mol/L}) + (4.3 \times 10^7 \text{ L/mol} \cdot \text{s})(5.89 \times 10^{-7} \text{ mol/L})}$$

$$= 0.0199$$

2. Determine Q_R for TCE and chloroacetic acid in tap water.
   a. Convert concentration of NOM in mg/L to mol/L:

$$C_{DOC} = \frac{3.0 \text{ mg C/L}}{(12 \text{ g/mol})(1000 \text{ mg/g})} = 2.5 \times 10^{-4} \text{ mol C/L}$$

   b. Q_R for TCE:

$$Q_R = \frac{(4.2 \times 10^9 \text{ L/mol} \cdot \text{s})(2.66 \times 10^{-7} \text{ mol/L})}{(3.0 \times 10^8 \text{L/mol} \cdot \text{s})(2.5 \times 10^{-4} \text{mol/L}) + (4.2 \times 10^9 \text{L/mol} \cdot \text{s})(2.66 \times 10^{-7} \text{mol/L})}$$

$$= 0.0147$$

c. Q_R for chloroacetic acid:

$$Q_R = \frac{(4.3 \times 10^7 \text{ L/mol} \cdot \text{s})(5.89 \times 10^{-7} \text{ mol/L})}{(3.0 \times 10^8 \text{ L/mol} \cdot \text{s})(2.5 \times 10^{-4} \text{ mol/L}) + (4.3 \times 10^7 \text{ L/mol} \cdot \text{s})(5.89 \times 10^{-7} \text{ mol/L})}$$
$$= 0.000338$$

Comment

The amount of quenching ($1/Q_R$) for TCE and CAA is about 32 (0.472/0.0147) and 59 (0.0199/0.000338) times higher in tap water than in distilled deionized water due to the quenching effect of the NOM. Hence, the results of tests in distilled water cannot be used directly to estimate oxidation rates in the field. Bench tests should be conducted in the same water that is to be treated in the field.

IMPACT OF REDUCED METAL IONS

Metal ions in reduced oxidation states such as Fe(II) and Mn(II) are often found in groundwater due to the anoxic conditions typically present in groundwater. These inorganic species can consume a significant quantity of chemical oxidants as well as scavenge HO• radicals. Consequently, the concentration of reduced metal ions should be measured as part of any treatability study, and the dosages of oxidants should be based on a consideration of the chemical oxygen demand (COD) of the reduced metal species. Similar to NOM, the amount of quenching can be estimated using the expression

$$Q_R = \frac{k_R C_R}{k_R C_R + k_{Fe(II)} C_{Fe(II)} + k_{Mn(II)} C_{Mn(II)}} \tag{18-8}$$

where $k_{Fe(II)}$ = second-order hydroxyl rate constants for Fe(II), L/mol · s
$k_{Mn(II)}$ = second-order hydroxyl rate constants for Mn(II), L/mol · s
$C_{Fe(II)}$ = concentration of Fe(II), mol/L
$C_{Mn(II)}$ = concentration of Mn(II), mol/L

The second-order rate constants reported for Fe(II) and Mn(II) are 3.3×10^8 and 3.0×10^7 M^{-1} s^{-1}, respectively (Buxton et al., 1988).

Example 18-3 Quenching the AOP reaction rate due to Fe(II)

The second-order hydroxyl rate constant for Fe(II) is 3.3×10^8 L/mol · s. For iron concentrations of 5.6 and 0.56 mg/L and a TCE concentration of 100 μg/L, calculate the quenching of the reaction.

Solution

1. Determine Q_R using Eq. 18-8 for an iron concentration of 5.6 mg/L:

$$C_R = \frac{100\ \mu g/L}{(131.39\ g/mol)(10^6\ \mu g/g)} = 7.611 \times 10^{-7}\ mol/L$$

$$C_{Fe} = \frac{5.6\ mg/L}{(55.84\ g/mol)(1000\ mg/g)} = 10^{-4}\ mol/L$$

$$Q_R = \frac{(4.2 \times 10^9\ L/mol \cdot s)(7.611 \times 10^{-7}\ mol/L)}{(3.3 \times 10^8\ L/mol \cdot s)(10^{-4}\ mol/L) + (4.2 \times 10^9\ L/mol \cdot s)(7.611 \times 10^{-7}\ mol/L)}$$
$$= 0.088$$

2. Determine Q_R for an iron concentration of 0.56 mg/L:

$$C_R = 7.611 \times 10^{-7}\ mol/L$$

$$C_{Fe} = \frac{(0.56\ mg/L)}{(55.84\ g/mol)(1000\ mg/g)} = 10^{-5}\ mol/L$$

$$Q_R = \frac{(4.2 \times 10^9\ L/mol \cdot s)(7.611 \times 10^{-7}\ mol/L)}{(3.3 \times 10^8\ L/mol \cdot s)(10^{-5}\ mol/L) + (4.2 \times 10^9\ L/mol \cdot s)(7.611 \times 10^{-7}\ mol/L)}$$
$$= 0.492$$

Comment

Although iron can also quench the hydroxyl radical reaction, NOM has a much greater impact, as shown in Example 18-2. The amount of quenching for TCE is about 33 (0.492/0.0147) times higher in tap water containing NOM as compared to water containing iron.

REACTIVITY OF PARENT COMPONENT WITH HYDROXYL RADICAL

Depending on the reactivity of the parent compound, the compound being oxidized can reduce the destruction rate of the parent compound significantly. For example, NOM can reduce the destruction rate of the parent compound by a factor of 100 for compounds with a second-order rate constant of 10^9 L/mol·s or 1000 for compounds with a second-order rate constant of 10^8 L/mol·s (Liao and Gurol, 1995; Westerhoff et al., 1999).

Assessing Feasibility of AOPs

To assess the feasibility of AOPs, the following parameters should be measured: (1) alkalinity, (2) pH, (3) COD, (4) TOC, (5) Fe, (6) Mn, and (7) light transmission. Once these parameters are known, this information can be used to interpret and plan treatability studies for AOPs and investigate pretreatment options that may be needed. In addition, these parameters can be used in the simple models presented later in this chapter to assess the feasibility of AOPs.

18-2 Ozonation as an Advanced Oxidation Process

Approximately one-third of the water treatment plants in the United States use ozone for disinfection, taste and odor control, and target compound destruction. The production of ozone, ozone contactor design, and disinfection using ozone are presented in Chap. 13. In this section, the focus is on the destruction of organoleptic and other target compounds including MIB, geosmin, and atrazine. The fundamental question is whether water utilities have received the maximum benefit for target compound destruction using ozone alone or whether they can receive additional benefit by adding hydrogen peroxide after their disinfection requirements (Ct) are met. The addition of hydrogen peroxide, which reacts with ozone to produce hydroxyl radical, is discussed in Sec. 18-3. Target compound destruction using ozone is discussed in this section.

Hydroxyl Radical Production from OH^-

High pH values (≈11) are thought to catalyze the formation of HO• radicals directly from O_3. The complete set of reactions for HO• production from OH^- is shown in Table 18-5 and listed in the following reaction sequence:

$$OH^- + O_3 \xrightarrow{k_2} HO_2^- + O_2 \tag{18-9}$$

$$HO_2^- + O_3 \xrightarrow{k_1} O_3^- \bullet + HO_2 \bullet \tag{18-10}$$

$$HO_2 \bullet \rightleftarrows O_2^- \bullet + H^+ \tag{18-11}$$

$$O_2^- \bullet + O_3 \xrightarrow{k_3} O_3^- \bullet + O_2 \tag{18-12}$$

$$O_3^- \bullet + H^+ \xrightarrow{k_4} HO_3 \bullet \tag{18-13}$$

$$HO_3 \bullet \xrightarrow{k_5} HO \bullet + O_2 \tag{18-14}$$

where k_1 = second-order rate constant between ozone and anion of hydrogen peroxide, L/mol·s ($M^{-1}\ s^{-1}$)

k_2 = second-order rate constant between ozone and hydroxyl ion, L/mol·s ($M^{-1}\ s^{-1}$)

k_3 = second-order rate constant between ozone and the ozonide ion radical, L/mol·s ($M^{-1}\ s^{-1}$)

Table 18-5
Important elementary reactions involved in H_2O_2/O_3 and H_2O_2/UV processes at near-neutral pH

Number	Reactions	Rate Constants at 25°C, $M^{-1}\ s^{-1}$	References
		Reactions Specifically for H_2O_2/O_3 Process	
1	$HO_2^- + O_3 \xrightarrow{k_1} O_3^- \bullet + HO_2 \bullet$	$k_1 = 2.8 \times 10^6$	Staehelip and Hoigné (1982)
2	$OH^- + O_3 \xrightarrow{k_2} HO_2^- + O_2$	$k_2 = 70$	Staehelip and Hoigné (1982)
3	$O_2^- \bullet + O_3 \xrightarrow{k_1} O_3^- \bullet + O_2$	$k_3 = 1.6 \times 10^9$	Buhler et al. (1984)
4	$O_3^- \bullet + H^+ \xrightarrow{k_4} HO_3 \bullet$	$k_4 = 5.2 \times 10^{10}$	Buhler et al. (1984)
5	$HO_3 \bullet \xrightarrow{k_5} HO \bullet + O_2$	$k_5 = 1.1 \times 10^5\ s^{-1}$	Buhler et al. (1984)
6	$O_3 + R \xrightarrow{k_6} \text{products}$	k_6^a	
7	$H_2O_2 + h\nu \rightarrow 2HO\bullet$	***Reactions Specifically for H_2O_2/UV Process***	
		$r_{UV,H_2O_2} = r_{HO}\bullet/2 = -\phi_{H_2O_2} P_{u\text{-}v} f_{H_2O_2}\left(1 - e^{-A}\right)$	
		$A = 2.303b\left(\varepsilon_{H_2O_2} C_{H_2O_2} + \varepsilon_{R1} C_{R1} + \varepsilon_{NOM} C_{NOM}\right)$	
		$f_{H_2O_2} = 2.303b\left(\varepsilon_{H_2O_2} C_{H_2O_2} + \varepsilon_{HO_2} - C_{HO_2^-}\right)/A$	
		$\varepsilon_{H_2O_2,254\ nm} = 17.9 \sim 19.6\ M^{-1}\ cm^{-1}$	
		$\phi_{H_2O_2} = \phi_{HO_2^-} = 0.5$	
8	$R + h\upsilon \rightarrow \text{products}$	$r_{UV,R} = -\phi_R P_{u-v}(1 - e^{-A})$ $f_R = 2.303b\, \varepsilon_R C_R / A$	

(*Continued*)

Table 18-5 *(Continued)*

Number	Reactions	Rate Constants at 25°C, $M^{-1}\ s^{-1}$	References
		Reactions Common for Both H_2O_2/O_3 and H_2O_2/UV Process	
9	$HO\bullet + HO_2^- \xrightarrow{k_9} OH^- + HO_2\bullet$	$k_9 = 7.5 \times 10^9$	Christensen et al. (1982)
10	$HO\bullet + H_2O_2 \xrightarrow{k_{10}} H_2O + HO_2\bullet$	$k_{10} = 2.7 \times 10^7$	Buxton and Greenstock (1988)
11	$HO\bullet + HCO_3^- \xrightarrow{k_{11}} CO_3^-\bullet + H_2O$	$k_{11} = 8.5 \times 10^6$	Buxton and Greenstock (1988)
12	$HO\bullet + R \xrightarrow{k_{12}} \text{products}$	k_{12}[b]	Buxton and Greenstock (1988)
13	$HO\bullet + NOM \xrightarrow{k_{13}} \text{products}$	$k_{13} = 3 \times 10^8$ to 4.5×10^8 (average 3.9×10^8) L/mol C^{-s}	Westerhoff et al. (1999)
		Acid Dissociation Constants	
14	$H_2CO_3^* \rightleftarrows H^+ + HCO_3^-$	$pk_{a1} = 6.3$	Stumm and Morgan (1981)
15	$HCO_3^- \rightleftarrows H^+ + CO_3^{2-}$	$pk_{a2} = 10.3$	Stumm and Morgan (1981)
16	$H_2O \rightleftarrows H^+ + OH^-$	$pK_{a3} = 14$	Stumm and Morgan (1981)
17	$H_2O_2 \rightleftarrows H^+ + HO_2^-$	$pK_{a5} = 11.75$	Behar et al. (1970)
18	$HO_2\bullet \rightleftarrows H^+ + O_2^-\bullet$	$pk_{a6} = 4.8$	Staehelip and Hoigné (1982)

[a]See electronic Table E-2 at the website listed in App.E.
[b]See electronic Table E-4 at the website listed in App. E.

k_4 = second-order rate constant between ozonide ion radical and hydrogen ion, L/mol·s ($M^{-1}\ s^{-1}$)
k_5 = first-order rate constant for ozonide radical, s^{-1}

The overall stoichiometry of the reaction is given by the following reaction:

$$3O_3 + OH^- + H^+ \rightarrow 4O_2 + 2HO\bullet \qquad (18\text{-}15)$$

Consequently, the high-pH ozone process is inefficient and requires 1.5 mol of ozone to produce 1 mol of HO•.

The first step in the reaction sequence, Eq. 18-9, is only fast at high pH. For example, the half-lives of ozone for the reaction in Eq. 18-9 at various pH values are (1) 1650 min at pH 7, (2) 165 min at pH 8, (3) 16.5 min at pH 9, (4) 1.65 min at pH 10, and (5) 0.165 min at pH 11. Consequently, the reaction does not proceed rapidly until the pH is 11.

Unfortunately, high pH values are detrimental to the production of HO•, as shown in Eq. 18-15, and carbonate species quench the hydroxyl radicals that are formed from subsequent reactions. As shown in Table 18-4, the relative reaction rates for TCE (100 μg/L) at a total concentration of all carbonate species ($C_{T,CO_3} = 2 \times 10^{-3}$ mol/L) decrease from a high of 5.78 percent at pH 7 to a low of 0.754 percent at pH 10. As a direct result of the low relative reaction rates at high pH, significant destruction rates of target compounds have not been observed with the high-pH O_3 process (Hoigné and Bader, 1976).

Hydroxyl Radical Production from NOM

When ozone reacts with NOM, it produces low levels of hydroxyl radical via the reaction

$$O_3 + NOM \rightarrow HO\bullet + \text{by-products} \qquad (18\text{-}16)$$

As discussed in Sec. 18-1, the hydroxyl radical that may be produced from Eq. 18-16 may also be quenched by the reaction with NOM, as shown in the reaction

$$HO\bullet + NOM \xrightarrow{k_{13}} \text{by-products} \qquad (18\text{-}17)$$

where k_{13} = second-order rate constant between hydroxyl radical and NOM, L/(mol NOM C·s)

The quenching of hydroxyl radical with NOM is usually more important than quenching by bicarbonate and carbonate or metal species (see Sec. 18-1 for details, including rate constants).

Accordingly, when it comes to target compound destruction, ozonation can destroy organic compounds by either direct reactions with O_3 or indirect reactions with HO• , as shown in the following:

$$O_3 \rightarrow \begin{cases} \xrightarrow[O_3]{\text{direct pathway}} & O_3 + R \rightarrow \text{product 1} \\ \xrightarrow[\text{NOM}]{\text{indirect pathway}} & HO\bullet + R \rightarrow \text{product 2} \end{cases} \qquad (18\text{-}18)$$

The rate of destruction of a target compound R is given by the equation

$$r_R = -k_{O_3}[O_3][R] - k_{HO\bullet}[HO\bullet][R] \quad (18\text{-}19)$$

where r_R = rate of disappearance of target compound R, mol/L·s
$[O_3]$ = concentration of ozone, mol/L
$[R]$ = concentration of target compound R, mol/L
$[HO\bullet]$ = concentration of hydroxyl radical, mol/L
$k_{HO\bullet}$ = second-order rate constants between hydroxyl radical and R, L/mol·s
kO_3 = second-order rate constants between ozone and R, L/mol·s

The relative importance of the direct reaction with ozone and the indirect reaction with HO• can be assessed using the expression

$$f_{HO\bullet} = \frac{k_{HO\bullet}[HO\bullet]}{k_{HO\bullet}[HO\bullet] + k_{O_3}[O_3]} = \frac{k_{HO\bullet}([HO\bullet]/[O_3])}{k_{HO\bullet}([HO\bullet]/[O_3]) + k_{O_3}}$$

$$= \frac{k_{HO\bullet}C_{[HO\bullet]/[O_3]}}{k_{HO\bullet}C_{[HO\bullet]/[O_3]} + k_{O_3}} \quad (18\text{-}20)$$

where $f_{HO\bullet}$ = fraction of target compound destruction due to indirect reaction with HO•, dimensionless
$C_{[HO\bullet]/[O3]}$ = ratio of hydroxyl radical concentration to ozone concentration, dimensionless

Second-order rate constants for ozone (listed in the electron Table E-2 at the website listed in App. E) are useful in assessing possible reactions and reaction kinetics. The rate constants for organics are highly dependent on the type of organic being oxidized. The indirect pathway reaction rate trend is from high for the hydroxyl radical with amine-substituted benzenes to low for aliphatics without nucleophilic sites. In contrast, many of the rate constants for the direct reaction with ozone appear to be low. It has been reported that the ratio of the concentration of the hydroxyl radical to the concentration of ozone $(C_{[HO\bullet]/[O_3]})$ was relatively constant during the decomposition of ozone in the presence of NOM, with typical values ranging from 10^{-7} to 10^{-10} (Elovitz and von Gunten, 1999).

The reaction of ozone with NOM to produce HO• is the most important mechanism to destroy target compounds (Elovitz and von Gunten, 1999; Westerhoff et al., 1999). For example, it has been demonstrated that 83 percent of MIB and 90 percent of geosmin were degraded by the hydroxyl radical for an ozonated natural water (Bruce et al., 2002). This finding is important because, if ozone is able to remove taste- and odor-causing organics, it is probably due to HO• production from the reaction of ozone with NOM. The following example will demonstrate when the indirect or direct reaction of target compounds with HO• is an important pathway for target compound destruction.

Example 18-4 Fraction of target compound destruction by direct and indirect reaction pathways

Determine the fraction of the reaction that is carried out by the indirect reaction with HO• for second-order HO• rate constants of 10^7, 10^8, and 10^9 L/mol · s. For the calculation, use $C_{[HO\bullet]/[O_3]}$ values of 10^{-7}, 10^{-8}, 10^{-9}, and 10^{-10} and a rate constant for the direct reaction with ozone of 10 L/mol · s.

Solution

1. Calculate the fraction of target compound destruction due to the indirect reaction with HO• for k_{O_3} of 10 L/mol · s using Eq. 18-20: For $k_{HO\bullet} = 10^7$ L/mol · s and $C_{[HO\bullet]/[O_3]} = 10^{-7}$,

$$f_{HO\bullet} = \frac{\left(10^7 \text{ L/mol}\cdot\text{s}\right)\times 10^{-7}}{\left(10^7 \text{ L/mol}\cdot\text{s}\right)10^{-7} + (10 \text{ L/mol}\cdot\text{s})} = 0.0909$$

2. Tabulate the results—fraction of target compound destruction due to indirect reaction with HO• for k_{O_3} of 10 L/mol · s:

	$f_{HO\bullet}$		
$C_{[HO\bullet]/[O_3]}$	$k_{HO\bullet} = 10^7\ M^{-1}\ s^{-1}$	$k_{HO\bullet} = 10^8\ M^{-1}\ s^{-1}$	$k_{HO\bullet} = 10^9\ M^{-1}\ s^{-1}$
10^{-7}	9.09×10^{-2}	5.00×10^{-1}	9.09×10^{-1}
10^{-8}	9.90×10^{-3}	9.09×10^{-2}	5.00×10^{-1}
10^{-9}	9.99×10^{-4}	9.90×10^{-3}	9.09×10^{-2}
10^{-10}	1.00×10^{-4}	9.99×10^{-4}	9.90×10^{-3}

Comment

The direct ozonation reaction is more important than the indirect reaction with HO• for $k_{HO\bullet} = 10^7\ M^{-1}\ s^{-1}$. For $k_{HO\bullet} = 10^8\ M^{-1}\ s^{-1}$, the direct ozonation reaction is less important than the indirect reaction with HO• for $C_{[HO\bullet]/[O_3]}$ greater than 10^{-7}. For $k_{HO\bullet} = 10^9\ M^{-1}\ s^{-1}$, the direct ozonation reaction is less important than the indirect reaction with HO• for $C_{[HO\bullet]/[O_3]}$ greater than 10^{-8}.

A simple model may be developed using Eq. 18-20. The loss of ozone and target compound may be described by the following pseudo-first-order reaction:

$$r_{O_3} = -k[O_3] \tag{18-21}$$

$$r_R = -(k_{O_3} + k_{HO\bullet}C_{[HO\bullet]/[O_3]})[O_3][R] \tag{18-22}$$

where r_{O_3} = rate of loss of ozone, mol/L · s
k = pseudo-first-order decay rate constant for ozone, s^{-1}

For a PFR with detention time τ or a completely mixed batch reactor with an elapsed time t, Eqs. 18-21 and 18-22 may be written and solved as shown below:

$$\frac{d[O_3]}{dt} = -k[O_3] \tag{18-23}$$

$$[O_3] = [O_3]_0 e^{-kt} \tag{18-24}$$

$$\frac{d[R]}{dt} = -\left(k_{HO\bullet}C_{[HO\bullet]/[O_3]}[R] + k_{O_3}[R]\right)[O_3]_0 e^{-kt} \tag{18-25}$$

$$[R] = [R]_0 \exp\left[\left([O_3]_0/k\right) + \left(k_{HO\bullet}C_{[HO\bullet]/[O_3]} + k_{O_3}\right)\left(e^{-kt} - 1\right)\right] \tag{18-26}$$

where $[O_3]_0$ = initial concentration of ozone, mol/L
$[R]_0$ = initial concentration of target compound R, mol/L

The following example is presented to illustrate how Eq. 18-26 can be used to predict the destruction of target compounds.

Example 18-5 Time required for destruction of target compound

Calculate the half-life and the time required for 95 percent destruction of geosmin and MIB in a batch reactor. The second-order rate constants with HO• for geosmin and MIB are $(1.4 \pm 0.3) \times 10^{10}$ and $(8.2 \pm 0.4) \times 10^{9}$, respectively. For the calculation, use $C_{[HO\bullet]/[O_3]}$ ranging from 10^{-9} to 10^{-7} and an initial ozone concentration of 3 mg/L.The rate constant for the direct reaction with ozone is 10 M^{-1} s^{-1}. Use a typical ozone pseudo-first-order decay rate constant of 0.1 min^{-1}.

Solution

1. Calculate the half-life for destruction of a target compound. Rearrange Eq. 18-26 to obtain an expression for $t_{1/2}$:

$$t_{1/2} = -\frac{\ln\left\{1 + \left[\ln\left(\frac{1}{2}\right)\right](k) / \left[[O_3]_0\left(k_{HO\cdot}C_{[HO\cdot]/[O_3]} + k_{O_3}\right)\right]\right\}}{k}$$

a. Convert initial ozone concentration from mg/L to mol/L:

$$[O_3]_0 = \frac{3\ \text{mg/L}}{(48\ \text{g/mol})(10^3\ \text{mg/g})} = 6.25 \times 10^{-5}\ \text{mol/L}$$

b. Calculate $t_{1/2}$ for MIB and geosmin when $C_{[HO\bullet]/[O_3]} = 10^{-7}$:

i. For MIB,

$$k_{HO\bullet}C_{[HO\bullet]/[O_3]} + k_{O_3} = \left\{\left[\left(8.2\times10^{9}\right)\left(10^{-7}\right)+10\right]\text{L/mol}\cdot\text{s}\right\}(60\text{ s/min})$$
$$= 49{,}800\text{ L/mol}\cdot\text{min}$$

$$t_{1/2} = -\frac{\ln\left\{1+\left[\ln\left(\frac{1}{2}\right)\right](0.1/\text{min})\,/\left[\left(6.25\times10^{-5}\text{ mol/L}\right)(49{,}800\text{ L/mol}\cdot\text{min})\right]\right\}}{0.1/\text{min}}$$
$$= 0.23\text{ min}$$

ii. For geosmin,

$$k_{HO\bullet}C_{[HO\bullet]/[O_3]} + k_{O_3} = \left\{\left[\left(1.4\times10^{10}\right)\left(10^{-7}\right)+10\right]\text{L/mol}\cdot\text{s}\right\}(60\text{ s/min})$$
$$= 84{,}600\text{ L/mol}\cdot\text{min}$$

$$t_{1/2} = -\frac{\ln\left\{1+\left[\ln\left(\frac{1}{2}\right)\right](0.1/\text{min})/\left[\left(6.25\times10^{-5}\text{ mol/L}\right)(84{,}600\text{ L/mol}\cdot\text{min})\right]\right\}}{0.1/\text{min}}$$
$$= 0.13\text{ min}$$

c. Tabulate the $t_{1/2}$ results for $10^{-9} \le C_{[HO\bullet]/[O_3]} \le 10^{-7}$:

	$t_{1/2}$, min	
$C_{[HO\cdot]/[O_3]}$	MIB	Geosmin
1.00×10^{-7}	0.23	0.13
5.00×10^{-8}	0.45	0.26
1.00×10^{-8}	2.24	1.32
5.00×10^{-9}	4.50	2.63
3.00×10^{-9}	7.64	4.39
1.50×10^{-9}	17.65	9.07
1.05×10^{-9}	49.95	13.80
1.00×10^{-9}	∞	14.70

2. Calculate the time required for 95 percent destruction of a target compound:

$$t = -\frac{\ln\left\{1+[\ln(1-0.95)](k)\,/\left[[O_3]_0\left(k_{HO\bullet}C_{[HO\bullet]/[O_3]}+k_{O_3}\right)\right]\right\}}{k}$$

$$= -\frac{\ln\left\{1+[\ln(1-0.95)](k)\,/\left[[O_3]_0\left(k_{HO\bullet}C_{[HO\bullet]/[O_3]}+k_{O_3}\right)\right]\right\}}{k}$$

a. Calculate t for MIB and geosmin when $C_{[HO\bullet]/[O_3]} = 10^{-7}$:

i. For MIB,

$$t = -\frac{\ln\left\{1 + [\ln(0.05)](0.1/\text{min}) / \left[(6.25 \times 10^{-5}\,\text{mol/L})(49{,}800\,\text{L/mol}\cdot\text{min})\right]\right\}}{0.1/\text{min}}$$

$= 1.01$ min

ii. For geosmin,

$$t = \frac{\ln\left\{1 + \left[\ln(0.05)\right](0.1/\text{min}) / [(6.25 \times 10^{-5}\,\text{mol/L})(84{,}600\,\text{L/mol}\cdot\text{min})]\right\}}{0.1/\text{min}}$$

$= 0.58$min

b. Tabulate the t results for $10^{-9} \le C_{[HO\bullet]/[O_3]} \le 10^{-7}$:

	t, min	
$C_{[HO\cdot]/[O_3]}$	MIB	Geosmin
1.00×10^{-7}	1.01	0.58
5.00×10^{-8}	2.11	1.19
1.00×10^{-8}	20.27	7.61
8.55×10^{-9}	58.8	9.57
8.50×10^{-9}	∞	9.66
5.00×10^{-9}	∞	65.55
4.95×10^{-9}	∞	∞

Comment

The destruction of geosmin and MIB is possible if $C_{[HO\cdot]/[O_3]}$ is greater than 5.00×10^{-8}. Many unsaturated compounds have a second-order rate constant for HO• less than $10^9\ M^{-1}\ s^{-1}$ and ozone alone would not be effective in destroying such compounds.

Determination of Destruction of Target Compounds from Bench-Scale Tests

While Eq. 18-26 can be used to describe the destruction rate of target compounds, there is no way to predict the value of $C_{[HO\cdot]/[O_3]}$ or the pseudo-first-order rate constant for ozone, k. Consequently, batch tests are required to determine $C_{[HO\cdot]/[O_3]}$ and k. The basics of the batch-testing method include the following steps:

1. Ozonate the water and measure the initial ozone concentration and the concentration of ozone as a function of time.
2. Determine the pseudo-first-order rate constant for ozone by fitting Eq. 18-24 to the ozone-versus-time data using commercially available software with a graphing function.

3. Measure the concentration of one or more target compounds as a function of time.
4. Determine the best-fit $C_{[HO\bullet]/[O_3]}$ value by fitting the target compound data using Eq. 18-26 and commercially available software with a graphing function. The suggested method is to define an objective function as shown below and to minimize an objective function by using a spreadsheet and making $C_{[HO\cdot]/[O_3]}$ the target cell, which is adjusted to find the minimum objective function or best fit:

$$\mathrm{OF} = \sqrt{\frac{1}{n-1}\sum_{i=1}^{n}\left(\frac{C_{\mathrm{data},i} - C_{\mathrm{model},i}}{C_{\mathrm{data},i}}\right)^2} \tag{18-27}$$

where
OF = objective function, dimensionless
n = number of data points, unitless
$C_{\mathrm{data},i}$ = measured concentration of data point i, mg/L
$C_{\mathrm{model},i}$ = predicted concentration of data point i, mg/L

The values for k_{O3} and $k_{HO\bullet}$ must be known to determine $C_{[HO\cdot]/[O_3]}$,and in most instances the contribution of k_{O_3} can be ignored. If the $k_{HO\bullet}$ is known for the destruction of the target compound, then the destruction of other target compounds can be predicted using the appropriate $k_{HO\bullet}$ and the fitted $C_{[HO\bullet]/[O_3]}$. If $k_{HO\bullet}$ is not known, then a fitted parameter ($k_{HO} \bullet C_{[HO\bullet]/[O_3]} + k_{O_3}$) can be determined. In either situation, the target compound destruction for different influent concentrations and ozone dosages can be predicted using the fitted parameter ($k_{HO} \bullet C_{[HO\bullet]/[O_3]} + k_{O_3}$) or $C_{[HO\bullet]/[O_3]}$. If acceptable destruction rates are not achieved by ozonation alone, then the addition of hydrogen peroxide should be considered. The O_3/H_2O_2 process is discussed in Sec. 18-3.

18-3 Hydrogen Peroxide/Ozone Process for Potable Water

The material in this section covers the benefits of adding hydrogen peroxide in conjunction with ozone and the basic reactions involved in the hydrogen peroxide/ozone process. The activation of NOM with ozone is ignored because it is insignificant when hydrogen peroxide is added. However, the quenching of hydroxyl radical by NOM is considered. At the end of the section, bench-scale testing that is used to determine process feasibility is discussed. The basic components of the hydrogen peroxide/ozone process with a single injection point for ozone is shown on Fig. 18-1. As discussed earlier, ozone can form bromate when bromide is present, and one way to reduce bromate formation is to reduce the ozone concentration. As discussed above, the use of a reactor with multiple injection points, lowering the pH, chlorine addition, and ammonia addition are methods that have been proposed to reduce bromate production.

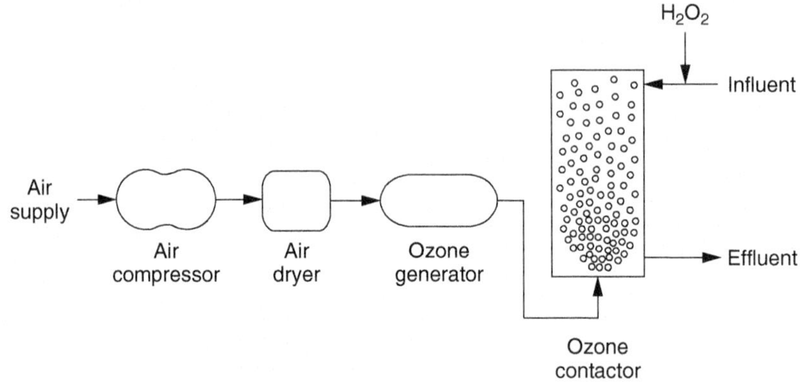

Figure 18-1
Basic components of hydrogen peroxide–ozone process. (Courtesy of Applied Process Technology).

Reaction Mechanisms

The elementary reactions that are involved in the production of HO• from H_2O_2/O_3 are listed in Table 18-5. The following discussion of the reaction mechanisms will cover the H_2O_2/O_3 process at neutral pH (values near 7). The H_2O_2/O_3 reaction sequence begins by forming the ozonide ion radical, O_3^{-}•, and the superoxide radical, HO_2•, which is the rate-limiting step (Buhler et al., 1984; Forni et al., 1982; Sehested et al., 1982):

$$H_2O_2 \rightleftarrows HO_2^{-} + H^{+} \qquad k_a = 1.6 \times 10^{-12} \quad \text{at } 25°C \qquad (18\text{-}28)$$

$$O_3 + HO_2^{-} \xrightarrow{k_1} O_3^{-} \bullet + HO_2 \bullet \qquad k_1 = 2.8 \times 10^{6}\, M^{-1}s^{-1} \quad \text{at } 25°C \qquad (18\text{-}29)$$

where K_a = equilibrium constant, dimensionless
k_1 = second-order rate constant between ozone and anion of hydrogen peroxide, L/mol·s

The rate-limiting step in the formation of HO• is the formation of O_3^{-}· and HO_2•, which is slow, especially at low pH. Consequently, low reaction rates have been observed at pH values of 5 or less and the H_2O_2/O_3 AOP may not be a viable option for the destruction of organics if the pH is 5 or less. Once the superoxide radical, HO_2•, has been formed, it can form the ozonide ion radical, O_3^{-}•, as follows:

$$HO_2 \bullet \rightleftarrows O_2^{-} \bullet + H^{+} \qquad k_a = 1.6 \times 10^{-5} \quad \text{at } 25°C \qquad (18\text{-}30)$$

$$O_2^{-} \bullet + O_3 \xrightarrow{k_3} O_3^{-} \bullet + O_2 \qquad k_3 = 1.6 \times 10^{9}\, M^{-1}s^{-1} \quad \text{at } 25°C \qquad (18\text{-}31)$$

where k_3 = second-order rate constant between ozonide ion radical and ozone, L/mol·s (M^{-1} s^{-1})

The ozonide ion radical forms HO• according to the following reactions:

$$O_3^{-} \bullet + H^{+} \xrightarrow{k_4} HO_3 \bullet \qquad k_4 = 5.2 \times 10^{10}\, M^{-1}s^{-1} \quad \text{at } 25°C \qquad (18\text{-}32)$$

$$HO_3 \bullet \xrightarrow{k_5} O_2 + HO \bullet \qquad k_5 = 1.1 \times 10^{5}\, s^{-1} \quad \text{at } 25°C \qquad (18\text{-}33)$$

where k_4 = second-order rate constant between ozonide ion radical and hydrogen ion, L/mol·s ($M^{-1}\ s^{-1}$)
k_5 = first-order rate constant for ozonide radical, s^{-1}

The overall reaction for HO• radical formation is

$$H_2O_2 + 2O_3 \rightarrow 2HO\bullet + 3O_2 \tag{18-34}$$

Proper Dosage of Hydrogen Peroxide and Ozone

According to Eq. 18-34, 0.5 mol H_2O_2 is needed for every mole of O_3 or a mass ratio of 0.354 kg H_2O_2 is needed for every kilogram of O_3. However, there are several issues that impact the proper dosages of H_2O_2 and O_3. First, O_3 tends to be more reactive with background organic matter and inorganic species than H_2O_2. (O_3 mass transfer efficiency is usually greater than 95 percent.) As a result, the applied O_3 dosage will have to be higher than estimated from stoichiometry to achieve the optimum ratio. However, an excess O_3 dosage has the potential of wasting O_3 and scavenging HO• radicals via the reaction

$$O_3 + HO\bullet \rightarrow HO_2\bullet + O_2 \tag{18-35}$$

The HO_2• radical, formed as shown in Eq. 18-35, may produce more HO• via reactions through Eqs. 18-30 to 18-33 assuming there is adequate ozone remaining in solution. Excess H_2O_2 is also detrimental to the H_2O_2/O_3 AOP because it may scavenge HO• via the following reactions:

$$HO\bullet + HO_2^- \xrightarrow{k_9} HO_2\bullet + OH^- \quad k_9 = 7.5\times10^9\ M^{-1}s^{-1} \quad \text{at } 25^\circ C \tag{18-36}$$

$$HO\bullet + H_2O_2 \xrightarrow{k_{10}} HO_2\bullet + H_2O \quad k_{10} = 2.7\times10^7\ M^{-1}s^{-1} \quad \text{at } 25^\circ C \tag{18-37}$$

where k_9 = second-order rate constant between hydroxyl radical and anion of hydrogen peroxide, L/mol·s
k_{10} = second-order rate constant between hydroxyl radical and hydrogen peroxide, L/mol·s

Further, the H_2O_2 residual can be more problematic than ozone because hydrogen peroxide is more stable than ozone. As pointed out previously, some vendors have attempted to overcome the problem of H_2O_2 quenching of HO• by adding H_2O_2 at multiple points in a single reactor or by using multiple reactors in series.

Elementary Reactions for H_2O_2/O_3 Process

The elementary reactions for the O_3/H_2O_2 process are listed in Table 18-5. The elementary reactions include the initiation (reactions 1, 3, 4, and 5 or Eqs. 18-29 and 18-31 to 18-33), propagation (reactions 9 and 10 or Eqs. 18-36 and 18-37), and termination reactions of the radical chain reaction. Termination reactions involve recombination of radical species and are not shown because they have a low probability of occurrence (e.g., $HO\bullet + HO\bullet \rightarrow H_2O_2$). The elementary reactions also include the oxidation of the target organic compound (R) and the scavenging of the hydroxyl radical by bicarbonate, carbonate, and NOM, as discussed in Sec. 18-1.

The net rates of formation of various radicals are given by the expressions

$$r_{HO\bullet} = k_5\left[HO_3\bullet\right] - k_9[HO\bullet]\left[HO_2^-\right] - k_{10}[HO\bullet][H_2O_2] \quad (18\text{-}38)$$

$$-k_{11}[HO\bullet]\left[HCO_3^-\right] - k_{12}[HO\bullet][R] - k_{13}[HO\bullet][NOM]$$

$$r_{HO_3\bullet} = k_4\left[O_3^{-}\bullet\right]\left[H^+\right] - k_5\left[HO_3\bullet\right] \quad (18\text{-}39)$$

$$r_{O_3^-\bullet} = k_1\left[O_3\right]\left[HO_2^-\right] + k_3\left[O_2^-\bullet\right]\left[O_3\right] - k_4\left[O_3^-\bullet\right]\left[H^+\right] \quad (18\text{-}40)$$

$$r_{HO_2\bullet/O_2^-\bullet} = k_1\left[HO_2^-\right]\left[O_3\right] + k_9[HO\bullet]\left[HO_2^-\right] + k_{10}[HO\bullet]\left[H_2O_2\right]$$

$$-k_3\left[O_3\right]\left[O_2^-\bullet\right] \quad (18\text{-}41)$$

where
$r_{HO\bullet}$ = rate of hydroxyl radical formation, mol/L·s
$r_{HO_3\bullet}$ = rate of ozonide radical formation, mol/L·s
$r_{O_3^-\bullet}$ = rate of ozonide ion radical formation, mol/L·s
$r_{HO_2\bullet/O_2^-\bullet}$ = total rate of superoxide radical formation, mol/L·s
k_{11} = second-order rate constant between hydroxyl radical and bicarbonate, L/mol·s
k_{12} = second-order rate constant between hydroxyl radical and target organic compound R, L/mol·s
k_{13} = second-order rate constant between hydroxyl radical and NOM, L/mol·s
$[HO_3\bullet]$ = concentration of ozonide radical, mol/L
$[HO\bullet]$ = concentration of hydroxyl radical, mol/L
$[HO_2^-]$ = concentration of conjugate base or anion of hydrogen peroxide, mol/L
$[H_2O_2]$ = concentration of hydrogen peroxide, mol/L
$[HCO_3^-]$ = concentration of bicarbonate, mol/L
[R] = concentration of target organic compound R, mol/L
[NOM] = concentration of NOM, mol/L
$[O_3^-\bullet]$ = concentration of ozonide ion radical, mol/L
$[H^+]$ = concentration of hydrogen ion, mol/L
$[O_2^-\bullet]$ = concentration of superoxide anion radical, mol/L
$[O_3]$ = concentration of ozone, mol/L

The net rate of formation of all radical species can be set to zero because their concentrations are small and change rapidly in response to changing solution conditions. This rapid response is known as the pseudo-steady-state approximation. Invoking the pseudo-steady-state approximation for the various radical intermediates, four algebraic equations are obtained and radical species other than HO• can be eliminated from Eq. 18-38.

Equation 18-38 can be rearranged to obtain the following expression for HO•:

$$[HO\bullet]_{ss} = \frac{2k_1\left[HO_2^-\right][O_3]}{k_{11}\left[HCO_3^-\right] + k_{12}[R] + k_{13}[NOM]} \tag{18-42}$$

where $[HO\bullet]_{ss}$ = pseudo-steady-state concentration of HO•, mol/L

When the H_2O_2/ozone ratio is close to the stoichiometric optimum, the liquid-phase reaction occurs so fast that the ozone transfer is the limiting factor in the reaction rate. The O_3 concentration can be assumed to be constant and is much lower than the saturation concentration, as follows:

$$\left[O_3\right]_s = \frac{P_{O_3}}{H_{O_3}} \tag{18-43}$$

where $[O_3]_s$ = saturation concentration of ozone, mol/L
P_{O_3} = partial pressure of ozone in inlet gas, atm
H_{O_3} = Henry's law constant for ozone, atm · L/mol

The reaction of O_3 with hydroxyl ion and the target compounds can be ignored because of the low O_3 concentration and the relatively low rate constant for the reaction between O_3 and the target compound. The resulting rate expression for O_3 formation is given by

$$r_{O_3} = (K_L a)_{O_3}\left(\frac{P_{O_3}}{H_{O_3}} - [O_3]\right) - k_1\left[HO_2^-\right][O_3] - k_3\left[O_2^-\bullet\right][O_3] \tag{18-44}$$

where $(K_L a)_{O_3}$ = overall mass transfer coefficient for ozone, s^{-1}

The pseudo-steady-state expression for the rate of formation of ozone is invoked and Eq. 18-44 may be rearranged to the form

$$(K_L a)_{O_3}\left(\frac{P_{O_3}}{H_{O_3}} - [O_3]\right) = k_1\left[HO_2^-\right][O_3] + k_3\left[O_2^-\bullet\right][O_3] \tag{18-45}$$

The pseudo-steady state for the rate of formation of $HO_2\bullet/O_2\bullet$ shown in Eq. 18-41 may be rearranged to the form

$$k_3[O_2^-\bullet][O_3] = k_1[HO_2^-][O_3] + k_9[HO_2^-][HO\bullet] + k_{10}[H_2O_2][HO\bullet] \tag{18-46}$$

Substituting Eq. 18-41 into Eq. 18-46 yields

$$2k_1\left[HO_2^-\right][O_3] = (K_L a)_{O_3}\left(\frac{P_{O_3}}{H_{O_3}} - [O_3]\right) - k_9\left[HO_2^-\right][HO\bullet] - k_{10}\left[H_2O_2\right][HO\bullet] \tag{18-47}$$

The following expression is obtained after substituting Eq. 18-47 into Eq. 18-42 and rearranging:

$$[\mathrm{HO}\bullet]_{ss} = \frac{(K_L a)_{\mathrm{O_3}}\left(P_{\mathrm{O_3}}/H_{\mathrm{O_3}} - [\mathrm{O_3}]\right)}{k_9\left[\mathrm{HO_2^-}\right] + k_{10}\left[\mathrm{H_2O_2}\right] + k_{11}\left[\mathrm{HCO_3^-}\right] + k_{12}[R] + k_{13}[\mathrm{NOM}]} \tag{18-48}$$

The initial pseudo-steady-state concentration of HO• is obtained by neglecting $[O_3]$ as compared to P_{O_3}/H_{O_3} and the reaction between O_3 and HO• (ignoring this reaction is reasonable because the concentrations of both O_3 and the hydroxyl radical are low), as shown in the expression

$$[\mathrm{HO}\bullet]_{ss,0} = \frac{(K_L a)_{\mathrm{O_3}}\left(P_{\mathrm{O_3}}/H_{\mathrm{O_3}}\right)}{k_9\left[\mathrm{HO_2^-}\right]_0 + k_{10}\left[\mathrm{H_2O_2}\right]_0 + k_{11}\left[\mathrm{HCO_3^-}\right]_0 + k_{12}[\mathrm{R}]_0 + k_{13}[\mathrm{NOM}]_0} \tag{18-49}$$

where $[\mathrm{HO}\bullet]_{ss,0}$ = initial steady-state concentration of HO•, mol/L
$[\mathrm{HO_2^-}]_0$ = initial concentration of anion of hydrogen peroxide, mol/L
$[\mathrm{H_2O_2}]_0$ = initial concentration of hydrogen peroxide, mol/L
$[\mathrm{HCO_3^-}]_0$ = initial concentration of bicarbonate, mol/L
$[\mathrm{R}]_0$ = initial concentration of target organic compound R, mol/L
$[\mathrm{NOM}]_0$ = initial concentration of NOM, mol/L

The initial steady-state O_3 concentration can be estimated from Eq. 18-44:

$$[\mathrm{O_3}]_{ss,0} = \frac{K_L a\left(P_{\mathrm{O_3}}/H_{\mathrm{O_3}}\right) - k_9[\mathrm{HO}\bullet]_{ss,0}\left[\mathrm{H_2O_2}\right]_0 \times 10^{(\mathrm{pH}-pK_{\mathrm{H_2O_2}})} - k_{10}[\mathrm{HO}\bullet]_{ss,0}\left[\mathrm{H_2O_2}\right]_0}{(K_L a)_{\mathrm{O_3}} + 2k_1\left[\mathrm{H_2O_2}\right]_0 \times 10^{(\mathrm{pH}-pK_{\mathrm{H_2O_2}})}} \tag{18-50}$$

where $pK_{\mathrm{H_2O_2}}$ = acid dissociation constant for hydrogen peroxide ($\mathrm{p}K_{a5}$ in Table 18-5)

The rate laws for the parent compound R and H_2O_2 are given by the equations

$$r_{\mathrm{R}} = -k_6[\mathrm{R}][\mathrm{O_3}] - \mathrm{k}_{12}[\mathrm{R}][\mathrm{HO}\bullet] \tag{18-51}$$

$$r_{\mathrm{H_2O_2}} = -k_1\left[\mathrm{HO_2^-}\right][\mathrm{O3}] - k_9\left[\mathrm{HO_2^-}\right][\mathrm{HO}\bullet] - k_{10}\left[\mathrm{H_2O_2}\right][\mathrm{HO}\bullet] \tag{18-52}$$

where r_{R} = rate of target compound R destruction, mol/L·s
$r_{\mathrm{H_2O_2}}$ = rate of hydrogen peroxide loss, mol/L·s
k_6 = second-order rate constant between target compound R and ozone, L/mol·s

For the situation where the direct ozonation rate of a target compound is much lower than the reaction rate with hydroxyl radicals (the most common situation), that is, $\mathrm{k}_6[\mathrm{O_3}] \ll k_{12}[\mathrm{HO}\bullet]$, the first term in Eq. (18-51) can be ignored.

Simplified Model for H_2O_2/O_3 Process

A simplified model of the H_2O_2/O_3 process can be developed for various cases to provide an estimate of the destruction rates of the parent compound and hydrogen peroxide. The following two cases are considered: (1) H_2O_2 and O_3 are added together and (2) H_2O_2 is added to water containing O_3.

H_2O_2 AND O_3 ARE ADDED SIMULTANEOUSLY

A simplified analysis for the H_2O_2/O_3 process can be obtained by assuming that the hydroxyl radical concentration does not change with time and is equal to the initial steady-state hydroxyl radical concentration. This assumption yields a pseudo-first-order rate law, which results in the prediction of reaction rates that are faster than would be observed. The pseudo-first-order rate law and coefficient are given by the expressions

$$k_R = k_{12}\left[\mathrm{HO}\bullet\right]_{\mathrm{ss},0} \tag{18-53}$$

$$r_\mathrm{R} = -k_\mathrm{R}\left[\mathrm{R}\right] \tag{18-54}$$

where k_R= pseudo-first-order rate constant for target compound R, s^{-1}

Other terms as defined preiouvsly.

The residual of the hydrogen peroxide concentration is of interest, and the following pseudo-first-order rate law and coefficient can be obtained by assuming that the hydroxyl radical and ozone concentrations do not change with time and are equal to their initial steady-state concentration, which are given by the following expressions, respectively:

$$k_{\mathrm{H_2O_2}} = k_1\left[\mathrm{O_3}\right]_{\mathrm{ss},0}\left(10^{\mathrm{pH}-pK_{\mathrm{H_2O_2}}}\right) + k_9[\mathrm{HO}\bullet]_{\mathrm{ss},0}\left(10^{\mathrm{pH}-pK_{\mathrm{H_2O_2}}}\right) + k_{10}[\mathrm{HO}\bullet]_{\mathrm{ss},0} \tag{18-55}$$

$$\mathrm{r}_{\mathrm{H_2O_2}} = -k_{\mathrm{H_2O_2}}\left[\mathrm{H_2O_2}\right] \tag{18-56}$$

where $k_{\mathrm{H_2O_2}}$ = pseudo-first-order rate constant for hydrogen peroxide, s^{-1}

The above model, termed the *simplified pseudo-steady-state model*, overestimates the destruction rates of the parent compound and hydrogen peroxide. Consequently, when these expressions are used to assess the feasibility of destroying organic compounds and to provide an estimate of the effluent concentration of hydrogen peroxide, the computed value will be lower than the observed value.

H_2O_2 IS ADDED TO A WATER CONTAINING O_3

Some utilities add ozone for disinfection, and, when *Ct* disinfection credit is obtained, it may be useful to estimate the potential benefit of adding hydrogen peroxide for the destruction of target micropollutants such as atrazine. In this situation, the residual ozone concentration $[O_3]_{res}$ is known, and hydrogen peroxide is added to produce the hydroxyl radical. The rate law for O_3 is given by the equation

$$r_{\mathrm{O_3}} = -k_1[\mathrm{HO_2^-}][\mathrm{O_3}] - k_3[\mathrm{O_2^-}\bullet][\mathrm{O_3}] \tag{18-57}$$

Substituting Eq. 18-46 into Eq. 18-57 yields

$$r_{O_3} = -(2k_1[HO_2^-][O_3] + k_9[HO\bullet][HO_2^-] + k_{10}[HO\bullet][H_2O_2]) \quad (18\text{-}58)$$

According to Eq. 18-58, the initial pseudo-steady-state concentration of HO• is given by the equation

$$[HO\bullet]_{ss,0} = \frac{2k_1[H_2O_2]_0(10^{pH-pK_{H_2O_2}})[O_3]_{res}}{k_{11}[HCO_3^-]_0 + k_{12}[R]_0 + k_{13}[NOM]_0} \quad (18\text{-}59)$$

The rate laws for the parent compound R and H_2O_2 are given by Eqs. 18-51 and 18-52. In most cases, because $k_6[O_3] \ll k_{12}[HO\bullet]$, the first term in Eq. 18-51 can be ignored. A simplified model can be obtained by assuming that the hydroxyl radical does not change with time and is equal to the initial steady-state hydroxyl radical concentration, which is given by Eq. 18-59. The pseudo-first-order rate law and coefficient are given by the expressions

$$k_R = k_{12}[HO\bullet]_{ss,0} \quad (18\text{-}60)$$

$$r_R = -k_R[R] \quad (18\text{-}61)$$

The pseudo-first-order rate law and coefficient for hydrogen peroxide can be obtained by assuming that the hydroxyl radical and ozone concentrations do not change with time and are equal to their initial steady-state concentration. The initial concentration of ozone is equal to $[O_3]_{res}$:

$$k_{H_2O_2} = k_1[O_3]_{res}(10^{pH-pK_{H_2O_2}}) + k_9[HO\bullet]_{ss,0}(10^{pH-pK_{H_2O_2}}) + k_{10}[HO\bullet]_{ss,0} \quad (18\text{-}62)$$

$$r_{H_2O_2} = -k_{H_2O_2}[H_2O_2] \quad (18\text{-}63)$$

Because the initial concentrations are used, the above model predicts reaction rates that are faster than would be observed.

MASS BALANCES USING SIMPLIFIED RATE LAWS

The steady-state mass balances for a CMFR, tanks in series (TIS), a PFR, and dispersed-flow reactor for a pseudo-first-order reaction are provided in Sec. 8-4. Another model for nonideal mixing, the segregated-flow model, is provided in Chap. 6.

Comparison of Simplified Model to Data and Its Limitations

The simple pseudo-steady-state (Sim-PSS) model was compared to the data that were provided by Glaze and Kang (1989). A comparison of the model predicted pseudo-first-order rate constants to the experimentally determined values is displayed on Fig. 18-2. The measured data are predicted well with the Sim-PSS model when the H_2O_2/O_3 mass ratio is from 0.3 to about 0.6, which is around the stoichiometric optimum of 0.35. For a mass ratio less than 0.3, the predicted rate constants are higher than the

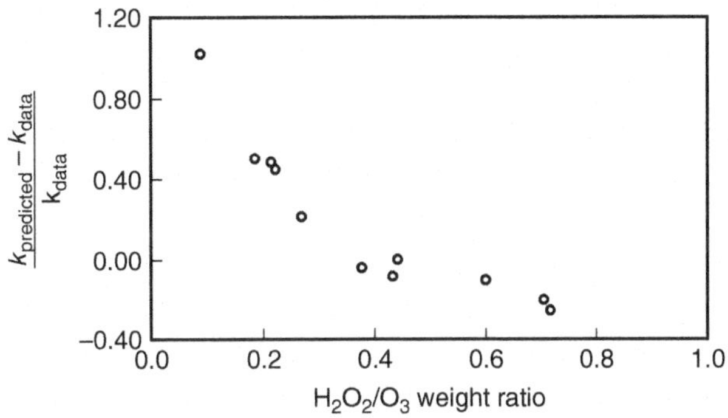

Figure 18-2
Comparison of model-predicted pseudo-first-order rate constants to experimental values.

measured values, and when the ratio exceeds 0.6, the predicted values are less than the measured values. The observed variations are due to the complexity of the H_2O_2/O_3 reaction system; in particular, different mechanisms control the overall reaction rate from O_3 control to H_2O_2 control as the H_2O_2/O_3 ratio changes. Consequently, to predict process performance more accurately, a sophisticated model is required. However, the Sim-PSS model is precise enough to examine the feasibility of the process. Moreover, pilot testing is necessary to evaluate the technology in the field once process feasibility has been assessed using the Sim-PSS model.

Numerous studies have been reported in the literature on the efficacy of the H_2O_2/O_3 process. When compared with the results of these studies, the simple model does a reasonable job of predicting process performance. The electrical efficiency per log order reduction (EE/O) (see Eq. 8-130 in Chap. 8) for the H_2O_2/O_3 process can also be calculated using the energy required to produce ozone and the observed first-order rate constants. These EE/O values are much lower than the values that are typically reported for the UV/hydrogen peroxide process for the destruction of TCE and PCE (0.53 to 2.64 kWh/m^3 per log order reduction for UV/hydrogen peroxide process vs. 0.11 kWh/m^3 per log order reduction for the H_2O_2/O_3 process). Thus, the H_2O_2/O_3 process appears to be a more energy-efficient method for producing hydroxyl radicals than the UV/hydrogen peroxide process.

Disadvantages of H_2O_2/O_3 Process

Several problems are associated with the hydrogen peroxide/ozone process. One problem is the stripping of volatile species into the off-gas from the ozone contactor. The stripping phenomenon is not important for the more reactive volatile species but can be for species that are less reactive with hydroxyl radical, such as carbon tetrachloride. Another problem with the use of the hydrogen peroxide/ozone process is the production of bromate when the water being treated contains significant amounts of bromide ion. The formation of bromate can occur with any advanced oxidation process, but the presence of ozone in the H_2O_2/O_3 process adds a second route of formation, namely direct oxidation by O_3 itself (see Chap. 8).

For California State Project water and Colorado River water (CRW), each containing about 90 μg/L bromide, it was found that the H_2O_2/O_3 process produced bromate concentrations in excess of the MCL of 10 μg/L (Liang et al., 1999). A number of researchers have investigated the mechanism of bromide oxidation by ozone alone (Buxton and Greenstock, 1988; Haag and Hoigné, 1983; Hoigné and Bader, 1976; Hoigné et al., 1985; Taube, 1942).

Models of bromate formation from bromide reactions with ozone alone, discussed under ozone reactions (see Chap. 19), have been developed (von Gunten and Hoigné, 1994; Westerhoff et al., 1994, 1998). A study of bromate formation in advanced oxidation processes in which the influence of H_2O_2 on the bromide oxidation by O_3 is considered is available in the literature (von Gunten and Hoigné, 1994). A model of the pathways of bromide oxidation reaction, which includes almost all of the known mechanisms, has been developed and tested (Westerhoff et al., 1998).

Advanced Oxidation plays a key disinfection role in San Diego's potable reuse program.
http://www.stantec.com/wt3r/18-1.html

It has been observed that lower bromate concentrations were produced when a hydrogen peroxide–ozone weight ratio of 1.0 was used as compared to ozone alone (Liang et al., 2001). It appears that bromate formation could be controlled effectively with higher hydrogen peroxide–ozone ratios because excess hydrogen peroxide accelerates ozone decay. However, a high hydrogen peroxide–ozone ratio might not be optimal for contaminant removal. As a result, removal efficiency and bromate formation have to be considered when using the hydrogen peroxide/ozone process. Other strategies, such as pH depression and ammonia addition (see Chap. 19), can also be used to control bromate formation.

In spite of the success of the bromate formation models and what is known about bromate formation, pilot studies are recommended for treatability assessment and process optimization. The simple models of the process presented in this section are only useful for preliminary process assessment and planning the scope of pilot investigations.

Example 18-6 Contaminant Effluent Concentration Using the Hydrogen peroxide–ozone process

A small city has recently discovered that one of its wells is contaminated with 200 μg/L (1.52 μmol/L) TCE. To continue using the well as a drinking water source, the TCE needs to be destroyed and the effluent concentration must be less than 5 μg/L. During normal pumping operations, the well produces water at about 0.025 m^3/s (400 gal/min). The HCO_3^-, pH, and DOC concentrations are 488 mg/L, 7.5, and 0.7 mg/L, respectively. The physicochemical properties of TCE and NOM are as follows:

Compound	MW, g/mol	HO· Rate Constant, $k_{HO\cdot}$, L/mol·s
TCE	131.389	4.20×10^9
NOM[a]	NA	3.90×10^8

[a]For NOM, the unit of $k_{HO\cdot}$ is L/mol NOM C·s.

For simplicity, a proprietary reactor will be used. It has been determined by conducting tracer studies on the reactor that its hydraulic performance can be described using four completely mixed flow reactors in series. Given the following information: (1) the H_2O_2 dosage is 0.8 mg/L, (2) O_3 is generated onsite and the ozone flow rate is 1 mg/L · min, (3) the partial pressure of ozone in the inlet gas is 0.07 atm, (4) the Henry's law constant for O_3 at 23°C is 83.9 atm · L/mol, (5) the overall mass transfer coefficient for 03, K_La, was measured to be $7 \times 10^{-4}\ s^{-1}$, and (6) the reactor volume is 5500 L (5.5 m^3). Determine the expected effluent concentration for TCE and the H_2O_2 residual concentration.

Solution

1. Calculate the hydraulic detention time (τ):

$$\tau = \frac{V}{Q} = \frac{5.5\ \text{m}^3}{\left(0.025\ \text{m}^3/\text{s}\right)\left(60\ \text{s/min}\right)} = 3.7\ \text{min}$$

2. Calculate the initial steady-state concentration of hydroxyl radical using Eq. 18-49:

$$[\text{HO}\bullet]_{ss,0} = \frac{K_La(P_{O_3}/H_{O_3})}{k_9[\text{HO}_2^-]_0 + k_{10}[\text{H}_2\text{O}_2]_0 + k_{11}[\text{HCO}_3^-]_0 + k_{12}[\text{R}]_0 + k_{13}[\text{NOM}]_0}$$

   a. Obtain the reaction rate constants and acid dissociation constants from Table 18-5:

$$k_9 = 7.5 \times 10^9\ \text{L/mol} \cdot \text{s}$$
$$k_{10} = 2.7 \times 10^7\ \text{L/mol} \cdot \text{s}$$
$$k_{11} = 8.5 \times 10^6\ \text{L/mol} \cdot \text{s}$$
$$k_{12} = 4.2 \times 10^9\ \text{L/mol} \cdot \text{s}$$
$$k_{13} = 3.9 \times 10^8\ \text{L/mol NOM C} \cdot \text{s}$$
$$pK_{H_2O_2} = 11.75$$

   b. Calculate the concentration of each component:

$$\left[\text{H}_2\text{O}_2\right]_0 = \frac{0.8\ \text{mg/L}}{(34\ \text{g/mol})\ (1000\ \text{mg/g})} = 2.35 \times 10^{-5}\ \text{mol/L}$$

$$\left[HO_2^-\right]_0 = [H_2O_2]_0\left(10^{pH-pK_{H_2O_2}}\right) = \left(2.35\times10^{-5}\text{ mol/L}\right)\left(10^{7.5-11.75}\right)$$
$$= 1.32\times10^{-9}\text{ mol/L}$$

$$\left[HCO_3^-\right]_0 = \frac{488\text{ mg/L}}{(61\text{ g/mol})(1000\text{ mg/g})} = 0.008\text{ mol/L}$$

$$[R]_0 = [TCE]_0 = \frac{1.52\ \mu\text{mol/L}}{10^6\ \mu\text{mol/mol}} = 1.52\times10^{-6}\text{ mol/L}$$

$$[NOM]_0 = \frac{0.7\text{ mg/L}}{(12\text{ g C/mol NOM C})(1000\text{ mg/g})}$$
$$= 5.83\times10^{-5}\text{ mol NOM C/L}$$

c. Calculate the product of the rate constant and initial concentration of each component needed in Eq. 18-49:

$$k_9\left[HO_2^-\right]_0 = \left(7.5\times10^9\text{ L/mol}\cdot\text{s}\right)\left(1.32\times10^{-9}\text{mol/L}\right) = 9.9\text{ s}^{-1}$$
$$k_{10}\left[H_2O_2\right]_0 = \left(2.7\times10^7\text{ L/mol}\cdot\text{s}\right)\left(2.35\times10^{-5}\text{ mol/L}\right) = 634.5\text{ s}^{-1}$$
$$k_{11}\left[HCO_3^-\right]_0 = \left(8.5\times10^6\text{ L/mol}\cdot\text{s}\right)(0.008\text{ mol/L}) = 68{,}000\text{ s}^{-1}$$
$$k_{12}[R]_0 = k_{12}[TCE]_0 = \left(4.2\times10^9\text{ L/mol}\cdot\text{s}\right)\left(1.52\times10^{-6}\text{ mol/L}\right)$$
$$= 6384\text{ s}^{-1}$$
$$k_{13}[NOM]_0 = \left(3.9\times10^8\text{ L/mol NOM C}\cdot\text{s}\right)\left(5.83\times10^{-5}\text{ mol NOM C/L}\right)$$
$$= 22737\text{ s}^{-1}$$

d. Calculate the initial steady-state concentration of the hydroxyl radical using Eq. 18-49:

$$[HO\bullet]_{ss,0} = \frac{\left(7\times10^{-4}\text{ s}^{-1}\right)[(0.07\text{ atm})/(83.9\text{ L}\cdot\text{atm/mol})]}{(9.9+634.5+68000+6384+22737)\text{ s}^{-1}}$$
$$= 5.97\times10^{-12}\text{ mol/L}$$

3. Calculate the TCE effluent concentration:
   a. Determine k_{TCE} using the pseudo-first-order rate law presented in Eq. 18-49:

$$k_R = k_{TCE} = k_{12}[HO\cdot]_{ss,0} = (4.2\times10^9\ L/mol\cdot s)\ (5.97\times10^{-12}\ mol/L)$$
$$= 0.025\ s^{-1}$$

   d. Determine the effluent TCE concentration using the TIS model presented in Sec. 8-4:

$$[TCE] = [R] = \frac{[R]_0}{(1+k_{TCE}\tau/n)^n} = \frac{200\ \mu g/L}{\left[1+(0.025\ s^{-1})(3.7\ min)(60\ s/min)/4\right]^4}$$
$$= 6.2\ \mu g/L$$

4. Estimate the initial steady-state concentration of O_3 using Eq. 18-50:

$$[O_3]_{ss,0} = \frac{K_La(P_{O_3}/H_{O_3}) - k_9[HO\cdot]_{ss,0}[H_2O_2]_0(10^{pH-pK_{H_2O_2}}) - k_{10}[HO\cdot]_{ss,0}[H_2O_2]_0}{K_La + 2k_1[H_2O_2]_0(10^{pH-pK_{H_2O_2}})}$$

   a. From Table 18-5,

$$k_1 = 2.8\times10^6\ L/mol\cdot s$$

   b. From steps 2a, 2b, and 2c,

$$k_9[H_2O_2]_0(10^{pH-pK_{H_2O_2}}) = k_9[HO_2^-]_0 = 9.9\ s^{-1}$$
$$k_{10}[H_2O_2]_0 = 634.5\ s^{-1}$$
$$[H_2O_2]_0(10^{pH-pK_{H_2O_2}}) = [HO_2^-]_0 = 1.32\times10^{-9}\ mol/L$$

   c. Solve for $K_La\ (P_{O_3}/H_{O_3})$:

$$K_La\frac{P_{O_3}}{H_{O_3}} = (7\times10^{-4}\ s^{-1})\left(\frac{0.07\ atm}{83.9\ L\cdot atm/mol}\right) = 5.84\times10^{-7} mol/L\cdot s$$

   d. Combine results of 4b with result of 2d:

$$k_9[H_2O_2]_0(10^{pH-pK_{H_2O_2}})\times[HO\cdot]_{ss,0} = (9.9\ s^{-1})(5.97\times10^{-12}\ mol/L)$$
$$= 5.9\times10^{-11}\ mol/L\cdot s$$

$$k_{10}[H_2O_2]_0[HO\cdot]_{ss,0} = (634.5\ s^{-1})(5.97\times10^{-12}\ mol/L)$$
$$= 3.79\times10^{-9}\ mol/L\cdot s$$

e. Solve for $[O_3]_{ss,0}$:

$$[O_3]_{ss,0} = \frac{\left[\left(5.84\times10^{-7}\right)-\left(5.9\times10^{-11}\right)-\left(3.79\times10^{-9}\right)\right]\text{mol/L}\cdot\text{s}}{\left(7\times10^{-4}\ \text{s}^{-1}\right)+\left[2\left(2.8\times10^{6}\ \text{L/mol}\cdot\text{s}\right)\left(1.32\times10^{-9}\ \text{mol/L}\right)\right]}$$

$$= 7.17\times10^{-5}\ \text{mol/L (3.44 mg/L)}$$

5. Estimate H_2O_2 residual:

   a. Estimate the pseudo-first-order rate constant for hydrogen peroxide using Eq. 18-55:

$$k_{H_2O_2} = k_1[O_3]_{ss,0}\left(10^{pH-pK_{H_2O_2}}\right)+k_9[HO\cdot]_{ss,0}\left(10^{pH-pK_{H_2O_2}}\right)+k_{10}[HO\cdot]_{ss,0}$$

   i. Determine the values of the rate constant times concentration needed in Eq. 18-55:

$$k_1[O_3]_{ss,0}\left(10^{pH-pK_{H_2O_2}}\right) = (2.8\times10^{6}\ \text{L/mol}\cdot\text{s})(7.17\times10^{-5}\ \text{mol/L})\times10^{(7.5-11.75)}$$

$$= 0.0112\ \text{s}^{-1}$$

$$k_9[HO\cdot]_{ss,0}\left(10^{pH-pK_{H_2O_2}}\right) = (7.5\times10^{9}\ \text{L/mol}\cdot\text{s})(5.97\times10^{-12}\ \text{mol/L})\times10^{(7.5-11.75)}$$

$$= 2.52\times10^{-6}\ \text{s}^{-1}$$

$$k_{10}[HO\cdot]_{ss,0} = (2.7\times10^{7}\ \text{L/mol}\cdot\text{s})(5.97\times10^{-12}\text{mol/L})$$

$$= 1.61\times10^{-4}\text{s}^{-1}$$

   ii. Determine $k_{H_2O_2}$:

$$k_{H_2O_2} = \left(0.0112\ \text{s}^{-1}\right)+\left(2.52\times10^{-6}\ \text{s}^{-1}\right)+\left(1.61\times10^{-4}\ \text{s}^{-1}\right)$$

$$= 0.0114\ \text{s}^{-1}$$

   b. Estimate the H_2O_2 residual using the TIS model (see Sec. 8-4):

$$[H_2O_2] = \frac{[H_2O_2]_0}{(1+k_{H_2O_2}\tau/n)^n} = \frac{0.8\ \text{mg/L}}{\left[1+(0.0114\ \text{s}^{-1})(3.7\ \text{min})(60\ \text{s / min})/4\right]^4}$$

$$= 0.11\ \text{mg/L}$$

Comment

The initial ozone concentration is only an approximate estimate because it was assumed that the reactor contents are mixed completely and the gas-phase ozone concentration does not change.The effluent hydrogen peroxide concentration is only an estimate, and, based on the reactions that were considered, it is the lowest expected effluent concentration. Measurements will have to be taken to ensure that this residual does not interfere with disinfection (e.g., consume chlorine) and is not transmitted to the distribution system.

18-4 Hydrogen Peroxide/UV Light Process

The UV/ hydrogen peroxide process includes hydrogen peroxide injection and mixing followed by a reactor that is equipped with UV lights. As shown on Fig. 18-3, a typical UV reactor is a stainless steel column that contains UV lights in a crisscrossing pattern. The details of the reactor are discussed in Secs. 8-4 and 13-9. The UV/H_2O_2 process cannot be used for potable water treatment because it has high effluent H_2O_2 concentrations. High effluent H_2O_2 concentrations are unavoidable, because high initial dosages of H_2O_2 are required in order to efficiently utilize the UV light and produce hydroxyl radical. Aside from the health issues associated with high effluent H_2O_2 concentrations in the finished water, the residual H_2O_2 will consume chlorine and interfere with disinfection. This challenge will have to be overcome before the UV/ H_2O_2 process is used in drinking water treatment.

Elementary Reactions for the Hydrogen Peroxide/UV Process

The complex elementary radical reactions that are involved with the H_2O_2/UV process have been discovered. It is now possible to predict the destruction of the target compound using these reactions and gain insight into the factors that impact the H_2O_2/UV process (Glaze et al., 1990; Liao and Gurol, 1995; Crittenden et al., 1999). The mechanisms that may be considered are (1) photolysis of H_2O_2 with a multichromatic light source, (2) UV absorption by the background components in the water matrix, (3) scavenging of hydroxyl radical by NOM and carbonate species, and (4) direct photolysis of NOM and the target compound. A rigorous AOP model was developed to predict the destruction of target compounds and the effluent H_2O_2 concentration using the complete radical reaction pathway presented by Crittenden et al. (1999).

However, reasonable predictions of target compound destruction and residual H_2O_2 can be obtained by using a simplified pseudo-steady-state model as shown below, although some accuracy will be lost (Crittenden et al., 1999). The most important elementary reactions in the H_2O_2/UV process at neutral pH are shown in Table 18-5. The reaction pathway is extremely simplified and ignores radical–radical reactions, the reactions between HO_2^- and CO_3^{2-} and other species (due to the large pK_a values) and unimportant radical species ($CO_3^-\cdot$; etc.).

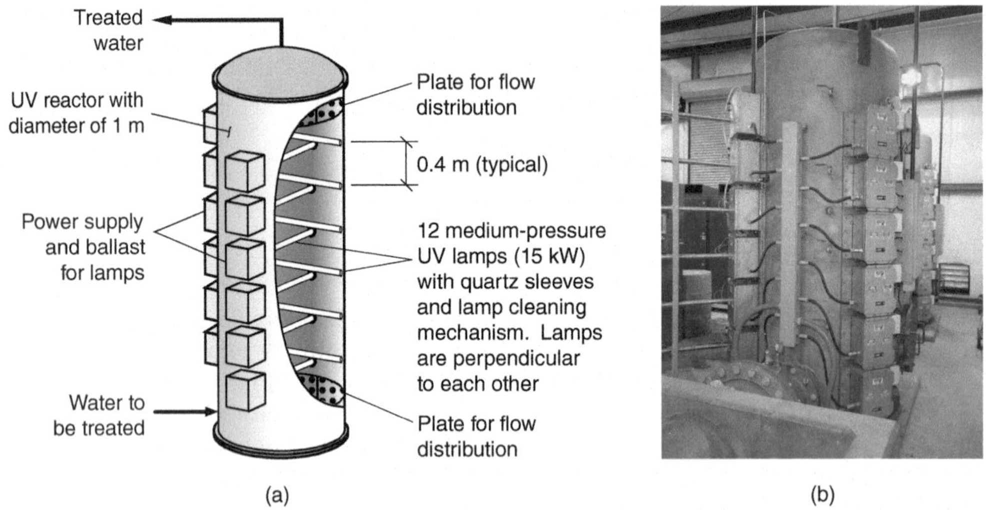

Figure 18-3
UV reactor used for advanced oxidation: (a) schematic and (b) photograph.

The elementary reactions that are involved in the H_2O_2/UV process include initiation (reaction 7), propagation (reactions 9, 10), and termination reactions of the radical chain reactions. (Termination reactions involve recombination of radical species and are not shown because they have a low probability of occurrence, e.g., $HO\bullet + HO\bullet \rightarrow H_2O_2$) The elementary reactions also include the oxidation of the target organic compound (R) and the scavenging of hydroxyl radical by bicarbonate, carbonate, and NOM.

PRODUCTION OF HYDROXYL RADICAL

Early investigations of hydrogen peroxide photolysis (Heidt, 1932; Volman, 1949; Volman and Chen, 1959; Hunt and Taube, 1952; Baxendale and Wilson, 1957) as well as more recent studies (Sehested et al., 1968; Bielski and Richter, 1977; Bielski et al., 1985) indicate that the following radical chain reactions occur in a hydrogen peroxide solution with UV light irradiation. As shown in Table 18-5, the production of hydroxyl radical is initiated via the following reaction:

$$H_2O_2 + h\nu \rightarrow 2HO\bullet \tag{18-64}$$

QUANTUM YIELD

The primary quantum yield for H_2O_2 is 0.5 for wavelengths in the UV region (Volman and Chen, 1959; Zellner et al., 1990), but the primary quantum yield of H_2O_2 depends slightly on temperature. For example, the quantum yield is 0.41 at 5°C. However, this is not important because the temperature in a UV reactor generally achieves room temperature due to heat produced during lamp illumination.

SIMPLIFIED PSEUDO-STEADY–STATE MODEL

Based on the reactions that are presented in Table 18-5, the rate expression for HO· is given by this expression.

$$r_{HO\bullet} = 2\phi_{H_2O_2}P_{u-v}f_{H_2O_2}(1-e^{-A}) - k_{10}[HO\bullet][H_2O_2] - k_{11}[HO\bullet][HCO_3^-] - k_{12}[HO\bullet][R] - k_{13}[HO\bullet][NOM] \quad (18\text{-}65)$$

where $r_{HO^\bullet}$ = rate of hydroxyl radical formation, mol/L · s
$\Phi_{H_2O_2}$ = quantum yield of hydrogen peroxide, mol/einstein
$P_{U\text{-}V}$ = photonic intensity per unit volume, einsteins/cm^3 · s
$f_{H_2O_2}$ = fraction of light absorbed by hydrogen peroxide, dimensionless
A = absorbance, dimensionless
k_{10} = second-order rate constant between hydroxyl radical and hydrogen peroxide, L/mol · s ($M^{-1}s^{-1}$)
k_{11} = second-order rate constant between hydroxyl radical and carbonate, L/mol · s ($M^{-1}s^{-1}$)
k_{12} = second-order rate constant between hydroxyl radical and target organic compound R, L/mol · s ($M^{-1}s^{-1}$)
k_{13} = second-order rate constant between hydroxyl radical and NOM, L/mol · s ($M^{-1}s^{-1}$)
[HO•] = concentration of hydroxyl radical, mol/L
$[H_2O_2]$ = concentration of hydrogen peroxide, mol/L
$[HCO_3^-]$ = concentration of carbonate, mol/L
[R] = concentration of target compound R, mol/L
[NOM] = concentration of NOM, mol carbon/L

The photonic intensity per unit volume of reactor, $P_{U\text{-}V,}$ can be calculated using the the following expression:

$$P_{U-V} = \frac{P\eta}{N_{av}Vh\nu} \quad (18\text{-}66)$$

where η = efficiency of the UV lamp, dimensionless
V = volume of reactor solution, L

According to the pseudo-steady-state assumption, the change of hydroxyl radical concentration with time is negligible because the rate of reactions involving HO• are very fast and HO• concentration is very small as compared to other compounds. Consequently, the formation rate of hydroxyl radical can be set equal to zero. By setting the formation rate of HO• equal to zero, the pseudo-steady-state concentration of hydroxyl radical can be determined.

$$[HO\bullet]_{ss} = \frac{2\phi_{H_2O_2}P_{u-v}f_{H_2O_2}\left(1-e^{-A}\right)}{k_{10}[H_2O_2] + k_{11}[HCO_3^-] + k_{12}[R] + k_{13}[NOM]} \quad (18\text{-}67)$$

where $[HO\bullet]_{ss}$ = pseudo-steady-state concentration of hydroxyl radical, mol/L

A further simplification of the UV/ H_2O_2 process model that can be used to show trends and estimate process feasibility is obtained by assuming that the NOM, R, and H_2O_2 concentrations are constant and equal to their initial concentration, when calculating the pseudo-steady-state concentration of hydroxyl radical. This version of the model is called the simple pseudo-steady-state (Sim-PSS) model and the hydroxyl radical concentration becomes the following expression:

$$[HO\bullet]_{ss,0} = \frac{2\phi_{H_2O_2} P_{u-v} f_{H_2O_2}\left(1-e^{-A}\right)}{k_{10}[H_2O_2]_0 + k_{11}[HCO_3^{-}]_0 + k_{12}[R]_0 + k_{13}[NOM]_0} \tag{18-68}$$

where $[HO\bullet]_{ss,0}$ = initial pseudo-steady-state concentration of hydroxyl radical, mol/L
$[H_2O_2]_0$ = initial concentration of hydrogen peroxide, mol/L
$[HCO_3^{-}]_0$ = initial concentration of carbonate, mol/L
$[R]_0$ = initial concentration of target compound R, mol/L
$[NOM]_0$ = initial concentration of NOM, mol/L

Accordingly, the rate law for the disappearance of the target compound and hydrogen peroxide are given by the following expressions:

$$r_R = -k_{12}[R][HO\bullet]_{ss,0} - \phi_R P_{u-v} f_R (1-e^{-A}) \tag{18-69}$$

$$r_{H_2O_2} = -\phi_{H_2O_2} P_{u-v} f_{H_2O_2}\left(1-e^{-A}\right) - k_{10}[HO\bullet]_{ss,0}\left[H_2O_2\right] \tag{18-70}$$

In many cases, the photolysis rate of the target compound is small and the second term in Eq. 18-69 can be neglected; and photoreactors are designed so all the light is absorbed. For this situation, Eqs. 18-69 and 18-70 simplify to the following equations:

$$r_R = -k'_{12}[R] \tag{18-71}$$

$$r_{H_2O_2} = -\phi_{H_2O_2} P_{u-v} f_{H_2O_2} - k_{10}[HO\bullet]_{ss,0}[H_2O_2] \tag{18-72}$$

where $k'_{12} = k_{12}[HO\bullet]_{ss,0}$ = pseudo-first-order rate constant, s^{-1}

Equation 18-71 may be further simplified by assuming that fH_2O_2 does not change with time and may be calculated from the following expression:

$$\phi_{H_2O_2} P_{u-v} f_{H_2O_2} = \frac{0.5 P_{u-v} \varepsilon_{H_2O_2}[H_2O_2]}{\sum \varepsilon_i C_i} \tag{18-73}$$

where $\varepsilon_{H_2O_2}$ = extinction coefficient for hydrogen peroxide, L/mol·cm
ε_i = extinction coefficient for component i, L/mol·cm
C_i = concentration of component i, mol/L

If the major background chromophores are Fe(II) and NOM, and their concentrations are assumed to be constant and equal to their initial concentration, Eq. 18-74 simplifies to the following expression:

$$\phi_{H_2O_2} P_{u-v} f_{H_2O_2} = \frac{0.5 P_{u-v} \varepsilon_{H_2O_2} [H_2O_2]}{\varepsilon_{H_2O_2} [H_2O_2]_0 + \varepsilon_{NOM}[NOM]_0 + \varepsilon_{Fe(II)}[Fe(II)]_0} \tag{18-74}$$

where $[Fe(II)]_0$ = initial concentration of ferrous ion, mol/L

The key assumption for Eq. 18-74 is that $\varepsilon_{H_2O_2}[H_2O_2]$ is a constant, which is equal to $\varepsilon_{H_2O_2}[H_2O_2]_0$, and this assumption will predict a lower photolysis rate. However, the effluent concentration that is predicted using the Sim-PSS is lower than that will be actually observed because the pseudo-steady-state concentration of hydroxyl radical is taken to be the initial value in the simplified pseudo-steady-state model. Accordingly, if the predicted concentration is too high, then the process may be considered infeasible.

The final rate expression for loss of H_2O_2 using the Sim-PSS model is given by the following expressions:

$$r_{H_2O_2} = -k'_{10}[H_2O_2] \tag{18-75}$$

$$k'_{10} = \frac{0.5 P_{u-v} \varepsilon_{H_2O_2}}{\varepsilon_{H_2O_2} [H_2O_2]_0 + \varepsilon_{NOM}[NOM]_0 + \varepsilon_{Fe(II)}[Fe(II)]_0} + k_{10}[HO\bullet]_{ss,0} \tag{18-76}$$

where k'_{10} = pseudo-first-order rate constant for the destruction of hydrogen peroxide, s^{-1}

Describing Reactor Performance

The steady-state mass balances for a completely mixed flow reactor (CMFR), CMFRs in series, a plug flow reactor (PFR), and a dispersed flow reactor for a pseudo-first-order reaction is provided in Sec. 8-4. The identical equations may be used. Another model for nonideal mixing, the segregated-flow model, is provided in Chap. 6.

UV LIGHT TRANSMISSION

The ability of H_2O_2 to absorb UV light and produce HO• via reaction shown in Eq. 18-76 is dependent upon the wavelength and quantum yield and the UV light absorbance of the background components in the water, as discussed in Sec. 8-4.

AOPs that utilize UV light for the production of HO• radicals must have reasonable light transmission in the UV region of the light because any light that is not absorbed by the oxidant is wasted, and the generation of UV light represents a significant operational cost. Accordingly, it is important to evaluate the influence of pretreatment effectiveness and cost (e.g., particle removal and the removal of certain UV absorbing species) on UV light transmission. For example, when considering the UV/ H_2O_2 process, a preliminary evaluation would include an estimate of the fraction of UV light that would be available to activate the H_2O_2 and the influence that pretreatment would have on the available light transmission. In a highly contaminated groundwater, an absorbance of 0.385 for a 1 cm depth at 254 nm was measured. The light absorption coefficient for H_2O_2 is about 19 $M^{-1}cm^{-1}$ at 254 nm and the quantity of light and the fraction of light that produces hydroxyl radical may be estimated from the following equation:

$$f_{\text{oxidant}} = \frac{\varepsilon_{H_2O_2} C_{H_2O_2} L}{\varepsilon_{H_2O_2} C_{H_2O_2} L + \varepsilon_{\text{bac}} C_{\text{bac}} L} = \frac{\varepsilon_{H_2O_2} C_{H_2O_2}}{\varepsilon_{H_2O_2} C_{H_2O_2} + \varepsilon_{\text{bac}} C_{\text{bac}}} \quad (18\text{-}77)$$

where f_{oxidant} = light absorbed by the oxidant, dimensionless
$C_{H_2O_2}$ = concentration of hydrogen peroxide, mol/L
L = reactor depth,cm
$\varepsilon_{H_2O_2}$ = extinction coefficients for hydrogen peroxide, L/mol · cm
ε_{bac} = extinction coefficients for background, L/mol · cm

For example, fraction of light absorbed by an H_2O_2 concentration of 80 mg/L is only 10.7 percent, or 90 percent of the light is wasted.

When considering pretreatment options, it is useful to know the light absorption of certain dissolved species in water because this can form the basis for pretreatment.

Example 18-7 Fraction of light absorbed by hydrogen peroxide for various Fe(II) solutions

Calculate the fraction of light absorbed by 1 mM H_2O_2 in solutions of 0.5 and 5 mg/L of Fe(II) at 254 nm. At 254 nm,the molar extinction coefficients of H_2O_2 and ferrous ion are 19 and 448 L/mol · cm, respectively.

Solution

1. Calculate the absorbance of H_2O_2:
   From the Beer–Lambert law presented in Sec. 8-4:

$$A = \varepsilon C x$$

The absorbance of H_2O_2 at 1 mM is

$$A_{H_2O_2} = \varepsilon_{H_2O_2} C_{H_2O_2} x = (19\ \text{L/mol}\cdot\text{cm})\left(1\times10^{-3}\ \text{mol/L}\right)x$$
$$= 0.019 \times \text{cm}^{-1}$$

2. Calculate the absorbance of Fe(II).
   a. Calculate the absorbance of Fe(II) at 0.5 mg/L:

$$A_{Fe(II)} = \varepsilon_{Fe(II)} C_{Fe(II)} x = (448\ \text{L/mol}\cdot\text{cm})\left[\frac{0.5\ \text{mg/L}}{(55.85\ \text{g/mol})(1000\ \text{mg/g})}\right]x$$
$$= (4.01\times10^{-3})x\ \text{cm}^{-1}$$

   b. Calculate the absorbance of Fe(II) at 5 mg/L:

$$A_{Fe(II)} = \varepsilon_{Fe(II)} C_{Fe(II)} x = (448\ \text{L/mol}\cdot\text{cm})\left[\frac{5\ \text{mg/L}}{(55.85\ \text{g/mol})(1000\ \text{mg/g})}\right]x$$
$$= (4.01\times10^{-2})x\,\text{cm}^{-1}$$

3. Calculate the fraction of light absorbed by H_2O_2.
   The total absorbance of the solution is

$$A_t = A_{H_2O_2} + A_{Fe(II)} = x\left(\varepsilon_{H_2O_2} C_{H_2O_2} + \varepsilon_{Fe(II)} C_{Fe(II)}\right)$$

   The fraction of light absorbed by H_2O_2 can be calculated using Eq. 18-77:

$$f_{H_2O_2} = \frac{A_{H_2O_2}}{A_t}\times100\% = \frac{\varepsilon_{H_2O_2} C_{H_2O_2}}{\varepsilon_{H_2O_2} C_{H_2O_2} + \varepsilon_{Fe(II)} C_{Fe(II)}}\times100\%$$

   a. For Fe(II) concentration of 0.5 mg/L:

$$f_{H_2O_2} = \frac{0.019x}{\left[0.019x + \left(4.01\times10^{-3}\right)x\right]}\times100\% = 82.6\%$$

   b. For Fe(II) concentration of 5 mg/L:

$$f_{H_2O_2} = \frac{0.019x}{0.019x + (4.01\times10^{-2})x}\times100\% = 32.1\%$$

Comment

Iron can absorb a significant amount of UV light and scavenges hydroxyl radical. Consequently, it may be worthwhile to remove iron from groundwaters that contain high concentrations, if the hydrogen peroxide/UV process is considered for organics destruction.

Example 18-8 Fraction of light adsorbed by hydrogen peroxide for various NOM concentrations

Calculate the fraction of light absorbed by 1 mM H_2O_2 for 0.5, 1, 2, 5, and 10 mg/L of NOM at 254 nm. At 254 nm, the extinction coefficient of NOM is 0.0196 L/mg NOM · cm. From Example 18-7 at 254 nm, the absorbance of H_2O_2 at 1 mM is 0.0196 cm^{-1}.

Solution

1. Calculate the fraction of light absorbed by H_2O_2 for 0.5 mg/L of NOM.
   a. Calculate the absorbance of NOM at a concentration of 0.5 mg/L:

$$A_{NOM} = \varepsilon_{NOM} C_{NOM} x = (0.0196 \text{ L/mg} \cdot \text{cm})(0.5 \text{ mg/L}) x = 0.01 \times \text{cm}^{-1}$$

   b. Calculate the fraction of light absorbed by H_2O_2:
   In a similar manner compared to the solution of Example 18-7, the fraction of light absorbed by H_2O_2 may be calculated from the following equation:

$$f_{H_2O_2} = \frac{\varepsilon_{H_2O_2} C_{H_2O_2}}{\varepsilon_{H_2O_2} C_{H_2O_2} + \varepsilon_{NOM} C_{NOM}}$$

   The fraction of light absorbed by H_2O_2 at 0.5 mg NOM/L is

$$f_{H_2O_2} = \left(\frac{0.019x}{0.019x + 0.01x}\right) \times 100\% = 65.5\%$$

2. Plot the fraction of UV light absorbed by H_2O_2 versus NOM concentrations (0.5, 1, 2, 5, and 10 mg/L).
   The following figure gives the fraction of UV light absorbed by H_2O_2 at different NOM concentrations:

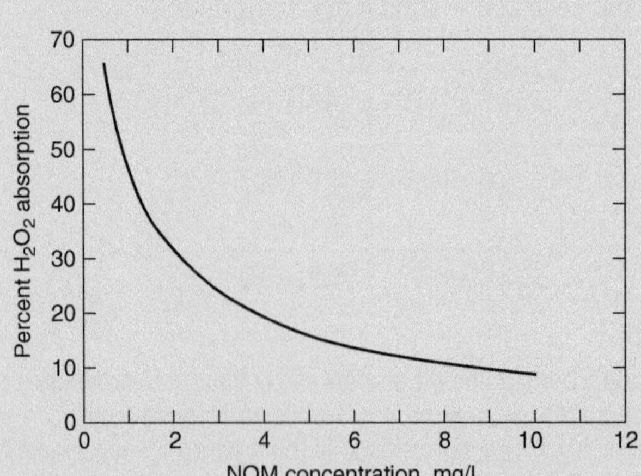

Comment

NOM can absorb a significant amount of UV light, and very high concentrations of hydrogen peroxide may be needed in such cases to increase the UV light absorption of hydrogen peroxide and hydroxyl radical concentration.

Comparison of the Simplified Model to Data and Its Limitations

Through numerous experiments with 1,2-dibromo-3-chloropropane (DBCP) it has been found that the data follows a pseudo-first-order reaction (Glaze et al., 1995). Consequently, the ability of the Sim-PSS and rigorous AOP models to predict process performance can be demonstrated by comparing the pseudo-first-order rate constants that were calculated with the various models to the rate constants that were determined from the data. The pseudo-first-order rate constants that are predicted from the various models are compared to the data in Table 18-6. The pseudo-steady-state model assumes that the concentration of hydrogen peroxide changes with time and Eqs. 18-67, 18-69, and 18-70 can be solved using MatLab or Mathcad, once they are used in appropriate reactor mass balances. The simplified pseudo-steady-state model assumes that the hydrogen peroxide concentration does not change and is equal to the initial concentration of hydrogen peroxide as shown in Eq. 18-68. Pseudo-first-order rate constants of 2.5×10^{-4} to 2.5×10^{-3} s^{-1}, which are reported in Table 18-6, would have a pseudo-steady-state hydroxyl concentrations that range from 5.95×10^{-14} and 5.95×10^{-13} M, respectively. This is 3 orders of magnitude lower than observed in full-scale processes because this was a small laboratory reactor. It appears that the pseudo-steady-state model and the simplified pseudo-steady-state model predict the DBCP destruction very well and follows most of the trends of the rigorous AOP model, which is termed AdOx (Li et al., 1999; Crittenden et al., 1999). Given the simplicity and accuracy of the simple pseudo-steady-state model, it can be used for preliminary design calculations. However, the rigorous model that solves all the equations is a much better design tool for detailed calculations because it can simulate pH change, and multiple wavelengths, and multiple components, and it has databases for physicochemical properties with graphical user interface (Li et al. 1999, Crittenden et al., 1999). And importantly as shown on Fig. 18-4, it predicts the experimental data more accurately (Crittenden et al., 1999).

Additional calculations that are not reported here show that the simplified PSS model can be used to evaluate the impact of pH and alkalinity on the TCE destruction. The simplified pseudo-steady-state model predicted similar results as AdOx (Li et al., 1999; Crittenden et al., 1999) except at high pH because the rate of photolysis of the hydrogen peroxide anion is much higher than the neutral species.

Table 18-6
A comparison of pseudo–first-order rate constants that are predicted from the various models and experimental data

$[DBCP]_0$ µM	$[H_2O_2]_0$ mM	$P_{u\text{-}v}$ eins./Ls	[TIC] mM	pH	k_0, s^{-1} Experiment	AdOx	PSS	Sim-PSS
Group I (changing $[H_2O_2]_0$)								
1.63	0.054	1.04×10^{-6}	4	8.4	26.5×10^{-5}	32.2×10^{-5}	7.6×10^{-5}	14.8×10^{-5}
1.43	0.096	1.04×10^{-6}	4	8.3	30.7×10^{-5}	37.7×10^{-5}	13.1×10^{-5}	24.9×10^{-5}
1.42	0.35	1.04×10^{-6}	4	8.4	61.9×10^{-5}	60.1×10^{-5}	41.9×10^{-5}	69.2×10^{-5}
1.55	0.56	1.04×10^{-6}	4	8.5	81.5×10^{-5}	74.6×10^{-5}	60.1×10^{-5}	89.7×10^{-5}
1.83	1	1.04×10^{-6}	4	8.4	106×10^{-5}	94.0×10^{-5}	90.6×10^{-5}	114.8×10^{-5}
1.77	1.5	1.04×10^{-6}	4	8.4	108×10^{-5}	103.3×10^{-5}	108.1×10^{-5}	120.2×10^{-5}
1.5	3	1.04×10^{-6}	4	8.4	107×10^{-5}	97.4×10^{-5}	116.8×10^{-5}	105.9×10^{-5}
1.52	4.4	1.04×10^{-6}	4	8.4	85.5×10^{-5}	82.0×10^{-5}	101.3×10^{-5}	87.0×10^{-5}
3.06	6.6	1.04×10^{-6}	4	8.5	70.9×10^{-5}	59.6×10^{-5}	73.7×10^{-5}	63.5×10^{-5}
Group II (changing $[H_2O_2]_0$)								
1.11	0.26	1.04×10^{-6}	0.1	8.1	258×10^{-5}	299.3×10^{-5}	271.3×10^{-5}	286.9×10^{-5}
1.32	1.5	1.04×10^{-6}	0.1	8.2	229×10^{-5}	260.7×10^{-5}	255.9×10^{-5}	217.8×10^{-5}
1.14	3	1.04×10^{-6}	0.1	8.2	160×10^{-5}	147.4×10^{-5}	189.9×10^{-5}	151.6×10^{-5}
Group III (changing $P_{u\text{-}v}$)								
1.38	1	2.6×10^{-7}	4	8.4	30.2×10^{-5}	25.5×10^{-5}	27.3×10^{-5}	28.4×10^{-5}
1.23	1	5.2×10^{-7}	4	8.4	52.5×10^{-5}	49.4×10^{-5}	51.7×10^{-5}	56.9×10^{-5}
1.16	1	7.7×10^{-7}	4	8.4	75.7×10^{-5}	71.0×10^{-5}	71.8×10^{-5}	84.2×10^{-5}
1.83	1	1.04×10^{-6}	4	8.4	106×10^{-5}	94.0×10^{-5}	90.6×10^{-5}	114.8×10^{-5}
Group IV (changing [TIC])								
1.91	1	1.04×10^{-6}	1	8.1	179×10^{-5}	165.7×10^{-5}	190.3×10^{-5}	199.7×10^{-5}
1.42	1	1.04×10^{-6}	2	8.4	150×10^{-5}	125.6×10^{-5}	136.2×10^{-5}	156.5×10^{-5}
1.83	1	1.04×10^{-6}	4	8.4	106×10^{-5}	94.0×10^{-5}	90.6×10^{-5}	114.8×10^{-5}
Group V (changing pH)								
1.01	1	1.04×10^{-6}	4	10.4	23.7×10^{-5}	27.5×10^{-5}	23.1×10^{-5}	22.8×10^{-5}
1.04	1	1.04×10^{-6}	4	9.4	36.1×10^{-5}	46.5×10^{-5}	63.3×10^{-5}	75.4×10^{-5}
1.83	1	1.04×10^{-6}	4	8.4	106×10^{-5}	94.0×10^{-5}	90.6×10^{-5}	114.8×10^{-5}
1.02	1	1.04×10^{-6}	4	7.4	132×10^{-5}	119.3×10^{-5}	99.6×10^{-5}	126.4×10^{-5}
1.15	1	1.04×10^{-6}	4	6.4	196×10^{-5}	160.4×10^{-5}	136.7×10^{-5}	161.7×10^{-5}
1.32	1	1.04×10^{-6}	4	5.4	219×10^{-5}	243.3×10^{-5}	247.9×10^{-5}	240.8×10^{-5}

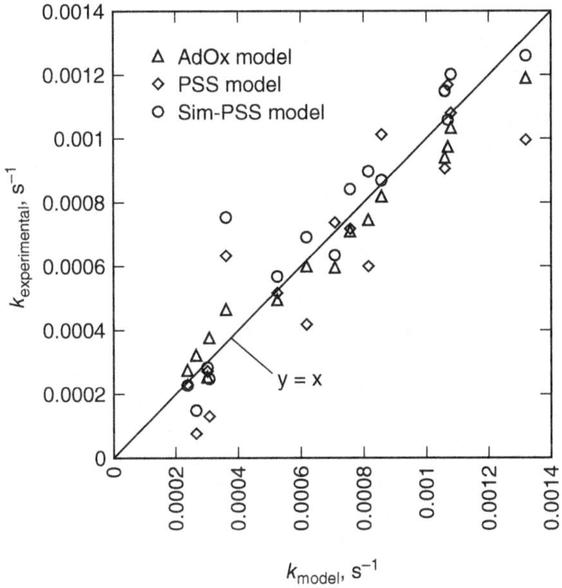

Figure 18-4
Comparison of pseudo-first-order rate constants of DBCP degradation from experimental data, the AdOx model, the pseudo-steady-state (PSS) model and the simplified pseudo-steady-state (Sim-PSS) model. $[DBCP]_0$= 1.42–1.83 μM, $[H_2O_2]_0$= 0.054 – 4.4 mM, I_0 = 1.04 × 10^{-6} einsteins/L · s, pH=8.3-8.4, total inorganic concentration = 4 mM.

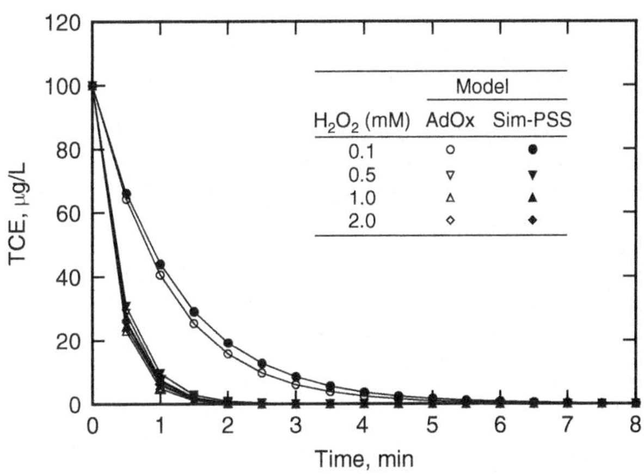

Figure 18-5
Comparison of predicted trichloroethylene concentration versus time for hydrogen peroxide dosages of 0.1, 0.5, 1.0, and 2.0 mM, alkalinity of 100 mg/L as $CaCO_3$, and a pH of 7 using AdOx and the simplified pseudo-steady-state model.

SELECTION OF HYDROGEN PEROXIDE DOSAGE

One of the most important design issues for the UV/H_2O_2 process is proper selection of the appropriate dosage of H_2O_2. The predicted trichloroethylene concentration versus time for hydrogen peroxide dosages of 0.1, 0.5, 1.0, and 2.0 mM, alkalinity of 100 mg/L as $CaCO_3$, and a pH of 7 using AdOx, and the simplified pseudo-steady-state model is shown on Fig. 18-5. The initial TCE concentration is 100 μg/L. The rate of destruction increases until the hydrogen peroxide concentration increases to 1 mM, and then it decreases slightly because of increased scavenging of hydroxyl radical by hydrogen peroxide. It appears that the optimum hydrogen peroxide dosage is in the range of 0.5 to 2 mM. The predicted results using simplified pseudo-steady-state model were very close to the fully dynamic model that does not invoke the pseudo-steady-state assumption (AdOx; Li et al., 1999); consequently, it could also be used to examine the impact of hydrogen peroxide dosage.

ELECTRICAL EFFICIENCY PER ORDER OF TARGET COMPOUND DESTRUCTION

Most photoreactors are designed to absorb all the UV light. For these reactors (see Sec. 8-4), the destruction of the target compound will only depend on the total radiant energy that is received by the reactor. Consequently, the EE/O is a very effective metric in evaluating the electrical efficiency of the UV/H_2O_2 process. The EE/O versus hydrogen peroxide concentration is plotted on Fig. 18-6, and the optimum hydrogen peroxide concentration can be determined. Predictions using the simplified pseudo-steady-state model and AdOx are identical. Accordingly, the optimum peroxide dosage is between 0.5 and 2 mM and about 0.50 kWh/m^3 of water treated (0.1 kwh/1000 gall) for an order of magnitude reduction in TCE concentration. This is a very low value but the influence of NOM has not been included.

Generally, EE/O values less than 0.265 kWh/m^3 (1.0 kWh/1000 gall) of water treated are considered favorable, but the process has been used in cases where much higher values have been observed when there are no other treatment options (Bolton and Carter, 1994). A value of 0.265 would correspond to electrical energy costs of 0.1325 dollars per cubic meter of water treated (50 cents per thousand gallons) for an order of magnitude reduction in concentration, assuming that electric power costs are 10 cents per kW h and the lamps have an electrical efficiency of 20 percent. Given the price of hydrogen peroxide versus the cost of electricity, EE/O is the most important design parameter, and the optimum hydrogen peroxide dosage must be selected on the basis of EE/O.

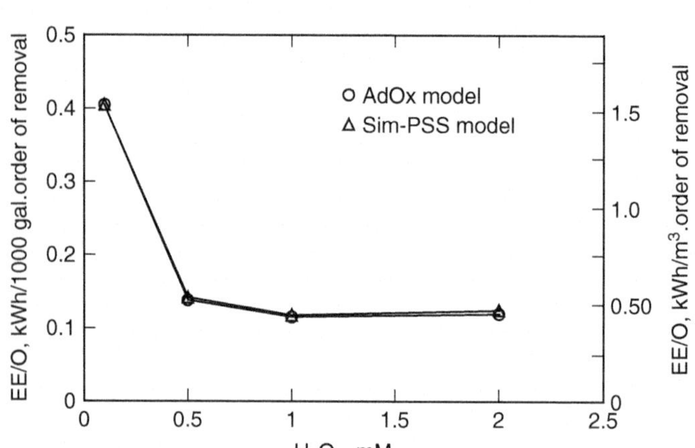

Figure 18-6
Impact of H_2O_2 dosage on EE/O for trichloroethylene (TCE) destruction using H_2O_2/UV process (operating conditions: $[TCE]_0 = 100$ μg/L, alkalinity = 100 mg/L $CaCO_3$, [NOM] = 0 mg/L, UV light intensity = 1.04×10^{-6} einstein L^{-1} s^{-1} at 254 nm, reactor size = 70 L with 15.8 cm of the effective path length, and the total lamp power is 160W (assuming 20 % efficiency).

Example 18-9 Lamp power requirement

Calculate the lamp wattage for a flow rate 0.03 m^3/s (500 gal/min), 1 order of magnitude of destruction, and an EE/O of 0.25 kWh/m^3 (0.95 kWh/1000 gal). The lamp efficiency is 30 percent.

Solution

1. Derive the equation for lamp power output based on the definition of EE/O (see Eq. 8-130 in Chap. 8):

$$\mathrm{EE/O} = \frac{P}{Q\log\left(\dfrac{C_i}{C_f}\right)}$$

$$P = (\mathrm{EE/O})(Q)\log\left(\frac{C_i}{C_f}\right)$$

2. Calculate lamp power output:

$$\begin{aligned} P &= (EE/O)(Q)\log\left(\frac{C}{C_0}\right) \\ &= \left(0.25\ \mathrm{kW\cdot h/m^3}\right)\left(0.03\ \mathrm{m^3/s}\right)(3600\ \mathrm{s/h}) \times \log(10) \\ &= 27\ \mathrm{kW} \end{aligned}$$

3. Calculate the lamp power requirement:

$$\text{Power requirement} = \frac{\text{power output}}{\text{efficiency}} = \frac{27\ \mathrm{kW}}{0.30} = 90\ \mathrm{kW}$$

Comment

High-output low-pressure lamps are more efficient than medium-pressure lamps. The high-output lamps are about 400 W and the medium-pressure lamps can be 15-kW. If 15-kW lamps are used, only 6 such lamps would be required. A reactor that uses 400-W lamps would need about 225 lamps.

Impact of NOM and Compound Type on Target Compound Destruction

Another important factor in the H_2O_2/UV AOP process is the reactivity of the compounds. Compounds with double bonds tend to react more quickly than saturated compounds because saturated compounds must undergo hydrogen abstraction, whereas double bonds only undergo addition. Consequently, more energy and hydrogen peroxide are required to destroy saturated compounds than compounds with double bonds. The EE/O for 1,1,1-trichloroethane (TCA), dibromochloropropane (DBCP), and trichloroethylene (TCE) are shown on Fig. 18-7 for the conditions of

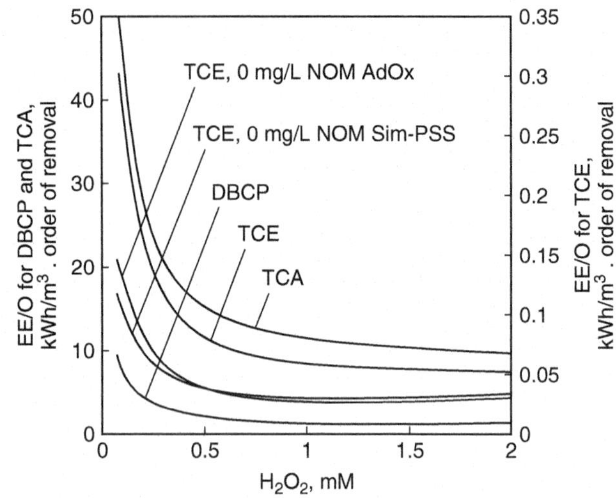

Figure 18-7
Comparison of EE/O values for 1,1,1-trichloroethane (TCA), dibromochloropropane (DBCP), and trichloroethylene (TCE) (initial concentrations = 100 μg/L, pH of 7, and alkalinity = 100 mg/L as $CaCO_3$). NOM = 1 mg/L except where noted. Results are both AdOX and Sim-PSS models except where noted.

initial concentrations of 100 μg/L, pH of 7, alkalinity of 100 mg/L CaCO3, and 1 mg/L NOM unless noted otherwise. The optimum EE/O for TCE, DBCP, and TCA are 0.052, 2.4, and 10.2 kW/m^3 order, respectively. The EE/O for DBCP is lower than TCA because there are more hydrogen atoms on the molecule for attack by hydrogen abstraction. As expected, TCA requires a great deal more radiant energy and hydrogen peroxide than does TCE. Further, the simplified pseudo-steady-state model can describe most situations at one wavelength and is useful to assess the feasibility of the process.

Of all of the factors, NOM has the greatest impact, and this is because it not only scavenges hydroxyl radical but it also absorbs UV that may otherwise be absorbed by the hydrogen peroxide and create hydroxyl radicals. Figure 18-7 shows the impact of NOM on the EE/O for TCE. As shown, the EE/O for TCE increases from 0.025 to 0.052 when 1 mg/L NOM is present.

Example 18-10 Using the simplified pseudo-steady-state model to estimate the effluent concentration

The city of Eagle River recently discovered that one of its wells was contaminated with 200 μg/L (1.52 μmol/L) TCE. Calculate the effluent concentration of TCE for H_2O_2 dosage of 2.5 mM (85 mg/L) and estimate the residual of H_2O_2 in the effluent. The treatment objective for TCE is 5.0 μg/L. During normal pumping of the well field, the flow rate is 0.20 m^3/s

(3200 gpm). The pH, HCO_3^-, and DOC concentrations are 6.8, 488 mg/L, and 0.7 mg/L, respectively. The following table shows some important physicochemical properties of H_2O_2, TCE, and NOM.

Compound	MW (g/mol)	OH Radical Rate Constant, k_{OH}, (L/mol · s)	Extinction Coefficient, ε, (L/mol · cm)	Quantum Yield, Φ (mol/einstein)
Trichloroethylene	131.389	4.20×10^9	Ignored	0
NOM[a]	NA	3.90×10^8	0.0196	0
H_2O_2	34.015	–	19.6	0.5

[a]For NOM, the units of k_{OH} is L/mol NOM carbon-s and the unit of ε is L/mg NOM • cm.

For simplicity, a proprietary reactor will be used. A dye study on the reactor has shown that four completely mixed flow reactors in series describes mixing that occurs in the reactor. The reactor size is 1 m in diameter by 3 m in height and the volume is approximately 2300 L with 12 × 15 kW medium-pressure lamp,as shown on Fig.18-8. To simplify the calculations, it can be assumed that the UV light intensity is monochromatic at 254 nm and that the lamps are 20 percent efficient. Assume that all the UV light is absorbed and $[HO_2^-]$ and $[CO_3^{2-}]$ can be neglected at pH 6.8.

Solution

1. Calculate the hydraulic detention time (τ):

$$\tau = \frac{\text{Total reactor volume}}{Q} = \left(\frac{2300\ \text{L}}{(0.20\ \text{m}^3/\text{s})(1000\ \text{L/m}^3)(60\text{s/min})}\right)$$

$$= 0.19\ \text{min}$$

2. Calculate the fraction of light absorbed by H_2O_2 according to Eq. 18-77: To simplify the calculation, it will be assumed that all the light is absorbed by the water matrix (the walls of the vessel reflect all light back into the water and absorb no light).

$$f_{H_2O_2} = \frac{\varepsilon_{H_2O_2} C_{H_2O_2}}{\left(\varepsilon_{H_2O_2} C_{H_2O_2} + \varepsilon_{NOM} C_{NOM}\right)}$$

$$= \frac{(19.6\ \text{L/mol}\cdot\text{cm})\left(2.5\times10^{-3}\ \text{mol/L}\right)}{(19.6\ \text{L/mol}\cdot\text{cm})\left(2.5\times10^{-3}\ \text{mol/L}\right) + (0.0196\ \text{L/mg}\cdot\text{cm})(0.7\ \text{mg/L})}$$

$$= 0.78$$

3. Determine the UV light intensity:
The light power is related to the UV intensity by Eq. 18-66:

$$P_{U-V} = \frac{P\eta}{N_{av} V h\nu}$$

a. Calculate the frequency of light using Eq. 8-94:

$$\nu = \frac{c}{\lambda} = \frac{(3 \times 10^8 \text{m/s})(10^9 \text{ nm/m})}{254 \text{ nm}} = 1.18 \times 10^{15} \text{ s}^{-1}$$

b. Calculate UV intensity: Assume 20 percent efficiency and 12 lamps turned on. The UV intensity can be calculated:

$$P_{U\text{-}V} = \frac{(180 \text{ kW})(1000 \text{ W/kW})[(1 \text{ J/s})/\text{W}](0.2)(1 \text{ einstein/mol})}{(6.023 \times 10^{23} \text{ photons/mol})(2300 \text{ L})(6.62 \times 10^{-34} \text{ J} \cdot \text{s})(1.18 \times 10^{15} \text{ s}^{-1})}$$
$$= 3.3 \times 10^{-5} \text{ einsteins / L} \cdot \text{s}$$

4. Calculate the effluent concentration of TCE.

a. Convert the concentration of each component from mg/L to mol/L:

$$[HCO_3^-]_0 = \frac{488 \text{ mg/L}}{(61 \text{ g/mol})(1000 \text{ mg/g})} = 0.008 \text{ mol/L}$$

$$[\text{NOM}]_0 = \frac{0.7 \text{ mg/L}}{(12 \text{ g C/mol NOM C})(1000 \text{ mg/g})} = 5.83 \times 10^{-5} \text{ mol NOM C/L}$$

b. From Table 18-5 and the problem statement:

$$k_{10} = 2.7 \times 10^7 \text{ L/mol} \cdot \text{s}$$
$$k_{11} = 8.5 \times 10^6 \text{ L/mol} \cdot \text{s}$$
$$k_{12} = 4.2 \times 10^9 \text{ L/mol} \cdot \text{s (see Table 18-2)}$$
$$k_{13} = 3.9 \times 10^8 \text{ L/mol NOM carbon} \cdot \text{s}$$

c. Determine values of the product of rate constant and concentration:

$$k_{10}[H_2O_2]_0 = (2.7 \times 10^7 \text{ L/mol} \cdot \text{s})(2.5 \times 10^{-3} \text{ mol/L} = 67{,}500 \text{ s}^{-1}$$
$$k_{11}[HCO_3^-]_0 = (8.5 \times 10^6 \text{ L/mol} \cdot \text{s})(0.008 \text{ mol/L}) = 68{,}000 \text{ s}^{-1}$$
$$k_{12}[R]_0 = k_{12}[\text{TCE}]_0 = (4.2 \times 10^9 \text{ L/mol} \cdot \text{s})(1.52 \times 10^{-6} \text{mol/L} = 6384 \text{ s}^{-1}$$

$$k_{13}[\text{NOM}]_0 = (3.9\times10^8\ \text{L/mol NOM C}\cdot\text{s})(5.83\times10^{-5}\text{mol NOM C/L}) = 22737\text{s}^{-1}$$

d. Calculate $[\text{HO}\bullet]_{ss,0}$ using Eq. 18-68:

Assuming that all the light was absorbed and $[\text{HO}_2^-]$ and $[\text{CO}_3^{2-}]$ can be neglected at pH6.8, the pseudo-steady-state concentration of hydroxyl radical is given by Eq.18-68

$$[\text{HO}\bullet]_{ss,0} = \frac{2\phi_{\text{H}_2\text{O}_2} I_0 f_{\text{H}_2\text{O}_2}\left(1-e^{-A}\right)}{k_{10}[\text{H}_2\text{O}_2]_0 + k_{11}\left[\text{HCO}_3^-\right] + k_{12}[\text{TCE}]_0 + k_{13}[\text{NOM}]_0}$$

$$= \frac{2(0.5\ \text{mol/einstein})\left(3.3\times10^{-5}\ \text{einstein/L}\cdot\text{s}\right)0.78}{(67{,}500+68{,}000+6384+22{,}737)\ \text{s}^{-1}}$$

$$= 1.58\times10^{-10}\ \text{mol/L}$$

e. Calculate pseudo-first-order rate constant for TCE:

$$k'_{12} = k_{12}[\text{HO}\bullet]_{ss,0} = \left(4.2\times10^9\ \text{L/mol}\cdot\text{s}\right)\left(1.58\times10^{-10}\ \text{mol/L}\right) = 0.66\ \text{s}^{-1}$$

f. Calculate TCE effluent concentration using the tanks in series model (see Sec. 8-4):

$$[\text{TCE}] = \frac{[\text{TCE}]_0}{\left(1+k'_{12}\tau/n\right)^n} = \frac{200\ \mu\text{g/L}}{\left[1+\dfrac{(0.661/\text{s})(0.19\ \text{min})(60\ \text{s/min})}{4}\right]^4}$$

$$= 2.9\ \mu\text{g/L}$$

5. Calculate the residual hydrogen peroxide concentration.
   a. Estimate pseudo-first-order rate constant for hydrogen peroxide assuming that NOM is the major background chromophore, using Eq. 18-76:

$$k'_{10} = \frac{0.5 I_0 p_{\text{u-v}}\varepsilon_{\text{H}_2\text{O}_2}}{\varepsilon_{\text{H}_2\text{O}_2}[\text{H}_2\text{O}_2]_0 + \varepsilon_{\text{NOM}}[\text{NOM}]_0} + k_{10}[\text{HO}\bullet]_{ss,0}$$

   i. Determine $0.5I_0\varepsilon_{\text{H}_2\text{O}_2}$, $\varepsilon_{\text{H}_2\text{O}_2}[\text{H}_2\text{O}_2]_0$, $\varepsilon_{\text{NOM}}[\text{NOM}]_0$, and $k_{10}[\text{HO}\bullet]_{ss,0}$:

$$0.5I_0\varepsilon_{\text{H}_2\text{O}_2} = (0.5\ \text{mol/einstein})(19.6\ \text{L/mol}\cdot\text{s})(3.3\times10^{-5}\ \text{einstein/L}\cdot\text{s}) = 3.23\times10^{-4}\ \text{s}^{-2}$$

$$\varepsilon_{H_2O_2}[H_2O_2]_0 = (19.6\ L/mol\cdot s)(2.5\times10^{-3}\ mol/L) = 0.049\ s^{-1}$$

$$\varepsilon_{NOM}[NOM]_0 = (0.7\ mg/L)(0.0196\ L/mg\cdot s) = 0.01372\ s^{-1}$$

$$k_{10}[HO\cdot]_{ss,0} = (2.7\times10^{7}\ L/mol\cdot s)(1.58\times10^{-10} mol/L) = 0.004266\ s^{-1}$$

ii. Determine k'_{10}:

$$\begin{aligned} k'_{10} &= \frac{3.23\times10^{-4}\ s^{-2}}{(0.049+0.01372)\ s^{-1}} + \left(0.004266\ s^{-1}\right) \\ &= 9.42\times10^{-3}\ s^{-1} \end{aligned}$$

b. Estimate H_2O_2 residual using the tanks in series model:

$$[H_2O_2] = \frac{[H_2O_2]_0}{\left[1+k'_{10}\tau/n\right]^n} = \frac{2.5\times10^{-3} mol/L}{\left[1+\dfrac{\left(9.42\times10^{-3} s^{-1}\right)(0.19\ min)(60\ s/min)}{4}\right]^4}$$

$$= 2.25\times10^{-3}\ mol/L (76.5\ mg/L)$$

Comment

While the treatment objective for TCE can be met, the residual hydrogen peroxide concentration is too high to use the process for water treatment. The residual hydrogen peroxide concentration is the lowest possible value because the pseudo-steady-state concentration of hydroxyl radical is taken to be the initial value in the simplified pseudo-steady-state model. However, this approach is still useful because it can be used to calculate the lowest expected residual hydrogen peroxide concentration, and, if the residual is unacceptable, then process is not a viable option. The effluent concentration of hydrogen peroxide predicted by the rigorous AOP model is 2.39×10^{-3} mol/L.

18-5 Other Advanced Oxidation Processes

The AOPs discussed in this section are not commonly used for water treatment or are only tested at a laboratory scale for feasibility. The AOPs considered are (1) the ozone/ UV light process, (2) the photocatalysis with titanium dioxide process, (3) the hydrogen peroxide/iron (Fenton's reaction) process, and (4) the sonolysis process.

Ozone/UV Light

The first step of the O_3/UV process is the formation of H_2O_2 by photolysis of ozone:

$$O_3 + H_2O + \text{UV light} \rightarrow O_2 + H_2O_2 \quad (18\text{-}78)$$

The reactions between H_2O_2 and O_3 species produce hydroxyl radical, as discussed in Sec. 18-3. It is also possible for UV light to split H_2O_2 into HO•; however, the extinction coefficient for O_3 is about 199 times greater than that for H_2O_2 at 254 nm. Using ozone to produce H_2O_2, which in turn reacts with O_3 to produce HO• , is a very inefficient way to produce HO• radicals because it takes a lot of energy to form ozone onsite. Although H_2O_2 is a relatively cheap and stable chemical that may be purchased in bulk, the net overall hydroxyl radical yield for the O_3/UV process approaches 1 mol HO• /mol O_3 (Peyton and Glaze, 1986). Due to the energy that is required to produce ozone and UV light, the O_3/UV process is most applicable where direct photolysis of the contaminants is significant (e.g., PCE and some aromatic halides) because the radical chain reactions can be initiated from reactions between photolytic by-products and ozone.

An additional complication for the reaction to proceed in the most efficient manner is the ratio of H_2O_2 to O_3. The proper ratio of H_2O_2 to O_3 must be maintained for compounds that do not undergo direct photolysis, and this ratio will be impossible to meet because of complexity of the competing reactions. Further discussion on the ratio that is required is provided in Sec 18-3. The one potential advantage is that one would not have to be concerned about how ozone mass transfer and mixing would affect the reactor performance.

Photocatalysis with Titanium Dioxide

When a photon with sufficient light energy is absorbed by titanium dioxide, an electron in the outer orbital of the valance band (VB) moves up to the conduction band (CB), as shown on Fig. 18-8. The difference in potential between the VB and the CB is called the band gap and the band gap of TiO_2 is 3.2 eV. The movement of an electron from the VB to the CB produces a hole in the valence band and an electron that can move freely in the conduction band, as shown in the expression

$$TiO_2 \xrightarrow{hv} h^+ + e_{cb} \quad (18\text{-}79)$$

where h^+= hole in valence band, electronic charge 1.6×10^{-19} C
e_{cb} = electron in conduction band, electronic charge 1.6×10^{-19} C

Unfortunately, most of the holes and conduction band electrons recombine before they undergo any chemical reactions on the surface and the light energy is simply wasted:

$$h^+ + e_{cb} \rightarrow \text{photocatalyst} + \text{heat and/or light (recombination)} \quad (18\text{-}80)$$

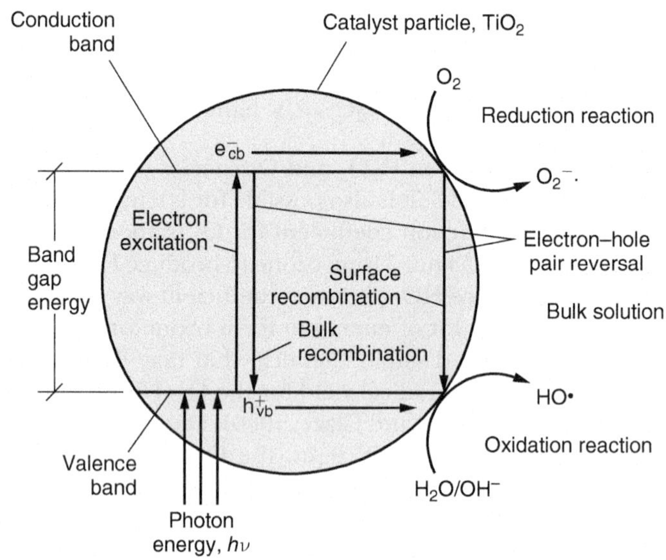

Figure 18-8
Schematic diagram of mechamisms of photocatalytic oxidation and reduction.

The electrochemical energy for the photocatalytic AOP comes from subsequent reactions with the hole and conduction band electron. At a pH value of 7, the hole has a reduction potential E_H° of approximately 2.9 V and the conduction band electron has a reduction potential E_H° of approximately −0.3 V (Hashimoto et al., 1984). Consequently, a hydroxyl radical is produced when the hole in the valence band reacts with water, as shown in the following reaction because the standard electrode potential for the formation of hydroxyl radical is 2.59 or 2.177 V at pH 7:

$$h^+ + H_2O \rightarrow H^+ + HO\bullet \qquad (18\text{-}81)$$

A conduction band electron can reduce the hydrogen ion and oxygen depending on the pH and oxygen concentration as follows:

$$H^+ + e_{cb} \rightarrow \tfrac{1}{2}H_2 \text{ (anaerobic)} \qquad (18\text{-}82)$$

$$O_2 + e_{cb} \rightarrow O_2^- \bullet \text{ (aerobic)} \qquad (18\text{-}83)$$

The addition of hydrogen peroxide has been shown to improve the reaction rate, which may occur because hydrogen peroxide has a higher reduction potential than oxygen. In addition, hydrogen peroxide photolysis can generate additional hydroxyl radicals.

The mechanisms of propagation and termination of the radical chain are the same as the UV/H_2O_2 process except that photolysis of hydrogen peroxide does not occur unless it is added. One advantage of the UV/TiO_2 process is that good light transmission is not as much of a factor because, if the TiO_2 is added as a slurry, then fluid mixing can transport the TiO_2 to the light and back into bulk solution. This will allow the chemical reactions to occur throughout the reactor depending on the level of fluid mixing.

As with the other AOPs, the primary reaction for oxidation of electron-rich organic pollutants begins with the attack of the organic pollutant by hydroxyl radicals that are produced on the surface. Subsequent organic radical chain reactions with oxygen and other species eventually will completely mineralize the parent compound, as discussed in Sec. 18-1.

Example 18-11 Titanium dioxide: band gap and wavelength for lowest energy

When an electron in the valence band absorbs a photon with energy greater than 3.2 eV, it can move up into the conduction band. Calculate the wavelength of light required to create a hole and conduction band electron in TiO_2 and prove that the hole can oxidize water to produce HO•. What band gap and wavelength corresponds to lowest energy input that can create HO•?

Solution

1. Calculate the frequency of light.
   The frequency of light can be calculated from the band gap energy for electron excitation that is required and Planck's constant as presented in Eq. 8.94 in Chap. 8.

$$\nu = \frac{E}{h} = \frac{(3.2\ \text{J/C})\left(1.60 \times 10^{-19}\text{C}\right)}{\left(6.62 \times 10^{-34}\ \text{J}\cdot\text{s}\right)} = 7.73 \times 10^{14}\ \text{s}^{-1}$$

2. Calculate the wavelength of light required.
   The wavelength of light can be calculated from the speed and frequency of light as presented in Eq. 8-95 in Chap. 8.

$$\lambda = \frac{c}{\nu} = \frac{\left(3.00 \times 10^{8}\text{m/s}\right)}{\left(7.73 \times 10^{14}\text{s}^{-1}\right)} = 3.88 \times 10^{-7}\text{m (388 nm)}$$

3. Prove that the hole can oxidize water to produce HO•

$$h^{+} + e \rightarrow TiO_2 \qquad E^{o} = 2.9V \text{ at pH} = 7$$

$$H_2O \rightarrow H^{+} + e + HO\bullet \qquad E^{o} = -2.177V \text{ at pH} = 7$$

   The hole in the valence band can power the oxidation of water to form hydroxyl radical because the hole has a reduction potential of 2.9 V and the oxidation reaction requires 2.177 V.

4. Determine band gap corresponding to the lowest energy input that can create HO•.
   The electrode potential for the formation of hydroxyl radical is 2.177 at pH of 7. Therefore, the lowest reduction potential in the valence band

is 2.177 V. The conduction band electron has a reduction potential of approximately − 0.3 V. The band gap corresponding to the lowest energy input that can create HO• is 2.477 eV.

5. Determine wavelength corresponding to the lowest energy input that can create HO•:
   a. Calculate the frequency of light:

$$\nu = \frac{E}{h} = \frac{(2.477\text{J/C})\left(1.60 \times 10^{-19}\text{C}\right)}{6.62 \times 10^{-34}\ \text{J} \cdot \text{s}} = 5.99 \times 10^{14}\text{s}^{-1}$$

   b. Calculate the wavelength of light required.

$$\lambda = \frac{c}{\nu} = \frac{3.00 \times 10^{8}\ \text{m/s}}{5.99 \times 10^{14}\ \text{s}^{-1}} = 5.01 \times 10^{-7}\ \text{m}(501\ \text{nm})$$

Comment

The band gap of TiO_2 has sufficient power to create hydroxyl radical from water. Further, the wavelength that is required to produce hydroxyl radical is fairly large, and it may be possible to harness sunlight to create an oxidizing environment for toxic organics destruction.

QUANTUM YIELD

The apparent quantum yield (moles of parent compound reacted per mole of photons illuminating the reactor) for most photocatalytic reactions is in the range of 0.1 to 3 percent depending on the reactant and its concentration, light intensity, and the constituents in the water matrix. There are strategies for increasing the quantum yield and this is an area of ongoing research. For example, it has been demonstrated that the catalyst recombination reaction is reduced by impregnating the catalyst with a small amount of platinum (1 percent) and platinum impregnation significantly increased the rate of reaction (Suri et al., 1993).

COMMERCIALLY AVAILABLE REACTORS

There are several vendors that provide TiO_2 treatment systems, and one of the most successful units is provided by Purifics, as shown in the photographs on Fig. 18-9a and b. A schematic of its system is displayed on Fig. 18-9c. The ceramic cross-flow membrane, which is shown on Fig. 18-9c, is used to recover the TiO_2. The quantity of TiO_2 is very small (~10 kg) and it can last several years before it has to be replaced. An interesting feature of their technology is that wipers are not required for their lamps because they do not develop fouling deposits when TiO_2 is present. They have many installations that are able to treat many kinds of organic contaminants and are resistant to alkalinity, pH, and turbidity and temperature changes.

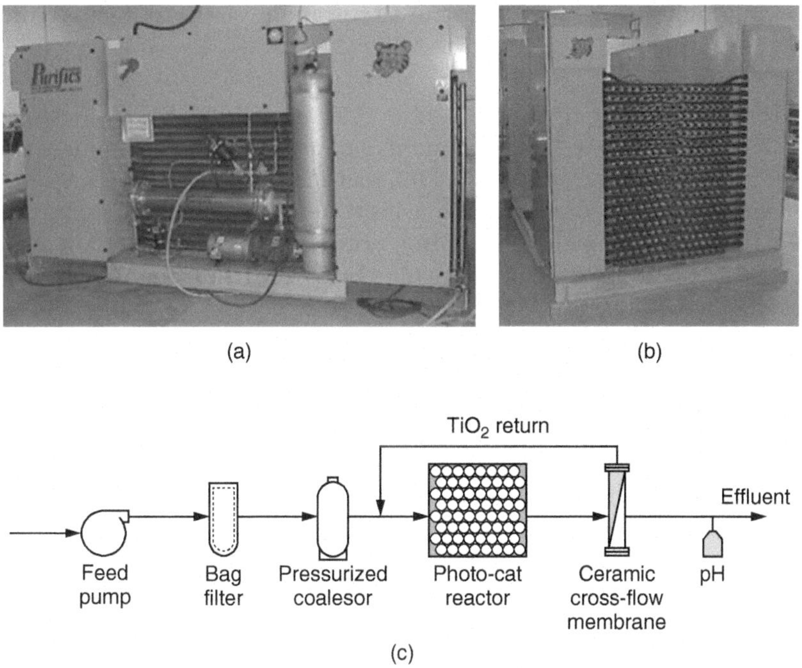

Figure 18-9
Purifics Photo-Cat System. (a) Photograph of front view, (b) Photograph of side view, and (c) Schematic of the Process.

Reactions with Iron (Fenton's Reactions)

Reductive metal ions can catalyze the hydrolysis of H_2O_2 to form hydroxyl radicals. Fenton's reagent, a mixture of ferrous ion and hydrogen peroxide, has been known as a powerful oxidant for organic contaminants. The principal mechanism for Fenton's reagent is as follows:

$$Fe^{2+} + H_2O_2 \xrightarrow{k} Fe^{3+} + OH^- + HO\cdot \qquad k = 76.5\ M^{-1}s^{-1} \quad \text{at } 25^\circ C \tag{18-84}$$

The ferric ions produced can then produce $HO_2\bullet$ through the following reactions (Barb et al., 1951):

$$Fe^{3+} + H_2O_2 \xrightarrow{k} Fe(OOH)^{2+} + H^+ \quad k = 2\times10^{-3} M^{-1}s^{-1} \quad \text{at } 25^\circ C \tag{18-85}$$

$$Fe^{3+} + H_2O_2 \rightarrow Fe^{2+} + HO_2\bullet + H^+ \tag{18-86}$$

Reaction 18-86 is much slower than reaction 18-85. As a result, the overall oxidation process slows down after the conversion of ferrous to ferric ion (Venkatadri and Peters, 1993), even though small concentrations of iron can generate hydroxyl radicals (as low as 0.05 mM). Reaction 18-85 is also the initial mechanism of the Fe(III) Fenton-like system.

The optimum pH for Fenton's reagent processes ranges from 2 to 4, which prevents its use for in situ treatment or in water treatment applications because pH adjustment before and after treatment is needed.

This would not only be expensive but also would raise the TDS above the secondary standard.

Sonolysis

Ultrasonic irradiation, or sonolysis, is an effective method for the destruction of many environmentally important and refractory contaminants, such as chlorinated hydrocarbons, pesticides, fuel additives, atrazine, parathion, aromatic compounds, or even humic acids (Cheung et al., 1991; Gondrexon et al., 1999; Nagata et al., 1996; Olson and Barbier, 1994; Schramm and Hua, 2001). Ultrasound destroys organics by creating small bubbles by cavitation, which occurs during the low-pressure portion of a wave cycle. Extremely high temperatures and high pressures occur during bubble collapse. This causes pyrolysis of organics and produces highly reactive chemical radicals. The estimated transient temperature is in the range of about 4000 to 10,000 K and the pressure is in the range of 300 to 975 bar (Hoffmann et al., 1996). This localized extreme condition is able to provide enough energy for the pyrolysis of most, if not all, contaminants, including water, to form radicals. As a result, two main mechanisms are responsible for the destruction of contaminants during the sonolysis process: (1) pyrolysis reactions at the cavitation bubble and (2) radical reactions by HO• and H• formed through the sonolysis of H_2O. The two mechanisms are as follows:

$$R \xrightarrow{\Delta+)))} \text{pyrolysis product} \tag{18-87}$$

$$H_2O \xrightarrow{\Delta+)))} H\bullet + HO\bullet \tag{18-88}$$

$$HO\bullet + R \xrightarrow{\Delta+)))} \text{product} \tag{18-89}$$

The concentration of HO· at a bubble interface can be as high as 4×10^{-3} M, which is 10^8 or 10^9 times higher than that in the other advanced oxidation process (Gutirrez et al., 1991).

The pyrolysis of contaminants can also form radicals and initiate chain reactions, such as the degradation of carbon tetrachloride (Fogler and Crittenden, 1970):

$$CCl_4 \xrightarrow{\Delta+)))} CCl_3\bullet + Cl\bullet \tag{18-90}$$

$$Cl\bullet + R(\text{or radical}) \xrightarrow{\Delta+)))} \text{product} \tag{18-91}$$

Although the technology has been shown to be feasible on a small scale, the commercialization of sonolysis is still a challenge, due to the high energy requirement of the process. Based on the observed destruction rates, the EE/O of sonolysis of organic compounds is on the order of 2600 kWh/m^3 per order of destruction.

The combination of sonolysis and other advanced oxidation options, such as ozone and elemental iron, can further improve the destruction of contaminants and process intermediates (Schramm and Hua, 2001). The enhancement has been attributed to the improvement of HO• and/or H• radical production.

Problems and Discussion Topics

18-1 The second-order rate constant of HO• for methyl ethyl ketone (MEK) is 9.0×10^8 L/mol·s. Calculate the half-life of MEK for a HO• concentration of 10^{-12} mol/L.

18-2 Calculate the quenching of the reaction for TCE and hydroxyl radical in a natural water with alkalinity of 75 mg/L as $CaCO_3$ assuming that all alkalinity is bicarbonate. The rate constant k_{TCE} is 4.2×10^9 L/mol·s and TCE concentration is 100 μg/L.

18-3 Examine the difference in the rate of oxidation of TCE and chloroacetic acid (CAA) in double-distilled water versus the rate in a typical natural water. Here, k_{NOM} is 3×10^8 L/mol C·s, k_{TCE} is 4.2×10^9 L/mol·s, and k_{CAA} is 4.3×10^7 L/mol·s. The NOM concentrations in distilled water and natural water are 0.05 and 1.4 mg/L, respectively. The initial concentrations of TCE and CAA are 100 and 60 μg/L. Estimate Q_R for each compound in distilled water and in the natural water.

18-4 The second-order hydroxyl rate constants for Fe(II) and Mn(II) are 2.3×10^8 and 1.4×10^8 L/mol·s, respectively. For iron concentration of 5.6 mg/L, manganese concentration of 5 mg/L, and TCE concentration of 100 μg/L, calculate the quenching of the reaction for TCE Assume k_{TCE} is 4.2×10^9 L/mol·s.

18-5 Determine the fraction of the reaction that is carried out by the indirect reaction with HO• versus the direct reaction with O_3 for the oxidation of geosmin and MIB. The second-order rate constants for HO• for geosmin and MIB are 1.4×10^{10} and 8.2×10^9 L/mol·s respectively. For the calculation, use $C_{[HO\bullet]/[O_3]}$ values of 10^{-7}, 10^{-8}, 10^{-9}, and 10^{-10} and a rate constant for the direct reaction with ozone of 10 L/mol·s.

18-6 Calculate the time required for 99 percent destruction of MIB in a batch reactor. The second-order rate constant with HO• is 8.2×10^9 L/mol·s. For the calculation, use $C_{[HO\bullet]/[O3]}$ ranging from 10^{-9} to 10^{-7} and an initial ozone concentration of 3 mg/L. The rate constant for the direct reaction with ozone is 10 L/mol·s. Use a typical ozone pseudo-first-order rate constant of 0.1 min^{-1}.

18-7 Derive the simple pseudo-steady-state model for the hydrogen peroxide/ozone processes.

18-8 A municipality recently discovered that one of its wells was contaminated with the compounds listed in the following:

Compound	Influent Concentration, C_0, μg/L	Treatment Objective, C_{TO}, μg/L
Trichloroethylene (TCE)	130	5.0
Tetrachloroethylene (PCE)	75	5.0
Vinyl chloride	15	2.0
Benzene	80	5.0

To continue using the well as a drinking water resource, the compounds need to be removed to meet the treatment objectives shown in the table. During normal pumping operations, the well produces about 400 gpm, and further expansion of the well field may be considered depending on the efficacy of the ozone/hydrogen peroxide process. The pH, alkalinity, and DOC concentrations are 7.5, 400 mg/L as $CaCO_3$, and 1.2 mg/L, respectively. Important physicochemical properties for the compounds that need to be removed are as follows:

Compound	MW g/mol	HO· Rate Constant, $k_{HO\cdot}$, L/mol · s
Trichloroethylene	131.4	4.20×10^9
Tetrachloroethylene	165.8	2.60×10^9
Vinyl chloride	62.5	1.20×10^{10}
Benzene[a]	78.1	7.80×10^9
NOM[b]	NA	17,666

[a] Molar extinction coefficient is high but quantum yield is very low; consequently, photolysis can be ignored.
[b] For NOM, the unit of k_{HO} • is L/mg·s.

For simplicity, a proprietary reactor will be used. Based on dye studies, it has been found that the reactor can be modeled as four completely mixed reactors in series. The reactor is 1 m in diameter and 3 m in height, and the volume is approximately 2300 L. For the given conditions, determine the optimum H_2O_2 dosage to achieve the treatment objectives based on the simplified model (Sim-PSS) for 0.025 m^3/s (400 gpm). Consider ozone dosages of 1, 3, and 5 mg/L.

18-9 Use the information given in Problem 18-8 with an H_2O_2 dosage of 0.8 mg/L. Ozone (O_3) is generated onsite and the ozone flow rate is 1 mg/L·min. The partial pressure of O_3 in the inlet gas is 0.07 atm. Henry's law constant for O_3 at 23°C is 83.9 atm·L/mol, and the overall mass transfer coefficient for O_3, $k_L a$, was measured to be $7 \times 10^{-4}\ s^{-1}$. Determine the expected effluent concentrations of all contaminants and H_2O_2 residual concentration.

18-10 Discuss the advantages and disadvantages of advanced oxidation processes that are not commonly used for water treatment.

References

Barb, W. G., Baxendale, J. H., George, P., and Hargrave, K. R. (1951) "Reactions of Ferrous and Ferric Ions with Hydrogen Peroxide," *Trans. Faraday Soc.*, **47**, 462–500, 591–616.

Baxendale, J. H.; and Wilson, J. A. (1957) "Photolysis of Hydrogen Peroxide at High Light Intensities," *Trans. Faraday SOC.*, **53**, 344–356.

Behar, D., Czapski, G., and Duchovny, I. (1970) "Carbonate Radical in Flash Photolysis and Pulse Radiolysis of Aqueous Carbonate Solutions," *J. Phys. Chem.*, **74**, 2206–2210.

Bielski, B. H. J., and Richter, H. W. (1977) "A Study of the Superoxide Radical Chemistry by Stopped-Flow Radiolysis and Radiation Induced Oxygen Consumption," *J. Am. Chem. Soc.*, **99**, 3019–3023.

Bielski, B. H. J. Cabelli, D. E., Arudi, R. L., Ross, (A.B.) (1985) "Reactivity of HO_2/O_2– Radicals in Aqueous Solution," *J. Phys. Chem. Ref. Data.*, **14**, 1041–1100.

Bolton, J. R., and Carter, S. R. (1994) Homogeneous Photodegradation of Pollutants in Contaminated Water: An Introduction, Chap. 33, in G. R. Helz, R. G. Zepp, and D. G. Crosby (eds.), *Aquatic and Surface Photochemistry*, CRC Press, Boca Raton, FL.

Bruce, D., Westerhoff, P., and Brawley-Chesworth, A. (2002) "Removal of 2- Methylisoborneol and Geosmin in Surface Water Treatment Plant in Arizona," *J. Water Supply: Res. Technol.—Aqua*, **51**, 4, 183–197.

Buhler, R. F., Staehelin, J., and Hoigné, J. (1984) "Ozonation Decomposition in Water Studied by Radiolysis," *J. Phys. Chem.*, **8**, 12, 2560–2564.

Buxton, G. V., and Greenstock, C. L. (1988) "Critical Review of Rate Constants for Reactions of Hydrated Electrons, Hydrogen Atoms and Hydroxyl Radicals (•OH/H•) in Aqueous Solution," *J. Phys. Chem. Ref. Data*, **17**, 2, 513–586.

Cheung, H. M., Bhatnagar, A., and Jansen, G. (1991) "Sonochemical Destruction of Chlorinated Hydrocarbons in Dilute Aqueous Solution," *Environ. Sci. Technol.*, **25**, 8, 1510–1512.

Christensen, H. S., Sehested, K., and Corftizan, H. (1982) "Reaction of Hydroxyl Radicals with Hydrogen Peroxide at Ambient Temperatures," *J. Phys. Chem.*, **86**, 15–88.

Crittenden, J., Hu, S., Hand, D., and Green, S., (1999) "A Kinetic Model for H_2O_2/UV Process in a Completely Mixed Batch Reactor," *Water Res.*, **33**, 10, 2315–2328.

Elovitz, M.S.,and von Gunten, U. (1999) "Hydroxyl Radical/Ozone Ratios During Ozonation Processes," *Ozone: Sci. Eng.*, **21**, 3, 239–260.

Fogler, H. S., and Crittenden, J. C. (1970) Feasibility of Using Ultrasonic Reactions to Degrade Recalcitrant Organic Compounds, internal report, University of Michigan, Ann Arbor, MI.

Forni, L., Bahnemann, D., and Hart, E. J. (1982) "Mechanism of the Hydroxide Ion Initiated Decomposition of Ozone in Aqueous Solution," *J. Phys. Chem.*, **86**, 2, 255–259.

Glaze, W. H., and Kang, J. (1988) "Advanced Oxidation Processes for Treating Groundwater Contaminated with TCE and PCE: Laboratory Studies," *J. AWWA*, **81**, 5, 57–63.

Glaze, W. H., and Kang, J.-W. (1989) "Advanced Oxidation Processes. Description of a Kinetic Model for the Oxidation of Hazardous Materials in Aqueous Media with Ozone and Hydrogen Peroxide in a Semibatch Reactor," *Ind. Eng. Chem. Res.*, **28**, 11, 1573–1580.

Glaze, W., Kang, J.-W., and Chapin, D. H. (1987) "Chemistry of Water Treatment Processes Involving Ozone, Hydrogen Peroxide and Ultraviolet Radiation," *Ozone: Sci. Eng.*, **9**, 4, 335–352.

Glaze, W. H., Lay, Y., Kang, J. W. (1995) "Advanced Oxidation Processes—A Kinetic Model for the Oxidation of 1,2-dibromo-3-chloropropane in Water by the Combination of Hydrogen Peroxide and UV Radiation," *Ind. Eng. Chem. Res.*, **34**, 7, 2314–2323.

Glaze, W. H., Schep, R., Chauncey, W., Ruth, E. C., Zarnoch, J. J., Aieta, E. M., Tate, C. H., and McGuire, M. J. (1990) "Evaluating Oxidants for the Removal of Model Taste and Odor Compounds from a Municipal Water Supply," *J. AWWA*, **82**, 5, 79–84.

Gondrexon, N., Renaudin, V., Petrier, C., Boldo, P., Bernis, A., and Gonthier, Y. (1999) "Degradation of Pentachlorophenol Aqueous Solutions Using a Continuous Flow Ultrasonic Reactor. Experimental Performance and Modeling," *Ultrason. Sonochem.*, **54**, 5, 125–131.

Gutiérrez, M., Henglein, A., and Ibanez, F. (1991) "Radical Scavenging in the Sonolysis of Aqueous Solution of I^-, Br^- and N_3^-," *J. Phys. Chem.*, **95**, 15, 6044–6047.

Haag, W. R., and Hoigné, J. (1983) "Ozonation of Bromide-Containing Waters: Kinetics of Formation of Hypobromous Acid and Bromate," *Environ. Sci. Technol.*, **17**, 5, 261 –267.

Hashimoto, K.,Kawai, T.,and Sakata, T. (1984) "Photocatalytic Reactions of Hydrocarbons and Fossil Fuels with Water. Hydrogen Production and Oxidation," *J. Phys. Chem.*, **88**, 18, 4083–4088.

Heidt, L. J. (1932) "The Photolysis of Hydrogen Peroxide in Aqueous Solution." *J. Am. Chem. Soc.*, **54**, 7, 2840–2843

Hoffmann, M. R., Hua, I., and Höchemer, R. (1996) "Application of Ultrasonic Irradiation for Degradation of Chemical Contaminants in Water," *Ultrason. Sonochem.*, **3**, S163–S172.

Hoigné, J., and Bader, H. (1976) "The Role of Hydroxyl Radical Reactions in Ozonation Process in Aqueous Solutions," *Water Res.*, **10**, 5, 377–386.

Hoigné, J., Bader, H., Haag, W. R., and Staehelin, J. (1985) "Rate Constants of Reactions of Ozone with Organic and Inorganic Compounds in Water—III: Inorganic Compounds and Radicals," *Water Res.*, **19**, 8, 993–1004.

Hunt, J.P.,and Taube, H. (1952) "The photochemical decomposition of Hydrogen Peroxide. Quantum Yields, Tracer and Fractionation Effects," *J. Am. Chem. Soc.*, **74**, 5999–6002.

Lal, M., Schöneich, C., Mönig, J., and Asmus, K.-D. (1988) "Rate Constants for the Reactions of Halogenated Organic Radicals," *Int. J. Radiation Biol.*, **54**, 5, 773–785.

Li, K., Crittenden, J. C., Hand, D. W., and Hokanson, D. R. (1999) Advanced Oxidation Process Simulation Software (AdOxTM) Version 1.0, Michigan Technological University, Houghton, MI.

Liang, S., Palencia, L. S., Yates, R. S., Davis, M. K., Bruno, J. M., and Wolfe, R. L. (1999) "Oxidation of MTBE by Ozone and Peroxone Processes," *J. AWWA*, **91**, 6, 104–114.

Liang, S., Yates, R. S., Davis, D. V., Pastor, S.J., Palencia, L. S., and Bruno, J. M. (2001) "Treatability of MTBE-Contaminated Groundwater by Ozone and Peroxone," *J. AWWA*, **93**, 6, 110–120.

Liao, C.H., and Gurol, M.D. (1995) "Chemical Oxidation by Photolytic Decomposition of Hydrogen Peroxide," *Environ. Sci. Technol.*, **29**, 12, 3007–3014.

Mao, Y., Schöneich, C., and Asmus, K.-D. (1991) "Identification of Organic Acids and Other Intermediates in Oxidative Degradation of Chlorinated Ethanes on TiO_2 Surfaces en Route to Mineralization. A Combined Photocatalytic and Radiation Chemical Study," *J. Phys. Chem.*, **95**, 24, 10080–10089.

Nagata,Y., Hirai, K., Bandow, H., and Maeda, Y. (1996) "Decomposition of Hydroxybenzoic and Humic Acids in Water by Ultrasonic Irradiation," *Environ. Sci. Technol.*, **30**, 4, 1133–1138.

Olson, T. M., and Barbier, P. F. (1994) "Oxidation Kinetics of Natural Organic Matter by Sonolysis and Ozone," *Water Res.*, **28**, 6, 1383–1391.

Peyton, G. R., and Glaze, W. H. (1986) Mechanism of Photolytic Ozonation, Chap. 6, in R. G. Zika and W. J. Cooper (eds.), *Photochemistry of Environmental Aquatic Systems*, ACS Symposium Series 327, American Chemical Society, Washington, DC.

Schramm, J. D., and Hua, I. (2001) "Ultrasonic Irradiation of Dichlorvos: Decom-position Mechanism," *Water Res.*, **35**, 3, 665–674.

Sehested, K., Rasmussen, O. L., and Fricke, H. (1968) "Rate Constants of OH with HO_2,O_2 and H_2O_2 from Hydrogen Peroxide Formation in Pulse-Irradiated Oxygenated Water," *J. Phys. Chem.*, **72**, 626–631.

Sehested, K., Holcman, J., Bjergbakke, E., and Hart, E. J. (1982) "Ultraviolet Spectrum and Decay of the Ozonide Ion Radical, O_3^-, in Strong Alkaline Solution," *J. Phys. Chem.*, **86**, 11, 2066–2069.

Staehelin, J., and Hoigné, J. (1982) "Decomposition of Ozone in Water: Rate of Initiation by Hydroxide Ions and Hydrogen Peroxide," *Environ. Sci. Technol.*, **16**, 10, 676–681.

Stefan, M.I., and Bolton, J.R. (1998) "Mechanism of the Degradation of 1,4–Dioxane in Dilute Aqueous Solution Using the UV/Hydrogen Peroxide Process," *Environ. Sci. Technol.*, **32**, 1588–1595.

Stefan, M. I., and Bolton, J. R. (1999) "Reinvestigation of the Acetone Degradation Mechanism in Dilute Aqueous Solution by the UV/H_2O_2 Process," *Environ. Sci. Technol.*, **33**, 870–873.

Stefan, M. I., and Bolton, J. R. (2002) Personal communication.

Stefan, M. I., Mack, J., and Bolton, J. R. (2000) "Degradation Pathways During the Treatment of Methyl *tert*-Butyl Ether by the UV/H_2O_2 Process," *Environ. Sci. Technol.*, **34**, 650–658.

Stumm, W., and Morgan, J. J. (1972) *Aquatic Chemistry*, Prentice-Hall, Engelwood Cliffs, NJ.

Stumm, W., and Morgan, J. J. (1981) *Aquatic Chemistry: An Introduction Emphasizing Chemical Equilibria in Natural Waters*, Wiley-Interscience, New York.

Suri, R. P. S., Liu, J., Hand, D.W., Crittenden, J. C., Perram, D. L. and Mullins, M. E. (1993) "Heterogeneous Photocatalytic Oxidation of Hazardous Organic Contaminants in Water," *Water Environ. Fed.*, **65**, 5, 665–673.

Taube, H. (1942) "Reactions of Solutions Containing Ozone, H_2O_2, H^+, and Br^-," *J. Am. Chem. Soc.*, **64**, 2468–2474.

Venkatadri, R., and Peters, R. W. (1993) "Chemical Oxidation Technologies: Ultraviolet Light/Hydrogen Peroxide, Fenton's Reagent, and Titanium Dioxide-Assisted Photocatalysis," *Hazard. Waste Hazard. Mat.*, **10**, 2, 107–149.

Volman, D. H. (1949) "The Vapor-Phase Photo Decomposition of Hydrogen Peroxide" *J. Chem. Phys.*, **17**, 10, 947–950.

Volman, D. H., and Chen, J. C. (1959) "The Photochemical Decomposition of Hydrogen Peroxide in Aqueous Solutions of Allyl Alcohol at 2537A," *JACS*, **81**, 16, 4141–4144.

von Gunten, U., and Hoigné, J. (1994) "Bromate Formation during Ozonation of Bromide-Containing Waters: Interaction of Ozone and Hydroxyl Radical Reactions," *Environ. Sci. Technol.*, **28**, 7, 1234–1242.

Westerhoff, P., Aiken, G., Amy, G., and Debroux, J. (1999) "Relationships between the Structure of Natural Organic Matter and Its Reactivity Towards Molecular Ozone and Hydroxyl Radicals," *Water Res.*, **33**, 10, 2265–2276.

Westerhoff, P., Mazyk, S. P., Cooper, W. J., and Minakata, D. (2007) "Electron Pulse Radiolysis Determination of Hydroxyl Radical Rate constants with Suwannee River Fulvic Acid and Other Dissolved Organic Matter Isolates," *Environ. Sci. Technol.*, **41**, 4640–4646.

Westerhoff, P., Ozekin, K., Siddiqui, M., and Amy, G. (1994) Kinetic Modeling of Bromate Formation in Ozonated Waters. Molecular Ozone versus Radical Pathways, paper presented at the American Water Works Association Annual Conference, New York.

Westerhoff, P., Song, R.,Amy, G., and Minear, R. (1998) "Numerical Kinetic Models for Bromide Oxidation to Bromine and Bromate," *Water Res.*, **32**, 5, 1687–1699.

Zellner, R., Exner, M., and Herrmann, H. (1990) Absolute OH Quantum Yields in the Laser Photolysis of Nitrate, Nitrite and Dissolved H_2O_2 at 308 and 351 nmin the Temperature Range 278–353 K. *J Atmos. Chem.*, **10**, 4, 411–425.

Zhang, Y., Crittenden, J. C., Hand, D. W., and Perram, D. L. (1994) "Fixed-Bed Photocatalysts for Solar Decontamination of Water," *Environ. Sci. Technol.*, **28**, 3, 435–442.

19 Disinfection/ Oxidation By-products

Stantec's Water Treatment: Principles and Design, Updated Third Edition. John C. Crittenden, R. Rhodes Trussell, David W. Hand, Kerry J. Howe, and George Tchobanoglous.

Terminology for Disinfection/Oxidation By-Products

Term	Definition
Assimilable organic carbon (AOC)	Measure of the biodegradability of the natural organic matter in a water sample, significant in that ozonation increases the AOC, which can lead to regrowth problems in distribution systems.
Biodegradable dissolved organic carbon (BDOC)	Measure of the biodegradability of the natural organic matter in a water sample.
Bromate	By-product of ozonation.
Disinfection by-product (DBP)	Product of a disinfection reaction other than the desired product. By-products are significant in disinfection because many are suspected to have negative human health consequences.
Formation potential	Test of the ability of DBPs to form in water. The test is conducted with predefined conditions that are expected to produce the maximum amount of the target DBP.
Haloacetic acid (HAA)	Class of DBPs caused by chlorination of water containing natural organic matter.
N-nitrosodimethylamine (NDMA)	By-product of chloramination.
Locational running annual average (LRAA)	Method of determining concentrations for compliance with DBP regulations in which running annual averages are calculated for every sample location.
Running annual average (RAA)	Method of determining concentrations for compliance with DBP regulations in which a year of data is averaged. For each sampling period, the new data is added and the oldest data is dropped, but the average always contains a year's worth of data.
Trihalomethane (THM)	Class of DBPs caused by chlorination of water containing natural organic matter.

Chemical disinfection became an integral part of municipal drinking water treatment over 100 years ago as a vital tool in achieving its principal objective: protection of public health. Oxidation, while not as vital to achieving public health objectives as disinfection, has also been accepted as an important part of drinking water treatment. Typically, oxidation is used to address aesthetic concerns such as color, taste, and odor, which impact consumer perception and acceptance of a water as fit for human consumption. Disinfection is discussed at length in Chap. 13 of this book and oxidation is presented in Chaps. 8 and 18. Unfortunately, disinfection and oxidation produce a variety of by-products, which is the focus of this chapter. The discussion of disinfection and oxidation by-products is presented in five sections. Following an introduction to the subject, separate sections are devoted to the by-products formed by chlorine, chloramines, chlorine dioxide, and ozone. For each disinfectant/oxidant, the chemistry of formation, means of controlling (i.e., reducing) their formation, and means of removing these by-products, if possible, after they form are presented and discussed in this chapter.

19-1 Introduction

The use of chemicals for disinfection and oxidation is common, occurring at nearly every drinking water treatment plant in the industrialized world. Familiarity with the by-products of the chemicals used for disinfection and oxidation is important because these by-products can impact consumer health. Some health effects are not fully understood, and some health effects have been identified only recently, making disinfection/oxidation by-products an ever-changing part of the drinking water treatment situation. An overview of disinfection by-products, including a historical perspective, some known by-products, regulatory requirements, and practical considerations regarding the balance between the need to prevent microbial contamination and minimizing exposure to by-products is presented in this section.

Historical Perspective

Since the introduction of chlorine as a disinfectant in drinking water treatment at the turn of the twentieth century, chemical disinfection has been an integral part of municipal drinking water treatment. In addition to use as a microbial disinfectant, chlorine—as well as other chemical disinfectants such as ozone and chlorine dioxide —had other benefits, including the ability to eliminate color and destroy many naturally occurring chemicals that cause objectionable taste and odor in the water. Consequently, water treatment plant operators commonly added as much disinfectant as necessary to achieve the desired aesthetic and microbial water quality.

In the early 1970s, researchers in the Netherlands and the United States were able to identify and quantify the formation of chloroform ($CHCl_3$) and other trihalomethanes (THMs) in drinking water and relate this formation to the use of chlorine during treatment (Rook, 1971, 1974; Bellar and Lichtenberg, 1974). These early findings led to a large number of studies in the United States on the formation of these "by-products" of chlorination. The studies included several monitoring surveys to assess the magnitude of the problem in drinking water treatment plants across the United States (i.e., occurrence studies), as well as studies to investigate how chlorination by-products were formed and what water quality and/or treatment conditions affected their formation.

It was also discovered that chloroform was not the only chemical formed as a result of the reaction of chlorine with natural organic matter (NOM) present in water and that in the presence of bromide ion (Br^-), the reaction between chlorine, bromide, and NOM resulted in the formation of a mix of chlorinated and brominated chemical by-products. Using mass balance calculations it has been shown that the known chlorination by-products constitute between 30 and 60 percent of the total organic halides (TOX) formed upon the reaction of chlorine with NOM and bromide (Singer and Chang, 1989). Additional work is needed to identify and characterize the unknown by-products of chlorination.

The formation of disinfection by-products (DBPs) is not limited to chlorine disinfection. Haag and Hoigné (1983) reported the formation of bromate ion (BrO_3^-) when ozone was added to waters containing bromide. Bromate was later identified as a suspected carcinogen (Kurokawa et al., 1990). Ozone addition to natural water was also implicated in the formation of numerous organic by-products, such as aldehydes (NAS, 1980). Bromate is now regulated, but thus far the presence of organic ozone by-products in drinking water has not been determined to be a public health concern at the typical levels at which they are formed.

Another disinfectant used in water treatment is chloramines (combined chlorine). In the 1970s, the U.S. EPA identified combined chlorine as an alternative to free chlorine because it was believed not to form THMs. Later research showed that while monochloramine, the principle component of combined chlorine, is less reactive than free chlorine, it does react to form DBPs but at much lower concentrations than are formed with free chlorine (Carlson and Hardy, 1998). The prominent category of DBPs formed during chloramination of potable water is haloacetic acids (HAAs), mostly the dihalogenated species (Cowman and Singer, 1996). The presence of bromide ion will both increase the production of HAAs and shift the speciation to the more brominated species, which are thought to present a greater carcinogen risk (Bull and Kopfler, 1991). In addition, Krasner et al. (1989) reported the results of a 35-utility study in which they identified several new by-products, including cyanogen halides (e.g., CNCl), as by-products of chloramination.

However, no chloramine by-product was believed to be a significant public health concern until Najm and Trussell (2001) reported that *N*-nitrosodimethylamine (NDMA) was a by-product of chloramination of drinking water and wastewater.

Chlorine dioxide forms by-products such as chlorite (ClO_2^-) and chlorate (ClO_3^-) ions, both of which have been suspected to cause health effects (Bull, 1982). While the health effects of chlorate and chlorite continue to be a topic of debate, there was sufficient concern that a maximum contaminant limit (MCL) for chlorite was adopted by the U.S. EPA (1998).

Some Known By-products

As described above, it has been well established over the last 35 years that all chemical disinfectants and oxidants currently used in water treatment form chemical by-products. Even among the known DBPs, the health effects of many are still uncertain. Some of the known DBPs that form from the use of disinfectants during drinking water treatment are summarized in Table 19-1. While many of the DBPs listed in this table have been detected in some treated waters, they are typically present at very low concentrations.

Regulatory Requirements

After the discovery that THMs can form during chlorination of drinking water and because of concerns about the health effects of chloroform (NCI, 1976), the U.S. EPA issued the THM Rule in 1979. The four regulated THMs were chloroform ($CHCl_3$), bromodichloromethane ($CHBrCl_2$), dibromochloromethane ($CHBr_2Cl$), and bromoform ($CHBr_3$). The THM Rule set an MCL of 0.10 mg/L (100 μg/L) for the total sum of these four THMs (on a mass basis) in the distribution system. The THM Rule required water systems to monitor a minimum of four locations throughout the distribution system on a quarterly basis. Three monitoring sites were to be located at average hydraulic travel time through the system and one site was to be located at the far reaches of the system, representing maximum THM formation due to long travel time. The running annual average (RAA) of all distribution system samples collected every quarter was not to exceed the MCL. The rationale for using the RAA for compliance instead of individual THM concentrations was that the adverse health effects of exposure to THMs were caused by long-term exposure not short-term effects.

In 1998, the U.S. EPA expanded the number of regulated DBPs by issuing Stage 1 of the Disinfectants/Disinfection By-products (D/DBP) Rule (U.S. EPA, 1998). This rule reduced the MCL for total THMs from 0.10 to 0.080 mg/L and added MCLs for additional DBPs, as listed in Table 19-2. Of nine HAAs formed with chlorine and bromine, only five were included in the D/DBP Rule because analysis of the remaining four HAAs was not practical at the time the regulation was promulgated. Compliance with the new MCLs began in January 2002. The Stage 1 D/DBP Rule also included a requirement to use chemical coagulation and filtration to maximize the removal of NOM through a conventional water treatment plant (i.e., enhanced coagulation, as discussed in Chap. 9). This requirement was based on the concept that minimizing the concentration of NOM would reduce the amount of DBPs formed with subsequent chlorination.

In January 2006, the U.S. EPA promulgated the Stage 2D/DBP Rule (U.S. EPA, 2006). The Stage 2 rule was designed to further reduce exposure to DBPs without undermining the control of microbial pathogens. A concern that led to passage of the rule was that some portions of distribution

Table 19-1
Some known by-products of chlorine, combined chlorine (chloramines), chlorine dioxide, and ozone application during drinking water treatment

Class of Compound	By-product Name	Chemical Formula	By-product of
Trihalomethanes	Chloroform	$CHCl_3$	Chlorine
	Bromodichloromethane	$CHBrCl_2$	Chlorine
	Dibromochloromethane	$CHBr_2Cl$	Chlorine
	Bromoform	$CHBr_3$	Chlorine, ozone
	Dichloroiodomethane	$CHICl_2$	Chlorine
	Chlorodiiodomethane	CHI_2Cl	Chlorine
	Bromochloroiodomethane	CHBrICl	Chlorine
	Dibromoiodomethane	$CHBr_2I$	Chlorine
	Bromodiiodomethane	$CHBrI_2$	Chlorine
	Triiodomethane	CHI_3	Chlorine
Haloacetic acids	Monochloroacetic acid	$CH_2ClCOOH$	Chlorine
	Dichloroacetic acid	$CHCl_2COOH$	Chlorine
	Trichloroacetic acid	CCl_3COOH	Chlorine
	Bromochloroacetic acid	CHBrClCOOH	Chlorine
	Bromodichloroacetic acid	$CBrCl_2COOH$	Chlorine
	Dibromochloroacetic acid	$CBr_2ClCOOH$	Chlorine
	Monobromoacetic acid	$CH_2BrCOOH$	Chlorine
	Dibromoacetic acid	$CHBr_2COOH$	Chlorine
	Tribromoacetic acid	CBr_3COOH	Chlorine
Haloacetonitriles	Trichloroacetonitrile	$CCl_3C \equiv N$	Chlorine
	Dichloroacetonitrile	$CHCl_2C \equiv N$	Chlorine
	Bromochloroacetonitrile	$CHBrClC \equiv N$	Chlorine
	Dibromoacetonitrile	$CHBr_2C \equiv N$	Chlorine
Haloketones	1,1-Dichloroacetone	$CHCl_2COCH_3$	Chlorine
	1,1,1-Trichloroacetone	CCl_3COCH_3	Chlorine
Aldehydes	Formaldehyde	HCHO	Ozone, chlorine
	Acetaldehyde	CH_3CHO	Ozone, chlorine
	Glyoxal	OHCCHO	Ozone, chlorine
	Methyl glyoxal	CH_3COCHO	Ozone, chlorine
Carboxylic acids	Formate	$HCOO^-$	Ozone
	Acetate	CH_3COO^-	Ozone
	Oxalate	$OOCCOO^{2-}$	Ozone
Ketoacids	Glyoxylic acid	OHCCOOH	Ozone
	Pyruvic acid	$CH_3COCOOH$	Ozone
	Ketomalonic acid	HOOCCOCOOH	Ozone
Oxyhalides	Chlorite	ClO_2^-	Chlorine dioxide
	Chlorate	ClO_3^-	Chlorine dioxide
	Bromate	BrO_3^-	Ozone
Nitrosamines	*N*-Nitrosodimethylamine	$(CH_3)_2NNO$	Chloramines

Table 19-1 *(Continued)*

Class of Compound	By-product Name	Chemical Formula	By-product of
Cyanogen halides	Cyanogen chloride	$ClCN$	Chloramines
	Cyanogen bromide	$BrCN$	Chloramines
Miscellaneous	Chloral hydrate	$CCl_3CH(OH)_2$	Chlorine
Trihalonitromethanes	Trichloronitromethane (Chloropicrin)	CCl_3NO_2	Chlorine
	Bromodichloronitromethane	$CBrCl_2NO_2$	Chlorine
	Dibromochloronitromethane	CBr_2ClNO_2	Chlorine
	Tribromonitromethane	CBr_3NO_2	Chlorine

Source: Krasner (1999); Krasner et al. (2001); Thibaud et al. (1987).

Table 19-2
Disinfection by-products regulated under D/DBP Rule

By-product	Regulatory Limit, mg/L	By-product of
Total THMs[a]	0.080	Chlorine
Five Haloacetic Acids (HAA5)[b]	0.060	Chlorine
Bromate (BrO_3^-)	0.010	Ozone
Chlorite (ClO_2^-)	1.0	Chlorine dioxide

[a]Sum of four THMs: chloroform, bromodichloromethane, dibromochloromethane, and bromoform.
[b]Sum of five HAAs: monochloroacetic acid, dichloroacetic acid, trichloroacetic acid, monobromoacetic acid, and dibromoacetic acid.

systems may have DBP concentrations considerably higher than the average and that some customers might be exposed to DBP concentrations above the MCLs on a consistent basis. The Stage 2 D/DBP Rule required water utilities to conduct an initial distribution system evaluation (IDSE) to identify the locations in the system with the highest DBP concentrations. Once suitable sampling locations were identified, compliance was evaluated via a locational running annual average (LRAA). The LRAA requires a running annual average at each sampling site, whereas the former RAA required only a running annual average over all sampling sites in the entire system. The IDSE and the LRAA provide increased assurance that customers are receiving more consistent protection against exposure to DBPs, even in areas of a distribution system that had typically had higher DBP concentrations. Furthermore, if a utility makes changes to treatment to reduce DBP concentrations in portions of the distribution system, concentrations may be reduced throughout the rest of the distribution system as well. Calculation of the RAA and LRAA are demonstrated in Example 19-1.

Example 19-1 Calculating compliance under the Stage 1 and Stage 2 D/DBP rules

Two years of quarterly THM data from sampling sites in a distribution system are listed in the following table, in μg/L. Calculate the RAA and the LRAA for THMs from this data, and determine if this distribution system would meet both the Stage 1 D/DBP Rule and the Stage 2 D/DBP Rule. Note that the THM data shows both spatial and temporal trends; THMs tend to be the highest in July in all locations, and the THMs at locations Distr-3 and Distr-4 are higher than at Distr-1 and Distr-2.

	Location			
Sample	Distr-1	Distr-2	Distr-3	Distr-4
Jan, 2009	31	35	55	42
Apr, 2009	35	41	57	51
Jul, 2009	62	78	110	91
Oct, 2009	45	55	90	78
Jan, 2010	27	23	45	37
Apr, 2010	39	41	72	72
Jul, 2010	78	94	131	105
Oct, 2010	51	65	104	73

Solution

1. Calculate the RAA for the distribution system data.
   The RAA is calculated by averaging the data from all locations for one sampling event, and then averaging the results from four consecutive sampling events (i.e, one year of data). As each new quarter of data is generated, the data from the quarter one year previous is dropped off and the new quarter is added.
   The RAA data are summarized in the following table:

Sample	RAA
Oct, 2009	60
Jan, 2010	58
Apr, 2010	60
Jul, 2010	65
Oct, 2010	66

2. Calculate the LRAA for the distribution system data.

The LRAA is calculated by averaging the results from four consecutive sampling events (i.e., one year of data) at each sample location. As each new quarter of data is generated, the data from the quarter one year previous is dropped and the new quarter is added.

The LRAA data are summarized in the following table:

	Location			
Sample	**Distr-1**	**Distr-2**	**Distr-3**	**Distr-4**
Oct, 2009	43	52	78	66
Jan, 2010	42	49	76	64
Apr, 2010	43	49	79	70
Jul, 2010	47	53	85	73
Oct, 2010	49	56	88	72

3. Evaluate compliance with the Stage 1 and Stage 2 D/DBP rules. The MCL for THMs is 80 μg/L. Several individual samples, particularly in July each year and particularly at sample location Distr-3, are above the MCL. However, the MCL does not apply to individual samples. Under the Stage 1 D/DBP Rule, compliance was calculated on the basis of the RAA. The calculated RAA values range from 58 to 66 μg/L. The system meets the Stage 1 D/DBP Rule for THMs every quarter. Under the Stage 2 D/DBP Rule, compliance is calculated on the basis of the LRAA. The LRAA values are below the MCL at locations Distr-1, Distr-2, and Distr-4, but the calculated LRAA values at Distr-3 are 85 μg/L in July 2010 and 88 μg/L in Oct 2010. The system is not in compliance with the Stage 2 D/DBP Rule.

Comment

Some water systems may require modifications to meet the Stage 2 D/DBP Rule. Operational changes may include (1) process modifications to reduce by-product precursor material, (2) removal of by-products after formation, (3) changes in the type of secondary disinfectants that are used, and (4) distribution system operation changes to minimize stagnant flow areas and reduce travel time to far reaches of the system to limit the time by-products may form in the distribution system.

The only ozone by-product currently regulated in the D/DBP Rule is bromate (BrO_3^-), with an MCL of 0.010 mg/L (10 μg/L). Some health effects data has indicated that the 10^{-6} cancer risk level for bromate may be as low as 0.05 μg/L (Bull et al., 2001). The U.S. EPA typically sets MCLs for carcinogens at a level between their 10^{-6} and 10^{-4} cancer risk levels, which suggests that the bromate MCL may be lowered to a level between 0.05 and 5 11 g/L at some time in the future. (The current practical quantification limit for bromate is 10 μg/L.) However, the bromate MCL was not identified as a candidate for revision in the most recent 6-year review of drinking water standards (U.S. EPA, 2010a), so action toward a revision to this MCL will probably not occur before the year 2016 when the next 6-year review will take place. A lower MCL for bromate will make it difficult to use ozone as a disinfectant when treating water containing measurable concentrations of bromide.

Practical Considerations

Minimizing DBP formation needs to be balanced with the need to maximize disinfection. Since DBPs are formed via the reaction of various water constituents with disinfectants, one way of reducing DBP formation is to reduce the concentration of the disinfectant in the water and/or the time it is present in that water. Reducing the disinfectant concentration and/or contact time will directly reduce disinfection efficiency, and thus reduce public protection against exposure to disease-causing microorganisms. Therefore, any effort to reduce the formation of DBPs through minimizing disinfectant concentration and/or contact time during water treatment must be balanced against the need to lower the microbial risk through adequate disinfection.

19-2 Free-Chlorine By-products

Chlorine is by far the most widely used disinfectant in drinking water treatment in the United States It is also the disinfectant that forms the greatest variety of known by-products, many of which are listed in Table 19-1. Of the known chlorination by-products, the primary by-products of concern in drinking water treatment are THMs and HAAs. A cumulative distribution profile of average concentrations of total THMs (TTHM) and five HAAs (HAA5) in the distribution systems of approximately 360 U.S. water treatment plants (McGuire, 1999) is shown on Fig. 19-1. This information was collected over a 9-month period spanning summer 1997 through winter 1998. As shown on Fig. 19-1, on average, 10 percent of U.S. distribution systems contained greater than 51 μg/L of HAA5 and greater than 73 μg/L of TTHM. Based on the data shown on Fig. 19-1, the majority of distribution systems in the United States contain HAAs and THMs, and slightly less than 10 percent of distribution systems may not be meeting the Stage 1 D/DBP Rule.

Chemistry of Formation

At an elementary level, chlorine reacts with NOM and bromide ions to form halogenated by-products, as shown in the following reaction:

$$\text{NOM with or without bromide} + \text{chlorine} \rightarrow \text{by-products} \qquad (19\text{-}1)$$

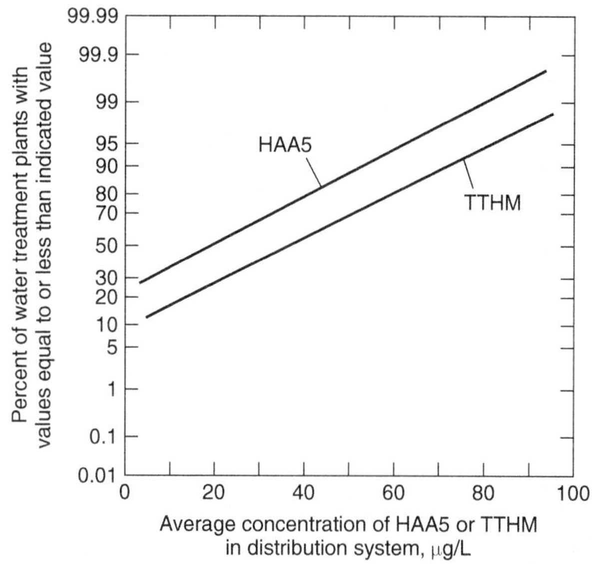

Figure 19-1
Cumulative distribution profile of average TTHMs and HAA5 in approximately 360 U.S. distribution systems monitored over a 9-month period beginning in summer 1997 and ending winter 1998 (adapted from McGuire, 1999).

According to Fuson and Bull (1934), the reaction between certain organic chemicals and chlorine to produce "haloforms" has been known since 1822. However, it was only after the discovery of chloroform formation upon chlorination of drinking water in the early 1970s that the occurrence of haloform reactions in drinking water treatment was recognized. Many researchers have explored the mechanisms of THM and other by-product formation upon chlorination of natural waters (Christman et al., 1978; Norwood et al., 1980, 1987; Reckhow and Singer, 1985; Rook, 1974, 1977). Because NOM in natural waters is complex and has unidentifiable structures, it has not been possible to identify and verify every specific reaction mechanism between NOM and chlorine. Regardless of the exact mechanisms involved in the formation of chlorination by-products, higher concentrations of chlorination by-products are formed with higher concentrations of organic precursor material, bromide ions (inorganic precursor), or chlorine, as shown in Eq. 19-1.

THMs and HAAs form simultaneously when chlorine reacts with NOM, but the ideal conditions for THM formation are different from the ideal conditions for HAA formation. It is generally believed that the reaction mechanism leading to the formation of THMs is base catalyzed, meaning the reaction is catalyzed by hydroxide ions (OH^-) present in the water, and therefore proceeds faster at more alkaline pH (Rook, 1977). Conversely, HAAs formation is enhanced under acidic conditions. Therefore, pH will directly influence whether THM formation is favored or HAA formation is favored.

The presence of bromide increases the mass concentration of THMs formed when chlorine reacts with NOM. Bromide ions (Br^-) participate in the reaction between NOM and chlorine to form various by-products that have a mix of chlorine and bromine substitutions (e.g., bromodichloromethane, bromochloroacetic acid). Brominated THMs form after hypochlorous acid oxidizes bromide ions to form hypobromous acid (HOBr) as follows:

$$HOCl + Br^- \rightarrow HOBr + Cl^- \tag{19-2}$$

Hypobromous acid is a weak acid that dissociates depending on pH:

$$HOBr \rightleftarrows H^+ + OBr^- \quad pK_a = 8.7 \tag{19-3}$$

HOBr and OBr^- are both capable of reacting with NOM during chlorination to form a mixture of chlorinated and brominated by-products. The increase in THM formation when bromide is present occurs because bromine has a molecular weight of 79 g/mol compared to 35.5 g/mol for chlorine. Thus, an equal molar concentration of a highly brominated DBP (e.g., bromoform) compared to a purely chlorinated DBP (e.g., chloroform) will result in a significantly higher mass-based concentration of THMs, as demonstrated in Example 19-2.

Example 19-2 Comparison of mass and molar concentrations of THMs chloroform and bromoform

Determine the mass concentrations of 0.5 μM solutions of (a) chloroform ($CHCl_3$) and (b) bromoform ($CHBr_3$) and compare the results.

Solution

1. Calculate the molecular weight of chloroform and bromoform:
   a. $CHCl_3 = [12 + 1 + 3(35.5)] = 119.5$ g/mol or 119.5 μg/μmol.
   b. $CHBr_3 = [12 + 1 + 3(79.9)] = 252.7$ g/mol or 252.7 μg/μmol.
2. Multiply the molar concentration (μM = μmol/L) by the molecular weight of each compound:
   a. Mass concentration of $CHCl_3$ = (0.5 μmol/L) (119.5 μg/μmol) = 59.8 μg/L = 0.0598 mg/L.
   b. Mass concentration $CHBr_3$ = (0.5 μmol/L) (252.7 μg/μmol) = 126 μg/L = 0.126 mg/L.
3. Compare the mass concentrations. The mass concentration of $CHBr_3$ is nearly double that of $CHCl_3$. If drinking water contained 0.5 μM of THMs, it would meet the MCL of 0.080 mg/L if only chloroform was present but would not if only bromoform was present.

Comment

Regulatory requirements for DBPs are based on mass. Therefore, waters with moderate organic carbon concentrations and high bromide ion concentrations produce higher levels of chlorination by-products on a mass concentration basis. There is some rationale for this approach, because brominated by-products are more toxic.

The effect of bromide increasing the mass concentration of by-products formed is demonstrated on Fig. 19-2. As shown on Fig. 19-2, the increase in bromide ion concentration from 0.12 to 0.44 mg/L increased the total THM levels formed in 3 h of free-chlorine contact time from 80 to 142 μg/L, a 70 percent increase. Increasing the bromide concentration resulted in a slight increase in the chlorine dose required to achieve the same 3-h residual (see values in parentheses on Fig. 19-2). Therefore, the increase in THM formation was caused by both the higher bromide concentration and the higher chlorine dose added. The linear relationship between bromide concentration and TTHMs shown on Fig. 19-2 is not universal, as other work has shown a more logarithmic relationship between bromide concentration and THM formation in natural waters (Trussell and Umphres, 1978).

The type of NOM present also influences the amount of DBP formation. Researchers have been able to fractionate NOM present in different waters according to its molecular size and/or chemical characteristics (Aiken et al., 1992; Croué et al., 1993; Leenheer and Croué, 2003; Reckhow et al., 1990). In natural waters, NOM consists of humic (hydrophobic) and nonhumic (hydrophilic/polar) fractions. In most, but not all, natural waters, hydrophobic NOM contributes more to THM and HAA formation than

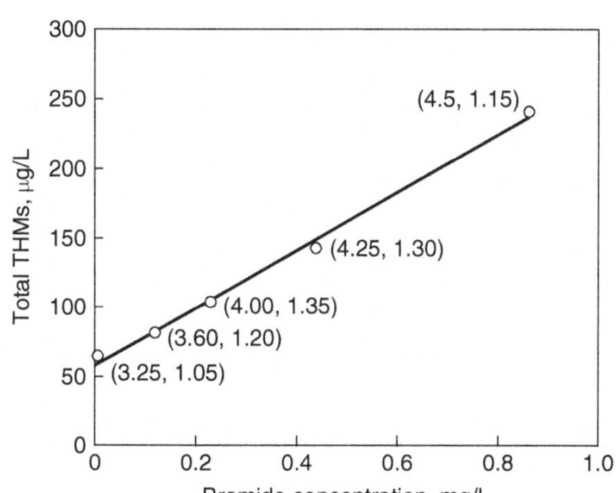

Figure 19-2
Impact of bromide on THM formation (adapted from Krasner et al., 1994) (Conditions: Temperature = 25°C, pH = 8.2, contact time = 3 h, TOC = 3.25 mg/L. Values in parentheses represent the chlorine dose and the 3-h chlorine residual for each data point.).

hydrophilic NOM (Kavanaugh, 1978; Croué et al., 2000). An example where the hydrophilic fraction has a higher THM and HAA yield compared to its hydrophobic fraction is Colorado River water (Hwang et al., 2000). Many conventional and advanced treatment processes preferentially remove hydrophobic NOM over hydrophilic NOM, so it may be important to characterize the source water NOM to assist in selecting the appropriate treatment process for the source water (see enhanced coagulation in Chap. 9). The specific UV absorbance (SUVA) is an important water quality indicator that correlates well with the amount of hydrophobic NOM in water; high SUVA values can be an indicator that a source water is predisposed to higher THM or HAA formation.

Estimating By-product Formation

THMs and HAAs formation has been studied extensively in the past 30 years. Formation data can be correlated to water quality conditions such as pH, alkalinity, and TOC concentration and operating parameters such as chlorine dose, chlorine residual, and time using multivariate regression analyses to determine the effect of each parameter on DBP formation. In an extensive review, Sadiq and Rodriguez (2004) identified 25 publications in peer-reviewed literature that presented correlations between THM and HAA concentrations and water quality and operating conditions. More recently, Obolensky and Singer (2008) used extensive field data from the Information Collection Rule to develop a similar correlation. The Obolensky and Singer regression considers temperature, alkalinity, TOC, UV absorbance, chlorine dose, chlorine residual, pH, and time as factors that affect THM formation. The THM formation model is

$$\begin{aligned}\log[\mathrm{THM}_4] = &-1.371 + 0.015\,(T) - 0.0015\,(\mathrm{ALK}) \\ &+ 0.188\,[\log(\mathrm{TOC})] + 0.326\,[\log(\mathrm{UV})] \\ &+ 0.291\,[\log(\mathrm{Cl}_2)] + 0.119\,[\log(t)] \\ &+ 0.087(\mathrm{pH}) + 0.167\,[\log(\mathrm{Cl}_{\mathrm{res}})]\end{aligned} \tag{19-4}$$

where THM_4 = THM concentration, μ mol/L
T = plant influent temperature, °C
ALK = plant influent alkalinity, mg/L as $CaCO_3$
TOC = total organic carbon conc. at point of Cl_2 addition, mg/L
UV = UV_{254} absorbance at point of Cl_2 addition, cm^{-1}
Cl_2 = chlorine consumed (dose minus residual), mg/L as Cl_2
t = Cl_2 contact time in treatment plant, h
Cl_{res} = chlorine residual in finished water, mg/L as Cl_2

Obolensky and Singer (2008) also presented model parameters for calculation of three individual THMs (excluding bormoform because low concentrations in the data set led to unacceptable model results) and five HAAs. Since the equations estimate DBPs in units of μmol/L, it is necessary to calculate each THM individually, multiply by the molecular weight, and sum the results to determine the THM concentration in μg/L.

Simpler models have been developed by Chen and Westerhoff (2010) that predict THM and HAA formation potentials. Formation potential tests evaluate the formation of DBPs under prescribed laboratory conditions so that factors such as time, temperature, and chlorine dose are constant across all tests. THM and HAA formation potentials are typically higher than those observed in distribution system and serve as an approximate upper limit on the expected DBP formation. The models developed by Chen and Westerhoff (2010) are

$$\text{THMFP}_4 = 1147\,(\text{UV})^{0.83}(\text{Br}+1)^{0.27} \tag{19-5}$$

$$\text{HAAFP}_9 = 1151\,(\text{DOC})^{0.17}(\text{UV})^{0.89}(\text{Br}+1)^{-0.60} \tag{19-6}$$

where THMFP_4 = THM formation potential for four THMs, μg/L
HAAFP_9 = THM formation potential for nine HAAs, μg/L
DOC = dissolved organic carbon conc., mg/L
UV = UV_{254} absorbance, cm^{-1}
Br = bromide conc., mg/L

Chen and Westerhoff (2010) also provided coefficients for the estimation of numerous DBPs individually.

Caution should be exercised when using these models to estimate DBP formation for design or modification of treatment facilities. While such correlations give an indication of DBP formation based on average water quality conditions, they are not necessarily accurate for any specific site. Nevertheless, they can provide general guidance for the level of DBPs that may be formed. An example of the use of these equations is demonstrated in Example 19-3.

Example 19-3 Estimation of THM and HAA formation potential

Estimate the THM and HAA formation potential for the following conditions: DOC = 4.0 mg/L, UV_{254} absorbance = 0.1 cm^{-1}, and bromide = 0.05 mg/L.

Solution

1. Estimate the THM formation potential using Eq. 19-5.

$$\text{THMFP}_4 = 1147\,(0.1)^{0.83}\,(0.05+1)^{0.27} = 172\ \mu\text{g/L}$$

2. Estimate the HAA formation potential using Eq. 19-6.

$$\text{HAAFP}_9 = 1151\,(4)^{0.17}\,(0.1)^{0.89}\,(0.05+1)^{-0.60} = 182\ \mu\text{g/L}$$

Formation Control

Since implementation of the THM rule in 1979, a large amount of work has been conducted to identify and evaluate alternatives for reducing DBP formation (primarily THMs and HAAs). The following are practical alternatives for reducing the formation of chorination by-products:

1. Use an alternate disinfectant/oxidant.
2. Reduce the free-chlorine contact time.
3. Reduce the concentration of NOM before chlorine addition.

USE OF ALTERNATE DISINFECTANTS/OXIDANTS

Chlorine is not the only oxidant/disinfectant available to treat water. To reduce concern about chlorination by-products, substitution of another oxidant/disinfectant for chlorine may be a viable alternative. For example, neither ozone (O_3) nor chlorine dioxide (ClO_2)—both of which are strong oxidants and disinfectants—produces measurable THMs or HAAs when applied to natural waters. If the source water contains a lot of humic acid that forms excessive amounts of chloroform, chlorine dioxide may be an appropriate alternative disinfectant to chlorine. The relative formation of chloroform with the addition of 10 mg/L of chlorine or chlorine dioxide to a solution containing 5 mg/L of humic acid is illustrated on Fig. 19-3 (Noack and Doerr, 1978). While 210 μg/L chloroform was formed with chlorination, the addition of 10 mg/L of chlorine dioxide formed no detectable levels of chloroform.

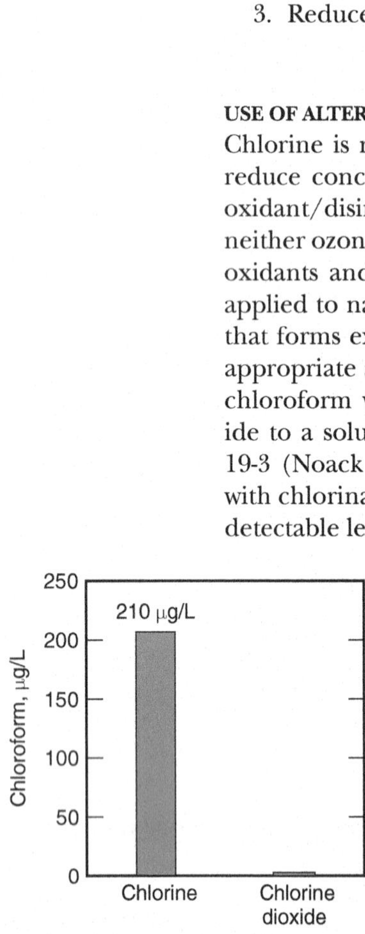

Figure 19-3
Relative formation of chloroform with 10 mg/L doses of chlorine and chlorine dioxide to water containing 5 mg/L humic acid at pH 7.0 (adapted from Noack and Doerr, 1978).

The implications of the use of alternative disinfectants should be thoroughly investigated. Disinfectants such as ozone and chlorine dioxide, while not forming THMs and HAAs, do form other by-products, some of which are regulated and of significant health concern. For instance, Trussell and Umphres (1978) showed that the addition of up to 10 mg/L ozone to a water containing 10 mg/L humic acid and 1 mg/L bromide ion formed less than 4 μg/L of THMs. However, Br^- originally present in the water could not be accounted for by summing the Br^- and HOBr present after ozonation. At the time it was not feasible to measure BrO_3^- at low levels. Since then it has been demonstrated that the addition of ozone to waters containing bromide ions results in the formation of bromate, which is now regulated.

Potassium permanganate ($KMnO_4$) is an alternative to chlorine for oxidation at water treatment plants. While permanganate is a weak disinfectant, it is a good oxidant for the control of iron, manganese, and sulfide. Iron is sometimes strongly complexed and hard to remove. Potassium permanganate forms less THMs and HAAs than chlorine. Potassium permanganate addition to drinking water is not known to produce any regulated by-products, but $KMnO_4$ is a weak disinfectant, so additional chemical disinfection is necessary and may result in DBP formation.

REDUCTION OF FREE-CHLORINE CONTACT TIME

Minimizing the free-chlorine contact time can minimize DBP formation. Reducing contact time may be achieved by placing the chlorine addition point near the end of the treatment train and adding ammonia to the treated water at the point where the chlorine disinfection requirements have been met. Ammonia reacts with free chlorine and forms combined chlorine, which is much less reactive with NOM. Experience from full-scale plants indicates that THMs and HAAs do continue to form under combined chlorine conditions, but at a much lower rate compared to their formation under free-chlorine conditions.

Ammonia added at a mass ratio of less than 1 : 5 (between the $NH_3 - N$ dose and the residual Cl_2 concentration at the point of ammonia addition) will convert the free chlorine residual to monochloramine according to the following reaction (see Eq. 13-17 in Chap. 13):

$$NH_3 + HOCl \rightarrow NH_2Cl + H_2O \tag{19-7}$$

The conversion of free chlorine to combined chlorine is an excellent strategy for limiting the formation of many chlorination by-products, especially THMs and HAAs. The extent of THM and HAA reduction is a function of how long the free chlorine is allowed to react with the NOM to form by-products before the ammonia is added to the water. Free chlorine contact time with the water is necessary to achieve the target disinfection goals, unless an alternative disinfectant is used, such as ozone or chlorine dioxide. When an alternate disinfectant is used for primary disinfection, little to no free-chlorine contact time may be required before ammonia addition to form chloramines for residual maintenance.

REDUCTION OF CONCENTRATION OF NOM BEFORE CHLORINE ADDITION

Based on Eq. 19-1, reducing the concentration of NOM will reduce the formation of chlorination by-products. The relationship between THM and HAA formation and NOM concentration was characterized in a 1994 study on Sacramento River water. As shown on Fig. 19-4, the concentration of THMs and HAAs increase with increasing the NOM concentration when using chlorination. Therefore, reducing the total organic carbon (TOC) concentration in natural water results in a corresponding decrease in the formation of by-products during exposure to chlorine. The relationship between NOM reduction and reduced DBP formation is the basis for the enhanced coagulation requirement in the D/DBP rule.

Several methods are available for removing NOM during drinking water treatment, including (1) enhanced coagulation, (2) adsorption on activated carbon, (3) ozone/biofiltration, (4) ion exchange, and (5) reverse osmosis. These treatment strategies are discussed in the following sections.

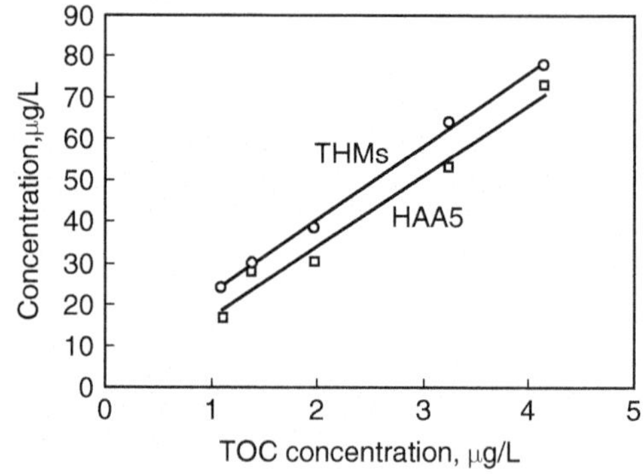

Figure 19-4
THM and HAA5 formation upon chlorination of Sacramento River water spiked with various doses of natural organic matter (adapted from Krasner et al., 1994). (Conditions: Temperature = 25°C, pH = 8.2; contact time = 3 h, $Br^- < 0.010$ mg/L. Chlorine dose was set to achieve a 3-h free-chlorine residual of 0.5 to 1.5 mg/L.).

Removal of NOM by enhanced coagulation

Chemical coagulation and precipitation is primarily achieved with the addition of inorganic coagulants such as aluminum or ferric salts. Coagulants are traditionally used for removing suspended material from water, but they can also adsorb NOM. When chemical precipitates are removed from the water through clarification and filtration, the adsorbed NOM is removed as well, reducing the concentration of NOM available to react with the chlorine added downstream for disinfection. The process of practicing coagulation specifically for NOM removal is now commonly practiced and is known as enhanced coagulation. TOC removal of 15 to 50 percent can be achieved with enhanced coagulation. Enhanced coagulation is discussed in Sec. 9-5 of Chap. 9.

Removal of NOM by adsorption on activated carbon

A less common method of NOM removal is adsorption on activated carbon. A large number of studies have evaluated the removal of NOM with adsorption on activated carbon (e.g., Crittenden et al., 1987; Hooper et al., 1996; Sontheimer et al., 1988; Wood and DeMarco, 1980) and have found that significant removal of NOM with activated carbon can be very costly. A significant fraction of NOM is comprised of large molecular weight organic molecules that are poorly adsorbed on activated carbon, resulting in the use of a large amount of activated carbon to remove a small amount of NOM. Adsorption is discussed in Chap. 15.

Removal of NOM by ozone/biofiltration

Frequently, ozone's use for DBP control is as an alternate disinfectant (i.e., replacing chlorine), but in some cases it can also be used for precursor removal prior to chlorination. Natural organic matter in water contains a significant amount of high-MW organic material that is relatively resistant to biodegradation. Ozone breaks carbon-carbon double bonds and, in doing so, breaks the organics into smaller, more readily biodegradable material.

The increase in biodegradable material is characterized by an increase in assimilable organic carbon (AOC) or biodegradable dissolved organic carbon (BDOC) following ozonation. There is essentially no decrease in DOC, however, because ozonation cannot completely mineralize the complex organics to inorganic products. Biologically active filtration is effective at degrading the smaller organics, so the combined process of ozone and filtration is capable of achieving as much as 35 to 40 percent DOC removal. Ozonation is presented in Chap. 13 and biological filtration is presented in Sec. 11-8 of Chap. 11.

Removal of NOM by ion exchange (MIEX)

NOM in water is negatively charged, so it could be removed using ion exchange (IX) technology using anion exchange resins (Fu and Symons, 1990; Kim and Symons, 1991), using a conventional packed-bed configuration. Depending on the water quality and resin parameters, TOC reduction of 50 percent can be achieved with run lengths ranging from 500 to 5000 bed volumes (BVs) between regenerations. A limitation of conventional IX technology for NOM removal, however, is slow process kinetics. The slow exchange process is caused by slow rate of mass transfer coupled with the size of the beads.

An alternative IX technology (MIEX) has been developed that uses a smaller size of resin bead (< 0.2 mm) (Singer and Bilyk, 2002) added to the water at a treatment plant as a slurry. The MIEX resin is allowed to contact the water for a short time and then settled to the bottom of the contactor, where it is collected, regenerated, and reused. To increase the bead settling velocity in the contactor, the beads are magnetized, which forces them to coalesce during settling to form larger masses, which then settle at a higher rate. The smaller bead size allows for faster removal kinetics.

A pilot study for the City of West Palm Beach, FL, found that THM formation potential was reduced by 70 percent using a process train that incorporated MIEX and coagulation, compared to water treated with enhanced softening (MWH, 2010). Use of MIEX also reduced coagulant use and sludge production by about 80 percent compared to enhanced coagulation alone. Significantly, the duration of free-chlorine contact time without excessive DBP formation was increased from about 3 min to nearly 4 h, long enough for the plant to get the disinfection CT credit it needed. As of the end of 2010, about 15 MIEX systems had been installed at treatment plants greater than 3785 m^3/d (1 mgd) in North America. Additional information on MIEX and ion exchange is presented in Chap. 16.

Removal of NOM by reverse osmosis

Reverse osmosis using nanofiltration (NF) or reverse osmosis (RO) membranes can be used to remove NOM. With pore sizes less than 1 nm, these membranes can reject more than 90 percent of NOM and DBP precursors (Fu et al., 1994). With this rate of removal, reverse osmosis is the most effective of the treatment processes for NOM removal discussed in this chapter. Reverse osmosis has been used effectively for NOM removal at a number of installations treating highly colored groundwater, in particular for treating Biscayne Aquifer water in South Florida.

Despite the high removal efficiency, other methods of NOM removal are frequently preferred because reverse osmosis has several disadvantages, including high capital cost, high operating cost, high energy consumption, and low recovery of product water compared to other treatment processes.

The low recovery of reverse osmosis is due to scaling caused by sparingly soluble salts that are concentrated during the reverse osmosis process. The cations that cause scaling are principally divalent cations (specifically Ca^{2+}, Ba^{2+}, and Sr^{2+}), thus, a membrane that can reject NOM without rejecting these cations would be able to operate at higher recovery. For instance, the Hydranautics NTR-7450 membrane has been reported to achieve 93 percent TOC removal but only 35 percent calcium removal when operating at 90 percent recovery (Fu et al., 1994). Reverse osmosis is discussed in additional detail in Chap. 17.

Removal of Chlorination By-products

Technologies exist to remove some chlorination by-products after they are formed, but they are seldom practical compared to the by-product reduction strategies discussed above.

Chlorination by-products are organic chemicals, so they can be removed from water via several trace-chemical removal technologies. These include adsorption on activated carbon (Chap. 15) or, for the more volatile byproducts, removal with aeration and air stripping (Chap. 14). Several researchers (e.g., Speth and Miltner, 1998; Crittenden et al., 1986) have evaluated the removal of chloroform with adsorption on GAC, but the adsorption capacity of GAC for chloroform is relatively low, requiring a high GAC replacement or regeneration frequency, possibly every few weeks. Similarly, air stripping is theoretically a viable approach for removing THMs from water because they are volatile chemicals (Umphres et al., 1983). However, DBPs, and THMs in particular, continue to be formed after they are removed as long as the water contains a chlorine residual. For this reason, coupled with the fact that HAAs are not removed, air stripping is a nonviable alternative for DBP removal.

19-3 Chloramine By-products

Converting free chlorine to combined chlorine (chloramines) via Eq. 19-4 after requirements for primary disinfection have been met has long been considered a cost-effective strategy to halt THM and HAA formation in chlorinated water while still complying with distribution system residual requirements. Most systems using chloramines engage in this practice. The practice is widely accepted in some countries, such as the United States, the United Kingdom, Denmark, and Australia. It is not as widely accepted in countries that do not make a practice of maintaining distribution system residuals, such as the Netherlands, Germany, and Switzerland. In the United States in particular, with tighter regulatory standards being implemented on DBPs (see Chap. 4), an increasing number of water utilities are converting the disinfectant residual in their distribution systems from free chlorine to combined chlorine (Jacangelo and Trussell, 2002). The addition of ammonia to water containing a free-chlorine residual to form chloramines will almost completely halt the formation of THMs, HAAs, and most other chlorinated DBPs with the exception of chloropicrin, although, in some cases, a gradual but modest increase has been observed. Until recently, there are were no known DBPs of significance formed by chloramine itself.

Around the year 2000, it was found that the reaction of combined chlorine with organic matter present in some waters may result in the formation of *N*-nitrosodimethylamine (NDMA) as a by-product (Najm and Trussell, 2000; Choi and Valentine, 2001, Mitch and Sedlak, 2002). The U.S. EPA currently classifies NDMA as a "probable human carcinogen," and has estimated its 10^{-6} cancer-risk level at 0.7 ng/L, which is well below the levels measured in chloraminated drinking water (Eaton and Briggs, 2000). No federal MCL has yet been established, but NDMA is on the Contaminant Candidate List (CCL3) released in 2010 and is being monitored as part of the Unregulated Contaminant Monitoring Regulation (UCMR). California has set an action limit of 10 ng/L (Cal DHS, 2002).

Chemistry of Formation

The chemical mechanism leading to the formation of NDMA is not completely resolved, but some principles have been established. NDMA formation in drinking water disinfection is a direct by-product of chloramine and not free chlorine, as shown on Fig. 19-5 (Najm and Trussell,

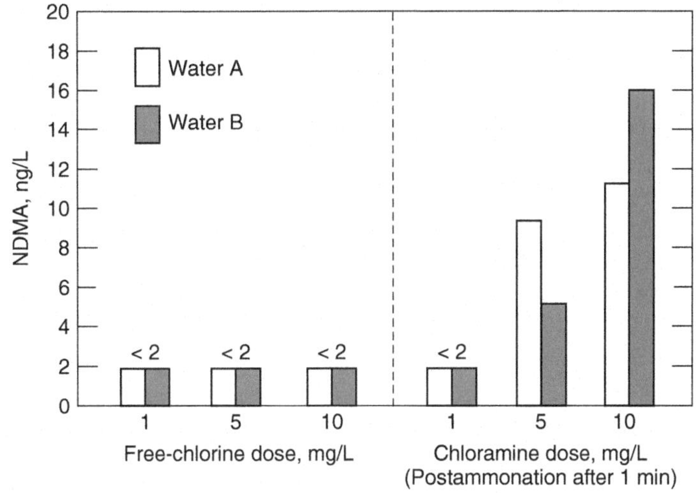

Figure 19-5
NDMA formation after chlorine and chloramine addition to two test waters (contact time = 24 h) (Adapted from Najm and Trussell, 2001).

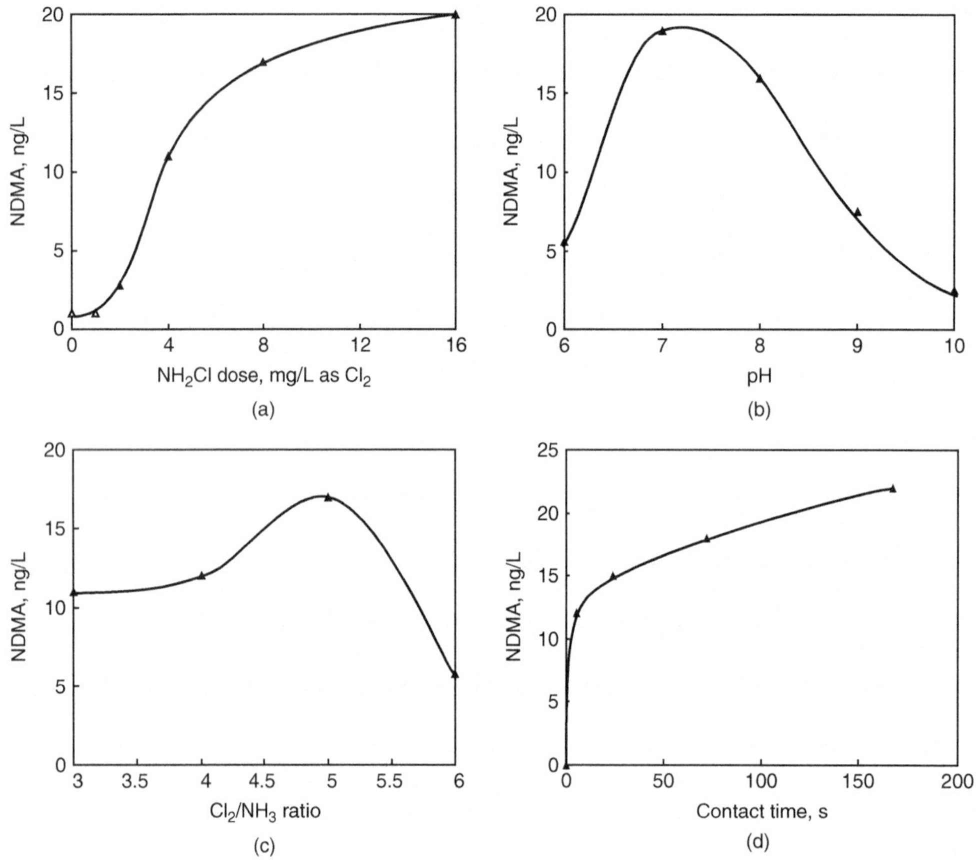

Figure 19-6
Factors influencing NDMA formation in the presence of chloramines: (a) NH_2Cl dose, (b) pH, (c) Cl_2:NH_3 ratio, and (d) contact time.

2001). Najm and Trussell (2001) found that the NDMA concentration increased with increasing chloramine dose, and the sequence of addition of chlorine and ammonia had no significant effect on the final levels of NDMA formed. Further unpublished research by MWH engineers on untreated water from the West Branch of the California State Aqueduct has shown that NDMA formation during chloramination of that supply peaks at a pH of about 7 and at a Cl_2:NH_3 ratio of approximately 5 : 1 on a weight basis (see Figs. 19-6a,b,c). Although significant NDMA formation occurs in the first few hours, it appears that the process continues at a slow pace for several days (see Fig. 19-6d). As a result of this prolonged formation time, the highest NDMA levels in the system are likely to be formed where the travel time is the longest.

Two pathways have been proposed for NDMA formation during chloramination. First, a mechanism based on the reaction of monochloramine

(NH_2Cl) with unsymmetrical dimethylhydrazine (UDMH) was proposed (Choi and Valentine, 2001, 2002a, 2002b; Choi et al. 2002; Mitch and Sedlak, 2002). This mechanism accounted for the fact that NDMA formation was observed during chloramination, but it predicted yields that were too low. In addition, other work showed that one of the reaction steps was much slower than had been estimated (Yagil and Anbar, 1962). Subsequently, another mechanism was proposed, one that requires the presence of dichloramine ($NHCl_2$) as well as oxygen (Schreiber and Mitch, 2005). The overall reaction takes on the following stoichiometry:

$$\underset{\text{DMA}}{NH(CH_3)_2} + NHCl_2 + O_2 \rightarrow \underset{\text{NDMA}}{(CH_3)_2 NNO} + HCl + HOCl \qquad (19\text{-}8)$$

DMA is dimethylamine.

The California Department of Health Services conducted a survey of water systems in 1999 and found that 50 percent of the system samples had no NDMA (method detection level (MDL) = 1 ng/L) and 5 percent of the system samples had NDMA levels above 20 ng/L. More recently, the UCMR2 data summary released by the EPA in July, 2009, showed NDMA in 179 of 729 systems sampled, suggesting most chloraminated systems had detections (U.S. EPA, 2010b). The average NDMA level in samples with detections was 14 ng/L and the maximum was 630 ng/L. Studies have shown that higher NDMA levels are generally observed downstream of ion exchange (Loveland, 2001; Najm and Trussell, 2001) and following the use of sorne cationic polymers (Neiss et al., 2003; Wilczak et al., 2003).

Formation Control

Alternatives for minimizing NDMA formation can be gleaned from an examination of the circumstances that promote its formation. Generally, reducing the dose of chloramines, increasing the pH, and minimizing the contact time in the distribution system are all practices that will reduce NDMA formation. It also seems likely that either minimizing the use of polyDADMAC cationic polymers (see Chap. 9) or selecting carefully among them will also reduce the levels formed (Wilczak et al., 2003). Care in recycling filter waste washwater has also been shown to be of significance (Neiss et al., 2003). Recent work in wastewater reuse suggests that, consistent with the Schreiber–Mitch mechanism shown in Eq. 19-8, NDMA formation can be reduced by avoiding dichloramine formation. This potential formation control strategy is illustrated on Fig. 19-7 in which data on NDMA formation with dichloramine and preformed monochloramine are compared. Preformed monochloramine resulted in significantly lower NDMA formation than dichloramine.

Removal of Chloramine By-products

NDMA can be removed by reverse osmosis (Chap. 17), photolysis, or advanced oxidation (Chap. 18). Although these alternatives may be practical for treating contaminated groundwater or water exiting ion exchange processes, they are not practical solutions when NDMA results from chloramine use for

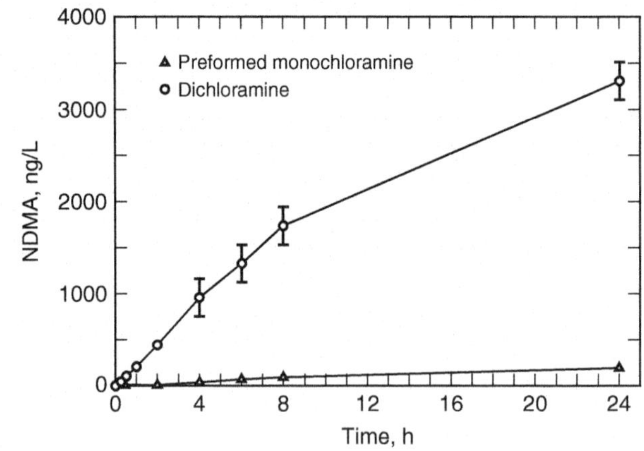

Figure 19-7
Formation of NDMA via dichloramine and monochloramine. (Data from Farré et al., 2011.)

disinfectant residual maintenance. As noted above, NDMA forms slowly in the distribution system, reaching a maximum concentration in the consumer's tap. As a result, any treatment for removal of NDMA would have to be applied at the consumer's tap. Preventing the formation of NDMA, as discussed above, may be the most economical approach.

19-4 Chlorine Dioxide By-products

Chlorine dioxide (ClO_2) is a unique oxidant because it is a stronger disinfectant than free chlorine (see Fig. 13-5 in Chap. 13) but, when added to water containing NOM, does not break the C-C bond and therefore does not form halogenated organic molecules. Chlorine dioxide does, however, produce two inorganic by-products: chlorite (ClO_2^-) and chlorate (ClO_3^-). The presence of chlorite in drinking water is of concern because it is believed to have serious adverse health effects. In the United States, the Stage 1 D/DBP Rule set the MCL for chlorite in water at 1 mg/L. Although no MCL for chlorate exists at the federal level in the United States, the state of California limits it to 0.8 mg/L.

Chemistry of Formation

There are several sources of chlorite and chlorate in drinking water, including some that are unrelated to the use of chlorine dioxide. Thus, the actual source needs to be identified before measures can be taken to reduce the concentrations. Chlorate can be present in the raw water entering a treatment plant due to agricultural use because sodium chlorate has been used as a herbicide for almost a century (Hayes and Laws, 1971). Chlorate salts are also used commercially as oxidizing agents in some industries (Clapper, 1979), which can result in significant levels of chlorate in raw waters. Chlorate is a degradation product of chlorine in liquid hypochlorite solutions (see Chap. 13); thus, the use of liquid hypochlorite at water treatment plants may add chlorate into the water.

CHLORITE AND CHLORATE FORMATION DURING CHLORINE DIOXIDE GENERATION

There are two sources of chlorite and chlorate related to the use of chlorine dioxide. The first is the process of chlorine dioxide generation itself. As shown in Eq. 13-37 in Chap. 13, chlorine dioxide can be generated by the controlled reaction between chlorine and sodium chlorite under very acidic conditions as follows:

$$2NaClO_2 + HOCl \rightarrow 2ClO_2(g) + 2Na^+ + Cl + OH^- \qquad (19\text{-}9)$$

When excess chlorine is present in the chlorine dioxide generator, however, chlorate can form according to the following simplified reaction:

$$NaClO_2 + HOCl \rightarrow ClO_3^- + Na^+ + H^+ + Cl^- \qquad (19\text{-}10)$$

In addition, residual chlorite may remain in the product solution if excessively high concentrations of sodium chlorite are used in the generator. The residual chlorite will then be injected into the process stream with the chlorine dioxide. New developments in chlorine dioxide generation technologies have improved the generator efficiency and greatly minimized the formation of chlorite and chlorate.

CHLORITE AND CHLORATE FORMATION AS BY-PRODUCTS

The second source of chlorite and chlorate in treatment plants using chlorine dioxide is formation as a by-product of the chlorine dioxide disinfection/oxidation reactions. Chlorine dioxide produces chlorite and chlorate as by-products in two ways. The first is through the oxidation of various water constituents such as reduced iron, manganese, or NOM. The reaction typically involves a one-electron transfer, resulting in the formation of chlorite as follows:

$$ClO_2 + e^- \rightarrow ClO_2^- \qquad (19\text{-}11)$$

Second, under high-temperature and/or high-pH conditions, chlorine dioxide disproportionates to form chlorite and chlorate. The oxidation state of chlorine in chlorine dioxide is (+4), which is between that in chlorite (+3) and that in chlorate (+5), which allows an electron transfer to produce both by-products. The overall reaction is

$$2ClO_2 + 2OH^- \rightarrow ClO_2^- + ClO_3^- + H_2O \qquad (19\text{-}12)$$

Researchers have estimated that 50 to 70 percent (by mass) of the chlorine dioxide applied during drinking water treatment is converted to chlorite (Werderhoff and Singer, 1987). With this formation rate, the formation of chlorite limits the chlorine dioxide dose that can be applied during drinking water treatment unless chlorite removal technologies are implemented downstream, since the MCL for chlorite is 1 mg/L.

Formation Control

There is currently no technology that can be implemented at a water treatment plant to reliably reduce the formation of chlorite as a decay product of chlorine dioxide (Eq. 19-11), except through reducing the chlorine dioxide demand. When the chlorine dioxide demand is reduced, a lower chlorine dioxide dose will have to be added, thus reducing the amount of chlorite formed. Reducing ClO_2 demand is especially effective when chlorine dioxide is used to oxidize reduced substances such as iron, manganese, and color present in the source water. Partial oxidation of iron and manganese with other oxidants (e.g., permanganate) is one approach. When chlorine dioxide is used for disinfection downstream of chemical precipitation, the chlorine dioxide demand can be reduced with NOM removal through coagulation.

Removal of Chlorine Dioxide By-products

The unpreventable production of chlorite is the primary obstacle to widespread use of chlorine dioxide in drinking water treatment. Because implementing mitigation measures to reduce chlorite formation is not feasible, a significant amount of work has been conducted to evaluate options for the destruction of chlorite after it is formed. While many options exist for chlorite destruction, the following discussion focuses on the most feasible options for a full-scale water treatment plant: (1) reduction with ferrous ion, (2) reduction with activated carbon, and (3) oxidation with ozone.

REDUCTION WITH FERROUS ION

Reduction with ferrous ion and activated carbon are based on the idea of reducing chlorite to chloride by the following half-reaction:

$$ClO_2^- + 4H^+ + 4e^- \rightarrow Cl^- + 2H_2O \qquad (19\text{-}13)$$

When ferrous (Fe^{2+}) ion is added to the water, it releases the needed electrons to form ferric (Fe^{3+}) ion with the following overall reaction:

$$ClO_2^- + 4Fe^{2+} + 4H^+ \rightarrow Cl^- + 4Fe^{3+} + 2H_2O \qquad (19\text{-}14)$$

The stoichiometric mass ratio of ferrous ion to chlorite in Eq. 19-14 is 3.3 : 1 mg Fe^{2+}/mg ClO_2^-, as demonstrated in Example 19-4. This stoichiometry was validated by Iatrou and Knocke (1992), who also showed that adding ferrous ions (as $FeSO_4 \bullet 7H_2O$) at a mass ratio of 3 : 1 (mg Fe^{2+}/mg ClO_2^-) resulted in virtually complete reduction of chlorite in less than 1 min of reaction time (pH between 5 and 7; temperature between 5 and 25°C). These researchers also verified that the ferrous ion was oxidized to ferric ion, which then enhanced downstream coagulation and flocculation by precipitating as $Fe(OH)_3(s)$. These findings were confirmed by Griese et al. (1992) who, along with Iatrou and Knocke (1992), showed that no chlorate was formed from the reaction between chlorite and ferrous ions. Subsequently, Hurst and Knocke (1997) verified this approach under alkaline conditions (pH between 8 and 10) and studied the effect of dissolved oxygen on the Fe^{2+} dose required. Based on these results, a mass ratio between 3.5 : 1 and 4 : 1 is more appropriate under high-O_2 conditions (>5 mg/L) to satisfy the added demand for Fe^{2+}.

REDUCTION WITH ACTIVATED CARBON

Reduction with activated carbon is based on the fact that the surface of activated carbon is a good reducing agent and, therefore, may be used to reduce chlorite to chloride. Chlorite is removed by GAC and the removal efficiency increases with increasing EBCT. Research has shown, however, that the removal efficiency decreases rapidly over time (Dixon and Lee, 1991; Karpel Vel Leitner et al., 1996).

OXIDATION WITH OZONE

When ozone is added to water containing chlorite, the chlorite is oxidized to chlorate as shown in the following overall reaction:

$$ClO_2^- + O_3 \rightarrow ClO_3^- + O_2 \qquad (19\text{-}15)$$

Water treatment plants that use chlorine dioxide as a raw-water preoxidant and ozonation for disinfection can rely on the ozonation process to oxidize the chlorite by-product to chlorate. When chlorine dioxide use is followed by ozone, the chlorine dioxide dose can be increased in response to changing raw-water quality conditions without violating the chlorite standard in the finished water. This strategy is of limited value, however, when chlorate is an issue.

Example 19-4 Calculating ferrous dosages for chlorite removal

Calculate the Fe^{2+} dose required to remove 1 mg/L of ClO_2^- using the stoichiometry of Eq. 19-14.

Solution

1. Calculate the molecular weight of ClO_2^-:

$$ClO_2^- + [35.5 + 2\,(16)] = 67.5 \text{ g/mol} = 67.5 \text{ mg/mmol}$$

2. Calculate the molar concentration of chlorite:

$$\left[ClO_2^-\right] = \frac{1 \text{ mg/L}}{67.5 \text{ mg/mmol}} = 0.0148 \text{ mmol/L}$$

3. Calculate the concentration of Fe^{2+}:
According to stoichiometric relationship given in Eq. 19-14, 4 mol Fe^{2+} are required to remove 1 mol chlorite.

$$\left[Fe^{2+}\right] = (0.0148 \text{ mmol/L})\,(4 \text{ mmol Fe/mmol})\,(56 \text{ mg/mmol})$$
$$= 3.3 \text{ mg/L}$$

19-5 Ozone By-products

Ozone is a very strong disinfectant, is commonly used in Europe, and is often evaluated as an alternative to free chlorine for primary disinfection when chlorinated DBPs are a problem. It is also one of the most effective oxidants for the destruction of chemicals that cause color, taste, and odor in drinking water. When applied to some raw waters, it has been shown to improve the downstream chemical coagulation, clarification, and granular filtration.

Ozone addition to water forms three types of by-products, as are listed in Table 19-1. The first type of by-product is inorganic bromate (BrO_3^-). Bromate, which forms when bromide ion (Br^-) is present during ozonation, is classified by the U.S. EPA as a "probable human carcinogen" with a 10^{-6} cancer risk level of 0.05 μg/L (U.S. EPA, 1998). Bromate is the most significant of the by-products formed by ozone.

The second type of by-product is low-molecular-weight (MW) organic molecules such as aldehydes, carboxylic acids, and ketoacids that form during reactions between ozone and NOM. It should be noted that by-products of the direct oxidation of NOM by ozone are not halogenated, that is, they do not contain chlorine or bromine. Ozone typically causes the partial oxidation of NOM to simpler organics and not the total oxidation of NOM to carbon dioxide and water (i.e., not complete mineralization of organic compounds to inorganic products). The low-MW organic by-products are not regulated by the U.S. EPA and are not believed to have significant adverse public health effects at the concentrations that occur in drinking water. Nevertheless, these by-products may still have negative impacts on municipal water systems because the by-products are more biodegradable than the high-molecular-weight "parent" NOM molecules. The increase in the biodegradability of the organic material present in the water is of concern because it can promote bacterial growth in the distribution system. The total concentration of the organic by-products is gauged by measuring the biodegradable dissolved organic carbon (BDOC) concentration or the assimilable organic carbon (AOC) concentration, which is believed to represent the more readily biodegradable fraction of the BDOC.

The third type of ozone by-product is brominated organics such as bromoform. As noted above, ozone does not react directly with NOM to produce halogenated by-products, but it does react with bromide. One product of the reaction with bromide is hypobromite ion (OBr^-). At neutral pH, hypobromite ion protonates to form hypobromous acid (HOBr). Hypobromous acid can, in turn, react with NOM to form brominated organics, as was discussed in Sec. 19-2. The concentration of bromide ion, NOM, and pH determine the quantity of brominated by-products formed. However, the concentrations of brominated by-products are at least one order of magnitude less than those formed by chlorination (Singer and Reckhow, 1999).

The following sections focus on the formation and control of these ozonation by-products.

Chemistry of Formation

The reaction between ozone and bromide forms bromate by three distinct pathways, as shown schematically on Fig. 19-8. Two of the pathways involve hydroxyl radicals (HO·) as well as molecular ozone (see Chaps. 13 and 18 for more discussion on hydroxyl radical formation during ozonation). The direct pathway, which involves only molecular ozone and is shown as pathway (a) on Fig. 19-8, involves three sequential ozonation reactions, forming hypobromite and bromite (BrO_2^-) before forming bromate, as given in the following reactions:

$$O_3 + Br^- \rightarrow OBr^- + O_2 \tag{19-16}$$

$$O_3 + OBr^- \rightarrow BrO_2^- + O_2 \tag{19-17}$$

$$O_3 + BrO_2^- \rightarrow BrO_3^- + O_2 \tag{19-18}$$

The kinetics of bromate formation with ozone are far more rapid (i.e., minutes) than those of THM formation with chlorine (i.e., hours to days). Equation 19-16 proceeds rapidly as a second-order reaction with a rate constant of 160 $M^{-1}\ s^{-1}$ at 25°C (see Table E-2 at the website listed in App. E).

$$r_{OBr} = k[Br^-][O_3] \tag{19-19}$$

As was shown in Eq. 19-3, hypobromite is the conjugate base of hypobromous acid. When the pH is below the pK_a value ($pK_a = 8.7$), the hypobromite is converted to hypobromous acid and the formation of bromite and bromate (as shown in Eqs. 19-17 and 19-18) will proceed more slowly, especially as the pH decreases.

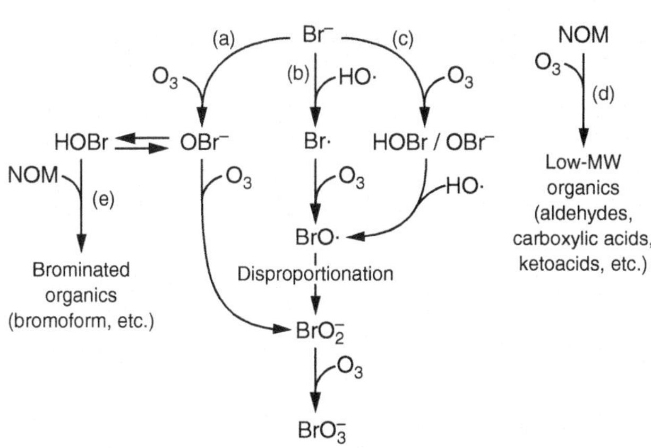

Figure 19-8
Pathways for ozone DBP formation: (a) molecular ozone pathway for bromate formation, (b) OH·/O_3 pathway for bromate formation, (c) O_3/OH· pathway for bromate formation, (d) low-MW organic formation, and (e) brominated organic formation. (Adapted from Song et al., 1997.)

Pathways (b) and (c) on Fig. 19-8 involve reactions with both ozone and hydroxyl radicals for bromate formation. Hydroxyl radicals can be produced by ozone reacting with NOM naturally present in water, or they can be generated in an advanced oxidation process by adding hydrogen peroxide or UV light with ozone. The hydroxyl radical pathways are complex and additional reaction steps not shown in Fig. 19-8 occur. The complete pathways as currently understood are available in the literature (Ozekin et al., 1998; von Gunten and Hoigné, 1994; von Gunten and Oliveras, 1998; Westerhoff et al., 1998).

In pathway (b) of Fig. 19-8, the reaction with hydroxyl radical occurs first to form bromide radical (Br·), followed by reaction with ozone to produce hypobromite radical (BrO·). In pathway (c), the reaction with ozone occurs first to form hypobromous acid or hypobromite, followed by the reaction with hydroxyl radical to produce the hypobromite radical. In both pathways, BrO· will disproportionate to form bromite via either of the following reactions:

$$BrO\cdot + H_2O \rightarrow BrO_2^- + 2H^+ \tag{19-20}$$

$$2BrO\cdot + H_2O \rightarrow OBr^- + BrO_2^- + 2H^+ \tag{19-21}$$

In natural waters without advanced oxidation (i.e., HO· only forms through reactions between NOM and O_3), the HO· concentration is relatively low after the first few seconds. Under these conditions, O_3 reacts with Br^- more rapidly than HO· does, so pathways (a) and (c) are the prime routes to bromate formation. In the next step, however, the pH is important. HO· reacts with both HOBr and OBr^-, but O_3 reacts only with OBr^-. At pH values of 7 to 8 (below the pK_a value of 8.7) the hypobromite is primarily in the form of HOBr, which slows the rate of pathway (a) and gives preference to pathway (c). Gillogly et al. (2001) reported that the HO· pathways contribute to bromate formation far more than the molecular ozone pathway. Subsequent studies found that the molecular ozone pathway contributed only 10 to 30 percent of the formation of bromate during ozonation (von Gunten and Hoigné, 1992; Yates and Stenstrom, 1993; Westerhoff et al., 1998).

Pathway (d) in Fig. 19-8 represents the degradation of NOM by ozone to form various low-MW organic by-products including aldehydes, carboxylic acids, and others as listed in Table 19-1. The low-MW organics are strictly the products of NOM oxidation and do not contain bromine.

During the formation of OBr^- in pathway (a), the OBr^- is also in equilibrium with its conjugate acid, HOBr, as shown in Eq. 19-3. The pK_a is 8.7 at 25°C, so HOBr will be the predominant form in most natural waters. The HOBr can react with NOM, as shown in pathway (e), to form brominated organic by-products such as bromoform. However, pathway (e) is usually not significant for two reasons. First, bromide levels in natural waters seldom exceed 0.5 mg/L. If 50 percent of the bromide (0.25 mg/L) is converted to OBr^- and subsequently to HOBr, only 0.3 mg/L HOBr will be formed (which is equivalent to about 0.22 mg/L as chlorine, a relatively low dose). Therefore, the amount of bromoform that would form is expected to be quite small; Trussell and Umphres (1978) showed that the addition of 10 mg/L ozone to a natural water containing 1 mg/L bromide formed less than 1 μg/L of bromoform.

The second reason for the low formation of brominated organics is, although decreasing water pH favors the conversion of OBr^- to HOBr, it hinders the base-catalyzed haloform reaction required to form bromoform. It has been shown that bromoform formation upon ozonation decreased from 37 μg/L at pH 6 to approximately 15 μg/L at pH 8.5 (bromide = 1 mg/L; DOC = 3.4 mg/L; ozone dose = 10.2 mg/L) (Siddiqui and Amy, 1993).

Estimating By-product Formation

As was described for THM and HAA formation with chlorine, empirical correlations have been developed to predict bromate formation based on measurable water quality parameters and operating conditions (Sadiq and Rodriguez, 2004; Song et al., 1997). The model by Song et al. (1997) considers bromate formation as a function of initial bromide concentration, ozone dose, pH, inorganic carbon concentration, DOC concentration, $NH_3 - N$ concentration, and time. The model is

$$[BrO_3^-] = \frac{10^{-6.11}[Br^-]_0^{0.88}[O_3]^{1.42}pH^{5.11}[IC]^{0.18}(t)^{0.27}}{[DOC]^{1.88}[NH_3 - N]^{0.18}} \qquad (19\text{-}22)$$

where $[BrO_3^-]$ = bromate conc. (for $[BrO_3^-] \geq 2$), μg/L
$[Br^-]$ = initial bromide conc. (for $100 \leq [Br^-] \leq 1000$), μg/L
$[O_3]$ = ozone dose (for $1.5 \leq [O_3] \leq 6$), mg/L
pH = pH (for $6.5 \leq pH \leq 8.5$)
$[IC]$ = inorganic carbon conc. (for $1 \leq [IC] \leq 216$), mg/L as $CaCO_3$
t = reaction time, (for $t \leq 30$), min
$[DOC]$ = DOC conc. (for $1.5 \leq [DOC] \leq 6$), mg/L
$[NH_3 - N]$ = $NH_3 - N$ conc. (for $0.005 \leq [NH_3 - N] \leq 0.7$), mg/L

The model can be applied similar to the models for THM and HAA formation shown in Example 19-3. As with other models for DBP formation based on multivariate regression, the results should be used with caution when predicting DBP formation for a specific site.

Formation Control

To minimize the formation of low-MW organic by-products such as aldehydes and carboxylic acids, the only viable options are to reduce the dose of ozone or remove NOM before adding ozone to the water. Reducing the ozone dose is usually not practical because the dose is determined by the oxidation or disinfection requirements. Options for the removal of NOM are the same as those discussed earlier in this chapter for minimizing the formation of chlorination by-products.

Similarly, bromate formation could be minimized by reducing the dose of ozone or removing bromide before adding ozone to the water. As noted above, reducing the ozone dose is usually not practical. Technologies exist to remove the bromide ion from water (ion exchange, reverse osmosis), but they are not viable options for minimizing DBP formation because of cost. Fortunately, several strategies can minimize bromate formation by chemically converting the bromide to a less reactive form or intercepting the reaction pathways to bromate formation. These measures include (1) pH depression, (2) ammonia addition, (3) conversion of bromide to bromamine by chlorine and ammonia addition, and (4) preoxidation with chlorine or chlorine dioxide.

pH DEPRESSION

Reducing the pH of water during ozonation is the most reliable and proven method for reducing bromate formation upon ozonation of bromide-containing waters. The rationale for this control strategy can be deduced from Fig. 19-8: The pK_a for the equilibrium reaction between HOBr and OBr^- is 8.7 at 25°C. Therefore, at lower pH values, a greater portion of the OBr^- is converted to HOBr, thus making it unavailable for the bromate formation reactions in the continuation of pathway (a), as shown on Fig. 19-9a. In addition, a reduction in pH reduces the ratio of HO· to O_3 in the water, decreasing the rate of all HO· reactions in water and thus reducing the rate of bromated formation by the other pathways as well. The published literature includes many examples of the effect of pH on bromate formation, one of which is shown on Fig. 19-10a (Krasner et al., 1993). At an ozone dose necessary to achieve 0.5-log inactivation of *Giardia* cysts in the test water (3.4 mg/L), the bromate level formed at pH 8.0 was 2.6 times the level formed at pH 7.0 (13 μg/L compared to 5 μg/L). With an ozone dose necessary for 2-log inactivation of *Giardia* cysts (6.0 mg/L), the bromate level formed at pH 8.0 was 3.6 times greater than the bromate level formed at pH 7.0 (58 μg/L compared to only 16 μg/L). In other studies it has been shown that further reduction in bromate formation can be achieved as the pH is decreased to 6.5 or lower, although acid addition for bromate minimization must be followed by caustic addition for corrosion control in the distribution system.

AMMONIA ADDITION

Because HOBr is a weak acid in equilibrium with OBr^-, a fraction of the oxidized bromide will exist as OBr^- even at low pH and still contribute to some bromate formation. To further minimize this fraction, ammonia can be added to serve as a "sink" for HOBr by transforming it to NH_2Br and creating a continuous driving force for the transformation of OBr^- to HOBr, as shown on Fig. 19-9b. The effect of ammonia addition on the formation of bromate in a natural water sample is shown on Fig. 19-10b (Glaze et al.,

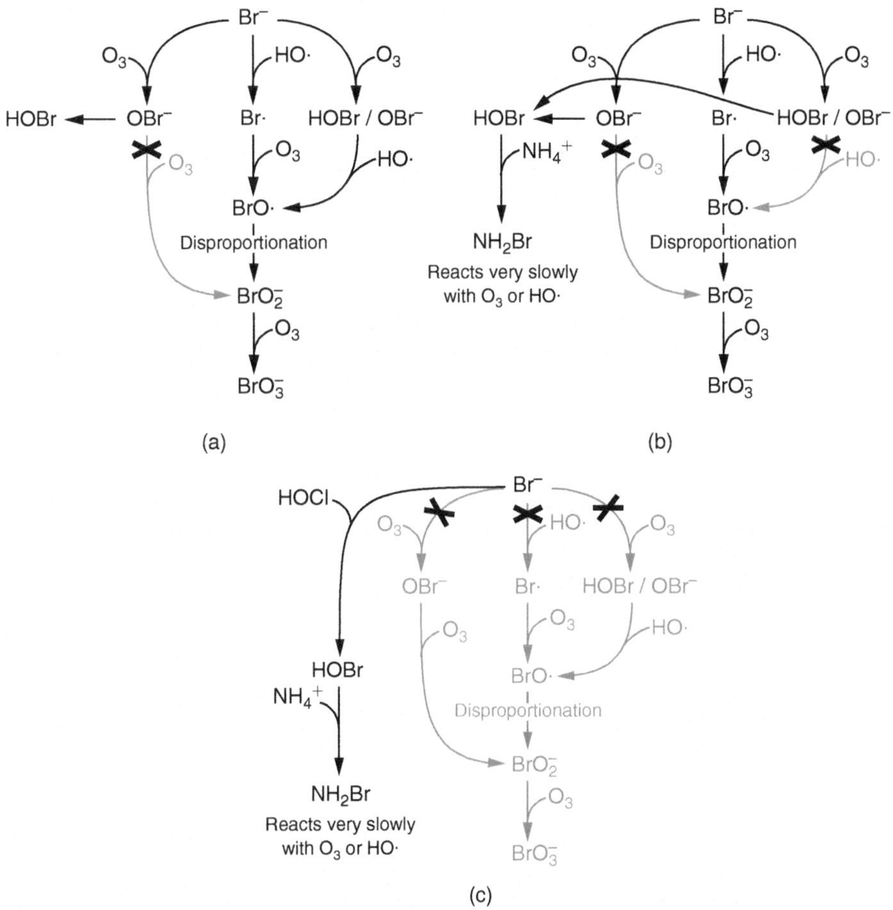

Figure 19-9
Effect of (a) pH adjustment, (b) ammonia addition, and (c) chlorine and ammonia addition on inhibition of bromate formation.

1993). Under ambient conditions (NH_3–N < 0.03 mg/L as N), adding 5 mg/L of ozone resulted in the formation of 26 μg/L bromate. Increasing the ammonia–nitrogen concentration to 0.7 mg/L (NH_3–N = 0.7 mg/L as N) decreased the bromate level formed to less than the detection limit of 5 μg/L, which is a greater than 500 percent reduction in bromate formation. Combining pH depression and ammonia addition may be an effective strategy for some waters.

Unfortunately, while pH depression has been demonstrated to reduce bromate formation, some studies do not show a significant impact of ammonia addition on bromate formation. The reason for the mixed ammonia results is not yet clear. For now, the impact of ammonia addition on bromate formation should be evaluated on a case-by-case basis.

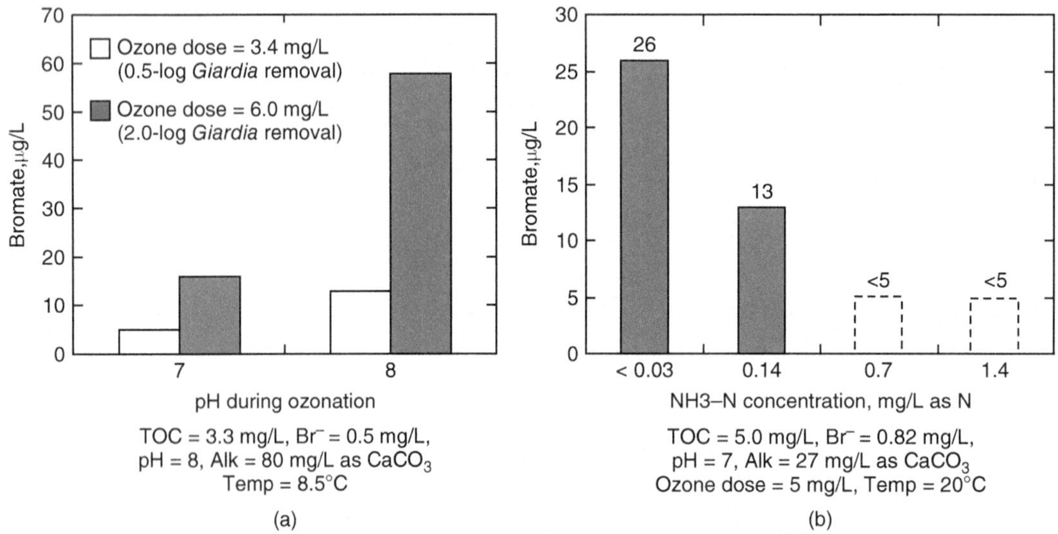

Figure 19-10
Effects of (a) pH (adapted from Krasner et al., 1993) and (b) ammonia (adapted from Glaze et al., 1993) on bromate formation during ozonation of natural water.

CONVERSION OF BROMIDE TO BROMAMINE

In 2004, the Southern Nevada Water Authority (SNWA) published the results of 3 years of studies it had been doing at the pilot scale (Neeman et al., 2004). After exploring preoxidation with chlorine and the addition of ammonia separately, the research moved to exploration of the addition of both chlorine and ammonia prior to ozonation. Pretreatment with 0.5 mg/L of chlorine followed by the addition of 0.1 mg/L of ammonia brought the bromate levels formed during ozonation from 20 to 25 μg/L to about 5 μg/L. SNWA patented the process and put the process in the public domain. Late that same year researchers at the Swiss Federal Institute of Aquatic Science and Technology (EAWAG) EAWAG in Switzerland published laboratory research designed to explain the results (Buffle et al., 2004). The bottom line is that the addition of the small amount of chlorine oxidizes the bromide to bromine; the subsequent addition of ammonia converts that bromine to bromamine, and, unlike the bromide ion, bromamine is not easily oxidized by ozone to bromate. As the EAWAG group describes it, the key reactions and their rate constants are

$$\mathrm{HOCl} + \mathrm{Br}^- \rightarrow \mathrm{HOBr} + \mathrm{Cl}^- \qquad k_1 = 1550\ \mathrm{M}^{-1}\ \mathrm{s}^{-1} \tag{19-23}$$

$$\mathrm{HOBr} + \mathrm{NH}_3 \rightarrow \mathrm{NH}_2\mathrm{Br} + \mathrm{H}_2\mathrm{O} \qquad k_2 = 8 \times 10^7\,\mathrm{M}^{-1}\,\mathrm{s}^{-1} \tag{19-24}$$

$$\mathrm{NH}_2\mathrm{Br} + 3\mathrm{O}_3 \rightarrow \mathrm{NO}_3^- + \mathrm{Br}^- + 3\mathrm{O}_2 + 2\mathrm{H}^+ \quad k_3 = 40\ \mathrm{M}^{-1}\ \mathrm{s}^{-1} \tag{19-25}$$

The oxidation of bromide is very rapid and the conversion to bromamine extremely fast. As a result, this method has the potential for preventing all mechanisms of bromate formation, as shown on Fig 19-9c. In the ozone contactor the oxidation of monobromamine is sluggish, masking the bromide from the ozone.

PREOXIDATION WITH CHLORINE OR CHLORINE DIOXIDE

SNWA studied preoxidation with chlorine to control bromate formation and found it unsatisfactory. A study at Contra Costa WD in Northern California explored the idea of preoxidation with chlorine dioxide with greater success (Zhou and Neeman, 2004). The requirements for success for this process remain somewhat elusive as substantial reductions have been achieved in some localities and not at others.

Removal of Ozonation By-products

As noted earlier, ozone produces three types of by-products. Bromonated organics are not produced in high enough concentration to warrant removal. Removal of the other two main types of ozonation by-products, bromate and low-MW organic by-products, are discussed below.

BROMATE

The chemistry of bromate (BrO_3^-) is quite similar to that of nitrate (NO_3^-): Both are monovalent anions, both are highly oxygenated, and both are at the top of their respective oxidation scales. Therefore, just like nitrate, bromate theoretically can be removed from water through the following water treatment technologies: (1) ion exchange, (2) reverse osmosis, (3) biological reduction, and (4) chemical reduction. However, these treatment technologies are not currently practical at full scale for bromate removal; the primary strategy for complying with the bromate MCL is to minimize its formation in the first place, as discussed previously. Nevertheless, the following discussion provides some insight into each option for process understanding and possible future use.

Most ion exchange (IX) resins have low selectivity for bromate that makes IX impractical for bromate removal from drinking water. Reverse osmosis (RO) can achieve greater than 90 percent removal of bromate. However, both RO and IX are more costly than any of the bromate formation control strategies discussed earlier in this chapter and produce a high-TDS residual stream that requires proper disposal. Because of these issues, RO and IX are generally impractical for bromate removal.

Bromate can be reduced biologically to bromide by denitrifying bacteria (Hijnen et al., 1995, 1999) when dissolved oxygen (DO) concentrations are low (<2.5 mg/L) and empty-bed contact times (EBCT) are high (>25 min) (Kirisits and Snoeyink, 1999; Kirisits et al., 2001). Similar to nitrification, bromate reduction is greatly inhibited by increased DO concentration. Unfortunately, this is a significant drawback to applying biological removal in a full-scale treatment plant because the DO concentration downstream of an ozonation process can be greater than 20 mg/L, especially when pure oxygen is used for ozone generation. The requirement for high EBCT is also a drawback as typical EBCT values in a water treatment plant filter are significantly lower than 25 min.

Chemical reduction of bromate to bromide can be achieved using reducing agents such as ferrous ions (Fe^{2+}) or the surface of activated carbon. Bromate can be reduced by Fe^{2+} under water treatment conditions according to the following reaction (Siddiqui et al., 1996):

$$BrO_3^- + 6Fe^{2+} + 6H^+ \rightarrow Br^- + 6Fe^{3+} + 3H_2O \qquad (19\text{-}26)$$

Bromate reduction through the reaction shown in Eq. 19-26 is typically 40 to 80 percent, depending upon the dose of Fe^{2+}. The unique aspect of using ferrous as a reducing agent is that it is oxidized to ferric (Fe^{3+}), which is used as a coagulant in water treatment. In plants that practice preozonation, a ferrous salt can be added to the rapid mix chamber downstream of the preozone contactor, which allows time for the reaction between ferrous ions and bromate to take place followed by the precipitation of the ferric coagulant formed. Even though the DO levels are high, because the rate of oxidation of Fe^{2+} with oxygen is quite slow, not all of the ferrous ions are oxidized. The residual iron passing through the treatment plant can cause aesthetic water quality problems, rendering this bromate removal option impractical unless subsequent iron removal strategies are employed (see Chap. 20).

LOW-MW ORGANIC BY-PRODUCTS

Ozonation of natural water forms various low-MW organic by-products such as aldehydes and carboxylic acids, all of which increase the concentration of the biodegradable organic matter (BOM) in the water. Due to the high biodegradation potential of these by-products, biologically active filtration downstream of ozonation has developed as the approach of choice for removing organic ozonation by-products. In typical dual-media filters (either anthracite–sand or GAC–sand), biological filtration will occur merely by allowing the water to pass through the filters without a disinfectant residual present. The effectiveness of BOM removal in a biological filter is affected by (1) BOM type and concentration, (2) filter media type (i.e., GAC, anthracite, and/or sand), (3) water temperature, and (4) EBCT through the filter. Additional details of biological filtration are presented in Sec. 11-8 of Chap. 11.

Problems and Discussion Topics

19-1 For each common disinfectant used in drinking water treatment, address the following issues: (a) list the common by-products of each disinfectant, (b) identify method(s) that may be used to reduce the production of these by-products, and (c) identify methods that may be used to remove the by-products. Which strategy is better for each by-product—removal of the by-product or preventing the by-product from forming in the first place?

19-2 What is the mass concentration of 1 μM concentrations of the following DBPs: chloroform, bromoform, dibromochloromethane, bromodichloromethane, monochloroacetic acid, and monobromoacetic acid?

19-3 What methods can be used to reduce the concentration of NOM before disinfectants are added? Discuss their effectiveness, advantages, and disadvantages.

19-4 How can the formation of by-products of chlorine including brominated by-products be reduced?

19-5 Describe a comprehensive approach that disinfects the water and reduces the production of disinfection by-products. Be sure to describe the jar testing and chemical analyses of the raw and finished water that would be required to develop this approach.

19-6 What is the impact of increasing bromide ion concentration on the TTHMs?

19-7 Estimate the THM and HAA formation potential for the following conditions: DOC = 6.4 mg/L, UV_{254} absorbance = 0.205 cm^{-1}, and bromide = 0.13 mg/L.

19-8 Estimate the THM and HAA formation potential for the following conditions: DOC = 2.1 mg/L, UV_{254} absorbance = 0.053 cm^{-1}, and bromide = 0.0 mg/L.

19-9 Estimate the bromate formation for the following conditions: bromide concentration = 0.35 mg/L, ozone dose = 1.7 mg/L, pH = 7.5, inorganic carbon concentration = 165 mg/L, DOC concentration = 3.2 mg/L, NH_3-N concentration = 0.005 mg/L, and time = 4 h.

19-10 Explain how ammonia addition can reduce the production of bromate when ozone is used to treat water.

References

Aiken, G. R., McKnight, D. M., Thorn, K. A., and Thurman, E. M. (1992) "Isolation of Hydrophilic Organic Acids from Water Using Nonionic Macroporous Resins," *Org. Geochem.*, **18**, 4, 567–573.

Bellar, T. A., and Lichtenberg, J. J. (1974) "Determining Volatile Organics at Microgram-per-Litre Levels by Gas Chromatography," *J. AWWA*, **66**, 12, 739–744.

Buffle, M., Galli, S., and von Gunten, U. (2004) "Enhanced Bromate Control During Ozonation: The Chlorine-Ammonia Process", *Environ. Sci. Technol.*, **38**, 5187–5195.

Bull, R. J. (1982) "Health Effects of Drinking Water Disinfectants and Disinfection By-Products," *Environ. Sci. Technol.*, **16**, 10, 554–559.

Bull, R., and Kopfler, F. (1991) *Health Effects of Disinfectants and Disinfection Byproducts*, American Water Works Association Research Foundation, Research Foundation, American Water Works Association, Denver, CO.

Bull, R. J., Krasner, S. K., Daniel, P., and Bull, R. D. (2001) *The Health Effects and Occurrence of Disinfection By-products*, American Water Works Association Research Foundation, American Water Works Association, Denver, CO.

Cal DHS (2002) *NDMA in California Drinking Water*, California Department of Health Services, Sacramento, CA.

Carlson, M., and Hardy, D. (1998) "Controlling DBPs with Monochloramine," *J. AWWA*, **90**, 2, 95–106.

Chen B. Y. and Westerhoff, P. (2010) "Predicting Disinfection By-product Formation Potential in Water," *Water Res.* **44**, 13, 3755–3762.

Choi, J., and Valentine, R. L. (2001) Formation of *N* -Nitrosodimethylamine (NDMA) from Reaction of Monochloramine: A New Disinfection By-product, paper presented at the American Water Works Association Annual Conference, Washington, DC.

Choi, J. and Valentine, R. (2002a) Formation of *N*-Nitrosodimethylamine (NDMA) from Reaction of Monochloramine: A New Disinfection By-product, "*Water Res.*, **36**, 817–824.

Choi, J. and Valentine, R. (2002b) A kinetic model of *N*-nitrosodimethylamine (NDMA) Formation During Water Chlorination/Chloramination, *Water Sci. Technol.* **46**, 65–71.

Choi, S. E. Duirk, R. L. and Valentine, R. (2002) "Mechanistic Studies of *N*-Nitrosodimethylamine (NDMA) Formation in Chlorinated Drinking Water," *J Environ Monit.* **4**, 249–252.

Christman, R. F., Johnson, J. D., Hass, J. R., Pfaender, F. K., Liao, W. T., Norwood, D. L., and Alexander, H. J. (1978) Natural and Model Aquatic Humics: Reaction with Chlorine, Chap. 2, in R. L. Jolley, H. Gorchev, and D. H. Hamilton, Jr. (eds.), *Water Chlorination: Environmental Impact and Health Effects*, Vol. 2, Ann Arbor Science, Ann Arbor, MI.

Clapper, T.W. (1979) Chloric Acid and Chlorates, in M. Grayson (ed.), *Kirk-Othmer Encyclopedia of Chemical Technology*, Vol. **5**, 3rd ed., John Wiley & Sons, New York.

Cowman, G. A., and Singer, P. C. (1996) "Effect of Bromide Ion on Haloacetic Acid Speciation Resulting from Chlorination and Chloramination of Aquatic Humic Substances," *Environ. Sci. Technol.*, **30**, 1, 16–24.

Crittenden, J. C., Berrigan, J. K., and Hand, D. W. (1986) "Design of Rapid Small-Scale Adsorption Tests for a Constant Diffusivity," *J WPCF*, **58**, 4, 312–319.

Crittenden, J. C., Hand, D. W., Arora, H., and Lykins Jr., B. W. (1987) "Design Considerations for GAC Treatment of Organic Chemicals," *J. AWWA*, **79**, 1, 74–82.

Croué, J.-P., Martin, B., Deguin, A., and Legube, B. (1993) Isolation and Characterization of Dissolved Hydrophobic and Hydrophilic Organic Substances of a Reservoir Water, paper presented at the Natural Organic Matter in Drinking Water: Origin, Characterization, and Removal Workshop, American Water Works Association Research Foundation and American Water Works Association, Chamonix, France.

Croué, J.-P., Violleau, D., and Labouyrie, L. (2000) DBP Formation Potentials of Hydrophobic and Hydrophilic NOM Fractions: A Comparison between a Low and a High-Humic Water, in S. E. Barrett, S. W. Krasner, and G. L Amy (eds.), *Natural Organic Matter and Disinfection By-products*, ACS Symposium Series 761, American Chemical Society, Washington, DC.

Dixon, K. L., and Lee, R. G. (1991) "The Effect of Sulfur-Based Reducing Agents and GAC Filtration on Chlorine Dioxide By-products," *J. AWWA*, **83**, 5, 48–55.

Eaton, A., and Briggs, M. (2000) NDMA—Analytical Methods Options for a New Dis-infection By-product, paper presented at the American Water Works Association, Water Quality Technology Conference, Salt Lake City, UT.

Farré, M.J, Döderer, K., Hearn, L., Poussade, Y., Keller, J., and Gernjak W. (2011) "Understanding the Operational Parameters Affecting NDMA Formation at Advanced Water Treatment Plants, *J. Hazard. Materials*, **185**, 2-3, 1575–1581.

Fu, P.L.-K., and Symons, J. M. (1990) "Removing Aquatic Organic Substances by Anion Exchange Resins," *J. AWWA*, **82**, 10, 70–77.

Fu, P., Ruiz, H., Thompson, K., and Spangenberg, C. (1994) "Selecting Membranes for Removing NOM and DBP Precursors," *J. AWWA*, **88**, 12, 55–72.

Fuson, R. C., and Bull, B. A. (1934) "The Haloform Reaction," *Chem. Rev.*, **15**, 3, 275–309.

Gillogly, T. E., Najm, I., Minear, R., Mariñas, B., Urban, M., Kim, J. H., Echigo, S., Amy, G., Douville, C., Daw, B., Andrews, R., Hofmann, R., and Croué, J.-P. (2001) *Bromate Formation and Control During Ozonation of Low-Bromide Waters*, American Water Works Association Research Foundation, Denver, CO.

Glaze, W. H., Weinberg, H. S., and Cavanagh, J. E. (1993) "Evaluating the Formation of Brominated DBPs During Ozonation," *J. AWWA* **85**, 1, 96–103.

Griese, M. H., Kaczur, J. J., and Gordon, G. (1992) "Combining Methods for the Reduction of Oxychlorine Residuals in Drinking Water," *J. AWWA*, **84**, 11, 69–77.

Haag, W. R., and Hoigné, J. (1983) "Ozonation of Bromide-Containing Waters: Kinetics of Formation of Hypobromous Acid and Bromate," *Environ. Sci. Technol.*, **17**, 5, 261.

Hayes, Jr., W. J., and Laws, Jr., E. R. (eds.) (1971) *Handbook of Pesticides Toxicology, Classes of Pesticides*, Vol. 2, Academic, San Diego.

Hijnen, W. A. M., Jong, R., and van der Kooij, D. (1999) "Bromate Removal in a Denitrifying Bioreactor Used in Water Treatment," *Water Res.*, **33**, 4, 1049–1053.

Hijnen, W. A. M., Voogt, R., Veenendaal, H. R., van der Jagt, H., and van der Kooij, D. (1995) "Bromate Reduction by Denitrifying Bacteria," *Appl. Environ. Microbiol.*, **61**, 2, 239–244.

Hooper, S. M., Summers, R. S., Solarik, G., and Hong S. (1996) GAC Performance for DBP Control: Effect of Influent Concentration, Seasonal Variation, and Pretreatment, pp. 21–40, in *Proceedings of AWWA Annual Conference*, Toronto, Ontario, Canada, American Water Works Association, Denver, CO.

Hurst, G. H., and Knocke, W. R. (1997) "Evaluating Ferrous Iron for Chlorite Ion Removal," *J. AWWA*, **89**, 8, 98–105.

Hwang, C. J., Sclimenti, M. J., and Krasner, S. W. (2000) DBP Formation Reactivities of NOM Fractions of a Low-Humic Water, pp. 173–187 in S. E. Barrett, S. W.

Kras- ner, and G. L. Amy (eds.), *Natural Organic Matterand Disinfection By-products*, ACS Symposium Series 761, American Chemical Society, Washington, DC.

Iatrou, A.,and Knocke, W. R. (1992) "Removing Chlorite by the Addition of Ferrous Iron," *J. AWWA*, **84**, 11, 63–68.

Jacangelo, J., and Trussell, R. (2002) "International Report: Waterand Waste water Disinfection—Trends Issues and Practices," *Water Sci. Technol.: Water Supply*, **2**, 3, 147–157.

Karpel Vel Leitner, N., De Laat, J., Doré, M., and Suty, H. (1996) The Use of ClO_2 in Drinking Water Treatment: Formation and Control of Inorganic By-products (ClO_2^-, ClO_3^-), in R. A. Minear and G. L. Amy (eds.), *Disinfection By-products in Water Treatment: The Chemistry of Their Formation and Control*, CRC Press, Lewis Publishers, New York.

Kavanaugh, M. (1978) "Modified Coagulation for Improved Removal of Tri- halomethane Precursors," *J. AWWA*, **70**, 11, 613–620.

Kim, P. H.-S.,and Symons, J. M. (1991) "Using Anion Exchange Resins to Remove THM Precursors," *J. AWWA*, **83**, 12, 61–68.

Kirisits, M. J., and Snoeyink, V. L. (1999) "Reduction of Bromate in a BAC Filter," *J. AWWA*, **91**, 8, 74–84.

Kirisits, M. J., Brown, J. C., Snoeyink, V. L., Raskin, L. M., Chee-Sanford, J. C., Liang, S., and Min, J. (2001) *Removal of Bromate and Perchlorate in Conventional Ozone/GAC Systems*, Final Report, American Water Works Association Research Foundation, Denver, CO.

Krasner, S. W. (1999) Chemistry of Disinfection By-product Formation, in P. C. Singer (ed.), *Formation and Control of Disinfection By-products in Drinking Water*, American Water Works Association, Denver, CO.

Krasner, S. W., Glaze, W. H., Weinberg, H. S., Daniel, P. A., and Najm, I. N. (1993) "Formation and Control of Bromate During Ozonation of Waters Containing Bromide," *J. AWWA*, **85**, 1, 73–81.

Krasner, S. W., McGuire, M. J., Jacangelo, J. G., Patania, N. I., Reagan, K. M., and Aieta, E. M. (1989) "The Occurrence of Disinfection By-products in U.S. Drinking Water," *J. AWWA*, **81**, 8, 41–53.

Krasner, S. W., Pastor, S., Chinn, R., Sclimenti, M. J., Wienberg, H. S., Richardson, S. D., and Thruston, Jr.,A. D. (2001) The Occurrence of a New Generation of DBPs (Beyond the ICR), paper presented at the AWWA Water Quality Technology Conference, Nashville, TN.

Krasner, S. W., Sclimenti, M. J., and Means, E. G. (1994) "Quality Degradation: Implications for DBP Formation," *J. AWWA*, **86**, 6, 34.

Kurokawa, Y., Maekawa, A., Takahashi, M., and Hayashi, Y. (1990) "Toxicity and Carcinogenecity of Potassium Bromate," *Environ. Health Perspectives*, **87**, 309.

Leenheer, J. A., and Croué, J.-P. (2003) "Characterizing Dissolved Aquatic Organic Matter," *Environ. Sci. Technol.*, **37**, 1, 18-26a.

Loveland, J. P. (2001) Emerging Water Quality Issues in Distribution Systems, paper presented at the AWWA Annual Conference, Washington, DC.

McGuire, M. J. (1999) Technical Working Group Presentation to Federal Advisory Committee (FACA) on M/DBP Regulations, Washington, DC.

Mitch, W., and Sedlak, D. (2002) "Formation of *N*-Nitrosodimethylamine (NDMA) from Dimethylamine During Chlorination," *Environ. Sci. Technol.*, **36**, 4, 588–959.

MWH (2010) Phase II Bench-Scale Evaluation of Post-membrane Treatment Alternatives Final Report submitted to City of West Palm Beach, March, 2010.

Najm, I., and Trussell, R. (2000) NDMA Formation in Water and Wastewater, paper presented at the American Water Works Association, Water Quality Technology Conference, Salt Lake City, UT.

Najm, I. N., and Trussell, R. R. (2001) "NDMA Formation in Water and Waste water," *J. AWWA*, **93**, 2, 92–99.

NAS (1980) The Chemistry of Disinfectants in Water: Reactions and Products, Chap. III, in *Drinking Water and Health*, Vol. 2, National Academy of Sciences, Washington, DC.

NCI(National Cancer Institute) (1976) *Report on Carcinogenesis Bioassay of Chloroform*, National Cancer Institute, NTIS PB 264018, Washington DC.

Neeman, J., Hulsey, R., Rexing, D., and Wert, E. (2004) "Controlling Bromate Formation with Chlorine and Ammonia," *J. AWWA*, **96**, 2, 26–28.

Neiss, L., Baird, R., Carr, S., Gute, J., Strand K., and Young C. (2003) Can *N*- Nitrosodimethylamine Formation Be Affected by Polymer Use During Advance Wastewater Treatment? paper presented at the Wateruse Association 18th Annual Symposium, Session A-2, Operational Issues and Disinfection, San Antonio, TX.

Noack, M. G., and Doerr, R. L. (1978) Reactions of Chlorine, Chlorine Dioxide, and Mixtures Thereof with Humic Acid: An Interim Report, in R. L. Jolley, H. Gorchev, and D. H. Hamilton, Jr. (eds.), *Water Chlorination: Environmental Impact and Health Effects*, Vol. 2, Ann Arbor Science, Ann Arbor, MI.

Norwood, D. L., Christman, R. F., and Hatcher, P. G. (1987) "Structural Characterization of Aquatic Humic Material. 2. Phenolic Content and Its Relationship to Chlorination Mechanism in an Isolated Aquatic Fulvic Acid," *Environ. Sci. Technol.*, **21**, 8, 791–798.

Norwood, D. L., Johnson, J. D., Christman, R. F., Hass, J. R., and Bobenrieth, M. J. (1980) "Reactions of Chlorine with Selected Aromatic Models of Aquatic Humic Material," *Environ. Sci. Technol.*, **14**, 2, 187–189.

Obolensky A., and Singer P. C., (2008) "Development and Interpretation of Disinfection Byproduct formation Models using the Information Collection Rule Database," *Environ. Sci. Technol.*, **42**, 15, 5654–5660.

Ozekin, K., Westerhoff, P., Amy, G. L., and Siddiqui, M. (1998) "Molecular Ozone and Radical Pathways of Bromate Formation during Ozonation," *J. Environ. Eng.*, **124**, 5, 456–462.

Reckhow, D. A., and Singer, P. C. (1985) Mechanisms of Organic Halide Formation During Fulvic Acid Chlorination and Implications with Respect to Preozonation, pp. 1229-1257, in R. L. Jolley, R. J. Bull, W. P. David, S. Katz, M. H. Roberts, Jr., and V. A. Jacobs (eds.), *Water Chlorination: Chemistry, Environmental Impact and Health Effects*, Vol. 5, Lewis, Chelsea, MI.

Reckhow, D. A., Singer, P. C., and Malcolm, R. L. (1990) "Chlorination of Humic Materials: By-product Formation and Chemical Interpretations," *Environ. Sci. Technol.*, **24**, 11, 1655–1664.

Rook, J. J. (1971) "Headspace Analysisin Water"[translated], *H_2O*, **4**, 17, 385–387.

Rook, J. J. (1974) "Formation of Haloforms During the Chlorination of Natural Water," *Water Treat. Examin.* **23**, 2, 234–243.

Rook, J. J. (1977) "Chlorination Reactions of Fulvic Acids in Natural Waters," *Environ. Sci. Technol.*, **11**, 5, 478–482.

Sadiq, R. and Rodriguez, M. J. (2004) "Disinfection By-products (DBPs) in Drinking Water and Predictive Models for Their Occurrence: A Review," *Sci. Tot. Environ.*, **321**, 1-3, 21–46.

Schreiber, M., and Mitch, W. (2005) "Influence of the Order of Reagent Addition on NDMA Formation During Chloramination, *Environ. Sci. Technol.* **39**, 3811–3818.

Siddiqui, M., and Amy, G. (1993) "Factors Affecting DBP Formation During Ozone-Bromide Reactions," *J. AWWA*, **85**, 1, 63–72.

Siddiqui, M., Zhai, W., Amy, G., and Mysore, C. (1996) "Bromate Ion Removal by Activated Carbon," *Water Res.*, **30**, 7, 1651–1660.

Singer, P. C., and Bilyk, K. (2002) "Enhanced Coagulation Using a Magnetic Ion Exchange Resin," *Water Res.*, **36**, 16, 4009–4022.

Singer, P. C., and Chang, S. D. (1989) "Correlations between Trihalomethanes and Total Organic Halides Formed During Water Treatment," *J. AWWA*, **81**, 8, 61–65.

Singer, P. C., and Reckhow, D. A. (1999) Chemical Oxidation, Chap. 12, in R. D. Letterman (ed.), *Water Quality and Treatment*, 5th ed., American Water Works Association, McGraw-Hill, New York.

Song, R., Westerhoff, P., Minear, R., and Amy, G. (1997) "Bromate Minimization During Ozonation," *J. AWWA*, **89**, 6, 69–78.

Sontheimer, H., Crittenden, J. C., and Summers, R. S. (1988) *Activated Carbon for Water Treatment*, DVGW-Forschungsstelle, Karlsruhe, Germany, distributed in the U.S. by the American Water Works Association, Denver, CO.

Speth, T. F., and Miltner, R. J. (1998) "Technical Note: Adsorption Capacity of GAC for Synthetic Organics," *J. AWWA*, **90**, 4, 171–174.

Thibaud, H., De Laat, J., and Dore M. (1987) Chlorination of Surface Waters: Effect of Bromide Concentration on the Chloropicrin Formation Potential, paper presented at the Sixth Conference on Water Chlorination: Environmental Impact and Health Effects, Oak Ridge, TN.

Trussell, R. R., and Umphres, M. D. (1978) "The Formation of Trihalomethanes," *J. AWWA*, **70**, 11, 604–612.

Umphres, M., Tate, C., Kavanaugh, M., and Trussell, R. (1983) "Trihalomethane Removal by Packed-Tower Aeration," *J. AWWA*, **75**, 8, 414–418.

U.S. EPA (1998) "National Primary Drinking Water Regulations: Disinfectants and Disinfection By-products; Final Rule," *Fed. Reg.*, **63**, 241, 69390.

U.S. EPA (2006) "National Primary Drinking Water Regulations: Stage 2 Disinfectant and Disinfection By-product Rule, Final Rule," *Fed. Reg.*, **71** 2, 388–493.

U.S. EPA (2010a) "Six-Year Review 2 of Drinking Water Standards," accessed on Nov 27, 2010 at: <http://water.epa.gov/lawsregs/rulesregs/regulatingcontaminants/sixyearreview/second_review/index.cfm>.

U.S. EPA (2010b) Unregulated Contaminant Monitoring Rule 2 (UCMR 2) Database, accessed Nov 17, 2010 at:<http://water.epa.gov/lawsregs/rulesregs/sdwa/ucmr/ucmr2/index.cfm>.

von Gunten, U., and Hoigné, J. (1992) "Factors Affecting the Formation of Bromate During Ozonation of Bromide-Containing Waters," *J. Water SRT-Aqua*, **41**, 5, 299–304.

von Gunten, U., and Hoigné, J. (1994) "Bromate Formation During Ozonation of Bromide-Containing Waters: Interaction of Ozone and Hydroxyl Radical Reactions," *Environ. Sci. Technol.*, **28**, 7, 1234–1242.

von Gunten, U., and Oliveras, Y. (1998) "Advanced Oxidation of Bromide- Containing Waters: Bromate Formation Mechanisms," *Environ. Sci. Technol.*, **32**, 1, 63–70.

Werderhoff, K. S., and Singer, P. C. (1987) "Chlorine Dioxide Effects on THMFP, TOXFP, and the Formation of Inorganic By-Products," *J. AWWA*, **79**, 9, 107–113.

Westerhoff, P., Song, R., Amy, G., and Minear, R. (1998) "Numerical Kinetic Models for Bromide Oxidation to Bromine and Bromate," *Water Res.*, **32**, 5, 1687–1699.

Wilczak, A., Assadi-Rad, A., Lai, H., Hoover, L., Smith, J., Berger, R., Rodigari, F., Beland, J., Lazzelle, L., Kincannon, E., Baker, H., and Heaney, T. (2003) "Formation of NDMA in Chloraminated Water Coagulated with DADMAC Cationic Polymer," *J. AWWA*, **95**, 9, 94–106.

Wood, P. R., and DeMarco, J. (1980) Removing Total Organic Carbon and Tri- halomethane Precursor Substances, in M. J. McGuire and I. H. Suffet (eds.), *Activated Carbon Adsorption of Organics from the Aqueous Phase*, Vol. **2**, Ann Arbor Science, Ann Arbor, MI.

Yagil, G., and Anbar, M. (1962) Kinetics of Hydrazine Formation From Chloramine and Ammonia,. *J. Am. Chem. Soc.* **84**, 1797.

Yates, R. S., and Stenstrom, M. K. (1993) Bromate Production in Ozone Contactors, paper presented at the America Water Works Association Annual Conference, San Antonio, TX.

Zhou, P., and Neeman, J. (2004) *Use of Chlorine Dioxide and Ozone for Control of Disinfection By-Products*, American Water Works Association Research Foundation, Denver, CO.

20 Removal of Selected Constituents

Stantec's Water Treatment: Principles and Design, Updated Third Edition. John C. Crittenden, R. Rhodes Trussell, David W. Hand, Kerry J. Howe, and George Tchobanoglous.

Terminology for Removal of Selected Constituents

Term	Definition
Carbonate hardness	Concentration of polyvalent ions that are associated with anions that comprise alkalinity (e.g. HCO_3^-, CO_3^{2-}).
Denitrification	Reduction or conversion of nitrate to nitrogen gas.
Enhanced softening	Process of removing hardness and TOC from water using the lime-softening process.
Greensand	Sand that contains a greenish colored mineral called glauconite (iron potassium phyllosilicate) and is typically coated with manganese oxide, known as manganese greensand, and is used to remove insoluble ferric iron and manganese.
Hardness	Sum of the soluble concentrations of polyvalent ions (e.g., Ca^{2+}, Mg^{2+}).
Natural radio-nucleotides	Molecules (e.g., radium-226, radon-222, uranium-238) that dissolve in groundwater as a result of radioactive gases and rock formations, and have unstable nuclei and emit energy in the form of ionizing radiation.
Noncarbonate hardness	Concentration of polyvalent ions in water that are associated with nonalkalinity anions (e.g., SO_4^{2-}, Cl^-).
Recarbonation	Process of adding carbon dioxide to water lower the pH after softening.
Single-stage softening	Process of adding lime to water to remove only the calcium hardness from the water, leaving the magnesium hardness.
Softening	Process of removing hardness from water.
Split-flow lime treatment	Process of separating the water stream into two or more streams, softening the streams to various degrees and then blending them to obtain the desired water quality.
Two-stage excess lime–soda treatment	Process of adding an excess of lime to water to bring about the removal of both calcium and magnesium followed by stabilization of the water with soda ash treatment.

Water purveyors are continuously striving to provide their communities with potable water that is safe for human consumption and aesthetically acceptable. The primary focus of the previous chapters of this textbook has been the removal of the traditional constituents found in most natural waters. There are, however, a number of nontraditional constituents that must also be removed from some natural waters to provide potable water. The purpose of this chapter is to introduce and discuss the removal of the most common of these nontraditional and emerging constituents, including arsenic, calcium, magnesium, nitrate, radionuclides, pharmaceutical and personal care products. However, before discussing the individual constituents, it will be helpful to review what constitutes traditional, nontraditional, and emerging constituents.

20-1 Traditional, Nontraditional, and Emerging Constituents

As discussed in Chap. 2, there are many constituents in water that impact potable water. The more common, also called traditional, constituents in most surface water supplies include turbidity, NOM, biological agents, taste and odor compounds, and low levels of synthetic organic compounds (SOCs). Low, but measurable, levels of SOCs are usually not specifically treated in present water treatment practices if their concentrations do not exceed their maximum contaminant level (MCL). In addition to the conventional constituents identified above, a variety of other constituents may be present in both surface waters and groundwaters. These other constituents are considered nontraditional in the sense that they are not encountered in most natural waters. For example, surface waters may contain inorganic constituents such as iron, manganese, hardness (Ca^{2+}, Mg^{2+}), and other trace dissolved metals (e.g., Ba, Se, Ra). Groundwater sources may also contain iron, manganese, hardness, and other dissolved metals such as those described above. These nontraditional constituents in one way or another also impact the provision of potable water.

In addition to the traditional and nontraditional constituents in water, water purveyors need to be aware of emerging constituents. For example, perchlorate, *N*-nitrosodimethylamine (NDMA), methyl tertiary butyl ether (MTBE), arsenic, radium, and disinfection by-products are now cause for health concern and some are being regulated. Many of these emerging constituents can be difficult and expensive to remove using conventional water treatment practices. New innovative water treatment practices are also being developed to meet regulatory demands for these emerging constituents. For example, new-iron-oxide-based adsorbents were developed to treat water sources containing arsenic, UV technology is being used to treat NDMA in some water supplies, and modifications to conventional water treatment practices are being employed to reduce disinfection by-products (see Chaps. 13 and 19).

The list of CECs is continually changing. Follow this link to see some recent recent chemicals of concern.

http://www.stantec.com/wt3r/20-1.html

Several constituents in water that impact potable water and treatment processes that may be used to remove them are reported in Table 20-1. The performance of these treatment processes for a given constituent is presented subjectively in terms of removal efficiencies reported in various studies. The information presented in Table 20-1 can be used as a general guideline when a particular constituent must be removed. However,

Table 20-1
General effectiveness of water treatment processes for selected nontraditional constituents[a]

		Water Treatment Processes												
			Coagulation Sedimentation,		Ion Exchange, Ch. 16		Membrane Processes		Chemical	Adsorption, Ch . 15				
Selected Contaminant	MCL, mg/L	Aeration, Stripping, Ch. 14	Granular Media Filtration, Ch. 9, 10, 11	Lime Softening Ch. 20	Anion	Cation	Reverse Osmosis Ch. 17	Membrane filtration Ch. 12	Oxidation Disinfection Ch. 8, 13, 18	GAC	PAC	Granular Ferric Hvdroxide	Activated Alumina	References
Inorganic Constituents														
Arsenic (+3)	0.010	P	G-E	F-E	G-E	P	E	P	P	F-G	P-F	E	F-E	1-22, 33, 36
Arsenic (+5)	0.010	P	G-E	F-E	G-E	P	E	F	P	F-G	P-F	E	F-E	1-22, 33, 36
Barium	2.0	P	P-F	G-E	P	E	E	NA	P	P	P	NA	P	31, 33, 35, 36
Chromium (+3)	0.10	P	G-E	G-E	P	E	E	NA	P	F-G	F	NA	P	31, 33, 35, 36
Chromium (+6)	0.10	P	P	P	E	P	E	NA	F	F-G	F	NA	P	32, 33, 35, 36, 39
Copper	1.3[b]	P	G	G-E	P	F-G	E	NA	P-F	F-G	P	NA	NA	31, 33, 35, 36
Fluoride	4.0	P	F-G	P-F	P-F	P	E	NA	P	G-E	P	NA	E	23, 24, 35, 36, 41
Hardness	NA	P	P	E	P	E	E	NA	P	P	P	NA	P	35, 36
Iron	0.30[c]	F-G	F-E	E	P	G-E	G-E	NA	G-E	P	P	NA	F-G	25, 26, 38
Lead	0.0[b]	P	E	E	P	F-G	F-G	NA	P	F-G	P-F	NA	P	31, 32, 33, 35, 36
Manganese	0.05[c]	P-G	F-E	E	P	G-E	G-E	NA	G-E	F-E	P	NA	P	26, 33, 36
Mercury (inorganic)	0.002	P	F-G	F-G	P	F-G	F-G	NA	P	F-G	F	NA	P	29, 36, 39
Nitrate	10.0	P	P	P	G-E	P	G	NA	P	P	P	NA	P	35, 36
Perchlorate	0.018[b,d]	P	NA	NA	G-E	P	G-E	NA	NA	F-G	NA	NA	NA	35, 36
Radium	5.0[e]	P	P-F	G-E	P	E	E	NA	P	P-F	P	NA	P-F	30, 35, 36, 41, 43
Uranium	0.030	P	G-E	G-E	E	G-E	E	NA	P	F	P-F	NA	G-E	42

Organic Constituents														
VOCs	NA	G-E	P	P-F	P	P	F-E	F-E	P-G	F-E	P-G	NA	P	36, 40
SOCs	NA	P-F	P-G	P-F	P	P	F-E	F-E	P-G	F-E	P-E	NA	P-F	36, 40, 50
Color	15[f]	P	F-G	F-G	P-G	NA	NA	NA	F-E	E	G-E	NA	P	35, 36
TTHMs	0.080	G-E	P	P	P	P	F-G	F-G	P-G	F-E	P-F	NA	P	27, 28, 37
MTBE	0.0208	G-E	P	P	P	P	F-E	F-E	P-G	F-E	P-E	NA	NA	44, 45
NDMA	0.02[h]	P	NA	NA	NA	NA	NA	NA	E[i]	NA	NA	NA	NA	46-48

[a]Abbreviations: P—poor (0-20% removal); F—fair (20-60% removal); G—good (60-90% removal); E—excellent (90-100% removal); NA, not applicable/insufficient data.

References: 1. Aus Planer-Friedrich, 2001; 2. Benjamin et al., 2000; 3. Brandhuber and Amy, 1998; 4. Chang et al., 1994; 5. Cheng et al., 1993; 6. Clifford, 1999; 7. Clifford and Lin, 1986; 8. Clifford and Ghurye, 1998a; 9. Driehaus et al., 1998; 10. Edwards, 1994; 11. Ferguson and Anderson, 1974; 12. Ferguson and Gavis, 1972; 13. Frey et al., 2000; 14. Gupta and Chen, 1978; 15. Hering and Elimelech, 1996; 16. Pontius et al., 1994; 17. Rubel and Williams, 1980; 18. Rubel and Hathaway, 1985; 19. Scott et al., 1995; 20. Simms and Azizian, 1977; 21. Smith et al., 1992; 22. Thompson and Chowdhury, 1993; 23. Sollo et al., 1984; 24. Rubel and Woosley, 1979; 25. Singer and Stumm, 1970; 26. Sly et al., 1990; 27. Blanck, 1979; 28. Weil, 1975; 29. Logsdon and Symons, 1979; 30. Brinck, 1976; 31. McRae and Parsi, 1974; 32. Mixon, 1973; 33. Logsdon et al., 1974; 34. Argo, 1984; 35. Sorg 1978a,b; 36. Sorg and Love, 1984; 37. Symons and Carswell, 1981; 38. Wood and DeMarco, 1980; 39. Zemansky, 1974; 40. Singley, 1979; 41. Sigworth and Smith, 1972; 42. Sorg, 1988; 43. Valentine et al., 1990; 44. Brown et al., 1997; 45. U.S. EPA, 1999; 46. Fleming et al., 1996; 47. Calgon, 1998; 48. Bolton et al., 1999; 49. Taylor and Jacobs, 1996; 50. Duranceau et al., 1992.

[b]Maximum contaminant limit goal.

[c]Secondary standard.

[d]Proposed standard.

[e]pCi/L.

[f]CFU (colony-forming units).

[g]Odor threshold.

[h]California Department of Health Services.

[i]Ultraviolet light.

constituents that comprise background water matrices can vary widely with respect to their number and concentration and can impact treatment process performance. Treatment processes that can be used in combination are not considered in Table 20-1.

20-2 Arsenic

Arsenic is widespread throughout our environment. For example, the crust of the Earth contains about 1.8 mg/kg of arsenic. Arsenic is typically composed of the minerals arsenopyrite (FeAsS), orpiment (As_2S_3), and realgar (AsS). The lithosphere varies from less than 0.2 to about 15 mg As/kg of soil. In addition, the atmosphere contains about 0.02 to 2.8 ng/m^3 while the aquatic environment contains less than 0.010 μg/L (Aus Planer-Friedrich, 2001). Most commercial arsenic is obtained by heating arsenopyrite. Because arsenic is tasteless and colorless, it was well known as a poison for humans in the Middle Ages and was used as far back as AD 55 in the poisoning by Nero of Britannicus. Arsenic has also been used for medicinal purposes. In the early 1800s, arsenic was used as a remedy for curing anorexia, rheumatism, asthma, tuberculosis, and diabetes and the treatment of malaria until the discovery of penicillin. Commercial uses of arsenic today include wood preservatives (accounts for 90 percent usage), paints, dyes, metals, drugs, soaps, rat poison, and semiconductors.

In the past few decades, the long-term exposure to arsenic has become a major health concern throughout the world. More than 100 million people worldwide are ingesting drinking water from wells that contain arsenic. Among other potential health risks, recent research shows a strong dose-dependent relationship between arsenic exposure and accelerated development of atherosclerosis in the arteries leading to the brain. Studies have suggested that arsenic may cause liver, kidney, and bladder cancer (Smith et al., 1992). Pontius et al. (1994) summarized the health implications of arsenic on humans. More recently, the water industry has focused some of its efforts on arsenic occurrence in water supplies and methodologies for its removal.

The concentrations of arsenic in water from various places throughout the world are summarized in Table 20-2. Perhaps the most well known area of the world noted for arsenic problems is in Bangladesh. It is estimated that 28 to 35 million people are exposed to arsenic in drinking water wells with arsenic concentrations exceeding 0.05 mg/L, and between 200,000 to 270,000 arsenic-related deaths from cancer are expected in this country alone (Smith et al., 1992). The U.S. EPA has estimated that some 13 million people in the United States, primarily in the western states, are exposed to arsenic in drinking water at levels greater than 0.01 mg/L.

The World Health Organization (WHO) proposed limit on arsenic in drinking water is 0.01 mg/L, which is largely based on analytical capability.

Table 20-2
Summary of waters with elevated arsenic concentrations throughout world

	Arsenic Concentration, μg/L	
Water Body and Location	**Range**	**Typical**
River water		
Baseline, various	0.13–2.1	0.83
Norway	<0.02–1.1	0.25
Southeast United States	0.15–0.45	0.30
Madison and Missouri Rivers, United States	10–370	
(geothermal influenced)	10–370	
United States		2.1
Dordogne, France		0.7
Po River, Italy		1.3
Polluted European rivers	4.5–45	
High-As groundwater influenced		
Northern Chile	190–21800	
Northern Chile	400–450	
Ron Phibun, Thailand (mining influenced)	4.8–583	218
Ashanti, Ghana	<2.0–7900	284
British Columbia, Canada	<0.2–556	17.5
Lake water		
Baseline	<0.2–0.42	0.28
British Columbia		
France	0.73–9.2	
Japan	0.38–1.9	
Sweden	0.06–1.2	
Western United States (geothermal influenced)	0.38–1000	
Estuarine water		
Oslofjord, Norway	0.7–2.0	
Saanich Inlet,	1.2–2.5	
British Columbia		
Rhone, France	1.1–3.8	2.2
Krka Estuary, Yugloslavia	0.13–1.8	
Seawater		
Deep Pacific and Atlantic Oceans	1.0–1.8	
Coastal Malaysia	0.7–1.8	1.0
Coastal Spain	0.5–3.7	1.5
Coastal Australia	1.1–1.6	1.3
Groundwater		
Baseline United Kingdom	<0.5–10	
As-rich provinces: Bengal Basin, Argentina, Mexico, northern China, Taiwan, Hungary	10–5000	

Source: Adapted from WHO (2001) Draft Report.

Developed countries such as the United States, European community, Japan, and Canada have adopted the 0.01-mg/L limit. The implications of these new regulations in terms of health and cost to society are discussed in several references (U.S. EPA, 2000, 2001; WHO, 2001; Health Canada, 2006). Presented below is a short summary of the forms of arsenic found in water followed by present treatment technologies for removing arsenic from water supplies.

Chemical Properties

A brief review of the chemical properties of arsenic in water is useful in understanding the treatment processes available for arsenic removal. Arsenic can occur in four oxidation states in water (+5, +3, 0, −3) but is usually found only in the trivalent [arsenite, As(III)] and pentavalent [arsenate, As(V)] states.

The predominance diagrams for arsenite and arsenate as a function of pH are shown on Fig. 20-1. In the pH range from 2 to 9, the undissociated form of arsenite (H_3AsO_3) is the predominate species. Most natural As(III)-containing surface and groundwaters in the pH range of 6.5 to 8.5 will have As in the H_3AsO_3 form. Arsenate ($HAsO_4^{2-}$) will be present in the pH range from 7 to 11.5, which will most likely occur in the normal pH range for most water supplies. For pH values less than 7.0, $H_2AsO_4^-$ will be the predominate species.

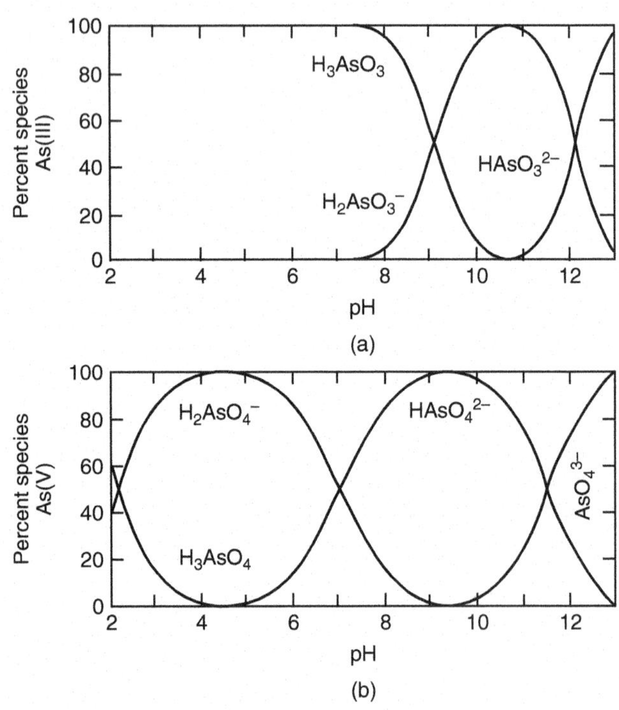

Figure 20-1
Predominance diagram for As(III) and As(V) as function of pH. (Adapted from Gupta and Chen, 1978.)

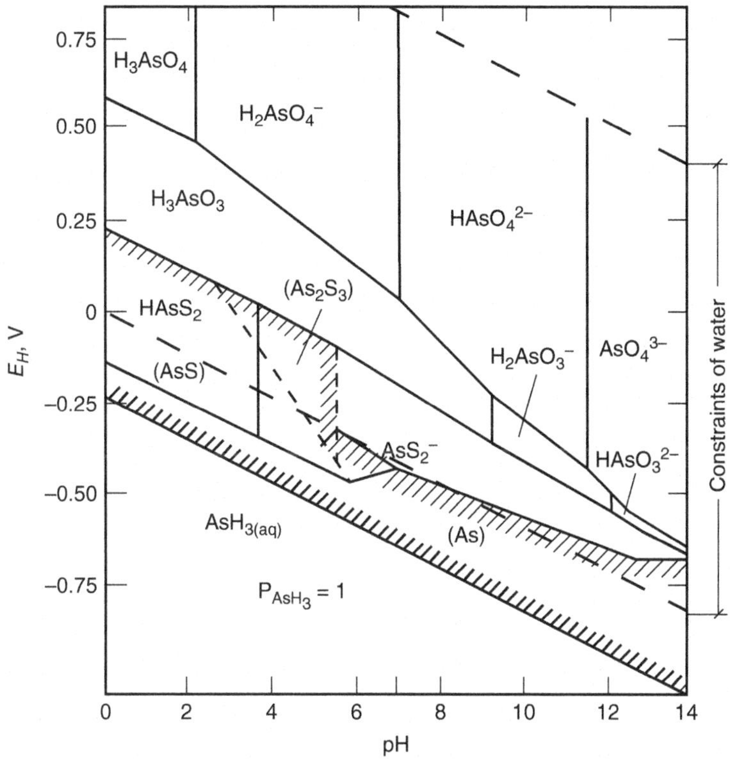

Figure 20-2
The E_H–pH diagram for arsenic at 25° C and 1 atm with total arsenic 10^{-5} mol/L and total sulfur 10^{-3} mol/L. Solid species are enclosed in parentheses in cross-hatched area, which indicates solubility less than $10^{-5.3}$ mol/L. (adapted From Ferguson and Gavis, 1972. Reprinted with permission.)

The E_H–pH diagram for arsenic in the presence of oxygen, sulfur, and water at 25°C and 1 atm is presented on Fig. 20-2. The E_H–pH diagram represents the equilibrium condition of arsenic under various redox potentials and pH conditions. For surface waters that are well aerated (oxidizing or high E_H conditions), $H_2AsO_4^-$ becomes the dominant species at pH values less than about 6.9 and $HAsO_4^{2-}$ becomes the dominant species at higher pH values. Usually little or no As(III) is present under oxidizing conditions. For well waters that contain little dissolved oxygen (mildly reducing or low E_H values), H_3AsO_3 should be the dominant species for pH values less than about 9.2.

In the presence of relatively high concentrations of reduced sulfur, dissolved arsenic–sulfide species can be present. Reducing, acidic conditions will favor the formation of reduced sulfur species such as orpiment (As_2S_3), realgar (AsS), or other sulfide–arsenic species. Waters high in arsenic are usually not expected where there is a high concentration of free sulfide. Thioarsenite species are more likely to be present at neutral and alkaline pH in the presence of very high sulfide concentrations. In addition, organic arsenic forms can be produced by biological activity in some surface waters but are not routinely measured.

Treatment Strategies

Treatment strategies for arsenic removal from water supplies include (1) conventional coagulation, (2) sorption with ferric and aluminum oxides, (3) ion exchange, and (4) membranes. More recently, the use of nonconventional adsorbents such as ferric and aluminum oxides have been evaluated as potential treatment alternatives for the removal of arsenic. A brief discussion of these treatment practices is presented and references are provided for further reading in this section.

COAGULATION PROCESSES

Traditionally, coagulation and filtration processes have been used to remove arsenic. Coagulants such as alum, ferric chloride, ferric sulfate, and lime have been used to remove arsenic to varying degrees. Coagulation processes usually require arsenic to be in the form of As(V); if arsenite, As(III), is present, it is usually oxidized first using chlorine, permanganate, or ozone. The mechanisms for As(V) removal by coagulation processes may be a combination of precipitation, coprecipitation, and adsorption. Edwards (1994) provides a discussion of the mechanisms associated with arsenic removal.

The performance results from several pilot- and full-scale studies using various coagulation processes are summarized in Table 20-3. Observed removals using ferric chloride have ranged from 81 to 100 percent using coagulant dosages from 5 to 304 mg/L. Typical doses for ferric salts are 5 to 30 mg/L and the pH is usually below 8.0. Alum coagulation has resulted in removals ranging from 23 to 100 percent with dosages from 6 to 50 mg/L. Typical dosages for alum are 10 to 50 mg/L and the pH is usually between 6 and 7.0. For Fe–Mn treatment processes, where the Fe^{2+} concentrations are greater than about 1.5 mg/L, around 80 to 90 percent of the As is removed. Plants that remove only Mn did not achieve significant removals of arsenic (McNeill and Edwards, 1997). Arsenic removals ranging from 60 to 90 percent have been observed in softening plants that use excess lime for Mg^{2+} treatment. For single-stage softening plants that only remove Ca^{2+}, arsenic removals of 0 to 40 percent have been observed. For source waters containing high levels of phosphate or silicate, coagulation can be less effective for arsenic removal. For waters containing sulfides, As(III) is precipitated as AsS and As_2S_3. Jar test screening is recommended at the bench level to evaluate different options for arsenic removal when using coagulants.

SORPTION PROCESSES

Because adsorption onto alum, ferric, and lime precipitates is a major arsenic removal mechanism, adsorbents composed of similar surface adsorption properties have been tested and also found to be promising for the removal of arsenic. Two promising granular adsorbents are activated alumina and granular ferric hydroxide (GFH) based adsorbents. The properties of these two granular adsorbents are reported in Table 20-4. In the fixed-bed

Table 20-3
Summary of some pilot- and full-scale results for As removal using coagulation processes

Type of Study	Water Type	Coagulant Conditions	Initial As Concentration, μg/L	Final As Concentration, μg/L	Percent As Removal, %	Reference
Pilot plant	Groundwater	Ferric hydroxide (2.4 - 2.7 mg/L Fe), pH 8.8	40-43	<10	75–82	Ghurye et al., (2004)
	Anoxic well water) (natural As)	Alum (7–46 mg/L), pH 7.4	640-830[a]	<10-470	38–100	Sorg and Logsdon (1978)
		Ferric chloride (18–77 mg/L) pH 7.0-8.5	560-920[a]	<10-290	60–100	Sorg and Logsdon (1978)
		Ferric chloride (17-304 mg/L) pH 7.0-7.6, prechlorinated	580-940[b]	<10–130	81–100	Sorg and Logsdon (1978)
	Surface water [spiked As(V)]	Alum (10–30 mg/L), pH 6.3–7.6, cationic	2.42-5.83[c]	0.26–3.50	13–94	Cheng et al. (1993)
		polymer 2 mg/L Ferric chloride (10–30 mg/L), pH 6.3–7.3, cationic polymer 3 mg/L	3.17-5.33[c]	0.02–0.43	91–98	Cheng et al. (1993)
	Surface water (natural As)	Ferric Hydroxide (2.7 mg/L Fe), pH 8.8	40–43	<10	>80	Ghurye et al. (2004)
Full-scale plant	Surface water (natural As)	Ferric chloride (3–10 mg/L), pH 7.18-7.8, chlorine 3–5 mg/L, cationic polymer 2–3 mg/L	1.2–1.7[a]	0.07–0.32	81–96	Scott et al. (1995)
		Alum (6–20 mg/L), pH 7.18–7.8, chlorine 3–5	2.1–2.2[c]	0.63–1.7	23–71	Scott et al. (1995)
		mg/L, cationic polymer 2–3 mg/L	2.1–2.2[c]	0.63–1.7	23–71	Scott et al. (1995)
	Groundwater (natural As)	Fe–Mn treatment (aeration, oxidation, filtration), pH 7.5	20.5[d]	2.7	86.7	McNeill and Edwards (1995)
		Fe–Mn treatment (aeration, oxidation, filtration), pH 7.5	5.2[d]	0.88	83.1	McNeill and Edwards (1995)

(*Continued*)

Table 20-3 *(Continued)*

Type of Study	Water Type	Coagulant Conditions	Initial As Concentration, μg/L	Final As Concentration, μg/L	Percent As Removal, %	Reference
	Blended surface and groundwater (natural As)	Single-stage calcium softening, pH 8.7	3.1[d]	2.9	6.1	McNeill and Edwards (1995)
	Surface water (natural As)	Excess lime enhanced softening, lime dose 120 mg/L,[e] pH 8.2	3.9[d]	1.26	67.7	McNeill and Edwards (1995)
	Groundwater (natural As)	Excess lime enhanced softening, pH 7.6	8.2[d]	1.75	78.6	McNeill and Edwards (1995)
		Excess lime enhanced softening, lime dose 450 mg/L[e], pH 7.1	32.4[d]	13.9	57.1	McNeill and Edwards (1995)
	Surface water (natural As)	Alum coagulation, alum dose 55 mg/L, pH 7.8	12.0[d]	3.4	71.3	McNeill and Edwards (1995)
		Alum coagulation, alum dose 10 mg/L, pH 7.8	4.3[d]	2.48	42.3	McNeill and Edwards (1995)
		Alum coagulation, alum dose 10 mg/L, pH 7.8	4.4[d]	3.06	30.4	McNeill and Edwards (1995)

[a]As(III) (assumed).
[b]As(V) (assumed).
[c]As(V) (known).
[d]As (unknown).
[e]As CaO.

Table 20-4
Properties of activated alumina and GFH adsorbents

Parameter	Unit	Activated Alumina, Alcoa F-1	GFH
Media size	mm	0.29–0.50	0.32–2.0
Grain density	g/cm^3	3.97	1.59
Bulk density	g/cm^3	0.641–0.960	1.22–1.29
Porosity of grains	%	—	72–77
Specific surface area	m^2/g (dry weight)	300–350	250–300

mode of operation, these adsorbents were tested successfully at both the pilot and full scale in the United States and Europe on natural waters. The results of some studies that have been performed on these adsorbents are given in Table 20-5. The reported range of capacities for activated alumina is from 1000 to 13,000 bed volumes (BV) but typically around 10,000 BV. The GFH adsorbent capacities have ranged from 32,000 up to 85,000 BV depending upon the water quality. The GFH adsorbents are very attractive from a performance standpoint; however, at this time the cost is rather prohibitive for large systems. For both adsorbents, the capacities will depend upon water quality parameters such as temperature, pH, and competing constituents in the water matrix, which may include NOM and competing ions (e.g., phosphate, silicate, sulfate). For example, activated

Table 20-5
Comparison of results for adsorbents in fixed beds in removing arsenic from water supplies

Adsorbent	Influent Arsenic Concentration, μg/L	Treatment Objective, μg/L	Bed Volumes Treated	Reference
Activated alumina	70	10	1,000	Benjamin et al. (2000)
	50	10	10,000	Benjamin et al. (2000)
	21	10	13,000	Clifford (1999)
	23	10	110,500	Simms and Azizian (1997)
	98	50	16,000	Clifford (1999)
	22	10	15,600	Clifford (1999)
	100	10	9,000	Rubel and Hathaway (1985)
Granular ferric hydroxide (GFH) based adsorbent	15–20	10	85,000	Driehaus et al. (1998)
	100	10	25,000	Thomson et al. (2005)
	100–180	—	34,000	Driehaus et al. (1998)
	21	—	37,000	Driehaus et al. (1998)
	16	—	32,000	Driehaus et al. (1998)

alumina performs best at pH values around 5.5 to 6.0 and drops off sharply above 7.0. It should also be noted that preoxidation of the water to convert any As(III) to As(V) is usually performed before the adsorption process.

In some cases, activated alumina can be regenerated with a strong base followed by a strong acid. Reported recoveries of arsenic from the regeneration process are about 75 percent. For small systems, one-time use may be more economical than regeneration, which will be required for large systems. A good discussion of the design and regeneration of activated alumina for fluoride removal is presented by Clifford (1999). The design parameters for arsenic removal may be similar; however, rapid small-scale and pilot testing may be used to develop site-specific design information. In addition, the use of powdered activated alumina coupled with membranes (microfiltration and ultrafiltration) may also be a promising treatment process.

Presently, the GFH adsorbent is used once and disposed of in a landfill. Regeneration techniques for GFH may be necessary to make it an economically viable process for large systems. Other adsorbents such as manganese greensand, manganese dioxide, hydrous iron oxide particles, and ironoxide-coated sand may be promising if verified through pilot testing and if it can be demonstrated that these processes are viable and economically feasible at full scale.

ION EXCHANGE

Ion exchange can be a viable process for the removal of arsenic from natural waters. Design considerations for arsenic removal by ion exchange include (1) oxidation state of arsenic, (2) resin type, (3) background ion concentrations and type of ion, (4) empty-bed contact time (EBCT), (5) regenerant strength and level, and (6) spent-brine reuse and treatment. Some of the design considerations for arsenic removal by ion exchange are discussed in this section, and an in-depth discussion is presented in Clifford (1999).

From the speciation diagram given on Fig. 20-1, As(V) is present as monovalent $H_2AsO_4^-$ and divalent $HAsO_4^{2-}$ in the pH range of natural waters, 6 to 9. If As(III) is present in the water, it usually exists as a neutral species, which cannot be removed by ion exchange and must be oxidized to As(V) prior to ion exchange treatment. For waters with total dissolved solids (TDS) concentrations less than about 500 mg/L and sulfate concentrations less than about 120 mg/L, anion ion exchange can be an economically attractive process for arsenic removal (Frey and Edwards, 1997). At low TDS and sulfate concentrations the competition for resin exchange sites with arsenic is low, and reasonable exchange capacity for arsenic can be achieved. However, for waters containing high sulfate and TDS levels, ion exchange may not be a viable process (Clifford and Ghurye, 1998b). For waters having high pH and alkalinity and low sulfate concentration, As(V) can be effectively treated with an anion exchange resin in the chloride form (Clifford and Lin, 1986).

Type of strong-base resin

With respect to the type of strong-base anion (SBA) exchange resin, no significant difference in performance has been observed among the various resins and both type 1 and type 2 polystyrene resins may provide slightly higher breakthrough capacities than other SBA resins. For waters containing arsenic and nitrate, $HAsO_4^{2-}$ has a poor affinity for monovalent nitrate-selective resins and should be avoided. When using conventional SBA resins, another concern is nitrate peaking or chromatographic overshoot due to the ions having different affinities for the resin. Nitrate has a slightly lower affinity than arsenic for standard SBA resins (see Chap. 16, Table 16-4) and will increase in liquid-phase concentration as it is pushed through the column and eventually will be present in the effluent at a much higher concentration than the influent concentration. The nitrate effluent concentration can sometimes exceed the effluent guidelines for nitrate in the water. Care should be taken to shorten the arsenic loading cycle time such that the nitrate standard is not exceeded. Nitrate will typically appear in the effluent just before arsenic breakthrough. In addition, lowering the pH to produce monovalent arsenic is not effective because $H_2AsO_4^-$ affinity is much less than $HAsO_4^{2-}$ (Clifford, 1999).

Based on pilot studies for arsenic removal, it has been shown that arsenic leakage (see Chap. 16) can develop during exhaustion cycles, but the observed concentrations (0.2 to 0.8 μg/L) are well below the proposed MCL of 10 μg/L. Arsenic leakage increases when particulate iron concentrations in the water increase due to the adsorption of the arsenic onto the iron particles. Steps should be taken to provide particulate filtration prior to the ion exchange process.

General design considerations

Typical EBCTs for arsenic removal range from 1.5 to 3.0 min and other operational parameters are similar to those presented in Table 16-1 in Chap. 16. Arsenic-loaded SBA resins can be regenerated easily with NaCl. Because arsenic is a divalent ion, it undergoes selectivity reversal in the presence of high-ionic-strength solutions and consequently is easy to remove from the resin during regeneration. In addition, regeneration superficial velocities greater than 0.02 m/h were found to work best for arsenic because they resulted in higher arsenic recoveries.

Down-flow co-current regeneration has been shown to be more effective for regenerating arsenic-laden resins than the conventional countercurrent mode of regeneration (Clifford and Ghurye, 1998a). Co-current is more effective because at the end of the exhaustion cycle the arsenic that is exchanged onto the resin bed is located near the feed end of the bed. Consequently, when the regenerant is passed through the bed in the down-flow mode, the regenerant is contacted with the highest arsenic resin concentration. Down-flow co-current regeneration will reduce arsenic leakage.

Brine management

Brine reuse and treatment are important in the removal of arsenic by ion exchange (see also Chap. 21). Field studies have demonstrated that arsenicladen resin could be regenerated successfully using spent brine for over 20 regeneration cycles before arsenic leakage exceeded 10 μg/L (Clifford, 1999). Sodium chloride was added to the spent-brine solution to maintain a 1.0N NaCl solution concentration. The concept of brine reuse should be investigated because it can provide significant cost savings with respect to lowering the salt requirements and brine disposal volume. Waste brine containing arsenic needs to be treated prior to disposal. Arsenic can be precipitated using $FeCl_3$, alum, or lime. For a brine waste containing 90 mg/L of As(V) and 50,000 mg/L TDS, about 12 times the stoichiometric quantity of $FeCl_3$ is required to reduce the As(V) concentration to less than 5 mg/L. Arsenic removal is very pH dependent, and greater than 99 percent As(V) can be removed from the brine solution using an iron-to-arsenic ratio of 20: 1 at a final pH of around 5.5. Consequently, pH adjustment should be included in brine disposal evaluations.

MEMBRANES

A number of studies have been conducted to evaluate the use of membranes for arsenic removal (Moore et al., 2008; Ghurye et al., 2004; Brandhuber and Amy, 1998; Chang et al., 1994; Hering and Elimelech, 1996; Thompson and Chowdhury, 1993). In one study, several ultrafiltration (UF), nanofiltration (NF), and reverse osmosis (RO) membranes were evaluated for arsenic removal from two spiked groundwater sources from Southern California (Brandhuber and Amy, 1998). Based on these studies, guidelines for arsenic removal using membrane treatment are summarized in Table 20-6. The combined treatment using coagulation and membranes can be used to effectively remove arsenic, and if arsenic is in particulate form, large-pore-size membranes may be effective in As removal. The As(III) form can only be removed by RO; otherwise it must be preoxidized, forming As(V) before membrane treatment. Reverse osmosis and NF can effectively remove the dissolved form of As(V), and tight UF membranes may be effective as well. Waters high in dissolved organic carbon (DOC) may inadvertently cause poor arsenic removal due to membrane fouling.

20-3 Iron and Manganese Removal

The fundamental concepts involved with removal of soluble iron and manganese are similar and the two often occur together in water supplies. These soluble species are typically unstable when exposed to oxidants such as dissolved oxygen and when processed in a water treatment plant will usually form precipitate during treatment, in the distribution piping system, or at the point of use. However, depending upon which treatment processes are used, it may be important to control these precipitation reactions. The

Table 20-6
Guidelines for the use of membranes for arsenic removal in water treatment[a]

Source Water Characteristic	Membrane Treatment Only				Oxidation Pretreatment
	RO	NF	UF	MF	
As speciation					
As(III)	R	PE	NR	NR	R
As(V)	R	R	NR	NR	NR
As size distribution					
Dissolved	R	PE	NR	NR	NR
Particulate	NR	NR	PE	PE	NR
Co-occurrence					
NOM	PE	NR	NR	NR	NR
Inorganic	R	NR	NR	NR	NR

[a]Removal of other forms possible with ferric coagulants.
[b]R—Recommended, NR—not recommended, PE—possibly effective.
Source: Brandhuber and Amy (1998).

most common treatment approach is to precipitate all the soluble forms of iron and manganese so that these constituents can be removed in other processes such as sedimentation and filtration. In addition to controlling the soluble species, colloidal and/or particulate iron and manganese are also important and must be considered in the overall removal. Consequently, the following sections provide a brief discussion of the background of these species, as well as their chemistry as related to redox properties, followed by control methods.

Iron

Iron is the fourth most abundant element in the Earth's crust, making up about 5.6 percent of the mass (McMurry and Fay, 2003). Common mineral sources (deposits) of iron include ferric oxides and hydroxides such as hematite (Fe_2O_3) and ferric hydroxide [$Fe(OH)_3$]. Ferric hydroxide gives rocks and soils their red and yellowish color. Sedimentary forms of iron may include sulfides, such as pyrite and marcasite; two minerals with identical composition (FeS_2) but different crystalline structures; carbonates such as siderite ($FeCO_3$); and mixed oxides such as magnetite (Fe_3O_4). The ferrous oxides and sulfides are the usual sources of dissolved iron in groundwaters. Weathering of iron silicates can produce dissolved iron in surface water; however, this is a relatively slow process.

OCCURRENCE AND IMPORTANCE IN WATER SUPPLIES

The interactions between iron-bearing soils or rock formations and water surrounding it can dissolve iron into the water. Iron is relatively soluble in a reducing environment or in natural waters, such as some low- oxygen-containing groundwaters and low-oxygen surface water, such as hypolimnetic waters of eutrophic lakes, large rivers, and reservoirs. In these waters,

Table 20-7
Maximum iron and manganese concentrations for selected industrial and commercial applications

	Maximum Concentration or Threshold Range (mg/L)		
Application	**Mn**	**Fe + Mn**	**Fe**
Air conditioning	0.5	0.5	—
Baking	0.2	0.2	0.2
Brewing	0.1	0.1	0.1–1.0
Canning	0.2	0.2	—
Carbonated beverages	0.2	0.1–0.2	0.1–0.2
Cooling water	0.2–0.5	0.2–0.5	0.5
Confectionary	0.2	0.2	0.2
Dyeing	0.0	0.0	—
Electroplating	—	—	Trace
Food processing, general	0.0	0.2	0.2
Ice	0.2	0.2	—
Laundering	—	—	0.2–1.0
Milk industry	0.03–0.1	—	—
Oil well flooding	—	—	0.1
Photographic processing	0.0	0.0	0.1
Pulp and paper			
Ground wood	0.5	1.0	0.3
Kraft pulp	0.1	0.2	—
Soda pulp	0.05	0.1	0.1
Kraft pulp, unbleached	0.5	—	—
Kraft pulp, bleached	0.1	—	0.05
Fine paper pulp	0.05	0.1	0.10
High-grade paper pulp	0.05	0.1	—
Plastics (clear)	0.02	0.02	—
Rayon pulp	0.03	0.05	—
Rayon manufacturing	0.0–0.02	0.0	0.05
Sugar manufacturing	—	—	0.1
Tanning	0.2	0.2	0.1–2.0

iron may be found in the reduced or ferrous form (Fe^{2+}) such as ferrous sulfate ($FeSO_4$) and ferrous bicarbonate [$Fe(HCO_3)_2$], hydroxide forms, or complexed with NOM. Low-alkalinity (<50 mg/L as $CaCO_3$) groundwaters may contain up to 10 mg/L of total iron. Also, some water treatment plant operators have observed high concentrations of iron in the raw water obtained from the hypolimnetic zone of reservoirs during periods of stagnation and stratification as it is seasonally mobilized from reduced lake sediments (Stumm and Lee, 1961).

Correspondingly, iron is very insoluble in an oxidizing environment or in natural waters containing sufficient quantities of dissolved oxygen. Depending upon the water quality, iron can exist in three physical forms.

as large oxidized particles, small oxidized colloidal particles, and the soluble reduced form. The smaller particles can often pass through filters of 0.45 μm pore size or smaller and are therefore classified as soluble in many operational schemes. For example, oxygenated surface waters (pH 5 to 8) typically have total iron concentrations in the range of 0.05 to 0.2 mg/L. In these waters, iron species present may consist of solids of large oxidized particles or small colloidal particles of ferric hydroxide sorbed onto clay particles, organic colloids, and other suspended solids and precipitates.

The U.S. EPA (1991) Secondary Drinking Water Regulations limit iron to 0.3 mg/L for taste and aesthetic reasons. The only advantage of having iron in water is for nutritional value, but only 1 to 2 mg/d is needed and most humans intake about 7 to 35 mg/d. Thus, daily consumption of water is not a major source of iron. There are several disadvantages of having iron in water supplies. Iron ions impart a metallic taste, and the taste, threshold is reported to be around 0.1 to 0.2 mg/L of ferrous sulfate or ferrous chloride as Fe^{2+}. Ferrous iron may precipitate as ferric hydroxide after oxidation and stain laundry and household fixtures such as bathtubs, porcelain basins, glassware, and dishes. Iron may discolor industrial products such as textiles and paper. Threshold values for industrial and commercial uses of iron are listed in Table 20-7. Iron precipitates can clog pipes and support the growth of iron bacteria (*Crenothrix* and *Gallionella*), which can cause taste and odor problems.

CHEMICAL PROPERTIES

While the chemistry of iron oxidation is complex and not clearly understood, there are some useful physical/chemical relationships that can be used to interpret observations made in water treatment facilities. Parameters that affect iron oxidation and its rate include water temperature and pH and constituents in the water such as dissolved oxygen (DO), bicarbonate, NOM, sulfate, dissolved silica, and particles. Many of these parameters are discussed below to help explain iron oxidation.

An E_H–pH diagram for iron is presented on Fig. 20-3. The equilibrium form of various iron species that would be expected under the conditions stated, at any specific E_H–pH combination, is shown with the boundary defined by the E_H–pH limit of water. The solid forms of iron are shown with an (s). Under reducing conditions ($E_H < 0$), over a wide range of pH, iron solubility is low and pyrite tends to precipitate. Under oxidizing conditions ($E_H > 0$) and pH values above 5.0, iron tends to precipitate to ferric hydroxide [$Fe(OH)_3$]. Between these two regions ferrous iron is quite soluble, and this region corresponds to typical E_H–pH conditions of groundwater (pH 5 to 9 and E_H 0.20 to –0.10 V).

The stoichiometric expressions for the oxidation of ferrous iron using DO and some commonly used oxidants are shown in Table 20-8. In addition, the quantity of oxygen required, the alkalinity consumed, and an estimate of the sludge produced are also provided in Table 20-8.

Table 20-8
Oxidation reactions for iron

Oxidant	Reaction	Equation Number	Oxidant Needed, mg/mg Fe^{2+}	Alkalinity Used, mg/mg Fe^{2+}	Sludge Produced,[a] kg/kg Fe^{2+}
Oxygen	$4Fe(HCO_3)_2 + O_2 + 2H_2O \rightarrow 4Fe(OH)_3 + 8CO_2$	20-1	0.14	1.80	1.9
Chlorine	$2Fe(HCO_3)_2 + Ca(HCO_3)_2 + Cl_2 \rightarrow 2Fe(OH)_3 + CaCl_2 + 6CO_2$	20-2	0.64	2.70	1.9
Chlorine dioxide	$Fe(HCO_3)_2 + NaHCO_3 + ClO_2 \rightarrow Fe(OH)_3 + NaClO_2 + 3CO_2$	20-3	1.21	2.70	1.9
Potassium permanganate	$3Fe(HCO_3)_2 + KMnO_4 + 2H_2O \rightarrow 3Fe(OH)_3 + MnO_2 + KHCO_3 + 5CO_2$	20-4	0.94	1.50	2.43

[a]Sludge weight is based on $Fe(OH)_3$ as the precipitate; however, it is likely that portions of the sludge will contain $FeCO_3$.
Source: Adapted from ASCE/AWWA (1990).

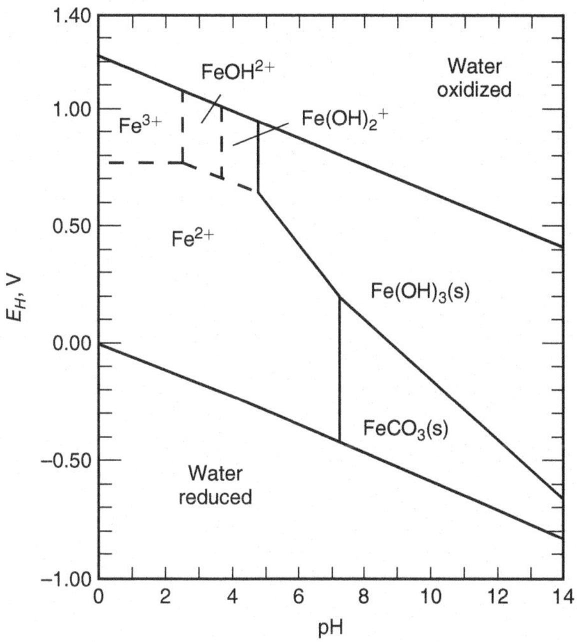

Figure 20-3
Forms of iron in water as function of redox potential versus pH constructed with total iron activity 10^{-7} M or 5.6 μg/L, 96 mg/L SO_4^{2-}, CO_2 species at 1000 mg/L HCO_3^-, temperature at 25°C, and pressure of 1 atm (adapted from Langmuir, 1997).

KINETICS OF IRON OXIDATION

In the absence of iron complexed with NOM and pH values greater than 5.5, the rate of oxygenation of Fe^{2+} iron was found to be first order with respect to Fe^{2+} and O_2 and second order with respect to OH^- ion (Stumm and Lee, 1961). Based on these experimental observations, the following expression was proposed:

$$-\frac{d\left[\mathrm{Fe}^{2+}\right]}{dt} = k\left[\mathrm{Fe}^{2+}\right]\left[\mathrm{OH}^-\right]^2 P_{\mathrm{O}_2} = k_r\left[\mathrm{Fe}^{2+}\right] \tag{20-5}$$

where $[Fe^{2+}]$ = aqueous-phase ferrous iron concentration, mol/L
k = rate constant, typically $8.0(\pm 2.5) \times 10^{13}$ $L^2/mol^2 \cdot min \cdot atm$ at 20°C (Stumm and Morgan, 1996)
$[OH^-]$ = aqueous-phase hydroxide ion concentration, mol/L
P_{O_2} = partial pressure of oxygen, atm
k_r = pseudo-first-order constant, $min^{-1} = k\,[OH]^2 P_{O_2}$

The oxygenation rate of $[Fe^{2+}]$ is very dependent upon pH, as shown in Eq. 20-5. The dependence of the oxygenation rate of Fe^{2+} on pH is shown on Fig. 20-4. The *y* axis represents the log of the rate of oxygenation of Fe^{2+} with respect to time and is obtained by rearrangement of Eq. 20-5 as

$$k\left[\mathrm{OH}^-\right]^2 P_{\mathrm{O}_2} = -\frac{d\left[\mathrm{Fe}^{2+}\right]}{\left[\mathrm{Fe}^{2+}\right]dt} = -\frac{d\ln\left[\mathrm{Fe}^{2+}\right]}{dt} \tag{20-6}$$

Example 20-1 Theoretical stoichiometric calculation

A 100,000-m^3/d raw-water source containing 5 mg/L ferrous iron is oxidized to ferric hydroxide with oxygen. Calculate the quantity of oxygen required, alkalinity consumed as $CaCO_3$, and quantity of sludge produced as ferric hydroxide.

Solution

1. Determine the quantity of oxygen required using Eq. 20-1 in Table 20-8:

$$\text{Dissolved oxygen required}$$
$$= \left(5\ Fe^{2+}\ mg/L\right)\left(0.14\ mg\ O_2/mg\ Fe^{2+}\right)$$
$$\left(100{,}000\ m^3/d\right)\left(1 kg/10^6\ mg\right)\left(10^3\ L/m^3\right)$$
$$= 70\ kg/d$$

2. Determine the quantity of alkalinity consumed as $CaCO_3$ using the values shown in Table 20-8:

$$\text{Quantity alkalinity consumed}$$
$$= \left(5\ mg\ Fe^{2+}/L\right)\left(1.80\ mg\ alkalinity/mg\ Fe^{2+}\right)$$
$$\left(100{,}000 m^3/d\right)\left(1 kg/10^6\ mg\right)\left(10^3\ L/m^3\right)$$
$$= 900\ kg/d$$

3. Estimate the quantity of sludge produced as ferric hydroxide using values from Table 20-8:

$$\text{Sludge produced as Fe } (OH)_3$$
$$= \left(5\ mg\ Fe^{2+}/L\right)\left(1.90\ mg\ sludge/mg\ Fe^{2+}\right)$$
$$= \left(100{,}000\ m^3/d\right)\left(1\ kg/10^6\ mg\right)\left(10^3\ L/m^3\right)$$
$$= 950 kg/d$$

Comment

Oxygen addition is seldom used to precipitate iron because iron is typically complexed to NOM and the oxidation kinetics are too slow.

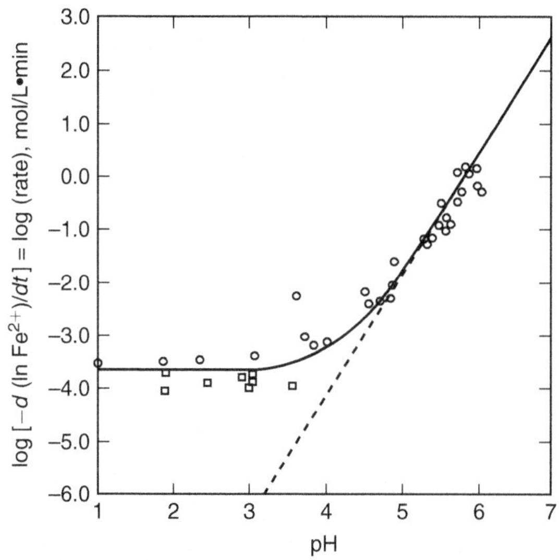

Figure 20-4
Oxidation rate of Fe^{2+} by oxygen for $P_{O_2} = 0.20$ atm and temperature = 25°C. (Adapted from from Singer and Stumm, 1970.)

Assuming the oxygen partial pressure is constant, Eq. 20-6 can be written as

$$-\frac{d\ln\left[Fe^{2+}\right]}{dt} = k'\left[OH^-\right]^2 \tag{20-7}$$

where k' = rate constant, L / mol · min, $= kP_{O_2}$

Taking the logarithms of both sides and substituting $K_w/[H^+]$ for $[OH^-]$, the following expression for the log of the rate of oxygenation of Fe^{2+} can be obtained (Snoeyink and Jenkins, 1980):

$$\log(\text{rate}) = \log(k'') + 2\,\text{pH} \tag{20-8}$$

where k'' = constant, $k'\ K_w^2$

Effect of pH

For pH values greater than 5.5, the experimental data and Eq. 20-5 are in agreement, and the rate of oxygenation of Fe^{2+} increases by 100-fold per pH unit. For pH values less than about 3.5, the oxygenation of Fe^{2+} is independent of pH.

Effect of temperature

The effect of temperature on the rate of oxygenation of Fe^{2+} appears to be large when data are plotted as shown on Fig. 20-5 (Sung and Morgan, 1980). However, when the data on Fig. 20-5 are normalized with respect to changes in K_w and O_2 solubility with temperature, the change in the value of the rate constants is small (Sung and Morgan, 1980).

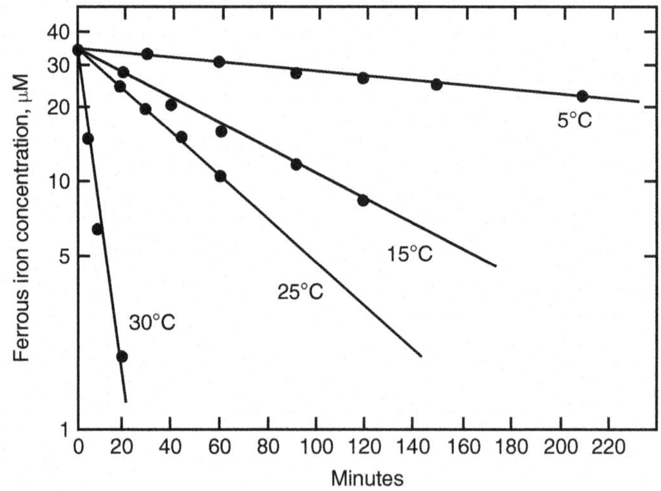

Figure 20-5
Impact of temperature on oxygen kinetics of Fe^{2+}. The experiments were conducted in 0.11 M ionic strength and adjusted with $NaClO_4$. The alkalinity was equal to 9.0 × 10^{-3} M as HCO_3, pH was equal to 6.82, P_{O_2} was 0.2 atm, and the initial Fe^{2+} concentration was 0.0347 M. (Adapted from Sung and Morgan, 1980.)

Effect of ionic strength

The impact of ionic strength on the oxygenation rate of Fe^{2+} is shown in Table 20-9. Increasing the ionic strength by a factor of about 10 from 0.009 to 0.11 M decreases the rate by about a factor of 4 from 4.0×10^{13} to 1.2×10^{13} $M^{-2} \cdot atm^{-1} \cdot min^{-1}$. The time required for 50 percent reduction of the Fe^{2+} concentration in seawater is nearly 100 times larger than values observed for freshwater (Kester et al., 1975). The presence of Cl^- and SO_4^{2-} is believed to form complexes with the Fe^{2+} ion and inhibit oxygenation (Tamura et al., 1976).

Table 20-9
Impact of ionic strength on oxygenation rate constant for Fe^{2+}

Ionic Strength, mol/L	*k*, $(mol/L)^{-2} \cdot atm^{-1} \cdot min^{-1}$
0.009	$4.0 \pm 0.6 \times 10^{13}$
0.012	$3.1 \pm 0.7 \times 10^{13}$
0.020	$2.9 \pm 0.6 \times 10^{13}$
0.040	$2.2 \pm 0.5 \times 10^{13}$
0.060	$1.8 \pm 0.3 \times 10^{13}$
0.110	$1.2 \pm 0.2 \times 10^{13}$

[a]$T = 25°C$; alkalinity = 9×10^{-3} M HCO_3^-; $[Fe2+]_0 = 34.7$ μM; $P_{O_2} = 0.20$ atm; pH = 6.84.
*Sources:*Adapted from Faust and Aly (1998).

Effect of complexing agents
Organic complexing agents can also impact the oxygenation of Fe^{2+}. Humic and tannic acids or other NOM can bind or complex iron and slow down the kinetics of oxidation. Iron complexed with NOM, often referred to as filterable iron because it is soluble, is usually associated with water sources high in organic color. The chemistry of NOM and its interactions with metals are very complex and considerable research has been performed to elucidate the chemical nature of these yellow-colored organic iron complexes (Liao et al., 1982; Thurman, 1985). Soluble Fe^{2+} has been shown to not complex in the presence of humic acid at low pH values (pH < 5), but increasing the pH (pH > 8.0) can result in the formation of a soluble Fe^{2+} complex. This complexing ability was attributed to greater dissociation of carboxyl groups of the humic acid. It was proposed that humic acids could chemically reduce Fe^{3+} to Fe^{2+} followed by complexation of the Fe^{2+} with the humic acids. The presence of humic and tannic acids has also been shown to inhibit the oxygenation of Fe^{2+} (Theis and Singer, 1974). In addition, organically bound Fe^{2+} cannot be effectively oxidized by the use of aeration (Kawamura, 2000).

Manganese

Even though manganese makes up a very small percentage of the Earth's crust (<0.1 percent), it is abundant in rocks and soils. Manganese is an essential nutrient for both humans and plants. Typical daily intake is about 10 mg, the majority of which comes from food sources. For plants, manganese moves as an enzyme activator and in animals it is important in growth and in nervous system functioning.

OCCURRENCE AND IMPORTANCE IN WATER

Manganese is similar to iron in that it is usually present in the +2 oxidation state (Mn^{2+}) in anoxic groundwaters and in the hypolimnion region of reservoirs and eutrophic lakes. When the groundwater is pumped to the surface and when the hypolimnetic waters are mixed, the Mn^{2+} is exposed to oxygen and begins to undergo a series of oxidation reactions to Mn^{4+}. This oxidation is accompanied by a decrease in pH and DO concentration and the formation of MnO_2 precipitate. Manganese may cause aesthetic problems such as laundry and fixture staining. Manganese concentrations around 0.2 to 0.4 mg/L may also impart an unpleasant taste to water and can promote the growth of microorganisms in reservoirs and distribution systems. Consumer complaints have been documented with manganese concentrations as low as 20 μg/L (Sly et al., 1990). The U.S. EPA (1991) specifies a secondary MCL for manganese of 50 μg/L.

CHEMICAL PROPERTIES

Typically manganese occurs in the form of oxides and hydroxides. Manganese has eight oxidation states [Mn^0, Mn^{2+}, $Mn_3O_4(s)$, $Mn_2O_3(s)$, MnO_2, MnO_4^{3-}, MnO_4^{2-}, and MnO_4^-]. At pH values of most natural waters,

Example 20-2 Time required for Fe^{2+} oxidation from a groundwater

Groundwater with a soluble Fe^{2+} concentration of 5.0 mg/L is to be oxidized by aeration to a Fe^{2+} concentration of 0.3 mg/L. The raw-water pH is 7.0 with a temperature of 10°C, and it is assumed that P_{O_2} is in equilibrium with the atmosphere. A typical pseudo-first-order rate constant for the oxygenation of Fe^{2+} is 0.168 min^{-1}. For steady-state operation and a flow rate of 10,000 m^3/d (2.64 mgd), calculate and compare the minimum hydraulic detention time and reactor volume for the oxidation of Fe^{2+} to Fe^{3+} for a completely mixed flow reactor (CMFR) and for a plug flow reactor (PFR).

Solution

1. Determine the steady-state residence time for the CMFR using Eq. 6-37:

$$\tau_{\text{CMFR}} = \frac{C_{A0} - C_A}{kC_A} = \frac{\text{Fe}_0^{2+} - \text{Fe}^{2+}}{k'\text{Fe}^{2+}} = \frac{(5.0 - 0.3)\ \text{mg/L}}{(0.168\ \text{min}^{-1})(0.3\ \text{mg/L})}$$

$$= 93\ \text{min}$$

2. Determine the CMFR volume:

$$V_{\text{CMFR}} = Q\tau_{\text{CMFR}} = (93\ \text{min})(10{,}000\ \text{m}^3/\text{d})(1\ \text{d}/1440\ \text{min})$$

$$= 645\ \text{m}^3$$

3. Determine the steady-state residence time for the PFR using Eq. 6-65:

$$\tau_{\text{PFR}} = \frac{1}{k}\ln\left(\frac{C_{A0}}{C_A}\right) = \frac{1}{0.168\ \text{min}^{-1}}\ln\left(\frac{5.0}{0.3}\right) = 16.7\ \text{min}$$

4. Determine the PFR volume:

$$V_{\text{PFR}} = (16.7\ \text{min})(10{,}000\,\text{m}^3/\text{d})(1\ \text{d}/1440\,\text{min}) = 116\ \text{m}^3$$

5. Compare the PFR and CMFR detention times and values. As discussed in Chap. 6, the PFR is much more efficient than a CMFR. In most cases bench and/or pilot tests will be used to determine the kinetics for the water source of interest because the impact of both colloidal and particulate iron will have to be evaluated. However, this calculation can be used to provide some preliminary insight into the minimum contact times required for Fe^{2+} removal.

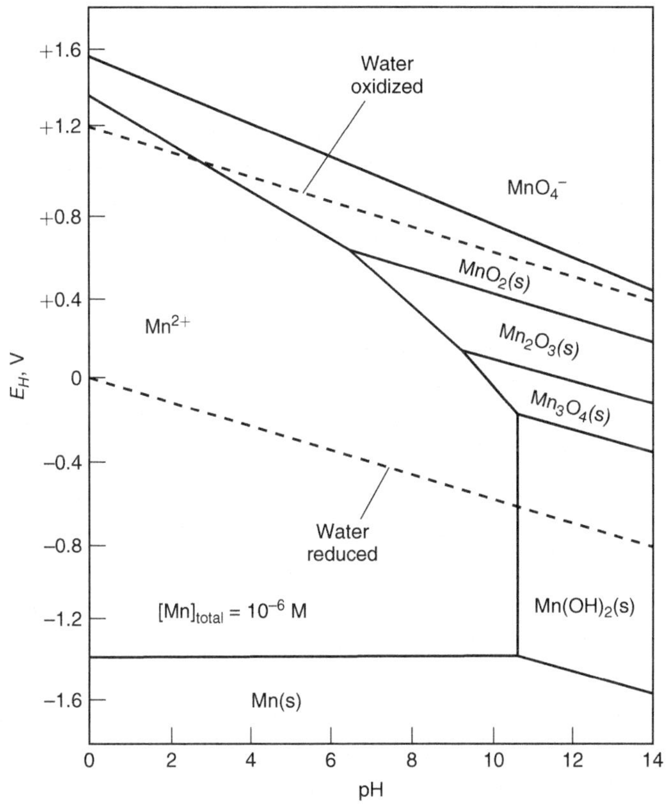

Figure 20-6
Forms of manganese in water as function of redox potential versus pH at a water temperature of 25°C. (Adapted from Pourbaix, 1966, 1974.)

aqueous Mn^{2+} is the predominant form of manganese, as shown on the E_H–pH diagram on Fig. 20-6. Concentrations of Mn^{2+} on the order of 0.1 to 1.0 mg/L are common, although in low-pH waters higher concentrations can occur. Oxidation of Mn^{2+} can thermodynamically lead to three different oxides of manganese depending upon the E_H and pH—MnO_2, Mn_2O_3, and Mn_3O_4, with the predominant form being MnO_2.

KINETICS OF OXIDATION

Three oxidation states of manganese important in drinking water are Mn^{2+} (soluble), Mn^{4+} (as MnO_2 precipitate), and Mn^{7+} (as MnO_4^- strong oxidant). The Mn^{2+} state can be removed from water via oxidation, but aeration is not a very effective treatment option unless the pH is greater than about 9.0. The reaction sequence for oxygenation of Mn^{2+} is suggested as (Faust and Aly, 1998)

$$Mn^{2+} + 0.5O_2 \xrightarrow{\text{slow}} MnO_2(s) \tag{20-9}$$

$$Mn^{2+} + MnO_2(s) \xrightarrow{\text{fast}} Mn^{2+} \cdot MnO_2(s) \tag{20-10}$$

$$Mn^{2+} \cdot MnO_2(s) + 0.5O_2 \xrightarrow{\text{slow}} 2MnO_2(s) \tag{20-11}$$

Table 20-10
Oxidation reactions for manganese

Oxidant	Reaction	Equation Number	Oxidant Needed, mg/mg Mn^{2+}	Alkalinity Used, mg/mg Mn^{2+}	Sludge Produced[a], kg/kg Mn^{2+}
Oxygen	$2MnSO_4 + 2Ca(HCO_3)_2 + O_2 \rightarrow 2MnO_2 + 2CaSO_4 + 2H_2O + 4CO_2$	20-12	0.29	1.80	1.58
Chlorine	$Mn(HCO_3)_2 + Ca(HCO_3)_2 + Cl_2 \rightarrow MnO_2 + CaCl_2 + 2H_2O + 4CO_2$	20-13	1.29	3.64	1.58
Chlorine dioxide	$Mn(HCO_3)_2 + 2NaHCO_3 + 2ClO_2 \rightarrow MnO_2 + 2NaClO_2 + 2H_2O + 4CO_2$	20-14	2.46	3.64	1.58
Potassium permanganate	$3Mn(HCO_3)_2 + 2KMnO_4 \rightarrow 5MnO_2 + 2KHCO_3 + 2H_2O + 4CO_2$	20-15	1.92	1.21	2.64

[a]Sludge weight is based on MnO_2 as the precipitate. It may be conservative in some cases.
Source: ASCE/AWWA (1990).

The rate of conversion of Mn^{2+} to MnO_2 also involves an autocatalytic process where the formation of MnO_2 solid provides for adsorption of Mn^{2+} and accelerates conversion of Mn^{2+} to MnO_2. Because of this catalytic effect, not all the Mn^{2+} that is removed from the process is converted to MnO_2 and may simply be adsorbed onto MnO_2. The products of manganese oxygenation appear to be nonstoichiometric and show various degrees of oxidation ranging from about $MnO_{1.3}$ to $MnO_{1.9}$ depending upon the pH (Stumm and Morgan, 1996). The general stoichiometric expressions for the oxidation of manganese using DO and some commonly used oxidants are displayed in Table 20-10 along with the quantity of oxygen required, the alkalinity consumed, and an estimate of the sludge produced. The values shown in Table 20-10 may be used to estimate the quantity of oxidant required, alkalinity consumed, and quantity of sludge produced for manganese oxidation. As discussed above, the extent of oxygenation of manganese is not accounted for by the stoichiometry of the oxidation alone and assuming that the sludge is MnO_2 will be conservative in terms of oxidant requirement.

Although the mechanism of this reaction is not understood completely, the following general expression may be used to describe the oxidation in a CMBR:

$$-\frac{d\left[Mn^{2+}\right]}{dt} = k_1\left[Mn^{2+}\right] + k_2\left[Mn^{2+}\right]\left[MnO_2(s)\right] \tag{20-16}$$

where k_1, k_2 = respective rate constants for oxidative and autocatalytic pathways

$[Mn^{2+}]$ = aqueous-phase manganese ion concentration, mol/L

$[MnO_2(s)]$ = manganese oxide precipitate concentration, mol/L

An alternative rate expression has been presented for the oxidation of Mn^{2+} to MnO_2 using potassium permanganate (Knocke et al., 1991):

$$\begin{aligned}-\frac{d\left[Mn^{2+}\right]}{dt} &= k_1\left[Mn^{2+}\right]\left[KMnO_4\right]\left[OH^-\right]^{1.1} \\ &\quad + k_2\left(\left[Mn^{2+}\right] - \left[Mn^{2+}\right]_e\right)\left[MnO_2(s)\right]\end{aligned} \tag{20-17}$$

where k_1 = rate constant for oxidative pathway, $9.55 \times 10^{12}\ s^{-1}\ (mol/L)^{-2.1}$

$[Mn^{2+}]$ = aqueous-phase Mn^{2+} ion concentration, mol/L

$[KMnO_4]$ = aqueous-phase $KMnO_4$ concentration, mol/L

$[OH^-]$ = aqueous-phase hydroxide ion concentration, mol/L

k_2 = rate constant for autocatalytic pathway, $8.7 \times 10^3\ s^{-1}\ (mol/L)^{-1}$

$[Mn^{2+}]_e$ = aqueous-phase Mn^{2+} ion concentration in finished water, mol/L

$[MnO_2(s)]$ = manganese oxide precipitate concentration, mol/L

Similar to Fe^{2+}, the rate of Mn^{2+} oxidation is dependent on P_{O_2} and $[OH^-]$, as shown in the equation

$$-\frac{d[Mn^{2+}]}{dt} = k_3 P_{O_2}[OH^-][Mn^{2+}] = k[Mn^{2+}] \tag{20-18}$$

where k_3 = rate constant, $L^2/mol^2 \cdot min \cdot atm$
k = pseudo-first-order constant, $min^{-1} = k_3 P_{O_2}[OH^-]$

The oxygenation rate dependence of both Fe^{2+} and Mn^{2+} as a function of pH is shown on Fig. 20-7 (Stumm and Morgan, 1996). The Fe^{2+} and Mn^{2+} states have the same slope but occur at much different pH ranges. The oxidation rate of Mn^{2+} is very low for pH values less than about 9.0. In the pH range encountered in water treatment, the use of oxygenation for the removal of manganese is not practical and alternative oxidants are typically used.

The presence of DOC in the water is not considered in Examples 20-2 and 20-3. Typically, the DOC in natural water will react with Fe^{2+} or Mn^{2+} to form organic complexes. Knocke et al. (1991) found that Mn^{2+} does not appear to be readily complexed by humic and fulvic acids, but Fe^{2+} is readily complexed. Complexed Fe^{2+} is not well oxidized by either $KMnO_4$ or ClO_2. However, preliminary studies indicated that alum coagulation may have an important role in the removal of complexed Fe^{2+}.

Treatment Strategies for Iron and Manganese

Several different treatment methods have been used to remove iron and manganese from drinking water supplies, including (1) oxidation using oxygen (aeration), chlorine, chlorine dioxide, potassium permanganate, or ozone followed by precipitate removal by sedimentation and filtration; (2) ion exchange; (3) lime softening; and (4) sequestering chemicals. Design considerations and performance information are provided for each process in the following discussion.

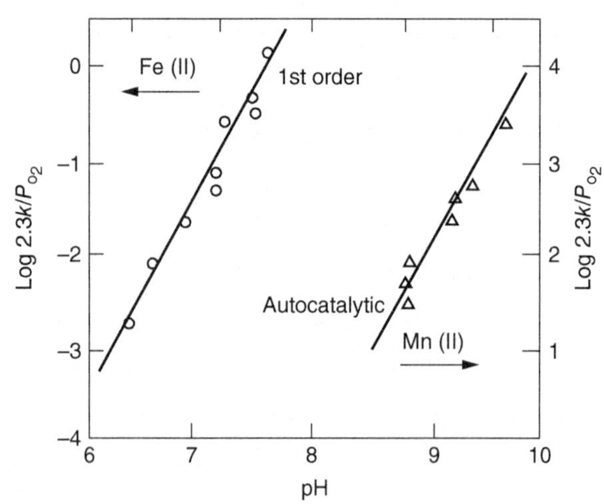

Figure 20-7
Comparison of impact of pH on oxidation rate of Fe^{2+} and Mn^{2+} by oxygen. (Adapted from Stumm and Morgan, 1996.).

Example 20-3 Time required for Mn^{2+} oxidation by potassium permanganate

The flow rate through a treatment process is 10,000 m^3/d (2.64 mgd). After removing the soluble iron from groundwater by aeration, 2 mg/L of Mn(II) ion is present. Potassium permanganate ($KMnO_4$) is added at a dose of 4.0 mg/L (2.53×10^{-5} mol/L) to oxidize Mn(II) ion (Mn^{2+}) to a concentration of 0.1 mg/L. The groundwater pH is 7.0 with a temperature of 10°C, and it is assumed that 5 mg/L (5.75×10^{-5} mol/L) of manganese dioxide (MnO_2) is in equilibrium with 0.1 mg/L of Mn^{2+}. For steady-state operation, calculate and compare the minimum hydraulic detention time and volume for the oxidation of Mn^{2+} to MnO_2 for a completely mixed flow reactor (CMFR) and for a plug flow reactor (PFR). Assume the $[Mn^{2+}]_e$ concentration is zero.

Solution

1. Calculate the pseudo-first-order rate constant.
   a. Write the rate equation for the oxidation of Mn^{2+}. The rate equation for the oxidation of Mn^{2+} to MnO_2 shown in Eq. 20-17 can be simplified to a pseudo-first-order reaction rate as follows:

$$-\frac{d[Mn^{2+}]}{dt} = k_1[Mn^{2+}][KMnO_4][OH^-]^{1.1} + k_2([Mn^{2+}] - [Mn^{2+}]_e)[MnO_2(s)]$$

$$-\frac{d[Mn^{2+}]}{dt} = k'[Mn^{2+}]$$

   where k' = pseudo-first-order rate constant

   b. Determine k' for the given reaction conditions:

$$\begin{aligned} k' &= k_1\,[KMnO_4][OH^-]^{1.1} + k_2[MnO_2(s)] \\ &= [9.55 \times 10^{12}\ s^{-1} \cdot (mol/L)^{2.1}] \\ &\quad \times (2.53 \times 10^{-5}\ mol/L)(10^{-7}\ mol/L)^{1.1} \\ &\quad + [8.7 \times 10^{3}\ s^{-1}\ (mol/L)^{-1}](5.75 \times 10^{-5}\ mol/L) \\ &= 4.82 + 0.50 = 5.3\ s^{-1} = 319\ min^{-1} \end{aligned}$$

2. Determine the steady-state residence time and volume for the CMFR using Eq. 6-24.
   a. Determine the steady-state residence time:

$$\tau_{CMFR} = \frac{[Mn^{2+}]_0 - [Mn^{2+}]}{k'[Mn^{2+}]} = \frac{(2-0.1)\ mg/L}{(319\ min^{-1})(0.1\ mg/L)}$$
$$= 5.96 \times 10^{-2}\ min$$

   b. Estimate the CMFR volume:

$$V_{CMFR} = \tau_{CMFR}\, Q$$
$$= (5.96 \times 10^{-2}\ min)(10{,}000\ m^3/d)(1\ d/1440\ min)$$
$$= 0.414\ m^3$$

3. Determine the steady-state residence time and volume for the PFR.
   a. Calculate the required residence time using Eq. 6-70:

$$\tau_{PFR} = \frac{1}{k'}\ln\left(\frac{\left[Mn^{2+}\right]_0}{\left[Mn^{2+}\right]}\right) = \frac{1}{319\ min}\ln\left(\frac{2}{0.1}\right) = 9.4 \times 10^{-3}\ min$$

   b. Estimate the required volume:

$$V_{PFR} = \tau_{PFR} Q$$
$$= (9.73 \times 10^{-3}\ min)(10{,}000\ m^3/d)(1 d/1440\ min)$$
$$= 0.065\ m^3$$

OXIDATION WITH AIR

Aeration can be used to provide DO to the water to convert Fe^2 and Mn^{2+} to $Fe(OH)_3$ and MnO_2, respectively. From a stoichiometric standpoint, 1 mg of oxygen can oxidize 7 mg of soluble Fe^{2+} and 3.4 mg soluble Mn^{2+} (see Tables 20-8 and 20-10). However, the rate of oxidation is slow and not practical for Mn^{2+} at typical pH values for natural waters. Even at a pH of 9.5, it takes about 1 h detention time for Mn^{2+} to oxidize. In comparison, iron can be completely oxidized in about 15 min at pH values around 7.5 to 8.0 when not complexed with NOM.

A number of aeration devices used to supply oxygen to water have been discussed in Chap. 14. Diffused aeration is one process where air diffusers are located along or near the bottom of a tank that is 3 to 5 m (12 to 15 ft) deep. The volumetric air-to-water ratio is typically around 0.75 to 1.0 and the average oxygen transfer efficiency is only around 5 to 10 percent. Diffused aeration is not used very often because it is not very effective in terms of oxygen transfer.

A more commonly used and effective aeration device is the coke tray aerator. This process contains a series of three to five perforated stainless steel trays 0.3 to 0.45 m (1.0 to 1.5 ft) apart that contain 5- to 10-cm-diameter crushed coke or limestone or plastic random packing. The packing provides air–water contact area as the water flows down through the trays while air flows across the trays providing oxygen to the water for oxidation to take place. The water-loading rate is typically around 600 to 800 $L/m^2 \cdot min$ (15 to 20 gpm/ft^2). After a short period of operation, the iron deposits will coat the surface of the coke and assist in the oxidation process. Eventually the oxidation products will build up on the coke and begin to clog the system and the coke must be cleaned or replaced (ASCE/AWWA, 1990). This process is effective for soluble iron removal but not for iron that is organically bound. A baffled basin, which will provide 15 to 30 min additional contact time for oxidation to take place, is used following the aeration device. The oxidized insoluble iron is then removed by filtration. When iron concentrations are greater than 5 mg/L, the addition of alum after aeration followed by flocculation and sedimentation may be required prior to filtration (Kawamura, 2000).

OXIDATION WITH CHLORINE

Traditionally, iron and manganese were controlled using chlorine alone or combined with potassium permanganate under alkaline pH conditions followed by alum coagulation, clarification, and filtration. Free-chlorine doses as high as 5 mg/L have been used effectively to oxidize soluble Fe^{2+} ions as well as organically bound iron. However, in the current regulatory climate using such high doses of chlorine may not be desirable from a disinfection by-product (DBP) formation control standpoint. The stoichiometric reactions for chlorination of Fe^{2+} and Mn^{2+} are shown in Tables 20-8 and 20-10, respectively. Oxidation of Fe^{2+} requires about 0.64 mg chlorine/mg Fe^{2+} while Mn^{2+} oxidation requires 1.29 mg chlorine/mg Mn^{2+}. Because the rate of oxidation is pH dependent, a pH of 8.0 to 8.5 is needed to provide Fe^{2+} oxidation times of about 15 to 30 min. Oxidation of Mn^{2+} requires 2 to 3 h and is not effective under these conditions. If ammonia is present in the water, it will consume chlorine and form chloramines, which will significantly reduce the rate of oxidation for both Fe^{2+} and Mn^{2+}.

A common process for iron and manganese removal incorporates prechlorination, alum coagulation, sedimentation, and filtration. In this process, the filter media is conditioned with permanganate to form a manganese oxide coating on the surface of the media. The Mn^{2+} and Fe^{2+} will readily adsorb onto the media but can desorb if not oxidized. Chlorine addition just prior to filtration can be used to oxidize the adsorbed Mn^{2+} and Fe^{2+} and convert them to oxides providing further adsorption sites for oxidation to take place. Any DO in the water can also oxidize adsorbed Fe^{2+}. This process has proven to be effective for Mn^{2+} removal to levels less than 0.02 mg/L and requires minimum process control. The advantage of this process is that the chlorine addition can be easily controlled depending upon the concentration of Mn^{2+} in the water, thus minimizing the formation of DBPs. When Mn^{2+} concentrations in the raw water are low and chlorination is shut off, the oxides will gradually be stripped off but can be reestablished when the chlorine is again added.

OXIDATION WITH CHLORINE DIOXIDE

Chlorine dioxide (ClO2) is a stronger oxidant than free chlorine and can effectively oxidize soluble Fe^{2+} and Mn^{2+} ions. The observed removal of soluble Mn^{2+} is slightly greater than predicted from the stoichiometry shown in Table 20-10, due to adsorption of Mn^{2+} on the MnO_x solids that are formed during oxidation (Knocke et al., 1991). When the ClO_2 dosages used are greater or equal to the theoretical stoichiometric quantity, the observed Mn^{2+} removal is very close to the theoretical prediction, which is attributed to the adsorbed Mn^{2+} on the MnO_x being oxidized. Overall, the observed oxidation reaction of Mn^{2+} by ClO_2 appears to be a one-electron transfer with chlorite being the oxidant by-product. Oxidation of Fe^{2+} by ClO_2 shows a five-electron transfer resulting in Cl^- being the oxidant by-product. Consequently, the dosage requirement for Fe^{2+} oxidation is about 10 times less than for Mn^{2+}. These results have been verified by tests using ClO_2 as the oxidant (Knocke et al., 1991).

The rate of reaction for Mn^{2+} and Fe^{2+} with ClO_2 is quite fast. In the absence of NOM and for pH values of 5.5 or greater, complete oxidation for Mn^{2+} can occur in about 20 s or less. In the presence of NOM and the same pH range, complete Mn^{2+} oxidation also requires about 20 s. The impact of NOM on Mn^{2+} oxidation is observed to be small, as it does not appear to complex with NOM (Knocke et al., 1991). Oxidation of soluble Fe^{2+} by ClO_2 requires about 5 s at a pH value of 5.5, and faster rates are possible at higher pH values. However, when Fe^{2+} is complexed with NOM, it becomes highly resistant to oxidation and must be removed by other processes that remove NOM, such as coagulation, or activated carbon treatment.

POTASSIUM PERMANGANATE AND GREENSAND FILTRATION

Soluble Fe^{2+} can also be oxidized using $KMnO_4$ at a similar rate, but the cost of $KMnO_4$ is higher than the cost of chlorine. Oxidation times for soluble and particulate Mn^{2+} in the presence of NOM are very fast (<20 s) at pH 5.5 and the rate increases as the pH increases. Soluble Fe^{2+} can also be oxidized using $KMnO_4$ at a similar rate, but because the cost of $KMnO_4$ is much higher than the cost of chlorine, the process is usually not practical. In applications where Fe^{2+} and Mn^{2+} are both present, Fe^{2+} is usually oxidized first using chlorine followed by the addition of $KMnO_4$ for Mn^{2+}

Example 20-4 Stoichiometric calculation for Fe^{2+} oxidation by chlorine

A process treating 100,000 m^3/d groundwater containing 5 mg/L of ferrous iron (Fe^{2+}) is being treated with chlorine. Calculate the quantity of chlorine required to oxidize ferrous to ferric hydroxide, alkalinity consumed as $CaCO_3$, and quantity of sludge produced given the oxidation reactions for iron using chlorine as shown in Table 20-8.

Solution

1. Determine the quantity of chlorine required using the values for the chlorine reaction shown in Table 20-8:

$$\begin{aligned}\text{Chlorine required} &= \left(5\ \text{mg Fe}^{2+}/\text{L}\right)\left(0.64\ \text{mg Cl}_2/\text{mg Fe}^{2+}\right)\\ &\quad\left(100{,}000\ \text{m}^3/\text{d}\right)\left(10^3\ \text{L/m}^3\right)\left(1\ \text{kg}/10^6\ \text{mg}\right)\\ &= 320\ \text{kg/d}\end{aligned}$$

2. Determine the quantity of alkalinity consumed for the chlorine reaction shown in Table 20-8:

$$\begin{aligned}\text{Alkalinity consumed} &= \left(5\ \text{mg Fe}^{2+}/\text{L}\right)\left(2.70\ \text{mg alkalinity/mg Fe}^{2+}\right)\\ &\quad\left(100{,}000\ \text{m}^3/\text{d}\right)\left(10^3\ \text{L/m}^3\right)\left(1\ \text{kg}/10^6\ \text{mg}\right)\\ &= 1350\ \text{kg/d}\end{aligned}$$

3. Determine the quantity of sludge produced for the chlorine reaction shown in Table 20-8:

$$\begin{aligned}&\text{Sludge produced as Fe(OH}_3)\\ &= \left(5\ \text{mg Fe}^{2+}/\text{L}\right)\left(1.90\ \text{mg sludge/mg Fe}^{2+}\right)\\ &\quad\left(100{,}000\ \text{m}^3/\text{d}\right)\left(10^3\ \text{L/m}^3\right)\left(1\ \text{kg}/10^6\ \text{mg}\right)\\ &= 950\ \text{kg/d}\end{aligned}$$

oxidation. The oxidation of Fe^{2+} complexed with NOM requires contact times greater than 1 h and oxidant dosages above the theoretical stoichiometric amount. As a result, oxidation with $KMnO_4$ is not a practical process for removal of Fe^{2+} complexed with NOM.

A typical process for the removal of soluble Fe^{2+} and Mn^{2+} involves adding $KMnO_4$ as a solution ahead of a filter. After the addition of $KMnO_4$ and an alkali (if required), the oxidized water is delivered to a specially prepared filter. The contact time after the oxidant addition is typically 5 min at 20°C or 10 min at 1°C which is more than enough time for Fe^{2+} and Mn^{2+} oxidation as mentioned above. The filter media may be natural greensand, but silica sand and/or anthracite may also be used. Silica sand or anthracite are first treated with $KMnO_4$ to provide a manganese oxide coating on the media. Under normal conditions, the coating can be applied by controlled operation for several days with optimum $KMnO_4$ feed rates. Partial or marginal treatment may occur during the coating process. Once the coating is applied completely, satisfactory removals are usually maintained. The process is more efficient at pH values above 7.5.

The filtration process generally used for iron and manganese removal is pressure filtration. Filtration and backwash rates typically range from 240 to 480 m/d (4 to 8 gpm/ft^2) and 480 to 1200 m/d (8 to 20 gpm/ft^2), respectively, depending on the media size, temperature, and supplemental scour. Greensand media require periodic regeneration with a $KMnO_4$ solution. Greensand filter depths are similar to those used in conventional filtration applications; however, greensand media usually have an effective size less than 0.3 mm.

The applied dose of $KMnO_4$ should be controlled carefully because permanganate gives an easily detectable pink color in water at concentrations in the 0.05-mg/L range. Controlling the dose range is critical in avoiding consumer complaints. Some waters, such as reservoirs, experiencing periodic hypolimnologic episodes, may have seasonal variations in Mn^{2+}. Consequently, bench-and pilot-scale studies are important in determining the required dosing rate associated with these variations.

OXIDATION WITH OZONE

Ozone can be used to oxidize Fe^{2+} and Mn^{2+} but is more costly than other oxidation methods and consequently is not practiced in the United States. However, ozonation has been used successfully in Europe for both Fe^{2+} and Mn^{2+} removal with conventional treatment processes combined with preozonation, and when preozonation is part of a process train for other treatment purposes, Fe^{2+} and Mn^{2+} oxidation occurs as incidental to the primary process purpose. The stoichiometric requirements for oxidation of Fe^{2+} and Mn^{2+} are 0.43 mg O_3/mg Fe^{2+} and 0.87 mg O_3/mg Mn^{2+}. Ozonation will not impact the removal of Fe^{2+} complexed with NOM in subsequent conventional processes. Based on practical experience, overdosing with ozone will lead to the formation of various forms of permanganate and, if present, result in a pink color.

ION EXCHANGE PROCESS

Ion exchange processes may be used for removing low concentrations (<0.5 mg/L) of Fe^{2+} and Mn^{2+} from groundwaters. The majority of ion exchange applications for Fe^{2+} and Mn^{2+} are limited to treatment of

Example 20-5 Theoretical stoichiometric calculation for Fe^{2+} and Mn^{2+} removal using $KMnO_4$

A groundwater containing 5 g/m^3 Fe^{2+} and 2 g/m^3 Mn^{2+} is processed at a flow rate of 100,000 m^3/d. Potassium permanganate ($KMnO_4$) is used to oxidize the Fe^{2+} and Mn^{2+}. Calculate the quantity of potassium permanganate required, alkalinity consumed as $CaCO_3$, and quantity of sludge produced. Use the oxidation reactions for iron and manganese using $KMnO_4$ as shown in Tables 20-8 and 20-10, respectively.

Solution

1. Determine the quantity of potassium permanganate required.
   a. Compute the amount of $KMnO_4$ needed for Fe^{2+} oxidation for the reaction shown with potassium permanganate in Table 20-8:

$$\begin{aligned} &KMnO_4 \text{ required due to } Fe^{2+} \\ &= \left(5 \text{ g } Fe^{2+}/m^3\right)\left(0.94 \text{ g } KMnO_4/\text{g } Fe^{2+}\right) \\ &\quad \left(100{,}000 \text{ m}^3/\text{d}\right)\left(1 \text{ kg}/10^3 \text{ g}\right) \\ &= 470 \text{ kg/d} \end{aligned}$$

   b. Calculate the $KMnO_4$ required due to Mn^{2+} for the reaction shown with potassium permanganate in Table 20-10:

$$\begin{aligned} &KMnO_4 \text{ required due to } Mn^{2+} \\ &= \left(2 \text{ g } Mn^{2+}/m^3\right)\left(1.92 \text{ g } KMnO_4/\text{g } Fe^{2+}\right) \\ &\quad \left(100{,}000 \text{ m}^3/\text{d}\right)\left(1 \text{ kg}/10^3 \text{g}\right) \\ &= 384 \text{ kg/d} \end{aligned}$$

   c. Calculate the total $KMnO_4$ required:

$$\text{Total } KMnO_4 \text{ required} = 470 + 384 = 854 \text{ kg/d}$$

2. Determine the quantity of alkalinity consumed as $CaCO_3$.
   a. Calculate the quantity of alkalinity consumed due to Fe^{2+} for the reaction shown with potassium permanganate in Table 20-8:

$$\begin{aligned} &\text{Alkalinity consumed due to } Fe^{2+} \\ &= \left(5 \text{ g } Fe^{2+}/m^3\right)\left(1.50 \text{ g alkalinity/g } Fe^{2+}\right) \\ &\quad \left(100{,}000 \text{ m}^3/\text{d}\right)\left(1 \text{ kg}/10^3 \text{ g}\right) \\ &= 750 \text{ kg/d} \end{aligned}$$

b. Calculate the quantity of alkalinity consumed due to Mn^{2+} for the reaction shown with potassium permanganate in Table 20-10:

$$\text{Alkalinity consumed due to } Mn^{2+}$$
$$= \left(2\ \text{g}\ Mn^{2+}/\text{m}^3\right)\left(1.21\ \text{g alkalinity/g}\ Fe^{2+}\right)$$
$$\left(100{,}000\ \text{m}^3/\text{d}\right)\left(1\ \text{kg}/10^3\ \text{g}\right)$$
$$= 242\ \text{kg/d}$$

c. Calculate the total quantity of alkalinity consumed:

$$\text{Total alkalinity consumed} = 750 + 242\ \text{kg/d} = 992\ \text{kg/d}$$

3. Determine the quantity of sludge produced.

a. Calculate the sludge produced from Fe^{2+} oxidation for the reaction shown with potassium permanganate in Table 20-8:

$$\text{Sludge produced due to } Fe^{2+}$$
$$= \left(5\ \text{g}\ Fe^{2+}/\text{m}^3\right)\left(2.43\ \text{g sludge/g}\ Fe^{2+}\right)$$
$$\left(100{,}000\ \text{m}^3/\text{d}\right)\left(1\ \text{kg}/10^3\ \text{g}\right)$$
$$= 1215\ \text{kg/d}$$

b. Calculate the sludge produced from Mn^{2+} oxidation for the reaction shown with potassium permanganate in Table 20-10:

$$\text{Sludge produced due to } Mn^{2+}$$
$$= \left(2\ \text{g}\ Mn^{2+}/\text{m}^3\right)\left(2.64\ \text{g sludge/g}\ Fe^{2+}\right)$$
$$\left(100{,}000\ \text{m}^3/\text{d}\right)(1\ \text{kg}/10^3\ \text{g})$$
$$= 528\ \text{kg/d}$$

c. Calculate the total sludge produced from Mn^{2+} and Fe^{2+} oxidation:

$$\text{Sludge produced due to } Mn^{2+} = 1215 + 528$$
$$= 1743\ \text{kg dry solids/d}$$

Comment

The values computed in this example were derived from a theoretical stoichiometric calculation point of view. In practical application, Fe^{2+} is usually oxidized first using chlorine followed by the addition of $KMnO_4$ for Mn^{2+} because the cost of $KMnO_4$ is much higher than the cost of chlorine.

industrial water and for point-of-use treatment systems for single-family dwellings using groundwater. Usually a strong-acid cation (SAC) exchange resin in the sodium form is used. The separation factors for several cations for a polystyrene resin sometimes found in groundwater are shown in Table 16-7. The separation factors for Fe^{2+} and Mn^{2+} are slightly lower than for Ca^{2+}, and the Mn^{2+} separation factor is slightly lower than Mg^{2+}. For waters containing moderate to high hardness, Fe^{2+} and Mn^{2+} may be removed, but the regeneration frequency will be higher than for waters that are low in hardness.

The use of SAC resins will not remove Fe^{2+} complexed with NOM; however, SBA resins have been reported to remove up to 95 percent of Fe^{2+} complexed with NOM (Clifford, 1999). The City of Santa Monica, California, removes about 1.0 mg/L of iron from a 30,000-m^3/d (7.5-mgd) groundwater flow by a process of aeration and filtration through 0.762 m (30 in.) of ion exchange media. The water is softened in the filtering stage. The principal design parameters for this system include aeration with 10min contact time and volumetric air-to-water ratio of 0.75; unbaffled contact basin with 60 min contact time; and 9.8 m/h (4 gpm/ft^2) filtration rate through a polystyrene resin bed with an effective size of 0.4 mm. The resin requires periodic acid treatment due to fouling of the surface by oxidation. Effective resin life is about 12 years.

MEMBRANE PROCESS

Reverse osmosis membranes can be very effective for the removal of soluble Fe^{2+} and Mn^{2+}. However, as discussed in Chap. 17, even a small amount of oxidized iron and manganese can foul membranes and cause a decrease in their effectiveness. Pretreatment systems are typically used to remove oxidized iron and manganese prior to treatment with RO. In recent work in the Netherlands, it was found that Fe^{2+} and Mn^{2+} were removed effectively when the feed water was under anaerobic conditions (Kartinen and Martin, 2001). Two full-scale RO plants were involved in these studies, with Fe^{2+} concentrations ranging from 11 to 25 mg/L Fe^{2+}. In addition, fouling of the membranes was found to be considerably less under anaerobic operating conditions.

STABILIZATION PROCESS

Stabilization of soluble Fe^{2+} and Mn^{2+} is the opposite of oxidation. The chemical used for stabilization in water treatment is sodium hexametaphosphate $(NaPO_3)_6$ (SHMP), commonly known as polyphosphate, glassy phosphate, or polysilicate. This chemical is available in crystal, granular, or liquid form and is highly soluble. Chemical addition should occur before the water has a chance to come in contact with air or chlorine to ensure that the Fe^{2+} and Mn^{2+} are still in the soluble state. However, stabilization agents can be aggressive compounds with respect to metals, and they may dissolve precipitated iron and manganese or promote dissolution of metallic pipe materials.

The high phosphate content (66 percent P_2O_5) of SHMP results in the formation of phosphate as a by-product of the reaction, which can promote biological growth. Where treated water is stored in open reservoirs, SHMP must be used with caution; otherwise, algal blooms and slimes may result. It should also be noted that stabilization does not remove Fe^{2+} and Mn^{2+} but merely holds it in an aesthetically acceptable condition that will degrade with time. Stabilized Fe^{2+} and Mn^{2+} will also appear on analytical tests. Feed rates for SHMP are typically less than 2 mg/L but should be determined by testing. Stabilization may be considered for waters if the Fe^{2+} is in the range of 0.3 to 1.0 mg/L and/or if the Mn^{2+} content is between 0.05 and 0.1 mg/L. The additive demand of polyvalent cations must be considered.

LIME TREATMENT

Lime treatment is effective in removing both Fe^{2+} and Mn^{2+}. Excellent removals are obtained if the water is preaerated, the pH exceeds 9.8 during the process, and sufficient alkalinity (>20 mg/L as $CaCO_3$) is present. Softening is more costly than other processes for Fe^{2+} and Mn^{2+} removal because of the high capital costs. High iron and manganese removals preclude recalcination of lime sludge due to impurity of the sludge.

20-4 Softening

Theoretically, the hardness of water is defined as the sum of the soluble concentrations of polyvalent cations—all expressed as equivalent concentrations of calcium carbonate. *Carbonate hardness* is defined as the concentration of Ca^{2+} and Mg^{2+} and other polyvalent cations in water that are associated with the anions that comprise alkalinity (e.g., HCO_3^-, CO_3^{2-}). Similarly, *noncarbonate hardness* is defined as the concentration of Ca^{2+} and Mg^{2+} and other polyvalent cations in water that are associated with nonalkalinity anions (e.g., SO_4^{2-}, Cl^-). For example, carbonate hardness would be present in water after dissolution of $CaCO_3$ and $MgCO_3$, whereas noncarbonate hardness would appear if $CaSO_4$, $CaCl_2$, $MgSO_4$, and $MgCl_2$ were dissolved in the water. In water treatment, total hardness is usually expressed as the sum of the carbonate hardness and noncarbonate hardness.

Water treatment issues related to hardness removal have traditionally been related to aesthetics, although there is some nutritional benefit. The presence of hardness causes scale in pipes and hot-water heaters, high soap consumption, and the deterioration of fabrics. Removing hardness, termed *softening*, may be accomplished either by chemical precipitation as insoluble compounds, ion exchange, or membrane processes. Hardness has been removed successfully using ion exchange or low-pressure softening membranes. These processes are discussed in Chaps. 16 and 17, respectively.

The purpose of this section is to present the removal of hardness by means of chemical precipitation. Topics to be considered include (1) sources of hardness, (2) occurrence of hardness, (3) chemical precipitation methods, and (4) types of process configurations.

Sources of Hardness

Hardness in natural waters usually comes from the dissolution of minerals from geologic formations that contain calcium and magnesium. Two of the most common forms are calcite ($CaCO_3$) and dolomite [$CaMg(CO_3)_2$]. Calcite, or *chalix*, the Greek word for "lime", is one of the most common minerals and makes up about 4 percent by weight of the Earth's crust. Calcite can dissolve in weak acidic environments such as in some groundwaters. Dolomite makes up approximately 2 percent by weight of the Earth's crust. The mineral was named after French geologist Deodat de Dolomieu (1750–1801). Dolomite does not easily dissolve in acidic environments as calcite and is more often found in tropical marine environments. Groundwaters rich in dissolved magnesium usually contain a significant quantity of salt, which is thought to be essential in the formation of dolomite.

The distribution of hard waters in the United States is shown on Fig. 20-8. Although waters above 150 mg/L hardness (as $CaCO_3$) are considered very

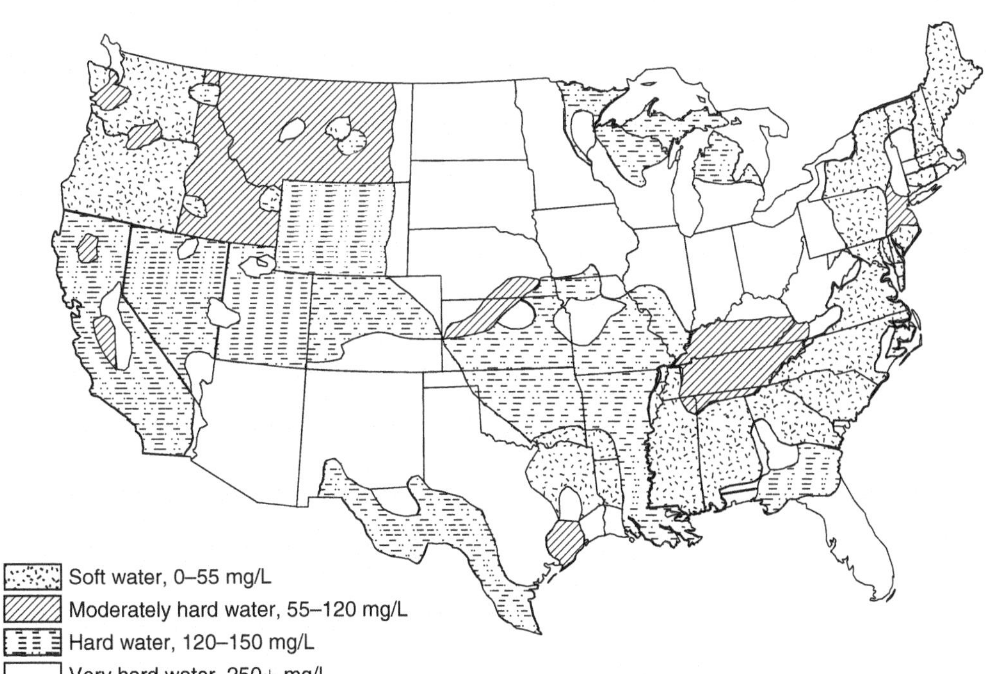

Figure 20-8
Distribution of hard water in United States. The areas shown define approximate hardness values for municipal water supplies. (Reprinted with permission from Ciaccio, 1971.)

hard, many utilities do not soften the water. Some utilities soften water when the total hardness exceeds 150 mg/L. The finished water hardness produced by a utility softening plant may vary from a low of 50 mg/L to a high of 150 mg/L depending upon the process configuration, economics, and public acceptance. A historical goal has been between 80 and 100 mg/L for both aesthetics and corrosion control (AWWA, 1969).

Softening by Chemical Precipitation

Precipitation softening relies on the relative insolubilities of calcium carbonate and magnesium hydroxide. The choice of precipitating chemicals depends upon the raw-water quality, which complicates the selection of the optimum treatment process, although the chemistry is straightforward. The equilibrium solubilities of calcium carbonate and magnesium hydroxide for typical water chemistry are shown on Figs. 20-9 and 20-10. As shown,

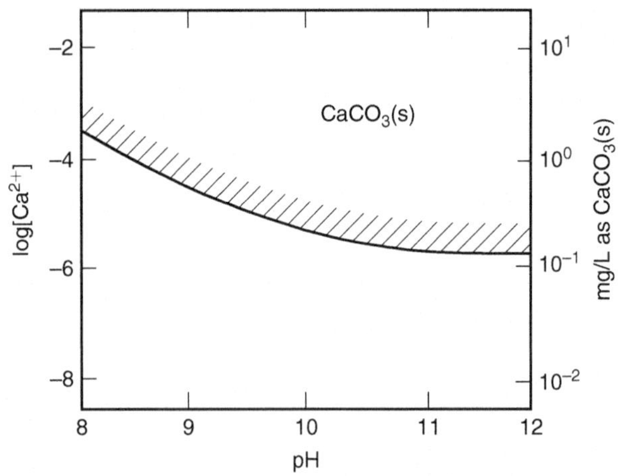

Figure 20-9
Solubility of $CaCO_3$ as function of pH ($K_{sp} = 4 \times 10^{-9}$); $C_T = 2 \times 10^{-3}$ mol/L.

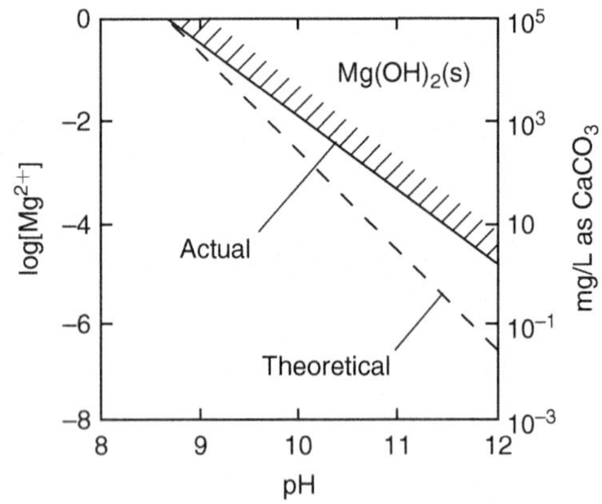

Figure 20-10
Solubility of $Mg(OH)_2$ as function of pH($K_{sp} = 2.5 \times 10^{-11}$). Solid line is for solubility in 23 waters determined by Thompson et al. (1972).

within the normal concentration range of each cation and pH found in water treatment, it is necessary to increase the pH for both calcium and magnesium precipitation. The chemicals most commonly used to precipitate calcium and magnesium are lime and caustic soda. The choice of the two depends upon the water quality and economics. When the carbonate hardness is adequate, the softening required can be accomplished by pH adjustment alone, and both calcium carbonate and magnesium hydroxide can be precipitated.

When carbonate hardness is too low, the carbonate content must be supplemented by soda ash (sodium carbonate). The minimum hardness that can be achieved depends on the solubilities of the calcium carbonate and magnesium hydroxide at the pH of softening as well as the subsequent processes. Lime–soda ash softening has been a traditional process for the removal of hardness, consisting of both carbonate and noncarbonate hardness, from water supplies.

Lime is sold commercially in the forms of quicklime and hydrated lime. Quicklime is granular and usually greater than 90 percent CaO with magnesium oxide the primary impurity. Although quicklime is less expensive than hydrated lime, it must be hydrated or slaked to $Ca(OH)_2$ before it is used for softening. Quicklime is usually crushed in a slaker and fed to form a slurry containing about 5 percent calcium hydroxide. Powdered, hydrated lime contains about 70 percent $Ca(OH)_2$ and is prepared for use in the softening process by fluidizing in a tank containing a turbine mixer. Soda ash is a grayish-white powder and is nearly 98 percent sodium carbonate. Soda ash may be added simultaneously with lime to the treatment train or it may be added sequentially, following the addition of lime.

Aside from eliminating the aesthetic problems associated with hardness, lime softening has many potential health benefits as well. The use of lime softening can effectively remove heavy metals (e.g., lead, chromium, mercury, arsenic), iron and manganese, turbidity, and some organic compounds including a substantial amount of NOM and kill algae, bacteria, and viruses. More recently, the U.S. EPA has termed a variation of the lime-softening process as *enhanced softening* because the TOC removals can be as high as 40 to 80 percent depending upon the pH. When TOC removals are high, significant control of THMs and HAAs can be achieved when using chlorine for disinfection.

Chemistry of Water Softening by Precipitation

The principal reactions involved in the precipitation of hardness depend on whether the hardness is carbonate or noncarbonate. Chemical reactions for the removal of carbonate and noncarbonate hardness are presented and discussed below, along with treatment process options.

LIME SOFTENING: CARBONATE HARDNESS

The chemical reactions for lime softening are

$$CO_2 + Ca(OH)_2 \rightarrow CaCO_3\,(s) + H_2O \tag{20-19}$$

$$Ca^{2+} + 2HCO_3^- + Ca(OH)_2 \rightarrow 2CaCO_3\,(s) + 2H_2O \tag{20-20}$$

$$Mg^{2+} + 2HCO_3^- + 2Ca(OH)_2 \rightarrow 2CaCO_3\,(s) + Mg(OH)_2\,(s) + 2H_2O \tag{20-21}$$

$$Mg^{2+} + SO4^{2-} + Ca(OH)_2 \rightarrow Mg(OH)_2\,(s) + Ca^{2+} + SO4^{2-} \tag{20-22}$$

When lime is added to water, it will first react with any free CO_2, forming a calcium carbonate precipitate as shown in Eq. 20-19, and does not reduce any hardness. This reaction will take place first as CO_2 is a stronger acid than HCO_3^-. The conversion of bicarbonate to carbonate as a function of pH is shown on Fig. 20-11. Complete conversion would require a pH value greater than 12 to attain complete utilization of the bicarbonate alkalinity for calcium precipitation, as shown on Fig. 20-11. In practice, when alkalinity is present as bicarbonate, the optimum pH for maximum calcium carbonate precipitation may be as low as 9.3 because a significant amount of carbonate is in equilibrium with bicarbonate and more carbonate is formed as precipitation occurs.

The overall reaction for the removal of magnesium when alkalinity is present as bicarbonate (i.e., magnesium carbonate hardness) is given by Eq. 20-21. Based on practical experience, it has been found that a pH value of at least 10.5 or greater is required for effective $Mg(OH)_2$ precipitation. The range of lime dosages in excess of the stoichiometric amount required to raise the pH for precipitation of $MgCO_3$ as $Mg(OH)_2$ has been reported to be from 30 to 70 mg/L as $CaCO_3$ (Degrémont, 2007; Faust and Aly, 1998; Hammer and Hammer, 2001). Jar testing is recommended to obtain

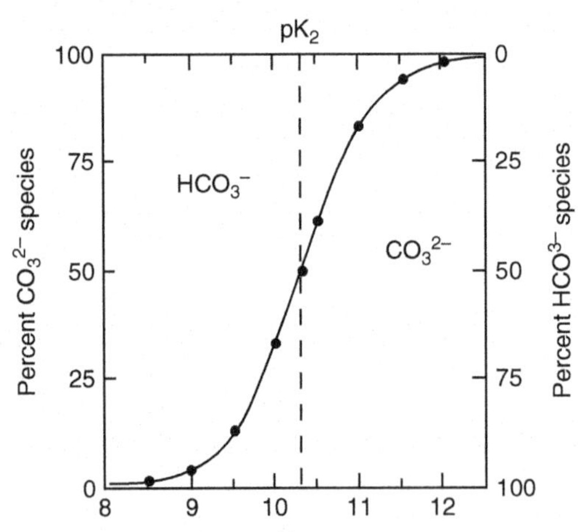

Figure 20-11
Distribution of carbonate and bicarbonate species as function of pH (25°C).

a more precise amount of excess lime required for effective $Mg(OH)_2$ precipitation for a given water source.

LIME–SODA SOFTENING: CARBONATE AND NONCARBONATE HARDNESS

Sometimes there may be a lack of carbonate alkalinity present to react with the lime and it becomes necessary to add an external source. Typically, soda ash, Na_2CO_3, is used, as shown in the reaction

$$Ca^{2+} + SO_4^{2-} + Na_2CO_3 \rightarrow CaCO_3\,(s) + 2Na^+ + SO_4^{2-} \tag{20-23}$$

The calcium noncarbonate hardness in Eq. 20-23 may be present in the untreated water or may result from the precipitation of magnesium noncarbonate hardness, as shown in Eq. 20-22. For both reactions, the amount of soda ash required depends upon the amount of noncarbonate hardness to be removed.

CAUSTIC SODA SOFTENING: CARBONATE AND NONCARBONATE HARDNESS

Caustic soda (NaOH) is an alternative to the use of lime softening when there is insufficient carbonate hardness present in the raw water to react with lime. The choice between using soda ash and caustic soda will depend upon the economics and other factors such as the ease of handling (NaOH is purchased as a 50 percent solution) and magnesium content. Some concern has been expressed about the use of sodium-containing compounds because, when noncarbonate hardness is removed, both use of soda ash and caustic soda result in the replacement of the divalent hardness ions with sodium.

The reactions for caustic soda are

$$CO_2 + 2NaOH \rightarrow Na_2CO_3 + H_2O \tag{20-24}$$

$$Ca^{2+} + 2HCO_3^{-} + 2NaOH \rightarrow CaCO_3\,(s) + 2Na^+ + CO_3^{2-} + 2H_2O \tag{20-25}$$

$$Mg^{2+} + 2HCO_3^{-} + 2NaOH \rightarrow Mg(OH)_2\,(s) + 2Na^+ + CO_3^{2-} + H_2O \tag{20-26}$$

$$Mg^{2+} + SO_4^{2-} + 2NaOH \rightarrow Mg(OH)_2\,(s) + 2Na^+ + SO_4^{2-} \tag{20-27}$$

$$Ca^{2+} + SO_4^{2-} + Na_2CO_3 \rightarrow CaCO_3\,(s) + 2Na^+ + SO_4^{2-} \tag{20-23}$$

In Eqs. 20-24, 20-25, and 20-26 the sodium carbonate that is formed is available for calcium carbonate precipitation. In situations where more sodium carbonate is formed than there is calcium noncarbonate hardness to remove, the excess must remain in solution. This presents a problem of very high carbonate alkalinity, which can be reduced by acidification, but none of the excess sodium will precipitate. In contrast, when lime is used, the excess calcium is precipitated as the carbonate. Calcium carbonate precipitation is particularly important in those cases where high pH (and the concomitant high caustic alkalinity) must be attained to precipitate magnesium as $Mg(OH)_2$.

RECARBONATION: PH ADJUSTMENT

When the pH of the water after softening is greater than the saturation pH (pH_s), it is necessary to reduce the pH. The addition of CO_2 is the most common and economical method for precipitation of excess calcium and pH reduction. When excess caustic (noncarbonate) alkalinity is present, the following reaction occurs:

$$2OH^- + CO_2 \rightleftarrows CO_3^{2-} + H_2O \tag{20-28}$$

When CO_3^{2-} is formed, it will react to precipitate any calcium that is present above the saturation of calcium carbonate. Adding additional CO_2 will then lower the pH to a point of saturation equilibrium, as shown in the reaction

$$CO_3^{2-} + CO_2 + H_2O \rightleftarrows 2HCO_3^- \tag{20-29}$$

Historically, CO_2 was produced by the combustion of diesel fuel or natural gas either under water or in external burners, but the combustion process only yields about 12 percent CO_2. Bulk pressurized/liquefied CO_2 is now commonly available and widely used because it is more convenient and the operation and maintenance problems associated with CO_2 generation are eliminated.

In the case of waters that only require selective calcium removal, the lime-treated water will be supersaturated with calcium carbonate and the pH will be between 10.0 and 10.6. The addition of carbon dioxide to the water will convert the carbonate ions, to bicarbonate ions, as shown by the reaction

$$Ca^{2+} + CO_3^{2-} + CO_2 + H_2O \rightleftarrows Ca^{2+} + 2HCO_3^- \tag{20-30}$$

For waters that require calcium and magnesium removal, excess lime is added to precipitate the magnesium carbonate as magnesium hydroxide at a pH above 11.0. Enough carbon dioxide is needed to convert the excess hydroxide ions to carbonate ions and then all the carbonate ions to bicarbonate ions. Conversion of the excess hydroxide ions to carbonate ions will drop the pH to about 10.0 to 10.5, and the calcium hydroxide is converted to calcium carbonate precipitate and the magnesium hydroxide is converted to soluble magnesium carbonate (if it has not been removed), as shown in the reactions (Benefield and Morgan, 1999)

$$Ca^{2+} + 2OH^- + CO_2 \rightleftarrows CaCO_3(s) + H_2O \tag{20-31}$$

$$Mg^{2+} + 2OH^- + CO_2 \rightleftarrows Mg^{2+} + CO_3^{2-} + H_2O \tag{20-32}$$

As stated above, additional carbon dioxide is required to lower the pH to about 8.4 to 8.6 (Benefield and Morgan, 1999). In this case, the calcium carbonate that was precipitated will dissolve in the bicarbonate form and the magnesium carbonate will also convert to the bicarbonate form, as shown by the reactions

$$CaCO_3(s) + CO_2 + H_2O \rightleftarrows Ca^+ + 2HCO_3^- \quad (20\text{-}33)$$

$$Mg^{2+} + CO_3{}^{2-} + CO_2 + H_2O \rightleftarrows Mg^{2+} + 2HCO_3{}^{2-} \quad (20\text{-}34)$$

In a water treatment plant, the recarbonation process may take place in single or double stages. The double-stage process, sometimes called the "split recarbonation" process, according to Eqs. 20-29 and 20-30, is only practiced in a few plants as it is much more mechanically and land intensive than a single-stage process. However, the water quality produced by the two-stage process is softer and lower in alkalinity than single-stage softening when magnesium reduction is desired.

When the softened water contains little or no excess caustic alkalinity, only single-stage recarbonation is possible. It is used to reduce the pH to the saturation pH. Usually the final pH is around 8.5 to 9.0.

Kinetics of Lime Softening and Recarbonation

As discussed in Chap. 10, reactor/clarifiers and sludge blanket clarifiers are typically used for lime-softening plants. For these systems, flocculation and sedimentation times are around 20 min and 1 to 2 h, respectively. Based on kinetic studies, it has been found that about 20 to 30 min of contact time is required for Ca^{2+} to precipitate as $CaCO_3(s)$ and approach equilibrium conditions (Alexander and McClanahan, 1975). The pH is also a factor in the kinetics of Ca^{2+} precipitation, as the kinetics of $CaCO_3$ formation increase with an increase in pH in the range of 9.0 to 12 (Faust and Aly, 1998). Recycling $CaCO_3$ sludge to provide a seeding effect for Ca^{2+} precipitation will increase the kinetics of $CaCO_3$ formation in a significant way.

The kinetics of recarbonation are more efficient than $CaCO_3$ precipitation and contact times in the CO_2 diffusion and recarbonation tanks are typically a minimum of 3 and 20 min, respectively. A typical basin has baffled channels and walls and is about 3.7 m (12 ft) deep.

Types of Softening Process Configurations

For the softening processes describe above, there are several process trains that are used in practice. Specific process train selection depends upon the raw-water quality and the treated-water quality objectives. Single-stage, two-stage, and split-stream treatment softening are discussed below.

SINGLE-STAGE SOFTENING

Single-stage softening, sometimes called *undersoftening*, is used for waters that do not require the removal of magnesium hardness. A typical process schematic of the single-stage process is shown on Fig. 20-12a. Lime is added to the raw water either upstream of the reactor/clarifier in a separate flash mix process or into the reactor/clarifier. Flash mixers are used because mixing with a conventional mechanical mixer will cause precipitation formation on the blades, resulting in the dislodging and release of $CaCO_3$. The pH of the water leaving the flash mixer is about 10.2 to 10.5. Soda ash can be added in the flash mix if noncarbonate hardness is present either with the lime or sequentially after the lime has been added.

As discussed in Chap. 10 (see Table 10-8), reactor/clarifiers and sludge blanket clarifiers are typically used for water-softening processes. Where reactor/clarifiers are used, lime sludge is frequently recycled to the head of the process train to improve the efficiency of the softening process. This recycle provides a seeding effect for the $CaCO_3$ particles to flocculate.

Following sedimentation, the lime-treated water will pass through a recarbonation reactor where the addition of CO_2 is used to reduce the pH from 10.2 to 10.5 to around 8.7 to 9.0. After recarbonation, particles that may form are removed by filtration. The water is usually softened to a final hardness of about 70 to 100 mg/L as $CaCO_3$ and it is very close to the pH_s value. Sometimes postchlorination is sufficient to reduce the pH to the pH_s, eliminating the need for the recarbonation step.

TWO-STAGE SOFTENING

Sometimes postchlorination is sufficient to reduce the pH to the saturation pH (pH_s), eliminating the need for the recarbonation step. A schematic of the split recarbonation process is shown on Fig. 20-12b. Excess lime is added with a flash mixer to raise the pH to 11.0 or higher to precipitate magnesium. After sedimentation, CO_2 recarbonation is used to reduce the pH to about 10.0 to 10.6 and soda ash is added to precipitate the excess lime added for magnesium removal. The second-stage precipitation step is followed by sedimentation and then the second stage of CO_2 recarbonation is used to lower the pH value around 8.3 to 8.5. Filtration is used to remove any particles formed from the second recarbonation step.

SPLIT-TREATMENT SOFTENING

Split-treatment configurations consist of separating the process water into two or more streams, treating the streams to various degrees with different process treatment techniques, and then blending them to obtain the desired effluent water quality. Three spilt-stream configurations that have been practiced are (1) parallel softening and coagulation, (2) parallel lime softening and ion exchange or reverse osmosis, and (3) split treatment with excess lime. The first two processes are discussed briefly first, as they are used less frequently, followed by a more detailed discussion of the split treatment with excess lime process.

PARALLEL SOFTENING AND COAGULATION

For waters high in magnesium, turbidity, and/or color, parallel softening and coagulation can sometimes be economical and cost effective. In this process, part of the water to be treated is softened with excess lime to remove calcium and magnesium hardness. The remaining part of the water to be treated is coagulated to remove turbidity and color and as a result will

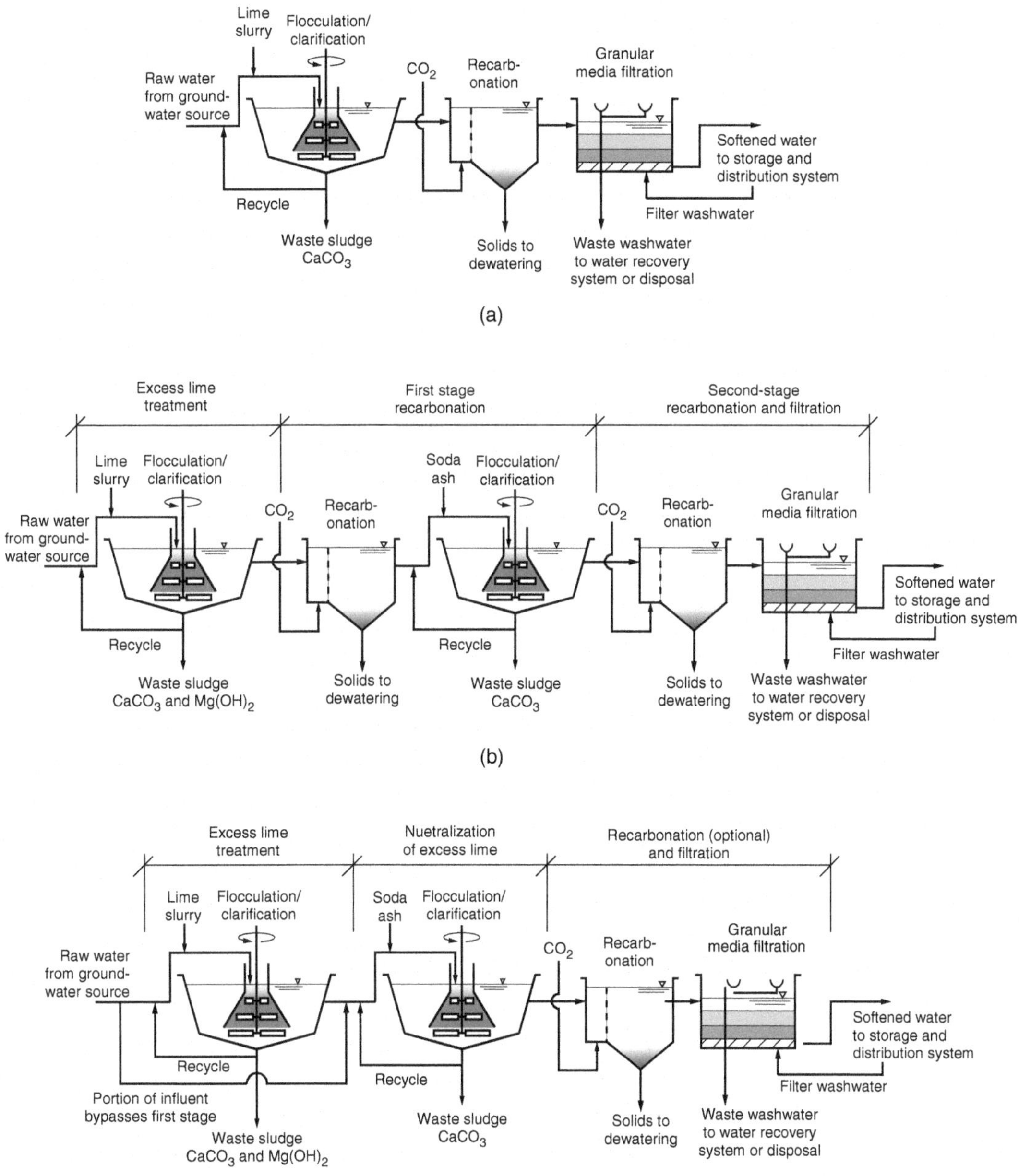

Figure 20-12
Process flow diagrams of common softening treatment techniques: (a) single-stage lime treatment; (b) two-stage excess lime–soda treatment, and (c) split-flow lime treatment.

be low in alkalinity. The waters are then blended to a balance of good color, turbidity, and hardness. A third stream has also been used in some cases to provide additional hardness for neutralization of excess caustic alkalinity in the softened water stream.

PARALLEL LIME SOFTENING AND ION EXCHANGE OR REVERSE OSMOSIS

This process modification can be used to soften raw waters with high noncarbonate and dissolved solids concentrations. Part of the water stream is treated with lime softening and the other part is softened with ion exchange or reverse osmosis. The use of ion exchange can be attractive when there is high noncarbonate hardness in the raw water, when the use of soda ash or caustic soda would not be economical, or in the case of an existing ion exchange plant. Reverse osmosis can be used instead of ion exchange for demineralization if the raw water requires additional reduction in dissolved solids.

SPLIT TREATMENT WITH EXCESS LIME

This process is used when magnesium must be reduced and the raw water contains very little noncarbonate hardness. A schematic of the excess lime split-stream process is shown on Fig. 20-12c. Part of the water is treated with excess lime softening for both calcium and magnesium hardness removal. In this part, the magnesium hardness can be removed down to its practical solubility limit of 10 mg/L. The other part of the water is bypassed and blended with the softened water prior to sedimentation. The alkalinity of the bypassed raw water is used to neutralize the excess caustic alkalinity required to reduce the magnesium in the treated water. Because the free carbon dioxide in the bypassed water is used to neutralize the excess lime of the processed stream, a recarbonation step is usually not needed. When the treatment objective of magnesium is chosen, a mass balance on the magnesium is used to determine the ratio of the bypass to treated flow, as given by the equation

$$X = \frac{\left[Mg^{2+}\right]_e - \left[Mg^{2+}\right]_t}{\left[Mg^{2+}\right]_0 - \left[Mg^{2+}\right]_t} \tag{20-35}$$

where X = ratio of bypassed flow

$[Mg^{2+}]_e$ = magnesium hardness concentration, mg/L as $CaCO_3$ in finished water (Typically magnesium hardness is less than 50 mg/L as $CaCO_3$ such that the level of total hardness in the finished water is less than 80 to 100 mg/L as $CaCO_3$)

$[Mg^{2+}]_0$ = magnesium hardness concentration (mg/L as $CaCO_3$) in raw water

$[Mg^{2+}]_t$ = magnesium hardness concentration (mg/L as $CaCO_3$) in excess lime treated water before blending

Split treatment works best for groundwaters. For surface waters, where color, taste, and odor may cause problems, the two-stage process is usually preferred. As discussed above, total hardness levels up to 120 mg/L as $CaCO_3$ are acceptable by some consumers so the calcium hardness in the finished water can be as high as 80 mg/L as $CaCO_3$.

Chemical Dose Calculations for Lime-Soda Ash Softening

An estimate of the chemical dosages for lime softening can be made using a number of different methods: (1) chemical stoichiometry, (2) the solution of simultaneous equilibria equations, (3) the use of softening diagrams, and (4) laboratory studies. All have applications in estimating chemical dosages with experimental methods providing results that most closely approximate plant performance. The chemical stoichiometry is typically used to obtain predesign information for planning and preliminary cost estimation. The solution of simultaneous equilibria equations is usually more rigorous as it involves numerical solution of a series of simultaneous equilibrium expressions. A number of programs such as RIDEQL, the MINEQL series, and the WATEQ series developed by the U.S. Geological Survey (Ball et al., 1979) can be used. These programs are typically used for research in the area of equilibrium and kinetics. The use of softening diagrams involves the use of Caldwell–Lawrence (CL) diagrams for estimating chemical dosages for the lime-softening process (Caldwell and Lawrence, 1953; Lowenthal and Marias, 1976; Merrill and Sanks, 1977). These diagrams were developed to enable equilibrium calculations to be made with less special training. A good discussion and application of this approach have been presented elsewhere (Benefield and Morgan, 1999; Sanks, 1978). The fourth approach, laboratory studies, is discussed in Chap. 9. Estimating chemical dosages for softening based on stoichiometry is illustrated below.

CHEMICAL DOSAGES BASED ON STOICHIOMETRY

The stoichiometric method is based on the assumption that all of the relevant reactions go to completion. Corrections can be made for the solubility of calcium carbonate and for the excess hydroxide alkalinity required for magnesium hydroxide precipitation. The chemical requirements can be estimated using the guidelines presented in Table 20-11. The guidelines were developed based on the stoichiometric expressions discussed previously (Eqs. 20-19 through 20-35) and the discussion of various lime-softening processes. The lime required as 100 percent CaO is calculated by noting that it serves the function of CO_2 removal, bicarbonate conversion to carbonate, and magnesium reduction. Use of the information in Table 20-11 for estimating the chemical dosages required for the various lime-softening processes is illustrated in Examples 20-6 and 20-7.

The soda ash requirement can be estimated from Eq. 20-23 by noting that soda ash is used only for noncarbonate hardness reduction because it requires about 1.9 times as much 100 percent $Na2CO_3$ as 100 percent CaO to remove the same amount of hardness. Further, because soda ash is more expensive on a mass basis than CaO, it is desirable to minimize the amount of soda ash used (i.e., the amount of noncarbonate hardness removed).

Table 20-11
Summary of chemical dosage calculations required for lime and lime–soda ash softening[a]

Process	Required Chemical Dosage Calculations
Single-stage lime: For waters with high calcium, low magnesium, and carbonate hardness	Lime addition for softening: $CaO = \{\text{carbonic acid concentration}\} + \{\text{calcium carbonate hardness}\}$ Soda ash addition for softening: $Na_2CO_3 = \text{none}$ Carbon dioxide for pH adjustment after softening: $CO_2 = \begin{Bmatrix}\text{estimated}\\\text{carbonate alkalinity}\\\text{of softened water}\end{Bmatrix} = \begin{Bmatrix}\text{source}\\\text{water}\\\text{alkalinity}\end{Bmatrix} - \begin{Bmatrix}\text{source}\\\text{water calcium}\\\text{hardness}\end{Bmatrix} + \begin{Bmatrix}\text{estimated residual}\\\text{calcium hardness}\\\text{of softened water}\end{Bmatrix}$
Excess lime: for waters with high calcium, high magnesium, and process may be one or two stages	Lime addition for softening: $CaO = \begin{Bmatrix}\text{carbonic acid}\\\text{concentration}\end{Bmatrix} + \begin{Bmatrix}\text{total}\\\text{alkalinity}\end{Bmatrix} + \begin{Bmatrix}\text{magnesium}\\\text{hardness}\end{Bmatrix} + \begin{Bmatrix}\text{excess}\\\text{lime dose}\end{Bmatrix}$ Soda ash addition for softening: $Na_2CO_3 = \text{none}$ Carbon dioxide for pH adjustment after softening: $CO_2 = \begin{Bmatrix}\text{source}\\\text{water}\\\text{alkalinity}\end{Bmatrix} - \begin{Bmatrix}\text{source}\\\text{water total}\\\text{hardness}\end{Bmatrix} - \begin{Bmatrix}\text{excess}\\\text{lime}\\\text{dose}\end{Bmatrix} + \begin{Bmatrix}\text{estimated residual}\\\text{calcium hardness}\\\text{of softened water}\end{Bmatrix} + 2\begin{Bmatrix}\text{excess}\\\text{lime}\\\text{dose}\end{Bmatrix}$ $+ \begin{Bmatrix}\text{estimated residual}\\\text{magnesium hardness}\\\text{of softened water}\end{Bmatrix}$
Single-stage lime-soda ash: for water with high calcium, low magnesium, and carbonate and noncarbonate hardness	Lime addition for softening: $CaO = \{\text{carbonic acid concentration}\} + \{\text{calcium carbonate hardness}\}$ Soda ash addition for softening: $Na_2CO_3 = \{\text{calcium noncarbonate hardness}\}$ and/or $\{\text{magnesium noncarbonate hardness}\}$ Carbon dioxide for pH adjustment after softening: $CO_2 = \begin{Bmatrix}\text{source}\\\text{water}\\\text{alkalinity}\end{Bmatrix} + \begin{Bmatrix}\text{soda}\\\text{ash}\\\text{dose}\end{Bmatrix} - \begin{Bmatrix}\text{source water}\\\text{calcium}\\\text{hardness}\end{Bmatrix} + \begin{Bmatrix}\text{estimated residual}\\\text{calcium hardness}\\\text{of softened water}\end{Bmatrix}$

Process	Required Chemical Dosage Calculations
Excess lime-soda ash: for waters with high calcium high magnesium, and carbonate and noncarbonate hardness; process may be one or two stages	Lime addition for softening: $CaO = \begin{Bmatrix}\text{carbonic}\\ \text{acid}\\ \text{concentration}\end{Bmatrix} + \begin{Bmatrix}\text{calcium}\\ \text{carbonate}\\ \text{concentration}\end{Bmatrix} + 2\begin{Bmatrix}\text{magnesium}\\ \text{carbonate}\\ \text{hardness}\end{Bmatrix} + \begin{Bmatrix}\text{magnesium}\\ \text{noncarbonate}\\ \text{hardness}\end{Bmatrix} + \begin{Bmatrix}\text{excess}\\ \text{lime}\\ \text{requirement}\end{Bmatrix}$ Soda ash addition for softening: $Na_2CO_3 = \begin{Bmatrix}\text{calcium}\\ \text{noncarbonate}\\ \text{hardness}\end{Bmatrix} + \begin{Bmatrix}\text{magnesium}\\ \text{noncarbonate}\\ \text{hardness}\end{Bmatrix}$ Carbon dioxide for pH adjustment after softening: $CO_2\text{, first stage} = \begin{Bmatrix}\text{estimated}\\ \text{hydroxide alkalinity}\\ \text{of softened water}\end{Bmatrix} = \begin{Bmatrix}\text{excess}\\ \text{lime}\\ \text{dose}\end{Bmatrix} + \begin{Bmatrix}\text{estimated residual}\\ \text{magnesium hardness}\\ \text{of softened water}\end{Bmatrix}$ $CO_2\text{ second stage} = \begin{Bmatrix}\text{estimated}\\ \text{hydroxide alkalinity}\\ \text{of softened water}\end{Bmatrix} = \begin{Bmatrix}\text{source}\\ \text{water}\\ \text{alkalinity}\end{Bmatrix} + \begin{Bmatrix}\text{soda}\\ \text{ash}\\ \text{dose}\end{Bmatrix} - \begin{Bmatrix}\text{source}\\ \text{water total}\\ \text{hardness}\end{Bmatrix} + \begin{Bmatrix}\text{estimated residual}\\ \text{hardness of}\\ \text{softened water}\end{Bmatrix}$

[a]All quantities are expressed as mg/L as $CaCO_3$.

Minimizing the use of soda ash can be done by reducing the residual carbonate hardness to a minimum or allowing the hardness of the finished water to be higher or both.

Example 20-6 Single-stage selective calcium removal

A 50-ML/d raw-water source is to be softened to reduce the hardness. The mineral analysis of the raw water is given below. The average raw-water temperature and pH were found to be 10°C and pH 7.0, respectively. Using the given information, determine the total, carbonate, and noncarbonate hardness present in the raw and finished waters; the kilograms per day of lime, soda ash, and CO_2 needed for selective calcium softening; and the kilograms per day of $CaCO_3$ solids produced. Draw the initial and final bar diagrams of the raw and softened water. Assume the residual calcium hardness in the softened water is 30 mg/L as $CaCO_3$.

Constituent	Unit	Value
H_2CO_3*	mg/L	72
Ca^{2+}	mg/L	75
Mg^{2+}	mg/L	6.1
Na^{+}	mg/L	36.8
Alkalinity	mg/L as $CaCO_3$	195
SO_4^{2-}	mg/L	60
Cl^{-}	mg/L	25

Solution

1. Develop a summary table for the chemical constituents and the conversion of all the concentrations to meq/L and mg/L as $CaCO_3$:

Chemical Constituent	Concentration, mg/L	Equivalent	Molecular Weight	Equivalent Weight	Concentration, meq/L	mg/L as $CaCO_3$
H_2CO_3*	72.0	2	62.0	31.0	2.32	116.0
Ca^{2+}	75.0	2	40.0	20.0	3.75	187.5
Mg^{2+}	6.1	2	24.4	12.2	0.50	25.0
Na^{+}	36.8	1	23.0	23.0	1.60	80.0
Total cation					5.85	292.5
Alk (HCO_3)	195.0	2	100.0	50.0	3.90	195
SO_4^{2-}	60.0	2	96.0	48.0	1.25	62.5
Cl^{-}	25.0	1	35.5	35.5	0.70	35
Total anion					5.85	292.5

As a check, the constituent anions and cations should balance as shown.

2. Construct a bar diagram of the raw water that includes the chemical constituents that are important for softening. In the development of the bar diagrams, the cation constituents are placed on the top of the diagram and the anions are placed on the bottom. In relationship to the order of cations to anions on the bar diagram, they are placed according to their reactivity to lime. For example, as stated above, lime will first react with $H_2CO_3^*$ ($[CO_2]_{aq}$) followed by HCO_3^- and then the rest of the nonreacting anions. The order for the cations is Ca^{2+} followed by Mg^{2+} and then the rest of the cations as shown below.

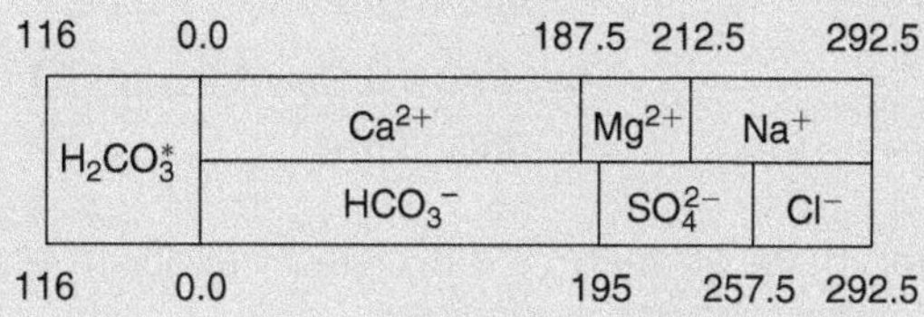

3. Determine the total hardness, carbonate hardness,and noncarbonate hardness.
   a. The total hardness can be calculated from the table as the sum of the calcium and magnesium ions as $CaCO_3$ or taken directly from the bar chart:

$$\text{Total hardness} = (187.5 + 25) = 212.5 \text{ mg/L as } CaCO_3$$

   b. The carbonate hardness is simply the sum of the calcium and magnesium ions associated with bicarbonate ions. From the bar diagram of the raw water, all the calcium is associated with bicarbonate and only a small amount of the magnesium is associated with the rest of the bicarbonate. Because all the bicarbonate is associated with calcium and magnesium, the carbonate hardness is simply equal to bicarbonate alkalinity as $CaCO_3$ as shown by the following:

$$\text{Carbonate hardness} = 187.5 + (195 - 187.5)$$
$$= 195 \text{ mg/L as } CaCO_3$$

   c. From the bar chart, the noncarbonate hardness is simply the magnesium ions not associated with carbonate hardness:

$$\text{Noncarbonate hardness} = 212.5 - 195.0$$
$$= 17.5 \text{ mg/L as } CaCO_3$$

4. Calculate the lime, soda ash, and carbon dioxide dosages required for selective calcium softening.
   a. From Table 20-11, the lime requirement for single-stage treatment is given as

$$\text{CaO required} = \text{carbonic acid concentration} + \text{calcium carbonate hardness concentration}$$

   or from the bar diagram:

$$\begin{aligned}\text{CaO required} &= [(116 \text{ mg CaCO}_3/\text{L}) + (187.5 \text{ mg CaCO}_3/\text{L})] \\ &\quad \times (28 \text{ mg CaO}/50 \text{ mg CaCO}_3) \\ &= 170 \text{ mg/L} \\ &= (170 \text{ mg CaO/L})(1 \text{ kg}/10^6 \text{ mg})(50 \times 10^6 \text{ L/d}) \\ &= 8500 \text{ kg/d}\end{aligned}$$

   b. Because there is sufficient alkalinity to precipitate calcium, no soda ash is required.
   c. From Table 20-11, the carbon dioxide requirement for selective calcium removal is equal to the estimated carbonate alkalinity of the softened water:

$$\begin{aligned}&\text{Estimated carbonate alkalinity of softened water} \\ &= \text{source water alkalinity} \\ &\quad - \text{source water calcium hardness} \\ &\quad + \text{estimated residual calcium hardness of softened water}\end{aligned}$$

   Assuming the residual calcium hardness in the softened water is 30 mg/L as $CaCO_3$, the CO_2 can be calculated as

$$\begin{aligned}&\text{Estimated carbonate alkalinity of softened water} \\ &\quad = (195 - 187.5 + 30.0) \text{ mg CaCO}_3/\text{L} \\ &\quad = 37.5 \text{ mg CaCO}_3/\text{L}\end{aligned}$$

$$\begin{aligned}&\text{CO}_2 \text{ required} \\ &\quad = (37.5 \text{ mg CaCO}_3/\text{L}) \\ &\qquad \times (22.0 \text{ mg CO}_2/50 \text{ mg CaCO}_3) = 16.5 \text{ mg/L} \\ &\quad = (16.5 \text{ mg CO}_2/\text{L})(1 \text{ kg}/10^6 \text{ mg})(50 \times 10^6 \text{ L/d}) \\ &\quad = 825 \text{ kg/d}\end{aligned}$$

5. Determine the quantity of precipitated solids. The quantity of precipitated solids will be equal to the sum of the $H_2CO_3{}^*$ precipitated as $CaCO_3$ (see Eq. 20-19) and two times the calcium carbonate hardness precipitated as $CaCO_3$ (see Eq. 20-20) because an equivalent amount of lime was added to remove the calcium as bicarbonate, minus the solubility of $CaCO_3$:

$$\begin{aligned} CaCO_3 \text{ solids produced} &= [(116.0 + 2 \times 187.5 - 30.0) \text{ mg } CaCO_3/L] \\ &\quad \times 1\left(1 \text{ kg}/10^6 \text{ mg}\right)\left(50 \times 10^6 \text{L/d}\right) \\ &= 23{,}050 \text{ kg/d} \end{aligned}$$

6. Calculate the total, carbonate, and noncarbonate hardness present in finished water. In this water, the final hardness will be the sum of the soluble calcium and the magnesium carbonate hardness plus the magnesium noncarbonate hardness:

$$\begin{aligned} &\text{Total final hardness of softened water} \\ &\quad = [(30 + 7.5) + (17.5)] \text{ mg/L} \\ &\quad = 55 \text{ mg/L as } CaCO_3 \\ &\text{Final noncarbonate hardness of softened water} \\ &\quad = 17.5 \text{ mg/L as } CaCO_3 \end{aligned}$$

7. Draw a bar diagram of the finished water.

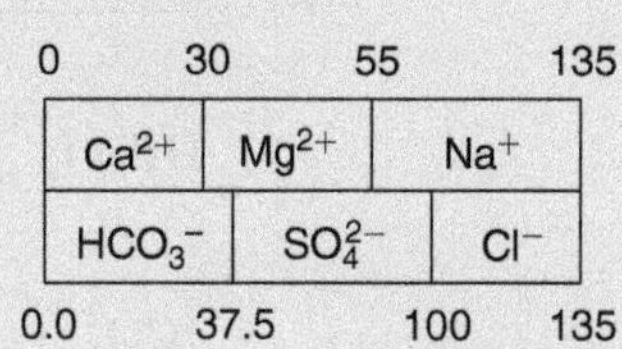

DETERMINATION OF CARBONIC ACID CONCENTRATION FOR SOFTENING CALCULATIONS

Some laboratories may provide the carbonic acid concentration, $H_2CO_3{}^*$. If the concentration of $H_2CO_3{}^*$ is not given, then it can be calculated from carbonic acid equilibria. The applicable equilibrium equations for the carbonic acid system are

$$CO_2 + H_2O \rightleftarrows H_2CO_3 \rightleftarrows H^+ + HCO_3^- \tag{20-36}$$

$$HCO_3^- \rightleftarrows H^+ + CO_3^{2-} \tag{20-37}$$

Because the dissolved fraction of the total CO_2 in water is small and hydrolyzes to H_2CO_3, the $H_2CO_3{}^*$ concentration is taken as the sum of CO_2 and

H_2CO_3 concentrations. The equilibrium constants for the stoichiometric equations (Eqs. 20-36 and 20-37) are

$$K_1 = \frac{[H^+][HCO_3^-]}{[H_2CO_3^*]} \quad (20\text{-}38)$$

$$K_2 = \frac{[H^+][CO_3^{2-}]}{[HCO_3^-]} \quad (20\text{-}39)$$

where K_1 = first dissociation constant for carbonic acid
K_2 = second dissociation constant for carbonic acid

Both K_1 and K_2 have been correlated to temperature and can be calculated from the expressions (Rossum and Merrill, 1983)

$$K_1 = 10^{14.8435-3404.71/T-0.032786T} \quad (20\text{-}40)$$

$$K_2 = 10^{6.498-2909.39/T-0.02379T} \quad (20\text{-}41)$$

where T = water temperature, K

The total carbonic species concentration, C_T, in water is represented by the expression

$$C_T = [H_2CO_3^*] + [HCO_3^-] + [CO_3^{2-}] \quad (20\text{-}42)$$

where C_T = total carbonic species concentration, mol/L

The ionization fractions of these species defined for each species are

$$\alpha_0 = \frac{[H_2CO_3^*]}{C_T} \quad (20\text{-}43)$$

$$\alpha_1 = \frac{[HCO_3^-]}{C_T} \quad (20\text{-}44)$$

$$\alpha_2 = \frac{[CO_3^{2-}]}{C_T} \quad (20\text{-}45)$$

Based on Eqs. 20-42 through 20-45, the following expressions can be used to calculate the ionization fractions based on acid reactions in natural waters (i.e., alkalinity is in the form of bicarbonate alkalinity):

$$\alpha_0 = \frac{1}{1 + K_1/[H^+] + K_1K_2/[H^+]^2} \quad (20\text{-}46)$$

$$\alpha_1 = \frac{1}{[H^+]/K_1 + 1 + K_2/[H^+]} \quad (20\text{-}47)$$

$$\alpha_2 = \frac{1}{[H^+]^2/(K_1K_2) + [H^+]/K_2 + 1} \quad (20\text{-}48)$$

Given the alkalinity, pH, and temperature of the water, Eqs. 20-42 through 20-48 can be used to calculate the H_2CO_3* concentration (procedure adapted from Benefield and Morgan, 1999).

Example 20-7 Single-stage excess lime softening

A 50-ML/d raw-water source is to be softened to reduce the hardness. The mineral analysis of the raw water is given below. The average raw-water temperature and pH were found to be 15°C and 7.2, respectively. Using the given information, determine the carbonic acid concentration in the raw water; the total carbonate and noncarbonate hardness present in the raw and finished water; the kilograms per day of lime, soda ash, and CO_2 needed for selective calcium softening; and the kilograms per day of $CaCO_3$ solids produced. Draw the initial and final bar diagrams of the raw and softened water. The residual calcium and magnesium hardnesses in the softened water are 30 and 20 mg/L as $CaCO_3$, respectively. Use an excess lime dose of 30 mg/L as $CaCO_3$.

Constituent	Unit	Value
Ca^{2+}	mg/L	60
Mg^{2+}	mg/L	20
Na^{+}	mg/L	15.9
Alk(HCO_3^-)	mg/L as $CaCO_3$	240
SO_4^{2-}	mg/L	12
Cl^-	mg/L	10

Solution

1. Develop a summary table for the chemical constituents and the conversion of all the concentrations to meq/L and mg/L as $CaCO_3$. Because the H_2CO_3* was not given, it must first be determined from the alkalinity, pH, and temperature.
   a. The carbonic acid concentration can be determined from Eq. 20–42 if it is assumed that at pH = 7.2 the alkalinity is primarily made up of bicarbonate alkalinity:

$$C_T = \left[H_2CO_3^*\right] + \left[HCO_3^-\right] + \left[CO_3^{2-}\right] = \left[H_2CO_3^*\right] + \left[HCO_3^-\right] + 0$$

   b. The C_T value can be determined from the pH, temperature, and bicarbonate alkalinity using Eqs. 20-44, 20-47, 20-40, and 20-41 as shown:

$$\alpha_1 = \frac{1}{\left[H^+\right]/K_1 + 1 + K_2/\left[H^+\right]}$$

$$C_T = \frac{\left[HCO_3^-\right]}{\alpha_1}$$

$$K_1 = 10^{14.8435-3404.71/T(K)-0.032786T(K)}$$
$$= 10^{14.8435-3404.71/288\text{ K}-0.032786\times 288\text{ K}} = 3.79\times 10^{-7}$$
$$K_2 = 10^{6.498-2909.39/T(K)-0.02379T(K)}$$
$$= 10^{6.498-2909.39/288\text{ K}-0.02379\times 288\text{ K}} = 3.50\times 10^{-11}$$
$$\alpha_1 = \frac{1}{(1.0\times 10^{-7.2})/3.79\times 10^{-7}+1+3.5\times 10^{-11}/(1.0\times 10^{-7.2})}$$
$$= 0.86$$
$$C_T = \frac{[HCO_3^-]}{\alpha_1} = \frac{(240\text{ mg/L})[1/(100\text{ g/mol})](1\text{ g}/10^3\text{ mg})}{0.86}$$
$$= 2.79\times 10^{-3}\text{ mol/L}$$

c. The carbonic acid concentration can be determined as

$$[H_2CO_3^*] = C_T - [HCO_3^-] + [CO_3^{2-}]$$
$$= 2.79\times 10^{-3} - 2.4\times 10^{-3} - 0$$
$$= (3.9\times 10^{-4}\text{ mol/L})(100\text{ g/mol})(10^3\text{ mg/g})$$
$$= 39.0\text{ mg/L as }CaCO_3$$

d. A summary of the chemical constituents in terms of $CaCO_3$ is shown in the following table:

Chemical Constituent	Concentration mg/L	Equivalents	Molecular Weight	Equivalent Weight	meq/L	mg/L as $CaCO_3$
H_2CO_3*	39	2	62.0	31.0	1.26	62.9
Cations						
Ca^{2+}	60	2	40.0	20.0	3.00	150
Mg^{2+}	20	2	24.4	12.2	1.64	82
Na^+	15.9	1	23.0	23.0	0.69	34.5
Total					5.33	266.5
Anions						
Alk as $CaCO_3$	240	2	100	50.0	4.8	240
SO_4^{2-}	12	2	96.0	48	0.25	12.5
Cl^-	10	1	35.5	35.5	0.28	14
Total					5.33	266.5

2. Construct a bar diagram of the raw water that includes the chemical constituents (as $CaCO_3$) that are important for softening (see step 2 in Example 20-6 for a discussion of the development of the bar diagram).

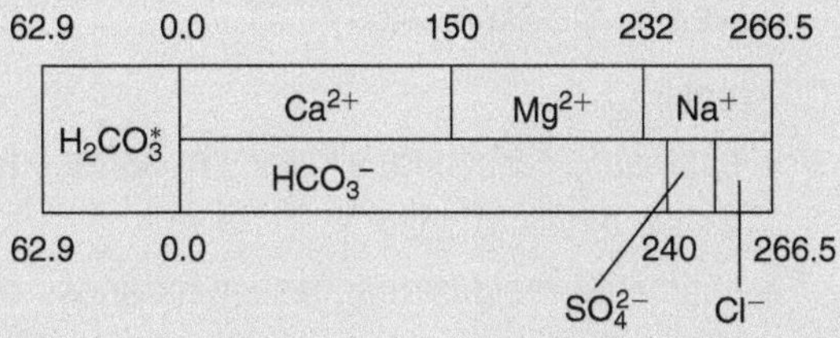

3. Determine the total hardness, carbonate hardness, and noncarbonate hardness.
   a. The total hardness can be calculated from the summary table as the sum of the calcium and magnesium ions as $CaCO_3$ or taken directly from the bar chart:

$$\text{Total hardness} = (150 + 82)\ \text{mg/L as CaCO}_3$$
$$= 232\ \text{mg/L as CaCO}_3$$

   b. The carbonate hardness is the same as the total hardness because there is no noncarbonate hardness present:

$$\text{Carbonate hardness} = 232\ \text{mg/L as CaCO}_3$$

   c. In this case the noncarbonate hardness is zero.
4. Calculate the lime, soda ash, and carbon dioxide dosages required for excess lime softening.
   a. From Table 20-11 for excess lime treatment, the following expression can be used to calculate the lime dose:

$$\begin{aligned}\text{CaO required} &= \text{carbonic acid concentration} + \text{calcium hardness} \\ &\quad + \text{magnesium hardness} + \text{excess lime dose} \\ &= [(62.9 + 240 + 82 + 30)\ \text{mg CaCO}_3/\text{L}] \\ &\quad \times (28\ \text{mg CaO}/50\ \text{mg CaCO}_3) \\ &= 232\ \text{mg/L} \\ &= (232\ \text{mg CaO/L})(1\ \text{kg}/10^6\ \text{mg})(50 \times 10^6\ \text{L/d}) \\ &= 11{,}600\ \text{kg/d}\end{aligned}$$

   b. Because all the hardness is in the carbonate form, no soda ash dose is required.

c. The carbon dioxide requirement for selective calcium removal is equal to the estimated carbonate alkalinity of the softened water plus two times the excess lime dose and the estimated residual magnesium hardness of the softened water or as given by the expression

$$\begin{aligned}CO_2 \text{ required} &= \text{estimated carbonate alkalinity of softened water}\\ &+ 2 \times \text{excess lime dose}\\ &+ \text{estimated residual magnesium hardness of softened water}\end{aligned}$$

d. Carbonate alkalinity of the softened water is calculated as

$$\begin{aligned}&\text{Estimated carbonate alkalinity of softened water}\\ &\quad= \text{source water alkalinity} - \text{source water total hardness}\\ &\qquad - \text{excess lime dose}\\ &\qquad + \text{estimated residual calcium hardness of softened water}\\ &\quad= (240 - 232 - 30 + 30)\text{ mg } CaCO_3/L = 8 \text{ mg } CaCO_3/L\end{aligned}$$

e. Consequently, the CO_2 dosage requirement is calculated as

$$\begin{aligned}CO_2 \text{ required} &= [(8 + 2 \times 30 + 20)\text{ mg } CaCO_3/L](22 \text{ mg } CO_2/50 \text{ mg } CaCO_3)\\ &= 39 \text{ mg } CO_2/L\\ &= (39 \text{ mg } CO_2/L)(1 \text{ kg}/10^6 \text{ mg})(50 \times 10^6 \text{ L/d}) = 1{,}950 \text{ kg/d}\end{aligned}$$

f. The $CaCO_3$ solids precipitated can be calculated from the table where it will be equal to the sum of the $H_2CO_3{}^*$ precipitated as $CaCO_3$, 2 times the calcium carbonate hardness precipitated as $CaCO_3$ (see Eq. 20-20), 2 times the magnesium hardness precipitated, precipitation of the excess lime added, and minus the solubility of $CaCO_3$:

$$\begin{aligned}CaCO_3 \text{ solids produced}\\ &= [(62.9 + 2 \times 150 + 2 \times 82 - 30)\text{ mg } CaCO_3/L]\\ &\quad (1 \text{ kg}/10^6 \text{ mg})(50 \times 10^6 \text{ L/d})\\ &= 24{,}845 \text{ kg/d}\end{aligned}$$

5. Calculate the total, carbonate, and noncarbonate hardness present in finished waters. The carbonate hardness will be the sum of the

soluble calcium and the magnesium carbonate hardness and the non-carbonate hardness is equal to zero. Therefore, the total hardness is equal to the total carbonate hardness:

$$\text{Total carbonate hardness} = (30 + 20) = 50 \text{ mg L as } CaCO_3$$

6. The final bar diagram of the finished water in terms of $CaCO_3$ is given below.

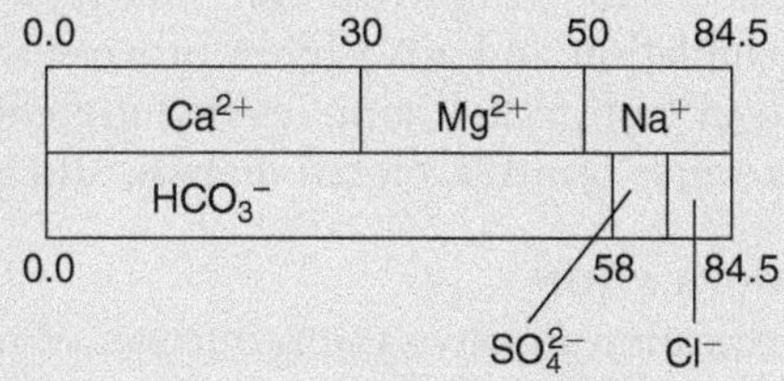

20-5 Nitrate

The principal sources of nitrogen are from (1) nitrogeneous compounds produced by plant and animals, (2) the mining of sodium nitrate for use in fertilizers, and (3) the atmosphere. The most oxidized form of nitrogen is nitrate (NO_3^-). In the United States, the average dietary intake of nitrate is about 75 to 100 mg/d, of which approximately 80 to 90 percent comes from vegetables. Vegetables with high nitrate levels are lettuce, beets, celery, and spinach. It is interesting that people on a vegetarian diet may consume as much as 250 mg/d of nitrate. Accordingly, drinking water accounts for only 5 to 10 percent of nitrates consumed. However, if the nitrate levels in the water are five times the MCL (10 mg/L), water may supply a person about half the daily diet requirements.

Occurrence and Importance in Water Supplies

Although nitrate occurs naturally in drinking water sources, elevated levels usually come from human activity such as municipal and industrial wastes, unmanaged solid waste landfills, onsite wastewater systems, runoff or groundwater from manured and fertilized agricultural lands, stormwater drainage, and animal feed lots. Because nitrates are extremely soluble in water, they can easily seep through the soil and into drinking water supplies. Surface waters such as lakes, reservoirs, and rivers are also susceptible to nitrate contamination from these sources. Nitrate concentrations in surface and groundwaters may be increasing worldwide (Nixon, 1992; Spalding and Exner, 1993).

Nitrate is of primary concern for infants younger than 6 months of age. Infants are very susceptible to methemoglobinemia, a condition known as "blue baby syndrome". High nitrate levels that are reduced in the stomach and/or the saliva of an infant to nitrite cause blue baby syndrome. Nitrite in the blood combines with hemoglobin to form methemoglobin, which reduces the capability of the blood to transport oxygen throughout the body. This results in the skin of a baby turning blue and can be fatal. The present MCL in the United States is 10 mg/L as nitrate and Canada has established a maximum acceptable concentration (MAC) of 10 mg NO_3^- – N/L. The European Union recommends a level of 11.3 mg NO_3^-/L.

Treatment Strategies

Because nitrate is a stable, highly soluble ion, it is difficult to remove by conventional coagulation and adsorption processes. Present technologies for nitrate removal from water supplies include chemical and biological denitrification, reverse osmosis, electrodialysis, and ion exchange.

CHEMICAL DENITRIFICATION

Chemical denitrification involves the reduction of nitrate to nitrogen gas using metals such as iron and aluminum. This process was evaluated and found to be very expensive due to the high metal dosages and other costs associated with the process (Murphy, 1991; Sorg, 1978a,b).

BIOLOGICAL DENITRIFICATION

Biological denitrification takes place in an anoxic environment where nitrate is converted to nitrogen gas through the following series of steps:

$$NO_3^- \rightarrow NO_2^- \rightarrow NO \rightarrow N_2O \rightarrow N_2 \qquad (20\text{-}49)$$

To provide an anoxic environment, the DO concentration in the water must be less than about 0.1 mg/L. Both heterotrophic and autotrophic bacteria can be used to denitrify water. Heterotrophic bacteria require an organic carbon source to be used as an electron acceptor for the denitrification reaction. For drinking water, the microorganisms can utilize NOM if the concentration is high enough and the form of NOM is amenable to biodegradation. In most cases, it is necessary to add in an organic source such as methanol, ethanol, or acetic acid. The carbon-to-nitrogen ratios required for methanol, ethanol, and acetic acid are reported to be 0.93, 1.05, and 1.32, respectively (Mateju et al., 1992).

Autotrophic bacteria are also capable of denitrification. Bacteria such as *Thiobacillus denitrificans* can use hydrogen or reduced sulfur species as substrate and carbon dioxide or bicarbonate as carbon sources for cell synthesis. Autotrophic bacteria require a minimum sulfur-to-nitrogen ratio of 4.3 when using thiosulfate and a minimum hydrogen-to-nitrogen ratio of 0.38 when using hydrogen as the substrate. Autotrophic bacteria have also been used in in situ denitrification in groundwater (Gayle et al., 1989).

Attached and suspended growth

Pilot studies have been performed to evaluate the feasibility and performance of several biological denitrification processes. The results from a number of biological denitrification pilot- and some full-scale studies that were designed to evaluate the biological denitrification process are reported

Table 20-12
Summary of several reported biological denitrification studies

Study	Reactor Type(s)	Water Source	Performance	Comments
1	1. Packed bed 2. Suspended growth 3. Fluidized sand bed (heterotrophic)	Thames water with methanol added	Reported denitrification rate 1. 12 g N/m^3 · h 2. 12-160 g N/m^3 · h 3. 160 g N/m^3 · h	Fluidized-bed reactor reduced nitrate by 45 mg/L at 2°C using upflow velocity of 12 m/h
2	Packed sand bed (heterotrophic)	Groundwater with sucrose added	With initial nitrate concentration of 22.6 mg/L, 100% removal was obtained with carbon-to-nitrogen ratio of 2.0	Vacuum used to remove trapped nitrogen gas in packed column
3	Static bed upflow reactor with spherical support medium (heterotrophic)	Groundwater with acetic acid added	With initial nitrate concentration of 100 mg/L, 100% removal was obtained with carbon-to-nitrogen ratio of 1.5	Retention time 9 h, effluent high in suspended solids and turbidity
4	Fluidized sand bed (heterotrophic)	Groundwater with methanol added	With initial nitrate concentration of 75 mg/L, 100% removal was observed for EBCT of 15 min	Flow rate 0.95 ML/d (0.25 mgd), reactor loading was 9.0 kg NO_3^-/m^3 d, methanol added 20–25% higher than stoichiometric requirement
5	DENITROPUR process, fixed-bed reactor (heterotrophic)	Groundwater with hydrogen added	With initial nitrate concentration of 75 mg/L, 99% removal observed	Flow rate 2.4 ML/d (0.63 mgd) with loading rate of 0.12 kg N/m^3 · d
6	DENIPOR process, fixed-bed reactor with polystyrene medium (autotrophic)	Groundwater with ethanol and phosphate added	Removal rate of 95% reported	Process loading rate 0.7–1.0 kg N/m^3 · d

Source: Adapted from Kapoor and Viraraghavan (1997).

in Table 20-12. Two processes developed in Germany are the DENIPOR and DENITROPUR processes. A schematic of the DENIPOR process used for full-scale microbiological denitrification in Mönchengladbach, Germany, is shown on Fig. 20-13. As shown, buoyant polystyrene spherical beads are used to support the biomass in the fixed-bed reactors. A periodic downward flushing is used to remove excess biofilm that can cause plugging.

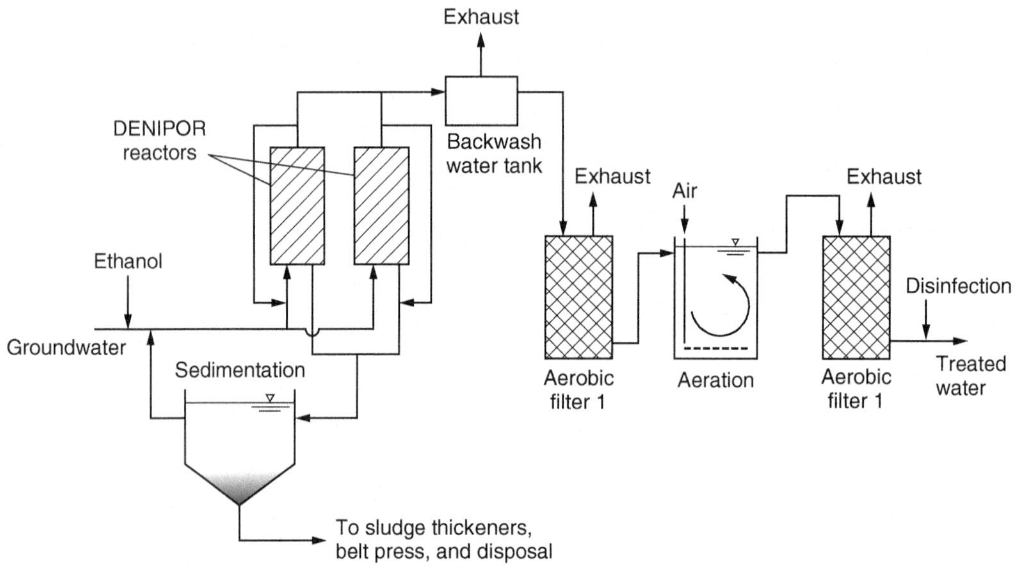

Figure 20-13
Schematic of DINIPOR process for biological denitrification.

The water from the denitrification reactors is treated further through two aerobic filters. The advantages of this process are (1) an organic substrate is not needed and (2) as the biomass production is low the sludge production is also low (a sludge production rate of only 0.2 kg organic matter/kg N was observed). The disadvantage of this process is that the kinetics of autotrophic bacteria is much slower than heterotrophic bacteria.

Hollow-fiber membrane bioreactor

An emerging autotrophic biological process for denitrification is called the hollow-fiber membrane bioreactor (HFMB) process, as shown on Fig. 20-14 (Lee and Rittmann, 1999). The HFMB process consists of membrane modules supporting several confined hollow-fiber membranes that contain a thin film of autotrophic bacteria on the outer portion of the hollow-fiber membranes. The inside of the confined hollow-fiber membranes contains hydrogen gas that diffuses outward in the radial direction through the membrane, providing dissolved hydrogen gas to the biofilm and creating a hydrogen atmosphere surrounding the biofilm. Biological denitrification is carried out by the autotrophic biofilm as NO_3^- is used as a terminal electron acceptor for respiration while hydrogen is used as the electron donor under anoxic conditions.

Results of pilot plant studies show the HFMB process is effective for the removal of nitrate from groundwater and is summarized in Table 20-13 (Falk and Ergas, 2002). Removal efficiencies of greater than 90 percent can be achieved with the HFMB process.

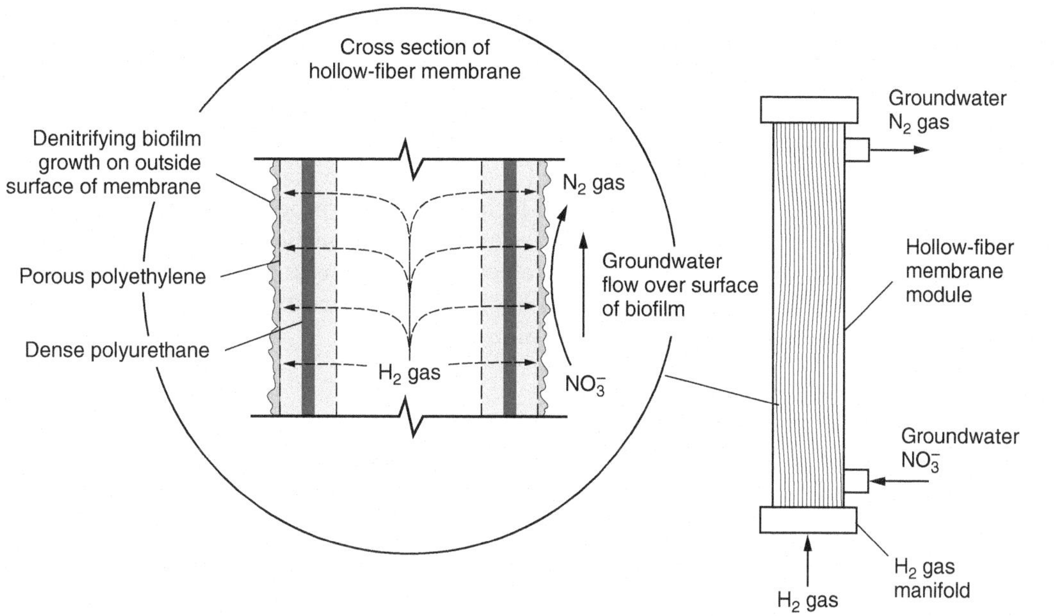

Figure 20-14
Schematic of bench-scale hollow-fiber membrane bioreactor process. (Adapted from Lee and Rittmann, 1999.)

Table 20-13
Performance data from studies evaluating HFMB process

		Study[a]			
		Falk and Ergas (2002)		Lee and Rittmann, 2002	
Parameter	**Unit**	**Influent**	**Effluent**	**Influent**	**Effluent**
Nitrate	mg NO_3^- –N/L	76	28	12.5	<1
Turbidity	NTU	2.4	2.9	nd	nd
Nitrite	mg NO_2^- –N/L	0.08	0.75	0.9	0.74
TOC	mg/L	0.7	3.4	nd	nd
DOC	mg/L	0.7	2.7	1.4	2.3
pH	Unitless	7.5	7.8	7	7.2
Heterotrophic plate count	CFU/100 mL	0	2.3×10^5	0	1.4×10^3
Nitrate utilization rate	g NO_3^- –N/m^2	Maximum = 1.4 Average = 0.8		Maximum = 1.4 Average = 0.8	
NO_3^- –N removal rate	%	63		92	
Hydrogen utilization efficiency	%	>94		>96	

[a]nd = nondetect.
Source: Adapted from Falk and Ergas (2002).

REVERSE OSMOSIS

Reverse osmosis and electrodialysis can be used to reduce nitrate levels in drinking water but is primarily used to treat high total dissolved solids and salt water. The cost of RO for treatment of nitrate only is much more expensive than the ion exchange process. Reverse osmosis is usually cost effective for nitrate removal if there are other water quality issues such as high TDS concentrations. Electrodialysis is a process that involves passing electric current through a series of semipermeable membranes to remove nitrate and other ions. The electrodialysis process provides about the same removal as reverse osmosis but is limited to treating soft waters, is expensive, and requires full-time monitoring.

ION EXCHANGE

Ion exchange is an effective treatment process for removal of nitrate. As discussed in Chap. 16, SBA exchange resins are used to remove anions. Typically major anions present in water supplies are sulfate, bicarbonate, and chloride. When nitrate is present in the water, these ions along with nitrate will be removed on SBA resins. The preference for anion exchange onto standard type 1 and type 2 SBA resins is sulfate > nitrate > bicarbonate > chloride. Because sulfate is preferred over nitrate, the impact of sulfate on nitrate exchange is very important and needs to be considered (see Chap. 16). For both type 1 and type 2 polystyrene resins and polyacrylic SBA resins, the quantity of nitrate removed in a given exchange cycle will depend upon the sulfate concentration, TDS, and nitrate concentration.

Resin types

There are also nitrate-specific resins that are specifically designed for nitrate removal. Nitrate-selective resins are usually used when chromatographic peaking in standard resins may impact the process performance (Clifford, 1999). Chromatographic peaking occurs when the more preferred sulfate ion migrates through the resin bed displacing the nitrate and concentrating it toward the end of the bed. When sulfate migrates to the end of the bed, it displaces the concentrated nitrate off the resin, resulting in a higher concentration of nitrate in the effluent than in the influent.

When treating a typical groundwater, the ion exchange effluent nitrate concentration can reach as high as 130 percent of the influent concentration. Consequently, resins used to treat nitrate are designed such that nitrate is preferred over sulfate. Some commercially available nitrate-selective resins are described in Table 20-14. Nitrate-selective resins are similar to standard type 1 resins but have ethyl, propyl, or butyl groups substituted for the methyl group on the trimethyl amine functionality [$RN(CH_3)_3$]. These substitutions separate the charged exchange sites along the backbone of the resins far enough apart so that the sulfate ion, which requires two charged sites close to each other, will not readily exchange. The nitrateselective resins also have a higher hydrophobicity than the standard SBA

Table 20-14
Summary of commercially available SBA nitrate-selective resins

Resin	Functional Groups	Ionic Form	Total Volume Capacity Minimum, meq/mL	Shipping Weight (approx.), kg/m^3	Water Retention, %
Purolite A520E	Quaternary ammonium	Cl^-	1.0	684	52–56
SR-6 Symbron	Trimethyl amine	Cl^-	0.85	668	42–45
Amberlite-996, Rohm & Haas	Trimethyl amine	Cl^-	1.0	—	50–56

resins. Nitrate-selective resins are also required when brine reuse by denitrification is implemented when the sulfate concentration in the feed water is greater than 200 mg/L (Liu and Clifford, 1996). Nitrate-selective resins and their technical information can be obtained from several manufacturers (e.g., Purolite, London, England; Rohm and Haas, Philadelphia, PA; and Sybron/Bayer Corporation, Pittsburgh, PA).

Ion exchange process
A typical ion exchange process for nitrate removal is shown on Fig. 20-15. Strong-base anion resins are usually in the chloride form so that NaCl can

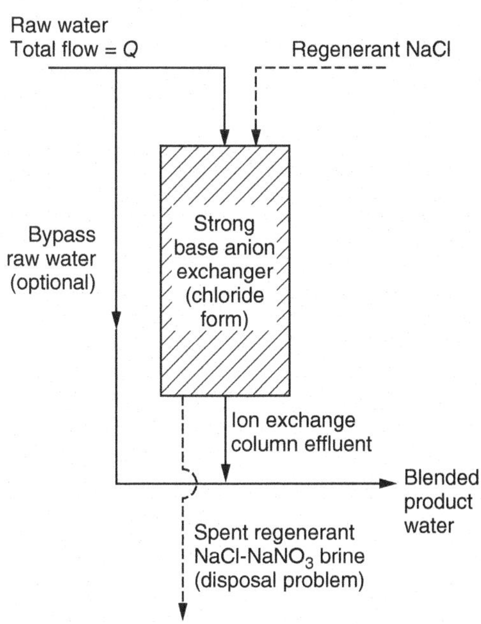

Figure 20-15
Schematic of typical ion exchange process using NaCl regeneration.

be used for regeneration of the resin. The fixed beds operate in the down-flow mode and co-current regeneration is usually used to reduce leakage of nitrate. An optional bypass of the water is sometimes used so that, when blended with the ion exchange process water, the nitrate concentration in the water is around 7 to 8 mg/L and is below the MCL. Depending on the water quality, nitrate breakthrough typically occurs around 100 to 500 BV. For example, field experiments showed that for an influent nitrate concentration of 21.5 mg/L and sulfate concentrations of 43 and 310 mg/L the nitrate breakthrough decreased from 400 to 180 BV (Clifford, 1999). Consequently, pilot studies are usually carried out to evaluate the breakthrough performance for a particular raw-water source.

Regeneration

Two types of regeneration techniques, partial and complete, can be used. However, for economic reasons, partial regeneration is used because significant leakage can be tolerated up to 10 mg/L of nitrate. For partial regeneration, it is important to completely mix the resin after each regeneration cycle to ensure leakage is not excessive. Performance for two full-scale plants using partial regeneration with NaCl is presented in Table 20-15. Both plants are used to treat 3.8 ML/d(1 mgd) using NaCl for resin regeneration. The water treated in the McFarland, California, plant contains 15 mg/L NO_3^-, 100 mg/L SO_4^{2-}, 90 mg/L Cl^-, and 100 mg/L HCO_3^-. In the plant in Binic, France, the surface water is pretreated with coagulation, sedimentation, and filtration ahead of the ion exchange process. The raw water treated in the Binic, France, plant contains 10 to 23 mg/L NO_3^-, 50 mg/L SO_4^{2-}, 50 mg/L Cl^-, and 85 mg/L HCO_3^-. This water is also

Table 20-15
Full-scale nitrate ion exchange plant performance

Description	Unit	McFarland, CA	Binic, France
Source water	—	Well water	Surface water
Polystyrene–DVB SBA Type I resin	—	Duolite A101D	Dowex SBRP
Regenerant concentration	N (%)	1.0 (6%)	—
Resin capacity, meq/L	meq/L	1.3	1.2
Brine utilization factor	eq Cl^-/eq	10	7.6
Run length to breakthrough	BV	260	400
Wastewater volume	% of blended product water	3.4	1.4
Bypass flow	% of blended product water	24	0
NaCl consumption for 3.8 ML/d blended product water	kg/d	1130	1150

Source: Clifford and Liu (1995).

high in TOC and the resin is replaced every 3 years due to organic fouling. Because the Binic plant had lower TDS and sulfate concentrations than the McFarland plant, it could treat up to 135 more BVs. It should also be noted that the Binic plant was unable to utilize a bypass stream and, thus, required higher salt usage than the McFarland plant. If the Binic plant were allowed to have a leakage of 7.9 mg/L nitrate, then the salt consumption in that plant would have only been 530 kg/d (Clifford and Liu, 1995).

Brine disposal

Nitrate-containing brine cannot be directly discharged into lakes and rivers due to its high salinity and nitrate, which causes eutrophication. For ion exchange plants near the coastal regions, ocean disposal may be an alternative if the conveyance system is available. For both the cities of McFarland, California, and Binic, France, the brine is disposed of in the sanitary sewer system as the brine waste is only a small portion of the total wastewater sent to the treatment plant and the biological treatment is not impacted. However, the McFarland plant, which uses aerated lagoons, discharges the water for irrigation of cotton crops. Unfortunately, salt concentrations are beginning to accumulate in the soil. While the exchange process is a cost-effective treatment process for nitrate removal from water, the issues of brine disposal must be carefully considered.

Recent studies have focused on brine treatment by denitrification and brine reuse to reduce the cost and problems associated with salt usage and brine disposal. The two processes shown on Fig. 20-16 are the biological denitrification of brine wastes using an upflow sludge blanket (USB) reactor (Hoek and Klapwijk, 1987, 1988) and a sequencing batch reactor (SBR) process (Liu and Clifford, 1996). The USB reactor is used to denitrify the spent nitrate-containing brine. An equalization tank is used to provide a constant nitrate influent in the USB and filtration is required for the denitrified brine before reuse. An advantage of the SBR process is that regeneration of an ion exchange column is also a batch process so the SBR process can also act as an equalization tank. Based on the results of bench and pilot studies for a methanol-to-nitrate ratio of 2.2, the nitrate removal efficiency is above 95 percent per batch. With the use of nitrateselective resins, the brine has been reused over 38 times without significant sulfate buildup problems in the brine. With the SBR process there were no significant problems associated with resin performance. Compared to the conventional partial-regeneration process, with reuse of the denitrified brine it was possible to treat 70 percent more water per run, use 23 percent less salt, and produce 60 percent less wastewater (Clifford and Liu, 1995). While this process shows much promise, it has yet to be tested as a full-scale application.

A comparison of the various nitrate removal processes is presented in Table 20-16, including (1) conventional ion exchange, (2) direct biological

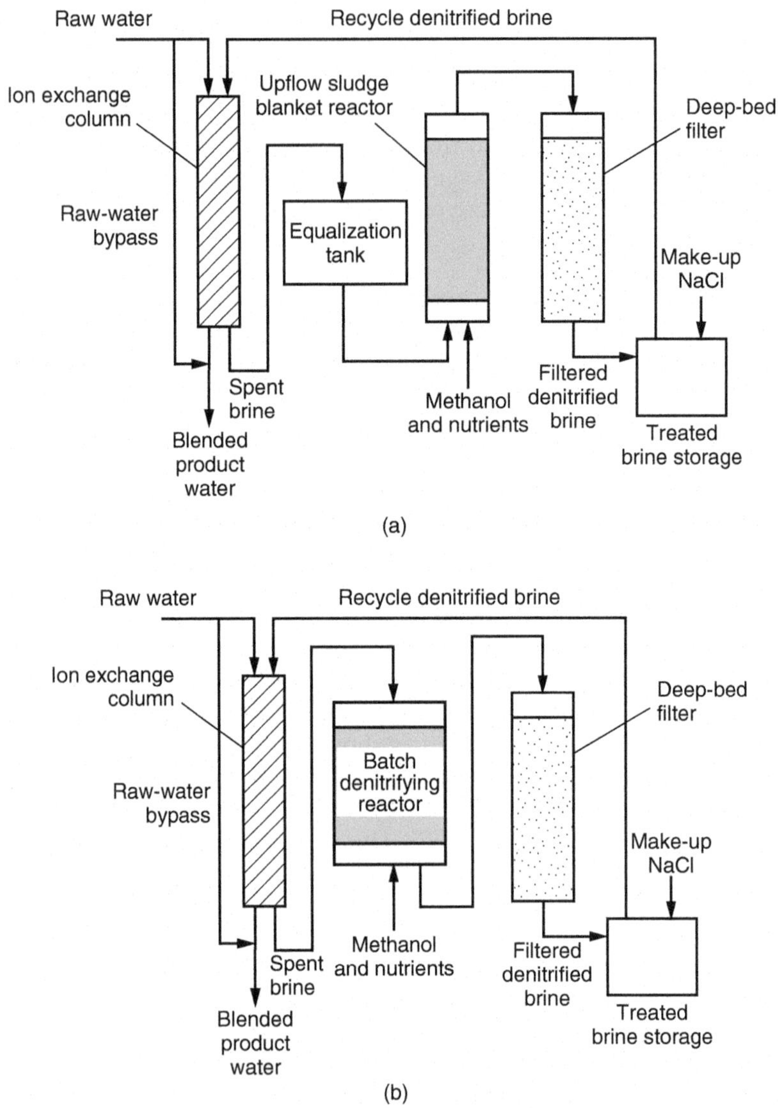

Figure 20-16
Dutch and American nitrate removal processes that incorporate biological denitrification into ion exchange process for spent-brine treatment: (a) Dutch ion exchange process with biological denitrification using upflow sludge blanket reactor (adapted from van der Hoek and Klapwijk, 1987) and (b) American ion exchange process with biological denitrification using sequencing batch reactor (adapted from Clifford and Liu, 1995).

denitrification, and (3) ion exchange with biological denitrification and reuse of the brine (Clifford and Liu, 1995). Both ion exchange and biological denitrification have been operated on a full-scale level, and the combined ion exchange and spent-brine denitrification process has only been tested on the pilot scale. If the nitrate-laden brine can be disposed of in a safe and cost-effective way, then the chloride ion exchange process may be the simplest and most cost-effective process. However, if there are problems with brine disposal, then direct biological denitrification or ion exchange with biological denitrification and brine reuse may be required.

Table 20-16
Comparison of nitrate removal techniques

Parameter	Ion Exchange	Direct Biological Heterotrophic Denitrification	Ion Exchange with Biological Denitrification and Brine Reuse
Raw-water quality (typically groundwater)	High TDS and sulfate reduce resin capacity for NO_3^-	Not influenced by water quality	Same as ion exchange
Treated-water quality	Increased chloride and corrosivity, variable anion concentrations	Posttreatment required to increase DO and eliminate TOC, NO_2^-, and turbidity	Same as ion exchange
Posttreatment required	Usually none	Biological oxidation, filtration, disinfection	Filtration of brine and possible disinfection required
Process complexity	Simple	Complex	Complex
Process control	Run length based on cumulative flow	Must control rate of ethanol and nutrient addition	Same as ion exchange plus addition of ethanol and nutrients[a]
Monitoring of treated water	NO_3^-	NO_3^-, NO_2^-, TOC, and bacteria	NO_3^- brine conductivity
Startup and shutdown	Operates on demand	Weeks to months required for startup; maintain viable denitrifying organisms	Maintain viable denitrifying organisms
Waste disposal	Large volume of brine containing NO_3^- and excess NaCl	Small volume of biomass sludge[b]	Small amount of nitrate-free brine and biomass sludge[b]

[a]For batch denitrification of brine, only batch addition of substrate and nutrients is required.
[b]Nitrate is converted to nitrogen gas by bacteria that grow slowly and produce a small amount of biomass for disposal.

20-6 Radionuclides

Natural radionuclides are the most common source of radioactivity in the environment. Natural radionuclides are formed from the dissolution of rock formations containing uranium ore and gases released from deep in the Earth's crust. The principal source of natural radionuclides is uranium ore (U_3O_8), and its abundance in the Earth's crust is only about 1 part per 10^{12} parts. The anthropogenic radionuclides come from sources such as nuclear power plants used to supply electrical energy, medical facilities that provide nuclear medicines and x-ray services, academic and research facilities using nuclear materials for research, commercial products such as televisions and smoke detectors, and nuclear weapons for national defense.

Occurrence and Importance in Water Supplies

Most drinking water sources contain very low levels of radioactive nuclei, or radionuclides, which is not usually a public health concern. However, there are some groundwater sources, primarily areas of the Midwest and western United States, with radionuclide concentrations that exceed present drinking water standards. Presently, radionuclide contamination of drinking water has not been a concern for large water utilities, but for small water utilities using groundwater in some areas of the country treatment is required. Use of radionuclide materials by industry may become greater due to increases in energy demand and possible medical applications, and the increase in demand may impact some water utilities. In addition, the present threat of terrorist activities related to the use of nuclear material in the United States and abroad may also be a future source of contamination in water supplies.

Chemistry and Removal

The most common radionuclide of concern in water treatment are radium-226, radon-222, radium-228, uranium-234, which are ionic decay products of uranium-238 and thorium-232. With respect to aqueous systems, radium-226 and radium-228 exist in natural groundwaters primarily as divalent cations. Radon-222 is a gaseous radioactive element that can exist as 25 different isotopes. Radon-222 is highly volatile and has a Henry's law constant of 1.69 (L_{Water}/L_{Air}) at 20°C. Uranium has four oxidation states, U(III), U(IV), U(V), and U(VI). The oxidation states U(III) and U(V) are unstable in both air and water, U(IV) is stable in air, and U(VI) is stable in water. In aqueous solution, uranium exists as uranyl ion and readily complexes primarily with carbonate and hydroxide, as shown on Fig. 20-17. In the pH range of most natural waters, the uranyl ion complexes primarily with the carbonate and bicarbonate anions to form uranyl carbonate complexes. At pH values between 5.0 and 6.5, the primary species is $UO_2CO_3^0$, and between pH values of 6.5 and about 7.5 the primary species is $UO_2(CO_3)_2^{2-}$. Small uranyl hydroxide complexes [UO_2OH^+ and $(UO_2)_3(OH)_5^+$] are also formed as shown on Fig. 20-17.

Treatment Methods

Several conventional water treatment technologies are capable of removing radionuclides. The U.S. EPA's best available technology (BAT) for radon-222 removal is aeration; for combined radium-226 and radium-228 it is coagulation-filtration; for radium-226 and radium-228 separately it is ion exchange, reverse osmosis, or lime softening; and for uranium it is ion exchange. Performance data for various processes used to remove radionuclides are reported in Table 20-17. The highest removal efficiencies

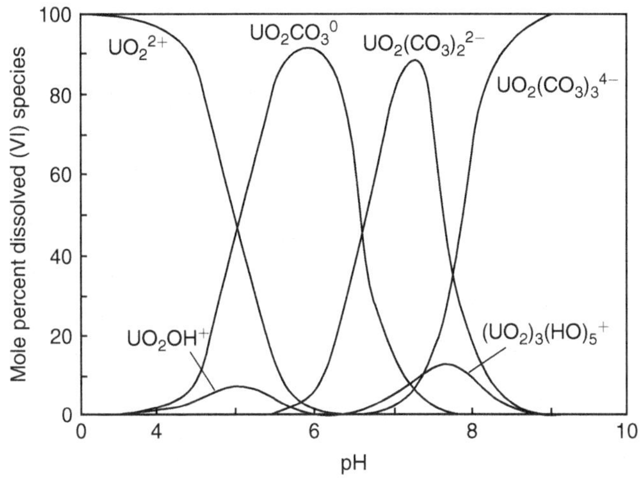

Figure 20-17
Distribution of uranium species in water.

Table 20-17
Summary of technology and performance of processes for removing radionuclides from drinking water

	Removal Efficiency, %		
Method	**Radon**	**Radium**	**Uranium**
Activated alumina			90
Aeration, packed tower	To 99+		
Aeration, diffused bubble	To 99+		
Aeration, spray	70–95+		
Coagulation–filtration			80–98
GAC adsorption–decay	62–99+		
Electrodialysis		90	
Greensand		25–50	
Hydrous manganese oxide filter		90	
Ion exchange		81–99	90–100
Lime softening		80–92	85–99
Reverse osmosis		90–95+	90–99

[a]The highest removal efficiencies for some technologies are associated with point-of-entry and point-of-use devices.
Source: Adapted from Lowery and Lowery (1988).

observed are for point-of-entry (POE) and point-of-use (POU) treatment systems. Water treatment processes for radionuclides are further discussed in the following sections.

AERATION

Radon-222 gas has a very high Henry's law constant and is very amenable to aeration. As discussed in Chap. 14, aeration methods such as spray, bubble, and packed-tower systems are very effective methods for removing radon-222 from water. In packed towers, reported radon-222 removal efficiencies have been as high as 98 percent using an air-to-water ratio of only 8 : 1 (Dixon et al., 1991). Using higher air-to-water ratios can achieve removals greater than 99 percent. Design procedures discussed in Chap. 14 can be used to design aeration devices to remove radon-222.

ION EXCHANGE

Strong-acid cation resins in the sodium form and weak acid cation (WAC) resins in the hydrogen form can be used to remove radium-226 and radium-228 from aqueous systems. For normal sodium ion exchange softening, both radium isotopes can be removed completely. The advantage of using WAC resins is that they are easier to regenerate and require less regenerant per unit volume treated than SAC resins. However, the disadvantages of using WACs are that they swell during exhaustion, acid-resistant materials are required to prevent corrosion when using HCl as a regenerant solution, noncarbonate hardness is not removed, it will be necessary to strip the CO_2 from the treated water, and pH adjustment is needed (Faust and Aly, 1998).

For combined radium and uranium removal, a mixed bed containing SBA and SAC resins can be used. A mixture of 10 percent SBA and 90 percent SAC resins in a small fixed bed was able to treat water containing 25 pCi/L radium and 120 μg/L uranium to less than 1 pCi/L radium and 20 μg/L uranium (Clifford and Zhang, 1994). Potassium chloride was found to be a better regenerant than sodium chloride for radium removal. Design concepts and procedures discussed in Chap. 16 can be used to provide design guidance for ion exchange systems. With respect to disposal of contaminated brines, present practice is disposal to municipal sanitary sewers.

COAGULATION–FILTRATION

Conventional coagulation–filtration plants can be effective in removing uranium from drinking water supplies. Based on laboratory studies using jar tests, it has been found that uranium can be removed effectively using iron and aluminum coagulants (Lee and Bondietti, 1983; Lee et al., 1982).

The full-scale Moffat CO water treatment plant has an influent uranium concentration of 15 μg/L, and 75 percent removal was achieved at a raw-water pH of 7.5 using alum, lime, and polymer (Hanson, 1987; Lee et al., 1982). However, there was wide variation in the percent removals. For the same water treated at the Moffat plant, Hanson (1987) also reported that the alum plant in Arvada, Colorado, which uses the microfloc system, was able to achieve removals of 18 to 90 percent, and an average efficiency of 67±15 percent. High removals have also been observed using chemical clarification and dissolved air flotation.

LIME SOFTENING

In the same jar test study described above, Lee et al. (1982) showed the addition of lime (50 to 250 mg/L) raised the pH to 10.6 to 11.5 and obtained uranium removals of 85 to 90 percent. The impact of pH between 10.6 and 11.5 did not impact the removals. Jar tests were also performed with various doses of lime and magnesium carbonate ($MgCO_3$). At pH values between 9.8 and 10.6 $MgCO_3$ dosages reduced the effectiveness of lime softening on uranium removal. However, at pH values greater than 10.6, uranium removal increased with increasing $MgCO_3$ dosages. The critical pH value was determined to be 10.6. Above this value the addition of $MgCO_3$ with lime increased uranium removals to 93 to 99 percent. Further experiments showed that magnesium hydroxide precipitate plays an important role in the removal of uranium.

REVERSE OSMOSIS

Reverse osmosis provides excellent removal of radium and uranium isotopes. Operating results for several membranes used to remove natural uranium from a groundwater in Florida with an influent concentration of 300 μg/L are displayed in Table 20-18. Removal efficiencies of 98 percent and greater were observed for each membrane. High removal efficiencies were also reported for radium-226. These results have been confirmed on similar waters containing uranium and radium (Sorg, 1988).

ADSORPTION

A promising adsorbent for the removal of radium is hydrous manganese oxides (HMOs), which has been successfully used in bench and pilot studies. The HMOs are preformed by the addition of potassium permanganate ($KMnO_4$), which brings about the oxidation of manganese sulfate ($MnSO_4$). Radium will adsorb onto the preformed HMOs and be removed by filtration. For conventional water treatment plants, this process can be incorporated

Table 20-18
Radionuclide removal using membranes[a]

		Contaminant Removal, %			
Radionuclide	Feed Concentration Range	Dow CTA, HF	DuPont ARAMIDHF	Filmtec TFC, SW	Hydranautics MCA, SW
Radium, pCi/L	2.2–9.8	97	96	—	97
Uranium(IV), μg/L	103–1650	99	98	99	99

[a]CA = cellulose acetate; CTA = cellulose triacetate; MCA = modified cellulose acetate; HF = hollow fiber; SW = spiral wound; TFC = thin-film composite.
Source: Adapted from Faust and Aly (1998).

easily into the process flow stream without significant plant modifications if radium removal is required. Periodic acid wash of the sand filters may be required to remove excess MnO_2. Preformed HMO was successfully used to remove radium from drinking water (Valentine et al., 1990).

20-7 Pharmaceuticals and Personal Care Products

Increasing interconnectedness between receiving waters for treated wastewater and source waters for potable water systems has created concern about whether trace contaminants can pass through wastewater treatment systems and enter the water supply. Many recent investigations have found evidence of low concentrations of pharmaceuticals and personal care products (PPCPs) and endocrine disrupting compounds (EDCs) in the source water for many communities throughout the United States and other developed nations. PPCPs are not currently regulated drinking water contaminants in the United States, but there is broad concern about their presence in drinking water supplies.

Pharmaceuticals include antibiotics, analgesics [painkillers such as aspirin, ibuprofen (e.g., Advil), acetaminophen (e.g., Tylenol)], lipid regulators (e.g., atorvastatin, the active ingredient in Lipitor), mood regulators (e.g., fluoxetine, the active ingredient in Prozac), antiepileptics (e.g., carbamazepine, the active ingredient in many epilepsy and bipolar disorder medications), and many other medications. Personal care products can include cosmetics and fragrances, acne medication, insect repellants, lotions, detergents, and other products. Ingested pharmaceuticals can be excreted with human waste and enter the wastewater system. Additional pharmaceuticals can enter the wastewater system because of the common practice of flushing unused medication down the toilet. Personal care products can be washed from the skin and hair during washing or showering. Discharge from wastewater treatment plants (WWTPs) have been shown to be a major source of many PPCPs (Wintgrens et al., 2004; Snyder et al., 2003) in the environment. Untreated animal waste, manufacturing residues, pesticides, and agricultural runoff are other sources of PPCPs (Kolpin et al., 2002).

Endocrine disrupting chemicals are chemicals that have the capability to interfere with the function of the human endocrine system (either stimulating or repressing hormonal function). EDCs can interfere with female sex hormones (estrogenic EDCs), male sex hormones (androgenic EDCs), or hormones that control metabolism and many other systems in the body (thyroidal EDCs). EDCs include natural hormones excreted from humans, ingested hormones such as estrogens subsequently excreted from females after use of birth-control pills, or synthetic compounds that mimic the function of hormones, such as bisphenol A.

Occurrence and Significance in Water Supplies

PPCPs have been detected in wastewater in the United States. since the 1960s and 1970s, and recent studies have detected them in a significant number of surface waters and even in treated drinking water supplies (Halling-Sorensen et al., 1998; Kolpin et al., 2002; Barnes et al., 2008; Focazio et al., 2008; Benotti et al., 2009). Analytical technologies have improved to the extent that the detection limit for many PPCPs is in the range of 5 to 10 ng/L, and these analytical capabilities have led to increased incidence of detection. In most cases, the concentrations of PPCPs and EDCs detected in source waters are below 1 μg/L. These concentrations are orders of magnitude below the therapeutic doses of many medications, which are typically tens to hundreds of milligrams per day. Example 20-8 demonstrates a comparison between level of exposure between concentrations found in water supplies and in therapeutic doses.

Example 20-8 Comparison of therapeutic and environmental exposures to pharmaceuticals

A single adult tablet of a pain medication contains 200 mg of ibuprofen. Compare the amount of ibuprofen consumed in one tablet to the amount consumed in drinking water in (1) one day and (2) over a lifetime, if the drinking water contains 1 μg/L of ibuprofen.

Solution

1. Daily exposure. A typical human consumes about 2 L of water per day. Comparing one day of drinking water consumption to one tablet is

$$\frac{200\ \text{mg}\left(10^3\ \mu\text{g/mg}\right)}{(1\ \mu\text{g/L})(2\ \text{L})} = 100{,}000 = 10^5$$

One tablet of pain medication contains 100,000 times more ibuprofen than the amount consumed by drinking water for 1 day.

2. Lifetime exposure. The average life expectancy in several countries is just over 80 years. Consumption of water containing 1 μg/L of ibuprofen for 80 years would expose a person to:

$$(1\ \mu\text{g/L})(2\ \text{L/d})(365\ \text{d/yr})(80\,\text{yr})\left(10^{-3}\ \text{mg}/\mu\text{g}\right) = 58.4\ \text{mg}$$

$$\frac{58.4\ \text{mg}}{200\ \text{mg}} \times 100 = 29\%$$

The amount of ibuprofen consumed in drinking water over a lifetime is less than 30 percent of the amount in one tablet.

Example 20-8 demonstrates that environmental exposure to PPCPs can be low compared to exposure for medical purposes. Nevertheless, exposure to PPCPs and EDCs through drinking water remains an area of potential concern. Concern because of the increased incidence of ssdetection that has been caused by improvements in analytical technologies has already been noted. Other possible reasons for recent concern about PPCPs and EDCs in water supplies include (1) increased interest in indirect potable reuse and increased awareness of the interconnections between wastewater discharge and water supplies, (2) the possibility of synergistic effects from exposure to trace concentrations of multiple PPCPs, and (3) the possibility that trace amounts of antibiotics in the environment may lead to the formation of resistant strains of bacteria.

Regardless of the public's concern regarding PPCPs and EDCs, it is important for the water treatment and regulatory communities to assess the concentration at which PPCPs and EDCs may pose health threats to consumers. An AWWARF study (Bruce et al., 2010) evaluated the toxicological relevance of PPCPs and indicated, for the compounds evaluated, the concentrations found in drinking water supplies were significantly below the concentrations considered to be significant from a toxicological perspective. Additional studies are ongoing to assess the human health significance of PPCPs and EDCs in drinking water.

Chemical Properties

Separation of constituents from water or wastewater is accomplished by exploiting differences in physical, chemical, and biological properties between the contaminants and water. These properties include molecular weight, solubility, charge, polarity, volatility, chemical reactivity, biodegradability, and others. Often, groups of compounds with similar properties can be removed by a single treatment process that exploits a specific property. However, there are thousands of different drugs and chemical compounds in use today that can, and do, end up in water with a correspondingly large variation in their physical, chemical, and biological properties (Dalton, 2004). The variability in properties means that these compounds will respond differently to different treatment techniques, so no treatment process will be effective for all PPCPs and EDCs.

Many PPCPs are organic compounds with relatively low molecular weight (< 1000 Da). However, they have differences in charge, polarity, volatility, and other properties that will make generalized treatment strategies more difficult. In addition, many pharmaceuticals are designed with specific properties to enhance their function as pharmacological agents that may interfere with treatment objectives. For instance, PPCPs may be designed with high chemical stability, high water solubility, low biodegradability, and low adsorbancy to nonpolar adsorbants, which may make treatment more difficult.

Treatment Strategies

Despite the wide range of chemical properties that may be expressed by PPCPs and EDCs, there are several treatment strategies that work well for many compounds. The most practical for the widest range of compounds are advanced oxidation, reverse osmosis, and adsorption onto activated carbon. The application of these treatment processes to PPCPs is described in the following sections.

ADVANCED OXIDATION

Oxidation and advanced oxidation processes (AOPs) achieve removal by chemical destruction. Advanced oxidation processes use combinations of chemical oxidants or combine a chemical oxidant with UV radiation to increase the rate of oxidation through generation of highly reactive free radicals such as the hydroxyl radical (OH·). Common AOPs include ozone/hydrogen peroxide, UV/ozone (UV/O_3), and UV/hydrogen peroxide (UV/H_2O_2). Advanced oxidation is discussed in more detail in Chap. 18.

Most conventional chemical oxidation processes are not very effective at removing many PPCPs at the doses used for disinfection (Snyder et al., 2008; Okuda et al., 2008). The same is generally true for UV light oxidation. UV irradiation alone achieves limited degradation of many PPCPs, particularly at doses used for disinfection (Kim et al., 2007, 2008; Canonica et al., 2008; Kruithof et al., 2007). However, doses at least 5 to 10 times higher than the typical doses for disinfection can achieve better removal of some PPCPs (Snyder et al., 2003; Kruithof et al., 2007).

Ozone, ozone-based, and UV-based AOPs can effectively degrade most PPCPs. Ozonation by itself can reduce both the concentration and number of compounds detected after treatment (Snyder et al., 2006; Andreozzi et al., 2004; Vieno et al., 2007). Okuda et al. (2008) found that ozone coupled with a biological activated carbon process reduced all residual pharmaceuticals to below quantification limits. As with other uses of ozonation, the formation of bromate is a consideration when using ozone for PPCP removal. In addition, ozonation of PPCPs may require longer contact times and/or higher doses than that used for disinfection (Andreozzi et al., 2004; Ternes et al., 2003). The combination of hydrogen peroxide and UV light has also been shown to be highly effective at degrading many PPCPs (Pereira et al., 2007a, 2007b; Chen et al., 2007).

Because of the differences in chemical properties, some compounds are slowly oxidized or poorly degraded by advanced oxidation processes. Cyclophosphamide, 2-quinoxaline carboxylic acid (2-QCA) and N,N- diethyl-m-toluavin (DEET) were poorly removed in one study (Kim et al., 2008). Other studies have found that clofibric acid (Westerhoff et al., 2005; Ternes et al., 2003) and ciprofloxacin (Vieno et al., 2007) are difficult to remove by advanced oxidation processes.

An important consideration in the use of advanced oxidation for PPCP removal is that the reaction products are almost certainly not fully mineralized to H_2O and CO_2. While an oxidation process may destroy the parent compound, it may produce degradation products with unknown biological activity. A subsequent process, such as biofiltration following ozone, may be appropriate to remove the oxidation products. Biofiltration is discussed in Sec. 11-8 of Chap. 11.

REVERSE OSMOSIS

Reverse osmosis is a membrane-based treatment process that separates contaminants from water by forcing water through the membrane under pressure. Dissolved contaminants are separated from the water as the water passes through the membrane. RO can effectively remove most PPCPs; removal efficiencies depend on properties of the feed water, membranes, and compounds to be removed. Reverse osmosis is discussed in Chap 17.

Many studies that have evaluated reverse osmosis for PPCP removal have found that RO can achieve excellent removal for most compounds (Snyder et al., 2003, 2007; Drewes et al., 2006). However, a number of factors can affect the level of removal. The most important parameters include the molecular weight or size, polarity, hydrophobicity, and charge of the compound, the membrane's surface charge and molecular weight cut-off (MWCO), and the fractional water recovery (Bellona et al., 2004; Kimura et al., 2003; Verliefde et al., 2007). The relationship between compound physicochemical properties and removal efficiency are qualitatively summarized in Fig. 20-18 (Drewes et al., 2006). Important compound properties reflected in this diagram are MWCO, charge, and hydrophobicity. As shown on the right side of Fig. 20-18, when the MWCO of the membrane is smaller than the MW of the compound, high rejection can be achieved for charged and hydrophobic [$\log(K_{ow}) > 2$] compounds, and lower rejection might be observed for small, neutral, hydrophilic compounds depending on the shape of the molecule. A number of researchers have shown that NF membranes do not achieve as good a rejection of PPCPs as tighter RO membranes (Xu et al., 2005; Yoon et al., 2002).

An operating parameter that can have a significant effect on PPCP removal is the feed water recovery. As noted in Chap. 17, osmotic pressure, concentration polarization, and the solubility of sparingly soluble salts can limit the recovery of water from an RO system. Higher recovery increases permeate volume but decreases its quality. Verliefde et al. (2007) showed that at a recovery of 10 percent, an NF membrane was able to remove >75 percent of all target compounds with most achieving >90 percent removal and a few compounds being removed at >99 percent. At 80 percent recovery, the same compounds were removed less effectively with one compound dropping to ~10 percent removal.

PPCP removal can be maximized by selecting membranes with a lower MWCO (i.e., seawater RO in lieu of brackish water RO membranes).

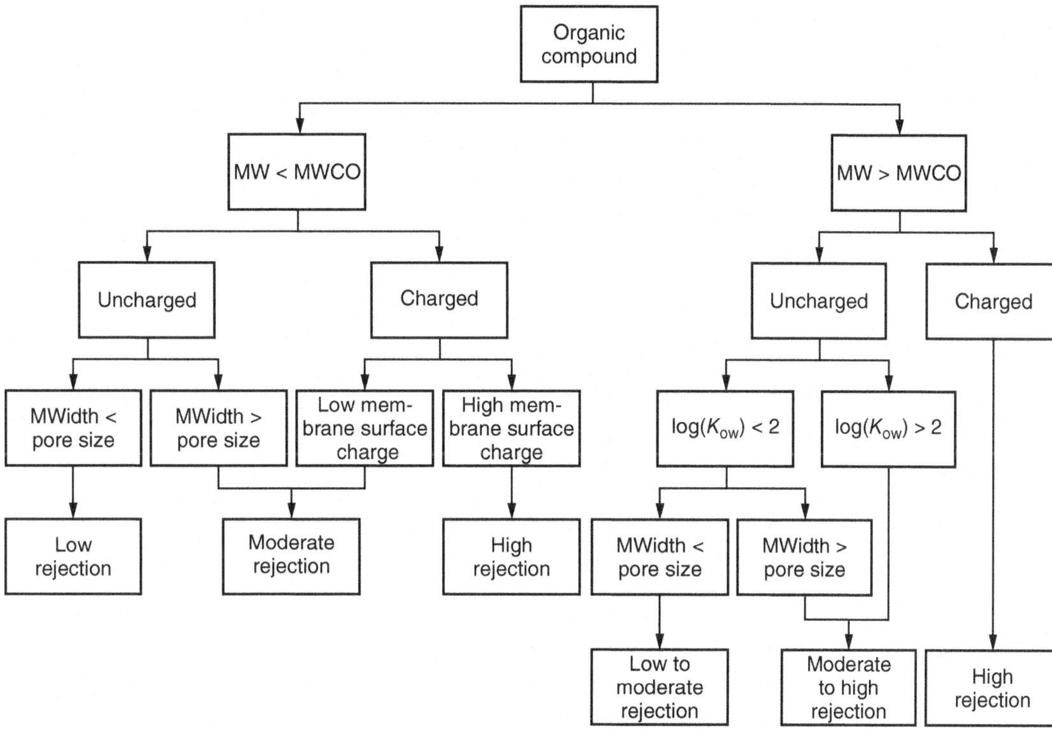

Figure 20-18
Rejection diagram for microconstituents using membrane processes as functions of both solute and membrane properties. (Adapted from Drewes et al., 2006.)

Membranes with a lower MWCO, however, typically operate at lower water flux rates. As a result, it would be necessary to increase the size of the system or increase the feed pressure, which increases capital and operating costs.

As a treatment process, reverse osmosis has several negative aspects. These include (1) high loss of product water because of low recovery, (2) high energy consumption, (3) large volume waste stream, which increases disposal costs. These negative aspects should be considered when comparing reverse osmosis to other treatment processes for PPCP treatment.

ADSORPTION ONTO ACTIVATED CARBON

Activated carbon is an effective adsorbent that is used for removing many dissolved compounds from water. Granular activated carbon (GAC) and powdered activated carbon (PAC) have both been evaluated for their effectiveness in removing PPCPs and found to be effective for many PPCPs.

The compounds most effectively removed by activated carbon include the more nonpolar, more hydrophobic, lower MW, uncharged, and lower solubility compounds. Lower MW compounds are more efficiently removed because of increased accessibility to inner pores of the carbon. The pH is also important for PPCPs that are weak acids because the pH affects the charge of the species.

Adsorption capacity affects the use of PAC and GAC differently. For PAC, a dose of carbon is added to the water and adsorption occurs until the capacity is reached, with the remaining pollutant staying in the water. Westerhoff et al. (2005) showed that protonated bases are well removed by PAC. Compounds with low K_{ow} values and deprotonated acid functional groups were the most difficult to remove. Increased removal efficiency for many compounds can be achieved by increasing the PAC dose and/or the contact time (Snyder et al., 2007; Baumgarten et al., 2007; Westerhoff et al., 2005).

For GAC, pollutants adsorb to the carbon bed and the pollutant concentration in the effluent can be unmeasurable until the capacity (measured as bed volumes) is reached, at which time the pollutant passes through the bed and the influent concentration of the pollutant is measured in the effluent. Studies have concluded that hydrophilic compounds break through the column sooner than the hydrophobic compounds (Snyder et al., 2007; Vieno et al., 2007). Vieno et al. (2007) found that the hydrophobic compound carbamazepine could be effectively removed by GAC even after treatment of >70,000 bed volumes of water, but that the more hydrophilic compounds could pass GAC treatment after only 2000 to 3000 bed volumes of water.

Once the adsorption capacity is reached, the media must be replaced or regenerated to restore removal effectiveness. Snyder et al. (2007) identified a facility with onsite and regular regeneration as having minimal breakthrough of organic contaminants and improved removal efficiency of selected PPCPs. In contrast, the study found little removal of trace organics in a facility with high levels of TOC that did not provide regular replacement/regeneration.

One parameter that affects both GAC and PAC is the NOM concentration (measured as DOC) in the feed water (Snyder et al., 2007). The presence of NOM can reduce the removal efficiency of PPCPs due to competition for adsorption sites. The NOM can block the pores within the activated carbon structure, leaving less opportunity for the PPCPs to be adsorbed. The quantity and characteristics of DOC in the feed water is an important parameter that can influence removal efficiencies for activated carbon (Westerhoff et al., 2005).

Studies have found that combined use of GAC or PAC with membrane processes is highly effective at removing PPCPs. One of the reported advantages is that this combination of processes effectively removes both DOC and DBPs. Verliefde et al. (2007) reported the combination of NF and GAC can provide a robust dual barrier for the removal of organic PPCPs. This is attributed to the NF membrane's ability to effectively remove high-molecular weight polar solutes, while activated carbon is more effective at removing nonpolar solutes. Similarly, use of RO to remove NOM would reduce the competition between NOM and PPCPs in a subsequent activated carbon process.

Problems and Discussion Topics

20-1 A 100 ML/d raw-water source containing 10-mg/L ferrous iron is oxidized to ferric hydroxide. Calculate the quantity of oxygen required, alkalinity consumed as $CaCO_3$ and quantity of sludge produced as ferric hydroxide.

20-2 A water supply containing a soluble Fe^{2+} concentration of 8.0 mg/L is to be oxidized by aeration to a concentration of 0.5 mg/L. The raw-water pH is 6.0, the temperature is 12°C, and it is assumed that P_{O_2} is in equilibrium with the atmosphere. Based on the results of laboratory studies, the pseudo-first-order rate constant for the oxygenation of Fe^{2+} is 0.17 min^{-1}. For steady-state operation and a flow rate of 40 ML/d (10.5 mgd), calculate and compare the minimum hydraulic detention time and reactor volume for the oxidation of Fe^{2+} to Fe^{3+} for plug flow and complete mixed flow reactors.

20-3 A 500-ML/d (132-mgd) flow of groundwater contains 2 mg/L of Mn(II) ion after aeration. A 5.0-mg/L (3.16×10^{-5} mol/L) dose of potassium permanganate ($KMnO_4$) is added to oxidize Mn(II) ion (Mn^{2+}) to a concentration of 0.1 mg/L. The groundwater pH is 6.5, the temperature is 15°C, and it is assumed that 5 mg/L (5.75×10^{-5} mol/L) of manganese dioxide (MnO_2) is in equilibrium with 0.1 mg/L of Mn^{2+}. For steady-state operation in a CMFR and plug flow reactor, calculate and compare the minimum hydraulic detention times for the oxidation of Mn^{2+} to MnO_2.

20-4 A 20-ML/d flow of groundwater contains 5 mg/L of ferrous iron (Fe^{2+}). Calculate the quantity of chlorine required to oxidize ferrous to ferric hydroxide, alkalinity consumed, and the quantity of sludge produced.

20-5 A groundwater contains 8 mg/L of ferrous iron (Fe^{2+}) and 2 mg/L of Mn (II) ion (Mn^{2+}). Potassium permanganate ($KMnO_4$) is used to oxidize ferrous iron and Mn(II). If the well pumping rate is 9.81 ML/d (1500 gpm), calculate the quantity of potassium permanganate required, alkalinity consumed, and quantity of sludge produced.

20-6 The results of a mineral analysis of a raw water are as follows: $H_2CO_3^* = 72$ mg/L, $Ca^{2+} = 100$ mg/L, $Mg^{2+} = 15$ mg/L, $Na^+ =$ 20 mg/L, Alk(HCO_3^-) =220 mg/L as $CaCO_3$, $SO_4^{2-} = 60$ mg/L, Cl^- = 5.15 mg/L. If 50 ML/d of water from this source is to be softened to reduce the hardness, calculate the total, carbonate, and noncarbonate hardness present in the raw and finished waters; the kg/d of lime, soda ash, and CO_2 needed for selective calcium softening; and the kg/d of $CaCO_3$ solids produced. Draw bar diagrams of the raw and softened water. Assume the residual hardness in the softened water is 30 mg/L as $CaCO_3$.

20-7 If the water in Problem 20-6 had an alkalinity of 100 mg/L as $CaCO_3$ and an SO_4^{2-} concentration of 151 mg/L, calculate the total, carbonate, and noncarbonate hardness present in the raw and finished waters; the kg/d of lime, soda ash, and CO_2 needed for selective calcium softening; and the kg/d of $CaCO_3$ solids produced. Draw bar diagrams of the raw and softened water.

20-8 A two-state excess lime-softening plant is designed to treat 57 ML/d (15 mgd) of a groundwater that contains the following constituents: Ca^{2+} = 80 mg/L, Mg^{2+} = 48.8 mg/L, Na^+ = 23 mg/L, Alk (HCO_3^-) =270 mg/L as $CaCO^3$, SO_4^{2-} =125 mg/L, and Cl^- = 35 mg/L. The water is to be softened by excess lime treatment. The average raw-water temperature and pH were found to be 10°C and 7.0, respectively. Draw a meq/L bar diagram, and determine the lime and soda ash dosages necessary in kg/d needed for softening. Assume that the soda ash is pure sodium carbonate and the lime is 85 percent CaO by weight. Also, calculate the kg/d of precipitated solids produced. Draw a meq/L bar graph of the water after the first stage of softening that includes the excess lime. Assume the practical limit of hardness removal for $CaCO_3$ is 30 mg/L, and that of Mg $(OH)_2$ is 10 mg/L as $CaCO_3$. Determine the kg/d of CO_2 required. Draw a meq/L bar diagram for the softened water after the second stage.

20-9 A 10-ML/d raw-water source is to be softened by two-stage lime softening to reduce the hardness. The results of a mineral analysis of the raw water are as follows: Ca^{2+} = 112 mg/L, Mg^{2+} = 20 mg/L, Na^+ = 11 mg/L, Alk(HCO_3^-) =260 mg/L as $CaCO_3$, SO_4^{2-} = 80.6 mg/L, Cl^- = 38.0 mg/L. Calculate the total, carbonate, noncarbonate hardness present in the raw finished waters; the kg/d of lime, soda ash, and CO_2 needed for selective calcium softening; and the kg/d of $CaCO_3$ solids produced. Assume the residual calcium hardness in the softened water is 30 mg/L as $CaCO_3$. Draw the initial and final bar diagrams of the raw and softened water. The average raw-water temperature and pH were found to be 15°C and 7.2, respectively.

20-10 A 50-ML/d raw-water source is to be softened using excess lime treatment to reduce the hardness. The results of a mineral analysis of the raw water are as follows: Ca^{2+} = 70 mg/L, Mg^{2+} = 15.9 mg/L, Na^+ = 23 mg/L, Alk(HCO_3^-) = 250 mg/L as $CaCO_3$ and SO_4^{2-} = 38.4 mg/L. Calculate the total, carbonate, and noncarbonate hardness present in the raw finished waters; the kg/d of lime, soda ash, and CO_2 needed for selective calcium softening; and the kg/d of $CaCO_3$ solids produced. Assume the residual calcium hardness in the softened water is 30 mg/L as $CaCO_3$. Draw bar diagrams of the raw and softened water.

20-11 For the groundwater given in Problem 20-10, determine the lime dose required for softening by split treatment assuming the magnesium concentration in the finished water does not exceed 40 mg/L as $CaCO_3$ and the total hardness does not exceed 155 mg/L as $CaCO_3$. Determine the final hardness of the finished water.

20-12 A200-ML/d (53 mgd) raw-water source is to be softened using excess lime treatment to reduce the hardness. The results of a mineral analysis of the raw water are as follows: $H_2CO_3{}^* = 65.1$, $Ca^{2+} = 80$ mg/L, $Mg^{2+} = 19.5$ mg/L, $Na^+ = 23$ mg/L, Alk($HCO_3{}^-$) = 280 mg/L as $CaCO_3$, $SO_4{}^{2-} = 28.8$ mg/L, and $Cl^- = 14.2$ mg/L. Calculate the total, carbonate, and noncarbonate hardness present in the raw and finished waters; the kg/d of lime, soda ash, and CO_2 needed for selective calcium softening; and the kg/d of $CaCO_3$ solids produced. Assume the residual calcium hardness in the softened water is 30 mg/L as $CaCO_3$. Draw bar diagrams of the raw and softened water.

20-13 Consider the precipitation softening by split-stream treatment of the raw water in Problem 20-12. Assume that only 75 percent of the water is treated by excess lime treatment and the other 25 percent bypasses the first stage and is mixed in the second stage. Compute the kg/d of chemicals required and the hardness of the water.

References

Alexander, H. J., and McClanahan, M. A. (1975) "Kinetics of Calcium Carbonate Precipitation in Lime-Soda Ash Softening," *J. AWWA*, **67**, 1, 618–621.

Andreozzi, R., Campanella, L., Fraysse, B., Garric, J., Gonnella, A., Lo Giudice, R., Marotta, R., Pinto, G., and Pollio, A. (2004) "Effects of advanced Oxidation Processes (AOPs) on the Toxicity of a Mixture of Pharmaceuticals," *Water Sci. Technol.*, **50**, 5, 23–28.

Argo, D. C. (1984) "Use of Lime Clarification and Reverse Osmosis in Water Treatment," *J. WPCF*, **56**, 10, 1238–1245.

ASCE/AWWA (1990) *Water Treatment Plant Design*, 3rd ed., McGraw-Hill, New York.

Aus Planer-Friedrich, B. (2001) Natural Attenuation Processes Regulating Arsenic Concentrations in Aquatic Environments, pp. 165–177 in M. Fall & B. Merkel (eds.), DFG-Graduiertenkolleg "Geowissenschaftliche und Geotechnische Umweltforschung".—Wissenschaftliche Mitteilungen des Institutes für Geologie, TU Bergakademie Freiberg, Heft 16, Germany.

AWWA (1969) *Water Treatment Plant Design Manual*, American Water Works Association, Denver, CO.

Ball, J. W., Jeanne, E. A., and Nordstrom, D. K. (1979) WATEQ2—A Computerized Chemical Model for Trace and Major Element Speciation and Mineral Equilibria of Natural Waters, pp. 815–835, in E. A Jeanne (ed.), *Chemical Modeling in Aqueous Systems,* ACS Symposium Series 93, American Chemical Society, Washington, DC.

Barnes, K. K., Kolpin, D. W., Furlong, E. T., Zaugg, S. D., and Meyer, M. T. (2008) "A National Reconnaissance of Pharmaceuticals and Other Organic Wastewater Contaminants in the United States. I. Groundwater," *Sci. Total Environ.*, **402**, 2–3, 192–200.

Baumgarten, S., Schroder, H. F., Charwath, C., Lange, M., Beier, S., and Pinnekamp, J. (2007) "Evaluation of Advanced. Treatment Technologies for the Elimination of Pharmaceutical Compounds," *Water Sci. Technol.*, **56**, 5, 1–8.

Bellona, C., Drewes, J. E., Xu, P., and Amy, G. (2004) "Factors Affecting the Rejection of Organic Solutes During NF/RO Treatment—A Literature Review," *Water Res.*, **38**, 12, 2795–2809.

Benefield, L. D., and Morgan, J. M. (1999) Chemical Precipitation, Chap. 10, in R. D. Letterman (ed.), *Water Quality and Treatment,* 5th ed., McGraw-Hill, New York.

Benjamin, M., Amy, G., Edwards, M., Carlson, K., Chwirka, J., Brandhuber, P., McNeill, L., and Vagliasindi, F. (2000) *Arsenic Treatability Options and Evaluation of Residuals Handling, AWWARF Final Report,* American Water Works Association, Denver, CO.

Benotti, M. J., Trenholm, R. A., Vanderford, B. J., Holady, J. C., Stanford, B. D., and Snyder, S. A. (2009) "Pharmaceuticals and Endocrine Disrupting Compounds in US Drinking Water," *Environ. Sci. Technol.*, **43**, 3, 597–603.

Blanck, C. A. (1979) "Trihalomethane Reduction in Operating Water Treatment Plants," *J. AWWA,* **71**, 9, 525–528.

Bolton, J. R., Stefan, M. I., and Cater, S. R. (1999) UV Light-Driven Degradation on Nitrosodimethylamine, paper presented at the Fifth International Conference on Advanced Oxidation Technologies for Water Remediation, Albuquerque, NM.

Brandhuber, P., and Amy, G. (1998) "Alternative Methods for Membrane Filtration of Arsenic from Drinking Water," *Desalination,* **117**, 3, 1–10.

Brinck, W. L. (1976) *Determination of Radium Removal Efficiencies in Water Treatment Processes,* USEPA, ORP/TAD-76-5, U.S. Environmental Protection Agency, Washington, DC.

Brown, A., Devinny, J. S., Browne, T. E., and Chitwood, D. (1997) A Review of Technologies for Removal of Methyl Tertiary Butyl Ether (MTBE) from Drinking Water, paper presented at the National Groundwater Association/American Petroleum Institute Petroleum Hydrocarbons and Organic Chemicals in Groundwater, Houston, TX.

Bruce, G. M., Pleus, R. C., and Snyder, S. A. (2010) "Toxicological Relevance of Pharmaceuticals in Drinking Water," *Environ. Sci. Technol.*, **44**, 14, 5619–5626.

Caldwell, D. H., and Lawrence, W. B. (1953) "Diagrams for Water Treatment Calculations," *Ind. Eng. Chem.*, **45**, 4, 523–555.

Calgon (1998) *UV Photolysis of NDMA in Contaminated Groundwater,* ABOT-007-8/98, Calgon Carbon Corporation Bulletin, Pittsburgh, PA.

Canonica, S., Meunier, L., and von Gunten, U. (2008). "Phototransformation of Selected Pharmaceuticals During UV Treatment of Drinking Water," *Water Res.*, **42**, 1–2, 121–128.

Chang, S., Ruiz, W., Bellamy, C., Spangenberg, C., and Clark, D. (1994) Removal of Arsenic by Enhanced Coagulation and Membrane Technology, in *Critical Issues in Water and Waste Water Treatment*, Boulder, CO, American Society of Civil Engineers, New York.

Chen, P.-J., Rosenfeldt, E. J., Kullman, S. W., Hinton, D. E., and Linden, K. G. (2007) "Biological Assessments of a Mixture of Endocrine Disruptors at Environmentally Relevant Concentrations in Water Following UV/H2O2 Oxidation," *Sci. Total Environ.*, **376**, 1–3, 18–26.

Cheng, R. C., Wang, H. C., and Liang, S. (1993) Metropolitan's Experience with Arsenic Removal by Enhanced Coagulation, paper presented at the American Water Works Association Water Quality Technology Conference, Miami, FL.

Ciaccio, L. (ed.) (1971) *Water and Water Pollution Handbook*, Marcel Dekker, New Yark.

Clifford, D. A. (1999) Ion Exchange and Inorganic Adsorption, Chap. 9, in R. D. Letterman (ed.), *Water Quality and Treatment: A Handbook of Community Water Supplies*, 5th ed., American Water Works Association, McGraw-Hill, New York.

Clifford, D. A., and Ghurye, G. (1998a) Arsenic Ion Exchange Process with Reuse of Spent Brines, paper presented at the American Water Works Association Annual Conference, Dallas, TX.

Clifford, D. A., and Ghurye, G. (1998b) *Final Report: Phase 3, City of Albuquerque Arsenic Study*, University of Houston, Department of Civil and Environmental Engineering, Houston, TX.

Clifford, D. A., and Lin C. C. (1986) *Arsenic Removal from Groundwater in Hanford, California— A Summary Report*, University of Houston, Texas, Houston, TX.

Clifford, D. A., and Liu, C. X. (1995) A Review of Processes for Removing Nitrate from Drinking Water, pp. 551–582, in *Proceeding AWWA Water Research Annual Conference*, Anheim, CA. American Water Works Association, Denver, CO.

Clifford, D. A., and Zhang, Z. (1994) "Combined Uranium and Radium Removal by Ion Exchange," *J. AWWA*, **86**, 4, 214–217.

Dalton, L. W. (2004). "After Chemicals Go down the Drain," *Chem. Eng. News*, **82**, 40, 44–45.

Degrémont (2007) *Water Treatment Handbook*, Vols. 1 and 2, 7th ed., Lavoisier, Paris.

Dixon, K. L, Lee, R. G., Smith, J., and Zielinski, P. (1991) "Evaluating Aeration Technology for Radon Removal," *J. AWWA*, **83**, 4, 141–148.

Drewes, E. J., Xu, P., Bellona, C., Oedekoven, M., and Macalady, D. (2006) *Rejection of Wastewater-Derived Micropollutants in High-Pressure Membrane Applications Leading to Indirect Potable Reuse*, WateReuse Foundation, Alexandria, VA.

Driehaus, W., Jekel, M., and Hildebrandt, U. (1998) "Granular Ferric Hydroxide— A New Adsorbent for the Removal of Arsenic from Natural Water," *J. Water Sci. Res. Technol.*, **47**, 1, 30–35.

Duranceau, S. J., Taylor, J. S., and Mumford, L. A. (1992) "SOC Removal in a Membrane Softening Process," *J. AWWA*, **84**, 4, 68–78.

Edwards, M. (1994) "Chemistry of Arsenic Removal During Coagulation and Fe-Mn Oxidation," *J. AWWA*, **86**, 9, 64–78.

Falk, M. W., and Ergas, S. J. (2002) Hydrogenotrophic Dentrification Using a Dead End Hollow Fiber Membrane Bioreactor, pp. 429–447 in *Proceedings Annual Conference, AWWA*, Washington DC, American Water Works Association, Denver, CO.

Faust, S. D., and Aly, O. M. (1998) *Chemistry of Water Treatment*, 2nd ed., Ann Arbor Press, Chelsea, MI.

Ferguson, J. F., and Anderson, M. A. (1974) Chemical Forms of Arsenic in Water Supplies and Their Removal, Chap. 3 in A. J. Rubin (ed.), *Chemistry of Water Supply Treatment and Distribution*, Vol. 1, Ann Arbor Science, Ann Arbor, MI.

Ferguson, J. F., and Gavis, J. (1972) "Review of Arsenic Cycle in Natural Waters," *Water Res.*, **6**, 11, 1259–1274.

Fleming, E. C., Pennington, J. C., Howe, R. A., Colsman, M. R., Garrett, K. E., and Wachob, B. (1996) "Removal of *N*-Nitrosodimethylamine from Waters Using Physical-Chemical Techniques," *Hazardous Ind. Wastes*, **51**, 1, 151–164.

Focazio, M. J., Kolpin, D. W., Barnes, K. K., Furlong, E. T., Meyer, M. T., Zaugg, S. D., Barber, L. B., and Thurman, M. E. (2008) "A National Reconnaissance for Pharmaceuticals and Other Organic Wastewater Contaminants in the United States. II.) Untreated Drinking Water Sources," *Sci. Total Environ.*, **402**, 2–3, 201–216.

Frey, M. M., and Edwards, M. A. (1997) "'Surveying Arsenic Occurrence," *J. AWWA*, **89**, 3, 105–117.

Frey, M., Chwirka, J., Narasimhan, R, Kommineni, S., and Chowdhury, Z. (2000) *Cost Implications of a Lower Arsenic MCL*, American Water Works Association Research Foundation, Denver, CO.

Gayle, B. P., Boardman, G. D., Sherreard, J. H., and Benoit, R. E. (1989) "Biological Denitrification of Water," *J. ASCE Environ. Div.*, **115**, 5, 930–940,

Ghurye, G., Clifford, D., and Tripp, T. (2004) "Iron Coagulation and Direct Microfiltration to Remove Arsenic from Groundwater", *J. AWWA*, **96**, 4, 143–152.

Gupta, S. K., and Chen, K. Y. (1978) "Arsenic Removal by Adsorption," *J. WPCF*, **50**, 3, 493–506.

Halling-Sorensen, B., Nielsen, S. N., Lanzky, P. F., Ingerslev, F., Lutzhoft, H. C. H., and Jorgensen, S. E. (1998) "Occurrence, Fate and Effects of Pharmaceutical Substances in the Environment — A Review," *Chemosphere*, **36**, 2, 357–394.

Hammer, M. J., and Hammer, Jr., M. J. (2001) *Water and Wastewater Technology*, 4th ed., Prentice Hall, Upper Saddle River, NJ.

Hanson, S. W. (1987) *Removal of Uranium from Drinking Water by Ion Exchange and Chemical Clarification*, EPA/600/S2-87/076, U.S. Environmental Protection Agency, Cincinnati, OH.

Health Canada (2006) Guidelines for Canadian Drinking Water Quality, available at http://www.hc-sc.gc.ca/ewh-semt/pubs/water-eau/sum_guide-res_recom/chemical-chimiques-eng.php.

Hering, J., and Elimelech, M. (1996) *Arsenic Removal by Enhanced Coagulation and Membrane Processes*, American Water Works Association Research Foundation, Denver, CO.

Hoek, J. P., and Klapwijk, A. (1987) Nitrate Removal from Ground Water, *Water Res.*, **21**, 8, 989–997.

Hoek, J. P., and Klapwijk, A. (1988) "Nitrate Removal from Ground Water—Use of a Nitrate Selective Resin and a Low Concentrated Regenerant," *J. Water Air Soil Pollut.*, **37**, 1, 41–53.

Kapoor, A., and Viraraghavan, T. (1997) "Nitrate Removal from Drinking Water—Review," *J. Environ. Engr., ASCE*, **123**, 4, 371–380.

Kartinen, E. O., and Martin, J. M. (2001) Selection of a Nitrate Removal Process for the City of Seymour, Texas, Chap. 21, in S. J. Duranceau (ed.), *Membrane Practices for Water Treatment*, American Water Works Association, Denver, CO.

Kawamura, S. (2000) *Integrated Design and Operation of Water Treatment Facilities*, 2nd ed., John Wiley & Sons, New York.

Kester, D. R., Byrne, Jr., R. H., and Liang, Y-J. (1975) Redox Reactions and Solution Complexes of Iron in Marine Systems, in T. M. Church (ed.), *Chemistry in the Coastal Environment*, ACS Symposium Series 18, American Chemical Society, Washington, DC.

Kim, S. D., Cho, J., Kim, I. S.; Vanderford, B. J., and Snyder, S. A. (2007). "Occurrence and Removal of Pharmaceuticals and Endocrine Disruptors in South Korean Surface, Drinking, and Waste Waters. *Water Res.* **41**, 5, 1013–1021.

Kim, I. H, Tanaka, H., Iwasaki, T., Takubo, T., Morioka, T., and Kato, Y. (2008). "Classification of the Degradability of 30 Pharmaceuticals in Water with Ozone, UV and H_2O_2". *Water Sci. Technol.* **57**, 2, 195–200.

Kimura, K., Amy, G., Drewes, J. E., Heberer, T., Kim, T.-U., and Watanabe, Y. (2003) "Rejection of Organic Micropollutants (Disinfection By-products, Endocrine Disrupting Compounds, and Pharmaceutically Active Compounds) by NF/RO Membranes" *J. Memb. Sci.*, **227**, 1–2, 113–121.

Knocke, W. R., Van Benschoten, J. E., Kearney, M. J., Soborski, A. W., and Reckhow, D. A. (1991) "Kinetics of Manganese and Iron Oxidation by Potassium Permanganate and Chlorine Dioxide," *J. AWWA*, **83**, 8, 80–87.

Kolpin, D. W., Furlong, E. T., Meyer, M. T., Thurman, E. M., Zaugg, S. D.; Barber, L. B., and Buxton, H. T. (2002) "Pharmaceuticals, Hormones, and Other organic Wastewater Contaminants in U.S. Streams, 1999–2000: A National Reconnaissance," *Environ. Sci. Technol.*, **36**, 6, 1202–1211.

Kruithof, J. C., Kamp, P. C., and Martijn, B. J. (2007) "UV/H_2O_2 treatment: A Practical Solution for Organic Contaminant Control and Primary Disinfection," *Ozone Sci. Eng.* **29**, 4, 273–280.

Langmuir, D. (1997) *Aqueous Environmental Geochemistry*, Prentice-Hall, Englewood Cliffs, NJ.

Lee, S. Y., and Bondietti, E. A. (1983) "Removing Uranium by Current Municipal Water Treatment Processes," *J. AWWA*, **75**, 7, 374.

Lee, K. C., and Rittmann, B. E. (1999) "A Novel Hollow-Fiber Membrane Biofilm Reactor for Autohydrogenotrophic Denitrification of Drinking Water," *Water Sci. Technol.*, **41**, 4, 219–226.

Lee, K. C., and Rittmann, B. E. (2002) "Applying a Novel Autohydrogenotrophic Hollow-Fiber Membrane Biofilm Reactor for Denitrification of Drinking Water," *Wates. Res.*, **368**, 2040–2052.

Lee, S. Y., Hall, S. K., and Bondietti, E. A. (1982) *Methods of Removing Uranium from Drinking Water: II. Present Municipal Treatment and Potential Removal Methods,* EPA 570/9-82-003, U.S. Environmental Protection Agency, Washington, DC.

Liao, W., Christman, R. F., Johnson, J. D., Millington, D. S., and Hass, R. J. (1982) "Structural Characterization of Humic Material," *Environ. Sci. Technol.*, **16**, 7, 403–410.

Liu, C. X., and Clifford, D. A. (1996) "Ion Exchange with Denitrified Brine Reuse," *J. AWWA*, **88**, 11, 88–99.

Logsdon, G. S., and Symons, J. M. (1979) "Mercury Removal by Conventional Water Treatment Techniques," *J. AWWA*, **71**, 9, 454–466.

Logsdon, G. S., Sorg, T. J., and Symons, J. M. (1974) Removal of Heavy Metals by Conventional Treatment, pp. 11–133, in *Proceedings of the 16th Water Quality Conference: Trace Metals in Water Supplies: Occurrence, Significance and Control,* February 13–14, 1974, University of Illinois, Champaign-Urbana, IL.

Lowenthal, R. E., and Marias, G. V. R. (1976) *Carbonate Chemistry of Aquatic Systems: Theory and Application,* Ann Arbor Science, Ann Arbor, MI.

Lowery, J. D., and Lowery, S. B. (1988) "Radionuclides in Drinking Water," *J. AWWA*, **80**, 7, 50–64.

Mateju, V., Cizinska, S., Krejci, J., and Janoch, T. (1992) "Biological Water Denitrification. A Review," *Enz. Microb. Technol.*, **14**, 3, 170–183.

McMurry, J., and Fay, R. C. (2003) *Chemistry,* 4nd ed., Prentice-Hall, Upper Saddle River, NJ.

McNeill, L. S., and Edwards, M. (1995) "Soluble Arsenic Removal at Water Treatment Plants," *J. AWWA*, **87**, 4, 105–113.

McNeill, L. S., and Edwards, M. (1997) "Predicting as Removal During Metal Hydroxide Precipitation," *J. AWWA*, **89**, 1, 75–86.

McRae, W. A., and Parsi, E. J. (1974) Removal of Trace Heavy Metal Ions from Water by Electrodialysis, in J. E. Sabadell (ed.), *Traces of Heavy Metals in Water Removal Processes and Monitoring,* Proc. Symp. Conducted by the Center for Environmental Studies and the Water Resources Program, Princeton University, EPA-902/9-74-001, U.S. Environmental Protection Agency, Washington, DC.

Merrill, D. T., and Sanks, R. L. (1977) "Corrosion Control by Deposition of Calcium Carbonate Films, Part I, A Practical Approach for Plant Operators," *J. AWWA*, **69**, 5, 592–599.

Mixon, F. O. (1973) *Removal of Toxic Metals from Water by Reverse Osmosis,* R&D Prog. Rep. No. 889, U.S. Department of Interior, Office of Saline Water, Washington, DC.

Moore, K. W., Huck, P. M., and Siverns, S. (2008) "Arsenic Removal Using Oxidative Media and Nanfiltration," *J. AWWA*, **100**, 12, 74–83.

Murphy, A. P. (1991) "Chemical Removal of Nitrate from Water," *Nature*, **350**, 6315, 223–225.

Nixon, N. (1992) "English Water Utility Tackles Nitrate Removal," *Water Eng. Management*, **139**, 3, 27–28.

Okuda, T., Kobayashi, Y., Nagao, R., Yamashita, N., Tanaka, H., Tanaka, S., Fujii, S., Konishi, C., and Houwa, I. (2008) "Removal Efficiency of 66 Pharmaceuticals During Wastewater Treatment Process in Japan," *Water Sci.*, **57**, 1, 65–71.

Pereira, V. J., Weinberg, H. S., Linden, K. G., and Singer, P. C. (2007a) "UV Degradation Kinetics and Modeling of Pharmaceutical Compounds in Laboratory Grade and Surface Water via Direct and Indirect Photolysis at 254 nm," *Environ. Technol.*, **41**, 5, 1682–1688.

Pereira, V. J., Linden, K. G., and Weinberg, H. S. (2007b) "Evaluation of UV irradiation for Photolytic and Oxidative Degradation of Pharmaceutical Compounds in Water," *Water Res.*, **41**, 19, 4413–4423.

Pontius, F. W., Brown, K. G., and Chen, C .J. (1994) "Health Implications of Arsenic in Drinking Water," *J. AWWA*, **86**, 9, 52–63.

Pourbaix, M. (1966) *Atlas of Electrochemical Equilibria in Aqueous Solutions*, Pergamon, Oxford, England.

Pourbaix, M. (1974) *Atlas of Electrochemical Equilibria in Aqueous Solutions*, 2nd ed., National Association of Corrosion Engineers, Houston, TX.

Rossum, J. R., and Merrill, D. T. (1983) "Valuation of the Calcium Carbonate Saturation Indexes," *J. AWWA*, **75**, 2, 95–100.

Rubel, F. R., and Hathaway, S. W. (1985) *Pilot Study for Removal of Arsenic from Drinking Water, Fallon, NV Naval Air Station*, Project Summary EPA-600/S2-35/094, U.S. Environmental Protection Agency, Washington, DC.

Rubel, F. R., and Williams, F. S. (1980) *Pilot Study of Flouride and Arsenic Removal from Potable Water*, EPA-600/2-80-100, U.S. Environmental Protection Agency, Washington, DC.

Rubel, F., Jr., and Woosley, R. D. (1979) "The Removal of Excess Fluoride from Drinking Water by Activated Alumina," *J. AWWA*, **71**, 1, 45–49.

Sanks, R. L. (1978) *Water Treatment Plant Design*, Ann Arbor Science, Ann Arbor, MI.

Scott, K. N., Greeen, J. F., Hoang, D., and McLean, S. J. (1995) "Arsenic Removal by Coagulation," *J. AWWA*, **87**, 4, 114–126.

Sigworth, E. A., and Smith, S. B. (1972) "Adsorption of Inorganic Compounds by Activated Carbon," *J. AWWA*, **64**, 6, 386.

Simms, J., and Azizian, F. (1977) Pilot Plant Trials on Removal of Arsenic from Potable Water Using Activated Alumina, paper presented at the American Water Works Association Water Quality Technology Conference, Denver, CO.

Simms, J., and Azizian, F. (1997) Pilot Plant Trials on Removal of Arsenic from Potable Water Using Activated Alumina, paper presented at the AWWA Water Quality Technology Conference, American Water Works Association, Denver, CO.

Singer, P. C., and Stumm, W. (1970) *Oxygenation of Ferrous Iron*, Water Pollution Control Research Series Report 14010-06/69, U.S. Department of Interior, Federal Water Quality Administration, Washington, DC.

Singley, J. E. (1979) Use of Powdered Activated Carbon for Removal of Specific Organic Compounds, paper presented at the American Water Works Association Annual Conference in Controlling Organics in Drinking Water, San Francisco, CA.

Sly, L. I., Hodgkinson, M. C., and Arunpairojana, V. (1990) "Deposition of Manganese in a Drinking Water Distribution Systems," *Appl. Environ. Microbiol.*, **56**, 3, 628–639.

Smith, A. H., Hopenhayn-Rich, C., Bates, M. N., Goeden, H. M., Hertz-Picciotto, I., Duggan, H. M., Wood, R., Kosnett, M. J., and Smith, M. T. (1992) "Cancer Risks from Arsenic in Drinking Water," *Environ. Health Perspective*, **97**, 3, 259–264.

Snoeyink, V. L., and Jenkins, D. (1980) *Water Chemistry*, John Wiley & Sons, New York.

Snyder, S. A., Westerhoff, P., Yoon, Y., and Sedlak, D. L. (2003) "Pharmaceuticals, Personal Care Products, and Endocrine Disruptors in Water: Implications for the Water Industry," *Environ. Eng. Sci.*, **20**, 5, 449–469.

Snyder, S. A., Wert, E. C., Rexing, D. J., Zegers, R. E., Drury, D. D. (2006) "Ozone Oxidation of Endocrine Disruptors and Pharmaceuticals in and Surface Water and Wastewater," *Ozone: Sci. Eng.*, **28**, 6, 445–460.

Snyder, S. A., Adham, S., Redding, A. M., Cannon, F. S., DeCarolis, J., Oppenheimer, J., Wert, E. C., and Yoon, Y. (2007) "Role of Membranes and Activated Carbon in the Removal of Endocrine Disruptors and Pharmaceuticals," *Desalination*, **202**, 1–3, 156–181.

Snyder, S. A., Vanderford, B. J., Drewes, J. E., Dickenson, E., Snyder, E. M., Bruce, G. M., and Pleus, R. C. (2008) *State of Knowledge of Endocrine Disruptors and Pharmaceuticals in Drinking Water*, Awwa Research Foundation, Denver, CO.

Sollo, F. W., Jr., Thurston, E. L., and Mueller, H. F. (1978) *Fluoride Removal from Potable Water Supplies*, Res. Rep. No. 136, Illinois State Water Survey, Urbana, IL.

Sorg, T. J. (1978a) "Treatment Technology to Meet the Interim Primary Drinking Water Regulations for Inorganics: Part 2," *J. AWWA*, **70**, 7, 379–393.

Sorg, T. J. (1978b) "Treatment Technology to Meet the Interim Primary Drinking Water Regulations for Inorganics, Part 3," *J. AWWA*, **70**, 12, 680–691.

Sorg, T. J. (1988) "Methods for Removing Uranium from Drinking Water," *J. AWWA*, **80**, 7, 105–111.

Sorg, T. J., and Logsdon, G. S. (1978) "Treatment Technology to Meet the Interim Primary Drinking Water Regulations for Inorganics, Part 5," *J. AWWA*, **72**, 7, 411.

Sorg, T. J., and Love, O. T. (1984) Reverse Osmosis Treatment to Control Inorganic and Volatile Organic Contamination, paper presented at the American Water Works Association Annual Conference in Experiences with Ground Water Contamination, Dallas, TX.

Spalding, R. F., and Exner, M. E. (1993) "Occurrence of Nitrate in Groundwater—A Review," *J. Environ. Quality*, **22**, 3, 392–402.

Stumm, W., and Lee, G. F. (1961) "Oxygenation of Ferrous Iron," *Ind. Eng. Chem.*, **53**, 4, 143–151.

Stumm, W., and Morgan, J. J. (1996) *Aquatic Chemistry*, Wiley-Interscience, New York.

Sung, W., and Morgan, J. J. (1980) "Kinetics and Product of Ferrous Iron Oxygenation in Aqueous Systems," *Environ. Sci. Technol.*, **14**, 5, 561–568.

Symons, J. M., and Carswell, J. K. (1981) *Treatment Techniques for Controlling Trihalomethanes in Drinking Water*, EPA-600/2-81-156, U.S. Environmental Protection Agency, Cincinnati, OH.

Tamura, H., Goto, K., and Nagayama, M. (1976), "Effect of Anions on the Oxygenation of Ferrous Ion in Neutral Solutions," *J. Inorgan. Nucl. Chem.*, **38**, 1, 113–117.

Taylor, J. S., and Jacobs, E. P. (1996) Reverse Osmosis and Nanofiltration, Chap. 9, in P. E. Mallevialle, J. Odendaal, and M. R. Wiesner (eds.), *Water Treatment Membrane Processes*, " McGraw-Hill, New York.

Ternes, T. A., Stuber, J., Herrmann, N., McDowell, D., Ried, A., Kampmann, M., and Teiser, B. (2003) "Ozonation: A Tool for removal of Pharmaceuticals, Contrast Media and Musk Fragrances from Wastewater?" *Water Res.*, **37**, 8, 1976–1982.

Theis, T. L., and Singer, P. C. (1974) "Complexation of Iron (II) by Organic Matter and Its Effect on Iron(II) Oxygenation," *Environ. Sci. Technol.*, **8**, 6, 569–573.

Thompson, M., and Chowdhury, Z. (1993) Evaluating Arsenic Removal Technologies, paper presented at the American Water Works Association Annual Conference, San Antonio, TX.

Thompson, C. G., Singley, J. E., and Black, A. P. (1972) "Magnesium Carbonate. A Recycled Coagulant," *J. AWWA*, **64**, 1, 11–19.

Thomson, B., Aragon, A., Anderson, J., Chwirka, J., and Brady P. (2005) *Rapid Small-Scale Column Testing for Evaluating Arsenic Adsorbents, AWWARF Final Report*, American Water Works Association, Denver, CO.

Thurman, E. M. (1985) *Organic Geochemistry of Natural Waters*, Martinus Nijhoff, Boston.

U.S. EPA (1991) "National Secondary Drinking Water Regulations; Final Rule," *Fed. Reg*, **56**, 20, 3526.

U.S. EPA (1999) *Achieving Clean Air and Clean Water: The Report of the Blue Ribbon Panel on Oxygenates in Gasoline*, EPA420-R-99-021, U.S. Environmental Protection Agency, Washington, DC.

U.S. EPA (2000) *Arsenic Removal from Drinking Water by Ion Exchange and Activated Alumina Plants*, EPA 600/R-00/088, U.S. Environmental Protection Agency, Washington, DC.

U.S. EPA (2001) "National Primary Drinking Water Regulations; Arsenic and Clarification to Compliance and News Source Contaminants Monitoring," 40CFR, Parts 9, 141, 142, *Fed. Reg.*, **66**, 141 6975.

Valentine, R. L., Spangler, K. M., and Meyer, J. (1990) "Removing Radium by Adding Preformed Hydrous Manganese Oxides," *J. AWWA*, **82**, 4, 66–71.

Van Der Hoek, J. P., and Klapwijk, A. (1987). "Nitrate Removal from Ground Water," *Waer. Res.*, **21**, 8, 989–997.

Verliefde, A. R. D., Heijman, S. G. J., Cornelissen, E. R., Amy, G., Van der Bruggen, B., and van Dijk, J. C. (2007) "Influence of Electrostatic Interactions on the Rejection with NF and Assessment of the Removal Efficiency During. NF/GAC Treatment of Pharmaceutically Active Compounds in Surface Water," *Water Res.*, **41**, 15, 3227–3240.

Vieno, N. M., Harkki, H., Tuhkanen, T., and Kronberg, L. (2007). "Occurrence of Pharmaceuticals in River Water and Their Elimination in a Pilot-Scale Drinking Water Treatment Plant," *Environ. Sci. Technol.*, **41**, 14, 5077–5084.

Weil, J. B. (1975) Aeration and Powdered Activated Carbon Adsorption for the Removal of Trihalomethanes from Drinking Water, Master of Engineering Thesis, University of Louisville, Louisville, KY.

Westerhoff, P., Yoon, Y., Snyder, S., and Wert, E. (2005) "Fate of Endocrine-Disruptor, Pharmaceutical, and Personal Care Product Chemicals During Simulated Drinking Water Treatment Processes," *Environ. Sci. Technol.*, **39**, 17, 6649–6663.

WHO (2001) *United Nations Synthesis Report on Arsenic in Drinking Water*, World Health Organization, Geneva, Switzerland.

Wintgens, T., Gallenkemper, M., and Melin, T. (2004) "Removal of Endocrine Disrupting Compounds with Membrane Processes in Wastewater Treatment and Reuse, *Water Sci. Technol.*, **50**, No. 5, 1–8.

Wood, P. R., and DeMarco, J. (1980) Effectiveness of Various Adsorbents in Removing Organic Compounds from Water. Part I: Removing Purgeable Halogenated Organics, in M. J. McGuire and I. H. Suffet (eds.), *Activated Carbon Adsorption of Organics from Aqueous Phase*, Vol. II, Ann Arbor Science, Ann Arbor, MI.

Xu, P., Drewes, J. E., Bellona, C., Amy, G., Kim, T.-U., Adam, M., and Heberer, T. (2005) "Rejection of Emerging Organic Micropollutants in Nanofiltrationreverse Osmosis Membrane Applications," *Water Environ. Res.*, **77**, 1, 40–48.

Yoon, Y., Westerhoff, P., Snyder, S., Song, R., and Levine, B. (2002) In A Review on Removal of Endocrine-Disrupting Compounds and Pharmaceuticals by Drinking Water Treatment Processes, Water Quality Technology Conference, 2002 American Water Works Association, p. 20.

Zemansky, G. M. (1974) "Removal of Trace Metals During Conventional Water Treatments," *J. AWWA*, **66**, 10, 606–618.

21 Residuals Management

Stantec's Water Treatment: Principles and Design, Updated Third Edition. John C. Crittenden, R. Rhodes Trussell, David W. Hand, Kerry J. Howe, and George Tchobanoglous.

Terminology for Residuals Managemnt

Term	Definition
Brine, ion exchange	Waste resulting from the regeneration of ion exchange resins including concentrated brine (high salinity solutions) and washwater.
Brine, sorbent	Waste resulting from the regeneration and conditioning of solid sorbents including activated aluminum and granular ferric hydroxide.
Chemical conditioning	Addition of chemicals to improve the dewaterability of treatment plant sludges.
Conditioning	Techniques and processes used improve the physical properties of the sludge so that it will dewater more easily.
Concentrate reverse osmosis	High salinity solution produced by the concentration of salts removed from brackish or saline waters during treatment.
Crystallization	Process of converting thickened concentrate and brine into mineral crystals that can be dewatered with a centrifuge or belt press.
Deep-well injection	Process of discharging membrane concentrate or ion exchange brine by injection into deep brackish or saline aquifers.
Dewatering	Process of removing excess water from sludge.
Flotation	Process of using air to float coagulated particles for thickening and removal.
Freezing	Freezing of sludge to enhance its dewaterability.
Heat treatment	Process of using heat to condition sludge for subsequent processing.
Leachate	Liquid underflow (percolate) from sand and other types of drying beds.
Membrane washwater	Waste stream resulting from the backwashing of the membranes.
Return flows	Flows returned to the plant influent or to separate treatment facilities resulting from operations such as centrifugal thickening of sludge (centrate), belt press thickening (filtrate or pressate), and plate and frame thickening of sludge (filtrate).
Residuals	Liquid, semisolid, solid, and gaseous phase by-products removed during the water treatment process, along with any transport water that is removed.

Term	Definition
Residuals management	Planning, designing, and operating of facilities to reuse and/or dispose of water treatment residuals.
Sorbent	Any material used to remove constituents from solution by sorption such as activated aluminum and granular ferric hydroxide.
Sorbent, spent	Sorbents that have lost significant adsorptive capacity and cannot be reactivated effectively.
Sterilization	Total destruction of disease-causing and other organisms.
Storage lagoons	Lined earthen basin used to store various sludges before processing.
Supernatant flow	Clear water decanted off residual solids resulting from the gravity and flotation thickening of sludge.
Underflows	See *Return flows*.
Waste washwater	Water from the backwashing of granular medium filters, other packed beds.

In most water treatment processes the objective is to remove certain materials from the water, to purify it. These materials are referred to as *residuals* and consist of the liquid-, solid-, semisolid, and gaseous-phase by-products removed during the water treatment process along with any transport water that is removed with them. These residuals include the turbidity-causing materials in raw water, organic and inorganic solids, algae, bacteria, viruses, colloids, precipitated from the raw water and those added in treatment, and dissolved salts. *Sludge* is the term used to refer to the solid, or liquid–solid, portion of some types of water plant residuals such as the underflow from sedimentation basins. In addition, some solid materials used to remove specific contaminants by sorption can become a *solid waste* when they can no longer be regenerated cost effectively or their removal (sorption) capacity has been reached when operated in a single-pass mode.

Residuals management is a term used to describe the planning, design, and operation of facilities to reuse or dispose of water treatment residuals. From a technical standpoint the objective in residuals management is usually to minimize the amount of material that must ultimately be disposed of by (1) recovering recyclable materials and (2) reducing the water content of the residuals. In most cases, the cost of transporting and ultimately disposing of the residuals makes up the major fraction of residuals management costs, and the most economical solution is to reduce the quantity of material for ultimate disposal. Other considerations include minimizing environmental impacts and meeting discharge requirements established by regulatory agencies.

Residuals management can have an important impact on the design and operation of many water treatment plants. For existing plants, residuals management systems may limit overall plant capacity if not designed and operated properly. Frequently, residuals are stored temporarily in the process train before removal for treatment, recycle, and/or disposal. Residual removal must be optimized for the process train and coordinated with the residuals management systems to maintain water quality.

Given the many issues associated with and the importance of residuals management, the purpose of this chapter is to (1) define the nature of the problem, including the sources of residuals; (2) review the physical and chemical properties used to characterize water treatment plant residuals; (3) consider the residuals and their properties, produced by the principal treatment processes; (4) review options available for the management of residual liquid streams; (5) review the options available for the management of residual concentrates and brines; (6) review the options available for the management of residual sludges; and (7) review options available for the ultimate reuse and/or disposal of residuals after processing.

21-1 Defining the Problem

The problem of residuals management can be quantified with respect to (1) the quantities and costs of handling residuals, (2) the constituents of concern, and (3) the environmental and regulatory constraints any engineered solution must meet. Before considering these topics, it is appropriate to consider the sources of residuals in water treatment processes.

Sources of Residuals

The principal residuals generated from the treatment of water can be classified as (1) sludges from water treatment processes, (2) liquid wastes from water treatment processes, (3) liquid wastes resulting from processes used to thicken process sludges and to treat liquid wastes, and (4) gaseous wastes from specialized water treatment processes. The sources of these residuals and a brief description are presented in Table 21-1. The specific types of sludges and liquid waste streams will depend on the type of treatment train as illustrated on Fig. 21-1.

Magnitude of the Problem

As much as 3 to 5 percent of the volume of the raw water entering a conventional water treatment plant may end up as solid, semisolid, and liquid residuals. The bulk of that volume will be the filter waste washwater, which typically contains less than 10 percent of the removed solids in a conventional treatment plant. Underflow from sedimentation basins typically contain on the order of 0.1 to 0.3 percent of the plant flow but contain most of the removed solids. In a direct or in-line filtration plant, however, all solids removal is accomplished in the filters. Typical values for the quantities of

Table 21-1
Sources of semisolid, solid, liquid, and gaseous residuals from the treatment of water

Source of Residual	Description
	Treatment Process Semisolid and Solid Residuals
Chemical precipitation with alum and iron	Sludges resulting from the chemical precipitation of surface waters that may contain clay, silt, colloidal material, and microorganisms with coagulant chemicals and polymers.
Coarse screens	Coarse screens prevent the entry of debris and fish into the intake structure. The coarse solids retained on the screens include rags, stringy material, and large wood pieces.
Flotation	Float, which in time thickens to sludge, resulting from the flotation process. Sludge, which settles, is removed periodically in small plants and continuously in large plants.
Presedimentation	Sludge resulting from presedimentation to remove gross amounts of sediment prior to conventional treatment.
Slow sand filter scrapings	Semisolid material resulting from the scraping of the surface of slow sand filters.
Spent sorbents	Solid material used to sorb constituents from solution such as hardness, arsenic, fluoride, phosphorus, and selected organic constituents, which have lost significant adsorptive capacity and/or cannot be reactivated effectively.
Traveling screens	Traveling screens are used to prevent grit, sand, and small rocks that have come through the intake from continuing into the treatment facility. Screenings include grit, sand, and small rocks.
Water softening	Lime sludge resulting from the removal of calcium and magnesium from hard waters during precipitation softening.
	Treatment Process Liquid Wastes
Brines and waste washwater from solid sorbents	Brine and rinse water from the reactivation of sorbents, principally ion exchange, along with waste washwater used to clean the beds.
Electrodialysis concentrate (brine)	The water that contains the dissolved constituents removed by the electrodialysis membranes used for softening and desalination.
Filter waste washwater	Waste washwater from backwashing filters to remove residual solids. Waste washwater is high in turbidity and may contain pathogenic organisms such as *Giardia* and *Cryptosporidium*.
Filter-to-waste water	Water used to condition filters after backwashing that has particles and turbidity above regulatory action levels.
Ion exchange brines and washwater	Brine and rinse water from the reactivation of ion exchange resins whose exchange capacity has been exhausted. Brines from resins used for softening typically containing sodium, chloride, and hardness ions; they are high in TDS, but low in suspended solids.
Membrane concentrate	The water that contains the dissolved constituents removed by reverse osmosis membranes.

Table 21-1 *(Continued)*

Source of Residual	Description
Membrane (mico- and ultrafiltration) washwater	Waste washwater from backwashing membranes to remove accumulated solids.
Slow sand filter washwater	Washwater high in turbidity that may contain pathogenic organisms such as *Giardia* and *Cryptosporidium* resulting from the cleaning of slow sand filter scrapings. (Note in many facilities the scraped sand is not washed for reuse on the filter beds but is used for other purposes.)
	Thickening/Dewatering Process Liquid Wastes
Centrate	Liquid resulting from centrifugal thickening of sludge.
Drying bed decant and underflow	Decant (supernatant) liquid from the surface and underflow (percolate) from sand and other types of drying beds.
Filtrate	Liquid resulting from plate and frame thickening of sludge.
Filtrate (pressate)	Liquid resulting from belt press thickening of sludge.
Supernatant flow	Clear water decanted off residual solids resulting from the gravity and flotation thickening of sludge.
	Treatment Process Gaseous Wastes
Stripping towers (not discussed in this chapter)	Off-gas from stripping operations contains contaminants that may need to be removed before gas can be discharged.

residuals produced by various treatment processes are summarized in Table 21-2. The costs of handling these residuals are dependent on the type of handling provided and the nature of the residuals. In general, the major portion of the cost with residuals management is associated with transport and ultimate disposal.

Constituents of Concern

Residual constituents of concern contained in the sludges and liquid wastes from treatment processes and thickening operations may include the following:

- ❑ Pathogenic microorganisms
- ❑ *Giardia* cysts and *Cryptosporidium* oocysts
- ❑ Turbidity/particles
- ❑ Disinfection by-products (DBPs)
- ❑ Precursors in the formation of DBPs (natural organic matter)
- ❑ Total organic carbon (TOC)
- ❑ Assimilative organic carbon (AOC)
- ❑ Taste- and odor-causing compounds

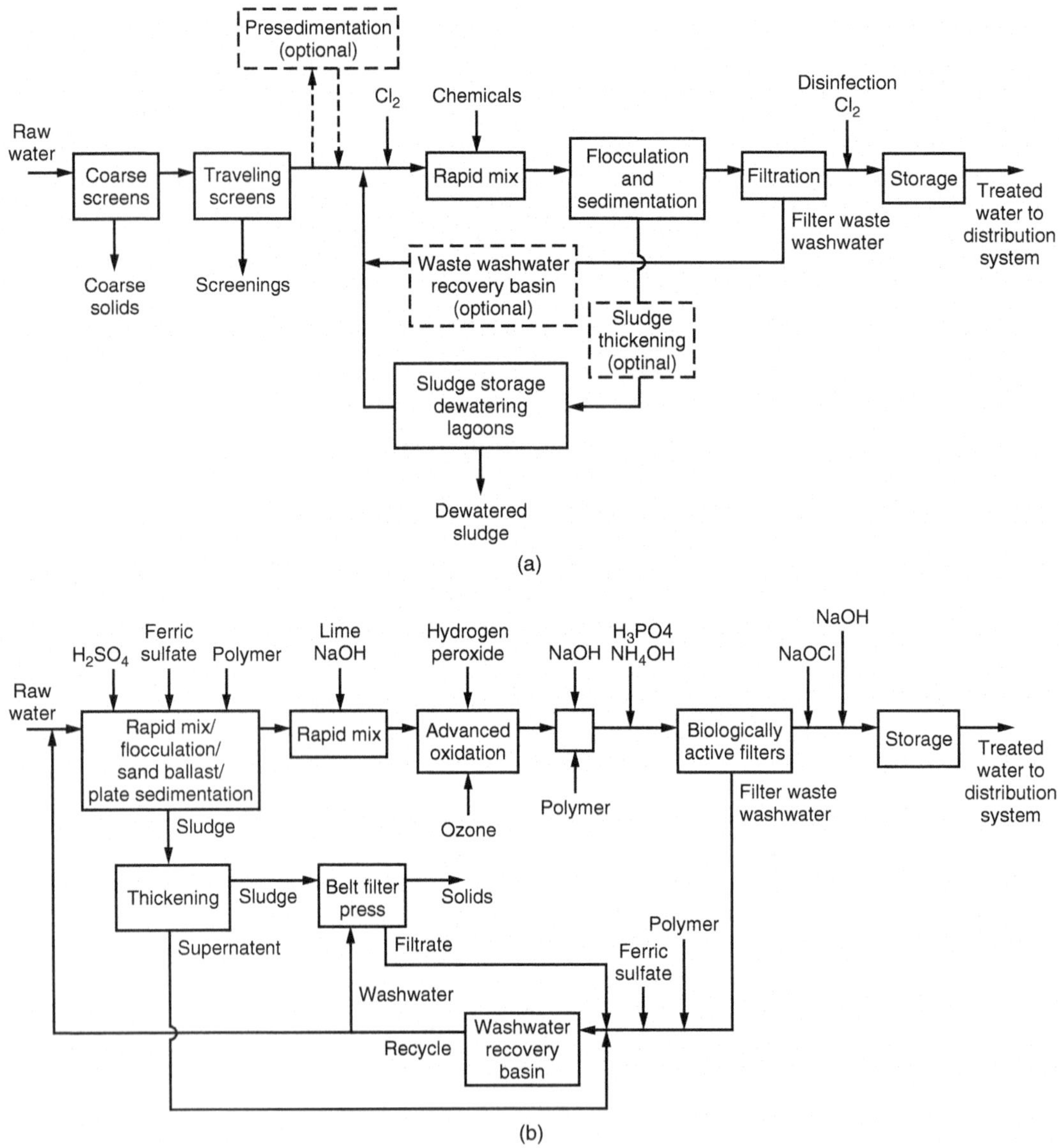

Figure 21-1
Typical water treatment process flow diagrams: (a) small plant with sludge storage lagoons. Future options include the addition of a waste washwater recovery basin and sludge thickening before discharge to sludge lagoons: (b) large plant with mechanically intensive sludge-processing facilities.

Table 21-2
Typical production of residuals in water treatment facilities as percent of plant flow

	Percent of Plant Flow	
Type of Residual	**Range**	**Typical**
Alum coagulation sludge	0.08–0.3	0.1
Direct filtration backwash water	4–8	
Filter backwash water	2–5	2[a], 3[b]
Flotation sludge (from reactor surface)	0.01–0.1	0.06
Flotation sludge (from reactor bottom)	0.001–0.04	
Ion exchange brine	1.5–10	5–8
Iron coagulation sludge	0.08–0.3	0.1
Lime-softening sludge	0.3–6	4
Microfiltration backwash water	2–8	6
Reverse osmosis concentrate	10–50	20–30

[a]During warn months.
[b]During cold months.

Coarse screen at headworks of water treatment plant

http://www.stantec.com/wt3r/21-1.html

- ❑ Synthetic organic compounds (SOCs)
- ❑ Manganese and iron
- ❑ Arsenic or other toxic compounds
- ❑ Radioactive materials
- ❑ Dissolved solids/salt

A variety of other compounds may also be of concern depending on the source of the water. Where liquid wastes and return flows from dewatering and thickening operations are recycled, these flows must, as noted previously, be returned to the headworks of the process train. However, concern over the presence of one or more of the above constituents has led, in some cases, to the use of separate treatment facilities for these liquid wastes. The use of separate treatment facilities is considered in Sec. 21-9.

Environmental Constraints

In the past, treatment plant residuals were often discharged to nearby streams, stored in lagoons, or spread on land with little or no processing, which created both negative aesthetic and environmental impacts. Aesthetic impacts include discoloration or increased turbidity in receiving waters and buildup of sludge deposits in waterways and occupying large land areas with lagoons. Impacts on the biota are, for sludges and waste washwaters, related primarily to the impact(s) on fish from increased water turbidity, pH, and hardness. Redissolved iron and aluminum may also pose a problem. In the cases of brine, there may be toxic effects caused by the high salt concentrations, especially in localized areas around the discharge. Most sludges, if spread on land to any depth, will prevent or inhibit plant growth; however, if mixed adequately into the soil, sludge may have little or no impact on plant growth. Lime sludges may have beneficial impacts on the soil, if used in appropriate amounts.

Regulatory Constraints

Regulatory constraints on residuals disposal have become increasingly severe in recent years. Prior to the late 1960s there was little concern for disposal of water treatment residuals. In most cases residuals were returned to the nearest receiving water, usually the source of the water supply. In the late 1960s some states began considering these residuals as pollutants and began establishing treatment or discharge standards for them.

The 1972 Federal Water Pollution Control Act classified water treatment plant residuals as pollutants and categorized them as industrial waste. As such, they are now required to meet standards for best practicable control technology (BPT) currently available and best available technology (BAT) economically achievable. There has also been legislation, both federal and state, to control toxic and hazardous substances. Such regulations, while protecting public and environmental health, can severely limit the available residuals management options and add to the cost of disposal.

21-2 Physical, Chemical, and Biological Properties of Residuals

An understanding of the physical, chemical, and biological properties of the residuals produced by treatment processes is fundamental to determining appropriate management techniques and to design facilities to implement those techniques. The physical, chemical, and biological properties used to characterize water treatment plant residuals are reviewed in this section. Additional information on the individual residuals is presented in the following sections dealing with coagulation sludges, lime sludges, filter waste washwater, and softening and demineralization concentrates.

Physical Properties

The physical properties of water treatment plant residuals are important for sizing and design of residuals management facilities. The physical properties used most commonly to characterize residuals are summarized in Table 21-3. Total solids is one of the most important physical parameters. Sludge density is dependent on the moisture content, varying from the density of water (1000 kg/m^3) for sludges below about 1 percent to 1100 kg/m^3 for a 15 percent sludge and higher for relatively dry sludges. A reasonable estimate of the wet density of inorganic sludges, typical of alum or iron salts, can be made by assuming the dry density of the solids is about 2300 kg/m^3 (see Example 2-1).

Table 21-3
Physical, chemical, and biological properties used to characterize water treatment plant residuals

Parameter	Unit of Expression	Description
Physical		
Total solids	%	Measure of total mass of material that must be handled on dry basis as percent of combined mass of solute and material
Dry density	kg/m^3	Measure of mass per unit volume on dry basis
Wet density	kg/m^3	Measure of mass per unit volume on wet basis
Specific gravity of dry solids	Unitless	Mass relative to mass of water
Specific resistance	m/kg	Measure of rate at which sludge can be dewatered (see Eq. 21-6)
Dynamic viscosity	$N \cdot s/m^2$	Measure of resistance to tangential or shear stress
Initial settling velocity	mm/s	Initial settling rate of a water–solids suspension
Chemical		
BOD	mg/L	Estimate of readily biodegradable organic content
COD	mg/L	Measure of oxygen equivalent of organic matter determined by chemical oxidation
pH	Unitless	Measure of effective acidity or alkalinity of solution
Alum content	% or mg/L	Derived from addition of coagulating chemical
Calcium, magnesium content	% or mg/L	Derived from addition of lime for water softening
Iron content	% or mg/L	Derived from addition of coagulating chemical
Silica and inert material	% or mg/L	Material present in surface water supplies
Trace constituents	μg/L or ng/L	Detection of specific constituents of concern
Biological		
Bacteria	no./100 mL	Variable depending on source of water and season
Protozoan cysts and oocysts	no./100 mL	Variable depending on source of water and season
Helminths	no./100 mL	Variable depending on source of water and season
Viruses	no./100 mL	Variable depending on source of water and season

SPECIFIC GRAVITY OF SLUDGE

The volume of sludge depends primarily on its water content and only slightly on the character of the solid matter. For example, a 5 percent sludge contains 95 percent water by weight. If the solid matter is composed of fixed (mineral) solids and volatile (organic) solids, the specific gravity of all the solid matter can be computed as

$$\frac{W_s}{S_s \rho_w} = \frac{W_f}{S_f \rho_w} + \frac{W_v}{S_v \rho_w} \tag{21-1}$$

where W_s = weight of total dry solids, kg
S_s = specific gravity of total solids, unitless

ρ_w = density of water, kg/m^3
W_f = weight of fixed solids (mineral matter), kg
S_f = specific gravity of fixed solids, unitless
W_v = weight of volatile solids, kg
S_v = specific gravity of volatile solids, unitless

Thus, if 90 percent by weight of the solid matter in a sludge containing 95 percent water is composed of fixed mineral solids with a specific gravity of 2.5 and 10 percent is composed of volatile solids with a specific gravity of 1.0, then the specific gravity of all solids, S_s, would be equal to 2.17, computed using Eq. 21-1:

$$\frac{1}{S_s} = \frac{0.90}{2.5} + \frac{0.10}{1.0} = 0.46$$

$$S_s = \frac{1.0}{0.46} = 2.17$$

If the specific gravity of the water is taken to be 1.00, the specific gravity of the sludge, S_{sl}, is 1.03, as follows:

$$\frac{1}{S_{sl}} = \frac{0.05}{2.17} + \frac{0.95}{1.00} = 0.97$$

$$S_{sl} = \frac{1.0}{0.97} = 1.03$$

DENSITY OF SLUDGE

The density of wet sludge, which is a mixture of solid matter and water, can be determined using the following expression:

$$\text{Density of wet sludge, } \rho_{sl} = \rho_w S_{sl} \tag{21-2}$$

where ρ_{sl} = density of sludge, kg/m^3
ρ_w = density of water, kg/m^3
S_{sl} = specific gravity of the sludge, unitless

VOLUME OF SLUDGE

The volume of a wet sludge may be computed with the following expression:

$$V = \frac{W_s}{\rho_w S_{sl} P_s} \tag{21-3}$$

where V = volume of wet sludge, m^3
W_s = weight of total dry solids, kg
ρ_w = density of water, kg/m^3
S_{sl} = specific gravity of sludge, unitless
P_s = percent solids expressed as a decimal

For approximate sludge volume calculations for a given solids content, it is simple to remember that the volume varies inversely with the percent of solid matter contained in the sludge, as given by

$$\frac{V_1}{V_2} = \frac{P_2}{P_1} \quad \text{(approximate)} \tag{21-4}$$

where V_1, V_2 = sludge volumes
P_1, P_2 = percent solid matter

Application of the above relationships is illustrated in Examples 21-1 and 21-2.

Example 21-1 Estimating density and volume of alum sludge

Determine the density and liquid volume of 1000 kg of dry alum sludge with the following characteristics:

Item	Unit	Value
Solids	%	15
Volatile matter	%	6
Specific gravity of fixed solids	Unitless	2.65
Specific gravity of volatile solids	Unitless	1.0
Temperature	°C	20

Solution

1. Compute the specific gravity of all the solids in the sludge using Eq. 21-1:

$$\frac{1}{S_s} = \frac{0.94}{2.65} + \frac{0.06}{1.0} = 0.41$$

$$S_s = \frac{1.0}{0.41} = 2.44$$

2. Compute the specific gravity of the wet alum sludge:

$$\frac{1}{S_{sl}} = \frac{0.15}{2.44} + \frac{0.85}{1.00} = 0.91$$

$$S_{sl} = \frac{1.0}{0.91} = 1.10$$

3. Compute the density of the wet alum sludge using Eq. 21-2:

$$\text{Density of wet sludge, } \rho_{sl} = \rho_w S_{sl}$$

The density of water at 15°C from App. C = 998.2 kg/m^3
Density of wet sludge, ρ_{sl} = (998.2 kg/m^3)(1.10) = 1098 kg/m^3

4. Compute the volume of wet sludge at 20°C using Eq. 21-2. The density of water at 15°C from App. C = 998.2 kg/m³.

$$V = \frac{W_s}{\rho_w S_{s1} P_s}$$

$$= \frac{1000 \text{ kg}}{\left(998.2 \text{ kg/m}^3\right)(1.10)(0.15)} = 6.07 \text{ m}^3$$

SPECIFIC RESISTANCE

Specific resistance, dynamic viscosity, initial settling velocity, and other physical properties are dependent on solids concentrations and the relative proportions of coagulant and other materials in the sludge. Specific resistance is a measure of the rate at which a sludge can be dewatered. Although developed for the vacuum filtration process, specific resistance has been found to be a useful parameter for assessing the dewaterability of sludges by gravity settling, centrifugation, belt filtration, plate and frame pressure filtration, and sand beds.

The concept of specific resistance is derived from the basic theory of filtration as developed by Carmen (1933, 1934) and extended by Coackley and Jones (1956)for conditions of streamline flow by the application of Poiseuille's and Darcy's law. The basic filtration equation is

$$\frac{dV}{dt} = \frac{PA^2}{\mu\left(rWV + R_m A\right)} \tag{21-5}$$

where V = volume of filtrate, m³

t = time, s

P = pressure, N/m²

A = area, m²

μ = viscosity of filtrate, N · s/m²

r = specific resistance of sludge cake, m/kg

W = mass of dry solids per unit volume of filtrate, kg/m³

V = volume, m³

R_m = resistance of filter medium, m^{-1}

For constant pressure, integration of Eq. 21-5 yields

$$\frac{t}{V} = \frac{\mu rWV}{2PA^2} \frac{\mu R_m}{PA} \tag{21-6}$$

The specific resistance of a sludge can be determined from laboratory data on the time for a given volume of water to be filtered, obtained using a Buchner funnel (see Fig. 21-2) or other filters specifically designed for the

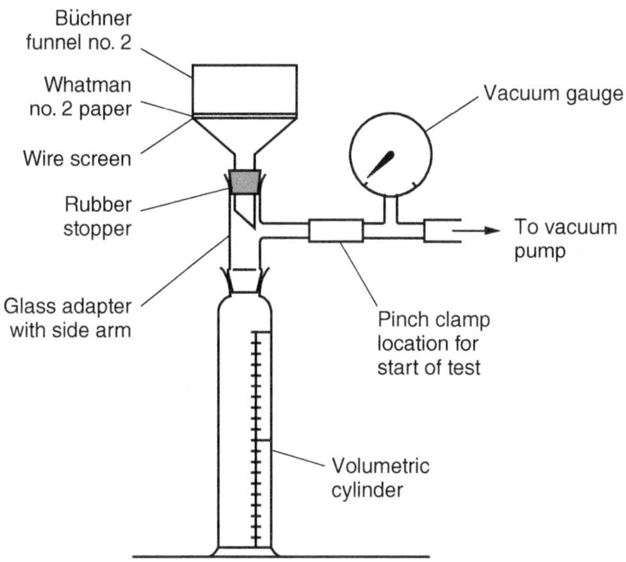

Figure 21-2
Buchner funnel test apparatus used for determination of specific resistance of sludge.

purpose. If the measured data are plotted in terms of t/V versus V, then the specific resistance can be determined from the slope of the line:

$$r = \frac{2PA^2m}{\mu W} \tag{21-7}$$

where m = slope of line

Typical specific resistance values for various sludges range from 5×10^{10} to 100×10^{10} m/kg. Sludges with specific resistance values below 10×10^{10} m/kg are considered to be readily dewaterable. Sludges with values greater than 100×10^{10} m/kg are considered to be difficult to dewater. Specific resistance is sometimes expressed in units of seconds squared per gram, which is not dimensionally correct for use in Eq. 21-7. To compare values expressed in meters per kilogram to values expressed in seconds squared per gram, the values in meters per kilogram must be divided by the gravitational constant (9.81 m/s^2) and converted from grams to kilograms (10^3 g/kg). Typically, specific resistance increases at solids concentrations below about 2 percent but is relatively constant above that concentration, as shown on Fig. 21-3. Shear strength and viscosity increase as solids concentrations increase, as shown on Figs. 21-4 and 21-5.

OTHER PHYSICAL PROPERTIES

The variation in physical properties among sludges of various compositions is due to the physical structure of the sludge. A coagulation sludge is made up of the suspended material in the raw water, metal hydroxides added during coagulation treatment, and a large amount of bound and entrapped

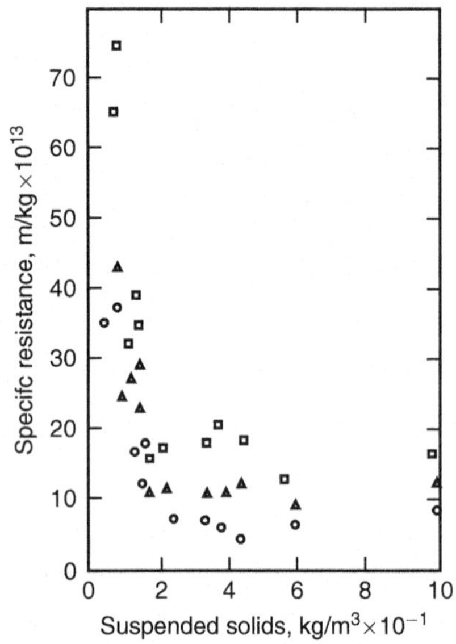

Figure 21-3
Relationship between suspended solids and specific resistance of alum sludges. (Adapted from Hawkins et al., 1974.)

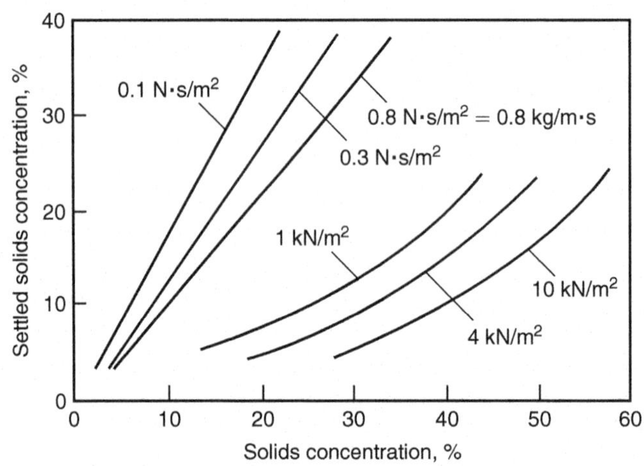

Figure 21-4
Variation in sludge physical properties. (Adapted from Novak and Calkins, 1975.)

water in a loose structure. The suspended materials are clay and sediment particles, color-causing colloids, algae, and other similar materials. Clays and sediments are structurally solid and have a specific gravity of around 2.6; the other materials are agglomerations of individual metal hydroxide molecules with various ions and water molecules all loosely held together by electrostatic bonds. Metal hydroxides become attached to the suspended materials by electrostatic bonds and also physically entrap suspended materials as well as water molecules. In the coagulation–flocculation process the

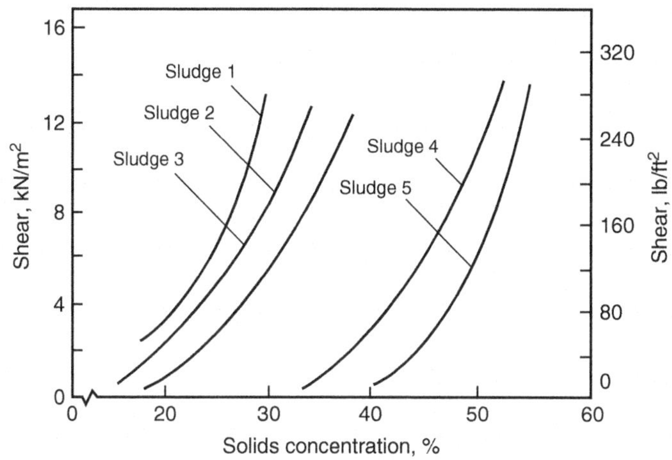

Figure 21-5
Effect of concentration on shear of five different sludges as they are dried. (Adapted from Novak and Calkins, 1975.) Because of the wide variation between the five different plants, generalized information such as presented in this graph should be used with caution.

suspended particles and metal hydroxides are brought together to form the flocs that then settle and make sludges. When the individual flocs come in contact with each other in the sludge, they become loosely bound by the same electrostatic forces that hold the individual flocs together.

The extent of the bonding depends on the extent of the contact between the flocs, which is limited by the entrapped water that separates the flocs. As more water is removed by draining, pressure, or other means, the particles contact more and the sludge becomes increasingly solid. Because the metal hydroxides have water molecules in their structure, direct particle contact is more difficult than for other suspended materials that do not have water in their structure. Therefore, sludges with high proportions of metal hydroxides are not as dewatered easily as are sludges that have higher proportions of other materials unless the sludge is conditioned with large amounts of lime or an appropriate polymer dose. The polymer molecules form bridges between floc particles and improve the bond between particles. If polymers are added to coagulation sludges, either as sludge conditioners or as a part of the coagulation process, the sludge produced will have a more solid structure.

Chemical Properties

The chemical properties of residual sludges are related directly to the chemical content of the raw water and the coagulant chemicals. Important chemical characteristics are summarized in Table 21-3. The BOD, COD, TOC, and related organic content are representative of the dissolved and suspended organic materials and algae removed from the water. The inorganic solids are derived from the coagulant chemicals and the clay and sediments removed from the raw water. The pH and dissolved solids in the liquid portion of the sludge are about the same as those in the water being treated. In a complete chemical analysis of an alum sludge, as reported by Schmitt and Hall (1975), a total of 72 elements were detected.

The major elements found were (in order of decreasing predominance by weight) silicon, aluminum, iron, titanium, calcium, potassium, magnesium, and manganese.

Biological Properties

Water treatment plant residuals may contain a variety of microorganisms, depending on the source, the quality of the raw water, the treatment process employed (e.g., prechlorination), and the time of year. Coagulation sludges, filter waste washwater, and membrane concentrates will contain bacteria, protozoan cysts and oocysts, and viruses removed during treatment. It is not possible to generalize on the number of microorganisms that may be present per unit mass or volume for the reasons cited above.

21-3 Alum and Iron Coagulation Sludges

Coagulation sludges are produced by the coagulation and settling of natural turbidity by added coagulant chemicals. In water treatment plants, coagulation sludges are collected in the sedimentation basins and on the filters. The amount and properties of the sludge collected in the basins and the filters depends upon the water quality, type and dose of coagulant used, efficiency of operation, plant design, and other factors. For typical plants using alum as the coagulant, between 60 and 90 percent of the total residuals will be collected in the sedimentation basins with the remainder in the filters. The residuals collected on the filters are removed from the filters during backwashing and, if the waste washwater is recovered, are removed from the waste washwater by settling.

Sedimentation facilities for the removal of turbidity

http://www.stantec.com/wt3r/21-2.html

Sludge from the sedimentation basins may be removed continuously or, more commonly, on an intermittent basis. Sludge removal may be accomplished using various mechanical devices (see Chap. 10) or by draining and manually washing down the basin. If basins are manually cleaned, the frequency may be once every 3 months or more. Mechanical cleaning equipment is usually designed to operate between once a week and once every few hours or continuously. Sludge removal frequency is decided based on balancing the competing interests of maintaining water quality in the process train, available sludge storage in the process train, the ability of the residuals management system to accept additional inflows, and disposal costs.

Filters are typically backwashed every 24 to 72 h, resulting in a relatively large volume of waste washwater produced in a short time. However, some proprietary filter types (such as automatic backwash filters) may backwash as frequently as every 2 to 6 h. Waste washwater recovery facilities must be designed to accept high, intermittent flows.

Coagulation sludges are grouped according to the type of primary coagulant employed. The principal types of coagulant employed are

(1) hydrolyzing metal salts of alum and iron, (2) prehydrolyzed metal salts such as polyaluminum chloride (PACl) and polyiron chloride (PICl), and (3) synthetic organic polymers.

Estimating Quantities of Coagulant Sludges

Typical overall values for the quantities of coagulant sludge produced were summarized previously in Table 21-2. For design purposes, the amount of sludge anticipated at a plant can be estimated based on the quality of the raw water and the type of chemical treatment. The suspended solids fraction of the sludge may be safely assumed to be equal to the suspended solids of the raw water, or, if total suspended solids (TSS) data are not available, it may be estimated from turbidity data. It is important to note, however, that there is great variability in the TSS/turbidity ratio depending on the organic content of the water source. The ratio for most water sources will vary between 1 and 2, with a typical value being about 1.4. For turbidities less than 10 NTU the ratio is nearly equl to 1.

PRECIPITATION REACTIONS

Typical precipitation reactions for alum and iron when used as coagulants, as shown in Eqs. 9-11 to 9-13 in Chap. 9, are as follows:

$$Al_2(SO_4)_3 \cdot 14H_2O \rightarrow 2Al(OH)_{3\downarrow} + 6H^+ + 3SO_4^{2-} + 8H_2O \quad (21\text{-}8)$$

$$FeCl_3 + 3H_2O \rightarrow Fe(OH)_{3\downarrow} + 3H^+ + 3Cl^- \quad (21\text{-}9)$$

$$Fe_2(SO_4)_3 \cdot 9H_2O \rightarrow 2Fe(OH)_{3\downarrow} + 6H^+ + 3SO_4^{2-} + 3H_2O \quad (21\text{-}10)$$

For alum or iron sludges, the precipitates are largely aluminum and iron hydroxide, respectively, and the quantity precipitated can be calculated from the stoichiometry. Using Eq. 21-8, it can be calculated that a total of 0.26 kg of sludge on a dry-solids basis will be produced for each kilogram of alum [$Al_2(SO_4)_3 \bullet 14H_2O$] added [e.g., 78 g/mol $Al(OH)_3$ × (2 mol/ mol)/(594 g/mol alum)]. The corresponding values for ferric coagulants are 0.53 kg sludge/kg ferric sulfate [$Fe_2(SO_4)_3$] added, and 0.66 kg sludge/kg ferric chloride (FeCl3) added. Typical values that can be used to estimate the quantity of alum and iron sludges are given in Table 21-4.

Prehydrolized PACl [$Al_nCl_{(3n-m)}(OH)_m$] is supplied in solution form containing varying amounts of aluminum. The amount of sludge produced can be estimated using the relationship

$$\text{mgAl}(OH)_3/\text{mg PACl added} = (\text{mg PACl/L})(\%\text{Al in PACl}/100)\left[\text{mg Al}(OH)_3/\text{mgAl}\right] \quad (21\text{-}11)$$

Typical values that can be used to estimate the quantity of PACl sludge are given in Table 21-4.

For polymer sludges or sludges with polymer used as coagulant aid, the amount of polymer added should also be included in the calculation of the total amount of sludge produced. Other coagulant aids, such as bentonite

Table 21-4
Typical values that can be used to estimate quantities of sludge resulting from addition of coagulating chemicals and polymers, turbidity removal, and softening in water treatment processes

Process	Unit	Range	Typical Value
Coagulation			
Alum, $Al_2(SO_4)_3{\bullet}14H_2O$	kg dry sludge/kg coagulant	0.2–0.33[a]	0.26
Ferric sulfate, $Fe_2(SO_4)_3$	kg dry sludge/kg coagulant	0.5–0.53[a]	0.53
Ferric chloride, $FeCl_3$	kg dry sludge/kg coagulant	0.6–0.66[a]	0.66
PACl	kg dry sludge/kg PACl	(0.0372–0.0489) × Al (%)	(0.0489) × (Al, %)
Polymer addition	kg dry sludge/kg coagulant	1	1
Turbidity removal	mg TSS/NTU removed	1.0–2	1.4
Softening			
Ca^{2+}[b]	kg dry sludge/kg Ca^{2+} removed	2.0	2.0
Mg^{2+}[c]	kg dry sludge/kg Mg^{2+} removed	2.6	2.6

[a]Value without bound water.
[b]Sludge is expressed as $CaCO_3$.
[c]Sludge is expressed as $Mg(OH)_2$.

or activated silica, should also be considered in the calculation as well as any other chemicals or materials, such as activated carbon, that may be collected in the basins or filters.

ESTIMATING SLUDGE MASS

The total sludge mass and volume produced can be calculated as

$$\begin{aligned}\text{Total sludge} &= \text{sludge from chemical coagulant}\\ &\quad+\text{sludge from suspended solids}\\ &\quad+\text{sludge from other chemicals or materials}\end{aligned} \tag{21-12}$$

Application of Eq. 21-12 is illustrated in Example 21-2, following the discussion of the relationships for alum, ferric sulfate, and ferric chloride.

For example, for alum with 14 bound water:

$$M = [D_{\text{Al}}(0.26) + \text{TSS} + X](10^{-3}\,\text{kg/g}) \tag{21-13}$$

where M = total sludge produced, kg/m^3
D_{Al} = alum dose, g/m^3
TSS = total dissolved solids, g/m^3
X = other coagulant aids added to enhance precipitation, g/m^3

Similarly for ferric sulfate and ferric chloride without bound water:

$$M = [D_{\text{FeSO4}}(0.53) + \text{TSS} + X](10^{-3}\ \text{kg/g}) \tag{21-14}$$

$$M = [D_{FeCl3}(0.66) + TSS + X](10^{-3}\ kg/g) \tag{21-15}$$

where D_{FeSO4} = ferric sulfate dose, g/m^3
D_{FeCl3} = ferric chloride dose, g/m^3

For the enhanced coagulation and precipitation process, which results in the production of additional sludge, the following relationship has been recommended by the U.S. EPA (1996):

$$M = [D_{Al}(0.36) + TOC + X](10^{-3} kg/g) \tag{21-16}$$

where TOC = total organic carbon, kg/m^3

The additional sludge that is produced is accounted for primarily by the factor by which the alum dose is multiplied (0.36) versus the factor (0.26) used in Eq. 21-12.

Example 21-2 Determination of quantity (mass and volume) of sludge from coagulant addition and the volume of sludge as percentage of the total flow

Determine the mass and volume of sludge produced from alum or PACl precipitation for the removal of turbidity. Assume the following conditions apply: (1) flow rate is 0.5 m^3/s (11.4 mgd), (2) average raw-water turbidity is 25 NTU, and (3) average alum or PACl dose is 30 mg/L, the sludge solids concentration is 5 percent with a corresponding specific gravity of 1.05, and the temperature is 15°C. Assume the ratio between TSS and turbidity expressed as NTU is 1.4, the ratio of alum sludge produced per kilogram of alum added is 0.26 (see Table 21-4), and that 1 mg/L of a polymer will be used. Assume the PACl contains 13 percent Al by weight. See also Example 9-2 for calculation of stoichiometry of the sludge from the coagulant.

Solution

Part A Mass and volume of alum sludge

1. Determine the amount of dry sludge produced from the addition of alum:

   Sludge from alum, kg/d

   $= (0.5\ m^3/s)(30\ mg/L)(0.26)(86{,}400\ s/d)(10^3\ L/m^3)(1\ kg/10^6\ mg)$

   $= 337\ kg/d$

2. Determine the amount of dry sludge produced from the removal of turbidity:

   Sludge from turbidity, kg/d

   $= (0.5\ m^3/s)(25\ NTU)(1.4\ g/m^3 \cdot NTU)(86{,}400\ s/d)(1\ kg/10^3\ g)$

   $= 1512\ kg/d$

3. Determine the amount of dry sludge produced from the addition of a polymer:

   Sludge from polymer addition

   $= (0.5\ m^3/s)(1\ mg/L)(86{,}400\ s/d)(10^3\ L/m^3)(1\ kg/10^6\ mg)$

   $= 43\ kg/d$

4. Determine the total amount of sludge produced using Eq. 21-12:

   $$\text{Total sludge} = (337 + 1512 + 43)\ kg/d = 1892\ kg/d$$

5. Estimate the volume of the sludge using the mass determined in step 4 and the given sludge characteristics using Eq. 21-3. The density of water at 15°C from App. C = 999.1 kg/m³.

   $$\text{Sludge volume} = \frac{1892\ kg/d}{\left(999.1\ kg/m^3\right)(0.05)(1.05)} = 36.1\ m^3/d$$

6. Estimate the volume of the sludge as a percent of the total flow:

   Sludge volume, % of total flow

   $= [(36.1\ m^3/d)/(0.5\ m^3/s)(86{,}400\ s/d)] \times 100$

   $= 0.08\%$

Part B Mass and Volume of PACl sludge

7. Determine the amount of dry sludge produced from the addition of PACl using the typical value given in Table 21-4:

   Sludge, kg/d

   $= (0.5\ m^3/s)(30\ mg/L)(0.0489)(13)(86{,}400\ s/d)$

   $\times (10^3\ L/m^3)(1\ kg/10^6\ mg)$

   $= 824\ kg/d$

8. Determine the total amount of sludge produced using Eq. 21-12 and the data from Part A:

   $$\text{Total sludge} = (824 + 1512 + 43)\ kg/d = 2379\ kg/d$$

9. Estimate the volume of the sludge using the mass determined in step 3 and the given sludge characteristics using Eq. 21-3.

$$\text{Sludge volume} = \frac{2379\ \text{kg/d}}{\left(999.1\ \text{kg/m}^3\right)(0.05)(1.05)} = 45.4\ \text{m}^3/\text{d}$$

10. Estimate the volume of the sludge as a percent of the total flow:

$$\begin{aligned}&\text{Sludge volume, \% of total flow}\\&= [(45.4\ \text{m}^3/\text{d})/(0.5\ \text{m}^3/\text{s})(86{,}400\ \text{s/d})] \times 100\\&= 0.11\%\end{aligned}$$

Physical Properties of Coagulant Sludges

The physical properties of alum and iron sludges are summarized in Table 21-5. The solids concentrations and physical properties are the most important properties for sizing and design of residuals management facilities. Solids concentrations depend on the design and operation of the sedimentation basins in addition to the type of sludge and its composition.

Table 21-5
Typical physical properties and chemical constituents of alum and iron sludges from chemical precipitation

		Type of Sludge	
Item	**Unit**	**Alum**	**Iron**
Physical properties			
Volume	% water treated	0.05–0.15	0.06–0.15
Total solids	%	0.1–4	0.25–3.5
Dry density	kg/m^3	1200–1500	1200–1800
Wet density	kg/m^3	1025–1100	1050–1200
Specific resistance[a]	m/kg	10–50×1011	40–150×1011
Viscosity at 20°C	$N \bullet s/m^2$	$2–4\times10^{-3}$	$2–4\times10^{-3}$
Initial settling velocity	M/h	2.2-5.5	1-5
Chemical constituents			
BOD	mg/L	30–300	30–300
COD	mg/L	30–5000	30–5000
pH	Unitless	6–8	6–8
Solids			
$Al_2O_3 \cdot 5.5H_2O$	%	15–40	
Fe	%		4–21
Silicates and inert materials	%	35–70	35–70
Organics	%	10–25	5–15

[a]Values of specific resistance reported in literature in units of s^2/g must be multiplied by 9.81×10^3 [$(s^2/g)(9.81\ m/s^2)(10^3\ g/kg)$ = m/kg] to obtain units of m/kg.

For example, alum sludge from an upflow clarifier would typically be drawn off at a concentration of 0.1 to 0.3 percent solids, compared to sludge from a horizontal-flow basin at 0.2 to 1.0 percent or more. Sludge may thicken to 4 to 6 percent solids if it is allowed to accumulate for a month or longer in a horizontal-flow sedimentation basin. Sludges that have relatively high proportions of alum or iron coagulant, as would result from treating low-turbidity water, will have lower solids concentrations than will those with relatively higher proportions of turbidity or silt. Coagulation of waters having substantial algae concentrations will also result in light, low-solids-concentration sludges. The addition of polymers generally tends to produce higher solids concentrations.

As coagulation sludges are dewatered and dried, there is a gradual transformation from a liquid to a semisolid to a solid. For the purposes of designing residuals management facilities, it is important to know when that transformation occurs as it will determine the type of equipment required to handle the sludge. As a liquid, the sludge can be pumped, piped, and transported in tank trucks, while as a semisolid or solid it must be shoveled and transported on a conveyor or in open trucks.

Unfortunately, the transition is not sharply defined and the transition point is not the same for all sludges. Coagulation sludges that have high proportions of gelatinous aluminum or iron hydroxides will act as liquids at higher solids concentrations than will those that contain more clay and sediments. These sludges are also thixotropic; that is, on standing they will seem to solidify, but when disturbed with a sudden jolt will revert to a liquid state. The minimum concentration at which sludge can be considered a solid is about 16 percent solids.

Chemical Properties of Coagulant Sludges

The chemical characteristics of coagulant sludges are directly related to the chemical content of the raw water and the coagulant chemicals. Typical data on the chemical characteristics of coagulant sludges are given in Table 21-5.

21-4 Lime Precipitation Sludges

Lime sludges are produced from the precipitation of calcium carbonate and magnesium hydroxide in the lime–soda softening process. Lime sludges may be essentially pure chemical sludges or they may include suspended materials from the raw water if turbidity removal is combined with softening. Similar sludges are produced in the magnesium carbonate softening process.

Estimating Quantities of Lime Sludges

Typical quantities of lime sludge produced from water softening are reported in Table 21-2. For design purposes, the amount of sludge anticipated at a plant can be estimated based on the chemical treatment and raw-water quality. The sludge is essentially composed of precipitated calcium carbonate and magnesium hydroxide, any turbidity or suspended solids that are removed during softening, and any insoluble impurities present in the treatment chemicals, such as lime grit. The suspended contribution can be estimated from turbidity data, as discussed above for coagulation sludges.

The amounts of precipitated calcium carbonate and magnesium hydroxide can be estimated directly from the anticipated calcium and magnesium removals. The pertinent precipitation reactions for lime softening are given in Chap. 20. Based on the reactions given in Chap. 20, the total lime sludge can be estimated with the following expression (Kawamura, 2000):

$$M = [(2.0\ \mathrm{Ca}^{2+} + 2.58\ \mathrm{Mg}^{2+}) + \mathrm{TSS} + X](10^{-3}\ \mathrm{kg/g}) \qquad (21\text{-}17)$$

where M = total sludge produced, kg/m^3
Ca^{2+} = calcium removed, g/m^3 as $CaCO_3$
Mg^{2+} = magnesium removed, g/m^3 as $CaCO_3$
TSS = total suspended solids, g/m^3
X = other coagulant aids added to enhance precipitation, g/m^3

Thus, the removal of 1.0 mg of calcium (expressed as $CaCO_3$) results in 2.0 mg of $CaCO_3$ in the sludge. Similarly, removal of 1.0 mg of magnesium (expressed as $CaCO_3$) results in 2.58 mg of sludge [$CaCO_3$+ $Mg(OH)_2$]. A graphical means of estimating sludge production on the total hardness removed is given on Fig 21-6.

Physical Properties of Lime Sludges

The physical properties of lime sludges, as reported in Table 21-6, are the most important factors in sizing treatment facilities. In larger plants, it is sometimes economical to recover lime from the sludge, in which case the chemical content of the sludge also becomes important. Solids concentrations

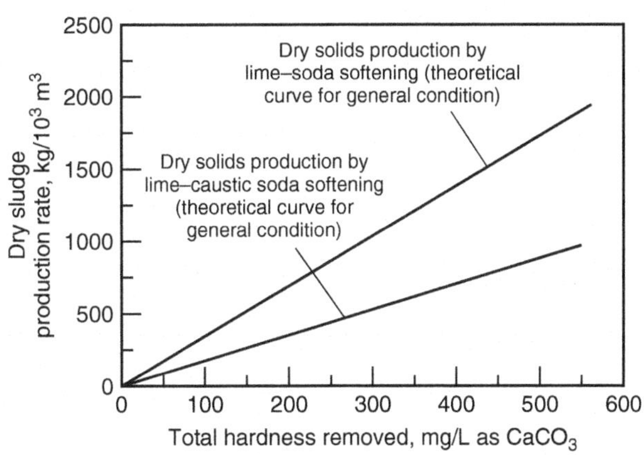

Figure 21-6
Dry sludge production rate versus total hardness removed.

Table 21-6
Typical physical properties and chemical constituents of lime-softening sludge

Item	Unit	Range of Values
Physical properties		
Volume	% water treated	0.3–6
Total solids	%	2–15
Dry density	kg/m^3	1100
Wet density	kg/m^3	1920
Specific resistance	m/kg	12×10^{10}
Viscosity	$N \cdot s/m^2$	$5–7 \times 10^{-3}$
Initial settling velocity	m/h	0.4–3.6
Chemical constituents		
BOD	mg/L	0–low
COD	mg/L	0–low
pH	Unitless	10.5–11.5
Total dissolved solids	%	2–15
Solids		
$CaCO_3$	%	85–94
Mg(OH)2	%	0.5–8
Silicates and inert materials	%	2–6
Organics	%	5–8

are dependent on the treatment process and on the proportions of the various chemical precipitates in the sludge. Sludges high in $CaCO_3$ typically have higher solids concentrations than sludges with more $Mg(OH)_2$ because $CaCO_3$ is a fine-grained, dense precipitate while $Mg(OH)_2$ is a more gelatinous material. The typical treatment process utilizing either upflow or horizontal-flow sedimentation basins following chemical addition and reaction produces fine-grained precipitates similar in nature to mud or silt deposits.

Chemical Properties of Lime Sludges

Lime sludges typically have a high pH (10.5 to 11.5) and are white, unless colored by turbidity, iron, or manganese. Generally, lime sludges are odorless, with little or no organic matter. Because of the high pH, lime sludges do not contain significant numbers of viable microorganisms. Typical chemical characteristics are summarized in Table 21-6. The specific chemical content of sludge from any given plant can be determined from the raw-water quality and the quantity of chemicals used.

21-5 Diatomaceous Earth Sludges

Sludges are produced during backwash of diatomaceous earth filters. Generally, the diatomaceous earth (DE) filter process is operated until the filter cake contains about two parts of DE for each part of impurities removed from the water. As a result, the sludge characteristics are predominantly those of the DE. The volume of waste washwater will vary from 2 to 5 percent of the plant flow. The corresponding solids concentration will typically vary from 6000 to 8000 mg/L, comprised primarily of DE. The total dry sludge is equivalent to the DE added in the process plus the suspended materials removed in the filter. The dry density of DE is about 160 kg/m^3.

21-6 Granular and Membrane Filter Waste Washwater

Waste washwater from the cleaning of granular or membrane filters is the most common type of liquid waste produced at water treatment plants.

Estimating Quantities of Filter Waste Washwater

It is difficult to estimate the quantity of waste washwater because it will depend on the raw-water quality, the degree and effectiveness of the treatment processes preceding the filtration step, the duration of the filter run, and the duration and type of backwash cycle employed. Based on the operating experience from a variety of water treatment plants,the quantity of waste washwater for both granular and membrane filters will typically comprise from 2 to 5 percent of the total amount of water processed. Some designers use 5 percent as a design value for the quantity of waste washwater. The volume of washwater to be handled depends on the frequency and duration of the backwash cycle. Because the backwash cycle from membrane filters is shorter and more frequent than that from granular filters, the volume of water to be handled from each backwash will be smaller.

Filter backwashing photos and video.

http://www.stantec.com/wt3r/21-3.html

Estimates of filter waste washwater quantities and frequencies may be obtained from pilot studies and filter design criteria. Pilot studies can be used to obtain critical information on backwashing rates and frequencies, which may be scaled up to address backwash duration at full scale. Using the information from pilot plant studies, the frequency, volume, and flow rate of waste washwater may be estimated.

Filter design criteria that are relevant to determining waste washwater frequency in granular filters are the unit filter run volume (UFRV) and the unit backwash volume (UBWV). These concepts are introduced in Chap. 11 and are used to determine the effective filtration rate (v_{EFF}) and the recovery or production efficiency ($\text{Rec} = v_{EFF}/v$) for a filter. The design criterion for production efficiency is typically 95 percent or greater. Typically, waste washwater quantities are 8 m^3/m^2 (200 gal/ft^2). To achieve a filter production efficiency of 95 percent, the UFRV would have to be at least 200 m^3/m^2 (5000 gal/ft^2) a run. At a filtration rate of 12.2 m/h (5 gpm/ft^2), a filter run would have to last at least 1000 min between backwash cycles.

The calculation for recovery in membrane filters is the same as for granular filters, but the terminology is different. Recovery (r) is the ratio of net-to-gross water production over a filter run,which is the volume of water fed to the membrane over a filter run (V_F) less the volume of water used during a backwash (V_{BW}) quantity divided by V_F. Recovery in membrane filtration is typically 95 to 98 percent, which is comparable to granular filters.

Another liquid waste stream from granular filters occurs when a filter is initially brought online after backwashing and the initial filter effluent is wasted, called *filter to waste*. During this initial period of operation, the filter is clean and does not have the same ability to remove particles as it does when fully ripened. The initial flow from a clean filter is typically diverted from the filter effluent to reduce the chance of undesirable constituents passing through the filter outlet and on through the process train. Filter-to-waste flow typically occurs for 15 min to 1 h after a filter is backwashed, but the specific time a filter operates in a filter-to-waste mode is based on the filter effluent quality.

The filter-to-waste flow may be captured and recycled through the treatment plant headworks or, in some cases,directly upstream of the filters. Filter-to-waste water quality is different than both filter waste washwater and supernatant from dewatering processes so it may need to be separated from these other waste streams.

Physicochemical Properties of Waste Washwater

The physicochemical properties of waste washwater are reported in Table 21-7. As reported in Table 21-7, the average total suspended solids concentration is typically on the order 100 to 1000 mg/L. Thus, the physical properties of waste washwater are similar to those for water. Because of the low concentration of solids and their settleability, waste washwater has historically been (1) returned directly to the headworks of the treatment plant when comprising less than 10 percent of the plant flow; (2) discharged to a flow equalization basin and then returned to the headworks of the treatment plant; (3) discharged to waste washwater recovery ponds, basins, or lagoons where it is allowed to settle for 24 h or more before being decanted and returned to the headworks of the treatment plant; or (4) discharged to surface waters with the appropriate National Pollutant Discharge Elimination System (NPDES) permit in place. The current trend for handling waste washwater is to have a separate treatment facility, especially in larger plants, because of concern over the presence and recycling of microorganisms, potential increases in the concentration of disinfection by-products, as well as other concerns such as taste and odor.

Table 21-7
Typical physical properties and chemical constituents of granular filter and membrane filter waste washwater

Item	Unit	Range of Values: Granular Filter	Range of Values: Microfiltration
Physical properties			
Volume	% water treated	1–5	2–8
Total solids	mg/L	100–1000	100–1000
Specific gravity	Unitless	1.00–1.025	1.00–1.025
Specific resistance	m/kg	$11–120 \times 10^{10}$	$11–120 \times 10^{10}$
Viscosity at 20°C	$N \cdot s/m^2$	$1–1.2 \times 10^{-3}$	$1–1.2 \times 10^{-3}$
Initial settling velocity	m/h	0.06–0.15	0.06–0.15
Chemical constituents			
BOD	mg/L	2–10	2–10
COD	mg/L	20–200	20–200
pH	Unitless	7.2–7.8	7.2–7.8
Solids			
Al_2O_3 or Fe	%	20–50	20–50
Silicates and inert materials	%	30–40	30–40
Organics	%	15–22	15–22

21-7 Reverse Osmosis Concentrate

Increasingly, greater use is being made of membrane processes for the treatment of water. As noted in Chap. 17, nanofiltration and reverse osmosis membranes are used for the removal of dissolved constituents from water. Nanofiltration is typically used for water softening, whereas reverse osmosis is used most commonly for demineralization of brackish water and seawater. In general, these processes produce wastes that are high in dissolved solids but low in suspended solids.

Estimating Quantities of Membrane Concentrate

Typical quantities of concentrate produced in reverse osmosis were shown in Table 21-2. The quantity of concentrate produced depends on the operating characteristics of the membrane and the water and solute mass transfer coefficients.

The solvent (i.e., water) recovery rate r is defined as (Eq. 17-12, reprinted here)

$$r = \frac{Q_P}{Q_F} \tag{21-18}$$

where Q_P = permeate stream flow, m^3/s
Q_F = feed stream flow, m^3/s

The rate of rejection Rej, which is used to describe the removal of the solute from the permeate, is calculated as (Eq. 17-1, reprinted here)

$$\text{Rej} = \frac{C_F - C_P}{C_F} = \left(1 - \frac{C_P}{C_F}\right) \qquad (21\text{-}19)$$

where C_F = concentration of solute in feed, g/m^3
C_P = concentration of solute in permeate, g/m^3

To determine the quantity and concentration of the concentrate, equations for recovery and rejection are combined with mass balances for water and solute (Eqs. 17-13 and 17-14, reprinted here):

$$Q_F = Q_P + Q_C \qquad (21\text{-}20)$$

$$Q_F C_F = Q_P C_P + Q_C C_C \qquad (21\text{-}21)$$

where Q_C = concentrate stream flow, m^3/s
C_C = concentration in concentrate, g/m^3

Combining Eqs. 21-20 and 21-18 results in the following expression for the concentrate stream flow rate:

$$Q_C = \frac{Q_P(1-r)}{r} \qquad (21\text{-}22)$$

The application of the above equations is illustrated in Example 21-3.

Example 21-3 Estimating quantity and quality of waste stream from a reverse osmosis facility

Estimate the quantity and quality of the waste stream and the total quantity of water that must be processed from a reverse osmosis facility that is to produce 4600 m^3/d of demineralized water. Assume that the recovery and rejection rates are 86 and 92 percent, respectively, and that the concentration of total dissolved solids in the feed steam is 2200 mg/L.

Solution

1. Determine the flow rate of the concentrated waste stream and the total amount of water that must be processed.

a. Determine the concentrate stream flow rate using Eq. 21-22:

$$Q_C = \frac{Q_P(1-r)}{r}$$

$$= \frac{(4600\ \text{m}^3/\text{d})(1-0.86)}{0.86} = 748.8\ \text{m}^3/\text{d}$$

b. Determine the total amount of water that must be processed to produce 4600 m^3/d of permeate. Using Eq. 21-20, the required amount of water is given as

$$Q_F = Q_P + Q_C = (4600 + 748.8)\ \text{m}^3/\text{d} = 5348.8\ \text{m}^3/\text{d}$$

2. Determine the concentration of the permeate stream. The permeate concentration is obtained by writing Eq. 21-19 as follows:

$$C_P = C_F(1 - \text{Rej}) = (2200\ \text{mg/L})(1 - 0.92) = 176\ \text{mg/L}$$

3. Determine the concentration of the concentrated waste stream using Eq. 21-21:

$$C_C = \frac{Q_F C_F - Q_P C_P}{Q_C}$$

$$= \frac{(5348.8\ \text{m}^3/\text{d})(2200\ \text{mg/L}) - (4600\ \text{m}^3/\text{d})(176\ \text{mg/L})}{748.8\ \text{m}^3/\text{d}}$$

$$= 14{,}634\ \text{mg/L}$$

Physicochemical Properties of Membrane Concentrate

Concentrate may be clear or colored, with the specific gravity being dependent on the salt concentration but typically in the range of 1.02 to 1.035. Any residual suspended material present in the water to be treated would also be included in the waste concentrate. The detailed chemical content of a waste concentrate can be determined from a mass balance on the process and, as discussed above, depends on the quality of the water to be treated, the water and solute mass transfer coefficients (specific for the membrane), and the detailed design and operation of the system.

21-8 Ion Exchange Brine

Ion exchange brines resulting from the softening of water are considered in this section. Specific quantities and characteristics of the brine from a particular softening plant will depend on the resin selected, the raw-water characteristics, and the operation of the regeneration process.

Table 21-8
Typical chemical properties of ion exchange brine from softening[a]

Item	Unit	Range of Values
BOD	mg/L	30–300
COD	mg/L	30–5000
pH	Unitless	6–8
Total dissolved solids	mg/L	15,000–30,000
Solids		
Ca^{2+}	mg/L	3000–6000
Mg^{2+}	mg/L	1000–2000
Na^{+}	mg/L	2000–5000
Cl^{-}	mg/L	9000–22,000
SO_4^{2-}	%	5–15

[a]The volume of brine as a function of the water treated will vary from 1.5 to 10%.

Estimating Quantities of Ion Exchange Brine

Typical quantities of brine produced in ion exchange were given previously in Table 21-2 and will vary from 1.5 to 10 percent of the plant flow. Quantities of brine required for regeneration are determined during design of an ion exchange facility, as discussed in Chap. 16. The brine to be disposed of is equal in volume to that used for regeneration, but instead of being entirely composed of water and regeneration salts, it will be a mixture of the excess sodium required to drive the regeneration process and the ions removed by the ion exchange process. In addition, there will be rinse water used to flush the brine out of the resin between the regeneration cycle and operation, and there may be waste washwater from an initial backwash cycle to remove any suspended materials collected by the bed. Depending on the design of the facilities, the various waste streams may be separated or combined. Ideally, the brine and the freshwater streams are segregated, as are less contaminated portions of the regenerant brine, so as to allow some brine recovery.

Chemical Properties of Ion Exchange Brine

The chemical properties of typical ion exchange brines are summarized in Table 21-8. Physically, these waste brines are clear, with the specific gravity being dependent on the salt concentration but typically in the range of 1.02 to 1.035. If waste washwater is included, any suspended material present in the raw water would also be included in the waste. The detailed chemical content of a waste brine can be determined from a mass balance on the process and, as discussed above, depends on the raw-water quality and the detailed design and operation of the system.

21-9 Solid Sorbent Brines and Washwater

Brines and other liquid wastes resulting from the regeneration of sorption media are reviewed briefly in this section. At the present time a variety of

Table 21-9
Common solid sorbents used for the removal of specific constituents in water

Sorbent	Application
Activated carbon, granular (GAC)	Control of toxic organic compounds; barrier to occasional spikes of toxic organics in surface waters; control of taste and odor compounds; control of disinfection by-product precursors or DOC
Activated carbon, powdered (PAC)	Used primarily in the treatment of taste and odor compounds and the treatment of low concentrations of pesticides and other organic micropollutants
Ion exchange resins, mixed bed resins	Specific inorganic and organic constituents, organic compounds
Activated alumina (AA)	Used for the removal of fluoride, arsenic (see also Table 4-7)
Granular ferric oxide (GFO), granular ferric hydroxide (GFH)	Used for the removal of arsenic (III and V), phosphate, and metals such as chromium, selenium, antimony, and copper
Fibroid chemosorbents, various	Used for the removal of radionuclides, heavy metal ions, and organic contaminants from drinking water

sorption processes are used for the removal of specific constituents. The most common solid sorbents are listed in Table 21-9. The use of granular and powdered activated carbon is discussed in Chap. 15. Ion exchange is considered in Chap. 16. The other two commonly used sorbents are activated alumina (AA) for fluoride and arsenic removal and granular ferric hydroxide (GFH) for the removal of phosphorus and arsenic.

Use of fibroid chemosorbents is not common in the water industry. With the exception of ion exchange resins, discussed previously in Sec. 21-8, these sorbents are typically not regenerated but used in a singlepass mode until the adsorptive capacity has been reached and then disposed of. Once-through use is advantageous because there is no need for onsite storage regeneration chemicals and solid waste disposal issues are minimized. Where these sorbents are regenerated, the specific quantities and characteristics of the brine and wash waters will depend on the sorbent, the raw water characteristics, and the specific regeneration process selected. The management of brines is considered in Sec. 21-13; the disposal of solid sorbents is considered in Sec. 21-16.

Estimating Quantities of Brines and Washwater from Sorption Processes

The focus of the following discussion is on the use of AA and GFH solid sorbents. Typical quantities of brines and washwater will vary with each sorbent and the specific chemicals used in the regeneration process. For example, where AA is used for fluoride removal, three methods have been used for regeneration including (1) $NaOH/H_2SO_4$, (2) $Al_2(SO_4)_3$, and H_2SO_4. If the $NaOH/H_2SO_4$ method is used, the four-step process described in Table 21-10 is used.

Table 21-10
Regeneration sequence for activated alumina using $NaOH/H_2SO_4$

Step	Description	Waste	Approximate volume, mL/100 g Al_2O_3
1	Backwash with raw water[a]	Washwater	*a*
2	Regeneration with 1% NaOH	Spent regenerant containing removed constituents	1000–1200
3	Rinse with raw water	Washwater	700–800
4	Neutralization with 0.05% H_2SO_4	Spent acid solution	1000–1200

[a]Backwash opreation takes about 10 min to expand bed and remove accumulated solids before regeneration.
Source: Adapted from Tramfloc, Inc., Tempe, Arizona.

The quantities of brine and washwater must be determined by laboratory and pilot plant testing because of the variability of local operating conditions. Typically, the amount of backwash water will be about 2 to 4 percent of the total throughput, depending on the solids accumulation. When used for fluoride removal, on a once-through basis, a dilute aluminum sulfate solution $[Al_2(SO_4)_3 \cdot 18H_2O]$ is used to contact the AA as a pretreatment step to enhance performance. This pretreatment step is not needed when the AA is regenerated as described above. In a similar manner, GFH can be regenerated using either NaOH or H_2O_2.

Depending on the operating conditions, NaOH concentrations between 0.1 to 1 M are required for regeneration following a backwash cycle to clean the bed. Minimum regenerant volumes are on the order of 4 to 6 bed volumes. The backwash rate is in the range from 20 to 30 m/h. Operationally, the GFH sorbent bed is backwashed every 2–6 weeks to remove accumulated solids and to prevent compaction of the filter bed.

Chemical Properties of Brines and Washwater from Sorption Processes

The chemical properties of typical waste washwater and spent regenerant brines will depend on the specific chemicals used for regeneration. For the AA example given above, the regeneration process results in waste washwater, spent caustic brine solution containing a high concentration of the removed constituents, and a spent acid solution. Similarly, the brine resulting from the GFH regeneration is caustic with a high concentration of phosphorus. As discussed in the section on ion exchange brine, the detailed chemical content of brine solutions is estimated based on a mass balance on the process and will depend on the raw-water quality and the detailed design and operation of the system. Physically, the waste brines are clear, with the specific gravity typically in the range of 1.01 to 1.035. The management of these brines in considered in Sec. 21-13.

21-10 Management of Residual Liquid Streams

In addition to sludges, a number of residual liquid streams result from the treatment of water. As identified in Table 21-1, the principal liquid waste streams, excluding membrane concentrates and ion exchange brines, are filter waste washwater and filter-to-waste water for water treatment plants with granular filters and filter waste washwater for water treatment plants that use membrane filtration. Other waste streams are comprised of recycle flows from sludge-processing operations, including centrate, filtrate, pressate, supernatant flow, and leachate. The combined volume of these waste streams may approach 4 to 5 percent of the total water treated, depending on the processes employed. In the past, these streams were returned to the headworks, discharged to nearby water bodies, land applied, or discharged to wastewater collection systems. Because of new regulations, many of these past practices are no longer acceptable. As a result, the management of these liquid waste streams is a major issue in the design and operation of most water treatment plants.

Concerns with Recycle Waste Streams

As noted in the introduction to this chapter, the concerns with recycle flows are related to the constituents contained in them. The principal constituents of concern are listed in Sec. 21-1 and the options for dealing with these constituents are considered below.

Flow Equalization Lagoons or Basins

Flow equalization is used to reduce the impact of the intermittent high-volume flows from backwashing operations (see Fig. 21-1b). By returning the waste washwater at a more constant rate, the impact on treatment process performance is minimized.

When the equalization basin also functions as a settling basin, the impact of the return flow is further mitigated (see Fig. 21-7). When the suspended material in the raw water has been effectively coagulated and flocculated prior to filtration, the solids in the waste washwater generally settle rapidly. To achieve a supernatant turbidity of about 5 NTU with this type of waste washwater, the equalization basin should provide a minimum detention time of 1 to 2 h. Coagulants and coagulant aids such as alum and cationic polymer may be added to improve the settling characteristics of the solids in the waste washwater.

Treatment of Recycle Waste Streams

Because of concern over the constituents in the return flows, separate treatment facilities are now used at some water treatment plants to process recycle flows. Treatment options include

- ❑ Flow equalization without or with chemical addition
- ❑ Lagoons without or with chemical addition
- ❑ Batch sedimentation without or with chemical addition

Figure 21-7
Typical waste washwater basins used for flow equalization and as settling basins.

- High-rate sedimentation without or with chemical addition and preflocculation
- Dissolved air flotation
- Granular filtration
- Membrane filtration
- Disinfection
- UV oxidation.

Because of the larger area required for waste washwater storage basins, the use of high-rate sedimentation (see Fig. 21-8) has become common in larger water treatment plants. The sludge resulting from the high-rate sedimentation process as well as from the treatment options identified above is typically combined with other plant sludges for further treatment. Some innovative applications of treatment processes for filter waste washwater are reviewed and discussed in Cornwell et al. (2010).

Disposal of Liquid Streams

In some cases, residual liquid waste streams have been discharged to surface waters and/or to wastewater collection systems. The ability to use either of these options is site specific.

DISPOSAL TO SURFACE WATERS

Surface water discharges are regulated under the Clean Water Act through the National Pollutant Discharge Elimination System (NPDES). These laws consider water treatment and supply to be an industry, and, therefore, consider water treatment residuals, such as concentrate, an industrial waste. The NPDES permits can specify a variety of water quality requirements,

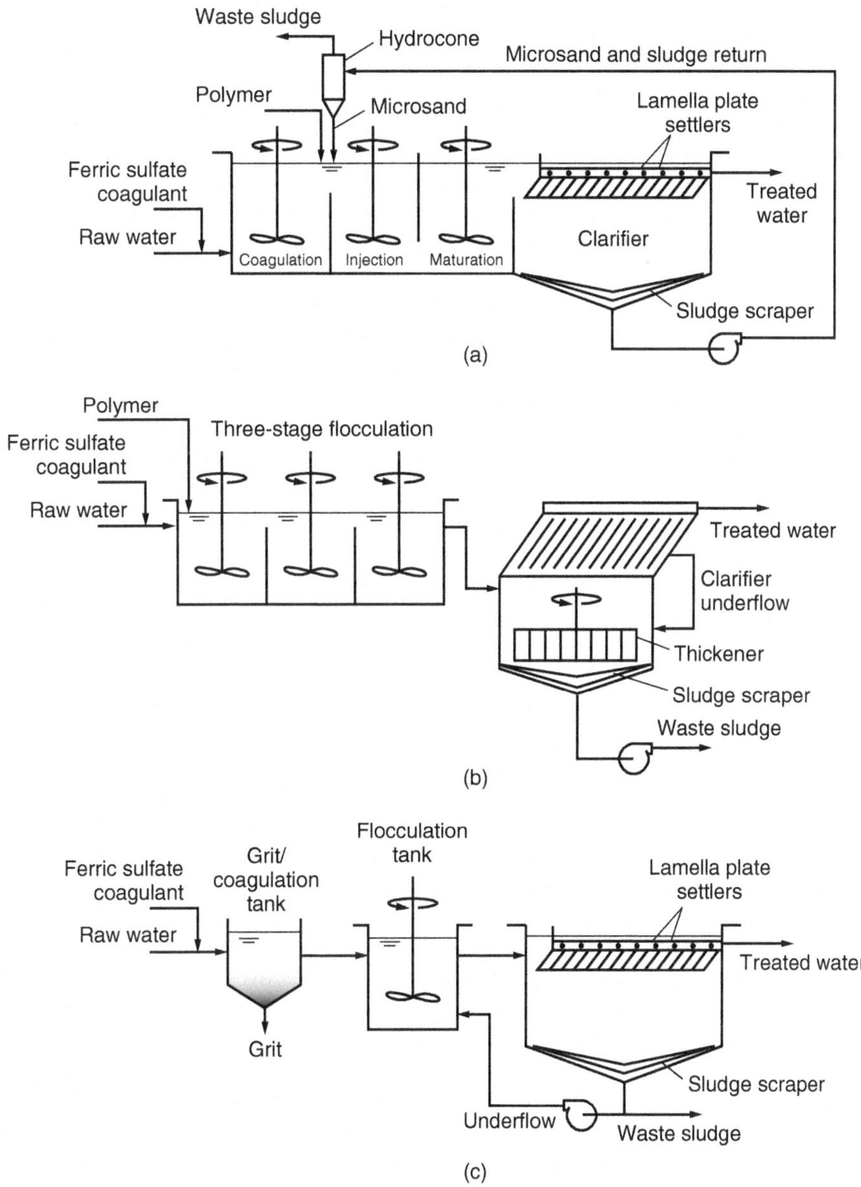

Figure 21-8
High-rate clarification processes: (a) ballasted flocculation, (b) lamella plate clarification and (c) dense sludge. (Adapted from Tchobanoglous et al., 2003.)

depending on classification of the receiving water body (e.g., potable water source, trout stream). State and local governments may impose additional restrictions on surface water discharges.

DISCHARGE TO WASTEWATER COLLECTION SYSTEM

The same laws that govern surface water discharge apply to wastewater collection system disposal. Pretreatment of the residual prior to discharge to the wastewater plant may be required because of state regulations or conditions imposed by the wastewater plant. In general, local pretreatment guidelines will cover the discharge from a water treatment plant to the wastewater collection system.

The capacity of the collection system or the wastewater treatment plant and the types of processes and operations at the wastewater facility may limit the amounts and type of liquids and/or solids that may be added to the wastewater system. The viability of sewer discharge is affected by the chemical characteristics of the residual stream, particularly with respect to whether high TDS, low dissolved oxygen, or high metals content may be toxic to the biological process at the wastewater plant.

Direct discharge to a wastewater collection system has a low capital cost and may also have a low operation and maintenance cost depending on monitoring requirements and sewer use fees. An advantage of this method is simpler permitting requirements. Discharge to a collection system is the easiest disposal method if a local wastewater treatment plant is willing to accept the waste, an issue that is often facilitated when a municipality operates both the water and wastewater systems.

A condition of discharge may be continuous monitoring of the organic strength and solids content of the residual flow. An attempt should be made to assess the impact of the residuals on the wastewater treatment facility prior to the selection of this alternative. It is possible that disposal of alum sludge may enhance phosphate removal at the wastewater treatment plant if any of the alum activity remains. Residual coagulant activity may also enhance primary sedimentation.

21-11 Management of Membrane Concentrates and Cleaning Solutions

Concentrate is produced when nanofiltration (NF) and reverse osmosis (RO) membranes are utilized in the treatment process. As noted previously, these processes produce a concentrate that is high in TDS but low in suspended solids. Concerns that must be addressed in the management of membrane concentrates include (1) the volume of concentrate (typically 15 to 50 percent) and (2) environmental classification and regulations, as discussed in Chap. 17. The management of cleaning solutions is also an important consideration. The management of membrane concentrates and cleaning solutions is considered in the following discussion. Ion exchange brines are considered in the following section. Additional details on the management of membrane concentrates may be found in Fox et al. (2009) and Mackey and Seacord (2008).

Concentrate Treatment Methods

Because of the interest in the application of RO, a variety of methods have been developed to treat the concentrate streams to both recover water and reduce the quantity of waste that must be processed further (see also Chap. 17). Some of the proven treatment processes for RO concentrate are summarized in Table 21-11. Of the methods listed in Table 21-11 precipitation, as discussed in Chap. 12, is used most commonly.

Methods of Thickening Concentrates

Because the volume of the concentrate stream from membrane processes that must be disposed of is larger than the waste stream from virtually all other water treatment processes, a number of alternative processes have been developed to further thicken (concentrate) the concentrate. Included among the thickening methods that have been developed are (1) membrane concentration, (2) evaporation/distillation, (3) crystallization, (4) solar evaporation, and (5) crystallization.

MEMBRANE CONCENTRATION

Two- and three-stage RO membrane concentration steps (see Fig. 21-9) have been used to increase the concentration of the brine to TDS values greater

Table 21-11
Treatment methods for RO concentrate

Treatment method	Description
Chemical precipitation	Precipitation to remove sparingly soluble salts and hardness. Precipitation has been used as a pretreatment step for conventional RO treatment to reduce scaling and for the pretreatment of brine before RO concentration to improve recovery. Reducing scaling is more critical in RO concentration operations because of enhanced scaling potential at high brine concentrations.
Electrodialysis reversal (EDR)	Use of a semipermeable ion exchange, ion selective, or electrodialysis membranes to separate the positively and negatively charged ions. Typically used as a pretreatment process.
High-efficiency reverse osmosis (HERO)	Involves specific pretreatment steps with RO to improve the recovery. Ions that cause scaling are removed and the pH is raised above 10 to limit the precipitation of silica on the membranes.
Natural treatment systems	Rely on naturally occurring physical, chemical, and biological processes to treat and assimilate the constituents in the brine. Free surface flow, subsurface flow, combined free and subsurface flow, and natural wetlands have been used and/or studied.

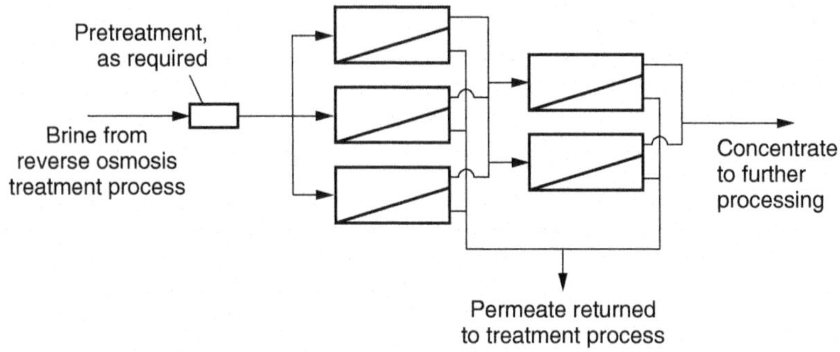

Figure 21-9
Two-stage reverse osmosis process.

than 35,000 mg/L. The concentrated brine can then be processed further by crystallization and solar evaporation. Details concerning membrane concentrate may be found in Chap. 17.

EVAPORATION/DISTILLATION

Residual concentrates and ion exchange brines (discussed in the following section) can be concentrated further by evaporation/distillation. Evaporation/distillation technologies that potentially could be used to concentrate residual brines include (1) boiling with submerged tube heating surface, (2) boiling with long-tube vertical evaporator, (3) flash evaporation, (4) forced circulation with vapor compression, (5) solar evaporation, (6) rotating-surface evaporation, (7) wiped-surface evaporation, (8) vapor reheating process, (9) direct heat transfer using an immiscible liquid, and (10) condensing-vapor heat transfer by vapor other than steam. Of these types of evaporation/distillation processes, multistage flash evaporation, multiple-effect evaporation, and vapor compression distillation appear most feasible for the processing of residual concentrates. A vertical-tube falling-film evaporator is illustrated on Fig. 21-10.

SOLAR EVAPORATION

Where climatic conditions are favorable, the use of evaporation ponds may be feasible. Important factors that affect the performance of evaporation ponds include relative humidity, wind velocity, barometric pressure, water temperature, and the salt content of the brine. In some locations, glass-covered solar ponds similar to those used for desalination in many of the dry Mediterranean countries are used to further concentrate brines by evaporation (see Fig. 21-11).

CRYSTALLIZATION

The crystallization process involves the conversion of thickened concentrate and brine into crystals that can be dewatered with a centrifuge or belt press. A typical brine crystallizer is illustrated on Fig. 21-12. The disposal of brine crystals is by landfilling.

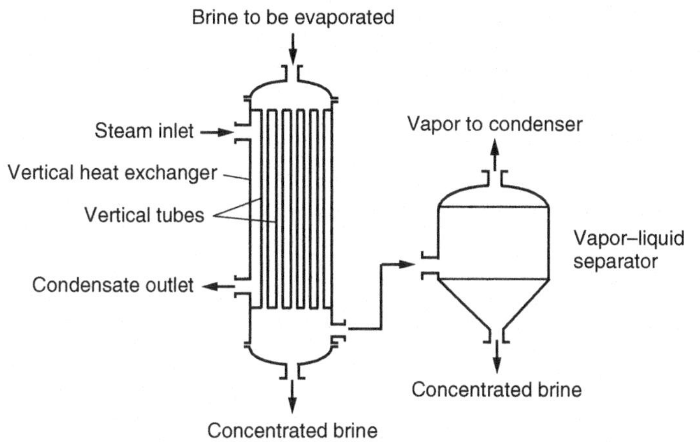

Figure 21-10
Schematic diagram of vertical-tube falling-film evaporator that can be used as a brine concentrator. Brine to be concentrated flows downward in a thin film along the inner walls of the vertical tubes and is partially evaporated. The evaporated vapor and concentrated brine flow into a vapor–liquid separator where the vapor is separated from the concentrated brine.

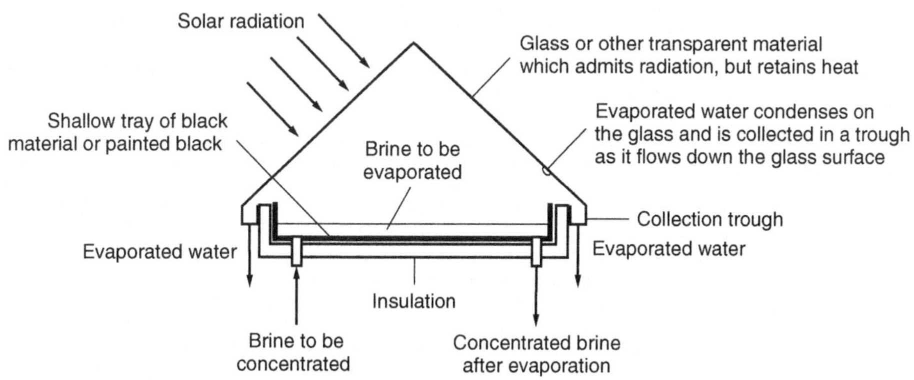

Figure 21-11
Schematic of solar brine evaporator. Solar radiation absorbed on dark tray heats the brine to be evaporated.

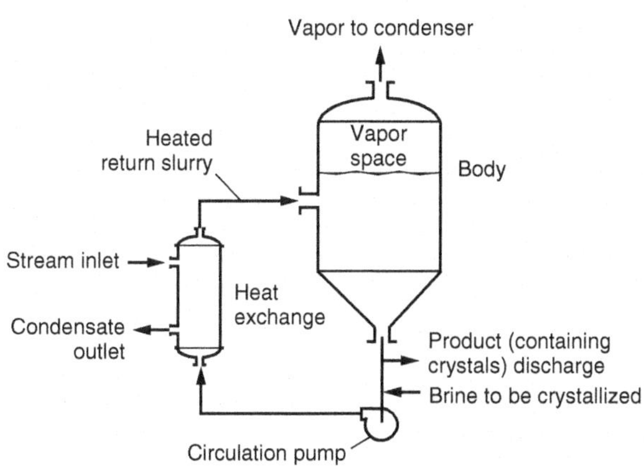

Figure 21-12
Schematic of forced circulation brine crystallizer. The heated return slurry causes the liquid in the body to boil. As the liquid boils and vapor is released, the concentration of the constituents in the liquid increases until supersaturation occurs and crystals form.

Ultimate Disposal Methods for Membrane Concentrates

Conventional methods used for the disposal of membrane concentrates include (1) disposal to surface waters, (2) discharge to wastewater collection systems, and (3) deep-well injection. Each of these methods is considered below.

DISPOSAL TO SURFACE WATERS

Despite complex regulatory requirements discussed previously in connection with the management of liquid waste streams, surface water discharge is often the most cost-effective disposal option for RO plants, especially for those plants located in coastal areas, where brackish or saline receiving waters are a viable discharge option. More than half of the plants in the United States now use this method of disposal. The advantage of a surface water disposal system is the relatively low capital and operation and maintenance costs. Disadvantages include the uncertainty of continued allowance of this practice in the future and the potential for creating a water pollution problem. Under the NPDES, extensive monitoring of the concentrate and the discharge water body is required.

Water quality issues

Surface water discharge is dependent on the quality of the concentrate. Nominally, the waste stream is comprised of inorganic solutes from the source water that have been concentrated. The degree of concentration is typically expressed as a concentration factor, which is estimated using a (worst-case) assumption of 100 percent rejection. With this approximation, the concentration factor (CF) is simply

$$CF = \frac{1}{1 - r} \tag{21-23}$$

where r = recovery rate expressed as a fraction

For instance, a low-pressure RO treating a brackish water (feed water TDS = 5000 mg/L) at a recovery of 85 percent produces a concentrate TDS of 33,300 mg/L, roughly the salinity of ocean water.

Thus, a coastal plant drawing water from a brackish aquifer and discharging concentrate to the ocean would appear to have little environmental impact. In addition to TDS, however, it is important to consider the toxicity of individual heavy metals, whose concentration is increased by the same factor. Additionally, many concentrate streams are anaerobic, which can be toxic to fish in the receiving water without sufficient dilution. Toxicity can initially be assessed by comparing predicted heavy-metal concentrations to regulated limits, but bioassays are often required before permits are issued. In many cases, economic considerations favor discharge to a brackish river or bay near the plant over an ocean outfall. However, the difference in salinity between the concentrate and the receiving water is more important in these locations.

Design considerations

Considerations for the design of a surface water disposal system include quality of the concentrate, pumping requirements, flow equalization, and outfall location and design. Outfall location is also an extremely important concern. The outfall should be located such that it discharges to a point of maximum dispersion. Similarly, the outfall should be designed to disperse the concentrate across the well-mixed zone of the water body. The location and design of the outfall significantly impact the pumping requirements. Consideration should be given to equalizing the residual flow to minimize pump and motor sizing and limit discharge of large residual slugs to the receiving water.

DISCHARGE TO WASTEWATER COLLECTION SYSTEM

As with the disposal of liquid streams, local pretreatment guidelines will cover the discharge of membrane concentrates to the wastewater collection system. In the case of RO plant residuals, which are primarily concentrated inorganic solutes, the biological process provides little treatment for the concentrate stream. In some situations it may be advantageous to discharge the concentrate to the wastewater plant effluent rather than the influent. Blending concentrate with wastewater effluent avoids toxicity issues in the wastewater process while still using the effluent to dilute the concentrate prior to discharge to a surface water. Dilution before discharge is a legitimate treatment strategy for a waste consisting primarily of concentrated inorganic solutes from a natural source water.

Design of a wastewater disposal system must provide for controlled discharges to eliminate the possibility of large slugs of residuals upsetting the wastewater treatment facility. Discharge should be coordinated with the wastewater treatment plant operators so that they may optimize the performance of their process units.

DEEP-WELL INJECTION OF MEMBRANE CONCENTRATE

Discharge of clear membrane concentrate and ion exchange brine by means of deep-well injection into a brackish or saline aquifer is regulated by federal and local environmental regulations and is dependent on the geology and groundwater hydrology of the area. Deep-well injection involves pumping the concentrate or brine stream into an injection well, typically thousands of meters deep. A typical injection well is shown on Fig. 21-13. The injection zone is typically a brackish or saline aquifer with no potential for use as a potable water supply, which is overlain by thick layers of impermeable rock that prevent contamination of shallower freshwater aquifers. Deep-well injection is used by about 10 percent of RO plants, although its use is becoming more common, particularly in Florida. Preference for deep-well disposal in Florida has arisen because of the existence of a reliable injection zone and public and regulatory resistance to surface water discharge.

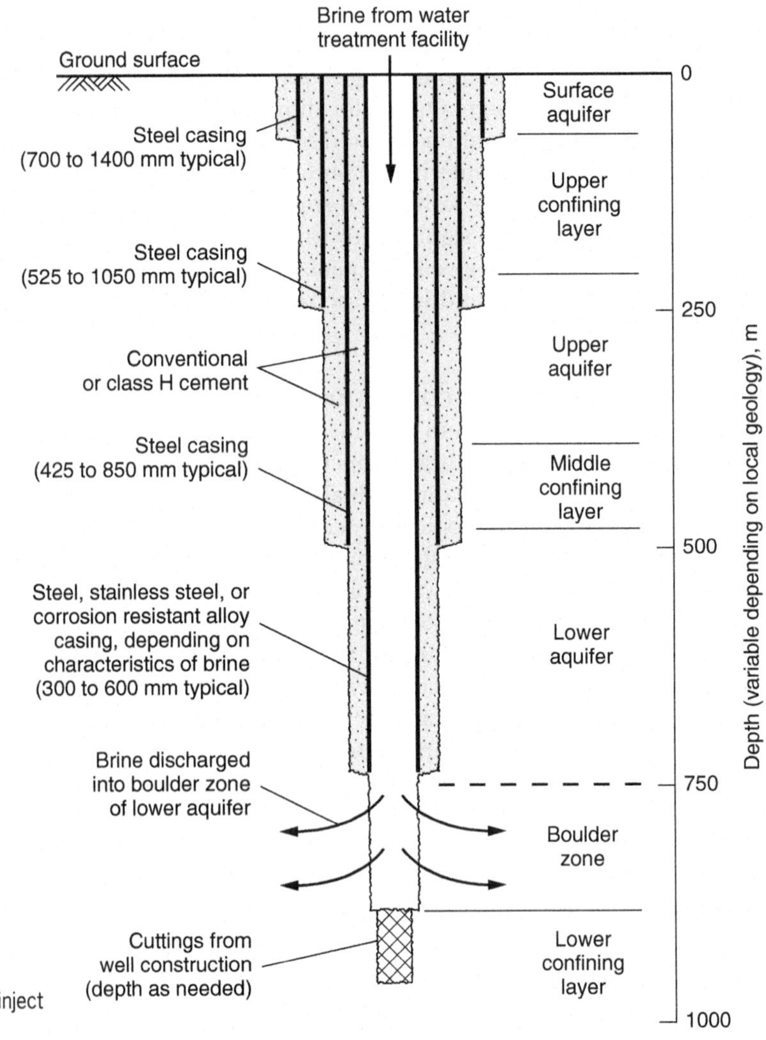

Figure 21-13
Schematic of typical well used to inject brine into subsurface aquifers.

Well construction is governed by regulations for deep-well injection of industrial wastes. Wells are constructed of three to four casings, with the space between each casing filled with cement grout. Each casing typically ends at a different depth. Depending on the local groundwater hydrology, there may be a significant potential for groundwater contamination and multiple casings are designed to prevent any leakage from one aquifer to the next (Cornwell and Roth, 2011). Deep-well injection systems tend to be fairly expensive due to well-drilling cost and maintenance costs. The high pressure at the bottom of the injection well and the saline solution tend to enhance the corrosion potential of the well screen and casing. Selection of materials resistant to corrosion under those conditions may prolong the operating life of an injection facility.

Management of Cleaning Solutions

Although the concentrate is by far the most voluminous waste stream, RO plants must also dispose of spent cleaning solutions. Frequently, the cleaning solutions are acidic or basic solutions with added detergents or surfactants. In many cases, the cleaning solution volume is so small compared to the concentrate stream that the cleaning solution is diluted into and disposed with the concentrate. In some cases, treatment of the cleaning solution may be required prior to disposal, but treatment may consist only of pH neutralization. Detergents and surfactants should be selected with disposal issues in mind.

21-12 Management of Ion Exchange Brines

The principal source of ion exchange brines is from the softening of hard waters. The characteristics of ion exchange brines were considered previously in Sec. 21-8. In general, large ion exchange water-softening plants have been located on or near coastal areas so that the resulting brines can be discharged to the ocean.

Processing of Brines

The management of brines will often involve some form of thickening before disposal. The thickening methods discussed previously in Sec. 21-11 in connection with management of membrane concentrates are also used for brines.

Ultimate Disposal of Brines

The principal methods used for the disposal of brines involves discharge to brackish or saline receiving waters, deep-well injection, and in the case of small facilities to wastewater collection systems. The same considerations discussed previously in Sec. 21-11 for the ultimate disposal of membrane concentrate also apply to ion exchange brines.

21-13 Management of Brines and Washwater from Sorption Processes

Because most solid sorbents are used on a single-pass (once-through) mode, the waste volumes are relatively small. Further, because the spent regenerant brine solutions contain contaminants of concern (e.g., arsenic, fluoride, etc.), ocean discharge and deep-well injection are not viable options for disposal as was the case with the ion exchange brines discussed in the previous section. The principal methods used to process wastes from the regeneration of AA and GFH are reported in Table 21-12. As reported

Table 21-12
Management of sorbents used in water treatment or the removal of specific constituents

Sorbent	Source of Waste	Disposal Method
Activated alumina (AA)	Backwash water	Where allowed, the backwash water can be discharged to wastewater collection system, if not it must be processed with the regenerant and neutralization solutions.
	Spent regenerant	In arid regions, the spent regenerant is concentrated in evaporation ponds and ultimately disposed of to a hazardous waste processing facility, otherwise it must be disposed of to a hazardous waste facility.
	Washwater	Where allowed, the washwater can be discharged to a wastewater collection system, if the residual constituent concentration is low. If not, it must be processed with the regenerant and neutralization solutions.
	Spent acid	In arid regions, the spent acid brine solution is concentrated in evaporation ponds and ultimately disposed of to a hazardous waste processing facility; otherwise it must be disposed of to a hazardous waste facility.
	Pretreatment	Because of the relatively small volume the dilute alum sulfate pretreatment solution is discharged to the wastewater collection system.
Granular ferric hydroxide (GFH)	Backwash water to clean and restore bed porosity	Where allowed, the backwash water can be discharged to wastewater collection system. If not, it must be processed with the regenerant solution, if used.
	Spent regenerant	In arid regions, the spent regenerant solution is concentrated in evaporation ponds and ultimately disposed of to a hazardous waste processing facility; otherwise it must be disposed of to a hazardous waste facility.

in Table 21-12 there are basically two methods for the management of these waste solutions: (1) discharge to a wastewater collection system and (2) evaporation (in arid locations) and ultimate transfer to a hazardous processing facility.

21-14 Management of Residual Sludges

Based on an understanding of the characteristics of the residuals, development of a complete residuals process treatment train requires an understanding of the technologies that are available for processing the residuals. The major unit operations and processes that are employed for the management of residuals will be reviewed in this section. A generalized

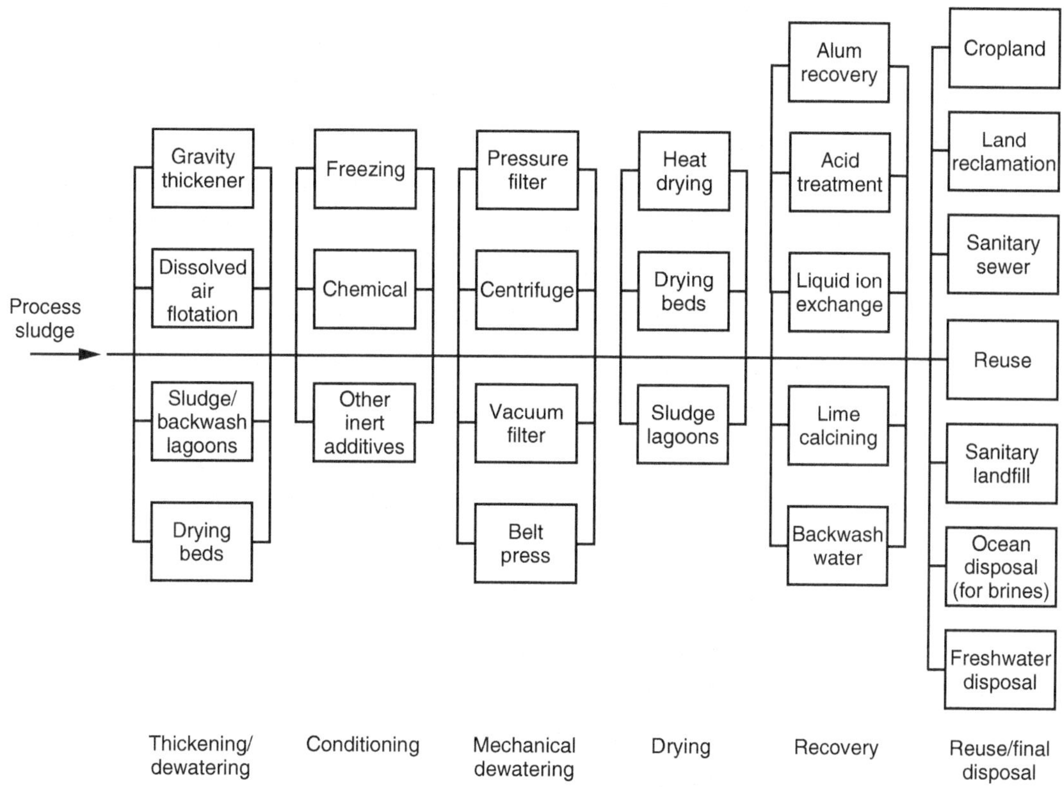

Figure 21-14
Unit operations and processes for management of water treatment plant sludges.

process diagram showing the various unit processes that may be used in residuals management and the sequence in which they may be assembled to form complete treatment systems are shown on Fig. 21-14. Some of the processes shown on Fig. 21-14 are omitted in the following discussion because either they are seldom used or insufficient data are available on their application.

A complete residuals management system is made up of one unit process from one or more of the process steps shown (e.g., thickening/dewatering, conditioning) and must include one of the unit processes from the final reuse and/or disposal step. Some typical residuals management processes are as follows:

- For alum sludge, gravity thickening, chemical conditioning, centrifugation, and final disposal to sanitary or monofill landfill
- For alum sludge, sludge lagoons, decant recovery and recycle, and final disposal to a sanitary or monofill landfill or wastewater collection system

- For lime sludge, gravity thickening, filter press dewatering, heat drying, lime calcining, and reuse
- For lime sludge, sludge lagoons, drying beds, cropland application, or monofill landfill
- For membrane concentrate, final disposal directly to brackish surface water, the ocean, deep-well injection, or wastewater collection system
- For ion exchange brines, membrane concentration, thermal brine concentration, and evaporation ponds

Unit operations that have proven to be the most successful and to have significant capabilities for dewatering sludges from water treatment plants are drying beds, vacuum filtration, pressure filter press, belt filter press, centrifugation, alum and lime recovery, and pellet flocculation. The above technologies are considered in the following discussion.

Thickening/ Dewatering

Thickening to increase the solids content of sludge involves the removal of excess water by decanting and the concentration of the solids by settling. The decanted water is usually recovered unless the water contains objectionable tastes or odors or large numbers of algae or other microorganisms while the solids are processed further or disposed of. Gravity thickening is used most commonly as the first step in the residuals management process. The most common methods of gravity thickening for sludges are lagoon settling with a decantation operation and conventional gravity thickening in specifically designed reactors. For coagulation or softening sludges, the primary process involved is compaction thickening of the sludge, while for filter waste washwater the processes of settling and hindered settling are most important.

MECHANICAL GRAVITY THICKENING

Gravity thickening is typically accomplished in a circular tank designed and operated similarly to a solids-contact clarifier or sedimentation tank (see Fig. 21-15). Sludge is introduced into the tank and allowed to settle and compact. Gentle agitation of the sludge prior to settling creates channels in the sludge matrix for water to escape and promote densification of the solids. The thickened sludge is collected and withdrawn at the bottom of the tank. A properly designed and operated gravity thickening system can produce softening sludges in excess of the 2 to 6 percent associated with alum sludge, depending upon the calcium carbonate and magnesium hydroxide content in the sludge. Lime sludges that are predominantly calcium carbonate can be thickened to 30 percent solids and higher. Typical performance and design data on gravity thickening are presented in Table 21-13.

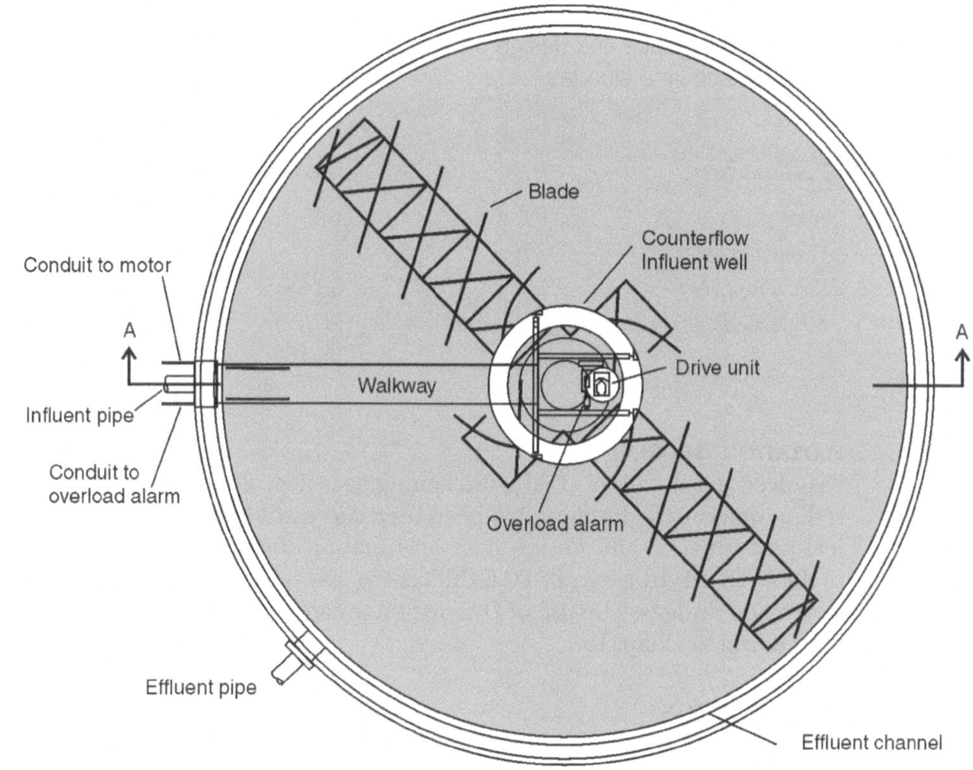

(a)

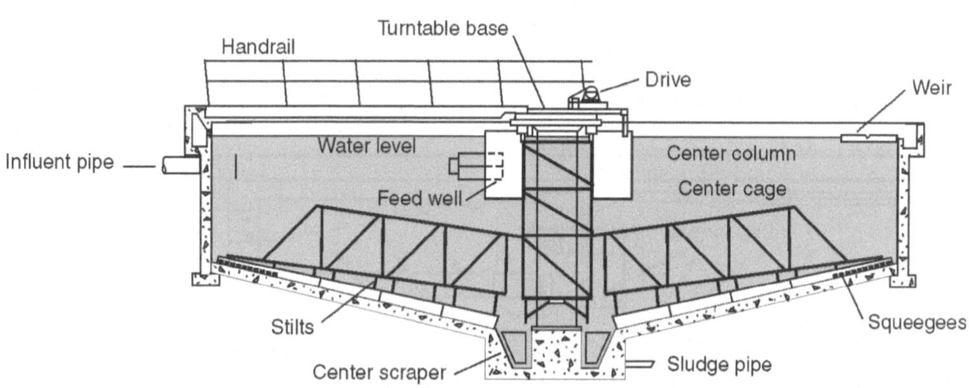

(b)

Figure 21-15
Typical mechanical gravity thickener for water treatment plant sludges: (a) plan view and (b) section through thickener.

Table 21-13
Typical performance and design data for gravity mechanical thickening of coagulant and lime sludges

		Type of Sludge	
Parameter	Unit	Coagulant	Lime Softening
Feed solids	%	0.2–1	1–4
Thickened solids	%	2–3	>5
Solids recovery	%	80–90	80–90
Solids loading	$kg/m^2 \cdot d$	20–80	100–200
	$lb/ft^2 \cdot d$	4–16	20–40

FLOTATION THICKENING

Dissolved air flotation (DAF) thickening (see Fig. 21-16) involving both sedimentation and flotation has been used successfully for dewatering. Typical performance and design data on flotation thickening are presented in Table 21-14. In general, DAF thickening has been most successful with hydroxide sludges. The use of DAF for the treatment of return flows is also considered in Chap. 10.

SLUDGE LAGOONS

A nonmechanical means of handling water treatment plant sludges consists of dewatering in sludge lagoons or drying beds. If land is readily available and inexpensive, the use of sludge lagoons is a cost-effective means of

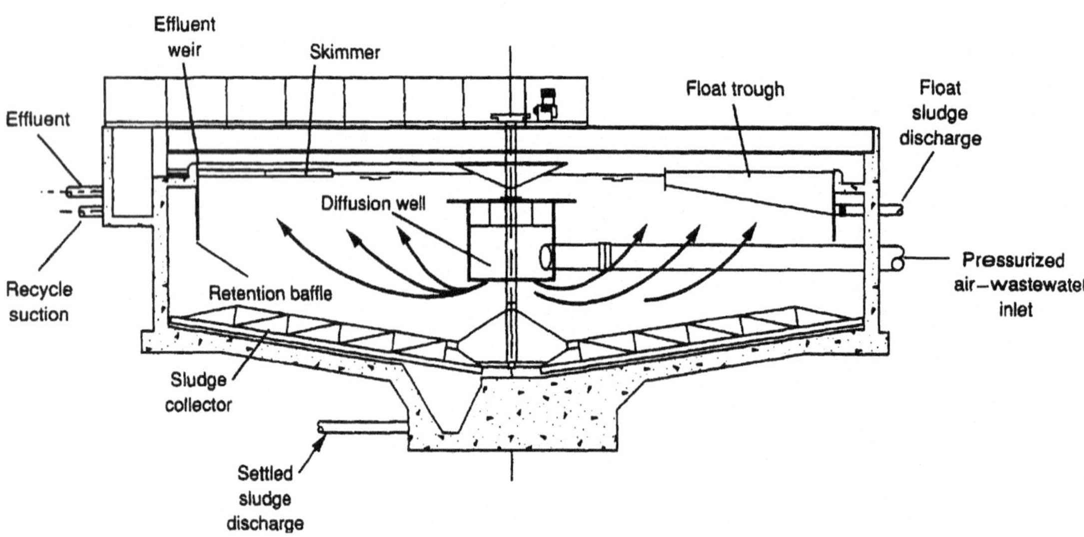

Figure 21-16
Section through typical flotation thickener for water treatment plant sludges (see also Fig. 10-22, Chap. 10).

Table 21-14
Typical performance and design data for dissolved air flotation thickening for coagulant and lime-softening sludges

		Type of Sludge	
Parameter	**Unit**	**Coagulant**	**Lime Softening**
Feed solids	%	0.5–1	0.5–1
Thickened solids	%	3–5	3–5
Solids recovery	%	80–90	80–90
Solids loading	$kg/m^2 \cdot d$	48–120	48–120
	$lb/ft^2 \cdot d$	10–24	10–24
Volumetric loading	$m^3/m^2 \cdot d$	110–150	110–150
	$gal/ft^2 \cdot d$	2800–3600	2800–3600

storing and thickening residuals. Lagoons are commonly lined earthen basins equipped with inlet control devices and overflow structures (see Fig. 21-17). Wastes with settleable solids are discharged into the lagoons from which the solids are separated by gravity sedimentation. Sludge lagoons can be classified by their mode of operation: permanent lagoons and dewatering lagoons. Permanent lagoons act as a final disposal site for settled water solids, whereas dewatering lagoons are cleaned periodically. Lime sludges have been dewatered to 50 percent solids concentration by the sludge lagoon handling method.

Lime-softening sludges dewater more readily than alum sludges. In addition, most softening plant sludges will air dry well in lagoons; therefore, it is important to design the lagoon so that the sludge does not remain submerged after initial filling, as lime sludge does not compact well when under water. Typically, sludge lagoons for lime wastes should be such that they can be filled and allowed to dry before being refilled. Typical values to which lime sludges can be dewatered (in percent solids concentration) vary from 30 to more than 50 percent. It should be noted that the final solids concentrations obtained by the various dewatering methods will vary depending on the type of sludge and sludge conditioning employed.

A common approach used at many water treatment plants in the United States is to use lagoons not only as thickeners (with continuous decanting) but also as drying beds after a predetermined filling period. Three months of filling and an average drying cycle of 3 months are the most common design parameters used. The required lagoon area can be determined using a sludge loading rate of 40 and 80 kg dry solids/m^2 of lagoon area (8.2 to 16.4 lb/ft^2) for wet and dry regions, respectively.

For example, based on a loading rate of 80 kg/m^2, the effective area of lagoons required to handle alum sludge from the 0.5-m^3/s (11.4-mgd) water treatment plant of Example 21-2, Part A, can be approximated as follows. Assuming a total of four lagoons and an average of 100 days filling

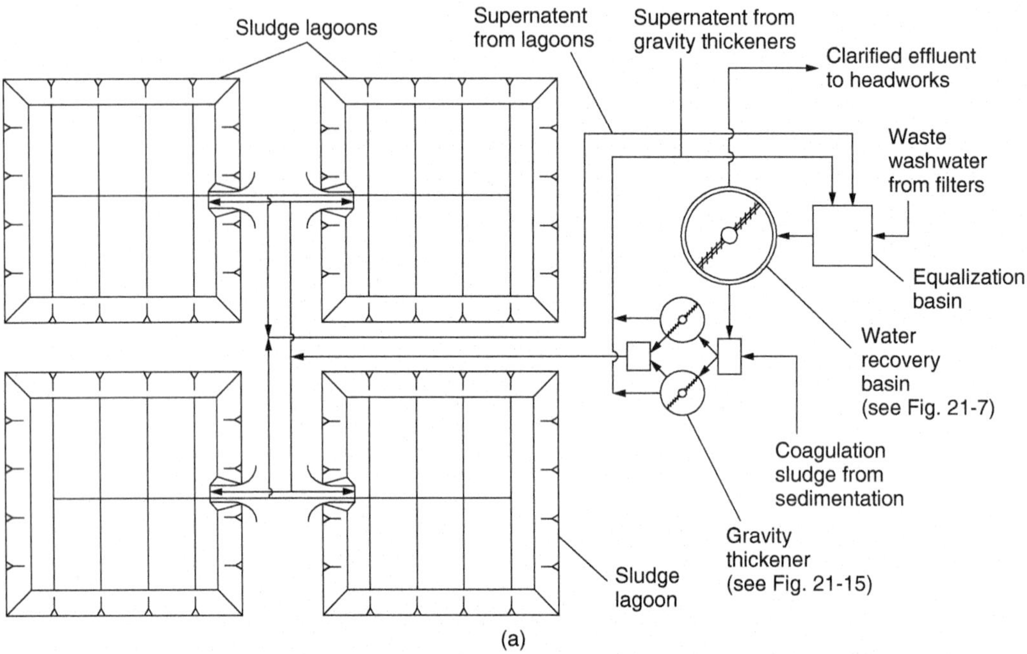

Figure 21-17
Typical sludge storage lagoons: (a) schematic (adapted from Qasim et al., 2000) and (b) view of large sludge lagoon.

cycle (detention) per lagoon in a dry region, an effective lagoon area for each lagoon is given as

$$\text{Effective lagoon area} = \frac{(1892 \text{ kg/d})(100 \text{ d})}{80 \text{ kg/m}^2} = 2365 \text{ m}^2 = 0.24 \text{ ha}$$

The actual area required for a lagoon would be at least 1.5 times the area computed because of the additional area required for berms and access roads.

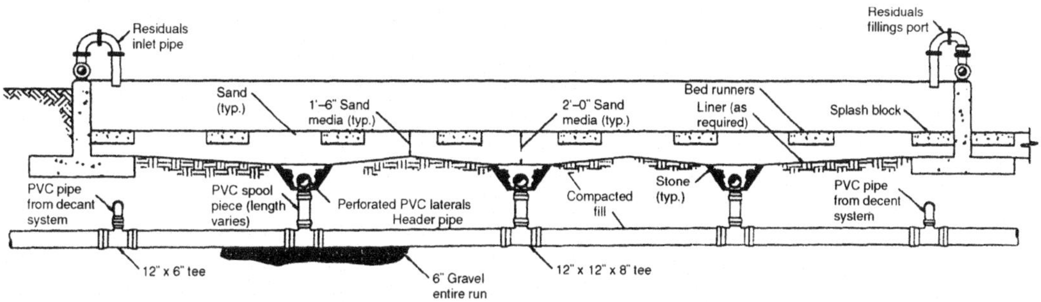

(a)

(b)

Figure 21-18
Typical sludge-drying beds for water treatment plant sludges: (a) section through sludge-drying bed and (b) view of sand-drying beds.

GRAVITY DEWATERING ON DRYING BEDS

Gravity dewatering involves placement of the sludge to be dewatered on a sand (see Fig. 21-18) or wedge wire filter surface and the subsequent drainage of water from the sludge through the filter material. The process will produce a relatively dry, solid sludge for further treatment or disposal. Gravity dewatering may be combined with other drying and dewatering operations to produce a sludge of any desired dryness. Gravity dewatering is applicable to dewatering of sludge discharged directly from sedimentation basins or following thickening.

Bed area

The size of the drying bed required is usually the factor that determines the feasibility of gravity dewatering at a given site. If land is readily available, gravity dewatering is the method of choice; otherwise a more sophisticated mechanical system will be required. Multiple drying beds must be sized to allow spreading of a relatively thin layer of sludge 0.15 to 0.3 m (6 to 12 in.) and sufficient time between spreading cycles to permit drainage, drying, and removal of the sludge. At least three and preferably four or more beds should be provided to allow discharge of sludge to one bed while the other beds are draining, drying, and being cleaned.

Underdrains and decanters

An underdrain system or decanter system must be provided to remove water from the sludge if the drying beds are constructed in wet regions. Underdrains are used to collect the water drained from the sludge and decanters are used to collect the water off the top of the sludge. Underdrains are not required in most dry regions, but decanters are helpful in any type of climate. The underflow or decanted water can then be either recycled to the plant inlet, if the quality is good, or discharged. Underdrains typically consist of gravel and perforated clay or PVC (polyvinyl chloride) pipes.

Cycle time, weather, and conditioning

As indicated above when discussing bed area, cycle time includes time for filling the bed, sludge draining and drying, and cleaning the bed. The major portion of the cycle time is the drainage and drying time. Ideally, the time required for draining and drying should be determined by bench or pilot testing, with due consideration given to variations in climate and sludge characteristics. In the absence of actual testing, the engineer must estimate the extent to which the sludge will drain. The use of polymers to condition alum sludge can reduce draining time but will probably not substantially increase the drained solids concentration.

After draining is completed, further dewatering occurs by evaporation. The time required to reach the desired dryness can be calculated from evaporation/precipitation data. Once the sludge is drained, rainfall will drain through or be decanted from the surface of the bed, rather than rewetting the sludge. For conservative design the net evaporation rate (evaporation minus precipitation) should be used as a reference for sizing drying beds.

Conditioning

As in thickening, successful dewatering often depends on proper conditioning of the sludge in advance. The objectives of conditioning are to improve the physical properties of the sludge so that water will be released easily from the sludge, improve the structural properties of the sludge to allow free draining of the released water, improve the solids recovery of the process (i.e., to reduce the fraction of solids lost in the removed water), and minimize dewatering process cycle times.

CHEMICAL ADDITION

Polymers are the most commonly used conditioners for dewatering water treatment sludges. Based on full-scale operating experience, it has been found that most types of polymers will improve the dewatering characteristics of sludges. The selection of a polymer for a given application should be based on bench tests or, preferably, pilot- or full-scale tests, as described in Chap. 9. Also, in general, higher molecular weight polymers are more effective, except that the viscosity of very high molecular weight polymers may cause handling problems.

Successful use of polymers is dependent on good dispersion of the polymer into the sludge to be conditioned. A typical blending device for polymers is shown on Fig. 21-19. As with any chemical addition, provision must be made for good initial mixing. Sizing of polymer feeding equipment should be based on bench-scale determination of dosage requirements. If that is not practical (such as in the design of a new plant), facilities can be sized based on estimated sludge solids concentrations and quantities. Polymer doses required are typically in the range of 10×10^{-4} to 100×10^{-4} kg polymer/kg sludge solids for metal hydroxide sludges.

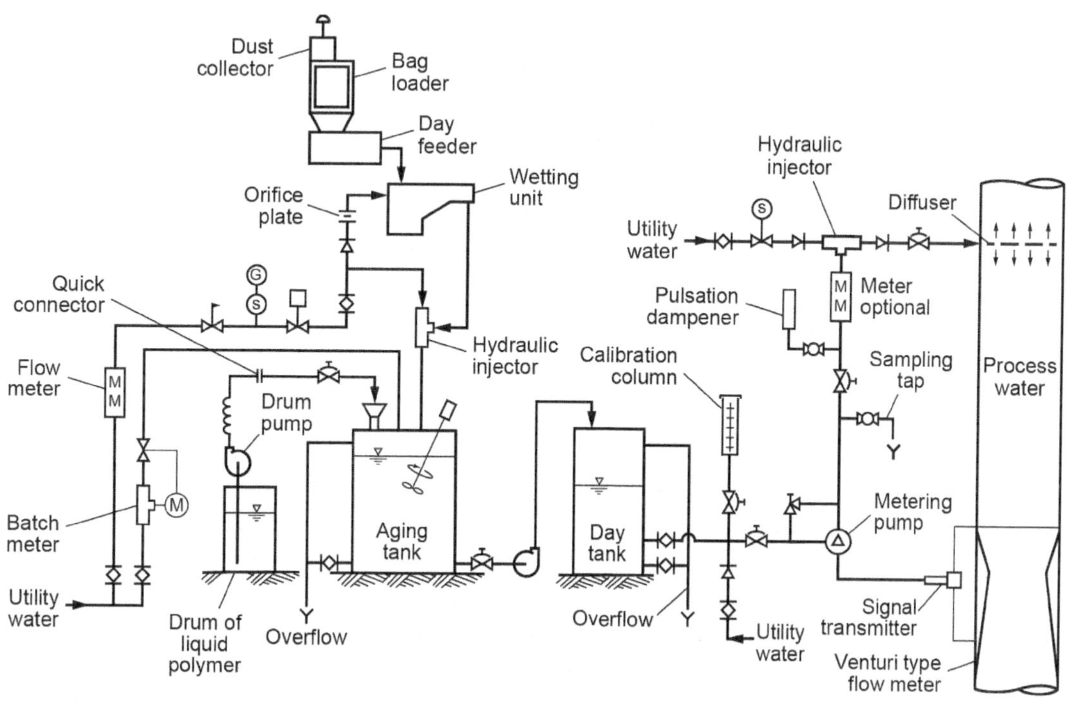

Figure 21-19
Blending diagram for organic polymers used to condition sludge. (Adapted From Kawamura, 2000.)

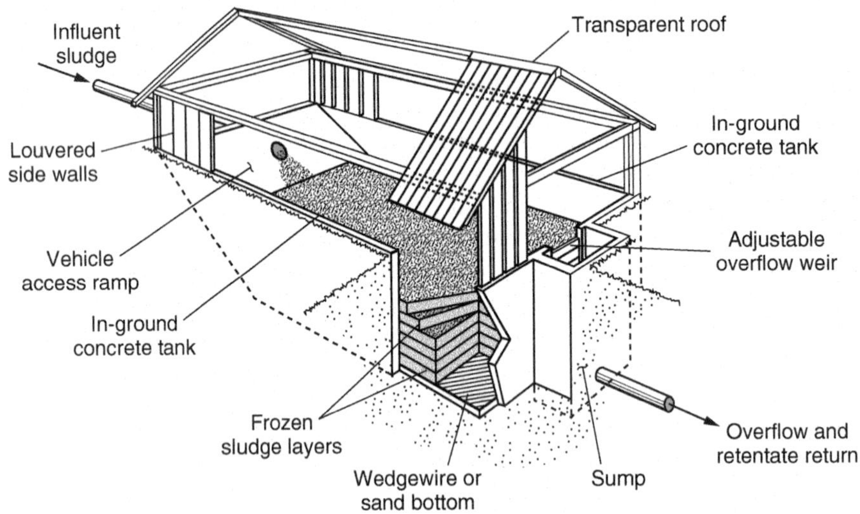

Figure 21-20
Typical installation for conditioning of sludge by freezing.

FREEZING

Freezing is very effective for metal hydroxide sludges such as alum and iron sludges (see Fig. 21-20). The effect is to destroy completely the gelatinous structure, leaving the sludge (after thawing) in the form of a fairly coarse granular material such as sand or coffee grounds. The process is irreversible. Unfortunately, the mechanical efficiencies of equipment for freezing and thawing sludge are low, so this process is usually applied only where natural freezing will occur in a lagoon. The lagoon must have sufficient capacity to allow the sludge to sit over the winter. With natural freezing of alum sludge, a 2 percent solids sludge can be converted to a 20 percent solids granular slurry that will readily drain to over 30 percent solids and can be easily handled. Additional details on the design of freeze-thaw sludge dewatering beds may be found in Martel (1989).

HEAT TREATMENT

Although heat treatment has been investigated as a sludge-conditioning process, results are not as dramatic as with freezing. Heat treatment of storage is not being employed on a full scale. With rising energy costs, heat treatment is not an attractive alternative for sludge conditioning.

ADDITION OF INERT MATERIALS

Another conditioning step often applied in pressure filtration of alum sludges is the addition of lime or inert granular materials such as fly ash or diatomaceous earth. Relatively high proportions of these materials are required.

Mechanical Dewatering

Dewatering includes all those processes intended to remove free water from sludges beyond that which can be removed by decanting from a thickener. The objective is to reduce the sludge volume and produce a sludge that can be easily handled for further processing. As the use of open storage lagoons and drying beds becomes less feasible for dewatering, some form of mechanical dewatering is now used at most large treatment plants. The principal types of mechanical dewatering devices now used are (1) vacuum filtration, (2) plate and frame filter presses, (3) belt filter presses, and (4) centrifuges.

VACUUM FILTRATION

The type of vacuum filter employed almost exclusively is the rotary drum vacuum filter. There are two basic types of rotary drum filters: (1) traveling media and (2) precoat media filters. The precoat filter is used mainly for dewatering coagulated sludges such as alum sludge (see Fig. 21-21). For alum and ferric hydroxide sludges, successful operation of vacuum filters requires the use of polymer or lime as sludge-conditioning chemicals. Lime sludges, however, generally do not require conditioning prior to vacuum filtration.

The variables to be considered in designing a vacuum filter system are the size and type of filter, the cake discharge mechanism, the filter media, vacuum level, cycle time, and sludge conditioning. For a specific sludge, filter leaf tests are used to help determine the appropriate conditioner and dose, filter media, and cycle time. Typical performance and design data for sludge dewatering with vacuum filtration are reported in Table 21-15.

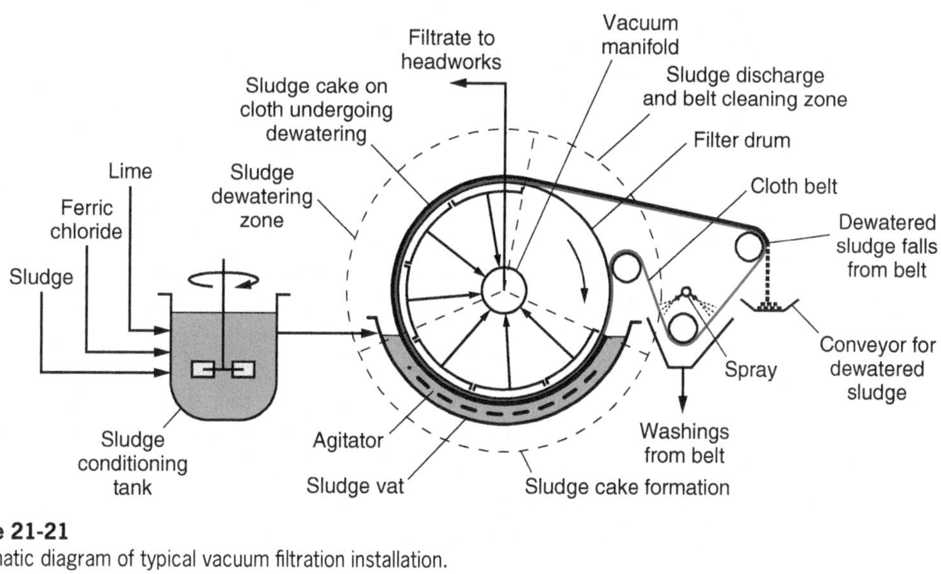

Figure 21-21
Schematic diagram of typical vacuum filtration installation.

Table 21-15
Typical performance and design data for precoat rotary vacuum filter for dewatering water treatment plant sludges

Parameter	Unit	Range of Values
Feed solids	%	2–6
Feed rate	$L/m^2 \cdot h$	0.7–2.1
	$gal/ft^2 \cdot h$	2–6
Solids recovery	%	96–99+
Dry-solids yield	$kg/m^2 \cdot h$	0.2–0.3
	$lb/ft^2 \cdot h$	1.0–1.5
Thickened solids		
Alum sludges	%	15–25
Lime sludges	%	20–40
Filtrate suspended solids	mg/L	10–20
Precoat recovery	%	30–35
Precoat rate	$kg/m^2 \cdot h$	0.02–0.04
	$lb/ft^2 \cdot h$	0.1–0.2
Precoat thickness	mm	38.1–63.5
	in.	1.5–2.5
Drum speed	rev/min	0.2–0.3
Operating vacuum	mm Hg	127–508
	in. Hg	5–20

PLATE AND FRAME FILTER PRESSES

Filter press dewatering is achieved by forcing the water from the sludge under high pressure. Although the filter press produces high solids concentration and low chemical consumption, its disadvantages include high labor costs and limitations on filter cloth life. A filter press consists of a number of plates or trays supported in a common frame (see Fig. 21-22). During sludge dewatering, these frames are pressed together either electromechanically or hydraulically between a fixed and moving end. A filter cloth is mounted on the face of each plate. Sludge is pumped into the press until the cavities or chambers between the trays are completely filled. Pressure is then applied, forcing the liquid through the filter cloth and plate outlet. The plates are then separated and the sludge removed.

Conditioning of the sludge prior to filtration is required and the degree of conditioning dictates the performance. In general, a filter cake of about 30 to 40 percent solids concentration is expected after pressure filtration with lime and polymers as sludge-conditioning chemicals. Lime sludges have been reported to readily dewater to above 50 percent solids without sludge conditioning. The filtrate may contain less than 10 mg/L of suspended solids if the sludge is conditioned properly. Both capital and operation and maintenance (O&M) costs for this process are high.

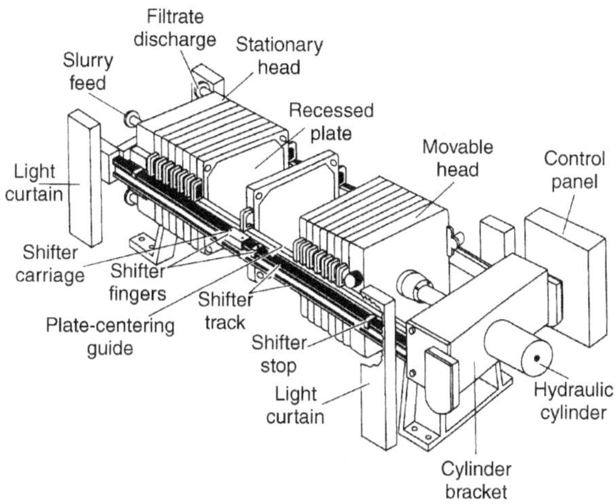

Figure 21-22
Schematic view of plate and frame filter press. (Adapted from Tchobanoglous, et al., 2003.)

GRAVITY AND PRESSURE BELT FILTERS

The application of belt filters for water sludge thickening and dewatering is a relatively recent application dating back to the early 1970s, although belt filters were developed much earlier. Thickening with a gravity belt filter involves two operational steps: (1) chemical conditioning of the sludge and (2) gravity drainage using a single belt as illustrated on Fig. 21-23a. In some designs a vacuum is applied to the under side of the belt to enhance dewatering. Sludge dewatering with a belt filter press involves three operational steps: (1) chemical conditioning of the sludge, (2) gravity drainage, and (3) mechanical application of pressure. To accomplish the application of pressure, two or more belts are used, depending on the manufacturer. A belt filter press employing two belts for dewatering is illustrated schematically on Fig. 21-23b and pictorially on Fig. 21-23c. For both types of belt filters, the key to successful performance is the sludge chemical conditioning step.

In thickening dilute sludges, both coagulant and polymer addition is employed. Coagulant addition is used to concentrate the solids. Polymer addition is used to coagulate and flocculate the sludge before it is applied to the gravity belt thickener. Once applied to the belt thickener, the sludge is distributed uniformly across the width of the belt and moves with the belt. Fixed guide veins or plows located just above the surface of the moving belt create clear zones for free water released from the sludge to drain through the belt. Typically, from 70 to 80 percent of the free water is drained within the first meter. Thickened solids, scraped from the belt, are collected in a hopper for further processing, transport, or disposal (see Fig. 21-13a). Thickened sludge cake with up to 20 percent solids is possible with proper conditioning.

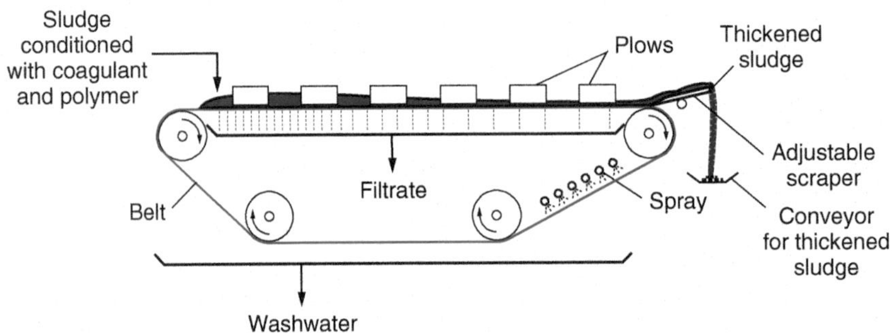

(a)

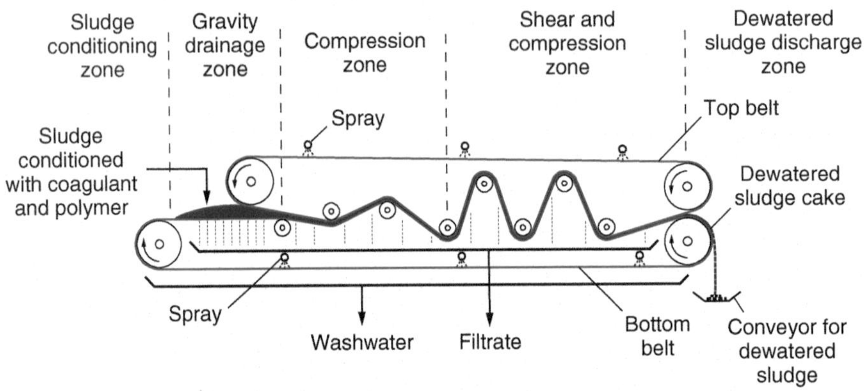

(b)

(c)

Figure 21-23
Belt filters for sludge thickening and dewatering: (a) schematic gravity belt thickener, (b) schematic belt filter press for dewatering sludge, and (c) view of typical belt filter press used for dewatering alum sludge.

Sludge dewatering as illustrated on Fig. 21-23b involves subjecting chemically conditioned sludge to gravity drainage, mechanical pressing, and shearing. Often shear and compression dewatering are shown as occurring together. Chemical conditioning typically involves the use of organic polymers. Complete and thorough mixing of the polymer and the sludge is the key to successful dewatering with belt presses. Once the coagulated solids are applied to the filter, gravity drainage occurs. The partially dewatered sludge then moves into the compression zone where it is squeezed between two moving belts. Additional dewatering occurs by shearing as the sludge moves to the outlet. Dewatered sludge, which falls off the belt, is transported by a conveyor belt to storage facilities for further processing or disposal (see Fig. 21-26). The capital cost and space requirements for pressure belt filtration sludge dewatering are generally significantly lower than for plate and frame pressure filter sludge dewatering systems. Typical performance and design data for sludge dewatering with belt filter presses are reported in Table 21-16.

Belt filters for sludge thickening and dewatering.

http://www.stantec.com/wt3r/21-4.html

CENTRIFUGES

Centrifuges are used to both thicken and dewater sludges. The centrifuge is basically a sedimentation device in which the solids/liquid separation is improved by rotating the liquid at high speeds to increase the gravitational forces applied on the sludge. There are two basic types of centrifuges: (1) solid-bowl and (2) basket centrifuges. The two principal elements of centrifuges are the rotating bowl, which is the settling vessel, and the conveyor discharge of the settled solids (see Fig. 21-24).

Application of centrifuges in the water treatment field has normally been in dewatering lime-softening sludges. Solids concentrations of 35 to

Table 21-16

Typical performance and design data for belt filter press dewatering of water treatment plant sludges

Parameter	Unit	Range of Values
Feed solids	%	4–30
Thickened sludge		
Alum sludges	%	15–30
Lime sludges	%	25–60
Solids recovery	%	95–99+
Cake yield	$kg/m^2 \cdot h$	0.8–4.0
	$lb/ft^2 \cdot h$	4–20
Filtrate solids	mg/L	950–1500
Filter speed	rev/min	0.2–0.5
Operating pressure	kPa	550–830
	$lb/in.^2$	80–120

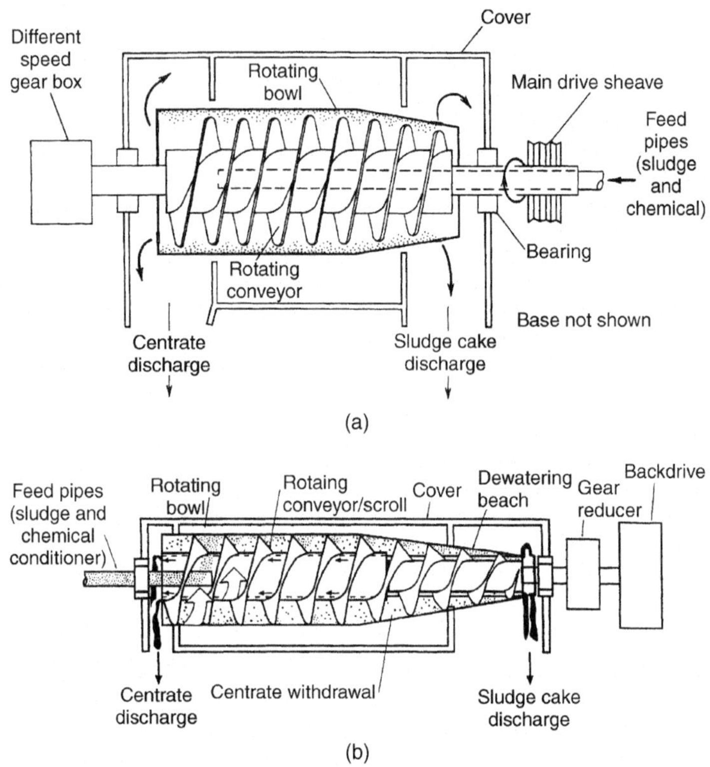

Figure 21-24
Typical centrifuges used for dewatering of water treatment plant sludges: (a) continuous countercurrent solid bowl and (b) continuous concurrent solid bowl.

50 percent are typical with lime sludges, although higher values have been reported. In centrifugation of predominantly alum sludges, solids concentrations of 12 to 15 percent have been obtained. Effective dewatering of alum sludge by centrifugation requires conditioning of the sludge with polymers and lime. Twenty to 25 percent solids may be obtained from 3 to 4 percent solids alum sludge. Polymer doses of approximately 1 to 2 g/kg (2 to 4 lb/ton) of feed solids are typical. Feed solids concentration for alum sludge centrifugation is in the range of 1 to 6 percent and 10 to 25 percent for lime sludge. Typical performance and design data for centrifuge thickening of coagulant and lime-softening sludges are reported in Table 21-17. Both capital and operation and maintenance costs for this process are relatively high.

Lime Sludge Pelletization

Taking advantage of the observation that sludge pelletization occurred during the suspended-bed cold-softening water treatment process led to the development of the lime sludge pelletizer shown on Fig. 21-25. Operationally, the conical vessel is charged with a silica catalyst. Lime sludge injected into the reactor reacts with calcium bicarbonate (hardness) in the carrier water

Table 21-17
Typical performance and design data for centrifuge thickening of coagulant and lime-softening sludges

		Type of Sludge	
Parameter	Unit	Coagulant	Lime Softening
Feed solids	%	1–6	10–25
Thickened solids	%	12–15[a]	35–50
Solids recovery	%	90–96	90–97
Polymer dosage	g/kg	1–2	na
	lb/ton	2–4	na

[a]Up to 25% solids has been achieved with the use of conditioning chemicals.

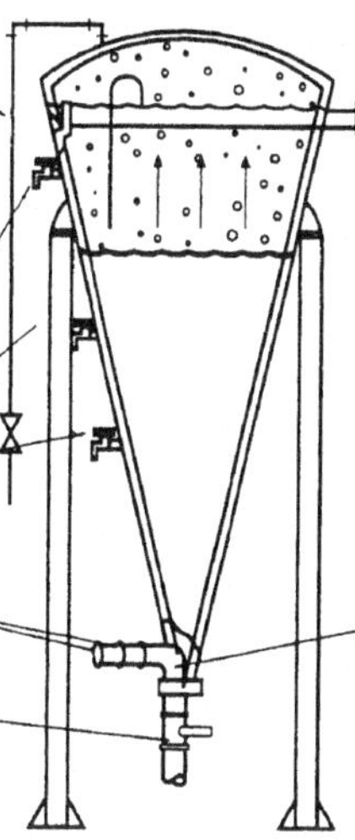

Figure 21-25
Schematic diagram of reactor for lime sludge pelletization. (Courtesy of US Filter.)

to form calcium carbonate precipitates on the silica catalyst. The softened water is removed from the top of the reactor. The calcium pellets removed from the reactor are discharged to a storage and drainage facility. The solids content of the calcium carbonate pellets is about 60 percent.

Aqua Pellet System

The Aqua Pellet system employs proprietary equipment of a Japanese firm and has been used successfully to treat alum sludge in Japan. The process consists of multistage thickening of the sludge using sodium silicate and polymer and a dewatering process using a large horizontal rotating drum called a Dehydrum. Based on a full-scale operating system, it appears that 20 to 25 percent solids can be obtained by the Dehydrum starting with a 0.5 to 2 percent solids alum sludge.

Recovery of Coagulant

Aluminum and iron recovery can be accomplished by adding acid (normally sulfuric acid) to sludges to solubilize the metal ion salts. Lime recovery from lime sludge has also been practiced by the recalcination process.

ALUM AND IRON RECOVERY

Alum and iron recovery can be accomplished by acidification with sulfuric acid. In simplified form,the reactions involved are

$$2Al\,(OH)_3 + 3H_2SO_4 \rightarrow Al_2\,(SO_4)_3 + 6H_2O \qquad (21\text{–}24)$$

$$2Fe\,(OH)_3 + 3H_2SO_4 \rightarrow Fe_2\,(SO_4)_3 + 6H_2O \qquad (21\text{-}25)$$

Normally, over 80 percent of alum and iron recovery is achieved at a pH of about 2.5. Unfortunately, heavy metals, manganese, and other organic compounds are often found in the recovered alum and iron. The presence of potential contaminants, as well as rising costs, has limited the recovery of alum and iron as a viable processing alternative.

LIME RECOVERY

Lime recovery by recalcination has been practiced in a number of locations in the United States. Quicklime (CaO) can be recovered from softening sludges after purification and dewatering. To recover CaO, water-softening sludges are burned at a temperature of 1010°C (1850°F). The pertinent reactions are

$$CaCO_3 \rightarrow CaO + CO_2 \qquad (21\text{-}26)$$

$$Mg\,(OH)_2 \rightarrow MgO + H_2O \qquad (21\text{-}27)$$

The types of furnaces that have been used for recalcining include the rotary kiln, flash calciner, fluidized-bed calciner, and multiple-hearth calciner. The primary consideration in selecting the recalcination process is the overall economics of the process because the process is energy intensive (10×10^9 to 15×10^9 J/kg of sludge).

MAGNESIUM BICARBONATE RECOVERY

The magnesium bicarbonate recovery process was developed as a part of an overall treatment process that would use magnesium bicarbonate (precipitating magnesium hydroxide and calcium carbonate) as the primary coagulant. The pertinent reactions are

Precipitation

$$Mg(HCO_3)_2 + 2Ca\,(OH)_2 \rightarrow Mg(OH)_2 + 2CaCO_3 \qquad (21\text{-}28)$$

Magnesium Bicarbonate Recovery

$$Mg(OH)_2 + 2CO_2 \rightarrow Mg(HCO_3)_2 \qquad (21\text{-}29)$$

During the late 1960s and the early 1970s, recycling magnesium carbonate was a developing and promising technology. Although the magnesium bicarbonate recovery process is technically feasible, based on pilot plant studies, the economics could not justify scale-up to a full-size facility. Presently, there are no known existing full-scale water treatment facilities that employ the magnesium bicarbonate recovery process; however, recovery of magnesium bicarbonate is being practiced in the paper and pulp industry.

21-15 Ultimate Reuse and Disposal of Semisolid Residuals

Several alternatives are available for the disposal or reuse of water treatment plant residuals. In practice, the options available for ultimate disposal or reuse of water treatment plant residuals frequently dictate the type of inplant handling system necessary. Selection of an alternative should be based on economic as well as regulatory considerations. The type of sludge and sludge characteristics are also important criteria to be used in developing disposal or reuse alternatives. It is critical that the ultimate solids disposal or reuse program be a reliable, environmentally sound practice to ensure that it does not affect the primary goal of the treatment plant—the production of potable water. Alternatives available for disposal or reuse of water treatment plant residuals include

- Landfilling
- Disposal on land (reuse as a soil amendment)
- Discharge to a wastewater collection system
- Codisposal with wastewater biosolids
- Reuse in building or fill materials

Landfilling, land spreading, and lagoon storage followed by landfilling or spreading are typical land disposal options. Residuals disposed of in a wastewater collection system end up in the wastewater treatment plant, where they are removed and disposed of with wastewater sludge. Codisposal involves the mixing of water treatment plant residuals with wastewater treatment plant sludges followed by disposal or reuse. Reuse as building or fill material is site specific. However, before discussing the various disposal methods, it is appropriate to consider the impacts of arsenic in residuals.

Arsenic in Residuals

As the regulations for arsenic in drinking water have become more stringent, water treatment processes have shifted to include greater arsenic removal from raw water, increasing arsenic concentrations in residuals. Arsenic in water treatment plant residuals comes from two sources: the raw water and the chemicals used for treating the raw water. The amount of arsenic contributed from treatment chemicals may be considered minor (Cornwell et al., 2003), but earlier research (Cornwell and Koppers, 1990) found up to 40 percent of the arsenic in water treatment plant residuals was contributed by iron coagulants.

The U.S. EPA has evaluated water treatment plants specifically designed to remove arsenic and the residuals in these plants (U.S. EPA, 2000a, 2000b). Adsorption and coprecipitation of As(V) with iron and aluminum flocs are believed to be the primary arsenic removal mechanisms in the water treatment plants in these studies.

The effectiveness of arsenic removal from water is dependent upon both the pH of the water and the oxidation state of the arsenic. Once arsenic is in the treatment plant residuals, changes in pH or changes that result in a reducing environment may cause the arsenic to resolubilize. Processes commonly used for water treatment plant residuals that may cause pH or oxidation state changes include dewatering, lagooning, and landfilling.

Landfilling

The most common disposal method for water treatment plant sludge in the United States is landfilling (see Fig. 21-26) in a commercial nonhazardous landfill (or monofill that receives only drinking water treatment plant residuals) or a hazardous waste landfill, which is regulated by the U.S. federal government. Water treatment plant sludge is tested to determine if it is a hazardous (Resource Conservation and Recovery Act, RCRA, subtitle C) or nonhazardous (RCRA subtitle D) waste to determine which type of landfill is appropriate for final disposal.

Figure 21-26
Dewatered sludge is placed in large storage containers that are hauled to a landfill, emptied, and returned. The elevated conveyor belt system used to transport the dewatered sludge to the storage containers is used in conjunction with the belt press shown in Fig. 21-23c.

Water treatment plant sludge testing is performed to meet the U.S. EPA requirement for solid waste characterization by the toxicity characteristic leaching procedure (TCLP). The TCLP test exposes a waste to a mildly acidic solution similar to what might be found in a municipal landfill (U.S. EPA, 1992). If the waste leachate contains any of the regulated compounds at or above the minimum concentration in leachate for toxicity characteristics, it is considered to be toxic and, therefore, a hazardous waste.

California has more stringent regulations than the U.S. EPA and requires solid waste to be tested according to the California waste extraction test (WET) (State of California, 2005). The WET uses a slightly more aggressive leaching procedure than is used by the TCLP test, as shown in Table 21-18. Both the TCLP test and the WET are designed to simulate landfill leaching. If the leachate contains any of the regulated compounds on the *List of Inorganic Persistent and Bioaccumulative Toxic Substances and Their Soluble Threshold Limit Concentration* (U.S. EPA, 1992) and the concentration of the compound is equal to or exceeds its listed soluble threshold limit concentration (STLC) or total threshold limit concentration (TTLC), the waste is considered toxic and therefore a hazardous waste.

A study of water treatment plant sludge leachate from plants that use either alum or iron as the primary coagulant was done by the American Water Works Research Foundation (Cornwell et al., 1992). The sludges were analyzed using the TCLP test, and all were found to be nonhazardous. Thus, landfilling of coagulant sludges in nonhazardous waste landfills is, in general, an appropriate disposal method.

Table 21-18
Comparison of toxicity characteristic leaching procedure (TCLP) and the waste extraction test (WET)

	Test Procedure	
Parameter	**TCLP**[a]	**WET**[b]
Extraction fluid	Acetic acid	Citric acid
Extraction fluid pH	4.93	5.00
Extraction duration, h	18	48
Dilution of waste to extraction fluid of solid portion of waste	20-fold	10-fold
Anaerobic conditions	No	Yes, by purging with N_2 gas prior to agitation
Inorganic constituents measured	8	19
Organic constituents measured	23	18
Aggressiveness for inorganic constituents	Less	More

[a]U.S. EPA (1992).
[b]State of California (2005).

There is concern that the current testing procedure, the TCLP, is not adequate for determining the long-term leachability of a material in its final disposal site. A variety of conditions can exist (wet, dry, acidic, basic) at a disposal site, making it unlikely that one test can assess all possible conditions. Research by the Department of Energy's Mixed Waste Focus Area and the U.S. EPA Office of Solid Waste evaluated six different testing protocols using a tiered approach. The conclusion reached in this study was that a better picture of how waste would behave at its final disposal site was obtainable but was expensive to achieve (Hulet et al., 2000).

Land Application

Land application of water treatment plant residuals is a disposal method that is regulated in the United States by the federal government under the Resource Conservation and Recovery Act (RCRA) as well as state and local governments. Sludges to be spread on land must be tested to determine if they are a hazardous (RCRA subtitle C) or nonhazardous (RCRA subtitle D) waste by either the TCLP or WET, which are compared in Table 21-18.

Residuals that have been land applied include coagulant sludges, lime-softening sludges, reverse osmosis concentrate, and slow sand filter washings (Novak, 1993). Benefits from land application of coagulant sludges have not been clearly demonstrated, while concerns have been reported (Gendebien et al., 2001). Specific concerns raised include aluminum having a negative impact on barley growth in soils where the pH is below 5.5; high levels of aluminum reducing the availability of phosphorus and increasing soil compaction; and iron becoming concentrated in grazing land resulting in a negative effect on copper metabolism, especially in sheep (Gendebien et al., 2001; Marshall, 2002).

Depending on local soil conditions, the spreading of lime sludges may have beneficial impacts on the soil and crop yields when used in appropriate amounts (see Fig. 1-3d, Chap. 1). Nitrogen fertilizers typically lower soil pH, resulting in a decrease in calcium availability and reduced crop production (Marshall, 2002). The addition of lime sludge raises the soil pH comparable to commercially available agricultural limestone materials (Bly et al., 2001). The effectiveness of lime sludge, measured in terms of the total neutralizing power (TNP), is typically around 100, which is comparable to the TNP of commercial agricultural lime products (Marshall, 2002). A small number of facilities use land application or irrigation for concentrate disposal. Land application is often limited by salinity, which can accumulate in soil and prevent plants from growing or leach into underlying freshwater aquifers. Land application disposal methods are more appropriate for low-pressure systems that primarily remove hardness or NOM; the concentrate from these systems have lower salinity.

Lagooning Prior to Disposal

As noted above, lagooning of water treatment plant residuals is typically an intermediate step prior to final disposal in a landfill. Lagooned residuals separate into sludge and supernatant. The sludge settles to the bottom of the lagoon and becomes more compressed, and less porous, anoxic and anaerobic conditions may develop leading to a reducing environment in the settled sludge layer. Under reducing conditions, arsenic may gain electrons, going from As(III) to As(IV), and resolublize into the lagoon supernatant.

Based on recent research, it has been found that the release of arsenic from settled sludge in lagoons into the supernatant is associated with the release of iron and generally follows a lowering of the redox potential (Cornwell et al., 2003). Arsenic release being associated with the release of iron was correlated with the change in iron concentration, not the total iron concentration, and was found to occur for both ferric and alum sludges. Results from lagoon simulations were used to conclude that reduced pH and biodegradable organic matter cause an increase in arsenic release from sludge.

Seven different sludges in a simulated lagoon study were tested at periodic intervals for toxicity using both the TCLP and the WET, and the test results were compared to the arsenic levels measured in the lagoon supernatant (Cornwell et al., 2003). In all cases, the WET results were much higher, up to 700 times higher, than the measured arsenic levels in the supernatant. The TCLP results tracked more closely to the actual measured concentrations, exceeding the measured concentrations four times and understating the measured concentrations three times.

The conclusion drawn from TCLP testing on sludges from settling ponds and lagoons in arsenic removal studies (Cornwell et al., 2003; U.S. EPA, 2000a, 2000b) is that these sludges did not qualify as hazardous waste as the arsenic levels were under 5.0 mg/L. However, some of the same sludges were also tested using the WET and were found to be hazardous because arsenic levels above 5.0 mg/L were measured. In California, where hazardous waste guidelines are stricter than the federal requirements, lagoon sludges with high arsenic concentrations may not pass the WET, so individual evaluation should be performed to confirm the type of waste that is present and to determine the appropriate disposal method.

21-16 Management of Spent Solid Sorbents

With the exception of activated carbon most solid sorbents such as AA and GFH are used in a once-through basis. As with granular activated carbon, regeneration of these sorbents is only economical for large installations, if then. The most common reuse and disposal methods for spent solid sorbents are given in Table 21-19. If the spent solid sorbent can pass the TCLP test (see Sec. 21-15), conventional landfilling in a certified landfill is the most cost-effective method for disposal. Alternatively, the solid sorbents must be disposed of in a hazardous waste landfill.

Table 21-19
Management of sorbents used in water treatment or the removal of specific constituents

Sorbent	Reuse or Disposal Method
Activated carbon, granular (GAC)	Spent GAC is not reactivated in water treatment applications unless the usage is greater than 150,000 kg/yr (see Chap. 15). If usage is less than 150,000 kg/yr, the spent GAC is shipped to a central facility for reactivation. Because reactivated GAC cannot be used again for potable water treatment, it is typically stored for use in other applications such as industrial wastewater treatment. In some parts of the country, the demand for reactivated carbon is lower than the quantity available, and as a result some reactivated GAC is ultimately disposed of in landfills.
Activated carbon, powdered (PAC)	Contained in sludge from a coagulation process, PAC is not reactivated and is disposed of along with the sludge (see Sec. 21-3 and 21-4).
Ion exchange resins, mixed bed resins	Resins whose useful life has been reached are typically disposed of in lined municipal or hazardous waste landfills or hazardous waste processing facilities. Depending on their use, some resins are destroyed by combustion or pyrolysis.
Activated alumina (AA)	Spent adsorbents (sorbents that have lost significant adsorptive capacity and cannot be reactivated effectively) must be disposed in a lined municipal or hazardous waste landfills or hazardous waste processing facilities, depending on the constituent sorbed and local regulations.
Granular ferric oxide (GFO), Granular ferric hydroxide (GFH)	Spent adsorbents (see above) that are not reactivated must be disposed in a lined municipal or hazardous waste landfills or hazardous waste processing facilities, depending on the constituent sorbed and local regulations.
Fibroid chemosorbents, various	Spent adsorbents (see above) are typically not reactivated and must be disposed in a lined municipal or hazardous waste landfills or hazardous waste processing facilities, depending on the constituent sorbed and local regulations.

21-17 Process Selection

The selection of process steps and unit processes for a specific installation depends on factors such as availability and cost of raw water, space available at the site, local weather conditions, type of residual, cost of reuse versus disposal, sophistication of the plant operations personnel, distance to an ultimate disposal site, and types of ultimate disposal available. In most cases it will be possible to rule out most of the options by inspection (e.g., a small site surrounded by residential development would not have space available for sludge lagoons or drying beds; an inland site could not consider ocean disposal), leaving a limited number of options that can be evaluated in more depth to determine the least cost alternative. With most water treatment plant residuals, the general treatment objective is to recover as much of the usable liquid as possible and reduce the volume as much as possible to reduce the costs of subsequent recovery or disposal steps. Because most of the residuals from the treatment process are fairly dilute, the simplest and most economical first step in liquid recovery and reducing volume is gravity thickening.

Operational Factors

Operational factors that must be evaluated are the plant location, size, and reliability. An extremely complicated system would probably not be either economical or effective for a small, simple plant. Similarly, the plant reliability history must be evaluated to determine if a complex system would be well operated and maintained at the particular facility.

Economic Factors

Economic factors that must be considered are capital, operation and maintenance, and solids disposal costs. Capital costs should include such items as construction costs, trucks, and special equipment needed for the process. Operation and maintenance costs should include power, chemicals, labor, parts replacement, and equipment repair costs. Disposal costs are typically such items as fees at the landfill or wastewater collection system discharge fees.

Problems and Discussion Topics

21-1 An alum sludge contains 10 percent solids. If the density of alum is 2400 kg/m^3, estimate the density of the wet sludge at 25°C.

21-2 A ferric iron sludge contains 15 percent solids. If the density of iron is 2500 kg/m^3, estimate the density of the wet sludge at 11°C.

21-3 Determine the total kilograms of sludge on a dry-solids basis that will be produced for each kilogram of alum [$Al_2(SO_4)_3 \cdot 6$ or 12 or $18H_2O$] added. Value of bound water to be selected by instructor.

21-4 Determine the total kilograms of sludge on a dry-solids basis that will be produced for each kilogram for ferric sulfate [$Fe_2(SO_4)_3 \cdot 9H_2O$] and ferric chloride [$FeCl_3 \cdot 6H_2O$] added. How do the computed values compare to the values given in Table 21-4.

21-5 Determine the total kilograms of sludge on a dry-solids basis that will be produced for each kilogram for ferrous sulfate [$FeSO_4 \cdot 7H_2O$] added. How does the computed values compare to the values given in Table 21-4 for ferric sulfate and ferric chloride.

21-6 Determine the mass and volume of sludge produced and the volume of sludge as percentage of the total flow from the use of alum [$Al_2(SO_4)_3 \cdot 14H_2O$] for the removal of turbidity. Assume the following conditions apply: (1) flow rate is 0.05 m^3/s, (2) average raw-water turbidity is 45 NTU, and (3) average alum dose is 40 mg/L, sludge solids concentration is 5 percent with a corresponding specific gravity of 1.05, and temperature is 10°C. Assume the ratio between total suspended solids and turbidity expressed as NTU is 1.33 and 0.3 kg of alum sludge is produced per kg of alum added. Assume 1 mg/L of a coagulant aid will also be used.

21-7 Determine the mass and volume of sludge produced and the volume of sludge as a percentage of the total flow from the use of PACl for the removal of turbidity. Assume the following conditions apply: (1) flow rate is 0.1 m^3/s, (2) average raw-water turbidity is 20 NTU, and (3) average alum or PACl dose is 45 mg/L, sludge solids concentration is 5 percent with a corresponding specific gravity of 1.05, and temperature is 15°C. Assume the ratio between total suspended solids and turbidity expressed as NTU is 1.5, the PACl contains 13 percent Al by weigh and that 1.25 mg/L of a coagulant aid will also be used.

21-8 Referring to Chap. 20, demonstrate the correctness of the coefficients for Ca^{2+} and Mg^{2+} in Eq., 21-17 which is used to determine the amount of lime sludge.

21-9 Using the following information, estimate the yearly volume of waste washwater relative to the throughput for a granular ferric hydroxide (GFH) filter for arsenic. The amount of water processed on a yearly basis is 40,000 bed volumes. The filter is backwashed every 4 weeks for 10 min at a rate of 22 m/h. The cross-sectional area of the filter is 1 m^2, the depth of the filter is 1 m, and the porosity of the filter bed is 0.33.

21-10 Estimate the quantity and quality of the waste stream and the total quantity of water that must be processed from a reverse osmosis facility that is to produce 48,000 m^3/d of demineralized water. Assume that the recovery and rejection rates are 86 and 92 percent, respectively, and that the concentration of total dissolved solids in the feed stream is 400 mg/L.

21-11 Estimate the quantity and quality of the waste stream and the total quantity of water that must be processed from a reverse osmosis facility that is to produce 3800 m^3/d of demineralized water. Assume that the recovery and rejection rates are 80 and 85 percent, respectively, and that the concentration of total dissolved solids in the feed stream is 500 mg/L.

21-12 Using a loading rate of 60 kg/m^2, estimate the effective area of lagoons required to handle alum sludge from a water treatment plant with a flow rate of 0.35 m^3/s. Assume that solids and alum dose are as described in Problem 21-6 and that at least two lagoons will be used and that the filling cycle will be 120 days. Allow an additional area of 40 percent times the lagoon area for berms and access roads.

21-13 Review and cite three articles dealing with the disposal of water treatment plant sludges on land. What general conclusions can you draw regarding this practice based on your review? Are there any critical issues that stand out in your mind?

21-14 Review and cite three articles dealing with the processing of alum sludge. What are the similarities and differences between the recommended processing technologies?

References

Bly, A., Woodard, H. J., and Winther, D. (2001) "Lime Application Effects on Corn and Soybean Grain Yield at Aurora, SD, in 2001," *Soil/Water Research*, South Dakota State University, Progress Report.

Carmen, P. C. (1933, 1934) "Study of the Mechanism of Filtration, Parts I-III," *J. Soc. Chem. Ind.*, **52**, **53**.

Coackley, P., and Jones, B. R. S. (1956) "Vacuum Sludge Filtration. I. Interpretation of Results by the Concept of Specific Resistance," *Sewage Ind. Wastes*, **28**, 6, 963–976.

Cornwell, D. A., and Roth, D. K. (2011) Water Treatment Plant Residuals Management, Chap. 22, in J. K. Edzwald (ed.), *Water Quality and Treatment: A Handbook on Drinking Water*, 6th ed., American Water Works Association, McGraw-Hill, New York.

Cornwell, D. A., Tobiason, J. E., and Brown, R. (2010) *Innovative Applications of Treatment Processes for Spent Filter Backwash*, Water Research, Foundation. Denver, CO.

Cornwell, D. A., MacPhee, M. J., Muter, R., Novak, J., Edwards, M., Parks J. L., and Itle, C. (2003) *Disposal of Waste Resulting from Arsenic Removal Processes*, AWWA Research Foundation and American Water Works Association, Washington, DC.

Cornwell, D. A., Vandermeyden, C., Dillow, G., and Wang, M. (1992) *Landfilling of Water Treatment Plant Coagulant Sludges*, Environmental Engineering & Technology, American Water Works Association Research Foundation, Denver, CO.

Cornwell, D. A., and Koppers H. M. M. (eds.) (1990) *Slib, Schlamm, Sludge*, American Water Works Association Research Foundation and KIWA, Denver, CO.

Fox, P., Abbaszadegan, M., Mohammadesmaeili, F., and Kabiri-Badr, M. (2009) *Inland Membrane Concentrate Treatment Strategies for Water Reclamation Systems*, Water Research Foundation, Denver, CO.

Gendebien, A., Ferguson, R., Brink, J., Horth, H., Sullivan, M., Davis, R., Brunet, H., Dalimier, F., Landrea, B., Krack, D., Perot, J., and Orsi, C. (2001, July) *Survey of Wastes Spread on Land—Final Report*,WRC Study Contract B4-3040/99/110194/MAR/E3, European Commission-Directorate-General for Environment, European Communities, Luxembourg.

Hawkins, F. C., Judkins, Jr., J. F., and Morgan, J. M. (1974) "Water Treatment Sludge Filtration Studies," *J. AWWA*, **66**, 11, 653–658.

Hulet, G. A., Kosson, D., Conley T. B., and Morris, M. I. (2000) Evaluation of Waste-Form Analysis Protocols That May Replace TCLP, paper presented at the Waste Management Conference, Tucson, AZ.

Kawamura, S. (2000) *Integrated Design and Operation of Water Treatment Facilities*, 2nd ed., John Wiley & Sons, New York.

Mackey, E. D., and Seacord, T. (2008) *Regional Solutions f'or Concentrate Management*, WateReuse Foundation, Alexandria, VA.

Marshall, T. (2002) "Sweeter Soil with Substantial Savings," *Ohio's Country J.* **11**, 5, 21–22.

Martel, C. J. (1989) "Development and Design of Sludge Freezing Beds," *J. EED*, **115**, 4, 99–808.

Novak, J. T. (1993) *Demonstration of Cropland Application of Alum Sludges*, American Water Works Association Research Foundation, Denver, CO.

Novak, J. T., and Calkins, D. D. (1975) "Sludge Dewatering and Its Physical Properties," *J. AWWA*, **67**, 1, 42–45.

Qasim, S. R., Morley, E. M., and Zhu, G. (2000) *Water Works Engineering: Planning, Design and Operation*, Prentice-Hall, Upper Saddle River, NJ.

Schmitt, C. R., and Hall, J. E. (1975) "Analytical Characterization of Water-Treatment-Plant Sludge," *J. AWWA*, **67**, 1, 40–42.

State of California (2005) *Waste Extraction Test (WET) Procedure*, California Code of Regulations, Title 22, Division 4.5, Chapter 11, Appendix II, Sacramento, CA.

Tchobanoglous, G., Burton, F. L., and Stensel, H. D. (2003) *Wastewater Engineering: Treatment and Reuse*, 4th ed., Metcalf and Eddy, McGraw-Hill, New York.

U.S. EPA (1992) *Test Methods for Evaluating Solid Waste, Physical/Chemical Methods*, EPA Publication SW-846 [Third Edition (September, 1986), as amended by Update I (July 1992)], U.S. Environmental Protection Agency, Washington, DC.

U.S. EPA, AWWA, ASCE (1996) *Technology Transfer Handbook Management of Water Treatment Plant Residuals*, U.S. Environmental Protection Agency, EPA/635/R-95/006, Office of Research and Development, Washington DC.

U.S. EPA (2000a) *Arsenic Removal from Drinking Water by Iron Removal Plants*, EPA/600/R-00/086, U.S. Environmental Protection Agency, Washington, DC.

U.S. EPA (2000b) *Arsenic Removal from Drinking Water by Coagulation/Filtration and Lime Softening Plants*, EPA/600/R-00/063, U.S. Environmental Protection Agency, Washington, DC.

22 Internal Corrosion of Water Conduits

Stantec's Water Treatment: Principles and Design, Updated Third Edition. John C. Crittenden, R. Rhodes Trussell, David W. Hand, Kerry J. Howe, and George Tchobanoglous.

Terminology for Internal Corrosion of Water Conduits

Term	Definition
Calcium carbonate scale	Natural scale that sometimes forms on the surface of materials used for conduits when the concentration of calcium carbonate in solution exceeds solubility.
Cathodic protection	A technique to control the corrosion of metal pipes using a sacrificial electrode.
Corrosion cell	The components of a system necessary to cause aqueous corrosion of metals: namely, an anode, a cathode, a conductor, and conducting electrolyte (the water).
Corrosion	Erosion of a structural material through chemical transformation—for metals, through oxidation.
Corrosion inhibition	Actions taken to reduce the rate of corrosion by inhibiting the rate of cathodic and anodic reactions.

Term	Definition
Corrosion, localized	Crevice corrosion and pitting are two types of localized corrosion as contrasted to uniform corrosion (see below).
Corrosion, uniform	Corrosion that occurs uniformly over the surface of a pipe as contrasted to localized corrosion.
Crevice corrosion	Localized corrosion that occurs in locations where conditions are stagnant with respect to water flow.
Current density	Rate of corrosion by oxidation, expressed as the flow of an electrical current per unit of surface area.
Electrolytic corrosion	Corrosion caused by stray electrical currents, which occurs at the point where the electrical current exits the pipe.
Galvanic corrosion	Corrosion that occurs when dissimilar metals or alloys are placed in contact with each other in the presence of an electrolyte (e.g., water).
Langelier saturation index (LSI)	An estimate of the thermodynamic driving force for either the precipitation or dissolution of calcium carbonate.
Mixed potential theory	Theory in which the concepts of thermodynamics, exchange current, and Tafel slopes are combined to provide one integrated model of the corrosion process.
Passivity	Formation of a nearly “perfect” protective oxide coating on transition metals, which reduces the rate of corrosion significantly.
Pitting	Localized corrosion in which small pits, holes, or cavities develop on the wall of a pipe.
Pourbaix diagram, E_H–pH diagram	Diagram drawn to display three regions for a metal with respect to corrosion: (1) immunity, (2) corrosion, and (3) passivity.
Tafel slope	Slope of the linear portion of plot of the logarithm of the current versus electrode potential.
Tubercles	Mounds of corrosion products on the inside of iron pipe by localized corrosion.

The interactions of water with the materials used to transport, distribute, and store it are presented and discussed in this chapter. Where mineral materials such as rock or concrete are concerned, the interaction is generally a question of the ability of water to dissolve them directly (e.g., solubility). Although metals do not dissolve directly in water, the metals used in water conduits are generally not stable in water and, as a result, they oxidize or corrode. In the case of these metals it is important to understand the phenomena that govern the underlying rate of oxidation as well as the solubility of the oxidation products. The plastic materials used as water conduits are generally stable in water. With these materials the issues are the leaching of organic contaminants in the plastic matrix or the leaching of solvents used to bind one plastic surface to another and the penetration of the pipe by organic solvents from the exterior environment. For all materials, microbiological activity can be important as well.

Recently, concern about the impact of metals on water quality has led to an increased emphasis on the release of metals from the pipe surface into solution. Although corrosion and the release of metals to solution are related, they are not the same. *Corrosion* is the oxidation process by which the native metal is converted to an oxidized species. If corrosion takes place slowly, the metal is often converted directly to an oxide of the metal on the metal surface. When corrosion takes place rapidly, as it often does in a pit, in crevice corrosion, or in galvanic corrosion, the oxidized metal is often released directly to solution in the form of ions. Once corrosion has taken place, other processes control the concentration of metal in solution. Prominent among these other processes are precipitation and dissolution. With some metals, particularly lead, iron, and copper, abrasion and erosion can also be important. Metals such as iron and copper, which exist in more than one oxidized state, are subject to particularly complex processes. In the case of both of these metals, understanding multiple oxidation states is important to understanding the dynamics between the primary surface oxide (the oxide formed immediately against the metal surface) and the rest of the surface scale. The relationship between corrosion of metals and the release of metal contaminants into the bulk water solution is conceptually illustrated on Fig. 22-1.

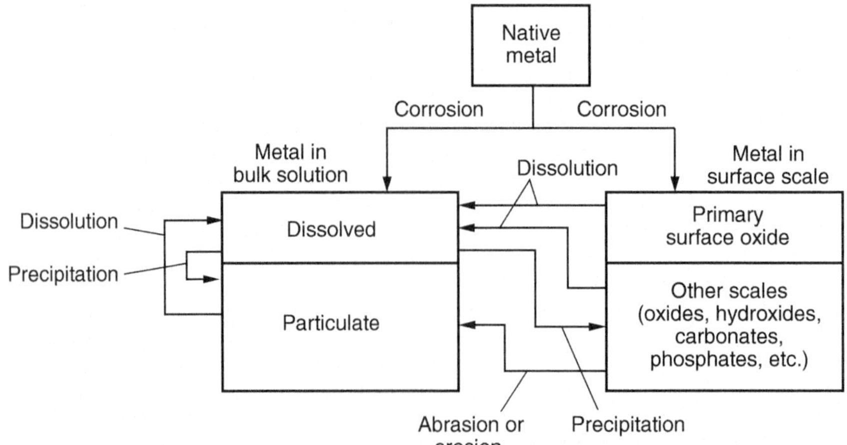

Figure 22-1
Dynamics of corrosion and metals release.

22-1 Materials Used to Transport, Distribute, and Store Water

Over the years water has been transported in conduits made of many materials. The earliest records are of bamboo pipes in China and of earthenware pipes used in early Greek and Middle-Eastern civilizations (Baker, 1948). The Roman aqueducts were primarily composed of local natural materials, and in many cases they were just carved in the local rock. An ancient Roman version of concrete was used to line and cover the canal, making the structure long-lasting and impervious. Ancient Roman concrete was prepared by mixing hydrated lime, pozzolan ash, and rock with minimal water. The mix was then pounded into place to produce a concrete surface superior to most produced today (Moore, 1993). Many early water systems in the United States were ditches or canals constructed of local materials; some were also constructed using concrete or wooden pipes and wooden flumes. In fact, wooden materials continued to be used, worldwide, until the end of the nineteenth century. Unfortunately, most of these systems had a useful life of 20 years or less; nevertheless, an occasional wooden pipe is still uncovered today.

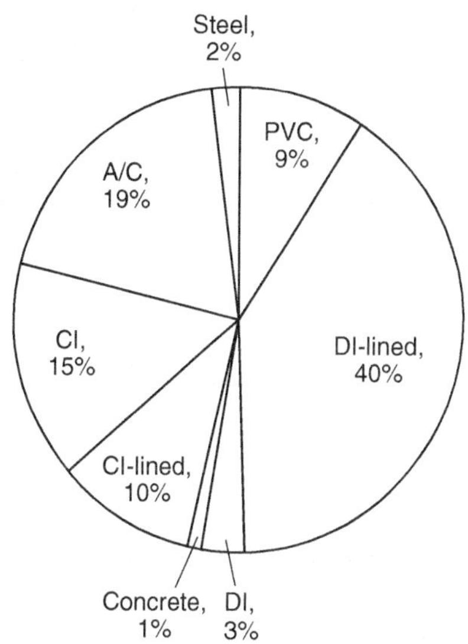

Figure 22-2
Pipe material in place in North America in 2002. (*Source*: water:\stats 2002 distribution survey based on 542,600 km of pipe as reported by 337 utilities.)

Based on a recent inventory of the materials used in public water supply distribution systems in the United States (see Fig. 22-2) the principal materials in place are ductile iron (DI), cast-iron (CI), and asbestos–cement (AC) pipe. In fact, together, ductile iron and cast iron account for nearly 70 percent of the total. Conduits lined with a cementaceous surface account for an even greater share of present installations (mortar-lined cast iron, ductile iron and steel, AC, and concrete pressure pipe). The primary uses, advantages, and disadvantages of these and other pipe materials are summarized in Table 22-1.

Cast and Ductile Iron

Gray cast iron was used for cannons as early as 1313, but it was eventually adapted for water systems as well. The first recorded use was at Dillenburg Castle in Germany in AD 1455. From the start, the use of cast iron has a pretty good record in water supply systems. In 1664, King Louis XIV of France had a cast-iron pipe installed to bring water to his palace in Versailles. This same cast-iron line was still functioning after 335 years of service. Once methods for refining iron improved enough, every progressive city was using cast iron for water mains. By the middle of the nineteenth century, gray cast iron had become the dominate choice for new water pipelines, worldwide. In the last half of the twentieth century ductile iron pipe began to displace gray cast iron in developed countries.

Table 22-1
Types of pipe, primary usage, and advantages and disadvantages when used in water supply systems

Pipe Material	Standard	Primary Use	Advantages	Disadvantages/ Limitations
Asbestos cement (AC) pipe (ACP)	AWWA C400-03	Distribution systems	Rigid light weight in long lengths	No longer used in new construction or manufactured in North America Contains asbestos
Cast iron (CI)	AWWA 106, 108	Transmission lines, distribution systems	Inexpensive, durable	Heavy, brittle Corrodes in soft water, water with high chloride or sulfate Tuberculation reduces carrying capacity No longer manufactured in North America: replaced by ductile iron
Cast iron—mortar lined	AWWA C151	Transmission lines, distribution systems	Inexpensive, durable, corrosion resistant	Heavy, brittle No longer manufactured in North America: replaced by ductile iron
Concrete cylinder pipe (CCP)	AWWA C303	Transmission lines	Adaptability of steel with the corrosion resistance of concrete and cement mortar	Limited pressure range Used in very large, gravity flow pipe
Copper tubing	ASTM B88	Consumer plumbing	Easy to work with	Can corrode in aggressive water; also due to poor workmanship during installation
Ductile iron pipe (DIP)	AWWA C150	Transmission lines, distribution systems	High strength for supporting earth loads Less brittle than CI Lighter than CI	May require wrapping or cathodic protection in corrosive soils or water. Typically lined to limit corrosion
Ductile iron pipe, cement lined (DIP)	AWWA C150, C104	Transmission lines, distribution systems	High strength for supporting earth loads Less brittle than CI Lighter than CI Extremely resistant to internal corrosion	May require wrapping or cathodic protection in corrosive soils
Fiberglass-reinforced plastic pipe (FPR)	AWWA C950 ASTM D2996	Transmission lines, water treatment systems	Light weight Corrosion resistance Long service life (durable)	Pipe can be coated for specific applications for piping

Table 22-1 *(Continued)*

Pipe Material	Standard	Primary Use	Advantages	Disadvantages/ Limitations
Galvanized steel pipe	ASTM53/ 53M-07	Consumer plumbing used in homes before 1960	Inexpensive Zinc protective coating reduces tendency to corrode	Prone to tuberculation and corrosion (severe in many cases) Corrodes from the inside Limited service life (~40 years)
High-density polyethylene pipe (HDPE)	AWWA C901-02 ASTM D2513, D3350	Distribution systems, consumer plumbing	Corrosion resistance Reduced formation of chemical scales Chlorine free Long service life (durable)	Limited structural properties (not very rigid) Thick walls in large-diameter pipe
Polypropylene (PP)	ASTM D2146, D4101	Water mains, consumer plumbing	Light weight Corrosion resistance Long service life (durable)	Protective coating required if installed outside
Chlorinated polyvinyl chloride pipe (CPVC)	ASTM D1784	Consumer plumbing	Light weight Corrosion resistance Long service life (durable)	Can have problems with chemical permeation (solvents like TCE, gasoline, etc)
Polyvinyl chloride pipe (PVC)	AWWA C900, C905	Consumer plumbing, water treatment plants	Light weight, high strength Corrosion resistance Long service life (durable)	Can have problems with chemical permeation (solvents such as TCE, gasoline, etc). Used for cold water
Prestressed concrete, steel cylinder pipe (PCCP)	AWWA C301	Transmission lines, distribution systems	Corrosion resistance	Often used in large pipe Corrosion of reinforcement can be a problem Expensive
Lead		Service lines and internal plumbing	Easily installed (malleable) Banned by U.S. EPA in 1986	No longer used in the U.S. in new construction for water service. A legacy material
Reinforced concrete pipe (RCP)	ASTM C76	Transmission lines, distribution systems	High strength for support of earth loads	Attacked by soft water May require protective coating Water hammer can rupture outer shell
Reinforced concrete pressure pipe (RCPP)	AWWA C300, 302	Transmission lines, distribution systems	Different types are available for different soil conditions High strength for support of earth loads Variety of sizes available Relatively inexpensive	Attacked by soft water May require protective coating Water hammer can rupture outer shell

(Continued)

Table 22-1 *(Continued)*

Pipe Material	Standard	Primary Use	Advantages	Disadvantages/ Limitations
Steel pipe	AWWA C200	Transmission lines, distribution systems	High strength for support of earth loads	Poor corrosion resistance Must be lined and protected on the outside against corrosion Corrosion resistant when used with both internal an external mortar lining
Stainless steel (SS) tubing	ASTM A312	Water treatment plants, building water supply lines	Corrosion resistance	Relatively expensive

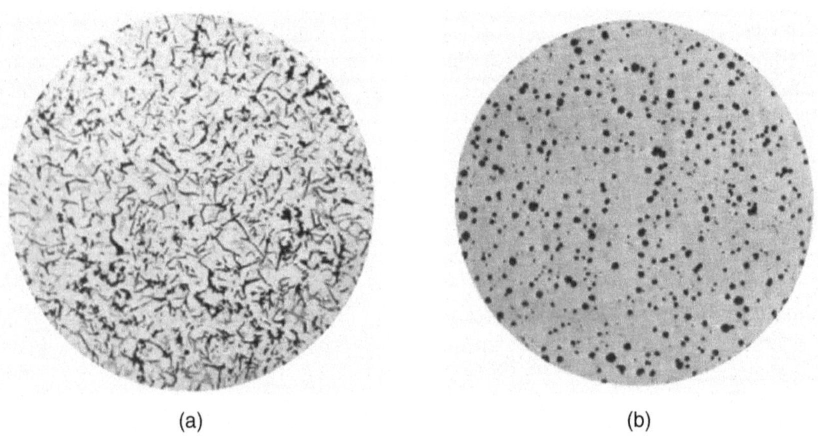

Figure 22-3 Photomicrographs of (a) gray cast iron and (b) ductile iron.

The difference between these two types of cast irons is evident in the photomicrographs displayed on Fig. 22-3. All cast irons include certain additives that make them easy to manufacture; prominent among these are carbon and silicon. The carbon in gray cast iron takes the form of graphite flakes, as may be observed in the photomicrograph on Fig. 22-3a. Unfortunately, although the graphite makes manufacturing and machining easier, when it takes the form of flakes, it also makes the pipe brittle, reducing its useful strength.

In 1948, German metallurgists discovered that the introduction of a small amount of magnesium or cerium in the molten cast iron could prevent the full development of the traditional graphite flakes, forcing the formation of round nodules instead (see photomicrograph on Fig. 22-3b). When the graphite in the final cast takes this form, ductility, toughness, and useful strength are substantially increased. As a result, whereas a 20-ft segment of 24-in. gray cast-iron pipe weighs 4065 lb (200 psi Class), the equivalent ductile iron pipe weighs 1710 lb, less than half as much. The approximate metric equivalents are 6 m, 600 mm, and 1850 kg for gray cast iron and 780 kg for ductile iron. Ductile iron almost completely displaced gray cast iron in the United States within a decade and a half after it was introduced in 1955. Its principal competitors are now plastic materials. Gray cast iron, which is easier to manufacture, still sees significant use on a worldwide basis. In fact, gray cast iron is still the dominant pipe material in many countries where ductile iron is not yet readily available.

Other Materials in Distribution Systems

Most modern water systems are made of cast iron, ductile iron, steel, galvanized steel, AC, polyvinyl chloride (PVC), lead, and copper (see Table 22-1). Of these, AC and lead are no longer used in new pipe, and lead and copper tubing are found almost exclusively in buildings and related connections. Galvanized iron is found both in residences and in smaller distribution system piping. However, since approximately 1980, it has seen limited use in new domestic plumbing in developed countries. Both ductile iron and steel are typically lined with a bitumastic seal coat, cement mortar, or both, to reduce corrosion. In older pipe, coal tar coatings were common and the pipe was often installed with no lining at all. Recent surveys show that approximately half the inventory of installed water pipe in the United States today is gray cast iron and half of that is unlined (Kirmeyer et al., 1994). Because it is stronger and more malleable, steel has also been employed as an alternative to cast iron or ductile iron, although most steels are more susceptible to corrosion than ductile iron. Steel is relatively long-lived in most systems when a cement–mortar lining is used. Asbestos–cement pipe, very popular between 1955 and 1970, is no longer used in new construction.

Materials in Consumer Plumbing

Lead was once used extensively in service lines and interior plumbing because of its ductility and its relatively long life. Its longevity is due to a low corrosion rate and the fact that it does not form the sort of thick surface scale and encrustation that often reduces the carrying capacity of ferrous pipe. Lead service connections between distribution mains and consumer plumbing were often used until lead materials were prohibited in the United States by the Lead Contamination Control Act of 1988 (Public Law 100-572).

Today galvanized pipe is probably the most common material in existing consumer plumbing worldwide. However, since 1950 copper tubing has gradually displaced it in most new consumer plumbing in the United States and Western Europe. Plastic materials are used for consumer plumbing with increasing frequency and also for many distribution mains.

Reservoir Materials

The materials used in the construction of water reservoirs have also changed with time. Many older facilities are made of wood or have wooden surfaces, which are often associated with bacterial contamination problems because they serve as a reservoir for bacteria. Today, new water storage tanks are made of either concrete or steel, and the issue of concern is the leaching of contaminants from materials used to coat the surface or seal joints.

22-2 Thermodynamics of Metallic Corrosion

Corrosion of metallic water conduits is of concern for three reasons: (1) it can reduce the wall thickness, (2) it can cause encrustation, reducing carrying capacity, and (3) it contributes to water contamination. These are separate but related phenomena. Understanding one of the three does not necessarily lead to understanding the others.

Most metals used to construct water conduits corrode when exposed to water, particularly when oxidants such as dissolved oxygen or chlorine are present. Thermodynamics can be used to find the free energy of the corrosion reaction. *If the free energy of the reaction is positive, then corrosion will not occur. If the free energy of the reaction is negative, then corrosion may occur.* The fact that the corrosion reaction is favored thermodynamically is a necessary condition but not a sufficient condition for corrosion to proceed. Sometimes thermodynamics favor corrosion, but the rate of the reaction is so slow that no significant corrosion occurs. This low rate occurs because the rate at which corrosion takes place is not just a question of thermodynamics. It is also a question of kinetics. The type of kinetics that control the rate of a corrosion reaction is called *electrode kinetics.* Electrode kinetics are influenced by thermodynamic conditions but also by the condition of the metal's surface, limitations of mass transport, as well as other factors. Thermodynamics cannot be used to determine the rate of the reaction, but thermodynamics can be used to determine if corrosion is possible. The relevant thermodynamics, discussed in Chap. 5, are half reactions, Faraday's law, standard electrode potentials, and E_H – pH diagrams. Selected standard electrode potentials of interest in solving problems in this chapter are listed in Table 22-2.

Components of Corrosion Cell

It is important to understand the basic components of a corrosion cell. An idealized diagram of a corrosion cell is illustrated on Fig. 22-4. The four necessary components of a corrosion cell are illustrated based on the corrosion of iron. These components are (1) the anode, (2) the cathode, (3) a conductor, and (4) a conducting electrolyte. At the anode the iron metal loses two electrons and goes into solution as a ferrous iron. At the cathode, hydrogen ions near the metal surface (probably absorbed to it) accept the electrons generated by the corroding iron, become reduced to nascent hydrogen, and, eventually, combine to form hydrogen molecules

Table 22-2
Selected standard oxidation–reduction (redox) potentials

Reaction		Reaction	
		Positive Redox Potentials	
Noble Metals	***E*°,mV**	**Oxidizers**	***E*°,mV**
$Au^{3+} + 3e^- \leftrightarrows Au(s)$	+1500	$HO\cdot + e^- \leftrightarrows OH^-$	+2590
$Pt^{2+} + 2e^- \leftrightarrows Pt(s)$	+1200	$O_3(g) + 2H^+ + 2e^- \leftrightarrows O_2 + H_2O$	+2080
$Pd^{2+} + 2e^- \leftrightarrows Pd$	+920	$ClO_2(g) + 5e^- + 2H_2O \leftrightarrows Cl^- + 4OH^-$	+1910
$Hg^{2+} + 2e^- \leftrightarrows Hg$	+851	$H_2O_2 + 42H^+ + 2e^- \leftrightarrows 2H_2O$	+1780
$Ag^+ + e^- \leftrightarrows Ag(s)$	+800	$2HOCl + 2H^+ + 2e^- \leftrightarrows Cl_2(aq) + 2H_2O$	+1610
$Cu^{2+} + 2e^- \leftrightarrows Cu(s)$	+340	$HOCl + H^+ + 2e^- \leftrightarrows H_2O + Cl^-$	+1500
		$Cl_2(g) + 2e^- \leftrightarrows 2Cl^-$	+1390
		$O_2(g) + 4H^+ + 4e^- \leftrightarrows 2H_2O$	+1230
		$NH_2Cl + H_2O + 2e^- \leftrightarrows NH_3 + Cl^- + OH^-$	+1200[a]
		$Fe(OH)_3(s) + 3H^+ + e^- \leftrightarrows Fe^{3+} + 3H_2O$	+1060
		$ClO_2(aq) + e^- \leftrightarrows ClO_2^-$	+950
		$NO_3^- + 2H^+ + 2e^- \rightleftarrows NO_2^- + H_2O$	+840
		$Fe^{3+} + e^- \leftrightarrows Fe^{2+}$	+770
		$O_2 + 2H_2O + 4e^- \leftrightarrows 4OH^-$	+410
		$S(s) + 2H^+ + 2e^- \leftrightarrows H_2S(g)$	+170
		$Cu^{2+} + e^- \leftrightarrows Cu^+$	+160
Reference for redox potentials: $2H^+ + 2e^- \leftrightarrows H_2(g)$ 0.00 mV			
		Negative Redox Potentials	
Active Metals		**Reducers**	
$Pb^{2+} + 2e^- \leftrightarrows Pb(s)$	−126	$SO_4^{2-} + 2H^+ + 2e^- \leftrightarrows SO_3^{2-} + H_2O$	−40
$Sn^{2+} + 2e^- \leftrightarrows Sn(s)$	−140	$SO_3^{2-} + 3H_2O + 4e^- \leftrightarrows S + 6OH^-$	−660
$Ni^{2+} + 2e^- \leftrightarrows Ni(s)$	−250	$SO_4^{2-} + H_2O + 2e^- \leftrightarrows SO_3^{2-} + 2OH^-$	−930
$Fe^{2+} + 2e^- \leftrightarrows Fe(s)$	−440		
$Cr^{3+} + 3e^- \leftrightarrows Cr(s)$	−740		
$Zn^{2+} + 2e^- \leftrightarrows Zn(s)$	−763		
$Ti^{2+} + 2e^- \leftrightarrows Ti(s)$	−1630		
$Al^{3+} + 3e^- \leftrightarrows Al(s)$	−1660		
$Mg^{2+} + 2e^- \leftrightarrows Mg(s)$	−2370		
$Na^+ + e^- \leftrightarrows Na(s)$	−2710		

[a]Approximation from Oxidation/reduction potential (ORP) data.
Source: Adapted from Lide (2000) and Lange (1985). (See also www.webelements.com.)

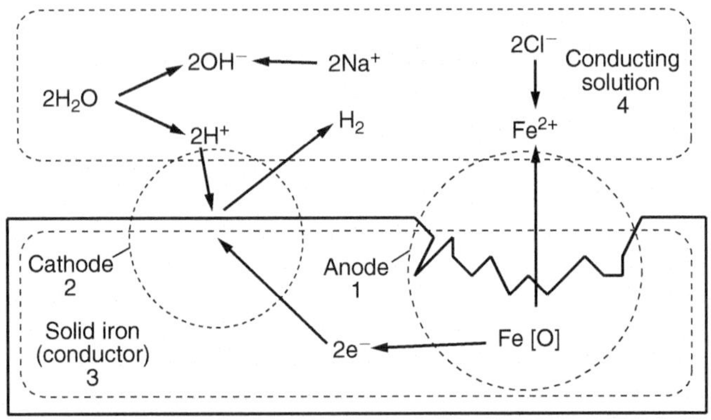

Figure 22-4 Idealized corrosion cell for iron. The four components involved in the corrosion of iron are identified as 1 through 4.

and leave the surface. The electrons travel from one site to another through the conductor. A conducting solution is also necessary to complete the electrical circuit. The movement of electrons is illustrated on Fig. 22-4 by the transport of positively charged sodium ions toward the cathode and negatively charged chloride ions toward the anode. The reaction couple for metallic iron corroding in an acid solution is shown below:

Reaction	E°(V)		
$Fe \leftrightarrows Fe^{2+} + 2e-$	−(−0.44)	Oxidation	(22-1)
$2H^+ + 2e- \leftrightarrows H_2$	0.00	Reduction	(22-2)
$2H^+ + Fe \leftrightarrows Fe^{2+} + H_2$	+0.44		(22-3)

It should be noted that the overall potential of this reaction must be obtained by the simple addition of the two half reactions as both reactions have the same number of electrons. When this is not the case, the reader should follow the procedure dictated by the Faraday equation and described in Chap. 5. The oxidative half reaction and the reductive half reaction do not need to take place at the same physical location, as the free electrons generated at the oxidative site can migrate through the conductor (and also through some metal oxides if they are semiconductors) to find a reductive half reaction.

Pourbaix Diagrams

The E_H–pH diagrams are shown for the four metals of greatest concern in drinking water on Fig. 22-5. These diagrams provide a pictorial view of the chief thermodynamic forces acting on the metal. The principles behind these diagrams are introduced in Chap. 5. The region where water is stable is depicted in each diagram by the dashed line. At potentials above the upper dashed line, water is converted to oxygen. At potentials below the

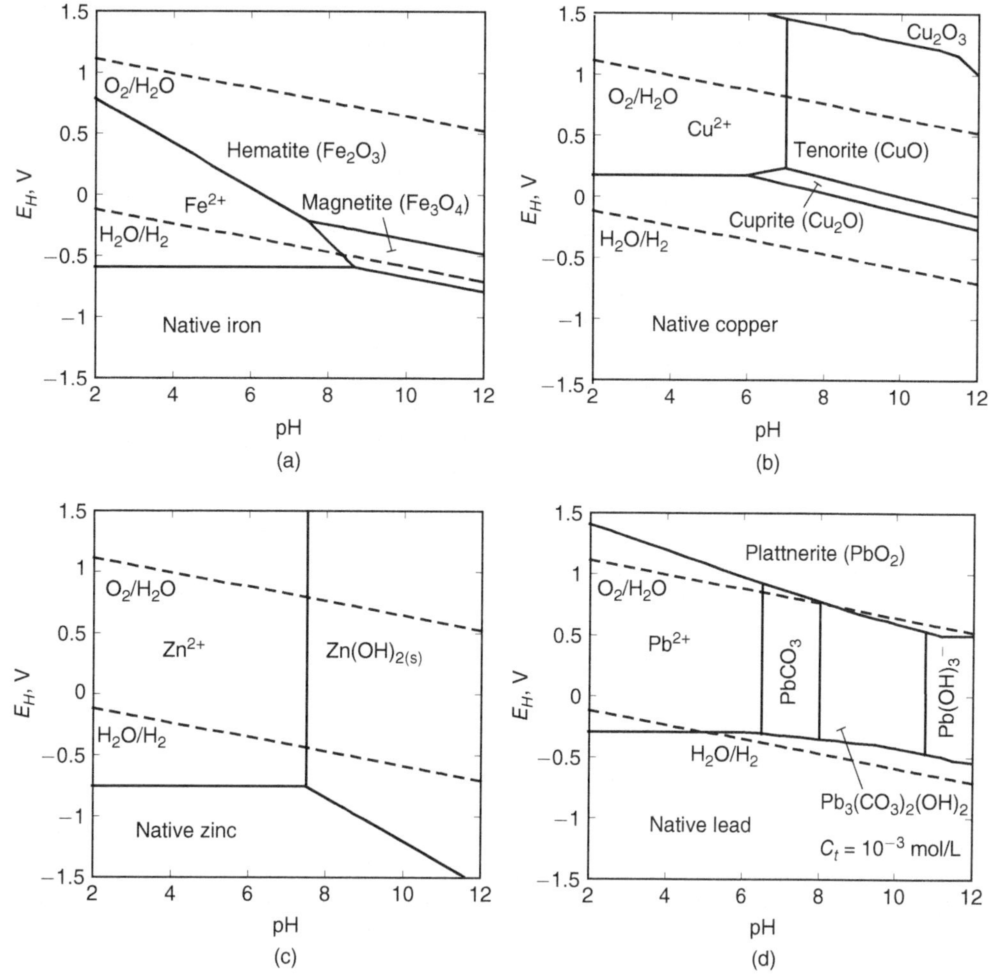

Figure 22-5
The E_H–pH diagrams for four metals common in water conduits: (a) iron, (b) copper, (c) zinc, and (d) lead.

lower dashed line, water is converted to hydrogen gas. It should be noted that iron, the most common material used for water conduits, is unstable throughout this region. Similarly, zinc, the material used to increase the life of iron pipe, is also unstable. Lead, probably the most corrosion resistant of the three, is only stable at very low redox potentials and alkaline conditions. Copper is the only conduit metal that is thermodynamically stable in a significant part of this range. Nevertheless, all of these materials have been known to serve for several decades as water conduits. Clearly, thermodynamic stability is not the only prerequisite for the proper choice of materials.

Expanding on his practical knowledge of corrosion phenomena, Pourbaix edited the E_H–pH diagrams on Fig. 22-5, and those for other metals, to better portray the metal's corrosion behavior, producing a new diagram that had both thermodynamic and practical meaning. In deference to Pourbaix's contribution, diagrams of this kind have become known in the corrosion industry as *Pourbaix diagrams.* Pourbaix diagrams are used to identify three regions for each metal: (1) immunity, (2) corrosion, and (3) passivity (Pourbaix, 1949). The region labeled "immunity" corresponds to that region of the E_H – pH diagram where corrosion of the metal is not thermodynamically possible. The label "corrosion" is applied to conditions where the stable metal species are highly soluble and, as a result, no natural protective layer can form. The region labeled "passivation" is applied to conditions where the stable species of the metal can form insoluble products that might retard corrosion to a significant degree. Thus, Pourbaix's use of the term *passivation* is broader than the meaning used elsewhere in this chapter.

Pourbaix diagrams for the same four metals analyzed on Fig. 22-5 are shown on Fig. 22-6. These diagrams are useful for discussing the influence of both pH and solution redox potential on corrosion, but they must be used with care because they are one step removed from the basic thermodynamic diagrams, which already suffer from an inability to reflect kinetic considerations. Such diagrams are most useful as a tool for communicating the influence of pH and redox conditions on the corrosion of a particular metal among practitioners who already have broad experience with a metal's behavior in corrosion.

Limitations of Pourbaix Diagrams

The position a sample of metal corroding in water occupies on these diagrams may be approximated directly by measuring the pH of the solution and the potential of the metal with respect to the solution, usually referred to as the metal's corrosion potential, E_{corr}. The corrosion potential can be approximated by measuring the potential of the corroding specimen with respect to a reference electrode, while the freely corroding specimen is immersed in and is at steady-state with the solution of interest. These E_{corr} measurements are subject to error and must be made with care. For example, the distance between the reference electrode and the specimen must be minimized to avoid interference from solution resistance. One of the most difficult aspects is obtaining a specimen without altering its conditions in a fashion that will significantly change its corrosion rate and, hence, its corrosion potential. Any E_{corr} measurements made in this fashion must be made with a particular specimen and are not a general property of the solution itself.

The unknowns do not end with the *E*corr measurement. Any E_H–pH diagram is necessarily constructed at a particular temperature, a particular ionic strength, and with specific assumptions about the concentration

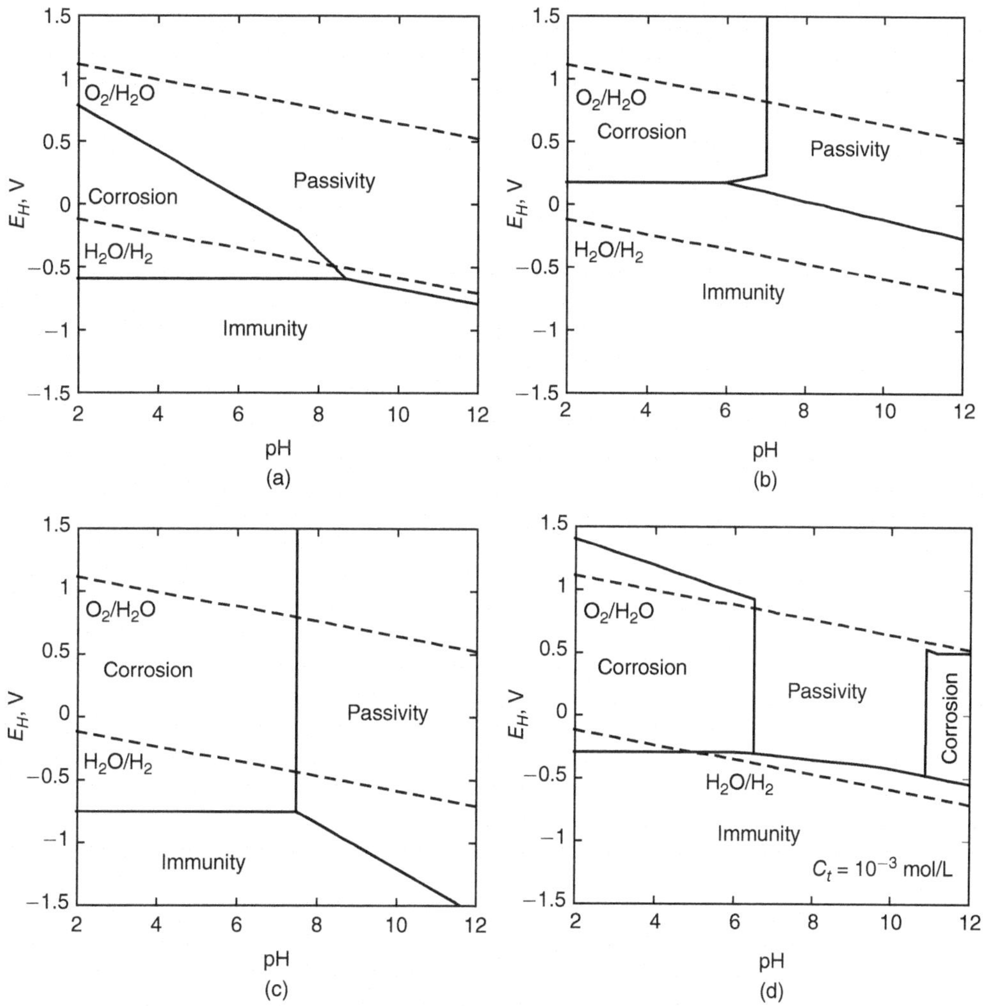

Figure 22-6
Pourbaix diagrams for four metals common in water conduits: (a) iron; (b) copper; (c) zinc; and (d) lead.

of species relevant to the metal's behavior. The E_H–pH diagrams are particularly useful as a didactic tool, illustrating the relationship between pH, potential, and other solution properties and the stable species of the material of concern. Even if the E_H is correctly measured and all the appropriate species were considered in drawing the diagram, the result is based on equilibrium concepts and can only be used to determine feasibility of corrosion. It cannot be used to predict the rate of corrosion. To pursue these answers, electrokinetics must be considered.

22-3 Electrokinetics of Metallic Corrosion

Environmental engineers tend to view the corrosion process as an interaction of cathodic and anodic sites on the macroscale and often analyze corrosion control through the solubility of surface scales and occasionally through some of the thermodynamic properties discussed in the preceding section. Through the first six decades of the twentieth century, the corrosion profession as a whole took a similar view. In the late 1950s, Stern and Geary (1957) presented a new and powerful description of the mixed-potential theory of corrosion that had been advanced nearly two decades earlier (Wagner and Traud, 1938) but had received little attention. Mixed-potential theory provides a rigorous connection between the fundamentals of corrosion and the principles of electrochemistry. The theory began to appear in corrosion texts in the late 1960s (Donahue, 1972; Fontana and Greene, 1967; Uhlig, 1967) and is now generally accepted (Fontana, 1985; Jones, 1996; Roberge, 1999; Revie and Uhlig, 1985). Mixed-potential theory is introduced here because it is the most powerful conceptual tool available for understanding the dynamics of the corrosion process. Before describing mixed-potential theory, three related concepts must be introduced: (1) corrosion as an electrical process, (2) the relationship between polarization and current flow, and (3) exchange current density.

Corrosion as an Electrical Process

Corrosion involves oxidation and reduction reactions and electrons as well. In fact, the process of corrosion can be viewed in electrical terms. Consider the oxidation of a metal to metal ions:

$$\mathrm{Me} \rightleftarrows \mathrm{Me}^{n+} + ne^{-} \tag{22-4}$$

The rate of this reaction is characterized by weight loss on the left and by electricity on the right. One can be converted to the other using Faraday's law:

$$M_w = \frac{Ia_M}{nF} \tag{22-5}$$

where M_w = rate of weight loss, g/s
a_M = atomic weight of metal, g/mol
I = electrical current, A, C/s (coulombs per second)
n = equivalents of metal in reaction, eq/mol
F = Faraday's constant, 96,500 C/eq

Although corrosion measurements are often made by measuring the weight lost by test coupons, the results are normally expressed as a rate of uniform attrition of the metal surface: mpy or mm/y. The acronym "mpy" stands for "mils per year," where 1 mil = 1/1000 in. The acronym "mm/y" stands for "mm per year." The conversion to these more conventional measures

Table 22-3
Converting corrosion rates and current densities

Metal	μA/ cm^2	eq/mol	Density, g/cm^2	Atomic Mass	mpy	mm/y	$mg/dm^2 \cdot d$
Iron	1	2	7.87	55.8	0.458	0.0116	2.50
Copper	1	2	8.96	63.5	0.457	0.0116	2.85
Lead	1	2	11.34	207	1.178	0.0299	9.28
Zinc	1	2	7.13	65.4	0.591	0.0150	2.93
Aluminum	1	3	2.7	27.0	0.430	0.0109	0.81
Tin	1	2	7.3	118.7	1.049	0.0266	5.32
Nickel	1	2	8.9	59.2	0.429	0.0109	2.65

of corrosion rate is made by introducing the density of the metal and the concept of current density. In this manner Eq. 22-5 becomes

$$\text{Rate (mpy)} = 0.129\left(\frac{a_M i}{n D_M}\right) \tag{22-6}$$

$$\text{Rat e(mm/y)} = 0.00327\left(\frac{a_M i}{n D_M}\right) \tag{22-7}$$

where i = corrosion current density, $\mu A/cm^2$
DM = density of metal, g/cm^3

Conversions for most metals encountered in water supply are summarized in Table 22-3.

Relationship between Polarization and Electrical Current

Mixed-potential theory begins with the relationship between electrode polarization and current flow. Perhaps the best way to develop an understanding of this aspect of electrode kinetics is by examining a simple polarization experiment. To make polarization measurements requires a three-electrode setup like the one illustrated on Fig. 22-7. In the sketch shown, the auxiliary or "working" electrode is platinum, the reference electrode is a Ag–AgCl electrode, and the test electrode is made of iron. A potentiometer is used to measure the potential of the test electrode with respect to the reference electrode. A source of current is placed between the test electrode and the working electrode along with a current meter so that the electrical current flowing into or out of the test electrode can be measured. The current of the working electrode divided by its surface area represents the net current produced by the electrokinetic reactions taking place on its surface. When the electrode is being oxidized, this is also the corrosion rate. It is usually expressed in A/cm^2 or mA/cm^2.

The apparatus includes an iron test or "working" electrode, a platinum auxiliary electrode, and a silver–silver chloride reference electrode

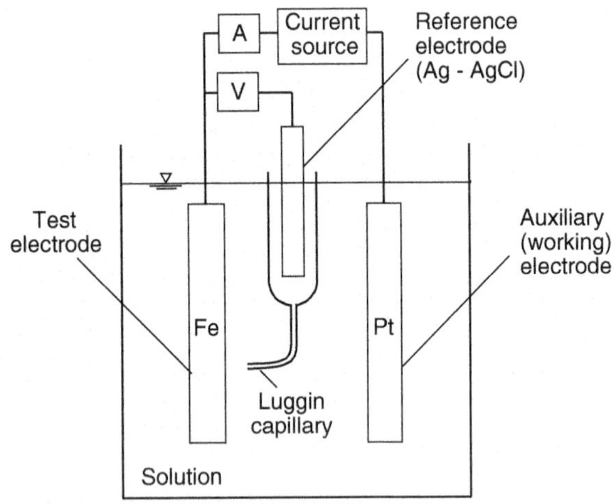

Figure 22-7
Apparatus used to measure polarization effects in metals.

immersed in the same acid solution. A high-impedance voltmeter (V) is connected between the reference electrode and the iron electrode to measure the potential between the iron test electrode and the solution. A Luggin capillary is used for the reference electrode to reduce solution resistance between the reference and test electrodes. A variable emf (current source) is connected between the strip of iron and the platinum electrode. Using the platinum electrode as a working electrode, a current is passed between the iron strip and the solution, polarizing the iron electrode. A low-resistance ammeter (A) is connected between the iron and platinum electrodes so that the current flow between them can be measured. The apparatus shown on Fig. 22-7 is conceptual only. A more detailed description of this experiment and apparatus may be found in Reiber et al. (1986).

Understanding electrode kinetics on the surface of an iron electrode corroding in acid solution begins with the realization that both cathodic and anodic reactions take place on the surface of the iron simultaneously. The cathodic reaction is the reduction of hydrogen ion ($2H^+ + 2e^- \leftrightarrows H_2$) and the anodic reaction is the oxidation of the iron itself ($Fe \rightleftarrows Fe^{2+} + 2e^-$). The polarization experiment is conducted by imposing an electrical current on the iron electrode, raising its potential and accelerating the density (or the rate) of anodic reactions taking place on its surface. By the convention established by Benjamin Franklin, electrical current flows in the opposite direction as the electrons. As a result, imposing a current on the iron electrode means drawing electrons from it. Corrosion engineers refer to this as "polarizing" the electrode. The increase in the rate of this reaction on the surface of the electrode corresponds to the increase in current observed. The experiment is then repeated, but with the current flowing in the opposite direction. In this second case, the potential of the iron electrode is reduced, and the current flow is generated by acceleration of the cathodic reaction on the surface of the iron electrode surface ($2H^+ + 2e^- \rightleftarrows H_2$).

When such polarization experiments are conducted, the slope of the plot of the logarithm of the current versus electrode potential quickly becomes linear. As Tafel was the first to make this observation (Tafel, 1905), the slope of this straight line is commonly referred to as the Tafel slope. In equation form, Tafel's concept is as follows:

$$\eta = \alpha \pm \beta \log(i) \tag{22-8}$$

where η = electrode overpotential = $E_H - E_0$
α = constant
E_H = electrode potential, mV
E_0 = electrode potential at equilibrium, mV
β = Tafel slope, mV/decade
i = current density, $\mu A/cm^2$
$\pm$ is + for the anodic reaction and – for the cathodic reaction

Tafel's concept applied to the polarization data from a polarization experiment conducted on iron in 1 N H_2SO_4 is illustrated on Fig. 22-8. The data are plotted in the form of an Evans diagram (Evans, 1948) where the logarithm of the absolute value of the current is plotted versus the electrode potential. To create the plot, the apparatus on Fig. 22-7 is used to impose a current on the electrode and to increase that current in a stepwise

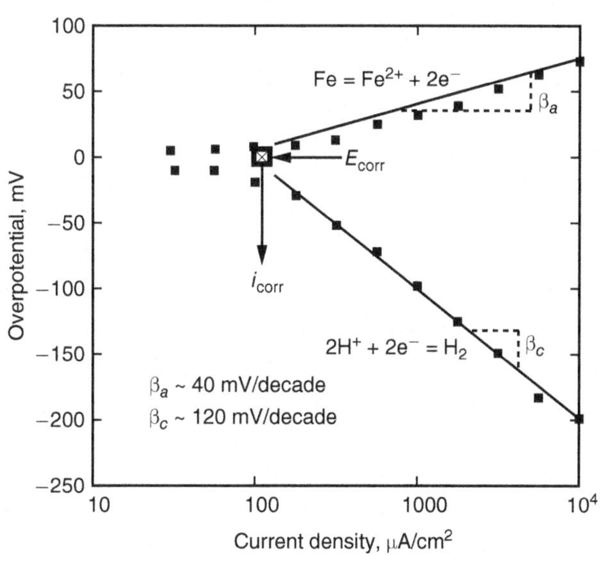

Figure 22-8
Polarization plot for iron corroding in acid solution. *Note:* In this plot the Evans convention is used in which the current is given no sign and both the anodic and cathodic currents are treated in the same manner (as if the absolute value of the current were being plotted).

Example 22-1 Estimating corrosion rate

Estimate the corrosion rate of the iron coupon on which the polarization test was being conducted on Fig. 22-8. Provide an answer in both mpy and mm/y.

Solution

1. Estimate the corrosion current density. From Fig. 22-8, the corrosion current density, I_{corr}, is estimated to be 110 μA/cm².
2. Estimate the corrosion rate. Using the conversion coefficients provided for iron in Table 22-3, the i_{corr} value corresponds to a corrosion rate of

$$\text{mpy} = 110\ \mu\text{A/cm}^2 \times 0.46 = 51\ \text{mpy}$$

$$\text{mm/y} = 110\ \mu\text{A/cm}^2 \times 0.012 = 1.3\ \text{mm/y}$$

fashion. The result is an *acceleration of the anodic reaction* on the surface of the electrode *and an increase in its potential.* The apparatus is then used to run the current in the opposite direction. Now the *cathodic reaction is accelerated and the potential is reduced.*

In 1905, Tafel observed that such curves become a straight line as they are extrapolated. As a result, the slopes of the lines bear his name. The anodic and cathodic slopes (β_a and β_c) are often different as different reactions generate their behavior. These slopes can be extrapolated to estimate the corrosion current density (i_{corr}) and, hence, the corrosion rate. At rest, the electrode is at its corrosion potential (E_{corr}).

Exchange Current Density

Another important concept in mixed-potential theory is the exchange current density (i_0). The exchange current density is a fundamental kinetic parameter and is characteristic of a particular reduction–oxidation half reaction on a particular surface. Consider the redox reaction for hydrogen:

$$2H^+ + 2e^- \rightleftarrows H_2 \tag{22-9}$$

The reversible potential for this reaction, E_{H2}, can be determined by the Nernst equation. When the potential of the metal coupon with respect to the solution equals E_{H_2}, the hydrogen redox reaction on its surface is presumed to be at reversible equilibrium. That is, the rate of the reaction in the forward direction (r_f) is equal to its rate in the backward direction (r_b). Using Faraday's law, the rate of this reaction can also be expressed as a current. The following expression describes the exchange current:

$$I = r_f nF \tag{22-10}$$

or

$$I = r_b n F \tag{22-11}$$

where I = exchange current, A (C/s)
r_f = rate of forward reaction, mol/s
r_b = rate of backward reaction, mol/s
n = electrons transferred per mol of reaction, eq/mol
F = Faraday constant, 96,500 C/eq

Arranging these rates as the sum of all the couples of that reaction reacting on the metal's surface divided by its area A produces the exchange current density i_0:

$$i_0 = \frac{\sum (r_f n_{H_2} F)}{A} = \frac{\sum (r_b n_{H_2} F)}{A} \tag{22-12}$$

Thus, the exchange current is the current equivalent of the rate at which the reaction goes in both the forward and reverse directions on a surface divided by the area of that surface—when that surface is maintained at the reaction's reversible potential. Whereas the potential of the half reaction is its fundamental thermodynamic parameter, the exchange current density is the fundamental kinetic parameter for that reaction when it occurs on a particular surface. It can be shown that the exchange current density is determined by the activation energy of the reaction (Jones, 1996), but it is not usually determined from first principles, rather it is measured. Also, whereas the equilibrium potential for a particular reaction does not depend on the nature of the surface upon which the reaction occurs, the exchange current density is specific to a particular surface.

The exchange current density for the hydrogen half reaction on a number of surfaces is displayed on the scale shown in Table 22-4. It is evident from these data that the exchange current density is extremely sensitive to the surface involved. The difference between the rate on platinum and on lead in acid conditions is 10 orders of magnitude. Not only is the particular metal important, but the condition of its surface is also important. The condition of the surface is influenced by solution conditions and other environmental factors such as temperature and velocity. For example, returning to Table 22-4, the difference between the hydrogen exchange current on a mercury surface in acid and basic conditions is 2 orders of magnitude. Finally, the condition of the surface changes with time and the exchange current density changes with it. Usually the exchange current density decreases as the film on the metal's surface continues to develop.

Table 22-4
Selected exchange current densities and Tafel slopes

Electrode Surface	Solution Condition	i_0, μA/cm^2	β, mV/decade
***Cathodic Reaction*: $2H^+ + 2e^- \leftrightarrows H_2$**			
Pt	1N HCl	10^4	110–130
	0.1N NaOH	7×10^2	110
Fe	1N HCl	10	150
	0.52 N H_2SO_4	20	110
	2 N H_2SO_4, 4%NaCl	1	120
Cu	0.1 N HCl	2	150
	1 N HCl	2	120
	0.15 N NaOH	10	120
Zn	1 N H_2SO_4	2×10^{-4}	120
Pb	0.01–8 N HCl	2×10^{-6}	150
Hg	0.1N HCl	7×10^{-6}	120
	0.1 N H_2SO_4	2×10^{-6}	120
	0.1N NaOH	3×10^{-8}	100
Sb	2 N H_2SO_4	10^{-2}	100
Al	2 N H_2SO_4	10^{-3}	100
Sn	1 N HCl	10^{-1}	150
Cd	1 N HCl	1	200
***Cathodic Reaction*: $O_2 + 4H^+ + 4e^- \leftrightarrows 2H_2O$**			
Pt	0.1 N H_2SO_4	9×10^{-5}	100
	0.1 N NaOH	4×10^{-6}	50
	1 N HCl	5×10^3	110
Zn	2 N H_2SO_4	10^{-1}	300
***Cathodic Reaction*: $Cl_2 + 2e^- \leftrightarrows 2Cl^-$**			
Pt	1 N HCl	5×10^3	110
***Cathodic Reaction*: $Fe^{3+} + e^- \leftrightarrows Fe^{2+}$**			
Pt	2 N H_2SO_4	10^3	150
***Anodic Reaction*: $Fe \leftrightarrows Fe^{2+} + 2e^-$**			
Fe	0.5 N H_2SO_4	10^{-4}	40
Fe	0.6 N $FeSO_4$	3×10^{-3}	60
Fe	0.5 N $FeSO_4$ +0.3 N $NaHSO_4$	3×10^{-4}	30

Example 22-2 Exchange current for zinc in an acid solution saturated with oxygen

A zinc electrode is immersed in 2 N H_2SO_4 solution that is saturated with an overlying atmosphere of pure oxygen at 25°C. The electrode is observed to be at the reversible potential for the oxygen reduction half reaction (+1230 mV). For these conditions, estimate the rate at which oxygen is converted to water. Express the answer in atoms/cm^2 · s. (*Hint*: At the given potential it is the same rate at which water is converted back to oxygen again.)

Solution

1. Estimate the exchange current. From Table 22-4, the exchange current for oxygen reduction on a zinc surface at these conditions is 0.1 μA/cm^2.
2. Convert μA/cm^2 to coulombs/cm^2 · s:

$$(0.1\ \mu\text{A/cm}^2)(10^{-6}\text{A}/\mu\text{A})(1\ \text{C/s}\cdot\text{A}) = 10^{-7}\ \text{C/cm}^2\cdot\text{s}$$

3. Use Faraday's law to convert coulombs/cm^2 · s to moles of oxygen/cm^2 · s:

$$\text{mol O}_2/\text{cm}^2\cdot\text{s} = (1\ \text{eq Zn}/96{,}500\ \text{C})(1\text{mol Zn}/2\ \text{eq Zn})(10^{-7}\ \text{C/cm}^2\cdot\text{s})$$
$$= 5.2\times10^{-13}\ \text{mol/cm}^2\cdot\text{s}$$

4. Estimate the rate in atoms/cm^2 · s. Using Avagadro's number, convert moles to atoms:

$$\text{Rate} = (5.2\times10^{-13}\ \text{mol/cm}^2\cdot\text{s})(6.02\times10^{23}\ \text{atoms/mol})$$
$$= 3.1\times10^{11}\ \text{atoms/cm}^2\cdot\text{s}$$

Mixed-Potential Diagram

The mixed-potential diagram provides a useful view of the overall corrosion process and the principles that control it from the perspective of mixed-potential theory. Construction of a mixed-potential diagram from the polarization plot on Fig. 22-8 begins with the extrapolation of the polarization curves to their reversible potential as determined by the Nernst equation. The current density at which each of these two reactions reaches its reversible potential is the exchange current density for that half reaction on the surface of iron corroding in acid conditions. The construction of the diagram is illustrated in the following example problem.

Example 22-3 Construction of a mixed-potential diagram

Construct a mixed-potential diagram for the results displayed in Fig. 22-8. Both [Fe^{2+}] and [H^+] are present at 1 mol/L and that the partial pressure of hydrogen, $pH_2 = 1$ atm. The corrosion potential, E_{corr}, was measured to be –270 mV. *Hint*: Use the slope of the anodic reaction (β_a) of 40 mV/decade and the slope of the cathodic reaction (β_c) of 120 mV/decade.

Solution

A. The anodic reaction, $Fe \leftrightarrows Fe^{2+} + 2e^-$

1. Overview: Start at (i_{corr}, E_{corr}) and extrapolate at the given anodic slope, β_a, until E_H is equal to the reversible potential for the anodic reaction, $E_{Fe/Fe^{2+}}$. The current at this potential is the exchange current for this reversible reaction on an iron surface.
2. Find i_{corr}: From Example 22-1, $i_{corr} = 110\ \mu A/cm^2$.
3. Find E_{corr}: E_{corr} is obtained by measuring the potential of the corroding iron electrode when no current is imposed from the working electrode. In this problem, E_{corr} has been given as –270 mV.
4. Therefore (i_{corr},E_{corr}) = (110 $\mu A/cm^2$, –270 mV).
5. By inspection of Table 22-4, the slope of the anodic reaction is found to be 40 mV/decade.
6. Fe^{2+} is given as 1 mol/L. Thus the reversible potential for the anodic reaction can be directly obtained from Table 22-2: $E_{Fe^{2+}Fe^{3+}} = -440$ mV.
7. Rearranging Eq. 22-8 for the anodic reaction:

$$\log(i_0) = \log(i_{corr}) + \frac{E_{Fe} + E_{corr}}{\beta}$$

$$\log(i_0) = \log(110) + \frac{270 - 440}{40} = -2.2$$

$$i_0 = 10^{-2.2}\ \mu A / cm^2$$

B. The cathodic reaction, $2H^+ + 2e^- = H_2(g)$

1. Overview: Start at (i_{corr}, E_{corr}) and extrapolate at the cathodic slope of β_c until E_H is equal to the reversible potential for the cathodic reaction, E_{H^+/H_2}.
   The current at this potential is the exchange current for the reaction.

2. The starting point has already been determined in step 4 above (i_{corr}, E_{corr}) = (110 μA/cm², −270 mV).
3. The slope of the cathodic reaction is given as 120 mV/decade.
4. For the purposes of these estimates, $H^+ = 1\ \text{mol}/\text{L}$ and $p_{H_2} = 1$ atm. Thus the reversible potential for the cathodic reaction, E_{H_2} is 0 mV (Table 22-2).
5. Rearranging Eq. 22-8, the equation for the anodic reaction is

$$\log(i_0) = \log(i_{corr}) - \frac{E_{H_2} + E_{corr}}{\beta}$$

$$\log(i_0) = \log(110) - \frac{0 + 270}{120}$$

$$\log(i_0) = 0$$

$$i_0 = 1\ \mu\text{A}/\text{cm}^2$$

6. The completed mixed-potential diagram for iron corroding in acid solution is shown below.

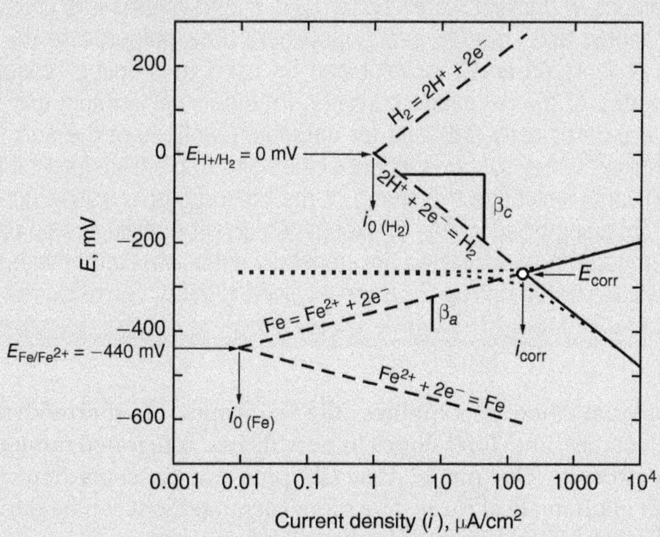

Mixed potential diagram for iron corroding in acid solution.

In the diagram constructed in the preceding example, the information shown on Fig. 22-8 has been augmented so that the important relationships between the anodic and cathodic reactions can be better understood. Diagrams of this kind, often called Evans diagrams in honor of their originator, Dr. Evans of the United Kingdom, can be used to illustrate most of the important concepts in the corrosion of metals.

The shape of the plot is determined by three factors, each for the anodic and cathodic reactions: (1) the Nernst potential at which the reaction is at equilibrium (E_H), (2) the exchange current of the reaction on the electrode surface (i_0), and (3) the reaction's characteristic Tafel slope on the electrode surface (β_a or β_c). The potential of the corroding electrode and the corrosion current are then determined as the point where the cathodic and anodic reactions are at equilibrium. As can be seen by inspection of the diagram, the reversible potentials of the cathodic and anodic reactions have important influence on the corrosion potential and on the corrosion rate. Note that the exchange current of the cathodic reaction for iron is on the order of a few nanoamperes per square centimeter, much lower than that of the anodic reaction, which is on the order of a few microamperes per square centimeter.

The preceding example (22-3) was prepared for the purpose of illustrating the dynamics of mixed-potential theory. Simple extrapolations such as those used in Example 22-3, while effective for iron immersed in sulfuric acid, are generally not practical under the environmental conditions in drinking water systems. In such systems it is unusual to find anodic and cathodic reactions where data gathered to the right side of *i*corr on an Evans plot can be used to produce reasonable estimates of the exchange currents. Problems with simple extrapolation are particularly the case for cathodic reactions of the sort found in drinking water systems where scales and mass transfer limitations significantly influence the shape of the cathodic polarization curve. In fact obtaining good values of exchange current densities is so difficult that most important values are generally unknown. Nevertheless, the forces illustrated on Fig. 22-8 are involved in every corrosion reaction.

Mixed-potential Corrosion Model

Mixed-potential theory combines the concepts of thermodynamics, exchange current, and Tafel slopes to provide one integrated model of the corrosion process. The model basically puts the concepts depicted on Fig. 22-8 in mathematical form. The model arrangement can be derived by restructuring the Tafel equation in the following manner:

$$E_H = E_a^0 + \beta_a \log\left(\frac{i}{i_a^0}\right) \quad \text{(for the anode)} \tag{22-13}$$

$$E_H = E_c^0 - \beta_c \log\left(\frac{i}{i_c^0}\right) \quad \text{(for the cathode)} \tag{22-14}$$

Where E_H = potential of electrode, mV
E_a^0 = reversible equilibrium potential for anodic half reaction, mV
E_c^0 = reversible equilibrium potential for cathodic half reaction, mV
β_c = Tafel slope of cathodic reaction, mV/decade

β_a = Tafel slope of anodic reaction, mV/decade
i = measured current density at E_H, $\mu A/cm^2$
i_a^0, i_c^0 = current density at $E_H = E_a^0$ and E_c^0, respectively

Measurements of electrode kinetics take advantage of the fact that the measured potential of the electrode with respect to the solution, E_H is well represented by Eq. 22-13, if its potential is depressed, and by Eq. 22-14, if its potential is elevated. If the reversible potential, the exchange current, and the Tafel slopes are known for both the cathodic and anodic reactions, then it is possible to combine Eqs. 22-13 and 22-14 to prepare an estimate of the corrosion current density (i_{corr}):

$$\log(i_{corr}) = \frac{\beta_c \log(i_c^0) - E_a^0 + \beta_a \log\left(i_a^0\right) + E_c^0}{\beta_a + \beta_c} \qquad (22\text{-}15)$$

At the corrosion current (i_{corr}) the electrode potential (E_H) is equal to the corrosion potential (E_{corr}); therefore E_{corr} can be determined by either Eq. 22-13 or 22-14, once Eq. 22-15 is solved.

Estimating corrosion rates using these equations is not the most powerful thing about mixed-potential theory. In fact too many empirical parameters must be determined $\left(i_c^0, i_a^0, \beta_a, \beta_c\right)$ before Eq. 22-15 can be used. It is easier to just measure the corrosion rate. The more significant use of mixed-potential theory is to conceptualize the forces that drive the corrosion process and to anticipate the outcome of contemplated system changes.

The reversible potential of a cathodic or anodic reaction on the surface of a corroding electrode can be estimated using the Nernst equation as shown below:

$$E_H = E_H^0 + \left(\frac{2303RT}{nF}\right)\log\left(\frac{\text{oxidized}}{\text{reduced}}\right) \qquad (22\text{-}16)$$

where E_H = reversible potential of reaction, mV
E_H^0 = reversible potential of reaction under standard temperature and pressure, mV
R = universal gas constant, 8.314 J/K · eq
T = temperature, K
n = equivalents of electrons per mole in reaction, eq/mol
F = Faraday's constant, 96,500 C/eq

22-4 Application of Electrokinetics

In the preceding sections the thermodynamics of metallic corrosion have been reviewed and an electrokinetic model of the corrosion process has been introduced. In the following discussion, some aspects of the application of these principles to understanding the corrosion of metallic conduits in water systems are reviewed.

Applying Mixed-Potential Corrosion Model

A simple way of examining the implications of the mixed-potential model is to examine the effect of the strength of the acid solution on the corrosion rate. In the examples so far, it has been assumed that all ions are at a concentration of 1 mol/L. The consequences of a milder acid solution are examined in the following example using mixed-potential theory.

Example 22-4 Corrosion of iron in a mild acid solution

Construct a mixed-potential diagram to examine and compare the corrosion rate of iron at pH 0 and at pH 2 for the following conditions:

1. For the cathodic reaction: $2H^+ + 2e^- \leftrightarrows H_2$

   Exchange current = 1 μA/cm²

   Tafel slope = 120 mV/decade

2. For the anodic reaction: $Fe \leftrightarrows Fe^{2+} + 2e^-$

   Exchange current = 0.01 μA/cm²

   Tafel slope = 40 mV/decade

For purposes of this example, assume that Fe^{2+} is present at a concentration of 1 mol/L, that the partial pressure of H_2 is 1 atm, and that the temperature is 25°C. The effect of any change in ionic strength may be neglected.

Solution

1. Determine the reversible potential for the anodic reaction. Because Fe^{2+} remains at 1 mol/L, the reversible potential of the anodic reaction remains at –440 mV (Table 22-2).
2. Find the reversible potential of the cathodic reaction:
   a. At pH 0, $H^+ = 1$ mol/L and $pH_2 = 1$ atm, thus from Table 22-2, $EH = 0$ mV.
   b. At pH 2, Eq. 22-16 can be used:

$$\begin{aligned} E_H &= 0 \text{ mV} + \frac{59}{1}\log\left(\frac{1}{10^{-2}}\right) \\ &= 0 - 59 \times 2 = -118 \text{ mV} \end{aligned}$$

3. Compare the corrosion rates at pH 0 and 2. A mixed-potential diagram in which the two conditions described is presented in the diagram below. Using both the diagram and Eq. 22-16, a reduction in corrosion rate of more than fivefold is estimated when the pH is changed from 0 to 2. All this change results from the reduction in the reversible potential of the cathodic reaction.

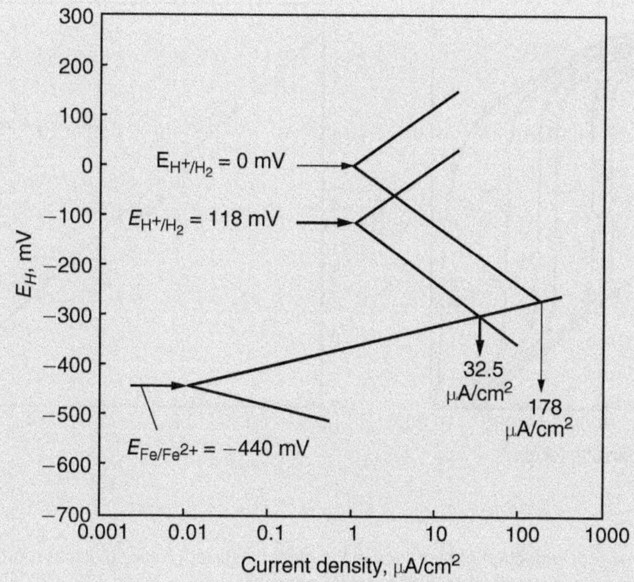

Concentration Polarization

So far, the examination of mixed-potential theory has been limited to *activation polarization.* In activation polarization, the change in electrode potential is governed by the activation energy of the redox reaction on the electrode surface. Although common with iron in acid solution and useful for purposes of illustration, activation polarization does not occur under the conditions present in drinking water systems. A more common rate-controlling phenomenon in such systems is concentration polarization. In concentration polarization, the rate of the redox reaction (and hence the current flow) is not governed by the activation energy but by mass transport considerations.

Mass transfer limitations are particularly important with respect to the cathodic reactions where a chemical species in the water is to be reduced (H^+, O2, Cl_2, etc.) and must be transported from the bulk solution to the electrode interface. Under these circumstances, concentration polarization results in a maximum or limiting current that cannot be exceeded. Above the limiting current, current flow is relatively insensitive to electrode potential. The extensive literature on mass transport can be used to examine the impact of flow rate, shape, and other factors on limiting current, but this is beyond the scope of this discussion. In simple terms, the diffusion-limited current density may be characterized in the following manner:

$$i_L = nF\ k_f C_b \tag{22-17}$$

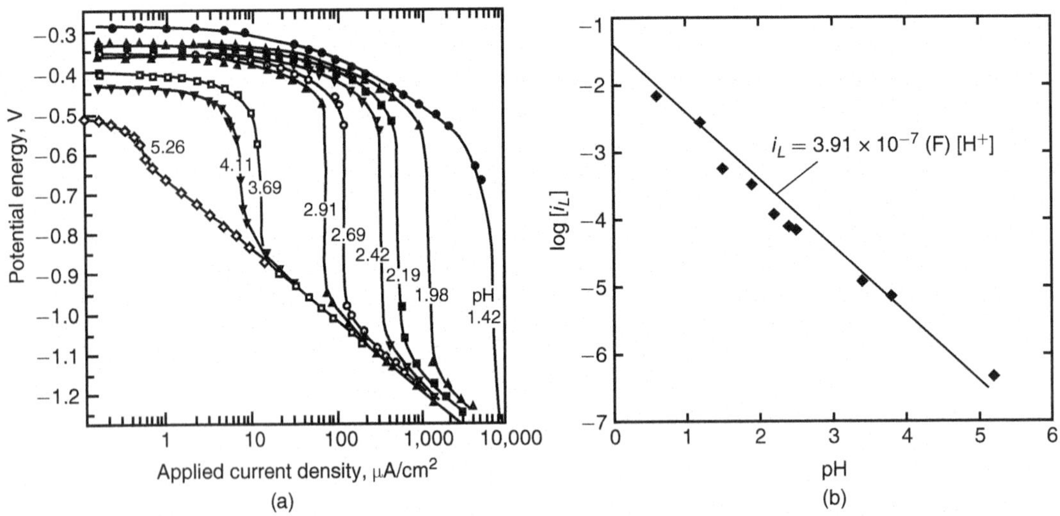

Figure 22-9
Stern's experiments on limiting current for reduction of hydrogen ion in quiet (unstirred) NaCl solutin at room temperature: (a) cathodic polarization curves at various pHs and (b) limiting current versus pH.

where i_L = limiting current density, A/cm^2
k_f = mass transfer constant for reacting ions, $L/cm^2 \cdot s$
C_b = concentration of ions in bulk solution, mol/L
n = equivalents of ions per mole in reaction, eq/mol
F = Faraday's constant (96,500 C/eq)

In an unstirred solution with dissolved oxygen as the cathodic reactant (electron acceptor), the value of k_f is approximately $9.2 \times 10^{-7}\ L/cm^2 \cdot s$. Stern (1955) conducted a series of experiments examining the limiting cathodic current for the reduction of hydrogen ion on pure iron at several different pH values in a 4 percent NaCl solution at room temperature. The diffusion-limited current found in Stern's experiments and the data fitted to Eq. 22-17 are shown on Figs. 22-9a and 22-9b, respectively. Based on Fig. 22-9a, in an unstirred solution with hydrogen ion as the electron acceptor, the value of k_f is approximately $3.91 \times 10^{-7}\ L/cm^2 \cdot s$.

Limitations of Mixed-Potential Theory

Mixed-potential theory is a powerful way of looking at the fundamental driving forces for the corrosion process, and useful perspectives can be gained by examining the corrosion of water conduits using the mixed-potential approach. At the present time, with the exception of the work of Reiber (1989, 1991) on copper and lead solder, the power of mixed-potential theory in exploring the internal corrosion of drinking water systems remains largely unexplored.

Example 22-5 Further examination of effect of pH on rate of iron corrosion in acid solution

Review Example 22-4 and determine if the conclusion derived from the analysis can be altered in light of the concept of limiting current. Also examine the fundamental changes that take place as higher pH values are considered and discuss the implications.

Solution

1. Using Eq. 22-17, determine the limiting current. From the paragraph preceding this example, $k_f = 3.91 \times 10^{-7}$ L/cm$^2 \cdot$ s.
   a. At pH 0, $C_b = C_{H+} = 10^0$ mol/L:

   $$i_L = (1 \text{ eq/mol})(3.91 \times 10^{-7} \text{ L/cm}^2 \cdot \text{s})(96{,}500 \text{ C/eq})(10^0 \text{ mol/L})(1 \text{ A/C} \cdot \text{s})$$

   $$= (3.77 \times 10^{-2} \text{ A/cm}^2 \cdot \text{s})(10^6\ \mu\text{A /A}) = 3.77 \times 10^4\ \mu\text{A/cm}^2 \cdot \text{s}$$

   b. At pH 2, $C_b = C_{H+} = 10^{-2}$ mol/L:

   $$i_L = (3.77 \times 10^4\ \mu\text{A/cm}^2 \cdot \text{s})(10^{-2}/10^0) = 377\ \mu\text{A/cm}^2 \cdot \text{s}$$

2. Compare the limiting current computed in step 1 to the corrosion current predicted earlier. For both pH 0 and pH 2, the corrosion rates estimated in Example 22-4 are lower. Thus, the conclusions reached in Example 22-4 do not change.
3. Examine the effect of higher pH values on the limiting current.
   a. Using the procedure from step 1, compute the limiting value as a function of pH. The computed value are summarized in the following table:

pH	i_L, μA/cm^2
0	37,770
1	3,777
2	378
3	37.8
4	3.78
5	0.378

   b. Between a pH of 4 and 5, the limiting current for hydrogen reduction falls below the exchange current density for hydrogen ion reduction on an iron surface as reported in Table 22-4. Under these conditions, corrosion is no longer possible via this mechanism.

Based on a limited exploration of mixed-potential theory, it appears that the determination of a few critical constants for certain materials of importance in water systems could lead to a powerful system for comprehensive modeling of corrosion behavior. In fact, the corrosion process is complex, and mixed-potential theory is not well suited for making direct predictions of corrosion rates under the conditions in which drinking water systems operate. As a result, it is important to understand the limitations of mixed potential theory as well as its strengths.

Most shortcomings have to do with the parameters the model uses. In summary the parameters required for the model are:

1. The reversible potential for the cathodic and anodic half reactions.
2. The exchange currents for the cathodic half reactions on the materials of interest. For drinking water systems, these would include cast iron, ductile iron, mild steel, galvanized iron, lead, copper, solder, and brass, among others. Electron acceptors of interest as cathodic reactants include the hydrogen ion, oxygen, and all the oxidants being used in drinking water systems. These include free chlorine, combined chlorine, chlorine dioxide, ozone, hydrogen peroxide, hydroxyl radical, and permanganate.
3. The Tafel slopes for the anodic reactions for the conduit materials.
4. The Tafel slopes for the various electron acceptors on these same materials.

Some of the limitations of each of these critical parameters are summarized in Table 22-5. The discussion of these limitations should not be taken to suggest that the principles of mixed-potential theory do not apply to corrosion of water conduits. Rather, it is important to understand that much needs to be learned before the corrosion process in water systems can be modeled adequately and the nature of further research determined. The relationship between scale formation and these electrokinetic behavior is particularly important.

One of the more important limitations identified in Table 22-5 is understanding the concentration of oxidized ions immediately adjacent to the corroding metal surface. It should be noted that in Table 22-4 the anodic constants are only for iron. Constants for copper, lead, and zinc are not provided, even in acid solution where, presumably, they are soluble as well. In fact, iron is the only common metal that corrodes in acid solution at a potential that corresponds to reasonably high levels of oxidized ion (e.g., near its standard potential). Recently published data on the polarization curves for copper, zinc, and brass in a solution of 2 N H_2SO_4 (Jinturkar et al., 1998) are presented on Fig. 22-10. It should be noted that the corrosion potential of these metals is substantially below their standard potential, suggesting that, despite the strong acid conditions, the surface of

Table 22-5
Parameters of mixed-potential model and their limits

Parameter	Comment on Limitations
E_a, reversible potential for anodic reaction	For conceptual purposes, the potentials in Table 22-2 are often used directly. But these values are only valid at unimolar concentrations and at standard temperature and pressure. Correction for pressure is generally not significant and corrections for temperature, while complicated, are straightforward. These can be accomplished using the same principles as are used for correction of other chemical equilibria. Where the reversible potential at the anode is concerned, the big difficulty is that the concentration of the oxidized form of the metal at the electrode surface (e.g., Fe^{2+}) is not ordinarily available. In these examples the ion is often assumed to be at unimolar concentration (allowing the values in Table 22-2 to be used directly). Generally, the potentials of corroding surfaces are at or below this reversible potential itself, suggesting that the concentration of oxidized ions at the anode surface is very low. Assuming that the concentration of these ions is controlled by solubility phenomena at the electrode surface is a promising approach that needs to be explored further.
E_c, reversible potential for cathodic reaction	Perhaps the reversible potential of the cathodic reactions is the parameter that can be most easily estimated with what we know today. This reversible potential is not dependent on the material corroding. Normally, it is just dependent on the properties of the solution, such as the concentration of the oxidizing agent of interest, and it can be determined using the Nernst equation.
i_a^0 and i_C^0, exchange current densities for anodic and cathodic reactions at their reversible potential	The exchange current densities for metals of interest in water systems cannot be directly measured. Moreover, they change as the film on the metal's surface changes. These two parameters are probably the most inaccessible, and we are forced to make some generous assumptions to estimate them.
β_a and β_c, Tafel slopes of anodic and cathodic reactions	Polarization experiments can be conducted and the corresponding Tafel slopes measured, but, except for acid conditions, the factors that control these slopes are poorly understood. This is particularly true of cathodic Tafel slopes on metals with the sort of scales that normally develop after a period of sustained service in drinking water systems. It is not alway so easy to determine whether the slope observed is under activation control or is controlled by some other process, such as diffusion.

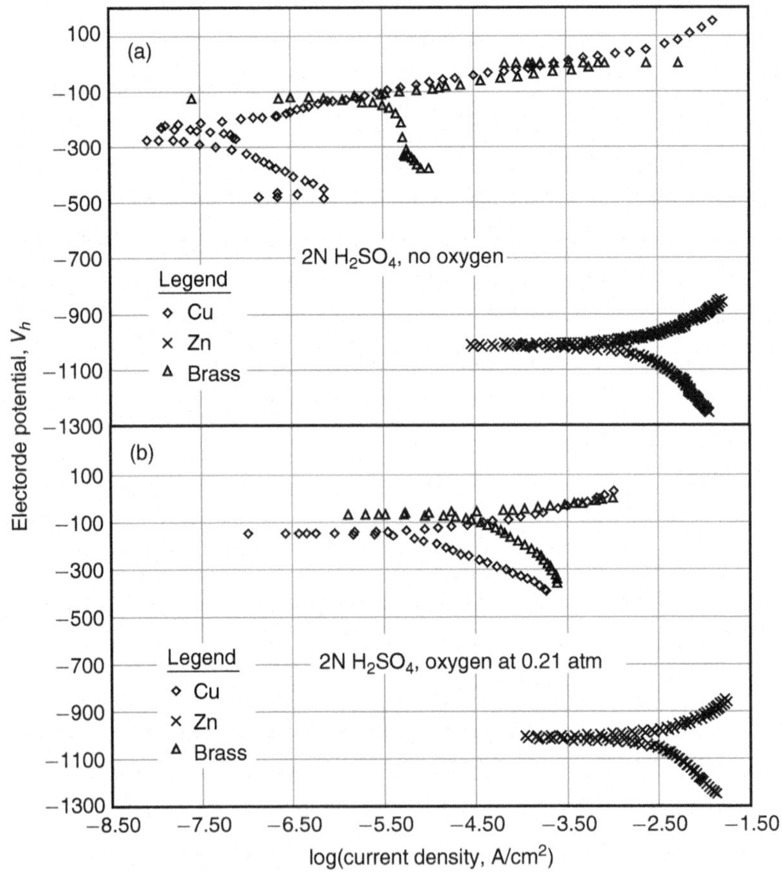

Figure 22-10
Polarization curves for copper, zinc, and brass: (a) in acid solution with no oxygen and (b) in acid solution with oxygen at 0.21 atm (data from Jinturkar, et al., 1998).

the metal is not exposed to high concentrations of the Me^{2+} ion. In fact, the levels that are implicated are far below solubility. It is likely that the data can only be explained through detailed knowledge of the microstructure and surface chemistry of the corroding surface.

Example 22-6 Estimation of Me^{2+} concentrations near the surface of zinc and copper corroding in acid solution

Using the data on Fig. 22-10 for acid solution in the absence of air, estimate how low the concentration of Zn(II) or Cu(II) ions would have to be to bring the reversible potential 50 mV below the corrosion potentials observed.

Solution

1. The corrosion potentials, estimated using Fig. 22-10, are as follows:

Copper	–235 mV
Zinc	–1010 mV

2. The reversible potential at standard conditions are found in Table 22-2:

Copper	+340 mV
Zinc	–763 mV

3. According to mixed-potential theory, the reversible potential of the anode must be *below* the metal's corrosion potential. According to the Nernst equation, this is possible if the oxidized ion is at a low enough concentration:

$$E_H = E_H^0 + \frac{59}{2} \log (\text{Me}^{2+})$$

4. Determine the concentration of the metal ions necessary to bring the reversible potential 50 mV below the corrosion potential.
   a. The required values are:

$$\text{For Cu(II): } E_H = -235 \text{ mV} - 50 \text{ mV} = -285 \text{ mV}$$

$$\text{For Zn(II): } E_H = -1010 \text{ mV} - 50 \text{ mV} = -1060 \text{ mV}$$

   b. Using the Nernst equation to estimate the level of metal ions:
   For copper,

$$-285 \text{ mV} = +340 \text{ mV} + 29.5 \log[\text{Cu}^{2+}]$$

   Rearranging,

$$\log\left[\text{Cu}^{2+}\right] = \frac{-285 \text{ mV} - 340 \text{ mV}}{29.5} = -21.2$$

$$\left[\text{Cu}^{2+}\right] = 10^{-21.2} \text{ mol/L}$$

   For zinc,

$$-1060 \text{ mV} = -760 \text{ mV} + 29.5 \log[\text{Zn}^{2+}]$$

   Rearranging,

$$\log\left[\text{Zn}^{2+}\right] = \frac{-1060 \text{ mV} + 760 \text{ mV}}{29.5} = -10.2$$

$$\left[\text{Zn}^{2+}\right] = 10^{-10.2} \text{ mol/L}$$

Because these reactions take place in a solution of 2 N H_2SO_4, the metal concentrations are extraordinarily low when compared to traditional solubility equilibria for metal hydroxides. The current tools available are not sufficient to characterize these electrode phenomena. Current understanding is limited to observations that can be made from the figure, namely: (1) brass corrodes at a potential that is higher than that for either of its alloy components, copper and zinc, especially zinc, (2) copper corrodes at an extremely low rate in acid solutions, (3) zinc, on the other hand, corrodes much more rapidly, and (4) the presence or absence of oxygen seems much more significant to copper and 60/40 brass (60 percent copper, 40 percent zinc) than it does to zinc.

Changes in the condition of the corroding metal surface with time can also have important impacts on the exchange current that the surface can support. As the condition of the surface changes, the Tafel slopes change, and it is likely that the exchange currents change as well. For example, while examining the behavior of copper tubing exposed to Seattle tap water over a 100-day period, Reiber (1989) observed a shift in the anodic Tafel slope from 55 to 95 and in the cathodic slope from 145 to 245.

Inhibition of Corrosion

The inhibition of corrosion involves actions that can be taken to reduce the rate of corrosion by cathodic and anodic reactions.There are several modes of corrosion inhibition that are implicated by the mixed-potential theory of corrosion. In the future, it seems likely that inhibitors will be designed to target each of the parameters that control the mixed-potential plot: the exchange current, the reversible potentials, and the response of both of the anodic and cathodic reactions to changes in polarization. Current understanding is not adequate to support interpretations that are too complex.

It is useful to think of inhibitors as being either anodic or cathodic in their action. These two modes of corrosion control are illustrated on Fig. 22-11. The original corrosion reaction is illustrated on Fig. 22-11a. The sketch displayed on Fig. 22-11b illustrates the effect of actions taken to

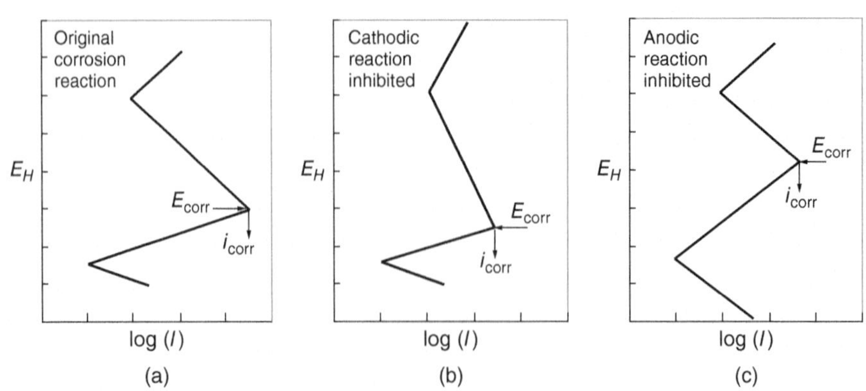

Figure 22-11 Modes of corrosion inhibition: (a) original corrosion reaction, (b) cathodic reaction inhibited, and (c) anodic reaction inhibited.

inhibit the cathodic reaction, for example, limiting the mass transfer of the electron acceptor (like H^+, O_2, or HOCl) to the cathode. Figure 22-11c is drawn to show the effect of actions taken to inhibit the anodic reaction. Reducing the solubility of the oxidized metal, causing the reversible potential of the anodic reaction to increase is such an action. Note that actions taken to inhibit the anodic reaction raise the potential of the corroding electrode, whereas actions taken to inhibit the cathodic reaction lower it. Thus, some basic conclusions about the mechanism of the action of a corrosion inhibitor can be drawn from the impact it has on the potential of the corroding specimen.

Insights about the mechanism of inhibition can be gained from measurements of the potential of the corroding surface. There are complications in making this measurement as the potential is not always stable, but insight can be gained by looking at the change in potential before and after the inhibitor is applied and over time as changes in the surface scales take place. Nevertheless, in an effort to gather data to understand the mechanisms underlying the corrosion of drinking water systems, the water industry has not gathered enough data on these potential measurements. As a result, the potentials that typically occur are not available. The following approximate potentials were reported after 96 h of exposure to Seattle water that had been adjusted to pH 8 and saturated with dissolved oxygen (Reiber et al., 1986):

Galvanized iron	–800 mV
Black iron	–200 mV
Copper	0 mV

The corrosion potential of 50/50 Pb–Sn solder was measured over a period of time:

Time of Exposure, Days	Corrosion Potential, mV
1	–345
14	–200
180	20

These and other observed corrosion potentials are summarized and compared with the standard electrode potentials of the relevant metals in Table 22-6. These corrosion potentials are given for purposes of illustration, and it is important to understand that the corrosion potentials observed in other water supplies will be different. Nevertheless, some important observations can be made. Once again, with the exception of iron, the corrosion potentials are lower than the standard electrode potentials. Second, it should be noted that the corrosion potential of copper is substantially higher than that of either galvanized iron or lead–tin solder. The higher potential for copper has important implications where galvanic corrosion is concerned.

Table 22-6
Comparison of observed corrosion potentials and standard electrode potentials

Metal	Potential in Seattle Water after Few Days, mV	Potentials Observed by Jones in Seawater, mV	Metals in Alloy	Standard Electrode Potential, mV
Galvanized iron[a]	−800	—	Fe[a]	−440
			Zn[a]	−760
Steel	−200	−603	Fe	−440
Copper	0	−350	Cu	+340
Lead–tin solder	−200	—	Pb	−130
			Sn	−140
Zinc	—	−1100	Zn	−760
Brass 70% Cu, 30% Zn	—	−378	Cu	+340
Brass 50% Cu, 50% Zn	—	−410	Cu	+340
Brass 30% Cu, 70% Zn	—	−988	Zn	−760
Brass 15% Cu, 85% Zn	—	−999	Zn	−760

[a]A new hot-dip galvanized pipe actually presents an alloy surface to the water that is 95% zinc (see Fig. 22-22). As the pipe ages and this nearly pure zinc layer is corroded away, alloy layers that are increasingly iron rich are exposed. Thus the corrosion potential of a galvanized pipe can be expected to shift gradually from the corrosion potential of nearly pure zinc to nearly pure iron over time. The reported potential for Seattle water is representative of fairly new pipe.

22-5 Microbiologically Induced Corrosion

Microorganisms can play a significant role in fostering corrosion of pipe materials. Bacteria have the ability, in certain situations, to:

1. Form microzones of high acidity or high concentrations of corrosive species.
2. Increase electrolytic concentration at surface sites.
3. Favor electron transfers.
4. Mediate the oxidation of reduced chemical species.
5. Disrupt the protective influence of surface films.
6. Mediate the removal of corrosion reaction products, enhancing the corrosion kinetics.
7. Take advantage of local gradients in redox potential to obtain energy for their metabolic needs.

Thus, bacteria are able to facilitate corrosion kinetics by accelerating the rate of redox reactions. Even anaerobic bacteria may thrive in otherwise aerobic water because of the formation of microbiological consortia in the biofilms on the surface of pipes, especially when the surface of the pipe has the heterogeneous surface associated with pitting or other microscopic irregularities. The most significant bacteria involved in mediation of corrosion reactions are sulfate reducers, methane producers, nitrate reducers, sulfur bacteria, and iron bacteria. In particular, the sulfate-reducing and iron bacteria groups are nuisances in corrosive behavior.

Under anaerobic environments, such as those that may occur within a corrosion pit, sulfate-reducing bacteria, such as *Desulfovibrio desulfuricans*, obtain their energy needs by reducing sulfate to sulfite, elemental sulfur, or other reduced forms. Bacteria in general can thrive at an interface between oxygenated and anaerobic areas because of the energetic difference associated with the oxygen gradient and the presence of oxidized species such as sulfate in an anaerobic region where sulfate is not thermodynamically stable. Sulfate reducers require ferrous iron and hydrogen gas as substrates. The uptake of Fe^{2+} and H_2, which are reaction products of the corrosion of iron metal, serves to lower the concentration of Fe^{2+} and H_2 in the microzone and thereby to enhance the kinetics of the forward direction of the corrosion reaction.

Hydrogen-oxidizing bacteria promote corrosion by removing the hydrogen gas produced at the cathode and, thus, depolarizing it. Iron bacteria such as *Gallionella* and *Sphaerotilus lepothrix* are aerobic organisms that mediate the oxidation of Fe(II) to Fe(III), reducing Fe(II) at the anode surface and depolarizing it. Iron bacteria may also be involved in the formation of FeOOH (goethite), which can result in the mineralization of *Gallionella* colonies and the formation of encrustations (Kolle and Rosch, 1980). Nitrate-reducing bacteria and methane-producing bacteria may play a role in altering the pH of a surface site, affecting the driving force for a corrosion reaction.

Current understanding of the interactions of microbiological and electrochemical reactions does not permit quantitative prediction of the role of microorganisms in enhancing corrosion reactions. It appears likely that complex ecosystems involving a variety of genera can arise in microzones on the surface of pipes where corrosion reactions take place. Empirical evidence suggests a major role for microbiological reactions in some systems. For example, in many instances, the chlorination of well water supplies has reduced corrosion rates, suggesting that disinfection can inhibit the role of bacteria. Laboratory studies of the corrosion of cast iron shows that unsterilized water can lead to significantly more rapid corrosive attack than sterilized water. Similarly, aeration of otherwise anaerobic or microaerobic water can reduce the activity of anaerobic bacteria.

A practical consequence of the effect of microorganisms in promoting corrosion is the relationship between disinfection and corrosion. Water utilities can experience relatively severe corrosion in parts of the distribution system where water stagnates. The decline in chlorine residual and lack of scouring action in distribution dead ends result in increased growth of microorganisms on surfaces, especially where pitting has taken place or

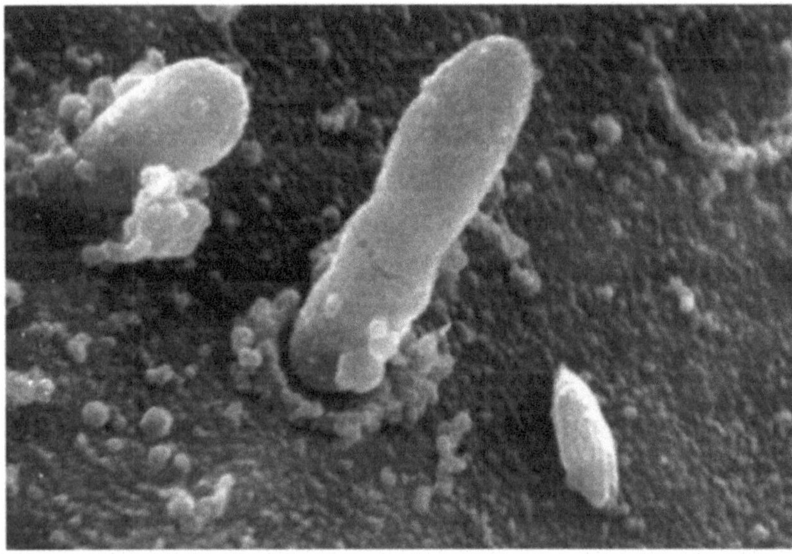

Figure 22-12
Bacteria attacking copper surface.

organic content is high. As a result, corrosive failures and main breaks tend to be more common in these locations. It is common practice for utilities to flush such dead ends routinely to minimize bacterial growth by scouring and by lowering the detention time of the chlorinated water.

Bacteria have even been found to attack copper surfaces (Lin and Olson, 1995). In the SEM photomicrograph shown on Fig. 22-12, from unpublished work of B. Olson (UC Irvine) and M. Keitz (U. Minnesota) gram-negative bacteria are shown growing on the surface of a copper ring. Copper is often thought to be toxic, but a number of instances have been reported where biological-mediated corrosion has occurred in copper tubing, particularly pitting. This photo is a particularly striking example demonstrating how some microorganisms can attack unlikely surfaces.

22-6 Surface Films and Surface Scales

When they are not lined, the conduits used to transport water appear to operate with the water in direct contact with the metal. In reality, all metals form a film on their surface upon contact with water, and these films (or scales) have important influences on the metal's interaction with the water. Zinc is a good illustration of this occurrence. Based on thermodynamics, it would appear that zinc, with a reduction potential of –760 mV, should be oxidized rapidly in water, especially if oxygen (+1230 mV) is present. In fact, in many domestic waters, the zinc coating on galvanized pipe is quite stable. The stability of zinc is, in large part, due to the fact that zinc forms a stable oxide scale that interferes with the corrosion process.

In the middle of the nineteenth century Faraday performed an experiment that illustrates how important these surface films can be. Faraday observed that if he immersed iron in dilute nitric acid, it reacted immediately, reducing hydrogen ion to emit hydrogen gas while the metal was rapidly oxidized. On the other hand, if he immersed the iron in concentrated nitric acid, it did not react. Moreover, even after the concentrated acid solution was subsequently replaced with a more dilute one, the iron continued in a passive state for quite some time. On the other hand, in the dilute acid solution, the iron specimen began to react violently if it was scratched. Faraday suggested that these properties were imparted to iron by an invisible surface oxide film, formed upon exposure to the concentrated nitric acid. Now, many decades later, it is known that he was right. Faraday had identified the special role that surface films can play in the corrosion process. It turns out that the films that form on the surfaces of metals when they are submerged in water play a critical role in the corrosion process, particularly oxide films.

Passivity

Passivity involves the formation of a nearly "perfect" protective oxide coating on transition metals, which reduces the rate of corrosion significantly. Since Faraday's time, the term *passivation* has been used in the corrosion literature to describe the phenomena he observed. Passivity is used in reference to a special condition that can be achieved only with transition metals and only when these are elevated to a fairly high anodic potential, either by means of an external electromotive force or by the means of an oxidizing agent in solution, such as the fuming nitric acid that Faraday used. Under these conditions, transition metals form a nearly "perfect" protective oxide film, and their corrosion rate drops several orders of magnitude (one order is usual, three orders is common, six orders has been demonstrated). Also since Faraday's experiment, it has been demonstrated that other oxidants besides nitric acid can form passivating films on the surface of iron (e.g., salts of chromate and nitrite). It has also been demonstrated that passivation can be accomplished by adjustment of the metal's electrical potential by means of an external electromotive force. More than that, it is now possible to make stainless steels, steels that are able to form such films in the presence of the oxidants present in natural environments.

The passivation phenomenon is illustrated in the Evans diagrams on Fig. 22-13 in which the different behavior of active and active–passive metals is contrasted. Ordinary, active metals exhibit the behavior shown on Fig. 22-13a where, as the oxidizing power of the environment is increased, the potential of the metal surface is increased and the metal corrodes more rapidly. More specifically, the corrosion current density increases logarithmically with an increase in the electrode potential. The logarithmic increase corresponds to the anodic reaction line in the mixed-potential diagrams shown earlier in this chapter.

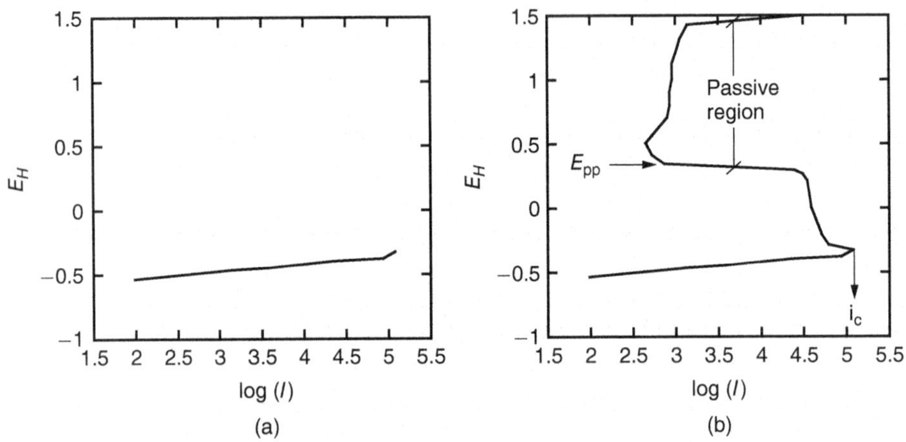

Figure 22-13
Polarization diagrams for different metals: (a) active and (b) active–passive.

With certain transition metals, however, once the potential of the metal surface exceeds a certain critical *passivation potential* (E_{pp}) (sometimes just called *critical potential*), the corrosion rate suddenly decreases. As potential is further increased, the corrosion rate remains low over a considerable range. In the zone where the corrosion rate remains low, the metal is described as passivated. The behavior shown on Fig. 22-13b is representative of the behavior of an active–passive metal such as iron. Once the potential of the electrode is driven past a certain critical potential (E_{pp}), the anodic current (corrosion rate) drops dramatically. Among materials used in water systems, iron and all its alloys are active–passive, whereas the others, copper, zinc, and lead, are active only. This behavior has important implications for corrosion control.

Passivation behavior is observed with transition metals such as iron, nickel, silicon, chromium, titanium, and their alloys. For a transition metal to become passivated, two conditions must be met. The potential of the metal must be raised above its E_{pp}, and the cathodic reaction must support a current density greater than the *critical current density*, i_{cc}. If the current density cannot reach this level, the cathodic reaction will never drive the reaction into the passive zone.

Although the subject of a great deal of research over the 16 decades since Faraday conducted his experiment, the exact nature of the passivating film and the mechanism by which such films protect the surface of metals on which they form is only just becoming understood. Most models of the passivation process envision the formation of a surface film, which is stable over a considerable range of potentials. In the case of iron, the passive layer appears to consist of an oxide of $Fe_{3-x}O_4$ with a spinal structure, varying in composition from Fe_3O_4 (magnetite), in oxygen-free solutions, to $Fe_{2.67}O_4$

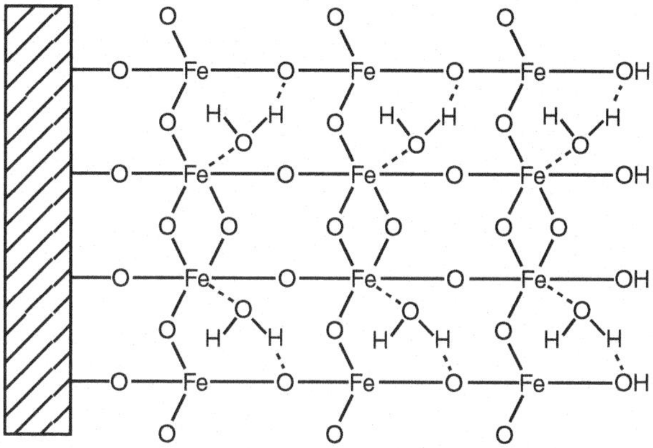

Figure 22-14
Formation of hydrated passive film on iron (Adapted from Pou et al., 1984).

in the presence of oxygen but which can, under certain conditions, transmit electrons from the metal surface to electron acceptors on the film surface (Newman, 2001). A schematic representation proposed by Bockris and collaborators (Pou et al., 1984) is shown on Fig. 22-14.

Stumm argues that it is the kinetics of the dissolution of this oxide film that controls the rate of corrosion. Perhaps more important, Stumm argues that dissolution also controls that rate of corrosion for the protective oxide films on copper and zinc as well (Grauer and Stumm, 1982; Stumm, 1990, 1998). If this view is correct, then understanding the phenomona that influence these kinetics may offer important insight into understanding corrosion kinetics as well. Moreover, passivity may just be an extreme case along a continuum of relative protection of metals by oxide films.

Stainless Steels

All the iron alloys used to transport drinking water exhibit active/passive behavior, though for many alloys, the conditions in drinking water rarely raise the surface of the metal to the potential where it can be passivated. Oxygen is the principal oxidant available in these systems, and, in water at room temperature, at equilibrium with oxygen in the atmosphere, the cathodic current is limited to about 100 $\mu A/cm^2$. For iron to be passivated in tap water at room temperature, its critical current (i_c on Fig. 22-13) must be less than 100 $\mu A/cm^2$. Based on experience, the critical current density for pure iron under near neutral conditions is on the order of 500 $\mu A/cm^2$ or more. As a result true passivation is not likely with a normally aerated solution.

Stainless steels have a composition designed to address this issue. There are two parts to the corrosion process, the cathodic process and the anodic process. If it is not possible to change the cathodic process, perhaps it is possible to change the anodic process. Changing the anodic process can be accomplished through alloying of the metal. Certain elements, when alloyed with iron, lower the critical current density of the metal (i_c). Alloying with chromium significantly reduces the passivation potential (E_{pp}) as well. Thus, the shape of the anodic polarization curve can be modified by using the alloying process.

The polarization curves for iron, 430 SS, 304L SS, and 316 SS in unagitated 1 N H_2SO_4 are displayed on Fig. 22-15. Also shown are the passivation potential and the critical current. All the stainless steels have a passivation potential substantially below that of iron. As a result they can be passivated in milder environments. On the other hand, the critical current required for passivation ranges from 104.3 μA/cm^2 in the case of 430 SS to as low as 10 μA/cm^2 in the case of 316 SS. As a result, only 316 SS is passivated in most aerated tap water, with 304L SS being passivated when mass transfer is optimized.

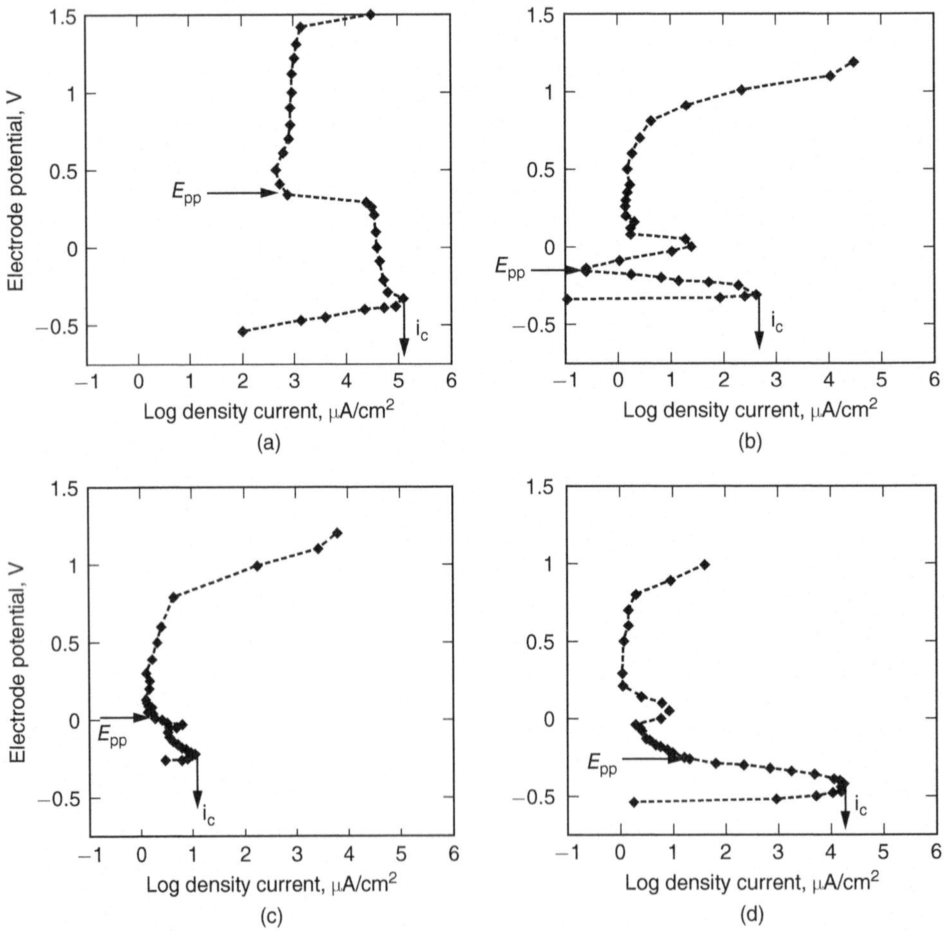

Figure 22-15
Polarization curves for iron and some common stainless steels in 1 N H_2SO_4: (a) iron, (b) 304L SS, (c) 316 SS, and (d) 430 SS.

In air-saturated solutions of 1 N sulfuric acid, with no agitation, the limiting current density for cathodic reduction of oxygen is about 100 μA/cm^2 and the reversible potential for this reaction is about –0.06 V. As a result both iron and 430 SS are vulnerable in such a solution because of their high critical current densities. Type 304L may be stable under some conditions and type 316 SS should be fairly resistant at any reasonable dissolved oxygen (DO) levels because its critical current density is substantially less than 100 μA/cm^2. These tests were all conducted in 1 N sulfuric acid. Consequently, the results do not directly translate to experience in domestic water supplies (lowering the pH significantly raises the required critical current by increasing the rate of dissolution of the passive layer). Nevertheless the behaviors illustrated on Fig. 22-15 demonstrate an important lesson about why stainless steels behave as they do.

Other Surface Scales

Thus, the protection afforded by surface oxides extends beyond the narrow definition of passivation alone. Understanding and controlling the formation and behavior of surface films and scales on metal water pipes is the key to controlling both corrosion and metals release. Virtually all these scales are protective to some degree. As a result, the corrosion rate of fresh metal coupons is almost always higher than the corrosion rate observed after a period of exposure. The scale found on water pipes can be thought of as organizing into three broad categories: passive films, thin scales, and thick scales (see Fig. 22-16). In drinking water systems thin and thick scales are the most common. The thin scales, which are tightly adherent, usually form in a short time, and they can often be quite protective, which is certainly the case with the oxide scale that forms on zinc and also on copper. It appears that it is also true of the carbonate and hydroxycarbonate scales that form on lead. All these materials usually achieve equilibrium with the water they are exposed to in a few weeks. The most common examples of thick scales occur on iron conduits. In many cases these scales grow to such great thickness that they seriously compromise the conduit's capacity to carry water. Obviously, such scales also continue to form over a long period of time. Under some circumstances, iron water conduits require more than a year or two before their corrosion rates begin to stabilize.

Earlier in the chapter it was noted that none of the metals used to construct water conduits are thermodynamically stable in water. Thus, it is logical that the scales that form on the surface of the metals make the metals last long enough to be useful for transporting water. Thus it appears that the minerals that form on the surfaces of these metals are so inert and insoluble that they can protect the surface of the metal. The solubility of the common oxidation states of iron, zinc, copper, and lead in a solution having a total dissolved inorganic carbon (DIC) content of 12 mg C/L (C_t = 1 mmol/L) over a wide range of pHs is shown on Fig. 22-17. Of all these metals, only the copper oxide, tenorite, and the ferric oxide have solubility low enough in the pH range of interest to support an argument

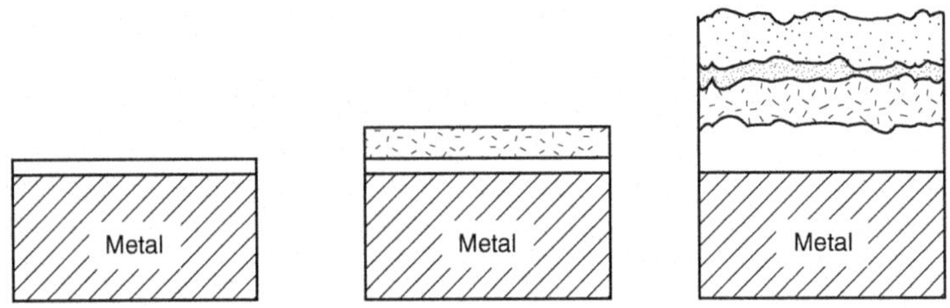

Film	Passive film	Thin scale	Thick scale
Typical thickness	< 10 nm	μm to mm	mm to several cm
Typical composition	Chromium oxides[a]	Mostly oxides, some hydrous oxides, some carbonates and hydroxycarbonates	Mostly oxides, some hydrous oxides, some carbonates and hydroxy-carbonates, often multiple oxidation states
Metals of interest in drinking water	Stainless steels[a]	Copper, galvanized iron, lead, cast iron and ductile iron	Iron and its alloys, including cast, ductile and galvanized iron
Some important characteristics	Semiconductive, selfhealing, tightly adherent, continuous coverage, anodic inhibitor	Often semiconductive, tightly adherent, usually continuous, usually cathodic inhibitors	Sometimes contains semi-conducting components, usually porous, often not adherent

[a] The film formed in Faraday's experiment is thought to be a spiral hydrous ferrous oxide (Figure 22-14) and it can form on most ferrous alloys used as drinking water conduits, but the types and concentrations of oxidants used in water systems are generally not sufficient to form a truly passive surface.

Figure 22-16
Types of surface scale that can form on metals.

that the reversible potential might be greatly depressed as a result of the low concentration of the oxidized species.

Neither of these oxides is thought to form a continuous, impervious film of the sort necessary to prevent corrosion, particularly, ferric oxide. Moreover, because iron has two oxidation states, it can form many complex oxides with different behaviors. Based on data presented on Fig. 22-17, it appears that, with the possible exception of copper, the solubility of these minerals provides limited insight into corrosion behavior. There is also some evidence that most surface scales exhibit much more limited solubility when they are in the form of a thin film on the surface of the metal. For example, copper, lead, and zinc are almost never observed at the levels suggested in these solubility diagrams in water transported by metal conduits, particularly zinc.

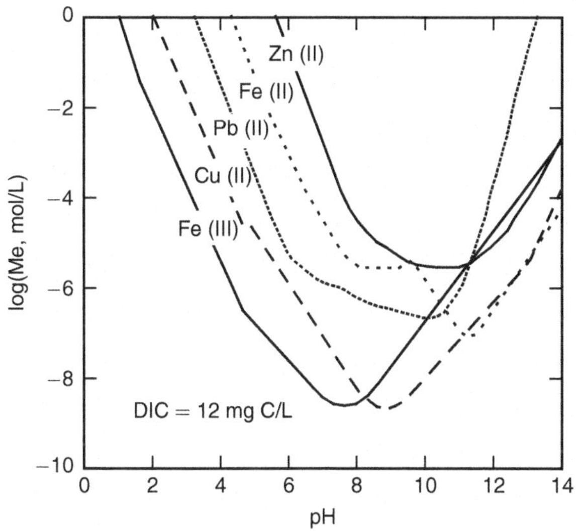

Figure 22-17
Solubility of principal corrosion cations.

Nevertheless, natural scales formed by the metals themselves are often important. These stable species are formed by the oxidation products from corrosion. These scales can serve to protect the metal itself from further corrosion by inhibiting transport to the metal surface of species necessary for the corrosion reaction. The most common example cited is the inhibition of the transport of electron acceptors, such as hydrogen ion, oxygen, and chlorine, but, as will be seen in later discussion, when tubercles are present, other forms of inhibition can also be important.

The semiconducting properties of some oxides are also thought to be important. Snoeyink and Wagner (1996) illustrate a model as to how semiconducting oxide scales could be one of the principal mechanisms of uniform corrosion. Others argue that semiconducting properties of oxides are an important mechanism of corrosion inhibition for zinc (Gilbert, 1948), copper (Ives and Rawson, 1962a– d), and iron (Stumm, 1995). Ideas common to all these models is that the cathodic reaction occurs at the oxide–water interface, whereas the corrosion reaction takes place at the metal–oxide interface; that electrons generated by the oxidation of the metal are transmitted through the semiconducting oxide layer. Those arguing inhibition suggest that the rate of the reaction is no longer controlled by activation polarization of the metal surface or by mass transfer of the electron acceptor, rather the rate is controlled by the diffusion of electron holes through the semiconductor or by the activities that must take place on the oxide surface. As mentioned earlier, Stumm (1998) argued that it is controlled by the dissolution kinetics of the oxide itself.

Example 22-7 Surface scale on lead–tin solder

Although a great deal of discussion continues regarding the mechanism of protection, there is a general consensus that these surface scales afford corrosion protection. In a study of the galvanic stimulation of corrosion on lead–tin solder joints in Seattle water, Reiber (1991) found that the nature of the corrosion reaction on the surface of lead–tin solder changed substantially with time, beginning with a corrosion potential (E_{corr}) of −345 mV, rising to −200 mV after 2 weeks of exposure and settling at +25 mV after 6 months. Using Fig. 22-11 as a guide, what sort of inhibition is occurring as the natural scale forms on the solder?

Solution

During the 6 months while the scale accumulated, E_{corr} increased from −345 mV to +25 mV and the corrosion rate dropped substantially. From Fig. 22-11 this behavior is consistent with an inhibitor that addresses (changes the slope of) the anodic reaction.

22-7 Common Forms of Corrosion

Up to this point, the discussion of corrosion has been approached broadly, viewing the process as being one where water conduits, made of metal materials that are not stable in water, react with the water and certain oxidizing agents to corrode or become oxidized. When corrosion occurs evenly across the metal surface, it is usually referred to as "uniform" corrosion, uniform in the sense that the metal surface is eaten away in a uniform manner. Corrosion of metal water conduits is universal, and uniform corrosion is the most common form, particularly where copper tubing is concerned. Uniform corrosion can also be a major contributor to the release of metal ions to the drinking water, an issue that became especially important in the last two decades of the twentieth century.

There are, however, other forms of corrosion that are also important, including (1) localized corrosion (pitting corrosion), (2) galvanic corrosion, and (3) concentration corrosion (crevice corrosion). All of these forms of corrosion are associated with local corrosion cells of a more permanent nature and all can result in important reductions in the useful life of the pipe.

Localized Corrosion

In the case of uniform corrosion as discussed above, it is generally assumed that anodic and cathodic sites move around on the metal's surface so that the entire surface is corroded at a more or less uniform rate. Another, particularly destructive form that corrosion can take is localized corrosion, or pitting. Pitting is a situation where one small part of the pipe surface becomes a permanent anode and much of the surrounding surface serves as the cathode. Formation of a pit in the interior surface of a pipe is accompanied by conditions that further the corrosion of the material at a specific site. Under pitting conditions the corrosion reaction rapidly eats its way through the pipe wall and may lead to pipe failure. Pitting failures in consumer plumbing typically occur in new pipe between 1 and 3 years after installation.

Several good reviews have been written on the subject of pitting (Jones, 1996; Szlarska-Smialowska, 1986). The principal factors in the pitting process are (1) the oxidizing potential of the solution, (2) the presence of aggressive ions (particularly chloride), and (3) the condition of the metal surface (films, scales, etc.). In the case of passivated iron surfaces Jones (1996) postulates the following:

1. In the absence of chloride, the passive film (represented as FeOOH) dissolves slowly, releasing ferric ions (see earlier discussion regarding Stumm model of passive film behavior):

$$FeOOH + H_2O \rightarrow Fe^{3+} + 3OH^- \tag{22-18}$$

2. When chloride is present, it substitutes for OH^- in the passive film, creating "salt islands" (represented below as FeOCl) in the outer layer of the passive film:

$$FeOOH + Cl^- \rightarrow FeOCl + OH^- \tag{22-19}$$

3. These salt islands presumably dissociate and lead to the liberation of Fe^{3+}:

$$FeOCl + H_2O \rightarrow Fe^{3+} + Cl^- + 2OH^- \tag{22-20}$$

The pitting process is autocatalytic in nature. Once a pit is started, rapid dissolution of metal at the site produces an excess of positive charge within the pit. The positive charge attracts negatively charged chloride ions. Because hydrochloric acid is highly ionized ($K_a \sim 10^6$), the presence of pits allows the hydrolysis of ferrous iron to drive the pH very low and the increase in acidity accelerates the corrosion rate. Pits normally expand in the direction of gravity because of the important role the dense, concentrated chloride solution plays in the pitting process. Oxygen reduction takes place on surfaces adjacent to the pit; however, the high salinity of the solution in the pit precludes significant reduction of oxygen there. High levels of chloride accelerate pit growth and seem to be associated with more frequent pit formation.

In addition to influencing the frequency and rate of growth of pits, aggressive ions, particularly chloride, also affect the induction time that precedes the initiation of the first pit. It has been shown (Engell and

Stolica, 1959; Szlarska-Smialowska, 1971) that, all else being equal, the induction time (T_i) can be predicted by the following relation:

$$T_i = \frac{1}{k(\text{Cl}^-)} \tag{22-21}$$

where T_i = induction time before first pits begin to appear, d
(Cl^-) = chloride concentration, mol/L
k = constant, d · L/mol

Unfortunately, for the waterworks industry, no values of the constant, k, are available for Eq. 22-21 for the conditions of importance in water supply. The equation is displayed because it suggests the principle that, for a given metal surface, the induction time before pitting is inversely related to the concentration of chloride ion.

Once a pit begins to form, its growth is often autocatalytic in nature. The sketch shown on Fig. 22-18 is modeled after the model Jones proposed for a pit on stainless steel (Jones, 1996), but it is analogous to the process of pitting in many other pits as well. Once the pit begins, metal ions released into the pit undergo hydrolysis, releasing hydrogen ions. These hydrogen ions create an electrostatic charge in the pit that attracts anions into the pit. The hydrolysis process also reduces the pH inside the pit, accelerating the anodic dissolution reaction. Certain anions, referred to as aggressive anions, are known to exacerbate the pitting process. Chloride is the most notorious of these because of its high mobility and because hydrochloric acid is highly ionized ($K_a \sim 10^6$ compared to $\sim 10^3$ for sulfuric acid). In the case of iron, as the ferrous hydroxide formed diffuses toward the exit from the pit, it meets the oxygen in the bulk solution and is further oxidized to ferric hydroxide. The ferric hydroxide forms a corrosion product cap on the top of the pit that is porous in nature. The shell blocks the transfer of Fe^{2+} but is porous enough so that chloride ions can migrate easily into the

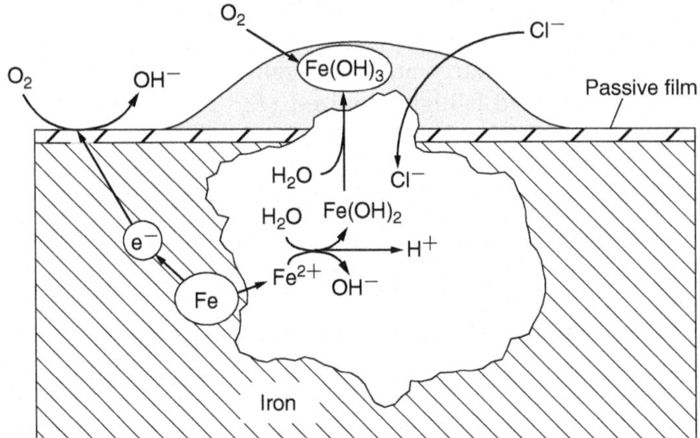

Figure 22-18
Formation of autocatalytic pit in iron pipe.

pit. The diffusion of chloride is necessary to complete the solution circuit as demonstrated on Fig. 22-4.

It is important to understand that a cathodic reaction is also required to support pit growth. As a result, pits are often widely spaced high-salinity solutions because oxygen has limited solubility and a large cathodic area is required to support each pit. Formation of new pits in the vicinity of an active existing pit will be suppressed by "cathodic protection," that is, by the availability of a surplus of electrons in the pit's immediate vicinity.

Corrosion Due to Potential Differences

There are two types of corrosion that occur as a result of potential differences: (1) electrolytic corrosion and (2) galvanic corrosion. Often these two types of corrosion are confused. Electrolytic corrosion occurs when an external emf is imposed on the water conduit, and galvanic corrosion occurs when plumbing materials made of dissimilar metals are coupled to each other. It is important to understand that although a very small insulating coupling can eliminate galvanic corrosion, the same coupling can aggravate electrolytic corrosion, causing the corrosion to occur at new locations.

ELECTROLYTIC CORROSION

Whenever an electrical current travels along a water conduit, corrosion occurs. As mentioned earlier, according to the convention established by Benjamin Franklin, electrical current and electrons flow in opposite directions (Franklin, 1769). As a result, the point where corrosion will occur corresponds to the point where the electrical current exits the pipe (compare Figs. 22-4 and 22-19). External corrosion occurs when the current exits the outside of the pipe (usually in the ground), but internal corrosion can also occur when the current exits the inside of the pipe to jump an insulating coupling. The ordinary insulating couplings used to prevent galvanic corrosion are often not adequate to address problems of electrolytic corrosion. The length of the insulating coupling required to control a problem of electrolytic corrosion is greater for water supplies with higher conductivities (Sutherland and Tekippe, 1972).

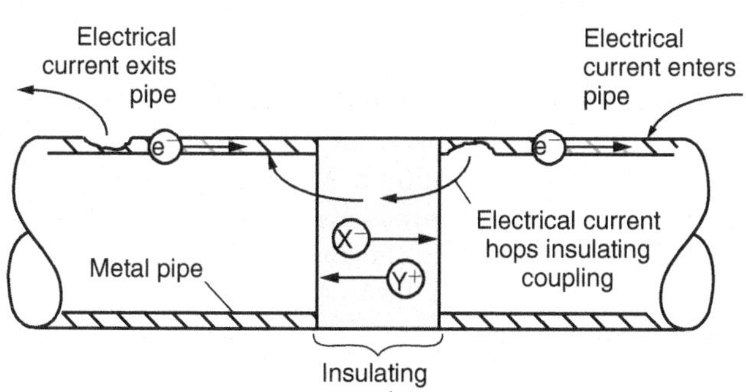

Figure 22-19
Definition sketch for electrolytic corrosion. The circled ions complete the circuit (see also Fig. 22-4).

GALVANIC CORROSION

Galvanic corrosion occurs when two different metals or metal alloys are placed in contact with each other in a water system. The most common examples are probably brass fittings coupled to galvanized pipe and solder in contact with copper tubing. Occasionally, insulating couplings are used to separate such materials and prevent galvanic corrosion. Whenever a system for water conveyance is designed with different metals or different alloys, each of these materials will have a different corrosion potential than the other and a corrosion cell will result. The severity of the corrosion that results in these circumstances depends on many things, but foremost among them are (1) the difference in the independent corrosion potential of the two metals or alloys, (2) the difference in the relative areas of the two metals, and (3) the conductivity of the water itself. The geometry and proximity of the two dissimilar surfaces is also important, though rigorous consideration of these effects requires the use of field theory and is beyond the resources of most water utility staff and their consultants as well.

Differences in corrosion potential

From the earlier sections of this chapter, it is clear that the corrosion potential of each metal in a particular water quality depends not only on that metal's place in the redox potentials displayed in the table of standard reduction–oxidation potentials (Table 22-2) but also on other factors that influence all the parameters of the mixed-potential phenomena summarized in Table 22-5. Important among these is the nature of the scale that forms on that metal in the particular water. Once these potentials have been determined, a list can be prepared, ranking these materials with respect to their relative corrosion potential. Such tables are referred to typically as the galvanic series.

A galvanic series prepared for seawater is provided in Table 22-7. This particular series has been widely studied and is fairly practical because seawater has a similar composition throughout the world. The information in Table 22-7 is not accurate for waters of the composition of most domestic drinking waters. In fact, no universal table exists for drinking water use, and it is unlikely there will ever be one because the series is likely to be somewhat unique for each individual water quality.

Unfortunately, most utilities are not aware of the significance of these data, and standard procedures for making these measurements are not available. On the other hand, the standard redox potentials are much of what underlies the galvanic series, and many of the other variables are not a great deal different from one site to another. Also the metals grouped together in Table 22-7 generally have potentials that are different from the potentials in the groups above and below. In the absence of a local galvanic series, the rankings for seawater found in Table 22-7 are often used as a place to start. Water utilities would be wise to conduct testing to determine the corrosion potential of metals commonly used with the water they serve.

Table 22-7
Galvanic series of metals and alloys in seawater[a]

Anodic (least noble, most likely to corrode)
Magnesium
Magnesium alloys
Zinc
Aluminum
Cadmium
Steel or iron
Cast iron
Iron alloys
Lead–tin solders
Lead
Tin
Nickel
Brasses
Copper
Bronzes
Titanium
Monel
Silver solder
Silver
Carbon (graphite)
Gold
Cathodic (most noble, least likely to corrode)

[a]Metals within each of the groupings have similar corrosion potentials (Jones, 1996).

Information of this kind would prove valuable to the utility as well as to other corrosion engineers working on the design of systems that use that drinking water.

Differences in area

The relative area of the two dissimilar surfaces is also important to corrosion behavior as well as whether most of the area is in the form of the cathodic surface or the anodic surface. Consider a brass fitting coupled to galvanized pipe. These two metals have considerably different corrosion potentials, the brass being the more noble of the two. The more noble metal is the metal with the higher potential and, hence, the one that will become the cathode. Moreover, connecting the brass to the galvanized pipe will significantly reduce the corrosion potential of the brass, accelerating the cathodic processes there. On the other hand, the area of the brass is much smaller than the area of the galvanized iron connected to it. As a result, the cathodic processes promoted by the brass generally result in a small acceleration of anodic (corrosion) processes along a considerable length of galvanized pipe.

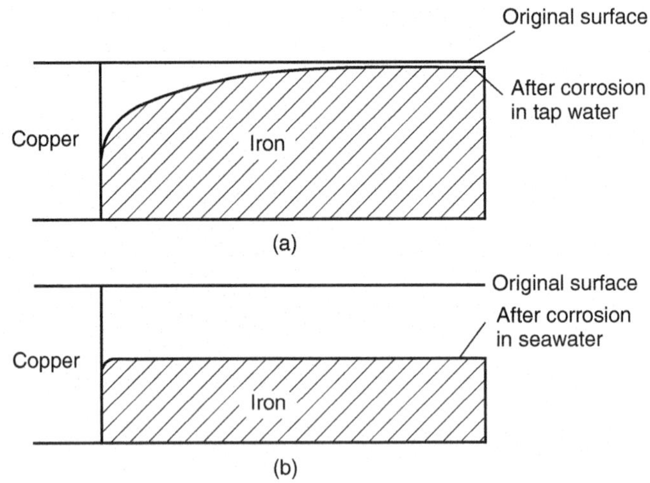

Figure 22-20
Effect of solution conductivity on nature of galvanic corrosion (Adapted from Uhlig, 1948): (a) corrosion of iron/copper galvanic cell in low-conductivity tap water and (b) corrosion of iron/copper galvanic cell in high-conductivity seawater.

Conductivity of water
Materials in waters with higher conductivity gain more benefit from the effects just discussed than those with lower conductivity. The extremes are best illustrated by comparing galvanic corrosion in the presence of a typical tap water (low conductivity) with galvanic corrosion in the presence of a typical seawater. The influence of conductivity on the behavior of corrosion near galvanic couples is illustrated with these two extremes on Fig. 22-20. With a low-conductivity tap water, the influence of the galvanic cell is intense but only in the immediate vicinity of the two metals. In seawater, with very high conductivity, the effect of the galvanic cell is spread widely across the surface of the less noble metal.

22-8 Metals of Interest in Domestic Drinking Water Systems

The metals of most interest in water systems are cast iron and ductile iron, galvanized iron, copper tubing, lead pipe, solders, and brass fittings. Each of these are addressed in the following discussion.

Iron Pipe

Because of their low cost and a history of relatively good service, iron alloys have been the principal component of modern water systems. Today most new installations are (or should be) provided with a mortar lining. In this event, understanding corrosion becomes a question of understanding the behavior of the mortar lining. But iron, particularly cast iron and galvanized iron, has been around for a long time and, as a result, unlined iron alloy piping represents a significant part of the conduit in almost every active water system. There are a lot of problems that can develop with this iron pipe, particularly loss of carrying capacity and red water. As a result, it is important to discuss iron and its alloys.

The reaction chemistry of domestic waters is extremely complex; attempting to characterize the effect in the immediate vicinity of a corroding iron surface is even more difficult. Long-term corrosion tests conducted in extremely soft waters show an increasing rate of corrosion with increasing pH (Larson and Skold, 1958a,b; Skold and Larson, 1957). Stumm (1960) made similar observations concerning the corrosion of iron in waters of various hardnesses. Both Stumm and Larson and Skold documented the effect of pH on the character of a corroding cast-iron surface, noting uniform corrosion at lower pH, and increased unevenness and tuberculation at higher pH (Larson and Skold, 1958a; Stumm, 1960). The photos taken by Werner Stumm in the late 1950s (see Fig. 22-21) illustrate the difference between corrosion behavior of cast-iron surfaces as a function of pH. Larson and Skold (1958a) produced similar photographs and made similar observations. Below pH 7, corrosion was found to be relatively uniform. Above pH 8, corrosion becomes increasingly more localized and at pH 9.5 serious tuberculation was observed. Stumm argued that it is likely that the change in the nature of the corroding surface is due in part to the influence of pH on the charge of ferric hydroxide micelles formed during the corrosion process. The isoelectric point of ferric hydroxide is slightly above a neutral pH. Thus at low pH, the micelles have a positive charge and will migrate toward the cathode. The change in charge reduces the rate of mass transport of other ions to the cathode surface and alters the distribution of anodic and cathodic areas. At more alkaline pH the $Fe(OH)_3$ micelles take on a more negative charge, causing them to remain at the anode; their presence at the anode increases the potential difference between the cathode and the anode, increases the rate of corrosion, and increases the heterogeneity of the corroding surface. Another argument can also be

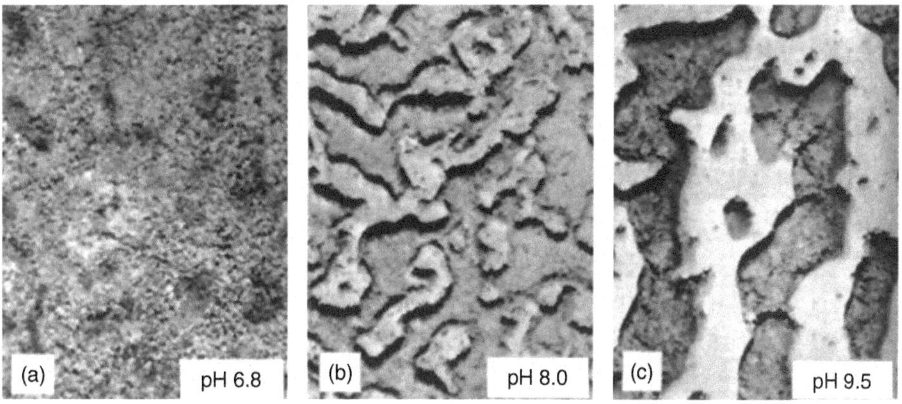

Figure 22-21
Effect of pH on uniformity of corrosion on cast iron: (a) pH 6.8, (b) pH 8.0, and (c) pH 9.5. (Adapted from Stumm, 1960.)

made that at high pH ferrous corrosion products are more likely to precipitate and stay at the anode, whereas at low pH, they would tend to remain dissolved and disperse. Similar arguments can be made regarding the rate of oxidation from Fe(II) to Fe(III).

Galvanized Pipe

Galvanized pipe is an old and common material for plumbing used to transport domestic drinking water. For most of the twentieth century, it was the dominant plumbing material in the United States and Europe. However, since 1970, copper tubing has gradually displaced galvanized pipe as the leading material for domestic plumbing in most developed nations. The use of copper pipe for domestic plumbing is driven by the rising cost of labor, higher consumer standards for reliability, and more liberal standards on the thickness of copper tubing. Galvanized pipe remains the dominant plumbing material in much of the developing world.

To understand the corrosion of galvanized plumbing it is useful to understand how it is made and how a proper galvanized layer should appear. The structure of the alloy layers that are to be found on a normal galvanized surface is given on Fig. 22-22. This photomicrograph shows the five layers in a proper galvanized surface as identified in ASTM A385. According to this standard, a hot-dip zinc coating should show the following separate metallurgical layers: (1) the base steel, (2) the γ layer at the steel-coating interface, (3) the δ layer, which is iron-rich zinc, (4) the ζ layer, which can be characterized as the iron–zinc diffusion zone, and (5) the η layer, which is virtually pure zinc in composition. Impurities, such as cadmium, are usually at their highest level in the η layer. Common problems are in contamination of the interface between the γ layer and the base steel due to poor preparation and "galvanealing" or diffusion of iron too deeply in the zinc coating as a result of running the dip at too high a temperature or providing too much time at high temperatures.

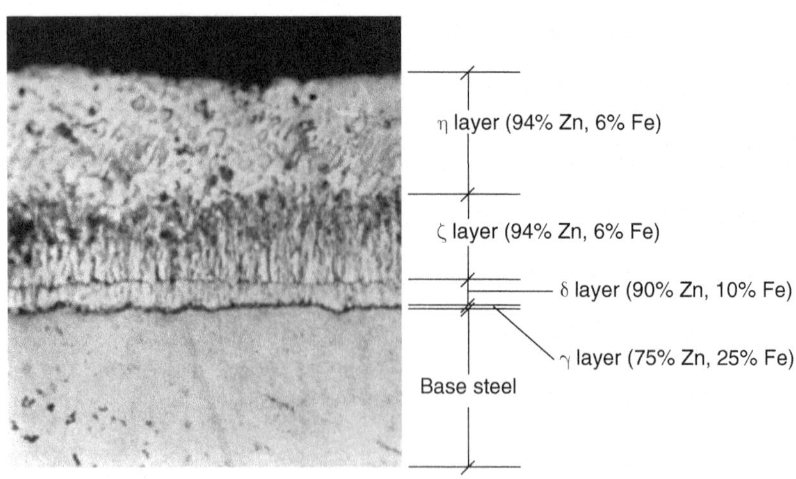

Figure 22-22
Metallurgical layers in galvanized surface.

When galvanized pipe is to be used, its quality should be examined using a qualitative visual examination as well as tests of coating thickness, coating uniformity, and microscopic examination of the layers shown on Fig. 22-22. Details of such examination are given in the AWWARF book on internal corrosion (Trussell and Wagner, 1996). A normal galvanized coating should have an average weight of 550 g of zinc per m^2 of pipe surface. An abbreviated description of the pipe manufacturing process is also given in Trussell and Wagner (1996). A more detailed description can be found in the earlier edition of the same reference (Trussell and Wagner, 1985). Manufacturing defects can be an important cause of a shortened lifetime for galvanized pipe, particularly inadequacies in welding or in the galvanizing process itself (Trussell and Wagner, 1996).

Galvanized pipe is not always that much better in service as compared to iron pipe. Once the zinc layer has corroded away, galvanized pipe actually does become iron pipe, and, of course, it then behaves as if it were. As a result, galvanized pipe gives its best service in regions where the zinc layer, itself, is stable. Generally, this good service is obtained in hard water with a pH between 7.5 and 8.5 and low chloride levels.

The method of protection provided by galvanized pipe is one of its important assets. The behavior near an imperfection in the tin layer in tin-plated steel with the behavior near an imperfection in the zinc layer in galvanized steel are illustrated on Fig. 22-23. Whereas tin is more noble than the steel, zinc is less so. Because of this, imperfections in the zinc coating are not of great significance. The nature of the protection of the galvanized layer allows for service to continue as the zinc layer is corroded away. Depending on the quality of the water, a substantial amount of the zinc layer can be corroded away before the galvanizing ceases to have its protective effect. Imperfections in tin plating can result in rapid failure by pitting. As will be seen later in the chapter, one of the more important causes of failure of galvanized pipe results from conditions where the potential between the zinc and steel layers becomes reversed.

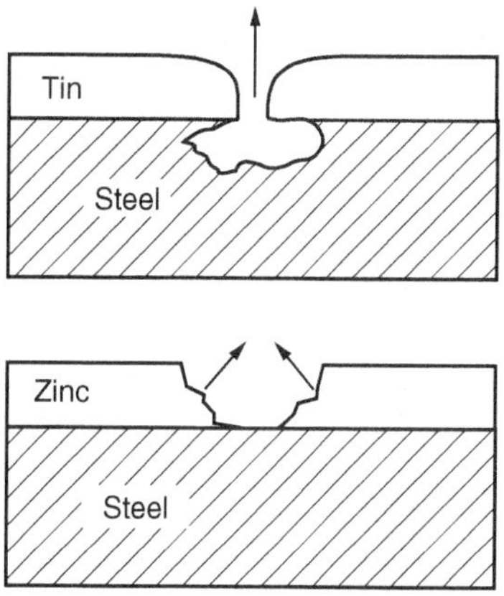

Figure 22-23
Illustration of difference between protection provided between zinc and tin coating on iron.

Returning to the corrosion of galvanized pipe, the transition from zinc to iron occurs, the behavior of the pipe can change materially, tuberculation is often observed, and the pipe's carrying capacity becomes reduced. Tuberculation refers to the formation of large, thick protrusions of oxide product on the inside of the corroding pipe, a phenomenon that is much more pronounced with the corrosion of iron. Some time after tuberculation develops fully, the consumer is often motivated to replace the pipe. The time sequence of the corrosion process in galvanized pipe is demonstrated on Fig. 22-24 in which the results of work by Bächle et al. (1981) are summarized. Bächle and co-workers examined the nature of the corrosion scale on the surface of galvanized pipe before and

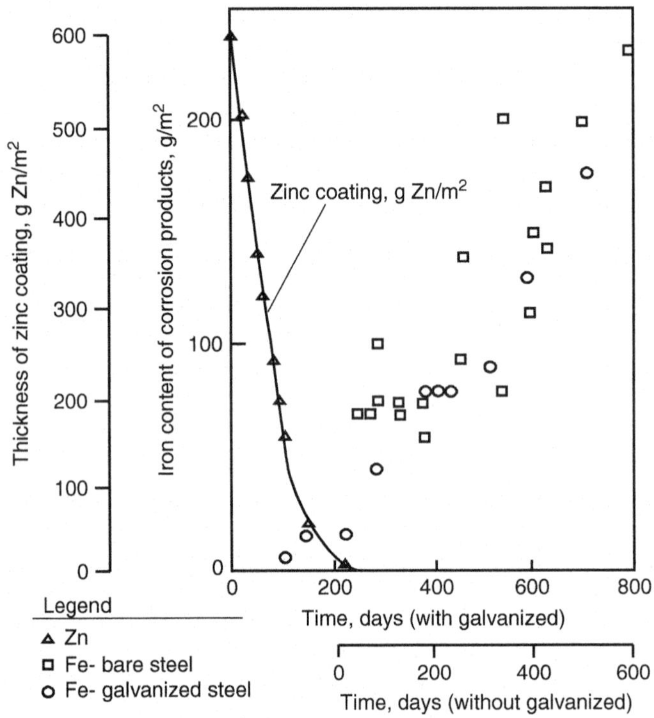

Figure 22-24
Loss of zinc layer. The corrosion of a galvanized pipe is compared with the corrosion of a bare steel pipe. For the galvanized pipe, for the first 200 days the zinc disappears (triangles) and then the iron corrodes (circles). The bare steel pipe, started 200 days into the test, corrodes the same (squares) (Adapted from Bächle et al., 1981)

after the loss of the zinc layer and compared it to the behavior of steel pipe that had not been galvanized.

Bächle et al., (1981) studied the deposits on the surface of iron pipe and zinc pipe corroding in a local groundwater. In this particular case, the water was rather aggressive to zinc, and the zinc coating was almost entirely corroded away in less than a year. Here the iron content of the scale formed by the corrosion products is displayed in a way that demonstrates that the galvanized pipe behaves much like bare iron pipe once the zinc has been removed. As noted on Fig. 22-24, a rapid development of scale mass occurs once this process begins. Because iron scales are usually voluminous, this development often results in failure of the pipe to transport an adequate volume of water.

The single most important water quality parameter influencing the stability of the zinc coating on galvanized pipe is probably the pH. Work on the effect of pH has been conducted by Sontheimer and associates (Bächle et al., 1981; Kruse, 1983; Pisigan and Singley, 1985; Werner et al., 1973). From the results of Bächle et. al (1981) (see Fig. 22-25), it can be concluded that the corrosion behavior of galvanized pipe consistently improves with increasing pH. Bächle et al., studied the effect of pH on the corrosion of galvanized pipe in seven different water supplies. The issue of pH is more thoroughly discussed by Trussell and Wagner (1996).

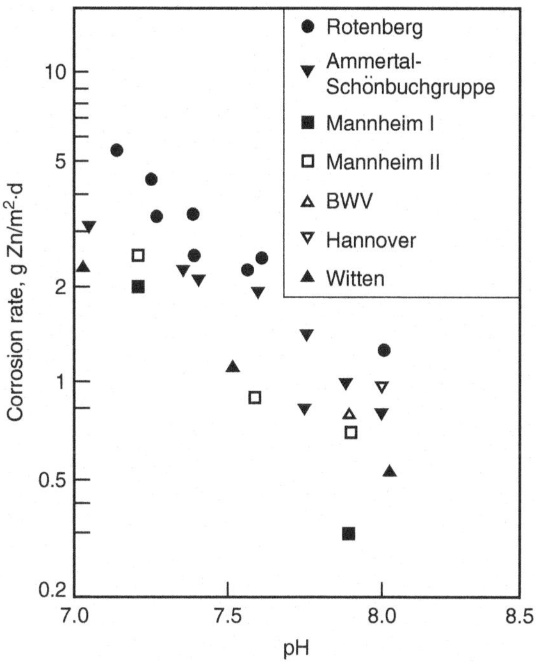

Figure 22-25
Corrosion of galvanized pipe as function of pH. (Adapted from Bächle et al., 1981.)

Trussell and Wagner (1996) also conducted a thorough examination of the chemistry of zinc ion and found that, under the conditions of importance in drinking water systems, zinc is typically coated with hydrozincite [$Zn_5(OH)_6(CO_3)_2$], zinc oxide (ZnO), and zinc hydroxide [$Zn(OH)_2$]. Occasionally, some calcium carbonate has also been observed on the zinc surface, usually in the form of aragonite, a form having slightly greater solubility than calcite. The solubility of zinc ion seems to be controlled by the oxide and the hydrozincite; but, as shown on Fig. 22-17, these zinc minerals do not have exceptionally low solubility. As a result, it is likely that there are conditions near the pipe surface that influence behavior to keep zinc levels in the bulk water matrix below those expected by consideration of mineral solubility.

Where satisfactory service is concerned, pitting is the mode of corrosion that has the most dramatic impact reducing the useful life of the pipe. There are three causes of pitting that are important to galvanized pipe: (1) pitting associated with manufacturing defects (discussed earlier), (2) pitting associated with copper in the water, and (3) pitting resulting from potential reversal between the steel and zinc.

PITTING ASSOCIATED WITH COPPER IN WATER

McKee (1932) reported that the use of copper algacides resulted in the pitting of galvanized pipe in cooling water equipment. Since that time, most incidents of pitting due to copper have been associated with recirculating water systems (Kenworthy, 1943), but others have reported copper-induced pitting of galvanized pipe with copper from a variety of sources (Cruse, 1971; Treweek et al., 1978). The principle involved is that copper, being the more noble metal, plates out over much of the surface of the zinc, reducing the anodic zones on the surface of the pipe and increasing the risk of pitting. On the other hand, the Metropolitan Water District of Southern California sponsored extensive studies on the role of copper on the pitting of galvanized pipe (Grasha et al., 1981; Treweek et al., 1978; Fox et al., 1983), and these studies came up with equivocal answers, researchers being unable to duplicate copper-induced pitting after extensive efforts. Corrosion was accelerated, but pitting was not evident.

Based on available evidence, it appears that copper-induced pitting of galvanized pipe is a real phenomenon, but the circumstances when it will and will not occur are not understood adequately. Perhaps the presence of copper can exacerbate the conditions associated with the potential reversal phenomenon discussed below.

PITTING ASSOCIATED WITH POTENTIAL REVERSAL

Under certain conditions, the electrochemical potential between zinc and iron can be reversed, so that iron becomes the sacrificial metal in place of zinc. Schikorr was the first to publish clear evidence of potential reversal (1939). Schikorr connected iron and zinc electrodes and then watched their relative potential in a hot water bath maintained between 50 and 80°C. Potential reversal was observed after 4 h of contact time (Fig. 22-26). Beyond increased temperatures, bicarbonate concentration and/or nitrate concentration have also been noted as possible causes of potential reversal, while increased chloride, sulfate, calcium, and silicate levels will mitigate against potential reversal (Hoxeng, 1950; Hoxeng and Prutton, 1949). Perhaps the most thoughtful

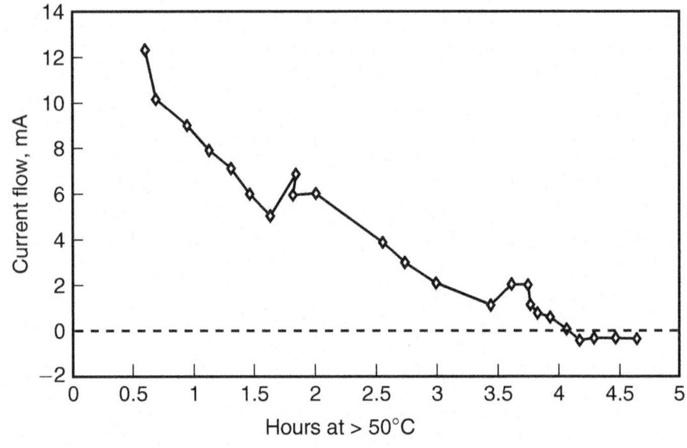

Figure 22-26
Shikorr experiment. After 4 h in hot water, the current flow between iron and zinc electrodes is reversed. (Data from Schikorr, 1939.)

examination of the subject of potential reversal between iron and zinc was conducted by Gilbert (1948). Gilbert hypothesized that, at elevated temperatures (and perhaps elevated pH values), scale on the zinc surface is transformed from the zinc hydroxide normally present to zinc oxide. This zinc oxide is more effective in "healing" anodic sites (reducing their area) while at the same time, as a semiconductor, it is also effective in depolarizing cathodic sites (reducing β_c in Eq. 22-4). The depolarization of the cathodic sites reduces the potential of the corroding zinc with respect to the solution, thus reducing the overall rate of corrosion, but accelerating the rate of corrosion at the few anodic sites remaining.

TREATMENT TO REDUCE CORROSION OF GALVANIZED PIPE

Much can be done in the design and installation of galvanized plumbing systems to reduce corrosion. Specifically, testing should be done to assure the quality of installed pipe and appropriate design and construction of galvanized systems. Once the system has been built, there are three approaches to water treatment that have been demonstrated to be successful under specific circumstances: pH adjustment, silicate addition, and the addition of orthophosphate. Adjustment of calcium carbonate saturation is also often advocated, but its role in controlling the corrosion of galvanized pipe has not been demonstrated as clearly.

Silicates have long been advocated for controlling corrosion in galvanized systems (Cox, 1934; Speller, 1926; Thresh, 1922); however, modern research has focused on their application in circumstances where pitting due to potential reversal in hot water systems is the principal issue (Lane et al., 1973, 1977; Lehrman and Schuldener, 1951, 1952). While it has been demonstrated in this more recent work that silicates can be successful in preventing the pitting of galvanized pipe, the dose of silicate typically required for successful treatment is rather high (8 to 25 mg/L as SiO_2). As a result the use of this treatment is limited to industrial applications.

Murray (1971), working for the City of Long Beach, California, was the first to report successfully employing orthophosphates to control the corrosion of galvanized pipe. Murray's inhibitor actually included zinc sulfate, orthophosphate, and sulfamic acid. Since that time, most studies examining Murray's formulation, or simpler mixtures without sulfamic acid, have focused on controlling the corrosion of iron pipe (Kelly et al., 1978; Mullen, 1974). Later Nancollas (1983) proposed a theoretical basis for inhibition of corrosion of zinc surfaces by orthophosphate.

Since that time, German researchers have demonstrated that orthophosphates alone can successfully control the corrosion of both iron and galvanized pipe. Werner did extensive work examining the influence of orthophosphate addition on the corrosion of galvanized pipe (Trussell and Wagner, 1985). Werner found that the corrosion rate is inversely proportional to the orthophosphate dose and to the ratio of the hydrogen ion concentration. The results from the Werner studies are reported on

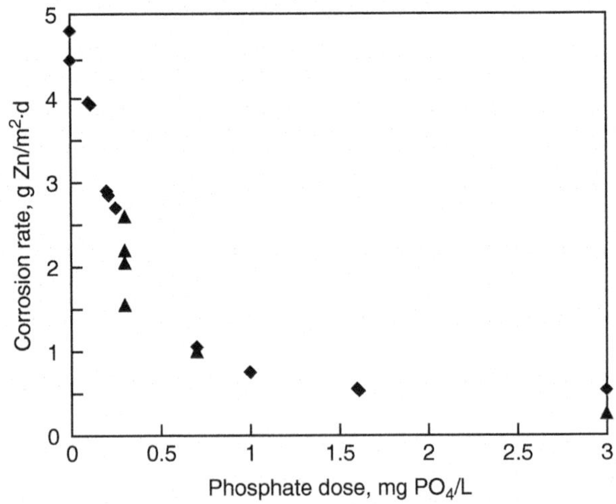

Figure 22-27
Effect of orthophosphate on corrosion of galvanized pipe. (Data from Trussell and Wagner, 1985.)

Fig. 22-27. Examination of the figures confirms that the corrosion rate is inversely proportional to the orthophosphate dose. The technique of phosphate addition has not been shown to be as successful in waters with very low pHs or very low alkalinities.

Copper Tubing

Copper tubing in water systems is subject to three forms of corrosion: (1) general corrosion, (2) impingement attack, and (3) pitting attack (Cohen and Lyman, 1972; Hatch, 1961). General corrosive attack is rare and generally occurs at low rates so that the service life of the tubing is not reduced significantly. Impingement and pitting attack can be both rapid and severe.

UNIFORM CORROSION

General corrosive attack of copper, which can produce "green" or "blue" water if the copper concentration is particularly high, is most often associated with soft, acid waters. General corrosion proceeds at a low rate and is characterized by a gradual buildup of corrosion products, generally basic cupric salts.

As described earlier, the oxide scale on the surface of copper tubing is understood to play an important role in the corrosion process. Evidence identified by Ives and Rawson (1962a–d) suggests that, when copper is immersed in water saturated with oxygen, the copper is rapidly oxidized to form a compact, adherent film of cuprite (cuprous oxide) and that this cuprite, a semiconductor, is in good electrical contact with the metal itself. When uniform corrosion occurs, the underlying copper is oxidized and the oxide formed remains in immediate contact with the metal's surface. Oxygen is reduced to water at the exterior of this tight layer of cuprite. Ives and Rawson have argued that the tight adherent layer can only reach a certain

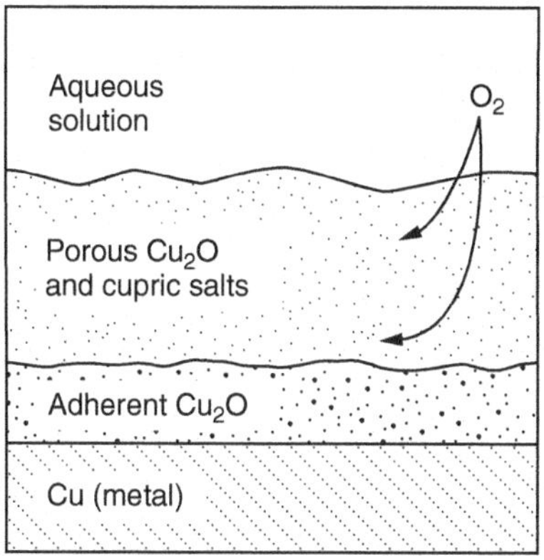

Figure 22-28
Ives–Rawson duplex film model for copper (adapted from Ives and Rawson, 1962c). The Ives–Rawson duplex film model for copper corrosion. Copper corrodes to form a compact, adherent film of cuprite (CuO), a semiconductor, that remains in electrical contact with the base metal. Over time, a more porous outer layer of cupric salts also forms. Oxygen from the aqueous solution can penetrate the outer layer, but when it reaches the cuprite, it does not penetrate but is reduced to water at the surface via electrons transported from the surface of the corroding base metal.

thickness and still maintain its structural integrity. Beyond that thickness, a second, outer oxide layer also develops (see Fig. 22-28). This outer layer becomes a mixture of cuprous and cupric salts, and the difference in the coordination requirements of these two oxidation states of copper results in the formation of a porous scale layer. The formation of the two layers might also be a result of two processes, oxidation of the metal at the Cu–CuO interface and penetration of oxygen to form the CuO–Cu_2O interface. In any case, there is wide agreement that the cuprite scale (CuO) is tight and continuous and that the tenorite scale (Cu_2O) is not.

The most important factors influencing general corrosion are (1) pH, (2) hardness, (3) temperature, (4) age of the pipe, and (5) oxygen. As with most metal corrosion, low pH waters attack the protective oxide layer. Waters with pH values lower than 6.5 are aggressive to copper. Corrosive attack is more severe in hot water systems. Obrecht and Quill (1960, 1961) report that temperatures exceeding 60°C (140°F) cause an increased rate of corrosive attack. Most serious cases of corrosion in copper tubing occur in new installations where the protective oxide layer has not yet formed.

The impacts of general corrosion are mostly of a nuisance nature. Green and blue water are caused by dispersion of copper corrosion products into the water. A related problem is blue or green staining of plumbing fixtures. Such water often exhibits a rather unpleasant taste due to high concentrations of dissolved copper. Increased copper concentrations in water and sludge discharged from wastewater plants is also an issue. Uniform corrosion can be controlled by raising the pH to 7.5.

IMPINGEMENT ATTACK

Impingement attack is the result of excessive flow velocities [greater than 1.25 m/s (4 ft/s)] and was at one time thought to be purely mechanical in nature (Hatch, 1961). It is now believed that high velocities disrupt formation of protective films, allowing electrochemical attack to proceed more rapidly. Factors besides velocity that aggravate impingement attack are soft water, high temperature, and low pH. For hard, cold water, velocities of 2.5 m/s (8 ft/s) can be tolerated. Soft waters, temperatures above 60°C, and pH below 6.5 all contribute to destruction of the protective film.

Impingement attack is characterized by a rough surface, often accompanied by horseshoe or U-shaped pits. In severe cases, perforation of the tube wall occurs in as little as 6 months. It is most severe at points of turbulence, such as sites downstream of fittings, and is most prevalent in recirculating systems.

PITTING ATTACK

Pitting of copper tubing takes many forms and has many causes. Edwards et al. (1994) summarized the causes, characteristics, and remedies for the most common forms of copper corrosion, including pitting. The pitting of copper tubing is most commonly associated with hard well waters and most often occurs in cold water piping. Usually, dissolved carbon dioxide exceeds 5 mg/L and dissolved oxygen is 10 to 12 mg/L or more (Cohen and Lyman, 1972; Rambow and Holmgren, 1965). The water quality parameters typical for water systems having copper pitting problems are not the typical soft, low-pH waters normally associated with corrosion of copper. Surface waters containing organic or humic substances are not associated with pitting attack (Campbell, 1954; Lucey, 1967). In England, the presence of a carbon film due to residues of drawing lubricants from manufacturing processes was found to exacerbate problems of cold water pitting in copper pipe (Campbell, 1950).

Pitting occurs most often in horizontal runs of piping with the deepest pits concentrated in the bottom of the pipe. It appears that gravity holds dense solutions of copper salts in the pit, sustaining the corrosion reaction (Cruse and Pomeroy, 1974). Horizontal surfaces are more vulnerable to attack (Lucey, 1967), suggesting that corrosion products stream from the vertical surface and concentrate gravitationally.

Pitting attack is most common in new installations, with 80 to 90 percent of the reported failures occurring within the first 2 to 3 years. In some extreme cases, failure occurs in as little as 3 months (Cruse and Pomeroy, 1974). Pitting occurs in soft, annealed tubing rather than in hard-drawn tubing (Ferguson et al., 1996). Typically, hard copper tubing (types K, L, and M) is used inside structures and soft copper tubing underground and inside slabs. As the name implies, soft copper tubing is more malleable and hard copper tubing is more rigid. According to Cohen and Lyman (1972), the relative frequency of failure does not appear to vary among types of tubing, even though type M has a thinner wall. If unfavorable water quality conditions occur before the protective coat has formed, then serious pitting attack may occur. After 3 or 4 years, the incidence of failure drops significantly even in systems having serious incidents of pitting.

In the diagram on Fig. 22-29, developed originally by Lucey (1967), the profile of a typical cold water pit in copper tubing is illustrated. Lucey

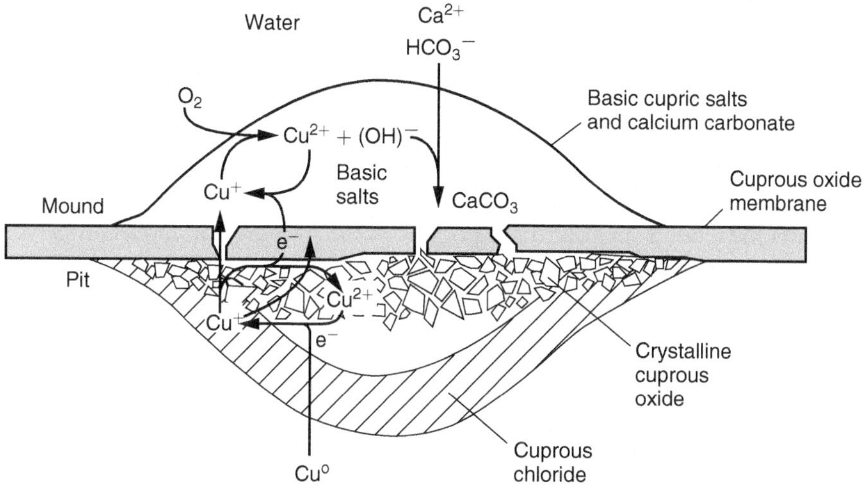

Figure 22-29
Profile of a cold-water pit in copper tubing (Adapted from Lucey, 1967.)

has proposed a mechanism for copper pitting in which the formation of a membrane cell is assumed, as shown on Fig. 22-29. The cuprous oxide membrane covers the pit cavity, which contains cuprous chloride and cuprous oxide crystals. The cuprous oxide membrane acts as a bipolar electrode with oxidation taking place on the inside face and reduction taking place on the outside face. There are reactions simultaneously with the calcium bicarbonate in the waters, which result in the precipitation of calcium carbonate and basic cupric salts such as malachite [$CuCO_3Cu(OH)_2$] in a mound above the membrane.

Lead Tubing

Pipes made of lead (*plumbum*) were first used for plumbing by the Romans. Lead is ductile and can be bent easily into desired shapes. Use of lead in plumbing declined beginning in the nineteenth century as iron became competitive as a pipe material. Later, health concerns about lead poisoning from corrosion by-products in drinking water further reduced the use of lead in water conduits, although household lead plumbing is still in place in some older cities. Lead "pigtails" and "goosenecks" were also in use as an easily deformable connection between distribution mains and household connections. Relatively little is known about corrosion of lead except that its corrosion rate is relatively low. The primary issue with lead has not been that it corrodes but rather that lead released to the water being carried in the conduit can be toxic. The release of lead to the solution in the pipe is discussed later in the chapter.

Lead Solder

Throughout the developed world, solders used to bond copper fittings together had long been lead–tin alloys. In the United States a 50 percent lead, 50 percent tin alloy had been common, while in Europe a similar 60 percent lead, 40 percent tin alloy was dominant. Tin has always been the key component in solders, maintaining their low melting point and providing effective bonding. Lead was used as a component in the interest of reducing cost, improving workability, and maintaining a low melting point. During the past two decades the use of lead-containing solders has been banned in the United States, the European Union, and Japan. In the United States, a solder that is 95 percent tin and 5 percent antimony has been the dominant replacement. In Europe, the replacements have been silver and copper alloys (95 percent tin, 5 percent silver and 95 percent tin, 5 percent copper). As a result, corrosion of lead solders is a somewhat temporary problem in the developed world. The use of lead solders is important because of the significant contribution they make toward lead at the tap, but of decreasing importance with time because the lead content in these solders will decrease over time. Only the corrosion of the lead–tin combinations is considered in this chapter, as these combinations are of principal interest, even at the present time.

As demonstrated by Oliphant and Schock (1996) and Reiber (1991), corrosion of lead-containing solders is fundamentally different from the corrosion of lead pipe, because these solders are lower on the galvanic series than is the copper pipe to which they are bound. For example, as reported in Table 22-6, the normal potential of lead–tin solder is about 200 mV lower than that of a copper surface in Seattle water. This finding would seem to be contraindicated by the standard electrode potentials of lead, tin, and copper. *The fact that the relative potential these metals exhibit in practice is the opposite of that expected by their standard electrode potentials illustrates a point made earlier*. Namely, the nature of the scale formed on a metal and the specifics of the mixed potentials acting on the metal surface in solution can result in a potential with respect to the solution that is different than might be expected from standard tables.

In the discussion of galvanic corrosion earlier in this chapter, the ratio of the area of the cathode and anode in a galvanic cell was of great importance. Unfortunately, this ratio is unfavorable for the lead–tin solder/copper couple in a soldered fitting. The ratio is unfavorable because the lead–tin solder presents only a small anodic surface; its oxidation by corrosion is driven by the cathodic reaction on a substantial copper surface.

In other work Oliphant has demonstrated that because the corrosion of the solder is driven by the galvanic cell rather than by a mixed-potential reaction on the solder surface, decreasing the solubility of the corrosion products from the solder did not result in a reduction in the rate of the corrosion reaction (Oliphant and Schock, 1996). As a result, it seems likely that the reduction in solder corrosion with increasing pH is a result of a reduction in the rate of the cathodic reaction, namely the reduction of oxygen to water.

Oliphant and Schock (1996) examined the impact of adding zinc on the galvanic corrosion of lead–tin solder. As shown on Fig. 22-30, the addition of zinc resulted in insignificant change in the behavior of the solder anode, but

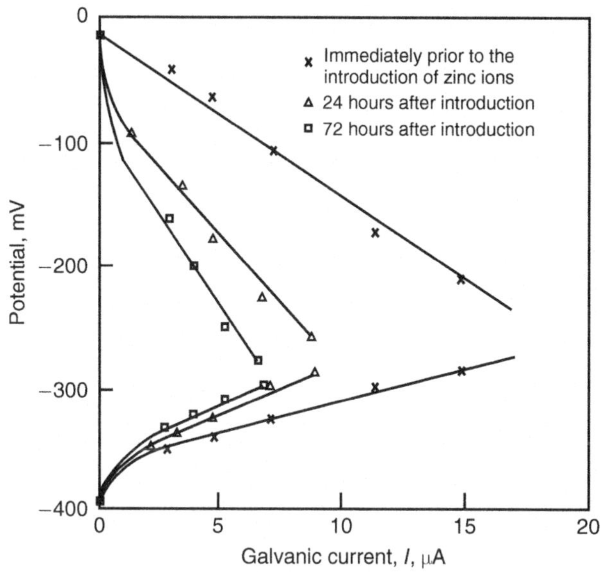

Figure 22-30
Impact of zinc on corrosion of solder. (Adapted from Oliphant and Schock, 1996.)

a significant reduction of the rate of the cathodic reaction taking place on the surface of the copper. Once again, zinc appears to be an effective cathodic inhibitor. Note that the estimated corrosion potential of the galvanic couple drops from approximately −260 mV to between −280 and −300 mV. Strategies that reduce the cathodic reaction appear to be the principal tools suitable for reducing solder corrosion.

Brasses and Gun Metals

Valves and fittings for domestic systems are made from brasses and gun metals. Both are copper-dominated alloys and their composition and use are summarized in Table 22-8. In the United States, the 1988 Lead Contamination Control Act limited the lead content of these materials to 8 percent (U.S. EPA, 1991). Nevertheless, these fittings have been implicated as an important potential source of lead in consumer plumbing. The corrosion of these fittings depends primarily on water quality, but also on the details of the plumbing. Comparing the data in Table 22-8 and Fig. 22-31, brass fittings have copper contents high enough that their corrosion potentials should be very close to that of copper. As a result, brass fittings, when used in copper-plumbed systems, will be at much the same potential as the rest of the system, perhaps, slightly anodic. On the other hand, brass fittings employed in a galvanized system should be strongly cathodic. As a result, the brass fittings, themselves, are not likely to fail as a result of galvanic action.

The principal issue where corrosion of brass and gun metal fittings is concerned is dezincification. As the name implies, dezincification is a process whereby the zinc in the alloy is corroded selectively out of the

Table 22-8
Brasses and gun metals used in drinking water systems

Alloy	Relevant Standard	Composition, % Cu	Zn	Sn	Pb[a]	Common Applications
Gun metal	C83600	85	5	5	5	Casting alloys used for domestic plumbing valves in United States
	C84400	81	9	3	7	
	LG2	84	5	5	5	Casting alloys used for domestic plumbing valves in United Kingdom
	LG4	88	2	5	4	
Brass	α Brass	64	30	—[b]	6	Used for tube and sand cast fittings
	α-β Brass	54	40	—[b]	6	Used for hot stamping

[a]In the United States the 1988 Lead Contamination Control Act limits lead content to no more than 8%.
[b]Up to 1% tin is sometimes used to improve corrosion resistance.

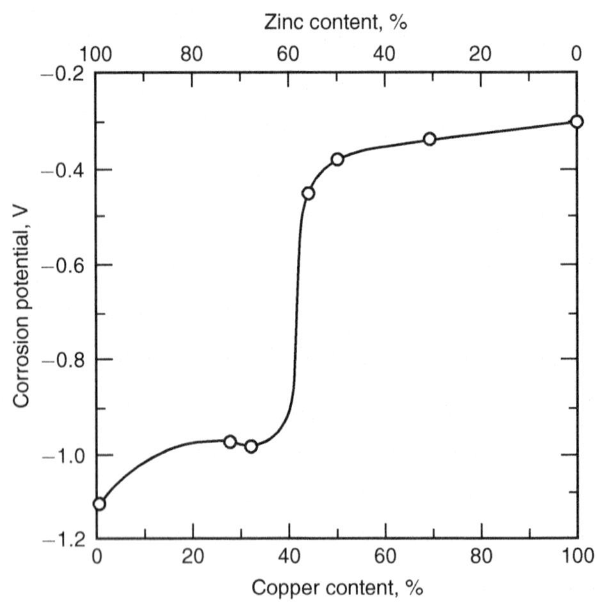

Figure 22-31
Corrosion potential versus copper content. (Adapted from Jones, 1984.)

matrix. A brass surface that has suffered from dezincification is distinctive in its reddish appearance in contrast to the more yellow appearance of a brass surface that remains intact. Alloy composition is important in dezincification. It has been found that zinc-rich alloys (β brasses) are more sensitive to the problem and that traces of arsenic in the alloy are known to improve resistance to dezincification. Nevertheless, water quality is also an important consideration as well. In general, waters with higher concentrations of chloride ion and lower alkalinities are more likely to suffer from dezincification.

22-9 Release of Contaminants

So far the discussion of corrosion has been focused on water conduits, specifically processes that change the oxidation state or, in some other manner, attack the structural integrity of the material, such as a water conduit. Processes that cause the release of contaminants to the water or cause deposition of undesirable scales on the surfaces of water conduits are considered in this section and the one to follow.

Red Water and Iron Pipe

Much of the following discussion is based on the work of Clement and coworkers (Clement et al., 2002). From the beginning of its use as a material for water conduits, iron pipe has been associated with the formation of discolored water. In some circumstances such water is only a minor inconvenience, in others it is a major irritation. The scale that develops on the surface of the pipe as a result of its corrosion is thought to be a key component in the development of red water. The scale that forms on iron pipe is far more complex than that found on water conduits constructed from other metals. The rather loose, porous, and bulky character of many ferric oxides and hydroxides is thought to be a root cause of these unique scales, which can become quite thick with time. It is common for the corrosion on iron pipe to form tubercles as well. Iron pipe with significant tuberculation is particularly noted for problems with discolored water, high disinfectant demand, bacterial problems, and hydraulic restrictions (reduced carrying capacity). The following discussion is focused on the mechanisms behind the formation of discolored, red water. Problems of discoloration are associated with all types or iron, steel, cast iron, and galvanized iron water conduits.

The scales formed by both ferric and ferrous iron are important. Ferric hydroxide, ferrous hydroxide, and ferrous carbonate are formed from corroding iron. Both of the iron hydroxides form rather porous oxide scales that do little to reduce corrosion. The ferrous carbonate scale is a much denser, more protective scale. As a consequence, in the last quarter of the twentieth century, quite a bit of effort was made to understand the role that ferrous carbonate (siderite) might have in the corrosion of iron pipe.

A number of researchers in the water field have tried to establish the significance of siderite in the formation of protective scales in water distribution pipes (Baylis, 1926; Larson and King, 1954). German investigators (Kolle and Rosch, 1980) reported the existence of a "shell-like layer" in scales found in cast-iron pipes. This layer seems to be important in the formation of corrosion resistance of iron in many waters. Kolle and Rosch speculated that siderite is important as an intermediate in the formation of the shell-like layer but may not be part of the layer itself.

A model for scale formation proposed by German researchers is based on the concept that siderite formation is a key step in the formation of corrosion-resistant scales (Sontheimer et al., 1979). The formation of goethite (FeOOH) in the shell-like layer may involve the conversion of siderite while retaining its dense crystalline structure (Kolle and Rosch, 1980). The formation of calcite also appears to be necessary to form a good protective scale.

To test the siderite model, experiments were designed to determine the influence of various factors related to scale formation on the corrosion rate (Sontheimer et al., 1981). It was shown that the corrosion rate decreases as the proportion of Fe(II) in the scale increases, tending to support the model. Increasing calcium concentration also showed a protective relationship.

Further research has led to a conceptual model of a tubercle pit similar to that shown on Fig. 22-18 (Clement et al., 2002; Herro and Port, 1993). Based on this work, it is suggested that tubercles take a form that includes four elements: (1) a corroded floor where the original metal has been partially removed, (2) a hard shell-like but porous cap made of a mixture of magnetite (Fe_2O_3), a semiconducting iron oxide, and geothite (α-FeOOH), (3) a porous interior cavity composed mostly of ferrous compounds in a water matrix (often including chloride and or sulfate), and (4) a soft external crust that is mostly dominated by ferric oxides and hydroxides.

Processes Associated with Tubercles

The processes associated with tubercles are thought to play an important role in the formation of red water (Sarin et al., 2001). Two conditions thought to be particularly important in red water and corrosion are illustrated on Fig. 22-32 using the model described above. The first condition occurs when water is flowing and oxygen is plentiful. Under these conditions, the native iron corrodes inside the tubercle, at its base. As this iron corrodes, it becomes an electron donor. The electrons generated travel through the metal, but they can also travel through the hard shell of the tubercle, as much of this hard shell is made of magnetite, a semiconductor. On the outside of the hard shell of the tubercle, oxygen, after diffusing through the soft crust, becomes the electron acceptor. In doing so, the oxygen is reduced to water, consuming hydrogen ions. As a result, the pH is elevated, and calcium carbonate may precipitate in the vicinity, *even if the bulk solution is not saturated with $CaCO_3$*. As Fe^{2+} cations are being generated inside the tubercle, anions must diffuse inside the tubercle so that electroneutrality can be maintained (shown as X^- on Fig. 22-32). To the extent that small anions, such as chloride, are present, they can accelerate the process. The deposition of $CaCO_3$ reduces the porosity of the hard shell and can inhibit the process, reducing the rate of the corrosion.

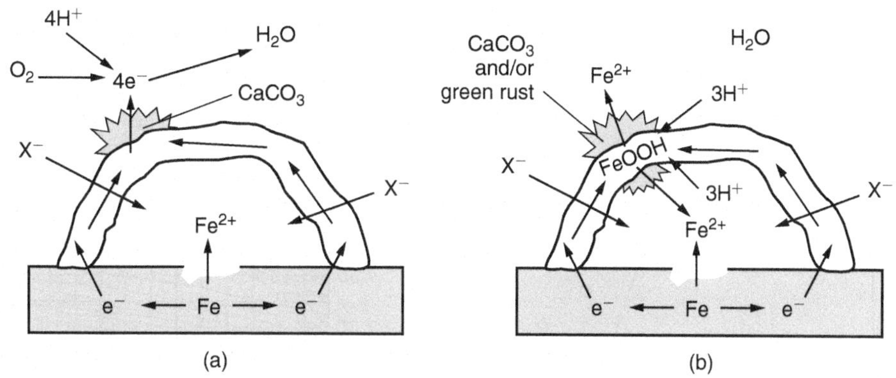

Figure 22-32
Processes associated with tubercles: (a) flowing water with oxygen (O_2 as electron acceptor), and (b) stagnant water with no oxygen (FeOOH as electron acceptor).

Drawing from the earlier discussion on electrode kinetics, the potential of the corroding electrode will be that at which the anodic and cathodic reactions develop the same current density. The potential–current relationship for the anodic reaction in the tubercle is defined by the reversible potential of corroding iron, by the exchange current and Tafel slope of that reaction on the iron surface, and by the rate of diffusion of anions into the tubercle. The potential–current relationship for the cathodic reaction in the tubercle is defined by (1) the reversible potential of oxygen redox reaction, (2) the exchange current and Tafel slope of that reaction on the surface of the hard shell, and (3) the rate of diffusion of oxygen through the solution and the soft crust to the surface of the hard shell. If $CaCO_3$ deposition is successful in slowing the corrosion reaction by inhibiting the influx of anions into the tubercle, then the potential of the electrode will rise and it will become more noble, resulting in anodic inhibition. This conclusion is consistent with Stumm's argument that $CaCO_3$ behaves as an anodic inhibitor (Stumm, 1960).

Blue-Green Water with Copper Pipe

The stability of the oxide layer on the surface of copper pipe is sensitive to pH. As a result, undesirably high levels of copper in solution are commonly associated with low pH water. As a broad rule of thumb, problems of blue and or green water (copper in the water) are common for water supplies with a pH below 7 and uncommon for water supplies with a pH greater than 8. The results of experiments conducted by Shull and Becker (1960) in which they examined the effect of pH on the copper released during a period of stagnation in copper tubing are shown on Fig. 22-33a. The experiments were conducted with tap water that had been adjusted to a particular pH and then placed in a section of copper tubing. After a 24-h period of stagnation the water in the copper tube was analyzed for

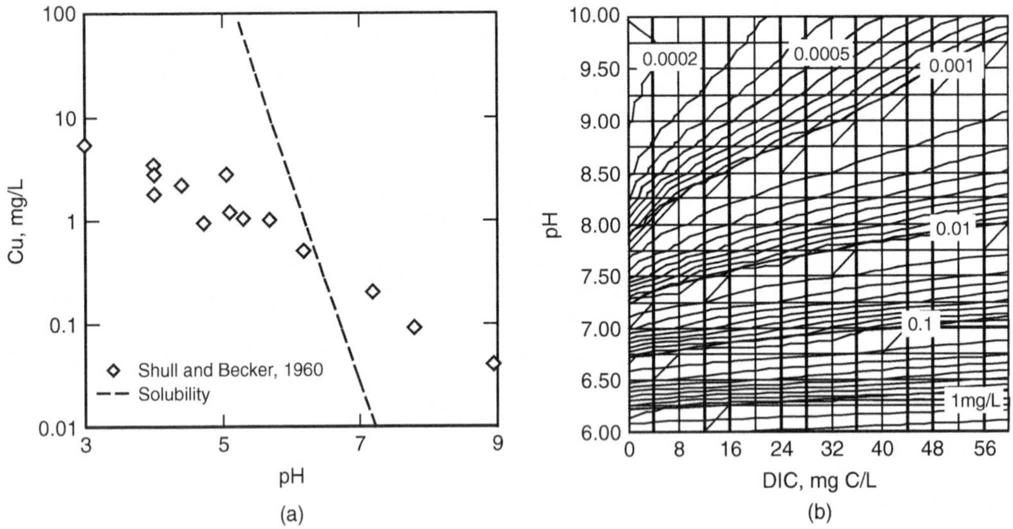

Figure 22-33
Copper solubility as function of (a) pH and (b) dissolved inorganic carbon (DIC). (Adapted from Shull and Becker, 1960).

copper. The experiments found a strong correlation with pH. Also shown on Fig. 22-33b the dashed line is used to delineate the maximum thermodynamic solubility of copper as a function of pH for a water having 1 mmol/L of dissolved inorganic carbon. The change in solubility has an even stronger connection with pH, in fact showing theoretical solubility below the observations of Shull and Becker for pH values above approximately 6.5.

Constructed with the equilibrium constants published by Ferguson et al. (1996), an overview of copper solubility over the entire range of water conditions normally encountered in drinking water is shown on Fig. 22-34. Figure 22-34 is useful in making water conditioning decisions designed to address the solubility of copper. While it cannot be expected that the copper levels in any standing water will match the conditions in this figure exactly, actions taken to move a water's pH–DIC to a point on the diagram with lower copper solubility are likely to reduce copper at the tap. The reader should be cautioned that, while pH is often the cause of excess copper levels, this is not always so. If the water supply having high copper levels is at a pH of 8 or higher, other mechanisms should be considered, particularly mechanisms that might be related to an unusually high corrosion rate.

Lead Release

The stable solid phase of lead also depends strongly on pH and much more strongly on the alkalinity. At lower pH, the dissolved plumbic ion (Pb^{2+}) is the most stable form, while lead carbonate, or cerussite ($PbCO_3$), is favored at neutral pH, and the hydroxycarbonate $Pb_3(OH)_2(CO_3)_2$ or

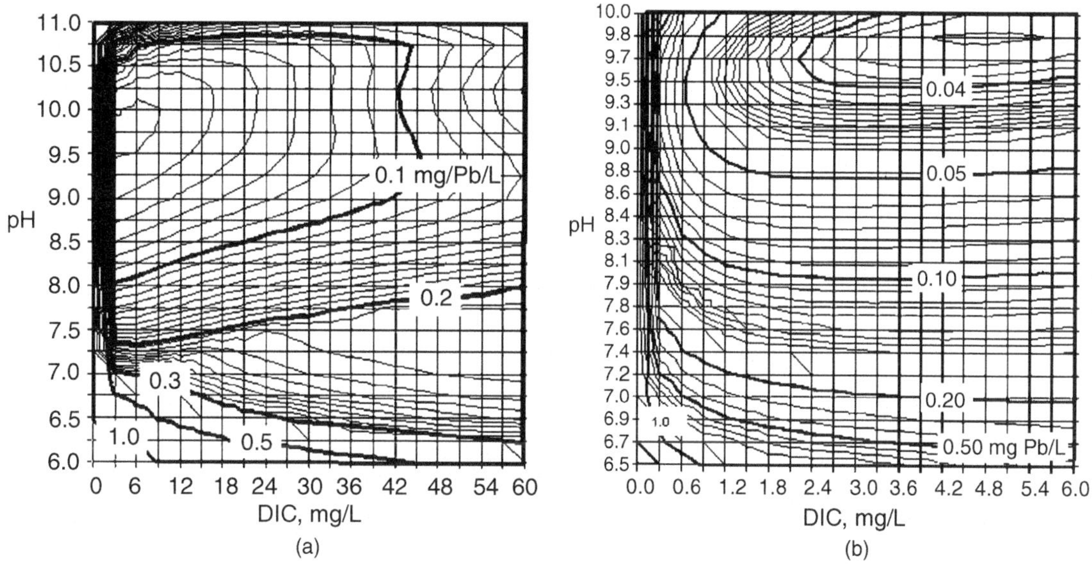

Figure 22-34
Contour plots of lead solubility as function of dissolved inorganic carbon (DIC): (a) for DIC concentrations from 0 to 6 mg/L and (b) DIC concentrations from 0 to 60 mg/L.

the hydroxide $Pb(OH)_2$ are favored at higher pH. Because the equilibrium between Pb^{2+} and $PbCO_3$ tends to govern the distribution of oxidized forms of lead, the solubility of lead increases dramatically as pH decreases below pH 8 (for fixed alkalinity). The equilibrium concentration of carbonate decreases by about two orders of magnitude for each decrease of one pH unit when H_2CO_3 is the predominating carbonate species. Also, as Pb^{2+} is a dissolved ion, while $PbCO_3$ is a solid scale, the rate of dissolution of lead tends to be accelerated by low pH. Lead solubility in natural waters is controlled by pH and by the concentration of dissolved inorganic carbon.

Constructed with data from Schock and Wagner (1985), the regions of lead solubility on a pH–DIC plane are shown on Fig. 22-34. Two diagrams are provided: one to enable an overview throughout the entire range of pH and DIC normally encountered in drinking water and the second to expand the region where problems normally occur, namely low levels of dissolved inorganic carbon. The shape of the contours on these diagrams are controlled by the soluble complexes that lead forms in water, but the dominant influence on the actual concentrations shown is the solubility of the solid phases. The concentrations of lead shown on these diagrams are higher than the concentrations that are ordinarily observed in real systems, suggesting that kinetic effects dominate, that the solid phases are less soluble than available equilibria would estimate, that important solid phases have not been identified, or that electrokinetic phenomena influence solubility. Work has been done to confirm that the crystalline minerals in

Update on lead and copper rule regulations.

http://www.stantec.com/wt3r/22-1.html

place are those used in the model, but this does not eliminate the possibility of amorphous solid phases that cannot be identified by crystallography (Schock and Wagner, 1985; Schock et al., 1996).

In any case, Fig. 22-34 has been found useful in examining alternate water conditioning strategies and their impact on lead levels. Movement from one place in the diagram to another with lower lead solubility usually results in reduced concentrations of soluble lead.

22-10 Formation of Treatment-Related Scales on Water Conduits

Important natural scales form on the surface of all the metals used for water conduits. In addition to these scales, water is often conditioned in the water treatment process to manage the development of other scales, particularly calcium carbonate. The formation of these scales is discussed below.

Calcium Carbonate Scale

During the first half of the twentieth century, one of the most prominent developments in waterworks-related chemistry was the calcium carbonate saturation index developed by Wilfred F. Langelier of the University of California at Berkeley (Langelier, 1936). Langelier's work was timely because it combined knowledge of chemical equilibria (Talbot and Blanchard, 1905) with the relatively recent availability of practical pH measurement devices (Baylis, 1929) to provide a practical means for evaluating calcium carbonate saturation. Langelier named the index the *saturation index.*

At the time, achieving calcium carbonate saturation was thought to be the principal means for controlling corrosion of iron distribution piping. If a solution were appropriately supersaturated with calcium carbonate, it would deposit protective calcium carbonate on the inside of the pipe, protecting it from the water. Earlier work had been conducted on the subject of $CaCO_3$ and corrosion, notably by a German chemist named Tillmans (Tillmans, 1913; Tillmans and Heublein, 1913). Langelier proposed a "saturation pH" (pH_s) at which the alkalinity and the calcium hardness would be at equilibrium with each other and with solid calcium carbonate. Without any simplifying assumptions, the saturation pH may be found with the following relationships:

- Definition of alkalinity in terms of molar quantities:

$$\text{Alk} = [HCO_3^-] + 2[CO_3^{2-}] + [OH^-] - [H^+] \tag{22-22}$$

- Equilibrium for dissolution of calcium carbonate:

$$K'_{so} = [Ca^{2+}][CO_3^{2-}] \tag{22-23}$$

where K'_{so} = mixed solubility constant for $CaCO_3$

- Equilibrium for dissociation of bicarbonate:

$$K_2' = \frac{a_{H^+}\left[CO_3^{2-}\right]}{\left[HCO_3^-\right]} \qquad (22\text{-}24)$$

where K_2' = acidity constant for dissociation of bicarbonate
a_{H^+} = activity of hydrogen ($a_{H^+} = a_{H_s^+}$ for pH = pH$_s$)

- Equilibrium for dissociation of water:

$$K_w' = a_{H^+}\,[OH^-] \qquad (22\text{-}25)$$

where K_w' = mixed dissociation constant of water

In this context the term "mixed constant" refers to an equilibrium constant that has been corrected for ionic strength of all ions except for the hydrogen ion. Correction for the hydrogen ion is not necessary because the pH measurement is a measure of activity, that is, pH = $-\log(a_{H+})$.

Rearranging these:

$$\left[CO_3^{2-}\right] = \frac{K_{so}'}{\left[Ca^{2+}\right]} \quad \left[HCO_3^-\right] = \frac{a_{H_s} + K_{so}'}{K_2'\left[Ca^{2+}\right]} \quad \left[OH^-\right] = \frac{K_w'}{a_{H_s^+}} \quad \left[H^+\right] \sim a_{H_s^+}$$

Substituting,

$$\text{Alk} = \frac{a_{H_s} + K_{so}'}{K_2'\left[Ca^{2+}\right]} + 2\times\frac{K_{so}'}{\left[Ca^{2+}\right]} + \frac{K_w'}{a_{H_s}^+} - a_{H_s^+} \qquad (22\text{-}26)$$

Rearranging,

$$\left(a_{H_s^+}\right)^2\left(K_{so}' - K_2'\left[Ca^{2+}\right]\right) + \left(a_{H_s}+\right)\left(2K_2'K_{so}' - K_2'\left[Ca^{2+}\right]\text{Alk}\right) + K_2'\left[Ca^{2+}\right]K_w' = 0 \qquad (22\text{-}27)$$

Equation 22-27 may be solved as a quadratic equation:

$$a_{H_s+} = \frac{-b \pm \sqrt{b^2 - 4ac}}{2a} \qquad (22\text{-}28)$$

where $a = K_{so}' - K_2'\left[Ca^{2+}\right]$

$b = 2K_2'K_{so}' - K_2'\left[Ca^{2+}\right]\text{Alk}$

$c = K_2'\left[Ca^{2+}\right]K_w'$

Taking the logarithm,

$$\text{pH}_s = -\log a_{H_s} + \qquad (22\text{-}29)$$

The necessary data for solving the equation are provided in Table 22-9. A simplified form of these equations has traditionally been used:

$$\text{pH}_s = \text{p}K_2' - \text{p}K_{so}' - \log\left[Ca^{2+}\right] - \log \text{Alk} \qquad (22\text{-}30)$$

Table 22-9
Thermodynamic data for the carbonate system

The following expressions can be used to determine equilibrium and solubility products as a function of temperature (K):

- Plummer and Busenberg (1982):

$$\log K_1 = -356.3094 - 0.06091964 \times T + \frac{21{,}834.37}{T} + 126.8339 \times \log T - \frac{1{,}684{,}915}{T^2}$$

$$\log K_2 = -107.8871 - 0.03252849 \times T + \frac{5151.79}{T} + 38.92561 \times \log T - \frac{56{,}713.9}{T^2}$$

$$\log K_{so}\ \text{(calcite)} = -171.9065 - 0.077993 \times T + \frac{2839.319}{T} + 71.595 \times \log T$$

- Harned and Owen (1958):

$$\log K_W = 6.008 - 0.017060 \times T - \frac{4471}{T}$$

For nonideal solutions (i.e., with an ionic strength above zero) the nonideality can be corrected using activity coefficients:

$$\gamma_B = \frac{\{B\}}{[B]}$$

The activity coefficient is given by the Davies equation (Eq. 5–43):

$$\log \gamma_B = -AZ_B^2\left(\frac{\sqrt{I}}{1+\sqrt{I}} - 0.3 \times I\right)$$

where Z_B = charge on species B
A = Debye–Hückel constant, 0.5 at 25°C
I = ionic strength, $= \frac{1}{2}\sum C_i Z_i^2$
C_i = concentration of species i
Z_i = ionic charge of species i

Using Eq. 22-30, pH_s is a function of the same variables. In Eq. 22-30, the influence of pH on alkalinity is neglected. When the pH is between 6.5 and 8.5, the contribution of alkalinity may be neglected. Langelier then defined the $CaCO_3$ or Langelier saturation index (LI) as the difference between the measured pH and the pH_s, thus

$$LI = pH - pH_s \tag{22-31}$$

The state of saturation with respect to calcium carbonate, therefore, depends on the value of the Langelier index:

$LI < 0$ solution is undersaturated with $CaCO_3$ (will dissolve $CaCO_3$)
$LI = 0$ solution at equilibrium with $CaCO_3$
$LI > 0$ solution supersaturated with $CaCO_3$ (will precipitate $CaCO_3$) It is a common practice for utilities to add lime or caustic soda to their treated

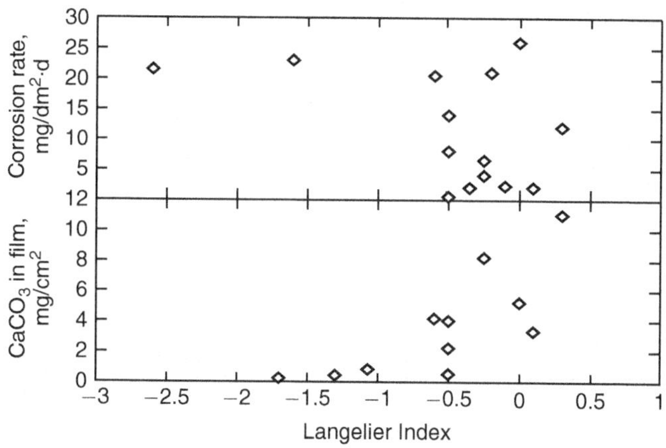

Figure 22-35
Correlation of calcium and carbonate ($CaCO_3$) deposition and corrosion of gray cast iron (Adapted from Stumm, 1956): (a) corrosion rate versus Langelier saturation index, and (b) $CaCO_3$ deposition versus Langelier saturation index.

water to achieve a saturation index that is near neutral or slightly positive (0 to +0.2), and there is some evidence that this chemical addition helps to reduce red water complaints (Clement et al., 2002; DeMartini, 1938). The appropriateness of the saturation index for controlling corrosion or even the deposition of a $CaCO_3$ scale is subject to contradictory evidence (Larson, 1975; Stumm, 1956, 1960). The problem of connecting $CaCO_3$ deposition and corrosion is illustrated by the data on Fig. 22-35 in which there is a poor correlation between Langelier's index and the rate of corrosion. Although the saturation pH has a rational basis, the definition of the saturation index itself is empirical. Nevertheless, calcium carbonate chemistry is a subject of broad interest.

The results summarized above were developed by Stumm (1960) in a study examining the corrosion of cast-iron pipe during exposure to several different water qualities. Based on these results, no correlation was found between the Langelier index ($CaCO_3$ saturation) and corrosion even though $CaCO_3$ deposition did increase substantially as the Langelier index increased.

Since Langelier's original research, there have been a number of attempts to improve upon the index (Larson and Buswell, 1942), to facilitate its determination (Dye, 1944, 1952, 1958; Hoover, 1938), or to demonstrate its utility (DeMartini, 1938). Others have also proposed alternate indices associated with calcium carbonate saturation. Notable among these are the driving force index (McCauley, 1960), the momentary excess (Dye, 1952), the Ryznar stability index (Ryznar, 1944), and the aggressiveness index (AWWA, 1977). These indices represent concepts that are basically the same as Langelier's original principle.

Perhaps the calcium carbonate chemistry of a water and its relationship to stability and corrosion is best characterized by three factors: (1) the free-energy driving force of the precipitation reaction (ΔG), (2) the calcium carbonate precipitation potential (CCPP), and (3) the calcium carbonate saturation buffer intensity $\left(\beta_{\text{Alk}}^{S}\right)$. A brief discussion of each of these is useful.

FREE-ENERGY DRIVING FORCE

The free energy (ΔG) is a quantitative measure of the energy available to drive the precipitation reaction. It can be shown that ΔG influences the rate of the precipitation reaction and that a certain ΔG threshold must be reached before crystal nucleation will ensue (Nielsen, 1964; Stumm and Morgan, 1996). It can be shown that the ΔG for $CaCO_3$ precipitation can be defined as follows:

$$\Delta G = \mathrm{RT}\ln\frac{\left[Ca^{2+}\right]\left[CO_3^{2-}\right]}{K_{so}^{'}} \tag{22-32}$$

Trussell et al. (1976) demonstrated that ΔG and the Langelier index are related in the following manner:

$$\Delta G = -2.3\mathrm{RT}\ (\mathrm{LI}) \tag{22-33}$$

Thus, the LI is a *driving force index.* Calcium carbonate will spontaneously precipitate from solution at a ΔG of less than about –5 to –7 kcal/mol (Trussell, 1972). When calcium saturation is used as a red water control technique, a ΔG of about –0.5 kcal/mol is probably sufficient. The calcium saturation technique should be used with caution for waters of low alkalinity and/or hardness because, even though the water is supersaturated, its capacity to precipitate protective calcium carbonate is limited. This illustrates the need for a capacity index.

CALCIUM CARBONATE PRECIPITATION POTENTIAL

The calcium carbonate precipitation potential, or the CCPP, is the amount of calcium carbonate that will precipitate or dissolve from the solution as it comes to equilibrium with solid $CaCO_3$ (Merrill and Sanks, 1977a, b, 1978). The magnitude of this *capacity index* can be derived as follows:

If the amount of $CaCO_3$ that will precipitate from a solution to achieve equilibrium is given by X mol/L, then the equilibrium condition is

$$\left[Ca^{2+}\right]_{\text{final}}\left[CO_3^{2-}\right]_{\text{final}} = K_{\text{so}}^{'} = \left(\left[Ca^{2+}\right]-X\right)\left(C_T - X\right)\alpha_2 \tag{22-34}$$

where $K_{\text{so}}^{'}$ = mixed solubility product for $CaCO_3$

X = amount of $CaCO_3$ that will precipitate, mol/L

$\alpha_2 = \left(K_1^{'}K_2^{'}\right)/\left([\mathrm{H}]^2 + K_1^{'}[\mathrm{H}] + K_1^{'}K_2^{'}\right)$

Alkalinity in the equilibrated system is given by

$$\mathrm{Alk} - 2X \simeq \left(C_T - X\right)\left(\alpha_1 + 2\alpha_2\right) + \frac{K_w^{'}}{a_{\mathrm{H}^+}} - a_{\mathrm{H}^+} \tag{22-35}$$

where $\alpha_1 = \alpha_2[\mathrm{H}]/K_2^{'}$

$K_w^{'}$ = mixed dissociation constant for water

These two equations can then be solved by the methods described by Trussell (1998) or by the computer programs identified in *Standard Methods* (APHA, 2000).

CALCIUM CARBONATE SATURATION BUFFER INTENSITY

The calcium carbonate buffer intensity $\left(\beta_{\text{Alk}}^{S}\right)$ is a measure of the sensitivity of the calcium carbonate saturation of the solution to changes in alkalinity (Trussell et al., 1976). This *sensitivity index* is of interest because it measures the sensitivity of the calcium carbonate saturation of the solution to changes in alkalinity near the cathodic reaction. The buffer intensity can be estimated from the following equation:

$$\beta_{\text{Alk}}^{S} = \left(\frac{\partial \text{Alk}}{\partial S}\right)_{C_t} = \left(\frac{\partial \text{Alk}}{\partial\left[\text{H}^{+}\right]}\right)\left(\frac{\partial\left[\text{H}^{+}\right]}{\partial\left[\text{CO}_3^{2-}\right]}\right)\left(\frac{\partial\left[\text{CO}_3^{2-}\right]}{\partial S}\right) \qquad (22\text{-}36)$$

where S is the saturation with respect to calcium carbonate:

$$S = \frac{\left[\text{Ca}^{2+}\right]\left[\text{CO}_3^{2-}\right]}{K_{\text{so}}^{\prime}} \qquad (22\text{-}37)$$

Higher values of β_{Alk}^{S} are associated with smaller changes in calcium carbonate saturation, for a given exogenous change in alkalinity, for example, by local inhomogeneities in solution associated with the reduction of oxygen at a cathodic site. Thus local supersaturation of $CaCO_3$ would be expected at cathodic sites where hydrogen ion is being withdrawn from solution, when β_{Alk}^{S} is larger, resulting in the deposition of $CaCO_3$ scale at the cathode, but not the anode. Therefore, it is desirable to enhance scale formation by operating in a region where β_{Alk}^{S} is small, all else being equal. When C_t is larger or $[Ca^{2+}]$ is larger, β_{Alk}^{S} is smaller. Higher pH generally favors corrosion control by increasing ΔG and CCPP and lowering β_{Alk}^{S}; however, not all of these fundamental indices can be optimized under the same conditions. Also, there may be practical limitations to pH control. There is probably no simple optimum for all three of these parameters. Rather it is important to consider each of them and understand their significance.

Finally, consideration must be given to the fact that calcium carbonate is a salt with inverse solubility, *that is, its solubility decreases as the temperature of the water increases.* Temperature is an important consideration because waters that appear to have satisfactory conditioning at the temperature in the water system often have problems of calcium carbonate deposition in hot water systems, resulting in undesirable deposition and sludging, reducing the useful life of water heaters.

Alumino-Silicate Scales

In addition to calcium carbonate, aluminum salts can also result in undesirable deposition downstream of water treatment plants (Snoeyink, 2002). Examples of such deposits are shown on Fig. 22-36. Alum is, by far, the most

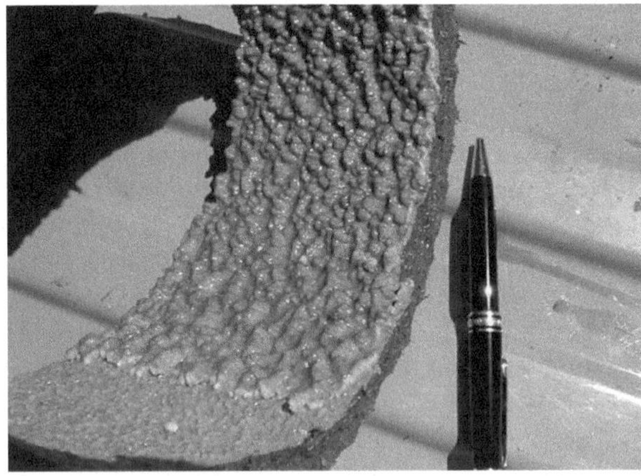

Figure 22-36
Alumino-silicate deposits on surface of asbestos–cement water main downstream of water treatment plant.

frequently used coagulant used in U.S. water treatment practice today. As it takes quite some time for aluminum hydrolysis to reach final equilibrium. As a result, virtually all water treatment plant effluents are supersaturated with aluminum. This can result in direct precipitation of aluminum hydroxide or the formation of other aluminum precipitates in downstream pipelines. When this precipitation occurs, it typically takes the form of a "washboard" surface that can significantly increase head loss. Sometimes this head loss increase can cause important problems for the distribution system.

When any metal ion is added to water, it first forms a hydroxide precipitate that is metastable in nature, meaning that in time it will hydrolyze further to form a more stable solid species. Metal hydrolysis also occurs when copper sulfate is added to water for algae control as well as when a metal ion coagulant is added to water. In the case of aluminum, which is particularly amphoteric, the amount of this unstable material can result in direct precipitation of aluminum hydroxide or the formation of other aluminum precipitates in downstream pipelines. To date, there is limited literature on this subject and all the conditions that control these problems are not well understood (Price, 1997).

22-11 Dissolution of Cement-Based Materials

Most of the largest structures and conduits used to transport, treat, and contain drinking water have surfaces of concrete or cement mortar. Use of cementaceous materials for this purpose dates back to Roman times, and evidence of the good service that concrete can provide is that some of the conduits built of these materials by the Romans are still in service today. Nevertheless, there are certain conditions when concrete will not provide good service. It is important to know what these conditions are and what can be done to mitigate them.

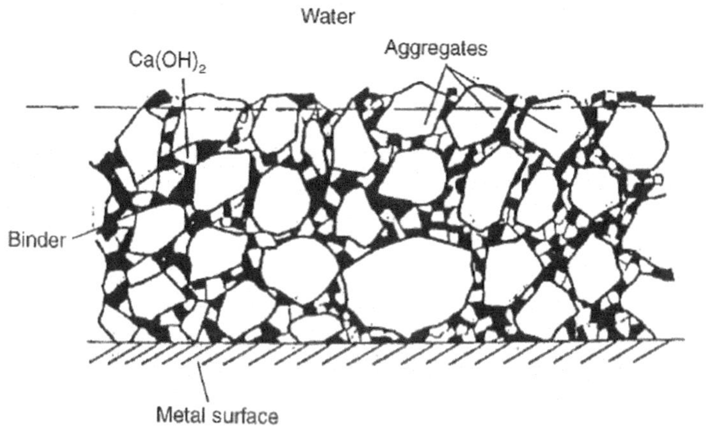

Figure 22-37
Schematic structure of cement-based material.

Composition of Cement and Corrosion Resistance

Concrete is composite material consisting of a cement binder in which inert filler materials called aggregate are imbedded (see Fig. 22-37). The cement binder is made through the hydration of a cement powder. Cement powders are primarily a mixture of calcium silicates, calcium aluminates, and other components, mixed in various proportions.

Cement-based materials, such as concrete and mortar are made from three components: inert aggregates, a cement powder, and water. Once water is added and the mixture is blended, the cement powder hydrates to form a binder with significant compressive strength and resistance to erosion. When mortar is applied to the surface of a metal water conduit, it not only provides physical protection but also chemical protection as a result of its basic alkaline character.

Generally, the cement powder used to make concrete for water-carrying structures (filters, basins, reservoirs, and pipes) is Portland cement, the cement with the lowest aluminum content. Cement mortar linings on water pipe are made from a variety of cements, generally with standard composition, shown in Table 22-10 (Leroy et al., 1996). The composition of each type of cement varies somewhat from one country to the next.

Reaction of Cementaceous Surfaces with Water

Once the cement powder comes in contact with water, it begins to undergo hydration, hydrolysis, and precipitation reactions that result in hardening of the material (Leroy et al., 1996). At first, the hydration process affects only the outer layer of the particles of cement powder. After hydration has occurred the process slows down and various solid phases formed come to a metastable equilibrium with the interstitial water. In most cements, this interstitial water is rich in calcium and potassium hydroxide. Typically, this pore water has a pH value of 13 to 13.5. The metastable phases formed in these early stages continue a process of slow hydrolysis over a period of years until the cement is completely cured. These changes also result in

Table 22-10
Standard composition of primary types of cements used for the manufacture of concrete pipe

Parameter	Unit	Blast Furnace Cement	Pozzolanic Metallurgical Cement	High-Alumina Cement	Portland Cement
SiO_2	%	27	29	3.5	22
CaO	%	48	44	38	65
Free CaO	%	0.5	1.5	0.5	2.5
Al_2O_3	%	13	13	36	5
Fe_2O_3	%	1.5	3	18	2.5
SO_3	%	3.8	2.2	0.1	1.9
Na_2O	%	0.4	0.3	0.1	0.1
K_2O	%	0.8	0.7–1.7	0.05	0.8
MgO	%	2.8	3	0.4	0.5
Ignition loss	%	0.2	1.2–2.4	—	2
Density	g/cm^3	2.9	2.8	3.2	3.1
Specific surface	cm^2/g	3350–3450	3650	2750	3870

some changes in the crystalline structure as well. As a result, the hardened material develops a heterogeneous structure that can be porous both at the microscale and the macroscale. Generally, cements with less calcium oxide and more aluminum and iron oxides are more resistant to attack from low-pH, low-alkalinity waters.

Aggregate used in concrete ranges from fine sand used in mortars to large stones 75 to 100 mm (3 to 4 in.) in diameter. Reactive aggregates may cause durability problems in water pipe and are to be avoided. The cement mortar used as a protective liner in cast-iron or steel water pipe is made by using a fine aggregate of silica sand. When feasible, it is applied using the centrifuged spinning method, designed to assure a dense, even coating.

Important Cementaceous Materials

For many years asbestos– cement was used to make a light, strong, low-cost water pipe (AC pipe). An additional advantage of AC pipe is its ability to maintain a smooth surface when used for waters that are corrosive to cast iron. AC pipe, made by using asbestos fibers in place of sand or aggregate, has a relatively strong, dense pipe wall. In the 1970s concerns about the release of asbestos fibers from asbestos–cement pipe subject to corrosion by low-pH, low-alkalinity waters began to surface. Worker exposure to asbestos fibers during pipe manufacture also raised additional concerns about AC pipe. As a result of these concerns AC pipe is not available on the market today, but there are still many thousands of miles of this pipe in water systems today.

The cement in reinforced concrete water pipe, cement mortar lining of steel and cast-iron water pipe, and asbestos–cement water pipe is subject to deterioration when exposed to prolonged contact with aggressive waters. There are at least two mechanisms involved in the dissolution of cements. The first is the dissolution of free lime and other compounds when in contact with low-pH, low-alkalinity waters. The second is chemical attack by aggressive ions such as sulfate and chloride.

Understanding Concrete Corrosion

Corrosion of cement pipe materials in domestic water systems is governed principally by solubility considerations. This statement suggests that simple chemical equilibrium could be applied to address the problem, but cement is not one simple compound, but a mixture of different solid phases. As a result, the chemistry of its interaction with water depends on the composition of the specific cement matrix as well as the composition of the water.

Leroy et al., (1996) outlined a simple conceptual model that is useful in understanding the role of $CaCO_3$ in the deterioration of cement-based materials exposed to water. The idea is that, after they cure, most cements have a certain amount of free lime in their pore structure. Two extremes can then be envisioned. On the one hand a pore-blocking condition can occur (Fig. 22-38a). Such a condition occurs when the cement-based material is exposed to a water that is hard, alkaline, and supersaturated with $CaCO_3$. Under these conditions, lime attempting to leach from the pores in the cement structure immediately reacts with the hardness and the alkalinity in the water to form $CaCO_3$, and the $CaCO_3$ formed seals off the pore. Because the water is supersaturated with $CaCO_3$, the $CaCO_3$ plug that is formed is thermodynamically stable and the concrete is protected. The

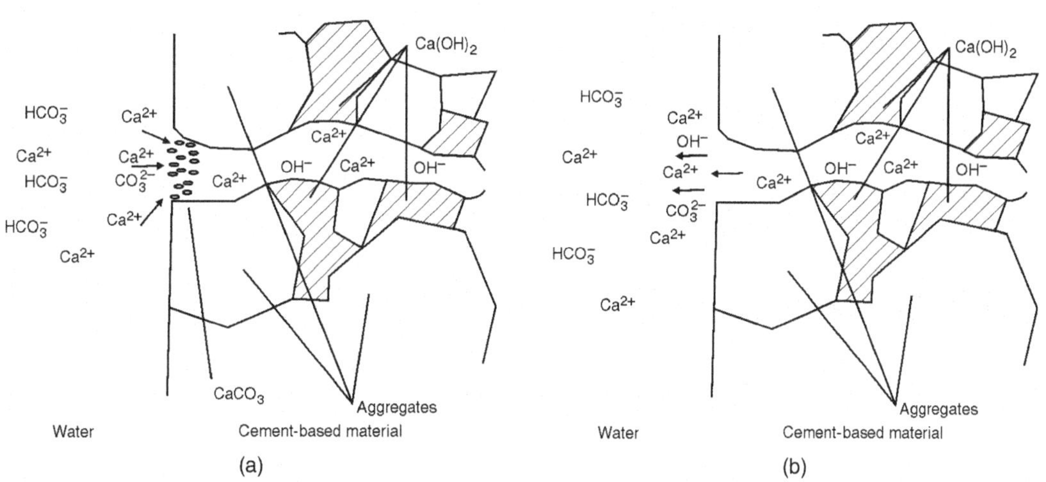

Figure 22-38
Leroy model for pore blocking: (a) blocking condition, and (b) nonblocking condition.

other condition (Fig. 22-38b) occurs when the cement-based material is exposed to a soft, low-pH water that is undersaturated with $CaCO_3$. Under these conditions, although some $CaCO_3$ may form, lime is able to leach out of the pore and, with time, it becomes empty.

The Leroy et al. (1996) model is also consistent with certain observations about the impact of cement chemistry on the deterioration of the cement matrix in low-alkalinity waters. For example, the general observation is that waters with pH values below 7 are generally regarded as aggressive to concrete surfaces. Coagulation with alum or ferric chloride can result in significant pH reduction. Consequently, concrete surfaces in water plants downstream of coagulation often show etching and other signs of deterioration before pH adjustment. As mentioned earlier in this section, cements with less lime and more aluminum and iron are less prone to forming deposits of free lime that might be leached out in the manner shown on Fig. 22-39. Soukatchoff (1990) conducted a 14-year study where samples of cement mortar, made with three different types of cement, were exposed to a soft water with the following characteristics:

pH 5.0

Alkalinity, 0.5 mg/L as $CaCO_3$

Hardness, 2.6 mg/L as $CaCO_3$ ($Ca = 0.45$ mg/L)

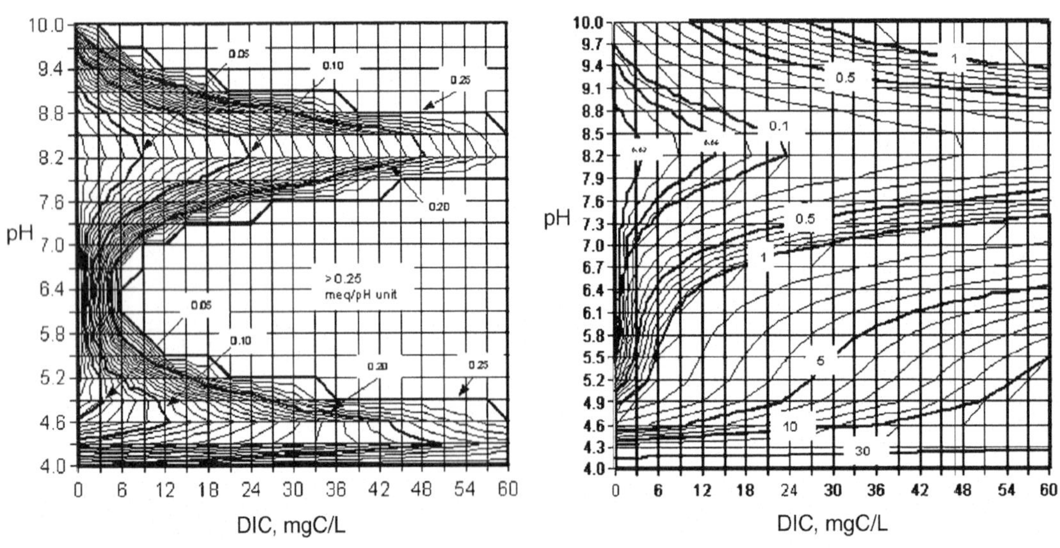

Figure 22-39
Isopleths of buffer intensity versus pH and dissolved inorganic carbon (DIC or C_t): (a) buffer intensity with respect to changes in alkalinity [∂ Alk/∂pH]DIC, and (b) buffer intensity with respect to changes in DIC [∂DIC/∂ pH]Alk. (Adapted from Leroy et al., 1996.)

Table 22-11
Cement character and impact of exposure to low-pH, low-alkalinity water22

Item		Composition or Value, %		
		Portland Cement	Blast Furnace Cement	High-Alumina Cement
Chemical composition	CaO	65	48	38
	SiO_2	22	27	3.5
	Al_2O_3	5	13	36
	Fe_2O_3	2.5	1.50	18
	Other	6	10.50	4.5
Change after 14 years of soft water	Change in porosity	+33	+22	+10
	Change in weight	−10	−6.5	−2.5
	Change in compressive strength	−62	−32	+134

[a]Chemical composition from Table 22-10; porosity and strength data from Soukatchoff (1990).

From the results of Soukatchoff's experiment, summarized in Table 22-11, it appears that cements with greater aluminum and iron content are more resistant than are traditional Portland cements.

Swelling, due to the use of poor-quality aggregates or because of excessive sulfates in the water, is also a problem with cement-based materials immersed in water. Waters with high levels of sulfate are thought to react with the calcium aluminates in the concrete to form ettringite, a compound that takes up a significantly larger volume, resulting in swelling. The problem of swelling has been associated with waters with sulfate levels above 400 mg/L (Leroy et al., 1996).

22-12 Treatment for Corrosion Control

Because of economic losses associated with corrosion of water conduits, much research has been devoted to developing effective corrosion control measures. General categories of such approaches to corrosion control include (1) control of environmental parameters affecting the corrosion rate, (2) addition of chemical inhibitors, (3) electrochemical measures, and (4) considerations of system design (Fontana, 1985; Revie and Uhlig, 1985; Jones, 1996). Because the focus of this chapter is primarily with internal corrosion of water distribution conduits, the following discussion will be limited to those control measures applicable to drinking water distribution systems. Applicable control measures include pH adjustment and addition of chemical inhibitors such as polyphosphates, zinc orthophosphate, and silicates.

Adjustment of pH–Alkalinity–Dissolved Inorganic Carbon

Generally, pH adjustment is made in an attempt to manage the deposition of calcium carbonate. Methods for estimating the state of calcium carbonate saturation have been discussed earlier.

In addition to affecting the carbonate system, pH is a key variable in the solubility of conduit materials such as lead, copper, and zinc. The thermodynamic stability of oxidized species such as metal oxides, hydroxides, and carbonates is strongly influenced by pH also, so pH may govern whether an insoluble natural protective scale can form or whether soluble species are favored as a metal corrodes. Therefore, pH adjustment can play a major role in stabilizing a pipe material such as lead: low pH favors dissolution into plumbic ions, whereas high pH favors formation of solid scales that deter further corrosion.

Water treatment practice to adjust pH has typically involved the addition of lime [as CaO or $Ca(OH)_2$] or sodium hydroxide to achieve a positive value of the Langelier index. The addition of these materials also increases alkalinity, which, in turn, has the side effect of decreasing the solubility of corrosion products such as lead carbonate and enhancing the formation of solid metal carbonate scales. Generally, pH adjustment may not be a satisfactory corrosion control measure for waters of very low alkalinity. In such cases, the buffer intensity is low and the amount of lime necessary to raise pH is small, but the resulting high-pH water does not have much resistance to pH changes. Therefore, it is easier for low-pH microsites to form near the surface (pits) even though the overall thermodynamic tendency is for calcium carbonate scale to form.

In waters of high alkalinity and hardness, it becomes more difficult to adjust the pH to a value above 8.0 because of the rapid precipitation of calcium carbonate in distribution lines. The interaction of pH change and alkalinity change has a complicated effect on buffer intensity, which may result in a lowered buffer intensity following treatment if not monitored carefully (Stumm, 1960).

When examining which chemicals might be used to address the solubility of copper and lead, it is often convenient to refer to solubility on contour diagrams of solubility on a pH–DIC plane. Such diagrams are provided on Figs. 22-33 and 22-39. Further insight as to which chemicals will drive the water in which direction on the pH–DIC plane is provided in Table 22-11 (Trussell, 1998).

It is also useful to consider the stability of water in three-dimensional space defined by the pH, alkalinity, and DIC. Stumm (1960) first introduced this concept by appropriating the parameter ∂ Alk/∂ pH from the field of medical science. Stumm demonstrated that the concept could be used to examine the impact of the water's resistance to pH changes on corrosion. At first he appropriated the name, buffer capacity, from the medical community as well. Later on, recognizing that the alkalinity is really the capacity factor, he renamed it the buffer intensity. Of the three components that define the system—pH, alkalinity, and DIC—the pH is

Table 22-12
Effect of common chemicals on DIC, Alk, and pH

Chemical	Change in DIC, mol DIC/mol X	Change in Alk, eq Alk/mol X	pH
CaO	0	2	Increase
NaOH	0	1	Increase
Na_2CO_3	1	2	Increase
$NaHCO_3$	1	1	Tends to pH 8.3
CO_2	1	0	Decrease
H_2SO_4	0	−2	Decrease
HCl	0	−1	Decrease
$CaCO_3$	1	2	Increase

Source: Trussell (1998).

the only intensive variable. There are several ways to look at buffer intensity. The buffer intensity most often referred to in connection with corrosion is a measure of the degree to which the water resists changes in pH due to reactions that change its alkalinity (the reduction of water, the hydrolysis of metal ions, etc.). The other major factor that can change the pH is exchange of CO_2 with the atmosphere. Thus, there are really two "buffer intensities" that characterize the system:

$$\beta_{\text{pH}}^{\text{Alk}} = \left[\frac{\partial \text{Alk}}{\partial \text{pH}}\right]_{\text{DIC}} \tag{22-38}$$

$$\beta_{\text{pH}}^{\text{DIC}} = \left[\frac{\partial \text{DIC}}{\partial \text{pH}}\right]_{\text{Alk}} \tag{22-39}$$

The parameter $\beta_{\text{pH}}^{\text{Alk}}$ provides information about the water's resistance to changes in pH that result from actions that consume or produce alkalinity, and $\beta_{\text{pH}}^{\text{DIC}}$ provides information about the resistance of the water to changes in pH that result from actions that remove or add CO_2 to the water. Isopleths of $\beta_{\text{pH}}^{\text{Alk}}$ and $\beta_{\text{pH}}^{\text{DIC}}$ on a DIC–pH plane are shown on Fig. 22-39. Using Fig. 22-39, it is possible to gain an understanding of how adjustments will affect the stability of the pH of a water. Generally, both indices are minimal in the pH range of 8 to 8.5.

Addition of Phosphates

Two fundamental types of phosphate compounds are commonly used in water conditioning: orthophosphates and polyphosphates. These two compounds, however, are presented in a number of different ways depending on the purpose for which they are designed.

All the forms of phosphate used in water treatment can be characterized by the formula $P_nO_{3n+1}{}^{-(n+2)}$. The most well-understood forms are those having n values of 1, 2, and 3, respectively. Phosphates with $n = 1$ (e.g., $PO_4{}^{3-}$)

are called orthophosphates and phosphates with $n > 1$ are called polyphosphates. The most well understood of the polyphosphates are pyrophosphate ($P_2O_7^{4-}$) and tripolyphosphate ($P_3O_{10}^{5-}$) and the least understood are probably the "glassy polyphosphates." It should be noted that the polyphosphates, not the orthophosphates, were first used in water treatment. Polyphosphates were introduced in the late 1930s for prevention of calcium carbonate scale formation (Rice and Hatch, 1939). It was observed subsequently that polyphosphates could also act as corrosion inhibitors.

Polyphosphates, generally linear chains of phosphorous atoms held together by a –PO–P– linkeage, are formed via the dehydration of orthophosphates using caustic. The chain length of the product is dependent on the ratio of Na_2O and P_2O_5 used in its preparation. The "hexametaphosphate" commonly used in water treatment is actually a mixture of polyphosphates having n values ranging from 5 to 22, and is not a true hexametaphosphate (Lytle and Snoeyink, 2002). A true hexametaphosphate would have the formula $Na_6(PO_3)_6$ and would be a cyclic compound rather than a polyphosphate.

When introduced to water, all polyphosphates hydrolyze with time, shortening their chain length until they eventually become orthophosphate. The rate at which this reversion process proceeds has been determined for certain concentrated solutions but not in conditions applicable to drinking water treatment (Green, 1950; Morgen and Swoope, 1943).

Both orthophosphates and polyphosphates are thought to behave as anodic inhibitors. It is possible that this behavior is related to the inhibition of the dissolution of the oxide layer in conformance with the theory mentioned earlier in connection with passive films on iron surfaces (Stumm, 1998). Such inhibition may be due to the formation of binuclear complexes on the oxide surface that compete with ligands attempting to promote dissolution protective scale (Stumm, 1995).

In water treatment, phosphates are employed as orthophosphates, polyphosphates, and bimetallic phosphates. Orthophosphates are most effective in slightly alkaline conditions, and they have been found to be effective with iron, galvanized, and lead pipe as well as solder, although the conditions that make orthophosphates most effective require further investigation.

Polyphosphates are most widely used for their properties in "sequestering" the color that iron and manganese impart to water. This sequestering action has been demonstrated to be due to the impact of polyphosphates on the size of particles formed in solution rather than sequestration in the chemical sense (Lytle and Snoeyink, 2002). Polyphosphates are not that effective in controlling color when Fe(III) solids are already present. Rather, they seem to be more successful at controlling color formation when they are added to water while the iron is still in the Fe(II) form. Thus polyphosphates are best added before chlorination or oxygen oxidize Fe(II) to Fe(III). The required dose is also influenced by hardness. In the absence of hardness Kleuh and Robinson (1988) observed that about 1 mg/L of polyphosphate as PO_4^{2-} was required to sequester 2 mg Fe/L. In the presence of a calcium hardness of 100 mg/L as $CaCO_3$ as much as 5 mg PO_4^{2-}/L was required. Polyphosphates are also widely used to inhibit the precipitation of calcium carbonate and calcium sulfate. Mixtures of medium-molecular-weight polyphosphates, such as "hexametaphosphate," are generally more effective than high-molecular-weight glassy polyphosphates.

Polyphosphates were once the most widely used corrosion inhibitor for recirculating cooling towers (Butler and Ison, 1966), and in that capacity they served as an effective anodic inhibitor when used at fairly high doses (20 to 30 mg/L as PO_4) and at low pH ($pH < 6$). In drinking water applications polyphosphates have been used as corrosion inhibitors also, but results have been more mixed. In some cases the addition of polyposphates has actually increased the rate of pitting. The observed increase in the pitting rate may be due to the rather weak properties of anodic inhibition they sometimes exhibit at neutral and alkaline pHs and low doses. Polyphosphates are ineffective for corrosion inhibition in stagnant water; protection for corrosion increases with velocity (Larson, 1957).

Blends of ortho- and polyphosphates have been used successfully in situations where both corrosion inhibition and iron sequestration or hardness stabilization are objectives of the same treatment program.

In 1970, Murray, at Long Beach, California, developed a zinc orthophosphate combination that has been used successfully in many drinking water applications (Murray, 1970). It appears that Murray's formulation takes advantage of the combination of zinc as a cathodic inhibitor and orthophosphate as an anodic inhibitor (Kelly et al., 1978).

The addition of zinc orthophosphate or zinc chloride, in combination with pH adjustment, can also reduce the deterioration of asbestos–cement pipe exposed to soft waters of low alkalinity (Buelow et al., 1980; Schock and Buelow, 1981). The addition of zinc orthophosphate at dosages of 0.3 to 0.6 mg/L as Zn, with pH adjusted up to 8.2, resulted in extremely low rates of leaching of calcium from the pipe. A dense protective coating of hydrozincite [$Zn_5(OH)_6(CO_3)_2$] may have played a major role in reducing loss of calcium from the concrete. Orthophosphate was believed not to have played a role in this protective mechanism; therefore, Schock and Buelow (1981) recommended that zinc chloride be used instead.

Reiber (1989) found that orthophosphate forms a very thin, labile, protective surface that serves to inhibit the anodic reaction on copper surfaces. After 2 weeks of exposure, the anodic Tafel slope increased substantially and corrosion current dropped 80 percent.

Addition of Silicates

Sodium silicates have been studied as an alternative corrosion inhibitor. They were first employed in the 1920s to reduce corrosion of lead piping (Butler and Ison, 1966). As with polyphosphates, the molecular composition of silicates tends to be indeterminate, with a formula of $Na_2O : n(SiO_2)$, where n is a variable ratio. The neutral silicate commonly used for treatment of alkaline waters is a silicate with n of 3.2 (Vic et al., 1996). Typical dosages required for protection range from 4 to 30 mg/L as SiO_2, the higher doses being required for waters with higher hardness, higher chlorides, and/or higher dissolved solids (Lane et al., 1977).

When used at low doses, silicates, like polyphosphates, tend to operate as anodic inhibitors. As a consequence, if doses are insufficient, anodic area is reduced and pitting is exacerbated (Vic et al., 1996). Recently, the usage of silicates increased in the United States, and it appears that one of the major benefits being gained is the pH increase associated with silicate addition. Because silicates are relatively expensive, lime and sodium hydroxide should be compared as alternatives.

Based on surface analysis, it appears that silicates form a thin layer over the corroded metal that consists of ferric or other metal oxides and silicates. Therefore, some corrosion must take place for the metal surface to be protected. The scales that form are self-limiting and do not appear to build to thick layers. If silica dosage is ended, the protective films begin to break down and gradually disappear (Vic et al., 1996).

The degree of effectiveness of silicate as a corrosion inhibitor depends on the characteristics of the water. Based on extensive tests by the Illinois State Water Survey (ISWS) (Lane et al., 1977), it was found that pH controls the silicate dosage required for effective control, with higher doses needed at pH values lower than 8.5. The concentration of calcium, magnesium, chloride, and other constituents affects the optimal silicate dosage. The presence of calcium may assist inhibition by the scale that forms, while high magnesium concentrations may cause deposits of magnesium and decrease effectiveness. The ISWS concluded that silicates are the best means of inhibiting corrosion of galvanized steel and copper-based metals in domestic hot water systems, especially recirculating systems such as those used in commercial buildings. It is clear that for any particular water, detailed testing of the effect of silicate at different dosages is necessary. When added at too low a dosage, silicate may intensify the corrosion rate in some waters, while overdosage can affect the taste of the supply and cause discoloration of food.

22-13 Corrosion Testing

Test loops are often a useful way to explore treatment alternatives and to gain a better understanding of alternatives for controlling corrosion, the release of metals, and deposition of undesirable scales. The design and operation of such test loops is beyond the scope of this book but can be found in the AWWARF book on internal corrosion of water distribution systems (Snoeyink and Wagner, 1996).

Because of the potentially high costs of corrosion damage, corrosion measurement techniques have been developed to determine corrosion rates in both the field and the laboratory. Techniques have been developed to assess quantitatively the effectiveness of various corrosion control measures and the corrosion resistance of new pipe and lining materials. Until the early 1950s, corrosion measurements were almost exclusively made using weight loss methods, which are inconvenient and time consuming. Subsequently, electrochemical techniques were developed that can be used to measure instantaneous corrosion rates and can be used to continuously monitor both laboratory corrosion studies and corrosion processes in operating water systems. A good survey of corrosion monitoring methods has been published by Moreland and Hines (1979). The principal methods of corrosion measurements currently in use in the water supply field are described below.

Weight Loss Methods

There are two weight loss corrosion rate measurement methods in general use in the water field. The older is the coupon method and the newer is the machined nipple method developed by the ISWS. Both of these methods are described in ASTM Standard D2688 (ASTM, 1983).

COUPON TEST

The coupon method employs a flat metal coupon. The coupon is located in the center of the pipe where the velocity is higher than at the pipe wall. At the pipe wall the coupon may be subject to turbulence induced by the coupon holder, and this can affect scale and oxide film formation and the action of corrosion inhibitors. Coupon holders are normally installed in a bypass pipe loop to permit removal and checking of specimens without upsetting normal system operation. Multiple specimens are usually installed to allow determination of the corrosion rate as a function of time. Coupons are 13 × 102 × 0.8 mm (0.5 × 4.0 × 0.032 in.) for sheet metals and 13 × 102 × 3 mm (0.5 × 4.0 × 0.125 in.) for cast metals.

Coupons are then installed in a test loop with flow velocities adjusted to match velocities in the system under consideration. Two time series of coupons are recommended. The first set of coupons should be removed at 4- to 7-day intervals to determine initial corrosion rates. The second set is used to determine the long-term steady-state corrosion rate. The first coupon of the long-term set should be removed after 1 month with the remainder removed at 1- to 3-month intervals as the test progresses.

MACHINED NIPPLE TEST

The machined nipple test employs a short length of actual pipe material machined on the outer surface. Machining permits insertion in a PVC pipe sleeve that is connected into the pipe system under test with pipe unions. The specimen holder is designed so that a smooth flow line is maintained through the pipe specimen so as to simulate actual flow conditions. The inner surface of the pipe nipple is not altered or machined.

When monitoring corrosion in an existing system, inserts are made from the same material as used in the system being observed. A bypass pipe loop is constructed to permit removal and inspection of the pipe inserts without disturbing normal operations. A minimum upstream straight run of 1 m (3.25 ft) should be provided ahead of the ISWS tester to minimize non-uniform flow or turbulence. One insert should be exposed a minimum of 120 days before evaluation, and the second should be exposed at least 12 months if possible. The corrosion rate is then computed according to ASTM D2668.

Electrical Resistance Method

The electrical resistance method of corrosion measurement is based on the principle that the rate of change of electrical resistance of a piece of metal is directly proportional to its rate of change of cross-sectional area, which is in turn a function of the corrosion rate (Bovankovich, 1973). Electrical resistance also changes with temperature so some means of temperature compensation must be provided. A reference element is provided by connecting in a Wheatstone bridge circuit in parallel with the measuring element. The reference element is of the same material as the measuring element but is protected from the corrosive environment. The reference element is enclosed in the body of the probe to assure that it is maintained at the same temperature as the measuring element.

Linear Polarization Resistance Method

The linear polarization resistance (LPR) method is based on the principle that at low corrosion potentials, corrosion rate is essentially a linear function of polarization resistance. Commercially manufactured equipment based on this principle is available. It is normally calibrated to read corrosion rate directly as a function of corrosion current, usually at a fixed corrosion potential of $\pm$ 10 mV. The advantage of this method is that corrosion rates can be determined instantaneously.

The basic equation on which the linear polarization method is based is (Stern and Geary, 1957)

$$i_{\mathrm{corr}} = \frac{1}{2.3}\left(\frac{\Delta i}{\Delta E}\right)\left(\frac{\beta_a \beta_c}{\beta_a + \beta_c}\right) \tag{22-40}$$

where i_{corr} = corrosion current, $\mu A/cm^2$
ΔE = change in emf, mV
Δi = change in current density, $\mu A/cm^2$
β_a, β_c = Tafel slopes for anodic and cathodic reactions, mV/decade

The LPR is defined as the change in polarization with changes in current at the corrosion potential:

$$\mathrm{LPR} = \left[\frac{\Delta E}{\Delta i}\right]_{i=i_{\mathrm{corr}}} \tag{22-41}$$

$$i_{\mathrm{corr}} = \frac{1}{2.3}\left[\frac{\beta_a\beta_c/(\beta_a+\beta_c)}{\mathrm{LPR}}\right] \tag{22-42}$$

or

$$i_{\mathrm{corr}} = \frac{B}{\mathrm{UPR}} \tag{22-43}$$

where

$$B = \frac{1}{2.3}\left[\frac{\beta_a\beta_c}{\beta_a+\beta_c}\right] \tag{22-44}$$

and the units are

$$i_{\mathrm{corr}} = \mu\mathrm{A}/\mathrm{cm}^2$$
$$\mathrm{LPR} = \frac{\mathrm{mV}}{\mu\mathrm{A}/\mathrm{cm}^2}$$
$$B = \mathrm{mV}$$

When corrosion is controlled by concentration polarization at the cathodic reaction as when oxygen depolarization is controlling, β_c becomes very large and Eq. 22-44 becomes

$$B \sim \frac{1}{2.3}\beta_a \tag{22-45}$$

Direct-reading corrosion meters based on the LPR principle use this relationship. The value of E is held at a constant value, usually ± 10 mV, and the instrument is calibrated to read corrosion rate directly as a function of current.

Various researchers have presented data relating corrosion rate to LPR. The principle behind the method was actually demonstrated before Stern and Geary explained the principles behind it (Skold and Larson, 1957; Stumm, 1959). A plot amalgamating the results of both these efforts is shown on Fig. 22-40. The results in the figure suggest that, for cast iron, $B = 35.2$ mV. Values for the constant, B for a variety metals of interest in water systems, are summarized in Table 22-13.

In 1957, Stern and Geary demonstrated theoretically that polarization resistance could be used to measure corrosion rates (Stern and Geary, 1957). Unaware of this work and working in another field, Larson and Skold published a study reporting an empirical relationship between polarization resistance and the corrosion rate of cast iron (Larson and Skold, 1958a). In 1959, Stumm published an even broader set of data demonstrating that the Stern and Geary method could be used successfully to measure

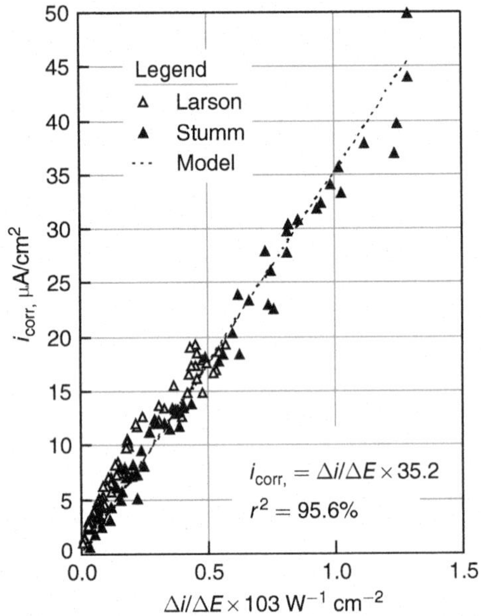

Figure 22-40
Measuring corrosion of cast iron with linear polarization resistance. (Data from Larson and Skold, 1958a; Stumm, 1959.)

Table 22-13
Summary of polarization resistance coefficients

Solution	Iron	Steel	Cast Iron	Stainless	Copper Alloys
Seawater	22	20 ± 9	29	—	22 ± 11
Miscelaneous water (includes tap water)	50 ± 28	43 ± 19	35 ± 4	—	—
Acids	15	20 ± 9	24	22 ± 6	—
Chlorides	22 ± 15	22 ± 10	60	31 ± 11	20 ± 9
Means	31	32	38	25 ± 7	20

Source: Grauer and Stumm (1982).

the corrosion rate of cast iron (Stumm, 1959). The data from both Larson and Skold (1958a) and Stumm (1959) are displayed together in Fig. 22-40, with good agreement.

Problems and Discussion Topics

22-1 Describe the differences in composition of gray cast iron and ductile iron. What are the advantages of ductile iron? What is its disadvantage?

22-2 What are the properties of lead that led to its early use as a plumbing material?

22-3 Why does corrosion not necessarily result in an increase in metal concentration in the bulk solution? Under what circumstances would the concentration of a metal increase in the water being served?

22-4 What is the potential of the O_2/H_2O half reaction at pH 7? Calculate the potential using the Nernst equation and the data in Table 22-2.

22-5 What behavior would be expected from iron, copper, zinc, or lead materials at pH 7 and E_H −250 mV? What about pH 6 and E_H +250 mV?

22-6 In an American study, it is reported that iron is corroding at a rate of 10 mpy in the water system being studied. Electrochemists like to work with the corrosion rate in A/m^2, and European corrosion engineers are comfortable with units like mm/y. What is the corrosion rate of the iron in these units?

22-7 Using Fig. 22-10, estimate the corrosion rate of copper and zinc in 2 N H_2SO_4 with and without exposure to air. Express the corrosion rate in mm/y.

22-8 Using Eq. 22-17, estimate the limiting current density for iron corroding in water containing oxygen at a concentration of 9 mg/L. What would the limiting corrosion rate be if the corroding metal were copper instead of iron? If it were lead?

22-9 Why are exchange currents hard to measure?

22-10 In Fig. 22-10, the E_{corr} for copper appears to be below its standard electrode potential. Explain why this might be true.

22-11 Zinc generally operates as a cathodic inhibitor. Will the addition of zinc raise or lower the potential of a corroding sample?

22-12 The standard electrode potential for copper is approximately 340 mV, yet Reiber (1989) observed an *E*corr of approximately 0 mV in Seattle tap water, Jones (1996) observed an E_{corr} of −350 mV in seawater, and Jinturkar et al. (1998) observed an E_{corr} of approximately −120 mV in aerated 1 N H_2SO_4. What is a possible explanation for these observations?

22-13 Zinc and steel specimens are both suspended as test electrodes in a polarization apparatus like that shown on Fig. 22-7 and held at a potential of approximately −400 mV. If the apparatus were adjusted to a potential of −600 mV, would the behavior of the two metals differ? If so, why?

22-14 A hydrogeologist is having trouble with the corrosion of the casing made of 304L stainless steel in a deep well. Could the situation be improved by replacing the existing casing with a new one made of 316 SS? The available analytical data are shown below.

Well water analysis

Constituent	Units	Concentration	Constituent	Units	Concentration
Ca	mg/L as Ca	20	Alkalinity	mg/L as $CaCO_3$	80
Mg	mg/L as Mg	18	Chloride	mg/L as Cl^-	35
Na	mg/L as Na	23	Sulfate	mg/L as $SO_4{}^{2-}$	48
K	mg/L as K	4	Si	mg/L as SiO_2	8
Iron	mg/L as Fe	2	Ammonia	mg/L as N	1
pH	Unitless	8	TDS	mg/L	150
H_2S	mg/L as H_2S	2			

22-15 Over what range of pH is the solubility of most scales minimized? Recognizing that most water systems have a pH between 6.5 and 9.5, at which extreme of these pH values is a corrosion scale most likely to form?

22-16 Which anion is thought to be the most aggressive where corrosion is concerned? What properties of this ion contribute to its aggressive behavior?

22-17 In all models of the pitting process it is assumed that a porous layer or cap will develop over the top of the pit. Why is it necessary that this cap be porous?

22-18 Explain the difference between galvanic and electrolytic corrosion.

22-19 Why are dielectric couplings often used between dissimilar metals in home construction? Under what circumstances can these couplings contribute to internal corrosion problems?

22-20 Which couple is more likely to cause galvanic corrosion? Galvanized pipe and brass or 90/10 tin–antimony solder and copper? Silver solder and copper or 90/10 tin–antimony solder and copper?

22-21 Both Stumm (1960) and Larson (1957) demonstrated that high pH results in rougher corroded surface on gray cast iron. What might be some reasons behind these observations?

22-22 Describe the five layers that form on the surface of a hot-dipped galvanized pipe. Why does the protection of the galvanizing diminish as each of these layers gradually corrodes away? Zinc is much less noble than iron, so why does it provide any protection at all. Does it not just corrode faster than the iron?

22-23 Explain the difference in protection provided by galvanizing and by tin plating when an imperfection or a place where the coating is absent is present.

22-24 Brasses exhibit a certain critical composition. Beneath the critical composition they tend to behave as pure zinc. Above that, they behave more like copper. What is that critical composition?

22-25 A water system exhibits signs of microbiologically induced corrosion. What can be done?

22-26 A water system is being designed that must handle a rather soft water with a low pH. What sort of cement will give best service under these circumstances?

22-27 A water distribution system has red water complaints and tuberculation in portions of the system. What are some alternatives that might be considered to remedy these problems?

22-28 A water distribution system with lots of new copper-plumbed homes has been receiving a number of complaints about blue water. The chemical characteristics of the water are given below. What can be done?

Water analysis

Constituent	Units	Concentration	Constituent	Units	Concentration
Ca	mg/L as Ca	5	Alkalinity	mg/L as $CaCO_3$	15
Mg	mg/L as Mg	0	Chloride	mg/L as Cl^-	2
Na	mg/L as Na	2	Sulfate	mg/L as SO_4^{2-}	0
K	mg/L as K	0	Si	mg/L as SiO_2	6
Iron	mg/L as Fe	0	Ammonia	mg/L as N	0
pH	Unitless	6.5	TDS	mg/L	25

22-29 The utility in Problem 22-28 is also having problems with the lead rule in the same homes. What might be done to mitigate this issue?

References

APHA (2000) *Standard Methods for the Examination of Water and Wastewater*, A. Eaton, L. Clesceri, and A. Greenberg (eds.), American Public Health Association, Washington, DC.

ASTM (1983) *Standard Test Methods for the Corrosivity of Water in the Absence of Heat Transfer (Weight Loss Protocol), Standard D2688-83, Method B*, American Society for Testing and Materials, Philadelphia, PA.

AWWA (1977) *AWWA Standard for Asbestos-Cement Pressure Pipe, 4 in. through 24 in., for Water and Other Liquids*, AWWA C400-77, Rev of C400-75, American Water Works Association, Denver, CO.

Bächle, A., Deschner, E., Weiss, H., and Wagner, I. (1981) "Das Korrosionsverthalten von verzinkten und unlegierten Stahlrohenren in trinkwässern mit unterschiedlicher härte und erhöhtem neutralsalzgehalt [Corrosion of Galvanized and Unalloyed Steel Tubes in Drinking Water of Different Hardness and Increased Neutral Salt Content]," *Werkstoffe und Korrosion,* **32**, 3, 435–442.

Baker, M. (1948) *The Quest for Pure Water,* American Water Works Association, Denver, CO.

Baylis, R. (1926) "Prevention of Corrosion and Red Water," *J. AWWA,* **15**, 598.

Baylis, R. (1929) Tungsten Electrode for Determining Hydrogen-Ion Concentration, U.S. Patent 7,727,094.

Bovankovich, J. (1973) "On-Line Corrosion Monitoring," *Mat. Protection Performance,* **7**, 1, 20–23.

Buelow, R.W., Millette, J. R., McFarren, E. F., and Symons, J.M. (1980) "Behavior of Asbestos-Cement Pipe under Various Water Quality Conditions: Progress Report," *J. AWWA,* **72**, 2, 91–102.

Butler, G., and Ison, H. (1966) *Corrosion and Its Prevention in Waters,* Reinhold, New York.

Campbell, H. (1950) "Pitting Corrosion of Copper Pipes Caused by Films of Carbonaceous Material Produced During Manufacture," *J. Inst. Metals,* **77**, 345–356.

Campbell, H. (1954) "A Natural Inhibition of Pitting of Copper in Tap Waters," *J. Appl. Chem.,* **4**, 633–647.

Clement, J., Hayes, M., Sarin, P., Kriven, W., Bebee, J., Jim, K., Beckett, M., Snoeyink, V., Kirmeyer, G., and Pierson, G. (2002) *Development of Red Water Control Strategies,* American Water Works Association Research Foundation, Denver, CO.

Cohen, M., and Lyman, W. (1972) "Service Experience with Copper Plumbing Tube," *Mat. Protection Performance,* **11**, 48–53.

Cox, C. (1934) "Equipment for the Chlorination of Small Water Supplies," *J. AWWA,* **26**, 11, 1587–1601.

Cruse, H. (1971) "Dissolved-Copper Effect on Iron Pipe," *J. AWWA,* **63**, 2, 79–81.

Cruse, H., and Pomeroy, R. (1974) "Corrosion of Copper Pipes," *J. AWWA,* **66**, 2, 479–483.

DeMartini, F. (1938) "Practical Application of the Langelier Method," *J. AWWA* **30**, 1, 85–111.

Donahue, F. (1972) Corrosion and Corrosion Control, Chap. 10, in W. Weber (ed.), *Physiochemical Processes for Water Quality Control,* Wiley-Interscience, New York.

Dye, J. (1944) "The Calculation of Alkalinities and Free Carbon Dioxide in Water by the Use of Nomographs," *J. AWWA,* **36**, 6, 895–900.

Dye, J. (1952) "Calculation of Effects of Temperature and pH, Free Carbon Dioxide and the Three Forms of Alkalinity," *J. AWWA,* **44**, 4, 356–372.

Dye, J. (1958) "Correlation of the Two Principal Methods of Calculating the Three Forms of Alkalinity," *J. AWWA,* **50**, 12, 800–820.

Edwards, M., Meyer, T., and Rehring, J. (1994) "Effect of Selected Anions on Copper Corrosion Rates," *J. AWWA,* **86**, 3, 73–78.

Engell, H., and Stolica, N. (1959) "Die kinetik der entstehung und des wachs-tums von lochfraßstedllen afu passiven eisenelektroden," *Z. Phys. Chem.*, **20**, S. 113–120.

Evans, U. (1948) *Introduction to Metallic Corrosion*, Edmund Arnold, London.

Ferguson, J., van Franqué, O., and Schock, M. (1996) Corrosion of Copper in Potable Water Systems, in V. Snoeyink and I. Wagner (eds.), *Internal Corrosion of Water Distribution Systems*, 2nd ed., American Water Works Association Research Foundation and DVGW, Denver, CO.

Fontana, M. (1985) *Corrosion Engineering*, 3rd ed., McGraw-Hill, New York.

Fontana, M., and Greene, N. (1967) *Corrosion Engineering*, McGraw-Hill, New York.

Fox, K., Tate, C., Treweek, G., and Bowers, G. (1983) "Interior Surface of Galvanized Steel Pipe: A Potential Factor in Corrosion Resistance," *J. AWWA*, **75**, 2, 84–86.

Franklin, B. (1769) *Experiments and Observations on Electricity, Made at Philadelphia in America ... to Which Are Added, Letters and Papers on Philosophical Subjects. The Whole Corrected, Methodized, Improved, and Now First Collected into One Volume, and Illustrated with Copper Plates*, David Henry, London.

Gilbert, P. (1948) "An Investigation into the Corrosion of Zinc and Zinc-Coated Steel in Hot Waters," *Sheet Metal Industries*, **10**, 2003–2012, 11, 2243–2254, and 12, 2441–2460.

Grasha, L., Risner, W., and Davis, M. (1981) Corrosive Effects of Dissolved Copper on Galvanized Steel, Phase II, Internal Report, Metropolitan Water District of Southern CA, Laverne, CA.

Grauer, R., and Stumm, W. (1982) "Die koordinationschemie oxidischer grenzflächen un ihre auswirkung auf die auflösungskinetik oxidischer festphasen in wäßrigen lösungen," *Colloid Polym. Sci.*, **260**, 10, 959–970.

Green, J. (1950) "Reversion of Molecularly Dehydrated Sodium Phosphates," *Ind. Eng. Chem.*, **42**, 8, 1542–1551.

Harned, H., and Owen, B. (1958) *Physical Chemistry of Electrolyte Solutions*, 3rd ed., Reinhold, New York.

Hatch, G. (1961) "Unusual Cases of Copper Corrosion," *J. AWWA*, **53**, 11, 1417–1428.

Herro, H. M., and Port, R. D. (1993) *The Nalco Guide to Cooling Water System Failure Analysis*, 1st ed., McGraw-Hill, New York.

Hoover, C. (1938) "Practical Application of the Langelier Method," *J. AWWA*, **30**, 11, 1802–1807.

Hoxeng, R. (1950) "Electrochemical Behavior of Zinc and Steel in Aqueous Media—Part II," *Corrosion*, **6**, 10, 308–312.

Hoxeng, R., and Prutton, C. (1949) "Electrochemical Behavior of Zinc and Steel in Aqueous Media," *Corrosion*, **5**, 10, 330–338.

Ives, D., and Rawson, A. (1962a) "Copper Corrosion I: Thermodynamic Aspects," *J. Electrochem. Soc.*, **109**, 6, 447–451.

Ives, D., and Rawson, A. (1962b) "Copper Corrosion II: Kinetic Studies," *J. Electrochem. Soc.*, **109**, 6, 452–457.

Ives, D., and Rawson, A. (1962c) "Copper Corrosion III: Electrochemical Theory of General Corrosion," *J. Electrochem. Soc.*, **109**, 6, 458–461.

Ives, D., and Rawson, A. (1962d) "Copper Corrosion IV: The Effect of Saline Additions," *J. Electrochem. Soc.*, **109**, 6, 462–465.

Jinturkar, P., Guan, Y., and Han, K. (1998) "Dissolution and Corrosion Inhibition of Copper, Zinc, and Their Alloys," *Corrosion*, **54**, 12, 106–114.

Jones, D. (1984) "Polarization Studies of Brass-Stell Galvanic Couples," *Corrosion*, **40**, 2, 181–185.

Jones, D. (1996) *Principles and Prevention of Corrosion*, 2nd ed., Prentice Hall, Upper Saddle River, NJ.

Kelley, M., Kise, M., and Steketee, F. (1978) "Zinc/Phosphate Combinations Control Corrosion in Potable Water Distribution Systems," *Mat. Protection Performance*, **12**, 1, 28–30.

Kenworthy, L. (1943) "The Problem of Copper and Galvanized Iron in the Same Water System," *J. Inst. Metals*, **69**, 1, 67–90.

Kirmeyer, G., Richards, W., and Smith, C. (1994) *An Assessment of Water Distribution Systems and Associated Research Needs*, American Water Works Association Research Foundation, Denver, CO.

Kleuh, K., and Robinson, R. (1988) "Sequestration of Iron in Groundwater by Polyphosphates," *J. Environ. Eng.*, **114**, 5, 1192–1199.

Kolle, W., and Rosch, H. (1980) "Untersuchungen an rohnetzinkrustierungen unter mineralogischen Gesichtspunkten," *Vom Wasser*, **55**, 159–163.

Kruse, C. (1983) Korrosionsverhalten von Zink und Feuerverzinktem Stahl in Kaltem und ewärmtem Trinkwasser unter besonderer Berücksichtigung des Transportres in Rohrleitungen, Ph.D. Dissertation, Technische Univedrsität, München.

Lane, R., Larson, T., and Neff, R. (1973) "Silicate Treatment Inhibits Corrosion of Galvanized Steel and Copper Alloys," *Mat. Protection Performance*, **12**, 1, 32–37.

Lane, R., Larson, T., and Schilsky, S. (1977) "The Effect of pH on the Silicate Treatment of Hot Water in Galvanized Piping," *J. AWWA*, **69**, 8, 457–460.

Lange, N. (1985) *Lange's Handbook of Chemistry*, J., Dean, (ed.), McGraw-Hill, New York.

Langelier, W. (1936) "The Analytical Control of Anti-Corrosion Water Treatment," *J. AWWA*, **28**, 11, 1500–1522.

Larson, T. (1957) "Evaluation of the Use of Polyphosphates in the Water Industry," *J. AWWA*, **49**, 12, 1581–1586.

Larson, T. E. (1975) *Corrosion by Domestic Waters*, ISWS 75 Bulletin 59, Illinois State Water Survey, Urbana, IL.

Larson, T., and Buswell, A. (1942) "Calcium Carbonate Saturation Index and Alkalinity Interpretations," *J. AWWA*, **34**, 11, 1667–1684.

Larson, T., and King, R. (1954) "Corrosion by Water at Low Flow Velocity," *J. AWWA*, **46**, 1, 1–9.

Larson, T., and Skold, R. (1958a) "Laboratory Studies Relating Mineral Quality of Water to Corrosion of Steel and Cast Iron," *Corrosion*, **13**, 8, 285–289.

Larson, T., and Skold, R. (1958b) "Current Research on Corrosion and Tuberculation of Cast Iron," *J. AWWA*, **50**, 11, 1429–1432.

Lehrman, L., and Schuldener, H. (1951) "The Role of Sodium Silicate in Inhibiting Corrosion by Film Formation on Water Piping," *J. AWWA*, **43**, 3, 175–179.

Lehrman, L., and Schuldener, H. (1952) "Action of Sodium Silicate as a Corrosion Inhibitor in Water Piping," *Ind. Eng. Chem.*, **44**, 12, 1765–1769.

Leroy, P., Schock, M., Wagner, I., and Holtschutte, H. (1996) Cement-Based Materials, in V. Snoeyink and I. Wagner (eds.), *Internal Corrosion of Water Distribution Systems*, 2nd ed., American Water Works Association Research Foundation and DVGW, Denver, CO.

Lide, D. (ed.) (2000) *Handbook of Chemistry and Physics*, 81st ed., CRC Press, New York.

Lin, C., and Olson, B. (1995) "Occurrence of *cop*-like Copper Resistance Genes among Bacteria Isolated from a Water Distribution System," *Can. J. Microbiol.*, **41**, 7, 643–646.

Lucey, V. (1967) "Mechanism of Pitting Corrosion of Copper in Supply Waters," *Br. Corrosion J.*, **2**, 9, 175–175.

Lytle, D., and Snoeyink, V. (2002) "The Effect of Ortho- and Polyphosphates on the Properties of Iron Particles and Suspensions," *J. AWWA*, **94**, 10, 87–99.

McCauley, R. (1960) "Controlled Deposition of Protective Calcite Coatings on Water Mains," *J. AWWA*, **52**, 11, 1386–1396.

McKee, C. (1932) "Effect of Copper Sulphate on Pipe Materials," *Power Plant Eng.*, **36**, 1, 658–659.

Merrill, D., and Sanks, R. (1977a) "Corrosion Control by Deposition of $CaCO_3$ Films, Part I," *J. AWWA*, **69**, 11, 592–599.

Merrill, D., and Sanks, R. (1977b) "Corrosion Control by Deposition of $CaCO_3$ Films, Part 2," *J. AWWA*, **69**, 12, 634–640.

Merrill, D., and Sanks, R. (1978) "Corrosion Control by Deposition of $CaCO_3$ Films, Part 3," *J. AWWA*, **70**, 1, 12–18.

Moore, D. (1993) "The Riddle of Ancient Roman Concrete," *The Spillway*, Newsletter of the U.S. Dept. of Interior, Bureau of Reclamation, Upper Colorado Region, Denver, CO.

Moreland, P., and Hines, J. (1979) "Corrosion Monitoring, Select the Right System," *Hydrocarbon Process.*, **57**, 11, 251–255.

Morgen, R. A., and Swoope, R. L. (1943) "The Useful Life of Pyro-, Meta-, and Tetraphosphates," *Ind. Eng. Chem.*, **35**, 7, 822–825.

Mullen, E. (1974) "Potable Water Corrosion Control," *J. AWWA*, **66**, 8, 474–479.

Murray, B. (1970) "A Corrosion Inhibitor Process for Domestic Water," *J. AWWA*, **62**, 10, 659–662.

Murray, B. (1971) "A Corrosion Inhibitor Process for Domestic Water," *J. AWWA*, **63**, 10, 659–662.

Nancollas, G. (1983) "Phosphate Precipitation in Corrosion Protection, Reaction Mechanisms," *Corrosion*, **39**, 3, 77–82.

Newman, R. (2001) "W.R. Whitney Award Lecture: Understanding the Corrosion of Stainless Steels," *Corrosion*, **57**, 12, 1030.

Nielsen, A. (1964) *Kinetics of Precipitation, Pergamon*, MacMillan, New York.

Obrecht, M., and Quill, L. (1960) "How Temperature and Velocity of Potable Water Affect Corrosion of Copper and Copper Alloys," etc. (several articles), *Heating Plumbing Air Conditioning*, **35**, 1, 165–169; 3, 109–116; 4, 131–137; 5, 105–113; 7, 115–122; 9, 125–133.

Obrecht, M., and Quill, L. (1961) "How Temperature, Velocity of Potable Water Affects Corrosion of Copper and Its Alloys," *Heating Plumbing Air Conditioning*, **36**, 4, 129–134.

Oliphant, R., and Schock, M. (1996) Copper Alloys and Solders, in V. Snoeying and I. Wagner (eds.), *Internal Corrosion of Water Distribution Systems*, 2nd ed,American Water Works Association Research Foundation and DVGW, Denver, CO.

Pisigan, R., and Singley, E. (1985) "Effects of Water Quality Parameters on the Corrosion of Galvanized Steel," *J. AWWA*, **77**, 11, 76–82.

Plummer, L., and Busenberg, E. (1982) "The Solubilities of Calcite Aragonite and Vaterite in CO_2-H_2O Solutions between 0 and 90 Degrees C, and an Evaluation of the Aqueous Model for the System $CaCO_3$-CO_2-H_2O," *Geochim. Cosmochim. Acta*, **46**, 1, 1011–1023.

Pou, T., Murphy, P., Young, V., and Bockris, J. (1984) "Passive Film on Iron: The Mechanism of Breakdown in Chloride Containing Solutions," *J. Electrochem. Soc.*, **131**, 6, 1243–1252.

Pourbaix, M. (1949) *Thermodynamics of Dilute Aqueous Solution, with Applications to Electrochemistry and Corrosion*, Edward Arnold, London.

Price, M. (1997) Control of Magnesium Silicate Scaling in Austin's Lime-Softened Water, in *Proceedings of AWWA WQTC (on CD)*, American Water Works Association, Denver, CO.

Rambow, C., and Holmgren, R. (1965) "Technical and Legal Aspects of Copper Tube Corrosion," *J. AWWA*, **58**, 347–353.

Reiber, S. (1991) "Galvanic Stimulation of Corrosion on Lead-Tin Solder-Sweated Joints," *J. AWWA*, **83**, 7, 83–91.

Reiber, S. H. (1989) "Copper Plumbing Surfaces: An Electrochemical Study," *J. AWWA*, **81**, 4, 114–122.

Reiber, S., Ferguson, J., and Benjamin, M. (1986) An Innovative Technique for Corrosion Rate Measurements within the Distribution System, in *Proc. AWWA WQTC*, Portland, OR, American Water Works Association, Denver, CO.

Revie, R., and Uhlig, H. (1985) *Corrosion and Corrosion Control*, 3rd ed, Wiley, New York.

Rice, O., and Hatch, G. (1939) "Threshold Treatment of Municipal Water Supplies," *J. AWWA*, **31**, 7, 1171–1185.

Roberge, P. (1999) *Handbook of Corrosion Engineering*, McGraw-Hill, New York.

Ryznar, T. (1944) "A New Index for Determining the Amount of Calcium Carbonate Scale Formed by a Water," *J. AWWA*, **36**, 4, 472–486.

Sarin, P., Snoeyink., V., Bebee, J., Krivenk, W., and Clement, J. (2001) "Physiochemical Characteristics of Scales in Old Iron Pipes," *Water Res.*, **35**, 12, 2961–2969.

Schikorr, G. (1939) "The Cathodic Behavior of Zinc versus Iron in Hot Tap Water," *Trans. Electrochem Soc.*, **76**, 247–258.

Schock, M., and Buelow, R. (1981) "The Behavior of Asbestos-Cement Pipe under Various Water Quality Conditions: Part 2, Theoretical Considerations," *J. AWWA*, **73**, 636–651.

Schock, M., and Wagner, I. (1985) Corrosion and Solubility of Lead in Drinking Water, Chap. 4, in *Internal Corrosion of Water Distribution Systems, a Cooperative Research Report,* AWWARF, FVGW-Forschungsstelle, American Water Works Research Foundation, Denver, CO.

Schock, M., Wagner, I., and Oliphant, R. (1996) Corrosion and Solubility of Lead in Drinking Water, in V. Snoeyink and I. Wagner (eds). *Internal Corrosion of Water Distribution Systems,* 2nd ed., a Cooperative Research Report, AWWARF, FVGW—Technologiezentrum Wasser, American Water Works Research Foundation, Denver, CO.

Shull, K., and Becker, R. (1960) "Cold Water Corrosion of Copper Tubing," *J. AWWA*, **59**, 8, 1033–1040.

Skold, R., and Larson, T. (1957) "Measurement of Instantaneous Corrosion Rate by Means of Polarization Data," *Corrosion,* **13**, 3, 139–142.

Snoeyink, V. (2002) *Aluminum Magnesium and Iron Scales in Distribution Systems,* Association of Environmental Engineering and Sciences Professors Lecture, American Water Works Association Annual Conference, New Orleans, LA, available on CD, American Water Works Association, Denver, CO.

Snoeyink, V., and Wagner, I. (1996) "Corrosion and Solubility of Lead in Drinking Water," Chap. 4, in V. Snoeyink and I. Wagner (eds.) *Internal Corrosion of Water Distribution Systems,* 2nd ed., a Cooperative Research Report, AWWARF, FVGW—Technologiezentrum Wasser, American Water Works Association Research Foundation, Denver, CO.

Sontheimer, H., Kolle, W., and Rudek, R. (1979) "Aufgaven und Methoden der Wasser-chemie—dargestellt an der Entwidklung der Erkenntnisse zur bildung von Korrosions-schutzschichten auf Metallen," *Vom Wasser,* **52**, 1–12.

Sontheimer, H., Kolle, W., and Snoeyink, V. (1981) "The Siderite Model of the Formation of Corrosion-Resistant Scales," *J. AWWA,* **73**, 11, 572–579.

Soukatchoff, P. (1990) "Résistance des Mortiers aux Eaux Agressives," *TSM,* **86,4**, 197–201.

Speller, F. (1926) *Corrosion, Causes and Prevention,* McGraw-Hill, New York.

Stern, M. (1955) "The Electrochemical Behavior Including Hydrogen over Voltage of Iron in Acid Environment," *J. Electrochem. Soc.*, **102**, 11, 609–616.

Stern, M., and Geary, A. (1957) "Electrochemical Polarization—A Theoretical Analysis of the Shape of the Polarization Curves," *J. Electrtochem. Soc.*, **104**, 1, 56–63.

Stumm, W. (1956) "Calcium Carbonate Deposition at Iron Surfaces," *J. AWWA,* **48**, 3, 300–310.

Stumm, W. (1959) "Evaluation of Corrosion in Water by Polarization Data," *Ind. Eng. Chem.*, **51**, 12, 1487–1490.

Stumm, W. (1960) "Investigations of Corrosive Behavior of Waters," *J. ASCE, San Div.*, **86**, A-6, 27–45.

Stumm, W. (1990) "The Coordination Chemistry of the Oxide-Electrolyte Interface: The Dependence of Surface Reactivity (Dissolution, Redox Reactions) on Surface Structure," *Croatica Chem. Acta,* **63**, 277–312.

Stumm, W. (1995) The Inner-Sphere Surface Complex, pp. 1–32, in *Aquatic Chemistry: Interfacial and Interspecies Processes,* American Chemical Society, Washington, DC.

Stumm, W. (1998) *Corrosion of Metals in Aqueous Systems,* Schriftenreihe der EAWAG NR, 12, 39. Swiss Federal Institute for Environmental Science and Technology, EAWAG, Zurich, Switzerland.

Stumm, W., and Morgan, J. (1996) *Aquatic Chemistry,* 3rd ed., Wiley, New York.

Sutherland, A., and Tekippe, R. (1972) "Effects of Insulating Coupling Lengths for Corrosion Control," *Mat. Protection Performance,* **11**, 5, 31–34.

Szlarska-Smialowska, Z. (1971) "Review of Literature on Pitting Corrosion Published Since 1960," *Corrosion,* **27**, 5, 222–233.

Szlarska-Smialowska, Z. (1986) *Pitting Corrosion of Metals,* National Association of Corrosion Engineers (NACE), Houston, TX.

Tafel, J. (1905) "über de polarisation bei kathodischer wasserstoffentwicklung," *Zeitschrift Physik. Chemie.,* **50**, 641–712.

Talbot, H., and Blanchard, A. (1905) *Electrolytic Dissociation Theory and Some of Its Applications,* MacMillan, New York.

Thresh, J. (1922) "The Action of Natural Waters on Lead, Part I. Effect of Various Saline Constituents" *Analyst — Proceedings of the Society of Public Analysts and Other Analytical Chemists (Br.)* **47**, 11, 459–469.

Tillmans, J. (1913) *Water Purification and Sewage Disposal,* Constable, London.

Tillmans, J., and Heublein, O. (1913) "Investigation of the Carbon Dioxide Which Attacks Calcium Carbonate in Natural Waters," *Gesundh. Ing.,* **35**, 8, 669–677.

Treweek, G., Trussell, R., and Pomeroy, R. (1978) Copper-induced Corrosion of Galvanized Steel Pipe, paper presented at the Sixth Annual AWWA WQTC, Denver, CO.

Trussell, R. (1972) *Systematic Aqueous Chemistry of Engineers,* Ph.D. Dissertation, University of California, Berkeley, CA.

Trussell, R., Russell, L., and Thomas, J. (1976) The Langelier Index, paper presented at the Sixth AWWA WQTC, Kansas City, MO.

Trussell, R., and Wagner, I. (1985) Corrosion of Galvanized Pipe, in V. Snoeyink and H. Sontheimer (eds.), *Internal Corrosion of Water Distribution Systems,* American Water Works Association Research Foundation, Denver, CO.

Trussell, R. R. (1998) "Spreadsheet Water Conditioning," *J. AWWA,* **90**, 6, 70–81.

Trussell, R., and Wagner, I. (1996) Corrosion of Galvanized Pipe, in V. Snoeyink and I. Wagner (eds.), *Internal Corrosion of Water Distribution Systems,* 2nd ed, a Cooperative Research Report, AWWARF, FVGW—Technologiezentrum Wasser, American Water Works Association Research Foundation, Denver, CO.

Uhlig, H. (1948) *Corrosion Handbook,* Wiley, New York.

Uhlig, H. (1967) *Corrosion and Corrosion Control,* Wiley, New York.

U.S. EPA (1991) "Maximum Contaminant Level Goals and National Primary Drinking Water Regulations for Lead and Copper, Final Rule," *Fed. Reg.*, **56**, 110, 26460–26564.

Vic, E., Ryder, R., Wagner, I., and Ferguson, J. (1996) Mitigation of Corrosion Effects, in V. Snoeyink and I. Wagner (eds.), *Internal Corrosion of Water Distribution Systems*, 2nd ed., a Cooperative Research Report, AWWARF, FVGW—Technologiezentrum Wasser, American Water Works Association Research Foundation, Denver, CO.

Wagner, C., and Traud, W. (1938) "Zeitschrift für Elektrochemie und Angewandte Physikalische Chemie," *Z. Elektrocherm.*, **44**, 7, 391–454.

Werner, G., Wurster, E., and Sontheimer, H. (1973) "Korrosionsversuche des Zweckverbandes Landeswasserversorgung mit Feuerverzinkten. Stahlrohren," *Gwf-Das Gas Wasserfach*, **114**, 105–113.

23 Potable Reuse

Stantec's Water Treatment: Principles and Design, Updated Third Edition. John C. Crittenden, R. Rhodes Trussell, David W. Hand, Kerry J. Howe, and George Tchobanoglous.

Terminology for Potable Reuse

Term	Definition
Advance treated water (ATW)	Municipal wastewater that has been treated by conventional wastewater treatment followed by advanced water treatment processes for the removal of pathogens and chemicals, in preparation for its use as a drinking water supply.
Advanced water treatment facility (AWTF)	The treatment facility where ATW is produced. The specific combination of treatment technologies employed will depend on the quality of the treated wastewater and the type of potable reuse (i.e., IPR or DPR).
Constituents of emerging concern (CECs)	Constituents in water that may cause ecological or human health impacts that are typically not regulated and/or not routinely monitored but may be candidates for study and future regulation.
Constituents of concern (COCs)	Constituents that have an adverse effect on human health that are regulated in drinking water or under consideration for regulation.
De facto potable reuse	The use of a natural water source that contains treated wastewater from other communities as a source of drinking water, without explicitly acknowledging the interconnection between the wastewater and water supply.

Term	Definition
Direct potable reuse (DPR)	Potable reuse, in which ATW is piped directly from the AWTF to a drinking water treatment facility or distribution system without passing through an intervening environmental buffer.
Engineered storage buffer (ESB)	An enclosed water storage facility used to provide retention time before ATW is introduced into the drinking water treatment facility or distribution system to (1) conduct testing to evaluate water quality or (2) hold the water if it does not meet specifications.
Environmental buffer	A groundwater aquifer or surface water body into which ATW is introduced before being withdrawn for potable reuse. In some cases, environmental buffers allow for (1) response time if the ATW does not meet specifications and (2) time for natural processes to affect water quality.
Indirect potable Reuse (IPR)	Potable reuse, in which ATW is discharged to and stored in an environmental buffer for a specified period prior to its use as a drinking water supply.
Log (base 10) reduction value (LRV)	A reduction in the concentration of a constituent or microorganism expressed on a logarithmic scale, in which the value indicates the reduction in factors of 10. In other words, an LRV of 1 corresponds to a reduction of 90% from the original concentration, an LRV of 2 corresponds to a reduction of 99%, etc. See Sec. 4-5 for the LRV equation.
Nonpotable reuse	A general term for all water reuse applications except those related to potable reuse.
Potable reuse (PR)	The augmentation of a drinking water supply with ATW.
Source control	The control or elimination of the discharge of constituents into a wastewater collection system that can impact wastewater treatment, are difficult to treat, or can impair the quality of the secondary effluent being delivered to an AWTF.

Potable reuse is the term used for the deliberate recycling of treated municipal wastewater into a municipal potable water supply system. Recycling of wastewater is becoming an important water supply option for communities as traditional sources of fresh water become less reliable because of increasing demand and uncertainty about future supply and as appropriate treatment technologies become available. Potable reuse can provide unique advantages to communities but must be properly implemented to ensure protection of public health.

In the modern urban water cycle, water is withdrawn from the environment, treated to compliance with the Safe Drinking Water Act (SDWA), used throughout the community, discharged to wastewater collection systems, treated in a wastewater treatment facility (WWTF) to compliance with the Clean Water Act (CWA), and then dispersed back into the environment. Drinking water and wastewater systems are regulated, designed, and operated separately. In potable reuse, this historical decoupling of wastewater and drinking water is eliminated.

Use of treated wastewater as a deliberate component of a city's water supply involves a myriad of complex issues. The objective of this chapter is to introduce issues that must be considered for the use of wastewater as a source of potable water. Because the field of potable reuse is evolving rapidly, the material presented in this chapter is meant to serve an as introduction to the subject.

23-1 Introduction to Potable Reuse

The purpose of this introductory section is to: (1) introduce the principal components of a potable reuse program, (2) introduce the various types of potable reuse, (3) identify and discuss the driving forces for potable reuse, and (4) present the evolution of potable reuse and the implications for the management of public water supplies.

Components of a Potable Reuse Program

A successful potable reuse program must effectively integrate three principal components: (1) technical aspects related to the production of safe drinking water, (2) regulatory considerations, and (3) effective public outreach and communication.

The technical aspects arise from the specific characteristics of wastewater as a source water. Treating wastewater to potable quality involves a multibarrier approach that includes source control programs to ensure difficult-to-treat harmful substances are not discharged to the wastewater collection system, modifying existing biological wastewater treatment process to optimize performance for subsequent additional treatment, providing advanced water treatment to remove constituents that are not removed in the biological wastewater treatment plant, providing posttreatment to prevent corrosion in the distribution system and ensure compatibility for blending with other water sources, employing an environmental or engineered buffer to provide additional treatment and time to detect and respond to system malfunction, and advanced monitoring to detect treatment that does not meet specifications and mount an appropriate response. These technical subjects are considered in Sec. 23-2 and Secs. 23-4 through 23-9.

Regulatory considerations are discussed in Sec. 23-3 and include the development of new regulations and standards for potable reuse that are protective of public health. Public outreach, discussed in Sec. 23-10, deals with how support is gained with the public for the implementation of a potable reuse program. Because the success of a potable reuse program depends on these three components, it is important to allocate appropriate resources to each of these components in the project planning phase.

Types of Potable Reuse

Potable reuse can be unplanned or planned, and when it is planned the use can be indirect or direct. These various uses of wastewater are described below.

UNPLANNED POTABLE REUSE

If a downstream community withdraws water from the same water body used for wastewater discharge by an upstream community, the downstream community is clearly reusing some portion of the upstream community's treated wastewater. This default scenario is known as *de facto* potable reuse. *De facto* potable reuse involves compliance with the SDWA and CWA but does not consider how much dilution might have occurred, whether additional degradation of constituents may have occurred in the natural water body, and the extent to which the upstream treated wastewater discharges may influence the water quality at the downstream withdrawal point (Asano et al., 2007; NRC, 2012). It is important to note that *de facto* reuse is the norm; for example, in 2002 the USGS documented the presence of pharmaceuticals in surface waters throughout the United States (Kolpin et al., 2002), and Rice et al. found that wastewater flow upstream of some water treatment facilities increased by 68% between 1980 and 2008 (Rice et al., 2013).

PLANNED POTABLE REUSE

Planned potable reuse projects start with the acknowledgement that treated wastewater will be part of the water supply. The advanced treatment processes and other features incorporated in planned potable reuse projects provide water of higher quality than water supplied by the *de facto* reuse scenario. Potable reuse is described as indirect potable reuse (IPR) or direct potable reuse (DPR). The defining difference between IPR and DPR is that the advance treated water (ATW) in an IPR system is discharged into an environmental buffer before being withdrawn and used for potable purposes.

INDIRECT POTABLE REUSE

IPR systems employ an environmental buffer that comprises either groundwater replenishment or surface water augmentation as illustrated on Fig. 23-1. Groundwater replenishment, as discussed in Sec. 23-8, can be achieved by injecting ATW directly into an aquifer via an injection well, or by placing the water in spreading basins and allowing it to percolate

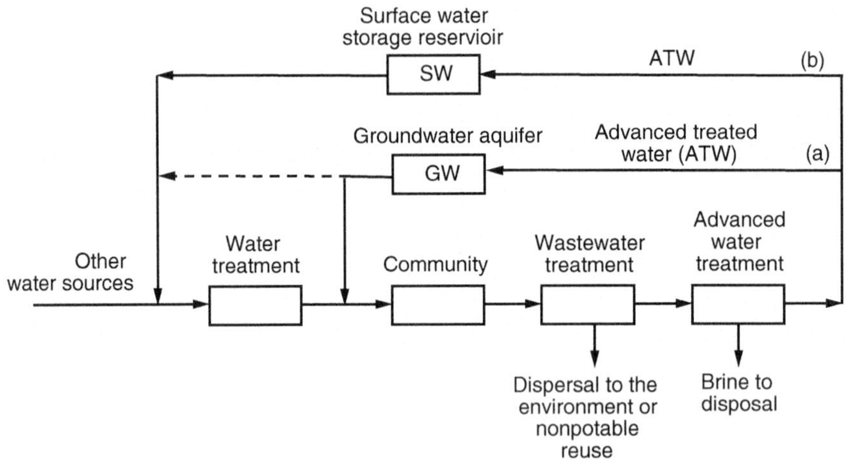

Figure 23-1
Flow diagram for indirect potable reuse, (a) with groundwater replenishment and (b) with surface water augmentation.

into the aquifer. Percolation through the overlying soil and alluvium allows an opportunity for soil aquifer treatment (SAT), an additional level of treatment discussed in Sec. 23-6.

Surface water augmentation involves discharge to a terminal reservoir, meaning a reservoir that is used as a water supply for a water treatment plant but does not have other outlets. Water is open to the atmosphere, such that birds and animals can interact with the water and constituents can enter the water through rainfall, atmospheric deposition, or being carried with the wind. Water from a surface water augmentation reservoir must be treated in a surface water treatment plant in compliance with the surface water treatment rule (SWTR) prior to being distributed to the community.

Discharge of ATW into an environmental buffer serves several purposes. First, additional degradation of constituents may occur by natural processes (e.g. photolysis, hydrolysis, sedimentation, and biodegradation) in the environmental system. Second, mixing the ATW with naturally occurring water provides some dilution of short-term quality variations by dispersion. Third, and perhaps most importantly, the environmental buffer serves as storage that provides a significant time delay between the treatment of the wastewater and delivery of potable water to customers. Thus, if a treatment failure occurs, the water system operators have time to detect and correct the failure to minimize risk to public health. A potential downside of IPR is that, because the water is discharged into the environment, it has an opportunity to be exposed to new contamination and may need to be treated in a conventional water treatment plant (in compliance with SWDA regulations) before being distributed to customers. Furthermore, unregulated contaminants in the advanced treated water may be regulated in the future, leading to expensive mitigation or treatment options.

DIRECT POTABLE REUSE

A DPR system does not use an environmental buffer. As illustrated on Fig. 23-2, a DPR system can be designed with or without an engineered storage buffer (ESB). Typically, an ESB will be a tank or covered reservoir. DPR eliminates the potential for recontamination that may occur in an IPR system, but it also eliminates the potential for beneficial natural treatment. The ATW can be blended with other water directly before treatment in a water treatment facility or, if the AWTF is permitted as a drinking water facility, the ATW can be added directly to the water distribution system.

The ESB must have enough residence time to allow monitoring of key performance indicators to ensure that the treatment system is operating properly, so that operators are able to halt the AWTF if off-spec water is detected. Monitoring becomes much more critical in a DPR system, as the residence time of the ESB in an DPR system will be measured in hours compared to a residence time measured in months for an environmental buffer of an IPR system.

THE TRANSITION BETWEEN IPR AND DPR

For an IPR project as on Fig. 23-1, the scale of the environmental buffer is an important design consideration. As the scale of the environmental buffer diminishes so does its contribution to project performance. Eventually, as that scale diminishes, the project is treated as DPR as shown on Fig. 23-2. As of 2021, it is difficult to identify objective, scientific measures that can be used to justify the classification as IPR or DPR, as a result that

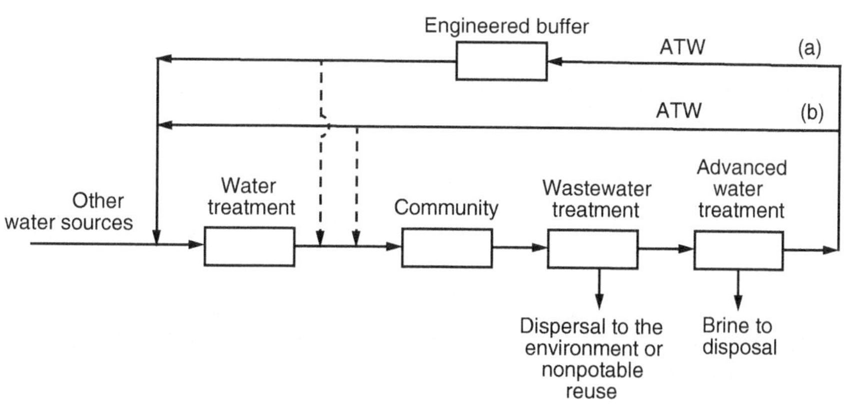

Figure 23-2
Flow diagrams for direct potable reuse, (a) with engineered buffer and (b) without engineered buffer.

designation is generally a decision by local regulators and practice varies from one jurisdiction to the next.

Driving Forces for Potable Reuse

A variety of forces are contributing to the increasing interest in potable reuse. First, water demand is increasing in urban areas because of population growth and migration patterns. From 1900 to 2020, the United States population increased by more than a factor of 4, from 76 million to 330 million people. During that same time, the fraction of the population in urban areas increased from 40 to more than 80%. Conservation can reduce per-capita demand; the USGS has published data indicating that total water use peaked in 1975 even though population has continued to grow (Dieter et al., 2017). Nevertheless, this increased population density in urban areas has created numerous areas in which the water demand is greater than the local supply.

Drought and climate change are contributing to water shortages.
http://www.stantec.com/wt3r/23-1.html

DECLINE IN SOURCE WATER SUPPLY

Multiple factors are contributing to a decline in supply. In some areas, decades of over pumping of groundwater have left aquifers depleted. Elsewhere, surface water availability is being reduced because of conflicts between municipalities and other users over water rights, reductions in diversions to preserve in-stream flows for environmental protection, and overall reductions due to long-term drought and climate change. As an example of the simultaneous forces of increasing demand and declining supply, 85–90% of the municipal water supply in San Diego, CA, is imported hundreds of miles, from Northern California or the Colorado River (CDM Smith, 2021). The Colorado River is overallocated and California may face additional shortfalls when other states claim their full allocations. Aqueducts used to import water are susceptible to leakage and evaporation. The water is used a single time by homeowners, industry, and commercial business, discharged to the municipal sewer system, treated in wastewater treatment plants, and discharged into the Pacific Ocean. Recycling of wastewater would allow a reduction of the amount of water to be imported on a gallon-for-gallon basis.

ENERGY USAGE AND ENVIRONMENTAL IMPACTS

Depending on elevation and pumping requirements, importing water large distances can be expensive and energy intensive. An alternative source in coastal areas is seawater desalination. The prevalent seawater desalination process in the United States is reverse osmosis (see Chap. 17), which is also expensive and energy intensive. The comparison between importation, desalination, and water reuse is site specific, but several studies have shown that potable reuse can have lower costs and environmental impacts than importing water or desalinating seawater (Chamberlain et al., 2020; Stokes and Horvath, 2006).

IMPROVED TREATMENT CAPABILITIES

Improvements in wastewater and advanced water treatment technologies, analytical capabilities, and process monitoring have been drivers for increased interest in potable reuse. It is now possible to measure a wide variety of constituents in water at concentrations of nanograms/liter and to design treatment processes to reliably and cost-effectively remove constituents to below detection. Improvements in pathogen removal are particularly notable. More sophisticated monitoring and control can give the public confidence that treatment processes work as intended. The net result of this improved ability to detect and effectively treat constituents and ensure that they are being removed to levels with no known health concerns is that a properly designed potable reuse system provides a safe, reliable, less expensive alternative to other water sources.

NO NEED FOR ALTERNATIVE DISTRIBUTION SYSTEM

An advantage of potable reuse over nonpotable reuse (e.g., for irrigation or industrial uses) is that potable reuse water can be delivered to the community in the same distribution system as the municipal water supply, whereas nonpotable reuse water must be delivered in a separate distribution system (known as a purple pipe distribution system for the pipe color used to identify the water as nonpotable). The cost of constructing and maintaining two parallel water distribution systems may be considerably more expensive than the cost of advanced treatment. In most highly developed urban environments, the installation of a parallel distribution system is not feasible.

ABILITY TO STORE ADVANCE TREATED WATER

A final advantage of potable reuse is the ability to use ATW to defer the consumption of traditional water supplies, or store ATW in reservoirs or aquifers for future use, which increases the reliability of the water supply. As an example, Albuquerque, NM, has excess water rights to surface water in winter months that goes unused and insufficient water rights to surface water in summer months. Practicing potable reuse during winter and storing the advance treated water in the aquifer would allow winter water rights to be used more fully and reduce demand for surface water in the summer, when river flows are often at their lowest. Storing water can also provide resiliency against declining supply due to climate change and future droughts.

Evolution of Potable Reuse

Despite the recent rising interest in water reuse, it is not a new concept. Domestic and municipal wastewater has been used for irrigation and fertilization of agriculture for thousands of years (Angelakis et al., 2019). In 1930, R.F. Goudey, a sanitary engineer in the Los Angeles Department of Water and Power, wrote "Reclamation of sewage in Southern California is coming whether we like it or not. It is not a question of sewage disposal, but one directly related to a legitimate increase in water supply." Goudey proposed an advanced wastewater treatment system, but the city voted to import water.

The first potable reuse installation in the United States, known as the Montebello Forebay, was installed in the 1960s by the County Sanitation Districts of Los Angeles County (NRC, 2012). The Montebello Forebay consists of 560 acres of spreading basins that allow treated wastewater to recharge the local aquifer, which is then withdrawn by wells. Another groundwater replenishment in southern California started in 1976 with Water Factory 21, operated by the Orange County Water District (NRC, 2012). Water Factory 21 was designed with groundwater injection wells spaced to create a mound in the groundwater table to prevent seawater intrusion from impacting the local aquifer. In addition to preventing seawater intrusion, the injected treated wastewater flowed toward potable water supply wells.

The first surface water augmentation IPR project in the United States was constructed in 1978 by the Upper Occoquan Service Authority in Fairfax County, VA (NRC, 2012). This system discharged advanced treated wastewater into a water supply reservoir, which is then used as a source for a drinking water treatment facility.

Internationally, the first DPR project was implemented in the city of Windhoek, Namibia, in 1968. Source control (see Sec. 23-3) is practiced by diverting industrial wastewater from the treatment facility, and then the treated wastewater is treated in an advanced treatment facility that includes, among other processes, ozone/biofiltration, advanced oxidation with ozone and hydrogen peroxide, and ultrafiltration. Numerous upgrades have occurred, including a new treatment facility in 2002 (NRC, 2012). Reclaimed water is blended with other sources and is up to 35 percent of the water supply in normal periods and up to 50 percent when water supplies are limited (NRC, 2012).

Interest in potable reuse has increased dramatically since 2000, with numerous potable reuse systems planned, in construction, or in operation in multiple states and internationally. The U.S. Environmental Protection Agency (U.S.EPA) released a National Water Reuse Action Plan in 2020 (U.S. EPA, 2020). The Potable Reuse Compendium (U.S. EPA, 2017) contains a list of operating potable reuse systems in the United States along with several detailed case studies in the appendices.

23-2 Wastewater Sources, Quantities, and Characteristics

Important characteristics of wastewater that affect its conversion to potable water include the (1) sources; (2) quantities; and (3) physical, chemical, and biological properties of wastewater. Changes in any of these characteristics will have ripple effects on both wastewater treatment and advanced water treatment.

Sources of Wastewater

The principal sources of municipal wastewater include residences, commercial establishments, institutional facilities, industrial facilities, and public facilities. Other sources can include infiltration/inflow and stormwater. Stormwater usually has lower total dissolved solids (TDS) than wastewater, resulting in lower treatment costs, so stormwater use can be optimized for potable reuse by incorporating inlet controls and structures that permit "first flush" into the stormwater system with subsequent flows into the reuse treatment system. The relative contribution of each source will vary with the location in the collection system, the age and condition of the collection system, the geographic location, the season of the year, and the social characteristics of the community (Tchobanoglous et al., 2014). The portion of the wastewater from each source will affect the development and implementation of an effective enhanced source control program, as described in Sec. 23-4, and may influence the design and operation of the wastewater and advanced water treatment facilities.

Quantities of Wastewater

The municipal uses of water, typical wastewater flowrates, variations in flowrates, and impacts of water conservation and climate change will impact the design and operation of wastewater treatment facilities.

MUNICIPAL WATER USE

The uses of water in a typical municipality are reported in Table 23-1 for the years 2000 and 2020 and projected for the year 2030. The large range in the values for each use is reflective of different water-use patterns in various parts of the country. Because outdoor water use varies so widely, caution is required in using published average outdoor water-use data.

Table 23-1
Municipal uses of water and typical quantities in the United States[a]

	2000		2020		2030	
	Flow, L/capita·d		Flow, L/capita·d		Flow, L/capita·d	
Use	Range	Typical	Range	Typical	Range	Typical
Residential indoor use	150–300	230	130–245	210	110–220	170
Residential outdoor use	60–190	130	60–190	130	60–190	130
Commercial	40–285	150	40–265	130	35–245	115
Public	60–100	75	55–95	70	55–95	55
Loss and waste	60–100	75	55–95	70	55–95	55
Total	370–970	660	340–990	890	315–845	525

[a]Data developed from the author's experience and numerous sources including Tchobanoglous et al. (2014).

The data show a trend of declining per-capita water use, particularly for residential indoor use. If the trends in commercial and public water use follow that trend, less wastewater may be available for planned potable reuse projects.

WASTEWATER FLOWRATES

In 2020, the typical total influent flow to wastewater treatment facilities was about 380 L/capita·d (100 gal/capita·d), but the flow can vary substantially from one community to another. When planning for potable reuse projects, local measurements must be made to assess wastewater flowrates. Also, the impact of future water conservation measures on wastewater flowrates must be considered when planning potable reuse projects. In southern California, conservation has caused flowrates to some upstream satellite WWTFs to fall considerably below the flowrate the facility was designed for. In some cases, it has been necessary to divert flow from one sewer shed to another to make up for reduced wastewater flowrates.

VARIATIONS IN WASTEWATER FLOWRATES

Wastewater flowrates vary during the time of day, day of the week, season of the year, and from year to year depending upon the nature of the discharges to the collection system, as discussed above. During dry weather periods, the influent flowrate to most WWTFs varies diurnally as shown on Fig. 23-3(a). During wet weather periods, the variation in flow during storm events often increases significantly as shown on Fig. 23-3(b) and discussed below. Variations in flowrate can degrade the performance of WWTFs. Flow equalization, as discussed in Sec. 23-4, can mitigate the impacts of flowrate variations.

IMPACTS OF WATER CONSERVATION ON WASTEWATER QUANTITIES AND CHARACTERISTICS

In the early 1990s, conservation measures began to be implemented in earnest in the United States. As a result, indoor water use has decreased significantly. As reported in Table 23-1, indoor water use has decreased from 150 to 300 L/capita·d (40–80 gal/capita·d) in the year 2000 to about 130–245 L/capita·d (35–65 gal/capita·d) in 2020. Further decreases are projected for the year 2030. The decrease in water use can affect potable reuse projects due to (1) reduced quantities of wastewater available for reuse; (2) the development of anaerobic conditions which can alter the characteristics of the wastewater; (3) low flow conditions which exacerbate the flow rate variations observed at WWTFs, especially in smaller plants; and (4) increases in wastewater constituent concentrations, even though the mass of constituents [expressed in kilograms per day (kg/d)] that must be treated has remained the same or has increased consistent with population growth (Tchobanoglous and Leverenz, 2019).

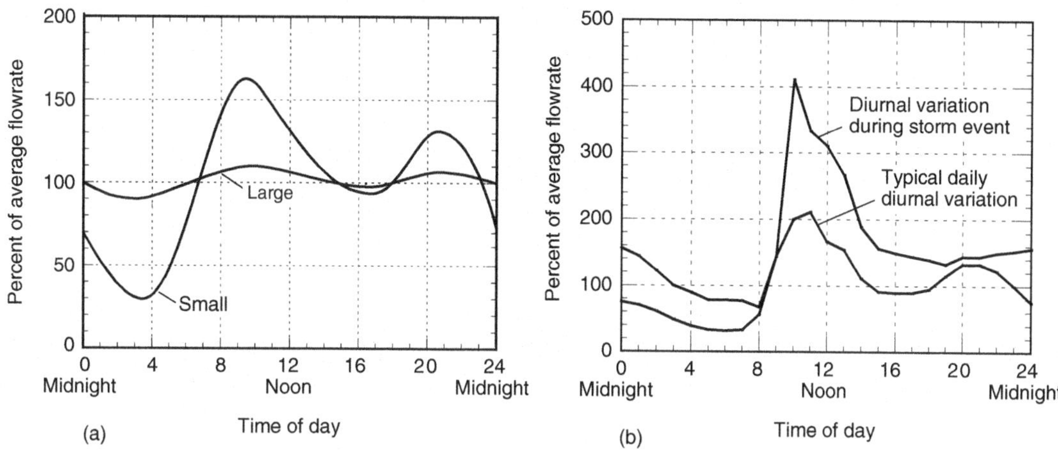

Figure 23-3
Typical variations in normalized influent flowrates at municipal WWTFs: (a) daily flowrate variations for small (4000–40,000 m^3/d) and large (>400,000 m^3/d) communities and (b) when collection systems are subject to stormwater infiltration (note change in vertical scale). (Adapted from Tchobanoglous et al., 2014).

IMPACTS OF CLIMATE CHANGE ON WASTEWATER QUANTITIES

Climate change is causing wet regions to get wetter and dry regions to get dryer. More importantly, climate change is making storm events more intense. The biological process in a WWTF can be washed out by the rapid increase in wastewater flowrates resulting from sudden intense rainfall events. In addition, the dilution effects of stormwater can degrade wastewater treatment plant performance (Plósz et al., 2009). In combined wastewater collection systems (i.e., collection systems used to transport both wastewater and stormwater), increased flowrates are primarily from direct inflow from street and other drains. In separate wastewater collection systems, increased flowrates are from illicit drains, inflow from surface access ports, and infiltration due to high groundwater levels. Historically, peak wet-weather flowrates have often exceeded the average flowrate by a factor of five or more. With climate change, peak to average flowrate ratios as high as 10 are now being observed, depending on the condition of the collection system (Tchobanoglous et al., 2021). Unfortunately, most WWTFs are not designed for such flowrate fluctuations. In some parts of the country, wastewater flowrates have decreased due to increased exfiltration from collection systems caused by the lowering of the water table due to drought conditions.

Wastewater Characteristics

The quality of wastewater can be significantly different from traditional freshwater sources and can vary significantly from one plant to the next, so the selection and design of wastewater and advanced water treatment processes will be site-specific. The physical and chemical quality of various source waters is discussed in Chap. 2 and microbiological quality in Chap. 3. Typical treatment trains for surface water and groundwater sources are presented in Sec. 4-5 in Chap. 4. Wastewater characteristics that affect process selection and the design and operation of wastewater treatment and advanced water treatment facilities are considered below. Wastewater treatment and advanced water treatment are considered in Secs. 23-5 and 23-6, respectively.

WASTEWATER CHARACTERISTICS OF IMPORTANCE FOR WASTEWATER TREATMENT

Important physical, chemical, and biological characteristics of wastewater with respect to treatment are given in Table 23-2. The definitions of the individual constituents listed as representative examples have been presented previously in Chapters 2 and 3. Additional details may be found

Table 23-2
Principal physical, chemical, and microbial constituents of concern (COC) in wastewater treatment

Characteristic or Constituent	Representative Examples or Units	Reason for Concern
	Physical Characteristics (Aggregate Properties)	
Solids concentration	TS, TVS, TFS, TSS, VSS, FSS, TDS, VDS, FDS, SS	Used in the design of WWTFs to assess most suitable treatment processes.
Particle size	μm	Affects removal by sedimentation and filtration processes.
Turbidity	NTU	Used to assess the clarity of treated wastewater. High values can reduce performance of disinfection, particularly when UV is used or LRV goals are high.
Temperature	°C or °F	Affects chemical and biological process kinetics. Also important in direct potable reuse where ATW is added directly to water distribution system.
Odor	TON	Often a source of complaints from the surrounding community. Also used in source control programs.
Transmittance (UVT)	%T	Used to assess applicability of treated effluent for UV disinfection.
	Inorganic Chemical Characteristics	
Dissolved inorganics	H^+, Na^+, Ca^{2+}, Mg^{2+}, Cl^-, SO_4^{2-}, alkalinity	Inorganic constituents are present in most public water supplies. Constituents such as calcium, sodium, and sulfate are added to the original domestic water supply because of water use and may have to be removed for some reuse applications. Alkalinity is a measure of buffering capacity of wastewater. Low alkalinity may limit nitrification.

(Continued)

Table 23-2 (*Continued*)

Characteristic or Constituent	Representative Examples or Units	Reason for Concern
Heavy metals	As, Cd, Cr, Co, Cu, Pb, Hg, Mo, Ni, Se, Zn	Heavy metals can be added to wastewater from commercial and industrial activities and may have to be removed if the wastewater is to be reused. Source control programs are designed to limit the discharge of these constituents.
Nutrients	NH_4^+-N, TKN (Org N + NH_4^+-N), NO_2^-, NO_3^-, Inorg P, PO_4, org P	Used in the design of WWTFs. The presence of nitrogen compounds can affect performance of membrane filtration. Nitrogen and phosphorus, along with carbon, are essential nutrients for growth. When discharged to the aquatic environment, these nutrients can lead to the growth of undesirable aquatic life.
pH	$pH = -\log \{H^+\}$	A measure of the acidity or basicity of an aqueous solution. Affects biological activity and other chemical reactions in wastewater treatment.
	Organic Chemical Characteristics	
Biodegradable organics	$CBOD_5$, UBOD (also BOD_u, BOD_L), COD, TOC, NOD	Used in the design of WWTFs. Composed principally of proteins, carbohydrates, and fats, biodegradable organics are measured most commonly in terms of BOD and COD.
Refractory organics	Surfactants, phenols, PFAS, and agricultural pesticides	Used in source control programs. These organics tend to resist conventional methods of wastewater treatment.
Priority pollutants	Defined under Section 307 of the Clean Water Act and the Code of Federal Regulations at 40 CFR 401.15	Compounds that are known or suspected to be carcinogenic, mutagenic, teratogenic, or have high acute toxicity. Many of these compounds are found in wastewater. Their measurement often incorporated into source control programs.
	Biological Characteristics	
Coliform organisms	*E. coli*	Used as a surrogate to assess potential presence of pathogenic bacteria and effectiveness of disinfection.
Pathogens	Enteric virus, *Cryptosporidium, Giardia*	Are a primary cause of acute waterborne illness.
Toxicity	Toxic unit acute (TU_a) or chronic (TU_c)	Can be used in source control programs.

Source: Adapted from Tchobanoglous et al. (2014).

in standard references (Tchobanoglous et al., 2014; WEF/ASCE, 2010). The total organic content (TOC) of wastewater is compared to other water sources on Fig. 2-12 in Chap. 2. The removal of organic matter is a key objective of wastewater treatment and typical numerical values for untreated and treated wastewater constituents are given in Table 23-9 in Sec. 23-5.

WASTEWATER CHARACTERISTICS OF IMPORTANCE FOR ADVANCED WASTEWATER TREATMENT

For potable reuse projects, the wastewater characteristics of greatest importance are the types and concentrations of microorganisms and chemicals.

Microbial Characteristics

The major classes of microorganisms to be considered in potable water projects are bacteria, viruses, and protozoa. Representative members of each class that potentially may be in wastewater are presented in Table 23-3. Extensive details on microorganisms are presented in Chap. 3. It is important to note that the analytical techniques used to assess the presence of many of these organisms are continually evolving and improving, especially with respect to the degree of microorganism recovery. As a result, caution should be used when comparing older reported values with more recent data. Pathogenic microorganisms used for regulatory compliance are considered in Sec. 23-3.

Chemical Characteristics

Chemicals in treated wastewater can include disinfectants, disinfection by-products, household chemicals, pharmaceuticals, personal care products (e.g. cosmetics, fragrances, and mosquito repellant), industrial chemicals, solvents, pesticides, herbicides, and other constituents of emerging concern (CEC), as

Constituents of emerging concern are continually changing.
http://www.stantec.com/wt3r/23-2.html

Table 23-3
Microbial groups and representative microorganisms of concern in potable reuse that could potentially be found in untreated wastewater

Microbial Group	Representative Members
Bacteria[a]	Atypical mycobacteria, *Campylobacter*, Enteropathogenic *Escherichia coli* (many other types of *E. coli* are not harmful), *Legionella, Leptospira, Pseudomonas, Salmonella, Shigella, Staphylococcus, Vibrio cholera, Yersinia.*
Viruses[b]	Adenovirus, Astrovirus, Caliciviruses (including Norovirus and Sapovirus), Coronavirus, Enteroviruses (polio, echo, coxsackie, new enteroviruses, serotype 68 to 71), Hepatitis A and E virus, Parvovirus, Rotavirus.
Protozoa[c]	*Cryptosporidium, Giardia, Cyclospora, Entamoeba, Microsporidia, Toxoplasma.*

[a]Refer to Tables 3-3 through 3-5 for additional details.
[b]Refer to Table 3-7.
[c]Refer to Table 3-8.

Table 23-4
Categories and representative examples of natural and synthetic CECs of relevance in potable reuse

Category	Representative Examples
Food additives	Bisphenol A (BPA), phthalates, sucralose
Household chemicals	Detergents, disinfectants, surfactants
Industrial and household chemicals	1,4-Dioxane, methyl tertiary butyl ether (MTBE), flame retardants, poly and perfluoroalkyl substances (PFAS)
Personal care products	Fragrances, pigments, sunscreen, triclosan
Pesticides/herbicides	Atrazine, DEET, diuron, fipronil, lindane
Pharmaceuticals and metabolites	Antibacterial, antidepressants, anti-inflammatory, analgesics, beta blockers, carbamazepine, lipid regulators
Synthetic and naturally occurring compounds	Hormones (ethinyl estradiol, 17 β-estradiol, estrone), geosmin, phytoestrogens
Transformation products (e.g., disinfection by-products)	Bromoform, chloroform, *N*-nitrosodimethylamine (NDMA), trihalomethanes

reported in Table 23-4. While many of these compounds are biodegradable in the wastewater treatment, others are not and will require advanced water treatment to ensure that they meet all regulatory and public health requirements. Because of concerns with these compounds, enhanced source control programs have been developed to control their discharge to wastewater collections systems to improve the quality of untreated wastewater for potable reuse.

Disinfection to inactivate pathogens is the final step in conventional wastewater treatment. Two common chemical disinfectants during wastewater treatment are chlorine and chloramines (see Chap. 13). Chlorine may be detrimental to advanced wastewater treatment processes for potable reuse, particularly reverse osmosis because RO membranes can be damaged by chlorine. Chloramines can form N-nitrosodimethylamine (NDMA) during interaction with organic matter in wastewater effluent. NDMA is a disinfection by-product that must be removed during advanced treatment. Disinfection by-products (DBPs) and the mechanisms for their formation are presented in Chap. 19. Chlorine DBPs include THMs, which are low molecular-weight, neutral halogens that are poorly removed by several common advanced treatment processes (Bellona et al., 2004; Breitner et al., 2019).

23-3 Regulatory Requirements

Wastewater discharge limits are defined by permit under the CWA, and potable water distributed to a community, regardless of source, must meet the requirements in the SDWA. Federal regulations for conventional water sources are discussed in Chap. 4. Where wastewater is the source of the water supply, additional requirements are needed to protect public health because of the diversity of constituents that may be present. Additional regulatory requirements for potable reuse of wastewater are considered in this section.

As of 2021, national regulations for potable reuse have not been developed. However, several states have regulations or guidelines, including California, Florida, North Carolina, Virginia, Texas, and Washington (U.S. EPA, 2017). Public health regulations and guidelines are based on a consideration of acute (short-term) toxicity, typically caused by pathogenic microorganisms, and chronic (long-term) toxicity, caused by chemical constituents.

Pathogens

Links to federal and state laws relevant to treatment of WW for potable reuse.
http://www.stantec.com/wt3r/23-3.html

The SDWA regulates the microbial quality of water by setting log reduction requirements for enteric virus, *Giardia*, and *Cryptosporidium*, with the reasoning that these are the most challenging to remove by various treatment processes and achieving removal of these will assure removal of all other important classes of pathogens. Treatment processes that meet performance requirements are given credit for achieving a specific log reduction value (LRV). The total LRV achieved by the treatment train is simply the sum of the LRVs achieved by the individual processes. States that have developed potable reuse guidelines have followed the same approach, using the same three target pathogens. The regulations and guidance adopted in California and Texas are described below as examples of the approach taken by various states.

CALIFORNIA LOG REDUCTION VALUES

For IPR, the Division of Drinking Water (DDW) in California requires the potable reuse system to achieve at least 12-log reduction of enteric virus, 10-log reduction of *Giardia* cysts, and 10-log reduction of *Cryptosporidium* oocysts from the raw wastewater. The derivation of the DDW LRVs is given in Table 23-5. These LRV were based on: (1) an annual risk of infection of 10^{-4} per person per year, which is the widely accepted goal as articulated by the U.S. EPA during the development of the SDWA; (2) acceptable microorganism concentrations in finished water based on dose–response studies; and (3) worst-case microorganism concentrations measured in untreated wastewater (Olivieri et al., 2016). Because the worst-case values were derived from data reported in the literature using a variety of analytical methods, future assessment of the actual range of values in untreated wastewater may increase or decrease as analytical methods become more robust. Furthermore, the DDW does not grant any LRV credit to the WWTF; thus, the LRVs must be achieved in the AWTF.

Table 23-5
Development of required log reduction values for IPR in California

Description	Enteric Virus	*Giardia* Cysts	*Cryptosporidium* Oocysts
Untreated wastewater maximum concentration	10^5 virus/L	10^5 cysts/L	10^4 oocysts/L
Tolerable drinking water concentration (TDWC)	2.2×10^{-7} virus/L	6.8×10^{-6} cysts/L	1.7×10^{-6} oocysts/L
Ratio of TDWC to wastewater concentration	2.2×10^{-12}	6.8×10^{-11}	1.7×10^{-10}
Required log reduction value	12	10	10

Source: Adapted from Olivieri et al. (2016).
Note: In the original report, the term "density" is used rather than "concentration," as used in this adapted table. Often, the term "density" is used in place in "concentration."

California, by law, must develop standards for DPR by the end of 2023 and draft legislation released in spring 2021. The draft legislation proposed higher LRVs to provide redundancy and reduce the possibility of a process failure causing the treatment train to fall below the minimum LRV requirement. The higher values will require more unit processes in the ATWF flow diagram. To achieve resiliency, the draft regulation requires the treatment train to contain at least four unit processes including at least one physical separation mechanism, one chemical disinfection mechanism, and one UV disinfection mechanism. Reliability of the process train is achieved by requiring continuous online monitoring of a surrogate for each unit process.

TEXAS LOG REDUCTION VALUES

The Texas Commission on Environmental Quality (TCEQ) has established guidance (not regulations) for potable reuse. IPR is, in general, governed by existing water quality regulations in the CWA and SDWA. In the face of severe drought and water shortages, the TCEQ has approved DPR projects on a case-by-case basis in accordance with an innovative treatment clause in state regulations (Steinle-Darling, 2015). The TCEQ requires DPR systems to demonstrate that they will achieve finished water goals for virus, *Giardia*, and *Cryptosporidium* that are similar in magnitude to the TDWC in Table 23-5, using the same LRV approach as the SDWA and the State of California. Baseline (or minimum) LRV targets have been developed based on the expected pathogen concentrations in wastewater effluent instead of the raw wastewater as adopted in California (Alan Plummer Associates, 2015). As a result, the LRVs for an AWTF in Texas are 4.0 to 4.5-log lower than in California, as shown in Table 23-6.

Table 23-6
Comparison of LRV targets for pathogen reduction in Texas DPR and California IPR

Pathogen Group	LRV for California IPR	LRV for Texas DPR
Enteric viruses	12	8
Giardia cysts	10	6
Cryptosporidium oocysts	10	5.5

Chemical Constituents

Constituents of emerging concern (CECs) can cause both acute and chronic illness and are problematic for regulators because of the enormous number of chemicals used in modern society that may end up in the wastewater supply. As analytical methods become more sensitive, hundreds of organic compounds are being detected in untreated wastewater. A list of classes of natural and synthetic CECs of relevance in potable reuse was presented in Table 23-4. Any list of chemicals is necessarily incomplete because new CECs are continually being identified and synergistic effects and degradation products must also be considered. Thus, the regulatory approach in the SDWA of identifying MCLs for specific constituents is likely inadequate. In some studies, however, it has been found that the occurrence of measured CECs in ATW is equivalent or lower than in drinking water from conventional sources, especially drinking water from streams subject to upstream discharges of treated wastewater effluent (Tchobanoglous et al., 2014; Mosher et al., 2016; Olivieri et al., 2016; U.S. EPA, 2017).

California has adopted an approach using treatment techniques (like filtration requirements in the SWTR) and specific chemicals as indicators (like the approach of using target pathogens as indicative of the removal of microorganisms broadly). For example, California requires that the AWTF contain reverse osmosis and an oxidation process, with specific criteria for the design (such as membrane selection, see Sec. 23-6) and operation of the processes. Examples of chemical surrogates in the regulations are a minimum 0.5 LRV of 1,2-dioxane by the oxidation process and finished water TOC concentration below 0.5 mg/L (CSWRCB, 2018). Texas guidance recommends that DPR projects include a monitoring program for indicators and surrogates (Alan Plummer Associates, 2015). To assess public safety of water containing unidentified chemicals, future regulations may consider bioassays or whole-water toxicity tests (Olivieri et al., 2016).

23-4 Source Control

The principal sources of municipal wastewater were identified in Sec. 23-2. Because of the diversity and variability of these sources, the number and concentration of constituents in wastewater can be quite variable.

Wastewater utilities can control the discharge of dangerous substances into the wastewater collection system by requiring industrial users to pretreat their waste before discharging it into a sewer, which is known as source control. Because of the concern for public health in potable reuse, source control programs for potable reuse must be more comprehensive than conventional source control programs. The evolution of enhanced and comprehensive source control programs for wastewater treatment and potable reuse are considered in the following discussion.

Conventional Source Control Programs for Wastewater Treatment Facilities

Pretreatment programs for industrial discharges are mandated under the CWA for publicly owned treatment works (POTWs) that have a design flow greater than 19,000 m^3/d (5 mgd) (U.S. EPA, 2011). The principal objectives of the National Pretreatment Program are to prevent the introduction of pollutants into a POTW that

- interfere with treatment operations (e.g., toxic substances that disrupt the biological process),
- pass through the treatment process and are discharged to receiving waters, and
- limit opportunities to recycle and reclaim municipal and industrial wastewaters and biosolids.

Based on these objectives, the primary focus of the National Pretreatment Program is on meeting discharge limits and nonpotable reuse criteria (U.S. EPA, 2011). The principal elements of the National Pretreatment Program are summarized in Table 23-7. In practice, it has been demonstrated that a well-designed pretreatment program can help a POTW operate more effectively and meet its discharge requirements more reliably.

For most POTWs, the primary focus of the source control program is on minimizing the discharge of potentially harmful or difficult-to-treat chemical and biological constituents to the wastewater collection system from industries, health care facilities, commercial businesses, and other nonresidential sources. Some communities have instituted special programs to eliminate fats, oils, and grease to prevent sewer blockages. Other communities have instituted household hazardous waste collection programs and facilities. Typically, source control programs vary from city to city, from minimal to extensive, depending primarily on the number of industrial, health care facilities, and commercial establishments in the community. However, when considering potable reuse, source control programs, as discussed below, must be far more rigorous and comprehensive.

Enhanced Source Control Programs for Potable Reuse

For potable reuse, conventional source control programs as discussed above are not sufficient. *Enhanced* source control programs are used to assess the presence of a greater range of compounds, including precursor

Table 23-7
Principal elements of the National Pretreatment Program for source control

Element	Description
Legal authority	The legal authority to enforce pretreatment standards is provided by state law, local codes, or sewer use ordinances.
Procedures	The POTW must develop procedures for: (1) receiving and analyzing self-monitoring reports and other notices submitted by industrial users (IUs); (2) random sampling and analysis of effluent from IUs; and (3) conducting surveillance activities to identify compliance or noncompliance independently from information supplied by IUs.
Funding	The POTW must dedicate a portion of its budget and personnel time to implementing the pretreatment program.
Discharge limits	The POTW must develop site-specific discharge limits for pollutants from IUs that address the specific needs and concerns of the POTW.
Enforcement response plan	The POTW must develop and implement an enforcement response plan that contains detailed procedures indicating how the POTW will investigate and respond to instances of industrial noncompliance.
List of industrial users	The POTW must maintain a list of all IUs.

Notes: IU = industrial user; POTW = publicly owned treatment work.
Sources: U.S. EPA (2011) and Alan Plummer Associates (2015).

compounds, specific constituents that have health implications, and constituents that are difficult to treat in AWTFs (Rimer et al., 2017a,b,c). Details of the source control program should be integrated with the WWTF optimization, AWTF design, and other elements of the potable reuse program.

The key elements of an enhanced source control program are identified in Table 23-8 (Tchobanoglous et al., 2015). The more extensive requirements for an enhanced source control program for potable reuse are evident comparing the many steps identified in Table 23-8 to those identified in Table 23-7. Because the elements of enhanced source control are continually changing as more experience with potable reuse is gained, it is important to follow the current literature on the subject. Two examples of enhanced source control programs for potable reuse include those developed by the Orange County Sanitation District (OCSD, 2018) and the City of San Diego (SDPUD, 2018).

Comprehensive Source Control for Potable Reuse

While conventional and enhanced source control measures deal primarily with interventions within the sewershed, comprehensive source control, as illustrated on Fig. 23-4, encompasses both conventional and enhanced source control, but also includes measures that can be taken

Table 23-8
Principal elements of an Enhanced Source Control Program for direct potable reuse

Element	Description
	Regulatory Authority
Legal authority	The POTW should evaluate whether the legal authority for conventional source control programs will cover its expanded programs.
Discharge limits	The POTW should expand the list of pollutants covered by discharge limits to include constituents that can have potential human health effects if not effectively removed by the WWTF and AWTF.
Enforcement	Ensure that the enforcement response program is effective for minimizing discharges of COCs.
Alternative control programs	Alternatives to discharge limits, such as best management practices (BMPs) or self-certification for zero discharge of pollutants, may be appropriate for classes of industries or commercial businesses.
	Monitoring and Assessment of the Wastewater Collection System Service Area (Sewershed)
Routine monitoring program	The influent to the WWTF and secondary or tertiary effluent sent to the AWTF are monitored routinely for regulated constituents and other COCs that may be discharged into the collection system service area.
Constituent prioritization program	COCs are identified and short-listed using results from the routine monitoring program. It may be necessary to develop separate monitoring programs for the constituents of greatest concern.
Evaluation of technically based site-specific limits	Regulated constituents and other COCs are evaluated for their potential to cause interference, pass through an AWTF, or affect human and environmental health and safety. For the development of site-specific limits, a broader spectrum of COCs, such as (1) regulated and nonregulated constituents that are relevant for DPR (e.g., drinking water constituents) or (2) CECs are considered.
	Source Investigations
Industrial and commercial business inventory	Develop and maintain a frequently updated, comprehensive inventory of industries and businesses that may use products or chemicals containing COCs or generate intermediate COCs. For agencies with large service areas, multiple communities, or industrial flows coming from other wastewater entities, it may be desirable to link the inventory to a service area mapping tool such as a geographic information system network.
WWTF–AWTF joint response plan	The response plan includes a flow chart showing key responsibilities and decision points to either investigate or mitigate COCs being discharged into the collection system.
	Maintenance of Current Inventory of Chemicals and Constituents
Chemical inventory program	Develop and maintain a database of the chemicals stored and inventory volumes used annually by industrial and commercial producers and manufacturers in the service area. Potential sources of this type of information include the industries themselves, State Emergency Response Commission, Local Emergency Response Commission, or local fire departments.
Waste hauler monitoring program	Develop a program to monitor and track discharges of septic wastes or other wastewater delivered to the collection system by truck. Haulers should be permitted and required to provide chemical inventory and discharge information to the wastewater treatment authority before being allowed to discharge. Consideration should be given to requiring waste haulers to deliver to a different treatment facility.

(Continued)

Table 23-8 *(Continued)*

Element	Description
Chemical fact sheets	Maintain a database of fact sheets for COCs and CECs encountered within the service area.
	Public Outreach Program
Industrial discharges	Provide industries with (1) public outreach information on potable reuse; (2) source control practices; and (3) compliance assistance and permit assistance to support the potable reuse program. Develop a program that encourages commercial and industrial dischargers to be partners in protecting the sewershed, such as environmental stewardship programs or award programs for consistent compliance. Assist and encourage industries and businesses that use chemicals that contain COCs to identify source control options, such as chemical substitution.
Service area pollution prevention partnership program	Develop a cooperative program with cities, counties, or other jurisdictions within the WWTF service area to disseminate information to the public about COCs and acceptable discharges to the sewer.
Public education and outreach program	Provide outreach to the public regarding the proper disposal of pharmaceuticals and household products containing chemicals that may be difficult to treat (e.g., what to flush and not flush). Consider developing a household hazardous waste collection program.
Education program	Develop school educational programs for grades 1 through 12 that address source control issues related to potable reuse.
	Response Plan for Identified Constituents
Interagency collaboration	The success of a source control program will depend on strong interagency cooperation and responsiveness between the WWTF and AWTF. For potable reuse projects that receive industrial waste from outside the service area, ensure that the agreement to accept the waste is consistent with source control program requirements. For projects where the agency that administers the source control program is not the agency that operates the AWTF, consider entering into a memorandum of understanding or other contractual agreement so that appropriate source control actions can be taken, if necessary, to protect water quality.
Response to water quality deviations	Develop an action plan for responding to water quality deviations. For example, if a specific chemical constituent is detected at the AWTF, review operation and calibration records for online meters and any analytical methods that may be involved. If a problem is not identified, then notify the WWTF to initiate a review and inspection of the WWTF for possible sources of the constituent. If no source is found at the WWTF, then initiate a wastewater collection system sampling program. If a problem is identified, the action plan should include procedures for the operations staff to notify the source control staff to respond to and correct the issue and, if necessary, procedures for bypassing or shutting down the facility.

Notes: AWTF = advanced water treatment facility; BMP = best management practice; CEC = constituent of emerging concern; COC = constituent of concern; DPR = direct potable reuse; WWTF = wastewater treatment facility.
Sources: U.S. EPA (2011), Alan Plummer Associates (2015), and Tchobanoglous et al. (2015).

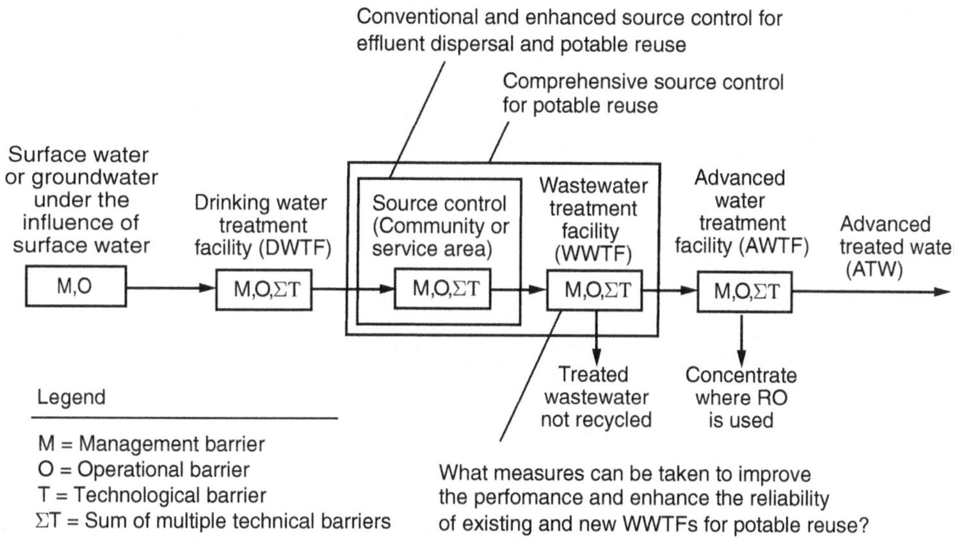

Figure 23-4
Definition sketch for the boundaries of conventional, enhanced, and comprehensive source control programs for potable reuse applications. (Adapted from Tchobanoglous et al., 2014).

within the wastewater treatment facility to further optimize the effluent for advanced water treatment (Tchobanoglous and Leverenz, 2019). Measures that can be taken to enhance wastewater treatment are considered in the following section. Comprehensive source control could also include separating the residential and industrial collection systems (as is done in Windhoek, see Sec. 23-6), collecting wastewater only from residential areas, or requiring specific industries to divert toxic or problematic waste streams away from the collection system into separate storage or holding facilities. The stored waste would be collected, hauled, and treated separately in a centralized facility.

23-5 Wastewater Treatment

Today, the principal objective of modern wastewater treatment is to remove constituents so that the treated effluent can be safely dispersed to the environment or used for nonpotable purposes. Although many types of WWTFs meet current discharge requirements, they are not equally effective at removing constituents that can impact the performance of AWTFs. As potable reuse becomes more common, WWTFs should be designed to produce effluent suitable for influent to AWTFs. The purpose of this section is to present a brief introduction to wastewater treatment, a description of the levels of treatment, some representative process flow diagrams along with corresponding performance data, a review of strategies to enhance the performance of WWTFs, and a discussion of the future of wastewater treatment for potable reuse.

Methods Used to Treat Wastewater

Wastewater treatment includes physical, chemical, and biological unit processes. As noted in Sec. 4-4 of Chap. 4, selection of unit processes depends on the source water quality and treated water goals. Many of the unit processes presented in Table 4-7 for water treatment are also applicable for wastewater treatment. The most important unit process in a WWTF, however, is the biological treatment process designed to remove dissolved and colloidal biodegradable organic substances found in wastewater. Textbooks are available on wastewater treatment (e.g., Tchobanoglous et al., 2014).

Levels of Wastewater Treatment

The treatment train in a WWTF is divided into treatment levels (or steps) identified as primary, secondary, tertiary, and advanced treatment as described in Table 23-9. In general, the term "primary" involves the application of physical unit processes, "secondary" involves chemical and biological unit processes to remove biodegradable organic matter, and "tertiary" involves physical unit processes to remove residual fine solids following secondary treatment. In some parts of the country, biological nutrient removal and/or advanced disinfection are included within the definition of tertiary treatment. Advanced treatment is typically used to remove specific constituents to meet requirements when the water is reused rather than returned to the environment.

For potable reuse, the selection of unit processes for the wastewater treatment train should consider the constituents that must be removed and the degree of constituent removal that is required so that treated wastewater can be processed effectively in an AWTF.

Typical Examples of Wastewater Treatment Processes

To illustrate how unit processes are combined into a wastewater treatment train, three WWTFs process flow diagrams are presented with corresponding performance data.

PROCESS FLOW DIAGRAMS

Three typical WWTF process flow diagrams are shown on Fig. 23-5. On Fig. 23-5(a), the conventional secondary treatment process can be either one of many variants of the suspended growth process (the most common being activated sludge) or an attached growth (biofilm) process. The optional

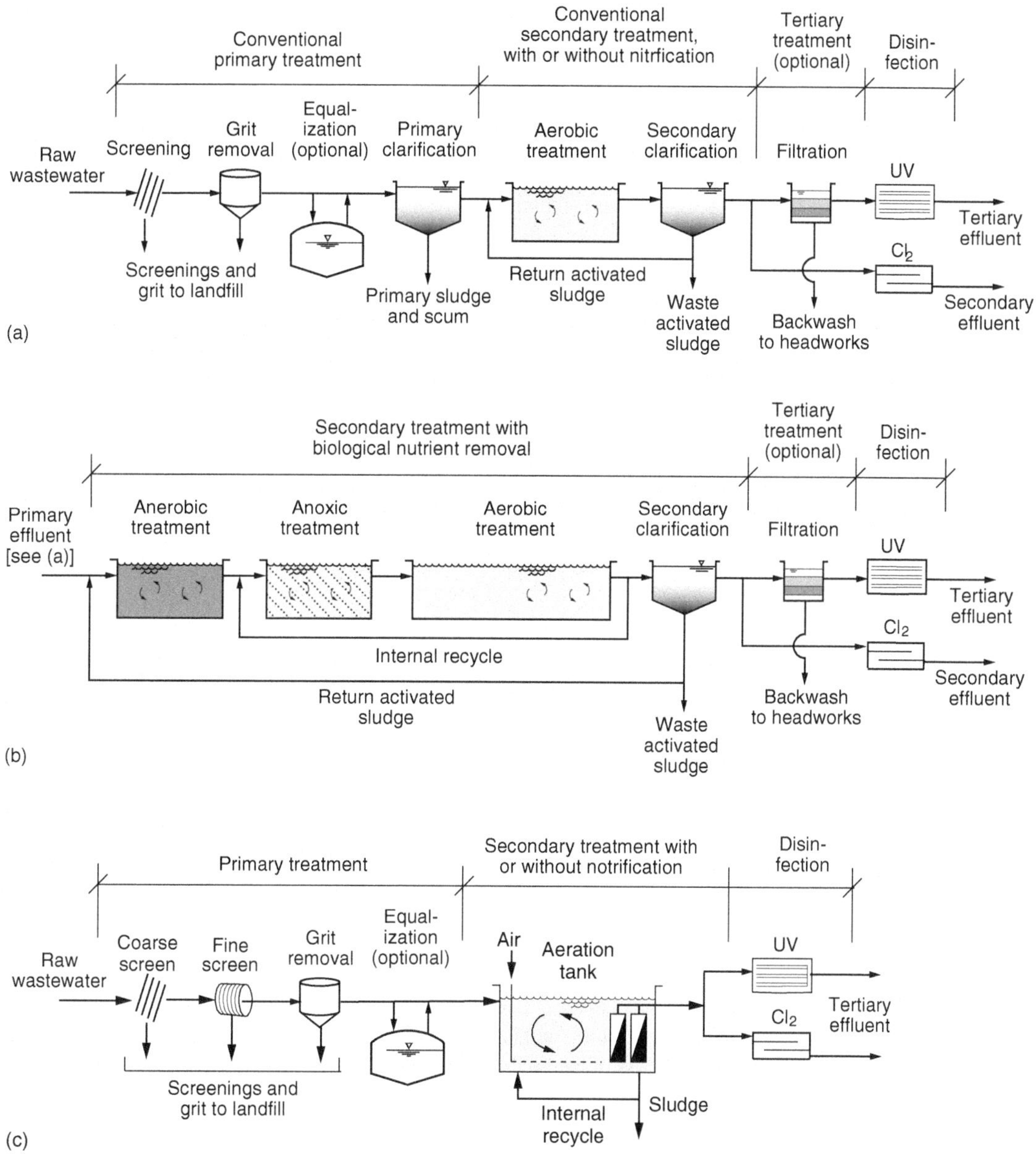

Figure 23-5
Generalized process flow diagrams for wastewater treatment: (a) activated sludge for TSS and BOD removal with or without nitrification; (b) suspended growth biological treatment for TSS, BOD, and biological nutrient removal; and (c) membrane bioreactor with or without nitrification. (Adapted from Tchobanoglous et al., 2014).

Table 23-9
Levels of wastewater treatment in a WWTF treatment train

Treatment Level	Description
Preliminary	Removal of wastewater constituents (such as rags, sticks, floatables, grit, and grease) that may cause maintenance or operational problems with the treatment operations, processes, and ancillary systems.
Primary	Removal of a portion of the suspended solids and organic matter from the wastewater. Typically accomplished by sedimentation. Primary filtration has been demonstrated as a replacement for primary sedimentation. In many references, preliminary treatment is included as part of primary treatment.
Advanced primary	Enhanced removal of suspended solids and organic matter from the wastewater typically accomplished by filtration of primary effluent sedimentation, chemical addition, or by replacing the primary clarifier with a primary filter.
Secondary	Removal of suspended and dissolved biodegradable organic matter (in solution or suspension) and suspended solids. In Federal regulations, secondary treatment is defined as meeting minimum removal standards for BOD and TSS and pH limits in effluents discharged from municipal WWTFs.
Secondary with nutrient removal	Removal of biodegradable organics, suspended solids, and nutrients (nitrogen, phosphorus, or both nitrogen and phosphorus).
Tertiary	Removal of residual suspended solids (after secondary treatment), usually by granular or cloth media filtration. Disinfection is often a part of tertiary treatment, although disinfection can be applied after any level of treatment prior to discharge. In some parts of the country, nutrient removal and/or more advanced disinfection are included under this definition.
Advanced	Removal of dissolved, colloidal, and suspended materials remaining after conventional biological or tertiary treatment to enhance the performance of AWTFs. Advanced treatment is often included as part of the AWTF design.

Source: Adapted from Tchobanoglous et al. (2014).

flow equalization can, as discussed later in this section, be used to achieve constant flow, which will enhance the performance of the biological treatment process. The biological treatment process shown on Fig. 23-5(b) is an example that achieves removal of nutrients in addition to the conventional parameters of BOD, TSS, and pathogenic microorganisms needed to meet federal and state discharge permit requirements. The membrane bioreactor shown on Fig. 23-5(c) has become a popular option for the treatment of wastewater for advanced water treatment.

The successful design and operation of the process trains illustrated on Fig. 23-5 requires an understanding of the types of microorganisms involved, the specific reactions involved, reaction kinetics, environmental factors that affect microbial performance, nutritional needs, and how these variables relate to the effluent quality that enhances the performance of

AWTFs. Details on these processes may be found in literature on wastewater treatment (Tchobanoglous et al., 2014).

Effective wastewater treatment is essential for producing advance-treated water.

http://www.stantec.com/wt3r/23-4.html

PERFORMANCE DATA FOR TYPICAL PROCESS FLOW DIAGRAMS

Typical effluent quality from the wastewater treatment trains shown on Fig. 23-5 is presented in Table 23-10. The differences in final water quality include nutrients, metals, microorganisms, organics, and solids concentrations. The effluent quality will affect the performance of the AWTF; for instance, higher organic concentrations in the WWTF effluent can increase membrane fouling in the AWTF. The range in values presented in this table are illustrative of typical performance and not guarantees for performance at any particular facility. Performance depends not only on the unit processes present but also the design criteria for those processes, how well they are operated, and the variability and stability resulting from diurnal flowrate conditions. The key message from the performance data presented in Table 23-10 is that the unit processes in the WWTF can have a dramatic effect on the quality of effluent that will be delivered to an AWTF and that certain processes can produce higher-quality effluent. For instance, a membrane bioreactor or an activated sludge plant containing biological nutrient removal and filtration can achieve substantially lower nitrogen content than a conventional activated sludge plant, which will reduce the fouling of MF and RO membranes, allow the membranes to operate at higher flux, and reduce the TOC concentration in the AWTF product water (Tchobanoglous et al., 2015).

Limitations of Existing WWTFs for Potable Reuse

A challenge for potable reuse projects is that most conventional WWTFs were not designed to provide source water for an AWTF and, as a result, plants that meet current discharge standards can be inadequate for feeding an AWTF. While there are numerous examples of existing AWTFs that produce ATW with varying degrees of secondary or tertiary wastewater quality, the role of the WWTF in a potable reuse system should be to provide a consistent, high-quality effluent to maximize treatment effectiveness and the protection of public health. Also, certain constituents such as pathogens, CECs, and disinfection by-product (DBP) precursors may be removed more cost-effectively through biological treatment. Although these removals have not been well-documented in the past, studies are underway, so it is expected that a wastewater treatment system used in conjunction with an AWTF will be upgraded as necessary to optimize consistency, reliability, and product water quality.

Enhancing the Performance of Existing and Proposed WWTFs

To produce the highest quality effluent for AWT, process considerations described in Table 23-11 can be evaluated to enhance the performance of a WWTF. The potential impacts of changing wastewater characteristics and climate change were considered previously in Sec. 23-2. The impact that

Table 23-10
Typical effluent quality from various wastewater treatment process trains

			Range of Effluent Quality after Indicated Treatment				
Constituent	**Unit**	**Untreated Wastewater**	**Conventional Activated Sludge**[a]	**Conventional Activated Sludge with Filtration**[a]	**Activated Sludge with BNR**[b]	**Activated Sludge with BNR and Filtration**[b]	**Membrane Bioreactor**[g]
Total suspended solids (TSS)	mg/L	130–400	5–25	2–8	5–20	1–4	<1–5
Turbidity	NTU	80–150	2–15	1–5	1–5	1–5	<1–2
Biochemical oxygen demand (BOD)	mg/L	130–400	5–25	<5–20	5–15	1–5	<1–5
Chemical oxygen demand (COD)	mg/L	340–1100	40–100	30–90	20–60	20–40	<10–40
Total organic carbon (TOC)	mg/L	110–340	20–40	15–30	10–20	1–5	<0.5–5
Ammonia nitrogen	mg N/L	12–45	1–10	1–6	1–3	1–2	<1–5
Nitrate nitrogen	mg N/L	0–trace	5–35	5–35	<2–8	1–8	<8[c]
Nitrite nitrogen	mg N/L	0–trace	0–trace	0–trace	0–trace	0.001–0.1	0–trace
Total nitrogen	mg N/L	25–70	15–35	15–35	3–8	2–5	<10[d]
Total phosphorus	mg P/L	4–12	3–10	3–8	1–2	≤2	<0.3[d]–5
Volatile organic compounds (VOCs)	μg/L	<100–>400	10–40	10–40	10–20	10–20	10–20
Iron and manganese	mg/L	1–2.5	1–1.5	1–1.4	1–1.5	1–1.5	Trace
Surfactants	mg/L	4–10	0.5–2	0.5–1.5	0.1–1	0.1–1	0.1–0.5
Total dissolved solids (TDS)	mg/L	375–1125	350–1050	350–1050	350–1050	350–1050	350–1000
Trace constituents[e]	μg/L	10–50	5–40	5–30	5–30	5–30	0.5–20
Total coliform	No./100 mL	10^6–10^{10}	10^4–10^5	10^3–10^5	10^4–10^5	10^4–10^5	<100
Protozoan cysts and oocysts	No./100 mL	10^1–10^5	10^1–10^3	0–10^2	0–10^2	0–1	0–1
Viruses	PFU/100 mL[f]	10^1–10^8	10^1–10^4	10^1–10^3	10^1–10^3	10^1–10^3	1–10^3

[a]Conventional secondary is defined as activated sludge treatment with nitrification as shown in Fig. 23-5(a).
[b]BNR = biological nutrient removal [see Fig. 23-5(b)].
[c]With anoxic stage.
[d]With coagulant addition.
[e]Fire retardants, personal care products, and prescription and nonprescription drugs.
[f]Plaque forming units.
[g]See Fig. 23-5(c).
Source: Adapted from Tchobanoglous et al. (2014).

Table 23-11

Factors, measures, and operational strategies that should be considered in optimizing the performance of wastewater treatment facilities for potable reuse[a]

Process Consideration	Principal Impact(s)
Changing wastewater characteristics	Reduced flowrates, increased constituent concentrations, changing chemical characteristics due to decomposition in collection system, decreased effluent quality, as discussed in Sec. 23-2.
Climate change	Peak-flow events, surge flows, decreased effluent quality without flow equalization, washout of biological treatment process, flows beyond capacity of disinfection facilities, as discussed in Sec. 23-2.
Screening and grit removal (preliminary treatment)	Improved treatment performance and effluent quality.
Flow equalization	Improved treatment performance and effluent quality, improved process reliability, reduced biological reactor size, ability to predict stable pathogen log reduction values for WWTFs. Similar benefits for AWTFs.
Filtration of primary effluent or primary filtration as a replacement for primary sedimentation	Improved treatment performance and effluent quality, reduced energy usage in biological treatment, more effective denitrification.
Elimination of return flows discharged to treatment plant headworks	Improved effluent quality and process reliability. Removal of stable nitrogenous compounds in return flows improves the performance of membrane filtration in AWTFs.
Equalization and pretreatment of return flows before discharge to headworks	Improved performance of biological treatment processes, especially when return flows are introduced during low load periods in the evening hours. May impact the removal of specific constituents.
Modification of biological treatment process operational mode	Improved treatment performance for specific constituents, effluent quality, and process reliability.
Implementation of new biological treatment process(es)	Improved treatment performance and effluent quality, process reliability.
Improved process monitoring	Improved process performance, process reliability, identification of constituents that may not be treated that can impact AWTFs.
Effluent filtration	Improved effluent water quality, minimizes impact of WWTF upsets on AWTF.
Effluent disinfection method	Use of chlorine compounds, ozone, peracetic acid or UV, consistent with AWTF needs (e.g., chlorine compounds can damage some membranes). In some cases, effluent disinfection may not be required.

[a]Adapted, in part, from Tchobanoglous et al. (2015).

the remaining process considerations will have on the effluent quality from an WWTF will vary depending on the design and operation of existing facilities. Because of its singular importance to the performance of both WWTFs and AWTFs, flow equalization is considered further in the following discussion.

BENEFITS OF FLOWRATE EQUALIZATION FOR WWTPs

Water conservation and climate change are causing more pronounced variations in flow in small and medium wastewater systems where storage capacity in the collection system does not provide damping. Because of these factors and exfiltration under drought conditions, the influent flowrates in the early morning hours approach zero in many wastewater treatment plants. Without flow equalization, these diurnal variations in flowrate and constituent concentrations can degrade plant performance, especially in plants with limited (or excessive) volumetric capacity. Reduced WWTF performance is characterized by greater variability in the constituent concentrations in the treated effluent and reduced removal of trace organic constituents (Gujer and Erni, 1978; Niku et al., 1981a,b; Plósz et al., 2009).

The benefits of constant flow for WWTFs include (1) smaller reactor size, (2) improved performance and increased process stability, (3) reduced energy usage, (4) more predictable performance, (5) ability to implement new biological treatment processes, (6) ability to optimize performance of existing facilities for the removal of specific constituents, and (7) the ability to assign pathogen LRV credits (Foess, 1977; Gujer and Erni, 1978; Niku et al., 1981a,b; Giokas et al., 2002; Tchobanoglous et al., 2014, 2021). Benefits for AWTFs are considered below.

BENEFITS OF FLOWRATE EQUALIZATION FOR AWTFs

Important benefits of constant flow for AWTFs include: (1) improved performance and increased process stability, (2) reduced capital and operational and maintenance costs, and (3) reduced energy usage (LaGrega and Keenan, 1974; Tchobanoglous et al., 2021). In water short areas, flow equalization also has the benefit of increasing the amount of wastewater that can be captured for potable reuse where diurnal variations in flowrate are significant.

The Future of Wastewater Treatment

With an appreciation of the impacts of water conservation and climate change, new operating strategies will have to be developed as existing WWTFs are upgraded and new potable reuse facilities are planned. New strategies to be examined will include the use of colocated satellite WWTFs and AWTFs, the use of flow equalization, the use of divided treatment trains at existing facilities, and diversion of fixed constant flow for large WWTFs, which will typically undergo additional treatment before advanced water treatment. As new WWTFs and AWTFs are planned, integrated treatment process design will become standard practice (Tchobanoglous, 2019).

23-6 Advanced Water Treatment

The objective of advanced water treatment is to remove regulated and unregulated constituents that cannot be removed by in the WWTF. Typical unit processes and ancillary systems employed in AWTFs are presented in Table 23-12. The core unit processes are each described in a dedicated chapter or section earlier in this book, and those chapters should be studied

Table 23-12
Common AWTF unit processes and ancillary systems. Chapters and sections in this book where unit processes are presented are included in parentheses

Treatment Process	Description
Filter screens	Removes large, suspended solids that may be present in the source water to protect downstream processes.
Flow equalization	Reduces diurnal flowrate and load variations to allow processes to operate at a constant rate, which can reduce unit process capacity, increase the reliability, and reduce wear on equipment.
Ozone/biologically active filtration (Ozone/BAF)	Achieves significant removal of both pathogens and chemicals. Ozone chemically degrades organic chemicals (see Chaps. 13 and 18) and makes them more biodegradable in the biological filter (see Sec. 12-8). Provides excellent pathogen removal. When used upstream of MF and RO, can increase the flux and reduce fouling in those processes. Ozone/BAF has higher water recovery and lower capital cost and energy consumption than RO.
Granular activated carbon (GAC)	Removes dissolved compounds via absorption (see Chap. 15). GAC is often combined with Ozone/BAF and referred to as biological activated carbon (BAC).
Membrane filtration (MF)	Achieves excellent pathogen removal, removes suspended particles (turbidity) by mechanical sieving (see Chap. 12). When used upstream of RO, the turbidity removal minimizes RO membrane fouling. When used upstream of Ozone/BAF, the turbidity removal can reduce ozone doses.
Cartridge filtration	Pretreatment step for RO, protects RO membranes from particles present in pretreatment chemicals. Some cartridge filters can receive protozoa removal credit.
Reverse osmosis (RO)	Removes salts and dissolved solids, including trace organics, CECs, and TOC, by means of size exclusion and solution/diffusion (see Chap. 17). Recognized as a significant barrier for all pathogens. Provides reliability for unregulated contaminants. When used upstream of UV, the high transmittance of RO permeate decreases required UV doses.
Advanced oxidation processes (AOP)	Chemically degrades pathogens and dissolved organic chemicals via a nonspecific oxidation reaction involving hydroxyl radicals (see Chap. 18). A common application is high UV doses coupled with a strong oxidant (hydrogen peroxide or chlorine).
UV photolysis	Removes photosensitive chemicals like NDMA and iopromide (see Sec. 8-4). Achieves excellent pathogen inactivation (see Sec. 13-9). Often used with AOP.
Stabilization	Stabilizes finished water to reduce its corrosive properties by adjusting pH, alkalinity, and hardness (see Chap. 22).
Disinfection	Achieves pathogen inactivation (see Chap. 13).
Soil aquifer treatment	Achieves substantial removal of pathogens and organic chemicals through extended contact with natural soils.

to understand the principles and design concepts for each process. The focus of this section is to describe the process for selecting the AWTF treatment, typical design considerations, operating conditions, and process performance for each of these processes when used for potable reuse, and to present AWTF case studies.

Treatment Train Selection

Treatment train selection must consider removal capabilities with respect to pathogens and chemicals, the effect of processes on each other, reliability and redundancy, and pilot testing.

PATHOGEN REMOVAL

The first consideration in process selection for an AWTF is to identify a series of processes that will achieve the required removal of pathogens (LRV credit) and chemicals (treatment techniques and surrogate parameters) specified in the regulations or guidance that the plant will have to meet. Capabilities for common AWTF processes to remove pathogens are summarized in Table 23-13.

The observed LRVs for each process span a broad range because many factors affect removal. Earlier studies used older test methodologies and may have reported lower pathogen removal because of poor sensitivity for pathogen detection. Process design and operation (chemical doses, etc.) affect removal, as does feed water characteristics. Equipment failures, such as broken fibers in an MF system, damaged O-rings in an RO system, or burnt-out bulbs in a UV reactor, will dramatically reduce removal. Pilot testing is an essential step to demonstrate the LRV that will be achieved in site-specific conditions.

Regulations and guidance vary from state to state, and some states will grant less LRV credit than other states relative to the capability of each process. An important factor in granting LRV credit is monitoring results that can be used to validate that processes are operating correctly. Because equipment or operating failures can degrade performance, regulators will only allow as much credit as can be demonstrated through validation testing. As an example, challenge testing has demonstrated that RO can achieve greater than 6-log removal of all pathogens. However, measurement of common parameters like conductivity or TOC can only be used to demonstrate 2-log removal; as a result, equipment damage or process degradation that reduces removal below 6-log can go undetected. Thus, regulators will only allow credit for removal that can be validated through continuous monitoring (2-log) instead of the process capabilities under ideal conditions (>6-log).

CHEMICAL REMOVAL

Capabilities for common AWTF processes to remove chemicals are summarized in Table 23-14. Several processes are capable of excellent removal of a wide range of CECs, but unfortunately each has specific groups of

Table 23-13
Pathogen removal capabilities for common AWTF processes

Treatment Process	Observed Pathogen Removal (LRV)			Comments
	Virus	Giardia	Crypto	
Ozonation	3–>6	2–4	1–2	Removal depends on the concentration *x* time (CT) value.
BAF and GAC	0	0–2	0–2	Physical removal if the biological filter is operated like a conventional granular media filter.
Membrane filtration	0–>6	4–>6	4–>6	Challenge tests with UF membranes have consistently demonstrated high virus removal but regulatory credit is limited because of an inability to validate performance at full-scale.
Reverse osmosis	2–>6	2–>6	2–>6	Ability to obtain regulatory credit for high pathogen removal is limited because of an inability to validate performance at full-scale.
UV/AOP	>6	>6	>6	UV doses for AOP are typically 800 mJ/cm^2 or higher.
UV photolysis	0–1	>3	>3	UV doses for disinfection are typically in the range of 40 mJ/cm^2.
Chlorination	1–4	0–0.5	0	LRV credit depends on the CT value.
Soil aquifer treatment	1–>6	1–>6	1–>6	California regulations allow 1-log virus removal for every month that the water spends underground.

Sources: Trussell et al. (2013); U.S. EPA (2017); Sharma and Kennedy (2017); Soller et al. (2017).

chemicals for which it is less effective. Thus, redundancy is an important factor in selecting the treatment train for a potable reuse system.

INTERACTION BETWEEN UNIT PROCESSES

In addition to their treatment capabilities, upstream advanced treatment processes can affect downstream processes and the order of processes should be selected to optimize their collective performance. For instance, it has been demonstrated that Ozone/BAF can reduce the fouling of downstream membrane processes and allow them to operate at a higher flux and with less frequent cleaning (Trussell et al., 2015). Reverse osmosis can

Table 23-14
Chemical removal capabilities for common AWTF processes

Treatment Process	Chemical Removal
Membrane filtration	No removal of dissolved chemicals.
Reverse osmosis	Excellent removal of many inorganic and organic compounds; low removal of low-MW, neutral compounds depending on functional chemistry (e.g., alkenes, ketones, and chlorinated compounds). Greater than 95% TOC removal.
Advanced oxidation	Excellent removal of many organic compounds; low removal of compounds with $k_{OH} < 10^8\ M^{-1}\ s^{-1}$ (e.g., alkanes, peroxides, and nitriles).
Ozone/BAF	Excellent removal of many organic compounds; Less effective for TOC removal than RO (40–70%).
Granular activated carbon	Excellent removal of many organic compounds; although some compounds may reach breakthrough rapidly (e.g., chloroalkanes). Less effective for TOC removal than RO.
Soil aquifer treatment	Excellent removal of many organic compounds, including NDMA. TOC removal of 65–75%, although some projects report higher removal.

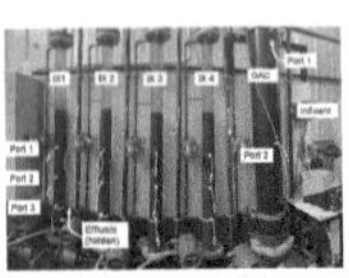

Pilot testing is essential for selecting the treatment train, obtaining design criteria, and communicating with the public about potable reuse.
http://www.stantec.com/wt3r/23-5.html

improve the UV transmittance so that downstream UV-AOP operates with higher efficiency.

MULTIPLE-BARRIER CONCEPT, RELIABILITY, AND REDUNDANCY

The multiple-barrier approach to water treatment and its ability to improve reliability and redundancy was presented in Sec. 4-6. The multiple-barrier approach is particularly important in potable reuse because of the direct link between municipal wastewater and the potable water supply. Source control, biological treatment, advanced water treatment, and the environmental or engineered buffer provide multiple barriers within the entire potable reuse system. The multiple-barrier concept is also used within the AWTF to improve reliability and redundancy, and the selected treatment train will often exceed the required pathogen LRV credit and chemical removal requirements to ensure that removal requirements are met even when a failure occurs within one of the processes. A framework that considers reliability, redundancy, robustness, and resilience can help protect public health (Pecson et al., 2015).

PILOT TESTING AND VALIDATION TESTING

Because of wide range of removal capabilities shown in Tables 23-12 and 23-13, pilot testing is an essential step in selecting the treatment train and designing an AWTF. Pilot testing is used to identify performance criteria such as flux rates, run time duration, water recovery, fouling potential, cleaning frequency, and other parameters. Pilot testing will also be necessary to validate performance for regulatory agencies. Finally, pilot testing can be an integral part of a public acceptance and communication plan, as discussed in Sec. 23-10.

Examples of AWTF Treatment Trains

While many treatment train configurations are possible, examples of process sequences are presented in Fig. 23-6. A process train consisting of membrane filtration, reverse osmosis, and UV/AOP, as presented in Fig. 23-6(a), is considered by some to be the "gold standard" for advanced water treatment and is known as "full advanced treatment" (FAT). The addition of ozone and biofiltration upstream of the MF, as presented in Fig. 23-6(b), can improve the performance of the MF and RO systems (higher flux, less fouling) while providing an additional barrier for pathogen and chemical removal. In inland areas, disposing of the RO concentrate can be an obstacle to implementation of potable reuse. Using ozone/biofiltration in lieu of RO and AOP, as presented in Fig. 23-6(c) can provide excellent

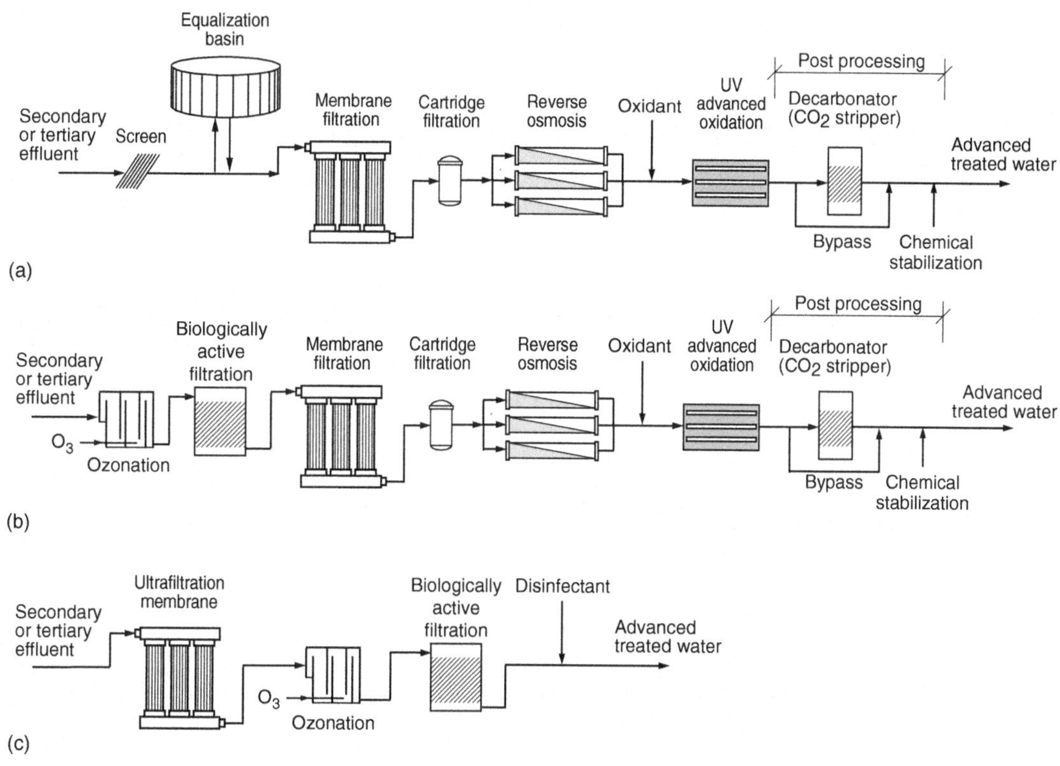

Figure 23-6
Examples of AWTF treatment trains: (a) full advanced treatment as described in California regulations, (b) full advanced treatment with Ozone/BAF pretreatment to improve MF and RO process performance, and (c) a treatment train that achieves pathogen and CEC removal with Ozone/BAF instead of RO and UV/AOP.

pathogen and CEC removal, eliminate the RO concentration disposal challenge, and provide higher overall water recovery (an advantage because potable reuse projects are developed in response to water scarcity).

Process Design Considerations

The use, performance, and limitations of the treatment processes used for AWT are highlighted in the following discussion. For additional information, the corresponding chapters or sections in the text related to each of the unit processes is also identified.

Potable reuse relies on treatment technologies that are commonly used for other water treatment purposes.
http://www.stantec.com/wt3r/23-6.html

MEMBRANE FILTRATION

Membrane filtration is often the first process in an advanced treatment train. Membrane filtration can be included as an integral part of the wastewater treatment activated sludge process when the membranes replace the secondary clarifiers; this configuration is known as a membrane bioreactor. Alternatively, MF can be located after the secondary clarifiers as part of the advanced treatment train. Typical MF design criteria for potable reuse applications are presented in Table 23-15, and Chap. 12 can be reviewed for general principles and design concepts.

The higher concentration of organic and biological matter in treated wastewater as compared to conventional source waters can lead to increased membrane fouling. Chloramines (see Chap. 13) are commonly used to control fouling. Research has demonstrated that ozone and biofiltration upstream of membrane filtration can reduce biological fouling and allow higher permeate flux (Trussell et al., 2015). Chemically enhanced backwash (CEB) cycles with the addition of 30–50 mg/L of sodium hypochlorite once per day or 100–500 mg/L one to three times per week can help minimize organic fouling (WEF, 2021). Clean-in-place cycles typically use sodium hypochlorite to remove organic and biological foulants. Typical cleaning frequency is every 1–3 months. Caustic soda to raise the pH above 10 may improve cleaning when heavy organic fouling is present (WEF, 2021). Hypochlorite concentrations for CIP cleaning are specific to the membrane manufacturer recommendations, but typically range from 1000 to 5000 mg/L.

Table 23-15
Typical design criteria for membrane filtration for potable reuse applications

Parameter	Units	Value
Membrane type	–	Microfiltration or ultrafiltration
Permeate flux	$L/m^2 \cdot h$ ($gal/ft^2 \cdot d$)	34–100 (20–60)
Feed pressure	bar (psi)	0.21–2 (3–30)
Filter run duration	min	15–90
Water recovery	%	90–95

REVERSE OSMOSIS

Typical RO design criteria for potable reuse applications are presented in Table 23-16, and Chap. 17 can be reviewed for general principles and design concepts. Brackish water RO (BWRO) membranes are used for potable reuse systems because seawater RO operate at higher pressure and nanofiltration (NF) membranes do not achieve enough removal of low-MW anthropogenic organic compounds. In California, regulations stipulate membrane products that achieve at least 99.2% NaCl rejection (Bernados, 2018). Membrane products that meet this criterion, however, can have widely varying rejection of low-MW uncharged organics. Fujioka et al. (2013) measured rejection of NDMA ranging from 19 to 82% among membranes that had greater than 99.2% NaCl rejection, and a similarly wide range in rejection has been observed for many other organic compounds (Breitner et al., 2018). It has been demonstrated that boron rejection is better than NaCl rejection for predicting the rejection of low-MW neutral compounds (Tu et al., 2013; Breitner et al., 2018).

Reverse osmosis can achieve excellent removal of a wide variety of inorganic and organic compounds, except some low-MW, hydrophilic, uncharged compounds. As examples, poor removal (<80%) has been reported for ethanol, acetone, chloroform, bromobenzene, NDMA, and vinyl chloride (Drewes et al., 2013; Breitner et al., 2019). Generally, rejection increases as molecular weight and hydrophobicity increase (Bellona et al., 2004; Drewes et al., 2013), but many exceptions to these trends are observed. An additional factor of importance is membrane–solute interactions resulting from functional chemistry, with decreased rejection caused by the presence of aromaticity, carbon–carbon double bonds, carbonyl functional groups, and halogen substituents (Breitner et al., 2019). Important constituents that are poorly rejected by RO include some industrial solvents (e.g. acetone, methylene chloride, trichloroethylene) and disinfection by-products (e.g., THMs, NDMA).

In additional to individual organic compounds, RO can achieve excellent removal of TOC, with permeate typically below 0.2 mg/L. In the

Table 23-16
Typical design criteria for reverse osmosis for potable reuse applications

Parameter	Units	Value
Membrane type	–	Brackish water RO
Number of stages	–	2 or more
Permeate flux	$L/m^2 \cdot h$ ($gal/ft^2 \cdot d$)	14–24 (8–14)
Feed pressure	bar (psi)	5–10 (73–145)
Water recovery	%	>75
Cartridge prefilter nominal pore size	μm	1–5

absence of membranes imperfections and damaged seals, RO membranes should achieve total removal of pathogens, but monitoring and verification remain an obstacle to achieving high LRV credit. With better surrogates, it should be possible for RO system to receive 4-log LRV credit for viruses (Trussell et al., 2020).

RO membranes are susceptible to organic and biological fouling from the organic matter in WWTF effluent. A chloramine residual of 2–3 mg/L can be used to minimize fouling but if chlorine is used, dechlorination is necessary to avoid damage to the membrane material. RO systems need pretreatment with cartridge filters even if they are preceded by membrane filtration because RO pretreatment chemicals (like antiscalants) may contain particulate matter that could clog RO element feed channels. When considering scale formation by limiting salts to determine allowable recovery, design calculations need to consider the saturation limit of calcium phosphate in addition to traditional limiting salts of RO systems when phosphate is present in the WWTF effluent.

Periodic CIP cleaning must be done to remove scaling, typically monthly to annually. Cleaning regimes will be developed in consultation with the membrane manufacturer, but typically involves citric acid to remove inorganic scalants and sodium hydroxide to remove organic and biological foulants.

UV PHOTOLYSIS AND ADVANCED OXIDATION

Both UV photolysis and advanced oxidation can remove pathogens and chemicals, but the higher UV dose and addition of strong oxidants in UV/AOP make it substantially more effective. Typical design criteria for UV photolysis and UV-AOP systems for potable reuse are presented in Table 23-17. Photolysis is presented in Sec. 8-4, disinfection with UV light is presented in Sec. 13-9, advanced oxidation is presented in Chap. 18, and UV-AOP is presented in Sec. 18-4.

Hydrogen peroxide is the typical oxidant in AOP, but free chlorine can have several advantages. First, the residual hydrogen peroxide must be removed before delivering water to the distribution system, which is typically

Table 23-17
Typical design criteria for UV photolysis and advanced oxidation for potable reuse applications

Parameter	Units	Value
UV dose for photolysis	mJ/cm^2	40–100
UV dose for UV/AOP	mJ/cm^2	800–1600
UV transmittance (UVT)	%	95–98
Oxidant dose for AOP	mg/L	2–7 (H_2O_2) 1–4 (chlorine residual)

done during the final disinfection process by adding additional chlorine to react with the hydrogen peroxide. Stoichiometry indicates that 2.1 mg/L as Cl_2 of chlorine is needed to quench each 1 mg/L of H_2O_2 residual. Second, the residual chlorine present after UV-AOP is directly available for final disinfection, reducing chlorine requirements for final disinfection. Finally, production of HO · may be more efficient with chlorine than H_2O_2, reducing the UV dose and energy requirements (Watts and Linden, 2007).

Like RO, AOP is effective at removing both regulated and unregulated chemicals but not all chemicals are removed. The ability of AOP to remove organic constituents depends on the reactivity of the chemical with the hydroxyl radical, as described in Chap. 18, which depends on the second-order rate constant, k_{OH}. The hydroxyl radical targets electron-rich areas of molecules, such as aromatic rings and double bonds. Advanced oxidation will effectively remove compounds with $k_{OH} > 10^8\ M^{-1}\ s^{-1}$. Advanced oxidation is less effective for simple alkanes (e.g., chloroform, methylene chloride), peroxides, and nitriles (see the hydroxyl rate constants in Appendix E4). UV-AOP is effective at removing NDMA, a compound that is poorly removed by RO. The California Department of Public Health regulations require that the AOP achieve 0.5-log destruction of 1,4-dioxane, so that requirement will often set the target UV dose (CSWRCB, 2018).

OZONE AND BIOLOGICALLY ACTIVE FILTRATION

Typical design criteria for ozone/BAF systems are presented in Table 23-18. Ozonation is presented in Sec. 13-7, AOP with ozone is presented in Sec. 18-2, and biofiltration is described in Sec. 11-8. GAC is frequently used as the biofilter media and referred to as biological activated carbon (BAC). While a wide range of Ozone/TOC values have been observed in literature, increasing the Ozone/TOC ratio can increase CEC removal (Sari et al., 2020). The biofilter is typically designed in a pressure–filter configuration.

The primary reactive site for ozone oxidation is the carbon–carbon double bond, and compounds will be well-removed if the ozone rate constant $k_{O3} > 10^3\ M^{-1}\ s^{-1}$. For many compounds, ozone/BAF can provide similar removal to RO. Recalcitrant compounds are those with $k_{O3} < 10\ M^{-1}\ s^{-1}$

Table 23-18
Typical design criteria for ozone/BAF for potable reuse applications

Parameter	Units	Value
Ozone/TOC ratio	mg O_3/mg TOC	0.2–2.0
Ozone contactor HRT	Min	5–15
BAF media	–	GAC, anthracite, or sand
BAF EBCT	Min	10–20
BAF loading rate	m/h (gpm/ft^2)	5–15 (2–6)

and include sucralose, PFAS, tris-(2-carboxyethyl)phosphine (TCEP), NDMA, meprobamate, primidone, and iohexol (Sari et al., 2020; Lee et al., 2012; Vaidya et al., 2020). Performance for removing trace organics can be affected by species that scavenge radicals, including carbonate, ammonia, sulfide, iron(II), and manganese(II). The formation of ozonation byproducts such as bromate must be evaluated (see Sec. 13-7), and NDMA has also been observed to form after ozonation (Sari et al., 2020). Hydrogen peroxide can be effective at increasing removal of compounds that have low ozone reaction rates while reducing the formation of bromate. BAF is effective at removing many compounds that form during ozonation, including aldehydes, glyoxals, and NDMA.

Ozone/BAF can have advantages over RO that include lower capital cost, higher water recovery, lower energy demand, and no concentrate disposal (Lee et al., 2012). Concentrate disposal is particularly challenging in inland areas and can be an obstacle to potable reuse. Disadvantage of ozone/BAF relative to RO is that it cannot achieve the same level of TOC removal and may be inadequate if treatment requirements require a low (<0.5 mg/L) finished water TOC concentration, it is not as effective for PFAS and some other specific constituents, and it may not be suitable for high-salinity wastewaters because of its inability to remove dissolved ions.

When RO is required because of TOC removal requirements, ozone/BAF may be advantageous as a first step in treatment because of its ability to effectively control biological and organic fouling of MF and RO membranes.

ADSORPTION WITH GRANULAR ACTIVATED CARBON

Typical design criteria for granular activated carbon (GAC) adsorption systems are presented in Table 23-19. Adsorption is described in Chap. 15. Design of GAC contactors for potable reuse is like GAC contactors for other applications. Adsorption of CECs onto GAC has been extensively studied and is well understood. Higher MW and more hydrophobic compounds are more strongly adsorbed. In a recent study, GAC was able to remove 96 CECs to below the detection limit up to 10,000 bed volumes (BV), after which 7 CECs were detected: sucralose, iohexol, acesulfame-k, meprobamate, cotinine, primidone, and acetaminophen (Vaidya et al., 2020). GAC has been coupled with ozone/BAF to meet lower TOC requirements and to remove refractory organics like PFAS (Vaidya et al., 2020).

Table 23-19
Typical design criteria for GAC adsorption for potable reuse applications

Parameter	Units	Value
EBCT	min	10–20
Loading rate	m/h (gpm/ft^2)	5–15 (2–6)

SOIL AQUIFER TREATMENT

Regulations in California for recharge of potable aquifers with recycled water allow the recycled water to be introduced by either surface spreading and subsurface injection. Projects employing surface spreading may rely on soil aquifer treatment (SAT) to remove pathogens and organic constituents. The term "soil aquifer treatment" describes the biological and physiochemical processes that collectively remove constituents from wastewater as it passes through the soil column and subsequently through aquifer systems beneath the soil (Idelovitch, 1978). Numerous studies, both at field sites and in pilot and bench-scale soil columns, have been conducted to improve understanding of the SAT process, how it works, what it can do, notably those in the Dan region of Israel (Idelovitch and Michail, 1984), the Tucson area (Fox et al., 2001), at Berlin (Drewes and Jekel, 1998), and at Montebello, CA (Laws et al., 2011). Recent studies have established the average TOC reduction for the process under aerobic conditions to range from 65 to 70 percent, with ozonation enhancing degradation during the first several days (Trussell et al., 2017, 2018a,b; Qu et al., 2020; Shabani et al., 2020).

AWTF Treatment Train Case Studies

Two case studies are presented to illustrate IPR and DPR operating systems. The first, the Orange County Water District (OCWD) Groundwater Replenishment System (GWRS), is the largest advanced water purification system for IPR in the world. The second, the Goreangab Water Reclamation Plant in Windhoek Namibia, is first and longest operating DPR system in the world. The following discussion is focused on the advanced water treatment facilities in these systems. Additional case studies are available in the U.S. EPA Potable Reuse Compendium (U.S. EPA, 2017).

ORANGE COUNTY WATER DISTRICT INDIRECT POTABLE REUSE

The GWRS is a joint project between the OCWD and the Orange County Sanitation District (OCSD). The AWTF has a capacity of 380,000 m^3/d (100 mgd) and replenishes groundwater that is an essential part of the local drinking water supply for more than 850,000 residents in Orange County.

Historical Development

To combat falling groundwater levels, seawater intrusion, and limitations in the available water sources for groundwater replenishment in the Orange County groundwater basin, OCWD built Water Factory 21 with a capacity to inject a blend of 37,000 m^3/d (15 mgd) of advanced treated wastewater and 37,000 m^3/d (15 mgd) of desalinated seawater to create a groundwater mound barrier to halt seawater intrusion. The seawater desalination facility used vertical tube evaporators and multistage flash distillation and the wastewater reclamation facility used lime coagulation, clarification, ammonia stripping, recarbonation, mixed media filtration, GAC adsorption, and chlorination. Construction on the plant started in 1971 and it went into operation in 1975. The seawater desalination portion of the plant ceased operation a year later, when federal government support for the facility was withdrawn. OCWD added reverse osmosis to the advanced wastewater process train in 1977. The OCWD started studies for the GWRS in 1995, which replaced Water Factory 21 in 2008. The original GWRS AWTF had a capacity of 265,000 m^3/d (70 mgd). It was expanded to 380,000 m^3/d (100 mgd) in 2015 and will be expanded to 490,000 m^3/d (130 mgd) to come on line in 2023.

Treatment Process Flow Diagram

The source of water for the GWRS is undisinfected secondary effluent from the OCSD. The GWRS AWTF process flow diagram shown on Fig. 23-7 includes flow equalization, microfiltration, cartridge filtration, reverse osmosis, advanced oxidation (UV photolysis and hydrogen peroxide), partial carbon dioxide stripping, and lime addition (Scott-Roberts, 2016; OCWD, 2021), which meets the criteria for full advanced treatment in California. The product water is discharged to spreading basins and seawater barrier injection wells. Purified water, along with rainwater, water from the Santa Ana River, and imported water, are delivered to the spreading basins and percolate into the aquifer, where the blended water eventually becomes part of Orange County's drinking water supply. Water pumped to the injection wells creates a barrier to seawater intrusion and becomes part of the drinking water supply. New mid-basin injection wells directly augment the groundwater basin. The product water from the GWRS meets and/or exceeds all DDW requirements for potable drinking water and the Santa Ana Regional Water Quality Control Board requirements for IPR (Burris, 2020).

WINDHOEK NAMIBIA DIRECT POTABLE REUSE

Windhoek is the capital of Namibia, one of the most arid countries in sub-Saharan Africa. Windhoek, which has a population of about 450,000, has been adding highly treated reclaimed water to the municipal water supply system since 1968. The blending of reclaimed water with potable water takes place directly in the pipeline that feeds its potable water distribution network (du Pisani, 2005).

Historical Development

Direct potable reuse was considered as early as 1954 because of chronic drought. Pilot-scale testing was performed from 1962 to 1965, the full-scale facility was designed based on the pilot results, plant construction started in 1967, and the plant started operation in October 1968. The initial treatment process has undergone extensive modifications considering new scientific and technological developments and actual operating experience. The DPR program incorporates the following features: (1) only domestic wastewater is used for DPR; (2) monitoring from the point of wastewater discharge to the point of water usage; (3) effective biological treatment; (4) multiple barrier treatment; (5) blending of final effluent with water from natural sources; and (6) continuous education and communication.

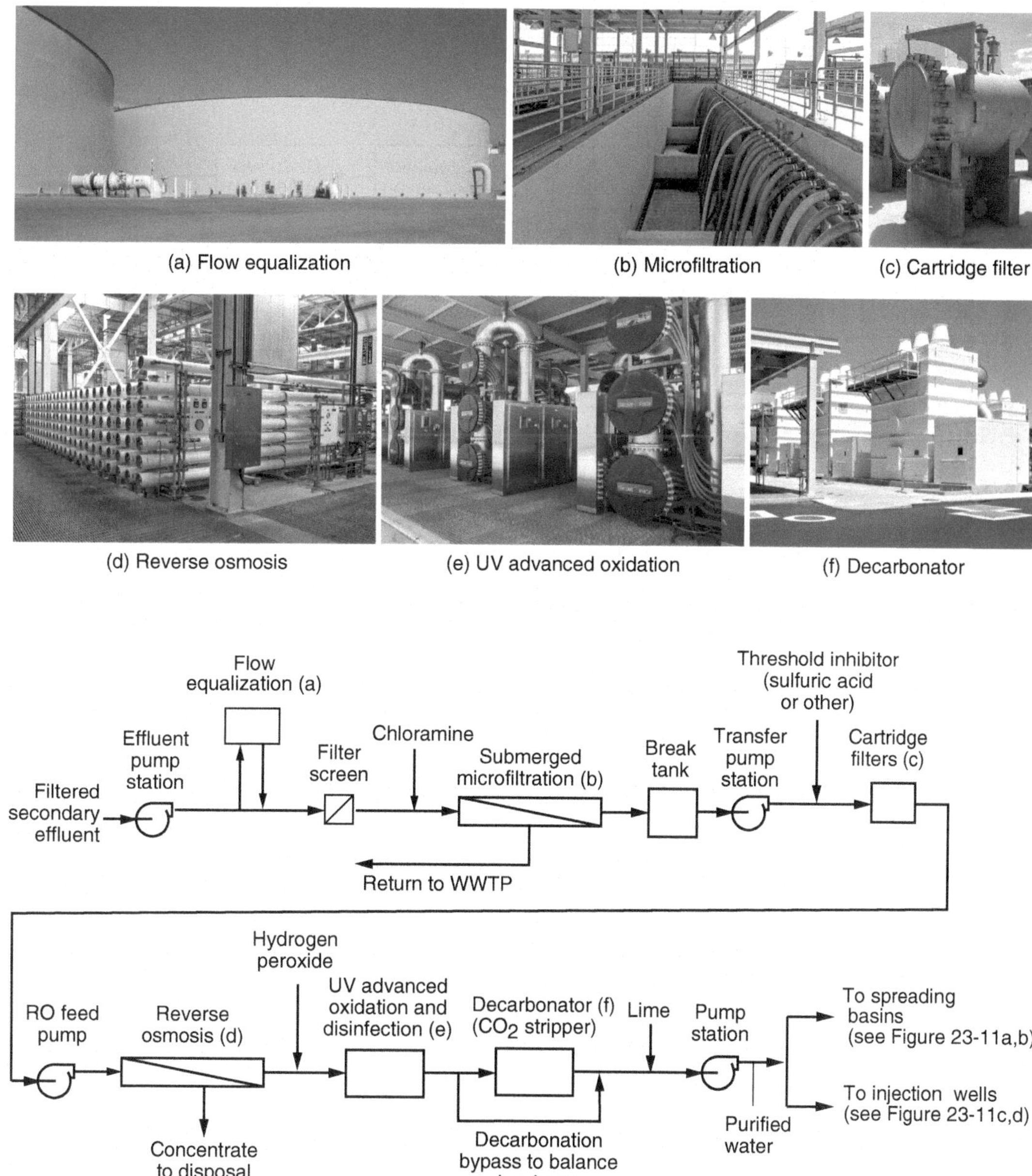

Figure 23-7
Process flow diagram for the OCWD GWRS advanced water treatment facility (Coordinates 33.692 N, 117.942 W).

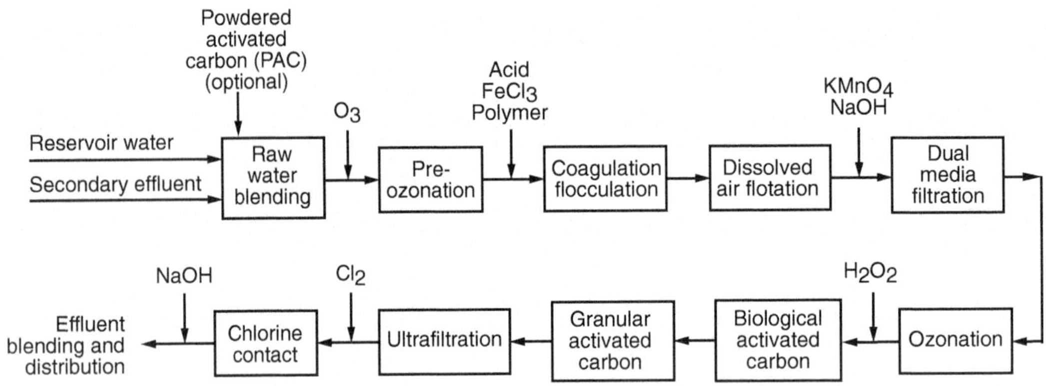

Figure 23-8
Process flow diagram for the Goreangab Mark IV Water Reclamation Plant (Coordinates 17.006E, 22.528S).

Treatment Process Flow Diagram
The current treatment plant, known as Goreangab Mark VI, was built on a new site in 2002. The process flow diagram is shown on Fig. 23-8. The source water is wastewater effluent from the Gammams Water Care Works, which only treats domestic wastewater, and raw water from Goreangab Dam. The design of the new plant is based on the experience gained from operating the various treatment processes and ongoing pilot-plant studies. The process flow diagram includes powdered activated carbon on a standby basis, pre-ozonation, ozonation followed by biological activated carbon, granular activated carbon, and membrane filtration (ultrafiltration). The latter two processes were pilot tested over a 30-month period to verify performance with this specific raw water. The Goreangab Water Reclamation Plant does not use reverse osmosis and UV advanced oxidation, which avoids the challenges associated with disposal of RO concentrate (Wingoc, 2021).

The long history of the Windhoek DPR experience demonstrates that highly treated municipal wastewater can be reused successfully for potable purposes. In the case of Windhoek, a combination of factors, with the lack of feasible alternative water sources being the most compelling, made DPR a viable option, even in financial terms.

23-7 Posttreatment

As described in the previous section, RO is a common component of many AWTF treatment trains because of its high efficacy at removing organic chemicals and pathogens. Reverse osmosis is also highly effective at removing nearly all inorganic species, producing permeate that is typically low in pH, alkalinity, and TDS. Typical quality of RO permeate from an AWTF is

Table 23-20
Typical RO permeate quality

Parameter	Unit	Values
pH	Unitless	4.8–6.0
TDS	mg/L	15–25
Calcium hardness	mg/L as $CaCO_3$	0.75–1.75
Alkalinity	mg/L as $CaCO_3$	8–15
CO_2(aq)	mg/L	25–400

reported in Table 23-20. This water can be aggressive to the metallic and cement-lined surfaces of conventional water conduits and structures. Furthermore, when this water is introduced into an existing distribution system accustomed to water of different quality, it can destabilize the existing scale on surfaces and mobilize the constituents in the scale, seriously degrading the quality of the highly purified water from the AWTF. The degradation can range from increases in turbidity and colored water to, in some cases, contamination with lead and copper that is of health significance. As a result, posttreatment is usually applied to stabilize (remineralize) the water and adjust its composition to be closer to that of a conventional water supply.

Remineralization of Advanced Treated Water

Remineralization is accomplished by increasing the hardness, alkalinity, and pH of the water so that it is slightly saturated with calcium carbonate ($CaCO_3$) and at a pH comparable to existing supplies. A common goal is to achieve a slight oversaturation of calcium carbonate and a pH of 8.5 or less. As described in Chap. 22 (Sec. 22-10) calcium carbonate saturation is evaluated using the Langelier Saturation Index (LSI), and the goal is to achieve a slightly positive LSI (LSI = 0–0.2 is a common target). The pH and $CaCO_3$ targets are often accomplished by adding lime and managing the concentration of CO_2. When this approach is used, the alkalinity and calcium hardness of the resulting water are nearly identical because the addition of lime increases both equally and the addition or removal of CO_2 changes neither. Neglecting temperature, ionic strength, and other minor corrections, the relationship between the saturation pH for calcium carbonate and the alkalinity of the treated water can be characterized using Eq. 23-1, which is a simplification of an equation derived by Langelier (see Eq. 22-30) (Langelier, 1936):

$$pH_s = 11.81 - 2\log(\text{Alk}) \tag{23-1}$$

where pH_s = saturation pH for calcium carbonate
Alk = alklinity, mg/L as $CaCO_3$

The Langelier saturation index is the difference between the measured pH and the saturation pH:

$$LSI = pH - pH_s \tag{23-2}$$

This relationship between LSI and alkalinity is illustrated on Fig. 23-9. As shown, an alkalinity of about 57 mg/L or greater is required before an LSI of 0.2 can be achieved at a pH of 8.5 or less. Typically, the alkalinity of the water following posttreatment ranges from 60 to 100 mg/L as $CaCO_3$.

The goals of remineralization and adjusting calcium carbonate saturation are often achieved by adding lime and managing the dissolved inorganic carbon (DIC) through the removal and/or addition of CO_2 as shown on Fig. 23-10. Other alternatives are used and the design and operation of even the seemingly simple lime/CO_2 strategy can have many nuances.

Typically, posttreatment will not include both CO_2 removal and CO_2 addition; one or the other is usually required, although it is possible for a system to include both. Mathematical modeling of the water chemistry before, during, and after the RO process can be used to determine the appropriate design for a combination of source water quality and finished water quality goals. Factors that can affect the addition or removal or CO_2 include the phosphorus concentration in the treated wastewater and the target alkalinity in the product water. High phosphorus concentrations can lead to increased acid addition ahead of RO for scale control

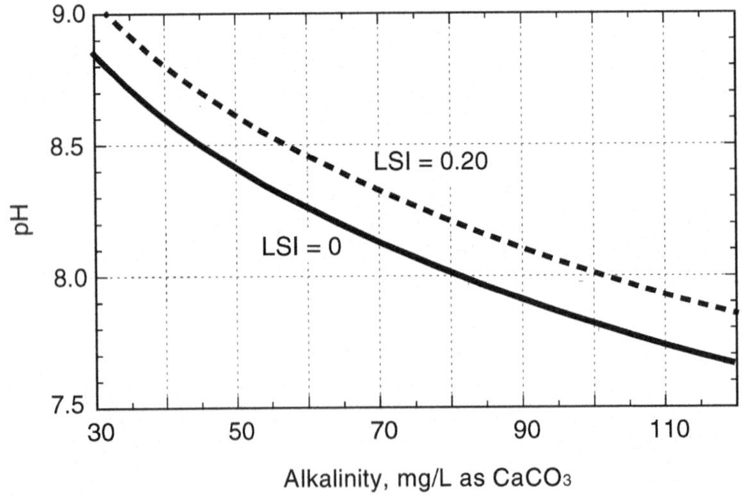

Figure 23-9
Relationship between pH and alkalinity for achieving slight supersaturation of calcium carbonate, based on LSI values from 0 to 0.2 and equal alkalinity and calcium hardness.

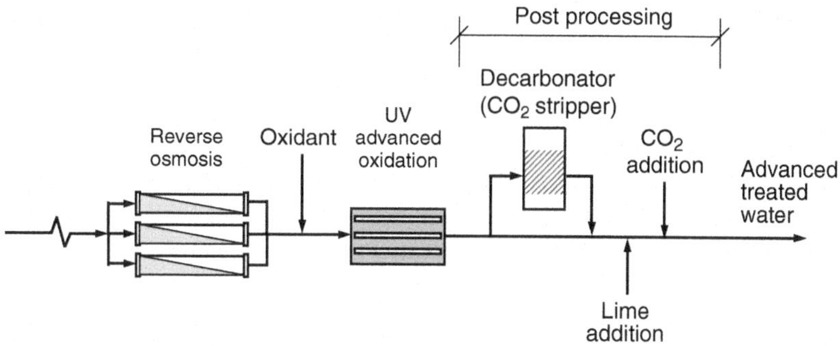

Figure 23-10
Posttreatment with lime addition and CO_2 management to stabilize advance treated water.

and, consequently, lower pH and higher CO_2 levels in the RO permeate. High alkalinity goals increase lime addition, necessitating correspondingly more CO_2. The latter conditions result in a lower pH goal, as shown in Fig. 23-9.

Lime Addition

Lime is produced by calcining limestone; that is, heating limestone to a temperature where it releases CO_2 [$CaCO_3 + \text{heat} \rightarrow CaO + CO_2\uparrow$]. The use of lime is complicated by its relatively low solubility in water, roughly 1.5 g/L as $CaCO_3$. As a result, lime is often added to the process stream as a slurry called milk of lime.

Manufacturers deliver lime in three forms; as quicklime pebbles or powder [CaO], hydrated lime powder [$Ca(OH)_2$], and hydrated lime slurry. Before quicklime can be used in water treatment, it must be hydrated by adding water, a process called slaking [$CaO + H_2O \rightarrow Ca(OH)_2 + \text{heat}$]. Treatment plants using quicklime must have facilities to slake the lime. Slaking ordinarily involves not only hydrating the lime, but also removing inert materials, or grit, from the product while forming the milk of lime. Treatment plants purchasing hydrated lime powder must have facilities for mixing the hydrated lime powder with water to make milk of lime. Treatment plants using a delivered lime slurry can use it directly, though sometimes further dilution is required to ensure accurate feeding.

Post treatment is essential for producing a stable product water.

http://www.stantec.com/wt3r/23-7.html

Although some purification is possible during the manufacturing process, the purity of lime is heavily dependent on the purity of the original limestone source (i.e., the purity of the mineral in the mine). The purity of commercial lime varies widely depending on the location, ranging from a low of about 85 percent to a high of about 96 percent. Impurities are also common in hydrated lime and delivered lime slurry.

PARTICLES IN LIME

The addition of lime slurry into the process stream can introduce extraneous particulate matter into the treated water. In addition to the grit in the lime, these particles can include $CaCO_3$ particles that form in high pH "hot spots" created as the $Ca(OH)_2$ particles in the slurry react with dissolved inorganic carbon in the water. When lime is added early in a treatment train, such as it is in lime softening, these particles are removed in downstream sedimentation and filtration processes. In potable reuse, however, no downstream separation processes are present and the particles can increase the turbidity of the finished water, which can exceed regulatory limits or contribute to clogging of injection wells for groundwater replenishment.

To prevent lime from contributing to increased finished water turbidity, the milk of lime can be purified prior to being used for treatment. A lime saturator, which is an upflow reactor-clarifier dedicated to producing saturated lime water, can produce a saturated lime solution with little or no extraneous particulate matter. Unfortunately, lime saturators are large, expensive, difficult to operate, and require a lot of maintenance. Developing better ways of introducing lime into treated water is an active area of investigation.

LIMESTONE CONTACTORS

Limestone contactors can be a cost-effective alternative for adding lime to water if sufficiently pure limestone is available at an affordable price. Considerations in the design of limestone contactors are the contact time, achieving slightly supersaturated conditions, achieving specific hardness/alkalinity goals, and maintaining low turbidity in the effluent. The appropriate contact time is a tradeoff between the size of the limestone pebbles in the contactor and the size of the contactor. Limestone contactors can closely approach saturation with calcium carbonate in the finished water but cannot achieve a positive LSI; thus, they are usually followed by a low dose of caustic soda.

Depending on the quality of the RO permeate (pH and alkalinity), CO_2 can be added or removed to manage the alkalinity and hardness of the final product water. Sulfuric acid (H_2SO_4) is sometimes added preceding the limestone contactor, which increases the calcium hardness without increasing the alkalinity. Reducing the influent pH with H_2SO_4 reduces the contact time needed to achieve a specific calcium hardness goal and increases the caustic soda dose needed to achieve a positive LSI. A design contact time of 15 minutes is common. The purity of the limestone is important in maintaining a low turbidity in the final effluent. Managing remaining inert materials released from previously dissolved limestone while recharging contactors can also be an important issue.

Table 23-21
Chemicals used in posttreatment and their impact on pH, calcium hardness, and alkalinity

Chemical	pH	Hardness	Alkalinity
CO_2	Reduce	No change	No change
$CaCO_3$	Increase	Increase	Increase
$Ca(OH)_2$	Increase	Increase	Increase
NaOH	Increase	No change	Increase
$CaCl_2$	No change	Increase	No change

Other Chemical Addition Options

While most common, lime, CO_2, and limestone are not the only chemicals that can be used for posttreatment. The combination of calcium chloride, sodium hydroxide, and carbon dioxide requires a low capital cost, achieves very low levels of particles in the treated water, and has been used in several projects with groundwater injection. Unfortunately, the chemical cost of this alternative is significantly higher than the use of lime. Table 23-21 characterizes the impact of posttreatment chemicals on the key components of calcium carbonate saturation.

23-8 Environmental Buffers and Engineered Storage Buffers

Planned potable reuse projects have been in operation since 1962 and, with few exceptions, have taken the form of indirect potable reuse, which means they have an environmental buffer designed into the system. Potential benefits of environmental buffers were identified by the National Research Council as: "(1) an opportunity to further reduce contaminants through natural processes, (2) a substantial lag time between the exit of the advanced wastewater treatment system and entrance into the potable system, and (3) the opportunity for the water of wastewater origin to blend with natural waters in the environment" (NRC, 1998).

The reduction in contaminants in environmental buffers depends on many site-specific characteristics and, as a result, can be difficult to quantify. IPR projects may not be able to rely on or receive credit for contaminant reduction in natural systems unless the performance can be reliably demonstrated. The lag time is also site-specific but can often be quantified. Some recent regulations have related the lag time to performance expectations and to the time available to respond to failures in upstream treatment processes. Blending can be quantified but may be of less value because the required reduction in contaminant concentrations typically cannot be accomplished with blending, so blending will generally not reduce treatment requirements.

Considering these factors, lag time may be the most significant benefit of an environmental buffer. An engineered storage buffer provides a similar benefit. Thus, a storage buffer may be defined as an intermediate location (either an environmental or engineered facility) where ATW can be held (stored) for some period before being withdrawn for either indirect or direct potable reuse. Environmental buffers can either be groundwater aquifers or surface water reservoirs or lakes. Engineered storage buffers (ESBs) can be aboveground storage tanks, covered and lined surface or underground storage reservoirs, or large diameter surface and subsurface pipelines. The implementation of these buffer facilities is considered in this section.

Groundwater Storage Buffers

An additional benefit of the use of environmental buffers is the concept of "loss of identity." The recycled water enters the environmental buffer as treated wastewater but exits it as "waters of the state," having lost the tie to its wastewater origin. This loss of identity is like the concept of *de facto* reuse. The regulatory requirement, the methods of application, and the modeling requirements are presented below.

REGULATORY REQUIREMENTS

No national regulations for groundwater recharge for IPR exist. Some states have developed guidelines. California, for example, requires a six-month retention time in the subsurface before the water can be withdrawn for IPR, though that time may be reduced to a minimum of two months if additional treatment is provided and it is satisfactorily demonstrated that response to treatment failures can be addressed in that time. If six months is provided, the project is credited with 10-log removal of protozoa and 6-log removal of viruses. Validation of the six-month residence time requires extensive modeling of the aquifer and subsequent confirmation with tracer tests. The ratio of the ATW flow to the natural groundwater base flow must also meet requirements. California regulations for groundwater augmentation may be found in Title 22, California Code of Regulations (DPH, 2014).

METHOD OF APPLICATION

Advanced treated water can be applied to groundwater aquifers by spreading in open recharge basins or injected directly into the aquifer with recharge wells. Surface spreading will benefit from soil aquifer treatment. The ATW must be stabilized to minimize contamination from the dissolution of constituents present in the aquifer such as arsenic and metals. Examples of surface spreading basins and injection wells are shown on Fig. 23-11. ATW from the Orange County Water District GWRS is delivered to spreading

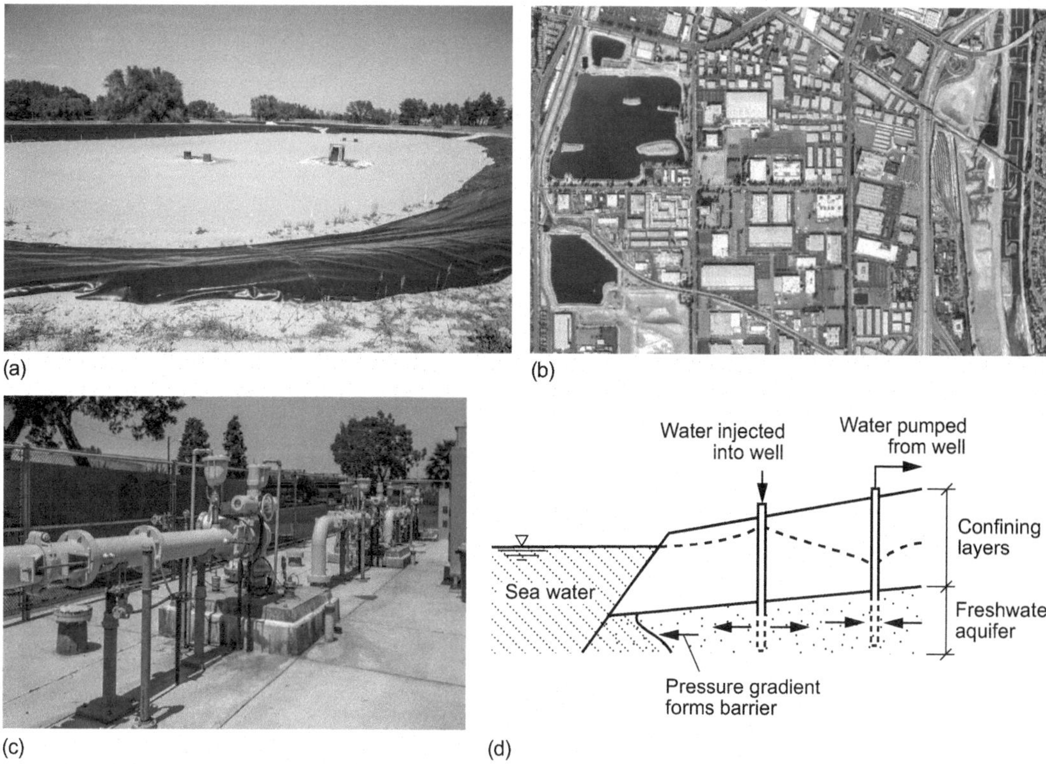

Figure 23-11
Methods of recharging groundwater aquifers: (a) spreading basins in Lakeland FL (Coordinates: 28.493 N, 81.620 W); (b) four spreading basins shown below Anaheim Lake, upper left, in Orange County, CA (Coordinates: 33.862 N, 117.857 W, view at altitude 6 km). River infiltration basins shown on right; (c) injection wells in Orange County, CA; and (d) in coastal regions, injection wells are commonly used for the control of sea water intrusion and for aquifer recharge.

basins [see Fig. 23-11(b)] to augment local groundwater and to injection wells [see Fig. 23-11(c)] to form the Talbert Seawater Intrusion Barrier to limit the intrusion of seawater into the freshwater groundwater basin.

GROUNDWATER MODELING

Extensive groundwater modeling is necessary to identify the flow patterns in an aquifer and confirm that the six-month residence time criterion is being met. The Orange County Groundwater Basin (OCGB) will be used to illustrate the approach to groundwater modeling.

Groundwater from the OCGB serves about 2.4 million residents in Orange County using about 300 water supply wells. These wells, along with about 400 monitoring wells installed by OCWD are used to monitor groundwater levels and to collect water quality samples. OCWD used well logs and pumping records to develop a groundwater model of the complex basin, as shown on Fig. 23-12.

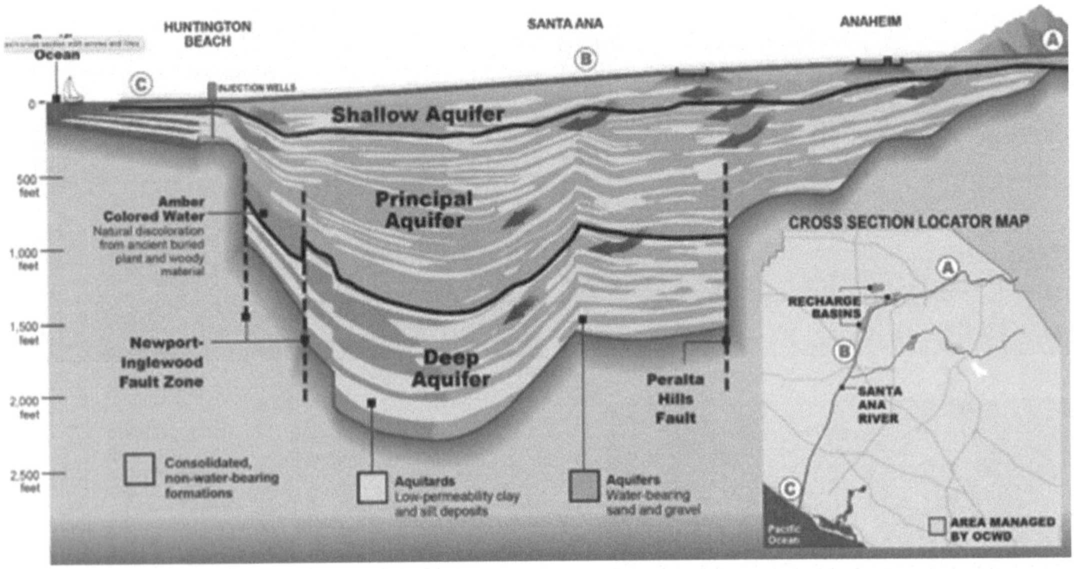

Figure 23-12
Geologic cross section through OCGB depicting the shallow, principal, and deep aquifers. The spreading basins and barrier wells shown in Fig. 23-11 are shown in the upper right and upper left side of the diagram, respectively. (Courtesy of the Orange County Water District).

The model is used to evaluate basin production capacity, recharge requirements, and groundwater flow dynamics. It is also used to estimate annual changes in groundwater storage, the effects of future pumping, and the effects of groundwater augmentation on water levels, storage, and overall water budget. With respect to regulatory requirements, the groundwater model has been used to predict underground travel times. Monitoring wells have validated the water retention and travel times (OCWD, 2015). Because of the time, effort, and expense involved, groundwater modeling studies should be initiated early in the planning process if groundwater augmentation is being considered. The OCWD had the benefit of numerous past studies on which to build its successful groundwater model and augmentation program.

Surface Water Storage Buffers

The regulatory requirement, the methods of application, and the required modeling for surface water augmentation are considered below.

REGULATORY REQUIREMENTS

As with groundwater, no national regulations for the implementation of surface water augmentation for IPR exist. Some states have developed guidelines. California has adopted regulations for surface water augmentation that requires that the reservoir has been in operation as an approved water supply for a sufficiently long period to establish the reservoir's baseline raw water quality (generally 5 years), have an approved minimum theoretical retention time (generally 6 months), that tracer studies and hydrodynamic modeling be performed to demonstrate that the extracted water contain no more than 1% of the wastewater delivered to the reservoir during any 24-hour period (10% if additional treatment is provided). California's regulations for surface water augmentation may be found in Title 22, California Code of Regulations (CSWRCB, 2018).

METHOD OF APPLICATION

Examples of reservoirs used for surface water augmentation projects are presented in Fig. 23-13). The most common method of introducing ATW into the reservoir is with a system of submerged diffusers so that effective initial mixing of the ATW with the resident dilution water can be achieved.

SURFACE WATER MODELING

To comply with the regulations for the discharge of ATW into an existing water supply reservoir, the hydrodynamics and the biochemistry of the reservoir must be modeled. To illustrate, two comprehensive models were used to study San Vincente Reservoir [see Fig. 23-13(a)]. The three-dimensional hydrodynamic Estuary Lake and Coastal Ocean Model (ELCOM) used was to study water velocities, temperatures, salinity (i.e., conductivity), and tracer simulations. A biochemistry model called the Computational Aquatic Ecosystem Dynamics Model (CAEDYM) was used to study dissolved oxygen (DO), nutrients, organic matter, pH, and chlorophyll. The coupled models were used to study the relationships between physical, biological, and chemical variables in the reservoir (Flow Science, 2012).

(a) (b)

Figure 23-13
Surface water storage reservoirs: (a) San Vicente Reservoir, San Diego County (Coordinates: 32.9310 N, 116.9134 W) and (b) Miramar Reservoir, San Diego County (Coordinates: 32.9169 N, 117.0985 W).

For the Miramar Reservoir, the ELCOM model was replaced with the newer Aquatic Ecosystem Model 3 Dimensions (AEM3D). User inputs to AEM3D include boundary conditions, bathymetry, initial reservoir conditions, physical inputs, and meteorological inputs; the outputs from the AEM3D model include temperature, salinity (i.e., conductivity), water velocities, and concentrations of conservative tracers. To illustrate the computational complexity involved, the AEM3D Model Grid for Miramar Reservoir is shown on Fig. 23-14 (Stantec MWH, 2020).

Studying the water quality effects of reservoir augmentation with ATW using computer hydrodynamic and water quality models typically involves the following steps.

- Select appropriate hydrodynamic and water quality computer models (e.g., AEM3D or CAEDYM),
- Obtain and assemble data for calibration and validation of the reservoir model,
- Set up the reservoir model and associated input data files,
- Perform model simulations to calibrate and validate the reservoir model,
- Perform model simulations to calibrate the water quality part of the reservoir model,
- Determine the future conditions and associated input data,
- Apply the calibrated reservoir model to different future operating conditions and evaluate water quality changes (adapted from Flow Science, 2012).

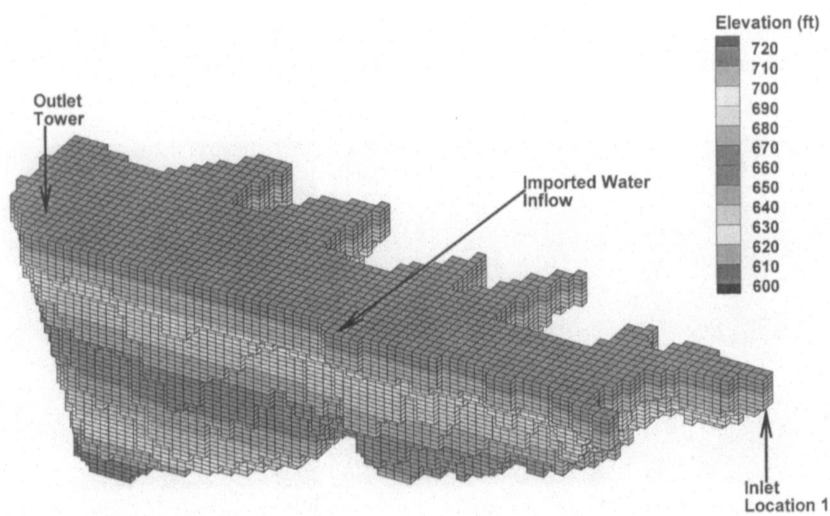

Figure 23-14
The AEMED Model Computational Grid for Miramar Reservoir. (Courtesy of the City of San Diego, CA, and Stantec MWH, 2020).

Because of the time, effort, and expense involved in reservoir modeling, modeling studies should be initiated early in the project planning process if reservoir augmentation is being considered.

Engineered Storage Buffers

In DPR, the ATW can be blended with other water directly before treatment in a water treatment facility or, if the AWTF is permitted as a drinking water facility, the ATW can be added directly to the water distribution system (see Fig. 23-2). As shown in Fig. 23-2(a), an engineered storage buffer may be used to store water for a specified period. The hydraulic retention time required should be sufficient to provide flow continuity and for the measurement and reporting of specific constituents or operational surrogates at critical control points to ensure the water quality meets all standards or goals before being delivered to the water distribution system (Tchobanoglous et al., 2011). When the AWTF has proven performance, reliability, and redundancy and the performance can be monitored rapidly, a relatively small ESB, if any, may be sufficient. In some cases, the storage capacity in the pipeline from the AWTF to the water treatment plant or distribution system may obviate the need for an ESB. In such a case, there must be the capability and time to divert the flow of recycled water before contamination reaches the drinking water system.

STORAGE REQUIREMENTS FOR ENGINEERED STORAGE BUFFERS

In general, the storage requirements will be controlled by the time required for constituent analysis and confirmation of the overall reliability of the monitoring system. Thus, key monitoring parameters to verify system performance that can be evaluated expediently must be identified. If off-spec product water is detected entering the buffer, it is necessary to discharge the off-spec batch to a predetermined alternate location or recycle it back to an acceptable point in the AWTF (Tchobanoglous et al., 2011).

DESIGN FEATURES OF ENGINEERED STORAGE BUFFERS

Important features of a well-defined engineered storage buffer include:

- Fully controlled environment
- Contained to prevent contamination and evaporative losses
- No source of contaminants from within the buffer itself
- Ability to divert flow going into or out of the buffer as needed
- Accommodation of monitoring and sampling equipment
- Sufficient residence time to allow failures in performance at critical points in the process train to be detected and a response to be initiated
- Well-characterized and optimized hydraulics
- High level of security (Tchobanoglous et al., 2011)

The design of the storage buffer will include site-specific constraints, capabilities of the monitoring systems, flow rate and flow equalization requirements, and safety factors.

23-9 Monitoring and Response

Although the ability of treatment processes to remove constituents to below regulated concentrations and produce water safe to drink is well established, process performance can degrade over time, equipment can be operated incorrectly or break down, and chemicals can be discharged into the wastewater collection system at concentrations that the treatment system cannot manage. The purpose of monitoring is to verify process performance, demonstrate compliance with regulations, document treated water quality, and assure regulators and the public that the water is safe to drink. If a potential failure is detected, a response by operators or control systems is necessary to prevent off-spec water from reaching customers. Nothing can assure absolute safety of treated water from a potable reuse system, but the same is true of any water supply. The monitoring and response plan should assure that water from a potable reuse system is as safe or safer than conventional drinking water supplies. This section presents the basic requirements of a monitoring and response plan.

Monitoring Response Time

As the distance and time between the point of treatment and point of consumption becomes shorter, the time for monitoring and response also become shorter. The total time for a response includes time to collect a sample and send it to analysis, time to analyze and report the results, time to assess the meaning of results, and time to execute a response such as diverting the off-spec water, adjusting the treatment process (e.g., increasing chemical or UV doses), or changing to an alternative water supply. When the time from treatment to consumption is measured in months, daily tests to confirm performance are more than adequate, but when the time proximity decreases to a few hours, a continuous method of performance testing may be more appropriate.

The frequency and rapidity of monitoring and response are affected by the type of risk to public health. Periodic (daily, monthly, or even annually) measurements may be appropriate for chemicals of chronic toxicity, but continuous measurements may be necessary for pathogens and chemicals of acute toxicity when the time for delivery of water to customers is also short.

A final factor affecting the frequency and rapidity of monitoring and response is the potential for dilution. The ability of environmental buffers to provide substantial dilution of off-spec water can increase public safety when monitoring is less frequent, but more frequent monitoring is essential for DPR because both the residence time and the potential for dilution are smaller.

Types and Methods of Monitoring

The sampling and monitoring methods used to produce advanced treated water are considered below.

SAMPLING METHODS

The purpose of water sampling is to obtain representative, reproducible, defensible, and useful data for monitoring programs. The types of samples used in monitoring programs are summarized in Table 23-22.

MONITORING PROGRAMS

Potable reuse systems incorporate various types of monitoring programs, as summarized in Table 23-23. The most common monitoring programs are regulatory compliance, routine monitoring of performance indicators, and long-term periodic sampling for specific constituents. The evaluation of new technologies and improvements in existing technologies

Table 23-22
Types of samples used for monitoring potable reuse systems

Type of Sampleor Test	Description	Uses
Grab (discrete)	An individual sample collected at a given time and a given location.	To measure constituents that can only be analyzed in offline analytical instruments (e.g., chemicals, pathogenic microorganisms).
Grab (composite)	Discrete grab samples collected at one location that are combined for a single analysis either by the sampling instrument or in the laboratory.	To characterize the composition of the process flow averaged over time or reduce analytical costs.
Grab (integrated)	Discrete grab samples collected at the same time at different locations that are combined for a single analysis.	To support special studies to identify the effectiveness of treatment processes.
Online	Probes or instruments used to monitor process parameters (e.g., pressure, flowrate) or constituents (e.g., pH, individual chemicals) continuously or at specified time intervals.	To control processes (e.g., measure pH to control acid feed), monitor and document process operation, or monitor critical control points that may indicate a loss of treatment effectiveness.
Offline	Testing that can only be accomplished by taking equipment out of service.	To assess whether equipment is functioning properly (e.g., for MF membrane integrity, which modules may have broken fibers).

Table 23-23
Types of monitoring programs in potable reuse systems

Monitoring Program	Types of Samples	Purpose
Basis of design	Online, grab	To evaluate technology alternatives, evaluate performance under varying operating conditions, and develop process design criteria.
Calibration during startup	Online, grab	To document baseline performance for process acceptance before becoming operational.
Regulatory compliance	Online, grab	To verify performance with respect to drinking water MCLs and other permit requirements.
Routine monitoring of performance indicators (including surrogates and indicator compounds)	Online, offline, grab	To verify performance with respect to priority pollutants, selected chemicals, and pathogenic microorganisms.
Long-term periodic performance monitoring	Online, offline, grab	To assess long-term aging, degradation, and performance decline of processes and equipment.
Evaluation of new technologies	Online, offline, grab	To identify and test new technologies for comparison to existing technologies.
Profiling	Online, grab	To assess the progress in removing a constituent as it travels through the process train.

can be conducted at laboratory or pilot scale using separate equipment, or parallel components in the full-scale system as illustrated on Fig. 23-15. Profiling is performed in the full-scale system to evaluate the reduction of a specific target constituent (including new emerging contaminants) by each process in the overall process train.

The Critical Control Point Concept

The term Critical Control Points (CCP) refers to the key unit operations that are critical for the control of pathogens and chemicals. The use of CCP is a derivative of efforts to apply the hazard analysis and critical control point (HACCP) methodology to potable reuse. The food industry developed HACCP methodology in the early 1990s to address food safety through the analysis and control of biological, chemical, and physical hazards from raw material production, procurement and handling, manufacturing, distribution, and consumption of the finished product. The HACCP methodology has been adapted for potable reuse with the identification of barriers in the process train where control can be applied to ensure a specified performance and monitoring of control parameters that can be used to confirm that performance (Halliwell et al., 2014; Walker et al., 2016).

Normally bench, pilot, and full-scale studies are used to establish the removals that can be expected and identify the operating conditions and performance measures that will establish the performance that has been achieved. CCPs

Figure 23-15
The upper RO pressure vessel is used to evaluate the performance of new RO membranes relative to the performance of existing membranes. The test vessel runs in parallel with the operational RO vessels.

often target an individual pathogen or chemical. Operating conditions might include measures as SRT and dissolved oxygen levels for biological treatment, O_3:TOC ratio and UVT reduction for ozonation, CT for disinfection processes, and filtration rate and effluent turbidity for filtration. Performance measurements might include monitoring surrogates whose removal has been shown to be predictive of the removal of the target constituent.

A surrogate parameter is an easily measured physical or chemical property that correlates directly with the concentration of an indicator compound, can be used to assess the removal of trace organic compounds, and/or provides an indication of a treatment process failure (CSWRCB, 2018). An indicator compound is a chemical or organism that represents the characteristics of a family of trace organic compounds or pathogens and can be monitored to assess the removal of those chemicals. *E. coli* is an indicator organism for disinfection (NRC, 2012).

Monitoring for pathogen removal in membrane processes generally follows the processes outlined in the Membrane Filtration Guidance Manual (U.S. EPA, 2005). Monitoring for disinfection generally follows the CT concept (see Sec. 13-3).

APPLICATION OF CCP METHODOLOGY

The general application of CCP methodology for an AWTF involves the identification of the CCP locations in the process train that are critical for

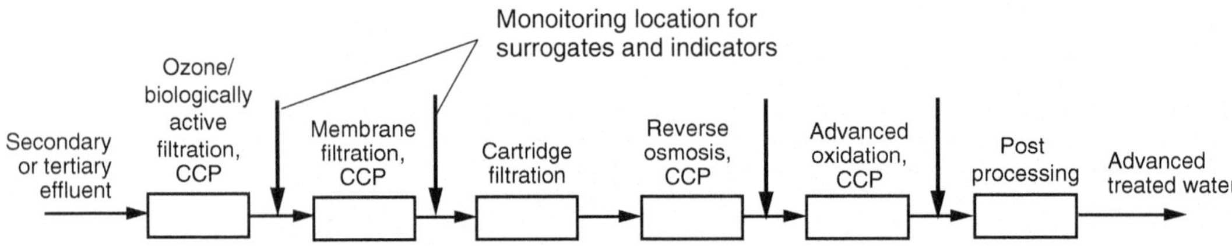

Figure 23-16
Examples of CCPs and corresponding monitoring points for surrogates and indicators in the application of CCP methodology in an AWTF.

the removal of specific constituents of concern. CCPs and their monitoring locations in a typical AWTF are identified on Fig. 23-16. Parameters monitored for the CCPs such as those shown on Fig. 23-16 are identified in Table 23-24.

Generally, CCPs are separated between CCPs for pathogen control and CCPs for chemical control. Although unit operations can serve both purposes, the monitoring and control criteria are usually different. For example, ozone system performance for disinfection is assessed using the CT value, whereas for organics removal it is assessed using the O_3:TOC ratio, the ratio of the applied ozone dose to the influent TOC, corrected for the presence of nitrite.

PROCESS CONTROL AND RESPONSE MEASURES

An essential step in monitoring is to establish critical limits for each CCP parameter, which establishes the range of allowable values. When the measured value falls outside the allowable range, the utility must respond to prevent out-of-spec water from reaching consumers. In some cases, the required response may be straightforward (e.g., if ozone residual drops too low, the operators can increase the ozone dose), but in other cases the treated water will need to be diverted or the plant shut down until the failure can be rectified. Response and corrective measures will be site-specific based on specific processes and design of an AWTF.

For pathogens, AWTFs include a series of processes designed to achieve an overall removal goal that cannot be achieved by any of the processes individually, and often the LRV of each process is determined separately (with pressure decay test for MF, constituent removal for RO, and CT for disinfection). Critical control limits are applied to the overall performance so an evaluation for meeting control limits is determined by summing the overall LRV of the process train. Under these circumstances, a process train with redundant overall LRV performance will exhibit greater reliability (Pecson et al., 2015).

Table 23-24
Typical critical control points and monitoring parameters for processes used in an AWTF[a]

CCP	Monitoring Parameters	Comments
Ozone/biologically-active filtration	Ozone dose (O_3:TOC ratio), ozone residual, CT value, change in UVA, TOC removal	Control of chemicals and pathogenic microorganisms. Measured values of UVA and TOC correspond to the destruction of residual organic carbon. CT value corresponds to pathogen reduction.
Membrane filtration	Pressure decay testing, effluent turbidity, particle counting	Control of pathogenic microorganisms. Pressure decay is related to membrane integrity. If fibers are broken, removal performance will deteriorate. Turbidity is not sensitive enough to demonstrate membrane integrity.
Reverse osmosis	Electrical conductivity (EC), TOC, specific surrogates (strontium, sulfate, etc.)	Control of chemicals and pathogenic microorganisms. Measured values of EC and TOC underestimate reduction in pathogens.
UV disinfection and advanced oxidation	UV dose, UV intensity, UV transmittance, lamp power utilization, oxidant dose	Control of pathogenic microorganisms and chemicals (e.g., NDMA, 1,4-dioxane). Few concerns with respect to pathogenic microorganisms if the sensors are calibrated properly.
Engineered storage buffer	Disinfectant residual, CT value	Control of pathogenic microorganisms.

[a]Adapted from Tchobanoglous et al. (2014).

The Future of Process Monitoring

Extensive experimental work is underway to develop improved surrogates and rapid biological test methods to enhance performance validation at ATWFs.

NEW AND IMPROVED SURROGATES

In some cases, the ability of processes to remove specific constituents exceeds the capability of existing surrogate parameters. As an example, studies have shown that RO can achieve greater than 4-log LRV of pathogens, but measurement of EC or TOC can only validate 2-log LRV (Trussell et al., 2013). Other surrogates such as fluorescent compounds and dyes (e.g., rhodamine WT, uranine, and fluorescein) are being evaluated, and it is anticipated that better surrogates will be developed for demonstrating higher LRV credit for the removal of pathogens by RO. In addition, as chemical detection limits become more sensitive, other surrogates will become available.

RAPID BIOLOGICAL TEST METHODS

The availability of rapid-response bioassays that eliminate need for monitoring individual compounds, give quick measure of overall "safety" of product water would further improve and enhance the ability to monitor the performance of AWTFs. While some instrumental online biological test units have been developed, they do not provide sufficient resolution to establish LRV credit, but future advancements may make this possible.

23-10 Public Outreach

Public outreach is essential in building and sustaining community support for a potable reuse program. The purpose here is to highlight key elements of public outreach program including establishing the need for potable reuse, identifying the challenges with implementation of a potable reuse program, developing an ongoing communication plan, and developing effective communications materials. A public outreach case study is also presented along with a consideration of future challenges.

Communicating the Need for Potable Reuse

The first step in developing a public outreach program is a clear statement of why potable reuse is needed. The responsible agency, having considered all other possibilities, must demonstrate that the added water source is essential for the community's long-term water supply. A clear rationale for why a particular potable reuse program was selected relative to other options must be articulated. A clear vision and rationale for the project will enhance public confidence in its merits.

Challenges in Implementing a Potable Reuse Program

In planning for potable reuse, a public outreach program must be initiated early in the planning process. Once the need for potable reuse has been established, a public outreach program that launches early on (i.e., before and during project planning) and is maintained throughout the lifetime of the project should be developed. In developing such a program, it is possible to benefit from previous experience about the challenges in developing and implementing a potable reuse program. Examples of communication and outreach challenges are reported in Table 23-25.

Developing an Effective Communication Plan

A communication plan provides a roadmap so that clear and consistent messaging is achieved when dealing with the public. For example, it is important that everyone involved in the outreach program use a consistent set of facts and terminology when speaking and engaging

Table 23-25
Examples of communication and outreach challenges

Type of Challenge	Specific Concerns
Concerns about current water supply	❑ Mistrust of drinking water quality. ❑ Lack of knowledge about current water supply. ❑ Increased use of point-of-use filters and bottled water. ❑ Distrust of water supplier and governmental agencies.
Perceptions about potable reuse	❑ The "yuck factor," which is a strong (but natural) reaction that can cause low initial acceptance of DPR projects. ❑ Water quality concerns, including those in the media, such as MTBE, perchlorate, chromium-6, and CECs (i.e., pharmaceuticals, hormones, endocrine disrupting compounds, and industrial chemicals). ❑ Potable reuse viewed as a "last resort" rather than a sustainable alternative that benefits the community. ❑ Impacts on the environment from managing residuals and/or reducing water released to the environment. ❑ Businesses (such as soft drink bottlers) viewing recycled water as negatively affecting their brands.
Community opposition to project	❑ Environmental justice issues (e.g., only certain demographics receive the water). ❑ Opponents concealing true motivations (such as antigrowth). ❑ Opposition occurring late in project development. ❑ Costs of water and/or increasing rates.

Source: Millan et al. (2015).

with the public. Key factors in a communications plan are summarized in Table 23-26.

Once a communication plan has been developed, the next step is developing outreach strategies for communicating with the public. Examples of outreach strategies are summarized in Table 23-27.

Developing Communication Materials

The final element of an effective communications program is the availability of clear written materials that can be left with the public for further study, exhibits about the project, and a list of speakers who are available to speak about the project. Examples of communication plan tools and materials are summarized in Table 23-28.

Public Outreach: A Case Study

To illustrate the importance and elements of public outreach, the "Pure Water San Diego" (PWSD) program will be used as a case study. The PWSD recycling program is multiyear phased project which, at buildout in 2035,

Table 23-26
Key factors in a communication plan

Factor	Significance
Schedule and duration	Communication should start early in the process, before project selection, and must continue through the design, construction, start-up phases, and throughout the operation of the system.
Purpose of communication	Communication activities should have a clearly stated purpose and be used to support decisions.
Messages	Messages should provide a framework for understanding the need for the project, including a narrative to engage the public, raise awareness, and gain acceptance. Messages should be consistent, accurate, and understandable to a nontechnical audience.
Terminology	Uniform terminology has not been developed for potable reuse, but specific projects have produced terminology that has been effective on a local level. Accessible terms like "advanced purified water" and "purified water" are more effective with the public than jargon like "potable reuse" and "IPR." Technical terms not understood by the public may not resonate well even when explained.
Problem solving	A clearly articulated problem will help the public better understand and support the need for potable reuse; therefore, define the water supply condition that will be resolved by the project. Another best practice is to focus on improvement (i.e., the project is improving the quality of life and making things better for the community).
Anticipated outcomes	The outcomes of the outreach program should be broad and include public agreement that wastewater is a resource and should be recycled; community trust in the utility to implement potable reuse that produces safe, high-quality water and a reliable water supply; and that the agency publicly commits to being transparent and seeks community engagement and involvement in projects.
Costs and benefits	Financial considerations may be the primary concern of some communities, and clear and transparent explanation of the costs is necessary to gain public confidence, especially if the potable reuse project is not the least costly option. The conversation on project costs should include a discussion of benefits (e.g., water reliability and the sustainability of the water supply option). Economic development may be an important benefit to some stakeholders.
Competing issues	Communities face many important issues, ranging from education to the economy. Water reliability and sustainability are part of the community's discussion, and consideration is needed as to how to illustrate the link between water supply and other important community topics.
Demographics and environmental justice	Because certain demographic groups are less likely to support potable reuse, attention should be given to communicating with these groups. Groups that tend to be less supportive of potable reuse include women (with children at home), minorities, less educated/less affluent individuals, non-English speakers, and senior citizens. These groups may also be concerned with environmental justice issues.

Adapted from Ruetten (2004), Millan et al. (2015), and Tchobanoglous et al. (2015).

Table 23-27
Examples of potable reuse outreach strategies

Element/Tactic	Details
Research on public perception	Telephone surveys, one-on-one stakeholder meetings, focus groups, and other research activities can be used to assess community concerns and gain an understanding of public perception and acceptance. Results can inform the development of the outreach strategy.
Audience identification	Communication with diverse audiences is needed because each group may present potential challenges to effective outreach. Specific audiences that require distinct outreach efforts include opinion leaders, community leaders, community organizations, and youth. Maintaining a database of individuals categorized by audience can be helpful in organizing outreach efforts.
Internal communication	Include an internal outreach component to educate agency staff members. Customers or friends and neighbors may approach staff members with questions about potable reuse; it is important the agency provides a consistent message.
Outreach to opinion leaders	Opinion leaders influence the attitudes and behaviors of others. This group should be made aware of the need to increase water supply sources and educated about potable reuse as an option. Effort is needed to identify opinion leaders in individual communities.
Outreach to other important leaders	These leaders include academic and educational staff; business organizations; civic groups; environmental organizations; water wholesalers and retailers; state and local elected officials and staff; and medical, public health, and water quality experts.
Written materials	Written materials must present confident messages. Possible formats include briefing binders, e-mail listservs, event invitations, brochures and flyers, e-newsletter articles, press releases, direct mail, websites, and social media.
Personal interaction	Meetings and presentations are needed to reach out to all identified stakeholders. Personal interactions can be in the form of one-on-one meetings, town hall meetings, neighborhood gatherings, formal presentations, speaker's bureau, and civic meetings. Board members and elected officials should be involved, in addition to agency staff.
Identify supporters and champions	Agencies can use these individuals or organizations as project partners. In addition to voicing support, these partners can become advocates for the project.
Message plan	Messages can be developed for both a general audience and specific audiences affected by an individual project. Messages should describe potable reuse technology, underscore the safety of the water and protection of public health, and provide the costs and benefits (e.g., increased water reliability and sustainability) of the project. Informational material needs to be developed and spokespersons need to be trained to deliver this information.
Communicate effectively	Guidance for effective communication includes training all project spokespeople; reviewing messages for consistency in presentations, interviews, and meetings with stakeholders; and ensuring written materials reflect the same information provided in oral communications.
Letters of support	After a presentation or meeting, agency staff should request a letter of support for the project from appropriate audience members. Post these letters on the project website and use them to reach out to other stakeholders.

(Continued)

Table 23-27 *(Continued)*

Element/Tactic	Details
Common questions and answers	Be prepared with answers to common questions. Information can be provided in written format, as well as in briefing materials.
Address difficult issues	Be prepared to address difficult topics raised by stakeholders. If needed, seek assistance from experts in the field, including academics, medical doctors, public health officials, and other credible individuals.
Other specialized needs	Other issues may need to be fulfilled as part of the DPR project. For example, risk communication and risk management expertise may be needed in response to stakeholder questions. These fields of study are often used in communication and managing risks associated with public health.

Adapted from Ruetten (2004), Millan et al. (2015), and Tchobanoglous et al. (2015).

Table 23-28
Examples of communication plan tools and materials

Tools/Materials	Purpose/Examples
Written materials	Examples: Fact sheets, frequently asked questions, brochures, bill inserts, posters and banners, materials for youth and children, white papers.
Digital materials	Examples: Project website, slide presentations, e-newsletters, videos.
Mailing lists	To communicate to different groups for different purposes. Mailing lists can be electronic and physical.
Centralized internal information system	To catalogue and store materials.
Speakers bureau	To facilitate opportunities to speak at group meetings, including business leaders, civic groups, environmental, multicultural, and other community groups.
Demonstration facility/visitor center	To provide a positive learning experience for participants. Visitor centers involve educational displays and materials, while demonstration facilities show the treatment processes and treated water for examination. If possible, allow visitors to taste test the product water.
Media outreach	To provide timely information and ensure the media is informed, as well as to quickly address misinformation in the media. Examples: spokespeople, media training, contacts, articles, tours, and response to media requests.
Social media	To reach certain segments of the population and provide information on a timely basis.
Independent advisory panels	To provide credibility and validation of a project. Local physicians and local and national experts in health, water quality, and technology can provide an independent viewpoint and make recommendations for improvement.
Stakeholder groups	To provide a process for input and feedback from interested parties within a community. Stakeholder group members can become important supporters of the project.
Rapid response plan	To prepare the agency to quickly address unexpected events related to the project.
Monitoring and evaluation	To provide measurable objectives that can be reviewed periodically. The results of the review will provide feedback for adapting or changing the communications plan and/or tools and materials.

Source: Millan et al. (2015).

will provide about 33 percent of the city's water needs. In the first phase of the PWSD program, 114,000 m^3/d (30 mgd) will be added to the Miramar Reservoir. Water from the Miramar reservoir will be treated in the Miramar Drinking Water Facility (MDWF) before being added to the City's water distribution system (PWSD, 2017).

EVOLUTION OF PWSD PROGRAM

The PWSD program has evolved from reuse studies involving aquatic treatment followed by advanced treatment dating back to the 1970s (WCPH, 1996). The current PWSD program began in 2004 when the City Council authorized a study to evaluate options to maximize the use of recycled water. Completed in 2006, the Water Reuse Study was the first phase of the authorized study. The second phase involved the construction of a 3800 m^3/d (1.0 mgd) advanced water treatment demonstration project. The results of the demonstration program were adopted by the City Council in 2013. In 2014, the City Council approved the proceed with the implementation of the program (City of San Diego, 2020).

PUBLIC OUTREACH PROGRAM

Public outreach has been the key factor in gaining public support for adding purified water to the City's drinking water supply. In 2004, public support for the program was 23 percent. By 2015, public support had increased to 73 percent. The outreach program was also successful in overcoming the stigma associated with the phrase toilet-to-tap.

The activities of the outreach team included: (1) development of outreach materials to inform the residents about the program and the science behind it; (2) participation in community events, (3) facility tours, (4) youth outreach programs, (5) Pure Water Day Open House, (6) social media and online engagement, (7) reaching out with all forms of media, and (8) sponsorship of an active speakers series. In 2019, more than 1900 people toured the facility, presentations were given to more than 2900 community members, City of San Diego staff participated in 40 community events, 30 stakeholder interviews, more than 21,600 website visits, and more than 220 support pledges were received from San Diego residents (City of San Diego, 2020). As of February 2021, over 18,665 people have toured the Pure Water Demonstration Facility and more than 157,250 people have visited the website. These statistics demonstrate that a comprehensive and all-encompassing outreach program can be effective in engaging the public.

Future Public Outreach Needs

In the future, as the need for additional water becomes more critical, more municipalities will consider the benefits of potable reuse. To avoid misperceptions of the role of potable reuse, it will be important to develop informational and resource materials for the public to introduce them to the subject and to address their concerns.

USE OF STANDARD TERMINOLOGY

Use of standard terminology is key to the development of effective resource materials and presentations. In developing these resource materials, the use of jargon must be avoided. When speaking with the public about potable reuse, it is important to develop clear and understandable terminology. For example, the distinctions between the various forms of IPR and DPR has been a continuing source of confusion with the public.

ANTICIPATING PUBLIC CONCERNS

In addition to developing standardized terminology, anticipating public concerns is critical for the success of a potable reuse program. For example, concern about CECs and other chemicals that may be found in water is an ongoing issue that must be addressed on a continuing basis in resource materials. As potable use becomes more commonplace, new techniques must be developed to anticipate and communicate with the public about their concerns with potable reuse.

Problems and Discussion Topics

23-1 If potable water reuse was being considered for your community, discuss the feasibility of indirect and direct potable reuse (IPR and DPR). In your judgment, which of the two options would be the easiest to promote and implement? What are the principal reasons for your answer?

23-2 What are some of the arguments for and against the use of a separate purple pipe system for reclaimed water versus either IPR or DPR?

23-3 For your community, discuss the feasibility of implementing the following options for source control for troublesome industrial constituents that are now discharged to the municipal wastewater collection system:

a. providing a separate pressurized small diameter wastewater collection system.

b. require industry to use holding tanks with separate collection and treatment and/or disposal via tanker trucks.

23-4 Discuss the advantages and disadvantages of including reverse osmosis in a potable reuse treatment train.

23-5 Discuss the advantages and disadvantages of including ozone and biofiltration in a potable reuse treatment train.

23-6 Propose an AWTF process train that does not include reverse osmosis that can achieve 12-log virus, 10-log Giardia, and 10-log Cryptosporidium reduction. Identify how much log removal is achieved by each process in your treatment train.

23-7 Calculate the required alkalinity to achieve an LSI of 0.1 at pH = 7.8.

23-8 Considering advances in monitoring technology, discuss whether it is necessary for operators to be on site 24 hours per day at advanced water treatment facilities. Explain your reasoning.

23-9 Some opponents of potable reuse say that it is prudent to wait until everything is known about every constituent in wastewater. What are the key points of a counterargument for moving ahead with a potable reuse project?

23-10 The need for transparency and public trust is especially important for potable reuse projects. Cite three recent literature references on potable reuse implementation and discuss how the need for transparency and public trust were dealt with.

References

Alan Plummer Associates (2015) *Direct Potable Reuse Resource Document – Final Report*, Report prepared for the Texas Water Development Board. Fort Worth, TX.

Angelakis, A.N., Asano, T., Bahri, A., Jimenez, B.E., and Tchobanoglous, G. (2019) "Water Reuse: From Ancient to Modern Times and the Future," *Front. Environ. Sci.*, **6**, 26, https://doi.org/10.3389/fenvs.2018.00026.

Asano, T., Burton, F.L., Leverenz, H., Tsuchihashi, R., and Tchobanoglous, G. (2007) *Water Reuse: Issues, Technologies, and Applications*, McGraw-Hill Book Company, New York.

Bellona, C., Drewes, J.E., Xu, P., and Amy, G. (2004) "Factors Affecting the Rejection of Organic Solutes during NF/RO Treatment—a Literature Review," *Water Res.*, **38**, 12, 2795–2809.

Bernados, B. (2018) "Reverse Osmosis for Direct Potable Reuse in California," *J. Am. Water Works Ass.*, **110**, 1, 28–36.

Breitner, L.N., Howe, K.J., and Minakata, D. (2018) "Boron can be Used to Predict Trace Organic Rejection through Reverse Osmosis Membrane for Potable Reuse," *Environ. Sci. Technol.*, **52**, 13871–13878.

Breitner, L.N., Howe, K.J., and Minakata, D. (2019) "Effect of Functional Chemistry on the Rejection of Low-Molecular Weight Neutral Organics through Reverse Osmosis Membranes for Potable Reuse," *Environ. Sci. Technol.*, **53**, 9, 11401–11409.

Burris, D.L. (2020) Groundwater Replenishment System 2019 Annual Report, Prepared for the California Regional Water Quality Control Board, Santa Ana Region, DDB Engineering, Inc.

CDM Smith (2021) City of San Diego 2020 Urban Water Management Plan (Draft), Report prepared for City of San Diego Public Utilities Department, CDM Smith, Carlsbad, CA.

Chamberlain, J.F., Tromble, E., Graves, M., and Sabatini, D. (2020) "Water Reuse versus Water Conveyance for Supply Augmentation: Cost and Carbon Footprint," *AWWA Water Science*, **2**, 1, https://doi.org/10.1002/aws2.1170.

City of San Diego (2020) 2019 Year in Review, City of San Diego https://www.sandiego.gov/sites/default/files/pure_water_2019_year_in_review_report.pdf

CSWRCB (2018) *Regulations Related to Recycled Water, October 1, 2018*, California State Water Resources Control Board, Sacramento, CA.

Dieter, C.A., Maupin, M.A., Caldwell, R.R., Harris, M.A., Ivahnenko, T.I., Loverlace, J.K., Barber, N.L., and Linsey, K.S. (2017) *Estimated Use of Water in the United States in 2015 - Circular 1441*, US Geological Survey, Reston, VA.

DPH (2014) GW Replenishment Using Recycled Water, DPH-14-003E, Title 22, California Code of Regulations State of California, Sacramento, CA. https://www.waterboards.ca.gov/drinking_water/certlic/drinkingwater/documents/gwreplenishmentregulation/DPH-14-003EFinalText.pdf.

Drewes, J.E., Bellona, C., Eberhart, M., and Chellam, S. (2013) *Predictive Models to Aid in Design of Membrane Systems for Organic Micropollutant Removal*, WateReuse Research Foundation, Alexandria, VA.

Drewes, J.E., and Jekel, M. (1998) "Behavior of DOC and AOX Using Advanced Treated Wastewater for Groundwater Recharge," *Water Res.*, **32**, 10, 3125–3133, https://doi.org/10.1016/S0043-1354(98)00064-5.

du Pisani, P.L. (2005) "Direct Reclamation of Potable Water at Windhoek's Goreangab Reclamation Plant," pp. 193–202, in S.J. Khan, M.H. Muston, and A.I. Schafer (Eds.) *Integrated Concepts in Water Recycling 2005*, University of Wollongong, Wollongong, Australia.

Flow Science. (2012) *Reservoir Augmentation Demonstration Project; Limnology and Reservoir Detention Study of San Vicente Reservoir Calibration of the Water Quality Model, Report prepared for City of San Diego*, Flow Science Incorporated, Harrisburg, VA.

Foess, G.W., Meenahan, J.G., and Blough, D. (1977) "Evaluation of In-Line and Side-Line Flow Equalization Systems," *J WPCF*, **49**, 1, 120–130.

Fox, P., Houston, S., Westerhoff, P., Drewes, J.E., Nellor, M., Yanko, W., Baird, R., Rincon, M., Arnold, R., Lansey, K., Bassett, R., Gerba, C., Karpiscak, M., Amy, G., and Reinhard, M. (2001) *Soil Aquifer Treatment for Sustainable Water Reuse*, American Water Works Association Research Foundation, Denver, CO.

Fujioka, T., Khan, S.J., McDonald, J.A., Roux, A., Poussade, Y., Drewes, J.E., and Nghiem, L.D. (2013) ""N-nitrosamine Rejection by Nanofiltration and Reverse Osmosis Membranes the Importance of Membrane Characteristics," *Desalination*, **316**, 67–75.

Giokas, D., Vlessidis, A., Angelídis, M., Tsímarakis, G., and Karayannis, M. (2002) "Systematic Analysis of the Operational Response of Activated Sludge Process to Variable Wastewater Flows: A Case Study," *Clean Technol and Envir.*, **4**, 3, 183–190.

Gujer, W., and Erni, P. (1978) The Effect of Diurnal Ammonium Load Variation on the Performance of Nitrifying Activated Sludge Processes. *Proceedings of the 9th International Conference on Water Pollution Research*, Stockholm, Sweden, 391–407.

Halliwell, D., Burris, D., Deere, D., Leslie, G., Rose, J., and Blackbeard, B. (2014) *Utilization of Hazard Analysis and Critical Control Points Approach for Evaluating Integrity of Treatment Barriers for Reuse, WRRF-09-03,* WateReuse Research Foundation, Alexandria, VA.

Idelovitch, E. (1978) ""Wastewater Reuse by Biological-chemical Treatment and Groundwater Recharge," *J. Water Pollut. Contr. Fed.*, **50**, 12, 2723–2740.

Idelovitch, E., and Michail, M. (1984) "Soil-Aquifer Treatment: A New Approach to an Old Method of Wastewater Reuse," *J. Water Pollut. Control Fed.*, **56**, 8, 936–943, http://www.jstor.org/stable/25042398.

Kolpin, D.W., Furlong, E.T., Meyer, M.T., Thurman, E.M., Zaugg, S.D., Barber, L.B., and Buxton, H.T. (2002) "Pharmaceutical, Hormones, and Other Organic Wastewater Contaminants in U.S. Streams, 1999-2000: A National Reconnaissance," *Environ. Sci. Technol.*, **36**, 6, 1202–1211.

LaGrega, M., and Keenan, J. (1974) "Effects of Equalizing Wastewater Flows," *J. WPCF*, **46**, 1, 123–132, http://www.jstor.org/stable/25038099.

Langelier, W. (1936) "The Analytical Control of Anti-corrosion Water Treatment," *J. AWWA*, **28**, 11, 1500–1522.

Laws, B.V., Dickenson, E.R.V., Johnson, T.A., Snyder, S.A., and Drewes, J.E. (2011) "Attenuation of Contaminants of Emerging Concern during Surface-spreading Aquifer Recharge," *Sci. Total Environ.*, **409**, 1087–1094, https://doi.org/10.1016/j.scitotenv.2010.11.021.

Lee, C.O., Howe, K.J., and Thomson, B.M. (2012) "Ozone and Biofiltration as an Alternative to Reverse Osmosis for Removing PPCPs and Micropollutants from Treated Wastewater," *Water Res.*, **46**, 4, 1005–1014.

Millan, M., Tennyson, P.A., and Snyder, S. (2015) *Model Communication Plans for Increasing Awareness and Fostering Acceptance of Potable Reuse, WRRF 13-02,* WateReuse Research Foundation, Alexandria, VA.

Mosher, J., Vartanian, G.M., and Tchobanoglous, G. (2016) *Potable Reuse Research Compilation: Synthesis of Findings,* Water Environment and Reuse Foundation, Washington, DC.

Niku, S., Schroeder, E.D., Tchobanoglous, G., and Samaniego, F.J. (1981a) *Performance of Activated Sludge Processes: Reliability, Stability and Variability, R805097-01,* U.S. Environmental Protection Agency, Washington, DC.

Niku, S., Schroeder, E.D., Tchobanoglous, G., and Samaniego, F.J. (1981b) *Project Summary: Performance of Activated Sludge Processes: Reliability, Stability and Variability, EPA-600/S2-81-227,* U.S. Environmental Protection Agency, Cincinnati, OH.

NRC (1998) *Issues in Potable Reuse: The Viability of Augmenting Drinking Water Supplies with Reclaimed Water,* National Research Council, National Academies Press, Washington, D.C.

NRC (2012) *Water Reuse: Potential for Expanding the Nation's Water Supply through Reuse of Municipal Wastewater,* National Research Council, National Academies Press, Washington, D.C.

OCSD (2018) *2017-18 Annual Report, Resource Protection Division, Pretreatment Program,* Orange County Sanitation District, Fountain Valley, CA. https://www.ocsd.com/Home/ShowDocument?id=26330

OCWD (2015) *OCWD Groundwater Management Plan 2015 Update,* Orange County Water District, Fountain Valley, CA.

OCWD (2021) *GWRS Groundwater Replenishment System Technical Brochure*, Orange County Water District, Fountain Valley, CA, Available at: https://www.ocwd.com/media/4267/gwrs-technical-brochure-r.pdf (assessed on May 20, 2021).

Olivieri, A.W., Crook, J., Anderson, M.A., Bull, R.J., Drewes, J.E., Haas, C.N., Jakubowski, W., McCarty, P.L., Nelson, K.L., Rose, J.B., Sedlak, D.L., and Wade, T.J. (2016) *Expert Panel Final Report: Evaluation of the Feasibility of Developing Uniform Water Recycling Criteria for Direct Potable Reuse*, Prepared by the National Water Research Institute for the State Water Resources Control Board, Sacramento, CA.

Pecson, B.M., Trussell, R.S., Pisarenko, A., and Trussell, R.R. (2015) "Achieving Reliability in Potable Reuse: The Four Rs," *J. Am. Water Works Ass.*, **107**, 3, 48–58.

Plósz, B.G., Liltved, H., and Ratnaweera, H. (2009) "Climate Change Impacts on Activated Sludge Wastewater Treatment: A Case Study from Norway," *Water Sci. Technol.*, **60**, 2, 533–541.

PWSD (2017) North City Pure Water Project Title 22 Engineering Report, Pure Water San Diego, San Diego, CA.

Qu, Y., Trussell, B., Olivas, L., Monroy, I., and Trussell, S. (2020) *Soil Aquifer Treatment Characterization with Soil Columns for Groundwater Recharge in the San Fernando Valley-Phase IV Microorganisms Challenge Testing, Project #4600*, Water Research Foundation, Denver, CO.

Rice, J., Wutich, A., and Westerhoff, P. (2013) "Assessment of De Facto Wastewater Reuse across the U.S.: Trends between 1980 and 2008," *Environ. Sci. & Technol.*, **47**, 11099–11105, https://dx.doi.org/10.1021/es402792s.

Rimer, A., DeCarolis, J., and Sathyamoorthy, S. (2017a) *Guidelines for Source Water Control Options and the Impact of Selected Strategies on Direct Potable Reuse*, Water Environment & Research Foundation, Washington, DC.

Rimer, A., DeCarolis, J., and Sathyamoorthy, S. (2017b) *Guidelines for Source Water Control Options and the Impact of Selected Strategies on Direct Potable Reuse: Literature Compendium*, Water Environment & Research Foundation, Washington, DC.

Rimer, A., DeCarolis, J., and Sathyamoorthy, S. (2017c) *Guidelines for Source Water Control Options and the Impact of Selected Strategies on Direct Potable Reuse: Case Studies*, Water Environment & Research Foundation, Washington, DC.

Ruetten, J. (2004) *Best Practices for Developing Indirect Potable Reuse Project: Phase 1 Report, WRF-01-004*, WateReuse Research Foundation. Alexandria, VA.

Sari, M.A., Oppenheimer, J., Robinson, K., Drewes, J.E., Pisarenko, A.N., Sundaram, V., and Jacangelo, J.G. (2020) "Persistent Contaminants of Emerging Concern in Ozone-biofiltration Systems: Analysis from Multiple Studies," *AWWA Water Science*, **2**, 5, https://doi.org/10.1002/aws2.1193.

Scott-Roberts, S. (2016) *Engineer's Report for the Water Production Flow Enhancement Project*, Orange County Water District, Fountain Valley, CA.

SDPUD (2018) *Draft Title 22 Engineering Report, North City Pure Water Project*, San Diego Public Utilities Department. Available at: https://www.sandiego.gov/sites/default/files/ncpw_project_draft_title_22_engineering_report_0.pdf

Shabani, F., Aflaki, R., Minamide, T., Venezia, T., and Stenstrom, M.K. (2020) "Soil Aquifer Treatment to Meet Reclaimed Water Requirements," *Water Environ. Res.*, **92**, 2, 266–277, https://doi.org/10.1002/wer.1245.

Sharma, S.K., and Kennedy, M.D. (2017) "Soil Aquifer Treatment for Wastewater Treatment and Reuse," *Int. Biodeterior. Biodegradation*, **119**, 671–677, http://dx.doi.org/10.1016/j.ibiod.2016.09.013.

Soller, J.A., Eftim, S.E., Warren, I., and Nappier, S.P. (2017) "Evaluation of Microbiological Risks Associated with Direct Potable Reuse," *Microb. Risk Anal.*, **5**, 3–14, http://dx.doi.org/10.1016/j.mran.2016.08.003.

Stantec MWH (2020) *Limnology Study of Miramar Reservoir*, Report prepared for City of San Diego Public Utilities Department, Stantec MWH, San Diego, CA.

Steinle-Darling, E. (2015) "The Many Faces of DPR in Texas," *J. AWWA*, **107**, 3, 16–20, https://doi.org/10.5942/jawwa.2015.107.0050.

Stokes, J., and Horvath, A. (2006) "Life Cycle Energy Assessment of Alternative Water Supply Systems," *Int. J. Life Cycle Assess.*, **11**, 335–343, http://dx.doi.org/10.1065/lca2005.06.214.

Tchobanoglous, G. (2019) "Integrated Wastewater Management: The Future of Water Reuse in Large Metropolitan Areas," *Integr. Environ. Assess. Manag.*, **15**, 1, 160–164, https://doi.org/10.1002/ieam.4103.

Tchobanoglous, G., Cotruvo, J., Crook, J., McDonald, E., Olivieri, A., Salveson, A., and Trussell, R.S. (2015) *Framework for Direct Potable Reuse, WRRF 14–20*, WateReuse Research Foundation, American Water Works Association, Water Environment Federation and National Water Research Institute, Washington, DC.

Tchobanoglous, G., Kenny, J., and Leverenz, H. (2021) "Rationale for Constant Flow to Optimize Wastewater Treatment and Advanced Water Treatment Performance for Potable Reuse Applications," *Water Environ. Res.*, https://doi.org/10.1002/wer.1531.

Tchobanoglous, G., and Leverenz, H. (2019) "Comprehensive Source Control for Potable Reuse," *Front. Environ. Sci.*, **7**, 95, 1–12, https://doi.org/10.3389/fenvs.2019.00095.

Tchobanoglous, G., Leverenz, H., Nellor, M.H., and Crook, J. (2011) *Direct Potable Reuse: A Path Forward*, WateReuse Research Foundation, Alexandria, VA.

Tchobanoglous, G., Stensel, H.D., Tsuchihashi, R., and Burton, F.L. (2014) *Wastewater Engineering: Treatment and Resource Recovery*, 5th ed., McGraw-Hill Education, New York, NY.

Trussell, B., Trussell, S., Qu, Y., Gerringer, F., Stanczak, S., Venezia, T., Monroy, I., Bacaro, F., and Trussell, R. (2018b) "A Four-year Simulation of Soil Aquifer Treatment Using Columns Filled with San Gabriel Valley Sand," *Water Res.*, **144**, 26–35, https://doi.org/10.1016/j.watres.2018.07.012.

Trussell, R.R., Kenny, J.D., Tackaert, R.A., Pisarenko, A.N., and Trussell, R.S. (2020) "Mass and Flow Balances for Reverse Osmosis Virus Removal Crediting with Tracers," *AWWA Water Sci.*, **2**, 6, https://doi.org/10.1002/aws2.1206.

Trussell, R.R., Salveson, A., Snyder, S.A., Trussell, R.S., Gerrity, D., and Pecson, B.M. (2013) *Potable Reuse: State of the Science Report and Equivalency Criteria for Treatment Trains, WateReuse-11–02*, WateReuse Research Foundation, Alexandria, VA.

Trussell, R.S., Pisarenko, A.N., and Chen, E. (2015) *Implementation of Advanced Water Purification Facility Extended Testing, Project Report for City of San Diego*, Available at: https://www.sandiego.gov/sites/default/files/implementation_of_awp_facility_extended_testing_project_report_july_2015.pdf

Trussell, S., Trussell, B., Qu, Y., and Trussell, R. (2017) *Soil Aquifer Treatment Characterization with Soil Columns for Groundwater Recharge in the San Fernando Valley, Project #4600,* Water Research Foundation, Denver, CO.

Trussell, S., Trussell, B., Qu, Y., and Trussell, R. (2018a) *Soil Aquifer Treatment Characterization with Soil Columns for Groundwater Recharge in the San Fernando Valley-Extended Testing Report, Project #4600,* Water Research Foundation, Denver, CO.

Tu, K.L., Fujioka, T., Khan, S.J., Poussade, Y., Roux, A., Drewes, J.E., Chivas, A.R., and Nghiem, L.D. (2013) "Boron as a Surrogate for N-nitrosodimethylamine Rejection by Reverse Osmosis Membranes for Potable Water Reuse Applications," *Environ. Sci. Technol.*, **47**, 6425–6430.

U.S. EPA (2005) *Membrane Filtration Guidance Manual, EPA-815-R-06-009,* U.S. Environmental Protection Agency, Washington, DC.

U.S. EPA (2011) *Introduction to the National Pretreatment Program, EPA-833-B-11-001,* U.S. Environmental Protection Agency, Washington, DC.

U.S. EPA (2017) *2017 Potable Reuse Compendium,* U.S. Environmental Protection Agency, Washington, DC.

U.S. EPA (2020) *National Water Reuse Action Plan, EPA-820-R-20-001,* U.S. Environmental Protection Agency, Washington, D.C.

Vaidya, R., Wilson, C.A., Salazar-Benites, G., Pruden, A., and Bott, C. (2020) "Implementing Ozone-BAC-GAC in Potable Reuse for Removal of Emerging Contaminants," *AWWA Water Science,* **2**, 5, https://doi.org/10.1002/aws2.1203.

Walker, T., Stanford, B.D., Khan, S., Robillot, C., Snyder, S., Valerdi, R., Dwivedi, S., and Vickers, J. (2016) Critical Control Point Assessment to Quantify Robustness and Reilability of Multiple Treatment Barriers of a DPR Scheme, Reuse-13-03, Water Environment & Reuse Foundation, Alexandria, VA.

Watts, M.J., and Linden, K.G. (2007) "Chlorine Photolysis and Subsequent OH Radical Production during UV Treatment of Chlorinated Water," *Water Res.*, **41**, 2871–2878, https://doi.org/j.watres.2007.03.032.

WCPH (1996) *Total Resource Recovery Project Final Report,* Report prepared for City of San Diego Water Utilities Department, Western Consortium for Public Health, Oakland, CA.

WEF (2021) *Wastewater Treatment Fundaments: Book 3 – Advanced Treatment,* Water Environment Federation, WEF Press, Alexandria, VA.

WEF/ASCE (2010) *Design of Municipal Wastewater Treatment Plants, WEF Manual of Practice No. 8,* ASCE Manuals and Reports on Engineering Practice No.76 (5th ed.), Water Environment Federation, WEF Press, Alexandria, VA.

Wingoc (2021) *50 Years of Direct Potable Reuse: The Windhoek Experience,* Windhoek Goreangab Operation Company, Windhoek, Namibia. Available at: http://documents.windhoekcc.org.na/Content/Documents/CoW50yrDPR/Day (accessed April 30, 2021).

APPENDIX A

Conversion Factors

Table A-1
Unit conversion factors, SI units to U.S. customary units and U.S. customary units to SI units

		To convert, multiply in direction shown by arrows			
SI Unit Name	**Symbol**	→	←	**Symbol**	**U.S. Customary Unit Name**
Acceleration					
Meters per second squared	m/s^2	3.2808	0.3048	ft/s^2	Feet per second squared
Meters per second squared	m/s^2	39.3701	0.0254	$in./s^2$	Inches per second squared
Area					
Hectare (10,000 m^2)	ha	2.4711	0.4047	ac	Acre
Square centimeter	cm^2	0.1550	6.4516	$in.^2$	Square inch
Square kilometer	km^2	0.3861	2.5900	mi^2	Square mile
Square kilometer	km^2	247.1054	4.047×10^{-2}	ac	Acre
Square meter	m^2	10.7639	9.2903×10^{-2}	ft^2	Square foot
Square meter	m^2	1.1960	0.8361	yd^2	Square yard
Energy					
Kilojoule	kJ	0.9478	1.0551	Btu	British thermal unit
Joule	J	2.7778×10^{-7}	3.6×10^{6}	kWh	Kilowatt-hour
Joule	J	0.7376	1.356	$ft \cdot lb_f$	Foot-pound (force)
Joule	J	1.0000	1.0000	$W \cdot s$	Watt-second
Joule	J	0.2388	4.1876	cal	Calorie
Kilojoule	kJ	2.7778×10^{-4}	3600	kWh	Kilowatt-hour
Kilojoule	kJ	0.2778	3.600	$W \cdot h$	Watt-hour
Megajoule	kJ	0.3725	2.6845	$hp \cdot h$	Horsepower-hour
Force					
Newton	N	0.2248	4.4482	lb_f	Pound force
Flow rate					
Cubic meters per day	m^3/d	264.1720	3.785×10^{-3}	gal/d	Gallons per day
Cubic meters per day	m^3/d	2.6417×10^{-4}	3.7854×10^{3}	Mgal/d (mgd)	Million gallons per day
Cubic meters per second	m^3/s	35.3147	2.8317×10^{-2}	ft^3/s (cfs)	Cubic feet per second
Cubic meters per second	m^3/s	22.8245	4.3813×10^{-2}	Mgal/d (mgd)	Million gallons per day
Cubic meters per second	m^3/s	15,850.3	6.3090×10^{-5}	gal/min (gpm)	Gallons per minute
Million liters per day	ML/d	0.26417	3.7854	Mgal/d (mgd)	Million gallons per day
Liters per second	L/s	22,824.5	4.3813×10^{-2}	gal/d	Gallons per day

Liters per second	L/s	2.2825×10^{-2}	43.8126	Mgal/d (mgd)	Million gallons per day
Liters per second	L/s	15.8508	6.3090×10^{-2}	gal/min (gpm)	Gallons per minute
Length					
Centimeter	cm	0.3937	2.540	in.	Inch
Kilometer	km	0.6214	1.6093	mi	Mile
Meter	m	39.3701	2.54×10^{-2}	in.	Inch
Meter	m	3.2808	0.3048	ft	Foot
Meter	m	1.0936	0.9144	yd	Yard
Millimeter	mm	0.03937	25.4	in.	Inch
Mass					
Gram	g	0.0353	28.3495	oz	Ounce
Gram	g	0.0022	4.5359×10^{2}	lb	Pound
Kilogram	kg	2.2046	0.45359	lb	Pound
Megagram (10^3 kg)	Mg	1.1023	0.9072	ton	Ton (short: 2000 lb)
Megagram (10^3 kg)	Mg	0.9842	1.0160	ton	Ton (long: 2240)
Power					
Kilowatt	kW	0.9478	1.0551	Btu/s	British thermal units per second
Kilowatt	kW	1.3410	0.7457	hp	Horsepower
Watt	W	0.7376	1.3558	ft-lb_f/s	Foot-pounds (force) per second
Pressure (force/area)					
Pascal (newtons per square meter)	Pa (N/m^2)	1.4504×10^{-4}	6.8948×10^{3}	lb_f/in.^{2} (psi)	Pounds (force) per square inch
Pascal (newtons per square meter)	Pa (N/m^2)	2.0885×10^{-2}	47.8803	lb_f/in.^{2} (psi)	Pounds (force) per square foot
Pascal (newtons per square meter)	Pa (N/m^2)	2.9613×10^{-4}	3.3768×10^{3}	in. Hg	Inches of mercury
Pascal (newtons per square meter)	Pa (N/m^2)	4.0187×10^{-3}	2.4884×10^{2}	in. H_2O	Inches of water
Kilopascal (kilonewtons per square meter)	kPa (kN/m^2)	0.1450	6.8948	lb_f/in.^{2} (psi)	Pounds (force) per square inch
Kilopascal (kilonewtons per square meter)	kPa (kN/m^2)	0.0099	1.0133×10^{2}	atm	Atmosphere (standard)

(*Continued*)

Table A-1 *(Continued)*

		To convert, multiply in direction shown by arrows			
SI Unit Name	**Symbol**	→	←	**Symbol**	**U.S. Customary Unit Name**
Temperature					
Degree Celsius (centigrade)	°C	1.8(°C) +32	0.0555(°F) – 32	°F	Degree Fahrenheit
Degree kelvin	K	1.8(K) – 459.67	0.0555(°F) + 459.67	°F	Degree Fahrenheit
Velocity					
Kilometers per second	km/s	2.2369	0.44704	mi/h	Miles per hour
Meters per second	m/s	3.2808	0.3048	ft/s	Feet per second
Volume					
Cubic centimeter	cm^3	0.0610	16.3781	$in.^3$	Cubic inch
Cubic meter	m^3	35.3147	2.8317×10^{-2}	$in.^3$	Cubic foot
Cubic meter	m^3	1.3079	0.7646	yd^3	Cubic yard
Cubic meter	m^3	264.1720	3.7854×10^{-3}	gal	Gallon
Cubic meter	m^3	8.1071×10^{-4}	1.2335×10^3	ac · ft	Acre · foot
Liter	L	0.2642	3.7854	gal	Gallon
Liter	L	0.0353	28.3168	ft^3	Cubic foot
Liter	L	33.8150	2.9573×10^{-2}	oz	Ounce (U.S. fluid)

Table A-2
Conversion factors for commonly used water treatment plant design parameters

To convert, multiply in direction shown by arrows			
SI Units	→	←	**U.S. Units**
g/m^3	8.3454	0.1198	lb/Mgal
kg	2.2046	0.4536	lb
kg/ha	0.8922	1.1209	lb/acre
kg/kWh	1.6440	0.6083	lb/hp · h
kg/m^2	0.2048	4.8824	lb/ft^2
kg/m^3	8345.4	1.1983×10^{-4}	lb/Mgal
$kg/m^3 \cdot d$	62.4280	0.0160	$lb/ft^3 \cdot d$
$kg/m^3 \cdot h$	0.0624	16.0185	$lb/ft^3 \cdot h$
kJ	0.9478	1.0551	Btu
kJ/kg	0.4303	2.3241	Btu/lb
kPa (gage)	0.1450	6.8948	$lbf/in.^2$ (gage)
kPa Hg	0.2961	3.3768	in. Hg
kW/m^3	5.0763	0.197	$hp/10^3 gal$
$kW/10^3\ m^3$	0.0380	26.3342	$hp/10^3\ ft^3$
L	0.2642	3.7854	gal
L	0.0353	28.3168	ft^3
$L/m^2 \cdot d$	2.4542×10^{-2}	40.7458	$gal/ft^2 \cdot d$ (gfd)
$L/m^2 \cdot min$	0.0245	40.7458	$gal/ft^2 \cdot min$
$L/m^2 \cdot min$	35.3420	0.0283	$gal/ft^2 \cdot d$ (gfd)
m	3.2808	0.3048	ft
m/h	3.2808	0.3048	ft/h
m/h	0.0547	18.2880	ft/min
m/h	0.4090	2.4448	$gal/ft^2 \cdot min$
$m^2/10^3\ m^3 \cdot d$	0.0025	407.4611	$ft^2/Mgal \cdot d$
m^3	1.3079	0.7646	yd^3
m^3/capita	35.3147	0.0283	ft^3/capita
m^3/d	264.1720	3.785×10^{-3}	gal/d (gpd)
m^3/d	2.6417×10^{-4}	3.7854×10^3	Mgal/d (mgd)
m^3/h	0.5886	1.6990	ft^3/min
$m^3/ha \cdot d$	106.9064	0.0094	$gal/ac \cdot d$
m^3/kg	16.0185	0.0624	ft^3/lb
$m^3/m \cdot d$	80.5196	0.0124	$gal/ft \cdot d$
$m^3/m \cdot min$	10.7639	0.0929	$ft^3/ft \cdot min$
$m^3/m^2 \cdot d$	24.5424	0.0407	$gal/ft^2 \cdot d$ (gfd)
$m^3/m^2 \cdot d$	0.0170	58.6740	$gal/ft^2 \cdot min$
$m^3/m^2 \cdot d$	1.0691	0.9354	$Mgal/ac \cdot d$
$m^3/m^2 \cdot h$	3.2808	0.3048	$ft^3/ft^2 \cdot h$
$m^3/m^2 \cdot h$	589.0173	0.0017	$gal/ft^2 \cdot d$
m^3/m^3	0.1337	7.4805	ft^3/gal

(Continued)

Table A-1 (*Continued*)

To convert, multiply in direction shown by arrows			
SI Units	→	←	**U.S. Units**
$m^3/10^3 m^3$	133.6805	7.04805×10^{-3}	ft^3/Mgal
$m^3/m^3 \cdot$ min	133.6805	7.04805×10^{-3}	$ft^3/10^3$ gal · min
$m^3/m^3 \cdot$ min	1000.0	0.001	$ft^3/10^3\ ft^3 \cdot$ min
Mg/ha	0.4461	2.2417	ton/ac
mm	3.9370×10^{-2}	25.4	in.
ML/d	0.2642	3.785	Mgal/d (mgd)
ML/d	0.4087	2.4466	ft^3/s

APPENDIX B

Physical Properties of Selected Gases and Composition of Air

Table B-1
Molecular weight, specific weight, and density of gases found in water at standard conditions (0°C, 1 atm)

Gas	Formula	Molecular Weight, g/mol	Density, g/L
Air	—	28.97[a]	1.2928
Ammonia	NH_3	17.03	0.7708
Carbon dioxide	CO_2	44.00	1.9768
Carbon monoxide	CO	28.00	1.2501
Hydrogen	H_2	2.016	0.0898
Hydrogen sulfide	H_2S	34.08	1.5392
Methane	CH_4	16.03	0.7167
Nitrogen	N_2	28.02	1.2507
Oxygen	O_2	32.00	1.4289

Source: Adapted from R. H. Perry, D. W. Green, and J. O. Maloney (1984) *Chemical Engineers' Handbook*, 6th ed., McGraw-Hill, New York.

[a]Value reported in the literature vary depending on the standard conditions. *Note*: (0.7803 × 28.02) + (0.2099 × 32.00) + (0.0094 × 39.95) + (0.0003 × 44.00) = 28.97.

Table B-2
Composition of dry air at 0°C and 1.0 atm

Gas	Formula	Percent by Volume[a]	Percent by Weight
Nitrogen	N_2	78.03	75.47
Oxygen	O_2	20.99	23.18
Argon	Ar	0.94	1.30
Carbon dioxide	CO_2	0.039	0.05
Other[d]	—	0.01	—

[a]Adapted from *North American Combustion Handbook*,2nd ed., North American Mfg., Cleveland, OH.
[b]Hydrogen, neon, helium, krypton, xenon.

Table B-3
Density and viscosity of air (SI units)

Temperature T (°C)	Density ρ (kg/m³)	Dynamic Viscosity[a,b] μ ($\times 10^{-5}$ kg/m · s)	Kinematic Viscosity ν ($\times 10^{-5}$ m²/s)
0	1.293	1.736	1.343
5	1.269	1.762	1.388
10	1.247	1.787	1.433
15	1.225	1.812	1.479
20	1.204	1.837	1.525
25	1.184	1.862	1.572
30	1.165	1.886	1.619
35	1.146	1.910	1.667
40	1.127	1.934	1.716
45	1.110	1.958	1.765
50	1.093	1.982	1.814
60	1.060	2.029	1.915
70	1.029	2.075	2.017
80	1.000	2.121	2.121
90	0.972	2.166	2.228
100	0.946	2.210	2.336

[a]Dynamic viscosity can also be expressed in units of N · s/m².
[b]Dynamic viscosity calculated at <http://www.lmnoeng.com/Flow/GasViscosity.htm>

B-1 Density of Air at Other Temperatures

The following relationship can be used to compute the density of air, ρ_a, at other temperatures at atmospheric pressure:

$$\rho_a = \frac{PM}{RT}$$

where ρ_a = density of air, g/m^3

P = atmospheric pressure, $1.01325 \times 10^5\ N/m^2$

M = molecular weight of air (see Table B-1), 28.97 g/mol

R = universal gas constant, $8.314\ N \cdot m/(mol \cdot K)$

T = temperature, K (273.15 + °C)

For example, at 20°C, the density of air is

$$\rho_{a,20^\circ C} = \frac{(1.01325 \times 10^5\ N/m^2)(28.97\ g/mol)}{[8.314\ N{\cdot}m/(mol{\cdot}K)][(273.15+20)K]}$$
$$= 1204\ g/m^3 = 1.204\ kg/m^3$$

B-2 Change in Atmospheric Pressure with Elevation

The following relationship can be used to compute the change in atmospheric pressure with elevation:

$$\frac{P_b}{P_a} = \exp\left[-\frac{gM(z_b - z_a)}{RT}\right]$$

where P_b = pressure at elevation z_b, N/m^2

P_a = atmospheric pressure at sea level, $1.01325 \times 10^5\ N/m^2$

g = acceleration due to gravity, $9.81\ m/s^2$

M = molecular weight of air (see Table B-1), 28.97 g/mol

z = elevation, m

R = universal gas constant, $8.314\ N \cdot m/(mol \cdot K)$

T = temperature, K (273.15 + °C)

APPENDIX C

Physical Properties of Water

Table C-1
Physical properties of water (SI units)

Temperature T (°C)	Specific Weight γ (kN/m^3)	Density[a] ρ (kg/m^3)	Dynamic Viscosity[b] μ ($\times 10^{-3}$ kg/m·s)	Kinematic Viscosity ν ($\times 10^{-6}$ m^2/s)	Surface Tension[c] σ (N/m)	Modulus of Elasticity[a] E ($\times 10^9$ N/m^2)	Vapor Pressure P_v (kN/m^2)
0	9.805	999.8	1.781	1.785	0.0765	1.98	0.61
5	9.807	1000.0	1.518	1.519	0.0749	2.05	0.87
10	9.804	999.7	1.307	1.306	0.0742	2.10	1.23
15	9.798	999.1	1.139	1.139	0.0735	2.15	1.70
20	9.789	998.2	1.002	1.003	0.0728	2.17	2.34
25	9.777	997.0	0.890	0.893	0.0720	2.22	3.17
30	9.764	995.7	0.798	0.800	0.0712	2.25	4.24
40	9.730	992.2	0.653	0.658	0.0696	2.28	7.38
50	9.689	988.0	0.547	0.553	0.0679	2.29	12.33
60	9.642	983.2	0.466	0.474	0.0662	2.28	19.92
70	9.589	977.8	0.404	0.413	0.0644	2.25	31.16
80	9.530	971.8	0.354	0.364	0.0626	2.20	47.34
90	9.466	965.3	0.315	0.326	0.0608	2.14	70.10
100	9.399	958.4	0.282	0.294	0.0589	2.07	101.33

Source: Adapted from J. K. Venard and R. L. Street (1975). *Elementary Fluid Mechanics*, 5th ed., Wiley, New York.
[a]At atmospheric pressure.
[b]Dynamic viscosity can also be expressed in units of N·s/m^2.
[c]In contact with air.

Table C-2
Physical properties of water (U.S. customary units)

Temperature T (°F)	Specific Weight γ (lb/ft^3)	Density[a] ρ (slug/ft^3)	Dynamic Viscosity μ ($\times 10^{-5}$ lb·s/ft^2)	Kinematic Viscosity ν ($\times 10^{-5}$ ft^2/s)	Surface Tension[b] σ (lb/ft)	Modulus of Elasticity[a] E (10^3 lb$_f$/in.^{2})	Vapor Pressure p_v (lb$_f$/in.^{2})
32	62.42	1.940	3.746	1.931	0.00518	287	0.09
49	62.43	1.940	3.229	1.664	0.00614	296	0.12
50	62.41	1.940	2.735	1.410	0.00509	305	0.18
60	62.37	1.938	2.359	1.217	0.00504	313	0.26
70	62.30	1.936	2.050	1.059	0.00498	319	0.36
80	62.22	1.934	1.799	0.930	0.00492	324	0.51
90	62.11	1.931	1.595	0.826	0.00486	328	0.70
100	62.00	1.927	1.424	0.739	0.00480	331	0.95
110	61.86	1.923	1.284	0.667	0.00473	332	1.27
120	61.71	1.918	1.168	0.609	0.00467	332	1.69
130	61.55	1.913	1.069	0.558	0.00460	331	2.22
140	61.38	1.908	0.981	0.514	0.00454	330	2.89
150	61.20	1.902	0.905	0.476	0.00447	328	3.72
160	61.00	1.896	0.838	0.442	0.00441	326	4.74
170	60.80	1.890	0.780	0.413	0.00434	322	5.99
180	60.58	1.883	0.726	0.385	0.00427	318	7.51
190	60.36	1.876	0.678	0.362	0.00420	313	9.34
200	60.12	1.868	0.637	0.341	0.00413	308	11.52
212	59.83	1.860	0.593	0.319	0.00404	300	14.70

Source: Adapted from J. K. Venard and R. L. Street (1975). *Elementary Fluid Mechanics*, 5th ed., Wiley, New York.
[a]At atmospheric pressure.
[b]In contact with the air.

The following equations (R. C. Weast, 1983, *CRC Handbook of Chemistry and Physics*, 64th edition, CRC Press, Boca Raton, FL) can be used to compute the density ρ_w (kg/m^3) and dynamic viscosity μ_w (kg/ m·s) at other temperatures:

$$\rho_w = \frac{\left[\begin{array}{c} 999.83952 + 16.945176(T) - 7.9870401\times 10^{-3}(T)^2 \\ -46.170461\times 10^{-6}(T)^3 + 105.56302\times 10^{-9}(T)^4 - 280.54253\times 10^{-12}(T)^5 \end{array}\right]}{1 + 16.879850\times 10^{-3}(T)}$$

For $0 < T < 20°C$, $\mu_w = 10^{-3}(10^A)$

$$\text{Where} \quad A = \frac{1301}{998.333 + 8.1855(T-20) + 0.00585(T-20)^2} - 1.30223$$

For $20 < T < 100°C$, $\mu_w = (1.002 \times 10^{-3})(10^B)$

$$\text{Where} \quad B = \frac{1.3272(20-T) - 0.001053(T-20)^2}{T+105}$$

APPENDIX D

Standard Atomic Weights 2001[a]

The atomic weights of many elements are not invariant but depend on the origin and treatment of the material. The standard values of $A_r(E)$ and the uncertainties (in parentheses, following the last significant figure to which they are attributed) apply to elements of natural terrestrial origin. The footnotes to this table elaborate the types of variation that may occur for individual elements and that may be larger than the listed uncertainties of values of $A_r(E)$. Names of elements with atomic numbers 110 to 116 are provisional.

Alphabetical order in English

Name	Symbol	Number	Atomic weight	Footnotes
Actinium*	Ac	89		
Aluminum	Al	13	26.981 538(2)	
Americium*	Am	95		
Antimony (Stibium)	Sb	51	121.760(1)	g
Argon	Ar	18	39.948(1)	g, r
Arsenic	As	33	74.921 60(2)	
Astatine*	At	85		
Barium	Ba	56	137.327(7)	
Berkelium*	Bk	97		
Beryllium	Be	4	9.012 182(3)	
Bismuth	Bi	83	208.980 38(2)	
Bohrium*	Bh	107		
Boron	B	5	10.811(7)	g, m, r
Bromine	Br	35	79.904(1)	

(Continued)

[a]Scaled to $A_r(^{12}C) = 12$, where ^{12}C is a neutral atom in its nuclear and electronic groundstate.

Alphabetical order in English				
Name	Symbol	Number	Atomic weight	Footnotes
Cadmium	Cd	48	112.411(8)	g
Caesium (Cesium)	Cs	55	132.905 45(2)	
Calcium	Ca	20	40.078(4)	g
Californium*	Cf	98		
Carbon	C	6	12.0107(8)	g, r
Cerium	Ce	58	140.116(1)	g
Chlorine	Cl	17	35.453(2)	g, m, r
Chromium	Cr	24	51.9961(6)	
Cobalt	Co	27	58.933 200(9)	
Copper (Cuprum)	Cu	29	63.546(3)	r
Curium*	Cm	96		
Dubnium*	Db	105		
Dysprosium	Dy	66	162.500(1)	g
Einsteinium*	Es	99		
Erbium	Er	68	167.259(3)	g
Europium	Eu	63	151.964(1)	g
Fermium*	Fm	100		
Fluorine	F	9	18.998 4032(5)	
Francium*	Fr	87		
Gadolinium	Gd	64	157.25(3)	g
Gallium	Ga	31	69.723(1)	
Germanium	Ge	32	72.64(1)	
Gold (Aurum)	Au	79	196.966 55(2)	
Hafnium	Hf	72	178.49(2)	
Hassium*	Hs	108		
Helium	He	2	4.002 602(2)	g, r
Holmium	Ho	67	164.930 32(2)	
Hydrogen	H	1	1.007 94(7)	g, m, r
Indium	In	49	114.818(3)	
Iodine	I	53	126.904 47(3)	
Iridium	Ir	77	192.217(3)	
Iron (Ferrum)	Fe	26	55.845(2)	
Krypton	Kr	36	83.798(2)	g, m
Lanthanum	La	57	138.9055(2)	g
Lawrencium*	Lr	103		
Lead (Plumbum)	Pb	82	207.2 (1)	g, r
Lithium	Li	3	[6.941(2)]†	g, m, r
Lutetium	Lu	71	174.967(1)	g
Magnesium	Mg	12	24.3050(6)	
Manganese	Mn	25	54.938 049(9)	
Meitnerium*	Mt	109		
Mendelevium*	Md	101		

Alphabetical order in English				
Name	Symbol	Number	Atomic weight	Footnotes
Mercury (Hydrargyrum)	Hg	80	200.59(2)	
Molybdenum	Mo	42	95.94(2)	g
Neodymium	Nd	60	144.24(3)	g
Neon	Ne	10	20.1797(6)	g, m
Neptunium*	Np	93		
Nickel	Ni	28	58.6934(2)	
Niobium	Nb	41	92.906 38(2)	
Nitrogen	N	7	14.0067(2)	g, r
Nobelium*	No	102		
Osmium	Os	76	190.23(3)	g
Oxygen	O	8	15.9994(3)	g, r
Palladium	Pd	46	106.42(1)	g
Phosphorus	P	15	30.973 761(2)	
Platinum	Pt	78	195.078(2)	
Plutonium*	Pu	94		
Polonium*	Po	84		
Potassium (Kalium)	K	19	39.0983(1)	
Praseodymium	Pr	59	140.907 65(2)	
Promethium*	Pm	61		
Protactinium*	Pa	91	231.035 88(2)	
Radium*	Ra	88		
Radon*	Rn	86		
Rhenium	Re	75	186.207(1)	
Rhodium	Rh	45	102.905 50(2)	
Rubidium	Rb	37	85.4678(3)	g
Ruthenium	Ru	44	101.07(2)	g
Rutherfordium*	Rf	104		
Samarium	Sm	62	150.36(3)	g
Scandium	Sc	21	44.955 910(8)	
Seaborgium*	Sg	106		
Selenium	Se	34	78.96(3)	r
Silicon	Si	14	28.0855(3)	r
Silver (Argentum)	Ag	47	107.8682(2)	g
Sodium (Natrium)	Na	11	22.989 770(2)	
Strontium	Sr	38	87.62(1)	g, r
Sulfur	S	16	32.065(5)	g, r
Tantalum	Ta	73	180.9479(1)	
Technetium*	Tc	43		
Tellurium	Te	52	127.60(3)	g
Terbium	Tb	65	158.925 34(2)	
Thallium	Tl	81	204.3833(2)	

(*Continued*)

Alphabetical order in English				
Name	Symbol	Number	Atomic weight	Footnotes
Thorium*	Th	90	232.0381(1)	g
Thulium	Tm	69	168.934 21(2)	
Tin (Stannum)	Sn	50	118.710(7)	g
Titanium	Ti	22	47.867(1)	
Tungsten (Wolfram)	W	74	183.84(1)	
Ununbium*	Uub	112		
Ununhexium*	Uuh	116		
Ununnilium*	Uun	110		
Ununquadium*	Uuq	114		
Unununium*	Uuu	111		
Uranium*	U	92	238.028 91(3)	g m
Vanadium	V	23	50.9415(1)	
Xenon	Xe	54	131.293(6)	g m
Ytterbium	Yb	70	173.04(3)	g
Yttrium	Y	39	88.905 85(2)	
Zinc	Zn	30	65.409(4)	
Zirconium	Zr	40	91.224(2)	g

*Element has no stable nuclides.

†Commercially available Li materials have atomic weights that range between 6.939 and 6.996; if a more accurate value is required, it must be determined for the specific material.

gGeological specimens are known in which the element has an isotopic composition outside the limits for normal material. The difference between the atomic weight of the element in such specimens and that given in the table may exceed the stated uncertainty.

mModified isotopic compositions may be found in commercially available material because it has been subjected to an undisclosed or inadvertent isotopic fractionation. Substantial deviations in atomic weight of the element from that given in the table can occur.

rRange in isotopic composition of normal terrestrial material prevents a more precise A_r (E) being given; the tabulated A_r (E) value should be applicable to any normal material.

APPENDIX E

Electronic Resources Available on the Stantec Website for This Textbook

Table or Resource	Filename	Description
E1	Standard_Reduction_Potentials.pdf	Selected standard reduction potentials for inorganic compounds at 25°C.
E2	Ozone_Reactions.pdf	Reactions of ozone with inorganic and organic compounds.
E3	Extinction_Coefficients.pdf	Extinction coefficients for common inorganic chemicals.
E4	Hydroxyl_Rate_Constants.pdf	Second-order rate constants between hydroxyl radical and various species in water.
E5	Freundlich_Isotherm_Parameters.xlsx	Freundlich isotherm parameters K and $1/n$ for various organic compounds in aqueous and gaseous phases.

(*Continued*)

These and other resources are available online.

http://www.stantec.com/wt3r/AppendixE.html

Table or Resource	Filename	Description
E6	HSDM_Solutions_for_PAC.pdf	Parameters for the empirical equation that describes solutions to the HSDM for PAC in a batch or plug flow reactor.
E7	HSDM_Solutions_for_GAC.pdf	Parameter values used in Equations 15-173 and 15-178 for constant pattern solutions to the plug-flow homogeneous surface diffusion model.
E8	HSDM_Solutions_for_GAC.pdf	Minimum Stanton number for which constant pattern calculations can be used with a 10% error in calculated breakthrough times.
E9	HSDM_Solutions_for_GAC.pdf	Values of γ and St_{min}, at various $1/n$ values that are required for constant pattern and external mass transfer control.
E10	AirStripCalc.xlsx	Spreadsheet designed to facilitate the calculations for countercurrent packed tower design.

Index

Stantec's Water Treatment: Principles and Design, Updated Third Edition. John C. Crittenden, R. Rhodes Trussell, David W. Hand, Kerry J. Howe, and George Tchobanoglous.

D

E

N

O

P

Q

T